MW01591212

COLD SPRING HARBOR SYMPOSIA ON QUANTITATIVE BIOLOGY

VOLUME LXXIII

http://symposium.cshlp.org

Institutions that have purchased the hardcover edition of this book are entitled to online access to the Symposium website. Please contact your institution's library to gain access to the website. The site contains the full-text articles from the 2008 Symposium and previous Symposia from 2004 onward.

If you received this book from Cold Spring Harbor Laboratory Press directly, the packing slip included with this book contains your activation instructions. If you received this book from your agent, please have your agent contact us for your activation instructions.

If you do not have an account number or are experiencing access problems, please contact Kathy Cirone, CSHL Press Subscription Manager, at 1-800-843-4388, extension 4044 (Continental U.S. and Canada), 516-422-4100 (all other locations), cironek@cshl.edu, or sqb-feedback@ highwire.stanford.edu.

COLD SPRING HARBOR SYMPOSIA ON QUANTITATIVE BIOLOGY

VOLUME LXXIII

Control and Regulation of Stem Cells

http://symposium.cshlp.org

Meeting Organized by Bruce Stillman, David Stewart, and Terri Grodzicker
COLD SPRING HARBOR LABORATORY PRESS
2008

COLD SPRING HARBOR SYMPOSIA ON QUANTITATIVE BIOLOGY VOLUME LXXIII

©2008 by Cold Spring Harbor Laboratory Press
International Standard Book Number 978-0-87969-861-4 (cloth)
International Standard Book Number 978-0-87969-862-1 (paper)
International Standard Serial Number 0091-7451
Web International Standard Serial Number 1943-4456
Library of Congress Catalog Card Number 34-8174

Printed in the United States of America
All rights reserved
COLD SPRING HARBOR SYMPOSIA ON QUANTITATIVE BIOLOGY
Founded in 1933 by
REGINALD G. HARRIS
Director of the Biological Laboratory 1924 to 1936
Previous Symposia Volumes

Front Cover (*Paperback*): Human stem cells differentiating to neurons. Staining for β-tubulin and DAPI. Courtesy of Juan Carlos Izpisúa Belmonte at the Center for Regenerative Medicine, Barcelona. *Back Cover* (*Paperback*): A brain lobe from a *Drosophila* third-instar larva labeled with Dlg (*red*), BrdU (*green*), and DAPI (*blue*). The large red cells are the neural stem cells of the central brain. (Courtesy of Jakob von Trotha and Andrea H. Brand.)

All Cold Spring Harbor Laboratory Press publications may be ordered directly from Cold Spring Harbor Laboratory Press, 500 Sunnyside Boulevard, Woodbury, NY 11797-2924. Phone: 1-800-843-4388 in Continental U.S. and Canada. All other locations: (516) 422-4100. FAX: (516) 422-4097. E-mail: cshpress@cshl.edu. For a complete catalog of all Cold Spring Harbor Laboratory Press publications, visit our World Wide Web site http://www.cshlpress.com/

Website Access: Institutions that have purchased the hardcover edition of this book are entitled to online access to the companion website at http://symposium.cshlp.org. For assistance with activation, please contact Kathy Cirone, CSHL Press Subscription Manager, at cironek@cshl.edu.

Symposium Participants

ABRANCHES, ELSA, Dept. of Developmental Biology, Institute of Molecular Medicine, Lisbon, Portugal

ADAMS, JERRY, Div. of Molecular Genetics of Cancer, The Walter and Eliza Hall Institute of Medical Research, Melbourne, Australia

ADOLPHE, CHRISTELLE, Swiss Institute for Cancer Research, Ecole Polytechnique Fédéral de Lausanne, Epalinges, Switzerland

AIGNER, STEFAN, Lab. of Genetics, Biterials, Seoul, La Jolla, California

AKALA, OMOBOLAJI, Institute for Stem Cell Biology and Regenerative Medicine, School of Medicine, Stanford University, Stanford, California

ALCANTARA LLAGUNO, SHEILA. Dept. of Developmental Biology, Southwestern Medical Center, University of Texas, Dallas

ALFRED, JANE, *Development*, Company of Biologists, Cambridge, United Kingdom

AL-HAJJ, MUHAMMAD, Dept. of Cancer Biology, AstraZeneca, Gaithersburg, Maryland

ALVAREZ-BUYLLA, ARTURO, Dept. of Neurological Surgery, Program in Developmental and Stem Cell Biology, University of California, San Francisco

AMARIGLIO, NINETTE, Cancer Research Center, Sheba Medical Center, Tel Hashomer, Israel

ANDERS, DAVID, New York State Dept. of Health, New York State Stem Cell Science, Wadsworth Center, Albany, New York

ANDROUTSELLIS-THEOTOKIS, ANDREAS, National Institute of Neurological Disorders and Stroke, National Institutes of Health, Bethesda, Maryland

ARANDA, VICTORIA, Cold Spring Harbor Laboratory, Cold Spring Harbor, New York

AUSONI, SIMONETTA, Dept. of Biomedical Sciences, University of Padua, Padua, Italy

BAGCHI, ANINDYA, Cold Spring Harbor Laboratory, Cold Spring Harbor, New York

BAI, LIXIA, Dept. of Basic Sciences, Fred Hutchinson Cancer Research Center, Seattle, Washington

BAKER, MONYA, *Nature Reports Stem Cells*, Nature Publishing Group, San Francisco, California

BALASUBRAMANIAN, D., Kallam Anji Reddy Campus, L.V. Prasad Eye Institute, Hyderabad, India

BALINT, BALINT, Dept. of Biochemistry and Molecular Biology, University of Debrecen, Debrecen, Hungary

BANK, ARTHUR, Dept. of Genetics and Development, Columbia University, New York, New York

BARON, MARGARET, Dept. of Medicine, Black Family Stem Cell Institute, Mt. Sinai School of Medicine, New York, New York

BEAUSÉJOUR, CHRISTIAN, Dept. of Pharmacology, CHU Ste-Justine, Université de Montréal, Montréal, Québec, Canada

BERNARDOS, REBECCA, Dept. of Neurobiology, Medical School, University of Massachusetts, Worcester

BERTA, MELANIE, Lab. of Epithelial Cell Biology, Cancer Research UK, Cambridge, United Kingdom

BEYRET, ERGIN, Dept. of Cell Biology, Stem Cell Center, Yale University, New Haven, Connecticut

BHATTACHARYA, PARAMITA, Lab. of Metabolism, National Cancer Institute, National Institutes of Health, Bethesda, Maryland

BLIN, GUILLAUME, Dept. of Cardiomyopathy, Institut National de la Santé et de la Recherche Médicale, Université d'Evry, Evry, France

BOGESTAL, YALDA, Dept. of Biomedicine, Gothenburg University, Gothenburg, Sweden

BORGHESI, LISA, Dept. of Immunology, School of Medicine, University of Pittsburgh, Pittsburgh, Pennsylvania

BOTELHO, JOAO, Champalimaud Foundation, Lisbon, Portugal

BRAND, ANDREA, Dept. of Physiology, Development, and Neuroscience, Gurdon Institute, University of Cambridge, Cambridge, United Kingdom

BROWNE, CATHERINE, Institute of Molecular Bioscience, University of Queensland, Brisbane, Australia

BUAC, KRISTINA, Genetics Disease Research Branch, National Human Genome Institute, National Institutes of Health, Bethesda, Maryland

BUCKINGHAM, MARGARET, Dept. of Developmental Biology, Centre National de la Recherche Scientifique, Pasteur Institute, Paris, France

BUSHELL, WILLIAM, Dept. of Anthropology, Massachusetts Institute of Technology, Cambridge, Massachusetts

CALS-GRIERSON, MARIE-MADELEINE, Dept. of Life Sciences, L'Oreal Research, Clichy, France

CARBONNEAU, CYNTHIA, Dept. of Pharmacology, Centre de Recherche, CHU Ste-Justine, Université de Montréal, Montréal, Québec, Canada

CAROTTA, SEBASTIAN, Div. of Immunology, The Walter and Eliza Hall Institute of Medical Research, Melbourne, Australia

CARROLL, PATRICK, Dept. of Basic Sciences, Fred Hutchinson Cancer Research Center, Seattle, Washington

CELLOT, SONIA, Dept. of Medicine, Institute for Research in Immunology and Cancer, University of Montréal, Montréal, Québec, Canada

CESARI, FRANCESCA, *Nature Reviews Molecular Cell Biology*, Nature Publishing Group, London, United Kingdom

CHAI, HYUN JONG, Biterials, Seoul, Korea

CHALAMALASETTY, RAVINDRA B., Lab. of Cancer and Developmental Biology, National Cancer Institute, Frederick, Maryland

CHANG, DAVID, Dept. of Bone Marrow Transplant, Div. of Research Immunology, Children's Hospital Los Angeles, Los Angeles, California

CHATTERJEE, SAMIT, Cold Spring Harbor Laboratory, Cold Spring Harbor, New York

CHEN, CHUN-MING, Dept. of Life Sciences, National Yang-Ming University, Taipei, Taiwan

CHEN, JIAN, Dept. of Developmental Biology, Southwestern Medical Center, University of Texas, Dallas

CHEN, JUN AN, Dept. of Pathology, Columbia University, New York, New York

CHEN, LEI, Dept. of Pathology, Stanford University, Stanford, California

CHEN, MUHAN, Cold Spring Harbor Laboratory, Cold Spring Harbor, New York

CHEN, THERESA, Dept. of Medicine, VA Palo Alto Health Care System, Palo Alto, California

CHEN, TING, Lab. of Mammalian Cell Biology and Development, Howard Hughes Medical Institute, The Rockefeller University, New York, New York

CHEN, XI, Dept. of Biological Sciences, National University of Singapore, Singapore

CHENG, LI-CHUN, Dept. of Pathology, Columbia University, New York, New York

CHEN-KIANG, SELINA, Dept. of Pathology, Weill-Cornell Medical College, New York, New York

CHENOWETH, JOSHUA, National Institute of Neurological Disorders and Stroke, National Institutes of Health, Bethesda, Maryland

CHI, HUNG YUAN, Lab. of Mammalian Cell Biology and Development, Howard Hughes Medical Institute, The Rockefeller University, New York, New York

CHIEN, KENNETH, Cardiovascular Research Center, Massachusetts General Hospital, Boston, Massachusetts

CHILDERS, MARTIN, Dept. of Neurology, School of Medicine, Wake Forest University, Winston-Salem, North Carolina

CHIN, BENNETT, Dept. of Radiology, Duke University, Durham, North Carolina

CHOI, JINHYANG, Dept. of Biomedical Sciences, Cornell University, Ithaca, New York

CHOI, KYUNG-DAL, Dept. of Pathology, Lab. of Medicine, University of Wisconsin, Madison

CHOU, RUEY-HWANG, Center for Molecular Medicine, Medical University of China, Tai-Chung, Taiwan

CHRISTOPHERSEN, NICOLAJ, Centre for Epigenetics, University of Copenhagen, Copenhagen, Denmark

CHUANG, AARON T.T., Dept. of Neurodegeneration Research, GlaxoSmithKline, Harlow, United Kingdom

CIMADAMORE, FLAVIO, Stem Cells and Regeneration Program, Burnham Institute for Medical Research, La Jolla, California

CLEVERS, HANS, Dept. of Immunology and Molecular Genetics, Hubrecht Institute for Developmental Biology, Utrecht, The Netherlands

CONNELL, LAUREEN, *Genes & Development*, Cold Spring Harbor Laboratory Press, Woodbury, New York

CORMIER, CATHERINE, Institute of Proteomics, Harvard Medical School, Cambridge, Massachusetts

CORONAS, VALÉRIE, Institut de Physiologie et Biologie Cellulaires, Centre National de la Recherche Scientifique, University of Poitiers, Poitiers, France

COTTRELL, TRICIA, Div. of Rheumatology, School of Medicine, Johns Hopkins University, Baltimore, Maryland

CRAWFORD, SARAH, Dept. of Biology, Southern Connecticut State University, New Haven

CROSIER, PHILIP, Dept. of Molecular Medicine, School of Medicine, University of Auckland, Auckland, New Zealand

CURTIS, STEPHEN, Dept. of Genetics, Stem Cell Program, Children's Hospital, Harvard Medical School, Boston, Massachusetts

DALERBA, PIERO, Institute for Stem Cell Biology and Regenerative Medicine, Stanford University, Stanford, California

DALEY, GEORGE, Div. of Hematology and Oncology, Children's Hospital, Boston, Massachusetts

DAMELIN, MARC, Dept. of Oncology, Wyeth Research, Pearl River, New York

DANIELSON, LAURA, Dept. of Pathology, New York University School of Medicine, New York, New York

DAVIES, ERIN, Dept. of Developmental Biology, Stanford University, Stanford, California

DAVIS, MATHEW, Dept. of Research, OncoMed Pharmaceuticals, Redwood City, California

DECATHELINEAU, AIMEE, *Journal of Cell Biology*, The Rockefeller University Press, New York, New York

DE JONG, PIETER, BACPAC Resources, Children's Hospital and Research Center Oakland, Oakland, California

DENG, YIBIN, Dept. of Cancer Genetics, M.D. Anderson Cancer Center, University of Texas, Houston

DENLI, AHMET, Lab. of Genetics, The Salk Institute for Biological Studies, La Jolla, California

DEPINHO, RONALD, Dept. of Medical Oncology, Dana-Farber Cancer Institute, Boston, Massachusetts

DE SOUZA, NATALIE, *Nature Methods*, Nature Publishing Group, New York, New York

DOETSCH, FIONA, Dept. of Pathology, Neurology, and Neuroscience, Columbia University, New York, New York

DONOVAN, SHANE, Dept. of BioScience, Baxter Healthcare, Round Lake, Illinois

DOSS, MICHAEL, Dept. of Medicine, Weill Cornell Medical College, New York, New York

DOULATOV, SERGEI, Dept. of Molecular and Medical Genetics, University of Toronto, Toronto, Ontario, Canada

DU, MIN, Dept. of Animal Science, University of Wyoming, Laramie

DUBOWITZ, VICTOR, Institute of Child Health, University College, University of London, London, United Kingdom

DUPIN, ELISABETH, Institute of Neurobiology, Centre National de la Recherche Scientifique, Gif-sur-Yvette, France

EBERT, JOAN, Cold Spring Harbor Laboratory Press, Woodbury, New York

ECONOMIDES, ARIS, Dept. of Genome Engineering and Technology Development, Regeneron Pharmaceuticals, Tarrytown, New York

EDOFF, KARIN, Gurdon Institute, University of Cambridge, Cambridge, United Kingdom

EGGER, BORIS, Dept. of Physiology, Development, and Neuroscience, Wellcome Trust Cancer Research, U.K. Gurdon Institute, University of Cambridge, Cambridge, United Kingdom

EISENBERGER, TOBIAS, Developmental Biology Unit, European Molecular Biology Laboratory, Heidelberg, Germany

ENCINAS, JUAN MANUEL, Cold Spring Harbor Laboratory, Cold Spring Harbor, New York

ENGLE, SANDRA, GeMM CoE Database, Pfizer, Groton, Connecticut

ENIKOLOPOV, GRIGORI, Cold Spring Harbor Laboratory, Cold Spring Harbor, New York

EPPERT, KOLJA, Div. of Cell and Molecular Biology, University Health Network, Toronto, Ontario, Canada

EPSTEIN, DAVID, Dept. of Oncology Research, OSI Pharmaceuticals, Farmingdale, New York

ERIKSSON, MALIN, Dept. of Cell and Molecular Biology, Karolinska Institute, Stockholm, Sweden

EYLER, CHRISTINE, Dept. of Pharmacology and Cancer Biology, Duke University, Durham, North Carolina

FALKOWSKA-HANSEN, BERIT, Div. of Genetics of Skin Carcinogenesis, German Cancer Research Center, Heidelberg, Germany

FAZZIO, THOMAS, Hooper Foundation, University of California, San Francisco

FERREIRA, MIGUEL, Gulbenkian Institute of Science, Lisbon, Portugal

FLEMING, WILLIAM, Dept. of Hematology, Oncology, and Medicine, Oregon Health and Sciences University, Portland, Oregon

FRANZ, WOLFGANG, Medical Clinic, University of Munich-Grosshadern, Munich, Germany

FRIAS, ANA, University of Minho, Braga, Portugal

FUCHS, ELAINE, Lab. of Mammalian Cell Biology and Development, Howard Hughes Medical Institute, The Rockefeller University, New York, New York

FUENTE MORA, CRISTINA, School of Biological Sciences, University of Liverpool, Liverpool, United Kingdom

FULLER, MARGARET, Dept. of Developmental Biology, School of Medicine, Stanford University, Stanford, California

GAIANO, NICHOLAS, Institute for Cell Engineering, School of Medicine, Johns Hopkins University, Baltimore, Maryland

GALLIOT, BRIGITTE, Dept. of Zoology and Animal Biology, University of Geneva, Geneva, Switzerland

GAVRILIUC, OANA, Dept. of Immunology and Physiology, University of Medicine and Pharmacy Victor Babes, Timisoara, Romania

GAVRILOV, SVETLANA, Dept. of Genetics and Development, Columbia University, New York, New York

GENANDER, MARIA, Dept. of Cell and Molecular Biology, Karolinska Institute, Stockholm, Sweden

GIANGRECO, ADAM, Keratinocyte Laboratory, Cancer Research U.K., University of Cambridge, Cambridge, United Kingdom

GIFFORD, DAVID, Dept. of Electrical Engineering and Computer Science, Massachusetts Institute of Technology, Cambridge, Massachusetts

GINESTIER, CHRISTOPHE, Dept. of Internal Medicine, Comprehensive Cancer Center, University of Michigan, Ann Arbor

GLUDISH, DAVID, Dept. of Hematology and Oncology, Children's Hospital, Boston, Massachusetts

GOLD, KATRINA, Dept. of Physiology, Development, and Neuroscience, Wellcome Trust Cancer Research, U.K. Gurdon Institute, University of Cambridge, Cambridge, United Kingdom

GOTHARD, DAVID, Dept. of Tissue Engineering and Modelling, Wolfson Centre, University of Nottingham, Nottingham, United Kingdom

GRECO, VALENTINA, Lab. of Mammalian Cell Biology and Development, Howard Hughes Medical Institute, The Rockefeller University, New York, New York

GREENE, MICHELLE, Scientific Collaborations, Millipore Corporation, West Palm Beach, Florida

GREENFIELD, RAZI, Dept. of Cellular Biochemistry and Human Genetics, Hebrew University, Jerusalem, Israel

GREVE, JEFFREY, Dept. of Biopharmaceuticals, Exelixis, South San Francisco, California

GRODZICKER, TERRI, Cold Spring Harbor Laboratory Press, Woodbury, New York

GROSSNIKLAUS, UELI, Institute of Plant Biology, University of Zürich, Zürich, Switzerland

GUIJARRO, MARIA, Dept. of Pathology, New York University Medical Center, New York, New York

GUO, BING, Dept. of Oncology, Wyeth, Pearl River, New York

HALEY, JOHN, Translational Research, OSI Pharmaceuticals, Farmingdale, New York

HAMRA, F. KENT, Dept. of Pharmacology, Southwestern Medical Center, University of Texas, Dallas

HAN, JONG JIN, Cold Spring Harbor Laboratory, Cold Spring Harbor, New York

HANNON, GREGORY, Cold Spring Harbor Laboratory, Cold Spring Harbor, New York

HAO, HSIAO-NAN, Dept. of Orthopaedic Surgery, School of Medicine, Wayne State University, Detroit, Michigan

HARTER, M.L. (NIKKI), Dept. of Biochemistry, Case Western Reserve University, Cleveland, Ohio

HASHIMOTO, SHUICHI, Dept. of Pulmonary, Allergy, and Critical Care, Duke University Medical Center, Durham, North Carolina

HASS, MATTHEW, Dept. of Developmental Biology, Washington University, St. Louis, Missouri

HE, WEI, Dept. of Discovery Oncology, Hoffmann La Roche, Nutley, New Jersey

HE, YE, Dept. of Neuroscience and Cell Biology, University of Medicine and Dentistry of New Jersey, Robert Wood Johnson Medical School, Piscataway, New Jersey

HECHT, MERAV, Dept. of Cell Biochemistry and Human Genetics, Hebrew University, Jerusalem, Israel

HELIN, KRISTIAN, Copenhagen Biocenter, Biotech Research and Innovation Centre, Copenhagen, Denmark

HENSELEIT, KORINNA, Centre for Cutaneous Research, Queen Mary University, London, United Kingdom

HERRERA, FRANCISCO, Dept. of Molecular and Cell Biology, University of California, Berkeley

HO, LENA, Dept. of Immunology, Stanford University, Stanford, California

HOCHEDLINGER, KONRAD, Center for Regenerative Medicine, Cancer Center, Massachusetts General Hospital, Boston, Massachusetts

HOEPPNER, DANIEL, National Institute of Neurological Disorders and Stroke, National Institutes of Health, Bethesda, Maryland

HOGAN, BRIGID, Dept. of Cell Biology, Duke University Medical Center, Durham, North Carolina

HOLOWACZ, TAMARA, Canadian Centre for Biomolecular Research, University of Toronto, Toronto, Canada

HOPE, KRISTIN, Dept. of Medicine, Institute for Research in Cancer and Immunology, Montréal, Canada

HRECKA, KASIA, Cold Spring Harbor Laboratory, Cold Spring Harbor, New York

HUA, KATE, Institute of Pharmacology, National Yang Ming University, Taipei, Taiwan

HUEBNER, MICHAEL, Cold Spring Harbor Laboratory, Cold Spring Harbor, New York

HUR, EUN MI, Dept. of Biology, California Institute of Technology, Pasadena, California

IBARRA, INGRID, Cold Spring Harbor Laboratory, Cold Spring Harbor, New York

ICHIMURA, MICHIO, BioFrontier Laboratories, Kyowa Hakko Kogyo, Machidashi, Japan

INGLIS, JOHN, Cold Spring Harbor Laboratory Press, Woodbury, New York

IZPISÚA BELMONTE, JUAN CARLOS, Lab. of Gene Expression, The Salk Institute for Biological Studies, La Jolla, California

JACOB, BINDYA, Div. of Genetics and Genomics, Institute of Molecular and Cell Biology, Singapore

JAENISCH, RUDOLF, Dept. of Biomedical Research, Whitehead Institute for Biomedical Research, Massachusetts Institute of Technology, Cambridge, Massachusetts

JAKIMO, ALAN, Dept. of Law, Hofstra University, Hempstead, New York

JANSSEN, KAAREN, Cold Spring Harbor Laboratory Press, Woodbury, New York

JENSEN, UFFE, Dept. of Clinical Genetics, Vejle Hospital, Vejle, Denmark

JIANG, XIN, Dept. of Cardiology, Massachusetts General Hospital, Boston, Massachusetts

JOHANSSON, HELENA, Dept. of Medical Biochemistry and Cell Biology, University of Gothenburg, Gothenburg, Sweden

JONES, DEBORAH, Oncology Discovery Technology Group, GlaxoSmithKline, Collegeville, Pennsylvania

JULIANO, CELINA, Dept. of Molecular Biology, Cell Biology, and Biochemistry, Brown University, Providence, Rhode Island

JUNG, JIN, Dept. of Physiology, School of Medicine, Pusan National University, Pusan, South Korea

KADYK, LISA, Dept. of Biopharmaceuticals, Exelixis, South San Francisco, California

KALABIS, JIRI, Dept. of Gastroenterology, University of Pennsylvania, Philadelphia

KANG, KYUNG-SUN, Adult Stem Cell Research Center, College of Veterinary Medicine, Seoul National University, Seoul, South Korea

KANG, SOO, Dept. of Physiology, School of Medicine, Pusan National University, Pusan, South Korea

KARETA, MICHAEL, Dept. of Molecular and Cellular Biology, University of California, Davis

KAUSHAL, SHALESH, Dept. of Ophthalmology, University of Florida, Gainesville

KEHLER, JAMES, Lab. of Molecular Biology, National Institute of Neurological Disorders and Stroke, National Institutes of Health, Bethesda, Maryland

KELLER, GORDON, McEwen Centre for Regenerative Medicine, Toronto, Ontario, Canada

KESHET, GILMOR, Cancer Research Center, Sheba Medical Center, Tel Hashomer, Israel

KEYES, WILLIAM, Cold Spring Harbor Laboratory, Cold Spring Harbor, New York

KIKYO, NOBUAKI, Stem Cell Institute, University of Minnesota, Minneapolis

KIM, CARLA, Dept. of Genetics, Stem Cell Program, Children's Hospital, Harvard Medical School, Boston, Massachusetts

KIM, HANA, Dept. of Biological Sciences, Louisiana State University, Baton Rouge

KIM, JUN SUNG, Dept. of Research, Biterials, Seoul, South Korea

KIM, MIN-JU, Dept. of Research and Development, Stemgent, San Diego, California

KIM, TAE-SHIN, Dept. of Biological Sciences, Korea Advanced Institute of Science and Technology, Daejeon, South Korea

KINOSHITA, MASAKI, Lab. for Stem Cell Biology, RIKEN Center for Developmental Biology, Kobe, Japan

KLASEN, CHRISTIAN, Transgenic Service, European Molecular Biology Laboratory, Heidelberg, Germany

KLOC, ANNA, Cold Spring Harbor Laboratory, Cold Spring Harbor, New York

KORKAYA, HASAN, Dept. of Internal Medicine, Comprehensive Cancer Center, University of Michigan, Ann Arbor

KOST, THOMAS, Molecular Discovery Research, GlaxoSmithKline, Research Triangle Park, North Carolina

KOTVAL, JEROO, Dept. of Health of New York State, Wadsworth Center, Albany, New York

KOUADIO, TANOH, Dept. of Cellular Biology and Development, Institut de Génétique et de Biologie Moléculaire et Cellulaire, Illkirch, France

KOULAKOV, ALEX, Cold Spring Harbor Laboratory, Cold Spring Harbor, New York

KRISHNAMURTHY, SUDHA, Dept. of Restorative Sciences, School of Dentistry, University of Michigan, Ann Arbor

KRUIJER, WIEBE, Dept. of Molecular Cell Biology, University of Twente, Enschede, The Netherlands

KUIPERS, HEDWICH, Dept. of Pathology and Neuropathology, VU University Medical Center, Amsterdam, The Netherlands

KUMARAN, ILENG, Cold Spring Harbor Laboratory, Cold Spring Harbor, New York

KUO, CHAY, Dept. of Cell Biology, Duke University Medical Center, Durham, North Carolina

LAN, FEI, Constellation Pharmaceuticals, Cambridge, Massachusetts

LANDRY, YANNICK, Dept. of Pharmacology, CHU Ste-Justine Research Center, Université de Montréal, Montréal, Québec, Canada

LARSEN, ELISABETH, Dept. of Molecular Biology, Massachusetts General Hospital, Boston, Massachusetts

LAWSON, DEVON, Dept. of Microbiology, Immunology, and Molecular Genetics, University of California, Los Angeles

LAWTON, LEE, Whitehead Insitute for Biomedical Research, Cambridge, Massachusetts

LAZETIC, SASHA, Dept. of Molecular Cell Biology, OncoMed Pharmaceuticals, Redwood City, California

LECHMAN, ERIC, Dept. of Cell and Molecular Biology, University Health Network, Toronto, Ontario, Canada

LEE, CHANG GEUN, Dept. of Molecular and Cellular Biology, School of Medicine, Sungkyunkwan University, Suwon, South Korea

LEE, JOO-HYEON, Dept. of Biological Sciences, Korea Advanced Institute of Science and Technology, Daejeon, South Korea

LEE, YOONSUNG, Dept. of Cell Biology, Duke University Medical Center, Durham, North Carolina

LEE, YOUNG, Dept. of Regenerative Medicine, Maria Biotech, Seoul, South Korea

LEHMANN, RUTH, Skirball Institute, New York University School of Medicine, New York, New York

LEMISCHKA, IHOR, Dept. of Gene and Cell Medicine, Mount Sinai School of Medicine, New York, New York

LI, BINGBING, Dept. of Biology, Indiana University-Purdue University, Indianapolis, Indiana

LI, CHUNG, Institute of Cellular and Organismic Biology, Genomics Research Center, Academia Sinica, Taipei, Taiwan

LI, LIQI, Lab. of Mammalian Genes and Development, National Institute of Child Health and Human Development, National Institutes of Health, Bethesda, Maryland

LI, XIN, Dept. of Molecular Biology and Genetics, Cornell University, Ithaca, New York

LI, YINA, Dept. of Cell and Developmental Biology, School of Medicine, Vanderbilt University, Nashville, Tennessee

LI, ZHIZHONG, Dept. of Pharmacology and Cancer Biology, Duke University, Durham, North Carolina

LIANG, CHIH-CHIA, Dept. of Life Sciences, National Yang-Ming University, Taipei, Taiwan

LIAO, JUN, Dept. of Molecular, Cellular, and Developmental Biology, Yale University, New Haven, Connecticut

LIM, DAE-SIK, Dept. of Biological Science, Korea Advanced Institute for Science and Technology, Daejeon, South Korea

LIN, HAIFAN, Dept. of Cell Biology and Genetics, Stem Cell Center, Yale University, New Haven, Connecticut

LIU, CHUNG HSIEN, Plastic Surgery and Stem Cell Therapy Center, American Academy of Cosmetic Surgery, Chia-Yi City, Taiwan

LIU, XIAOFEI, Cold Spring Harbor Laboratory, Cold Spring Harbor, New York

LIU, YAN, Dept. of Medicine, Memorial Sloan-Kettering Cancer Center, New York, New York

LIU, YANG, Dept. of Research and Development, Stemgent, San Diego, California

LORENZO, LAUREANNE, Dept. of Molecular Microbiology and Immunology, University of Maryland, Baltimore

LOWE, SCOTT, Howard Hughes Medical Institute, Cold Spring Harbor Laboratory, Cold Spring Harbor, New York

LU, BAISONG, Institute for Regenerative Medicine, Wake Forest University, Winston-Salem, North Carolina

LU, WANGE, Dept. of Biochemistry and Molecular Biology, University of Southern California, Los Angeles, California

LUKACS, RITA, Dept. of Microbiology, Immunology, and Molecular Genetics, University of California, Los Angeles

LUPO, GIUSEPPE, Dept. of Physiology, Development, and Neuroscience, University of Cambridge, Cambridge, United Kingdom

MACFARLAN, TODD, Lab. of Gene Expression, The Salk Institute for Biological Studies, La Jolla, California

MACK, DAVID, Mammary Biology and Tumorigenesis Laboratory, National Cancer Institute, National Institutes of Health, Bethesda, Maryland

MAEDGE, BRITTA, *Cell*, Cell Press, Cambridge, Massachusetts

MAENG, SUNGHO, Dept. of Regenerative Medicine, Maria Biotech, Seoul, South Korea

MAERKI, DAVID, Dept. of Anatomy, University of Zürich, Zürich, Switzerland

MAHERALI, NIMET, Center for Regenerative Medicine, Cancer Center, Massachusetts General Hospital, Boston, Massachusetts

MAHONY, SHAUN, Dept. of Electrical Engineering and Computer Science, Massachusetts Institute of Technology, Cambridge, Massachusetts

MAJUMDER, SADHAN, Dept. of Cancer Genetics, M.D. Anderson Cancer Center, University of Texas, Houston

MALETIĆ-SAVATIĆ, MIRJANA, Dept. of Neurology, Stony Brook University Medical Center, Stony Brook, New York

MANTESSO, ANDREA, Dept. of Craniofacial Development, Kings College, London, United Kingdom

MAROLT, DARJA, Dept. of Biomedical Engineering, Columbia University, New York, New York

MARSHAK, DANIEL, Corporate Headquarters, Perkin-Elmer, Waltham, Massachusetts

MARTELLO, GRAZIANO, Dept. of Histology, Microbiology, and Medical Biotechnology, University of Padua, Padova, Italy

MARTIENSSEN, ROBERT, Cold Spring Harbor Laboratory, Cold Spring Harbor, New York

MATSAS, REBECCA, Lab. of Cellular and Molecular Neurobiology, Hellenic Pasteur Institute, Athens, Greece

MAZUREK, ANTHONY, Cold Spring Harbor Laboratory, Cold Spring Harbor, New York

MAZZONI, ESTEBAN, Dept. of Pathology, Columbia University, New York, New York

MCDONNELL, KEVIN, Cold Spring Harbor Laboratory, Cold Spring Harbor, New York

MCKAY, RONALD, Lab. of Molecular Biology, National Institute of Neurological Disorders and Stroke, National Institutes of Health, Bethesda, Maryland

MCQUALTER, JONATHAN, Adult Lung Stem Cell Laboratory, Australian Stem Cell Centre, Monash University, Clayton, Australia

MELTON, COLLIN, Dept. of Urology, University of California, San Francisco

MELTON, DOUGLAS, Dept. of Molecular and Cellular Biology, Howard Hughes Medical Institute, Harvard University, Cambridge, Massachusetts

MERRILL, BRAD, Dept. of Biochemistry and Molecular Genetics, University of Illinois, Chicago

MEUWISSEN, RALPH, Institut National de la Santé et de la Recherche Médicale, Institut Albert Bonniot, Grenoble, France

MEYERROSE, TODD, Dept. of Biomedical Research, Whitehead Institute for Biomedical Research, Cambridge, Massachusetts

MEYN, STEPHEN, Program in Genetics and Genome Biology, The Hospital for Sick Children, Toronto, Ontario, Canada

MITTAL, VIVEK, Cold Spring Harbor Laboratory, Cold Spring Harbor, New York

MOLYNEUX, GEMMA, Dept. of Mammary Stem Cells, Institute of Cancer Research, London, United Kingdom

MORONI, MARIA CRISTINA, Dept. of Experimental Oncology, European Institute of Oncology, Milan, Italy

MORRISON, SEAN, Dept. of Cell and Developmental Biology, University of Michigan, Ann Arbor

MURATANI, MASAFUMI, Transcription Laboratory, Cancer Research UK, London, United Kingdom

MURR, RABIH, International Agency for Research on Cancer, Lyon, France

MURTUZA, BARI, Dept. of Medicine, Imperial College, Northwood, Middlesex, United Kingdom

NAKANISHI, ATSUSHI, Frontier Research Laboratories, Takeda Pharmaceutical Company, Tsukuba, Ibaraki, Japan

NAKANO, ATSUSHI, Cardiovascular Research Center, Massachusetts General Hospital, Boston, Massachusetts

NAPPER, JENNIFER, Dept. of Biochemistry and Microbiology, School of Medicine, Marshall University, Huntington, West Virginia

NARKIS, GINAT, National Institute for Child Health and Human Development, National Institutes of Health, Bethesda, Maryland

NAVARRO, PABLO, Dept. of Developmental Biology, Institut Pasteur, Paris, France

NEMAJEROVA, ALICE, Dept. of Pathology, Stony Brook University, Stony Brook, New York

NEWMARK, PHILLIP, Dept. of Cell and Developmental Biology, University of Illinois, Urbana-Champaign

NG, HUCK-HUI, Dept. of Stem Cell Biology, Genome Institute of Singapore, Singapore

NGUYEN, HOANG, Dept. of Molecular and Cellular Biology, Baylor College of Medicine, Houston, Texas

NI, HSIAO-TZU, Dept. of Antibody Application, R&D Systems, Minneapolis, Minnesota

NIKI, MASARU, Dept. of Internal Medicine, University of Iowa, Iowa City

NIMER, STEVEN, Dept. of Hematology, Memorial Sloan-Kettering Cancer Center, New York, New York

NISHIKAWA, SHIN-ICHI, Dept. of Stem Cell Biology, RIKEN Center for Developmental Biology, Kobe, Japan

NOLDE, MONA, Dept. of Cell Biology, Stem Cell Center, Yale University, New Haven, Connecticut

NOWAK, JONATHAN, Lab. of Mammalian Cell Biology and Development, The Rockefeller University, New York, New York

NURY, DAVID, ISTEM, Institut National de la Santé et de la Recherche Médicale, Unité Mixte de Recherche, University of Evry, Evry, France

NUSSE, ROEL, Dept. of Developmental Biology, Howard Hughes Medical Institute, Stanford University, Stanford, California

O'BRIEN, LUCY ERIN, Dept. of Molecular and Cell Biology, University of California, Berkeley

ORKIN, STUART, Howard Hughes Medical Institute, Children's Hospital, Harvard Medical School, Boston, Massachusetts

ORLANDO, VALERIO, Dept. of Epigenetics and Genome Reprogramming, Dulbecco Telethon Institute, Fondazione Telethon, Naples, Italy

OSENBERG, SIVAN, Dept. of Genetics, Tel Aviv University, Tel Aviv, Israel

OUGLAND, RUNE, Dept. of Molecular Biology, Massachusetts General Hospital, Harvard Medical School, Boston, Massachusetts

OVIEDO, NESTOR, Dept. of Developmental Biology, Center for Regenerative and Developmental Biology, Forsyth Institute, Harvard Medical School, Boston, Massachusetts

OWUSU-ANSAH, EDWARD, Dept. of Molecular Cell and Developmental Biology, University of California, Los Angeles

PAIN, BERTRAND, School of Medecine, Institut National de la Santé et de la Recherche Médicale, Unité Mixte de Recherche, Centre National de la Recherche Scientifique, Clermont-Ferrand, France

PALAKODETI, DASARADHI, Dept. of Genetics and Developmental Biology, Health Science Center, University of Connecticut, Farmington

PAN, HAIYAN, Dept. of Microbiology, Medical Center, Columbia University, New York, New York

PAN, YI, Dept. of Immunology, Merck, Rahway, New Jersey

PANKRATZ, MATTHEW, Lab. of Gene Expression, The Salk Institute for Biological Studies, La Jolla, California

PARADA, LUIS, Dept. of Developmental Biology, Southwestern Medical Center, University of Texas, Dallas

PARIENTE, NONIA, *EMBO Reports*, Nature Publishing Group, Heidelberg, Germany

PARK, ANGIE INKYUNG, Dept. of Cancer Biology, Oncomed Pharmaceuticals, Redwood City, California

PARK, JUNE-HEE, Cold Spring Harbor Laboratory, Cold Spring Harbor, New York

PARK, SAEYOUNG, Dept. of Regenerative Medicine, Maria Biotech, Seoul, South Korea

PATEL, PARTHIVE, Dept. of Basic Sciences, Fred Hutchinson Cancer Research Center, Seattle, Washington

PATEL, SHETAL, Abramson Family Cancer Research Institute, University of Pennsylvania, Philadelphia

PELICCI, GIULIANA, Dept. of Experimental Oncology, European Institute of Oncology, Milan, Italy

PERSAUD, TRIKALDARSHI, Dept. of Genome Engineering Technology, Regeneron Pharmaceuticals, Tarrytown, New York

PERVIN, SHEHLA, Dept. of Obstetrics and Gynecology, University of California, Los Angeles

PESTELL, RICHARD, Kimmel Cancer Center, Jefferson University, Philadelphia, Pennsylvania

PING, YUEH-HSIN, Institute of Pharmacology, National Yang-Ming University, Taipei, Taiwan

POLLOCK, MILA, Cold Spring Harbor Laboratory Library, Cold Spring Harbor, New York

PREMSRIRUT, PREM, Cold Spring Harbor Laboratory, Cold Spring Harbor, New York

PUCÉAT, MICHEL, Institut National de la Santé et de la Recherche Médicale, Unité Mixte de Recherche, University of Evry, Evry, France

QIAN, DALONG, Institute for Stem Cell Biology and Regenerative Medicine, Stanford University, Stanford, California

QYANG, YIBING, Medicine, Cardiovascular Research Center, Massachusetts General Hospital, Harvard Medical School, Boston, Massachusetts

RAISER, DAVE, Dept. of Genetics, Stem Cell Program, Children's Hospital, Harvard Medical School, Boston, Massachusetts

RALSTON, AMY, Dept. of Developmental and Stem Cell Biology, Hospital for Sick Children, Toronto, Ontario, Canada

RAMALHO-SANTOS, MIGUEL, Institute for Regeneration Medicine, University of California, San Francisco

RANDO, THOMAS, Dept. of Neurology and Neurological Sciences, School of Medicine, Stanford University, Stanford, California

RANGAN, PRASHANTH, Dept. of Developmental Genetics, Howard Hughes Medical Institute, School of Medicine, New York University, New York

RAO, NANDINI, Dept. of Biology, Indiana University-Purdue University, Indianapolis, Indiana

RAPP, ULF, Institut für Medizinische Strahlenkunde und Zellforschung, University of Würzburg, Würzburg, Germany

RAVIN, REA, National Institute of Neurological Disorders and Stroke, National Institutes of Health, Bethesda, Maryland

REBATCHOUK, DMITRI, Dept. of Biological Sciences, Sanofi-Aventis, Bridgewater, New Jersey

RECHAVI, GIDEON, Cancer Research Center, Sheba Medical Center, Tel Hashomer, Israel

REID, LOLA, Institute for Regenerative Medicine, School of Medicine, Wake Forest University, Winston-Salem, North Carolina

REINER, STEVEN, Dept. of Medicine, Abramson Family Cancer Research Institute, University of Pennsylvania, Philadelphia

REINKE, VALERIE, Dept. of Genetics, School of Medicine, Yale University, New Haven, Connecticut

RENDL, MICHAEL, Dept. of Developmental and Regenerative Biology, Mount Sinai School of Medicine, New York, New York

RICH, JEREMY, Dept. of Medicine, Duke University Medical Center, Durham, North Carolina

RICUPERO, CHRISTOPHER, Dept. of Cell Biology and Neuroscience, Rutgers University, Piscataway, New Jersey

RIO, CARLOS, Dept. of Regenerative Medicine, Fundación Caubet, Centro Internacional de Medicina Respiratoria Avanzada, Bunyola, Spain

ROACH, REBECCA, Dept. of Genetics, Stem Cell Program, Harvard Stem Cell Institute, Children's Hospital, Harvard Medical School, Boston, Massachusetts

RODRIGUES, CLAUDIA, Dept. of Molecular and Cellular Pharmacology, Leonard M. Miller School of Medicine, University of Miami, Miami, Florida

ROHRSCHNEIDER, LARRY, Dept. of Basic Sciences, Fred Hutchinson Cancer Research Center, Seattle, Washington

ROLLINS, FRED, Cold Spring Harbor Laboratory, Cold Spring Harbor, New York

ROSARIO, LUIS, Gulbenkian Institute of Science, Lisbon, Portugal

ROSSMANN, MARLIES, Cold Spring Harbor Laboratory, Cold Spring Harbor, New York

ROTHSTEIN, RODNEY, Dept. of Genetics and Development, Columbia University Medical Center, New York, New York

ROY, RAHUL, Dept. of Chemistry and Chemical Biology, Harvard University, Cambridge, Massachusetts

RUARO, MARIA ELISABETTA, Dept. of Neurobiology, International School for Advanced Studies, Trieste, Italy

RUDNICKI, MICHAEL, Sprott Centre for Stem Cell Research, Ottawa Health Research Institute, Ottawa, Ontario, Canada

RUDOLPH, K. LEONARD, Institute of Molecular Medicine, Ulm University, Ulm, Germany

SALVAGIOTTO, GIORGIA, Dept. of Blood Research, WiCell Reseach Institute, Madison, Wisconsin

SÁNCHEZ ALVARADO, Alejandro, Dept. of Neurobiology and Anatomy, Howard Hughes Medical Institute, University of Utah, Salt Lake City

SARMENTO, OLGA, Dept. of Cell Biology, Health Science Center, University of Virginia, Charlottesville

SAUNDERS, LAURA, Gladstone Institute of Virology and Immunology, University of California, San Francisco

SAUVAGEAU, GUY, Institute for Research in Immunology and Cancer, University of Montréal, Montréal, Québec, Canada

SCHERES, BEN, Dept. of Molecular Cell Biology, Utrecht University, Utrecht, The Netherlands

SCHNAPP, ESTHER, Stem Cell Research Institute, DiBiT, Centro San Raffaele, Milan, Italy

SCIMONE, MARIA LUCILA, Dept. of Biology, Whitehead Institute for Biomedical Research, Massachusetts Institute of Technology, Cambridge, Massachusetts

SEITA, JUN, Institute for Stem Cell Biology and Regenerative Medicine, Stanford University, Stanford, California

SEO, SACHIKO, Dept. of Hematology and Oncology, University of Tokyo, Tokyo, Japan

SESHI, BEERELLI, Dept. of Pathology, Los Angeles Biomedical Research Institute at Harbor-UCLA, Torrance, California

SEVER, RICHARD, Cold Spring Harbor Laboratory Press, Woodbury, New York

SHARMA, MANJU, Dept. of Research, The Jackson Laboratory, Bar Harbor, Maine

SHARMA, PRANAV, Cold Spring Harbor Laboratory, Cold Spring Harbor, New York

SHCHERBATA, HALYNA, Dept. of Biochemistry, University of Washington, Seattle

SHEN, MICHAEL, Depts. of Medicine and Genetics and Development, Columbia University Medical Center, New York, New York

SHERR, CHARLES, Dept. of Tumor Cell Biology, Howard Hughes Medical Institute, St. Jude Children's Research Hospital, Memphis, Tennessee

SHEU, YI-JUN, Cold Spring Harbor Laboratory, Cold Spring Harbor, New York

SHINOHARA, TAKASHI, Dept. of Molecular Genetics, Kyoto University, Kyoto, Japan

SHIRAS, ANJALI, Dept. of Molecular Biology, National Centre for Cell Science, Pune, India

SIDDALL, NICOLE, Dept. of Anatomy and Cell Biology, University of Melbourne, Parkville, Australia

SIEGRIST, SARAH, Dept. of Molecular and Cell Biology, University of California, Berkeley

SIERRA, AMANDA, Dept. of Neurology, Stony Brook University, Stony Brook, New York

SINGH, SANJAY, Dept. of Cancer Genetics, M.D. Anderson Cancer Center, University of Texas, Houston

SINKEVICIUS, KERSTIN, Children's Hospital Stem Cell Program, Children's Hospital, Boston, Massachusetts

SITTAMPALAM, SITTA, Dept. of Therapeutics, Discovery, and Development, Cancer Center, University of Kansas, Kansas City

SLATER, JILL, Dept. of Physiology, Wayne State University, Detroit, Michigan

SMITH, STEPHEN, Dept. of BioScience and BioSurgery, Baxter Healthcare, Round Lake, Illinois

SOLTER, DAVOR, Agency for Science, Technology and Research, Institute of Medical Biology, Immunos, Singapore

SONG, KIWON, Dept. of Biochemistry, Yonsei University, Seoul, South Korea

SONG, YAN, Dept. of Pharmacology, Stony Brook University, Stony Brook, New York

SOUTHALL, TONY, Dept. of Physiology, Development and Neuroscience, Wellcome Trust Cancer Research, U.K. Gurdon Institute, University of Cambridge, Cambridge, United Kingdom

SPARMANN, ANKE, Div. of Molecular Genetics, Netherland Cancer Institute, Amsterdam, The Netherlands

SPECTOR, DAVID, Cold Spring Harbor Laboratory, Cold Spring Harbor, New York

SPEES, JEFF, Dept. of Medicine, University of Vermont, Colchester

SPRADLING, ALLAN, Dept. of Embryology, Carnegie Instition of Washington, Baltimore, Maryland

STEFANOVIC, SONIA, Institut National de la Santé et de la Recherche Médicale, Unité Mixte de Recherche, University of Evry, Evry, France

STEWART, DAVID, Meetings and Courses Program, Cold Spring Harbor Laboratory, Cold Spring Harbor, New York

STILLMAN, BRUCE, President, Cold Spring Harbor Laboratory, Cold Spring Harbor, New York

STRIPP, BARRY, Dept. of Pulmonary, Allergy, and Critical Care, Duke University Medical Center, Durham, North Carolina

STUDER, LORENZ, Dept. of Developmental Biology, Memorial Sloan-Kettering Cancer Center, New York, New York

SUBRAMANIAN, VASANTA, Dept. of Biology and Biochemistry, University of Bath, Bath, United Kingdom

SUETTERLIN, PHILIPP, Wolfson Centre for Age Related Diseases, King's College, London, United Kingdom

SULLIVAN, GARETH, Center of Regenerative Medicine, Edinburgh University, Edinburgh, Scotland, United Kingdom

SUN, XUYANG ALFRED, Dept. of Pathology, Stanford University, Stanford, California

SURANI, AZIM, Wellcome Trust Cancer Research, U.K. Gurdon Institute, University of Cambridge, Cambridge, United Kingdom

SURZENKO, NATALIA, Dept. of Genetics, University of North Carolina, Chapel Hill

SUTHERLAND, KATE, Div. of Molecular Genetics, Netherlands Cancer Institute, Amsterdam, The Netherlands

SWEET, DEBORAH, Developmental Cell, Cell Press, Cambridge, Massachusetts

SWINDLE, SCOTT, Dept. of Microbiology, University of Alabama, Birmingham

SZEKELY, ANNA, Dept. of Genetics and Neurology, School of Medicine, Yale University, New Haven, Connecticut

TAKACOVA, SYLVIA, Dept. of Biology, Faculty of Medicine, Palacky University, Olomouc, Czech Republic

TALIB, SOHEL, California Institute for Regenerative Medicine, San Francisco, California

TALOS, FLAMINIA, Dept. of Pathology, Stony Brook University, Stony Brook, New York

TAM, PATRICK, Embryology Unit, Children's Medical Research Institute, Westmead, Australia

TAN, KAH YONG, Dept. of Stem Cells and Development, Joslin Diabetes Center, Boston, Massachusetts

TANAKA, ELLY, Max-Planck-Institute of Molecular Cell Biology and Genetics, Dresden, Germany

TAUCHI, TETSUZO, Dept. of Internal Medicine, Tokyo Medical University, Tokyo, Japan

TAVAZOIE, MASOUD, Dept. of Pathology, Columbia University, New York, New York

TEE, WEE-WEI, Gurdon Institute, University of Cambridge, Cambridge, United Kingdom

TEISANU, ROXANA, Dept. of Pulmonary, Allergy, and Critical Care, Duke University Medical Center, Durham, North Carolina

TERSKIKH, ALEXEY, Dept. of Neuroscience, Burnham Institute for Medical Research, La Jolla, California

TESAR, PAUL, Dept. of Molecular Biology, National Institute of Neurological Disorders and Stroke, National Institutes of Health, Bethesda, Maryland

THAL, MELISSA, Dept. of Microbiology, University of Alabama, Birmingham

THEUNISSEN, JAN-WILLEM, Nature Biotechnology, Nature Publishing Group, New York, New York

THOMPSON, ALYSON, Wellcome Trust Cancer Research, U.K. Gurdon Institute, University of Cambridge, Cambridge, United Kingdom

THORPE, PETER, Dept. of Genetics and Development, Columbia University Medical Center, New York, New York

TIMMERMANS, MARJA, Cold Spring Harbor Laboratory, Cold Spring Harbor, New York

TING, STEPHEN, Dept. of Molecular Genetics of Stem Cells, Institute for Research in Immunology and Cancer, Montréal, Québec, Canada

TJIAN, ROBERT, Dept. of Molecular and Cell Biology, University of California, Berkeley

TROTTIER, MAGAN, Dept. of Genetics and Genome Biology, Hospital for Sick Children, University of Toronto, Toronto, Ontario, Canada

TZENG, RUEI-YING, Cold Spring Harbor Laboratory, Cold Spring Harbor, New York

UNGEWITTER, ERICA, Dept. of Molecular Cell and Developmental Biology, University of Virginia, Charlottesville

VALDIMARSDOTTIR, GUDRUN, Dept. of Biochemistry and Molecular Biology, University of Iceland, Reykjavik, Iceland

VAN ZOELEN, E. JOOP, Dept. of Cell Biology, Radboud University of Nymegen, Nymegen, The Netherlands

VEIT, BRUCE, Dept. of Forage Biotechnology, AgResearch, Palmerston North, New Zealand

VENERE, MONICA, Dept. of Urology, Institute for Regeneration Medicine, University of California, San Francisco

VERDIN, ERIC, Gladstone Institute of Virology and Immunology, University of California, San Francisco

VINSON, CHARLES, Lab. of Metabolism, National Institutes of Health, Bethesda, Maryland

VISVADER, JANE, Royal Melbourne Hospital, The Walter and Eliza Hall Institute of Medical Research, Parkville, Victoria, Australia

VOGT, THOMAS, Dept. of Genetically Engineered Models, Merck Research Laboratories, Rahway, New Jersey

VOLANAKIS, EMMANUEL, Dept. of Oncology, St. Jude Children's Research Hospital, Memphis, Tennessee

VOOG, JUSTIN, Lab. of Genetics, The Salk Institute for Biological Studies, La Jolla, California

VULLIET, RICHARD, Dept. of Veterinary Molecular Biosciences, University of California, Davis

WADA, NAOKO, Pharmacology and Pathology Research Center, Chugai Research Institute for Medical Science, Gotemba, Japan

WAGERS, AMY, Dept. of Developmental and Stem Cell Biology, Joslin Diabetes Center, Boston, Massachusetts

WANG, JIALIANG, Dept. of Surgery, Duke University, Durham, North Carolina

WANG, XIUPING, Brigham and Women's Hospital, Harvard Medical School, Boston, Massachusetts

WANG, ZHONG, Cardiovascular Research Center, Massachusetts General Hospital, Harvard Medical School, Boston, Massachusetts

WATT, FIONA, Li Ka Shing Centre, Cancer Research UK, Cambridge Research Institute, Cambridge, United Kingdom

WEBB, CAROL, Dept. of Immunobiology and Cancer, Oklahoma Medical Research Foundation, Oklahoma City, Oklahoma

WEISSMAN, IRVING, Dept. of Pathology, School of Medicine, Stanford University, Stanford, California

WELNER, ROBERT, Dept. of Immunobiology and Cancer, Oklahoma Medical Research Foundation, Oklahoma City, Oklahoma

WENEMOSER, DANIELLE, Whitehead Institute for Biomedical Research, Cambridge, Massachusetts

WICHA, MAX, Comprehensive Cancer Center, University of Michigan, Ann Arbor

WICHTERLE, HYNEK, Dept. of Pathology, Columbia University, New York, New York

WIENHOLDS, ERNO, Dept. of Cellular and Molecular Biology, University Health Network, Toronto, Ontario, Canada

WILLIAMS, RICHARD, Dept. of Oncology, St. Jude Children's Research Hospital, Memphis, Tennessee

WILSCHUT, KARLIJN, Dept. of Farm Animal Health, Faculty of Veterinary Medicine, Utrecht University, Utrecht, The Netherlands

WITKOWSKI, JAN, Banbury Center, Cold Spring Harbor Laboratory, Cold Spring Harbor, New York

WITTE, OWEN, Dept. of Microbiology, Immunology, and Molecular Genetics, Howard Hughes Medical Institute, University of California, Los Angeles

WONG, CHRISTINE, Dept. of Obstetrics and Gynecology, Samuel Lunenfeld Research Insitute, Toronto, Ontario, Canada

WOOD, STEPHEN, National Centre for Adult Stem Cell Research, Griffith University, Nathan, Australia

WU, JIANG, Dept. of Pathology, Howard Hughes Medical Institute, Stanford University, Stanford, California

WU, PAO-SHU, Wellcome Trust Cancer Research, U.K. Gurdon Institute, University of Cambridge, Cambridge, United Kingdom

XI, RONGWEN, National Institute of Biological Sciences, Beijing, China

XIE, TING, Stowers Institute for Medical Research, Kansas City, Missouri

XU, JIAN, Molecular Biology Institute, University of California, Los Angeles

XU, NA, Neuroscience Research Institute, University of California, Santa Barbara

XUAN, ZHENYU, Cold Spring Harbor Laboratory, Cold Spring Harbor, New York

YAMAGUCHI, TERRY, Lab. of Cancer and Developmental Biology, National Cancer Institute, Frederick, Maryland

YANG, PETER, Regeneron Pharmaceuticals, Tarrytown, New York

YANG, ZIPING, Dept. of BioScience, Baxter Healthcare, Round Lake, Illinois

YOSHIDA, SHOSEI, Dept. of Pathology and Tumor Biology, Graduate School of Medicine, Kyoto University, Kyoto, Japan

YOUNG, RICHARD, Dept. of Biology, Whitehead Institute for Biomedical Research, Massachusetts Institute of Technology, Cambridge, Massachusetts

YU, YUNG-LUEN, Graduate Institute of Cancer Biology, China Medical University, Taichung, Taiwan

YUAN, SHAUNA, Dept. of Neuroscience, University of California at San Diego, La Jolla

ZABALA UGALDE, MAIDER, Institute for Stem Cell Biology and Regenerative Medicine, Stanford University, Stanford, California

ZACHAREK, SIMA, Dept. of Hematology and Oncology, Stem Cell Program, Children's Hospital, Harvard Medical School, Boston, Massachusetts

ZARET, KENNETH, Dept. of Cell and Developmental Biology, Fox Chase Cancer Center, Philadelphia, Pennsylvania

ZATZ, MARION, Div. of Genetics and Developmental Biology, National Instute of General Medical Sciences, National Institutes of Health, Bethesda, Maryland

ZENTAR, MARC, Wolfson Centre for Age Related Diseases, King's College London, London, United Kingdom

ZHANG, GUANGJUN, Koch Institute, Massachusetts Institute of Technology, Cambridge, Massachusetts

ZHANG, HAIYING, Lab. of Mammalian Cell Biology and Development, Howard Hughes Medical Institute, The Rockefeller University, New York, New York

ZHANG, SUI, Dept. of Cardiology, M.D. Anderson Cancer Center, University of Texas, Houston

ZHANG, YANPING, Dept. of Radiation Oncology, University of North Carolina, Chapel Hill

ZHOU, YAN, Dept. of Molecular, Cellular, and Developmental Biology, Yale University, New Haven, Connecticut

ZON, LEONARD, Dept. of Hematology, Howard Hughes Medical Institute, Children's Hospital, Boston, Harvard Medical School, Massachusetts

ZUNINO, FEDERICA, Dept. of Experimental Oncology, European Institute of Oncology, Milan, Italy

ZWAKA, THOMAS, Dept. of Molecular and Cellular Biology, Baylor College of Medicine, Houston, Texas

Row 1: R. McKay, A. Androutsellis-Theotokus; H. Lin; M. Wicha; B. Stillman, M. Timmermans
Row 2: Wine and Cheese party; A. Spradling, A. Sánchez Alvarado
Row 3: T. Grodzicker, C. Sherr; C. Kim, B. Stripp; R. Tjian
Row 4: E. Fuchs; M. Tavazoie; A Spradling, J. Greve; J. Adams
Row 5: Poster Session; Get-together on the beach

Row 1: Participant happy with her caricature; D. Hoeppner, G. Enkilopov; L. Bar, L. Rohrschneicer
Row 2: T. Grodzicker, R. McKay; J. Silveira Botelho, D. Stewart; J. Seita, S. Nishikawa
Row 3: Picnic
Row 4: A. Surani, I. Weissman; R. Jaenisch, K. Zaret; B. Hogan, D. Solter

Row 1: K. Janssen, V. Mittal; C. Li, G. Sullivan; K. Zaret, R. Young
Row 2: Symposium Interview—J. Witkowski, B. Hogan; D. Balasubramanian
Row 3: T. Rando, M. Childers; J. Inglis, J. Alfred; S. Aigner, A. Denli
Row 4: T. Vogt, M. Shen; S. Ting, K. Hope
Row 5: Poster Session; I. Ibarra; S. Morrison, K. Helin

Row 1: B. Hogan, E. Fuchs; L.E. O'Brien; A. Economides
Row 2: V. Dubowitz; N. Harter, R. Rothstein; H. Johanssen
Row 3: M. Niki; A. Nakano, W. Lu; B. Seeshi; P. Carroll
Row 4: Walking Tour

Row 1: M. Baron, A. Bank; R. Vulliet; B. Hogan, J. Alfred
Row 2: L. Zon; N. Siddall, V. Coronas; M. Berta, M. Cals-Grierson
Row 3: C. Wong, M. Trottier; Symposium Interview—C. Cormier, L. Parada
Row 4: J. Williams (*right*) with Professor for a Day high school students

Row 1: R. Nusse, T. Zwaka, C. Scherr, M. Fuller; J. Witkowski, E. Tanaka
Row 2: U. Grossniklaus; R. Nusse, T. Yamaguchi; F. Watt; M. Buckingham
Row 3: R. Doerge, R. Martienssen; R. Jaenisch, K. Hochedlinger

Foreword

The roots of stem cell research can be traced back to classical work in embryology and regeneration performed in the 19th century. By the early 1900s, researchers in Europe had come to understand that various types of blood cells were derived from a particular "stem cell," resulting in physicians' attempts to administer bone marrow by mouth to patients suffering from anemia and leukemia. Although such therapy was unsuccessful, laboratory experiments eventually demonstrated that mice with defective marrow could be restored to health with infusions into the bloodstream of marrow taken from other mice, stimulating physicians to speculate whether it was feasible to transplant bone marrow from one human to another. By the late 1960s, pioneering work by Till and McCulloch demonstrated the existence of hematopoietic stem cells. Later, bone marrow transplantation (a stem cell transplant) was dogged by problems of histocompatability, but as the basis of the HLA system became increasingly understood, successful bone marrow transplantation between unrelated individuals was first demonstrated in 1973.

Stem cell research has since exploded, with the derivation of mouse embryonic stem cells in 1981, the culturing of neural stem cells as neurospheres in 1992, and the establishment of the first human embryonic stem cell lines in 1998. A definitive link between stem cells and cancer was established when certain leukemias were shown to originate from hematopoietic stem cells. During the last decade, the concept of adult stem cell plasticity has gained credence, although many findings have proved controversial, and the ethical arguments for and against the use of human embryonic stem cells have been widely debated on national and international levels. Within the last 2 years, it has been reproducibly established that an embryonic stem cell–like state, previously achieved only by somatic cell nuclear transfer into enucleated oocytes ("cloning") or by fusion with embryonic stem cells, can be induced by reprogramming differentiated adult cells using a simple combination of key transcription factors.

Progress in stem cell research is now extremely intense, with more than 5000 research papers on embryonic and adult stem cells published in reputable scientific journals every year. It therefore seemed appropriate to focus the 73rd Symposium on this important and rapidly developing field, providing a unique synthesis of the exciting progress being made in the field of stem cell biology, not only for the Symposia attendees, but for a wider global audience via interviews freely available on the world wide web, and, we anticipate, for readers of these Proceedings.

In organizing this Symposium, we relied on the assistance of Elaine Fuchs, Rusty Gage, Greg Hannon, Ron McKay, Davor Solter, and Allan Spradling for suggestions for speakers. Opening night speakers included Rudolph Jaenisch, Minx Fuller, Arturo Alvarez-Buylla, and Max Wicha, and Irving Weissman presented the Reginald Harris Lecture on normal and neoplastic stem cells. Elaine Fuchs enlightened a mixed audience of scientists, lay friends, and neighbors with her Dorcas Cummings Lecture on skin stem cells, and Brigid Hogan ended the meeting with a masterful and eloquent Summary.

This Symposium was attended by 465 scientists from more than 30 countries, and the program included 68 oral presentations and 177 poster presentations. We particularly thank the Lisbon-based Champalimaud Foundation for their extensive and generous support of this year's Symposium. Additional funds to run this meeting were obtained from the National Institutes of Health. Financial support of our meetings program from corporate patrons, benefactors, sponsors, associates, affiliates, contributors, and foundations is essential for these Symposia to remain a success and we are most grateful for their continued support.

We wish to thank Val Pakaluk and Mary Smith in the Meetings and Courses Program office for their efficient help in organizing the Symposium. Joan Ebert and Rena Steuer in the Cold Spring Harbor Laboratory Press, headed by John Inglis, ensured that this volume would be produced. We thank them for their dedication to producing high-quality publications.

Bruce Stillman
David Stewart
Terri Grodzicker
March 2009

Sponsors

The 73rd Symposium was generously supported by the **Champalimaud Foundation**, a Lisbon-based international organization with an ongoing interest in basic and clinical aspects of stem cells.

This meeting was funded in part by the **National Cancer Institute,** a branch of the **National Institutes of Health.**

Contributions from the following companies provide core support for the Cold Spring Harbor meetings program.

Corporate Patron

Pfizer, Inc.

Corporate Benefactors

Amgen, Inc.
GlaxoSmithKline

Merck Research Laboratories
Novartis Institutes for BioMedical Research

Corporate Sponsors

Agilent Technologies
Applied Biosystems
AstraZeneca
BioVentures, Inc.
Bristol-Myers Squibb Company
Genentech, Inc.
Hoffmann-La Roche, Inc.

Invitrogen Corporation
IRX Therapeutics, Inc.
Kyowa Hakko Kogyo Co., Ltd.
New England BioLabs, Inc.
OSI Pharmaceuticals, Inc.
Sanofi-Aventis
Schering-Plough Research Institute

Plant Corporate Associates

Monsanto Company
Pioneer Hi-Bred International, Inc.

Corporate Affiliates

Affymetrix

Corporate Contributors

Epicentre Biotechnologies
Illumina

Foundations

Hudson-Alpha Institute for Biotechnology

Contents

Germ Cells and Totipotency

Niches and Asymmetry

Embryonic Development and Multipotent Progenitors

Reprogramming Somatic Cells

Gene Expression and Transcriptional Networks

Epigenetics

Adult Stem Cells

Neural Stem Cells and Brain Tumors

Regulating Gene Expression in the *Drosophila* Germ Line

P. RANGAN, M. DeGENNARO, AND R. LEHMANN

HHMI and Kimmel Center for Biology and Medicine of the Skirball Institute,
Department of Cell Biology, New York University School of Medicine, New York, New York 10016

Germ cells are the ultimate stem cells because they have the potential to give rise to a new organism. Specified during early embryogenesis in most species, germ cells evade somatic differentiation by using mechanisms such as transcriptional silencing and translational control (Seydoux and Braun 2006; Cinalli et al. 2008). To identify germ-line targets of translational regulation and to understand their mechanism of regulation, we used publicly available databases to identify RNAs localized to germ plasm. Using a transgenic reporter assay, we find that these germ-line RNAs are both spatially and temporally regulated during both oogenesis and embryogenesis by their 3′-untranslated regions (3′UTRs) (Rangan et al. 2008). We find that many RNAs that are spatially and temporally regulated in the early embryo are also translationally regulated during oogenesis. However, RNAs that are similarly regulated during oogenesis are no longer coregulated during embryogenesis, demonstrating that *cis*-acting sequences within a single RNA are used differentially during the life cycle of the germ line. Our study emphasizes a multifaceted role of translational regulation in germ cells. Many aspects of cellular behavior are shared between germ cells and other stem cells; thus, analysis of the translational regulatory networks controlling translation during the germ-line life cycle may reveal important general features of RNA regulation in stem cells.

Germ cells allow for the continuity of life in all sexually reproducing organisms by providing a link among generations and creating genetic diversity. In most organisms studied, germ cells are first to form and are set aside from somatic cells during early embryogenesis (Hayashi et al. 2007; Strome and Lehmann 2007; Cinalli et al. 2008). There are two main modes of specifying germ cells. First, as in *Drosophila*, germ cells arise from a maternally deposited cytoplasm called germ plasm that contains specialized RNA–protein particles, referred to as germ granules, P granules, or polar granules (Seydoux and Braun 2006). Second, in particular in mammals, germ cells are formed at the interface between the embryo and the extraembryonic tissue as the result of an inductive event requiring bone morphogenetic proteins 4 and 8 (BMP4/8) signaling (Hayashi et al. 2007). In mammals, primordial germ cells have the potential to differentiate into all germ layers and resemble pluripotent embryonic stem cells (Surani et al. 2007). Thus, understanding the mechanisms that primordial germ cells use to avoid somatic differentiation may provide more general insight into the control of pluripotency.

Although germ cells are specified by distinct mechanisms in different organisms, they share several common characteristics. One common characteristic is that in contrast to other tissues that develop in the early embryo, no conserved master transcriptional regulator has yet been identified for germ cells (Cinalli et al. 2008). Instead, one of the hallmarks of early germ cells is the fact that they are transcriptionally silent (Hayashi et al. 2007; Strome and Lehmann 2007; Cinalli et al. 2008; Nakamura and Seydoux 2008). This is achieved by direct repression of the function of RNA polymerase II (RNA Pol II), as was shown in *Caenorhabditis elegans* and in *Drosophila*, where the PIE-1 and Pgc proteins, respectively, interfere with phosphorylation and activation of the carboxy-terminal tail of RNA Pol II (Seydoux et al. 1996; Martinho et al.

2004; Hanyu-Nakamura et al. 2008; Nakamura and Seydoux 2008). In mouse, transcriptional repression in germ cells is chromatin mediated and requires Blimp1, a SET/PR domain and zinc-finger-containing protein (Ohinata et al. 2005; Vincent et al. 2005; Ancelin et al. 2006). Irrespective of the specific mechanism used, the outcome of this repression is that somatic differentiation programs are repressed in germ cells, allowing germ cells to develop. Mutations in any of these transcriptional regulators cause loss of germ cell specification or germ cell death (Nakamura et al. 1996; Seydoux et al. 1996; Martinho et al. 2004). Transcriptional repression is, however, not sufficient for germ cell specification because ectopic expression of these repressors independent of other factors that promote germ cell development does not lead to ectopic germ cell formation (Cinalli et al. 2008).

Another common characteristic of germ cells is the role that RNA regulators have in germ cell biology. RNA regulators are present throughout germ-line development and have important roles in both germ-line specification and maintenance (Seydoux and Braun 2006; Cinalli et al. 2008). In fact, two of these conserved RNA regulators, the RNA helicase Vasa and the translational repressor Nanos, are universal germ cell markers (Seydoux and Braun 2006). Functional analysis of mutations in RNA regulators suggests that they fulfill essential roles during germ-line development by preventing early differentiation of germ cells into gametes or preventing somatic dedifferentiation, such as in the case of *nanos* (Forbes and Lehmann 1998; Subramaniam and Seydoux 1999; Gilboa and Lehmann 2004; Hayashi et al. 2004; Wang and Lin 2004; Sato et al. 2007). These RNA regulators are thought to work by specifying the timing and spatial parameters of expression of target genes. Although several of the RNA regulators have been characterized and are known to affect all aspects of germ-line biology, little is known about the specific RNA targets and their regulation.

In this study, we have identified targets of germ-line RNA regulators by compiling a list of RNAs localized to the germ plasm in the early embryo. We demonstrate that 3'UTRs of these germ-line RNAs are sufficient to localize RNAs to germ plasm by making reporter constructs. We investigated the translational state of these reporter constructs in the embryo and during oogenesis. Although all of the RNAs involved in this study show similar localization to the germ plasm, they are translated at different times in the germ line, showing that 3'UTRs encode not only spatial information but also temporal information for translation. Moreover, when the expression pattern of RNAs in the embryo was compared to the pattern during oogenesis, we observed that RNAs that were similarly regulated in embryos are not necessarily regulated in similar pattern during oogenesis, suggesting that the information encoded by 3'UTRs is dynamic and differentially used during the germ-line life cycle.

RESULTS

Identification of Germ-line RNA

To identify translational targets of conserved RNA regulators, we focused on RNAs present in the germ line throughout its life cycle. We used published reports to assemble a list of germ-cell-specific RNAs (Table 1) (Lecuyer et al. 2007; Tomancak et al. 2007). We focused on RNAs with an expression pattern similar to that previously described for *nanos*, *germ-cell-less* (*gcl*), and *polar granule component* (*pgc*) (Wang and Lehmann 1991; Jongens et al. 1992; Nakamura et al. 1996). These RNAs are synthesized during oogenesis and become localized to the germ plasm at the posterior pole of the oocyte at the end of oogenesis. Within the germ plasm, these RNAs are found enriched in polar granules which are RNA–protein particles specific for germ cells (Nakamura et al. 1996; Amikura et al. 2001). At egg deposition, the germ plasm containing these RNAs forms a crescent at the posterior pole of the embryo (stages 1–2) before it is incorporated into the developing germ cells (stages 3–4).

We created a reporter cassette to systematically test 3'UTRs of selected RNAs for their ability to recapitulate the endogenous RNA localization and translation pattern. The 3'UTRs of selected localized RNAs were added to this reporter cassette in which the maternally active *nanos* promoter and its 5'UTR was fused to green fluorescent protein (GFP), which was flanked by HA sequences at both the carboxyl and amino termini and transgenic lines (*pnos::HA-GFP-HA-geneX3'UTR*) were generated (Table 1; Fig. 1A). We found that the 3'UTRs of the selected RNAs were sufficient to localize the reporter RNA to the germ plasm (Rangan et al. 2008). We tested the translational state of these germ plasm RNAs and found that unlocalized RNA is not translated and RNA localized to the posterior pole is specifically translated (Rangan et al. 2008). We identified several patterns of translational regulation in the embryo (Fig. 2A). As described previously, *nanos* RNA is translated upon localization. We also observed a similar pattern for *orb* (class I; Fig. 2A); *gcl* RNA is translated as germ cell nuclei reach the germ plasm and germ cells begin to bud (class II; Fig. 2A); five RNAs (*pgc, sra, CG5292, CG18446*, and *rapgap1*) are translated when germ cells are formed (class III; Fig. 2A); three RNAs (*bruno, CG2774*, and *cyclinB*) are not translated in embryonic germ cells; only *cyclinB* is translated in germ cells at the end of embryogenesis (classes IV and V; Fig. 2A) (Rangan et al. 2008). These results suggest that localized RNAs are, in general, translationally regulated and that RNA localization does not necessarily act as a trigger for translational activation.

MATERNALLY SUPPLIED GERM PLASM RNAS ARE DIFFERENTIALLY REGULATED DURING OOGENESIS

To determine whether RNAs localized to the embryonic germ plasm are also translationally regulated during oogenesis and whether this regulation is mediated by their respective 3'UTRs, we assayed the transgenic lines (*pnos::HA-GFP-HA-geneX3'UTR*) described above for expression during oogenesis. For each construct, we analyzed the pattern of GFP or HA-tag expression during all stages of oogenesis. *Drosophila* ovary development is divided into 14 stages. *Drosophila egg* chambers of different stages mature in one of 16 ovarioles in each ovary (see Fig. 1B). At the tip of the ovariole, germ-line stem cells regenerate and produce a differentiating daughter cell called the cystoblast. The cystoblast undergoes four synchronous incomplete cell divisions, giving rise to a 16-cell interconnected cyst. Only one of the 16 cells of a cyst becomes the oocyte, whereas the others become nurse cells that are connected to the oocyte by ring canals that allow the passage of proteins and RNAs from the nurse cells into the oocyte. Somatic follicle cells surround the 16-cell cyst. During oogenesis, nurse cells and follicle cells endoreplicate and the oocyte grows in volume. At the end of oogenesis, the nurse cells dump their contents into the oocyte and degenerate.

We observed three principle patterns of RNA regulation during oogenesis (Fig. 2B; Table 2). RNAs in the first

Table 1. RNAs localized to germ plasm

RNA	Function	Germ plasm[a]	3'UTR sufficiency
nos	translational control	+	+
bruno	translational control	+	+
pgc	transcriptional silencing	+	+
gcl	germ cell formation	+	+
CG5292	RNA binding[b]	+	+
sra	Ca²⁺ signaling	+	+
CG18446	zinc ion binding[b]	+	+
CG2774	endocytosis[b]	+	+
cyclin B	cell cycle	+	+
orb	translational control	+	+
rapgap1	GTPase	+	n.d.

Table lists the RNAs used in this study (column 1), their established or predicted ([b]) function (column 2), localization to germ plasm (column 3), and sufficiency of 3'UTR to localize reporter constructs to germ plasm (column 4).
n.d., not determined.
[a]Localization data from the Berkeley *Drosophila* Genome Project (Lecuyer et al. 2007; Tomancak et al. 2007).
[b]Predicted function.

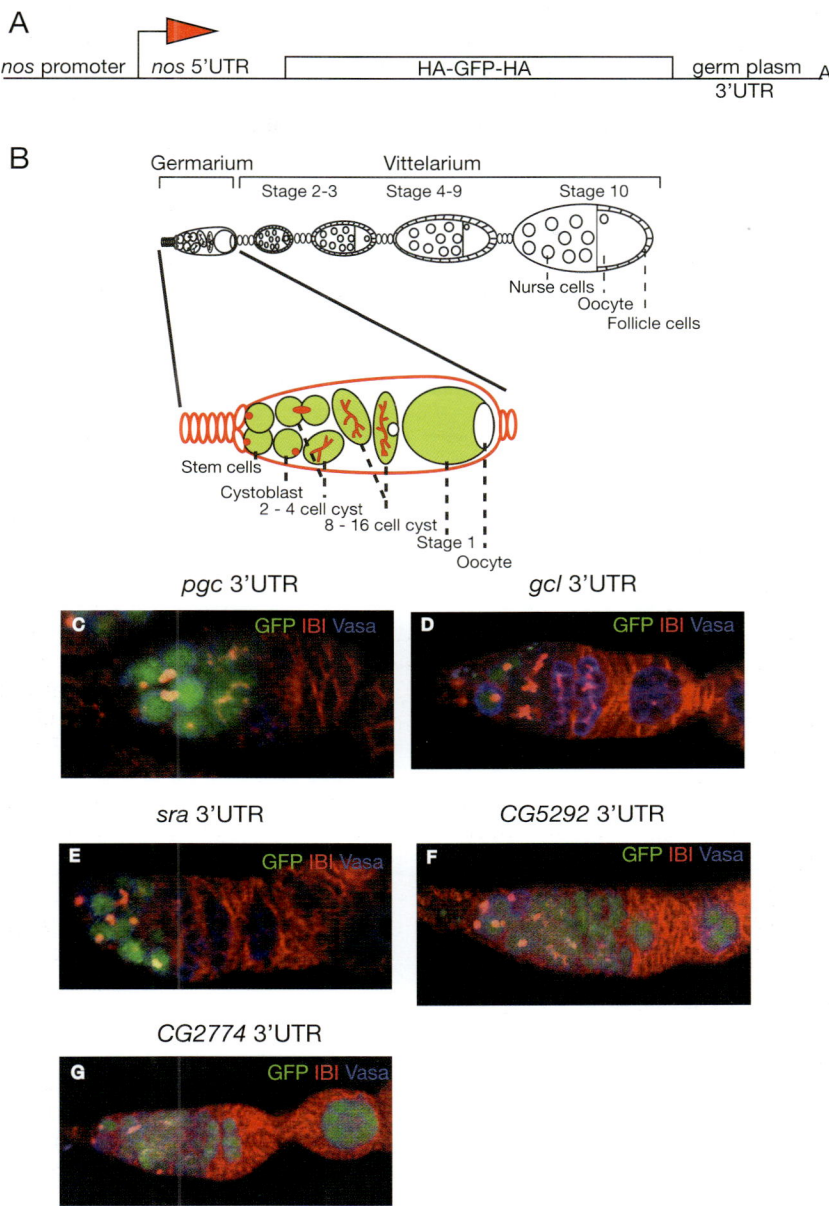

Figure 1. Germ plasm RNAs are regulated in the germarium by their 3′UTRs. (*A*) Reporter construct that was used to test various germ plasm 3′UTRs used in this study. (*B*) Summary of *Drosophila* oogenesis. Each ovary is comprised of 16–18 ovarioles (*top*). Each ovariole contains 2–3 stem cells in the germarium (*below*). (*C–G*) Expression patterns of various transgenes under control of indicated 3′UTR monitored by GFP expression (*green*). Anti-Vasa (*blue*) marks the germ line. IB1 (*red*) marks spectrosome, fusome, and the somatic cell membranes. Reporter constructs fused to *pgc*, *gcl*, and *sra* 3′UTRs show expression in anteriormost part of the germarium and are then silenced. Germaria are oriented with the niche containing the stem cells to the left.

class were expressed strongly in the early stages of oogenesis in the germarium. These included *pgc, CG18446, gcl,* and *sra* (Table 2; Fig. 1C–E). *pgc* and *CG18446* expression was only observed in the germarium and we detected very little expression during later stages, whereas *gcl* and *sra* expression was detected throughout oogenesis (Table 2; Fig. 3A–C). The second pattern displayed by *CG5292* and *CG2774* showed more uniform expression throughout all stages of oogenesis (Table 2; Figs. 1F,G and 3D,E). The third pattern was exemplified

by *bruno* and *orb* 3′UTRs and showed low expression in early germarial stages and up-regulation during later stages of oogenesis (Table 2). Indeed this pattern very closely resembled the pattern of expression of the endogenous Bruno and Orb proteins (Table 2) (Lantz et al. 1994; Webster et al. 1997). These results show that 3′UTRs also carry information about translational control during oogenesis. Most interestingly, RNAs sharing an expression pattern during oogenesis are not necessarily coregulated during embryogenesis.

A. Embryo

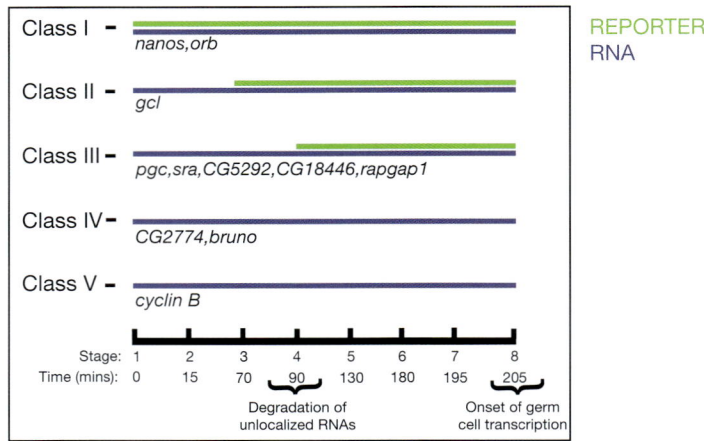

B. Oogenesis

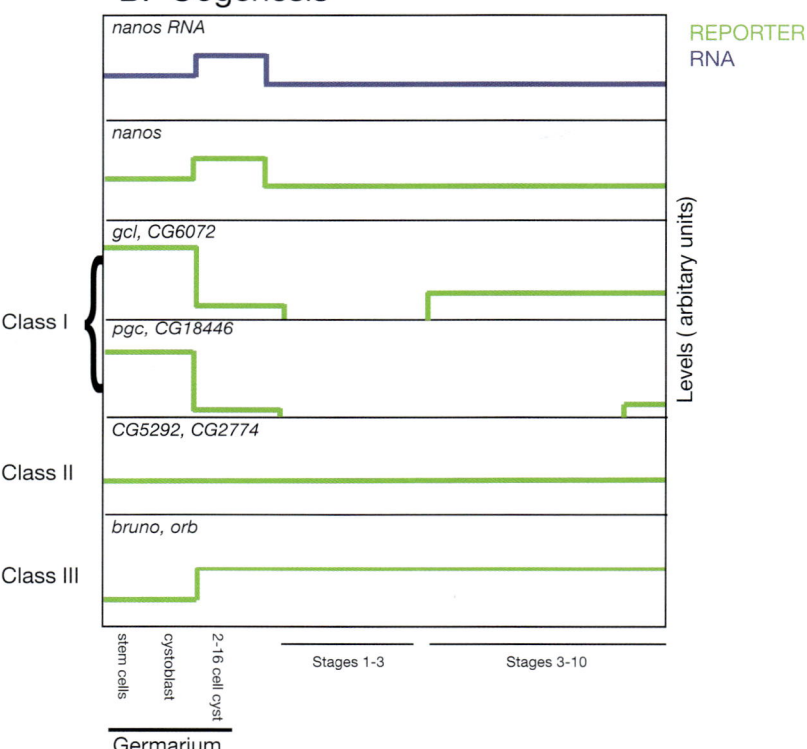

Figure 2. RNAs localized to germ plasm are spatially and temporally regulated during germ-line development. (*A*) Germ plasm RNAs are classified into five classes according to the timing of onset of translation of the reporter or endogenous protein. The RNAs that belong to each of these classes are shown below the corresponding classes. (*Green line*) Expression of the reporter construct under control of the respective 3′UTRs; (*blue line*) endogenous localized RNA; (*black line*) time line of embryonic development after egg deposition. Class IV showed no expression of reporter construct in the germ line. Class V was tested for expression of endogenous protein and showed no protein expression in the germ line, but protein was expressed in the soma. (*B*) (*Blue*) Expression profile of the *nanos* promoter used for all of the reporter constructs; (*green line*) expression of the reporter construct under control of the respective 3′UTRs as indicated. The reporters carrying germ plasm RNA 3′UTRs have at least three classes of expression patterns, showing that despite carrying the same promoter and 5′UTR, the 3′UTRs dictate translation of these RNAs. The time line of oogenesis is indicated at the bottom of the panel.

Table 2. Regulation of germ plasm RNA during oogenesis

Promoter	Protein 3′UTR	Stem cells	Cysto-blast	4-cell cyst	16-cell cyst	Stages 2a–3	Stages 3–10	Stage 10	Reference
nanos	Gal4-tublin 3′UTR	+/–	+/–	+	+	+	+	+	Swantek and Gergen (2004); this study
nanos	NosGFP-*nos* 3′UTR	+	+	++	++	+	+	+	Forrest and Gavis (2003)
nanos	moeRFP-*nos* 3′UTR	+	+	++	++	+	+	+	this study
nanos	NANOS	+	+/–	++	++	+	+	+	Wang and Lehmann (1994)
nanos	GFP-*pgc* 3′UTR	++	++	+	+/–	–	–	+/–	this study
nanos	GFP-CG18446 3′UTR	++	++	+	+/–	–	–	+/–	this study
nanos	GFP-*gcl* 3′UTR	++	++	+/-	–	–	+	+	this study
nanos	GFP-*sra* 3′UTR	++	++	+/–	–	–	+	+	this study
hsp 83	lacZ-*orb* 3′UTR	–	–	–	+	+	+	+	Lantz and Schedl (1994)
Orb	ORB	–	–	–	+	+	+	+	Lantz et al. (1994)
nanos	GFP-*bru* 3′UTR	+	+	+	+	+	+	+	this study
Bruno	BRUNO	–	–	–	+	+	+	+	Webster et al. (1997)
nanos	GFP-CG5292 3′UTR	+	+	+	+	+	+	+	this study
nanos	GFP-CG2774 3′UTR	+	+	+	+	+	+	+	this study

Table lists the promoters for the reporter constructs used in this study (column 1), reporters used in this study fused to the 3′UTRs as indicated (column 2), and expression in different stages of oogenesis as indicated (column 3–7). (*Dark gray*) Regions of highest reporter expression (++); (*light gray*) regions of intermediate reporter expression (+/–); (*white*) regions of no reporter expression (–). Expression of endogenous proteins is shown below the reporter constructs.

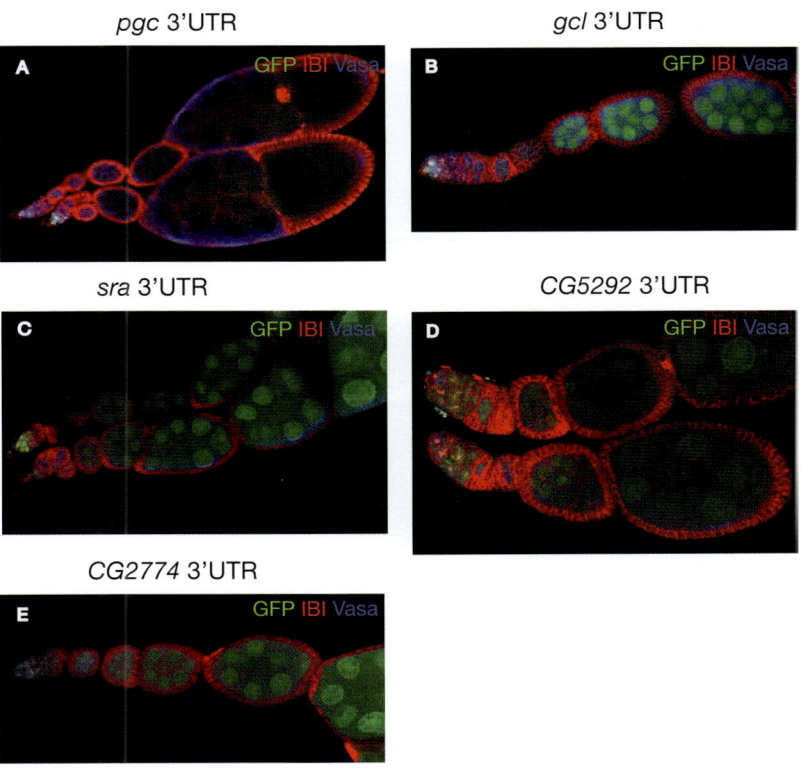

Figure 3. Germ plasm RNAs are regulated throughout oogenesis by their 3′UTRs. Expression patterns of various transgenes under control of the indicated 3′UTRs monitored by GFP expression (*green*). Anti-Vasa (*blue*) marks the germ line. IB1 (*red*) marks spectrosomes, fusomes, and the somatic cell membranes. Reporter fused to *pgc* 3′UTR is completely repressed in later stages. Reporter fused to *gcl* and *sra* 3′UTR is repressed in early stages of vitellogenesis but is expressed in later stages. Reporter fused to *CG5292* and *CG2774* 3′UTRs is uniformly expressed. Egg chambers are oriented with the earlier stages to the left.

TRANSCRIPTIONAL REGULATION OF *NANOS* IN THE GERMARIUM

The pattern of Nanos protein during oogenesis has been well studied (Wang et al. 1994; Forbes and Lehmann 1998). Nanos protein is expressed at low level in stem cells and cystoblasts and is up-regulated in 2–16 cell cysts and then expressed at moderate levels in the nurse cells throughout oogenesis (Forbes and Lehmann 1998). None of the heterologous 3′UTR reporters exhibited this pattern. To distinguish between genuine regulation of translation or stability and the transcriptional aspects conveyed by the *nanos* promoter, we analyzed the expression pattern generated by a transgene that contained the *nanos* promoter, the GAL4 coding sequence, and an "inert" tubulin 3′UTR (*pnos::GAL4-tub3′UTR*), which is unlikely to confer specific translational regulatory function. Expression of this construct was analyzed in females carrying the *pnos::GAL4-tub3′UTR* and UAS-GFP-*tubulin3′UTR*. GFP was expressed weakly in stem cells and dividing cystoblasts and more strongly in 16 cell cysts. Strongest expression was observed in oocytes from stages 2–3 onward (Table 2; Fig. 4) (Wang et al. 1994). This pattern was not observed with any of the 3′UTRs used, suggesting that the respective 3′UTRs recapitulate diverse patterns of translational control during oogenesis, largely independent of transcriptional regulation (Fig. 4C). These results suggest that during oogenesis, *nos* is regulated at the transcriptional and translational levels.

MATERIALS AND METHODS

The following stocks were used in this study: w^{118}, *nos*::Gal4-VP16-*nos3′UTR*, *nos*::Gal4-VP16-*tub3′UTR*, UAS-GFP-*k103′UTR*, and UAS-moeRFP-*nos3′*UTR. Fixation and immunostaining of ovaries were carried out as previously described (Forbes and Lehmann 1998). Rabbit anti-Vasa antibody was used at a dilution of 1:5000. IB1 monoclonal supernatant was used at 1:25 (Developmental Studies Hybridoma Bank). Secondary antibodies (Invitrogen) were used at 1:500. The reporters were monitored by direct fluorescence microscopy for GFP and RFP, and LacZ was monitored by antibody staining. Immunostainings was visualized using a Zeiss LSM 510 Meta confocal microscope using 10x, 20x, and oil-immersed 40x lenses.

CONCLUSIONS

Translational regulation is a widespread phenomenon and has been observed in particular for RNAs, such as *oskar, nanos, gurken*, and *orb*, during oogenesis (Johnstone and Lasko 2001). These RNAs are specifically transported into the oocyte and are localized to a particular region of the developing egg. In these cases, translational regulation is closely linked to localization, because RNAs are translationally repressed during transport and are translated upon localization. Our analysis of 3′UTR reporter constructs revealed that a significant number of maternal RNAs, that become localized to the posterior pole late during oogenesis and are part of the germ plasm in the early embryo, are also regulated at the translational level during the early stages of oogenesis. We observe translational regulatory patterns during oogenesis that are apparently independent of a particular localization pattern and the timing of translation during embryogenesis.

Posttranscriptional regulation of RNA is a widespread phenomenon in germ cells but also in other stem cells (Chen and Daley 2008; Kohlmaier and Edgar 2008). Our systematic study emphasizes the regulatory role that 3′UTRs have in this regulation. *Trans*-acting factors such

pnos::nos-GFP-nos3′UTR

pnos::GAL4-tub3′UTR;
UASp-GFP-k10 3′UTR

pnos::GFP-pgc 3′UTR;
pnos::RFP-nos 3′UTR

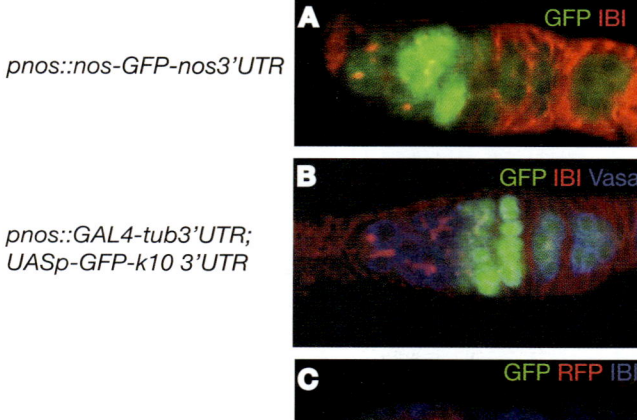
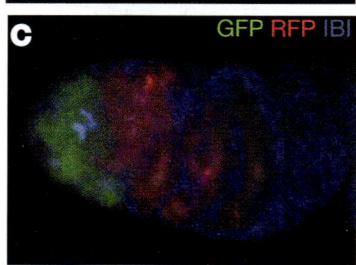

Figure 4. *Nanos* is transcriptionally regulated in the germarium. (*A*) Nanos-GFP fusion protein (*green*) expressed under the control of its endogenous promoter and 5′ and 3′UTR shows an up-regulation of protein expression in the 4–16-cell cyst. IB1 (*red*) marks spectrosomes, fusomes, and the somatic cell membranes. (*B*) GFP (*green*) expressed under control of the *nanos* promoter and *tubulin* 3′UTR also shows an up-regulation of expression in the 16-cell cyst and a down-regulation in later stages, demonstrating the role of the promoter and on nos regulation. Anti-Vasa (*blue*) marks the germ line. IB1 (*red*) marks spectrosomes, fusomes, and the somatic cell membranes. (*C*) Red fluorescent protein (RFP) under the control of the *nanos* promoter and *nanos* 3′UTR has an up-regulation in the 16-cell cyst, showing that the up-regulation is not due to protein stability. GFP (*green*) under the control of the *nanos* promoter and *pgc* 3′UTR shows expression in the germarium and down-regulation in 16-cell cyst, showing the effect of the *pgc* 3′UTR. Germaria are oriented with the niche containing the stem cells to the left.

as specific RNA-binding proteins or microRNAs (miRNAs) have been shown to regulate RNA translational repression, often by interfering with the ability of the poly(A) tail to stimulate translational initiation (Besse and Ephrussi 2008; Kohlmaier and Edgar 2008). Less is known about the mechanisms that activate translationally repressed RNAs. It has been suggested that either competition among binding factors for the RNA or signaling events such as phosphorylation inactivate repressors or target components of the translational activation machinery (Besse and Ephrussi 2008). How this regulatory information is encoded within 3′UTRs and is integrated over space and time remains largely unclear. By comparing translation patterns of the same RNAs during embryogenesis and oogenesis, we found that RNAs coregulated at one stage are not necessarily coregulated at another stage (Table 3). Thus, different *trans*-acting factors or combinations of factors must control the response of the 3′UTR. Our results suggest that the repertoire of *trans*-acting factors may itself be regulated by translation. For example, *bruno*, which encodes a translational repressor, and *orb*, which encodes the *Drosophila* CPEB homolog that regulates poly(A) elongation, are coregulated during oogenesis, but *bruno* RNA is completely silenced during embryogenesis, whereas *orb* is translated in the germ plasm (Fig. 2) (Kim-Ha et al. 1995; Webster et al. 1997; Chang et al. 1999; Castagnetti et al. 2000). Further systematic analysis of RNA localization and translation within its developmental context should reveal a code for 3′UTR-mediated RNA regulation that promises to be as complex and multileveled as the much better known transcriptional control mechanisms. Importantly, because some germ cells share the ability to self-propagate with other stem cells, we expect that principles discovered by the analysis of RNA regulation in germ cells will also provide insight into the regulation and maintenance of other stem cell populations (Surani et al. 2007; Chen and Daley 2008; Kohlmaier and Edgar 2008; Sampath et al. 2008).

Table 3. Regulation of germ plasm RNAs in the ovary does not parallel regulation in the embryo

Classes of similarly regulated RNAs in oogenesis (stem cells–stage 10)		Classes of similarly regulated RNAs in embryogenesis (stages 1–5)	
	nanos	Class 1	*nanos* *orb*
Class I	*pgc* CG18846 *gcl* *sra*	Class 2	*gcl*
Class II	CG2774 CG5292	Class 3	*pgc* CG18846 CG5292 *sra* *rapgap1*
Class III	*bru* *orb*	Class 4	*bru* CG2774
		Class 5	*cyclin B*

ACKNOWLEDGMENTS

We thank Paul Schedl and Iswar Hariharan for flies and antibodies. We also thank Alexey Arkov for the gift of the plasmid containing the *nanos* promoter with the *nanos* 5′UTR linked to GFP. P.R. is an HHMI Research Associate. R.L. is an HHMI investigator and a member of the Kimmel Center for Stem Cell Biology at NYULMC.

REFERENCES

Amikura, R., Kashikawa, M., Nakamura, A., and Kobayashi, S. 2001. Presence of mitochondria-type ribosomes outside mitochondria in germ plasm of *Drosophila* embryos. *Proc. Natl. Acad. Sci.* **98:** 9133–9138.

Ancelin, K., Lange, U.C., Hajkova, P., Schneider, R., Bannister, A.J., Kouzarides, T., and Surani, M.A. 2006. Blimp1 associates with Prmt5 and directs histone arginine methylation in mouse germ cells. *Nat. Cell Biol.* **8:** 623–630.

Besse, F. and Ephrussi, A. 2008. Translational control of localized mRNAs: Restricting protein synthesis in space and time. *Nat. Rev. Mol. Cell Biol.* **9:** 971–980.

Castagnetti, S., Hentze, M.W., Ephrussi, A., and Gebauer, F. 2000. Control of *oskar* mRNA translation by Bruno in a novel cell-free system from *Drosophila* ovaries. *Development* **127:** 1063–1068.

Chang, J.S., Tan, L., and Schedl, P. 1999. The *Drosophila* CPEB homolog, orb, is required for oskar protein expression in oocytes. *Dev. Biol.* **215:** 91–106.

Chen, L. and Daley, G.Q. 2008. Molecular basis of pluripotency. *Hum. Mol. Genet.* **17:** R23–R27.

Cinalli, R.M., Rangan, P., and Lehmann, R. 2008. Germ cells are forever. *Cell* **132:** 559–562.

Forbes, A. and Lehmann, R. 1998. Nanos and Pumilio have critical roles in the development and function of *Drosophila* germline stem cells. *Development* **125:** 679–690.

Forrest, K.M. and Gavis, E.R. 2003. Live imaging of endogenous RNA reveals a diffusion and entrapment mechanism for nanos mRNA localization in *Drosophila*. *Curr. Biol.* **13:** 1159–1168.

Gilboa, L. and Lehmann, R. 2004. Repression of primordial germ cell differentiation parallels germ line stem cell maintenance. *Curr. Biol.* **14:** 981–986.

Hanyu-Nakamura, K., Sonobe-Nojima, H., Tanigawa, A., Lasko, P., and Nakamura, A. 2008. *Drosophila* Pgc protein inhibits P-TEFb recruitment to chromatin in primordial germ cells. *Nature* **451:** 730–733.

Hayashi, Y., Hayashi, M., and Kobayashi, S. 2004. Nanos suppresses somatic cell fate in *Drosophila* germ line. *Proc. Natl. Acad. Sci.* **101:** 10338–10342.

Hayashi, K., de Sousa Lopes, S.M., and Surani, M.A. 2007. Germ cell specification in mice. *Science* **316:** 394–396.

Johnstone, O. and Lasko, P. 2001. Translational regulation and RNA localization in *Drosophila* oocytes and embryos. *Annu. Rev. Genet.* **35:** 365–406.

Jongens, T.A., Hay, B., Jan, L.Y., and Jan, Y.N. 1992. The germ cell-less gene product: A posteriorly localized component necessary for germ cell development in *Drosophila*. *Cell* **70:** 569–584.

Kim-Ha, J., Kerr, K., and Macdonald, P.M. 1995. Translational regulation of *oskar* mRNA by Bruno, an ovarian RNA-binding protein, is essential. *Cell* **81:** 403–412.

Kohlmaier, A. and Edgar, B.A. 2008. Proliferative control in *Drosophila* stem cells. *Curr. Opin. Cell Biol.* **20:** 699–706.

Lantz, V. and Schedl, P. 1994. Multiple *cis*-acting targeting sequences are required for orb mRNA localization during *Drosophila* oogenesis. *Mol. Cell. Biol.* **14:** 2235–2242.

Lantz, V., Chang, J.S., Horabin, J.I., Bopp, D., and Schedl, P. 1994. The *Drosophila* orb RNA-binding protein is required for the formation of the egg chamber and establishment of polarity. *Genes Dev.* **8:** 598–613.

Lecuyer, E., Yoshida, H., Parthasarathy, N., Alm, C., Babak, T.,

Cerovina, T., Hughes, T.R., Tomancak, P., and Krause, H.M. 2007. Global analysis of mRNA localization reveals a prominent role in organizing cellular architecture and function. *Cell* **131:** 174–187.

Martinho, R.G., Kunwar, P.S., Casanova, J., and Lehmann, R. 2004. A noncoding RNA is required for the repression of RNApolII-dependent transcription in primordial germ cells. *Curr. Biol.* **14:** 159–165.

Nakamura, A. and Seydoux, G. 2008. Less is more: Specification of the germline by transcriptional repression. *Development* **135:** 3817–3827.

Nakamura, A., Amikura, R., Mukai, M., Kobayashi, S., and Lasko, P.F. 1996. Requirement for a noncoding RNA in *Drosophila* polar granules for germ cell establishment. *Science* **274:** 2075–2079.

Ohinata, Y., Payer, B., O'Carroll, D., Ancelin, K., Ono, Y., Sano, M., Barton, S.C., Obukhanych, T., Nussenzweig, M., Tarakhovsky, A., Saitou, M., and Surani, M.A. 2005. Blimp1 is a critical determinant of the germ cell lineage in mice. *Nature* **436:** 207–213.

Rangan, P., DeGennaro, M., Jaime-Bustamante, K., Coux, R., Martinho, R., and Lehmann, R. 2008. Temporal and spatial control of germ plasm RNAs. *Curr. Biol.* **19:** 72–77.

Sampath, P., Pritchard, D.K., Pabon, L., Reinecke, H., Schwartz, S.M., Morris, D.R., and Murry, C.E. 2008. A hierarchical network controls protein translation during murine embryonic stem cell self-renewal and differentiation. *Cell Stem Cell* **2:** 448–460.

Sato, K., Hayashi, Y., Ninomiya, Y., Shigenobu, S., Arita, K., Mukai, M., and Kobayashi, S. 2007. Maternal Nanos represses *hid/skl*-dependent apoptosis to maintain the germ line in *Drosophila* embryos. *Proc. Natl. Acad. Sci.* **104:** 7455–7460.

Seydoux, G. and Braun, R.E. 2006. Pathway to totipotency: Lessons from germ cells. *Cell* **127:** 891–904.

Seydoux, G., Mello, C.C., Pettitt, J., Wood, W.B., Priess, J.R., and Fire, A. 1996. Repression of gene expression in the embryonic germ lineage of *C. elegans*. *Nature* **382:** 713–716.

Strome, S. and Lehmann, R. 2007. Germ versus soma decisions: Lessons from flies and worms. *Science* **316:** 392–393.

Subramaniam, K. and Seydoux, G. 1999. *nos-1* and *nos-2*, two genes related to *Drosophila nanos*, regulate primordial germ cell development and survival in *Caenorhabditis elegans*. *Development* **126:** 4861–4871.

Surani, M.A., Hayashi, K., and Hajkova, P. 2007. Genetic and epigenetic regulators of pluripotency. *Cell* **128:** 747–762.

Swantek, D. and Gergen, J.P. 2004. Ftz modulates Runt-dependent activation and repression of segment-polarity gene transcription. *Development* **131:** 2281–2290.

Tomancak, P., Berman, B.P., Beaton, A., Weiszmann, R., Kwan, E., Hartenstein, V., Celniker, S.E., and Rubin, G.M. 2007. Global analysis of patterns of gene expression during *Drosophila* embryogenesis. *Genome Biol.* **8:** R145.

Vincent, S.D., Dunn, N.R., Sciammas, R., Shapiro-Shalef, M., Davis, M.M., Calame, K., Bikoff, E.K., and Robertson, E.J. 2005. The zinc finger transcriptional repressor Blimp1/Prdm1 is dispensable for early axis formation but is required for specification of primordial germ cells in the mouse. *Development* **132:** 1315–1325.

Wang, C. and Lehmann, R. 1991. Nanos is the localized posterior determinant in *Drosophila*. *Cell* **66:** 637–647.

Wang, Z. and Lin, H. 2004. Nanos maintains germline stem cell self-renewal by preventing differentiation. *Science* **303:** 2016–2019.

Wang, C., Dickinson, L.K., and Lehmann, R. 1994. Genetics of nanos localization in *Drosophila*. *Dev. Dyn.* **199:** 103–115.

Webster, P.J., Liang, L., Berg, C.A., Lasko, P., and Macdonald, P.M. 1997. Translational repressor bruno plays multiple roles in development and is widely conserved. *Genes Dev.* **11:** 2510–2521.

Germ Line, Stem Cells, and Epigenetic Reprogramming

M.A. Surani, G. Durcova-Hills, P. Hajkova, K. Hayashi, and W.W. Tee

Wellcome Trust Cancer Research UK Gurdon Institute, University of Cambridge,
Cambridge CB2 1QN, United Kingdom

The germ cell lineage has the unique attribute of generating the totipotent state. Development of blastocysts from the totipotent zygote results in the establishment of pluripotent primitive ectoderm cells in the inner cell mass of blastocysts, which subsequently develop into epiblast cells in postimplantation embryos. The germ cell lineage in mice originates from these pluripotent epiblast cells of postimplantation embryos in response to specific signals. Pluripotent stem cells and unipotent germ cells share some fundamental properties despite significant phenotypic differences between them. Additionally, early primordial germ cells can be induced to undergo dedifferentiation into pluripotent embryonic germ cells. Investigations on the relationship between germ cells and pluripotent stem cells may further elucidate the nature of the pluripotent state. Furthermore, comprehensive epigenetic reprogramming of the genome in early germ cells, including extensive erasure of epigenetic modifications, is a critical step toward establishment of totipotency. The mechanisms involved may be relevant for gaining insight into events that lead to reprogramming of somatic cells into pluripotent stem cells.

The maternal inheritance of key pluripotency transcription factors and epigenetic modifiers in the mammalian oocyte are important components of the totipotent zygote; they contribute to the foundation of the pluripotent primitive ectoderm cells in the inner cell mass (ICM) of blastocysts (Surani et al. 2007). The latter are the precursors of all somatic tissues and germ cells in vivo, as well as the pluripotent embryonic stem (ES) cells in vitro. The germ cell lineage in mice originates from the pluripotent epiblast cells of the early postimplantation embryo commencing at embryonic day (E) 6.25 (Hayashi et al. 2007). In the mouse, there is no compelling evidence for the existence of preformed germ cell determinants in the zygote or early embryo, as is the case in some other model organisms (Strome and Lehmann 2007); instead, specification of primordial germ cells (PGCs) occurs from pluripotent cells according to the "stem cell model" in response to signals from the extraembryonic tissues (Lawson and Hage 1994; McLaren 2003; Surani et al. 2007).

In postimplantation embryos, development of pluripotent epiblast cells from the ICM is accompanied by a series of epigenetic changes, such as silencing of specific genes including *stella*, a marker of pluripotency and germ cells, as well as random X inactivation in cells of female embryos (McMahon et al. 1981). These pluripotent epiblast cells are the immediate precursors of primordial germ cells, as well as of the in-vitro-derived pluripotent epiblast stem cells (EpiSCs), which differ from pluripotent ES cells derived from the ICM (Brons et al. 2007; Tesar et al. 2007). The unipotent germ cells are the only cells during postimplantation development to reacquire some characteristics that are closer to those of the ICM following epigenetic reprogramming of early germ cells (Hayashi et al. 2007; Surani et al. 2007). This is reflected in the properties of the embryonic germ (EG) cells derived from early PGCs (Matsui et al. 1992; Resnick et al. 1992) that are more like ES cells, and not like EpiSCs despite their origin from postimplantation epiblast cells (Fig. 1).

The extensive epigenetic reprogramming of the genome in the germ cell lineage includes comprehensive erasure of epigenetic modifications (Hajkova et al. 2002, 2008). The underlying mechanisms, including reprogramming of somatic cells, are of wide interest. Analysis of the germ cell lineage may also provide a deeper understanding of the pluripotent state itself, as well as the relationship between germ cells and pluripotent stem cells. In this regard, our recent studies on the arginine methylase Prmt5 may provide insights into the link between pluripotent stem cells and the establishment of germ cells at the molecular level (W.W. Tee, unpubl.). Indeed, all pluripotent stem cells generated and maintained in vitro by definition are potential germ cells.

THE STEM CELL MODEL FOR PGC SPECIFICATION IN MICE

PGC specification in vivo follows after pluripotent epiblast cells gain competence in response to signals from the extraembryonic tissues, notably BMP4 and the Smad1/5 signaling pathways (Lawson et al. 1999; Tremblay et al. 2001; Hayashi et al. 2002). This triggers expression of Blimp1/Prdm1, the key determinant of PGC specification, which is initially detected in a few cells present within the proximal epiblast cells (Ohinata et al. 2005; Vincent et al. 2005). Expression of Blimp1/Prdm1 initiates a cascade of events, including extensive epigenetic reprogramming of the germ cell lineage and reactivation of the inactive X chromosome (Surani et al. 2004; de Napoles et al. 2007; Sugimoto and Abe 2007; Chuva de Sousa Lopes et al. 2008). Reactivation of the inactive X is also seen earlier during the establishment of pluripotent cells in the ICM (Mak et al. 2004; Okamoto et al. 2004). As the germ cells migrate into the gonads, this lineage is maintained in the unipotent state probably by the Blimp1–Prmt5 complex (Ancelin et al. 2006). Notably, dedifferentiation of PGCs into EG cells is coupled with

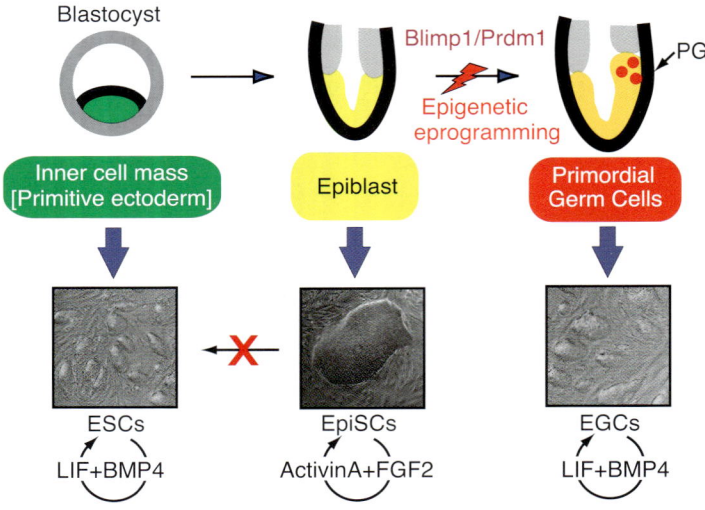

Figure 1. Establishment of pluripotent cells and the emergence of the germ cell lineage. The primitive ectoderm cells develop into pluripotent epiblast cells of postimplantation embryos. Specification of primordial germ cells (PGCs) occurs in response to signals, including BMP4. Induction of Blimp1/Prdm1 expression in PGC precursors causes an exit from the pluripotent state, initiates a germ-cell-specific program, represses the somatic program, and initiates an extensive epigenetic program in nascent PGCs. Pluripotent stem cells derived from the ICM (ESC: embryonic stem cells) are more like those derived from PGCs (EGC: embryonic germ cells), and these differ from pluripotent epiblast stem cells (EpiSCs) derived from the epiblast cells.

the early down-regulation of Blimp1, which is accompanied by the expression of some of the targets of Blimp1 (Durcova-Hills et al. 2008). During this process, Prmt5 translocates to the cytoplasm where it has a distinct and critical role in pluripotency (see below).

Initiation of Blimp1/Prdm1 expression apparently induces an exit from the very rapidly dividing state seen in pluripotent epiblast cells toward a unipotent germ cell lineage characterized by a significantly increased cell cycle time (Ohinata et al. 2005; K. Hayashi, unpubl.). How this is accomplished precisely is not yet clear, but it is likely that some direct targets of Blimp1, such as c-*Myc*, become repressed. Profound epigenetic changes also ensue in nascent PGCs, which culminate in the reexpression of *stella* as well as of some key pluripotency-specific genes such as *Esg1, Sox2*, and *nanog* (Yamaguchi et al. 2005; Yamaji et al. 2008). More importantly, the early epigenetic changes lead to the repression of the *Xist* gene and reactivation of the inactive X chromosome (de Napoles et al. 2007; Sugimoto and Abe 2007; Chuva de Sousa Lopes et al. 2008), as well as of the extensive erasure of epigenetic modifications and DNA methylation (Hajkova et al. 2002, 2008).

Blimp1 also causes the initiation of the PGC-specific program in precursor cells, as is evident from the initiation of expression of PGC-specific genes including *Dnd1, Nanos3*, and *Prdm14* (Kurimoto et al. 2008). Notably, Blimp1 also has a critical role in the repression of the somatic program, which continues in the neighboring cells that share common ancestry (Ohinata et al. 2005; Kurimoto et al. 2008). Repression of the somatic program is a key event during specification of germ cells also in other organisms, although the precise mechanisms differ (Blackwell 2004; Strome and Lehmann 2007). Compared to PGC specification in other model organisms such as worms and flies where germ cell fate is initiated through inheritance of germ plasm at the start of development, the epiblast cells in vivo are fast dividing and transcriptionally active. This

makes it necessary to convert these cells, which already show initiation of the somatic program, into PGCs. Blimp1 has the potential to form a repressive complex, with Groucho, HDACs, and diverse histone modifiers such as G9a and Prmt5 (Gyory et al. 2004; Ancelin et al. 2006). The precise role of Blimp1 during the very early stages of PGC specification remains to be fully elucidated. Similarly, the precise role of Prmt5, which is detected in all epiblast cells, is also unclear (W.W. Tee, unpubl.). However, the Blimp1–Prmt5 complex is detected in the nucleus of nascent PGCs after E7.5 (Ancelin et al. 2006), where it probably has a role in maintaining early germ cells as they migrate into the developing gonads.

In this context, it will be of interest to determine the relative potential of ES cells and EpiSCs for generating germ cells in vitro. EpiSCs are a stage closer to the cell type from which PGCs emerge during normal development in vivo, assuming that they inherit key properties of postimplantation epiblast cells. Our evidence suggests that EpiSCs can produce PGC precursor cells that show expression of Blimp1, and some of these go onto PGCs and show expression of Stella (K. Hayashi , unpubl.). Note that the culture conditions for ES (LIF and BMP4) differ from the culture conditions for EpiSCs (activin + FGF-2) (Fig. 1) (Brons et al. 2007; Tesar et al. 2007). Importantly, however, these observations demonstrate that PGC specification in mice indeed conforms to the stem cell model. It is noteworthy that human ES cells (hES), which resemble EpiSCs, also show expression of germ-cell-specific markers such as *Blimp1* and *Prdm14* (Clark and Reijo Pera 2006; Tsuneyoshi et al. 2008).

EPIGENETIC REPROGRAMMING IN EARLY GERM CELLS

PGCs emerge from the postimplantation pluripotent epiblast cells that have already undergone X inactivation in female embryos and where some of the pluripotency-specific genes, including *stella, Sox2*, and *nanog*, are repressed (Surani et al. 2007). However, not only does PGC specification involve activation of the germ-cell-specific genes, but these cells also undergo epigenetic changes that in some respects bring them closer to the pluripotent primitive ectoderm cells in the ICM. The importance of these reprogramming changes in germ cells remains to be fully elucidated, but they could be prerequisite for the eventual establishment of totipotency.

Epigenetic reprogramming in nascent PGCs occurs in a step-wise manner. The initial changes at specification in chromatin modifications in PGCs begin to reinstate the epigenetic status that is closer to the pluripotent stem cells (Ancelin et al. 2006; Seki et al. 2007; Hajkova et al. 2008). The earliest changes include the loss of the repressive mark of dimethylation of lysine 9 on histone H3 (H3K9me2) (Seki et al. 2005; Hajkova et al. 2008). This epigenetic modification is attributed to the histone methyltransferase G9a (also called Ehmt2). Although G9a itself is not down-regulated in early PGCs, its binding partner GLP (or Ehmt1) clearly is (Yabuta et al. 2006), and this affects the activity of the G9a–GLP complex, which may account for the loss of H3K9me2 modification from PGCs (Seki et al. 2007). However, the involvement of a specific histone demethylase in removing this modification could not be excluded. It is also noteworthy that the inhibition of G9a by a specific inhibitor enhances reprogramming of somatic cells toward the induced pluripotent stem (iPS) cells (Kubicek et al. 2007; Shi et al. 2008). At the same time, there is enhancement of trimethylation of lysine 27 of histone H3 (H3K27me3), concomitant with an increase in the polycomb group enzyme Ezh2 (Hajkova et al. 2008). There is also an enhancement of lysine 4 of histone H3 (H3K4me2 and H3K4me3) and of histone acetylation marks (such as H3K9ac), which are chromatin modifications associated with gene activation. Notably, these chromatin modifications occur in conjunction with the expression of pluripotency-specific genes *Sox2, Oct4, nanog,* and *stella* (Hajkova et al. 2008; Kurimoto et al. 2008). These chromatin modifications also provide permissive conditions for reprogramming and dedifferentiation of early PGCs into pluripotent EG cells (see below). However, in vivo, the unipotent germ cell lineage is probably maintained and propagated by the Blimp1–Prmt5 complex through symmetrical methylation of arginine 3 on histone H2A and H4 (H2A/H4R3me2s) presumably through repression of target genes (Ancelin et al. 2006).

RESETTING THE GENOME FOR TOTIPOTENCY IN GONADAL PGCs

One of the hallmarks of epigenetic reprogramming in germ cells is the extensive erasure of the existing epigenetic modifications, including genomic imprints when PGCs enter into the developing gonads (Hajkova et al. 2002, 2008). This includes genome-wide DNA demethylation accompanied by extensive changes in chromatin modifications. The underlying mechanism responsible for DNA demethylation remains to be elucidated, but recent studies have advanced our knowledge of the accompanying chromatin changes. The Blimp1–Prmt5 complex, that is evidently detected in the nucleus during migration of PGCs into the gonads, translocates to the cytoplasm (Ancelin et al. 2006), but whether this has any significance for the extensive epigenetic changes in germ cells or merely reflects a major phenotypic change in the lineage is unclear.

The entry of PGCs into the gonads is associated with marked changes in nuclear morphology and a significant increase in nuclear size, which to an extent is reminiscent of changes observed in somatic nuclei during the initiation of reprogramming when transplanted into oocytes (Hajkova et al. 2008). This nuclear enlargement is accompanied by a rapid loss of linker histone H1 and the loss of chromocenters. There are other extensive changes in histone modifications, some of which are lost transiently, including H3K27me3 and H3K9me3, whereas others are erased more permanently, including H3K9ac and H2A/H4R3me2s. The latter modification is associated with the Blimp1–Prmt5 complex, and the translocation of this complex to the cytoplasm potentially explains the persistent loss of this epigenetic mark, concomitant with the derepression of a target gene *Dhx38.*

The sequence of events associated with the erasure of epigenetic modifications and the large-scale chromatin changes, which most likely involves histone replacement, is of particular interest (Hajkova et al. 2008). Histone variants such as H2A.Z and H3.3, unlike canonical histones H2A, H2B, H3.1, H3.2, and H4, can be incorporated independently of DNA replication at S phase (Henikoff et al. 2004). Indeed, although H2A.Z was present at significant levels in PGCs early at E10.5, this signal diminished dramatically in PGCs at E11.5–12.5. More significantly however, under appropriate conditions, PGCs from E11.5 could be separated by FACS (fluorescence-activated cell sorting) into two distinct groups, A and B, due to some intrinsic changes in their physical attributes (Fig. 2). Notably, group B is highly transient and detected only during a 3–4-hour duration. These two groups of PGCs exhibit distinct chromatin modifications because PGCs in group A show the presence of the linker histone H1 as well as high levels of H2A.Z, whereas PGCs in group B are devoid of both. Other modifications associated with histone H3, such as H3K27me3 and H3K9ac, are also lost in PGCs in group B, including the Blimp1–Prmt5-associated H2A/H4R3me2s. These observations indicate that PGCs in group B are at a more advanced stage of development compared to those in group A. Furthermore, these chromatin modifications occur in these PGCs in both groups in G_2 phase of the cell cycle and therefore must occur independently of DNA replication.

Histone replacement and eviction require deployment of specific histone chaperones, depending on whether the process occurs in conjunction with DNA replication. For example, introduction of canonical histones, such as H3.1, are coupled to DNA replication and require the chaperone CAF-1, but the noncanonical histones, such as H3.3, are associated with HIRA (Tagami et al. 2004). In agreement with the predicted active histone replacement occurring in these PGCs, CAF-1 is not used because it is

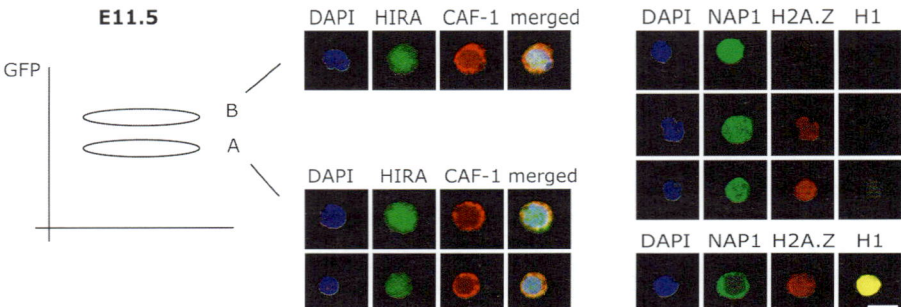

Figure 2. Epigenetic reprogramming in gonadal germ cells at E11.5. In the course of epigenetic reprogramming, PGCs undergo profound changes in nuclear structure and chromatin modifications. Using an Oct4-GFP transgene and based on their physical properties, PGCs can be separated into two distinct subpopulations. PGCs in these populations differ in the presence of linker histone H1, multiple histone modifications, and the histone variant H2A.Z (for details, see the text). This is due to the large-scale histone replacement as documented by the change in the localization of NAP-1 and high levels of nuclear HIRA. The images represent immunofluoresce stainings of fluorescence-activated cell-sorted PGCs at E11.5. Bar, 10 μm.

detected in the cytoplasm of PGCs, which is rarely seen in most cases (Fig. 2). In contrast, HIRA is detected in the nuclei of PGCs in both groups, consistent with its role in the deposition of noncanonical histones outside the S phase of the cell cycle. Of great interest is the localization of a histone chaperone, nucleosome assembly protein-1 (NAP-1), which by forming a complex with H2A-H2B dimers can cause release of H3 and H4 during transcription (Lorch et al. 2006). Furthermore NAP-1 has the potential to extract linker histones from chromatin (Kepert et al. 2005). Notably, NAP-1 is detected in the cytoplasm of PGCs in group A, but it translocates to the nucleus in PGCs of group B, concomitant with the loss of H1 and H2A.Z from these cells (Fig. 2). These observations suggest a very significant role for NAP-1 in epigenetic reprogramming in gonadal PGCs.

How the chromatin changes are linked to the erasure of DNA methylation and its underlying mechanism is not yet clear, except that DNA demethylation is already evident in PGCs from group A and therefore precedes the extensive chromatin modifications that follow in PGCs in group B (Hajkova et al. 2008). This being the case, it appears that the chromatin changes and histone replacement observed in PGCs may be a response to DNA demethylation, perhaps because this event compromises the integrity of the DNA.

Deeper insights into the mechanism of epigenetic reprogramming in early PGCs in vivo may provide valuable mechanistic insights on reprogramming of somatic nuclei in the oocyte, which is likely to involve extensive erasure of existing epigenetic modifications involving chromatin modifications and DNA demethylation. Some of this knowledge may also help to unravel the mechanism that underlies the formation of iPS cells from somatic cells.

FROM PGCs TO PLURIPOTENT STEM CELLS

Although germ cells share some of the core properties of pluripotency with stem cells, this is a unipotent lineage, and early PGCs do not dedifferentiate or participate in the formation of chimeric embryos if transplanted into blastocysts. However, in vitro, E8.5–11.5 PGCs can be induced to undergo dedifferentiation into pluripotent stem cells called embryonic germ (EG) cells when cultured in the presence of leukemia inhibitory factor (LIF), fibroblast growth factor-2 (FGF-2), and stem cell factor (SCF) (Matsui et al. 1992; Resnick et al. 1992). EG cells closely resemble ES cells and not the EpiSCs. This is presumably because specification of PGCs entails epigenetic reprogramming that eventually leads to reactivation of the inactive X chromosome as well as other important epigenetic modifications (de Napoles et al. 2007; Sugimoto and Abe 2007; Chuva de Sousa Lopes et al. 2008). Reactivation of the inactive paternal X chromosome is also observed in the ICM of blastocyst from which ES cells are derived (Mak et al. 2004; Okamoto et al. 2004). It is possible that these epigenetic modifications reflect other underlying similarities between ICM and PGCs, which may account for the similarities between pluripotent ES and EG cells.

There are, however, also differences between EG and ES cells, the primary difference being that the DNA methylation marks associated with the imprinted genes are predominantly erased during the formation of EG cells (Tada et al. 1997, 1998), an event that is also observed in PGCs in vivo when the germ cells enter into the developing gonads (see above). Other differences between ES and EG cells are evident in cell fusion experiments with somatic cells. Although both ES and EG somatic cell hybrids exhibit reprogramming of the somatic nucleus to pluripotency, there is an additional effect observed with EG cells because these hybrids additionally reveal erasure of the DNA methylation of imprinted genes from somatic nuclei (Tada et al. 1997). Furthermore, hybrids between EG and ES cells result in the erasure of imprints from the ES cells, suggesting that EG cells are dominant over ES cells in this test (Tada et al. 2001). However, the precise nature and mechanism underlying DNA demethylation of imprinted genes are not yet known.

Several mutations, including some in germ-cell-specific genes such as *Dnd1*, as well as in *Pten, Pgct1*, and *Akt* signaling, increase the efficiency of EG derivation from PGCs (Muller et al. 2000; Kimura et al. 2003, 2008; Youngren et al. 2005). This process, to some extent, represents a reversal of the events associated with the specification of PGCs from pluripotent cells. EG cells exhibit the ability to undergo differentiation into a variety of cell types. Dedifferentiation of PGCs to pluripotent EG cells requires FGF-2, but only for the first 24-hour period of a process that takes 7–10 days to be completed (Durcova-Hills et al. 2006). Addition of FGF-2 after the first 24 hours does not lead to the dedifferentiation process, but the reasons for this are unknown. Recent study of differential gene expression has shown that although there is expression of some of the key pluripotency-specific genes in both PGCs and EG cells, including *Oct4, Sox2,* and *nanog,* other genes such as c-*Myc, Stat-3, Klf4, Dnmt3l,* and *Eras* are not expressed in E8.5 PGCs (Durcova-Hills et al. 2008), which must be up-regulated during the derivation of EG cells (Fig. 3).

Because Blimp1–Prmt5 is potentially the key complex involved in the maintenance of early germ cell lineage, we anticipated a change in their expression during dedifferentiation of PGCs into pluripotent EG cells. Indeed, among an early event during the derivation of EG cells is the down-regulation of Blimp1 (but not of Prmt5; see below) within 1–2 days of culture (Fig. 3) (Durcova-Hills et al. 2008). Down-regulation of Blimp1 does not occur in the absence of FGF-2 nor if this factor is added 24 hours after the start of the PGC culture, and consequently no EG cells are detected. In contrast, Prmt5 is detected in the nuclei of PGCs in culture for up to 7 days, but it undergoes translocation into the cytoplasm after this time (see below for further discussion). Loss of Blimp1 leads to the up-regulation of some of its direct and indirect targets. Among these targets is *Dhx38* (Ancelin et al. 2006), which is detectable as PGCs dedifferentiate into EG cells. However, among the more significant direct targets of

Blimp1 are c-*Myc* and *Klf4*, which begin to be detectable after 7 and 4 days of culture, respectively (Fig. 3) (Durcova-Hills et al. 2008). It is also very likely that *Eras* may also be up-regulated during dedifferentiation of PGCs. It is possible that during PGC specification from epiblast cells, these are among the initial and key genes to be repressed in response to the expression of Blimp1. They are also among the key genes that along with *Sox2* and *Oct4* have a critical role in reprogramming of somatic cells into pluripotent cells.

These studies also reveal the importance of the LIF/STAT3 signaling pathway during reprogramming of PGCs to EG cells. STAT3 is undetectable in E8.5 PGCs, but it is clearly present in EG cells, and it starts to be detected predominantly in the cytoplasm when PGCs undergo dedifferentiation at about day 4 of culture; however, it is also occasionally seen in the nuclei of these cells and in older cultures when EG cells become established. LIF is known to enhance the survival and proliferation of PGCs. Unlike FGF-2, LIF is apparently not required during the first 24 hours of culture, but the addition of LIF after a 48-hour culture of PGCs improves the derivation of EG cells. Furthermore, the addition of WHI-P131 (1–5 μM), an inhibitor of the Jak/STAT3 signaling pathway to the culture medium, also abrogated the derivation of EG cells (Durcova-Hills et al. 2008).

Further investigations have shown that Trichostatin A (TSA; 5 ng/ml), an inhibitor of histone deacetylases (HDACs), can substitute for FGF-2 (Durcova-Hills et al. 2008). TSA proved to be very effective for the derivation of EG cells, and the process of dedifferentiation was also accelerated by TSA when the EG colonies were detected 1–2 days earlier. There was also a quicker down-regulation of Blimp1 from PGCs, which was already evident within 1 day after culture. The mechanism through which TSA action manifests itself and accelerates the down-regulation of Blimp1 from PGCs in culture is unknown, but this could involve epigenetic modifications of certain loci during this process. Because there is the presence of feeder cells in culture, it is also possible that TSA may sensitize PGCs to some of the cytokines that are produced by these cells to induce reprogramming of PGCs.

It is of interest to compare reprogramming of PGCs with that of neural progenitor cell lines. The latter are self-renewing cells when cultured under appropriate conditions but they are not pluripotent. In contrast, PGCs show expression of key pluripotency-specific genes, but early PGCs in vivo, unlike EG cells and neural progenitors, are not self-renewing. However, both cell types can be converted into pluripotent stem cells.

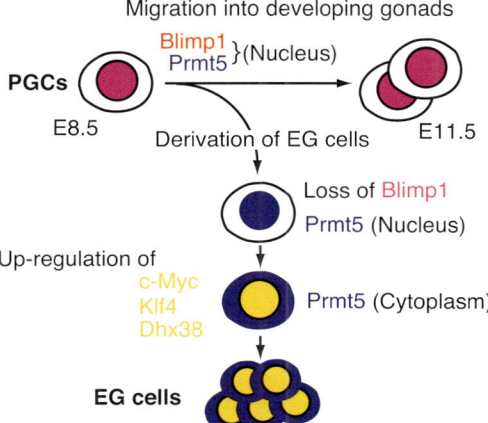

Figure 3. Derivation of pluripotent embryonic germ (EG) cells from primordial germ cells (PGCs). EG cells can be derived from E8.5–11.5 PGCs in vitro when cultured in the presence of FGF-2, LIF, and SCF. Early loss of Blimp1 is followed by translocation of Prmt5 to the cytoplasm and up-regulation of direct targets of Blimp1, including c-*Myc, Klf4,* and *Dhx38.* Many epigenetic changes, including erasure of epigenetic modifications and reactivation of the inactive X chromosome, occur in EG cells as seen in PGCs in vivo.

MULTIPLE ROLES OF PRMT5: THE LINK BETWEEN PLURIPOTENT STEM CELLS AND GERM CELLS

A potentially important observation that emerges from studies on reprogramming of PGCs to EG cells is the apparent multiple roles of Prmt5, an enzyme with specificity for symmetrical dimethylation of arginine 3 of histones H2A and H4. Additionally, Prmt5 has been shown to methylate both histones and nonhistone substrates (Bedford and

Richard 2005). In conjunction with its complex with Blimp1 detected in the nuclei of early PGCs (Ancelin et al. 2006), Prmt5 is responsible for the symmetrical dimethylation of arginine 3 of histone H2A/H4 (H2A/H4R3me2s). This repressive modification is presumably responsible for silencing target genes and maintaining the early germ cell lineage. However, whereas Blimp1 is rapidly down-regulated during the derivation of pluripotent EG cells from PGCs, Prmt5 translocates to the cytoplasm (Fig. 3), where it is detected in the absence of Blimp1 in pluripotent stem cells (W.W. Tee, unpubl.). Indeed, Prmt5 is predominantly detected in the cytoplasm of all pluripotent stem cells, including ES cells, as well as in the pluripotent epiblast cells in vivo. Our ongoing studies indicate that Prmt5 may be crucial for pluripotency and for germ cell specification. This is in addition to the role of the Blimp1–Prmt5 complex in the maintenance of the germ cell lineage during the migration of PGCs into the gonads. The apparent involvement of Prmt5 in pluripotency and in the germ cell lineage provides a major opportunity to explore the links between pluripotent cells and the germ cell lineage.

Preliminary studies indicate that Prmt5 is maternally inherited protein in the oocyte, and its loss of function results in very early embryonic lethality. Loss of Prmt5 abrogates pluripotency and induces differentiation of ES cells. Prmt5 is detected predominantly in the cytoplasm of ES cells, and we have identified several novel interactors of Prmt5 in ES cells, where presumably some of the components serve as the substrates of this enzyme. Notably, some of these interactors have previously been demonstrated to be important for ES pluripotency (W.W. Tee, unpubl.). We are currently investigating how Prmt5 potentially integrates into the pluripotency circuitry.

Prmt5 is also present in the cytoplasm of pluripotent epiblast cells, and here it may have an additional role during the formation of PGC precursors and in events leading to the specification of germ cells. This possibility is further suggested by the observation that in *Drosophila*, a Prmt5 homolog (Capsuleen/Dart5) forms together with MEP50, a methylosome that is important for the localization of Tudor to the pole plasm (Anne and Mechler 2005). Loss of function of any of these components in flies results in the loss of germ cells. Recent evidence suggests the presence of a Tudor domain protein, Tdrd5, and MEP50 in the cytoplasm of ES cells, which interact with Prmt5 (W.W. Tee, unpubl.). Notably, Tdrd5 levels increase when Blimp1 expression commences in PGC precursors. This raises the possibility that Prmt5–Tdrd5 interaction in the cytoplasm of PGC precursors may have a role during PGC specification in mice. Following PGC specification, Tdrd5 is down-regulated and Prmt5 translocates to the nucleus, where as part of a complex with Blimp1, it performs the role of maintaining the early germ cell lineage as described above. Detailed analysis of the multiple roles of Prmt5 in pluripotency and germ cells should further reveal how the two phenotypic states are linked and regulated.

PERSPECTIVE

Evidence shows that the germ cell lineage and pluripotent stem cells are intimately linked entities, and they share many core properties despite germ cells being unipotent. This link can be attributed to the stem cell mode of germ cell specification in mice. Recent derivation of pluripotent iPS cells directly from adult somatic cells, which have the potential to generate germ cells, suggests that once pluripotency is restored to somatic cells, they also acquire the potential for specification into germ cells. This can occur even when such cells are introduced directly into blastocysts, thus bypassing preimplantation development. Further studies may lead to the derivation of germ cells from fully or partially reprogrammed somatic cells directly in vitro without reintroduction into blastocysts.

The molecular basis for the similarities and differences between PGCs and pluripotent stem cells is also becoming more evident. It is possible that detailed investigations on the role of Prmt5 will enhance our understanding of the properties of pluripotent stem cells and unipotent germ cells, and indeed on reprogramming of somatic cells to a pluripotent state.

ACKNOWLEDGMENTS

We thank all members of the lab for their valuable contributions. The work described in the paper was funded by the Medical Research Council (G.D.-H.) and the Wellcome Trust (M.A.S.).

REFERENCES

Ancelin, K., Lange, U.C., Hajkova, P., Schneider, R., Bannister, A.J., Kouzarides, T., and Surani, M.A. 2006. Blimp1 associates with Prmt5 and directs histone arginine methylation in mouse germ cells. *Nat. Cell Biol.* **8:** 623–630.

Anne, J. and Mechler, B.M. 2005. Valois, a component of the nuage and pole plasm, is involved in assembly of these structures, and binds to Tudor and the methyltransferase Capsuléen. *Development* **132:** 2167–2177.

Bedford, M.T. and Richard, S. 2005. Arginine methylation an emerging regulator of protein function. *Mol. Cell* **18:** 263–272.

Blackwell, T.K. 2004. Germ cells: Finding programs of mass repression. *Curr. Biol.* **14:** R229–R230.

Brons, I.G., Smithers, L.E., Trotter, M.W., Rugg-Gunn, P., Sun, B., Chuva de Sousa Lopes, S.M., Howlett, S.K., Clarkson, A., Ahrlund-Richter, L., Pedersen, R.A., and Vallier, L. 2007. Derivation of pluripotent epiblast stem cells from mammalian embryos. *Nature* **448:** 191–195.

Chuva de Sousa Lopes, S.M., Hayashi, K., Shovlin, T.C., Mifsud, W., Surani, M.A., and McLaren, A. 2008. X chromosome activity in mouse XX primordial germ cells. *PLoS Genet.* **4:** e30.

Clark, A.T. and Reijo Pera, R.A. 2006. Modeling human germ cell development with embryonic stem cells. *Regen. Med.* **1:** 85–93.

de Napoles, M., Nesterova, T., and Brockdorff, N. 2007. Early loss of *Xist* RNA expression and inactive X chromosome associated chromatin modification in developing primordial germ cells. *PLoS ONE* **2:** e860.

Durcova-Hills, G., Adams, I.R., Barton, S.C., Surani, M.A., and McLaren, A. 2006. The role of exogenous fibroblast growth factor-2 on the reprogramming of primordial germ cells into pluripotent stem cells. *Stem Cells* **24:** 1441–1449.

Durcova-Hills, G., Tang, F., Doody, G., Tooze, R., and Surani, M.A. 2008. Reprogramming primordial germ cells into pluripotent stem cells. *PLoS ONE* (in press).

Gyory, I., Wu, J., Fejér, G., Seto, E., and Wright, K.L. 2004. PRDI-BF1 recruits the histone H3 methyltransferase G9a in transcriptional silencing. *Nat. Immunol.* **5:** 299–308.

Hajkova, P., Erhardt, S., Lane, N., Haaf, T., El-Maarri, O., Reik, W., Walter, J., and Surani, M.A. 2002. Epigenetic reprogramming in mouse primordial germ cells. *Mech. Dev.* **117:** 15–23.

Hajkova, P., Ancelin, K., Waldmann, T., Lacoste, N., Lange, U.C., Cesari, F., Lee, C., Almouzni, G., Schneider, R., and Surani, M.A. 2008. Chromatin dynamics during epigenetic reprogramming in the mouse germ line. *Nature* **452**: 877–881.

Hayashi, K., Kobayashi, T., Umino, T., Goitsuka, R., Matsui, Y., and Kitamura, D. 2002. SMAD1 signaling is critical for initial commitment of germ cell lineage from mouse epiblast. *Mech. Dev.* **118**: 99–109.

Hayashi, K., de Sousa Lopes, S.M., and Surani, M.A. 2007. Germ cell specification in mice. *Science* **316**: 394–396.

Henikoff, S., Furuyama, T., and Ahmad, K. 2004. Histone variants, nucleosome assembly and epigenetic inheritance. *Trends Genet.* **20**: 320–326.

Kepert, J.F., Mazurkiewicz, J., Heuvelman, G.L., Toth, K.F., and Rippe, K. 2005. NAP1 modulates binding of linker histone H1 to chromatin and induces an extended chromatin fiber conformation. *J. Biol. Chem.* **280**: 34063–34072.

Kimura, T., Suzuki, A., Fujita, Y., Yomogida, K., Lomeli, H., Asada, N., Ikeuchi, M., Nagy, A., Mak, T.W., and Nakano, T. 2003. Conditional loss of PTEN leads to testicular teratoma and enhances embryonic germ cell production. *Development* **130**: 1691–1700.

Kimura, T., Tomooka, M., Yamano, N., Murayama, K., Matoba, S., Umehara, H., Kanai, Y., and Nakano, T. 2008. AKT signaling promotes derivation of embryonic germ cells from primordial germ cells. *Development* **135**: 869–879.

Kubicek, S., O'Sullivan, R.J., August, E.M., Hickey, E.R., Zhang, Q., Teodoro, M.L., Rea, S., Mechtler, K., Kowalski, J.A., Homon, C.A., Kelly, T.A., and Jenuwein, T. 2007. Reversal of H3K9me2 by a small-molecule inhibitor for the G9a histone methyltransferase. *Mol. Cell* **25**: 473–481.

Kurimoto, K., Yabuta, Y., Ohinata, Y., Shigeta, M., Yamanaka, K., and Saitou, M. 2008. Complex genome-wide transcription dynamics orchestrated by Blimp1 for the specification of the germ cell lineage in mice. *Genes Dev.* **22**: 1617–1635.

Lawson, K.A. and Hage, W.J. 1994. Clonal analysis of the origin of primordial germ cells in the mouse. *Ciba Found. Symp.* **182**: 68–91.

Lawson, K.A., Dunn, N.R., Roelen, B.A., Zeinstra, L.M., Davis, A.M., Wright, C.V., Korving, J.P., and Hogan, B.L. 1999. Bmp4 is required for the generation of primordial germ cells in the mouse embryo. *Genes Dev.* **13**: 424–436.

Lorch, Y., Maier-Davis, B., and Kornberg, R.D. 2006. Chromatin remodeling by nucleosome disassembly in vitro. *Proc. Natl. Acad. Sci.* **103**: 3090–3093.

Mak, W., Nesterova, T.B., de Napoles, M., Appanah, R., Yamanaka, S., Otte, A.P., and Brockdorff, N. 2004. Reactivation of the paternal X chromosome in early mouse embryos. *Science* **303**: 666–669.

Matsui, Y., Zsebo, K., and Hogan, B.L. 1992. Derivation of pluripotential embryonic stem cells from murine primordial germ cells in culture. *Cell* **70**: 841–847.

McLaren, A. 2003. Primordial germ cells in the mouse. *Dev. Biol.* **262**: 1–15.

McMahon, A., Fosten, M., and Monk, M. 1981. Random X-chromosome inactivation in female primordial germ cells in the mouse. *J. Embryol. Exp. Morphol.* **64**: 251–258.

Muller, A.J., Teresky, A.K., and Levine, A.J. 2000. A male germ cell tumor-susceptibility-determining locus, pgct1, identified on murine chromosome 13. *Proc. Natl. Acad. Sci.* **97**: 8421–8426.

Ohinata, Y., Payer, B., O'Carroll, D., Ancelin, K., Ono, Y., Sano, M., Barton, S.C., Obukhanych, T., Nussenzweig, M., Tarakhovsky, A., Saitou, M., and Surani, M.A. 2005. Blimp1 is a critical determinant of the germ cell lineage in mice. *Nature* **436**: 207–213.

Okamoto, I., Otte, A.P., Allis, C.D., Reinberg, D., and Heard, E. 2004. Epigenetic dynamics of imprinted X inactivation during early mouse development. *Science* **303**: 644–649.

Resnick, J.L., Bixler, L.S., Cheng, L., and Donovan, P.J. 1992. Long-term proliferation of mouse primordial germ cells in culture. *Nature* **359**: 550–551.

Seki, Y., Hayashi, K., Itoh, K., Mizugaki, M., Saitou, M., and Matsui, Y. 2005. Extensive and orderly reprogramming of genome-wide chromatin modifications associated with specification and early development of germ cells in mice. *Dev. Biol.* **278**: 440–458.

Seki, Y., Yamaji, M., Yabuta, Y., Sano, M., Shigeta, M., Matsui, Y., Saga, Y., Tachibana, M., Shinkai, Y., and Saitou, M. 2007. Cellular dynamics associated with the genome-wide epigenetic reprogramming in migrating primordial germ cells in mice. *Development* **134**: 2627–2638.

Shi, Y., Do, J.T., Desponts, C., Hahm, H.S., Scholer, H.R., and Ding, S. 2008. A combined chemical and genetic approach for the generation of induced pluripotent stem cells. *Cell Stem Cell* **2**: 525–528.

Strome, S. and Lehmann, R. 2007. Germ versus soma decisions: Lessons from flies and worms. *Science* **316**: 392–393.

Sugimoto, M. and Abe, K. 2007. X chromosome reactivation initiates in nascent primordial germ cells in mice. *PLoS Genet.* **3**: e116.

Surani, M.A., Ancelin, K., Hajkova, P., Lange, U.C., Payer, B., Western, P., and Saitou, M. 2004. Mechanism of mouse germ cell specification: A genetic program regulating epigenetic reprogramming. *Cold Spring Harbor Symp. Quant. Biol.* **69**: 1–9.

Surani, M.A., Hayashi, K., and Hajkova, P. 2007. Genetic and epigenetic regulators of pluripotency. *Cell* **128**: 747–762.

Tada, M., Tada, T., Lefebvre, L., Barton, S.C., and Surani, M.A. 1997. Embryonic germ cells induce epigenetic reprogramming of somatic nucleus in hybrid cells. *EMBO J.* **16**: 6510–6520.

Tada, T., Tada, M., Hilton, K., Barton, S.C., Sado, T., Takagi, N., and Surani, M.A. 1998. Epigenotype switching of imprintable loci in embryonic germ cells. *Dev. Genes Evol.* **207**: 551–561.

Tada, M., Takahama, Y., Abe, K., Nakatsuji, N., and Tada, T. 2001. Nuclear reprogramming of somatic cells by in vitro hybridization with ES cells. *Curr. Biol.* **11**: 1553–1558.

Tagami, H., Ray-Gallet, D., Almouzni, G., and Nakatani, Y. 2004. Histone H3.1 and H3.3 complexes mediate nucleosome assembly pathways dependent or independent of DNA synthesis. *Cell* **116**: 51–61.

Tesar, P.J., Chenoweth, J.G., Brook, F.A., Davies, T.J., Evans, E.P., Mack, D.L., Gardner, R.L., and McKay, R.D. 2007. New cell lines from mouse epiblast share defining features with human embryonic stem cells. *Nature* **448**: 196–199.

Tremblay, K.D., Dunn, N.R., and Robertson, E.J. 2001. Mouse embryos lacking Smad1 signals display defects in extra-embryonic tissues and germ cell formation. *Development* **128**: 3609–3621.

Tsuneyoshi, N., Sumi, T., Onda, H., Nojima, H., Nakatsuji, N., and Suemori, H. 2008. PRDM14 suppresses expression of differentiation marker genes in human embryonic stem cells. *Biochem. Biophys. Res. Commun.* **367**: 899–905.

Vincent, S.D., Dunn, N.R., Sciammas, R., Shapiro-Shalef, M., Davis, M.M., Calame, K., Bikoff, E.K., and Robertson, E.J. 2005. The zinc finger transcriptional repressor Blimp1/Prdm1 is dispensable for early axis formation but is required for specification of primordial germ cells in the mouse. *Development* **132**: 1315–1325.

Yabuta, Y., Kurimoto, K., Ohinata, Y., Seki, Y., and Saitou, M. 2006. Gene expression dynamics during germline specification in mice identified by quantitative single-cell gene expression profiling. *Biol. Reprod.* **75**: 705–716.

Yamaguchi, S., Kimura, H., Tada, M., Nakatsuji, N., and Tada, T. 2005. Nanog expression in mouse germ cell development. *Gene Expr. Patterns* **5**: 639–646.

Yamaji, M., Seki, Y., Kurimoto, K., Yabuta, Y., Yuasa, M., Shigeta, M., Yamanaka, K., Ohinata, Y., and Saitou, M. 2008. Critical function of Prdm14 for the establishment of the germ cell lineage in mice. *Nat. Genet.* **40**: 1016–1022.

Youngren, K.K., Coveney, D., Peng, X., Bhattacharya, C., Schmidt, L.S., Nickerson, M.L., Lamb, B.T., Deng, J.M., Behringer, R.R., Capel, B., Rubin, E.M., Nadeau, J.H., and Matin, A. 2005. The Ter mutation in the dead end gene causes germ cell loss and testicular germ cell tumours. *Nature* **435**: 360–364.

Brief History, Pitfalls, and Prospects of Mammalian Spermatogonial Stem Cell Research

M. Kanatsu-Shinohara,* M. Takehashi,*† and T. Shinohara*‡

*Department of Molecular Genetics, Graduate School of Medicine, Kyoto University, Kyoto
606-8501, Japan; ‡Japan Science and Technology Agency, CREST, Kyoto 606-8501, Japan

Spermatogonial stem cells (SSCs) provide the foundation for spermatogenesis. During the last decade, several techniques for the manipulation of this cell type have been developed; as a result, SSCs can now be subjected to long-term in vitro expansion and genetically manipulated for knockout mouse production. These techniques have allowed SSCs to serve as a new target for animal transgenesis, which may provide an alternative to embryonic stem (ES) cells. Furthermore, SSCs may be converted into ES-like cells, demonstrating that the postnatal testis is a source of pluripotent stem cells. These techniques were first established in mice, but they are currently being extended to other animal species. SSC-based technologies will be useful in agriculture and medicine and will also provide valuable opportunities to study SSC biology. The mechanisms of self-renewal division and differentiation and the regulation of pluripotency in SSCs are now being studied at the molecular level. However, some technical and conceptual pitfalls must be kept in mind when designing and analyzing experimental results. Nevertheless, these advances in SSC research will provide valuable insight into the study of mammalian stem cell systems.

Unlike other stem cells, SSCs are distinct in that they can transmit genetic information to the next generation. Although the number of SSCs is very small (2×10^4 to 3×10^4 cells per mouse testis) (Meistrich and van Beek 1993; de Rooij and Russell 2000), they continue to proliferate throughout life and provide the foundation for spermatogenesis. Although most previous studies of SSCs have involved morphological analyses, the detailed behavior of this rare cell population has been determined precisely. Many basic concepts in SSC behavior were elaborated through these rigorous analyses and have survived the test of time. However, by definition, stem cells are identified only by their ability to undergo self-renewal, which becomes evident only retrospectively after the production of two daughter cells. The lack of such a functional assay has limited studies on SSCs.

DEVELOPMENT OF A GERM CELL TRANSPLANTATION TECHNIQUE: FUNCTIONAL ANALYSIS OF SSCS

In 1994, Dr. Ralph Brinster developed a germ cell transplantation technique, in which dissociated donor testicular cells colonized and underwent spermatogenesis in the empty seminiferous tubules of infertile recipient mice (Brinster and Zimmermann 1994). The recipient animals were eventually able to produce offspring. This experiment had two important implications. First, it provided a functional assay for SSCs. Only stem cells can proliferate over the long term and undergo spermatogenesis after transplantation; other differentiated progenitor cells disappear after 35 days, or one cycle of mouse spermatogenesis (Meistrich and van Beek 1993; de Rooij and Russell

2000). This technique is somewhat similar to the transplantation assay for hematopoietic stem cells (HSCs), in which stem cells are transplanted into irradiated recipients (Till and McCulloch 1961). Second, this technique allowed for the first manipulation of the male germ line for transgenic animal production. By genetically manipulating SSCs and transplanting these cells into recipient animals, donor stem cells may be used to produce transgenic animals, thereby providing a competitive approach for ES-cell-based mutagenesis.

Since the development of this germ cell transplantation technique, significant progress has been made in several areas of SSC biology; however, the most direct application of this technique was the phenotypic characterization of SSCs. By taking advantage of the concepts of HSC purification, cell surface SSC markers were identified systematically by fractionating and transplanting testicular cells (Shinohara et al. 1999, 2000; Kubota et al. 2003). Second, it became possible to analyze whether specific types of spermatogenic defects were caused by the stem cells or their microenvironment during reciprocal transplantation (Ogawa et al. 2000). Third, xenogeneic spermatogenesis was achieved by interspecific germ cell transplantation. Remarkably, rat spermatogenesis occurred in immunodeficient nude mouse testis, indicating significant flexibility in germ cell–Sertoli cell interactions (Clouthier et al. 1996; Shinohara et al. 2006).

Although germ cell transplantation was very useful in analyzing the biology of SSCs, little progress was made in the genetic manipulation of the male germ line. In fact, the germ cell transplantation technique was originally developed for the genetic manipulation of animal species, an application for which current ES-cell-based gene targeting technique have not been applied. Although ES cells are available for several animal species, only mouse ES cells can produce germ cells, and it has been impossible to produce knockout animals in most species. Although knockout

---

†Present address: Laboratory of Pathophysiology and Pharmacotherapeutics, Faculty of Pharmacy, Osaka Ohtani University, Tondabayashi, Osaka 584-8540, Japan.

animals can be produced by the nuclear transplantation of genetically engineered somatic cells (Denning and Priddle 2003), the limited proliferative potential of somatic cells and inefficient production of offspring has restricted its practical applications. Therefore, it is important to establish a novel method for the production of knockout animals that may be used in a wide range of species. However, although transgenic mice and rats were produced by transfecting SSCs with retroviral vectors, the efficiency was low, and it was not possible to produce knockout animals (Nagano et al. 2001b; Hamra et al. 2002).

THE DEVELOPMENT OF A LONG-TERM CULTURE SYSTEM: GERM-LINE STEM CELLS

The biggest hurdle for knockout animal production was the inability to culture SSCs. Although SSCs can increase in number in vivo (Kanatsu-Shinohara et al. 2003a; Ogawa et al. 2003), more than 90% of cultured SSCs undergo apoptosis within 1 week, probably because of a lack of a self-renewal factor (Orwig et al. 2002). In 2000, glial-cell-line-derived neurotrophic factor (GDNF) was found to induce the proliferation of spermatogonia (Meng et al. 2000). Whereas homozygous GDNF knockout mice die perinatally, heterozygous knockout mice exhibit reduced spermatogenesis and eventually become infertile because of germ cell depletion. In contrast, GDNF-transgenic mice possess clumps of undifferentiated spermatogonia, suggesting that GDNF stimulates the self-renewal division of SSCs.

Taking advantage of this finding, our group succeeded in the long-term culture of SSCs in 2003 (Kanatsu-Shinohara et al. 2003b). In the presence of GDNF, SSCs produced uniquely shaped germ cell colonies that were apparently different from ES cell colonies (see Fig. 1a). Unlike ES cells, the cells in the germ cell colony were loosely attached to one another and could be easily dissociated without trypsin. We designated these cells as germ-line stem (GS) cells. Although GS cells were originally established from neonatal testis, similar cells were subsequently established from adult testis, demonstrating that GS cells can be derived from SSCs at various stages (Kanatsu-Shinohara et al. 2004b; Kubota et al. 2004; Ogawa et al. 2004). Subsequent studies have shown that GS cells are very stable in terms of their germ-line potential and can produce normal fertile offspring even after 2 years of culture (Fig. 1c) (Kanatsu-Shinohara et al. 2005c). Surprisingly, the cultured cells retain a normal karyotype and DNA methylation pattern. This is in contrast to ES cells, which often lose their germ cell potential due to trisomy (Liu et al. 1997; Longo et al. 1997). Similar to ES cells, GS cells can be used to produce transgenic and knockout animals via genetic transduction and drug selection (Kanatsu-Shinohara et al. 2005a, 2006c). However, the efficiency of transgenesis is about 50%, which reflects the fact that the transgene is transmitted to half of the haploid cells. The level of efficiency is five to ten times higher than that achieved by conventional methods using eggs or oocytes (Nagano et al. 2001b). Moreover, the frequency of homologous recombination is comparable to that achieved in ES cells (Kanatsu-Shinohara et al. 2006c). Thus, SSCs may be used as a vehicle for gene targeting.

An unexpected finding that developed from these experiments was the pluripotency of SSCs. Although primordial germ cells (PGCs), the fetal precursors of SSCs, can give rise to ES-like pluripotent cells (Matsui et al. 1992; Resnick et al. 1992), it was thought that germ-line cells are fully committed to the germ line by the middle of gestation and that such ES-like potential is missing from postnatal germ cells (Labosky et al. 1994). In the course of our gene-targeting experiments, we detected abnormal colonies in the GS cell culture that subsequently transformed into ES-like cells (Fig. 1b) (Kanatsu-Shinohara et al. 2004b). These cells, referred to as multipotent GS (mGS) cells, differentiate not only into somatic cells, but also into germ cells, and they are capable of producing knockout animals in a manner similar to that of ES cells (Fig. 1d) (Takehashi et al. 2007b). The pluripotency of SSCs was subsequently confirmed by several other groups (Guan et al. 2006; Seandel et al. 2007; Izadyar et al. 2008; Ogawa 2008). However, there are pronounced differences in the phenotypic and functional characteristics of these ES-like cells, which will be the focus of future studies. Although the ori-

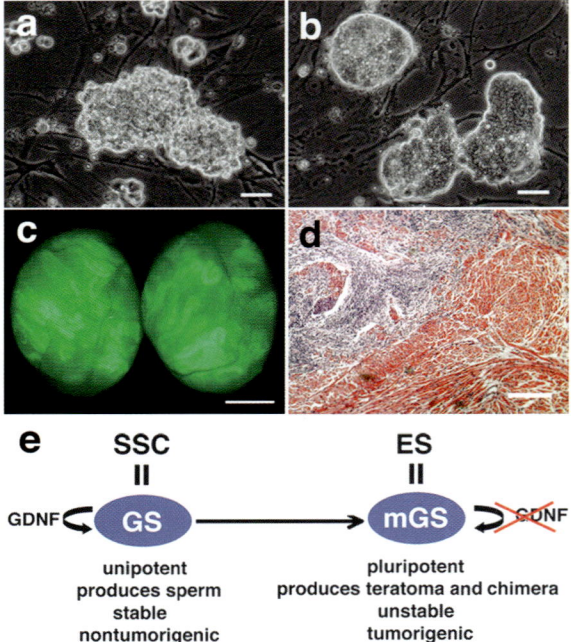

Figure 1. GS and mGS cell characteristics. (*a*) GS cell culture; (*b*) mGS cell culture; (*c*) germ cell colonies formed from GS cells by transplantation into testes. Green fluorescence indicates germ cell colonies from enhanced green fluorescent protein (EGFP)-expressing GS cells. (*d*) A teratoma produced from mGS cells by subcutaneous transplantation. (*e*) Schema showing the features of GS cells and mGS cells. GS cells are unipotent and undergo spermatogenesis in the testes. In contrast, mGS cells are pluripotent, like ES cells, and produce teratomas. Although GS cells maintain a normal karyotype, genomic imprinting, and SSC activity during long-term culture, mGS cells are unstable and prone to accumulate genomic and epigenomic abnormalities. A recent report demonstrated that mGS cells are derived from the conversion of GS cells in culture. Bars, (*a,b*) 50 μm; (*c*) 1 mm; (*d*) 200 μm.

gin of mGS cells is unclear, we recently found that GS cells may be converted directly into mGS cells in vitro (Fig. 1e) (Kanatsu-Shinohara et al. 2008b); however, the process is very rare (mGS cells were established in only 1 of 30 testes), and attempts at increasing the conversion rate have been unsuccessful. Although the activation of Akt enhances ES-like cell formation from PGCs (Kimura et al. 2008), the same treatment only stimulates GS cell self-renewal (Lee et al. 2007); i.e., it does not enhance their conversion into mGS cells. Therefore, SSCs may be pluripotent, but their conversion into ES-like cells occurs in a manner different from that of PGCs.

PROBLEMS AND PITFALLS IN SSC RESEARCH

The development of germ cell transplantation and culture techniques has enabled functional analyses and the manipulation of SSCs. No other self-renewing tissues have these advantages; thus, spermatogenesis is an ideal model system for analyzing the regulation of stem cells. However, there are several pitfalls that may confuse the novice in the field.

Use of Functional Analyses to Define SSCs

The identification of SSC markers has facilitated the identification of SSCs; however, no SSC-specific marker allows the definitive identification of SSCs in situ. For example, not all markers expressed in germ cells from neonatal or immature pup testes should be considered bona fide SSC markers. In fact, only a small fraction of these gonocytes or spermatogonia act as SSCs on the basis of functional criteria (Shinohara et al. 2001). Likewise, the genes that are expressed in primitive germ cell colonies (e.g., chain or network) may not always be appropriate as SSC markers. For example, *Neurog3*, which is expressed in a chain of spermatogonia (Yoshida et al. 2004), is now known to be more strongly expressed after SSCs start to differentiate (Oatley et al. 2006; Lee et al. 2007). SSC markers should be determined only by prospective isolation and functional transplantation assay.

Although the transplantation assay is the only reliable established method for the identification of SSCs, it should be kept in mind that the criteria for SSCs determined by transplantation are also flexible. For example, the choice of recipient animals can influence the outcome of an experiment. Although approximately 10% of transplanted As (single) cells undergo spermatogenesis in adult recipients (Nagano et al. 1999), the transplantation of the same cell population into pup recipients results in approximately ten times greater colonization (Shinohara et al. 2001). This result suggests that most of the As cells can colonize pup recipients and that As cells and SSCs, as defined by transplantation assay in the adult recipients, are not equivalent. Hormonal environment in the recipient testis also influences the level of colonization significantly (Ogawa et al. 1999a). Thus, the outcome of a transplantation assay depends on the recipient animal, and great care must be taken to use these definitions when interpreting experimental data.

Not All Cultured Cells Are Bona Fide SSCs

Another potential source of confusion in SSC research is the distinction between proliferation and self-renewal. For example, SSCs proliferate in response to GDNF in vitro. Without GDNF, no growth occurs, and all of the cells undergo apoptosis within weeks. However, this does not necessarily mean that all cells that respond to GDNF are SSCs. Our analyses have shown that only 1–2% of GS cells have the potential to colonize seminiferous tubules (Kanatsu-Shinohara et al. 2005c). Therefore, the remaining cells should be defined as differentiated progenitor cells. The distinction between SSCs and progenitor cells is particularly important in the analysis of in vitro data. Because the definition of a stem cell is based on functional criteria, it is impossible to define cultured cells as stem cells simply because they proliferate. For example, imagine an experiment aimed at identifying a molecule that enhances the proliferation of the cultured cells. Even if one of the test molecules efficiently enhances the proliferation of the cultured cells, it does not necessarily mean that it is enhancing self-renewal division. Cultured cells should be assessed for the number of stem cells by transplantation before and after the addition of a test molecule to determine its effect on SSC proliferation. This will determine whether the molecule is involved in self-renewal, proliferation, or apoptosis and cancel its potential effect in colonization.

Definition of a Germ Cell Colony

Determining the number of colonies in a recipient testis sounds easy, but there are several potential problems. When SSCs are transplanted, donor cells produce colonies of various sizes (Nagano et al. 1999). Because a single SSC produces a single colony (Kanatsu-Shinohara et al. 2006b), this difference in size indicates that the proliferative potential of individual SSCs is variable. The regulatory mechanism underlying this process awaits further study, but it poses a practical problem for the functional analysis of SSCs. In fact, it is impossible to count all of the blue cell clusters in many germ cell transplantation experiments; thus, specific criteria to define a germ cell colony must be set to estimate the number of SSCs in a transplanted cell population. In our work, we have slightly modified the criterion of Nagano et al. (1999), and a cluster of germ cells are defined as a colony when it is at least 0.1 mm in length and occupies the entire circumference of the seminiferous tubule.

Conversely, individual stem cells may not produce single colonies under other experimental situations. The clonal origin of a germ cell colony may be confirmed only via transplantation experiments and cannot be applied to other cases. For example, when SSCs are labeled in situ (Kanatsu-Shinohara et al. 2004a), the germ cell colonies are generally smaller and have a morphology different from those that develop after germ cell transplantation, which indicates that germ cell colony formation has different kinetics during normal spermatogenesis. In such cases, each colony is not necessarily derived from a single stem cell. It is possible that several clusters of SSCs may reside in a niche and only appear to form a single colony; however, no study has addressed this point.

Technical Problems

A well-known problem in SSC research is transplantation-induced inflammation. Recipient testes often suffer severe inflammation starting several weeks after transplantation. Inflammation occurs most frequently in adult recipients and is rarely observed in immature pup testes (T. Shinohara, unpubl.). The mechanism underlying this response has not been clarified; however, trypan blue may be a causative factor (K. Jahnukainen, pers. comm.). Trypan blue is very useful for examining the success of microinjection and is often used in germ cell transplantation. However, it is advisable to avoid using trypan blue if signs of inflammation are observed after transplantation.

Another relatively frequent problem is the failure to initiate and maintain GS cells. The initiation of GS cell culture is somewhat tricky because they usually divide very slowly in vivo. These cells are probably unable to proliferate quickly when they are placed in culture. Instead, somatic cells often grow faster in vitro to overcome GS cell growth. This problem can be alleviated, in part, by reducing the serum concentration and increasing the GDNF concentration. The growth of somatic cells is effectively blocked by reducing the serum concentration without disturbing GS cell growth. However, SSCs cannot grow in the complete absence of serum at present. Although two different serum-free culture systems were reported (Kubota et al. 2004; Kanatsu-Shinohara et al. 2005b), the culture was maintained using serum to stop trypsin reaction at each passage in one study (Kubota and Brinster 2008). Moreover, we have not been able to use the serum-free medium to culture GS cells under feeder-free conditions using laminin or suspension cultures (Kanatsu-Shinohara et al. 2005b, 2006a), suggesting that a residual amount of serum included with the feeder cells may promote GS cell propagation. In contrast, low concentrations of GDNF are detrimental to GS cell culture; GS cells do not increase in number in the absence of GDNF.

The culture of GS cells may be initiated from both neonatal and adult testes (Kanatsu-Shinohara et al. 2003b, 2004b; Kubota et al. 2004; Ogawa et al. 2004). Immature testes are useful because the germ cell/somatic cell ratio is relatively high, and germ cells are easily separated from somatic cells based on their differential ability to attach to a gelatin-coated dish. In contrast, it is a prerequisite to purify SSCs from adult testes because only 0.02–0.03% of the cells are stem cells. In both cases, it is advisable to remove as many somatic cells as possible.

Finally, unlike ES cells, GS cells do not grow robustly and their genetic manipulation is still difficult. For example, because the proliferation of GS cells is influenced by their density, clonal expansion from a single transfected GS cell requires mixing with nontransfected cells to maintain proliferation at low densities following drug selection (Kanatsu-Shinohara et al. 2005a). In addition, although the genetic manipulation of GS cells offers the possibility of examining the function of specific genes in SSCs, care must be taken to confirm that the manipulation will not

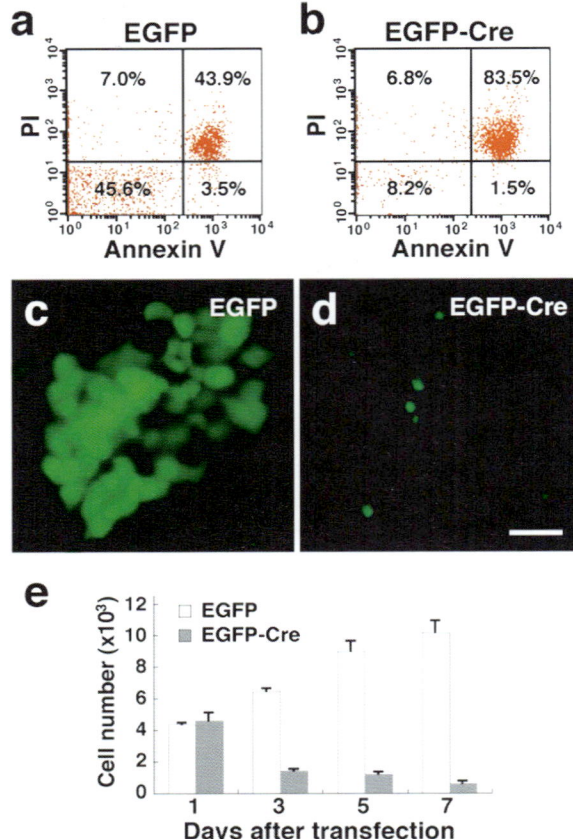

Figure 2. Toxic effects of the overexpression of Cre in GS cells. (*a,b*) GS cells were transfected with a vector expressing enhanced green fluorescent protein (EGFP, control) (*a*) or an EGFP-Cre fusion (*b*) under control of the CAG promoter. To detect cells that were dead or undergoing apoptosis, the cells were stained with annexin V and propidium iodide (PI) after 24 hours and analyzed by flow cytometry. Widespread cell death in response to Cre overexpression was observed. (*c,d*) Growth of a GS cell colony 9 days after transfection of the EGFP (*c*) or EGFP-Cre (*d*) transgene. GS cell colony growth was suppressed when EGFP-Cre was overexpressed in GS cells. Similar results were obtained following the overexpression of a nonfusion Cre transgene by lipofection, electroporation, and retroviral infection. (*e*) GS cell survival after transfection. GS cells (2.5×10^5 cells) were transfected with the plasmids, and the number of EGFP-expressing germ cells was determined at 1, 3, 5, and 7 days after transfection. The values are the mean ± S.E.M. (*n* = 3). Bar, 50 μm for both *c* and *d*.

have a toxic effect on the GS cells. We found that GS cells are extremely responsive to Cre expression and that stable Cre expression results in apoptosis (Fig. 2). However, this problem was overcome by using an adenovirus that can transiently express Cre in GS cells (Takehashi et al. 2007a). These technical limitations must be considered when designing experiments using GS cells.

PERSPECTIVES IN SSC RESEARCH

The transplantation and culture techniques described above were originally developed in mice. However, ES-cell-based technologies are already available for mouse mutagenesis, in which more than 70% of the total genes are mutated. Therefore, the real value of SSCs lies in their ability to induce mutagenesis in other animal species in which ES-cell-based techniques are not applicable. The use of SSCs has several advantages over conventional methods based on eggs/oocytes. First, GS cells are established from postnatal testes, whereas ES cells are derived from embryos. Unlike mice, most animals do not ovulate large numbers of oocytes and they require a long period of time to reach sexual maturity. These factors limit the genetic manipulation of such animal species. Therefore, GS cell technology has an advantage in many animal species that produce small numbers of offspring. Second, GS cells have very stable genetic and epigenetic properties, probably reflecting the features of SSCs as committed stem cells. This is in contrast to ES cells. Because they are uncommitted to a specific lineage, ES cells can easily differentiate into other lineages, but they may lose their germ-line potential. Therefore, we have good reason to think that GS cells will become a target for mutagenesis in many animal species.

Following its development in 1994, germ cell transplantation has been applied to several animal species, including rats, pigs, bulls, goats, monkeys, and humans (Jiang and Short 1995; Ogawa et al. 1999b; Schlatt et al. 1999; Brook et al. 2001; Honoramooz et al. 2002, 2003). In contrast, the culture of SSCs is only now being developed. The culture conditions for rat and hamster GS cells have been established (Hamra et al. 2005; Ryu et al. 2005; Kanatsu-Shinohara et al. 2008a). Although these results suggest a promising future for the manipulation of SSCs, their application remains limited because of factors such as low fertility (Brinster and Avarbock 1994; Ogawa et al. 1999a). Outside of rodents, the transplantation of SSCs has been used to produce offspring only in goats (Honoramooz et al. 2003). Although the problem of low fertility was successfully overcome by the transplantion of SSCs into immature pup testes, the same treatment may not be applicable to other animal species because of differences in testicular anatomy and biology. In vitro microinsemination may help in some, but apparently not all, cases of infertility.

GS cell culture techniques should also be improved. Although SSCs from rats and hamsters have been cultured successfully, their growth is significantly slower than that of SSCs from mice, which limits the possibility of genetic manipulation. Because SSCs from many animal species grow efficiently in mouse testis (Dobrinski et al. 1999, 2000; Ogawa et al. 1999b; Nagano et al. 2001a), these germ cells could be cultured under conditions similar to those of mouse GS cells.

CONCLUSIONS

To solve these problems, two issues must be addressed: the improvement of culture conditions and the establishment of an in vitro differentiation technique for GS cells. First, to improve the culture conditions, efforts to identify additional self-renewal factors and establish complete serum- and feeder-free culture systems should be continued. Given the fact that xenogeneic donor stem cells from several animal species can proliferate in mouse testes, there is no doubt that this will become possible in the near future. Second, although germ cell transplantation has been attempted in large animal species and higher primates, low transplantation efficiency and complicated recipient preparation make it unlikely for practical applications. However, with the development of an in vitro sperm differentiation technique, there will be no need to keep animals for germ cell transplantation. The best-case scenario would be to recover SSCs from a piece of testis for GS cell establishment and direct the cells in vitro for sperm production after necessary genetic maniuplations. Those sperm that develop in vitro would in turn be used to fertilize eggs for transgenic animal production. In this way, there will be no need to maintain numerous males for transgenesis, and such a scheme will make mammalian mutagenesis a reality.

ACKNOWLEDGMENTS

Financial support for this study was provided by the Ministry of Education, Culture, Sports, Science, and Technology of Japan. This work was also supported by the Genome Network Project and by the Program for Promotion of Fundamental Studies in Health Sciences of the National Institute of Biomedical Innovation (NIBIO), and JST, CREST.

REFERENCES

Brinster, R.L. and Avarbock, M.R. 1994. Germline transmission of donor haplotype following spermatogonial transplantation. *Proc. Natl. Acad. Sci.* **91:** 11303–11307.

Brinster, R.L. and Zimmermann, J.W. 1994. Spermatogenesis following male germ-cell transplantation. *Proc. Natl. Acad. Sci.* **91:** 11298–11302.

Brook, P.F., Radford, J.A., Shalet, S.M., Joyce, A.D., and Gosden, R.G. 2001. Isolation of germ cells from human testicular tissue for low temperature storage and autotransplantation. *Fertil. Steril.* **75:** 269–274.

Clouthier, D.E., Avarbock, M.R., Maika, S.D., Hammer, R.E., and Brinster, R.L. 1996. Rat spermatogenesis in mouse testis. *Nature* **381:** 418–421.

Denning, C. and Priddle, H. 2003. New frontiers in gene targeting and cloning: Success, application and challenges in domestic animals and human embryonic stem cells. *Reproduction* **126:** 1–11.

de Rooij, D.G. and Russell, L.D. 2000. All you wanted to know about spermatogonia but were afraid to ask. *J. Androl.* **21:** 776–798.

Dobrinski, I., Avarbock, M.R., and Brinster, R.L. 1999. Transplantation of germ cells from rabbits and dogs into mouse testes. *Biol. Reprod.* **61:** 1331–1339.

Dobrinski, I., Avarbock, M.R., and Brinster, R.L. 2000. Germ cell transplantation from large domestic animals into mouse testes. *Mol. Reprod. Dev.* **57:** 270–279.

Guan, K., Nayernia, K., Maier, L.S., Wagner, S., Dressel, R., Lee, J.H., Nolte, J., Wolf, F., Li, M., Engel, W., and Hasenfuss, G. 2006. Pluripotency of spermatogonial stem cells from adult mouse testis. *Nature* **440:** 1199–1203.

Honaramooz, A., Megee, S.O., and Dobrinski, I. 2002. Germ cell transplantation in pigs. *Biol. Reprod.* **66:** 21–28.

Honaramooz, A., Behboodi, E., Megee, S.O., Overton, S.A., Galantino-Homer, H., Echelard, Y., and Dobrinski, I. 2003. Fertility and germline transmission of donor haplotype following germ cell transplantation in immunocompetent goats. *Biol. Reprod.* **69:** 1260–1264.

Hamra, F.K., Chapman, K.M., Nguyen, D.M., Williams-Stephens, A.A., Hammer, R.E., and Garbers, D.L. 2005. Self-renewal, expansion, and transfection of rat spermatogonial stem cells in culture. *Proc. Natl. Acad. Sci.* **102:** 17430–17435.

Hamra, F.K., Gatlin, J., Chapman, K.M., Grellhesl, D.M., Garcia, J.V., Hammer, R.E., and Garbers, D.L. 2002. Production of transgenic rats by lentiviral transduction of male germ-line stem cells. *Proc. Natl. Acad. Sci.* **99:** 14931–14936.

Izadyar, F., Pau, F., Marh, J., Slepko, N., Wang, T., Gonzalez, R., Ramos, T., Howerton, K., Sayre, C., and Silva, F. 2008. Generation of multipotent cell lines from a distinct population of male germ line stem cells. *Reproduction* **135:** 771–784.

Jiang, F.X. and Short, R.V. 1995. Male germ cell transplantation in rats: Apparent synchronization of spermatogenesis between host and donor seminiferous epithelia. *Int. J. Androl.* **18:** 326–330.

Kanatsu-Shinohara, M., Toyokuni, S., and Shinohara, T. 2004a. Transgenic mouse produced by retroviral transduction of male germline stem cells in vivo. *Biol. Reprod.* **71:** 1202–1207.

Kanatsu-Shinohara, M., Toyokuni, S., and Shinohara, T. 2005a. Genetic selection of mouse male germline stem cells in vitro: Offspring from single stem cells. *Biol. Reprod.* **72:** 236–240.

Kanatsu-Shinohara, M., Toyokuni, S., Morimoto, T., Matsui, S., Honjo, T., and Shinohara, T. 2003a. Functional assessment of self-renewal activity of male germline stem cells following cytotoxic damage and serial transplantation. *Biol. Reprod.* **68:** 1801–1807.

Kanatsu-Shinohara, M., Miki, H., Inoue, K., Ogonuki, N., Toyokuni, S., Ogura, A., and Shinohara, T. 2005b. Long-term culture of mouse male germline stem cells under serum- or feeder-free conditions. *Biol. Reprod.* **72:** 985–991.

Kanatsu-Shinohara, M., Ogonuki, N., Inoue, K., Miki, H., Ogura, A., Toyokuni, S., and Shinohara, T. 2003b. Long-term proliferation in culture and germline transmission of mouse male germline stem cells. *Biol. Reprod.* **69:** 612–616.

Kanatsu-Shinohara, M., Inoue, K., Lee, J., Miki, H., Ogonuki, N., Toyokuni, S., Ogura, A., and Shinohara, T. 2006a. Anchorage-independent growth of mouse male germline stem cells in vitro. *Biol. Reprod.* **74:** 522–529.

Kanatsu-Shinohara, M., Inoue, K., Miki, H., Ogonuki, N., Takehashi, M., Morimoto, T., Ogura, A., and Shinohara, T. 2006b. Clonal origin of germ cell colonies after spermatogonial transplantation in mice. *Biol. Reprod.* **75:** 68–74.

Kanatsu-Shinohara, M., Muneto, T., Lee, J., Takenaka, M., Chuma, S., Nakatsuji, N., Horiuchi T., and Shinohara, T. 2008a. Long-term culture of male germline stem cells from hamster testes. *Biol. Reprod.* **78:** 611–617.

Kanatsu-Shinohara, M., Ikawa, M., Takehashi, M., Ogonuki, N., Miki, H., Inoue, K., Kazuki, Y., Lee, J., Toyokuni, S., Oshimura, M., et al. 2006c. Production of knockout mice by random and targeted mutagenesis in spermatogonial stem cells. *Proc. Natl. Acad. Sci.* **103:** 8018–8023.

Kanatsu-Shinohara, M., Inoue, K., Lee, J., Yoshimoto, M., Ogonuki, N., Miki, H., Baba, S., Kato, T., Kazuki, Y., Toyokuni, S., et al. 2004b. Generation of pluripotent stem cells from neonatal mouse testis. *Cell* **119:** 1001–1012.

Kanatsu-Shinohara, M., Lee, J., Inoue, K., Ogonuki, N., Miki, H., Toyokuni, S., Ikawa, M., Nakamura, T., Ogura, A., and Shinohara, T. 2008b. Pluripotency of a single spermatogonial stem cell in mice. *Biol. Reprod.* **78:** 681–687.

Kanatsu-Shinohara, M., Ogonuki, N., Iwano, T., Lee, J., Kazuki, Y., Inoue, K., Miki, H., Takehashi, M., Toyokuni, S., Oshimura, M., et al. 2005c. Genetic and epigenetic properties of mouse male germline stem cells during long-term culture. *Development* **132:** 4155–4163.

Kimura, T., Tomooka, M., Yamano, N., Murayama, K., Matoba, S., Umehara, H., Kanai, Y., and Nakano, T. 2008. AKT signaling promotes derivation of embryonic germ cells from primordial germ cells. *Development* **135:** 869–879.

Kubota, H. and Brinster, R.L. 2008. Culture of rodent spermatogonial stem cells, male germline stem cells of the postnatal animal. *Methods Cell Biol.* **86:** 59–84.

Kubota, H., Avarbock, M.R., and Brinster, R.L. 2003. Spermatogonial stem cells share some, but not all, phenotypic and

functional characteristics with other stem cells. *Proc. Natl. Acad. Sci.* **100:** 6487–6492.

Kubota, H., Avarbock, M.R., and Brinster, R.L. 2004. Growth factors essential for self-renewal and expansion of mouse spermatogonial stem cells. *Proc. Natl. Acad. Sci.* **101:** 16489–16494.

Labosky, P.A., Barlow, D.P., and Hogan, B.L.M. 1994. Mouse embryonic germ (EG) cell lines: Transmission through the germline and differences in the methylation imprint of insulin-like growth factor 2 receptor (*Igf2r*) gene compared with embryonic stem (ES) cell lines. *Development* **120:** 3197–3204.

Lee, J., Kanatsu-Shinohara, M., Inoue, K., Ogonuki, N., Miki, H., Toyokuni, S., Kimura, T., Nakano, T., Ogura, A., and Shinohara, T. 2007. Akt mediates self-renewal division of mouse spermatogonial stem cells. *Development* **134:** 1853–1859.

Liu, X., Wu, H., Loring, J., Hormuzdi, S., Disteche, C.M., Bornstein, P., and Jaenisch, R. 1997. Trisomy eight in ES cells is a common potential problem in gene targeting and interferes with germ line transmission. *Dev. Dyn.* **209:** 85–91.

Longo, L., Bygrave, A., Grosveld, F.G., and Pandolfi, P.P. 1997. The chromosome make-up of mouse embryonic stem cells is predictive of somatic and germ cell chimaerism. *Transgenic Res.* **6:** 321–328.

Matsui, Y., Zsebo, K., and Hogan, B.L. 1992. Derivation of pluripotential embryonic stem cells from murine primordial germ cells in culture. *Cell* **70:** 841–847.

Meistrich, M.L. and van Beek, M.E.A.B. 1993. Spermatogonial stem cells. In *Cell and molecular biology of the testis* (ed. C. Desjardins and L.L. Ewing), pp. 266–295. Oxford University Press, New York.

Meng, X., Lindahl, M., Hyvönen, M.E., Parvinen, M., de Rooij, D.G., Hess, M.W., Raatikainen-Ahokas, A., Sainio, K., Rauvala, H., Lakso, M., et al. 2000. Regulation of cell fate decision of undifferentiated spermatogonia by GDNF. *Science* **287:** 1489–1493.

Nagano, M., Avarbock, M.R., and Brinster, R.L. 1999. Pattern and kinetics of mouse donor spermatogonial stem cell colonization in recipient testes. *Biol. Reprod.* **60:** 1429–1436.

Nagano, M., McCarrey, J.R., and Brinster, R.L. 2001a. Primate spermatogonial stem cells colonize mouse testes. *Biol. Reprod.* **64:** 1409–1416.

Nagano, M., Brinster, C.J., Orwig, K.E., Ryu, B.Y., Avarbock, M.R., and Brinster, R.L. 2001b. Transgenic mice produced by retroviral transduction of male germ-line stem cells. *Proc. Natl. Acad. Sci.* **98:** 13090–13095.

Oatley, J.M., Avarbock, M.R., Telaranta, A.I., Fearon, D.T., and Brinster, R.L. 2006. Identifying genes important for spermatogonial stem cell self-renewal and survival. *Proc. Natl. Acad. Sci.* **103:** 9524–9529.

Ogawa, T. 2008. Reproductive stem cell research and its application to urology. *Int. J. Urol.* **15:** 121–127.

Ogawa, T., Dobrinski, I., and Brinster, R.L. 1999a. Recipient preparation is critical for spermatogonial transplantation in the rat. *Tissue Cell* **31:** 461–472.

Ogawa, T., Dobrinski, I., Avarbock, M.R., and Brinster, R.L. 1999b. Xenogeneic spermatogenesis following transplantation of hamster germ cells to mouse testes. *Biol. Reprod.* **60:** 515–521.

Ogawa, T., Dobrinski, I., Avarbock, M.R., and Brinster, R.L. 2000. Transplantation of male germ line stem cells restores fertility in infertile mice. *Nat. Med.* **6:** 29–34.

Ogawa, T., Ohmura, M., Yumura, Y., Sawada, H., and Kubota, Y. 2003. Expansion of murine spermatogonial stem cells through serial transplantation. *Biol. Reprod.* **68:** 316–322.

Ogawa, T., Ohmura, M., Tamura, Y., Kita, K., Ohbo, K., Suda, T., and Kubota, Y. 2004. Derivation and morphological characterization of mouse spermatogonial stem cell lines. *Arch. Histol. Cytol.* **67:** 307–314.

Orwig, K.E., Avarbock, M.R., and Brinster, R.L. 2002. Retrovirus-mediated modification of male germline stem cells in rats. *Biol. Reprod.* **67:** 874–879.

Resnick, J.L., Bixler, L.S., Cheng, L., and Donovan, P.J. 1992.

Long-term proliferation of mouse primordial germ cells in culture. *Nature* **359:** 550–551.

Ryu, B.Y., Kubota, H., Avarbock, M.R., and Brinster, R.L. 2005. Conservation of spermatogonial stem cell self-renewal signaling between mouse and rat. *Proc. Natl. Acad. Sci.* **102:** 14302–14307.

Schlatt, S., Rosiepen, G., Weinbauer, G.F., Rolf, C., Brook, P.F., and Nieschlag E. 1999. Germ cell transfer into rat, bovine, monkey and human testes. *Hum. Reprod.* **14:** 144–150.

Seandel, M., James, D., Shmelkov, S.V., Falciatori, I., Kim, J., Chavala, S., Scherr, D.S., Zhang, F., Torres, R., Gale, N.W., et al. 2007. Generation of functional multipotent adult stem cells from GPR125$^+$ germline progenitors. *Nature* **449:** 346–350.

Shinohara, T., Avarbock, M.R., and Brinster, R.L. 1999. β1- and α6-integrin are surface markers on mouse spermatogonial stem cells. *Proc. Natl. Acad. Sci.* **96:** 5504–5509.

Shinohara, T., Orwig, K.E., Avarbock, M.R., and Brinster, R.L. 2000. Spermatogonial stem cell enrichment by multiparameter selection of mouse testis cells. *Proc. Natl. Acad. Sci.* **97:** 8346–8351.

Shinohara, T., Orwig, K.E., Avarbock, M.R., and Brinster, R.L. 2001. Remodeling of the postnatal mouse testis is accompa-nied by dramatic changes in stem cell number and niche accessibility. *Proc. Natl. Acad. Sci.* **98:** 6186–6191.

Shinohara, T., Kato, M., Takehashi, M., Lee, J., Chuma, S., Nakatsuji, N., Kanatsu-Shinohara, M., and Hirabayashi, M. 2006. Rats produced by interspecies spermatogonial trans-plantation in mice and in vitro microinsemination. *Proc. Natl. Acad. Sci.* **103:** 13624–13628.

Takehashi, M., Kanatsu-Shinohara, M., Inoue, K., Ogonuki, N., Miki, H., Toyokuni, S., Ogura, A., and Shinohara, T. 2007a. Adenovirus-mediated gene delivery into mouse spermatogo-nial stem cells. *Proc. Natl. Acad. Sci.* **104:** 2596–2601.

Takehashi, M., Kanatsu-Shinohara, M., Miki, H., Lee, J., Kazuki, Y., Inoue, K., Ogonuki, N., Toyokuni, S., Oshimura, M., Ogura, A., and Shinohara, T. 2007b Production of knock-out mice by gene targeting in multipotent germline stem cells. *Dev. Biol.* **312:** 344–352.

Till, J.E. and McCulloch, E.A. 1961. A direct measurement of the radiation sensitivity of normal mouse bone marrow cells. *Radiat. Res.* **14:** 213–222.

Yoshida, S., Takakura, A., Ohbo, K., Abe, K., Wakabayashi, J., Yamamoto, M., Suda, T., and Nabeshima, Y. 2004. Neuro-genin3 delineates the earliest stages of spermatogenesis in the mouse testis. *Dev. Biol.* **269:** 447–458.

Spermatogenic Stem Cell System in the Mouse Testis

S. YOSHIDA

Division of Germ Cell Biology, National Institute for Basic Biology,
Higashiyama, Myodaiji, Okazaki 444-8787, Okazaki, Japan

Mouse spermatogenesis represents a highly potent and robust stem cell system. Decades of research have made it one of the most intensively studied mammalian tissue stem cell systems. These studies include detailed morphological examinations, posttransplantation colony formation, and in vitro culture of the stem cells; however, the nature of the stem cells as well as their niche are mostly to be elucidated in the context of homeostatic spermatogenesis. Our group has been challenging this issue by means of transgenic and live-imaging approaches that enable the investigation of live behaviors of "undifferentiated spermatogonia," the candidate stem cell population. A pulse-label experiment has suggested a hierarchical composition of the stem cell functional compartments, unlike the general idea. In addition, live imaging revealed the preferential localization of undifferentiated spermatogonia in the area adjacent to the blood vessel, leading to the proposal of a vasculature-associated niche. These results have suggested the idea of "flexibility" in the mouse spermatogenic stem cell system, which makes a good contrast to the "strict" stem-cell-niche system observed, for example, in the *Drosophila* germ line. This flexible nature seems to be advantageous for mammalians.

AN ANATOMICAL OVERVIEW OF MAMMALIAN SPERMATOGENESIS

In mammals, the male gonad (i.e., the testis) carries a highly efficient stem cell system that continuously produces numerous differentiating cells (i.e., sperm) during the reproduction period (Russell et al. 1990; Meistrich and van Beek 1993; de Rooij and Russell 2000). Figure 1 represents the anatomical basis of mouse spermatogenesis: Spermatogenesis proceeds inside the seminiferous tubule, a convoluted tubular structure with a diameter of about 200 µm that connects to the common outlet of the mature sperm (rete testes) with both ends to form loops. Each mouse testis contains about 20 tubules that are highly convoluted and tightly packed inside the testicular capsule (tunica albuginea). Their total length is up to 2 m, and spermtaogenesis occurs evenly throughout the inner surface of the tubules, the seminiferous epithelium. Therefore, in the mouse testis, an overall "polarity" that covers the entire organ cannot be recognized, as is clear in the gonads of several other "model organisms" such as *Drosophila* or *Caenorhabditis elegans* (Fig. 1A,B) (Decotto and Spradling 2005).

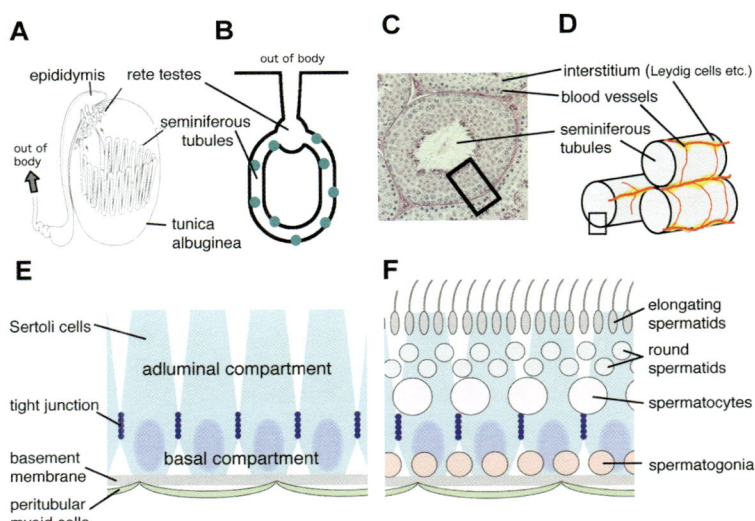

Figure 1. Anatomy of the mouse testis and seminiferous tubules. (*A*) Schematic overview of the mouse testis. Seminiferous tubules, the spermatogenic center of the testis, are highly convoluted and tightly packed in the tunica albuginea. A single tubule out of approximately 20 is shown. Individual tubules form loops with both ends open to the rete testis. (*B*) Diagram of mouse seminiferous tubule topology. As shown by *green dots*, stem cells are scattered throughout the tubule loops, which do not show apparent overall polarity. (*C,D*) Seminiferous tubules and the surrounding network of vasculature and interstitium. The blood vessels (*rea*) never penetrate the seminiferous tubule, but instead they run through the interstitial space and form a network among the seminiferous tubules. Vessels are surrounded by Leydig and other types of interstitial cells (*yellow*). (*E,F*) Scheme of the seminiferous epithelium and spermatogenesis, corresponding to the area shown by *rectangles* in *C* and *D*. (*E*) Anatomical framework composed of somatic components; (*F*) spermatogenic cells. See text for details. (Modified from Yoshida 2008 [© Kyoritsu Shuppan].)

Seminiferous tubules show a simple structural frame-work composed of Sertoli and peritubular myoid cells, the two somatic cell types that cover the inside and outside of the basement membrane, respectively (Fig. 1E). Sertoli cells show clear polarity and form a typical epithelium with tight junctions between them. The tight junction is the anatomical basis of the blood testis barrier and separates the tubules into basal and adluminal compartments. The basal compartment (i.e., between the junction and basement membrane) is occupied with spermatogonia (i.e., spermatogenic cells in mitotic stages) that contain stem cells and their differentiating progeny. Then, germ cells translocate to the adluminal compartment when entering meiosis, somehow through the tight junction. Subsequently, postmeiotic round and elongating spermatids are pushed up toward the lumen, which results in the beautifully arranged organization of the seminiferous epithelium (Fig. 1F) (Russell et al. 1990). The matured sperm are released into the lumen and ejaculated outside the body via the rete testes, epididymis, and vas deferens. The seminiferous tubules represent a common structure throughout the longitudinal and perpendicular; no specialized substructures or subsets of somatic cells that suggest a stem cell niche have been described. Blood vessels nourish the tubules but never penetrate them and run in the triangular intertubular interstitial space to form a network (Fig. 1C,D). Leydig cells (the main producer of testicular testosterone), lymphatetic epithelium, and macrophages surround the vessels to form interstitium.

Mammalian spermatogenesis therefore progresses in an apparently different anatomical context from that elucidated in other organisms. To my understanding, this has made the mammalian spermatogenic stem cell system a big challenge: Which germ cell population acts as stem cells? How do they behave in the testis to achieve stem cell functions?

In this chapter, I overview the research of the mammalian spermatogenic stem cell system (mainly in the mouse system) from a historical point of view and summarize the essential achievements as well as their potential drawbacks that allow us to recognize the remaining essential questions. I then discuss our recent work with this important and attractive system. I hope that this chapter provides a direction for a fuller understanding of mammalian spermatogenesis.

UNDIFFERENTIATED SPERMATOGONIA AND THE "A_S MODEL"

The detailed morphological observations of testis sections and whole-mount seminiferous tubule specimens in the 1950s through the 1970s established the backbone of spermatogenesis research (Russell et al. 1990; Meistrich and van Beek 1993; de Rooij and Russell 2000). There is no doubt that spermatogenic stem cells consist of only a tiny fraction of spermatogonia; however, strictly speaking, it is still to be elucidated which fraction of the numerous spermatogonia contains the "stem cells" that support homeostatic spermatogenesis.

The morphologically most primitive spermatogonia found in the adult mouse testis are A_s or A_{single} spermatogonia (i.e., single, isolated spermatogonia) (Fig. 2)

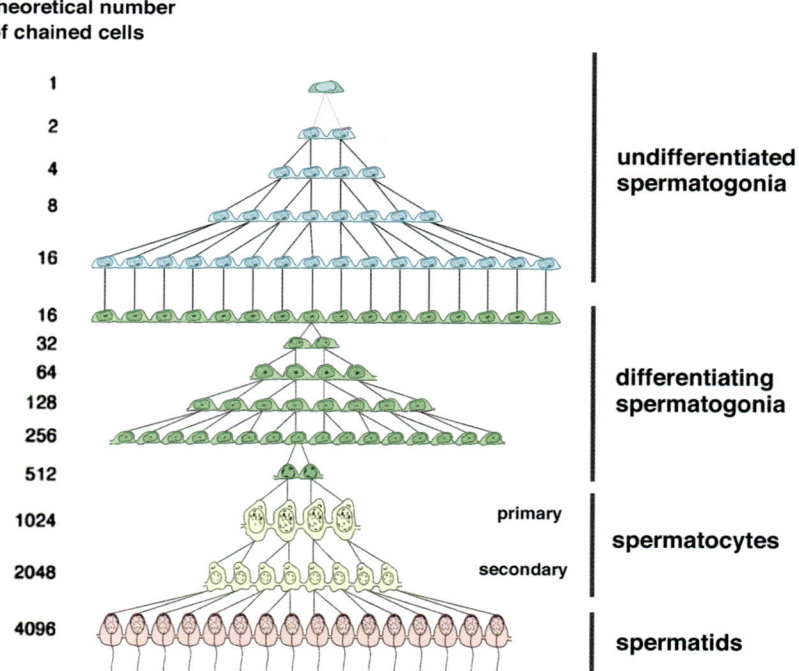

theoretical number of chained cells

1
2
4
8
16

16
32
64
128
256

512
1024
2048
4096

undifferentiated spermatogonia

differentiating spermatogonia

primary

secondary

spermatocytes

spermatids

Figure 2. Spermatogenic differentiation process occurring in the mouse testis. The most primitive germ cell found in the adult mouse testis is A_s. As a result of subsequent incomplete cell division, syncytial cysts of 2, 4, 8, 16.... cells form. A_{undiff} (undifferentiated spermatogonia) consists of A_s, A_{pr} (connected pairs of spermatogonia), and A_{al} (chains of 4, 8, 16, or occasionally 32 spermatogonia). Differentiation of A_{undiff} into A_1 differentiating spermatogonia is followed by a highly organized synchronized program leading to mature spermatozoa. See text for details. (Modified from Russell et al. 1990 [© Cache River Press].)

(Russell et al. 1990; de Rooij and Russell 2000). Their progeny remain interconnected by intercellular bridges due to incomplete cytokinesis, forming syncytial chains of 2^n cells (2, 4, 8, 16, etc.). It has been experimentally established that "undifferentiated spermatogonia" (or "A_{undiff}" hereafter)—which consist of the most primitive set of spermatogonia with minimal heterochromatin condensation, including A_s, A_{pr} (A_{paired}; interconnected pairs), and A_{al} ($A_{aligned}$; chains of 4, 8, 16, or occasionally 32 cells)—contain stem cells. A_{undiff} consists of less than 1% of all testicular cells. Note that this anatomical entity is defined based on nuclear morphology and a lack of synchronicity with the surrounding differentiating spermatogonia (see next paragraph), rather than the number of syncytial cells.

In mouse seminiferous tubules, spermatogenesis proceeds as a cyclic program that takes 8.6 days, known as the seminiferous epithelial cycle (Leblond and Clermont 1952; Russell et al. 1990; de Rooij 2001). A_{undiff} spermatogonia persist throughout the cycle and give rise to A_1 differentiating spermatogonia once every cycle. A_1 spermatogonia subsequently go through six mitoses (each forming A_2, A_3, A_4, In and B spermatogonia, and preleptotene primary spermatocytes) and two meiotic divisions before forming haploid spermatids, in a highly synchronous manner within a particular seminiferous tubule segment; therefore, A_{undiff} as a population behaves as the stem cell compartment. Compatible with this idea, A_{undiff} is often found as the only germ cell type that survives after insult caused by chemicals, radiation, or high temperature, which is enough for complete regeneration of spermatogenesis. However, A_{undiff} is a heterogeneous population and it is unlikely that all A_{undiff} act equivalently as the stem cells.

Which fraction of A_{undiff} consists of the actually self-renewing stem cell compartment in homeostasis and how does it behave (proliferate, self-renew, or die) in the testis? Decades ago, several models were proposed for this issue (Meistrich and van Beek 1993). Among them, the "A_s model" (Huckins 1971; Oakberg 1971) is cur-

rently most widely considered to be true (Meistrich and van Beek 1993; de Rooij and Russell 2000). This model proposes that A_s is the only cell type that can act as stem cells, whereas the interconnected population of A_{undiff} (A_{pr} and A_{al}) is devoid of stem cell capacity (Fig. 3, left). This comprehensive model is persuasive and attractive and is found frequently in the literature; however, it is theoretically impossible to be entirely conclusive regarding stem cell function based on "snapshots" from fixed specimens. Given that "stem cells" are defined as cells that maintain themselves while producing differentiating progeny for a long period, an experimental strategy that enables long-term analyses is warranted.

POSTTRANSPLANTATION SPERMATOGENIC COLONY FORMATION

A great breakthrough was brought about by intratubular stem cell transplantation developed by Brinster and colleagues in 1994 (Brinster and Avarbock 1994; Brinster and Zimmermann 1994; Brinster 2002). After a single cell suspension of the donor testis is transplanted into the recipient's seminiferous tubules, stem cells in the suspension reach and settle in the basal compartment (by an unknown mechanism) and proliferate to form colonies showing persisting spermatogenesis. This system has made mammalian spermatogenesis today's invaluable tissue stem cell system in which quantitative analyses by posttransplantation colony formation have been achieved, like mammalian hematopoiesis.

Taking advantage of stem cell detection by transplantation, a number of cell surface markers have been identified to enrich colony-forming stem cell activity. These and other experiments support that colony-forming activity is enriched in the A_{undiff} population (Shinohara et al. 2000; Ohbo et al. 2003; Tokuda et al. 2007); however, further purification of A_{undiff} subfractions has not been done. This system has also led to the establishment by Shinohara and colleagues and Brinster and coworkers of

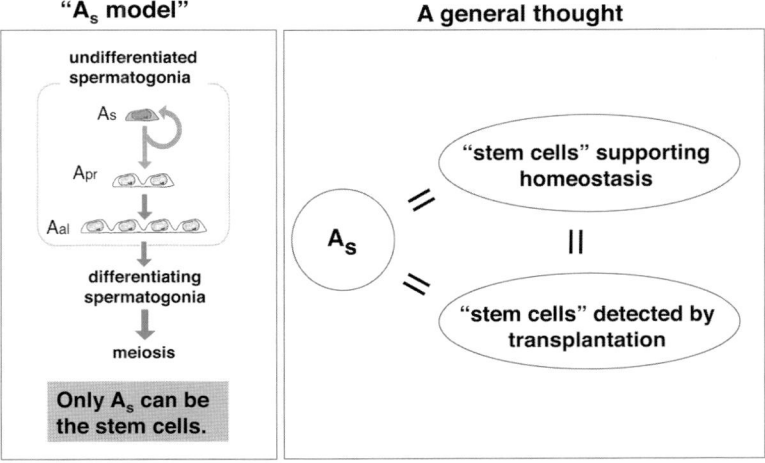

Figure 3. (*Left*) Schematic representation of the A_s model. See text for details. (*Right*) General thoughts about the mouse spermatogenic stem cell system.

long-term spermatogonial cultures that retains colony-forming stem cell activity (GS or germ-line stem cells; Kanatsu-Shinohara et al. 2003; Kubota et al. 2004). Strikingly, GS cell culture stably retains colony-forming activities for at least 2–3 years. On the other hand, only a small portion (at most several percent) of cells in these cultures exhibit colony-forming activity (Kanatsu-Shinohara et al. 2003). Compatible with this, a significant portion of cells exhibits differentiating characteristics. Further optimization might increase the stem cell content up to, theoretically, 100%, or this might reflect an unknown important property of stem cell maintenance, such as the "population effect." Again, the establishment of spermatogonial cultures that retain stem cell activity is a breakthrough that has enabled the investigation of stem cell characteristics and/or mechanisms of stem cell maintenance in vitro; this has not been achieved for hematopoietic stem cells. For example, stem cell control by the GDNF (glial-cell-line-derived neurotrophic factor) signaling pathway, which has an essential role in stem cell maintenance in vivo (Meng et al. 2000; Jijiwa et al. 2008), has been investigated extensively (Braydich-Stolle et al. 2005; Sariola and Immonen 2008).

GENERAL THOUGHTS ON SPERMATOGENIC STEM CELLS

To my understanding, the original A_s model claimed that stem cell activity resides in the A_s compartment but not in other morphological entities (Huckins 1971; Oakberg 1971) and did not consider whether all A_ss act equivalently as stem cells; however, without evidence of heterogeneity, A_s has often been considered to be a synonym of stem cells, raising the idea that all A_ss equally act as stem cells. Indeed, in some of the literature, A_s has been designated as A_{stem}. Furthermore, this idea has been easily combined with posttransplantation colony-forming stem cells. It is frequently considered to be true that all A_s spermatogonia are equivalent and act as stem cells that support both homeostatic spermatogenesis and posttransplantation colony formation (Fig. 3, right).

It is acknowledged that this is a reasonable consequence in the absence of experimental links between these different means of stem cell recognition; however, it is also clear that a fuller understanding of mammalian spermatogenesis warrants experiment-based evaluation. In particular, intratubular transplantation has been designed to achieve maximum sensitivity in detecting self-renewing potential: Typically, a single cell suspension is prepared from the donor testis (an artificial breakdown of syntytia into single cells) and germ cells are depleted from the host seminiferous tubules before transplantation (this is thought to empty the stem cell niche). Therefore, it is still to be evaluated whether the "stem cells" detected by transplantation are identical to the "stem cells" that actually self-renew in homeostasis.

GENETIC LABELING OF A_{UNDIFF}

The ultimate goal of our group is to fully understand the nature of the mouse spermatogenic stem cell system in the context of testicular tissue. For that purpose, we have been investigating the behavior and function of A_{undiff}. Authentic identification of A_{undiff} was performed on whole-mount specimens (Clermont and Bustos-Obregon 1968; Huckins and Oakberg 1978) and/or based on the nuclear morphology judged from electron microscopy or high-resolution light microscopy of plastic-embedded sections (Chiarini-Garcia and Russell 2001, 2002). These strategies inevitably require fixation, making it impossible to address the live behavior of A_{undiff}. In addition, the reliability of identification largely depends on the skill and experience of the researchers.

On the other hand, genes that delineate this population have long been unknown. We identified that Ngn3 (neurogenin3), a bHLH (basic helix-loop-helix) transcription factor, is expressed in the A_{undiff} population, by means of yeast two-hybrid screening of a spermatogonia-derived cDNA library (Yoshida et al. 2004). In transgenic mice in which Ngn3[+] cells were labeled with green fluorescent protein (GFP), isolated and interconnected spermatogonia that fulfill the authentic criteria for A_{undiff} were visualized (Fig. 4) (Yoshida et al. 2004). Other than Ngn3, A_{undiff}-specific expression of genes has been reported (Buaas et al. 2004; Costoya et al. 2004; Yoshida et al. 2004; Hofmann et al. 2005; Tokuda et al. 2007) that is making the heterogeneous nature of A_{undiff} population apparent.

Taking advantage of the genetic labeling of A_{undiff} by means of the Ngn3 regulatory sequence, we have established experimental systems to investigate their live behavior without disturbing normal tissue architecture.

FUNCTIONAL HIERARCHY IN THE STEM CELL SYSTEM, SUGGESTED BY PULSE-LABEL EXPERIMENTS

We first asked whether "stem cells" detected by transplantation are identical to "stem cells" that actually self-renew in homeostasis. To address this question, it was necessary to establish an experimental strategy to identify actual self-renewing stem cells without disturbing the homeostatic testicular architecture. For that purpose, a tamoxifen-dependent Cre recombinase (CreER™; Hayashi and McMahon 2002) was expressed in Ngn3[+] spermatogonia (Yoshida et al. 2006). In double transgenic mice with a CAG-CAT-Z reporter (Araki et al. 1995), Ngn3[+] spermatogonia and their progeny were irreversibly labeled with the constitutive expression of LacZ in a tamoxifen-dependent manner.

This enabled the first quantitative detection of "actual stem cells" (i.e., a cell population that persists for a long time while producing differentiating progeny and that supports tissue homeostasis; after the definition by Potten and Loeffler [1990]). Intriguingly, contribution of the pulse-labeled subpopulation of A_{undiff} to "actual stem cells" and "posttransplantation colony-forming stem cells" represent a great difference (~40 times higher in the latter than the former (for details, see Nakagawa et al. 2007; Yoshida et al. 2007a). Therefore, these two "stem cells" represent different subpopulations of A_{undiff}. We concluded that in addition to actual stem cells, an extended population exits that does not self-renew but retains the potential of self-renewal, which was defined as

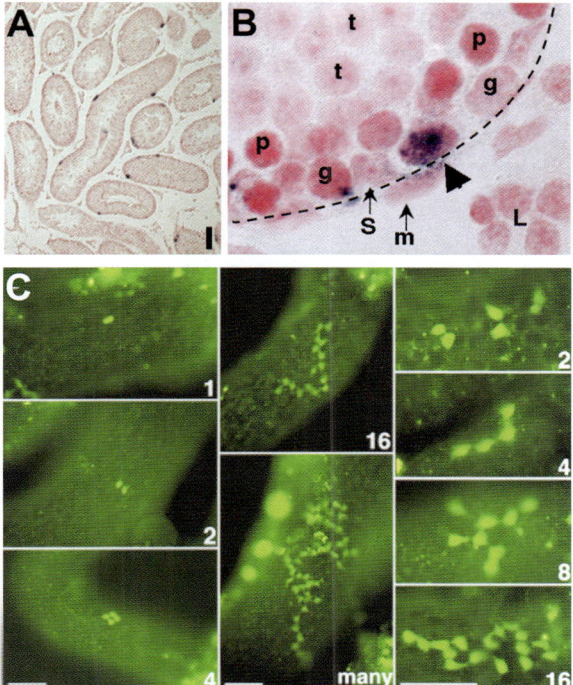

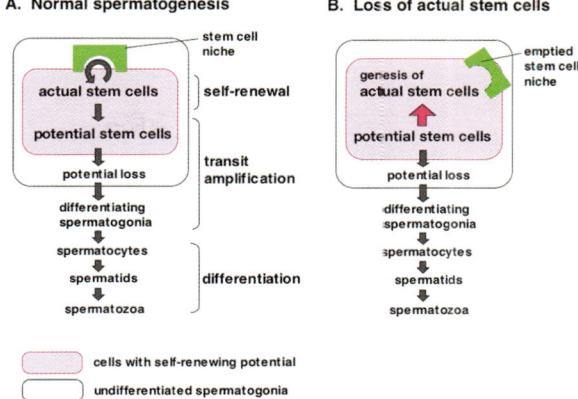

Figure 5. Proposed model of the functional compartments in mouse spermatogenesis. See text for details. (Reprinted, with permission, from Nakagawa et al. 2007 [© Elsevier].)

Figure 4. Ngn3-positive spermatogonia in the testis. (*A,B*) Ngn3 expression in adult mouse testis, revealed by in situ hybridization. Ngn3$^+$ cells (*purple*) are sparsely observed on the periphery of seminiferous tubules, counterstained with nuclear fast red (*A*). At a higher magnification (*B*), the signal is localized to spermatogonia with flattened nuclear morphology (*arrowhead*) on the basement membrane (*dotted line*). (g) Ngn3-negative spermatogonia; (p) pachytene spermatocytes; (t) spermatids; (S) Sertoli cells; (m) peritubular myoid cells. Bar, 100 µm. (*C*) Live visualization of Ngn3$^+$ spermatogonia by GFP expression driven by the regulatory genomic sequence of the *Ngn3* gene. A small number of spermatogonia with characteristic morphology for A$_{undiff}$ (i.e., isolated cells [A$_s$], or chains of 2, 4, 8, 16 cells [A$_{pr}$ and A$_{al}$]) were visualized in seminiferous tubules of the resultant transgenic mice. (Modified, with permission, from Yoshida et al. 2004 [© Elsevier].)

"potential stem cells" (see also Potten and Loeffler 1990). The "potential stem cells" were shown to rapidly turn over in homeostasis, suggesting that they consist of a transit-amplifying compartment.

Figure 5 shows a model for a hierarchical composition of the mouse spermatogenic stem cell system proposed as the simplest interpretation (Nakagawa et al. 2007). In case of actual stem cell loss, potential stem cells might revert to the self-renewing mode and replenish actual stem cells. Indeed, we also observed that actual stem cells are sometimes lost during a long period and are substituted by new actual stem cells supplied by neighboring actual stem cells (Nakagawa et al. 2007). We suppose that potential stem cells may have active roles in such normal stem cell turnover.

These results have raised a number of questions (Nakagawa et al. 2007; Yoshida et al. 2007a). Most important is the function-morphology relationship. Given that actual and colony-forming stem cells are different populations, the general thought that these two "stem cells" and A$_s$ spermatogonia are all identical (Fig. 3, right) needs to be reconsidered. Do actual and potential stem cells consist of different subsets of A$_s$? If so, A$_s$ must be heterogeneous. Are all A$_s$ homogeneous and do they equivalently act as actual stem cells, as the A$_s$ model may suggest? If so, A$_{pr}$ or A$_{al}$ must include potential stem cells, which may suggest their fragmentation in homeostasis or regeneration, as observed in the *Drosophila* germ line (Brawley and Matunis 2004; Kai and Spradling 2004). Addressing these questions as well as challenging the actual and potential stem cell model (Fig. 5) experimentally will elucidate a fuller understanding of the stem cell system.

A VASCULATURE-ASSOCIATED NICHE FOR A$_{UNDIFF}$, REVEALED BY LIVE IMAGING AND THREE-DIMENSIONAL RECONSTRUCTION

Here, we discuss the microenvironmental niche for stem cells in the mouse testis. Most mammalian spermatogenetic stem cell research, including our work described above, does not involve localization and movement of cells. However, transplantation and/or regeneration experiments have suggested an intimate relationship between stem cells and the niche microenvironment (Shinohara et al. 2002; Hess et al. 2006). Therefore, we also aim to identify the nature and function of the mouse spermatogenic stem cell niche; however, this is difficult because seminiferous tubules do not exhibit suspicious substructures. Moreover, actual stem cells can be identified only functionally, and their histological detection has not yet been achieved. Therefore, our current aim is to clarify the niche of A$_{undiff}$.

We have developed a live-imaging system during which GFP-labeled Ngn3$^+$ A$_{undiff}$ and their progeny (based on the residual GFP signal after down-regulation of the GFP transgene transcription) can be continuously filmed in undisturbed testes (Yoshida et al. 2007b). As a result, A$_{undiff}$ showed preferential localization to the area adjacent to blood vessels and interstitial cells that surround the seminiferous tubules. This is compatible with preceding observations from mouse and rat testis sections that A$_{undiff}$ shows a significant biased localization to the area facing the interstitium (Chiarini-Garcia et al. 2001, 2003). In addition, the dynamic migration of spermatogo-

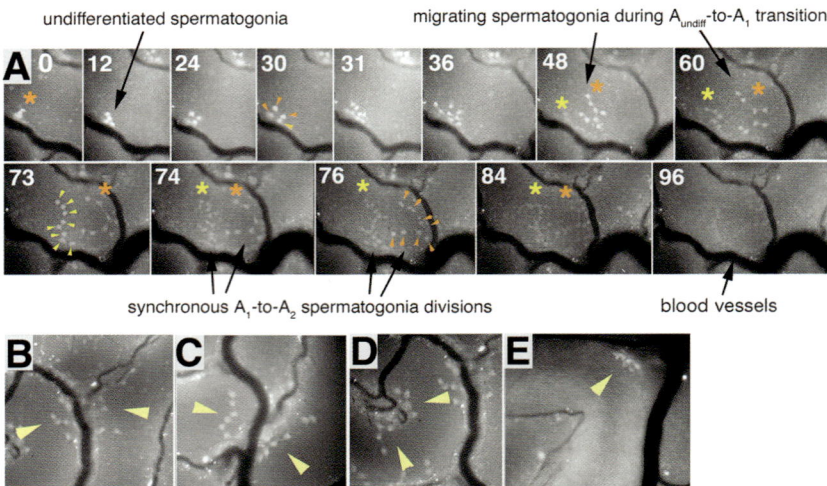

Figure 6. Localization of GFP-labeled Ngn3$^+$ A$_{undiff}$ and their relocation upon transition into differentiating spermatogonia. (*A*) Behavior of spermatogonia upon A$_{undiff}$-to-A$_1$ transition, revealed by live imaging. Before transition (0 hour; the elapsed time is indicated in each panel in hours), labeled A$_{undiff}$ preferentially localized to the area adjacent to the blood vessels (seen as a *black line*) and surrounding interstitium. Upon transition into A$_1$, two chains of eight-cell cysts (A$_{al-8}$; indexed in *yellow* and *orange*) migrated from this position to spread all over the basal compartment of the tubule (~36–60 hours). Subsequently, the two cysts underwent synchronous mitotic division with as short as a 2–3 hour interval, resulting in the formation of two 16-cell cysts of A$_2$ differentiating spermatogonia (73–74 hours). Stability of the GFP protein enabled us to follow differentiating spermatogonia even after Ngn3 (enhanced GFP) transcription was shut down during the transition process. For details, see Yoshida et al. (2007b). (*B–E*) Examples of the vasculature-proximal localization of A$_{undiff}$. A$_{undiff}$ (*arrowhead*) preferentially localized to area adjacent to blood vessels, more characteristically to their branch points. In *B* and *C*, A$_{undiff}$ in neighboring seminiferous tubules shows back-to-back localization over branching vessels. (Modified from Yoshida et al. 2007b [© AAAS].)

nia from the vasculature proximity to spread throughout the tubules was also observed upon A$_{undiff}$-to-A$_1$ transition (Fig. 6) (Yoshida et al. 2007b). The same relocation was also supported by a three-dimensional reconstruction based on authentic morphological identification of A$_{undiff}$ on serial sections (Fig. 7) (Yoshida et al. 2007b). On the basis of these observations, we proposed the area of the basal compartment of seminiferous tubules adjacent to the blood vessels as the niche for A$_{undiff}$ (Fig. 8). It is also suggested that changes of the vasculature pattern may accompany niche rearrangement (Yoshida et al. 2007b).

These observations provided the idea of a "flexible" niche for the spermatogenic stem cells, which may be reversibly specified in accordance with the vasculature pattern and its reorganization. This makes a good contrast to the *Drosophila* germ-line stem cell niche, which is specified after a highly programmed developmental process (Kitadate et al. 2007) and, once damaged, never regenerates. However, identification of the actual stem cells in the tissue and/or live imaging of their in vivo behaviors are warranted before final identification of the "spermatogenic stem cell niche," in its real meaning of the words. Another challenge is the mechanism by which vessels and/or interstitial cells specify the niche region. Further investigations are expected to resolve these essential questions.

CONCLUSIONS

The current status of the study of mammalian (mouse) spermatogenic stem cells, including our own works, was

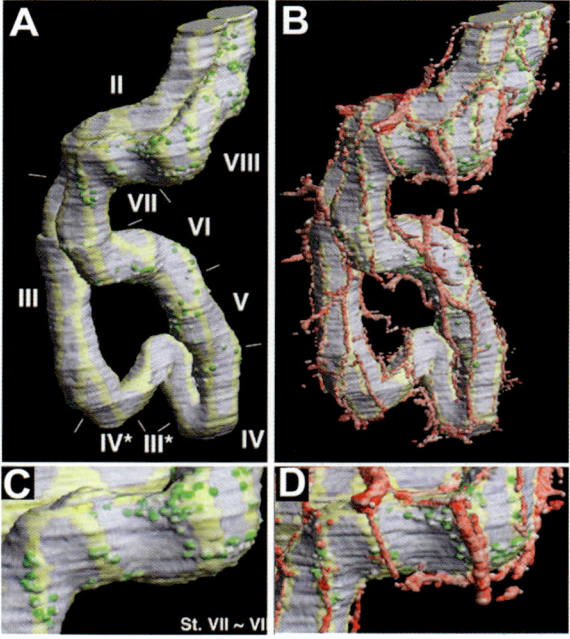

Figure 7. Localization of A$_{undiff}$ by three-dimensional reconstruction. Computationally reconstituted three-dimensional image of the seminiferous tubules based on 280 serial sections. A$_{undiff}$ (*green*) shows biased localization to the blood vessel network (*red*) and the area adjacent to the interstitium (*yellow*). (*A,C* and *B,D*) Images without or with blood vessels. Roman numerals indicate the stage of the seminiferous epithelium. (Reprinted from Yoshida et al. 2007b [© AAAS].)

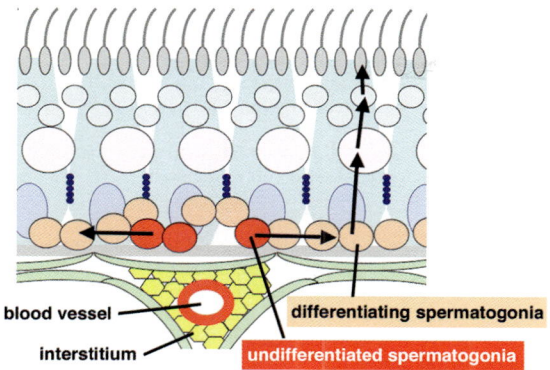

Figure 8. Schematic model of the niche microenvironment for A_{undiff}. Although seminiferous tubules do not harbor specialized structures (see Fig. 1), the niche region may be specified based on the spatial relationship with the surrounding vasculature network. Within the basal compartment of the tubules, A_{undiff} localized the adjacent region to the blood vessels. Upon transition into A_1, they migrate horizontally to spread throughout the basal compartment, followed by six mitotic divisions in the basal compartment and subsequent vertical translocation into the adluminal compartment upon entering meiosis. (Modified from Yoshida 2008 [© Kyoritsu Shuppan].)

reviewed. The mouse spermatogenic stem cell system involving a niche may be characterized by its "flexibility," which, I believe, can offer "robustness" to the entire system. This may be advantageous for mammals, which have a far larger body (i.e., organs harbor many more cells) and live much longer (i.e., the stem cell system needs to persist much longer). Further investigations will reveal more about the "flexible" mouse spermatogenic stem cell system.

ACKNOWLEDGMENTS

The work of our group introduced in this manuscript was performed in the Department of Pathology and Tumor Biology, Graduate School of Medicine, Kyoto University. I am deeply grateful to the tolerant and continuous support of Professor Yo-ichi Nabeshima. I also thank my colleagues, especially Dr. Toshinori Nakagawa, for his involvement in the pulse-label experiments, and Ms. Mamiko Sukeno and Mr. Tsutomu Obata for their excellent technical assistance. These studies were financially supported by grants-in-aid for Scientific Research from MEXT (Ministry of Education, Culture, Sports, Science and Technology) and JSPS (Japan Society for the Promotion of Science), the PRESTO (Precursory Research for Embryonic Science and Technology) program of the JST (Japan Science and Technology Agency), The Naito Foundation, and The Uehara Memorial Foundation.

REFERENCES

Araki, K., Araki, M., Miyazaki, J., and Vassalli, P. 1995. Site-specific recombination of a transgene in fertilized eggs by transient expression of Cre recombinase. *Proc. Natl. Acad. Sci.* **92:** 160–164.

Brawley, C. and Matunis, E. 2004. Regeneration of male germ-line stem cells by spermatogonial dedifferentiation in vivo. *Science* **304:** 1331–1334.

Braydich-Stolle, L., Nolan, C., Dym, M., and Hofmann, M.C. 2005. Role of glial cell line-derived neurotrophic factor in germ-line stem cell fate. *Ann. N.Y. Acad. Sci.* **1061:** 94–99.

Brinster, R.L. 2002. Germline stem cell transplantation and transgenesis. *Science* **296:** 2174–2176.

Brinster, R.L. and Avarbock, M.R. 1994. Germline transmission of donor haplotype following spermatogonial transplantation. *Proc. Natl. Acad. Sci.* **91:** 11303–11307.

Brinster, R.L. and Zimmermann, J.W. 1994. Spermatogenesis following male germ-cell transplantation. *Proc. Natl. Acad. Sci.* **91:** 11298–11302.

Buaas, F.W., Kirsh, A.L., Sharma, M., McLean, D.J., Morris, J.L., Griswold, M.D., de Rooij, D.G., and Braun, R.E. 2004. Plzf is required in adult male germ cells for stem cell self-renewal. *Nat. Genet.* **36:** 647–652.

Chiarini-Garcia, H. and Russell, L.D. 2001. High-resolution light microscopic characterization of mouse spermatogonia. *Biol. Reprod.* **65:** 1170–1178.

Chiarini-Garcia, H. and Russell, L.D. 2002. Characterization of mouse spermatogonia by transmission electron microscopy. *Reproduction* **123:** 567–577.

Chiarini-Garcia, H., Raymer, A.M., and Russell, L.D. 2003. Non-random distribution of spermatogonia in rats: Evidence of niches in the seminiferous tubules. *Reproduction* **126:** 669–680.

Chiarini-Garcia, H., Hornick, J.R., Griswold, M.D., and Russell, L.D. 2001. Distribution of type A spermatogonia in the mouse is not random. *Biol. Reprod.* **65:** 1179–1185.

Clermont, Y. and Bustos-Obregon, E. 1968. Re-examination of spermatogonial renewal in the rat by means of seminiferous tubules mounted "in toto." *Am. J. Anat.* **122:** 237–247.

Costoya, J.A., Hobbs, R.M., Barna, M., Cattoretti, G., Manova, K., Sukhwani, M., Orwig, K.E., Wolgemuth, D.J., and Pandolfi, P.P. 2004. Essential role of Plzf in maintenance of spermatogonial stem cells. *Nat. Genet.* **36:** 653–659.

Decotto, E. and Spradling, A.C. 2005. The *Drosophila* ovarian and testis stem cell niches: Similar somatic stem cells and signals. *Dev. Cell* **9:** 501–510.

de Rooij, D.G. 2001. Proliferation and differentiation of spermatogonial stem cells. *Reproduction* **121:** 347–354.

de Rooij, D.G. and Russell, L.D. 2000. All you wanted to know about spermatogonia but were afraid to ask. *J. Androl.* **21:** 776–798.

Hayashi, S. and McMahon, A.P. 2002. Efficient recombination in diverse tissues by a tamoxifen-inducible form of Cre: A tool for temporally regulated gene activation/inactivation in the mouse. *Dev. Biol.* **244:** 305–318.

Hess, R.A., Cooke, P.S., Hofmann, M.C., and Murphy, K.M. 2006. Mechanistic insights into the regulation of the spermatogonial stem cell niche. *Cell Cycle* **5:** 1164–1170.

Hofmann, M.C., Braydich-Stolle, L., and Dym, M. 2005. Isolation of male germ-line stem cells: influence of GDNF. *Dev. Biol.* **279:** 114–124.

Huckins, C. 1971. The spermatogonial stem cell population in adult rats. I. Their morphology, proliferation and maturation. *Anat. Rec.* **169:** 533–557.

Huckins, C. and Oakberg, E.F. 1978. Morphological and quantitative analysis of spermatogonia in mouse testes using whole mounted seminiferous tubules. I. The normal testes. *Anat. Rec.* **192:** 519–528.

Jijiwa, M., Kawai, K., Fukihara, J., Nakamura, A., Hasegawa, M., Suzuki, C., Sato, T., Enomoto, A., Asai, N., Murakumo, Y., and Takahashi, M. 2008. GDNF-mediated signaling via RET tyrosine 1062 is essential for maintenance of spermatogonial stem cells. *Genes Cells* **13:** 365–374.

Kai, T. and Spradling, A. 2004. Differentiating germ cells can revert into functional stem cells in *Drosophila melanogaster* ovaries. *Nature* **428:** 564–569.

Kanatsu-Shinohara, M., Ogonuki, N., Inoue, K., Miki, H., Ogura, A., Toyokuni, S., and Shinohara, T. 2003. Long-term proliferation in culture and germline transmission of mouse male germline stem cells. *Biol. Reprod.* **69:** 612–616.

Kitadate, Y., Shigenobu, S., Arita, K., and Kobayashi, S. 2007. Boss/Sev signaling from germline to soma restricts germline-stem-cell-niche formation in the anterior region of *Drosophila* male gonads. *Dev. Cell.* **13:** 151–159.

Kubota, H., Avarbock, M.R., and Brinster, R.L. 2004. Growth factors essential for self-renewal and expansion of mouse spermatogonial stem cells. *Proc. Natl. Acad. Sci.* **101:** 16489–16494.

Leblond, C.P. and Clermont, Y. 1952. Definition of the stages of the cycle of the seminiferous epithelium in the rat. *Ann. N.Y. Acad. Sci.* **55:** 548–573.

Meistrich, M.L. and van Beek, M.E. 1993. Spermatogonial stem cells. In *Cell and molecular biology of the testis* (ed. C. Desjardins and L.L. Ewing), pp. 266–295. Oxford University Press, New York.

Meng, X., Lindahl, M., Hyvönen, M.E., Parvinen, M., de Rooij, D.G., Hess, M.W., Raatikainen-Ahokas, A., Sainio, K., Rauvala, H., Lakso, M., et al. 2000. Regulation of cell fate decision of undifferentiated spermatogonia by GDNF. *Science* **287:** 1489–1493.

Nakagawa, T., Nabeshima, Y., and Yoshida, S. 2007. Functional identification of the actual and potential stem cell compartments in mouse spermatogenesis. *Dev. Cell* **12:** 195–206.

Oakberg, E.F. 1971. Spermatogonial stem-cell renewal in the mouse. *Anat. Rec.* **169:** 515–531.

Ohbo, K., Yoshida, S., Ohmura, M., Ohneda, O., Ogawa, T., Tsuchiya, H., Kuwana, T., Kehler, J., Abe, K., Schöler, H.R., and Suda, T. 2003. Identification and characterization of stem cells in prepubertal spermatogenesis in mice. *Dev. Biol.* **258:** 209–225.

Potten, C.S. and Loeffler, M. 1990. Stem cells: Attributes, cycles, spirals, pitfalls and uncertainties. Lessons for and from the crypt. *Development* **110:** 1001–1020.

Russell, L., Ettlin, R., Sinha Hikim, A., and Clegg, E. 1990. *Histological and histopathological evaluation of the testis.* Cache River Press, Clearwater, Florida.

Sariola, H. and Immonen, T. 2008. GDNF maintains mouse spermatogonial stem cells in vivo and in vitro. *Methods Mol. Biol.* **450:** 127–135.

Shinohara, T., Orwig, K.E., Avarbock, M.R., and Brinster, R.L. 2000. Spermatogonial stem cell enrichment by multiparameter selection of mouse testis cells. *Proc. Natl. Acad. Sci.* **97:** 8346–8351.

Shinohara, T., Orwig, K.E., Avarbock, M.R., and Brinster, R.L. 2002. Germ line stem cell competition in postnatal mouse testes. *Biol. Reprod.* **66:** 1491–1497.

Tokuda, M., Kadokawa, Y., Kurahashi, H., and Marunouchi, T. 2007. CDH1 is a specific marker for undifferentiated spermatogonia in mouse testes. *Biol. Reprod.* **76:** 130–141.

Yoshida, S. 2008. "Flexible" stem cell-niche system in mouse spermatogenesis (transl.). *Tanpakushitsu Kakusan Koso* (Protein, Nucleic Acid and Enzyme) **53:** 1125–1132.

Yoshida, S., Nabeshima, Y., and Nakagawa, T. 2007a. Stem cell heterogeneity: Actual and potential stem cell compartments in mouse spermatogenesis. *Ann. N.Y. Acad. Sci.* **1120:** 47–58.

Yoshida, S., Sukeno, M., and Nabeshima, Y. 2007b. A vasculature-associated niche for undifferentiated spermatogonia in the mouse testis. *Science* **317:** 1722–1726.

Yoshida, S., Sukeno, M., Nakagawa, T., Ohbo, K., Nagamatsu, G., Suda, T., and Nabeshima, Y. 2006. The first round of mouse spermatogenesis is a distinctive program that lacks the self-renewing spermatogonia stage. *Development* **133:** 1495–1505.

Yoshida, S., Takakura, A., Ohbo, K., Abe, K., Wakabayashi, J., Yamamoto, M., Suda, T., and Nabeshima, Y. 2004. Neurogenin3 delineates the earliest stages of spermatogenesis in the mouse testis. *Dev. Biol.* **269:** 447–458.

Reprogramming and Differentiation in Mammals: Motifs and Mechanisms

W.N. DE VRIES, A.V. EVSIKOV, L.J. BROGAN, C.P. ANDERSON, J.H. GRABER,
B.B. KNOWLES, AND D. SOLTER*

*The Jackson Laboratory, Bar Harbor, Maine 04609; *Institute of Medical Biology,
A*STAR, Immunos, Singapore 138648*

The natural reprogramming of the mammalian egg and sperm genomes is an efficient process that takes place in less than 24 hours and gives rise to a totipotent zygote. Transfer of somatic nuclei to mammalian oocytes also leads to their reprogramming and formation of totipotent embryos, albeit very inefficiently and requiring an activation step. Reprogramming of differentiated cells to induced pluripotent stem (iPS) cells takes place during a period of time substantially longer than reprogramming of the egg and sperm nuclei and is significantly less efficient. The stochastic expression of endogenous proteins during this process would imply that controlled expression of specific proteins is crucial for reprogramming to take place. The fact that OCT4, NANOG, and SOX2 form the core components of the pluripotency circuitry would imply that control at the transcriptional level is important for reprogramming to iPS cells. In contradistinction, the much more efficient reprogramming of the mammalian egg and sperm genomes implies that other levels of control are necessary, such as chromatin remodeling, translational regulation, and efficient degradation of no longer needed proteins and RNAs.

Reprogramming of differentiated cells into stem cells occurs at the outset of each new generation. Sperm and egg nuclei become reprogrammed in the cytoplasm of the fertilized oocyte to form the totipotent blastomeres of the newly formed embryo. J.B. Gurdon was the first to report that adult *Xenopus* could be derived by transferring a single somatic cell nucleus into the cytoplasm of a *Xenopus* oocyte (Gurdon et al. 1958). It was not until almost 40 years later that reprogramming of a mammalian somatic cell nucleus to totipotency in a mammalian oocyte was demonstrated by the birth of Dolly the sheep (Campbell et al. 1996). Reports of reprogramming of differentiated cells into induced pluripotent stem (iPS) cells have again challenged the existing knowledge of how reprogramming is accomplished (Jaenisch and Young 2008). Determining the molecular circuitry of pluripotency by looking at iPS gave some insight into this problem. However, another approach would be to expand the knowledge base of how two differentiated cells, the sperm and egg, are reprogrammed to give rise to the totipotent embryo. To this day, the maternal messages, and the mechanisms controlling their translation, that accomplish this natural embryonic reprogramming in mammals are only loosely understood.

EMBRYOGENESIS IN METAZOANS

Early embryogenesis in metazoans varies at the anatomical level, but in each model organism, the ooplasm contents control oocyte maturation, reprogramming of the sperm and egg genomes, activation of the embryonic genome, and one or more mitotic divisions. Reprogramming of the sperm and egg genomes in metazoans has one purpose: to set up totipotent stem cells to give rise to the new organism. Information gleaned from determining the regulation of maternal gene expression in the model organisms *Drosophila, Caenorhabditis elegans, Xenopus, Danio rerio*, and mouse have shed some light on the molecules potentially involved in reprogramming, but by and large, they are still unknown.

In the invertebrates *Drosophila* and *C. elegans* and in the vertebrates *Xenopus* and *D. rerio*, transcription from the embryonic genome is not initiated until some number of nuclear or cell divisions have taken place (Fig. 1). Embryonic genome activation takes place in *Drosophila* 2 hours after egg deposition and a number of nuclear divisions, in *Xenopus* during the micblastula transition (~4000 cells) 7 hours following fertilization, and in *D. rerio* at the 512-cell stage 3–4 hours after fertilization. In mammals, however, the embryonic genome is activated during cleavage, at the two-cell stage in mouse about 30 hours after fertilization, at the four–eight-cell stage in humans, and between the eight-cell and 12-cell stage in the cow (Telford et al. 1990).

The asymmetric *Drosophila, Xenopus,* and *D. rerio* oocytes exhibit clear localization of maternal transcripts, and these localized maternal mRNAs and proteins have a crucial role in establishing the embryonic axes. In contrast, mammalian oocytes appear to be radially symmetrical and are unlikely to be patterned (Motosugi et al. 2005). Mammalian oocytes are postnatally recruited into the growth phase, and at puberty, the full-grown oocyte (FGO) completes the first meiotic division, progresses into the second meiotic cell cycle, and arrests at meiotic metaphase II until ovulation and fertilization (Eppig 2001; Matzuk et al. 2002). Transcription decreases dramatically in the FGO and does not recommence until after the mature, ovulated oocytes are fertilized and the first cleavage division takes place (Bevilacqua et al. 1992; Worrad et al. 1995).

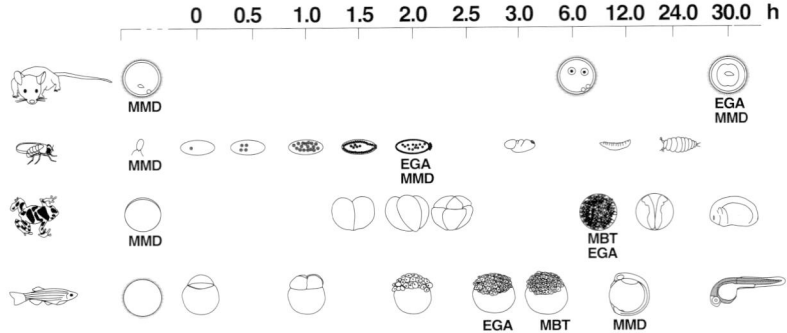

Figure 1. Time line for the development of *Drosophila, Xenopus,* and *Danio rerio* with respect to the oocyte-to-embryo transition in the mouse. The time line is given in hours, with 0 hour fixed at fertilization. The time points when maternal message degradation (MMD), the midblastula transition (MBT), and embryonic genome activation (EGA) take place are indicated.

PROGRAMMED DEGRADATION OF MATERNAL MRNAS

A critical role in embryonic development in metazoans is fulfilled by mechanisms that degrade maternal RNAs once they are no longer required. In *Drosophila, Xenopus*, and *D. rerio,* activation of the embryonic genome and degradation of maternal transcripts occur during the midblastula transition (MBT) when the embryo is already a multicellular organism. In *C. elegans*, the destruction of maternal transcripts takes place over a longer period of time. Although embryonic genome activation (EGA) takes place at the four-cell stage in somatic cells, embryonic genome control of development does not dominate until after gastrulation. In addition, transcription in the germ-line blastomeres is repressed until gametogenesis in the larval stage, with degradation of maternal messages taking place only at that point. In the mouse, the majority of maternal transcripts are degraded during oocyte maturation (Su et al. 2007), with the remainder being degraded at the two-cell stage upon embryonic genome activation.

Although the degradation of maternal mRNAs has been well documented, the mechanisms controlling it have not. In *D. rerio*, microRNA miR-430 is expressed right after the onset of MBT until gastrulation (Giraldez et al. 2006). In *Drosophila*, SMAUG regulates destabilization of maternal mRNAs (Tadros et al. 2007) and microRNAs (miRNAs) encoded from the miR-309 cluster are involved in degradation of a subset of maternal mRNAs (Bushati et al. 2008). Only recently has the involvement of pseudogene-derived endo-siRNAs (small interfering RNAs) and piRNAs (Piwi-interacting RNAs) in the regulation of gene expression in mouse oocytes been determined (Tam et al. 2008). These interfering RNAs (iRNAs) appear to be solely necessary for regulation of up-regulated endogenous retroviruses.

STABILIZATION OF MATERNAL mRNAs

In the mouse, maternal mRNAs are degraded between the FGO and ovulated oocyte stages, but some are stabilized for use following fertilization and are later degraded in the embryo. Indeed, mammalian maternal transcripts must be stabilized for quite a long time to guide embryo development until embryonic genome activation. Maternal transcripts must also be protected from degradation and activated for translation at specific times to enable successful nuclear reprogramming.

The primary knowledge base for control of maternal mRNA expression and degradation has been gleaned from study of meiotic maturation of the large and easily accessed oocytes of *Xenopus*. Sequences in the 3′ untranslated region (3′UTR) of maternal transcripts, and binding of their cognate *trans*-factor proteins, are central to the control of maternal mRNA polyadenylation and translation. For example, the polyadenylation signal (PAS) AAUAAA, which binds the cleavage and polyadenylation-specific factor, and the cytoplasmic polyadenylation element (CPE) UUUUUAU and variants thereof, which bind the cytoplasmic polyadenylation element-binding protein, are known to control translation (Richter 2007). When a CPE-binding protein is bound to a CPE-containing mRNA, a complex containing Maskin is recruited to the 3′UTR of mRNAs, which represses polyadenylation and translation by preventing entry of the mRNA into the ribosome. Phosphorylation of the CPE-binding protein by Aurora A kinase results in release of the Maskin-containing inhibitory complex from the mRNA, binding of the cleavage and polyadenylation specificity factor to the mRNA, entry into the ribosome, and initiation of translation. A pumilio-binding element (PBE) in the 3′UTR of mRNAs, activated/repressed by members of the PUF protein family (Nakahata et al. 2003), can also control the timing of cytoplasmic polyadenylation and translation of maternal transcripts.

Genome-wide analysis of the expression patterns and regulatory mechanisms responsible for these patterns of expression has been undertaken in *Xenopus*. Piqué et al. (2008), by determining the translational control of the cyclins B1–B5 during *Xenopus* oocyte maturation, established how a combinatorial code for the relationship between the number of, and physical distances among, PAS, CPE, and PBE in a given mRNA 3′UTR determines the time of activation of specific maternal mRNAs. The AU-rich elements (AREs) are also involved in the regulation of *Xenopus* maternal transcripts during oocyte maturation. When bound by the protein C3H-4, which recruits the CCR4 deadenylase complex to the 3'UTR, the mRNA's poly(A) tail is shortened, thereby repressing translation (Belloc and Méndez 2008). An embryonic deadenylation element (EDEN) has also been described, which, upon binding of its cognate binding protein in *Xenopus*, prevents translation of c-Mos and Aurora A mRNA translation after fertilization (Paris and Richter 1990; Paillard and Osborne 2003; Graindorge et al. 2006).

Many of these mechanisms governing maternal mRNA stability and translation in *Xenopus* appear to be conserved in the mouse. The stability and degradation of maternal transcripts after fertilization are controlled by ARE characterized by AUUUA motifs repeated in an AU-rich region (Chen and Shyu 1995; Voeltz and Steitz 1998). These elements are bound by poly(A)-binding proteins that confer stability to mRNAs (Wagner et al. 2001). Many maternal mRNAs present in the FGO contain CPEs in their 3'UTRs and a limited number of maternal transcripts in mouse under translational control have been identified (Oh et al. 2000; Sakurai et al. 2005a,b). Maternal mRNAs containing a PBE motif in their 3'UTR were found to be overrepresented among the stable mouse maternal transcripts (Evsikov et al. 2006).

THE CPE IN STABILITY AND TRANSLATION OF MOUSE MATERNAL MESSAGES

Maternal messages containing CPEs are cytoplasmically polyadenylated and translated during the oocyte-to-embryo transition (Oh et al. 2000). There is a bias toward longer 3'UTRs in stable maternal messages, i.e., in mRNAs present in both the FGO and the two-cell-stage embryo (Evsikov et al. 2006). Furthermore, a CPE and/or PBE is also present in the majority of the stable transcripts. This suggested that such motifs might contribute to regulation of polyadenylation and translational activation of messages as well as to their stability.

To test this hypothesis in the mouse, we turned to the *Spin* gene, which encodes an abundant protein in the oocytes and preimplantation embryos and has three transcripts of 0.8, 1.7, and 4.1 kb (Oh et al. 2000). These three transcripts arise as a consequence of differential PAS usage resulting in different 3'UTRs, each containing the same open reading frame. Although the 0.8-kb mRNA quickly degrades at the onset of maturation, the two larger transcripts are relatively stable and both are present at the two-cell stage. The 1.7-kb transcript contains a UA-rich CPE-like sequence, whereas the 4.1-kb transcript contains two CPE sequences. The 1.7-kb transcript is polyadenylated in the ovulated oocyte and zygote, and the 4.1-kb transcript is deadenylated in the oocyte and then readenylated in the zygote. The CPE located 37 nucleotides from the PAS of the 4.1-kb *Spin* transcript was mutated (TTTTTAT to CACGCGT) and inserted into a previously described reporter construct pBlueβgalSpin (4.1 UTR vs. 4.1 UTR-CPE; Fig. 2, upper panel). Both transcripts were transcribed in vitro, injected into full-grown or ovulated oocytes, and the effect of the mutation on translation of the reporter construct was measured at various times after its introduction. Neither tran-

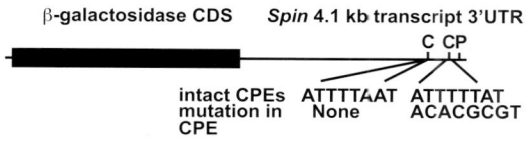

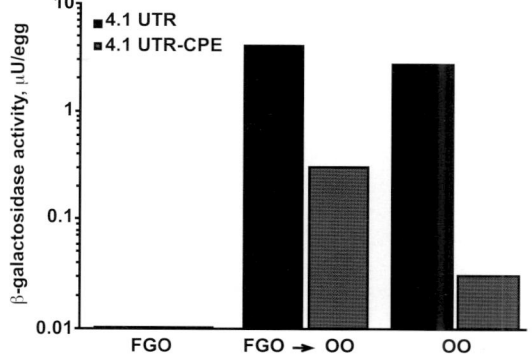

Figure 2. Mutation of the CPE in the *Spin* 4.1-kb 3'UTR leads to ineffective translation. (*Top panel*) Schematic representation of the *Spin* 4.1-kb reporter constructs used to determine the effect of mutation of the CPE on message stability in the FGO and ovulated oocyte. The CPE sequences in the *Spin* 4.1-kb transcript are indicated next to "intact CPEs." The 3'CPE is 37 nucleotides from the PAS, and the 5'CPE is 119 nucleotides away from the PAS. The sequences replacing the 3'CPE sequence in the *Spin* 4.1-kb transcript are indicated next to "mutation in CPE." P indicates the polyadenylation signal sequence. (*Lower panel*) Graphic representation of the decrease in translation of the reporter construct when the CPE sequence in the *Spin* 4.1-kb transcript is mutated. (FGO) Full-grown oocyte; (FGO → OO) transition from FGO to ovulated oocyte; (OO) ovulated oocyte. (*Black*) Graph of reporter construct containing the intact CPE sequence; (*gray*) graph of reporter construct containing the mutated CPE sequence.

script was translated in injected FGOs kept under meiotic arrest by addition of IBMX in the culture medium. The pBlueβgalSpin4.1UTR transcript was efficiently translated in injected FGOs that were allowed to mature in vitro for 12 hours, as well as in ovulated metaphase II oocytes cultured for 3 hours after injection (Fig. 2, lower panel). Conversely, there was a marked decrease in the translation of the pBlueβgalSpin4.1 UTR-CPE transcript in both in-vitro-matured FGOs and ovulated oocytes. Thus, mutation of the CPE proximal to the PAS most likely decreases the efficiency of reporter mRNA translation, suggesting an important role of the CPE in regulation of translational efficiency.

Spin serves as a prototype of the kind of regulation expected for genes involved in the establishment of the totipotent embryo: It is a stable maternal transcript that undergoes translational activation at specific times during the oocyte-to-embryo transition and the message is degraded when no longer needed.

PLURIPOTENCY REGULATORY CIRCUITRY

Transfection of human and mouse fibroblasts and other differentiated somatic cells with viral vectors containing coding sequences for *Lin28, Sox2, Klf4, Myc,* and *Pou5f1*

(*Oct4*) leads to reprogramming of these cells to a pluripotent state, i.e., iPS cells (Meissner et al. 2007; Okita et al. 2007; Takahashi et al. 2007; Wernig et al. 2007; Yu et al. 2007; Hanna et al. 2008; Jaenisch and Young 2008; Nakagawa et al. 2008). This reprogramming takes place in a stochastic fashion, where partially reprogrammed cells are first established, which then over time develop into fully reprogrammed iPS cells. Specific molecules, AP1, SSEA1, OCT4, and NANOG are sequentially expressed during the transformation of differentiated cells into iPS cells. Three transcription factors, OCT4, NANOG, and SOX2, are responsible for activating genes involved in maintaining the ES cell state, as well as for repressing genes that would lead to differentiation. TCF3, a transcription factor responsible for activating genes in the presence of WNT signaling and repressing genes in the absence of WNT signaling, also seems to have a central role in the maintenance of the stem cell state. Interestingly, TCF3 may have different roles in reprogramming in the oocyte-to-embryo transition versus iPS reprogramming. During the oocyte-to-embryo transition, TCF3 is expressed in an environment devoid of Wnt signaling (de Vries et al. 2004); thus, it would have a gene-silencing function, whereas during iPS reprogramming, it would be expressed in the presence of Wnt signaling and would thus have an activating function. The expression pattern of the genes used to induce iPS cells during the oocyte-to-embryo transition gives a clue as to the sequential expression of genes needed to trigger efficient reprogramming of differentiated genomes.

EMBRYONIC EXPRESSION OF GENES THAT EFFECT REPROGRAMMING OF DIFFERENTIATED CELLS TO IPS CELLS

We determined the timing of expression of *Nanog, Klf4, Myc, Lin28,* and *Pou5f1* (*Oct4*) during the oocyte-to-embryo transition in our oocyte and two-cell stage library databases (Evsikov et al. 2004, 2006) and also by performing reverse transcriptase–polymerase chain reaction (RT-PCR) on samples obtained throughout oocyte maturation, ovulation, and fertilization and in preimplantation embryos (Fig. 3). *Lin28* and *Pou5f1* (*Oct4*) were represented by two and one expressed sequence tags (EST), respectively, in the oocyte and two-cell-stage cDNA libraries, i.e., at low levels, whereas *Nanog, Klf4,* and *Myc* were not found in either of them. RT-PCR revealed that *Lin28* and *Oct4* are indeed expressed throughout the oocyte-to-embryo transition and at later preimplantation stages (Fig. 3), whereas *Nanog, Klf4,* and *Myc* only begin to be expressed at later stages, i.e., the late two-cell stage, the morula, and the blastocyst, respectively. These data suggest that genes crucial for reprogramming to iPS cells are different from those that reprogram the egg and sperm nucleus.

OOCYTE-TO-EMBRYO TRANSITION AND STEM CELL GENES

Analysis of human embryonic stem cells and their differentiating derivatives identified a set of genes whose transcript abundance is significantly reduced upon differ-

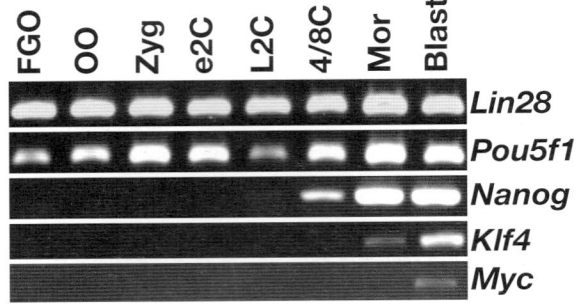

Figure 3. Expression of "iPS" genes in oocytes and preimplantation embryos. RT-PCR, using cDNA from two oocyte or embryo equivalents, was used to determine the expression of iPS genes during the oocyte-to-embryo transition and in later preimplantation stages. The names of the genes are indicated on the right; oocyte and embryo stages are indicated at the top. (FGO) Full-grown oocyte; (OO) ovulated oocyte; (Zyg) zygote; (e2C) early two-cell stage; (L2C) late two-cell stage; (4/8C) four–eight-cell stage; (Mor) morula; (Blast) blastocyst. Primer information, PCR conditions, and results are available in the Gene Expression Database (GXD) (http://www.informatics.jax.org; GXD ID = J:138685).

entiation. These are *Tgf1, Gabrb3, Dnmt3b, Gdf3, Pou5f1, Fgf4, Gal, Lefty1, Ifitm1, Nodal, Tert, Utf1, Foxd3, Lin28, Grb7, Podxl, Cd9, Bxdc2, Sox2, Klf4,* and *Nanog* (Adewumi et al. 2007). Microarray expression data (Wang et al. 2004), obtained from ArrayExpress (http://www.ebi.ac.uk/arrayexpress), which were quantile-normalized and corrected for background noise using the standard statistical RMA package and further analyzed using the R/MAANOVA software package (Wu et al. 2003), were searched to determine expression of these genes during the mouse oocyte-to-embryo transition and in later preimplantation stages. According to this public data set (Wang et al. 2004) *Cd9* is degraded during oocyte maturation and *Tgf1, Gabrb3, Dnmt3b, Gdf3, Oct4, Fgf4, Gal, Lefty1, Ifitm1, Nodal, Tert, Utf1, Foxd3, Lin28, Grb7, Podxl,* and *Bxdc2* are expressed during the oocyte-to-embryo transition. *Sox2, Klf4,* and *Nanog* were only expressed after activation of the embryonic genome. In contrast, a search of our own full-grown oocyte EST data set (Evsikov et al. 2006) revealed ESTs corresponding to only *Oct4, Tert, Dnmt3b, Bxdc2,* and *Lin28,* whereas the two-cell stage EST database (Evsikov et al. 2004) revealed, in addition, ESTs corresponding to *Utf1* and *Sox2.* The discrepancy in results obtained using these two different approaches (*Tgf1, Gabrb3, Gdf3, Fgf4, Gal, Lefty1, Ifitm1, Nodal, Utf1, Foxd3, Grb7, Podxl,* and *Bxdc2* are present in the microarray data set and absent in the EST databases) may reflect the different shortcomings of the two assays. Microarray analysis is very sensitive, but in some cases, absence of expression could be interpreted as presence if the expression level is close to background. In contrast, ESTs in libraries represent only those genes sequenced. Because saturation sequencing is impractical, some genes that are expressed at a very low level can be missed. Thus, the microarray approach can produce false positives and EST analysis can produce false-negative results. Nevertheless, when we analyzed the expression of *Nanog, Klf4,* and *Myc,* all of which are predicted by microarray data to be expressed during the oocyte-to-embryo transition, RT-PCR confirmed our negative EST data.

To determine if there is a possibility that any of those human "stemness" genes expressed throughout the oocyte-to-embryo transition are activated for translation at specific time points during this developmental period, we searched the 3'UTR of the transcripts for known motifs. Three genes (*Lin28, Grb7,* and *Tert*) contain CPE sequences in their 3'UTRs. This implies that, similar to *Spin*, these messages could be activated for translation at different times during the oocyte-to-embryo transition.

CONCLUSION

Genomic reprogramming is a very complex process involving numerous and possibly alternative controlling mechanisms. Natural reprogramming, i.e., events following fertilization, is regulated by maternally inherited molecules, proteins, and mRNAs synthesized and stored during oogenesis. These molecules mediate rapid reprogramming of the egg and sperm genomes likely through chromatin remodeling, differential mRNA use, and directed mRNA and protein degradation. Reprogramming following somatic cell nuclear transfer is mechanistically similar if not identical, with the low efficiency most likely arising due to the specific state of the genome of the transferred nuclei. In contrast, artificial reprogramming to iPS cells, induced by transfection, is obviously under transcriptional control and is a much slower and less efficient process. Evolution has perfected the just in time supply of specific proteins to effect swift natural reprogramming, which would be very difficult to mimic in the iPS type of reprogramming. Comparing and contrasting these two models of reprogramming will help us to identify relevant molecules and mechanisms.

ACKNOWLEDGMENTS

This work was supported by the U.S. Public Health Service/National Institutes of Health grants 1RO1 HD37102 (B.B.K.) and 1R01 GM072706 (J.H.G.). The authors thank Jesse Hammer of Multi Media Services at The Jackson Laboratory for assistance with figures.

REFERENCES

Adewumi, O., Aflatoonian, B., Ahrlund-Richter, L., Amit, M., Andrews, P.W., Beighton, G., Bello, P.A., Benvenisty, N., Berry, L.S., Bevan, S., et al. 2007. Characterization of human embryonic stem cell lines by the International Stem Cell Initiative. *Nat. Biotechnol.* **25:** 803–816.

Belloc, E. and Méndez, R. 2008. A deadenylation negative feedback mechanism governs meiotic metaphase arrest. *Nature* **452:** 1017–1021.

Bevilacqua, A., Kinnunen, L.H., and Mangia, F. 1992. Genetic manipulation of mammalian dictyate oocytes: Factors affecting transient expression of microinjected DNA templates. *Mol. Reprod. Dev.* **33:** 124–130.

Bushati, N., Stark, A., Brennecke, J., and Cohen, S.M. 2008. Temporal reciprocity of miRNAs and their targets during the maternal-to-zygotic transition in *Drosophila. Curr. Biol.* **18:** 501–506.

Campbell, K.H., McWhir, J., Ritchie, W.A., and Wilmut, I. 1996. Sheep cloned by nuclear transfer from a cultured cell line. *Nature* **380:** 64–66.

Chen, C.Y. and Shyu, A.B. 1995. AU-rich elements: Characterization and importance in mRNA degradation. *Trends Biochem. Sci.* **20:** 465–470.

de Vries, W.N., Evsikov, A.V., Haac, B.E., Fancher, K.S., Holbrook, A.E., Kemler, R., Solter, D., and Knowles, B.B. 2004. Maternal β-catenin and E-cadherin in mouse development. *Development* **131:** 4435–4445.

Eppig, J.J. 2001. Oocyte control of ovarian follicular development and function in mammals. *Reproduction* **122:** 829–838.

Evsikov, A.V., de Vries, W.N., Peaston, A.E., Radford, E.E., Fancher, K.S., Chen, F.H., Blake, J.A., Bult, C.J., Latham, K.E., Solter, D., and Knowles, B.B. 2004. Systems biology of the 2-cell mouse embryo. *Cytogenet. Genome Res.* **105:** 240–250.

Evsikov, A.V., Graber, J.H., Brockman, J.M., Hampl, A., Holbrook, A.E., Singh, P., Eppig, J.J., Solter, D., and Knowles, B.B. 2006. Cracking the egg: Molecular dynamics and evolutionary aspects of the transition from the fully grown oocyte to embryo. *Genes Dev.* **20:** 2713–2727.

Giraldez, A.J., Mishima, Y., Rihel, J., Grocock, R.J., Van Dongen, S., Inoue, K., Enright, A.J., and Schier, A.F. 2006. Zebrafish MiR-430 promotes deadenylation and clearance of maternal mRNAs. *Science* **312:** 75–79.

Graindorge, A., Thuret, R., Pollet, N., Osborne, H.B., and Audic, Y. 2006. Identification of post-transcriptionally regulated *Xenopus tropicalis* maternal mRNAs by microarray. *Nucleic Acids Res.* **34:** 986–995.

Gurdon, J.B., Elsdale, T.R., and Fischberg, M. 1958. Sexually mature individuals of *Xenopus laevis* from the transplantation of single somatic nuclei. *Nature* **182:** 64–65.

Hanna, J., Markoulaki, S., Schorderet, P., Carey, B.W., Beard, C., Wernig, M., Creyghton, M.P., Steine, E.J., Cassady, J.P., Foreman, R., Lengner, C.J., Dausman, J.A., and Jaenisch, R. 2008. Direct reprogramming of terminally differentiated mature B lymphocytes to pluripotency. *Cell* **133:** 250–264.

Jaenisch, R. and Young, R. 2008. Stem cells, the molecular circuitry of pluripotency and nuclear reprogramming. *Cell* **132:** 567–582.

Matzuk, M.M., Burns, K.H., Viveiros, M.M., and Eppig, J.J. 2002. Intercellular communication in the mammalian ovary: Oocytes carry the conversation. *Science* **296:** 2178–2180.

Meissner, A., Wernig, M., and Jaenisch, R. 2007. Direct reprogramming of genetically unmodified fibroblasts into pluripotent stem cells. *Nat. Biotechnol.* **25:** 1177–1181.

Motosugi, N., Bauer, T., Polanski, Z., Solter, D., and Hiiragi, T. 2005. Polarity of the mouse embryo is established at blastocyst and is not prepatterned. *Genes Dev.* **19:** 1081–1092.

Nakagawa, M., Koyanagi, M., Tanabe, K., Takahashi, K., Ichisaka, T., Aoi, T., Okita, K., Mochiduki, Y., Takizawa, N., and Yamanaka, S. 2008. Generation of induced pluripotent stem cells without Myc from mouse and human fibroblasts. *Nat. Biotechnol.* **26:** 101–106.

Nakahata, S., Kotani, T., Mita, K., Kawasaki, T., Katsu, Y., Nagahama, Y., and Yamashita, M. 2003. Involvement of *Xenopus* Pumilio in the translational regulation that is specific to cyclin B1 mRNA during oocyte maturation. *Mech. Dev.* **120:** 865–880.

Oh, B., Hwang, S., McLaughlin, J., Solter, D., and Knowles, B.B. 2000. Timely translation during the mouse oocyte-to-embryo transition. *Development* **127:** 3795–3803.

Okita, K., Ichisaka, T., and Yamanaka, S. 2007. Generation of germline-competent induced pluripotent stem cells. *Nature* **448:** 313–317.

Paillard, L. and Osborne, H.B. 2003. East of EDEN was a poly(A) tail. *Biol. Cell* **95:** 211–219.

Paris, J. and Richter, J.D. 1990. Maturation-specific polyadenylation and translational control: Diversity of cytoplasmic polyadenylation elements, influence of poly(A) tail size, and formation of stable polyadenylation complexes. *Mol. Cell. Biol.* **10:** 5634–5645.

Piqué, M., López, J.M., Foissac, D., Guigó, R., and Méndez, R. 2008. A combinatorial code for CPE-mediated translational control. *Cell* **132:** 434–448.

Richter, J.D. 2007. CPEB: A life in translation. *Trends Biochem. Sci.* **32:** 279–285.

Sakurai, T., Sato, M., and Kimura, M. 2005a. A novel method for constructing murine cDNA library enriched with maternal mRNAs exhibiting de novo independent post-fertilization polyadenylation. *Biochem. Biophys. Res. Commun.* **327:** 688–699.

Sakurai, T., Sato, M., and Kimura, M. 2005b. Diverse patterns of poly(A) tail elongation and shortening of murine maternal mRNAs from fully grown oocyte to 2-cell embryo stages. *Biochem. Biophys. Res. Commun.* **336:** 1181–1189.

Su, Y.Q., Sugiura, K., Woo, Y., Wigglesworth, K., Kamdar, S., Affourtit, J., and Eppig, J.J. 2007. Selective degradation of transcripts during meiotic maturation of mouse oocytes. *Dev. Biol.* **302:** 104–117.

Tadros, W., Goldman, A.L., Babak, T., Menzies, F., Vardy, L., Orr-Weaver, T., Hughes, T.R., Westwood, J.T., Smibert, C.A., and Lipshitz, H.D. 2007. SMAUG is a major regulator of maternal mRNA destabilization in *Drosophila* and its translation is activated by the PAN GU kinase. *Dev. Cell* **12:** 143–155.

Takahashi, K., Tanabe, K., Ohnuki, M., Narita, M., Ichisaka, T., Tomoda, K., and Yamanaka, S. 2007. Induction of pluripotent stem cells from adult human fibroblasts by defined factors. *Cell* **131:** 861–872.

Tam, O.H., Aravin, A.A., Stein, P., Girard, A., Murchison, E.P., Cheloufi, S., Hodges, E., Anger, M., Sachidanandam, R., Schultz, R.M., and Hannon, G.J. 2008. Pseudogene-derived small interfering RNAs regulate gene expression in mouse oocytes. *Nature* **453:** 534–538.

Telford, N.A., Watson, A.J., and Schultz, G.A. 1990. Transition from maternal to embryonic control in early mammalian development: A comparison of several species. *Mol. Reprod. Dev.* **26:** 90–100.

Voeltz, G.K. and Steitz, J.A. 1998. AUUUA sequences direct mRNA deadenylation uncoupled from decay during *Xenopus* early development. *Mol. Cell. Biol.* **18:** 7537–7545.

Wagner, M.J., Gogela-Spehar, M., Skirrow, R.C., Johnston, R.N., Riabowol, K., and Helbing, C.C. 2001. Expression of novel ING variants is regulated by thyroid hormone in the *Xenopus laevis* tadpole. *J. Biol. Chem.* **276:** 47013–47020.

Wang, Q.T., Piotrowska, K., Ciemerych, M.A., Milenkovic, L., Scott, M.P., Davis, R.W., and Zernicka-Goetz, M. 2004. A genome-wide study of gene activity reveals developmental signaling pathways in the preimplantation mouse embryo. *Dev. Cell.* **6:** 133–144.

Wernig, M., Meissner, A., Foreman, R., Brambrink, T., Ku, M., Hochedlinger, K., Bernstein, B.E., and Jaenisch, R. 2007. In vitro reprogramming of fibroblasts into a pluripotent ES-cell-like state. *Nature* **448:** 318–324.

Worrad, D.M., Turner, B.M., and Schultz, R.M. 1995. Temporally restricted spatial localization of acetylated isoforms of histone H4 and RNA polymerase II in the 2-cell mouse embryo. *Development* **121:** 2949–2959.

Wu, H., Kerr, M.K., Cui, X., and Churchill, G.A. 2003. MAANOVA: A software package for the analysis of spotted cDNA experiments. In *The analysis of gene expression data: Methods and software* (ed. G. Parmigiani et al.), pp. 313–341. Springer-Verlag, New York.

Yu, J., Vodyanik, M.A., Smuga-Otto, K., Antosiewicz-Bourget, J., Frane, J.L., Tian, S., Nie, J., Jonsdottir, G.A., Ruotti, V., Stewart, R., Slukvin, I.I., and Thomson, J.A. 2007. Induced pluripotent stem cell lines derived from human somatic cells. *Science* **318:** 1917–1920.

Interactions between Stem Cells and Their Niche in the *Drosophila* Ovary

T. XIE, X. SONG, Z. JIN, L. PAN, C. WENG, S. CHEN, AND N. ZHANG

Stowers Institute for Medical Research, Kansas City, Missouri 64110

The *Drosophila* ovary contains at least three types of active stem cells, namely, germ-line stem cells (GSCs), escort stem cells (ESCs), and follicular stem cells (FSCs), which work together to efficiently assemble egg chambers. Among the three stem cell types, the GSC is among the first shown to be controlled by the niche due to its easy identification and well-defined surrounding cells. We have shown that the niche controls GSC self-renewal, anchorage, aging, and competition, and the GSC also signals back to the niche for its maintenance. The FSC is an attractive model for studying epithelial stem cell regulation and signal integration because we have shown that it resembles mammalian epithelial stem cells and requires multiple signaling pathways for its self-renewal. In this chapter, we have highlighted the findings of our studies on interactions between *Drosophila* ovarian stem cells and their niches during normal development and aging and on stem cell competition for niche occupancy. We further discuss their implications in general stem cell biology and future directions in this exciting area.

Stem cell self-renewal and differentiation have been shown to be controlled by the regulatory microenvironment or niche in many different systems (Li and Xie 2005; Morrison and Spradling 2008). However, it remains largely unclear how niche signals collaborate with intrinsic factors to control stem cell self-renewal and differentiation. In the past 8 years, our laboratory has been using *Drosophila* ovary as a model system to identify niche signals and intrinsic factors that are essential for stem cell self-renewal and to study the molecular mechanisms underlying stem cell/niche interactions, niche formation and maintenance, stem cell aging, and competition.

Each *Drosophila* female carries 30–40 individual egg assembly lines known as ovarioles, where the anterior tip has the germarium in which three types of active stem cells reside (Fig. 1A,B). The terminal filament (TF) is positioned at the most anterior end of the germarium, whereas the cap cells, located posterior to the TF, contact GSCs posteriorly and ESCs laterally (Fig. 1B–D). GSCs can be identified by their location (in contact with cap cells) and anteriorly localized spectrosome, which can be labeled with Hts staining (Fig. 1C,D). We have shown that cap cells are necessary and sufficient for maintaining GSCs (Xie and Spradling 2000; Song et al. 2007). Interestingly, when both of the daughters of a GSC are in contact with cap cells, they both become stem cells, representing the first demonstration of the existence of the stem cell niche (Xie and Spradling 2000). Consistently, cap cells express Yb and Piwi, which are required for maintaining GSCs (Cox et al. 1998, 2000; King and Lin 1999; King et al. 2001). In addition, cap cells also express secreted bone morphogenetic protein (BMP)-like growth factors, which are essential for GSC self-renewal (Xie and Spradling 1998, 2000; Song et al. 2004). Intrinsically, two

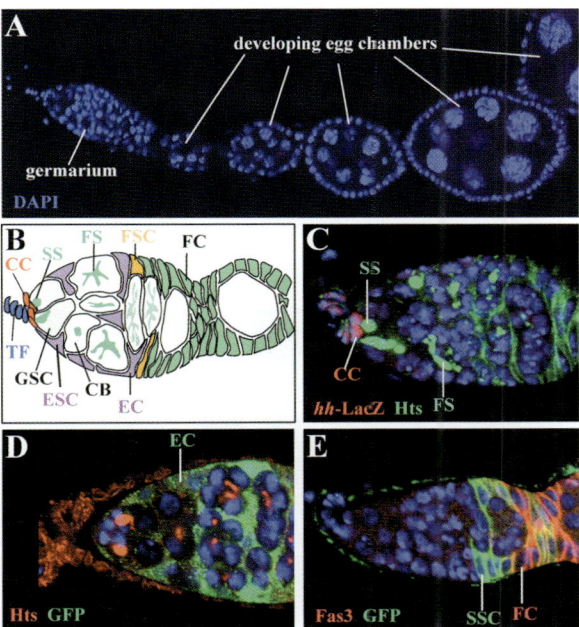

Figure 1. Structures of *Drosophila* ovariole and germarium. (*A*) A part of an ovariole, which is stained by DAPI (*blue*), contains a germarium and five developing egg chambers. (*B*) Diagram of a germarium: TF (terminal filament, *blue*); CC (cap cell, *red*); ESC (escort stem cell, *purple*); EC (escort cell, *purple*); SS (spectrosome, *green*); FS (fusome, *green*); FSC (follicular stem cell, *brown*); and FC (follicle cell, *green*). (*C*) A germarium, labeled for LacZ (TF and CCs, *red*) and Hts (SS and FS, *green*), shows cap cells and two GSCs. (*D*) A part of a germarium, labeled for Hts (*red*) and GFP (*green*), shows the escort cells (*green*). (*E*) A germarium, labeled for Fas3 (FCs, *red*) and GFP (*green*), shows a GFP-marked FSC clone.

classes of intrinsic factors control the balance between GSC self-renewal and differentiation: self-renewing factors such as Pumilio and Nanos and differentiation-promoting factors such as Bam and Bgcn (McKearin and Spradling 1990; Lin and Spradling 1997; Forbes and Lehmann 1998; Ohlstein et al. 2000; Wang and Lin 2004). The interplay between niche signals and intrinsic factors that control *Drosophila* ovarian GSCs has just begun to be revealed (Chen and McKearin 2003a; Song et al. 2004; Xi and Xie 2005).

Two FSCs, previously also known as somatic stem cells (SSCs), which can be followed by green fluorescent protein (GFP) or LacZ markers, are located on the opposite sides of the germarium to produce the follicular epithelium that wraps around germ cell cysts produced by GSCs (Fig. 1E) (Margolis and Spradling 1995; Zhang and Kalderon 2001; Song and Xie 2002; Nystul and Spradling 2007). Cap-cell-expressing Hedgehog (Hh) is required for maintaining FSC self-renewal (Forbes et al. 1996; King et al. 2001; Zhang and Kalderon 2001). Using genetically marked mutant FSCs, we have shown that cap-cell-expressing Wingless and escort-cell-expressing BMP-like Gbb directly control FSC self-renewal (Song and Xie 2003; Kirilly et al. 2005). In addition, E-cadherin-mediated cell adhesion is also required for maintaining FSCs by keeping them in the proximity of signaling resources (Song and Xie 2002). The adhesion may help early FSC progeny to migrate from one side of the germarium to the other for stem cell replacement (Nystul and Spradling 2007). In this chapter, we summarize the findings from our studies on *Drosophila* ovarian GSCs and FSCs and discuss their potential implications in other stem cell systems.

SHORT-RANGE BMP NICHE SIGNALS CONTROL GSC SELF-RENEWAL BY DIRECTLY REPRESSING DIFFERENTIATION

Although it is known that Dpp (Decapentaplegic)/BMP is a niche-derived growth factor essential for GSC self-renewal (Xie and Spradling 1998), it remains unclear how this BMP-like factor mechanistically controls stem cell self-renewal and whether other BMP-like molecules also participate in GSC regulation. We have found that similar to *dpp*, another *Drosophila* BMP-like gene *gbb* is expressed specifically in the somatic cells surrounding GSCs, including cap cells, and *gbb* mutant females lose their GSCs prematurely, indicating that Gbb is also an essential niche signal (Song et al. 2004). To further our understanding of how BMP-like signals control GSC self-renewal, we have shown that only GSCs, but not cystoblasts (immediate differentiating GSC daughters lying one cell away from cap cells), express pMad and *Dad*, two BMP activity indicators, indicating that BMPs only function in one cell diameter (Fig. 2A) (Song et al. 2004). Expressed in cystoblasts, *bam* is necessary and sufficient for their differentiation because mutations in *bam* can completely block cystoblast differentiation, and its forced expression in GSCs can sufficiently cause their differentiation (McKearin and Ohlstein 1995; Ohlstein and McKearin 1997). Although it is normally repressed in wild-type GSCs, *bam* transcription is up-regulated in the GSCs that are in the *dpp* and *gbb* mutant niche or are mutant for BMP downstream transducers, indicating that BMP signaling is essential for repressing *bam* expression and thus maintaining GSC self-renewal (Fig. 2C,D) (Song et al. 2004). Dpp overexpression can completely repress

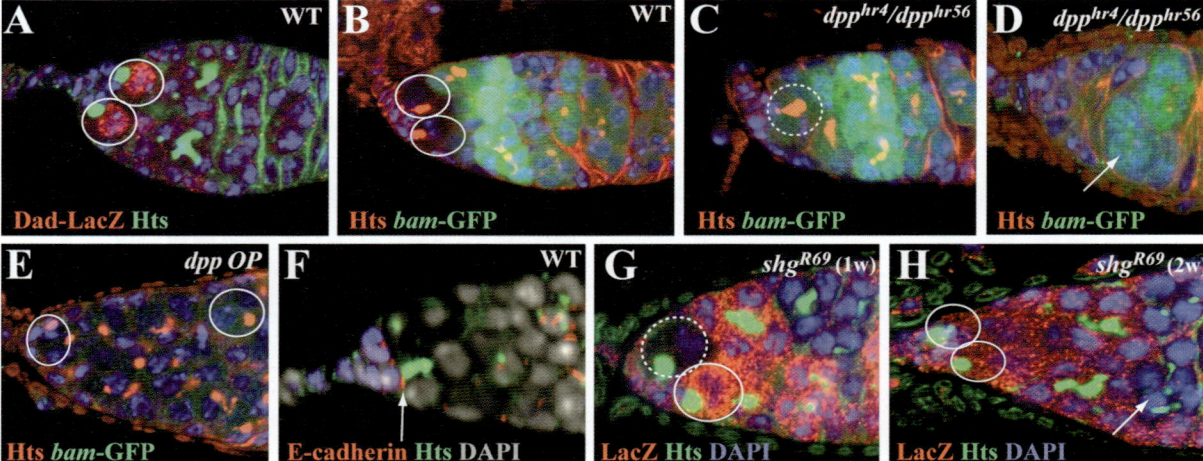

Figure 2. Niche-derived BMP signaling and E-cadherin-mediated GSC-niche adhesion are required for GSC maintenance. (*A*) A germarium shows the restricted BMP signaling activity indicated by *Dad-lacZ* expression (*red*) in GSCs (*solid circles*). (*B*) A germarium shows the repression of *bam-GFP* expression (*green*) in GSCs (*solid circles*). (*C*) A *dpp* mutant germarium shows the up-regulated expression of *bam-GFP* (*green*) in the remaining GSC (*dashed circle*). (*D*) A *dpp* mutant germarium shows the differentiation of a *bam GFP*-expressing GSC into a cyst (arrow) and thus GSC loss. (*E*) The tip of a *dpp*-overexpressing (*dpp OP*) germarium shows the absence of *bam-GFP* expression (*green*) in single germ cells (*circles*). (*F*) The tip of a wild-type germarium shows the accumulation of E-cadherin (*red*) in the GSC-niche junction (*arrow*). (*G*) The tip of a wild-type germarium carries one marked week-old *shg* mutant GSC and an unmarked wild-type GSC. (*H*) The tip of a wild-type germarium carries two unmarked wild-type GSCs and marked *shg* mutant cysts, indicating the loss of a marked mutant GSC (near *arrow*).

bam expression in germ cells by activating and forming the Mad-Medea transcriptional complex, which can directly bind to a *bam* silencer (Fig. 2E) (Chen and McKearin 2003a,b; Song et al. 2004). Similarly, we have demonstrated that niche-derived BMP signaling controls GSC self-renewal by repressing *bam* expression in the *Drosophila* testis (Kawase et al. 2004). Therefore, BMP signals maintain the self-renewal of GSC daughters remaining in the niche by directly repressing *bam* expression and allow other daughters outside the niche to differentiate due to their short-range function. These findings have, for the first time, offered a simple model to explain the stem cell dogma: A stem cell divides to generate a self-renewing stem cell in the niche and a differentiating daughter staying out of the niche.

E-CADHERIN-MEDIATED CELL ADHESION ANCHORS GSCS IN THE NICHE FOR THEIR SELF-RENEWAL

Because BMP niche signals only function as one cell diameter, the temporary departure of GSCs from the niche can potentially jeopardize their self-renewal potential. Thus, it is imperative to know how self-renewing GSCs are maintained in the niche constantly. We have shown that E-cadherin and its associated protein Armadillo (β-catenin) are expressed in GSCs and cap cells and accumulated in the stem cell/niche junction to form adherens junctions, indicating that GSCs are anchored to their niche (Fig. 2F) (Song et al. 2002). Removal of E-cadherin or Armadillo specifically from GSCs themselves sufficiently disrupts adherens junctions between GSCs and cap cells due to homophilic interactions and causes rapid GSC loss, further indicating that the niche anchorage through E-cadherin-mediated cell adhesion is essential for GSC self-renewal (Fig. 2G,H). In addition, E-cadherin is required for recruiting GSCs to their niche during niche and GSC formation (Song et al. 2002). Therefore, these findings demonstrate that E-cadherin-mediated niche anchorage is important for GSC maintenance (Song et al. 2002). Intriguingly, N-cadherin has been shown to be required for keeping mouse hematopoietic stem cells in the niche, indicating that cadherin-mediated stem cell anchorage is a conserved mechanism for anchoring stem cells in the niche (Haug et al. 2008).

INTRINSIC FACTORS ARE ALSO REQUIRED FOR CONTROLLING GSC SELF-RENEWAL BY REPRESSING DIFFERENTIATION

Although we have shown that BMP signaling can directly repress *bam* expression in GSCs and thereby maintain self-renewal, it remains unclear how the BMP-regulated Medea-Mad (SMAD) transcriptional complex represses *bam* transcription, because SMADs are known to be transcriptional activators in mammals (Kretzschmar and Massagué 1998). We have recently shown that an ATP-dependent chromatin-remodeling protein ISWI, a member of the SNF2/SWI protein family, is required in GSCs for their self-renewal, because marked mutant *iswi* GSCs are rapidly lost from the niche (Fig. 3A) (Xi and Xie 2005).

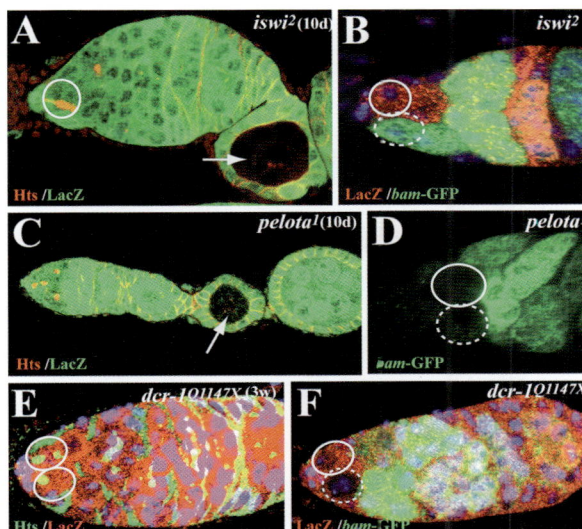

Figure 3. Intrinsic factors ISWI, Pelota, and Dcr-1 are required for GSC self-renewal. (*A*) A germarium only contains wild-type GSCs (*solid circle*) with the marked *iswi* mutant GSC differentiating into a cyst in an egg chamber (*arrow*). (*B*) *bam*-GFP expression is up-regulated in the marked *iswi* mutant GSC (*broken circle*) but not in the unmarked wild-type GSC (*solid circle*). (*C*) A part of an ovariole shows that the lost marked *pelota* mutant GSC has differentiated into a cyst in an egg chamber (*arrow*). (*D*) In the unmarked wild-type GSC (*solid circle*) and the marked *pelota* mutant GSC (*dashed circle*), *bam*-GFP expression is still repressed. (*E*) A germarium only contains wild-type GSCs (*solid circles*) with a lost marked *dcr-1* mutant GSC moving out of the germarium (not shown). (*F*) In the marked *dcr-1* mutant GSC and the unmarked wild-type GSC, *bam*-GFP expression remains repressed.

Interestingly, in the *iswi* mutant GSCs, *bam* transcription is elevated, and BMP signaling activity is down-regulated, indicating that ISWI is required for BMP-signaling-mediated transcriptional repression for *bam* in GSCs (Fig. 3B). Therefore, this study demonstrates that chromatin remodeling or epigenetic control has a critical role in controlling GSC self-renewal (Xi and Xie 2005). Because ISWI is a catalytic subunit of three different chromatin-remodeling complexes, namely, CHRAC (chromatin accessibility complex), NURF (nucleosome-remodeling factor), and ACF (ATP-utilizing chromatin assembly and remodeling factor) (Deuring et al. 2000), it would be interesting to investigate which ISWI-containing complex is involved in GSC regulation and if and how the ISWI-containing complex interacts with the Mad-Medea complex in repressing *bam* expression.

To further understand how BMP signaling controls GSC self-renewal, *pelota* (*pelo*), a gene known only to be required for *Drosophila* male meiosis (Eberhart and Wasserman 1995), was identified genetically as a dominant suppressor of the *dpp* overexpression-induced GSC tumor phenotype (Xi et al. 2005). In *pelo* mutant ovaries, GSCs are lost rapidly owing to differentiation, and our genetic results show that it functions inside GSCs to control self-renewal (Fig. 3C). In those *pelo* mutant GSCs, *bam* expression is still repressed, and *bam* mutant germ cells are still able to differentiate into cystocytes without *pelo* function, indicating that *pelo* controls GSC self-renewal by repressing a *bam*-independent differentiation pathway (Fig. 3D). Because Pelo is shown to function as an RNA endonuclease (Lee et al. 2007), our findings suggest that Pelo is involved in degrading mRNAs encoding proteins important for GSC differentiation (Xi et al. 2005). It is important to identify Pelo target mRNAs in GSCs to obtain a better understanding of how Pelo controls GSC self-renewal. Because Pelo is highly conserved from *Drosophila* to mammals, it may also be involved in the regulation of adult stem cell self-renewal in mammals, including humans.

microRNAs (miRNAs) regulate gene expression to modulate a variety of cellular events such as cell-fate determination and maintenance by controlling the turnover and/or translation of specific mRNAs in different cell types and organisms (Valencia-Sanchez et al. 2006). In *Drosophila*, Dicer-1 (Dcr-1) and Loquacious (Loqs) form a protein complex essential for generating mature miRNAs from their corresponding precursors. In the *Drosophila* ovary, Dcr-1 was first shown to be required for controlling GSC division only (Hatfield et al. 2005). Surprisingly, our results show that *dcr-1* mutant GSCs are lost rapidly from the niche without discernible features of cell death, indicating that Dcr-1 controls GSC self-renewal (Fig. 3E) (Jin and Xie 2007). *bam* transcription and protein expression, however, are not up-regulated in *dcr-1* mutant GSCs, and their removal does not slow down *dcr-1* mutant GSC loss, suggesting that Dcr-1 controls GSC self-renewal by repressing a Bam-independent differentiation pathway (Fig. 3F) (Jin and Xie 2007). In addi-

tion, Loqs is also shown to be required in GSCs for their maintenance (Park et al. 2007). Therefore, the miRNA pathway is required not only for regulating the GSC division rate, but also for controlling GSC self-renewal. In the future, it will be exciting to investigate functions of individual miRNAs and their targets in GSC regulation.

NOTCH SIGNALING CONTROLS GSC-NICHE FORMATION AND MAINTENANCE

Although a number of stem cell niches have been defined, it remains largely unknown how niche formation and maintenance are controlled. As mentioned earlier, cap cells are a key cellular component of the GSC niche (Xie and Spradling 2000). During the late third-instar larval stage, the TFs in the developing female gonad form before the emergence of cap cells (Zhu and Xie 2003). Interestingly, these newly formed TF cells express high levels of Delta that can only activate Notch signaling in adjacent Notch-expressing somatic precursor cells in the posterior because Delta is a transmembrane ligand (Song et al. 2007). Indeed, Notch signaling, detected by an *E(spl)* reporter line, is active in the cells that lie adjacent to the TF posteriorly and are destined to form cap cells (Fig. 4A) (Song et al. 2007). Expanded Notch activation causes the formation of more cap cells, which support more GSCs, whereas compromising Notch signaling during niche formation decreases the cap cell number and consequently the GSC number (Fig. 4B) (Song et al. 2007). Furthermore, the niches located away from their normal location can still sufficiently sustain GSC self-renewal by maintaining high local BMP signaling and repressing *bam* in

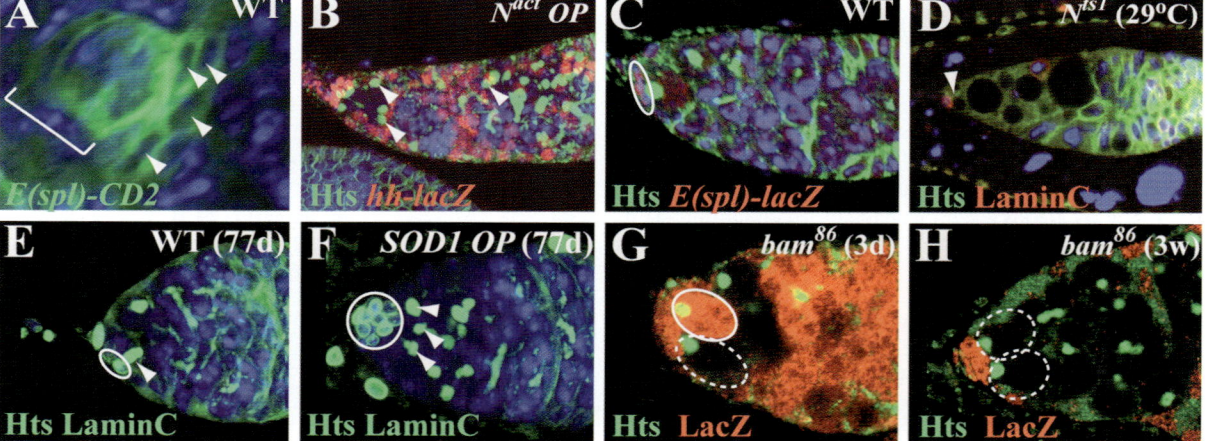

Figure 4. Notch signaling controls the maintenance and formation of the GSC niche, which further contributes to GSC aging and competition. (*A*) In the larval-pupal transitional stage, newly formed TFs (*bracket*) and cap cells (*arrowheads*) have active Notch signaling indicated by *E(spl)-CD2* expression (*green*). (*B*) Ectopic activation of Notch signaling during the third-instar larval stage can lead to formation of excessive cap cells (*red*) that in turn support more GSCs (*arrowheads*). (*C*) The cap cells (*oval*) in the adult ovary retain active Notch signaling activity indicated by *E(spl)-lacZ* (*red*). (*D*) A germarium (its tip indicated by *arrowhead*) shows the loss of cap cells and GSCs 2 weeks after inactivation of Notch signaling. (*E*) A 77-day-old germarium carries two cap cells (*oval*) and one GSC (its spectrosome indicated by *arrowhead*). (*F*) A 77-day-old germarium, in which *SOD1* is overexpressed in the cap cells, still has eight cap cells (*circle*) and three GSCs (*arrowheads*), like a young germarium. (*G*) A germarial tip contains unmarked wild-type GSC (*circle*) and a 3-day-old marked *bam* mutant GSC (*dashed circle*). (*H*) A germarial tip contains two marked *bam* mutant GSCs (*dashed circles*) 3 weeks after generation of a marked GSC. In the same ovariole, the outcompeted wild-type GSC has differentiated into a normal cyst in an old egg chamber.

ectopic GSCs. Similar findings on formation of extra cap cells induced by expanded Notch signaling have also been made by the Ruohola-Baker group (Ward et al. 2006). Finally, Delta starts to gain its expression in adult GSCs and maintains active Notch signaling in the cap cells of the adult ovary, and loss of Notch function results in rapid loss of the adult GSC niche, including cap cells (Fig. 4C,D). These findings demonstrate that Notch signaling is important for formation and maintenance of the GSC niche and that cap cells help to determine niche size and function (Song et al. 2007).

We have also shown that following the niche formation, anterior primordial germ cells (PGCs) adjacent to cap cells can develop into GSCs at the early pupal stage while the rest differentiate directly (Zhu and Xie 2003). The anterior PGCs are very mitotically active and exhibit two division patterns with respect to the niche: asymmetric division (one daughter contacting cap cells and the other daughter away from cap cells) and symmetric division (both daughters contacting cap cells). Indeed, our genetic lineage analysis results indicate that newly formed GSCs can divide asymmetrically and symmetrically (Zhu and Xie 2003). At the third-instar larval stage, *dpp* overexpression promotes PGC proliferation and causes the accumulation of more PGCs in the gonad. The PGCs mutant for *tkv*, encoding an essential *dpp* receptor, lose their ability to divide symmetrically. Therefore, *dpp* is probably one of the mitotic signals that promote the symmetric division and expansion of GSCs during niche and GSC formation (Zhu and Xie 2003).

NICHE AGING AND INTRINSIC GSC AGING COLLECTIVELY CONTRIBUTE TO OVERALL GSC AGING

It is widely postulated that tissue aging could be, at least partially, caused by reduction of stem cell number, activity, or both. However, it largely remains a mystery as to how stem cells and their niche deal with the aging process. Indeed, the number of GSCs and cap cells in old *Drosophila* females decreases in comparison with that in young females, and the GSC proliferation rate also declines with age (Fig. 4E) (Pan et al. 2007). As expected, BMP signaling activity and E-cadherin accumulation in the stem cell/niche junction undergo an age-dependent decline. In females heterozygous for mutations in the BMP signaling and E-cadherin-mediated adhesion pathways, the age-dependent decline in the number of GSCs and cap cells is accelerated, whereas providing more BMP in the niche or E-cadherin in GSCs can significantly rescue the age-dependent decline in GSC number and proliferation rate, demonstrating that the age-dependent decline in BMP signaling and E-cadherin-mediated cell adhesion contribute to GSC aging (Pan et al. 2007).

Reactive oxygen species (ROS)-induced cellular damage is known to cause cellular and organismal aging, whereas overexpression of superoxide dismutase (SOD), an enzyme that helps remove ROS from the cell, can prolong life span in *Drosophila* (Tower 2000). Interestingly, overexpression of SOD specifically in cap cells can prolong the life span of cap cells and GSCs and promote proliferation of aged

GSCs, indicating that the aged niche contributes to overall stem cell aging (Fig. 4F) (Pan et al. 2007). Similarly, SOD overexpression in GSCs alone can prolong GSC life span and promote GSC proliferation, indicating that GSCs also undergo intrinsic aging. Therefore, our study demonstrates that ROS-induced cellular damage causes niche aging and intrinsic stem cell aging, which collectively contribute to overall stem cell aging (Pan et al. 2007).

COMPETITION IS USED BY DIFFERENTIATION-DEFECTIVE STEM CELLS FOR THEIR EXPANSION AND MAY ALSO SERVE AS A STEM CELL QUALITY CONTROL MECHANISM

Although much progress has recently been made regarding how the niche controls stem cell function, little is yet known about how stem cells in the same niche or tissue interact with one another. In addition, cancer stem cells (CSCs) have recently been proposed to be rare, self-renewing mitotic cells for driving tumor growth, but it remains unclear how CSCs interact with normal stem cells. We have used *Drosophila* ovarian GSCs as a model system to investigate how stem cells interact with one another in the same niche or tissue. *bam* or *bgcn* mutant GSCs can continuously self-renew and generate a large number of differentiation-defective stem cells, resembling mammalian CSCs (McKearin and Ohlstein 1995; Kai et al. 2005). The differentiation-defective *bam* or *bgcn* mutant GSCs gradually push neighboring wild-type GSCs out of the niche, indicating that these CSC-like *bam* or *bgcn* mutant GSCs are more competitive than wild-type ones (Fig. 4G,H) (Jin et al. 2008). Furthermore, *bam* or *bgcn* mutant GSCs up-regulate E-cadherin expression in the stem cell/niche junction, and different E-cadherin levels can sufficiently stimulate GSC competition (Jin et al. 2008). Therefore, our findings demonstrate that differentiation-defective stem cells can outcompete normal stem cells for niche occupancy through up-regulation of E-cadherin and further suggest that CSCs may use competition to expand themselves and invade normal tissues (Jin et al. 2008).

What is the biological function of competition in stem cell regulation? We have shown that when GSCs differentiate and up-regulate *bam* expression, they are detached from the niche and are lost rapidly, although there are adherens junctions between GSCs and their niche (Song et al. 2002, 2004; Xi and Xie 2005). Stem cell competition offers important insight into how differentiated stem cells are expelled out of the niche. The following is our current working model: A differentiating GSC up-regulates *bam* expression and consequently down-regulates E-cadherin expression; over time, the mutant GSC has less E-cadherin than its wild-type neighboring stem cell in the same niche and is then pushed out of the niche by a wild-type neighboring stem cell, which then generates a new stem cell to replace the lost one. Thus, we propose that GSCs have a competitive relationship for niche occupancy, which likely serves as a quality control mechanism to ensure that accidentally differentiated stem cells are rapidly removed from the niche and replaced by functional ones.

FSCs CONTROL THEIR SELF-RENEWAL USING A DIFFERENT COMBINATION OF NICHE SIGNALS FROM GSCs

To investigate if and how different stem cell types in the same tissue are regulated differently, we have chosen to study FSC regulation in the *Drosophila* ovary. We have shown that Wingless (Wg) is also expressed in cap cells and is required for follicle cell production (Song and Xie 2003). Down-regulation of Wg signaling through removal of its positive regulators *disheveled* (*dsh*) and *armadillo* (*arm*) results in rapid FSC loss (Fig. 5A,B). Surprisingly, constitutive Wg signaling in FSCs through the removal of its negative regulators *Axin* and *shaggy* also causes their loss, suggesting that appropriate levels of Wg signaling are critical for FSC self-renewal (Song and Xie 2003). In addition, constitutive Wg signaling causes overproliferation and abnormal differentiation of early FSC progeny. These findings demonstrate that *wg* signaling regulates FSC maintenance and also influences follicle cell differentiation (Song and Xie 2003). In mammals, Wnt signaling is important for maintaining the epithelial stem cell compartment in the intestine (Korinek et al. 1998; Alonso and Fuchs 2003), whereas constitutive Wnt signaling causes overproliferation and abnormal differentiation of epithelia stem cells in the skin and the intestine, resulting in cancer formation (Korinek et al. 1997; DasGupta and Fuchs 1999). Possibly, mechanisms regulating proliferation and differentiation of epithelial stem cells are conserved from *Drosophila* to humans.

dpp is restricted to cap cells and follicle cells, whereas *gbb* is expressed in all of the somatic cells including cap cells and escort cells covering differentiated germ cells, but their role in FSC regulation remains unclear (Song et al. 2004). We have shown that although *dpp* mutant ovaries show a weak follicle cell production defect, *gbb* mutant ovaries exhibit a severe deficiency in follicle cell production (Kirilly et al. 2005). In addition, the marked

FSC clones mutant for BMP downstream components have retarded proliferation and are lost rapidly, indicating that BMP signaling is necessary for promoting FSC self-renewal and proliferation (Fig. 5C,D). Moreover, constitutive BMP signaling prolongs the FSC life span. Therefore, BMP signaling directly promotes SSC self-renewal and proliferation in the *Drosophila* ovary (Kirilly et al. 2005). In the future, it will be critical to understand how these different signaling pathways are integrated in FSCs to control their self-renewal.

As one would have expected, FSCs must stay close to sources for self-renewing signals, including cap cells and escort cells, in order to maintain their self-renewal. We have shown that E-cadherin-mediated cell adhesion is required for keeping FSCs in the proximity to signaling sources for their self-renewal (Song and Xie 2002). The marked *shg* or *arm* mutant FSC clones are rapidly lost from germaria (Fig. 5E). In addition, we have identified Domino (Dom), an SNF2-related ATP-dependent chromatin remodeling factor, as an essential intrinsic factor for controlling FSC self-renewal because the marked *dom* mutant FSCs are lost rapidly from germaria (Fig. 5F) (Xi and Xie 2005). Finally, we have also shown that the miRNA pathway is needed in FSCs for their self-renewal (Jin and Xie 2007). In the future, it will be of great interest to study the relationships between intrinsic pathways and niche-signal-mediated signaling pathways.

CONCLUSIONS AND FUTURE DIRECTIONS

Along with the findings from other colleagues in the field, our findings on *Drosophila* ovarian GSC and FSC regulation have produced several general principles for stem cell biology as described below.

First, the stem cell is anchored to its niche through cadherin-mediated cell adhesion to ensure long-term self-renewal. Both GSCs and FSCs require E-cadherin-mediated

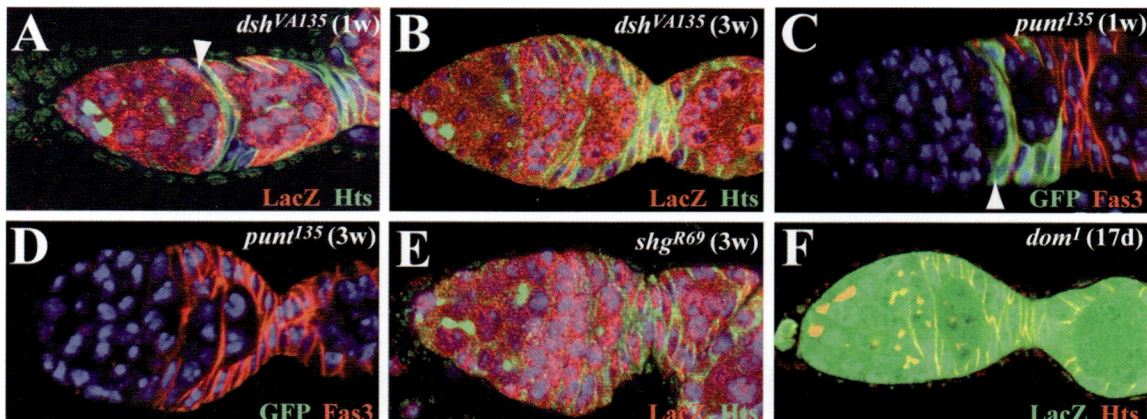

Figure 5. Wingless signaling, BMP signaling, E-cadherin-mediated adhesion, and Dom are required for maintaining FSCs. (*A*) A germarium carries a LacZ-negative marked *dsh* mutant FSC (*arrowhead*) 1 week after clone induction. (*B*) A germarium has lost a LacZ-negative marked *dsh* mutant FSC clone 3 weeks after clone induction. (*C*) A germarium carries a GFP-positive marked *punt* mutant FSC (*arrowhead*) 1 week after clone induction. *punt* encodes a BMP type II receptor. (*D*) A germarium has lost a GFP-positive marked *punt* mutant FSC clone 3 weeks after clone induction. (*E*) A germarium has lost a LacZ-negative marked *shg* mutant FSC clone 3 weeks after clone induction. (*E*) A germarium has lost a LacZ-negative marked *dom* mutant FSC clone 3 weeks after clone induction.

cell adhesion to keep them in the proximity of signals from the niche (Fig. 6) (Song and Xie 2002; Song et al. 2002). Such anchorage guarantees that the future stem cell will continuously self-renew and that the unanchored daughter that leave the niche and differentiate.

Second, different stem cell niches can have distinct structures and use different combinations of signals (Fig. 6). The GSC niche is composed of cap cells and possibly ESCs, and produces BMPs and the unidentified Piwi-regulated signal controlling GSC self-renewal and proliferation (Xie and Spradling 1998, 2000; King et al. 2001; Song et al. 2004, 2007; Decotto and Spradling 2005), whereas the FSC niche is composed of at least cap cells and escort cells and produces BMP, Hh, and Wg signals for promoting self-renewal and proliferation (Forbes et al. 1996; Zhang and Kalderon 2001; Song and Xie 2003; Kirilly et al. 2005). Interestingly, cap cells produce self-renewing signals for both GSCs and FSCs, likely representing a common niche component for coordinating the activities of both stem cells. In addition, BMPs from cap cells only function in one cell diameter to control GSC self-renewal (Kai and Spradling 2003; Song et al. 2004), whereas Wg and Hh in cap cells function in many cell diameters to control FSC self-renewal (Forbes et al. 1996; King et al. 2001; Song and Xie 2003).

Third, different classes of intrinsic factors are required for GSC or FSC self-renewal (Fig. 6). In addition to previously identified Pumilio/Nanos, cyclin B, and Stonewall (Lin and Spradling 1997; Forbes and Lehmann 1998; Wang and Lin 2005; Maines et al. 2007), we have shown that Dcr-1, Pelota, and ISWI are required intrinsically for GSC self-renewal, and Dcr-1 and Dom are required intrinsically for FSC self-renewal (Xi et al. 2005; Xi and Xie 2005; Jin and Xie 2007). These intrinsic factors are involved in different biological pathways, such as mRNA translation and degradation, chromatin remodeling, miRNAs, and the cell cycle.

Fourth, intrinsic factors and niche signals control self-renewal by repressing expression or functions of differentiation-promoting genes. BMP niche signaling, Piwi-mediated niche signaling (Fig. 6), and ISWI repress expression of the differentiation-promoting gene *bam* in GSCs and thus maintain their undifferentiated state (Chen and McKearin 2003a, 2005; Song et al. 2004; Szakmary et al. 2005; Xi and Xie 2005), whereas the intrinsic factors Pumilio/ Nanos, Pelota, Stonewall, and Dcr-1/Loqs are required for maintaining GSC self-renewal by repressing a *bam*-independent pathway (Gilboa and Lehmann 2004; Wang and Lin 2004; Xi et al. 2005; Jin and Xie 2007; Maines et al. 2007; Park et al. 2007).

Fifth, the stem cell niche requires signals for maintenance from its resident stem cells (Fig. 6). We have shown that Notch signaling is required for GSC-niche formation during early development, and GSC-expressing Delta can signal back to cap cells for their maintenance and integrity in the adult ovary (Ward et al. 2006; Song et al. 2007). These findings indicate that stem cells and their niche are mutually dependent.

Sixth, stem cells and their niche undergo intrinsic aging, collectively contributing to overall stem cell aging (Fig. 6). ROS-induced cellular damage leads to niche and GSC aging in the *Drosophila* ovary (Pan et al. 2007).

Finally, competition serves as a quality control mechanism. We have found that *bam* and *bgcn* mutant GSCs can outcompete wild-type GSCs by up-regulating E-cadherin expression (Jin et al. 2008). This finding also suggests the existence of a quality control mechanism to ensure that only undifferentiated stem cells stay in the niche. So far, some of the generalizations of the control mechanisms for *Drosophila* ovarian stem cells have been confirmed and also have provided guidance for studying stem cells in mammalian systems; others await further verification.

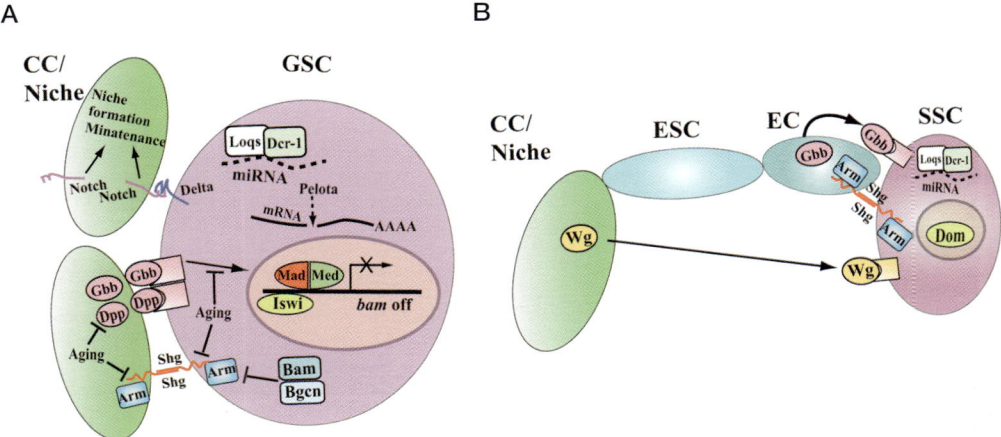

Figure 6. Schematic diagrams summarizing our contributions to understanding *Drosophila* ovarian GSC and FSC regulation. (*A*) For GSCs, we have identified niche-derived BMP signaling for maintaining self-renewal, E-cadherin for anchorage in the niche, Notch signaling for controlling niche formation and maintenance, intrinsic factors Dcr-1, Pelota, and ISWI for controlling self-renewal, and Bam/Bgcn for controlling stem cell competitiveness. In addition, we have also shown that ROS-induced cellular damage leads to niche aging and intrinsic GSC aging, collectively contributing to overall GSC aging. (*B*) For FSCs, we have identified BMP and Wingless signaling pathways and chromatin remodeling factors Dom and Dcr-1 for controlling self-renewal and E-cadherin for anchorage in the niche.

Although much progress has been made in studying *Drosophila* ovarian stem cells, such as defining stem cell niche structures and signals and identifying self-renewal-promoting intrinsic factors, many important questions still remain to be answered. In the future, we will continue to use a combination of genetic, molecular, biochemical, genomic, and cell biological approaches and *Drosophila* ovarian stem cells as a model system to address the following important stem-cell-related questions, including: How are GSC and FSC niches built in the *Drosophila* ovary? How do shared niche components contribute to functions of different niches and coordination of these stem cell activities? What are additional intrinsic factors and extrinsic signals from the niche for controlling GSC and FSC self-renewal? How are multiple pathways integrated in stem cells to control self-renewal? How do intrinsic factors interact with extrinsic signals to control stem cell self-renewal, proliferation, and differentiation? The answers to these questions will provide further insights into how *Drosophila* ovarian stem cells are regulated and how stem cells are controlled in general.

ACKNOWLEDGMENTS

Our special thanks go to all the former Xie laboratory members for their outstanding scientific contributions. The work in the Xie lab is supported by the Stowers Institute for Medical Research and the National Institutes of Health (GM64428).

REFERENCES

Alonso, L. and Fuchs, E. 2003. Stem cells in the skin: Waste not, Wnt not. *Genes Dev.* **17:** 1189–1200.

Chen, D. and McKearin, D. 2003a. Dpp signaling silences *bam* transcription directly to establish asymmetric divisions of germline stem cells. *Curr. Biol.* **13:** 1786–1791.

Chen, D. and McKearin, D.M. 2003b. A discrete transcriptional silencer in the *bam* gene determines asymmetric division of the *Drosophila* germline stem cell. *Development* **130:** 1159–1170.

Chen, D. and McKearin, D. 2005. Gene circuitry controlling a stem cell niche. *Curr. Biol.* **15:** 179–184.

Cox, D.N., Chao, A., Baker, J., Chang, L., Qiao, D., and Lin, H. 1998. A novel class of evolutionarily conserved genes defined by *piwi* are essential for stem cell self-renewal. *Genes Dev.* **12:** 3715–3727.

Cox, D.N., Chao, A., and Lin, H. 2000. *piwi* encodes a nucleoplasmic factor whose activity modulates the number and division rate of germline stem cells. *Development* **127:** 503–514.

DasGupta, R. and Fuchs, E. 1999. Multiple roles for activated LEF/TCF transcription complexes during hair follicle development and differentiation. *Development* **126:** 4557–4568.

Decotto, E. and Spradling, A.C. 2005. The *Drosophila* ovarian and testis stem cell niches: Similar somatic stem cells and signals. *Dev. Cell* **9:** 501–510.

Deuring, R., Fanti, L., Armstrong, J.A., Sarte, M., Papoulas, O., Prestel, M., Daubresse, G., Verardo, M., Moseley, S.L., Berloco, M., et al. 2000. The ISWI chromatin-remodeling protein is required for gene expression and the maintenance of higher order chromatin structure in vivo. *Mol. Cell* **5:** 355–365.

Eberhart, C.G. and Wasserman, S.A. 1995. The *pelota* locus encodes a protein required for meiotic cell division: An analysis of G2/M arrest in *Drosophila* spermatogenesis. *Development* **121:** 3477–3486.

Forbes, A. and Lehmann, R. 1998. Nanos and Pumilio have critical roles in the development and function of *Drosophila* germline stem cells. *Development* **125:** 679–690.

Forbes, A.J., Lin, H., Ingham, P.W., and Spradling, A.C. 1996. *hedgehog* is required for the proliferation and specification of ovarian somatic cells prior to egg chamber formation in *Drosophila*. *Development* **122:** 1125–1135.

Gilboa, L. and Lehmann, R. 2004. Repression of primordial germ cell differentiation parallels germ line stem cell maintenance. *Curr. Biol.* **14:** 981–986.

Hatfield, S.D., Shcherbata, H.R., Fischer, K.A., Nakahara, K., Carthew, R.W., and Ruohola-Baker, H. 2005. Stem cell division is regulated by the microRNA pathway. *Nature* **435:** 974–978.

Haug, J.S., He, X.C., Grindley, J.C., Wunderlich, J.P., Gaudenz, K., Ross, J.T., Paulson, A., Wagner, K.P., Xie, Y., Zhu, R., et al. 2008. N-cadherin expression level distinguishes reserved versus primed states of hematopoietic stem cells. *Cell Stem Cell* **2:** 367–379.

Jin, Z. and Xie, T. 2007. Dcr-1 maintains *Drosophila* ovarian stem cells. *Curr. Biol.* **17:** 539–544.

Jin, Z., Kirilly, D., Weng, C., Kawase, E., Song, X., Smith, S., Schwartz, J., and Xie, T. 2008. Differentiation-defective stem cells outcompete normal stem cells for niche occupancy in the *Drosophila* ovary. *Cell Stem Cell* **2:** 39–49.

Kai, T. and Spradling, A. 2003. An empty *Drosophila* stem cell niche reactivates the proliferation of ectopic cells. *Proc. Natl. Acad. Sci.* **100:** 4633–4638.

Kai, T., Williams, D., and Spradling, A.C. 2005. The expression profile of purified *Drosophila* germline stem cells. *Dev. Biol.* **283:** 486–502.

Kawase, E., Wong, M.D., Ding, B.C., and Xie, T. 2004. Gbb/Bmp signaling is essential for maintaining germline stem cells and for repressing *bam* transcription in the *Drosophila* testis. *Development* **131:** 1365–1375.

King, F.J. and Lin, H. 1999. Somatic signaling mediated by *fs(1)Yb* is essential for germline stem cell maintenance during *Drosophila* oogenesis. *Development* **126:** 1833–1844.

King, F.J., Szakmary, A., Cox, D.N., and Lin, H. 2001. *Yb* modulates the divisions of both germline and somatic stem cells through *piwi-* and *hh*-mediated mechanisms in the *Drosophila* ovary. *Mol. Cell* **7:** 497–508.

Kirilly, D., Spana, E.P., Perrimon, N., Padgett, R.W., and Xie, T. 2005. BMP signaling is required for controlling somatic stem cell self-renewal in the *Drosophila* ovary. *Dev. Cell* **9:** 651–662.

Korinek, V., Barker, N., Morin, P.J., van Wichen, D., de Weger, R., Kinzler, K.W., Vogelstein, B., and Clevers, H. 1997. Constitutive transcriptional activation by a β-catenin-Tcf complex in APC$^{-/-}$ colon carcinoma. *Science* **275:** 1784–1787.

Korinek, V., Barker, N., Moerer, P., van Donselaar, E., Huls, G., Peters, P.J., and Clevers, H. 1998. Depletion of epithelial stem-cell compartments in the small intestine of mice lacking Tcf-4. *Nat. Genet.* **19:** 379–383.

Kretzschmar, M. and Massagué, J. 1998. SMADs: Mediators and regulators of TGF-β signaling. *Curr. Opin. Genet. Dev.* **8:** 103–111.

Lee, H.H., Kim, Y.S., Kim, K.H., Heo, I., Kim, S.K., Kim, O., Kim, H.K., Yoon, J.Y., Kim, H.S., Kim do, J., et al. 2007. Structural and functional insights into Dom34, a key component of no-go mRNA decay. *Mol. Cell* **27:** 938–950.

Li, L. and Xie, T. 2005. Stem cell niche: Structure and function. *Annu. Rev. Cell Dev. Biol.* **21:** 605–631.

Lin, H. and Spradling, A.C. 1997. A novel group of *pumilio* mutations affects the asymmetric division of germline stem cells in the *Drosophila* ovary. *Development* **124:** 2463–2476.

Maines, J.Z., Park, J.K., Williams, M., and McKearin, D.M. 2007. Stonewalling *Drosophila* stem cell differentiation by epigenetic controls. *Development* **134:** 1471–1479.

Margolis, J. and Spradling, A. 1995. Identification and behavior of epithelial stem cells in the *Drosophila* ovary. *Development* **121:** 3797–3807.

McKearin, D.M. and Ohlstein, B. 1995. A role for the *Drosophila* Bag-of-marbles protein in the differentiation of cystoblasts from germline stem cells. *Development* **121:** 2937–2947.

McKearin, D.M. and Spradling, A.C. 1990. *bag-of-marbles*: A *Drosophila* gene required to initiate both male and female gametogenesis. *Genes Dev.* **4:** 2242–2251.

Morrison, S.J. and Spradling, A.C. 2008. Stem cells and niches:

Mechanisms that promote stem cell maintenance throughout life. *Cell* **132:** 598–611.

Nystul, T. and Spradling, A. 2007. An epithelial niche in the *Drosophila* ovary undergoes long-range stem cell replacement. *Cell Stem Cell* **1:** 277–285.

Ohlstein, B. and McKearin, D. 1997. Ectopic expression of the *Drosophila* Bam protein eliminates oogenic germline stem cells. *Development* **124:** 3651–3662.

Ohlstein, B., Lavoie, C.A., Vef, O., Gateff, E., and McKearin, D.M. 2000. The *Drosophila* cystoblast differentiation factor, *benign gonial cell neoplasm,* is related to DExH-box proteins and interacts genetically with *bag-of-marbles. Genetics* **155:** 1809–1819.

Pan, L., Chen, S., Weng, C., Call, G.B., Zhu, D., Tang, H., Zhang, N., and Xie, T. 2007. Stem cell aging is controlled both intrinsically and extrinsically in the *Drosophila* ovary. *Cell Stem Cell* **1:** 458–469.

Park, J.K., Liu, X., Strauss, T.J., McKearin, D.M., and Liu, Q. 2007. The miRNA pathway intrinsically controls self-renewal of *Drosophila* germline stem cells. *Curr. Biol.* **17:** 533–538.

Song, X. and Xie, T. 2002. DE-cadherin-mediated cell adhesion is essential for maintaining somatic stem cells in the *Drosophila* ovary. *Proc. Natl. Acad. Sci.* **99:** 14813–14818.

Song, X. and Xie, T. 2003. *wingless* signaling regulates the maintenance of ovarian somatic stem cells in *Drosophila. Development* **130:** 3259–3268.

Song, X., Zhu, C.H., Doan, C., and Xie, T. 2002. Germline stem cells anchored by adherens junctions in the *Drosophila* ovary niche. *Science* **296:** 1855–1857.

Song, X., Wong, M.D., Kawase, E., Xi, R., Ding, B.C., McCarthy, J.J., and Xie, T. 2004. Bmp signals from niche cells directly repress transcription of a differentiation-promoting gene, *bag of marbles,* in germline stem cells in the *Drosophila* ovary. *Development* **131:** 1353–1364.

Song, X., Call, G.B., Kirilly, D., and Xie, T. 2007. Notch signaling controls germline stem cell niche formation in the *Drosophila* ovary. *Development* **134:** 1071–1080.

Szakmary, A., Cox, D.N., Wang, Z., and Lin, H. 2005. Regulatory relationship among *piwi, pumilio,* and *bag-of-marbles* in *Drosophila* germline stem cell self-renewal and differentiation. *Curr. Biol.* **15:** 171–178.

Tower, J. 2000. Transgenic methods for increasing *Drosophila* life span. *Mech. Ageing Dev.* **118:** 1–14.

Valencia-Sanchez, M.A., Liu, J., Hannon, G.J., and Parker, R. 2006. Control of translation and mRNA degradation by miRNAs and siRNAs. *Genes Dev.* **20:** 515–524.

Wang, Z. and Lin, H. 2004. *Nanos* maintains germline stem cell self-renewal by preventing differentiation. *Science* **303:** 2016–2019.

Wang, Z. and Lin, H. 2005. The division of *Drosophila* germline stem cells and their precursors requires a specific cyclin. *Curr. Biol.* **15:** 328–333.

Ward, E.J., Shcherbata, H.R., Reynolds, S.H., Fischer, K.A., Hatfield, S.D., and Ruohola-Baker, H. 2006. Stem cells signal to the niche through the Notch pathway in the *Drosophila* ovary. *Curr. Biol.* **16:** 2352–2358.

Xi, R. and Xie, T. 2005. Stem cell self-renewal controlled by chromatin remodeling factors. *Science* **310:** 1487–1489.

Xi, R., Doan, C., Liu, D., and Xie, T. 2005. Pelota controls self-renewal of germline stem cells by repressing a Bam-independent differentiation pathway. *Development* **132:** 5365–5374.

Xie, T. and Spradling, A.C. 1998. *decapentaplegic* is essential for the maintenance and division of germline stem cells in the *Drosophila* ovary. *Cell* **94:** 251–260.

Xie, T. and Spradling, A.C. 2000. A niche maintaining germ line stem cells in the *Drosophila* ovary. *Science* **290:** 328–330.

Zhang, Y. and Kalderon, D. 2001. Hedgehog acts as a somatic stem cell factor in the *Drosophila* ovary. *Nature* **410:** 599–604.

Zhu, C.H. and Xie, T. 2003. Clonal expansion of ovarian germline stem cells during niche formation in *Drosophila. Development* **130:** 2579–2588.

Stem Cells and Their Niches: Integrated Units That Maintain *Drosophila* Tissues

A.C. Spradling, T. Nystul, D. Lighthouse, L. Morris, D. Fox, R. Cox, T. Tootle, R. Frederick, and A. Skora

Howard Hughes Medical Institute Research Laboratories, Department of Embryology, Carnegie Institution of Washington, Baltimore, Maryland 21218

The genetic analysis of four distinct *Drosophila* stem cells and their niches has revealed principles of stem cell biology that are likely to apply widely. A stem cell and its niche act together as integral parts of a system that supplies replacement cells when and where they are needed within a tissue. Stem cell/niche units are highly regulated and continue to operate despite the periodic turnover and replacement of all of their component cells. To successfully respond to tissue needs, these units receive and process a wide range of local and systemic information. A stem cell alone would be no more use at this task than an isolated neuron. It is only when integrated into a system of multiple interacting cells (the niche) that stem cells achieve the capacity to serve as the fundamental units of tissue homeostasis and repair.

Adult metazoan tissues maintain a high level of functionality by routinely replacing their component cells. Stably differentiated cells may temporarily transition to a state where they can divide and produce daughters like themselves, without reiterating their entire process of developmental specification. More commonly, regenerative potential is based on tissue stem cells. During development, many tissues set aside a relatively small number of specific cells that maintain themselves in a less differentiated state while generating daughter cells that complete development in an adult milieu. Some of the most accessible and best-understood adult metazoan stem cells are found in the gonads and gut of *Drosophila*. The powerful genetics and relatively simple tissue architecture of this model organism has made it possible to identify stem cells in situ and to analyze their behavior and molecular regulation, providing useful insights for stem cell biology in general (for review, see Morrison and Spradling 2008).

One of the most important contributions of *Drosophila* stem cell studies has been to document the central role of the stem cell niche. Stem cell behavior in vivo depends on the local microenvironment—the stem cell niche (Xie and Spradling 2000). Although some stem cells can be isolated (Bryder et al. 2006) and cultured in vitro (Kanatsu-Shinohara et al. 2003; Niki and Mahowald 2003), they must still be returned to a tissue niche for normal activity. The defining characteristic of niches is that they are required for stem cell maintenance, i.e., for self-renewing divisions that also generate new tissue cells. In the absence of this natural milieu, or the artificial mixes of growth factors found in stem cell culture media, stem cells begin to differentiate. Making stem cells dependent on factors that are not readily available throughout the tissue, but only found in isolated "niches," may be a strategy to tightly align stem cell activity with tissue needs and limit unregulated stem cell proliferation, which poses an obvious danger. Once a tissue's niches are filled, any additional potential stem cells that arise by somatic muta-

tion or failed differentiation would be unable to sustain themselves. Stem cell preservation provides the defining test for a candidate stem cell niche: It must maintain introduced competent cells as stem cells, cells that would differentiate or die at all other locations.

Here, we review recent progress and argue that stem cell/niche systems do more than store stem cells and limit overproliferation. They must determine the number of replacement cells and their states of differentiation needed at each location. Answering these questions requires the stem cell/niche unit to receive and process inputs from hormones, intercellular signals, mechanical stresses, sensory information, etc. Thus, the task of the stem cell and its niche is not unlike that of a neural ganglion. A stem cell alone would be of no more use than an isolated neuron. It is only when integrated into an information processing system comprising multiple linked cells and cell types (the niche) do stem cells achieve the capacity to serve as a fundamental unit of tissue homeostasis and repair.

METHODS

***Drosophila* strains.** Clonal analysis was performed essentially as described by Nystul and Spradling (2007). β-galactosidase (β-gal⁺)-positive clones (Harrison and Perrimon 1993) were generated using *yw, p[(hsFlp)¹², ry⁺]; X.15.29*, and *yw; X.15.33*. Dual-marked clones were induced using *y¹²², hsFlp; FRT 42D tub-lacZ/cyo; TM2/TM6B* and *w; and FRT 42D, Ubi-GFP/CyO* (Bloomington). For further information, see FlyBase (http://flybase.bio.indiana.edu).

Immunofluoresence microscopy. Immunofluoresence microscopy was performed essentially as described in Nystul and Spradling (2007). Ovaries dissected in Grace's medium were fixed in 4% paraformaldehyde for 10 minutes, rinsed with 0.2% Triton X-100 in phosphate-buffered saline (1x PBST) and 1x PBST + 5% normal goat serum, and then incubated with primary antibody (overnight at 4°C). After washing in 1x PBST (3x 20 minutes), tissue was incubated with secondary antibody (1x PBST, 0.5% bovine serum albumin [BSA], and 60 minutes), and washed in 1x PBST (3x 20 minutes) and 1x PBS (3x 20 minutes). Slides were stained with DAPI (4'-6-diamidino-2-phenylindole) and mounted with Vectashield (Vector).

Electron microscopy. Electron microscopy was performed essentially as described by Cox and Spradling (2003). Tissue isolated in Grace's media was fixed for 1 hour (1% gluteraldehyde, 1% OsO$_4$, 0.1 M cacodylate buffer, 2 mM Ca at pH 7.5). Following washes in cacodylate buffer (3x 5 minutes) the tissue was embedded in agarose at 55°C, rinsed in 0.05 M maleate (pH 6.5) (1x 5 minutes), and incubated in 0.5% uranyl acetate, 0.05 M maleate (1x 1.5 hours). Tissue was then dehydrated through ethanol (35% 2x 5 minutes, 50% 10 minutes, 75% 10 minutes, 95% 10 minutes, 100% 3x 10 minutes), incubated in propylene oxide (2x 10 minutes), and 1:1 propylene oxide:resin (Epon 812:Quetol 651(2:1):1% silicone 200, 2% benzyldimethylamine [BDMA]) (1x 1 hour). Resin was changed (3x 1 hour) and then allowed to polymerize at 50°C (>8 hours) and then at 70°C (>8 hours). Images were captured with a Phillips Tecnai 12 microscope and recorded with a GATAN multiscan CCD (charged-coupled device) camera using Digital Micrograph software.

Lineage analysis. Lineage analysis, as described by Nystul and Spradling (2007), was performed by generating either β-gal[+] clones or dual-marked clones (see *Drosophila* strains). F$_1$ flies that contained hsFlp and the two appropriate FRT-containing chromosomes were maintained with fresh food at 25°C, and mitotic clones were generated by heat shocks (37°C x 60 minutes). To assay for nondividing FSC niche cells, this treatment was repeated once a day for 3–5 days.

RESULTS

Anatomically Fixed Niches: The Germ-line Stem Cell Niche

Early studies of *Drosophila* niches focused on the female and male germ-line stem cells (GSCs) (Xie and Spradling 2000; for review, see Fuller and Spradling 2007), because these stem cells and niches are among the simplest and most accessible in all of biology. The location of the female GSC niche at the blind end of each tubular ovariole fulfilled many preconceptions of what a niche should be like (Fig. 1). Here, the terminal filament and cap cells define a unique tissue microenvironment that activates the Dpp (Decapentaplegic) signaling pathway within resident germ cells to much higher levels than cells just one cell diameter away (Chen and McKearin 2003; Song et al. 2004). No other region of the normal

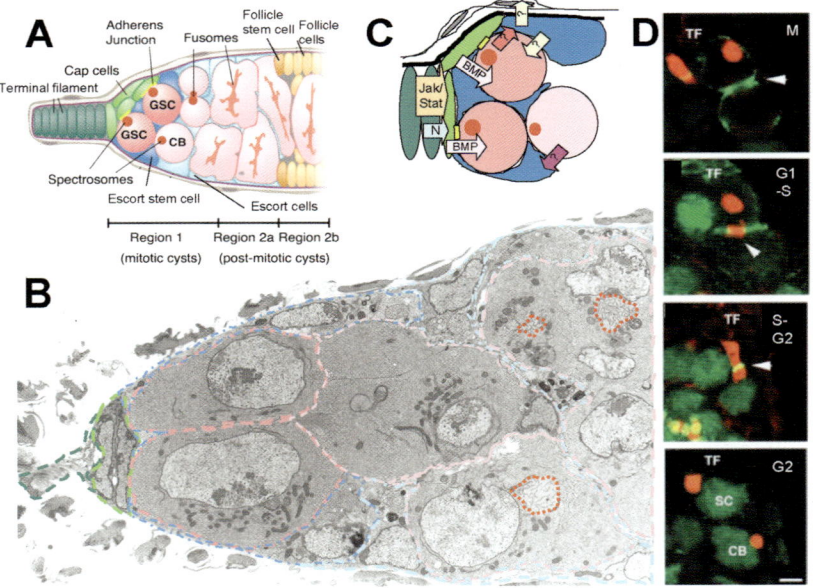

Figure 1. The germ-line stem cell niche. (*A*) Diagram of a *Drosophila* germarium (Nystul and Spradling 2007) showing the germ-line stem cells (GSC), escort stem cells, and stromal cells that make up the GSC niche located just posterior to the cap cells (*light green*). Follicle stem cells (*gold*) are found at the junction of regions 2a and 2b, where cysts (*pink*) begin to move in single file. (*B*) Electron micrograph of the GSC niche region. The cells have been outlined in colors that correspond to the diagram above. Cross sections of the fusome are outlined in *red*. (*C*) Known and proposed signals that mediate niche/stem cell regulation. Cells are colored as in *A*. (*Thick black line*) Basement membrane; (*gray*) ovarian (muscle) sheath. (*D*) Fusome behavior during the GSC cell cycle. At M phase, the spectrosome is entirely within the anterior cell (closer to the terminal filament, TF). The cytokinesis furrow is beginning to form (*arrowhead*). During G$_1$-S, new fusome material grows within the arrested cytokinesis furrow (*arrowhead*). During late S-G$_2$, the two segments of fusome material fuse; however, more always is found within the GSC than in the cystoblast. During G$_2$, the cytokinesis furrow separating the GSC from the cystoblast (CB) finally closes, leaving the two cells with differing amounts of fusome. (*Red*) Hts protein; (*green*) Anilin. (Modified from deCuevas and Spradling 1998.)

germarium induces such strong pathway activation (Kai and Spradling 2003), and only this high level is sufficient to repress the *bag-of-marbles* (*bam*) gene and suppress GSC differentiation. Remarkably, a germ cell daughter derived from the division of a nearby GSC or a cystocyte up to the eight-cell stage, cells that would differentiate outside of the niche, can dedifferentiate and become functional GSCs if they gain niche access (Brawley and Matunis 2004; Kai and Spradling 2004).

If the niche was simply a small zone capable of holding a GSC in place and blocking its differentiation by regulating *bam*, it could not adequately function. A major role of the niche is to regulate cell production, not simply generate daughters continuously. Stem cell division responds directly to nutritional status (Drummond-Barbosa and Spradling 2001) and the actual rate of egg deposition. Nutritional levels are sensed by GSCs and their progeny using the insulin-signaling pathway (Drummond-Barbosa and Spradling 2001; LeFever and Drummond-Barbosa 2005). How oviposition is sensed and whether oviposition from specific ovarioles can be distinguished remain unknown, but they may involve prostaglandin signaling (Tootle and Spradling 2008). The niche probably also mediates interactions among individual GSCs and facilitates the replacement of some GSCs by the daughters of others (Xie and Spradling 2000; Yamashita et al. 2007; Jin et al. 2008). The functioning of both male and female niches and their stem cells declines with age, because of changes in niche cells, niche signals, and stem cell responsiveness (Wallenfang et al. 2006; Boyle et al. 2007; Pan et al. 2007).

Given this sophistication, it is not surprising that more cells and signals participate in niche operation than originally appreciated. A second type of stem cell in females, the escort stem cell (ESC), also contacts cap cells and covers most of the GSC surface (Decotto and Spradling 2005). ESCs maintain the population of escort cells (also known as inner germarium sheath cells) that surround developing germ cells before they acquire a follicle cell layer. JAK/STAT signaling has an important role in the female niche, as well as the male niche (Decotto and Spradling 2005; López-Oneiva et al. 2008; Wang et al. 2008). JAK/STAT ligand levels are high in the terminal filament, and these cells signal to cap cells and ESCs, which relay the critical bone morphogenetic protein (BMP) signals, i.e., dpp for GSC division and maintenance and possibly other signals to GSCs and muscle sheath cells as well (Fig. 1C).

Even after the niche has determined that it is time to stimulate GSC division and the distal daughter has separated and up-regulated *bam* transcription, it remains unclear how this leads to cystoblast differentiation. GSCs express transcripts required for growth, but they repress nearly all genes associated with embryonic and tissue differentiation (Kai et al. 2005), much like ES cells. Premature expression of differentiation genes is repressed at multiple levels, including extensive translational control mediated by products of the *pumilio*, *nanos*, *CPEB*, and microRNA (miRNA) genes (Spradling et al. 1997; Jin and Xie 2007; Park et al. 2007; Neumüller et al. 2008). One theory is that Bam binds to differentiation gene

mRNAs to counteract the translational repression mediated by Nanos and Pumilio (Chen and McKearin 2005; Szakmary et al. 2005). However, it remains possible that translational up-regulation is an effect of cyst differentiation, rather than a direct cause, and that Bam acts by a different mechanism.

The Role of the Fusome

A better understanding of the fusome (Fig. 1B), an organelle rich in endoplasmic reticulum (ER)-like tubules that is found in early germ cells (Spradling et al. 1997; Snapp et al. 2004), might help to resolve how *bam* controls GSC differentiation. A specific subset of Bam protein appears to reside within the fusome, and a *bam* null allele greatly reduces the number of fusome vesicles (McKearin and Ohlstein 1995). Moreover, fusome protein Ter94, related to the yeast ER vesicle fusion mediator Cdc48, interacts with Bam and depends on Bam for its normal fusome enrichment (León and McKearin 1999). It is plausible that after separation from the GSC is complete, Bam catalyzes a process of ER maturation within the fusome that initiates cystoblast development. So far, however, the nature of any molecular differences between the round fusome within the GSC (which is called the spectrosome), and the cystoblast fusome remains unclear (Spradling et al. 1997). To learn more about the role of the fusome in GSC maintenance and cystoblast specification, we developed a large collection of protein trap strains (Buszczak et al. 2007) and used them to identify new fusome components (Lighthouse et al. 2008).

The fusome synchronizes the cystocyte cell cycles and induces cleavage-like divisions (Spradling et al. 1997; Lilly et al. 2000), but a function in GSCs has not been determined previously. We analyzed GSC clones of mutations in many of the newly identified fusome components to determine whether fusome structure, GSC behavior, or cystoblast development was defective. Loss of several genes, including *tmod*, encoding the cytoskeletal protein tropomodulin; *Fer1HCH*, encoding the iron-binding protein ferritin (heavy chain); and *scrib*, encoding an epithelial polarity protein, has no effect on fusome morphology (not shown) or on GSC lifetime (Fig. 2A). In contrast, clones disrupting production of the small GTPase Rab11, a key component of the recycling endosome, drastically alter fusome morphology, disrupt normal cyst formation, and greatly accelerate GSC loss (Fig. 2B). Fusomes within the rab11 mutant GSCs contain greatly reduced numbers of ER-like tubules (Fig. 2C). Because we could find no changes in the adherens junctions between GSCs and cap cells, we suggest that the alterations in the fusome disrupt a germ cell signal whose production depends on fusome vesicle maturation, leading to changes that disrupt cyst development and destabilize the GSCs (Lighthouse et al. 2008). The nature of this fusome-dependent signal is currently under investigation.

Dynamic Niches: The Follicle Stem Cell Niche

Studies of GSC niches have provided significant insight into the development, function, and aging of stem cells

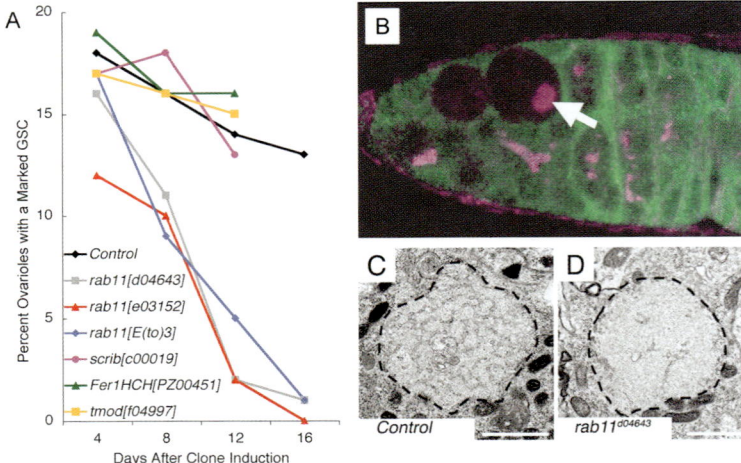

Figure 2. Rab11 is required to maintain GSCs and for normal cyst formation. (*A*) Rab11 mutations accelerate GSC loss. The graph shows the rate of loss of marked GSCs bearing the indicated mutations or no mutation (control). All three tested rab11 mutations greatly accelerate the rate of GSC loss, compared to wild type or the other gene mutations tested. (*B*) Germarium containing germ-line clones rab11[d04643] indicated by the absence of LacZ staining (*green*). One mutant GSC is present, along with a cyst that has developed aberrantly, and contains a large mass of fusome material in one oversized cell (*arrow*). (*C*) Electron micrograph of a normal round fusome from wild-type GSC, showing the presence of ER-like vesicles. (*D*) Electron micrograph of round fusome from a rab11[d04643] GSC. The number of ER-like vesicles is greatly reduced. (Modified from Lighthouse et al. 2008.)

interacting with their niches in an intact tissue. Niches based on similar specialized stroma may also occur in mammals, for example, within intestinal crypts, the bulge region of hair follicles, or along the surface of muscle fibers and seminiferous tubules. Some tissues, such as mammalian interfollicular epidermis, lack obvious candidate regions, however. It remains controversial whether specific stem cells and niches even exist in this tissue (Clayton et al. 2007). The *Drosophila* ovary provides a potential model for understanding mammalian epithelial stem cells and their niches. Each developing *Drosophila* follicle arises from an epithelial layer in the germarium produced by two epithelial stem cells (FSCs) that so closely resemble their daughters that they were never distinguished by classical histology (Spradling et al. 1997).

Each germarium contains exactly two FSCs and they each reside in a separate small niche (Nystul and Spradling 2007). However, FSC niches differ sharply from those previously characterized for GSCs. First, they lack permanent, nondividing stromal partner cells, the cells around which GSC niches are built. We demonstrated this using a *Drosophila* strain in which all cells in the germarium are marked with both green fluorescent protein (GFP) and LacZ (Fig. 3A'). Upon heat shock, stem cells undergo recombination and produce daughters that are either LacZ+ or GFP+ but not both. Following multiple heat shocks, we found that all of the cells surrounding the FSCs could become unilabeled, whereas the cap cells of the GSC niche always retained both markers (Fig. 3A). Therefore, all of the FSC's neighbors are transient, moving cells that will either die or depart to join new follicles. There are no cap cell equivalents. Yet despite this lack of permanent cells, FSCs remain in the same relative location, at the position where cysts form a single file, lose their escort cells, and begin to acquire fol-

licle cell replacements (known as the region 2a/2b border). Further analyses show that FSCs in this region contact the basement membrane, the escort cells covering approaching cysts, and several FSC daughter cells, but they never touch germ cells (Fig. 3B).

Not only are the cellular components of the FSC niche transitory, but its location is not anatomically fixed. The exact number of cysts in each region can vary with time, but there is always a characteristic region 2a/2b border. Cysts upstream of this boundary are covered with escort cells, whereas region 2b cysts span the width of the germarium and are covered with follicle cells. The two FSCs are always found in the same relative position at this boundary (Fig. 3C). We propose that this type of "dynamic niche" in which stem cells are maintained in a defined but constantly changing microenvironment represents a common and widespread type of tissue niche.

The dynamic niche is our term for this cyclically changing microenvironment amid the moving, developing cells that surround the FSC. Rather than residing in a fixed microenvironment and responding to a relatively small number of signals, FSCs likely experience a 24-hour cycle of microenvironments. By making appropriate choices at each juncture, the sequence eventually repeats. This looping sequence of microenvironments would be generated by the changing contacts between the approaching and leaving cyst and its escort cells, apoptosis of escort cells, and inward migration of young follicle cells adjacent to the FSCs. Generating and maintaining such a dynamic signaling milieu seems complex and subject to perturbations that might cause the resident stem cells to be lost or to be programmed incorrectly. However, such a system may also offer advantages over static niches in assuring that the resident stem cell remains perfectly in tune with the needs of the tissue it serves.

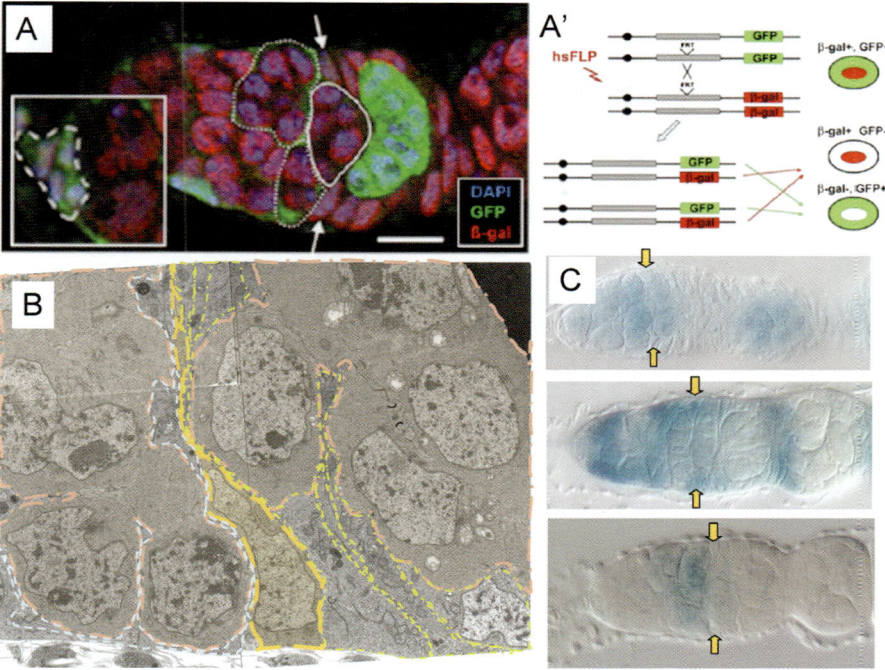

Figure 3. The dynamic FSC niche lacks fixed stromal cells. (*A*) A strain bearing a dual clone-marking system (*A′*) was subjected to multiple heat shocks to label all cells that are cycling within the anterior germarium. The arrows point to the two FSCs, and the absence of any cells bearing both the LacZ and GFP markers (such as the nondividing cap cells, *inset*) shows that no nondividing cells border the FSCs. (*B*) An electron micrograph of an FSC (*gold outline*) in its normal microenvironment shows that it contacts only escort cells on upstream cysts, its own follicle cell progeny (*yellow outlines*), and those of its partner stem cell on the other side of the germarium (not visible). The FSC does contact a region of basement membrane, but it apparently does not directly touch any germ cells. (*C*) Three micrographs of wild-type *Drosophila* germaria are shown to illustrate the variation in the position of the FSCs (*arrows*) relative to the germaria as a whole. Some variation is due to differences in the number of 2a and 2b cysts between germaria; however, additional variation is intrinsic due to the movements that take place over a 24-hour cycle as cysts move into single file at the region 2a/2b border. X-gal staining (*blue*) is not comparable between panels. (*A*, Modified from Nystul and Spradling 2007.)

Stem Cell Dynamics: Both Stem Cells and Niche Cells Are Frequently Replaced

The dynamic FSC niche maintains stem cell activity, despite undergoing continuous morphological change: the loss of old niche cells downstream and the addition of new ones upstream. This is reminiscent of the stem cell replacement observed previously within the stable GSC niche. Although individual cap cells and terminal filament cells do not change with time, individual GSCs frequently turn over and are replaced (Margolis and Spradling 1995; Xie and Spradling 2000). There appears to be no loss of stem cell/niche functionality associated with this stem cell replacement, and indeed, ongoing replacement has been postulated to depend on competition and to ensure that highly functional cells serve as stem cells (Nystul and Spradling 2007; Jin et al. 2008). Thus, the maintenance of functionality despite frequent substitution of constituent cells is a prominent aspect of all studied stem cell/niche systems.

In the dynamic niche, both niche cells and stem cells are subject to replacement. In fact, in the FSC niche, external niche cells are undergoing cyclic replacement with the passage of each cyst, whereas individual FSCs turn over about once every 30 divisions (Nystul and Spradling 2007). Thus, stem cell/niche systems are self-correcting systems that maintain their critical functions even though all of their constituent cells are ultimately dispensable. They are like the wave, not the moving particles of the underlying medium.

Talking Back: Intestinal Stem Cells and Daughter Cell Programming

We have emphasized the importance of the niche in controlling stem cell behavior and adapting their activity to the needs of the tissue and organism. However, study of another *Drosophila* stem cell provides clear indications that stem cells as well as niche cells participate in regulatory interactions. The *Drosophila* posterior midgut is maintained throughout adulthood by 800–1000 intestinal stem cells (ISCs) (Michelli and Perrimon 2006; Ohlstein and Spradling 2006) that resemble mouse intestinal stem cells (Barker et al. 2007) in many important respects. Like mammalian ISCs, *Drosophila* stem cells are multipotent, giving rise to both enterocytes and enteroendocrine cells (Fig. 4A). In addition, ISCs in both flies and mice require Notch signaling to produce enterocytes; when Notch is mutant, an excess number of secretory (enteroendocrine) cells result. Finally, ISCs even show morphological similarities (Fig. 4B); in both species, they reside against the

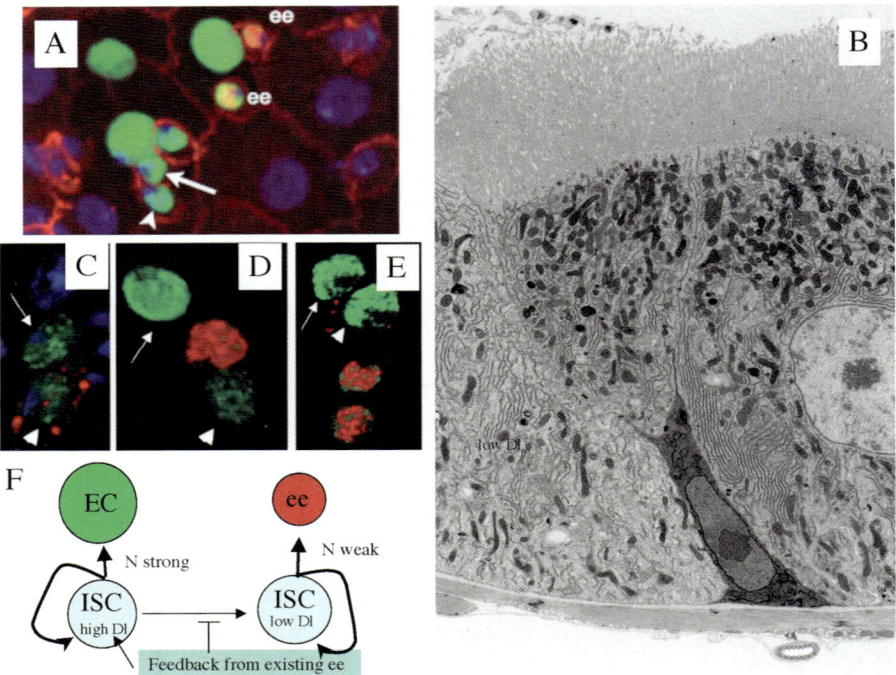

Figure 4. ISCs program their daughters by differential Notch signaling. (*A*) A clonally labeled ISC (*arrowhead*) has given rise to seven progeny (*green*): five enterocytes and two enteroendocrine cells (*red*), proving that ISCs are multipotent (from Ohlstein and Spradling 2006). (*B*) An electron micrograph of the adult midgut showing that like the FSC, ISCs (*dark cell*) contact only the basement membrane (*bottom*) and downstream cells such as the enterocytes shown. A cross section of a trachiole lies below the basement membrane under the ISC. (*C–E*) ISCs program progeny cell type via Notch signaling. (*C*) A two-cell clone (*green*) showing that ISCs (*arrowhead*) giving rise to enterocytes (*arrow*) are always rich in Delta-containing vesicles (*red*, cytoplasmic). (*D*) A three-cell clone (*green*) showing that when an ISC (*arrowhead*) switches from enterocyte (*arrow*) to enteroendocrine cell (*red*, nuclear) production, the level of Delta (*red*, cytoplasmic) falls drastically. (*E*) A four-cell clone (*green*) showing that when an ISC (*arrowhead*) switches from enteroendocrine cell production to enterocyte production, Delta-rich vesicles return to the cytoplasm. (*F*) Model of ISC programming: A feedback signal from existing enteroendocrine cells maintains high Delta levels in ISCs that causes a strong signal to be sent specifying enterocyte fate. In the absence of nearby enteroendocrine cells, Delta levels fall, and the resulting low-level Notch signal specifies enteroendocrine cell fate. (*A–F*, Modified from Ohlstein and Spradling 2007.)

basement membrane but extend thin cytoplasmic extensions up to the vilar surface where they may receive information from the gut lumen (Ohlstein and Spradling 2006; Barker et al. 2007).

The ISC niche is less well characterized than that of the FSC, but there appear to be some interesting similarities. ISCs are not found in fixed positions within the midgut, but it remains to be determined if they are in a predictable position relative to the surrounding cells, i.e., if there is a gut equivalent to the 2a/2b boundary. They contact the basement membrane and differentiating or differentiated ISC daughters but not any permanent niche cells. One of the most interesting aspects of ISCs is their multipotency, which depends on Notch signals sent by ISCs shortly after division that program daughter cell fate. A strong Notch signal specifies the daughter to become an enterocyte (Fig. 4C,E), whereas a weak signal results in enteroendocrine differentiation (Fig. 4D) (Ohlstein and Spradling 2007). The observation that enteroendocrine cells are made in pairs and that each ISC makes no more than one pair if existing enteroendocrine cells remain nearby lead to a simple feedback model for cell-fate specification by a multipotent stem cell (Fig. 4F). Thus, in this tissue, sig-

nals reporting a tissue's status may be sensed and interpreted directly by the stem cell, which then responds by specifying an appropriate type of daughter cell.

DISCUSSION

Niches and Stem Cells Work as Dynamic Integrated Units

Our studies demonstrate that both stem cells and niche cells are essential components of a regulatory system that produces new cells of the appropriate types at the times and locations needed to maintain tissue health. None of the individual cells, whether stem cells or niche cells, on their own appears to be essential for stem cell/niche function. The capacity for stem cell/niche function is not maintained due to the long life or unusual properties of any individual cell. Rather, tissue maintenance is based on the regular production of interchangeable parts and a stable program of cellular interactions, governed by reciprocal signals highly stabilized to perturbations. These stem cell/niche systems are capable of reacting to diverse situations, calculating and putting into effect appropriate responses including changes in division rate, movement, and cell specification. We probably need new terminology to describe this type of biological system, which likely constitutes a fundamental aspect of multicellularity.

The Dynamic Niche and the Importance of Studying Stem Cells In Vivo

There are several practical corollaries to these findings. It has too often been assumed that stem cells are sufficiently autonomous that properties such as developmental potency and division potential will be intrinsically maintained even during culture in the absence of normal neighbors and in the presence of very different signals and factors than they experience in vivo. However, this belief has yet to be documented in the case of a stem cell whose behavior in vivo has been independently determined. Stem cells have also been thought to behave in such a uniform fashion that they could be distinguished from other cells in diverse tissues using simple general properties such as a slow cell cycle. However, we see that true stem cells observed in vivo interact extensively with their neighboring cells and microenvironment in ways that differ among particular stem cells, niches, and physiological situations.

The more we know about stem cell/niche systems, the more we realize how difficult they will be to reconstruct in culture. Not only might it be difficult to find a feeder cell or soluble factors that can mimic the surface interactions between stem cell and niche, but our findings suggest that the microenvironment may need to cycle in a complex three-dimensional sequence. After in vivo study to understand the cycle, simulation might be possible, but the chances seem low that a solution can be found empirically by guesswork. For these reasons, it is now clear that stem cells should initially be identified and studied in vivo to establish their behavioral parameters. Fortunately, an ever-widening array of noninvasive cell-marking techniques are becoming available (for review, see Fox et al. 2009), as are methods for the direct observation of stem cell/niche systems in living tissue.

Implications of Stem Cell/Niche Units as Dynamic Entities Mediating Homeostatic Responses

Sustained tissue function over an extended period requires many decisions. For example, *Drosophila* can lay eggs at a high rate under ideal conditions, but egg production slows in many situations: circadian day, limited diet (especially protein), lack of attractive oviposition sites, etc. Thus, sensory and nutritional inputs are essential. Insulin levels are read directly by the stem cell. Neuronal input regarding oviposition sites is likely relayed to the oviducts, either by direct neuronal connection to oviduct musculature or via neurosecretion. It is not known how oviposition controls follicle development all the way back to the stem cells, but hormonal signals such as prostaglandins that have been implicated in oviposition in many species are possible candidates.

Previously, intercellular signaling and asymmetric division mechanisms have been recognized as being par-

ticularly important for understanding stem cell behavior. However, if niches must cycle through a series of particular states, each inducing a response in the stem cell, then many additional cell biological mechanisms will likely affect stem cell/niche function. Cell migration, apoptosis, shape change, extracellular matrix characteristics, and many other mechanisms might all affect stem cell/niche choreography.

The Dynamic Niche and the Role of Stroma in Cancer

The idea that epithelial stem cells read and react to complex sequences of neighboring cell activity has several potentially significant implications. Abnormalities in stromal cells contribute to uncoordinated growth and may act as a stepping-stone to cancer. Repeated wounding increases the chance of cancer development, but this is often ascribed to the elevated number of divisions that stem cells and early progenitors undergo to repair these insults. However, repeated wounding and scar formation may also disrupt the ability of cells to move normally and might change the way moving cells contact other cells in the affected region. Our findings suggest that this alone might affect stem cell programming, leading to inappropriate cycles of activity and metaplasia. To test these ideas, it will be essential to understand in detail the complete cycle of a normal dynamic niche and to study the effects of perturbations at each point in the program. In addition, we will need to learn whether these interactions are unique to the niche and stem cell or whether nonstem cell populations are guided by similar cues and might be capable of correcting or interfering with normal stem cell/niche programming.

Implications of Stem Cell/Niche Interactions for Stem Cell Therapy

Our growing realization that stem cells act in conjunction with tissue niches as integrated units for tissue maintenance has implications for stem cell therapy using cells grown in vitro. Unless the target tissue has a large supply of empty and compatible niches, it is unlikely that the introduced cells will persist or function as stem cells or that any therapeutic effect will be more than short-lived. Niches sometimes appear to be available when external factors such as radiation, infection, or autoimmunity preferentially destroy stem cells. However, as normal *Drosophila* adults age, their niches remain filled with stem cells, despite regular replacement of individual cells. In very old animals, niches may no longer produce high enough levels of signal to remain functional (Boyle et al. 2007; Pan et al. 2007). However, the condition most amenable to stem cell therapy, i.e., the presence of functional niches unoccupied by stem cells, has not been observed during normal *Drosophila* development.

What about the possibility of supplying cells farther downstream in the lineage? Here again, our studies suggest that knowledge of normal tissue biology will likely be needed to guide cell transplantation. In the stem cell lineages we studied, cell specification often begins close to or within the niche; ectopic cells may not differentiate or function properly without such a start. In a tissue such as the intestine with its common, dispersed stem cells, replacement cells are probably generated almost on site from the stem cells closest to the lesion and differentiate as part of an interacting group of cells. Cell therapy would be most promising in a tissue where replacement cells travel some distance from their site of origin and have evolved an ability to find, identify, and repair damage. Added cells will likely have to closely match some normal process that occurs in vivo, and it will be crucially important to identify those situations and understand the conditions required for repair to take place.

Consequently, much more emphasis in stem cell research is needed on the tissue niches and on the behavior of downstream cells in normal tissue repair. Medical conditions need to be evaluated in terms of the type of defects in cell replacement that have caused them and the subsequent consequences: loss of stem cells, loss of niches, disordering of tissue architecture, etc. Many conditions may be treatable by addressing changes in the niche, rather than by adding stem cells. Tissue homeostasis is an equal partnership between stem cells and their niches. Consequently, stem cell therapy needs to become a more balanced field in which the state of the host tissue is given the same consideration as the nature of the cells to be added.

ACKNOWLEDGMENTS

We thank many former members of the Spradling lab including Drs. Ben Ohlstein and Michael Buszczak, whose work contributed greatly to the development of these ideas. The authors are also grateful to Dianne Williams, Shelley Paterno, Megan Kutzer, and Vanessa Damm for technical assistance. Mike Sepanski provided expert help with electron microscopy, and Mamuhd Siddiqi assisted with 3D image reconstruction and time-lapse micrography. This work was supported by the Howard Hughes Medical Institute and the Carnegie Institution of Washington.

REFERENCES

Barker, N., van Es, J.H., Kuipers, J., Kujala, P., van den Born, M., Cozijnsen, M., Haegebarth, A., Korving, J., Begthel, H., Peters, P.J., and Clevers, H. 2007. Identification of stem cells in small intestine and colon by marker gene *Lgr5*. *Nature* **449:** 1003–1007.

Boyle, M., Wong, C., Rocha, M., and Jones, D.L. 2007. Decline in self-renewal factors contributes to aging of the stem cell niche. *Cell Stem Cell* **1:** 458–469.

Brawley, C. and Matunis, E. 2004. Regeneration of male germline stem cells by spermatogonial dedifferentiation in vivo. *Science* **304:** 1331–1334.

Bryder, D., Rossi, D.J., and Weissman, I.L. 2006. Hematopoietic stem cells: The paradigmatic tissue-specific stem cell. *Am. J. Pathol.* **169:** 338–346.

Buszczak, M., Paterno, S., Lighthouse, D., Bachman, J., Plank, J., Owen, S., Skora, A., Nystul, T., Ohlstein, B., Allen, A., et al. 2007. The Carnegie protein trap library: A versatile tool for *Drosophila* developmental studies. *Genetics* **175:** 1505–1531.

Chen, D. and McKearin, D. 2003. Dpp signaling silences *bam* transcription directly to establish asymmetric divisions of germline stem cells. *Curr. Biol.* **13:** 1786–1791.

Chen, D. and McKearin, D. 2005. Gene circuitry controlling a stem cell niche. *Curr. Biol.* **15:** 179–184.

Clayton, E., Doupe, D.P., Klein, A.M., Winton, D.J., Simons, B.D., and Jones, P.H. 2007. A single type of progenitor cell maintains normal epidermis. *Nature* **446:** 185–189.

Cox, R. and Spradling, A.C. 2003. A Balbiani body and the fusome mediate mitochondrial inheritance during *Drosophila* oogenesis. *Development* **130:** 1579–1590.

Decotto, E. and Spradling, A.C. 2005. The male and female *Drosophila* germline stem cell niche: Similar cells and signals. *Dev. Cell* **9:** 501–510.

deCuevas, M. and Spradling, A.C. 1998. Morphogenesis of the fusome and its implications for oocyte specification. *Development* **125:** 2781–2789.

Drummond-Barbosa, D. and Spradling, A.C. 2001. Stem cells and their progeny respond to nutritional changes during *Drosophila* oogenesis. *Dev. Biol.* **231:** 265–278.

Fox, D., Morris, L., Nystul, D., and Spradling, A.C. 2009. Analysis of stem cells by lineage analysis. *StemBook* (in press).

Fuller, M.T. and Spradling, A.C. 2007. Male and female *Drosophila* germline stem cells: Two versions of immortality. *Science* **316:** 402–404.

Harrison, D. and Perrimon, N. 1993. A simple and efficient generation of marked clones in *Drosophila*. *Curr. Biol.* **3:** 424–433.

Jin, Z. and Xie, T. 2007. Dcr-1 maintains *Drosophila* ovarian stem cells. *Curr. Biol.* **17:** 539–544.

Jin, Z., Kirilly, D., Weng, C., Kawase, E., Song, X., Smith, S., Schwartz, J., and Xie, T. 2008. Differentiation-defective stem cells outcompete normal stem cells for niche occupancy in the *Drosophila* ovary. *Cell Stem Cell* **10:** 39–49.

Kai, T. and Spradling, A. 2003. An empty *Drosophila* stem cell niche reactivates the proliferation of ectopic cells. *Proc. Natl. Acad. Sci.* **100:** 4633–4638.

Kai, T. and Spradling, A.C. 2004. Differentiating germ cells can revert into functional stem cells in *Drosophila melanaogaster* ovaries. *Nature* **428:** 564–569.

Kai, T., Williams, D., and Spradling, A.C. 2005. The expression profile of purified *Drosophila* germline stem cells. *Dev. Biol.* **283:** 486–502.

Kanatsu-Shinohara, M., Ogonuki, N., Inoue, K., Miki, H., Ogura, A., Toyokuni, S., and Shinohara, T. 2003. Long-term proliferation in culture and germline transmission of mouse male germline stem cells. *Biol. Reprod.* **69:** 612–616.

LaFever, L. and Drummond-Barbosa, D. 2005. Direct control of germline stem cell division and cyst growth by neural insulin in *Drosophila*. *Science* **309:** 1071–1073.

León, A. and McKearin, D. 1999. Identification of TER94, an AAA ATPase protein as a Bam-dependent component of the *Drosophila* fusome. *Mol. Biol. Cell* **10:** 3825–3834.

Lighthouse, D., Buszczak, M., and Spradling, A.C. 2008. New components of the *Drosophila* fusome suggest it plays novel roles in signaling and transport. *Dev. Biol.* **217:** 59–71.

Lilly, M., deCuevas, M., and Spradling, A.C. 2000. Cyclin A associates with the fusome during germline cyst formation in the *Drosophila* ovary. *Dev. Biol.* **218:** 53–63.

López-Oneiva, L., Fernandez-Minan, A., and Gonzalez-Reyes, A. 2008. Jak/Stat signalling in niche support cells regulates dpp transcription to control germline stem cell maintenance in the *Drosophila* ovary. *Development* **135:** 533–540.

Margolis, J. and Spradling, A. 1995. Identification and behaviour of epithelial stem cells in the *Drosophila* ovary. *Development* **121:** 3797–3807.

McKearin, D.M. and Ohlstein, B. 1995. A role for the *Drosophila* Bag-of-marbles protein in the differentiation of cystoblasts from germline stem cells. *Development* **121:** 2937–2947.

Michelli, C.A. and Perrimon, N. 2006. Evidence that stem cells reside in the adult *Drosophila* midgut epithelium. *Nature* **439:** 475–479.

Morrison, S. and Spradling, A.C. 2008. Stem cells and niches: Mechanisms that promote stem cell maintenance throughout life. *Cell* **132:** 598–611.

Neumüller, R.A., Betschinger, J., Fischer, A., Bushati, N., Poernbacher, I., Mechtler, K., Cohen, S.M., and Knoblich, J.A. 2008. Mei-P26 regulates microRNAs and cell growth in the *Drosophila* ovarian stem cell lineage. *Nature* **454:** 241–245.

Niki, Y. and Mahowald, A.P. 2003. Ovarian cystocytes can repopulate the embryonic germ line and produce functional gametes. *Proc. Natl. Acad. Sci.* **100:** 14042–14047.

Nystul, T. and Spradling, A.C. 2007. An epithelial niche in the *Drosophila* ovary undergoes long range stem cell replacement. *Cell Stem Cell* **1:** 277–285.

Ohlstein, B. and Spradling, A.C. 2006. The adult *Drosophila* posterior midgut is maintained by pluripotent stem cells. *Nature* **439:** 470–474.

Ohlstein, B. and Spradling, A.C. 2007. Multipotent *Drosophila* intestinal stem cells specify daughter cell fates by differential Notch signaling. *Science* **315:** 988–992.

Fan, L., Chen, S., Weng, C., Call, G., Zhu, D., Tang, H., Zhang, N., and Xie, T. 2007. Stem cell aging is controlled both intrinsically and extrinsically in the *Drosophila* ovary. *Cell Stem Cell* **1:** 458–469.

Park, J.K., Liu, X., Strauss, T.J., McKearin, D.M., and Liu, Q. 2007. The miRNA pathway intrinsically controls self-renewal of *Drosophila* germline stem cells. *Curr. Biol.* **17:** 533–538.

Snapp, E.L., Iida, T., Frescas, D., Lippincott-Schwartz, J., and Lilly, M.A. 2004. The fusome mediates intercellular endoplasmic reticulum connectivity in *Drosophila* ovarian cysts. *Mol. Biol. Cell* **15:** 4512–4521.

Song, X., Wong, M.D., Kawase, E., Xi, R., Ding, B.C., McCarthy, J.J., and Xie, T. 2004. Bmp signals from niche cells directly repress transcription of a differentiation-promoting gene, *bag of marbles*, in germline stem cells in the *Drosophila* ovary. *Development* **131:** 1353–1364.

Spradling, A.C., de Cuevas, M., Drummond-Barbosa, D., Keyes, L., Lilly, M., Pepling, M., and Xie, T. 1997. The *Drosophila* germarium: Stem cells, germ line cysts and oocytes. *Cold Spring Harbor Symp. Quant. Biol.* **62:** 25–34.

Szakmary, A. Cox, D.N., Wang, Z., and Lin, H. 2005. Regulatory relationships among *piwi*, *pumilio*, and *bag-of-marbles* in germline stem cells in the *Drosophila* ovary. *Curr. Biol.* **15:** 171–178.

Tootle, T.L. and Spradling, A.C. 2008. *Drosophila* Pxt: A cyclooxygenase-like facilitator of follicle maturation. *Development* **135:** 839–847.

Wallenfang, M.R., Nayak, R., and DiNardo, S. 2006. Dynamics of the male germline stem cell population during aging of *Drosophila melanogaster*. *Aging Cell* **5:** 297–304.

Wang, L., Li, Z., and Cai, Y. 2008. The JAK/STAT pathway positively regulates DPP signaling in the *Drosophila* germline stem cell niche. *J. Cell Biol.* **180:** 721–728.

Xie, T. and Spradling, A.C. 2000. A stem cell niche maintaining germ cell production in the *Drosophila* ovary. *Science* **290:** 328–330.

Yamashita, Y.M., Mahowald, A.P., Perlin, J.R., and Fuller, M.T. 2007. Asymmetric inheritance of mother versus daughter centrosome in stem cell division. *Science* **315:** 518–521.

Wnt Signaling and Stem Cell Control

R. Nusse, C. Fuerer, W. Ching, K. Harnish, C. Logan, A. Zeng,
D. ten Berge, and Y. Kalani

Howard Hughes Medical Institute, Department of Developmental Biology, Stanford University
School of Medicine, Stanford, California 94305

In many contexts, self-renewal and differentiation of stem cells are influenced by signals from their environment, constituting a niche. It is postulated that stem cells compete for local growth factors in the niche, thereby maintaining a balance between the numbers of self-renewing and differentiated cells. A critical aspect of the niche model for stem cell regulation is that the availability of self-renewing factors is limited and that stem cells compete for these factors (Fig. 1). Consequently, the range and concentrations of the niche factors are of critical importance. Now that some of the few self-renewing factors have become identified, aspects of the niche models can be tested experimentally. In this chapter, we address mechanisms of signal regulation that take place at the level of signal-producing cells, constituting a niche for stem cells. We emphasize the biochemical properties and posttranslational modifications of the signals, all in the context of Wnt signaling. We propose that these modifications control the range of Wnt signaling and have critical roles in establishing niches for stem cells in various tissues.

SIGNALING BY THE WNT PATHWAY

Wnt proteins comprise one family of secreted signaling molecules that is conserved in a wide range of species, from the cnidarian *Hydra* to humans. Members of the Wnt family are classified based on sequence homology with the first-identified Wnt proteins: mouse *Wnt-1* and *Drosophila wingless*. Signaling by Wnt ligands is used in a variety of contexts, having key roles in both embryonic development and adult homeostasis (for review, see Logan and Nusse 2004). Misregulation of Wnt signaling has been implicated in developmental abnormalities and cancer.

Most Wnt signaling events are transduced by what is referred to as the canonical Wnt pathway (Fig. 2). In this setting, Wnt ligands are secreted and bind to their cognate receptor complex on the surface of receiving cells. The receptor complex consists of Frizzled, a 7-transmembrane domain receptor, and a low-density lipoprotein (LDL) receptor-related protein (LRP) coreceptor (Bhanot et al. 1996; Wehrli et al. 2000). The signal is relayed through the intracellular protein Dishevelled (Dsh/Dvl), which, in the presence of Wnt ligand binding, promotes inhibition of a destruction complex consisting of Axin, APC (adenomatous polyposis coli), and GSK3 (glycogen synthase kinase 3) (for review, see Cadigan and Liu 2006). In the absence of a Wnt signal, the destruction complex phosphorylates β-

catenin, targeting it for degradation by the proteasome. Inhibition of the destruction complex leads to an accumulation of cytoplasmic β-catenin and the subsequent translocation of β-catenin into the nucleus. In the nucleus, β-catenin interacts with T-cell factor/lymphoid enhancer factor

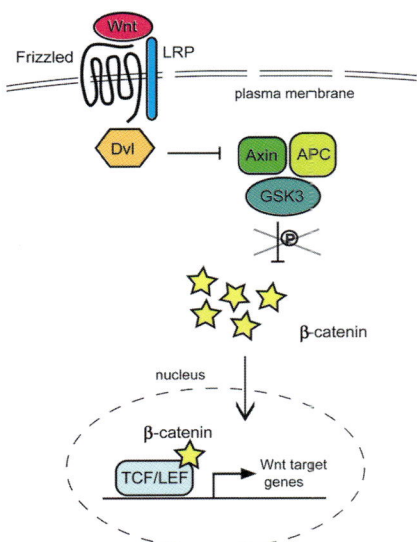

Figure 2. Wnt signal transduction. Wnt ligands are secreted and bind to their cognate receptor complex on the surface of receiving cells. The receptor complex consists of Frizzled, a 7-transmembrane domain receptor, and a low-density lipoprotein (LDL) receptor-related protein (LRP) coreceptor. The *Drosophila* homolog of this LRP coreceptor is called Arrow. The signal is relayed through the intracellular protein Dishevelled (Dsh/Dvl), which, in the presence of Wnt ligand binding, promotes inhibition of a destruction complex consisting of Axin, APC, and GSK3 (glycogen synthase kinase 3; Zeste-white 3/Shaggy in *Dro-sophila*). In the absence of a Wnt signal, the destruction complex phosphorylates β-catenin, targeting it for degradation by the proteasome. Inhibition of the destruction complex leads to an accumulation of cytoplasmic β-catenin and the subsequent translocation of β-catenin into the nucleus. In the nucleus, β-catenin interacts with TCF/LEF transcription factors to regulate transcription of Wnt target genes.

niches regulate stem cell behavior

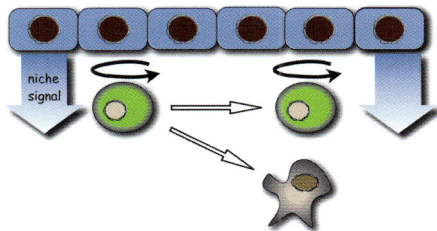

Figure 1. Niches are involved in stem cell control by secreting short-range signals. Cells that are outside of the range of the niche signal will differentiate.

(TCF/LEF) transcription factors to regulate transcription of Wnt target genes (Clevers and Van de Wetering 1997). In some instances, Wnt signals are transduced independently of β-catenin, for example, in planar polarity in *Drosophila* and during morphogenetic movements in vertebrate gastrulation (Veeman et al. 2003).

BIOCHEMICAL PROPERTIES OF WNT PROTEINS

Wnt proteins are characterized by the presence of a hydrophobic signal sequence followed by an invariant spacing pattern of 22 highly conserved cysteine residues. Many Wnt proteins are known to be *N*-glycosylated or possess potential glycosylation sites. An unusual characteristic of Wnt family members is posttranslational lipid modification. The specific lipid modifications observed on Wnt proteins include a palmitic acid (a saturated 16-carbon fatty acid chain) attachment on the most amino-terminal-conserved cysteine residue and a palmitoleic acid (monounsaturated 16-carbon fatty acid chain) attachment on a highly conserved serine residue (Willert et al. 2003; Takada et al. 2006; Komekado et al. 2007; Kurayoshi et al. 2007). These sites correspond to cysteine 77 (Cys-77) and serine 209 (Ser-209), respectively, in mouse Wnt3a. Lipid modification, or the covalent attachment of fatty acids, has been shown in different instances to regulate the biological activity of signaling proteins by affecting structural stability, membrane targeting, and protein–protein interactions (Smotrys and Linder 2004; Miura and Treisman 2006). Although it is more common for intracellular proteins to be lipid-modified,

secreted signaling proteins can also be regulated in this way. Hedgehog is one example of an important signaling molecule that is both secreted and lipid-modified. Like Wnt, Hedgehog is used for cell-to-cell communication in a variety of processes during embryonic and larval development. Hedgehog is modified by the addition of cholesterol and palmitic acid (Porter et al. 1996; Pepinsky et al. 1998). Loss of the cholesterol attachment affects the cell surface localization, range of activity, and asymmetrical trafficking of Hedgehog (Porter et al. 1995; Gallet et al. 2003), whereas loss of the palmitic acid group reduces the signaling activity of the protein without affecting its processing, secretion, or distribution (Chamoun et al. 2001; Lee et al. 2001; Gallet et al. 2003). Another more recent example of a secreted signaling protein that undergoes lipid modification is the *Drosophila* EGFR (epidermal growth factor receptor) ligand Spitz. Spitz is modified by the attachment of palmitic acid, which functions to enhance its association with the plasma membrane, thereby restricting its range of activity (Miura et al. 2006). These examples highlight the fact that altering the biochemical properties of a signaling protein by posttranslational lipid modification can impact both its potency and its range of signaling activity. This raises interesting questions about the role of lipid modification in Wnt signaling, with respect to both the mechanisms of signal transduction taking place at Wnt-receiving cells and the mechanisms of signal regulation taking place in Wnt-producing cells.

The Wnt protein structure is likely governed by intramolecular disulfide bonds formed among conserved cysteine residues, along with a number of posttranslational modifications (Fig. 3). The posttranslational glyco-

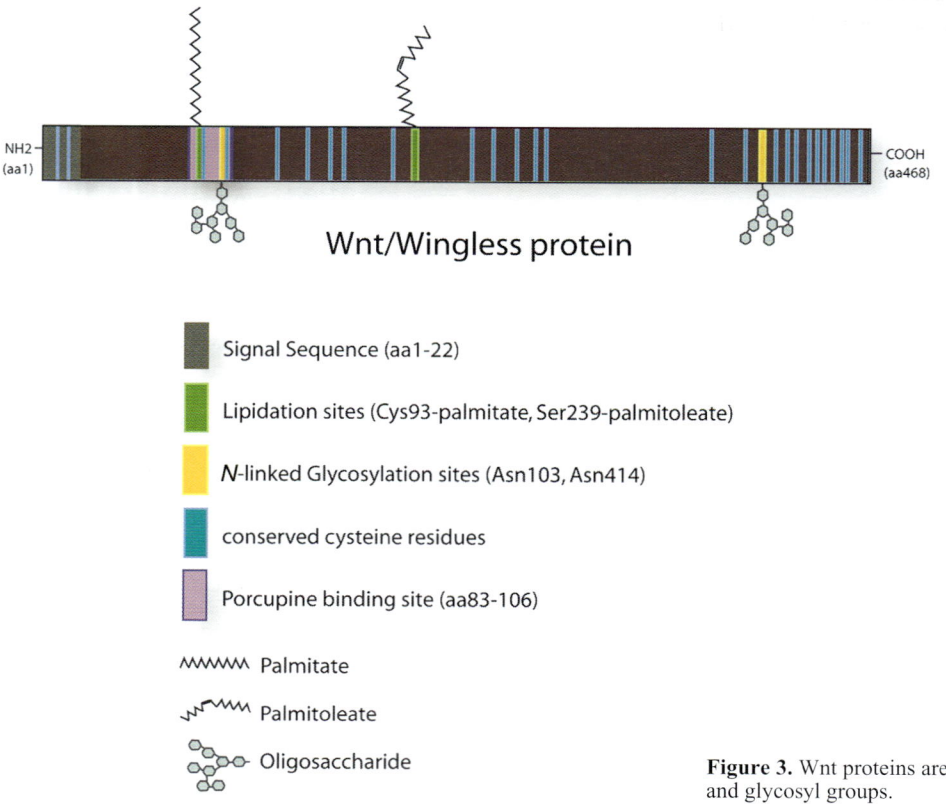

Wnt/Wingless protein

■ Signal Sequence (aa1-22)

■ Lipidation sites (Cys93-palmitate, Ser239-palmitoleate)

■ *N*-linked Glycosylation sites (Asn103, Asn414)

■ conserved cysteine residues

■ Porcupine binding site (aa83-106)

〜〜〜 Palmitate

〜〜〜 Palmitoleate

Oligosaccharide

Figure 3. Wnt proteins are modified by several lipids and glycosyl groups.

sylation and lipid modifications of Wnts begin once the nascent proteins reach the endoplasmic reticulum (ER). A member of the membrane-bound O-acyltransferase (MBOAT) family Porcupine (Porc) is a multipass transmembrane ER protein that has been shown to have a crucial role in the posttranslational processing of Wnt proteins (Tanaka et al. 2000, 2002; Zhai et al. 2004). Loss of *porc* leads to an accumulation of Wg in *porc* mutant cells, indicating that Wg cannot be properly secreted (van den Heuvel et al. 1993; Siegfried et al. 1994; Tanaka et al. 2002). There is evidence that Porc has a direct role in lipid modification of Wnt proteins. Beyond belonging to an acyltransferase protein family and thus being capable of catalyzing an acyl bond formation between palmitic acid and a cysteine residue, overexpression of *porc* increases the hydrophobicity of mWnt3a (Komekado et al. 2007), whereas loss of *porc* decreases the hydrophobicity of Wg (Zhai et al. 2004). In addition, the Porc-binding site of Wg includes Cys-93, the cysteine residue thought to be the site of palmitic acid attachment (Tanaka et al. 2002).

WNT SECRETION

Once Wnt proteins have been posttranslationally modified, they are ready to enter the secretory pathway. In Wnt-producing cells, Wg appears to move from the ER to the *trans*-Golgi network (TGN) and then on to different subcellular compartments, including endosomes (González et al. 1991). The Wnt secretory pathway has proved to be highly specialized, requiring the assistance of dedicated proteins to ensure delivery of Wnts to the plasma membrane (Fig. 4). The multipass transmembrane protein Wntless/Evi/Sprinter (Wls) was initially identified as a key component of the Wnt secretory pathway in *Drosophila* (Banziger et al. 2006; Bartscherer et al. 2006; Goodman et al. 2006). Wls is specifically required in Wnt-producing cells where, in the absence of Wls, Wnt will accumulate intracellularly and not be secreted (Banziger et al. 2006; Bartscherer et al. 2006; Franch-Marro et al. 2008; Port et al. 2008). This secretion defect is specific to Wnts, because loss of Wls has no effect on other proteins dependent on the secretory pathway, such as Hedgehog (Hh) (Banziger et al. 2006; Bartscherer et al. 2006; Goodman et al. 2006). Studies of the Wls ortholog in *Caenorhabditis elegans*, MOM-3/MIG-14, revealed that the function of Wls is conserved across species and necessary for the signaling activity of all Wnts examined (Harris et al. 1996; Thorpe et al. 1997; Eisenmann and Kim 2000; Banziger et al. 2006). Wls does not affect Wnt glycosylation or lipid modification, although it has been shown to directly interact with Wnt and is localized to the TGN and late endosomes of Wnt-producing cells (Banziger et al. 2006; Franch-Marro et al. 2008; Port et al. 2008). Research to date has led to the currently accepted model of Wls as an intracellular chaperone for Wnt, which enables trafficking from the TGN to the plasma membrane of Wnt-producing cells.

The retromer complex was recently established as a necessary component for Wnt signaling in *C. elegans*, where it is also specifically required in Wnt-producing cells (Coudreuse et al. 2006; Prasad and Clark 2006). Initial studies linking the retromer complex to Wnt signaling in *C. elegans* showed a breakdown in the establishment of the anteroposterior gradient of Wnt/EGL-20 in the absence of retromer activity and a subsequent impairment in long-range, but not short-range, signaling events (Coudreuse et al. 2006). Although retromer activity is required in Wnt-producing cells, preliminary biochemical analysis indicated that Wnt secretion was unaffected in cultured mammalian cells lacking retromer activity (Coudreuse et al. 2006). These data are consistent with a model of Wnt signaling that involves specific secretory routes of Wnts destined for short- or long-range signaling. In this model, the retromer complex is required for the intracellular trafficking events necessary for proper secretion of Wnts intended for long-range signaling events.

Two recent studies have found evidence that challenge this model of retromer activity. They show that, similar to a loss of Wls, loss of the retromer complex leads to a complete block in Wnt secretion and consequent accumulation of Wnt in producing cells of the *Drosophila* imaginal wing disc (Franch-Marro et al. 2008; Port et al. 2008). Interestingly, overexpression of *wls* can overcome this block in Wnt secretion induced by the loss of the retromer

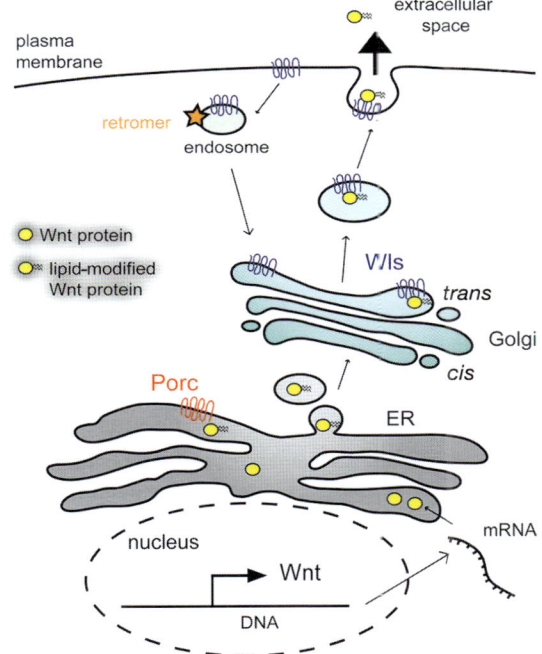

Figure 4. Wnt processing and secretion. The processing and secretion of Wnt proteins is a highly regulated process with several dedicated factors. Because Wnts are secreted proteins with a hydrophobic amino-terminal signal sequence, they are translated on the rough endoplasmic reticulum (ER). Once in the ER, many Wnt proteins are posttranslationally lipid-modified by the covalent attachment of fatty acids. The multipass transmembrane protein Porcupine (Porc), which resides in the ER and is homologous to a family of *O*-acyltransferases, is required for lipid modification of Wnt proteins. After transiting through the Golgi apparatus (Golgi) from the *cis* to the *trans* face, Wnt proteins are escorted to the plasma membrane by Wntless, another multipass transmembrane protein that is specifically required for Wnt secretion. At the plasma membrane, Wnt proteins disassociate with Wntless. Wntless is then recycled back to the *trans*-Golgi network through an endosomal pathway via the action of the retromer complex, where it can once again assist in the transport of Wnt proteins to the plasma membrane.

complex (Port et al. 2008). This raises the possibility that the primary function of the retromer complex is to regulate the intracellular transit of Wls. Indeed, upon close inspection of retromer-deficient cells using electron microscopy, Wls was found to be diverted from the plasma membrane to degradative mulitvesicular bodies as opposed to being transported back to the TGN (Franch-Marro et al. 2008).

Biochemical analysis revealed that the retromer complex binds directly to Wls (Franch-Marro et al. 2008). Taken together, these data suggest that Wnt uses a specialized secretory pathway dependent on its interaction with Wls in the TGN that facilitates endosomal transit to the plasma membrane, whereupon Wls is recycled back to the TGN in a retromer-dependent fashion.

The controversy regarding the sole requirement of the retromer complex for secretion of Wnts targeted for long-range, but not short-range, signaling events as opposed to enabling secretion of all Wnt proteins will require further analysis. However, both models agree that the retromer complex is required for a specific secretion route used by Wnts in producing cells. How Wnts are released into the extracellular space from the plasma membrane of Wnt-producing cells remains unknown.

WNTS AS MORPHOGENS

Wnts belong to a class of secreted proteins called morphogens (Strigini and Cohen 2000). Morphogens are secreted molecules that influence the movement and organization of cells by providing spatial information in the form of a concentration gradient. For example, in the developing *Drosophila* wing imaginal disc, Wg affects target gene transcription in a graded manner from its sight of production at the dorsoventral (DV) boundary out to the periphery of the wing pouch (Zecca et al. 1996). Different concentrations of Wg lead to induction of different target genes (Cadigan et al. 1998). High-threshold target genes, such as *achaete* and *senseless* (*sens*), require high levels of Wg to be induced and are therefore only up-regulated in cells directly adjacent to Wg-producing cells (Zecca et al. 1996; Nolo et al. 2000). Low-threshold target genes, such as *vestigial* and *distalless* (*dll*), can be turned on in response to much lower amounts of Wg and are consequently induced in the entire pouch of the imaginal wing disc (Zecca et al. 1996; Cadigan et al. 1998).

The dually lipid-modified Wnt proteins are highly hydrophobic and are presumably associated with cell membranes after secretion (Zhai et al. 2004). These observations raise several interesting biological questions: If lipid modification is indispensable for signaling (Willert et al. 2003), how is it possible for Wnts to affect cells at a distance if this lipid moiety confers high affinity for cell membranes? And if Wnts are released from membranes, how is it possible for a hydrophobic protein to move through the aqueous extracellular environment without aggregating? One possibility is that Wnt proteins are always chaperoned while in the extracellular space.

Previous studies have identified a number of factors that contribute to the extracellular stabilization of Wg. In *Drosophila*, two membrane-tethered heparin sulfate proteoglycan (HSPG) molecules, Dally (division abnormally delayed) and Dlp (Dally-like protein), are essential in shaping the Wg gradient (Lin and Perrimon 1999; Tsuda et al. 1999; Franch-Marro et al. 2005; Han et al. 2005). In addition, lipoprotein particles composed of lipophorin have also been suggested to have a role in the transport of two lipid-linked morphogens in *Drosophila*, Wg and Hh (Panakova et al. 2005).

A number of different mechanisms are used by morphogens to achieve their graded extracellular distribution. The different models include passive diffusion, active transport via planar transcytosis, the "bucket brigade" relay mechanism, cellular protrusion-mediated transport, and carrier-assisted transport (Strigini and Cohen 1999; Christian 2000; Tabata 2001; Cadigan 2002; Gonzalez-Gaitan 2003; Zhu and Scott 2004). It is certainly possible for a given signaling molecule to use a combination of these mechanisms.

Passive diffusion is the simplest mechanism of extracellular distribution of a signaling molecule. In this model, the morphogen freely diffuses away from its site of production, forming a concentration gradient. Although it is widely accepted that Wg somehow manages the feat of extracellular diffusion, it is unlikely that Wg can freely diffuse through the epithelia, because the extracellular space is an aqueous environment making the free diffusion of a hydrophobic protein such because Wg impossible. Planar transcytosis is a model of extracellular transport that involves the movement of a signal through cells via multiple rounds of receptor-mediated endocytosis and subsequent exocytosis. This would prevent Wg from exposure to the aqueous extracellular environment, but it would also take an unrealistic amount of time because the estimated time for crossing just one cell is between 20 minutes and 4 hours (Lander et al. 2002).

Another mechanism that has been challenged as being too slow by theoretical assessment is the "bucket brigade" relay mechanism (Kerszberg and Wolpert 1998). In this model, extracellular matrix proteins or cell surface receptor proteins, such as HSPGs, bind to the signaling molecules and pass them along in a "bucket brigade" fashion to adjacent cells.

The final model of diffusion is carrier-assisted diffusion. In this model, another diffusible molecule or molecules bind to the morphogen to aid in its movement. A carrier protein could function to solubilize Wg, thereby preventing aggregation in the aqueous extracellular environment. We have recently found that a novel Wg-binding protein, Swim, functions in this capacity to maintain solubility and activity, thereby mediating the long-range action of Wg. We identified *Drosophila* Swim by mass spectrometry as a trace component of the purified Wg preparation. This was a particularly interesting finding because preparations of mammalian Wnt3a and Wnt5a purified in parallel from mouse L cells contained variable quantities of the mammalian homolog of Swim. We have shown that Swim is both necessary and sufficient for purified Wg activity in cultured Schneider2 (S2) cells and that purified Swim can bind to Wg with nanomolar affinity. Purified Swim can maintain Wg solubility in an aqueous environment. Interestingly, in vivo overexpression of *swim* leads to a wg loss-of-function phenotype, whereas RNA interference (RNAi) of *swim* also appears to decrease the range of Wg signaling in the wing. These data suggest that the in vivo role of Swim is to mediate extracellular Wg transport and that this interaction is sensitive to dose. Swim is a putative member of the Lipocalin superfamily of extracellular transport molecules. Lipocalins are typically small (averaging 20 kD), extracellular proteins that bind hydrophobic molecules (Bratt 2000; Flower et al. 2000; Ganfornina et al. 2000). The hallmark of the Lipocalin family is the highly conserved tertiary structure formed by these proteins, an eight-stranded antiparallel β-barrel with a hydrophobic binding pocket (Flower et al. 2000).

WNT SIGNALING AND STEM CELL CONTROL

Wnt signaling and Wnt proteins are important for the maintenance of stem cells of various lineages. One example is in the digestive tract, where in the crypt of the colon, the loss of transcription factor TCF4 leads to depletion of stem cells (Van de Wetering et al. 2002). More recently, lineage labeling based on the Wnt target gene *LGR5* has shown conclusive evidence for stem cells in the crypts (Barker et al. 2007). The Wnt pathway has also been implicated as a self-renewal signal in the hematopoietic system (Willert et al. 2003). In the nervous system, the anatomical phenotypes of mouse Wnt mutants suggest that Wnts are involved in regulating neural stem and progenitor cell activity. Wnt1 knockout results in loss of midbrain, and Wnt3a mutant mice exhibit underdevelopment of the hippocampus due to lack of proliferation (Lee et al. 2000). Recent work demonstrating enhanced neurogenesis in vivo via exogenous expression of Wnt3a via lentiviral vectors strengthens the model that the Wnt signaling pathway is a major regulator of adult stem cell activity and fate in the hippocampus (Lie et al. 2005). A β-catenin gain-of-function study shows that continuous Wnt signaling results in marked and generalized hypercellularity of the brain (Chenn and Walsh 2002).

WNT SIGNALING AND EMBRYONIC STEM CELLS

The hallmark of embryonic stem (ES) cells is the ability to give rise to progenitors that can either self-renew or differentiate into all cell lineages. One key question is how these cells decide which path to follow, and consequently, how these two pathways (i.e., self-renewal vs. differentiation) are controlled. Mouse ES cells placed in culture were found to spontaneously differentiate unless they were grown on top of a feeder layer of fibroblast or in presence of a differentiation inhibitory activity, which was later found to be the cytokine leukemia inhibitory factor (LIF) (Smith et al. 1988; Williams et al. 1988). In presence of serum, LIF prevents differentiation by triggering phosphorylation of STAT3 through the activation of the receptor tyrosine kinase gp130 (Niwa et al. 1998). In the absence of serum, LIF is not sufficient to maintain mouse ES cells in culture unless BMP4 is added to prevent differentiation through SMAD phosphorylation and transcription of inhibitors of differentiation (Id) genes (Ying et al. 2003). Activation of the Wnt signaling pathway has been shown to maintain ES cells in a pluripotent stage. ES cells

that harbor inactivating mutations in the negative regulator APC or activating mutations in β-catenin have a profound reduction in their ability to form teratomas or to differentiate following LIF withdrawal (Kielman et al. 2002). Addition of Wnt-conditioned medium to ES cell cultures also sustained their pluripotency, but Wnt proteins alone were not sufficient to inhibit ES cell differentiation in absence of LIF. This apparent discrepancy was explained by the detection of LIF activity in the Wnt-conditioned medium and the subsequent discovery of a synergistic effect between LIF and Wnt signaling pathways (Ogawa et al. 2006). This effect might be due at least in part by Wnt-triggered transcription of the STAT3 gene itself (Hao et al. 2006). Activation of the Wnt signaling pathway through inhibition of the negative regulator GSK-3β with the chemical compound 6-bromo-indirubin-3′-oxime (BIO) was sufficient to maintain both mouse and human ES cells in their pluripotent stage (Sato et al. 2004), but BIO was found to also activate STAT3 through an unknown mechanism (Ogawa et al. 2006). Thus, Wnt signaling was only able to maintain pluripotent ES cells in conjunction with activation of the LIF/STAT3 pathway.

In culture, mouse ES cell differentiation is triggered by the spontaneous expression of fibroblast growth factor 4 (FGF4) and the subsequent activation of the mitogen-activated protein kinase/extra-cellular signal-related kinase (MAPK/ERK) signaling pathway (Kunath et al. 2007; Stavridis et al. 2007), but neither LIF nor BMP4 inhibit ERK phosphorylation. Instead, inhibition of the MAPK/ERK pathway using two small molecules was found to be sufficient to expand pluripotent ES cells, and concomitant inhibition of GSK-3 using the small-molecule CHIR99021 enabled highly efficient expansion of undifferentiated ES cells, bypassing the requirement for LIF/STAT3 (Ying et al. 2008). Genome-wide chromatin immunoprecipitation (ChIP) experiments revealed that promoters bound by the ES cell transcription factors Oct4, Nanog, and Sox2 also bind Tcf3. This is true for both the active class of genes (which encode ES cell transcription factors and proliferation genes) and the inactive class of genes (which are mostly coding for developmental regulators and are silenced by bound Polycomb proteins). In absence of a Wnt signal, Tcf3 represses the active gene set, but activation of the pathway relieves this repression and enables expression of Oct4, Nanog, and Sox2 target genes, including the ES cell regulators themselves (Cole et al. 2008). Altogether, these experiments suggest that activation of the Wnt signaling pathway helps to maintain ES cells in their undifferentiated state by affecting the balance between pluripotency and differentiation and that in absence of any differentiation signal or in presence of inhibitors of differentiation, the Wnt signaling pathway will sustain ES cell growth and self-renewal.

ONGOING RESEARCH ON WNT SIGNALING AND STEM CELLS

Our current research is based on methods we developed to purify Wnt proteins in an active form (Willert et al. 2003). With the purified active Wnt in hand, we have extensively tested how the protein can be used to manipulate the behavior of stem cells in culture. We have also used various Wnt reporter mice to identify Wnt-responsive cells in vivo. These reporters are based on a multimerized TCF-binding site, driving expression of LacZ. Because the expression of Axin2 is under the control of Wnt signaling in many tissues and Axin2 may even be a universal Wnt target, the reporter line made by Lustig et al. (2002) based on the Axin2 gene is very useful to visualize expression of the Wnt target in animals.

These studies have borne fruit in two different stem cell areas: mammary stem cells and neural stem cells. In all cases, we established that we can expand stem cells in an undifferentiated state, with full retention of specific differentiation capacities. The latter is revealed when the Wnt protein is withdrawn and the stem cells are transplanted in vivo. Despite the different origins of these stem cells, there is a common theme: The Wnt protein works in combination with other signals to promote stem cell expansion, these other molecules often being growth factors that activate a tyrosine kinase pathway. These findings have significant fundamental and practical implications, because one of the key questions in the stem cell field is how to control the decisions that these cells make to stay undifferentiated or to become committed. Using the Axin2–LacZ Wnt reporter mouse, we identified Wnt-responsive cells in the subventricular zone of the developing mouse brain as well as in known neurogenic zones in the adult brain such as the dentate gyrus of the hippocampus. We tested directly whether Wnt signals act as a stem cell factor by adding purified Wnt protein to neural cells in culture, finding that the protein causes a clonogenic outgrowth of neural stem cells. These cells self-renew in culture for several passages, each time starting from single cells, in a Wnt-protein-dependent manner. By itself, the Wnt protein is poorly mitogenic, but it appears to act mainly by blocking the differentiation of the cells, whereas FGF must be added as a mitogen. When Wnt is withdrawn, neural stem cell colonies are multipotential and can form the three cell types of the central nervous system. Blocking the Wnt signaling pathway with the soluble Wnt inhibitor Dickkopf (Dkk) results in a depletion of stem cell populations. To demonstrate that Wnt reporter-positive cells from mouse embryos have stem cell properties, we isolated the cells and cultured them in vitro. Wnt-responsive cells from these embryos exhibit enhanced colony-forming potential similar to cultures of whole populations of neural stem cells treated with Wnt. Our data show that Wnt proteins not only are important regulators of neurogenesis in vivo, but can be used in vitro for the clonal expansion of neural stem cells in an undifferentiated state.

Another example of the critical role of Wnt signals and stem cell behavior comes from the mammary gland. This tissue is known to contain a population of multipotent mammary stem cells. The existence of mammary stem cells was previously established by the fact that the mammary gland can be regenerated by transplantation. Elucidation of the cellular signals that maintain mammary stem cells is of broad interest and will lead to the design of more effective treatments for breast cancer, because mutations of Wnt pathway components have been impli-

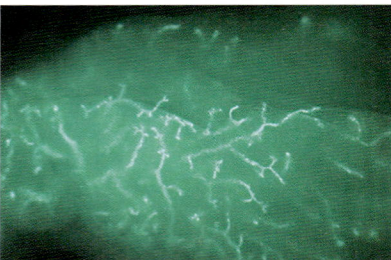

Figure 5. Mammary stem cells grown for several passages in cell culture in the presence of Wnt protein have retained normal differentiation properties when transplanted into a mouse. The cells are labeled with green fluorescent protein (GFP).

cated in this disease. Using reporter mice that visualize Wnt signaling in intact animals, we found that Wnt-responsive cells reside in the basal epithelial layer of the ducts coinciding with stem cell locations. In mutant mice that have slightly elevated Wnt signaling, the stem cells have a selective growth advantage when transplanted. When we use purified Wnt protein in mammary epithelium cell culture, we found that it promotes the maintenance of the stem cells. These Wnt-responsive stem cells are able to regenerate an entire mammary gland in transplantation assays (Fig. 5). Our data suggest that normal stem cells in the mammary gland are under the control of the Wnt pathway in the sense that Wnt signals promote self-renewal.

CONCLUSIONS

The extensive characterization of Wnt signaling and Wnt proteins has led to an increased understanding of how this pathway is involved in stem cell control. The distribution of the Wnt proteins, an important aspect of their function as stem cell niche factors, is controlled by specific modifications, including acylation by multiple lipids. Wnts can bind to several extracellular proteins, another level of regulation of availability for target cells. By purifying active Wnt proteins, we have been able to examine directly how they influence stem cell fate, using cell culture assays. These studies contribute to stem cell biology not only by generating deeper insight into the stem cell control mechanism, but also in the practical sense of being able to manipulate stem cells for therapeutic purposes.

ACKNOWLEDGMENTS

Studies in our lab are supported by the Howard Hughes Medical Institute and a grant from the California Institute of Regenerative Medicine.

REFERENCES

Banziger, C., Soldini, D., Schutt, C., Zipperlen, P., Hausmann, G., and Basler, K. 2006. Wntless, a conserved membrane protein dedicated to the secretion of Wnt proteins from signaling cells. *Cell* **125:** 509–522.
Barker, N., van Es, J.H., Kuipers, J., Kujala, P., van den Born,

M., Cozijnsen, M., Haegebarth, A., Korving, J., Begthel, H., Peters, P.J., and Clevers, H. 2007. Identification of stem cells in small intestine and colon by marker gene *Lgr5*. *Nature* **449:** 1003–1007.

Bartscherer, K., Pelte, N., Ingelfinger, D., and Boutros, M. 2006. Secretion of Wnt ligands requires Evi, a conserved transmembrane protein. *Cell* **125:** 523–533.

Bhanot, P., Brink, M., Samos, C.H., Hsieh, J.C., Wang, Y., Macke, J.P., Andrew, D., Nathans, J., and Nusse, R. 1996. A new member of the *frizzled* family from *Drosophila* functions as a Wingless receptor. *Nature* **382:** 225–230.

Bratt, T. 2000. Lipocalins and cancer. *Biochim. Biophys. Acta* **1482:** 318–326.

Cadigan, K.M. 2002. Regulating morphogen gradients in the *Drosophila* wing. *Semin. Cell Dev. Biol.* **13:** 83–90.

Cadigan, K.M. and Liu, Y.I. 2006. Wnt signaling: Complexity at the surface. *J. Cell Sci.* **119:** 395–402.

Cadigan, K.M., Fish, M.P., Rulifson, E.J., and Nusse, R. 1998. Wingless repression of *Drosophila frizzled 2* expression shapes the Wingless morphogen gradient in the wing. *Cell* **93:** 767–777.

Chamoun, Z., Mann, R.K., Nellen, D., von Kessler, D.P., Bellotto, M., Beachy, P.A., and Basler, K. 2001. Skinny hedgehog, an acyltransferase required for palmitoylation and activity of the hedgehog signal. *Science* **293:** 2080–2084.

Chenn, A. and Walsh, C.A. 2002. Regulation of cerebral cortical size by control of cell cycle exit in neural precursors. *Science* **297:** 365–369.

Christian, J.L. 2000. BMP, Wnt and Hedgehog signals: How far can they go? *Curr. Opin. Cell Biol.* **12:** 244–249.

Clevers, H. and Van de Wetering, M. 1997. TCF/LEF factors earn their wings. *Trends Genet.* **13:** 485–489.

Cole, M.F., Johnstone, S.E., Newman, J.J., Kagey, M.H., and Young, R.A. 2008. Tcf3 is an integral component of the core regulatory circuitry of embryonic stem cells. *Genes Dev.* **22:** 746–755.

Coudreuse, D.Y., Roel, G., Betist, M.C., Destree, O., and Korswagen, H.C. 2006. Wnt gradient formation requires retromer function in Wnt-producing cells. *Science* **312:** 921–924.

Eisenmann, D.M. and Kim, S.K. 2000. Protruding vulva mutants identify novel loci and Wnt signaling factors that function during *Caenorhabditis elegans* vulva development. *Genetics* **156:** 1097–1116.

Flower, D.R., North, A.C., and Sansom, C.E. 2000. The lipocalin protein family: Structural and sequence overview. *Biochim. Biophys. Acta* **1482:** 9–24.

Franch-Marro, X., Marchand, O., Piddini, E., Ricardo, S., Alexandre, C., and Vincent, J.P. 2005. Glypicans shunt the Wingless signal between local signalling and further transport. *Development* **132:** 659–666.

Franch-Marro, X., Wendler, F., Guidato, S., Griffith, J., Baena-Lopez, A., Itasaki, N., Maurice, M.M., and Vincent, J.P. 2008. Wingless secretion requires endosome-to-Golgi retrieval of Wntless/Evi/Sprinter by the retromer complex. *Nat. Cell Biol.* **10:** 170–177.

Gallet, A., Rodriguez, R., Ruel, L., and Therond, P.P. 2003. Cholesterol modification of hedgehog is required for trafficking and movement, revealing an asymmetric cellular response to hedgehog. *Dev. Cell* **4:** 191–204.

Ganfornina, M.D., Gutierrez, G., Bastiani, M., and Sanchez, D. 2000. A phylogenetic analysis of the lipocalin protein family. *Mol. Biol. Evol.* **17:** 114–126.

González, F., Swales, L., Bejsovec, A., Skaer, H., and Martinez Arias, A. 1991. Secretion and movement of wingless protein in the epidermis of the *Drosophila* embryo. *Mech. Dev.* **35:** 43–54.

Gonzalez-Gaitan, M. 2003. Signal dispersal and transduction through the endocytic pathway. *Nat. Rev. Mol. Cell Biol.* **4:** 213–224.

Goodman, R.M., Thombre, S., Firtina, Z., Gray, D., Betts, D., Roebuck, J., Spana, E.P., and Selva, E.M. 2006. Sprinter: A novel transmembrane protein required for Wg secretion and signaling. *Development* **133:** 4901–4911.

Han, C., Yan, D., Belenkaya, T., and Lin, X. 2005. *Drosophila*

glypicans Dally and Dally-like shape the extracellular Wingless morphogen gradient in the wing disc. *Development* **132:** 667–679.

Hao, J., Li, T.G., Qi, X., Zhao, D.F., and Zhao, G.Q. 2006. WNT/β-catenin pathway up-regulates Stat3 and converges on LIF to prevent differentiation of mouse embryonic stem cells. *Dev. Biol.* **290:** 81–91.

Harris, J., Honigberg, L., Robinson, N., and Kenyon, C. 1996. Neuronal cell migration in *C. elegans*: Regulation of Hox gene expression and cell position. *Development* **122:** 3117–3131.

Kerszberg, M. and Wolpert, L. 1998. Mechanisms for positional signalling by morphogen transport: A theoretical study. *J. Theor. Biol.* **191:** 103–114.

Kielman, M.F., Rindapaa, M., Gaspar, C., van Poppel, N., Breukel, C., van Leeuwen, S., Taketo, M.M., Roberts, S., Smits, R., and Fodde, R. 2002. Apc modulates embryonic stem-cell differentiation by controlling the dosage of β-catenin signaling. *Nat. Genet.* **32:** 594–605.

Komekado, H., Yamamoto, H., Chiba, T., and Kikuchi, A. 2007. Glycosylation and palmitoylation of Wnt-3a are coupled to produce an active form of Wnt-3a. *Genes Cells* **12:** 521–534.

Kunath, T., Saba-El-Leil, M.K., Almousailleakh, M., Wray, J., Meloche, S., and Smith, A. 2007. FGF stimulation of the Erk1/2 signalling cascade triggers transition of pluripotent embryonic stem cells from self-renewal to lineage commitment. *Development* **134:** 2895–2902.

Kurayoshi, M., Yamamoto, H., Izumi, S., and Kikuchi, A. 2007. Post-translational palmitoylation and glycosylation of Wnt-5a are necessary for its signalling. *Biochem. J.* **402:** 515–523.

Lander, A.D., Nie, Q., and Wan, F.Y. 2002. Do morphogen gradients arise by diffusion? *Dev. Cell* **2:** 785–796.

Lee, J.D., Kraus, P., Gaiano, N., Nery, S., Kohtz, J., Fishell, G., Loomis, C.A., and Treisman, J.E. 2001. An acylatable residue of Hedgehog is differentially required in *Drosophila* and mouse limb development. *Dev. Biol.* **233:** 122–136.

Lee, S., Tole, S., Grove, E., and McMahon, A. 2000. A local Wnt-3a signal is required for development of the mammalian hippocampus. *Development* **127:** 457–467.

Lie, D.C., Colamarino, S.A., Song, H.J., Desire, L., Mira, H., Consiglio, A., Lein, E.S., Jessberger, S., Lansford, H., Dearie, A.R., and Gage, F.H. 2005. Wnt signalling regulates adult hippocampal neurogenesis. *Nature* **437:** 1370–1375.

Lin, X. and Perrimon, N. 1999. Dally cooperates with *Drosophila* Frizzled 2 to transduce Wingless signalling. *Nature* **400:** 281–284.

Logan, C.Y. and Nusse, R. 2004. The Wnt signaling pathway in development and disease. *Annu. Rev. Cell Dev. Biol.* **20:** 781–810.

Lustig, B., Jerchow, B., Sachs, M., Weiler, S., Pietsch, T., Karsten, U., Van de Wetering, M., Clevers, H., Schlag, P.M., Birchmeier, W., and Behrens, J. 2002. Negative feedback loop of Wnt signaling through up-regulation of conductin/axin2 in colorectal and liver tumors. *Mol. Cell. Biol.* **22:** 1184–1193.

Miura, G.I. and Treisman, J.E. 2006. Lipid modification of secreted signaling proteins. *Cell Cycle* **5:** 1184–1188.

Miura, G.I., Buglino, J., Alvarado, D., Lemmon, M.A., Resh, M.D., and Treisman, J.E. 2006. Palmitoylation of the EGFR ligand Spitz by Rasp increases Spitz activity by restricting its diffusion. *Dev. Cell* **10:** 167–176.

Niwa, H., Burdon, T., Chambers, I., and Smith, A. 1998. Self-renewal of pluripotent embryonic stem cells is mediated via activation of STAT3. *Genes Dev.* **12:** 2048–2060.

Nolo, R., Abbott, L.A., and Bellen, H.J. 2000. Senseless, a Zn finger transcription factor, is necessary and sufficient for sensory organ development in *Drosophila*. *Cell* **102:** 349–362.

Ogawa, K., Nishinakamura, R., Iwamatsu, Y., Shimosato, D., and Niwa, H. 2006. Synergistic action of Wnt and LIF in maintaining pluripotency of mouse ES cells. *Biochem. Biophys. Res. Commun.* **343:** 159–166.

Panakova, D., Sprong, H., Marois, E., Thiele, C., and Eaton, S. 2005. Lipoprotein particles are required for Hedgehog and Wingless signalling. *Nature* **435:** 58–65

Pepinsky, R.B., Zeng, C., Wen, D., Rayhorn, P., Baker, D.P.,

Williams, K.P., Bixler, S.A., Ambrose, C.M., Garber, E.A., Miatkowski, K., et al. 1998. Identification of a palmitic acid-modified form of human Sonic hedgehog. *J. Biol. Chem.* **273:** 14037–14045.

Port, F., Kuster, M., Herr, P., Furger, E., Banziger, C., Hausmann, G., and Basler, K. 2008. Wingless secretion promotes and requires retromer-dependent cycling of Wntless. *Nat. Cell Biol.* **10:** 178–185.

Porter, J.A., Young, K.E., and Beachy, P.A. 1996. Cholesterol modification of hedgehog signaling proteins in animal development. *Science* **274:** 255–259.

Porter, J.A., von Kessler, D.P., Ekker, S.C., Young, K.E., Lee, J.J., Moses, K., and Beachy, P.A. 1995. The product of hedgehog autoproteolytic cleavage active in local and long-range signalling. *Nature* **374:** 363–366.

Prasad, B.C. and Clark, S.G. 2006. Wnt signaling establishes anteroposterior neuronal polarity and requires retromer in *C. elegans. Development* **133:** 1757–1766.

Sato, N., Meijer, L., Skaltsounis, L., Greengard, P., and Brivanlou, A.H. 2004. Maintenance of pluripotency in human and mouse embryonic stem cells through activation of Wnt signaling by a pharmacological GSK-3-specific inhibitor. *Nat. Med.* **10:** 55–63.

Siegfried, E., Wilder, E.L., and Perrimon, N. 1994. Components of wingless signalling in *Drosophila. Nature* **367:** 76–80.

Smith, A.G., Heath, J.K., Donaldson, D.D., Wong, G.G., Moreau, J., Stahl, M., and Rogers, D. 1988. Inhibition of pluripotential embryonic stem cell differentiation by purified polypeptides. *Nature* **336:** 688–690.

Smotrys, J.E. and Linder, M.E. 2004. Palmitoylation of intracellular signaling proteins: Regulation and function. *Annu. Rev. Biochem.* **73:** 559–587.

Stavridis, M.P., Lunn, J.S., Collins, B.J., and Storey, K.G. 2007. A discrete period of FGF-induced Erk1/2 signalling is required for vertebrate neural specification. *Development* **134:** 2889–2894.

Strigini, M. and Cohen, S.M. 1999. Formation of morphogen gradients in the *Drosophila* wing. *Semin. Cell Dev. Biol.* **10:** 335–344.

Strigini, M. and Cohen, S.M. 2000. Wingless gradient formation in the *Drosophila* wing. *Curr. Biol.* **10:** 293–300.

Tabata, T. 2001. Genetics of morphogen gradients. *Nat. Rev. Genet.* **2:** 620–630.

Takada, R., Satomi, Y., Kurata, T., Ueno, N., Norioka, S., Kondoh, H., Takao, T., and Takada, S. 2006. Monounsaturated fatty acid modification of Wnt protein: Its role in Wnt secretion. *Dev. Cell* **11:** 791–801.

Tanaka, K., Kitagawa, Y., and Kadowaki, T. 2002. *Drosophila* segment polarity gene product porcupine stimulates the posttranslational N-glycosylation of wingless in the endoplasmic reticulum. *J. Biol. Chem.* **277:** 12816–12823.

Tanaka, K., Okabayashi, K., Asashima, M., Perrimon, N., and Kadowaki, T. 2000. The evolutionarily conserved porcupine gene family is involved in the processing of the Wnt family. *Eur. J. Biochem.* **267:** 4300–4311.

Thorpe, C.J., Schlesinger, A., Carter, J.C., and Bowerman, B. 1997. Wnt signaling polarizes an early *C. elegans* blastomere to distinguish endoderm from mesoderm. *Cell* **90:** 695–705.

Tsuda, M., Kamimura, H., Nakato, M., Archer, W., Staatz, B., Fox, M., Humphrey, S., Olson, T., Futch, V., Kaluza, E., et al. 1999. The cell-surface proteoglycan Dally regulates Wingless signalling in *Drosophila. Nature* **400:** 276–280.

Van de Wetering, M., Sancho, E., Verweij, C., de Lau, W., Oving, I., Hurlstone, A., van der Horn, K., Batlle, E., Coudreuse, D., Haramis, A.P., et al. 2002. The β-catenin/TCF-4 complex imposes a crypt progenitor phenotype on colorectal cancer cells. *Cell* **111:** 241–250.

van den Heuvel, M., Harryman-Samos, C., Klingensmith, J., Perrimon, N., and Nusse, R. 1993. Mutations in the segment polarity genes *wingless* and *porcupine* impair secretion of the wingless protein. *EMBO J.* **12:** 5293–5302.

Veeman, M.T., Axelrod, J.D., and Moon, R.T. 2003. A second canon. Functions and mechanisms of β-catenin-independent Wnt signaling. *Dev. Cell* **5:** 367–377.

Wehrli, M., Dougan, S.T., Caldwell, K., O'Keefe, L., Schwartz, S., Vaizel-Ohayon, D., Schejter, E., Tomlinson, A., and DiNardo, S. 2000. *arrow* encodes an LDL-receptor-related protein essential for Wingless signalling. *Nature* **407:** 527–530.

Willert, K., Brown, J.D., Danenberg, E., Duncan, A.W., Weissman, I.L., Reya, T., Yates III, J.R., and Nusse, R. 2003. Wnt proteins are lipid-modified and can act as stem cell growth factors. *Nature* **423:** 448–452.

Williams, R.L., Hilton, D.J., Pease, S., Willson, T.A., Stewart, C.L., Gearing, D.P., Wagner, E.F., Metcalf, D., Nicola, N.A., and Gough, N.M. 1988. Myeloid leukaemia inhibitory factor maintains the developmental potential of embryonic stem cells. *Nature* **336:** 684–687.

Ying, Q.L., Nichols, J., Chambers, I., and Smith, A. 2003. BMP induction of Id proteins suppresses differentiation and sustains embryonic stem cell self-renewal in collaboration with STAT3. *Cell* **115:** 281–292.

Ying, Q.L., Wray, J., Nichols, J., Batlle-Morera, L., Doble, B., Woodgett, J., Cohen, P., and Smith, A. 2008. The ground state of embryonic stem cell self-renewal. *Nature* **453:** 519–523.

Zecca, M., Basler, K., and Struhl, G. 1996. Direct and long-range action of a wingless morphogen gradient. *Cell* **87:** 833–844.

Zhai, L., Chaturvedi, D., and Cumberledge, S. 2004. *Drosophila* wnt-1 undergoes a hydrophobic modification and is targeted to lipid rafts, a process that requires porcupine. *J. Biol. Chem.* **279:** 33220–33227.

Zhu, A.J. and Scott, M.P. 2004. Incredible journey: How do developmental signals travel through tissue? *Genes Dev.* **18:** 2985–2997.

Niche Required for Inducing Quiescent Stem Cells

S.-I. Nishikawa, M. Osawa, S. Yonetani, S. Torikai-Nishikawa, and R. Freter

Laboratory for Stem Cell Research, RIKEN Center for Developmental Biology, Kobe, Japan

Quiescence is an important feature distinguishing stem cells (SCs) from other compartments for most SC systems. Evidence suggests that the quiescent state is directed by external cues expressed in the presumptive microenvironment, the niche, although the cellular and molecular nature of the niche remains obscure in most SC systems. Our group has been addressing this question using the melanocyte (MC) as a model, because MC SCs (MSCs) and other compartments are distinguished by their location in the hair follicle, the former in the bulge and the other in the hair matrix. On the basis of the gene expression profiles of MSCs, we developed a method to distinguish MSCs from other compartments by using their own characteristics. Using the new criterion for MSCs, we investigated the molecular cues that induce the quiescent MSCs from proliferating melanoblasts. Our study showed that fibroblast growth factor-2 (FGF-2), or an equivalent signal, is essential for inducing a set of MSC signatures, although additional signals required for inducing the ultimate MSCs remain to be identified.

Niche is an important issue common to all types of SCs (Spradling et al. 2001; Ohlstein et al. 2004; Moore and Lemischka 2006; Scadden 2006). To the best of our knowledge, however, in no mammalian SC system have cells responsible for niche function ever been specified in such an exact manner as has been shown in the gonads of *Drosophila* and *Caenorhabditis elegans*. Indeed, the niche cells for germ cells of these species (such as tip cell and cap cell) have been fully defined (Spradling et al. 2001; Ohlstein et al. 2004). In contrast, the SCs in mammalian tissues have been characterized to a considerable extent by their own features, and they are now able to be specified in a relatively precise manner (Fuchs et al. 2004; Joseph et al. 2004). Together, SC research in mammals remains in a situation where the niche can be defined no more than the presumptive area in the vicinity of SCs.

Although the situation for the MSC system is not different from that of other SC systems, there are appreciable advantages in MCs for studying the niche. In this chapter, we summarize our recent studies on MSCs in the hair follicle, particularly the quiescent SCs, and present our views on the microenvironment required for inducing the quiescent SCs.

MSCs SERVE AS AN IDEAL MODEL TO UNDERSTAND QUIESCENCE OF SCs

Two types of SC systems exist in the vertebrate body: a constitutive type, such as gut epithelium in which SCs undergo continuous self-renewal, and the other, a regenerative system in which SCs repeat the quiescence and regeneration cycles with a variable length of the quiescent stage. The MSC system is a typical example of the regenerative-type system, and its regeneration cycle is linked to that of follicular keratinocytes (hair cycle). MSCs are functionally distinguished from other compartments of the MSC system by a number of features (Nishimura et al. 2002). First, in mature hair follicles, MSCs are distinguished by their

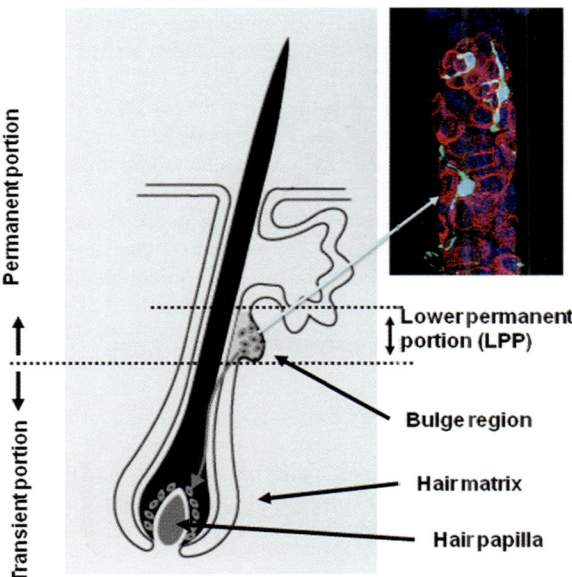

Figure 1. Architecture of the mouse hair follicle. (*Left*) A hair follicle is divided into permanent and transient portions. Only a transient portion undergoes apoptosis/regeneration cycles. Previous studies demonstrated that SCs for follicular keratinocytes are present in the bulge region. Most MCs in hair follicles are present in the hair matrix at the bottom of follicles. MCs are also found in the most lower portion of the permanent region, which corresponds to MSCs. (*Right*) Whole-mount staining of the lower permanent portion (LPP) of a hair follicle. MSCs are visualized by anti-LacZ staining. MSCs exist solitarily.

location in the hair follicles. As shown in Figure 1, a hair follicle is divided into two parts: the upper permanent portion and the lower transient portion. Only the transient portion is renewed in each hair cycle. The hair cycle is mainly directed by the hair papilla that localizes at the bottom of the follicles and emanates a set of molecules regulating the activities of follicular keratinocytes and the MC. Hence, all proliferating and differentiating MCs need to localize in the hair matix in the vicinity of the hair papilla. In mouse skin, MCs cannot survive outside the hair matrix. However, the small region at the bottom of the permanent portion that includes a region called the bulge is the exception, because this region has a distinctive capacity to serve as the niche capable of supporting the survival of the quiescent MSCs. In this chapter, we designate this region as the lower permanent portion (LPP). This invariable definition of MSCs in terms of its location allows further characterization of MSCs. In fact, we have shown that MSCs in the LPP correspond to quiescent SCs, because they do not proliferate until a new hair regeneration cycle is initiated and they become resistant to deprivation of the c-Kit signal.

Another unique feature of MSCs, as compared with other SC systems, is that they can exist solitarily separated from other components of the MSC system. The LPP usually contains multiple MSCs, but it is possible, under certain experimental conditions, to prepare hair follicles that contain only a single MSC in the LPP. Moreover, Figure 1 shows that, even in the LPP containing multiple MSCs, each MSC exists solitarily without notable connections to one another (Nishimura et al. 2002; Mak et al. 2006). Hence, if there exists a distinctive niche for MSC, it should comprise heterologous cells, most likely keratinocytes. On the other hand, in many SC systems including the hair follicle, SCs are integrated in a continuous cluster together with other compartments of the SC systems. For instance, SCs of the gut epithelium are wedged by their immediate progeny (Potten et al. 2003). In such a situation, it is difficult to rule out the possibility that the progeny of SCs themselves are involved in the function of the niche, thereby making the cellular composition of the niche more complicated. Indeed, Hans Clevers (this volume) demonstrated that gut epithelial cells autonomously generate an SC system in the absence of heterologous cell lineages.

In conclusion, an explicit definition of SCs and the simplicity of the cellular components of the niche point to the MSC as an ideal model for studying the SC niche.

CELLULAR COMPARTMENTS IN THE MSC SYSTEM

All skin MCs are derived from neural crest cells that spread over the whole body surface through the subcutaneous mesenchyme, enter the developing epidermis, and eventually settle in hair follicles. In mouse skin, because of the absence of stem cell factor (SCF), MCs cannot survive in the dermis or interfollicular epidermis. Hence, mouse MCs are present exclusively in the hair follicle, except for regions such as the auricle and palm where MCs are found in dermis. Interestingly, when the *scf* transgene is expressed in the epidermis, MCs also become distributed in the interfollicular epidermis (Kunisada et al. 1998). As a result, localization of MCs is directed extrinsically, rather than according to the cell-autonomous program. This passivity is another unique characteristic of the MSC system. Figure 2 summarizes our current understanding of the fate divergence of the embryonic melanoblast (MB) in postnatal

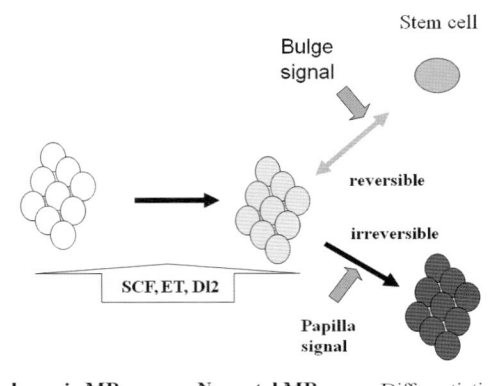

Figure 2. Divergence of the MB in neonatal skin. During embryogenesis, MBs expand and migrate to hair follicles. This process is regulated by various molecules including SCF and endothelin expressed in embryonic dermal tissues. After birth, the MB undergoes fate divergence. Some that migrate to the hair matrix undergo irreversible commitment to differentiated MCs under the control of molecules emanated from hair papilla. The rest migrate to the LPP and differentiate to MSCs. MSCs are reactivated by yet unknown mechanisms. Some of the reactivated MBs can maintain the capacity to revert to MSCs, when they meet the niche in the LPP. However, most of reactivated MBs migrate to the hair matrix and give rise to differentiated MCs.

tal life. During embryogenesis, the MB maintains self-renewal with only a limited rate of differentiation. After entering developing hair follicles, those migrating into the hair matrix undergo terminal differentiation, whereas those trapped in the LPP are induced to become quiescent MSCs that can be reactivated in the next hair cycle (Nishimura et al. 2002). Of note is that some of progeny of the reactivated MSCs can maintain the ability to revert to MSCs. Nonetheless, all of the states are regulated by extrinsic cues present in distinct environments.

MSC SIGNATURE

As described above, LPP localization is the most reliable mark for MSC signature. Thus, we first isolated individual MSCs by manual dissection of hair follicles. Using transgenic mice that were engineered to express green fluorescent protein (GFP) specifically in the MC lineage, we could distinguish MSCs from other cells in the LPP. We next prepared a single-cell cDNA library of harvested MSCs, analyzed gene expression profiles, and compared them with those prepared from MBs in the embryonic epidermis and MCs in the hair matrix. Figure 3 presents characteristic gene expression profiles in these three MC populations. Despite unavoidable variation due to technical difficulties in analyzing gene expression at the single-cell level, MCs collected from the LPP are significantly similar in gene expression (Osawa et al. 2005). Moreover, the overall gene expression profiles of the three populations are different from one another, although some variations are observed among individual cells in the same group.

Taken together, this single-cell analysis defines two molecular features of MSCs: a low level of housekeeping gene expression and down-regulation of MC-specific genes.

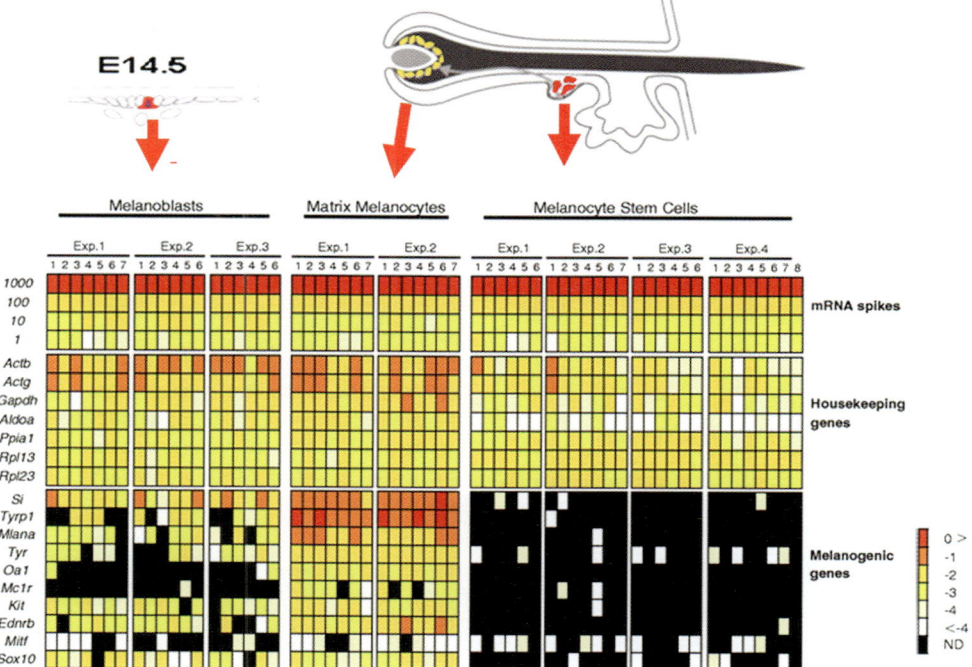

Figure 3. Gene expression profiles of MCs in embryos, hair matrix, and LPP. (*Top*) Individual MCs are isolated from embryonic skin, hair matrix, and LPP. RNA was prepared from each individual cell and amplified. Expression levels of genes in single-cell cDNA libraries were analyzed by quantitative polymerase chain reaction. Expression levels were classified into five grades and expressed as black, white, yellow, light orange, and orange. Each column represents the data of each cell. (Modified from Osawa et al. 2005.)

USE OF FACS FOR ANALYZING MSC FEATURES

The aforementioned data on gene expression of each individual MSC suggest additional methods to specify MSCs. The first method is to use the low expression level of MSC housekeeping genes. Because the method that we used for distinguishing MCs from other lineages is the chicken γ-actin (CAG) promoter-driven *gfp* gene, we expected that the expression level of the housekeeping genes could be monitored by measuring GFP expression level. As expected, the expression level of *gfp* in MC lineage cells in hair follicles from P10 skin varies to a considerable extent (Fig. 4). To verify whether MSC is included in the GFP^{low} population, we took advantage of another feature of MSCs in neonatal skin: the resistance against starvation of SCF. Previously, we showed that treatment of neonatal mice by ACK2, an antagonistic monoclonal antibody to c-Kit, depletes nearly all MCs from the skin except for those that colonize to the LPP. Thus, MCs surviving after ACK2 treatment are likely to contain MSCs and are expected to be GFP^{low}. In fact, ACK2 treatment depleted all of the GFP^{high} population, whereas the GFP^{low} population survived. This GFP^{low} population is easily purified by fluorescence-activated cell sorting (FACS), and we were able to confirm that the GFP^{low} cells share the many characteristics of MSCs such as down-regulation of a group of MC-specific genes such as *trp2* and *sox10*.

The other method for distinguishing MSCs by FACS is to use the *gfp* gene driven by promoters of MC-specific

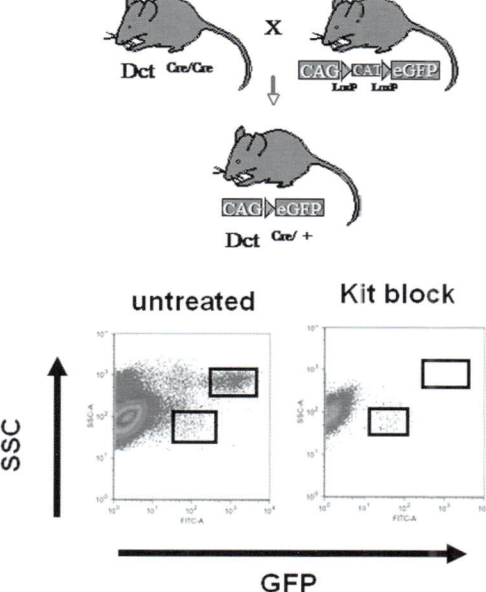

Figure 4. Low activity of the chicken γ-actin (CAG) promoter in MSCs. (*Top*) Generation of mice that express GFP driven by the CAG promoter in MCs. The Dct-Cre mouse was a kind gift from Dr. F. Beermann (Guyonneau et al. 2002). Mice were mated with a transgenic strain that bears the CAG-driven GFP gene activated by Cre recombinase. (*Bottom*) FACS analysis of cells dissociated from neonatal skin. (*Right panel*) Analysis of untreated mice. Two GFP$^+$ populations are found. (*Left panel*) Analysis of mice injected with anti-c-Kit monoclonal antibodies (ACK2) at birth. Only GFP^{low} cells remain in the skin preparation.

genes. We have been trying to confirm this possibility by generating two mouse strains harboring the *gfp* gene, one driven by the *dct* promoter and the other driven by the *sox10* promoter. Up to now, we have completed analysis of the *dct-gfp* transgenic mouse and confirmed that the MC of P10 mouse skin contains both GFP[high] and GFP[low] populations. Analysis of *sox10-gfp* mice is currently in progress.

These results clearly indicate that the two features of MSCs defined originally by analysis at the single-cell level can now be used for defining MSCs at the population level by using FACS. This progress is essential for distinguishing MSCs from other compartments without using histological localization.

GENETIC APPROACH TO INVESTIGATE THE MOLECULAR MECHANISMS UNDERLYING MSC MAINTENANCE

This progress also allows us to use FACS to sort the MSC population en mass. Comparing CAG-driven GFP high and low populations, we were able to list genes that are expressed at higher levels in MSCs. Of importance is that despite low expression of some housekeeping genes and MC-specific genes, we found genes that were expressed at higher levels in MSCs. Molecules involved in Notch signaling are an example that were listed to be expressed at higher levels. An advantage of MC lineage is that complete absence of MCs does not have much of an effect on the life of the animal. This unique feature of MCs allows us to evaluate the role of a particular molecule on MC activity during a long period of time. As is the case for some of the other cell lineages, a number of MC-specific gene expression systems have been developed that are used for MC-specific gene knockouts. Using these advantages, we have investigated the function of the Notch signal in the development and maintenance of the MSC system (Moriyama et al. 2006). Our results unequivocally demonstrate the role of Notch in the development and maintenance of MSCs.

INDUCTION OF QUIESCENT MSCs

In the mouse, the MB undergoes the final differentiation to MSCs as late as about P4, because this is the stage when the functional niche is formed in the LPP. Despite this, the cellular identity of niche has not yet been clarified; the timing of the niche formation was able to be estimated through the analysis of the *bcl2*[−/−] mouse. In the *bcl2*[−/−] mouse, the MB undergoes faster apoptosis upon SCF starvation. Thus, neonatal ACK2 injection resulted in the complete absence of MSCs in hair follicles due to rapid depletion of MBs before the formation of the niche (Mak et al. 2006). On the other hand, in the wild-type mouse, a portion of the MB can survive for more than 3 days against SCF starvation and is able to meet the niche that is newly formed in the LPP of guard hairs. Nonetheless, all key events required for induction of MSCs should occur during these neonatal 4 days.

To dissect this process, we analyzed the timing of the appearance of distinct MSC features. Our histological analysis showed that Sox10 expression in MCs is down-regulated at about P2 in the upper area of the developing

hair follicle. FACS analysis showed that cells that are low in *dct*-promoter-driven *gfp* are first detected at E18.5. These results suggests that, even among MC-specific genes that are down-regulated in MSCs, the timing of when each individual feature appears is variable. Likewise, cells that show low expression of CAG-promoter-driven GFP were first detected at E18.5. Concerning the resistance to ACK2 treatment (resistance against SCF starvation), which we are using to specify MSCs in neonatal hair follicles, our previous studies showed that such a state is also observed in embryonic stages. Indeed, MBs in E15.5–17.5 embryos are ACK2-resistant (Yoshida et al. 1993). As such, it is likely that each feature associated with the MSCs in the LPP is acquired through multiple and probably independent steps, rather than a single step. If so, it is important to define each distinct MSC signature in terms of function significance. In this respect, it must be emphasized that the functional definition of the ultimate SCs has not yet been fully attained.

ATTEMPT TO INDUCE QUIESCENT MSCs IN VITRO

As described above, genetic methods are available for manipulating genes in the MC lineage. Thus, determining the functional role of each individual molecule in MSC activity may be feasible. Moreover, there is a definite conception about the phenotype resulting from the defect in this process. Because MSC differentiation was completed later than the colonization of MBs to the hair matrix, defects specifically affecting the MSC should be expressed such that mice born pigmented lose pigmentation in the next hair cycle. Thus, a continuous effort to generate conditional knockout strains of MSC signature genes is important for understanding molecular mechanisms involved in induction and maintenance of MSCs.

What might be difficult with this genetic approach, although not impossible, is to determine a minimum requirement for a given process, e.g., the molecular requirement for inducing MSCs. As shown in the preceding sections, the induction of MSCs is likely to be regulated by multiple extrinsic signals. We think that in vitro culture of embryonic MCs may address this issue more effectively, because in vitro culture per se is a constitutive approach to recapitulate the process. With this in mind, we decided to use MB culture to search for the minimum requirements for inducing quiescent MSCs.

Although we have established a method for distinguishing MSCs from other compartments by FACS, culturing FACS-purified MBs is a new challenge that has not yet been reported. In fact, all previous culture conditions for MBs were optimized for MBs mixed with other skin cells. We have tested those culture conditions reported previously (Hirobe 1991; Sviderskaya et al. 1995), but none of them were able to support the proliferation of FACS-purified MBs. On the other hand, culture with the XB2 (Rheinwald and Green 1975) feeder layer originally developed by Bennett's group (Sviderskaya et al. 1995) turned out to be capable of supporting clonogenic proliferation of FACS-purified MCs (Yonetani et al. 2008). The proliferation of MCs under these conditions is c-Kit-dependent, and its differentiation is promoted by endothelin. More importantly, MBs proliferating in this culture are able to reconstitute the whole MSC system in hair follicles regenerated from dissociated keratinocytes. This result shows unequivocally that at least a portion of the cells in this culture can replenish MSCs in vivo. This culture can now be used to explore extrinsic signals that induce various features of MSCs and eventually the quiescent MSCs.

Using this culture system, we have been searching for molecules that are able to induce SC signature. XB2+SCF supports sustained proliferation of MBs, but the MSC signature was not induced. Among the molecules that we have tested in this culture, basic FGF showed a striking activity to induce CAG-GFP^{low} as well as Dct-GFP ^{low} populations. Those two populations are indeed overlapped (S. Torikai-Nishikawa et al., unpubl.). We are currently examining whether other features of MSCs such as low-Sox10 activity are associated with this state, but our preliminary analysis suggests that the Sox10 expression level in the CAG-GFP^{low} population is comparable to that of GFP^{high} MBs, even though they are in GFP^{low} populations. Hence, the final state of the quiescent MSCs is not attained by basic FGF alone. On the other hand, cells that are slowly cycling and ACK2-resistant are indeed induced in this FGF-induced GFP^{low} population. Further attempts to define molecules required for inducing all MSC features are currently in progress.

CONCLUSION

The MSC system has a unique advantage for studying the niche for the quiescent MSCs. Our study suggests, however, that the quiescent MSCs are induced not by a single encounter with the niche, but rather through multiple steps acquiring each distinct MSC feature. After extensive analyses of in vivo MSCs, we are now able to investigate the signals involved in this process by using in vitro culture. Although the molecular mechanisms inducing and maintaining the ultimate MSCs remain totally obscure, some signals such as basic FGF that are involved in inducing quiescence, as well as some features of MSCs, are being uncovered.

ACKNOWLEDGMENTS

This study is supported by the Leading Project for Realization of Regenerative Medicine.

REFERENCES

Fuchs, E., Tumbar, T., and Guasch, G. 2004. Socializing with the neighbors: Stem cells and their niche. *Cell* **116**: 769–778.

Guyonneau, L., Rossier, A., Richard, C., Hummler, E., and Beermann, F. 2002. Expression of Cre recombinase in pigment cells. *Pigment Cell Res.* **15**: 305–309.

Hirobe, T. 1991. Selective growth and serial passage of mouse melanocytes from neonatal epidermis in a medium supplemented with bovine pituitary extract. *J. Exp. Zool.* **257**: 184–194.

Joseph, N.M., Mukouyama, Y.S., Mosher, J.T., Jaegle, M., Crone, S.A., Dormand, E.L., Lee, K.F., Meijer, D., Anderson, D.J., and Morrison, S.J. 2004. Neural crest stem cells undergo multilineage differentiation in developing peripheral nerves to generate endoneurial fibroblasts in addition to Schwann cells. *Development* **131**: 5599–5612.

Kunisada, T., Yoshida, H., Yamazaki, H., Miyamoto, A., Hemmi, H., Nishimura, E., Shultz, L.D., Nishikawa, S., and Hayashi, S. 1998. Transgene expression of steel factor in the basal layer of epidermis promotes survival, proliferation, differentiation and migration of melanocyte precursors. *Development* **125**: 2915–2923.

Mak, S.S., Moriyama, M., Nishioka, E., Osawa, M., and Nishikawa, S. 2006. Indispensable role of Bcl2 in the development of the melanocyte stem cell. *Dev. Biol.* **291**: 144–153.

Moore, K.A. and Lemischka, I.R. 2006. Stem cells and their niches. *Science* **311**: 1880–1885.

Moriyama, M., Osawa, M., Mak, S.S., Ohtsuka, T., Yamamoto, N., Han, H., Delmas, V., Kageyama, R., Beermann, F., Larue, L., and Nishikawa, S. 2006. Notch signaling via Hes1 transcription factor maintains survival of melanoblasts and melanocyte stem cells. *J. Cell Biol.* **173**: 333–339.

Nishimura, E.K., Jordan, S.A., Oshima, H., Yoshida, H., Osawa, M., Moriyama, M., Jackson, I.J., Barrandon, Y., Miyachi, Y., and Nishikawa, S. 2002. Dominant role of the niche in melanocyte stem-cell fate determination. *Nature* **416**: 854–860.

Ohlstein, B., Kai, T., Decotto, E., and Spradling, A. 2004. The stem cell niche: Theme and variations. *Curr. Opin. Cell Biol.* **16**: 693–699.

Osawa, M., Egawa, G., Mak, S.S., Moriyama, M., Freter, R., Yonetani, S., Beermann, F., and Nishikawa, S. 2005. Molecular characterization of melanocyte stem cells in their niche. *Development* **132**: 5589–5599.

Potten, C.S., Booth, C., Tudor, G.L., Booth, D., Brady, G., Hurley, P., Ashton, G., Clarke, R., Sakakibara, S., and Okano, H. 2003. Identification of a putative intestinal stem cell and early lineage marker; musashi-1. *Differentiation* **71**: 28–41.

Rheinwald, J.G. and Green, H. 1975. Formation of a keratinizing epithelium in culture by a cloned cell line derived from a teratoma. *Cell* **6**: 317–330.

Scadden, D.T. 2006. The stem-cell niche as an entity of action. *Nature* **441**: 1075–1079.

Spradling, A., Drummond-Barbosa, D., and Kai, T. 2001. Stem cells find their niche. *Nature* **414**: 98–104.

Sviderskaya, E.V., Wakeling, W.F., and Bennett, D.C. 1995. A cloned, immortal line of murine melanoblasts inducible to differentiate to melanocytes. *Development* **121**: 1547–1557.

Yonetani, S., Moriyama, M., Nishigori, C., Osawa, M., and Nishikawa, S. 2008. In vitro expansion of immature melanoblasts and their ability to repopulate melanocyte stem cells in the hair follicle. *J. Invest. Dermatol.* **128**: 408–420.

Yoshida, H., Nishikawa, S., Okamura, H., Sakakura, T., and Kusakabe, M. 1993. The role of c-*kit* proto-oncogene during melanocyte development in mouse. In vivo approach by the in utero microinjection of anti-c-kit antibody. *Dev. Growth Differ.* **35**: 209–220.

Asymmetric Division and Stem Cell Renewal without a Permanent Niche: Lessons from Lymphocytes

J.T. CHANG AND S.L. REINER

Abramson Family Cancer Research Institute and Department of Medicine,
University of Pennsylvania, Philadelphia, Pennsylvania 19104

Numerous tissues in long-lived organisms are composed of short-lived cells. The continual regeneration of some barrier surfaces, for example, relies on adult stem cells that have the capacity to divide and produce one daughter cell destined for terminal differentiation and function and another daughter cell that renews the stem cell fate. The immune system of higher animals possesses a cellular component called lymphocytes, which face a similar need for regeneration. A lymphocyte that is recruited during an infection must give rise to cellular progeny that undergo terminal differentiation to eliminate an invading microbe, yet retain progeny that replace the recruited cell in order to maintain immunity to reinfection. Emerging evidence suggests that specifying the divergent cell fates necessary for immunity relies on the ability of the lymphocyte to exploit an evolutionarily conserved strategy for making kindred cells different—asymmetric cell division. Although the lymphocyte does not possess constitutive polarity, it appears to use a facultative interaction with another cell to nucleate unequal segregation of fate determinants relative to its plane of division. Herein, we propose that other mobile and nonadherent cells, such as blood and cancer stem cells, might exploit provisional interactions with their niche or microenvironment to achieve diversity among their daughter cells.

GENERATING CELL FATE DIVERSITY, WITH OR WITHOUT A PERMANENT NICHE

Life is a succession of cell divisions. For multicelled beings, diversity in the fate of a cell's progeny is essential for orderly tissue formation and specialized function. Knowledge of how kindred cells diverge in fate is essential for understanding organ formation and function, stem cell and tissue regeneration, immunity, and cancer. At least two distinct mechanisms, representing the cellular versions of "nature" versus "nurture," can promote daughter cell diversity (Fig. 1). After inception, identically born daughter cells can be nurtured to adopt different fates by encountering distinct signals in their environments. There also exists an evolutionarily conserved mechanism, called asymmetric cell division, whereby a dividing cell imparts unequal inheritance of its components to its two daughter cells, making them different from inception (Betschinger and Knoblich 2004; Knoblich 2008).

Scattered instances of asymmetric cell division have been described across evolution for generating specialized cells within an organ and for enabling the regeneration of progenitor cells in tissues that require continuous production of differentiated progeny, such as skin (Lechler and Fuchs 2005). Asymmetric cell division has typically been associated with tissues that already possess intrinsic apical-basal polarity. In those situations, the constitutive attachment of a cell to the basement membrane provides a framework that allows the daughter attached to the basement membrane to retain the stem cell fate,

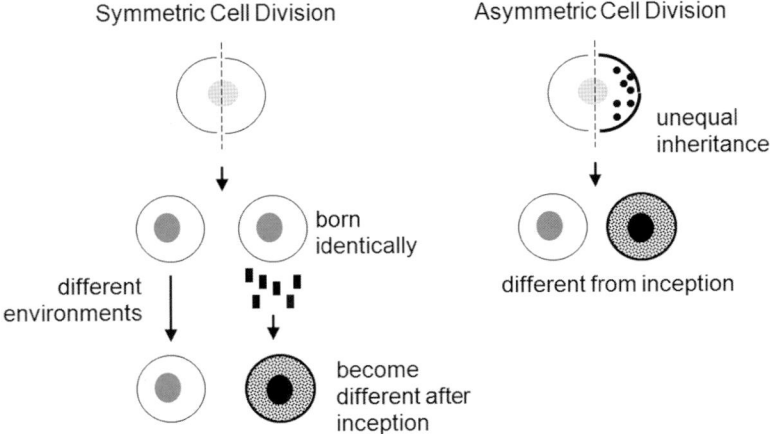

Figure 1. Alternative mechanisms for generating cell-fate diversity. (*Left*) Two identically born daughter cells encounter a disparity in their extrinsic signaling (*black rectangles*), resulting in their subsequent adoption of distinct cell fates. (*Right*) A cell is signaled in such a way that it organizes asymmetric localization of a determinant (*black circles*) relative to the plane of division. The two daughters receive an unequal share of the determinant, making them different from inception.

whereas the daughter that detaches initiates the process of terminal differentiation.

Not all stem cells, however, possess a constitutive niche. Nonadherent or migratory cells such as blood stem cells and metastatic cancer-initiating cells are motile and may have only facultative encounters with their niche or microenvironment. We have recently begun to explore the possibility that extrinsic signaling emanating from a point on a cell's surface could act as a guiding principle to orient the machinery necessary for the execution of cell division and simultaneously reorganize the preformed components of the cell (Chang et al. 2007). The coordination of these events could enable a dividing cell whose external surface may have only provisionally communicated with the microenvironment or niche to create two daughter cells that are different from inception. Such a mechanism may offer a solution to a long-standing puzzle of vertebrate immunity and provide a framework to study how self-renewal or specialization is achieved in other mobile cells.

REGENERATING LYMPHOCYTES FOR IMMUNITY

Most of the mature blood lineages, red or white, have a finite and relatively short life span. Attrition of these lineages during homeostasis is continually offset by the production and subsequent maturation of lineage-committed progenitors from a blood stem cell. Each of the mature blood lineages is postmitotic and does not proliferate in response to stress. The need for excess numbers to respond to acute changes (hemorrhage or infection) is accommodated by mobilization and differentiation from progenitors and sequestration or migration to inflamed sites.

The most recent evolutionary addition to the immune armamentarium of vertebrates is a set of cells called lymphocytes. The major lineages of lymphocytes are B cells, the producers of antibodies, and $CD4^+$ and $CD8^+$ T lymphocytes. $CD4^+$ T cells orchestrate the selection of the type of defense mechanisms that are appropriate to fend off the different classes of microbial pathogens (Reiner 2007; Reiner et al. 2007). $CD8^+$ T cells participate in the defense against intracellular microbes, viruses, and tumors by killing target cells with cytolytic granules and secreting cytokines that activate intracellular microbicides. Successful habitation of a microbe-filled planet is dependent on the lymphocyte system of defense that arose late in evolution in an ancestor of jawed fish. Lack of normal gene products that are required for lymphocyte development results in a severe combined immunodeficiency (SCID) phenotype.

The way lymphocytes recognize foreign intruders is different from that of all other types of white blood cells. All other mature white blood cells are clonally identical among their lineage with regard to the receptors that bind foreign ligands. The lymphocyte system of recognition, however, is predicated on an unprecedented diversity of recognition. The foreign ligand-binding proteins (antigen receptor) of each lymphocyte within a given lineage are clonally distinct or bar-coded from one another (Chang and Reiner 2007). During lymphocyte development, random somatic recombination occurs at the multiple poly-

morphic gene segments encoding the potential antigen receptor polypeptides. After its acquisition of a unique antigen receptor and successful screening to ensure that it will not be overtly self-destructive to the host, which occurs in specialized developmental sites (bone marrow and thymus), a nascent lymphocyte is exported to the peripheral lymphoid tissues for circulation through lymph nodes, spleen, and blood. This army of lymphocytes carries such tremendous diversity in antigen recognition that it is now likely that the host could be infected with virtually any microbe and mount a specific response against it.

Lymphocytes that have not yet encountered their foreign antigen are called naïve T cells and B cells. Naïve lymphocytes are highly motile and are continually patrolling peripheral lymphoid tissues for evidence of microbial invasion. Although they do not divide during their surveillance, they have a much longer life span than that of the other blood cells, many months if not a year. During an immune response to pathogen, a naïve T cell patrolling peripheral lymphoid tissues encounters an antigen-presenting cell displaying microbial components on its surface (Reiner 2007; Reiner et al. 2007). Because of its sheer rarity in recognizing the microbial invader, the lymphocyte faces the challenge of producing large numbers of terminally differentiated progeny needed for acute defense (called effector cells). Whereas other blood lineages are postmitotic, naïve lymphocytes indeed have the capacity to undergo massive clonal expansion in response to the antigen trigger in order to accommodate the need for large numbers of effector cells to match the microbial burden. The progeny of a naïve lymphocyte undergoing massive cellular expansion are also empowered with new programs of gene expression that make them effective as an effector cell (Bird et al. 1998; Reiner 2007; Reiner et al. 2007). Antigen-experienced T cells heritably remodel chromatin at genes encoding cytokines and cytolytic materials. The progeny of the naïve cell that have undergone transit amplification and effector cell differentiation leave the lymph node and migrate to the tissue site of microbial invasion. Upon encounter with pathogen in skin or lung, these effector cells now release their cytokines or destructive granules at target cells harboring intracellular infection or those cells displaying evidence of pathogen interaction.

In contrast to other white blood cells, replenishment of a useful naïve lymphocyte cannot be derived from a blood stem cell because it is impossible for the blood stem cell to specify that the correct antigen receptor would eventually be expressed (Chang and Reiner 2007). The effector cell progeny of a naïve lymphocyte that migrate from lymph node to peripheral tissues to do battle with the invading microbe are themselves now postmitotic, terminally differentiated, and short-lived. If all progeny of a naïve lymphocyte underwent terminal differentiation into short-lived effector cells, it might be expected that the host would be left with a gaping hole in its defense because the first encounter has consumed or depleted the cells specific for this microbe. Centuries of experience, however, have taught that subsequent encounters with a microbe are rarely met by ignorance or an unresponsive state. Instead, subsequent encounters with a given microbe are often heralded with a faster and stronger lymphocyte response than the first battle, a phenomenon known as immunity. This is thought to be possible owing to the ability of some antigen-experienced progeny of the naïve lymphocyte to undergo a program of regeneration and self-renewal (so-called memory cells).

ASYMMETRIC CELL DIVISION AS A POTENTIAL SOLUTION FOR THE PARADOX OF IMMUNITY

How does a thin layer of defense repair itself when punctured, leaving behind a thicker segment of armor instead of a gaping hole? In other words, how does a naïve T cell accommodate the mutually exclusive needs for terminally differentiated effector cells and self-renewing memory cells? One potential solution to this conundrum is for the lymphocyte to use a mechanism that is a frequent solution for accommodating lifelong function in long-lived beings possessing short-lived cells. The adult somatic stem cell is a paradigm for tissue regeneration (Morrison and Kimble 2006; Knoblich 2008). In skin, for example, a stem cell residing in the basement membrane can periodically divide asymmetrically, producing one daughter that is amplified and growing apically to eventually give rise to the mature tegument and another daughter that retains its attachment to the basal surface and maintains the identity of the original stem cell (Lechler and Fuchs 2005). In this way, the stem cell uses asymmetric division to continually regenerate cells destined for acute function and cells that maintain the lineage over a lifetime.

Is it possible for nonadherent cells, such as members of the blood lineages that lack constitutive attachment and apparent polarity, to undergo an asymmetric division? We began to entertain the possibility that lymphocytes might exploit the principle of asymmetric cell division in order to achieve essential heterogeneity in the fates of daughter cells (Chang et al. 2007). Our hypothesis arose because of ample literature describing facultative polarity in interphase T cells undergoing stimulation by antigen (Monks et al. 1998; Huppa and Davis 2003; Lin et al. 2005; Cemerski and Shaw 2006). The hypothesis became further embellished based on in vivo time-lapse imaging, documenting an unusually prolonged contact between the naïve T cell and the cell that is decorated with foreign antigen before the T cell's first division (Miller et al. 2002; Stoll et al. 2002; Bousso and Robey 2003; Mempel et al. 2004).

LYMPHOCYTES ESTABLISH FACULTATIVE POLARITY BEFORE MITOSIS

When a T cell (naïve, effector, or memory) is engaged by an antigen-presenting cell, ligation of the antigen receptor propagates a signal transduction cascade that rapidly activates or represses the transcription of multiple genes. It is also recognized that antigen-receptor-induced signals do not solely transmit information to the nucleus. In particular, antigen receptor signaling, probably in cooperation with T-cell integrin signaling, results in the coalescence of signaling and adhesive components at the site of contact between the T cell and antigen-presenting cell. This structure, termed the immunological synapse, is composed of an inner confluence of antigen receptor and associated polypeptides encircled by an outer ring composed of an adhesive integrin receptor (Monks et al. 1998). Other components of the immunological synapse include the receptors for certain cytokines, such as interferon-γ (IFN-γ) and interleukin-2 (IL-2), as well as cytoplasmic signaling intermediates associated with the antigen receptor (Monks et al. 1998; Maldonado et al. 2004).

The formation of the immunological synapse is accompanied by a substantial reorganization of several cellular components. The microtubule-organizing center and Golgi apparatus move to a site just below the immunological synapse. Recruitment of mitochondria to the synapse has been suggested to regulate local concentrations of calcium and allow its influx across plasma membrane channels, which is needed to activate signaling components (Quintana et al. 2007). Signaling through the T-cell antigen receptor also communicates with an evolutionarily conserved network of proteins with roles in coordinating polarity and asymmetric cell division (Betschinger and Knoblich 2004; Knoblich 2008). In lymphocytes, the partitioning-defective (PAR) proteins appear to be reorganized both during migration and following antigen receptor signaling (Ludford-Menting et al. 2005). A mammalian homolog of atypical protein kinase C (aPKC), an essential component of a complex containing the PAR proteins Par-3 and Par-6, has further been implicated in T-cell function (Ludford-Menting et al. 2005; Martin et al. 2005; Real et al. 2007). A role for two other conserved polarity proteins, Scribble (Scrib) and discs large (Dlg), has also been suggested during T-cell activation and migration (Xavier et al. 2004; Ludford-Menting et al. 2005; Round et al. 2005; Stephenson et al. 2007).

The initial descriptions of the immunological synapse were made in effector T cells, and its formation was understood as an adaptive mechanism to ensure that cytokine and cytolytic granules could be directed judiciously to the site of intercellular communication with a target cell (Huse et al. 2006; Stinchcombe et al. 2006). Although such facultative polarity has also been observed in naïve T cells, the reasons for such behavior remained controversial and ambiguous. As the field began to image the dynamics of intercellular communication between the naïve T cell and its antigen-presenting cell in vivo, it became tempting to hypothesize that the facultative polarity of naïve cells could serve a purpose different from that of effector cells, wherein it literally guides the execution of their acute function. Intravital, time-lapse imaging of a naïve T cell during its first day of the immune response indicated that the T cell forms a prolonged conjugate with its antigen-presenting cell that lasts until the first T-cell division (Miller et al. 2002; Stoll et al. 2002; Bousso and Robey 2003; Mempel et al. 2004). We therefore hypothesized that synapse-like polarity in a naïve cell might not be forming solely to instruct the acute behavior of the interphase lymphocyte, but rather to establish an organization of cellular components in the mitotic cell preparing for its first division (Fig. 2).

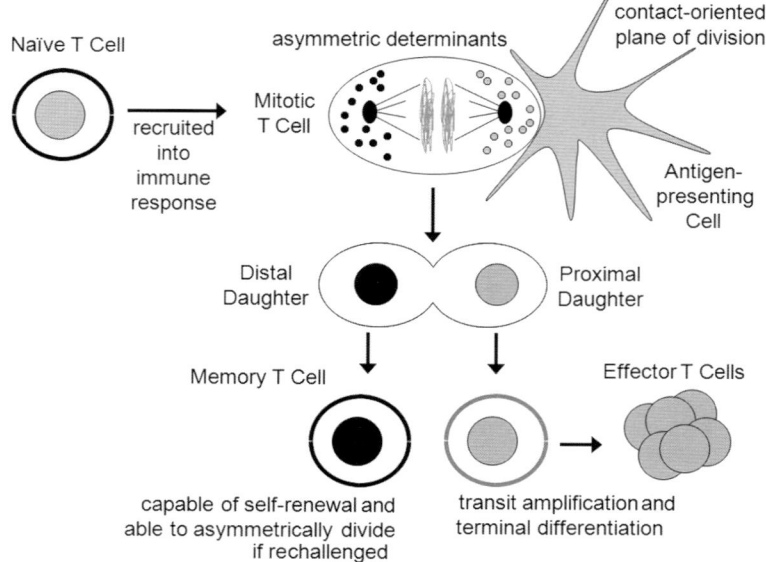

Figure 2. Asymmetric cell division initiating the immune response. A migrating naïve T cell is recruited into the immune response when its unique antigen receptor recognizes microbial components presented by an antigen-presenting cell in the draining lymph node. The activation induced by antigen receptor ligation results in facultative polarity of the lymphocyte. The provisional but prolonged interaction between the T cell and antigen-presenting cell lasts until mitosis. The contact surface allows the T cell to orient its plane of division as well as to segregate asymmetric determinants relative to the plane of division. The proximal daughter undergoes vigorous proliferation and terminal differentiation to create terminally differentiated effector cells that leave the lymph node to perform acute function at the barrier surfaces where pathogens invade. The distal daughter differentiates into a memory cell, which serves as the long-term replacement of this antigen-specific clone. The stem-cell-like memory cell is capable of self-renewal by slowly dividing symmetrically in the absence of recurrent infection. In the face of reinfection, the newly activated memory cell would be capable of asymmetrically dividing to produce more effector cell progeny from its proximal daughter as well as a distal daughter that would renew the memory cell.

ASYMMETRIC DIVISION OF CLONALLY SELECTED T CELLS

Examination of a naïve T lymphocyte responding to a pathogen suggested that it indeed exploits contact with its antigen-presenting cell as a provisional surface on which to organize polarity and subsequent asymmetric division (Chang et al. 2007). Interphase T-cell blasts were found to contain synapse-like organization during the in vivo immune response. This morphology persisted into mitosis and it was contingent on prolonged contact between the T cell and antigen-presenting cell (Fig. 2). One centrosome remains proximal to the contact, whereas the other is positioned at the distal pole of the mitotic T-cell blast. In so doing, the plane of division of the T cell is oriented parallel to the site of contact. In addition, key components are segregated to one or the other side of the impending plane of division. Several classes of proteins appear to be asymmetrically distributed in this manner, including immune receptors, signaling molecules, polarity proteins, and potential fate determinants. We also observed complementary localization of members of the aPKC-Par3-Par6 complex distally and the Scrib-Dlg-Lgl family of polarity proteins proximally, which was evident in both interphase and mitotic lymphocyte blasts (Chang et al. 2007).

As naïve T cells underwent their initial cytokinesis, the incipient daughters inherited unequal shares of several of these proteins, which was also contingent on prolonged contact with the antigen-presenting cell (Chang et al.

2007). Phenotypic and functional analyses indicated that the daughter "proximal" to the provisional contact with the antigen-presenting cell undergoes transit amplification and terminal differentiation as an effector cell, whereas the daughter "distal" to the contact site eventually gives rise to a self-renewing memory cell (Fig. 2).

In a sense, neither of the first two daughter lymphocytes is identical to the original naïve mother cell. The terminally differentiated effector branch (proximal daughter) appears to be destined for acute function, whereas the self-renewing memory branch (distal daughter) carries a distinctive marking of its lack of naïveté and is also empowered with a capacity for slow proliferative turnover in the absence of antigen to offset the attrition of a finite life span. The self-renewing memory lineage, however, does appear to accomplish the fundamental need of barcoded lymphocytes for a system of intrinsic regeneration: These memory cells retain the same anatomic and migratory behavior of the original naive lymphocyte, patrolling peripheral lymphoid tissue in anticipation of a recurrent invasion by the microbe that their naïve progenitor was suited to recognize. Indeed, preliminary studies suggest that these memory cells are stem-cell-like, able to maintain their numbers by periodic division in the absence of the threat and to undergo conservative, asymmetric division in the face of reinvasion by their microbial nemesis (Fig. 2). In the case of reinfection, memory cells appear to yield proximal daughters that undergo transit amplification and terminal differentiation to effector cells, whereas the distal daughters appears to retain the identity of the parental memory cell.

It might be considered unusual for the daughter proximal to the contact with the antigen-presenting cell to abandon the conservative role of replacing the mother cell while the distal daughter becomes the long-term replacement for the selected naïve clone. In typical stem cell arrangements, the daughter cell closer to the niche retains the mother cell's less-differentiated identity, whereas the distal daughter cell is liberated to undergo terminal differentiation (Morrison and Kimble 2006; Knoblich 2008). It is possible that the explanation for the apparent upside-down behavior of the T cell is that the antigen-presenting cell is not really a niche but is instead an "anti-niche." The pathogen-associated antigen-presenting cell has been induced to provide signals that are necessary to initiate rapid cell division and terminal effector differentiation of lymphocytes. Proximity to the differentiative information is thus likely to specify greater change, whereas arising from the distal pole is more likely to foster the less-differentiated state of the memory cell.

ASYMMETRIC CELL DIVISION IN OTHER BLOOD LINEAGES AND CANCER?

That the process of asymmetric T-cell division involves a core set of evolutionarily conserved regulators of cell polarity raises the possibility that other lymphocytes having provisional contacts may also be capable of exploiting this strategy to generate cell-fate diversity. Selected CD4$^+$ T cells give rise to daughter cell fates that have differing roles in the host defense, suggesting the possibility that asymmetric cell division can create diversity among effector cells in addition to enabling terminal differentiation versus self-renewal (Reiner 2007; Reiner et al. 2007). A potential mechanism for more than two branches in the fate map of a lymphocyte is suggested by the findings that the daughter cells of a naïve T cell may also partake in cell-to-cell communication with antigen-presenting cells (Celli et al. 2005; Obst et al. 2005).

Like T lymphocytes, B lymphocytes divide and diversify their fates upon encountering pathogen. The progeny of a selected naïve B cell undergo further somatic recombination to generate antibody molecules that have selective functions in defense against microbes. In addition, an activated naïve B lymphocyte must give rise to terminally differentiated, antibody-producing plasma cells as well as self-renewing memory cells. During pathogen invasion, B cells migrate within a lymph node to an area where they can interact with helper T cells. Such interactions precede B-cell division, raising the possibility that B cells could exploit their interaction with helper T cells as a provisional surface on which to organize polarity and subsequent asymmetric divisions.

Seminal experiments performed in the 1950s and 1960s demonstrated that a single hematopoietic stem cell has the capacity to reconstitute all of the terminally differentiated, mature blood lineages while still yielding progeny with an undifferentiated fate (Becker et al. 1963). This is compatible with the hypothesis that hematopoietic stem cells might undergo conservative cell divisions, producing one differentiating progeny and one undifferentiated, self-renewing progeny. It has remained unclear, however, whether hematopoietic stem cells accomplish asymmetric cell division and, if so, how.

Hematopoietic stem cells within the bone marrow have been observed in close proximity to stromal cells, osteoblasts, and endothelial cells of blood vessels (Kiel et al. 2005). These observations have led to the hypothesis that one or more of these cells might constitute the stem cell niche in the bone marrow. Contact with the niche could direct orientation of the mitotic spindle of a hematopoietic stem cell in such a way that the daughter cell remaining attached to the niche would retain the ability to self-renew, whereas the daughter dividing away from the niche would undergo terminal differentiation. Other cells lining the bone marrow cavity could also serve to direct asymmetric cell division in later branches of the differentiating progeny in order to parse two distinct lineage branches from a common progenitor. Improvements in time-lapse intravital imaging will be of great value in determining whether any of these hypothetical scenarios are operative (Sipkins et al. 2005).

One of the fundamental questions of cancer biology relates to the mechanism by which tumors are able to maintain themselves. One possibility is that each cell within a tumor has the capacity to maintain itself. Alternatively, it is possible that self-renewal properties might be a characteristic of a small fraction of cells within the tumor, so-called cancer stem cells (Wang and Dick 2005). How might cancer stem cells self-renew? The tumor microenvironment, including stromal cells and extracellular matrix proteins, is thought to have a profound influence on cancer development (Bissell and Labarge 2005; Engler et al. 2006; Karnoub et al. 2007). It is therefore tempting to speculate that the tumor microenvironment might promote the self-renewal of cancer stem cells by serving as the provisional surface to organize cellular components that might enable asymmetric cell division.

A UNIFYING MODEL FOR CELL-FATE DIVERSITY IN MULTICELLULAR LIFE

Signal transduction is frequently considered to result in information transfer to the nucleus, reprogramming the gene expression profile needed to specify a given cell fate. During development, immunity, or cancerous growth, cell fate may not always be adopted or at least finalized until after a cell has divided. In some settings, the progeny of a cell, rather than the original cell that received the signal, must alter their gene expression profile. It is known that signaling can also communicate to the preformed components of the cell within the cytoplasm, resulting in spatial reorganization that can direct the cytoskeleton and other elements that may impact a cell that is preparing to divide.

Although we have recently provided evidence that an antigen-presenting cell might provide a provisional surface for lymphocytes to organize polarity and asymmetric cell division, it is also possible that other cell types, and even noncellular surfaces, such as extracellular matrix protein, could serve the same function (Fig. 3). In this

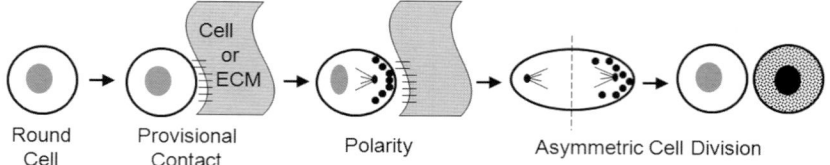

Figure 3. A unifying model of asymmetric division in multicelled life. A cell may encounter a provisional interaction with some cellular surface or extracellular matrix (ECM) that may transduce a signal to its nucleus but also organize polarity in its cytoplasmic contents and its machinery that execute cell division. If the cell is preparing for division, this could lead to unequal information transfer to its daughter cells. It is speculated that such provisional polarity may explain how mobile cells (lymphocytes, blood stem cells, metastatic cancer stem cells) achieve asymmetric cell division in the absence of a constitutive niche.

way, cells that do not exist in a conventionally stratified apical-basal tissue could have provisional communications before they divide to coordinate cell polarity and subcellular segregation. By signaling to cellular components other than the nucleus, information transfer to daughter cells can thus be rendered unequal. Some well-characterized examples of asymmetric cell division have been described in varied aspects of multicellular life. It now remains to be determined how frequently in life a signal delivered to a point on a cell that is preparing to divide may actually result in an asymmetric cell division.

ACKNOWLEDGMENTS

The authors are grateful to members of the Reiner laboratory for helpful discussion, Jiyeon Kim for critical suggestions, and the National Institutes of Health, Abramson Family, the American Gastroenterological Association, and Howard Hughes Medical Institute for financial support.

REFERENCES

Becker, A.J., McCulloch, E.A., and Till, J.E. 1963. Cytological demonstration of the clonal nature of spleen colonies derived from transplanted mouse marrow cells. *Nature* **197:** 452–454.

Betschinger, J. and Knoblich, J.A. 2004. Dare to be different: Asymmetric cell division in *Drosophila, C. elegans* and vertebrates. *Curr. Biol.* **14:** R674–R685.

Bird, J.J., Brown, D.R., Mullen, A.C., Moskowitz, N.H., Mahowald, M.A., Sider, J.R., Gajewski, T.F., Wang, C.R., and Reiner, S.L. 1998. Helper T cell differentiation is controlled by the cell cycle. *Immunity* **9:** 229–237.

Bissell, M.J. and Labarge, M.A. 2005. Context, tissue plasticity, and cancer: Are tumor stem cells also regulated by the microenvironment? *Cancer Cell* **7:** 17–23.

Bousso, P. and Robey, E. 2003. Dynamics of CD8[+] T cell priming by dendritic cells in intact lymph nodes. *Nat. Immunol.* **4:** 579–585.

Celli, S., Garcia, Z., and Bousso, P. 2005. CD4 T cells integrate signals delivered during successive DC encounters in vivo. *J. Exp. Med.* **202:** 1271–1278.

Cemerski, S. and Shaw, A. 2006. Immune synapses in T-cell activation. *Curr. Opin. Immunol.* **18:** 298–304.

Chang, J.T. and Reiner, S.L. 2007. Protection one cell thick. *Immunity* **27:** 832–834.

Chang, J.T., Palanivel, V.R., Kinjyo, I., Schambach, F., Intlekofer, A.M., Banerjee, A., Longworth, S.A., Vinup, K.E., Mrass, P., Oliaro, J., et al. 2007. Asymmetric T lymphocyte division in the initiation of adaptive immune responses. *Science* **315:** 1687–1691.

Engler, A.J., Sen, S., Sweeney, H.L., and Discher, D.E. 2006. Matrix elasticity directs stem cell lineage specification. *Cell* **126:** 677–689.

Huppa, J.B. and Davis, M.M. 2003. T-cell-antigen recognition and the immunological synapse. *Nat. Rev. Immunol.* **3:** 973–983.

Huse, M., Lillemeier, B.F., Kuhns, M.S., Chen, D.S., and Davis, M.M. 2006. T cells use two directionally distinct pathways for cytokine secretion. *Nat. Immunol.* **7:** 247–255.

Karnoub, A.E., Dash, A.B., Vo, A.P., Sullivan, A., Brooks, M.W., Bell, G.W., Richardson, A.L., Polyak, K., Tubo, R., and Weinberg, R.A. 2007. Mesenchymal stem cells within tumour stroma promote breast cancer metastasis. *Nature* **449:** 557–563.

Kiel, M.J., Yilmaz, O.H., Iwashita, T., Yilmaz, O.H., Terhorst, C., and Morrison, S.J. 2005. SLAM family receptors distinguish hematopoietic stem and progenitor cells and reveal endothelial niches for stem cells. *Cell* **121:** 1109–1121.

Knoblich, J.A. 2008. Mechanisms of asymmetric stem cell division. *Cell* **132:** 583–597.

Lechler, T. and Fuchs, E. 2005. Asymmetric cell divisions promote stratification and differentiation of mammalian skin. *Nature* **437:** 275–280.

Lin, J., Miller, M.J., and Shaw, A.S. 2005. The c-SMAC: Sorting it all out (or in). *J. Cell Biol.* **170:** 177–182.

Ludford-Menting, M.J., Oliaro, J., Sacirbegovic, F., Cheah, E.T., Pedersen, N., Thomas, S.J., Pasam, A., Iazzolino, R., Dow, L.E., Waterhouse, N.J., et al. 2005. A network of PDZ-containing proteins regulates T cell polarity and morphology during migration and immunological synapse formation. *Immunity* **22:** 737–748.

Maldonado, R.A., Irvine, D.J., Schreiber, R., and Glimcher, L.H. 2004. A role for the immunological synapse in lineage commitment of CD4 lymphocytes. *Nature* **431:** 527–532.

Martin, P., Villares, R., Rodriguez-Mascarenhas, S., Zaballos, A., Leitges, M., Kovac, J., Sizing, I., Rennert, P., Marquez, G., Martinez, A.C., Diaz-Meco, M.T., and Moscat, J. 2005. Control of T helper 2 cell function and allergic airway inflammation by PKCζ. *Proc. Natl. Acad. Sci.* **102:** 9866–9871.

Mempel, T.R., Henrickson, S.E., and Von Andrian, U.H. 2004. T-cell priming by dendritic cells in lymph nodes occurs in three distinct phases. *Nature* **427:** 154–159.

Miller, M.J., Wei, S.H., Parker, I., and Cahalan, M.D. 2002. Two-photon imaging of lymphocyte motility and antigen response in intact lymph node. *Science* **296:** 1869–1873.

Monks, C.R., Freiberg, B.A., Kupfer, H., Sciaky, N., and Kupfer, A. 1998. Three-dimensional segregation of supramolecular activation clusters in T cells. *Nature* **395:** 82–86.

Morrison, S.J. and Kimble, J. 2006. Asymmetric and symmetric stem-cell divisions in development and cancer. *Nature* **441:** 1068–1074.

Obst, R., van Santen, H.M., Mathis, D., and Benoist, C. 2005. Antigen persistence is required throughout the expansion phase of a CD4[+] T cell response. *J. Exp. Med.* **201:** 1555–1565.

Quintana, A., Schwindling, C., Wenning, A.S., Becherer, U., Rettig, J., Schwarz, E.C., and Hoth, M. 2007. T cell activation requires mitochondrial translocation to the immunological synapse. *Proc. Natl. Acad. Sci.* **104:** 14418–14423.

Real, E., Faure, S., Donnadieu, E., and Delon, J. 2007. Cutting edge: Atypical PKCs regulate T lymphocyte polarity and scanning behavior. *J. Immunol.* **179:** 5649–5652.

Reiner, S.L. 2007. Development in motion: Helper T cells at

work. *Cell* **129:** 33–36.

Reiner, S.L., Sallusto, F., and Lanzavecchia, A. 2007. Division of labor with a workforce of one: Challenges in specifying effector and memory T cell fate. *Science* **317:** 622–625.

Round, J.L., Tomassian, T., Zhang, M., Patel, V., Schoenberger, S.P., and Miceli, M.C. 2005. Dlgh1 coordinates actin polymerization, synaptic T cell receptor and lipid raft aggregation, and effector function in T cells. *J. Exp. Med.* **201:** 419–430.

Sipkins, D.A., Wei, X., Wu, J.W., Runnels, J.M., Cote, D., Means, T.K., Luster, A.D., Scadden, D.T., and Lin, C.P. 2005. In vivo imaging of specialized bone marrow endothelial microdomains for tumour engraftment. *Nature* **435:** 969–973.

Stephenson, L.M., Sammut, B., Graham, D.B., Chan-Wang, J., Brim, K.L., Huett, A.S., Miletic, A.V., Kloeppel, T., Landry, A., Xavier, R., and Swat, W. 2007. DLGH1 is a negative reg-

ulator of T lymphocyte proliferation. *Mol. Cell. Biol.* **27:** 7574–7581.

Stinchcombe, J.C., Majorovits, E., Bossi, G., Fuller, S., and Griffiths, G.M. 2006. Centrosome polarization delivers secretory granules to the immunological synapse. *Nature* **443:** 462–465.

Stoll, S., Delon, J., Brotz, T.M., and Germain, R.N. 2002. Dynamic imaging of T cell-dendritic cell interactions in lymph nodes. *Science* **296:** 1873–1876.

Wang, J.C. and Dick, J.E. 2005. Cancer stem cells: Lessons from leukemia. *Trends Cell Biol.* **15:** 494–501.

Xavier, R., Rabizadeh, S., Ishiguro, K., Andre, N., Ortiz, J.B., Wachtel, H., Morris, D.G., Lopez-Ilasaca, M., Shaw, A.C., Swat, W., and Seed, B. 2004. Discs large (Dlg1) complexes in lymphocyte activation. *J. Cell Biol.* **166:** 173–178.

Modeling Stem Cell Asymmetry in Yeast

P.H. Thorpe,* J. Bruno,† and R. Rothstein*

*Department of Genetics and Development, Columbia University Medical Center, New York,
New York 10032; †Massachusetts Institute of Technology, Cambridge, Massachusetts 02139

For adult stem cells to both self-renew and give rise to differentiating progenitors, they must undergo an inherently asymmetric division. This defining model of asymmetric cell division requires either that stem cells preferentially distribute internal factors, thereby maintaining a stem cell phenotype in one lineage, or that extrinsic signals determine the fate of daughter cells, allowing the maintenance of one stem cell lineage. Although microbial systems are often used to model asymmetry, lineage-specific asymmetry has not been characterized in these organisms. Recently, we identified a stem-cell-like lineage-specific pattern of kinetochore asymmetry in postmeiotic yeast spores. Because the function of the kinetochore is to segregate chromosomes, this asymmetry has the potential to segregate sister chromatids nonrandomly. This may be relevant to stem cells because more than 30 years ago, it was proposed that stem cells selectively segregate one strand of their chromosomes into the self-renewing stem cell lineage (Cairns 1975). Although advanced labeling methods have provided evidence to both support and refute this hypothesis, it remains unclear how nonrandom sister-chromatid segregation might be achieved in a stem cell lineage. We have identified a kinetochore-specific mechanism in yeast that could support lineage-specific nonrandom sister-chromatid segregation and we discuss the implications of this observation.

Adult stem cells exhibit a lineage-specific asymmetric division pattern. When each stem cell divides, it generates a new stem cell and a progenitor cell, the latter ultimately giving rise to new cells in the tissue. This standard model of adult stem cell division does not preclude symmetric division, for example, to expand a stem cell pool. However, the asymmetric division provides a means for a stable population of stem cells to be maintained while enabling the creation of progenitor cells to replace damaged or aging cells within a tissue. One interesting consequence of this model, illustrated in Figure 1 (top), is that the stem cell lineage, represented by shaded cells, is a unique single lineage within a dividing population.

There are a number of microbial models of cellular asymmetry, but in these, the asymmetries are not confined to a single lineage and are found throughout the population. In *Saccharomyces cerevisiae*, every cell division is asymmetric, producing a mother and bud that have distinct fates. This inequality allows factors (e.g., alleles or protein complexes) to go into either the mother or the bud. An example of this asymmetry is illustrated by the RNA encoding Ash1, an inhibitor of the *HO* endonuclease, which is specifically transported to the bud rather than the mother (Long et al. 1997). This results in buds being unable to switch mating type, unlike mother cells that can switch. This pattern of asymmetry is illustrated in Figure 1 (bottom), with cells capable of mating-type switching indicated by shading.

Using a strategy that involves fluorescently tagged proteins, we identified a lineage-specific asymmetry in the segregation of yeast kinetochore proteins. The asymmetric pattern is strikingly similar to the asymmetric model of adult stem cell division (Fig. 1, top). This phenotype only occurs immediately after meiosis in the mother lineage

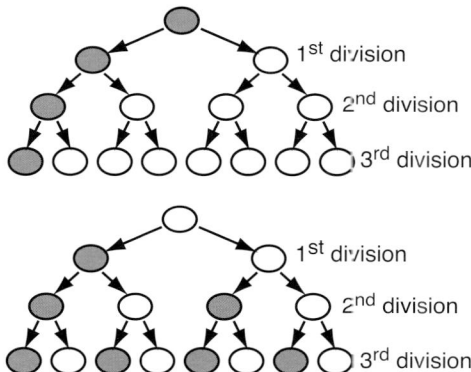

Figure 1. Two kinds of asymmetric divisions. (*Top*) Lineage-specific pattern of asymmetry seen in adult stem cells. Each division of the stem cell (*shaded*) gives rise to both a stem cell and progenitor cell (*white*). In contrast, progenitors give rise only to progenitors. This results in a single lineage of the stem cell in a given pedigree. This lineage contrasts with other patterns of asymmetry such as mating-type switching of *S. cerevisiae* shown below (*bottom*). In mating-type switching, a cell gives rise to both a switchable mother cell (*shaded*) and a nonswitching bud (*white*); however, this pattern is repeated at each division and therefore does not establish a single lineage.

that is derived from each spore (P. Thorpe et al., unpubl.). We have extended these studies and report the asymmetric segregation of a protein component of the spindle pole body, the yeast microtubule-organizing center. The kinetochore and spindle proteins direct segregation of sister chromatids during meiosis, and we discuss the implications of this phenotype with respect to potential nonrandom segregation of sister chromatids.

A NEW METHOD TO TRACK NONENCODED PROTEIN

Because it is an excellent model for many aspects of basic cellular functions, we set out to discover a lineage-specific asymmetric program in yeast. We reasoned that such a program may be established when a new yeast strain is generated. This happens only twice in the normal yeast life cycle: when two haploid cells fuse to form a diploid cell and when a diploid cell undergoes meiosis to generate four new haploid spores. We examined meiosis because the transition from a diploid to haploid cell allowed us to use a straightforward strategy to follow protein segregation. We created haploid strains that have a given gene fused to a sequence encoding either yellow fluorescent protein (YFP) or cyan fluorescent protein (CFP). Two haploids were mated to form a diploid strain that was "heteroallelically" tagged for individual genes (YFP/CFP). This diploid strain was induced to undergo meiosis, generating haploid spores that inherited either the YFP- or CFP-tagged allele. Importantly, however, these spores also inherited both of the tagged proteins from the diploid parent, one of which is no longer genetically encoded (the method is illustrated in Fig. 2). Using time-lapse fluorescence microscopy, we specifically assayed the fluorescent protein that was no longer genetically encoded. By assaying the levels of nonencoded fluorescent protein, we avoid the confounding effects of transcription in these studies.

SEGREGATION OF THE HISTONE HTA1

We first examined histone H2A, encoded by *HTA1*. For this analysis, we created a diploid strain that is heteroallelically tagged, *HTA1-YFP/HTA1-CFP*, and induced this strain to undergo meiosis. Figure 3 shows an example of the resulting time-lapse sequences; levels of fluorescence within the nucleus at each division were assessed for each cell and the dilution of the nonencoded (YFP-tagged) pro-

tein is evident. We found that the amount of nonencoded (or indeed total) fluorescence in the nucleus of the cells was equivalent for the mother and bud at each cell division. For nonencoded protein, we calculated the ratio of the fluorescence in the mother versus the bud—the mother-to-bud ratio (m/b)—to indicate if there was any deviation from parity. An m/b ratio of 1 indicates equal fluorescence in the mother and bud after cell division. We found that the m/b ratio of nonencoded Hta1 is 0.9 (standard error of the mean [S.E.M.] = 0.05, standard deviation [S.D.] = 0.27), indicating that approximately equal amounts of protein segregate to the mother and bud during cell division. Using the time-lapse sequences, we monitored all of the divisions from a single spore, allowing us to create a pedigree and examine each separate cell division in the lineage. Figure 4 shows the first three divisions of spores and includes, with each, the mean m/b ratio for Hta1. It is clear that the m/b ratio for Hta1 is the same in each of the cell divisions examined.

We found that other proteins also show the same pattern of segregation in the spore and its progeny. These include Rad52, a DNA-repair and recombination protein, and an artificially engineered construct that consists of the tetracycline repressor fused to red fluorescent protein (TetR-RFP), which binds to an array of Tet operator sequences located on yeast chromosome III (data not shown). The TetR-RFP fusion protein forms a focus specifically at a tandem array of operator sequences, and in this case, we specifically measured the fluorescence of each focus, rather than that of the whole nucleus.

ASYMMETRIC SEGREGATION OF KINETOCHORE PROTEINS

Interestingly, we found a different pattern of protein segregation when we examined four different kinetochore proteins. Ask1, Ctf19, Ndc10, and Mtw1 are each members of separate kinetochore subcomplexes (Cheeseman et al.

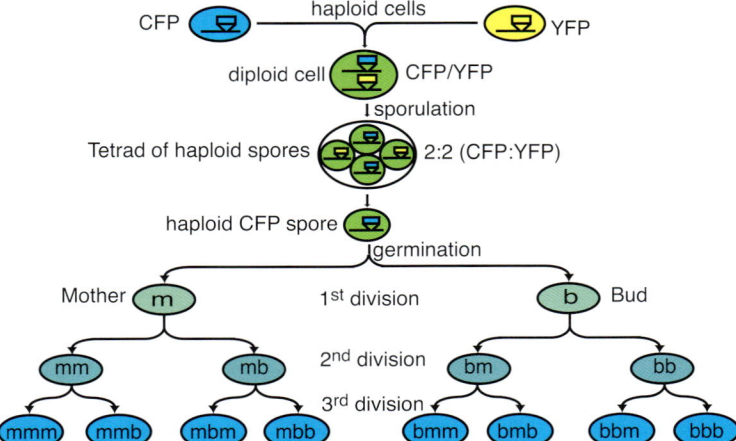

Figure 2. A labeling strategy for assaying nonencoded protein segregation. Two haploid strains containing the same gene tagged with different fluorescent proteins (YFP and CFP) are mated together to create a heteroallelically tagged strain. Because this diploid expresses both yellow- and blue-tagged protein, the cell is illustrated as *green*. This diploid is induced to undergo meiosis, producing four haploid spores that contain only one of the tagged alleles, but both proteins. An example of a spore inheriting the CFP-tagged allele is shown. As this cell undergoes successive divisions, the nonencoded yellow-tagged protein is diluted away and replaced with exclusively blue-tagged protein, illustrated as a transition from *green* to *blue*.

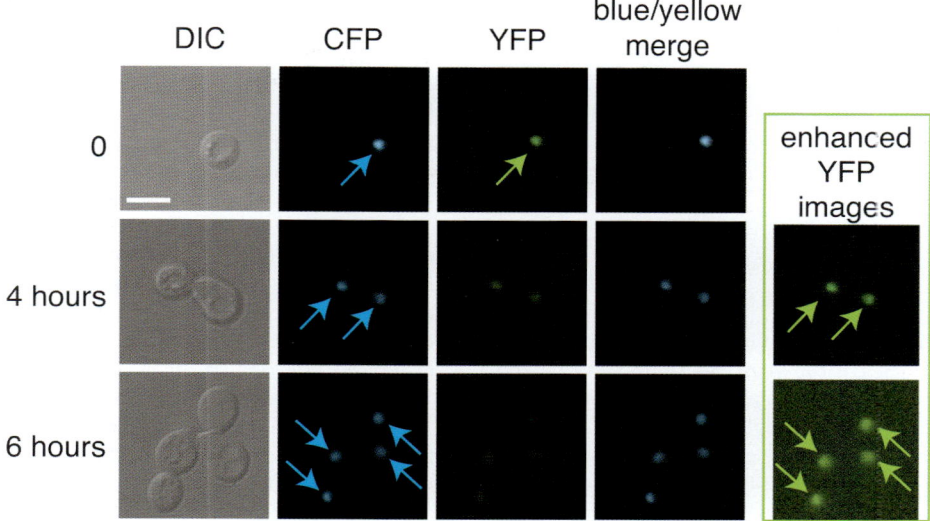

Figure 3. Microscopy images of a yeast spore encoding the tagged histone Hta1-CFP. Successive time-lapse images are arranged from top to bottom. The images in the first column are differential interference contrast (DIC) showing the spore and its progeny (bar, 5 μm). The next two columns, CFP and YFP, show the blue and yellow fluorescent channels, respectively, with the tagged histones in the nuclei of the cells indicated with arrows. These fluorescent images were deconvolved from a stack of vertically separated fluorescent exposures and are equivalently and uniformly enhanced. The fourth column shows a merge of the two fluorescent images. In the later time points (4 and 6 hours), the yellow-tagged histone becomes diluted, because it is no longer genetically encoded. However, to show that the protein is still detectable, in the fifth column, the YFP images have been artificially enhanced and the YFP-tagged histone indicated with arrows.

2001; De Wulf et al. 2003; Nekrasov et al. 2003; Pinsky et al. 2003; Bouck and Bloom 2005) and their concentrations can be quantified using fluorescent imaging (Joglekar et al. 2006). These proteins all localize to a single kinetochore focus in yeast, and it is the fluorescence of these foci that we used to calculate m/b ratios (as for the TetR-RFP analysis described above). All four of these proteins show an asymmetric pattern of segregation that is confined to the spore and the mother lineage derived from it (Fig. 4) (P.

Thorpe et al., unpubl.). On average, we see twice as much protein in the mother cell versus the bud in this asymmetric mother lineage. Bud cells do not show this asymmetry nor do mother cells that are not directly descended from the spore. For example, nonencoded Ndc10 has a mean m/b ratio of 2.5 in the mother lineage (S.E.M. = 0.31, S.D. = 1.8) compared with a mean m/b ratio of 1.2 in the other lineages (S.E.M. = 0.1, S.D. = 0.7). To assess whether this phenotype was specific to postmeiotic cells, we used time-lapse fluo-

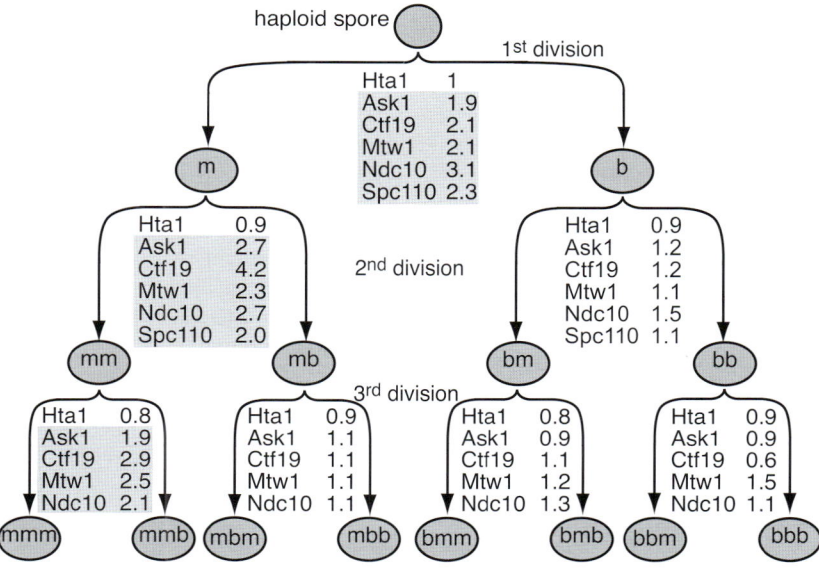

Figure 4. The mean mother-to-bud ratios (m/b) for different proteins at different postmeiotic cell divisions. The histone Hta1 divides symmetrically at each division, and the m/b ratios are approximately 1. In contrast, the kinetochore proteins (Ask1, Ctf19, Mtw1, and Ndc10) segregate asymmetrically only in a single mother-specific lineage derived from the spore, indicated as shaded values. Spc110, a spindle pole body protein, also shows this same pattern of asymmetry in the first two postmeiotic divisions that were measured.

rescence microscopy to monitor protein segregation in vegetatively growing haploid and diploid cells and did not see any asymmetry within the population. Consequently, this kinetochore asymmetry defines a single postmeiotic lineage originating at the spore and maintained within the mother lineage (Fig. 4). This phenomenon bears a striking resemblance to the asymmetric division pattern of adult stem cells (see Fig. 1, top).

ASYMMETRY OF SPINDLE POLE BODY SEGREGATION

Kinetochores are attached via microtubules to the spindle pole body (SPB), the yeast microtubule-organizing center, which is a structure equivalent to that of the metazoan centrosome. Interestingly, in *Drosophila*, centrosome asymmetry has been characterized in specific stem cell lineages (Rebollo et al. 2007; Rusan and Peifer 2007; Yamashita et al. 2007). In the latter study, Fuller and colleagues found that by labeling old pericentrin-AKAP450 centrosomal targeting (PACT) domain protein, mother centriole pairs tend to remain in male *Drosophila* germ cells during cell division. Therefore, we asked whether Spc110, a PACT domain SPB protein, segregates asymmetrically in the postmeiotic lineage in a manner similar to that seen in kinetochore proteins. In yeast, components of the SPB normally segregate asymmetrically in vegetatively growing haploid cells via a Cdc28-dependent mechanism (Pereira et al. 2001; Liakopoulos et al. 2003). SPBs replicate conservatively before mitosis, with the old SPB usually segregating to the bud during cell division. However, instead of the asymmetric pattern of segregation normally found in vegetatively growing cells, we see a pattern of segregation in spores that mirrors that of the kinetochore proteins (Fig. 4). So far, we have only examined the first two postmeiotic divisions for this protein and thus we do not yet know when the standard asymmetric pattern becomes established. However, it is interesting to note that the only time in the yeast life cycle when the SPB divides without an accompanying cycle of DNA replication is during meiosis II. The SPB asymmetry seen in haploid vegetatively growing cells is associated with Cdc28-dependent cell cycle progression (Liakopoulos et al. 2003). Thus, the atypical cell cycle during meiosis II may account for a disruption in the normal pattern of SPB asymmetry. It should also be noted that the W303 yeast strain used for these studies carries a mutation in *BUD4*, a gene involved in bud site selection (Voth et al. 2005) that may also disrupt the normal SPB segregation phenotype.

A MECHANISM TO SEGREGATE IMMORTAL DNA STRANDS

From our studies, we have uncovered a lineage-specific pattern of asymmetry in yeast. This phenotype mimics the divisions seen in adult stem cells and thus provides a new model with which to study asymmetric stem cell divisions. Perhaps more importantly, the asymmetric segregation of kinetochore proteins offers a possible mechanism to direct the nonrandom segregation of sister centromeres (and perhaps chromatids) to a single lineage.

It is clear that the two complementary strands of a DNA molecule are not equivalent. For example, only one strand of the DNA serves as a template for transcription of a given gene. We arbitrarily refer to the two strands of the DNA molecule as Watson and Crick. After DNA replication, the old (parental) Watson and Crick strands segregate into different daughter cells (Taylor et al. 1957; Meselson and Stahl 1958). The two newly synthesized DNA strands are complementary to the old strands; hence, the two sister chromatids have largely identical sequences. The segregation of the resulting sister chromatids is therefore unimportant because no genetic differences exist between the two chromatids; i.e., unless the old strands of DNA are different from the new, or to put it another way, the old Watson is different from the new Watson.

The distinction between old and new DNA strands, based on epigenetic marks, is well established. The dramatic importance of DNA methylation first became apparent through studies done with bacteriophage. Bacterial DNA methylation protects the "old" bacterial genome from restriction endonucleases that recognize and cleave, infecting "new" unmethylated phage DNA. Phage become resistant to restriction if their DNA is methylated at the appropriate sequences. Both restricted and resistant phage share the same DNA sequence, but epigenetic methylation determines their ability to infect their host (Arber and Dussoix 1962; Arber 1965; Murray 2000).

Methylation provides a mechanism by which the cell can distinguish between old and new DNA strands after DNA replication. The newly synthesized strands are unmethylated until a methylase acts upon them. This difference in the DNA strands is exploited in bacteria to correct errors that are introduced by polymerases into the newly synthesized strands during replication. After replication, the cells are able to differentiate between the new and old strands, allowing them to correct errors using only the old strand as a template. In bacteria, the temporary hemimethylated pattern of newly synthesized DNA signals that the methylated strand is the appropriate template for mismatch repair (Glickman and Radman 1980; Modrich and Lahue 1996). In eukaryotes, the nicks created by discontinuous DNA synthesis, particularly on the lagging strand, likely serve as a guide for the mismatch repair machinery (Kunkel and Erie 2005; Modrich 2006).

Another instance of DNA strand distinction occurs in fission yeast, because the pattern of mating-type switching is determined by an epigenetic mark on one strand of the DNA (Arcangioli et al. 2007; Klar 2007). This DNA strand difference is established by the direction of progression of the replication fork (which itself is asymmetric, i.e., leading- and lagging-strand synthesis), resulting in only one of the daughter cells containing the strand imprint. Only at the next division does one of the granddaughter cells switch mating type via a recombination-dependent mechanism.

Despite mismatch repair activity, it is possible for replication errors to persist and give rise to mutations within the genome. To counter such an accumulation of mutations in a defined lineage of stem cells, John Cairns (1975) hypothesized that stem cells could retain the old

Watson strand of DNA from each chromosome at each cell division (illustrated in Fig. 5, top). In this way, an ancestral strand of DNA from each chromosome would be retained in a single cell from one generation to the next; hence, the term for this model: the "immortal strand hypothesis." On the basis of this mechanism, replication errors would be passed on to progenitor cells, but the stem cell would always retain the original, unaltered Watson strand of DNA.

For such a model to be correct, a number of cellular requirements would have to be met. First, the cell would need a way to mark and detect the old DNA strand. Second, the cells would have to undergo an asymmetric cell division in which each chromatid that contains an ancestral DNA strand is segregated to the stem cell lineage. Finally, genetic recombination, primarily sister-chromatid exchange (SCE), would have to be suppressed to prevent the old and new DNA strands from becoming mixed.

TESTING THE IMMORTAL STRAND HYPOTHESIS

An obvious approach to test this hypothesis is to label the DNA (akin to the 1958 Meselson and Stahl experiment) and assay for retention of label in a single lineage. At the time when the immortal strand hypothesis was initially proposed, however, there was no method available to isolate pure populations of adult stem cells. Nevertheless, there was some evidence to support selective strand segregation in a number of organisms (Lark et al. 1966; Rosenberger and Kessel 1968; Priest and Shikes 1970). For example, using tritiated thymidine labeling in mouse epithelia of the tongue or intestine, it was possible to find retention of the labeled DNA in daughter cells in regions where stem cells resided (Potten et al. 1978). Experiments in budding yeast also identified nonrandom retention of labeled DNA in daughter cells (Williamson and Fennell 1981). However, a later experiment using the thymidine

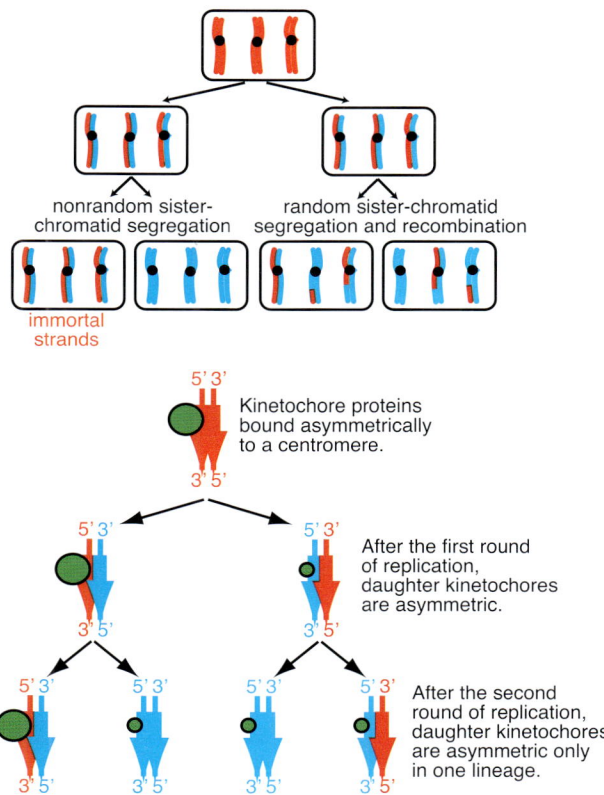

Figure 5. The immortal strand hypothesis and model for kinetochore asymmetry. (*Top*) The immortal strand hypothesis proposes that one original DNA strand of each chromosome segregates in a single lineage. A hypothetical haploid cell is displayed with three chromosomes each shown as double-stranded DNA. (*Red*) Initial DNA strands; (*blue*) newly synthesized DNA. If the hypothesis is correct, the original (*red*) strands will be cosegregated to a single lineage of (stem) cells (*lower left*). If not, unbiased chromatid segregation and recombination will randomly dilute DNA strands into progeny cells (*lower right*). (*Bottom*) Kinetochore sequences have the potential to direct strand asymmetry, because yeast centromeres can be considered to be unidirectional (*red arrows*). If a unique kinetochore structure (*large green circle*) was laid down on centromeres in meiosis, all of the chromosomes containing this structure could segregate to a single lineage in subsequent postmeiotic divisions. New kinetochore structures are indicated as *small green circles* bound to newly synthesized centromeric DNA (*blue arrows*) and these segregate randomly. This model provides an explanation for the pattern of kinetochore asymmetry that we find in yeast spores. Furthermore, this model could provide a means to segregate original centromeric DNA (*red*) to a single lineage, whereas newly synthesized DNA (*blue*) segregates randomly.

analog, bromodeoxyuridine as a DNA label, did not identify any asymmetry (Neff and Burke 1991).

Along with the ability to identify and study individual stem cell lines have come new tests of the immortal strand hypothesis, although these are not without their own shortcomings. One of the problems with any labeling strategy is that if cells do not divide, they will retain the DNA label irrespective of the segregation pattern of their DNA strands. Some recent studies have used independent means to verify that label-retaining cells are undergoing division (Conboy et al. 2007; Kiel et al. 2007). To date, the results of these studies have been mixed, with some groups reporting asymmetry, whereas others have found evidence to support a symmetric pattern of segregation; these data are reviewed elsewhere (Lansdorp 2007; Rando 2007). Alternatively, it is possible that not all of the chromosomes show selective strand segregation, in which case, a DNA-labeling strategy would be inconclusive. Indeed, there is genetic evidence in mouse stem cells for a chromosome-7-specific segregation pattern (Armakolas and Klar 2006).

Our observation of an asymmetrically dividing kinetochore in a single lineage of cells provides a means to test the selective segregation of sister chromatids. An underlying assumption is that centromere sequences direct strand asymmetry by binding to kinetochore proteins unidirectionally. We suspect that once the determinant of the asymmetry is established, it persists through subsequent cell divisions as illustrated in Figure 5 (bottom). The asymmetry may be determined by specific epigenetic marks on the centromeric DNA or it could be due to strand-specific binding of the kinetochore proteins themselves. It is possible that the direction of DNA replication through the centromere establishes a strand-specific imprint (Lew et al. 2008). A precedent for this latter type of imprinting exists in fission yeast mating-type switching (Dalgaard and Klar 1999). In any event, we predict that the cell "senses" orientation of the centromeres to facilitate asymmetric segregation. We are currently testing this model by engineering a dicentric chromosome. In yeast, dicentric chromosomes can be created and maintained by using a conditional centromere that can be reversibly activated (Hill and Bloom 1989). In vegetatively growing haploid cells, the two centromeres act independently: They either cosegregate to one daughter, resulting in a normal mitosis, or, they segregate to separate daughter cells, resulting in anaphase bridges and chromosome breakage (Brock and Bloom 1994). If the orientation of the centromere is the critical determinant of asymmetry, then in the postmeiotic mother lineage, cooriented centromeres on a dicentric chromosome will always cosegregate and not result in chromosome breakage (Fig. 6, left). Whereas if the orientation of one centromere is reversed resulting in opposing orientations, then chromosome breakage should occur in each asymmetric division (Fig. 6, right).

It is worth mentioning that the asymmetric segregation of kinetochore proteins in the postmeiotic mother lineage is not found in 100% of yeast spores. The frequency of kinetochore/spindle asymmetry is 45–69% in the mother lineage compared with 7–15% in the other lineages (Fig. 7). This variation may indicate that not all yeast spores behave the same way in terms of postmeiotic asymmetry. Indeed, it is possible that only two yeast spores from each tetrad show the asymmetric phenotype. This could be related to an unusual pattern of SPB asymmetry that is found during meiosis and results in the four spores of each tetrad having nonequivalent SPBs (Taxis et al. 2005).

IMMORTAL STRANDS ARE UNLIKELY TO BE FOUND IN YEAST

At this time, we have no evidence to suggest that the postmeiotic mother lineage selectively segregates DNA strands. Indeed, even if it did, the mother lineage in yeast is mortal; it ages and senesces after approximately 30–40 divisions (Mortimer and Johnston 1959). Furthermore, SCE would need to be suppressed in the mother lineage because SCE would result in crossing-over of the two sis-

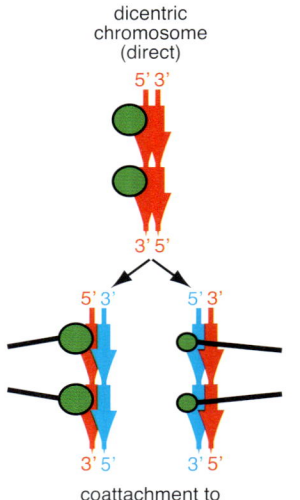

coattachment to
the same spindle pole

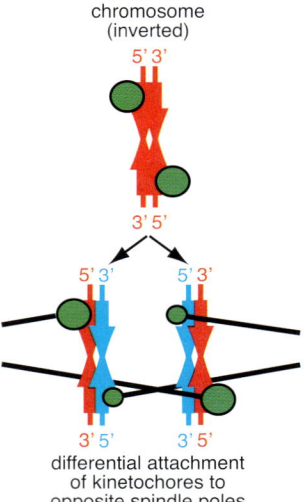

differential attachment
of kinetochores to
opposite spindle poles
results in chromosome
breakage

Figure 6. A dicentric chromosome test of nonrandom segregation. As in Figure 5 (bottom), a unique kinetochore complex is established on the centromeres unidirectionally in meiosis (*large green circle*). (*Left*) DNA replication of a direct dicentric chromosome (*top*) results in two chromatids, each of which containing both old (*red*) and new (*blue*) DNA strands (*below*). The meiotically induced kinetochore complexes are on the same sister chromatid after replication. Cosegregation of these complexes to a single postmeiotic mother lineage results in cosegregation of co-aligned centromeres (indicated by the orientation of spindle microtubules; *black lines*). (*Right*) However, if the centromere sequences are in inverted orientations, the meiosis-induced kinetochore structure will be bound to separate sister chromatids after DNA replication. In this case, if these kinetochore complexes cosegregate, both of the dicentric chromatids will likely break.

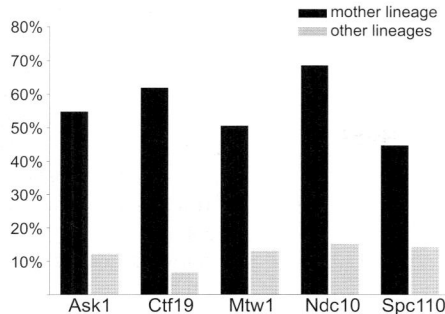

Figure 7. Proportion of asymmetric divisions for each of the asymmetrically dividing proteins that we studied. The proportion (%) of asymmetric divisions (defined as an m/b ratio greater than 1.6) is shown for the mother lineage (*black*) and the other lineages (*gray*).

ter-chromatid arms, thereby scrambling the old and new DNA strands in each chromatid (Fig. 5, top). Although SCE is difficult to measure, it is thought to occur frequently during G_2 and M phases in yeast (Kadyk and Hartwell 1992; Zou and Rothstein 1997; Gonzalez-Barrera et al. 2003). A reasonable surrogate for measuring SCE is the appearance of a DNA-repair focus that arises spontaneously during S phase (Lisby et al. 2001, 2003). We reasoned that if SCE were repressed, we might see suppression of DNA recombination foci in dividing spores and the mother cells derived from them. However, Rad52 foci do appear in these cells, indicating that recombination is not completely inactive (P. Thorpe et al., unpubl.). Together, these data argue against an immortal strand in postmeiotic yeast. However, kineotchore-driven asymmetry could selectively segregate centromeres and their closely linked sequences in the single postmeiotic lineage.

SELECTIVE CENTROMERE SEGREGATION

The selective segregation of centromere sequences could have an important role in speciation. For example, in metazoan diploids where the two parental species are diverging, each set of centromeres may have different efficiencies of kinetochore binding (Henikoff and Malik 2002). This difference is due to the coevolution of the centromere sequence together with its kinetochore-binding protein, thereby necessitating cosegregation of these components. During female meiosis, only one of the four products forms a gamete, and the other three polar bodies are lost. Hence, there is an opportunity to select for the cosegregation of one parental set of centromeres and its kinetochore-binding protein, a process akin to meiotic drive (Sandler and Novitski 1957; Henikoff et al. 2001; Pardo-Manuel de Villena and Sapienza 2001; Copenhaver 2004). Surprisingly, it appears that yeast centromeres are also evolving rapidly despite the absence of meiotic drive, because all four meiotic products are viable in yeast (Bensasson et al. 2008). Thus, the kinetochore asymmetry described here may enable a single lineage derived from a spore to select for "preferred" centromere sequences with a high affinity for their kinetochore-binding proteins. This in turn may explain the

divergence of yeast centromere sequences. Therefore, it is possible that the kinetochore asymmetry described here is involved in promoting evolutionary divergence. Yeast provides one of the few organisms in which such a model could be tested practically.

CONCLUSIONS AND PERSPECTIVES

Yeast is a powerful model system due in part to a large array of experimental techniques and has consequently been used to characterize many basic cell biological pathways, such as those involved in cell cycle control and genetic recombination. We have characterized a lineage-specific pattern of asymmetry that mimics that seen in adult stem cells. This type of asymmetric division is now amenable to study in yeast. A first step toward further study of this phenomenon will be a high-throughput microscopy screen to identify the genes that are required for this postmeiotic asymmetric pattern. This set of genes is likely to include those that control the lineage-specific asymmetric segregation pattern of the yeast kinetochore. We predict that these will be conserved and will be key regulators of the pattern of asymmetric division that occurs during development and in adult stem cells.

ACKNOWLEDGMENTS

We thank Argiris Efstratiadis and Rudy Leibel for motivating our interest in cellular asymmetry. Phil Heiter, Trisha Davis, Kerry Bloom, Tim Bestor, Kara Bernstein, Michael Lisby, Rebecca Burgess, Bob Reid, Qi Feng, and Jackie Barlow provided strains and/or helpful advice and suggestions on this manuscript. This work was supported by the Russ Berrie Diabetes Foundation and National Institutes of Health grants GM50237 and GM67055.

REFERENCES

Arber, W. 1965. Host specificity of DNA produced by *Escherichia coli*. V. The role of methionine in the production of host specificity. *J. Mol. Biol.* **11:** 247–256.

Arber, W. and Dussoix, D. 1962. Host specificity of DNA produced by *Escherichia coli*. I. Host controlled modification of bacteriophage λ. *J. Mol. Biol.* **5:** 18–36.

Arcangioli, B., Roseaulin, L., and Holmes, A. 2007. Mating-type switching in *S. pombe*. In *Molecular genetics of recombination* (ed. A. Aguilera and R. Rothstein), pp. 251–275. Spinger-Verlag, Berlin.

Armakolas, A. and Klar, A.J. 2006. Cell type regulates selective segregation of mouse chromosome 7 DNA strands in mitosis. *Science* **311:** 1146–1149.

Bensasson, D., Zarowiecki, M., Burt, A., and Koufopanou, V. 2008. Rapid evolution of yeast centromeres in the absence of drive. *Genetics* **178:** 2161–2167.

Bouck, D. and Bloom, K. 2005. The role of centromere-binding factor 3 (CBF3) in spindle stability, cytokinesis, and kinetochore attachment. *Biochem. Cell Biol.* **83:** 696–702.

Brock, J.A. and Bloom, K. 1994. A chromosome breakage assay to monitor mitotic forces in budding yeast. *J. Cell Sci.* **107:** 891–902.

Cairns, J. 1975. Mutation selection and the natural history of cancer. *Nature* **255:** 197–200.

Cheeseman, I.M., Brew, C., Wolyniak, M., Desai, A., Anderson, S., Muster, N., Yates, J.R., Huffaker, T.C., Drubin, D.G., and Barnes,

G. 2001. Implication of a novel multiprotein Dam1p complex in outer kinetochore function. *J. Cell Biol.* **155:** 1137–1145.

Conboy, M.J., Karasov, A.O., and Rando, T.A. 2007. High incidence of nonrandom template strand segregation and asymmetric fate determination in dividing stem cells and their progeny. *PLoS Biol.* **5:** e102.

Copenhaver, G.P. 2004. Who's driving the centromere? *J. Biol.* **3:** 17.

Dalgaard, J.Z. and Klar, A.J. 1999. Orientation of DNA replication establishes mating-type switching pattern in *S. pombe*. *Nature* **400:** 181–184.

De Wulf, P., McAinsh, A.D., and Sorger, P.K. 2003. Hierarchical assembly of the budding yeast kinetochore from multiple subcomplexes. *Genes Dev.* **17:** 2902–2921.

Glickman, B.W. and Radman, M. 1980. *Escherichia coli* mutator mutants deficient in methylation-instructed DNA mismatch correction. *Proc. Natl. Acad. Sci.* **77:** 1063–1067.

Gonzalez-Barrera, S., Cortes-Ledesma, F., Wellinger, R.E., and Aguilera, A. 2003. Equal sister chromatid exchange is a major mechanism of double-strand break repair in yeast. *Mol. Cell* **11:** 1661–1671.

Henikoff, S. and Malik, H.S. 2002. Centromeres: Selfish drivers. *Nature* **417:** 227.

Henikoff, S., Ahmad, K., and Malik, H.S. 2001. The centromere paradox: Stable inheritance with rapidly evolving DNA. *Science* **293:** 1098–1102.

Hill, A. and Bloom, K. 1989. Acquisition and processing of a conditional dicentric chromosome in *Saccharomyces cerevisiae*. *Mol. Cell. Biol.* **9:** 1368–1370.

Joglekar, A.P., Bouck, D.C., Molk, J.N., Bloom, K.S., and Salmon, E.D. 2006. Molecular architecture of a kinetochore-microtubule attachment site. *Nat. Cell Biol.* **8:** 581–585.

Kadyk, L.C. and Hartwell, L.H. 1992. Sister chromatids are preferred over homologs as substrates for recombinational repair in *Saccharomyces cerevisiae*. *Genetics* **132:** 387–402.

Kiel, M.J., He, S., Ashkenazi, R., Gentry, S.N., Teta, M., Kushner, J.A., Jackson, T.L., and Morrison, S.J. 2007. Haematopoietic stem cells do not asymmetrically segregate chromosomes or retain BrdU. *Nature* **449:** 238–242.

Klar, A.J. 2007. Lessons learned from studies of fission yeast mating-type switching and silencing. *Annu. Rev. Genet.* **41:** 213–236.

Kunkel, T.A. and Erie, D.A. 2005. DNA mismatch repair. *Annu. Rev. Biochem.* **74:** 681–710.

Lansdorp, P.M. 2007. Immortal strands? Give me a break. *Cell* **129:** 1244–1247.

Lark, K.G., Consigli, R.A., and Minocha, H.C. 1966. Segregation of sister chromatids in mammalian cells. *Science* **154:** 1202–1205.

Lew, D.J., Burke, D.J., and Dutta, A. 2008. The immortal strand hypothesis: How could it work? *Cell* **133:** 21–23.

Liakopoulos, D., Kusch, J., Grava, S., Vogel, J., and Barral, Y. 2003. Asymmetric loading of Kar9 onto spindle poles and microtubules ensures proper spindle alignment. *Cell* **112:** 561–574.

Lisby, M., Rothstein, R., and Mortensen, U.H. 2001. Rad52 forms DNA repair and recombination centers during S phase. *Proc. Natl. Acad. Sci.* **98:** 8276–8282.

Lisby, M., Mortensen, U.H., and Rothstein, R. 2003. Colocalization of multiple DNA double-strand breaks at a single Rad52 repair centre. *Nat. Cell Biol.* **5:** 572–577.

Long, R.M., Singer, R.H., Meng, X., Gonzalez, I., Nasmyth, K., and Jansen, R.P. 1997. Mating type switching in yeast controlled by asymmetric localization of *ASH1* mRNA. *Science* **277:** 383–387.

Meselson, M. and Stahl, F.W. 1958. The replication of DNA in *Escherichia coli*. *Proc. Natl. Acad. Sci.* **44:** 671–682.

Modrich, P. 2006. Mechanisms in eukaryotic mismatch repair. *J. Biol. Chem.* **281:** 30305–30309.

Modrich, P. and Lahue, R. 1996. Mismatch repair in replication fidelity, genetic recombination, and cancer biology. *Annu. Rev. Biochem.* **65:** 101–133.

Mortimer, R.K. and Johnston, J.R. 1959. Life span of individual yeast cells. *Nature* **183:** 1751–1752.

Murray, N.E. 2000. Type I restriction systems: Sophisticated molecular machines (a legacy of Bertani and Weigle). *Microbiol. Mol. Biol. Rev.* **64:** 412–434.

Neff, M.W. and Burke, D.J. 1991. Random segregation of chromatids at mitosis in *Saccharomyces cerevisiae*. *Genetics* **127:** 463–473.

Nekrasov, V.S., Smith, M.A., Peak-Chew, S., and Kilmartin, J.V. 2003. Interactions between centromere complexes in *Saccharomyces cerevisiae*. *Mol. Biol. Cell* **14:** 4931–4946.

Pardo-Manuel de Villena, F. and Sapienza, C. 2001. Female meiosis drives karyotypic evolution in mammals. *Genetics* **159:** 1179–1189.

Pereira, G., Tanaka, T.U., Nasmyth, K., and Schiebel, E. 2001. Modes of spindle pole body inheritance and segregation of the Bfa1p-Bub2p checkpoint protein complex. *EMBO J.* **20:** 6359–6370.

Pinsky, B.A., Tatsutani, S.Y., Collins, K.A., and Biggins, S. 2003. An Mtw1 complex promotes kinetochore biorientation that is monitored by the Ipl1/Aurora protein kinase. *Dev. Cell* **5:** 735–745.

Potten, C.S., Hume, W.J., Reid, P., and Cairns, J. 1978. The segregation of DNA in epithelial stem cells. *Cell* **15:** 899–906.

Priest, J.H. and Shikes, R.H. 1970. Distribution of labeled chromatin. I. M 1 and M 2 anaphases of diploid and tetraploid cultured mammalian cells. *J. Cell Biol.* **47:** 99–106.

Rando, T.A. 2007. The immortal strand hypothesis: Segregation and reconstruction. *Cell* **129:** 1239–1243.

Rebollo, E., Sampaio, P., Januschke, J., Llamazares, S., Varmark, H., and Gonzalez, C. 2007. Functionally unequal centrosomes drive spindle orientation in asymmetrically dividing *Drosophila* neural stem cells. *Dev. Cell* **12:** 467–474.

Rosenberger, R.F. and Kessel, M. 1968. Nonrandom sister chromatid segregation and nuclear migration in hyphae of *Aspergillus nidulans*. *J. Bacteriol.* **96:** 1208–1213.

Rusan, N.M. and Peifer, M. 2007. A role for a novel centrosome cycle in asymmetric cell division. *J. Cell Biol.* **177:** 13–20.

Sandler, I. and Novitski, E. 1957. Meiotic drive as an evolutionary force. *Am. Nat.* **91:** 105–110.

Taxis, C., Keller, P., Kavagiou, Z., Jensen, L.J., Colombelli, J., Bork, P., Stelzer, E.H., and Knop, M. 2005. Spore number control and breeding in *Saccharomyces cerevisiae*: A key role for a self-organizing system. *J. Cell Biol.* **171:** 627–640.

Taylor, J.H., Woods, P.S., and Hughes, W.L. 1957. The organization and duplication of chromosomes as revealed by autoradiographic studies using tritium-labeled thymidine. *Proc. Natl. Acad. Sci.* **43:** 122–128.

Voth, W.P., Olsen, A.E., Sbia, M., Freedman, K.H., and Stillman, D.J. 2005. *ACE2*, *CBK1*, and *BUD4* in budding and cell separation. *Eukaryot. Cell* **4:** 1018–1028.

Williamson, D.H. and Fennell, D.J. 1981. Nonrandom assortment of sister chromatids in yeast mitosis. *Alfred Benzon Symp.* **16:** 89–107.

Yamashita, Y.M., Mahowald, A.P., Perlin, J.R., and Fuller, M.T. 2007. Asymmetric inheritance of mother versus daughter centrosome in stem cell division. *Science* **315:** 518–521.

Zou, H. and Rothstein, R. 1997. Holliday junctions accumulate in replication mutants via a RecA homolog-independent mechanism. *Cell* **90:** 87–96.

The Maternal to Zygotic Transition in Animals and Plants

C. Baroux,*[†] D. Autran,‡[†] C.S. Gillmor,§ D. Grimanelli,‡ and U. Grossniklaus*

*Institute of Plant Biology & Zürich-Basel Plant Science Center, University of Zürich, CH-8008 Zürich,
Switzerland; ‡IRD, Institut de Recherche pour le Développement, UMR 5096, BP 56501, 34394 Montpellier
Cedex 5, France; §Department of Biology, University of Pennsylvania, Philadelphia, Pennsylvania 19104

In the animal kingdom, maternal control of early development is a common feature. The onset of zygotic control over early development, defined as the maternal to zygotic transition (MZT), follows fertilization with a delay of a variable number of cell divisions, depending on the species. The MZT has been well defined in animals, but investigations remain in their infancy in plants. Recent evidence suggests, however, that in plants as in animals, the MZT also occurs several division cycles after fertilization. The likely convergent evolution of the MZT in the animal and plant kingdoms is fascinating and raises major questions regarding its biological significance, particularly with regard to its importance in genome reprogramming and the acquisition of totipotency by the embryo.

Plants and animals have each evolved different reproductive strategies that determine the extent to which the parents, usually the female, exert control over the developing offspring. Fertilization can be external or internal. In the latter case, the fertilized egg can be laid on a nutritive substrate (e.g., insects) or it develops within a protective and nurturing matrix (the placenta of mammals, the egg of birds and reptiles, the seed of seed plants). The MZT marks the end of maternal control and the onset of zygotic control over embryo development. It is characterized by the simultaneous degradation of stored maternal products and zygotic genome activation (Andéol 1994; Schultz 2002; Tadros et al. 2003; Stitzel and Seydoux 2007). In animals, the MZT occurs several cell cycles after fertilization. As a consequence, the first stages of embryonic development essentially rely on maternally deposited products. Despite divergent reproductive strategies and differences in the parental provisioning for offspring development, current evidence indicates that plants and animals share a marked maternal control of early postfertilization development. We review this evidence and our knowledge regarding the mechanisms controlling the onset of the MZT in both animals and plants. We hypothesize that shared biological constraints on early embryo development explain the existence of convergent mechanisms for maintaining maternal predominance.

EARLY EMBRYOGENESIS IN ANIMALS AND HIGHER PLANTS

Embryogenesis in flowering plants and animals has in common the union of a male and a female gamete that produces the zygote and the sequential events of cell proliferation, morphogenesis, and organogenesis that follow (Fig. 1). In animals, a single female gamete, the oocyte, fuses with a motile sperm to produce the zygote. Cell proliferation generates a multicellular morula of 16–32 cells

in mammals, a blastula of approximately 30,000 cells in *Xenopus,* and a syncytial blastoderm of about 6000 nuclei in *Drosophila*. Cellularization (for the syncytial blastoderm of insects) and asymmetric divisions define the onset of morphogenesis, and cellular migration marks the beginning of gastrulation and organogenesis (Browder et al. 1991).

In flowering plants, two pairs of gametes fuse, a process termed double fertilization. The two female gametes, the egg and central cell, are each fertilized by a sperm to produce the embryo and the endosperm, respectively. The endosperm is a protective and nurturing tissue that has a role similar to that of the placenta in eutherian mammals (Harper et al. 1970). The female gametes are produced together with "accessory" cells (the synergids and antipodals) within the female gametophyte (embryo sac), which is enclosed in the ovule that, after fertilization, forms the seed. Ovule and seed development occurs in the gynoecium in the center of the flower. The two sperms are produced by the male gametophyte (pollen), which germinates and grows through tissues of the gynoecium to deliver the sperm cells to the embryo sac. In plants, the early events of embryo development and pattern formation have been best described in the model plant *Arabidopsis thaliana*. In contrast to the situation in animals, the first division of the zygote is asymmetric and produces an apical cell, giving rise to the embryo proper, and a basal cell, forming the embryonic suspensor and the hypophysis, which will contribute to the root meristem of the embryo (Park and Harada 2008). The apical and basal cells are distinguished not only by their shape, but also by the differential expression of key regulatory genes (Breuninger et al. 2008). The apical cell then divides symmetrically until the fourth, asymmetric cleavage, which establishes the protoderm, the precursor of the epidermis. Further asymmetrical divisions will set up the apical-basal and radial polarity of the globular-stage embryo that contains about 100 cells. The transition from radial to bilateral symmetry occurs later with the formation of the early heart-stage embryo (Jürgens 1992; Park and Harada 2008). The embryo and endosperm of flowering

---

[†]These authors contributed equally to this work.

mouse

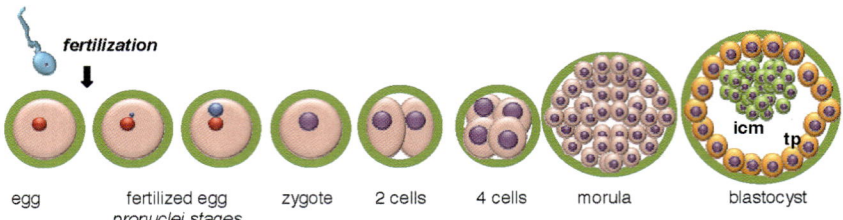

egg | fertilized egg *pronuclei stages* | zygote | 2 cells | 4 cells | morula | blastocyst

Arabidopsis

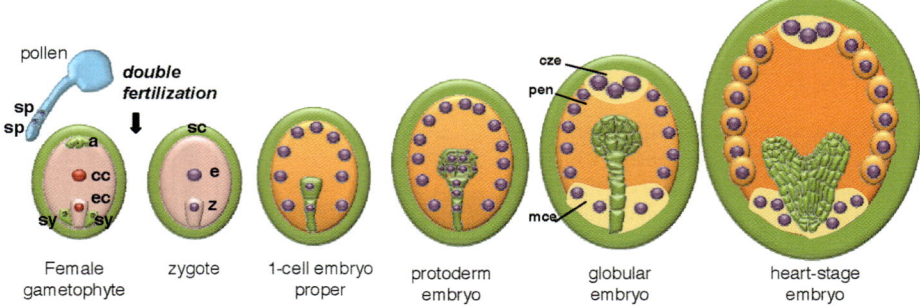

Female gametophyte | zygote | 1-cell embryo proper | protoderm embryo | globular embryo | heart-stage embryo

Figure 1. Early development in animals and plants using the mouse and *Arabidopsis* as models. Sexual reproduction involves dimorphic gametes in both animals and plants, which are highly differentiated. The products of fertilization will undergo several cleavage cycles before morphogenesis. These stages of early development are largely under maternal control. In mammals (e.g., the mouse) symmetrical cleavages form a morula, whereas further asymmetrical cleavages and positioning of daughter cells separate the inner cell mass (icm) from the peripheral trophectoderm (tp), being precursors of the embryo and the placenta, respectively. In flowering plants (e.g., *Arabidopsis*), double fertilization takes place. It involves two sperm cells (sp) delivered by the pollen to the female gametophyte. The latter consists of seven cells, two of which are the female gametes (the egg [ec] and the central cell [cc]) and the others (synergids [syn] and antipodals [a]) are accessory cells. Double fertilization produces the zygote (z) and the primary endosperm nucleus (e), precursors of the embryo (*green*) and the endosperm (*orange*), respectively. Seed stages are defined according to the embryo stage (protoderm, globular, and heart stages only are represented here). The endosperm develops initially as a syncytium. Three mitotic domains are established at the globular stage and are positioned along the anteroposterior axis (chalazal endosperm [cze], peripheral endosperm [pen], micropylar endosperm [mce]). The peripheral endosperm becomes cellularized when the embryo reaches the early heart stage.

plants develop simultaneously and in coordination but along distinct developmental pathways. In *Arabidopsis*, the endosperm undergoes about ten nuclear division cycles within a syncytium before cellularization occurs and three endosperm domains differentiate along the anteroposterior axis (Brown et al. 1999; Boisnard-Lorig et al. 2001).

MATERNAL CONTROL OF EARLY DEVELOPMENT AND THE MATERNAL TO ZYGOTIC TRANSITION

Defining the Maternal to Zygotic Transition

Early development in animals is under maternal control. This was dramatically illustrated by enucleation experiments first conducted on sea urchins by Harvey in 1936. When induced by seawater, enucleated eggs are able to undergo normal cleavages and form plutei, their free swimming larval form (Harvey 1936). Thus, the cleavage stage of the sea urchin relies solely on maternally stored products and does not require the expression of the zygotic genome until the larvae metamorphose into the adult form.

Since Harvey's experiments, it has been found that early development of many animal species—evolutionarily as divergent as echinoderms, amphibians, fishes, worms, insects, birds, and mammals—also relies on maternally deposited products (Andéol 1994). The stage at which maternal reliance ends and control of embryo development is transferred to the zygotic genome is referred to as the MZT. The MZT, under this definition, is distinct from the stage of zygotic genome activation (ZGA), which corresponds to the onset of de novo transcription from the zygotic genome. ZGA and MZT can coincide, but this is not always the case (Table 1). The MZT is also distinct from the midblastula transition (MBT), a developmental transition in amphibians and fishes associated with cell cycle lengthening, asynchronous cleavage, and the acquisition of cell motility, which, however, does not require zygotic transcription (Newport and Kirschner 1982a). The distinction is not always clear in the literature, and the MBT and ZGA are often referred to as marking the MZT, although this is clearly not true in all organisms. In this review, we adhere to a strict distinction as defined above.

Importantly, ZGA is not sufficient to ensure control by

Table 1. Maternal to zygotic transition and zygotic gene activation in plants and animals

Organism	ZGA first detection	major ZGA	MZT	References
Animals				
Sea urchin	zygote, paternal pronucleus	zygote	pluteus (larvae)	Harvey (1936); Poccia et al. (1985)
C. elegans	four cells	90–125 cells	gastrulation (~28 cells)	Schauer and Wood (1990); Edgar et al. (1994); Seydoux and Fire (1994)
Xenopus	pre-MBT (~256 cells)	MBT (~4000 cells)	blastula (~30,000 cells)	Newport and Kirschner (1982b); Yang et al. (2002)
Drosophila	cleavage 8	cleavage 14 (~6000 cells)	blastoderm	Robbins (1980); Pritchard and Schubiger (1996)
Zebra fish	pre-MBT (~64/128 cells)	MBT	at epiboly	Kane et al. (1996); Mathavan et al. (2005)
Mouse	zygote, paternal pronucleus	two-cell embryo	two-cell embryo	Schultz (2002); Zeng et al. (2004)
Plants				
Arabidopsis	two-cell embryo	globular embryo	globular embryo?	Vielle-Calzada et al. (2000); Baroux et al. (2001); Weijers et al. (2001)
Maize	zygote	globular embryo	n.d.	Grimanelli et al. (2005); Meyer and Scholten (2007)
Maize IVF	zygote	n.d.	n.d.	Dresselhaus et al. (1996, 1999); Leduc et al. (1996) Scholten et al. (2002)
Wheat	two-cell proembryo	n.d.	n.d.	Sprunck et al. (2005)
Tobacco	zygote	n.d.	n.d.	Ning et al. (2006)

ZGA is gradual in animals and plants and can occur concomitantly or earlier than the MZT, as assessed by enucleation experiments (sea urchin), pharmacological treatments (inhibition of RNA pol II with α-amanitin), mutant analyzes, or both (other organisms). n.d. indicates not determined; IVF, in vitro fertilization.

the zygotic genome. Although ZGA is a necessary condition, it is not sufficient because the MZT also requires that the maternally stored products no longer influence the further development of the embryo. Regulation of the MZT must thus depend on an appropriate balance between maternal mRNA clearance and ZGA.

Timing of the Maternal to Zygotic Transition in Plants and Animals

The MZT was originally defined using α-amanitin, an inhibitor of RNA polymerase II and III. Application on fertilized eggs or later stages blocks zygotic transcription and the induced developmental arrest marks the MZT. In worms, echinoderms, amphibians, fishes, and insects, the MZT takes place after the entire cleavage period is completed. In particular, in amphibians and fishes, the MZT occurs after the MBT, only at the onset of the first morphogenetic event (Newport and Kirschner 1982a; Stroband et al. 1992; Kane et al. 1996). In contrast, mammalian embryos require zygotic transcription already at the two-cell stage (Andéol 1994). The MZT can take place after as few as one cycle (two-cell mouse embryo) or about 15 cell cycles (Xenopus blastula). On an absolute timescale, the MZT occurs between 2 hours (Drosophila nuclear blastoderm) and 2 days (four- to eight-cell human embryo) after fertilization. Thus, the extent of maternal control and the timing of the MZT vary greatly among species.

The timing of the MZT can also be defined genetically. Because the MZT occurs a certain number of division cycles after fertilization, mutations affecting early development in animals often display maternal effects. Mutations showing a maternal effect can affect a gene product stored in the egg whose function is required between fertilization and the MZT. Alternatively, maternal effects can result from genomic imprinting, as is the case for genes that are zygotically expressed but with only the maternal allele being active and the paternal allele remaining silent. Genetic characterization coupled with parental allele-specific expression analyses is thus required to distinguish the different mechanisms underlying maternal effects. Maternal-effect mutants with a developmental defect before the MZT are informative for the role of maternal factors in early embryo development, for example, in chromatin remodeling and pronuclear congression (Loppin et al. 2001; Dekens et al. 2003) or the definition of body axes (Moody et al. 1996; Pelegri 2003; Mtango et al. 2008). The mutant class informative for zygotic gene contribution after the MZT consists of zygotic maternal mutants, which display a maternal phenotype that is either rescuable by the wild-type paternal allele (recessive mutant) or is enhanced by a mutant paternal allele (dominant mutant). In zebra fish, zygotic maternal mutants have been isolated that show a developmental arrest or delay of epiboly, the first morphogenetic event and the stage of the MZT (Kane et al. 1996), or they affect cell-fate determination processes at later stages of development (Pelegri 2003). Similarly in Drosophila, cell division and patterning genes acting maternally and zygotically have been uncovered in genetic screens (Garcia-Bellido and Robbins 1983; Perrimon et al. 1989, 1996).

Because of their inaccessibility, the plant embryo and

endosperm are not amenable to pharmacological experiments, such as inhibition of transcription by α-amanitin. Thus, genetic approaches have been used to investigate the MZT. However, the analysis of maternal-effect mutants in plants is complicated by the fact that maternal influences on seed development can stem from several possible sources (for review, see Grossniklaus and Schneitz 1998; Chaudhury et al. 2001; Baroux et al. 2002): (1) the intimate relationship of the fertilization products with the surrounding maternal tissues (sporophytic effects), (2) the dominant cytoplasmic contribution of female versus male gametes (gametophytic effects), (3) non-cell-autonomous effects of the endosperm on embryo development, and (4) genomic imprinting as observed in mammals. Thus, genetic characterization, together with thorough expression analyses, has to dissect the contribution of each effect to the observed mutant phenotype.

In plants, very few maternal-effect mutations are informative for the MZT, mostly because relatively few experiments have been conducted to specifically target maternal effects. Thus, very little is known about the nature and role of maternal factors stored in the female gametes. PROLIFERA, an MCM7 protein, is encoded by a maternal-effect gene important for the cell divisions following fertilization (Springer et al. 1995, 2000), but the low penetrance of the mutation suggests a redundant function with other factors. Interestingly, mutations in ORC2, a subunit of the origin recognition complex that, like MCM7, participates in the formation of the prereplicative complex, are suppressed by the maternal-effect mutation *medea*, further illustrating the interplay of maternal and zygotic factors early in development (Collinge et al. 2004). *Arabidopsis* MSI1 (MULTISUPPRESSOR of IRA1), which, among others, interacts with the RETINOBLASTOMA-RELATED (RBR) protein regulating the initiation of replication (Ach et al. 1997), is required in the female gametophyte, and not zygotically, to maternally control embryo and endosperm proliferation (Leroy et al. 2007).

Two genetic screens aimed at the isolation of gametophytic mutations identified a number of maternal-effect mutants affecting embryo development at various, often early, stages (Moore 2002; Pagnussat et al. 2005). As outlined above, it is not easy to determine the underlying mechanisms of a maternal effect in plants, and the mode of action of the genes affected in these mutants is currently not known. Large numbers of recessive mutations with zygotic effects have also been isolated, with the embryo phenotype only observed when both parental alleles are defective (Tzafrir et al. 2004). Some of them can affect embryo development at very early stages, including the zygote (Weijers et al. 2001; Lukowitz et al. 2004; Wu et al. 2007; Ronceret et al. 2008). This observation indicates that the paternal genome can complement for a deficient maternal contribution. Although it is generally assumed that such a paternal rescue stems from transcription of the paternally contributed genome, it could also be contributed by the sperm's cytoplasm. In fact, *Arabidopsis* sperm cells were recently shown to contain a distinct and diverse transcript profile (Borges et al. 2008). Moreover, zygotically acting mutants provide limited information regarding the MZT, because the timing of the paternal rescue is generally unknown. This has rarely been investigated but is illustrated by the mutants *gnom/emb30* and *vacuoleless1* (*vcl1*), which were both classified as recessive zygotic embryo lethals. Both heterozygous mutants produce approximately 25% aborted seeds when screened at a late stage of fruit development (Mayer et al. 1993; Rojo et al. 2001). However, at early stages of embryo development (e.g., two- to four-cell stage), mutant phenotypes are observed at similar frequencies whether self-fertilized or crossed to wild-type pollen (Table 2) (Vielle-Calzada et al. 2000). This indicates that at the two- to four-cell stage, a paternally inherited wild-type allele cannot provide enough activity to rescue the mutant phenotype. Thus, paternal rescue of the maternal mutation occurs later during embryo development. The time course analysis performed for the *vcl1*

Table 2. Delayed paternal rescue of *vacuoleless1* and *gnom/emb30*

	Two–four cells (2 dap)	Globular (3 dap)	Heart (4 dap)	Late heart (5 dap)	Mature seed
vacuoleless1/vcl1					
vcl1/VCL1 selfed	53% (n = 243)	25% (n = 271)	16% (n = 290)	20% (n = 287)	25% (n = 999)
vcl1/VCL1 × VCL1/VCL1	38% (n = 258)	12% (n = 232)	1% (n = 210)	0% (n = 318)	
VCL1/VCL1 × vcl1/VCL1	3% (n = 137)	1% (n = 171)	1% (n = 119)	0% (n = 137)	
gnom/emb30					
emb30/EMB30 selfed	12% (n = 131)				25% (n = ?)*
emb30/EMB30 × EMB30/EMB30	13% (n = 116)				0% (n = 157)
EMB30/EMB30 × emb30/EMB30	0% (n = 53)				

vcl1 and wild-type plants were pollinated 2 days after emasculation, and embryos were scored at 2, 3, 4, and 5 days after pollination (dap); seeds were cleared as described previously (Rojo et al. 2001). The frequency of *emb30-3* mutant embryos is cited from Vielle-Calzada et al. (2000) except for (*), which is cited from www.arabidopsis.org (lethal phenotype curated by ABRC). The percentage of mutant embryos is shown, with the total number examined in parentheses. For each cross, the female genotype is listed first.

mutation (Table 2) suggests that the maternally driven mutant phenotype is gradually rescued by the paternal allele and that the MZT in *Arabidopsis* takes place between the two- to four-cell stage and the globular stage. Additional reassessments of existing recessive embryo lethal mutants arrested at late stages would allow defining the MZT with more precision.

ZYGOTIC GENE ACTIVATION

Zygotic Genome Activation Precedes the Maternal to Zygotic Transition and Is Gradual

ZGA was initially defined as the stage where a large increase in de novo RNA synthesis could be measured, for instance, following the incorporation of radioactive-labeled uridine or adenosine (Zalokar 1976; Clegg and Piko 1977). However, it was early recognized that specific transcripts such as α-histone mRNA, selective species of rRNA or tRNA could be synthesized de novo before the major ZGA (Anderson and Lengyel 1980). Furthermore, microinjection of reporter genes, as done in *Xenopus* and mouse embryos (Newport and Kirschner 1982b; Telford et al. 1990), elegantly demonstrated the developmental acquisition of transcriptional capacity before the major ZGA. Since then, additional examples of zygotic genes expressed before the major ZGA in *Drosophila*, but also in *Xenopus*, zebra fish, mouse, and *Caenorhabditis elegans,* established that ZGA is a gradual process (see Table 1). Recent profiling experiments comparing α-amanitin-sensitive and α-amanitin-insensitive genes (differentiating de novo from maternally deposited transcripts) confirmed that a large number of zygotic genes are already active at the one-cell stage in the mouse embryo with a selective, although abundant, activation of genes involved in transcription and RNA processing at the two-cell stage (Zeng et al. 2004). Such early expressed genes are likely to have a specific role in the MZT itself. This view is supported by the observation that in *Drosophila*, but also in *Xenopus* and *C. elegans*, early inhibition of zygotic gene transcription (using α-amanitin) prolongs maternally driven embryonic development in comparison to blocking the major ZGA by a treatment just before the MZT (for review, see Andéol 1994). It was suggested that early expressed genes, such as the *Drosophila string* gene, may also have an important role in controlling the degradation of maternally stored transcripts (Edgar and Lehner 1996). The recent identification of the zinc finger protein Zelda, a regulator of ZGA in *Drosophila*, confirms this scenario (Liang et al. 2008).

ZGA coincides with the MZT in mouse and *Drosophila* embryos (Andéol 1994). However, ZGA precedes the MZT in amphibians or fishes by several cell cycles (Table 1). In the sea urchin, where the zygote has no transcriptional quiescence, this difference is even more dramatic. Importantly, even in these animal species, ZGA is a gradual process with a subset of genes being activated early before the dramatic increase in transcription at the major ZGA.

Similar experiments of ZGA have not been performed in plants, where the developing embryo is deeply embedded in maternal tissues that act as a barrier for pharmacological treatments. It is also difficult, as in animals, to discriminate maternally stored from de-novo–synthesized zygotic transcripts in fertilization products. This could be done by looking at nascent transcripts performing RNA-FISH (fluorescence in situ hybridization) on developing embryos, as was done in *Drosophila,* for instance (Ronshaugen and Levine 2004). To date, similar techniques have only been applied to the *Arabidopsis* endosperm, where the maternally expressed, imprinted *MEDEA* locus was found to be actively transcribed immediately following fertilization (Vielle-Calzada et al. 1999). One route around this problem is to analyze the activation of paternally inherited alleles as a substitute for zygotic gene expression, albeit with provisions regarding the synchronous activation of both parental genomes (see later in the text). In *Arabidopsis* and maize, the activation of paternal genes has been followed using reporter transgenes, allele-specific reverse transcriptase–polymerase chain reaction (RT-PCR), or both, for a discrete number of loci. It was found that most paternal loci remained silent or were expressed at very low levels until the globular embryo stage (Vielle-Calzada et al. 2000; Baroux et al. 2001; Grimanelli et al. 2005). Importantly, the verification of paternal activation of endogenous genes using allele-specific RT-PCR excluded transgene-specific paternal silencing effects. This finding was corroborated in maize, where transcript profiles from sexually produced seeds were compared with profiles from seeds of exclusively maternal origin (generated through asexual reproduction) (Grimanelli et al. 2005). The results indicate major changes in the transcript profile only around the globular stage of embryogenesis. A ZGA only after several division cycles was confirmed in *Arabidopsis* by the observation of delayed expression for several other paternally inherited transgenes, or endogenous genes, in unrelated reports (Table 3). Importantly, a delay in paternal gene expression is observed in both the embryo and endosperm.

Interestingly, as in animals, several genes are zygotically active before the major stage of paternal genome activation as defined above. A few paternally inherited alleles were found to be expressed already in two-cell *Arabidopsis* embryos (Weijers et al. 2001), albeit at low levels (Baroux et al. 2001). Early activation is more prominent in maize. Embryo-expressed genes were shown to be expressed biallelically in the zygote using allele-specific assays on manually dissected maize zygotes and embryos (Meyer and Scholten 2007). Although for 13 of the 25 genes tested, there was no significant difference in the levels of maternally and paternally derived transcripts in the zygote, paternal activation is not complete at this stage for other loci. At the zygote stage, 10 of 25 genes showed predominantly maternal expression, decreasing to 8 of 25 loci at 3 days after pollination and to 5 of 25 loci at 6 days after pollination (Meyer and Scholten 2007). Thus, maternal predominance decreases gradually because either paternal alleles are increasingly more transcribed or maternal transcripts are degraded, or both.

Therefore, as in animals, ZGA seems to be a gradual process in higher plants. Differences between maize and *Arabidopsis* seem to exist, however, up to the earliest stage at which zygotic genes can be detected. This discrepancy may relate to different reproductive strategies. Out-crossing plants, such as maize, may benefit significantly from heterosis effects related

Table 3. Reported examples of predominant maternal expression in the early embryo and/or endosperm of higher plants

Locus (gene or transgene)	Method of investigation	Reference
Arabidopsis		
AGP18	allele-specific RT-PCR/ enhancer detector	our unpublished observations
AtLPT1:GUS, AtLTP1:LhG4)	reporter expression	Baroux et al. (2001
AtSUC5	allele-specific RT-PCR/ reporter expression	our unpublished observations
CYCB1;1:GUS, CBCB1;1:LhG4	reporter expression	Baroux et al. (2001)
DCL1:GUS	reporter expression	Golden et al. (2002)
DD36	reporter expression	our unpublished observations
ET346	enhancer detector	Vielle-Calzada et al. (2000)
ET552	enhancer detector	Vielle-Calzada et al. (2000)
ET1041	enhancer detector	Vielle-Calzada et al. (2000)
ET1051	enhancer detector	Vielle-Calzada et al. (2000)
ET1119	enhancer detector	Vielle-Calzada et al. (2000)
ET1275	enhancer detector	Vielle-Calzada et al. (2000)
ET1278	enhancer detector	Vielle-Calzada et al. (2000)
ET1811	enhancer detector	Vielle-Calzada et al. (2000)
ET1849	enhancer detector	Vielle-Calzada et al. (2000)
ET2209	enhancer detector	Vielle-Calzada et al. (2000)
ET2567	enhancer detector	Vielle-Calzada et al. (2000)
ET2612	enhancer detector	Vielle-Calzada et al. (2000)
ET2634	enhancer detector	Vielle-Calzada et al. (2000)
ET3536	enhancer detector	Vielle-Calzada et al. (2000)
ET3757	enhancer detector	Vielle-Calzada et al. (2000)
ET3988	enhancer detector	Vielle-Calzada et al. (2000)
ET3992	enhancer detector	Vielle-Calzada et al. (2000)
ET4320	enhancer detector	Vielle-Calzada et al. (2000)
ET4336	enhancer detector	Vielle-Calzada et al. (2000)
ET4563	enhancer detector	Vielle-Calzada et al. (2000)
FIE:GUS, FIE:GFP	reporter expression	Yadegari et al. (2000)
GNOM/EMB30	allele-specific RT-PCR	Vielle-Calzada et al. (2000)
KS117	reporter expression	Sørensen et al. (2001)
LACHESIS:GUS	reporter expression	our unpublished observations
MSI1	allele-specific RT-PCR	Leroy et al. (2007)
pOp/LhG4 components	reporter expression	Baroux et al. 2001)
PROLIFERA	allele-specific RT-PCR/ enhancer detector/gene trap	Springer et al. (2000); Vielle-Calzada et al. (2000)
Zea mays		
AB073081	allele-specific RT-PCR	Grimanelli et al. (2005)
AF371278	allele-specific RT-PCR	Grimanelli et al. (2005)
AI670662	allele-specific RT-PCR	Grimanelli et al. (2005)
AI677212	allele-specific RT-PCR	Grimanelli et al. (2005)
AI677270	allele-specific RT-PCR	Grimanelli et al. (2005)
AI745997	allele-specific RT-PCR	Grimanelli et al. (2005)
AI746088	allele-specific RT-PCR	Grimanelli et al. (2005)
AI746192	allele-specific RT-PCR	Grimanelli et al. (2005)
AI833700	allele-specific RT-PCR	Grimanelli et al. (2005)
AI854929	allele-specific RT-PCR	Grimanelli et al. (2005)
AW066244	allele-specific RT-PCR	Grimanelli et al. (2005)
AW066927	allele-specific RT-PCR	Grimanelli et al. (2005)
AW091461	allele-specific RT-PCR	Grimanelli et al. (2005)
AW181192	allele-specific RT-PCR	Grimanelli et al. (2005)
AW216004	allele-specific RT-PCR	Grimanelli et al. (2005)
AW216025	allele-specific RT-PCR	Grimanelli et al. (2005)
AW216194	allele-specific RT-PCR	Grimanelli et al. (2005)
DW475554	MS on RT-PCR products	Meyer and Scholten (2007)
EH038205	MS on RT-PCR products	Meyer and Scholten (2007)
EH038208	MS on RT-PCR products	Meyer and Scholten (2007)
EH038209	MS on RT-PCR products	Meyer and Scholten (2007)
EH038210	MS on RT-PCR products	Meyer and Scholten (2007)
EH038211	MS on RT-PCR products	Meyer and Scholten (2007)
EH038212	MS on RT-PCR products	Meyer and Scholten (2007)
EH038213	MS on RT-PCR products	Meyer and Scholten (2007)
EH038215	MS on RT-PCR products	Meyer and Scholten (2007)
EH038218	MS on RT-PCR products	Meyer and Scholten (2007)
Fie2	allele-specific RT-PCR	Danilevskaya et al. (2003)
Meg1	allele-specific RT-PCR	Gutiérrez-Marcos et al. (2004)

Genes or transgenes (enhancer detectors, reporter gene fusions, components of gene *trans*-activation systems) are listed, together with methods used to discriminate the expression of parental alleles for the two species *Arabidopsis* and maize. Putative or known functions assigned to these genes do not suggest any common trend with respect to their cellular functions. (MS) Mass spectrometry.

to early paternal genome activation (Meyer and Scholten 2007), but no heterosis effects are observed in embryos of self-fertilizing species, such as *Arabidopsis*, at this early stage.

Mechanisms of Zygotic Genome Activation

Establishment of a permissive chromatin state. Following fertilization in mammals, reprogramming of chromatin occurs on a large scale by rapid and active demethylation of the paternal genome, whereas the maternal genome is progressively and passively demethylated (Santos et al. 2002). Imprinted genes, however, escape these demethylation processes (Branco et al. 2008). Genome-wide demethylation reflects the release of a global, silent chromatin state, a prerequisite for transcriptional activation. Furthermore, the apparent increase in histone acetylation at the one- to two-cell transition in the mouse may provide the basis for a permissive transcription state (Sarmento et al. 2004). In favor of this argument, depletion of maternal BRG1, a catalytic subunit of SWI/SNF-related chromatin remodeling complexes, does not affect global levels of histone acetylation, but affects levels of H3K4me2, a mark of active chromatin (Bultman et al. 2006). Maternal depletion of BRG1 causes embryos to arrest at the MZT (two-cell arrest) and results in down-regulation of 30% of the genes that are normally expressed at this stage. Maternal mutations in the mouse homolog of *Xenopus* nucleoplasmin 2 (NPM2), which induces sperm DNA decondensation in vitro, lead to a loss of heterochromatin and deacetylated histone H3 associated with nucleoli (Burns et al. 2003). However, the exact role of maternal NPM2 in regulating zygotic gene expression levels is not known.

During this reprogramming process, repressive mechanisms also act to ensure relative embryonic quiescence. The role of transcriptional repressors has been uncovered by conditional inhibition of protein synthesis during embryo development. This is the case for the *Xenopus* homolog of the mammalian DNA-methyltransferase Dnmt1 (xDnmt1), where embryos deficient in xDnmt1 exhibit premature gene expression at least two cell cycles earlier than normal (Stancheva and Meehan 2000; Stancheva et al. 2002). Repression by xDnmt1 is independent of its catalytic activity, and it may act as a general DNA-binding transcriptional repressor (Dunican et al. 2008). Similarly, the methyl-CpG-binding protein KAISO was identified as a global transcriptional repressor of early transcription in *Xenopus* (Ruzov et al. 2004). In KAISO-depleted embryos, 35S-UTP incorporation was detected two cell cycles earlier than in mock injected embryos, which was associated with a developmental arrest similar to that observed in embryos depleted for xDnmt1.

The timing of ZGA results, therefore, from a fine-tuned balance between chromatin-based repressive mechanisms and the establishment of a chromatin state permissive for transcription. Silencing and activating epigenetic pathways acting at the genome-wide scale are well described in plants (for review, see Vaillant and Paszkowski 2007). In *Arabidopsis*, these include DNA methylation at symmetric CG sites controlled by the maintenance of methyltransferase MET1, DNA methylation at non-CG sites (a plant-specific modification) controlled by CMT3 (CHROMOMETHYLASE3), which involves RNA-dependent DNA methylation linking the chromatin small interfering RNA (siRNA)-dependent pathway with DNA methylation, and histone H3 methylation on lysine 9 (H3K9me2). The potential role of these pathways in early zygote transcriptional silencing and the MZT remains to be determined.

Transcriptional activation of zygotic genes. Genome-wide studies identified *cis*-regulatory elements in the 5′ region of zygotically transcribed genes in *Drosophila*, which may prime them for expression during early cleavage stages (ten Bosch et al. 2006; De Renzis et al. 2007). The genes possessing the heptamer motif "CAGGTAG" in their 5′-regulatory regions are referred to as "TAG genes" and suggest a collective control of their expression during ZGA by a sequence-specific transcriptional activator(s). This motif was used as an entry point to identify the Zelda (Zld) transcription factor, in a one-hybrid screen (Liang et al. 2008). Zld is maternally stored in the zygote and is required for normal cell division and patterning of the embryo. The broad range of phenotypes observed in *zld* mutants indicates that *zld* embryos fail to express genes essential for cellular blastoderm formation. This was confirmed for several patterning genes by in situ hybridization, and microarray analyses detected at least 279 genes controlled by Zld. Among these, 82% were zygotically active genes. Surprisingly, an equal amount of genes are up-regulated in *zld* mutants and correspond to maternal genes. This effect can be explained by a lack of maternal transcript turnover due to the expression of miR309, a target of Zld that is derepressed in mutant *zld* embryos. Zld therefore provides a mechanistic link between ZGA and maternal transcript degradation (see below and Fig. 2).

Parent-specific mechanisms. Differences in parental genome activation have been observed and can be related to the distinct epigenetic control of maternal and paternal chromosomes. In the mouse, the paternal pronucleus shows transcriptional activity as early as the one-cell stage, before the maternal pronucleus, based on BrUTP incorporation (Aoki et al. 1997). Differential DNA methylation profiles have been found between the two parental genomes in mammals: The maternal genome undergoes a stepwise passive demethylation (see above), whereas the paternal genome is rapidly demethylated before the first cell division (Reik 2007). Moreover, genome-wide analysis of DNA methylation in promoters using mouse embryonic stem cells, embryonic germ cells, and sperm cells shows that their DNA methylation patterns are surprisingly similar. This suggests that although the sperm is a highly specialized and differentiated cell type, its epigenome is already largely reprogrammed before fertilization, resembling that of a pluripotent state (Farthing et al. 2008). Moreover, in contrast to the female genome, the male genome must undergo drastic chromatin remodeling after fertilization. The protamines, which are required for tight chromatin packaging in the sperm, have to be replaced by histones, including H3.3 variants. These histone variants have been associated with transcriptionally active chromatin in animals. The distinct chromatin composition of male and female genomes at

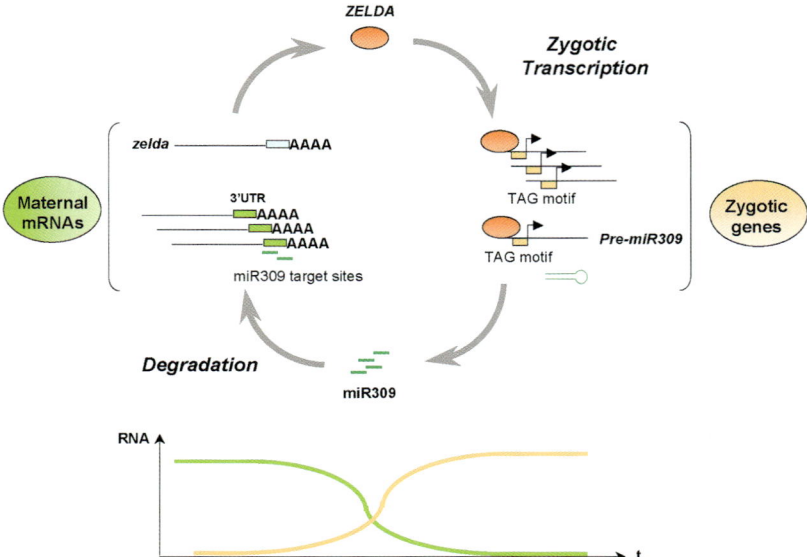

Figure 2. Mechanisms of zygotic genome activation and maternal transcript degradation in *Drosophila*. Early development is driven by maternal transcripts deposited in the egg. *Zelda* transcripts are inherited maternally and encode a global transcription factor that recognizes a specific regulatory motif called "TAG." Genes containing the TAG motif in their regulatory regions are transcribed actively in the early embryo after fertilization. Several hundred TAG genes have been identified. Among these, the precursor of microRNA-miR309 (pre-miR309) allows the coupling of zygotic transcriptional activation to the large-scale degradation of maternal transcripts. miR309 targets hundreds of maternally deposited transcripts, inducing their translational inhibition and destabilization. This coupling results in the coordinated decrease and increase in levels of maternal and zygotic transcripts, respectively.

fertilization may provide a mechanistic basis for parentally distinct transcriptional activation mechanisms.

As in animals, the plant sperm chromatin is highly compacted at fertilization. Sperm-specific histone variants have been identified that may package plant sperm genomes similar to the animal protamins (Okada et al. 2005). In *Arabidopsis*, karyogamy is quickly followed by the removal of at least one of these histones, an H3.3 variant (Ingouff et al. 2007). Interestingly, this removal follows distinct dynamics in the two fertilization products, with a rapid elimination in the zygote and a progressive dilution through successive replication rounds in the endosperm (Ingouff et al. 2007). These findings suggest that changes in core nucleosome composition occur in plants after fertilization, indicating a converging role in plants and animals for histone H3 and its variants in enabling transcription from paternally inherited chromatin.

OVERCOMING MATERNAL DOMINANCE: DEGRADATION OF MATERNAL FACTORS DURING THE MATERNAL TO ZYGOTIC TRANSITION

The establishment of the zygotic transcriptional program requires the degradation of maternally contributed RNAs. Although some of the maternal mRNAs are stable and continue to contribute to development long after ZGA, degradation mechanisms start to act early after fertilization, creating a mixed maternal/zygotic gene expression profile in the young embryo. Genome-wide profiling in *Drosophila* showed that 33% of the maternally

deposited transcripts are degraded in the embryo (De Renzis et al. 2007). Two RNA degradation pathways are used to promote turnover of maternal transcripts during the MZT, and both target 3′UTR (untranslated region) sequence motifs (for review, see Stitzel and Seydoux 2007; Tadros et al. 2007a). The first pathway is driven by maternally encoded factors, whereas the second coincides with the onset of zygotic transcription.

A survey of 1095 genes known to be maternally deposited before the *Drosophila* MZT identified two families of motifs in their 3′UTR, based on sequence similarity (De Renzis et al. 2007). The first family contains a UUGUU core, which resembles the target site for the PUF family of RNA-binding proteins, represented by the Pumilio translational regulator (Wharton et al. 1998). Pumilio controls the expression of the maternally encoded SMAUG protein, which recognizes stem-loop structures present in 3′UTR or coding sequences of maternal mRNAs (Tadros et al. 2007a; Semotok et al. 2008). Microarray analysis showed that Smaug is a general activator of RNA degradation (Tadros et al. 2007b). Sequences from the second family match the AU-rich element (AREs, canonically defined as UAUUUAU), a mediator of mRNA degradation (Shaw and Kamen 1986). The existence of the zygotic degradation pathway was first demonstrated for the maternal mRNAs of the string/Cdc25 and cyclinA1 cell cycle regulators. Inhibition of transcription in the early embryo by α-aminitin treatment inhibited mRNA degradation, thus showing that ZGA and maternal mRNA degradation are coupled during the MZT (Anderson et al. 2001; Audic et al. 2002). It was proposed that a ribonuclease activity was responsible for this coupling (Andéol 1994). The regulatory pathway became more intricate with the discovery that regulation by microRNAs (miRNAs) is involved in mRNA degradation during early embryogenesis. miRNAs are small noncoding RNAs, produced from precursors by the Dicer ribonuclease. They silence gene expression by repressing translation or promoting mRNA turnover, via 3′UTR sequence-specific recognition of their target mRNAs. Zebra fish *dicer* mutants, which lack mature miRNAs, show a maternal zygotic effect (Mishima et al. 2006). In zebra fish, a single miRNA family (miR430) drives the repression of several hundreds of maternal mRNAs (Giraldez et al. 2006). miR430 is expressed zygotically shortly after fertilization and is required for embryogenesis to be completed, as shown by reversion of the *dicer* embryo phenotype by miR430 injection. miR430 promotes target mRNA clearance by accelerating deadenylation (Giraldez et al. 2006). Use of miRNAs to promote mRNA turnover during the MTZ appears to be a conserved phenomenon because a comparable role of miRNAs was recently reported in *Drosophila* (Bushati et al. 2008). A cluster of zygotically expressed miRNAs (miR309), activated about 2 hours after fertilization, targets maternal mRNAs for turnover as part of the zygotic degradation pathway. Interestingly, it was recently shown that miR309 zygotic expression is driven by the general ZGA activator Zelda (Fig. 2), providing a link between both the genome activating pathway and the mRNA degrading pathway.

In plants, nothing is known about the mechanisms that lead to a reduction of the maternal dominance at the MZT. A genome-wide analysis of maternal transcripts and their dynamic levels following fertilization would be a prerequisite to identify potential motifs and regulators of their degradation, as was done in *Drosophila* (Liang et al. 2008). Such studies are challenging because female gametes and fertilization products are embedded in maternal tissues and are difficult to access. This may, however, be overcome using manual or laser-assisted dissection of eggs and young embryos (Dresselhaus et al. 1996, 1999; Day et al. 2005; Sprunck et al. 2005; Meyer and Scholten 2007) in combination with transcript profiling or deep sequencing approaches.

BIOLOGICAL SIGNIFICANCE

Gametes are highly differentiated cells and perform unique tasks associated with reproduction. At the same time, they can rapidly lose gametic cell fate following fertilization and endow the zygote with a totipotent state. Performing both functions implies a dramatic reprogramming of the zygotic genome in order to erase the marks associated with gametic cell fate and to establish the transcriptional status associated with totipotency. The consensus today is that the potential for totipotency in both animals and plants is maternally controlled, with most of the control over genome structure and function residing in the egg cytoplasm (Stitzel and Seydoux 2007). The most striking evidence is the ability of parthenogenetic plants and animals to develop functional organisms without any paternal contribution. That male contributions to the zygote are fully dispensable in these cases illustrates that the female genome maintains enough flexibility to compensate for any essential male contribution, whenever missing. Thus, the oocyte probably drives the events required for totipotency, including the reprogramming of the male genome following fertilization. This is well illustrated by the replacement of paternal histone components by maternally provided histone variants immediately following fertilization.

In animals, this model appears to be true whether the germ line is "predetermined" as in *C. elegans* or *Drosophila* (primordial germ cells are formed early through inheritance of maternal germplasm) or formed by inductive signals later during development, such as in mice. In both systems, transcriptional repression in the germ line is apparently decisive to totipotency (Stitzel and Seydoux 2007). In addition, gene regulation is dependent on cytoplamic posttranscriptional mechanisms in both of these systems. Although the underlying processes are different in *C. elegans*, *Drosophila*, or mice, shared developmental constraints associated with genome reprogramming and the acquisition of totipotency have led to a set of convergent mechanisms. Transcriptional repression and extensive chromatin remodeling in the gametes are followed by the establishment of a totipotent zygotic program after fertilization through another wave of chromatin reprogramming, and finally, at the MZT, the maternally stored gene products are cleared in the developing embryo.

It is tempting to speculate that the transfer of gene expression to the cytoplasm and the maintenance of relative quiescence in the early embryo represent a buffering mechanism that protects the embryo against uncoordinated genic activity during this dramatic phase of reprogramming. Likely, all living organisms share this constraint, and most animals have apparently responded by evolving remarkably similar mechanisms. Plants more resemble mice than flies, in the sense that they differentiate a germ line very late during development in an inductive manner, often many years after the formation of the embryo. Although the life cycles of plants and animals differ in important ways, the requirement for a reprogramming phase after fertilization remains. Although the data currently available remain inconclusive, we anticipate that the coming years will provide ample opportunities for enriching comparisons between the earliest phase of postfertilization development in animals and that in plants.

ACKNOWLEDGMENTS

We thank Stephanie Meyer and Stefan Scholten (Universität Hamburg) for their help in identifying loci showing bi-parental expression but maternal dominance in maize zygotes and Chris Somerville (Carnegie Institution) in whose laboratory C.S.G. collected the data on *vcl1* shown in Table 2. C.B. and U.G. are supported by the Swiss National Science Foundation and the University of Zürich; D.A. and D.G. are supported by the Institut de Recherche pour le Développement and Agence National de la Recherche.

REFERENCES

Ach, R.A., Taranto, P., and Gruissem, W. 1997. A conserved family of WD-40 proteins binds to the retinoblastoma protein in both plants and animals. *Plant Cell* **9:** 1595–1606.

Andéol, Y. 1994. Early transcription in different animal species: Implication for transition from maternal to zygotic control in development. *Roux Arch. Dev. Biol.* **204:** 3–10.

Anderson, J.E., Matteri, R.L., Abeydeera, L.R., Day, B.N., and Prather, R.S. 2001. Degradation of maternal cdc25c during the maternal to zygotic transition is dependent upon embryonic transcription. *Mol. Reprod. Dev.* **60:** 181–188.

Anderson, K.V. and Lengyel, J.A. 1980. Changing rates of histone mRNA synthesis and turnover in *Drosophila* embryos. *Cell* **21:** 717–727.

Aoki, F., Worrad, D.M., and Schultz, R.M. 1997. Regulation of transcriptional activity during the first and second cell cycles in the preimplantation mouse embryo. *Dev. Biol.* **181:** 296–307.

Audic, Y., Garbrecht, M., Fritz, B., Sheets, M.D., and Hartley, R.S. 2002. Zygotic control of maternal cyclin A1 translation and mRNA stability. *Dev. Dyn.* **225:** 511–521.

Baroux, C., Blanvillain, R., and Gallois, P. 2001. Paternally inherited transgenes are down-regulated but retain low activity during early embryogenesis in *Arabidopsis*. *FEBS Lett.* **509:** 11–16.

Baroux, C., Spillane, C., and Grossniklaus, U. 2002. Genomic imprinting during seed development. *Adv. Genet.* **46:** 165–214.

Boisnard-Lorig, C., Colon-Carmona, A., Bauch, M., Hodge, S., Doerner, P., Bancharel, E., Dumas, C., Haseloff, J., and Berger, F. 2001. Dynamic analyses of the expression of the HISTONE::YFP fusion protein in *Arabidopsis* show that syncytial endosperm is divided in mitotic domains. *Plant Cell* **13:** 495–509.

Borges, F., Gomes, G., Gardner, R., Moreno, N., McCormick, S., Feijó, J.A., and Becker, J.D. 2008. Comparative transcriptomics of *Arabidopsis* sperm cells. *Plant Physiol.* **148:** 1168–1181.

Branco, M.R., Oda, M., and Reik, W. 2008. Safeguarding parental identity: Dnmt1 maintains imprints during epigenetic reprogramming in early embryogenesis. *Genes Dev.* **22:** 1567–1571.

Breuninger, H., Rikirsch, E., Hermann, M., Ueda, M., and Laux, T. 2008. Differential expression of *WOX* genes mediates apical-basal axis formation in the *Arabidopsis* embryo. *Dev. Cell* **14:** 867–876.

Browder, L.W., Erickson, C.A., and Jeffery, W.R. 1991. *Developmental biology*. Saunders, Philadelphia.

Brown, R.C., Lemmon, B.E., Nguyen, H., and Olsen, O.-A. 1999. Development of endosperm in *Arabidopsis thaliana*. *Sex. Plant Reprod.* **12:** 32–42.

Bultman, S.J., Gebuhr, T.C., Pan, H., Svoboda, P., Schultz, R.M., and Magnuson, T. 2006. Maternal BRG1 regulates zygotic genome activation in the mouse. *Genes Dev.* **20:** 1744–1754.

Burns, K.H., Viveiros, M.M., Ren, Y., Wang, P., DeMayo, F.J., Frail, D.E., Eppig, J.J., and Matzuk, M.M. 2003. Roles of NPM2 in chromatin and nucleolar organization in oocytes and embryos. *Science* **300:** 633–636.

Bushati, N., Stark, A., Brennecke, J., and Cohen, S.M. 2008. Temporal reciprocity of miRNAs and their targets during the maternal-to-zygotic transition in *Drosophila*. *Curr. Biol.* **18:** 501–506.

Chaudhury, A.M., Koltunow, A., Payne, T., Luo, M., Tucker, M.R., Dennis, E.S., and Peacock, W.J. 2001. Control of early seed development. *Annu. Rev. Cell Dev. Biol.* **17:** 677–699.

Clegg, K.B. and Piko, L. 1977. Size and specific activity of the UTP pool and overall rates of RNA synthesis in early mouse embryos. *Dev. Biol.* **58:** 76–95.

Collinge, M.A., Spillane, C., Köhler, C., Gheyselinck, J., and Grossniklaus, U. 2004. Genetic interaction of an origin recognition complex subunit and the *Polycomb* group gene *MEDEA* during seed development. *Plant Cell* **16:** 1035–1046.

Danilevskaya, O.N., Hermon, P., Hantke, S., Muszynski, M.G., Kollipara, K., and Ananiev, E.V. 2003. Duplicated *fie* genes in maize: Expression pattern and imprinting suggest distinct functions. *Plant Cell* **15:** 425–438.

Day, R.C., Grossniklaus, U., and Macknight, R.C. 2005. Be more

specific! Laser-assisted microdissection of plant cells. *Trends Plant Sci.* **10:** 397–406.

Dekens, M.P., Pelegri, F.J., Maischein, H.M., and Nüsslein-Volhard, C. 2003. The maternal-effect gene *futile cycle* is essential for pronuclear congression and mitotic spindle assembly in the zebrafish zygote. *Development* **130:** 3907–3916.

De Renzis, S., Elemento, O., Tavazoie, S., and Wieschaus, E.F. 2007. Unmasking activation of the zygotic genome using chromosomal deletions in the *Drosophila* embryo. *PLoS Biol.* **5:** e117.

Dresselhaus, T., Cordts, S., and Lörz, H. 1999. A transcript encoding translation initiation factor eIF-5A is stored in unfertilized egg cells of maize. *Plant Mol. Biol.* **39:** 1063–1071.

Dresselhaus, T., Hagel, C., Lörz, H., and Kranz, E. 1996. Isolation of a full-length cDNA encoding calreticulin from a PCR library of in vitro zygotes of maize. *Plant Mol. Biol.* **31:** 23–34.

Dunican, D.S., Ruzov, A., Hackett, J.A., and Meehan, R.R. 2008. xDnmt1 regulates transcriptional silencing in pre-MBT *Xenopus* embryos independently of its catalytic function. *Development* **135:** 1295–1302.

Edgar, B.A. and Lehner, C.F. 1996. Developmental control of cell cycle regulators: A fly's perspective. *Science* **274:** 1646–1652.

Edgar, L.G., Wolf, N., and Wood, W.B. 1994. Early transcription in *Caenorhabditis elegans* embryos. *Development* **120:** 443–451.

Farthing, C.R., Ficz, G., Ng, R.K., Chan, C.F., Andrews, S., Dean, W., Hemberger, M., and Reik, W. 2008. Global mapping of DNA methylation in mouse promoters reveals epigenetic reprogramming of pluripotency genes. *PLoS Genet.* **4:** e1000116.

Garcia-Bellido, A. and Robbins, L.G. 1983. Viability of female germ-line cells homozygous for zygotic lethals in *Drosophila melanogaster*. *Genetics* **103:** 235–247.

Giraldez, A.J., Mishima, Y., Rihel, J., Grocock, R.J., Van Dongen, S., Inoue, K., Enright, A.J., and Schier, A.F. 2006. Zebrafish MiR-430 promotes deadenylation and clearance of maternal mRNAs. *Science* **312:** 75–79.

Golden, T.A., Schauer, S.E., Lang, J.D., Pien, S., Mushegian, A.R., Grossniklaus, U., Meinke, D.W., and Ray, A. 2002. *Short Integuments1/suspensor1/carpel Factory*, a Dicer homolog, is a maternal effect gene required for embryo development in *Arabidopsis*. *Plant Physiol.* **130:** 808–822.

Grimanelli, D., Perotti, E., Ramirez, J., and Leblanc, O. 2005. Timing of the maternal-to-zygotic transition during early seed development in maize. *Plant Cell* **17:** 1061–1072.

Grossniklaus, U. and Schneitz, K. 1998. The molecular and genetic basis of ovule and megagametophyte development. *Semin. Cell Dev. Biol.* **9:** 227–238.

Gutiérrez-Marcos, J.F., Costa, L.M., Biderre-Petit, C., Khbaya, B., O'Sullivan, D.M., Wormald, M., Perez, P., and Dickinson, H.G. 2004. *maternally expressed gene1* is a novel maize endosperm transfer cell-specific gene with a maternal parent-of-origin pattern of expression. *Plant Cell* **16:** 1288–1301.

Harper, J.L., Lovell, P.H., and Moore, K.G. 1970. The shapes and sizes of seeds. *Annu. Rev. Ecol. Syst.* **1:** 327–356.

Harvey, E.B. 1936. Parthenogenetic merogony or cleavage without nuclei in *Arbacia punctulata*. *Biol. Bull.* **71:** 101–121.

Ingouff, M., Hamamura, Y., Gourgues, M., Higashiyama, T., and Berger, F. 2007. Distinct dynamics of HISTONE3 variants between the two fertilization products in plants. *Curr. Biol.* **17:** 1032–1037.

Jürgens G. 1992. Pattern formation in the flowering plant embryo. *Curr. Opin. Genet. Dev.* **2:** 567–570.

Kane, D.A., Hammerschmidt, M., Mullins, M.C., Maischein, H.M., Brand, M., van Eeden, F.J., Furutani-Seiki, M., Granato, M., Haffter, P., Heisenberg, C.P., et al. 1996. The zebrafish epiboly mutants. *Development* **123:** 47–55.

Leduc, N., Matthys-Rochon, E., Rougier, M., Mogensen, L., Holm, P., Magnard, J.L., and Dumas, C. 1996. Isolated maize zygotes mimic in vivo embryonic development and express microinjected genes when cultured in vitro. *Dev. Biol.* **177:** 190–203.

Leroy, O., Hennig, L., Breuninger, H., Laux, T., and Kohler, C. 2007. Polycomb group proteins function in the female gametophyte to determine seed development in plants. *Development*

134: 3639–3648.

Liang, H.L., Nien, C.Y., Liu, H.Y., Metzstein, M.M., Kirov, N., and Rushlow, C. 2008. The zinc-finger protein Zelda is a key activator of the early zygotic genome in *Drosophila*. *Nature* **456:** 400–403.

Loppin, B., Berger, F., and Couble, P. 2001. The *Drosophila* maternal gene *sesame* is required for sperm chromatin remodeling at fertilization. *Chromosoma* **110:** 430–440.

Lukowitz, W., Roeder, A., Parmenter, D., and Somerville, C. 2004. A MAPKK kinase gene regulates extra-embryonic cell fate in *Arabidopsis*. *Cell* **116:** 109–119.

Mathavan, S., Lee, S.G., Mak, A., Miller, L.D., Murthy, K.R., Govindarajan, K.R., Tong, Y., Wu, Y.L., Lam, S.H., Yang, H., et al. 2005. Transcriptome analysis of zebrafish embryogenesis using microarrays. *PLoS Genet.* **1:** 260–276.

Mayer, U., Büttner, G., and Jürgens, G. 1993. Apical-basal pattern formation in the *Arabidopsis* embryo: Studies on the role of the *gnom* gene. *Development* **117:** 149–162.

Meyer, S. and Scholten, S. 2007. Equivalent parental contribution to early plant zygotic development. *Curr. Biol.* **17:** 1686–1691.

Mishima, Y., Giraldez, A.J., Takeda, Y., Fujiwara, T., Sakamoto, H., Schier, A.F., and Inoue, K. 2006. Differential regulation of germline mRNAs in soma and germ cells by zebrafish miR-430. *Curr. Biol.* **16:** 2135–2142.

Moody, S.A., Bauer, D.V., Hainski, A.M., and Huang, S. 1996. Determination of *Xenopus* cell lineage by maternal factors and cell interactions. *Curr. Top. Dev. Biol.* **32:** 103–138.

Moore, J.M. 2002. "Isolation and characterization of gametophytic mutants in *Arabidopsis thaliana*." Ph.D. thesis, State University of New York, Stony Brook.

Mtango, N.R., Potireddy, S., and Latham K.E. 2008. Oocyte quality and maternal control of development. *Int. Rev. Cell Mol. Biol.* **268:** 223–290.

Newport, J. and Kirschner, M. 1982a. A major developmental transition in early *Xenopus* embryos. I. Characterization and timing of cellular changes at the midblastula stage. *Cell* **30:** 675–686.

Newport, J. and Kirschner, M. 1982b. A major developmental transition in early *Xenopus* embryos. II. Control of the onset of transcription. *Cell* **30:** 687–696.

Ning, J., Peng, X.B., Qu, L.H., Xin, H.P., Yan, T.T., and Sun, M. 2006. Differential gene expression in egg cells and zygotes suggests that the transcriptome is restructed before the first zygotic division in tobacco. *FEBS Lett.* **580:** 1747–1752.

Okada, T., Endo, M., Singh, M.B., and Bhalla, P.L. 2005. Analysis of the histone H3 gene family in *Arabidopsis* and identification of the male-gamete-specific variant *AtMGH3*. *Plant J.* **44:** 557–568.

Pagnussat, G.C., Yu, H.J., Ngo, Q.A., Rajani, S., Mayalagu, S., Johnson, C.S., Capron, A., Xie, L.F., Ye, D., and Sundaresan, V. 2005. Genetic and molecular identification of genes required for female gametophyte development and function in *Arabidopsis*. *Development* **132:** 603–614.

Park, S. and Harada, J.J. 2008. *Arabidopsis* embryogenesis. *Methods Mol. Biol.* **427:** 3–16.

Pelegri, F. 2003. Maternal factors in zebrafish development. *Dev. Dyn.* **228:** 535–554.

Perrimon, N., Engstrom, L., and Mahowald, A.P. 1989. Zygotic lethals with specific maternal effect phenotypes in *Drosophila melanogaster*. I. Loci on the X chromosome. *Genetics* **121:** 333–352.

Perrimon, N., Lanjuin, A., Arnold, C., and Noll, E. 1996. Zygotic lethal mutations with maternal effect phenotypes in *Drosophila melanogaster*. II. Loci on the second and third chromosomes identified by P-element-induced mutations. *Genetics* **144:** 1581–1692.

Poccia, D., Wolff, R., Kragh, S., and Williamson, P. 1985. RNA synthesis in male pronuclei of the sea urchin. *Biochim. Biophys. Acta* **824:** 349–356.

Pritchard, D.K. and Schubiger, G. 1996. Activation of transcription in *Drosophila* embryos is a gradual process mediated by the nucleocytoplasmic ratio. *Genes Dev.* **10:** 1131–1142.

Reik, W. 2007. Stability and flexibility of epigenetic gene regulation in mammalian development. *Nature* **447:** 425–432.

Robbins, L.G. 1980. Maternal-zygotic lethal interactions in *Drosophila melanogaster:* The effects of deficiencies in the zeste-white region of the *X* chromosome. *Genetics* **96:** 187–200.

Rojo, E., Gillmor, C.S., Kovaleva, V., Somerville, C.R., and Raikhel, N.V. 2001. *VACUOLELESS1* is an essential gene required for vacuole formation and morphogenesis in *Arabidopsis*. *Dev. Cell* **1:** 303–310.

Ronceret, A., Gadea-Vacas, J., Guillemirot, J., Lincker, F., Delorme, V., Lahmy, S., Pelletier, G., Chabouté, M.E., and Devic, M. 2008. The first zygotic division in *Arabidopsis* requires de novo transcription of thymidylate kinase. *Plant J.* **53:** 776–789.

Ronshaugen, M. and Levine, M. 2004. Visualization of *trans*-homolog enhancer-promoter interactions at the *Abd-B* Hox locus in the *Drosophila* embryo. *Dev. Cell* **7:** 925–932.

Ruzov, A., Dunican, D.S., Prokhortchouk, A., Pennings, S., Stancheva, I., Prokhortchouk, E., and Meehan, R.R. 2004. Kaiso is a genome-wide repressor of transcription that is essential for amphibian development. *Development* **131:** 6185–6194.

Santos, F., Hendrich, B., Reik, W., and Dean. W. 2002. Dynamic reprogramming of DNA methylation in the early mouse embryo. *Dev. Biol.* **241:** 172–182.

Sarmento, O.F., Digilio, L.C., Wang, Y., Perlin, J., Herr, J.C., Allis, C.D. and Coonrod, S.A. 2004. Dynamic alterations of specific histone modifications during early murine development. *J. Cell Sci.* **117:** 4449–4459.

Schauer, I.E. and Wood, W.B. 1990. Early *C. elegans* embryos are transcriptionally active. *Development* **110:** 1303–1317.

Scholten, S., Lörz, H., and Kranz, E. 2002. Paternal mRNA and protein synthesis coincides with male chromatin decondensation in maize zygotes. *Plant J.* **32:** 221–231.

Schultz, R.M. 2002. The molecular foundations of the maternal to zygotic transition in the preimplantation embryo. *Hum. Reprod. Update* **8:** 323–331.

Semotok, J.L., Luo, H., Cooperstock, R.L., Karaiskakis, A., Vari, H.K., Smibert, C.A., and Lipshitz, H.D. 2008. *Drosophila* maternal *Hsp83* mRNA destabilization is directed by multiple SMAUG recognition elements in the open reading frame. *Mol. Cell. Biol.* **28:** 6757–6772.

Seydoux, G. and Fire, A. 1994. Soma-germline asymmetry in the distributions of embryonic RNAs in *Caenorhabditis elegans*. *Development* **120:** 2823–2834.

Shaw, G. and Kamen, R. 1986. A conserved AU sequence from the 3′ untranslated region of GM-CSF mRNA mediates selective mRNA degradation. *Cell* **46:** 659–667.

Sørensen, M.B., Chaudhury, A.M., Robert, H., Bancharel, E., and Berger, F. 2001. Polycomb group genes control pattern formation in plant seed. *Curr. Biol.* **11:** 277–281.

Springer, P.S., McCombie, W.R., Sundaresan, V., and Martienssen, R.A. 1995. Gene trap tagging of *PROLIFERA*, an essential *MCM2-3-5*-like gene in *Arabidopsis*. *Science* **268:** 877–880.

Springer, P.S., Holding, D.R., Groover, A. Yordan, C., and Martienssen, R.A. 2000. The essential Mcm7 protein PROLIFERA is localized to the nucleus of dividing cells during the G_1 phase and is required maternally for early *Arabidopsis* development. *Development* **127:** 1815–1822.

Sprunck, S., Baumann, U., Edwards, K., Langridge, P., and Dresselhaus, T. 2005. The transcript composition of egg cells changes significantly following fertilization in wheat (*Triticum aestivum* L.). *Plant J.* **41:** 660–672.

Stancheva, I. and Meehan, R.R. 2000. Transient depletion of xDnmt1 leads to premature gene activation in *Xenopus* embryos. *Genes Dev.* **14:** 313–327.

Stancheva, I., El-Maarri, O., Walter, J., Niveleau, A., and Meehan, R.R. 2002. DNA methylation at promoter regions regulates the timing of gene activation in *Xenopus laevis* embryos. *Dev. Biol.* **243:** 155–165.

Stitzel, M.L. and Seydoux, G. 2007. Regulation of the oocyte-to-zygote transition. *Science* **316:** 407–408.

Stroband, H.W.J., te Krounie, G., and van Gestel, W. 1992. Differential susceptibility of early steps in carp (*Cyrinus carpio*) development to α-amanitin. *Dev. Genes Evol.* **202:** 61–65.

Tadros, W., Westwood, J.T., and Lipshitz, H.D. 2007a. The mother-to-child transition. *Dev. Cell* **12:** 847–849.

Tadros, W., Houston, S.A., Bashirullah, A., Cooperstock, R.L., Semotok, J.L., Reed, B.H., and Lipshitz, H.D. 2003. Regulation of maternal transcript destabilization during egg activation in *Drosophila*. *Genetics* **164:** 989–1001.

Tadros, W., Goldman, A.L., Babak, T., Menzies, F., Vardy, L., Orr-Weaver, T., Hughes, T.R., Westwood, J.T., Smibert, C.A., and Lipshitz, H.D. 2007b. SMAUG is a major regulator of maternal mRNA destabilization in *Drosophila* and its translation is activated by the PAN GU kinase. *Dev. Cell* **12:** 143–155.

Telford, N.A., Watson, A.J., and Schultz, G.A. 1990. Transition from maternal to embryonic control in early mammalian development: A comparison of several species. *Mol. Reprod. Dev.* **26:** 90–100.

ten Bosch, J.R., Benavides, J.A., and Cline, T.W. 2006. The TAGteam DNA motif controls the timing of *Drosophila* preblastoderm transcription. *Development* **133:** 1967–1977.

Tzafrir, I., Pena-Muralla, R., Dickerman, A., Berg, M., Rogers, R., Hutchens, S., Sweeney, T.C., McElver, J., Aux, G., Patton, D., and Meinke, D. 2004. Identification of genes required for embryo development in *Arabidopsis*. *Plant Physiol.* **135:** 1206–1220.

Vaillant, I. and Paszkowski, J. 2007. Role of histone and DNA methylation in gene regulation. *Curr. Opin. Plant Biol.* **10:** 528–533.

Vielle-Calzada, J.P., Baskar, R., and Grossniklaus, U. 2000. Delayed activation of the paternal genome during seed development. *Nature* **404:** 91–94.

Vielle-Calzada, J.P., Thomas, J., Spillane, C.S., Coluccio, A.,

Hoeppner, M.A., and Grossniklaus, U. 2000. Maintenance of genomic imprinting at *Arabidopsis MEDEA* locus requires zygotic DDMI activity. *Genes Dev.* **13:** 2971–2982.

Weijers, D., Geldner, N., Offringa, R., and Jürgens, G. 2001. Seed development: Early paternal gene activity in *Arabidopsis*. *Nature* **414:** 709–710.

Wharton, R.P., Sonoda, J., Lee, T., Patterson, M., and Murata, Y. 1998. The Pumilio RNA-binding domain is also a translational regulator. *Mol. Cell* **1:** 863–872.

Wu, X., Chory, J., and Weigel, D. 2007. Combinations of WOX activities regulate tissue proliferation during *Arabidopsis* embryonic development. *Dev. Biol.* **309:** 306–316.

Yadegari, R., Kinoshita, T., Lotan, O., Cohen, G., Katz, A., Choi, Y., Nakashima, K., Harada, J.J., Goldberg, R.B., Fischer, R.L., and Ohad, N. 2000. Mutations in the *FIE* and *MEA* genes that encode interacting polycomb proteins cause parent-of-origin effects on seed development by distinct mechanisms. *Plant Cell* **12:** 2367–2382.

Yang, J., Tan, C., Darken, R.S., Wilson, P.A., and Klein, P.S. 2002. β-Catenin/Tcf-regulated transcription prior to the midblastula transition. *Development* **129:** 5743–5752.

Zalokar, M. 1976. Autoradiographic study of protein and RNA formation during early development of *Drosophila* eggs. *Dev. Biol.* **49:** 425–437.

Zeng, F., Baldwin, D.A., and Schultz, R.M. 2004. Transcript profiling during preimplantation mouse development. *Dev. Biol.* **272:** 483–496.

Directed Differentiation of Pluripotent Stem Cells: From Developmental Biology to Therapeutic Applications

S. Irion, M.C. Nostro, S.J. Kattman, and G.M. Keller

McEwen Centre for Regenerative Medicine, University Health Network, Toronto, Ontario M5G 1L7, Canada

The discovery of human pluripotent stem cells has laid the foundation for an emerging new field of biomedical research that holds promise to develop models of human development and disease, establish new strategies for discovering and testing drugs, and provide systems for the generation of cells and tissues for transplantation for the treatment of disease. The remarkable potential of pluripotent stem cells has sparked interest and excitement in academia, the biotechnology and pharmaceutical industries, as well as the lay public. Although the potential of human pluripotent stem cells is truly outstanding, fulfilling this potential is solely dependent on our ability to efficiently generate functional cell types from them. Some of the most successful approaches in this area to date are those that have applied the principles of developmental biology to stem cell differentiation. In this chapter, we review these concepts and highlight specific examples demonstrating that pluripotent stem cell differentiation in culture recapitulates the key aspects of early embryonic development. By continuing to translate insights from embryology to stem cell biology, progress in our ability to generate specific cell types from pluripotent stem cells will advance, yielding enriched populations of human cell types, including cardiomyocytes, hematopoietic cells, hepatocytes, pancreatic β cells, and neural cells, for drug discovery, functional evaluation in preclinical models of human disease, and ultimately clinical applications.

Although early insights into pluripotent stem cells (PSCs) were provided by studies using embryonal carcinoma cells isolated from teratocarcinomas (Stevens and Little 1954), realization of the true potential of such stem cells came with the discovery of mouse embryonic stem cells (mESCs) (Evans and Kaufman 1981; Martin 1981). The initial studies with mESCs rapidly established the defining characteristics of PSCs, namely, their ability to proliferate for extended periods of time in an undifferentiated state in culture and their capacity to generate derivatives of the three primary germ layers. Within a decade of the discovery of mESCs, technologies for their genetic alteration through homologous recombination (HR) were developed, enabling the routine generation of strains of mice carrying specific mutations (Glaser et al. 2005). The potential of mESCs was also tested in vitro in studies that demonstrated that these stem cells were able to generate a variety of mature cell types following differentiation in culture (Keller 2005). Although most early experiments used crude culture conditions that resulted in the generation of mixed lineage populations, they provided a clear demonstration that the in vitro differentiation system could be used as a model for studies on mammalian development and provided the basis for the widely held notion that PSCs may some day represent a novel source of cell types and tissues for future pharmaceutical and clinical applications (Fig. 1).

These theoretical applications moved a significant step closer to practice with the isolation of human embryonic stem cells (hESCs) (Thomson et al. 1998). Although hESCs appear to represent a slightly different stage of development and respond to different signaling pathways than their mouse counterparts, they do display the key characteristics of PSCs. hESCs can be maintained and expanded in an undifferentiated state in culture and are able to generate progeny of the three primary germ layers when induced to differentiate in culture or to form teratomas in vivo (Thomson et al. 1998). The fact that hESCs display pluripotency was critical because it fulfilled the requirement of a stem cell population that could potentially give rise to unlimited supplies of functional human cells appropriate for transplantation, drug discovery, and developmental biology. A major hurdle that remained with the use of hESCs as a source of "donor" tissue for transplantation was allogeneic graft rejection. Efforts to overcome this obstacle included establishing banks of hESCs and generating patient-specific PSCs through somatic cell nuclear transfer or genetic reprogramming of somatic cells.

The landmark studies of Takahashi et al. (Takahashi and Yamanaka 2006) demonstrated the feasibility of the latter approach and showed that transient expression of four transcription factors normally found in ESCs led to the reprogramming of mouse embryonic fibroblasts (MEFs) to a pluripotent state. These reprogrammed cells were known as induced PSCs (iPSCs). Mouse iPSCs (miPSCs) have molecular profiles similar to those of mESCs and display pluripotent potential and the capacity to proliferate and expand in an undifferentiated state in culture (Takahashi and Yamanaka 2006). Following the discovery of miPSCs, human iPSCs (hiPSCs) were generated from fibroblasts using a similar strategy of transient expression of pluripotency genes (Takahashi et al. 2007; Yu et al. 2007; Park et al. 2008). hiPSCs could be maintained in a undifferentiated state in culture and were able to generate progeny of the three primary germ layers, documenting their pluripotency. Since the initial discovery of miPSCs and hiPSCs, significant advances have been made in expanding the target cell types that can be reprogrammed as well as in developing nonviral approaches for delivery

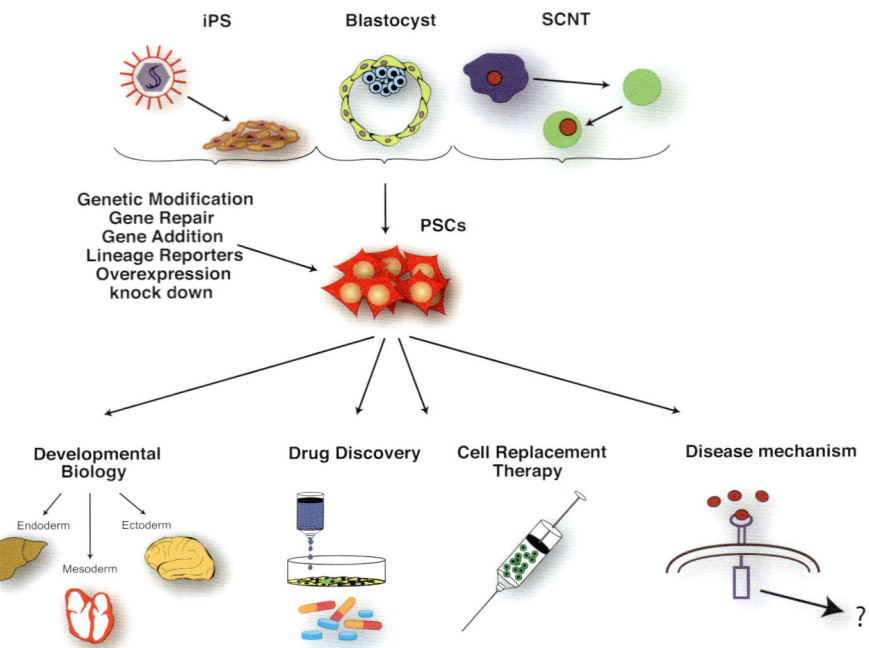

Figure 1. Pluripotent stem cell sources and their application. PSCs can be generated from the inner cell mass of the mammalian blastocyst from a normal fertilized egg, from an egg that has undergone somatic cell nuclear transfer, or by reprogramming somatic cells with pluripotency factors. Unmanipulated PSCs or those that have been genetically modified can be used for developmental biology studies, drug discovery and drug testing, cell replacement therapy, or modeling human diseases in culture.

of the reprogramming factors (Li et al. 2009). Given the pace of progress in this field, it is likely that technologies for reprogramming somatic cells in a safe and controlled manner will be developed in the near future. Combined with the establishment of animal-free culture systems, the creation of clinically compliant patient-specific PSCs could soon be a reality (Fig. 1).

With the availability of hESCs, mESCs, and iPSCs (hereafter referred to as mouse and human pluripotent stem cells; mPSCs and hPSCs), the challenge facing the field is the development of strategies for the generation of homogeneous populations of functional cell types from them. Such strategies must be reliable, scalable, and applicable to a broad range of stem cells. As different cell types are generated, appropriate assays need to be developed to evaluate their function. For cell-based therapy, predictive preclinical models of human disease must be used and assays need to be in place to demonstrate that the transplanted population has integrated into the graft site and that the cells are functioning as expected. Drug discovery and screening applications will require a constant supply of relatively pure cell populations at comparable stages of development to provide reproducibility in the various assays. When establishing models of human disease from patient-specific iPSCs, the generation of the appropriate lineages possibly in the context of a three-dimensional organ-like structure will be required. For all applications, the maturation stage of the target population will be an important factor in determining the success of the approach. The demands of these different applications

highlight the need for studies aimed at understanding the mechanisms that control the induction, expansion, and maturation of specific cell types from PSCs.

In this chapter, we review the early events regulating mammalian development, including germ-layer specification, and illustrate how the principles of developmental biology have been successfully applied to strategies for the directed differentiation of PSCs. In the final section, we conclude with a discussion of applications for the genetic modification of PSCs.

MOUSE GASTRULATION: FORMATION OF THE PRIMITIVE STREAK AND THE PRIMARY GERM LAYERS

The primary germ layers, the founder populations of all cell types in the body, are established early in embryonic development, during a process known as gastrulation (Tam and Behringer 1997). At the onset of gastrulation, a transient structure known as the primitive streak (PS) forms in the region of the epiblast that will ultimately give rise to the posterior part of the embryo (Fig. 2). As gastrulation proceeds, epiblast cells migrate through the PS and exit as populations fated to either the mesoderm or definitive endoderm cell lineages. Gene expression and lineage-tracing studies have been used to define three main regions of the PS: the posterior, mid, and anterior regions. With respect to gene expression patterns, *brachyury* (*t*) (Kispert and Herrmann 1994) and *mixl1* (Hart et al. 2002) are found throughout the entire PS, *foxa2* and *goosecoid* (Sasaki and Hogan 1993; Kinder et al. 2001) mark the anterior region, and *hoxb1* and *evx1* define the posterior region (Dush and Martin 1992; Forlani et al. 2003). The movement of uncommitted epiblast cells through the PS is controlled both spatially and temporally. The first cells to move through this structure do so in the posterior region and give rise to the extraembryonic mesoderm that forms the allantois and amnion as well as the hematopoietic, endothelial, and vascular smooth muscle cells of the yolk sac (Parameswaran and Tam 1995; Kinder et al. 1999). The next populations of epiblast cells traverse the PS in progressively more anterior sites, contributing to cranial, cardiac, paraxial, and axial mesoderm, respectively. Finally, cells that migrate through the most anterior part of the PS give rise to the definitive endoderm. Epiblast cells fated to the third germ layer, ectoderm, lie in the anterior region of the early embryo and do not migrate through the PS.

Although the mechanisms controlling PS formation and germ-layer induction are not fully understood, targeting studies and expression analysis have identified transforming growth factor-β (TGF-β) (BMP4 and nodal) (Conlon et al. 1994; Hogan 1996; Schier and Shen 2000) and Wnt (Yamaguchi 2001) signaling pathways as key regulators of this process. Moreover, they have shown that the gradient of these pathways established in the embryo likely has a critical role in the spatiotemporal allocation of cells to different fates. For instance, bone morphogenetic protein 4 (BMP4) expressed from the extraembryonic ectoderm is found at the highest levels in the proximal or posterior part of the embryo and at the lowest levels in a distal region that contains the anterior PS. Inhibitors of BMP signaling, including noggin and chordin, present in the anterior region contribute to the formation of this gradient (McMahon et al. 1998; Bachiller et al. 2000). Nodal forms a reverse gradient to that of BMP4. The highest concentrations of nodal are found in the node near the distal tip of the PS, whereas the lowest are detected in the posterior region of the embryo. During gastrulation, Wnt expression is restricted to the posterior region of the PS as well as to the posterior and lateral proximal epiblast. Collectively, these gradients establish signaling domains that in part regulate the establishment of specific fates. Recreating these precise signaling domains is an important consideration for generating comparable cell types from PSCs in culture (Gadue et al. 2005).

RECREATING KEY ASPECTS OF GASTRULATION IN PSC DIFFERENTIATION CULTURES

Strategies for PSC Differentiation

To achieve reproducible and efficient differentiation from PSCs, it is necessary to develop approaches to study the early stages of lineage commitment in a quantitative fashion. Reporter mESC lines in which enhanced green fluorescent protein (EGFP) cDNA has been targeted to either the *t* or *mixl1* locus (Fehling et al. 2003; Ng et al. 2005) have been generated to enable the investigator to monitor PS formation in differentiation cultures and to isolate cells that represent these early developmental stages. A similar *MIXL1-EGFP* cell line has also been generated for hESCs (Davis et al. 2008). To segregate the in-vitro-generated PS into posterior and anterior populations, a truncated human CD4 (hCD4) cDNA was targeted to the anterior PS gene *foxa2* in the mESC line carrying the *t-EGFP* knockin (*t-EGFP/foxa2-hCD4* ESCs) (Gadue et al. 2006). Following PS formation, mesoderm induction can be monitored by expression of *flk-1* (Ema et al. 2006), and the combination of Cxcr4 and c-Kit can be used successfully to identify early endoderm (Gouon-Evans et al. 2006).

Although early studies demonstrated the in vitro potential of ESCs using fetal bovine serum (FBS) as a differentiation inducer, these conditions were not optimal for induction of a broad range of lineages. In addition, consistency and reproducibility of differentiation often relied on specific batches of selected serum. During the past 5 years,

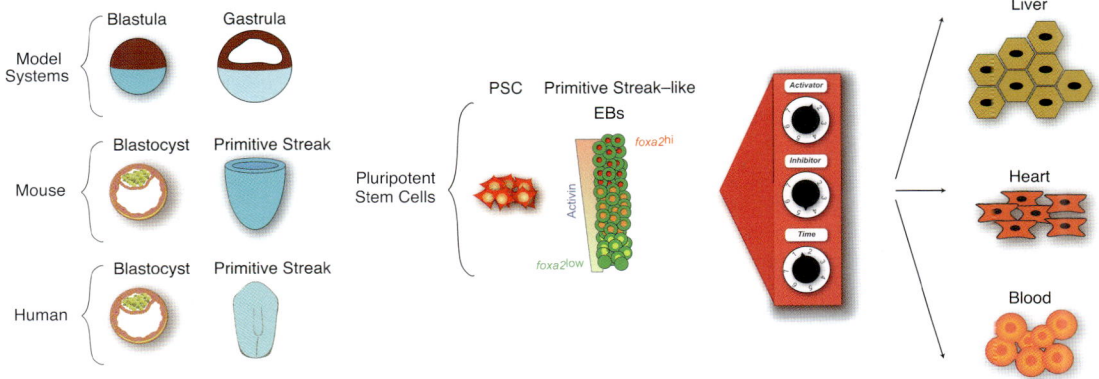

Figure 2. Translating developmental biology to pluripotent stem cell differentiation. By translating principles of development to the directed differentiation of PSCs, it is now possible to generate highly enriched populations of different cell types including immature hepatocytes, cardiomyocytes, and hematopoietic cells. The dial on the right of the figure highlights the importance of applying appropriate concentrations of agonists (activators) and antagonists (inhibitors) of key signaling pathways at the correct stage (time) of development.

FBS has been successfully replaced with specific agonists of signaling pathways, known to function in early development. With appropriate combinations, concentrations, and timing of addition of both agonists and antagonists of these pathways, it is possible to recapitulate many of the early events associated with gastrulation in a culture dish (Fig. 2).

PS Formation from PSCs

Using the dual reporter *t-EGFP/foxa2-hCD4* ESC line, it has been possible to investigate the signaling requirements for generation of a PS-like population in culture. By monitoring EGFP and hCD4 expression, we demonstrated that both Wnt and nodal signaling are simultaneously required for the development of the population that coexpresses *t* and *foxa2* (Gadue et al. 2006). The effects of Wnt3A and activin A/nodal are synergistic and result in the generation of a PS-like population composed of T-EGFP^{pos}/Foxa2-hCD4^{high}, T-EGFP^{pos}/Foxa2-hCD4^{medium}, and T-EGFP^{pos}/ Foxa2-hCD4^{low} subpopulations that display characteristics of the anterior, mid, and posterior streak, respectively (Monaghan et al. 1993). The anterior T-EGFP^{pos}/Foxa2-hCD4^{high} population gives rise to definitive endoderm and derivative cell types, whereas the mid and posterior T-EGFP^{pos}/Foxa2-hCD4^{medium} and T-EGFP^{pos}/Foxa2-hCD4^{low} populations generate different subsets of mesoderm.

The relative size of these subpopulations can be influenced by levels of nodal signaling, with high activin-A concentrations preferentially inducing the T-EGFP^{pos}/ Foxa2-hCD4^{high} subpopulation and low concentrations promoting the formation of the T-EGFP^{pos}/Foxa2-hCD4$^{medium/low}$ subpopulations. Although Wnt and nodal are required for PS formation, BMP signaling appears to be dispensable at this stage because blocking the pathway with soluble BMP receptors had no effect on the development of the Wnt/activin A–induced T-EGFP^{pos}/Foxa2-hCD4^{pos} population (Nostro et al. 2008). Interestingly, when added alone, BMP4 does induce a T-EGFP^{pos}/ Foxa2-hCD4^{low} PS population; however, the effect appears to be indirect and likely mediated through the induction of the Wnt and nodal pathways as development of the population was blocked when Wnt and nodal signaling was inhibited and both Wnt3A and nodal messages were detected in the embryoid body cells (EBCs) following BMP4 induction. When added together with Wnt3A and activin A, BMP4 appears to have a dominant "posteriorizing" effect, resulting in the generation of a T-EGFP^{pos}/Foxa2-hCD4^{low} PS population.

Studies with hESCs suggest that the early development of a human PS is regulated in a similar fashion. BMP4 alone appears to induce the equivalent of a posterior PS population able to generate hematopoietic cells, whereas high doses of activin A promote the formation of anterior PS cells that give rise to definitive endoderm and derivative cell types (D'Amour et al. 2005; Kennedy et al. 2007; Basma et al. 2008; Davis et al. 2008). Detailed analysis revealed that activin A/nodal and canonical Wnt function in a synergistic fashion to induce the anterior PS population (Sumi et al. 2008), similar to the activity observed with the mESCs. Collectively, these observations with mouse and hESCs highlight two important aspects of ESC

differentiation. First, they clearly demonstrate that cells from both species respond to the same signaling pathways that regulate gastrulation and PS formation in the early mouse embryo, strongly supporting the notion that PSC differentiation in culture recapitulates the key events in normal embryonic development. Second, they highlight the absolute requirement for selecting appropriate combinations and concentrations of cytokines during the initial stages of differentiation (Fig. 2).

Endoderm Formation from PSCs

Endoderm forms from epiblast cells that move through the anterior PS, in a region adjacent to the node that expresses high levels of nodal. As anterior PS cells migrate out of the streak, they give rise to a sheet of endodermal cells that subsequently forms the gut tube. Cells within the gut tube grow into a columnar epithelium that is specified by the surrounding tissue into the different organs including the thymus, thyroid, lung stomach, liver, pancreas, and intestines (Wells and Melton 1999). Studies using different animal model systems have highlighted the critical role of nodal signaling for both PS formation and endoderm development (Zhou et al. 1993; Conlon et al. 1994; Jones et al. 1995; Feldman et al. 1998; Osada and Wright 1999; Gritsman et al. 2000; Lowe et al. 2001).

A major hurdle in studying definitive endoderm differentiation in vitro has been the paucity of markers that distinguish it from visceral endoderm. Given that visceral endoderm is not derived from T$^+$ cells, we used the *t-EGFP/foxa2-hCD4* ESC line to follow the progression of the T-EGFP^{pos}/Foxa2-hCD4^{high} anterior PS population to a definitive endoderm population characterized by a loss of T expression and the retention of high levels of Foxa2-hCD4. This developmental step, from anterior PS to definitive endoderm, is dependent on sustained activin A/nodal signaling, clearly demonstrating that the signaling pathways that regulate the induction of this germ layer in the early embryo also function at this stage in mESC differentiation cultures.

Formation of endoderm-derived tissues and organs in vivo is dependent on complex cellular interactions that need to be recapitulated in culture. During embryogenesis, the newly formed endoderm folds anteriorly and posteriorly to generate three distinct areas: the foregut, midgut, and hindgut. BMP4 and FGF signals, provided by the septum transversum and cardiac mesoderm, respectively, are required to pattern the foregut endoderm to a liver phenotype (Zaret 2008). Both factors have been shown to function at the level of hepatic specification of ESC-derived endoderm, indicating that the requirement of these pathways is recapitulated in the in vitro model (Gouon-Evans et al. 2006).

In addition to cardiac mesoderm and the septum transversum, the endothelial lineage also has a pivotal role in the establishment of endoderm-derived tissues. In the absence of Flk-1$^+$ cells, the liver and pancreatic buds do not expand and thus are unable to form mature organs, suggesting that endothelial cells may have a pivotal role in proliferation and differentiation of early hepatic and pancreatic progenitors (Matsumoto et al. 2001; Yoshitomi and Zaret

2004). Observations from ESC differentiation studies suggest that a similar interaction between endothelial cells and endoderm-derived populations may be required in vitro. When induced to differentiate to a hepatic fate, the T-EGFP^{pos}/Foxa2-hCD4^{high}/c-Kit^{high} PS cells generate a population consisting of large distinct clusters or colonies of α-fetoprotein(afp)^{pos}, albumin(alb)^{pos}, and Foxa2-hCD4^{pos} hepatic-like cells surrounded by CD31^{+} endothelial cells (Gouon-Evans et al. 2006). Fluorescence-activated cell sorting (FACS) analysis revealed that the endothelial and endodermal cells represent approximately 40% and 50% of the population, respectively (Fig. 3, top). The close spatial development of the endothelial cells and maturing hepatocytes suggests some interaction between the two lineages. Evidence in support of this was provided by additional studies which showed that anterior PS cells isolated from the activin-A–induced EBCs on the basis of c-Kit and Cxcr4 gave rise to the endodermal population (Foxa2-hCD4^{pos}) but did not generate the endothelial cells (CD31^{pos}) (Fig. 3, bottom). Maturation along the hepatic lineage did proceed in the absence of the endothelial cells, but the total cell yield in these cultures was significantly less than in the cultures containing both lineages generated from the T-EGFP^{pos}/Foxa2-hCD4^{high}/c-Kit^{high} population (Gouon- Evans et al. 2006). These findings clearly document a role for endothelial cells in the development of the ESC-derived hepatic lineage in ESC cultures. Access to these populations at the earliest stages of induction in the ESC model provides an unprecedented opportunity to further investigate the interactions between the endothelial cells and the maturing hepatic lineage.

The developmental biology approach has also been successfully applied to the differentiation of hESCs to endoderm and derivative cell types. Using a step-wise differentiation approach aimed at recreating specific developmental stages within the pancreatic lineage, D'Amour et al. (2006) were able to generate insulin-producing β

cells from hESCs. Whereas the cells formed in the cultures did not display all of the characteristics of functional β cells, the study further highlights the importance of translating findings from developmental biology to the in vitro differentiation of PSCs.

Mesoderm

Understanding the specific stages of mesoderm development and the signaling pathways that regulate them is a prerequisite for generating appropriate derivative cell types from PSCs in vitro. Subpopulations of mesoderm fated to form different tissues are generated by distinct subsets of epiblast cells that egress through the posterior, mid, and anterior regions of the primitive streak. Access to markers that distinguish these early populations will be important for isolating specific mesoderm-derived lineages from PSCs. Although extensive marker sets do not yet exist to measure mesoderm induction and specification, the receptor tyrosine kinase Flk-1 (vascular endothelial growth factor receptor-2, VEGFR-2) has been used to monitor some of the early differentiation stages. Flk-1 was originally identified as an endothelial-cell-specific receptor (Millauer et al. 1993; Quinn et al. 1993; Yamaguchi et al. 1993). However, expression analyses of early embryo and lineage-tracing studies uncovered a broader pattern before endothelial specification and demonstrated expression in mesoderm fated to the hematopoietic, vascular, cardiac, and somitic lineages (Millauer et al. 1993; Quinn et al. 1993; Yamaguchi et al. 1993; Motoike et al. 2003; Ema et al. 2006).

To formally demonstrate that Flk-1 is a valid marker for mesoderm induction, we evaluated the potential of the population to give rise to nonmesoderm cell types, specifically endoderm following exposure to high concentrations of activin A. Posterior PS cells that have not yet up-regulated Flk-1 (Flk-1^{neg}/T-EGFP^{pos}/Foxa2-hCD4^{neg}) will differentiate to Foxa2-hCD4^{pos} endoderm when cultured in activin A for 24 hours, indicating that their fate is not yet fixed to the mesoderm lineage. In contrast, the Flk-1^{pos}/T-EGFP^{pos}/Foxa2-hCD4^{neg} population isolated from the same stage EBCs appears to have lost the potential to generate endoderm, because it does not give rise to Foxa2-hCD4^{pos} cells following culture with activin A. Rather, this population up-regulated expression of the hematopoietic-vascular marker CD31 under these conditions, strongly suggesting that its mesodermal fate is fixed and that Flk-1 is a valid marker for this developmental step (Nostro et al. 2008).

Hematopoietic Mesoderm

The blood islands of the yolk sac represent the onset of hematopoiesis in the mammalian embryo and derive from the first mesodermal cell population to egress from the PS (Russell 1979). These structures are composed of clusters of primitive erythroblasts surrounded by developing endothelial cells and are present in the presumptive yolk sac by E7.5 of mouse development (Haar and Ackerman 1971). The close temporal and spatial development of the hematopoietic and endothelial lineages in blood islands provided the basis for the hypothesis that these lineages

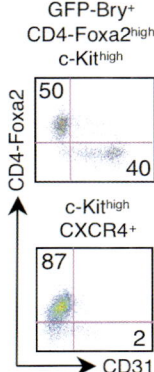

Figure 3. Endothelial potential of GFP-Bry^{+}/CD4-Foxa2^{high}/c-Kit^{high} (T-GFP^{pos}/Foxa2-hCD4^{pos}/c-Kit^{pos}) and c-Kit^{high}/CXCR4+ (Cxcr4^{pos}/c-Kit^{pos}) populations. T-GFP^{pos}/Foxa2-hCD4^{pos}/c-Kit^{pos} (*top*) and Cxcr4^{pos}/c-Kit^{pos} (*bottom*) were isolated by cell sorting from 4-day-old EBCs induced with high concentrations of activin A. The sorted cells were plated on gelatin and analyzed 3 days later for the presence of endothelial CD31^{+} and endodermal (Foxa2-hCD4^{pos}) cells. (Reprinted, with permission, from Gouon-Evans et al. 2006 [© Nature Publishing Group].)

develop from a common ancestor, a cell known as the hemangioblast (Sabin 1917; Murray 1932). Although gene expression and targeting studies in different model systems provided evidence that supported the concept of the hemangioblast (Keller 2005), it was studies with the mESC model that demonstrated the existence of this progenitor (Choi et al. 1998; Nishikawa et al. 1998). The mESC-derived hemangioblasts, also known as blast colony-forming cells (BL-CFCs), express Flk-1 and T, develop early in differentiation cultures, and are able to generate blast cell colonies that contain hematopoietic, endothelial, and vascular smooth-muscle potential (Choi et al. 1998; Fehling et al. 2003). The emergence of this progenitor within the differentiation cultures, before the establishment of all other blood cell lineages, indicates that it defines the earliest stage of hematopoietic commitment. On the basis of findings from the ESC studies, we were able to identify a comparable progenitor in the early embryo (Huber et al. 2004). This cell also expresses Flk-1 and T, develops in the posterior PS of the E7.0 embryo, and is able to generate progeny of the hematopoietic, endothelial, and vascular smooth-muscle lineages. The potential of this progenitor strongly suggests that it represents the in vivo hemangioblast, the progenitor of the yolk sac hematopoietic system. A progenitor with hemangioblast characteristics has also been identified in hESC differentiation cultures (Kennedy et al. 2007). The human hemangioblast also expresses KDR (Flk-1), displays the potential to generate both hematopoietic and endothelial progeny, and emerges within the cultures before the establishment of other hematopoietic lineages. Collectively, these studies highlight two important aspects of PSC differentiation. First, they demonstrate the power of the in vitro system for studying early developmental stages that are difficult to access in the mammalian embryo. Second, they provide strong evidence that lineage development in the ESC model follows that of the early embryo and that information gained from the mESC model can be translated to the mouse embryo as well as to the human system.

The identification of the Flk-1[pos]/T-EGFP[pos] hemangioblast as the earliest stage of hematopoietic commitment in the ESC system was important because it provided a readily detectable marker for quantifying blood cell commitment. By monitoring T-EGFP, Flk-1, and the hemato-

poietic-specific marker CD41 (Ferkowicz et al. 2003; Mikkola et al. 2003), we were able to map the stage-specific signaling requirements in the developmental pathway from ESCs to blood (Nostro et al. 2008). The outcome of this study showed that a combination of Wnt and activin/nodal signaling were required for the initial step, the formation of the T-EGFP[pos] PS population. Wnt, activin/nodal, and BMP4 were essential for induction of Flk-1[pos] hematopoietic mesoderm from the PS, whereas VEGF/Flk-1 signaling was required to specify this population to a hematopoietic fate. Further detailed analysis revealed that Wnt, produced endogenously by the Flk-1[pos] mesoderm, is essential for specification of the primitive erythroid lineage. This effect of Wnt appeared to be specific to primitive erythropoieis because the macrophage lineage, which also develops at this early stage, was not affected (Fig. 4).

More recent studies using the ESC model have shown that primitive erythroid specification is more complex and is regulated by cross-talk between the Wnt and Notch signaling pathways. Notch signaling inhibits primitive erythroid specification in vivo as well as in ESC differentiation cultures in vitro and therefore must be inhibited during primitive erythroid development (Fig. 4) (Hadland et al. 2004; Cheng et al. 2008). Numb, a Notch inhibitor, present in the early yolk sac as well as in developing hemangioblast colonies in culture likely functions in this capacity. The demonstration that the combination of enforced expression of numb together with activation of Wnt signaling displayed a synergistic effect on primitive erythroid development supports this interpretation (Cheng et al. 2008). Collectively, these findings provide the first detailed insights into the regulation of the primitive erythroid lineage, the first blood lineage to develop, and demonstrate that Wnt signaling has a pivotal role at this developmental step.

Cardiac Mesoderm

Cardiac mesoderm is derived from epiblast cells that move through the PS at a more distal site and at a slightly later time point than the hemangioblast mesoderm fated to colonize the yolk sac (Parameswaran and Tam 1995). Following induction, the cardiac mesodermal population migrates to the anterior region of the embryo where it is

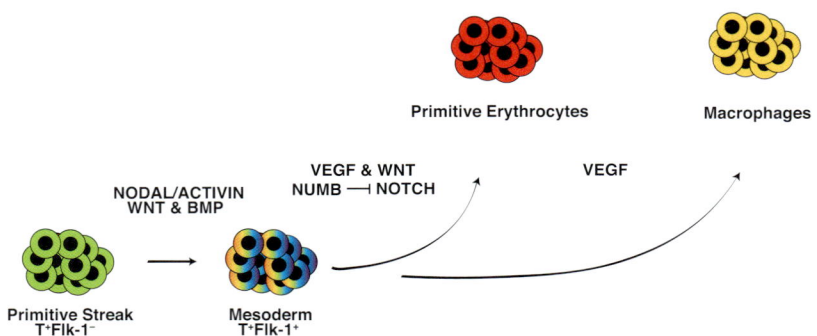

Figure 4. Model of primitive hematopoiesis during ESC differentiation. BMP4, nodal/activin A, and Wnt signaling are all necessary for the induction of a mesoderm population (T[+]Flk-1[+]) from primitive streak-like cells (T[+]Flk[−]). Wnt, VEGF, and numb function in concert to induce the primitive erythroid lineage from the Flk-1[+] mesoderm. VEGF alone is sufficient to induce the macrophage lineage from the same population.

exposed to cytokines of the BMP (Schultheiss et al. 1997) and FGF (Reifers et al. 2000) families as well as inhibitors of the Wnt pathway (Marvin et al. 2001; Schneider and Mercola 2001; Tzahor and Lassar 2001). This signaling environment specifies the mesodermal cells to a cardiac fate, leading to the formation of the cardiac crescent that gives rise to the primitive heart tube consisting of cardiac (myocardium) and endothelial (endocardium) cells.

To study the earliest stages of cardiac development in the ESC differentiation system, we developed a strategy to recreate in vitro the temporal allocation of mesoderm fates observed in the early embryo (Kouskoff et al. 2005). Using the t-EGFP reporter cell line, we identified the emergence of two distinct Flk-1^{pos}/T-EGFP^{pos} populations at different times in the ESC cultures (Kattman et al. 2006). The first Flk-1^{pos}/T-EGFP^{pos} population to develop contained hemangioblasts, whereas the second population was enriched for cardiac potential. Detailed analyses of the second cardiogenic Flk-1^{pos}/T-EGFP^{pos} population revealed that it contained cardiovascular progenitors that displayed endothelial, cardiomyocyte, and vascular smooth-muscle development. A comparable progenitor population was identified in the early mouse embryo, suggesting that the cardiovascular lineages in vivo and in vitro derive from a Flk-1^{pos} progenitor. The concept of a common Flk-1^{pos} cardiovascular progenitor is supported by the studies of Moretti et al. (2006), who described a Flk-1^{pos}/Isl1^{pos} population with similar potential, and by gene targeting and lineage-tracing studies showing that cells of the cardiac lineage express the receptor or are derived from cells that had expressed it (Motoike et al. 2003; Yamashita et al. 2005; Ema et al. 2006). The cardiovascular progenitors develop before the establishment of cardiac lineage, indicating that they define the onset of cardiovascular development in this model.

The identification of the cardiovascular progenitor in mESC cultures provided the basis for the identification of a comparable human progenitor (Yang et al. 2008). By recapitulating the signaling environment of the cardiac region of the mouse embryo in hESC differentiation cultures, we were able to establish conditions that promoted the efficient and reproducible development of the cardiac lineage from the hESCs. Analysis of the EBCs at different stages led to the identification of a distinct KDR^{pos}(Flk-1) population between day five and six of differentiation that displayed cardiovascular potential. The cardiogenic KDR^{pos} population was easily distinguished from the hematopoietic/vascular KDR^{pos} population by lack of c-KIT expression. As observed in the mouse system, the human cardiogenic KDR^{pos} population contained cardiovascular progenitors that displayed cardiac, endothelial, and vascular smooth-muscle potential.

The identification of cardiovascular progenitors in both mouse and human cultures highlights important similarities between these species in the establishment of lineages that participate in the formation of the first organs to develop in the early embryo. Access to these progenitors provides an enriched source of cardiomyocytes for in vitro and in vivo studies as well as a unique opportunity for investigating the molecular pathways that regulate the specification of the cardiac, endothelial, and vascular smooth-muscle lineages that comprise three of the major cell types of the adult heart.

Genetic Modification of PSCs

The ability to genetically alter hPSCs is a prerequisite for creating reporter cell lines to monitor specific developmental steps, for establishing gain- and loss-of-function approaches to investigate the molecular mechanisms regulating early human development, and ultimately for repairing genetic mutations in patient-specific iPSCs. With mESCs, HR has been the most widely used method for genetic modification because it enables the investigator to either create or repair mutations as well as to generate reporter ESC lines and mice with confidence that the transgene will faithfully recapitulate expression of the endogenous locus. In contrast to the routine use of this technology in mESCs, its application to hESCs has been limited (Zwaka and Thomson 2003; Irion et al. 2007; Davis et al. 2008). Although detailed protocols of the entire strategy have been published addressing some of the technical differences between mouse and human ESCs, the methodology has not translated well to the human system (Zwaka and Thomson 2003; Costa et al. 2007).

The identification of specific loci that allow broad expression of targeted genetic material (reporter genes, cDNAs) has had an important role in a wide variety of developmental biology and genetic studies in the mouse. The most notable of these loci is Rosa26, identified originally through a gene-trap approach (Zambrowicz et al. 1997). Rosa26 is a preferred site for genetic modification because it is expressed in most tissues throughout development and in adult life and is relatively easy to target through homologous recombination. Using a bioinformatics approach, we identified a locus in the human genome that shared marked homology with the mouse Rosa26 sequence (Irion et al. 2007). The locus is broadly expressed in fetal and adult tissues and can be targeted through HR using a gene-trap approach, comparable to the one routinely used to target the mouse locus (Soriano 1999; Luche et al. 2007). Importantly, the tandem-dimer red fluorescent protein (tdRFP) (Campbell et al. 2002) reporter introduced into human ROSA26 was expressed in the undifferentiated hESCs as well as in derivatives of all three germ layers generated during the in vitro differentiation of these cells. There was no evidence of silencing of the locus following long periods of culture of the undifferentiated cells or following their differentiation to derivative lineages (Fig. 5). The targeting vector contained two loxP sites (one wild type and one mutant) that created a "docking" site in the ROSA26 locus for introduction of new genetic material through recombinase-mediated cassette exchange (RMCE) (Bouhassira et al. 1997). Using RMCE, it was possible to replace the tdRFP cDNA with other sequences including the EGFP and puromycin resistance cDNAs. Strategies to introduce vectors that allow the inducible expression of transgenes from the ROSA26 locus, now under way, will provide a model system for using gain- and loss-of-gene-function approaches to study human development.

Given that the standard approaches for HR can be slow and have met with limited success in the human system, other state-of-the-art technologies for genetic modification of hPSCs must be considered. One interesting approach is the use of zinc finger nucleases (ZFNs) that can dramati-

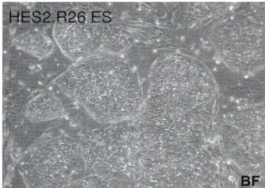
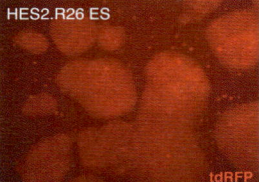
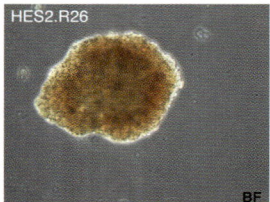
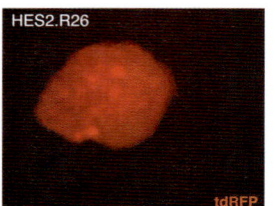

Figure 5. Genetic modification of human embryonic stem cells. HES2 hESCs were targeted at the *ROSA26* locus with a targeting construct harboring a red fluorescent protein (tdRFP) cDNA. Under bright-field microscopy (BF, *upper left*) these cells show normal morphology and display robust expression of tdRFP when examined by fluorescent microscopy (tdRFP, *upper right*). The targeted cells are able to differentiate into the hematopoietic lineage, as demonstrated by the development of an erythroid colony in methylcellulose cultures (BF, *lower left*). The cells within the erythroid colony express tdRFP (tdRFP, *lower right*), demonstrating that expression of the reporter is maintained following differentiation of the targeted hESCs.

cally increase the efficiency of HR through directing double-strand breaks to the desired target gene (Urnov et al. 2005; Lombardo et al. 2007). When ZFNs are delivered to the cells together with a conventional targeting construct, HR is significantly more efficient compared to a random double-strand break, possibly alleviating the need for drug selection (Santiago et al. 2008). Other approaches that may improve the efficiency of HR in hESCs include the use of adeno-associated viruses to target a specific sequence in the human genome. Proof-of-principle studies in the mouse have demonstrated that this strategy can be used to correct a small deletion in the *Rosa26* locus (Miller et al. 2004, 2006).

In addition to HR, other approaches that rely on random integration of the introduced genetic material in hPSCs are being developed and optimized. These include the use of lentiviral vectors (Lombardo et al. 2007), bacterial artificial chromosomes (Placantonakis et al. 2008), the sleeping beauty transposon system (Wilber et al. 2007), and the φC31 integrase (Thyagarajan et al. 2008). All have been successfully used to genetically modify mammalian cells and all have their own unique advantages and disadvantages. As the field matures and as these and other technologies yet to be developed are evaluated, the efficient and genetic modification of hPSCs will be become a routine practice in most labs.

ACKNOWLEDGMENTS

We thank members of the Keller lab for many helpful discussions and a critical reading of the manuscript. This work was supported by grants from the National Institutes of Health (G.M.K., S.J.K.) and the Leopoldina Stiftung der Naturforscher (S.I.).

REFERENCES

Bachiller, D., Klingensmith, J., Kemp, C., Belo, J.A., Anderson, R.M., May, S.R., McMahon, J.A., McMahon, A.P., Harland, R.M., Rossant, J., et al. 2000. The organizer factors Chordin and Noggin are required for mouse forebrain development. *Nature* **403:** 658–661.

Basma, H., Soto-Gutierrez, A., Yannam, G.R., Liu, L., Ito, R., Yamamoto, T., Ellis, E., Carson, S.D., Sato, S., Chen, Y., et al. 2008. Differentiation and transplantation of human embryonic stem cell-derived hepatocytes. *Gastroenterology* **3:** 990–999.

Bouhassira, E.E., Westerman, K., and Leboulch, P. 1997. Transcriptional behavior of LCR enhancer elements integrated at the same chromosomal locus by recombinase-mediated cassette exchange. *Blood* **90:** 3332–3344.

Campbell, R.E., Tour, O., Palmer, A.E., Steinbach, P.A., Baird, G.S., Zacharias, D.A., and Tsien, R.Y. 2002. A monomeric red fluorescent protein. *Proc. Natl. Acad. Sci.* **99:** 7877–7882.

Cheng, X., Huber, T.L., Chen, V.C., Gadue, P., and Keller, G.M. 2008. Numb mediates the interaction between Wnt and Notch to modulate primitive erythropoietic specification from the hemangioblast. *Development* **135:** 3447–3458.

Choi, K., Kennedy, M., Kazarov, A., Papadimitriou, J.C., and Keller, G. 1998. A common precursor for hematopoietic and endothelial cells. *Development* **125:** 725–732.

Conlon, F.L., Lyons, K.M., Takaesu, N., Barth, K.S., Kispert, A., Herrmann, B., and Robertson, E.J. 1994. A primary requirement for nodal in the formation and maintenance of the primitive streak in the mouse. *Development* **120:** 1919–1928.

Costa, M., Dottori, M., Sourris, K., Jamshidi, P., Hatzistavrou, T., Davis, R., Azzola, L., Jackson, S., Lim, S.M., Pera, M., Elefanty, A.G., and Stanley, E.G. 2007. A method for genetic modification of human embryonic stem cells using electroporation. *Nat. Protoc.* **2:** 792–796.

D'Amour, K.A., Agulnick, A.D., Eliazer, S., Kelly, O.G., Kroon, E., and Baetge, E.E. 2005. Efficient differentiation of human embryonic stem cells to definitive endoderm. *Nat. Biotechnol.* **23:** 1534–1541.

D'Amour, K.A., Bang, A.G., Eliazer, S., Kelly, O.G., Agulnick, A.D., Smart, N.G., Moorman, M.A., Kroon, E., Carpenter, M.K., and Baetge, E.E. 2006. Production of pancreatic hormone-expressing endocrine cells from human embryonic stem cells. *Nat. Biotechnol.* **24:** 1392–1401.

Davis, R.P., Ng, E.S., Costa, M., Mossman, A.K., Sourris, K., Elefanty, A.G., and Stanley, E.G. 2008. Targeting a GFP reporter gene to the *MIXL1* locus of human embryonic stem cells identifies human primitive streak-like cells and enables isolation of primitive hematopoietic precursors. *Blood* **111:** 1876–1884.

Dush, M.K. and Martin, G.R. 1992. Analysis of mouse *Evx* genes: *Evx-1* displays graded expression in the primitive streak. *Dev. Biol.* **151:** 273–287.

Ema, M., Takahashi, S., and Rossant, J. 2006. Deletion of the selection cassette, but not *cis*-acting elements, in targeted *Flk1-lacZ* allele reveals *Flk1* expression in multipotent mesodermal progenitors. *Blood* **107:** 111–117.

Evans, M. and Kaufman, M. 1981. Establishment in culture of pluripotent cells from mouse embryos. *Nature* **292:** 154–156.

Fehling, H.J., Lacaud, G., Kubo, A., Kennedy, M., Robertson, S., Keller, G., and Kouskoff, V. 2003. Tracking mesoderm induction and its specification to the hemangioblast during embryonic stem cell differentiation. *Development* **130:** 4217–4227.

Feldman, B., Gates, M.A., Egan, E.S., Dougan, S.T., Rennebeck, G., Sirotkin, H.I., Schier, A.F., and Talbot, W.S. 1998. Zebrafish organizer development and germ-layer formation require nodal-related signals. *Nature* **395:** 181–185.

Ferkowicz, M.J., Starr, M., Xie, X., Li, W., Johnson, S.A., Shelley, W.C., Morrison, P.R., and Yoder, M.C. 2003. CD41 expression defines the onset of primitive and definitive hematopoiesis in the murine embryo. *Development* **130:** 4393–4403.

Forlani, S., Lawson, K.A., and Deschamps, J. 2003. Acquisition of Hox codes during gastrulation and axial elongation in the mouse embryo. *Development* **130:** 3807–3819.

Gadue, P., Huber, T.L., Paddison, P.J., and Keller, G.M. 2006. Wnt and TGF-β signaling are required for the induction of an in vitro model of primitive streak formation using embryonic stem cells. *Proc. Natl. Acad. Sci.* **103:** 16806–16811.

Gadue, P., Huber, T.L., Nostro, M.C., Kattman, S., and Keller, G.M. 2005. Germ layer induction from embryonic stem cells. *Exp. Hematol.* **33:** 955–964.

Glaser, S., Anastassiadis, K., and Stewart, A.F. 2005. Current issues in mouse genome engineering. *Nat. Genet.* **37:** 1187–1193.

Gouon-Evans, V., Boussemart, L., Gadue, P., Nierhoff, D., Koehler, C.I., Kubo, A., Shafritz, D.A., and Keller, G. 2006. BMP-4 is required for hepatic specification of mouse embryonic stem cell-derived definitive endoderm. *Nat. Biotechnol.* **24:** 1402–1411.

Gritsman, K., Talbot, W.S., and Schier, A.F. 2000. Nodal signaling patterns the organizer. *Development* **127:** 921–932.

Haar, J.L. and Ackerman, G.A. 1971. A phase and electron microscopic study of vasculogenesis and erythropoiesis in the yolk sac of the mouse. *Anat. Rec.* **170:** 199–223.

Hadland, B.K., Huppert, S.S., Kanungo, J., Xue, Y., Jiang, R., Gridley, T., Conlon, R.A., Cheng, A.M., Kopan, R., and Longmore, G.D. 2004. A requirement for Notch1 distinguishes 2 phases of definitive hematopoiesis during development. *Blood* **104:** 3097–3105.

Hart, A.H., Hartley, L., Sourris, K., Stadler, E.S., Li, R., Stanley, E.G., Tam, P.P., Elefanty, A.G., and Robb, L. 2002. Mixl1 is required for axial mesendoderm morphogenesis and patterning in the murine embryo. *Development* **129:** 3597–3608.

Hogan, B.L. 1996. Bone morphogenetic proteins in development. *Curr. Opin. Genet. Dev.* **6:** 432–438.

Huber, T.L., Kouskoff, V., Fehling, H.J., Palis, J., and Keller, G. 2004. Haemangioblast commitment is initiated in the primitive streak of the mouse embryo. *Nature* **432:** 625–630.

Irion, S., Luche, H., Gadue, P., Fehling, H.J., Kennedy, M., and Keller, G.M. 2007. Identification and targeting of the ROSA26 locus in human embryonic stem cells. *Nat. Biotechnol.* **25:** 1477–1482.

Jones, C.M., Kuehn, M.R., Hogan, B.L., Smith, J.C., and Wright, C.V. 1995. Nodal-related signals induce axial mesoderm and dorsalize mesoderm during gastrulation. *Development* **121:** 3651–3662.

Kattman, S.J., Huber, T.L., and Keller, G.M. 2006. Multipotent flk-1+ cardiovascular progenitor cells give rise to the cardiomyocyte, endothelial, and vascular smooth muscle lineages. *Dev. Cell* **11:** 723–732.

Keller, G. 2005. Embryonic stem cell differentiation: Emergence of a new era in biology and medicine. *Genes Dev.* **19:** 1129–1155.

Kennedy, M., D'Souza, S.L., Lynch-Kattman, M., Schwantz, S., and Keller, G. 2007. Development of the hemangioblast defines the onset of hematopoiesis in human ES cell differentiation cultures. *Blood* **109:** 2679–2687.

Kinder, S.J., Tsang, T.E., Quinlan, G.A., Hadjantonakis, A.K., Nagy, A., and Tam, P.P. 1999. The orderly allocation of mesodermal cells to the extraembryonic structures and the anteroposterior axis during gastrulation of the mouse embryo. *Development* **126:** 4691–4701.

Kinder, S.J., Tsang, T.E., Wakamiya, M., Sasaki, H., Behringer, R.R., Nagy, A., and Tam, P.P. 2001. The organizer of the mouse gastrula is composed of a dynamic population of progenitor cells for the axial mesoderm. *Development* **128:** 3623–

3634.

Kispert, A. and Herrmann, B.G. 1994. Immunohistochemical analysis of the *Brachyury* protein in wild-type and mutant mouse embryos. *Dev. Biol.* **161:** 179–193.

Kouskoff, V., Lacaud, G., Schwantz, S. Fehling, H.J., and Keller, G. 2005. Sequential development of hematopoietic and cardiac mesoderm during embryonic stem cell differentiation. *Proc. Natl. Acad. Sci.* **102:** 13170–13175.

Li, W., Wei, W., Zhu, S., Zhu, J., Shi, Y., Lin, T., Hao, E., Hayek, A., Deng, H., and Ding, S. 2009. Generation of rat and human induced pluripotent stem cells by combining genetic reprogramming and chemical inhibitors. *Cell Stem Cell* **4:** 16–19.

Lombardo, A., Genovese, P., Beausejour, C.M., Colleoni, S., Lee, Y.L., Kim, K.A., Ando, D., Urnov, F.D., Galli, C., Gregory, P.D., Holmes, M.C., and Naldini, L. 2007. Gene editing in human stem cells using zinc finger nucleases and integrasedefective lentiviral vector delivery. *Nat. Biotechnol.* **25:** 1298–1306.

Lowe, L.A., Yamada, S., and Kuehn, M.R. 2001. Genetic dissection of nodal function in patterning the mouse embryo. *Development* **128:** 1831–1843.

Luche, H., Weber, O., Nageswara Rao, T., Blum, C., and Fehling, H.J. 2007. Faithful activation of an extra-bright red fluorescent protein in "knock-in" Cre-reporter mice ideally suited for lineage tracing studies. *Eur. J. Immunol.* **37:** 43–53.

Martin, G.R. 1981. Isolation of a pluripotent cell line from early mouse embryos cultured in medium conditioned by teratocarcinoma stem cells. *Proc. Natl. Acad. Sci* **78:** 7634–7638.

Marvin, M.J., Di Rocco, G., Gardiner, A., Bush, S.M., and Lassar, A.B. 2001. Inhibition of Wnt activity induces heart formation from posterior mesoderm. *Genes Dev.* **15:** 316–327.

Matsumoto, K., Yoshitomi, H., Rossant, J., and Zaret, K.S. 2001. Liver organogenesis promoted by endothelial cells prior to vascular function. *Science* **294:** 559–563.

McMahon, J.A., Takada, S., Zimmerman, L.B., Fan, C.M., Harland, R.M., and McMahon, A.P. 1998. Noggin-mediated antagonism of BMP signaling is required for growth and patterning of the neural tube and somite. *Genes Dev.* **12:** 1438–1452.

Mikkola, H.K., Fujiwara, Y., Schlaeger, T.M., Traver, D., and Orkin, S.H. 2003. Expression of CD41 marks the initiation of definitive hematopoiesis in the mouse embryo. *Blood* **101:** 508–516.

Millauer, B., Wizigmann-Voos, S., Schnurch, H., Martinez, R., Moller, N.P., Risau, W., and Ullrich, A. 1993. High affinity VEGF binding and developmental expression suggest Flk-1 as a major regulator of vasculogenesis and angiogenesis. *Cell* **72:** 835–846.

Miller, D.G., Petek, L.M., and Russell, D.W. 2004. Adeno-associated virus vectors integrate at chromosome breakage sites. *Nat. Genet.* **36:** 767–773.

Miller, D.G., Wang, P.R., Petek, L.M., Hirata, R.K., Sands, M.S., and Russell, D.W. 2006. Gene targeting in vivo by adeno-associated virus vectors. *Nat. Biotechnol.* **24:** 1022–1026.

Monaghan, A.P., Kaestner, K.H., Grau, E., and Schutz, G. 1993. Postimplantation expression patterns indicate a role for the mouse *forkhead*/HNF-3 α, β and γ genes in determination of the definitive endoderm, chordamesoderm and neuroectoderm. *Development* **119:** 567–578.

Moretti, A., Caron, L., Nakano, A., Lam, J.T., Bernshausen, A., Chen, Y., Qyang, Y., Bu, L., Sasaki, M., Martin-Puig, S., et al. 2006. Multipotent embryonic *isl1+* progenitor cells lead to cardiac, smooth muscle, and endothelial cell diversification. *Cell* **127:** 1151–1165.

Motoike, T., Markham, D.W., Rossant, J., and Sato, T.N. 2003. Evidence for novel fate of Flk1+ progenitor: Contribution to muscle lineage. *Genesis* **35:** 153–159.

Murray, P.D.F. 1932. The development of in vitro of the blood of the early chick embryo. *Proc. Roy. Soc. Lond.* **11:** 497–521.

Ng, E.S., Azzola, L., Sourris, K., Robb, L., Stanley, E.G., and Elefanty, A.G. 2005. The primitive streak gene *Mixl1* is required for efficient haematopoiesis and BMP4-induced ventral mesoderm patterning in differentiating ES cells. *Development* **132:** 873–884.

Nishikawa, S.I., Nishikawa, S., Hirashima, M., Matsuyoshi, N., and Kodama, H. 1998. Progressive lineage analysis by cell sorting and culture identifies FLK1$^+$VE-cadherin$^+$ cells at a diverging point of endothelial hematopoietic lineages. *Development* **125:** 1747–1757.

Nostro, M.C., Cheng, X., Keller, G.M., and Gadue, P. 2008. Wnt, activin, and BMP signaling regulate distinct stages in the developmental pathway from embryonic stem cells to blood. *Cell Stem Cell* **2:** 60–71.

Osada, S.I. and Wright, C.V. 1999. *Xenopus* nodal-related signaling is essential for mesendodermal patterning during early embryogenesis. *Development* **126:** 3229–3240.

Parameswaran, M. and Tam, P.P. 1995. Regionalisation of cell fate and morphogenetic movement of the mesoderm during mouse gastrulation. *Dev. Genet.* **17:** 16–28.

Park, I.H., Zhao, R., West, J.A., Yabuuchi, A., Huo, H., Ince, T.A., Lerou, P.H., Lensch, M.W., and Daley, G.Q. 2008. Reprogramming of human somatic cells to pluripotency with defined factors. *Nature* **451:** 141–146.

Placantonakis, D.G., Tomishima, M.J., Lafaille, F., Desbordes, S.C., Jia, F., Socci, N.D., Viale, A., Lee, H., Harrison, N., Tabar, V., et al. 2008. BAC transgenesis in human ES cells as a novel tool to define the human neural lineage. *Stem Cells* (in press).

Quinn, T., Peters, K., De Vries, C., Ferrara, N., and Williams, L. 1993. Fetal liver kinase 1 is a receptor for vascular endothelial growth factor and is selectively expressed in vascular endothelium. *Proc. Natl. Acad. Sci.* **90:** 7533–7537.

Reifers, F., Walsh, E.C., Leger, S., Stainier, D.Y., and Brand, M. 2000. Induction and differentiation of the zebrafish heart requires fibroblast growth factor 8 (*fgf8/acerebellar*). *Development* **127:** 225–235.

Russell, E. 1979. Heriditary anemias of the mouse: A review for geneticists. *Adv. Genet.* **20:** 357–459.

Sabin, F.R. 1917. Origin and development of the primitive vessels of the chick and of the pig. *Contrib. Embryol. Carnegie Inst.* **226:** 61–124.

Santiago, Y., Chan, E., Liu, P.Q., Orlando, S., Zhang, L., Urnov, F.D., Holmes, M.C., Guschin, D., Waite, A., Miller, J.C., et al. 2008. Targeted gene knockout in mammalian cells by using engineered zinc-finger nucleases. *Proc. Natl. Acad. Sci.* **105:** 5809–5814.

Sasaki, H. and Hogan, B.L. 1993. Differential expression of multiple forkhead related genes during gastrulation and axial pattern formation in the mouse embryo. *Development* **118:** 47–59.

Schier, A.F. and Shen, M.M. 2000. Nodal signalling in vertebrate development. *Nature* **403:** 385–389.

Schneider, V.A. and Mercola, M. 2001. Wnt antagonism initiates cardiogenesis in *Xenopus laevis*. *Genes Dev.* **15:** 304–315.

Schultheiss, T.M., Burch, J.B., and Lassar, A.B. 1997. A role for bone morphogenetic proteins in the induction of cardiac myogenesis. *Genes Dev.* **11:** 451–462.

Soriano, P. 1999. Generalized *lacZ* expression with the ROSA26 Cre reporter strain. *Nat. Genet.* **21:** 70–71.

Stevens, L.C. and Little, C.C. 1954. Spontaneous testicular teratomas in an inbred strain of mice. *Proc. Natl. Acad. Sci.* **40:** 1080–1087.

Sumi, T., Tsuneyoshi, N., Nakatsuji, N., and Suemori, H. 2008. Defining early lineage specification of human embryonic stem cells by the orchestrated balance of canonical Wnt/β-catenin, activin/nodal and BMP signaling. *Development* **135:** 2969–2979.

Takahashi, K. and Yamanaka, S. 2006. Induction of pluripotent stem cells from mouse embryonic and adult fibroblast cultures by defined factors. *Cell* **126:** 663–676.

Takahashi, K., Tanabe, K., Ohnuki, M., Narita, M., Ichisaka, T.,

Tomoda, K., and Yamanaka, S. 2007. Induction of pluripotent stem cells from adult human fibroblasts by defined factors. *Cell* **131:** 861–872.

Tam, P.P. and Behringer, R.R. 1997. Mouse gastrulation: The formation of a mammalian body plan. *Mech. Dev.* **68:** 3–25.

Thomson, J.A., Itskovitz-Eldor, J., Shapiro, S.S., Waknitz, M.A., Swiergiel, J.J., Marshall, V.S., and Jones, J.M. 1998. Embryonic stem cell lines derived from human blastocysts. *Science* **282:** 1145–1147.

Thyagarajan, B., Liu, Y., Shin, S., Lakshmipathy, U., Scheyhing, K., Xue, H., Ellerstrom, C., Strehl, R., Hyllner, J., Rao, M.S., and Chesnut, J.D. 2008. Creation of engineered human embryonic stem cell lines using phiC31 integrase. *Stem Cells* **26:** 119–126.

Tzahor, E. and Lassar, A.B. 2001. Wnt signals from the neural tube block ectopic cardiogenesis. *Genes Dev.* **15:** 255–260.

Urnov, F.D., Miller, J.C., Lee, Y.L., Beausejour, C.M., Rock, J.M., Augustus, S., Jamieson, A.C., Porteus, M.H., Gregory, P.D., and Holmes, M.C. 2005. Highly efficient endogenous human gene correction using designed zinc-finger nucleases. *Nature* **435:** 646–651.

Wells, J.M. and Melton, D.A. 1999. Vertebrate endoderm development. *Annu. Rev. Cell Dev. Biol.* **15:** 393–410.

Wilber, A., Linehan, J.L., Tian, X., Woll, P.S., Morris, J.K., Belur, L.R., McIvor, R.S., and Kaufman, D.S. 2007. Efficient and stable transgene expression in human embryonic stem cells using transposon-mediated gene transfer. *Stem Cells* **25:** 2919–2927.

Yamaguchi, T.P. 2001. Heads or tails: Wnts and anterior-posterior patterning. *Curr. Biol.* **11:** R713–R724.

Yamaguchi, T.P., Dumont, D.J., Conlon, R.A., Breitman, M.L., and Rossant, J. 1993. *flk-1*, an *flt*-related receptor tyrosine kinase is an early marker for endothelial cell precursors. *Development* **118:** 489–498.

Yamashita, J.K., Takano, M., Hiraoka-Kanie, M., Shimazu, C., Peishi, Y., Yanagi, K., Nakano, A., Inoue, E., Kita, F., and Nishikawa, S. 2005. Prospective identification of cardiac progenitors by a novel single cell-based cardiomyocyte induction. *FASEB J.* **19:** 1534–1536.

Yang, L., Soonpaa, M.H., Adler, E.D., Roepke, T.K., Kattman, S.J., Kennedy, M., Henckaerts, E., Bonham, K., Abbott, G.W., Linden, R.M., Field, L.J., and Keller, G.M. 2008. Human cardiovascular progenitor cells develop from a KDR$^+$ embryonic-stem-cell-derived population. *Nature* **453:** 524–528.

Yoshitomi, H. and Zaret, K.S. 2004. Endothelial cell interactions initiate dorsal pancreas development by selectively inducing the transcription factor Ptf1a. *Development* **131:** 807–817.

Yu, J., Vodyanik, M.A., Smuga-Otto, K., Antosiewicz-Bourget, J., Frane, J.L., Tian, S., Nie, J., Jonsdottir, G.A., Ruotti, V., Stewart, R., et al. 2007. Induced pluripotent stem cell lines derived from human somatic cells. *Science* **318:** 1917–1920.

Zambrowicz, B.P., Imamoto, A., Fiering, S., Herzenberg, L.A., Kerr, W.G., and Soriano, P. 1997. Disruption of overlapping transcripts in the ROSA βgeo 26 gene trap strain leads to widespread expression of β-galactosidase in mouse embryos and hematopoietic cells. *Proc. Natl. Acad. Sci.* **94:** 3789–3794.

Zaret, K.S. 2008. Genetic programming of liver and pancreas progenitors: Lessons for stem-cell differentiation. *Nat. Rev. Genet.* **9:** 329–340.

Zhou, X., Sasaki, H., Lowe, L., Hogan, B.L., and Kuehn, M.R. 1993. *Nodal* is a novel TGF-β-like gene expressed in the mouse node during gastrulation. *Nature* **361:** 543–547.

Zwaka, T.P. and Thomson, J.A. 2003. Homologous recombination in human embryonic stem cells. *Nat. Biotechnol.* **21:** 319–321.

Regulation of Stem Cells in the Zebra Fish Hematopoietic System

H.-T. Huang and L.I. Zon

Harvard Medical School, Stem Cell Program and Division of Hematology/Oncology,
Children's Hospital and Dana Farber Cancer Institute, Howard Hughes Medical Institute,
Harvard Stem Cell Institute, Boston, Massachusetts 02115

Hematopoietic stem cells (HSCs) have been used extensively as a model for stem cell biology. Stem cells share the ability to self-renew and differentiate into multiple cell types, making them ideal candidates for tissue regeneration or replacement therapies. Current applications of stem cell technology are limited by our knowledge of the molecular mechanisms that control their proliferation and differentiation, and various model organisms have been used to fill these gaps. This chapter focuses on the contributions of the zebra fish model to our understanding of stem cell regulation within the hematopoietic system. Studies in zebra fish have been valuable for identifying new genetic and signaling factors that affect HSC formation and development with important implications for humans, and new advances in the zebra fish toolbox will allow other aspects of HSC behavior to be investigated as well, including migration, homing, and engraftment.

Stem cells and early progenitors are important for organ formation during development and support tissue function throughout an organism's lifetime. These principles are illustrated in the blood system, where HSCs are needed to maintain a constant pool of progenitors committed to the various blood lineages to replenish the mature blood cells that turn over. Consequently, HSCs, like other stem cells, have the ability to self-renew in order to generate more HSCs and to differentiate along multiple lineage pathways to make erythrocytes, megakaryocytes, monocytes/macrophages, neutrophils, or lymphocytes. Because of their potential therapeutic value, HSCs have been the subject of intense study for many years, but methods for maintaining them in vitro and differentiating them into specific cell types are limited. In addition, HSCs remain difficult to study because their ontogeny is tightly regulated in a spatial and temporal manner, with progenitors of varying potentials arising from different sites.

Research using the zebra fish (*Danio rerio*) model has provided new insights into some of the major issues regarding the regulation and function of HSCs (Davidson and Zon 2004; Carradice and Lieschke 2008; Orkin and Zon 2008). The advantages of using zebra fish include its high fecundity, rapid growth, and external development of transparent embryos, which facilitates visualization of early embryonic processes. Most importantly, the genes controlling hematopoiesis are highly conserved in fish and mammals. Many blood mutants have been isolated from large-scale genetic screens (Ransom et al. 1996; Weinstein et al. 1996), thereby providing powerful tools for dissecting the molecular aspects of HSC control.

EMBRYONIC HEMATOPOIESIS: PRIMITIVE AND DEFINITIVE WAVES

The onset of hematopoiesis in the early embryo is characterized by the induction of distinct progenitors at various anatomical sites. Hematopoiesis subsequently shifts location during the course of development, as depicted in Figure 1. Blood formation occurs in two major waves. In zebra fish, the primitive wave consists of the formation of primitive erythrocytes (*gata1*⁺) in the intermediate cell

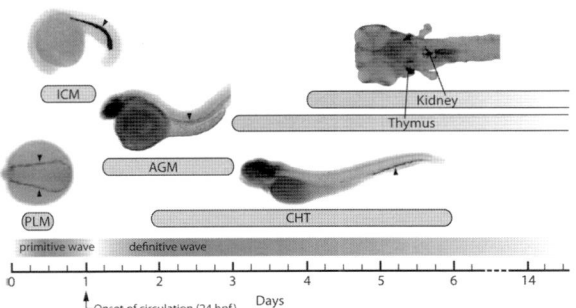

Figure 1. Embryonic hematopoiesis in zebra fish. Hematopoiesis occurs in two waves. In zebra fish, the primitive wave begins with the induction of precursors in the bilateral stripes of the posterior lateral mesoderm (PLM), which converge medially to form the intermediate cell mass (ICM) region where primitive erythrocytes are formed. After the onset of circulation (about 24 hours postfertilization [hpf]), definitive HSCs appear in the aorta-gonad mesonephros (AGM) region. These cells subsequently migrate and colonize the caudal hematopoietic tissue (CHT), thymus, and kidney. Each hematopoietic site is identified by in situ hybridization for *gata1* at 14 hpf (PLM), *gata1* at 20 hpf (ICM), c-*myb* at 36 hpf (AGM), *scl* at day 4 (CHT), and c-*myb* at day 6 (thymus and kidney). (Photos courtesy of X. Bai and T.V. Bowman.)

mass (ICM) region from posterior lateral mesoderm (PLM) (Detrich et al. 1995) and the emergence of primitive myeloid cells (*pu.1*[+]) from anterior lateral mesoderm (ALM) (Herbomel et al. 1999; Bennett et al. 2001). This initial wave of blood production is equivalent to blood formation on the extraembryonic yolk sac of other vertebrates. The primary function of primitive hematopoiesis is to provide red blood cells to deliver oxygen within the rapidly developing embryo. Although dispensable in zebra fish, this property allows primitive blood mutants to be studied in the fish because early blood defects in most other vertebrates are embryonic-lethal.

Soon after the formation of the ICM, erythromyeloid progenitor cells (*lmo2-gata1*[+]) appear in the posterior blood island (PBI) located in the tail region (Bertrand et al. 2007), and a wave of definitive HSC (*runx1*[+], c-*myb*[+], *ikaros*[+], *lmo2*[+], *scl*[+], and *CD41*[+]) production ensues in the aorta-gonad-mesonephros (AGM) region along the ventral wall of the dorsal aorta (Liao et al. 1998; Thompson et al. 1998; Willett et al. 2001; Kalev-Zylinska et al. 2002; Bertrand et al. 2008; Kissa et al. 2008). These AGM cells subsequently colonize the caudal hematopoietic tissue (CHT), an expansion of the PBI, as well as the thymus and kidney (Murayama et al. 2006; Jin et al. 2007; Kissa et al. 2008).

These secondary hematopoietic tissues provide a niche for blood progenitors to expand and begin to differentiate. The CHT is similar to mouse placenta or fetal liver, and developing lymphoid and myeloid cells in this region can be identified by markers such as c-*myb*, *scl*, *runx1*, and *ikaros* (Murayama et al. 2006; Zhang and Rodaway 2007). Thymic immigrants differentiate into *rag1*[+] lymphoid cells (Murayama et al. 2006; Jin et al. 2007; Kissa et al. 2008). The kidney is the adult hematopoietic organ in zebra fish equivalent to mammalian bone marrow, and like the CHT, c-*myb*, *scl*, *runx1*, and *ikaros* are expressed in this region (Murayama et al. 2006; Jin et al. 2007). Within the kidney marrow, HSCs reside adjacent to renal tubule epithelial cells (Kobayashi et al. 2008).

ORIGIN OF BLOOD CELLS: HEMANGIOBLASTS AND HEMOGENIC ENDOTHELIUM

Currently, it is thought that blood and vascular cells derive from a common progenitor based on shared marker expression and physical proximity between the two cell types during development. The lack of blood and vascular markers in developing zebra fish *cloche* (*clo*) mutants provides evidence for the existence of hemangioblasts (Strainier et al. 1995; Thompson et al. 1998), further supported by recent fate-mapping experiments in the early embryo. Photoactivation of fluorescein dextran in single cells within the ventral mesoderm at shield stage (6 hours postfertilization [hpf]) labeled blood and vascular cells later at 30 hpf (Vogeli et al. 2006).

Definitive HSCs are believed to arise from the hemogenic endothelium (Jaffredo et al. 1998; de Brujin et al. 2002) and are transplantable and capable of multilineage differentiation (Cumano et al. 1996; Medvinsky and Dzierzak 1996). In addition, other data suggest that mesenchymal cells ventral to the dorsal aorta have HSC poten-

tial (North et al. 2002). By analogy, HSCs produced within the zebra fish AGM region are believed to be equivalent to those found in the same region in mice, supported by expression of homologous markers and lineage-tracing data. Furthermore, hematopoietic mutants such as *clo* and *spadetail* (*spt*) that disrupt formation of the dorsal aorta show loss of HSC induction (Thompson et al. 1998).

TRANSCRIPTIONAL REGULATORS OF HSC FORMATION

Specification of hematopoietic cells involves both the action of master blood transcriptional regulators and signaling molecules from the surrounding tissues. The transcription factors important for hematopoiesis in zebra fish are listed in Table 1. Many of them have been implicated in blood development by the blood-specific roles these factors have in other vertebrates or by genetic analysis of mutants isolated from large-scale screens (Fig. 2). Together, they

Table 1. Hematopoietic transcription factors in zebra fish

Gene	Family	Loss of function	Reference
scl	basic helix loop helix	morphant	Dooley et al. (2005)
c/ebp1	bZIP	morphant	Su et al. (2007)
c/ebpa	bZIP	–	Lyons et al. (2001)
c/ebpb	bZIP	–	Lyons et al. (2001)
tif1y	B box, PHD, Bromo	*moonshine*	Ransom et al. (2004)
cbfb	core-binding factor	–	Blake et al. (2000)
ets1	ETS	–	Zhu et al. (2005)
etsrp	ETS	*y11*	Pham et al. (2007)
mef	ETS	–	Zhu et al. (2005)
pu.1	ETS	morphant	Rhodes et al. (2005)
fli1a	ETS	–	Brown et al. (2000)
cdx1a	homeobox	morphant	Davidson and Zan (2006)
cdx4	homeobox	*kugelig*	Davidson et al. (2003)
hhex	homeobox	deletion b16	Liao et al. (2000)
ldb1	LIM domain binding	–	Toyama et al. (1998)
lmo2	LIM domain	morphant	Patterson et al. (2007)
runx1	runt	morphant	Kalev-Zylinska et al. (2002)
runx3	runt	morphant	Kalev-Zylinska et al. (2003)
stat5	STAT	morphant	Paffett-Lugassy et al. (2007)
tbx16	T box	*spadetail*	Thompson et al. (1998)
trf3	TBP	morphant	Hart et al. (2007)
c-myb	zinc finger	deletion b316	Thompson et al. (1998)
draculin	zinc finger	–	Herbomel et al. (1999)
fog1	zinc finger	–	Nishikawa et al. (2003)
gata1	zinc finger	*vlad tepes*	Detrich et al. (1995)
gata2	zinc finger	morphant	Galloway et al. (2005)
gfi1.1	zinc finger	morphant	Wei et al. (2008)
ikaros	zinc finger	–	Willett et al. (2001)
klfd	zinc finger	–	Oates et al. (2001)
klf4	zinc finger	morphant	Gardiner et al. (2007)
klf12	zinc finger	–	Oates et al. (2001)
nfe2	zinc finger	–	Pratt et al. (2002)
ZBP-89	zinc finger	morphant	Li et al. (2006)

Adapted from Davidson and Zon (2004).

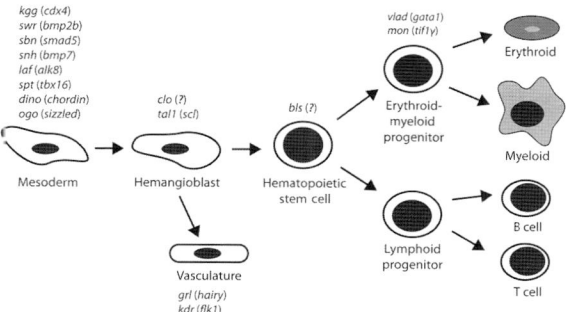

Figure 2. Mutants affecting blood development in zebra fish. Corresponding genes are in parenthesis.

form a transcriptional network that directs various aspects of the blood program from specification to differentiation.

scl and *lmo2* are expressed in hematopoietic progenitors and endothelial cells, possibly acting together to specify the hemangioblast (Patterson et al. 2007). *scl* encodes a basic helix-loop-helix transcription factor, whereas *lmo2* is a LIM-domain transcription factor. *scl* expression is initiated at 10 hpf and is coexpressed with *lmo2*, *gata2*, and *flk1* in the ALM and PLM (Gering et al. 1998; Thompson et al. 1998). A subset of these *scl*⁺ cells in the PLM becomes *gata1*⁺, committing to the erythroid lineage, whereas the *scl*⁺/*gata1*⁻ population is believed to form endothelial cells (Gering et al. 1998). In the ALM, *pu.1* expression is initiated with the differentiation of myeloid cells (Bennett et al. 2001). Overexpression of *scl* and *lmo2* expands the formation of blood and vascular progenitors along the anteroposterior (AP) axis (Gering et al. 2003), although *scl* itself is capable of activating hematopoietic genes independently of *lmo2* (Dooley et al. 2005). Once primitive erythroblasts from the ICM enter the circulation, additional markers appear that define the definitive wave.

Targeted disruption of *runx1* in mice has demonstrated its requirement for HSC induction in the AGM (Okuda et al. 1996; North et al. 2002). In zebra fish, *runx1* and c-*myb* are expressed in the AGM as well (Thompson et al. 1998; Burns et al. 2002; Gering and Patient 2005). *runx1* is a member of the runt family of transcriptional regulators involved in many developmental processes including blood and bone development in mammals. In zebra fish, *runx1* is first expressed in the bilateral stripes of the PLM that then migrate to the midline to form the ICM. It subsequently appears during the definitive wave in the ventral wall of the dorsal aorta (Burns et al. 2002). *runx1* overexpression causes HSC expansion in embryos, whereas morpholino knockdown causes defects in definitive hematopoiesis and vasculoangiogenesis (Burns et al. 2005). Furthermore, c-*myb* expression is reduced in *runx1* morphants (Kalev-Zylinska et al. 2002).

runx3 was found to regulate primitive and definitive hematopoietic cells but not vascular cells. Depletion of *runx3* resulted in decreased *runx1* expression in the ventral wall of the dorsal aorta. Conversely, overexpression of *runx3* increased primitive *scl*⁺ cell numbers and *runx1*⁺ cells in the aorta, suggesting that *runx3* regulates hematopoietic pro-

genitor numbers and cooperates with *runx1* to regulate HSC formation in the AGM (Kalev-Zylinska et al. 2003).

c-*myb* encodes a proto-oncogene that marks the initiation of HSCs in the ventral wall of the dorsal aorta. In zebra fish, it is expressed in primitive erythroid cells in the ICM region, but it is not required for the primitive wave (Thompson et al. 1998). Expression then shifts to the ventral wall of the dorsal aorta, presumably marking the definitive HSCs in the AGM region. Although these c-*myb*⁺ cells have not been transplanted, the lack of c-*myb* expression in *clo* and *spt* mutants provides further evidence for c-*myb* as a marker of definitive HSCs (Thompson et al. 1998).

The contribution of the *cdx-hox* pathway to specification of hematopoietic cell fate has been elucidated in zebra fish. *cdx* genes belong to the *caudal* family of homeobox transcription factors implicated in the regulation of *hox* gene expression and in AP patterning. Both *cdx1a* and *cdx4* establish the correct *hox* expression domains necessary for blood development in zebra fish (Davidson and Zon 2006). Loss of *cdx4* gene function in homozygous *kugelig* (*kgg*) mutants results in severe anemia with embryos having few *scl*⁺ and *gata1*⁺ cells, although the number of *flk1*⁺ angioblasts appears to be normal. All *hox* genes examined (*hoxb4*, *hoxb5a*, *hoxb6b*, *hoxb7a*, *hoxb8a*, *hoxb8b*, and *hoxa9a*) displayed altered expression patterns in *kgg* mutants, but overexpression of *hoxb7a* and *hox9a* could rescue erythropoiesis (Davidson et al. 2003). This pathway regulating blood specification was recently found to be conserved in mouse embryonic stem (ES) cells (Lengerke et al. 2008).

Additional transcription factors important for definitive hematopoiesis include *fli1a*, *hhex*, and *tbx16*. *fli1a* is an ETS-domain transcription factor implicated in proliferation or differentiation of hematopoietic precursors. It is coexpressed in the hemangioblasts of the PLM with *gata2*, diverging later to mark only the endothelial cells (Brown et al. 2000). Given that its initial expression is normal in *clo* mutants, *fli1a* may be the earliest marker of hemangioblasts. *hhex* encodes a homeobox-containing protein whose expression begins about 12 hpf in the ALM and PLM. Overexpression enhances blood and endothelial markers but is not essential for their development, which can be compensated by *scl* (Liao et al. 2000). *tbx16* encodes a T-box transcription factor that regulates mesodermal cell migration, which is defective in *spt* mutants. As described previously, abnormal somite patterning and accumulation of mesodermal cells perturb vessel formation, which subsequently leads to defective HSC formation as demonstrated by loss of hematopoietic markers *gata2*, *gata1*, and *runx1* in the PLM (Ho and Kane 1990; Thompson et al. 1998). Overexpression of *scl* rescues blood formation in *spt* mutants, indicating that *tbx16* is upstream of *scl* in directing HSC formation (Dooley et al. 2005).

The analogous expression of these different markers in zebra fish and other vertebrate models suggests that the molecular mechanisms are highly conserved. Once the hematopoietic precursors have been specified, additional blood transcription factors such as *gata1*, *pu.1*, and *ikaros* direct the lineage-specific differentiation of these progenitors into erythroid, myeloid, and lymphoid cell types, respectively. Some of the factors required for HSC formation (*runx1*, *scl*, and *lmo2*) reappear later within the differentiation of individual blood lineages, and conversely, factors that have more lineage-restricted roles (*pu.1*) can also be found in HSCs (Orkin and Zon 2008). How the same transcription factors are used at various stages of hematopoietic development is not well understood, but the fact that many of these genes are mutated or translocated in human hematopoietic malignancies underscores the importance of these transcription factors in blood development throughout the life of an animal.

PATHWAYS IN THE INDUCTION OF HSCS

In addition to transcriptional regulators, signal transduction pathways are also important for modulating blood formation. Although distinct hematopoietic precursors are generated in different anatomical sites, common signaling events at each site are expected to lead to blood formation. How these different signaling pathways are coupled to control stem cell induction and development is a subject of ongoing research.

The family of Hedgehog (Hh) proteins are known to be involved in embryonic patterning and cell-fate specification. Based on murine mutants, there was no role for Hh signaling in hematopoiesis (Dyer et al. 2001; Byrd et al. 2002). In zebra fish, Hh was found to be required for definitive but not primitive hematopoiesis (Gering and Patient 2005). It is secreted by midline structures (floor plate, notochord, and hypochord) in the developing embryo. When Hh signaling is inhibited by chemicals or in genetic mutants, embryos showed normal numbers of β-*globinE1*+ primitive erythrocytes but reduced *runx1*+ definitive stem cells and *rag1*+ thymocytes, suggesting that Hh is required only for induction of definitive HSCs. Impaired medial migration of *flk1*+ angioblasts was also observed, indicating the possibility that improper patterning of the aorta is the cause of HSC loss. These effects were similar to those seen with vascular endothelial growth factor (VEGF) and Notch inhibition (Gering and Patient 2005).

Notch signaling has been previously shown to be required for induction of HSCs during embryogenesis in mice (Okuda et al. 1996). The Notch pathway is highly conserved throughout evolution, regulating cell-fate decisions in a wide range of biological processes. Using zebra fish *mind bomb* mutants that lack an E3 ubiquitin ligase essential for Notch signaling, *runx1* was identified as a downstream effector (Burns et al. 2005). Overexpression of *notch1a* intracellular domain (NICD) expanded the population of c-*myb*+ and *runx1*+ cells in the AGM region, and this increase was not due to proliferation or conversion of vein-to-artery identity. This phenotype was recapitulated in *runx1* overexpressing embryos (Burns et al. 2005). Given that it rescues the *mind bomb* phenotype, *runx1* may be acting in parallel or downstream from Notch signaling. In addition, *runx1* morphants show reduced c-*myb*+ and *ikaros*+ cells at 50 hpf and loss of *rag1*+ thymocytes at 6 days (Gering and Patient 2005).

A new pathway modulating HSC formation by prostaglandins was recently identified in the zebra fish. Prostaglandins are part of the eicosanoid signal transduction pathway, with prostaglandin E2 (PGE2) being the main effector prostanoid produced in the zebra fish. They are regulated by cyclooxygenases Cox1 and Cox2 (Grosser et al. 2002). When treated with PGE2, zebra fish embryos showed increased *runx1*+ and *cmyb*+ cells in the AGM region by in situ hybridization, confocal microscopy, and quantitative polymerase chain reaction (PCR). Chemical or morpholino inhibition of the pathway reduced HSC formation. These results were verified in both colony-forming and -limiting dilution competitive transplantation assays in the mouse, demonstrating a functional conservation of prostaglandin signaling not only in inducing HSCs, but also in adult maintenance (North et al. 2007). 16,16-Dimethyl-PGE2 (dmPGE2), a stable derivative of PGE2, will be tested in a human phase I clinical trial to determine whether it can improve the efficiency of cord blood transplantations (Lord et al. 2007).

MIGRATORY ROUTES OF HSCs TO SECONDARY HEMATOPOIETIC SITES

The ontogeny of HSCs in the AGM region is followed by subsequent colonization of secondary hematopoietic sites, presumably as different niches become available to support HSC growth within the constantly evolving microenvironment of the developing embryo. Lineage tracing in the mouse has been complicated by the inability to stage embryos precisely in utero and to determine the kinetics of conditional recombination activity, making it difficult to identify clearly the anatomic origins of adult HSCs. One advantage of performing in vivo fate mapping in zebra fish is the optical transparency and external development of its embryos.

Live imaging of cells labeled with green fluorescent protein (GFP) driven by HSC-specific promoters, such as *CD41* and c-*myb*, and caged fluorescein-dextran cell-tracing experiments have shown that HSCs from the AGM region migrate to colonize the CHT, thymus, and pronephros (Murayama et al. 2006; Jin et al. 2007; Zhang and Rodaway 2007; Bertrand et al. 2008; Kissa et al. 2008). *CD41* marks HSCs in the mouse (Mitjavila-Garcia et al. 2002; Ferkowicz et al. 2003; Mikkola et al. 2003). In zebra fish, both *CD41*-GFP+ and c-*myb*-GFP+ cells were observed in the AGM region, consistent with *runx1* expression at the same site (North et al. 2007; Kissa et al. 2008). Knockdown of *runx1* suppressed the appearance of *CD41*-GFP+ cells in this region, the CHT, and thymus (Kissa et al. 2008). Transplantation of *CD41*-GFP+ AGM cells into sibling embryonic recipients demonstrated colonization of the thymus and CHT (Bertrand et al. 2008), although assays have yet to be developed in the zebra fish that can support long-term reconstitution of embryonic donor cells.

Unlike chick and mouse, where AGM HSCs presumably bud off into circulation from intra-aortic clusters, AGM cells in zebra fish enter the circulation through the cardinal vein (CV) to seed the CHT (Kissa et al. 2008). Migration to the CHT requires circulation because *CD41*-GFP+ and c-*myb*+ cells were not found in the CHT of *silent heart* (*sih*) morphants (Murayama et al. 2006), which lack blood flow due to disruption of cardiac tropomyosin. Seeding of the thymus occurs by both circulation and migration through the mesenchyme from either the AGM or the CHT. These cells then proliferate and generate *rag1*[+] T-lymphocyte precursors (Murayama et al. 2006; Kissa et al. 2008).

Recently, a novel route was found for HSCs to seed the kidney. Using time-lapse imaging of *CD41*-GFP and c-*myb*-GFP transgenic animals, hematopoietic cells were observed crawling along the pronephric tubules from the AGM toward the anterior glomeruli (Bertrand et al. 2008). The migration appears to be circulation-independent given that it remains intact in *sih* morphants. These migrating cells were found to express *runx1* and the pan-hematopoietic marker *CD45* as well, suggesting that they are stem cells (Bertrand et al. 2008).

ADVANCES IN ZEBRA FISH FOR THE STUDY OF HSCs

The power of the zebra fish model lies in the ability to perform large-scale genetic screens. Both forward genetic screens, using *N*-ethyl-*N*-nitrosourea (ENU) (Haffter et al. 1996; Weinstein et al. 1996) or insertional mutagenesis (Amsterdam et al. 1999), and reverse genetic screens, using targeted induced local lesions in genomes (TILLING) (Wienholds et al. 2002) and morpholinos, have been described. Zebra fish also provide a unique platform for conducting in vivo whole-animal chemical screening to identify novel compounds of therapeutic value. Although a number of hematopoietic mutants have been isolated from previous screens, a small number affect HSCs (Fig. 2); continued screening thus has the potential to generate new mutations that affect other pathways or regulators of HSC development. More precise genetic manipulations in zebra fish can be accomplished with transgenic fish, for example, using the heat shock Cre/*lox* system to induce tissue-specific gene expression (Feng et al. 2007). Recently, zinc finger nucleases have been used successfully for inducing targeted mutations in the germ line (Meng et al. 2008). The combination of these molecular methods makes it feasible to perform very precise genetic manipulations in zebra fish.

Assays that test stem cell function have also been developed in zebra fish. The major blood lineages (erythroid, lymphoid, myeloid, and precursors) can be segregated by flow cytometry using only the forward scatter and side scatter profiles (Traver et al. 2003); thus, multilineage reconstitution can be measured in irradiation-recovery or transplantation assays (Fig. 3, top left and bottom) (Traver et al. 2004). This method was used to determine the contribution of Notch and prostaglandin signaling to adult hematopoiesis, because these were both identified initially as regulators of HSC induction during embryogenesis (Burns et al. 2005; North et al. 2007). A brief dose of Notch activation or treatment with PGE2 enhanced marrow recovery postirradiation by expanding early multilineage precursors. *runx1*, *lmo2*, *scl*, and even *fli1* expression were significantly up-regulated in these cells (Burns et al. 2005; North et al. 2007). The effects of PGE2 were verified by limiting dilution transplantation assays of HSCs in the mouse (North et al. 2007). Although the mechanism by which these pathways enhance recovery remains unknown, they are ideal candidates for clinical use

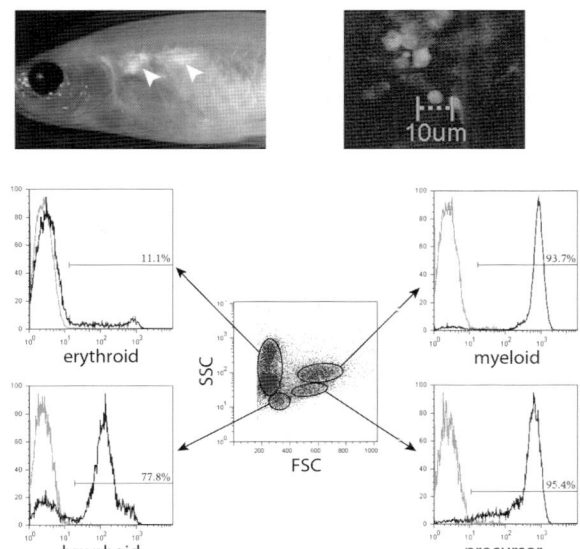

Figure 3. Multilineage reconstitution posttransplantation of transparent *casper* fish. (*Top left*) β-actin:GFP cells transplanted into irradiated *casper* fish show engraftment at 4 weeks posttransplantation. (*White arrowheads*) GFP[+] cells observed in the kidney by fluorescent microscopy. (*Top right*) In vivo visualization of recipient kidney marrow by confocal laser scanning shows single GFP[+] cells. Bar, 10 μm. (*Bottom*) Whole kidney marrow isolated from transplanted fish can be sorted into erythroid, lymphoid, myeloid, and precursor cells by flow cytometry based on forward scatter (FSC) and side scatter (SSC). Histograms of GFP[+] cells show reconstitution in the lymphoid, myeloid, and precuror populations. (*Top right,* Reprinted, with permission, from White et al. 2008 [© *Cell Press*]; *bottom*, images courtesy of J. de Jong.)

because any chemical that can enhance their signaling may have the potential to improve patient recovery posttransplantation by stimulating hematopoietic stem and progenitor cells.

Finally, visualization of HSCs in adult fish is now facilitated due to the recent development of a transparent zebra fish called *casper* (White et al. 2008). *casper* fish are doubly mutant for the nacre allele (encoding the *mitfa* gene) and the roy allele (encoding an unknown pigment gene), which blocks the development melanocytes and iridophores, respectively. As a result, the lack of pigmentation allows internal organs such as the heart, intestinal tube, liver, and gallbladder to be observed in vivo with the naked eye (White et al. 2008). HSC homeostasis could be studied in the *casper* fish within the context of the endogenous marrow niche, which has been a traditionally difficult process to observe. The ability to use these fish for examining the kinetics of stem cell homing and engraftment in the transplantation setting with resolution down to the single-cell level is currently unparalleled in other model systems (Fig. 3, top left and right).

CONCLUSIONS

Studies of HSCs in zebra fish have complemented investigations in other model organisms and have advanced our understanding of hematopoiesis. Much of what has been learned about the signaling pathways and transcription factors involved in the development of HSCs in zebra fish and other vertebrates is highly conserved. Given that zebra fish are amenable to large-scale screens, future genetic screens will continue to uncover new mutants with interesting hematopoietic phenotypes, and whole-animal chemical screening will identify new compounds of clinical value. In vivo screens are feasible using reporter lines and mutant and transgenic strains. As zebra fish methods advance, new opportunities for revealing more details of the molecular mechanisms of stem cell regulation will arise. Finally, the knowledge gained about HSCs will likely be applicable to stem cell biology in general.

ACKNOWLEDGMENTS

We thank X. Bai, J. de Jong, and R.M. White for a critical reading of the manuscript.

REFERENCES

Amsterdam, A., Burgess, S., Golling, G., Chen, W., Sun, Z., Townsend, K., Farrington, S., Haldi, M., and Hopkins, N. 1999. A large-scale insertional mutagenesis screen in zebrafish. *Genes Dev.* **13:** 2713–2724.

Bennett, C.M., Kanki, J.P., Rhodes, J., Liu, T.X., Paw, B.H., Kieran, M.W., Langenau, D.M., Delahaye-Brown, A., Zon, L.I., Fleming, M.D., and Look, A.T. 2001. Myelopoiesis in the zebrafish, *Danio rerio. Blood* **98:** 643–651.

Bertrand, J.Y., Kim, A.D., Ten, S., and Traver, D. 2008. CD41⁺ cmyb⁺ precursors colonize the zebrafish pronephros by a novel migration route to initiate adult hematopoiesis. *Development* **135:** 1853–1862.

Bertrand, J.Y., Kim, A.D., Violette, E.P., Stachura, D.L., Cisson, J.L., and Traver, D. 2007. Definitive hematopoiesis initiates through a committed erythromyeloid progenitor in the zebrafish embryo. *Development* **134:** 4147–4156.

Blake, T., Adya, N., Kim, C.H., Oates, A.C., Zon, L., Chitnis, A., Weinstein, B.M., and Liu, P.P. 2000. Zebrafish homolog of the leukemia gene *CBFB:* Its expression during embryogenesis and its relationship to *scl* and *gata-1* in hematopoiesis. *Blood* **96:** 4178–4184.

Brown, L.A., Rodaway, A.R., Schilling, T.F., Jowett, T., Ingham, P.W., Patient, R.K., and Sharrocks, A.D. 2000. Insights into early vasculogenesis revealed by expression of the ETS-domain transcription factor Fli-1 in wild-type and mutant zebrafish embryos. *Mech. Dev.* **90:** 237–252.

Burns, C.E., DeBlasio, T., Zhou, Y., Zon, L.I., and Nimer, S.D. 2002. Isolation and characterization of *runxa* and *runxb,* zebrafish members of the runt family of transcriptional regulators. *Exp. Hematol.* **30:** 1381–1389.

Burns, C.E., Traver, D., Mayhall, E., Shepard, J.L., and Zon, L.I. 2005. Hematopoietic stem cell fate is established by the Notch-Runx pathway. *Genes Dev.* **19:** 2331–2342.

Byrd, N., Becker, S., Maye, P., Narasimhaiah, R., St-Jacques, B., Zhang, X., McMahon, J., McMahon, A., and Grabel, L. 2002. Hedgehog is required for murine yolk sac angiogenesis. *Development* **129:** 361–372.

Carradice, D. and Lieschke, G.J. 2008. Zebrafish in hematology: Sushi or science? *Blood* **111:** 3331–3342.

Cumano, A., Dieterlen-Lièvre, F., and Godin, I. 1996. Lymphoid potential, probed before circulation in mouse, is restricted to caudal intraembryonic splanchnopleura. *Cell* **86:** 907–916.

Davidson, A.J. and Zon, L.I. 2004. The "definitive" (and "primitive") guide to zebrafish hematopoiesis. *Oncogene* **23:** 7233–7246.

Davidson, A.J. and Zon, L.I. 2006. The *caudal*-related homeobox genes *cdx1a* and *cdx4* act redundantly to regulate *hox* gene expression and the formation of putative hematopoietic stem cells during zebrafish embryogenesis. *Dev. Biol.* **292:** 506–518.

Davidson, A.J., Ernst, P., Wang, Y., Dekens, M.P.S., Kingsley, P.D., Palis, J., Korsmeyer, S.J., Daley, G.Q., and Zon, L.I. 2003. *cdx4* mutants fail to specify blood progenitors and can be rescued by multiple *hox* genes. *Nature* **425:** 300–306.

de Bruijn, M.F., Ma, X., Robin, C., Ottersbach, K., Samchez, M.J., and Dzierzak, E. 2002. HSCs localize to the endothelial layer in the midgestation mouse aorta. *Immunity* **16:** 673–683.

Detrich III, H.W., Kieran, M.W., Chan, F.Y., Barone, L.M., Yee, K., Rundstadler, J.A., Pratt, S., Ransom, D., and Zon, L.I. 1995. Intraembryonic hematopoietic cell migration during vertebrate development. *Proc. Natl. Acad. Sci.* **92:** 10713–10717.

Dooley, K.A., Davidson, A.J., and Zon, L.I. 2005. Zebrafish *scl* functions independently in hematopoietic and endothelial development. *Dev. Biol.* **277:** 522–536.

Dyer, M.A., Farrington, S.M., Mohn, D., Munday, J.R., and Baron, M.H. 2001. Indian hedgehog activates hematopoiesis and vasculogenesis and can respecify prospective neuroectodermal cell fate in the mouse embryo. *Development* **128:** 1717–1730.

Feng, H., Langenau, D.M., Madge, J.A., Quinkertz, A., Gutierrez, A., Neuberg, D.S., Kanki, J.P., and Look, T.A. 2007. Heat-shock induction of T-cell lymphoma/leukaemia in conditional Cre/lox-regulated transgenic zebrafish. *Br. J. Haematol.* **138:** 169–175.

Ferkowicz, M.J., Starr, M., Xie, X., Li, W., Johnson, S.A., Shelley, W.C., Morrison, P.R., and Yoder, M.C. 2003. CD41 expression defines the onset of primitive and definitive hematopoiesis in the murine embryo. *Development* **130:** 4393–4403.

Galloway, J.L., Wingert, R.A., Thisse, C., Thisse, B., and Zon, L.I. 2005. Loss of gata1 but not gata2 converts erythropoiesis to myelopoiesis in zebrafish embryos. *Dev. Cell* **8:** 109–116.

Gardiner, M.R., Gongora, M.M., Grimmond, S.M., and Perkins, A.C. 2007. A global role for zebrafish *klf4* in embryonic erythropoiesis. *Mech. Dev.* **124:** 762–774.

Gering, M. and Patient, R. 2005. Hedgehog signaling is required for adult blood stem cell formation in zebrafish embryos. *Dev. Cell* **8:** 389–400.

Gering, M., Yamada, Y., Rabbitts, T.S., and Patient, R.K. 2003. Lmo2 and Scl/Tal1 convert non-axial mesoderm into haemangioblasts which differentiate into endothelial cells in the absence of Gata1. *Development* **130:** 6187–6199.

Gering, M., Rodaway, A.R.F., Gottgens, B., Patient, R.K., and Green, A.R. 1998. The *SCL* gene specifies haemangioblast development from early mesoderm. *EMBO J.* **17:** 4029–4045.

Grosser, T., Yusuff, S., Cheskis, E., Pack, M.A., and FitzGerald, G.A. 2002. Developmental expression of functional cyclooxygenases in zebrafish. *Proc. Natl. Acad. Sci.* **99:** 8418–8423.

Haffter, P., Granato, M., Brand, M., Mullins, M.C., Hammerschmidt, M., Kane, D.A., Odenthal, J., van Eeden, F.J., Jiang, Y.J., Heisenberg, C.P., et al. 1996. The identification of genes with unique and essential functions in the development of the zebrafish, *Danio rerio. Development* **123:** 1–36.

Hart, D.O., Raha, T., Lawson, N.D., and Green, M.R. 2007. Initiation of zebrafish haematopoiesis by the TATA-box-binding protein-related factor Trf3. *Nature* **450:** 1082–1085.

Herbomel, P., Thisse, B., and Thisse, C. 1999. Ontogeny and behavior of early macrophages in the zebrafish embryo. *Development* **126:** 3735–3745.

Ho, R.K. and Kane, D.A. 1990. Cell-autonomous action of zebrafish *spt-1* mutation in specific mesodermal precursors. *Nature* **348:** 728–730.

Jaffredo, T., Gautier, R., Eichmann, A., and Dieterlen-Lièvre, F. 1998. Intraaortic hemopoietic cells are derived from endothelial cells during ontogeny. *Development* **125:** 4575–4583.

Jin, H., Xu, J., and Wen, Z. 2007. Migratory path of definitive hematopoietic stem/progenitor cells during zebrafish development. *Blood* **109:** 5208–5214.

Kalev-Zylinska, M.L., Horsfield, J.A., Flores, M.V.C., Postlethwait, J.H., Vitas, M.R., Baas, A.M., Crosier, P.S., and Crosier,

K.E. 2002. Runx1 is required for zebrafish blood and vessel development and expression of a human RUNX1-CBF2T1 transgene advances a model for studies of leukemogenesis. *Development* **129:** 2015–2030.

Kalev-Zylinska, M.L., Horsfield, J.A., Flores, M.V.C., Postlethwait, J.H., Chau, J.Y.M., Cattin, P.M., Vitas, M.R., Crosier, P.S., and Crosier, K.E. 2003. Runx3 is required for hematopoietic development in zebrafish. *Dev. Dyn.* **228:** 323–336.

Kissa, K., Murayama, E., Zapata, A., Cortes, A., Perret, E., Machu, C., and Herbomel, P. 2008. Live imaging of emerging hematopoietic stem cells and early thymus colonization. *Blood* **111:** 1147–1156.

Kobayashi, I., Saito, K., Moritomo, T., Araki, K., Takizawa, F., and Nakanishi, T. 2008. Characterization and localization of side population (SP) cells in zebrafish kidney hematopoietic tissue. *Blood* **111:** 1131–1137.

Lengerke, C., Schmitt, S., Bowman, T.V., Jang, I.H., Maouche-Chretien, L., McKinney-Freeman, S., Davidson, A.J., Hammerschmidt, M., Rentzsch, F., Green, J.B.A., Zon, L.I., and Daley, G.Q. 2008. BMP and Wnt specify hematopoietic fate by activation of the *Cdx-Hox* pathway. *Cell Stem Cell* **2:** 72–82.

Li, X., Xiong, J.W., Shelley, C.S., Park, H., and Arnaout, M.A. 2006. The transcription factor ZBP-89 controls generation of the hematopoietic lineage in zebrafish and mouse embryonic stem cells. *Development* **133:** 3641–3650.

Liao, E.C., Paw, B.H., Oates, A.C., Pratt, S.J., Postlethwait, J.H., and Zon, L.I. 1998. SCL/Tal-1 transcription factor acts downstream of *cloche* to specifiy hematopoietic and vascular progenitors in zebrafish. *Genes Dev.* **12:** 621–626.

Liao, W., Ho, C.Y., Yan, Y.L., Postlethwait, J., and Stainier, D.Y. 2000. Hhex and scl function in parallel to regulate early endothelial and blood differentiation in zebrafish. *Development* **127:** 4303–4313.

Lord, A.M., North, T.E., and Zon, L.I. 2007. Prostaglandin E2: Making more of your marrow. *Cell Cycle* **6:** 3054–3057.

Lyons, S.E., Shue, B.C., Lei, L., Oates, A.C., Zon, L.I., and Liu, P.P. 2001. Molecular cloning, genetic mapping, and expression analysis of four zebrafish *c/ebp* genes. *Gene* **281:** 43–51.

Medvinsky, A. and Dzierzak, E. 1996. Definitive hematopoiesis is autonomously initiated by the AGM region. *Cell* **86:** 897–906.

Meng, X., Noyes, M.B., Zhu, L.J., Lawson, N.D., and Wolfe, S.A. 2008. Targeted gene inactivation in zebrafish using engineered zinc-finger nucleases. *Nat. Biotechnol.* **26:** 695–701.

Mikkola, H.K., Fujiwara, Y., Schlaeger, T.M., Traver, D., and Orkin, S.H. 2003. Expression of CD41 marks the initiation of definitive hematopoiesis in the mouse embryo. *Blood* **101:** 508–516.

Mikkavila-Garcia, M.T., Cailleret, M., Godin, I., Nogueira, M.M., Cohen-Solal, K., Schiavon, V., Lecluse, Y., Le Pesteur, F., Lagrue, A.H., and Vainchenker, W. 2002. Expression of CD41 on hematopoietic progenitors derived from embryonic hematopoietic cells. *Development* **129:** 2003–2013.

Murayama, E., Kissa, K., Zapata, A., Mordelet, E., Briolat, V., Lin, H.F., Handin, R.I., and Herbomel, P. 2006. Tracing hematopoietic precursor migration to successive hematopoietic organs during zebrafish development. *Immunity* **25:** 963–975.

Nishikawa, K., Kobayashi, M., Masumi, A., Lyons, S.E., Weinstein, B.M., Liu, P.P., and Yamamoto, M. 2003. Self-association of Gata1 enhances transcriptional activity in vivo in zebra fish embryos. *Mol. Cell Biol.* **23:** 8295–8305.

North, T.E., de Bruijn, M.F., Stacey, T., Talebian, L., Lind, E., Robin, C., Binder, M., Dzierzak, E., and Speck, N.A. 2002. Runx1 expression marks long-term repopulating hematopoietic stem cells in the midgestation mouse embryo. *Immunity* **16:** 661–672.

North, T.E., Goessling, W., Walkley, C.R., Lengerke, C., Kopani, K.R., Lord, A.M., Weber, G.J., Bowman, T.V., Jang, I.H., Grosser, T., et al. 2007. Prostaglandin E2 regulates vertebrate haematopoietic stem cell homeostasis. *Nature* **447:** 1007–1011.

Oates, A.C., Pratt, S.J., Vail, B., Yan, Y.I., Ho, R.K., Johnson, S.L., Postlethwait, J.H., and Zon, L.I. 2001. The zebrafish *klf* gene family. *Blood* **98:**1792–1801.

Okuda, T., van Deursen, J., Hiebert, S.W., Grosveld, G., and Downing, J.R. 1996. AML1, the target of multiple chromosomal translocations in human leukemia, is essential for normal fetal liver hematopoiesis. *Cell* **84:** 321–330.

Orkin, S.H. and Zon, L.I. 2008. Hematopoiesis: An evolving paradigm for stem cell biology. *Cell* **132:** 631–644.

Patterson, L.J., Gering, M., Eckfeldt, C.E., Green, A.R., Verfaillie, C.M., Ekker, S.C., and Patient, R. 2007. The transcription factors Scl and Lmo2 act together during development of the hemangioblast in zebrafish. *Blood* **109:** 2389–2398.

Paffett-Lugassy, N., Hsia, N., Fraenkel, P.G., Paw, B., Leshinsky, I., Barut, B., Bahary, N., Caro, J., Handin, R., and Zon, L.I. 2007. Functional conservation of erythropoietin signaling in zebrafish. *Blood* **110:** 2718–2726.

Pham, V.N., Lawson, N.D., Mugford, J.W., Dye, L., Castranova, D., Lo, B., and Weinstein, B.M. 2007. Combinatorial function of ETS transcription factors in the developing vasculature. *Dev. Biol.* **303:** 772–783.

Pratt, S.J., Drejer, A., Foott, H., Barut, B., Brownlie, A., Postlethwait, J., Kato, Y., Yamamoto, M., and Zon, L.I. 2002. Isolation and characterization of zebrafish NFE2. *Physiol. Genomics* **11:** 91–98.

Ransom, D.G., Bahary, N., Niss, K., Traver, D., Burns, C. Trede, N.S., Paffett-Lugassy, N., Saganic, W.J., Lim, C.A., Hersey, C., et al. 2004. The zebrafish *moonshine* gene encodes transcriptional intermediary factor 1γ, an essential regulator of hematopoiesis. *PloS Biol.* **2:** E237.

Ransom, D.G., Haffter, P., Odenthal, J., Brownlie, A., Vogelsang, E., Kelsh, R.N., Brand, M., van Eeden, F.J., Furutani-Seiki, M., Granato, M., et al. 1996. Characterization of zebrafish mutants with defects in embryonic hematopoiesis. *Development* **123:** 311–319.

Rhodes, J., Hagen, A., Hsu, K., Deng, M., Liu, T.X., Look, A.T., and Kanki, J.P. 2005. Interplay of *pu.1* and *gata1* determines myelo-erythroid progenitor cell fate in zebrafish. *Dev. Cell* **8:** 97–108.

Stainier, D.Y.R., Weinstein, B.M., Detrich III, H.W., Zon, L.I., and Fishman, M.C. 1995. *cloche*, an early acting zebrafish gene, is required for both the endothelial and hematopoietic lineages. *Development* **121:** 3141–3150.

Su, F., Juarez, M.A., Cooke, C.L., Lapointe, L., Shavit, J.A., Yamaoka, J.S., and Lyons, S.E. 2007. Differential regulation of primitive myelopoiesis in the zebrafish by Spi-1/Pu.1 and C/ebp1. *Zebrafish* **4:** 187–199.

Thompson, M.A., Ransom, D.G., Pratt, S.J., MacLennan, H., Kieran, M.W., Detrich III, H.W., Vail, B., Huber, T.L., Paw, B., Brownlie, A.J., et al. 1998. The *cloche* and *spadetail* genes differentially affect hematopoiesis and vasculogenesis. *Dev. Biol.* **197:** 248–269.

Toyama, R., Kobayashi, M., Tomita, T., and Dawid, I.B. 1998. Expression of LIM-domain binding protein (ldb) genes during zebrafish embryogenesis. *Mech. Dev.* **71:** 197–200.

Traver, D., Paw, B.H., Poss, K.D., Penberthy W.T., Lin, S., and Zon, L.I. 2003. Transplantation and in vivo imaging of multilineage engraftment in zebrafish bloodless mutants. *Nat. Immunol.* **4:** 1238–1246.

Traver, D., Winzeler, A., Stern, H.M., Mayhall, E.A., Langenau, D.M., Kutok, J.L., Look, A.T., and Zon, L.I. 2004. Effects of lethal irradiation in zebrafish and rescue by hematopoietic cell transplantation. *Blood* **104:** 1298–1305.

Vogeli, K.M., Jin, S.W., Martin, G.R., and Stainier, D.Y.R. 2006. A common progenitor for hematopoietic and endothelial lineages in the zebrafish gastrula. *Nature* **443:** 337–339.

Wei, W., Wen, L., Huang, P., Zhang, Z., Chen, Y., Xiao, A., Huang, H., Zhu, Z., Zhang, B., and Lin, S. 2008. Gfi1.1 regulates hematopoietic lineage differentiation zebrafish embryogenesis. *Cell Res.* **18:** 677–685.

Weinstein, B.M., Schier, A.F., Abdelilah, S., Malicki, J., Solnica-Krezel, L., Stemple, D.L., Stainier, D.Y., Zwartkruis, F., Driever, W., and Fishman, M.C. 1996. Hematopoietic mutations in the zebrafish. *Development* **123:** 303–309.

White, R.M., Sessa, A., Burke, C., Bowman, T., LeBlanc, J., Ceol, C., Bourque, C., Dovey, M., Goessling, W., Burns, C.E., and Zon, L.I. 2008. Transparent adult zebrafish as a tool for in vivo transplantation analysis. *Cell Stem Cell* **2:** 183–189.

Wienholds, E., Schulte-Merker, S., Walderich, B., and Plasterk, R.H.A. 2002. Targeted-selected inactivation of the zebrafish *rag1* gene. *Science* **297:** 99–102.

Willett, C.E., Kawasaki, H., Amemiya, C.T., Lin, S., and Steiner, L.A. 2001. *Ikaros* expression as a marker for lymphoid progenitors during zebrafish development. *Dev. Dyn.* **222:** 694–698.

Zhang, X.Y. and Rodaway, A.R.F. 2007. *SCL*-GFP transgenic zebrafish: In vivo imaging of blood and endothelial development and identification of the initial site of definitive hematopoiesis. *Dev. Biol.* **307:** 179–194.

Zhu, H., Traver, D., Davidson, A.J., Dibiase, A., Thisse, C., Thisse, B., Nimer, S., and Zon, L.I. 2005. Regulation of the *lmo2* promoter during hematopoietic and vascular development in zebrafish. *Dev. Biol.* **281:** 256–269.

Pioneer Factors, Genetic Competence, and Inductive Signaling: Programming Liver and Pancreas Progenitors from the Endoderm

K.S. Zaret,* J. Watts,* J. Xu,[†] E. Wandzioch,* S.T. Smale,[†] and T. Sekiya*

*Epigenetics and Progenitor Cells Program, Fox Chase Cancer Center, Philadelphia, Pennsylvania 19111;
[†]Molecular Biology Institute, University of California, Los Angeles, California 90095

The endoderm is a multipotent progenitor cell population in the embryo that gives rise to the liver, pancreas, and other cell types and provides paradigms for understanding cell-type specification. Studies of isolated embryo tissue cells and genetic approaches in vivo have defined fibroblast growth factor/mitogen-activated protein kinase (FGF/MAPK) and bone morphogenetic protein (BMP) signaling pathways that induce liver and pancreatic fates in the endoderm. In undifferentiated endoderm cells, the FoxA and GATA transcription factors are among the first to engage silent genes, helping to endow competence for cell-type specification. FoxA proteins can bind their target sites in highly compacted chromatin and open up the local region for other factors to bind; hence, they have been termed "pioneer factors." We recently found that FoxA proteins remain bound to chromatin in mitosis, as an epigenetic mark. In embryonic stem cells, which lack FoxA, FoxA target sites can be occupied by FoxD3, which in turn helps to maintain a local demethylation of chromatin. By these means, a cascade of Fox factors helps to endow progenitor cells with the competence to activate genes in response to tissue-inductive signals. Understanding such epigenetic mechanisms for transcriptional competence coupled with knowledge of the relevant signals for cell-type specification should greatly facilitate efforts to predictably differentiate stem cells to liver and pancreatic fates.

The activation of a particular cell-type program within multipotent progenitor and stem cells is perhaps the most dramatic of gene regulatory events: It enables all subsequent gene regulatory events specific to a lineage but generally excludes all other cell-type programs available to the progenitor cell. Although cells within a blastula or embryonic stem (ES) cells are pluripotent and thus have all embryological fates available to them, after gastrulation, cells of ectoderm, endoderm, and mesoderm lineages are more restricted in their potential fates, and derivatives of each of these germ layers have successively fewer fates choices available. Nonetheless, any cell with an alternate fate choice has at least two parameters governing the cell-type decision: (1) signals that provide a "go" to make or allow a decision and (2) the intrinsic competence of the genome, in terms of its chromatin state, to respond to the signal. Our laboratory investigates both of these areas for initiation of liver and pancreatic programs from the endoderm. Understanding the basis for cell-type specification will provide insight into normal development, homeostatic self-renewal within adult tissues, regeneration upon tissue damage, and the prospective programming of stem cells and other progenitor cells to these biomedically relevant tissue types.

MULTIPLE EMBRYONIC ORIGINS OF THE LIVER AND PANCREAS

Liver and pancreas cells are derived from the foregut endoderm. Our fate-mapping studies demonstrated that the liver bud is derived from paired lateral domains of foregut endoderm, as well as from a physically separated domain of ventromedial endoderm (Tremblay and Zaret 2005). Although both the lateral and ventromedial domains give rise to liver bud cells that express early liver genes, including *Alb1*, *afp*, and *Hnf4*, it remains to be determined whether descendants from the different progenitor domains have different functions or regenerative capabilities in adult tissues. The earliest cells to express liver genes are hepatoblasts; later, they differentiate into hepatocytes and cholangiocytes (bile duct cells) (Shiojiri 1981; Zaret 2008). Similarly, the pancreas is derived from two domains of endoderm. In this case, the caudal portion of the paired lateral prospective liver domains of endoderm also give rise to the ventral pancreatic bud (Tremblay and Zaret 2005), and a separate domain of dorsal endoderm, positioned near the notochord, gives rise to the dorsal pancreatic bud (Slack 1995). Later in development, both pancreatic buds merge to create the gland, and descendants of both embryonic origins give rise to exocrine and endocrine cell types. Endocrine cells differentiate into five different cell types that are each specialized to express a single hormone. β cells are the most abundant and important pancreatic endocrine cell type; they secrete insulin into the bloodstream in response to high blood glucose concentrations that cause body tissues to store glucose after a meal. In contrast, each hepatocyte in the liver has many functions, including the secretion of hormones, serum proteins, and bile salts; the metabolism of nutrients and toxicants; and the storage of glucose.

MESODERMAL SIGNALS THAT SPECIFY LIVER AND PANCREAS FATES IN THE ENDODERM

Tissue explant studies in the the mouse and chick demonstrated that the foregut endoderm cells are a multipotent population, in that the cells can be induced to initiate gene expression programs for the liver or pancreas fate in response to FGF, BMP, and Shh signaling from different overlying types of mesodermal cells (Kim et al. 1997; Hebrok et al. 1998; Jung et al. 1999; Deutsch et al. 2001; Rossi et al. 2001). Specifically, cardiogenic mesoderm cells secrete FGFs as a hepatogenic signal (Jung et al. 1999), which activates liver genes in the adjacent ventral foregut endoderm via MAPK pathway signaling (Calmont et al. 2006); such signaling suppresses the ventral pancreatic program (Fig. 1) (Deutsch et al. 2001). These findings provide a molecular explanation for the original discovery with chick embryos that the cardiogenic mesoderm induces hepatogenesis in the endoderm (Le Douarin 1975).

We found that septum transversum mesenchyme cells in the vicinity of cardiogenic mesoderm cells secrete BMPs; this also promotes hepatic gene induction in the endoderm (Fig. 1) (Rossi et al. 2001). Such BMP signaling also promotes ventral pancreas induction (E. Wandzioch and K. Zaret, in prep.). Lateral plate mesoderm cells also promote ventral pancreatic induction (Kumar et al. 2003). Recent data from *Xenopus* indicates that Wnt signaling must be suppressed in the foregut endoderm to allow hepatic specification (McLin et al. 2007). Later, Wnt signaling promotes the outgrowth of the liver bud (Ober et al. 2006). The dorsal endoderm expresses Shh, which in turn is inhibitory to the dorsal pancreatic program; signals from the notochord suppress endodermal Shh expression and allow dorsal pancreas specification (Kim et al. 1997; Hebrok et al. 1998). The liver and pancreas both have endocrine function, and their development is tightly coordinated with that of the vasculature; more specifically, after the initial specification of liver and pancreas progenitors, their morphogenetic bud development is dependent on signals from adjacent endothelial cells (Lammert et al. 2001; Matsumoto et al. 2001; Yoshitomi and Zaret 2004). In summary, diverse signals and cell interactions induce the initial genetic programs for the liver and pancreas. Indeed, this information has been used to prospectively differentiate embryonic stem cells to liver and pancreas fates (Fair et al. 2003; Teratani et al. 2005; D'Amour et al. 2006; Gouon-Evans et al. 2006).

MULTIPOTENCY OF FOREGUT ENDODERM

Evidence that foregut endoderm cells are truly multipotent, capable of initiating either liver or pancreas fates, comes from tissue explant studies as well as genetic studies of the *Hex* mutation in mice. Isolated foregut endoderm, along with associated septum transversum mesenchyme cells, readily induce early pancreatic genes in culture (Deutsch et al. 2001). However, inclusion of cardiac mesoderm in the endoderm explants, or treatment of the explants with low concentrations of FGF-2, induces liver genes in the explants and suppresses pancreatic gene induction. Changes in proliferation or cell death are not observed. Thus, the default program for foregut endoderm explants is to initiate the pancreatic program, and cardiac–FGF signals seem to divert the cells to a hepatic fate.

In a different line of research, homozygous null *Hex* mutants exhibit a defect in liver development after the initiation of the hepatic program and formation of the liver bud (Bort et al. 2004). Interestingly, the liver bud cells fail to continue their differentiation and revert to a gut-like fate (Bort et al. 2006). However, in the *Hex* null embryos, ventral pancreas genes exhibit a complete failure to be activated (Bort et al. 2004). Further studies showed that the *Hex* mutation causes cell morphogenetic and movement defects, so that the prospective ventral pancreatic endoderm domain fails to move beyond the cardiogenic domain, which, in turn, normally induces the liver (see above). We found that isolation of the foregut endoderm from *Hex* mutant embryos and culturing it in vitro, in the absence of cardiogenic mesoderm, allowed the normal induction of early pancreatic genes in the mutant endoderm (Bort et al. 2006). Differences in growth or cell apoptosis were not observed. It thus appears that in *Hex* null embryos, the ventral pancreatic fate is suppressed in the endoderm by cardiac, hepatogenic signaling, but the endoderm cells retain the competence to initiate the pancreatic program. Thus, foregut endoderm cells are bipotential with regard to liver and pancreas fates, and in *Hex* mutant embryos, the nascent liver cells later revert to a gut fate, indicating further multipotency. These findings raise the question of how the cells gain the potential to activate the different cell fates.

PIONEER FACTORS AND THE DEVELOPMENTAL COMPETENCE OF THE ENDODERM

Upon discovering that the *Alb1* locus in mouse embryos is activated in the endoderm by the earliest hepatogenic signals (Gualdi et al. 1996; Jung et al. 1999), we used regulatory sequences of *Alb1* as sentinels of transcription factor occupancy during hepatic cell-type specification. The *Alb1* gene contains an upstream enhancer sequence that binds numerous liver-enriched transcription factors and confers liver-specific transcription upon linked genes in transgenic mice and transfected cells

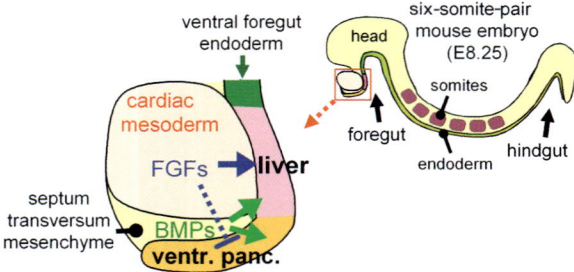

Figure 1. Cell interactions and signals that pattern the ventral foregut endoderm into liver and pancreas progenitors. (*Right*) Parasagital section of a mouse embryo at 8.25 days gestation (E8.25) (containing six somite pairs), which is the time of liver and pancreas induction (Gualdi et al. 1996; Jung et al. 1999; Deutsch et al. 2001). (*Left*) Expansion of the ventral foregut area and depiction of different tissue domains and inductive (*arrow*) and repressive (*dotted line*) signals that pattern the local endoderm. See text for details.

(Pinkert et al. 1987; DiPersio et al. 1991; Liu et al. 1991; Jackson et al. 1993). Our laboratory performed in vivo footprinting studies on the enhancer in adult liver cells, embryonic liver buds, and undifferentiated endoderm, as well as in control tissues in which *Alb1* is not expressed. We found that most of the key transcription-factor-binding sites in adult hepatocytes are occupied all the way back, temporally, to the embryonic liver bud stage (McPherson et al. 1993; Gualdi et al. 1996). Extending further to the endoderm, where the *Alb1* gene is silent, we discovered only two of the binding sites occupied, for FoxA (formerly HNF3) and GATA factors (Fig. 2) (Bossard and Zaret 1998).

Mammals have three unlinked FoxA genes (*FoxA1*, *FoxA2*, and *FoxA3*), of which *FoxA2* is expressed in and necessary for endoderm development in apparently all metazoans (Zaret 1999; Davidson and Erwin 2006). Indeed, ectopic *FoxA2* expression promotes endoderm development in ES cells (Ishizaka et al. 2002), and endogenous *FoxA2* expression is used to efficiently monitor for endoderm formation from ES cells (Kubo et al. 2004). These results underscore the importance of developmental gene regulation studies for being able to predictably manipulate the fates of stem cells. Furthermore, after endoderm formation, *FoxA1* and *FoxA2* are redundantly necessary in the endoderm for liver induction (Lee et al. 2005). FoxA factors represent a subclass of the Fox transcription factor family; together, Fox factors function in diverse developmental and signaling contexts (Katoh 2004). GATA-4 and GATA-6 factors are expressed in the endoderm as well as other developing tissues and are redundantly necessary for early liver development (Holtzinger and Evans 2005; Zhao et al. 2005; Watt et al. 2007). Notably, throughout the developmental period in which the *Alb1* enhancer sites for FoxA and GATA are occupied in the endoderm, the endoderm remains competent to activate the *Alb1* gene (Bossard and Zaret 2000). The chromatin occupancy studies, together with the genetics, indicate that FoxA and GATA either mark or help to maintain the competence of the endoderm to activate liver genes in response to inductive signals. The early engagement of these factors in progenitor cell chromatin, before target gene activation, along with the factors' ability to help open chromatin structure (see below), has led them to be termed "competence" or "pioneer" factors (Zaret 1999). Furthermore, the mere engagement of target genes by a subset of enhancer-binding factors may facilitate a more rapid and homogeneous activation of a genetic program in a field of progenitor cells in development, in response to inductive signals.

CHROMATIN OPENING BY DIVERSE FOX PIONEER FACTORS

The crystal structure of the FoxA DNA-binding domain immediately suggested a role for FoxA factors in chromatin, in that the fold of the protein was found to highly resemble that for linker histone (Clark et al. 1993; Ramakrishnan et al. 1993). Further studies showed that purified FoxA1 protein binds its target sites on reconstituted nucleosome core particles in vitro (Cirillo et al. 1998) and MNase mapping, and sequential chromatin immunoprecipitation studies of liver chromatin showed that FoxA1 occupies its *Alb1* enhancer target sequence on a nucleosome in vivo (McPherson et al. 1993; Chaya et al. 2001). Purified FoxA1 binding to compacted polynucleosome array templates in vitro creates a nuclease-hypersensitive site where FoxA1 binds the DNA (Cirillo et al. 2002), showing that the protein opens up the local chromatin (Fig. 3). Binding of FoxA1 protein to nucleosomal templates in vitro enables binding by GATA-4 and NF1 at adjacent sites (Cirillo and Zaret 1999; Cirillo et al. 2002). FoxA1 binding to chromatin in vivo also creates hypersensitive sites and enables estrogen receptor binding (Carroll et al. 2005). Nucleosome binding by FoxA1 and FoxA2 in vitro occurs equally well on substrates made with mouse histones or recombinant histones expressed in *Escherichia coli* (Sekiya and Zaret 2007); thus, FoxA1 does not require a prior histone modification for its intrinsic chromatin-binding capacity. Yet, in vivo, FoxA1 does not bind most of its consensus sites in chromatin, and certain histone modifications may exclude or enhance FoxA1 binding to potential targets (Lupien et al. 2008). The ability of FoxA proteins to engage target sites in nucleosomal DNA and enable other factors to bind is consistent with its early pioneer function in development.

Various Fox factors have now been found to possess chromatin opening activity and enable gene expression. Pha4, a FoxA homolog, enables the activation of the pharyngeal program in *Caenorhabditis elegans* embryos, with the strongest Pha4-binding genes turning on earliest in pharyngeal development and genes that bind Pha4 weaker turning on later in pharyngeal development, as Pha4 levels continue to rise (Gaudet and Mango 2002). These findings would suggest that simple protein concentration in the nucleus is a major determinant of Pha4 binding. FoxI1 creates hypersensitive sites at its genetic targets in vivo (Yan et al. 2006), FoxE1 binds compacted chromatin in vitro and in vivo and generates local hypersensitivity (Cuesta et al. 2007), and FoxO1 binds and opens chromatin in a phosphorylation-sensitive fashion

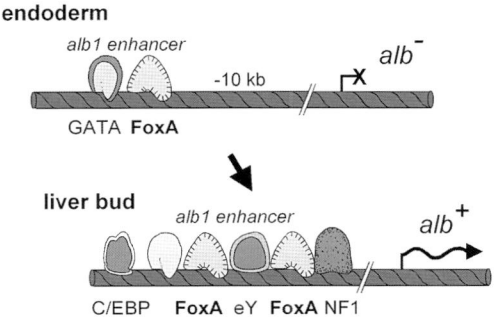

Figure 2. Chromatin occupancy at the *Alb1* enhancer in endoderm and liver bud cells. The blobs represent inferred protein occupancies at DNA target sequences for the designated transcription factors from in vivo footprinting experiments cited in the text. These data from mouse embryonic tissues led to the hypothesis that FoxA and GATA factors can serve as pioneer factors in the endoderm, helping to enable liver gene induction under the appropriate developmental signals.

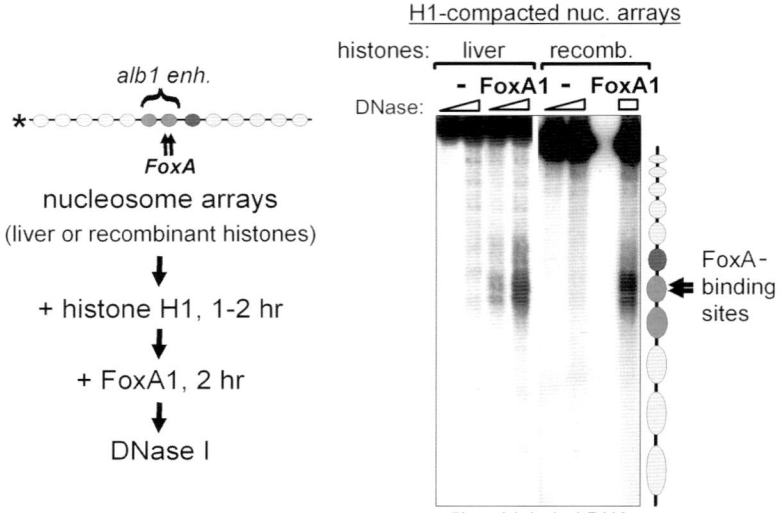

Figure 3. Chromatin opening in vitro by FoxA1 with purified proteins. The *Alb1* enhancer sequence was inserted into an array of 5S rRNA gene repeats that promote naturally phased nucleosome arrays upon chromatin assembly (for details, see Cirillo et al. 2002). Nucleosome arrays were assembled onto end-labeled DNA with histones purified from liver cells or in recombinant form from *E. coli*. The arrays were compacted by the addition of linker histone at a unit molar ratio to nucleosomes. Aliquots of the arrays were incubated with purified FoxA1 protein for 2 hours, then treated briefly with DNase I. DNA was purified, run on an agarose gel, and autoradiographed. The data show that FoxA1 induces the nucleosome underlying its enhancer-binding sites to become hypersensitive to nuclease. The data also show that FoxA binding and generation of hypersensitivity do not require a mammalian modification on the template histones.

(Hatta and Cirillo 2007). Together, these findings indicate that Fox factors generally appear high in the hierarchy of the transcriptional regulatory mechanism, exposing target DNA sequences in chromatin for gene activity.

TRANSCRIPTION FACTOR BINDING AS AN EPIGENETIC MARK IN MITOTIC CHROMATIN

When a cell divides, in mitosis or meiosis, the extreme state of chromatin compaction excludes most regulatory proteins and leads to gene silencing (Gottesfeld and Forbes 1997). After mitosis, regulatory proteins must once again engage their genomic target sites and reestablish appropriate gene expression states. The means by which mitotic and meiotic chromatin are marked so that genes can be appropriately reactivated is a central aspect of epigenetic regulation. There has been intense focus on DNA methylation and certain histone modifications as being retained through mitosis and hence serving as epigenetic marks (Jenuwein and Allis 2001; Turner 2002; Bird and Macleod 2004; Goll and Bestor 2005; Kouzarides 2007). Yet, cytology, fluorescence recovery after photobleaching, and in vivo footprinting studies indicate that a small subset of transcription factors are also retained on mitotic chromatin (Martinez-Balbas et al. 1995; Michelotti et al. 1997; Chen et al. 2002; Christova and Oelgeschlager 2002). Such factors would be predicted to be able to bind highly compacted chromatin in other contexts, and this is exactly what has been found for Fox factors (Fig. 3). Indeed, FoxI1 has been directly visualized on metaphase chromosomes, suggesting that the factor could serve as an epigenetic mark (Yan et al. 2006).

To investigate whether FoxA factors could be maintained in mitotic chromatin, we created green fluorescent protein (GFP) fusion proteins with FoxA1 and with C/EBPα and the nuclear localization signal (NLS) of SV40 large T antigen, as controls. Plasmids encoding the constructs were transiently transfected into HUH7 hepatoma

cells and the cells were treated with aphidicolin (1 μg/ml) for 14 hours and then with nocodozole (0.2 μg/ml) for 16 hours, to enrich for cells in mitosis. The resultant cells were treated with DAPI (4′-6-diamidino-2-phenylindole) and viewed under fluorescence optics to visualize the nuclear DNA state and location of GFP fusion proteins.

As seen in Figure 4, A and B, cells that expressed the GFP-NLS and GFP-C/EBPα control proteins contained GFP fluorescence in nuclei that were decondensed and hence in metaphase. Cells that contained highly condensed chromatin, indicative of mitosis, had GFP-NLS and GFP-C/EBPα excluded from the bulk DNA (Fig. 4A,B). In contrast, the GFP-FoxA1 protein was readily visualized in nuclei that were decondensed, as well as in cells containing highly compacted mitotic chromatin (Fig. 4C). As a secondary assessment of the mitotic state of the FoxA1-GFP-transfected cells, we treated cultures with an antibody to histone H3 phosphorylated on serine 10 (Ser-10), a mitotic marker. Cells containing highly compacted chromatin with bound FoxA-GFP also stained for the H3 Ser-10 marker, demonstrating FoxA1 occupancy in mitotic chromatin (Fig. 4C). Taken together, these studies demonstrate that FoxA1 can bind to mitotic chromatin, and hence its binding to DNA could serve as an epigenetic mark. We speculate that FoxA binding to mitotic chromatin could enable target genes to be more rapidly or synchronously activated as they progress toward interphase.

FOXD3 AND A DEMETHYLATION MARK AT THE *ALB1* ENHANCER IN EMBRYONIC STEM CELLS

DNA methylation constitutes a major epigenetic mark that is typically associated with inactive genes (Bird and Macleod 2004). Like many genes, *Alb1* is heavily methylated during early development, where the gene is silent, but becomes less methylated as the gene becomes activated, during hepatocyte differentiation (Kunnath and Locker 1983). Despite otherwise heavy methylation in

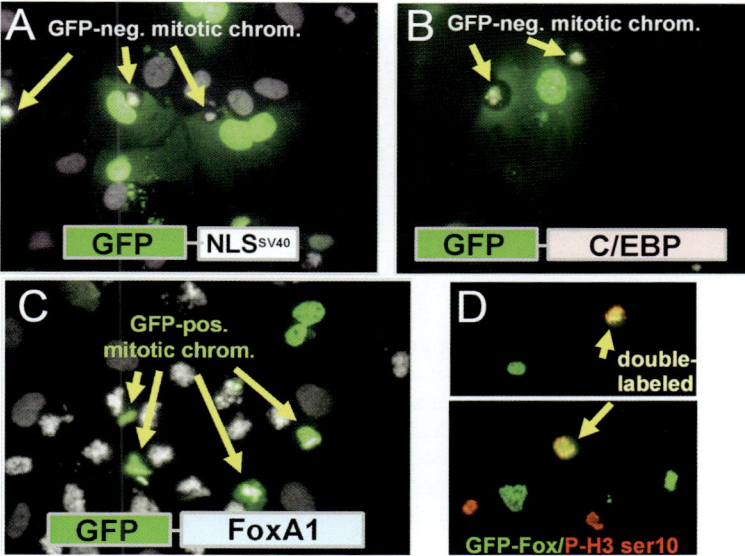

Figure 4. Engagement of FoxA1 in mitotic chromatin. HUH7 hepatoma cells were transfected with plasmids encoding GFP-NLS^{SV40} (*A*), GFP-C/EBPα (*B*), or GFP-FoxA1 (*C,D*) and subjected to a protocol to enrich for cells in mitosis (see text for details). GFP-NLS^{SV40} and GFP-C/EBPα were present in interphase nuclei, but not in highly compacted mitotic chromatin (*A,B*). GFP-FoxA1 was readily detected in highly compacted mitotic chromatin (*C*) and colabeled cells enriched for histone H3 that was enriched for phosphorylated Ser-10, a mitotic marker (*D*).

early development, a single unmethylated CpG was discovered at the *Alb1* enhancer in various ES cell lines, underlying one of the FoxA-binding sites (Xu et al. 2007). Given that ES cell lines do not express FoxA proteins (data not shown), we investigated the potential role of FoxD3, which is expressed in ES cells (Sutton et al. 1996). FoxD3 is necessary to maintain ES cell pluripotency (Hanna et al. 2002; Tompers et al. 2005; Pan et al. 2006) and it functions in various other progenitor cell contexts in the embryo (Guo et al. 2002; Perera et al. 2006). Interestingly, chromatin immunoprecipitation (ChIP) studies revealed that FoxD3 engages the *Alb1* enhancer in ES cells, where *Alb1* is silent (Xu et al. 2007). This resembles the aforementioned occupancy by FoxA

in endoderm cells, where *Alb1* is also silent, but competent to be activated. Recent studies indicate that FoxD3 is necessary to maintain demethylation of the *Alb1* enhancer site in ES cells and that ectopic FoxD3 expression in fibroblasts induces *Alb1* enhancer site demethylation (J. Xu et al., in prep.). In summary, it appears that a cascade of Fox factors engage the *Alb1* gene, starting with *FoxD3* before gastrulation and continuing with FoxA in the endoderm and its liver descendants (Fig. 5). Coordinated with this cascade is the striking and selective demethylation at the *Alb1* enhancer Fox-binding site.

The combined chromatin mark consisting of a local CpG demethylation amid highly methylated chromatin and its attendant bound transcription factor extends be-

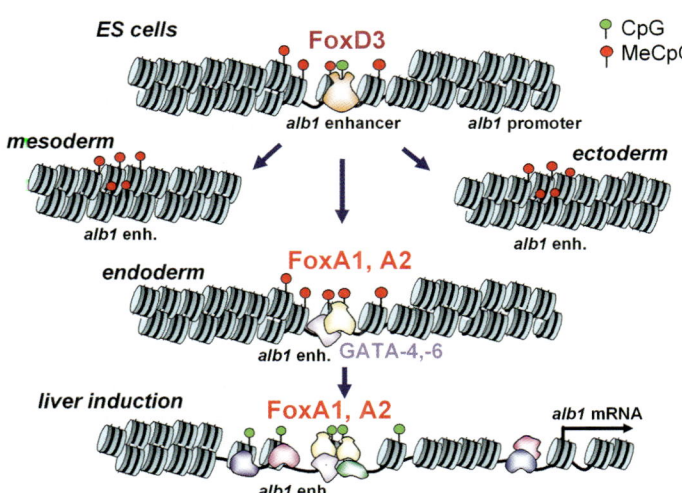

Figure 5. A cascade of Fox factors in multipotent progenitor cells in development. The diagram depicts the chromatin structure, DNA methylation state, and designated bound transcription factors at the *Alb1* enhancer at different stages of development. These include ES cells, which presumably represent the chromatin state in blastocyst cells, the three germ layers arising during gastrulation, and the early liver bud in the embryo. FoxD3 and FoxA transcription factors appear to sequentially occupy the *Alb1* enhancer throughout the ES to liver bud lineage. Occupancy at the endoderm stage correlates with (Gualdi et al. 1996; Bossard and Zaret 1998, 2000) and is necessary for (Lee et al. 2005) hepatic induction. Occupancy by FoxD3 correlates with (Xu et al. 2007) and is necessary for (J. Xu et al., in prep.) a demethylation mark at the *Alb1* enhancer in ES cells.

yond FoxD3 and its targets. Studies of the *Ptcra* and *Il12* genes, which are normally expressed in the thymocyte and macrophage lineages, show that their respective enhancers are also demethylated in ES cells, whereas other regions of these genes are methylated (Xu et al. 2007). Introduction of bacterial artificial chromosomes (BACs) harboring the *Il12* gene into ES cells, where CpG sequences within the BAC were methylated before transduction, showed that the enhancer sequences became demethylated selectively (Xu et al. 2007). In vivo footprinting reveals diverse factors bound to the demethylated enhancers in ES cells (Xu et al. 2007). Although the *trans*-acting factors that elicit demethylation are not yet known, mutagenesis of clusters of transcription-factor-binding sites causes the failure of transfected methylated *Ptcra* and *Il12* BACs to be demethylated at the enhancers (J. Xu et al., in prep.). The binding sites that were mutated do not appear to be Fox targets. These findings suggest that distinct *trans*-acting factors can elicit the local demethylation mark in ES cells.

FUTURE DIRECTIONS

Nature has evolved detailed cellular mechanisms by which multipotent cells come to possess particular developmental competencies as well as by which inductive signals and response networks elicit cell-type specification. Further elucidation of the marks of developmental competence and their potential mechanisms of action will ultimately allow for the prediction of differentiation capacity of a given progenitor or stem cell population. Thus, we wish to better understand chromatin states in embryonic endoderm cells, their progenitors, and in different stages of their descendants to the liver and pancreatic fates. To this end, we are currently using fluorescence-activated cell sorting (FACS) of different endodermal and embryonic liver and pancreas cell populations to isolate progenitors for detailed chromatin analysis. Given that a major difficulty with current stem cell differentiation protocols is to develop cells that fully express hepatocyte and β-cell phenotypes, we anticipate that the information we obtain can be used to assess whether cells at early stages in the differentiation protocol have been properly programmed, for example, by possession of the appropriate chromatin competence marks. This could provide a novel dimension to prospectively programming stem cells to desired fates.

We are also taking a different perspective on how the engagement of pioneer transcription factors at silent genes may mark or endow the potential for gene activity. We have performed a sub-genome-wide location analysis of FoxA2-bound sites in the adult mouse liver and focused on the approximately one third of FoxA2 targets that occur at silent genes (J. Watts and K. Zaret, in prep.). Typically, these silent FoxA2-bound genes are active in other endoderm-derived cell types. Notably, at least one of the targets is a regulatory gene whose activation can cause cell-type conversion, or metaplasia, among gut tissues. Further analysis has revealed a network of repressive transcription factors that help to keep the FoxA2 target gene silent in liver cells and that may be disrupted during gut pathologies associated with metaplasia. These studies reveal how pioneer factors enable dysregulated cell differentiation events that may underlie disease.

With regard to our fate mapping of the foregut endoderm, revealing different endoderm domains that together give rise to the embryonic liver bud, we have performed laser-capture microdissection of different endodermal domains, followed by RNA isolation, amplification, and microarray analysis. With the resulting RNA expression profiles, we subtracted the expression profiles of adjacent mesodermal tissue, yolk sac, and the total embryo, providing us with lists of genes whose expression is highly enriched in different domains of the foregut endoderm. We are now using BACs harboring these genes to drive the expression of Cre recombinase in different endodermal domains of transgenic mice. This will allow us to perform genetic lineage-marking analysis and determine whether descendants of different endodermal progenitors have different growth, regenerative, and stem cell capacities in the adult liver and pancreas.

A major application of developmental biology studies is to use the tissue-inductive signals that were identified from studies of embryos to prospectively program stem cells. Furthermore, understanding the signal transduction pathways that mediate tissue induction events, and how the pathways interact to form a network, can reveal agonist, antagonist, and other small-molecule targets to promote efficient stem cell differentiation. To this end, we are investigating the signal transduction pathways and interactions within endoderm cells during the period preceding and culminating in liver and pancreas induction, as well as within cells at the subsequent steps of differentiation. Understanding how such pathways converge on pioneer factors at target genes and other chromatin parameters and induce new regulatory events leading to cell-type specification will provide a cohesive view of how to control cell fates at will.

ACKNOWLEDGMENTS

Research in the Zaret laboratory described in this review has been supported by the National Institutes of Health (R01-GM36477, R01-GM74903, U01-DK072503, CA-06927), the Searle Scholars Program, the Human Frontiers Science Program, and the Mathers Charitable Foundation.

REFERENCES

Bird, A. and Macleod, D. 2004. Reading the DNA methylation signal. *Cold Spring Harbor Symp. Quant. Biol.* **69:** 113–118.

Bort, R., Martinez-Barbera, J.P., Beddington, R.S., and Zaret, K.S. 2004. *Hex* homeobox gene-dependent tissue positioning is required for organogenesis of the ventral pancreas. *Development* **131:** 797–806.

Bort, R., Signore, M., Tremblay, K., Martinez Barbera, J.P., and Zaret, K.S. 2006. Hex homeobox gene controls the transition of the endoderm to a pseudostratified, cell emergent epithelium for liver bud development. *Dev. Biol.* **290:** 44–56.

Bossard, P. and Zaret, K.S. 1998. GATA transcription factors as potentiators of gut endoderm differentiation. *Development* **125:** 4909–4917.

Bossard, P. and Zaret, K.S. 2000. Repressive and restrictive mesodermal interactions with gut endoderm: Possible relation to Meckel's Diverticulum. *Development* **127:** 4915–4923.

Calmont, A., Wandzioch, E., Tremblay, K.D., Minowada, G.,

Kaestner, K.H., Martin, G.R., and Zaret, K.S. 2006. An FGF response pathway that mediates hepatic gene induction in embryonic endoderm cells. *Dev. Cell.* **11:** 339–348.

Carroll, J.S., Liu, X.S., Brodsky, A.S., Li, W., Meyer, C.A., Szary, A.J., Eeckhoute, J., Shao, W., Hestermann, E.V., Geistlinger, T.R., et al. 2005. Chromosome-wide mapping of estrogen receptor binding reveals long-range regulation requiring the forkhead protein FoxA1. *Cell* **122:** 33–43.

Chaya, D., Hayamizu, T., Bustin, M., and Zaret, K.S. 2001. Transcription factor FoxA (HNF3) on a nucleosome at an enhancer complex in liver chromatin. *J. Biol. Chem.* **276:** 44385–44389.

Chen, D., Hinkley, C.S., Henry, R.W., and Huang, S. 2002. TBP dynamics in living human cells: Constitutive association of TBP with mitotic chromosomes. *Mol. Biol. Cell* **13:** 276–284.

Christova, R. and Oelgeschlager, T. 2002. Association of human TFIID-promoter complexes with silenced mitotic chromatin in vivo. *Nat. Cell Biol.* **4:** 79–82.

Cirillo, L., Lin, F.R., Cuesta, I., Jarnik, M., Friedman, D., and Zaret, K. 2002. Opening of compacted chromatin by early developmental transcription factors HNF3 (FOXA) and GATA-4. *Mol. Cell* **9:** 279–289.

Cirillo, L.A. and Zaret, K.S. 1999. An early developmental transcription factor complex that is more stable on nucleosome core particles than on free DNA. *Mol. Cell* **4:** 961–969.

Cirillo, L.A., McPherson, C.E., Bossard, P., Stevens, K., Cherian, S., Shim, E.-Y., Clark, E.A., Burley, S.K., and Zaret, K.S. 1998. Binding of the winged-helix transcription factor HNF3 to a linker histone site on the nucleosome. *EMBO J.* **17:** 244–254.

Clark, K.L., Halay, E.D., Lai, E., and Burley, S.K. 1993. Co-crystal structure of the HNF3/*fork head* DNA recognition motif resembles histone H5. *Nature* **364:** 412–420.

Cuesta, I., Zaret, K.S., and Santisteban, P. 2007. The forkhead factor FoxE1 binds to the thyroperoxidase promoter during thyroid cell differentiation and modifies compacted chromatin structure. *Mol. Cell. Biol.* **27:** 7302–7314.

D'Amour, K.A., Bang, A.G., Eliazer, S., Kelly, O.G., Agulnick, A.D., Smart, N.G., Moorman, M.A., Kroon, E., Carpenter, M.K., and Baetge, E.E. 2006. Production of pancreatic hormone-expressing endocrine cells from human embryonic stem cells. *Nat. Biotechnol.* **24:** 1392–1401.

Davidson, E.H. and Erwin, D.H. 2006. Gene regulatory networks and the evolution of animal body plans. *Science* **311:** 796–800.

Deutsch, G., Jung, J., Zheng, M., Lóra, J., and Zaret, K.S. 2001. A bipotential precursor population for pancreas and liver within the embryonic endoderm. *Development* **128:** 871–881.

DiPersio, C.M., Jackson, D.A., and Zaret, K.S. 1991. The extracellular matrix coordinately modulates liver transcription factors and hepatocyte morphology. *Mol. Cell. Biol.* **11:** 4405–4414.

Fair, J.H., Cairns, B.A., Lapaglia, M., Wang, J., Meyer, A.A., Kim, H., Hatada, S., Smithies, O., and Pevny, L. 2003. Induction of hepatic differentiation in embryonic stem cells by co-culture with embryonic cardiac mesoderm. *Surgery* **134:** 189–196.

Gaudet, J. and Mango, S.E. 2002. Regulation of organogenesis by the *Caenorhabditis elegans* FoxA protein PHA-4. *Science* **295:** 821–825.

Goll, M.G. and Bestor, T.H. 2005. Eukaryotic cytosine methyltransferases. *Annu. Rev. Biochem.* **74:** 481–514.

Gottesfeld, J.M. and Forbes, D.J. 1997. Mitotic repression of the transcriptional machinery. *Trends Biochem. Sci.* **22:** 197–202.

Gouon-Evans, V., Boussemart, L., Gadue, P., Nierhoff, D., Koehler, C.I., Kubo, A., Shafritz, D.A., and Keller, G. 2006. BMP-4 is required for hepatic specification of mouse embryonic stem cell-derived definitive endoderm. *Nat. Biotechnol.* **24:** 1402–1411.

Gualdi, R., Bossard, P., Zheng, M., Hamada, Y., Coleman, J.R., and Zaret, K.S. 1996. Hepatic specification of the gut endoderm in vitro: Cell signaling and transcriptional control. *Genes Dev.* **10:** 1670–1682.

Guo, Y., Costa, R., Ramsey, H., Starnes, T., Vance, G.,

Robertson, K., Kelley, M., Reinbold, R., Scholer, H., and Hromas, R. 2002. The embryonic stem cell transcription factors Oct-4 and FoxD3 interact to regulate endodermal-specific promoter expression. *Proc. Natl. Acad. Sci.* **99:** 3663–3667.

Hanna, L.A., Foreman, R.K., Tarasenko, I.A., Kessler, D.S., and Labosky, P.A. 2002. Requirement for Foxd3 in maintaining pluripotent cells of the early mouse embryo. *Genes Dev.* **16:** 2650–2661.

Hatta, M. and Cirillo, L.A. 2007. Chromatin opening and stable perturbation of core histone:DNA contacts by FoxO1. *J. Biol. Chem.* **282:** 35583–35593.

Hebrok, M., Kim, S.K., and Melton, D.A. 1998. Notochord repression of endodermal sonic hedgehog permits pancreas development. *Genes Dev.* **12:** 1705–1713.

Holtzinger, A. and Evans, T. 2005. Gata4 regulates the formation of multiple organs. *Development* **132:** 4005–4014.

Ishizaka, S., Shiroi, A., Kanda, S., Yoshikawa, M., Tsujinoue, H., Kuriyama, S., Hasuma, T., Nakatani, K., and Takahashi, K. 2002. Development of hepatocytes from ES cells after transfection with the HNF-3β gene. *FASEB J.* **16:** 1444–1446.

Jackson, D.A., Rowader K.E., Stevens, K.A., Jiang, C., Milos, P., and Zaret K.S. 1993. Modulation of liver-specific transcription by interactions between HNF3 and NF1/CTF proteins binding in close apposition. *Mol. Cell. Biol.* **13:** 2401–2410.

Jenuwein, T. and Allis, C.D. 2001. Translating the histone code. *Science* **293:** 1074–1080.

Jung, J., Zheng, M., Goldfarb, M., and Zaret, K.S. 1999. Initiation of mammalian liver development from endoderm by fibroblast growth factors. *Science* **284:** 1998–2003.

Katoh, M. 2004. Human FOX gene family (review). *Int. J. Oncol.* **25:** 1495–1500.

Kim, S.K., Hebrok, M., and Melton, D.A. 1997. Notochord to endoderm signaling is required for pancreas development. *Development* **124:** 4243–4252.

Kouzarides, T. 2007. Chromatin modifications and their function. *Cell* **128:** 693–705.

Kubo, A., Shinozaki, K., Shannon, J.M., Kouskoff, V., Kennedy, M., Woo, S., Fehling, H.J., and Keller, G. 2004. Development of definitive endoderm from embryonic stem cells in culture. *Development* **131:** 1651–1662.

Kumar, M., Jordan, N., Melton, D., and Grapin-Botton, A. 2003. Signals from lateral plate mesoderm instruct endoderm toward a pancreatic fate. *Dev. Biol.* **259:** 109–122.

Kunnath, L. and Locker, J. 1983. Developmental changes in the methylation of the rat albumin and α-fetoprotein genes. *EMBO J.* **2:** 317–324.

Lammert, E., Cleaver, O., and Melton, D. 2001. Induction of pancreatic differentiation by signals from blood vessels. *Science* **294:** 564–567.

Le Douarin, N.M. 1975. An experimental analysis of liver development. *Med. Biol.* **53:** 427–455.

Lee, C.S., Friedman, J.R., Fulmer, J.T., and Kaestner, K.H. 2005. The initiation of liver development is dependent on Foxa transcription factors. *Nature* **435:** 944–947.

Liu, J.K., DiPersio, C.M., and Zaret, K.S. 1991. Extracellular signals that regulate liver transcription factors during hepatic differentiation in vitro. *Mol. Cell. Biol.* **11:** 773–784.

Lupien, M., Eeckhoute, J., Meyer, C.A., Wang, Q., Zhang, Y., Li, W., Carroll, J.S., Liu, X.S., and Brown, M. 2008. FoxA1 translates epigenetic signatures into enhancer-driven lineage-specific transcription. *Cell* **132:** 958–970.

Martinez-Balbas, M.A., Dey, A., Rabindran, S.K., Ozato, K., and Wu, C. 1995. Displacement of sequence-specific transcription factors from mitotic chromatin. *Cell* **83:** 29–38.

Matsumoto, K., Yoshitomi, H., Rossant, J., and Zaret, K.S. 2001. Liver organogenesis promoted by endothelial cells prior to vascular function. *Science* **294:** 559–563.

McLin, V.A., Rankin, S.A., and Zorn, A.M. 2007. Repression of Wnt/β-catenin signaling in the anterior endoderm is essential for liver and pancreas development. *Development* **134:** 2207–2217.

McPherson, C.E., Shim, E.-Y., Friedman, D.S., and Zaret, K.S. 1993. An active tissue-specific enhancer and bound transcription factors existing in a precisely positioned nucleosomal array. *Cell* **75:** 387–398.

Michelotti, E.F., Sanford, S., and Levens, D. 1997. Marking of active genes on mitotic chromosomes. *Nature* **388:** 895–899.

Ober, E.A., Verkade, H., Field, H.A., and Stainier, D.Y. 2006. Mesodermal Wnt2b signalling positively regulates liver specification. *Nature* **442:** 688–691.

Pan, G., Li, J., Zhou, Y., Zheng, H., and Pei, D. 2006. A negative feedback loop of transcription factors that controls stem cell pluripotency and self-renewal. *FASEB J.* **20:** 1730–1732.

Perera, H.K., Caldwell, M.E., Hayes-Patterson, D., Teng, L., Peshavaria, M., Jetton, T.L., and Labosky, P.A. 2006. Expression and shifting subcellular localization of the transcription factor, Foxd3, in embryonic and adult pancreas. *Gene Expr. Patterns* **6:** 971–977.

Pinkert, C.A., Ornitz, D.M., Brinster, R.L., and Palmiter, R.D. 1987. An albumin enhancer located 10 kb upstream functions along with its promoter to direct efficient, liver-specific expression in transgenic mice. *Genes Dev.* **1:** 268–276.

Ramakrishnan, V., Finch, J.T., Graziano, V., Lee, P.L., and Sweet, R.M. 1993. Crystal structure of globular domain of histone H5 and its implications for nucleosome binding. *Nature* **362:** 219–224.

Rossi, J.M., Dunn, N.R., Hogan, B.L.M., and Zaret, K.S. 2001. Distinct mesodermal signals, including BMP's from the septum transversum mesenchyme, are required in combination for hepatogenesis from the endoderm. *Genes Dev.* **15:** 1998–2009.

Sekiya, T. and Zaret, K.S. 2007. Repression by Groucho/TLE/Grg proteins: Genomic site recruitment generates compacted chromatin in vitro and impairs activator binding in vivo. *Mol. Cell* **28:** 291–303.

Shiojiri, N. 1981. Enzymo- and immunocytochemical analyses of the differentiation of liver cells in the prenatal mouse. *J. Embryol. Exp. Morphol.* **62:** 139–152.

Slack, J.M.W. 1995. Developmental biology of the pancreas. *Development* **121:** 1569–1580.

Sutton, J., Costa, R., Klug, M., Field, L., Xu, D., Largaespada, D.A., Fletcher, C.F., Jenkins, N.A., Copeland, N.G., Klemsz, M., et al. 1996. Genesis, a winged helix transcriptional repressor with expression restricted to embryonic stem cells. *J. Biol. Chem.* **271:** 23126–23133.

Teratani, T., Yamamoto, H., Aoyagi, K., Sasaki, H., Asari, A., Quinn, G., Terada, M., and Ochiya, T. 2005. Direct hepatic fate specification from mouse embryonic stem cells. *Hepatology* **41:** 836–846.

Tompers, D.M., Foreman, R.K., Wang, Q., Kumanova, M., and Labosky, P.A. 2005. Foxd3 is required in the trophoblast progenitor cell lineage of the mouse embryo. *Dev. Biol.* **285:** 126–137.

Tremblay, K.D. and Zaret, K.S. 2005. Distinct populations of endoderm cells converge to generate the embryonic liver bud and ventral foregut tissues. *Dev. Biol.* **280:** 87–99.

Turner, B.M. 2002. Cellular memory and the histone code. *Cell* **111:** 285–291.

Watt, A.J., Zhao, R., Li, J., and Duncan, S.A. 2007. Development of the mammalian liver and ventral pancreas is dependent on GATA4. *BMC Dev. Biol.* **7:** 37.

Xu, J., Pope, S.D., Jazirehi, A.R., Attema, J.L., Papathanasiou, P., Watts, J.A., Zaret, K.S., Weissman, I.L., and Smale, S.T. 2007. Pioneer factor interactions and unmethylated CpG dinucleotides mark silent tissue-specific enhancers in embryonic stem cells. *Proc. Natl. Acad. Sci.* **104:** 12377–12382.

Yan, J., Xu, L., Crawford, G., Wang, Z., and Burgess, S.M. 2006. The forkhead transcription factor FoxI1 remains bound to condensed mitotic chromosomes and stably remodels chromatin structure. *Mol. Cell. Biol.* **26:** 155–168.

Yoshitomi, H. and Zaret, K.S. 2004. Endothelial cell interactions initiate dorsal pancreas development by selectively inducing the transcription factor Ptf1a. *Development* **131:** 807–817.

Zaret, K. 1999. Developmental competence of the gut endoderm: Genetic potentiation by GATA and HNF3/fork head proteins. *Dev. Biol.* **209:** 1–10.

Zaret, K.S. 2008. Genetic programming of liver and pancreas progenitors: Lessons for stem-cell differentiation. *Nat. Rev. Genet.* **9:** 329–340.

Zhao, R., Watt, A.J., Li, J., Luebke-Wheeler, J., Morrisey, E.E., and Duncan, S.A. 2005. GATA6 is essential for embryonic development of the liver but dispensable for early heart formation. *Mol. Cell. Biol.* **25:** 2622–2631.

Generation of Cardiomyocytes from New Human Embryonic Stem Cell Lines Derived from Poor-quality Blastocysts

A. RAYA,*†‡+ I. RODRÍGUEZ-PIZÀ,*+ B. ARÁN,* A. CONSIGLIO,*§ P.N. BARRI,¶
A. VEIGA,*¶ AND J.C. IZPISÚA BELMONTE*#

*Center for Regenerative Medicine in Barcelona, 08003 Barcelona, Spain; †Institució Catalana de Recerca
i Estudis Avançats (ICREA); ‡Networking Center of Biomedical Research in Bioengineering, Biomaterials
and Nanomedicine (CIBER-BBN); §Department of Biomedical Sciences and Biotechnology, Unit of
Biochemistry, University of Brescia, 25123 Brescia, Italy; ¶Departament of Obstetrics, Gynecology, and
Reproduction, Institut Universitari Dexeus, Barcelona, Spain; #Gene Expression Laboratory,
Salk Institute for Biological Studies, La Jolla, California 92037

Human embryonic stem (hES) cells represent a potential source for cell replacement therapy of many degenerative diseases. Most frequently, hES cell lines are derived from surplus embryos from assisted reproduction cycles, independent of their quality or morphology. Here, we show that hES cell lines can be obtained from poor-quality blastocysts with the same efficiency as that obtained from good- or intermediate-quality blastocysts. Furthermore, we show that the self-renewal, pluripotency, and differentiation ability of hES cell lines derived from either source are comparable. Finally, we present a simple and reproducible embryoid body-based protocol for the differentiation of hES cells into functional cardiomyocytes. The five new hES cell lines derived here should widen the spectrum of available resources for investigating the biology of hES cells and advancing toward efficient strategies of regenerative medicine.

hES cells are permanent cell lines derived from preimplantation human embryos, most frequently from the inner cell mass (ICM) of human blastocysts. Since 1998, when the first lines of hES cells were derived (Thomson et al. 1998), numerous hES cell lines have been produced by different laboratories (Allegrucci and Young 2007; Veiga et al. 2007). The two main defining features of these cells are their self-renewal capacity and pluripotency (i.e., their ability to differentiate into cell types of the three embryonic germ layers). Because of these properties, hES cells are thought to hold great potential as a source of cells for therapeutic use. Parkinson's disease, spinal cord injury, diabetes, heart failure, and bone marrow failure are examples of pathological conditions amenable to being treated by means of stem cell transplantation (Gerecht-Nir and Itskovitz-Eldor 2004; Liew et al. 2005; Menasché 2005; Semb 2005). In addition to their potential value as therapeutic agents, hES cells also appear to be a powerful experimental model for studying early human development, a platform to develop and test new drugs and treatment protocols (Pera and Trounson 2004), and an aid for research on human monogenic diseases (Pickering et al. 2003; Verlinsky et al. 2005; Mateizel et al. 2006).

Typically, hES cell lines originate from surplus embryos created during in vitro fertilization (IVF) treatments. Couples reaching an end to their IVF treatments with surplus embryos have a decision to make. These frozen surplus embryos can remain in cryostorage, be made available for adoption, be thawed and discarded, or be donated to research (Lyerly and Faden 2007). It is from

this latter option that most hES cell lines to date have been derived. IVF programs often use in vitro culture of preimplantation embryos to the blastocyst stage of development to achieve better implantation rates in specific groups of patients (Ménézo et al. 1992; Veiga et al. 1995). Approximately 35% of cultured human embryos develop successfully to the blastocyst stage but not all of these show good quality morphology. The remainder show retarded or arrested development as well as abnormal morphology due to unequal cell division or cellular fragmentation (Ménézo et al. 1998; Gardner et al. 2000).

Today, the methodology of hES cell derivation is still highly empirical, and various protocols are used in the different steps of the process, including feeder cell preparation (if feeder cells are used), embryo culture, ICM isolation, and initial steps of derivation. It is worth mentioning that an important goal in derivation attempts is to achieve derivation and culture of hES cells under feeder-free (Amit et al. 2004; Klimanskaya et al. 2005) and animal-free (Genbacev et al. 2005; Li et al. 2005; Ellerström et al. 2006; Rajala et al. 2007) conditions, as well as under chemically defined culture conditions (Li et al. 2005; Lu et al. 2006; Ludwig et al. 2006). Moreover, the entire protocol is performed following good manufacturing practice (GMP) conditions necessary for the safe clinical use of hES cells in human therapy. Recent publications have revealed differences among hES cell lines related to the environment to which the cells have been exposed after embryo culture and derivation (Adewumi et al. 2007; Allegrucci and Young 2007).

Here, we present our recent experience at the Center for Regenerative Medicine in Barcelona (CMRB) and describe the successful derivation of five new hES cell lines. Interestingly, three out of the five new hES cell lines were derived from poor-quality blastocysts, underscoring

---

+These authors contributed equally to this work.

the fact that not only embryos of good quality, but those of poor quality can be used for hES cell derivation. Moreover, one of these lines has been adapted to enzymatic passaging, displays extremely high plating efficiency from single cells, and can be routinely maintained on Matrigel- or gelatin-coated plates for extended periods of time (>30 passages) using fibroblast-conditioned media or serum-free chemically defined media without loss of pluripotency or accumulation of karyotypic abnormalities. Finally, we show that functional cardiomyocytes can be reproducibly differentiated from hES cell lines using a modified embryoid body-based protocol, irrespective of the quality of the blastocyst used for their derivation.

DERIVATION OF FIVE NEW hES CELL LINES

A total of 61 human preembryos, which had been frozen at different stages of development (pronuclear-20, cleavage-35, and blastocyst-6) were used in this study (Table 1). The survival rate after thawing ranged from 80% for cleavage stage to 50% for pronuclear stage and averaged 69% overall. In vitro culture of pronuclear and cleavage-stage embryos resulted in 13 reaching blastocyst stage from the 38 that survived thawing (34%); 4 of the 6 embryos from blastocyst-stage thaws survived, yielding 17 blastocyst-stage cultures overall. These relatively low rates of survival and blastocyst formation are likely attributed to the substandard freezing protocols that were used at the time that these preembryos were placed into cryostorage. In addition, most supernumerary embryos were frozen irrespective of their quality. If an adequate selection had been performed before freezing, it is probable that derivation efficiency would have achieved higher rates of success (Sjögren et al. 2004).

In total, we obtained 17 blastocysts that were classified according to the criteria (put forth by Stephenson et al. 2006) into three groups: good, intermediate, and poor quality (Table 2). Four of these were graded as good-quality blastocysts, showing expanded morphology, an ICM with compacted cells, and a trophectoderm forming a continuous or almost continuous layer. Four blastocysts were classified as intermediate because they had an ICM with few cells visible and were not compacted. Nine embryos did not have a distinguishable ICM or it appeared degenerated, and these were classified as poor-quality blastocysts. Five of these poor-quality preembryos did not show distinct blastocyst morphology; the ICM was indistinguishable from the trophectoderm and consisted of only a few surviving cells.

Table 1. Summary of results of embryo thawing and derivation

	Pronuclear	Cleavage	Blastocyst	Total
Thawed embryo	20	35	6	61
Embryo survival	10 (50%)	28 (80%)	4 (67%)	42 (69%)
Blastocyst rate	3 (30%)	10 (36%)		
Blastocyst/ICM seeded	3	10	4	17
Outgrowths	–	3	2	5
hES cell lines	–	3	2	5

Table 2. Scoring of blastocyst quality

Embryo number	Expansion	ICM	Trophectoderm	Quality score	Lines
1	1	D	C	poor	
2	2	D	C	poor	ES[2]
3	2	D	C	poor	
4	3	C	C	intermediate	ES[3]
5	2	D	C	poor	
6	3	B	B	intermediate	
7	2	B	A	good	
8	3	B	A	good	
9	3	C	A	intermediate	
10	1	D	C	poor	
11	2	D	C	poor	
12	2	A	B	good	
13	2	C	B	intermediate	
14	2	B	A	good	ES[5]
15	2	D	B	poor	ES[6]
16	1	D	C	poor	ES[4]
17	1	D	C	poor	

Blastocyst quality was scored following the criteria of Stephenson et al. (2006) with respect to blastocyst expansion: (1) no expansion in overall size, zona pellucida still thick; (2) some expansion in overall size, zona pellucida beginning to thin; (3) full expansion, zona pellucida very thin. ICM appearance ([A] cells compacted, tightly adhered together, and indistinguishable as individual cells; [B] cells less compacted so larger in size, loosely adhered together, some visible as individual cells; [C] very few cells visible, either compacted or loose, may be difficult to completely distinguish from trophectoderm; [D] cells of ICM appear degenerate) and trophectoderm appearance ([A] many small identical cells forming a continuous trophectoderm layer; [B] fewer, larger cells, may not form continuous trophectoderm layer; [C] sparse cells, may be very large, very flat, or appear degenerate).

During the derivation process, the ICM could not be distinguished in some of the blastocysts and thus it was not possible to isolate it. As a consequence, in a first series of experiments, we performed immunosurgery to isolate the ICM only in good-quality blastocysts, whereas those scored as intermediate or poor quality were seeded whole. In a second series of derivations, we avoided ICM isolation altogether for all blastocysts in order to eliminate the use of antibodies and complement from animal origin (Heins et al. 2004; Suss-Toby et al. 2004).

After 4–5 days in culture, a small clump of cells with hES cell morphology (a compact colony structure, a high nuclear-to-cytoplasmic ratio, and prominent nucleoli) (Thomson et al. 1998; Reubinoff et al. 2000) appeared among other cell types of unspecific morphology (Fig. 1A). These compact cell clumps were mechanically dissociated and replated onto fresh feeders after 10–12 days. During that time, the outgrowth of such structures was carefully monitored and the culture media changed on a daily basis. Overall, from the 17 blastocysts seeded, we obtained five hES cell lines (ES[2]–ES[6]; Fig. 1B–C). One line (ES[5]) was derived from one of the four good-quality blastocysts (25%); one hES cell line (ES[3]) was derived from the four intermediate-quality blastocysts (25%), and the remaining three lines (ES[2], ES[4], and ES[6]) were derived from the nine poor-quality blastocysts (33%). The overall derivation efficiency was 29% and, surprisingly, did not depend on the quality of blastocysts used. This contrasts with the widely held notion that

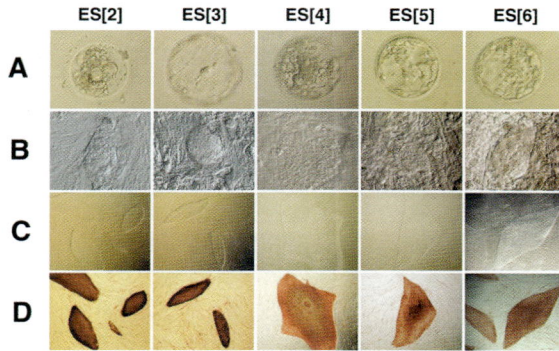

Figure 1. Derivation of five new human embryonic stem cell lines. (*A*) Blastocysts used for successful hES cell derivations. (*B*) Initial cellular outgrowth of the successful hES cell derivations, surrounded by trophectoderm derivatives. Note the atypical initial flat growth of the ES[4] line. (*C*) Morphology of hES cell colonies after 15–20 passages. (*D*) Analysis of alkaline phosphatase activity on the same colonies as those shown in *C*.

embryo quality is correlated with developmental competence and embryo viability (Alikani et al. 1999) and with previous evidence suggesting that hES cells are more efficiently derived from good-quality embryos (Stojkovic et al. 2004; Oh et al. 2005; Simon et al. 2005; Zhang et al. 2006). Nevertheless, our efficiency in the derivation of hES cell lines from poor-quality blastocycts is consistent with previous studies reporting the derivation of four lines from 19 discarded embryos (21% derivation efficiency) (Mitalipova et al. 2003) and two lines from 19 blastocysts derived from day-3 embryos with low morphological scores (10% derivation efficiency) (Chen et al. 2005).

CHARACTERIZATION OF THE NEW hES CELL LINES

Karyotype analysis, human leukocyte antigen (HLA) typing, expression of pluripotency-associated markers, and evaluation of pluripotency in vitro and in vivo were used to characterize all five hES cell lines. A summary of the results obtained from these analyses is presented in Table 3.

The five hES cell lines were positive for all pluripotency markers tested, displaying high levels of alkaline phosphatase (Fig. 1D) and telomerase activity (data not shown) and high levels of Oct4, Nanog, Sox2, SSEA-3, SSEA-4, TRA-1-60, and TRA-1-81 immunoreactivity (Fig. 2 and data not shown). All five lines displayed a normal karyotype (46, XY; data not shown). We are currently investigating the possible significance of a male karyotype for all five lines derived in this study. Four hES cell lines (ES[2]–[4] and ES[6]) have been continuously kept in culture for more than 50 passages (>150 population doublings; Table 3) without evident changes in their growth characteristics, expression of pluripotency-associated markers, or karyotype (data not shown).

All five hES cell lines formed teratomas in severe combined immunodeficient (SCID) beige mice. The tumors contained derivatives from the three embryonic germ layers, such as respiratory epithelium (Fig. 3A–B and data not shown), cartilage (Fig. 3C–D and data not shown), or organized structures that stained positive for neural β-tubulin III, α-fetoprotein, or α-actinin (Fig. 3E–H and data not shown).

To ascertain the ability of hES cell lines to differentiate in vitro, we initially induced the formation of embryoid bodies from colony fragments maintained in suspension for 3–4 days in hES cell media. Under these conditions, embryoid bodies were formed in a reproducible manner, and we could overcome the characteristic difficulty of hES cells to grow as aggregates (Reubinoff et al. 2000). Withdrawal of basic fibroblast growth factor (bFGF) during this phase resulted in extensive cell death and failure to maintain initial cell aggregates, as has been previously reported (Reubinoff et al. 2000). After 3–4 days of growth in hES cell media, embryoid bodies generated in this way were plated onto gelatin-coated plates and allowed to undergo further differentiation by removal of bFGF and the addition of serum. Endoderm derivatives displaying strong α-fetoprotein immunoreactivity were readily observed after 2–3 weeks in all five hES cell lines (Fig. 4A and data not shown). However, neuroectoderm derivatives were obtained at a very low frequency under these conditions. To promote differentiation of hES cells toward neuronal fates, we adapted a protocol of coculture with the stromal cell line PA6 that has been described to enhance neural differentiation of mouse ES cells (Kitajima et al. 2005). All five hES cell lines cocultured with PA6 cells gave rise to differentiated cells with the morphology of mature neurons that expressed high levels of β-tubulin III (Fig. 4B and data not shown).

Table 3. Summary of characterization analyses of the five new hES cell lines

	ES[2]	ES[3]	ES[4]	ES[5]	ES[6]
Karyotype	46, XY	46, XY	46, XY	46, XY	46, XY
Alkaline phosphatase	+	+	+	+	+
Telomerase	+	+	+	+	+
HLA typing	done[a]	done[a]	done[a]	done[a]	done[a]
MDF	done[a]	done[a]	done[a]	done[a]	done[a]
Number of passages	50	65	51	50	78
Pluripotency markers					
SSEA-3	+	+	+	+	+
SSEA-4	+	+	+	+	+
TRA-1-60	+	+	+	+	+
TRA-1-81	+	+	+	+	+
Oct-4	+	+	+	+	+
NANOG	+	+	+	+	+
Sox-2	+	+	+	+	+
SSEA-1	–	–	–	–	–
Freezing/thawing	+	+	+	+	+
Pluripotency					
In vitro					
Endoderm	+	+	+	+	+
Ectoderm	+	+	+	+	+
Mesoderm	+	+	+	+	+
In vivo					
Endoderm	+	+	+	+	+
Ectoderm	+	+	–	+	+
Mesoderm	+	+	–	+	+

Definitions: (HLA) Human leukocyte antigen; (MDF) microsatellite DNA fingerprinting; (+) positive/present; (–) negative.
[a]Full details are available upon request.

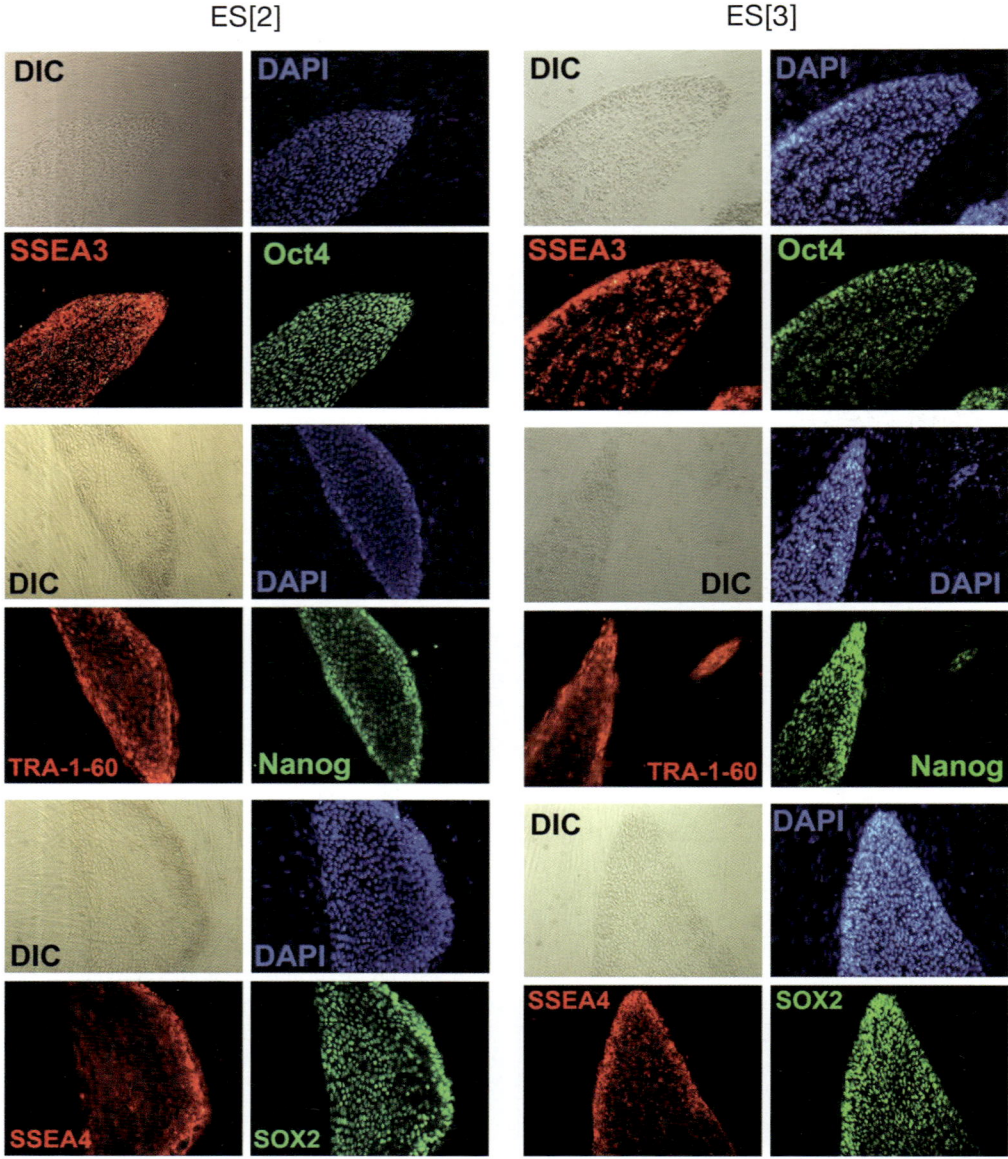

Figure 2. Expression of pluripotency-associated markers. Morphology (as shown by differential interference contrast [DIC]) and immunofluorescence localization of SSEA3, Oct4, TRA-1-60, Nanog, SSEA4, and SOX2 in colonies of ES[2] and ES[3] lines. Cell nuclei were counterstained with DAPI.

EMBRYOID BODY-BASED CARDIAC DIFFERENTIATION OF hES CELLS

In contrast to mouse ES cells, the differentiation of hES cells toward cardiomyocytes is notoriously inefficient (Reubinoff et al. 2000), particularly more so when using embryoid body-based differentiation protocols (Kehat et al. 2001; Laflamme et al. 2005, 2007). Alternatively, coculture of hES cells with mouse visceral endoderm-like END-2 cells has been shown to result in robust induction of cardiomyocyte differentiation (Mummery et al. 2003). To characterize the ability of the new hES cell lines to differentiate into cardiomyocytes in vitro, and in an attempt to establish differentiation protocols that do not rely on coculture systems, we initially set out to analyze the pres-

ence of heart muscle derivatives in embryoid bodies allowed to differentiate in serum-containing media for 2–4 weeks. In contrast to previous studies (Xu et al. 2002), but in agreement with observations from different laboratories using a variety of hES cell lines (Kehat et al. 2001; Laflamme et al. 2005, 2007), we did not observe rhythmically beating cells under these conditions in any of the five new hES cell lines (Fig. 5A and data not shown). We next analyzed the effect of ascorbic acid, because it has been shown that this compound enhances the cardiac differentiation of mouse ES cells (Takahashi et al. 2003) and potentiates the cardiogenic effect of END-2–hES cell cocultures (Passier et al. 2005). The supplementation of the differentiation medium with 100 μM ascorbic acid resulted in approximately 20% of the embryoid bodies generated

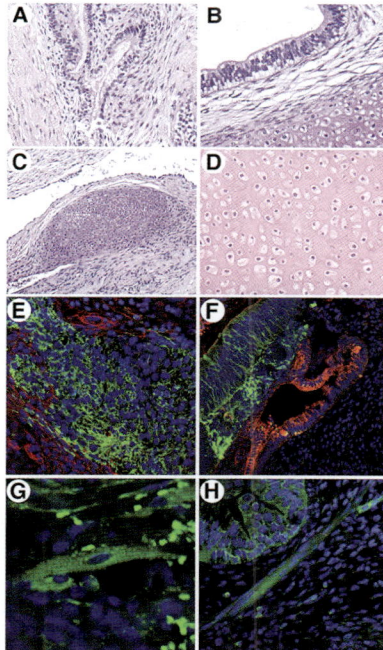

Figure 3. Pluripotency of hES cells in vivo. Hematoxylin and eosin staining (*A–D*) and immunofluorescence localization (*E–H*) in paraffin sections of teratomas induced by injecting ES[2] (*A,C,E,G*) or ES[3] (*B,D,F,H*) cells into severe combined immunodeficient (SCID) beige mice. Histologically recognizable structures include respiratory epithelia (*A,B*) and cartilage formations (*C,D*). (*E,F*) Neuroectoderm and endoderm derivatives were detected by strong immunoreactivity for β-tubulin III (*green channel*) and α-fetoprotein (*red channel*), respectively. (*G,H*) Mesoderm derivatives were detected by strong immunoreactivity for α-actinin (*green channel*). Cell nuclei were counterstained with DAPI in *E–H*.

from any of the five hES cell lines displaying rhythmically beating areas (Fig. 5B and data not shown) that stained strongly positive for cardiac α-actinin (Fig. 5C and data not shown). Our results show that reproducible cardiac differentiation of hES cells can be obtained using relatively simple in vitro protocols and provide a valuable experi-

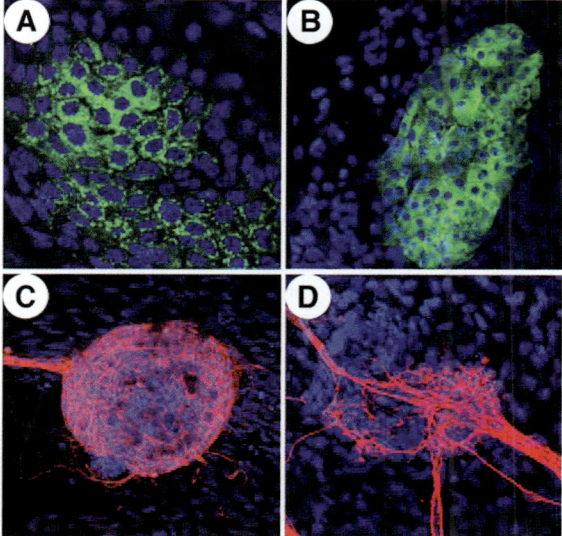

Figure 4. In vitro differentiation of endoderm and neuroectoderm derivatives. ES[2] (*A,C*) and ES[3] (*B,D*) cells readily generate endoderm derivatives after embryoid body-based differentiation protocols (*A,B*), as evidenced by glandular structures immunoreactive for α-fetoprotein (*green channel*) and differentiated cells with morphology of mature neurons that stain positive for β-tubulin III (*red channel*) after coculture with PA6 cells (*C,D*). Cell nuclei were counterstained with DAPI.

mental platform to further optimize such protocols for the directed generation of large quantities of human cardiomyocytes as well as for exploring the mechanisms that control cardiomyocyte differentiation.

ESTABLISHMENT OF ENZYMATICALLY DISPERSABLE hES CELL LINES

Human ES cells typically display very low plating efficiency when passaged as single cells (Amit et al. 2000) and therefore require mechanical or mild enzymatic dissociation of cell clumps. These procedures are time-con-

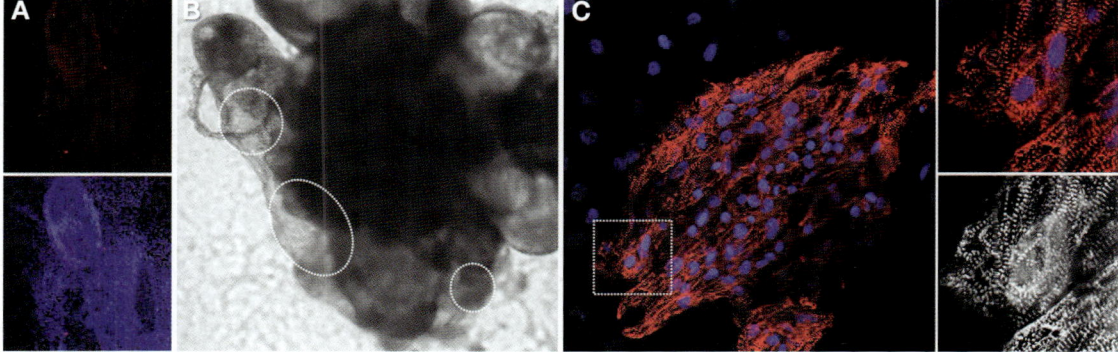

Figure 5. Embryoid body-based in vitro differentiation of hES cells into cardiomyocytes. (*A*) Embryoid bodies of ES[3] cells after 3 weeks of differentiation in serum-containing media fail to differentiate into cardiomyocytes, as evidenced by the absence of α-actinin immunoreactivity (*red channel*). Cell nuclei were counterstained with DAPI in the lower panel. (*B,C*) Addition of ascorbic acid to the differentiation medium results in cardiomyocyte differentiation of ES[3]-derived embryoid bodies, as shown by the appearance of rhythmically beating areas, encircled in *B*, and strong immunoreactivity for α-actinin (*red channel* in *C*). A higher magnification of the area marked in *C* is shown in the right panels to better appreciate the sarcomeric striations of the α-actinin staining. Cell nuclei were counterstained with DAPI in *C*.

suming, result in variable degrees of cell differentiation during hES cell culture maintenance, and make the establishment of genetically modified hES cell clones difficult. The derivation of hES cell lines that can be routinely passaged as single cells (Cowan et al. 2004) and the adaptation to single-cell dissociation of preexisting hES cell lines (Ellerström et al. 2007) are thus likely to facilitate the maintenance of undifferentiated hES cells and their experimental manipulation.

For these reasons, we attempted to establish subclones of hES cells that displayed high plating efficiency after single-cell trypsinization. Subclones were particularly easily adapted from early-passage (passage 3 or 4) ES[4] cells and have been maintained for more than 40 passages on feeders after single-cell dissociation (scES[4]). Furthermore, scES[4] cells can also be propagated for extended periods of time in fibroblast-free cultures using feeder-conditioned hES cell media and Matrigel- or gelatin-coated plates. It should be noted that, even though it has been suggested that such culture adaptations may increase the propensity of hES cells to accumulate karyotypic abnormalities (Enver et al. 2005; Baker et al. 2007), scES[4] cells continuously maintained in our laboratory for more than 30 passages under feeder-free conditions using Matrigel retain a normal karyotype and pluripotency in vivo and in vitro (data not shown). Finally, scES[4] cells have also been easily adapted to commercially available serum-free, chemically defined media (mTeSR™1 or STEMPRO® hESC SFM) and have been maintained undifferentiated for more than 10 passages on Matrigel- or gelatin-coated plates.

METHODS

Source of human preembryos. Human preimplantation embryos were specifically donated for this research project by couples undergoing IVF cycles at the Institut Universitari Dexeus (Barcelona). Embryos were used for the present study with the informed consent of the couples, following the protocol previously approved by the Institutional Ethics Committee on Clinic Investigation and the Spanish competent authorities (Comisión de Seguimiento y Control de la Donación de Células y Tejidos Humanos del Instituto de Salud Carlos III). Derivations were performed at the Stem Cell Bank of the Center of Regenerative Medicine in Barcelona.

Embryos were thawed using Vitrolife thawing media, Thaw-kit 1, or Thaw-kit Blast (Vitrolife, Göteborg, Sweden), depending on the embryo stage and according to manufacturer instructions.

Derivation of hES cell lines. Early-cleavage embryos were cultured in G1.2 medium (Vitrolife, Göteborg, Sweden) until day 3 and then in G2.2 medium (Vitrolife) to the blastocyst stage. Embryos thawed at the blastocyt stage were cultured in G2.2 medium overnight to allow for reexpansion. Blastocysts were classified according to criteria reported elsewhere (Stephenson et al. 2006).

The zona pellucida was removed with 5 mg/ml pronase (Roche) or with Tyrode's acid (Medicult). Fragments and degenerated cells were gently removed with a pulled Pasteur pipette (Humagen Fertility Diagnostics). When performed, immunosurgery for ICM isolation was performed according to published protocols (Solter and Knowles 1975). Whole blastocysts or isolated ICMs were plated on top of irradiated (55Gy) human foreskin fibroblasts (CCD1112Sk ATCC), seeded at a density of 7×10^4 cells/cm^2 on IVF dishes (Falcon, Becton Dickinson) coated with 0.1% gelatin (Chemicon) and cultured further at 37°C and 5% CO_2 in hES cell media consisting of KO–Dulbecco's modified Eagle's medium (DMEM) supplemented with 20% KO–Serum Replacement (Invitrogen), 2 mM GlutaMAX (Invitrogen), 50 μM 2-mercaptoethanol (Invitrogen), nonessential amino acids (Cambrex), 8 ng/ml bFGF (Invitrogen), and penicillin-streptomycin (Invitrogen). Outgrowths were mechanically dissociated into small clumps after 10–12 days using a 150-μm-diameter plastic pipette (The Stripper, Midatlantic Diagnostics) and replated onto a fresh feeder layer.

Propagation of hES cell lines. Individual undifferentiated colonies were mechanically dissociated into small clumps using a 150-μm-diameter plastic pipette and replated every 5–7 days. In some cases, single-cell suspensions were prepared after enzymatic dispersion (0.05% Trypsin/EDTA, Invitrogen) and plated at 1:5-1:10 dilutions onto feeder layers or on plates coated with Matrigel (Becton Dickinson) diluted 1:15 with KO–DMEM or 0.1% gelatin. In feeder-free cultures, we used hES cell media preconditioned by feeders. The following commercially available serum-free, chemically defined media were also assayed following manufacturer directions: mTeSR™1 (StemCell Technologies) and STEMPRO® hESC SFM (Invitrogen).

Karyotyping was performed after incubating colonies for 30 minutes in a 1:500 dilution of Colcemid (Invitrogen) by the G-banding method and imaged and processed using the software CytoVision (Applied Imaging, Olympus). A minimum of 15 metaphases were analyzed for each line.

Colonies were also periodically selected and cryopreserved in cryotube vials (Nunc, Roskilde) in 90% fetal bovine serum (FBS) and 10% dimethylsulfoxide (DMSO) (Sigma-Aldrich) at a cooling rate of 1°C per minute and stored in liquid nitrogen. After thawing at 37°C, some colonies were plated on new feeder layers in hES medium to assess freezing/thawing survival.

In vitro differentiation. Embryoid bodies (EBs) were generated from large colony fragments obtained by mechanical splitting with a finely drawn Pasteur pipette or by treatment of confluent hES cell cultures with 1 mg/ml collagenase IV (Invitrogen) for 10 minutes at 37ºC. Colony fragments were maintained in nonadherent dishes for 3–4 days in hES medium. Embryoid bodies were then plated in 0.1% gelatin-coated glass chamber slides and cultured in differentiation medium (DMEM supplemented with 20% fetal bovine serum, 2 mM GlutaMAX, 100 μM 2-mercaptoethanol, nonessential amino acids, and penicillin-streptomycin) for 2–3 weeks. Neuronal differentiation of hES cells was induced by coculture with the stromal cell line PA6 for 3–5 weeks (Kitajima et al. 2005). Differentiation into mesoderm derivatives was enhanced by supplementing the differentiation medium with 100 μM ascorbic acid (Sigma-Aldrich).

Teratoma formation. Approximately 10^6 hES cells were resuspended in 20–40 μl of hES cell media and injected intramuscularly into the gastrocnemius of SCID beige mice (Charles River Laboratories). Eight weeks after cell injection, mice were sacrificed and tumors were processed and analyzed following conventional immunohistochemistry protocols. All animal experiments were conducted following experimental protocols previously approved by the Institutional Ethics Committee on Experimental Animals, in full compliance with Spanish and European laws and regulations.

Immunofluorescence. Cells grown on chamber slides (LabTek, Nunc) were washed with PBS and fixed with 4% paraformaldehyde for 15 minutes at 4°C. We used the following primary antibodies for immunofluorescence: mouse anti-Oct4 (1:500, Santa Cruz Biotechnology), rabbit anti-Sox2 (1:100, Chemicon), rabbit anti-Nanog (1:500, Abcam), rat anti-SSEA-3, mouse anti-SSEA-4, mouse anti-TRA-1-60, mouse anti-TRA-1-81 (1:10, Chemicon), mouse anti-α-actinin (1:100, Sigma-Aldrich), mouse anti-β-tubulin III (1:1000, Covance), and rabbit anti-α-fetoprotein (1:400, Dako). Incubation with primary antibody was for 24 hours at 4°C. Incubation with FITC-, Cy2-, or Cy3-conjugated secondary antibodies (1:200, Jackson ImmunoResearch) was for 2 hours at room temperature followed by counterstaining with 4′-6-diamidino-2-phenylindole (DAPI) (10 mg/ml, Sigma-Aldrich).

Alkaline phosphatase activity was measured using the Alkaline Phosphatase Red Membrane Substrate (Sigma-Aldrich) after fixation with 4% paraformaldehyde for 2 minutes. Telomerase activity was detected using the TRAPeze Telomerase Detection Kit (Chemicon International) following manufacturer directions.

Molecular typing. HLA typing of hES cell lines was performed by sequence-based typification (SBT) with the AlleleSEQR HLA Sequencing Kit (Atria Genetics). Microsatellite DNA fingerprinting was performed using multiplex polymerase chain reaction of nine microsatellites or short tandem repeats (STRs) plus amelogenine gene using AmplFlSTR Profiler Plus Kit (Applied Biosystems).

CONCLUSIONS

The five new hES cell lines derived in this study provide valuable tools for investigating the mechanisms of ES cell self-renewal, pluripotency, and differentiation toward specific cell types. ES[2] and ES[3] are currently available to interested researchers through the Spanish Stem Cell Bank (http://www.isciii.es/htdocs/terapia/terapia_lineas.jsp). The remaining three lines (ES[4], ES[5], and ES[6]) are in the process of being registered and deposited and will soon be similarly available.

ACKNOWLEDGMENTS

We are grateful to Joana Visa for help with teratoma formation assays, Vicenç Català and Esther Cuatrecases for help and advice with karyotyping, Ricardo Pujol and Eduard Palou for HLA typing, and the personnel of the Institut Universitari Dexeus, especially Marta Luna, for assistance with embryo handling. We are indebted to Travis Berggren and members of the Belmonte laboratory for comments on the manuscript. We thank Meritxell Carrió, Yolanda Muñoz, and Yvonne Richaud for excellent technical assistance in hES cell culture, and Mercé Martí for expert assistance in confocal microscopy. I.R.-P. is a recipient of a MEC predoctoral fellowship. This work was partially supported by Fondo de Investigaciones Sanitarias grants RETIC-RD06/0010/0016 (to J.C.I.B.), PI052847 (to A.V.), and CP05/00294 and PI061897 (to A.C.), Ministerio de Educación y Ciencia grants BFU2006-12247 (to J.C.I.B.) and BFU2006-12251 (to A.R.), AGAUR grant 2005SGR00331 (to J.C.I.B.), European Commission "Marie-Curie Reintegration Grant" MIRG-CT-2007-046523 (to A.R.), Marató de TV3 grant 063430 (to J.C.I.B.), and Fundación Cellex grant (to J.C.I.B.).

REFERENCES

Adewumi, O., Aflatoonian, B., Ahrlund-Richter, L., Amit, M., Andrews, P.W., Beighton, G., Bello, P.A., Benvenisty, N., Berry, L.S., Bevan, S., et al. 2007. Characterization of human embryonic stem cell lines by the International Stem Cell Initiative. *Nat. Biotechnol.* **25:** 803–816.

Alikani, M., Cohen, J., Tomkin, G., Garrisi, G.J., Mack C., and Scott, R.T. 1999. Human embryo fragmentation in vitro and its implications for pregnancy and implantation. *Fertil. Steril.* **71:** 836–842.

Allegrucci, C. and Young, L.E. 2007. Differences between human embryonic stem cell lines. *Hum. Reprod. Update* **13:** 103–120.

Amit, M., Shariki, C., Margulets, V., and Itskovitz-Eldor, J. 2004. Feeder layer- and serum-free culture of human embryonic stem cells. *Biol. Reprod.* **70:** 837–845.

Amit, M., Carpenter, M.K., Inokuma, M.S., Chiu, C.P., Harris, C.P., Waknitz, M.A., Itskovitz-Eldor, J., and Thomson, J.A. 2000. Clonally derived human embryonic stem cell lines maintain pluripotency and proliferative potential for prolonged periods of culture. *Dev. Biol.* **227:** 271–278.

Baker, D.E., Harrison, N.J., Maltby, E., Smith, K., Moore, H.D., Shaw, P.J., Heath, P.R., Holden, H., and Andrews, P.W. 2007. Adaptation to culture of human embryonic stem cells and oncogenesis in vivo. *Nat. Biotechnol.* **25:** 207–215.

Chen, H., Qian, K., Hu, J., Liu, D., Lu, W., Yang, Y., Wang, D., Yan, H., Zhang, S., and Zhu, G. 2005. The derivation of two additional human embryonic stem cell lines from day 3 embryos with low morphological scores. *Hum. Reprod.* **20:** 2201–2206.

Cowan, C.A., Klimanskaya, I., McMahon, J., Atienza, J., Witmyer, J., Zucker, J.P., Wang, S., Morton, C.C., McMahon, A.P., Powers, D., and Melton, D.A. 2004. Derivation of embryonic stem-cell lines from human blastocysts. *N. Engl. J. Med.* **350:** 1353–1356.

Ellerström, C., Strehl, R., Noaksson, K., Hyllner, J., and Semb, H. 2007. Facilitated expansion of human embryonic stem cells by single-cell enzymatic dissociation. *Stem Cells* **25:** 1690–1696.

Ellerström, C., Strehl, R., Moya, K., Andersson, K., Bergh, C., Lundin, K., Hyllner, J., and Semb, H. 2006. Derivation of a xeno-free human embryonic stem cell line. *Stem Cells* **24:** 2170–2176.

Enver, T., Soneji, S., Joshi, C., Brown, J., Iborra, F., Orntoft, T., Thykjaer, T., Maltby, E., Smith, K., Dawud R.A., et al. 2005.

Cellular differentiation hierarchies in normal and culture-adapted human embryonic stem cells. *Hum. Mol. Genet.* **14:** 3129–3140.

Gardner, D.K., Lane, M., Stevens, J., Schlenker, T., and Schoolcraft, W.B. 2000. Blastocyst score affects implantation and pregnancy outcome: Towards a single blastocyst transfer. *Fertil. Steril.* **73:** 1155–1158.

Genbacev, O., Krtolica, A., Zdravkovic, T., Brunette, E., Powell, S., Nath, A., Caceres, E., McMaster, M., McDonagh, S., Li, Y., et al. 2005. Serum-free derivation of human embryonic stem cell lines on human placental fibroblast feeders. *Fertil. Steril.* **83:** 1517–1529.

Gerecht-Nir, S. and Itskovitz-Eldor, J. 2004. Human embryonic stem cells: A potential source for cellular therapy. *Am. J. Transplant.* (suppl. 6) **4:** 51–57.

Heins, N., Englund, M.C., Sjoblom, C., Dahl, U., Tonning, A., Bergh, C., Lindahl, A., Hanson, C., and Semb, H. 2004. Derivation, characterization, and differentiation of human embryonic stem cells. *Stem Cells* **22:** 367–376.

Kehat, I., Kenyagin-Karsenti, D., Snir, M., Segev, H., Amit, M., Gepstein, A., Livne, E., Binah, O., Itskovitz-Eldor, J., and Gepstein, L. 2001. Human embryonic stem cells can differentiate into myocytes with structural and functional properties of cardiomyocytes. *J. Clin. Invest.* **108:** 407–414.

Kitajima, H., Yoshimura, S., Kokuzawa, J., Kato, M., Iwama, T., Motohashi, T., Kunisada, T., and Sakai, N. 2005. Culture method for the induction of neurospheres from mouse embryonic stem cells by coculture with PA6 stromal cells. *J. Neurosci. Res.* **80:** 467–474.

Klimanskaya, I., Chung, Y., Meisner, L., Johnson, J., West, M.D., and Lanza, R. 2005. Human embryonic stem cells derived without feeder cells. *Lancet* **365:** 1636–1641.

Laflamme, M.A., Gold, J., Xu, C., Hassanipour, M., Rosler, E., Police, S., Muskheli, V., and Murry, C.E. 2005. Formation of human myocardium in the rat heart from human embryonic stem cells. *Am. J. Pathol.* **167:** 663–671.

Laflamme, M.A., Chen, K.Y., Naumova, A.V., Muskheli, V., Fugate, J.A., Dupras, S.K., Reinecke, H., Xu, C., Hassanipour, M., Police, S., et al. 2007. Cardiomyocytes derived from human embryonic stem cells in pro-survival factors enhance function of infarcted rat hearts. *Nat. Biotechnol.* **25:** 1015–1024.

Li, T., Zhou, C.Q., Mai, Q.Y., and Zhuang, G.L. 2005. Establishment of human embryonic stem cell line from gamete donors. *Chin. Med. J.* **118:** 116–122.

Liew, C.G., Moore, H., Ruban, L., Shah, N., Cosgrove, K., Dunne, M., and Andrews, P. 2005. Human embryonic stem cells: Possibilities for human cell transplantation. *Ann. Med.* **37:** 521–532.

Lu, J., Hou, R., Booth, C.J., Yang, S.H., and Snyder, M. 2006. Defined culture conditions of human embryonic stem cells. *Proc. Natl. Acad. Sci.* **103:** 5688–5693.

Ludwig, T.E., Levenstein, M.E., Jones, J.M., Berggren, W.T., Mitchen, E.R., Frane, J.L., Crandall, L.J., Daigh, C.A., Conard, K.R., Piekarczyk, M.S., Llanas, R.A., and Thomson, J.A. 2006. Derivation of human embryonic stem cells in defined conditions. *Nat. Biotechnol.* **24:** 185–187.

Lyerly, A.D. and Faden, R.R. 2007. Embryonic stem cells. Willingness to donate frozen embryos for stem cell research. *Science* **317:** 46–47.

Mateizel, I., De Temmerman, N., Ullmann, U., Cauffman, G., Sermon, K., Van de Velde, H., De Rycke, M., Degreef, E., Devroey, P., Liebaers, I., and Van Steirteghem, A. 2006. Derivation of human embryonic stem cell lines from embryos obtained after IVF and after PGD for monogenic disorders. *Hum. Reprod.* **21:** 503–511.

Menasché, P. 2005. The potential of embryonic stem cells to treat heart disease. *Curr. Opin. Mol. Ther.* **7:** 293–299.

Ménézo, Y., Veiga, A., and Benkhalifa, M. 1998. Improved methods for blastocyst formation and culture. *Hum. Reprod.* (suppl. 4) **13:** 256–265.

Ménézo, Y., Hazout, A., Dumont, M., Herbaut, N., and Nicollet, B. 1992. Coculture of embryos on Vero cells and transfer of blastocysts in humans. *Hum. Reprod.* (suppl. 1) **7:** 101–106.

Mitalipova, M., Calhoun, J., Shin, S., Wininger, D., Schulz, T., Noggle, S., Venable, A., Lyons, I., Robins, A., and Stice, S. 2003. Human embryonic stem cell lines derived from discarded embryos. *Stem Cells* **21:** 521–526.

Mummery, C., Ward-van Oostwaard, D., Doevendans, P., Spijker, R., van den Brink, S., Hassink, R., van der Heyden, M., Opthof, T., Pera, M., de la Riviere, A.B., Passier, R., and Tertoolen, L. 2003. Differentiation of human embryonic stem cells to cardiomyocytes: Role of coculture with visceral endoderm-like cells. *Circulation* **107:** 2733–2740.

Oh, S.K., Kim, H.S., Ahn, H.J., Seol, H.W., Kim, Y.Y., Park, Y.B., Yoon, C.J., Kim, D.W., Kim, S.H., and Moon, S.Y. 2005. Derivation and characterization of new human embryonic stem cell lines: SNUhES1, SNUhES2, and SNUhES3. *Stem Cells* **23:** 211–219.

Passier, R., Oostwaard, D.W., Snapper, J., Kloots, J., Hassink, R.J., Kuijk, E., Roelen, B., de la Riviere, A.B., and Mummery, C. 2005. Increased cardiomyocyte differentiation from human embryonic stem cells in serum-free cultures. *Stem Cells* **23:** 772–780.

Pera, M.F. and Trounson, A.O. 2004. Human embryonic stem cells: Prospects for development. *Development* **131:** 5515–5525.

Pickering, S.J., Braude, P.R., Patel, M., Burns, C.J., Trussler, J., Bolton, V., and Minger, S. 2003. Preimplantation genetic diagnosis as a novel source of embryos for stem cell research. *Reprod. Biomed. Online* **7:** 353–364.

Rajala, K., Hakala, H., Panula, S., Aivio, S., Pihlajamaki, H., Suuronen, R., Hovatta, O., and Skottman, H. 2007. Testing of nine different xeno-free culture media for human embryonic stem cell cultures. *Hum. Reprod.* **22:** 1231–1238.

Reubinoff, B.E., Pera, M.F., Fong, C.Y., Trounson, A., and Bongso, A. 2000. Embryonic stem cell lines from human blastocysts: Somatic differentiation in vitro. *Nat. Biotechnol.* **18:** 399–404.

Semb, H. 2005. Human embryonic stem cells: Origin, properties and applications. *APMIS* **113:** 743–750.

Simon, C., Escobedo, C., Valbuena, D., Genbacev, O., Galan, A., Krtolica, A., Asensi, A., Sanchez, E., Esplugues, J., Fisher, S., and Pellicer, A. 2005. First derivation of human embryonic stem cell lines: Use of long-term cryopreserved embryos and animal-free conditions. *Fertil. Steril.* **83:** 246–249.

Sjögren, A., Hardarson, T., Andersson, K., Caisander, G., Lundquist, M., Wikland, M., Semb, H., and Hamberger, L. 2004. Human blastocysts for the development of embryonic stem cells. *Reprod. Biomed. Online* **9:** 326–329.

Solter, D. and Knowles, B.B. 1975. Immunosurgery of mouse blastocyst. *Proc. Natl. Acad. Sci.* **72:** 5099–5102.

Stephenson, E.L., Braude, P.R., and Mason, C. 2006. Proposal for a universal minimum information convention for the reporting on the derivation of human embryonic stem cell lines. *Regen. Med.* **1:** 739–750.

Stojkovic, M., Lako, M., Stojkovic, P., Stewart, R., Przyborski, S., Armstrong, L., Evans, J., Herbert, M., Hyslop, L., Ahmad, S., Murdoch, A., and Strachan, T. 2004. Derivation of human embryonic stem cells from day-8 blastocysts recovered after three-step in vitro culture. *Stem Cells* **22:** 790–797.

Suss-Toby, E., Gerecht-Nir, S., Amit, M., Manor, D., and Itskovitz-Eldor, J. 2004. Derivation of a diploid human embryonic stem cell line from a mononuclear zygote. *Hum. Reprod.* **19:** 670–675.

Takahashi, T., Lord, B., Schulze, P.C., Fryer, R.M., Sarang, S.S., Gullans, S.R., and Lee, R.T. 2003. Ascorbic acid enhances differentiation of embryonic stem cells into cardiac myocytes. *Circulation* **107:** 1912–1916.

Thomson, J.A., Itskovitz-Eldor, J., Shapiro, S.S., Waknitz, M.A., Swiergiel, J.J., Marshall, V.S., and Jones, J.M. 1998. Embryonic stem cell lines derived from human blastocysts. *Science* **282:** 1145–1147.

Veiga, A., Camarasa, M.V., Aran, B., Raya, A., and Izpisúa Belmonte, J.C. 2007. Selection of embryos for stem cell derivation: Can we optimize the process? In *Stem cells in reproductive medicine: Basic science and therapeutic potential* (ed. C.

Simon and A. Pellicer), pp. 133–145. Informa, London.

Veiga, A., Torello, M.J., Boiso, I., Sandalinas, M., Busquets, A., Calderon, G., and Barri, P.N. 1995. Optimization of implantation in the in-vitro fertilization laboratory. *Hum. Reprod.* (suppl. 2) **10:** 98–106.

Verlinsky, Y., Strelchenko, N., Kukharenko, V., Rechitsky, S., Verlinsky, O., Galat, V., and Kuliev, A. 2005. Human embryonic stem cell lines with genetic disorders. *Reprod. Biomed.*

Online **10:** 105–110.

Xu, C., Police, S., Rao, N., and Carpenter, M.K. 2002. Characterization and enrichment of cardiomyocytes derived from human embryonic stem cells. *Circ. Res.* **91:** 501–508.

Zhang, X., Stojkovic, P., Przyborski, S., Cooke, M., Armstrong, L., Lako, M., and Stojkovic, M. 2006. Derivation of human embryonic stem cells from developing and arrested embryos. *Stem Cells* **24:** 2669–2676.

Regulation of Self-renewal and Differentiation in Adult Stem Cell Lineages: Lessons from the *Drosophila* Male Germ Line

E.L. Davies and M.T. Fuller

Departments of Developmental Biology and Genetics, Stanford University School of Medicine, Stanford, California 94305-5329

The ability to identify stem cells and trace their descendants in vivo has yielded insights into how self-renewal, proliferation, and differentiation are regulated in adult stem cell lineages. Analysis of male germ-line stem cells in *Drosophila* has revealed the importance of local signals from the microenvironment, the stem cell niche, in controlling stem cell behavior. Germ-line stem cells physically attach to the niche via localized adherens junctions that provide a polarity cue for orientation of centrosomes in interphase and the spindle in mitosis. As a result, stem cells divide asymmetrically: One daughter inherits attachment to the niche and remains within its embrace, whereas the other is displaced away and initiates differentiation. Strikingly, much as leukemia inhibitory factor (LIF) and transforming growth factor-β (TGF-β) signaling maintain mouse embryonic stem (ES) cells, maintenance of stem cell state in the *Drosophila* male germ line is regulated by cytokine-like signals from hub cells that activate the transcription factor STAT (signal transducer and activator of transcription) and TGF-β class signals from surrounding support cells that repress expression of a key differentiation factor. Surprisingly, transit-amplifying cells can revert to the stem cell state if they reoccupy the niche. Upon cessation of mitosis and the switch to terminal differentiation, germ cells express cell-type- and stage-specific transcription machinery components that drive expression of terminal differentiation genes, in part by removing Polycomb transcriptional silencing machinery.

Regulated activity of adult stem cells is essential for tissue homeostasis. Adult stem cell lineages are responsible for the long-term maintenance and repair of tissues containing highly specialized, short-lived cell types, including blood, sperm, skin, and intestinal epithelium. Recent studies suggest that differentiated cells in other tissues, including breast, lung, skeletal muscle, and prostate, are also produced from adult stem cells in response to physiological changes or damage. Hallmarks of adult stem cells are a committed but relatively undifferentiated state, long-term ability to proliferate, and ability to produce both new stem cells (self-renewal) and differentiating progeny (Fig. 1). Understanding the mechanisms that regulate adult stem cell self-renewal, as well as those that regulate the proliferation and differentiation of their progeny, will be key for harnessing the potential of adult stem cells for regenerative medicine and may also suggest new strategies for the war on cancer. Many common cancers arise in adult stem cell lineages, and there is increasing evidence that defects in the mechanisms that regulate self-renewal, proliferation, and differentiation in adult stem cell lineages can contribute to tumorigenesis (Clarke and Fuller 2006).

KEY REGULATORY POINTS IN ADULT STEM CELL LINEAGES

Tissue homeostasis in adult stem cell lineages requires the proper execution and coordination of several key switches in cell state (Fig. 1A). When adult stem cells

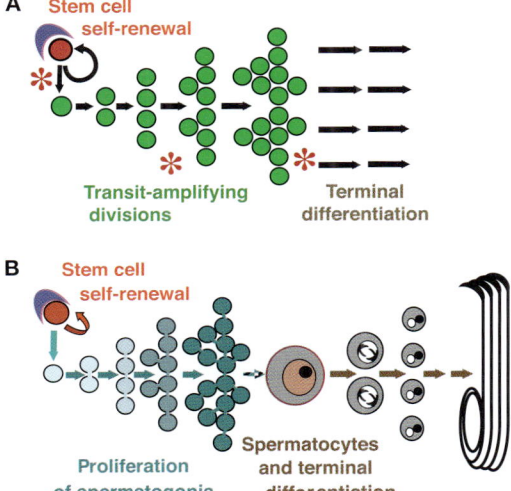

Figure 1. Key decisions in adult stem cell lineages. (*A*) An archetypical adult stem cell lineage. A stem cell (*red*) in the niche (*blue crescent*) divides to both self-renew and produce a daughter that initiates differentiation (*green*). After a limited series of transit-amplifying mitotic divisions (*green*), the cells cease proliferation and turn on the terminal differentiation program (*bank of arrows*). Asterisks mark key regulatory points: (*Left to right*) Self-renewal versus onset of differentiation; limiting the number of transient-amplifying divisions; switch to terminal differentiation. (*B*) Male germ-line stem cell lineage in *Drosophila*. A stem cell (*red*) in the niche at the tip of the testis divides asymmetrically, producing a new stem cell and a gonialblast, which initiates four rounds of synchronous spermatogonial mitotic divisions with incomplete cytokinesis (*green arrows*). The resulting 16 interconnected germ cells undergo premeiotic DNA replication in synchrony and turn on the spermatocyte program of cell growth, meiotic prophase, and transcription of terminal differentiation genes (*large cell with brown nucleus:* All 16 cells become spermatocytes, but only one is shown for simplicity). After the two meiotic divisions, the 64 haploid spermatids undergo complete remodeling to generate mature sperm.

divide, the daughter cells must first choose between maintenance of stem cell identity or initiation of differentiation. Specialized microenvironments, known as stem cell niches, regulate the outcome of this critical cell-fate decision in vivo, ensuring that populations of stem cells and differentiating progeny are properly maintained (Watt and Hogan 2000; Spradling et al. 2001; Yamashita et al. 2005; Fuller and Spradling 2007). In several well-studied adult stem cell model systems, stem cell self-renewal relies on close-range signals from a support cell niche; failure to maintain an intimate association with the niche causes cells to adopt differentiated fates. Discrete anatomical niches provide a strategy for regulating stem cell number and may defend against cancer by preventing undifferentiated precursors from self-renewing outside of the niche. In most systems, however, the difficulty in identifying stem cells in situ has precluded investigations of the relationship between stem cells and their natural environment.

In many adult stem cell lineages, stem cell daughters that initiate differentiation execute a series of transit-amplifying divisions before terminal differentiation, thereby allowing a large number of differentiated cells to be produced from a single stem cell division (Fig. 1A). Mechanisms that limit the number of transit-amplifying cell divisions comprise another key regulatory point: They ensure that mutations arising in precursor cells are not maintained in the pool of long-term proliferating cells. Third, the switch from proliferation to terminal differentiation entails dramatic changes in the cell cycle and gene expression programs, culminating in expression of large numbers of terminal differentiation genes that were either silenced or expressed at only low levels in the precursor cells. Additionally, many adult stem cell lineages are multipotent, and precursor cells must choose to differentiate along a single, context-appropriate developmental pathway.

Investigation of the male and female germ-line stem cell (GSC) lineages in *Drosophila* has provided insight into the types of mechanisms that regulate these critical cell-fate switches in vivo and paradigms for researchers working in mammalian stem cell systems. In *Drosophila*, GSC self-renewal depends on short-range signals from a support cell niche (Xie and Spradling 1998; Kiger et al. 2001; Tulina and Matunis 2001). Other key features include physical attachment to the niche, orientation of stem cell divisions with respect to the niche (Xie and Spradling 2000; Yamashita et al. 2003, 2007), and the ability of transit-amplifying cells that reoccupy the niche to revert to stem cell function (Brawley and Matunis 2004; Kai and Spradling 2004; Cheng et al. 2008). For a review of research on the female GSC niche, see Spradling et al. (this volume). Here, we discuss the mechanisms that regulate self renewal and differentiation in the *Drosophila* male GSC niche. Comparison of these two systems has revealed key conserved features and also interesting differences.

SPERMATOGENESIS AS A MODEL ADULT STEM CELL SYSTEM

Male gametes are produced in a unipotent adult stem cell lineage (Fig. 1B). In *Drosophila*, spermatogenesis initiates with the asymmetric division of a male GSC. One daughter maintains stem cell identity and the other initiates differentiation as a gonialblast (Gb), the founder of a clone of transit-amplifying spermatogonia (Figs. 1A and 2C). Transit-amplifying spermatogonia can be clearly distinguished from GSCs on the basis of cell behavior, structure, and gene expression. GSCs divide asynchronously, execute complete cytokinesis, and contain a ball-shaped spectrin-rich intracellular membranous organelle called the spectrosome (Fig. 2C). In contrast, transit-amplifying spermatogonia descended from a Gb divide synchronously with incomplete cytokinesis, producing a cyst of interconnected germ cells joined by cytoplasmic bridges. As spermatogonial cysts grow with each round of mitosis, the ball-shaped spectrosome elongates and threads through the ring canals that connect germ cells within a cyst, creating a branched organelle called the fusome (Fig. 2C). After undergoing a genetically predetermined number of transit-amplifying divisions (four in *Drosophila melanogaster*), the resulting 16 germ cells together exit the mitotic program and commit to terminal differentiation. The cells synchronously execute premeiotic DNA synthesis, adopt primary spermatocyte fate, grow dramatically in volume, and turn on an extremely active cell-type-specific transcription program in preparation for terminal differentiation (Fig. 1A). Many genes are transcribed for the first time in development in primary spermatocytes, and many widely expressed genes are transcribed from alternate spermatocyte-specific promoters (for review, see Fuller 1993; Hecht 1993). The 16 spermatocytes then together undergo the two meiotic divisions, and the resulting 64 haploid spermatids execute a dramatic remodeling program to generate mature sperm.

THE TESTIS STEM CELL NICHE MAINTAINS TWO ADULT STEM CELL POPULATIONS

Drosophila male GSCs reside in a niche that instructs their self-renewal. Approximately eight to ten GSCs lie in a rosette around a cluster of somatic support cells called the hub at the apical tip of each testis (Fig. 2). The hub supports stem cell self-renewal by serving as the exclusive source of the signaling ligand *Unpaired* (*Upd*) (Fig. 2A). In a cytokine-like signaling pathway, *Upd* produced by hub cells locally activates the Janus kinase–signal transducer and activator of transcription (JAK-STAT) pathway in neighboring cells. Action of STAT is required cell autonomously in male GSCs for stem cell maintenance: Male GSCs made homozygous mutant for a null allele of *stat* fail to maintain stem cell identity and instead differentiate (Kiger et al. 2001; Tulina and Matunis 2001). Age-dependent decrease in *Upd* expression in the hub may contribute to GSC loss with advanced age (Boyle et al. 2007).

The apical hub serves as a niche for two adult stem cell populations: male GSCs and somatic cyst stem cells (CySCs) (Figs. 2C and 3), allowing spatially coordinated production of progeny cells that interact and codifferentiate. The cell bodies of CySCs sit distal to the GSCs, and they maintain direct contact with the hub via narrow cytoplasmic processes that intercalate between the GSCs.

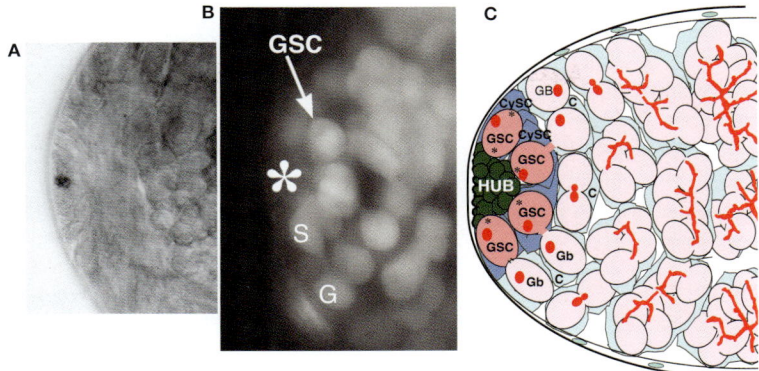

Figure 2. Stem cells in the niche at the apical tip of the *Drosophila* testis. (*A*) Hub cells at the apical tip of the testis express the signaling ligand *Upd.* In situ hybridization shows local expression of *Upd* mRNA. (*B*) Male germ cells at the apical tip of the testis viewed live by germ-line-specific expression of UAS-GFP under control of *nos-GAL4.* Asterisk indicates the apical hub composed of somatic cells. (S) GSC next to the hub; (G) gonialblast displaced away from the hub. Note the transient cytoplasmic bridge connecting the stem cell and the gonialblast, indicating that the two arose recently from division of a stem cell. (*C*) Diagram of the testis stem cell niche and germinal proliferation center. GSCs (*dark pink*) attach to hub cells (*green*) and are flanked by somatic cyst stem cells (CySC) (*dark blue*). Gonialblasts (Gb) produced by GSCs are displaced away from the hub and enveloped by two somatic cyst cells (C) (*light blue*) produced by the CySCs. Each gonialblast finds a mitotic clone of spermatogonia (*light pink*) that divide in synchrony, forming cysts of 2, 4, 8, and 16 interconnected germ cells. (*Red balls*) Spectrosomes; (*red lines*) fusomes. The germ cell cysts execute premeiotic S and become spermatocytes (not shown) (*A,* Reprinted, with permission, from Kiger et al. 2001 [©AAAS]; *C,* reprinted, with permission, from Fuller and Spradling 2007 [©AAAS].)

CySCs self-renew and give rise to daughters that differentiate into somatic cyst cells (Gönczy and DiNardo 1996). Cyst cells occasionally undergo one round of transit-amplifying division, but they usually adopt a postmitotic fate and initiate differentiation immediately after the CySC division (Voog et al. 2008). Two cyst cells associate with and envelop each gonialblast, forming a discrete packet, or cyst (Fig. 3), that remains intact throughout the rest of spermatogenesis. The mechanisms that specify the ratio of one Gb to two cyst cells are not yet understood. The cyst cells codifferentiate with the germ cells that they enclose (Gönczy et al. 1992), and continued association and communication between the germ line and the enveloping somatic cyst cells is essential for proper differentiation of both cell types. CySCs are bipotential: CySCs or their daughters occasionally contribute to the hub (Voog et al. 2008), suggesting a novel mechanism for maintaining the hub. This relationship may reflect the common segmental origin of hub and CySCs in the embryonic gonad (Le Bras and Van Doren 2006). The choice between cyst cell versus hub cell identity may be influenced by the presence of germ cells, because the hub expands and changes morphology in agametic testes (Gönczy and DiNardo 1996; Voog et al. 2008). The picture emerging is that stem cells and their niches are mutually interdependent self-sustaining systems.

The same cytokine-like signal from the hub that maintains GSCs also maintains the somatic CySCs (Fig. 3). Like GSCs, the CySCs overproliferate when *Upd* is ectopically expressed in early germ cells (paracrine stimulation) or in early cyst cells (autocrine stimulation) (Kiger et al. 2001; Tulina and Matunis 2001; Leatherman and DiNardo 2008). Upd activates the JAK-STAT cascade in CySCs to specify their self-renewal. Action of *stat* is required cell autonomously for CySC maintenance: Somatic CySCs

made homozygous mutant for *stat* fail to maintain stem cell identity and instead commit to differentiation (Leatherman and DiNardo 2008). The transcriptional repressor *zinc finger-homeodomain transcription factor 1* (*zfh-1*) is likely to be a critical downstream target of the JAK-STAT pathway in CySCs. Zfh-1 protein accumulates in CySCs, and its levels drop precipitously as cyst cells codifferentiate with

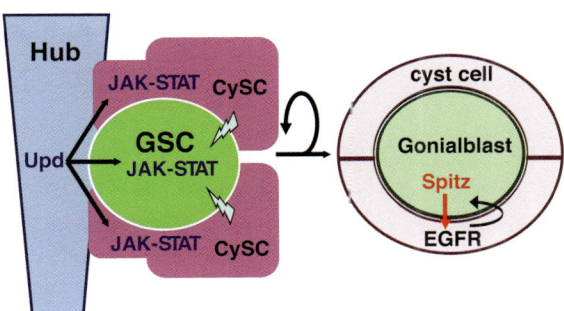

Figure 3. GSCs and CySCs require the *Upd* signal from the hub, and the hub and CySCs both contribute to the GSC niche. (*Left*) Diagram of the stem cell niche. *Upd* secreted by hub cells (*blue*) activates the JAK-STAT pathway, which is required cell autonomously in both the germ line (GSC) and cyst stem cells (CySC) for stem cell self-renewal. However, activation of STAT is not sufficient to instruct GSC self-renewal, which also requires as yet unknown signal(s) (*lightning bolt*) from the CySCs. (*Right*) Interactions between the progeny of somatic and germ line stem cells set up the functional unit of differentiation, the cyst. A Gb (*light green*), produced by asymmetric division of a GSC, signals via the epidermal growth factor receptor (EGFR) ligand *Spitz* to cyst cells (*light pink*) produced by asymmetric division of CySCs. Activation of EGFR triggers the somatic cyst cells to envelop the Gb and may also induce changes in gene expression in the cyst cells required for the germ cells to properly differentiate.

spermatogonia. *Zfh-1* function is required cell autonomously for CySC maintenance. Importantly, ectopic expression of *Zfh-1* in the soma is sufficient to promote CySC self-renewal and to block differentiation, even without concomitant up-regulation of JAK-STAT pathway activity in CySCs (Leatherman and DiNardo 2008).

MULTIPLE CELL TYPES CONTRIBUTE TO NICHE FUNCTION

Recent studies suggest that the niche that maintains male GSCs is complex, consisting of conversations among three different cell types: the GSCs, CySCs, and hub cells. Complex niches may be a feature of systems where multiple stem cell types must be maintained in close proximity, and their activities coordinated to produce progeny that form functional units and differentiate together. These mechanisms may be especially important in organs composed of cell types from different developmental origins.

The somatic CySCs contribute to the niche that maintains stem cell state in the germ line. Although activation of STAT in GSCs by *Upd* secreted from the hub is required for GSC self-renewal, it is not sufficient. Forced expression of a constitutively active allele of the JAK tyrosine kinase in early germ cells, leading to STAT activation, did not cause massive proliferation of GSCs at the expense of differentiation (Leatherman and DiNardo 2008; A.A. Kiger and M.T. Fuller, unpubl.). Strikingly, however, forced expression of constitutively active JAK in the cyst cell lineage resulted in overproliferation of GSCs or their immediate progeny and inhibited germ cell differentiation (Leatherman and DiNardo 2008). Similar results were seen when *Zfh-1* was forcibly expressed in the somatic cyst cell lineage. Taken together, these data suggest that GSCs are maintained in the niche and instructed to self-renew by a combination of cytokine-like signals from the hub and as yet unknown *Zfh-1*-dependent signal(s) from the neighboring somatic CySCs (Fig. 3).

STEM CELLS ORIENT TOWARD THE NICHE TO PROGRAM ASYMMETRIC DIVISION

Drosophila male GSCs normally divide asymmetrically: The mitotic spindle is set up perpendicular to the hub–GSC interface so that, upon cytokinesis, one daughter remains in the niche and maintains stem cell identity, whereas the other daughter is displaced out of the niche and initiates differentiation (Fig. 2B,C). Thus, the mechanisms that polarize and orient the cytoskeleton in GSCs are key regulators of cell fate.

GSCs are physically attached to the hub by localized adherens junctions. *Drosophila* E-cadherin concentrates at the GSC cortex adjacent to the hub (Fig. 4A). β-catenin and APC2, a homolog of the adenomatous polyposis coli tumor suppressor, colocalize with E-cadherin at the adherens junctions, which provide a polarity cue toward which GSCs orient throughout the cell cycle (Yamashita et al. 2003). In G_1, the single centrosome in each GSC localizes near the cell cortex where the germ cell attaches to the hub (Fig. 4B, arrows). When the duplicated centro-

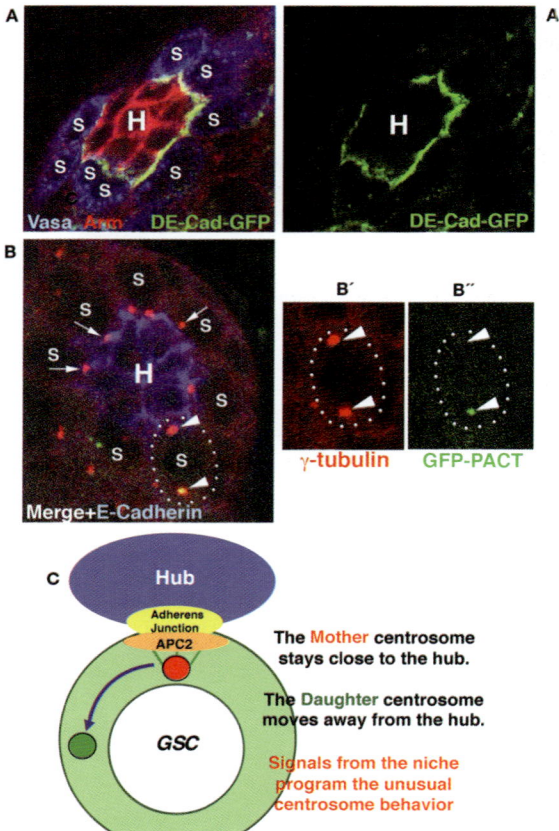

Figure 4. GSCs attach to and orient their centrosomes toward the hub. (*A*) Apical tip of a *Drosophila* testis expressing UAS-DE-cadherin-GFP (*green*) exclusively in germ cells, under control of *nos-GAL4*. Note the localization of E-cadherin-GFP to the GSC cortex closest to the hub. GSCs (S) are clustered around the hub (germ cells are stained *blue* with anti-Vasa). Hub cells (H) are stained with anti-Armadillo (*red*) but not anti-Vasa (*blue*). (*B*) End view of testis tip, showing hub cells (H) outlined by anti-DE-cadherin staining (*blue*), surrounded by GSCs (S). (*Arrows*) Centrosomes in the interphase GSCs, visualized by staining with anti-γ-tubulin (*red*), lie next to the interface with the hub. (*Dotted outline*) Germ cell in G_2 with two centrosomes (*arrowheads*), separated to opposite sides of the nucleus. The younger of the two centrosomes, pulse-labeled with GFP-PACT (visualized in *B* and *B"*), has moved away from the hub, whereas the mother centrosome (not labeled with GFP-PACT) remains next to the hub. (*C*) Model for how GSCs attach to and orient toward the hub. (Figure courtesy of Y. Yamashita.)

somes separate in G_2, one stays next to the hub and the other migrates to the opposite side of the cell (Fig. 4B, arrowheads). The separation of centrosomes occurs unusually early in male GSCs: A substantial fraction of GSCs have centrosomes on opposite sides of the nucleus but have not yet built a spindle. APC2 function is required for GSCs to maintain proper orientation of the centrosomes toward the hub. The stereotyped position of the centrosomes in turn orients the mitotic spindle perpendicular to the GSC–hub cell interface, ensuring that the outcome of GSC divisions is asymmetric: One daughter inherits the attachment to the niche, remains subject to its signals, and renews stem cell identity, whereas the other s displaced away from the niche and initiates differentiation (Yamashita et al. 2003, 2007). Surprisingly, centrosomes are not oriented toward the niche throughout the cell cycle in female GSCs, indicating that spindle orientation is set up by a different mechanism in the germarium.

The two centrosomes in male GSCs have different characters and fates. Differential labeling of mother versus daughter centrosomes by transient expression of a green fluorescent protein (GFP)-tagged centriolar protein fragment during centrosome duplication revealed that the mother centrosome normally remains adjacent to the hub and is inherited by the GSC, whereas the daughter centrosome migrates to the opposite side of the cell and is inherited by the Gb (Fig. 4B). As a result, male GSCs maintain a centriolar Eve, a centriole that was assembled many cell generations earlier (Yamashita et al. 2007). The underlying cellular mechanism for the asymmetric inheritance of centrosomes during male GSC division may involve asymmetric maintenance of astral microtubule arrays. The mother centrosome appears to retain a robust astral microtubule array throughout the cell cycle, whereas the centrosome migrating to the opposite side of the cell had few associated astral microtubules until late in G_2, near the onset of mitosis. Consistent with this model, loss of attachment of astral microtubules to centrosomes in *cnn* mutant germ cells resulted in misoriented centrosomes and spindles (Yamashita et al. 2003) and randomized the positions of mother and daughter centrosomes (Yamashita et al. 2007).

The cellular mechanisms that maintain centrosome position and orient the mitotic spindle in stem cells are important because they ensure the asymmetric outcome of stem cell divisions within the niche. In the male germ line, Yamashita and colleagues hypothesize that a checkpoint mechanism blocks progression into mitosis unless a centrosome is properly situated next to the attachment to the hub (Cheng et al. 2008). Accumulation of GSCs that are arrested by this orientation checkpoint may contribute to the decrease in production of differentiating germ cells observed with aging.

MANY INTERCELLULAR CONVERSATIONS REGULATE SELF-RENEWAL AND DIFFERENTIATION

Two somatic cyst cells enclose each gonialblast to form a cyst (Fig. 3), the functional unit that initiates differentiation, starting with the onset of the spermatogonial transit-amplifying divisions. The cyst cells codifferentiate with the germ cells that they enclose (Gönczy et al. 1992). Bidirectional intercellular signaling between the germ cells and cyst cells is required to set up the cyst and to orchestrate codifferentiation of the two cell types.

Intercellular signaling via the epidermal growth factor receptor (EGFR) helps set up proper interactions between early germ cells and cyst cells. Early germ cells express the EGFR ligand *Spitz* that activates the EGFR in adjacent cyst cells (Schulz et al. 2002; Sarkar et al. 2007). Production of secreted *Spitz* by early germ cells requires the activity of *stet*, a germ-line-specific *rhomboid* class protease (Schulz et al. 2002; Sarkar et al. 2007). Activation of the EGFR

along the Gb–cyst cell interface locally activates Vav, a guanine nucleotide exchange factor (GEF) for the small GTPase Rac1, which promotes local rearrangements in the cytoskeleton that induce cyst cells to send out processes and envelop the Gb (Sarkar et al. 2007). EGFR stimulation in cyst cells also activates a mitogen-activated protein kinase (MAPK)-dependent cascade that ultimately promotes differentiation of the enclosed germ cells (Kiger et al. 2000; Tran et al. 2000). One possibility is that enveloping cyst cells physically insulate early germ cells from self-renewal signals from the hub or CySCs. Another possibility is that EGFR activation modulates expression in cyst cells of secreted factors that nonautonomously influence germ cell differentiation.

Signals from CySCs and early cyst cells may prevent premature differentiation of transit-amplifying germ cells by blocking expression of *bag of marbles* (*bam*), a key regulator of germ cell differentiation. In male germ cells, expression of *bam* normally initiates midway through the transit-amplification divisions, and wild-type function of *bam* is required for spermatogonia to cease mitotic divisions and initiate spermatocyte development (Gönczy et al. 1997). In the female germ line, *bam* transcription is silenced in GSCs by TGF-β signaling from the niche (Xie and Spradling 2000; Spradling et al., this volume), which causes an activated Smad-CoSmad complex containing Mad and Medea to bind *cis*-acting control elements just downstream from the *bam* transcription start site, blocking *bam* expression in female GSCs (Chen and McKearin 2003a,b). Expression of bam commences as female germ cells exit the niche and is both necessary and sufficient for the cells to initiate transit-amplifying divisions and differentiation (McKearin and Ohlstein 1995). In contrast, *bam* function is not required for male GSCs to divide asymmetrically or for spermatogonia to undergo transit-amplifying divisions (Gönczy et al. 1997). However, *bam* expression must be repressed in GSCs and early germ cells to allow survival and prevent premature differentiation. Forced high-level expression of *bam* in early male germ cells caused arrest and cell death (Schulz et al. 2004).

In males, TGF-β signaling from soma to germ line appears to inhibit precocious expression of *bam* in GSCs and spermatogonial cysts. Intracellular components of the TGF-β signal transduction machinery, including the receptors *punt* and *thick veins*, are required cell autonomously for male GSC maintenance (Shivdasani and Ingham 2003; Bunt and Hime 2004; Kawase et al. 2004). Hub and early cyst cells express the TGF-β class signaling ligand *glass bottom boat* (*gbb*), which is required for GSC survival (Shivdasani and Ingham 2003). A likely scenario is that the hub, CySCs, and early cyst cells secrete Gbb and potentially other TGF-β ligands, activating the TGF-β signal transduction pathway in neighboring germ cells and causing transcriptional repression of *bam*. The resulting silencing of *bam* in early germ cells allows GSC and Gb survival. The mechanisms that downregulate *gbb* expression as cyst cells differentiate to allow expression of *bam* in late spermatogonia are not known, but it is tempting to speculate that a germ line to soma signal, perhaps germ-line-derived Spitz activating EGFR in cyst cells, might be involved.

TRANSIT-AMPLIFYING CELLS CAN REVERT TO STEM CELL IDENTITY

Remarkably, male and female germ cells undergoing transit-amplifying divisions can and indeed often do reoccupy the niche and revert back to stem cell identity (Brawley and Matunis 2004; Kai and Spradling 2004; Cheng et al. 2008). Reversion of transit-amplifying cells to stem cell state may provide a mechanism to replace stem cells lost from the niche by normal turnover or damage and may be especially important in cases where stereotyped spindle orientation normally prevents a symmetric outcome of stem cell divisions in the niche.

Although transit-amplifying cells are clearly morphologically and behaviorally different from GSCs, they are not yet irreversibly committed to differentiation. Dedifferentiation of spermatogonia requires STAT activity. Adult males conditionally mutant for *stat* maintained stem cells at permissive temperature, but they rapidly lost both GSCs and CySCs to differentiation when shifted to nonpermissive temperature. However, if the flies were returned to permissive conditions within a few days, transit-amplifying spermatogonial cysts adjacent to the hub broke apart and provided germ cells that regained stem cell behavior, reconstituting the vacated niche (Brawley and Matunis 2004). Unlike spermatogonia, spermatocytes did not appear to be able to revert to stem cell identity.

Genetic marking experiments demonstrated that transit-amplifying germ cells that have progressed far enough to turn on *bam* expression can repopulate the niche and revert to stem cell behavior, even without drastic clearing of the niche and manipulations of niche signaling pathways (Cheng et al. 2008). The frequency of stem cell replacement by transit-amplifying cell reversion increased when flies were irradiated, suggesting that dedifferentiation of transit-amplifying cells may be an important mechanism for restoring stem cells to the niche after damage. Interestingly, dedifferentiated GSCs exhibit misoriented centrosomes and undergo cell cycle arrest until proper centrosome orientation toward the hub is reestablished (Cheng et al. 2008). Thus, it may take some time under the influence of niche signals for returning germ cells to express the molecules responsible for the unusual cytoskeletal behavior that anchors the oldest centrosome near the junction between the GSC and the hub.

THE SWITCH TO TERMINAL DIFFERENTIATION REPROGRAMS THE TRANSCRIPTION MACHINERY

Once the germ cells have completed four rounds of transit-amplifying divisions, the resulting 16 cells exit the mitotic division program and initiate spermatocyte growth and differentiation. This cell state change involves dramatic remodeling of the transcriptional landscape: Many genes expressed in mitotic precursor cells are turned off and many new genes are expressed, some for the first time in the development of the animal. Quite frequently, genes normally thought to serve a general or housekeeping function have homologs in the genome that are expressed in spermatocytes but not in the proliferating precursor cells. These include homologs encoding sub-

units of the proteosome (Yuan et al. 1996), mitochondrial import complex (Hwa et al. 2004), and translation initiation machinery (Baker and Fuller 2007). The emerging picture is that the switch to the spermatocyte state initiates a multistep transcriptional hierarchy that controls spermatocyte growth and development, the meiotic cell cycle, and eventually the expression of hundreds of genes required for spermatid differentiation.

A key question in stem cell lineages is how terminal differentiation genes kept silent in precursor cells are activated in the correct cell type and at the right time during differentiation. Studies on the meiotic arrest genes in *Drosophila* have provided insight into the underlying regulatory mechanisms at work in the male germ line. In meiotic-arrest mutant males, GSCs, transit-amplifying cells, and spermatocytes are formed. Although they grow to normal size and have normal morphology, the mutant spermatocytes arrest at the G_2/M transition of meiosis I and fail to initiate spermatid differentiation. The phenotype of the *Drosophila* meiotic-arrest mutants is strikingly similar to the clinical description of meiosis I maturation arrest azoospermia, a common form of human idiopathic male infertility (Meyer et al. 1992). Wild-type function of the meiotic arrest genes is required for normal expression of hundreds of genes involved in postmeiotic spermatid differentiation. The defect in differentiation is independent of cell cycle arrest, because spermatid differentiation genes that are direct targets of the meiotic-arrest genes are normally transcribed starting at the onset of the primary spermatocyte period, days before the G_2/M transition of meiosis I.

Phenotypic analysis and molecular cloning revealed that the *Drosophila* meiotic-arrest genes fall into two functional classes, encoding components of two separate regulatory complexes. The first, represented by *aly*, encodes components of tMAC, a testis-specific variant of MIP/dREAM (White-Cooper et al. 2000; Perezgasga et al. 2004; Beall et al. 2007; Jiang et al. 2007), a multisubunit complex conserved from worms (the SynMuv pathway) to mammals (E2F/Rb/Myb interacting proteins). MIP/dREAM complexes have an important but not yet understood role in cell cycle control and expression of developmentally controlled genes (Korenjak et al. 2004). Strikingly, the *aly* class meiotic-arrest genes encode spermatocyte-specific homologs of generally expressed MIP/dREAM subunits, presumably endowing the tMAC complex with unique properties not shared by the MIP/dREAM transcriptional regulator. tMAC function is required for transcription in primary spermatocytes of meiotic cell cycle control genes such as *cyclin B, boule*, and the *CDC25* cell cycle phosphatase *twine*, as well as spermatid differentiation genes (White-Cooper et al. 1998).

The second class of meiotic arrest genes, represented by *can, nht, rye, sa,* and *mia*, all encode spermatocyte-specific TBP-associated factor (TAF) homologs (Hiller et al. 2001, 2004). TAFs interact with the TATA-box binding protein (TBP) to form the general transcription factor TFIID that interprets and integrates molecular signals regulating the core RNA polymerase II (Pol II) machinery for transcriptional initiation (for review, see Hochheimer and Tjian 2003). Certain TAFs are also core components of histone acetylase complexes such as SAGA (Spt-Ada-

Gcn5-acetyltransferase) (Struhl et al. 1998) or bind stoichiometrically to the Polycomb transcriptional silencing complex in some tissues (Saurin et al. 2001). Wild-type function of the testis TAFs (tTAFs) is required for robust transcription in spermatocytes of many of the same spermatid differentiation genes that require tMAC action (White-Cooper et al. 1998; Hiller et al. 2004). Strikingly, regulation of transcription by the tTAFs and tMAC is gene selective: Although transcription of spermatid differentiation genes is severely affected in mutant spermatocytes, many other genes are expressed normally in the same cells. Thus, tissue-specific homologs of general Pol II transcription machinery components regulate gene-selective, cell-type-specific transcriptional programs for terminal differentiation in the *Drosophila* male germ cell lineage (White-Cooper et al. 1998; Hiller et al. 2001).

Emerging evidence indicates that tissue-specific TAFs and other core promotor recognition machinery components control cell-type-specific gene expression programs in a variety of developmental contexts. Incorporation of tissue-specific TAFs may alter the properties of TFIID, enabling the modified complex to direct widely expressed transcription factors to activate specific target genes in certain tissues. In mice, TAF4b, a tissue-specific alternate form of TAF4, is required for maintenance of spermatogenesis (Falender et al. 2005) and differentiation of mouse ovarian granulosa cells (Freiman et al. 2001). Substitution of TAF4b for one of the two TAF4 subunits in TFIID strongly facilitates transcriptional activation by c-Jun at selected promoters (Liu et al. 2008), activating tissue-specific gene-selective transcription programs in cells where both c-Jun and TAF4b are expressed. In a more dramatic core machinery switch, during muscle development, TFIID is down-regulated and instead TAF3 and the TBP homolog TRF3 turn on transcription of myogenin and other genes required for differentiation of myoblasts into myotubes (Deato and Tjian 2007 and this volume).

In the *Drosophila* male GSC lineage, expression of terminal differentiation genes in spermatocytes depends on cell-type-specific modification of core promotor recognition machinery components and chromatin regulatory complexes. These in turn selectively promote expression of genes required for meiosis and spermiogenesis in germ cells committed to terminal differentiation. The cell-type- and stage-specific expression of tTAFs and tMAC components in spermatocytes begs the question of how expression of these factors is regulated by the stem cell lineage program. We imagine a transcriptional cascade where regulatory factor(s) expressed as part of the switch to spermatocyte state induce expression of the tTAFs and spermatocyte-specific tMAC components. These in turn act together to allow transcription of terminal differentiation genes.

TERMINAL DIFFERENTIATION INVOLVES CELL-TYPE-SPECIFIC MECHANISMS TO COUNTERACT REPRESSION BY POLYCOMB

The testis TAFs and tMAC may facilitate expression of terminal differentiation genes by reversing repression by *Polycomb* epigenetic transcriptional silencing machinery. Chromatin immunoprecipitation (ChIP) experiments on

tTAF mutant testes showed Polycomb protein at the promoter regions of all three of the terminal differentiation genes assayed. In wild-type testes, these promoters were instead occupied by the tTAFs, and the differentiation genes were expressed (Chen et al. 2005). The tTAFs also appear to control subnuclear localization of the Polycomb complex PRC1. In wild-type spermatocytes, tTAF proteins accumulate in a subdomain of the nucleolus. PRC1 components localized to the same nucleolar subcompartment in a tTAF-dependent manner (Chen et al. 2005). Taken together, these results suggest that the tTAFs might bind PRC1 and sequester the complex to the nucleolus. Sequestering PRC1 may promote expression of spermatid differentiation genes by preventing rebinding of free PRC1 to target gene promoters. Alternatively, PRC1 may have a novel function in the nucleolus.

Recent studies in mammalian embryonic stem cells suggest that many genes required for lineage commitment and tissue-specific differentiation programs are silenced in embryonic stem cells by Polycomb group (PcG) machinery. Loss of PcG function in ES cells causes inappropriate expression of developmental regulators and defects in differentiation of particular cell types (Boyer et al. 2006; Lee et al. 2006; Chamberlain et al. 2008). PcG machinery appears to have critical roles in decisions between proliferating precursor cell fate and terminal differentiation in mammalian adult stem cell lineages as well. In mice, loss of function of the PcG protein Bmi-1 causes loss of long-term self-renewing capacity in hematopoietic and neural stem cells. Additionally, Bmi1 expression is up-regulated in certain leukemias (Lessard and Sauvageau 2003; Molofsky et al. 2003; Park et al. 2003). If PcG activity blocks expression of differentiation genes in precursor cells, a major question is how this epigenetically silenced state is normally reversed to allow expression of differentiation genes appropriate to distinct cell types and developmental stages. Studies on the role of the testis TAFs and tMAC in the *Drosophila* male GSC lineage provide a model for how cell-type-specific forms of core transcriptional regulatory machinery might have a role in reversing epigenetic silencing to activate cell-type-specific transcription programs as precursor cells initiate and execute terminal differentiation.

DISCUSSION

Studies of model stem cell lineages in vivo are yielding important lessons for understanding tissue homeostasis, aging, regenerative medicine, and cancer. The importance of the niche in controlling stem cell behavior is paramount. However, we must understand the complexities of natural stem cell niches before we can hope to mimic them for expansion of adult stem cells in vitro, for construction of artificial niches for therapeutic applications, or even for understanding the relationship between stem cell lineages and cancer. Indeed, recent work on basal cell carcinoma indicates that tumor cells are maintained by close-range signals from accompanying stromal cells, which provide an ersatz niche (Sneddon et al. 2006). The picture emerging from work on GSC niches in *Drosophila* points toward a self-sustaining and mutually reinforcing community of several cell types engaged in close-range signaling conversations that influence the fate choices of neighbors within the niche (see also Spradling et al., this volume).

A second important lesson is the ability of cells to embark on transit-amplifying divisions but still be able to revert to the stem cell state, given access to an empty niche. Replacement of stem cells by dedifferentiation of transit-amplifying cells occurs in the female (Kai and Spradling 2004) as well as the male germ line in *Drosophila* and also appears to occur during mammalian spermatogenesis (Nakagawa et al. 2007; also see Yoshida, this volume). This ability means that transplantation assays may detect not only the most fundamental stem cells, but also possibly early transit-amplifying cells that have initiated but are not yet committed to differentiation. It will be interesting to find out if this underlies the recent discovery of two interconverting populations of hematopoietic stem cells, one deeply quiescent and one more rapidly proliferating (Wilson et al. 2008).

As we understand more clearly the signaling pathways and regulatory strategies that maintain the stem cell state in adult tissues, it will be interesting to compare these to the signals that maintain pluripotency of embryonic stem cells. In the case of the *Drosophila* male germ line, stem cell self-renewal and maintenance requires activation of STAT by a cytokine-like signaling ligand as well as repression of differentiation by TGF-β class signals from neighboring cells. Without the TGF-β signal, GSCs precociously express the lineage-specific differentiation factor *bam*. The cooperative action of JAK-STAT and TGF-β pathways in male GSC self-renewal is reminiscent of the dependence of murine embryonic stem cells on LIF/STAT3 and BMP/Smad signaling where, again, these pathways sustain the stem cell state by inhibiting cells from adopting lineage-specific fates (Ying et al. 2003). A second intriguing parallel between adult and embryonic stem cell lineages may lie in a need for mechanisms to reverse Polycomb-mediated repression by allowing context-appropriate expression of terminal differentiation genes.

ACKNOWLEGMENTS

We thank many present and former members of the Fuller lab, including Alexis Bailey, Catherine Baker, Xin Chen, Edith Glusman, Megan Insco, D. Leanne Jones, Amy Kiger, Chenggang Lu, Anthony Mahowald, Jose Morillo Prado, Alicia Shields, Allyson Campbell Spence, Vidhya Srinivasan, and Yukiko Yamashita, whose scientific discussions and work contributed greatly to the development of these ideas. E.D. was supported by the Smith Stanford Graduate Fellowship and predoctoral fellowships from the National Science Foundation and the American Heart Association Western Affiliate. M.T.F. is the Reed-Hodgson Professor of Human Biology at Stanford. This work was supported by National Institutes of Health grants 1 R01 GM080501 and 3 RO1 GM061986 to M.T.F.

REFERENCES

Baker, C.C. and Fuller, M.T. 2007. Translational control of meiotic cell cycle progression and spermatid differentiation in male germ cells by a novel eIF4G homolog. *Development*
134: 2863–2869.
Beall, E.L., Lewis, P.W., Bell, M., Rocha, M., Jones, D.L., and Botchan, M.R. 2007. Discovery of tMAC: A *Drosophila* testis-specific meiotic arrest complex paralogous to Myb-Muv B. *Genes Dev.* 21: 904–919.
Boyer, L.A., Plath, K., Zeitlinger, J., Brambrink, T., Medeiros, L.A., Lee, T.I., Levine, S.S., Wernig, M., Tajonar, A., Ray, M.K., et al. 2006. Polycomb complexes repress developmental regulators in murine embryonic stem cells. *Nature* 441: 349–353.
Boyle, M., Wong, C., Rocha, M., and Jones, D.L. 2007. Decline in self-renewal factors contributes to aging of the stem cell niche in the *Drosophila* testis. *Cell Stem Cell* 1: 470–478.
Brawley, C. and Matunis, E. 2004. Regeneration of male germline stem cells by spermatogonial dedifferentiation in vivo. *Science* 304: 1331–1334.
Bunt, S.M. and Hime, G.R. 2004. Ectopic activation of Dpp signalling in the male *Drosophila* germline inhibits germ cell differentiation. *Genesis* 39: 84–93.
Chamberlain, S.J., Yee, D., and Magnuson, T. 2008. Polycomb repressive complex 2 is dispensable for maintenance of embryonic stem cell pluripotency. *Stem Cells* 26: 1496–1505.
Chen, D. and McKearin, D. 2003a. Dpp signaling silences *bam* transcription directly to establish asymmetric divisions of germline stem cells. *Curr. Biol.* 13: 1786–1791.
Chen, D. and McKearin, D.M. 2003b. A discrete transcriptional silencer in the *bam* gene determines asymmetric division of the *Drosophila* germline stem cell. *Development* 130: 1159–1170.
Chen, X., Hiller, M., Sancak, Y., and Fuller, M.T. 2005. Tissue-specific TAFs counteract Polycomb to turn on terminal differentiation. *Science* 310: 869–872.
Cheng, J., Turkel, N., Hemati, N., Fuller, M.T., Hunt, A.J., and Yamashita, Y.M. 2008. Centrosome misorientation reduces stem cell division during ageing. *Nature* 456: 599–604.
Clarke, M.F. and Fuller, M. 2006. Stem cells and cancer: Two faces of Eve. *Cell* 124: 1111–1115.
Deato, M.D. and Tjian, R. 2007. Switching of the core transcription machinery during myogenesis. *Genes Dev.* 21: 2137–2149.
Falender, A.E., Freiman, R.N., Geles, K.G., Lo, K.C., Hwang, K., Lamb, D.J., Morris, P.L., Tjian, R., and Richards, J.S. 2005. Maintenance of spermatogenesis requires TAF4b, a gonad-specific subunit of TFIID. *Genes Dev.* 19: 794–803.
Freiman, R.N., Albright, S.R., Zheng, S., Sha, W.C., Hammer, R.E., and Tjian, R. 2001. Requirement of tissue-selective TBP-associated factor TAF$_{II}$105 in ovarian development. *Science* 293: 2084–2087.
Fuller, M.T. 1993. Spermatogenesis. In *The development of Drosophila melanogaster* (ed. M. Bate and A. Martinez-Arias), pp. 71–147. New York, Cold Spring Harbor Laboratory Press, Cold Spring Harbor, New York.
Fuller, M.T. and Spradling, A.C. 2007. Male and female *Drosophila* germline stem cells: Two versions of immortality. *Science* 316: 402–404.
Gönczy, P. and DiNardo, S. 1996. The germ line regulates somatic cyst cell proliferation and fate during *Drosophila* spermatogenesis. *Development* 122: 2437–2447.
Gönczy, P., Matunis, E., and DiNardo, S. 1997. *bag-of-marbles* and *benign gonial cell neoplasm* act in the germline to restrict proliferation during *Drosophila* spermatogenesis. *Development* 124: 4361–4371.
Gönczy, P., Viswanathan, S., and DiNardo, S. 1992. Probing spermatogenesis in *Drosophila* with P-element enhancer detectors. *Development* 114: 89–98.
Hecht, N.B. 1993. Gene expression during male germ cell development. In *Cell and molecular biology of the testis* (ed C. Desjardins and L.L. Ewing), pp. 400–432. Oxford University Press, New York.
Hiller, M., Chen, X., Pringle, M.J., Suchorolski, M., Sancak, Y., Viswanathan, S., Bolival, B., Lin, T.Y., Marino, S., and Fuller, M.T. 2004. Testis-specific TAF homologs collaborate to control a tissue-specific transcription program. *Development* 131: 5297–5308.
Hiller, M.A., Lin, T.Y., Wood, C., and Fuller, M.T. 2001. Developmental regulation of transcription by a tissue-specific TAF homolog. *Genes Dev.* 15: 1021–1030.

Hochheimer, A. and Tjian, R. 2003. Diversified transcription initiation complexes expand promoter selectivity and tissue-specific gene expression. *Genes Dev.* **17:** 1309–1320.

Hwa, J.J., Zhu, A.J., Hiller, M.A., Kon, C.Y., Fuller, M.T., and Santel, A. 2004. Germ-line specific variants of components of the mitochondrial outer membrane import machinery in *Drosophila*. *FEBS Lett.* **572:** 141–146.

Jiang, J., Benson, E., Bausek, N., Doggett, K., and White-Cooper, H. 2007. Tombola, a tesmin/TSO1-family protein, regulates transcriptional activation in the *Drosophila* male germline and physically interacts with Always early. *Development* **134:** 1549–1559.

Kai, T. and Spradling, A. 2004. Differentiating germ cells can revert into functional stem cells in *Drosophila melanogaster* ovaries. *Nature* **428:** 564–569.

Kawase, E., Wong, M.D., Ding, B.C., and Xie, X. 2004. Gbb/Bmp signaling is essential for maintaining germline stem cells and for repressing bam transcription in the *Drosophila* testis. *Development* **131:** 1365–1375.

Kiger, A.A., White-Cooper, H., and Fuller, M.T. 2000. Somatic support cells restrict germline stem cell self-renewal and promote differentiation. *Nature* **407:** 750–754.

Kiger, A.A., Jones, D.L., Schulz, C., Rogers, M.B., and Fuller, M.T. 2001. Stem cell self-renewal specified by JAK-STAT activation in response to a support cell cue. *Science* **294:** 2542–2545.

Korenjak, M., Taylor-Harding, B., Binne, U.K., Satterlee, J.S., Stevaux, O., Aasland, R., White-Cooper, H., Dyson, N., and Brehm, A. 2004. Native E2F/RBF complexes contain Myb-interacting proteins and repress transcription of developmentally controlled E2F target genes. *Cell* **119:** 181–193.

Leatherman, J.L. and DiNardo, S. 2008. *Zfh-1* controls somatic stem cell self-renewal in the *Drosophila* testis and nonautonomously influences germline stem cell self-renewal. *Cell Stem Cell* **3:** 44–54.

Le Bras, S. and Van Doren, M. 2006. Development of the male germline stem cell niche in *Drosophila*. *Dev. Biol.* **294:** 92–103.

Lee, T.I., Jenner, R.G., Boyer, L.A., Guenther, M.G., Levine, S.S., Kumar, R.M., Chevalier, B., Johnstone, S.E., Cole, M.F., Isono, K., et al. 2006. Control of developmental regulators by Polycomb in human embryonic stem cells. *Cell* **125:** 301–313.

Lessard, J. and Sauvageau, G. 2003. *Bmi-1* determines the proliferative capacity of normal and leukaemic stem cells. *Nature* **423:** 255–260.

Lu, W.L., Coleman, R.A., Grob, P., King, D.S., Florens, L., Washburn, M.P., Geles, K.G., Yang, J.L., Ramey, V., Nogales, E., and Tjian, R. 2008. Structural changes in TAF4b-TFIID correlate with promoter selectivity. *Mol. Cell* **29:** 81–91.

McKearin, D. and Ohlstein, B. 1995. A role for the *Drosophila* Bag-of-marbles protein in the differentiation of cystoblasts from germline stem cells. *Development* **121:** 2937–2947.

Meyer, J.M., Maetz, J.L., and Rumpler, Y. 1992. Cellular relationship impairment in maturation arrest of human spermatogenesis: An ultrastructural study. *Histopathology* **21:** 25–33.

Molofsky, A.V., Pardal. R., Iwashita, T., Park, I.K., Clarke, M.F., and Morrison, S.J. 2003. *Bmi-1* dependence distinguishes neural stem cell self-renewal from progenitor proliferation. *Nature* **425:** 962–967.

Nakagawa, T., Nabeshima, Y., and Yoshida, S. 2007. Functional identification of the actual and potential stem cell compartments in mouse spermatogenesis. *Dev. Cell* **12:** 195–206.

Park, I.K., Qian, D., Kiel, M., Becker, M.W., Pihalja, M., Weissman, I.L., Morrison, S.J., and Clarke, M.F. 2003. Bmi-1 is required for maintenance of adult self-renewing haematopoietic stem cells. *Nature* **423:** 302–305.

Perezgasga, L., Jiang, J., Bolival, Jr., B., Hiller, M., Benson, E., Fuller, M.T., and White-Cooper, H. 2004. Regulation of transcription of meiotic cell cycle and terminal differentiation genes by the testis-specific Zn-finger protein *matotopetli*. *Development* **131:** 1691–1702.

Sarkar, A., Parikh, N., Hearn, S.A., Fuller, M.T., Tazuke, S.I., and Schulz, C. 2007. Antagonistic roles of Rac and Rho in organizing the germ cell microenvironment. *Curr. Biol.* **17:** 1253–1258.

Saurin, A.J., Shao, Z., Erdjument-Bromage, H., Tempst, P., and

Kingston, R.E. 2001. A *Drosophila* Polycomb group complex includes Zeste and dTAFII proteins. *Nature* **412:** 655–660.

Schulz, C., Wood, C.G., Jones, D.L., Tazuke, S.I., and Fuller, M.T. 2002. Signaling from germ cells mediated by the *rhomboid* homologue stet organizes encapsulation by somatic support cells. *Development* **129:** 4523–4534.

Schulz, C., Kiger, A.A., Tazuke, S.I., Yamashita, Y.M., Pantalena-Filho, L.C., Wood, C.G., Jones, D.L., and Fuller, M.T. 2004 A misexpression screen reveals effects of *bag-of-marbles* and TGFβ class signaling on the *Drosophila* male germ line stem cell lineage. *Genetics* **167:** 707–723.

Shivdasani, A.A. and Ingham, P.W. 2003. Regulation of stem cell maintenance and transit amplifying cell proliferation by TGF-β signaling in *Drosophila* spermatogenesis. *Curr. Biol.* **13:** 2065–2072.

Sneddon, J.B., Zhen, H.H., Montgomery, K., van de Rijn, M., Tward, A.D., West, R., Gladstone, H., Chang, H.Y., Morganroth, G.S., Oro, A.E., and Brown, P.O. 2006. Bone morphogenetic protein antagonist gremlin 1 is widely expressed by cancer-associated stromal cells and can promote tumor cell proliferation. *Proc. Natl. Acad. Sci.* **103:** 14842–14847.

Spradling, A., Drummond-Barbosa, D., and Kai, T. 2001. Stem cells find their niche. *Nature* **414:** 98–104.

Struhl, K., Kadosh, D., Keaveney, M., Kuras, L., and Moqtaderi, Z. 1998. Activation and repression mechanisms in yeast. *Cold Spring Harbor Symp. Quant. Biol.* **63:** 413–421.

Tran, J., Brenner, T.J., and DiNardo, S. 2000. Somatic control over the germline stem cell lineage during *Drosophila* spermatogenesis. *Nature* **407:** 754–757.

Tulina, N. and Matunis, E. 2001. Control of stem cell self-renewal in *Drosophila* spermatogenesis by JAK-STAT signaling. *Science* **294:** 2546–2549.

Voog, J., D'Alterio, C., and Jones, D.L. 2008 Multipotent somatic stem cells contribute to the stem cell niche in the *Drosophila* testis. *Nature* **454:** 1132–1136.

Watt, F.M. and Hogan, B.L. 2000. Out of Eden: Stem cells and their niches. *Science* **287:** 1427–1430.

White-Cooper, H., Leroy, D., MacQueen, A., and Fuller, M.T. 2000. Transcription of meiotic cell cycle and terminal differentiation genes depends on a conserved chromatin associated protein, whose nuclear localisation is regulated. *Development* **127:** 5463–5473.

White-Cooper, H., Schafer, M.A., Alphey, L.S., and Fuller, M.T. 1998. Transcriptional and post-transcriptional control mechanisms coordinate the onset of spermatid differentiation with meiosis I in *Drosophila*. *Development* **125:** 125–134.

Wilson, A., Laurenti, E., Oser, G., van der Wath, R.C., Blanco-Bose, W., Jaworski, M., Offner, S., Dunant, C.F., Eshkind, L., Bockamp, E., et al. 2008. Hematopoietic stem cells reversibly switch from dormancy to self-renewal during homeostasis and repair. *Cell* **135:** 1118–1129.

Xie, T. and Spradling, A.C. 1998. *decapentaplegic* is essential for the maintenance and division of germline stem cells in the *Drosophila* ovary. *Cell* **94:** 251–260.

Xie, T. and Spradling, A.C. 2000. A niche maintaining germ line stem cells in the *Drosophila* ovary. *Science* **290:** 328–330.

Yamashita, Y.M. and Fuller, M.T. 2005. Asymmetric stem cell division and function of the niche in the *Drosophila* male germ line. *Int. J. Hematol.* **82:** 377–380.

Yamashita, Y.M., Fuller, M.T., and Jones, D.L. 2005. Signaling in stem cell niches: Lessons from the *Drosophila* germline. *J. Cell Sci.* **118:** 665–672.

Yamashita, Y.M., Jones, D.L., and Fuller, M.T. 2003. Orientation of asymmetric stem cell division by the APC tumor suppressor and centrosome. *Science* **301:** 1547–1550.

Yamashita, Y.M., Mahowald, A.P., Perlin, J.R., and Fuller, M.T. 2007. Asymmetric inheritance of mother versus daughter centrosome in stem cell division. *Science* **315:** 518–521.

Ying, Q.L., Nichols, J., Chambers, I., and Smith, A. 2003. BMP induction of Id proteins suppresses differentiation and sustains embryonic stem cell self-renewal in collaboration with STAT3. *Cell* **115:** 281–292.

Yuan, X., Miller, M. and Belote, J.M. 1996 Duplicated proteasome subunit genes in *Drosophila melanogaster* encoding testes-specific isoforms. *Genetics* **144:** 147–157.

Reprogramming of Somatic Cell Identity

J. HANNA,* B.W. CAREY,*† AND R. JAENISCH*†

*The Whitehead Institute for Biomedical Research, Cambridge, Massachusetts 02142;
†Massachusetts Institute of Technology, Department of Biology, Cambridge, Massachusetts 02142

All mammalian somatic cells originate from a single fertilized cell, the zygote, and share identical genetic information despite the dramatic changes in cell structure and function that accompany organismal development. The genome is subjected to a wide array of epigenetic modifications during lineage specification, a process that contributes to the implementation and maintenance of specific gene expression programs in somatic cells. Nuclear transfer and cell-fusion experiments demonstrate that the epigenetic signature directing a cell identity can be erased and modified into that of another cell type. Furthermore, in the case of cloning, differentiated cells can be reprogrammed back to pluripotency to support the reexpression of all developmental programs. Recent breakthroughs highlight the importance of transcription factors as well as epigenetic modifiers in the establishment, maintenance, and rewiring of cell identity. By focusing on reprogramming of terminally differentiated lymphocytes, we review and highlight recent insights into the molecular mechanisms and cellular events potentially underlying programming and reprogramming of somatic cell identity in mammals.

During embryonic development, pluripotent cells undergo cellular differentiation by the gradual acquisition of complex epigenetic modifications, including changes in DNA methylation and histone marks, to establish heritable gene expression programs concordant with a specific cell identity (Bernstein et al. 2007; Meissner et al. 2008). Somatic cells of the adult body are composed of both unipotent and multipotent cell types, many of which exist to support growth and repair of tissues and organs throughout the lifetime of an organism. The successful generation of animals by somatic cell nuclear transfer (SCNT) experiments in amphibians (Gurdon 1999, 2006) and subsequently mammals (Wilmut et al. 1997; Wakayama et al. 1998; Lee et al. 2005; Byrne et al. 2007) proved that cellular differentiation depends on epigenetic differences between genomes of differentiated and embryonic cells that can be reset by exposure to the oocyte cytoplasm.

For more than 20 years since the initial isolation of pluripotent embryonic stem (ES) cells, with the ability to differentiate into all cell types in vitro, many labs have used these cells to study the molecular events and genetic circuitry underlying pluripotency and cellular differentiation. In the field of regenerative medicine, ES cells have garnered much attention because the possibility of generating "patient-specific and genetically identical" ES cells by nuclear transfer holds great medical promise (Jaenisch 2004; Lerou and Daley 2005). However, technical and ethical considerations regarding the nuclear transfer procedure have hampered the application of this technique to the development of "customized" human pluripotent stem cells (Hochedlinger and Jaenisch 2006). This has stimulated efforts to better understand the molecular basis of reprogramming and to devise alternative strategies for reprogramming somatic cells to pluripotency that do not depend on nuclear transfer into oocytes. This chapter focuses on recent strategies for reprogramming somatic cells into pluripotent ES-like cells by using defined factors, with a specific focus on direct reprogramming of terminally differentiated lymphocytes as well as the use of a drug-inducible transgenic system to study the reprogramming process. We review current knowledge of potential pathways and principles underlying resetting somatic cell identity.

DEFINING CELLULAR REPROGRAMMING

Cellular reprogramming refers to the concept of rewiring the epigenetic and transcriptional network of one cell state to that of a different cell type (Egli et al. 2008). In general, reprogramming can be induced either by molecularly undefined means or by molecularly defined means (here, referred to as "direct reprogramming"). Molecularly undefined reprogramming methods, such as SCNT, fusion of somatic cells with ES cells, or explantation of cells in tissue culture, use a milieu of components or elements that are largely unknown to achieve cellular reprogramming (Fig. 1) (Hochedlinger and Jaenisch 2006; Rodolfa and Eggan 2006; Yamanaka 2008). For instance, in SCNT, undefined factors present in the oocyte cytoplasm reprogram the injected nucleus to pluripotency (e.g., transcription factors, histone-modifying and chromatin-remodeling enzymes, and DNA demethylases). In contrast, direct reprogramming methods use defined genetic or nongenetic elements to induce rewiring of the cell state. This is best illustrated by recent work describing transcription-factor-induced reprogramming of somatic cells in which defined genes are overexpressed, initiating a series of largely undefined events that eventually lead to the acquisition of a pluripotent state (Takahashi and Yamanaka 2006; Maherali et al. 2007; Takahashi et al.

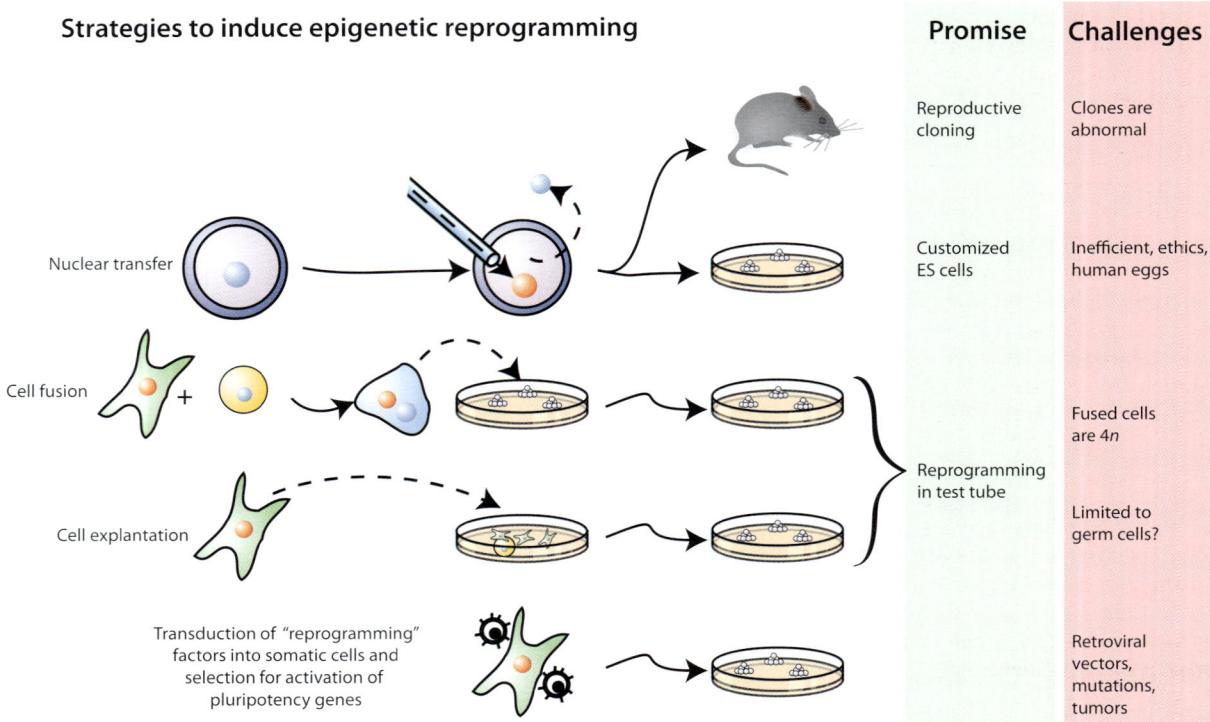

Figure 1. Methods for reprogramming somatic cells to pluripotency and schematic representation of strategies to induce reprogramming of somatic cells. Nuclear transfer involves the injection of a somatic nucleus into an enucleated oocyte, which, upon transfer into a surrogate mother, can give rise to a clone ("reproductive cloning"), or, upon explantation in culture, can give rise to genetically identical ES cells ("somatic cell nuclear transfer," SCNT). Cell fusion of ES cells with somatic cells results in the generation of hybrids and pluripotent ES-like cells. Explantation of somatic cells in culture selects for immortal cell lines that may be pluripotent or multipotent. At present, spermatogonial stem cells are the only reproducible source of pluripotent cells that can be derived from postnatal animals (Kanatsu-Shinohara et al. 2004; Guan et al. 2006; Seandel et al. 2007). Transduction of somatic cells with defined factors can initiate reprogramming to a pluripotent state and generate induced pluripotent stem (iPS) cells.

2007; Wernig et al. 2007). It should be noted that reprogramming not only encompasses dedifferentiation of somatic cells to a pluripotent ES-cell-like state, but also includes the conversion of one somatic cell type to another cell type, also referred to as "*trans*-differentiation." For example, Weintraub et al. (1989) have demonstrated reprogramming of fibroblasts into myoblast-like cells upon ectopic expression of the MyoD transcription factor. More recent work has shown that lineage-committed B and T cells can be converted into macrophage-like cells by overexpressing a myeloid transcription factor CCAAT/enhancer-binding protein (C/EBPα) and by growth of the transduced cells in lymphoid growth-promoting cell culture conditions (Xie et al. 2004; Laiosa et al. 2006). These initial experiments were important because they paved the way for using ectopic expression of transcription factors to reprogram cells back to a pluripotent state.

REPROGRAMMING SOMATIC CELLS BY MOLECULARLY UNDEFINED MECHANISMS

Following the pioneering work of Briggs and King, John Gurdon (2006) first demonstrated that adult somatic nuclei can be reprogrammed to pluripotency when injected into enucleated oocytes. Cloned mouse blasto-

cysts either can complete development in vivo and generate live adult mice or can be used to generate established ES lines in culture (Fig. 1) (Hochedlinger and Jaenisch 2006). Recent work has extended this technology to primates since the generation of blastocyst-derived pluripotent ES lines from adult somatic nuclei of rhesus macaques has been reported (Byrne et al. 2007). Despite major efforts, attempts to generate human ES cells by SCNT have so far failed, highlighting the difficulties of using nuclear transfer to generate patient-specific human ES cells. However, groundbreaking work from Eggan and colleagues demonstrates that mouse zygotes can be used as recipients in nuclear transfer (NT) experiments for somatic nuclei contingent upon temporary drug-mediated mitotic arrest of the recipient zygote (Egli et al. 2007), providing hope that discarded human in vitro fertilization (IVF)-derived zygotes could potentially be used instead of oocytes for generating human NT-ES cells. Currently, this has not been successful in humans and it is unclear whether this approach can be used to generate patient-specific ES cells.

An alternative strategy for reprogramming to pluripotency is fusion of somatic cells with mouse or human ES cells. Cell fusion has been shown to produce pluripotent hybrid tetraploid (4n) cells that have successfully reactivated key pluripotency genes as well as the inactive X chromosome in the somatic nucleus and can generate mature differentiated teratomas when injected in immunodeficient mice, thus providing evidence for epigenic reprogramming of these cells (Fig. 1) (Tada et al. 2001; Cowan et al. 2005). It will be important to map the extent of reprogramming during cell fusion by genome-wide epigenetic mapping of chromatin marks of these cells in comparison to ES cells and characterize the extent of cell identity erasure that the somatic epigenome experiences upon the cell-fusion event. From a mechanistic standpoint, convincing evidence is lacking and it remains to be determined whether fusion of the nuclei and/or cell proliferation are absolute requirements for reprogramming the somatic epigenome by cell fusion or whether factors found in the ES cytoplasm would be sufficient to induce reprogramming (Jaenisch and Young 2008). Although cell fusion might prove to be useful for studying mechanisms of reprogramming, the tetraploidy of the genome in these pluripotent cells constitutes a major shortcoming of this method for generating cells that would be useful for therapy (Tada et al. 2001), because reseparating the two genomes after fusion and completion of reprogramming have not been achieved.

DIRECT IN VITRO REPROGRAMMING OF SOMATIC CELLS INTO ES-LIKE CELLS BY DEFINED FACTORS

Takahashi and Yamanaka (2006) showed that pluripotent ES-like cells can be derived from mouse fibroblast cultures upon ectopic viral transduction of the four transcription factors Oct4, Sox2, Klf4, and c-Myc, which are normally expressed in ES cells, followed by drug selection for reactivation of the *Fbx15* gene (Fig. 1). Such cells, termed induced pluripotent stem (iPS) cells, were pluripotent by some criteria because they were able to form teratomas and generate early chimeric embryos. However, the initially described Fbx15 iPS cells were different from ES cells in two respects: The endogenous pluripotency genes were not fully reactivated and the pluripotent state was strictly dependent on continuous expression of the transgenic copies of Oct4 and Sox2. Moreover, Fbx15 iPS lines failed to generate postnatal chimeras, suggesting that these cells were not fully reprogrammed to pluripotency (Takahashi and Yamanaka 2006). Oct4 and Nanog gene reactivation offered a more stringent selection criteria because these proved to be functionally relevant components of the core ES cell transcriptional circuitry (Boyer et al. 2005; Jaenisch and Young 2008). Subsequent experiments introducing the same four factors into fibroblasts, followed by selection for reactivation of Oct4 and Nanog, generated fully reprogrammed iPS cells that were genetically, epigenetically, and developmentally indistinguishable from blastocyst-derived ES cells (Maherali et al. 2007; Okita et al. 2007; Wernig et al. 2007). Futhermore, iPS cells showed reactivation of the inactive X chromosome in female iPS lines, generated live adult chimeras with germ-line contribution, and, more stringently, generated "all iPS cell" embryos by tetrapolid complementation assay. Also, in addition to the complete demethylation and reactivation of

key endogenous pluripotency factors such as Oct4 and Nanog, the ectopic viral transgenes were silenced in iPS cells (Maherali et al. 2007; Okita et al. 2007; Wernig et al. 2007), consistent with well-established evidence that Moloney viruses are efficiently silenced in pluripotent ES cells (Jahner et al. 1982). This showed that following induction of pluripotency in somatic fibroblasts, the pluripotent state can be maintained by the endogenous pluripotency circuitry. This concept was further substantiated in iPS cells induced by doxycycline (Dox)-inducible lentiviral vectors that were shown to maintain the pluripotent state independently of Dox in the culture media (Brambrink et al. 2008; Stadtfeld et al. 2008a).

Importantly, our group demonstrated that pluripotent iPS cells can be established from genetically unmodified fibroblasts without drug selection for reactivation of pluripotency genes (Meissner et al. 2007). Similarly, human iPS lines have been derived from human fibroblasts upon transduction of different combinations of reprogramming factors (Oct4, Sox2, Klf4, and c-Myc or Oct4, Sox2, Nanog, and lin28) without the need for drug selection (Takahashi et al. 2007; Yu et al. 2007; Lowry et al. 2008; Park et al. 2008b). Proof-of-principle experiments demonstrated that mouse iPS cells can be used in combined gene and cellular therapy for alleviation of diseases in vivo (Hanna et al. 2007; Wernig et al. 2008a). Finally, pluripotent lines derived from human patients with specific diseases constitute invaluable tools and an unlimited source for biological material that can be used to study these complex diseases in the Petri dish (Dimos et al. 2008; Park et al. 2008a).

REPROGRAMMING OF TERMINALLY DIFFERENTIATED LYMPHOCYTES TO PLURIPOTENCY

Calculations for reprogramming efficiency performed in various studies on mouse cells have indicated that approximately 0.01–0.1% of cells infected with all four factors are reprogrammed into iPS cells (Meissner et al. 2007; Okita et al. 2007; Wernig et al. 2007). The low reprogramming efficiency can be explained by a number of nonmutually exclusive possibilities. One possibility is that reprogrammed cells originate from rare less-differentiated cells or adult stem cells present in the donor population, a question raised in the initial cloning experiments. It was not clear whether Dolly the sheep was generated by reprogramming of a rare somatic stem cell present in the heterogeneous donor cell population, rather than a differentiated mammary gland cell as assumed, because no genetic marker was available that would allow the retrospective identification of the donor nucleus (Wilmut et al. 1997). Similarly, it could not be excluded that "adult stem cells" could have been the source for iPS lines derived from somatic fibroblast cultures, insulin-expressing pancreatic β cells, or albumin-expressing liver cell cultures (Aoi et al. 2008; Sridharan and Plath 2008; Stadtfeld et al. 2008b). In these experiments, it is difficult to exclude that a rare stem cell, instead of a differentiated cell, was reprogrammed because no unambiguous genetic marker was used to retrospectively identify the differentiation state of the donor nucleus. To resolve these issues, we reprogrammed mature B and T lymphocytes and used the genetic rearrangements in the endogenous immunoglobulin or T-cell receptor (TCR) genes as markers to identify the differentiation state of the donor cell.

NUCLEAR TRANSFER OF MATURE MOUSE B AND T LYMPHOCYTES

Initial attempts to clone nuclei from mouse lymphocytes by SCNT techniques failed, suggesting that cloning these specialized cells might be impossible or highly inefficient. We therefore devised a two-step strategy to generate adult mice from mature B and T lymphocytes by first deriving ES cells from cloned blastocysts, followed by generating adult monoclonal mice using the tetraploid embryo complementation technique. Two ES lines were derived from approximately 1000 eggs injected with mature B- and T-cell nuclei. The NT-ES cell lines and the derived monoclonal mice carried rearrangements of the IgH and IgL immunoglobulin loci or TCRα and TCRβ chain rearrangements, demonstrating their origin from a mature B or T cell, respectively. These results provided unequivocal proof that terminally differentiated cells can be reprogrammed to a pluripotent state following exposure of the nucleus to the egg cytoplasm (Hochedlinger and Jaenisch 2002; Inoue et al. 2005). This experiment suggested that the two-step approach could overcome technical obstacles faced when cloning somatic cells. It is possible that prolonged in vitro culture of the cloned embryo in the two-step approach allowed faithful reactivation of pluripotency genes and/or selection for rare cells that acquired an ES-like phenotype. However, adult mice were derived from mature NK-T lymphocytes in a one-step cloning approach, suggesting that reprogramming these cells might be more efficient than reprogramming mature B and T cells (Inoue et al. 2005).

DIRECT REPROGRAMMING OF MATURE B LYMPHOCYTES TO PLURIPOTENCY BY DEFINED FACTORS

The ability to derive iPS cells from fibroblasts by defined factors and the success of SCNT experiments performed on mature lymphocytes to generate pluripotent lines suggested that direct reprogramming of terminally differentiated lymphocytes can, in principle, be achieved. Moreover, the hematopoietic system provides the opportunity to compare the reprogramming efficiency of cells at different stages of differentiation in the same lineage. However, the difficulty of culturing mouse lymphocytes for prolonged periods in vitro and achieving sufficiently high transduction rates of each ectopic factor constituted major obstacles to reprogramming B or T cells. To overcome these obstacles, we devised a drug-inducible transgenic system, termed "secondary system," that supports the reprogramming of multiple somatic cell types carrying identical Dox-inducible viral transgenes encoding Oct4, Sox2, Klf4, and c-Myc (Fig. 2) (Hanna et al. 2008; Wernig et al. 2008b). This was achieved by infecting fibroblasts

with Dox-inducible lentiviral vectors carrying the four reprogramming factors. When cultured in the presence of Dox, multiple iPS lines were generated that could be propagated independently of Dox. As a next step, "primary" iPS lines were injected into blastocysts to generate embryonic or adult mouse chimeras, thus allowing clonal expansion and their redifferentiation into multiple somatic cells types in vivo. Because the injected iPS lines carried a constitutively expressed antibiotic resistance gene, homogeneous iPS-derived somatic cell populations such as embryonic fibroblasts, mesenchymal stem cells, neural precursors, and lymphocytes could be isolated that carried provirus integration patterns identical to those in the primary iPS cell line (Hanna et al. 2008; Wernig et al. 2008b). Cultivation of these "secondary" somatic cells in the presence of Dox efficiently generated "secondary" iPS cells, which grew independently of Dox and were shown to be pluripotent by stringent criteria (Hanna et al. 2008; Wernig et al. 2008b). Two major advantages are offered by this strategy. First, because secondary somatic cells do not require new vector-mediated factor transduction, cells that are difficult to infect can be reprogrammed. Second, the approach avoids the genetic heterogeneity produced by direct viral infection of somatic cells (Hanna et al. 2008; Wernig et al. 2008b). More recently, we have also established a secondary system for human cells (Hockemeyer et al. 2008). Human fibroblasts carrying Dox-inducible vectors encoding the human Oct4, Sox2, and Klf4 cDNAs generated secondary iPS cells with an efficiency similar to that of the mouse secondary somatic cells, which was several orders of magnitude higher than that of newly infected human fibroblasts (Meissner et al. 2007; Okita et al. 2007; Wernig et al. 2007).

Pro-, pre-, and mature B-cell populations were derived from adult "secondary" chimeras and cultured in the presence of Dox. As expected, the Oct4, Sox2, KLf4, and c-Myc vectors were highly induced, but the cells failed to proliferate and most died within 5 days in culture. This suggested that initial propagation of the cells under B-cell-compatible culturing conditions might be required during the initial steps of reprogramming. Therefore, optimized culture conditions were devised to allow growth of B lymphocytes at different stages of differentiation. Cells were grown on OP9 bone marrow stromal cells in media supplemented with leukemia inhibitory factor (LIF), which is required for ES cell growth, with interleukin-7 (IL-7), stem cell factor (SCF), and Flt-3L, which promote bone-marrow-derived B-cell development (Milne et al. 2004), and with IL-4, anti-CD40, and lipopolysaccharide (LPS), which induce robust expansion of mature B cells in vitro.

Under these conditions, we were able to derive multiple iPS lines from purified pro- and pre-B-cell subpopulations that were cultured in the presence of Dox for 14 days and subsequently subcloned and cultured in normal ES cell growth conditions independent of Dox. These lines, termed iB-iPS lines (immature B-cell-derived iPS), were indistinguishable from ES cells or mouse embryonic fibroblast (MEF)-derived iPS cells, because they expressed all known pluripotency markers and generated teratomas in vivo and live adult chimeras with contribu-

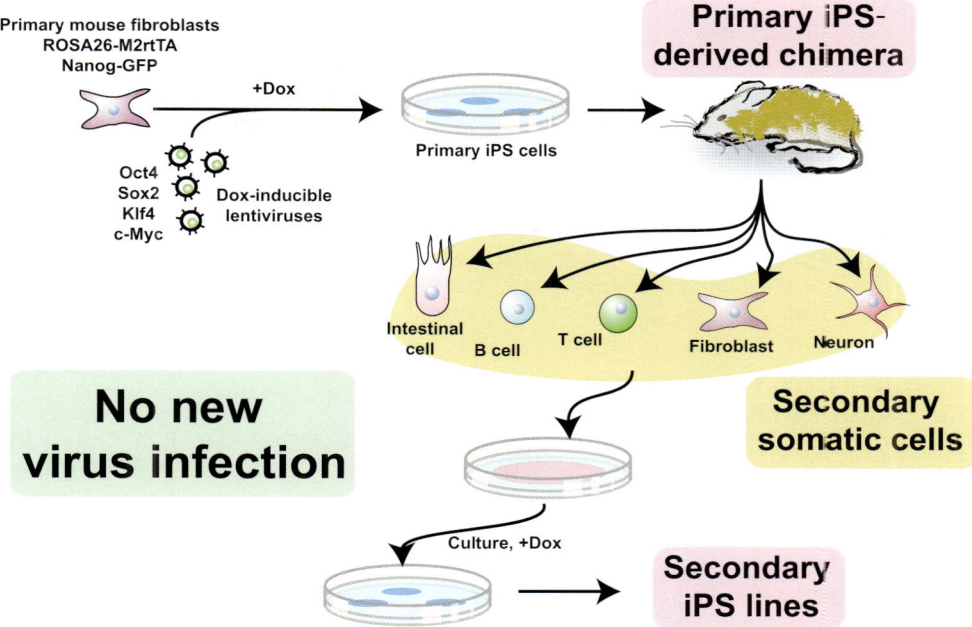

Figure 2. "Reprogrammable mouse model" for studying induction of pluripotency in homogeneous transgenic somatic cells. This mouse model constitutes an experimental system that circumvents genetic heterogeneity obtained when infecting somatic cell cultures with viruses and allows for the induction of iPS cells without requiring virus infection. In this "secondary system" strategy, primary iPS cells are generated from fibroblasts infected with Dox-inducible lentiviral vectors carrying the reprogramming factors Oct4, Sox2, Klf4, and c-Myc. Single iPS clones are injected into mouse blastocysts to generate chimeras and contribute to a variety of genetically homogeneous "secondary" somatic cells that carry the same Dox-inducible factors in their genome as their primary iPS cells. Cells from chimeric mice created using genetically reprogrammed cells can be triggered via drug administration to enter an ES-cell-like state without the need for additional direct genetic manipulation. This technical advancement enables creation of large numbers of genetically identical somatic cells that can be reprogrammed to an ES-cell-like state simply by exposure to a drug.

tions to the germ line (Hanna et al. 2008). Importantly, polymerase chain reaction (PCR) and southern blot analysis on genomic DNA obtained from iB-iPS lines demonstrated different patterns of incomplete (D-J) and complete (V-DJ) heavy-chain genetic rearrangements in the cell lines obtained, providing definitive proof that cells committed to the B-cell lineage were reprogrammed to pluripotency in our experiment.

In contrast to pro- or pre-B cells, mature B cells did not yield iPS cells when cultured in the presence of Dox. Although the cells were able to proliferate in conditioned media with Dox for a relatively extended period, no reactivation of any pluripotency-associated markers was observed, and ectopic expression of other pluripotency-related transcription factors did not facilitate reprogramming (Hanna et al. 2008). It appeared possible that reprogramming mature B cells required the disruption of the transcriptional circuitry that maintains the B-cell state. Indeed, studies by the Busslinger, Graf, and Weissman groups defined the role of Pax5, which is a transcriptional master regulator, in inducing and maintaining mature B-cell identity (Mikkola et al. 2002; Xie et al. 2004; Matthias and Rolink 2005; Delogu et al. 2006; Hsu et al. 2006; Cobaleda et al. 2007; Schebesta et al. 2007). Conditional deletion of Pax5 in mature B cells causes loss of B-cell markers and dedifferentiation to a lymphoid pro-

genitor-like state (Cobaleda et al. 2007). In vivo, these cells can be redifferentiated into the T-cell lineage. Moreover, Pax5 null mature B cells can be differentiated into macrophages when cultured on OP-9 stromal cells under myeloid-promoting culture conditions (Mikkola et al. 2002). In addition, trans-differentiation of B cells into macrophages can be induced by transduction of the C/EBPα transcription factor, which is a major player in inducing myeloid lineage differentiation and inhibits Pax5 (Xie et al. 2004).

These observations suggested to us that interfering with the transcriptional network maintaining B-cell identity might sensitize transgenic mature B lymphocytes to overexpress Oct4, Sox2, Klf4, and c-Myc and result in robust reprogramming into pluripotent ES-like cells. Indeed, overexpression of the C/EBPα or knockdown of the Pax5 transcription factor resulted in the efficient reprogramming of mature B cells to iPS cells upon Dox-induced vector expression (Hanna et al. 2008). The derived B-iPS cells carried at least one productive heavy-chain and light-chain rearrangement each, providing unequivocal proof that they were derived from fully differentiated adult mature B cells. The derived iPS cells were proven to be pluripotent, as evidenced by their ability to generate adult live chimeras with germ-line contribution and embryonic day 14.5 (E14.5) (4n) tetraploid chimeras (Hanna et al. 2008). The efficiency of reprogramming secondary mature B cells following additional overexpression of C/EBPα was determined to be approximately 3%, similar to that of secondary embryonic or adult fibroblasts (Hanna et al. 2008; Wernig et al. 2008b). Our results suggest that reprogramming some cell types can only be achieved by sensitizing the cells to the expression of the four factors Oct4, Sox2, c-Myc, and Klf4 through interfering with their somatic cell identity. Importantly, the novel drug-inducible transgenic system allowed us to gain insights into key events and principles underlying the reprogramming process, as discussed in the following section.

INSIGHTS INTO THE MECHANISM OF DIRECT REPROGRAMMING

Reprogramming the somatic epigenome to a pluripotent ES-like cell through ectopic expression of transcription factors is a process characterized by being (1) slow, (2) gradual, and (3) inefficient, with most cells in a given population ultimately failing to reprogram (Jaenisch and Young 2008). In both mouse and human fibroblast cultures, regardless of the combination of reprogramming factors used, formation of iPS cell colonies requires at least 2–3 weeks of continuous culture and proliferation (Takahashi and Yamanaka 2006; Wernig et al. 2007). It is remarkable that reprogramming terminally differentiated B lymphocytes, as well as less-differentiated neural precursor cells, occurs within a similar time frame (Eminli et al. 2008; Kim et al. 2008), arguing that the epigenetic resetting of the genome from different cell types follows similar kinetics. Unlike in SCNT experiments, where the differentiation state influences the efficiency of reprogramming (Blelloch et al. 2006), experimental evidence that somatic stem cells are easier to reprogram than differentiated by factor transduction is still lacking.

One possibility to explain the low efficiency of reprogramming is the activation of cellular genes by insertional mutagenesis. For example, proviral insertions may activate certain regulatory pathways that enhance the action of the four defined factors in resetting the epigenome. Yamanaka and colleagues (Aoi et al. 2008) failed to detect common integration sites in four iPS lines tested based on the mapping of about 20 insertion sites. However, this result does not exclude the potential role of insertional mutagenesis. For example, it may be that several alternative genes, not just one, may enhance reprogramming. It is also possible that insertional mutagenesis may not be strictly required for reprogramming but that the activation of certain cellular genes could increase the efficiency of the process.

Another possibility influencing the efficiency of reprogramming is the requirement for an optimal stoichiometry of factor expression transduced by the viral vectors. Because iPS cells carry multiple proviral copies, it may be that an optimal expression ratio of the different factors is achieved by selection for the "right" number of proviruses integrated at different chromosomal locations. Neural precursor cells, which already express the endogenous *Sox2* gene, can be reprogrammed without addition of *Sox2* (Eminli et al. 2008; Kim et al. 2008). Interestingly, infection with a *Sox2* virus reduced the number of iPS cells,

suggesting that high levels of *Sox2* inhibit the reprogramming process (Eminli et al. 2008).

Reprogramming secondary somatic cells is consistent with the conclusion that the level of factor expression and/or the ratio between the expression levels of different factors may influence the efficiency of the process. Thus, Dox addition to secondary somatic cells derived from different primary iPS cells generated secondary iPS cells with efficiencies ranging from .1 % to 4% (Hanna et al. 2008; Wernig et al. 2008b). Because the primary iPS cells carried different and distinct proviral integrations, it is possible that those cells which were reprogrammed with high efficiency expressed the factors in a more optimal ratio than those with low reprogramming efficiency.

STOCHASTIC EVENTS DURING DIRECT REPROGRAMMING

Because secondary somatic cells are genetically homogeneous, carrying identical proviral copies, one might have expected that most if not all cells can be reprogrammed to secondary iPS cells. However, as discussed above, only a small fraction of secondary somatic cells (at best, 4%) can be reprogrammed by Dox-induced factor activation (Hanna et al. 2008; Wernig et al. 2008b). This is consistent with the notion that stochastic epigenetic events such as demethylation of pluripotency genes or chromatin remodeling are important for the reprogramming process and restrict the fraction of cells that can ultimately be converted to secondary iPS cells. These considerations also raise the question of whether intermediate stages in the reprogramming process can be defined. Indeed, clonal cell lines that are partially reprogrammed have been recently derived from secondary fibroblasts and B lymphocytes (Meissner et al. 2007; Mikkelsen et al. 2008). These cells display a morphology different from that of iPS cells and can be extensively expanded in culture. Occasional interspersed ES-like cells grow within these cultures, which prove to be pluripotent and molecularly indistinguishable from ES cells.

Genomic analyses helped to define molecular differences between iPS cells and partially reprogrammed cells. In fully reprogrammed iPS cells, total demethylation of pluripotency genes and concurrent silencing of developmental regulators were observed, whereas only partial demethylation of pluripotency genes (e.g., *Fgf4*, *Rex1*, and *GDF3*) and only limited silencing of developmental regulators occurred in the partially reprogrammed cells (Mikkelsen et al. 2008). The importance of these molecular differences was underlined by the observations that DNA demethylation induced by treatment with 5-Aza or knockdown of Dnmt1 facilitated complete reprogramming of the partially reprogrammed cells into iPS lines. In addition, inhibition of a single partially silenced developmental regulator such as Sox9 or Pax7 enhanced complete reprogramming (Mikkelsen et al. 2008). Future studies aimed at defining the nature of additional secondary events that occur during reprogramming by either characterizing genetic or other epigenetic modifications that significantly increase efficiency will significantly enhance our understanding of the complex reprogramming process.

TWO SIDES OF THE SAME COIN? NUCLEAR CLONING VERSUS DIRECT REPROGRAMMING

Reprogramming by NT sparked major interest into specific factors that could induce reprogramming of somatic nuclei. Although NT-ES cells and iPS cells have highly similar, if not identical, epigenetic and developmental potential characteristics, it is interesting to explore whether the mechanisms for inducing reprogramming by a defined factor or the oocyte cytoplasm are similar. An important observation from NT is that the key pluripotency gene *Oct4* is reactivated in the somatic nucleus at the two to four cell stage in cloned embryos, suggesting that major reprogramming events can occur following one to two cell divisions (Boiani et al. 2005). In contrast, direct in vitro reprogramming of different somatic cell types proceeds during a period of at least 3 weeks in mice and even longer in humans (Fig. 3) (Jaenisch and Young 2008). These data suggest that the mechanism for reprogramming could be dramatically different. However, one cannot exclude the possibility that the oocyte cytoplasm contains a larger repertoire of reprogramming factors and therefore accomplishes reprogramming during a much shorter period. Defined transcription factor combinations can be thought of as the "minimal ingredients" recipe and most likely

achieve reprogramming by induction and/or recruitment of additional targets. However, in this case, proliferation and the sequence of stochastic events may be required to make these components available. Indeed, there are a number of certain similarities in the somatic nucleus response to the four factors or oocyte cytoplasm. Both processes are highly inefficient, with most cells failing to reprogram or reactivate endogenous Oct4 and other pluripotent components. Genomic analysis studying the response of fibroblasts and B cells to induction of the four factors demonstrated strong up-regulation of lineage-specific genes from unrelated lineages not normally expressed in ES cells (e.g., factors required for axon guidance, renal glomerular proteins, and epidermal-specific proteins) (Mikkelsen et al. 2008). Similarly, Gurdon and colleagues observed a similar pattern for promiscuous induction of gene expression in cloned somatic nuclei in amphibians (Trendelenburg et al. 1978). These results suggest that in both reprogramming methods, molecules supporting promiscuous transcription, possibly reflecting global changes in chromatin state, may be a crucial step in mediating successful reprogramming. Although this gene activation may act as transcriptional "noise," creating a bottleneck in the activation of pluripotency circuitry (as suggested by Mikkelsen et al. 2008), it remains possible that this phenomenon is a prerequisite

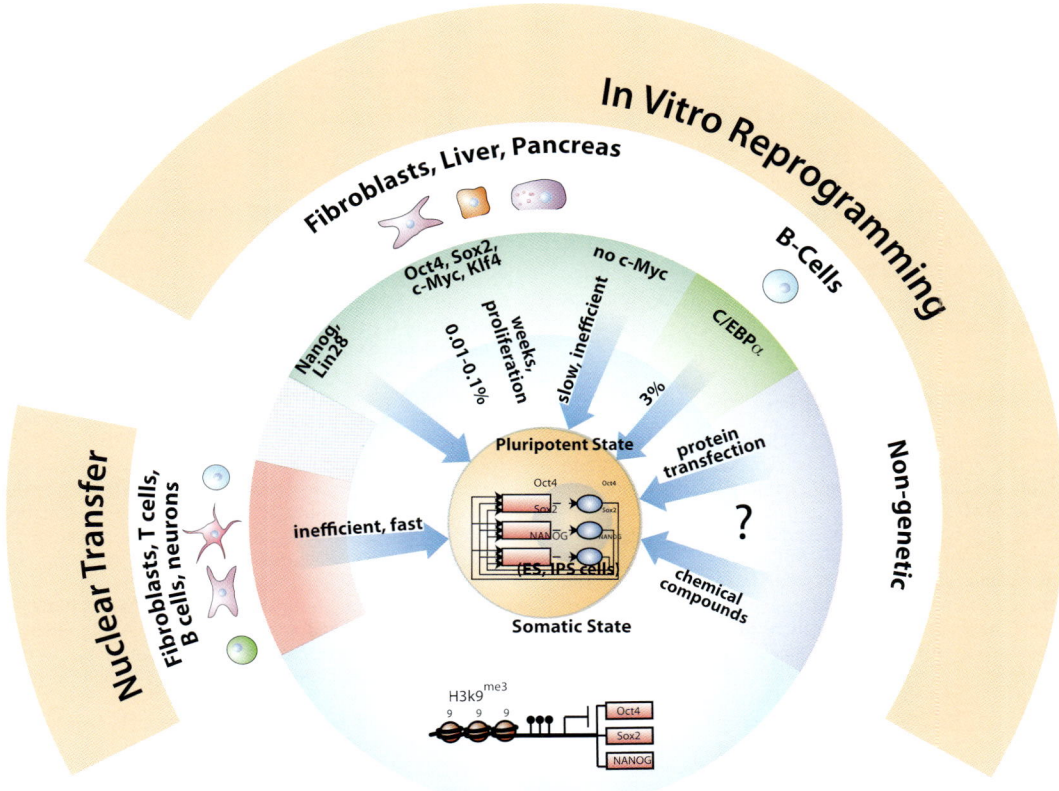

Figure 3. Reprogramming to pluripotency: past, present, and future. Shown is a schematic representation of induction of pluripotency in somatic cells by nuclear transfer and direct in vitro reprogramming. The figure highlights examples of somatic cell types used in these experiments and their efficiency and kinetics as donor nuclei for generating pluripotent ES-like cells. We also point out potential strategies that might be used in the future to generate genetically unmodified pluripotent cells in vitro.

and precipitates the activation of specific pluripotency genes (e.g., *Oct4*). We predict that in the near future, many cellular events and pathways used in direct reprogramming will be unraveled, and it would be highly informative to contrast the involvement of such mediators in resetting the epigenome from different reprogramming approaches. Understanding such events will also be instrumental for devising alternative strategies to achieve safe and efficient reprogramming, i.e., by using small molecules, cytokines, or protein transfections to replace the action of some of the reprogramming factors and generate genetically unmodified iPS cells (Fig. 3).

OUTLOOK

Understanding the molecular circuitry that dictates the identity of different cell types has greatly aided our efforts toward reprogramming different somatic cells, including terminally differentiated cells, to a pluripotent state. An unresolved question is whether one somatic cell type can be converted into another cell type, without prior dedifferentiation to a pluripotent state, by direct *trans*-differentiation. Recently, the in vivo conversion of exocrine pancreas cells to endocrine insulin-producing cells has been achieved by expression of three transcription factors (Zhou et al. 2008). It will be a major challenge for future work to use our current knowledge of transcriptional networks active in different somatic cell types to achieve the direct reprogramming of somatic cells to cells of a different germ layer in the Petri dish.

ACKNOWLEDGMENTS

We thank L. Boyer, J. Cassady, S. Markoulaki, and members of the Jaenisch lab for excellent assistance and helpful comments. We also thank Tom DiCesare for generating illustrations. R.J. is supported by grants from the National Institutes of Health (5-RO1-HDO45022, 5-R37-CA084198, and 5-RO1-CA087869). J.H. is a Novartis Fellow of the Helen Hay Whitney Foundation.

REFERENCES

Aoi, T., Yae, K., Nakagawa, M., Ichisaka, T., Okita, K., Takahashi, K., Chiba, T., and Yamanaka, S. 2008. Generation of pluripotent stem cells from adult mouse liver and stomach cells. *Science* **321:** 699–702.

Bernstein, B.E., Meissner, A., and Lander, E.S. 2007. The mammalian epigenome. *Cell* **128:** 669–681.

Blelloch, R., Wang, Z., Meissner, A., Pollard, S., Smith, A., and Jaenisch, R. 2006. Reprogramming efficiency following somatic cell nuclear transfer is influenced by the differentiation and methylation state of the donor nucleus. *Stem Cells* **24:** 2007–2013.

Boiani, M., Gentile, L., Gambles, V.V., Cavaleri, F., Redi, C.A., and Schöler, H.R. 2005. Variable reprogramming of the pluripotent stem cell marker Oct4 in mouse clones: Distinct developmental potentials in different culture environments. *Stem Cells* **23:** 1089–1104.

Boyer, L.A., Lee, T.I., Cole, M.F., Johnstone, S.E., Levine, S.S., Zucker, J.P., Guenther, M.G., Kumar, R.M., Murray, H.L., Jenner, R.G., et al. 2005. Core transcriptional regulatory circuitry in human embryonic stem cells. *Cell* **122:** 947–956.

Brambrink, T., Foreman, R., Welstead, G.G., Lengner, C.J.,

Wernig, M., Suh, H., and Jaenisch, R. 2008. Sequential expression of pluripotency markers during direct reprogramming of mouse somatic cells. *Cell Stem Cell* **2:** 151–159.

Byrne, J.A., Pedersen, D.A., Clepper, L.L., Nelson, M., Sanger, W.G., Gokhale, S., Wolf, D.P., and Mitalipov, S.M. 2007. Producing primate embryonic stem cells by somatic cell nuclear transfer. *Nature* **450:** 497–502.

Cobaleda, C., Schebesta, A., Delogu, A., and Busslinger, M. 2007. Pax5: The guardian of B cell identity and function. *Nat. Immunol.* **8:** 463–470.

Cowan, C.A., Atienza, J., Melton, D.A., and Eggan, K. 2005. Nuclear reprogramming of somatic cells after fusion with human embryonic stem cells. *Science* **309:** 1369–1373.

Delogu, A., Schebesta, A., Sun, Q., Aschenbrenner, K., Perlot, T., and Busslinger, M. 2006. Gene repression by Pax5 in B cells is essential for blood cell homeostasis and is reversed in plasma cells. *Immunity* **24:** 269–281.

Dimos, J.T., Rodolfa, K.T., Niakan, K.K., Weisenthal, L.M., Mitsumoto, H., Chung, W., Croft, G.F., Saphier, G., Leibel, R., Goland, R., et al. 2008. Induced pluripotent stem cells generated from patients with ALS can be differentiated into motor neurons. *Science* **321:** 1218–1221.

Egli, D., Rosains, J., Birkhoff, G., and Eggan, K. 2007. Developmental reprogramming after chromosome transfer into mitotic mouse zygotes. *Nature* **447:** 679–685.

Egli, D., Birkhoff, G., and Eggan, K. 2008. Mediators of reprogramming: Transcription factors and transitions through mitosis. *Nat. Rev. Mol. Cell Biol.* **9:** 505–516.

Eminli, S., Utikal, J.S., Arnold, K., Jaenisch, R., and Hochedlinger, K. 2008. Reprogramming of neural progenitor cells into iPS cells in the absence of exogenous Sox2 expression. *Stem Cells* (in press).

Guan, K., Nayernia, K., Maier, L.S., Wagner, S., Dressel, R., Lee, J.H., Nolte, J., Wolf, F., Li, M., Engel, W., and Hasenfuss, G. 2006. Pluripotency of spermatogonial stem cells from adult mouse testis. *Nature* **440:** 1199–1203.

Gurdon, J.B. 1999. Genetic reprogramming following nuclear transplantation in Amphibia. *Semin. Cell Dev. Biol.* **10:** 239–243.

Gurdon, J.B. 2006. From nuclear transfer to nuclear reprogramming: The reversal of cell differentiation. *Annu. Rev. Cell Dev. Biol.* **22:** 1–22.

Hanna, J., Wernig, M., Markoulaki, S., Sun, C.W., Meissner, A., Cassady, J.P., Beard, C., Brambrink, T., Wu, L.C., Townes, T.M., et al. 2007. Treatment of sickle cell anemia mouse model with iPS cells generated from autologous skin. *Science* **318:** 1920–1923.

Hanna, J., Markoulaki, S., Schorderet, P., Carey, B.W., Beard, C., Wernig, M., Creyghton, M.P., Steine, E.J., Cassady, J.P., Foreman, R., et al. 2008. Direct reprogramming of terminally differentiated mature B lymphocytes to pluripotency. *Cell* **133:** 250–264.

Hochedlinger, K. and Jaenisch, R. 2002. Monoclonal mice generated by nuclear transfer from mature B and T donor cells. *Nature* **415:** 1035–1038.

Hochedlinger, K. and Jaenisch, R. 2006. Nuclear reprogramming and pluripotency. *Nature* **441:** 1061–1067.

Hockemeyer, D., Soldner, F., Cook, E., Gao, X., Mitalipova, M., and Jaenisch, R. 2008. A drug inducible system for direct reprogramming of human somatic cells to pluripotency. *Cell Stem Cells* **3:** 346–353.

Hsu, C.L., King-Fleischman, A.G., Lai, A.Y., Matsumoto, Y., Weissman, I.L., and Kondo, M. 2006. Antagonistic effect of CCAAT enhancer-binding protein-α and Pax5 in myeloid or lymphoid lineage choice in common lymphoid progenitors. *Proc. Natl. Acad. Sci.* **103:** 672–677.

Inoue, K., Wakao, H., Ogonuki, N., Miki, H., Seino, K., Nambu-Wakao, R., Noda, S., Miyoshi, H., Koseki, H., Taniguchi, M., et al. 2005. Generation of cloned mice by direct nuclear transfer from natural killer T cells. *Curr. Biol.* **15:** 1114–1118.

Jaenisch, R. 2004. Human cloning—The science and ethics of nuclear transplantation. *N. Engl. J. Med.* **351:** 2787–2791.

Jaenisch, R. and Young, R. 2008. Stem cells, the molecular circuitry of pluripotency and nuclear reprogramming. *Cell* **132:** 567–582.

Zahner, D., Stuhlmann, H., Stewart, C.L., Harbers, K., Lohler, J., Simon, I., and Jaenisch, R. 1982. De novo methylation and expression of retroviral genomes during mouse embryogenesis. *Nature* **298:** 623–628.

Kanatsu-Shinohara, M., Inoue, K., Lee, J., Yoshimoto, M., Ogonuki, N., Miki, H., Baba, S., Kato, T., Kazuki, Y., Toyokuni, S., et al. 2004. Generation of pluripotent stem cells from neonatal mouse testis. *Cell* **119:** 1001–1012.

Kim, J.B., Zaehres, H., Wu, G., Gentile, L., Ko, K., Sebastiano, V., Arauzo-Bravo, M.J., Ruau, D., Han, D.W., Zenke, M., et al. 2008. Pluripotent stem cells induced from adult neural stem cells by reprogramming with two factors. *Nature* **454:** 646–650.

Laiosa, C.V., Stadtfeld, M., Xie, H., de Andres-Aguayo, L., and Graf, T. 2006. Reprogramming of committed T cell progenitors to macrophages and dendritic cells by C/EBPα and PU.1 transcription factors. *Immunity* **25:** 731–744.

Lee, B.C., Kim, M.K., Jang, G., Oh, H.J., Yuda, F., Kim, H.J., Hossein, M.S., Kim, J.J., Kang, S.K., Schatten, G., et al. 2005. Dogs cloned from adult somatic cells. *Nature* **436:** 641.

Lerou, P.H. and Daley, G.Q. 2005. Therapeutic potential of embryonic stem cells. *Blood Rev.* **19:** 321–331.

Lowry, W.E., Richter, L., Yachechko, R., Pyle, A.D., Tchieu, J., Sridharan, R., Clark, A.T., and Plath, K. 2008. Generation of human induced pluripotent stem cells from dermal fibroblasts. *Proc. Natl. Acad. Sci.* **105:** 2883–2888.

Maherali, N., Sridharan, R., Xie, W., Utikal, J., Eminli, S., Arnold, K., Stadtfeld, M., Yachechko, R., Tchieu, J., Jaenisch, R., et al. 2007. Directly reprogrammed fibroblasts show global epigenetic remodeling and widespread tissue contribution. *Cell Stem Cell* **1:** 55–70.

Matthias, P. and Rolink, A.G. 2005. Transcriptional networks in developing and mature B cells. *Nat. Rev.* **5:** 497–508.

Meissner, A., Wernig, M., and Jaenisch, R. 2007. Direct reprogramming of genetically unmodified fibroblasts into pluripotent stem cells. *Nat. Biotechnol.* **25:** 1177–1181.

Meissner, A., Mikkelsen, T.S., Gu, H., Wernig, M., Hanna, J., Sivachenko, A., Zhang, X., Bernstein, B.E., Nusbaum, C., Jaffe, D.B., et al. 2008. Genome-scale DNA methylation maps of pluripotent and differentiated cells. *Nature* **454:** 766–770.

Mikkelsen, T.S., Hanna, J., Zhang, X., Ku, M., Wernig, M., Schorderet, P., Bernstein, B.E., Jaenisch, R., Lander, E.S., and Meissner, A. 2008. Dissecting direct reprogramming through integrative genomic analysis. *Nature* **454:** 49–55.

Mikkola, I., Heavey, B., Horcher, M., and Busslinger, M. 2002. Reversion of B cell commitment upon loss of Pax5 expression. *Science* **297:** 110–113.

Milne, C.D., Fleming, H.E., and Paige, C.J. 2004. IL-7 does not prevent pro-B/pre-B cell maturation to the immature/sIgM$^+$ stage. *Eur. J. Immunol.* **34:** 2647–2655.

Okita, K., Ichisaka, T., and Yamanaka, S. 2007. Generation of germline-competent induced pluripotent stem cells. *Nature* **448:** 313–317.

Park, I.H., Arora, N., Huo, H., Maherali, N., Ahfeldt, T., Shimamura, A., Lensch, M.W., Cowan, C., Hochedlinger, K., and Daley, G.Q. 2008a. Disease-specific induced pluripotent stem cells. *Cell* **134:** 877–886.

Park, I.H., Zhao, R., West, J.A., Yabuuchi, A., Huo, H., Ince, T.A., Lerou, P.H., Lensch, M.W., and Daley, G.Q. 2008b. Reprogramming of human somatic cells to pluripotency with defined factors. *Nature* **451:** 141–146.

Rodolfa, K.T. and Eggan, K. 2006. A transcriptional logic for nuclear reprogramming. *Cell* **126:** 652–655.

Schebesta, A., McManus, S., Salvagiotto, G., Delogu, A., Busslinger, G.A., and Busslinger, M. 2007. Transcription factor Pax5 activates the chromatin of key genes involved in B cell signaling, adhesion, migration, and immune function. *Immunity* **27:** 49–63.

Seandel, M., James, D., Shmelkov, S.V., Falciatori, I., Kim, J., Chavala, S., Scherr, D.S., Zhang, F., Torres, R., Gale, N.W., et al. 2007. Generation of functional multipotent adult stem cells from GPR125$^+$ germline progenitors. *Nature* **449:** 346–350.

Sridharan, R. and Plath, K. 2008. Illuminating the black box of reprogramming. *Cell Stem Cell* **2:** 295–297.

Stadtfeld, M., Brennand, K., and Hochedlinger, K. 2008a. Reprogramming of pancreatic β cells into induced pluripotent stem cells. *Curr. Biol.* **18:** 890–894.

Stadtfeld, M., Maherali, N., Breault, D.T., and Hochedlinger, K. 2008b. Defining molecular cornerstones during fibroblast to iPS cell reprogramming in mouse. *Cell Stem Cell* **2:** 230–240.

Tada, M., Takahama, Y., Abe, K., Nakatsuji, N., and Tada, T. 2001. Nuclear reprogramming of somatic cells by in vitro hybridization with ES cells. *Curr. Biol.* **11:** 1553–1558.

Takahashi, K. and Yamanaka, S. 2006. Induction of pluripotent stem cells from mouse embryonic and adult fibroblast cultures by defined factors. *Cell* **126:** 663–676.

Takahashi, K., Tanabe, K., Ohnuki, M., Narita, M., Ichisaka, T., Tomoda, K., and Yamanaka, S. 2007. Induction of pluripotent stem cells from adult human fibroblasts by defined factors. *Cell* **131:** 861–872.

Trendelenburg, M.F., Zentgraf, H., Franke, W.W., and Gurdon, J.B. 1978. Transcription patterns of amplified *Dytiscus* genes coding for ribosomal RNA after injection in *Xenopus* oocyte nuclei. *Proc. Natl. Acad. Sci.* **75:** 3791–3795.

Wakayama, T., Perry, A.C., Zuccotti, M., Johnson, K.R., and Yanagimachi, R. 1998. Full-term development of mice from enucleated oocytes injected with cumulus cell nuclei. *Nature* **394:** 369–374.

Weintraub, H., Tapscott, S.J., Davis, R.L., Thayer, M.J., Adam, M.A., Lassar, A.B., and Miller, A.D. 1989. Activation of muscle-specific genes in pigment, nerve, fat, liver, and fibroblast cell lines by forced expression of MyoD *Proc. Natl. Acad. Sci.* **86:** 5434–5438.

Wernig, M., Meissner, A., Foreman, R., Brambrink, T., Ku, M., Hochedlinger, K., Bernstein, B.E., and Jaenisch, R. 2007. In vitro reprogramming of fibroblasts into a pluripotent ES-cell-like state. *Nature* **448:** 318–324.

Wernig, M., Zhao, J.P., Pruszack, J., Hedlund, E., Fu, D., Soldner, F., Broccoli, V., Constantine-Paton, M., Isacson, O., and Jaenisch, R. 2008a. Neurons derived from reprogrammed fibroblasts functionally integrate into the fetal brain and improve symptoms of adult rats with Parkinson's. *Proc. Natl. Acad. Sci.* **105:** 5856–5861.

Wernig, M., Lengner, C.J., Hanna, J., Lodato, M.A., Steine, E., Foreman, R., Staerk, J., Markoulaki, S., and Jaenisch, R. 2008b. A drug-inducible transgenic system for direct reprogramming of multiple somatic cell types. *Nat. Biotechnol.* **26:** 916–924.

Wilmut, I., Schnieke, A.E., McWhir, J., Kind, A.J., and Campbell, K.H. 1997. Viable offspring derived from fetal and adult mammalian cells. *Nature* **385:** 810–813.

Xie, H., Ye, M., Feng, R., and Graf, T. 2004. Stepwise reprogramming of B cells into macrophages. *Cell* **117:** 663–676.

Yamanaka, S. 2008. Pluripotency and nuclear reprogramming. *Philos. Trans. R. Soc. Lond. B Biol. Sci.* **363:** 2079–2087.

Yu, J., Vodyanik, M.A., Smuga-Otto, K., Antosiewicz-Bourget, J., Frane, J.L., Tian, S., Nie, J., Jonsdottir, G.A., Ruotti, V., Stewart, R., et al. 2007. Induced pluripotent stem cell lines derived from human somatic cells. *Science* **318:** 1917–1920.

Zhou, Q., Brown, J., Kanarek, A., Rajagopal, J., and Melton, D.A. 2008. In vivo reprogramming of adult pancreatic exocrine cells to β-cells. *Nature* **455:** 627–632.

Induced Pluripotency of Mouse and Human Somatic Cells

N. Maherali*† and K. Hochedlinger*

*Massachusetts General Hospital Cancer Center and Center for Regenerative Medicine;
Harvard Stem Cell Institute; Department of Stem Cell and Regenerative Biology; Boston, Massachusetts 02114;
†Department of Molecular and Cellular Biology, Harvard University, Cambridge, Massachusetts 02138

The identification of transcription factors to induce pluripotency directly in somatic cells has given researchers a unique platform on which to dissect the mechanisms underlying epigenetic reprogramming. In addition, induced pluripotent stem (iPS) cells have enabled the derivation of patient-specific cells for the study and potential treatment of a variety of diseases. Here, we discuss recent discoveries in the reprogramming field including work from our own laboratory.

Reprogramming denotes the experimentally induced dedifferentiation of somatic cells into pluripotent cells, and pluripotency defines the ability of cells to give rise to all embryonic cell types including the germ line. Two major aims of reprogramming research are to understand the underlying mechanims at a biochemical level and to generate custom-tailored cells for studying and treating degenerative diseases. Experiments in mice have shown that iPS cells can indeed alleviate the disease phenotypes of sickle cell anemia (Hanna et al. 2007) and Parkinson's disease (Wernig et al. 2008b). This, and the recent derivation of patient-specifc iPS cell lines from 11 different genetic disorders (Dimos et al. 2008; Park et al. 2008a), demonstrates the potential utility of reprogramming research in a therapeutic setting.

Several different approaches have been developed to study the reprogramming of somatic cells into pluripotent cells, including nuclear transfer (NT), cell fusion, and direct reprogramming (Fig. 1). During NT, the nucleus of a somatic cell is injected into an enucleated oocyte, which then develops into a cloned embryo. When transferred into a recipient female, the cloned embryo can develop into a live animal (so-called reproductive cloning) (Fig. 1A) (Wilmut et al. 1997; Wakayama et al. 1998). When explanted in vitro, NT-derived embryonic stem (ES) cells can be derived, which may be useful for custom-tailored cell therapy (so-called therapeutic cloning) (Rideout et al. 2002). The process of NT is extremely inefficient (on average, only 1–3% of cloned blastocysts develop to term), and many of the cloned animals are abnormal, likely due to faulty epigenetic reprogramming of the genome (Hochedlinger and Jaenisch 2002b). Interestingly, the derivation of ES cells from cloned blastocysts seems to select for fully reprogrammed cells because NT ES cells are molecularly and functionally indistinguishable from fertilization-derived ES cells, including their ability to give rise to entirely ES-cell-derived mice (Brambrink et al. 2006). Thus, the abnormalities seen in cloned animals should not impede the therapeutic use of NT technology. However, ethical and legal constraints surrounding NT make alternative approaches to reprogramming desirable. Moreover, NT does not provide sufficient amounts of material to biochemically dissect the process of reprogramming.

An alternative approach to studying reprogramming is cell fusion, which involves the fusion of somatic cells with ES cells, thus generating pluripotent hybrids in which the somatic genome acquires epigenetic marks of ES cells (Fig. 1B) (Tada et al. 2001; Cowan et al. 2005). Cell fusion is also quite inefficient, making it difficult to perform biochemical analyses. Nevertheless, cell fusion has been helpful in determining the effects of individual genes on reprogramming efficiency. For example, overexpression of the transcription factor Nanog in ES cells has been shown to result in an up to 200-fold increase in the number of reprogrammed hybrids upon fusion with neural stem cells (NSCs) (Silva et al. 2006). Cell fusion is not a viable approach in cell therapy, however, because the genomes of two different individuals are combined in hybrid cells, resulting in tetraploid cells that are prone to chromosomal abnormalities (Fujiwara et al. 2005).

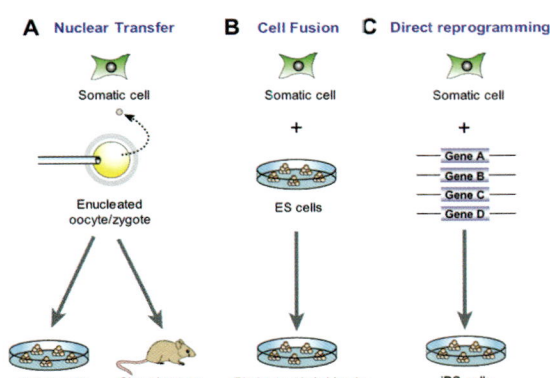

Figure 1. Different approaches for reprogramming somatic cells to pluripotency. (*A*) Nuclear transfer (NT) involves injecting a somatic nucleus into an enucleated oocyte or zygote, resulting in the formation of cloned blastocysts, which can be transferred into recipient females to produce cloned mice ("reproductive cloning") or explanted in culture to derive NT ES cells for therapeutic purposes ("therapeutic cloning"). (*B*) Cell fusion between somatic cells and pluripotent cells generated hybrid cells, in which the somatic genome becomes reprogrammed by the pluripotent cell. (*C*) Direct or in vitro reprogramming requires ectopic expression of defined genes in somatic cells, which converts them into a pluripotent state.

Cell-fusion experiments indicated that ES cells, like oocytes, must contain factors that mediate reprogramming to pluripotency and that these molecules should be identifiable to directly reprogram somatic cells into a pluripotent state (Fig. 1C). This notion prompted Yamanaka and colleagues to identify genes that are predominantly expressed in ES cells (Mitsui et al. 2003) and test the effect of their overexpression on somatic cells (Takahashi and Yamanaka 2006). Indeed, they identified four transcription factors, Oct4, Sox2, Klf4, and c-Myc, out of an initial group of 24 candidate genes, which when overexpressed in fibroblasts, gave rise to pluripotent cells called iPS cells (Fig. 2) (Takahashi and Yamanaka 2006). This chapter is aimed at summarizing the current knowledge on iPS cell research with an emphasis on work from our laboratory.

TRANSCRIPTION-FACTOR-INDUCED REPROGRAMMING

Previous work had indicated that individual transcription factors, when overexpressed or deleted, could induce cell-fate changes in somatic cells. Classical experiments in the 1980s showed that overexpression of the myogenic transcription factor MyoD was sufficient to convert fibroblasts into myogenic cells (Davis et al. 1987). Similarly, elimination of Pax5 from B cells results in their dedifferentiation into progenitors that can give rise to multiple hematopoietic lineages (Nutt et al. 1999), and overexpression of the transcription factor CEBPα has been shown to reprogram B and T cells into macrophages (Xie et al. 2004; Laiosa et al. 2006). Together, these experiments provided the rationale for attempts to reprogram somatic cells directly into iPS cells.

iPS cells were initially obtained using drug selection for the ES-cell-specific, but nonessential, gene *Fbx15*

(Takahashi and Yamanaka 2006). These first-generation iPS cells were similar but not identical to ES cells. They appeared to be transcriptionally and epigenetically intermediate between ES cells and fibroblasts, and although they did give rise to teratomas, they could not support the development of viable mice. Subsequent studies, however, showed that identifying iPS cells based on drug selection for the ES-cell-specific genes *Oct4* or *Nanog* (Maherali et al. 2007; Okita et al. 2007; Wernig et al. 2007), or simply based on morphological criteria alone (Blelloch et al. 2007; Maherali et al. 2007; Meissner et al. 2007), was sufficient to generate iPS cells that were highly similar to ES cells (Fig. 2). At the molecular level, iPS cells showed demethylation of the *Oct4* and *Nanog* promoter regions and transcriptional patterns akin to ES cells (Maherali et al. 2007; Okita et al. 2007; Wernig et al. 2007). Global analysis of histone methylation patterns including histone H3 lysine 4 (K4) and H3 lysine 27 (K27) methylation indicated that iPS cells were indistinguishable from ES cells (Maherali et al. 2007). In addition, the somatically silenced X chromosome in female cells became reactivated and underwent random inactivation upon differentiation (Maherali et al. 2007), similar to ES cells derived by NT from female somatic cells (Eggan et al. 2000). At the functional level, iPS cells produced viable chimeras that showed contribution to the germ line (Maherali et al. 2007; Okita et al. 2007; Wernig et al. 2007) and even supported the development of fetuses that were derived entirely from iPS cells (Wernig et al. 2007; Hanna et al. 2008; Kim et al. 2008).

iPS cell have since been generated from multiple tissues, including blood, liver, stomach, pancreas, brain, skin, and adrenals (Fig. 2) (Aoi et al. 2008; Eminli et al. 2008; Hanna et al. 2008; Kim et al. 2008; Stadtfeld et al. 2008a; Wernig et al. 2008a). Moreover, human fibroblasts (Takahashi et al. 2007; Yu et al. 2007; Lowry et al. 2008; Maherali et al. 2008; Park et al. 2008b) and keratinocytes (Maherali et al. 2008) have been converted into iPS cells by the same or a different combination of factors including OCT4, SOX2, LIN28, and NANOG. These results suggest that in vitro reprogramming is a universal process that functions in cell types derived from all three germ layers and in different species. However, the mechanisms underlying reprogramming remain largely unknown.

MOLECULAR CORNERSTONES OF DIRECT REPROGRAMMING

In vitro reprogramming is a gradual process that takes between 1 and 2 weeks in murine fibroblasts to generate pluripotent cells from somatic cells (Takahashi and Yamanaka 2006; Maherali et al. 2007; Okita et al. 2007; Wernig et al. 2007). To dissect the mechanism of reprogramming, it has been informative to study partially reprogrammed cells that have failed to silence the retroviral transgenes and regain expression of many important pluripotency regulators. Partially reprogrammed cells may have been generated when Fbx15 selection was initially used to identify iPS cells (Takahashi and Yamanaka 2006) and are frequently obtained when morphological

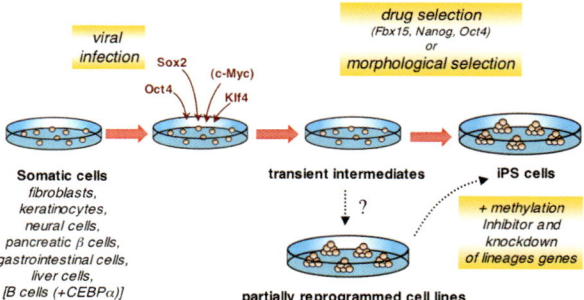

Figure 2. Direct reprogramming of somatic cells into induced pluripotent stem (iPS) cells. Flowchart of events involved in the direct reprogramming of somatic cells into iPS cells. Introduction of the Yamanaka factors into somatic cells results in the generation of iPS cells via defined intermediate steps. Drug selection or morphological criteria have been used to identify successfully reprogrammed colonies. Partially reprogrammed cell lines are thought to originate from transient intermediate cell populations and can be converted into iPS cells upon knockdown of lineage-specific genes and treatment with DNA methylation inhibitors. B lymphocytes could only be reprogrammed by additional overexpression of CEBPα and by use of a "secondary" system.

criteria are used to isolate iPS cells (Mikkelsen et al. 2008). These cell lines show incomplete demethylation and reactivation of pluripotency genes (Takahashi and Yamanaka 2006; Mikkelsen et al. 2008). Interestingly, genome-wide expression analyses showed that partially reprogrammed cell lines derived from B cells and fibroblasts are more similar to one another than to their cells of origin, suggesting that there could be one or several common intermediate state(s) in which somatic cells get trapped, irrespective of the cell of origin (see Fig. 2) (Mikkelsen et al. 2008).

Interestingly, partially reprogrammed cell lines show activation of lineage-specific genes not normally expressed in the starting cell population or in pluripotent cells (Mikkelsen et al. 2008). Knockdown of these genes, combined with the treatment with a DNA methylation inhibitor, resulted in a more efficient transition from the partially to a fully reprogrammed state, which suggests that ectopic expression of lineage-specific transcription factors and hypermethylation may prevent conversion into a pluripotent state (Fig. 2).

Although the analysis of partially reprogrammed cell states has been informative for understanding the "trapped" intermediate stages, a more detailed analysis of earlier and later stages of reprogramming will be critical for establishing the sequence of transcriptional and epigenetic events that lead to a pluripotent state. In attempts to define such early intermediates, two independent studies have shown that in vitro reprogramming of fibroblasts follows a defined sequence of molecular events, beginning with the down-regulation of somatic markers such as Thy1 and collagens, followed by the reactivation of the embryonic marker SSEA1 (Fig. 3) (Brambrink et al. 2008; Stadtfeld et al. 2008b). SSEA1-positive cells then gradually reactivate other markers associated with pluripotency including Oct4, Sox2, Nanog, telomerase, and the silent X chromosome in female cells. Reactivation of these late markers correlates with the time window when cells become independent of viral transgene expression and enter a self-sustaining pluripotent state. Importantly, sorting and plating of these rare intermediate cell populations result in a significant increase in the number of successfully reprogrammed iPS colonies, thus validating the functional importance of the identified biomarkers (Stadtfeld et al. 2008b). The observation that somatic markers are down-regulated before progressing into a pluripotent state further supports the notion that silencing of cell-type-specific programs is an important initial step toward reestablishing pluripotency. It also suggests that the differentiation state of the starting cell may affect the efficiency and kinetics of in vitro reprogramming.

EFFECT OF CELL TYPE AND DIFFERENTIATION STATE ON REPROGRAMMING EFFICIENCY

The derivation of iPS cells remains an extremely inefficient process ranging from 0.01% to 0.1%. This low efficiency of iPS cell derivation may be due to rare cells within the starting population, which serve as selective cells of origin. For example, adult stem cells are present in many tissues at about the same frequency as the success rate of reprogramming. Consistent with this notion, NSCs have been suggested to give rise to iPS cells 30–50 times more efficiently than fibroblasts (Kim et al. 2008), although another report came to different conclusions (Eminli et al. 2008).

The identity of the starting cells that give rise to iPS cells remains controversial. Two recent experiments addressed the cell-of-origin question in different cellular systems and came to different conclusions. In the first set of experiments, Hanna et al. (2008) attempted reprogramming of B lymphocytes into iPS cells. B cells carry differentiation-associated DNA rearrangements, which serve as an unambiguous genetic marker for their differentiation state (Hochedlinger and Jaenisch 2002a). Interestingly, ectopic expression of Oct4, Sox2, c-Myc, and Klf4 alone was insufficient to reprogram B lymphocytes into iPS cells, even when using a "secondary" system, in which most, if not all, cells express the four factors homogeneously (Maherali et al. 2008; Wernig et al. 2008a). The authors had to either overexpress the transcription factor CEBPα or knock down its suppressed target gene *Pax5* in addition to overexpressing the four factors to generate iPS cells from B cells. In contrast, progenitor B cells (pro-B cells) were permissive for reprogramming by the four factors alone, consistent with the notion that the differentiation state of the starting cell may affect reprogramming efficiency. Although this experiment indicated that defined terminally differentiated cells remain susceptible for reprogramming by defined factors, it is unclear, at this point, whether the difficulty in reprogramming lymphocytes is due to a biological barrier or merely reflects technical limitations.

In another set of experiments, Stadtfeld et al. (2008a) used genetically marked, terminally differentiated pancreatic β cells for reprogramming into iPS cells. β cells gave rise to iPS cells at a frequency comparable to fibroblasts (0.1–0.2%), demonstrating that this terminally differentiated cell type can be reprogrammed into iPS cells by just four factors and that adult stem cells are unlikely

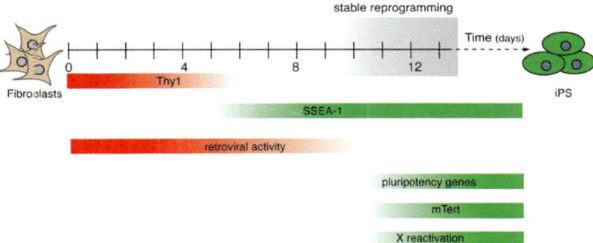

Figure 3. Molecular cornerstones of fibroblast into iPS cell reprogramming. Timescale of in vitro reprogramming into iPS cells with a summary of the major molecular events. (*Gray box*) Time window when cells convert into a pluripotent state and no longer depend on exogenous factor expression. (*Red*) Time frame of down-regulation of the fibroblast marker Thy1 and the silencing of retroviruses; (*green*) time frame of activation of the stem cell marker SSEA1 and other pluripotency associated events (reactivation of telomerase, pluripotency genes, and silent X chromosome in female cells). (Reprinted, with permission, from Stadtfeld et al. 2008b [© Cell Press].)

the selective cell type in successful reprogramming experiments. There are several explanations for the different outcomes of reprogramming lymphocytes and β cells. First, lymphocytes belong to the mesodermal lineage, whereas β cells are derived from endoderm. Liver and stomach cells, which are also endodermal derivatives, have recently been shown to be more amenable to reprogramming (Aoi et al. 2008) than fibroblasts (mesoderm). Alternatively, β cells could be more easily reprogrammed than lymphocytes because the pancreas does not contain stem, progenitor, and differentiated cells, as seen in the hematopoietic system, but rather a pool of β cells that continuously replicate (Dor et al. 2004). It will undoubtedly be interesting to compare the reprogramming efficiencies between undifferentiated and fully differentiated cells within a given cell lineage, for example, in the hematopoietic system.

REPROGRAMMING AND INSERTIONAL MUTAGENESIS

Another possible explanation for the low efficiency of reprogramming is the potential requirement for insertional mutagenesis by the viral transgenes (Hawley 2008). It has been previously shown that the retroviral infection of explanted blood stem cells selects for clones in which viruses inserted into genes conferring self-renewal to cells (Kustikova et al. 2005). Similarly, one or several of the viral copies present in iPS cells could have integrated into and activated a gene(s) that facilitates the reacquisition of a pluripotent, self-renewing state. The sequencing of 34 viral insertion sites in iPS cells derived from liver and stomach cells did not reveal any common integration sites (Aoi et al. 2008). However, sequencing has not yet been performed at saturation and it was still possible that insertional mutagenesis has a role during in vitro reprogramming (Hawley 2008). To ultimately exclude the possibility that there may be "hidden" reprogramming genes that become activated or inactivated as a result of viral integration, iPS cells needed to be generated without any genetic manipulation, for example, by using small chemical compounds, direct protein transduction, or nonintegrating viruses. Indeed, we have recently generated iPS cells without integrating viruses by infecting tail fibroblasts and hepatocytes with adenoviruses transciently expressing Oct4, Sox4, Kef4, and c-Myc (Stadtfeld et al. 2008).

A HIGH EFFICIENCY SYSTEM TO STUDY THE MECHANISMS OF DIRECT REPROGRAMMING

A third possibility for the low efficiency of reprogramming is the idea that the factors can only achieve reprogramming when expressed in precise relative amounts. Fibroblast-derived iPS cells carry on average 10–20 proviral transgenes expressing Oct4, Sox2, Klf4, and c-Myc, and the four transgenes are found at different copy numbers per cell (Takahashi and Yamanaka 2006; Maherali et al. 2007; Wernig et al. 2007; Eminli et al. 2008), suggesting that stoichiometry may be important.

This is consistent with observations in ES cells, where the levels of Oct4 and Sox2 are critical for maintaining a self-renewing pluripotent state (Niwa et al. 2000; Kopp et al. 2008). In further agreement, the reprogramming of Sox2-expressing NSCs into iPS cells in the absence of exogenous Sox2 expression results in an approximately fourfold increase in overall efficiency (Eminli et al. 2008). Thus, it is conceivable that the frequency at which a single somatic cell receives the four viral transgenes at the appropriate stoichiometry is extremely low.

If viral infection is indeed the rate-limiting step, one would predict that cells that can reactivate all four factors at the correct stoichiometry should give rise to iPS cells at an efficiency close to 100%. To address this question, we generated differentiated fibroblast-like "secondary" cells from human-induced pluripotent stem (hiPS) cells produced with doxycycline-inducible viral transgenes (Fig. 4) (Maherali et al. 2008). In these cells, most of the cells express the four factors together and likely at the correct stoichiometry upon exposure to doxycycline. Indeed, secondary hiPS cells were obtained up to 100 times more efficiently compared with iPS cells produced by direct viral infection, indicating that viral infection and the correct stoichiometry of the four factors are parameters that limit reprogramming efficiency. Experiments with mouse cells came to similar conclusions (Hanna et al. 2008; Wernig et al. 2008a). However, the overall efficiency was still quite low in all of these experiments (1–3% for secondary cells compared with 0.01–0.1% for primary infected cells), suggesting that additional events need to take place in order to reprogram somatic cells to pluripotency. These events likely involve stochastic epigenetic alterations including changes in DNA and histone methylation. Consistent with this notion, the treatment of somatic cells with compounds that inhibit DNA or histone methylation or histone deacetylation enhances the recovery of iPS cells significantly (Meissner et al. 2007; Huangfu et al. 2008; Shi et al. 2008).

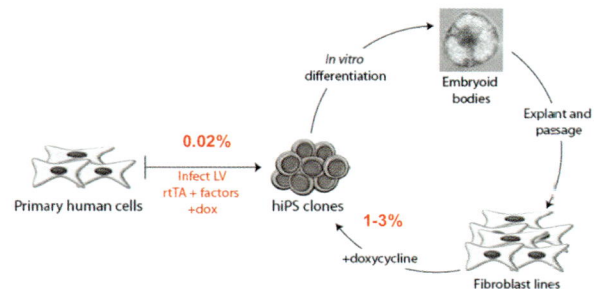

Figure 4. A high-efficiency system for the study of human-induced pluripotency. Outline for the generation of "secondary cells" from hiPS cells. Infection of fibroblasts or keratinocytes with doxycycline-inducible lentiviruses (LV) and a reverse tetracycline *trans*-activator (rtTA) in the presence of doxycycline results in the formation of primary hiPS cells at low frequency (~0.02%). Upon differentiation of clonal hiPS cell lines into fibroblast-like cells in vitro, followed by exposure to doxycycline, secondary hiPS cells are generated at high frequency (1–3%). The secondary system eliminates variability in infection rate efficiencies and factor expression and provides a platform for screening for drugs and genes that affect reprogramming. (Reprinted, with permission, from Maherali et al. 2008 [© Cell Press].)

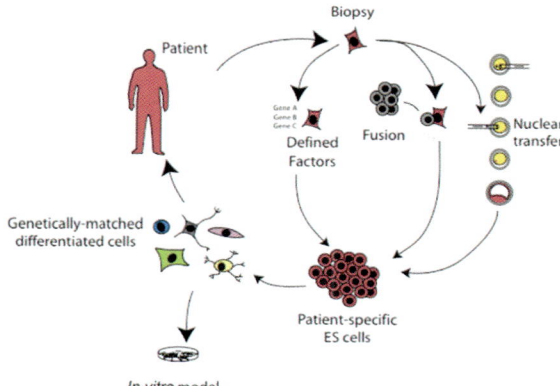

Figure 5. Therapeutic potential of reprogramming technology. Reprogramming by nuclear transfer, cell fusion, or defined factors may allow for the generation of patient-specific pluripotent cell lines. Pluripotent cells could be differentiated in vitro into the desired cell types and then transplanted into patients (cell therapy approach) or, alternatively, be used to establish in vitro models of diseases for drug screens and toxicology testing (disease modeling approach).

CONCLUSIONS

The study of induced pluripotency by defined transcription factors has yielded some important insights into the mechanisms underlying epigenetic reprogramming. Moreover, it has facilitated the derivation of patient-specific iPS cells from a variety of diseases (Dimos et al. 2008; Park et al. 2008a) to establish in vitro models for these disorders (Fig. 5). A major challenge before translating iPS technology into a therapeutic setting will be to induce pluripotency without genetic manipulation. Another open question is whether iPS cells are truly identical to ES cells or whether they retain a molecular memory of their cell of origin, as has been seen in some NT experiments (Hochedlinger and Jaenisch 2003). Finally, iPS technology has sparked interest in attempts to *trans*-differentiate specialized cells directly into other types of specialized cells by using alternative sets of transcription factors, for example from fibroblasts into myogenic cells (Davis et al. 1987), B lymphocytes into macrophages (Xie et al. 2004), or pancreatic exocrine cells into insulin-producing cells (Zhou et al. 2008). The next few years will undoubtedly bring exciting new discoveries on the role of transcription factors in cellular differentiation and epigenetic reprogramming.

ACKNOWLEDGMENTS

N.M. is supported by a graduate scholarship from the Natural Sciences and Engineering Research Council of Canada and a Sir James Lougheed Award from Alberta Scholarships. Support to K.H. is from the National Institutes of Health Director's Innovator Award, the Harvard Stem Cell Institute, the Kimmel Foundation, and the V Foundation. We thank members of the Hochedlinger lab for their scientific contributions and Chad Cowan for sharing graphic images.

REFERENCES

Aoi, T., Yae, K., Nakagawa, M., Ichisaka, T., Okita, K., Takahashi, K., Chiba, T., and Yamanaka, S. 2008. Generation of pluripotent stem cells from adult mouse liver and stomach cells. *Science* **321:** 699–702

Blelloch, R., Venere, M., Yen, J., and Ramalho-Santos, M. 2007. Generation of induced pluripotent stem cells in the absence of selection. *Cell Stem Cell* **1:** 245–247.

Brambrink, T., Hochedlinger, K., Bell, G., and Jaenisch, J. 2006. ES cells derived from cloned and fertilized blastocysts are transcriptionally and functionally indistinguishable. *Proc. Natl. Acad. Sci.* **103:** 933–938.

Brambrink, T., Foreman, R., Welstead, G.G., Lengner, C.J., Wernig, M., Suh, H., and Jaenisch, R. 2008. Sequential expression of pluripotency markers during direct reprogramming of mouse somatic cells. *Cell Stem Cell* **2:** 151–159.

Cowan, C.A., Atienza, J., Melton, D.A., and Eggan, K. 2005. Nuclear reprogramming of somatic cells after fusion with human embryonic stem cells. *Science* **309:** 1369–1373.

Davis, R.L., Weintraub, H., and Lassar, A.B. 1987. Expression of a single transfected cDNA converts fibroblasts to myoblasts. *Cell* **51:** 987–1000.

Dimos, J.T., Rodolfa, K.T., Niakan, K.K., Weisenthal, L.M., Mitsumoto, H., Chung, W., Croft, G.F., Saphier, G., Leibel, R., Goland, R., et al. 2008. Induced pluripotent stem cells generated from patients with ALS can be differentiated into motor neurons. *Science* **321:** 1218–1221.

Dor, Y., Brown, J., Martinez, O.I., and Melton, D.A. 2004. Adult pancreatic β-cells are formed by self-duplication rather than stem-cell differentiation. *Nature* **429:** 41–46.

Eggan, K., Akutsu, H., Hochedlinger, K., Rideout III, W., Yanagimachi, R., and Jaenisch, R. 2000. X-Chromosome inactivation in cloned mouse embryos. *Science* **290:** 1578–1581.

Eminli, S., Utikal, J.S., Arnold, K., Jaenisch, R., and Hochedlinger, K. 2008. Reprogramming of neural progenitor cells into iPS cells in the absence of exogenous Sox2 expression. *Stem Cells* (in press).

Fujiwara, T., Bandi, M., Nitta, M., Ivanova, E.V., Bronson R.T., and Pellman, D. 2005. Cytokinesis failure generating tetraploids promotes tumorigenesis in p53-null cells. *Nature* **437:** 1043–1047.

Hanna, J., Wernig, M., Markoulaki, S., Sun, C.W., Meissner, A., Cassady, J.P., Beard, C., Brambrink, T., Wu, L.C., Townes, T.M., and Jaenisch, R. 2007. Treatment of sickle cell anemia mouse model with iPS cells generated from autologous skin. *Science* **318:** 1920–1923.

Hanna, J., Markoulaki, S., Schorderet, P., Carey, B.W., Beard, C., Wernig M., Creyghton, M.P., Steine E.J., Cassady, J.P., Foreman, R., et al. 2008. Direct reprogramming of terminally differentiated mature B lymphocytes to pluripotency. *Cell* **133:** 250–264.

Hawley, R.G. 2008. Does retroviral insertional mutagenesis play a role in the generation of induced pluripotent stem cells? *Mol. Ther.* **16:** 1354–1355.

Hochedlinger, K. and Jaenisch, R. 2002a. Monoclonal mice generated by nuclear transfer from mature B and T donor cells. *Nature* **415:** 1035–1038.

Hochedlinger, K. and Jaenisch, R. 2002b. Nuclear transplantation: Lessons from frogs and mice. *Curr. Opin. Cell Biol.* **14:** 741–748.

Hochedlinger, K. and Jaenisch, R. 2003 Nuclear transplantation, embryonic stem cells, and the potential for cell therapy. *N. Engl. J. Med.* **349:** 275–286

Huangfu, D., Maehr, R., Guo, W., Eijkelenboom, A., Snitow, M., Chen A.E., and Melton, D.A. 2008. Induction of pluripotent stem cells by defined factors is greatly improved by small-molecule compounds. *Nat. Biotechnol.* **26:** 795–797.

Kim, J.B., Zaehres, H., Wu, G., Gentile, L., Ko, K., Sebastiano, V., Arauzo-Bravo, M.J., Ruau, D., Han, D.W., Zenke, M., and Scholer, H.R. 2008. Pluripotent stem cells induced from adult neural stem cells by reprogramming with two factors. *Nature* **454:** 646–650.

Kopp, J.L., Ormsbee, B.D., Desler, M., and Rizzino, A. 2008. Small increases in the level of Sox2 trigger the differentiation of mouse embryonic stem cells. *Stem Cells* **26:** 903–911.

Kustikova, O., Fehse, B., Modlich, U., Yang, M., Dullmann, J., Kamino, K., von Neuhoff, N., Schlegelberger, B., Li, Z., and Baum, C. 2005. Clonal dominance of hematopoietic stem cells triggered by retroviral gene marking. *Science* **308:** 1171–1174.

Laiosa, C.V., Stadtfeld, M., Xie, H., de Andres-Aguayo, L., and Graf, T. 2006. Reprogramming of committed T cell progenitors to macrophages and dendritic cells by C/EBPα and PU.1 transcription factors. *Immunity* **25:** 731–744.

Lowry, W.E., Richter, L., Yachechko, R., Pyle, A.D., Tchieu, J., Sridharan, R., Clark, A.T., and Plath, K. 2008. Generation of human induced pluripotent stem cells from dermal fibroblasts. *Proc. Natl. Acad. Sci.* **105:** 2883–2888.

Maherali, N., Sridharan, R., Xie, W., Utikal, J., Eminli, S., Arnold, K., Stadtfeld, M., Yachechko, R., Tchieu, J., Jaenisch, R., Plath, K., and Hochedlinger, K. 2007. Directly reprogrammed fibroblasts show global epigenetic reprogramming and widespread tissue contribution. *Cell Stem Cell* **1:** 55–70.

Maherali, N., Ahfeldt, T., Rigamonti, A., Utikal, J., Cowan, C.A., and Hochedlinger, K. 2008. A high-efficiency system for the generation and study of human induced pluripotent stem cells. *Cell Stem Cell* **3:** 340–345.

Meissner, A., Wernig, M., and Jaenisch, R. 2007. Direct reprogramming of genetically unmodified fibroblasts into pluripotent stem cells. *Nat. Biotechnol.* **25:** 1177–1181.

Mikkelsen, T.S., Hanna, J., Zhang, X., Ku, M., Wernig, M., Schorderet, P., Bernstein, B.E., Jaenisch, R., Lander, E.S., and Meissner, A. 2008. Dissecting direct reprogramming through integrative genomic analysis. *Nature* **454:** 794.

Mitsui, K., Tokuzawa, Y., Itoh, H., Segawa, K., Murakami, M., Takahashi, K., Maruyama, M., Maeda, M., and Yamanaka, S. 2003. The homeoprotein Nanog is required for maintenance of pluripotency in mouse epiblast and ES cells. *Cell* **113:** 631–642.

Niwa, H., Miyazaki, J., and Smith, A.G. 2000. Quantitative expression of Oct-3/4 defines differentiation, dedifferentiation or self-renewal of ES cells. *Nat. Genet.* **24:** 372–376.

Nutt, S.L., Heavey, B., Rolink, A.G., and Busslinger, M. 1999. Commitment to the B-lymphoid lineage depends on the transcription factor Pax5. *Nature* **401:** 556–562.

Okita, K., Ichisaka, T., and Yamanaka, S. 2007. Generation of germline-competent induced pluripotent stem cells. *Nature* **448:** 313–317.

Park, I.H., Arora, N., Huo, H., Maherali, N., Ahfeldt, T., Shimamura, A., Lensch, M.W., Cowan, C., Hochedlinger, K., and Daley, G.Q. 2008a. Disease-specific induced pluripotent stem cells. *Cell* **134:** 877–886.

Park, I.H., Zhao, R., West, J.A., Yabuuchi, A., Huo, H., Ince, T.A., Lerou, P.H., Lensch, M.W., and Daley, G.Q. 2008b. Reprogramming of human somatic cells to pluripotency with defined factors. *Nature* **451:** 141–146.

Rideout III, W.M., Hochedlinger, K., Kyba, M., Daley, G.Q., and Jaenisch, R. 2002. Correction of a genetic defect by nuclear transplantation and combined cell and gene therapy. *Cell* **109:** 17–27.

Shi, Y., Do, J.T., Desponts, C., Hahm, H.S., Scholer, H.R., and Ding, S. 2008. A combined chemical and genetic approach for the generation of induced pluripotent stem cells. *Cell Stem Cell* **2:** 525–528.

Silva, J., Chambers, I., Pollard, S., and Smith, A. 2006. Nanog promotes transfer of pluripotency after cell fusion. *Nature* **441:** 997–1001.

Stadtfeld, M., Brennand, K., and Hochedlinger, K. 2008a. Reprogramming of pancreatic β cells into induced pluripotent stem cells. *Curr. Biol.* **18:** 890–894.

Stadtfeld, M., Maherali, N., Breault, D.T., and Hochedlinger, K. 2008b. Defining molecular cornerstones during fibroblast to iPS cell reprogramming in mouse. *Cell Stem Cell* **2:** 230–240.

Stadtfeld, M., Nagaya, M., Utikal, J., Weir, G., and Hochedlinger, K. 2008. Induced pluripotent stem cells generated without viral integration. *Science* (in press).

Tada, M., Takahama, Y., Abe, K., Nakatsuji, N., and Tada, T. 2001. Nuclear reprogramming of somatic cells by in vitro hybridization with ES cells. *Curr. Biol.* **11:** 1553–1558.

Takahashi, K. and Yamanaka, S. 2006. Induction of pluripotent stem cells from mouse embryonic and adult fibroblast cultures by defined factors. *Cell* **126:** 663–676.

Takahashi, K., Tanabe, K., Ohnuki, M., Narita, M., Ichisaka, T., Tomoda, K., and Yamanaka, S. 2007. Induction of pluripotent stem cells from adult human fibroblasts by defined factors. *Cell* **131:** 861–872.

Wakayama, T., Perry, A.C., Zuccotti, M., Johnson, K.R., and Yanagimachi, R. 1998. Full-term development of mice from enucleated oocytes injected with cumulus cell nuclei. *Nature* **394:** 369–374.

Wernig, M., Meissner, A., Foreman, R., Brambrink, T., Ku, M., Hochedlinger, K., Bernstein, B.E., and Jaenisch, R. 2007. In vitro reprogramming of fibroblasts into a pluripotent ES-cell-like state. *Nature* **448:** 318–324.

Wernig, M., Lengner, C.J., Hanna, J., Lodato, M.A., Steine, E.J., Foreman, R., Staerk, J., Markoulaki, S., and Jaenisch, R. 2008a. A drug-inducible transgenic system for direct reprogramming of multiple somatic cell types. *Nat. Biotechnol.* **26:** 916–924.

Wernig, M., Zhao, J.P., Pruszak, J., Hedlund, E., Fu, D., Soldner, F., Broccoli, V., Constantine-Paton, M., Isacson, O., and Jaenisch, R. 2008b. Neurons derived from reprogrammed fibroblasts functionally integrate into the fetal brain and improve symptoms of rats with Parkinson's disease. *Proc. Natl. Acad. Sci.* **105:** 5856–5861.

Wilmut, I., Schnieke, A.E., McWhir, J., Kind, A.J., and Campbell, K.H. 1997. Viable offspring derived from fetal and adult mammalian cells. *Nature* **385:** 810–813.

Xie, H., Ye, M., Feng, R., and Graf, T. 2004. Stepwise reprogramming of B cells into macrophages. *Cell* **117:** 663–676.

Yu, J., Vodyanik, M.A., Smuga-Otto, K., Antosiewicz-Bourget, J., Frane, J.L., Tian, S., Nie, J., Jonsdottir, G.A., Ruotti, V., Stewart, R., Slukvin, I.I., and Thomson, J.A. 2007. Induced pluripotent stem cell lines derived from human somatic cells. *Science* **318:** 1917–1920.

Zhou, Q., Brown, J., Kanarek, A., Rajagopal, J., and Melton, D.A. 2008. *In vivo* reprogramming of adult pancreatic exocrine cells to β-cells. *Nature* **455:** 627–632.

Ronin and Caspases in Embryonic Stem Cells: A New Perspective on Regulation of the Pluripotent State

T.P. ZWAKA

Baylor College of Medicine, Houston, Texas 77030

Described here are recent discoveries in my laboratory which suggest that the complement of factors needed to direct ES cell pluripotency may be considerably larger than originally thought and may include proteins that act independently of the canonical factors described thus far. They also provide insight into a novel screening method that could be used to accelerate the identification and characterization of such factors.

Embryonic stem (ES) and ES-like cells afford attractive model systems for unraveling the complex molecular networks that give rise to cellular identity (Evans and Kaufman 1981; Martin 1981; Thomson et al. 1998). They are relatively easy to generate and maintain in culture and are amenable to a broad range of manipulations, including the creation of site-specific mutations in particular genes, with the use of a growing repertoire of molecular research tools now in the hands of cell biologists (Thomas and Capecchi 1987; Zwaka and Thomson 2003). More importantly, ES cells have the property of pluripotency, meaning that they can differentiate to any cell type in the body, even to germ cells (Pedersen 1986). Several groups now believe that a small core set of regulatory factors, including Pou5f1 (encoding the Oct4 protein), Sox2, and Nanog, maintain ES cells in the pluripotent state by acting on a limited number of target genes (for review, see Jaenisch and Young, 2008). This fundamental concept has generated enormous excitement in the biomedical research community, leading to successful reprogramming of somatic cells to an ES-like state (Takahashi and Yamanaka 2006; Yu et al. 2006; Okita et al. 2007; Takahashi et al. 2007; Wernig et al. 2007; Lowry et al. 2008; Park et al. 2008). This advancement was not serendipitous but represents the culmination of a long series of studies, such as somatic cell nuclear transfer (SCNT), to understand how the body's different cell lineages emerge from embryonic tissue (for review, see Boyer et al. 2006). Given the complexity of any cellular state, it is difficult to comprehend in detail even a single genetic network that is needed to guide cells smoothly from one cellular state to another. Thus, novel strategies must be identified that will allow identification of the pivotal components that maintain or alter stem cell states at one of more levels. Here, I describe recent discoveries in my laboratory which suggest that the complement of factors needed to direct ES cell pluripotency may be considerably larger than originally thought and may include proteins that act independently of the canonical factors described thus far (Dejosez et al. 2008; Fujita et al. 2008). They also provide insight into a novel

screening method that could be used to accelerate the identification and characterization of such factors.

CASPASES AS MEDIATORS OF ES CELL DIFFERENTIATION

We have discovered that members of a specific family of enzymes (caspases), previously thought to be involved only in programmed cell death (Thornberry and Lazebnik 1998; Earnshaw et al. 1999), have an unexpected role in ES cell differentiation by disabling one of the transcriptional pathways responsible for ES cell pluripotency (Fujita et al. 2008). Thus, ES cells may use apoptotic elements not only to regulate their pool size and maintain their genomic integrity, but also to induce a shift from self-renewal to differentiation in response to specific extrinsic cues (Fig. 1). This hypothesis is of fundamental importance, because programmed cell death is typically viewed as an isolated process restricted to the removal of damaged or unnecessary cells. One of the reasons for this separation is that most cell biologists have focused on differentiated cell types, including tumor cell lines with only one lineage choice, in contrast to stem cells, with multiple lineage options.

With the notable exceptions of caspases 1 and 11, which are involved in inflammation, caspases have been assigned primary functions in apoptosis, with scant attention paid to their possible involvement in nonapoptotic pathways (Thornberry and Lazebnik 1998). Recently, however, investigators have identified contexts in which caspase activity is associated with processes other than cell death. In mouse erythroid precursor cells, for example, the activation of cell death receptors triggers caspase activation and subsequent inhibition of differentiation via cleavage of the differentiation-promoting transcription factor Gata-1, and treatment of these cells with caspase inhibitors promotes differentiation (De Maria et al. 1999a,b). Another report links caspases directly to the differentiation decisions of hematopoietic stem cells via effects on pathways such as that driven by the interleukins (Janzen et al. 2008).

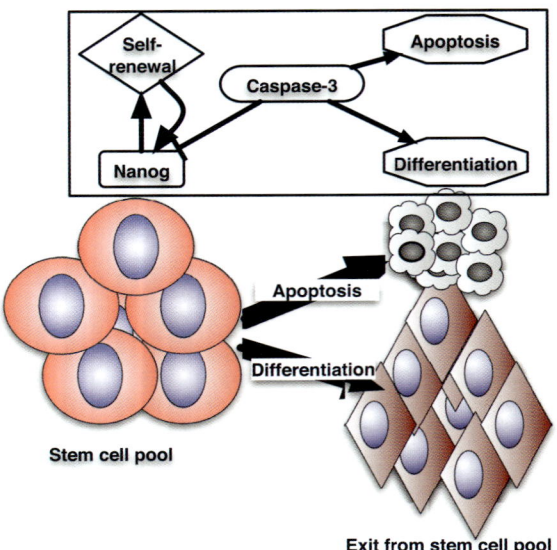

Figure 1. Stem cells can exit the stem cell pool either by differentiation or apoptosis. We found that caspase-3, previously known to be only essential for apoptosis, has a role in differentiation by cleaving Nanog and therefore allowing ES cells to exit the self-renewal cycle.

Further examples of processes that recruit caspases for purposes other than apoptosis include cell degradation or fusion during the differentiation of specific blood lineages (e.g., erythrocytes and megakaryocytes), the fusion of trophoblast-derived cells, the enucleation of keratinocytes and the lens epithelium, and sperm maturation (Ishizaki et al. 1998; De Botton et al. 2002; Arama et al. 2003; Fernando and Megeney 2007). Together, these observations support the notion that caspases can affect key developmental decisions, although the mechanisms and the nuances of context that induce these proteases into non-apoptotic pathways remain unclear.

Given the emerging reputation of caspases as post-translational modifiers of mammalian cell development, we considered them to be attractive candidates for one of the mechanistic elements that allow ES cells to escape their self-renewal constraints and rapidly begin to generate developmentally committed progeny. To accommodate the capacity of these cells to undergo germ-layer-specific differentiation, we predicted that caspase-3 may negatively regulate core self-renewal factors such as Oct4, Sox2, and Nanog, based on a surprising increase in the activity of this enzyme after induction of differentiation. Indeed, further experiments demonstrated caspase-induced cleavage of Nanog in differentiating ES cells, whereas stem cells lacking the *Casp3* gene displayed a marked defect in differentiation, and forced expression of a caspase-cleavage-resistant Nanog mutant in ES cells strongly promoted self-renewal (Fujita et al. 2008). We also found that in certain contexts, caspase-3 cooperates with caspase-9 to deactivate Nanog (Fujita et al. 2008). These findings suggest that ES cells exploit caspase-3 and -9 and perhaps others for rapid and spe-

cific deactivation of Nanog, thus disrupting the autoregulatory circuit that preserves the pluripotent state in these cells (Boyer et al. 2005).

RONIN, A CANDIDATE PLURIPOTENCY FACTOR WITH DISTINCT FUNCTIONS

Despite recent remarkable progress in reprogramming somatic cells to an ES-like state (induced pluripotent stem or iPS cells) via the manipulation of several key transcription factors (Takahashi and Yamanaka 2006; Yu et al. 2006; Okita et al. 2007; Takahashi et al. 2007; Wernig et al. 2007; Lowry et al. 2008; Park et al. 2008), the molecular mechanisms underlying the pluripotent state have only begun to emerge. The key regulators of ES cell pluripotency appear to be a small set of specific transcription factors able to promote self-renewal by repressing the transcription factors that initiate differentiation (Boyer et al. 2005; Bernstein et al. 2006; Lee et al. 2006). Prominent among these factors are Oct4, Sox2, and Nanog (Nichols et al. 1998; Avilion et al. 2003; Chambers et al. 2003). Although each of these proteins has been described by different authors as a "master regulator" of pluripotency, only Oct4 appears to be absolutely essential, whereas both Sox2 and Nanog are dispensable, at least in certain molecular contexts (Chambers et al. 2007; Masui et al. 2007). A second tier of pluripotency control is likely achieved via enzyme-mediated modification of chromatin (Klochendler-Yeivin et al. 2000; Loh et al. 2007; Fazzio et al. 2008). Thus, maintenance of the pluripotent state could require several different mechanisms acting in parallel and, potentially, even independently of one another. Given this background, and our finding that caspase-3 targets and cleaves Nanog upon induction of ES cell differentiation, we devised a yeast two-hybrid screen for other caspase-3 targets in ES cells, whose functions might augment the current repertoire of factors known to regulate the pluripotent state of ES cells. Constitutively modified caspase-3 spontaneously folds into its active conformation and recognizes and binds to target proteins, but no longer cleaves them due to a C163S substitution that was used as bait, in our two-hybrid screen because (1) unmodified procaspase-3 is functionally inactive and (2) actual cleavage of targets would render the interaction screening analysis useless (Kamada et al. 1998). After screening approximately 32 million clones, we identified 556 yeast wells that were positive for interaction with caspase-3 and selected a representative set of 286 positive wells for further study (Dejosez et al. 2008). Cognizant of the overwhelming number of proteins that could at least distantly be associated with pluripotency, we performed confirmation experiments that included an in vitro transcription/translation caspase-3 cleavage assay yielding cDNAs coding for proteins that were specific targets of caspase-3 or at least were associated with the protease. Among the inner circle of resultant genes was a cDNA encoding a protein that bore a striking resemblance to the *Drosophila* P-element transposase. Named Ronin (a masterless Japanese Samurai) to indicate its lack of any apparent relationship to known pluripotency factors, this protein possessed a THAP (thanatos-associated protein) domain at the amino terminus (Roussigne et al. 2003a,b; Macfarlan 2005; Quesneville et al. 2005) and therefore could be expected to participate in sequence-specific DNA binding and epigenetic silencing of gene expression, rendering it an attractive candidate for a role in ES cell development.

STRUCTURAL FEATURES OF RONIN

Ronin's THAP domain is remarkable in that is appears to be part of a "domesticated" DNA transposon. The Ronin gene located on chromosomes 16q22.1 and 8D3 in humans and mice, respectively, contains only one exon, suggesting that there is strong selective pressure against its interruption with introns. First characterized in 2003, the THAP domain is shared by numerous proteins across all animal species (~200 proteins with this domain have been identified so far) (Roussigne et al. 2003a,b). The THAP domain zinc-finger-containing DNA-binding protein motif is characterized by a C2CH signature (Cys-Xaa_{2-4}-Cys-Xaa_{33-50}-Cys-X-aa_2). Other essential residues are P26, W36, F38, and P78. The three cysteines and the histidine bind a central zinc atom, with the two β-sheets folded against one another while being separated by an amino acid chain that is unusually long for zinc fingers and contains an α-helix that is partly responsible for the sequence specificity of DNA binding. In contrast to other zinc finger motifs, the THAP domain contains a second region of amino acids that is also important for DNA sequence-specific binding. It is located toward the end of the domain, which lies in the three-dimensional structure exactly parallel to the α-helix (the sequence is AVPTIF) (Liew et al. 2007; Bessière et al. 2008). One of the most widely known members of the THAP domain family, the *Drosophila* P-element transposase, exhibits site-specific DNA binding to a consensus sequence at both the 5' and 3' regions of the transposase. The human genome contains 12 THAP-domain-containing proteins, of which only DAP4/p52rIPK, THAP1, and THAP7 have been characterized in detail (the mouse genome contains five THAP-domain-containing proteins). *Caenorhabditis elegans* expresses several important THAP proteins, including Lin-15b (Chesney et al. 2006) and Lin-36 (encoded by the class-B synthetic multivulva genes) (Clouaire et al. 2005), CDC-14B and CTB-1 (an ortholog of CtBP-1), all of which are involved in transcriptional control, and the protein HIM-17 (Reddy and Villeneuve 2004), which is involved in meiotic recombination and recruitment of the methyltransferase activity of histone H3 at lysine 9. Evidence for active DNA transposons containing a THAP domain has been obtained in zebra fish, and it is clear that this class of transposons was "domesticated" in a common ancestor of birds and mammals (Hammer et al. 2005; Quesneville et al. 2005). The most likely explanation for the wide distribution of THAP-containing protein domains across animal species is that they initially formed as a result of exon shuffling after DNA transposition of the first exon (containing the THAP domain) into other genes. These genes then acquired the DNA-binding activity of the DNA transposon and evolved further. It should be noted that this "domestication" of selfish DNA elements has precedence in evolutionary history; the most famous example is a DNA transposase that became the RAG-1/RAG-2 DNA recombination protein in mammals (Hammer et al. 2005; Quesneville et al. 2005).

Like most other THAP-domain-containing proteins, Ronin contains this motif at the amino terminus, followed by a polyglutamine stretch of 29 glutamine residues. These residues are encoded by CAGs occasionally interrupted by CAAs, suggesting that this repetitive sequence possesses at least some stability. Because glutamine chains of this size (~ 30 amino acids) are known to be relatively stiff, this conformation suggests that the polyglutamine tract separates the DNA-binding activity from the function of the carboxyl terminus.

RONIN EXPRESSION PATTERN DURING DEVELOPMENT

We searched for expression of Ronin in mouse and human ES cells and were able to detect its expression at the RNA level in the form of an approximately 1.8-kb transcript. Using a series of antibodies, we were successful in detecting Ronin protein in tissue sections and in cells stained in vitro. In the adult mouse, Ronin expression was mostly restricted to ovary and testes. Antibody staining of the ovary revealed high levels of Ronin in the ooplasm, although equal distribution of the protein was found across the entire oocyte after ovulation. We could not identify *lacZ*-positive cells in any of the adult organs tested (including brain, lung, thymus, heart, liver, gastrointestinal tract, bone marrow, skeletal muscle, and skin). We therefore concluded that Ronin expression is likely restricted to the earliest stages of development. Indeed, further antibody testing detected significant amounts of Ronin in the fertilized zygote. The *lacZ* reporter mouse lines showed increased Ronin expression in the inner cell mass of the blastocyst, whereas analysis of E8.5 and E10.5 embryos revealed substantial staining in specific organs such as the genital ridge, the heart, and the central nervous system, suggesting a specific role for Ronin during these phases of development. Importantly, gene expression studies with an inducible cell line showed that Ronin is exclusively a nuclear protein. It did not colocalize with DAPI (4'-6-diamidino-2-phenylindole)-stained areas in the nucleus or with polymerase II, indicating that it is primarily localized in silent regions of the genome. These findings contrast with the detection of Ronin in the cytoplasm during other stages of development, underscoring the complexity of its expression pattern and the need for a more detailed analysis of this feature in the future.

RONIN KNOCKOUT INDUCES LETHAL DEFECTS IN PRE- AND PERI-IMPLANTATION EMBRYOS

We created a Ronin knockout mouse by gene targeting. Although none of the pups or E10.5 and E7.5 embryos were *Ronin*[−/−], we did find a substantial number of empty deciduas on day E7.5, suggesting that a decidua formation was initiated but did not result in an embryo. In contrast, a proportion of the E3.5 embryos (blastocysts) had a *Ronin*[−/−] genotype. These blastocysts lacked obvious morphologic defects, and TUNEL (terminal deoxynucleotidyl-mediated nick-end labeling) staining did not detect an increase in apoptotic cells. To test the outgrowth potential of the inner cell mass, we placed the embryos in plastic dishes with medium derived from ES cell cultures. Whereas a significant majority of *Ronin*[+/-] and *Ronin*[+/+] embryos were able to attach and formed the typical flattened trophoblast epithelium and the inner cell mass, *Ronin*[-/-] blastocyst formed only the trophectoderm layer, suggesting a severe defect in outgrowth of the inner cell mass. Although this deficiency could be attributed to an abnormality in the trophoblastic cells leading to failure to support the inner cell mass, we argue against this interpretation because *Ronin* is not expressed in trophoblastic cells and its knockout in ES cells is lethal. Thus, Ronin appears to be essential for normal mammalian development but only during its earliest stages.

ECTOPIC EXPRESSION OF *RONIN* IN ES CELLS INDUCES A DIFFERENTIATION DEFECT AND GLOBAL TRANSCRIPTIONAL REPRESSION

We also generated a novel mouse ES cell line that ectopically expresses *Ronin* under a constitutive promoter. We noticed that such ES cells possess essentially the same morphology as that of normal ES cells but typically failed to differentiate in vitro and in immunocompromised mice. This antidifferentiation effect of Ronin is analogous to the ability of Nanog to promote self-renewal under unfavorable conditions (Chambers et al. 2003) and is observed in vitro only when ES cells are grown in the absence of leukemia-inhibitory factor (LIF), an essential factor for self-renewal that triggers signaling through the STAT (signal transducer and activator of transcription) factor. Ronin can substitute for the LIF-STAT pathway in ES cell cultures, even over two passages when the ES cells are propagated at clonal densities, similar to Nanog.

The nature of the differentiation defect in ES cells overexpressing *Ronin* was unclear, prompting us to perform microarray studies of ES cells with only transient expression of *Ronin*. This experiment revealed a striking shift in the expression of many genes, some of which have recognized functions in development, whereas measurement of newly synthesized RNA levels after induction of Ronin expression indicated pronounced decreases in the expression of key developmental genes. Thus, Ronin could be involved in the control of RNA transcription of multiple genes with roles in cell differentiation. This repressive activity could be exerted by binding directly to DNA or, as indicated by our protein–protein interaction data, by directly recruiting HCF-1 and subsequently other proteins with the ability to modify chromatin, such as mixed-lineage leukemia (MLL), Set1 (histone H3K4 methyltransferase), Sin3, histone deacetylase (HDAC), and histone acetyltransferase (HAT) (previously associated with chromatin modification) (Wysocka et al. 2003; Yokoyama et al. 2004), to sites of specific genes contributing to ES cell differentiation. The exact manner in which this putative protein complex suppresses gene transcription is unclear. In addition to its role as a transcriptional repressor, HCF-1 has been linked to regulation of the cell cycle; however, both acute and chronic expression of *Ronin* in ES cells as well as

somatic cells have at most only a marginal effect on the proliferation of ES cells, making it unlikely that the interaction of Ronin with HCF-1 contributes to cell cycle control.

WHAT DOES RONIN TEACH US ABOUT PLURIPOTENCY?

A very elegant concept of pluripotency is that of a default state whereby a core set of transcription factors controls both the transcriptional and epigenetic landscape (Boyer et al. 2006). This would imply that pluripotency occurs by default and in a hierarchical fashion when these "core" transcription factors activate other transcription factors, as well as epigenetic modifiers that control the pluripotent state. This theory is supported by the finding that specific epigenetic factors, such as the histone demethylases Jmjd1a and Jmjd2c, are essential for pluripotency and transcriptionally regulated by Oct4 (Loh et al. 2007). This transcription-factor-centric hierarchical model is further corroborated by the recent discovery that forced expression of *Oct4, Sox2, Klf4*, and *c-Myc* in differentiated cells leads to a pluripotent state similar to that in ES cells. Although this concept is attractive, it focuses entirely on Oct4/Sox2 as the main axis of pluripotency and discounts numerous biological examples showing that particular states are often established and maintained by independent pathways in parallel. Thus, one could predict the existence of pluripotency factors (either transcription factors or epigenetic modulators) that are (1) not directly downstream from the known core pluripotency factor genes (*Oct4, Sox2*, and *Nanog*) and do not positively affect their expression and (2) still essential and even sufficient to sustain pluripotency. We suggest that our newly identified factor, Ronin, belongs in this category and may represent an entirely different mechanism by which pluripotency could be achieved and maintained (Fig. 2).

The expression pattern of Ronin shows some major differences compared to those of the other pluripotency factors, including Oct4, Nanog, and Sox2. Ronin is clearly expressed during the earliest stages of development, but it does not disappear rapidly at the blastocyst stage. Indeed, we detected Ronin expression in the adult ovary and, more interestingly, in the hippocampus, olfactory region, subventricular zone, and cerebellum of the adult brain, suggesting specific roles in these regions. Thus, we propose that Ronin is expressed in specific cells during development and in the adult, where it is either required to fulfill roles analogous to those during early embryonic development or has distinct functions that remain undefined. Expression of Ronin, at least in the brain, indicates that the protein participates in other stem cell/progenitor systems, a concept that has been applied to the zinc finger protein Zfx (Galan-Caridad et al. 2007; Chen et al. 2008). The hypothesis that Ronin is expressed in stem cell progenitor populations in the developing embryo requires further investigation.

Numerous issues concerning the functional roles of Ronin remain unresolved. For instance, the protein appears to recognize and bind to DNA in a specific manner, but its mechanism of action is still undefined. Several members of the THAP protein family regulate cells at the

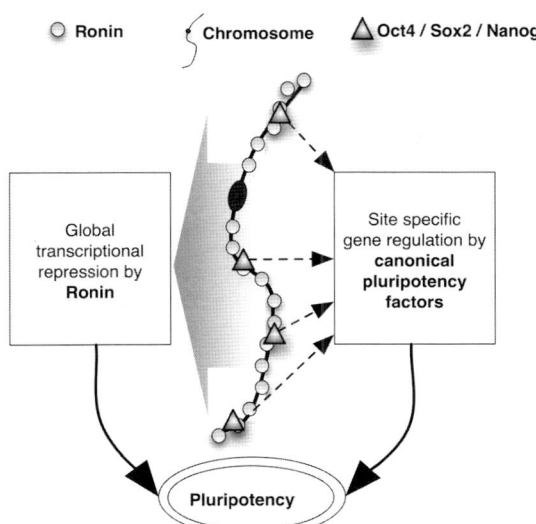

Figure 2. Proposed model for the mechanism of Ronin function. In contrast to the site-specific gene modulatory activity of canonical pluripotency factors (e.g., Oct4, Sox2, and Nanog), Ronin appears to repress transcription more broadly to ensure the maintenance of ES cell pluripotency. (Reprinted from Dejosez et al. 2008 [© Cell Press].)

epigenetic level (Reddy and Villeneuve 2004; Macfarlan 2005; Macfarlan et al. 2005), suggesting that epigenetic silencing of gene expression is a function of Ronin as well. It will be important to distinguish between these two functions in future efforts to explain the mechanisms by which Ronin sustains pluripotency. Finally, because our discovery of Ronin has several implications for studies of pluripotency, as mentioned earlier, the full repertoire of factors involved in establishing or maintaining pluripotency, or in reprogramming somatic cells to an ES-like state, is unknown. In our opinion, focusing exclusively on gene expression profiles is unlikely to close this gap in the near future, whereas screening cDNA libraries for candidate factors, using a fundamental strategy as outlined in this chapter, could yield much more rapid insights into the constituent molecules of pluripotency pathways.

TOWARD UNDERSTANDING CELLULAR IDENTITY

Deciphering the transcription factor code underlying particular cellular states will be essential for experimentally changing these states and attempting to understand naturally occurring transitions during development and in the adult organism. Perhaps the most significant impact of identifying new caspase targets such as Nanog and Ronin is that it provides a much-needed window into the complexity of transcriptional networks that give rise to cellular identity and increases the set of cell-fate-determining transcription factors that could be introduced into mammalian cells to reprogram them to the phenotype of another cell type. The most extreme form of such cellular "shape shifting" was the recent demonstration that well-

differentiated cells, such as fibroblasts, could be reprogrammed to a state akin to that of ES cells, the most primitive cell type sustainable in cell culture. The practical implications of using *trans*-differentiation or reprogramming factors to change cellular identities in any direction are enormous and most likely will revise the research directions of many different fields of biology. Equally critical is the potential impact of these discoveries on our comprehension of how cells differ from one another and why most cells retain their identity for very long periods of time without appreciable changes. It is not unreasonable to predict that the new understanding of cellular identity will enhance our ability to produce therapeutically relevant cell types that could be administrated to a broad range of patients whose diseases reflect malfunctions in the cellular circuitry regulating large blocks of gene expression. An obvious question that must be addressed before the promises of the genetic reprogramming can be realized is whether the emerging observations on control of the pluripotent state are in fact relevant to the developing embryo and the adult organism or merely phenomena induced by the introduction of artificial factors.

At the conceptional level, it is intriguing to consider transcription-factor-mediated shifts in cell lineages in the context of another compelling theory, epigenetic ground states. The epigenetic ground state has its roots in experimental observations of epigenetic marks, in particular K4 and K27 histone methylation and DNA methylation, and the observation that some of these epigenetic marks can be found in association with specific DNA sequences in undifferentiated ES cells (Bernstein et al. 2006). The prediction is that discrete genomic DNA sequences specifically encode signatures that can be recognized by epigenetic factors that then impose specific epigenetic alterations (Bernstein et al. 2007). An obvious extension of this idea is that there are specific determinants in the genome not only for pluripotent undifferentiated cells, but eventually also for every cellular state in the body. Specific subsets of cell-type-specific transcription factors could therefore be expected to initiate the transcriptional program that imposed the epigenetic changes associated with a particular cellular state or developmental stage. This model is analogous to a holographic image where every point of the image contains the information of the entire image, even though only one perspective is revealed to the viewer, depending on the angle of the laser light. Thus, just as a change in light angle produces a change in the holograph, different sets of transcription factors could be expected to induce a different phenotype, even though all cells store the same genetic information. If multiple ground states are indeed responsible for cellular identities, and if these states are determined by specific transcriptional programs, then one could reasonably postulate the existence of master regulators that dictate cellular identity. Logically then, such regulators would be capable of overriding preexisting cellular states at least under certain circumstances. Even though such transitions are still purely theoretical, their experimental confirmation would challenge the fundamental dogma of developmental biology: lineage linearity throughout development, whereby cells imperatively transit through specific stages that dictate subsequent cell fates through the expression of morphogens or

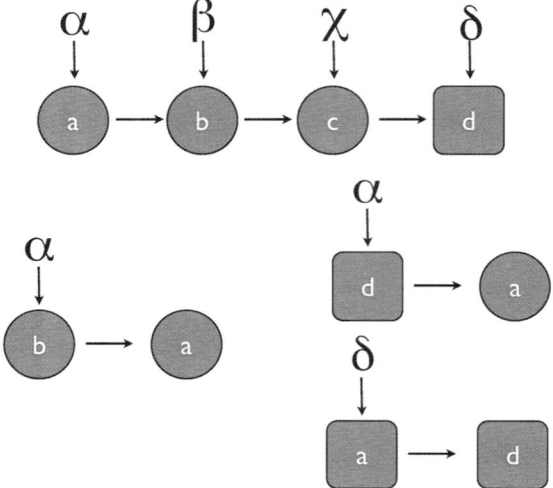

Figure 3. Until now, we have assumed that a developing and differentiating cell goes through different and defined steps to terminally differentiate (*a–d*), that a specific cascade of transcription and epigenetic factors allows progression to the next step, and that the order of steps is critical for successful differentiation. However, the fact that cells can "jump" from a state of terminal differentiation to one of complete undifferentiation indicates that the concept of lineage linearity can be bypassed, at least artificially in the laboratory. According to this concept, master regulators (α, β, χ, and δ) can dictate the cellular state.

other developmental factors. We would argue that developmental linearity and sequentiality, while appearing critical for morphogenesis, are not necessarily a requirement for cellulogenesis, in which master transcriptional regulators and epigenetic ground states could have the central role (Fig. 3). Hence, the most challenging task for the future will be to devise novel strategies that could be used to identify critical regulatory factors at any given step in the acquisition of cellular identity. We believe that our discovery of caspases as regulatory elements specifically targeting transcription factors involved in the shaping of cellular identities is an important first step in this direction.

ACKNOWLEDGMENTS

The author is supported by the Diana Helis Henry Medical Research Foundation, the Huffington Foundation, and the National Institutes of Health.

REFERENCES

Arama, E., Agapite, J., and Steller, H. 2003. Caspase activity and a specific cytochrome C are required for sperm differentiation in *Drosophila*. *Dev. Cell* **4:** 687–697.

Avilion, A.A., Nicolis, S.K., Pevny, L.H., Perez, L., Vivian, N., and Lovell-Badge, R. 2003. Multipotent cell lineages in early mouse development depend on SOX2 function. *Genes Dev.* **17:** 126–140.

Bernstein, B.E., Mikkelsen, T.S., Xie, X., Kamal, M., Huebert, D.J., Cuff, J., Fry, B., Meissner, A., Wernig, M., Plath, K., et al. 2006. A bivalent chromatin structure marks key developmental genes in embryonic stem cells. *Cell* **125:** 315–326.

Bernstein, B.E., Meissner, A., and Lander, E.S. 2007. The mammalian epigenome. *Cell* **128:** 669–681.

Bessière, D., Lacroix, C., Campagne, S., Ecochard, V., Guillet, V., Mourey, L., Lopez, F., Czaplicki, J., Demange, P., Milon, A., et al. 2008. Structure-function analysis of the THAP zinc finger of THAP1, a large C2CH DNA-binding module linked to Rb/E2F pathways. *J. Biol. Chem.* **283:** 4352–4363.

Boyer, L.A., Lee, T.I., Cole, M.F., Johnstone, S.E., Levine, S.S., Zucker, J.P., Guenther, M.G., Kumar, R.M., Murray, H.L., Jenner, R.G., et al. 2005. Core transcriptional regulatory circuitry in human embryonic stem cells. *Cell* **122:** 947–956.

Boyer, L.A., Mathur, D., and Jaenisch, R. 2006. Molecular control of pluripotency. *Curr. Opin. Genet. Dev.* **16:** 455–462.

Chambers, I., Colby, D., Robertson, M., Nichols, J., Lee, S., Tweedie, S., and Smith, A. 2003. Functional expression cloning of nanog, a pluripotency sustaining factor in embryonic stem cells. *Cell* **113:** 643–655.

Chambers, I., Silva, J., Colby, D., Nichols, J., Nijmeijer, B., Robertson, M., Vrana, J., Jones, K., Grotewold, L., and Smith, A. 2007. Nanog safeguards pluripotency and mediates germline development. *Nature* **450:** 1230–1234.

Chen, X., Xu, H., Yuan, P., Fang, F., Huss, M., Vega, V.B., Wong, E., Orlov, Y.L., Zhang, W., Jiang, J., et al. 2008. Integration of external signaling pathways with the core transcriptional network in embryonic stem cells. *Cell* **133:** 1106–1117.

Chesney, M.A., Kidd III, A.R., and Kimble, J. 2006. gon-14 functions with class B and class C synthetic multivulva genes to control larval growth in *Caenorhabditis elegans*. *Genetics* **172:** 915–928.

Clouaire, T., Roussigne, M., Ecochard, V., Mathe, C., Amalric, F., and Girard, J.P. 2005. The THAP domain of THAP1 is a large C2CH module with zinc-dependent sequence-specific DNA-binding activity. *Proc. Natl. Acad. Sci.* **102:** 6907–6912.

De Botton, S., Sabri, S., Daugas, E., Zermati, Y., Guidotti, J.E., Hermine, O., Kroemer, G., Vainchenker, W., and Debili, N. 2002. Platelet formation is the consequence of caspase activation within megakaryocytes. *Blood* **100:** 1310–1317.

Dejosez, M., Krumenacker, J.S., Zitur, L.J., Passeri, M., Chu, L.F., Songyang, Z., Thomson, J.A., and Zwaka, T.P. 2008. Ronin is essential for embryogenesis and the pluripotency of mouse embryonic stem cells. *Cell* **133:** 1162–1174.

De Maria, R., Testa, U., Luchetti, L., Zeuner, A., Stassi, G., Pelosi, E., Riccioni, R., Felli, N., Samoggia, P., and Peschle, C. 1999a. Apoptotic role of Fas/Fas ligand system in the regulation of erythropoiesis. *Blood* **93:** 796–803.

De Maria, R., Zeuner, A., Eramo, A., Domenichelli, C., Bonci, D., Grignani, F., Srinivasula, S.M., Alnemri, E.S., Testa, U., and Peschle, C. 1999b. Negative regulation of erythropoiesis by caspase-mediated cleavage of GATA-1. *Nature* **401:** 489–493.

Earnshaw, W.C., Martins, L.M., and Kaufmann, S.H. 1999. Mammalian caspases: Structure, activation, substrates, and functions during apoptosis. *Annu. Rev. Biochem.* **68:** 383–424.

Evans, M.J. and Kaufman, M.H. 1981. Establishment in culture of pluripotential cells from mouse embryos. *Nature* **292:** 154–156.

Fazzio, T.G., Huff, J.T., and Panning, B. 2008. An RNAi screen of chromatin proteins identifies Tip60-p400 as a regulator of embryonic stem cell identity. *Cell* **134:** 162–174.

Fernando, P. and Megeney, L.A. 2007. Is caspase-dependent apoptosis only cell differentiation taken to the extreme? *FASEB J.* **21:** 8–17.

Fujita, J., Crane, A.M., Souza, M.K., Dejosez, M., Kyba, M., Flavell, R.A., Thomson, J.A., and Zwaka, T.P. 2008. Caspase activity mediates the differentiation of embryonic stem cells. *Cell Stem Cell* **2:** 595–601.

Galan-Caridad, J.M., Harel, S., Arenzana, T.L., Hou, Z.E., Doetsch, F.K., Mirny, L.A., and Reizis, B. 2007. Zfx controls the self-renewal of embryonic and hematopoietic stem cells. *Cell* **129:** 345–357.

Hammer, S.E., Strehl, S., and Hagemann, S. 2005. Homologs of *Drosophila* P transposons were mobile in zebrafish but have been domesticated in a common ancestor of chicken and human. *Mol. Biol. Evol.* **22:** 833–844.

Ishizaki, Y., Jacobson, M.D., and Raff, M.C. 1998. A role for caspases in lens fiber differentiation. *J. Cell Biol.* **140:** 153–158.

Jaenisch, R. and Young, R. 2008. Stem cells, the molecular circuitry of pluripotency and nuclear reprogramming. *Cell* **132:** 567–582.

Janzen, V., Fleming, H.E., Riedt, T., Karlsson, G., Riese, M.J., Lo Celso, C., Reynolds, G., Milne, C.D., Paige, C.J., Karlsson, S., et al. 2008. Hematopoietic stem cell responsiveness to exogenous signals is limited by caspase-3. *Cell Stem Cell* **2:** 584–594.

Kamada, S., Kusano, H., Fujita, H., Ohtsu, M., Koya, R.C., Kuzumaki, N., and Tsujimoto, Y. 1998. A cloning method for caspase substrates that uses the yeast two-hybrid system: Cloning of the antiapoptotic gene *gelsolin*. *Proc. Natl. Acad. Sci.* **95:** 8532–8537.

Klochendler-Yeivin, A., Fiette, L., Barra, J., Muchardt, C., Babinet, C., and Yaniv, M. 2000. The murine SNF5/INI1 chromatin remodeling factor is essential for embryonic development and tumor suppression. *EMBO Rep.* **1:** 500–506.

Lee, T.I., Jenner, R.G., Boyer, L.A., Guenther, M.G., Levine, S.S., Kumar, R.M., Chevalier, B., Johnstone, S.E., Cole, M.F., Isono, K., et al. 2006. Control of developmental regulators by Polycomb in human embryonic stem cells. *Cell* **125:** 301–313.

Liew, C.K., Crossley, M., Mackay, J.P., and Nicholas, H.R. 2007. Solution structure of the THAP domain from *Caenorhabditis elegans* C-terminal binding protein (CtBP). *J. Mol. Biol.* **366:** 382–390.

Loh, Y.H., Zhang, W., Chen, X., George, J., and Ng, H.H. 2007. Jmjd1a and Jmjd2c histone H3 Lys 9 demethylases regulate self-renewal in embryonic stem cells. *Genes Dev.* **21:** 2545–2557.

Lowry, W.E., Richter, L., Yachechko, R., Pyle, A.D., Tchieu, J., Sridharan, R., Clark, A.T., and Plath, K. 2008. Generation of human induced pluripotent stem cells from dermal fibroblasts. *Proc. Natl. Acad. Sci.* **105:** 2883–2888.

Macfarlan, T. 2005. "Identification and initial characterization of thanatos associated protein 7 (THAP7), a putative transcriptional repressor." Ph.D. thesis. University of Pennsylvania, Philadelphia.

Macfarlan, T., Kutney, S., Altman, B., Montross, R., Yu, J., and Chakravarti, D. 2005. Human THAP7 is a chromatin-associated, histone tail-binding protein that represses transcription via recruitment of HDAC3 and nuclear hormone receptor corepressor. *J. Biol. Chem.* **280:** 7346–7358.

Martin, G.R. 1981. Isolation of a pluripotent cell line from early mouse embryos cultured in medium conditioned by teratocarcinoma stem cells. *Proc. Natl. Acad. Sci.* **78:** 7634–7638.

Masui, S., Nakatake, Y., Toyooka, Y., Shimosato, D., Yagi, R., Takahashi, K., Okochi, H., Okuda, A., Matoba, R., Sharov, A.A., et al. 2007. Pluripotency governed by Sox2 via regulation of Oct3/4 expression in mouse embryonic stem cells. *Nat. Cell Biol.* **9:** 625–635.

Nichols, J., Zevnik, B., Anastassiadis, K., Niwa, H., Klewe-Nebenius, D., Chambers, I., Scholer, H., and Smith, A. 1998. Formation of pluripotent stem cells in the mammalian embryo depends on the POU transcription factor Oct4. *Cell* **95:** 379–391.

Okita, K., Ichisaka, T., and Yamanaka, S. 2007. Generation of germline-competent induced pluripotent stem cells. *Nature* **448:** 313–317.

Park, I.H., Zhao, R., West, J.A., Yabuuchi, A., Huo, H., Ince, T.A.,

Lerou, P.H., Lensch, M.W., and Daley, G.Q. 2008. Reprogramming of human somatic cells to pluripotency with defined factors. *Nature* **451:** 141–146.

Pedersen, R.A. 1986. Potency, lineage and allocation in preimplantation mouse embryos. In *Experimental approaches to mammalian embryonic development* (ed. J. Rossant and R.A. Pedersen), pp. 3–33. Cambridge University Press, New York.

Quesneville, H., Nouaud, D., and Anxolabehere, D. 2005. Recurrent recruitment of the THAP DNA-binding domain and molecular domestication of the P-transposable element. *Mol. Biol. Evol.* **22:** 741–746.

Reddy, K.C. and Villeneuve, A.M. 2004. *C. elegans* HIM-17 links chromatin modification and competence for initiation of meiotic recombination. *Cell* **118:** 439–452.

Roussigne, M., Cayrol, C., Clouaire, T., Amalric, F., and Girard, J.P. 2003a. THAP1 is a nuclear proapoptotic factor that links prostate-apoptosis-response-4 (Par-4) to PML nuclear bodies. *Oncogene* **22:** 2432–2442.

Roussigne, M., Kossida, S., Lavigne, A.C., Clouaire, T., Ecochard, V., Glories, A., Amalric, F., and Girard, J.P. 2003b. The THAP domain: A novel protein motif with similarity to the DNA-binding domain of P element transposase. *Trends Biochem. Sci.* **28:** 66–69.

Takahashi, K. and Yamanaka, S. 2006. Induction of pluripotent stem cells from mouse embryonic and adult fibroblast cultures by defined factors. *Cell* **126:** 663–676.

Takahashi, K., Tanabe, K., Ohnuki, M., Narita, M., Ichisaka, T., Tomoda, K., and Yamanaka, S. 2007. Induction of pluripotent stem cells from adult human fibroblasts by defined factors. *Cell* **131:** 861–872.

Thomas, K.R. and Capecchi, M.R. 1987. Site-directed mutagenesis by gene targeting in mouse embryo-derived stem cells. *Cell* **51:** 503–512.

Thomson, J.A., Itskovitz-Eldor, J., Shapiro, S.S., Waknitz, M.A., Swiergiel, J.J., Marshall, V.S., and Jones, J.M. 1998. Embryonic stem cell lines derived from human blastocysts. *Science* **282:** 1145–1147.

Thornberry, N.A. and Lazebnik Y. 1998. Caspases: Enemies within. *Science* **281:** 1312–1316.

Wernig, M., Meissner, A., Foreman, R., Brambrink, T., Ku, M., Hochedlinger, K., Bernstein, B.E., and Jaenisch, R. 2007. In vitro reprogramming of fibroblasts into a pluripotent ES-cell-like state. *Nature* **448:** 318–324.

Wysocka, J., Myers, M.P., Laherty, C.D., Eisenman, R.N., and Herr, W. 2003. Human Sin3 deacetylase and trithorax-related Set1/Ash2 histone H3-K4 methyltransferase are tethered together selectively by the cell-proliferation factor HCF-1. *Genes Dev.* **17:** 896–911.

Yokoyama, A., Wang, Z., Wysocka, J., Sanyal, M., Aufiero, D.J., Kitabayashi, I., Herr, W., and Cleary, M.L. 2004. Leukemia proto-oncoprotein MLL forms a SET1-like histone methyltransferase complex with menin to regulate *Hox* gene expression. *Mol. Cell. Biol.* **24:** 5639–5649.

Yu, J., Vodyanik, M.A., He, P., Slukvin, I.I., and Thomson, J.A. 2006. Human embryonic stem cells reprogram myeloid precursors following cell-cell fusion. *Stem Cells* **24:** 168–176.

Zwaka, T.P. and Thomson, J.A. 2003. Homologous recombination in human embryonic stem cells. *Nat. Biotechnol.* **21:** 319–321.

Common Themes of Dedifferentiation in Somatic Cell Reprogramming and Cancer

G.Q. DALEY

Division of Pediatric Hematology and Oncology, Children's Hospital Boston and Dana-Farber Cancer Institute; Department of Biological Chemistry and Molecular Pharmacology, Harvard Medical School; Division of Hematology, Brigham and Women's Hospital; Harvard Stem Cell Institute; Howard Hughes Medical Institute, Karp Family Research Building 7214, Boston, Massachusetts 02115

With its hallmarks of unregulated cell proliferation and compromised differentiation, cancer represents a derangement of normal tissue homeostasis. A common set of pathways are activated in the transformed state, through either mutation or altered epigenetic regulation, and both heritable effects sustain the tumor. Classical views of cancer have invoked tissue dedifferentiation in the oncogenic process, whereas modern views embodied in the cancer stem cell hypothesis hold that cancer emerges from primitive tissue stem cells or specific progenitor populations that through mutations assume the self-renewal properties of stem cells. Recently, somatic tissues have been reprogrammed to a pluripotent state resembling embryonic stem (ES) cells by ectopic expression of a cocktail of transcription factors. The factors that drive reprogramming are oncogenes or have been linked to cellular transformation, suggesting that tumorigenesis and somatic cell reprogramming might indeed share common mechanisms of dedifferentiation.

Cancer is the quintessential clonal disorder, arising in a single cell that over time sustains a series of mutations in genes that govern cellular growth control, survival, and cell–cell interactions. During the last several decades, cancer researchers have focused on the nature of oncogenic mutations and their effects on the function of signal transduction pathways and states of cellular differentiation. The assembly of a catalog of oncogenic mutations has led to a host of effective anticancer agents directed against specific targets essential to the cancer cell.

More recently, a major goal of cancer research has been to define the specific cell of origin of cancer among the hierarchy of cells that constitute a specific tissue. It is well documented that tissues such as the skin, gut, and blood arise from a rare stem cell that self-renews and whose daughters undergo differentiation into a variety of specialized progeny that constitute the tissue mass. Cancers in these tissues might thus arise from mutations in the tissue stem cell itself, which already possess the machinery for continuous proliferation, or by alterations in the physiology of various stages of progenitor cells that have limited self-renewal potential and would otherwise be committed to differentiation. The "cancer stem cell hypothesis" suggests that the cancer is sustained by a cell with stem-like properties of self-renewal and multilineage differentiation (Reya et al. 2001). The implication is that only rare cells within the tumor are critical to the maintenance of the tumor, and this has considerable implications for understanding the origins and treatment of malignancy.

The disease chronic myelogenous leukemia (CML) exemplifies many aspects of the cancer stem cell hypothesis (Daley 2004). The initial phase of the disease arises in the hematopoietic stem cell itself, by virtue of sustaining the Philadelphia chromosome translocation, but the more aggressive blast crisis stage appears to arise by mutation in more differentiated cells such as the granulocyte-macrophage progenitor (GMP) (Jamieson et al. 2004). Whereas normal GMPs lack self-renewal potential, the leukemic GMPs of blast crisis CML have activated a gene expression program that mimics that of hematopoietic stem cells, which entail heightened function of β-catenin. Analysis of cases of human acute myeloid leukemia (AML) identifies the hematopoietic stem cell as the likely cell of origin of the initiating oncogenic translocations, but this population is essentially nonleukemic; the leukemic population is a more committed progenitor that has acquired additional cooperating oncogenic lesions (Miyamoto et al. 2000). Transformation by the MLL-AF9 oncogene associated with AML represents another model wherein expression in GMPs induces a self-renewal gene expression signature typical of hematopoietic stem cells, meaning that the tissue progenitors assume the replicative properties of the tissue stem cells (Krivtsov et al. 2006). Accumulating evidence in animal models of solid tumors suggests that the transit-amplifying cells of a tissue may frequently be the target population for oncogenic mutations, and the resulting cancer stem cell acquires features not only of self-renewal, but of a dedifferentiated phenotype. Unraveling the molecular mechanisms of tumorigenesis, in the context of the specific target cells involved, will be critical for developing therapies that target cancer stem cells while salvaging the function of normal tissue stem cells.

The traditional notion that differentiated cells become "dedifferentiated" in cancer has been controversial, even more so since the wider appreciation that rare stem cells persist in many adult tissues and that these stem cells can act as a reservoir of oncogenic mutations. A rich literature has documented the reactivation of primitive programs of tissue gene expression, most notably the "cancer-testis antigens" that characterize a number of human tumors and that represent novel targets for cancer immunotherapy (Simpson et al. 2005). Recent studies of a number of highly malignant, poorly differentiated solid tumors have identified similarities in gene expression programs with embryonic stem cells (Ben-Porath et al. 2008), highlighting the fact that tumorigenesis can hijack embryonic pathways of tissue development. Indeed, morphogenetic changes exemplified by the epithelial-mesenchymal transition (EMT) of gastrulation are replayed in the evolution of primary tumors toward metastatic phenotypes and typically invoke the same players: the transcription factors snail, slug, and twist (Yang et al. 2004; Huber et al. 2005; Baum et al. 2008; Natalwala et al. 2008). In the context of mammary epithelial cells, EMT is associated with acquisition of stem cell markers and stem cell properties, including enhanced mammosphere formation and the capacity to form tumors, establishing a link between the EMT and acquisition of epithelial stem cell qualities (Mani et al. 2008).

Pathologists examining tumor tissue have long described the histologic appearance of various tumors as "dedifferentiated," reflecting the expansion of primitive less-specialized malignant cell types. In cases where cancers arise from primitive tissue stem cells that are otherwise rare but become morphologically apparent after extensive proliferation or from immature transit-amplifying progenitors, the description of dedifferentiation in fact reflects expansion of an undifferentiated cell population, rather than a dedifferentiated cell population that resides normally within the tissues. Dedifferentiation would apply only in cases where a more specialized cell type literally extinguishes expression of lineage-specific genes of specialized tissue function, in favor of expression of the more primitive program of tissue development, perhaps accompanied by a block in differentiation that makes the transition apparent. Indeed, it is likely, at least in some cancers, that alterations in differentiated progenitor populations result in reversion to the stem cell phenotype. Evidence for such a tissue transition must come from experimental systems that prospectively identify the cell of origin of the tumor as having already acquired markers of tissue specialization.

Insights into common mechanisms of dedifferentiation and cancer have arisen unexpectedly from studies of direct cellular reprogramming to the pluripotent state. Yamanaka's pioneering work has demonstrated that ectopic expression of a set of transcription factors (Oct4, Sox2, Myc, and Klf4) can reset the epigenetic state of differentiated somatic cells to pluripotency (Takahashi and Yamanaka 2006; Takahashi et al. 2007). The alternative pluripotency factors Nanog and Lin28 can be substituted for Myc and KLF4 to reprogram human somatic cells (Yu et al. 2007b). Of great interest, these reprogramming factors have all been implicated in tumorigenesis, raising the provocative hypothesis that cellular transformation and reprogramming entail common pathways and thus represent variations on similar biological themes. Examining the factors that mediate somatic cell reprogramming shows how dedifferentiation may be a more

prominent element of tumorigenesis than previously appreciated.

Somatic cell reprogramming entails the erasure of gene expression programs characteristic of differentiated somatic cells and the reactivation of embryonic patterns of gene expression that are characteristic of the pluripotent state (Boyer et al. 2005, 2006). This genome-wide epigenetic transformation appears to be initiated by the Oct4 and Sox2 transcription factors, which bind to a large common set of target genes and alter chromatin structure. Aberrant expression of either Oct4 or Sox2 is associated with deranged tissue homeostasis and tumorigenesis (Hochedlinger et al. 2005; Chen et al. 2008), and nanog misexpression is characteristic of germ cell tumors (Clark 2007). Myc is among the most well-documented oncogenes, known for its role in promoting immortalization of tumor cells in part through direct actions on telomerase (Wang et al. 1998). Klf4 is associated with colorectal tumors (Wei et al. 2006), and a close relative of Lin28, Lin28B, is associated with hepatocellular cancer (Guo et al. 2006). Activation targets of Oct4, Sox2, Nanog, and Myc are frequently expressed in poorly differentiated tumors (Ben-Porath et al. 2008), again revealing a common molecular circuitry in reprogramming and cancer.

The case of *lin-28* in particular highlights a common pathway of reprogramming, dedifferentiation, and cancer. *lin-28* was originally identified in *C. elegans* as a gene regulating the developmental timing of larval stages (Ambros and Horvitz 1984). Later, *lin-28* was shown to reverse cell fates during vulval development, directly linking it to reprogramming (Euling and Ambros 1996). Recently, in studies aimed at understanding why certain microRNAs (miRs) were transcribed in embryonic stem cells but not expressed as functional miRs, we and other investigators showed that a prominent function of *lin-28* is to inhibit the *let-7* family of miRs by binding to the loop domain and inhibiting processing to the mature form (Newman et al. 2008; Piskounova et al. 2008; Rybak et al. 2008; Viswanathan et al. 2008). *let-7* has been shown to function as a tumor suppressor in a variety of cancers, including breast, lung, and colon (Takamizawa et al. 2004; Akao et al. 2006; Yu et al. 2007a), likely by inhibiting the expression of known oncogenes such as Ras, Myc, and HMGA2 (Johnson et al. 2005; Sampson et al. 2007; Yu et al. 2007a). To date, only a closely related homolog of *lin-28* (*lin-28B*) has been directly linked to hepatocellular carcinoma (Guo et al. 2006), a tumor type that has long been described as showing histopathologic dedifferentiation (Abelev 1971; Abelev and Lazarevich 2006). In unpublished studies, we have considerably extended the association of *lin-28* and *lin-28B* with a number of human malignancies, strongly suggesting that modulation of processing of the *let-7* family of tumor suppressors is a common feature of malignancy (S.R. Viswanathan et al., in prep.). *lin-28* is the only nontranscription factor to date to be implicated in reprogramming of human somatic cells to pluripotency (Yu et al. 2007b). Understanding precisely how *lin-28* functions in somatic cell reprogramming and tumorigenesis, whether by direct effects on *let-7* processing or perhaps other mechanisms such as mRNA stability (Bagga et al. 2005; Balzer and Moss 2007; Chendrimada et al. 2007), will shed light on how the dedifferentiation that is so evident in somatic cell reprogramming might also underlie tissue metaplasia, dysplasia, and tumor formation.

The predominant target cell for reprogramming studies has been the fibroblast, a motile mesenchymal cell type. Following ectopic expression of reprogramming factors, fibroblasts assume properties of epithelial cells, with tight cell–cell association through gap junctions and expression of E-cadherin reminiscent of mesenchymal-epithelial transitions in embryonic organ development. Interestingly, cells derived from gastric epithelia and keratinocytes appear to be reprogrammed more efficiently than fibroblasts (Aoi et al. 2008; Maherali et al. 2008). It will be revealing to probe how mediators of mesenchymal-epithelial transitions influence the reprogramming process.

Although there are common pathways activated during reprogramming and tumorigenesis, pluripotent stem cells and tumorigenic cells have important differences. A cardinal feature of pluripotent stem cells is the capacity to form teratomas, a distinct type of highly differentiated benign tumor that differs in fundamental ways from a malignancy (Lensch and Ince 2007; Lensch et al. 2007). Teratomas, by virtue of their predisposition to differentiation and tissue complexity, remain encapsulated and do not metastasize. Thus, whereas pluripotent cells are immortal, self-renew, and can form tumors, they fundamentally remain dependent on mitogenic signals in their environment and responsive to exogenous cues to differentiate. Teratocarcinomas, the malignant counterparts of teratoma, typically retain a degree of pluripotency but show more limited degrees of differentiation and behave more like malignant tumors in their potential for metastasis and tissue invasion. The critical distinctions between true cancer cells and reprogrammed somatic cells may be that reprogrammed cells remain genetically intact. Unraveling the critical distinctions—both genetic and epigenetic—between the pathways of tumorigenesis and somatic cell reprogramming promises to illuminate fundamental aspects of dedifferentiation that will be essential to understanding the safety of cell replacement products derived from pluripotent stem cells.

ACKNOWLEDGMENTS

Research in the author's laboratory was supported by grants from the National Institutes of Health (NIH), the NIH Director's Pioneer Award of the NIH Roadmap for Medical Research, and private funds contributed to the Harvard Stem Cell Institute and the Children's Hospital Stem Cell Program. G.Q.D. is a recipient of Clinical Scientist Awards in Translational Research from the Burroughs Wellcome Fund and the Leukemia and Lymphoma Society and is an Investigator of the Howard Hughes Medical Institute.

REFERENCES

Abelev, G.I. 1971. α-Fetoprotein in ontogenesis and its association with malignant tumors. *Adv. Cancer Res.* **14:** 295–358.
Abelev, G.I. and Lazarevich, N.L. 2006. Control of differentiation in progression of epithelial tumors. *Adv. Cancer Res.* **95:** 61–113.
Akao, Y., Nakagawa, Y., and Naoe, T. 2006. *let-7* microRNA functions as a potential growth suppressor in human colon cancer cells. *Biol. Pharm. Bull.* **29:** 903–906.
Ambros, V. and Horvitz, H.R. 1984. Heterochronic mutants of the nematode *Caenorhabditis elegans*. *Science* **226:** 409–416.
Aoi, T., Yae, K., Nakagawa, M., Ichisaka, T., Okita, K., Takahashi, K., Chiba, T., and Yamanaka, S. 2008. Generation of pluripotent stem cells from adult mouse liver and stomach cells. *Science* **321:** 699–702.
Bagga, S., Bracht, J., Hunter, S., Massirer, K., Holtz, J., Eachus, R., and Pasquinelli, A.E. 2005. Regulation by *let-7* and *lin-4* miRNAs results in target mRNA degradation. *Cell* **122:** 553–563.
Balzer, E. and Moss, E.G. 2007. Localization of the developmental timing regulator Lin28 to mRNP complexes, P-bodies and stress granules. *RNA Biol.* **4:** 16–25.
Baum, B., Settleman, J., and Quinlan, M.P. 2008. Transitions between epithelial and mesenchymal states in development and disease. *Semin. Cell Dev. Biol.* **19:** 294–308.
Ben-Porath, I., Thomson, M.W., Carey, V.J., Ge, R., Bell, G.W., Regev, A., and Weinberg, R.A. 2008. An embryonic stem cell-like gene expression signature in poorly differentiated aggressive human tumors. *Nat. Genet.* **40:** 499–507.
Boyer, L.A., Lee, T.I., Cole, M.F., Johnstone, S.E., Levine, S.S., Zucker, J.P., Guenther, M.G., Kumar, R.M., Murray, H.L., Jenner, R.G., et al. 2005. Core transcriptional regulatory circuitry in human embryonic stem cells. *Cell* **122:** 947–956.
Boyer, L.A., Plath, K., Zeitlinger, J., Brambrink, T., Medeiros, L.A., Lee, T.I., Levine, S.S., Wernig, M., Tajonar, A., Ray, M.K., et al. 2006. Polycomb complexes repress developmental regulators in murine embryonic stem cells. *Nature* **441:** 349–353.
Chen, Y., Shi, L., Zhang, L., Li, R., Liang, J., Yu, W., Sun, L., Yang, X., Wang, Y., Zhang, Y., et al. 2008. The molecular mechanism governing the oncogenic potential of SOX2 in breast cancer. *J. Biol. Chem.* **283:** 17969–17978.
Chendrimada, T.P., Finn, K.J., Ji, X., Baillat, D., Gregory, R.I., Liebhaber, S.A., Pasquinelli, A.E., and Shiekhattar, R. 2007. MicroRNA silencing through RISC recruitment of eIF6. *Nature* **447:** 823–828.
Clark, A.T. 2007. The stem cell identity of testicular cancer. *Stem Cell Rev.* **3:** 49–59.
Daley, G.Q. 2004. Chronic myeloid leukemia: Proving ground for cancer stem cells. *Cell* **119:** 314–316.
Euling, S. and Ambros, V. 1996. Reversal of cell fate determination in *Caenorhabditis elegans* vulval development. *Development* **122:** 2507–2515.
Guo, Y., Chen, Y., Ito, H., Watanabe, A., Ge, X., Kodama, T., and Aburatani, H. 2006. Identification and characterization of lin-28 homolog B (LIN28B) in human hepatocellular carcinoma. *Gene* **384:** 51–61.
Hochedlinger, K., Yamada, Y., Beard, C., and Jaenisch, R. 2005. Ectopic expression of *Oct-4* blocks progenitor-cell differentiation and causes dysplasia in epithelial tissues. *Cell* **121:** 465–477.
Huber, M.A., Kraut, N., and Beug, H. 2005. Molecular requirements for epithelial-mesenchymal transition during tumor progression. *Curr. Opin. Cell Biol.* **17:** 548–558.
Jamieson, C.H., Ailles, L.E., Dylla, S.J., Muijtjens, M., Jones, C., Zehnder, J.L., Gotlib, J., Li, K., Manz, M.G., Keating, A., Sawyers, C.L., et al. 2004. Granulocyte-macrophage progenitors as candidate leukemic stem cells in blast-crisis CML. *N. Engl. J. Med.* **351:** 657–667.
Johnson, S.M., Grosshans, H., Shingara, J., Byrom, M., Jarvis, R., Cheng, A., Labourier, E., Reinert, K.L., Brown, D., and Slack, F.J. 2005. *RAS* is regulated by the *let-7* microRNA family. *Cell* **120:** 635–647.
Krivtsov, A.V., Twomey, D., Feng, Z., Stubbs, M.C., Wang, Y., Faber, J., Levine, J.E., Wang, J., Hahn, W.C., Gilliland, D.G., et al. 2006. Transformation from committed progenitor to leukaemia stem cell initiated by MLL-AF9. *Nature* **442:** 818–822.
Lensch, M.W. and Ince, T.A. 2007. The terminology of teratocarcinomas and teratomas (author reply). *Nat. Biotechnol.* **25:** 1211–1212.
Lensch, M.W., Schlaeger, T.M., Zon, L.I., and Daley, G.Q. 2007. Teratoma formation assays with human embryonic stem cells:

A rationale for one type of human-animal chimera. *Cell Stem Cell* **1:** 253–258.

Maherali, N., Ahfeldt, T., Rigamonti, A., Utikal, J., Cowan, C., and Hochedlinger, K. 2008. A high-efficiency system for the generation and study of human induced pluripotent stem cells. *Cell Stem Cell* **3:** 340–345.

Mani, S.A., Guo, W., Liao, M.J., Eaton, E.N., Ayyanan, A., Zhou, A.Y., Brooks, M., Reinhard, F., Zhang, C.C., Shipitsin, M., et al. 2008. The epithelial-mesenchymal transition generates cells with properties of stem cells. *Cell* **133:** 704–715.

Miyamoto, T., Weissman, I.L., and Akashi, K. 2000. AML1/ETO-expressing nonleukemic stem cells in acute myelogenous leukemia with 8;21 chromosomal translocation. *Proc. Natl. Acad. Sci.* **97:** 7521–7526.

Natalwala, A., Spychal, R., and Tselepis, C. 2008. Epithelial-mesenchymal transition mediated tumourigenesis in the gastrointestinal tract. *World J. Gastroenterol.* **14:** 3792–3797.

Newman, M.A., Thomson, J.M., and Hammond, S.M. 2008. Lin-28 interaction with the Let-7 precursor loop mediates regulated microRNA processing. *RNA* **14:** 1539–1549.

Piskounova, E., Viswanathan, S.R., Janas, M., LaPierre, R.J., Daley, G.Q., Sliz, P., and Gregory, R.I. 2008. Determinants of microRNA processing inhibition by the developmentally regulated RNA-binding protein Lin28. *J. Biol. Chem.* **283:** 21310–21314.

Reya, T., Morrison, S.J., Clarke, M.F., and Weissman, I.L. 2001. Stem cells, cancer, and cancer stem cells. *Nature* **414:** 105–111.

Rybak, A., Fuchs, H., Smirnova, L., Brandt, C., Pohl, E.E., Nitsch, R., and Wulczyn, F.G. 2008. A feedback loop comprising *lin-28* and *let-7* controls pre-*let-7* maturation during neural stem-cell commitment. *Nat. Cell Biol.* **10:** 987–993.

Sampson, V.B., Rong, N.H., Han, J., Yang, Q., Aris, V., Soteropoulos, P., Petrelli, N.J., Dunn, S.P., and Krueger, L.J. 2007. MicroRNA let-7a down-regulates MYC and reverts MYC-induced growth in Burkitt lymphoma cells. *Cancer Res.* **67:** 9762–9770.

Simpson, A.J., Caballero, O.L., Jungbluth, A., Chen, Y.T., and Old, L.J. 2005. Cancer/testis antigens, gametogenesis and cancer. *Nat. Rev. Cancer* **5:** 615–625.

Takahashi, K. and Yamanaka, S. 2006. Induction of pluripotent stem cells from mouse embryonic and adult fibroblast cultures by defined factors. *Cell* **126:** 663–676.

Takahashi, K., Tanabe, K., Ohnuki, M., Narita, M., Ichisaka, T., Tomoda, K., and Yamanaka, S. 2007. Induction of pluripotent stem cells from adult human fibroblasts by defined factors. *Cell* **131:** 861–872.

Takamizawa, J., Konishi, H., Yanagisawa, K., Tomida, S., Osada, H., Endoh, H., Harano, T., Yatabe, Y., Nagino, M., Nimura, Y., et al. 2004. Reduced expression of the *let-7* microRNAs in human lung cancers in association with shortened postoperative survival. *Cancer Res.* **64:** 3753–3756.

Viswanathan, S.R., Daley, G.Q., and Gregory, R.I. 2008. Selective blockade of microRNA processing by Lin28. *Science* **320:** 97–100.

Wang, J., Xie, L.Y., Allan, S., Beach, D., and Hannon, G.J. 1998. Myc activates telomerase. *Genes Dev.* **12:** 1769–1774.

Wei, D., Kanai, M., Huang, S., and Xie, K. 2006. Emerging role of KLF4 in human gastrointestinal cancer. *Carcinogenesis* **27:** 23–31.

Yang, J., Mani, S.A., Donaher, J.L., Ramaswamy, S., Itzykson, R.A., Come, C., Savagner, P., Gitelman, I., Richardson, A., and Weinberg, R.A. 2004. Twist, a master regulator of morphogenesis, plays an essential role in tumor metastasis. *Cell* **117:** 927–939.

Yu, F., Yao, H., Zhu, P., Zhang, X., Pan, Q., Gong, C., Huang, Y., Hu, X., Su, F., Lieberman, J., et al. 2007a. *let-7* regulates self renewal and tumorigenicity of breast cancer cells. *Cell* **131:** 1109–1123.

Yu, J., Vodyanik, M.A., Smuga-Otto, K., Antosiewicz-Bourget, J., Frane, J.L., Tian, S., Nie, J., Jonsdottir, G.A., Ruotti, V., Stewart, R., et al. 2007b. Induced pluripotent stem cell lines derived from human somatic cells. *Science* **318:** 1917–1920.

Pathways to New β Cells

Q. Zhou and D.A. Melton

Department of Stem Cell and Regenerative Biology, Howard Hughes Medical Institute,
Harvard Stem Cell Institute, Harvard University, Cambridge, Masssachusetts 02138

Diabetes is a leading health problem of the world and its prevalence continues to rise. With Type I diabetes, and in some patients with Type II, the lack of insulin can be counterbalanced by providing new β (insulin-producing) cells. For Type I diabetes, treating the autoimmune attack remains a serious challenge. Several strategies to produce new β cells have been proposed. These include differentiation from embryonic stem cells, proliferation of existing adult β cells, derivation from putative adult progenitors/stem cells, and reprogramming of non-β cells to β cells. Each of these strategies has distinct merits and risks, and they are at different stages of understanding and development. In particular, the approach based on differentiation from embryonic stem cells has had strong support and in recent years has made notable progress. Nevertheless, significant hurdles remain to transform the current research into future therapies. To expedite this transformation, we believe particular emphasis should be placed on overcoming key knowledge gaps in β-cell biology, developing strategies that produce patient-specific β cells, and carefully addressing potential treatment-related complications or limitations.

Many aspects of human physiology are regulated by insulin, a hormone produced exclusively by β cells of the pancreas. Insulin insufficiency is the cause of type I diabetes, where β cells are selectively destroyed by autoimmune attack, and type II diabetes, where normal insulin supplies fail to meet demand due to impaired insulin sensitivity in peripheral organs and β-cell failure (Bell and Polonsky 2001). Although both types of diabetes can be treated effectively by insulin injections, the long-term health of diabetic patients appears to critically depend on how insulin is delivered (Kahn 2004; Daneman 2006). Continuous release in a measured fashion relative to physiological glucose levels is clearly beneficial. So far, attempts to mechanically mimic the action of β cells have fallen short; evolution has produced a cell with exquisite sensitivity, response time, and capacity to measure glucose and release insulin in a physiological manner (Kahn 2004). The best way to supply insulin is clearly to use β cells.

Type I diabetes patients could to benefit from a supply of new β cells. Although the cause of type I diabetes is autoimmune attack on β cell and a cure requires understanding and stopping autoimmunity, transplanted cadaveric islets have demonstrated impressive relief from diabetic symptoms (Lakey et al. 2006). The demand for transplantable new β cells, however, far outstrips available supplies. In addition, it would be ideal to supply autologous β cells to eliminate the need for immunosuppression (in the case of Type II diabetes; for Type I, blocking the autoimmune attack remains a key hurdle). Over the years, several strategies to produce new β cells have been proposed. These include directed differentiation from embryonic stem (ES) cells, proliferation of existing β cells, derivation from putative adult endocrine progenitor cells, and more recently, reprogramming of non-β cells to β cells. Each of these strategies exploits a different aspect of β-cell biology. Although notable progress has been made, significant challenges remain and discoveries will be needed in order to translate this research into therapy. Here, we briefly review

key areas of β-cell development and critically evaluate the current approaches to produce new β cells.

EMBRYONIC DEVELOPMENT OF β CELLS

Pancreatic Morphogenesis

β cells first arise in the pancreas during embryogenesis (Slack 1995; Edlund 2001). During early embryogenesis, it is thought that multiple signals, including some from blood vessels, induce part of the early endoderm to form the future pancreas (Lammert et al. 2001; Zaret 2008). From this domain, two pancreatic buds arise. The epithelium of the buds later sprouts finger-like protrusions into the surrounding mesenchyme to initiate branching morphogenesis. Several rounds of branching results in a tree-like epithelial structure surrounded by mesenchyme. Further growth and elaboration of this structure over time eventually gives rise to the adult pancreas with exocrine tissues located at the tip of the duct system and interspersed islets of Langerhans that harbor β cells (Fig. 1).

β-cell Lineage

Cell-type specification in the pancreas is thought to proceed in a stepwise fashion similar to the hematopoietic system where multipotent progenitors give rise to cell-type-specific precursor cells that subsequently produce mature cell types (Murtaugh 2007). Genetic lineage-tracing studies provide evidence for the existence of such a pancreatic multipotent progenitor (Zhou et al. 2007). These cells appear to reside in the early pancreatic buds and later, when the branching begins, at the tip of the branches. They give rise to all pancreatic cell types, including exocrine, endocrine, and duct cells.

A unique aspect of the early pancreatic progenitor cells is that they do not appear to have significant compensatory growth in response to a reduction in number (Stanger et al. 2007). Unlike organs such as liver, where reduction of progenitor numbers results in growth compensation and a

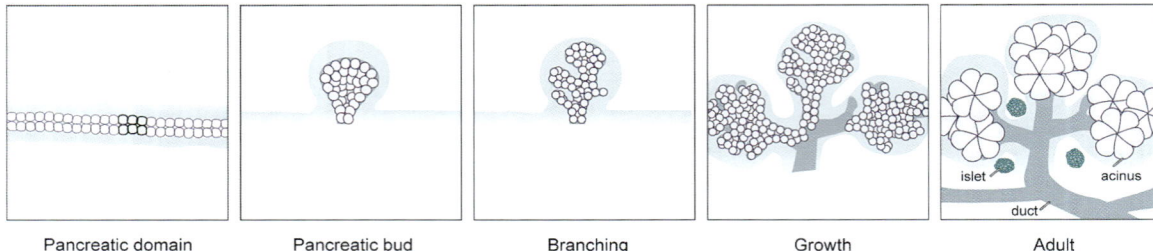

Figure 1. Pancreatic development. Schematic drawings of different stages of pancreatic development. At all stages, the epithelium is surrounded by a layer of mesenchyme. Pancreatic bud formation and branching are shown for dorsal pancreas only.

final organ of normal size, reduction of pancreatic progenitors results in a proportionately smaller pancreas.

Details about lineage-restricted precursor cells produced from multipotent pancreatic progenitors remain unresolved. What is clear, however, is that a population of endocrine precursors, recognized by the marker gene *Ngn3*, appears at midgestation among developing epithelial cells. Genetic lineage-tracing and knockout studies demonstrated convincingly that *Ngn3*[+] cells give rise to all pancreatic endocrine cells including β cells and that they are the only source of β cells during embryogenesis (Gradwohl et al. 2000; Schwitzgebel et al. 2000; Gu et al. 2002). Newly generated endocrine cells migrate out of epithelium into the surrounding mesenchyme and coalesce into clusters. These are precursors of the islets of Langerhans.

Genes in Development

Many genes associated with embryonic development of the pancreas and β cells have been identified (Wilson et al. 2003; Jensen 2004). These include transcription factors such as Pdx1, Ptf1a, and Ngn3, and secreted molecules of the Hedgehog (HH), fibroblast growth factor (FGF), retinoic acid (RA), Notch, and Wnt families. Their importance is largely demonstrated through gene knockout studies, where phenotypes involving pancreas and/or β cells were observed. These studies mostly establish the necessity of the genes in pancreas and β-cell development, but few have addressed the issue of whether any of them are sufficient to instruct the generation of particular pancreatic cell types. For example, it is not clear what signals would be sufficient to direct early endoderm to adopt the pancre-

atic cell fate or to promote Ngn3[+] precursor cells to become β cells. Nevertheless, a number of factors can strongly bias pancreatic fate choices. For example, SHH inhibits pancreatic domain formation from early endoderm (Hebrok et al. 2000); expression of activated Pdx1 under certain conditions can convert part of the liver domain to pancreas (Horb et al. 2003); *ngn3* expression in early pancreatic progenitors appears to be sufficient to promote endocrine cell fates (Johansson et al. 2007); and Notch signaling functions to maintain progenitor cells in undifferentiated states (Apelqvist et al. 1999; Jensen et al. 2000; Murtaugh et al. 2003). These factors have subsequently been used in various approaches to generate new β cells.

RECAPITULATING DEVELOPMENT: DIRECTED DIFFERENTIATION OF B CELLS FROM EMBRYONIC STEM CELLS

Recent Progress

Many studies have been carried out in recent years to generate β cells by differentiation from ES cells (Spence and Wells 2007). This is based on the premise that pluripotent ES cells have the ability to generate any cell type in the body and that key steps of β-cell development may be recapitulated in culture. Indeed, as we discussed above, the key steps for β-cell development are known: from definitive endoderm to pancreatic progenitors, endocrine precursors, and finally mature β cells (Fig. 2). Some factors that promote the progression from one step to another are also known, and cells that reach each developmental stage can be recognized by their expression of unique gene products (Fig. 2).

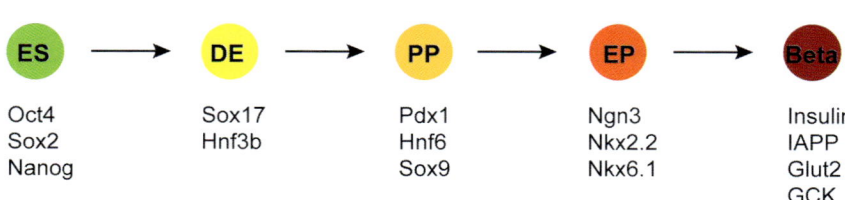

Figure 2. Directed differentiation from embryonic stem cells. There are four major steps that lead from ES cells to β cells. These are definitive endoderm (DE), pancreatic progenitor cells (PP), endocrine progenitors (EP), and mature β cells (Beta). Some markers that define each cell type are listed.

Using a protocol of stepwise differentiation based on conclusions from other studies of normal pancreatic development, significant advances have been made to produce β cells from human ES cells (D'Amour et al. 2005, 2006; Kroon et al. 2008). This protocol uses Activin/Nodal to convert a monolayer culture of ES cells into cells that resemble definitive endoderm. Suppression of SHH signaling and addition of other permissive factors allow some of the cells to further develop into pancreatic and endocrine progenitor-like cells in culture. These cells are subsequently transplanted into mice where they appear to mature into various endocrine cell types including β cells, organize into islet-like structures, and gain the ability to ameliorate hyperglycemia.

Issues That Remain

The encouraging results from human ES differentiation, while validating the ES cell approach, also point to several remaining hurdles. So far, little is known about the factors that promote pancreatic endocrine precursors to become β cells and even less is known about the factors that induce maturation of β cells. These two steps are now accomplished by transplanting progenitor cells into host animals where the signals and mechanisms remain opaque. The size, composition, and organization of the resulting cellular mass from progenitor cell differentiation in vivo are unpredictable. Therefore, although directly supplying pancreatic progenitor cells to patients may be clinically possible, it is more desirable to derive mature β cells for transplantation and that requires finding a way to produce mature β cells in culture in a controlled manner.

Part of the difficulty in deriving fully mature β cells in culture may reflect an inadequate in vitro environment. In vivo β cells develop in a three-dimensional structure with extensive interactions with a diverse array of cell types that includes mesenchymal and endothelial cells (Rutter et al. 1978; Nikolova et al. 2006). The current ES derivation protocol, however, uses a two-dimensional culture system that is largely devoid of other cell types. Reintroducing mesenchymal and endothelial cells or the signals they produce into the culture system may promote further β-cell differentiation.

It has been shown recently that different human ES cell lines exhibit marked differences in differentiation potential into different cell types, with variations as large as 100-fold (Adewumi et al. 2007; Osafune et al. 2008). It is therefore necessary to optimize the human ES lines used for each derivation protocol by screening through multiple human ES cell lines.

Each adult human has about 1 billion β cells. Producing such large numbers of cells requires a highly efficient process. Current methods of ES differentiation rely on protein factors that are expensive and varying in quality. To improve consistency and efficiency of β-cell derivation, it will likely be advantageous to identify small molecules that can replace the function of these protein factors.

Finally, it has long been known that islet structure is important for β-cell function and that intra-islet β-cell inter-

actions serve to suppress basal insulin secretion and enhance glucose-stimulated insulin release (Konstantinova et al. 2007). Any effort to produce fully functional β cells would likely benefit from proper reconstruction of the three-dimensional islet structure with the associated endocrine cell types in vitro.

Patient-specific β cells: Combining iPS Technology with Directed Differentiation

In the past, human ES cell lines were derived from embryos. Recent advances in stem cell reprogramming raised the possibility of generating patient-specific induced pluripotent stem (iPS) cells from skin cells of individual patients (Takahashi and Yamanaka 2006; Takahashi et al. 2007; Yu et al. 2007), which can be redifferentiated to produce patient-specific β cells. Although there are some remaining concerns with stem cell reprogramming, notably the use of viruses that may cause cancer, these problems are expected to be overcome in the near future. The iPS approach also carries some risks. Extensive cellular proliferation in culture during skin-to-stem-cell reprogramming and the subsequent differentiation of iPS cells may encourage accumulation of genetic mutations that allow the cells to survive and proliferate better in culture. This may increase the likelihood that differentiated cell types are predisposed for malignant transformation. In addition, contamination of differentiated cells by undifferentiated stem cells may also lead to teratoma formation. Properly addressing these risks would be necessary before clinical applications.

STRATEGIES BASED ON β-CELL REPLICATION

The formation of β cells during embryogenesis is followed by a period of rapid proliferation during early postnatal life when β-cell number increases to meet physiological demand. During adult life, β-cell mass is maintained at a constant level with very slow cellular turnover (Teta et al. 2005). Recent genetic experiments demonstrate that adult β cells derive from self-duplication of existing β cells (Dor et al. 2004). Clonal analysis and lineage-tracing studies further suggest that all β cells have equal potential for replication (Brennand et al. 2007; Teta et al. 2007). Stimulating proliferation of existing adult β cells is a very attractive approach for supplying new β cells as the β cells are immunologically fully matched with individual patients.

Signals That Influence β-cell Replication

Many intracellular and extracellular factors are known to influence proliferation and apoptosis of β cells (Heit et al. 2006). For example, direct manipulation of cell cycle regulators, including cyclins, cyclin-dependent kinases (CDK), and CDK inhibitors (CDI), changes β-cell proliferation, with some resulting in β-cell hyperplasia (Heit et al. 2006). However, direct manipulations of cell cycle regulators often lead to cancer (Sherr and Roberts 2004).

In addition, cell cycle regulators are normally expressed in multiple cell types, and changes in their levels may cause growth and apoptosis changes of an array of cell types. These concerns have led to searches for extracellular signals that function under physiological conditions to promote β-cell replication.

Despite the typical slow growth rate of adult β cells, β-cell mass can expand dramatically during pregnancy (Van Assche et al. 1978; Scaglia et al. 1995; Kahn et al. 2006). There is evidence that prolactin and placental lactogen, two maternal hormones, can directly regulate β-cell proliferation during pregnancy through the intracellular factor Menin (Parsons et al. 1992; Brelje et al. 1993; Sorenson and Brelje 1997; Karnik et al. 2007). However, the actions of these hormones are complicated and the prospect of using them as a therapy to increase β-cell number is unclear.

β-cell number also increases in obesity or genetic insulin resistance in both human and animals (Kahn et al. 2006). Strikingly, in some genetic models of peripheral insulin resistance, such as liver-specific deletion of insulin receptor (LIRKO), an almost tenfold islet hyperplasia is observed (Michael et al. 2000). Under insulin resistance conditions, blood glucose levels are elevated and insulin secretion is enhanced. Interestingly, both glucose and insulin have been suggested to be β-cell mitogens (Chick 1973; Kulkarni 2005). Indeed, infusion of either glucose or insulin into animal models results in increased β-cell proliferation and mass (Bonner-Weir et al. 1989; Paris et al. 2003). On the other hand, the actions of glucose and insulin are tightly linked. Glucose stimulates insulin release from β cells, which can subsequently trigger autocrine insulin signaling (Kulkarni 2002). It has proved to be difficult to separate the effect of glucose from insulin, and it remains unclear whether glucose/insulin are primarily responsible for increased β-cell proliferation under insulin-resistant conditions or whether other factors primarily drive this process.

Issues That Remain

Factors that are currently known for β-cell proliferation, such as prolactin, glucose, and insulin, are not suitable for direct application in therapies. Nevertheless, studies of them and the associated signaling pathways may eventually yield better targets for therapeutic interventions. It is also possible to directly screen for factors and chemical compounds that stimulate β-cell replication. One way to achieve this is to develop an in vitro culture system for islets or isolated β cells that allows rapid screening of large numbers of factors and compounds.

To achieve patient-specific therapies, it is important that the growth-stimulating factors can be administered directly in vivo and that their actions are specific to β cells or whole islets but not to other cell types. This specificity may be accomplished by exploiting mechanisms such as differential expression of certain glucose transporters in β cells that selectively import specific molecules into β cells.

The major risk associated with this strategy of increasing β-cell mass is the danger of cancer, especially if the agents used are not highly specific and induce changes in proliferation and apoptosis of pancreatic duct or exocrine cells. This may lead to pancreatic cancers, a particularly deadly class of tumors.

New Injury Model to Study β-cell Replication

In some organ systems such as the liver, studies of injury models have revealed important factors and pathways that mediate compensatory cellular proliferation (Taub 2004). Adult pancreas, however, appears to have very limited compensatory growth in various injury models (Tanigawa et al. 1997). Recently, a new study with inducible genetic ablation of β cells provided evidence for robust compensatory β-cell growth (Nir et al. 2007). The development of this model system provides a new avenue of investigating β-cell proliferation.

Dedifferentiation from Adult Islet Cells

It was suggested recently that β cells in cultured islets may dedifferentiate into proliferating progenitor-like cells (Gershengorn et al. 2004). Some of these fibroblast-like cells can be induced to redifferentiate and reexpress insulin. Whether these insulin-expressing cells represent bona fide β cells, however, has yet to be established. In addition, a recent study showed that the proliferating fibroblast-like cells do not derive from existing β cells (Morton et al. 2007). Their origin remains mysterious. Even if functional β cells can be produced with islet dedifferentiation, it would not be suitable for patient-specific therapies because it would be difficult to collect islets from individuals.

STRATEGIES BASED ON PUTATIVE ADULT PROGENITOR CELLS OR FACULTATIVE STEM CELLS

Basis on certain histological observations, it has long been hypothesized that adult endocrine progenitor cells or facultative stem cells may exist in adult pancreas (Bonner-Weir and Weir 2005). A recent study showed that a small number of Ngn3[+] cells arise in adult mouse pancreas after partial duct ligation (PDL) and some of them are capable of further development into endocrine cells that include β cells (Xu et al. 2008). Nevertheless, it is not clear whether this event represents the mobilization of existing adult endocrine progenitor cells that already express low levels of Ngn3, activation of dormant stem-like cells, or conversion of other cells into endocrine cells under the PDL conditions. Moreover, the factors involved in this process remain unknown and the number of Ngn3[+] progenitor cells produced is very small. In addition, the adult pancreas may be a nonpermissive environment for differentiation of progenitor cells. These are formidable challenges and major progress is clearly needed to make this approach a viable strategy for producing new β cells.

REPROGRAMMING NON-β CELLS TO β CELLS

In rare cases, cells of one lineage can be converted into cells of another lineage (Slack 2007). For example, embryonic dermal fibroblasts and pigmented epithelial cells can

Table 1. Comparison of current approaches to produce new β cells

	ES differentiation	β-cell replication	Adult progenitor	Reprogramming
Instructive factors	some known	some known	unknown	known
Patient specificity	possible with iPS	possible with direct in vivo replication	possible with direct in vivo induction	possible with in vivo or in vitro conversion
Starting material	abundant skin cells	existing β cells	unknown	abundant exocrine and liver cells
β-like cells produced	yes, large number	yes, large number	yes, very small number	yes, large number

be converted to contracting myocytes, whereas mature β cells can be reprogrammed into macrophages (Choi et al. 1990; Xie et al. 2004). These examples have led to research into potential ways to reprogram non-β cells, such as pancreatic exocrine cell, duct cells, and liver cells, into β cells (Ferber et al. 2000; Heremans et al. 2002; Gasa et al. 2004; Kaneto et al. 2005). Most of these efforts, however, have resulted in gene expression changes but not cell-type conversions. Interestingly, mature exocrine cells of the pancreas can turn on endocrine cell programs in culture (Minami et al. 2005). Dissociation itself is apparently sufficient to initiate the endocrine programs, whereas the addition of growth factors improves cell survival (Baeyens et al. 2005; Minami and Seino 2008). The molecular mechanism of this phenomenon, however, remains largely unknown.

Recent experiments in our laboratory showed that reprogramming of non-β cells to β cells can be achieved in an instructive manner. We demonstrated that a simple combination of three transcription factors is sufficient to convert fully differentiated pancreatic exocrine cells in adult mice into cells that closely resemble β cells in morphology, ultrastructure, molecular signatures, and function.

A key challenge of the reprogramming approach is that induced β cells persist as individual cells or small clusters and do not organize into islets. In addition, viruses are currently used to express the reprogramming factors. They would need to be replaced by safer reagents such as chemical compounds.

There are several potential ways to achieve patient specificity with the reprogramming approach. For example, given the difficulty of biopsying human pancreas, reagents can be developed to convert pancreatic exocrine cells directly in vivo. Alternatively, adult liver cells can be relatively easily harvested from individuals and used for reprogramming in vitro to produce β cells for subsequent transplants.

CONCLUSIONS

Different strategies have been proposed in recent years to produce new β cells. Each of these strategies has merits and drawbacks (Table 1), and so far, the most promising strategy appears to be the one based on differentiation of human ES cells.

Before transferring the ongoing research into clinics, several common issues face all of these approaches. Diabetes, after all, can be treated effectively with insulin for long term. Any cell therapies based on new β cells will be scrutinized and stringent criteria will be demanded. First, new β cells are expected to resemble endogenous β cells in function and that requires islet structure formation. Second, new β cells are expected to be patient-specific. Third, they must be safe to use for the long term. Various risk factors, particularly cancers, should be properly addressed before clinical applications.

Current approaches for producing β cells still remain at early research and development stages; it is difficult to foresee which approach may eventually yield β cells that are therapeutically useful. Given the severity and increasing prevalence of diabetes, each of these strategies, and potential new innovative approaches, should be pursued in earnest. From a broader perspective, a major goal of regenerative medicine is to generate new cells to replace those lost due to disease or damage. Successes in regenerating new β cells should inform other cell-based therapeutic strategies.

ACKNOWLEDGMENTS

We thank many of the past and present members of the Melton laboratory for stimulating discussions on pancreas and β cells. We apologize to colleagues whose work we could not cite due to space limitations. We thank Drs. Justin Annes and Jay Rajagopal for critical readings of this manuscript. Q.Z. was supported by a Damon-Runyon Cancer Research Foundation postdoctoral fellowship and a Pathway to Independence (PI) Award from the National Institutes of Health (NIH). D.A.M. is a Howard Hughes Medical Institute (HHMI) investigator, and work in the laboratory was supported in part by the Harvard Stem Cell Institute, HHMI, JDRF, and the NIH.

REFERENCES

Adewumi, O., Aflatoonian, B., Ahrlund-Richter, L., Amit, M., Andrews, P.W., Beighton, G., Bello, P.A., Benvenisty, N., Berry, L.S., Bevan, S., et al. 2007. Characterization of human embryonic stem cell lines by the International Stem Cell Initiative. *Nat. Biotechnol.* **25:** 803–816.

Apelqvist, A., Li, H., Sommer, L., Beatus, P., Anderson, D.J., Honjo, T., Hrabe de Angelis, M., Lendahl, U., and Edlund, H. 1999. Notch signalling controls pancreatic cell differentiation. *Nature* **400:** 877–881.

Baeyens, L., De Breuck, S., Lardon, J., Mfcpou, J.K., Rooman, I., and Bouwens, L. 2005. In vitro generation of insulin-producing β cells from adult exocrine pancreatic cells. *Diabetologia* **48:** 49–57.

Bell, G.I. and Polonsky, K.S. 2001. Diabetes mellitus and genetically programmed defects in β-cell function. *Nature* **414:** 788–791.

Bonner-Weir, S. and Weir, G.C. 2005. New sources of pancreatic β cells. *Nat. Biotechnol.* **23:** 857–861.

Bonner-Weir, S., Deery, D., Leahy, J.L., and Weir, G.C. 1989.

Compensatory growth of pancreatic β cells in adult rats after short-term glucose infusion. *Diabetes* **38:** 49–53.

Brelje, T.C., Scharp, D.W., Lacy, P.E., Ogren, L., Talamantes, F., Robertson, M., Friesen, H.G., and Sorenson, R.L. 1993. Effect of homologous placental lactogens, prolactins, and growth hormones on islet β-cell division and insulin secretion in rat, mouse, and human islets: Implication for placental lactogen regulation of islet function during pregnancy. *Endocrinology* **132:** 879–887.

Brennand, K., Huangfu, D., and Melton, D. 2007. All β cells contribute equally to islet growth and maintenance. *PLoS Biol.* **5:** e163.

Chick, W.L. 1973. β Cell replication in rat pancreatic monolayer cultures. Effects of glucose, tolbutamide, glucocorticoid, growth hormone and glucagon. *Diabetes* **22:** 687–693.

Choi, J., Costa, M.L., Mermelstein, C.S., Chagas, C., Holtzer, S., and Holtzer, H. 1990. MyoD converts primary dermal fibroblasts, chondroblasts, smooth muscle, and retinal pigmented epithelial cells into striated mononucleated myoblasts and multinucleated myotubes. *Proc. Natl. Acad. Sci.* **87:** 7988–7992.

D'Amour, K.A., Agulnick, A.D., Eliazer, S., Kelly, O.G., Kroon, E., and Baetge, E.E. 2005. Efficient differentiation of human embryonic stem cells to definitive endoderm. *Nat. Biotechnol.* **23:** 1534–1541.

D'Amour, K.A., Bang, A.G., Eliazer, S., Kelly, O.G., Agulnick, A.D., Smart, N.G., Moorman, M.A., Kroon, E., Carpenter, M.K., and Baetge, E.E. 2006. Production of pancreatic hormone-expressing endocrine cells from human embryonic stem cells. *Nat. Biotechnol.* **24:** 1392–1401.

Daneman, D. 2006. Type 1 diabetes. *Lancet* **367:** 847–858.

Dor, Y., Brown, J., Martinez, O.I., and Melton, D.A. 2004. Adult pancreatic β cells are formed by self-duplication rather than stem-cell differentiation. *Nature* **429:** 41–46.

Edlund, H. 2001. Developmental biology of the pancreas. *Diabetes* (suppl. 1) **50:** S5–S9.

Ferber, S., Halkin, A., Cohen, H., Ber, I., Einav, Y., Goldberg, I., Barshack, I., Seijffers, R., Kopolovic, J., Kaiser, N., et al. 2000. Pancreatic and duodenal homeobox gene 1 induces expression of insulin genes in liver and ameliorates streptozotocin-induced hyperglycemia. *Nat. Med.* **6:** 568–572.

Gasa, R., Mrejen, C., Leachman, N., Otten, M., Barnes, M., Wang, J., Chakrabarti, S., Mirmira, R., and German, M. 2004. Proendocrine genes coordinate the pancreatic islet differentiation program in vitro. *Proc. Natl. Acad. Sci.* **101:** 13245–13250.

Gershengorn, M.C., Hardikar, A.A., Wei, C., Geras-Raaka, E., Marcus-Samuels, B., and Raaka, B.M. 2004. Epithelial-to-mesenchymal transition generates proliferative human islet precursor cells. *Science* **306:** 2261–2264.

Gradwohl, G., Dierich, A., LeMeur, M., and Guillemot, F. 2000. *neurogenin3* is required for the development of the four endocrine cell lineages of the pancreas. *Proc. Natl. Acad. Sci.* **97:** 1607–1611.

Gu, G., Dubauskaite, J., and Melton, D.A. 2002. Direct evidence for the pancreatic lineage: NGN3+ cells are islet progenitors and are distinct from duct progenitors. *Development* **129:** 2447–2457.

Hebrok, M., Kim, S.K., St-Jacques, B., McMahon, A.P., and Melton, D.A. 2000. Regulation of pancreas development by hedgehog signaling. *Development* **127:** 4905–4913.

Heit, J.J., Karnik, S.K., and Kim, S.K. 2006. Intrinsic regulators of pancreatic β-cell proliferation. *Annu. Rev. Cell Dev. Biol.* **22:** 311–338.

Heremans, Y., Van De Casteele, M., in't Veld, P., Gradwohl, G., Serup, P., Madsen, O., Pipeleers, D., and Heimberg, H. 2002. Recapitulation of embryonic neuroendocrine differentiation in adult human pancreatic duct cells expressing neurogenin 3. *J. Cell Biol.* **159:** 303–312.

Horb, M.E., Shen, C.N., Tosh, D., and Slack, J.M. 2003. Experimental conversion of liver to pancreas. *Curr. Biol.* **13:** 105–115.

Jensen, J. 2004. Gene regulatory factors in pancreatic development. *Dev. Dyn.* **229:** 176–200.

Jensen, J., Pedersen, E.E., Galante, P., Hald, J., Heller, R.S., Ishibashi, M., Kageyama, R., Guillemot, F., Serup, P., and Madsen, O.D. 2000. Control of endodermal endocrine development by Hes-1. *Nat. Genet.* **24:** 36–44.

Johansson, K.A., Dursun, U., Jordan, N., Gu, G., Beermann, F., Gradwohl, G., and Grapin-Botton, A. 2007. Temporal control of neurogenin3 activity in pancreas progenitors reveals competence windows for the generation of different endocrine cell types. *Dev. Cell* **12:** 457–465.

Kahn, S.E. 2004. Engineering a new β-cell: A critical venture requiring special attention to constantly changing physiological needs. *Semin. Cell Dev. Biol.* **15:** 359–370.

Kahn, S.E., Hull, R.L., and Utzschneider, K.M. 2006. Mechanisms linking obesity to insulin resistance and type 2 diabetes. *Nature* **444:** 840–846.

Kaneto, H., Nakatani, Y., Miyatsuka, T., Matsuoka, T.A., Matsuhisa, M., Hori, M., and Yamasaki, Y. 2005. PDX-1/VP16 fusion protein, together with NeuroD or Ngn3, markedly induces insulin gene transcription and ameliorates glucose tolerance. *Diabetes* **54:** 1009–1022.

Karnik, S.K., Chen, H., McLean, G.W., Heit, J.J., Gu, X., Zhang, A.Y., Fontaine, M., Yen, M.H., and Kim, S. K. 2007. Menin controls growth of pancreatic β cells in pregnant mice and promotes gestational diabetes mellitus. *Science* **318:** 806–809.

Konstantinova, I., Nikolova, G., Ohara-Imaizumi, M., Meda, P., Kucera, T., Zarbalis, K., Wurst, W., Nagamatsu, S., and Lammert, E. 2007. EphA-Ephrin-A-mediated β cell communication regulates insulin secretion from pancreatic islets. *Cell* **129:** 359–370.

Kroon, E., Martinson, L.A., Kadoya, K., Bang, A.G., Kelly, O.G., Eliazer, S., Young, H., Richardson, M., Smart, N.G., Cunningham, J., et al. 2008. Pancreatic endoderm derived from human embryonic stem cells generates glucose-responsive insulin-secreting cells in vivo. *Nat. Biotechnol.* **26:** 443–452.

Kulkarni, R.N. 2002. Receptors for insulin and insulin-like growth factor-1 and insulin receptor substrate-1 mediate pathways that regulate islet function. *Biochem. Soc. Trans.* **30:** 317–322.

Kulkarni, R.N. 2005. New insights into the roles of insulin/IGF-I in the development and maintenance of β-cell mass. *Rev. Endocr. Metab. Disord.* **6:** 199–210.

Lakey, J.R., Mirbolooki, M., and Shapiro, A.M. 2006. Current status of clinical islet cell transplantation. *Methods Mol. Biol.* **333:** 47–104.

Lammert, E., Cleaver, O., and Melton, D. 2001. Induction of pancreatic differentiation by signals from blood vessels. *Science* **294:** 564–567.

Michael, M.D., Kulkarni, R.N., Postic, C., Previs, S.F., Shulman, G.I., Magnuson, M.A., and Kahn, C.R. 2000. Loss of insulin signaling in hepatocytes leads to severe insulin resistance and progressive hepatic dysfunction. *Mol. Cell* **6:** 87–97.

Minami, K. and Seino, S. 2008. Pancreatic acinar-to-β cell transdifferentiation in vitro. *Front. Biosci.* **13:** 5824–5837.

Minami, K., Okuno, M., Miyawaki, K., Okumachi, A., Ishizaki, K., Oyama, K., Kawaguchi, M., Ishizuka, N., Iwanaga, T., and Seino, S. 2005. Lineage tracing and characterization of insulin-secreting cells generated from adult pancreatic acinar cells. *Proc. Natl. Acad. Sci.* **102:** 15116–15121.

Morton, R.A., Geras-Raaka, E., Wilson, L.M., Raaka, B.M., and Gershengorn, M.C. 2007. Endocrine precursor cells from mouse islets are not generated by epithelial-to-mesenchymal transition of mature β cells. *Mol. Cell. Endocrinol.* **270:** 87–93.

Murtaugh, L.C. 2007. Pancreas and β-cell development: From the actual to the possible. *Development* **134:** 427–438.

Murtaugh, L.C., Stanger, B.Z., Kwan, K.M., and Melton, D.A. 2003. Notch signaling controls multiple steps of pancreatic differentiation. *Proc. Natl. Acad. Sci.* **100:** 14920–14925.

Nikolova, G., Jabs, N., Konstantinova, I., Domogatskaya, A., Tryggvason, K., Sorokin, L., Fassler, R., Gu, G., Gerber, H.P., Ferrara, N., et al. 2006. The vascular basement membrane: A niche for insulin gene expression and β cell proliferation. *Dev. Cell* **10:** 397–405.

Nir, T., Melton, D.A., and Dor, Y. 2007. Recovery from diabetes in mice by β cell regeneration. *J. Clin. Invest.* **117:** 2553–2561.

Osafune, K., Caron, L., Borowiak, M., Martinez, R.J., Fitz-Gerald, C.S., Sato, Y., Cowan, C.A., Chien, K.R., and Melton, D.A. 2008. Marked differences in differentiation propensity among human embryonic stem cell lines. *Nat. Biotechnol.* **26:** 313–315.

Paris, M., Bernard-Kargar, C., Berthault, M.F., Bouwens, L., and Ktorza, A. 2003. Specific and combined effects of insulin and glucose on functional pancreatic β-cell mass in vivo in adult rats. *Endocrinology* **144:** 2717–2127.

Parsons, J.A., Brelje, T.C., and Sorenson, R.L. 1992. Adaptation of islets of Langerhans to pregnancy: Increased islet cell proliferation and insulin secretion correlates with the onset of placental lactogen secretion. *Endocrinology* **130:** 1459–1466.

Rutter, W.J., Pictet, R.L., Harding, J.D., Chirgwin, J.M., MacDonald, R.J., and Przybyla, A.E. 1978. An analysis of pancreatic development: Role of mesenchymal factor and other extracellular factors. *Symp. Soc. Dev. Biol.* **1978:** 205–227.

Scaglia, L., Smith, F.E., and Bonner-Weir, S. 1995. Apoptosis contributes to the involution of β cell mass in the post partum rat pancreas. *Endocrinology* **136:** 5461–5468.

Schwitzgebel, V.M., Scheel, D.W., Conners, J.R., Kalamaras, J., Lee, J.E., Anderson, D.J., Sussel, L., Johnson, J.D., and German, M.S. 2000. Expression of neurogenin3 reveals an islet cell precursor population in the pancreas. *Development* **127:** 3533–3542.

Sherr, C.J. and Roberts, J.M. 2004. Living with or without cyclins and cyclin-dependent kinases. *Genes Dev.* **18:** 2699–2711.

Slack, J.M. 1995. Developmental biology of the pancreas. *Development* **121:** 1569–1580.

Slack, J.M. 2007. Metaplasia and transdifferentiation: From pure biology to the clinic. *Nat. Rev. Mol. Cell Biol.* **8:** 369–378.

Sorenson, R.L. and Brelje, T.C. 1997. Adaptation of islets of Langerhans to pregnancy: β-cell growth, enhanced insulin secretion and the role of lactogenic hormones. *Horm. Metab. Res.* **29:** 301–307.

Spence, J.R. and Wells, J.M. 2007. Translational embryology: Using embryonic principles to generate pancreatic endocrine cells from embryonic stem cells. *Dev. Dyn.* **236:** 3218–3227.

Stanger, B.Z., Tanaka, A.J., and Melton, D.A. 2007. Organ size is limited by the number of embryonic progenitor cells in the pancreas but not the liver. *Nature* **445:** 886–891.

Takahashi, K. and Yamanaka, S. 2006. Induction of pluripotent stem cells from mouse embryonic and adult fibroblast cultures by defined factors. *Cell* **126:** 663–676.

Takahashi, K., Tanabe, K., Ohnuki, M., Narita, M., Ichisaka, T., Tomoda, K., and Yamanaka, S. 2007. Induction of pluripotent stem cells from adult human fibroblasts by defined factors. *Cell* **131:** 861–872.

Tanigawa, K., Nakamura, S., Kawaguchi, M., Xu, G., Kin, S., and Tamura, K. 1997. Effect of aging on β-cell function and replication in rat pancreas after 90% pancreatectomy. *Pancreas* **15:** 53–59.

Taub, R. 2004. Liver regeneration: From myth to mechanism. *Nat. Rev. Mol. Cell Biol.* **5:** 836–847.

Teta, M., Long, S.Y., Wartschow, L.M., Rankin, M.M., and Kushner, J.A. 2005. Very slow turnover of β cells in aged adult mice. *Diabetes* **54:** 2557–2567.

Teta, M., Rankin, M.M., Long, S.Y., Stein, G.M., and Kushner, J.A. 2007. Growth and regeneration of adult β cells does not involve specialized progenitors. *Dev. Cell.* **12:** 817–826.

Van Assche, F.A., Aerts, L., and De Prins, F. 1978. A morphological study of the endocrine pancreas in human pregnancy. *Br. J. Obstet. Gynaecol.* **85:** 818–820.

Wilson, M.E., Scheel, D., and German, M.S. 2003. Gene expression cascades in pancreatic development. *Mech. Dev.* **120:** 65–80.

Xie, H., Ye, M., Feng, R., and Graf, T. 2004. Stepwise reprogramming of β cells into macrophages. *Cell* **117:** 663–676.

Xu, X., D'Hoker, J., Stange, G., Bonne, S., De Leu, N., Xiao, X., Van de Casteele, M., Mellitzer, G., Ling, Z., Pipeleers, D., et al. 2008. β Cells can be generated from endogenous progenitors in injured adult mouse pancreas. *Cell* **132:** 197–207.

Yu, J., Vodyanik, M.A., Smuga-Otto, K., Antosiewicz-Bourget, J., Frane, J.L., Tian, S., Nie, J., Jonsdottir, G.A., Ruotti, V., Stewart, R., et al. 2007. Induced pluripotent stem cell lines derived from human somatic cells. *Science* **318:** 1917–1920.

Zaret, K.S. 2008. Genetic programming of liver and pancreas progenitors: Lessons for stem-cell differentiation. *Nat. Rev. Genet.* **9:** 329–340.

Zhou, Q., Law, A.C., Rajagopal, J., Anderson, W.J., Gray, P.A., and Melton, D.A. 2007. A multipotent progenitor domain guides pancreatic organogenesis. *Dev. Cell* **13:** 103–114.

Mapping Key Features of Transcriptional Regulatory Circuitry in Embryonic Stem Cells

M.F. COLE AND R.A. YOUNG

Whitehead Institute for Biomedical Research, Cambridge, Massachusetts 02142;
Department of Biology, Massachusetts Institute of Technology, Cambridge, Massachusetts 02139

The process by which a single fertilized egg develops into a human being with more than 200 cell types—each with a distinct gene expression pattern controlling its cellular state—is poorly understood. Knowledge of the transcriptional regulatory circuitry that establishes and maintains gene expression programs in mammalian cells is fundamental to understanding development and should provide the foundation for improved diagnosis and treatment of disease. Although it is not yet feasible to map the entirety of this circuitry in vertebrate cells, recent work in embryonic stem (ES) cells has demonstrated that core features of the circuitry can be discovered through studies involving selected regulators. Here, we highlight the fundamental insights that have emerged from studies that examined the role of transcription factors, chromatin regulators, signaling pathways, and noncoding RNAs in the regulatory circuitry of ES cells. Maps of regulatory circuitry and the insights that have emerged from these studies have improved our understanding of global gene expression and are facilitating efforts to reprogram cells for disease therapeutics and regenerative medicine.

More than 200 different cell fates are generated during vertebrate development, each with a gene expression program unique to that cell type (Su et al. 2004; Shyamsundar et al. 2005; Vickaryous and Hall 2006). The expression programs are controlled by transcription factors, chromatin modifiers, and regulatory RNAs that can be influenced by signals from the extracellular environment. The subset of regulators that are expressed in each cell type have a key role in establishing and maintaining cell state. How these regulators control global gene expression programs is almost entirely unmapped in vertebrates because of limitations in our knowledge of the transcriptional, chromatin, and signaling components that are key to the control of each cell type. Nonetheless, multiple groups have begun to tackle the challenge of mapping transcriptional regulatory circuitry, particularly in ES cells, and have demonstrated that important insights into the global control of cell state can be obtained from these efforts (Boyer et al. 2005, 2006; Chew et al. 2005; Bernstein et al. 2006; Lee et al. 2006; Loh et al. 2006; Wu et al. 2006; Pan et al. 2007; Zhao et al. 2007; Chen et al. 2008; Cole et al. 2008; Endoh et al. 2008; Jiang et al. 2008; Kim et al. 2008a; Tam et al. 2008).

Mapping transcriptional regulatory circuitry is important because it provides insights into the regulators and mechanisms that control global gene expression, cell state, and development. Because deficiencies in control of gene expression can contribute to many human diseases such as cancer, immune disease, and diabetes (Villard 2004; Kloosterman and Plasterk 2006; Latchman 2008), knowledge of normal and abnormal transcriptional regulatory circuitries may provide new approaches to disease diagnosis and therapy. Here, we describe initial efforts to dissect core features of the ES cell transcriptional regulatory network and highlight the key insights that have emerged from such studies.

MOLECULAR MECHANISMS CONTROLLING EUKARYOTIC TRANSCRIPTION

Transcription factors, chromatin regulators, signaling pathways, and noncoding RNAs are among the key components that control mRNA gene expression (Fig. 1). The molecular mechanisms by which these regulators control expression of individual genes have been studied extensively and are reviewed elsewhere (Lee and Young 2000; Orphanides and Reinberg 2002; Berger 2007; Li et al. 2007; Core and Lis 2008; Hobert 2008) Our understanding of these mechanisms suggests how to organize models of the transcriptional regulatory circuitry, as described below.

DNA-binding transcription factors recognize sequence motifs, are key to specific gene regulation, and can thus be used to anchor transcriptional regulatory networks (Harrison 1991; Pabo and Sauer 1992; Kadonaga 2004; Remenyi et al. 2004). Transcription factors are also the single largest protein family encoded in the human genome, where they account for approximately 10% of protein-coding genes (Lander et al. 2001; Levine and Tjian 2003). They bind to both promoter-proximal and -distal regulatory DNA sequences and can aid or inhibit recruitment of the transcription apparatus at target genes (Latchman 1997; Blackwood and Kadonaga 1998; Ogata et al. 2003; West and Fraser 2005). At most well-studied promoters, it is evident that multiple transcription factors are bound, which allows for the combinatorial control of gene expression (Evans et al. 1990; Greene 1990; Harbison et al. 2004; Panne et al. 2004; Remenyi et al. 2004).

Chromatin regulators are often recruited to specific portions of the genome by DNA-binding transcription factors or the transcription apparatus where they act to augment gene expression or repression through their effects on chromatin state (Berger 2007; Kouzarides 2007; Li et al. 2007). Chromatin regulators that methylate

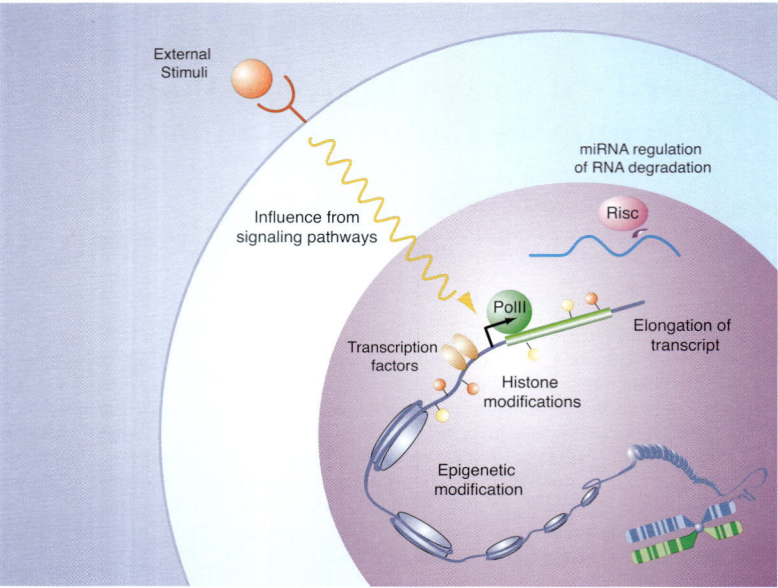

Figure 1. Mechanisms controlling eukaryotic gene expression: Examples of various types of inputs contributing to the control of mRNA levels within a cell. Site-specific transcription factors, recruitment and control of RNA polymerase initiation and elongation, DNA and histone modifications, input from signal transduction pathways, and the actions of miRNAs on mRNA levels can influence the gene expression program of a cell. Knowledge of the molecular mechanisms involved in control of gene expression can guide efforts to elucidate the transcriptional regulatory circuitry that produces a specific gene expression program. For further details and references, see text.

DNA and certain nucleosomal histone residues have been implicated in heritable chromatin states and thus have important roles in developmental control. Methylation of CpG islands in the promoter regions of some genes by DNA methyltransferases contributes to their repression and is maintained during development by maintenance methylases (Goll and Bestor 2005; Turek-Plewa and Jagodziński 2005; Klose and Bird 2006). Methylation of nucleosomal histones by Trithorax and Polycomb group protein complexes is also thought to be important for maintaining gene expression programs associated with specific cell states; Trithorax group complexes are associated with actively transcribed genes, whereas Polycomb group regulators are associated with repression of most genes that they occupy (Pirrotta 1998; Orlando 2003; Ringrose and Paro 2004; Schuettengruber et al. 2007; Schwartz and Pirrotta 2007).

Signaling pathways act to maintain or initiate changes in the regulatory circuitry in response to environmental or developmental cues. The terminal components of signaling pathways are often protein kinases that can phosphorylate and activate transcriptional regulators or are themselves transcription factors and chromatin modifiers (Hunter 2000; Brivanlou and Darnell 2002; Yang et al. 2003; Pokholok et al. 2006). Knowledge of the target genes of each of the signaling pathways that contribute to control of cell state is critical to understanding how these pathways control the gene expression program associated with such states.

Noncoding RNAs can influence gene expression and chromatin state (Goodrich and Kugel 2006; Amaral et al. 2008; Hawkins and Morris 2008). For example, a large class of noncoding RNAs, termed microRNAs (miRNAs),

modify gene expression by regulating translation and degradation of mRNA transcripts (Ambros 2004; Bartel 2004; Valencia-Sanchez et al. 2006; Meister 2007; Makeyev and Maniatis 2008). Noncoding RNA species have also been implicated in control of chromatin state (Verdel et al. 2004; Moazed et al. 2006; Grewal and Elgin 2007; Rinn et al. 2007; Zaratiegui et al. 2007). We have limited understanding of the regulation of expression of noncoding RNA species, and in most cases, we have yet to identify the specific set of genes that are under the control of these noncoding RNA species.

CONCEPT OF CORE TRANSCRIPTIONAL REGULATORY CIRCUITRY

Hundreds of gene expression regulators are present in each cell, making it a challenge to map the regulatory network that they form in even one cell type, much less in 200 cell types (Lander et al. 2001; Brivanlou and Darnell 2002). For this reason, even the most ambitious global studies have examined only a handful of transcriptional regulators and then in only a few cell types (Cawley et al. 2004; Odom et al. 2004, 2006; Boyer et al. 2005; Rada-Iglesias et al. 2005, 2008; Loh et al. 2006; Barski et al. 2007; Mikkelsen et al. 2007; Chen et al. 2008; Cole et al. 2008; Jaenisch and Young 2008; Jiang et al. 2008; Kim et al. 2008a; Komashko et al. 2008; Marson et al. 2008b; Park et al. 2008; Reed et al. 2008; Wang et al. 2008). However, several lines of evidence argue that a small subset of transcription factors and other regulators have a key role in the control of cell state. Cells can be reprogrammed into other cell states through forced expression of a very small number of transcription factors. For example, fibroblasts can be reprogrammed into induced pluripotent stem cells upon forced expression of four transcription factors (Takahashi and Yamanaka 2006; Okita et al. 2007; Takahashi et al. 2007; Wernig et al. 2007; Yu et al. 2007). Similarly, fibroblasts and other cells can take on a skeletal muscle state when the myogenic transcription factor MyoD is expressed (Davis et al. 1987; Weintraub et al. 1989, 1991; Choi et al. 1990). Screens to identify genes that are key to maintaining the ES cell state have identified only a small number of all of the transcription factor genes that are expressed in these cells (Ivanova et al. 2006; Zhang et al. 2006; Fazzio et al. 2008). Furthermore, several studies have shown that many transcription factors can be eliminated without dire consequences for the cell (Winzeler et al. 1999; Giaever et al. 2002; Kemphues 2005). The small set of transcription factors that have been demonstrated to be important for establishment or maintenance of a cell state will henceforth be termed "key regulators."

A simplified version of the transcriptional regulatory circuitry of a cell can thus be deduced by discovering the population of genes that are occupied and controlled by the key regulators for that cell type. We call this simplified network the "core transcriptional regulatory circuitry." Given current experimental limitations to elucidating complete vertebrate circuitry, we propose that the mapping of core regulatory circuitry provides a shortcut to discovering key network themes, a concept that we believe has been validated with the study of ES cells.

KEY ES CELL TRANSCRIPTION FACTORS ESTABLISH A CORE REGULATORY CIRCUITRY

Initial studies of transcription factors in the ES cell transcriptional regulatory network focused on the key regulators Oct4, Sox2, and Nanog (Boyer et al. 2005; Loh et al. 2006). Knowledge of genetic phenotypes, expression profiles, and molecular relations was leveraged to identify these factors as key components of the ES cell network. Genetic studies demonstrated functional consequences in ES cells of inappropriate expression of these factors, the expression of Oct4 and Nanog was found to be specific to pluripotent cells, and Sox2 was known to form a heterodimer with Oct4 (Schöler et al. 1990; Ambrosetti et al. 1997; Nichols et al. 1998; Avilion et al. 2003; Chambers et al. 2003; Mitsui et al. 2003; Hart et al. 2004). Because of the overwhelming evidence for key roles for these regulators in the ES cell network, multiple groups have mapped their target genes in human and murine ES cells (Boyer et al. 2005; Loh et al. 2006).

Several important themes emerged from the study of target genes for Oct4, Sox2, and Nanog (Fig. 2). The key regulators clearly prefer to cooccupy their target genes, thus forming a network structure called a multi-input motif (Fig. 2A) (Lee et al. 2002; Boyer et al. 2005; Loh et al. 2006; Alon 2007). Because they form a heterodimer, Oct4 and Sox2 were expected to bind the same target genes, but Nanog was also found to occupy a large percentage of the Oct4-Sox2 bound genes. More recent studies have mapped additional transcription factors in ES cells and have found that they also follow the theme of target gene cooccupancy (Wu et al. 2006; Chen et al. 2008; Jiang et al. 2008; Kim et al. 2008a). These studies, and similar investigations of key regulators in other cell types (Odom et al. 2004, 2006), suggest that the multi-input motif is an important theme in the transcriptional regulatory circuitry of vertebrate cells.

A related feature or theme that emerged from these global binding studies is that the key regulators tend to occupy DNA sequences in very close proximity to one another (Fig. 2B) (Boyer et al. 2005; Loh et al. 2006; Cole et al. 2008; Marson et al. 2008b). Oct4, Sox, and Nanog were often found to bind within 25 bp of one another at target genes (Marson et al. 2008b). This proximity suggests that these factors are forming tightly associated complexes on DNA to coordinately affect transcription. Some of these transcription factors are competing for binding to overlapping or similar DNA sequences, and because the data come from a population of cells, it is also possible that the complete set of transcription factors is not simultaneously bound at these sites in individual cells. Further studies into the biochemical nature of these binding events are needed to test these possibilities.

One of the more important themes that emerged from the initial studies of the key regulators of ES cells was that Oct4, Sox2, and Nanog together cooccupy their own promoter regions and thus form a network structure called an interconnected autoregulatory loop (Fig. 2C) (Boyer et al. 2005; Loh et al. 2006). This network motif may have two purposes: Feedback gene regulation by these transcription factors may contribute to the stability of the core ES cell transcriptional regulatory network, yet this network structure may also allow for a rapid change in core regulatory circuitry if one regulator is eliminated upon receipt of differentiation signals. Indeed, the circuit formed by Oct4, Sox2, and Nanog could apparently act as a bistable switch controlling ES cell maintenance versus differentiation (Chickarmane et al. 2006).

The early global studies also revealed that the key regulators occupy and control genes encoding many other transcriptional regulators that are expressed in ES cells (Boyer et al. 2005; Loh et al. 2006), forming a hierarchical regulatory network structure (Fig. 2D). The target genes of Oct4, Sox2, and Nanog were significantly enriched for transcription factors and developmental regulators (Boyer et al. 2005). The control of these secondary regulators allows key transcription factors to indirectly control a much larger set of genes. This hierarchical network structure has been described in model organisms and will likely prove to be a common vertebrate network architecture (Martinez-Antonio and Collado-Vides 2003; Ma et al. 2004; Farkas et al. 2006). This network structure may allow for rapid large-scale changes in the transcription program in response to signals that may only directly target a handful of key regulators.

One additional fundamental theme that has emerged from global binding studies is that Oct4, Sox2, and Nanog occupy both actively transcribed genes encoding ES cell transcription factors and repressed genes encoding lineage-specific developmental regulators (Fig. 2E) (Boyer et al. 2005; Loh et al. 2006). This observation suggested that reg-

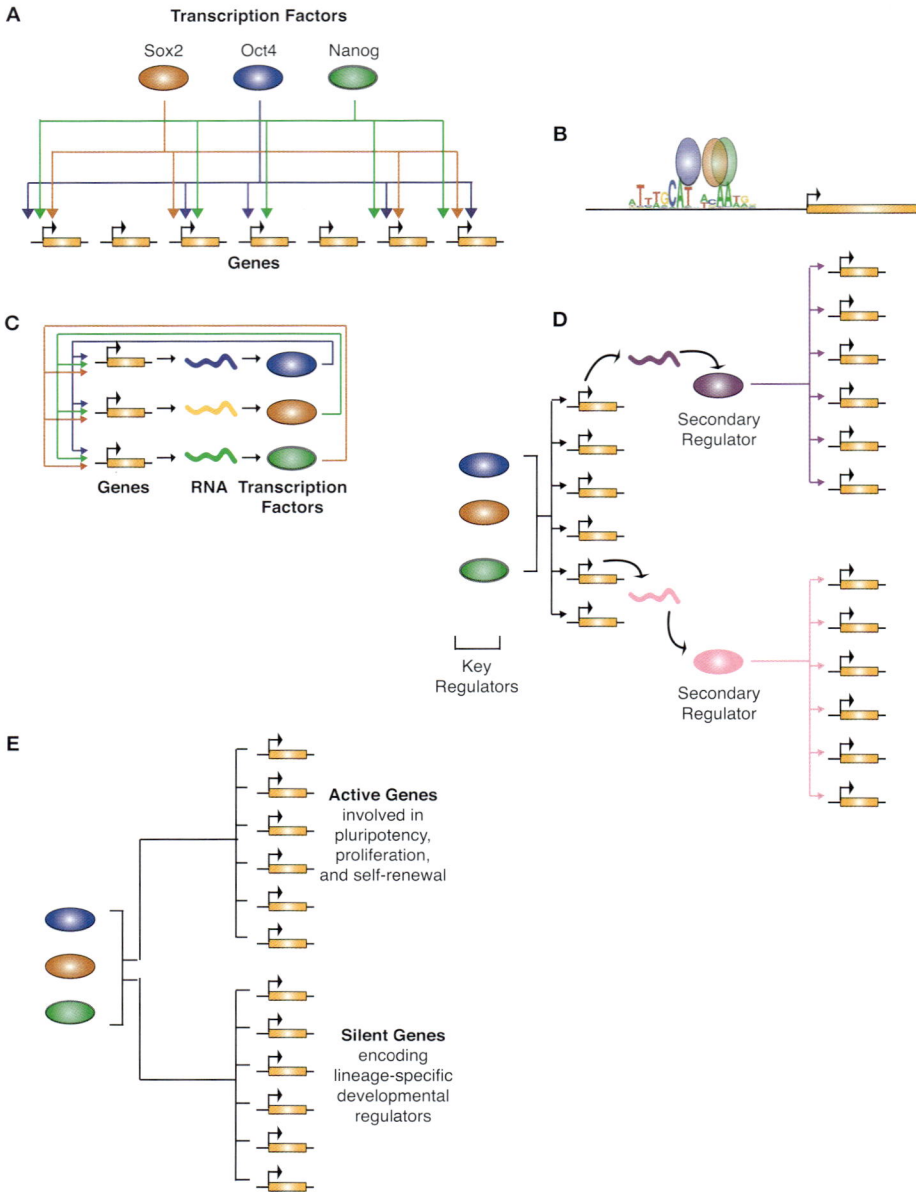

Figure 2. Major themes emerging from studies of key transcription factors in ES cells. (*A*) Key regulators largely target the same set of genes and thus form a multi-input motif. Multi-input motifs consist of a set of regulators that bind to a common set of genes. (*Colored ovals*) Proteins; (*orange rectangles*) genes. An arrow from a protein to a gene indicates direct regulation of the gene by the protein. (*B*) Key transcription factors often bind in extremely close proximity to one another on DNA. The motif associated with the key transcription factors in ES cells is depicted in the promoter region for a gene targeted by all three factors. The average location of binding with respect to this motif as determined by large-scale binding profiles is shown and demonstrates the proximity of binding. (*Colored ovals*) Proteins; (*orange rectangle*) gene. (*C*) Key transcription factors often form interconnected autoregulatory loops where the transcription factors together bind to and regulate each of their promoters. (*Colored ovals*) Proteins; (*orange rectangles*) genes; (*squiggly lines*) their resulting transcripts. An arrow from a protein to a gene indicates direct regulation of the gene by the protein. (*D*) Key transcription factors regulate genes encoding secondary regulators, thus allowing the key regulators ultimate control over many more genes and therefore a larger effect on the complete transcriptional program. (*Colored ovals*) Proteins; (*orange rectangles*) genes; (*squiggly lines*) their resulting transcripts. An arrow from a protein to a gene indicates direct regulation of the gene by the protein. (*E*) Key transcription factors in ES cells bind to the promoters of both active genes involved in pluripotency and self-renewal and to the promoters of inactive genes encoding lineage-specific developmental regulators that are silent in ES cells. (*Colored ovals*) Proteins; (*orange rectangles*) genes. A line from a protein to a gene indicates direct regulation of the gene by the protein.

ulation of silent developmental regulators in ES cells might be critical for pluripotency. Inappropriate expression of lineage-specific developmental regulators could initiate gene expression programs for other cell states, and thus it appears to be important to maintain these key regulators of other cell types in a repressed state in ES cells. Precisely how Oct4, Sox2, and Nanog contribute to repression of lineage-specific developmental regulators is not known.

POLYCOMB AND TRITHORAX CHROMATIN REGULATORS IN THE CORE CIRCUITRY

The genomic locations and functions of Polycomb and Trithorax group (PcG and TrxG) proteins, along with the histone modifications catalyzed by these chromatin regulators, have been the subject of much study in ES cells (Bernstein et al. 2006; Boyer et al. 2006; Lee et al. 2006; Pan et al. 2007; Zhao et al. 2007; Endoh et al. 2008). There is considerable genetic evidence that these chromatin regulators have an important role in early development (Faust et al. 1998; O'Carroll et al. 2001; Pasini et al. 2004; Breiling et al. 2007). Studies of their role in the ES cell network have revealed several important insights that are likely to become general themes of vertebrate transcriptional regulatory networks.

One key insight that emerged from studying these chromatin regulators is that the previous assumption that TrxG complexes are associated with actively transcribed genes, whereas PcG regulators are associated with repressed genes, is imperfect (Bernstein et al. 2006; Boyer et al. 2006; Lee et al. 2006; Pan et al. 2007; Zhao et al. 2007; Endoh et al. 2008). The set of silent genes encoding lineage-specific developmental regulators that are occupied by Oct4, Sox2, and Nanog was also occupied by both PcG and TrxG complexes and contained nucleosomes trimethylated at both histone H3 lysine (K)4 and H3K27. These silent genes encoding developmental regulators

were therefore described as being bivalently marked by both activating and repressive marks.

Further studies revealed that the transcription apparatus was recruited to the promoters of these bivalently marked genes encoding developmental regulators and that transcription was initiated but full-length transcript was not produced (Guenther et al. 2007; Stock et al. 2007). Studies in *Drosophila* suggest that transcriptional pausing is a conserved regulatory feature at genes encoding silent developmental regulators in embryonic tissues (Muse et al. 2007; Zeitlinger et al. 2007; Hendrix et al. 2008; Nechaev and Adelman 2008). This regulatory feature may keep genes encoding developmental regulators in a poised expression state, allowing rapid transcription of certain genes upon induction of differentiation. How only a specific subset of these Pc-repressed genes is induced to overcome transcriptional pausing to permit lineage-specific differentiation is not yet understood.

SIGNALING PATHWAYS BRING DEVELOPMENTAL CUES TO THE NETWORK

Recent studies have revealed how some signal transduction pathways contribute to control of the ES cell transcriptional regulatory network (Pereira et al. 2006; Chen et al. 2008; Cole et al. 2008; Tam et al. 2008; Yi et al. 2008). The Wnt signaling pathway has important roles throughout development and can influence ES cell state (Logan and Nusse 2004; Reya and Clevers 2005). A terminal component of this pathway, the transcription factor Tcf3, was identified as a likely key regulator in ES cells due to its genetic and expression phenotypes (Korinek et al. 1998; Merrill et al. 2004; Pereira et al. 2006).

Subsequent genome-wide studies of Tcf3 in ES cells revealed that the Wnt signaling pathway is intimately connected to the core transcriptional circuitry of ES cells (Fig. 3) (Cole et al. 2008; Marson et al. 2008b; Tam et al.

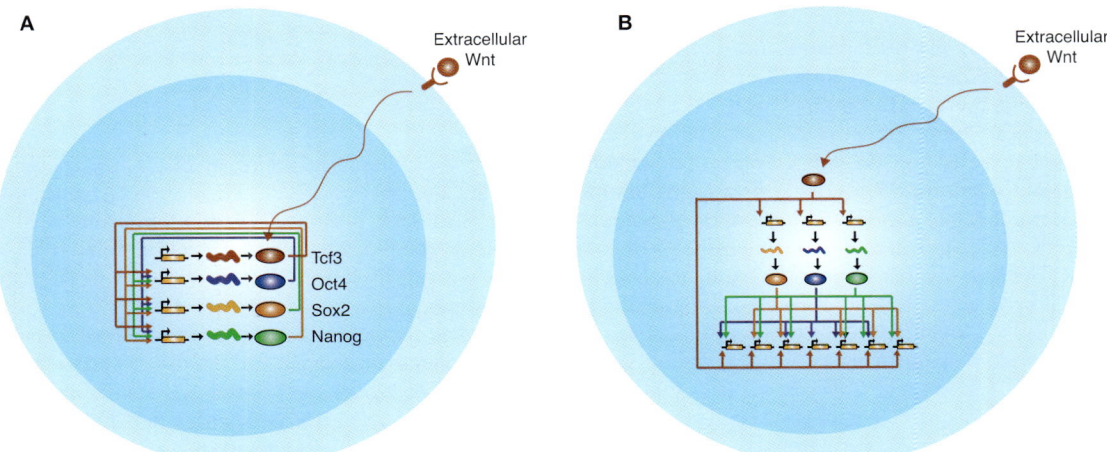

Figure 3. Key themes from studies of signaling pathways in ES cells. (*A*) Signaling pathways can directly connect to key transcription factors in the core interconnected autoregulatory loop through terminal components of their signal transduction pathway. (*Colored ovals*) Proteins; (*orange rectangles*) genes; (*squiggly lines*) their resulting transcripts. An arrow from a protein to a gene indicates direct regulation of the gene by the protein. (*Red Y shape* and *long squiggly arrow*) Wnt signal transduction pathway. (*B*) The terminal component of signaling pathways such as the Wnt pathway can form a feedforward loop by regulating both the key regulators and their target genes. Proteins, genes, and regulation are depicted as in *A*.

2008; Yi et al. 2008). Tcf3 occupies promoters of the key transcription factors Oct4, Sox2, and Nanog, and these factors, together with Tcf3, occupy the Tcf3 promoter (Fig. 3A). Thus, Tcf3 is a component of the interconnected autoregulatory loop that is at the core of ES cell transcriptional regulatory circuitry. These studies also revealed that Tcf3 cooccupied the genome with the key transcription factors, suggesting that the Wnt signaling pathway can affect cellular state by directly connecting to the core circuitry. In this manner, cells could respond to Wnt signaling through a feedforward loop where the key ES cell regulators as well as their targets are immediately targeted by Tcf3 (Fig. 3B). This network structure would allow for both a rapid and stable response to environmental stimuli.

Manipulation of the canonical Wnt pathway through Tcf3 can affect the balance between pluripotency and differentiation in ES cells (Cole et al. 2008). High Wnt pathway activity favors pluripotency, whereas low activity favors differentiation (Sato et al. 2004; Ogawa et al. 2006; Singla et al. 2006; Miyabayashi et al. 2007). Tcf3 and its associated proteins apparently contribute to gene activation when the Wnt pathway is activated and to repression when the pathway is not (Cole et al. 2008; Tam et al. 2008). This suggests that under conditions of high Wnt activity, Tcf3 and the key transcription factors Oct4, Sox2, and Nanog generally function to activate target gene expression (although such activity can be overridden by PcG proteins and other repressors). In contrast, under conditions of low Wnt activity, Tcf3 acts to repress target gene expression and may thus counter the activating functions of Oct4, Sox2, and Nanog. These opposing inputs thus allow ES cells to modulate the level of target gene expression in the core circuitry on the basis of the cell's external environment, which in turn influences the balance between pluripotency and differentiation.

NONCODING RNAS ADD ANOTHER LAYER OF REGULATION TO THE CIRCUITRY

miRNAs are critical for normal ES cell self-renewal and differentiation and have demonstrated roles in early development (Bernstein et al. 2003; Kanellopoulou et al. 2005; Murchison et al. 2005; Wang et al. 2007; Sinkkonen et al. 2008; Stefani and Slack 2008). Recent studies have revealed how the miRNA class of noncoding RNAs is controlled in ES cells, and this information has been incorporated into a model of the core regulatory circuitry of ES cells (Marson et al. 2008b). This class of noncoding RNAs adds another layer to the regulation of gene expression because the RNAs act posttranscriptionally to influence mRNA stability and translation. miRNAs can regulate the expression of many protein-coding genes (Farh et al. 2005; Krek et al. 2005; Lewis et al. 2005; Lim et al. 2005) and thus form a number of interesting control circuits in cells where they are expressed.

miRNA gene expression is regulated in a manner similar to that of regulation of protein-coding genes in ES cells (Fig. 4A) (Marson et al. 2008b). The key transcription factors Oct4, Sox2, Nanog, and Tcf3 occupy and positively regulate the promoters of miRNA genes that are

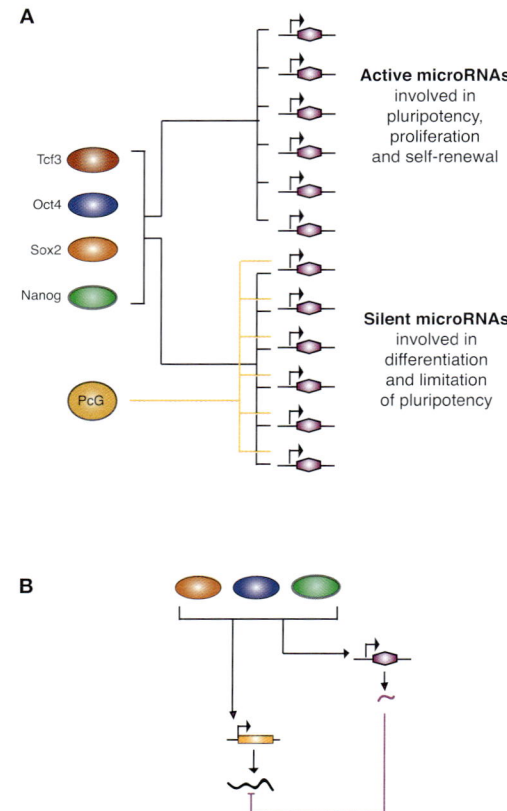

Figure 4. Key themes from studies of miRNAs in ES cells. (*A*) miRNA genes are regulated in a manner similar to that of protein-coding genes. Key transcription factors cooccupy the promoters of active miRNAs and, along with Pc proteins, cooccupy the promoters of silent miRNAs. (*Colored ovals*) Proteins; (*purple hexagons*) miRNA genes. A line from a protein to a gene indicates direct regulation of the gene by the protein. (*B*) miRNAs can participate in an incoherent feedforward loop with key transcription factors to modulate the level of target gene expression. Key transcription factors bind and activate an miRNA gene along with another target gene that specifies an mRNA; the miRNA represses by reducing translation or stability of the mRNA. (*Colored ovals*) Proteins; (*orange rectangle*) protein-coding gene; (*black squiggly line*) its mRNA; (*purple hexagon*) miRNA gene; (*squiggly lines*) resulting transcripts. Regulation is depicted by arrows and t bars.

actively expressed in ES cells. These key transcription factors also occupy a set of silent miRNA genes that are expressed later during differentiation. This set of silent miRNA genes is occupied by PcG proteins in ES cells, thus poising these miRNA genes for expression during development in a lineage-specific fashion.

Studies of miRNAs and the core circuitry of ES cells also revealed recognizable network motifs that provide insights into how networks can control cell state (Marson et al. 2008b). Certain miRNAs, such as mir-290-295, form a common network motif termed an incoherent feed-forward loop with the key transcription factors in ES cells (Fig. 4B) (Alon 2007). This network architecture may allow ES cells to fine-tune gene expression levels of important target genes and facilitate removal of certain ES-cell-specific mRNAs when cells are stimulated to differentiate.

MODEL OF ES CELL CORE REGULATORY CIRCUITRY

A model for ES cell core regulatory circuitry has recently been described that incorporates key transcription factors, chromatin regulators, the Wnt signaling pathway, and miRNAs (Fig. 5) (Marson et al. 2008b). This model represents only a portion of the available data, but it serves to illustrate several important features of ES cell regulatory circuitry. The transcription factors Oct4, Sox2, Nanog, and Tcf3 form an interconnected autoregulatory loop, to which the Wnt signaling pathway connects. The key transcription factors occupy and regulate a set of actively transcribed protein-coding and noncoding genes whose functions contribute to the ES cell state. The products of some of these genes add another layer of regulation; for example, the miRNAs fine-tune the levels of mRNAs for certain protein-coding genes. The key transcription factors also occupy silent protein-coding and noncoding genes involved in lineage-specific functions, and these genes appear to experience transcription initiation, but transcript completion is prevented by PcG proteins and perhaps additional repressors (Guenther et al. 2007; Stock et al. 2007).

REPROGRAMMING TRANSCRIPTIONAL REGULATORY CIRCUITRIES TO THE ES CELL STATE

Somatic cells can be reprogrammed into induced pluripotent stem (iPS) cells by ectopic expression of four or fewer transcriptional regulators (Takahashi and Yamanaka 2006; Meissner et al. 2007; Okita et al. 2007; Takahashi et al. 2007; Wernig et al. 2007, 2008; Yu et al. 2007; Aoi et al. 2008; Jaenisch and Young 2008; Kim et al. 2008b; Nakagawa et al. 2008; Park et al. 2008). The transcription factors that have been used for iPS cell generation have typically included a combination of Oct4, Sox2, Klf4, and cMyc or a mix of Oct4, Sox2, Nanog, and Lin28. Knowledge of the transcriptional regulatory circuitry has already provided insights into the mechanisms by which forced expression of these transcription factors leads to reprogramming of somatic cells (Jaenisch and Young 2008). For example, the interconnected autoregulatory loop of ES cells—composed of genes encoding the transcription factors Oct4, Sox2, Nanog, and Tcf3—can be jump-started by transient expression of these reprogramming factors.

Knowledge of the transcriptional regulatory circuitry

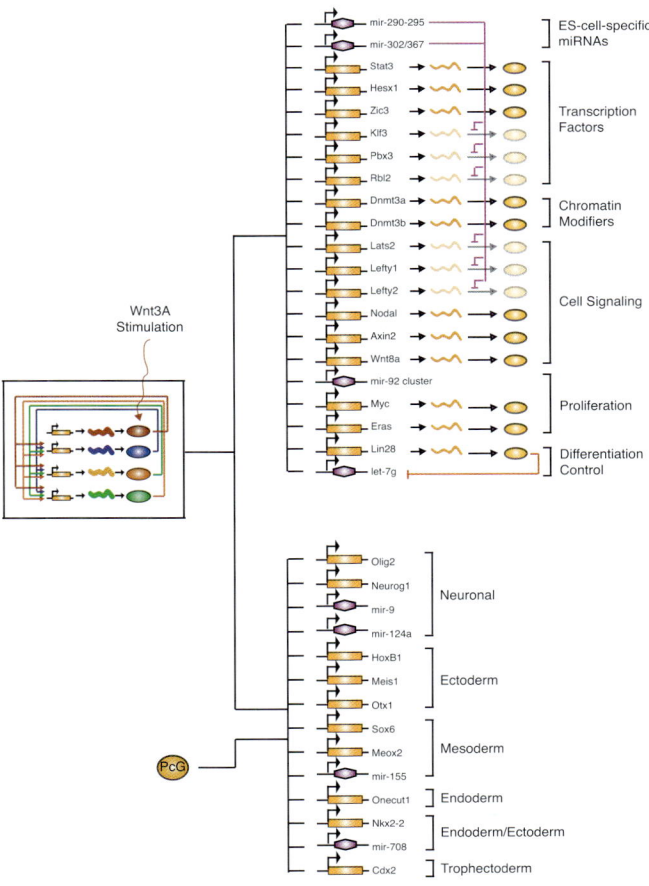

Figure 5. Model of the ES cell transcriptional regulatory circuitry. The interconnected autoregulatory loop formed by Oct4, Sox2, Nanog, and Tcf3, with input from the Wnt pathway, is shown to the left. (*Top right*) Active transcripts; (*bottom right*) inactive transcripts. (*Colored ovals*) Proteins; (*orange rectangles*) protein-coding genes; (*squiggly lines*) their resulting transcripts; (*purple hexagons*) miRNA genes. Regulation is depicted by lines, arrows, or t bars.

has recently been used to improve the reprogramming process (Marson et al. 2008a; Mikkelsen et al. 2008). The discovery that the Wnt pathway is connected directly to ES cell core regulatory circuitry (Cole et al. 2008; Tam et al. 2008; Yi et al. 2008) suggested that manipulation of the Wnt pathway might facilitate reprogramming. Indeed, addition of the Wnt3a ligand allows efficient reprogramming even in the absence of c-Myc (Marson et al. 2008a), which is important because the presence of the exogenous c-Myc oncogene in iPS cells leads to tumors in animals derived from these cells (Okita et al. 2007; Jaenisch and Young 2008). For somatic cells to adopt ES cell transcriptional regulatory circuitry, it is thought that they must silence the expression of key regulators of the somatic cell state. This was confirmed by experiments revealing that repression of key regulators of the somatic cell circuitry using RNA inhibition can substantially improve reprogramming efficiency (Mikkelsen et al. 2008).

CONCLUDING REMARKS

Knowledge of the transcriptional regulatory circuitry is important because it provides insights into the mechanisms by which key regulators control global gene expression, cell state, and development. Despite having identified only core elements of the ES cell transcriptional regulatory circuitry, important insights have been gained into the control of pluripotency and self-renewal in these cells. This knowledge has also provided insights into the mechanisms involved in reprogramming of cell state (Jaenisch and Young 2008) and has led to improved methods for reprogramming (Marson et al. 2008a; Mikkelsen et al. 2008). The new understanding of the core transcriptional regulatory circuitry in ES cells is also likely to shed light on key aspects of cancer, because many features of the ES cell gene expression program are recapitulated in cancer cells (Ben-Porath et al. 2008; Wong et al. 2008). These advances highlight the importance of future work to further map the regulatory circuitry of a wide range of cells, particularly those of medical importance.

ACKNOWLEDGMENTS

We are grateful for the contributions made by many members of the Young and Jaenisch labs to the insights and concepts described in this paper.

REFERENCES

Alon, U. 2007. Network motifs: Theory and experimental approaches. *Nat. Rev. Genet.* **8:** 450–461.

Amaral, P.P., Dinger, M.E., Mercer, T.R., and Mattick, J.S. 2008. The eukaryotic genome as an RNA machine. *Science* **319:** 1787–1789.

Ambros, V. 2004. The functions of animal microRNAs. *Nature* **431:** 350–355.

Ambrosetti, D.C., Basilico, C., and Dailey, L. 1997. Synergistic activation of the fibroblast growth factor 4 enhancer by Sox2 and Oct-3 depends on protein-protein interactions facilitated by a specific spatial arrangement of factor binding sites. *Mol. Cell. Biol.* **17:** 6321–6329.

Aoi, T., Yae, K., Nakagawa, M., Ichisaka, T., Okita, K.,

Takahashi, K., Chiba, T., and Yamanaka, S. 2008. Generation of pluripotent stem cells from adult mouse liver and stomach cells. *Science* **321:** 699–702.

Avilion, A.A., Nicolis, S.K., Pevny, L.H., Perez, L., Vivian, N., and Lovell-Badge, R. 2003. Multipotent cell lineages in early mouse development depend on SOX2 function. *Genes Dev.* **17:** 126–140.

Barski, A., Cuddapah, S., Cui, K., Roh, T.Y., Schones, D.E., Wang, Z., Wei, G., Chepelev, I., and Zhao, K. 2007. High-resolution profiling of histone methylations in the human genome. *Cell* **129:** 823–837.

Bartel, D.P. 2004. MicroRNAs: Genomics, biogenesis, mechanism, and function. *Cell* **116:** 281–297.

Ben-Porath, I., Thomson, M.W., Carey, V.J., Ge, R., Bell, G.W., Regev, A., and Weinberg, R.A. 2008. An embryonic stem cell-like gene expression signature in poorly differentiated aggressive human tumors. *Nat. Genet.* **40:** 499–507.

Berger, S.L. 2007. The complex language of chromatin regulation during transcription. *Nature* **447:** 407–412.

Bernstein, B.E., Mikkelsen, T.S., Xie, X., Kamal, M., Huebert, D.J., Cuff, J., Fry, B., Meissner, A., Wernig, M., Plath, K., et al. 2006. A bivalent chromatin structure marks key developmental genes in embryonic stem cells. *Cell* **125:** 315–326.

Bernstein, E., Kim, S.Y., Carmell, M.A., Murchison, E.P., Alcorn, H., Li, M.Z., Mills, A.A., Elledge, S.J., Anderson, K.V., and Hannon, G.J. 2003. Dicer is essential for mouse development. *Nat. Genet.* **35:** 215–217.

Blackwood, E.M. and Kadonaga, J.T. 1998. Going the distance: A current view of enhancer action. *Science* **281:** 60–63.

Boyer, L.A., Lee, T.I., Cole, M.F., Johnstone, S.E., Levine, S.S., Zucker, J.P., Guenther, M.G., Kumar, R.M., Murray, H.L., Jenner, R.G., et al. 2005. Core transcriptional regulatory circuitry in human embryonic stem cells. *Cell* **122:** 947–956.

Boyer, L.A., Plath, K., Zeitlinger, J., Brambrink, T., Medeiros, L.A., Lee, T.I., Levine, S.S., Wernig, M., Tajonar, A., Ray, M.K., et al. 2006. Polycomb complexes repress developmental regulators in murine embryonic stem cells. *Nature* **441:** 349–353.

Breiling, A., Sessa, L., and Orlando, V. 2007. Biology of Polycomb and Trithorax group proteins. *Int. Rev. Cytol.* **258:** 83–136.

Brivanlou, A.H. and Darnell, Jr., J.E. 2002. Signal transduction and the control of gene expression. *Science* **295:** 813–818.

Cawley, S., Bekiranov, S., Ng, H.H., Kapranov, P., Sekinger, E.A., Kampa, D., Piccolboni, A., Sementchenko, V., Cheng, J., Williams, A.J., et al. 2004. Unbiased mapping of transcription factor binding sites along human chromosomes 21 and 22 points to widespread regulation of noncoding RNAs. *Cell* **116:** 499–509.

Chambers, I., Colby, D., Robertson, M., Nichols, J., Lee, S., Tweedie, S., and Smith, A. 2003. Functional expression cloning of Nanog, a pluripotency sustaining factor in embryonic stem cells. *Cell* **113:** 643–655.

Chen, X., Xu, H., Yuan, P., Fang, F., Huss, M., Vega, V.B., Wong, E., Orlov, Y.L., Zhang, W., Jiang, J., et al. 2008. Integration of external signaling pathways with the core transcriptional network in embryonic stem cells. *Cell* **133:** 1106–1117.

Chew, J.L., Loh, Y.H., Zhang, W., Chen, X., Tam, W.L., Yeap, L.S., Li, P., Ang, Y.S., Lim, B., Robson, P., and Ng, H.H. 2005. Reciprocal transcriptional regulation of Pou5f1 and Sox2 via the Oct4/Sox2 complex in embryonic stem cells. *Mol. Cell. Biol.* **25:** 6031–6046.

Chickarmane, V., Troein, C., Nuber, U.A., Sauro, H.M., and Peterson, C. 2006. Transcriptional dynamics of the embryonic stem cell switch. *PLoS Comput. Biol.* **2:** e123.

Choi, J., Costa, M.L., Mermelstein, C.S., Chagas, C., Holtzer, S., and Holtzer, H. 1990. MyoD converts primary dermal fibroblasts, chondroblasts, smooth muscle, and retinal pigmented epithelial cells into striated mononucleated myoblasts and multinucleated myotubes. *Proc. Natl. Acad. Sci.* **87:** 7988–7992.

Cole, M.F., Johnstone, S.E., Newman, J.J., Kagey, M.H., and Young, R.A. 2008. Tcf3 is an integral component of the core regulatory circuitry of embryonic stem cells. *Genes Dev.* **22:** 746–755.

Core, L.J. and Lis, J.T. 2008. Transcription regulation through promoter-proximal pausing of RNA polymerase II. *Science* **319:** 1791–1792.

Davis, R.L., Weintraub, H., and Lassar, A.B. 1987. Expression of a single transfected cDNA converts fibroblasts to myoblasts. *Cell* **51:** 987–1000.

Endoh, M., Endo, T.A., Endoh, T., Fujimura, Y., Ohara, O., Toyoda, T., Otte, A.P., Okano, M., Brockdorff, N., Vidal, M., and Koseki, H. 2008. Polycomb group proteins Ring1A/B are functionally linked to the core transcriptional regulatory circuitry to maintain ES cell identity. *Development* **135:** 1513–1524.

Evans, T., Felsenfeld, G., and Reitman, M. 1990. Control of globin gene transcription. *Annu. Rev. Cell Biol.* **6:** 95–124.

Farh, K.K., Grimson, A., Jan, C., Lewis, B.P., Johnston, W.K., Lim, L.P., Burge, C.B., and Bartel, D.P. 2005. The widespread impact of mammalian microRNAs on mRNA repression and evolution. *Science* **310:** 1817–1821.

Farkas, I.J., Wu, C., Chennubhotla, C., Bahar, I., and Oltvai, Z.N. 2006. Topological basis of signal integration in the transcriptional-regulatory network of the yeast, *Saccharomyces cerevisiae*. *BMC Bioinformatics* **7:** 478.

Faust, C., Lawson, K.A., Schork, N.J., Thiel, B., and Magnuson, T. 1998. The *Polycomb*-group gene *eed* is required for normal morphogenetic movements during gastrulation in the mouse embryo. *Development* **125:** 4495–4506.

Fazzio, T.G., Huff, J.T., and Panning, B. 2008. An RNAi screen of chromatin proteins identifies Tip60-p400 as a regulator of embryonic stem cell identity. *Cell* **134:** 162–174.

Giaever, G., Chu, A.M., Ni, L., Connelly, C., Riles, L., Veronneau, S., Dow, S., Lucau-Danila, A., Anderson, K., Andre, B., et al. 2002. Functional profiling of the *Saccharomyces cerevisiae* genome. *Nature* **418:** 387–391.

Goll, M.G. and Bestor, T.H. 2005. Eukaryotic cytosine methyltransferases. *Annu. Rev. Biochem.* **74:** 481–514.

Goodrich, J.A. and Kugel, J.F. 2006. Non-coding-RNA regulators of RNA polymerase II transcription. *Nat. Rev. Mol. Cell Biol.* **7:** 612–616.

Greene, W.C. 1990. Regulation of HIV-1 gene expression. *Annu. Rev. Immunol.* **8:** 453–475.

Grewal, S.I. and Elgin, S.C. 2007. Transcription and RNA interference in the formation of heterochromatin. *Nature* **447:** 399–406.

Guenther, M.G., Levine, S.S., Boyer, L.A., Jaenisch, R., and Young, R.A. 2007. A chromatin landmark and transcription initiation at most promoters in human cells. *Cell* **130:** 77–88.

Harbison, C.T., Gordon, D.B., Lee, T.I., Rinaldi, N.J., Macisaac, K.D., Danford, T.W., Hannett, N.M., Tagne, J.B., Reynolds, D.B., Yoo, J., et al. 2004. Transcriptional regulatory code of a eukaryotic genome. *Nature* **431:** 99–104.

Harrison, S.C. 1991. A structural taxonomy of DNA-binding domains. *Nature* **353:** 715–719.

Hart, A.H., Hartley, L., Ibrahim, M., and Robb, L. 2004. Identification, cloning and expression analysis of the pluripotency promoting Nanog genes in mouse and human. *Dev. Dyn.* **230:** 187–198.

Hawkins, P.G. and Morris, K.V. 2008. RNA and transcriptional modulation of gene expression. *Cell Cycle* **7:** 602–607.

Hendrix, D.A., Hong, J.W., Zeitlinger, J., Rokhsar, D.S., and Levine, M.S. 2008. Promoter elements associated with RNA pol II stalling in the *Drosophila* embryo. *Proc. Natl. Acad. Sci.* **105:** 7762–7767.

Hobert, O. 2008. Gene regulation by transcription factors and microRNAs. *Science* **319:** 1785–1786.

Hunter, T. 2000. Signaling—2000 and beyond. *Cell* **100:** 113–127.

Ivanova, N., Dobrin, R., Lu, R., Kotenko, I., Levorse, J., DeCoste, C., Schafer, X., Lun, Y., and Lemischka, I.R. 2006. Dissecting self-renewal in stem cells with RNA interference. *Nature* **442:** 533–538.

Jaenisch, R. and Young, R. 2008. Stem cells, the molecular circuitry of pluripotency and nuclear reprogramming. *Cell* **132:** 567–582.

Jiang, J., Chan, Y.S., Loh, Y.H., Cai, J., Tong, G.Q., Lim, C.A., Robson, P., Zhong, S., and Ng, H.H. 2008. A core Klf circuitry

regulates self-renewal of embryonic stem cells. *Nat. Cell Biol.* **10:** 353–360.

Kadonaga, J.T. 2004. Regulation of RNA polymerase II transcription by sequence-specific DNA binding factors. *Cell* **116:** 247–257.

Kanellopoulou, C., Muljo, S.A., Kung, A.L., Ganesan, S., Drapkin, R., Jenuwein, T., Livingston, D.M., and Rajewsky, K. 2005. Dicer-deficient mouse embryonic stem cells are defective in differentiation and centromeric silencing. *Genes Dev.* **19:** 489–501.

Kemphues, K. 2005. Essential genes. In *WormBook* (The *C. elegans* Research Community, WormBook), pp. 1–7 (http://www.wormbook.org).

Kim, J., Chu, J., Shen, X., Wang, J., and Orkin, S.H. 2008a. An extended transcriptional network for pluripotency of embryonic stem cells. *Cell* **132:** 1049–1061.

Kim, J.B., Zaehres, H., Wu, G., Gentile, L., Ko, K., Sebastiano, V., Araúzo-Bravo, M.J., Ruau, D., Han, D.W., Zenke, M., and Schöler, H.R. 2008b. Pluripotent stem cells induced from adult neural stem cells by reprogramming with two factors. *Nature* **454:** 646–650.

Kloosterman, W.P. and Plasterk, R.H. 2006. The diverse functions of microRNAs in animal development and disease. *Dev. Cell* **11:** 441–450.

Klose, R.J. and Bird, A.P. 2006. Genomic DNA methylation: The mark and its mediators. *Trends Biochem. Sci.* **31:** 89–97.

Komashko, V.M., Acevedo, L.G., Squazzo, S.L., Iyengar, S.S., Rabinovich, A., O'Geen, H., Green, R., and Farnham, P.J. 2008. Using ChIP-chip technology to reveal common principles of transcriptional repression in normal and cancer cells. *Genome Res.* **18:** 521–532.

Korinek, V., Barker, N., Willert, K., Molenaar, M., Roose, J., Wagenaar, G., Markman, M., Lamers, W., Destree, O., and Clevers, H. 1998. Two members of the Tcf family implicated in Wnt/β-catenin signaling during embryogenesis in the mouse. *Mol. Cell. Biol.* **18:** 1248–1256.

Kouzarides, T. 2007. Chromatin modifications and their function. *Cell* **128:** 693–705.

Krek, A., Grun, D., Poy, M.N., Wolf, R., Rosenberg, L., Epstein, E.J., MacMenamin, P., da Piedade, I., Gunsalus, K.C., Stoffel, M., and Rajewsky, N. 2005. Combinatorial microRNA target predictions. *Nat. Genet.* **37:** 495–500.

Lander, E.S., Linton, L.M., Birren, B., Nusbaum, C., Zody, M.C., Baldwin, J., Devon, K., Dewar, K., Doyle, M., FitzHugh, W., et al. 2001. Initial sequencing and analysis of the human genome. *Nature* **409:** 860–921.

Latchman, D.S. 1997. Transcription factors: An overview. *Int. J. Biochem. Cell Biol.* **29:** 1305–1312.

Latchman, D.S. 2008. Transcription factors and human disease. In *Eukaryotic transcription factors,* 5th ed., chap. 9, pp. 373–448. Elsevier/Academic, New York.

Lee, T.I. and Young, R.A. 2000. Transcription of eukaryotic protein-coding genes. *Annu. Rev. Genet.* **34:** 77–137.

Lee, T.I., Jenner, R.G., Boyer, L.A., Guenther, M.G., Levine, S.S., Kumar, R.M., Chevalier, B., Johnstone, S.E., Cole, M.F., Isono, K., et al. 2006. Control of developmental regulators by Polycomb in human embryonic stem cells. *Cell* **125:** 301–313.

Lee, T.I., Rinaldi, N.J., Robert, F., Odom, D.T., Bar-Joseph, Z., Gerber, G.K., Hannett, N.M., Harbison, C.T., Thompson, C.M., Simon, I., et al. 2002. Transcriptional regulatory networks in *Saccharomyces cerevisiae*. *Science* **298:** 799–804.

Levine, M. and Tjian, R. 2003. Transcription regulation and animal diversity. *Nature* **424:** 147–151.

Lewis, B.P., Burge, C.B., and Bartel, D.P. 2005. Conserved seed pairing, often flanked by adenosines, indicates that thousands of human genes are microRNA targets. *Cell* **120:** 15–20.

Li, B., Carey, M., and Workman, J.L. 2007. The role of chromatin during transcription. *Cell* **128:** 707–719.

Lim, L.P., Lau, N.C., Garrett-Engele, P., Grimson, A., Schelter, J.M., Castle, J., Bartel, D.P., Linsley, P.S. and Johnson, J.M. 2005. Microarray analysis shows that some microRNAs downregulate large numbers of target mRNAs. *Nature* **433:** 769–773.

Logan, C.Y. and Nusse, R. 2004. The Wnt signaling pathway in development and disease. *Annu. Rev. Cell Dev. Biol.* **20:** 781–810.

Loh, Y.H., Wu, Q., Chew, J.L., Vega, V.B., Zhang, W., Chen, X., Bourque, G., George, J., Leong, B., Liu, J., et al. 2006. The Oct4 and Nanog transcription network regulates pluripotency in mouse embryonic stem cells. *Nat. Genet.* **38:** 431–440.

Ma, H.W., Kumar, B., Ditges, U., Gunzer, F., Buer, J., and Zeng, A.P. 2004. An extended transcriptional regulatory network of *Escherichia coli* and analysis of its hierarchical structure and network motifs. *Nucleic Acids Res.* **32:** 6643–6649.

Makeyev, E.V. and Maniatis, T. 2008. Multilevel regulation of gene expression by microRNAs. *Science* **319:** 1789–1790.

Marson, A., Foreman, R., Chevalier, B., Bilodeau, S., Kahn, M., Young, R., and Jaenisch, R. 2008a. Wnt signaling promotes reprogramming of somatic cells to pluripotency. *Cell Stem Cell* **3:** 132–135.

Marson, A., Levine, S.S., Cole, M.F., Frampton, G.M., Brambrink, T., Johnstone, S., Guenther, M.G., Johnston, W.K., Wernig, M., Newman, J., et al. 2008b. Connecting microRNA genes to the core transcriptional regulatory circuitry of embryonic stem cells. *Cell* **134:** 521–533.

Martinez-Antonio, A. and Collado-Vides, J. 2003. Identifying global regulators in transcriptional regulatory networks in bacteria. *Curr. Opin. Microbiol.* **6:** 482–489.

Meissner, A., Wernig, M., and Jaenisch, R. 2007. Direct reprogramming of genetically unmodified fibroblasts into pluripotent stem cells. *Nat. Biotechnol.* **25:** 1177–1181.

Meister, G. 2007. miRNAs get an early start on translational silencing. *Cell* **131:** 25–28.

Merrill, B.J., Pasolli, H.A., Polak, L., Rendl, M., Garcia-Garcia, M.J., Anderson, K.V., and Fuchs, E. 2004. Tcf3: A transcriptional regulator of axis induction in the early embryo. *Development* **131:** 263–274.

Mikkelsen, T.S., Hanna, J., Zhang, X., Ku, M., Wernig, M., Schorderet, P., Bernstein, B.E., Jaenisch, R., Lander, E.S., and Meissner, A. 2008. Dissecting direct reprogramming through integrative genomic analysis. *Nature* **454:** 49–55.

Mikkelsen, T.S., Ku, M., Jaffe, D.B., Issac, B., Lieberman, E., Giannoukos, G., Alvarez, P., Brockman, W., Kim, T.K., Koche, R.P., et al. 2007. Genome-wide maps of chromatin state in pluripotent and lineage-committed cells. *Nature* **448:** 553–560.

Mitsui, K., Tokuzawa, Y., Itoh, H., Segawa, K., Murakami, M., Takahashi, K., Maruyama, M., Maeda, M., and Yamanaka, S. 2003. The homeoprotein Nanog is required for maintenance of pluripotency in mouse epiblast and ES cells. *Cell* **113:** 631–642.

Miyabayashi, T., Teo, J.L., Yamamoto, M., McMillan, M., Nguyen, C., and Kahn, M. 2007. Wnt/β-catenin/CBP signaling maintains long-term murine embryonic stem cell pluripotency. *Proc. Natl. Acad. Sci.* **104:** 5668–5673.

Moazed, D., Buhler, M., Buker, S.M., Colmenares, S.U., Gerace, E.L., Gerber, S.A., Hong, E.J., Motamedi, M.R., Verdel, A., Villen, J., and Gygi, S.P. 2006. Studies on the mechanism of RNAi-dependent heterochromatin assembly. *Cold Spring Harbor Symp. Quant. Biol.* **71:** 461–471.

Murchison, E.P., Partridge, J.F., Tam, O.H., Cheloufi, S., and Hannon, G.J. 2005. Characterization of Dicer-deficient murine embryonic stem cells. *Proc. Natl. Acad. Sci.* **102:** 12135–12140.

Muse, G.W., Gilchrist, D.A., Nechaev, S., Shah, R., Parker, J.S., Grissom, S.F., Zeitlinger, J., and Adelman, K. 2007. RNA polymerase is poised for activation across the genome. *Nat. Genet.* **39:** 1507–1511.

Nakagawa, M., Koyanagi, M., Tanabe, K., Takahashi, K., Ichisaka, T., Aoi, T., Okita, K., Mochiduki, Y., Takizawa, N., and Yamanaka, S. 2008. Generation of induced pluripotent stem cells without Myc from mouse and human fibroblasts. *Nat. Biotechnol.* **26:** 101–106.

Nechaev, S. and Adelman, K. 2008. Promoter-proximal pol II: When stalling speeds things up. *Cell Cycle* **7:** 1539–1544.

Nichols, J., Zevnik, B., Anastassiadis, K., Niwa, H., Klewe-Nebenius, D., Chambers, I., Schöler, H., and Smith, A. 1998. Formation of pluripotent stem cells in the mammalian embryo depends on the POU transcription factor Oct4. *Cell* **95:** 379–391.

O'Carroll, D., Erhardt, S., Pagani, M., Barton, S.C., Surani, M.A., and Jenuwein, T. 2001. The Polycomb-group gene *Ezh2* is required for early mouse development. *Mol. Cell. Biol.* **21:** 4330–4336.

Odom, D.T., Dowell, R.D., Jacobsen, E.S., Nekludova, L., Rolfe, P.A., Danford, T.W., Gifford, D.K., Fraenkel, E., Bell, G.I., and Young, R.A. 2006. Core transcriptional regulatory circuitry in human hepatocytes. *Mol. Syst. Biol.* **2:** 0017.

Odom, D.T., Zizlsperger, N., Gordon, D.B., Bell, G.W., Rinaldi, N.J., Murray, H.L., Volkert, T.L., Schreiber, J., Rolfe, P.A., Gifford, D.K., Fraenkel, E., Bell, G.I. and Young, R.A. 2004. Control of pancreas and liver gene expression by HNF transcription factors. *Science* **303:** 1378–1381.

Ogata, K., Sato, K., and Tahirov, T.H. 2003. Eukaryotic transcriptional regulatory complexes: Cooperativity from near and afar. *Curr. Opin. Struct. Biol.* **13:** 40–48.

Ogawa, K., Nishinakamura, R., Iwamatsu, Y., Shimosato, D., and Niwa, H. 2006. Synergistic action of Wnt and LIF in maintaining pluripotency of mouse ES cells. *Biochem. Biophys. Res. Commun.* **343:** 159–166.

Okita, K., Ichisaka, T., and Yamanaka, S. 2007. Generation of germline-competent induced pluripotent stem cells. *Nature* **448:** 313–317.

Orlando, V. 2003. Polycomb, epigenomes, and control of cell identity. *Cell* **112:** 599–606.

Orphanides, G. and Reinberg, D. 2002. A unified theory of gene expression. *Cell* **108:** 439–451.

Pabo, C.O. and Sauer, R.T. 1992. Transcription factors: Structural families and principles of DNA recognition. *Annu. Rev. Biochem.* **61:** 1053–1095.

Pan, G., Tian, S., Nie, J., Yang, C., Ruotti, V., Wei, H., Jonsdottir, G.A., Stewart, R., and Thomson, J.A. 2007. Whole-genome analysis of histone H3 lysine 4 and lysine 27 methylation in human embryonic stem cells. *Cell Stem Cell* **1:** 299–312.

Panne, D., Maniatis, T., and Harrison, S.C. 2004. Crystal structure of ATF-2/c-Jun and IRF-3 bound to the interferon-β enhancer. *EMBO J.* **23:** 4384–4393.

Park, I.H., Zhao, R., West, J.A., Yabuuchi, A., Huo, H., Ince, T.A., Lerou, P.H., Lensch, M.W. and Daley, G.Q. 2008. Reprogramming of human somatic cells to pluripotency with defined factors. *Nature* **451:** 141–146.

Pasini, D., Bracken, A.P., Jensen, M.R., Lazzerini Denchi, E., and Helin, K. 2004. Suz12 is essential for mouse development and for EZH2 histone methyltransferase activity. *EMBO J.* **23:** 4061–4071.

Pereira, L., Yi, F., and Merrill, B. J. 2006. Repression of Nanog gene transcription by Tcf3 limits embryonic stem cell self-renewal. *Mol. Cell. Biol.* **26:** 7479–7491.

Pirrotta, V. 1998. Polycombing the genome: PcG, trxG, and chromatin silencing. *Cell* **93:** 333–336.

Pokholok, D.K., Zeitlinger, J., Hannett, N.M., Reynolds, D.B., and Young, R.A. 2006. Activated signal transduction kinases frequently occupy target genes. *Science* **313:** 533–536.

Rada-Iglesias, A., Ameur, A., Kapranov, P., Enroth, S., Komorowski, J., Gingeras, T.R., and Wadelius, C. 2008. Whole-genome maps of USF1 and USF2 binding and histone H3 acetylation reveal new aspects of promoter structure and candidate genes for common human disorders. *Genome Res.* **18:** 380–392.

Rada-Iglesias, A., Wallerman, O., Koch, C., Ameur, A., Enroth, S., Clelland, G., Wester, K., Wilcox, S., Dovey, O.M., Ellis, P.D., et al. 2005. Binding sites for metabolic disease related transcription factors inferred at base pair resolution by chromatin immunoprecipitation and genomic microarrays. *Hum. Mol. Genet.* **14:** 3435–3447.

Reed, B.D., Charos, A.E., Szekely, A.M., Weissman, S.M., and Snyder, M. 2008. Genome-wide occupancy of SREBP1 and its partners NFY and SP1 reveals novel functional roles and combinatorial regulation of distinct classes of genes. *PLoS Genet.* **4:** e1000133.

Remenyi, A., Schöler, H.R., and Wilmanns, M. 2004. Combinatorial control of gene expression. *Nat. Struct. Mol. Biol.* **11:** 812–815.

Reya, T. and Clevers, H. 2005. Wnt signalling in stem cells and cancer. *Nature* **434:** 843–850.

Ringrose, L. and Paro, R. 2004. Epigenetic regulation of cellular memory by the Polycomb and Trithorax group proteins. *Annu. Rev. Genet.* **38:** 413–443.

Rinn, J.L., Kertesz, M., Wang, J.K., Squazzo, S.L., Xu, X., Brugmann, S.A., Goodnough, L.H., Helms, J.A., Farnham, P.J., Segal, E., and Chang, H.Y. 2007. Functional demarcation of active and silent chromatin domains in human *HOX* loci by noncoding RNAs. *Cell* **129:** 1311–1323.

Sato, N., Meijer, L., Skaltsounis, L., Greengard, P., and Brivanlou, A.H. 2004. Maintenance of pluripotency in human and mouse embryonic stem cells through activation of Wnt signaling by a pharmacological GSK-3-specific inhibitor. *Nat. Med.* **10:** 55–63.

Schöler, H.R., Dressler, G.R., Balling, R., Rohdewohld, H., and Gruss, P. 1990. Oct-4: A germline-specific transcription factor mapping to the mouse t-complex. *EMBO J.* **9:** 2185–2195.

Schuettengruber, B., Chourrout, D., Vervoort, M., Leblanc, B., and Cavalli, G. 2007. Genome regulation by polycomb and trithorax proteins. *Cell* **128:** 735–745.

Schwartz, Y.B. and Pirrotta, V. 2007. Polycomb silencing mechanisms and the management of genomic programmes. *Nat. Rev. Genet.* **8:** 9–22.

Shyamsundar, R., Kim, Y.H., Higgins, J.P., Montgomery, K., Jorden, M., Sethuraman, A., van de Rijn, M., Botstein, D., Brown, P.O., and Pollack, J.R. 2005. A DNA microarray survey of gene expression in normal human tissues. *Genome Biol.* **6:** R22.

Singla, D.K., Schneider, D.J., LeWinter, M.M., and Sobel, B.E. 2006. wnt3a but not wnt11 supports self-renewal of embryonic stem cells. *Biochem. Biophys. Res. Commun.* **345:** 789–795.

Sinkkonen, L., Hugenschmidt, T., Berninger, P., Gaidatzis, D., Mohn, F., Artus-Revel, C.G., Zavolan, M., Svoboda, P., and Filipowicz, W. 2008. MicroRNAs control de novo DNA methylation through regulation of transcriptional repressors in mouse embryonic stem cells. *Nat. Struct. Mol. Biol.* **15:** 259–267.

Stefani, G. and Slack, F.J. 2008. Small non-coding RNAs in animal development. *Nat. Rev. Mol. Cell Biol.* **9:** 219–230.

Stock, J.K., Giadrossi, S., Casanova, M., Brookes, E., Vidal, M., Koseki, H., Brockdorff, N., Fisher, A.G., and Pombo, A. 2007. Ring1-mediated ubiquitination of H2A restrains poised RNA polymerase II at bivalent genes in mouse ES cells. *Nat. Cell Biol.* **9:** 1428–1435.

Su, A.I., Wiltshire, T., Batalov, S., Lapp, H., Ching, K.A., Block, D., Zhang, J., Soden, R., Hayakawa, M., Kreiman, G., Cooke, M.P., Walker, J.R., and Hogenesch, J.B. 2004. A gene atlas of the mouse and human protein-encoding transcriptomes. *Proc. Natl. Acad. Sci.* **101:** 6062–6067.

Takahashi, K. and Yamanaka, S. 2006. Induction of pluripotent stem cells from mouse embryonic and adult fibroblast cultures by defined factors. *Cell* **126:** 663–676.

Takahashi, K., Tanabe, K., Ohnuki, M., Narita, M., Ichisaka, T., Tomoda, K., and Yamanaka, S. 2007. Induction of pluripotent stem cells from adult human fibroblasts by defined factors. *Cell* **131:** 861–872.

Tam, W.L., Lim, C.Y., Han, J., Zhang, J., Ang, Y.S., Ng, H.H., Yang, H., and Lim, B. 2008. T-cell factor 3 regulates embryonic stem cell pluripotency and self-renewal by the transcriptional control of multiple lineage pathways. *Stem Cells* **26:** 2019–2031.

Turek-Plewa, J. and Jagodziński, P.P. 2005. The role of mammalian DNA methyltransferases in the regulation of gene expression. *Cell. Mol. Biol. Lett.* **10:** 631–647.

Valencia-Sanchez, M.A., Liu, J., Hannon, G.J., and Parker, R. 2006. Control of translation and mRNA degradation by miRNAs and siRNAs. *Genes Dev.* **20:** 515–524.

Verdel, A., Jia, S., Gerber, S., Sugiyama, T., Gygi, S., Grewal, S.I., and Moazed, D. 2004. RNAi-mediated targeting of heterochromatin by the RITS complex. *Science* **303:** 672–676.

Vickaryous, M.K. and Hall, B.K. 2006. Human cell type diver-

sity, evolution, development, and classification with special reference to cells derived from the neural crest. *Biol. Rev. Camb. Philos. Soc.* **81:** 425–455.

Villard, J. 2004. Transcription regulation and human diseases. *Swiss Med. Wkly.* **134:** 571–579.

Wang, Y., Medvid, R., Melton, C., Jaenisch, R., and Blelloch, R. 2007. DGCR8 is essential for microRNA biogenesis and silencing of embryonic stem cell self-renewal. *Nat. Genet.* **39:** 380–385.

Wang, Z., Zang, C., Rosenfeld, J.A., Schones, D.E., Barski, A., Cuddapah, S., Cui, K., Roh, T.Y., Peng, W., Zhang, M.Q., and Zhao, K. 2008. Combinatorial patterns of histone acetylations and methylations in the human genome. *Nat. Genet.* **40:** 897–903.

Weintraub, H., Tapscott, S.J., Davis, R.L., Thayer, M.J., Adam, M.A., Lassar, A.B., and Miller, A.D. 1989. Activation of muscle-specific genes in pigment, nerve, fat, liver, and fibroblast cell lines by forced expression of MyoD. *Proc. Natl. Acad. Sci.* **86:** 5434–5438.

Weintraub, H., Davis, R., Tapscott, S., Thayer, M., Krause, M., Benezra, R., Blackwell, T.K., Turner, D., Rupp, R., Hollenberg, S., et al. 1991. The myoD gene family: Nodal point during specification of the muscle cell lineage. *Science* **251:** 761–766.

Wernig, M., Meissner, A., Cassady, J.P., and Jaenisch, R. 2008. c-Myc is dispensable for direct reprogramming of mouse fibroblasts. *Cell Stem Cell* **2:** 10–12.

Wernig, M., Meissner, A., Foreman, R., Brambrink, T., Ku, M., Hochedlinger, K., Bernstein, B.E., and Jaenisch, R. 2007. In vitro reprogramming of fibroblasts into a pluripotent ES-cell-like state. *Nature* **448:** 318–324.

West, A.G. and Fraser, P. 2005. Remote control of gene transcription. *Hum. Mol. Genet.* (spec. no. 1) **14:** R101–R111.

Winzeler, E.A., Shoemaker, D.D., Astromoff, A., Liang, H., Anderson, K., Andre, B., Bangham, R., Benito, R., Boeke, J.D., Bussey, H., et al. 1999. Functional characterization of the *S. cerevisiae* genome by gene deletion and parallel analysis. *Science* **285:** 901–906.

Wong, D.J., Liu, H., Ridky, T.W., Cassarino, D., Segal, E., and Chang, H.Y. 2008. Module map of stem cell genes guides creation of epithelial cancer stem cells. *Cell Stem Cell* **2:** 333–344.

Wu, Q., Chen, X., Zhang, J., Loh, Y.H., Low, T.Y., Zhang, W., Sze, S.K., Lim, B., and Ng, H.H. 2006. Sall4 interacts with Nanog and co-occupies Nanog genomic sites in embryonic stem cells. *J. Biol. Chem.* **281:** 24090–24094.

Yang, S.H., Sharrocks, A.D., and Whitmarsh, A.J. 2003. Transcriptional regulation by the MAP kinase signaling cascades. *Gene* **320:** 3–21.

Yi, F., Pereira, L. and Merrill, B.J. 2008. Tcf3 functions as a steady-state limiter of transcriptional programs of mouse embryonic stem cell self-renewal. *Stem Cells* **26:** 1951–1960.

Yu, J., Vodyanik, M.A., Smuga-Otto, K., Antosiewicz-Bourget, J., Frane, J.L., Tian, S., Nie, J., Jonsdottir, G.A., Ruotti, V., Stewart, R., Slukvin, I.I., and Thomson, J.A. 2007. Induced pluripotent stem cell lines derived from human somatic cells. *Science* **318:** 1917–1920.

Zaratiegui, M., Irvine, D.V., and Martienssen, R.A. 2007. Noncoding RNAs and gene silencing. *Cell* **128:** 763–776.

Zeitlinger, J., Stark, A., Kellis, M., Hong, J.W., Nechaev, S., Adelman, K., Levine, M., and Young, R.A. 2007. RNA polymerase stalling at developmental control genes in the *Drosophila melanogaster* embryo. *Nat. Genet.* **39:** 1512–1516.

Zhang, J.Z., Gao, W., Yang, H.B., Zhang, B., Zhu, Z.Y., and Xue, Y.F. 2006. Screening for genes essential for mouse embryonic stem cell self-renewal using a subtractive RNA interference library. *Stem Cells* **24:** 2661–2668.

Zhao, X.D., Han, X., Chew, J.L., Liu, J., Chiu, K.P., Choo, A., Orlov, Y.L., Sung, W.K., Shahab, A., Kuznetsov, V.A., et al. 2007. Whole-genome mapping of histone H3 Lys4 and 27 trimethylations reveals distinct genomic compartments in human embryonic stem cells. *Cell Stem Cell* **1:** 286–298.

The Transcriptional Network Controlling Pluripotency in ES Cells

S.H. Orkin, J. Wang, J. Kim, J. Chu, S. Rao, T.W. Theunissen, X. Shen, and D.N. Levasseur

Department of Pediatric Oncology, Dana Farber Cancer Institute and Children's Hospital Boston,
Howard Hughes Medical Institute, Harvard Stem Cell Institute, Boston, Maassachusetts 02115

Embryonic stem (ES) cells are capable of continuous self-renewal and pluripotential differentiation. A "core" set of transcription factors, Oct4, Sox2, and Nanog, maintains the ES cell state, whereas various combinations of factors, invariably including Oct4 and Sox2, reprogram somatic cells to pluripotency. We have sought to define the transcriptional network controlling pluripotency in mouse ES cells through combined proteomic and genomic approaches. We constructed a protein interaction network surrounding Nanog and determined gene targets of the core and reprogramming factors, plus others. The expanded transcriptional network we have constructed forms the basis for further studies of directed differentiation and lineage reprogramming, and a paradigm for comprehensive elucidation of regulatory pathways in other stem cells.

Stem cells are distinguished by two properties: self-renewal and the potential to differentiate. Our understanding of the mechanisms underlying these features in vertebrate stem cells is incomplete. Whereas powerful genetic approaches in invertebrates permit delineation of the cellular components required for self-renewal and differentiation, as well as for interactions of stem cells with their niches, analogous strategies are not readily applicable to vertebrate stem cells. In considering this landscape, we chose to focus on mouse ES cells (Evans and Kaufman 1981; Martin 1981) as a tractable system in which to apply biochemical and genetic methods with the goal of providing a comprehensive description of the networks controlling self-renewal and pluripotency. Although ES cells may represent a "special case" given their origin and the artificial manner in which they are maintained in culture, they provide a convenient source of unlimited, quite homogeneous, self-renewing stem cells for biochemical studies. Moreover, the facility with which ES cells may be modified by gene-targeting or other loss-of-function approaches (e.g., si/shRNA inhibition) permits functional assessment of the contribution of specific components to the pluripotent state. Hence, in the work summarized here, our goal has been to identify in a comprehensive, relatively unbiased manner the proteins critical for maintenance of pluripotency and self-renewal and to delineate how they act individually and together in these processes. These approaches should lead to general concepts that may be applied to other stem cell systems and suggest methods for improved reprogramming of somatic cells to pluripotency (Takahashi and Yamanaka 2006; Takahashi et al. 2007) or to alternative fates.

ITERATIVE AFFINITY PROTEIN PURIFICATION FOR GENERATION OF A PLURIPOTENCY PROTEIN–PROTEIN INTERACTION NETWORK: STRATEGIC CONSIDERATIONS

To provide an initial point of reference in our analysis, we elected to focus on the transcription factor Nanog in light of its capacity to drive mouse ES cell self-renewal in the absence of leukemia inhibitory factor (LIF) in the medium (Chambers et al. 2003; Mitsui et al. 2003) and facilitate fusion-induced cellular reprogramming (Silva et al. 2006). In addition, loss of Nanog leads to cellular differentiation, specifically along the primitive endodermal pathway. As an initial hypothesis, we entertained the possibility that Nanog, as a critical regulatory factor, might interact physically with other proteins that participate in maintenance of pluripotency (Fig. 1). If so, purification of Nanog with its associated proteins would serve as a tool for the discovery of novel proteins involved in pluripotency and/or connect Nanog to already recognized proteins (such as Oct4 or Sox2). At one extreme, proteins involved in pluripotency might be "concentrated" within a subnetwork among all cellular proteins (Fig. 2) (Dezso et al. 2003). In this manner, the proteins would "talk" to

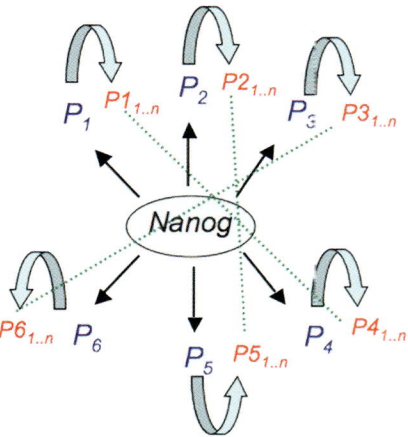

Figure 1. Iterative identification of Nanog-associated proteins. Proteins that associate with Nanog in protein complexes are depicted as P1....P6. Subsequent isolation of complexes containing P1....P6 in an iterative fashion leads to identification of secondary interaction proteins, e.g., $P1_{1...n}$. In some instances (*dashed lines*), the same protein will be recovered in different protein complexes.

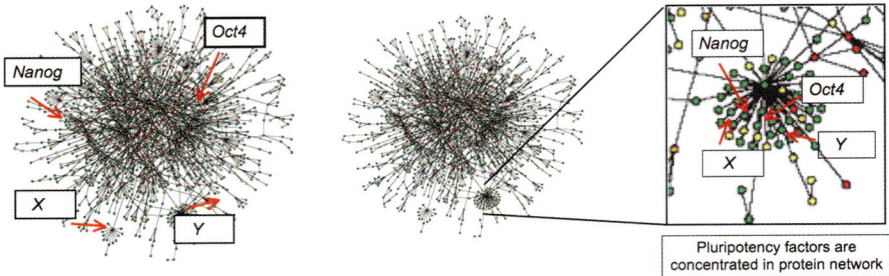

Figure 2. Alternative organization of pluripotency factors within the cellular proteome. (*Left*) Pluripotency factors Nanog and Oct4, as well as unknown factors X and Y, do not associate in shared protein complexes within the cellular proteome (indicated by the "*hairball*"). (*Right*) Pluripotency factors are concentrated within shared protein complexes.

one another to create a highly interactive network. At the other extreme, however, Nanog or other pluripotency factors would not associate within shared protein complexes but rather converge downstream in the control of critical target genes (Fig. 2).

We considered several alternative experimental strategies for identifying Nanog-associated proteins. Yeast two-hybrid screening using an ES cDNA library would provide a convenient, high-throughput discovery approach (Li et al. 2004). However, the nonphysiologic nature of this method, taken together with underrepresentation of DNA-binding transcription factors in large-scale two-hybrid screens, argued against this strategy. Instead, we favored purification of protein complexes under relatively physiologic conditions to approximate the in vivo setting as closely as possible. At the time we initiated our experiments, antibodies to the Nanog protein were not readily available. Hence, we chose to engineer into expressed Nanog protein an affinity tag suitable for protein purification. As part of this strategy, we also envisioned iterative purification of protein complexes, because Nanog-associated proteins could be used as "hooks" for the isolation of their associated proteins. In this iterative fashion, a protein network could be extended from a central point, the Nanog protein, to many other proteins. Hence, the ease with which affinity purification could be applied to purification of numerous proteins was given high priority.

With these issues in mind, we chose to use in vivo biotinylation of proteins, coupled with streptavidin affinity capture (Wang et al. 2006). Although the method was described more than 10 years ago, Strouboulis and colleagues revived its use (de Boer et al. 2003), as illustrated by one-step purification of complexes containing the GATA-1 transcription factor (Rodriguez et al. 2005). In this approach, the *Escherichia coli* biotin ligase (*BirA*) gene is stably introduced in a suitable expression vector into a host cell. Subsequently, an expression vector harboring the cDNA of interest with a short biotin ligase substrate tag is also stably transferred to the *BirA*-expressing host cell. Following expression of polypeptide, the substrate tag is biotinylated by *BirA*. To facilitate tandem affinity purification, we engineered a FLAG epitope tag into the substrate. After either streptavidin bead capture or tandem FLAG-antibody immunoprecipitation followed by streptavidin capture, samples are subjected to whole-

lane liquid chromatography–tandem mass spectrometry (LC-MS/MS). Putative associated proteins are revealed by the peptide sequences obtained. Validation of protein association can be accomplished by conventional immunoprecipitation experiments.

In the application of this method, we have been careful to choose ES cell clones that express low levels of the exogenous cDNA, because high-level expression of critical proteins could affect the protein network itself and the behavior of ES cells (Wang et al. 2006). Indeed, in most instances, we have expressed exogenous proteins at well below their endogenous level. As noted below, ES cells expressing tagged proteins may also be used to identify DNA targets of the respective proteins. Hence, a single cellular platform can be used for protein interaction and target gene analyses (Kim et al. 2008).

A PROTEIN INTERACTION NETWORK SURROUNDING NANOG

We first isolated protein complexes containing Nanog and then proceeded to tag several of the partner proteins in an iterative fashion, each time microsequencing the recovered proteins (Wang et al. 2006). Large-scale purifications were performed with both one-step (streptavidin capture alone) or tandem (FLAG-immunoprecipation followed by streptavidin capture) in order to identify as many associated proteins as possible. Stringent criteria were applied for selection of candidate interacting proteins (Wang et al. 2006).

The protein interaction network surrounding Nanog is depicted in Figure 3. Tagged proteins used as "baits" are shown in red. Lines connect proteins that were present together in isolated complexes. Given that proteins may be brought into complexes through secondary protein interactions, rather than by direct interaction with the tagged protein itself, this representation does not specify the number or variety of actual protein complexes. We suspect that there are many protein complexes, often containing shared components.

Several features of the protein interaction network are notable. First, Nanog is, indeed, connected to other critical pluripotency factors through its associated proteins. For example, Oct4 is recovered in Nanog-associated complexes. Concurrent or subsequent studies have also identified additional associated proteins, including Dax1, Sall4,

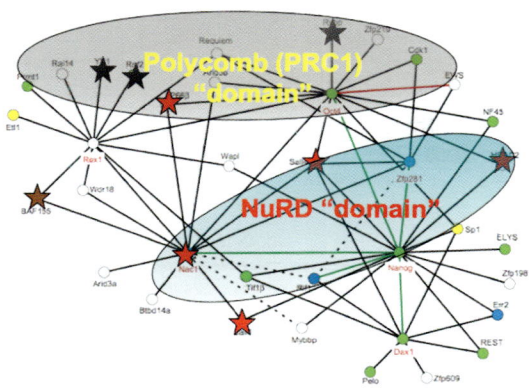

Figure 3. Protein interaction network surrounding Nanog. (*Red*) Proteins tagged by biotin and used for protein purification and peptide microsequencing, including Nanog, Oct4, Nac1, Dax1, Zfp281, and Rex1. (*Green circles*) Genes that are essential for early mouse development or maintenance of the ES cell state. (*Yellow circles*) Proteins that are dispensable. (*Red, black, brown stars*) HDAC/NuRD, PRC1, and Swi/Snf components, respectively. (Modified from Wang et al. 2006.)

Rif1, REST, Zfp281, Nac1, and Errb, among others (Loh et al. 2006; Niakan et al. 2006; Wang et al. 2006; Wu et al. 2006; Zhang et al. 2006; Singh et al. 2008), as being important within an ES cell context. Thus, our findings argue that pluripotency factors are highly concentrated within a subdomain of all proteins within ES cells (Fig. 2). This organization provides a simple rationale for the dose-dependent action of many of these factors, as higher-order protein complexes assemble, disassemble, and presumably compete for shared components.

Second, proteins within the Nanog interaction network associate with various components of chromatin remodeling or corepression complexes. Nanog and its immediate partners associate with histone deacetylases (HDACs) and the NuRD remodeling complex. Oct4 and its more immediate partners associate with components of the PRC1 polycomb complex. In addition, at least one connection to the Swi/Snf ATPase-dependent remodeling complex is evident. These relationships provide the means by which cell-specific factors within ES cells are linked to rather general modulators of transcription, largely implicated in transcriptional repression. Silencing differentiation-promoting genes is an essential role of pluripotency factors. The protein interaction network suggests pathways through which repression is maintained.

Third, an unexpectedly high fraction of proteins within the network are essential either for early mouse development and/or for maintenance of ES cell pluripotency. The vast majority of proteins are required based on either knockout or knockdown studies. Thus, the network, although highly interconnected, is also susceptible to breakdown through loss of any one of many components.

Fourth, consistent with this last point, our studies (as well as concurrent work by others) establish several "new" proteins beyond the pluripotency core factors Oct4, Sox2, and Nanog that must be considered in a broader accounting of transcriptional control in ES cells. These proteins include Dax1, Sall4, REST, Rif1, Zfp281,

Errb, and Nac1. For example, depletion of these components by short hairpin RNA (shRNA) leads to differentiation of ES cells. In the studies leading to construction of the initial Nanog protein interaction network, neither the core factor Sox2 nor the reprogramming factor Klf4 was identified. Subsequent work in our laboratory has linked these proteins to the Nanog network through affinity purification of Sox2 and Klf4 complexes (J. Chu and S.H. Orkin, unpubl.). Thus, the majority of transcription factors contributing to the maintenance of pluripotency in ES cells are contained within the broad Nanog protein interaction network. Consistent with this finding is the observation that most of the proteins within the network are down-regulated upon differentiation of ES cells.

Finally, taken together, the proteomic data suggest that the Nanog protein interaction network operates as a cellular "module" dedicated to pluripotency in ES cells. A priori, this might not have been anticipated. On reconsideration, however, the connection of the critical regulators within a subdomain of cellular proteins reflects parsimony in the evolution of pluripotency control.

IDENTIFICATION OF TARGET GENES OF THE PLURIPOTENCY NETWORK

The advent of chromatin immunoprecipitation (ChIP)-on-Chip, ChIP-PET, and ChIP-Sequencing (ChIP-Seq) methods now permits global identification of DNA targets of transcription factors. Initial work by other investigators indicated that the core pluripotency factors Nanog, Oct4, and Sox2 each bind several hundred putative target loci and also cooccupy many gene promoters (Boyer et al. 2005; Loh et al. 2006). For example, Boyer et al. (2005) reported that Nanog, Oct4, and Sox2 cooccupy approximately 350 target genes in human ES cells. In addition, each binds to its own regulatory sequences and those of other core members, leading to feed-forward and autoregulatory circuits. In our studies, we sought to develop a more comprehensive view of the transcriptional circuitry by determining the putative direct targets of additional pluripotency factors from the Nanog interaction network.

In an effort to maximize consistency in the experimental platform, we first evaluated the suitability of biotinylated factors expressed in our bank of tagged ES cell lines for ChIP-on-Chip analyses using streptavidin bead capture in place of conventional immunoprecipitation (Kim et al. 2008). Standard ChIP-on-Chip and biotin–ChIP-on-Chip analyses of promoter arrays for Nanog and c-Myc were comparable. Moreover, we demonstrated that low-level expression of tagged Nanog does not affect the sensitivity or range of targets identified by conventional ChIP-on-Chip. Because it is often challenging to identify suitable quality antibodies for ChIP-on-Chip studies, use of the biotin–ChIP-on-Chip method provides a convenient alternative to the conventional approach. The extraordinary avidity of streptavidin–biotin interactions also allows for the use of more stringent washing conditions. As a consequence, we believe that the sensitivity and specificity of biotin–ChIP-on-Chip should exceed that of other methods. The platform should be readily applicable to the emerging ChIP-Seq strategies that are especially promising.

In our studies, we initially determined promoter-binding targets for nine proteins within the ES cell protein network. This set includes the core pluripotency factors (Nanog, Oct4, and Sox2), the "Yamanaka reprogramming set" (Oct4, Sox2, c-Myc, and Klf4) (Takahashi and Yamanaka 2006), and others (Dax1, Rex1, Zfp281, and Nac1). We chose to use promoter arrays, rather than whole-genome arrays, because most binding within the +8- to –2-kb window of the Affymetrix array (relative to the transcriptional start site [TSS]) was observed within a few hundred base pairs of the TSS. Moreover, although characterization of distant binding sites might eventually be of interest, it is difficult to assign specific target genes with far-distant binding events. Stringent threshold criteria were applied to choose putative targets. Direct ChIP analyses demonstrated a false-positive assignment rate of <5%. Table 1 lists the numbers of putative target promoters for each of the factors studied.

Several general conclusions emerge from review of the data. First, the target loci shared by the principal pluripotency factors (excluding Rex1) are largely distinct from those bound by c-Myc. Second, c-Myc targets tend to be largely expressed in ES cells, rather than either expressed or "off" (or repressed), as is the case for pluripotency factor targets. Consistent with this observation, we find that targets of c-Myc are highly associated with the active H3K4me3 chromatin mark. As such, we speculate that c-Myc binding is associated with global effects on chromatin accessibility, a finding that may account for its role in facilitating somatic cell reprogramming to a pluripotent state. Third, closer inspection of target loci of the pluripotency factors reveals a striking association between the number of factors bound to a promoter region and the likelihood of target gene expression in undifferentiated ES cells. Remarkably, approximately 800 targets are bound by four or more of the nine factors we analyzed, and 450 targets are bound by five or more (Fig. 4). These "multifactor" binding loci tend to be expressed rather than "off" or repressed in ES cells, and then turned off on differentiation. In marked contrast, loci that are bound by less than four factors tend to be silent in ES cells and then expressed upon differentiation. This correlation is particularly striking for target loci bound by only a single factor. Thus, the extent of factor binding appears to correlate with gene expression in ES cells and

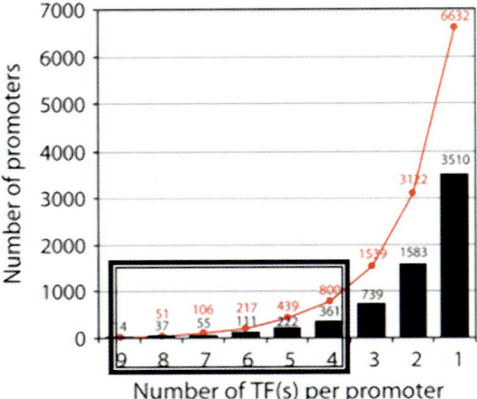

Figure 4. Distribution of numbers of factors bound to target proteins. Multifactor target genes (*boxed*) are defined as those promoters bound by more than four transcription factors among the nine tested. Approximately 800 target loci are represented in the multifactor category. (Modified from Kim et al. 2008 [© Elsevier].)

also segregates target loci into two broad classes.

Finally, the merging of the protein interaction network with target gene identification highlights "hubs" (Dezso et al. 2003), including Nanog, Oct4, Sox2, REST, Sall4, Rif1 among others (Fig. 5). The regulatory circuit is highly intertwined and the effectors of the network (i.e., the pluripotency factors) are themselves direct targets. A full understanding of pluripotency necessitates consideration of more than the "core" set of factors (Oct4, Nanog, and Sox2).

DUAL ROLE OF PLURIPOTENCY FACTORS

Our analysis of target loci of the pluripotency factors provides a logic for their dual action in maintaining the ES cell state (Fig. 6) (Orkin 2005). On the one hand, the pluripotency factors must prevent expression of differentiation-promoting genes. For example, it appears that GATA-6, which is essential for primitive endoderm gene expression, is under direct repression by Nanog, likely in concert with two of its partners (Zfp281 and Nac1) (Wang et al. 2006). The differentiation-promoting targets tend to be occupied by a limited number of the pluripotency factors. Indeed, the extent to which the lack of target gene expression is due to active repression versus insufficient factor loading to achieve transcriptional activation is uncertain. In parallel, pluripotency factors provide a positive stimulus for self-renewal and pluripotency, in part through maintenance of their own expression by autoregulatory and cross-regulatory interactions, but also through activation of additional targets. The multifactor gene targets fall within this broad category and reflect a dominant action of the pluripotency factors. Among this class are numerous transcription factors, the majority of which have not been studied in an ES cell context (Table 2). We speculate that the set of multifactor target loci is highly enriched for additional proteins that participate in maintenance of pluripotency. Further work will be required to validate this prediction.

Table 1. Numbers of promoters occupied by transcription factors in ES cells

Protein	Number of promoters
Nanog	1284
Sox2	819
Dax1	1754
Nac1	804
Oct4	783
Klf4	1790
Zfp281	601
Rex1	1543
Myc	3542

Data from Kim et al. (2008 [© Elsevier]).

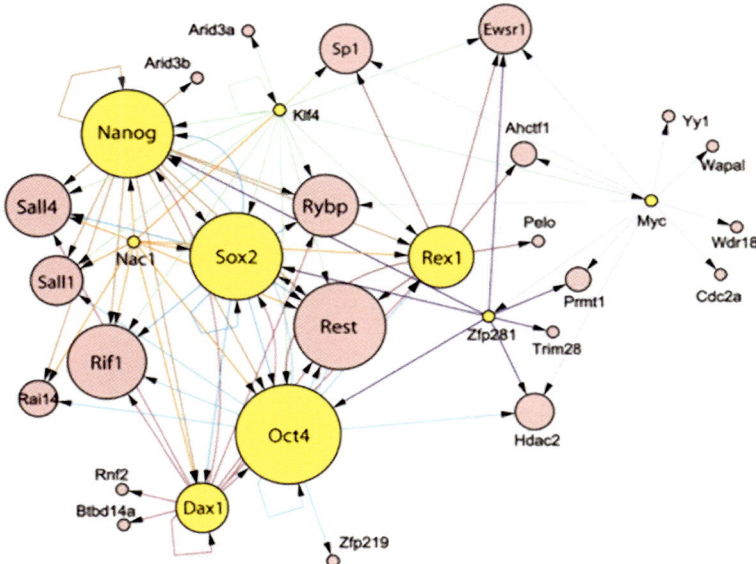

Figure 5. Expanded transcriptional regulatory network showing target hubs of multiple factors within the protein interaction network (Fig. 3). (*Yellow circles*) Nine transcription factors for which target loci were identified. The size of each circle reflects the extent of factor cooccupancy. (*Arrowheads*) Directions of transcriptional regulation. (Reprinted, from Kim et al. 2008 [© Elsevier].)

INSIGHTS INTO NANOG PROTEIN FUNCTION

Besides identifying new pluripotency regulators, it is of particular interest to characterize how the known factors participate in transcriptional control. To this end, we have examined how the Nanog protein functions. Among homeodomain proteins, Nanog is most closely related to the Nk2 subfamily. Members of this class often homodimerize through the homeodomain (HD). Outside of the HD, Nanog bears little resemblance to other homeodomain proteins. In size fractionation of nuclear extracts, we had observed that Nanog polypeptide (34 kD) distributes into two broad regions, one corresponding to large protein complexes (–2 MDa) and another approximating the mass of a Nanog dimer (Wang et al. 2006). Through study of tagged forms of Nanog, we demonstrated that Nanog monomers assemble into dimers (Wang et al. 2008). However, dimerization is mediated through a tryptophan-rich (WR) subregion of the carboxy-terminal CD domain, which had previously been associated with *trans*-activation potential. In contrast to the NK2 proteins, the HD of Nanog does not mediate dimer formation. To assess the role of dimer for-

mation in protein interactions and the function of Nanog, we generated tethered dimers, based on the pioneering studies of Wold and her colleagues (Neuhold and Wold 1993). We established that Nanog interacts with other pluripotency proteins (e.g., Oct4, Sall4, Zfp281, and Dax1) principally as a dimer. Furthermore, Nanog dimers promote self-renewal of ES cells in the absence of LIF. These findings suggest that Nanog dimer formation constitutes a critical point of control in ES cell pluripotency. We anticipate mechanisms that might shift the equilibrium of dimers to monomers and thereby inactivate Nanog. Recently, caspase cleavage of Nanog has been described as a control point for ES cell differentiation (Fujita et al. 2008). Thus, as supported by independent studies of Chambers (Mullin et al. 2008), Nanog serves to fine-tune the pluripotent state.

OCT4 DEPENDENCE OF CHROMATIN STRUCTURE WITHIN THE NANOG LOCUS

Multiple lines of evidence indicate that Oct4 has a central role in establishing and maintaining the pluripotent state. Besides the recognized dose-dependent role of Oct4 in ES cells, its inclusion in all somatic cell reprogramming "cocktails" to date suggests that its functions cannot be readily replaced by other factors. Moreover, the consensus binding motif predicted for multifactor binding targets very closely resembles an Oct4-binding sequence (Kim et al. 2008). Thus, it is likely that Oct4 protein bound to DNA provides a docking site for other pluripotency factors. The *Nanog* gene lies within a phylogenetically conserved chromosomal region that encodes several other genes that are expressed in early development, including *Aicda, Apobec1, GDF3*, and *Dppa3* (also known as *Stella* and *PGC7*). We have hypothesized that this extended *Nanog* locus may provide a window into regulation of pluripotency-related genes. Using a high-throughput quantitative chromatin profiling approach,

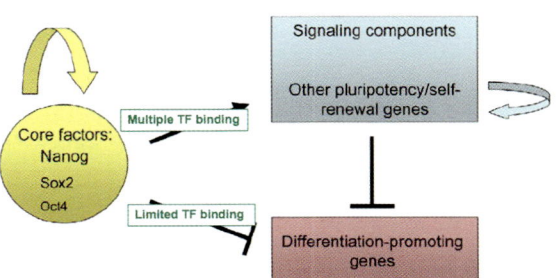

Figure 6. Dual roles of pluripotency factors in ES cells. (Modified from Orkin 2005 [© Elsevier].)

Table 2. Examples of DNA-binding proteins that are common targets of multiple transcription factors
(at least five of six factors: Nanog, Dax1, Sox2, Nac1, Oct4, and Klf4)

Symbol	Accession no.	Gene name	Symbol	Accession no.	Gene name
6030445D17Rik	NM_177079	Riken cDNA 6030445d17 gene	Nkx2-2	NM_010919	Nk2 transcription-factor-related, locus 2 (*Drosophila*)
Ankrd10	NM_133971	Ankyrin repeat domain 10			
Asxl1	NM_001039939	Additional sex combs like 1 (*Drosophila*)	Otx2	NM_144841	Orthodenticle homolog 2 (*Drosophila*)
Cbx1	NM_007622	Chromobox homolog 1 (*Drosophila* hp1 β)	Pax6	NM_013627	Paired box gene 6
			Phc1	NM_007905	Polyhomeotic-like 1 (*Drosophila*)
Cbx7	NM_144811	Chromobox homolog 7	Pou5f1	NM_013633	Pou domain, class 5, transcription factor 1
Cdx1	NM_009880	Caudal type homeobox 1			
Chd9	NM_177224	Chromodomain helicase DNA-binding protein 9	Rarg	NM_001042727	Retinoic acid receptor, γ
			Rax	NM_013833	Retina and anterior neural fold homeobox
Dido1	NM_175551	Death inducer-obliterator 1			
E2f4	NM_148952	E2F transcription factor 4	Rbbp5	NM_172517	Riken cDNA 4933411j24 gene
Evx1	NM_007966	Even-skipped homeotic gene 1 homolog	Rest	NM_011263	Re1-silencing transcription factor
			Rnf12	NM_011276	Ring finger protein 12
Fubp3	NM_001033389	Far upstream element (fuse) binding protein 3	Sall4	NM_175303	Testis expressed gene 20
			Sox13	NM_011439	Sry-box containing gene 13
Gbx2	NM_010262	Gastrulation brain homeobox 1	Sox2	NM_011443	Sry-box containing gene 2
Grhl3	NM_001013756	Grainyhead-like 3 (*Drosophila*)	Spic	NM_011461	Spi-c transcription factor (spi-1/pu.1-related)
H2afx	NM_010436	H2A histone family, member X			
Hist1h2an	NM_178184	Hypothetical protein 1190022l06	T	NM_009309	Brachyury
			Tbx3	NM_198052	T-box 3
Hist1h3i	NM_178207	Histone 1, h3g	Tcea3	NM_011542	Transcription elongation factor a (sii), 3
Hnrpdl	NM_016690	Heterogeneous nuclear ribonucleoprotein d-like			
			Tcfap2c	NM_009335	Transcription factor ap-2, γ
Hoxb13	NM_008267	Homeobox b13	Tcfcp2l1	NM_023755	Riken cDNA 4932442m07 gene
Jarid2	NM_021878	Jumonji, at rich interactive domain 2	Tgif	NM_009372	TG interacting factor
			Trib3	NM_144554	Induced in fatty liver dystrophy 2
Klf2	NM_008452	Krüppel-like factor 2 (lung)	Trib3	NM_175093	Induced in fatty liver dystrophy 2
Klf9	NM_010638	Krüppel-like factor 9	Trp53bp1	NM_013735	Transformation-related protein-53-binding protein 1
Max	NM_008558	Max protein			
Mllt6	NM_139311	Myeloid/lymphoid or mixed lineage-leukemia translocation to 6 homolog (*Drosophila*)	Zfp13	NM_011747	Zinc finger protein 13
			Zfp206	NM_001033425	Zinc finger protein 206
			Zfp36l1	NM_007564	Zinc finger protein 36, c3h type-like 1
Msh6	NM_010830	Muts homolog 6 (*E. coli*)	Zfp42	NM_009556	Zinc finger protein 42
Msx2	NM_013601	Homeobox, msh-like 2	Zfp704	NM_133218	Zinc finger protein 704
Mybl2	NM_008652	Myeloblastosis oncogene-like 2	Zic2	NM_009574	Zinc finger protein of the cerebellum 2
Myst2	NM_177619	Myst histone acetyltransferase 2			
Mzf1	NM_145819	Myeloid zinc finger 1	Zic5	NM_022987	Zinc finger protein of the cerebellum 5
Nanog	NM_028016	Nanog homeobox			

Reprinted from Kim et al. (2008 [© Elsevier]).

we identified multiple potential regulatory elements over more than 160 kb, as reflected by DNase-I-hypersensitive sites (Levasseur et al. 2008). ChIP assays reveal the cooccupancy of DNase-I-hypersensitive regions by Oct4 and other pluripotency factors, including Nanog and Zfp281. Activity of these regions in conventional enhancer assays suggests that they are likely to function as authentic regulatory elements in situ. Chromatin conformation capture (3C) assay also indicates that the Nanog proximal promoter contacts hypersensitivity sites as far as 150 kb away. Importantly, these long-range interactions are sensitive to depletion of Oct4, indicating that Oct4 is critical for maintenance of the structure of the extended Nanog chromatin region (Fig. 7). We speculate that Oct4 serves an analogous role at many other critical gene targets in ES cells.

CONCLUSIONS

Through our studies (Wang et al. 2006; Kim et al. 2008) and those of others (Boyer et al. 2005; Ivanova et al. 2006; Loh et al. 2006), the transcription factors and their direct targets responsible for maintaining ES cells in a self-renewing, pluripotent state are being uncovered in a comprehensive manner. The panoply of factors individually required for pluripotency is remarkable. Although Oct4, Sox2, and Nanog have earned respect as "core" factors, it is still unknown how many of the other factors that are just being studied can drive LIF-independent self-renewal and/or substitute to other factors in somatic cell reprogramming experiments. Although the identification of transcription factors and targets critical for pluripotency is a powerful strategy for discovery of biologically relevant genes and proteins for more in-depth analysis, the use of ES cell protein or transcriptional networks as tools for prediction of reprogramming factors or how networks change on cellular differentiation has yet to be fully exploited. Realization of the value of these networks may require development of new computational methods to model changes on a global scale. The current efforts are a first step in that direction.

ACKNOWLEDGMENTS

J.W. was supported in part by a pilot grant from the Harvard Stem Cell Institute. S.H.O. is an Investigator at the Howard Hughes Medical Institute.

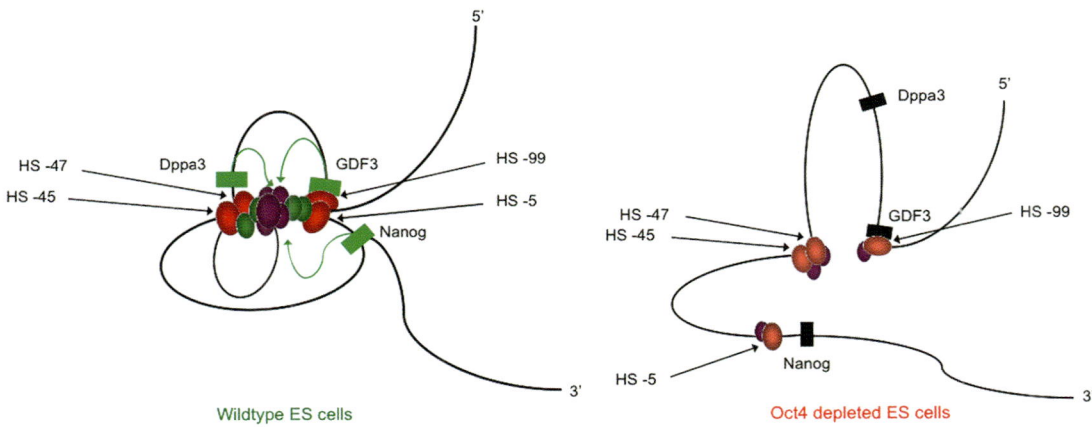

Figure 7. Proposed conformation of the extended Nanog locus as determined by 3C analysis. (*Red ovals*) Distant regulatory elements. (*Green and black rectangles*) Nanog locus genes indicating active and repressed states, respectively. The model depicts DNA-bound factors within the proximal promoters of GDF3, Dppa3, and Nanog initiating contact (*green arrows*) with an active transcriptional node formed by RNA polymerase II (*large purple oval*); accessory DNA-binding or bridging transcription factors p300, zfp281, Nac1, and CTCF (*smaller purple ovals*); and essential DNA-bound transcription factors Oct4 and Nanog (*green ovals*). (Reprinted from Levasseur et al. 2008.)

REFERENCES

Boyer, L.A., Lee, T.I., Cole, M.F., Johnstone, S.E., Levine, S.S., Zucker, J.P., Guenther, M.G., Kumar, R.M., Murray, H.L., Jenner, R.G., et al. 2005. Core transcriptional regulatory circuitry in human embryonic stem cells. *Cell* **122:** 947–956.

Chambers, I., Colby, D., Robertson, M., Nichols, J., Lee, S., Tweedie, S., and Smith, A. 2003. Functional expression cloning of Nanog, a pluripotency sustaining factor in embryonic stem cells. *Cell* **113:** 643–655.

de Boer, E., Rodriguez, P., Bonte, E., Krijgsveld, J., Katsantoni, E., Heck, A., Grosveld, F., and Strouboulis, J. 2003. Efficient biotinylation and single-step purification of tagged transcription factors in mammalian cells and transgenic mice. *Proc. Natl. Acad. Sci.* **100:** 7480–7485.

Dezso, Z., Oltvai, Z.N., and Barabasi, A.L. 2003. Bioinformatics analysis of experimentally determined protein complexes in the yeast *Saccharomyces cerevisiae*. *Genome Res.* **13:** 2450–2454.

Evans, M.J. and Kaufman, M.H. 1981. Establishment in culture of pluripotential cells from mouse embryos. *Nature* **292:** 154–156.

Fujita, J., Crane, A.M., Souza, M.K., Dejosez, M., Kyba, M., Flavell, R.A., Thomson, J.A., and Zwaka, T.P. 2008. Caspase activity mediates the differentiation of embryonic stem cells. *Cell Stem Cell* **2:** 595–601.

Ivanova, N., Dobrin, R., Lu, R., Kotenko, I., Levorse, J., DeCoste, C., Schafer, X., Lun, Y., and Lemischka, I.R. 2006. Dissecting self-renewal in stem cells with RNA interference. *Nature* **442:** 533–538.

Kim, J., Chu, J., Shen, X., Wang, J., and Orkin, S.H. 2008. An extended transcriptional network for pluripotency of embryonic stem cells. *Cell* **132:** 1049–1061.

Levasseur, D.N., Wang, J., Dorschner, M.O., Stamatoyannopoulos, J.A., and Orkin, S.H. 2008. Oct4 dependence of chromatin structure within the extended Nanog locus in ES cells. *Genes Dev.* **22:** 575–580.

Li, S., Armstrong, C.M., Bertin, N., Ge, H., Milstein, S., Boxem, M., Vidalain, P.O., Han, J.D., Chesneau, A., Hao, T., et al. 2004. A map of the interactome network of the metazoan *C. elegans*. *Science* **303:** 540–543.

Loh, Y.H., Wu, Q., Chew, J.L., Vega, V.B., Zhang, W., Chen, X., Bourque, G., George, J., Leong, B., Liu, J., et al. 2006. The Oct4 and Nanog transcription network regulates pluripotency in mouse embryonic stem cells. *Nat. Genet.* **38:** 431–440.

Martin, G.R. 1981. Isolation of a pluripotent cell line from early mouse embryos cultured in medium conditioned by teratocarcinoma stem cells. *Proc. Natl. Acad. Sci.* **78:** 7634–7638.

Mitsui, K., Tokuzawa, Y., Itoh, H., Segawa, K., Murakami, M., Takahashi, K., Maruyama, M., Maeda, M., and Yamanaka, S. 2003. The homeoprotein Nanog is required for maintenance of pluripotency in mouse epiblast and ES cells. *Cell* **113:** 631–642.

Mullin, N.P., Yates, A., Rowe, A.J., Nijmeijer, B., Colby, D., Barlow, P.N., Walkinshaw, M.D., and Chambers, I. 2008. The pluripotency rheostat Nanog functions as a dimer. *Biochem. J.* **411:** 227–231.

Neuhold, L.A. and Wold, B. 1993. HLH forced dimers: Tethering MyoD to E47 generates a dominant positive myogenic factor insulated from negative regulation by Id. *Cell* **74:** 1033–1042.

Niakan, K.K., Davis, E.C., Clipsham, R.C., Jiang, M., Dehart, D.B., Sulik, K.K., and McCabe, E.R. 2006. Novel role for the orphan nuclear receptor Dax1 in embryogenesis, different from steroidogenesis. *Mol. Genet. Metab.* **88:** 261–271.

Orkin, S.H. 2005. Chipping away at the embryonic stem cell network. *Cell* **122:** 828–830.

Rodriguez, P., Bonte, E., Krijgsveld, J., Kolodziej, K.E., Guyot, B., Heck, A.J., Vyas, P., de Boer, E., Grosveld, F., and Strouboulis, J. 2005. GATA-1 forms distinct activating and repressive complexes in erythroid cells. *EMBO J.* **24:** 2354–2366.

Silva, J., Chambers, I., Pollard, S., and Smith, A. 2006. Nanog promotes transfer of pluripotency after cell fusion. *Nature* **441:** 997–1001.

Singh, S.K., Kagalwala, M.N., Parker-Thornburg, J., Adams, H., and Majumder, S. 2008. REST maintains self-renewal and pluripotency of embryonic stem cells. *Nature* **453:** 223–227.

Takahashi, K. and Yamanaka, S. 2006. Induction of pluripotent stem cells from mouse embryonic and adult fibroblast cultures by defined factors. *Cell* **126:** 663–676.

Takahashi, K., Tanabe, K., Ohnuki, M., Narita, M., Ichisaka, T., Tomoda, K., and Yamanaka, S. 2007. Induction of pluripotent stem cells from adult human fibroblasts by defined factors. *Cell* **131:** 861–872.

Wang, J., Rao, S., Chu, J., Shen, X., Levasseur, D.N., Theunissen, T.W., and Orkin, S.H. 2006. A protein interaction network for pluripotency of embryonic stem cells. *Nature* **444:** 364–368.

Wang, J., Levasseur, D.N., and Orkin, S.H. 2008. Requirement of Nanog dimerization for stem cell self-renewal and pluripotency. *Proc. Natl. Acad. Sci.* **105:** 6326–6331.

Wu, Q., Chen, X., Zhang, J., Loh, Y.H., Low, T.Y., Zhang, W., Zhang, W., Sze, S.K., Lim, B., and Ng, H.H. 2006. Sall4 interacts with Nanog and co-occupies Nanog genomic sites in embryonic stem cells. *J. Biol. Chem.* **281:** 24090–24094.

Zhang, J., Tam, W.L., Tong, G.Q., Wu, Q., Chan, H.Y., Soh, B.S., Lou, Y., Yang, J., Ma, Y., Chai, L., et al. 2006. Sall4 modulates embryonic stem cell pluripotency and early embryonic development by the transcriptional regulation of Pou5f1. *Nat. Cell Biol.* **8:** 1114–1123.

Transcriptional Regulatory Networks in Embryonic Stem Cells

X. Chen,*† V.B. Vega,‡ and H.-H. Ng*†

*Gene Regulation Laboratory, Genome Institute of Singapore, Singapore 138672; †Department of
Biological Sciences, National University of Singapore, Singapore 117543; ‡Computational &
Mathematical Biology, Genome Institute of Singapore, Singapore 138672

Embryonic stem (ES) cells are characterized by their ability to self-renew and remain pluripotent. Transcription factors have critical roles in the maintenance of ES cells through specifying an ES-cell-specific gene expression program. Deciphering the transcriptional regulatory network that describes the specific interactions of these transcription factors with the genomic template is crucial for understanding the design and key components of this network. Recent advances in genomic technologies have facilitated genome-wide disclosure of the repertoire of transcription-factor-binding sites. Transcription factor colocalization hot spots targeted by multiple transcription factors have been identified. These are sites that integrate the external signaling pathways to the transcriptional regulatory circuitry governed by Oct4, Sox2, and Nanog. In addition, these sites may serve as focal points for the assembly of nucleoprotein complexes known as enhanceosomes. Studying the properties of ES-cell-specific enhanceosomes in different pluripotent cells will shed light on the composition and regulation of their activity. Knowledge of the transcriptional regulatory networks in different pluripotent cells will also help to distinguish the core and peripheral parts of the networks. Collectively, these studies will facilitate the understanding of molecular mechanisms behind transcription-factor-mediated regulation of pluripotent stem cells.

Mouse ES cells are derived from the inner cell mass (ICM) of preimplantation embryos. These cells are pluripotent because they have the potential to differentiate into any cell type given the right signals. Previous work during the past decades has increased our understanding of the molecular mechanisms underlying pluripotency (Rossant 2008; Silva and Smith 2008). Several signaling and intrinsic pathways involving transcription factors have been shown to have key roles in maintaining the self-renewal and pluripotent state of ES cells (Boiani and Schöler 2005; Jaenisch and Young 2008). Leukemia inhibitory factor (LIF) has long been known to be essential for the derivation and maintenance of mouse ES cells in the presence of serum. The binding of LIF to the LIF receptor (LIFR)–gp130 heterodimer receptor on the cell membrane activates STAT3 (signal transducer and activator of transcription 3) by phosphorylation, resulting in its subsequent dimerization, nuclear translocation, and target gene activation (Niwa et al. 1998; Matsuda et al. 1999; Raz et al. 1999). However, LIF alone is not sufficient to sustain the self-renewal of ES cells in the absence of serum. Ying et al. (2003) has identified bone morphogenetic proteins (BMPs) as the growth factors that work in conjunction with LIF to promote self-renewal in the absence of serum (Ying et al. 2003). Addition of BMP4 to chemically defined media leads to the phosphorylation of Smad1 and the activation of members of the *Id* (*inhibitor of differentiation*) gene family, which are effectors of the BMP pathway (Ying et al. 2003).

Besides these signaling pathways, which sense the presence of extrinsic growth factors in the environment, additional transcription factors are also essential for sustaining the undifferentiated state of ES cells. Oct4, Sox2, and Nanog are key components of the core regulatory network that governs ES cell pluripotency (Boyer et al. 2005; Loh et al. 2006). Other more recently identified transcriptional regulators, such as Tbx3, Esrrb, and Zfx, are also required to maintain ES cells (Ivanova et al. 2006; Loh et al. 2006; Galan-Caridad et al. 2007). Despite the critical roles of transcriptional regulators in the maintenance of mouse ES cells, in-depth knowledge of their in vivo targets is still lacking. Moreover, it is still not clear how the different transcription circuitries are connected or integrated in ES cells. Elucidation of the transcription networks that are operating in ES cells is fundamental to understand the molecular nature of pluripotency of ES cells, self-renewal of ES cells, and reprogramming of somatic cells to acquire a pluripotent state.

In this chapter, we discuss the recent advances in dissecting the transcriptional regulatory circuitries underlying self-renewal and pluripotency of mouse ES cells, with an emphasis on the emerging transcriptional regulatory network linking the key regulators.

OCT4/SOX2 TRANSCRIPTION CIRCUIT

The POU family transcription factor Oct4, which is encoded by *Pou5f1* and specifically expressed in the pluripotent cells of the ICM and epiblast, acts as a gatekeeper to prevent ES cell differentiation (Nichols et al. 1998). Oct4 has been reported to regulate a set of downstream targets by forming heterodimers with Sox2 (Boiani and Schöler 2005; Chew et al. 2005; Kuroda et al. 2005; Okumura-Nakanishi et al. 2005; Rodda et al. 2005; Jaenisch and Young 2008; Silva and Smith 2008). Sox2 is a high mobility group (HMG) domain-containing transcription factor and has an expression pattern similar to that of Oct4 through mouse preimplantation development. Both *Oct4* and *Sox2* null animals have primary defects in the pluripotent epiblast and both *Oct4* and *Sox2* null blastocysts are incapable of giving rise to pluripotent ES cells (Nichols et al. 1998; Avilion et al. 2003).

The fine-tuning of an appropriate level of Oct4 is crucial for determining the development of distinct cell fates from ES cells. A greater than twofold increase of Oct4 causes ES cells to differentiate into primitive endoderm and mesoderm, whereas a reduction to less than 50% of normal level triggers differentiation into the trophectoderm (Niwa et al. 2000). Thus, the precise control of Oct4 levels is critical not only for the maintenance of pluripotency, but also to cell differentiation decisions.

Oct4 and Sox2 bind to and regulate the respective regulatory regions of their own genes (Chew et al. 2005). Disruption of the circuit by short hairpin RNA (shRNA)-mediated knockdown leads to the reduction of both genes' enhancer activities and endogenous expression levels and results in ES cell differentiation. Two simple network motifs are proposed to account for the Oct4-Sox2 transcription circuit: autoregulation and multicomponent loops (Boyer et al. 2005; Chew et al. 2005). The autoregulation loop is characterized by binding of the gene product to its own regulatory element, which provides a positive feedback loop and allows for self-perpetuation and enhanced stability of gene expression. The multicomponent loop is characterized by the binding of one regulator to the regulatory element of another regulator in a closed loop. This closed-circuit loop of Oct4 and Sox2 could efficiently generate a bistable system that confers on the network the capability of maintaining options between both stability and developmental switching (Boyer et al. 2005; Chew et al. 2005).

NANOG/SALL4 TRANSCRIPTION CIRCUIT

Nanog, an NK-2 class homeobox transcription factor, is another core regulator predominantly expressed in the ICM and epiblast (Chambers et al. 2003; Mitsui et al. 2003). Nanog knockout embryos fail to form epiblasts and are mostly composed of disorganized extraembryonic tissue (Chambers et al. 2003; Mitsui et al. 2003). More interestingly, overexpression of *Nanog* could bypass the need for LIF to maintain pluripotency in culture, suggesting a potential cross-talk between Nanog and the LIF pathway (Chambers et al. 2003).

Using affinity purification coupled to liquid chromatography–tandem mass spectrometry, Wu et al. (2006) identified Sall4, a spalt-like zinc finger protein, as an interacting partner of Nanog. The finding that Sall4 resides in the Nanog complex was also independently discovered by two other laboratories (Wang et al. 2006; Liang et al. 2008). Nanog and Sall4 were found to cooccupy common targets including the genes encoding *Pou5f1*, *Sox2*, *Nanog*, and *Sall4* and several other important pluripotency genes (Wu et al. 2006; Zhang et al. 2006). Disruption of this circuit through shRNA-mediated knockdown causes the loss of self-renewal in ES cells, indicating that Sall4 is indispensable for the maintenance of ES cells (Zhang et al. 2006). Wu et al. (2006) also identified the autoregulatory and feedforward network motifs in the Nanog/Sall4 circuit, similar to that of the Oct4/Sox2 circuit. Although the genome-wide cooc-

cupancy of Nanog and Sall4 has not been demonstrated, it is conceivable that such cotargeting by these two factors could be extended to other target genes.

Oct4, Sox2, Nanog, and Sall4 are important components of the transcriptional regulatory network in ES cells and may function in a feedforward manner that maintains the appropriate levels of pluripotency-associated genes. Such a tightly integrated transcription circuit involving Oct4, Sox2, Nanog, and Sall4 may confer ES cells stability in resisting transient environmental noise and plasticity in a rapid response to differentiation cues. Elucidation of the downstream targets of these principle regulators is fundamental for dissecting their molecular mechanism in controlling pluripotency.

CORE TRANSCRIPTIONAL REGULATORY NETWORK IN ES CELLS

The core transcriptional regulatory circuitry consisting of Oct4, Nanog, and Sox2 has been mapped in both human and mouse ES cells using ChIP coupled with either microarray (ChIP-on-Chip) or paired end tag sequencing (ChIP-PET) (Boyer et al. 2005; Loh et al. 2006; H.H. Ng et al., unpubl.). A striking observation that emerged from these studies is the extent of cooccupancy by all three key factors. Both studies have found that Nanog and Oct4 cooccupy considerable genomic sites. In mouse ES cells, 345 (44.5%) Oct4 target genes are bound by Nanog (Loh et al. 2006), whereas 433 (70%) Oct4 and Nanog cobound sites are detected in human ES cells (Boyer et al. 2005). Gene ontology analysis reveals that the cooccupied targets are enriched in both self-renewal genes and differentiation genes, suggesting that the maintenance of pluripotency could be mediated by promoting the expression of self-renewal regulators and repressing lineage-specific developmental genes. These interesting observations suggest that these core factors converge on a set of common targets, rather than working independently, to perform their regulatory roles. The assembly of multiple factors on selective genomic fragments could assure steady gene expression of important downstream targets because transient fluctuation of any input signal would be neutralized by the other two sibling partners.

It is, however, imperative to note that protein occupancy is not necessarily linked to the regulation of the presumptive target gene. A substantial fraction of the binding sites is probably nonfunctional and may be the consequence of biological noise (Struhl 2007). It is difficult to tease apart noise from genuine regulation, but one approach is to deplete the particular gene product and attempt to infer regulation from the consequent response of the known bound genes. Loh et al. (2006) showed that a minor portion of Oct4- and Nanog-bound targets were responsive to the depletion of *Oct4* and *Nanog*, suggesting that only a fraction of the total binding sites are functional (Loh et al. 2006). Alternatively, there could be other redundant molecules that mask the effect of depleting a single factor.

To identify the key nodes of the Oct4 and Nanog transcriptional regulatory networks, genetic manipulation is required to assess the functional roles of the candidate target genes. Mouse ES cells depleted of *Esrrb, Sall4*, and *Rif1*, which are Oct4 and Nanog cobound targets, have been shown to lose pluripotency and undergo lineage commitment (Ivanova et al. 2006; Loh et al. 2006; Wu et al. 2006; Zhang et al. 2006). Developmental genes are preferentially up-regulated, whereas pluripotency genes are down-regulated in these knockdown cells. In another study, by coupling gene expression profiles with a ChIP data set, Matoba et al. (2006) has defined a list of 372 genes directly regulated by Oct4 and showed by an RNA interference (RNAi) experiment that Tcl1 is required for the self-renewal of ES cells (Matoba et al. 2006). In a large-scale RNAi screening study, Ivanova et al. (2006) examined the effects of RNAi-induced knockdown of 70 genes that are preferentially up-regulated in mouse ES cells. The self-renewal capacity of ES cells was compromised after depletion of 8 of the 70 genes (*Nanog, Oct4, Sox2, Tbx3, Esrrb, Tcl1, Dppa4*, and *Mm.343880*) (Ivanova et al. 2006). More importantly, the effects could be rescued by the ectopic expression of these genes, excluding the possibility of off-target effects.

These genome-wide location analyses coupled with RNAi-mediated functional studies provide a candidate list of important genes and assist in the identification of novel regulators of self-renewal and pluripotency (Ivanova et al. 2006; Loh et al. 2006; Matoba et al. 2006). Loh et al. (2007) characterized two histone H3 lysine 9 (H3K9) demethylases of the JmjC family that are direct targets of Oct4 and unraveled the roles of these two H3K9 demethylases in the maintenance of ES cell pluripotency (Loh et al. 2007). Jmjd2c, a histone demethylase that converts H3K9 trimethylation to dimethylation, was found to regulate the H3K9Me3 status of *Nanog*. Specific demethylation of H3K9Me3 by Jmjd2c at the *Nanog* promoter prevents the binding of transcription corepressors such as HP1 and KAP1 and sustains *Nanog* expression. Jmjd1a demethylates H3K9Me2 at the *Tcl1* promoter, a regulator of stem cell proliferation (Ivanova et al. 2006; Matoba et al. 2006). Depletion of Jmjd1a resulted in the induction of H3K9Me2, and the altered chromatin state obstructs Oct4 from binding to the *Tcl1* promoter. These results reveal a novel pathway in which Oct4 regulates pluripotency by activating histone modifiers that modify and maintain the expression of other Oct4 downstream target genes.

More recently, new insights have been uncovered by associating the Wnt signaling pathway with the core transcription circuitry through the terminal component Tcf3 in mouse ES cells (Pereira et al. 2006; Cole et al. 2008; Tam et al. 2008; Yi et al. 2008). *Tcf3* null ES cells are unable to undergo effective differentiation in the absence of LIF (Pereira et al. 2006; Yi et al. 2008). Interestingly, Tcf3 is shown to bind to the *Nanog* promoter and repress its expression. Hence, elevated levels of Nanog in the absence of Tcf3 may promote the maintenance of the undifferentiated state of ES cells (Chambers et al. 2003). Genome-wide mapping of Tcf3-binding sites using ChIP-on-Chip analysis revealed its global association with Oct4, Sox2, and Nanog targets with extensive colocalization. By down-regulating *Oct4, Sox2*, and *Nanog* expression, Tcf3 is suggested to strike a balance between lineage

commitment and maintenance of an undifferentiated state (Cole et al. 2008; Tam et al. 2008). Tcf3 is shown to bind to two categorically diverse sets of regulators, designated as pluripotency-associated genes and developmental genes (Cole et al. 2008; Tam et al. 2008). It is postulated that Wnt signaling impacts directly to the core regulatory network to affect the balance between pluripotency and differentiation through Tcf3 (Cole et al. 2008).

Transcription factors often recruit cofactors to modulate gene expression. The mapping of the protein interactome that defines pluripotency is pivotal for the identification of protein–protein connectivities within the transcriptional regulatory network. Two groups have explored the protein networks centered on Oct4 and Nanog using a proteomics approach (Wang et al. 2006; Liang et al. 2008). Both studies have revealed that the protein network was highly enriched for factors known to be critical for the self-renewal of ES cells and appeared to function as a module for pluripotency. Dax1, Sall4, Nac1, and Zfp281 are implicated in the maintenance of ES cell pluripotency (Wang et al. 2006). Besides transcription factors, this protein network includes cofactors found in other transcriptional repression complexes including histone deacetylase (HDAC2) and the polycomb group (YY1, Rnf2, and Rybp) (Wang et al. 2006; Liang et al. 2008). Interestingly, a unique HDAC1/2- and Mta1/2-containing complex NODE (for Nanog- and Oct4-associated deacetylase) has been identified (Liang et al. 2008). This NODE complex differs from the well-known NuRD complex, with Mbd3 and Rbbp7 either absent or present at substoichiometric levels (Liang et al. 2008). Depletion of certain NODE subunits by RNAi led to increased expression of developmentally regulated genes and ES cell differentiation, highlighting the importance of these multiprotein complexes (Liang et al. 2008).

EXPANDING THE TRANSCRIPTIONAL REGULATORY NETWORK IN ES CELLS

Initial studies on the ES cell transcriptional regulatory network have revealed new insights into the structure and key nodes of this network. However, many important questions still remain unanswered. First, given the essential roles of the LIF and BMP pathways in maintaining mouse ES cell pluripotency in culture, how they are integrated into the core transcriptional network is largely unknown. Second, how the other transcription factors implicated in ES cell biology are wired into the ES cell genome is also not clear. Third, transcriptional regulation is dependent not only on transcription factors, but also on a host of coregulators. How the different coregulators are recruited to the ES cell genome to exert their roles in regulating ES-cell-specific gene expression remains unknown. Finally, the previous genome-wide mapping studies were mainly based on ChIP-on-Chip (promoter array) or ChIP-PET (paired end ditags). Due to the limited genome coverage of promoter arrays and the inadequate depth of sequence sampling by ChIP-PET, considerable binding sites of transcription factors may be missed in the previous studies.

Recent advances in ultra-high-throughput sequencing technology through massively parallel short read sequencing (Solexa technology) have allowed the implementation of comprehensive and unbiased large-scale genome-wide mapping of transcription factors. ChIP coupled with Solexa sequencing technology (ChIP-Seq) is advantageous compared to other methods because it is simple, involves direct sequencing of small amounts of ChIP DNA, and has deeper sequencing coverage. Chen et al. (2008) have used ChIP-Seq to map the locations of 13 sequence-specific transcription factors (Nanog, Oct4, STAT3, Smad1, Sox2, Zfx, c-Myc, n-Myc, Klf4, Esrrb, Tcfcp2l1, E2f1, and CTCF) and two coregulators (p300 and Suz12) in ES cells. By mapping the binding sites, this study investigated the binding behavior of these factors and uncovered insights into how they are wired into the ES cell genome. More binding sites were identified by ChIP-Seq as compared to ChIP-on-Chip or ChIP-PET. For example, ChIP-Seq uncovered more than 10,000 putative binding loci for Nanog, whereas ChIP-PET identified only about 3000 binding sites, with a significant fraction of the ChIP-PET identified sites inclusive in the ChIP-Seq data set (Loh et al. 2006; Chen et al. 2008). Collectively, these data sets serve as a valuable resource for understanding gene regulation in mammalian cells.

Cluster analysis of the binding sites reveals two major clusters of binding activity (Fig. 1). They are the Oct4-centric and Myc-centric clusters. The Oct4-centric clusters encompass Sox2, Nanog, and Smad1. This result is consistent with the previous findings showing that they either reside in the same complex or are known to interact with one another (Boiani and Schöler 2005; Suzuki et al. 2006; Wang et al. 2006). STAT3, Esrrb, Klf4, and Tcfcp2l1 are also associated with Oct4-centric clusters.

On the other hand, Myc-centric clusters consist of c-Myc, n-Myc, E2f1, Zfx, and CTCF. Another interesting finding is the discovery of transcription factor colocalization hot spots cooccupied by multiple transcription factors. A total of 3583 multiple transcription-factor-binding loci (MTL) were bound by four or more transcription factors. It is also of interest to note that transcription factor colocalization hot spots have also been found in other higher eukaryotes such as *Drosophila* (Moorman et al. 2006; Li et al. 2008). By testing 25 MTL from the Oct4-centric cluster and 8 MTL from the Myc-centric cluster for enhancer activity, Nanog-Oct4-Sox2–bound regions were found to exhibit ES-cell-specific enhancer activity, whereas Myc MTL were not active or showed very weak ES-cell-specific enhancer activity. p300 is a coactivator commonly found at enhancer regions. ChIP-Seq analysis of p300 showed that it is preferentially localized to the Oct4-centric clusters, indicative of a close relationship of p300 with the ES-cell-specific transcription circuit.

Despite the critical roles of LIF and BMP in the maintenance of mouse ES cells, the targets of the downstream effectors of these two key signaling pathways are poorly defined. It remains unresolved how the two pathways are integrated with the core transcription network. It was reported that 87.4% of Smad1-binding and 56.8% of STAT3-binding sites within MTL are associated with the Oct4-centric MTL (Chen et al. 2008). Depletion of Oct4 leads to a reduction of STAT3 and Smad1 binding on these sites, indicating that the convergence of LIF and BMP pathways with the core transcription network is mediated by an Oct4-dependent binding of STAT3 and Smad1. However, perturbation of either the LIF or BMP pathways has no effect on Oct4 binding, suggesting that Oct4 itself has a pivotal role in stabilizing this nucleoprotein complex (Chen et al. 2008).

An enhanceosome is a nucleoprotein complex composed of multiple transcription factors binding directly or indirectly to DNA (Thanos and Maniatis 1995). The enhancer of the interferon-β gene is a prototypical enhanceosome. This 55-bp enhancer is bound by the p50 and p65 subunits of the NF-κB, ATF-2, IRF-3, IRF-7, and c-Jun and the architectural transcription factor HMGA (high mobility group A). The cobinding of these transcription factors is proposed to create an interface for the recruitment of coactivators such as p300. Hence, the densely occupied Oct4-centric clusters exhibit features of enhanceosomes. The proposed enhanceosomes may serve as anchor points for the relay of signaling information to the core ES cell circuitry (Fig. 2). The model of ES-cell-specific enhanceosome needs to be validated through future studies that define the DNA elements to which the transcription factors bind and ultimately confirmed by structural studies to construct the atomic model of this multiprotein/DNA complex.

The regulatory network model inferred from this study reveals both anticipated and novel links among these key transcription factors (Chen et al. 2008). Besides the regulatory feedback loops for Oct4, Sox2, Nanog, and STAT3, extensive interconnectivity was uncovered among 11 of the 13 transcription factors profiled. Moreover, several network motifs, such as the feedforward loop, biparallel motif,

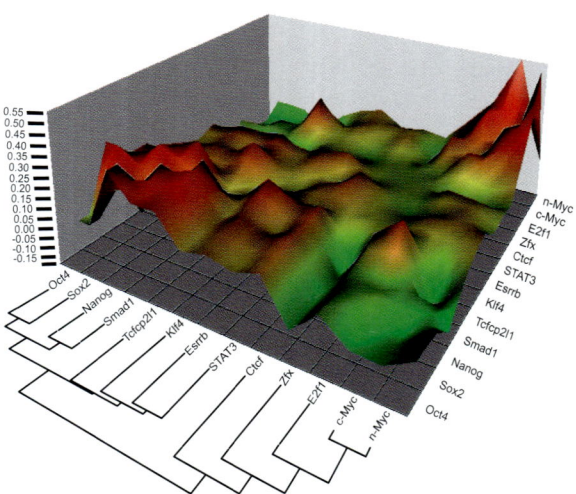

Figure 1. Cluster analysis of transcription-factor-binding sites based on 13 ChIP-Seq data sets. The transcription factor location data sets were obtained from Chen et al. (2008). The transcription factors were grouped through hierarchical clustering of the binding intensity correlation coefficient matrix calculated using the Gamma-rank correlation that measures the overall pairwise concordance and discordance over all possible pairings. The *y* axis represents the binding intensity correlation. (*Red, green*) Positive and negative coefficients, respectively. The diagonal was set to zero to avoid obscuring nonidentical pairs.

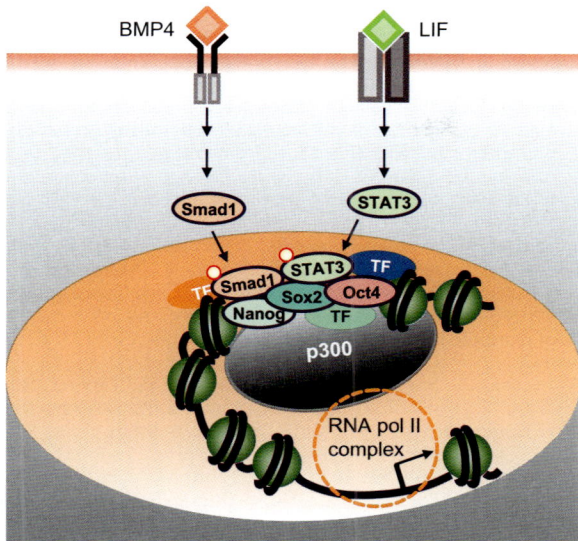

Figure 2. Integration of external signaling pathways with the core transcription network in maintaining pluripotency of mouse ES cells: A model for ES-cell-specific enhanceosome. LIF and BMP signaling pathways phosphorylate and activate nuclear translocation of STAT3 and Smad1, respectively. In the nucleus, STAT3 and Smad1 cobind specific genomic regions with Oct4, Sox2, and Nanog to form ES-cell-specific enhanceosomes. Other novel transcription factors may also be present at these sites. Coactivators such as p300 are selectively recruited to the enhanceosomes and facilitate the transcription of ES-cell-specific genes. Certain genes under the influence of these enhanceosomes may, however, be silenced by other nearby repressive *cis*-regulatory elements.

fully connected triad motif, and multiple input motif, were identified in the ES cell regulatory network. These building blocks of complex architecture may serve to stabilize gene expression patterns in undifferentiated ES cells.

In an independent study, Kim et al. (2008) reported the mapping of nine biotin-tagged transcription factors including Oct4, Sox2, Nanog, Klf4, c-Myc, Dax1, Rex1, Zfp281, and Nac1 using ChIP-on-Chip methodology (Kim et al. 2008). Similarly, the authors found cobinding of at least four transcription factors at 800 gene promoters. The difference in the number of multiple transcription-factor-bound loci identified could be attributed to the different factors studied and the different methodology used (ChIP-Seq versus ChIP-on-Chip). However, both groups found similar wiring patterns of the ES cell genome: the extensive colocalization of multiple transcription factors; distinct binding behavior of c-Myc from Oct4, Sox2, and Nanog; and an association of multiple binding loci to gene expression in ES cells (Chen et al. 2008; Kim et al. 2008). With these large-scale transcription factor/cofactor location maps, it is imperative to integrate these data with expression data and arrive at a regulatory code behind ES-cell-specific gene expression.

The knowledge gained in ES cells may be useful for understanding the mechanisms behind reprogramming whereby somatic cells can be reverted to pluripotent cells via coexpression of Oct4, Sox2, c-Myc, and Klf4 (Takahashi and Yamanaka 2006). Klf4 is extensively connected with Oct4, Sox2, or Nanog. It binds to the key regulatory regions of *Pou5f1* and *Nanog*. More than 40% of Klf4 sites within the MTL clusters are also bound by Oct4, Sox2 or Nanog. During the process of reprogramming, the wiring of the transcriptional regulatory network of a pluripotent cell needs to be reformed. It is possible that Oct4, Sox2, and Klf4 are able to jump-start the partial or entire ES cell expression program in somatic cells through targeting key nodes of the pluripotency network. Although Myc is not necessary for reprogramming, it is of interest to note that Oct4, Sox2, Klf4, and five other transcription factors bind to n-Myc, and r-Myc can replace c-Myc in reprogramming somatic cells to induced pluripotent cells (Blelloch et al. 2007). Therefore, it is conceivable that the reprogramming factors are able to activate endogenous n-Myc to form part of the transcriptional regulatory network unique to pluripotent cells. Embedded within this network is an additional level of complexity due to the presence of transcription factors exhibiting redundant functions. This is best exemplified by the finding that Klf4 alone is not important for the maintenance of ES cells (Nakatake et al. 2006). Instead, the concurrent depletion of Klf4 and two other Krüppel-like factors (Klf2 and Klf5) is required for the differentiation phenotype to be manifested (Jiang et al. 2008). Hence, there is functional redundancy for the Krüppel-like factors in maintaining ES cell properties.

CONCLUSIONS

Transcription factors are key for specifying gene expression programs and imparting distinct cellular phenotypes. Recent advancements in genomic technologies provide powerful platforms for a comprehensive mapping of chromatin modification profiles and binding sites of transcription factors and coregulators in ES cells (Mikkelsen et al. 2007; Chen et al. 2008). These efforts, coupled with previous knowledge of the ES cell transcriptome, will provide a framework for understanding the ES cell transcriptional regulatory networks in higher eukaryotes and facilitate a more precise association, hopefully prediction, of gene expression through the transcription-factor-binding pattern. In addition, these studies will greatly assist in the identification of novel self-renewal pathways, factors, and reprogramming regulators. The rapid progress made in tackling the transcriptome and genome use of ES cells will serve as a framework for further investigation to understand the biology of stem cells. With the derivation of different pluripotent cells (Rossant 2008), it will be of interest to delineate the core transcriptional regulatory network common in these cells.

ACKNOWLEDGMENTS

This work is supported by the Agency for Science, Technology, and Research (A*STAR) of Singapore. X.C. is supported by the Singapore Millennium Foundation scholarship and National University of Singapore graduate scholarship. We thank Andrew Hitchins, Jia-Hui Ng, and Wai-Leong Tam for critical comments on this manuscript.

REFERENCES

Avilion, A.A., Nicolis, S.K., Pevny, L.H., Perez, L., Vivian, N., and Lovell-Badge, R. 2003. Multipotent cell lineages in early mouse development depend on SOX2 function. *Genes Dev.* **17:** 126–140.

Blelloch, R., Venere, M., Yen, J., and Ramalho-Santos, M. 2007. Generation of induced pluripotent stem cells in the absence of drug selection. *Cell Stem Cell* **1:** 245–247.

Boiani, M. and Schöler, H.R. 2005. Regulatory networks in embryo-derived pluripotent stem cells. *Nat. Rev. Mol. Cell Biol.* **6:** 872–884.

Boyer, L.A., Lee, T.I., Cole, M.F., Johnstone, S.E., Levine, S.S., Zucker, J.P., Guenther, M.G., Kumar, R.M., Murray, H.L., Jenner, R.G., et al. 2005. Core transcriptional regulatory circuitry in human embryonic stem cells. *Cell* **122:** 947–956.

Chambers, I., Colby, D., Robertson, M., Nichols, J., Lee, S., Tweedie, S., and Smith, A. 2003. Functional expression cloning of Nanog, a pluripotency sustaining factor in embryonic stem cells. *Cell* **113:** 643–655.

Chen, X., Xu, H., Yuan, P., Fang, F., Huss, M., Vega, V.B., Wong, E., Orlov, Y.L., Zhang, W., Jiang, J., et al. 2008. Integration of external signaling pathways with the core transcriptional network in embryonic stem cells. *Cell* **133:** 1106–1117.

Chew, J.L., Loh, Y.H., Zhang, W., Chen, X., Tam, W.L., Yeap, L.S., Li, P., Ang, Y.S., Lim, B., Robson, P., and Ng, H.H. 2005. Reciprocal transcriptional regulation of *Pou5f1* and *Sox2* via the Oct4/Sox2 complex in embryonic stem cells. *Mol. Cell. Biol.* **25:** 6031–6046.

Cole, M.F., Johnstone, S.E., Newman, J.J., Kagey, M.H., and Young, R.A. 2008. Tcf3 is an integral component of the core regulatory circuitry of embryonic stem cells. *Genes Dev.* **22:** 746–755.

Galan-Caridad, J.M., Harel, S., Arenzana, T.L., Hou, Z.E., Doetsch, F.K., Mirny, L.A., and Reizis, B. 2007. Zfx controls the self-renewal of embryonic and hematopoietic stem cells. *Cell* **129:** 345–357.

Ivanova, N., Dobrin, R., Lu, R., Kotenko, I., Levorse, J., DeCoste, C., Schafer, X., Lun, Y., and Lemischka, I.R. 2006. Dissecting self-renewal in stem cells with RNA interference. *Nature* **442:** 533.

Jaenisch, R. and Young, R. 2008. Stem cells, the molecular circuitry of pluripotency and nuclear reprogramming. *Cell* **132:** 567–582.

Jiang, J., Chan, Y.S., Loh, Y.H., Cai, J., Tong, G.Q., Lim, C.A., Robson, P., Zhong, S., and Ng, H.H. 2008. A core Klf circuitry regulates self-renewal of embryonic stem cells. *Nat. Cell Biol.* **10:** 353–360.

Kim, J., Chu, J., Shen, X., Wang, J., and Orkin, S.H. 2008. An extended transcriptional network for pluripotency of embryonic stem cells. *Cell* **132:** 1049–1061.

Kuroda, T., Tada, M., Kubota, H., Kimura, H., Hatano, S.Y., Suemori, H., Nakatsuji, N., and Tada, T. 2005. Octamer and Sox elements are required for transcriptional *cis* regulation of *Nanog* gene expression. *Mol. Cell. Biol.* **25:** 2475–2485.

Li, X.Y., MacArthur, S., Bourgon, R., Nix, D., Pollard, D.A., Iyer, V.N., Hechmer, A., Simirenko, L., Stapleton, M., Luengo Hendriks, C.L., et al. 2008. Transcription factors bind thousands of active and inactive regions in the *Drosophila* blastoderm. *PLoS Biol.* **6:** e27.

Liang, J., Wan, M., Zhang, Y., Gu, P., Xin, H., Jung, S.Y., Qin, J., Wong, J., Cooney, A.J., Liu, D., and Songyang, Z. 2008. Nanog and Oct4 associate with unique transcriptional repression complexes in embryonic stem cells. *Nat. Cell Biol.* **10:** 731–739.

Loh, Y.H., Wu, Q., Chew, J.L., Vega, V.B., Zhang, W., Chen, X., Bourque, G., George, J., Leong, B., Liu, J., et al. 2006. The Oct4 and Nanog transcription network regulates pluripotency in mouse embryonic stem cells. *Nat. Genet.* **38:** 431–440.

Loh, Y.H., Zhang, W., Chen, X., George, J., and Ng, H.H. 2007. Jmjd1a and Jmjd2c histone H3 Lys 9 demethylases regulate self-renewal in embryonic stem cells. *Genes Dev.* **21:** 2545–2557.

Matoba, R., Niwa, H., Masui, S., Ohtsuka, S., Carter, M.G.,

Sharov, A.A., and Ko, M.S. 2006. Dissecting Oct3/4-regulated gene networks in embryonic stem cells by expression profiling. *PLoS ONE* **1:** e26.

Matsuda, T., Nakamura, T., Nakao, K., Arai, T., Katsuki, M., Heike, T., and Yokota, T. 1999. STAT3 activation is sufficient to maintain an undifferentiated state of mouse embryonic stem cells. *EMBO J.* **18:** 4261–4269.

Mikkelsen, T.S., Ku, M., Jaffe, D.B., Issac, B., Lieberman, E., Giannoukos, G., Alvarez, P., Brockman, W., Kim, T.K., Koche, R.P., et al. 2007. Genome-wide maps of chromatin state in pluripotent and lineage-committed cells. *Nature* **448:** 553–560.

Mitsui, K., Tokuzawa, Y., Itoh, H., Segawa, K., Murakami, M., Takahashi, K., Maruyama, M., Maeda, M., and Yamanaka, S. 2003. The homeoprotein Nanog is required for maintenance of pluripotency in mouse epiblast and ES cells. *Cell* **113:** 631–642.

Moorman, C., Sun, L.V., Wang, J., de Wit, E., Talhout, W., Ward, L.D., Greil, F., Lu, X.J., White, K.P., Bussemaker, H.J., and van Steensel, B. 2006. Hotspots of transcription factor colocalization in the genome of *Drosophila melanogaster*. *Proc. Natl. Acad. Sci.* **103:** 12027–12032.

Nakatake, Y., Fukui, N., Iwamatsu, Y., Masui, S., Takahashi, K., Yagi, R., Yagi, K., Miyazaki, J., Matoba, R., Ko, M.S., and Niwa, H. 2006. Klf4 cooperates with Oct3/4 and Sox2 to activate the *Lefty1* core promoter in embryonic stem cells. *Mol. Cell. Biol.* **26:** 7772–7782.

Nichols, J., Zevnik, B., Anastassiadis, K., Niwa, H., Klewe-Nebenius, D., Chambers, I., Schöler, H., and Smith, A. 1998. Formation of pluripotent stem cells in the mammalian embryo depends on the POU transcription factor Oct4. *Cell* **95:** 379–391.

Niwa, H., Burdon, T., Chambers, I., and Smith, A. 1998. Self-renewal of pluripotent embryonic stem cells is mediated via activation of STAT3. *Genes Dev.* **12:** 2048–2060.

Niwa, H., Miyazaki, J., and Smith, A.G. 2000. Quantitative expression of Oct-3/4 defines differentiation, dedifferentiation or self-renewal of ES cells. *Nat. Genet.* **24:** 372–376.

Okumura-Nakanishi, S., Saito, M., Niwa, H., and Ishikawa, F. 2005. Oct-3/4 and Sox2 regulate *Oct-3/4* gene in embryonic stem cells. *J. Biol. Chem.* **280:** 5307–5317.

Pereira, L., Yi, F., and Merrill, B.J. 2006. Repression of Nanog gene transcription by Tcf3 limits embryonic stem cell self-renewal. *Mol. Cell. Biol.* **26:** 7479–7491.

Raz, R., Lee, C.K., Cannizzaro, L.A., d'Eustachio, P., and Levy, D.E. 1999. Essential role of STAT3 for embryonic stem cell pluripotency. *Proc. Natl. Acad. Sci.* **96:** 2846–2851.

Rodda, D.J., Chew, J.L., Lim, L.H., Loh, Y.H., Wang, B. Ng, H.H., and Robson, P. 2005. Transcriptional regulation of *Nanog* by OCT4 and SOX2. *J. Biol. Chem.* **280:** 24731–24737.

Rossant, J. 2008. Stem cells and early lineage development *Cell* **132:** 527–531.

Silva, J. and Smith, A. 2008. Capturing pluripotency. *Cell* **132:** 532–536.

Struhl, K. 2007. Transcriptional noise and the fidelity of initiation by RNA polymerase II. *Nat. Struct. Mol. Biol.* **14:** 103–105.

Suzuki, A., Raya, A., Kawakami, Y., Morita, M., Matsui, T., Nakashima, K., Gage, F.H., Rodríguez-Esteban, C., and Izpisúa Belmonte, J.C. 2006. Nanog binds to Smad1 and blocks bone morphogenetic protein-induced differentiation of embryonic stem cells. *Proc. Natl. Acad. Sci.* **103:** 10294–10299.

Takahashi, K. and Yamanaka, S. 2006. Induction of pluripotent stem cells from mouse embryonic and adult fibroblast cultures by defined factors. *Cell* **126:** 663–676.

Tam, W.L., Lim, C.Y., Han, J., Zhang, J., Ang, Y.S., Ng, H.H., Yang, H., and Lim, B. 2008. Tcf3 regulates embryonic stem cell pluripotency and self-renewal by the transcriptional control of multiple lineage pathways. *Stem Cells* **26:** 2019–2031.

Thanos, D. and Maniatis, T. 1995. Virus induction of human IFNβ gene expression requires the assembly of an enhanceosome. *Cell* **83:** 1091–1100.

Wang, J., Rao, S., Chu, J., Shen, X., Levasseur, D.N., Theunissen, T.W., and Orkin, S.H. 2006. A protein interaction network for pluripotency of embryonic stem cells. *Nature* **444:** 364–368.

Wu, Q., Chen, X., Zhang, J., Loh, Y.H., Low, T.Y., Zhang, W., Zhang, W., Sze, S.K., Lim, B., and Ng, H.H. 2006. Sall4 interacts with Nanog and co-occupies Nanog genomic sites in embryonic stem cells. *J. Biol. Chem.* **281:** 24090–24094.

Yi, F., Pereira, L., and Merrill, B.J. 2008. Tcf3 functions as a steady state limiter of transcriptional programs of mouse embryonic stem cell self renewal. *Stem Cells* **26:** 1951–1960.

Ying, Q.L., Nichols, J., Chambers, I., and Smith, A. 2003. BMP induction of Id proteins suppresses differentiation and sustains embryonic stem cell self-renewal in collaboration with STAT3. *Cell* **115:** 281–292.

Zhang, J., Tam, W.L., Tong, G.Q., Wu, Q., Chan, H.Y., Soh, B.S., Lou, Y., Yang, J., Ma, Y., Chai, L., et al. 2006. Sall4 modulates embryonic stem cell pluripotency and early embryonic development by the transcriptional regulation of *Pou5f1*. *Nat. Cell Biol.* **8:** 1114–1123.

Toward Stem Cell Systems Biology: From Molecules to Networks and Landscapes

B.D. MacArthur,*[†] A. Ma'ayan,*[‡] and I.R. Lemischka*

*Department of Gene and Cell Medicine, The Black Family Stem Cell Institute, †Department of
Pharmacology and Systems Therapeutics, ‡Systems Biology Center New York,
Mount Sinai School of Medicine, New York, New York 10029

The last few years have seen significant advances in our understanding of the molecular mechanisms of stem-cell-fate specification. New and emerging high-throughput techniques, as well as increasingly accurate loss-of-function perturbation techniques, are allowing us to dissect the interplay among genetic, epigenetic, proteomic, and signaling mechanisms in stem-cell-fate determination with ever-increasing fidelity (Boyer et al. 2005, 2006; Ivanova et al. 2006; Loh et al. 2006; Cole et al. 2008; Jiang et al. 2008; Johnson et al. 2008; Kim et al. 2008; Liu et al. 2008; Marson et al. 2008; Mathur et al. 2008). Taken together, recent reports using these new techniques demonstrate that stem-cell-fate specification is an extremely complex process, regulated by multiple mutually interacting molecular mechanisms involving multiple regulatory feedback loops. Given this complexity and the sensitive dependence of stem cell differentiation on signaling cues from the extracellular environment, how are we best to develop a coherent *quantitative* understanding of stem cell fate at the systems level? One approach that we and other researchers have begun to investigate is the application of techniques derived in the computational disciplines (mathematics, physics, computer science, etc.) to problems in stem cell biology. Here, we briefly sketch a few pertinent results from the literature in this area and discuss future potential applications of computational techniques to stem cell systems biology.

FROM MOLECULES TO NETWORKS

Modern stem cell studies now typically use a variety of different high-throughput techniques to deconstruct the molecular basis of cell-fate specification. Nevertheless, individual studies inevitably only focus on one chosen aspect of stem cell self-renewal, fate specification, or reprogramming. Consequently, although they typically produce a wealth of data, each individual study still only represents a small aspect of our collective knowledge of stem cell behavior. Therefore, it is useful for information from a large number of individual studies to be collated and cataloged into structured *meta*-data sets representing the collective knowledge about the molecular regulatory mechanisms that control stem cell self-renewal and differentiation. However, the task of constructing and maintaining such collective knowledge data sets is computationally and biologically challenging because different experimental studies consider different types of stems cells, under different culture conditions, using different experimental techniques that may naturally produce biased results due to inherent limitations of experimental techniques. For example, proteomic experiments are known to enrich for highly abundant proteins, whereas gene expression microarrays are noisy and mRNA levels often only partially correlate with protein expression and function. To tackle the data integration and knowledge accumulation challenge, applications of techniques from the mathematical field of graph theory (Ma'ayan 2009) have been particularly successful. The realization that complex biological systems can be conceptually represented as networks (also known as graphs in the mathematical literature) has revolutionized our approach to exploring complex biochemical systems. To construct a biological regulatory network, elements such as genes, proteins,

mRNAs, microRNAs (miRNAs), or any other kind of molecular species are represented as nodes, whereas the biochemical interactions between species, for example, protein–protein interactions or transcription factor regulation of gene expression, are represented as edges or links. Because a variety of different types of regulatory mechanisms can be represented as networks (Ma'ayan et al. 2005a), representing complex biochemical systems as networks allows the merging of different types of experimental data into a single conceptual framework (Ma'ayan 2008). An example of the successful application of graph-theoretic techniques to data integration in stem cell biology was recently given by Franz-Josef Müller, Jeanne Loring, and coworkers (Müller et al. 2008). They first classified different types of human stem cells on the basis of their genome-wide mRNA expression signatures (Müller et al. 2008) and identified a set of genes that are specifically up-regulated across a variety of different types of stem cells. Then, using available mammalian protein–protein interaction databases, they "connected" their identified stem cell gene set into a network of protein interactions, naming this integrated network PluriNet. To build PluriNet, the authors made use of a graph-theoretic algorithm and software package called Matisse (Ulitsky and Shamir 2007) to identify modules in gene expression data using background knowledge about known protein–protein interactions. In general, algorithms such as Matisse can be used to identify functional modules in complex data sets (Berger et al. 2007), whereas statistical tools can be used to characterize the functional theme of such modules (Subramanian et al. 2005). Alternatively, protein–protein interaction networks can be readily reconstructed experimentally using proteomic techniques such as immunoprecipitation-based "pull-downs" followed by mass spectrometry (IP-MS) (Gygi and Aebersold 2000).

For example, a protein–protein interaction network centered around the transcription factor Nanog was recently constructed by Jianlong Wang, Stuart Orkin, and coworkers using a set of serial IP-MS experiments in which they pulled down different components of the Nanog interaction complex one at a time (Wang et al. 2006). Resources such as the PluriNet and empirically constructed interaction networks are useful because they can be used as a reference upon which to "project" future data and interpret new findings within the context of known biology.

Another source of data for building regulatory networks comes from high-throughput chromatin immunoprecipitation (ChIP)-chip (Kidder et al. 2008), ChIP-seq (Chen et al. 2008), and ChIP-PET (Loh et al. 2006) experiments. These techniques are commonly used to identify transcription factor—DNA interactions and thereby connect transcription factors to the putative sets of genes that they regulate. These techniques can also be used to identify a broad range of epigenetic chromatin modifications such as methylation/acetylation status of histone proteins. Several studies have used these techniques to identify targets of a number of the core pluripotency transcription factors. To clarify the nature of the observed binding events, high-throughput transcription-factor-binding studies are often coupled to loss-of-function experiments and genome-wide mRNA expression profiling to assess the functionality of any identified putative regulatory interactions (e.g., whether observed transcriptional binding induces activation or repression of the target gene). In addition to transcriptional regulation of stem cell fate, accumulating evidence, first in *Drosophila* (Hatfield et al. 2005) and more recently in mammalian stem cells (Houbaviy et al. 2003; Tay et al. 2008), suggests that miRNAs are also intimately involved in the regulation of stem-cell-fate decisions (Gangaraju and Lin 2009). For example, it has recently been shown that mir-21 suppresses a set of core pluripotency genes and is itself transcriptionally suppressed by the pluripotency factor Rest (Singh et al. 2008). Consequently, databases and network analyses deconstructing the place of miRNAs in the regulation of mammalian cells are also rapidly emerging (Altuvia et al. 2005; Griffiths-Jones et al. 2006). Although transcriptional regulation of stem cell fate is now being dissected with increasing detail, the complex signaling network (Ma'ayan et al. 2005b) upstream of the transcriptional network is less well understood. Even less information is available that sheds light on how signaling networks converge on events that occur in the nucleus. However, we anticipate that emerging high-throughput phosphoproteomics and RNA interference experiments will provide insights into the structure and function of stem cell signaling networks and their relationship to the core transcriptional network. Indeed, some progress has already been made toward this end (Chen et al. 2008).

Taken together, these reports suggest that stem-cell-fate determination is an intrinsically complex process, regulated by a dynamic interplay among genetic, proteomic, miRNA, and epigenetic mechanisms. To begin to make sense of this complexity, it is useful to cast this multiplicity of biochemical interactions in the form of networks that encode the architecture of the regulatory

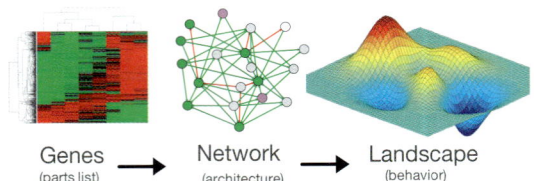

Figure 1. Understanding stem cells at the systems level: From genes to networks and landscapes.

mechanisms of stem-cell-fate specification at the system level (Fig. 1). However, this approach inevitably leads to a paradox: As our understanding of the molecular basis of cell fate becomes more detailed, the networks that arise from these integrative studies become correspondingly more complex and difficult to interpret. For this reason, there is now a pressing need to generate new tools to make sense of these complex networks to understand how internal molecular circuitry defines cell fate at the systems level. In a sense, new tools are needed to "see the forest and not just the trees." In the following section, we discuss ways in which mathematical models may be fruitfully used to make sense of this complexity.

FROM NETWORKS TO ATTRACTORS

A mentioned above, a number of recent reports have begun to reconstruct the transcriptional circuitry underpinning the maintenance of stem cell pluripotency and self-renewal (Boyer et al. 2005; Ivanova et al. 2006; Kim et al. 2008). Taken together, these studies report a complex transcriptional regulatory circuit centered around a set of core pluripotency factors (including Oct4, Sox2, Nanog, Esrrb, Tbx3, Tcl1, Dppa4, Tcf3, and others) connected to an extended set of lineage-specifying factors. Crucially, this extended circuit appears to have a highly enriched feedback loop structure, in which the core pluripotency factors regulate the expression of their target genes in a highly combinatorial manner and are themselves regulated in a coordinated way. The multiplicity of positive and negative feedback loops present in this core circuit makes determination and prediction of cell behavior from regulatory architecture intrinsically difficult. To tackle this problem, it is conceptually convenient to take a physical approach and think of transcriptional activation and inhibition as forces that "push" and "pull" the cell's internal transcriptional state in different directions: some synergistically, pushing the cell in the same genetic direction; others competitively, pushing the cell in divergent directions. Within this framework, cell-fate determination may be seen as resulting from the sum of the internal forces that the cell experiences in response to environmental signaling cues, and cell "types" as equilibrium states in which the core transcription factors are expressed at a level that balances the system. Within the mathematical literature, such balanced configurations are referred to as *attractors* because if perturbed away, the system is attracted back to the balanced state over time. A useful analogy is that of a marble perturbed from the bottom of a bowl that tracks out a transient trajectory around the sides of the bowl only to eventually return to rest again at its bottom. A system that supports the existence of multiple different attractor states is said to exhibit *multistability*.

The notion that different cell types may result from multistability of an underlying dynamical system was first suggested by the Nobel-Prize-winning physicist Max Delbrück in the late 1940s (Delbrück 1949; Thomas 1998) and has been developed extensively in a theoretical context by Stuart Kauffman and other researchers since the late 1960s (Kauffman 1969, 1993; Thomas 1998; MacArthur et al. 2008). However, although this notion has received much attention in the theoretical literature, experimental evidence that distinct mammalian cell fates may correspond to attractors of underlying *high-dimensional* regulatory networks has only recently been provided by Sui Huang, Donald Ingber, and coworkers (Huang et al. 2005). To do so, these authors made use of the experimental observation that similar in vitro cellular responses can often be induced by disparate chemical stimuli. In particular, they used the fact that human promyelocytic HL60 cells may be triggered to neutrophil differentiation in vitro, either by treatment with retinoic acid (RA) or by treatment with dimethylsulfoxide (DMSO). By taking time courses of microarrays during differentiation, they found that RA and DMSO initially triggered widely divergent patterns of gene expression; however, although genome-scale patterns of expression were initially divergent, they found that, over time, the patterns of gene expression induced by RA and DMSO ultimately converged to a common end point. The fact that alternative perturbations affect a common response through divergent routes is characteristic of an attracting state, and their results therefore suggest that the HL60 neutrophil state is an attractor of a (as yet undefined) complex regulatory network. Since this initial report, further evidence that other mammalian cell types may be high-dimensional attractors has been provided (Chang et al. 2006, 2008; Ying et al. 2008). For instance, by blocking fibroblast growth factor (FGF) receptor and extracellular signal-regulated kinase (ERK) signaling, Qi-Long Ying, Austin Smith, and coworkers demonstrated that, if protected from external inductive differentiation stimuli, mouse embryonic stem cells may be maintained in culture in a self-renewing state without the need for the additional culture stimuli usually required for their maintenance. This result suggests that in the mouse, the core self-renewing pluripotent state is internally stable and self-sustaining, indicating that it may correspond to an attractor of the complex pluripotency circuit. Within this context, recent reports that fully differentiated cell types may be reprogrammed to a primitive pluripotent state by a variety of different means (Yu et al. 2007; Nakagawa et al. 2008; Feng et al. 2009) are also indicative of the presence of a core pluripotent attracting state.

Taken together, these experimental reports are consistent with the notion that cell fates, including the primitive pluripotent self-renewing state, may correspond to different high-dimensional attractors of the cell's internal regulatory circuitry. However, current reports have generally only provided indirect evidence of attractors or evidence

for cellular attractors at the RNA level. Because cell fate is controlled by complex feedback among genetic, epigenetic, and proteomic mechanisms, a current challenge in stem cell systems biology is to extend these initial reports, to map not only the genetic profile of cellular attractors, but also the proteomic and epigenetic profiles of cellular attractors. For example, from a biological point of view, it is usual to think of cell types as characterized by fixed molecular signatures (Ivanova et al. 2002); however, from a mathematical point of view it is also natural to suspect that the complex circuitry at the core of cell-fate specification may allow not just static "fixed-point" attractors, but also stable self-sustaining oscillatory states, in which transcriptional forces balance in a dynamic manner. Oscillators are ubiquitous in complex systems containing feedback loops, and many biochemical oscillators have correspondingly been described (Winfree 2001). In the context of stem cell differentiation, recent data indicating that Nanog expression fluctuates in murine embryonic stem cells (Chambers et al. 2007) are possibly indicative of a dynamic, rather than static, attracting state. The notion of dynamic stem cell attractors is intuitively appealing because, if present, they may allow individual cells to be *dynamically* primed: At the Nanog high-expression phase, cells are resistant to inductive stimuli, whereas at the Nanog low-expression phase, cells are sensitive to inductive differentiation stimuli. Evidence for dynamic stem cell attractors is currently lacking; however, we anticipate that this may be a fruitful area for future stem cell systems biology research.

FROM ATTRACTORS TO LANDSCAPES

In an attempt to understand the robustness of cellular differentiation, Conrad Waddington suggested his now famous *epigenetic landscape* (Waddington 1957). His view was that development occurs rather like a marble rolling down a tilted, funneled landscape containing multiple "hills" and "valleys": As differentiation progresses, the cell adopts a more and more specific state, corresponding to a deeper valley in the landscape, and is barred from spontaneous movement between states by the hills that split the landscape into discrete valleys. Crucially, within Waddington's view, cell types are not terminally fixed, but rather, they are maintained by "epigenetic" barriers that can, given sufficient perturbation, be overcome. Recent demonstrations that cells can be reprogrammed from one type to another (Jaenisch and Young 2008; Zhou et al. 2008) suggest that this is indeed the case, and these reports have correspondingly led to a revived interest in Waddington's ideas (Goldberg et al. 2007). The notion that cell fate is guided by an underlying regulatory landscape is also appealing from a theoretical point of view because, for many complex systems, attractors may be directly associated in a precise way with local minima of an appropriately defined potential energy (or energy-like). This observation has led other authors to conjecture that Waddington's epigenetic landscape may, in fact, correspond to the "energy" landscape of a cell's underlying regulatory architecture (Huang and Ingber 2007). Energy landscapes have proven to be successful in helping to understand many other complex phenomena (such as the protein folding problem, for example [Wales 2003; Janke 2007]), and we therefore anticipate that applications of energy landscape theory will be useful in addressing the relationship between internal regulatory circuitry and cell-fate determination. In particular, by determining the topology of cellular "energy" landscapes, it may be possible to understand not just the nature of individual cellular attractors, but also the ways in which individual attractors relate to one another (e.g., the heights of the barriers separating them). In the context of cellular reprogramming, such information would be particularly useful because it would provide a means to determine how efficiently different cell types may be reprogrammed, either to the pluripotent state or to alternative differentiated or multipotent states.

CONCLUSIONS

These are exciting times for stem cell biology. New and emerging high-throughput technologies are allowing us to deconstruct the mechanisms of cell-fate determination with ever-increasing detail. By representing the multiplicity of regulatory interactions underpinning stem cell fate as networks, we are beginning to dissect stem-cell-fate specification at the systems level. However, it is becoming clear that cell-fate specification is a fundamentally complex process and this complexity makes it intrinsically difficult to determine and predict cell behavior from regulatory network architecture. One potential way to connect cell fate to regulatory circuitry is by using regulatory architecture to define a cellular "energy" landscape—in which valleys are associated with different cell types and hills are associated with the barriers between them—and computationally explore the topology of this landscape. This approach is conceptually reminiscent of Waddington's epigenetic landscape but has, until recently, been hampered by lack of data. However, with the advent of high-resolution high-throughput techniques, we are now beginning to accumulate sufficient data at multiple molecular and biochemical levels to make Waddington's vision quantitative. Doing so will require interdisciplinary collaboration among experimentalists, mathematicians, physicists, and computer scientists. Thus, this is not only an outstanding problem in stem cell systems biology, but also an area rich in collaborative opportunities between experimentalists and theoreticians. Consequently, developing a rigorous understanding of stem-cell-fate determination at the systems level is a significant challenge as well as a great opportunity.

REFERENCES

Altuvia, Y., Landgraf, P., Lithwick, G., Elefant, N., Pfeffer, S., Aravin, A., Brownstein, M.J., Tuschl, T., and Margalit, H. 2005. Clustering and conservation patterns of human microRNAs. *Nucleic Acids Res.* 33: 2697–2706.

Berger, S., Posner, J., and Ma'ayan, A. 2007. Genes2Networks: Connecting lists of gene symbols using mammalian protein interactions databases. *BMC Bioinformatics* 8: 372.

Boyer, L.A., Lee, T.I., Cole, M.F., Johnstone, S.E., Levine, S.S., Zucker, J.P., Guenther, M.G., Kumar, R.M., Murray, H.L.,

Jenner, R.G., et al. 2005. Core transcriptional regulatory circuitry in human embryonic stem cells. *Cell* 122: 947–956.

Boyer, L.A., Plath, K., Zeitlinger, J., Brambrink, T., Medeiros, L.A., Lee, T.I., Levine, S.S., Wernig, M., Tajonar A., Ray, M.K., et al. 2006. Polycomb complexes repress developmental regulators in murine embryonic stem cells. *Nature* 441: 349–353.

Chambers, I., Silva, J., Colby, D., Nichols, J., Nijmeijer, B., Robertson, M., Vrana, J., Jones, K., Grotewold, L., and Smith, A. 2007. Nanog safeguards pluripotency and mediates germline development. *Nature* 450: 1230–1234.

Chang, H., Oh, P., Ingber, D., and Huang, S. 2006. Multistable and multistep dynamics in neutrophil differentiation. *BMC Cell Biol.* 7: 11.

Chang, H.H., Hemberg, M., Barahona, M., Ingber, D.E., and Huang, S. 2008. Transcriptome-wide noise controls lineage choice in mammalian progenitor cells. *Nature* 453: 544–547.

Chen, X., Xu, H., Yuan, P., Fang, F., Huss, M., Vega, V.B., Wong, E., Orlov, Y.L., Zhang, W., Jiang, J., et al. 2008. Integration of external signaling pathways with the core transcriptional network in embryonic stem cells. *Cell* 133: 1106–1117.

Cole, M.F., Johnstone, S.E., Newman, J.J., Kagey, M.H., and Young, R.A. 2008. Tcf3 is an integral component of the core regulatory circuitry of embryonic stem cells. *Genes Dev.* 22: 746–755.

Delbrück, M. 1949. Discussion. In *Unités Biologiques Douées de Continuité Génétique* (Editions du Centre National de la Recherche Scientifique, Paris), pp. 33–35.

Feng, B., Jiang, J., Kraus, P., Ng, J.-H., Heng, J.-C.D., Chan, Y.-S., Yaw, L.-P., Zhang, W., Loh, T.-H., Han, J., et al. 2009. Reprogramming of fibroblasts into induced pluripotent stem cells with orphan nuclear receptor Esrrb. *Nat. Cell Biol.* 11: 197–203.

Gangaraju, V.K. and Lin, H. 2009. microRNAs: Key regulators of stem cells. *Nat. Rev. Mol. Cell Biol.* 10: 116–125.

Goldberg, A.D., Allis, C.D., and Bernstein, E. 2007. Epigenetics: A landscape takes shape. *Cell* 128: 635–638.

Griffiths-Jones, S., Grocock, R.J., van Dongen, S., Bateman, A., and Enright, A.J. 2006. miRBase: microRNA sequences, targets and gene nomenclature. *Nucleic Acids Res.* 34: D140–D144.

Gygi, S.P. and Aebersold, R. 2000. Mass spectrometry and proteomics. *Curr. Opin. Chem. Biol.* 4: 489–494.

Hatfield, S.D., Shcherbata, H.R., Fischer, K.A., Nakahara, K., Carthew, R.W., and Ruohola-Baker, H. 2005. Stem cell division is regulated by the microRNA pathway. *Nature* 435: 974–978.

Houbaviy, H.B., Murray, M.F., and Sharp, P.A. 2003. Embryonic stem cell-specific microRNAs. *Dev. Cell* 5: 351–358.

Huang, S. and Ingber, D.E. 2007. A non-genetic basis for cancer progression and metastasis: Self-organizing attractors in cell regulatory networks. *Breast Dis.* 26: 27–54.

Huang, S., Eichler, G., Bar-Yam, Y., and Ingber, D.E. 2005. Cell fates as high-dimensional attractor states of a complex gene regulatory network. *Phys. Rev. Lett.* 94: 128701-1–128701-4.

Ivanova, N., Dobrin, R., Lu, R., Kotenko, I., Levorse, J., DeCoste, C., Schafer, X., Lun, Y., and Lemischka, I.R. 2006. Dissecting self-renewal in stem cells with RNA interference. *Nature* 442: 533–538.

Ivanova, N.B., Dimos, J.T., Schaniel, C., Hackney, J.A., Moore, K.A., and Lemischka, I.R. 2002. A stem cell molecular signature. *Science* 298: 601–604.

Jaenisch, R. and Young, R. 2008. Stem cells, the molecular circuitry of pluripotency and nuclear reprogramming. *Cell* 132: 567–582.

Janke, W., ed. 2007. *Rugged free energy landscapes: Common computational approaches to spin glasses, structural glasses and biological macromolecules.* Springer, New York.

Jiang, J., Chan, Y.S., Loh, Y.H., Cai, J., Tong, G.Q., Lim, C.A., Robson, P., Zhong, S., and Ng, H.H. 2008. A core Klf circuitry regulates self-renewal of embryonic stem cells. *Nat. Cell Biol.* 10: 353–360.

Johnson, R., Teh, C.H., Kunarso, G., Wong, K.Y., Srinivasan, G.,

Cooper, M.L., Volta, M., Chan, S.S., Lipovich, L., Pollard, S.M., et al. 2008. REST regulates distinct transcriptional networks in embryonic and neural stem cells. *PLoS Biol.* **6:** e256.

Kauffman, S. 1969. Homeostasis and differentiation in random genetic control networks. *Nature* **224:** 177–178.

Kauffman, S. 1993. *The origins of order: Self-organization and selection in evolution.* Oxford University Press, New York.

Kidder, B.L., Yang, J., and Palmer, S. 2008. Stat3 and c-Myc genome-wide promoter occupancy in embryonic stem cells. *PLoS ONE* **3:** e3932.

Kim, J., Chu, J., Shen, X., Wang, J., and Orkin, S.H. 2008. An extended transcriptional network for pluripotency of embryonic stem cells. *Cell* **132:** 1049–1061.

Liu, X., Huang, J., Chen, T., Wang, Y., Xin, S., Li, J., Pei, G., and Kang J. 2008. Yamanaka factors critically regulate the developmental signaling network in mouse embryonic stem cells. *Cell Res.* **18:** 1177–1189.

Loh, Y.-H., Wu, Q., Chew, J.L., Vega, V.B., Zhang, W., Chen, X., Bourque, G., George, J., Leong, B., Liu, J., et al. 2006. The Oct4 and Nanog transcription network regulates pluripotency in mouse embryonic stem cells. *Nat. Genet.* **38:** 431–440.

Ma'ayan, A. 2008. Network integration and graph analysis in mammalian molecular systems biology. *IET Syst. Biol.* **2:** 206–221.

Ma'ayan, A. 2009. Insights into the organization of biochemical regulatory networks using graph theory analyses. *J. Biol. Chem.* **284:** 5451–5455.

Ma'ayan, A., Blitzer, R.D., and Iyengar, R. 2005a. Toward predictive models of mammalian cells. *Annu. Rev. Biophys. Biomol. Struct.* **34:** 319–349.

Ma'ayan, A., Jenkins, S.L., Neves, S., Hasseldine, A., Grace, E., Dubin-Thaler, B., Eungdamrong, N.J., Weng, G., Ram, P.T., Rice, J.J., et al. 2005b. Formation of regulatory patterns during signal propagation in a mammalian cellular network. *Science* **309:** 1078–1083.

MacArthur, B.D., Please, C.P., and Oreffo, R.O.C. 2008. Stochasticity and the molecular mechanisms of induced pluripotency. *PLoS ONE* **3:** e3086.

Marson, A., Levine, S.S., Cole, M.F., Frampton, G.M., Brambrink, T., Johnstone, S., Guenther, M.G., Johnston, W.K., Wernig, M., Newman, J., et al. 2008. Connecting microRNA genes to the core transcriptional regulatory circuitry of embryonic stem cells. **134:** 521–533.

Mathur, D., Danford, T.W., Boyer, L.A., Young, R.A., Gifford, D.K., and Jaenisch, R. 2008. Analysis of the mouse embryonic stem cell regulatory networks obtained by ChIP-chip and ChIP-PET. *Genome Biol.* **9:** R126.

Müller, F.-J., Laurent, L.C., Kostka, D., Ulitsky, I., Williams, R., Lu, C., Park, I.H., Rao, M.S., Shamir, R., Schwartz, P.H., et al. 2008. Regulatory networks define phenotypic classes of human stem cell lines. *Nature* **455:** 401–405.

Nakagawa, M., Koyanagi, M., Tanabe, K., Takahashi, K., Ichisaka, T., Aoi, T., Okita, K., Mochiduki, Y., Takizawa, N., and Yamanaka, S. 2008. Generation of induced pluripotent stem cells without Myc from mouse and human fibroblasts. *Nat. Biotechnol.* **26:** 101–106.

Singh, S.K., Kagalwala, M.N., Parker-Thornburg, J., Adams, H., and Majumder, S. 2008. REST maintains self-renewal and pluripotency of embryonic stem cells. *Nature* **453:** 223–227.

Subramanian, A., Tamayo, P., Mootha, V.K., Mukherjee, S., Ebert, B.L., Gillette, M.A., Paulovich, A., Pomeroy, S.L., Golub, T.R., Lander, E.S., and Mesirov J.P. 2005. Gene set enrichment analysis: A knowledge-based approach for interpreting genome-wide expression profiles. *Proc. Natl. Acad. Sci.* **102:** 15545–15550.

Tay, Y., Zhang, J., Thomson, A.M., Lim, B., and Rigoutsos, I. 2008. microRNAs to Nanog, Oct4 and Sox2 coding regions modulate embryonic stem cell differentiation. *Nature* **455:** 1124–1128.

Thomas, R. 1998. Laws for the dynamics of regulatory networks. *Int. J. Dev. Biol.* **42:** 479–485.

Ulitsky, I. and Shamir, R. 2007. Identification of functional modules using network topology and high-throughput data. *BMC Systems Biol.* **1:** 8.

Waddington, C.H. 1957. *The strategy of the genes.* Allen and Unwin, London.

Wales, D.J. 2003. *Energy landscapes: Applications to clusters, biomolecules and glasses.* Cambridge University Press, Cambridge.

Wang, J., Rao, S., Chu, J., Shen, X., Levasseur, D.N., Theunissen, T.W., and Orkin, S.H. 2006. A protein interaction network for pluripotency of embryonic stem cells. *Nature* **444:** 364–368.

Winfree, A. 2001. *The geometry of biological time,* 2nd ed. Springer, New York.

Ying, Q.-L., Wray, J., Nichols, J., Batlle-Morera, Doble, B., Woodgett, J., Cohen, P., and Smith A. 2008. The ground state of embryonic stem cell self-renewal. *Nature* **453:** 519–523.

Yu, J., Vodyanik, M.A., Smuga-Otto, K., Antosiewicz-Bourget, J., Frane, J.L., Tian, S., Nie, J., Jonsdottir, G.A., Ruotti, V., Stewart, R., et al. 2007. Induced pluripotent stem cell lines derived from human somatic cells. *Science* **318:** 1917–1920.

Zhou, Q., Brown, J., Kanarek, A., Rajagopal, J., and Melton, D.A. 2008. In vivo reprogramming of adult pancreatic exocrine cells to β-cells. *Nature* **455:** 627–632.

An Unexpected Role of TAFs and TRFs in Skeletal Muscle Differentiation: Switching Core Promoter Complexes

M.D.E. DEATO,*‡ AND R. TJIAN*†

*Howard Hughes Medical Institute, Department of Molecular and Cell Biology,
University of California, Berkeley, California 94720; †Li Ka Shing Center for Biomedical
and Health Sciences, University of California, Berkeley, California 94720

Sequence-specific enhancer-binding transcription factors and chromatin-modifying proteins are well recognized for their potential contributions to cell-type-specific gene regulation. In contrast, the role of core promoter recognition factors, such as TFIID in modulating gene- and cell-type-specific programs of transcription has been less understood. In general the so-called basal factors have largely been relegated to a supporting role as invariant components of the preinitiation complex. To dissect the potential contributions of TFIID to cell-type-specific transcription, we have studied the developmental process of skeletal myogenesis. Terminal differentiation during myogenesis involves an intricate reprogramming of transcription that is thought to be directed by cell-type-specific transcription regulatory factors. Here, we summarize our findings that the canonical TFIID complex must first be dismantled as a requisite step during the differentiation of myoblasts into myotubes and subsequently substituted by a novel core transcription complex composed of TAF3 and TRF3. Although this remarkable mechanism of completely switching core promoter recognition complexes to drive terminal differentiation has not been previously documented, it may eventually prove to be the rule rather than the exception as we learn more about cell-type-specific gene regulation.

Tissue formation during metazoan development requires an exquisitely controlled process of cellular proliferation and differentiation. Cell-type specialization ultimately involves a multistep terminal differentiation pathway wherein proliferating cells permanently exit the cell cycle. Driving this process requires spatial and temporal programs of gene regulation to achieve specialization and maintenance of tissue-specific functions. Because proliferating and differentiated cell types are sustained by distinct transcriptional programs, different batteries of genes must be kept off while others must be activated during the lifetime of a terminally differentiated cell type. The process of skeletal myogenesis represents a well-established and useful example of cellular differentiation. Skeletal muscle formation is a multistep process wherein predetermined muscle precursor myoblast cells undergo a series of differentiation steps to form mature myofibers that comprise the contractile muscle. This developmental process is responsible for skeletal muscle formation both in developing embryos and in postnatal muscle regeneration upon injury (Chargé and Rudnicki 2004).

Myogenesis requires a carefully orchestrated program of transcriptional activation and repression events. Key sequence-specific DNA-binding transcription factors that include Myf5, MyoD, Myogenin, and Mrf4 have key roles in directing specific transcriptional programs to form skeletal muscle (Berkes and Tapscott 2005). These myogenic factors have been characterized both in cell culture and in genetic loss-of-function mouse models that establish their important regulatory roles within two distinct stages of skeletal myogenesis: specification or determination of a myogenic state and differentiation (Pownall et al. 2002). The recruitment of these factors to regulatory enhancer sequences found in most skeletal muscle genes is one important step necessary to achieve the appropriate temporal gene expression patterns for muscle differentiation. Indeed, the primary transcriptional regulatory mechanism that is thought to be responsible for governing skeletal myogenesis has been attributed to these myogenic regulatory factors (MRFs). In conjunction with various chromatin-modifying activities, MRFs have been postulated to be necessary and sufficient to drive the divergent programs of cellular proliferation and terminal differentiation (Berkes and Tapscott 2005; Sartorelli and Caretti 2005; Palacios and Puri 2006). In contrast, the core promoter recognition apparatus required for transcription initiation, also commonly referred to as the "basal machinery," has largely been relegated to a supporting role that was not expected to play a key part in differential gene or cell-type-selective programs of transcription. Indeed, many studies performed out in the past 20 years have been based on the assumption that the composition of the core transcriptional machinery (TFIIA, B, D, E, F, and H and RNA polymerase II [pol II]) remains largely invariant in different cell types, and thus components of the preinitiation complex (PIC) were not expected to contribute significantly to the promoter or cell-specific activation responsible for cellular differentiation.

An integral and key component of the prototypic transcription machinery that comprises the PIC is the core promoter recognition complex, TFIID. The elements of this multisubunit complex (composed of TATA-box-

‡Present address: NGM Biopharmaceuticals Inc., 630 Gateway Boulevard, South San Francisco, California 94080.

binding protein [TBP] and its associated factors, TAFs) have generally been thought to remain invariant in different metazoan cell types. Targeted depletion of TAFs in various organisms such as yeast, worms, flies, frogs, and mice have all exhibited lethal phenotypes often associated with cell cycle arrest (Karim et al. 1996; Soldatov et al. 1999; Albright and Tjian 2000; Voss et al. 2000; Wassarman et al. 2000; Mohan et al. 2003). In addition, expression analyses of TAFs revealed a ubiquitous pattern in different cell types and tissues consistent with the phenotypes of null TAF mutations. However, the discovery of TAF homologs within different tissues (Freiman et al. 2001; Hiller et al. 2001, 2004; Pointud et al. 2003; Falender et al. 2005) as well as TBP-related factors (TRFs) (Crowley et al. 1993; Rabenstein et al. 1999; Persengiev et al. 2003) raised the possibility of at least some altered core promoter recognition complexes being differentially used in specific tissues and cell types. Indeed, conditional knockout approaches to bypass the lethal phenotype associated with null mutations in the TAFs have suggested a differential requirement for certain TAFs within specified tissue types (Indra et al. 2005). These early studies thus opened the possibility that some diversification in the core promoter recognition complexes may accommodate the necessary refinements in promoter selectivity and temporal transcriptional regulation in multicellular organisms (Hochheimer and Tjian 2003).

Here, we present evidence for an emerging new paradigm that relies on the use of distinct core promoter recognition complexes to regulate the process of cellular differentiation. By using skeletal myogenesis as a prototypic differentiation model system, we dissect the transcriptional consequences wherein the prototypic TFIID complex is eliminated early during myoblast differentiation and subsequently replaced by a second distinct core promoter recognition complex to selectively turn on one developmental program while effectively shutting off another.

DIFFERENTIAL EXPRESSION OF CORE PROMOTER RECOGNITION COMPLEX COMPONENTS DURING MYOBLAST DIFFERENTIATION

In our efforts to investigate the potential cell-type-specific functions of ubiquitously expressed TAFs, we have observed a surprisingly high level of TAF3 transcripts in skeletal muscle and testes but very low levels in other cell types (M. Deato, unpubl.). Furthermore, in our subsequent analysis of several mouse cell lines derived from different origins, we found that TAF3 levels are highly elevated in myoblast C2C12 cells when compared to fibroblast cells, whereas a testes cell line exhibited an intermediate level of TAF3. Thus, although low levels of TAF3 are detected in most mouse cell types, our analysis revealed a differential expression pattern in which levels of TAF3 are particularly highly expressed in myoblasts.

The observations that TAF3 is highly enriched in the myoblast cell line C2C12 provided us with a useful tool to study the potential function of TAF3 and TFIID in myo-

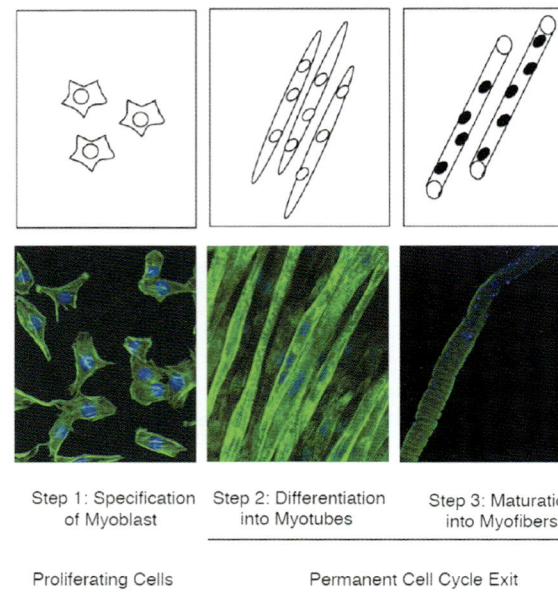

Figure 1. An overview of the key steps involved during skeletal myogenesis. Myogenesis involves multistep processes that are distinctly separated into two phases: proliferating and differentiated stages. Myoblasts are proliferating muscle-precursor cells. Upon induction of differentiation, these myoblast cells permanently exit the cell cycle and enter the differentiation stage to form multinucleated myotubes. These cells further mature by fusion to form myofibers that comprise the contractile unit of the skeletal muscle tissue. Each step in this process is characterized by the distinct expression of key myogenic transcription factors.

genesis when myoblasts differentiate into multinucleated myotubes (Fig. 1). This myoblast in vitro model system has been used extensively, and changes in various key transcription factors have previously been reported (Perletti et al. 1999; Guermah et al. 2003). However, key to our analyses of TAF3 and TFIID expression profiles in C2C12 cells is our efforts to separate myotubes from reserve cells. These reserve cells retain Myf5 expression under differentiation conditions and remain serum-responsive to reenter the cell cycle as myoblasts (Kitzmann et al. 1998). To circumvent the complications associated with mixed cell types, we have isolated nearly homogeneous (~95%) populations of myotubes and reserve cells derived from differentiated C2C12 cells by taking advantage of the selective capacity of reserve cells to remain attached to plates while myotubes remain largely nonadherent upon trypsinization.

We have used these highly purified myotubes as well as myoblasts and reserve cells to analyze the expression levels and cellular localization of TAF3 and the other TFIID subunits by quantitative polymerase chain reaction (PCR) (Deato and Tjian 2007), immunoblot, and immunofluorescence (Fig. 2). We find that in dividing cells, myoblasts contain a normal component of all the prototypic TFIID as well as TAF3 (Fig. 2A,B). Surprisingly, we found that myotubes contain dramatically reduced levels of TAF1,

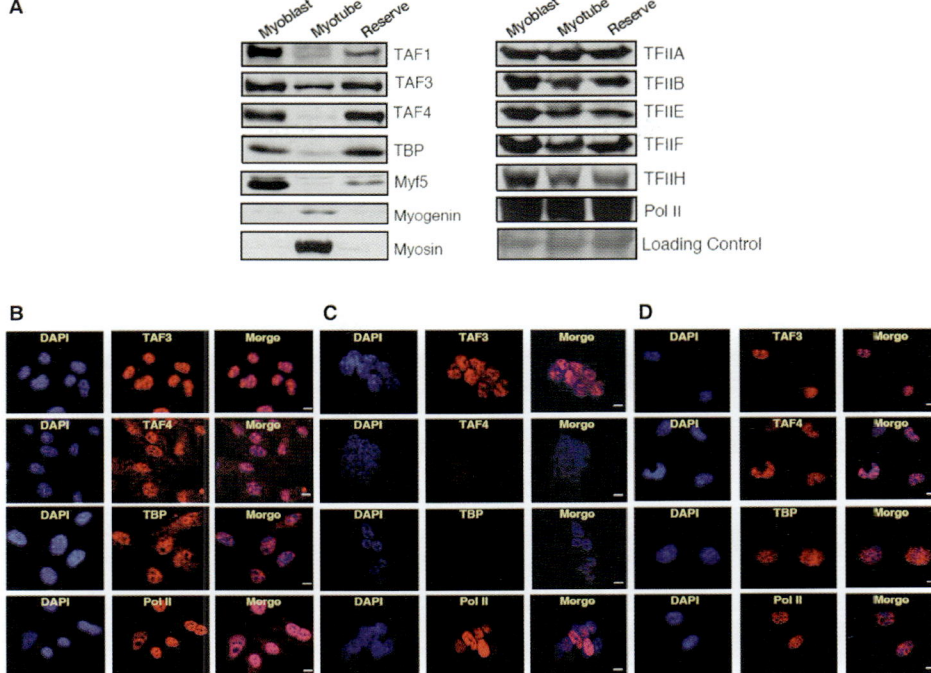

Figure 2. Differential expression of TAF3, TFIID, and other components of the preinitiation complex in myotubes. (*A*) Homogeneous populations of myoblast, myotube, and reserve cells derived from C2C12 cells were analyzed for TAF1, TAF3, TAF4, and TBP expression by immunoblot assay. The expression of other general transcription factors (TFIIA, TFIIB, TFIIE, TFIIF, TFIIH, and RNA pol II) were similarly analyzed in various C2C12 cell types. The expression of various myogenic transcription factors (Myf5 and Myogenin), as well as the muscle-specific protein myosin, also showed a differential pattern consistent with their expected profile. (*B–D*) Cellular localization analyses for TAF3 and TFIID subunits in myoblasts, myotubes, and reserve cells. (Reprinted, with permission, from Deato and Tjian 2007.)

TAF4, and TBP proteins and transcript levels, three canonical subunits of TFIID. In contrast, TAF3 protein levels in differentiated myotubes remained high (Fig. 2A,C). The levels of RNA pol II subunits and other basal factors remained unchanged during myoblast differentiation into myotubes. As a control, we monitored various myogenic transcription factors and other muscle-specific markers to gauge the efficiency of our myotube differentiation and isolation (Fig. 2A). Thus, by isolating purer populations of myotubes, we have uncovered a dramatic and severe loss of TFIID subunits (including TAF1, TAF4, and TBP) during differentiation of muscle cells with a concomitant and striking retention of TAF3 in differentiated myotubes. Taken together, our observations suggest that the very basic composition of the basal transcription machinery is critically altered with perhaps various core components (i.e., TFIID), even discarded as the cells transition into a postmitotic state of terminal differentiation. One possible mechanistic explanation for this observation may be the targeted degradation of the TFIID subunit TAF4, which was recently reported to be a keystone subunit in TFIID complex stability (Wright et al. 2006). Thus, it is possible that the loss of TAF4 during myoblast differentiation has severely compromised the stability of the TFIID complex. Furthermore, these findings suggest that TAF3 may emerge to have a regulatory role in directing transcription of muscle-specific genes in differentiated myotubes possibly replacing the canonical TFIID complex.

DIFFERENTIAL EXPRESSION OF CORE PROMOTER RECOGNITION COMPLEX COMPONENTS IN SKELETAL MUSCLE

Although C2C12 cells represent a useful in vitro model cell line to study certain stages of muscle development, there are limitations and differences relative to in vivo muscle differentiation (Pownall et al. 2002). For example, C2C12 cells can be induced to form the intermediate differentiated-stage myotubes but fail to progress further into myofibers that subsequently organize into bundles of the contractile muscle. During mouse skeletal muscle development, the myoblast cells that differentiate into myotubes retain a small pool of quiescent satellite cells (reminiscent of reserve cells) that are thought to potentiate future muscle regeneration (Chargé and Rudnicki 2004; Collins et al. 2005; Holterman and Rudnicki 2005). To investigate the relevance of the observed differential expression of TAF3 versus TFIID made in C2C12 cells, we analyzed various cell types that are representative of both proliferating and terminally differentiated states within bona fide skeletal muscle. First, we isolated primary myoblast cells (i.e., activated satellite cells) from newborn mice (Rando and Blau 1994) and analyzed for TAF3, TAF4, TBP, and pol II expression (Fig. 3A). We observed that the expression patterns for these factors are indistinguishable from those observed in both C2C12 myoblasts and reserve cells (Figs. 3E and 2A–B,D). Next,

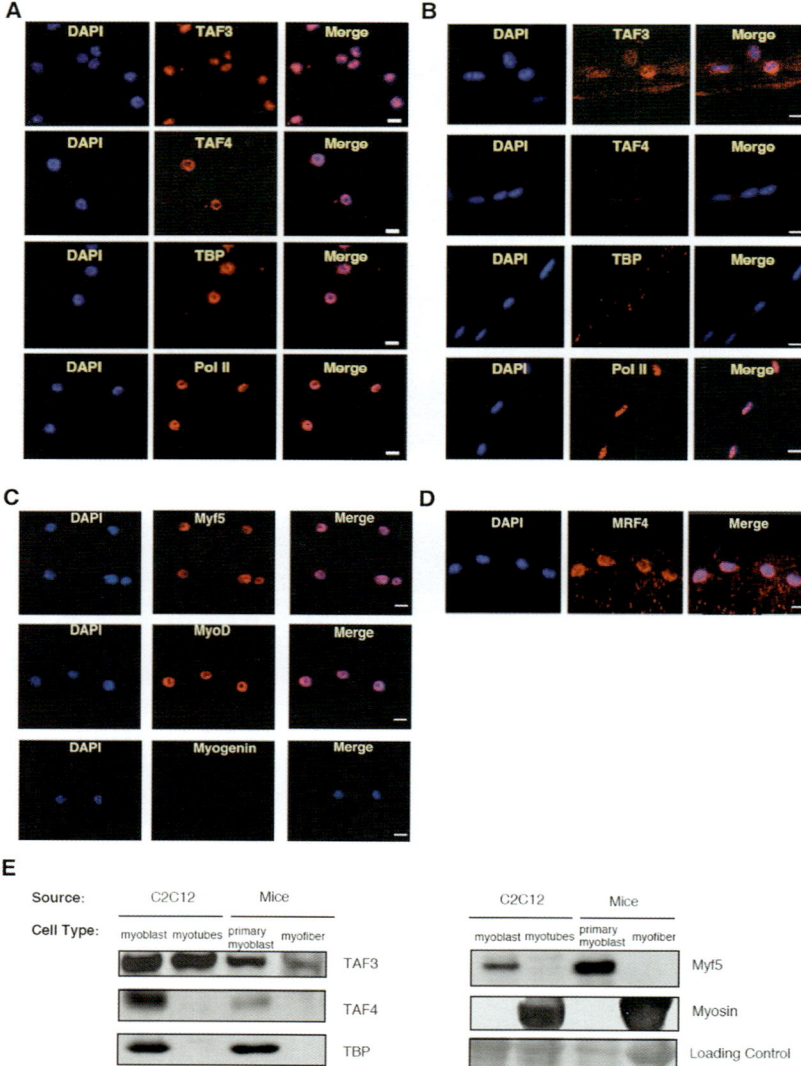

Figure 3. Differential expression of TAF3 and TFIID subunits in skeletal muscle. (*A,B*) Primary myoblasts and myofibers were isolated from newborn and adult mice, respectively, and analyzed by immunofluorescence. TAF3 is selectively expressed and retained in differentiated myofibers. In contrast, TAF3, TAF4, TBP, and RNA pol II expression in primary myoblast cells showed a profile similar to that previously observed in C2C12 myoblasts. (*C,D*) Expression patterns for various myogenic transcription factors Myf5, MyoD, Myogenin, and Mrf4 were also used as controls. (*E*) Myoblast and myotube cells derived from C2C12 cells were compared to primary myoblasts and myofibers isolated from mice for TAF3, TAF4, and TBP expression by immunoblot assay. (Reprinted, with permission, from Deato and Tjian 2007.)

individual myofibers from adult mice were isolated through a series of enzymatic digestion and manual trituration steps (Rosenblatt et al. 1995). We find that in contrast to primary myoblast cells, myofibers revealed the dramatic absence of TAF4 and TBP but retained TAF3 and pol II expression (Fig. 3B,E). Thus, the pattern of TAF3 levels in bona fide myofibers closely mimics our findings in cell culture myotubes (Figs. 3E and 2A,C). Indeed, our cell-staining and protein analyses of various muscle cell types strongly suggest that loss of TFIID subunits is likely a postmitotic event specific to terminally differentiated cell types such as myotubes/myofibers.

Furthermore, in cell types that are actively dividing or have the capacity to reenter the cell cycle (i.e., reserve cells, satellite, or primary myoblast cells), the canonical TFIID complex remains intact and presumably active. This suggests that loss of the TFIID complex may be part of an important and dramatic regulatory switch in the transcription mode as myoblast cells undergo terminal differentiation. This striking switch in basal factor components underscores the significance of TAF3 expression in differentiated skeletal muscle cell types and suggests a potential for TAF3 protein to regulate muscle-specific gene expression.

EXPRESSION OF A TBP-RELATED FACTOR IN MYOTUBES AND MYOFIBERS

TBP is conserved from yeast to humans. The recruitment of TBP on target genes is an essential step in the recruitment of other general transcription factors and subsequently the formation of the preinitiation complex (Thomas and Chiang 2006). In addition, TBP is also an essential gene required for metazoan viability. The targeted inactivation of TBP in various model organisms such as fish, worms, frogs, and mice all collectively point to the requirement for TBP in embryonic development (Veenstra et al. 2000; Martianov et al. 2001; Müller et al. 2001). However, the identification of TRFs in various metazoan genomes raised the possibility that TBP is not universally required for all eukaryotic gene expression (Hochheimer and Tjian 2003).

With the loss of conventional TFIID in differentiated myotubes, TAF3 may switch to a primary role in the transcriptional regulation of muscle-specific gene expression through the recruitment of TAF3 to the core promoters of these genes. However, all of our attempts to detect any sequence-specific activity attributed to TAF3 have failed. We have therefore hypothesized that TAF3 alone likely has no intrinsic sequence-specific DNA-binding capacity or core promoter recognition properties, characteristic of most TAFs (Albright and Tjian 2000). Rather, we infer that TAF3 may work in conjunction with a partner or partners to perform core promoter recognition of target genes. One possibility is that in the absence of TBP, as previously documented during embryonic stages in several model systems such as *Xenopus laevis* and *Danio renio*, cell-type-specific transcription may instead enlist the function of TRFs (Bartfai et al. 2004; Jallow et al. 2004). Our survey for various TRF expression revealed that TRF3 is highly expressed in myotube cells and in myofibers, whereas no TBP and other TAFs were detected. Thus, both TRF3 and TAF3 protein levels remain high in differentiated cell types such as myotubes and myofibers, concomitant with a dramatic loss of TBP and TFIID (Fig. 4).

TAF3 AND TRF3 ARE REQUIRED FACTORS FOR MYOGENIC DIFFERENTIATION

We have recently established that TAF3 and TRF3 form a complex in both myoblasts and myotubes (Deato and Tjian 2007). Next, we investigated the potential functional relevance of the TAF3 and TRF3 complex during myotube formation by generating stable RNA interference (RNAi) lines of both TAF3 and TRF3 in C2C12 cells. To assess the effect of depleting TAF3 and TRF3 in C2C12 cells, we analyzed these cell lines using several parameters that include cell morphology and the expression patterns of myogenic transcription factors during myoblast to myotube differentiation (Fig. 5). We find that reducing the levels of TAF3 and/or TRF3 in C2C12 cells results in a dramatic disruption of myoblast differentiation into myotubes (Fig. 5A,B). This phenotype is apparently not due to a failure to exit the cell cycle because both cell lines exhibit the prototypic loss of cell cycle markers

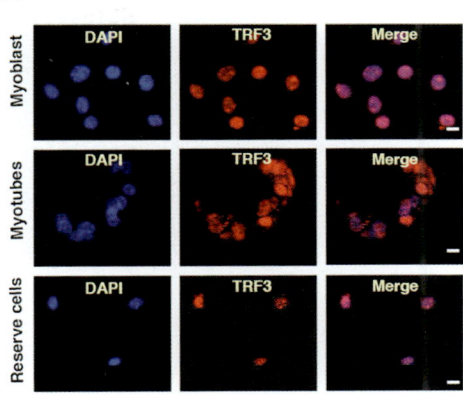

A

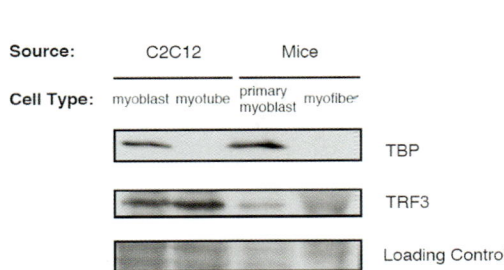

B

Figure 4. TRF3 expression in differentiated skeletal muscle cell types. (*A*) Immunofluorescence analysis for TRF3 localization in various C2C12 cell types showed its retention in myotubes. (*B*) Immunoblot analyses for TRF3 in myoblasts and myotubes derived from C2C12 cells and primary myoblasts and myofibers. In the absence of TBP expression in both myotubes and myofibers, TRF3 expression was observed. (Reprinted, with permission, from Deato and Tjian 2007.)

such as cyclin B1 expression when induced to differentiate (M. Deato, unpubl.). We also found that the levels of TAF4 and TBP remained unchanged when either TAF3 or TRF3 was targeted for depletion, suggesting that TFIID subunit expression in these stable RNAi cell lines remains largely unaffected and the complex presumably remains functional (Fig. 5B).

Because the depletion of TAF3 and TRF3 disrupted the process of myoblast differentiation into myotubes, we next set out to determine the stage at which this differentiation block may have occurred by analyzing the expression patterns of key myogenic transcription factors. Our studies found that cell lines lacking TAF3 and TRF3 are unable to express key regulators of myogenic differentiation, such as Myogenin, a transcription factor that specifies the differentiated state (Fig. 5C). Interestingly, a significant level of TAF4 and TBP proteins was detected even after treatment to induce differentiation in TAF3 and TRF3 RNAi knockdown cells (Fig. 5C). It is therefore possible that the loss of TAF3 or TRF3 not only blocks transcription of key myogenic factors (i.e., myogenin),

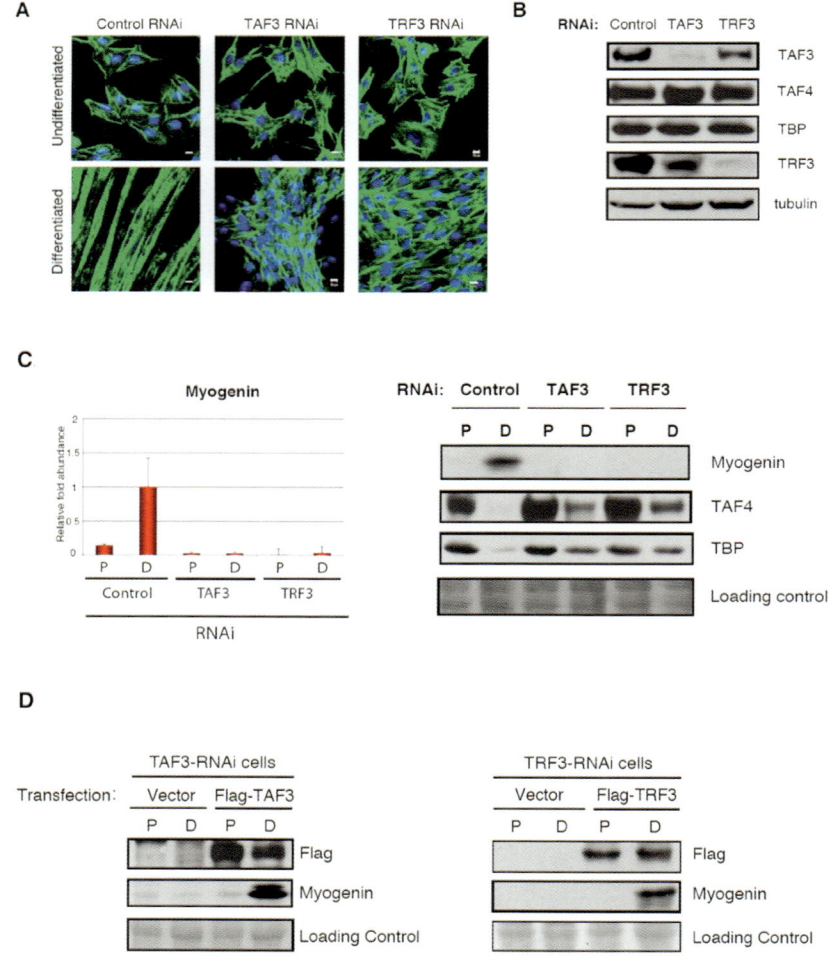

Figure 5. TAF3 and TRF3 are required factors for myoblast to myotube differentiation. (*A*) DAPI and phalloidin staining of control RNAi, TAF3 RNAi, and TRF3 RNAi cell lines to highlight the morphology of myoblast cells depleted of these factors. (*B*) Protein level analyses of control, TAF3, and TRF3 depleted cells. (*C*) Quantitative PCR analyses of control, TAF3 RNAi, and TRF3 RNAi cells grown under either proliferating (P) or differentiation (D) conditions for myogenin transcript levels. Each data bar shown is normalized to the abundance of U6 RNA in each sample and is represented as means plus or minus standard deviation from four independent samples. Immunoblot analyses of control, TAF3, and TRF3 RNAi cell lines grown under proliferating (P) or differentiated (D) conditions for the expression patterns of the key myogenic transcription factor myogenin. In addition, protein levels of TAF4 and TBP were analyzed. (*D*) Ectopic expression of full-length TAF3 or TRF3 into a TAF3-RNAi cell line and TRF3-RNAi cell line, respectively, rescues the differentiation defect that includes the expression of endogenous Myogenin. (Reprinted, with permission, from Deato and Tjian 2007.)

but also leads to the persistent (ectopic) expression of TBP and TFIID subunits, which may contribute to the disruption of myoblast to myotube differentiation. Importantly, the presence of the ectopic TBP and TFIID levels in myotubes appears to be insufficient to drive the expression of genes that are necessary for muscle differentiation (i.e., myogenin), providing further support that TAF3 and TRF3 likely have primary regulatory roles in muscle-specific gene expression (Fig. 5C).

The RNAi-mediated depletion experiments with TAF3 and TRF3 in myoblast cells thus provide strong evidence of a functional requirement for these factors during myoblast differentiation. Moreover, we find that forced expression of TAF3 and TRF3 back into RNAi cell lines rescues the observed differentiation defect. Importantly, the rescued cell lines were able to form robust myotubes

upon differentiation, and expression of myogenin was observed (Fig. 5D).

Our loss- and gain-of-function studies suggest that TAF3 and TRF3 are key transcription factors that direct cell-type-specific transcription of the myogenin gene during C2C12 differentiation. Indeed, the myogenin core promoter appears to be a direct regulatory target of both TAF3 and TRF3 during differentiation. To further establish this functional relationship, we performed chromatin immunoprecipitation (ChIP) studies and found that TAF3 and TRF3 are both highly enriched at the myogenin promoter in differentiated samples but not at intergenic control sites or in undifferentiated cells (Fig. 6). As a control, we also included pol II Ser-5 and TBP ChIP in our analyses and found that, as expected, pol II Ser-5 is efficiently recruited to the myogenin promoter along with TAF3 and

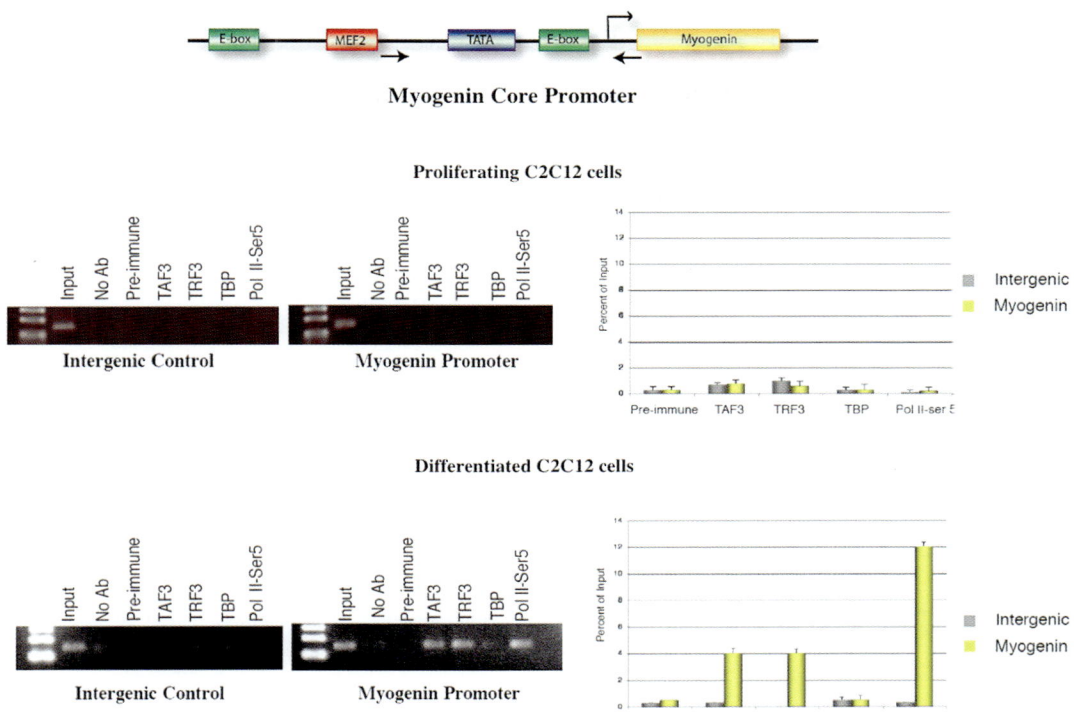

Figure 6. Enrichment of TAF3 and TRF3 on myogenin core promoter in differentiated C2C12 cells. The mouse myogenin core promoter consists of a TATA box flanked by several activator sites (E boxes and MEF2). *Arrowheads* correspond to the region within the myogenin core promoter wherein primer sequences were designed for chromatin immunoprecipitation (ChIP) studies. Intergenic primers were used as a control to amplify a region about 10 kb upstream of the myogenin transcription start site. ChIP samples from both proliferating and differentiated C2C12 cells were analyzed by quantitative PCRs using primer pairs directed against the intergenic control region or myogenin core promoter. The values shown were normalized as percent of input and represented as mean plus or minus standard deviation from four independent reactions. (Reprinted, with permission, from Deato and Tjian 2007.)

TRF3 in differentiated samples but not in proliferating C2C12 cells. In contrast, no TBP enrichment was observed on the myogenin promoter (Fig. 6). To eliminate the possibility that this is a nonspecific effect, we also analyzed the actin promoter, a gene expressed in both proliferating and differentiated cells. We observed that TBP is enriched on this promoter when cells are undifferentiated but not in differentiated C2C12 cells. In contrast, pol II Ser-5 is efficiently recruited to the actin promoter in both undifferentiated and differentiated cells (M. Deato, unpubl.). Taken together, our loss-of-function, rescue experiments and ChIP analyses all strongly point to an important role for TAF3 and TRF3 during muscle differentiation.

CONCLUSION

Coordinated Switching of TFIID, TAF3/TRF3 Complex in Skeletal Muscle Specialization

Terminal differentiation during myogenesis involves an intricate transcriptional reprogramming directed by cell-type-specific regulatory factors. Here, we highlighted our findings that a novel core transcription complex composed of TAF3 and TRF3 replaces the canonical TFIID complex, which is discarded as a requisite step during the differentiation of myoblasts into myotubes. We

further proposed that an unexpected critical component for directing cell-type specification involves dramatic changes to the composition of the core promoter complex (Fig. 7). Although a surprising observation, the complete loss of the prototypic holo-TFIID and the wholesale replacement by an alternative core promoter complex may help to explain a previous report that in C2C12 cells, TAF4 and TBP levels are reduced and may be targeted for proteolytic degradation during myoblast differentiation (Perletti et al. 1999). Because we recently established that TAF4 is a linchpin subunit of TFIID required for its stability (Wright et al. 2006), we hypothesize that the targeted degradation of one or few cornerstone subunits could be sufficient to severely compromise the stability of the entire TFIID complex and thus trigger the destruction of the holo-complex. Our studies suggest that this costly removal of functional TFIID may be a necessary step during differentiation of myoblasts into myotubes. Consistent with previous reports that TFIID is required for cellular proliferation and cell cycle regulation, we speculate that the loss of a key core promoter recognition complex such as TFIID could provide an efficient mechanism to drive cells out of the cell cycle and into a terminally differentiated state. In short, the loss of TFIID would provide a quick and relatively easy way to ensure keeping the proliferation transcription program shut off in terminally differentiated muscle cells.

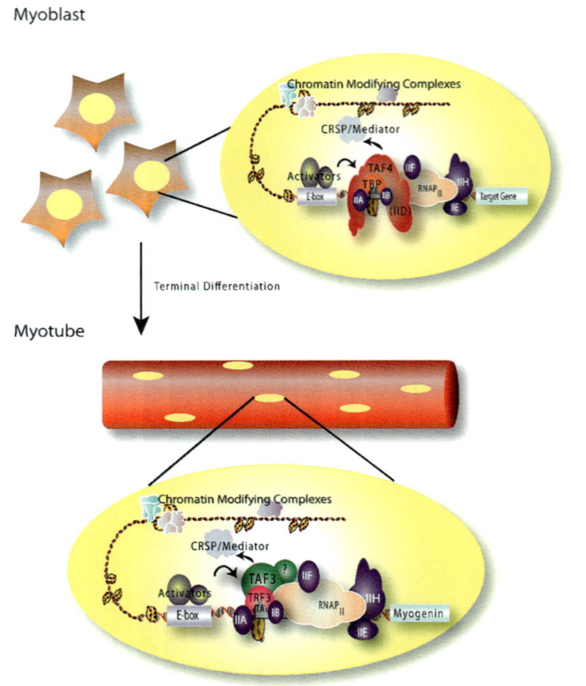

Figure 7. Proposed model for transcription initiation in myoblast and myotube cells. The novel core transcription initiation complex composed of TAF3 and TRF3 functionally replaces the canonical TFIID complex as a requisite step during the differentiation of myoblasts into myotubes. This mechanism of switching the core transcription machinery may regulate the unique transcriptional program during cell-type-specific terminal differentiation. (Reprinted, with permission, from Deato and Tjian 2007.)

Importantly, we have discovered that this degradation of TFIID occurs in a coordinated manner with a seamless replacement by a newly discovered TAF3 and TRF3 complex. Consistent with this mutually dependent switching of core promoter recognition complexes, we observed an enhanced retention of TAF4 and TBP levels in differentiation-defective cells depleted of TAF3 and TRF3. The recovery of myotube formation and myogenin expression observed in TAF3 and TRF3 rescue experiments further strengthens the correlation between the loss of TFIID and its replacement by the TAF3/TRF3 complex during myotube formation.

In light of our findings, it would be of interest to determine whether a similar switch in the core transcription machinery also occurs during the differentiation of other cell types. Another equally interesting question is whether other "prototypic" components of the preinitiation complex (i.e., CRSP/mediator, SAGA, etc.) might also be switched or discarded during differentiation. Perhaps it is possible that the initiation and maintenance of a specific terminally differentiated state are most economically achieved by a complete overhaul of certain basal transcription components, and such a novel mechanism may be a required step in cellular differentiation. Thus, the diversification of both sequence-specific enhancers and the core promoter recognition apparatus would significantly enhance the ability of metazoan organisms to accommodate tissue-specific patterns of gene expression by essentially adapting a more elaborate eukaryotic version of the "sigma factor switching" model for reprogramming transcription during development and differentiation (Gross et al. 1998).

REFERENCES

Albright, S.R. and Tjian, R. 2000. TAFs revisited: More data reveal new twists and confirm old ideas. *Gene* **242:** 1–13.

Bartfai, R., Balduf, C., Hilton, T., Rathmann, Y., Hadzhiev, Y., Tora, L., Orban, L., and Muller, F. 2004. TBP2, a vertebrate-specific member of the TBP family, is required in embryonic development of zebrafish. *Curr. Biol.* **14:** 593–598.

Berkes, C.A. and Tapscott, S.J. 2005. MyoD and the transcriptional control of myogenesis. *Semin. Cell Dev. Biol.* **16:** 585–595.

Chargé, S.B. and Rudnicki, M.A. 2004. Cellular and molecular regulation of muscle regeneration. *Physiol. Rev.* **84:** 209–238.

Collins, C.A., Olsen, I., Zammit, P.S., Heslop, L., Petrie, A., Partridge, T.A., and Morgan, J.E. 2005. Stem cell function, self-renewal, and behavioral heterogeneity of cells from the adult muscle satellite cell niche. *Cell* **122:** 289–301.

Crowley, T.E., Hoey, T., Liu, J.K., Jan, Y.N., Jan, L.Y., and Tjian, R. 1993. A new factor related to TATA-binding protein has highly restricted expression patterns in *Drosophila*. *Nature* **361:** 557–561.

Deato, M.D. and Tjian, R. 2007. Switching of the core transcription machinery during myogenesis. *Genes Dev.* **21:** 2137–2149.

Falender, A.E., Freiman, R.N., Geles, K.G., Lo, K.C., Hwang, K., Lamb, D.J., Morris, P.L., Tjian, R., and Richards, J.S. 2005. Maintenance of spermatogenesis requires TAF4b, a gonad-specific subunit of TFIID. *Genes Dev.* **19:** 794–803.

Freiman, R.N., Albright, S.R., Zheng, S., Sha, W.C., Hammer, R.E., and Tjian, R. 2001. Requirement of tissue-selective TBP-associated factor TAF$_{II}$105 in ovarian development. *Science* **293:** 2084–2087.

Gross, C.A., Chan, C., Dombroski, A., Gruber, T., Sharp, M., Tupy, J., and Young, B. 1998. The functional and regulatory roles of σ factors in transcription. *Cold Spring Harbor Symp. Quant. Biol.* **63:** 141–155.

Guermah, M., Ge, K., Chiang, C.M., and Roeder, R.G. 2003. The TBN protein, which is essential for early embryonic mouse development, is an inducible TAFII implicated in adipogenesis. *Mol. Cell* **12:** 991–1001.

Hiller, M., Chen, X., Pringle, M.J., Suchorolski, M., Sancak, Y., Viswanathan, S., Bolival, B., Lin, T.Y., Marino, S., and Fuller, M.T. 2004. Testis-specific TAF homologs collaborate to control a tissue-specific transcription program. *Development* **131:** 5297–5308.

Hiller, M.A., Lin, T.Y., Wood, C., and Fuller, M.T. 2001. Developmental regulation of transcription by a tissue-specific TAF homolog. *Genes Dev.* **15:** 1021–1030.

Hochheimer, A. and Tjian, R. 2003. Diversified transcription initiation complexes expand promoter selectivity and tissue-specific gene expression. *Genes Dev.* **17:** 1309–1320.

Holterman, C.E. and Rudnicki, M.A. 2005. Molecular regulation of satellite cell function. *Semin. Cell Dev. Biol.* **16:** 575–584.

Indra, A.K., Mohan II, W.S., Frontini, M., Scheer, E., Messaddeq, N., Metzger, D., and Tora, L. 2005. TAF10 is required for the establishment of skin barrier function in foetal, but not in adult mouse epidermis. *Dev. Biol.* **285:** 28–37.

Jallow, Z., Jacobi, U.G., Weeks, D.L., Dawid, I.B., and Veenstra, G.J. 2004. Specialized and redundant roles of TBP and a vertebrate-specific TBP paralog in embryonic gene regulation in *Xenopus*. *Proc. Natl. Acad. Sci.* **101:** 13525–13530.

Karim, F.D., Chang, H.C., Therrien, M., Wassarman, D.A., Laverty, T., and Rubin, G.M. 1996. A screen for genes that function downstream of Ras1 during *Drosophila* eye development. *Genetics* **143:** 315–329.

Kitzmann, M., Carnac, G., Vandromme, M., Primig, M., Lamb, N.J., and Fernandez, A. 1998. The muscle regulatory factors MyoD and myf-5 undergo distinct cell cycle-specific expression in muscle cells. *J. Cell Biol.* **142:** 1447–1459.

Martianov, I., Fimia, G.M., Dierich, A., Parvinen, M., Sassone-Corsi, P., and Davidson, I. 2001. Late arrest of spermiogenesis and germ cell apoptosis in mice lacking the TBP-like *TLF/TRF2* gene. *Mol. Cell* **7:** 509–515.

Mohan, Jr., W.S., Scheer, E., Wendling, O., Metzger, D., and Tora, L. 2003. TAF10 (TAF$_{II}$30) is necessary for TFIID stability and early embryogenesis in mice. *Mol. Cell. Biol.* **23:** 4307–4318.

Müller, F., Lakatos, L., Dantonel, J., Strähle, U., and Tora, L. 2001. TBP is not universally required for zygotic RNA polymerase II transcription in zebrafish. *Curr. Biol.* **11:** 282–287.

Palacios, D. and Puri, P.L. 2006. The epigenetic network regulating muscle development and regeneration. *J. Cell. Physiol.* **207:** 1–11.

Perletti, L., Dantonel, J.C., and Davidson, I. 1999. The TATA-binding protein and its associated factors are differentially expressed in adult mouse tissues. *J. Biol. Chem.* **274:** 15301–15304.

Persengiev, S.P., Zhu, X., Dixit, B.L., Maston, G.A., Kittler, E.L., and Green, M.R. 2003. TRF3, a TATA-box-binding protein-related factor, is vertebrate-specific and widely expressed. *Proc. Natl. Acad. Sci.* **100:** 14887–14891.

Pointud, J.C., Mengus, G., Brancorsini, S., Monaco, L., Parvinen, M., Sassone-Corsi, P., and Davidson, I. 2003. The intracellular localisation of TAF7L, a paralogue of transcription factor TFIID subunit TAF7, is developmentally regulated during male germ-cell differentiation. *J. Cell Sci.* **116:** 1847–1858.

Pownall, M.E., Gustafsson, M.K., and Emerson, Jr., C.P. 2002. Myogenic regulatory factors and the specification of muscle progenitors in vertebrate embryos. *Annu. Rev. Cell Dev. Biol.* **18:** 747–783.

Rabenstein, M.D., Zhou, S., Lis, J.T., and Tjian, R. 1999. TATA box-binding protein (TBP)-related factor 2 (TRF2), a third member of the TBP family. *Proc. Natl. Acad. Sci.* **96:** 4791–4796.

Rando, T.A. and Blau, H.M. 1994. Primary mouse myoblast purification, characterization, and transplantation for cell-mediated gene therapy. *J. Cell Biol.* **125:** 1275–1287.

Rosenblatt, J.D., Lunt, A.I., Parry, D.J., and Partridge, T.A. 1995. Culturing satellite cells from living single muscle fiber explants. *In Vitro Cell. Dev. Biol. Anim.* **31:** 773–779.

Sartorelli, V. and Caretti, G. 2005. Mechanisms underlying the transcriptional regulation of skeletal myogenesis. *Curr. Opin. Genet. Dev.* **15:** 528–535.

Soldatov, A., Nabirochkina, E., Georgieva, S., Belenkaja, T., and Georgiev, P. 1999. TAF$_{II}$40 protein is encoded by the *e(y)1* gene: Biological consequences of mutations. *Mol. Cell. Biol.* **19:** 3769–3778.

Thomas, M.C. and Chiang, C.M. 2006. The general transcription machinery and general cofactors. *Crit. Rev. Biochem. Mol. Biol.* **41:** 105–178.

Veenstra, G.J., Weeks, D.L., and Wolffe, A.P. 2000. Distinct roles for TBP and TBP-like factor in early embryonic gene transcription in *Xenopus*. *Science* **290:** 2312–2315.

Voss, A.K., Thomas, T., Petrou, P., Anastassiadis, K., Schöler, H., and Gruss, P. 2000. Taube nuss is a novel gene essential for the survival of pluripotent cells of early mouse embryos. *Development* **127:** 5449–5461.

Wassarman, D.A., Aoyagi, N., Pile, L.A., and Schlag, E.M. 2000. TAF250 is required for multiple developmental events in *Drosophila*. *Proc. Natl. Acad. Sci.* **97:** 1154–1159.

Wright, K.J., Marr II, M.T., and Tjian, R. 2006. TAF4 nucleates a core subcomplex of TFIID and mediates activated transcription from a TATA-less promoter. *Proc. Natl. Acad. Sci.* **103:** 12347–12352.

Stemness Is Only a State of the Cell

M.N. KAGALWALA,*# S.K. SINGH,* AND S. MAJUMDER*†‡§¶

*Department of Genetics, †Department of Neuro-Oncology, ‡The Brain Tumor Center, §Center for Stem Cell
and Developmental Biology, The University of Texas M.D. Anderson Cancer Center, Houston, Texas 77030;
¶Program in Genes and Development, The University of Texas Graduate School of Biomedical Sciences
at Houston, Houston, Texas 77030

How the programming and reprogramming of stem/progenitor cells regulate normal cell development and cancer is still not well known. One of the tools that we have chosen to use to investigate stem cell regulation is the transcriptional repressor element 1-silencing transcription factor (REST). REST contains a DNA-binding domain and two repressor domains. Once bound to its target genes, REST can interact with several cellular corepressors to regulate epigenetic modifications. REST is expressed in most nonneural cells, including neural stem/progenitor cells (NSCs), but it is absent in most neuronal cells. REST was originally found to be a major transcriptional repressor of neural differentiation. Previously, we found that activation of REST target genes in NSCs was sufficient to cause neuronal differentiation. Furthermore, the activation of REST target genes in myoblasts was sufficient to override the muscle differentiation pathway and produce a physiologically active neuronal phenotype. Although REST is normally not expressed in most neural cells, we previously found that approximately 50% of human medulloblastomas, a malignant pediatric brain tumor, express REST and that abnormal expression of REST in NSCs causes medulloblastoma-like cerebellar tumors by blocking neuronal differentiation. Interestingly, REST is also expressed at high levels in mouse embryonic stem (mES) cells, but its role in these cells is not understood. Recently, we found that REST maintains self-renewal and pluripotency in mES cells through suppression of microRNA-21 (miRNA21). Thus, REST is a newly discovered element of the interconnected regulatory network that maintains the self-renewal and pluripotency of mES cells. Taken together, the results of several different studies indicate that stem/progenitor cells are more flexible than previously believed and that a simple alteration of transcriptional regulators in these cells can affect both normal cell development and cancer.

REST has been the focus of research in many laboratories since its discovery (Chong et al. 1995; Schoenherr and Anderson 1995), and our understanding of its regulation and roles has improved considerably. Initial studies concluded that REST was a transcriptional repressor of a group of genes primarily involved in terminal neuronal differentiation. In accordance with these findings, the REST expression pattern was shown to be present mainly in nonneuronal cells and rarely in neurons, with few exceptions (Koenigsberger et al. 2000; Griffith et al. 2001; Shimojo and Hersh 2004). REST was also found to regulate neurogenesis via miRNA-124a (miR-124a) (Conaco et al. 2006). These authors found that REST and miR-124a are expressed in a reciprocal manner and that miR-124a is a direct target of REST. It was no coincidence that many of the targets of miR-124a were shown to be nonneuronal transcripts that assist REST in efficient neuronal differentiation. Since that first report, the list of miRNAs that are predicted to be or are regulated by REST has increased (Wu and Xie 2006; Otto et al. 2007). Many outstanding reviews about REST regulation and its proposed functions have been written in recent years (Ballas and Mandel 2005; Coulson 2005; Hsieh and Gage 2005; Lunyak and Rosenfeld 2005; Cao et al. 2006; Majumder 2006; Weissman 2008). Here, we briefly describe the critical aspects of REST biology and the con-

tributions to our understanding of REST from work done in our laboratory.

REST IN ACTION

Same Lyrics

REST is a zinc finger protein (1097 amino acids long) with a DNA-binding domain sandwiched between two repressor domains, RD1 and RD2, on its amino terminus and carboxyl terminus, respectively (Ballas and Mandel 2005; Coulson 2005; Hsieh and Gage 2005; Lunyak and Rosenfeld 2005; Cao et al. 2006). A 21–23-bp DNA element known as RE-1 was shown to be necessary and sufficient for the recruitment of REST to specific gene loci, and about 2000 such elements are predicted to be present throughout the mammalian genome (Bruce et al. 2004). Recently, a novel bipartite REST-binding element was discovered (Johnson et al. 2007).

Different Rhythm, Different Melody

REST interacts with several cellular cofactors through its two repressor domains and brings about direct transcriptional repression or alters the chromatin to a repressed state (Ballas and Mandel 2005; Coulson 2005; Hsieh and Gage 2005; Lunyak and Rosenfeld 2005; Cao et al. 2006). The interactors of the RD1 domain include mSin3A, the histone deacetylase (HDAC) complex, and N-CoR; the RD2 domain interactors include CoREST, the mSin3A/HDAC

#Present address: Laboratory of Genetics, Salk Institute for Biological Studies, La Jolla, California 92037.

complex, histone H3-K9 methyltransferase G9a, and histone H3-K4 demethylase LSD1. Depending on the context, the REST complex can also function in conjunction with, for example, DNA methyltransferase 1, DNA-methyl-CpG-binding protein-2, the chromatin-remodeling complex, and SWI/SNF. In protein–protein interaction studies, REST has also been shown to interact with the TATA-box binding protein and small RNA polymerase II carboxyl-terminal domain phosphatases.

Different Music

Although REST is largely regarded as a repressor molecule in certain cells such as neural stem cells, the presence of double-stranded RNA molecules with RE-1-like sequences can modulate its function and turn it into an activator molecule (Kuwabara et al. 2004). Various isoforms (e.g., REST4) have been shown to act as dominant-negative molecules in REST function and compete with its interacting partner RILP (a LIM domain protein) for cellular localization (Shimojo et al. 1999; Shimojo and Hersh 2004). The REST gene was found to be transcriptionally activated by Wnt signaling in human embryonic carcinoma cells (Willert et al. 2002) and directly in the chick spinal cord (Nishihara et al. 2003). In contrast, REST gene transcription can be blocked by the binding of the unliganded retinoic acid receptor-repressor complex to the retinoic acid receptor element found on the REST gene promoter (Lunyak et al. 2002; Ballas et al. 2005).

Besides having a critical role in neurogenesis, REST has been found to regulate several genes that may affect important biological processes. For example, ischemic insults were found to derepress REST mRNA and protein in dying neurons, and this was suggested to be a critical mechanism in insult-induced neuronal death (Calderone et al. 2003). REST and huntingtin protein were found to form a complex that forced REST to localize in the cytoplasm in normal neuronal cells. In patients with Huntington's disease, this interaction was ablated by the huntington mutation, resulting in the translocation of REST to the nucleus and causing a blockade of the expression of neuronal genes such as *BDNF* (brain-derived neurotrophic factor) (Zuccato et al. 2003, 2007), suggesting that REST has a role in Huntington's disease. REST was found to repress the μ-opioid receptor in neuronal cells and thus may have a role in opium addiction (Kim et al. 2004; Formisano et al. 2007). REST was found to repress the serotonin 1A receptor, which is implicated in depression and anxiety (Lemonde et al. 2004). REST also may have a role in other mental disorders, such as X-linked mental retardation (Tahiliani et al. 2007; Ding et al. 2008) and Down syndrome (Canzonetta et al. 2008).

Among nonneuronal tissues, REST was found to be present in normal ventricular myocytes and to suppress the expression of multiple fetal cardiac genes (Kuwahara et al. 2003). In cardiac dysfunction and arrhythmogenesis, REST expression was inhibited, resulting in the expression of fetal cardiac genes. These results suggest that REST has a role in maintaining normal cardiac structure and function and that its deregulation may cause cardiac dysfunction (Bingham et al. 2007). REST was found to be expressed in vascular smooth muscle cells, where it suppressed the expression of a critical potassium channel gene and, thereby, regulated both normal and diseased states of the cell (Cheong et al. 2005).

REST shows both tumor suppressor and oncogenic effects, depending on the cellular context (Coulson 2005; Majumder 2006). The epithelial cells of tissues such as those of the lung, breast, and colon normally express REST and suppress expression of neuronal genes. In this context, REST functions as a tumor suppressor, and the abnormal lack of REST activity in these cells may lead to oncogenesis. In contrast, differentiating or differentiated neuronal cells generally do not express REST and do express neuronal genes. In this context, the suppressive activity of REST may have an oncogenic function. Proteasomal degradation is a major mechanism regulating the REST protein and, thereby, REST activity (Ballas et al. 2005; Westbrook et al. 2005, 2008; Guardavaccaro et al. 2008). It was found that the carboxyl terminus of the REST protein interacts with the F-box protein β-TrCP and is degraded in a ubiquitin ligase SCF–β-TrCP-dependent manner. Proteasomal degradation is critical during neuronal differentiation (Ballas et al. 2005; Westbrook et al. 2008). In addition, β-TrCP overexpression may cause oncogenic transformation of human mammary epithelial cells via REST degradation, which is consistent with the function of REST as a tumor suppressor (Westbrook et al. 2005, 2008; Weissman 2008). On the other hand, degradation of REST protein β-TrCP during the G_2 phase of the cell cycle has been shown to cause transcriptional activation of the REST target gene *Mad2*, an essential component required for proper mitotic spindle assembly. Thus, a REST mutant defective in binding to β-TrCP was unable to activate the spindle checkpoint and demonstrated various mitotic defects. These results suggest that high levels of unchecked REST could have a role in promoting genomic instability, thus resulting in the oncogenic function of REST. Previous studies from our laboratory and others have shown that REST misexpression or overexpression in neural cells causes medulloblastomas (see below). It would be worth investigating whether this process involves genomic instability.

Depending on the cellular context, the relative abundance of its interacting partners, and the specific interactions, a variety of combinatorial mechanisms can thus be used by REST to differentially influence the transcription of each of its many target genes (Andres et al. 1999; Jepsen et al. 2000; Ballas et al. 2001; Lunyak et al. 2002). Furthermore, the epigenetic marks caused by the REST–cosuppressor complex on its target genes were found to differ in differentiated fibroblasts, pluripotent embryonic stem (ES) cells, neural progenitors, mature neurons derived from ES cells, and mature neurons from the adult brain (Ballas and Mandel 2005; Hsieh and Gage 2005).

FLEXIBILITY OF NEURAL STEM/PROGENITOR CELLS: NEURONAL DIFFERENTIATION

In early embryogenesis (stages E11.5–E13.5), REST is expressed in most nonneuronal cells (Chong et al. 1995). Studies with REST$^{-/-}$ mice suggested that although REST is required for the suppression of its target genes, activation of many of its targets was not solely dependent on the absence of the REST protein (Chen et al. 1998), and lack of REST alone did not activate these REST target genes. Our work suggested that, depending on the requirements for their activation, REST target genes can be broadly categorized into two groups. The first group is suppressed in the presence of REST but is transcriptionally active when REST-mediated repression is absent (e.g., β-tubulin III gene); the second group of target genes is also suppressed by REST but require additional promoter/enhancer-specific activators/events apart from the absence of REST-mediated repression for their activation (e.g., synapsin and glutamate receptor genes) (Immaneni et al. 2000).

Mandel's group used the staged conversion of ES cells into neural stem cells (NSCs) and further differentiated the NSCs into mature neurons in the presence of retinoic acid (Ballas et al. 2005). Their work provided a mechanistic basis for the two classes of REST target genes. They found that REST and its cosuppressor complex use two distinct mechanisms to regulate the target genes. In the first mechanism, the RE-1 site of the promoter is occupied by the REST–cosuppressor complex, which suppresses chromatin expression such that the removal of REST from the chromatin results in the expression of class I neuronal genes. In the second mechanism, the REST–cosuppressor complex occupies the RE-1 site of the promoter; in addition, CoREST/HDAC and methyl-CpG-binding protein-2/mSin3/HDAC complexes occupy distinct but adjacent methylated CpG sites. In this scenario, removal of REST from the chromatin alone does not activate these genes (class II neuronal genes) because of the additional suppressor complexes located on the methylated CpG sites. However, the class II genes can be activated upon application of a specific stimulus, such as membrane depolarization, which relieves the additional chromatin suppression and provides neuronal plasticity. These findings were validated in cortical progenitors and cortical neurons isolated from mouse embryos.

To determine the impact of REST-mediated suppression on neuronal differentiation, we examined what happens when we force the REST target genes to be activated. For this purpose, we constructed a recombinant transcription factor, REST-VP16, in which both suppressor domains of REST were replaced with the strong activation domain of the herpes simplex virus protein VP16 (Immaneni et al. 2000; Lawinger et al. 2000; Watanabe et al. 2004; Su et al. 2006). We found that REST-VP16 binds to the same DNA-binding site as does REST, competes with REST for DNA binding, and activates its target genes even in the presence of REST in multiple cell types (Immaneni et al. 2000). When we expressed REST-VP16 in an NSC line generated from newborn mouse cerebellum, it rapidly converted the cells into a mature neuronal phenotype that included neurite outgrowth, expression of multiple neuronal genes (both target and nontarget REST genes), survival in the presence of mitotic inhibitors, and synaptic vesicle recycling and glutamate-induced calcium influx (Su et al. 2004). Such expeditious differentiation was not seen in NSCs that did not express REST-VP16. These results showed that direct activation of REST target genes in

NSCs is sufficient to cause rapid neuronal differentiation and indicated that these cells are more flexible than previously thought.

FLEXIBILITY OF MUSCLE PROGENITOR CELLS: NEURONAL DIFFERENTIATION

To examine the cellular flexibility of nonneuronal stem/progenitor cells, we expressed REST-VP16 in muscle progenitor cells (myoblasts) (Watanabe et al. 2004). These cells were not thought to be capable of converting to a neuronal phenotype. Surprisingly, our results showed that the expression of REST-VP16 in myoblasts grown under muscle differentiation conditions blocked entry into the muscle differentiation pathway, countered endogenous REST-dependent suppression, activated the REST target genes, and, surprisingly, activated other neuronal differentiation genes and converted the myoblasts into a physiologically active neuronal phenotype. Furthermore, these in-vitro–differentiated neuronal cells, when injected into the mouse brain, survived, incorporated into the normal brain, and did not form tumors.

Our further unpublished results indicated that when the myoblast-derived neuronal cells were transplanted into the cerebella of newborn mice, the cells integrated into the cerebellar environment, did not form tumors, exhibited neuronal properties, including action potential, and were capable of receiving glutamatergic synaptic input. This was the first study showing myoblasts being converted into a neuronal phenotype, which indicated a high degree of flexibility in terms of the cellular fate of these cells. This enormous cellular flexibility has been further demonstrated in recent studies by other investigators, indicating that mouse and human fibroblasts can be reprogrammed to an induced-pluripotent state by the transfer of a set of four transcription factors (Jaenisch and Young 2008; Lengner et al. 2008; Yamanaka 2008; Yu and Thomson 2008;).

FLEXIBILITY OF NEURAL STEM/PROGENITOR CELLS: MEDULLOBLASTOMA

Medulloblastoma is one of the most malignant brain tumors in children and is believed to arise from undifferentiated NSCs present in the cerebellum. Pathways involved in normal early development of the brain (such as Hedgehog and Wnt signaling) are implicated in medulloblastoma tumorigenesis (McMahon 2000; Eberhart and Burger 2003; Gilbertson 2004; Raffel 2004; Rutka et al. 2004; Fogarty et al. 2005; Romer and Curran 2005; Dellovade et al. 2006; Knoepfler and Kenney 2006). During the progression or initiation of medulloblastoma, the activities of these pathways are sustained at higher than normal levels or the pathways are abnormally activated. However, mutations activating these pathways have been documented in only a modest percentage of human medulloblastoma tumors.

We found that many human medulloblastoma samples and established human medulloblastoma cell lines had a much higher expression of REST than did neuronal cells or normal brain cells (Lawinger et al. 2000; Su et al. 2004; Fuller et al. 2005). As expected, the REST-positive human medulloblastoma tumor cells did not express REST target genes, such as synapsin. Our further work showed that transgenic mice expressing REST in neuronal cells appeared to develop normally, without tumor formation (S. Majumder, unpubl.), indicating that abnormal expression of REST in NSCs alone was not sufficient to cause tumorigenesis. Similarly, David Anderson's group found that neuronal cells constitutively expressing REST did not form tumors and appeared to acquire a normal neuronal morphology, except that they manifested axon pathfinding errors (Paquette et al. 2000). Reexamination of human medulloblastoma tumors indicated that many of these tumors expressed abnormally high levels of both REST and Myc (Su et al. 2006). Furthermore, NSCs engineered to overexpress activated Myc and REST, and not the control cells, gave rise to tumors in the mouse cerebellum, the site of human medulloblastoma formation. These tumors were blocked in neuronal differentiation and were morphologically similar to human medulloblastoma.

The cells that produced tumors in the cerebellum did not produce tumors in the cortex, indicating the critical role of the local brain environment in the formation of tumors (Su et al. 2006). Such site-specific tumor growth suggests the role of a niche where REST-MYC-driven tumors can originate. Thus, efficient medulloblastoma tumorigenesis occurs when NSCs are present in an appropriate environment/niche and are forced to express both Myc, causing overall increased proliferation, and REST, the suppressor function of which causes the blockade of differentiation (maintenance of "stemness"). We further found that countering the effects of REST with REST-VP16 in Myc- and REST-expressing NSCs counteracted the tumorigenic potential of the cells, in a manner similar to that observed in human medulloblastoma cell lines, indicating that REST has a critical role in medulloblastoma tumorigenesis.

The oncogenic role of the overexpression/misexpression of REST in neuronal cells is further supported by studies from other laboratories that found REST overexpression in several neuroblastoma cells, with concomitant suppression of neuronal differentiation genes (Nishimura et al. 1996; Higashino et al. 2003). Furthermore, when neuroblastoma cells were forced to differentiate, REST expression decreased and neuronal markers increased in these cells. Taken together, these results suggested that the abnormal maintenance of stemness and the resulting blockade of differentiation in combination with increased proliferation in NSCs causing a state of restricted cellular flexibility can impart oncogenic properties.

FLEXIBILITY OF ES CELLS: SELF-RENEWAL VERSUS DIFFERENTIATION

ES cells are pluripotent cells that show extreme cellular flexibility and have the potential for both indefinite self-renewal and differentiation into all three germ layers of the body (Pan and Thomson 2007). These properties of ES cells are regulated by a combination of core regulatory factors, such as Oct4, Sox2, Nanog, and Klf4 and Lin28 (Jaenisch and Young 2008; Yamanaka 2008; Yu and Thomson 2008). REST was found to be expressed at high levels in mouse ES (mES) cells (Ballas and Mandel 2005), but its role in these cells was not understood. By performing chromatin immunoprecipitation (ChIP)-on-Chip assays, researchers showed that Oct4, Sox2, and Nanog cooccupy the REST gene, suggesting that REST is in the expanded regulatory circuit that controls self-renewal and pluripotency in ES cells (Boyer et al. 2005). Indeed, we found that REST maintains self-renewal and pluripotency in mES cells through suppression of miR-21 (Singh et al. 2008). As with known self-renewal markers, REST expression is much higher in self-renewing mES cells than in differentiating mES (mEB) cells. The heterozygous deletion of REST ($REST^{+/-}$) and its small interfering RNA (siRNA)-mediated knockdown in mES cells—*both of which result in significantly lower levels of REST protein*—cause a loss of self-renewal, even when the cells are grown under self-renewal conditions, and lead to the expression of markers specific for multiple lineages. Conversely, exogenously added REST maintains self-renewal in mEB cells. In addition, $REST^{+/-}$ mES cells cultured under self-renewal conditions express substantially reduced levels of several self-renewal regulators, including Oct4, Nanog, Sox2, and c-Myc, and exogenously added REST in mEB cells maintains the self-renewal phenotypes and expression of these self-renewal regulators.

We further found that in mES cells, REST was bound to the gene chromatin of a set of miRNAs that potentially target self-renewal genes. Whereas mES cells and mEB cells containing exogenously added REST expressed lower levels of these miRNAs, wild-type mEB cells, $REST^{+/-}$ mES cells, and siREST-treated mES cells expressed higher levels of these miRNAs. At least one of these REST-regulated miRNAs, miR-21, specifically suppressed the self-renewal of mES cells, corresponding to the decreased expression of Oct4, Nanog, Sox2, and c-Myc. Thus, REST is a newly discovered element of the interconnected regulatory network that maintains the self-renewal and pluripotency of mES cells.

Our conclusions are not yet universally accepted; two recent papers contain somewhat contradictory results. In the first, the investigators report that REST ablation does not alter ES cell pluripotency (Sun et al. 2008). In the second, Johnson et al. (2008) report that although REST indeed has a key role in the maintenance of the ES cell phenotype, REST does not bind to the *miR-21* gene chromatin. Both our conventional results and the ChIP–quantitative polymerase chain reaction (qPCR) results suggested specific binding of REST to a particular site in the *miR-21* gene chromatin. Moreover, our data suggests that this binding is functional and REST directly regulates miR-21 expression. This was clearly demonstrated by the up-regulation of miR-21 upon depletion of REST. Unfortunately, the Johnson et al. study did not provide a REST-miR-21 functional assay. In contrast, two other recent papers support our work: Results from Stuart Orkin's laboratory (Kim et al. 2008) indicated that REST is indeed part of the network that regulates ES cell self-renewal and pluripotency, and Dean Nizetic's laboratory (Canzonetta et al. 2008) showed that mES cells with lower REST levels derived from a mouse model of Down syndrome have decreased levels of self-renewal markers and a higher propensity toward differentiation, even when cultured in the presence of leukemia inhibitory factor (LIF). In addition, Phillip Sharp's laboratory (Houbaviy et al. 2003) found that miR-21 expression was higher in differentiated mES cellsthan in undifferentiated mES cells, supporting our conclusions. Thus, future work is needed to clarify these contradictory observations in terms of REST's role in maintaining self-renewal and pluripotency of mES cells.

ROLE OF REST IN THE ICM OF BLASTOCYSTS

We found that the inner cell mass (ICM) of mouse blastocysts coexpressed REST and the self-renewal markers Oct4, Sox2, and Nanog. However, the role of REST in the ICM of the developing mouse blastocyst is still unclear. $REST^{+/-}$ haploinsufficient mice have an apparently normal phenotype (Chen et al. 1998), similar to that in mice haploinsufficient in most of the other known self-renewal factors or regulators, such as Oct4 (Nichols et al. 1998), Nanog (Mitsui et al. 2003), and Sox2 (Avilion et al. 2003), and the components of the LIF-STAT3-c-Myc pathway (LIF, Stewart et al. 1992; LIF-receptor β [LIFRβ], Li et al. 1995; gp130, Nichols et al. 2001; STAT3, Takeda et al. 1997; c-Myc, Davis et al. 1993). All of these haploinsufficient mice were apparently normal. However, the complete loss of the individual regulators had variable effects. $REST^{-/-}$ embryos could survive past the blastocyst stage but showed progressive embryonic lethality between embryonic days 9.5 and 11, an apparent result of the lack of REST's role in regulating self-renewal and pluripotency in the ICM cell population under normal conditions (Chen et al. 1998). In comparison, $Oct4^{-/-}$, $Nanog^{-/-}$, and $Sox2^{-/-}$ embryos developed to the blastocyst stage, but the ICM cells failed to develop normally (Nichols et al. 1998; Avilion et al. 2003; Chambers et al. 2003; Mitsui et al. 2003). In contrast, embryos deficient in the components of the LIF-STAT3 pathway showed an apparent lack of an immediate effect on the ICM cells. $LIF^{-/-}$ mice were viable and showed only retarded postnatal growth (Stewart et al. 1992), $LIFR\beta^{-/-}$ mice died only after birth (Li et al. 1995), $gp130^{-/-}$ mice died between 12.5 days postcoitum and term (Nichols et al. 2001); $STAT3^{-/-}$ embryos developed until embryonic day 6 and then degenerated between embryonic days 6.5 and 7.5 (Takeda et al. 1997); and $c\text{-}Myc^{-/-}$ mice developed until embryonic day 10.5 (Davis et al. 1993). Thus, in terms of ICM development, $REST^{-/-}$ embryos are more similar to those deficient in the LIF-STAT3 pathway than to those deficient in Oct4, Nanog, and Sox2.

Although the exact nature of REST's role in the pluripotent ICM of blastocysts is still unknown, there are several possible explanations. One is that the ICM cell population under normal conditions is transient, so that even the complete absence of REST does not produce its full detrimental effect. However, if the ICM cells were capable of slowed development ("diapause"), a phenomenon that has evolved in some mammals, including mice, to get around adverse conditions during pregnancy, the deficiency of REST would result in the developmental arrest of blastocysts. This is supported by the finding that the development of $gp130^{-/-}$ blastocysts was arrested only when they were subjected to diapause (Nichols et al. 2001).

REST AS A TOOL TO UNDERSTAND CELLULAR FLEXIBILITY

On the basis of our knowledge about REST and the mechanisms it uses to affect multiple cellular processes in different cellular contexts, we present a model of how REST may be able to accomplish all of these seemingly diverse functions (Fig. 1). The following are two crucial factors about REST: (1) It is normally expressed in multiple cell types ranging from embryonic stem cells to adult nonneuronal cells and it has an important role in each of these cell types, although expression levels vary between cell types. (2) Different cellular functions seem to be affected by REST levels/function. During early development, the key role of REST appears to be to restrict expression of lineage-specific genes and maintain self-renewal. In ES cells, where REST is highly expressed, REST maintains self-renewal by repressing its targets, mainly the expression of miR-21 (Fig. 1). As shown by other investigators, REST protein levels must decrease if ES cells are to undergo differentiation (Ballas et al. 2005), and this decrease is mediated by cytoplasmic proteasome machinery after the REST degron is marked for degradation by β-TrCP (Ballas et al. 2005; Westbrook et al. 2008).

In the course of normal development, the gradual decrease in REST levels thus has a critical role in the controlled loss of pluripotency and yet allows some cells (such as NSCs) to retain the capacity to differentiate into more restricted cell types. It is imperative to understand whether there is an internal hierarchy among the targets of REST (based on the affinity of REST for RE-1 sites and chromatin context) and their repression is thus a function of the level of REST (and/or its cofactors). This built-in affinity sensor between REST levels and suppressed targets would allow for the cell-type-specific manifestation of REST function, including maintenance of pluripotency in ES cells and differentiation of NSCs into neurons. The microenvironments (although not completely understood) of these cells may have a critical role in controlling REST levels and other factors needed for its functions.

The Wnt and Shh signaling pathways have been shown to regulate neurogenesis (Kenney et al. 2004; Lie et al. 2005), where Wnt and Shh signaling are important for the initial expansion of NSCs, and another set of signals then allows these cells to differentiate by further repressing REST levels such that terminally differentiated neurons do not have any detectable levels of REST. Finally, in nonneuronal cells, REST levels are sufficient to suppress neuronal differentiation and prevent any misexpression of REST target genes, thus helping to maintain the cellular identity. It is also noteworthy that REST function is dependent on the levels of cofactors and on the presence of cell-specific activator(s). Thus, there are two scenarios by which altered REST function may lead to cellular abnormalities and tumorigenesis. As discussed earlier, REST functions as a tumor suppressor or oncogene in a context-dependent manner.

CONCLUSIONS

The flexibility of stem/progenitor cells is increasingly found to regulate both normal and abnormal development

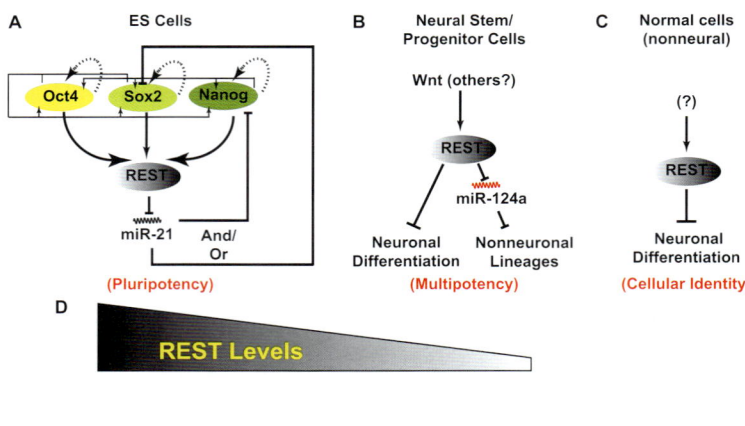

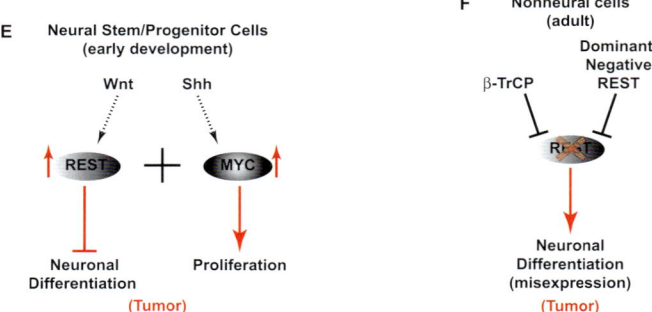

Figure 1. REST-directed cellular flexibility. (*A*) REST maintains self-renewal and pluripotency in ES cells via miRNA-21, which in turn could potentially target Sox2 and/or Nanog and the Oct4-Sox2-Nanog circuit. (*B*) REST has a critical role in blocking differentiation of neural stem/progenitor cells into neuronal lineage both by directly blocking expression of neuronal target genes and indirectly by preventing the expression of miR-124a, which destabilizes mRNAs of nonneuronal genes. (*C*) REST also maintains cellular identity of nonneural cells such as fibroblast cells by blocking neuronal genes. (*D*) Level of REST protein in embryonic stem cells compared with neural stem/progenitor cells. (*E*) Overexpression of both REST and Myc in neural stem/progenitor cells leads to tumor formation as a result of the block in neural differentiation (maintenance of "stemness") and a simultaneous increase in proliferation. (*F*) Loss of REST in nonneural cells, in which it is normally present, causes precocious expression of neuronal genes, leading to tumor formation in a context-dependent manner.

such as cancer and other diseases. Recent discoveries have shed light on REST's role in directing some important aspects of this process. As a regulator of a large number of genes, depending on the cellular context, on the amount of REST protein present in the cell, and on the affinity of the REST protein complex toward its specific target gene in the given cellular environment, including the cell's chromatin architecture, REST is likely to modulate the flexibility of stem/progenitor cells and, thus, impact health and disease in more ways than are currently known. Most of the exciting discoveries in that respect are yet to come.

ACKNOWLEDGMENTS

We thank the National Institutes of Health for funding (grants CA97124 and CA81255 to S.M.). M.N.K. and S.K.S. were both recipients of the Dodie Hawn Fellowship in Cancer Genetics.

REFERENCES

Andres, M.E., Burger, C., Peral-Rubio, M.J., Battaglioli, E., Anderson, M.E., Grimes, J., Dallman, J., Ballas, N., and Mandel, G. 1999. CoREST: A functional corepressor required

for regulation of neural-specific gene expression. *Proc. Natl. Acad. Sci.* **96:** 9873–9878.

Avilion, A.A., Nicolis, S.K., Pevny, L.H., Perez, L., Vivian, N., and Lovell-Badge, R. 2003. Multipotent cell lineages in early mouse development depend on SOX2 function. *Genes Dev.* **17:** 126–140.

Ballas, N. and Mandel, G. 2005. The many faces of REST oversee epigenetic programming of neuronal genes. *Curr. Opin. Neurobiol.* **15:** 500–506.

Ballas, N., Grunseich, C., Lu, D.D., Speh, J.C., and Mandel, G. 2005. REST and its corepressors mediate plasticity of neuronal gene chromatin throughout neurogenesis. *Cell* **121:** 645–657.

Ballas, N., Battaglioli, E., Atouf, F., Andres, M.E., Chenoweth, J., Anderson, M.E., Burger, C., Moniwa, M., Davie, J.R., Bowers, W.J., et al. 2001. Regulation of neuronal traits by a novel transcriptional complex. *Neuron* **31:** 353–365.

Bingham, A.J., Ooi, L., Kozera, L., White, E., and Wood, I.C. 2007. The repressor element 1-silencing transcription factor regulates heart-specific gene expression using multiple chromatin-modifying complexes. *Mol. Cell. Biol.* **27:** 4082–4092.

Boyer, L.A., Lee, T.I., Cole, M.F., Johnstone, S.E., Levine, S.S., Zucker, J.P., Guenther, M.G., Kumar, R.M., Murray, H.L., Jenner, R.G., et al. 2005. Core transcriptional regulatory circuitry in human embryonic stem cells. *Cell* **122:** 947–956.

Bruce, A.W., Donaldson, I.J., Wood, I.C., Yerbury, S.A., Sadowski, M.I., Chapman, M., Gottgens, B., and Buckley, N.J. 2004. Genome-wide analysis of repressor element 1 silencing transcription factor/neuron-restrictive silencing factor (REST/NRSF) target genes. *Proc. Natl. Acad. Sci.* **101:** 10458–10463.

Calderone, A., Jover, T., Noh, K.M., Tanaka, H., Yokota, H., Lin, Y., Grooms, S.Y., Regis, R., Bennett, M.V., and Zukin, R.S. 2003. Ischemic insults derepress the gene silencer REST in neurons destined to die. *J. Neurosci.* **23:** 2112–2121.

Canzonetta, C., Mulligan, C., Deutsch, S., Ruf, S., O'Doherty, A., Lyle, R., Borel, C., Lin-Marq, N., Delom, F., Groet, J., et al. 2008. DYRK1A-dosage imbalance perturbs NRSF/REST levels, deregulating pluripotency and embryonic stem cell fate in Down syndrome. *Am. J. Hum. Genet.* **83:** 388–400.

Cao, X., Yeo, G., Muotri, A.R., Kuwabara, T., and Gage, F.H. 2006. Noncoding RNAs in the mammalian central nervous system. *Annu. Rev. Neurosci.* **29:** 77–103.

Chambers, I., Colby, D., Robertson, M., Nichols, J., Lee, S., Tweedie, S., and Smith, A. 2003. Functional expression cloning of Nanog, a pluripotency sustaining factor in embryonic stem cells. *Cell* **113:** 643–655.

Chen, Z.F., Paquette, A.J., and Anderson, D.J. 1998. NRSF/REST is required in vivo for repression of multiple neuronal target genes during embryogenesis. *Nat. Genet.* **20:** 136–142.

Cheong, A., Bingham, A.J., Li, J., Kumar, B., Sukumar, P., Munsch, C., Buckley, N.J., Neylon, C.B., Porter, K.E., Beech, D.J., and Wood, I.C. 2005. Downregulated REST transcription factor is a switch enabling critical potassium channel expression and cell proliferation. *Mol. Cell* **20:** 45–52.

Chong, J.A., Tapia-Ramirez, J., Kim, S., Toledo-Aral, J.J., Zheng, Y., Boutros, M.C., Altshuller, Y.M., Frohman, M.A., Kraner, S.D., and Mandel, G. 1995. REST: A mammalian silencer protein that restricts sodium channel gene expression to neurons. *Cell* **80:** 949–957.

Conaco, C., Otto, S., Han, J.J., and Mandel, G. 2006. Reciprocal actions of REST and a microRNA promote neuronal identity. *Proc. Natl. Acad. Sci.* **103:** 2422–2427.

Coulson, J.M. 2005. Transcriptional regulation: Cancer, neurons and the REST. *Curr. Biol.* **15:** R665–R668.

Davis, A.C., Wims, M., Spotts, G.D., Hann, S.R., and Bradley, A. 1993. A null c-*myc* mutation causes lethality before 10.5 days of gestation in homozygotes and reduced fertility in heterozygous female mice. *Genes Dev.* **7:** 671–682.

Dellovade, T., Romer, J.T., Curran, T., and Rubin, L.L. 2006. The hedgehog pathway and neurological disorders. *Annu. Rev. Neurosci.* **29:** 539–563.

Ding, N., Zhou, H., Esteve, P.O., Chin, H.G., Kim, S., Xu, X., Joseph, S.M., Friez, M.J., Schwartz, C.E., Pradhan, S., and Boyer, T.G. 2008. Mediator links epigenetic silencing of neuronal gene expression with x-linked mental retardation. *Mol. Cell* **31:** 347–359.

Eberhart, C.G. and Burger, P.C. 2003. Anaplasia and grading in medulloblastomas. *Brain Pathol.* **13:** 376–385.

Fogarty, M.P., Kessler, J.D., and Wechsler-Reya, R.J. 2005. Morphing into cancer: The role of developmental signaling pathways in brain tumor formation. *J. Neurobiol.* **64:** 458–475.

Formisano, L., Noh, K.M., Miyawaki, T., Mashiko, T., Bennett, M.V., and Zukin, R.S. 2007. Ischemic insults promote epigenetic reprogramming of μ opioid receptor expression in hippocampal neurons. *Proc. Natl. Acad. Sci.* **104:** 4170–4175.

Fuller, G.N., Su, X., Price, R.E., Cohen, Z.R., Lang, F.F., Sawaya, R., and Majumder, S. 2005. Many human medulloblastoma tumors overexpress repressor element-1 silencing transcription (REST)/neuron-restrictive silencer factor, which can be functionally countered by REST-VP16. *Mol. Cancer Ther.* **4:** 343–349.

Gilbertson, R.J. 2004. Medulloblastoma: Signalling a change in treatment. *Lancet Oncol.* **5:** 209–218.

Griffith, E.C., Cowan, C.W., and Greenberg, M.E. 2001. REST acts through multiple deacetylase complexes. *Neuron* **31:** 339–340.

Guardavaccaro, D., Frescas, D., Dorrello, N.V., Peschiaroli, A., Multani, A.S., Cardozo, T., Lasorella, A., Iavarone, A., Chang, S., Hernando, E., and Pagano, M. 2008. Control of chromosome stability by the β-TrCP-REST-Mad2 axis. *Nature* **452:** 365–369.

Higashino, K., Narita, T., Taga, T., Ohta, S., and Takeuchi, Y. 2003. Malignant rhabdoid tumor shows a unique neural differentiation as distinct from neuroblastoma. *Cancer Sci.* **94:** 37–42.

Houbaviy, H.B., Murray, M.F., and Sharp, P.A. 2003. Embryonic stem cell-specific microRNAs. *Dev. Cell* **5:** 351–358.

Hsieh, J. and Gage, F.H. 2005. Chromatin remodeling in neural development and plasticity. *Curr. Opin. Cell. Biol.* **17:** 664–671.

Immaneni, A., Lawinger, P., Zhao, Z., Lu, W., Rastelli, L., Morris, J.H., and Majumder, S. 2000. REST-VP16 activates multiple neuronal differentiation genes in human NT2 cells. *Nucleic Acids Res.* **28:** 3403–3410.

Jaenisch, R. and Young, R. 2008. Stem cells, the molecular circuitry of pluripotency and nuclear reprogramming. *Cell* **132:** 567–582.

Jepsen, K., Hermanson, O., Onami, T.M., Gleiberman, A.S., Lunyak, V., McEvilly, R.J., Kurokawa, R., Kumar, V., Liu, F., Seto, E., et al. 2000. Combinatorial roles of the nuclear receptor corepressor in transcription and development. *Cell* **102:** 753–763.

Johnson, D.S., Mortazavi, A., Myers, R.M., and Wold, B. 2007. Genome-wide mapping of in vivo protein-DNA interactions. *Science* **316:** 1497–1502.

Johnson, R., Teh, C.H., Kunarso, G., Wong, K.Y., Srinivasan, G., Cooper, M.L., Volta, M., Chan, S.S., Lipovich, L., Pollard, S.M., et al. 2008. REST regulates distinct transcriptional networks in embryonic and neural stem cells. *PLoS Biol.* **6:** e256.

Kenney, A.M., Widlund, H.R., and Rowitch, D.H. 2004. Hedgehog and PI-3 kinase signaling converge on Nmyc1 to promote cell cycle progression in cerebellar neuronal precursors. *Development* **131:** 217–228.

Kim, C.S., Hwang, C.K., Choi, H.S., Song, K.Y., Law, P.Y., Wei, L.N., and Loh, H.H. 2004. Neuron-restrictive silencer factor (NRSF) functions as a repressor in neuronal cells to regulate the μ opioid receptor gene. *J. Biol. Chem.* **279:** 46464–46473.

Kim, J., Chu, J., Shen, X., Wang, J., and Orkin, S.H. 2008. An extended transcriptional network for pluripotency of embryonic stem cells. *Cell* **132:** 1049–1061.

Knoepfler, P.S. and Kenney, A.M. 2006. Neural precursor cycling at sonic speed: N-Myc pedals, GSK-3 brakes. *Cell Cycle* **5:** 47–52.

Koenigsberger, C., Chicca II, J.J., Amoureux, M.C., Edelman, G.M., and Jones, F.S. 2000. Differential regulation by multiple promoters of the gene encoding the neuron-restrictive silencer factor. *Proc. Natl. Acad. Sci.* **97:** 2291–2296.

Kuwabara, T., Hsieh, J., Nakashima, K., Taira, K., and Gage, F.H. 2004. A small modulatory dsRNA specifies the fate of adult neural stem cells. *Cell* **116:** 779–793.

Kuwahara, K., Saito, Y., Takano, M., Arai, Y., Yasuno, S., Nakagawa, Y., Takahashi, N., Adachi, Y., Takemura, G., Horie, M., et al. 2003. NRSF regulates the fetal cardiac gene program and maintains normal cardiac structure and function. *EMBO J.* **22:** 6310–6321.

Lawinger, P., Venugopal, R., Guo, Z.S., Immaneni, A., Sengupta, D., Lu, W., Rastelli, L., Marin Dias Carneiro, A., Levin, V., Fuller, G.N., et al. 2000. The neuronal repressor REST/NRSF is an essential regulator in medulloblastoma cells. *Nat. Med.* **6:** 826–831.

Lemonde, S., Rogaeva, A., and Albert, P.R. 2004. Cell type-dependent recruitment of trichostatin A-sensitive repression of the human 5-HT1A receptor gene. *J. Neurochem.* **88:** 857–868.

Lengner, C.J., Welstead, G.G., and Jaenisch, R. 2008. The pluripotency regulator Oct4: A role in somatic stem cells? *Cell Cycle* **7:** 725–728.

Li, M., Sendtner, M., and Smith, A. 1995. Essential function of LIF receptor in motor neurons. *Nature* **378:** 724–727.

Lie, D.C., Colamarino, S.A., Song, H.J., Desire, L., Mira, H., Consiglio, A., Lein, E.S., Jessberger, S., Lansford, H., Dearie, A.R., and Gage, F.H. 2005. Wnt signalling regulates adult hippocampal neurogenesis. *Nature* **437:** 1370–1375.

Lunyak, V.V. and Rosenfeld, M.G. 2005. No rest for REST: REST/NRSF regulation of neurogenesis. *Cell* **121:** 499–501.

Lunyak, V.V., Burgess, R., Prefontaine, G.G, Nelson, C., Sze, S.H., Chenoweth, J., Schwartz, P., Pevzner, P.A., Glass, C., Mandel, G., and Rosenfeld, M.G. 2002. Corepressor-dependent silencing of chromosomal regions encoding neuronal genes. *Science* **298:** 1747–1752.

Majumder, S. 2006. REST in good times and bad: Roles in tumor suppressor and oncogenic activities. *Cell Cycle* **5:** 1929–1935.

McMahon, A.P. 2000. More surprises in the Hedgehog signaling pathway. *Cell* **100:** 185–188.

Mitsui, K., Tokuzawa, Y., Itoh, H., Segawa, K., Murakami, M., Takahashi, K., Maruyama, M., Maeda, M., and Yamanaka, S. 2003. The homeoprotein Nanog is required for maintenance of pluripotency in mouse epiblast and ES cells. *Cell* **113**: 631–642.

Nichols, J., Chambers, I., Taga, T., and Smith, A. 2001. Physiological rationale for responsiveness of mouse embryonic stem cells to gp130 cytokines. *Development* **128**: 2333–2339.

Nichols, J., Zevnik, B., Anastassiadis, K., Niwa, H., Klewe-Nebenius, D., Chambers, I., Scholer, H., and Smith, A. 1998. Formation of pluripotent stem cells in the mammalian embryo depends on the POU transcription factor Oct4. *Cell* **95**: 379–391.

Nishihara, S., Tsuda, L., and Ogura, T. 2003. The canonical Wnt pathway directly regulates NRSF/REST expression in chick spinal cord. *Biochem. Biophys. Res. Commun.* **311**: 55–63.

Nishimura, E., Sasaki, K., Maruyama, K., Tsukada, T., and Yamaguchi, K. 1996. Decrease in neuron-restrictive silencer factor (NRSF) mRNA levels during differentiation of cultured neuroblastoma cells. *Neurosci. Lett.* **211**: 101–104.

Otto, S.J., McCorkle, S.R., Hover, J., Conaco, C., Han, J.J., Impey, S., Yochum, G.S., Dunn, J.J., Goodman, R.H., and Mandel, G. 2007. A new binding motif for the transcriptional repressor REST uncovers large gene networks devoted to neuronal functions. *J. Neurosci.* **27**: 6729–6739.

Pan, G. and Thomson, J.A. 2007. Nanog and transcriptional networks in embryonic stem cell pluripotency. *Cell Res.* **17**: 42–49.

Paquette, A.J., Perez, S.E., and Anderson, D.J. 2000. Constitutive expression of the neuron-restrictive silencer factor (NRSF)/REST in differentiating neurons disrupts neuronal gene expression and causes axon pathfinding errors in vivo. *Proc. Natl. Acad. Sci.* **97**: 12318–12323.

Raffel, C. 2004. Medulloblastoma: Molecular genetics and animal models. *Neoplasia* **6**: 310–322.

Romer, J. and Curran, T. 2005. Targeting medulloblastoma: Small-molecule inhibitors of the Sonic Hedgehog pathway as potential cancer therapeutics. *Cancer Res.* **65**: 4975–4978.

Rutka, J.T., Kuo, J.S., Carter, M., Ray, A., Ueda, S., and Mainprize, T.G. 2004. Advances in the treatment of pediatric brain tumors. *Expert Rev. Neurother.* **4**: 879–893.

Schoenherr, C.J. and Anderson, D.J. 1995. The neuron-restrictive silencer factor (NRSF): A coordinate repressor of multiple neuron-specific genes. *Science* **267**: 1360–1363.

Shimojo, M. and Hersh, L.B. 2004. Regulation of the cholinergic gene locus by the repressor element-1 silencing transcription factor/neuron restrictive silencer factor (REST/NRSF). *Life Sci.* **74**: 2213–2225.

Shimojo, M., Paquette, A.J., Anderson, D.J., and Hersh, L.B. 1999. Protein kinase A regulates cholinergic gene expression in PC12 cells: REST4 silences the silencing activity of neuron-restrictive silencer factor/REST. *Mol. Cell. Biol.* **19**: 6788–6795.

Singh, S.K., Kagalwala, M.N., Parker-Thornburg, J., Adams, H., and Majumder, S. 2008. REST maintains self-renewal and pluripotency of embryonic stem cells. *Nature* **453**: 223–227.

Stewart, C.L., Kaspar, P., Brunet, L.J., Bhatt, H., Gadi, I., Kontgen, F., and Abbondanzo, S.J. 1992. Blastocyst implantation depends on maternal expression of leukaemia inhibitory factor. *Nature* **359**: 76–79.

Su, X., Kameoka, S., Lentz, S., and Majumder, S. 2004. Activation of REST/NRSF target genes in neural stem cells is sufficient to cause neuronal differentiation. *Mol. Cell. Biol.* **24**: 8018–8025.

Su, X., Gopalakrishnan, V., Stearns, D., Aldape, K., Lang, F.F., Fuller, G., Snyder, E., Eberhart, C.G., and Majumder, S. 2006. Abnormal expression of REST/NRSF and Myc in neural stem/progenitor cells causes cerebellar tumors by blocking neuronal differentiation. *Mol. Cell. Biol.* **26**: 1666–1678.

Sun, Y.M., Cooper, M., Finch, S., Lin, H.H., Chen, Z.F., Williams, B.P., and Buckley, N.J. 2008. Rest-mediated regulation of extracellular matrix is crucial for neural development. *PLoS ONE* **3**: e3656.

Tahiliani, M., Mei, P., Fang, R., Leonor, T., Rutenberg, M., Shimizu, F., Li, J., Rao, A., and Shi, Y. 2007. The histone H3K4 demethylase SMCX links REST target genes to X-linked mental retardation. *Nature* **447**: 601–605.

Takeda, K., Noguchi, K., Shi, W., Tanaka, T., Matsumoto, M., Yoshida, N., Kishimoto, T., and Akira, S. 1997. Targeted disruption of the mouse *Stat3* gene leads to early embryonic lethality. *Proc. Natl. Acad. Sci.* **94**: 3801–3804.

Watanabe, Y., Kameoka, S., Gopalakrishnan, V., Aldape, K.D., Pan, Z.Z., Lang, F.F., and Majumder, S. 2004. Conversion of myoblasts to physiologically active neuronal phenotype. *Genes Dev.* **18**: 889–900.

Weissman, A.M. 2008. How much REST is enough? *Cancer Cell* **13**: 381–383.

Westbrook, T.F., Martin, E.S., Schlabach, M.R., Leng, Y., Liang, A.C., Feng, B., Zhao, J.J., Roberts, T.M., Mandel, G., Hannon, G.J., et al. 2005. A genetic screen for candidate tumor suppressors identifies REST. *Cell* **121**: 837–848.

Westbrook, T.F., Hu, G., Ang, X.L., Mulligan, P., Pavlova, N.N., Liang, A., Leng, Y., Maehr, R., Shi, Y., Harper, J.W., and Elledge, S.J. 2008. SCFβ-TRCP controls oncogenic transformation and neural differentiation through REST degradation. *Nature* **452**: 370–374.

Willert, J., Epping, M., Pollack, J.R., Brown, P.O., and Nusse, R. 2002. A transcriptional response to Wnt protein in human embryonic carcinoma cells. *BMC Dev. Biol.* **2**: 8.

Wu, J. and Xie X. 2006. Comparative sequence analysis reveals an intricate network among *REST, CREB* and miRNA in mediating neuronal gene expression. *Genome Biol.* **7**: R85.

Yamanaka, S. 2008. Induction of pluripotent stem cells from mouse fibroblasts by four transcription factors. *Cell Prolif.* (suppl. 1) **41**: 51–56.

Yu, J. and Thomson, J.A. 2008. Pluripotent stem cell lines. *Genes Dev.* **22**: 1987–1997.

Zuccato, C., Tartari, M., Crotti, A., Goffredo, D., Valenza, M., Conti, L., Cataudella, T., Leavitt, B.R., Hayden, M.R., Timmusk, T., et al. 2003. Huntingtin interacts with REST/NRSF to modulate the transcription of NRSE-controlled neuronal genes. *Nat. Genet.* **35**: 76–83.

Zuccato, C., Belyaev, N., Conforti, P., Ooi, L., Tartari, M., Papadimou, E., MacDonald, M., Fossale, E., Zeitlin, S., Buckley, N., and Cattaneo, E. 2007. Widespread disruption of repressor element-1 silencing transcription factor/neuron-restrictive silencer factor occupancy at its target genes in Huntington's disease. *J. Neurosci.* **27**: 6972–6983.

Stem Cell Factors in Plants: Chromatin Connections

N. KORNET AND B. SCHERES

Department of Molecular Cell Biology, Utrecht University, 3584 CH Utrecht, The Netherlands

The progression of pluripotent stem cells to differentiated cell lineages requires major shifts in cell differentiation programs. In both mammals and higher plants, this process appears to be controlled by a dedicated set of transcription factors, many of which are kingdom specific. These divergent transcription factors appear to operate, however, together with a shared suite of factors that affect the chromatin state. It is of major importance to investigate whether such shared global control mechanisms indicate a common mechanistic basis for preservation of the stem cell state, initiation of differentiation programs, and coordination of cell state transitions.

CELL STATE, TRANSCRIPTIONAL ACTIVITY, AND CHROMATIN

In the nucleus, DNA is compacted by chromatin, but it still remains dynamic and accessible for transcription (for review, see Luger and Hansen 2005; Mellor 2005). The nucleosome core particle is the base of the chromatin structure and consists of 147 bp of DNA wrapped around a histone octamer core (formed by two copies of histones H2A, H2B, H3, and H4).

The chromatin state can be influenced in four ways: (1) Methylation of the DNA can take place (Bernstein et al. 2007), (2) histone variants can be incorporated with properties different from those of the core histones (for review, see Kamakaka and Biggins 2005; Hake and Allis 2006; Loyola and Almouzni 2007), (3) sliding or replacement of nucleosomes can be facilitated by ATP-dependent chromatin remodelers (for review, see Cairns 2005; Kwon and Wagner 2007), and (4) both the nucleosomal core and the histone tails can be posttranslationally modified, including acetylation, methylation, phosphorylation, ubiquitination, and sumoylation (for review, see Peterson and Laniel 2004; Bannister and Kouzarides 2005; Margueron et al. 2005; Kouzarides 2007; Shahbazian and Grunstein 2007; Shi and Whetstine 2007).

Histone variants, remodelers, DNA methylation, and histone modifications regulate genes in many ways. Certain histone variants, such as H3.3, are found in transcriptionally active regions, whereas other H3 variants are found in silent chromatin (for review, see Kamakaka and Biggins 2005; Hake and Allis 2006; Loyola and Almouzni 2007). Chromatin remodelers can either activate or repress transcription by sliding, positioning, or exchanging nucleosomes (for review, see Cairns 2005; Kwon and Wagner 2007). Gene transcription is a tightly regulated process that involves many histone modifications (for review, see Workman 2006; Berger 2007; Kouzarides 2007; Li et al. 2007). In general, the promoter and the 5′ end of the coding region of active genes are enriched for acetylation of histone H3 or H4 and trimethylation of lysine 4 of histone 3 (H3K4me3). During transcription elongation, the histone variant H3.3 is incorporated and H3K36me3 is deposited. Inactive gene promoters generally contain DNA methyla-

tion, hypoacetylation of H3 and H4, H3K9me3, and/or H3K37me3 (for review, see Workman 2006; Berger 2007; Kouzarides 2007; Li et al. 2007).

Posttranslational modification of a histone tail not only influences other modifications on the same tail (in *cis*), but also affects the modifications on other tails (in *trans*) in the same or neighboring nucleosomes (for review, see Fischle et al. 2003; Fuks 2005; Margueron et al. 2005; Nightingale et al. 2006; Kouzarides 2007). This has led to the "histone code" hypothesis, which states that histone modifications, presumably in specific combinations, provide binding sites for chromatin factors that influence gene transcription (Strahl and Allis 2000; Jenuwein and Allis 2001; Turner 2002; Ruthenburg et al. 2007b).

Chromatin factors can be recruited to their targets by direct interactions with transcription factors or by association to the basal transcriptional machinery. Another targeting mechanism involves small heterochromatic RNAs (shRNAs) that are produced by transcription of centromeric repeats and form double-stranded RNAs. These are processed by the RNA interference (RNAi) machinery and the resulting small interfering RNAs (siRNAs) target chromatin factors to the centromere, leading to heterochromatin formation. Furthermore, noncoding RNAs can direct chromatin factors to specific locations on the DNA, for example, during X inactivation (for review, see Fischle et al. 2003; Peterson and Laniel 2004; Bernstein and Allis 2005; Wassenegger 2005; Ruthenburg et al. 2007a).

The histone code is read by proteins with specific chromatin-binding domains (for review, see de la Cruz et al. 2005; Kouzarides 2007; Ruthenburg et al. 2007a). Examples are the chromodomain and bromodomain that bind methylated lysines and acetylated lysines, respectively. These domains are found in histone acetyltransferases (HATs), histone methyltransferases (HMTs), and remodelers that in turn can modify chromatin and regulate gene transcription.

The histone code hypothesis implies that chromatin modifications are inherited through cell division to provide memory of a transcriptional state. But how are chromatin states maintained during replication and mitosis? Active chromatin has been proposed to be inherited by several mechanisms that still need validation. These mechanisms include the contribution of H3 variants to cellular memory (Hake and Allis 2006), the influence of the chromatin state on nucleosome stability (Henikoff 2008), and the reinstatement of active chromatin due to accessibility to transcription factors after replication (Groth et al. 2007). Accordingly, activating parental histone modifications, such as acetylation and methylation, have been shown to "survive" replication (Benson et al. 2006). Alternatively, transcription factors may also dictate the cell state and chromatin state after mitosis (for review, see Egli et al. 2008). Therefore, mitosis may provide a window of opportunity to change cell fate, because both chromatin and transcription factors dissociate and reassociate during mitosis.

Silent chromatin is thought to be reinstated quickly after replication (for review, see Wallace and Orr-Weaver 2005; Groth et al. 2007). DNA methyltransferase DNMT1 is recruited to the replicated DNA by the replication factor proliferating cell nuclear antigen (PCNA) and ensures the propagation of DNA methylation (for review, see Nightingale et al. 2006; Groth et al. 2007; Martin and Zhang 2007). In addition, DNMT1 recruits histone modifiers, such as histone deacetylases (HDACs) and HMTs, and in this way might help to replicate chromatin modifications (Fig. 1) (Groth et al. 2007). PCNA recruits another chromatin factor, the chromatin assembly factor 1 (CAF-1) complex (Shibahara and Stillman 1999; Zhang et al. 2000). The CAF-1 complex is involved in nucleosome assembly after replication by loading newly synthesized H3-H4 dimers onto the DNA. CAF-1 interacts with heterochromatin protein 1 (HP1) (Fig. 1), which is known to be important for heterochromatin stabilization (Murzina et al. 1999), and may in this way facilitate chromatin maturation after replication (Quivy et al. 2004). In addition to HP1,

CAF-1 binds a DNA methyl-binding domain (MBD) protein and an H3K9 HMT (Fig. 1) (Reese et al. 2003; Sarraf and Stancheva 2004). In turn, H3K9me3 is known to recruit HP1. Thus, both DNA methylation and histone methylation can immediately be reinstated after replication (for review, see Wallace and Orr-Weaver 2005; Groth et al. 2007).

CHROMATIN AND THE STEM CELL STATE

Stem cells replenish the cells present in an organism throughout its lifetime (for review, see Li and Xie 2005; Wong et al. 2005; Jones and Wagers 2008). They have two unique characteristics: the capability to self-renew and to differentiate into (several) cell types. The stem cell niche arose independently in animal and plant kingdoms and is defined by kingdom-specific transcription factors. Recently, it has become clear that chromatin factors support the unique features of mammalian stem cells (for review, see Niwa 2007a,b; Spivakov and Fisher 2007; Jaenisch and Young 2008). The role of chromatin factors in plant stem cell control is just starting to be revealed.

Mammalian Stem Cell Systems

Many stem cells are localized in stem cell niches, where a limited number of "organizer" cells maintain the stem cells by inhibiting differentiation and regulating their self-renewal. Besides the influence of the stem cell niche microenvironment, stem cells are thought to possess unique intrinsic properties. *Oct4, Nanog*, and *Sox2* genes are essential intrinsically for the pluripotency and self-renewal of mammalian embryonic stem (ES) cells. *Oct4* encodes a POU domain transcription factor, *Nanog* encodes a divergent homeodomain protein, and *Sox2* encodes a high mobility group (HMG)-box transcription factor. Analysis of the targets of Oct4, Nanog, and Sox2 revealed that they regulate themselves, one another, and key signaling pathways in stem cells and, on the other hand, repress key differentiation genes. These data suggest that complex regulatory mechanisms are at work in ES cells, probably providing stability within the system by positive and negative feedback loops (Boyer et al. 2005; Johnson et al. 2006; Loh et al. 2006; Niwa 2007a).

What makes stem cells unique? Not all progenitor cells are able to revert to a stem cell state, indicating that stem cells have unique features thought to be imposed by chromatin (for review, see Mikkers and Frisen 2005; Buszczak and Spradling 2006; Meshorer and Misteli 2006; Niwa 2007a; Spivakov and Fisher 2007). Accordingly, pluripotent ES cells contain hyperdynamic or "breathing" chromatin that is lost during differentiation. Furthermore, elevated levels of unbound histones in ES cells enhance differentiation, whereas ES cells expressing a strongly bound form of H1 show differentiation arrest. This suggests that "loose" chromatin not only is a unique trait of pluripotent cells, but also facilitates the structural chromatin changes occurring during the early stages of differentiation (Meshorer et al. 2006). That ES cells contain open chromatin is supported by the fact that ES cells are transcriptionally globally hyperactive, which is probably facil-

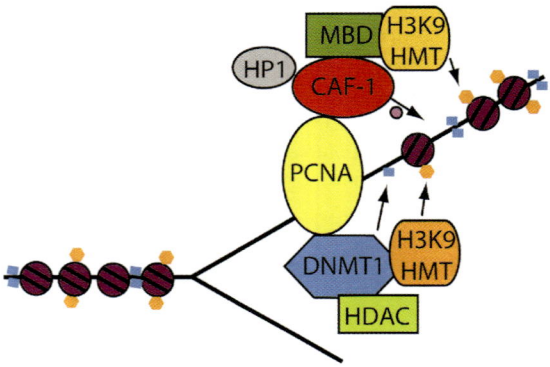

Figure 1. Inheritance of the histone code. Replication factor PCNA recruits DNA methyltransferase DNMT1 to ensure proper inheritance of DNA methylation. DNMT1 in turn is known to be associated with HDACs and an H3K9 HMT. This ensures proper inheritance of histone methylation through an epigenetic feedback loop. In addition, PCNA recruits the CAF-1 complex involved in H3-H4 loading (*small, light purple sphere*) to the newly replicated DNA. CAF-1 interacts with HP1 (which also can bind H3K9me3), a DNA methyl-binding domain (MBD) protein and an H3K9 HMT (which ensures propagation of histone methylation). (*Blue squares*) DNA methylation; (*orange pentagons*) histone methylation. For simplicity, only one strand is depicted.

tated by chromatin-remodeling factors (Efroni et al. 2008). Major changes in chromatin organization occur during early mouse embryogenesis. Heterochromatin is diffuse during the two-cell stage and is progressively assembled into condensed heterochromatin until the blastocyst stage. Condensation of chromatin is abolished by deletion of the p150 subunit of the CAF-1 complex and eventually leads to developmental arrest (Houlard et al. 2006). Thus, the CAF-1 complex and extensive chromatin reorganization are essential during the early steps of mouse embryogenesis and in ES cells (Houlard et al. 2006). Possibly, the CAF-1 complex is involved in the establishment of hyperdynamic or "breathing" chromatin in ES cells.

BMI1, a member of the Polycomb repressive complex 1 (PRC1), is essential for the self-renewal of hematopoietic, neural, and cancer stem cells (for review, see Valk-Lingbeek et al. 2004). BMI1 negatively regulates the *Ink4a/Arf* locus, encoding two cell cycle inhibitors, $p16^{ink4a}$ and $p19^{arf}$ (Valk-Lingbeek et al. 2004; Bruggeman et al. 2005; Molofsky et al. 2005), and thereby stimulates stem cell proliferation and prevents apoptosis (Fig. 2). Recently, it was shown that the PRC1 complex (containing BMI1) and the PRC2 complex bind to the $p16^{ink4a}$ locus in a retinoblastoma (pRb) family-dependent manner in differentiated fibroblasts (Fig. 2) (Bracken et al. 2007; Kotake et al. 2007). In turn, pRb family proteins are regulated by $p16^{ink4a}$, forming a regulatory loop (for review, see Lowe and Sherr 2003; Sharpless 2005). Possibly, elevated levels of BMI1 in stem cells ensure continuous proliferation by abolishing this feedback loop. BMI1 is mainly involved in the maintenance of adult stem cells, whereas other complexes are probably acting in ES cells (Valk-Lingbeek et al. 2004).

Recently, many genes involved in differentiation were

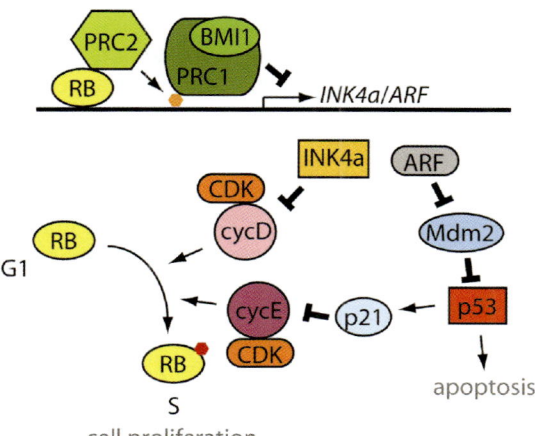

Figure 2. Cell cycle regulation by Polycomb proteins and Rb. At the *INK4a/ARF* locus, Rb probably recruits the PRC2 complex, which mediates H3K27me3 (*orange pentagon*). This in turn recruits the PRC1 complex containing BMI1 and represses transcription of the *Ink4a/Arf* locus. ARF inhibits degradation of p53 and thereby activates apoptosis and blocks cell cycle progression through p21 and cycE/CDK. INK4a inhibits cycD/CDK activity. The cycD/CDK complex promotes phosphorylation of Rb (*red pentagon*), which releases Rb binding and thereby promotes the transition through the G_1-S phase of the cell cycle.

shown to posses a special chromatin signature in mouse and human ES cells. They contain both active (H3K4me2/3 and H3K9 acetylation) and repressive (H3K27me3) modifications (for review, see Buszczak and Spradling 2006; Meshorer and Misteli 2006; Gan et al. 2007). These "bivalent" domains are found exclusively in ES cells and are lost upon differentiation by resolving into either active (H3K4me3) or repressive (H3K27me3) chromatin. The PRC2 complex is responsible for H3K27me3 and repression in those domains (Azuara et al. 2006; Boyer et al. 2006; Lee et al. 2006; Mikkelsen et al. 2007; Pasini et al. 2007). Interestingly, stem cell factors Oct4, Nanog, and Sox2 are found specifically at a large subset of these differentiation genes. Therefore, stem cell pluripotency might be imposed by the intrinsic stem cell factors Oct4, Nanog, and Sox2 through activation of a stem cell self-renewal program and a balance between activation and (a temporary) repression ("poised state") of differentiation factors by Polycomb complexes (Boyer et al 2005; Lee et al. 2006). Differentiation into a specific lineage could be accomplished by selective activation of a subset of genes by lineage-specific transcription factors, whereas other lineages of primed differentiation genes are permanently silenced. Although the theory of the poised state is appealing, the significance of bivalent domains must be further studied (for discussion, see Gan et al. 2007; Niwa 2007a). Another point of interest is that PRC2 members are required to prevent stem cell differentiation and do not seem to affect proliferation, in contrast to PRC1 member BMI1 (Azuara et al. 2006; Boyer et al. 2006; Niwa 2007a; Pasini et al. 2007). Recently, it was proposed that bivalent domains facilitate postponement of cell-fate decisions (Pietersen and van Lohuizen 2008).

Overexpression of Oct4, Sox2, Nanog, c-Myc, and Klf4 is sufficient to convert mouse and human somatic cells into induced pluripotent stem (iPS) cells, which show characteristics of stem cells, including chromatin features such as bivalent domains and proper genomic imprinting (Takahashi and Yamanaka 2006; Niwa 2007a; Okita et al. 2007; Takahashi et al. 2007; Wernig et al. 2007; Jaenisch and Young 2008). Oncogenes c-Myc and Klf4 are not essential, but they increase the frequency of iPS cells by unknown mechanisms (for review, see Jaenisch and Young 2008). c-Myc (important for proliferation and stem cells) is required for the maintenance of active chromatin in progenitor cells through regulation of histone acetyltransferase GCN5 (Knoepfler et al. 2006) and may in this way aid in the reprogramming process (Jaenisch and Young 2008). The induction of iPS cells is enhanced by inhibitors of both DNA methyltransferase and histone deacetylases (Huangfu et al. 2008). Recently, it was shown that the pluripotent transcription factor Oct4 regulates the expression of histone demethylases Jmjd1a and Jmjd2c, which may have a role in maintaining open chromatin in stem cells. In turn, Jmjd1a mediates the access of Oct4 to the promoter of the self-renewal regulator *Tcl1* (Loh et al. 2007; Niwa 2007b). Together, these data suggest that chromatin factors are important for the execution rather than the initiation of pluripotency.

In conclusion, a unique chromatin state is present in mammalian stem cells that appears to be imposed by transcription factors that program pluripotency through regulation of chromatin factors. Several classes of chromatin factors are important, including histone acetyltransferase GCN5, the CAF-1 complex, and histone demethylases. Polycomb proteins facilitate the unique characteristics of stem cells: They control the expression of genes mediating proliferation and differentiation into specific cell types.

Plant Stem Cell Systems

Similar to animals, plant stem cells are regulated by extrinsic signals from their niche (for review, see Stahl and Simon 2005; Vernoux and Benfey 2005; Williams and Fletcher 2005; Singh and Bhalla 2006; Scheres 2007). Therefore, equivalent mechanisms may be present to control stem cells, although multicellularity arose independently in both kingdoms. Postembryonically, *Arabidopsis* stem cells are located and maintained at two sites: the shoot apical meristem (SAM) and the root meristem (RM).

SAM stem cells produce all above-ground tissues. These stem cells are maintained at the population level and are controlled through two parallel pathways. The initiation and maintenance of the SAM and the suppression of differentiation are regulated by the homeodomain transcription factor SHOOT MERISTEMLESS (STM). In addition, stem cell maintenance is regulated by homeodomain transcription factor WUSCHEL (WUS) and the CLAVATA (CLV) receptor kinase pathway. WUS is expressed in the organizing center and induces the expression of the *CLV3* ligand in the overlying stem cells. CLV3 binds the CLV1-CLV2 receptor kinase complex, which represses the expression of *WUS*. This negative feedback loop regulates the size of the stem cell pool (for review, see Stahl and Simon 2005; Williams and Fletcher 2005; Tucker and Laux 2007).

The RM contains stem cells that produce all underground tissues. These stem cells surround and are physically linked to the organizing center or quiescent center (QC), which is required for their maintenance (van den Berg et al. 1995). The position of the root stem cell niche is controlled through two parallel plant-specific pathways. The stem cell niche is positioned in the apical-basal direction through the accumulation of the phytohormone auxin, which is achieved by the polar localization of PIN proteins (putative auxin efflux carriers) in the plasma membrane (Blilou et al. 2005). This auxin maximum regulates the expression of AP2 domain transcription factors *PLETHORA1* (*PLT1*) and *PLT2*, which are redundantly required for stem cell maintenance (Aida et al. 2004). In turn, the PLT proteins regulate the expression of *PIN* genes, revealing a feedback loop (Blilou et al. 2005). The stem cell niche is controlled in the radial direction through the GRAS family transcription factors SHORTROOT (SHR) and SCARECROW (SCR). *SHR* is expressed in the vascular tissue and moves into the outer layer (endodermis and QC), where it induces *SCR* expression. SCR is required cell-autonomously in the QC to maintain the surrounding stem cells (Helariutta et al. 2000; Nakajima et al. 2001; Sabatini et al. 2003; Heidstra et al. 2004). SCR appears to inhibit RETINOBLASTOMA RELATED1

(RBR1) activity in the QC or in the stem cells (Wildwater et al. 2005). RBR1 promotes the transition of stem cells to differentiating daughter cells without changes in cell cycle progression.

The position of the stem cell niche is defined by the overlap between the highest *PLT* and *SCR* expression (Aida et al. 2004). The combination of key transcription factor pathways to program stem cells resembles the combined action of pluripotent transcription factor Oct4, Nanog, and Sox2 in animal stem cells (Johnson et al. 2006; Niwa 2007a). Similarly, the PLT stem cell transcription factors are sufficient to induce ectopic root development including a functional root stem cell niche (Aida et al. 2004; Galinha et al. 2007). It is interesting to note that, for ectopic reprogramming, PLT genes require sites of active cell division.

There are two pools of root stem cells present that produce daughter cells by asymmetric divisions and have different characteristics. The daughters derived from the distal stem cells do not divide and differentiate immediately. The proximal stem cell daughters form a transit-amplifying cell population in which a few extra rounds of cell division take place. These cells contribute to the growth of the root by expansion and eventually differentiate into several cell types.

Chromatin factors are implicated in stem cell maintenance in plants. The SNF2 chromatin-remodeling factor SPLAYED (SYD) maintains the SAM during the reproductive stage through the WUS pathway. SYD directly interacts with the WUS promoter, regulates WUS transcription, and thereby promotes stem cell maintenance in the shoot (Kwon et al. 2005). This is similar to the role of remodeling factors in stem cell maintenance in the *Drosophila* ovary (Xi and Xie 2005) and in mammalian neural stem/progenitor cells (Lessard et al. 2007).

The PRC2 complex represses key differentiation genes in mammalian stem cells as discussed above (for review, see Buszczak and Spradling 2006; Meshorer and Misteli 2006; Gan et al. 2007). Similarly, members of *Arabidopsis* PRC2 complexes are required for root stem cell niche maintenance (N. Kornet and B. Scheres, unpubl.). Preliminary data indicate that RBR1 and PRC2 complexes may together control proliferation of transit-amplifying cells (our unpublished data).

Histone acetyltransferase GCN5 has been indirectly linked to expression regulation of transcription factor WUS in the floral meristem and embryo (Bertrand et al. 2003; Long et al. 2006). It is unknown whether shoot stem cells are affected directly. In addition, mutation of *GCN5* causes defects in root development (Vlachonasios et al. 2003). Our data show that GCN5 is required for root stem cell maintenance through regulation of *PLT* expression (N. Kornet and B. Scheres, unpubl.). Furthermore, GCN5 and its cofactor ADA2b influence proliferation of transit-amplifying cells (N. Kornet and B. Scheres, unpubl.).

Another chromatin factor implicated in both SAM and RM maintenance is the *Arabidopsis* CAF-1 complex. The mammalian CAF-1 complex subunits p150 and p60 are encoded by *Arabidopsis FAS1* and *FAS2* genes that are involved in nucleosome assembly and stable maintenance of epigenetic states (Kaya et al. 2001; Costa and Shaw

2006; Schonrock et al. 2006; Ramirez-Parra and Gutierrez 2007a,b). The SAM in *fas* mutants is broader and flatter, the organization is disrupted, and *WUS* expression is expanded. In the RM, the organization of the stem cell niche is perturbed in *fas* mutants (Kaya et al. 2001). CAF-1 activity appears to be linked to the role of the RETINOBLASTOMA RELATED1 (RBR1) protein in cell state progression. The CAF-1 complex and RBR1 synergistically repress stem cell proliferation in the root through regulation of stem cell transcription factors (N. Kornet and B. Scheres, unpubl.). The *Arabidopsis* CAF-1 complex is thought to be involved in the stable expression of key developmental genes through correct chromatin formation. The *Arabidopsis* CAF-1 complex was shown to influence chromatin conformation at the homeobox gene *GLABRA2* (*GL2*). GL2 is an atrichoblast cell-fate determinant that is expressed in nonhair root cells (atrichoblasts) and is repressed in hair root cells (trichoblasts). The chromatin conformation specifically at the *GL2* locus is "open" in atrichoblasts and "closed" in trichoblasts. In *fas* mutants, the chromatin conformation at the *GL2* locus is open, regardless of the cell type, and *GL2* is ectopically expressed (Costa and Shaw 2006). Furthermore, in *fas* mutants, the heterochromatin fraction is reduced (Kirik et al. 2006; Schonrock et al. 2006), suggesting that the CAF-1 complex is involved in heterochromatin compaction. Similarly, the largest subunit of CAF-1 (p150) was found to be essential for the assembly of the heterochromatin organization in mouse embryonic stem cells, and epigenetic marks at pericentric heterochromatin are affected (Houlard et al. 2006). However, histone modifications (DNA methylation, H3K9me2, H3, and H4 acetylation) are not affected globally in *fas* mutants (Schonrock et al. 2006; Ramirez-Parra and Gutierrez 2007a). Only a small fraction of heterochromatic genes are up-regulated in *fas* mutants and these were mainly expressed during late S phase (Schonrock et al. 2006). In addition, these genes (but not their neighboring genes) contained less H3K9me2 and more H3 and H4 acetylation (Ramirez-Parra and Gutierrez 2007a). This suggests that the CAF-1 complex ensures the proper epigenetic inheritance of specific genes in *Arabidopsis*, whereas in animals, the CAF-1 complex has a more general role in heterochromatin formation. It is noteworthy that the replication-associated function of the CAF-1 protein appears to be a plausible link between cell state transitions and cell cycle progression.

TOWARD ANALYSIS OF STATE TRANSITIONS IN PLANT STEM CELLS

Key transcription factors and several classes of chromatin modifying factors serve as stem cell factors in both animal and plant kingdoms. GCN5, the CAF-1 complex, and Polycomb proteins have been shown to be important in mammalian stem cells and the same is true for *Arabidopsis* root stem cells (our unpublished data).

In mammals, pluripotent stem cell transcription factors probably regulate chromatin factors to obtain a special chromatin state in ES cells. In *Arabidopsis*, stem cell transcription factors are also sufficient to induce root or shoot

stem cells, and it is tempting to speculate that they likewise regulate chromatin factors to stabilize cellular states. Conversely, stem-cell-associated chromatin factors (remodeler SYD and HAT GCN5) regulate key stem cell transcription factors (WUS and PLT) (Kwon et al. 2005; N. Kornet and B. Scheres, unpubl.). Feedback from these chromatin modifiers to transcription factors may serve to stabilize cellular states characterized by a defined transcriptional program. It remains to be seen whether feedback circuits between chromatin factors and transcription factors have been invented twice during evolution of multicellular plants and animals. Alternatively, chromatin factors might have been already at the nexus of proliferative and differentiation cell states of a common unicellular ancestor.

In the *Arabidopsis* root meristem, the dosage of the PLT stem cell regulators controls the transition from stem cell to transit-amplifying cells and the transition from transit-amplifying cell to terminal differentiation (Galinha et al. 2007). Part of the ability to shift expression and cell states can be explained by a feedback loop between PLT activity and the distribution of its upstream regulator, the plant hormone auxin, which establishes a PLT protein gradient. However, efficient blocking of state transitions (for example, from stem cell to transit-amplifying cell) requires high PLT levels and a simultaneous reduction of RBR1 levels (Galinha et al. 2007). Given the well-known connections between RBR1 and chromatin modification complexes in both kingdoms (Frolov and Dyson 2004; Hennig et al. 2005; Giacinti and Giordano 2006), it seems likely that the rapid shift between these states is facilitated by the interplay among transcription factors that instruct stem cell or transit-amplifying cell fates and chromatin modifiers. Our current data support a formal model for the requirement of chromatin modifiers and core transcription factors in this process (Fig. 3). It is possible to follow differentiation of root stem cell descen-

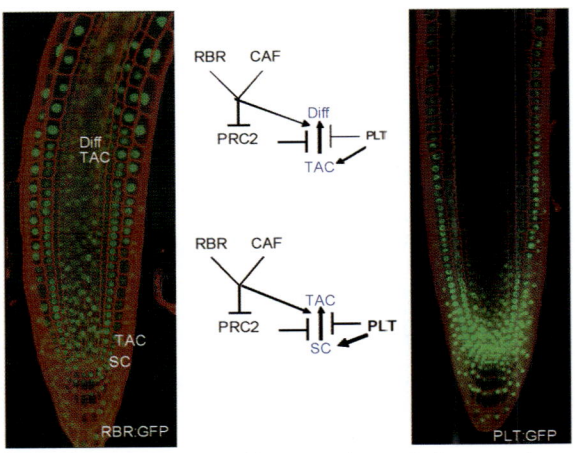

Figure 3. Interaction model between chromatin factors and core stem cell factors in root. Two state transitions are depicted: from stem cell (SC) to transit-amplifying cell (TAC) and from transit-amplifying cell to terminally differentiating cell (Diff). *Arrows* and *repression bars* depict a possible scenario by which RBR1, CAF-1 complex subunits, and PRC2 complex subunits may control state transitions.

dants in continuous cell files and it is now feasible to determine their gene expression state at selective stages of differentiation (Brady et al. 2007). This should allow us to dissect state transitions during the differentiation trajectory of pluripotent plant stem cells and to understand the hitherto unresolved role of chromatin factors and cell cycle regulators in this process.

REFERENCES

Aida, M., Beis, D., Heidstra, R., Willemsen, V., Blilou, I., Galinha, C., Nussaume, L., Noh, Y.S., Amasino, R., and Scheres, B. 2004. The *PLETHORA* genes mediate patterning of the *Arabidopsis* root stem cell niche. *Cell* **119:** 109–120.

Azuara, V., Perry, P., Sauer, S., Spivakov, M., Jorgensen, H.F., John, R.M., Gouti, M., Casanova, M., Warnes, G., Merkenschlager, M., and Fisher, A.G. 2006. Chromatin signatures of pluripotent cell lines. *Nat. Cell Biol.* **8:** 532–538.

Bannister, A.J. and Kouzarides, T. 2005. Reversing histone methylation. *Nature* **436:** 1103–1106.

Benson, L.J., Gu, Y., Yakovleva, T., Tong, K., Barrows, C., Strack, C.L., Cook, R.G., Mizzen, C.A., and Annunziato, A.T. 2006. Modifications of H3 and H4 during chromatin replication, nucleosome assembly, and histone exchange. *J. Biol. Chem.* **281:** 9287–9296.

Berger, S.L. 2007. The complex language of chromatin regulation during transcription. *Nature* **447:** 407–412.

Bernstein, B.E., Meissner, A., and Lander, E.S. 2007. The mammalian epigenome. *Cell* **128:** 669–681.

Bernstein, E. and Allis, C.D. 2005. RNA meets chromatin. *Genes Dev.* **19:** 1635–1655.

Bertrand, C., Bergounioux, C., Domenichini, S., Delarue, M., and Zhou, D.X. 2003. *Arabidopsis* histone acetyltransferase AtGCN5 regulates the floral meristem activity through the *WUSCHEL/AGAMOUS* pathway. *J. Biol. Chem.* **278:** 28246–28251.

Blilou, I., Xu, J., Wildwater, M., Willemsen, V., Paponov, I., Friml, J., Heidstra, R., Aida, M., Palme, K., and Scheres, B. 2005. The PIN auxin efflux facilitator network controls growth and patterning in *Arabidopsis* roots. *Nature* **433:** 39–44.

Boyer, L.A., Lee, T.I., Cole, M.F., Johnstone, S.E., Levine, S.S., Zucker, J.P., Guenther, M.G., Kumar, R.M., Murray, H.L., Jenner, R.G., et al. 2005. Core transcriptional regulatory circuitry in human embryonic stem cells. *Cell* **122:** 947–956.

Boyer, L.A., Plath, K., Zeitlinger, J., Brambrink, T., Medeiros, L.A., Lee, T.I., Levine, S.S., Wernig, M., Tajonar, A., Ray, M.K., et al. 2006. Polycomb complexes repress developmental regulators in murine embryonic stem cells. *Nature* **441:** 349–353.

Bracken, A.P., Kleine-Kohlbrecher, D., Dietrich, N., Pasini, D., Gargiulo, G., Beekman, C., Theilgaard-Mönch, K., Minucci, S., Porse, B.T., Marine, J.C., et al. 2007. The Polycomb group proteins bind throughout the *INK4A-ARF* locus and are disassociated in senescent cells. *Genes Dev.* **21:** 525–530.

Brady, S.M., Orlando, D.A., Lee, J.Y., Wang, J.Y., Koch, J., Dinneny, J.R., Mace, D., Ohler, U., and Benfey, P.N. 2007. A high-resolution root spatiotemporal map reveals dominant expression patterns. *Science* **318:** 801–806.

Bruggeman, S.W., Valk-Lingbeek, M.E., van der Stoop, P.P., Jacobs, J.J., Kieboom, K., Tanger, E., Hulsman, D., Leung, C., Arsenijevic, Y., Marino, S., and van Lohuizen, M. 2005. *Ink4a* and *Arf* differentially affect cell proliferation and neural stem cell self-renewal in *Bmi1*-deficient mice. *Genes Dev.* **19:** 1438–1443.

Buszczak, M. and Spradling, A.C. 2006. Searching chromatin for stem cell identity. *Cell* **125:** 233–236.

Cairns, B.R. 2005. Chromatin remodeling complexes: Strength in diversity, precision through specialization. *Curr. Opin. Genet. Dev.* **15:** 185–190.

Costa, S. and Shaw, P. 2006. Chromatin organization and cell fate switch respond to positional information in *Arabidopsis*.

Nature **439:** 493–496.

de la Cruz, X., Lois, S., Sanchez-Molina, S., and Martinez-Balbas, M.A. 2005. Do protein motifs read the histone code? *Bioessays* **27:** 164–175.

Efroni, S., Duttagupta, R., Cheng, J., Dehghani, H., Hoeppner, D.J., Dash, C., Bazett-Jones, D.P., Le Grice, S., McKay, R.D., Buetow, K.H., et al. 2008. Global transcription in pluripotent embryonic stem cells. *Cell Stem Cell* **2:** 437–447.

Egli, D., Birkhoff, G., and Eggan, K. 2008. Mediators of reprogramming: Transcription factors and transitions through mitosis. *Nat. Rev. Mol. Cell Biol.* **9:** 505–516.

Fischle, W., Wang, Y., and Allis, C.D. 2003. Histone and chromatin cross-talk. *Curr. Opin. Cell Biol.* **15:** 172–183.

Frolov, M.V. and Dyson, N.J. 2004. Molecular mechanisms of E2F-dependent activation and pRB-mediated repression. *J. Cell Sci.* **117:** 2173–2181.

Fuks F. 2005. DNA methylation and histone modifications: Teaming up to silence genes. *Curr. Opin. Genet. Dev.* **15:** 490–495.

Galinha, C., Hofhuis, H., Luijten, M., Willemsen, V., Blilou, I., Heidstra, R., and Scheres, B. 2007. PLETHORA proteins as dose-dependent master regulators of *Arabidopsis* root development. *Nature* **449:** 1053–1057.

Gan, Q., Yoshida, T., McDonald, O.G., and Owens, G.K. 2007. Concise review: Epigenetic mechanisms contribute to pluripotency and cell lineage determination of embryonic stem cells. *Stem Cells* **25:** 2–9.

Giacinti, C. and Giordano, A. 2006. RB and cell cycle progression. *Oncogene* **25:** 5220–5227.

Groth, A., Rocha, W., Verreault, A., and Almouzni, G. 2007. Chromatin challenges during DNA replication and repair. *Cell* **128:** 721–733.

Hake, S.B. and Allis, C.D. 2006. Histone H3 variants and their potential role in indexing mammalian genomes: The "H3 barcode hypothesis." *Proc. Natl. Acad. Sci.* **103:** 6428–6435.

Heidstra, R., Welch, D., and Scheres, B. 2004. Mosaic analyses using marked activation and deletion clones dissect *Arabidopsis* SCARECROW action in asymmetric cell division. *Genes Dev.* **18:** 1964–1969.

Helariutta, Y., Fukaki, H., Wysocka-Diller, J., Nakajima, K., Jung, J., Sena, G., Hauser, M.T., and Benfey, P.N. 2000. The *SHORT-ROOT* gene controls radial patterning of the *Arabidopsis* root through radial signaling. *Cell* **101:** 555–567.

Henikoff, S. 2008. Nucleosome destabilization in the epigenetic regulation of gene expression. *Nat. Rev. Genet.* **9:** 15–26.

Hennig, L., Bouveret, R., and Gruissem, W. 2005. MSI1-like proteins: An escort service for chromatin assembly and remodeling complexes. *Trends Cell Biol.* **15:** 295–302.

Houlard, M., Berlivet, S., Probst, A.V., Quivy, J.P., Hery, P., Almouzni, G., and Gerard, M. 2006. CAF-1 is essential for heterochromatin organization in pluripotent embryonic cells. *PLoS Genet.* **2:** e181.

Huangfu, D., Maehr, R., Guo, W., Eijkelenboom, A., Snitow, M., Chen, A.E., and Melton, D.A. 2008. Induction of pluripotent stem cells by defined factors is greatly improved by small-molecule compounds. *Nat. Biotechnol.* **26:** 795–797.

Jaenisch, R. and Young, R. 2008. Stem cells, the molecular circuitry of pluripotency and nuclear reprogramming. *Cell* **132:** 567–582.

Jenuwein, T. and Allis, C.D. 2001. Translating the histone code. *Science* **293:** 1074–1080.

Johnson, B.V., Rathjen, J., and Rathjen, P.D. 2006. Transcriptional control of pluripotency: Decisions in early development. *Curr. Opin. Genet. Dev.* **16:** 447–454.

Jones, D.L. and Wagers, A.J. 2008. No place like home: Anatomy and function of the stem cell niche. *Nat. Rev. Mol. Cell Biol.* **9:** 11–21.

Kamakaka, R.T. and Biggins, S. 2005. Histone variants: Deviants? *Genes Dev.* **19:** 295–310.

Kaya, H., Shibahara, K.I., Taoka, K.I., Iwabuchi, M., Stillman, B., and Araki, T. 2001. *FASCIATA* genes for chromatin assembly factor-1 in *Arabidopsis* maintain the cellular organization of apical meristems. *Cell* **104:** 131–142.

Kirik, A., Pecinka, A., Wendeler, E., and Reiss, B. 2006. The chromatin assembly factor subunit FASCIATA1 is involved in homologous recombination in plants. *Plant Cell* **18:** 2431–2442.

Knoepfler, P.S., Zhang, X.Y., Cheng, P.F., Gafken, P.R., McMahon, S.B., and Eisenman, R.N. 2006. Myc influences global chromatin structure. *EMBO J.* **25:** 2723–2734.

Kotake, Y., Cao, R., Viatour, P., Sage, J., Zhang, Y., and Xiong, Y. 2007. pRB family proteins are required for H3K27 trimethylation and Polycomb repression complexes binding to and silencing p16^{INK4}– tumor suppressor gene. *Genes Dev.* **21:** 49–54.

Kouzarides, T. 2007. Chromatin modifications and their function. *Cell* **128:** 693–705.

Kwon, C.S. and Wagner, D. 2007. Unwinding chromatin for development and growth: A few genes at a time. *Trends Genet.* **23:** 403–412.

Kwon, C.S., Chen, C., and Wagner, D. 2005. *WUSCHEL* is a primary target for transcriptional regulation by SPLAYED in dynamic control of stem cell fate in *Arabidopsis*. *Genes Dev.* **19:** 992–1003.

Lee, T.I., Jenner, R.G., Boyer, L.A., Guenther, M.G., Levine, S.S., Kumar, R.M., Chevalier, B., Johnstone, S.E., Cole, M.F., Isono, K., et al. 2006. Control of developmental regulators by Polycomb in human embryonic stem cells. *Cell* **125:** 301–313.

Lessard, J., Wu, J.I., Ranish, J.A., Wan, M., Winslow, M.M., Staahl, B.T., Wu, H., Aebersold, R., Graef, I.A., and Crabtree, G.R. 2007. An essential switch in subunit composition of a chromatin remodeling complex during neural development. *Neuron* **55:** 201–215.

Li, B., Carey, M., and Workman, J.L. 2007. The role of chromatin during transcription. *Cell* **128:** 707–719.

Li, L. and Xie, T. 2005. Stem cell niche: Structure and function. *Annu. Rev. Cell Dev. Biol.* **21:** 605–631.

Loh, Y.H., Zhang, W., Chen, X., George, J., and Ng, H.H. 2007. Jmjd1a and Jmjd2c histone H3 Lys 9 demethylases regulate self-renewal in embryonic stem cells. *Genes Dev.* **21:** 2545–2557.

Loh, Y.H., Wu, Q., Chew, J.L., Vega, V.B., Zhang, W., Chen, X., Bourque, G., George, J., Leong, B., Liu, J., et al. 2006. The Oct4 and Nanog transcription network regulates pluripotency in mouse embryonic stem cells. *Nat. Genet.* **38:** 431–440.

Long, J.A., Ohno, C., Smith, Z.R., and Meyerowitz, E.M. 2006. TOPLESS regulates apical embryonic fate in *Arabidopsis*. *Science* **312:** 1520–1523.

Lowe, S.W. and Sherr, C.J. 2003. Tumor suppression by *Ink4a-Arf*: Progress and puzzles. *Curr. Opin. Genet. Dev.* **13:** 77–83.

Loyola, A. and Almouzni, G. 2007. Marking histone H3 variants: How, when and why? *Trends Biochem. Sci.* **32:** 425–433.

Luger, K. and Hansen, J.C. 2005. Nucleosome and chromatin fiber dynamics. *Curr. Opin. Struct. Biol.* **15:** 188–196.

Margueron, R., Trojer, P., and Reinberg, D. 2005. The key to development: Interpreting the histone code? *Curr. Opin. Genet. Dev.* **15:** 163–176.

Martin, C. and Zhang, Y. 2007. Mechanisms of epigenetic inheritance. *Curr. Opin. Cell Biol.* **19:** 266–272.

Mellor, J. 2005. The dynamics of chromatin remodeling at promoters. *Mol. Cell* **19:** 147–157.

Meshorer, E. and Misteli, T. 2006. Chromatin in pluripotent embryonic stem cells and differentiation. *Nat. Rev. Mol. Cell Biol.* **7:** 540–546.

Meshorer, E., Yellajoshula, D., George, E., Scambler, P.J., Brown, D.T., and Misteli, T. 2006. Hyperdynamic plasticity of chromatin proteins in pluripotent embryonic stem cells. *Dev. Cell* **10:** 105–116.

Mikkelsen, T.S., Ku, M., Jaffe, D.B., Issac, B., Lieberman, E., Giannoukos, G., Alvarez, P., Brockman, W., Kim, T.K., Koche, R.P., et al. 2007. Genome-wide maps of chromatin state in pluripotent and lineage-committed cells. *Nature* **448:** 553–560.

Mikkers, H. and Frisen, J. 2005. Deconstructing stemness. *EMBO J.* **24:** 2715–2719.

Molofsky, A.V., He, S., Bydon, M., Morrison, S.J., and Pardal, R. 2005. Bmi-1 promotes neural stem cell self-renewal and

neural development but not mouse growth and survival by repressing the p16^{Ink4a} and p19^{Arf} senescence pathways. *Genes Dev.* **19:** 1432–1437.

Murzina, N., Verreault, A., Laue, E., and Stillman, B. 1999. Heterochromatin dynamics in mouse cells: Interaction between chromatin assembly factor 1 and HP1 proteins. *Mol. Cell* **4:** 529–540.

Nakajima, K., Sena, G., Nawy, T., and Benfey, P.N. 2001. Intercellular movement of the putative transcription factor SHR in root patterning. *Nature* **413:** 307–311.

Nightingale, K.P., O'Neill, L.P., and Turner, B.M. 2006. Histone modifications: Signalling receptors and potential elements of a heritable epigenetic code. *Curr. Opin. Genet. Dev.* **16:** 125–136.

Niwa, H. 2007a. How is pluripotency determined and maintained? *Development* **134:** 635–646.

Niwa, H. 2007b. Open conformation chromatin and pluripotency. *Genes Dev.* **21:** 2671–2676.

Okita, K., Ichisaka, T., and Yamanaka, S. 2007. Generation of germline-competent induced pluripotent stem cells. *Nature* **448:** 313–317.

Pasini, D., Bracken, A.P., Hansen, J.B., Capillo, M., and Helin, K. 2007. The polycomb group protein Suz12 is required for embryonic stem cell differentiation. *Mol. Cell. Biol.* **27:** 3769–3779.

Peterson, C.L. and Laniel, M.A. 2004. Histones and histone modifications. *Curr. Biol.* **14:** R546–R551.

Pietersen, A.M. and van Lohuizen, M. 2008. Stem cell regulation by polycomb repressors: Postponing commitment. *Curr. Opin. Cell Biol.* **20:** 201–207.

Quivy, J.P., Roche, D., Kirschner, D., Tagami, H., Nakatani, Y., and Almouzni, G. 2004. A CAF-1 dependent pool of HP1 during heterochromatin duplication. *EMBO J.* **23:** 3516–3526.

Ramirez-Parra, E. and Gutierrez, C. 2007a. E2F regulates FASCIATA1, a chromatin assembly gene whose loss switches on the endocycle and activates gene expression by changing the epigenetic status. *Plant Physiol.* **144:** 105–120.

Ramirez-Parra, E. and Gutierrez, C. 2007b. The many faces of chromatin assembly factor 1. *Trends Plant Sci.* **12:** 570–576.

Reese, B.E., Bachman, K.E., Baylin, S.B., and Rountree, M.R. 2003. The methyl-CpG binding protein MBD1 interacts with the p150 subunit of chromatin assembly factor 1. *Mol. Cell. Biol.* **23:** 3226–3236.

Ruthenburg, A.J., Allis, C.D., and Wysocka, J. 2007a. Methylation of lysine 4 on histone H3: Intricacy of writing and reading a single epigenetic mark. *Mol. Cell* **25:** 15–30.

Ruthenburg, A.J., Li, H., Patel, D.J., and Allis, C.D. 2007b. Multivalent engagement of chromatin modifications by linked binding modules. *Nat. Rev. Mol. Cell Biol.* **8:** 983–994.

Sabatini, S., Heidstra, R., Wildwater, M., and Scheres, B. 2003. SCARECROW is involved in positioning the stem cell niche in the *Arabidopsis* root meristem. *Genes Dev.* **17:** 354–358.

Sarraf, S.A. and Stancheva, I. 2004. Methyl-CpG binding protein MBD1 couples histone H3 methylation at lysine 9 by SETDB1 to DNA replication and chromatin assembly. *Mol. Cell* **15:** 595–605.

Scheres, B. 2007. Stem-cell niches: Nursery rhymes across kingdoms. *Nat. Rev. Mol. Cell Biol.* **8:** 345–354.

Schonrock, N., Exner, V., Probst, A., Gruissem, W., and Hennig, L. 2006. Functional genomic analysis of CAF-1 mutants in *Arabidopsis thaliana*. *J. Biol. Chem.* **281:** 9560–9568.

Shahbazian, M.D. and Grunstein, M. 2007. Functions of site-specific histone acetylation and deacetylation. *Annu. Rev. Biochem.* **76:** 75–100.

Sharpless, N.E. 2005. *INK4a/ARF*: A multifunctional tumor suppressor locus. *Mutat. Res.* **576:** 22–38.

Shi, Y. and Whetstine, J.R. 2007. Dynamic regulation of histone lysine methylation by demethylases. *Mol. Cell* **25:** 1–14.

Shibahara, K. and Stillman, B. 1999. Replication-dependent marking of DNA by PCNA facilitates CAF-1-coupled inheritance of chromatin. *Cell* **96:** 575–585.

Singh, M.B. and Bhalla, P.L. 2006. Plant stem cells carve their own niche. *Trends Plant Sci.* **11:** 241–246.

Spivakov, M. and Fisher, A.G. 2007. Epigenetic signatures of stem-cell identity. *Nat. Rev. Genet.* **8:** 263–271.

Stahl, Y. and Simon, R. 2005. Plant stem cell niches. *Int. J. Dev. Biol.* **49:** 479–489.

Strahl, B.D. and Allis, C.D. 2000. The language of covalent histone modifications. *Nature* **403:** 41–45.

Takahashi, K. and Yamanaka, S. 2006. Induction of pluripotent stem cells from mouse embryonic and adult fibroblast cultures by defined factors. *Cell* **126:** 663–676.

Takahashi, K., Tanabe, K., Ohnuki, M., Narita, M., Ichisaka, T., Tomoda, K., and Yamanaka, S. 2007. Induction of pluripotent stem cells from adult human fibroblasts by defined factors. *Cell* **131:** 861–872.

Tucker, M.R. and Laux, T. 2007. Connecting the paths in plant stem cell regulation. *Trends Cell Biol.* **17:** 403–410.

Turner, B.M. 2002. Cellular memory and the histone code. *Cell* **111:** 285–291.

Valk-Lingbeek, M.E., Bruggeman, S.W., and van Lohuizen, M. 2004. Stem cells and cancer: The Polycomb connection. *Cell* **118:** 409–418.

van den Berg, C., Willemsen, V., Hage, W., Weisbeek, P., and Scheres, B. 1995. Cell fate in the *Arabidopsis* root meristem determined by directional signalling. *Nature* **378:** 62–65.

Vernoux, T. and Benfey, P.N. 2005. Signals that regulate stem cell activity during plant development. *Curr. Opin. Genet. Dev.* **15:** 388–394.

Vlachonasios, K.E., Thomashow, M.F., and Triezenberg, S.J. 2003. Disruption mutations of *ADA2b* and *GCN5* transcriptional adaptor genes dramatically affect *Arabidopsis* growth, development, and gene expression. *Plant Cell* **15:** 626–638.

Wallace, J.A. and Orr-Weaver, T.L. 2005. Replication of heterochromatin: Insights into mechanisms of epigenetic inheritance. *Chromosoma* **114:** 389–402.

Wassenegger, M. 2005. The role of the RNAi machinery in heterochromatin formation. *Cell* **122:** 13–16.

Wernig, M., Meissner, A., Foreman, R., Brambrink, T., Ku, M., Hochedlinger, K., Bernstein, B.E., and Jaenisch, R. 2007. In vitro reprogramming of fibroblasts into a pluripotent ES-cell-like state. *Nature* **448:** 318–324.

Wildwater, M., Campilho, A., Perez-Perez, J.M., Heidstra, R., Blilou, I., Korthout, H., Chatterjee, J., Mariconti, L., Gruissem, W., and Scheres, B. 2005. The *RETINOBLAS-TOMA-RELATED* gene regulates stem cell maintenance in *Arabidopsis* roots. *Cell* **123:** 1337–1349.

Williams, L. and Fletcher, J.C. 2005. Stem cell regulation in the *Arabidopsis* shoot apical meristem. *Curr. Opin. Plant Biol.* **8:** 582–586.

Wong, M.D., Jin, Z., and Xie, T. 2005. Molecular mechanisms of germline stem cell regulation. *Annu. Rev. Genet.* **39:** 173–195.

Workman, J.L. 2006. Nucleosome displacement in transcription. *Genes Dev.* **20:** 2009–2017.

Xi, R. and Xie, T. 2005. Stem cell self-renewal controlled by chromatin remodeling factors. *Science* **310:** 1487–1489.

Zhang, Z., Shibahara, K., and Stillman, B. 2000. PCNA connects DNA replication to epigenetic inheritance in yeast. *Nature* **408:** 221–225.

Genetic and Epigenetic Regulation of Stem Cell Homeostasis in Plants

M. Lodha, C.F. Marco, and M.C.P. Timmermans

Cold Spring Harbor Laboratory, Cold Spring Harbor, New York 11724

Plants generate new organs through the activity of small populations of stem cells present in specialized niches called meristems. Stem cell homeostasis is attained by dynamic regulatory networks involving transcriptional regulators, hormones, and other intercellular signals that specify cell fate and convey positional information to the apical stem cells and the organizing center located immediately below. The balance between stem cell maintenance within the shoot apical meristem (SAM) and differentiation of cells that are displaced from the niche to form new organs involves the epigenetic silencing of stem cell regulatory genes. Recent advances have identified highly conserved chromatin remodeling factors as epigenetic regulators of stem cell fate that confer plasticity in plant development and ensure the stable inheritance of repressed expression states during organogenesis. These advances reveal that common mechanisms contribute to stem cell homeostasis in plants and animals.

One of the fundamental differences between plant and animal development is that plants produce new organs throughout their lifetime, which can span hundreds of years. This reiterative process of organogenesis depends on the activity of populations of pluripotent stem cells present in specialized niches, termed meristems, at the growing tip of the root and shoot (Fig. 1). Despite their common functions in stem cell maintenance and organ formation, the root and shoot meristems are organized differently. In the root, a single layer of stem cells surrounds an organizing center comprised of a small group of infrequently dividing cells that signal to the adjoining stem cells to inhibit their differentiation (van den Berg et al. 1997; Jiang and Feldman 2005). Each stem cell divides asymmetrically, giving rise to one daughter cell that maintains stem cell identity through contact with the organizing center and one daughter that is displaced from the niche and differentiates.

The angiosperm SAM consists of a small dome of cells that is organized into central and peripheral zones, distinguished by cell division rates and gene expression patterns. In the SAM of the model plant *Arabidopsis*, approximately three stem cells are maintained in each of the three most apical cell layers in the central zone (CZ) of the meristem (Fig. 1C) (see Bäurle and Laux 2003; Williams and Fletcher 2005). Their pluripotency is maintained by signals from an organizing center located immediately below the stem cell cluster. Thus, in contrast to the root, direct contact with the organizing center is not required to acquire or maintain shoot stem cell identity. Following stem cell divi-

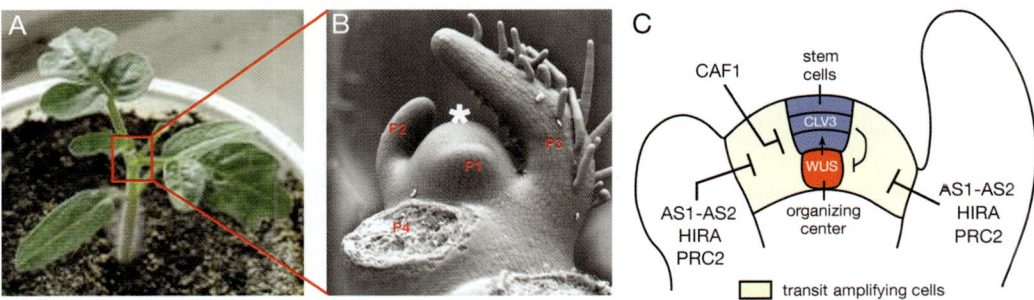

Figure 1. The shoot stem cell niche. (*A*) The shoot stem cell niche or meristem is located at the growing tip of the plant. The approximate position is indicated on this tomato seedling by the *red box*. (*B*) A scanning electron micrograph shows close-up of a tomato apex. (*Asterisk*) The shoot apical meristem. Stem cell activity leads to the periodic formation of new differentiating organs, such as the leaves shown here (marked P1–P4). (*C*) Diagram of the shoot stem cell niche. Stem cells are located in the central zone at the most apical tip (*blue*) overlaying the organizing center (*red*). *WUS* is expressed in the organizing center. *WUS* is required to maintain the pluripotent state of the overlying stem cells and to activate *CLV3* expression in these cells. CLV3 signaling in turn restricts the *WUS* expression domain. *WUS* expression is further regulated by the chromatin assembly factor CAF1. Stem cells give rise to a population of transit-amplifying cells (*yellow*) on the meristem periphery from which differentiating leaves are formed. *KNOX* genes are expressed in the central zone and transit-amplifying cells, but following organ initiation, *KNOX* genes are maintained in a repressed chromatin state through the action of AS1, AS2, HIRA, and PRC2. (*B*, Kindly provided by C. Kuhlemeier; *C*, modified from Tucker and Laux 2007.)

sion, one of the daughter cells typically enters a zone of indeterminate transit-amplifying cells at the meristem periphery (PZ), from which differentiating lateral organs are periodically initiated. Superimposed on the organization of the SAM into central and peripheral zones are discrete lineage layers. Cells in the outermost L1 divide anticlinally and give rise to the epidermal layer. Cells of the L2 also form a single layer that gives rise mostly to mesophyll tissue, and the innermost L3 comprises multiple cell layers that form the internal tissues, including mesophyll and vasculature. Predictably, the shoot stem cell cluster overlaps with each of these lineage layers (Fig. 1C).

During the past 5 years, mechanistic models for root and shoot meristem development have begun to emerge (see Sablowski 2007; Scheres 2007; Tucker and Laux 2007; Dinneny and Benfey 2008). Gene regulatory networks involved in stem cell specification and patterning of the root and shoot stem cell niches during embryogenesis have been identified. Important advances have also been reported on the signaling pathways that operate within these stem cell niches to maintain stem cell homeostasis and coordinate organogenesis. In this review, we describe regulatory networks that control stem cell activity within the vegetative shoot apex, focusing first on key genetic pathways involved in maintaining stem cell homeostasis within the SAM and subsequently on the molecular mechanisms that repress stem cell fate upon commitment to differentiation.

REGULATION OF STEM CELL HOMEOSTASIS BY THE WUS-CLV FEEDBACK LOOP

The homeodomain transcription factor WUSCHEL (WUS) is expressed exclusively in the organizing center of the SAM and is required to maintain the pluripotent state of the overlying stem cells, such that in plants that lack WUS activity, these stem cells undergo differentiation (Fig. 1C) (Laux et al. 1996; Mayer et al. 1998). In contrast to the root meristem, cell division patterns in and around the shoot organizing center are not predictable, and cells at the SAM tip are continuously displaced during shoot growth. Nonetheless, the position and size of the *WUS* expression domain remain remarkably constant throughout the life of the plant. *WUS* expression thus appears to be dynamically regulated and continuously redefined by positional information within the shoot. In support of this idea, a new organizing center can be established de novo after laser ablation of the CZ (Reinhardt et al. 2003a).

Not surprisingly, several regulatory genes that spatially define the domain of *WUS* expression during various stages of plant development have been identified, e.g., *HANABA TARANU, STIMPY/WOX9, HAIRY MERISTEM, AGAMOUS, ULTRAPETELA*, the *HD-ZIPIII* genes, and possibly *APETELA2* (Lenhard et al. 2001; Stuurman et al. 2002; Zhao et al. 2004; Carles et al. 2005; Prigge et al. 2005; Williams et al. 2005; Wu et al. 2005; Würschum et al. 2006). In addition to these transcription factors, *WUS* expression appears to be regulated epigenetically through recruitment of the SNF2 chromatin-remodeling factor SPLAYED and perhaps indirectly by small RNAs (Kwon et al. 2005). In the absence of functional PINHEAD/ZWILLE

(PNH/ZLL), a member of the ARGONAUTE family of small RNA-induced silencing factors, the *WUS* expression domain is enlarged (Moussian et al. 1998; Lynn et al. 1999; Tucker et al. 2008). Expression of *PNH/ZLL* from a vascular-specific promoter can rescue the shoot meristem phenotype of *pnh/zll* mutants, suggesting that the basal stem region may provide positional information to regulate *WUS* activity (Tucker et al. 2008).

The *WUS* expression domain is further defined by a positional signal from the overlying stem cells in the CZ via the CLAVATA (CLV) signaling pathway. In response to WUS-dependent signaling, the stem cells not only remain undifferentiated, but also induce expression of *CLV3* (Fig. 1C). The *CLV3* gene encodes a small protein that is processed into a secreted 12-amino-acid peptide (Fletcher et al. 1999; Kondo et al. 2006). Similar to plants overexpressing the *CLV3* gene, application of this 12-amino-acid peptide onto shoot apices terminates meristem activity, indicating that it constitutes the active CLV3 ligand. This ligand is perceived via the CLV1-CLV2 receptor kinase complex, which is expressed in a region that overlaps with the organizing center, and, via negative regulation of the POLTERGEIST (POL) and POL-like 1 phosphatases, down-regulates *WUS* expression (Fig. 1C) (Carles and Fletcher 2003; Song et al. 2006). This feedback loop between the organizing center and the stem cells creates a self-regulatory system to maintain a near constant size of the stem cell population (Brand et al. 2000; Schoof et al. 2000).

WUS induces stem cell identity in the overlying tissue layers in part by regulating cytokinin signaling. Many classical studies had demonstrated a role for cytokinin in promoting cell cycle activity and shoot identity. More recent studies provide the first insights into how cytokinin signaling is regulated in the SAM to control meristem function. *Arabidopsis* plants, in which cytokinin signaling is perturbed through mutation of the three known cytokinin receptors *CYTOKININ RESPONSE1* (*CRE1*), *ARABIDOPSIS HISTIDINE KINASE2* (*AHK2*), and *AHK3*, display a reduction in shoot meristem size (Higuchi et al. 2004). Likewise, mutations in the rice gene *LONELY GUY* (*LOG*), which encodes a cytokinin biosynthetic enzyme, reduce meristem size (Kurakawa et al. 2007). *LOG* is expressed near the tip of the SAM, suggesting that localized cytokinin production near the CZ is important for stem cell maintenance. One of the outcomes of cytokinin signaling is the activation of A-type ARABIDOPSIS RESPONSE REGULATORS (ARR) that provide negative feedback regulation and thus dampen the intracellular cytokinin response (for review, see Müller and Sheen 2007). Using chromatin immunoprecipitation (ChIP) coupled to microarray analyses, Leibfried et al. (2005) identified several A-type *ARR* genes, including *ARR5* and *ARR7*, as direct targets of repression by WUS. However, *ARR7* is expressed in the SAM, in a region that roughly coincides with the expression domain of *WUS*. This suggests that WUS modulates ARR7 levels and the cytokinin response. In plants that express a constitutively active from of ARR7, the stem cell population is depleted (Liebfried et al. 2005). In contrast, mutations in the maize A-type RESPONSE REGULATOR *abphyl1* lead to an increase in meristem size (Giulini et al. 2004). Together, these studies reveal that stem cell homeostasis requires a precisely defined cytokinin signaling response that is attained in part through the local repression of A-type ARR activity by WUS.

REGULATION OF STEM CELL ACTIVITY BY KNOX HOMEODOMAIN PROTEINS

In parallel to the WUS-CLV signaling pathway, class I KNOTTED1-like homeodomain (KNOX) proteins are key regulators of stem cell homeostasis in the plant shoot (Endrizzi et al. 1996; Long et al. 1996). In *Arabidopsis*, the class I *KNOX* gene family includes *SHOOT MERISTEM-LESS* (*STM*), *BREVIPEDICELLUS* (*BP*) (also known as *KNAT1*), *KNAT2*, and *KNAT6*. These genes have distinct but overlapping expression patterns within the SAM that reflect their diverse partially redundant activities in stem cell maintenance and organogenesis. *STM* is expressed in both the central and peripheral zones of the SAM, whereas expression of *BP* and *KNAT2* is confined to the periphery of the meristem (Lincoln et al. 1994; Long et al. 1996; Pautot et al. 2001). The expression domains reported for class I *KNOX* genes of other plant species similarly correspond to one of two general patterns. Like *STM*, maize *knotted1* (*kn1*), tomato *LeT6*, and tobacco *NTH1* are expressed in both the central and peripheral zones of the SAM, whereas expression of the class I *KNOX* genes *rough sheath1* (*rs1*) and *gnarly1* (*gn1*) of maize, *osh15* of rice, and *NTH20* of tobacco is limited to the basal region of the SAM's peripheral zone (for references, see Reiser et al. 2000; Scofield and Murray 2006).

Loss-of-function mutations in *STM* and *kn1* give rise to embryos that lack a SAM and so fail to develop any postembryonic vegetative tissues (Barton and Poethig 1993; Long et al. 1996; Vollbrecht et al. 2000). Even though *bp* single mutants have no obvious SAM defects, *BP* is conditionally redundant with *STM* and, in certain genetic backgrounds, can functionally replace STM in the maintenance of the shoot stem cell niche (Byrne et al. 2002). A role for other class I *KNOX* genes in stem cell homeostasis has been inferred from overexpression studies. Expression of all *KNOX* genes is down-regulated in leaf founder cells and, in simple leafed species, such as *Arabidopsis*, rice, and maize, typically remains repressed throughout organogenesis (see, e.g., Jackson et al. 1994; Lincoln et al. 1994; Long et al. 1996). Ectopic expression of *KNOX* genes in developing primordia interferes with organ differentiation and patterning. In *Arabidopsis*, *KNOX* misexpression causes the development of highly dissected or lobed leaves (Lincoln et al. 1994; Chuck et al. 1996; Brand et al. 2002). Depending on the severity of the phenotype, shoot meristems can form de novo in the sinuses between the lobes or on the upper leaf surface. In monocots, ectopic *KNOX* expression leads to the overproliferation of cells and the displacement of proximal tissue types into the distal leaf blade (see Hake et al. 2004). On the basis of these recessive and dominant phenotypes, *KNOX* genes have been implicated in maintaining proliferation of stem cells and transit-amplifying cells in the SAM and/or preventing their differentiation.

The activities of *KNOX* genes in stem cell homeostasis are closely entwined with several plant hormone pathways. A number of recent studies reveal a link also between *KNOX* activity and cytokinin. As mentioned above, the size of the shoot stem cell niche is reduced in mutants that perturb cytokinin biogenesis or signaling (Higuchi et al. 2004; Kurakawa et al. 2007). Conversely, the shoot meristem defects of *stm* mutants can be suppressed by increasing the level of cytokinin, whether through application of exogenous hormone or overexpression of *ISOPENTENYL TRANSFERASE* (*IPT*) genes, which encode an enzyme in the cytokinin biosynthetic pathway (Yanai et al. 2005). Moreover, *STM* and *KNAT2* were shown to positively regulate *IPT7*, and ectopic *KNOX* expression leads to increased accumulation of cytokinin and up-regulation of the cytokinin-induced response regulators *ARR5* and *ARR7* (Jasinski et al. 2005; Yanai et al. 2005). Because gibberelic acid (GA) signaling antagonizes cytokinin signaling, *KNOX* genes further promote the cytokinin pathway by negatively regulating the expression of GA-20 oxidase, a key GA biosynthetic enzyme, in the SAM (Sakamoto et al. 2001; Hay et al. 2002). Thus, it is conceivable that *KNOX* genes contribute to stem cell activity at least in part by promoting cytokinin accumulation in the meristem. However, the various cell types of the SAM may well respond differently to cytokinin. Whereas *KNOX* genes promote expression of the negative response regulators *ARR5* and *ARR7*, WUS represses their expression (Leibfried et al. 2005). Future studies will tell whether local differences in cytokinin signaling contribute to partitioning of the meristem into regions containing stem cells, the organizing center, or transit-amplifying cells.

Down-regulation of *KNOX* gene expression is required for the specification of organ initials from transit-amplifying cells on the flank of the SAM (Jackson et al. 1994; Long et al. 1996). Sites of organ initiation are further defined by local accumulation of the hormone auxin (Reinhardt et al. 2000, 2003b). An antagonistic relationship between auxin and *KNOX* activity could provide a mechanism for the required down-regulation in *KNOX* expression associated with organogenesis. Indirect evidence for such a link is presented by the misexpression of *BP* in *Arabidopsis* plants carrying mutations in genes involved in auxin transport or signaling (Hay et al. 2006). However, how the local accumulation of auxin might trigger the repression of *KNOX* genes is not currently understood.

NEGATIVE REGULATORS OF *KNOX* GENE EXPRESSION

As is evident from the marked developmental defects observed in simple leaves that ectopically express *KNOX* genes, their normal progression to differentiation requires that KNOX activity remains repressed throughout organogenesis. This process is mediated by the orthologous MYB-domain proteins ROUGH SHEATH2 (RS2), ASYMMETRIC LEAVES1 (AS1), and PHANTASTICA from maize, *Arabidopsis*, and *Antirrhinum*, respectively (Timmermans et al. 1999; Tsiantis et al. 1999; Byrne et al. 2000; Ori et al. 2000). These proteins are expressed in a pattern complementary to the *KNOX* genes, in organ founder cells and developing primordia, but not within the meristem proper. Loss-of-function mutations in *rs2* and *AS1* lead to perturbations in cell determination typical of ectopic *KNOX* accumulation (Fig. 2). However, the down-regulation of *KNOX* genes at the site of leaf initiation is unaffected in these mutants, indicating that RS2 and AS1 act after organ initiation to maintain *KNOX* genes silenced during subsequent leaf development. Expression studies have shown that multiple *KNOX* genes, including *kn1* and *rs1*, are misexpressed in leaves of *rs2* mutants (Schnee-berger et al. 1998; Scanlon et al. 2002; Alexander et al. 2005; Evans 2007). In *Arabidopsis*, *BP* and *KNAT2* are targets for negative regulation by *AS1* (Byrne et al. 2000; Ori et al. 2000; Semiarti et al. 2001). However, expression of *STM* is unaffected in *as1*. In fact, *STM* appears to repress *AS1* expression (Byrne et al. 2000).

Additional mutations that lead to *KNOX* misexpression in developing leaves have been identified. These include *semaphore1* and *corkscrew1* in maize and *asymmetric leaves2* (*as2*), *blade-on-petiole1*, and *yabby3 filamentous*

flower double mutants in *Arabidopsis*. Some of these mutants likely have an indirect effect on *KNOX* gene expression. For instance, mutations in *semaphore1* cause pleiotropic defects in the kernel and seedling that are asso-ciated with altered auxin transport (Scanlon et al. 2002), and ectopic *KNOX* expression in the *yabby3 filamentous flower* double mutant is associated with defects in adaxial-abaxial leaf polarity (Kumaran et al. 2002). AS2, on the other hand, functions with AS1 in a common genetic path-way. The *Arabidopsis as2* mutant has leaf phenotypes com-parable to *as1* that are associated with the misexpression of *BP* and *KNAT2*, and like *as1*, *as2* can suppress the *stm* mutant phenotype (Ori et al. 2000; Semiarti et al. 2001; Byrne et al. 2002; Lin et al. 2003). AS2 is a member of the LBD family of transcription factors, which contain a highly conserved amino-terminal zinc finger and leucine-zipper-like motif, and has been shown to physically inter-act with AS1 (Iwakawa et al. 2002; Shuai et al. 2002; Phelps-Durr et al. 2005; Husbands et al. 2007). Likewise, RS2 interacts physically with the maize homolog of AS2 (Phelps-Durr et al. 2005), presenting the possibility that *KNOX* gene silencing during organogenesis involves a mechanism that is conserved between monocot and dicot simple leafed species, which last shared a common ances-tor approximately 100 million years ago.

Insight into the mechanism by which AS1 and AS2 repress *KNOX* genes and establish determinacy came from a recent study by Guo et al. (2008), which showed that these proteins form a repressor complex that binds directly to two sites in the promoters of *BP* and *KNAT2*. A repressor activ-ity for AS1 was demonstrated by fusing its carboxy-termi-nal non-MYB domain to the DNA-binding domain of the floral regulator LEAFY (LFY). Expression of this fusion protein from the native *LFY* promoter leads to the repres-sion of LFY targets giving rise to characteristic *lfy* loss-of-function phenotypes. Binding of an AS1 complex to the promoters of *BP* and *KNAT2* was demonstrated using ChIP. Both *KNOX* promoters contain at least two AS1-complex-binding sites. Electrophoretic mobility-shift assays revealed that complex binding at each site is mediated by the regula-tory motif arrangement CWGTTD-KMKTTGAHW and requires interaction between AS1 and AS2. Interestingly, deletion of either AS1-complex-binding site in *BP* results in ectopic expression in developing leaves, indicating that the two sites act nonredundantly despite their analogous AS1-AS2-binding properties. Accordingly, interaction between AS1 complexes at each site appears to be required, suggest-ing that a repressive loop may be formed in the *KNOX* pro-moters that mediates stable *KNOX* gene silencing during organogenesis (Guo et al. 2008). The observation that AS1 can form homodimers (Theodoris et al. 2003; Phelps-Durr et al. 2005) presents a possible mechanism via which AS1 complexes can interact. Promoter deletion analysis further revealed that enhancer elements required for *BP* expression in the leaf are located between the AS1-AS2 complex bind-ing sites. On the basis of these observations, it was proposed that AS1 and AS2 maintain *KNOX* gene silencing via a mechanism that is conceptually similar to the action of insu-lators that form chromatin loop domains that sequester enhancer elements and block their action on promoters (Gaszner and Felsenfeld 2006).

Figure 2. *KNOX* expression in developing leaf primordia inter-feres with organ differentiation and patterning. (*A*) Image of a wild-type *Arabidopsis* plant. (*B*) Compared to wild type, *as1* plants are more compact and *as1* leaves are smaller, asymmetri-cally lobed, and have a rough leaf surface. (*C*) Mutants with reduced *HIRA* activity develop *as1*-like defects in cell prolifera-tion and differentiation. (*D*) Normal maize leaves develop a sharp boundary between proximal sheath (*arrow*) and distal blade regions, as marked by the *asterisk*. (*E*) Mutations in *rs2* affect proximodistal patterning of the leaf, as indicated by a dis-ruption of the sheath-blade boundary (*asterisks*) and sectors of sheath tissue extending distally into the leaf blade (*arrows*). (*F*) Mature *rs2* leaves develop a rough sheath due to the overprolif-eration of cells. *Arrows* mark sectors of sheath tissue in the leaf blade. (*G,H*) Immunolocalization of KNOX proteins in trans-verse sections of wild-type (*G*) and *rs2* null mutant apices (*H*). KNOX proteins accumulate in the SAM of both wild-type and *rs2* mutant plants (*arrows*). In addition, sectors expressing KNOX proteins are present in *rs2* leaves (*asterisks*).

EPIGENETIC REGULATION OF
PLANT STEM CELL FATE

Cellular memory is a general feature of development. Transcriptional outputs resulting from early cell-fate decisions need to be stably maintained throughout many rounds of cell division. In plants and animals, several epigenetic regulators have been identified that ensure the stable inheritance of activated or repressed expression states at developmental loci. The pattern of ectopic KNOX accumulation in rs2 leaves provided a first indication that the stable repression of KNOX genes during organogenesis involves an epigenetic cellular memory system. In rs2, KNOX genes become reactivated in a variable variegated pattern, such that rs2 null leaves are clonal mosaics of KNOX+ and KNOX- sectors (Fig. 2G,H) (Timmermans et al. 1999). This pattern of KNOX reactivation is reminiscent of the variegated expression patterns resulting from a failure to stably maintain a repressive chromatin state in all cells of a lineage. Sectors of KNOX reactivation are not observed in as1. Deletion of AS1-AS2-binding sites or loss of AS1 or AS2 function results in ectopic expression of BP and KNAT2 throughout young leaf primordia and in the petiole region and vasculature of older leaves (Ori et al. 2000; Guo et al. 2008). KNOX gene silencing in Arabidopsis thus has a stricter requirement for AS1. Nonetheless, AS1-AS2-mediated KNOX gene silencing may similarly involve an epigenetic maintenance mechanism. The expression domains of AS1 and AS2 overlap only in the very young leaf primordia (Iwakawa et al. 2007), indicating that KNOX misexpression in older leaves is unlikely a direct reflection of lost AS1-AS2-complex activity. An epigenetic mechanism for repressing stem cell fate in differentiating organs allows stem cell pluripotency to be reprogrammed during later stages of plant development and facilitates the somatic regeneration of plants upon wounding or in culture.

Consistent with an epigenetic mode of KNOX gene repression during leaf development, AS1 and RS2 interact with the chromatin-remodeling factor HIRA (Phelps-Durr et al. 2005). This WD-repeat protein is highly conserved throughout evolution. The yeast and animal HIRA proteins promote the DNA-synthesis-independent assembly of nucleosomes and can modulate chromatin structure both during heterochromatic gene silencing and to control the spatial and temporal expression of specific euchromatic genes (Spector et al. 1997; Magnaghi et al. 1998; Sharp et al. 2001; Ray-Gallet et al. 2002; Tagami et al. 2004; Zhang et al. 2005; Nakayama et al. 2007). Interestingly, in mouse, HIRA functions in the epigenetic regulation of stem cell pluripotency as well as in the reprogramming of expression states upon cellular differentiation (Meshorer et al. 2006; Hajkova et al. 2008). In Arabidopsis, hira null mutants are early embryonic lethal. However, in plants with reduced HIRA function, BP and KNAT2 are reactivated in developing leaves, and such plants develop as1-like defects in cell proliferation and differentiation (Fig. 2C) (Phelps-Durr et al. 2005). This presents the likely possibility that the AS1-AS2-mediated recruitment of HIRA to the BP and KNAT2 promoters leads to formation of a repressive chromatin state that is stably inherited throughout the many rounds of cell division associated with leaf development (Guo et al. 2008).

HIRA shares homology with the p60 subunit of the chromatin assembly factor 1 (CAF1) complex (Phelps-Durr et al. 2005). Interestingly, CAF1 also has an important role in maintaining stem cell homeostasis. Mutations in the Arabidopsis FASCIATA1 (FAS1) or FAS2 genes, which encode the p150 and p60 components of CAF1, respectively, lead to fasciation of the stem and enlargement and flattening of the SAM (Kaya et al. 2001). In addition, the functional organization of the SAM in distinct central and peripheral zones is severely disrupted in fas mutants. FAS1 and FAS2 affect meristem organization and function by regulating the size of the WUS expression domain, which in fas mutants is dispersed and expanded laterally and apically. Expression of WUS and CLV3 is unaffected in weak hira mutants (C. Fernandez-Marco and M. Timmermans, unpubl.), suggesting that HIRA and CAF1 regulate meristem function and organogenesis through independent pathways.

The involvement of CAF1, rather then HIRA, in regulating WUS expression is particularly intriguing considering that asymmetric cell divisions are key to defining the WUS expression domain and limiting the size of the stem cell population in the shoot (Mayer et al. 1998; see Scheres 2007). Both HIRA and CAF1 are nucleosome assembly factors. However, whereas HIRA can function independently of DNA synthesis, CAF1 is associated with proliferating cell nuclear antigen (PCNA) at the replication fork and assembles nucleosomes in a DNA-synthesis-dependent manner (Ray-Gallet et al. 2002; Tagami et al. 2004). Although chromatin states are normally inherited by both daughter cells, it has long been recognized that DNA replication provides a window of opportunity for changes in chromatin structure that might affect gene expression. The inherent asymmetry of the replication fork with a leading and lagging strand leads to a temporary asymmetric distribution of PCNA among the sister chromatids (Shibahara and Stillman 1999; Zhang et al. 2000). Because CAF1 binds to PCNA to direct chromatin assembly after DNA replication, this situation offers an opportunity for the establishment of an asymmetric chromatin structure on the two sister chromatids before division of proliferating cells (Shibahara and Stillman 1999; Zhang et al. 2000). Such an asymmetric distribution of epigenetic information, if inherited by the daughter cells, could provide the foundation for phenotypic asymmetry in sister cells during development that could perhaps allow one cell to maintain WUS expression or stem cell identity and the other cell to change cell fate.

As in animal systems, Polycomb group (PcG) proteins also have an important role in stem cell homeostasis in Arabidopsis (Katz et al. 2004; Boyer et al. 2006; Lee et al. 2006; Schubert et al. 2006; Pietersen and van Lohuizen 2008). These proteins form multiple evolutionary conserved complexes, termed Polycomb repressive complexes (PRCs). Animals contain two biochemically distinct PRCs. PRC2, which contains four core components—Suppressor-of-Zeste12 (Su[z]12), Enhancer-of-Zeste (E[z]), Extra-Sex-Comb (Esc), and p55/RbAp48—catalyzes the trimethylation of histone H3 lysine 27 (H3K27)

(Lund and van Lohuizen 2004; Schwartz and Pirrotta 2007). This chromatin mark provides a binding site for PRC1, which establishes a condensed chromatin structure that maintains genes stably silenced.

The PRC2 is structurally conserved in plants (see Makarevich et al. 2006; Schubert et al. 2006; Pien and Grossniklaus 2007). The *Arabidopsis CURLY LEAF* (*CLF*), *MEDEA* (*MEA*), and *SWINGER* (*SWN*) genes encode homologs of E[z]; *FERTILIZATION-INDEPENDENT SEED2* (*FIS2*), *VERNALIZATION2* (*VRN2*), and *EMBRYONIC FLOWER2* (*EMF2*) encode Su[z]12 homologs; FERTILIZATION-INDEPENDENT ENDOSPERM (FIE) is a homolog of Esc; and MULTICOPY SUPPRESSOR OF IRA1(MSI1) is homologous to p55/RbAp48. Data obtained from expression analyses, ChIPs, and protein–protein interaction studies indicate that multiple PRC2s contribute to the stable repression of *KNOX* genes during leaf development (Katz et al. 2004; Schubert et al. 2006; M. Lodha and M. Timmermans, unpubl.). CLF, MSI1, FIE, and EMF2 or VRN2 form a PRC2 that confines expression of *STM* to the SAM (Gendall et al. 2001; Chanvivattana et al. 2004; Schubert et al. 2006). CLF binds to *STM* in developing leaves and, similar to E[z], catalyzes the trimethylation of H3K27. H3K27me3 levels at *STM* are dramatically decreased in *clf* loss-of-function mutants and *STM* is misexpressed in differentiating *clf* leaves. CLF may act partially redundant with SWN, because the H3K27me3 marks at *STM* are lost completely in the *clf swn* double mutants. Interestingly, loss of this repressive histone mark is associated with gain of H3K4 dimethylation, which is typically associated with actively transcribed genes. Gain of this histone mark suggests the involvement of a Trithorax-like complex in the derepression of *STM*.

The PRC1 appears to be unique to animals. In plants, LIKE HETEROCHROMATIC PROTEIN1 (LHP1) is suggested to perform an analogous function (Turck et al. 2007; Zhang et al. 2007). LHP1 contains a chromodomain and chromoshadow domain and is homologous to yeast swi6 and Heterochromatic Protein1 (HP1) in animals. Similar to PRC1, LHP1 binds to trimethylated H3K27 (Turck et al. 2007; Zhang et al. 2007). We have recently shown that CLF-based PRCs also maintain the stable repression of

BP and *KNAT2* during leaf development (M. Lodha and M. Timmermans, unpubl.). LHP1 localizes to repressed *BP* and *KNAT2* loci and both genes are derepressed in leaves of *lhp1* mutants. This presents the possibility that LHP1 is recruited to these *KNOX* loci by H3K27 trimethylation and, like PRC1, functions in long-term memory of a repressed chromatin state.

One key outstanding question in the mammalian stem cell field is how PRCs are recruited to their target loci. Interestingly, our recent observations indicate that HIRA acts upstream of PRC2. In weak *hira* mutant leaves, nucleosomes at *BP* and *KNAT2* remain enriched for active chromatin marks and lack H3K27me3. These changes in chromatin modifications are not just observed around the AS1-complex-binding sites but occur throughout the *KNOX* loci, suggesting that stable *KNOX* gene silencing involves the spreading of epigenetic chromatin modifications. In chicken, HIRA interacts directly with p55/RbAp48 (Ahmad et al. 2003, 2005), suggesting the exciting possibility that HIRA can mediate the recruitment of PRC2 to target loci. If so, this presents the following working model for the stable repression of stem cell activity during organogenesis (Fig. 3): HIRA through interaction with AS1-AS2 binds the promoters of *KNOX* targets; this facilitates the recruitment of PRC2, perhaps through direct interaction between MSI1 and HIRA; PRC2-mediated trimethylation of H3K27 subsequently recruits LHP1, which through a currently unknown mechanism leads to a stable repressive chromatin state at the *KNOX* loci that allows cellular differentiation.

CONCLUSION

Networks of regulatory genes and intercellular signals maintain stem cell homeostasis in the plant shoot. The transcription factors that maintain the pluripotent state of stem cells function at least partly by regulating the activity of plant hormones, such as cytokinin. Local cell–cell communication among the organizing center, stem cells, and their immediate derivatives creates a self-regulatory system to maintain a near constant size of the stem cell population. Although the nature of the stem-cell-maintaining

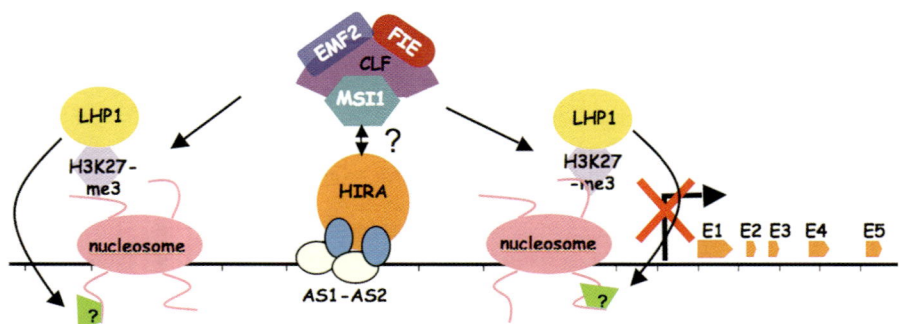

Figure 3. Model for stable *KNOX* gene silencing in differentiating leaves. AS1-AS2 binds to two sites in the promoters of *BP* and *KNAT2* and recruits HIRA to these *KNOX* targets. Recruitment of HIRA is required for the trimethylation of H3K27 in nucleosomes at *BP* and *KNAT2* by a CLF-containing PRC2. In chicken, HIRA interacts directly with the MSI1 homolog p55/Rbp48, presenting the possibility that HIRA functions to recruit PRC2 to *BP* and *KNAT2*. PRC2-mediated trimethylation of H3K27 subsequently recruits LHP1, which, through a currently unknown mechanism, leads to a stable repressive chromatin state at the *KNOX* loci that allows cellular differentiation.

signal from the organizing center remains elusive, recent advances have revealed an active peptide ligand that regulates stem cell number. On organ initiation, epigenetic silencing mechanisms suppress stem cell fate to allow the progression of normal development. As in animal systems, HIRA and PRC2 are part of a cellular memory system that facilitates differentiation through the stable repression of stem cell regulators. The possibility that the DNA-binding proteins AS1-AS2 via interaction with HIRA recruit PRC2 to the *KNOX* genes provides a molecular mechanism for silencing stem cell fate and could provide a framework for the recruitment of PcG complexes to target loci during the epigenetic regulation of pluripotency in animals.

ACKNOWLEDGMENTS

The authors thank past and present members of the Timmermans lab for discussions and input, especially Tara Phelps-Durr and Mengjuan Guo. Work on the epigenetic regulation of *KNOX* gene expression in the laboratory of M.T. is supported by grants from the National Science Foundation (MCB-0616114) and NYSTEM (C023044), and C.M. is funded by a postdoctoral fellowship from the Spanish Ministry of Education and Science (2007-0937).

REFERENCES

Ahmad, A., Takami, Y., and Nakayama, T. 2003. WD dipeptide motifs and LXXLL motif of chicken HIRA are necessary for transcription repression and the latter motif is essential for interaction with histone deacetylase-2 in vivo. *Biochem. Biophys. Res. Commun.* **312:** 1266–1272.

Ahmad, A., Kikuchi, H., Takami, Y., and Nakayama, T. 2005. Different roles of N-terminal and C-terminal halves of HIRA in transcription regulation of cell cycle-related genes that contribute to control of vertebrate cell growth. *J. Biol. Chem.* **280:** 32090–32100.

Alexander, D.L., Mellor, E.A., and Langdale, J.A. 2005. *CORKSCREW1* defines a novel mechanism of domain specification in the maize shoot. *Plant Physiol.* **138:** 1396–1408.

Barton, M. and Poethig, S. 1993. Formation of the shoot apical meristem in *Arabidopsis thaliana*: An analysis of development in the wild type and in the *shoot meristemless* mutant. *Development* **119:** 823–831.

Bäurle, I. and Laux, T. 2003. Apical meristems: The plant's fountain of youth. *Bioessays* **25:** 961–970.

Boyer, L.A., Plath, K., Zeitlinger, J., Brambrink, T., Medeiros, L.A., Lee, T.I., Levine, S.S., Wernig, M., Tajonar, A., Ray, M.K., et al. 2006. Polycomb complexes repress developmental regulators in murine embryonic stem cells. *Nature* **441:** 349–353.

Brand, U., Grünewald, M., Hobe, M., and Simon, R. 2002. Regulation of *CLV3* expression by two homeobox genes in *Arabidopsis*. *Plant Physiol.* **129:** 565–575.

Brand, U., Fletcher, J.C., Hobe, M., Meyerowitz, E.M., and Simon, R. 2000. Dependence of stem cell fate in *Arabidopsis* on a feedback loop regulated by *CLV3* activity. *Science* **289:** 617–619.

Byrne, M.E., Simorowski, J., and Martienssen, R.A. 2002. *ASYMMETRIC LEAVES1* reveals *knox* gene redundancy in *Arabidopsis*. *Development* **129:** 1957–1965.

Byrne, M.E., Barley, R., Curtis, M., Arroyo, J.M., Dunham, M., Hudson, A., and Martienssen, R.A. 2000. *ASYMMETRIC LEAVES1* mediates leaf patterning and stem cell function in *Arabidopsis*. *Nature* **408:** 967–971.

Carles, C.C. and Fletcher, J.C. 2003. Shoot apical meristem maintenance: The art of a dynamic balance. *Trends Plant Sci.* **8:** 394–401.

Carles, C.C., Choffnes-Inada, D., Reville, K., Lertpiriyapong, K., and Fletcher, J.C. 2005. *ULTRAPETALA1* encodes a SAND domain putative transcriptional regulator that controls shoot and floral meristem activity in *Arabidopsis*. *Development* **132:** 897–911.

Chanvivattana, Y., Bishopp, A., Schubert, D., Stock, C., Moon, Y.H., Sung, Z.R., and Goodrich J. 2004. Interaction of Polycomb-group proteins controlling flowering in *Arabidopsis*. *Development* **131:** 5263–5276.

Chuck, G., Lincoln, C., and Hake, S. 1996. *KNAT1* induces lobed leaves with ectopic meristems when overexpressed in *Arabidopsis*. *Plant Cell* **8:** 1277–1289.

Dinneny, J.R. and Benfey, P.N. 2008. Plant stem cell niches: Standing the test of time. *Cell* **132:** 553–557.

Endrizzi, K., Moussian, B., Haecker, A., Levin, J.Z., and Laux, T. 1996. The *SHOOT MERISTEMLESS* gene is required for maintenance of undifferentiated cells in *Arabidopsis* shoot and floral meristems and acts at a different regulatory level than the meristem genes *WUSCHEL* and *ZWILLE*. *Plant J.* **10:** 967–979.

Evans, M.M. 2007. The *indeterminate gametophyte1* gene of maize encodes a LOB domain protein required for embryo sac and leaf development. *Plant Cell* **19:** 46–62.

Fletcher, J.C., Brand, U., Running, M.P., Simon, R., and Meyerowitz, E.M. 1999. Signaling of cell fate decisions by *CLAVATA3* in *Arabidopsis* shoot meristems. *Science* **283:** 1911–1914.

Gaszner, M. and Felsenfeld, G. 2006. Insulators: Exploiting transcriptional and epigenetic mechanisms. *Nat. Rev. Genet.* **7:** 703–713.

Gendall, A.R., Levy, Y.Y., Wilson, A., and Dean, C.D. 2001. The *VERNALIZATION2* gene mediates the epigenetic regulation of vernalization in *Arabidopsis*. *Cell* **107:** 525–535.

Giulini, A., Wang, J., and Jackson, D. 2004. Control of phyllotaxy by the cytokinin-inducible response regulator homologue *ABPHYL1*. *Nature* **430:** 1031.

Guo, M., Thomas, J., Collins, G., and Timmermans, M.C. 2008. Direct repression of *KNOX* loci by the *ASYMMETRIC LEAVES1* complex of *Arabidopsis*. *Plant Cell* **20:** 48–58.

Hajkova, P., Ancelin, K., Waldmann, T., Lacoste, N., Lange, U.C., Cesari, F., Lee, C., Almouzni, G., Schneider, R., and Surani, M.A. 2008. Chromatin dynamics during epigenetic reprogramming in the mouse germ line. *Nature* **452:** 877–881.

Hake, S., Smith, H.M., Holtan, H., Magnani, E., Mele, G., and Ramirez, J. 2004. The role of *knox* genes in plant development. *Annu. Rev. Cell Dev. Biol.* **20:** 125–151.

Hay, A., Barkoulas, M., and Tsiantis, M. 2006. ASYMMETRIC LEAVES1 and auxin activities converge to repress *BREVIPEDICELLUS* expression and promote leaf development in *Arabidopsis*. *Development* **133:** 3955–3961.

Hay, A., Kaur, H., Phillips, A., Hedden, P., Hake, S., and Tsiantis, M. 2002. The gibberellin pathway mediates *KNOTTED1*-type homeobox function in plants with different body plans. *Curr. Biol.* **12:** 1557–1565.

Higuchi, M., Pischke, M.S., Mähönen, A.P., Miyawaki, K., Hashimoto, Y., Seki, M., Kobayashi, M., Shinozaki, K., Kato, T., Tabata, S., et al. 2004. In planta functions of the *Arabidopsis* cytokinin receptor family. *Proc. Natl. Acad. Sci.* **101:** 8821–8826.

Husbands, A., Bell, E.M., Shuai, B., Smith, H.M., and Springer, P.S. 2007. *LATERAL ORGAN BOUNDARIES* defines a new family of DNA-binding transcription factors and can interact with specific bHLH proteins. *Nucleic Acids Res.* **35:** 6663–6671.

Iwakawa, H., Iwasaki, M., Kojima, S., Ueno, Y., Soma, T., Tanaka, H., Semiarti, E., Machida, Y., and Machida, C. 2007. Expression of the *ASYMMETRIC LEAVES2* gene in the adaxial domain of *Arabidopsis* leaves represses cell proliferation in this domain and is critical for the development of properly expanded leaves. *Plant J.* **51:** 173–184.

Iwakawa, H., Ueno, Y., Semiarti, E., Onouchi, H., Kojima, S., Tsukaya, H., Hasebe, M., Soma, T., Ikezaki, M., Machida, C., and Machida, Y. 2002. The *ASYMMETRIC LEAVES2* gene of *Arabidopsis thaliana*, required for formation of a symmetric flat leaf lamina, encodes a member of a novel family of proteins characterized by cysteine repeats and a leucine zipper. *Plant Cell Physiol.* **43:** 467–478.

Jackson, D., Veit, B., and Hake, S. 1994. Expression of maize *KNOTTED1* related homeobox genes in the shoot apical meristem predicts patterns of morphogenesis in the vegetative shoot. *Development* **120:** 405–413.

Jasinski, S., Piazza, P., Craft, J., Hay, A., Woolley, L., Rieu, I., Phillips, A., Hedden, P., and Tsiantis, M. 2005. *KNOX* action in *Arabidopsis* is mediated by coordinate regulation of cytokinin and gibberellin activities. *Curr. Biol.* **15:** 1560–1565.

Jiang, K. and Feldman, L.J. 2005. Regulation of root apical meristem development. *Annu. Rev. Cell Dev. Biol.* **21:** 485–509.

Katz, A., Oliva, M., Mosquna, A., Hakim, O., and Ohad, N. 2004. FIE and CURLY LEAF polycomb proteins interact in the regulation of homeobox gene expression during sporophyte development. *Plant J.* **37:** 707–719.

Kaya, H., Shibahara, K.I., Taoka, K.I., Iwabuchi, M., Stillman, B. and Araki, T. 2001. *FASCIATA* genes for chromatin assembly factor-1 in *Arabidopsis* maintain the cellular organization of apical meristems. *Cell* **104:** 131–142.

Kondo, T., Sawa, S., Kinoshita, A., Mizuno, S., Kakimoto, T., Fukuda, H., and Sakagami, Y. 2006. A plant peptide encoded by *CLV3* identified by in situ MALDI-TOF MS analysis. *Science* **313:** 845–848.

Kumaran, M.K., Bowman, J.L., and Sundaresan, V. 2002. *YABBY* polarity genes mediate the repression of *KNOX* homeobox genes in *Arabidopsis*. *Plant Cell* **14:** 2761.

Kurakawa, T., Ueda, N., Maekawa, M., Kobayashi, K., Kojima, M., Nagato, Y., Sakakibara, H., and Kyozuka, J. 2007. Direct control of shoot meristem activity by a cytokinin activating enzyme. *Nature* **445:** 652–655.

Kwon, C.S., Chen, C., and Wagner, D. 2005. *WUSCHEL* is a primary target for transcriptional regulation by SPLAYED in dynamic control of stem cell fate in *Arabidopsis*. *Genes Dev.* **19:** 992–1003.

Laux, T., Mayer, K.F., Berger, J., and Jürgens, G. 1996. The *WUSCHEL* gene is required for shoot and floral meristem integrity in *Arabidopsis*. *Development* **122:** 87–96.

Lee, T.I., Jenner, R.G., Boyer, L.A., Guenther, M.G., Levine, S.S., Kumar, R.M., Chevalier, B., Johnstone, S.E., Cole, M.F., Isono, K., et al. 2006. Control of developmental regulators by Polycomb in human embryonic stem cells. *Cell* **125:** 301–313.

Leibfried, A., To, J.P., Busch, W., Stehling, S., Kehle, A., Demar, M., Kieber, J.J., and Lohmann, J.U. 2005. *WUSCHEL* controls meristem function by direct regulation of cytokinin-inducible response regulators. *Nature* **438:** 1172–1175.

Lenhard, M., Bohnert, A., Jürgens, G., and Laux, T. 2001. Termination of stem cell maintenance in *Arabidopsis* floral meristems by interactions between *WUSCHEL* and *AGAMOUS*. *Cell* **105:** 805–814.

Lin, W.C., Shuai, B., and Springer, P.S. 2003. The *Arabidopsis LATERAL ORGAN BOUNDARIES*-domain gene *ASYMMETRIC LEAVES2* functions in the repression of *KNOX* gene expression and in adaxial-abaxial patterning. *Plant Cell* **15:** 2241–2252.

Lincoln, C., Long, J., Yamaguchi, J., Serikawa, K., and Hake, S. 1994. A *knotted1*-like homeobox gene in *Arabidopsis* is expressed in the vegetative meristem and dramatically alters leaf morphology when overexpressed in transgenic plants. *Plant Cell* **6:** 1859–1876.

Long, J.A., Moan, E.I., Medford, J.I., and Barton, M.K. 1996. A member of the KNOTTED class of homeodomain proteins encoded by the *STM* gene of *Arabidopsis*. *Nature* **379:** 66–69.

Lund, A.H. and van Lohuizen, M. 2004. Polycomb complexes and silencing mechanisms. *Curr. Opin. Cell Biol.* **16:** 239–246.

Lynn, K., Fernandez, A., Aida, M., Sedbrook, J., Tasaka, M., Masson, P., and Barton, M.K. 1999. The *PINHEAD/ZWILLE* gene acts pleiotropically in *Arabidopsis* development and has overlapping functions with the *ARGONAUTE1* gene. *Development* **126:** 469–481.

Magnaghi, P., Roberts, C., Lorain, S., Lipinski, M., and Scambler, P.J. 1998. HIRA, a mammalian homologue of *Saccharomyces cerevisiae* transcriptional co-repressors, interacts with Pax3. *Nat. Genet.* **20:** 74–77.

Makarevich, G., Leroy, O., Akinci, U., Schubert, D., Clarenz, O., Goodrich, J., Grossniklaus, U., and Köhler, C. 2006. Different Polycomb group complexes regulate common target genes in *Arabidopsis*. *EMBO Rep.* **7:** 947–952.

Mayer, K.F., Schoof, H., Haecker, A., Lenhard, M., Jürgens, G., and Laux, T. 1998. Role of *WUSCHEL* in regulating stem cell fate in the *Arabidopsis* shoot meristem. *Cell* **95:** 805–815.

Meshorer, E., Yellajoshula, D., George, E., Scambler, P.J., Brown, D.T., and Misteli, T. 2006. Hyperdynamic plasticity of chromatin proteins in pluripotent embryonic stem cells. *Dev. Cell* **10:** 105–116.

Moussian, B., Schoof, H., Haecker, A., Jürgens, G., and Laux, T. 1998. Role of the *ZWILLE* gene in the regulation of central shoot meristem cell fate during *Arabidopsis* embryogenesis. *EMBO J.* **17:** 1799–1809.

Müller, B. and Sheen, J. 2007. Advances in cytokinin signaling. *Science* **318:** 68–69.

Nakayama, T., Nishioka, K., Dong, Y.X., Shimojima, T., and Hirose, S. 2007. *Drosophila* GAGA factor directs histone H3.3 replacement that prevents the heterochromatin spreading. *Genes Dev.* **21:** 552–561.

Ori, N., Eshed, Y., Chuck, G., Bowman, J.L., and Hake, S. 2000. Mechanisms that control *knox* gene expression in the *Arabidopsis* shoot. *Development* **127:** 5523–5532.

Pautot, V., Dockx, J., Hamant, O., Kronenberger, J., Grandjean, O., Jublot, D., and Traas, J. 2001. *KNAT2*: Evidence for a link between *knotted-like* genes and carpel development. *Plant Cell* **13:** 1719–1734.

Phelps-Durr, T.L., Thomas, J., Vahab, P., and Timmermans, M.C. 2005. Maize rough sheath2 and its *Arabidopsis* orthologue ASYMMETRIC LEAVES1 interact with HIRA, a predicted histone chaperone, to maintain *knox* gene silencing and determinacy during organogenesis. *Plant Cell* **17:** 2886–2898.

Pien, S. and Grossniklaus, U. 2007. Polycomb group and trithorax group proteins in *Arabidopsis*. *Biochim. Biophys. Acta* **1769:** 375–382.

Pietersen, A.M. and van Lohuizen, M. 2008. Stem cell regulation by polycomb repressors: Postponing commitment. *Curr. Opin. Cell Biol.* **20:** 201–207.

Prigge, M.J., Otsuga, D., Alonso, J.M., Ecker, J.R., Drews, G.N., and Clark, S.E. 2005. Class III homeodomain-leucine zipper gene family members have overlapping, antagonistic, and distinct roles in *Arabidopsis* development. *Plant Cell* **17:** 61–76.

Ray-Gallet, D., Quivy, J.P., Scamps, C., Martini, E.M., Lipinski, M., and Almouzni, G. 2002. HIRA is critical for a nucleosome assembly pathway independent of DNA synthesis. *Mol. Cell* **9:** 1091–1100.

Reinhardt, D., Mandel, T., and Kuhlemeier, C. 2000. Auxin regulates the initiation and radial position of plant lateral organs. *Plant Cell* **12:** 507–518.

Reinhardt, D., Frenz, M., Mandel, T., and Kuhlemeier, C. 2003a. Microsurgical and laser ablation analysis of interactions between the zones and layers of the tomato shoot apical meristem. *Development* **130:** 4073–4083.

Reinhardt, D., Pesce, E.R., Stieger, P., Mandel, T., Baltensperger, K., Bennett, M., Traas, J., Friml, J., and Kuhlemeier, C. 2003b. Regulation of phyllotaxis by polar auxin transport. *Nature* **426:** 255–260.

Reiser, L., Sánchez-Baracaldo, P., and Hake, S. 2000. Knots in the family tree: Evolutionary relationships and functions of *knox* homeobox genes. *Plant Mol. Biol.* **42:** 151–166.

Sablowski, R. 2007. The dynamic plant stem cell niches. *Curr. Opin. Plant Biol.* **10:** 639–644.

Sakamoto, T., Kamiya, N., Ueguchi-Tanaka, M., Iwahori, S., and Matsuoka, M. 2001. KNOX homeodomain protein directly suppresses the expression of a gibberellin biosynthetic gene in the tobacco shoot apical meristem. *Genes Dev.* **15:** 581–590.

Scanlon, M.J., Henderson, D.C., and Bernstein, B. 2002. *SEMAPHORE1* functions during the regulation of ancestrally duplicated *knox* genes and polar auxin transport in maize. *Development* **129**: 2663–2673.

Scheres, B. 2007. Stem-cell niches: Nursery rhymes across kingdoms. *Nat. Rev. Mol. Cell Biol.* **8**: 345–354.

Schneeberger, R., Tsiantis, M., Freeling, M., and Langdale, J.A. 1998. The *rough sheath2* gene negatively regulates homeobox gene expression during maize leaf development. *Development* **125**: 2857–2865.

Schoof, H., Lenhard, M., Haecker, A., Mayer, K.F., Jürgens, G., and Laux, T. 2000. The stem cell population of *Arabidopsis* shoot meristems is maintained by a regulatory loop between the *CLAVATA* and *WUSCHEL* genes. *Cell* **100**: 635–644.

Schubert, D., Primavesi, L., Bishopp, A., Roberts, G., Doonan, J., Jenuwein, T., and Goodrich, J. 2006. Silencing by plant Polycomb-group genes requires dispersed trimethylation of histone H3 at lysine 27. *EMBO J.* **25**: 4638–4649.

Schwartz, Y.B. and Pirrotta, V. 2007. Polycomb silencing mechanisms and the management of genomic programmes. *Nat. Rev. Genet.* **8**: 9–22.

Scofield, S. and Murray, J.A. 2006. *KNOX* gene function in plant stem cell niches. *Plant Mol. Biol.* **60**: 929–946.

Semiarti, E., Ueno, Y., Tsukaya, H., Iwakawa, H., Machida, C., and Machida, Y. 2001. The *ASYMMETRIC LEAVES2* gene of *Arabidopsis thaliana* regulates formation of a symmetric lamina, establishment of venation and repression of meristem-related homeobox genes in leaves. *Development* **128**: 1771–1783.

Sharp, J.A., Fouts, E.T., Krawitz, D.C., and Kaufman, P.D. 2001. Yeast histone deposition protein Asf1p requires Hir proteins and PCNA for heterochromatic silencing. *Curr. Biol.* **11**: 463–473.

Shibahara, K. and Stillman, B. 1999. Replication-dependent marking of DNA by PCNA facilitates CAF-1-coupled inheritance of chromatin. *Cell* **96**: 575–585.

Shuai, B., Reynaga-Peña, C.G., and Springer, P.S. 2002. The lateral organ boundaries gene defines a novel, plant-specific gene family. *Plant Physiol.* **129**: 747–761.

Song, S.K., Lee, M.M., and Clark, S.E. 2006. POL and PLL1 phosphatases are *CLAVATA1* signaling intermediates required for *Arabidopsis* shoot and floral stem cells. *Development* **133**: 4691–4698.

Spector, M.S., Raff, A., DeSilva, H., Lee, K., and Osley, M.A. 1997. Hir1p and Hir2p function as transcriptional corepressors to regulate histone gene transcription in the *Saccharomyces cerevisiae* cell cycle. *Mol. Cell. Biol.* **17**: 545–552.

Stuurman, J., Jäggi, F., and Kuhlemeier, C. 2002. Shoot meristem maintenance is controlled by a GRAS-gene mediated signal from differentiating cells. *Genes Dev.* **16**: 2213–2218.

Tagami, H., Ray-Gallet, D., Almouzni, G., and Nakatani, Y. 2004. Histone H3.1 and H3.3 complexes mediate nucleosome assembly pathways dependent or independent of DNA synthesis. *Cell* **116**: 51–61.

Theodoris, G., Inada, N., and Freeling, M. 2003. Conservation and molecular dissection of *ROUGH SHEATH2* and *ASYMMETRIC LEAVES1* function in leaf development. *Proc. Natl. Acad. Sci.* **100**: 6837–6842.

Timmermans, M.C., Hudson, A., Becraft, P.W., and Nelson, T. 1999. ROUGH SHEATH2: A Myb protein that represses *knox* homeobox genes in maize lateral organ primordia. *Science* **284**: 151–153.

Tsiantis, M., Schneeberger, R., Golz, J.F., Freeling, M., and Langdale, J.A. 1999. The maize *rough sheath2* gene and leaf development programs in monocot and dicot plants. *Science* **284**: 154–156.

Tucker, M.R. and Laux, T. 2007. Connecting the paths in plant stem cell regulation. *Trends Cell Biol.* **17**: 403–410.

Tucker, M.R., Hinze, A., Tucker, E.J., Takada, S., Jürgens, G., and Laux, T. 2008. Vascular signalling mediated by ZWILLE potentiates *WUSCHEL* function during shoot meristem stem cell development in the *Arabidopsis* embryo. *Development* **135**: 2839–2843.

Turck, F., Roudier, F., Farrona, S., Martin-Magniette, M.L., Guillaume, E., Buisine, N., Gagnot, S., Martienssen, R.A., Coupland, G., and Colot, V. 2007. *Arabidopsis TFL2/LHP1* specifically associates with genes marked by trimethylation of histone H3 lysine 27. *PLoS Genet.* **3**: e86.

van den Berg, C., Willemsen, V., Hendriks, G., Weisbeek, P., and Scheres, B. 1997. Short-range control of cell differentiation in the *Arabidopsis* root meristem. *Nature* **390**: 287–289.

Vollbrecht, E., Reiser, L., and Hake, S. 2000. Shoot meristem size is dependent on inbred background and presence of the maize homeobox gene, *knotted1*. *Development* **127**: 3161–3172.

Williams, L. and Fletcher, J.C. 2005. Stem cell regulation in the *Arabidopsis* shoot apical meristem. *Curr. Opin. Plant Biol.* **8**: 582–586.

Williams, L., Grigg, S.P., Xie, M., Christensen, S., and Fletcher, J.C. 2005. Regulation of *Arabidopsis* shoot apical meristem and lateral organ formation by microRNA *miR166g* and its *AtHD-ZIP* target genes. *Development* **132**: 3657–3668.

Wu, X., Dabi, T., and Weigel, D. 2005. Requirement of homeobox gene *STIMPY/WOX9* for *Arabidopsis* meristem growth and maintenance. *Curr. Biol.* **15**: 436–440.

Würschum, T., Gross-Hardt, R., and Laux, T. 2006. *APETALA2* regulates the stem cell niche in the *Arabidopsis* shoot meristem. *Plant Cell* **18**: 295–307.

Yanai, O., Shani, E, Dolezal, K., Tarkowski, P., Sablowski, R., Sandberg, G., Samach, A., and Ori, N. 2005. *Arabidopsis* KNOXI proteins activate cytokinin biosynthesis. *Curr. Biol.* **15**: 1566–1571.

Zhang, R., Poustovoitov, M.V., Ye, X., Santos, H.A., Chen, W., Daganzo, S.M., Erzberger, J.P., Serebriiskii, I.G., Canutescu, A.A., Dunbrack, R.L., et al. 2005. Formation of MacroH2A-containing senescence-associated heterochromatin foci and senescence driven by *ASF1a* and *HIRA*. *Dev. Cell* **8**: 19–30.

Zhang, X., Germann, S., Blus, B.J., Khorasanizadeh, S., Gaudin, V., and Jacobsen, S.E. 2007. The *Arabidopsis* LHP1 protein colocalizes with histone H3 Lys27 trimethylation. *Nat. Struct. Mol. Biol.* **14**: 869–871.

Zhang, Z., Shibahara, K., and Stillman, B. 2000. PCNA connects DNA replication to epigenetic inheritance in yeast. *Nature* **408**: 221–225.

Zhao, Y., Medrano, L., Ohashi, K., Fletcher, J.C., Yu, H., Sakai, H., and Meyerowitz, E.M. 2004. *HANABA TARANU* is a GATA transcription factor that regulates shoot apical meristem and flower development in *Arabidopsis*. *Plant Cell* **16**: 2586–2600.

Regulation of Stem Cell Differentiation by Histone Methyltransferases and Demethylases

D. Pasini, A.P. Bracken,* K. Agger, J. Christensen,
K. Hansen, P.A.C. Cloos, and K. Helin
*Biotech Research and Innovation Centre (BRIC) and Centre for Epigenetics,
University of Copenhagen, Ole Maaløes Vej 5, Copenhagen, Denmark*

The generation of different cell types from stem cells containing identical genetic information and their organization into tissues and organs during development is a highly complex process that requires defined transcriptional programs. Maintenance of such programs is epigenetically regulated and the factors involved in these processes are often essential for development. The activities required for cell-fate decisions are frequently deregulated in human tumors, and the elucidation of the molecular mechanisms that regulate these processes is therefore important for understanding both developmental processes and tumorigenesis.

Polycomb group (PcG) proteins control the expression of genes important for cell-fate decisions and are essential for embryogenesis, cell proliferation, and stem cell self-renewal. The PcG proteins form multiprotein complexes, called Polycomb repressive complexes (PRCs). The PRC2 complex contains the PcG proteins EZH2, SUZ12, and EED and is believed to repress transcription through methylation of lysine (K) 27 of histone H3 (H3). All three PcG components of the PRC2 complex are essential for the activity of the complex and they are required for early mouse embryogenesis.

In this chapter, we present some of the recent findings regarding the role of PcG proteins and lysine demethylases in embryonic stem cell differentiation, with the perspective of discussing models by which histone methyltransferases and demethylases are involved in regulating transcription during stem cell maintenance and cellular differentiation.

The development of a whole organism is a complex process that requires the precise regulation of transcription. Such regulation is essential to allow correct cellular commitment during development. Embryonic stem (ES) cells are pluripotent cells derived from the inner cell mass (ICM) of preimplantation embryos, which can be indefinitely maintained in tissue culture and differentiated into many different cell types (Keller 2005). Thus, ES cells are powerful as a tool, not only to generate genetic mouse models, but also to study the molecular mechanisms that regulate pluripotency and differentiation. Furthermore, loss of cellular commitment is a common feature of human cancers (Hanahan and Weinberg 2000), and the delineation of the mechanisms involved in regulating differentiation programs therefore becomes important for understanding the development of cancer.

THE POLYCOMB GROUP PROTEINS

The first PcG protein was discovered in *Drosophila* 60 years ago (Kennison 1995). The PcGs were defined as factors that, when mutated, give phenotypes similar to those of mutations in homeotic genes. This effect is counteracted by mutations in the Trithorax group (TrxG) proteins, which compete with the PcG proteins for binding to the same DNA elements (Ringrose and Paro 2004). In general, PcG proteins act as transcriptional repressors and are required for maintaining the repressive state of homeotic genes during fly development. The expression of homeotic genes is set up by the activity of segmentation proteins. This occurs transiently in early development, and the PcG and TrxG proteins maintain their repression or activation, respectively, during later development (Ringrose and Paro 2004). However, other results have shown that PcG proteins are expressed and are associated with homeotic genes earlier than the segmentation proteins (Orlando et al. 1998), suggesting that the mechanism by which fly development is regulated is more complex than current models indicate.

PcG proteins are highly conserved during evolution, and they execute their functions in multiprotein complexes. The best-characterized PcG complexes are the Polycomb repressive complexes 1 and 2 (PRC1 and PRC2) (Francis and Kingston 2001; Simon and Tamkun 2002; Pasini et al. 2004a). Orthologs of PRC1 members can be found in *Drosophila* to mammals, whereas orthologs of the PRC2 complex are also found in nematodes and plants (Brock and Fisher 2005). The PRC1 complex contains several different subunits, and its composition can change in different cell types (Pasini et al. 2007; Puschendorf et al. 2008). The four *Drosophila* proteins Pc, Psc, Ph, and dRING1 define the PRC1 core (Francis et al. 2001). The PRC1 complex contains two different enzymatic activities: *Drosophila* RING1 (RING1A and RING1B in mammals) catalyzes the ubiquitylation of histone H2A and its loss of function leads to

*Present address: Smurfit Institute of Genetics, Trinity College, Dublin 2, Ireland.

global loss of H2A K119 ubiquitylation in vivo (de Napoles et al. 2004; Wang et al. 2004; Cao et al. 2005; Stock et al. 2007; van der Stoop et al. 2008). Mammalian CBX4 (one of five homologs of *Drosophila* Pc) is an E3 SUMO ligase that has been shown to catalyze the sumoylation of the transcriptional repressor CtBP (Kagey et al. 2003, 2005). In vitro, the PRC1 complex retains the ability to bind in *trans* to nucleosomal arrays in a sequence-independent manner (Lavigne et al. 2004) and to induce compaction of nucleosomes (Francis et al. 2004). These different activities of PRC1 suggest various mechanisms by which PRC1 inhibits transcription.

The PRC2 complex is smaller than PRC1, and it is composed of the three PcG proteins EZH2, EED, and SUZ12 and the histone-binding proteins RbAp48/46. Through its catalytic subunit EZH2, the PRC2 complex preferentially methylates (me) K27 of histone H3 in vitro (Cao et al. 2002; Czermin et al. 2002; Kuzmichev et al. 2002; Müller et al. 2002). However, different groups have reported that PRC2 also can methylate histone H3K9 and histone H1K26 in vitro (Czermin et al. 2002; Kuzmichev et al. 2002, 2004; Erhardt et al. 2003; Su et al. 2003; Pasini et al. 2004b), but so far, genetic models only provide support for the involvement of EZH2 in catalyzing dimethylation and trimethylation of histone H3K27 in vivo (Cao and Zhang 2004; Pasini et al. 2007; Puschendorf et al. 2008; Riising et al. 2008). The activity of the PRC2 complex is required for the association of PRC1 to target genes (Cao et al. 2005; Pasini et al. 2007). This might involve a mechanism by which PRC1 is recruited to the H3K27me3 mark through direct binding of the chromodomain containing the CBX component of the PRC1 complex (Cao et al. 2002; Kuzmichev et al. 2002). Therefore, most current models of PcG-mediated transcriptional repression involve, first, the recruitment of the PRC2 complex to target genes, leading to methylation of H3K27 and the subsequent recruitment of PRC1 through binding to H3K27me3. PRC1 recruitment results in H2AK119 ubiquitylation (ubq) and chromatin compaction, resulting in a more stably transcriptional repressive state. Despite the fact that this model requires further experimental verification, the finding that PRC complexes share most if not all target genes in different cell types (Boyer et al. 2006; Bracken et al. 2006; Lee et al. 2006) strongly suggests that the two complexes have coordinated functions and supports the repressive model presented in Figure 1.

The components of the PRC2 complex are essential for mouse embryonic development. Knockout embryos for *Ezh2*, *Eed*, and *Suz12* die during early postimplantation stages during gastrulation (O'Carroll et al. 2001; Pasini et al. 2004b; Montgomery et al. 2005). The generation of mouse ES cell lines from *Eed* and *Suz12* knockout preimplantation embryos has shown that PRC2 activity is dispensable for the stem cell self-renewal but that *Eed* and *Suz12* knockout ES cells have an increased expression of differentiation-specific genes and a tendency to differentiate (Montgomery et al. 2005; Boyer et al. 2006; Lee et al. 2006; Pasini et al. 2007; Chamberlain et al. 2008). Attempts to derive *Ezh2* knockout ES cell lines were reported to be unsuccessful, suggesting that EZH2 is

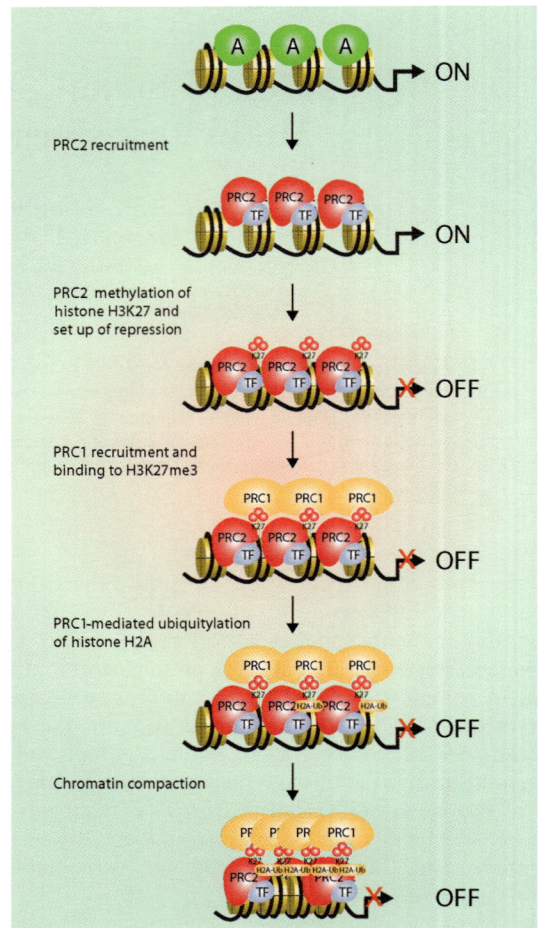

Figure 1. Proposed model for the mechanism by which Polycomb group proteins regulate their target genes. First, the PRC2 complex is recruited by ill-defined transcription factors, leading to the methylation of histone H3K27. This methylation results in the recruitment of the PRC1 complex through the binding to methylated K27. Subsequently, the RING1B protein of the PRC1 complex catalyzes the ubiquitylation of histone H2A K119, which may lead to chromatin compaction.

required for ES cell self-renewal (O'Carroll et al. 2001). The fact that both *Eed* and *Suz12* knockout ES cells globally lose H3K27me2/3 might suggest that Ezh2 has PRC2-independent functions that do not involve H3K27 methylation.

Interestingly, the analyses of ES cell differentiation of *Suz12* and *Eed* knockout ES cells have shown that the proteins are required for proper differentiation. The lack of differentiation correlates with failure to activate lineage-specific genes and to repress genes involved in stem cell self-renewal (Pasini et al. 2007; Chamberlain et al. 2008). These results are in agreement with the early developmental defects in PRC2 knockout embryos and show that the PRC2 proteins are critical for cell-fate determination during development.

Differently from PRC2, mice with mutations in genes of the PRC1 complex have less pronounced embryonic defects. For example, *Bmi1* (*Psc* in *Drosophila*) knockout mice die at birth with pronounced neural and hematopoietic defects (van der Lugt et al. 1994). This result could appear at odds with the proposed coordinated regulation of PRC2 and PRC1 repressive activities, but it is most likely a result of functional redundancy between the PRC1 subunits (Bmi1 has, for instance, five homologs in mice). In agreement with this, the inactivation of Rnf2 (Ring1B, dRING1 in *Drosophila*), which has only one homolog (Ring1a), leads to embryonic lethality in early postimplantation stages in mice, demonstrating that lack of PRC1 enzymatic activity is indeed essential for embryonic development (Voncken et al. 2003). Although *Rnf2* knockout ES cells show global loss of H2A ubiquitylation and have an increased expression of differentiation-specific genes, they can be established, similarly to ES cells devoid of PRC2 activity (de Napoles et al. 2004; Stock et al. 2007; Endoh et al. 2008; van der Stoop et al. 2008). Taken together, these data imply a crucial role for Rnf2 in PRC1 function and support the functional overlap between PRC1 and PRC2.

PCG TARGET GENES

Many aspects of PcG transcriptional regulation are still not fully understood. This includes the signaling pathways that control PRC2 activity, factors that mediate PcG recruitment to target genes, molecular mechanisms by which PcG inhibits transcription, and downstream pathways regulated by the PcG proteins. The identification of direct target genes of the PcG proteins within the last couple of years has significantly extended our knowledge regarding the downstream pathways regulated by the PcG proteins and, at the same time, has provided us with the tools to unravel the mechanism by which PcG proteins regulate transcription. Genome-wide location analyses using chromatin immunoprecipitation (ChIP) of PcG proteins and H3K27me3 in different cells lines, including mouse and human ES cells, have led to the following observations:

1. PcG and H3K27me3 are mainly associated with transcriptional start sites (TSS) of genes (Lee et al. 2006). In contrast to this, *Drosophila* PcG proteins are also found to be associated with Polycomb responsive elements (PREs) located several kilobases from the TSS (Ringrose and Paro 2004). This may suggest that the mechanism by which PcG proteins control transcription is not conserved between *Drosophila* and mammals.

2. The PcG proteins bind directly to the promoters of a large number of important regulators controlling cell-fate decisions during development in both human and mouse ES cells as well as in human embryonic fibroblasts (hEF) (Boyer et al. 2006; Bracken et al. 2006; Lee et al. 2006). The studies have shown that the PcGs are associated with the TSS of entire gene families, including homeobox, Gata, Pax, Sox, Wnt, Fgf, and T-box genes. These findings suggest a model by which the PcG proteins are displaced from their target genes when cells differentiate and that this displacement is required for the correct expression of lineage-specific genes.

3. PcG target genes are not fully conserved in different cell types. Comparison between cell lines such as ES, mouse embryonic fibroblast (MEF), C2C12 myoblasts, and hEF has shown that a proportion of PcG target genes are not conserved. Perhaps, not so surprisingly, PcG target genes show the largest degree of difference between normal and cancer cells (Bernstein et al. 2006a; Boyer et al. 2006; Bracken et al. 2006; Lee et al. 2006; Squazzo et al. 2006; Mikkelsen et al. 2007; Mohn et al. 2008). This difference could be partly flawed if the cancer cells are not compared to their normal counterparts, i.e., the cell of origin for the tumor cells. However, more intriguingly, the observed differences could be a result of the transformation process. Interestingly, despite the low degree of overlap between normal and tumor cells, PRC2 activity is essential for the proliferation of all cells tested so far, and it does not exclusively depend on the ability of PcG proteins to repress the *INK4A-ARF* locus (Bracken et al. 2003; Pasini et al. 2004b).

4. Different models describing how the PcG proteins regulate their target genes during differentiation have been proposed. We and other investigators have shown that PcG proteins are displaced from promoters when ES cells undergo differentiation. Moreover, the PcGs are also actively recruited to a substantial number of genes during differentiation (Bracken et al. 2006; Pasini et al. 2007; Mohn et al. 2008). Surprisingly, we have also found that the PcG proteins accumulate on some genes, even though they are activated, during differentiation (Pasini et al. 2007). Taking into consideration that ES cells do not differentiate synchronously, these results could suggest either that PcG activity is required for the transcriptional activation of some target genes or that, in some cases, the binding of the PcG proteins does not exclude transcription and may instead predispose the target genes for repression in later differentiation stages. Some support for the first hypothesis is provided by the fact that the activation of these genes does not occur in *Suz12* knockout ES cells (Pasini et al. 2007); however, this may also be due to the lack of proper differentiation of these cells. The second hypothesis is supported by the fact that transcribed genes such as *NEUROG2* and *OLIG2* are bound by PcGs and enriched for H3K27me3 in NT2 embryonic teratocarcinoma cells (Bracken et al. 2006).

We have summarized the different models for how the PcG proteins regulate transcription in Figure 2. Whereas solid evidence supports both the derepression and the recruitment models, further studies using differentiation systems with highly pure cellular populations combined with ChIP–re-ChIP (reverse ChIP) approaches are required to validate the activation model.

DOWNSTREAM FROM THE PCG PROTEINS

As previously mentioned, PcGs associate with the promoters of several genes whose activities are required for cell-fate determinations during development (Boyer et al.

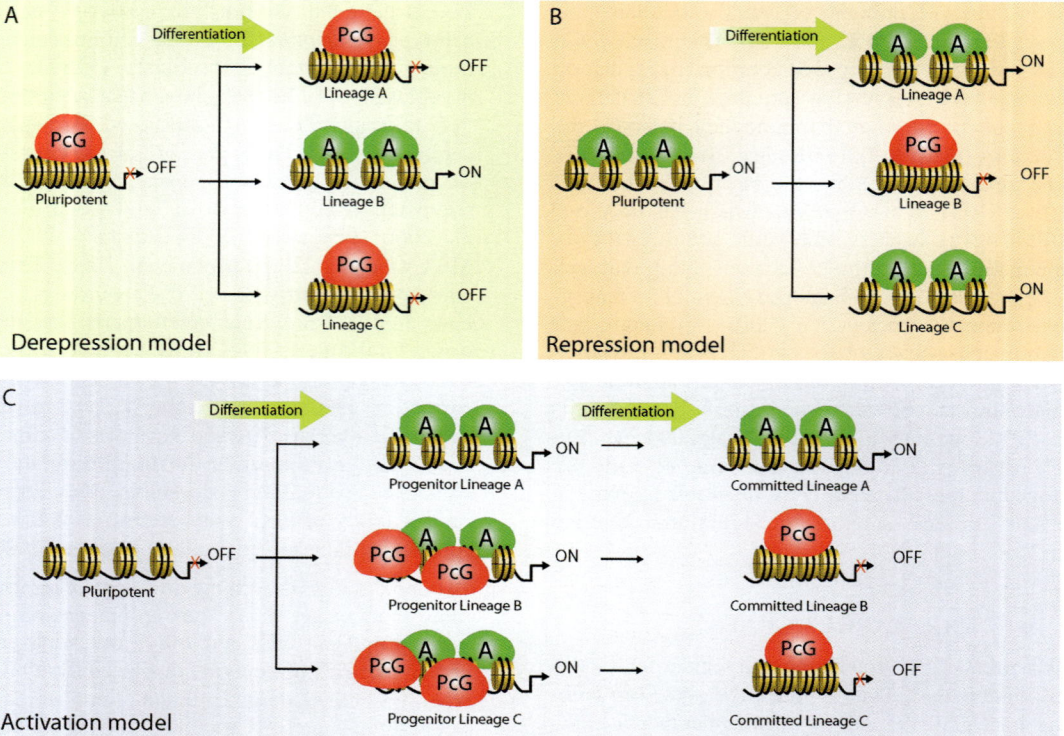

Figure 2. Model for the potential mechanisms by which the PcG proteins regulate transcription during differentiation. (*A*) Schematic representation of the derepression model, suggesting that displacement of PcG-binding contributes to the lineage-specific expression of target genes. (*B*) Schematic representation of the repression model, suggesting that PcGs can be actively recruited to the promoters of lineage-specific genes during differentiation. (*C*) Schematic representation of the activation model, showing that PcGs can be recruited to specific target genes that undergo transcriptional activation during differentiation. The model further highlights the fact that PcG recruitment may predispose these target genes for repression in later differentiation stages.

2006; Bracken et al. 2006; Lee et al. 2006). Among these, some deserve particular considerations.

The PcG proteins control expression of the *HOX* genes. They are bound to all four *HOX* gene clusters in ES cells (Boyer et al. 2006; Lee et al. 2006), and the use of tiling arrays in hES cells demonstrated that PcGs and H3K27me3 spread throughout the entire cluster (Lee et al. 2006). Interestingly, similar experiments in committed hEF cells showed that the 3′ end of all clusters lose PcG association and H3K27me3 (Bracken et al. 2006). Similar to their requirement in fly development, the HOX proteins are indispensable for mammalian development. Their spatial and temporal expression is essential and requires a tight regulation throughout development (Wellik 2007). Several results have suggested that PcGs are required for maintaining all *HOX* genes silenced in pluripotent cells and that differential PcG dissociation from *HOX* genes may be required for the coordinated regulation of *HOX* expression during development (Cao et al. 2005; Boyer et al. 2006; Bracken et al. 2006).

Another group of transcription factors regulated by the PcG proteins are the GATA factors. These transcription factors have different roles in development, and whereas GATA1, -2, and -3 are essential in hematopoiesis (Bresnick et al. 2005), GATA4, -5, and -6 seem to be required for myocardial development (Laverriere et al.

1994). Interestingly, the loss of PcGs in ES cells results in increased levels of GATA3, -4, and -6 (Boyer et al. 2006). This result further suggests an important role for PcG proteins in GATA repression, and consistent with this, the tissue-specific inactivation of Ezh2 leads to B- and T-cell developmental defects (Su et al. 2003, 2005).

PcG proteins directly associate with the promoters of several members of other transcription factor families such as the SRY box (SOX), T-box (TBX), forkhead box (FOX), and paired box (PAX) genes that have essential roles in regulating different developmental processes (Carlsson and Mahlapuu 2002; Plageman and Yutzey 2005; Kiefer 2007; Lang et al. 2007). Moreover, PcGs also associate with the promoters of gene families involved in different signaling pathways, including genes involved in transforming growth factor-β (TGF-β), WNT, and fibroblast growth factor (FGF) signaling. These signaling pathways regulate development by acting on cell-fate, patterning, and mitogenic signals (Kitisin et al. 2007; Cohen et al. 2008; Hayward et al. 2008; Itoh and Ornitz 2008).

Finally, our studies have demonstrated that PcG proteins directly bind and repress the *INK4A-ARF* tumor suppressor locus (Bracken et al. 2007). This locus codes for the expression of two negative regulators of the cell division cycle: p16^{INK4A} and p14^{ARF} (p19^{ARF} in mouse).

p16INK4A inhibits CDK4/CDK6 activity, and therefore phosphorylation of members of the retinoblastoma protein pRB family, whereas p14ARF inhibits the activity of MDM2, thereby leading to the accumulation of p53. Elevated levels of both p16INK4A and p14ARF expression lead to cell cycle arrest (Sherr 1998; Lowe and Sherr 2003). Importantly, activation of p16INK4A and p14ARF expression as a result of oncogene activation, oxidative stress, or other types of stress signals correlates with the displacement of the PcG proteins and decreased levels of H3K27me3 from the *INK4A-ARF* locus (Bracken et al. 2007). Consistent with a causal role of PcG proteins in the regulation of the locus, both genetic inactivation and RNA interference (RNAi)-based depletion of different PcG members induce premature senescence, whereas overexpression of PcG proteins such as BMI1, EZH2, CBX7, and CBX8 prevents stressed-induced senescence in an *INK4A-ARF*–dependent manner (Jacobs et al. 1999; Bracken et al. 2003, 2007; Gil et al. 2004; Dietrich et al. 2007). Importantly, the regulation of cell proliferation through binding to the *INK4A-ARF* locus is an essential feature of PcG proteins in mammalian cells, as first illustrated by the remarkable observation that several phenotypes of *Bmi1*−/− knockout mice are rescued by con- comitant deletion of the *Ink4a-Arf* locus (Jacobs et al. 1999; Bruggeman et al. 2005; Molofsky et al. 2005). These results also show that PcG proteins regulate embryonic development and normal differentiation by controlling both cell proliferation and the expression of a large number of genes involved in cell-fate decisions.

MECHANISMS FOR PCG TRANSCRIPTIONAL REGULATION

Although understanding of the downstream regulatory pathways regulated by PcG proteins is becoming clearer, two other aspects of PcG regulation are still poorly understood. First, the mechanisms by which PcGs are recruited to their target genes in mammalian cells remain elusive, and second, it is not fully understood how the PcG proteins repress transcription.

In *Drosophila*, the transcription factors Gaga, Psq, Zeste, Pho, and Pho-like (PhoL) mediate PcG recruitment to PREs (Brown et al. 2003). A PRE consists of multiple binding sites for these transcription factors, and it is believed that a PRE is defined by a combination of these binding sites. An algorithm to predict *Drosophila* PREs has been designed and has successfully predicted novel PREs (Ringrose et al. 2003; Ringrose and Paro 2007). This algorithm cannot predict any putative PREs from mammalian genomes and, together with the ChIP-chip data (Lee et al. 2006; Mikkelsen et al. 2007), suggests that these elements are not conserved.

Homologs for the transcription factors binding the *Drosophila* PREs are, with the exception of Pho, not found in mammals. Pho is homologous to the mammalian transcription factor Yin Yang 1 (YY1) (Brown et al. 1998). Consistent with this, it was shown that Yy1 and Ezh2 interact and colocalize to the *Mhc2b* promoter in undifferentiated C2C12 myoblasts (Caretti et al. 2004).

The fact that YY1 has not been reported to copurify with the PRC complexes could suggest that the association between YY1 and the PRCs is not stable (Cao et al. 2002; Czermin et al. 2002; Kuzmichev et al. 2002; Müller et al. 2002). Importantly, *Yy1* knockout mice are not viable and are embryonic-lethal in early postimplantation stages, similar to *Ezh2*, *Eed*, *Suz12*, and *Rnf2* knockout embryos (Donohoe et al. 1999). This result supports the possibility of a functional link between Yy1 and PRC complexes. Thus, it appears that the identification of target genes for YY1 in different cell types by genome-wide techniques is important for understanding the extent of overlap between YY1 and PcG proteins.

As described above, recent ChIP-chip and ChIP-sequencing (ChIP-seq) studies have led to the identification of PcG-associated DNA sequences (Boyer et al. 2006; Bracken et al. 2006; Lee et al. 2006; Mikkelsen et al. 2007). On the basis of these studies, it should be possible to identify DNA sequences or elements that function as PREs in mammalian cells. However, so far, no studies have been published in which a mammalian PRE has been identified. This could suggest that PREs do not exist or that there are many different PREs containing DNA-binding sites for a number of different transcription factors. The existence of a large number of PREs would allow the PcGs to differentially regulate gene expression depending on cell-type specification. Supporting a role for specific transcription factors in the regulation of PcG target genes is the recent demonstration that the transcription factor Snail1 recruits the PRC2 complex to the E-cadherin promoter to repress its expression in the mesoderm of developing mouse embryos (Herranz et al. 2008). Consistent with this, PcG target genes diverge when comparing different cell types (Squazzo et al. 2006; Mohn et al. 2008).

How do PcGs regulate transcription? As we have discussed above, PcG binding mainly correlates with transcriptional silencing (Boyer et al. 2006; Lee et al. 2006; Mikkelsen et al. 2007). Forced recruitment of the PRC2 complex to an artificial promoter induces transcriptional repression and suggests that PcGs "by default" act as transcriptional repressors (Pasini et al. 2008). However, as discussed previously, PcGs also associate with promoters that are actively transcribed, suggesting that PcG-mediated transcriptional regulation is a complex process that likely involves different mechanisms and factors. If we only consider the repressive activity of the PRC complexes, it is not fully understood how PcG recruitment can negatively influence transcription. First of all, it has not been formally established that the catalytic activity and therefore the H3K27me3 mark are required for PRC2 function. Results from several laboratories, including ours, have shown that the growth-promoting and oncogenic effects of EZH2 depend on the presence of a functional SET domain (Bracken et al. 2003; Kleer et al. 2003). This suggests that K27 methylation is essential for PcG activity, but the fact that these studies are based on the use of a truncated form of EZH2 still leaves the question open. Mechanistically, it has been proposed that H3K27me3 serves as a docking site for PRC1 recruitment and that this leads to ubiquitylation of H2AK119 and to compaction of chromatin (Francis et al. 2004; Cao et al.

2005). These observations may suggest that PRC complexes repress transcription by preventing the binding of FACT, thereby leading to a block of RNA-polymerase-II–dependent elongation (Zhou et al. 2008).

The physiological role of H3K27me3 in recruiting the PRC1 complex is also not fully understood. Even though PRC2 is required for the recruitment and maintenance of PRC1 on promoters (Cao et al. 2005; Boyer et al. 2006; Pasini et al. 2007), biochemical studies have shown that chromodomains of CBX proteins, which are part of different PRC1 complexes, have relatively low binding affinity for methylated lysines (K_d in a higher micromolar range) and, in addition, do not appear to discriminate between H3K9 and H3K27 trimethylation (Bernstein et al. 2006b). Thus, although these studies have been performed in vitro, they suggest that binding of CBX proteins to trimethyl lysines is unlikely to be the only mechanism by which PRC1 complexes are recruited to target genes. In agreement with this interpretation is the demonstration that the swap of chromodomains between HP1 and Pc is not sufficient to completely relocalize these proteins to other genes/structures in the cell (Platero et al. 1995). Taken together, these results suggest that the PRC1 complexes contain domains other than the chromodomains of the CBX proteins that are essential for their recruitment to PcG target genes.

Once PRC1 is recruited to its target gene, it is believed to maintain transcriptional repression. The question is how this is achieved. The PRC1 complex contains two different enzymatic activities. RING1B catalyzes the ubiquitylation of H2AK119 (Wang et al. 2004), whereas CBX4 contains E3 SUMO ligase activity (Kagey et al. 2003). How these two activities are related to the ability of PRC1 to repress transcription is currently unknown. However, the PRC1 complex has the ability to compact nucleosomes in vitro, suggesting that the complex could create a less-accessible chromatin environment at target genes (Francis et al. 2004). It is possible that such an effect will exclude RNA polymerase II from PcG target genes, but the fact that PcG binding in *Drosophila* does not exclude RNA polymerase II association to some promoters (Breiling et al. 2001) and that 23% of Suz12 target genes in hES cells have polymerase II associated with their promoter (Lee et al. 2006) does not support such a mechanism.

In this connection, it is important to note that loss of PcG activity only leads to the transcriptional activation of a small subset of target genes (Boyer et al. 2006; Bracken et al. 2006; Lee et al. 2006). Moreover, this activation is relatively low when compared with the expression levels that are achieved during normal activation of PcG target genes. An example of this is illustrated in Figure 3A. Here, the expression levels of *Gata4* and *Tbx1* are shown to be significantly higher during ES cell differentiation compared to their expression levels in *Suz12*$^{-/-}$ ES cells. These observations suggest that PcG displacement from promoters during differentiation is required, but not sufficient, for efficient activation of transcription and propose that in ES cells, the transcription factors required for this activation are either not expressed or their recruitment is impaired by a PcG-independent mechanism. This interpretation is further supported by genome-wide studies (summarized in Fig. 3B–D) showing that a large proportion of PcG target genes present higher transcript levels during differentiation when compared to loss of PcG activity. The existence of different layers of transcriptional regulators to control the expression of the same gene is in agreement with the complex spatial and temporal transcription network that must be established to regulate proper development of an organism.

HISTONE METHYLATION AND DEMETHYLATION

The recent discovery of lysine demethylases (KDM) that can actively remove methylation groups from lysine residues has increased the complexity of transcriptional regulation even further. Until the discovery of LSD1, histone lysine methylation was believed to be a stable mark that could only be lost during DNA replication and/or by histone replacement. LSD1 can catalyze the demethylation of me1 and me2 groups but not me3 groups (Shi et al. 2004). For this reason, me3 groups were considered a stable modification until the recent discovery that Jumonji C domain (JmjC)-containing proteins can actively remove me3 groups. This includes KDMs specific for H3K9me3/me2/me1, H3K36me3/me2, H3K27me3/me2, and H3K4me3/me2. Several of these KDMs are important for transcriptional regulation and are involved in different physiological processes, including differentiation and development (Agger et al. 2008; Cloos et al. 2008). For example, Jmjd1a (H3K9me2/me1-specific) and Jmjd2c (H3K9me3/me3- and H3K36me3/me2-specific) have been reported to be required for the pluripotency of mouse ES cells, potentially involving a mechanism removing H3K9me3 from the *Oct4* promoter and thereby preventing its transcriptional repression (Loh et al. 2007).

We and other investigators have shown that members of the JARID1 family specifically catalyze the demethylation of H3K4me3/me2 and that they are important for normal development (Christensen et al. 2007; Iwase et al. 2007; Klose et al. 2007; Lee et al. 2007a; Yamane et al. 2007). Recently, we have addressed the functional role of one member of the Jarid1 family (Rbp2/Jarid1a) in mouse ES cells (Pasini et al. 2008). By promoter location analysis, we have shown that Rbp2 is bound at the promoter of a large number of differentiation-related genes, suggesting that this activity may be involved in the repression of these genes in ES cells. In addition, we have demonstrated that Rbp2 and the PRC2 complex have a large number of common target genes, that Rbp2 and PRC2 interact, and that Rbp2 is required for the maintenance of the repression of PcG target genes (Pasini et al. 2008). These results show that the coordinated regulation of histone modifications by a repressive complex, containing KMT and KDM activities, may be important for the transcriptional regulation of different genes during differentiation and development. Interestingly, similar biochemical structures containing KMT and KDM activities have also been reported for the TrxG-like MLL complex, where the MLL2 and ASH2 H3K4me3 KMTs are associated in a complex containing

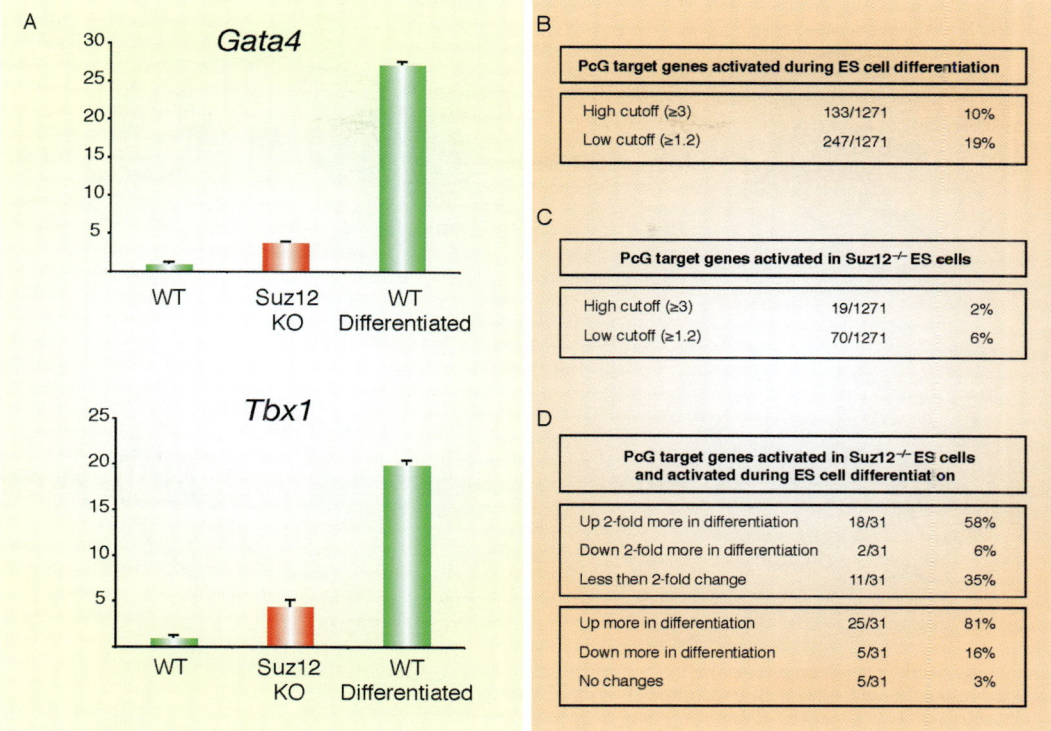

Figure 3. Different activities are required for the transcriptional regulation of PcG target genes during differentiation. (*A*) *Gata4* and *Tbx1* transcription in wild-type, *Suz12*[−/−], and differentiated wild-type mouse ES cells, showing that loss of PcG activity in ES cells is not sufficient for obtaining the full transcriptional activation achieved during ES cell differentiation (Pasini et al. 2007; D. Pasini, unpubl.). (*B*) Summary table obtained from genome-wide studies (Boyer et al. 2006; Lee et al. 2006), showing the proportion of PcG target genes that undergo transcriptional activation during mouse ES cell differentiation. (*C*) Summary table obtained from genome-wide studies (Boyer et al. 2006; Lee et al. 2006) showing the proportion of PcG target genes that undergo transcriptional activation in *Suz12*[−/−] ES cells. (*D*) Summary table obtained from genome-wide studies (Boyer et al. 2006; Lee et al. 2006), showing the proportion of PcG target genes that undergo transcriptional activation during mouse ES cell differentiation and in *Suz12*[−/−] ES cells. Similar to the data presented in *A*, this table shows that a large proportion of PcG target genes that are activated during differentiation and in *Suz12*[−/−] ES cells present higher transcript levels during differentiation relative to loss of Suz12 activity (58% with a stringent cutoff or 81% with a low cutoff). "*Up 2-fold more in differentiation*" includes genes that have a twofold higher expression level during differentiation versus *Suz12*[−/−] ES cells. "*Down 2-fold more in differentiation*" includes genes that have a twofold lower expression level during differentiation relative to Suz12[−/−] ES cells. "*Less than 2-fold change*" includes genes with expression changes lower than twofold during differentiation relative to *Suz12*[−/−] ES cells. "*Up more in differentiation*" includes genes that have a higher expression level during differentiation relative to *Suz12*[−/−] ES cells. "*Down more in differentiation*" includes genes that have a lower expression level during differentiation relative to *Suz12*[−/−] ES cells. "*No changes*" includes genes that have no expression changes during differentiation relative to *Suz12*[−/−] ES cells.

the H3K27me3-specific demethylase UTX (Agger et al. 2007; Issaeva et al. 2007; Lee et al. 2007b). These findings indicate that these different enzymatic activities function together to coordinately regulate lysine methylation and transcription. Moreover, they suggest that antagonistic activities act reciprocally during differentiation to regulate the "on" and "off" state of gene expression (Fig. 4).

BIVALENT DOMAINS:
AN ES CELL FEATURE?

Recent studies using genome-wide tiling ChIP-chip and ChIP-seq approaches for different methylated residues of histone tails have identified a modification pattern in ES cells termed "bivalent domains," consisting of H3K4me3 and H3K27me3 on the same nucleosomes (Bernstein et al. 2006a). Despite the considerable

number of H3K4me3-positive promoters in ES cells (78% of total annotated promoters), a large proportion of H3K27me3-positive promoters are also H3K4me3-positive (96% of Suz12 promoters) (Mikkelsen et al. 2007). This correlation is statistically significant (probability value = 1.4E[−18]), suggesting a functional role for the coexistence on these modifications. Interestingly, bivalent promoters correlate with low levels of expression, and bivalent domains are often lost in committed lineages (Mikkelsen et al. 2007; Mohn et al. 2008). Maintenance of H3K4me3, but not H3K27me3, correlates with activation of expression of bivalent genes in committed cell lines, suggesting that bivalency in ES cells has the function to maintain genes in a poised state for activation (Mikkelsen et al. 2007). This observation is extremely interesting and agrees with the transcriptional plasticity of pluripotent ES cells. Despite this, different questions in this regard remain unanswered. For

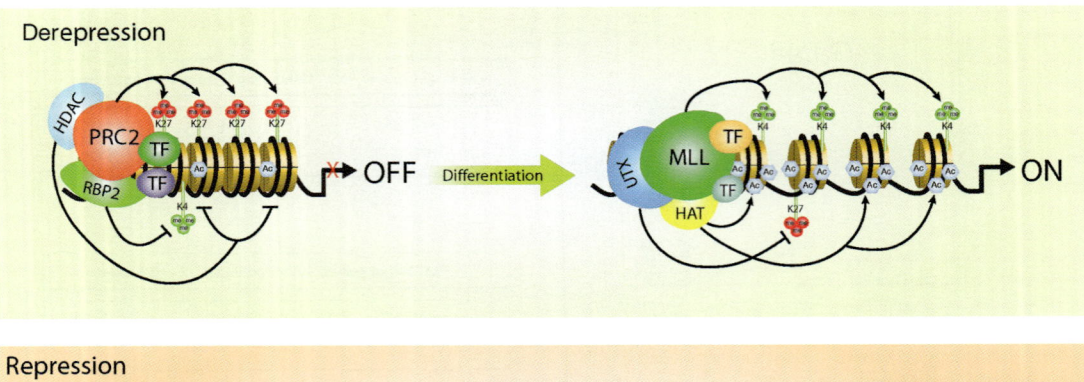

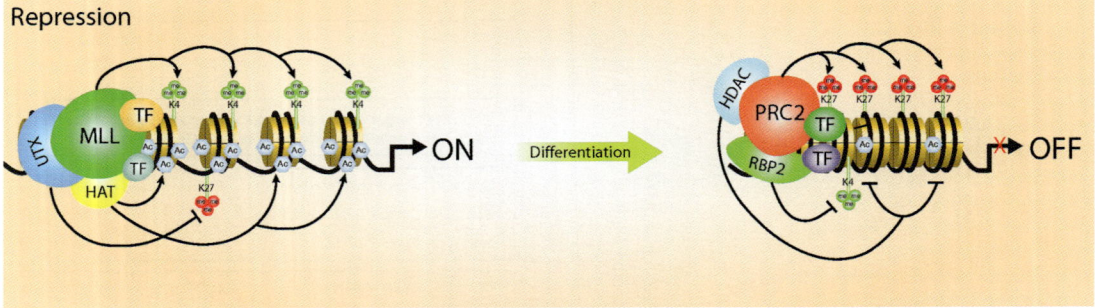

Figure 4. A model for the coordinated regulation of transcription by histone methyltransferases and demethylases during differentiation. The model highlights the antagonistic recruitment of repressive and activator complexes to regulate transcription during differentiation. The activator and repressive complexes contain opposite enzymatic activities. The repressive complex is formed by PRC2 H3K27 KMT activity, RBP2 H3K4 KDM activity, and HDAC de-acetylase activity, whereas the activator complex is formed by MLL H3K4 KMT activity, UTX H3K27 KDM activity, and HAT acetyltransferase activity. Furthermore, the model stresses that recruitment of the complexes could be mediated by multiple uncharacterized transcription factors.

example, is the bivalent mark important for transcriptional activation? What is the relevance for the bivalent mark at the CpG-rich promoter (HCG) if they are virtually all H3K4me3 (Mikkelsen et al. 2007)? Moreover, Bernstein and colleagues have suggested that most bivalent genes in ES cells resolve in committed cells, suggesting that bivalency is a specific feature of ES cells; however, recent data have shown that together with a proportion of genes that lose bivalency, there is a similar proportion of genes that gain bivalency during different stages of ES cell differentiation, demonstrating that bivalency is a feature not only of ES cells, but also of different committed cells types (Mohn et al. 2008). Additional work is required to better address the role of bivalent domains in regulating transcription during development.

The presence of bivalent domains may seem to be in contrast with our findings that PcG proteins coexist on a large number of PcG target genes with the H3K4 demethylase Rbp2 (Pasini et al. 2008). However, the association of Rbp2 with promoters does not appear to "erase" H3K4me3 from promoters but likely associates with the PRC2 complex to "fine-tune" the levels of H3K4me3 at TSS. Loss of Rbp2 activity leads to increased H3K4me3 levels, induces transcriptional activation, but does not affect PRC2 binding or H3K27me3 levels (Pasini et al. 2008). This interpretation is further supported by data that the H3K27me3 demethylase UTX is associated with promoters that are trimethylated on histone H3K27 (Agger et al. 2007; Lee et al. 2007b).

PCGS AND DNA METHYLATION

The methylation of a DNA cytosine residue within a CpG dinucleotide is an important mechanism for inherited epigenetic silencing of gene expression. CpG dinucleotides are found all over the genome, but they accumulate with high frequency in the proximity of TSS. These clusters of CpGs are defined as CpG islands. The hypermethylation of CpG islands leads to transcriptional repression (Bird 2002). DNA methylation is believed to be a stable modification that is inherited through many cell divisions, even a lifetime in the case of the inactivated X chromosome. In agreement with this, DNA methylation is faithfully inherited during DNA replication, and no bona fide DNA demethylase has so far been identified.

Recent data have shown that very few CpG islands are methylated in ES cells (Meissner et al. 2008). This is not surprising, because ES cells can differentiate into many different cell types and therefore require high transcriptional flexibility. Because DNA methylation is a stable modification, it will restrict the developmental potential of ES cells. In contrast, in terminally differentiated cells, which require low levels of plasticity and hence a stable expression program, many promoters are DNA-methylated. This DNA methylation, which is essential for differentiation, is acquired during the differentiation process. The enzymes that catalyze DNA methylation belong to the DNMT protein family (Bird 2002). Recently, it was shown that PcG proteins are able to interact with all members of the DNMT family, and it was suggested that PcGs and DNMTs are dependent on each other for their recruitment to target promoters (Vire et al. 2006). These results are also interesting in the perspective of tumor development. Promoter DNA hypermethylation frequently silences tumor suppressor genes, whereas members of the PRC2 complex are overexpressed in many human tumors. Taken together, these findings suggest a mechanism that links PcG overexpression to promoter hypermethylation in the development of human cancer. However, the mechanism by which PcGs set up and maintain DNA methylation is still under debate because other publications have suggested that RNAi-based PcG depletion does not reactivate the expression of fully methylated genes (McGarvey et al. 2007) and that trimethylation of histone H3K27 is an alternative mechanism for repressing tumor suppressor genes in prostate cancer, which does not require DNA methylation (Kondo et al. 2008). Nevertheless, genome-wide studies in primary and cancer cell lines have shown that there is a significant correlation between being a PcG- and H3K27me3-enriched gene in ES cells and differentiated cells and becoming DNA-hypermethylated in cancer cells (Schlesinger et al. 2007; Widschwendter et al. 2007; Mohn et al. 2008). Moreover, consistent with an essential role of PcGs to repress genes instructive for cell-fate decisions, recent results have demonstrated that low expression of PcG target genes contributes significantly to an embryonic stem-cell-like gene expression signature in poorly differentiated aggressive human tumors (Ohm et al. 2007).

CONCLUDING REMARKS

In this symposium review, we have summarized the recent advances in understanding how histone methylation regulates pluripotency. In particular, we have focused on the role of Polycomb group proteins in regulating transcription during development. We have discussed several models by which they control gene expression, how they are recruited to their sites of action, and a number of other unresolved questions that will be extremely interesting to address in the future. This is an exciting field that moves very quickly, and we expect that many of the questions we have posed will be solved within the next few years.

ACKNOWLEDGMENTS

We thank members of the Helin lab for discussions. D.P. was supported by a postdoctoral fellowship from the Danish Medical Research Council and P.A.C.C. by a grant from the Benzon Foundation. The work in the Helin lab was supported by grants from the Danish Cancer Society, the Danish National Research Foundation, the Novo Nordisk Foundation, the Lundbeck Foundation, the Danish Medical Research Council, the Danish Natural Science Research Council, the International Association for Cancer Research, and the European Union.

REFERENCES

Agger, K., Cloos, P.A., Christensen, J., Pasini, D., Rose, S., Rappsilber, J., Issaeva, I., Canaani, E., Salcini, A.E., and Helin, K. 2007. UTX and JMJD3 are histone H3K27 demethylases involved in *HOX* gene regulation and development. *Nature* **449:** 731–734.

Agger, K., Christensen, J., Cloos, P.A., and Helin, K. 2008. The emerging functions of histone demethylases. *Curr. Opin. Genet. Dev.* **18:** 159–168.

Bernstein, B.E., Mikkelsen, T.S., Xie, X., Kamal, M., Huebert, D.J., Cuff, J., Fry, B., Meissner, A., Wernig, M., Plath, K., et al. 2006a. A bivalent chromatin structure marks key developmental genes in embryonic stem cells. *Cell* **125:** 315–326.

Bernstein, E., Duncan, E.M., Masui, O., Gil, J., Heard, E., and Allis, C.D. 2006b. Mouse polycomb proteins bind differentially to methylated histone H3 and RNA and are enriched in facultative heterochromatin. *Mol. Cell. Biol.* **26:** 2560–2569.

Bird, A. 2002. DNA methylation patterns and epigenetic memory. *Genes Dev.* **16:** 6–21.

Boyer, L.A., Plath, K., Zeitlinger, J., Brambrink, T., Medeiros, L.A., Lee, T.I., Levine, S.S., Wernig, M., Tajonar, A., Ray, M.K., et al. 2006. Polycomb complexes repress developmental regulators in murine embryonic stem cells. *Nature* **441:** 349–353.

Bracken, A.P., Pasini, D., Capra, M., Prosperini, E., Colli, E., and Helin, K. 2003. *EZH2* is downstream of the pRB-E2F pathway, essential for proliferation and amplified in cancer. *EMBO J.* **22:** 5323–5335.

Bracken, A.P., Dietrich, N., Pasini, D., Hansen, K.H., and Helin, K. 2006. Genome-wide mapping of Polycomb target genes unravels their roles in cell fate transitions. *Genes Dev.* **20:** 1123–1136.

Bracken, A.P., Kleine-Kohlbrecher, D., Dietrich, N., Pasini, D., Gargiulo, G., Beekman, C., Theilgaard-Monch, K., Minucci, S., Porse, B.T., Marine, J.C., et al. 2007. The Polycomb group proteins bind throughout the INK4A-ARF locus and are disassociated in senescent cells. *Genes Dev.* **21:** 525–530.

Breiling, A., Turner, B.M., Bianchi, M.E., and Orlando, V. 2001. General transcription factors bind promoters repressed by Polycomb group proteins. *Nature* **412:** 651–655.

Bresnick, E.H., Martowicz, M.L., Pal, S., and Johnson, K.D. 2005. Developmental control via GATA factor interplay at chromatin domains. *J. Cell. Physiol.* **205:** 1–9.

Brock, H.W. and Fisher, C.L. 2005. Maintenance of gene expression patterns. *Dev. Dyn.* **232:** 633–655.

Brown, J.L., Mucci, D., Whiteley, M., Dirksen, M.L., and Kassis, J.A. 1998. The *Drosophila* Polycomb group gene *pleiohomeotic* encodes a DNA binding protein with homology to the transcription factor YY1. *Mol. Cell* **1:** 1057–1064.

Brown, J.L., Fritsch, C., Mueller, J., and Kassis, J.A. 2003. The *Drosophila pho-like* gene encodes a YY1-related DNA binding protein that is redundant with *pleiohomeotic* in homeotic gene silencing. *Development* **130:** 285–294.

Bruggeman, S.W., Valk-Lingbeek, M.E., van der Stoop, P.P., Jacobs, J.J., Kieboom, K., Tanger, E., Hulsman, D., Leung, C., Arsenijevic, Y., Marino, S., and van Lohuizen, M. 2005. *Ink4a* and *Arf* differentially affect cell proliferation and neural stem cell self-renewal in *Bmi1*-deficient mice. *Genes Dev.* **19:** 1438–1443.

Cao, R. and Zhang, Y. 2004. SUZ12 is required for both the histone methyltransferase activity and the silencing function of the EED-EZH2 complex. *Mol. Cell* **15:** 57–67.

Cao, R., Wang, L., Wang, H., Xia, L., Erdjument-Bromage, H., Tempst, P., Jones, R.S., and Zhang, Y. 2002. Role of histone H3 lysine 27 methylation in Polycomb-group silencing. *Science* **298:** 1039–1043.

Cao, R., Tsukada, Y., and Zhang, Y. 2005. Role of Bmi-1 and Ring1A in H2A ubiquitylation and Hox gene silencing. *Mol. Cell* **20:** 845–854.

Caretti, G., Di Padova, M., Micales, B., Lyons, G.E., and Sartorelli, V. 2004. The Polycomb Ezh2 methyltransferase regulates muscle gene expression and skeletal muscle differentiation. *Genes Dev.* **18:** 2627–2638.

Carlsson, P. and Mahlapuu, M. 2002. Forkhead transcription factors: Key players in development and metabolism. *Dev. Biol.* **250:** 1–23.

Chamberlain, S.J., Yee, D., and Magnuson, T. 2008. Polycomb repressive complex 2 is dispensable for maintenance of embryonic stem cell pluripotency. *Stem Cells* **26:** 1496–1505.

Christensen, J., Agger, K., Cloos, P.A., Pasini, D., Rose, S., Sennels, L., Rappsilber, J., Hansen, K.H., Salcini, A.E., and Helin, K. 2007. RBP2 belongs to a family of demethylases, specific for tri- and dimethylated lysine 4 on histone 3. *Cell* **128:** 1063–1076.

Cloos, P.A., Christensen, J., Agger, K., and Helin, K. 2008. Erasing the methyl mark: Histone demethylases at the center of cellular differentiation and disease. *Genes Dev.* **22:** 1115–1140.

Cohen, E.D., Tian, Y., and Morrisey, E.E. 2008. Wnt signaling: An essential regulator of cardiovascular differentiation, morphogenesis and progenitor self-renewal. *Development* **135:** 789–798.

Czermin, B., Melfi, R., McCabe, D., Seitz, V., Imhof, A., and Pirrotta, V. 2002. *Drosophila* enhancer of Zeste/ESC complexes have a histone H3 methyltransferase activity that marks chromosomal Polycomb sites. *Cell* **111:** 185–196.

de Napoles, M., Mermoud, J.E., Wakao, R., Tang, Y.A., Endoh, M., Appanah, R., Nesterova, T.B., Silva, J., Otte, A.P., Vidal, M., et al. 2004. Polycomb group proteins Ring1A/B link ubiquitylation of histone H2A to heritable gene silencing and X inactivation. *Dev. Cell* **7:** 663–676.

Dietrich, N., Bracken, A.P., Trinh, E., Schjerling, C.K., Koseki, H., Rappsilber, J., Helin, K., and Hansen, K.H. 2007. Bypass of senescence by the polycomb group protein CBX8 through direct binding to the *INK4A-ARF* locus. *EMBO J.* **26:** 1637–1648.

Donohoe, M.E., Zhang, X., McGinnis, L., Biggers, J., Li, E., and Shi, Y. 1999. Targeted disruption of mouse Yin Yang 1 transcription factor results in peri-implantation lethality. *Mol. Cell. Biol.* **19:** 7237–7244.

Endoh, M., Endo, T.A., Endoh, T., Fujimura, Y., Ohara, O., Toyoda, T., Otte, A.P., Okano, M., Brockdorff, N., Vidal, M., and Koseki, H. 2008. Polycomb group proteins Ring1A/B are functionally linked to the core transcriptional regulatory circuitry to maintain ES cell identity. *Development* **135:** 1513–1524.

Erhardt, S., Su, I.H., Schneider, R., Barton, S., Bannister, A.J., Perez-Burgos, L., Jenuwein, T., Kouzarides, T., Tarakhovsky, A., and Surani, M.A. 2003. Consequences of the depletion of zygotic and embryonic enhancer of zeste 2 during preimplantation mouse development. *Development* **130:** 4235–4248.

Francis, N.J. and Kingston, R.E. 2001. Mechanisms of transcriptional memory. *Nat. Rev. Mol. Cell Biol.* **2:** 409–421.

Francis, N.J., Saurin, A.J., Shao, Z., and Kingston, R.E. 2001. Reconstitution of a functional core polycomb repressive complex. *Mol. Cell* **8:** 545–556.

Francis, N.J., Kingston, R.E., and Woodcock, C.L. 2004. Chromatin compaction by a polycomb group protein complex. *Science* **306:** 1574–1577.

Gil, J., Bernard, D., Martínez, D., and Beach, D. 2004. Polycomb CBX7 has a unifying role in cellular lifespan. *Nat. Cell Biol.* **6:** 67–72.

Hanahan, D. and Weinberg, R.A. 2000. The hallmarks of cancer. *Cell* **100:** 57–70.

Hayward, P., Kalmar, T., and Martinez Arias, A. 2008. Wnt/Notch signalling and information processing during development. *Development* **135:** 411–424.

Herranz, N., Pasini, D., Díaz, V.M., Francí, C., Gutierrez, A., Dave, N., Escrivà, M., Hernandez-Muñoz, I., Di Croce, L., Helin, K., et al. 2008. Polycomb complex 2 is required for E-cadherin repression by the Snail1 transcription factor. *Mol. Cell. Biol.* **28:** 4772–4781.

Issaeva, I., Zonis, Y., Rozovskaia, T., Orlovsky, K., Croce, C.M., Nakamura, T., Mazo, A., Eisenbach, L., and Canaani, E. 2007. Knockdown of ALR (MLL2) reveals ALR target genes and leads to alterations in cell adhesion and growth. *Mol. Cell. Biol.* **27:** 1889–1903.

Itoh, N. and Ornitz, D.M. 2008. Functional evolutionary history of the mouse *Fgf* gene family. *Dev. Dyn.* **237:** 18–27.

Iwase, S., Lan, F., Bayliss, P., de la Torre-Ubieta, M., Qi, H.H., Whetstine, J.R., Bonni, A., Roberts, T.M., and Shi, Y. 2007. The X-linked mental retardation gene *SMCX/JARID1C* defines a family of histone H3 lysine 4 demethylases. *Cell* **128:** 1077–1088.

Jacobs, J.J., Kieboom, K., Marino, S., DePinho, R.A., and van Lohuizen, M. 1999. The oncogene and Polycomb-group gene *bmi-1* regulates cell proliferation and senescence through the

ink4a locus. *Nature* **397:** 164–168.

Kagey, M.H., Melhuish, T.A., and Wotton, D. 2003. The polycomb protein Pc2 is a SUMO E3. *Cell* **113:** 127–137.

Kagey, M.H., Melhuish, T.A., Powers, S.E., and Wotton, D. 2005. Multiple activities contribute to Pc2 E3 function. *EMBO J.* **24:** 108–119.

Keller, G. 2005. Embryonic stem cell differentiation: Emergence of a new era in biology and medicine. *Genes Dev.* **19:** 1129–1155.

Kennison, J.A. 1995. The Polycomb and trithorax group proteins of *Drosophila*: Trans-regulators of homeotic gene function. *Annu. Rev. Genet.* **29:** 289–303.

Kiefer, J.C. 2007. Back to basics: *Sox* genes. *Dev. Dyn.* **236:** 2356–2366.

Kitisin, K., Saha, T., Blake, T., Golestaneh, N., Deng, M., Kim, C., Tang, Y., Shetty, K., Mishra, B., and Mishra, L. 2007. Tgf-β signaling in development. *Sci. STKE* **2007:** cm1.

Kleer, C.G., Cao, Q., Varambally, S., Shen, R., Ota, I., Tomlins, S.A., Ghosh, D., Sewalt, R.G., Otte, A.P., Hayes, D.F., et al. 2003. EZH2 is a marker of aggressive breast cancer and promotes neoplastic transformation of breast epithelial cells. *Proc. Natl. Acad. Sci.* **100:** 11606–11611.

Klose, R.J., Yan, Q., Tothova, Z., Yamane, K., Erdjument-Bromage, H., Tempst, P., Gilliland, D.G., Zhang, Y., and Kaelin Jr., W.G. 2007. The retinoblastoma binding protein RBP2 is an H3K4 demethylase. *Cell* **128:** 889–900.

Kondo, Y., Shen, L., Cheng, A.S., Ahmed, S., Boumber, Y., Charo, C., Yamochi, T., Urano, T., Furukawa, K., Kwabi-Addo, B., et al. 2008. Gene silencing in cancer by histone H3 lysine 27 trimethylation independent of promoter DNA methylation. *Nat. Genet.* **40:** 741–750.

Kuzmichev, A., Nishioka, K., Erdjument-Bromage, H., Tempst, P., and Reinberg, D. 2002. Histone methyltransferase activity associated with a human multiprotein complex containing the Enhancer of Zeste protein. *Genes Dev.* **16:** 2893–2905.

Kuzmichev, A., Jenuwein, T., Tempst, P., and Reinberg, D. 2004. Different ezh2-containing complexes target methylation of histone h1 or nucleosomal histone H3. *Mol. Cell* **14:** 183–193.

Lang, D., Powell, S.K., Plummer, R.S., Young, K.P., and Ruggeri, B.A. 2007. PAX genes: Roles in development, pathophysiology, and cancer. *Biochem. Pharmacol.* **73:** 1–14.

Laverriere, A.C., MacNeill, C., Mueller, C., Poelmann, R.E., Burch, J.B., and Evans, T. 1994. GATA-4/5/6, a subfamily of three transcription factors transcribed in developing heart and gut. *J. Biol. Chem.* **269:** 23177–23184.

Lavigne, M., Francis, N.J., King, I.F., and Kingston, R.E. 2004. Propagation of silencing: Recruitment and repression of naive chromatin in *trans* by polycomb repressed chromatin. *Mol. Cell* **13:** 415–425.

Lee, T.I., Jenner, R.G., Boyer, L.A., Guenther, M.G., Levine, S.S., Kumar, R.M., Chevalier, B., Johnstone, S.E., Cole, M.F., Isono, K., et al. 2006. Control of developmental regulators by Polycomb in human embryonic stem cells. *Cell* **125:** 301–313.

Lee, M.G., Norman, J., Shilatifard, A., and Shiekhattar, R. 2007a. Physical and functional association of a trimethyl H3K4 demethylase and Ring6a/MBLR, a polycomb-like protein. *Cell* **128:** 877–887.

Lee, M.G., Villa, R., Trojer, P., Norman, J., Yan, K.P., Reinberg, D., Di Croce, L., and Shiekhattar, R. 2007b. Demethylation of H3K27 regulates polycomb recruitment and H2A ubiquitination. *Science* **318:** 447–450.

Loh, Y.H., Zhang, W., Chen, X., George, J., and Ng, H.H. 2007. Jmjd1a and Jmjd2c histone H3 Lys 9 demethylases regulate self-renewal in embryonic stem cells. *Genes Dev.* **21:** 2545–2557.

Lowe, S.W. and Sherr, C.J. 2003. Tumor suppression by *Ink4a-Arf*: Progress and puzzles. *Curr. Opin. Genet. Dev.* **13:** 77–83.

McGarvey, K.M., Greene, E., Fahrner, J.A., Jenuwein, T., and Baylin, S.B. 2007. DNA methylation and complete transcriptional silencing of cancer genes persist after depletion of EZH2. *Cancer Res.* **67:** 5097–5102.

Meissner, A., Mikkelsen, T.S., Gu, H., Wernig, M., Hanna, J., Sivachenko, A., Zhang, X., Bernstein, B.E., Nusbaum, C., Jaffe, D.B., et al. 2008. Genome-scale DNA methylation maps of pluripotent and differentiated cells. *Nature* **454:** 766–770.

Mikkelsen, T.S., Ku, M., Jaffe, D.B., Issac, B., Lieberman, E., Giannoukos, G., Alvarez, P., Brockman, W., Kim, T.K., Koche, R.P., et al. 2007. Genome-wide maps of chromatin state in pluripotent and lineage-committed cells. *Nature* **448:** 553–560.

Mohn, F., Weber, M., Rebhan, M., Roloff, T.C., Richter, J., Stadler, M.B., Bibel, M., and Schubeler, D. 2008. Lineage-specific polycomb targets and de novo DNA methylation define restriction and potential of neuronal progenitors. *Mol. Cell* **30:** 755–766.

Molofsky, A.V., He, S., Bydon, M., Morrison, S.J., and Pardal, R. 2005. Bmi-1 promotes neural stem cell self-renewal and neural development but not mouse growth and survival by repressing the p16^Ink4a and p19^Arf senescence pathways. *Genes Dev.* **19:** 1432–1437.

Montgomery, N.D., Yee, D., Chen, A., Kalantry, S., Chamberlain, S.J., Otte, A.P., and Magnuson, T. 2005. The murine polycomb group protein Eed is required for global histone H3 lysine-27 methylation. *Curr. Biol.* **15:** 942–947.

Müller, J., Hart, C.M., Francis, N.J., Vargas, M.L., Sengupta, A., Wild, B., Miller, E.L., O'Connor, M.B., Kingston, R.E., and Simon, J.A. 2002. Histone methyltransferase activity of a *Drosophila* Polycomb group repressor complex. *Cell* **111:** 197–208.

O'Carroll, D., Erhardt, S., Pagani, M., Barton, S.C., Surani, M.A., and Jenuwein, T. 2001. The *Polycomb*-group gene *Ezh2* is required for early mouse development. *Mol. Cell. Biol.* **21:** 4330–4336.

Ohm, J.E., McGarvey, K.M., Yu, X., Cheng, L., Schuebel, K.E., Cope, L., Mohammad, H.P., Chen, W., Daniel, V.C., Yu, W., et al. 2007. A stem cell-like chromatin pattern may predispose tumor suppressor genes to DNA hypermethylation and heritable silencing. *Nat. Genet.* **39:** 237–242.

Orlando, V., Jane, E.P., Chinwalla, V., Harte, P.J., and Paro, R. 1998. Binding of trithorax and Polycomb proteins to the bithorax complex: Dynamic changes during early *Drosophila* embryogenesis. *EMBO J.* **17:** 5141–5150.

Pasini, D., Bracken, A.P., and Helin, K. 2004a. Polycomb group proteins in cell cycle progression and cancer. *Cell Cycle* **3:** 396–400.

Pasini, D., Bracken, A.P., Jensen, M.R., Denchi, E.L., and Helin, K. 2004b. Suz12 is essential for mouse development and for EZH2 histone methyltransferase activity. *EMBO J.* **23:** 4061–4071.

Pasini, D., Bracken, A.P., Hansen, J.B., Capillo, M., and Helin, K. 2007. The polycomb group protein Suz12 is required for embryonic stem cell differentiation. *Mol. Cell. Biol.* **27:** 3769–3779.

Pasini, D., Hansen, K.H., Christensen, J., Agger, K., Cloos, P.A., and Helin, K. 2008. Coordinated regulation of transcriptional repression by the RBP2 H3K4 demethylase and Polycomb-Repressive Complex 2. *Genes Dev.* **22:** 1345–1355.

Plageman, Jr., T.F. and Yutzey, K.E. 2005. T-box genes and heart development: Putting the "T" in heart. *Dev. Dyn.* **232:** 11–20.

Platero, J.S., Hartnett, T., and Eissenberg, J.C. 1995. Functional analysis of the chromo domain of HP1. *EMBO J.* **14:** 3977–3986.

Puschendorf, M., Terranova, R., Boutsma, E., Mao, X., Isono, K., Brykczynska, U., Kolb, C., Otte, A.P., Koseki, H., Orkin, S.H., et al. 2008. PRC1 and Suv39h specify parental asymmetry at constitutive heterochromatin in early mouse embryos. *Nat. Genet.* **40:** 411–420.

Riising, E.M., Boggio, R., Chiocca, S., Helin, K., and Pasini, D. 2008. The polycomb repressive complex 2 is a potential target of SUMO modifications. *PLoS ONE* **3:** e2704.

Ringrose, L. and Paro, R. 2004. Epigenetic regulation of cellular memory by the Polycomb and Trithorax group proteins. *Annu. Rev. Genet.* **38:** 413–443.

Ringrose, L. and Paro, R. 2007. Polycomb/Trithorax response elements and epigenetic memory of cell identity. *Development* **134:** 223–232.

Ringrose, L., Rehmsmeier, M., Dura, J.M., and Paro, R. 2003.

Genome-wide prediction of Polycomb/Trithorax response elements in *Drosophila melanogaster*. *Dev. Cell* **5:** 759–771.

Schlesinger, Y., Straussman, R., Keshet, I., Farkash, S., Hecht, M., Zimmerman, J., Eden, E., Yakhini, Z., Ben-Shushan, E., Reubinoff, B.E., et al. 2007. Polycomb-mediated methylation on Lys27 of histone H3 pre-marks genes for de novo methylation in cancer. *Nat. Genet.* **39:** 232–236.

Sherr, C.J. 1998. Tumor surveillance via the ARF-p53 pathway. *Genes Dev.* **12:** 2984–2991.

Shi, Y., Lan, F., Matson, C., Mulligan, P., Whetstine, J.R., Cole, P.A., Casero, R.A., and Shi, Y. 2004. Histone demethylation mediated by the nuclear amine oxidase homolog LSD1. *Cell* **119:** 941–953.

Simon, J.A. and Tamkun, J.W. 2002. Programming off and on states in chromatin: Mechanisms of Polycomb and trithorax group complexes. *Curr. Opin. Genet. Dev.* **12:** 210–218.

Squazzo, S.L., O'Geen, H., Komashko, V.M., Krig, S.R., Jin, V.X., Jang, S.W., Margueron, R., Reinberg, D., Green, R., and Farnham, P.J. 2006. Suz12 binds to silenced regions of the genome in a cell-type-specific manner. *Genome Res.* **16:** 890–900.

Stock, J.K., Giadrossi, S., Casanova, M., Brookes, E., Vidal, M., Koseki, H., Brockdorff, N., Fisher, A.G., and Pombo, A. 2007. Ring1-mediated ubiquitination of H2A restrains poised RNA polymerase II at bivalent genes in mouse ES cells. *Nat. Cell Biol.* **9:** 1428–1435.

Su, I.H., Basavaraj, A., Krutchinsky, A.N., Hobert, O., Ullrich, A., Chait, B.T., and Tarakhovsky, A. 2003. Ezh2 controls B cell development through histone H3 methylation and Igh rearrangement. *Nat. Immunol.* **4:** 124–131.

Su, I.H., Dobenecker, M.W., Dickinson, E., Oser, M., Basavaraj, A., Marqueron, R., Viale, A., Reinberg, D., Wulfing, C., and Tarakhovsky, A. 2005. Polycomb group protein ezh2 controls actin polymerization and cell signaling. *Cell* **121:** 425–436.

van der Lugt, N.M., Domen, J., Linders, K., van Roon, M., Robanus-Maandag, E., te Riele, H., van der Valk, M., Deschamps, J., Sofroniew, M., van Lohuizen, M., et al. 1994. Posterior transformation, neurological abnormalities, and severe hematopoietic defects in mice with a targeted deletion of the *bmi-1* proto-oncogene. *Genes Dev.* **8:** 757–769.

van der Stoop, P., Boutsma, E.A., Hulsman, D., Noback, S., Heimerikx, M., Kerkhoven, R.M., Voncken, J.W., Wessels, L.F., and van Lohuizen, M. 2008. Ubiquitin E3 ligase Ring1b/Rnf2 of polycomb repressive complex 1 contributes to stable maintenance of mouse embryonic stem cells. *PLoS ONE* **3:** e2235.

Vire, E., Brenner, C., Deplus, R., Blanchon, L., Fraga, M., Didelot, C., Morey, L., Van Eynde, A., Bernard, D., Vanderwinden, J.M., et al. 2006. The Polycomb group protein EZH2 directly controls DNA methylation. *Nature* **439:** 871–874.

Voncken, J.W., Roelen, B.A., Roefs, M., de Vries, S., Verhoeven, E., Marino, S., Deschamps, J., and van Lohuizen, M. 2003. *Rnf2 (Ring1b)* deficiency causes gastrulation arrest and cell cycle inhibition. *Proc. Natl. Acad. Sci.* **100:** 2468–2473.

Wang, H., Wang, L., Erdjument-Bromage, H., Vidal, M., Tempst, P., Jones, R.S., and Zhang, Y. 2004. Role of histone H2A ubiquitination in Polycomb silencing. *Nature* **431:** 873–878.

Wellik, D.M. 2007. *Hox* patterning of the vertebrate axial skeleton. *Dev. Dyn.* **236:** 2454–2463.

Widschwendter, M., Fiegl, H., Egle, D., Mueller-Holzner, E., Spizzo, G., Marth, C., Weisenberger, D.J., Campan, M., Young, J., Jacobs, I., and Laird, P.W. 2007. Epigenetic stem cell signature in cancer. *Nat. Genet.* **39:** 157–158.

Yamane, K., Tateishi, K., Klose, R.J., Fang, J., Fabrizio, L.A., Erdjument-Bromage, H., Taylor-Papadimitriou, J., Tempst, P., and Zhang, Y. 2007. PLU-1 is an H3K4 demethylase involved in transcriptional repression and breast cancer cell proliferation. *Mol. Cell* **25:** 801–812.

Zhou, W., Zhu, P., Wang, J., Pascual, G., Ohgi, K.A., Lozach, J., Glass, C.K., and Rosenfeld, M.G. 2008. Histone H2A monoubiquitination represses transcription by inhibiting RNA polymerase II transcriptional elongation. *Mol. Cell* **29:** 69–80.

Epigenetic Inheritance and Reprogramming in Plants and Fission Yeast

R.A. Martienssen, A. Kloc, R.K. Slotkin, and M. Tanurdžić

Cold Spring Harbor Laboratory, Cold Spring Harbor, New York 11724

Plants and fission yeast exhibit a wealth of epigenetic phenomena, including transposon regulation, heterochromatic silencing, and gene imprinting. They provide excellent model organisms to address the question of how epigenetic information is propagated to daughter cells. We have addressed the questions of establishment, maintenance, and inheritance of heterochromatic silencing using the fission yeast *Schizosaccharomyces pombe* and the plant *Arabidopsis thaliana* by using a variety of genetic and genomic approaches. We present here results showing the cell cycle dependence of RNA in fission yeast RNA interference (RNAi), which is required for proper transcriptional silencing of the centromeric heterochromatin, and that this process occurs during S phase, allowing for precise copying and reestablishment of heterochromatic histone modifications following DNA replication and cell division. We also show that in plants, cells in culture and male germ-line cells undergo massive epigenomic changes correlated with the appearance of a novel class of 21-nucleotide small interfering RNA (siRNA) from transcriptionally reactivated transposable elements (TEs) following loss of heterochromatic DNA and histone methylation. We propose a model for the role of deliberate TE reactivation in germ-line companion cells as part of a developmental mechanism for first revealing and then silencing TEs via small RNA, which may contribute to reprogramming during early development in plants and animals.

Eukaryotic genomes are traditionally characterized as euchromatic or heterochromatic, originally based on their staining patterns (Heitz 1928). These patterns correspond to domains of different chromatin compaction levels, with euchromatin being the less compact of the two. Heterochromatin formation and maintenance have important roles in genome stability and centromere function as well as transcriptional silencing of various TEs and position effect variegation in eukaryotes, X chromosome inactivation in mammals, and gene imprinting in plants and animals (Lippman and Martienssen 2004; Henderson and Jacobsen 2007; Zaratiegui et al. 2007). The two chromatin states also differ in their content and transcriptional activity: Euchromatin is largely composed of genes, whereas heterochromatin is composed of various repeats, including all classes of TEs (Lippman and Martienssen 2004). The difference in composition of the two compartments is reflected in their transcriptional activity: Euchromatin is traditionally viewed as the active portion of the genome, transcribed in response to regulatory cues, whereas heterochromatin is considered transcriptionally silenced, although, paradoxically, it must be transcribed in order to be silenced (Volpe et al. 2002).

The molecular and cytological distinctions of the two chromatin compartments are established by a combination of DNA and histone modifications (Martienssen and Colot 2001; Henderson and Jacobsen 2007). In both plants and fission yeast, heterochromatic DNA is associated with a histone octamer carrying a dimethylation mark at lysine 9 on the amino terminus of histone H3 (H3K9me2). In plants, and *Arabidopsis* in particular, heterochromatic DNA is also heavily methylated, with 5-methyl cytosine (5mC) found in all three possible sequence contexts found in plants: CG, CHG, and CHH (Henderson and Jacobsen 2007). The targeting of heterochromatic DNA and histone modifications is achieved through the action of RNAi, which provides the sequence specificity necessary to guide the silencing complexes containing DNA and histone modifiers (Zaratiegui et al. 2007).

EPIGENETIC INHERITANCE IN FISSION YEAST

The three centromeres of the fission yeast *S. pombe* have been an excellent system to study the molecular mechanisms that establish heterochromatic silencing and its inheritance upon cell division. Their repetitive structure is very similar, but simplified, compared to higher eukaryotic centromeres (Smith and Corces 1995). The central centromeric region is the site of kinetochore binding that is flanked by innermost repeats which, in turn, are flanked by the outermost repeats consisting of *dg* and *dh* repeat sequences. The *dg* and *dh* repeats are enriched in H3K9me2 (Volpe et al. 2003), which is bound by the fission yeast homolog of the heterochromatic protein 1 (HP1) Swi6 (Holm et al. 2005). This, in turn, is necessary for proper attachments of the kinetochore (Nonaka et al. 2002). Targeting of the heterochromatic chromatin modifications is accomplished by RNAi (Volpe et al. 2002).

RNAi mechanisms take advantage of the *dh* and *dg* transcripts being generated from both DNA strands by RNA polymerase II (PolII) (Volpe et al. 2002). The sole fission yeast Argonaute protein Ago1 cotranscriptionally slices the nascent heterochromatic transcripts (Irvine et al. 2006). This is hypothesized to create sites for the RNA-dependent RNA polymerase (RDRC) complex, which results in formation of double-stranded RNA molecules. These are quickly processed into 21–24-nucleotide siRNA molecules by the Dicer protein Dcr1 (Colmenares et al. 2007). The siRNA molecules then serve as guides when they are bound by Ago1 and incorporated into the RNA-induced transcriptional silencing (RITS) complex (Verdel et al. 2004). This is required for the final step of heterochromatin formation, whereby the RITS complex interacts, through an unknown mechanism, with the Rik1 complex that contains the histone methyltransferase Clr4 responsible for H3K9me2 methylation (Horn et al. 2005). Ultimately, H3K9me2 is recognized by the chromodomain-containing Swi6, which itself is bound by cohesins (Holm et al. 2005), ensuring the proper functioning of fission yeast centromeres and chromosomal segregation during cell division.

Heterochromatin and the Cell Cycle in Fission Yeast

Given the basic premises of chromatin replication, it is clear that the need for the RNAi-targeted heterochromatin formation process differs between the stages of the cell cycle. It also poses a problem for heterochromatin inheritance, following cell division, because, in the case of fission yeast, the epigenetic information is carried by the histone marks that themselves are not inherited (Kloc and Martienssen 2008).

We have addressed the question of heterochromatin inheritance in fission yeast by profiling *dg* and *dh* transcripts, siRNAs, and histone modifications throughout the cell cycle. Contrary to most eukaryotes, fission yeast heterochromatin is known to replicate during early S phase (Kim and Huberman 2001). This is precisely the time in the cell cycle when the *dg* and *dh* centromeric transcripts can first be detected in cells arrested in early S phase (Kloc et al. 2008), concomitant with the localization of PolII in the same regions (Chen et al. 2008). The *dg* transcripts often appear slightly ahead of the *dh* transcripts, which may be due to the two repeats being located at different distances from the origins of replication. At the same time, H3K9me2 local concentration is at its lowest at the same time in the cell cycle. With the progression of heterochromatic DNA replication past early S phase, an increase in the RNAi factors can be seen, as well as the appearance of heterochromatic siRNA. This indicates that RNAi occurs specifically in the S phase (Kloc et al. 2008). Consequently, this is followed by an increase in H3K9me2, upon the finish of DNA replication, and the faithful reestablishment of heterochromatin. The H3K9me2 levels further increase, peaking during the G_2 phase of the cell cycle.

The loss of Swi6 binding in heterochromatic repeats is the earliest event detected so far that leads to cell-cycle-specific phasing in heterochromatin transcriptional activity, RNAi activity, and histone modification. Swi6 binding depends on the presence of H3K9me2; however, Swi6–H3K9me2 interaction is further dependent on the so-called phosphomethyl switch (Fischle et al. 2003). This term refers to the presence or absence of a phosphate group on histone H3 serine10 (H3S10p), located directly next to the lysine 9 residue carrying the dimethylation modification. In the presence of H3S10p, the Swi6–H3K9me2 interaction is abolished and the transcriptional silencing of the region is lost (Chen et al. 2008; Kloc et al. 2008). Phosphorylation of H3S10 is under the control of the aurora-B kinase Ark1.

Taken together, these results point to a cell-cycle-regulated and coordinated series of molecular events that ensure that epigenetic information contained within parental chromosomes is faithfully transmitted to daughter chromosomes, even in the absence of a directly heritable epigenetic mark such as DNA methylation.

EPIGENETIC INHERITANCE IN PLANTS

Plants offer the full palette of epigenetic modifications, both at the level of DNA methylation and numerous histone modifications. This potential for epigenetic control in plants impacts plant development, responses to biotic and abiotic stress, and integration of environmental and ontogenic cues in the transition to flowering (Henderson and Jacobsen 2007). Perhaps one of the best-studied epigenetic systems in plants is heterochromatin regulation (Lippman and Martienssen 2004).

Heterochromatic Changes in Proliferating Immortalized Plant Cells

Plant cell culture is a known source of induced genetic and epigenetic variation (Hirochika 1993; Kaeppler et al. 2000). A number of plant species when grown in culture undergo transposon reactivation and a number of genes undergo epigenetic changes that lead to epiallele formation (Grandbastien et al. 1989; Kaeppler et al. 2000). This represents a good model for studying the epigenetic mechanisms involved in establishment and maintenance of heterochromatin in a higher eukaryote such as *Arabidopsis*.

We first examined a compendium of gene expression profiles from various organs and cell types from *Arabidopsis*, including a long-term cell suspension culture (Schmid et al. 2005). All of the gene expression profiles contained within this set were created using the Affymetrix ATH1 gene expression platform. This array contains about 1200 specific oligonucleotide probes matching various copies of transposable elements in the *Arabidopsis* genome. This analysis showed that a large number of TEs (mainly those that represent relatively intact copies in the genome) become reactivated in cell suspension culture, whereas TEs remain transcriptionally silenced in *Arabidopsis* leaves and callus samples.

Transcriptional reactivation of TEs in plants is accompanied by changes in epigenetic chromatin modifications. This can be accomplished by down-regulation or by mutations in key regulators of heterochromatin: DNA methyltransferases (*MET1*, *CHROMOMETHYLTRANSFERASE3* [*CMT3*], and *DOMAINS REARRANGED METHYLATRANSFERASE1* and *2* [*DRM1* and *DRM2*], chromatin remodeler *DECREASE IN DNA METHYLATION1* [*DDM1*], histone methyltransferases, in particular the major H3K9me2 methyltransferase KRYPTONITE (KYP), as well as some components of the RNAi pathway (Lippman et al. 2003). The *ddm1* mutation has been particularly well studied in the context of TE regulation, because most TEs in the *Arabidopsis* genome become reactivated in this background (Singer et al. 2001; Lippman et al. 2003). This is accompanied by a nondiscriminating loss of heterochromatic DNA and H3K9me2 methylation (Lippman et al. 2004).

We therefore analyzed the chromatin states of immortalized *Arabidopsis* cells grown in a long-term suspension culture using a custom-made tiling microarray of *Arabidopsis* chromosome 4 to determine the DNA and histone methylation patterns (http://chromatin.cshl.edu/epiculture/). As predicted, we detected significant hypomethylation of heterochromatic DNA and loss of H3K9me2, often replaced by the euchromatic histone modification H3K4me2. Interestingly, these changes only affected some families of TEs, whereas other TEs, often spatially juxtaposed to the hypomethylated TEs, remained decorated with heterochromatic marks. Our experiments also revealed that these patterns are stable and mitotically heritable throughout many cell divisions. Because loss of heterochromatic marks was specific to some but not all TE families, we postulated that this was not an effect of transcriptional down-regulation of numerous chromatin regulators. Indeed, expression profiling revealed that all of the known chromatin regulators are overexpressed in cell culture, including all classes of DNA and histone methyltransferases and proteins involved in RNAi. The only protein whose expression was lost in cell culture was ROS1, a DNA demethylase (Zhu et al. 2007). We then used next-generation Illumina DNA sequencing to profile small RNA molecules in this sample and compared this profile to the small RNA profiles of young rosette leaves and freshly induced callus samples. This analysis revealed drastic shifts in siRNA abundance and length specific to those TE families that lost heterochromatic marks, with a shift from the predominant 24-nucleotide-size class to the 21-nucleotide size class whose production is under different genetic control (Fig. 1) (Brodersen and Voinnet 2006). This change was particularly striking for the demethylated and reactivated family of *Athila* retrotransposons (Fig 1). It therefore seems that this failure to specifically target appropriate heterochromatic marks upon cell division is due to perturbations in the RNAi systems used to target heterochromatic modifications. However, the fact that the heterochromatic state of other TE families, such as *Atlantys*, is faithfully reproduced over many cell divisions reflects the general availability of the core RNAi machinery and other chromatin regulators. The loss of heterochromatic modification was also detected in short regions of homology with TEs such as short interspersed repetitive elements (SINEs). This can have a particularly significant effect on gene regulation because SINEs are often found in the vicinity of genes. In the case of the imprinted gene *FWA* (Kinoshita et al. 2004), the SINE known to be the imprinting site in planta, becomes hypomethylated in cell culture, with a 25-fold increase in transcription at the *FWA* locus.

Concurrently with the loss of heterochromatic silencing in TEs, we also detected genic hypermethylation in these immortalized plant cells. Although the majority of hypermethylation is located in the coding regions and corresponds to an increase in the novel gene body CpG methylation (Vaughn et al. 2007), a number of genes undergo epiallele formation in these cells. This was evidenced by the concomitant appearance of novel DNA and H3K9me2 methylation. In addition, close to 90% of these new epialleles, including an epiallele at the *AGAMOUS* (*AG*) locus, a gene known to be prone to epiallele formation (Jacobsen et al. 2000), also have novel 24-nucleotide

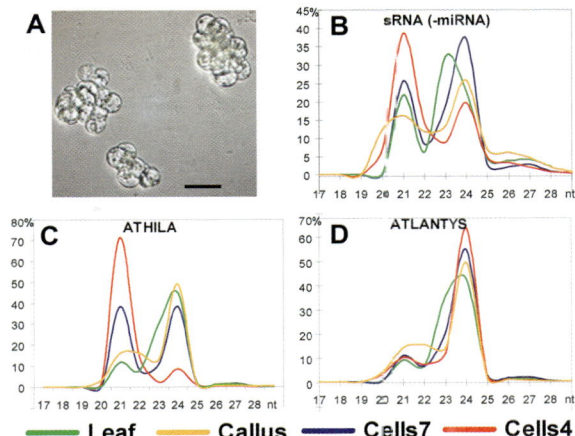

Figure 1. Size distribution of siRNA sequences in plant cell suspension culture (*A*), wild-type (Col) leaves, and wild-type (Col) callus. The relative frequencies of heterochromatic small RNAs (*B*) and those matching exclusively *Arabidopsis* retrotransposon families *Athila* (*C*) and *Atlantys* (*D*) were plotted according to size in cell culture 4 days and 7 days after transfer to fresh medium. The 21-nucleotide siRNAs accumulate only in cultured cells but not wild-type leaf or callus samples and correspond to *Athila* elements (which lose methylation), but not to *Atlantys* elements (which do not). (Modified from Tanurdžić et al. 2008.)

siRNA associated with them in cell culture. It is tempting to speculate that the appearance of novel epialleles in cell culture may be, at least partially, due to the inability of the cells to prune RNA-directed DNA methylation due to transcriptional down-regulation of *ROS1* (Zhu et al. 2007). Intriguingly, the proximal *ROS1* promoter contains a fragment of a helitron TE that becomes hypermethylated in cell culture, and thus, *ROS1* may itself be prone to epiallele formation.

Epigenetic Reprogramming and Plant Germ Cells

In addition to TE reactivation in cell culture, reexamination of previous *Arabidopsis* expression profiling data (Schmid et al. 2005) revealed that the transcriptional reactivation of TEs also occurs in mature wild-type pollen (Slotkin et al. 2009). Unlike animals, plant meiocytes differentiate from adult somatic cells late in development. The pollen grain is produced from two subsequent mitotic divisions that occur after meiosis, so that pollen contains three haploid cells. The largest is the pollen vegetative cell, controlled by the pollen vegetative nucleus, which contains decondensed chromatin. Enclosed within the pollen vegetative cell are two sperm cells with compact chromatin. The vegetative nucleus does not contribute genetic material to the next generation, whereas the two sperm cells are involved in the double fertilization process typical of most seed plants, providing paternal genetic and epigenetic contributions to the zygote and the endosperm.

Our analysis of TE transcriptional activity of pollen using *Arabidopsis* microarray data (Becker et al. 2003; Honys and Twell 2004; Pina et al. 2005; Schmid et al.

2005) showed that although TEs are generally silenced throughout the plant, they are up-regulated in pollen. This co-coordinated TE up-regulation was not dependent on the chromosomal position of TEs nor was it specific to particular families of TEs. Therefore, we concluded that TE transcriptional reactivation in pollen is due to a loss of particular *trans*-acting silencing factors.

Using gene-trap and enhancer-trap reporter lines inserted into TEs (Sundaresan et al. 1995), we assayed the precise location and timing of TE reactivation in the pollen grain. As expected, no GUS staining was detected in the plant body; however, strong GUS staining was detected in mature pollen for all 24 different reporter lines with insertion in various TEs throughout the *Arabidopsis* genome. Furthermore, upon closer inspection, we were able to confirm that the positive GUS-staining signal in pollen was in the pollen grain cytoplasm, whose content is under the control of the vegetative nucleus (VN). In contrast, no GUS accumulation was detected in the sperm cells (SCs). We further tested the hypothesis that TE reactivation is specific to the VN by looking at novel, heritable transposition events using transposon display. We could detect numerous transposition events in the pollen DNA itself, but we were not able to detect any novel transposition in 500 F_1 plants. This evidence also supports the notion that pollen TE reactivation is specific to the VN and does not occur in the SC. In addition, microarray data and reverse transcriptase–polymerase chain reaction (RT-PCR) both demonstrate that TEs are not expressed in SCs purified from the pollen grain using fluorescence-activated cell sorting (FACS) (Borges et al. 2008).

What is the extent, then, of epigenetic changes in the VN chromatin that allow transcriptional TE reactivation? We assayed DNA methylation by sequence analysis of bisulfite-converted DNA from pollen as well as purified SCs using TE-specific primers in PCR amplification. The results of this experiment showed that, on average, TEs are hypomethylated in the genomic DNA isolated from pollen, retaining some DNA methylation. In particular, methylation seems to be lost from the asymmetric methylation sites, which is targeted by siRNA. In contrast, SC DNA contains fully methylated cytosines in both symmetric and asymmetric sequence contexts. This result implicated a VN-specific down-regulation of one or more genes required for heterochromatic silencing in *Arabidopsis* and may be accompanied by active DNA demethylation specifically targeted to the VN. RT-PCR analysis showed that of several candidate genes known to regulate TE and heterochromatic silencing, genes involved in the production of the 24-nucleotide siRNA (particularly *RNA DEPENDENT RNA POLYMERASE2* [*RDR2*] and *DICER-LIKE3* [*DCL3*]) are down-regulated in pollen, whereas the *trans*-acting siRNA and microRNA (miRNA)-generating pathways (21-nucleotide *trans*-acting siRNA and miRNA) seemed to be fully functional. Importantly, we also showed that the master regulator of heterochromatic silencing, DDM1 protein, does not accumulate in the VN but does in the SCs.

We investigated the role of siRNA in targeting heterochromatic silencing in the male germ line by high-throughput sequencing of small RNA molecules from pollen and isolated SCs and compared them to small RNA data sets from wild-type *Arabidopsis* leaves and inflorescences as well as *ddm1* inflorescences. When the small RNA sequences are aligned to the reference genome, a striking similarity in the patterns and size of matching siRNAs is seen between the two samples with epigenetically misregulated TEs: *ddm1* plants and pollen (Fig. 2). Most TEs, such as *AtMu1*, seem to lose or have significantly reduced levels of the heterochromatic 24-nucleotide siRNA in both *ddm1* (Lippman et al. 2003) and pollen. In addition, the largest family of TEs in the *Arabidopsis* genome, the *Athila* retrotransposons, show a massive increase in the 21-nucleotide siRNA in both samples (Fig. 2). We therefore conclude that upon epigenetic reactivation of TEs, a novel class of siRNA, the 21-nucleotide in size, and matching *Athila* TEs, is massively up-regulated.

Interestingly, the 21-nucleotide siRNA from reactivated TEs accumulate in the SC small RNA library (Fig. 2), even though the primary TE transcripts are not found in the SC RNA. This result would imply that the 21-nucleotide siRNA can move between the VN and SC. We tested whether small RNAs could communicate from the VN to the SC by expressing an artificial miRNA (amiRNA) in the VN (Engel et al. 2005; Schwab et al. 2006) and targeting a green fluorescent protein (GFP)

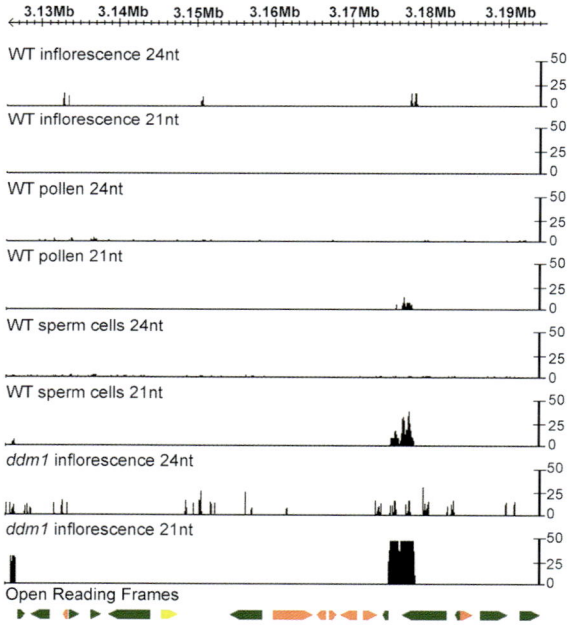

Figure 2. siRNA distribution according to size, frequency, and sequence homology in *Arabidopsis* pollen, sperm cells, and a *ddm1* mutant. Distributions of 21- and 24-nucleotide siRNAs are shown for a 65-kb section of pericentromeric heterochromatin from chromosome 4. Gene and ORF annotations include DNA transposons (*red*), retrotransposons (*green*), and gene (*yellow*). Bars represent copy-corrected counts per read of 21- and 24-nucleotide siRNAs in 100-bp windows for wild-type inflorescence, wild-type pollen, *ddm1* inflorescence, and wild-type sperm cells. In pericentromeric heterochromatin, large peaks of 21-nucleotide siRNAs in *ddm1*, sperm cells, and pollen match *Athila* family retrotransposons.

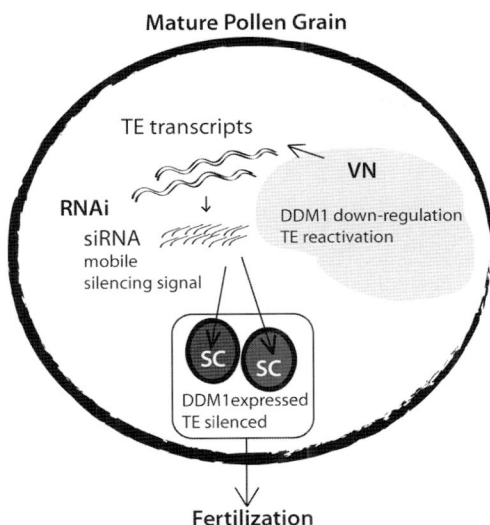

Figure 3. Model of TE reactivation and its role in the reprogramming of the germ line. The pollen VN and SCs differ in DDM1 localization and TE expression. Reactivated TEs, such as *Athila* retrotransposons, from the VN produce 21-nucleotide siRNAs via RNAi. This novel class of siRNA is mobile and accumulates in the SC. Thus, TEs are "revealed" by epigenetic reprogramming of the VN in order to target and silence themselves in the SC.

reporter expressed specifically in the SC (Engel et al. 2005). In the plants homozygous for both the amiRNA and GFP reporter, we could detect almost complete silencing of the GFP reporter in the SC. These data further corroborate our hypothesis that small RNA can move between the two nuclear compartments in pollen and communicate with the silencing signal generated in the VN to the SC to ensure TE silencing and both the genetic and epigenetic stability of germ cells (Fig. 3).

SMALL RNA-MEDIATED TRANSPOSON CONTROL IN ANIMALS AND PROTISTS

Our results may shed light on transposon regulation in animals as well as in plant germ cells and even in protists such as *Tetrahymena* and *Paramecium*. In plants and protists, genomic reprogramming seems to be confined to companion nuclei, such as the vegetative nucleus in pollen and the macronucleus in protists, which undergoes DNA elimination and RNAi and does not contribute to the next generation. In mammals, transposon silencing is transiently lost through extensive reprogramming, both in the germ line and in the early embryo. An ancient conserved mechanism of reprogramming may underlie transposon activation in each case.

Recently, there has been a great deal of interest in germ-line small RNA in *Drosophila* and the mouse. Much of this research has focused on Piwi-interacting RNA (piRNA), a 25–29-nucleotide small RNA associated with the PIWI class of Argonaute proteins (which are apparently not found in plants) (Aravin et al. 2007a; Gunawardane et al. 2007). Although their biogenesis is

still somewhat obscure, it depends on the PIWI proteins themselves (Aravin et al. 2007a; Gunawardane et al. 2007). Retrotransposon targeting is achieved by sequence specificity of piRNAs that arise from genomic locations with sequences containing antisense retrotransposon fragments (Aravin et al. 2007b; Brennecke et al. 2007). In the presence of active transposons, sense piRNAs are also found, which are thought to be derived from the transposon, and may help to amplify the more abundant antisense piRNAs (Gunawardane et al. 2007). It is generally assumed that transposons have evolved to infect the germ line, which is where piRNAs mount a host defense (Sarot et al. 2004; Brennecke et al. 2007).

On the basis of our results in plants, however, it is possible that not only are piRNA expressed in the germ line, but so are the transposons they target, deliberately released by the host to provide a small RNA signal and efficiently silence themselves. In *Drosophila*, somatic follicular cells adjacent to female germ-line stem cells are the site of long terminal repeat (LTR) retrotransposon *Gypsy* expression (Sarot et al. 2004; Brennecke et al. 2007). Furthermore, piRNA-mediated repression of hybrid dysgenesis by non-LTR retrotransposons depends on expression of full-length autonomous elements in nurse cell nuclei adjacent to the oocyte (Chambeyron et al. 2008). It is possible that piRNA or TE primary transcripts are generated in these companion cells and then translocated into the oocyte.

In mammals, deep sequencing of small RNA from oocytes has revealed abundant 25–29-nucleotide piRNA and similarly abundant 21-nucleotide endogenous siRNA (Tam et al. 2008; Watanabe et al. 2008). In these reports, the vast majority of 21-nucleotide siRNAs were from retrotransposons, as we have observed in plant sperm cells (Slotkin et al. 2009). It is not yet known at which stage of germ-line development these siRNAs arise; however, the transient loss of heterochromatin in germ cells around E11.5 (Hajkova et al. 2008) could provide the window of opportunity for retrotransposon transcripts and associated 21-nucleotide siRNA to be produced.

REPROGRAMMING AND CANCER

Reprogramming is not limited to the germ cells. Indeed, similarities in expression profiles among germ cells, stem cells, and cancer cells have, in part, led to the "cancer stem cell" hypothesis (Reya et al. 2001), and TEs are significantly up-regulated in these contexts. Cultured cells in general, and especially cancer cells, lose methylation genome-wide, largely from transposons and repeats. In addition, developmental reprogramming itself, in which a handful of transcription factors induce a pluripotent state, is greatly enhanced by inhibiting DNA methylation genome-wide (Meissner et al. 2008; Mikkelsen et al. 2008). In the developing embryo immediately after fertilization, DNA methylation is lost (passively) from the female pronucleus and (actively) from the male pronucleus, only being regained after differentiation of the trophoblast, which remains hypomethylated (Branco et al. 2008). In both the developing oocyte and the developing embryo, these cycles of demethylation and remethylation depend on down-regulation and mislocalization of the DNA methyltransferases Dnmt1, Dnmt3a, and Dnmt3b, resembling reprogramming in pollen in this respect.

The activation of transposable elements is thus programmed and "intentional" in somatic and germ cells in both plants and animals. Rather than representing a victory of parasite over host, we propose that the transposons are activated transiently at a time when they need to be silenced and that silencing occurs efficiently via RNAi. Importantly, when the RNAi machinery is inactivated in *piwi* mutants in animals or in certain *argonaute* mutants in plants (Nonomura et al. 2007), sterility ensues presumably because of the very significant transposon load. Although certainly speculative, it is hard to ignore the potential for selectively targeting cancer cells by loss of transposon RNAi.

CONCLUSIONS

The genetics and genomics of heterochromatin in fission yeast and plants have expanded our understanding of the establishment, maintenance, and inheritance of epigenetic marks that render heterochromatin transcriptionally silent. Heterochromatin inheritance in fission yeast is tightly coupled to the cell cycle via the phosphomethyl switch, which releases transcriptional repression from heterochromatic repeats. Their transcripts are then rapidly turned over during S phase by RNAi, ensuring that newly synthesized chromatin is once again fully decorated with heterochromatic marks and is returned to transcriptional silencing.

In long-term plant cell culture, similarly to immortalized animal cell lines and cancer cell lines, there is a substantial, yet very localized, loss of heterochromatic silencing specific to families of TEs, in parallel with hypermethylation of gene coding and regulatory regions. The latter represents epigenetic restructuring of the genome in immortalized cells and can lead to mitotically heritable changes in gene expression programs. A novel class of 21-nucleotide siRNA corresponding to the epigenetically reactivated repeats appears in plant cell culture as well as in mutants in *met1* (*Dnmt1*) and *ddm1* (*Lsh1*) and in wild-type pollen.

Down-regulation of chromatin regulators in the pollen VN represents a form of epigenetic reprogramming that leads to reactivation of TEs in companion cells that support the germ cells without contribution to the next generation, revealing the TE content of the parental genome and silencing them in the sperm via mobile siRNA. This likely ensures proper heterochromatic silencing following fertilization.

ACKNOWLEDGMENTS

We thank M.W. Vaughn and W.R. McCombie for their help with high-throughput sequencing and analysis. The work described here is funded by grants from the National Institutes of Health (GM076396) and the National Science Foundation project (DBI-042165) to R.A.M. and a National Institutes of Health postdoctoral fellowship (F32 CA125977) to R.K.S.

REFERENCES

Aravin, A.A., Hannon, G.J., and Brennecke, J. 2007a. The Piwi-piRNA pathway provides an adaptive defense in the transposon arms race. *Science* **318:** 761–764.

Aravin, A.A., Sachidanandam, R., Girard, A., Fejes-Toth, K., and Hannon, G.J. 2007b. Developmentally regulated piRNA clusters implicate MILI in transposon control. *Science* **316:** 744–747.

Becker, J.D., Boavida, L.C., Carneiro, J., Haury, M., and Feijo, J.A. 2003. Transcriptional profiling of *Arabidopsis* tissues reveals the unique characteristics of the pollen transcriptome. *Plant Physiol.* **133:** 713–725.

Borges, F., Gomes, G., Gardner, R., Moreno, N., McCormick, S., Feijo, J.A., and Becker, J.D. 2008. Comparative transcriptomics of *Arabidopsis thaliana* sperm cells. *Plant Physiol.* **148:** 1168–1181.

Branco, M.R., Oda, M., and Reik, W. 2008. Safeguarding parental identity: Dnmt1 maintains imprints during epigenetic reprogramming in early embryogenesis. *Genes Dev.* **22:** 1567–1571.

Brennecke, J., Aravin, A.A., Stark, A., Dus, M., Kellis, M., Sachidanandam, R., and Hannon, G.J. 2007. Discrete small RNA-generating loci as master regulators of transposon activity in *Drosophila*. *Cell* **128:** 1089–1103.

Brodersen, P. and Voinnet, O. 2006. The diversity of RNA silencing pathways in plants. *Trends Genet.* **22:** 268–280.

Chambeyron, S., Popkova, A., Payen-Groschene, G., Brun, C., Laouini, D., Pelisson, A., and Bucheton, A. 2008. piRNA-mediated nuclear accumulation of retrotransposon transcripts in the *Drosophila* female germline. *Proc. Natl. Acad. Sci.* **105:** 14964–14969.

Chen, E.S., Zhang, K., Nicolas, E., Cam, H.P., Zofall, M., and Grewal, S.I.S. 2008. Cell cycle control of centromeric repeat transcription and heterochromatin assembly. *Nature* **451:** 734–737.

Colmenares, S.U., Buker, S.M., Buhler, M., Dlakic, M., and Moazed, D. 2007. Coupling of double-stranded RNA synthesis and siRNA generation in fission yeast RNAi. *Mol. Cell* **27:** 449–461.

Engel, M.L., Holmes-Davis, R., and McCormick, S. 2005. Green sperm. Identification of male gamete promoters in *Arabidopsis*. *Plant Physiol.* **138:** 2124–2133.

Fischle, W., Wang, Y.M., and Allis, C.D. 2003. Binary switches and modification cassettes in histone biology and beyond. *Nature* **425:** 475–479.

Grandbastien, M.A., Spielmann, A., and Caboche, M. 1989. Tnt1, a mobile retroviral-like transposable element of tobacco isolated by plant-cell genetics. *Nature* **337:** 376–380.

Gunawardane, L.S., Saito, K., Nishida, K.M., Miyoshi, K., Kawamura, Y., Nagami, T., Siomi, H., and Siomi, M.C. 2007. A slicer-mediated mechanism for repeat-associated siRNA 5′ end formation in *Drosophila*. *Science* **315:** 1587–1590.

Hajkova, P., Ancelin, K., Waldmann, T., Lacoste, N., Lange, U.C., Cesari, F., Lee, C., Almouzni, G., Schneider, R., and Surani, M.A. 2008. Chromatin dynamics during epigenetic reprogramming in the mouse germ line. *Nature* **452:** 877–881.

Heitz, E. 1928. Das heterochromatin der Moose. *Jahrb. Wiss. Botanik.* **69:** 762–818.

Henderson, I.R. and Jacobsen, S.E. 2007. Epigenetic inheritance in plants. *Nature* **447:** 418–424.

Hirochika, H. 1993. Activation of tobacco retrotransposons during tissue-culture. *EMBO J.* **12:** 2521–2528.

Holm, T.M., Jackson-Grusby, L., Brambrink, T., Yamada, Y., Rideout III, W.M., and Jaenisch, R. 2005. Global loss of imprinting leads to widespread tumorigenesis in adult mice. *Cancer Cell* **8:** 275–285.

Honys, D. and Twell, D. 2004. Transcriptome analysis of haploid male gametophyte development in *Arabidopsis*. *Genome Biol.* **5:** R85.

Horn, P.J., Bastie, J.N., and Peterson, C.L. 2005. A Rik1-associated, cullin-dependent E3 ubiquitin ligase is essential for heterochromatin formation. *Gene Dev* **19:** 1705–1714.

Irvine, D.V., Zaratiegui, M., Tolia, N.H., Goto, D.B., Chitwood, D.H., Vaughn, M.W., Joshua-Tor, L., and Martienssen, R.A. 2006. Argonaute slicing is required for heterochromatic silencing and spreading. *Science* **313:** 1134–1137.

Jacobsen, S.E., Sakai, H., Finnegan, E.J., Cao, X., and Meyerowitz, E.M. 2000. Ectopic hypermethylation of flower-specific genes in *Arabidopsis*. *Curr. Biol.* **10:** 179–186.

Kaeppler, S.M., Kaeppler, H.F., and Rhee, Y. 2000. Epigenetic aspects of somaclonal variation in plants. *Plant Mol. Biol.* **43:** 179–188.

Kim, S.M. and Huberman, J.A. 2001. Regulation of replication timing in fission yeast. *EMBO J.* **20:** 6115–6126.

Kinoshita, T., Miura, A., Choi, Y., Kinoshita, Y., Cao, X., Jacobsen, S.E., Fischer, R.L., and Kakutani, T. 2004. One-way control of FWA imprinting in *Arabidopsis* endosperm by DNA methylation. *Science* **303:** 521–523.

Kloc, A. and Martienssen, R. 2008. RNAi, heterochromatin and the cell cycle. *Trends Genet.* **24:** 511–517.

Kloc, A., Zaratiegui, M., Nora, E., and Martienssen, R. 2008. RNA interference guides histone modification during the S phase of chromosomal replication. *Curr. Biol.* **18:** 490–495.

Lippman, Z. and Martienssen, R. 2004. The role of RNA interference in heterochromatic silencing. *Nature* **431:** 364–370.

Lippman, Z., May, B., Yordan, C., Singer, T., and Martienssen, R. 2003. Distinct mechanisms determine transposon inheritance and methylation via small interfering RNA and histone modification. *PLoS Biol.* **1:** E67.

Lippman, Z., Gendrel, A.V., Black, M., Vaughn, M.W., Dedhia, N., McCombie, W.R., Lavine, K., Mittal, V., May, B., Kasschau, K.D., et al. 2004. Role of transposable elements in heterochromatin and epigenetic control. *Nature* **430:** 471–476.

Martienssen, R.A. and Colot, V. 2001. DNA methylation and epigenetic inheritance in plants and filamentous fungi. *Science* **293:** 1070–1074.

Meissner, A., Mikkelsen, T.S., Gu, H.C., Wernig, M., Hanna, J., Sivachenko, A., Zhang, X.L., Bernstein, B.E., Nusbaum, C., Jaffe, D.B., et al. 2008. Genome-scale DNA methylation maps of pluripotent and differentiated cells. *Nature* **454:** 766–770.

Mikkelsen, T.S., Hanna, J., Zhang, X.L., Ku, M.C., Wernig, M., Schorderet, P., Bernstein, B.E., Jaenisch, R., Lander, E.S., and Meissner, A. 2008. Dissecting direct reprogramming through integrative genomic analysis. *Nature* **454:** 49–55.

Nonaka, N., Kitajima, T., Yokobayashi, S., Xiao, G.P., Yamamoto, M., Grewal, S.I.S., and Watanabe, Y. 2002. Recruitment of cohesin to heterochromatic regions by Swi6/HP1 in fission yeast. *Nat. Cell Biol.* **4:** 89–93.

Nonomura, K., Morohoshi, A., Nakano, M., Eiguchi, M., Miyao, A., Hirochika, H., and Kurata, N. 2007. A germ cell specific gene of the ARGONAUTE family is essential for the progression of premeiotic mitosis and meiosis during sporogenesis in rice. *Plant Cell* **19:** 2583–2594.

Pina, C., Pinto, F., Feijo, J.A., and Becker, J.D. 2005. Gene family analysis of the *Arabidopsis* pollen transcriptome reveals biological implications for cell growth, division control, and gene expression regulation. *Plant Physiol.* **138:** 744–756.

Reya, T., Morrison, S.J., Clarke, M.F., and Weissman, I.L. 2001. Stem cells, cancer, and cancer stem cells. *Nature* **414:** 105–111.

Sarot, E., Payen-Groschene, G., Bucheton, A., and Pelisson, A. 2004. Evidence for a *piwi*-dependent RNA silencing of the *gypsy* endogenous retrovirus by the *Drosophila melanogaster flamenco* gene. *Genetics* **166:** 1313–1321.

Schmid, M., Davison, T.S., Henz, S.R., Pape, U.J., Demar, M., Vingron, M., Scholkopf, B., Weigel, D., and Lohmann, J.U. 2005. A gene expression map of *Arabidopsis thaliana* development. *Nat. Genet.* **37:** 501–506.

Schwab, R., Ossowski, S., Riester, M., Warthmann, N., and Weigel, D. 2006. Highly specific gene silencing by artificial microRNAs in *Arabidopsis*. *Plant Cell* **18:** 1121–1133.

Singer, T., Yordan, C., and Martienssen, R.A. 2001. Robertson's *Mutator* transposons in *A. thaliana* are regulated by the chromatin-remodeling gene *decrease in DNA methylation* (*DDM1*). *Genes Dev.* **15:** 591–602.

Slotkin, R.K., Vaughn, M., Borges, F., Tanurdžić M., Becker, J.D., Feijó, J.A., and Martienssen, R.A. 2009. Epigenetic reprogramming and small RNA silencing of transposable elements in pollen. *Cell* **136:** 461–472.

Smith, P.A. and Corces, V.G. 1995. The *suppressor* of *hairy-wing* protein regulates the tissue-specific expression of the *Drosophila gypsy* retrotransposon. *Genetics* **139:** 215–228.

Sundaresan, V., Springer, P., Volpe, T., Haward, S., Jones, J.D., Dean, C., Ma, H., and Martienssen, R. 1995. Patterns of gene action in plant development revealed by enhancer trap and gene trap transposable elements. *Genes Dev.* **9:** 1797–1810.

Tam, O.H., Aravin, A.A., Stein, P., Girard, A., Murchison, E.P., Cheloufi, S., Hodges, E., Anger, M., Sachidanandam, R., Schultz, R.M., and Hannon, G.J. 2008. Pseudogene-derived small interfering RNAs regulate gene expression in mouse oocytes. *Nature* **453:** 534–538.

Tanurdžić, M., Vaughn, M.W., Jiang, H., Lee, T.-J., Slotkin, R.K., Sosinski, B., Thompson, W.F., Doerge, R.W., and Martienssen, R.A. 2008. Epigenomic consequences of immortalized plant cell suspension culture. *PLoS Biol.* **6:** e302.

Vaughn, M.W., Tanurdžić, M., Lippman, Z., Jiang, H., Carrasquillo, R., Rabinowicz, P.D., Dedhia, N., McCombie, W.R., Agier, N., Bulski, A., et al. 2007. Epigenetic natural variation in *Arabidopsis thaliana*. *PLoS Biol.* **5:** e174.

Verdel, A., Jia, S.T., Gerber, S., Sugiyama, T., Gygi, S., Grewal, S.I.S., and Moazed, D. 2004. RNAi-mediated targeting of heterochromatin by the RITS complex. *Science* **303:** 672–676.

Volpe, T., Schramke, V., Hamilton, G.L., White, S.A., Teng, G., Martienssen, R.A., and Allshire, R.C. 2003. RNA interference is required for normal centromere function in fission yeast. *Chromosome Res.* **11:** 137–146.

Volpe, T.A., Kidner, C., Hall, I.M., Teng, G., Grewal, S.I.S., and Martienssen, R.A. 2002. Regulation of heterochromatic silencing and histone H3 lysine-9 methylation by RNAi. *Science* **297:** 1833–1837.

Watanabe, T., Totoki, Y., Toyoda, A., Kaneda, M., Kuramochi-Miyagawa, S., Obata, Y., Chiba, H., Kohara, Y., Kono, T., Nakano, T., et al. 2008. Endogenous siRNAs from naturally formed dsRNAs regulate transcripts in mouse oocytes. *Nature* **453:** 539–543.

Zaratiegui, M., Irvine, D.V., and Martienssen, R.A. 2007. Noncoding RNAs and gene silencing. *Cell* **128:** 763–776.

Zhu, J.H., Kapoor, A., Sridhar, V.V., Agius, F., and Zhu, J.K. 2007. The DNA glycosylase/lyase ROS1 functions in pruning DNA methylation patterns in *Arabidopsis*. *Curr. Biol.* **17:** 54–59.

A Novel Epigenetic Mechanism in *Drosophila* Somatic Cells Mediated by PIWI and piRNAs

H. Lin and H. Yin*

Yale Stem Cell Center and Department of Cell Biology, Yale University School of Medicine, New Haven, Connecticut 06509

Small noncoding RNAs have emerged as key players in epigenetic regulation. Recently, a novel class of small RNAs that interact with Piwi proteins has been discovered in the mammalian and *Drosophila* germ line. These Piwi-interacting RNAs (piRNAs) represent a distinct small RNA pathway that is widely thought to function only in the germ line. In this chapter, we review our recent work with our collaborators on the epigenetic function of the *Drosophila* Piwi protein and its associated piRNAs in somatic cells. This work has revealed a novel epigenetic mechanism mediated by Piwi and its associated piRNAs in somatic cells that might also be applicable to the germ line. On the basis of these results, we propose a "Piwi-piRNA guidance hypothesis" for Piwi/piRNA-mediated epigenetic programming, in which the Piwi-piRNA complex serves as sequence-recognition machinery that recruits epigenetic effectors such as heterochromatin protein 1a (HP1a) to specific sites in the genome to execute epigenetic regulation.

It is an emerging theme that stem cell fate and other fundamental properties of development are not determined by turning on or off just a few genes. Instead, they are determined by modulating the transcriptional activity of the genome through changes in the local and global organization of chromatin in a heritable manner, a process called epigenetic programming (Surani 2001). Epigenetics refers to the regulation of gene expression by heritable but potentially reversible changes in chromatin structure and/or DNA methylation, but not in DNA sequences themselves. Regulators of epigenetic programming are called epigenetic factors. They include specific chromatin factors, histone modification proteins, and DNA methylation/demethylation enzymes.

In recent years, a role for noncoding small RNAs in the formation of epigenetic chromatin domains has been uncovered, owing mostly to exciting discoveries in the fission yeast *Schizosaccharomyces pombe* (Verdel et al. 2004). Endogeneous small interfering RNAs (siRNAs) in the nucleus have been proposed to provide a sequence-specific interface between a DNA sequence and its epigenetic state, presumably by their base pairing with genomic DNA or nascent RNA (Wassenegger 2005). Recent studies in fission yeast suggest that RNAi-mediated heterochromatin assembly occurs via a self-enforcing loop mechanism (Motamedi et al. 2004; Noma et al. 2004; Verdel et al. 2004; Sugiyama et al. 2005). A central player in this feed-forward loop is the RNAi-induced initiation of transcriptional gene silencing (RITS) complex, which contains a chromodomain-containing protein (Chp1), Argonaute 1 (Ago1), a novel protein Tas3, and siRNAs (Motamedi et al. 2004; Noma et al. 2004; Verdel et al. 2004; Sugiyama et al. 2005). Ago1 confers sequence specificity by binding to siRNAs and recruits other chro- matin proteins to initiate heterochromatization (Noma et al. 2004). Despite these exciting findings, the role of noncoding small RNAs in epigenetic regulation in higher organisms remains largely unexplored. In this chapter, we present a novel epigenetic mechanism in *Drosophila* somatic cells that is mediated by the Piwi protein and piRNAs.

ARGONAUTE/PIWI PROTEINS AND piRNAs

The *piwi/argonaute* (*ago*) gene family was first discovered for their evolutionarily conserved function for stem cell self-renewal (Cox et al. 1998). This gene family can be divided into *ago* and *piwi* subfamilies, herein referred to as *ago* and *piwi* genes, respectively. Both subfamilies encode highly basic proteins (pI~10) composed of four domains: the amino-terminal domain of variable length, the central PAZ (Piwi, Argonaute, and Zwille) domain that binds to a 3′ portion of single-stranded small RNAs without sequence specificity, the Mid domain that participates in small RNA binding, and the RNase H-like carboxy-terminal Piwi domain (Wang et al. 2008). The Piwi domain of some Ago proteins contains RNA slicing activity that is responsible for mRNA degradation mediated by siRNAs (Liu et al. 2004). However, such activity is not detected in all Ago or Piwi proteins (Liu et al. 2004), indicating that different Ago and Piwi proteins are biochemically distinct.

Piwi proteins are known to have diverse germ-line functions. For example, two Piwi proteins in *Drosophila*, Piwi and Aubergine, are initially required as maternal factors for germ-line establishment in early embryos (Harris and Macdonald 2001; Megosh et al. 2006). The dose of maternal Piwi determines the number of primordial germ cells (PGCs) in early embryos. Subsequently, Piwi is required for maintaining the transcriptional quiescence of PGCs (Cox 1999) and for germ-line stem cell self-renewal (Cox

*Present address: The Ottawa Health Research Institute, Ottawa, Canada.

et al. 1998). In mice, Piwi proteins Mili and Miwi are differentially expressed during spermatogenesis. Mili is expressed in spermatogonia (including male germ-line stem cells) spermatocytes and round spermatids, whereas Miwi is expressed from pachytene spermatocytes to elongating spermatids (Deng and Lin 2002; Kuramochi-Miyagawa et al. 2004; Unhavaithaya et al. 2009). Correspondingly, *mili*(–/–) mice are blocked in germ-line stem cell self-renewal, with a few escaping spermatogenic cells showing terminal arrest at the pachytene stage (Deng and Lin 2002; Kuramochi-Miyagawa et al. 2004; Unhavaithaya et al. 2009). This implicates a role for Mili in germ-line stem cell self-renewal and meiosis. In contrast, *miwi*(–/–) mice display uniform arrest at the beginning of the round spermatid stage, indicating its role as a key regulator of spermatid differentiation (Deng and Lin 2002; Kuramochi-Miyagawa et al. 2004; Unhavaithaya et al. 2009). A human *piwi* gene called *hiwi* is expressed in the testis, with its overexpression correlated to seminomas— common testicular cancers originating from malignant PGCs and/or germ-line stem cells (Qiao et al. 2002). In addition, *piwi* genes in both mammalian systems and in *Drosophila* have recently been implicated in transposon silencing in the germ line (Girard and Hannon 2008). Despite the well-recognized functions of Piwi proteins in the germ line, their underlying molecular mechanism remains elusive.

Although it is commonly believed that Piwi proteins only function in the germ line, the molecular mechanisms mediated by these proteins are actually better characterized in somatic cells, especially in somatic cells in *Drosophila*. Studies in *Drosophila* have provided several lines of evidence for a role of Piwi in epigenetic regulation in somatic cells. First, Piwi is a nuclear protein in both somatic and germ-line cells, except in early-cleavage-stage (i.e., syncytial) embryos (Cox et al. 2000; Megosh et al. 2006). Second, Piwi is a typical suppressor of position effect variegation in somatic cells (Pal-Bhadra et al. 2004), similar to other key epigenetic factors such as HP1 and HP2. Third, *piwi* deficiency results in loss of methylation of histone 3 at lysine 9 (H3K9me) and the delocalization of HP1 and HP2 from polytene chromosomes in the soma (Pal-Bhadra et al. 2004). Fourth, Piwi directly interacts with HP1, as described below. Thus, Piwi proteins, when in the nucleus, may behave as a key component of a complex required for heterochromatin assembly, just like Ago1 in fission yeast (Verdel et al. 2004). Interestingly, these studies highlight an ignored fact: Piwi proteins also have important functions in somatic cells, at least in *Drosophila*. The expression of *human Piwi* (*hiwi*) in human CD34[+] hematopoietic stem and progenitor cells (Sharma et al. 2001) and gastric epithelial cells (Liu et al. 2006) as well as the expression and function of Piwi proteins in neoblasts of planarians (Reddien et al. 2005; Rossi et al. 2006; Palakodeti et al. 2008) indicate that Piwi proteins have somatic functions in other organisms as well. The study of Piwi function in somatic cells should also shed light on their function in the germ line.

An important clue to the molecular function of Piwi proteins lies in their small RNA partners, the recently discovered piRNAs (Aravin et al. 2006; Girard et al. 2006; Grivna et al. 2006; Lau et al. 2006; Watanabe et al. 2006). piRNAs differ from siRNAs or microRNAs (miRNAs) in several ways. First, piRNAs interact with Piwi proteins but not argonautes. Mouse Piwi (Miwi) is required for piRNA biogenesis and/or stability (Grivna et al. 2006), whereas mouse Ago2 is required for the siRNA pathway (Liu et al. 2004). Second, piRNAs are mostly 24–31 nucleotides instead of approximately 21 nucleotides. Third, there are more than 50,000 cloned species of piRNAs, in contrast to several hundred species of miRNAs. Fourth, most piRNAs match to the genome in clusters of 20–90 kb in a strand-specific manner, with each cluster likely representing a long single-stranded RNA precursor or, more often, two nonoverlapping and divergently transcribed precursors (Kim 2006). In contrast, siRNAs and miRNAs are derived from double-stranded and short hairpin RNA precursors, respectively. Finally, some piRNAs may be involved in epigenetic regulation (see below). whereas siRNAs and miRNAs generally target mRNAs.

In *Drosophila*, piRNAs were first isolated as repeat-associated small interfering RNAs (rasiRNAs) (Saito et al. 2006; Vagin et al. 2006). Unlike siRNAs and miRNAs, these rasiRNAs bind to Piwi or Aubergine but not to Ago proteins (Vagin et al. 2006). Moreover, the production of these rasiRNAs requires neither Dicer-1 nor Dicer-2, which generate miRNAs and siRNAs, respectively. Thus, by definition, they are piRNAs. Despite this, the vast majority of piRNAs differ from previously identified rasiRNAs in that they show DNA strand specificity.

A PIWI-piRNA COMPLEX BINDS TO A piRNA CORRESPONDING SITE ON CHROMATIN IN SOMATIC CELLS

To explore the role of Piwi proteins and piRNAs in epigenetic regulation in somatic cells, we focused on Piwi and its interacting piRNAs in *Drosophila*. We cloned approximately 13,000 Piwi-associated piRNAs from *Drosophila* (Yin and Lin 2007). The size of these piRNAs shows a Gaussian distribution, with an average length of 25 nucleotides, shorter than mammalian piRNAs but longer than siRNA and miRNAs. Despite this large number, 82% of these piRNAs were sequenced only once, suggesting that this analysis is far from saturation and that our approximately 13,000 piRNAs are probably only a fraction of the entire pool. Consistent with this, our collection of Piwi-associated piRNAs overlaps with the collection in Gregory Hannon's lab (Brennecke et al. 2007) of 13,904 Piwi-associated piRNAs by only 10%. In addition, including the collection in Haru Siomi's lab of 330 Piwi-associated piRNAs (Saito et al. 2006), only two piRNAs are found in all three collections. Based on these data, we estimate that there are probably as many as 205,000 piRNAs in *Drosophila* that are associated with the Piwi protein alone.

Piwi-associated piRNAs correspond to all types of genomic sequences (Yin and Lin 2007). One third of them correspond to unique sequences in the genome, including gene-coding sequences and intergenic sequences. The remaining two thirds are transcribed from repetitive sequences. In fact, most of the repetitive piRNAs are derived from transposon sequences that account for only about 10% of the fully assembled euchromatic genome. These piRNAs are especially over-represented in retrotransposon encoding regions. Large numbers of Piwi-associated piRNAs are mapped into pericentromeric regions, subtelomeric regions, and telomeres. This broad distribution of piRNAs in nongene coding regions implicates their potential role in epigenetic regulation.

To explore the epigenetic function of piRNAs in the soma, we focused on a piRNA that uniquely corresponds to a repetitive sequence in the subtelomeric region (known as the telomere-associated sequence [TAS]) of the right arm of chromosome 3 (3R-TAS), a well-characterized heterochromatic region in *Drosophila*. This piRNA (called 3R-TAS1 piRNA) is expressed in the ovary and extra-ovarian somatic cells and is at least enriched, if not exclusively present, in the nucleus (Figs. 1B and 3B [below]). Keeping in mind that Piwi is also a nuclear protein in both somatic and germ-line cells (Cox et al. 2000), we postulated that the Piwi–TAS piRNA complex is associated with the piRNA-corresponding DNA sequence in the 3R-TAS genomic sequence in somatic and possibly germ-line cells. To test this possibility, we conducted chromatin immunoprecipitation (ChIP) experiments to precipitate Piwi from whole flies and quantified the amount of a 73-nucleotide region spanning the 3R-TAS1 piRNA-coding sequence (D, distal in Fig. 1A) that is coprecipitated with Piwi. Piwi is associated with TAS repeats 400-fold over the nonspecific control and 46-fold over the housekeeping gene *Rp49*. This specific association is further confirmed by ChIP in files carrying a functional *myc-piwi* transgene using an anti-Myc antibody. Myc-Piwi is enriched in 3R-TAS(D) 16.6-fold over an intergenic region on chromosome 2 and 11.4- and 11.5-fold over *succinate dehydrogenase B* and *actin88F* genes, respectively. In contrast, Piwi is not associated with any of the five piRNA-poor genomic regions examined or even with a proximal 3R-TAS sequence [TAS(P)] only 387 bp away (Fig. 1A). Thus, Piwi is strongly associated with the 3R-TAS(D) that encodes 3R-TAS1 piRNA.

We then confirmed Piwi binding to this genomic sequence in somatic cells by immumno-staining for Piwi protein in the polytene chromosomes from larval salivary glands. Piwi is indeed localized onto the chromosomes in many bands enriched in H3K9me, including the chromosome 3R-TAS region (Fig. 2).

Although it is technically challenging to directly visualize 3R-TAS1 piRNA binding to the 3R-TAS(D) sequence, this binding in complex with Piwi is strongly supported by six lines of evidence: First, this piRNA is at least enriched, if not specifically present, in the nucleus, just like Piwi; second, it binds to Piwi as shown both by coimmunoprecipitation and electrophoretic mobility shift (Yin and Lin 2007); third, only this piRNA corresponds to 3R-TAS(D) and is in fact transcribed from this sequence (Yin and Lin 2007); fourth, Piwi binds specifically to the 3R-TAS(D) site; fifth, Piwi binding to many sites in the polytene chromosome, including the 3R-TAS(D) site, is RNase sensitive (Brower-Toland et al. 2007); finally, the overexpression of 3R-TAS1 piRNA can largely rescue the epigenetic defects of the 3R-TAS(D) sequence in *piwi* mutants (see below). These data collectively implicate that Piwi and 3R-TAS1-piRNA form a complex that binds to the piRNA-matching DNA sequence.

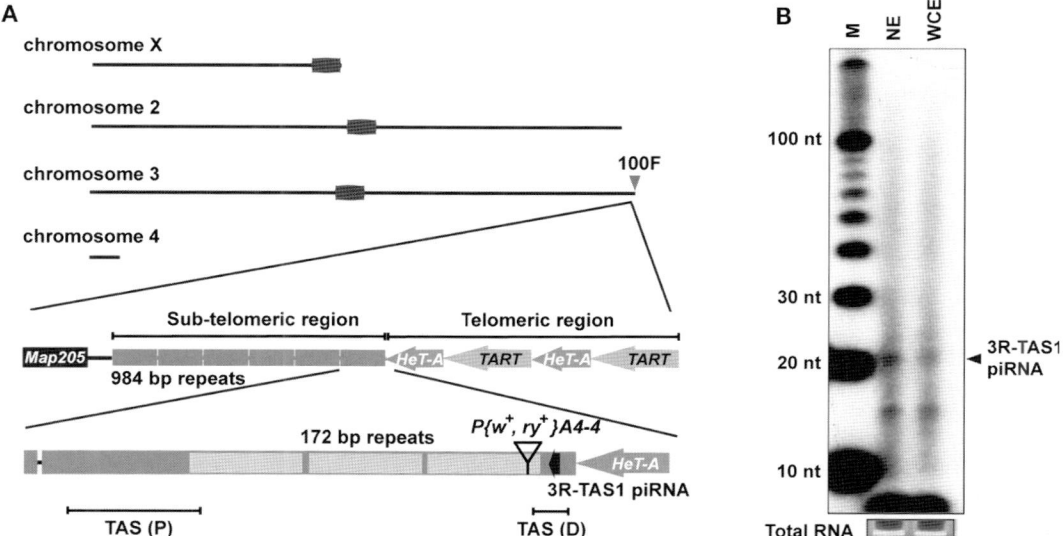

Figure 1. Expression of 3R-TAS1 piRNA that uniquely corresponds to the 3R-TAS(D) sequence. (*A*) Organization of 3R-TAS. (*Black arrow*) Position and direction of the unique mapped 3R-TAS1 piRNA. (*B*) Ribonuclease protection assay showing that 3R-TAS1 piRNA can protect a 597-bp antisense probe that spans the piRNA-coding region and covers 386–982 nucleotides of the 984-bp repeat unit in the antisense direction. The 3R-TAS1 piRNA is at least enriched, if not exclusively present, in nuclear extracts (NE) over whole-cell extracts (WCE). (M) 10-bp DNA marker. (Modified from Yin and Lin 2007 [© Nature Publishing Group].)

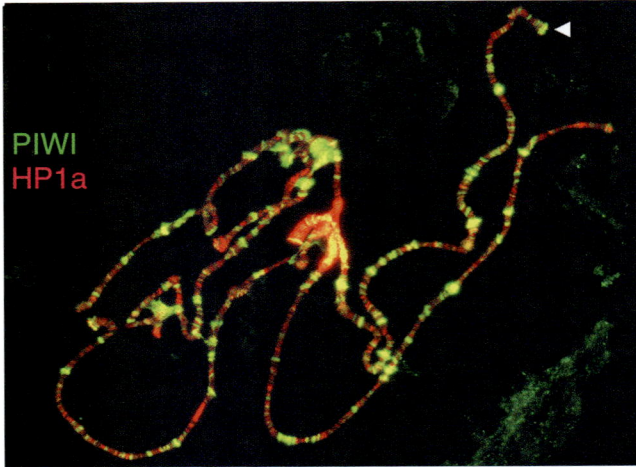

Figure 2. PIWI is associated with chromatin, where it colocalizes with HP1a in many regions. A *Drosophila* genome is represented by polytene chromosomes from a wild-type *Drosophila* third-instar larval salivary gland cell. PIWI (*green*) shows colocalization with HP1a (*red*) at multiple euchromatic bands along polytene chromosome arms, at telomeres, and in distinct regions of the chromocenter, including the 3R-TAS region (*white arrowhead*). HP1a is most concentrated in the chromocenter and along chromosome 4. (Adapted from Brower-Toland et al. 2007.)

PIWI-piRNA COMPLEX REGULATES THE EPIGENETIC STATE OF THE TARGET GENOMIC SITES IN SOMATIC CELLS

To determine whether the binding of Piwi to the TAS region has any effect on its chromatin status, we examined the histone modifications of the 73-bp 3R-TAS(D) sequence in wild-type versus *piwi* mutant flies by ChIP using antibodies specific against euchromatic or heterochromatic histone modifications. To our surprise, in wild-type flies, transcription activation markers, such as acetylated histone 3K9 and dimethylated and trimethylated histone 3K4, are abundantly associated with the TAS region, whereas heterochromatin markers, such as HP1a and methylated H3K9, are not as abundantly present as we expected (Fig. 3A). This indicates that this heterochromatin might be transcriptionally active. In *piwi* mutants, these euchromatic histone modifications are dramatically reduced from the TAS region. On the other hand, heterochromatic markers, such as HP1a and dimethylated and trimethylated histone 3K9, are dramatically accumulated on TAS genomic DNA in *piwi* mutants. This suggests that Piwi, possibly together with its associated 3R-TAS1 piRNA, promotes the euchromatic character of 3R-TAS(D).

To test whether the level of 3R-TAS1 piRNA has an effect on the epigenetic state of 3R-TAS(D), we isolated a P-element mutation *P{w⁺, ry⁺}A4-4* that is inserted in the 3R-TAS(D) sequence. This insertion causes the specific overexpression of 3R-TAS1 piRNA in wild-type flies (Fig. 3B) and rescues the expression of this piRNA in *piwi* mutants (data not shown). As a result, this rescue of piRNA expression restores the chromatin state of the 3R-TAS(D) sequence, with the euchromatic marker H3AcK4 increased to the wild-type level; heterochromatic markers also reduced to the wild-type level (Fig. 3A). This suggests the involvement of 3R-TAS1 piRNA in the Piwi-mediated epigenetic activation of 3R-TAS(D) sequence.

To determine whether the effect of Piwi on these chromatin modifications indeed has a functional consequence regarding the transcriptional activity of 3R-TAS(D) in somatic cells, we performed two experiments to assay the transcriptional activity of the sequence in somatic cells. First, we used an RNase protection assay to detect the transcription of TAS piRNA itself. 3R-TAS1 piRNA is indeed transcribed from this region, both in the ovary and, importantly, in extra-ovarian somatic cells (Fig. 3B). Second, we further inserted an eye color reporter gene into the 3R-TAS(D) sequence, only 128 bp downstream from the TAS piRNA coding sequence. In *piwi⁺/piwi⁺* flies, the eye-color reporter is moderately expressed in the eye. In *piwi⁺/piwi⁺* flies, the expression is significantly reduced, and in *piwi⁻/piwi⁻* mutants, the expression is barely detectable (Fig. 3C). Three independent quantifications of the eye pigment indicate that there is at least a fivefold decrease in reporter gene expression from heterozygous to homozygous mutants. These analyses further support the hypothesis that Piwi, presumably together with 3R-TAS1 piRNA, promotes the transcriptional activity of the 3R-TAS(D) sequence function in somatic cells.

The effect of Piwi on the transcriptional activity of 3R-TAS(D) in somatic cells is clearly inherited during cell division, because both the position effect of the *piwi* mutant and the transcriptional activation role of Piwi toward 3R-TAS(D) are variably expressed in a polyclonal fashion in fly eyes (Yin and Lin 2007). This indicates that Piwi-mediated transcriptional activation is an epigenetic phenomenon. To our knowledge, this is the first case illustrating that a small RNA-related mechanism has an epigenetic activation function.

PIWI DIRECTLY INTERACTS WITH HETEROCHROMATIN PROTEIN 1A

To further investigate the Piwi-mediated epigenetic mechanism in somatic cells, we sought to identify direct Piwi interactors and to determine their chromosomal localization. Yeast two-hybrid screens were conducted using full-length (FL), amino-terminal (NT) half, and carboxy-terminal (CT) half Piwi as baits (Fig. 4A) and recovered 102, 100, and 0 strong positives for these baits, respectively (Brower-Toland et al. 2007). Fourty four of 102 positives for Piwi-FL and 39 of 100 positives for

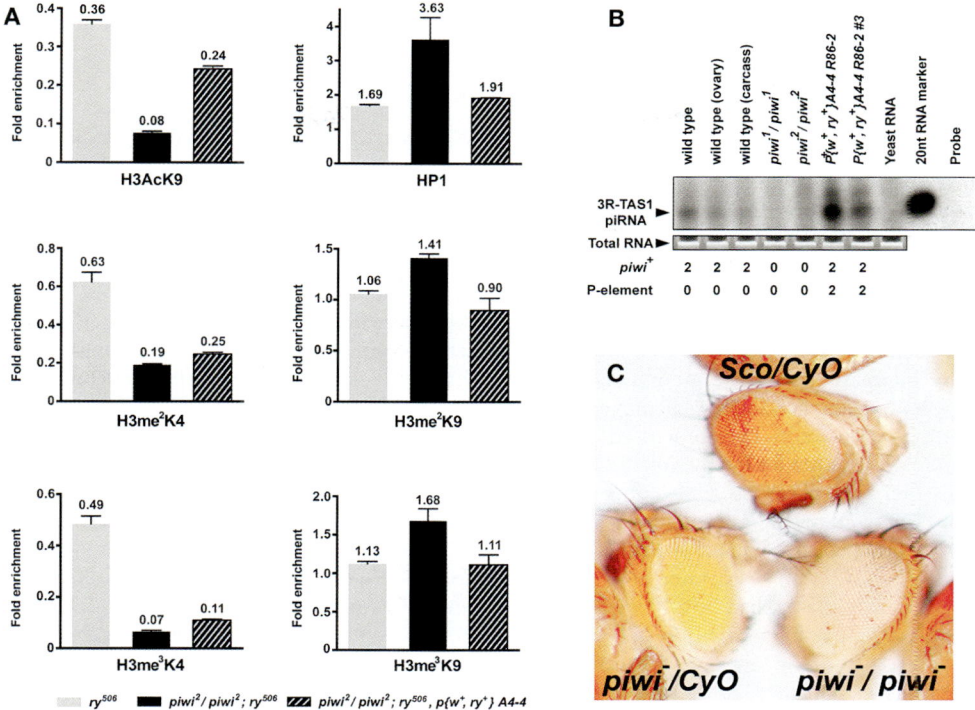

Figure 3. Piwi promotes the euchromatic features and transcriptional activity of 3R-TAS heterochromatin. (*A*) The association of modified histones H3Me^{2}K4, H3Me^{3}K4, H3AcK9, H3Me^{2}K9, H3Me^{3}K9, and HP1a with 3R-TAS(D) was assayed by ChIP and quantitative polymerase chain reaction (qPCR). The relative enrichment of modified histones is calculated by normalizing the quantity of 3R-TAS(D) DNA against the quantity of *Actin5C*. Each individual experiment was repeated at least three times. Standard deviations were used to indicate the error bars. (*B*) Ribonuclease protection assay showing that 3R-TAS1 piRNA is expressed in the wild-type adult fly, both in ovarian and extra-ovarian cells. Its expression is dramatically reduced in *piwi^1* and *piwi^2* mutants and drastically increased by the *P{w$^+$,ry$^+$}A4-4* insertion (four separately maintained lines, *221, 516, R86-2,* and *R86-2#3* were checked, with two lines shown in the figure). All lanes are loaded with equal amounts of total RNA. The copy numbers of *piwi* and *P{w$^+$,ry$^+$}A4-4* are noted at the bottom of *B*. (*C*) Piwi promotes the transcription of the eye-color reporter gene *white*, inserted into the 3R-TAS(D) sequence. Shown are eye colors of a *w^{1118};P{w$^+$,ry$^+$}A4-4* strain in wild-type, heterozygous, and homozygous *piwi^2* backgrounds. (Modified from Yin and Lin 2007 [© Nature Publishing Group].)

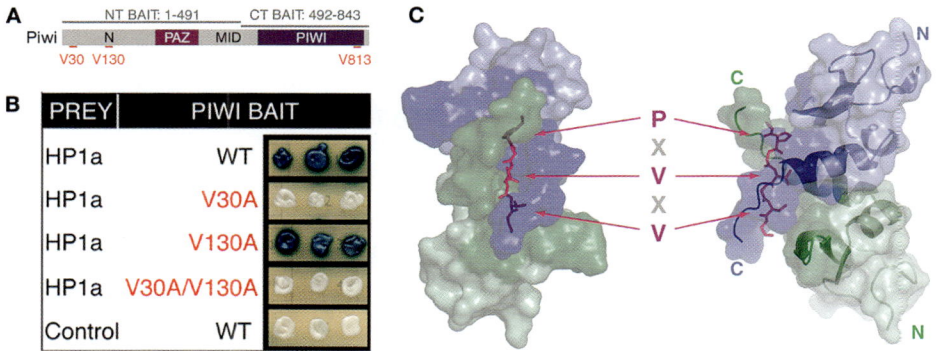

Figure 4. Piwi directly interacts with HP1a. (*A*) Baits used in PIWI Y2H screens. PIWI contains N, PAZ, MID, and PIWI domains. Y2H baits are PIWI-FL and residues 1–843, PIWI-NT and residues 1–491, and PIWI-CT and residues 492–843. The positions of the three "PXV" sequences, at V30, V130, and V813, are indicated. (*B*) V30 of PIWI but not V130 is required for HP1a interaction. Wild-type and V130A PIWI mutants produce comparably strong LacZ signals, whereas the signal in V30A, the V30A/V130A double mutant, or the negative control ("Control") is undetectable. (*C*) Model of the HP1a–chromoshadow domain (CSD) dimer in complex with the PIWI peptide (TSRGSGDGPRVKVFRGSSSGD). Top and side views are shown in left and right panels, respectively. Each monomer of the HP1a-CSD dimer is color-coded (*green* or *blue*). The CSD-binding motif (PXVXV) in the PIWI peptide is shown in stick models (*magenta*). Side chains of the conserved residues (P,V,V) are displayed. Chemical shift perturbations are calculated as $\delta_{CS} = (\delta_H^2 + (0.2\delta_N)^2)^{1/2}$. Residues of HP1a that experience significant resonance perturbation ($\delta_{CS} > 0.06$ ppm) during PIWI peptide titration are colored in *dark green* and *dark blue* and are mapped on the CSD surface and the ribbon diagram. The homology model of the complex was built using the XLOOK program and the figure was generated by PyMOL (Delano Scientific, South San Francisco, CA). (Modified from Brower-Toland et al. 2007.)

Piwi-NT encoded HP1a, a key component of epigenetic silencing systems (James et al. 1989; Hiragami and Festenstein 2005).

HP1a binding to Piwi is strong and highly specific: HP1a does not interact with a panel of unrelated baits. HP1a also fails to interact with any of the four Piwi homologs in *Drosophila* (Aubergine, Ago1, Ago2, and Ago3; Brower-Toland et al. 2007). Conversely, Piwi interacts with HP1a but not with two other HP1a-like chromatin proteins, HP1b and HP1c.

Piwi-HP1a interaction domains were then mapped using a yeast two-hybrid assay and nuclear magnetic resonance (NMR) analysis. HP1a is known to bind to heterochromatin by interacting with the methylated K9 of histone 3 (H3K9me). This interaction is achieved by the amino-terminal chromodomain. In addition, HP1a has a central hinge domain required for interaction with histone H1 and a carboxy-terminal chromoshadow domain (CSD) required for the dimerization of HP1a and interaction with many HP1 target proteins that contain a PXVXL motif (Smothers and Henikoff 2000; Lechner et al. 2005). Deletion analysis demonstrates that the HP1a CSD alone is necessary and sufficient for binding to NT- and FL-PIWI baits.

This interaction was further confirmed using two *Drosophila* HP1a point mutations that should disrupt the interaction between the CSD dimmer and its target (Brower-Toland et al. 2007). In mice, it is known that two CSDs symmetrically dimerize to form the interface for the asymmetrical binding of a single target peptide (Thiru et al. 2004). The W200A mutant does not affect dimerization but abolishes HP1 binding to the PXVXL motif in targets, and the I191E mutant disrupts HP1 dimerization. PIWI binding is lost in both of these HP1a mutants, suggesting that an intact CSD dimer interface is required for binding to PIWI. In contrast, PIWI interaction is maintained in the CD mutation V26M, which abolishes H3K9me binding. Thus, HP1a binding to methylated H3K9 is not required for PIWI binding.

Deletion analyses indicate that HP1a requires only the amino domain of PIWI for interaction. This domain contains two known vertebrate HP1-binding motifs: PRVKV centered on V30 and PRVRM centered on V130. Mutating V30 to alanine abolishes HP1a binding. In contrast, mutating V130 to alanine does not impact the interaction (Fig. 4B).

The direct interaction between the CSD of HP1a and the PRVKV motif centered on V30 was further confirmed by NMR titration of a 21-mer PIWI peptide centered on V30 with the HP1a CSD dimer (Brower-Toland et al. 2007). The exchange rate was slow on the NMR timescale, consistent with tight binding between the HP1a-CSD dimmer and the PIWI peptide. These analyses allowed the construction of a structural model of a dimmer of the *Drosophila* HP1a CSDs, where a single PIWI peptide binds across the dimer of an HP1a CSD. CSD residues that experience significant resonance perturbation during NMR titration are primarily distributed along the β strands close to the PIWI peptide, the carboxy-terminal portion of the central helices, and the carboxy-terminal extended loop. Together, these data strongly

suggest that PIWI binds specifically to HP1a, which represents a direct interaction between an RNAi component and a protein of the canonical epigenetic machinery in any eukaryote.

PIWI INTERACTS WITH HP1A AT MANY CHROMOSOMAL SITES IN SOMATIC CELLS

Because both Piwi and HP1a are chromatin-associated proteins in somatic cells, it is likely that their interaction occurs on chromosomes in these cells. To test this possibility and to assess the extent of Piwi-HP1a interaction on chromosomes, indirect immunofluorescence microscopy was conducted to examine Piwi-HP1a colocalization on *Drosophila* salivary gland polytene chromosomes. This experiment revealed a complex banded distribution of PIWI on polytenes that is partially coincident with HP1a (Brower-Toland et al. 2007). Although HP1a stained prominently localizes to the chromocenter (an agglomeration of pericentric heterochromatin) and to the largely heterochromatic fourth chromosome, HP1a is also found in all telomeres and in more than 200 bands along polytene chromosome arms. A prominent site of HP1a binding on chromosome arms is cytological region 31A, wherein HP1a localizes to eight discrete bands. PIWI colocalizes with HP1a in a complex pattern at multiple classes of chromatin loci, but it is completely absent from region 31A. Within the chromocenter, while HP1a is uniformly distributed, but PIWI has an irregular and particulate distribution within the pericentric heterochromatin of each chromosome. In chromosome 4, PIWI is restricted to a small number of distinct bands relative to HP1a, including one at or near the telomere. PIWI deposition also overlaps with HP1a in polytene telomeres. These data suggest that PIWI interacts with HP1 on chromosomes at many, but not all, genomic sites.

PIWI-HP1A INTERACTION IS REQUIRED FOR THE EPIGENETIC EFFECT OF PIWI IN SOMATIC CELLS

The direct interaction between Piwi and HP1a potentially represents a very direct means by which PIWI protein could act to recruit epigenetic effectors to specific sites in the *Drosophila* genome. To test the functional significance of PIWI-HP1a interaction in somatic cells, two assays were conducted (Brower-Toland et al. 2007). First, *piwi* transgenes encoding either the wild-type or V30A mutant form of *piwi* were introduced into a (lethal) *piwi*-null genetic background. The wild-type transgene was able to rescue viability, whereas the V30A mutant transgene failed to rescue viability. Second, it has been shown that hypomorphic alleles of *piwi* act as dominant suppressors of heterochromatic gene silencing on chromosome 4 in somatic cells, as revealed by the expression of transgenic *w*[+] reporter genes in the eye (Pal-Bhadra et al. 2004). This allowed the measurement of either transgene's ability to rescue dominant silencing defects produced in flies heterozygous for the protein-null *piwi*[2] allele. In this haploinsufficient *piwi* background, the presence of the wild-type transgene supported greater silencing of variegating *white* reporters than did the presence of the V30A transgene (see Fig. 6 below). These data strongly suggest that Piwi-HP1a interaction is directly involved in silencing *white* reporters embedded in constitutive heterochromatin in these somatic cells. In addition, they indicate that Piwi has opposing epigenetic effects toward different sites on the genome in somatic cells.

PIWI IS REQUIRED FOR HP1A BINDING AND H3K9 METHYLATION AT MANY GENOMIC SITES IN SOMATIC CELLS

We then investigated the relationship among PIWI, HP1a, and the epigenetic marker H3K9me. It has been long established that HP1a largely, but not exclusively, depends on the H3K9me marker provided by SUVAR3-9 family histone methyltransferases (HMTs) to bind to chromosomes. In addition, both HP1 targeting and H3K9me marking of many sites of polytene chromosomes have some dependence on *piwi* (Pal-Bhadra et al. 2004). This would suggest that Piwi acts upstream of H3K9 methylation and the binding of HP1a to these sites in these somatic cells. If so, we would expect H3K9me markers to colocalize with Piwi in polytene chromosomes, which indeed is the case (Brower-Toland et al. 2007). This, taken together with the fact that Piwi binds to HP1a and that Hp1a can bind to histone methylase, suggests the possibility that Piwi-piRNA recruits HP1a to these sites to initiate epigenetic marking. This then some-

how further recruits H3K9 histone methylase, resulting in H3K9 methylation to further heterochromatinize Piwi-piRNA target sites.

This mechanism would predict that PIWI localization to polytene chromosomes should not depend on the presence of HP1a and that PIWI binding to chromosomes is RNA dependent. Both predictions turn out to be true. In an *HP1a* mutant where HP1a is not detectable, Piwi localization in chromatin is not globally altered (Fig. 5B). However, when polytene chromosomes are pretreated with RNase A, PIWI staining on chromatin is greatly diminished (Fig. 5C).

PIWI/PIRNA-MEDIATED EPIGENETIC MECHANISM: THE PIWI-PIRNA GUIDANCE HYPOTHESIS

The results reviewed above revealed a novel epigenetic pathway in somatic cells that is mediated by Piwi and its associated piRNAs and allowed us to propose the following "Piwi-piRNA guidance hypothesis" (Fig. 6): A PIWI-piRNA complex binds to piRNA-corresponding genomic sequences via either piRNA-DNA base pairing (by either a conventional or unconventional paring mechanism) or by a piRNA forming a duplex with a nascent noncoding RNA transcript being transcribed from the target genomic sequence. This leads to the direct recruitment of a HP1a molecule to the genomic site, which initiates heterochromatization. The PIWI-piRNA-HP1a complex then

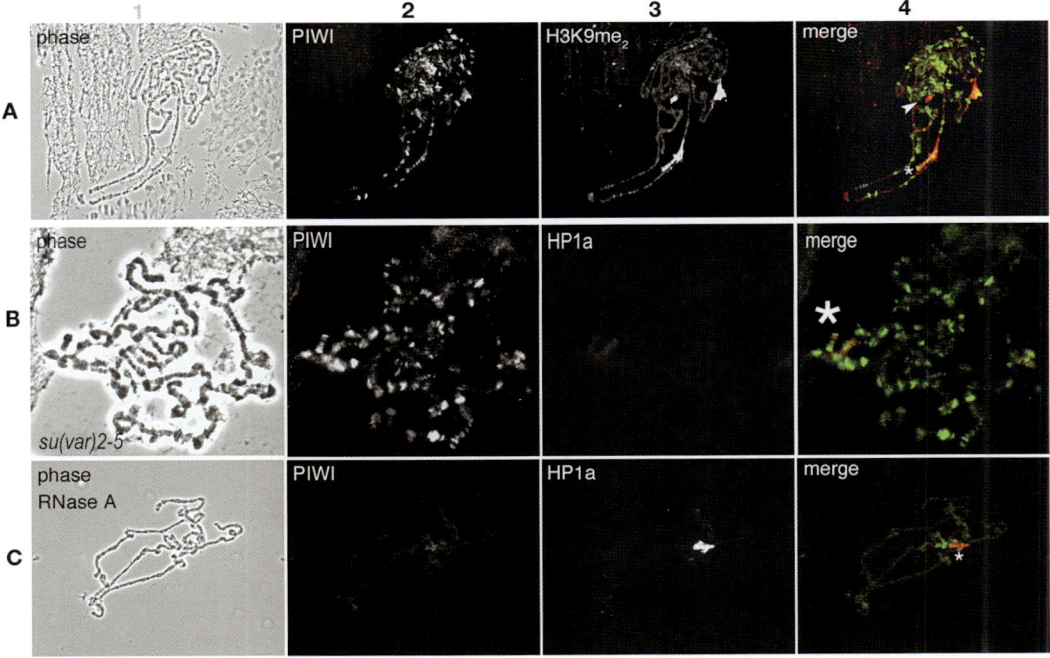

Figure 5. Chromatin binding of Piwi overlaps H3K9 methylation and depends on RNA hybrids but not HP1a. All images are of wild-type (Oregon R) chromosomes, except for *B*, which shows *Su(var)2-5* (the HP1a gene) mutant chromosomes. (*A*) PIWI shows significant overlap with dimethyl-H3K9 in the chromocenter and fourth chromosome but not region 31 (*white arrowhead*). (*B*) PIWI localization to polytene chromosomes is not globally perturbed in the absence of HP1a. Because of their small size, Su(var)2-5 chromosomes are shown at higher magnification by comparison with wild-type chromosomes. (*C*) Mild treatment with RNase A eliminates the PIWI-binding pattern seen on untreated chromosomes without perturbing HP1a. (Adapted from Brower-Toland et al. 2007.)

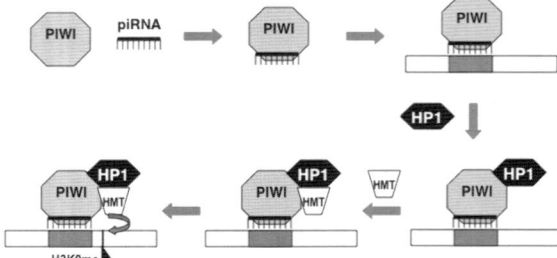

Figure 6. Novel epigenetic pathway mediated Piwi and piRNAs. For details, see text. Not shown are other factors possibly involved in the pathway.

further recruits HMTs such as SU(VAR)3-9 to methylate H3K9, which further leads to the stabilization and/or enhancement of heterochromatization.

This model is distinct from the RNAi model proposed in fission yeast (Volpe et al. 2002; Noma et al. 2004). Our model represents an H3K9me-independent mode for initial HP1 localization, an alternative but potentially effective means of triggering formation of heterochromatin at specific genomic sites. In this model, the epigenetic activation effect of Piwi at certain sites in the genomes could be achieved by either recruiting an epigenetic activator or having HP1a exhibiting epigenetic activation function at particular genomic sites. Such a function has been previously reported (Hediger and Gasser 2006).

It remains possible that heterochromatin formation is guided by a different mechanism. In this case, the presence of HP1a could allow stable binding of PIWI to heterochromatin for posttranscriptional gene silencing. It is also possible that the Piwi-piRNA–mediated mechanism acts at earlier developmental stages than the expressed epigenetic phenotype. In addition, there is evidence that Piwi proteins can also mediate other molecular mechanisms, especially when they are in the cytoplasm. Regardless of these possibilities, the Piwi-piRNA–mediated epigenetic mechanism, as revealed in somatic cells, may also be responsible for Piwi functions in the germ line, because Piwi and piRNAs are also localized in the nucleus in the germ line.

CONCLUDING REMARKS

Epigenetic programming of gene expression represents an exciting new frontier of stem cell research. A key yet essentially unexplored question in epigenetic programming is that how epigenetic regulators, most of which lack any DNA sequence recognition ability, are correctly guided to specific loci in the genome to exert their function. Recent studies suggest that transcriptional factors may have such a guidance role. Even so, this can only explain the targeting of a small fraction of the genome, because the vast majority of the genome is not gene coding. Our work in *Drosophila* somatic cells suggests that Piwi and its associated piRNAs guide epigenetic regulation to specific genomic sites. Indeed, this function of Piwi and its piRNAs may represent a major epigenetic guidance

mechanism in *Drosophila* somatic cells, given the association of Piwi with many regions of polytene chromosomes and the sufficient complexity of Piwi-associated piRNAs in targeting diverse regions throughout the genome. However, this mechanism alone is probably insufficient to account for the epigenetic programming of the entire genome, because Piwi does not bind to all HP1a sites and HP1a may not participate in epigenetic programming at all genomic sites. Other questions also remain: Which regions of the *Drosophila* genome are regulated by the Piwi-piRNA–mediated mechanism? How widely has the Piwi-piRNA mechanism been adopted during the course evolution? Does this mechanism also exist in the germ line? Answers to these questions should have tremendous impact on stem cell and developmental biology.

ACKNOWLEDGMENTS

We are grateful to Seth Findley, Brent Brower-Toland, Sarah Elgin, Pei Zhou, and Liang Jiang whose work with us on the Piwi-HP1a interaction represents an important part of this review. We thank Jonathan Saxe, Travis Thomson, and Vamsi Gangaraju for critically reading the manuscript on short notice. The research in the Lin lab is supported by the National Institutes of Health (grants HD33760, HD33760-S1, and HD42012), the Connecticut Stem Cell Research Fund, and the Mathers Foundation Award.

REFERENCES

Aravin, A., Gaidatzis, D., Pfeffer, S., Lagos-Quintana, M., Landgraf, P., Iovino, N., Morris, P., Brownstein, M.J., Kuramochi-Miyagawa, S., Nakano, T., et al. 2006. A novel class of small RNAs bind to MILI protein in mouse testes. *Nature* **442:** 203–207.

Brennecke, J., Aravin, A.A., Stark, A., Dus, M., Kellis, M., Sachidanandam, R., and Hannon, G.J. 2007. Discrete small RNA-generating loci as master regulators of transposon activity in *Drosophila*. *Cell* **128:** 1089–1103.

Brower-Toland, B., Findley, S.D., Jiang, L., Liu, L., Yin, H., Dus, M., Zhou, P., Elgin, S.C., and Lin, H. 2007. *Drosophila* PIWI associates with chromatin and interacts directly with HP1a. *Genes Dev.* **21:** 2300–2311.

Cox, D.N. 1999. "Function of the *Drosophila piwi* gene in the self-renewing division of germline stem cells and in germline development." Ph.D. thesis. Duke University, Durham, North Carolina.

Cox, D.N., Chao, A., and Lin, H. 2000. *piwi* encodes a nucleoplasmic factor whose activity modulates the number and division rate of germline stem cells. *Development* **127:** 503–514.

Cox, D.N., Chao, A., Baker, J., Chang, L., Qiao, D., and Lin, H. 1998. A novel class of evolutionarily conserved genes defined by *piwi* are essential for stem cell self-renewal. *Genes Dev.* **12:** 3715–3727.

Deng, W. and Lin, H. 2002. *miwi*, a murine homolog of *piwi*, encodes a cytoplasmic protein essential for spermatogenesis. *Dev. Cell* **2:** 819–830.

Girard, A. and Hannon, G.J. 2008. Conserved themes in small-RNA-mediated transposon control. *Trends Cell Biol.* **18:** 136–148.

Girard, A., Sachidanandam, R., Hannon, G.J., and Carmell, M.A. 2006. A germline-specific class of small RNAs binds mammalian Piwi proteins. *Nature* **442:** 199–202.

Grivna, S.T., Beyret, E., Wang, Z., and Lin, H. 2006. A novel class of small RNAs in mouse spermatogenic cells. *Genes Dev.* **20:** 1709–1714.

Harris, A.N. and Macdonald, P.M. 2001. *aubergine* encodes a

Drosophila polar granule component required for pole cell formation and related to eIF2C. *Development* **128:** 2823–2832.

Hediger, F. and Gasser, S.M. 2006. Heterochromatin protein 1: Don't judge the book by its cover! *Curr. Opin. Genet. Dev.* **16:** 143–150.

Hiragami, K. and Festenstein, R. 2005. Heterochromatin protein 1: A pervasive controlling influence. *Cell. Mol. Life Sci.* **62:** 2711–2726.

James, T.C., Eissenberg, J.C., Craig, C., Dietrich, V., Hobson, A., and Elgin, S.C. 1989. Distribution patterns of HP1, a heterochromatin-associated nonhistone chromosomal protein of *Drosophila. Eur. J. Cell Biol.* **50:** 170–180.

Kim, V.N. 2006. Small RNAs just got bigger: Piwi-interacting RNAs (piRNAs) in mammalian testes. *Genes Dev.* **20:** 1993–1997.

Kuramochi-Miyagawa, S., Kimura, T., Ijiri, T.W., Isobe, T., Asada, N., Fujita, Y., Ikawa, M., Iwai, N., Okabe, M., Deng, W., Lin, H., Matsuda, Y., and Nakano, T. 2004. *Mili*, a mammalian member of *piwi* family gene, is essential for spermatogenesis. *Development* **131:** 839–849.

Lau, N.C., Seto, A.G., Kim, J., Kuramochi-Miyagawa, S., Nakano, T., Bartel, D.P., and Kingston, R.E. 2006. Characterization of the piRNA complex from rat testes. *Science* **313:** 363–367.

Lechner, M.S., Schultz, D.C., Negorev, D., Maul, G.G., and Rauscher III, F.J. 2005. The mammalian heterochromatin protein 1 binds diverse nuclear proteins through a common motif that targets the chromoshadow domain. *Biochem. Biophys. Res. Commun.* **331:** 929–937.

Liu, J., Carmell, M.A., Rivas, F.V., Marsden, C.G., Thomson, J.M., Song, J.J., Hammond, S.M., Joshua-Tor, L., and Hannon, G.J. 2004. Argonaute2 is the catalytic engine of mammalian RNAi. *Science* **305:** 1437–1441.

Liu, X., Sun, Y., Guo, J., Ma, H., Li, J., Dong, B., Jin, G., Zhang, J., Wu, J., Meng, L., and Shou, C. 2006. Expression of *hiwi* gene in human gastric cancer was associated with proliferation of cancer cells. *Int. J. Cancer* **118:** 1922–1929.

Megosh, H.B., Cox, D.N., Campbell, C., and Lin, H. 2006. The role of PIWI and the miRNA machinery in *Drosophila* germline determination. *Curr. Biol.* **16:** 1884–1894.

Motamedi, M.R., Verdel, A., Colmenares, S.U., Gerber, S.A., Gygi, S.P., and Moazed, D. 2004. Two RNAi complexes, RITS and RDRC, physically interact and localize to noncoding centromeric RNAs. *Cell* **119:** 789–802.

Noma, K., Sugiyama, T., Cam, H., Verdel, A., Zofall, M., Jia, S., Moazed, D., and Grewal, S.I. 2004. RITS acts in *cis* to promote RNA interference-mediated transcriptional and post-transcriptional silencing. *Nat. Genet.* **36:** 1174–1180.

Palakodeti, D., Smielewska, M., Lu, Y.C., Yeo, G.W., and Graveley, B.R. 2008. The PIWI proteins SMEDWI-2 and SMEDWI-3 are required for stem cell function and piRNA expression in planarians. *RNA* **14:** 1174–1186.

Pal-Bhadra, M., Leibovitch, B.A., Gandhi, S.G., Rao, M., Bhadra, U., Birchler, J.A., and Elgin, S.C. 2004. Heterochromatic silencing and HP1 localization in *Drosophila* are dependent on the RNAi machinery. *Science* **303:** 669–672.

Qiao, D., Zeeman, A.M., Deng, W., Looijenga, L.H., and Lin, H. 2002. Molecular characterization of *hiwi*, a human member of the *piwi* gene family whose overexpression is correlated to seminomas. *Oncogene* **21:** 3988–3999.

Reddien, P.W., Oviedo, N.J., Jennings, J.R., Jenkin, J.C., and Sanchez Alvarado, A. 2005. SMEDWI-2 is a PIWI-like protein that regulates planarian stem cells. *Science* **310:** 1327–1330.

Rossi, L., Salvetti, A., Lena, A., Batistoni, R., Deri, P., Pugliesi, C., Loreti, E., and Gremigni, V. 2006. *DjPiwi-1*, a member of the *PAZ-Piwi* gene family, defines a subpopulation of planarian stem cells. *Dev. Genes Evol.* **216:** 335–346.

Saito, K., Nishida, K.M., Mori, T., Kawamura, Y., Miyoshi, K., Nagami, T., Siomi, H., and Siomi, M.C. 2006. Specific association of Piwi with rasiRNAs derived from retrotransposon and heterochromatic regions in the *Drosophila* genome. *Genes Dev.* **20:** 2214–2222.

Sharma, A.K., Nelson, M.C., Brandt, J.E., Wessman, M., Mahmud, N., Weller, K.P., and Hoffman, R. 2001. Human CD34$^+$ stem cells express the *hiwi* gene, a human homologue of the *Drosophila* gene *piwi*. *Blood* **97:** 426–434.

Smothers, J.F. and Henikoff, S. 2000. The HP1 chromo shadow domain binds a consensus peptide pentamer. *Curr. Biol.* **10:** 27–30.

Sugiyama, T., Cam, H., Verdel, A., Moazed, D., and Grewal, S.I. 2005. RNA-dependent RNA polymerase is an essential component of a self-enforcing loop coupling heterochromatin assembly to siRNA production. *Proc. Natl. Acad. Sci.* **102:** 152–157.

Surani, M.A. 2001. Reprogramming of genome function through epigenetic inheritance. *Nature* **414:** 122–128.

Thiru, A., Nietlispach, D., Mott, H.R., Okuwaki, M., Lyon, D., Nielsen, P.R., Hirshberg, M., Verreault, A., Murzina, N.V., and Laue, E.D. 2004. Structural basis of HP1/PXVXL motif peptide interactions and HP1 localisation to heterochromatin. *EMBO J.* **23:** 489–499.

Unhavaithaya, Y., Hao, Y., Beyret, E., Yin, H., Kuramochi-Miyagawa, S., Nakano, T., and Lin, H. 2009. MILI, a piRNA binding protein, is required for germline stem cell self-renewal and appears to positively regulate translation. *J. Biol. Chem.* (in press).

Vagin, V.V., Sigova, A., Li, C., Seitz, H, Gvozdev, V., and Zamore, P.D. 2006. A distinct small RNA pathway silences selfish genetic elements in the germline. *Science* **313:** 320–324.

Verdel, A., Jia, S., Gerber, S., Sugiyama, T., Gygi, S., Grewal, S.I., and Moazed, D. 2004. RNAi-mediated targeting of heterochromatin by the RITS complex. *Science* **303:** 672–676.

Volpe, T.A., Kidner, C., Hall, I.M., Teng, G., Grewal, S.I., and Martienssen, R.A. 2002. Regulation of heterochromatic silencing and histone H3 lysine-9 methylation by RNAi. *Science* **297:** 1833–1837.

Wang, Y., Juranek, S., Li, H., Sheng, G., Tuschl, T., and Patel, D.J. 2008. Structure of an argonaute silencing complex with a seed-containing guide DNA and target RNA duplex. *Nature* **456:** 921–926.

Wassenegger, M. 2005. The role of the RNAi machinery in heterochromatin formation. *Cell* **122:** 13–16.

Watanabe, T., Takeda, A., Tsukiyama, T., Mise, K., Okuno, T., Sasaki, H., Minami, N., and Imai, H. 2006. Identification and characterization of two novel classes of small RNAs in the mouse germline: Retrotransposon-derived siRNAs in oocytes and germline small RNAs in testes. *Genes Dev.* **20:** 1732–1743.

Yin, H. and Lin, H. 2007. An epigenetic activation role of Piwi and a Piwi-associated piRNA in *Drosophila melanogaster*. *Nature* **450:** 304–308.

Small RNA Silencing Pathways in Germ and Stem Cells

A.A. Aravin and G.J. Hannon

Watson School of Biological Sciences, Howard Hughes Medical Institute,
Cold Spring Harbor Laboratory, Cold Spring Harbor, New York 11724

During the past several years, it has become clear that small RNAs guard germ cell genomes from the activity of mobile genetic elements. Indeed, in mammals, a class of small RNAs, known as Piwi-interacting RNAs (piRNAs), forms an innate immune system that discriminates transposons from endogenous genes and selectively silences the former. piRNAs enforce silencing by directing transposon DNA methylation during male germ cell development. As such, piRNAs represent perhaps the only currently known sequence-specific factor for deposition of methylcytosine in mammals. The three mammalian Piwi proteins Miwi2, Mili, and Miwi are required at different stages of germ cell development. Moreover, distinct classes of piRNAs are expressed in developmental waves, with particular generative loci and different sequence content distinguishing piRNAs populations in embryonic germ cells from those that appear during meiosis. Although our understanding of Piwi proteins and piRNA biology have deepened substantially during the last several years, major gaps still exist in our understanding of these enigmatic RNA species.

GENERAL MECHANISMS OF RNA SILENCING

RNA interference (RNAi) was originally discovered as an artificial process of silencing cellular mRNAs in response to the addition of exogenous double-stranded RNAs (Fire et al. 1998). The power of this response strongly hinted at an endogenous function for such mechanisms, and work in the field quickly revealed that multiple, RNAi-related pathways formed a previously unrecognized layer of gene regulation and genome defense that is essential to virtually all eukaryotic organisms. The development of methods to characterize small RNA sequences opened previously unseen worlds of genomic output in organisms ranging from plants and fungi to protozoans and mammals. These fell into a number of classes based on their biogenesis mechanisms, genomic origins, and biological functions. Thus far, small RNAs that act in RNAi-related pathways fall mainly into these three classes.

In a classical RNAi experiment, exogenously provided, long double-stranded RNA is processed in the cell by an RNase III–family nuclease, Dicer, into mature small interfering RNAs (siRNAs) of 21–23 nucleotides (Bernstein et al. 2001; Elbashir et al. 2001). One strand of siRNA selected and incorporated in the RNA-induced silencing complex (RISC), which contains at its core a member of the Argonaute protein family (Hammond et al. 2001; Martinez et al. 2002; Liu et al. 2004). This protein directly interacts with both the 5′ and 3′ ends of the small RNA and uses it as a guide to recognize complementary targets. In classical RNAi, Argonaute also performs silencing by cleaving target transcripts using an endogenous, RNase H–related nuclease domain (Liu et al. 2004; Meister et al. 2004; Song et al. 2004). Endogenous siRNAs have revealed themselves in plants, fungi, and animals, wherein they have roles in gene regulation, transposon control, and chromosome organization.

microRNAs (miRNAs) share many similarities with siRNAs. They are processed from largely double-stranded precursors, produced from the genome by transcription of short, imperfect inverted repeats (Bartel 2004). The defining difference between miRNAs and siRNAs in animals is the addition of a nuclear processing step: the conversion of the primary miRNA (pri-miRNA) transcript into the pre-miRNA by Drosha, another RNase III family enzyme (Lee et al. 2003; Denli et al. 2004; Han et al. 2004). Following export to the cytoplasm (Kim 2004), maturation into an siRNA-like species is accomplished by Dicer (Grishok et al. 2001; Hutvagner et al. 2001), and the mature miRNAs then joins an Argonaute protein (Hutvagner and Zamore 2002; Mourelatos et al. 2002). In this case, target recognition does not depend on the entire extent of the small RNA but instead on a short guide sequence comprising roughly nucleotides 2–8 of the miRNA (Lewis et al. 2003, 2005; Stark et al. 2003). Most often, this region binds complementary sites in the 3′ UTRs of its regulatory targets, impacting their expression by regulation of translation and stability. In plants, relationships between siRNAs and miRNAs are less clear, because plants have no Drosha equivalent and appear to have evolved miRNA-mediated regulation separately. Plants do appear to have specialized different Dicer proteins for the production of miRNAs and siRNAs (Xie et al. 2004); however, a full understanding of the biochemistry of these pathways is still emerging.

The third class of small RNAs appears to be specific to animals and interacts with an animal-specific clade of Argonaute proteins, the Piwi family (Aravin et al. 2007a). These Piwi-interacting RNAs (piRNAs) have been found in *Caenorhabditis elegans* (Batista et al. 2008; Das et al. 2008), *Drosophila* (Saito et al. 2006; Brennecke et al. 2007), zebra fish (Houwing et al. 2007, 2008), and mammals (Aravin et al. 2006, 2007b; Girard et al. 2006; Grivna et al. 2006; Lau et al. 2006; Watanabe et al. 2006), where their expression is most prominent in male and female germ cells. In most organisms, piRNAs have clear roles in guarding germ cell genomes from the activity of mobile genetic elements (Aravin et al. 2007b; Brennecke et al.

2007; Houwing et al. 2007). They accomplish this as part of an elegantly constructed innate immune system that recognizes and silences transposons through both transcriptional and posttranscriptional mechanisms. However, it is clear that piRNAs must also have additional biological roles that are essential to germ cell maturation.

Although the roles of small RNAs are ubiquitous, their importance is particularly high in animal stem and germ cells. In stem cells, entire classes of miRNAs appear to be excluded from expression, instead defining a variety of differentiated cell fates. In germ cells, the broadest variety of small RNAs is elaborated, underscoring the special requirement for these pathways in reproductive tissues.

MiRNA AND siRNA PATHWAYS IN MAMMALIAN STEM AND GERM CELLS

Small RNA silencing pathways have important roles in controlling gene expression in many if not all of the cell types of multicellular organisms. Mice deficient in *Dicer*, a key enzyme required for miRNA and siRNA biogenesis, die before gastrulation (Bernstein et al. 2003), preventing analysis of Dicer function at later stages of development. Indeed, the ability of mice to sustain 7.5 days of development is probably due, at last in part, to the maternal contribution of Dicer and processed small RNAs (Tam et al. 2008; Watanabe et al. 2008). Production of conditional Dicer alleles by several groups indicated that small RNA pathways are required for the maintenance of embryonic stem (ES) cell potency and for the correct development of many different organ systems (Cobb et al. 2005; Harfe et al. 2005; Kanellopoulou et al. 2005; Murchison et al. 2005; Harris et al. 2006). Dicers have also proven essential for development in *Drosophila*, zebra fish, *Arabidopsis*, and many other systems in which mutant alleles have been analyzed.

Analysis of small RNA profiles in different tissue types revealed sets of ubiquitous as well as cell-type-specific miRNAs. For example, human and mouse ES cells express a group of miRNAs that are not present or exist only at very low levels in somatic cells (Houbaviy et al. 2003; Suh et al. 2004). One particular ES-cell-specific miRNA family, miR-290–295, was suggested to indirectly control the expression of de novo DNA methyltransferases by silencing the repressor protein Rbl2 (Benetti et al. 2008; Sinkkonen et al. 2008). A comprehensive investigation of small RNA profiles in adult stem cells is still underway, and it is currently not clear whether there is a common stem cell miRNA signature. However, a number of miRNAs that appear to be widely expressed in most differentiated cell types, such as the *let-7* family, are present at relatively low levels in tissue stem cells (Viswanathan et al. 2008). Overall patterns of miRNA expression have also been shown to track cell-fate specification in well-studied systems, such as hematopoietic development (Chen et al. 2004; Zhou et al. 2007).

Small RNA profiling in different tissues and cell types of mammals has revealed a plethora of miRNAs, with more than 800 species in humans that have substantial experimental support (Griffiths-Jones 2006; Landgraf et al. 2007). However, until recently such studies failed to reveal endogenous siRNAs (endo-siRNAs). The absence of abundant endo-siRNAs in mammals is likely explained by the activity of the interferon/protein kinase R (PKR) pathway that arrests translation in the presence of long double-stranded RNAs. Two groups of investigators, including our own, identified endogenous siRNAs in mouse oocytes, a cell type that demonstrably lacks a strong PKR pathway (Tam et al. 2008; Watanabe et al. 2008). One source of endogenous siRNAs is transposable elements (TEs); however, a small number of protein-coding genes also produce endo-siRNAs. In both cases, formation of double-stranded RNAs would require transcription of both strands of a particular locus. In the case of transposons, one can easily imagine that some copies might have integrated near a promoter that could produce antisense transcripts (but see below). However, the source of antisense information for protein-coding genes was mysterious. Deep sequencing and the analysis of polymorphic sense and antisense siRNAs corresponding to protein-coding genes allowed assignment of antisense information specifically to pseudogenes. Thus, the gene and pseudogene transcripts must form intermolecular hybrids that are substrates for Dicer. In part, the efficient interaction of transcripts from discrete genomic loci might hinge on the unique capacity of the oocyte to serve as a storehouse for RNAs that drive early development of the zygote.

The functional relevance of endo-siRNAs is supported by the observation that lesions in the RNAi pathway impact their target genes. For example, in Dicer-deficient oocytes, expression of endo-siRNA targets is increased (Tam et al. 2008; Watanabe et al. 2008). Genetic evidence also supports a role for Ago2, the only catalytically competent mammalian family member, in this regulatory circuit (Watanabe et al. 2008). Endo-siRNA targets are enriched for genes related to microtubule dynamics (Tam et al. 2008), and Dicer mutants show severe chromosome segregation and spindle defects during oocyte maturation (Murchison et al. 2007). This suggests that pseudogenes might not simply be molecular artifacts but might also have essential roles in gene regulation. There is also an intriguing possibility, fueled by recent studies of piRNAs in *Drosophila* (Brennecke et al. 2008), that inherited small RNA populations might have roles during early development.

Recently, endo-siRNAs have also been reported in ES cells (Babiarz et al. 2008), another context in which PKR-mediated nonspecific responses to dsRNA are not strong. In contrast, deep sequencing efforts have failed to reveal endo-siRNAs in a variety of somatic cell types. endo-siRNAs are seen in both somatic and germ cell compartments of *Drosophila* (Czech et al. 2008; Kawamura et al. 2008; Okamura et al. 2008) and *C. elegans* (Ambros et al. 2003), suggesting that there is no conserved restriction of these species to germ or multipotent cell types.

FUNCTION OF PIWI PROTEINS IN GERM CELLS

The third class of small silencing RNAs, piRNAs, and their protein partners, the Piwis, have expression patterns that are largely restricted to germ cells. In *Drosophila*, which has three members of the Piwi subfamily, these proteins are expressed in male and female germ cells (Brennecke et al. 2007). One family member, Piwi itself, is additionally expressed in follicular cells that have somatic origin but that tightly associated with and form a niche for germ-line cells. In *C. elegans* (Cox et al. 1998; Batista et al. 2008), zebrafish (Houwing et al. 2007, 2008), and mouse (Deng and Lin 2002; Kuramochi-Miyagawa et al. 2004; Aravin et al. 2008), the expression of Piwi family members is also restricted to germ-line cells. Interestingly, in more primitive animals, such as flatworms, Piwis are expressed in pluripotent neoblast cells that are responsible for the amazing regeneration potential of these animals and that can differentiate in any type of somatic or germ-line cell (Reddien et al. 2005).

The intracellular localization of Piwis also implicates them in germ-cell-specific processes. Most Piwi family members appear in cytoplasmic granular structures called nuage/germ plasm (*Drosophila*) or P granules (*C. elegans*) that mark germ cells throughout their development (Brennecke et al. 2007; Houwing et al. 2007; Aravin et al. 2008; Batista et al. 2008). Germ plasm/P granules have been strongly linked to germ cell-fate determination in multiple organisms. Interestingly, nuage-like granules containing Piwi proteins are present in both germ cells and neoblast cells in flatworms (Palakodeti et al. 2008). The localization of Piwi proteins seems linked to their function, and it likely that one of the major roles of nuage/germ plasm is related to their relationship to small RNA pathways. However, these structures also contain proteins that have not yet been linked to small RNA biology, suggesting that these locales likely have additional roles in RNA metabolism.

In mouse, determination of germ cells occurs relatively late in development and inheritance of nuage granules seems to have no role in this process. However, granules containing one of the Piwi proteins, MILI, are present in female germ cells from the arrested to the growing oocyte stage (Aravin et al. 2008). In embryos, the presence of MILI granules can be detected in both sexes soon after migrating primordial germ cells reach the somatic gonadal ridge (E12.5). Therefore, although Piwi-containing granules in mammals may not be directly involved in germ cell determination, even here they remain tightly associated with germ cells throughout almost all stages of gametogenesis.

Genetic analyses have also supported critical roles for Piwi proteins in germ cell development. Individual mutants for any of three Piwi family genes in mouse are male-sterile (Deng and Lin 2002; Kuramochi-Miyagawa et al. 2004, 2008; Carmell et al. 2007), whereas in zebra fish and *Drosophila* both male and female fertility is affected. No drastic defects in the development of somatic cells is observed upon Piwi mutation in any of these three organisms, although several reports indicate that Piwi family members in *Drosophila* might be involved in heterochromatin formation in somatic cells (Pal-Bhadra et al. 2004; Yin and Lin 2007). In flatworms, Piwi expression is critical for maintaining the regenerative potential of neoblasts (Reddien et al. 2005).

A detailed characterization of mutant phenotypes has revealed a number of specific developmental and molecular defects. In mouse, *Miwi*-mutant male germ cells arrest gametogenesis at the postmeiotic, round spermatid stage (Deng and Lin 2002). In *Mili* and *Miwi2* mutants, spermatogenic arrest occurs during meiosis with visible defects in chromosome alignment (Kuramochi-Miyagawa et al. 2004; Carmell et al. 2007). Deficiency in *Mili* and *Miwi2* also leads to defects in spermatogenic stem cell maintenance that manifests itself in a progressive loss of germ cells in adult animals (Carmell et al. 2007; Unhavaithaya et al. 2008). Analogously, mutation of Piwi family members in *Drosophila* causes defects at various stages of oogenesis. *Aub* mutants have defects in oocyte polarity and axis specification (Harris and Macdonald 2001). These defects are suppressed by mutations in *ATR* and *Chk2*, genes that encode checkpoint kinases that respond to nonrepaired double-stranded DNA breaks formed during meiosis (Klattenhoff et al. 2007). This result and the finding of unrepaired DNA breaks in *Miwi2*-deficient mouse cells (Carmell et al. 2007) suggest that the phenotype of at least some Piwi mutations in *Drosophila* and mouse might be secondary consequences of activating the DNA-damage checkpoint. Mutation of the founding member of the Piwi subfamily, Piwi itself, in *Drosophila*, causes depletion of stem cells in the ovary (Cox et al. 1998, 2000), similar to *Mili*- and *Miwi2*-mutant phenotypes in mouse. However, in contrast to mouse, the *Drosophila* Piwi defect seems to be noncell autonomous; lack of Piwi protein in somatic follicular cells results in defects in germ-line stem cell maintenance. Therefore, at least partially, the role of Piwi in *Drosophila* might be to maintain a proper somatic niche for germ-line stem cells.

Although analysis of Piwi-deficient mutants demonstrated links to germ cell maintenance, meiosis, and DNA-damage checkpoints, understanding the molecular mechanisms of Piwi function was impossible without identifying the Piwi small RNA binding partners, piRNAs.

piRNA POPULATIONS IN GERM CELLS

piRNAs derived from repetitive elements were initially identified in total RNA isolated from *Drosophila* germ cells and were termed repeat-associated small interfering RNAs (rasiRNAs) (Aravin et al. 2003). Subsequent studies showed that rasiRNAs significantly differed from siRNA and deserved separation in specific class (Vagin et al. 2006). piRNAs arise by a Dicer-independent mechanism that results in mature species that are larger than miRNAs and siRNAs. The precursors to piRNAs do not have the pronounced double-stranded structure that is a signature of miRNAs and siRNAs. Genomic mapping of piRNAs showed that multiple piRNAs mapped within discrete genomic intervals, forming so-called piRNA clusters that can exceed 100 kb. Each piRNA cluster produces numerous piRNAs that can have overlapping sequences, generating an amazing diversity of mature piRNA species. Although the majority of piRNA clusters in *Drosophila* produce piRNAs that match to both genomic strands, some piRNA clusters are single stranded (Brennecke et al. 2007). Similarly, piRNA clusters in mouse and zebra fish are either single stranded or consist of several segments in which the polarity of piRNA production switches between the plus and minus strands (Aravin et al. 2006, 2008; Girard et al. 2006; Houwing et al. 2007). Such a structure suggests that piRNAs are processed from long, single-stranded, precursor molecules that traverse the piRNA cluster. Indeed, insertional mutations in the putative promoter region of one of the *Drosophila* piRNA clusters, *flamenco*, eliminate piRNAs produced from the cluster as much as 100 kb away from the site of insertion (Brennecke et al. 2007).

Deep sequencing of piRNA libraries cloned from total cellular RNAs or immunoprecipitated Piwi complexes from *Drosophila*, *C. elegans*, zebra fish, and mammals revealed two types of piRNAs. Nonrepetitive piRNAs start to become expressed during the pachytene stage of meiosis in mouse germ cells and were accordingly termed pachytene piRNAs (Aravin et al. 2006; Girard et al. 2006). Similarly, piRNAs in *C. elegans*, called 21U RNAs, are derived from unique genomic regions and match only to the sites in the genome from which they are derived (Batista et al. 2008; Das et al. 2008). In contrast, a majority of piRNAs in *Drosophila* and zebra fish correspond to repetitive genomic elements, particularly different TEs. Whereas *Drosophila*, *C. elegans*, and zebra fish seem to have only one type of piRNA (repetitive or nonrepetitive), germ cells in mouse express both types at different stages of spermatogenesis (Aravin et al. 2007b, 2008; Kuramochi-Miyagawa et al. 2008). Expression of nonrepetitive, pachytene piRNAs is preceded by a population of piRNA derived from repeats. These initiate expression in germ cells during embryogenesis. Furthermore, expression of repetitive piRNAs during spermatogenesis is dynamic and regulated in a developmental fashion.

Nonrepetitive, pachytene piRNAs appear only at meiosis and interact with MIWI and MILI proteins (Aravin et al. 2006; Girard et al. 2006). Loss of these species is likely responsible for postmeiotic arrest of spermatogenesis observed in *Miwi*-deficient animals. Currently, the molecular function of pachytene piRNAs in mouse and of 21U piRNAs in *C. elegans* remains a matter of speculation, but it likely involves a germ-cell-specific process that occurs during and/or after meiosis.

The repetitive nature of the other class of piRNAs immediately suggested their function in the silencing of repetitive genomic elements, particularly transposons. This function is strongly supported by mutational analysis of Piwi proteins. In *Drosophila*, zebra fish, and mouse, deficiency for Piwi family members leads to overexpression of several types of repetitive elements in germ cells (Sarot et al. 2004; Kalmykova et al. 2005; Vagin et al. 2006; Aravin et al. 2007b; Carmell et al. 2007; Houwing et al. 2007). piRNA-producing loci have not yet been subjected to systematic genetic analysis, but mutations that lead to the perturbed function of one locus, *flamenco/COM*, were identified in *Drosophila* (Prud'homme et al. 1995; Desset et al. 2003). The *flamenco/COM* locus controls the expression of at least three different retrotransposons—*gypsy*, *Zam*, and *idefix*—and mutations that activate transposon expression also eliminate piRNA production from the locus (Brennecke et al. 2007).

Activation of TEs is likely responsible for the phenotypes of Piwi mutants in *Drosophila* and zebra fish and of *Mili* and *Miwi2* (but not Miwi) knockouts in mouse. It was proposed that active transposition in these mutant animals generates double-stranded breaks that are detected by the ATR/chk2 checkpoint and that, in turn, lead to meiotic arrest (Klattenhoff et al. 2007). Indeed, numerous Spo11-independent DNA breaks were detected in Piwi mutants in *Drosophila* (Klattenhoff et al. 2007) and mouse (Carmell et al. 2007).

It is less obvious how activation of TEs might affect the maintenance of germ-line stem cells, which are defective in *Piwi* mutants in *Drosophila* and in *Mili* and *Miwi2* mutants in mouse. In *Drosophila*, the *Piwi* defect seems to be caused by a lack of the Piwi protein in somatic follicular cells that in turn causes depletion of germ-line stem cells (Cox et al. 2000). The *flamenco* mutation, which activates transposons in follicular cells, has a phenotype very similar to that of *Piwi*, indicating that *Piwi*-induced stem cell phenotypes can be explained by transposon misregulation. The activation of TEs in follicular cells might perturb their function, preventing them from providing the niche necessary for maintaining germ-line stem cells.

In mouse, *Mili* and *Miwi2* are expressed exclusively in germ cells, and the defects caused by these mutations seem to be cell autonomous (Kuramochi-Miyagawa et al. 2004; Carmell et al. 2007). Activation of transposons in stem cells themselves might be inconsistent with the long-term maintenance of a multipotent state. Although piRNA profiles in pure stem cell populations have not yet been characterized, it is plausible that stem-cell-maintenance defects are caused by transposon activation either in the stem cell itself or its niche cells, rather than by another regulatory defect. Transposon activation can affect cellular physiology in many different ways. For example, transposition can generate dsDNA breaks that are sensed by DNA-repair machinery as described above. Alternatively, abundant TE transcripts might impact the expression of normal cellular mRNAs. Finally, as described below, activation of transposons in mouse correlates with demethylation of their genomic sequences. Therefore, it is plausible that failures in genome-wide methylation of TEs perturbs the normal chromatin landscape of stem cell.

THE PIWI/piRNA PATHWAY AS AN EPIGENETIC SENSOR AND MEMORY IN GERM CELLS

As compared to miRNAs and siRNAs, the Piwi/piRNA pathway has several features that make it suitable as a mediator of epigenetic memory in germ cells. First, the pathway is not linear but instead includes a self-perpetuating loop called the ping-pong cycle (Aravin et al. 2007a; Brennecke et al. 2007). In the ping-pong cycle, primary piRNAs, generated by an unknown mechanism, can be amplified if complementary transcripts are available. Amplification initiates when the endonucleolytic activity of Piwi proteins is used to cleave complementary transcripts. This induces the formation of a new, secondary piRNA with its 5′ end precisely at the cleavage site. The secondary piRNA can in turn regenerate the initial piRNA by cleavage of its complementary target. In *Drosophila*, the majority of primary piRNAs are derived from piRNA clusters and are antisense to transposon mRNAs (Brennecke et al. 2007). Therefore, on recognition of a transposon, they cleave its transcript and generate a sense secondary piRNA. Although the original silencing program is hardwired within the sequence of the piRNA clusters, amplification of each individual piRNA sequence depend on the existence of complementary transcripts, that is, on transcription of particular TEs. Thus, the system is able to sense expression of specific types of TEs and amplify piRNAs able to target those elements that are active within a given animal. In mouse, transcripts of TEs themselves are recognized as a source for primary processing, and therefore, primary piRNAs are sense and secondary are antisense, although the overall cycle operates similarly (Aravin et al. 2008).

Cycles that include production of secondary small RNAs are not restricted to the piRNA pathway (Baulcombe 2006). Indeed, situations in which production of secondary siRNAs depends on primary siRNAs or miRNAs have been described in fission yeast (Sugiyama et al. 2005), *C. elegans* (Pak and Fire 2007; Sijen et al. 2007), and plants (Allen et al. 2005; Axtell et al. 2006; Daxinger et al. 2009). Interestingly, in these organisms, piRNAs are either absent (yeast and plants) or produced exclusively by a primary processing mechanism without an amplification cycle (*C. elegans*). In all of these organisms production of secondary siRNAs depends on the activity of an RNA-dependent RNA polymerase (RdRP).

In plants, cleavage caused by primary siRNA or miRNA generates aberrant RNAs that are recognized as RdRP substrates (Allen et al. 2005; Axtell et al. 2006). In *C. elegans*, the primary siRNA somehow induces the RDRP to synthesize secondary siRNAs as direct transcription products (Pak and Fire 2007; Sijen et al. 2007). Secondary siRNAs can also recognize original transcripts, at least in some cases, closing the loop that amplifies the pool of small RNAs. In RdRP-dependent siRNA amplification mechanisms, secondary siRNAs spread from the site of primary siRNA, either in the 5′ direction (*C. elegans*) or both the 3′ and 5′ directions. Currently, there is no evidence of spreading during ping-pong amplification of piRNAs.

Amplification of piRNAs in the ping-pong cycle allows fine-tuning of piRNA populations to repetitive elements expressed in the cell and perpetuation of piRNA pools over cellular and even organism (in *Drosophila*) generations. Perpetuation of piRNA pools establishes a new system of epigenetic memory, completely independent of DNA/chromatin modifications (Brennecke et al. 2008). Furthermore, at least in mouse, there is clear evidence that piRNAs are linked to more classical epigenetic pathways that store information in the form of DNA methylation patterns (Aravin et al. 2007b, 2008; Carmell et al. 2007; Kuramochi-Miyagawa et al. 2008).

In mouse, it is well established that transposons are repressed transcriptionally, in a manner that depends on DNA methylation. Derepression of TEs in *Mili*- and *Miwi2*-deficient animals correlates with a failure of de novo methylation of these elements in the genome (Aravin et al. 2007b, 2008; Carmell et al. 2007; Kuramochi-Miyagawa et al. 2008). Several lines of evidence support a direct, functional link between the piRNA pathway and DNA methylation. The ping-pong cycle is most active during prenatal development, when patterns of DNA methylation on TEs are established (Aravin et al. 2008). At that time, two Piwi proteins are involved in the cycle: MILI and MIWI2. Expression of *Miwi2* precisely corresponds to the window of de novo DNA methylation in male germ cells, and this protein is present in the nucleus primed with small RNAs corresponding to mobile elements. Therefore, antisense piRNAs that are abundant in MIWI2 complexes might serve as sequence-specific guides that find and mark genomic sequences of transposons for DNA methylation. De novo DNA methyltransferases as well as *Miwi2* are not expressed after birth and the methylation pattern of TE sequences established in embryogenesis are maintained in germ-line stem cells. Similar patterns persist after fertilization and after the remodeling of the epigenome that occurs during early development. Because TEs constitute approximately 40% of genome in mouse, changes in their methylation patterns might have a substantial impact on the entire chromatin landscape of the cell, particularly if many components of heterochromatin normally exist in a tightly controlled balance with the sites that they regulate. Thus, although it is clear that loss of the piRNA pathway can impact DNA methylation states of transposons, this might also have secondary effects on broader gene expression patterns.

CONCLUSIONS

Small RNA pathways appear to be of particular importance in pluripotent and multipotent cells types. This relies on both inclusion and exclusion. Specific classes of miRNAs are often present in stem cell populations, and these may have critical roles in maintaining "stemness." Indeed, reprogramming experiments use lin-28, a regulator of miRNAs biogenesis, to aid in the conversion of fate-restricted to pluripotent cells (Yu et al. 2007; Darr and Benvenisty 2008). In this case, lin-28 is known to repress maturation of the let-7 family, a group of miRNAs that show widespread expression in differentiated cells (Heo et al. 2008; Newman et al. 2008; Viswanathan et al. 2008). Germ cells have greatly elaborated small RNA pathways, particularly in mammals. Germ cells are the only known cellular context in which all three extant RNAi-related pathways operate, the siRNA, miRNA, and piRNA pathways. Evidence from several animal models suggests that all three pathways are essential for germ cell integrity, with small RNAs serving to both control gene expression patterns and guard the integrity of the germ cell genome. Oocytes particularly are loaded with a variety of small RNAs during their maturation (Watanabe et al. 2006, 2008; Tam et al. 2008). At least in *Drosophila*, these maternally inherited species can have profound impacts on progeny, impacts that are also transmitted through subsequent generations via the maternal lineage (Brennecke et al. 2008). Thus, small RNAs may not only be critical for germ cell maintenance but may form a mechanism for epigenetic inheritance.

ACKNOWLEDGMENTS

We thank members of the Hannon lab for helpful discussions. This work was supported by grants from the National Institutes of Health (NIH) to G.J.H. and an NIH Pathway to Independence Award K99HD057233 to A.A.A.

REFERENCES

Allen, E., Xie, Z., Gustafson, A.M., and Carrington, J.C. 2005. microRNA-directed phasing during *trans*-acting siRNA biogenesis in plants. *Cell* **121:** 207–221.

Ambros, V., Lee, R.C., Lavanway, A., Williams, P.T., and Jewell, D. 2003. MicroRNAs and other tiny endogenous RNAs in *C. elegans*. *Curr. Biol.* **13:** 807–818.

Aravin, A., Gaidatzis, D., Pfeffer, S., Lagos-Quintana, M., Landgraf, P., Iovino, N., Morris, P., Brownstein, M.J., Kuramochi-Miyagawa, S., Nakano, T., et al. 2006. A novel class of small RNAs bind to MILI protein in mouse testes. *Nature* **442:** 203–207.

Aravin, A.A., Hannon, G.J., and Brennecke, J. 2007a. The Piwi-piRNA pathway provides an adaptive defense in the transposon arms race. *Science* **318:** 761–764.

Aravin, A.A., Sachidanandam, R., Girard, A., Fejes-Toth, K., and Hannon, G.J. 2007b. Developmentally regulated piRNA clusters implicate MILI in transposon control. *Science* **316:** 744–747.

Aravin, A.A., Sachidanandam, R., Bourc'his, D., Schaefer, C., Pezic, D., Toth, K.F., Bestor, T., and Hannon, G.J. 2008. A piRNA pathway primed by individual transposons is linked to de novo DNA methylation in mice. *Mol. Cell* **31:** 785–799.

Aravin, A.A., Lagos-Quintana, M., Yalcin, A., Zavolan, M., Marks, D., Snyder, B., Gaasterland, T., Meyer, J., and Tuschl, T. 2003. The small RNA profile during *Drosophila melanogaster* development. *Dev. Cell* **5:** 337–350.

Axtell, M.J., Jan, C., Rajagopalan, R., and Bartel, D.P. 2006. A two-hit trigger for siRNA biogenesis in plants. *Cell* **127:** 565–577.

Babiarz, J.E., Ruby, J.G., Wang, Y., Bartel, D.P., and Blelloch, R. 2008. Mouse ES cells express endogenous shRNAs, siRNAs, and other Microprocessor-independent, Dicer-dependent small RNAs. *Genes Dev.* **22:** 2773–2785.

Bartel, D.P. 2004. MicroRNAs: Genomics, biogenesis, mechanism, and function. *Cell* **116:** 281–297.

Batista, P.J., Ruby, J.G., Claycomb, J.M., Chiang, R., Fahlgren, N., Kasschau, K.D., Chaves, D.A., Gu, W., Vasale, J.J., Duan, S., et al. 2008. PRG-1 and 21U-RNAs interact to form the piRNA complex required for fertility in *C. elegans*. *Mol. Cell* **31:** 67–78.

Baulcombe, D.C. 2006. Short silencing RNA: The dark matter of genetics? *Cold Spring Harbor Symp. Quant. Biol.* **71:** 13–20.

Benetti, R., Gonzalo, S., Jaco, I., Munoz, P., Gonzalez, S., Schoeftner, S., Murchison, E., Andl, T., Chen, T., Klatt, P., et al. 2008. A mammalian microRNA cluster controls DNA methylation and telomere recombination via Rbl2-dependent regulation of DNA methyltransferases. *Nat. Struct. Mol. Biol.* **15:** 998.

Bernstein, E., Caudy, A.A., Hammond, S.M., and Hannon, G.J. 2001. Role for a bidentate ribonuclease in the initiation step of RNA interference. *Nature* **409:** 363–366.

Bernstein, E., Kim, S.Y., Carmell, M.A., Murchison, E.P., Alcorn, H., Li, M.Z., Mills, A.A., Elledge, S.J., Anderson, K.V., and Hannon, G.J. 2003. Dicer is essential for mouse

development. *Nat. Genet.* **35:** 215–217.

Brennecke, J., Malone, C.D., Aravin, A.A., Sachidanandam, R., Stark, A., and Hannon, G.J. 2008. An epigenetic role for maternally inherited piRNAs in transposon silencing. *Science* **322:** 1387–1392.

Brennecke, J., Aravin, A.A., Stark, A., Dus, M., Kellis, M., Sachidanandam, R., and Hannon, G.J. 2007. Discrete small RNA-generating loci as master regulators of transposon activity in *Drosophila*. *Cell* **128:** 1089–1103.

Carmell, M.A., Girard, A., van de Kant, H.J., Bourc'his, D., Bestor, T.H., de Rooij, D.G., and Hannon, G.J. 2007. MIWI2 is essential for spermatogenesis and repression of transposons in the mouse male germline. *Dev. Cell* **12:** 503–514.

Chen, C.Z., Li, L., Lodish, H.F., and Bartel, D.P. 2004. MicroRNAs modulate hematopoietic lineage differentiation. *Science* **303:** 83–86.

Cobb, B.S., Nesterova, T.B., Thompson, E., Hertweck, A., O'Connor, E., Godwin, J., Wilson, C.B., Brockdorff, N., Fisher, A.G., Smale, S.T., and Merkenschlager, M. 2005. T cell lineage choice and differentiation in the absence of the RNase III enzyme Dicer. *J. Exp. Med.* **201:** 1367–1373.

Cox, D.N., Chao, A., and Lin, H. 2000. *piwi* encodes a nucleoplasmic factor whose activity modulates the number and division rate of germline stem cells. *Development* **127:** 503–514.

Cox, D.N., Chao, A., Baker, J., Chang, L., Qiao, D., and Lin, H. 1998. A novel class of evolutionarily conserved genes defined by *piwi* are essential for stem cell self-renewal. *Genes Dev.* **12:** 3715–3727.

Czech, B., Malone, C.D., Zhou, R., Stark, A., Schlingeheyde, C., Dus, M., Perrimon, N., Kellis, M., Wohlschlegel, J.A., Sachidanandam, R., Hannon, G.J., and Brennecke, J. 2008. An endogenous small interfering RNA pathway in *Drosophila*. *Nature* **453:** 798–802.

Darr, H. and Benvenisty, N. 2008. Genetic analysis of the role of the reprogramming gene LIN-28 in human embryonic stem cells. *Stem Cells* (in press).

Das, P.P., Bagijn, M.P., Goldstein, L.D., Woolford, J.R., Lehrbach, N.J., Sapetschnig, A., Buhecha, H.R., Gilchrist, M.J., Howe, K.L., Stark, R., et al. 2008. Piwi and piRNAs act upstream of an endogenous siRNA pathway to suppress Tc3 transposon mobility in the *Caenorhabditis elegans* germline. *Mol. Cell* **31:** 79–90.

Daxinger, L., Kanno, T., Bucher, E., van der Winden, J., Naumann, U., Matzke, A.J., and Matzke, M. 2009. A stepwise pathway for biogenesis of 24-nt secondary siRNAs and spreading of DNA methylation. *EMBO J.* **28:** 48–57.

Deng, W. and Lin, H. 2002. *miwi*, a murine homolog of *piwi*, encodes a cytoplasmic protein essential for spermatogenesis. *Dev. Cell* **2:** 819–830.

Denli, A.M., Tops, B.B., Plasterk, R.H., Ketting, R.F., and Hannon, G.J. 2004. Processing of primary microRNAs by the Microprocessor complex. *Nature* **432:** 231–235.

Desset, S., Meignin, C., Dastugue, B., and Vaury, C. 2003. COM, a heterochromatic locus governing the control of independent endogenous retroviruses from *Drosophila melanogaster*. *Genetics* **164:** 501–509.

Elbashir, S.M., Lendeckel, W., and Tuschl, T. 2001. RNA interference is mediated by 21- and 22-nucleotide RNAs. *Genes Dev.* **15:** 188–200.

Fire, A., Xu, S., Montgomery, M.K., Kostas, S.A., Driver, S.E., and Mello, C.C. 1998. Potent and specific genetic interference by double-stranded RNA in *Caenorhabditis elegans*. *Nature* **391:** 806–811.

Girard, A., Sachidanandam, R., Hannon, G.J., and Carmell, M.A. 2006. A germline-specific class of small RNAs binds mammalian Piwi proteins. *Nature* **442:** 199–202.

Griffiths-Jones, S. 2006. miRBase: The microRNA sequence database. *Methods Mol. Biol.* **342:** 129–138.

Grishok, A., Pasquinelli, A.E., Conte, D., Li, N., Parrish, S., Ha, I., Baillie, D.L., Fire, A., Ruvkun, G., and Mello, C.C. 2001. Genes and mechanisms related to RNA interference regulate expression of the small temporal RNAs that control *C. elegans* developmental timing. *Cell* **106:** 23–34.

Grivna, S.T., Beyret, E., Wang, Z., and Lin, H. 2006. A novel class of small RNAs in mouse spermatogenic cells. *Genes Dev.* **20:** 1709–1714.

Hammond, S.M., Boettcher, S., Caudy, A.A., Kobayashi, R., and Hannon, G.J. 2001. Argonaute2, a link between genetic and biochemical analyses of RNAi. *Science* **293:** 1146–1150.

Han, J., Lee, Y., Yeom, K.H., Kim, Y.K., Jin, H., and Kim, V.N. 2004. The Drosha-DGCR8 complex in primary microRNA processing. *Genes Dev.* **18:** 3016–3027.

Harfe, B.D., McManus, M.T., Mansfield, J.H., Hornstein, E., and Tabin, C.J. 2005. The RNaseIII enzyme *Dicer* is required for morphogenesis but not patterning of the vertebrate limb. *Proc. Natl. Acad. Sci.* **102:** 10898–10903.

Harris, A.N. and Macdonald, P.M. 2001. *aubergine* encodes a *Drosophila* polar granule component required for pole cell formation and related to eIF2C. *Development* **128:** 2823–2832.

Harris, K.S., Zhang, Z., McManus, M.T., Harfe, B.D., and Sun, X. 2006. Dicer function is essential for lung epithelium morphogenesis. *Proc. Natl. Acad. Sci.* **103:** 2208–2213.

Heo, I., Joo, C., Cho, J., Ha, M., Han, J., and Kim, V.N. 2008. Lin28 mediates the terminal uridylation of let-7 precursor microRNA. *Mol. Cell* **32:** 276–284.

Houbaviy, H.B., Murray, M.F., and Sharp, P.A. 2003. Embryonic stem cell-specific microRNAs. *Dev. Cell* **5:** 351–358.

Houwing, S., Berezikov, E., and Ketting, R.F. 2008. Zili is required for germ cell differentiation and meiosis in zebrafish. *EMBO J.* **27:** 2702–2711.

Houwing, S., Kamminga, L.M., Berezikov, E., Cronembold, D., Girard, A., van den Elst, H., Filippov, D.V., Blaser, H., Raz, E., Moens, C.B., et al. 2007. A role for Piwi and piRNAs in germ cell maintenance and transposon silencing in zebrafish. *Cell* **129:** 69–82.

Hutvagner, G. and Zamore, P.D. 2002. A microRNA in a multiple-turnover RNAi enzyme complex. *Science* **297:** 2056–2060.

Hutvagner, G., McLachlan, J., Pasquinelli, A.E., Balint, E., Tuschl, T., and Zamore, P.D. 2001. A cellular function for the RNA-interference enzyme Dicer in the maturation of the *let-7* small temporal RNA. *Science* **293:** 834–838.

Kalmykova, A.I., Klenov, M.S., and Gvozdev, V.A. 2005. Argonaute protein PIWI controls mobilization of retrotransposons in the *Drosophila* male germline. *Nucleic Acids Res.* **33:** 2052–2059.

Kanellopoulou, C., Muljo, S.A., Kung, A.L., Ganesan, S., Drapkin, R., Jenuwein, T., Livingston, D.M., and Rajewsky, K. 2005. Dicer-deficient mouse embryonic stem cells are defective in differentiation and centromeric silencing. *Genes Dev.* **19:** 489–501.

Kawamura, Y., Saito, K., Kin, T., Ono, Y., Asai, K., Sunohara, T., Okada, T.N., Siomi, M.C., and Siomi, H. 2008. *Drosophila* endogenous small RNAs bind to Argonaute 2 in somatic cells. *Nature* **453:** 793–797.

Kim, V.N. 2004. microRNA precursors in motion: Exportin-5 mediates their nuclear export. *Trends Cell Biol.* **14:** 156–159.

Klattenhoff, C., Bratu, D.P., McGinnis-Schultz, N., Koppetsch, B.S., Cook, H.A., and Theurkauf, W.E. 2007. *Drosophila* rasiRNA pathway mutations disrupt embryonic axis specification through activation of an ATR/Chk2 DNA damage response. *Dev. Cell* **12:** 45–55.

Kuramochi-Miyagawa, S., Kimura, T., Ijiri, T.W., Isobe, T., Asada, N., Fujita, Y., Ikawa, M., Iwai, N., Okabe, M., Deng, W., et al. 2004. *Mili*, a mammalian member of *piwi* family gene, is essential for spermatogenesis. *Development* **131:** 839–849.

Kuramochi-Miyagawa, S., Watanabe, T., Gotoh, K., Totoki, Y., Toyoda, A., Ikawa, M., Asada, N., Kojima, K., Yamaguchi, Y., Ijiri, T.W., et al. 2008. DNA methylation of retrotransposon genes is regulated by Piwi family members MILI and MIWI2 in murine fetal testes. *Genes Dev.* **22:** 908–917.

Landgraf, P., Rusu, M., Sheridan, R., Sewer, A., Iovino, N., Aravin, A., Pfeffer, S., Rice, A., Kamphorst, A.O., Landthaler, M., et al. 2007. A mammalian microRNA expression atlas based on small RNA library sequencing. *Cell* **129:** 1401–1414.

Lau, N.C., Seto, A.G., Kim, J., Kuramochi-Miyagawa, S., Nakano, T., Bartel, D.P., and Kingston, R.E. 2006. Characterization of the piRNA complex from rat testes. *Science* **313:** 363–367.

Lee, Y., Ahn, C., Han, J., Choi, H., Kim, J., Yim, J., Lee, J., Provost, P., Radmark, O., Kim, S., and Kim, V.N. 2003. The nuclear RNase III Drosha initiates microRNA processing. *Nature* **425:** 415–419.

Lewis, B.P., Burge, C.B., and Bartel, D.P. 2005. Conserved seed pairing, often flanked by adenosines, indicates that thousands of human genes are microRNA targets. *Cell* **120:** 15–20.

Lewis, B.P., Shih, I.H., Jones-Rhoades, M.W., Bartel, D.P., and Burge, C.B. 2003. Prediction of mammalian microRNA targets. *Cell* **115:** 787–798.

Liu, J., Carmell, M.A., Rivas, F.V., Marsden, C.G., Thomson, J.M., Song, J.J., Hammond, S.M., Joshua-Tor, L., and Hannon, G.J. 2004. Argonaute2 is the catalytic engine of mammalian RNAi. *Science* **305:** 1437–1441.

Martinez, J., Patkaniowska, A., Urlaub, H., Luhrmann, R., and Tuschl, T. 2002. Single-stranded antisense siRNAs guide target RNA cleavage in RNAi. *Cell* **110:** 563.

Meister, G., Landthaler, M., Patkaniowska, A., Dorsett, Y., Teng, G., and Tuschl, T. 2004. Human Argonaute2 mediates RNA cleavage targeted by miRNAs and siRNAs. *Mol. Cell* **15:** 185–197.

Mourelatos, Z., Dostie, J., Paushkin, S., Sharma, A., Charroux, B., Abel, L., Rappsilber, J., Mann, M., and Dreyfuss, G. 2002. miRNPs: A novel class of ribonucleoproteins containing numerous microRNAs. *Genes Dev.* **16:** 720–728.

Murchison, E.P., Stein, P., Xuan, Z., Pan, H., Zhang, M.Q., Schultz, R.M., and Hannon, G.J. 2007. Critical roles for Dicer in the female germline. *Genes Dev.* **21:** 682–693.

Murchison, E.P., Partridge, J.F., Tam, O.H., Cheloufi, S., and Hannon, G.J. 2005. Characterization of Dicer-deficient murine embryonic stem cells. *Proc. Natl. Acad. Sci.* **102:** 12135–12140.

Newman, M.A., Thomson, J.M., and Hammond, S.M. 2008. Lin-28 interaction with the Let-7 precursor loop mediates regulated microRNA processing. *RNA* **14:** 1539–1549.

Okamura, K., Chung, W.J., Ruby, J.G., Guo, H., Bartel, D.P., and Lai, E.C. 2008. The *Drosophila* hairpin RNA pathway generates endogenous short interfering RNAs. *Nature* **453:** 803–806.

Pak, J. and Fire, A. 2007. Distinct populations of primary and secondary effectors during RNAi in *C. elegans*. *Science* **315:** 241–244.

Pal-Bhadra, M., Leibovitch, B.A., Gandhi, S.G., Rao, M., Bhadra, U., Birchler, J.A., and Elgin, S.C. 2004. Heterochromatic silencing and HP1 localization in *Drosophila* are dependent on the RNAi machinery. *Science* **303:** 669–672.

Palakodeti, D., Smielewska, M., Lu, Y.C., Yeo, G.W., and Graveley, B.R. 2008. The PIWI proteins SMEDWI-2 and SMEDWI-3 are required for stem cell function and piRNA expression in planarians. *RNA* **14:** 1174–1186.

Prud'homme, N., Gans, M., Masson, M., Terzian, C., and Bucheton, A. 1995. *Flamenco*, a gene controlling the *gypsy* retrovirus of *Drosophila melanogaster*. *Genetics* **139:** 697–711.

Reddien, P.W., Oviedo, N.J., Jennings, J.R., Jenkin, J.C., and Sanchez Alvarado, A. 2005. SMEDWI-2 is a PIWI-like protein that regulates planarian stem cells. *Science* **310:** 1327–1330.

Saito, K., Nishida, K.M., Mori, T., Kawamura, Y., Miyoshi, K., Nagami, T., Siomi, H., and Siomi, M.C. 2006. Specific association of Piwi with rasiRNAs derived from retrotransposon and heterochromatic regions in the *Drosophila* genome. *Genes Dev.* **20:** 2214–2222.

Sarot, E., Payen-Groschene, G., Bucheton, A., and Pelisson, A. 2004. Evidence for a *piwi*-dependent RNA silencing of the *gypsy* endogenous retrovirus by the *Drosophila melanogaster flamenco* gene. *Genetics* **166:** 1313–1321.

Sijen, T., Steiner, F.A., Thijssen, K.L., and Plasterk, R.H. 2007. Secondary siRNAs result from unprimed RNA synthesis and form a distinct class. *Science* **315:** 244–247.

Sinkkonen, L., Hugenschmidt, T., Berninger, P., Gaidatzis, D.,

Mohn, F., Artus-Revel, C.G., Zavolan, M., Svoboda, P., and Filipowicz, W. 2008. microRNAs control de novo DNA methylation through regulation of transcriptional repressors in mouse embryonic stem cells. *Nat. Struct. Mol. Biol.* **15:** 259–267.

Song, J.J., Smith, S.K., Hannon, G.J., and Joshua-Tor, L. 2004. Crystal structure of Argonaute and its implications for RISC slicer activity. *Science* **305:** 1434–1437.

Stark, A., Brennecke, J., Russell, R.B., and Cohen, S.M. 2003. Identification of *Drosophila* microRNA targets. *PLoS Biol.* **1:** E60.

Sugiyama, T., Cam, H., Verdel, A., Moazed, D., and Grewal, S.I. 2005. RNA-dependent RNA polymerase is an essential component of a self-enforcing loop coupling heterochromatin assembly to siRNA production. *Proc. Natl. Acad. Sci.* **102:** 152–157.

Suh, M.R., Lee, Y., Kim, J.Y., Kim, S.K., Moon, S.H., Lee, J.Y., Cha, K.Y., Chung, H.M., Yoon, H.S., Moon, S.Y., Kim, V.N., and Kim, K.S. 2004. Human embryonic stem cells express a unique set of microRNAs. *Dev. Biol.* **270:** 488–498.

Tam, O.H., Aravin, A.A., Stein, P., Girard, A., Murchison, E.P., Cheloufi, S., Hodges, E., Anger, M., Sachidanandam, R., Schultz, R.M., and Hannon, G.J. 2008. Pseudogene-derived small interfering RNAs regulate gene expression in mouse oocytes. *Nature* **453:** 534–538.

Unhavaithaya, Y., Hao, Y., Beyret, E., Yin, H., Kuramochi-Miyagawa, S., Nakano, T., and Lin, H. 2008. MILI, a piRNA binding protein, is required for germline stem cell self-renewal and appears to positively regulate translation. *J. Biol. Chem.* (in press).

Vagin, V.V., Sigova, A., Li, C., Seitz, H., Gvozdev, V., and Zamore, P.D. 2006. A distinct small RNA pathway silences selfish genetic elements in the germline. *Science* **313:** 320–324.

Viswanathan, S.R., Daley, G.Q., and Gregory, R.I. 2008. Selective blockade of microRNA processing by Lin28. *Science* **320:** 97–100.

Watanabe, T., Takeda, A., Tsukiyama, T., Mise, K., Okuno, T., Sasaki, H., Minami, N., and Imai, H. 2006. Identification and characterization of two novel classes of small RNAs in the mouse germline: Retrotransposon-derived siRNAs in oocytes and germline small RNAs in testes. *Genes Dev.* **20:** 1732–1743.

Watanabe, T., Totoki, Y., Toyoda, A., Kaneda, M., Kuramochi-Miyagawa, S., Obata, Y., Chiba, H., Kohara, Y., Kono, T., Nakano, T., et al. 2008. Endogenous siRNAs from naturally formed dsRNAs regulate transcripts in mouse oocytes. *Nature* **453:** 539–543.

Xie, Z., Johansen, L.K., Gustafson, A.M., Kasschau, K.D., Lellis, A.D., Zilberman, D., Jacobsen, S.E., and Carrington, J.C. 2004. Genetic and functional diversification of small RNA pathways in plants. *PLoS Biol.* **2:** E104.

Yin, H. and Lin, H. 2007. An epigenetic activation role of Piwi and a Piwi-associated piRNA in *Drosophila melanogaster.* *Nature* **450:** 304–308.

Yu, J., Vodyanik, M.A., Smuga-Otto, K., Antosiewicz-Bourget, J., Frane, J.L., Tian, S., Nie, J., Jonsdottir, G.A., Ruotti, V., Stewart, R., Slukvin, I.I., and Thomson, J.A. 2007. Induced pluripotent stem cell lines derived from human somatic cells. *Science* **318:** 1917–1920.

Zhou, B., Wang, S., Mayr, C., Bartel, D.P., and Lodish, H.F. 2007. miR-150, a microRNA expressed in mature B and T cells, blocks early B cell development when expressed prematurely. *Proc. Natl. Acad. Sci.* **104:** 7080–7085.

Epithelial Stem/Progenitor Cells in Lung Postnatal Growth, Maintenance, and Repair

E.L. RAWLINS,* T. OKUBO,[†] J. QUE,* Y. XUE,* C. CLARK,* X. LUO,* AND B.L.M. HOGAN*

*Department of Cell Biology, Duke University Medical Center, Durham, North Carolina 27710; [†]Center For Integrative Bioscience, National Institutes of Natural Sciences, Okazaki, Aichi 444-8787, Japan

The adult lung consists of a trachea leading into a system of branched airways ending in millions of alveolar sacs. It contains many different epithelial cell types arranged in precise patterns along the proximodistal axis. Each region of the lung has the capacity to repair through the proliferation of different epithelial cell types. However, the precise identity of the cells mediating repair is not fully resolved. To address this problem, we are using genetic lineage-labeling techniques in the mouse. The tools we have made will also be useful for understanding how progenitor cell behavior is regulated under normal and pathological conditions.

The lung is a complex and sophisticated organ. It is designed to function in gas exchange day in and day out over many decades while protecting itself against infectious agents, allergens, and toxic agents brought in with the air. Multiple mechanisms are used for this protection. Those initiated within the epithelium include the secretion of mucus and surfactant proteins, the mechanical activity of the ciliated cells, and the function of cells of the innate immune system, such as dendritic cells, embedded within the epithelial cells (Holt et al. 2008). In addition, the normal lung epithelium has a robust capacity to replace endogenous cells lost by "wear and tear" or specific injuries. Under normal circumstances, and particularly in experimental animals under laboratory conditions, cell turnover in the respiratory epithelium is very low (Kauffman 1980; Rawlins and Hogan 2008). This is in sharp contrast to the normal high turnover of epithelial cells in organs such as the stomach and intestine that, like the lung, are derived from the definitive endoderm during development (see below). However, if subsets of lung epithelial cells are killed experimentally, then under some circumstances, they can be replaced by the rapid proliferation and differentiation of cells from within the surviving population. Considerable effort has gone into identifying the stem and progenitor cells that affect this lung epithelial repair, and real progress has been made. One advance has been the generation of a series of gene-targeted and transgenic mouse lines that can be used in the living animal to trace the fate of specific populations of epithelial cells and their descendants during lung development and repair (Perl et al. 2002; Hong et al. 2004a, 2004b; Rawlins et al. 2007). Such experiments will allow lineage relationships between the different epithelial cell types to be established. In addition, these lines will be useful for understanding pathological processes. For example, they will likely have a key role in developing new insights into the cell of origin of different kinds of lung cancer. Another potential use for these lines is in studying "mucus metaplasia," a prominent feature of several lung diseases in which mucus-producing cells accumulate in the airways. One possibility is that these mucus-producing cells arise from differentiated secretory, and possibly ciliated, cells of the adult respiratory epithelium that change their program of gene expression ("transdifferentiate") without undergoing cell division (Evans et al. 2004; Tyner et al. 2006). This hypothesis could be tested by following the phenotype of lineage-traced secretory or ciliated cells under experimental conditions that promote mucus metaplasia.

MODELS FOR EPITHELIAL STEM/PROGENITOR CELLS OF THE MOUSE LUNG

Studies of the intestine and insulin-producing β cells of the pancreatic islets provide two very different models for how endodermal organs maintain their epithelial population throughout life. In the intestine, a small number of undifferentiated stem cells, residing at or near the base of the crypts, divide infrequently to give rise to transit-amplifying (TA) cells (Barker et al. 2007; Sangiorgi and Capecchi 2008). Classically, TA cells are defined as cells that proliferate rapidly, self-renew over the short term, and undergo lineage diversification, in this case, into goblet, absorptive, and enteroendocrine cells. In the mouse pancreas, in contrast, there is strong evidence that over several months at steady state, or after partial pancreatectomy, the insulin-producing cells of the islets self-duplicate and are not renewed from an undifferentiated, quiescent stem cell population (Dor et al. 2004; Teta et al. 2007). As an organ, the lung fits neither the intestinal nor β-cell model exactly. Rather, it contains different stem and progenitor cells, depending on the region examined (for review, see Borok et al. 2006; Rawlins and Hogan 2006). In the trachea and main bronchi, there has been good (but not watertight) evidence that during steady state, relatively undifferentiated, Trp63 (also known as p63)- and cytokeratin-5-positive basal cells function as classical stem cells (Fig. 1). Basal cells divide infrequently, self-renew over an extended period, and

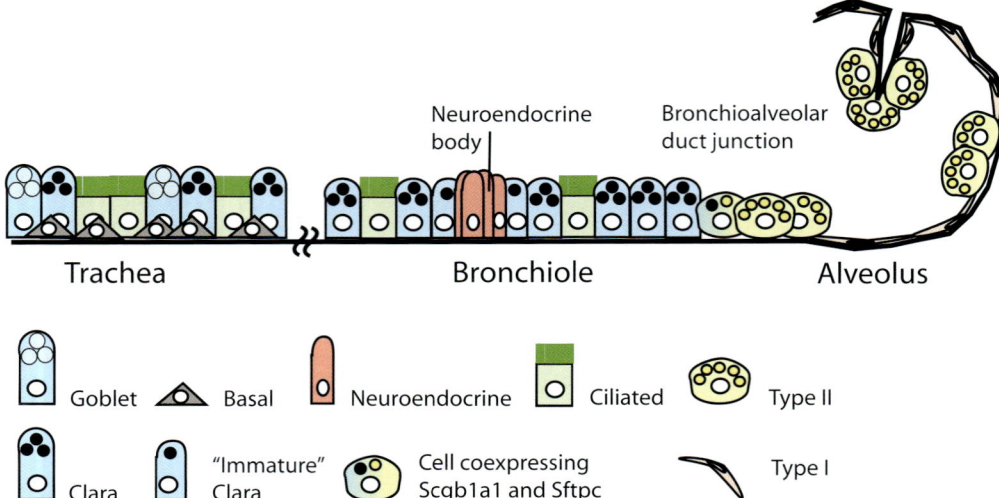

Figure 1. Schematic representation of the epithelial cells of the mouse lung. In the trachea, the main epithelial cell types are Clara-like secretory cells, mucus-producing goblet cells, ciliated cells, and undifferentiated basal cells that express p63 and cytokeratin Krt5. The upper regions also contain submucosal glands that open into the lumen through ducts (Rawlins and Hogan 2005). There are only a few neuroendocrine clusters in the trachea. In the interlobular airways (bronchi and bronchioles) of the mouse, there are no basal cells and more clusters of neuroendocrine cells. At the junction between the bronchioles and alveoli (known as the bronchioalveolar duct junction or BADJ), a very few cells coexpress proteins normally expressed at high levels by Clara cells (Scgb1a1) and alveolar type II cells (surfactant protein C or Sftpc). It has been proposed that these dual-positive cells are bronchioalveolar stem cells (BASCs), with the potential to give rise to descendants in both the airways and alveoli (see text). In the alveoli, there are two different epithelial cells types: the type II cells that produce large amounts of surfactant protein and highly attenuated type I cells across which gas exchange takes place. In the human lung, the BADJ is longer and contains cuboidal epithelial cells that have not yet been well defined. In addition, submucosal glands and p63[+ve] basal cells extend more distally into the lung than in the mouse, and even the bronchioles contain small numbers of basal cells.

give rise to differentiated descendants, in this case, ciliated and secretory cells (Borthwick et al. 2001; Hong et al. 2004a). However, it is unclear whether *all* basal cells have the same properties or whether there are subsets with higher proliferative capacity or different lineage potential (Hong et al. 2004b; Schoch et al. 2004). For example, the tracheal submucosal glands may be a protected niche environment for a subset of basal stem cells (Borthwick et al. 2001; Rawlins and Hogan 2005). In the intralobar airways (bronchioles) of the mouse lung, there are no basal cells (Fig. 1). Here, the evidence suggests that Clara cells expressing the secretoglobin gene *Scgb1a1* (also known as *CCSP* or *CC10*) self-renew over the long term and give rise to ciliated cells. Some debate exists, however, about whether *all* Clara cells have the same capacity to self-renew and give rise to ciliated cells or whether there are subsets with different properties. According to one theoretical model, all Clara cells have the same potential. In this case, the cells would be analogous to the population of "committed progenitors" that makes up the basal layer of the interfollicular epidermis of the mouse tail (Clayton et al. 2007; Jones and Simons 2008). According to other published models, at least two subpopulations of Clara cells are revealed during the response of the lung to treatment with naphthalene, which specifically kills differentiated Clara cells (Reynolds et al. 2000; Giangreco et al. 2002; Kim et al. 2005). One subpopulation has been called "variant

Clara" or Clara[V] cells. These are thought to be relatively undifferentiated (but still to express the Clara cell marker *Scgb1a1*), to be resistant to killing by naphthalene, and to reside in specific niches around the neuroendocrine bodies and at the bronchioalveolar duct junctions (BADJs). They are also thought to be relatively quiescent. However, when the majority of Clara cells are killed by naphthalene, the surviving Clara[V] cells rapidly proliferate and give rise to more Clara cells and to ciliated cells. According to this model, Clara[V] cells behave more as classical stem cells than committed progenitors because they are relatively undifferentiated (compared to most Clara cells) and are thought to be relatively quiescent at steady state. Another potential subclass of *Scgb1a1*-expressing epithelial cells has been termed bronchioalveolar stem cells or BASCs (Kim et al. 2005). These express both Scgb1a1 and SftpC (surfactant protein C, also known as SpC) and reside in small numbers exclusively in the BADJs. Like Clara[V] cells, BASCs are resistant to naphthalene and proliferate rapidly when most of the Clara cells are killed. However, unlike Clara[V] cells, it has been argued that BASCs are bipotential and their daughters can give rise not only to new Clara cells, but also to alveolar type II cells. Finally, in the alveoli of the lung (Fig. 1), there is evidence from injury models that type II cells can proliferate and give rise to type I cells, the flattened cells that are critical for gas exchange (Evans et al. 1973, 1975).

CELL LINEAGE ANALYSIS OF THE MOUSE LUNG EPITHELIUM

The only way to definitively test the models described above is to perform lineage-tracing experiments in vivo. This technique is based on the irreversible activation of a conditional reporter allele in the genome of specific cell types. This allows the cell, and any of its descendants, to express a reporter protein that can be detected histologically or by live imaging. The goal of these experiments is twofold: (1) to follow the fate of the labeled cells, and their descendants, over time and under different conditions (steady state or repair) and (2) to measure the extent to which the proportion of labeled cells in the population is maintained or diluted over time. From the first parameter, it is possible to define the developmental potential (potency) and fate of specific individual cells, or populations of cells, under the different conditions. From the second parameter, one can judge whether a population is self-maintaining over time or renewed from another, unlabeled, population. This approach is not without its pitfalls, and interpretation of results can be subject to many caveats. However, lineage tracing is generally accepted as the best way to follow the behavior of a cell and its descendants over time in vivo. In other words, the method allows us to determine whether cells are functioning as stem cells, TA cells, committed precursors, self-renewing differentiated cells, or terminally differentiated cells. During the past few years, we have generated a number of transgenic and gene-targeted mice for following cell fate and cell lineages in the embryonic and adult lung. These are all based on the CreER^{T2}/loxp system in which Cre recombinase activity in the cell nucleus is activated by the estrogen analog tamoxifen (Tmx) (Joyner and Zervas 2006). We have used a variety of conditional reporter alleles, including *Rosa26R-lacZ*, *Rosa26R-eYFP*, and *Rosa26R-CAG-farnesylatedGFP*. In all experiments, it is important to have controls to test for the potential toxicity of the Cre recombinase in different cell types and for the leakiness of the CreER allele (activity in the absence of exogenous Tmx) (Naiche and Papaioannou 2007).

In the first series of experiments, we used a *FOXJ1-CreER* transgenic mouse line to activate the reporter specifically in ciliated cells of the adult mouse lung (Fig. 2D,G) (Rawlins et al. 2007). We followed the behavior of labeled ciliated cells in response to two very different lung injuries in multiple experiments. We were never able to observe either the division or transdifferentiation of the labeled ciliated cells, strongly suggesting that ciliated cells are terminally differentiated. Subsequently, we observed that the labeled ciliated cell population was diluted by new, unlabeled ciliated cells over an 18-month period at steady state (Rawlins and Hogan 2008). This confirmed that ciliated cells are a terminally differentiated population and provided an estimate for rates of epithelial cell turnover in the steady-state lungs of the laboratory mouse. We have not yet used this mouse strain to test models of mucus cell metaplasia or cancer cell of origin.

More recently, we have generated Clara-cell- and basal-cell-specific CreER mouse strains: *Scgb1a1-CreER*

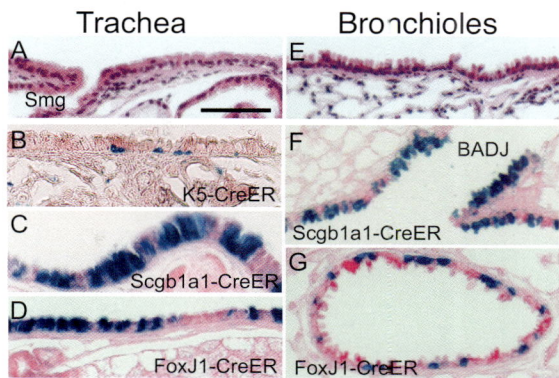

Figure 2. In vivo lineage-tracing of individual cell types in the mouse trachea and intralobular airways. Paraffin sections of adult mouse trachea (*A–D*) and bronchioles (*E–G*). (*A,E*) Hematoxylin and eosin staining show cell morphology; (*B*) lineage-labeled basal cells in section of X-gal-stained trachea from *K5-CreER;Rosa26R-LacZ* mouse; (*C,F*) lineage-labeled Clara and Clara-like cells from *Scgb1a1-CreER;Rosa26R-LacZ* mouse; (*D,G*) lineage-labeled ciliated cells from *FoxJ1-CreER;Rosa26R-LacZ* mouse. Bar, 200 μm in all sections.

(Fig. 2C,F) and *K5-CreER* (Fig. 2B). Our unpublished data confirm that in the intralobar airways of the adult mouse at steady state, Clara cells are able to self-renew and give rise to ciliated cells. However, we have not observed Clara cells, or cells coexpressing Clara and type II cell markers (putative BASCs), giving rise to type II cells under any of the conditions that we have tested. These include postnatal growth (embryonic day 18.5 to 1 year), exposure of neonatal and adult mice to hyperoxia (which damages alveoli and induces epithelial proliferation), and treatment with naphthalene. In contrast, in the trachea, basal cells indeed function as a classical stem cell, both self-renewing and generating new Clara cells. It is not yet clear whether a tracheal basal cell can give rise to a ciliated cell directly or whether Clara cells act as a TA intermediate between the basal cells and postmitotic ciliated cells. Certainly, our data suggest that Clara cells in the trachea do not have the capacity to self-renew extensively at steady state. This conclusion is based on the observation that the proportion of lineage-labeled Clara cells is gradually diluted by nonlabeled cells over time. We are currently testing whether all Clara and basal cells in the various regions of the mouse airways have the same properties. We are also studying the fate of Clara cells in the trachea after large numbers of the epithelial cells have been destroyed by exposure to sulfur dioxide.

CONCLUSIONS

In conclusion, our recent studies support the concept that in the mouse lung, the epithelial populations in each anatomical region—the trachea, the bronchioles, and the alveoli—are sustained and repaired by different kinds of stem and progenitor cells (Fig. 3). However, it should kept in mind that the mouse lung is not exactly anatomically comparable to the human lung. Our own studies, and reports in the literature, show that in the human lung,

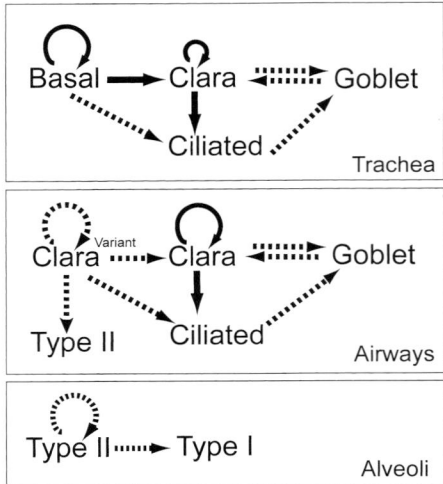

Figure 3. Summary of our current ideas about the stem/progenitor cell relationships in the trachea, intralobar airways, and alveoli of the adult mouse. (*Solid lines*) Evidence based on pulse-chase lineage-labeling experiments in the mouse. (*Curly arrows*) Self-renewal; (*dashed lines*) evidence based on experiments or hypotheses not yet validated by in vivo lineage tracing.

basal cells that express Trp63 and cytokeratin 5 extend much more distally than in the mouse (Sheikh et al. 2004). In other words, the distal mouse lung lacks a stem cell population that is present in the human. This means that caution must be exercised when extrapolating to the human the results, for example, of experiments in the mouse designed to identify the cell of origin of lung cancer. Finally, our most recent studies support the concept that the mesothelium of the mammalian lung functions as another stem cell population. In this case, our lineage-tracing studies demonstrate that mesothelial cells give rise to mesenchymal cells internal to the lung, including vascular smooth muscle. This brings the lung into line with other major visceral organs such as the heart and the gut, in which the outer mesothelial layer is a significant source of internal mesenchymal cells (Wilm et al. 2005; Zhou et al. 2008).

ACKNOWLEDGMENTS

Work from the Hogan lab described here was supported by grant HL071303. E.R. is supported by a Francis Family Foundation Parker B. Francis Fellowship.

REFERENCES

Barker, N., van Es, J.H., Kuipers, J., Kujala, P., van den Born, M., Cozijnsen, M., Haegebarth, A., Korving, J., Begthel, H., Peters, P.J., and Clevers, H. 2007. Identification of stem cells in small intestine and colon by marker gene *Lgr5*. *Nature* **449:** 1003–1007.

Borok, Z., Li, C., Liebler, J., Aghamohammadi, N., Londhe, V.A., and Minoo, P. 2006. Developmental pathways and specification of intrapulmonary stem cells. *Pediatr. Res.* **59:** 84R–93R.

Borthwick, D.W., Shahbazian, M., Krantz, Q.T., Dorin, J.R., and Randell, S.H. 2001. Evidence for stem-cell niches in the tracheal epithelium. *Am. J. Respir. Cell Mol. Biol.* **24:** 662–670.

Clayton, E., Doupe, D.P., Klein, A.M., Winton, D.J., Simons, B.D., and Jones, P.H. 2007. A single type of progenitor cell maintains normal epidermis. *Nature* **446:** 185–189.

Dor, Y., Brown, J., Martinez, O.I., and Melton, D.A. 2004. Adult pancreatic β-cells are formed by self-duplication rather than stem-cell differentiation. *Nature* **429:** 41–46.

Evans, M.J., Cabral, L.J., Stephens, R.J., and Freeman, G. 1973. Renewal of alveolar epithelium in the rat following exposure to NO2. *Am. J. Pathol.* **70:** 175–198.

Evans, M.J., Cabral, L.J., Stephens, R.J., and Freeman, G. 1975. Transformation of alveolar type 2 cells to type 1 cells following exposure to NO2. *Exp. Mol. Pathol.* **22:** 142–150.

Evans, C.M., Williams, O.W., Tuvim, M.J., Nigam, R., Mixides, G.P., Blackburn, M.R., DeMayo, F.J., Burns, A.R., Smith, C., Reynolds, S.D., Stripp, B.R., and Dickey, B.F. 2004. Mucin is produced by Clara cells in the proximal airways of antigen-challenged mice. *Am. J. Respir. Cell Mol. Biol.* **31:** 382–394.

Giangreco, A., Reynolds, S.D., and Stripp, B.R. 2002. Terminal bronchioles harbor a unique airway stem cell population that localizes to the bronchoalveolar duct junction. *Am. J. Pathol.* **161:** 173–182.

Holt, P.G., Strickland, D.H., Wikstrom, M.E., and Jahnsen, F.L. 2008. Regulation of immunological homeostasis in the respiratory tract. *Nat. Rev. Immunol.* **8:** 142–152.

Hong, K.U., Reynolds, S.D., Watkins, S., Fuchs, E., and Stripp, B.R. 2004a. Basal cells are a multipotent progenitor capable of renewing the bronchial epithelium. *Am. J. Pathol.* **164:** 577–588.

Hong, K.U., Reynolds, S.D., Watkins, S., Fuchs, E., and Stripp, B.R. 2004b. In vivo differentiation potential of tracheal basal cells: Evidence for multipotent and unipotent subpopulations. *Am. J. Physiol. Lung Cell. Mol. Physiol.* **286:** L643–L649.

Jones, P. and Simons, B.D. 2008. Epidermal homeostasis: Do committed progenitors work while stem cells sleep? *Nat. Rev.* **9:** 82–88.

Joyner, A.L. and Zervas, M. 2006. Genetic inducible fate mapping in mouse: Establishing genetic lineages and defining genetic neuroanatomy in the nervous system. *Dev. Dyn.* **235:** 2376–2385.

Kauffman, S.L. 1980. Cell proliferation in the mammalian lung. *Int. Rev. Exp. Pathol.* **22:** 131–191.

Kim, C.F., Jackson, E.L., Woolfenden, A.E., Lawrence, S., Babar, I., Vogel, S., Crowley, D., Bronson, R.T., and Jacks, T. 2005. Identification of bronchioalveolar stem cells in normal lung and lung cancer. *Cell* **121:** 823–835.

Naiche, L.A. and Papaioannou, V.E. 2007. Cre activity causes widespread apoptosis and lethal anemia during embryonic development. *Genesis* **45:** 768–775.

Perl, A.K., Wert, S.E., Nagy, A., Lobe, C.G., and Whitsett, J.A. 2002. Early restriction of peripheral and proximal cell lineages during formation of the lung. *Proc. Natl. Acad. Sci.* **99:** 10482–10487.

Rawlins, E.L. and Hogan, B.L. 2005. Intercellular growth factor signaling and the development of mouse tracheal submucosal glands. *Dev. Dyn.* **233:** 1378–1385.

Rawlins, E.L. and Hogan, B.L. 2006. Epithelial stem cells of the lung: Privileged few or opportunities for many? *Development* **133:** 2455–2465.

Rawlins, E.L. and Hogan, B.L. 2008. Ciliated epithelial cell lifespan in the mouse trachea and lung. *Am. J. Physiol. Lung Cell. Mol. Physiol.* **295:** L231–L234.

Rawlins, E.L., Ostrowski, L.E., Randell, S.H., and Hogan, B.L. 2007. Lung development and repair: Contribution of the ciliated lineage. *Proc. Natl. Acad. Sci.* **104:** 410–417.

Reynolds, S.D., Giangreco, A., Power, J.H., and Stripp, B.R. 2000. Neuroepithelial bodies of pulmonary airways serve as a reservoir of progenitor cells capable of epithelial regeneration. *Am. J. Pathol.* **156:** 269–278.

Sangiorgi, E. and Capecchi, M.R. 2008. Bmi1 is expressed in vivo in intestinal stem cells. *Nat. Genet.* **40:** 915–920.

Schoch, K.G., Lori, A., Burns, K.A., Eldred, T., Olsen, J.C., and Randell, S.H. 2004. A subset of mouse tracheal epithelial basal cells generates large colonies in vitro. *Am. J. Physiol. Lung Cell. Mol. Physiol.* **286:** L631–L642.

Sheikh, H.A., Fuhrer, K., Cieply, K., and Yousem, S. 2004. p63 expression in assessment of bronchioloalveolar proliferations of the lung. *Mod. Pathol.* **17:** 1134–1140.

Teta, M., Rankin, M.M., Long, S.Y., Stein, G.M., and Kushner, J.A. 2007. Growth and regeneration of adult β cells does not involve specialized progenitors. *Dev. Cell* **12:** 817–826.

Tyner, J.W., Kim, E.Y., Ide, K., Pelletier, M.R., Roswit, W.T., Morton, J.D., Battaile, J.T., Patel, A.C., Patterson, G.A., Castro, M., et al. 2006. Blocking airway mucous cell metaplasia by inhibiting EGFR antiapoptosis and IL-13 transdif-ferentiation signals. *J. Clin. Invest.* **116:** 309–321.

Wilm, B., Ipenberg, A., Hastie, N.D., Burch, J.B., and Bader, D.M. 2005. The serosal mesothelium is a major source of smooth muscle cells of the gut vasculature. *Development* **132:** 5317–5328.

Zhou, B., Ma, Q., Rajagopal, S., Wu, S.M., Domian, I., Rivera-Feliciano, J., Jiang, D., von Gise, A., Ikeda, S., Chien, K.R., and Pu, W.T. 2008. Epicardial progenitors contribute to the cardiomyocyte lineage in the developing heart. *Nature* **454:** 109–113.

Multipotent Islet-1 Cardiovascular Progenitors in Development and Disease

A. Nakano,* H. Nakano,* and K.R. Chien

Cardiovascular Research Center, Massachusetts General Hospital, Harvard Stem Cell Institute, Department of Stem Cell and Regenerative Biology, Harvard Medical School, Boston, Massachusetts 02114-2790

During the past several years, advances at the intersection of cardiovascular development and heart stem cell biology have begun to reshape our view of the fundamental logic that drives the formation of discrete tissue components in the mammalian heart. Although many of the critical genes that control cardiac myogenesis have been identified, our understanding of how a highly diverse and specialized subset of heart cell lineages arises from mesodermal precursors and is subsequently assembled into distinct muscle chambers, coronary arterial tree and large vessels, valvular tissue, and conduction system/pacemaker cells remains at a relatively primitive stage. Recent studies have uncovered a diverse group of closely related heart progenitors that are central in controlling and coordinating these complex steps of cardiogenesis. Understanding the pathways that control their formation, renewal, and subsequent conversion to specific differentiated progeny forms the underpinning for unraveling the pathways for congenital heart disease and has direct relevance to cardiovascular regenerative medicine. This current brief review highlights the discovery and delineation of the role of Islet-1 cardiovascular progenitors in the generation of diverse heart cell lineages and how the implications of these findings are revising our classification and thinking about congenital heart disease in general.

The mammalian heart is a complex multichambered organ consisting of two ventricles, two atria, outflow tract and inflow tract, conduction system, valves, endocardium, and epicardium. The proper function of the whole heart requires a highly coordinated process of differentiation and integration of all of the cellular components (Olson 2006; Srivastava 2006). One of the central questions in cardiovascular biology relates to how these various components are assembled from early mesodermal precursors located in the heart field in the early mammalian embryo.

The cardiac precursor cells are specified in the anterior part of the primitive streak shortly after gastrulation. These precursor cells migrate toward the anterior lateral plate mesoderm and form a pair of cardiac crescents, the first visible and morphologically distinguishable structure of the cardiac primordia. Endocardial tubes are already formed in each of cardiac crescents at this stage. As the neuroectoderm develops rapidly at this stage, the heart-forming region is pushed ventrally and fuses to form a single linear heart tube. It has been long thought that the identity of each heart chamber is already specified at an early stage within each segment of the linear heart tube—outflow tract, right ventricle, left ventricle, atria, and sinus venosus, from the cranial to caudal direction.

The discovery of the second progenitor source has begun to challenge this classical segmental prespecification view of cardiogenesis and is reshaping our view of congenital heart abnormalities (Buckingham et al. 2005). This second source of cardiac progenitors is formed in the dorsomedial part of the anterior lateral plate mesoderm adjacent to the cardiac crescent (Fig. 1). Recently, this extra-crescent heart field (second heart field; SHF) was revisited and characterized by several different techniques: dye labeling, detection of specific molecular markers, and retrospective single-cell-fate tracing experiments (Kelly et al. 2001; Mjaatvedt et al. 2001; Waldo et al. 2001; Cai et al. 2003; Dodou et al. 2004; Meilhac et al. 2004). Different methods and different markers have determined slightly different populations of the cardiac progenitors, which has led to the initial confusion of the definition of SHF and anterior heart field (AHF). The SHF is well delineated by the expression of Isl1, a member of the LIM homeodomain transcription factor family (Cai et al. 2003), whereas several other molecular markers, such as FGF8/10, Wnt11, and a specific Mef2c enhancer, label the anterior subset of SHF. In this chapter, "SHF" is used in a broader sense, including both the anterior heart field subset and the posterior subset that mainly contributes to the atria and inflow tract.

The dorsal mesocardium, the mesentery that suspends the linear heart tube, is initially continuous to the splanchnic mesoderm at any level at the linear heart tube stage. Cardiac progenitor cells in the splanchnic mesoderm continue to migrate through the dorsal mesocardium. As the dorsal mesocardium breaks down, the heart tube becomes connected to the body wall only at the arterial pole and the venous pole. From this point, the SHF progenitors migrate to the heart tube at either pole.

The rightward looping of the heart tube is then triggered by the addition of second heart field progenitors onto the scaffold of the primary heart tube and subsequent segmental ballooning of outer curvature of each chamber. The rapid growth of each segment places the forming atria and inflow portion above the ventricular chambers. The rightward looping is the first appearance of left-right asymmetry in the heart. Obviously, the rapid growth of the polar compart-

*Present address: Department of Molecular, Cell & Developmental Biology, Eli and Edythe Broad Center of Regenerative Medicine and Stem Cell Research, University of California, Los Angeles.

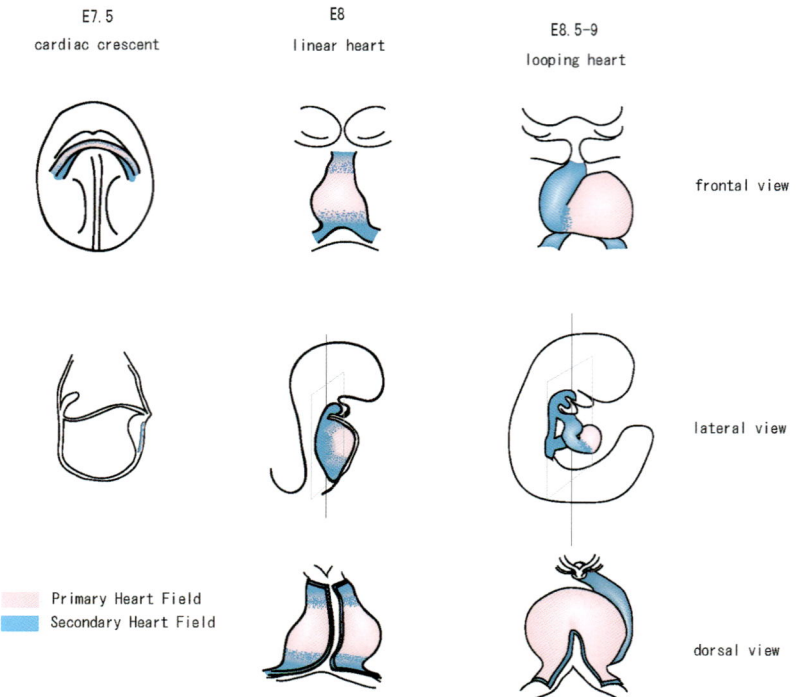

Figure 1. Morphogenesis of the mouse embryonic heart. The primordia of the heart are first morphologically recognizable as the cardiac crescent. Whereas the cells in the first heart field (FHF, *pink*) mainly contribute to the formation of the cardiac crescent, the cells in the second heart field (SHF, *blue*) appear in extra-crescent tissue. At around E8, bilateral crescents fuse in the midline to form a single linear heart tube. The primary heart tube is mainly composed of FHF cells, but SHF cells start to migrate toward the primary heart tube through the dorsal mesocardium, which is still continuous to the heart tube at any level at this stage. As the dorsal mesocardium dissolves, the heart tube become suspended by two poles, creating two subpopulations of SHF cells: anterior/arterial pole/outflow tract population and posterior/venous pole/inflow tract population. The anterior subpopulation gives rise to the myocardium of the outflow tract and right ventricle and acquires the ventricular phenotype. The posterior population contributes mainly to atrial myocytes and acquires the atrial phenotype. Concomitant with this, the heart tube starts to undergo rightward looping (d loop).

ments by itself is not the primary initiator of the asymmetry but an enhancer of preexisting laterality information.

After the cardiac chambers are positioned as seen in the postnatal heart, the heart undergoes a series of septation events. The formation of the conotruncal part of the heart requires the contribution of two distinct cell types: cardiac neural crest and second heart field progenitors. These two populations have distinct developmental origins, but they migrate through the pharyngeal mesoderm in a closely related pathway (Kelly and Buckingham 2002). Cardiac neural crest cells arise from the dorsal neural tube at the level of rhombomeres 6–8 and migrate into the arterial pole of the looping heart tube (Stoller and Epstein 2005; Hutson and Kirby 2007; Snider et al. 2007). This specific population of the neural crest contributes to the septum of the outflow tract and the smooth muscle wall of the ascending aorta, aortic arch, and orifice of the left and right coronary arteries. Simultaneously to this septation event, the outflow myocardium of the second heart field origin undergoes shortening. Ablation of either cell population results in the hypoplasia and rotation defects of the cardiac outflow tract (Yelbuz et al. 2002; Baldini 2005). Therefore, the interaction of these two populations has an important role in the morphogenesis of the cardiac outflow tract (Waldo et al. 2005a).

Endocardial cushions also contribute to the formation of septa as well as the valves. After the completion of the looping, endocardial cells at both ends of the heart tube—outflow tract and atrioventricular (AV) canal—transform into mesenchymal cell types and migrate into the cardiac jelly between endocardial and myocardial layers to form endocardial cushions. Cushions in the outflow portion are the primordia of the outflow septum and semilunar valves, and the cushion in the AV canal gives rise to atrial and ventricular septa and the AV valves.

CELL-FATE MAP

The earliest process of mammalian cardiogenesis has largely been a black box, partly because of the technical inaccessibility to the embryo at this stage. Instead, this step has been intensively studied in lower vertebrates (chick, amphibians, zebra fish) and nonvertebrate (fly) models. These nonmammalian models suggested that at least part of the endothelial/endocardial lineage shares a common origin with cardiac precursors. Recent advances in mouse genetics and embryonic stem (ES) cell technology have enabled us to study this process at the cellular level in mammals (Fig. 2). Mesodermal precursor cells expressing Flk1/Brachyury (T) and early cardiac progenitors labeled by Isl1/Nkx2.5/ Flk1 are shown to be capable of differentiating into cardiac, smooth muscle, and endothelial lineages from single cells (Garry and Olson 2006; Kattman et

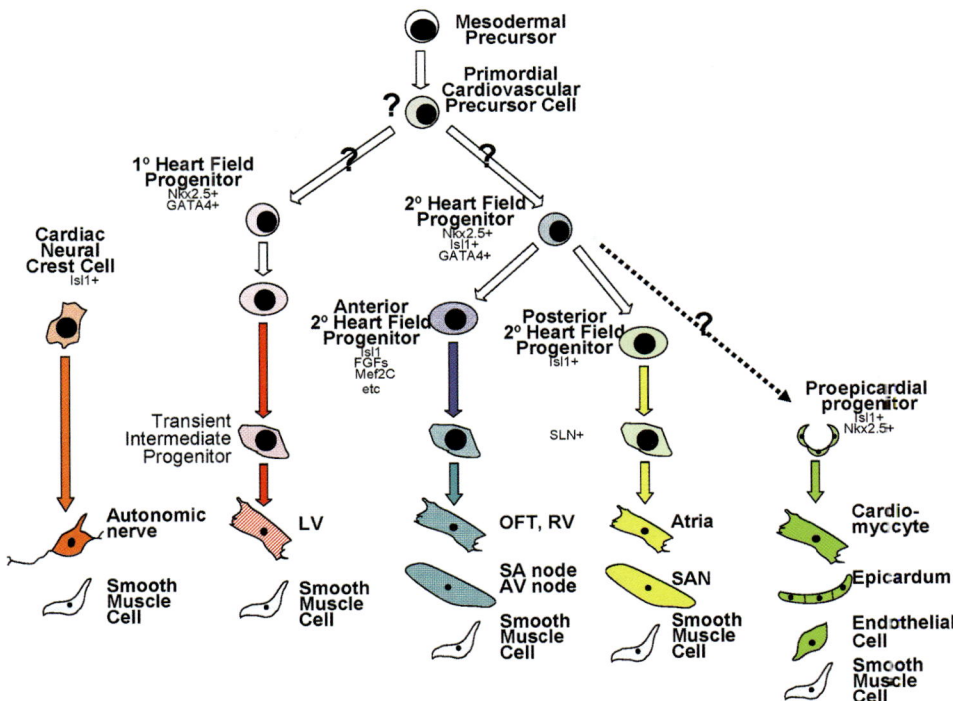

Figure 2. A model for cell-fate map of cardiac progenitors. Putative common precursor cells are specified from mesodermal precursors shortly after gastrulation. This common precursors give rise to first (Isl1$^-$) and second (Isl1$^+$) heart field progenitors (FHF and SHF). The SHF population can be further subdivided into anterior and posterior secondary heart field subpopulations. The FHF population gives rise to the left ventricle (*red*); the anterior SHF population gives rise to the outflow tract, right ventricle, and AV nodal cells; and the posterior SHF population contributes mainly to atrial myocytes and SA nodal cells. Proepicardial progenitor cells may share a common origin with SHF progenitors (they also give rise to cardiomyocytes, express early cardiac markers Isl1 and Nkx2.5, and arise in close vicinity to the posterior SHF). (LV) Left ventricle; (OFT) outflow tract; (RV) right ventricle; (SA) sinoatrial; (AV) atrioventricular; (SAN) sinoatrial node.

al. 2006, 2007; Moretti et al. 2006; Martin-Puig et al. 2008; Yang et al. 2008). Mesp1 acts as a master regulator during the commitment of mesodermal cells into cardiac lineages (Bondue et al. 2008; Lindsley et al. 2008; Wu 2008). Whereas endothelial lineages separate from cardiac and smooth muscle lineages at an early stage, cardiac and smooth muscle lineages are closely related until later stages of cardiogenesis (Wu et al. 2006). Smooth muscle cells appear to be recruited locally from a wide range of mesodermal tissues and differentiate in response to local inductive signals (Waldo and Kirby 1993; Topouzis and Majesky 1996; Waldo et al. 2005b; Majesky 2007).

Using in vivo retrospective clonal assays, Meilhac et al. (2004) showed that cardiac progenitors come from two major origins corresponding to the FHF and SHF. Both FHF and SHF progenitors contribute to most of the cardiac regions except for the left ventricle (mainly FHF) and outflow tract (mainly SHF), challenging the classical segmental view of cardiogenesis. Furthermore, they also speculate that these two lineages likely originate from putative common precursors. It remains to be answered whether there is a fundamental functional difference other than the marker expression between these two populations and whether these two are mutually exchangeable. Indeed, there are no morphological differences between the progenitors in FHF and SHF at a cellular level, nor are there

clear morphological boundaries at the cardiac crescent stage. Even in the postnatal heart, there are no obvious morphological differences among cardiomyocytes in OFT/RV and LV at the cellular level. Recent analysis suggests the possibility that Isl1 is transiently expressed in the FHF (Prall et al. 2007), although lineage-tracing studies support the concept of a distinct SHF subset of progenitors. Thus, the argument still remains as to whether it is relevant to define so-called SHF as another "organ field" of the heart (Abu-Issa et al. 2004; Moorman et al. 2007). Current studies designed to purify and tag specific subsets of FHF and SHF heart progenitors from ES cells and early-stage embryos should help to resolve this issue.

The breakdown of dorsal mesocardium further subdivides SHF into two major cardiac progenitor populations: anterior/arterial pole/outflow populations and posterior/venous pole/inflow populations (Fig. 3) The anterior population is labeled by several molecular markers, including fibroblast growth factors (FGFs), bone morphogenetic protein (BMPs), Wnt11, and a specific enhancer of Mef2c (Kelly et al. 2001; Cai et al. 2003; Dodou et al. 2004; Lin et al. 2007), and eventually acquires the ventricular phenotype contributing to the outflow tract and right ventricle. The posterior population eventually gives rise to atrial myocardium (Galli et al. 2008). The posterior population may also contribute to multiple lineages, such as the

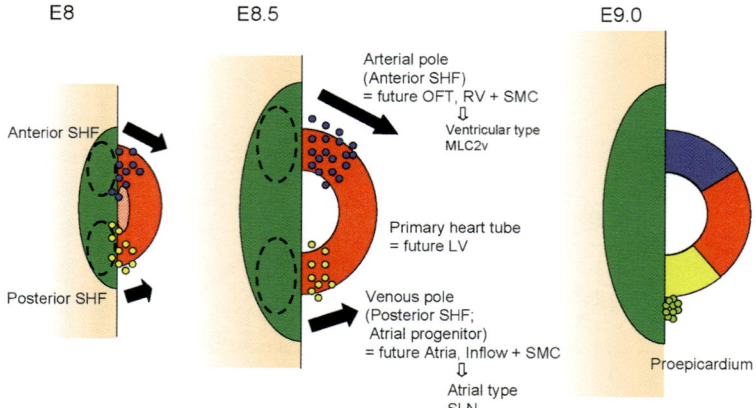

Figure 3. Migration of anterior and posterior population of SHF progenitors. Shortly after the formation of the linear heart tube, the anterior SHF progenitors migrate to the arterial pole of the primitive heart tube. This population is marked by FGF8/10 and the specific enhancer of Mef2c. The anterior population eventually acquires the ventricular phenotype and gives rise to the cardiomyocytes in the outflow tract and right ventricle, as well as smooth muscle cells in the base of the ascending aorta. The posterior subpopulation of the SHF migrates toward the venous pole and contributes to atrial myocytes and myocardial sleeves, as well as to smooth muscle cells in the cardiac inflow. Note that the proepicardium arises from the boundary between the embryonic liver and sinus venosus and thus spatially and temporarily continuous to the posterior SHF.

Tbx18$^+$ population (Christoffels et al. 2006) and mediastinal myocardial population (Anderson et al. 2006; Snarr et al. 2007a,b). Identification of additional specific molecular markers will further help to establish the cellular hierarchy of cardiac lineages (Laugwitz et al. 2008).

Recent findings suggest that the proepicardium may be another source of cardiomyocytes (Cai et al. 2008; Zhou et al. 2008). The proepicardium is a pair of grape-like clusters of epithelial cells that arise between the sinus venosus and liver primordium. Using retroviral tagging methods, Mikawa and Gourdie (1996) proved the speculation by Goor and Lillehei (1975) that the proepicardium is the source of the coronary vasculature. A combinatorial approach using mouse genetic models and the ES system showed that proepicardial cells also contribute to some of the cardiomyocytes (Cai et al. 2008; Zhou et al. 2008). These proepicardial progenitor cells are positive for both Isl1 and Nkx2.5. Given that the proepicardium is spatially and temporarily continuous to the posterior population of SHF, the proepicardial progenitors may share a common origin with cardiac progenitors in the SHF.

Apart from the argument regarding the expression of Isl1 in the FHF, Isl1 is expressed in most of the other cardiac lineages, as well as the proepicardial lineages. Hence, Isl1 labels a pool of various cardiac progenitor subpopulations. Although Isl1 is down-regulated as the cardiomyocytes mature, it stays on until a later stage in some subsets and turns off early in other subsets. Notably, these diverse progenitor populations arise from the spatiotemporal continuity of the Isl1-positive field of the primordial heart. The common function of the Isl1 transcription factor in these diverse populations is of interest. Obviously, Isl1 is expressed in proliferative cardiac progenitors. Our recent data show that Isl1 expression strongly correlates with cell cycle activity (A. Nakano et al., unpubl.). Therefore, Isl1 may well be regarded as a marker for proliferative progenitors during their transition to cardiomyocytes as well as for the cells early in the SHF lineage.

CONGENITAL HEART DISEASE AND ISL1 PROGENITOR

Many monogenic heart diseases result in closely related phenotypes, suggesting that abnormalities of cardiac progenitors resulting from these genetic deficiencies may be fundamental to the pathogenesis of congenital heart disease. Identification of the cellular level of the defects in the cardiac progenitor lineages in these diseases should link molecular function of genes to the pathogenesis of congenital heart defects. Figure 4 displays several of the human congenital heart defects associated with gene mutations. Although studies in model systems have shown that Isl1 is expressed in a variety of cardiac lineages and has a pivotal role in the transcription network during cardiogenesis (Black 2007), it is currently of major interest to determine if Isl1 genetic variation might relate to human congenital heart abnormalities.

CARDIAC OUTFLOW TRACT AND WNT/β-CATENIN SIGNALING

Many of the key major signaling molecules such as BMP, FGF, Shh, Wnt, and Notch are also implicated in multiple processes of cardiogenesis. Among these signals, the Wnt/β-catenin pathway is unique in that it has a highly context-dependent role during cardiogenesis. Early in avian cardiogenesis, canonical Wnt signals from the neuroectoderm are known to inhibit the specification of cardiac progenitors, resulting in a crescent-shaped heart-forming region (Marvin et al. 2001; Schneider and Mercola 2001; Tzahor and Lassar 2001). Mouse ES-based data mostly suggest a positive effect of Wnt/β-catenin signaling on the appearance of beating cardiomyocytes (49–52). Genetic loss- and gain-of-function experiments using various Cre mice mostly resulted in the conclusion that Wnt/β-catenin is involved in expansion of the cardiac progenitor population (Naito et al. 2006; Cohen et al. 2007; Klaus et al. 2007; Kwon et al.

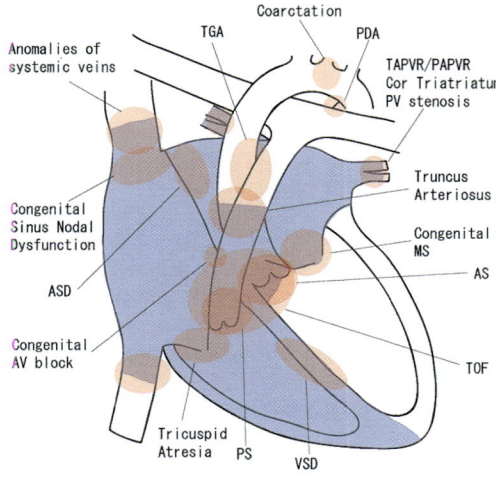

Figure 4. Distribution of Isl1-positive progenitor-derived cells and congenital heart disease. (AS) Aortic stenosis; (ASD) atrial septal defect; (AV) atrioventricular; (MS) mitral stenosis; (PDA) patent ductus arteriosus; (PS) pulmonary stenosis; (PV) pulmonary vein; (TAPVR/PAPVR) total/partial anomalous pulmonary venous return; (TGA) transposition of great arteries; (TOF) tetralogy of Fallot.

	Isl1-derived cells in the lesion	Congenital anomalies
Outflow tract defects		
Aortic stenosis		
Fibromuscular subvalvar	○	NKX2.5 (A)
Bicuspid aortic valve	○	Notch2 (Alagille) (B)
Supravalvar (SVAS)	○	Elastin (Williams-Beuren) (C)
Pulmonary stenosis	○	NF1 (D), Jag1 (Alagille) (E), Notch2 (Alagille) (B), PTPN11/Shp2 (Noonan) (F), MEK1/2 (G), B-Raf//H/K-Ras (H), GATA4 (I)
Tetralogy of Fallot		
RV/OFT obstruction, overriding aorta	○	Jag1 (Alagille) (E), Notch2 (Alagille) (B), NKX2.5 (A)
Double-outlet right ventricles	○	CFC1 (J), NKX2.5 (A), GATA4 (I)
Truncus arteriosus	○	TBX1 (DiGeorge) (K), VEGF promotor (DiGeorge) (L)
Transposition of great arteries	○	ZIC3 (M), CFC1 (J), THRAP2 (N)
Inflow tract defects		
Cor triatriatum	?	
TAPVR and PAPVR	?	
Anomalies of systemic veins	○	NKX2.5 (A)
Septation defects		
VSD	○	TBX1 (DiGeorge) (O), TBX5 (P), CHD7 (CHARGE) (Q), Sema3E (CHARGE) (R), NKX2.5 (A), GATA4 (I)
ASD	○	TBX5 (P), CHD7 (CHARGE) (Q), Sema3E (CHARGE) (R), NKX2.5 (A), GATA4 (I), TBX20 (S), MYH6 (T)
AVSD (endocardial cushion defect)	○	GJA1 (U), CRELD1 (V), FOG2 (W), GATA4 (I)
Valvular defects		
Congenital MS	?	
Tricuspid atresia	?	NKX2.5 (Epstein) (A)
Defects in aortic remodeling		
Patent ductus arteriosus (usually secondary)	○	TFAP2β (X)
Coarctation of the aorta	○	TBX1 (DiGeorge) (O)
Aortopulmonary septal defect (AP window)	○	
Coronary defects		
Anomalous origin of coronary arteries	○	
Congenital distal coronary defects (fistula, aneurysm, stenosis, hypoplasia)	○	
Congenital cardiac conduction system defects		
Congenital sinus nodal dysfunction	○	SCN5A (Y)
WPW syndrome	?	PRKAG2 (Z)
Congenital AV block	○	TBX5 (P), NKX2.5 (A)
Long QT syndrome	×	KCNQ1, KCNH2, KCNE1, KCNE2, CACNA1c, CAV3, SCN5A, SCN4B, HCN4
RBBB	?	TBX5 (P)
Left ventricular defects		
Hypoplastic left heart syndrome	×	GJA1 (U)
Left ventricular noncompaction syndrome	×	DTNA (ZZ), NKX2.5 (Epstein) (A)

References: (A) Schott et al. 1998; Benson et al. 1999; Kasahara et al. 2000; Goldmuntz et al. 2001; Gutierrez-Roelens et al. 2002; Ikeda et al. 2002; Watanabe et al. 2002; McElhinney et al. 2003; Kasahara and Benson 2004; Pashmforoush et al. 2004; (B) McDaniell et al. 2006 (C) Ewart et al. 1993; (D) Li et al. 1992; Bahuau et al. 1996, 1998; Lin et al. 2000; (E) Li et al. 1997a; Oda et al. 1997; (F) Tartaglia et al. 2001; (G) Rodriguez-Viciana et al. 2006; (H) Aoki et al. 2005; Niihori et al. 2006; Schubbert et al. 2006; (I) Pehlivan et al. 1999; Garg et al. 2003; Hirayama-Yamada et al. 2005; (J) Bamford et al. 2000; Goldmuntz et al. 2002; (K) Jerome and Papaioannou 2001; Lindsay et al. 2001; Merscher et al. 2001; Yagi et al. 2003; (L) Stalmans et al. 2003; (M) Gebbia et al. 1997; (N) Muncke et al. 2003; (O) Jerome and Papaioannou 2001; Lindsay et al. 2001; Merscher et al. 2001; (P) Basson et al. 1997; Li et al. 1997b; (Q) Vissers et al. 2004; (R) Lalani et al. 2004; (S) Kirk et al. 2007; (T) Ching et al. 2005; (U) Dasgupta et al. 2001; (V) Robinson et al. 2003; Zatyka et al. 2005; Maslen et al. 2006; (W) Pizzuti et al. 2003; (X) Satoda et al. 2000; Zhao et al. 2001; (Y) Benson et al. 2003; (Z) Gollob et al. 2001a,b; (ZZ) Ichida et al. 2001.

2007; Qyang et al. 2007; Ueno et al. 2007). These results reflect the difference in each experimental system. First, Wnt/β-catenin has various roles at different stages during cardiogenesis (stage dependency) (Tzahor 2007). Second, it is likely that the Wnt/β-catenin pathway has a different role in various subsets of the progenitors (sublineage dependency). Third, many of the key signaling molecules and transcription factors are known to have different roles at different dosages and different gradients (dosage and gradient dependency). Fourth, in some cases, it is difficult to segregate specification, migration, and proliferation of the Isl1 progenitors in some models. Thus, the Wnt/β-catenin pathway has a highly context-dependent role. However, despite all of the limitations of each experimental system, it is clear that the Wnt/β-catenin pathway is one of the most promising signaling molecules from the viewpoint of future therapeutic application.

ANCHORING ATRIAL CHAMBERS AND GREAT VEINS IN THE CARDIAC INFLOW TRACT

As discussed above, the SHF delineated by Isl1 can be subdivided into two major cardiac populations: the anterior subpopulation in the arterial pole and the posterior subpopulation in the venous pole of the heart tube (Fig. 3) (Galli et al. 2008). Isl1-positive cardiac progenitors that populate in the anterior SHF/arterial pole of the primitive heart tube give rise to both cardiomyocytes and smooth muscle cells in the cardiac outflow tract (Waldo et al. 2005b). Analogous to this observation, we have recently found that Isl1-positive cardiac progenitors in the posterior SHF/venous pole give rise to atrial cardiomyocytes as well as smooth muscle cells in the cardiac inflow region (A. Nakano et al., unpubl.). Single-cell analysis using a cardiac mesenchymal feeder system revealed that this lineage diversification is controlled by the bipotency of

atrial progenitors that express Isl1 and sarcolipin (SLN), a marker for committed atrial cells.

An important biological question is why these Isl1-positive cardiac progenitors show bipotency until a relatively late stage and how this bipotency serves to coordinate the morphogenesis of atrial chambers and the great veins. One possibility is that bipotency is required to connect the boundary between the atrial chambers and great veins and to maintain the cardiac syncytium. In fact, when the differentiation of atrial progenitors is inhibited, murine embryos overexpressing β-catenin in the atrial lineage (SLN[cre/+]; βcat[ex3/+]) showed significantly smaller atria and inflow tracts compared with control littermates (Fig. 5). Some of the mutants examined showed a massive hemorrhage, possibly leaking from the boundary between atrial chamber and inflow veins. These data suggest that bipotency of atrial progenitors is required for anchoring the atrial chamber to the great veins and forming functional syncytia that facilitate the dynamism of the central circulation system. From a developmental perspective, it implies that the cells in the boundary of two different tissues have to be phenotypically plastic in order to glue these tissues and maintain anatomical and functional continuity. It would be of interest to determine whether the anterior/arterial pole population of the cardiac progenitors displays similar bipotency during cardiogenesis.

CONGENITAL HEART DISEASE AFFECTING THE DEVELOPMENT OF CARDIAC INFLOW TRACT

The defect in the connection of the atrial chamber and great veins possibly has an important role in the pathogenesis of several particular congenital heart malformations (Fig. 4). Anomalous pulmonary venous return denotes a spectrum of malformations in which at least one pulmonary vein connects to the right atrium directly or

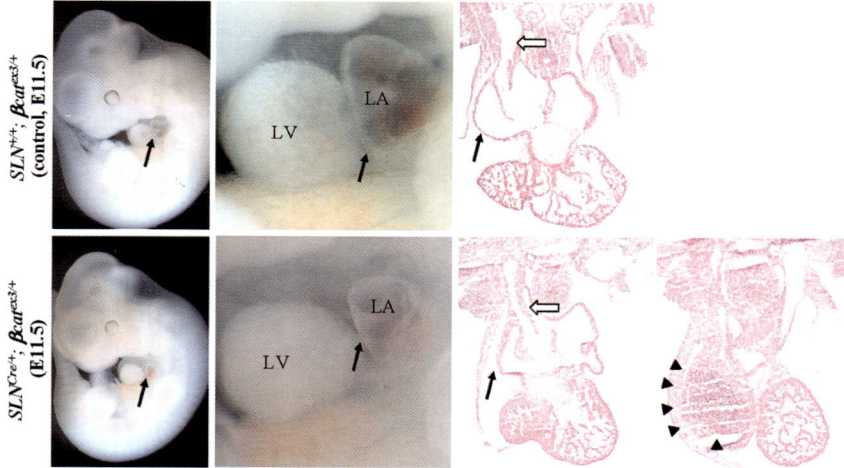

Figure 5. Inhibition of the differentiation of atrial progenitors results in the malformation of the boundary between the atrial chamber and great veins. The differentiation of atrial progenitors is inhibited by the overexpressing active form of β-catenin. The SLN[cre/+]; βcat[ex3/+] embryo showed significantly smaller atria (*black arrows*) and a narrower vena cava (*white arrows*). One of three mutants showed a massive hemorrhage, possibly due to the malconnection of the atrial chamber and great veins (*arrowheads*). These data suggest that the proper differentiation of atrial progenitors is required for anchoring the vascular smooth muscle wall to the myocardium of atria and for generating functional syncytium. (LV) Left ventricle; (LA) left atrium.

through a systemic vein(s). This anomaly can be asymptomatic depending on the severity of the left-to-right shunt, and the incidence may be more common than we find from clinical cases. Cor triatriatum is another congenital heart disease in which the connection of pulmonary veins and left atrium is affected. The inadequate integration of the common pulmonary vein results in the formation of two chambers: the accessory chamber (enlarged common pulmonary vein) and the true left atrium. Mice have a common pulmonary vein that collects blood from four lobar pulmonary veins. Pulmonary vein stenosis is a congenital heart disease characterized by recurrent obstruction of pulmonary veins by proliferation of spindle-shaped myofibroblasts of the junction between the pulmonary vein and the left atrial chamber (Sadr et al. 2000). Compared with the anomalies on the left side, the connection problem of the systemic veins to the right side of the heart is rare. Precise analysis of these human congenital diseases in the inflow tract will answer the questions of how the myocardial sleeves develop, how the boundary of atrial myocardium and vascular smooth muscle cells is formed during cardiogenesis, and the biological significance of the plasticity of atrial cells.

RETRODIFFERENTIATION

The Isl1/SLN atrial progenitors from the posterior/venous part of the SHF maintain their migratory and proliferative potential even until midgestation. Consistent with this, Isl1 expression continues at E13.5 in mouse atria (Sun et al. 2007). The late development of atrial progenitors is likely linked to the fact that atrial progenitors remain bipotent until later stages of heart development. Atrial cells can acquire a smooth muscle and ventricular phenotype even after birth. Interestingly, this phenotypical plasticity is at least partially acquired by a retrodifferentiation process. Upon culture on cardiac feeder or upon in vivo injury, atrial cells reenter the cell cycle and re-express the Isl1 that is not normally expressed in postnatal atrial myocytes. The cellular plasticity of the atrial cells in the inflow may be related to the pathogenesis of atrial fibrillation.

A number of stem/progenitor cell populations have been identified in the postnatal heart including c-kit-positive and Sca1-positive stem/progenitor cells and Isl1-positive progenitors that are embedded in cardiac mesenchyme and considered to be developmental remnants of the secondary heart field (Orlic et al. 2001; Beltrami et al. 2003; Oh et al. 2003; Laugwitz et al. 2005). Researchers speculate that regenerating zebra fish myocardium also arises from undifferentiated cardiac progenitors (Lepilina et al. 2006). The discovery of these preexisting stem/progenitor cells offers a great opportunity to regenerate the lost cardiomyocytes (Murry et al. 2005; Srivastava and Ivey 2006; Rubart and Field 2008). However, besides the activation of these residential undifferentiated stem/progenitor cells (dedifferentiation-independent regeneration), a trace of another mechanism of tissue regeneration may also take place in the heart: mature cells dedifferentiate, reenter the cell cycle, and redifferentiate another mature cell type (dedifferentia-

tion-dependent mechanism). Further analysis using defined markers will lead to a better understanding of the pathogenesis of the above-mentioned congenital heart diseases, the normal developmental process of the atrial lineage, the cellular mechanism of their plasticity, and eventually the clinical usage of Isl1-positive progenitor cells derived from atria.

ACKNOWLEDGMENTS

We thank all of our collaborators who contributed to the current work and to our ongoing research. Our studies are supported by LeDuq foundation, Harvard Stem Cell Institute, Massachusetts General Hospital, and the National Institutes of Health.

REFERENCES

Abu-Issa, R., Waldo, K., and Kirby, M.L. 2004. Heart fields: One, two or more? *Dev. Biol.* **272:** 281–285.

Anderson, R.H., Brown, N.A., and Moorman, A.F. 2006. Development and structures of the venous pole of the heart. *Dev. Dyn.* **235:** 2–9.

Aoki, Y., Niihori, T., Kawame, H., Kurosawa, K., Ohashi, H., Tanaka, Y., Filocamo, M., Kato, K., Suzuki, Y., Kure, S., et al. 2005. Germline mutations in *HRAS* proto-oncogene cause Costello syndrome. *Nat. Genet.* **37:** 1038–1040.

Bahuau, M., Flintoff, W., Assouline, B., Lyonnet, S., Le Merrer, M., Prieur, M., Guilloud-Bataille, M., Feingold, N., Munnich, A., Vidaud, M., et al. 1996. Exclusion of allelism of Noonan syndrome and neurofibromatosis-type 1 in a large family with Noonan syndrome-neurofibromatosis association. *Am. J. Med. Genet.* **66:** 347–355.

Bahuau, M., Houdayer, C., Assouline, B., Blanchet-Bardon, C., Le Merrer, M., Lyonnet, S., Giraud, S., Recan, D., Lakhdar, H., Vidaud, M., et al. 1998. Novel recurrent nonsense mutation causing neurofibromatosis type 1 (NF1) in a family segregating both NF1 and Noonan syndrome. *Am. J. Med. Genet.* **75:** 265–272.

Baldini, A. 2005. Dissecting contiguous gene defects: *TBX1*. *Curr. Opin. Genet. Dev.* **15:** 279–284.

Bamford, R.N., Roessler, E., Burdine, R.D, Saplakoglu, U., dela Cruz, J., Splitt, M., Goodship, J.A., Towbin, J., Bowers, P., Ferrero, G.B., et al. 2000. Loss-of-function mutations in the EGF-CFC gene *CFC1* are associated with human left-right laterality defects. *Nat. Genet.* **26:** 365–369.

Basson, C.T., Bachinsky, D.R., Lin, R.C., Levi, T., Elkins, J.A., Soults, J., Grayzel, D., Kroumpouzou, E., Traill, T.A., Leblanc-Straceski, J., et al. 1997. Mutations in human *TBX5* (corrected) cause limb and cardiac malformation in Holt-Oram syndrome. *Nat. Genet.* **15:** 30–35.

Beltrami, A.P., Barlucchi, L., Torella, D., Baker, M., Limana, F., Chimenti, S., Kasahara, H., Rota, M., Musso, E., Urbanek, K., et al. 2003. Adult cardiac stem cells are multipotent and support myocardial regeneration. *Cell* **114:** 763–776.

Benson, D.W., Wang, D.W., Dyment, M, Knilans, T.K., Fish, F.A., Strieper, M.J., Rhodes, T.H., and George, Jr., A.L. 2003. Congenital sick sinus syndrome caused by recessive mutations in the cardiac sodium channel gene (*SCN5A*). *J. Clin. Invest.* **112:** 1019–1028.

Benson, D.W., Silberbach, G.M., Kavanaugh-McHugh, A., Cottrill, C., Zhang, Y., Riggs, S., Smalls, O., Johnson, M.C., Watson, M.S., Seidman, J.G., et al. 1999. Mutations in the cardiac transcription factor *NKX2.5* affect diverse cardiac developmental pathways. *J. Clin. Invest.* **104:** 1567–1573.

Black, B.L. 2007. Transcriptional pathways in second heart field development. *Semin. Cell Dev. Biol.* **18:** 67–76.

Bondue, A., Lapouge, G., Paulissen, C., Semeraro, C., Iacovino, M., Kyba, M., and Blanpain, C. 2008. Mesp1 acts as a master regulator of multipotent cardiovascular progenitor specification. *Cell Stem Cell* **3:** 69–84.

Buckingham, M., Meilhac, S., and Zaffran, S. 2005. Building the mammalian heart from two sources of myocardial cells. *Nat. Rev. Genet.* **6:** 826–835.

Cai, C.L., Liang, X., Shi, Y., Chu, P.H., Pfaff, S.L., Chen, J., and Evans, S. 2003. Isl1 identifies a cardiac progenitor population that proliferates prior to differentiation and contributes a majority of cells to the heart. *Dev. Cell* **5:** 877–889.

Cai, C.L., Martin, J.C., Sun, Y., Cui, L., Wang, L., Ouyang, K., Yang, L., Bu, L., Liang, X., Zhang, X., et al. 2008. A myocardial lineage derives from *Tbx18* epicardial cells. *Nature* **454:** 104–108.

Ching, Y.H., Ghosh, T.K., Cross, S.J., Packham, E.A., Honeyman, L., Loughna, S., Robinson, T.E., Dearlove, A.M., Ribas, G., Bonser, A.J., et al. 2005. Mutation in myosin heavy chain 6 causes atrial septal defect. *Nat. Genet.* **37:** 423–428.

Christoffels, V.M., Mommersteeg, M.T., Trowe, M.O., Prall, O.W., de Gier-de Vries, C., Soufan, A.T., Bussen, M., Schuster-Gossler, K., Harvey, R.P., Moorman, A.F., et al. 2006. Formation of the venous pole of the heart from an Nkx2-5-negative precursor population requires *Tbx18*. *Circ. Res.* **98:** 1555–1563.

Cohen, E.D., Wang, Z., Lepore, J.J., Lu, M.M., Taketo, M.M., Epstein, D.J., and Morrisey, E.E. 2007. Wnt/β-catenin signaling promotes expansion of Isl-1-positive cardiac progenitor cells through regulation of FGF signaling. *J. Clin. Invest.* **117:** 1794–1804.

Dasgupta, C., Martinez, A.M., Zuppan, C.W., Shah, M.M., Bailey, L.L., and Fletcher, W.H. 2001. Identification of *connexin43* (α1) gap junction gene mutations in patients with hypoplastic left heart syndrome by denaturing gradient gel electrophoresis (DGGE). *Mutat. Res.* **479:** 173–186.

Dodou, E., Verzi, M.P., Anderson, J.P., Xu, S.M., and Black, B.L. 2004. Mef2c is a direct transcriptional target of ISL1 and GATA factors in the anterior heart field during mouse embryonic development. *Development* **131:** 3931–3942.

Ewart, A.K., Morris, C.A., Atkinson, D., Jin, W., Sternes, K., Spallone, P., Stock, A.D., Leppert, M., and Keating, M.T. 1993. Hemizygosity at the elastin locus in a developmental disorder, Williams syndrome. *Nat. Genet.* **5:** 11–16.

Galli, D., Dominguez, J.N., Zaffran, S., Munk, A., Brown, N.A., and Buckingham, M.E. 2008. Atrial myocardium derives from the posterior region of the second heart field, which acquires left-right identity as Pitx2c is expressed. *Development* **135:** 1157–1167.

Garg, V., Kathiriya, I.S., Barnes, R., Schluterman, M.K., King, I.N., Butler, C.A., Rothrock, C.R., Eapen, R.S., Hirayama-Yamada, K., Joo, K., et al. 2003. GATA4 mutations cause human congenital heart defects and reveal an interaction with TBX5. *Nature* **424:** 443–447.

Garry, D.J. and Olson, E.N. 2006. A common progenitor at the heart of development. *Cell* **127:** 1101–1104.

Gebbia, M., Ferrero, G.B., Pilia, G., Bassi, M.T., Aylsworth, A., Penman-Splitt, M., Bird, L.M., Bamforth, J.S., Burn, J., Schlessinger, D., et al. 1997. X-linked situs abnormalities result from mutations in ZIC3. *Nat. Genet.* **17:** 305–308.

Goldmuntz, E., Geiger, E., and Benson, D.W. 2001. NKX2.5 mutations in patients with tetralogy of fallot. *Circulation* **104:** 2565–2568.

Goldmuntz, E., Bamford, R., Karkera, J.D., dela Cruz, J., Roessler, E., and Muenke, M. 2002. *CFC1* mutations in patients with transposition of the great arteries and double-outlet right ventricle. *Am. J. Hum. Genet.* **70:** 776–780.

Gollob, M.H., Seger, J.J., Gollob, T.N., Tapscott, T., Gonzales, O., Bachinski, L., and Roberts, R. 2001a. Novel PRKAG2 mutation responsible for the genetic syndrome of ventricular preexcitation and conduction system disease with childhood onset and absence of cardiac hypertrophy. *Circulation* **104:** 3030–3033.

Gollob, M.H., Green, M.S., Tang, A.S., Gollob, T., Karibe, A., Ali Hassan, A.S., Ahmad, F., Lozado, R., Shah, G., Fananapazir, L., et al. 2001b. Identification of a gene responsible for familial Wolff-Parkinson-White syndrome. *N. Engl. J. Med.* **344:** 1823–1831.

Goor, D.A. and Lillehei, C.W. 1975. *Congenital malformations of the heart: Embryology, anatomy and operative considerations.* Grune and Stratton, New York.

Gutierrez-Roelens, I., Sluysmans, T., Gewillig, M., Devriendt, K., and Vikkula, M. 2002. Progressive AV-block and anomalous venous return among cardiac anomalies associated with two novel missense mutations in the *CSX/NKX2-5* gene. *Hum. Mutat.* **20:** 75–76.

Hirayama-Yamada, K., Kamisago, M., Akimoto, K., Aotsuka, H., Nakamura, Y., Tomita, H., Furutani, M., Imamura, S., Takao, A., Nakazawa, M., et al. 2005. Phenotypes with *GATA4* or *NKX2.5* mutations in familial atrial septal defect. *Am. J. Med. Genet. A* **135:** 47–52.

Hutson, M.R. and Kirby, M.L. 2007. Model systems for the study of heart development and disease. Cardiac neural crest and conotruncal malformations. *Semin. Cell Dev. Biol.* **18:** 101–110.

Ichida, F., Tsubata, S., Bowles, K.R., Haneda, N., Uese, K., Miyawaki, T., Dreyer, W.J., Messina, J., Li, H., Bowles, N.E., et al. 2001. Novel gene mutations in patients with left ventricular noncompaction or Barth syndrome. *Circulation* **103:** 1256–1263.

Ikeda, Y., Hiroi, Y., Hosoda, T., Utsunomiya, T., Matsuo, S., Ito, T., Inoue, J., Sumiyoshi, T., Takano, H., Nagai, R., et al. 2002. Novel point mutation in the cardiac transcription factor *CSX/NKX2.5* associated with congenital heart disease. *Circ. J.* **66:** 561–563.

Jerome, L.A. and Papaioannou, V.E. 2001. DiGeorge syndrome phenotype in mice mutant for the T-box gene, *Tbx1*. *Nat. Genet.* **27:** 286–291.

Kasahara, H. and Benson, D.W. 2004. Biochemical analyses of eight *NKX2.5* homeodomain missense mutations causing atrioventricular block and cardiac anomalies. *Cardiovasc. Res.* **64:** 40–51.

Kasahara, H., Lee, B., Schott, J.J., Benson, D.W., Seidman, J.G., Seidman, C.E., and Izumo, S. 2000. Loss of function and inhibitory effects of human CSX/NKX2.5 homeoprotein mutations associated with congenital heart disease. *J. Clin. Invest.* **106:** 299–308.

Kattman, S.J., Adler, E.D., and Keller, G.M. 2007. Specification of multipotential cardiovascular progenitor cells during embryonic stem cell differentiation and embryonic development. *Trends Cardiovasc. Med.* **17:** 240–246.

Kattman, S.J., Huber, T.L., and Keller, G.M. 2006. Multipotent flk-1+ cardiovascular progenitor cells give rise to the cardiomyocyte, endothelial, and vascular smooth muscle lineages. *Dev. Cell* **11:** 723–732.

Kelly, R.G. and Buckingham, M.E. 2002. The anterior heart-forming field: Voyage to the arterial pole of the heart. *Trends Genet.* **18:** 210–216.

Kelly, R.G., Brown, N.A., and Buckingham, M.E. 2001. The arterial pole of the mouse heart forms from *Fgf10*-expressing cells in pharyngeal mesoderm. *Dev. Cell* **1:** 435–440.

Kirk, E.P., Sunde, M., Costa, M.W., Rankin, S.A., Wolstein, O., Castro, M.L., Butler, T.L., Hyun, C., Guo, G., Otway, R., et al. 2007. Mutations in cardiac T-box factor gene TBX20 are associated with diverse cardiac pathologies, including defects of septation and valvulogenesis and cardiomyopathy. *Am. J. Hum. Genet.* **81:** 280–291.

Klaus, A., Saga, Y., Taketo, M.M., Tzahor, E., and Birchmeier, W. 2007. Distinct roles of Wnt/β-catenin and Bmp signaling during early cardiogenesis. *Proc. Natl. Acad. Sci.* **104:** 18531–18536.

Kwon, C., Arnold, J., Hsiao, E.C., Taketo, M.M., Conklin, B.R., and Srivastava, D. 2007. Canonical Wnt signaling is a positive regulator of mammalian cardiac progenitors. *Proc. Natl. Acad. Sci.* **104:** 10894–10899.

Lalani, S.R., Safiullah, A.M., Molinari, L.M., Fernbach, S.D., Martin, D.M., and Belmont, J.W. 2004. *SEMA3E* mutation in a patient with CHARGE syndrome. *J. Med. Genet.* **41:** e94.

Laugwitz, K.L., Moretti, A., Caron, L., Nakano, A., and Chien, K.R. 2008. Islet1 cardiovascular progenitors: A single source for heart lineages? *Development* **135:** 193–205.

Laugwitz, K.L., Moretti, A., Lam, J., Gruber, P., Chen, Y., Woodard, S., Lin, L.Z., Cai, C.L., Lu, M.M., Reth, M., et al.

2005. Postnatal isl1[+] cardioblasts enter fully differentiated cardiomyocyte lineages. *Nature* **433:** 647–653.

Lepilina, A., Coon, A.N., Kikuchi, K., Holdway, J.E., Roberts, R.W., Burns, C.G., and Poss, K.D. 2006. A dynamic epicardial injury response supports progenitor cell activity during zebrafish heart regeneration. *Cell* **127:** 607–619.

Li, L., Krantz, I.D., Deng, Y., Genin, A., Banta, A.B., Collins, C.C., Qi, M., Trask, B.J., Kuo, W.L., Cochran, J., et al. 1997a. Alagille syndrome is caused by mutations in human *Jagged1*, which encodes a ligand for Notch1. *Nat. Genet.* **16:** 243–251.

Li, Q.Y., Newbury-Ecob, R.A., Terrett, J.A., Wilson, D.I., Curtis, A.R., Yi, C.H., Gebuhr, T., Bullen, P.J., Robson, S.C., Strachan, T., et al. 1997b. Holt-Oram syndrome is caused by mutations in TBX5, a member of the *Brachyury* (*T*) gene family. *Nat. Genet.* **15:** 21–29.

Li, Y., Bollag, G., Clark, R., Stevens, J., Conroy, L., Fults, D., Ward, K., Friedman, E., Samowitz, W., Robertson, M., et al. 1992. Somatic mutations in the neurofibromatosis 1 gene in human tumors. *Cell* **69:** 275–281.

Lin, A.E., Birch, P.H., Korf, B.R., Tenconi, R., Niimura, M., Poyhonen, M., Armfield Uhas, K., Sigorini, M., Virdis, R., Romano, C., et al. 2000. Cardiovascular malformations and other cardiovascular abnormalities in neurofibromatosis 1. *Am. J. Med. Genet.* **95:** 108–117.

Lin, L., Cui, L., Zhou, W., Dufort, D., Zhang, X., Cai, C.L., Bu, L., Yang, L., Martin, J., Kemler, R., et al. 2007. β-Catenin directly regulates *Islet1* expression in cardiovascular progenitors and is required for multiple aspects of cardiogenesis. *Proc. Natl. Acad. Sci.* **104:** 9313–9318.

Lindsay, E.A., Vitelli, F., Su, H., Morishima, M., Huynh, T., Pramparo, T., Jurecic, V., Ogunrinu, G., Sutherland, H.F., Scambler, P.J., et al. 2001. *Tbx1* haploinsufficieny in the DiGeorge syndrome region causes aortic arch defects in mice. *Nature* **410:** 97–101.

Lindsley, R.C., Gill, J.G., Murphy, T.L., Langer, E.M., Cai, M., Mashayekhi, M., Wang, W., Niwa, N., Nerbonne, J.M., Kyba, M., et al. 2008. Mesp1 coordinately regulates cardiovascular fate restriction and epithelial-mesenchymal transition in differentiating ESCs. *Cell Stem Cell* **3:** 55–68.

Majesky, M.W. 2007. Developmental basis of vascular smooth muscle diversity. *Arterioscler. Thromb. Vasc. Biol.* **27:** 1248–1258.

Martin-Puig, S., Wang, Z., and Chien, K.R. 2008. Lives of a heart cell: Tracing the origins of cardiac progenitors. *Cell Stem Cell* **2:** 320–331.

Marvin, M.J., Di Rocco, G., Gardiner, A., Bush, S.M., and Lassar, A.B. 2001. Inhibition of Wnt activity induces heart formation from posterior mesoderm. *Genes Dev.* **15:** 316–327.

Maslen, C.L., Babcock, D., Robinson, S.W., Bean, L.J., Dooley, K.J., Willour, V.L., and Sherman, S.L. 2006. *CRELD1* mutations contribute to the occurrence of cardiac atrioventricular septal defects in Down syndrome. *Am. J. Med. Genet. A* **140:** 2501–2505.

McDaniell, R., Warthen, D.M., Sanchez-Lara, P.A., Pai, A., Krantz, I.D., Piccoli, D.A., and Spinner, N.B. 2006. *NOTCH2* mutations cause Alagille syndrome, a heterogeneous disorder of the notch signaling pathway. *Am. J. Hum. Genet.* **79:** 169–173.

McElhinney, D.B., Geiger, E., Blinder, J., Benson, D.W., and Goldmuntz, E. 2003. *NKX2.5* mutations in patients with congenital heart disease. *J. Am. Coll. Cardiol.* **42:** 1650–1655.

Meilhac, S.M., Esner, M., Kelly, R.G., Nicolas, J.F., and Buckingham, M.E. 2004. The clonal origin of myocardial cells in different regions of the embryonic mouse heart. *Dev. Cell* **6:** 685–698.

Merscher, S., Funke, B., Epstein, J.A., Heyer, J., Puech, A., Lu, M.M., Xavier, R.J., Demay, M.B., Russell, R.G., Factor, S., et al. 2001. *TBX1* is responsible for cardiovascular defects in velo-cardio-facial/DiGeorge syndrome. *Cell* **104:** 619–629.

Mikawa, T. and Gourdie, R.G. 1996. Pericardial mesoderm generates a population of coronary smooth muscle cells migrating into the heart along with ingrowth of the epicardial organ. *Dev. Biol.* 221–232.

Mjaatvedt, C.H., Nakaoka, T., Moreno-Rodriguez, R., Norris,

R.A., Kern, M.J., Eisenberg, C.A., Turner, D., and Markwald, R.R. 2001. The outflow tract of the heart is recruited from a novel heart-forming field. *Dev. Biol.* **238:** 97–109.

Moorman, A.F., Christoffels, V.M., Anderson, R.H., and van den Hoff, M.J. 2007. The heart-forming fields: One or multiple? *Philos. Trans. R. Soc. Lond. B Biol. Sci.* **362:** 1257–1265.

Moretti, A., Caron, L., Nakano, A., Lam, J.T., Bernshausen, A., Chen, Y., Qyang, Y., Bu, L., Sasaki, M., Martin-Puig, S., et al. 2006. Multipotent embryonic *isl1[+]* progenitor cells lead to cardiac, smooth muscle, and endothelial cell diversification. *Cell* **127:** 1151–1165.

Muncke, N., Jung, C., Rudiger, H., Ulmer, H., Roeth, R., Hubert, A., Goldmuntz, E., Driscoll, D., Goodship, J., Schon, K., et al. 2003. Missense mutations and gene interruption in *PROSIT240*, a novel *TRAP240*-like gene, in patients with congenital heart defect (transposition of the great arteries). *Circulation* **108:** 2843–2850.

Murry, C.E., Field, L.J., and Menasche, P. 2005. Cell-based cardiac repair: Reflections at the 10-year point. *Circulation* **112:** 3174–3183.

Naito, A.T., Shiojima, I., Akazawa, H., Hidaka, K., Morisaki, T., Kikuchi, A., and Komuro, I. 2006. Developmental stage-specific biphasic roles of Wnt/β-catenin signaling in cardiomyogenesis and hematopoiesis. *Proc. Natl. Acad. Sci.* **103:** 19812–19817.

Niihori, T., Aoki, Y., Narumi, Y., Neri, G., Cave, H., Verloes, A., Okamoto, N., Hennekam, R.C., Gillessen-Kaesbach, G., Wieczorek, D., et al. 2006. Germline *KRAS* and *BRAF* mutations in cardio-facio-cutaneous syndrome. *Nat. Genet.* **38:** 294–296.

Oda, T., Elkahloun, A.G., Pike, B.L., Okajima, K., Krantz, I.D., Genin, A., Piccoli, D.A., Meltzer, P.S., Spinner, N.B., Collins, F.S., et al. 1997. Mutations in the human *Jagged1* gene are responsible for Alagille syndrome. *Nat. Genet.* **16:** 235–242.

Oh, H., Bradfute, S.B., Gallardo, T.D., Nakamura, T., Gaussin, V., Mishina, Y., Pocius, J., Michael, L.H., Behringer, R.R., Garry, D.J., et al. 2003. Cardiac progenitor cells from adult myocardium: Homing, differentiation, and fusion after infarction. *Proc. Natl. Acad. Sci.* **100:** 12313–12318.

Olson, E.N. 2006. Gene regulatory networks in the evolution and development of the heart. *Science* **313:** 1922–1927.

Orlic, D., Kajstura, J., Chimenti, S., Jakoniuk, I., Anderson, S.M., Li, B., Pickel, J., McKay, R., Nadal-Ginard, B., Bodine, D.M., et al. 2001. Bone marrow cells regenerate infarcted myocardium. *Nature* **410:** 701–705.

Pashmforoush, M., Lu, J.T., Chen, H., Amand, T.S., Kondo, R., Pradervand, S., Evans, S.M., Clark, B., Feramisco, J.R., Giles, W., et al. 2004. Nkx2-5 pathways and congenital heart disease; loss of ventricular myocyte lineage specification leads to progressive cardiomyopathy and complete heart block. *Cell* **117:** 373–386.

Pehlivan, T., Pober, B.R., Brueckner, M., Garrett, S., Slaugh, R., Van Rheeden, R., Wilson, D.B., Watson, M.S., and Hing, A.V. 1999. *GATA4* haploinsufficiency in patients with interstitial deletion of chromosome region 8p23.1 and congenital heart disease. *Am. J. Med. Genet.* **83:** 201–206.

Pizzuti, A., Sarkozy, A., Newton, A.L., Conti, E., Flex, E., Digilio, M.C., Amati, F., Gianni, D., Tandoi, C., Marino, B., et al. 2003. Mutations of *ZFPM2/FOG2* gene in sporadic cases of tetralogy of Fallot. *Hum. Mutat.* **22:** 372–377.

Prall, O.W., Menon, M.K., Solloway, M.J., Watanabe, Y., Zaffran, S., Bajolle, F., Biben, C., McBride, J.J., Robertson, B.R., Chaulet, H., et al. 2007. An Nkx2-5/Bmp2/Smad1 negative feedback loop controls heart progenitor specification and proliferation. *Cell* **128:** 947–959.

Qyang, Y., Martin-Puig, S., Chiravuri, M., Chen, S., Xu, H., Bu, L., Jiang, X., Lin, L., Granger, A., Moretti, A., et al. 2007. The renewal and differentiation of *Isl1[+]* cardiovascular progenitors are controlled by a Wnt/β-catenin pathway. *Cell Stem Cell* **1:** 165–179.

Robinson, S.W., Morris, C.D., Goldmuntz, E., Reller, M.D., Jones, M.A., Steiner, R.D., and Maslen, C.L. 2003. Missense mutations in *CRELD1* are associated with cardiac atrioventricular septal defects. *Am. J. Hum. Genet.* **72:** 1047–1052.

Rodriguez-Viciana, P., Tetsu, O., Tidyman, W.E., Estep, A.L., Conger, B.A., Cruz, M.S., McCormick, F., and Rauen, K.A. 2006. Germline mutations in genes within the MAPK pathway cause cardio-facio-cutaneous syndrome. *Science* **311:** 1287–1290.

Rubart, M. and Field, L.J. 2008. Stem cell differentiation: Cardiac repair. *Cells Tissues Organs* **188:** 202–211.

Sadr, I.M., Tan, P.E., Kieran, M.W., and Jenkins, K.J. 2000. Mechanism of pulmonary vein stenosis in infants with normally connected veins. *Am. J. Cardiol.* **86:** 577–579.

Satoda, M., Zhao, F., Diaz, G.A., Burn, J., Goodship, J., Davidson, H.R., Pierpont, M.E., and Gelb, B.D. 2000. Mutations in *TFAP2B* cause Char syndrome, a familial form of patent ductus arteriosus. *Nat. Genet.* **25:** 42–46.

Schneider, V.A. and Mercola, M. 2001. Wnt antagonism initiates cardiogenesis in *Xenopus laevis*. *Genes Dev.* **15:** 304–315.

Schott, J.J., Benson, D.W., Basson, C.T., Pease, W., Silberbach, G.M., Moak, J.P., Maron, B.J., Seidman, C.E., and Seidman, J.G. 1998. Congenital heart disease caused by mutations in the transcription factor *NKX2-5*. *Science* **281:** 108–111.

Schubbert, S., Zenker, M., Rowe, S.L., Boll, S., Klein, C., Bollag, G., van der Burgt, I., Musante, L., Kalscheuer, V., Wehner, L.E., et al. 2006. Germline *KRAS* mutations cause Noonan syndrome. *Nat. Genet.* **38:** 331–336.

Snarr, B.S., Wirrig, E.E., Phelps, A.L., Trusk, T.C., and Wessels, A. 2007a. A spatiotemporal evaluation of the contribution of the dorsal mesenchymal protrusion to cardiac development. *Dev. Dyn.* **236:** 1287–1294.

Snarr, B.S., O'Neal, J.L., Chintalapudi, M.R., Wirrig, E.E., Phelps, A.L., Kubalak, S.W., and Wessels, A. 2007b. *Isl1* expression at the venous pole identifies a novel role for the second heart field in cardiac development. *Circ. Res.* **101:** 971–974.

Snider, P., Olaopa, M., Firulli, A.B., and Conway, S.J. 2007. Cardiovascular development and the colonizing cardiac neural crest lineage. *ScientificWorldJournal* **7:** 1090–1113.

Srivastava, D. 2006. Making or breaking the heart: From lineage determination to morphogenesis. *Cell* **126:** 1037–1048.

Srivastava, D. and Ivey, K.N. 2006. Potential of stem-cell-based therapies for heart disease. *Nature* **441:** 1097–1099.

Stalmans, I., Lambrechts, D., De Smet, F., Jansen, S., Wang, J., Maity, S., Kneer, P., von der Ohe, M., Swillen, A., Maes, C., et al. 2003. *VEGF:* A modifier of the del22q11 (DiGeorge) syndrome? *Nat. Med.* **9:** 173–182.

Stoller, J.Z. and Epstein, J.A. 2005. Cardiac neural crest. *Semin. Cell Dev. Biol.* **16:** 704–715.

Sun, Y., Liang, X., Najafi, N., Cass, M., Lin, L., Cai, C.L., Chen, J., and Evans, S.M. 2007. Islet 1 is expressed in distinct cardiovascular lineages, including pacemaker and coronary vascular cells. *Dev. Biol.* **304:** 286–296.

Tartaglia, M., Mehler, E.L., Goldberg, R., Zampino, G., Brunner, H.G., Kremer, H., van der Burgt, I., Crosby, A.H., Ion, A., Jeffery, S., et al. 2001. Mutations in *PTPN11,* encoding the protein tyrosine phosphatase SHP-2, cause Noonan syndrome. *Nat. Genet.* **29:** 465–468.

Topouzis, S. and Majesky, M.W. 1996. Smooth muscle lineage diversity in the chick embryo. Two types of aortic smooth muscle cell differ in growth and receptor-mediated transcriptional responses to transforming growth factor-β. *Dev. Biol.* **178:** 430–445.

Tzahor, E. 2007. Wnt/β-catenin signaling and cardiogenesis: Timing does matter. *Dev. Cell* **13:** 10–13.

Tzahor, E. and Lassar, A.B. 2001. Wnt signals from the neural tube block ectopic cardiogenesis. *Genes Dev.* **15:** 255–260.

Ueno, S., Weidinger, G., Osugi, T., Kohn, A.D., Golob, J.L., Pabon, L., Reinecke, H., Moon, R.T., and Murry, C.E. 2007. Biphasic role for Wnt/β-catenin signaling in cardiac specification in zebrafish and embryonic stem cells. *Proc. Natl. Acad. Sci.* **104:** 9685–9690.

Vissers, L.E., van Ravenswaaij, C.M., Admiraal, R., Hurst, J.A., de Vries, B.B., Janssen, I.M., van der Vliet, W.A., Huys, E.H., de Jong, P.J., Hamel, B.C., et al. 2004. Mutations in a new member of the chromodomain gene family cause CHARGE syndrome. *Nat. Genet.* **36:** 955–957.

Waldo, K.L. and Kirby, M.L. 1993. Cardiac neural crest contribution to the pulmonary artery and sixth aortic arch artery complex in chick embryos aged 6 to 18 days. *Anat. Rec.* **237:** 385–399.

Waldo, K.L., Hutson, M.R., Stadt, H.A., Zdanowicz, M., Zdanowicz, J., and Kirby, M.L. 2005a. Cardiac neural crest is necessary for normal addition of the myocardium to the arterial pole from the secondary heart field. *Dev. Biol.* **281:** 66–77.

Waldo, K.L., Kumiski, D.H., Wallis, K.T., Stadt, H.A., Hutson, M.R., Platt, D.H., and Kirby, M.L. 2001. Conotruncal myocardium arises from a secondary heart field. *Development* **128:** 3179–3188.

Waldo, K.L., Hutson, M.R., Ward, C.C., Zdanowicz, M., Stadt, H.A., Kumiski, D., Abu-Issa, R., and Kirby, M.L. 2005b. Secondary heart field contributes myocardium and smooth muscle to the arterial pole of the developing heart. *Dev. Biol.* **281:** 78–90.

Watanabe, Y., Benson, D.W., Yano, S., Akagi, T., Yoshino, M., and Murray, J.C. 2002. Two novel frameshift mutations in *NKX2.5* result in novel features including visceral inversus and sinus venosus type ASD. *J. Med. Genet.* **39:** 807–811.

Wu, S.M. 2008. Mesp1 at the heart of mesoderm lineage specification. *Cell Stem Cell* **3:** 1–2.

Wu, S.M., Fujiwara, Y., Cibulsky, S.M., Clapham, D.E., Lien, C.L., Schultheiss, T.M., and Orkin, S.H. 2006. Developmental origin of a bipotential myocardial and smooth muscle cell precursor in the mammalian heart. *Cell* **127:** 1137–1150.

Yagi, H., Furutani, Y., Hamada, H., Sasaki, T., Asakawa, S., Minoshima, S., Ichida, F., Joo, K., Kimura, M., Imamura, S., et al. 2003. Role of *TBX1* in human del22q11.2 syndrome. *Lancet* **362:** 1366–1373.

Yang, L., Soonpaa, M.H., Adler, E.D., Roepke, T.K., Kattman, S.J., Kennedy, M., Henckaerts, E., Bonham, K., Abbott, G.W., Linden, R.M., et al. 2008. Human cardiovascular progenitor cells develop from a KDR+ embryonic-stem-cell-derived population. *Nature* **453:** 524–528.

Yelbuz, T.M., Waldo, K.L., Kumiski, D.H., Stadt, H.A., Wolfe, R.R., Leatherbury, L., and Kirby, M.L. 2002. Shortened outflow tract leads to altered cardiac looping after neural crest ablation. *Circulation* **106:** 504–510.

Zatyka, M., Priestley, M., Ladusans, E.J., Fryer, A.E., Mason, J., Latif, F., and Maher, E.R. 2005. Analysis of *CRELD1* as a candidate 3p25 atrioventricular septal defect locus (*AVSD2*). *Clin. Genet.* **67:** 526–528.

Zhao, F., Weismann, C.G., Satoda, M., Pierpont, M.E., Sweeney, E., Thompson, E.M., and Gelb, B.D. 2001. Novel TFAP2B mutations that cause Char syndrome provide a genotype-phenotype correlation. *Am. J. Hum. Genet.* **69:** 695–703.

Zhou, B., Ma, Q., Rajagopal, S., Wu, S.M., Domian, I., Rivera-Feliciano, J., Jiang, D., von Gise, A., Ikeda, S., Chien, K.R., et al. 2008. Epicardial progenitors contribute to the cardiomyocyte lineage in the developing heart. *Nature* **454:** 109–113.

Regulation of Skeletal Muscle Stem Cell Behavior by Pax3 and Pax7

M. Lagha,* T. Sato,* L. Bajard,‡ P. Daubas,* M. Esner,‡
D. Montarras,* F. Relaix,† and M. Buckingham*
*Department of Developmental Biology, CNRS URA2578, Pasteur Institute, 75015 Paris, France

Pax genes have important roles in the regulation of stem cell behavior, leading to tissue differentiation. In the case of skeletal muscle, *Pax3* and *Pax7* perform this function both during development and on regeneration in the adult. The myogenic determination gene *Myf5* is directly activated by Pax3, leading to the formation of skeletal muscle. *Fgfr4* is also a direct Pax3 target and *Sprouty1*, which encodes an intracellular inhibitor of fibroblast growth factor (FGF) signaling, is under *Pax3* control. Orchestration of FGF signaling, through *Fgfr4/Sprouty1*, modulates the entry of cells into the myogenic program, thus controling the balance between stem cell self-renewal and tissue differentiation. This and other aspects of Pax3/7 function in regulating the behavior of skeletal muscle stem cells are discussed.

The entry of stem cells into the skeletal muscle program is characterized by the activation of genes of the *MyoD* family, which encode myogenic regulatory factors essential for the commitment to this cell fate, controling myogenic determination and subsequent differentiation (Fig. 1). However, more recently, it has become clear that members of the *Pax* gene family, *Pax3* and *Pax7*, lie upstream of these myogenic genes and that their expression marks the skeletal muscle stem cell. In this context, we present the cell populations that contribute to myogenesis in the mouse embryo and in the perinatal period during muscle growth and later regeneration.

Pax genes have important roles in tissue specification and organogenesis during development (Fig. 2). A classic example is provided by *Pax6* required for the formation of the eye, a function conserved from *Drosophila* to humans. Other cell types that depend on *Pax* genes include the B-lymphocyte lineage that depends on *Pax5* or the endocrine cells of the pancreas that depend on *Pax4* and *Pax6*. Many facets of Pax function have emerged from the different cell

systems in which they have been studied, notably maintenance of a multipotent state, direction into a differentiation program, cell migration, proliferation, and survival (see Buckingham and Relaix 2007). In skeletal muscle stem cells, Pax3 and Pax7 appear to be implicated in these various aspects of cell behavior. Identification of Pax3/7 targets is now beginning to clarify the role of these key regulators. Here, this is discussed mainly with reference to skeletal muscle stem cells during development, with an emphasis on the function of Pax3/7 in modulating stem cell renewal versus entry into the myogenic program.

THE CELLS THAT FORM SKELETAL MUSCLE

To understand the stem cells of adult tissues, and indeed to learn how to manipulate them for therapeutic purposes, it is important to know their embryological origin and how such cells contribute to the formation of the tissue during development (Buckingham and Montarras 2008).

Myogenesis in the Embryo

Skeletal muscles in the trunk and limbs are derived from paraxial mesoderm and from cells present in the dorsal compartment of the somite, an epithelial structure known as the dermomyotome (Fig. 3a). These cells express Pax3/7. At the onset of myogenesis in the mouse embryo, the linked myogenic determination genes *Myf5* and *Mrf4* are activated, without direct intervention of the Pax factors. Canonical Wnt (Borello et al. 2006) and Sonic Hedgehog (Shh) (Borycki et al. 1999; Gustafsson et al. 2002; Teboul et al. 2003) signaling, from the axial structures of the adjacent neural tube and notochord, directly regulate an early epaxial *Myf5* enhancer element, in cells at the epaxial (closest to the axis) edge of the dermomyotome. At later stages, *MyoD* is also activated independently in this domain by signals from the neural tube and notochord (Tajbakhsh et al. 1998). Activation of myogenic

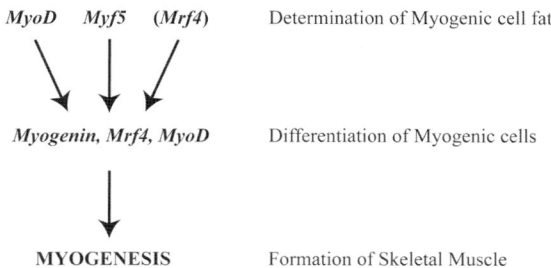

MyoD   Myf5   (Mrf4)   Determination of Myogenic cell fate

Myogenin, Mrf4, MyoD   Differentiation of Myogenic cells

MYOGENESIS   Formation of Skeletal Muscle

Figure 1. Myogenic regulatory factors. The functions of these skeletal-muscle-specific basic helix-loop-helix transcription factors have been determined by gene mutation in the mouse.

Present addresses: †Mouse Molecular Genetics group, UMR S787, Groupe Myologie, INSERM-UPMC-Paris VI, Faculté de Médecine Pitié-Salpétrière, 105 bd de l'Hôpital, 75634, Paris Cedex 13, France; ‡Max-Planck Institute of Molecular Cell Biology and Genetics, Pfotenhauerstr. 108, 01307 Dresden, Germany.

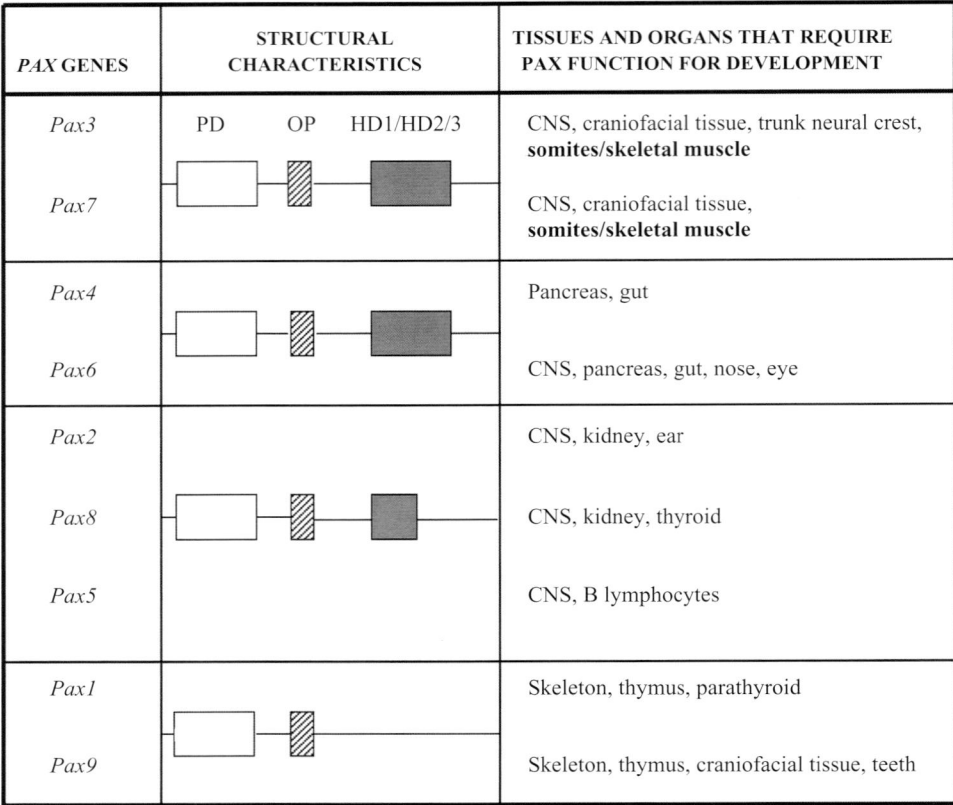

PAX GENES	STRUCTURAL CHARACTERISTICS	TISSUES AND ORGANS THAT REQUIRE PAX FUNCTION FOR DEVELOPMENT
Pax3	PD OP HD1/HD2/3	CNS, craniofacial tissue, trunk neural crest, **somites/skeletal muscle**
Pax7		CNS, craniofacial tissue, **somites/skeletal muscle**
Pax4		Pancreas, gut
Pax6		CNS, pancreas, gut, nose, eye
Pax2		CNS, kidney, ear
Pax8		CNS, kidney, thyroid
Pax5		CNS, B lymphocytes
Pax1		Skeleton, thymus, parathyroid
Pax9		Skeleton, thymus, craniofacial tissue, teeth

Figure 2. The Pax family of regulatory factors.

determination genes by axial signals leads to the formation of the early myotome, a differentiated muscle mass located underneath the dermomyotome. However, the activation of *Myf5* later in the hypaxial (furthest from the axis) dermomyotome depends on Pax3 (Bajard et al. 2006), and Pax3 transcriptional activity, which also controls *MyoD*, is regulated by noncanonical Wnt signaling from the overlying dorsal ectoderm, acting through a protein kinase C (PKC)-dependent pathway (Brunelli et al. 2007). Pax3 functions as a transcriptional activator in the myogenic context in vivo (Relaix et al. 2003), but it is not very effective on its own and probably requires cofactors and/or protein modifications such as phosphorylation, which may be affected by the PKC pathway. Pax3-positive cells that form the early myotome delaminate from the edges of the dermomyotomal epithelium as Myf5/Mrf4-positive cells and rapidly differentiate. Other cells delaminate and migrate from the somite to more distant sites of myogenesis, such as those in the limbs, before activating the myogenic regulatory factor genes (Fig. 3b). This process is Pax3-dependent, through direct regulation of *c-met* (see Buckingham and Relaix 2007).

Subsequent myogenesis in the trunk depends on Pax3/7. After the initial Myf5/Mrf4-dependent wave of myogenesis in the somite, the epithelial structure of the dermomyotome begins to disaggregate and Pax3/7-positive cells from the central dermomyotome are parachuted into the underlying myotomal muscle (Fig. 3c). These cells provide a source of myogenic progenitors for all subsequent muscle growth. They are a proliferative population that either self-renews or moves into the myogenic program with activation of *Myf5* and *MyoD*. Cells that express these myogenic determination factors undergo limited proliferation, as a transit-amplifying population, but then differentiate into skeletal muscle. It is therefore the upstream Pax-positive cells, in both the trunk and the limbs, that constitute a self-renewing muscle stem cell population. In double *Pax3/7* mutant embryos, the early myotome forms, but subsequent muscle development is compromised (Relaix et al. 2005).

The Muscle Satellite Cell

Pax-positive cells are initially intermingled with muscle fibers in the growing muscle masses. However, by late fetal stages, a basal lamina begins to be laid down and Pax3/7-positive cells are now found adjacent to the muscle fiber under the basal lamina. This is the characteristic position of the satellite cell, a quiescent cell that when activated contributes to postnatal muscle growth and to muscle regeneration (Montarras and Buckingham 2008). Satellite cells are marked by the expression of Pax7, and in many, but not all, muscles, *Pax3* is also transcribed in these cells (Relaix et al. 2006). Satellite cells are heterogeneous, also, in that many of them already transcribe *Myf5* (Beauchamp et al. 2000), and indeed most of them have done so at some stage in their history (Kuang et al. 2007), unlike the proliferating *Pax3/7* cells of prenatal develop-

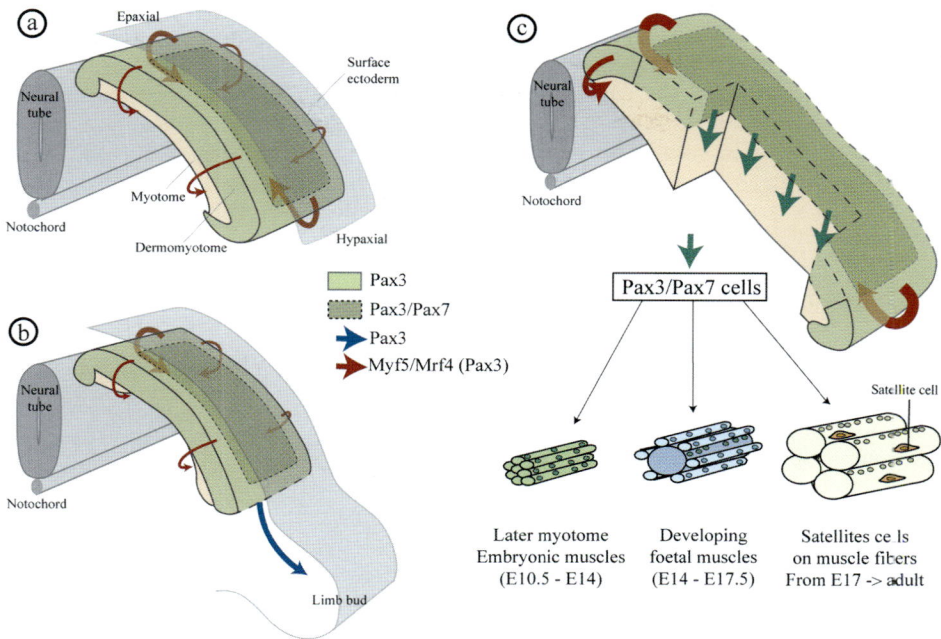

Figure 3. Successive waves of cells from the dermomyotome of the embryonic somite contribute to skeletal muscle formation. (*a*) Pax3-positive cells that express the myogenic determination factors Myf5 and Mrf4 and, later, MyoD, delaminate from the edges of this epithelium to form the early myotomal muscle. (*b*) At the limb level, cells delaminate and migrate from the hypaxial (further from the axis) dermomyotome into the limb bud where the myogenic determination genes *Myf5* and *MyoD* are subsequently activated. This migration is Pax3-dependent. (*c*) The central region of the dermomyotome, where both Pax3 and Pax7 are expressed, loses its epithelial structure so that Pax3/7-positive cells enter the underlying myotomal muscle. All subsequent muscle growth depends on this proliferating population of muscle stem cells. Before birth, these cells take up a satellite cell position under the basal lamina of the muscle fiber.

ing muscles in the mouse embryo, which either self-renew or differentiate (Fig. 4). Satellite cells that have activated myogenic determination genes either in vivo or after cell culture (Montarras et al. 2005) regenerate muscle much less efficiently, compared to the Pax-only satellite cells that also display more stem-cell-like properties, such as asymmetric cell division (Kuang et al. 2007).

The origin of satellite cells had been a subject of debate, but the recent identification of the dermomyotomal source of Pax3/7 stem cells for skeletal muscle development has led to the view that satellite cells are derived from this population. *Pax3-Cre/Rosa26* experiments suggested that this is the case for muscles such as those of the hind limbs in which satellite cells do not express *Pax3* (Schienda et

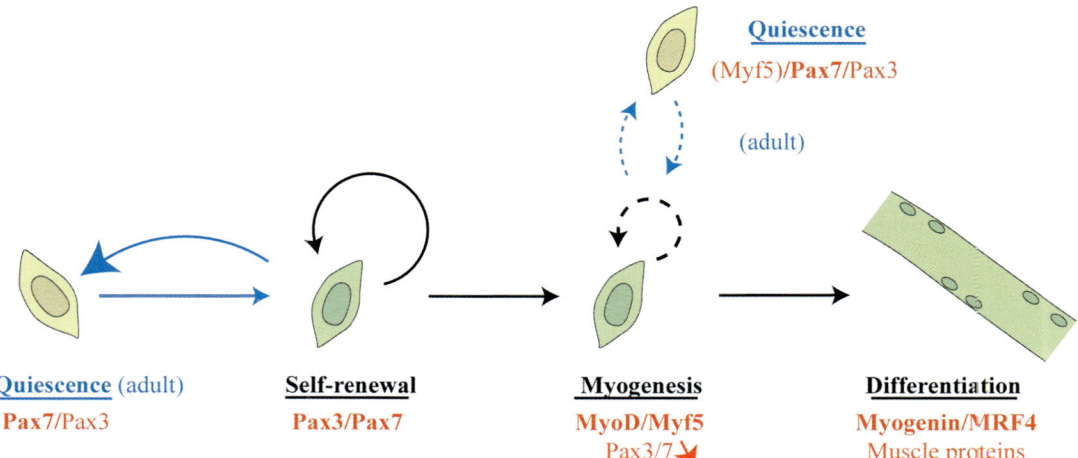

Figure 4. Skeletal muscle stem cell choices. In the embryo (*black arrows*), Pax3/Pax7-positive cells either self-renew or enter the myogenic program with activation of the myogenic determination genes *Myf5* and *MyoD*, accompanied by down-regulation of *Pax3/7*. In postnatal muscle, satellite cells participate in self-renewal and myogenesis, when activated in the context of postnatal growth and muscle repair, or remain quiescent (*blue arrows*) under the basal lamina of the muscle fiber. Many of these cells transcribe low levels of *Myf5*, indicating that they had already entered the myogenic program.

al. 2006), and experimental manipulation of chick/quail chimeras has demonstrated that satellite cells come from the dermomyotome, from similar Pax3/7-positive cells to those in the mouse (Gros et al. 2005). More than 90% of all satellite cells at birth derive from this source. This does not preclude that other stem cells, such as the mesoangioblast cells present in the walls of blood vessels (Sampaolesi et al. 2003), may not also contribute, especially after injury.

The Implications of Two Waves of Myogenesis

In conclusion, there is therefore an early wave of myogenesis in which myogenic determination genes are directly activated by axial signals in the somite. This leads to the formation of an early differentiated muscle, the myotome. However, a second wave of myogenesis depends on Pax3/7-positive cells, also derived from the dermomyotome, that initially invade the muscle scaffold provided by the early myotome. These cells, unlike those of the first wave, self-renew as well as participate in myogenesis. They contribute the satellite cells of postnatal muscle, some of which have retained the stem cell properties of Pax7$^+$/Myf5$^-$ cells, whereas others (Pax7$^+$, Myf5$^+$) have at some stage engaged the myogenic program before reverting to a quiescent satellite cell state (Fig. 4). The evolutionary implications of the two waves of myogenesis are interesting. It is possible that the first constitutes a more primitive and static way of making segmented muscles following their somitic origin, whereas the second wave of Pax-dependent myogenesis leads to flexibility of muscle location and muscle growth, retaining myogenic stem cell potential also for regeneration in the adult.

THE FUNCTION OF PAX3/7 IN SKELETAL MUSCLE STEM CELLS

Equivalence of Pax3 and Pax7

A first question is whether Pax3 and Pax7 are equivalent. Their transcription profiles differ and, notably, *Pax3* is more extensively expressed in the somite, whereas *Pax7* is restricted to the central domain of the dermomyotome. To address their relative function, we replaced the *Pax3* gene with a Pax7-coding sequence and showed that, in homozygote *Pax3$^{Pax7/Pax7}$* embryos, myogenesis in the trunk proceeds normally. However, in the limbs, this was not the case and muscle masses were deficient, notably distal muscles, with a more pronounced phenotype in the forelimbs. This deficit reflects reduced proliferation of Pax7 progenitor cells and probably also migration defects. We therefore suggest that Pax3 has acquired additional functions required for myogenesis in the limbs, at the time of tetrapod radiation, whereas functions in the trunk remain similar and probably reflect that of the ancestral *Pax3/7* gene, present in the cephalochordate *Amphioxus*.

In postnatal satellite cells, the lack of Pax7 is not compensated by Pax3 in those muscles in which the two *Pax* genes are coexpressed. It is not clear whether this reflects the level of Pax3. When Pax7 replaces Pax3, as mentioned above, *Pax3$^{Pax7/Pax7}$* phenotypes are milder than

Pax3^{Pax7}, and this, together with other observations (Zhou et al. 2008), illustrates the dose sensitivity of Pax function, and indeed, Pax proteins in other systems demonstrate threshold effects (Buckingham and Relaix 2007). In satellite cells, manipulation of constructs expressing dominant-negative Pax3- or Pax7-Engrailed fusion proteins suggests that Pax7 antiapoptotic targets are not affected by Pax3-Engrailed (Relaix et al. 2006). This is then different from the situation in the embryo, where Pax3 and Pax7 both have an antiapoptotic role (see next section).

ROLE OF *PAX3/7* IN MUSCLE STEM CELLS

In most of the following discussion of the role of Pax3/7, they are considered equivalent, which is the case during the development of trunk musculature.

Multipotency

Maintenance of multipotency is an important issue in the Pax-positive cells of the dermomyotome of the somite. This epithelium, as its name implies, gives rise to the dermis of the back and to skeletal muscle, as well as endothelial and smooth muscle cells of some blood vessels (Fig. 5). Cell-labeling experiments in the chick (Kardon et al. 2002; Ben-Yair and Kalcheim 2005, 2008) have shown that a single cell can give rise to dermis/skeletal muscle,

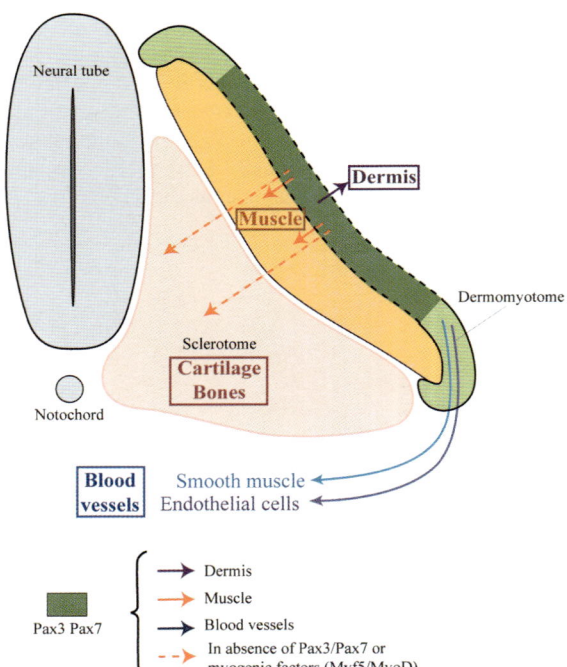

Figure 5. Pax-positive cells in the epithelial dermomyotome are multipotent, contributing to dermis and blood vessels as well as to skeletal muscle (*arrows*). In the absence of myogenic determination factors, cells that mislocate ventrally can adopt a cartilage/bone cell fate (*dashed arrows*). Normally, vertebrae and ribs are formed by Pax1/9-positive cells from the mesenchymal sclerotome compartment of the ventral somite.

endothelial cells/skeletal muscle, and smooth muscle/skeletal muscle, and indeed it is probable that a multipotent Pax-positive cell gives rise to multiple derivatives. In the mouse embryo, retrospective clonal analysis and Pax3-GFP (green fluorescent protein) perdurance also demonstrates a common Pax progenitor for smooth muscle/skeletal muscle (Esner et al. 2006). As nonskeletal muscle derivatives form, Pax3/7 are down-regulated and their maintenance marks the skeletal muscle stem cell. In *Pax3/7* double mutants, in the absence of myogenic determination factors, cells that would have formed skeletal muscle, depending on their environment, can also enter other differentiation pathways, such as that leading to bone and cartilage (Relaix et al. 2005; Tajbakhsh et al. 1996), which are derivatives of the ventral somite. Adipose tissue is also a default pathway when myogenic determination factors are absent (Rudnicki et al. 1993), and indeed, *Pax7*-positive satellite cells will form fat under some culture conditions (Shefer and Yablonka-Reuveni 2007).

It is not yet clear which Pax targets maintain multipotency. We have identified *Foxc2* as a gene that is negatively regulated by *Pax3* and have preliminary evidence that a *Pax3/Foxc2*–negative feedback loop may modulate cell-fate choices in the dermomyotome. We have also recently identified *Dmrt2* as a Pax3 target in the dermomyotome. This gene, via its regulation of laminin, a ligand for integrin receptors, is probably involved in coordinating the structural integrity of the somite and the dermomyotome epithelium (Seo et al. 2006). This, of course, is critical for the reception of signals that will decide the subsequent fate of Pax3/7-positive cells. Integrins are also involved in the correct location of myogenic cells to the myotome, before differentiation (Bajanca et al. 2006).

Migration

Pax3, expressed at the borders of the dermomyotome, has an essential role in the delamination of cells from this epithelium, via its target c-*met* (Epstein et al. 1996; Relaix et al. 2003), which encodes a tyrosine kinase receptor required for this process, as well as for subsequent migration to more distant sites of myogenesis in response to its ligand hepatocyte growth factor (HGF) (Bladt et al. 1995). Pax3-regulated c-Met function thus determines the location of the Pax3/7 cells in the embryo and hence their exposure to signals that promote skeletal muscle formation. During muscle regeneration, migration of satellite cells to sites of injury is also important; it is not yet clear to what extent this depends on c-Met or cytokine receptors such as CXCR4, also involved in early migration of cells from the somites to the limbs (Vasyutina et al. 2005), and probably also genetically downstream from *Pax3*.

Proliferation

Pax3/7 probably promote proliferation, although targets have not yet been identified in this context. Indicative of this is the generation of rhabdomyosarcomas, with overactivation of Pax3/7 targets (see Relaix et al. 2003) by PAX3- or PAX7-FKHR, in which a chromosome

translocation results in fusion of the PAX3/7 DNA-binding domain to the *trans*-activation domain of FKHR (FOXO1A). When Pax7 replaces Pax3, progenitor cells in the embryonic limb do not proliferate as well (Relaix et al. 2004). Futhermore, it has been suggested that the deficit in satellite cells in the *Pax7* mutant mouse may be due to compromised proliferation (Oustanina et al. 2004); the cell cycle is perturbated in satellite cells that lack Pax7 (Relaix et al. 2006). Expression of Pax3 during proliferation of activated satellite cells may be important for their amplification (Conboy and Rando 2002). This is seen in Pax7-positive satellite cells from the limb, which do not normally express *Pax3*. However, in vivo injury of limb muscle of *Pax3*[GFP+] mice did not result in detectable GFP (Montarras et al. 2005), as might be expected if *Pax3* is expressed by all activated satellite cells. Conditional deletion of *Pax3* postnatally (*Pax3* mutant mice die prenatally because of neural crest defects) should clarify the issue of Pax3 function in satellite cells.

Survival

Pax3/7 have an important role in the survival of the muscle stem cell population. In *Pax3* mutant embryos, apoptosis takes place in the hypaxial dermomyotome, where *Pax7* is not expressed, and indeed in looking at the myogenic function of Pax3 in these cells, it was necessary to use heterozygote *Pax3*[Pax3-En/+] mice, expressing a single allele encoding the dominant-negative Pax3-Engrailed protein, in which apoptosis is abrogated (Bajard et al. 2006). The Pax3/7-positive muscle stem cell population undergoes extensive apoptosis in the double-mutant mice, whereas this is not observed in the absence of Pax3 or Pax7 alone. Postnatally, the satellite cells of *Pax7* mutant mice undergo progressive cell death, and this antiapoptotic function of Pax7 is confirmed by the observation that introduction of a dominant-negative form of Pax7 (Pax7-Engrailed) into wild-type satellite cells leads to cell death (Relaix et al. 2006).

Differentiation into Skeletal Muscle

Pax3/7 expression maintains the myogenic potential of cells, and in the absence of these Pax factors, the myogenic determination genes *Myf5* and *MyoD* are not activated. *Myf5* and *MyoD* therefore lie genetically downstream from *Pax3/7*. This is observed for the major Pax3/7 population of muscle stem cells present in all developing muscle masses (Relaix et al. 2005) and also at earlier stages in the hypaxial somite where, when Pax3 activity is attenuated, but the cells have not undergone apoptosis, *Myf5* is not activated (Bajard et al. 2006).

Myf5 is regulated by a complex organization of *cis*-acting sequences more than 100 kb 5′ of the gene (Fig. 6). A sequence at −57.5 kb directs expression of *Myf5* in the maturing hypaxial somite and limb buds, and deletion of this region in a bacterial artificial chromosome (BAC) containing the *Mrf4-Myf5* locus had shown that it is essential for *Myf5* activation (Hadchouel et al. 2003). We have shown by in vivo chromatin immunoprecipitation

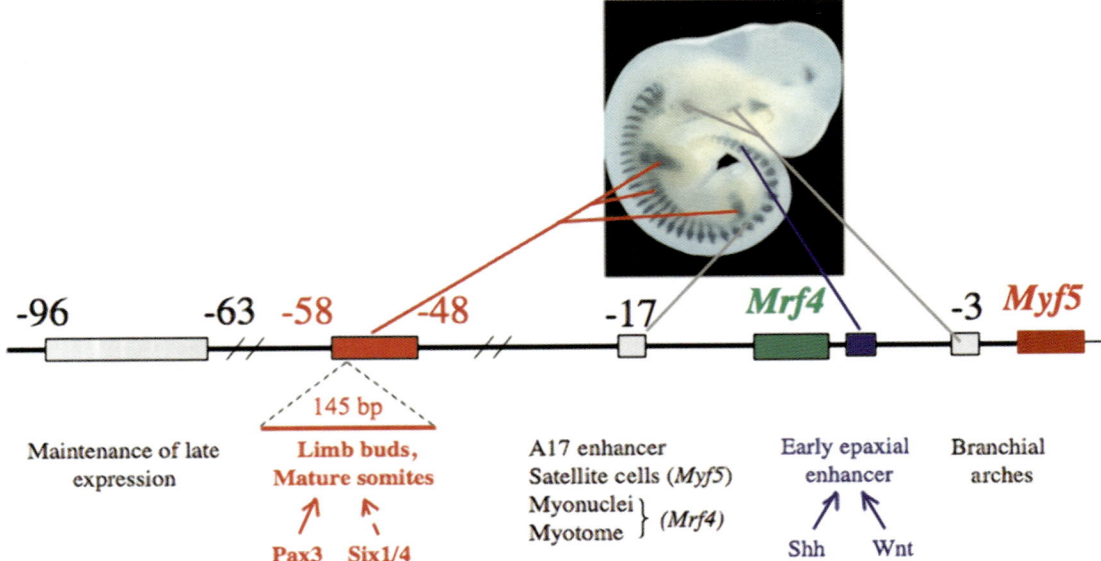

Figure 6. *cis*-regulatory elements of the *Mrf4/Myf5* locus. The interaction between *Mrf4* and *Myf5* promoters with intragenic and 5′ regulatory elements is complex (see Carvajal et al. 2008). *Myf5* activation during the first wave of myogenesis depends on an early epaxial enhancer, directly controlled by Wnt and Shh signaling. Most later embryonic expression depends on a region at –58/–48 kb. Within this sequence, a 145-bp element alone directs expression in myogenic cells of mature somites and limb buds. This activity depends directly on Pax3 and is modulated by Six1/4.

(ChIP) analysis that Pax3 binds to a site in this sequence, which when mutated, abolishes all activity, leading us to conclude that *Myf5* is a direct target of Pax3 (Bajard et al. 2006). This therefore illustrates direct regulation of entry into the myogenic program by Pax3/7. It is not yet clear whether *MyoD* is a direct Pax3/7 target; however, in the adult, as in the embryo, it lies genetically downstream from *Pax3/7*. In many satellite cells, *Myf5* is already activated and no longer dependent on Pax3/7, but this is not the case for *MyoD*. In *Myf5* mutant satellite cells, expression of a dominant-negative form of Pax3/7 prevents *MyoD* expression and subsequent myogenesis (Relaix et al. 2006).

Activation of myogenic determination genes not only depends on Pax3/7, but is also modulated by other factors. An example is provided by the Six homeodomain factors. Six1/4 are expressed at many sites of myogenesis, and notably, in the Pax3-positive cells of the hypaxial somite and limbs. The –57.5-kb *Myf5* regulatory sequence that is Pax3-dependent contains a Six site, also occupied in vivo. Mutation of this site results in partial loss of expression in transgenic mice (Fig. 6) (Giordani et al. 2007). In keeping with this observation, mutation of *Six1/4* affects *Myf5* activation (Grifone et al. 2005) and mutation of the genes encoding the Six cofactors Eya1/Eya2 also affects myogenesis (Grifone et al. 2007). It is notable that formation of the *Drosophila* eye depends not only on *Eyeless* and *Twin of Eyeless* (*Pax*), but also on *Sine oculis* (*Six*) and *Eyes absent* (*Eya*), acting in a regulatory cascade on downstream "*eye*" genes, thus illustrating the extent to which *Pax* gene regulatory networks are conserved across species and tissues (see Relaix and Buckingham 1999).

MUSCLE STEM CELL SELF-RENEWAL VERSUS DIFFERENTIATION

A major question in stem cell biology is how the balance between stem cell self-renewal versus differentiation is maintained. In the case of muscle stem cells, Pax3/7 activate myogenic determination genes leading to entry into the myogenic program. Other factors may also contribute to this step, such as Six1/4, which is also present in the progenitor cells. To examine further how Pax3/7 modulates muscle stem cell behavior, we have performed screens based on different *Pax3* alleles that we had engineered. They include a *Pax3*[GFP] allele, which permits purification of cells by flow cytometry (Montarras et al. 2005), and a *Pax3*[PAX3-FKHR] allele, which leads to overactivation of Pax3 targets. This allele encodes the DNA-binding domain of Pax3 fused to the transcriptional activation domain of FKHR (FOXO1A). In *Pax3*[PAX3-FKHR/–] embryos, the *Pax3* mutant phenotype is rescued, showing that Pax3 functions as a transcriptional activator in the myogenic context (Relaix et al. 2003). Comparative microarray analyses of cells from the limb buds or somites of *Pax3*[GFP/+] or *Pax3*[GFP/PAX3-FKHR] embryos led to the identification of potential Pax3 targets that were then tested by in situ hybridization on gain-of-function (*Pax3*[PAX3-FKHR/+]) (Relaix et al. 2003), partial loss-of-function (*Pax3*[Pax3-En/+]) (Bajard et al. 2006), or loss-of-function (*Pax3*[–/–]) embryos.

Fgfr4 was identified as a gene lying genetically downstream from Pax3, and ChIP-chip analysis, using embryo extracts on a tiling array across this locus, led to the identification of a Pax3-binding region 3′ of the *Fgfr4* gene. This 559-bp sequence directs reporter transgene expression at sites of myogenesis where *Fgfr4* is expressed. Mutation of the Pax3-binding sites abolished expression, and we therefore conclude that *Fgfr4* is a direct Pax3 target. Other components of the FGF-signaling pathway also show genetic dependence on Pax3, notably *Sprouty1*, which was up-regulated in the microarray screen. Manipulation of Sprouty, using a conditional *Sprouty*-expressing transgenic line (Basson et al. 2008) crossed to *Pax3^{Cre/+}* mice (Engleka et al. 2005), permitted in vivo modulation of FGF signaling in Pax3-expressing cells and their progeny. In this experiment, the ratio between Pax3/7-positive stem cells and myogenin-positive differentiating muscle cells was perturbed in favor of the former. We therefore think that Pax3 orchestrates maintenance of stem cells versus entry into the myogenic program by regulation of FGF signaling (Fig. 7) (Lagha et al. 2008). *Fgfr4* continues to be expressed in myogenic cells; E boxes, also present in the 3′ regulatory sequence, are implicated in its regulation, and transgenic analysis suggests that myogenic factors maintain expression of this FGF receptor in differentiated muscle.

In adult satellite cells, *Fgfr4* is also expressed, and *Fgfr4* mutants have regeneration defects (Zhao et al. 2006). *Sprouty1* is present in quiescent satellite cells and is down-regulated in cultured satellite cells that have activated *MyoD* and are progressing toward muscle differentiation (Fukada et al. 2007). Thus, Sprouty1/Fgfr4 modulation is also evident in postnatal muscle stem cells.

CONCLUSIONS

Pax3/7 orchestrate the entry of muscle stem cells into the myogenic program by direct activation of the myogenic determination gene *Myf5* and also by direct activation of the FGF receptor gene *Fgfr4*. Pax3 regulation of *Sprouty1*, which preliminary results suggest may also be direct, permits Pax3/7 control of the read-out of FGF signaling through this inhibitor. Sproutys affect tyrosine kinase receptor signaling in general so that Sprouty1 may also modulate signaling from other receptors such as PDGFRα, also potentially implicated in myogenesis

(Crosby et al. 1998). The Notch signaling pathway also regulates muscle stem cell entry into the myogenic program, both postnatally for satellite cells (Conboy and Rando 2002) and in the embryo, where mutation of *RBP-J* (Vasyutina et al. 2007) or *Delta1* (Schuster-Gossler et al. 2007) in Pax3/7-positive cells shifts them toward myogenesis, depleting the stem cell pool, with subsequent failure of muscle growth in the absence of Notch signaling. In keeping with this, interference with Notch signaling in the multipotent Pax3/7 cells of the dermomyotome promotes the muscle cell fate (Ben-Yair and Kalcheim 2008). Myostatin, a transforming growth factor-β (TGF-β) family member, and its inhibitor, follistatin, also affect the balance between self-renewal and myogenic differentiation, as shown by experimental manipulation in the chick embryo (Manceau et al. 2008). There is no evidence at present that Pax3/7 regulation intervenes in the Notch or Myostatin signaling pathways in muscle stem cells, and these therefore probably correspond to extrinsic controls of myogenic cell behavior.

Other genetic circuits linked to Pax3/7 also affect entry into the myogenic program. Six1/4 homeodomain proteins impact the Pax3-dependent *Myf5* enhancer (Giordani et al. 2007), and E-box motifs, through which myogenic regulatory factors act, are probably responsible for the maintenance of transcription by the Pax3-dependent *Fgfr4* enhancer (Lagha et al. 2008). These regulatory sequences provide examples of how Pax3/7-based genetic networks control skeletal myogenesis. Pax3 targets are also beginning to emerge from our screens, shedding light on upstream regulation of multipotent stem cells in the dermomyotome. As mentioned briefly, Pax3/7 also regulate other aspects of muscle stem cell behavior: migration, through the gene encoding the c-Met tyrosine kinase receptor, and also proliferation and cell survival, through as yet unidentified Pax target genes.

These multiple functions of Pax3/7 in the myogenic context reflect those of other Pax factors that control stem cell populations required for tissue differentiation (see Fig. 2). It is evident that maintenance of multipotency and of self-renewal versus differentiation, together with correct positioning of stem cells, are critical for tissue formation in the embryo and regeneration in the adult. Furthermore, Pax function in cell survival, illustrated by Pax3/7 in the muscle context, is also observed in different

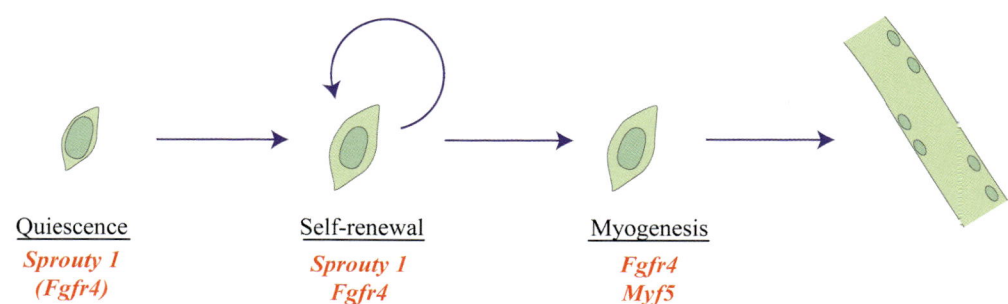

Quiescence	Self-renewal	Myogenesis
Sprouty 1	*Sprouty 1*	*Fgfr4*
(Fgfr4)	*Fgfr4*	*Myf5*

Figure 7. Stem cell choices are regulated by *Pax3* target genes; *Myf5* marks the entry of a cell into the myogenic program, also promoted by Fgfr4 signaling that is modulated by Sprouty1, expressed in Pax3/Pax7-positive muscle stem cells.

stem cell contexts. It is no doubt important that genes required for the normal tissue cell fate of a cell should also regulate its survival. Loss of stem cell regulators, such as Pax3/7, results in a cell that is out of control and potentially dangerous for the organism.

ACKNOWLEDGMENTS

Work on skeletal muscle stem cells in M.B.'s laboratory is supported by the Pasteur Institute and the Centre National de la Recherche Scientifique (CNRS), with additional grants from the Association Française contre les Myopathies (AFM); the European Union, through the Integrated Project "EuroSyStem"; and the Networks of Excellence "Cells into Organs" and "MYORES." M.L. is supported by a fellowship from the AFM and T.S. from the "Cells into Organs" network.

REFERENCES

Bajanca, F., Luz, M., Raymond, K., Martins, G.G., Sonnenberg, A., Tajbakhsh, S., Buckingham, M., and Thorsteinsdottir, S. 2006. Integrin α6β1-laminin interactions regulate early myotome formation in the mouse embryo. *Development* **133:** 1635–1644.

Bajard, L., Relaix, F., Lagha, M., Rocancourt, D., Daubas, P., and Buckingham, M.E. 2006. A novel genetic hierarchy functions during hypaxial myogenesis: Pax3 directly activates *Myf5* in muscle progenitor cells in the limb. *Genes Dev.* **20:** 2450–2464.

Basson, M.A., Echevarria, D., Petersen, Ahn, C., Sudarov, A., Joyner, A.L., Mason, I.J., Martinez, S., and Martin, G.R. 2008. Specific regions within the embryonic midbrain and cerebellum require different levels of FGF signaling during development. *Development* **135:** 889–898.

Beauchamp, J.R., Heslop, L., Yu, D.S., Tajbakhsh, S., Kelly, R.G., Wernig, A., Buckingham, M.E., Partridge, T.A., and Zammit, P.S. 2000. Expression of CD34 and Myf5 defines the majority of quiescent adult skeletal muscle satellite cells. *J. Cell Biol.* **151:** 1221–1234.

Ben-Yair, R. and Kalcheim, C. 2005. Lineage analysis of the avian dermomyotome sheet reveals the existence of single cells with both dermal and muscle progenitor fates. *Development* **132:** 689–701.

Ben-Yair, R. and Kalcheim, C. 2008. Notch and bone morphogenetic protein differentially act on dermomyotome cells to generate endothelium, smooth, and striated muscle. *J. Cell Biol.* **180:** 607–618.

Bladt, F., Riethmacher, D., Isenmann, S., Aguzzi, A., and Birchmeier, C. 1995. Essential role for the c-*met* receptor in the migration of myogenic precursor cells into the limb bud. *Nature* **376:** 768–771.

Borello, U., Berarducci, B., Murphy, P., Bajard, L., Buffa, V., Piccolo, S., Buckingham, M., and Cossu, G. 2006. The Wnt/β-catenin pathway regulates Gli-mediated *Myf5* expression during somitogenesis. *Development* **133:** 3723–3732.

Borycki, A.G., Li, J., Jin, F., Emerson, C.P., and Epstein, J.A. 1999. Pax3 functions in cell survival and in *pax7* regulation. *Development* **126:** 1665–1674.

Brunelli, S., Relaix, F., Baesso, S., Buckingham, M., and Cossu, G. 2007. β Catenin-independent activation of MyoD in presomitic mesoderm requires PKC and depends on Pax3 transcriptional activity. *Dev. Biol.* **304:** 604–614.

Buckingham, M. and Montarras, D. 2008. The origin and genetic regulation of myogenic cells: From the embryo to the adult. In *Advances in muscle research: Skeletal muscle repair and regeneration* (ed. S. Schiaffino and T. Partridge), vol. 3, pp. 19–44. Springer, Heidelberg.

Buckingham, M. and Relaix, F. 2007. The role of *Pax* genes in the development of tissues and organs: *Pax3* and *Pax7* regu-

late muscle progenitor cell functions. *Annu. Rev. Cell Dev. Biol.* **23:** 645–673.

Carvajal, J.J., Keith, A., and Rigby, P.W. 2008. Global transcriptional regulation of the locus encoding the skeletal muscle determination genes *Mrf4* and *Myf5. Genes Dev.* **22:** 265–276.

Conboy, I.M. and Rando, T.A. 2002. The regulation of notch signaling controls satellite cell activation and cell fate determination in postnatal myogenesis. *Dev. Cell* **3:** 397–409.

Crosby, J.R., Seifert, R.A., Soriano, P., and Bowen-Pope, D.F. 1998. Chimaeric analysis reveals role of Pdgf receptors in all muscle lineages. *Nat. Genet.* **18:** 385–388.

Engleka, K.A., Gitler, A.D., Zhang, M., Zhou, D.D., High, F.A., and Epstein, J.A. 2005. Insertion of Cre into the *Pax3* locus creates a new allele of *Splotch* and identifies unexpected *Pax3* derivatives. *Dev. Biol.* **280:** 396–406.

Epstein, J.A., Shapiro, D.N., Cheng, J., Lam, P.Y., and Maas, R.L. 1996. Pax3 modulates expression of the c-Met receptor during limb muscle development. *Proc. Natl. Acad. Sci.* **93:** 4213–4218.

Esner, M., Meilhac, S.M., Relaix, F., Nicolas, J.F., Cossu, G., and Buckingham, M.E. 2006. Smooth muscle of the dorsal aorta shares a common clonal origin with skeletal muscle of the myotome. *Development* **133:** 737–749.

Fukada, S., Uezumi, A., Ikemoto, M., Masuda, S., Segawa, M., Tanimura, N., Yamamoto, H., Miyagoe-Suzuki, Y., and Takeda, S. 2007. Molecular signature of quiescent satellite cells in adult skeletal muscle. *Stem Cells* **25:** 2448–2459.

Giordani, J., Bajard, L., Demignon, J., Daubas, P., Buckingham, M., and Maire, P. 2007. Six proteins regulate the activation of *Myf5* expression in embryonic mouse limbs. *Proc. Natl. Acad. Sci.* **104:** 11310–11315.

Grifone, R., Demignon, J., Houbron, C., Souil, E., Niro, C., Seller, M.J., Hamard, G., and Maire, P. 2005. Six1 and Six4 homeoproteins are required for Pax3 and Mrf expression during myogenesis in the mouse embryo. *Development* **132:** 2235–2249.

Grifone, R., Demignon, J., Giordani, J., Niro, C., Souil, E., Bertin, F., Laclef, C., Xu, P.X., and Maire, P. 2007. Eya1 and Eya2 proteins are required for hypaxial somitic myogenesis in the mouse embryo. *Dev. Biol.* **302:** 602–616.

Gros, J., Manceau, M., Thome, V., and Marcelle, C. 2005. A common somitic origin for embryonic muscle progenitors and satellite cells. *Nature* **435:** 954–958.

Gustafsson, M.K., Pan, H., Pinney, D.F., Liu, Y., Lewandowski, A., Epstein, D.J., and Emerson, Jr., C. 2002. *Myf5* is a direct target of long-range Shh signaling and Gli regulation for muscle specification. *Genes Dev.* **16:** 114–126.

Hadchouel, J., Carvajal, J.J., Daubas, P., Bajard, L., Chang, T., Rocancourt, D., Cox, D., Summerbell, D., Tajbakhsh, S., Rigby, P.W., et al. 2003. Analysis of a key regulatory region upstream of the *Myf5* gene reveals multiple phases of myogenesis, orchestrated at each site by a combination of elements dispersed throughout the locus. *Development* **130:** 3415–3426.

Kardon, G., Heanue, T.A., and Tabin, C.J. 2002. *Pax3* and *Dach2* positive regulation in the developing somite. *Dev. Dyn.* **224:** 350–355.

Kuang, S., Kuroda, K., Le Grand, F., and Rudnicki, M.A. 2007. Asymmetric self-renewal and commitment of satellite stem cells in muscle. *Cell* **129:** 999–1010.

Lagha, M., Kormish, J.D., Rocancourt, D., Manceau, M., Epstein, J.A., Zaret, K.S., Relaix, F., and Buckingham, M.E. 2008. Pax3 regulation of FGF signaling affects the progression of embryonic progenitor cells into the myogenic program. *Genes Dev.* **22:** 1828–1837.

Manceau, M., Gros, J., Savage, K., Thome, V., McPherron, A., Paterson, B., and Marcelle, C. 2008. Myostatin promotes the terminal differentiation of embryonic muscle progenitors. *Genes Dev.* **22:** 668–681.

Montarras, D. and Buckingham, M. 2008. Isolation, characterisation and origin of muscle satellite cells. In *Recent advances in skeletal myogenesis* (ed. K. Tsuchida). Research Signpost Series, Trivandrum, Kerala, India. (In press.)

Montarras, D., Morgan, J., Collins, C., Relaix, F., Zaffran, S., Cumano, A., Partridge, T., and Buckingham M. 2005. Direct isolation of satellite cells for skeletal muscle regeneration. *Science* **309:** 2064–2067.

Custanina, S., Hause, G., and Braun, T. 2004. Pax7 directs postnatal renewal and propagation of myogenic satellite cells but not their specification. *EMBO J.* **23:** 3430–3439.

Relaix, F. and Buckingham, M. 1999. From insect eye to vertebrate muscle: Redeployment of a regulatory network. *Genes Dev.* **13:** 3171–3178.

Relaix, F., Polimeni, M., Rocancourt, D., Ponzetto, C., Schafer, B.W., and Buckingham, M. 2003. The transcriptional activator PAX3-FKHR rescues the defects of *Pax3* mutant mice but induces a myogenic gain-of-function phenotype with ligand-independent activation of Met signaling in vivo. *Genes Dev.* **17:** 2950–2965.

Relaix, F., Rocancourt, D., Mansouri, A., and Buckingham, M. 2004. Divergent functions of murine Pax3 and Pax7 in limb muscle development. *Genes Dev.* **18:** 1088–1105.

Relaix, F., Rocancourt, D., Mansouri, A., and Buckingham, M. 2005. A Pax3/Pax7-dependent population of skeletal muscle progenitor cells. *Nature* **435:** 948–953.

Relaix, F., Montarras, D., Zaffran, S., Gayraud-Morel, B., Rocancourt, D., Tajbakhsh, S., Mansouri, A., Cumano, A., and Buckingham, M. 2006. Pax3 and Pax7 have distinct and overlapping functions in adult muscle progenitor cells. *J. Cell Biol.* **172:** 91–102.

Rudnicki, M.A., Schnegelsberg, P.N., Stead, R.H., Braun, T., Arnold, H.H., and Jaenisch, R. 1993. MyoD or Myf-5 is required for the formation of skeletal muscle. *Cell* **75:** 1351–1359.

Sampaolesi, M., Torrente, Y., Innocenzi, A., Tonlorenzi, R., D'Antona, G., Pellegrino, M.A., Barresi, R., Bresolin, N., De Angelis, M.G., Campbell, K.P., et al. 2003. Cell therapy of α-sarcoglycan null dystrophic mice through intra-arterial delivery of mesoangioblasts. *Science* **301:** 487–492.

Schienda, J., Engleka, K.A., Jun, S., Hansen, M.S., Epstein, J.A., Tabin, C.J., Kunkel, L.M., and Kardon, G. 2006. Somitic origin of limb muscle satellite and side population cells. *Proc. Natl. Acad. Sci.* **103:** 945–950.

Schuster-Gossler, K., Cordes, R., and Gossler, A. 2007. Premature myogenic differentiation and depletion of progenitor cells cause severe muscle hypotrophy in Delta1 mutants. *Proc. Natl. Acad. Sci.* **104:** 537–542.

Seo, K.W., Wang, Y., Kokubo, H., Kettlewell, J.R., Zarkower, D.A., and Johnson, R.L. 2006. Targeted disruption of the DM domain containing transcription factor *Dmrt2* reveals an essential role in somite patterning. *Dev. Biol.* **290:** 200–210.

Shefer, G. and Yablonka-Reuveni, Z. 2007. Reflections on lineage potential of skeletal muscle satellite cells: Do they sometimes go MAD? *Crit. Rev. Eukaryot. Gene Expr.* **17:** 13–29.

Tajbakhsh, S., Rocancourt, D., and Buckingham, M. 1996. Muscle progenitor cells failing to respond to positional cues adopt non-myogenic fates in myf-5 null mice. *Nature* **384:** 266–270.

Tajbakhsh, S., Borello, U., Vivarelli, E., Kelly, R., Papkoff, J., Duprez, D., Buckingham, M., and Cossu, G. 1998. Differential activation of *Myf5* and *MyoD* by different Wnts in explants of mouse paraxial mesoderm and the later activation of myogenesis in the absence of *Myf5*. *Development* **125:** 4155–4162.

Teboul, L., Summerbell, D., and Rigby, P.W. 2003. The initial somitic phase of *Myf5* expression requires neither Shh signaling nor Gli regulation. *Genes Dev.* **17:** 2870–2874.

Vasyutina, E., Stebler, J., Brand-Saberi, B., Schulz, S., Raz, E., and Birchmeier, C. 2005. *CXCR4* and *Gaδ1* cooperate to control the development of migrating muscle progenitor cells. *Genes Dev.* **19:** 2187–2198.

Vasyutina, E., Lenhard, D.C., Wende, H., Erdmann, B., Epstein, J.A., and Birchmeier, C. 2007. *RBP-J* (*Rbpsuh*) is essential to maintain muscle progenitor cells and to generate satellite cells. *Proc. Natl. Acad. Sci.* **104:** 4443–4448.

Zhao, P., Caretti, G., Mitchell, S., McKeehan, W.L., Boskey, A.L., Pachman, L.M., Sartorelli, V., and Hoffman, E.P. 2006. Fgfr4 is required for effective muscle regeneration in vivo. Delineation of a MyoD-Tead2-Fgfr4 transcriptional pathway. *J. Biol. Chem.* **281:** 429–438.

Zhou, H.M., Wang, J., Rogers, R., and Conway, S.J. 2008. Lineage-specific responses to reduced embryonic Pax3 expression levels. *Dev. Biol.* **315:** 369–382.

Regulation and Function of Skeletal Muscle Stem Cells

M. Cerletti,* J.L. Shadrach,* S. Jurga,* R. Sherwood,[†] and A.J. Wagers*

*Section on Developmental and Stem Cell Biology, Joslin Diabetes Center, Department of Stem Cell
and Regenerative Biology, Harvard University, and Harvard Stem Cell Institute, Boston, Massachusetts 02215;
[†]Department of Molecular and Cellular Biology, Harvard University, Cambridge, Massachusetts 02138

Skeletal muscle satellite cells, which reside beneath the basal lamina of mature muscle fibers, function as myogenic precursors and are required for normal muscle growth and repair. Satellite cells share a common anatomical localization, yet they exhibit substantial phenotypic and functional heterogeneity. Recent efforts in the field of adult myogenesis have been aimed at dissecting this heterogeneity and reveal the presence of discrete cell lineages within the muscle that function independently and interactively to maintain muscle homeostasis and to determine the outcome of muscle damage. Normal developmental regulation of the frequency and function of these distinct tissue precursors, and pathological deregulation of their activity, may have an important role in age- and disease-dependent loss of muscle regenerative activity.

DEVELOPMENTAL ORIGINS OF SKELETAL MUSCLE AND SATELLITE CELLS

Skeletal muscle is a highly specialized tissue composed of postmitotic, multinucleated muscle fibers that contract to generate force and movement. Skeletal muscle accounts for up to half the mass of human bodies, and, in addition to its role in voluntary and involuntary movement, has a critical role in maintaining metabolic health by regulating glucose uptake and insulin sensitivity.

During development, the cells that ultimately give rise to skeletal muscle are formed from paraxial mesoderm-derived cells of the dorsal somites (Buckingham et al. 2003). These cells are myogenically specified by signals emanating from neighboring cells of the notochord, neural tube, and dorsal ectoderm through the action of transcription factors such as the paired-box proteins Pax3 and Pax7 (Fig. 1) (Goulding et al. 1994; Cossu et al. 1996a,b; Borycki et al. 1999). Once committed to the muscle lineage, somite-derived cells migrate from the myotome to multiple sites of embryonic myogenesis, begin to express later markers of myogenic differentiation such as the basic helix-loop-helix transcription factors Myf-5 and MyoD (Birchmeier and Brohmann 2000) and the intermediate filament protein desmin, and differentiate into mature muscle fibers. Pax3 is required for appropriate establishment of limb muscles and regulates the migration of embryonic precursors from the somites to the limbs (Daston et al. 1996; Tremblay et al. 1998; Relaix et al. 2005). Studies of Pax7 "knockout" mice, on the other hand, suggest that Pax7 is dispensable for prenatal muscle formation but required for normal postnatal muscle growth (Seale et al. 2000; Oustanina et al. 2004; Relaix et al. 2005).

Because mature muscle fibers are postmitotic, postnatal muscle growth and repair after injury depend on the maintenance of a reservoir of mononuclear muscle-precursor cells, which appear to be descended from Pax3[+] Pax7[+] somite-derived progenitors that do not differentiate into myofibers during embryogenesis (Fig. 1) (Relaix et al. 2006). These progenitors form a pool of specialized muscle fiber-associated cells called satellite cells (Mauro 1961; Armand et al. 1983; Yablonka-Reuveni et al. 1987; Gros et al. 2005; Relaix et al. 2005) that contain the major regenerative activity for adult skeletal muscle (Mauro 1961; Collins et al. 2005). Satellite cells are found beneath the basal lamina and adjacent to the plasma membrane of mature muscle fibers (Fig. 2A) (Mauro 1961) and first appear in the limb muscles of mouse embryos between 16 and 18 days postconception (dpc). These cells

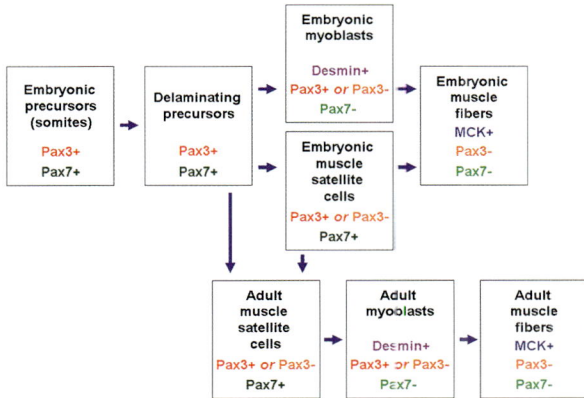

Figure 1. Myogenic gene expression in muscle precursors and differentiated muscle cells. The embryonic precursors of skeletal muscle originate in the somites and express Pax3 and Pax7. These precursors delaminate from the somites and migrate in a Pax3-dependent manner to multiple sites of myogenesis throughout the body. Embryonic myoblasts and satellite cells are formed from these precursors. Adult satellite cells arise either directly from the precursors or from embryonic satellite cells. Activation of satellite cells in adult muscle regenerates muscle fibers via production of more differentiated fusion-competent myoblasts. Cells at each stage of differentiation are specifically marked by unique expression of differentiation proteins; for example, expression of Pax proteins marks satellite cells and somitic precursors only, desmin marks myoblasts but not satellite cells, and muscle creatine kinase (MCK) marks fused myotubes and muscle fibers but not myoblasts or satellite cells.

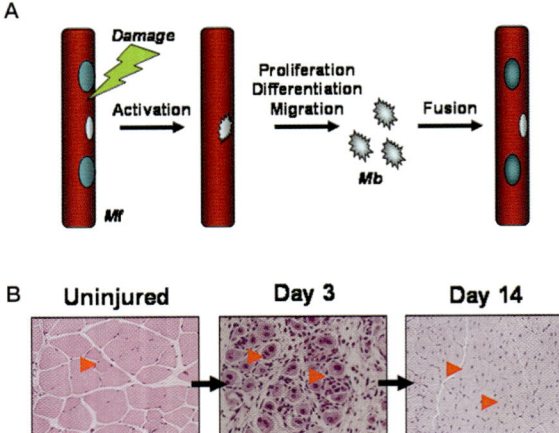

Figure 2. Normal muscle regeneration in young mice. (*A*) Muscle repair by endogenous myogenic satellite cells (*light blue*). Satellite cells reside normally beneath the basal lamina of myofibers (Mf). Satellite cells are activated by muscle damage to proliferate, differentiate and migrate. Differentiation of satellite cells generates fusion-competent myoblasts (Mb) that fuse with one another and with residual myofibers to repair muscle. Regenerated myofibers exhibit centrally located myonuclei (*dark blue*). Self-renewal of muscle stem cells reestablishes the satellite cell pool (*light blue*). (*B*) H&E staining shows uninjured and regenerating muscle at 3 and 14 days after intramuscular injection of the snake venom cardiotoxin. Uninjured muscle shows characteristic peripheral nuclei (*red arrowhead*). Three days after injury, muscle shows extensive inflammatory infiltrate and some small regenerating fibers with characteristic centrally localized nuclei (*red arrowheads*). By 14 days after injury, the muscle is fully repaired, but regenerated fibers are still distinguishable by their central nuclei (*red arrowheads*).

are most abundant early in life, such that in neonatal mice, satellite cell nuclei comprise approximately 30% of myofiber-associated nuclei. However, satellite cell number declines with age, and only about 5% of myofiber nuclei in the muscles of adult mice represent satellite cells (Bischoff 1994).

SATELLITE CELLS AND ADULT MYOGENESIS

Normal skeletal muscle possesses remarkable regenerative capacity (Fig. 2B), and can undergo multiple rounds of rapid repair in response to injury. This regeneration appears to be mediated primarily by satellite cells, which are activated by muscle damage. Activated satellite cells proliferate, differentiate, and migrate to form fusion-competent myoblasts that fuse with residual myofibers and with one another to regenerate the muscle (Fig. 2A) (for review, see Wagers and Conboy 2005). This differentiation process invokes a well-studied cascade of myogenic regulatory factors that, in some regards, recapitulate molecular events of embryonic myogenesis (Fig. 1). Although normally highly efficient, in certain disease states, muscle regenerative activity can be impaired and, together with a loss of homeostatic muscle maintenance, this loss of repair function can contribute to an overall loss of muscle mass and to progressive muscle weakness (see below).

SATELLITE CELL HETEROGENEITY

Satellite cells are so named because they associate very closely with muscle fibers, and in fact, satellite cells typically have been defined by their peculiar localization between the basal lamina and plasma membrane (sarcolemma) of muscle fibers (Mauro 1961). Although the adult satellite cell compartment is often thought of as homogeneous, we (Sherwood et al. 2004) and other investigators (Shefer et al. 2004; Kuang et al. 2007) have found that myofiber-associated cells are actually quite heterogeneous and contain both myogenic and nonmyogenic cells. As described in more detail below, our prospective fractionation of myofiber-associated cells, isolated from intact skeletal muscle by two-step enzymatic digestion, indicates that the muscle-forming and non-muscle-forming populations contained in adult muscle can be separated by virtue of their differential expression of the surface proteins CD45 and Sca-1 (data summarized in Table 1). In particular, in vitro differentiation and in vivo transplant assays demonstrate that the Sca-1⁻CD45⁻ subset of myofiber-associated cells contains the vast majority of autonomously myogenic cells (Sherwood et al. 2004; Montarras et al. 2005; Kuang et al. 2007; Cerletti et al. 2008; Sacco et al. 2008). These Sca-1⁻CD45⁻ cells exhibited no in vitro fibroblastic or adipocytic differentiation potential in any of the in vitro or in vivo studies that we have performed (Sherwood et al. 2004; Cerletti et al. 2008).

In contrast, consistent with recent studies describing the isolation of Sca-1⁺ mesenchymal precursor cells from adipose tissue (Rodeheffer et al. 2008; Tang et al. 2008), Sca-1⁺ cells purified from skeletal muscle exhibit fibroblastic and adipogenic differentiation upon culture (Sherwood et al. 2004) and after in vivo transplant (M. Cerletti et al., in prep.). Finally, in both resting and injured muscle, CD45⁺ blood-lineage cells can be detected in the myofiber-associated cell compartment (Sherwood et al. 2004; Cerletti et al. 2008). Like Sca-1⁺ cells, these CD45⁺ cells also exhibit no autonomously myogenic activity (Saba 2004; Sherwood et al. 2004; Montarras et al. 2005) and likely represent tissue-resident immune cells. Interestingly, CD45⁺ cells are replenished from the bloodstream (Table 1), but only slowly, and they remain predominantly host-derived in muscles of parabiotic mice, which exhibit approximately 50% chimerism of circulating blood leukocytes (Wright et al. 2001). Thus, the normal role of these cells may be to support rapid, local immune and inflammatory responses, equivalent to Langerhans cells in the skin (Merad et al. 2003) and microglia in the brain (Kennedy and Abkowitz 1997).

The close association of these many different cell types with normal muscle fibers suggests that complex interactions among these cells are likely important for normal muscle homeostasis and effective repair. Perturbation of any of these interactions might contribute to muscle dysfunction in disease or following myotrauma. In addition, because the relative representation of myogenic versus nonmyogenic cells within the myofiber-associated satellite cell compartment can change significantly during development, during the course of normal adult life, and in response to pathological muscle conditions, we pro-

Table 1. Summary of cell-surface-marker expression and functional characterization of myofiber-associated cells in adult, wild-type mice (Sherwood et al. 2004; Cerletti et al. 2008)

Cell-surface-marker phenotype	Autonomous in vitro myogenicity	In vivo myogenic activity	Formation of fibroblastic colonies in vitro	In vitro adipogenic activity	Can be seeded from circulation
(1) CD45$^+$ (mostly Mac-1$^+$)	No	No	No	N/D	Yes
(2) CD45$^-$Sca-1$^+$					
CD45$^-$Sca-1$^+$CD34$^+$	No	No	Yes	Yes	Yes
CD45$^-$Sca-1$^+$CD34$^-$	No	No	Yes	N/D	Yes
(3) CD45$^-$Sca-1Mac-1$^-$ CXCR4$^+$β1-integrin$^+$ (all are CD34$^+$)	Yes	Yes	No	No	No

From these data, we propose that the myofiber-associated cell compartment is heterogeneous and contains (1) CD45$^+$ hematopoietic cells, predominantly of the myeloid lineage (Mac-1$^+$), (2) CD45$^-$Sca-1$^+$ cells, including cells with differentiation potential for mesenchymal lineages (CD45$^-$Sca-1$^+$CD34$^+$), and (3) committed muscle stem cells (skeletal muscle precursors [SMPs], CD45$^-$Sca-1$^-$Mac-1$^-$CXCR4$^+$β1-integrin$^+$ [CSM4β]). N/D, not done.

pose that age- or disease-induced imbalance in the relative frequencies of myogenic versus nonmyogenic cells in the muscle could contribute to pathological changes in the regenerative responses of injured muscle tissue (Hidestrand et al. 2008).

PROSPECTIVE ISOLATION OF HIGHLY REGENERATIVE MUSCLE-FORMING STEM CELLS

The ability of skeletal muscle to undergo multiple rounds of regeneration throughout life while still maintaining the satellite cell pool suggests that at least a subset of satellite cells exhibits both self-renewal and differentiation capacities—hallmark properties of tissue stem cells (Collins and Partridge 2005; Wagers and Conboy 2005). To isolate such potential muscle stem cells, we have used cell-surface-marker profiling and fluorescence-activated cell sorting (FACS) to prospectively identify and separate distinct satellite cell populations (Sherwood et al. 2004). This strategy was based on the hypothesis that, as with blood-forming and neural stem cells (Wagers and Weissman 2004), primitive muscle stem cells would express on their surface particular combinations of marker proteins that specifically distinguish them from nonmyogenic cells and from their more differentiated progeny that also may be present in the satellite cell compartment (Wagers and Conboy 2005).

Consistent with this hypothesis, by systematic analysis of more than 40 antibodies recognizing individual cell surface markers, we identified a combination of five markers (CD45$^-$Sca-1$^-$Mac-1$^-$CXCR4$^+$β1-integrin$^+$ [CSM4β]) that distinguishes a population of myogenic satellite cells with muscle stem cell properties (Sherwood et al. 2004). These cells, which we call skeletal muscle precursors (SMPs), can be sorted from freshly dissociated mouse skeletal muscle tissue (Sherwood et al. 2004) and exhibit very efficient in vitro myogenic differentiation at the single-cell level (~50–80% colony formation from clonally sorted cells; Sherwood et al. 2004) and robust muscle regenerative activity in vivo (Fig. 1) (Sherwood et al. 2004). Significantly, when transferred into the muscle of *mdx* mice (Sicinski et al. 1989), which model several features of human Duchenne's muscular dystrophy (DMD), SMPs

contribute to hundreds of new muscle fibers and support donor cell-derived restoration of expression of the muscle-specific protein dystrophin (Cerletti et al. 2008), a protein that is normally absent in *mdx* animals (Sicinski et al. 1989). The mechanism by which these cells incorporate into muscle involves fusion with both endogenous fibers and myogenic precursors and de novo myogenesis (seen in ~2% of donor-marker-expressing myofibers) (Cerletti et al. 2008). Moreover, in highly engrafted recipients, incorporation of dystrophin-expressing donor nuclei supported functional improvement of muscle contraction in recipient mice (Cerletti et al. 2008). Finally, unlike differentiated muscle myoblasts, transplanted SMPs also re-seed a reserve pool of muscle precursors within the recipient's muscle (Cerletti et al. 2008). In addition, because this reserve of muscle stem cells remains in the transplanted tissue, it can be recruited again to mediate subsequent rounds of muscle regeneration, providing a renewing source of cells for muscle repair.

Extensive studies from numerous laboratories have defined characteristic properties of skeletal muscle satellite cells (including mitotic quiescence, localization beneath the basal lamina, and expression of the myogenic transcription factors Pax3 and Pax7) (for review, see Wagers and Conboy 2005). To evaluate the relationship of SMPs to canonically defined satellite cells and other described myogenic populations, we also have investigated their molecular and cell biological properties in adult skeletal muscle. Like bulk muscle satellite cells, SMPs are present at differing frequencies in various skeletal muscle beds but absent from cardiac muscle (S. Jurga and A.J. Wagers, unpubl.). When harvested from uninjured muscle, SMPs are largely quiescent (in the G_0/G_1 phase of the cell cycle) and express mRNA encoding markers of undifferentiated satellite cells, including the myogenic transcription factors Pax3 and Pax7 (Cerletti et al. 2008). They do not stain with antibodies against Flk1 or VE-cadherin (markers of vascular endothelial cells), but they do express CD34, which marks both endothelium and satellite cells (Cerletti et al. 2008; S. Jurga and A.J. Wagers, unpubl.). Like satellite cells, CSM4β SMPs do not express markers of differentiated myoblasts, including desmin and myosin heavy chain (MyHC) (Cerletti et al. 2008). Finally, immunostaining of isolated single myofibers indicates expression of the

CSM4β marker CXCR4 on a subset of sublaminar, Pax7[+] satellite cells, suggesting that CSM4β SMPs indeed localize to the canonical satellite cell niche (Cerletti et al. 2008).

Taken together, these data indicate that autonomously myogenic skeletal muscle stem cells can be distinguished from nonmyogenic myofiber-associated cells by virtue of their unique expression of the cell-surface-marker profile CSM4β. CSM4β SMPs exhibit unique myogenic and stem cell properties and robust muscle regenerative activities. These cells likely represent a major regenerative population in normal adult myogenesis and may be targeted in muscle disease (see below). Therefore, strategies to restore their normal function, either by transplantation (Cerletti et al. 2008) or manipulation of the endogenous SMP population, may provide promising new approaches for the treatment of muscle degenerative disease.

MUSCLE STEM CELLS AND MUSCLE DISEASE

A large number of genetically distinct, congenital diseases impair the structure and/or function of skeletal muscle, resulting in muscular dystrophy (MD), muscle weakness, and metabolic disorder. In Duchenne MD (DMD), the most common childhood form of MD, the impaired structural integrity of dystrophic myofibers necessitates repeated cycles of precursor-cell-mediated muscle regeneration, resulting in a continual need for satellite cell proliferation. It has been hypothesized that such repeated recruitment of satellite cells to participate in muscle repair may ultimately lead to premature loss or impairment of their myogenic activity (Blau et al. 1983; Wright et al. 1985; Luz et al. 2002). Consistent with this notion, several studies have documented alterations in satellite cell number and/or function associated with muscular dystrophy (Wakayama et al. 1979; Ishimoto et al. 1983; Reimann et al. 2000; Yablonka-Reuveni and Anderson 2006), including in vitro studies indicating an enhanced rate of differentiation among dystrophic (*mdx*) mouse satellite cells (Yablonka-Reuveni and Anderson 2006). We also have noted a significant age-dependent loss of phenotypically identified SMP muscle stem cells in *mdx* muscle, although clonal analysis indicated that the myogenic efficiency of SMPs sorted from *mdx* muscle is equivalent to that of wild-type SMPs (Cerletti et al. 2008). Interestingly, we have observed a similar numeric loss of SMPs in the affected muscles of mice deficient in expression of the membrane repair protein dysferlin (J.L. Shadrach and A.J. Wagers, in prep.). In both mice and humans, dysferlin deficiency causes a relatively late-onset dystrophy (known as limb girdle muscular dystrophy type 2B [LGMD2B] or Miyoshi myopathy in humans) that shows stereotypical involvement of particular muscle groups and is thought to arise from failure to efficiently repair tears in the sarcolemma of muscle fibers (Bashir et al. 1998; Liu et al. 1998; Bansal et al. 2003; Lennon et al. 2003; Bansal and Campbell 2004; Ho et al. 2004). These data, obtained in two different models of muscular dystrophy, suggest that chronic muscle damage can perturb the normal maintenance and regulation of muscle stem cells, leading to accelerated loss of these cells from the muscle and possibly accelerating disease progression. Although the specific molecular processes that induce such alterations in satellite cell homeostasis in dystrophic muscle have yet to be described, we believe that delineating these effects will likely be informative for understanding the progressive pathophysiology of dystrophic disease.

AGING OF SKELETAL MUSCLE STEM CELLS

In addition to congenital myopathies, the growing incidence of age-related loss of muscle mass and strength (also known as sarcopenia) presents an urgent health concern in both developed and developing countries. Defects in muscle maintenance and function contribute significantly to morbidity and loss of physical independence among the elderly and also predispose to the occurrence of additional age-associated conditions, including increased adiposity, insulin resistance, frailty fracture and falls. Thus, in light of current demographic trends that predict a near doubling of the number of individuals over the age of 65 in the next 20–25 years (U.S. Census Bureau 2004), it is more than ever imperative to develop a clear understanding of the consequences of physiological aging on muscle lineage cells.

In most tissues, aging involves a progressive decline in the ability to maintain homeostatic cell replacement and to regenerate after injury. How aging causes this widespread deterioration of tissue function is poorly understood, but several lines of evidence implicate loss or functional impairment of tissue-specific stem cells in these age-dependent failures (Rossi et al. 2008). For example, the aged hematopoietic system exhibits deregulation of hematopoietic stem cell self-renewal and differentiation, leading to an apparent phenotypic expansion of HSCs that nonetheless exhibit decreased hematopoietic activity and skewed cell-fate determination (Morrison et al. 1996; Liang et al. 2005; Rossi et al. 2005). Similarly, aging of the central nervous system (CNS) impairs the self-renewal activity of neural stem cells, leading to decreased production of new neurons in the neurogenic subventricular zone (SVZ) of aged animals (Enwere et al. 2004; Maslov et al. 2004; Molofsky et al. 2006).

Like the blood and nervous systems, skeletal muscle also appears to suffer deficiencies in precursor cell activity as a result of physiological aging. As muscle ages, its regenerative abilities decline, such that damage to old muscle tissue results less frequently in replacement by new muscle fibers and more frequently in a prolonged inflammatory response and ultimate replacement by fat and fibrous tissue (Conboy et al. 2003, 2005; Brack et al. 2007). This decline in myogenic function leads to progressive muscle wasting, weakness, and slow or absent recovery after injury in elderly individuals (Di Iorio et al. 2006). Although some inconsistencies exist in published literature (Conboy et al. 2003), several studies have reported decreases in the number of muscle satellite cells associated with aged muscle fibers (Shefer et al. 2006; Collins et al. 2007). Consistent with this, flow cytometric analysis indicates an approximately 50% reduction in the relative frequency of SMPs among myofiber-associated cells isolated from 2-year-old, as opposed to 2-month-old, mice (M. Cerletti et al., in prep.). In addition, satellite cell myogenic activity appears to be impaired in old muscle. Aged satellite cells fail to respond appropriately to muscle injury and as a result, may remain quiescent (Conboy et al. 2003) or exhibit perturbed differentiation potential (Brack et al. 2007) even in the face of strong regenerative signals that normally initiate myogenic repair in young tissue. Interestingly, in the skeletal muscle, age-related effects on regenerative cell function appear to arise largely from alterations in the aged muscle environment (Zacks and Sheff 1982; Carlson and Faulkner 1989; Conboy et al. 2005; Carlson et al. 2008) that appear to suppress normal stem cell activity in older animals and can be regulated by factors that circulate naturally in the bloodstream (Brack et al. 2007; Carlson et al. 2008). These findings have important implications for regenerative-medicine approaches in aged individuals. In particular, although identification of the relevant age-related factors that inhibit muscle regeneration in aged animals could provide new strategies to restore or maintain healthy muscle function in aging individuals, alterations in the aged environment could limit the function of young, healthy cells transplanted into aged muscle.

SUMMARY AND PERSPECTIVE

Adult skeletal muscle maintains remarkable regenerative potential throughout most of life, due predominantly to the presence of self-renewing myogenic stem cells within the canonically defined satellite cell pool. These muscle stem cells can be prospectively isolated, based on combinatorial cell-surface-marker expression, and display unique myogenic and regenerative capabilities. Distinct, nonmyogenic cell populations can also be isolated from the myofiber-associated cell compartment, and these may contribute to local inflammatory responses and support fibrogenic and adipogenic activities in the muscle. Changes in the relative representation of myogenic versus nonmyogenic populations are seen in the context of chronic muscle disease and sarcopenia, and they may reflect proliferative exhaustion of muscle stem cells or inhibitory signals emanating from the dystrophic or aged muscle environment. Future therapeutic strategies based on transplantation of myogenic precursor cells or targeting of endogenous regenerative cells must account for potential modulatory effects of the host environment on effective engraftment and activity of muscle-forming stem cells.

ACKNOWLEDGMENTS

This work was supported in part by a Burroughs Wellcome Fund career award, a Seed Grant from the Harvard Stem Cell Institute, and a grant from the Jain Foundation (A.J.W.).

REFERENCES

Armand, O., Boutineau, A.M., Mauger, A., Pautou, M.P., and Kieny, M. 1983. Origin of satellite cells in avian skeletal muscles. *Arch. Anat. Microsc. Morphol. Exp.* **72:** 163–181.

Bansal, D. and Campbell, K.P. 2004. Dysferlin and the plasma membrane repair in muscular dystrophy. *Trends Cell Biol.* **14:** 206–213.

Bansal, D., Miyake, K., Vogel, S.S., Groh, S., Chen, C.C., Williamson, R., McNeil, P.L., and Campbell, K.P. 2003. Defective membrane repair in dysferlin-deficient muscular dystrophy. *Nature* **423:** 168–172.

Bashir, R., Britton, S., Strachan, T., Keers, S., Vafiadaki, E., Lako, M., Richard, I., Marchand, S., Bourg, N., Argov, Z., et al., 1998. A gene related to *Caenorhabditis elegans* spermatogenesis factor *fer-1* is mutated in limb-girdle muscular dystrophy type 2B. *Nat. Genet.* **20:** 37–42.

Birchmeier, C. and Brohmann, H. 2000. Genes that control the development of migrating muscle precursor cells. *Curr. Opin. Cell Biol.* **12:** 725–730.

Bischoff, R. 1994. The satellite cell and muscle regeneration. In *Myogenesis* (ed. A.G. Engel and C. Franszini-Armstrong), vol. 2, pp. 97–118. McGraw-Hill, New York.

Blau, H.M., Webster, C., and Pavlath, G.K. 1983. Defective myoblasts identified in Duchenne muscular dystrophy. *Proc. Natl. Acad. Sci.* **80:** 4856–4860.

Borycki, A.G., Li, J., Jin, F., Emerson, C.P., and Epstein, J.A. 1999. Pax3 functions in cell survival and in *pax7* regulation. *Development* **126:** 1665–1674.

Brack, A.S., Conboy, M.J., Roy, S., Lee, M., Kuo, C.J., Keller, C., and Rando, T.A. 2007. Increased Wnt signaling during aging alters muscle stem cell fate and increases fibrosis. *Science* **317:** 807–810.

Buckingham, M., Bajard, L. Chang, T. Daubas, P., Hadchouel, J., Meilhac, S., Montarras, D., Rocancourt D., and Relaix, F. 2003. The formation of skeletal muscle: From somite to limb. *J. Anat.* **202:** 59–68.

Carlson, B.M. and Faulkner, J.A. 1989. Muscle transplantation between young and old rats: Age of host determines recovery. *Am. J. Physiol.* **256:** C1262–C1266.

Carlson, M.E., Hsu, M., and Conboy, I.M. 2008. Imbalance between pSmad3 and Notch induces CDK inhibitors in old muscle stem cells. *Nature* **454:** 528–532.

Cerletti, M., Jurga, S., Witczak, C.A., Hirshman, M.F., Shadrach, J.L., Goodyear, L.J., and Wagers, A.J. 2008. Highly efficient, functional engraftment of skeletal muscle stem cells in dystrophic muscles. *Cell* **134:** 37–47.

Collins, C.A. and Partridge, T.A. 2005. Self-renewal of the adult skeletal muscle satellite cell. *Cell Cycle* **4:** 1338–1341.

Collins, C.A., Zammit, P.S., Ruiz, A.P., Morgan, J.E., and Partridge, T.A. 2007. A population of myogenic stem cells that survives skeletal muscle aging. *Stem Cells* **25:** 885–894.

Collins, C.A., Olsen I., Zammit, P.S., Heslop, L., Petrie, A., Partridge, T.A., and Morgan, J.E. 2005. Stem cell function, self-renewal, and behavioral heterogeneity of cells from the adult muscle satellite cell niche. *Cell* **122:** 289–301.

Conboy, I.M., Conboy, M.J., Smythe, G.M., and Rando, T.A. 2003. Notch-mediated restoration of regenerative potential to aged muscle. *Science* **302:** 1575–1577.

Conboy, I.M., Conboy, M.J., Wagers, A.J., Girma, E.R., Weissman, I.L., and Rando, T.A. 2005. Rejuvenation of aged progenitor cells by exposure to a young systemic environment. *Nature* **433:** 760–764.

Cossu, G., Tajbakhsh, S., and Buckingham, M. 1996a. How is myogenesis initiated in the embryo? *Trends Genet.* **12:** 218–223.

Cossu, G., Kelly R., Tajbakhsh, S., Di Donna. S., Vivarelli, E., and Buckingham, M. 1996b. Activation of different myogenic pathways: myf-5 is induced by the neural tube and MyoD by the dorsal ectoderm in mouse paraxial mesoderm. *Development* **122:** 429–437.

Daston, G., Lamar, E., Olivier, M., and Goulding, M. 1996. *Pax-3* is necessary for migration but not differentiation of limb muscle precursors in the mouse. *Development* **122:** 1017–1027.

Di Iorio, A., Abate, M., Di Renzo, D., Russolillo, A., Battaglini, C., Ripari, P., Saggini, R., Paganelli, R., and Abate G. 2006. Sarcopenia: Age-related skeletal muscle changes from determinants to physical disability. *Int. J. Immunopathol. Pharmacol.* **19:** 703–719.

Enwere, E., Shingo, T., Gregg, C., Fujikawa, H., Ohta, S., and Weiss, S. 2004. Aging results in reduced epidermal growth factor receptor signaling, diminished olfactory neurogenesis, and deficits in fine olfactory discrimination. *J. Neurosci.* **24:** 8354–8365.

Goulding, M., Lumsden, A., and Paquette, A.J. 1994. Regulation of *Pax-3* expression in the dermomyotome and its role in muscle development. *Development* **120:** 957–971.

Gros, J., Manceau, M., Thomé, V., and Marcelle, C. 2005. A common somitic origin for embryonic muscle progenitors and satellite cells. *Nature* **435:** 954–958.

Hidestrand, M., Richards-Malcolm, S., Gurley, C.M., Nolen, G., Grimes, B., Waterstrat, A., Zant, G.V., and Peterson, C.A. 2008. Sca-1-expressing nonmyogenic cells contribute to fibrosis in aged skeletal muscle. *J. Gerontol. A Biol. Sci. Med. Sci.* **63:** 566–579.

Ho, M., Post, C.M., Donahue, L.R., Lidov, H.G., Bronson, R.T., Goolsby, H., Watkins, S.C., Cox, G.A., and Brown, Jr., R.H. 2004. Disruption of muscle membrane and phenotype divergence in two novel mouse models of dysferlin deficiency. *Hum. Mol. Genet.* **13:** 1999–2010.

Ishimoto, S., Goto, I., Ohta, M., and Kuroiwa, Y. 1983. A quantitative study of the muscle satellite cells in various neuromuscular disorders. *J. Neurol. Sci.* **62:** 303–314.

Kennedy, D.W. and Abkowitz, J.L. 1997. Kinetics of central nervous system microglial and macrophage engraftment: Analysis using a transgenic bone marrow transplantation model. *Blood* **90:** 986–993.

Kuang, S., Kuroda, K., Le Grand, F., and Rudnicki, M.A. 2007. Asymmetric self-renewal and commitment of satellite stem cells in muscle. *Cell* **129:** 999–1010.

Lennon, N.J., Kho, A., Bacskai, B.J., Perlmutter, S.L., Hyman, B.T., and Brown, Jr., R.H. 2003. Dysferlin interacts with annexins A1 and A2 and mediates sarcolemmal wound-healing. *J. Biol. Chem.* **278:** 50466–50473.

Liang, Y., Van Zant, G., and Szilvassy, S.J. 2005. Effects of aging on the homing and engraftment of murine hematopoietic stem and progenitor cells. *Blood* **106:** 1479–1487.

Liu, J., Aoki, M., Illa, I., Wu, C., Fardeau, M., Angelini, C., Serrano, C., Urtizberea, J.A., Hentati, F., Hamida, M.B., et al. 1998. *Dysferlin*, a novel skeletal muscle gene, is mutated in Miyoshi myopathy and limb girdle muscular dystrophy. *Nat. Genet.* **20:** 31–36.

Luz, M.A., Marques, M.J., and Santo Neto, H. 2002. Impaired regeneration of dystrophin-deficient muscle fibers is caused by exhaustion of myogenic cells. *Braz. J. Med. Biol. Res.* **35:** 691–695.

Maslov, A.Y., Barone, T.A., Plunkett, R.J., and Pruitt, S.C. 2004. Neural stem cell detection, characterization, and age-related changes in the subventricular zone of mice. *J. Neurosci.* **24:** 1726–1733.

Mauro, A. 1961. Satellite cells of muscle skeletal fibers. *J. Biophys. Biochem.* **9:** 493–495.

Merad, M., Manz, M.G., Karsunky, H., Wagers, A., Peters, W., Charo, I., Weissman, I.L., Cyster, J.G., and Engelman, E.G. 2002. Langerhans cells renew in the skin throughout life under steady-state conditions. *Nat. Immunol.* **3:** 1135–1141.

Molofsky, A.V., Slutsky, S.G., Joseph, N.M., He, S., Pardal, R., Krishnamurthy, J., Sharpless, N.E., and Morrison, S.J. 2006. Increasing *p16^{INK4a}* expression decreases forebrain progenitors and neurogenesis during ageing. *Nature* **443:** 448–452.

Montarras, D., Morgan, J., Collins, C., Relaix, F., Zaffran, S., Cumano, A., Partridge, T., and Buckingham, M. 2005. Direct isolation of satellite cells for skeletal muscle regeneration. *Science* **309:** 2064–2067.

Morrison, S.J., Wandycz, A.M., Akashi, K., Globerson, A., and Weissman, I.L. 1996. The aging of hematopoietic stem cells. *Nat. Med.* **2:** 1011–1016.

Oustanina, S., Hause, G., and Braun, T. 2004. Pax7 directs postnatal renewal and propagation of myogenic satellite cells but not their specification. *EMBO J.* **23:** 3430–3439.

Reimann, J., Irintchev, A., and Wernig, A. 2000. Regenerative capacity and the number of satellite cells in soleus muscles of normal and *mdx* mice. *Neuromuscul. Disord.* **10:** 276–282.

Relaix, F., Rocancourt, D., Mansouri, A., and Buckingham, M. 2005. A Pax3/Pax7-dependent population of skeletal muscle progenitor cells. *Nature* **435:** 948–953.

Relaix, F., Montarras, D., Zaffran, S., Gayraud-Morel, B., Rocancourt, D., Tajbakhsh, S., Mansouri, A., Cumano, A., and Buckingham, M. 2006. Pax3 and Pax7 have distinct and overlapping functions in adult muscle progenitor cells. *J. Cell Biol.* **172:** 91–102.

Rodeheffer, M.S., Birsoy, K., and Friedman, J.M. 2008. Identification of white adipocyte progenitor cells in vivo. *Cell* **135:** 240–249.

Rossi, D.J., Jamieson, C.H., and Weissman, I.L. 2008. Stems cells and the pathways to aging and cancer. *Cell* **132:** 681–696.

Rossi, D.J., Bryder, D., Zahn, J.M., Ahlenius, H., Sonu, R., Wagers, A.J., and Weissman, I.L. 2005. Cell intrinsic alterations underlie hematopoietic stem cell aging. *Proc. Natl. Acad. Sci.* **102:** 9194–9199.

Saba, J.D. 2004. Lysophospholipids in development: Miles apart and edging in. *J. Cell. Biochem.* **92:** 967–992.

Sacco, A., Doyonnas, R., Kraft, P., Vitorovic, S., and Blau, H.M. 2008. Self-renewal and expansion of single transplanted muscle stem cells. *Nature* **456:** 502–506.

Seale, P., Sabourin, L.A., Girgis-Gabardo, A., Mansouri A., Gruss, P., and Rudnicki, M.A. 2000. Pax7 is required for the specification of myogenic satellite cells. *Cell* **102:** 777–786.

Shefer, G., Wleklinski-Lee, M., and Yablonka-Reuveni, Z. 2004. Skeletal muscle satellite cells can spontaneously enter an alternative mesenchymal pathway. *J. Cell Sci.* **117:** 5393–5404.

Shefer, G., Van de Mark, D.P., Richardson, J.B., and Yablonka-Reuveni, Z. 2006. Satellite-cell pool size does matter: Defining the myogenic potency of aging skeletal muscle. *Dev. Biol.* **294:** 50–66.

Sherwood, R.I., Christensen, J.L., Conboy, I.M., Conboy, M.J., Rando, T.A., Weissman, I.L., and Wagers, A.J. 2004. Isolation of adult mouse myogenic progenitors: Functional heterogeneity of cells within and engrafting skeletal muscle. *Cell* **119:** 543–554.

Sicinski, P., Geng, Y., Ryder-Cook, A.S., Barnard, E.A., Darlison, M.G., and Barnard, P.J. 1989. The molecular basis of muscular dystrophy in the *mdx* mouse: A point mutation. *Science* **244:** 1578–1580.

Tang, W., Zeve, D., Suh, J.M., Bosnakovski, D., Kyba, M., Hammer, R.E., Tallquist, M.D., and Graff, J.M. 2008. White fat progenitor cells reside in the adipose vasculature. *Science* **322:** 583–586.

Tremblay, P., Dietrich, S., Mericskay, M., Schubert, F.R., Li, Z., and Paulin D. 1998. A crucial role for *Pax3* in the development of the hypaxial musculature and the long-range migration of muscle precursors. *Dev. Biol.* **203:** 49–61.

U.S. Census Bureau. 2004. *U.S. Interim Projections by Age, Race, and Hispanic Origin.* U.S. Government Printing Office, Washington, D.C.

Wagers, A.J. and Conboy, I.M. 2005. Cellular and molecular signatures of muscle regeneration: Current concepts and controversies in adult myogenesis. *Cell* **122:** 659–667.

Wagers, A.J. and Weissman, I.L. 2004. Plasticity of adult stem cells. *Cell* **116:** 639–648.

Wakayama, Y., Schotland, D.L., Bonilla, E., and Orecchio, E. 1979. Quantitative ultrastructural study of muscle satellite cells in Duchenne dystrophy. *Neurology* **29:** 401–407.

Wright, D.E., Wagers, A.J., Gulati, A.P., Johnson, F.L., and Weissman, I.L. 2001. Physiological migration of hematopoietic stem and progenitor cells. *Science* **294:** 1933–1936.

Wright, W.E. 1985. Myoblast senescence in muscular dystrophy. *Exp. Cell Res.* **157:** 343–354.

Yablonka-Reuveni, Z. and Anderson, J.E. 2006. Satellite cells from dystrophic (Mdx) mice display accelerated differentiation in primary cultures and in isolated myofibers. *Dev. Dyn.* **235:** 203–212.

Yablonka-Reuveni, Z., Quinn, L.S., and Nameroff, M. 1987. Isolation and clonal analysis of satellite cells from chicken pectoralis muscle. *Dev. Biol.* **119:** 252–259.

Zacks, S.I. and Sheff, M.F. 1982. Age-related impeded regeneration of mouse minced anterior tibial muscle. *Muscle Nerve* **5:** 152–161.

The Molecular Regulation of Muscle Stem Cell Function

M.A. Rudnicki,* F. Le Grand, I. McKinnell, and S. Kuang

*The Sprott Centre for Stem Cell Research, Regenerative Medicine Program,
Ottawa Health Research Institute, Ottawa, Ontario, Canada K1H 8L6*

Muscle satellite cells are responsible for the postnatal growth and robust regeneration capacity of adult skeletal muscle. A subset of satellite cells purified from adult skeletal muscle is capable of repopulating the satellite cell pool, suggesting that it has direct therapeutic potential for treating degenerative muscle disease. Satellite cells uniformly express the transcription factor Pax7, and Pax7 is required for satellite cell viability and to give rise to myogenic precursors that express the basic helix-loop-helic (bHLH) transcription factors Myf5 and MyoD. Pax7 activates expression of target genes such as *Myf5* and *MyoD* through recruitment of the Wdr5/Ash2L/MLL2 histone methyltransferase complex. Extensive genetic analysis has revealed that Myf5 and MyoD are required for myogenic determination, whereas myogenin and MRF4 have roles in terminal differentiation. Using a Myf5-Cre knockin allele and an R26R-YFP Cre reporter, we observed that in vivo about 10% of satellite cells only express Pax7 and have never expressed Myf5. Moreover, we found that Pax7$^+$/Myf5$^-$ satellite cells give rise to Pax7$^+$/Myf5$^+$ satellite cells through basal-apical asymmetric cell divisions. Therefore, satellite cells in skeletal muscle are a heterogeneous population composed of satellite stem cells (Pax7$^+$/Myf5$^-$) and satellite myogenic cells (Pax7$^+$/Myf5$^+$). Evidence is accumulating that indicates that satellite stem cells represent a true stem cell reservoir, and targeting mechanisms that regulate their function represents an important therapeutic strategy for the treatment of neuromuscular disease.

In adult skeletal muscle, satellite cells reside beneath the basal lamina of muscle closely juxtaposed to muscle fibers and comprise 2–7% of the nuclei associated with a particular fiber. Satellite cells are normally mitotically quiescent, but they are activated (i.e., enter the cell cycle) in response to stress induced by weight bearing or by trauma such as injury (Bischoff 1994). The descendants of activated satellite cells, called myogenic precursor cells (MPCs), undergo multiple rounds of division before fusion and terminal differentiation. Satellite cells are distinct from their daughter myogenic precursor cells as defined by biological, biochemical, and genetic criteria (Charge and Rudnicki 2004). Activated satellite cells also generate progeny that restore the pool of quiescent satellite cells (Kuang et al. 2008).

Our laboratory discovered that satellite cells express the transcription factor Pax7 (Seale et al. 2000). Satellite cells also express vascular cell adhesion molecule 1 (VCAM-1) (Rosen et al. 1992), c-*met* (receptor for hepatocyte growth factor [HGF]), M-cadherin protein (Irintchev et al. 1994; Cornelison and Wold 1997), neural cell adhesion molecule 1 (NCAM1) (Bischoff 1994), Foxk1 (Garry et al. 1997), CD34 (Beauchamp et al. 2000), and syndecans 3 and 4 (Cornelison et al. 2001). Our laboratory also recently identified several additional novel genes including *IgSF4, neuritin, Hoxc10, TcR*-β, *Klra18, Itm2*α, *G0S2*, and *MEGF10* that are expressed in satellite cells in vivo but are not expressed by primary myoblasts (Seale et al. 2004a). Additionally, numerous growth factors such as fibroblast growth factor 6 (FGF6), HGF, bone morphogenetic proteins (BMPs), and nitric oxide (NO) have been suggested to have roles in stimulating satellite cell activation (Charge and Rudnicki 2004). Nevertheless, the precise molecular mechanisms regulating satellite cell function remain poorly understood.

Early experiments using quail-chick chimeras suggested that satellite cells were derived from the somite (Armand et al. 1983). Recent experiments support this work and indicate that the progenitors of satellite cells originate in embryonic somites as Pax3/Pax7-expressing cells (Kassar-Duchossoy et al. 2005; Relaix et al. 2005). However, studies by De Angelis et al. (1999) provided evidence that satellite cells may also be derived from cells associated with the embryonic vasculature including the dorsal aorta. In the adult, results from several laboratories support the notion that under conditions of severe tissue damage and ischemia, satellite cells can be derived from so-called adult stem cells during regeneration (Asakura et al. 2002; LaBarge and Blau 2002; Polesskaya et al. 2003). However, under physiological conditions, satellite cells are the primary source of myogenic progenitors (Parise et al. 2008).

The maintenance of satellite cell numbers in aged muscle after repeated cycles of degeneration and regeneration has been interpreted to support the notion that satellite cells possess an intrinsic capacity for self-renewal (Bischoff 1994). Asymmetric distribution of Numb protein in daughters of satellite cells in cell culture has been implicated in the asymmetric generation of distinct daughter cells for self-renewal or differentiation (Conboy and Rando 2002). However, whether satellite cells are true stem cells or are dedifferentiated myoblasts (Zammit et al. 2004) has remained unresolved. Notably, recent work from our laboratory has defined a subset of satellite cells that form a stem cell reservoir for the satellite cell compartment (Kuang et al. 2007).

PAX7 AND REGENERATIVE MYOGENESIS

The Pax Family of Developmental Control Transcription Factors

The paired-box family of transcription factors (Pax1–9) has important functions in the regulation of the develop-

---

*Corresponding author.

ment and differentiation of diverse cell lineages during embryogenesis (Mansouri et al. 1999). *Pax7* and the closely related *Pax3* gene are paralogs with almost identical amino acid sequences and partially overlapping expression patterns during mouse embryogenesis (Jostes et al. 1990; Goulding et al. 1991). Notably, *Pax3* has an essential role in regulating the developmental program of *MyoD*-dependent migratory myoblasts during embryogenesis (Maroto et al. 1997; Tajbakhsh et al. 1997). More recently, Pax3[+]/Pax7[+] progenitors originating in the embryonic somite have been suggested to be the precursors of satellite cells in adult muscle (Kassar-Duchossoy et al. 2005; Relaix et al. 2005).

Pax7 and Pax3 proteins bind similar if not identical sequence-specific DNA elements, suggesting that they regulate similar sets of target genes (Schafer et al. 1994). Furthermore, increased expression and gain-of-function mutations in both Pax3 and Pax7 are associated with the development of alveolar rhabdomyosarcomas, indicating that both molecules regulate similar activities in the myogenic program (Bennicelli et al. 1999). Although the Pax3 and Pax7 proteins are structurally similar, analysis of null mutations in mice indicates that they are required for the development of a number of distinct cell lineages (Mansouri et al. 1996; Conway et al. 1997; Tremblay et al. 1998; Seale et al. 2000) and appear to have nonredundant roles in myogenesis (Seale et al. 2000, 2004b; Oustanina et al. 2004; Kassar-Duchossoy et al. 2005; Relaix et al. 2005).

Splotch (*Sp*) mice, lacking a functional *Pax3* gene, do not survive to term and fail to form limb muscles due to impaired migration of *Pax3*-expressing cells originating from the somite (Daston et al. 1996; Tremblay et al. 1998). Compound mutant *Sp/Myf5*[−/−] mice do not express *MyoD* in their somites, suggesting that *Myf5* and *Pax3* function upstream of *MyoD* in myogenic determination (Tajbakhsh et al. 1997). Forced expression of *Pax3* induces *MyoD* expression and subsequent myogenesis in nonmuscle tissues in avian embryos (Maroto et al. 1997). However, ectopic expression of *Pax3* in C2C12 myoblasts efficiently inhibits myogenic differentiation (Epstein et al. 1995). Coexpression of *MyoD* and *Pax3* is not observed in the mouse myotome (Williams and Ordahl 1994). Therefore, *Pax3* was suggested to function as an indirect upstream factor that induced migration or other cellular changes to facilitate subsequent induction of *MyoD* transcription (Borycki and Emerson 1997). Contrary to this notion, a *Pax3-FKHR* fusion was observed to activate many muscle regulatory genes including *Myf5* following expression in NIH-3T3 cells (Khan et al. 1999). Notably, Pax7 was recently demonstrated to bind a 57-kb regulatory element upstream of the Myf5 transcription start site (Bajard et al. 2006). These data together with the coexpression of Pax3 and Pax7 in somite-derived pro-satellite cells (Kassar-Duchossoy et al. 2005; Relaix et al. 2005) suggest the hypothesis that Pax3 mediates the migratory phase of the lineage, whereas Pax7 is required to achieve their myogenic potential.

Pax7 Is Required for the Myogenic Specification of Satellite Cells

Our laboratory discovered that satellite cells express the transcription factor Pax7 and that Pax7 has a critical role in regulating the function of satellite cells (Seale et al. 2000; Kuang et al. 2006). Pax7 is specifically expressed in satellite cells in adult muscle, and their daughter myogenic precursor cells in vivo, and primary myoblasts in vitro. Cell culture and electron microscopic analysis indicated ablation of satellite cells in *Pax7*[−/−] skeletal muscle. Fluorescence-activated cell sorting (FACS)/Hoechst analysis demonstrated that the proportion of muscle-derived side population (SP) cells, a putative adult stem cell population, was unaffected. These results demonstrate that satellite cells and muscle-derived SP cells represent distinct cell populations and revealed an essential role for Pax7 in specifying the satellite cell myogenic lineage functioning upstream of the MyoD-family of bHLH factors (Seale et al. 2000, 2004b; Polesskaya et al. 2003).

We performed an extensive analysis of *Pax7*[−/−] mice and have confirmed the progressive ablation of the satellite cell lineage in multiple muscle groups (Seale et al. 2000; Oustanina et al. 2004). Small *Pax7*-deficient cells do survive in the satellite cell position, but these cells arrest and die upon entering mitosis. *Pax7*[−/−] muscles are reduced in size, the fibers contain approximately 50% the normal number of nuclei, and fiber diameters are significantly reduced. Together, these data confirm an essential role for Pax7 in regulating the myogenic potential of satellite cells (Kuang et al. 2006).

In previous studies, we investigated the potential of atypical nonsatellite cell progenitors to participate in muscle regeneration. CD45[+]/Sca1[+] cells purified from regenerating wild-type muscle express Pax7 and give rise to skeletal myoblasts. In contrast, CD45[+]/Sca1[+] cells from regenerating *Pax7*[−/−] muscle do not undergo myogenic progression unless exposed to Wnts (Polesskaya et al. 2003). Retroviral expression of Pax7 in CD45[+]/Sca1[+] cells from uninjured muscle induced the formation of myogenic progenitors expressing Myf5 and MyoD, which differentiated into myogenin and myosin-heavy-chain–expressing myocytes (Seale et al. 2004b). Together, these results demonstrate that Pax7 is required for the myogenic specification of muscle-derived adult stem cells during regenerative myogenesis (Polesskaya et al. 2003; Seale et al. 2004b). It is important to note that although atypical myogenic cell progenitors have the proven potential to participate to some degree in muscle regeneration under conditions of severe trauma (Peault et al. 2007), our experiments strongly support the contention that under physiological conditions, the growth and regeneration of skeletal muscle are mediated largely if not exclusively by muscle satellite cells (Parise et al. 2008).

In summary, satellite cells arise from a novel population of muscle progenitor cells that originate in the central domain of the dermomyotome. These progenitors express Pax3 and Pax7 (Gros et al. 2005; Relaix et al. 2005), and although neither their emergence nor their maintenance requires Pax3 function (Kassar-Duchossoy et al. 2005), recent studies have demonstrated that Pax7 is uniquely indispensable (Bajard et al. 2006). In the absence of Pax7, satellite cells die and thus fail to repopulate their niche (Relaix et al. 2005, 2006; Kuang et al. 2006). Pax7 is therefore essential for the formation and maintenance of a population of functional satellite cells (Fig. 1).

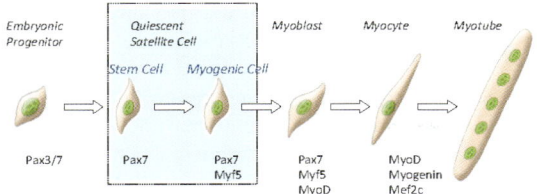

Figure 1. Transcriptional hierarchy regulating the developmental program of satellite cells in skeletal muscle. Progenitors of satellite stem cells originate in the somite as Pax3- and or Pax7-expressing progenitors. Satellite stem cells express Pax7, whereas satellite myogenic cells have additionally activated Myf5 transcriptional competence, as revealed by expression of Myf5-*lacZ* and Myf5-*cre* knockin alleles. Following activation and entrance into the cell cycle, myogenic precursor cells express Myf5 and MyoD. Induction of myogenin and Mef2c together with downregulation of Myf5 and then MyoD mark withdrawal from the cell cycle and entrance into the terminal differentiation program.

Molecular Regulation of Pax7 Function

The mechanisms by which Pax7 activates downstream target genes were difficult to address experimentally due to the relatively weak *trans*-activation properties of Pax7 (Bennicelli et al. 1999). We therefore undertook a multilevel approach to investigate the molecular determinants regulating Pax7 function in satellite cells (McKinnell et al. 2008). We used gene expression analysis to identify several novel and strongly regulated candidate target genes. Together with several other target genes (*PlagL1, Lix1, Sync2, Cipar1, Trim54,* and *Mest*), and *Myf5* was confirmed as an important Pax7 target gene using small interfering RNA (siRNA), FACS of Pax7-deficient satellite cells, and chromatin immunoprecipitation (ChIP) of Pax7 over regulatory sequences. Tandem affinity purification (TAP) and mass spectrometry were used to purify Pax7 together with associated protein complexes. This analysis revealed that Pax7 associates with the Wdr5/Ash2L/MLL2 histone methyltransferase (HMT) complex (McKinnell et al. 2008), which directs methylation of histone H3 lysine 4 (H3K4).

Pax3 binds at approximately 57.5 kb upstream of the transcriptional start of the *Myf5* gene (Bajard et al. 2006). Pax7 similarly binds and recruits the Wdr5/Ash2L/MLL2 HMT complex to binding sites in the regulatory sequences of target genes (McKinnell et al. 2008). Importantly, binding of the Pax7-HMT complex results in H3K4 trimethylation of surrounding chromatin (McKinnell et al. 2008). Together, these experiments indicate that Pax7 enforces satellite cell commitment by recruiting an HMT complex to Myf5, resulting in transcriptional activation. Notably, Pax family genes are essential for the embryonic specification of diverse tissues; thus, Pax recruitment of HMT complexes could be a conserved mechanism for seeding lineage-specific gene expression programs during development.

MYOGENIC bHLH REGULATORY FACTORS

The MyoD Family of Transcription Factors and Myogenesis

The myogenic regulatory factors (MRFs) form a group of bHLH transcription factors (MyoD, Myf5, myogenin,

and MRF4) that regulate the skeletal muscle developmental program (Perry and Rudnick 2000). MRF proteins contain a conserved basic DNA-binding domain that binds the E box, a DNA motif that contains the core E-box sequence CANNTG (Weintraub et al. 1991). The HLH domain mediates dimerization with other HLH-containing proteins (e.g., factors encoded by the E2-2 and E2-5 genes: E12, E47, HEB, and ITF2) (Barndt and Zhuang 1999). Myf5 and MyoD are expressed in proliferating myoblasts and are subject to distinct cell cycle regulation (Kitzmann et al. 1998). MyoD is up-regulated shortly after induction of differentiation followed by myogenin. Myf5 and MyoD levels progressively decrease after this point. The levels of myogenin increase through differentiation, followed by up-regulation of MRF4 several days after the induction of differentiation as myogenin levels decrease (Megeney and Rudnicki 1995).

The Mef2 class of transcription factors also has an important regulatory role in the control of muscle-specific transcription. There exists four alternatively spliced *Mef2* genes: *Mef2A, Mef2B,* and *Mef2D* are expressed ubiquitously, whereas *Mef2C* is restricted to muscle, brain, and spleen. Mef2 sites are found in the promoters of many muscle-specific genes including *myogenin,* suggesting that the MRFs and Mefs form an autoregulatory network (Berkes and Tapscott 2005). In addition, Mef2 proteins and MRFs synergistically coactivate E-box- and Mef2-site-containing promoters (Kaushal et al. 1994; Molkentin et al. 1995; Naidu et al. 1995). During development, Mef2 expression follows myogenin expression, suggesting that Mef2 proteins likely act following differentiation (Edmondson et al. 1994). Moreover, mice lacking myogenin contain virtually no mature myotubes but express normal levels of MyoD and Mef2C, suggesting that expression of Myf5 or MyoD induces myogenin and MEF2 expression upon initiation of terminal differentiation (Venuti et al. 1995). Importantly, Mef2d recruits the Ash2L histone methyltransferase complex to target genes, and this recruitment is regulated by p38 signaling (Rampalli et al. 2007).

The introduction of null mutations in the MyoD family into the germ line of mice revealed the hierarchical relationships existing among the MRFs and established that functional overlap is a feature of the MRF regulatory network. Newborn mice lacking a functional MyoD gene display no overt abnormalities in muscle but express about fourfold higher levels of Myf5 (Rudnicki et al. 1992). Newborn Myf5-deficient animals are also viable and display apparently normal muscle (Braun et al. 1992; Kaul et al. 2000). Muscle development in the trunk of embryos lacking Myf5 is delayed until the onset of MyoD expression, which occurs with somewhat delayed kinetics (Braun et al. 1992; Tajbakhsh and Cossu 1997). Strikingly, newborn mice deficient in both Myf5 and MyoD are totally devoid of myoblasts and myofibers. Thus, Myf5 and MyoD are required for the determination of myogenic precursors and act upstream of myogenin and MRF4 (Rudnicki et al. 1993).

Mice lacking myogenin are immobile and die perinatally due to deficits in myoblast differentiation, as evidenced by an almost complete absence of myofibers (Hasty et al. 1993; Nabeshima et al. 1993). However, normal numbers of MyoD-expressing myoblasts are present and these are organized in groups similar to wild-type muscle. Myogenin-deficient embryos form primary myofibers normally but appear unable to form secondary myofibers (Venuti et al. 1995). Therefore, myogenin has an essential in vivo role in the terminal differentiation of myoblasts.

Mice carrying different targeted MRF4 mutations display a range of phenotypes consistent with a late role for MRF4 in the myogenic pathway (Zhang et al. 1995; Rawls et al. 1998). Interestingly, mice lacking both MyoD and MRF4 display a phenotype similar to the myogenin-null phenotype (Rawls et al. 1998). Therefore, MRF4 function may be substituted by the presence of myogenin but only in the presence of MyoD. Notably, MRF4 appears to have a role as a determination factor in a subset of myocytes in the early somite and as a differentiation factor in later muscle fibers (Kassar-Duchossoy et al. 2004).

Gene targeting and expression analysis have therefore suggested the functional classification of the MRFs into two groups: In the first group, Myf5 and MyoD act as determination factors, and in the second group, myogenin and MRF4 act as differentiation factors.

Myf5 and MyoD Regulate the Development of Distinct Myogenic Lineages

The temporal-spatial patterns of myogenesis in Myf5- and MyoD-deficient embryos provide strong evidence for unique roles of Myf5 and MyoD in the development of epaxial and hypaxial musculature (Kablar et al. 1997, 1998, 2003; Kablar and Rudnicki 1999). Embryos lacking MyoD display normal development of paraspinal and intercostal muscles in the body proper, whereas muscle development in limb buds and branchial arches is delayed by about 2.5 days. In contrast, embryos lacking Myf5 display normal muscle development in limb buds and branchial arches and a marked delay in development of paraspinal and intercostal muscles. Although MyoD mutant embryos exhibit delayed development of limb musculature, the migration of Pax-3-expressing cells into the limb buds and subsequent induction of Myf5 in myogenic precursors occur normally. These results indicate that Myf5 expression in the limb is insufficient for the normal progression of myogenic development.

The phenotype of Myf5- and MyoD-deficient animals strongly supports the notion that the putative myogenic lineages that give rise to epaxial and hypaxial musculature have different requirements for Myf5 or MyoD for appropriate development. Importantly, analysis of primary myoblasts expressing either MyoD (wild type) or Myf5 (MyoD$^{-/-}$) has revealed striking differences in gene expression, morphology, and differentiation potential and indicate that Myf5 and MyoD are not functionally equivalent (Sabourin et al. 1999; Ishibashi et al. 2005). Indeed, recent studies using conditional ablation in mice have confirmed that Myf5 determines a distinct myogenic cell population (Gensch et al. 2008; Haldar et al. 2008).Therefore, these experiments clearly support the hypothesis that Myf5 and MyoD have important and unique roles in the development of epaxial and hypaxial musculature (Kablar et al. 1997, 1998, 1999, 2003).

Molecular Regulation of Myogenic Factor Function

Initiation of myogenic differentiation is characterized by cell cycle withdrawal, stimulation of MyoD transcriptional activity, and sequential induction of myogenin and Mef2 expression (Berkes and Tapscott 2005). However, at odds with this simplistic model are data from gene expression and ChIP-on-Chip studies revealing that MyoD directs multiple subprograms of gene expression, each of which is uniquely regulated (Bergstrom et al. 2002; Blais et al. 2005; Cao et al. 2006). For example, activation of a subset of late-activated MyoD target genes requires p38α/β kinases, whereas expression of an immediate-early target, myogenin, requires formation of a MyoD-Pbx1 complex (Berkes et al. 2004; Penn et al. 2004). Genes expressed early in differentiation, such as *myogenin*, are induced primarily by MyoD (Cao et al. 2006). In contrast, MyoD initiates regional histone modifications at late-expressed targets, such as *MyHC* and *MLC*. However, full expression of *MyHC* and *MLC* requires myogenin transcriptional activity. Thus, MyoD recruits different complexes to unique binding sites on specific genes at different times to orchestrate gene expression during myogenic differentiation.

MyoD specifically forms an E-box-associated complex with histone acetyltransferases (HATs) p300 and p300/CBP (CREB-binding protein)-associated factor (PCAF) in vitro (Puri et al. 1997). These observations suggest that MyoD recruits p300 and PCAF to muscle-specific promoters, which induces histone acetylation and transcriptional activation (Berkes and Tapscott 2005). The capacity of MyoD to stimulate myogenic differentiation genes is partially correlated with its ability to associate with acetyltransferases and to reorganize chromatin. In addition, an in vitro transcription system has identified specific roles for p300/CBP and PCAF (Dilworth et al. 2004).

HATs stimulate MyoD-dependent transcription by acetylating nucleosomal histones surrounding E boxes and directly acetylating MyoD. Nucleosomal histones H3 and H4 are acetylated following recruitment by MyoD to its promoter. Subsequently, p300/CBP recruits PCAF that then acetylates three MyoD lysine residues adjacent to its bHLH domain: Lys-99, -102, and -104 (Sartorelli et al. 1999; Polesskaya et al. 2000). This PCAF-mediated acetylation increases the interaction among these acetylated lysines of MyoD and the bromodomain of CBP, which ultimately enhances transcriptional activity (Polesskaya et al. 2001). By interacting simultaneously with the basal transcription machinery and with upstream transcription factors, p300/CBP functions as a physical bridge that stabilizes the transcription complex.

Histone deacetylases (HDACs) negatively regulate myogenic gene expression through interactions with MyoD and Mef2. HDAC-4/5 interact with Mef2 proteins to repress transcriptional activation from promoters containing Mef2 sites (McKinsey et al. 2001). In addition, these HDACs inhibit MyoD-activated transcription in promoters that contain an Mef2 site and an E box, suggesting that this class II histone deacetylase inhibition is mediated indirectly through Mef2 (Lu et al. 2000). In contrast, the class I deacetylase HDAC-1 is able to directly associate and deacetylate MyoD in vitro. Cell culture experiments demonstrate that in undifferentiated myoblasts, but not differentiating myotubes, HDAC-1 is complexed with MyoD and inhibits PCAF-permissive MyoD-dependent transcription (Mal et al. 2001). Following the induction of myogenic differentiation, HDAC-1 dissociates from MyoD following displacement by the retinoblastoma (Rb) protein, permitting MyoD to function in transcriptional activation and associate with PCAF (Puri et al. 2001; Mal and Harter 2003).

MyoD can also repress transcription in proliferating myoblasts by recruiting Suv39h1 to promoters of genes such as *myogenin* that are activated during differentiation (Mal 2006). Suv39h1 is a histone H3 lysine 9 (H3-K9)-specific methyltransferase (HMT) that silences genes by modifying chromatin (Schotta et al. 2003). Silencing of *myogenin* transcription requires sustained methylation of H3-K9 on the *myogenin* promoter as well as a stable interaction between Suv39h1 and MyoD (Mal 2006).

The MyoD family functions as heterodimers with members of the E-protein family to induce myogenic gene activation. Our laboratory found that the switch from α to β alternative splicing forms of HEB have an important role in regulating the switch from growth to differentiation. Upon induction of differentiation, a MyoD-HEBβ complex bound the E1 E box of the *myogenin* promoter, leading to transcriptional activation. Importantly, forced expression of HEBβ with MyoD synergistically led to precocious myogenin expression in proliferating myoblasts. However, after differentiation, HEBα and HEBβ synergized with myogenin, but not MyoD, to activate the *myogenin* promoter. Therefore, HEBα and HEBβ have novel and central roles in orchestrating the regulation of myogenic factor activity through myogenic differentiation (Parker et al. 2006).

During the induction of myoblast differentiation, p38α and p38β kinases are progressively phosphorylated, promoting the myogenic differentiation pathway (Cuenda and Cohen 1999; Zetser et al. 1999; Wu et al. 2000). Furthermore, myoblasts treated with the inhibitor SB203580, specific to the p38α and β isoforms, fail to fuse into myotubes or induce muscle-specific genes. Within the *myogenin* promoter, it has been proposed that MyoD interacts with the Pbx transcription factor in a p38α-dependent manner (Berkes et al. 2004; de la Serna et al. 2005). Constitutively bound Pbx associates with MyoD and recruits the SWI/SNF chromatin-remodeling complex, facilitating the direct binding of MyoD to an E box within the *myogenin* promoter. p38α/β phosphorylation permits targeting of SWI/SNF to muscle promoters, in addition to increased Mef2 transcriptional activity, whose interaction with MyoD contributes to the induction of muscle-specific gene transcription (Black and Olson 1998; Zetser et al. 1999; Wu et al. 2000; Simone et al. 2004). The mammalian SWI/SNF enzyme is a multiprotein chromatin-remodeling complex that actively alters nucleosomal structure by catalyzing a shift in the histone octamer along the DNA, thus exposing previously condensed DNA to transcriptional complexes (Simone 2006). These studies indicate a crucial and specific role for p38α/β in the recruitment of SWI/SNF to muscle gene promoters, providing an additional mechanism to account for the positive effect of p38α/β in myogenesis (Simone et al. 2004).

SATELLITE STEM CELLS AND REGENERATIVE MYOGENESIS

The Stem Cell Niche

Recent advances have provided important insights into the role that the microenvironment has in regulating stem cell function (Fuchs et al. 2004). The stem cell niche directs the maintenance of stem cell identity as well as the asymmetric generation and issue of committed daughter cells from the niche. The stem cell niche was originally described in studies of *Drosophila* oogenesis. Germ stem cells adhere to cap cells in contact with the basal lamina. Daughter cells that lose cap cell contact become cytoblasts that are destined to differentiate into oocytes (Song et al. 2002). In mammals, analogous stem cell niches have been described for neural stem cells in the brain (Zhong et al. 2000), hematopoietic stem cells in bone marrow (Calvi et al. 2003; Zhang et al. 2003), stem cells in the crypt of the intestinal villus (Sancho et al. 2003), and stem cells in the hair follicle (Tumbar et al. 2004).

Cell polarity has been hypothesized to be established within the niche by cell–cell interactions mediated by cadherins and cell–extracellular matrix interactions mediated by integrins (Fuchs et al. 2004). Stem cell polarity and spindle orientation relative to the basal lamina determine whether a stem cell division will be symmetric or asymmetric. Planar divisions (parallel to the basal lamina) are symmetrical, generating identical daughter cells. In contrast, apical-basal divisions (90° to the basal lamina) are asymmetrical, with one daughter cell remaining a stem cell at the basal surface and a committed daughter cell destined for differentiation on the apical surface (Fuchs et al. 2004; Tumbar et al. 2004). Therefore, the discovery of the stem cell niche provides an additional means to establish the identity of a stem cell population within a particular tissue.

Satellite Stem Cells Maintain the Satellite Cell Compartment

Previous studies of satellite cell activation have used individual myofibers in cell culture (Zammit et al. 2004). However, we have noted that in cultured myofibers, satellite cells uniformly and immediately melt through the basal lamina and either proliferate on the outside matrix of the myofiber or migrate onto the culture plate. In contrast, in vivo satellite cell divisions on a living fiber are uniformly beneath the basal lamina (our unpublished data). The existence of a polarized environment for muscle satellite cells is evidenced by expression of M-cadherin on the apical surface facing the myotube (Irintchev et al. 1994) and by expression of α7/β1-integrin on the basal surface facing the basal lamina (Blanco-Bose et al. 2001).

In recent work from our laboratory, we discovered that the satellite cell population is heterogeneous based on the

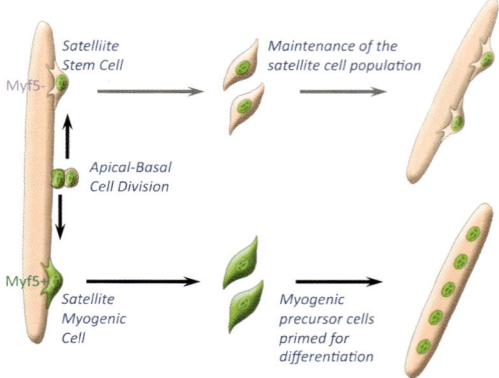

Figure 2. The satellite cell population is heterogeneously composed of satellite stem cells and satellite myogenic cells. Using *Myf5-Cre* and *ROSA26-YFP Cre* alleles, we observed that, in vivo, 10% of sublaminar Pax7-expressing satellite cells have never expressed *Myf5*. Moreover, we found that Pax7$^+$/Myf5$^-$ satellite cells gave rise to Pax7$^+$/Myf5$^+$ satellite cells through apical-basal–oriented divisions that asymmetrically generated a basal Pax7$^+$/Myf5$^-$ cell and an apical Pax7$^+$/Myf5$^+$ cell. Prospective isolation and transplantation into muscle revealed that, whereas Pax7$^+$/Myf5$^+$ cells exhibited precocious differentiation, Pax7$^+$/Myf5$^-$ cells extensively contributed to the satellite cell reservoir throughout the injected muscle. Therefore, satellite cells are a heterogeneous population composed of stem cells and committed progenitors.

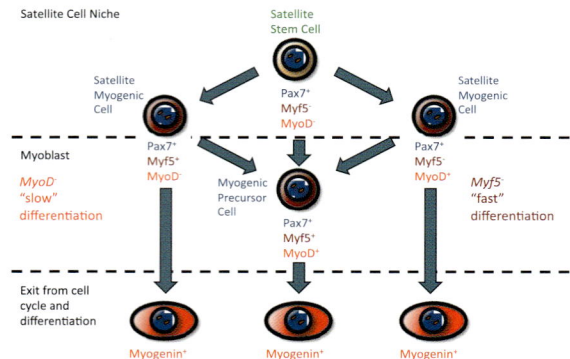

Figure 3. Hypothetical scheme describing the generation of different myogenic lineages from satellite stem cells during regenerative myogenesis. Satellite stem cells would enter different developmental programs depending on whether Myf5 or MyoD expression predominates. Predominance of MyoD would drive the program toward early differentiation, analogous to the behavior of *Myf5$^-$* myoblasts, whereas predominance of Myf5 would drive the program toward enhanced proliferation and delayed differentiation. Myoblasts coexpressing Myf5 and MyoD would exhibit the intermediate growth and differentiation program of primary cultures of typical satellite-cell-derived myoblasts.

expression of two transcriptional factors, Pax7 and Myf5 (Fig. 2) (Kuang et al. 2007). Genetic analysis using the Cre-LoxP system (*Myf5-Cre/R26R-YFP*) revealed that about 10% of satellite cells have never previously expressed *Myf5-cre*. However, time-lapse videography of cultured myofibers demonstrated up-regulation of *Myf5-Cre* and induction of yellow fluorescent protein (YFP) expression in satellite daughter cells. Moreover, we observed by reverse transcriptase–polymerase chain reaction (RT-PCR) analysis following prospective isolation that YFP$^+$ cells expressed Myf5 mRNA, whereas YFP$^-$ cells did not. We therefore hypothesized that the Pax7$^+$/YFP$^-$ satellite cells represent a novel stem cell population and that the majority of Pax7$^+$/YFP$^+$ satellite cells represent committed myogenic progenitors.

During muscle regeneration, both YFP$^+$ and YFP$^-$ satellite cells proliferate. Both types of satellite cells underwent planar divisions, with the polarity of cell division parallel to the basal lamina entirely symmetrical, generating two identical daughter cells that both remained either YFP$^+$ or YFP$^-$. Strikingly, basal-apical divisions, with the polarity of cell division at right angles to the basal lamina, were typically asymmetrical, with the basal cell remaining YFP$^-$ and the apical cell up-regulating YFP. Prospective isolation and transplantation into muscle revealed that whereas Pax7$^+$/Myf5$^+$ cells undergo terminal differentiation, Pax7$^+$/Myf5$^-$ cells extensively contributed to the satellite cell reservoir throughout the injected muscle. Repression of Notch signaling resulted in reduced numbers of satellite cells due to inhibition of satellite stem cell self-renewal. We therefore conclude that Pax7$^+$/Myf5$^-$ satellite cells represent a novel population of stem cells that actively maintain the homeostatic composition of

stem cells and committed progenitors within the satellite cell niche (Fig. 2) (Kuang et al. 2007).

The induction of expression of MyoD during satellite cell activation has not been well characterized relative to Myf5 induction. Interestingly, early studies suggested that Myf5 and MyoD were up-regulated independently of each other during satellite cell activation (Cornelison and Wold 1997). Notably, Myf5 and MyoD appear to direct the developmental program of distinct myogenic lineages during embryogenesis (Kablar et al. 1997, 1998, 1999, 2003; Gensch et al. 2008; Haldar et al. 2008). Therefore, these data suggest a hypothesis whereby satellite stem cells can enter different developmental programs depending on whether Myf5 or MyoD expression predominates (Fig. 3). Predominance of MyoD would drive the program toward early differentiation, analogous to the behavior of *Myf5$^{-/-}$* myoblasts (Montarras et al. 2000). In contrast, predominance of Myf5 would drive the program toward enhanced proliferation and delayed differentiation, analogous to the behavior of *MyoD$^{-/-}$* myoblasts (Sabourin et al. 1999). Myoblasts coexpressing Myf5 and MyoD would exhibit the intermediate growth and differentiation program of primary cultures of typical satellite-cell-derived myoblasts. This hypothesis also explains the spectrum of differentiation observed in clonal cultures of myogenic cells derived from skeletal muscle. Future studies using lineage markers should elucidate these phenomena as well as the molecular control of the lineage choice.

CONCLUSIONS

Recent experiments have established that a subset of satellite cells, termed satellite stem cells, can be purified from skeletal muscle and become capable of repopulating the satellite cell pool following transplantation. Significant challenges remain regarding transplantation as a therapeutic approach, including expansion of cells and delivery to affected muscles. Notably, studies are well under way to molecularly characterize satellite stem cells and define the regulatory pathways that govern their self-renewal and commitment to differentiate. Clearly, the development of drugs or biologics that specifically target these mechanisms represents an important therapeutic strategy for the treatment of neuromuscular disease.

ACKNOWLEDGMENTS

Studies from the laboratory of M.A.R. were supported by grants from the Canadian Institutes of Health Research, Muscular Dystrophy Association, National Institutes of Health, Howard Hughes Medical Institute, Canadian Stem Cell Network, and Canada Research Chair Program. M.A.R. holds the Canada Research Chair in Molecular Genetics and is an International Research Scholar of the Howard Hughes Medical Institute.

REFERENCES

Armand, O., Boutineau, A.M., Mauger, A., Pautou, M.P., and Kieny, M. 1983. Origin of satellite cells in avian skeletal muscles. *Arch. Anat. Microsc. Morphol. Exp.* **72:** 163–181.

Asakura, A., Seale, P., Girgis-Gabardo, A., and Rudnicki, M.A. 2002. Myogenic specification of side population cells in skeletal muscle. *J. Cell Biol.* **159:** 123–134.

Bajard, L., Relaix, F., Lagha, M., Rocancourt, D., Daubas, P., and Buckingham, M.E. 2006. A novel genetic hierarchy functions during hypaxial myogenesis: Pax3 directly activates *Myf5* in muscle progenitor cells in the limb. *Genes Dev* **20:** 2450–2464.

Barndt, R.J. and Zhuang, Y. 1999. Controlling lymphopoiesis with a combinatorial E-protein code. *Cold Spring Harbor Symp. Quant. Biol.* **64:** 45–50.

Beauchamp, J.R., Heslop, L., Yu, D.S., Tajbakhsh, S., Kelly, R.G., Wernig, A., Buckingham, M.E., Partridge, T.A., and Zammit, P.S. 2000. Expression of CD34 and myf5 defines the majority of quiescent adult skeletal muscle satellite cells. *J. Cell Biol.* **151:** 1221–1234.

Bennicelli, J.L., Advani, S., Schafer, B.W., and Barr, F.G. 1999. PAX3 and PAX7 exhibit conserved *cis*-acting transcription repression domains and utilize a common gain of function mechanism in alveolar rhabdomyosarcoma. *Oncogene* **18:** 4348–4356.

Bergstrom, D.A., Penn, B.H., Strand, A., Perry, R.L., Rudnicki, M.A., and Tapscott, S.J. 2002. Promoter-specific regulation of MyoD binding and signal transduction cooperate to pattern gene expression. *Mol. Cell* **9:** 587–600.

Berkes, C.A. and Tapscott, S.J. 2005. MyoD and the transcriptional control of myogenesis. *Semin. Cell Dev. Biol.* **16:** 585–595.

Berkes, C.A., Bergstrom, D.A., Penn, B.H., Seaver, K.J., Knoepfler, P.S., and Tapscott, S.J. 2004. Pbx marks genes for activation by MyoD indicating a role for a homeodomain protein in establishing myogenic potential. *Mol. Cell* **14:** 465–477.

Bischoff, R. 1994. The satellite cell and muscle regeneration. In *Myogenesis* (ed. A.G. Engel and C. Franzini-Armstrong), pp. 97–118. McGraw-Hill, New York.

Black, B.L. and Olson, E.N. 1998. Transcriptional control of muscle development by myocyte enhancer factor-2 (MEF2) proteins. *Annu. Rev. Cell Dev. Biol.* **14:** 167–196.

Blais, A., Tsikitis, M., Acosta-Alvear, D., Sharan, R., Kluger, Y., and Dynlacht, B.D. 2005. An initial blueprint for myogenic differentiation. *Genes Dev.* **19:** 553–569.

Blanco-Bose, W.E., Yao, C.C., Kramer, R.H., and Blau, H.M. 2001. Purification of mouse primary myoblasts based on α7 integrin expression. *Exp. Cell Res.* **265:** 212–220.

Borycki, A.G. and Emerson, C.P. 1997. Muscle determination: Another key player in myogenesis? *Curr. Biol.* **7:** R620–R623.

Braun, T., Rudnicki, M.A., Arnold, H.H., and Jaenisch, R. 1992. Targeted inactivation of the muscle regulatory gene *Myf-5* results in abnormal rib development and perinatal death. *Cell* **71:** 369–382.

Calvi, L.M., Adams, G.B., Weibrecht, K.W., Weber, J.M., Olson, D.P., Knight, M.C., Martin, R.P., Schipani, E., Divieti, P., Bringhurst, F.R., et al. 2003. Osteoblastic cells regulate the haematopoietic stem cell niche. *Nature* **425:** 841–846.

Cao, Y., Kumar, R.M., Penn, B.H., Berkes, C.A., Kooperberg, C., Boyer, L.A., Young, R.A., and Tapscott, S.J. 2006. Global and gene-specific analyses show distinct roles for Myod and Myog at a common set of promoters. *EMBO J.* **25:** 502–511.

Charge, S.B. and Rudnicki, M.A. 2004. Cellular and molecular regulation of muscle regeneration. *Physiol. Rev.* **84:** 209–238.

Conboy, I.M. and Rando, T.A. 2002. The regulation of Notch signaling controls satellite cell activation and cell fate determination in postnatal myogenesis. *Dev. Cell* **3:** 397–409.

Conway, S.J., Henderson, D.J., and Copp, A.J. 1997. *Pax3* is required for cardiac neural crest migration in the mouse: Evidence from the *splotch* (*Sp²ᴴ*) mutant. *Development* **124:** 505–514.

Cornelison, D.D. and Wold, B.J. 1997. Single-cell analysis of regulatory gene expression in quiescent and activated mouse skeletal muscle satellite cells. *Dev. Biol.* **191:** 270–283.

Cornelison, D.D., Filla, M.S., Stanley, H.M., Rapraeger, A.C., and Olwin, B.B. 2001. Syndecan-3 and syndecan-4 specifically mark skeletal muscle satellite cells and are implicated in satellite cell maintenance and muscle regeneration. *Dev. Biol.* **239:** 79–94.

Cuenda, A. and Cohen, P. 1999. Stress-activated protein kinase-2/p38 and a rapamycin-sensitive pathway are required for C2C12 myogenesis. *J. Biol. Chem.* **274:** 4341–4346.

Daston, G., Lamar, E., Olivier, M., and Goulding, M. 1996. *Pax-3* is necessary for migration but not differentiation of limb muscle precursors in the mouse. *Development* **122:** 1017–1027.

De Angelis, L., Berghella, L., Coletta, M., Lattanzi, L., Zanchi, M., Cusella-De Angelis, M.G., Ponzetto, C., and Cossu, G. 1999. Skeletal myogenic progenitors originating from embryonic dorsal aorta coexpress endothelial and myogenic markers and contribute to postnatal muscle growth and regeneration (comments). *J. Cell Biol.* **147:** 869–878.

de la Serna, I.L., Ohkawa, Y., Berkes, C.A., Bergstrom, D.A., Dacwag, C.S., Tapscott, S.J., and Imbalzano, A.N. 2005. MyoD targets chromatin remodeling complexes to the myogenin locus prior to forming a stable DNA-bound complex. *Mol. Cell. Biol.* **25:** 3997–4009.

Dilworth, F.J., Seaver, K.J., Fishburn, A.L., Htet, S.L., and Tapscott, S.J. 2004. In vitro transcription system delineates the distinct roles of the coactivators pCAF and p300 during MyoD/E47-dependent transactivation. *Proc. Natl. Acad. Sci.* **101:** 11593–11598.

Edmondson, D.G., Lyons, G.E., Martin, J.F., and Olson, E.N. 1994. *Mef2* gene expression marks the cardiac and skeletal muscle lineages during mouse embryogenesis. *Development* **120:** 1251–1263.

Epstein, J.A., Lam, P., Jepeal, L., Maas, R.L., and Shapiro, D.N. 1995. Pax3 inhibits myogenic differentiation of cultured myoblast cells. *J. Biol. Chem.* **270:** 11719–11722.

Fuchs, E., Tumbar, T., and Guasch, G. 2004. Socializing with the neighbors: Stem cells and their niche. *Cell* **116:** 769–778.

Garry, D.J., Yang, Q., Bassel-Duby, R., and Williams, R.S. 1997. Persistent expression of MNF identifies myogenic stem cells in postnatal muscles. *Dev. Biol.* **188:** 280–294.

Gensch, N., Borchardt, T., Schneider, A., Riethmacher, D., and Braun, T. 2008. Different autonomous myogenic cell populations revealed by ablation of Myf5-expressing cells during mouse embryogenesis. *Development* **135:** 1597–1604.

Goulding, M.D., Chalepakis, G., Deutsch, U., Erselius, J.R., and Gruss, P. 1991. Pax-3, a novel murine DNA binding protein expressed during early neurogenesis. *EMBO J.* **10:** 1135–1147.

Gros, J., Manceau, M., Thome, V., and Marcelle, C. 2005. A common somitic origin for embryonic muscle progenitors and satellite cells. *Nature* **435:** 954–958.

Haldar, M., Karan, G., Tvrdik, P., and Capecchi, M.R. 2008. Two cell lineages, *myf5* and *myf5*-independent, participate in mouse skeletal myogenesis. *Dev. Cell* **14:** 437–445.

Hasty, P., Bradley, A., Morris, J.H., Edmondson, D.G., Venuti, J.M., Olson, E.N., and Klein, W.H. 1993. Muscle deficiency and neonatal death in mice with a targeted mutation in the myogenin gene (comments). *Nature* **364:** 501–506.

Irintchev, A., Zeschnigk, M., Starzinski-Powitz, A., and Wernig, A. 1994. Expression pattern of M-cadherin in normal, denervated, and regenerating mouse muscles. *Dev. Dyn.* **199:** 326–337.

Ishibashi, J., Perry, R.L., Asakura, A., and Rudnicki, M.A. 2005. MyoD induces myogenic differentiation through cooperation of its NH$_2$- and COOH-terminal regions. *J. Cell Biol.* **171:** 471–482.

Jostes, B., Walther, C., and Gruss, P. 1990. The murine paired box gene, *Pax7*, is expressed specifically during the development of the nervous and muscular system. *Mech. Dev.* **33:** 27–37.

Kablar, B. and Rudnicki, M.A. 1999. Development in the absence of skeletal muscle results in the sequential ablation of motor neurons from the spinal cord to the brain. *Dev. Biol.* **208:** 93–109.

Kablar, B., Krastel, K., Tajbakhsh, S., and Rudnicki, M.A. 2003. *Myf5* and *MyoD* activation define independent myogenic compartments during embryonic development. *Dev. Biol* **258:** 307–318.

Kablar, B., Krastel, K., Ying, C., Asakura, A., Tapscott, S.J., and Rudnicki, M.A. 1997. MyoD and Myf-5 differentially regulate the development of limb versus trunk skeletal muscle. *Development* **124:** 4729–4738.

Kablar, B., Krastel, K., Ying, C., Tapscott, S.J., Goldhamer, D.J., and Rudnicki, M.A. 1999. Myogenic determination occurs independently in somites and limb buds. *Dev. Biol.* **206:** 219–231.

Kablar, B., Asakura, A., Krastel, K., Ying, C., May, L.L., Goldhamer, D.J., and Rudnicki, M.A. 1998. MyoD and Myf-5 define the specification of musculature of distinct embryonic origin. *Biochem. Cell Biol.* **76:** 1079–1091.

Kassar-Duchossoy, L., Giacone, E., Gayraud-Morel, B., Jory, A., Gomes, D., and Tajbakhsh, S. 2005. Pax3/Pax7 mark a novel population of primitive myogenic cells during development. *Genes Dev.* **19:** 1426–1431.

Kassar-Duchossoy, L., Gayraud-Morel, B., Gomes, D., Rocancourt, D., Buckingham, M., Shinin, V., and Tajbakhsh, S. 2004. Mrf4 determines skeletal muscle identity in *Myf5:Myod* double-mutant mice. *Nature* **431:** 466–471.

Kaul, A., Koster, M., Neuhaus, H., and Braun, T. 2000. Myf-5 revisited: Loss of early myotome formation does not lead to a rib phenotype in homozygous *Myf-5* mutant mice. *Cell* **102:** 17–19.

Kaushal, S., Schneider, J.W., Nadal-Ginard, B., and Mahdavi, V. 1994. Activation of the myogenic lineage by MEF2A, a factor that induces and cooperates with MyoD. *Science* **266:** 1236–1240.

Khan, J., Bittner, M.L., Saal, L.H., Teichmann, U., Azorsa, D.O., Gooden, G.C., Pavan, W.J., Trent, J.M., and Meltzer, P.S. 1999. cDNA microarrays detect activation of a myogenic transcription program by the *PAX3-FKHR* fusion oncogene. *Proc. Natl. Acad. Sci.* **96:** 13264–13269.

Kitzmann, M., Carnac, G., Vandromme, M., Primig, M., Lamb, N.J., and Fernandez, A. 1998. The muscle regulatory factors MyoD and myf-5 undergo distinct cell cycle-specific expression in muscle cells. *J. Cell Biol.* **142:** 1447–1459.

Kuang, S., Gillespie, M.A., and Rudnicki, M.A. 2008. Niche regulation of muscle satellite cell self-renewal and differentiation. *Cell Stem Cell* **2:** 22–31.

Kuang, S., Kuroda, K., Le Grand, F., and Rudnicki, M.A. 2007. Asymmetric self-renewal and commitment of satellite stem cells in muscle. *Cell* **129:** 999–1010.

Kuang, S., Charge, S.B., Seale, P., Huh, M., and Rudnicki, M.A. 2006. Distinct roles for Pax7 and Pax3 in adult regenerative myogenesis. *J. Cell Biol.* **172:** 103–113.

LaBarge, M.A. and Blau, H.M. 2002. Biological progression from adult bone marrow to mononucleate muscle stem cell to multinucleate muscle fiber in response to injury. *Cell* **111:** 589–601.

Lu, J., McKinsey, T.A., Zhang, C.L., and Olson, E.N. 2000. Regulation of skeletal myogenesis by association of the MEF2 transcription factor with class II histone deacetylases. *Mol. Cell* **6:** 233–244.

Mal, A. and Harter, M.L. 2003. MyoD is functionally linked to the silencing of a muscle-specific regulatory gene prior to skeletal myogenesis. *Proc. Natl. Acad. Sci.* **100:** 1735–1739.

Mal, A., Sturniolo, M., Schiltz, R.L., Ghosh, M.K., and Harter, M.L. 2001. A role for histone deacetylase HDAC1 in modulating the transcriptional activity of MyoD: Inhibition of the myogenic program. *EMBO J.* **20:** 1739–1753.

Mal, A.K. 2006. Histone methyltransferase Suv39h1 represses MyoD-stimulated myogenic differentiation. *EMBO J.* **25:** 3323–3334.

Mansouri, A., Goudreau, G., and Gruss, P. 1999. *Pax* genes and their role in organogenesis. *Cancer Res.* (suppl. 7) **59:** 1707s–1710s.

Mansouri, A., Stoykova, A., Torres, M., and Gruss, P. 1996. Dysgenesis of cephalic neural crest derivatives in *Pax7*$^{-/-}$ mutant mice. *Development* **122:** 831–838.

Maroto, M., Reshef, R., Munsterberg, A.E., Koester, S., Goulding, M., and Lassar, A.B. 1997. Ectopic Pax-3 activates MyoD and Myf-5 expression in embryonic mesoderm and neural tissue. *Cell* **89:** 139–148.

McKinnell, I.W., Ishibashi, J., Le Grand, F., Punch, V.G., Addicks, G.C., Greenblatt, J.F., Dilworth, F.J., and Rudnicki, M.A. 2008. Pax7 activates myogenic genes by recruitment of a histone methyltransferase complex. *Nat. Cell Biol.* **10:** 77–84.

McKinsey, T.A., Zhang, C.L., and Olson, E.N. 2001. Control of muscle development by dueling HATs and HDACs. *Curr. Opin. Genet. Dev.* **11:** 497–504.

Megeney, L.A. and Rudnicki, M.A. 1995. Determination versus differentiation and the MyoD family of transcription factors. *Biochem. Cell Biol.* **73:** 723–732.

Molkentin, J.D., Black, B.L., Martin, J.F., and Olson, E.N. 1995. Cooperative activation of muscle gene expression by MEF2 and myogenic bHLH proteins. *Cell* **83:** 1125–1136.

Montarras, D., Lindon, C., Pinset, C., and Domeyne, P. 2000. Cultured myf5 null and myoD null muscle precursor cells display distinct growth defects. *Biol. Cell* **92:** 565–572.

Nabeshima, Y., Hanaoka, K., Hayasaka, M., Esumi, E., Li, S., and Nonaka, I. 1993. Myogenin gene disruption results in perinatal lethality because of severe muscle defect (comments). *Nature* **364:** 532–535.

Naidu, P.S., Ludolph, D.C., To, R.Q., Hinterberger, T.J., and Konieczny, S.F. 1995. Myogenin and MEF2 function synergistically to activate the MRF4 promoter during myogenesis. *Mol. Cell. Biol.* **15:** 2707–2718.

Oustanina, S., Hause, G., and Braun, T. 2004. Pax7 directs postnatal renewal and propagation of myogenic satellite cells but not their specification. *EMBO J.* **23:** 3430–3439.

Parise, G., McKinnell, I.W., and Rudnicki, M.A. 2008. Muscle satellite cell and atypical myogenic progenitor response following exercise. *Muscle Nerve* **37:** 611–619.

Parker, M.H., Perry, R.L., Fauteux, M.C., Berkes, C.A., and Rudnicki, M.A. 2006. MyoD synergizes with the E-protein HEBβ to induce myogenic differentiation. *Mol. Cell. Biol.* **26:** 5771–5783.

Peault, B., Rudnicki, M., Torrente, Y., Cossu, G., Tremblay, J.P., Partridge, T., Gussoni, E., Kunkel, L.M., and Huard, J. 2007. Stem and progenitor cells in skeletal muscle development, maintenance, and therapy. *Mol. Ther.* **15:** 867–877.

Penn, B.H., Bergstrom, D.A., Dilworth, F.J., Bengal, E., and Tapscott, S.J. 2004. A MyoD-generated feed-forward circuit temporally patterns gene expression during skeletal muscle differentiation. *Genes Dev.* **18:** 2348–2353.

Perry, R.L. and Rudnick, M.A. 2000. Molecular mechanisms regulating myogenic determination and differentiation. *Front. Biosci.* **5**: D750–D767.

Polesskaya, A., Seale, P., and Rudnicki, M.A. 2003. Wnt signaling induces the myogenic specification of resident CD45[+] adult stem cells during muscle regeneration. *Cell* **113**: 841–852.

Polesskaya, A., Naguibneva, I., Duquet, A., Bengal, E., Robin, P., and Harel-Bellan, A. 2001. Interaction between acetylated MyoD and the bromodomain of CBP and/or p300. *Mol. Cell. Biol.* **21**: 5312–5320.

Polesskaya, A., Duquet, A., Naguibneva, I., Weise, C., Vervisch, A., Bengal, E., Hucho, F., Robin, P., and Harel-Bellan, A. 2000. CREB-binding protein/p300 activates MyoD by acetylation. *J. Biol. Chem.* **275**: 34359–34364.

Puri, P.L., Iezzi, S., Stiegler, P., Chen, T.T., Schiltz, R.L., Muscat, G.E., Giordano, A., Kedes, L., Wang, J.Y., and Sartorelli, V. 2001. Class I histone deacetylases sequentially interact with MyoD and pRb during skeletal myogenesis. *Mol. Cell* **8**: 885–897.

Puri, P.L., Sartorelli, V., Yang, X.J., Hamamori, Y., Ogryzko, V.V., Howard, B.H., Kedes, L., Wang, J.Y., Graessmann, A., Nakatani, Y., and Levrero, M. 1997. Differential roles of p300 and PCAF acetyltransferases in muscle differentiation. *Mol. Cell* **1**: 35–45.

Rampalli, S., Li, L., Mak, E., Ge, K., Brand, M., Tapscott, S.J., and Dilworth, F.J. 2007. p38 MAPK signaling regulates recruitment of Ash2L-containing methyltransferase complexes to specific genes during differentiation. *Nat. Struct. Mol. Biol.* **14**: 1150–1156.

Rawls, A., Valdez, M.R., Zhang, W., Richardson, J., Klein, W.H., and Olson, E.N. 1998. Overlapping functions of the myogenic bHLH genes *MRF4* and *MyoD* revealed in double mutant mice. *Development* **125**: 2349–2358.

Relaix, F., Rocancourt, D., Mansouri, A., and Buckingham, M. 2005. A Pax3/Pax7-dependent population of skeletal muscle progenitor cells. *Nature* **435**: 948–953.

Relaix, F., Montarras, D., Zaffran, S., Gayraud-Morel, B., Rocancourt, D., Tajbakhsh, S., Mansouri, A., Cumano, A., and Buckingham, M. 2006. Pax3 and Pax7 have distinct and overlapping functions in adult muscle progenitor cells. *J. Cell Biol.* **172**: 91–102.

Rosen, G.D., Sanes, J.R., LaChance, R., Cunningham, J.M., Roman, J., and Dean, D.C. 1992. Roles for the integrin VLA-4 and its counter receptor VCAM-1 in myogenesis. *Cell* **69**: 1107–1119.

Rudnicki, M.A., Braun, T., Hinuma, S., and Jaenisch, R. 1992. Inactivation of *MyoD* in mice leads to up-regulation of the myogenic HLH gene *Myf-5* and results in apparently normal muscle development. *Cell* **71**: 383–390.

Rudnicki, M.A., Schnegelsberg, P.N., Stead, R.H., Braun, T., Arnold, H.H., and Jaenisch, R. 1993. MyoD or Myf-5 is required for the formation of skeletal muscle. *Cell* **75**: 1351–1359.

Sabourin, L.A., Girgis-Gabardo, A., Seale, P., Asakura, A., and Rudnicki, M.A. 1999. Reduced differentiation potential of primary MyoD[−/−] myogenic cells derived from adult skeletal muscle. *J. Cell Biol.* **144**: 631–643.

Sancho, E., Batlle, E., and Clevers, H. 2003. Live and let die in the intestinal epithelium. *Curr. Opin. Cell Biol.* **15**: 763–770.

Sartorelli, V., Puri, P.L., Hamamori, Y., Ogryzko, V., Chung, G., Nakatani, Y., Wang, J.Y., and Kedes, L. 1999. Acetylation of MyoD directed by PCAF is necessary for the execution of the muscle program. *Mol. Cell* **4**: 725–734.

Schafer, B.W., Czerny, T., Bernasconi, M., Genini, M., and Busslinger, M. 1994. Molecular cloning and characterization of a human PAX-7 cDNA expressed in normal and neoplastic myocytes. *Nucleic Acids Res* **22**: 4574-4582.

Schotta, G., Ebert, A., and Reuter, G. 2003. SU(VAR)3-9 is a conserved key function in heterochromatic gene silencing. *Genetica* **117**: 149–158.

Seale, P., Ishibashi, J., Holterman, C., and Rudnicki, M.A. 2004a. Muscle satellite cell-specific genes identified by genetic profiling of MyoD-deficient myogenic cell. *Dev. Biol.* **275**: 287–300.

Seale, P., Ishibashi, J., Scime, A., and Rudnicki, M.A. 2004b. Pax7 is necessary and sufficient for the myogenic specification of CD45[+]:Sca1[+] stem cells from injured muscle. *PLoS Biol.* **2**: E130.

Seale, P., Sabourin, L.A., Girgis-Gabardo, A., Mansouri, A., Gruss, P., and Rudnicki, M.A. 2000. Pax7 is required for the specification of myogenic satellite cells. *Cell* **102**: 777–786.

Simone, C. 2006. SWI/SNF: The crossroads where extracellular signaling pathways meet chromatin. *J. Cell. Physiol.* **207**: 309–314.

Simone, C., Forcales, S.V., Hill, D.A., Imbalzano, A.N., Latella, L., and Puri, P.L. 2004. p38 pathway targets SWI-SNF chromatin-remodeling complex to muscle-specific loci. *Nat. Genet.* **36**: 738–743.

Song, X., Zhu, C.H., Doan, C., and Xie, T. 2002. Germline stem cells anchored by adherens junctions in the *Drosophila* ovary niches. *Science* **296**: 1855–1857.

Tajbakhsh, S. and Cossu, G. 1997. Establishing myogenic identity during somitogenesis. *Curr. Opin. Genet. Dev.* **7**: 634–641.

Tajbakhsh, S., Rocancourt, D., Cossu, G., and Buckingham, M. 1997. Redefining the genetic hierarchies controlling skeletal myogenesis: *Pax-3* and *Myf-5* act upstream of *MyoD. Cell* **89**: 127–138.

Tremblay, P., Dietrich, S., Mericskay, M., Schubert, F.R., Li, Z., and Paulin, D. 1998. A crucial role for *Pax3* in the development of the hypaxial musculature and the long-range migration of muscle precursors. *Dev. Biol.* **203**: 49–61.

Tumbar, T., Guasch, G., Greco, V., Blanpain, C., Lowry, W.E., Rendl, M., and Fuchs, E. 2004. Defining the epithelial stem cell niche in skin. *Science* **303**: 359–363.

Venuti, J.M., Morris, J.H., Vivian, J.L., Olson, E.N., and Klein, W.H. 1995. Myogenin is required for late but not early aspects of myogenesis during mouse development. *J. Cell Biol.* **128**: 563–576.

Weintraub, H., Davis, R., Tapscott, S., Thayer, M., Krause, M., Benezra, R., Blackwell, T.K., Turner, D., Rupp, R., Hollenberg, S., et al. 1991. The myoD gene family: Nodal point during specification of the muscle cell lineage. *Science* **251**: 761–766.

Williams, B.A. and Ordahl, C.P. 1994. *Pax-3* expression in segmental mesoderm marks early stages in myogenic cell specification. *Development* **120**: 785–796.

Wu, Z., Woodring, P.J., Bhakta, K.S., Tamura, K., Wen, F., Feramisco, J.R., Karin, M., Wang, J.Y., and Puri, P.L. 2000. p38 and extracellular signal-regulated kinases regulate the myogenic program at multiple steps. *Mol. Cell. Biol.* **20**: 3951–3964.

Zammit, P.S., Golding, J.P., Nagata, Y., Hudon, V., Partridge, T.A., and Beauchamp, J.R. 2004. Muscle satellite cells adopt divergent fates: A mechanism for self-renewal? *J. Cell Biol.* **166**: 347–357.

Zetser, A., Gredinger, E., and Bengal, E. 1999. p38 mitogen-activated protein kinase pathway promotes skeletal muscle differentiation. Participation of the Mef2c transcription factor. *J. Biol. Chem.* **274**: 5193–5200.

Zhang, J., Niu, C., Ye, L., Huang, H., He, X., Tong, W.G., Ross, J., Haug, J., Johnson, T., Feng, J.Q., et al. 2003. Identification of the haematopoietic stem cell niche and control of the niche size. *Nature* **425**: 836–841.

Zhang, W., Behringer, R.R., and Olson, E.N. 1995. Inactivation of the myogenic bHLH gene *MRF4* results in up-regulation of myogenin and rib anomalies. *Genes Dev.* **9**: 1388–1399.

Zhong, W., Jiang, M.M., Schonemann, M.D., Meneses, J.J., Pedersen, R.A., Jan, L.Y., and Jan, Y.N. 2000. Mouse *numb* is an essential gene involved in cortical neurogenesis. *Proc. Natl. Acad. Sci.* **97**: 6844–6849.

Building Epithelial Tissues from Skin Stem Cells

E. Fuchs and J.A. Nowak

Howard Hughes Medical Institute, Laboratory of Mammalian Cell Biology and Development,
The Rockefeller University, New York, New York 10065

The skin epidermis and its appendages provide a protective barrier that guards against loss of fluids, physical trauma, and invasion by harmful microbes. To perform these functions while confronting the harsh environs of the outside world, our body surface undergoes constant rejuvenation through homeostasis. In addition, it must be primed to repair wounds in response to injury. The adult skin maintains epidermal homeostasis, hair regeneration, and wound repair through the use of its stem cells. What are the properties of skin stem cells, when do they become established during embryogenesis, and how are they able to build tissues with such remarkably distinct architectures? How do stem cells maintain tissue homeostasis and repair wounds and how do they regulate the delicate balance between proliferation and differentiation? What is the relationship between skin cancer and mutations that perturbs the regulation of stem cells? In the past 5 years, the field of skin stem cells has bloomed as we and others have been able to purify and dissect the molecular properties of these tiny reservoirs of goliath potential. We report here progress on these fronts, with emphasis on our laboratory's contributions to the fascinating world of skin stem cells.

The mammalian skin epidermis has long been an archetype for exploring homeostasis and injury repair in a stratified epithelium. It maintains a single inner (basal) layer of proliferative cells that adhere to an underlying basement membrane (BM), rich in extracellular matrix (ECM) and growth factors, which separates the epidermis from the underlying dermis (Fig. 1A). Cells in the basal layer are responsible for generating the layers of nondividing cells that undergo a program of terminal differentiation as they move outward and are continually shed from the skin surface. To maintain this self-perpetuating barrier that keeps harmful microbes out and essential body fluids in, the basal epidermal layer must perfectly balance proliferation and differentiation. How it does this is still under investigation. Although the process of epidermal homeostasis might seem to be a simple one, the microenvironment of the basal stem cell niche is multifaceted and can change both suddenly due to injury and also progressively with cumulative damage from harmful UV rays and other external assaults that can eventually lead to cancer. As our molecular understanding of epidermal homeostasis expands, the layers of complexity are beginning to unravel.

THE EPIDERMIS: ARCHITECTURE

Basal epidermal cells must remain proliferative and also protect themselves from the physical traumas of our environment. To do so, they possess an abundant infrastructure of 10-nm intermediate filaments (IFs) composed of keratins 5 and 14 (K5 and K14). At the base of each cell, keratin IFs fasten through plakin proteins (BPAG1e and plectin) to hemidesmosomes that provide mechanical strength to the cytoplasm below the nucleus of the columnar basal cells (Fig. 1B). Hemidesmosomes also anchor the cells to their underlying BM through a transmembrane collagen (BPAG2) and an $\alpha6\beta4$ integrin core that adheres to laminin 5 in the BM. At intercellular borders, keratin filaments also bind to desmoplakin, a plakin component of

desmosomes, thereby completing the fortifying infrastructure of the IF cytoskeleton. Desmosomes also use their robust transmembrane core of desmosomal cadherins (desmogleins and desmocollins) to reinforce cell–cell adhesion. The physiological significance of this structural IF network is underscored by mutations in genes that perturb keratin IFs, hemidesmosomes, or desmosomes, which typically lead to cell degeneration and subsequent skin blistering (Fuchs and Cleveland 1998).

To maintain their proliferative capacity and their ability to migrate in response to injury, basal cells must also possess elaborate and dynamic actin and microtubule cytoskeletons. The actin cytoskeleton indirectly links to the underlying BM beneath the epidermis through $\alpha3\beta1$-rich focal adhesions (FAs) that also use laminin 5 as their ligand and to neighboring cells through E-cadherin-rich adherens junctions (AJs) (Fig. 1B). Intracellularly, the $\alpha3\beta1$ integrin core associates with a complex array of actin-binding proteins and regulatory kinases that in turn act on effectors of Rho-GTPases to orchestrate actin dynamics, focal adhesion, and migration (Watt 2002; Lorenz et al. 2007; Schober et al. 2007). Transmembrane E-cadherins form homotypic intercellular and intracellular interactions that are stabilized through binding of their cytoplasmic domain to p120-catenin. At a second site, the cytoplasmic domain of E-cadherin binds to β-catenin that in turn binds α-catenin, a mediator of actin dynamics (Perez-Moreno and Fuchs 2006). AJs act in conjunction with FAs to coordinate actin dynamics across the epithelial tissue and also organize the underlying cortical belt of actin fibers that reinforce the plasma membrane of each epidermal cell. Through a number of still unfolding mechanisms, microtubules associate with actin, FAs, and AJs to polarize the cytoskeleton, asymmetrically distribute proteins and organelles within the cell, and orchestrate adhesion and directed movements within the cell. Below, we highlight some of the ways in which these dynamics are important to epidermal stem cells and tissue homeostasis.

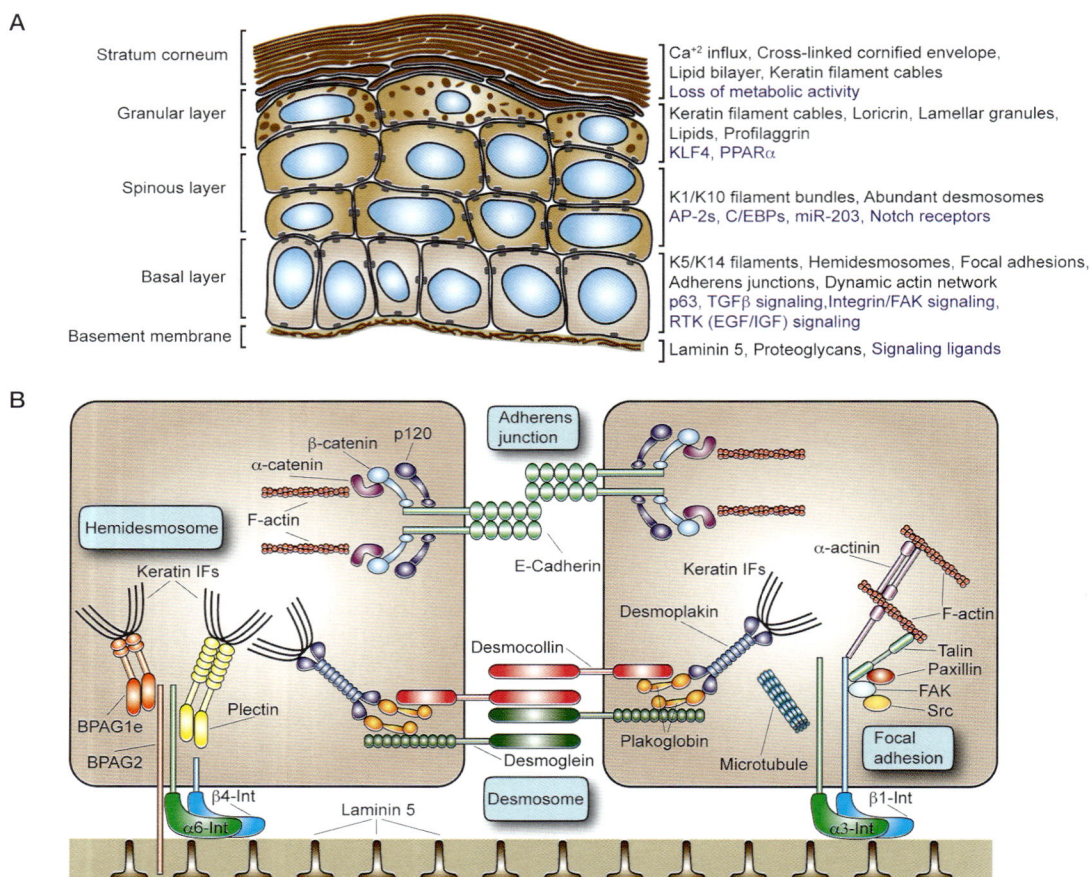

Figure 1. The cellular architecture of the epidermis and major components of the epidermal cytoskeleton. (*A*) Program of epidermal differentiation illustrating the BM at the base, the proliferative basal layer, and the three differentiation stages: spinous layer, granular layer, and outermost stratum corneum. Shown at the right are key molecular markers described in the text. (*Black text*) Structural and cytoskeletal components present in each layer; (*purple text*) regulatory mechanisms and signaling activity necessary for proper function of each layer. (*B*) Key components of the cytoskeleton in basal layer keratinocytes. Keratin IFs connect to the BM via hemidesmosomes, whose core components include homodimers of the plakin proteins plectin and BPAG1e, the transmembrane collagen BPAG2, and a heterodimer composed of α6 and β4 integrin. Intercellularly, the keratin network is also linked at adjacent cells by desmosomes that are composed of the desmosomal cadherins desmocollin and desmoglein, the linker protein plakoglobin, and a homodimer of the plakin protein desmoplakin. Analogously, the actin cytoskeleton attaches to the underlying BM through focal adhesions that are composed of an α3β1 integrin heterodimer core linked to actin filaments (F-actin) by talin and α-actinin. Paxillin, FAK, and Src are also associated with focal adhesions and help to orchestrate signaling events regulating cell adhesion and motility. Microtubules can also contact focal adhesions, helping to regulate their turnover. Between adjacent cells, the actin cytoskeleton is linked through adherens junctions that associate through homotypic interactions between E-cadherin molecules. Heterodimers composed of β-catenin and α-catenin regulate the link between E-cadherin and the actin cytoskeleton, whereas p120-catenin stabilizes E-cadherin interactions by binding to E-cadherin's cytoplasmic domain.

Intercellular and cell-substratum junctions must be robust enough to provide mechanical strength, but they must also be dynamic in order to allow for a flux of epidermal cells upward through the tissue during normal homeostasis and to orchestrate repair after wounding. A frequent paradigm not only in the regulation of normal homeostasis and migration but also in the epithelial-mesenchymal transition (EMT) process that can drive tumorigenesis is the inverse regulation of cadherin-dependent intercellular junctions and integrin-dependent cell motility (Frame et al. 2002). Although many details of the underlying regulatory mechanism remain elusive, the Src tyrosine kinase and small GTPases are central players in the process (Frame et al. 2002). Src is itself a downstream sub-

strate of focal adhesion kinase (FAK), which is activated by α3β1 integrin, and Src activation can result in activation of p190RhoGAP. This leads to reduced Rho-GTPase and ROCK activities as well as actin stress fibers, features that promote FA turnover and enhance cell migration in keratinocytes in vitro and that are required for normal wound repair in vivo (Lorenz et al. 2007; Schober et al. 2007). Activated Src is also known to phosphorylate E-cadherin and p120-catenin, and these marks appear to weaken intercellular junctions by promoting their endocytosis. It has been proposed that internalized E-cadherin results in the activation of an additional GTPase, Rap1, that in turn controls the polarized redistribution of integrins and/or integrin regulators to new adhesion sites, thereby enhancing integrin-mediated cell–matrix adhesion (Balzac et al. 2005).

Increasing evidence underscores the coordinate regulation not only of FA and AJ dynamics but also of actin and microtubule dynamics in epithelial cells. Microtubules contribute to adhesion dynamics by targeting and promoting FA turnover (Kaverina et al. 1999) and enhancing intercellular junction formation between keratinocytes (Kee and Steinert 2001). Although the underlying mechanisms remain to be elucidated, loss-of-function studies suggest a role for both α3β1 integrins and α-catenin in orienting the mitotic spindle properly in the epidermis (Lechler and Fuchs 2005), whereas p120-catenin appears to be involved in cadherin-independent stabilization of microtubules (Ichii and Takeichi 2007). Whether at cell-substratum junctions or cell–cell junctions, harmonizing actin and microtubule networks is critical for establishing cellular polarity, without which the epidermal basal layer cannot properly separate proliferative and differentiating compartments to generate correct tissue architecture.

Basal epidermal cells must integrate signals from multiple pathways in order to set their rate of proliferation. In addition to cell migration and adhesion, integrin signaling has a major role in basal layer growth control. In part, this is likely mediated through the ability of integrins to activate the Src family tyrosine kinases, which are potent activators of the Ras–mitogen-activated protein kinase (MAPK) signaling cascade (Lorenz et al. 2007; Schober et al. 2007). Additionally, transmembrane receptor tyrosine kinases (RTKs) for epidermal and insulin growth factors (EGFs and IGFs) have critical roles in stimulating basal cell proliferation (Barrandon and Green 1987a; Zenz and Wagner 2006; Scholl et al. 2007).

Whereas integrin and RTK signaling function as accelerators for epidermal proliferation, signaling through the transforming growth factor-β (TGF-β) pathway acts as the brake (Shi and Massagué 2003). Upon TGF-β ligand engagement, the TGF-β transmembrane receptor, which possesses serine/threonine kinase activity, becomes activated and subsequently phosphorylates Smad2 or Smad3 transcription factors. These phosphorylated Smads join larger transcription factor complexes to regulate downstream target genes, some of which encode cell cycle inhibitors. Intriguingly, when the TGF-β receptor II subunit is missing in the skin epidermis, the activity of αβ1 integrins and FAK are elevated, leading to a enhanced wound repair and cell migration (Guasch et al. 2007).

When coupled with prior studies identifying direct associations between RTKs and integrins (Lee and Juliano 2004), these data suggest a regulatory network where the activity of one type of transmembrane receptor will affect the activity of others. An additional twist comes from the fact that α-, β-, and p120 catenins each influence actin dynamics and/or proliferation in ways that appear to extend beyond their roles in cadherin-mediated intercellular adhesion. Counterbalancing these regulatory circuits are additional underlying mechanisms that can act to offset excessive proliferation with enhanced differentiation and/or apoptosis, thereby restoring normalcy to tissue morphology and minimizing the phenotypic consequences of genetic mutations.

As epidermal cells exit the basal layer and cease to proliferate, they progress upward through three distinct differentiation stages: spinous layer, granular layer, and stratum corneum (Fig. 1A). The major structural change at the basal-to-spinous-layer transition is the switch from K5 and K14 IFs in the basal layer to K1 and K10 suprabasally. Discovered three decades ago, this switch is transcriptionally controlled and remains among the most faithful of indicators that a cell has withdrawn from the cell cycle and is committed to terminally differentiate (Fuchs and Green 1980). Additional structural changes occurring at this time include switches in expression of desmosomal cadherins and the initial expression of a few components of the "cornified envelope" that are deposited beneath the plasma membrane for late-stage differentiation events.

To orchestrate these changes in structure and function, the basal/spinous transition is accompanied by dramatic changes in the expression of transcription factors. One transcription factor that is likely to have a key role in regulating the self-renewal and long-term proliferative capacity of the basal layer is p63, a member of the p53 family of proto-oncogenes. The ΔN isoform of p63 is preferentially expressed in basal epidermal cells (Laurikkala et al. 2006), and although the exact functions of p63 are still under investigation, it appears that p63 is necessary for basal cells to maintain proliferative potential. In its absence, the epidermis fails to stratify and differentiate properly during embryonic development (Mills et al. 1999; Yang et al. 1999; Senoo et al. 2007). In contrast to p63, which is down-regulated in the basal-to-spinous-layer switch, elevated levels of several AP-2 and C/EBP family members are associated with the terminal differentiation program (Maytin and Habener 1998; Wang et al. 2006). Loss-of-function studies also unveil an essential role for Notch signaling in governing the basal-to-spinous-layer fate switch, and in this regard, it is notable that Notch ligands are expressed both basally and suprabasally, whereas Notch receptors are suprabasal (Powell et al. 1998; Rangarajan et al. 2001; Pan et al. 2004; Blanpain et al. 2006; Nguyen et al. 2006a; Lee et al. 2007; Moriyama et al. 2008) Exactly how Notch, AP-2s, and C/EBPs cooperate to regulate the basal-to-spinous switch remains to be elucidated.

Adding another level of complexity to the process is the microRNA miR-203, which is expressed in suprabasal layers of the epidermis and was recently shown to directly target ΔNp63 mRNA for translational repression (Yi et al. 2008). Another miR-203 target is Zpf280, a nuclear protein expressed not only in epidermal stem cells, but also in embryonic stem cells (Yi et al. 2008). As the field continues to unfold, it will be interesting to see the extent to which miR-203 and other microRNAs function in stem cell fate commitment. If miR-203 targets are primarily basal genes, miR-203 might be viewed as a fine-tuner of commitment, accelerating repression of basal markers in suprabasal cells. A tantalizing possibility for future investigation is the notion that the gene encoding miR-203 could itself be a target for Notch signaling.

Although AP-2 and C/EBP families appear to function in conjunction with Notch signaling to regulate the basal-to-spinous switch early in suprabasal differentiation, the transcription factors KLF4 and PPARα seem to regulate genes expressed later in the process (Segre et al. 1999; Di-Poi et al. 2004). As cells enter the granular layer, the primary cornified envelope protein loricrin is expressed, and lamellar granules packed full of lipids appear. Profilaggrin is also expressed at this time, and soon afterward, it is processed to generate filaggrin, a protein that bundles keratin filaments into indestructible cables. As granular cells transit to the stratum corneum, all metabolic activity ceases, and an influx of calcium results in activation of transglutaminases that initiate γ-glutamyl-ε-lysine crosslinks to produce the cornified envelope characteristic of this layer. Lamellar granules are extruded onto this scaffold, where they form a lipid bilayer that temporarily seals the dead, protective cells at the body surface (de Guzman Strong et al. 2006; Elias 2007) .

The process of terminal differentiation is in a continual flux so that dead surface cells are continually sloughed off and replaced by inner cells differentiating and moving outward. In human epidermis, the self-renewing capacity of epidermal stem cells is enormous, and within 4 weeks, a basal cell has terminally differentiated and exited at the skin surface. In mice, the postnatal trunk epidermis becomes thinner and proliferation slows substantially as the hair coat develops and becomes the primary line of physical protection.

THE EPIDERMIS: HOMEOSTASIS

Researchers have known for decades that stem cells exist within the basal layer of adult epidermis, but it remains unknown whether all cells within the basal layer are stem cells. Early work on epidermal homeostasis defined the epidermal proliferative unit (EPU) as an architecturally discernible structure composed of a bed of approximately ten basal cells overlaid by a stack of increasingly larger and flatter cells (Potten 1974). EPU proliferation studies and genetic marking of epidermal clonal units have led to the hypothesis that there is one self-renewing, slower-cycling stem cell in the center of each EPU and that the other basal cells are so-called "transit-amplifying" (TA) cells, i.e., committed cells that divide several times and then exit the basal layer and terminally differentiate (Potten 1974; Mackenzie 1997). In support of this notion are in vitro studies that show that human epidermal cells with the highest level of surface β1 integrins give rise to the largest colonies (holoclones) that can be passaged long-term, whereas cells with lower levels of β1 integrins produce smaller meroclones that do not survive passaging (Barrandon and Green 1987b; Jones et al. 1995). Regional variations within the BM and microenvironment have been described to explain how distinct populations of integrin-rich stem cells and TA cells might arise within the basal layer if its residents divide symmetrically (Lavker and Sun 1982; Jensen et al. 1999). However, recent studies suggest that basal epidermal cells can divide asymmetrically, affording an alternative view of how one basal stem cell and one committed cell might arise (Lechler and Fuchs 2005; Clayton et al. 2007).

During embryogenesis in the mouse, some basal cells shift their spindle orientation from parallel to the BM to a more angled orientation, resulting in asymmetric divisions at the onset of stratification (Fig. 2A) (Lechler and Fuchs 2005). Approximately 70% of embryonic basal cell mitoses appear to be asymmetric relative to the underlying BM (Lechler and Fuchs 2005). As the animal matures, a marked reduction in basal cell divisions occurs that has posed technical difficulties in measuring division orientations in postnatal mice. However, asymmetric divisions have been detected in tongue, ear, and tail skin (Lechler and Fuchs 2005; Clayton et al. 2007). Intriguingly, mitotic basal cells about to undergo an asymmetric division display a cortical crescent of proteins that in lower eukaryotes are essential for proper spindle orientation (Lechler and Fuchs 2005). Additional experiments will be needed to ascertain whether these proteins function similarly in mammalian epidermis and whether asymmetric divisions are essential for epidermal homeostasis and/or wound repair.

Although the extent to which asymmetric divisions control epidermal dynamics remains unknown, it is nevertheless intriguing to consider the potential consequences of such divisions. If the divisions occur at an angle relative to the plane of the basement membrane, the basal daughter cell would inherit the majority of the growth-promoting RTKs and integrins (Fig. 2B). Such divisions appear to be prevalent in embryogenesis, where they occur concomitant with stratification (Lechler and Fuchs 2005). In contrast, lineage-tracing studies suggest that many of the asymmetric divisions in adult tail skin result in both daughters retaining contact with their substratum, reflecting a division lateral to the basement membrane (Fig. 2B) (Clayton et al. 2007). In this model, it was noted that the Notch antagonist Numb is asymmetrically distributed in cells dividing lateral to the BM (Clayton et al. 2007). Differential partitioning of Notch signaling between the daughter cells of a lateral divison could result in one daughter being primed to undergo a basal-to-spinous-layer transition. Consistent with this notion, activated Notch signaling results in a decrease in integrin expression in basal keratinocytes (Blanpain et al. 2006).

These different models also have distinct implications for the numbers of epidermal stem cells and their organization in the basal layer. The EPU model predicts a small number of basal stem cells that reside individually in spatially organized niche microenvironments and are surrounded by TA cells (Fig. 2B). In contrast, the asymmetric division models predict a large, homogeneous population of basal progenitors that give rise to spinous cells through differential partitioning of proteins between daughters. The lateral asymmetric division model implies that asymmetric divisions are intrinsic to the epidermal stem cell, alleviating the need for a microenvironmental change to trigger commitment. The perpendicular asymmetric division model suggests that the BM itself may constitute the epidermal stem cell niche, naturally positioning committed progeny upward in a column. This perpendicular asymmetric division model is analogous to that of *Drosophila* germ cell development, where preservation of contact with the niche maintains stemness, and perpendicular asymmetric divisions drive the fate determination of committed daughter cells that depart from the niche

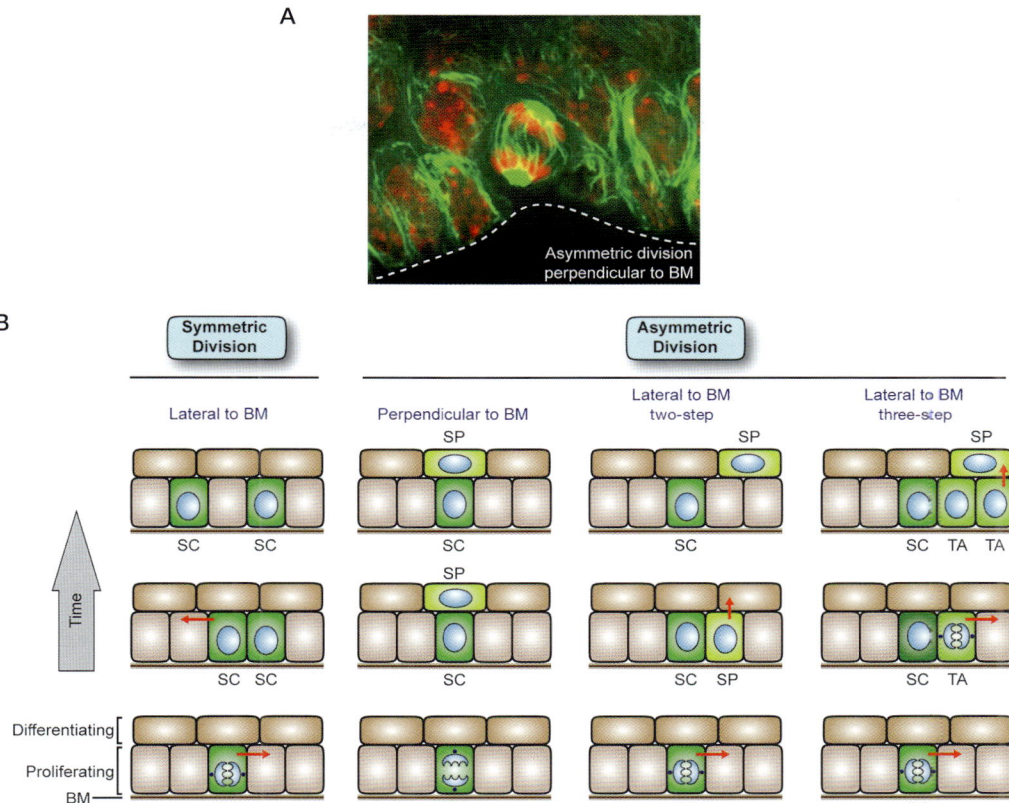

Figure 2. Roles for symmetric and asymmetric cell division in epidermal development and homeostasis. (*A*) Immunofluorescence image of a cell in the basal layer of an E15.5 embryonic tongue undergoing asymmetric division perpendicular to the BM (*dotted white line*). The microtubule network is marked by green fluorescent protein (GFP), and DNA is marked by red propidium iodide. (*B*) Self-renewing stem cells (SC) exist in the basal layer of the epidermis. Symmetrical divisions lateral to the BM produce two stem cells, a process that can serve to refill vacancies in the basal layer or increase the area of the epidermis during development. Asymmetric divisions can occur both laterally and perpendicular to the BM. In the two-step asymmetric division model, a stem cell divides asymmetrically to preferentially partition proliferation-associated factors into the stem cell daughter and provide differentiation-inducing components to the other daughter, fated to become a spinous (SP) cell. If the spindle orientation was perpendicular to the BM, the division could result in direct positioning of the SP daughter away from the BM, whereas lateral spindle orientation relative to the BM would then necessitate subsequent delamination of the committed SP daughter. In the three-step asymmetric division model, a transit-amplifying (TA) intermediate arises, which has been postulated to divide three to four times before delaminating (*arrows*) and entering into a terminal differentiation program. Once the spinous cells have separated from the BM, they enter a program of terminal differentiation as they move outward and are eventually sloughed from the skin surface (see Fig. 1). Differentiating cells are continually replaced by a flux of inner cells committing to terminally differentiate and move outward. Immunofluorescence image courtesy of Terry Lechler (when in the Fuchs lab).

(Fuller and Spradling 2007). Finally, lateral symmetric divisions would yield two stem cells and could provide a mechanism to replenish old or damaged basal stem cells or increase the area of the epidermis during development (Fig. 2B).

Irrespective of the role of the BM in governing basal cell division and the basal-to-suprabasal transition, its mechanophysical properties, along with those of the underlying dermis, are also likely to impact the behavior of basal cells (Dobereiner et al. 2005). The ECM polymers and growth factors of the BM also provide a complex repertoire of stimuli for basal cells. Among them is laminin 5, which, as outlined above, promotes anchorage and signaling/migration through its respective abilities to act as a ligand for both $\alpha6\beta4$ and $\alpha3\beta1$ integrins (Owens and Watt 2003; Raghavan et al. 2003; Manohar et al. 2004). The BM is also rich in ECM ligands for more

minor epidermal integrins, proteoglycans, and both positive and negative growth factors (Fuchs 2007). Together, these features of the BM create a microenvironment that enables basal epidermal stem cells to maintain homeostasis under normal conditions and to respond appropriately to injury.

The epidermis has a huge proliferative capacity, but the balance of proliferation and differentiation is easily perturbed. This feature quite possibly represents a necessary molecular trade-off for a tissue that has to be adaptable enough to quickly repair wounds without depleting proliferative capacity over time. Disruptions of even a single component of the proliferation regulatory network can have serious consequences for the epidermis, as evidenced by the fact that mice harboring loss-of-function mutations in TGF-β receptor II or gain-of-function mutations in TGF-α or integrins display an increased susceptibility to squamous cell carcinomas (SQCCs), whereas those with loss-of-function mutations in FAK are more resistant to tumorigenesis than normal (McLean et al. 2004; Janes and Watt 2006; Guasch et al. 2007; Marinkovich 2007). As additional regulators of basal cell activity are discovered and links between known pathways become clearer, our understanding of the interplay between the BM and microenvironment and how they cooperate to regulate epidermal stem cell biology should continue to deepen.

THE HAIR FOLLICLE: MORPHOGENESIS

One of the most extraordinary features of the vertebrate epidermis is its ability to generate highly specialized elaborate appendages, including the feathers of birds, scales on a snake, hoofs of a horse, wool of a sheep, and hairs, nails, and sweat (eccrine) and oil (sebaceous) glands of our skin. Notably, although all of these appendages seem to be quite distinct, they all begin development in a similar way (Mikkola 2007), and knowledge of this process has been gained by studying the embryonic skin at a stage at which it exists as a single-layered epithelium.

Before hair follicle morphogenesis, a uniform layer of epithelial cells overlies a disperse population of dermal cells. At about E14.5 of mouse development, mesenchymal-epithelial interactions result in the formation of the first wave of hair placodes that appear as small epidermal invaginations into the dermis (Fig. 3A). Once specified, signals from the epithelium cause dermal cells to aggregate and form a dermal condensate under each placode. This specification process occurs in four overlapping waves, with rare primary guard hairs being specified at E14.5 and the remainder of follicles being specified from E15.5 to P0. Once placodes have formed, they become highly proliferative and grow downward into the dermis, forming hair germs and then hair pegs. The most proliferative cells in peg-stage follicles are at the bottom leading edge of the follicle, and these cells give rise to the epithelial matrix as they surround the dermal condensate that then becomes the dermal papilla (DP). Although the matrix is transient, proliferating and differentiating only during the growth (anagen) phase of the hair cycle, the DP remains permanently associated with each follicle.

The matrix produces two differentiating structures: a three-layered inner root sheath (IRS) and a three-layered central hair shaft. The IRS forms first, providing the channel for the emerging hair. Both of these structures are internal to the outer root sheath (ORS), whose cells are in direct contact with the basement membrane and are topologically contiguous with the basal layer of the interfollicular epidermis. Reciprocal signaling between the DP and TA matrix cells in the bottom portion of the hair follicle, or bulb, allows the matrix progeny to engage in the distinct programs of gene expression that generate the full complement of differentiated cell lineages in the hair follicle.

At birth, sebaceous gland precursor cells appear in the upper portion of the ORS, and the sebaceous gland forms shortly thereafter (Horsley et al. 2006). For most backskin follicles, maturation is completed toward the end of the first postnatal week, when follicle downgrowth stops and the upward-moving, terminally differentiated hairs break through the skin surface (for review, see Schmidt-Ullrich and Paus 2005). As the hair shaft elongates, neural-crest-derived melanocytes, located on the epithelial side of the BM of the hair bulb, provide pigment to the differentiating cells of the hair shaft, giving them their color. In humans, melanocytes also disperse within the basal layer of the epidermis.

The developmental decision to form a hair follicle is the result of mesenchymal-epithelial cross-talk that integrates multiple instructive signals necessary to initiate hair follicle morphogenesis. These signaling cues include Wnt/β-catenin, sonic hedgehog (Shh), fibroblast growth factors (FGFs), and bone morphogenetic proteins (BMPs). Several lines of evidence suggest that activation of Wnt/β-catenin signaling in epithelial cells is a key initial step in placode formation. The bipartite transcription factor complex composed of the DNA-binding protein LEF1 and its activation partner (stabilized β-catenin) can be readily detected in the nuclei of developing placode cells, and the "Wnt reporter" gene TOPGAL, containing an enhancer element composed of multimerized LEF1 binding sites, confirms the specific transcriptional activity of these complexes in the placode (van Genderen et al. 1994; Zhou et al. 1995; DasGupta and Fuchs 1999; Millar 2002). Notably, when the Wnt inhibitor Dickoff 1 (Dkk1) was expressed ectopically or when β-catenin was conditionally targeted for ablation in epithelial cells, hair follicle morphogenesis was blocked altogether, whereas mice lacking LEF1 displayed a reduced number of follicles (van Genderen et al. 1994; Huelsken and Birchmeier 2001; Andl et al. 2002). Strong evidence for Wnt signaling as a sufficient, instructive cue for placode formation came from experiments expressing excessive stabilized β-catenin in the epithelium, which resulted in super-furry mice exhibiting ectopic hair follicles within their interfollicular epidermis (Gat et al. 1998). Recently, Cotsarelis and colleagues (Ito et al. 2007) showed that when the skin of mice is severely wounded, endogenous Wnt signaling is elevated in the regenerating epithelium and this leads to the induction of hair follicle formation from epidermal stem cells. Taken together, these findings underscore the role for Wnt signaling and stabilized β-catenin in governing the choice of whether to become epidermis or hair follicle.

In contrast to Wnt signaling, which is activated during placode formation, BMP signaling must be inhibited for placode morphogenesis to progress. One dermal cue that appears to be particularly critical is the BMP inhibitory protein Noggin, whose absence severely impairs hair follicle morphogenesis (Botchkarev et al. 1999) as well as hair cycling (Botchkarev et al. 2001). When skin is genetically engineered to overexpress Noggin, genes controlling cell cycle progression are enhanced (Sharov et al. 2006). Consistent with these findings, mice conditionally null for the BMPR1a receptor in the epithelium form the correct number of hair follicles, which are later blocked in differentiation of the IRS and hair shaft (Kobielak et al. 2003; Andl et al. 2004; Ming Kwan et al. 2004; Yuhki et al. 2004). The interplay between the Wnt and BMP pathways is highlighted by the requirement of BMP inhibition for proper LEF1 expression in the placode (Jamora et al.

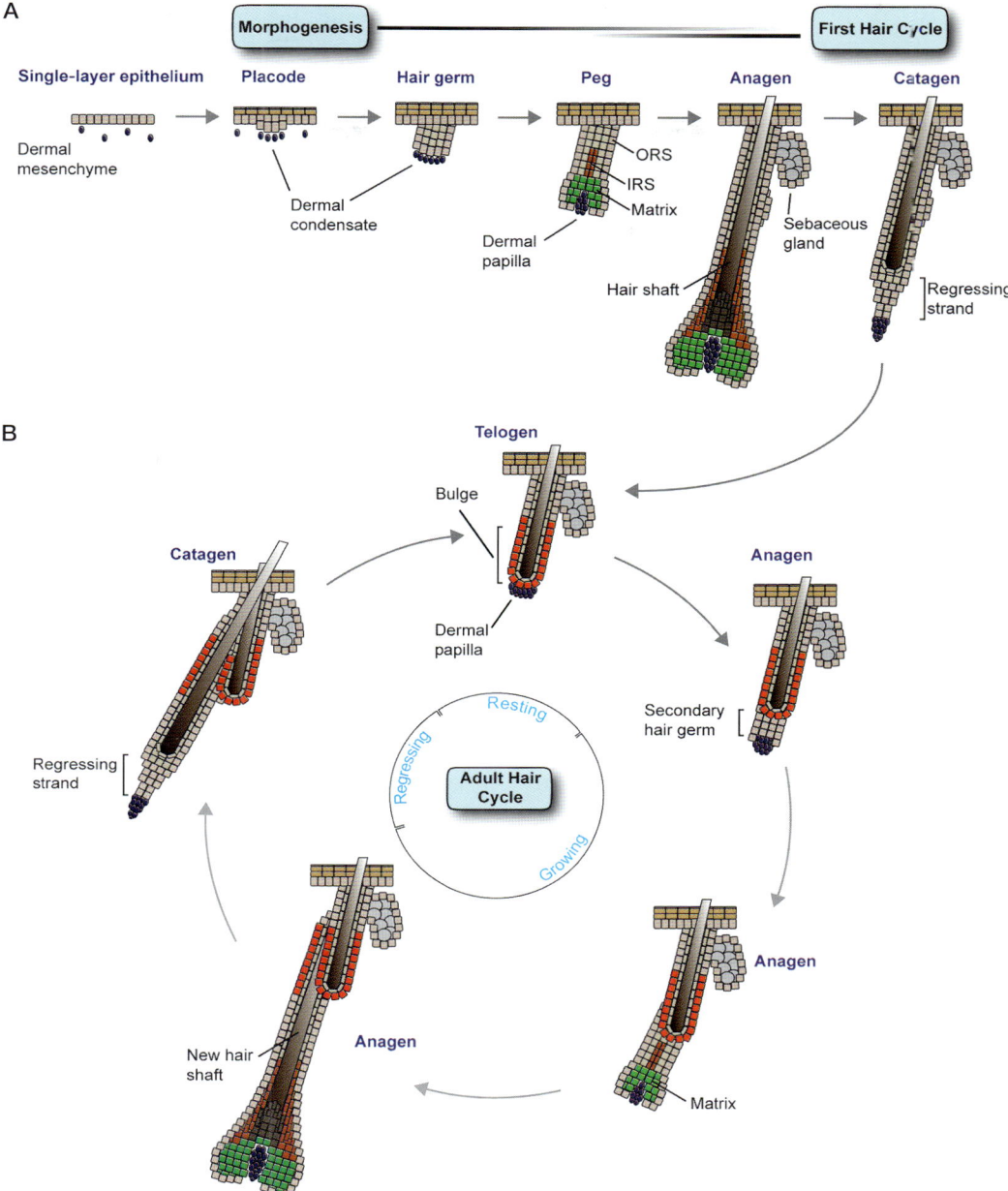

A

Morphogenesis — First Hair Cycle

Single-layer epithelium | Placode | Hair germ | Peg | Anagen | Catagen

Dermal mesenchyme

Dermal condensate

ORS
IRS
Matrix

Dermal papilla

Hair shaft

Sebaceous gland

Regressing strand

B

Telogen

Bulge

Dermal papilla

Catagen

Regressing strand

Anagen

Secondary hair germ

Resting

Regressing

Adult Hair Cycle

Growing

Anagen

Matrix

New hair shaft

Anagen

Figure 3. Embryonic hair follicle morphogenesis and the adult hair cycle. (*A*) The process of follicle morphogenesis occurs in several overlapping waves that begin at E14.5 when small invaginations termed placodes appear in the basal layer of the epidermis, accompanied by aggregations of dermal cells termed dermal condensates. Cells of the placode proliferate and grow downward into the dermis to form hair germs (~E16.5–E17.5). Next, during the peg stage (~E18.5–P0), transit-amplifying matrix cells appear at the base of the follicle and encapsulate the dermal papilla (DP). The upper portion of the follicle also becomes separated into the outer root sheath (ORS) and inner root sheath (IRS) at this time. Soon thereafter (~P1–P3), fully differentiated hair shafts and sebaceous glands (SGs) appear, and follicle morphogenesis reaches completion at about P9. At this time, follicles reach their maximum length, marking entrance into the anagen phase of the first hair cycle. After production of the first hair coat, follicles transition to the catagen phase of the hair cycle (~P16–P19), where the matrix undergoes apoptosis and degenerates into a regressing epithelial strand that draws the DP upward to rest just below the base of the first "club" hair, which is surrounded by the newly formed adult bulge niche. (*B*) Direct contact between the DP and adult bulge cells indicates the onset of telogen (~P20). Soon thereafter, at about P21, bulge stem cells are activated and a secondary hair germ grows downward from the base of the bulge niche, signaling entry into the anagen phase of the next hair cycle. The emergence of a new hair follicle from the side of the original club hair lends the stem cell niche its characteristic "bulge" morphology and divides the CD34-positive stem cells into two populations based on high or low integrin expression levels and adherence to the BM. In a process that has many similarities to that of initial hair follicle morphogenesis, the secondary germ expands and gives rise to a new matrix that begins to produce the differentiated lineages of the IRS and hair shaft. This new hair shaft then exits from the same channel as the existing club hair. After several weeks of growth, anagen ceases and follicles enter catagen, again drawing the DP upward to rest below the bulge stem cell niche. After a variable period of time, bulge stem cells are again activated to initiate a new hair growth cycle. For a comprehensive analysis of the details and classification of hair follicle morphogenesis and adult hair cycling, see Muller-Rover et al. (2001), Stenn and Paus (2001), and Schmidt-Ullrich and Paus (2005).

2003; Kobielak et al. 2003; Andl et al. 2004; Zhang et al. 2006a).

Interestingly, although this paradigm serves the bulk of follicle morphogenesis, the external cues that trigger these internal regulatory processes differ for the formation of the large primary guard hairs. The guards are the first hair follicles to appear in embryonic backskin, and they are uniquely dependent on a ligand–receptor complex composed of the TNF-related ectodysplasin (EDA) and the EDA receptor (EDAR) (Schmidt-Ullrich and Paus 2005). Notably, EDA/EDAR signaling results in induction of two BMP inhibitors different from Noggin (Pummila et al. 2007). *EDA* is itself a target of Wnt signaling (Laurikkala et al. 2001) that is already active in all other hair follicles at this stage, as judged from *TOPGAL* reporter activity and LEF1/β-catenin (DasGupta and Fuchs 1999). Thus, although the initiating events may differ, the recipe for hair follicle morphogenesis has many common ingredients across different hair types.

To drive hair follicle morphogenesis, signaling pathways ultimately have to elicit changes in the expression and dynamics of the ECM, cytoskeleton, and cell–matrix and cell–cell junction proteins in order to remodel the epithelium from its single layer to a hair placode. Several changes that accompany this process involve down-regulation of some adhesion proteins, such as E-cadherin, and up-regulation of others, such as P-cadherin (Jamora et al. 2003). One intriguing intersection between signaling and cytoskeletal organization is that the *E-cadherin* gene itself harbors a functional LEF1-binding site and is down-regulated concomitant with the appearance of Wnt reporter activity in developing placodes (Jamora et al. 2003). Additional links between signaling and cytoskeletal organization are Cdc42 and Rac1, both essential for maintaining follicle stem cells (Benitah et al. 2005; Chrostek et al. 2006; Wu et al. 2006). Although best understood for their roles in actin–cell junction dynamics and cell junction formation, Cdc42 and Rac1 also seem to act as effectors of Wnt signaling. It has been posited that Cdc42 functions in β-catenin stabilization (Wu et al. 2006) and that Rac1 functions in nuclear β-catenin localization (Wu et al. 2008). These new findings provide tantalizing glimpses as to how external cues received by the single layer of epidermal cells can translate into the early steps of hair follicle morphogenesis.

Once the placode forms, signaling events downstream from Wnts/BMPs drive the downgrowth and maturation of hair follicles. *Shh* is an early gene expressed downstream from Wnt/BMP receptor signaling (and EDA/EDARs in the case of guard hairs) once placodes have formed (Oro et al. 1997; Gat et al. 1998; Morgan et al. 1998; St-Jacques et al. 1998). Without *Shh*, hair follicles arrest at the placode stage and fail to form a dermal condensate, suggesting critical roles for *Shh* in proper epithelial-mesenchymal signaling and the dramatic expansion of cells involved in the transition from a placode to a mature follicle (Hardy 1992; St-Jacques et al. 1998; Oro and Higgins 2003; Levy et al. 2007). That said, many facets of *Shh*'s role in skin appendage formation remain mysterious. This is perhaps best exemplified by recent studies showing that abrogation of *Shh* responsiveness in the epithelium by conditional targeting of the transmembrane protein Smoothened leads to hair follicles that adopt features of mammary glands during development (Gritli-Linde et al. 2007).

THE HAIR FOLLICLE BULGE: A RESERVOIR OF SLOW-CYCLING MULTIPOTENT ADULT STEM CELLS

As a normal feature of skin homeostasis, the hair follicle undergoes cyclic bouts of degeneration and regeneration, producing a new hair with each cycle. During the hair growth phase of the hair cycle (anagen), the DP acts as a signaling center for the epithelial-mesenchymal cross-talk that regulates the balance between matrix cell proliferation and hair production (Schmidt-Ullrich and Paus 2005; Alonso and Fuchs 2006). TA matrix cells proliferate rapidly during anagen but then disappear when follicle growth ceases. After the anagen phase of the first hair cycle in postnatal mice, which is an extension of initial follicle morphogenesis, follicles enter a destructive phase (catagen) (Fig. 3A). Beginning at about P16, this stage is initially characterized by massive apoptosis of matrix cells. During the ensuing 3 days, the hair bulb of each follicle degenerates into an epithelial strand that contracts, dragging the DP upward to the base of the permanent, noncycling portion of the follicle.

In mice, these backskin follicles remain in a dormant resting phase (telogen) for several days before initiating the next anagen phase (Fig. 3B). The anagen and catagen phases are relatively constant in length, whereas the telogen phase of the second adult hair cycle lasts more than 3 weeks and the third is even longer. The ability of old club hairs to remain in their socket through several rounds of hair cycling means that only a portion of the hair coat is replaced during each cycle. Although the first several rounds of the hair cycle are synchronized, individual follicles become more asynchronous as animals age, concomitant with the extension of time spent in the telogen phase of the cycle.

The ability of the postnatal hair follicle to regenerate necessitates a reservoir of stem cells. Although long surmised, it has only recently been established that follicular stem cells reside in a niche within the ORS (Fig. 3B). This niche is located at the base of the telogen-phase follicle. In approximately 3-week-old mice, as the new follicle develops adjacent to the previous one, these stem cells reorganize, creating a bulge in the ORS that allows the new hair to share the same orifice as the club hair.

Lineage-tracing experiments identified the bulge as the source of cells that can give rise to new hair follicles in normal homeostasis, the interfollicular epidermis during wound repair, and the sebaceous gland when its own resident stem cells are defective (Fig. 4A) (Morris et al. 2004; Tumbar et al. 2004; Horsley et al. 2006). Clonal expansion of individually marked bulge cells, followed by engraftment, revealed that the bulge contains multipotent stem cells, rather than a heterogeneous mixture of epidermal, sebaceous gland, and hair follicle stem cells (Blanpain et al. 2004; Claudinot et al. 2005). The ability to culture and clonally expand bulge stem cells has also underscored their long-term potential for self-renewal and their intrinsic ability to maintain multipotency outside the niche. That said, the in vivo microenvironment of the stem cell niche is clearly important in directing cell fates, because bulge stem cells normally function in cyclic hair follicle growth and contribute to the interfollicular epidermis only when a wound creates a signal that recruits them to this lineage (Ito et al. 2005; Levy et al. 2005, 2007).

Nucleotide pulse-chase experiments show that the cells within the bulge proceed through the cell cycle, but they retain the label longer than other epithelial cells within the skin (Bickenbach and Mackenzie 1984; Cotsarelis et al. 1990; Potten 2004). Recent studies monitoring dilution of the histone label over multiple hair cycles in pulse-chased inducible histone H2B–GFP mice led to the conclusion that most if not all bulge stem cells divide an average of three times during the hair cycle (Waghmare et al. 2008). Both nucleotide and H2B–GFP pulse-chase experiments further show that at the start of each new hair cycle, some label-retaining cells exit from the base of the bulge, lose their label, and proliferate to form the new hair follicle (Taylor et al. 2000; Tumbar et al. 2004). Together, these studies support the pioneering observations of Barrandon and coworkers on whisker follicles (Oshima et al. 2001) that showed that stem cells migrate from the bulge along the ORS to the base of the follicle where they become rapidly proliferating TA matrix cells (Fig. 4B). The ability of bulge cells to periodically exit their niche throughout anagen provides an explanation for how the size of the

bulge remains constant during the hair cycle despite the continuous slow division of bulge cells during anagen.

When coupled with the view that stem cells divide asymmetrically (Morrison and Kimble 2006), the label-retaining feature of bulge cells has revived the hypothesis that a self-renewing daughter stem cell might be able to selectively retain the master DNA template strands, thereby minimizing the number of replication-associated DNA mutations it acquires (Cairns 1975). The notion is attractive, and a possible mechanistic basis has recently been proposed for how such segregation of DNA strands might occur within the existing framework for DNA replication (Lew et al. 2008). That said, pulse-chase experiments with nucleotides and histone H2B–GFP label the same bulge stem cells (Tumbar et al. 2004; Waghmare et al. 2008). Given that nucleosomal histones distribute evenly between the two DNA daughter strands, and H2A–H2B dimers exchange between different nucleosomes throughout the genome during interphase (Luger and Hansen 2005), if the immortal strand hypothesis is true, new biology would be needed to explain nonrandom segregation of nucleosomes in cycling bulge cells. At present, the simplest explanation for the existing data is that bulge cells divide cycle less frequently than their progeny but still segregate their DNA randomly.

ADULT FOLLICLE STEM CELLS: HOW DO THEY DIFFER FROM RAPIDLY PROLIFERATING EPIDERMAL CELLS?

To understand the special features of the bulge, we and other investigators have conducted microarray profiling on purified bulge cells. Approximately 150 genes are preferentially expressed in the bulge relative to the proliferating basal cells of the epidermis (Blanpain et al. 2004; Morris et al. 2004; Tumbar et al. 2004). The purification of bulge stem cells has been accomplished by fluorescence-activated cell sorting (FACS) based on either (1) bulge cell surface markers α6 integrin and CD34 coupled with *K14-GFP* transgene expression (Blanpain et al. 2004), (2) *K15-GFP* transgene expression (Morris et al. 2004), or (3) the H2B–GFP pulse-chase experiment outlined above (Tumbar et al. 2004). Although each of these different procedures marks slightly different cell populations, the array data are in quite good agreement, enabling researchers to exploit this information to learn more about follicle stem cells. Additionally, some of the markers, e.g., Lgr5, are upregulated not only in the bulge, but also in other stem cells (Barker et al. 2007), suggesting that common features of stemness are reflected in these transcriptional arrays and are intrinsic to the hair follicle stem cells.

Like the embryonic epidermis, bulge cells are Wnt-responsive, as judged from their expression of two LEF1-related DNA-binding proteins (TCF3 and TCF4) and several Wnt receptor proteins (Fzds). However, at least in telogen, most bulge cells appear to be in a state of Wnt inhibition, as judged from the array data and the lack of detectable nuclear β-catenin or Wnt reporter (TOPGAL) activity (Fig. 5) (DasGupta and Fuchs 1999). These findings are consistent with recent studies showing that TCF3 can function as a repressor when β-catenin is absent or

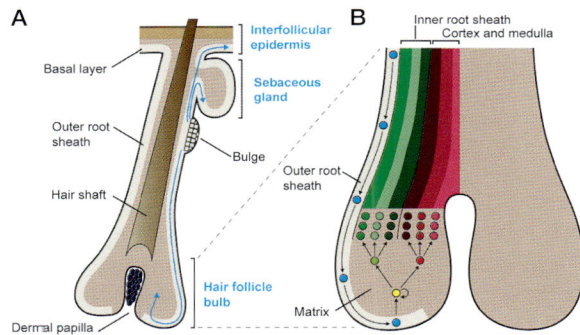

Figure 4. Multipotency of bulge epithelial stem cells and lineage decisions in the hair follicle. (*A*) Epithelial stem cells reside in a specialized niche in the upper ORS of each hair follicle and can give rise to all three epithelial lineages of the skin. During normal homeostasis, bulge stem cells are periodically activated to form a new hair follicle. During the hair follicle growth period (anagen), bulge cells migrate down the lower ORS toward the matrix, which is a specialized population of highly proliferative transit-amplifying cells responsible for producing a new hair. Bulge stem cells can also migrate to and differentiate along an SG lineage when sebaceous progenitors are absent or impaired. In a wound environment, bulge stem cells can also migrate upward and out of the hair follicle to contribute to regeneration of the interfollicular epidermis. (*B*) As bulge progeny migrate down the ORS, they subsequently enter the matrix. Matrix cells then detach from the BM and differentiate along one of six hair lineages, three of which comprise the IRS and three the cortex and medulla of the hair shaft. Intimate contact with the dermal papilla is essential for maintaining the high proliferative capacity of the matrix and driving lineage decisions.

underrepresented and that TCF3 on its own functions to repress the differentiation of skin stem cells (Nguyen et al. 2006b). Nevertheless, bulge stem cells seem to require at least a low level of Wnt signaling to maintain their identity, because conditional ablation of β-catenin in adult bulge cells results in their loss and subsequent hair follicle degeneration (Lowry et al. 2005). Coupled with the finding that transgenically elevating the levels of stabilized β-catenin results in early bulge cell activation and premature entry into anagen, it appears that bulge stem cells are responsive to a gradient of Wnt activity, with low levels required for self-renewal and higher levels driving activation (Van Mater et al. 2003; Lo Celso et al. 2004; Lowry et al. 2005).

In adult hair follicles, the role of Wnt signaling extends beyond stem cells to the TA matrix cells. In fact, our initial studies on LEF1/TCFs emanated from our identification of functional, conserved LEF1-binding sites in the promoters of the hair keratin gene family (Zhou et al. 1995). Moreover, the strongest Wnt reporter activity occurs in matrix cells as they withdraw from the cell cycle, initiate hair keratin gene expression, and commit to terminally differentiate (DasGupta and Fuchs 1999). Perhaps most intriguing is that in humans and mice, stabilizing mutations in β-catenin results in pilomatricomas, which are tumors composed of proliferating matrix-like cells and

differentiated hair shaft cells (Gat et al. 1998; Chan et al. 1999). Thus, although β-catenin stabilization and TCF3/4 are required to activate the follicle stem cells and initiate the follicle lineage, a much stronger signal involving Wnt3 (Millar et al. 1999) and LEF1 is required for hair shaft differentiation. Coupled with requirements of Wnt/β-catenin signaling in placode formation, these results highlight the manner in which a single signaling pathway functions at multiple times and locations to direct different aspects of hair follicle function. Interestingly, it was recently shown that human and mouse SQCCs are also dependent on β-catenin (Malanchi et al. 2008), raising the possibility that Wnts have even broader roles in skin tumorigenesis than previously thought.

BMP signaling is another pathway that exerts its effects at multiple points in follicle function, including the regulation of bulge stem cells (Fig. 5). In the absence of BMPr1a, otherwise quiescent bulge stem cells begin to proliferate rapidly and display markers of activated stem cells, e.g., Lhx2, Sox4, Sox9, and Shh, demonstrating that in contrast to Wnt signaling, active BMP signaling maintains bulge cell quiescence in vivo (Kobielak et al. 2007). However, these abnormally activated stem cells are later blocked in terminal differentiation stages because they can no longer receive the BMP signals necessary for IRS and hair shaft differentiation (Millar 2002; Kobielak et al. 2003; Andl et al. 2004; Yuhki et al. 2004). Although the BMPr1a-deficient bulge cells are no longer slow-cycling and do not express appreciable CD34, follicle cells lacking BMPR1a are still able to repair epidermis in a wound response and they generate what appear to be long-lived tumors (Zhang et al. 2006a; Kobielak et al. 2007). When taken together with the recent studies of Huelsken and colleagues (Malanchi et al. 2008), these studies are tantalizing in that they suggest that relative quiescence may not be an essential hallmark of follicle stem cells, at least in their tumorigenic state.

In vitro, bulge stem cells respond to BMP signaling by transient withdrawal from the cell cycle (Blanpain et al. 2004). In vivo, BMPs are made not only by telogen-phase bulge cells (Blanpain et al. 2004), but also by the surrounding dermis (Plikus et al. 2008). Interestingly, BMP expression in the dermis decreases in late telogen, coinciding with the timing of elevated WNT/β-catenin activity in follicle stem cells (Plikus et al. 2008). When coupled with the expression of multiple BMP antagonists by the DP (Rendl et al. 2005, 2008), these data suggest a model whereby the telogen-phase bulge niche receives multiple BMP signals that maintain its stem cells in a quiescent state, but as threshold levels of BMP inhibitors (and most likely additional stimulants) accumulate, the balance is tipped from stem cell quiescence to activation. How this mechanism might govern the age-related increase in telogen length still remains a mystery. However, the fact that bulge cells in the whisker follicle are able to initiate a new anagen without coming near the DP suggests that additional signaling pathways also operate to regulate bulge stem cell activity (Oshima et al. 2001).

The effects of BMPs on bulge stem cells appear to be mediated through the canonical pathway, based on the presence of phosphorylated Smad1 in the bulge and the

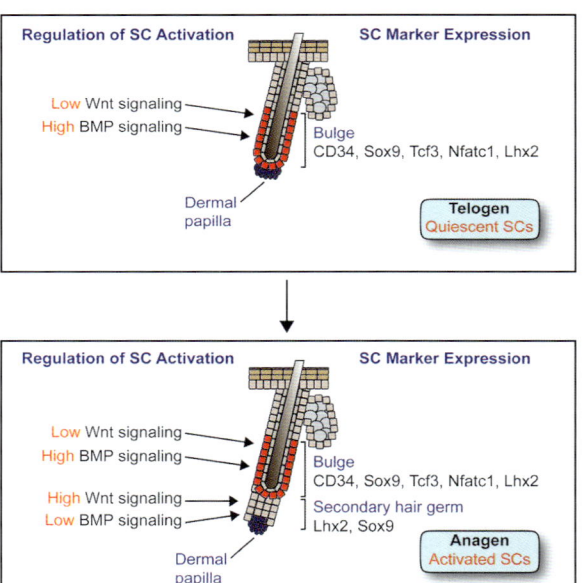

Figure 5. Regulation of stem cell identity and activity in the adult bulge stem cell niche. (*Top*) In telogen-phase hair follicles, stem cells residing in the bulge are quiescent and express the markers CD34, Sox9, Tcf3, Nfatc1, and Lhx2. High levels of BMP signaling maintain stem cells in a quiescent state, whereas low levels of Wnt signaling may help to maintain stem cell identity but are insufficient to drive SC activation. (*Bottom*) In early anagen-phase hair follicles, stem cells in the bulge proliferate and give rise to a secondary hair germ that loses most stem cell markers but still retains Sox9 and Lhx2 expression. In contrast to the bulge, BMP signaling is down-regulated and Wnt signaling is up-regulated in the germ, allowing cells to proliferate rapidly in order to produce a new hair follicle.

fact that conditional loss of its partner SMAD4 results in features that partially resemble those of the *BMPr1a* null state (Yang et al. 2005). There are likely to be many downstream effectors of BMP/SMAD signaling that maintain stem cell quiescence, but one key direct target appears to be *NFATc1*, encoding a transcription factor specifically expressed in the bulge (Tumbar et al. 2004; Horsley et al. 2008). Whereas BMP signaling dramatically enhances *NFATc1* transcription, NFATc1's nuclear localization and activity depend on a calcium/calcineurin-mediated mechanism. Although the underlying mechanisms remain to be elucidated, loss of NFATc1 leads to continuous cycling of hair follicles without resting in telogen. Moreover, the immunosuppressive drug cyclosporine A (CsA), which inhibits calcineurin, acts on the bulge to promote proliferation (Horsley et al. 2008). This finding is particularly intriguing in light of the well-known side effects of CsA on promoting hair growth and the established, often negative, effects of calcium on cell cycle control. From these data, a model emerges whereby BMP and calcium act coordinately to regulate NFATc1 and an additional downstream BMP effector PTEN (a dual specificity phosphatase implicated in β-catenin stabilization), which together function by holding bulge cells in a relatively quiescent, slow-cycling state (Fig. 5) (Zhang et al. 2006b; Kobielak et al. 2007; Horsley et al. 2008).

HOMEOSTASIS IN THE SEBACEOUS GLAND

An appendage of the hair follicle, sebaceous glands (SGs) are located above the bulge and just below the hair shaft orifice at the skin surface. The major role of the gland is to generate terminally differentiated sebocytes that degenerate to release lipids and sebum and lubricate the skin surface as they are expelled from the SG into the hair canal. Because of this holocrine manner of secretion, SG homeostasis necessitates a population of progenitor cells that can regenerate the differentiated cells constantly being lost. Lineage tracing by retrovirus-mediated gene transfer suggests that a small population of cells near or at the base of the SG might be stem cells (Ghazizadeh and Taichman 2001).

Recently, the transcriptional repressor protein Blimp1 was identified in a genetic screen for hair follicle transcription factors, and it was shown to mark a small population of cells at the SG base (Horsley et al. 2006). These Blimp1-positive cells appeared to be in close association with the BM that surrounds the gland and are contiguous with the BM underlying the epidermis and surrounding the hair follicle. Accordingly, the Blimp1-positive SG cells also express K5 and K14, markers of the basal layer in epidermis and follicles. Genetic lineage-tracing experiments revealed that Blimp1-positive SG cells can regenerate the entire gland, including the sebocytes. When Blimp1 was conditionally targeted for ablation, SGs became larger. The likely explanation for this alteration stems from Blimp1's ability, first detected in B lymphocytes (Chang et al. 2000), to transcriptionally repress c-*myc*, a gene known to induce SG hyperplasia and sebocyte differentiation at the expense of hair follicle differentiation (Arnold and Watt 2001; Waikel et al.

2001). However, if Blimp1 controls production of differentiated cells by SG progenitors, why don't the enlarged glands eventually degenerate after an initial period of hyperplasia?

Although further studies are needed, the underlying explanation could be rooted in the ability of bulge stem cells to become mobilized when the SG progenitor cell population is depleted. As judged from BrdU pulse and pulse-chase experiments, Blimp1-negative bulge cells show signs of active cycling and reduced label retention in the absence of Blimp1-marked SG progenitors (Horsley et al. 2006). In this regard, the behavior of bulge stem cells in response to a loss of Blimp1 in SG progenitors resembles their response to the epidermis lost upon injury in normal mice. Such a precursor–product relationship between the bulge and other skin stem cell populations has also been documented by engraftment experiments with isolated bulge stem cells (Fig. 4) (Blanpain et al. 2004; Claudinot et al. 2005).

TRACING THE EMBRYONIC ROOTS OF ADULT STEM CELLS IN THE FOLLICLE

Major questions in stem cell biology are where and when adult stem cell niches become established. We recently exploited the relatively slow-cycling behavior of adult follicle stem cells to trace their developmental origins (Nowak et al. 2008). Our embryonic pulse-chase studies revealed that, surprisingly, label-retaining cells are specified early in skin development and later become adult bulge stem cells. Moreover, we have discovered that these early label-retaining cells express a number of bulge-preferred transcription factors even though the widely used bulge marker CD34 (Trempus et al. 2003) is not expressed in these early bulge cells. Sox9 (Vidal et al. 2005) and Tcf3 (Nguyen et al. 2006b) emerged as key potential regulators of stem cell identity, marking both early and adult stem cells. Moreover, in contrast to CD34, these transcription factors also mark the trail of follicle stem cells that appear to migrate from the bulge to the TA matrix during anagen (Nowak et al. 2008).

By taking advantage of Sox9 as a marker of early bulge cells, we discovered that the hair placode that we recently transcriptionally profiled on the basis of P-cadherin and K14-GFP-actin expression (Rhee et al. 2006) is not a homogeneous cluster of cells but rather consists of two populations that can be distinguished by their differential expression of Lhx2 and Sox9. Our *Sox9* genetic marking studies suggest that the early basal placode cells that are marked by Lhx2 and high P-cadherin expression are transient rather than long-lived progenitors (Fig. 6). They give rise to the initial hair bulb but then seem to disappear entirely by approximately P7. In contrast, the Sox9-expressing suprabasal placode cells give rise not only to the long-lived self-renewing stem cells of the bulge, but also to the entire pilosebaceous unit. Moreover, the early emergence of a niche of Sox9-positive label-retaining cells is accompanied by their concomitant expression of Lhx2, Nfatc1, and Tcf3. Thereafter, these markers identify the multipotent stem cell niche of the hair follicle.

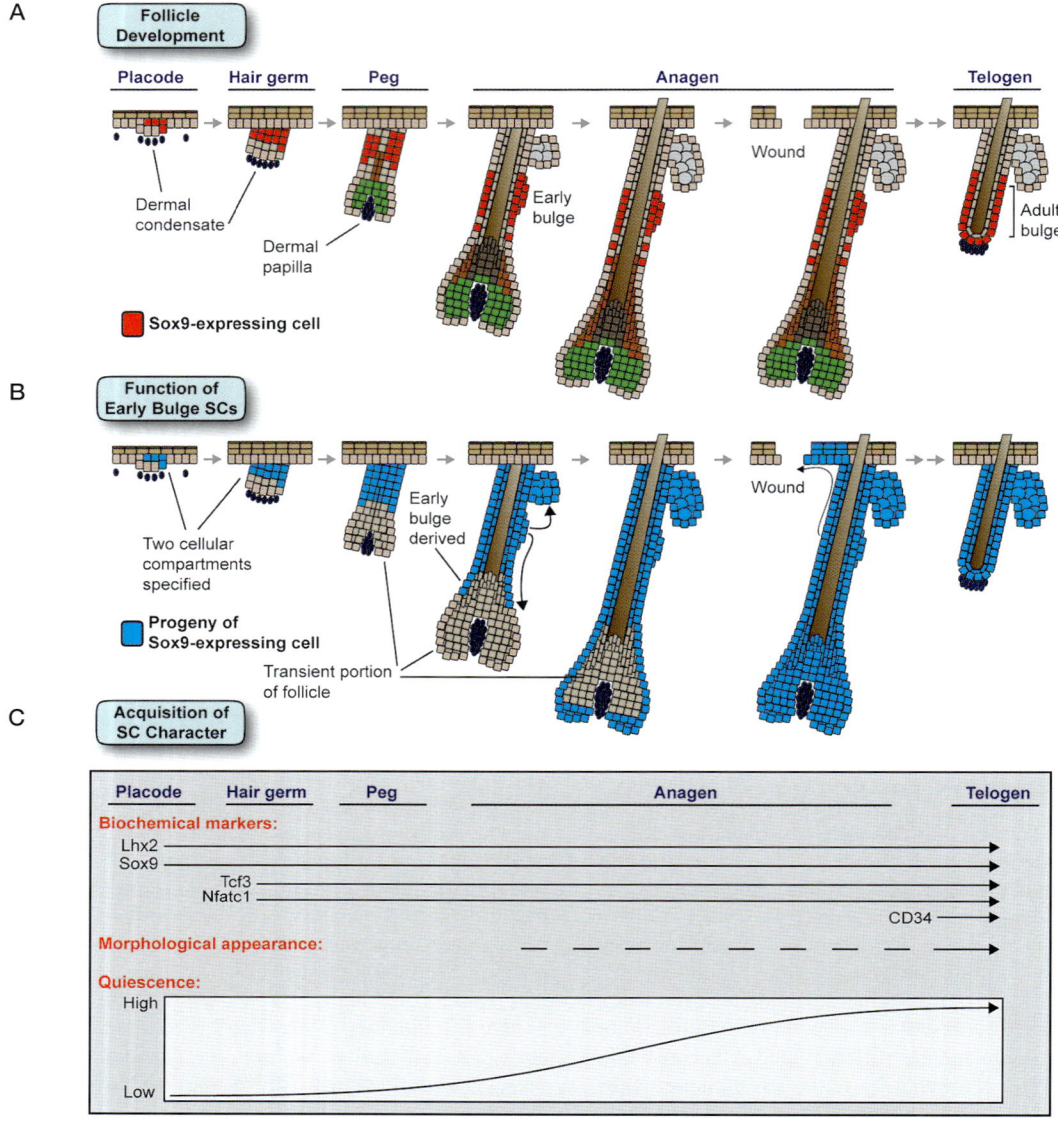

Figure 6. Embryonic specification of the hair follicle stem cell population. (*A*) *Sox9*, a gene essential for specifying the hair follicle stem cell population, is expressed in suprabasal cells of the placode at the first stage of hair follicle morphogenesis. As development proceeds, Sox9 remains in the upper portion of the ORS, marking the early follicle stem cell population, as well as in their progeny that migrate down the ORS toward the matrix. In adult follicles, Sox9 remains expressed in the bulge stem cell niche. (*B*) Genetic marking studies of Sox9-expressing cells show that hair follicles are initially composed of two cell populations, Sox9-expressing early stem cells and transient Sox9-negative cells that give rise to the initial matrix. As follicle morphogenesis proceeds, the progeny of Sox9-expressing cells move down the ORS and completely replace the initial matrix population. Additionally, progeny of Sox9-positive cells give rise to the SG lineage and can help repair the interfollicular epidermis in a wound environment. By the time hair follicle morphogenesis has finished, hair follicles are entirely derived from Sox9-expressing cells. (*C*) Markers and functional characteristics of the hair follicle stem cell population are acquired in a stepwise manner. The transcription factors Lhx2 and Sox9 are expressed at the first placode stage of morphogenesis, whereas Tcf3 and Nfatc1 appear later at the hair germ stage. Although all four of these genes mark both early and adult stem cells, CD34 is only up-regulated in adult follicle stem cells. Morphologically, the location of early stem cells in the follicle can be inferred by a thickening in the ORS that appears during late stages of morphogenesis and becomes much more pronounced when the adult bulge stem cell niche forms during the first telogen. Although all early stem cells divide at least several times during follicle morphogenesis, they gradually increase their slow-cycling character as the rate of hair follicle growth decreases. Adult stem cells in resting telogen hair follicles are highly quiescent, but they undergo periods of reduced quiescence in the growing anagen phase of the hair cycle.

Although the functional significance of separate TA and stem cell populations in development remains unclear, precedence exists in blood development, where a transient burst of primitive hematopoietic progenitors drives initial blood formation before being later replaced by definitive hematopoietic stem cells (Orkin and Zon 2008). The existence of a transient Sox9-independent population explains why in mouse embryos, conditionally targeted for *Sox9*, HF maturation and hair production begin normally but then halt midstream (Vidal et al. 2005; Nowak et al. 2008). These findings also provide compelling support for the notion that whether in the postnatal hair cycle or during morphogenesis, the matrix relies upon input from early Sox9-positive stem cells for its maintenance and cannot enhance its own self-renewal to compensate for a lack of bulge stem cell input. These studies further strengthen the evidence that stem cell migration occurs during hair follicle morphogenesis and that the fueling of the matrix by Sox9-positive bulge cells is functionally required for hair production (Fig. 6).

Another surprising finding that emerged from targeting *Sox9* ablation in the embryo was that the early bulge cells are required not only to complete hair follicle morphogenesis, but also to initiate SG morphogenesis. Our studies further revealed that without early bulge stem cells, wound repair of the IFE is markedly impaired and also that early Sox9-derived progeny contribute long-term to repopulating wounded IFE with close to 100% efficiency (Nowak et al. 2008). Prior lineage-tracing studies with bulge stem cells genetically marked by *K15-CrePR* suggest that adult bulge cells may contribute only transiently to wound repair (Ito et al. 2005). Our studies raise the possibility that early bulge stem cells may have greater potential than their adult counterparts. These studies provide major new insights into the existence and usage of a follicle stem cell population at a stage and for a purpose that has hitherto been unanticipated. Future exploration of the relationship between embryonic and adult skin stem cells will undoubtedly open new avenues for defining the features of stemness in the skin.

UNIFYING FEATURES OF SKIN STEM CELLS

An emerging view of skin stem cells is that there appears to be three different niches for skin stem cells: the follicle bulge, the base of the sebaceous gland, and the basal layer of the epidermis. It has not yet been resolved whether the basal layer contains a subset of stem cells as originally posited (Potten 1974) or whether the basal compartment is composed of a single progenitor population, as more recently proposed (Clayton et al. 2007). However, at least in the developing mouse skin, the division rates of cells within the basal layer of the epidermis are significantly more uniform than those in the hair follicle, where the emergence of stem cells is coincident with their adoption of slow-cycling behavior (Nowak et al. 2008).

Are there unifying features of keratinocyte stem cells and the activation mechanisms that guide them along specific lineages? Although the answers to these questions are still at the molecular drawing board, it seems reasonable to predict that, given the similarities in their niche

environments and their common developmental origin, certain features should be shared among all three stem cell populations. Notably, all three progenitor populations express K5, K14, and ΔNp63. Although ΔNp63 has also been implicated in differentiation, it remains one of the best candidates to date for a common gatekeeper for proliferation in epithelial stem cells (Truong et al. 2006; Senoo et al. 2007). Cells within all three progenitor pools also express E-cadherin and display elevated levels of adherens junctions. A reduction in E-cadherin appears to be an essential feature in mobilizing embryonic epidermal cells to invaginate to form a hair follicle (Jamora et al. 2003), and reductions in α-catenin and p120-catenin have also been linked to epithelial cancers, a less organized form of invagination (Scott and Yap 2006; Reynolds 2007). Such studies make it tempting to speculate that the mobilization of stem cells to exit their niches and either form a hair follicle or repair wounds may entail the remodeling of intercellular junctions.

The three progenitor compartments are also typified by their proximity to an underlying BM, and all three types of progenitors express integrins, including α6β4 and α3β1, as well as their associated proteins. The higher the integrin level, the greater the proliferative potential, as judged from the early studies of Jones and Watt on cultured epidermal stem cells (Jones et al. 1995). One intriguing aspect is the up-regulation of β6 integrin and the presence of the αvβ6 integrin ligand tenascin C that occurs in bulge stem cells as they transit from telogen to anagen (Tumbar et al. 2004). Similar changes are seen in the epidermal basal layer in response to injury (Fassler et al. 1996) and in tumorigenesis (Guasch et al. 2007), leading to the speculation that these changes might be important in understanding how stem cells become activated to migrate from their niche and differentiate along a specific lineage.

Although keratinocytes with proliferative capacity are often associated with an underlying BM, some bulge stem cells seem to be suprabasal. Thus, during the second postnatal anagen, the α6-integrin–low bulge stem cells are uniquely positioned between two differentiated hair shafts in mature follicles, yet they are nevertheless CD34-positive, slow-cycling, and transcriptionally and functionally similar to the α6-integrin–high bulge cells that appear earlier in the niche (Blanpain et al. 2004). How these α6-integrin–low cells manage to maintain their stem cell features and migrate downward to contribute to future hair follicle development will be a valuable area of future study. In a similar vein, the Sox9-positive cells that first appear in developing placodes appear to be primarily suprabasal, again hinting at the possibility of cells to possess stem cell character, at least transiently, in the absence of basement adherence.

The similarities in keratinocyte stem cell behavior may extend further to shared principles and pathways, even if the precise players differ. One example is Blimp1, which appears to mark SG progenitors and repress c-*myc* expression (Horsley et al. 2006), whereas Miz1 seems to have a similar c-*myc* repressive function in the epidermal basal layer (Gebhardt et al. 2006). The outcome of c-Myc expression is keratinocyte proliferation in both the SG and epidermis, leading to the postulate that c-Myc might control the conversion of stem cells to transit amplifying, i.e., committed cells (Waikel et al. 2001; Frye et al. 2003). In this regard, it is intriguing that later in the epidermal lineage, Ovo1, yet another transcriptional repressor of c-Myc, is expressed suprabasally, where it may act to prevent proliferation after TA cells commit to terminally differentiate (Nair et al. 2006).

Another signaling pathway likely to impact on more than one keratinocyte stem cell niche is Notch, which controls selective cell-fate determination through close-range interactions not only in the epidermis, but also in the hair follicle and the SG (Yamamoto et al. 2003; Pan et al. 2004; Vauclair et al. 2005; Blanpain and Fuchs 2006; Estrach et al. 2006; Nguyen et al. 2006a). The studies to date support a role for Notch in activating the keratinocyte switch from the undifferentiated to differentiated fate.

Although Notch signaling appears to have a similar role throughout different keratinocyte populations in the skin, other signaling pathways seem to have more potent effects on one lineage over another. Thus, in the follicle bulge, stem cells enter the new hair cycle when threshold levels of stabilized β-catenin are reached (Gat et al. 1998; Huelsken et al. 2001; Van Mater et al. 2003; Lo Celso et al. 2004; Lowry et al. 2005). Conversely, a negative role for Wnt signaling has been proposed for SG fate determination, because SG hyperplasia occurs when a truncated version of LEF1 is expressed that is unable to associate with β-catenin (Merrill et al. 2001; Niemann et al. 2002). Although epidermal stem cells are seemingly unaffected by loss of β-catenin (Huelsken and Birchmeier 2001), they adopt a hair follicle cell fate when β-catenin is stabilized through either genetic manipulation or wounding (Gat et al. 1998; Ito et al. 2007). When taken with the recent observation that the basal epidermal layer of chemically induced β-catenin dependent SQCCs is positive for a number of bulge markers, including CD34 and Sox9 (Malanchi et al. 2008), Wnt/β-catenin signaling emerges as a promoting force for the establishment of bulge SC.

Given that cell-fate signaling pathways are often interdependent on one another, subtle perturbances of a single pathway can exert a potent impact on lineage determination, as recently demonstrated by Watt and colleagues (Estrach et al. 2006). In the precortex of the hair follicle (DasGupta and Fuchs 1999), where Wnt signaling is particularly high, the Wnt target gene *Jag1* is expressed, leading to active Notch signaling in these cells and promoting proliferation and differentiation to produce hair (Estrach et al. 2006). Conversely, in the absence of *Jag1*, hair follicles fail to form, but the epidermis is unaffected. Thus, by activating expression of other signaling genes, Wnt signaling can accentuate its effects on a particular lineage within the skin.

CONCLUSIONS

In closing, the existence of different skin stem cell niches and the ability of cells in these niches to respond differentially to environmental cues has been an exciting new development in the skin field. Increasing knowledge about these stem cell populations during the past 5 years

has begun to shed light on the similarities and differences among these niches. Important questions for the future are now raised. Are these different stem cell populations multipotent as recent studies suggest (Ito et al. 2007)? Is the multipotency of stem cells in response to wounding a reflection of mechanical disruption of the niche or a response to growth/differentiation factors released in a wound response? To what extent do the gene expression programs of other progenitor populations overlap with that of follicle stem cells and do these similarities reflect their self-renewing undifferentiated features? When do these different proliferative epithelial compartments form during skin development and how are their differences in gene expression influenced by their distinct in vivo microenvironments? What factors control the differences in the slow-cycling ability of different stem cell populations within the skin? The answers to these questions will bring us closer to a complete understanding of how stem cells are maintained in their undifferentiated growth-restricted state and what prompts them to become activated, exit their niches, and embark upon distinct lineages.

ACKNOWLEDGMENTS

We are grateful to Howard Green, who many years ago inspired the field of skin stem cell biology with his pioneering work on human epidermal culture. Through his application of cultured keratinocytes to the treatment of burn patients, he taught the field the importance of bringing basic biology to a clinical setting. We thank former Green lab members James Rheinwald, Henry Sun, Fiona Watt, and Yann Barrandon, who helped to build the framework of molecular skin biology and who have provided valuable advice and insights to Elaine and her laboratory over the years. We are particularly grateful to the many members of the Fuchs lab, past and present, who contributed so mightily over the years to formulating our current knowledge of the molecular, cellular, and developmental biology of the skin and its associated genetic disorders. Finally, we thank our many colleagues in the field, both friends and competitors, for giving us the constructive criticism and motivation to do the best science possible and for making the field an enjoyable and interactive one. E.F. is an Investigator of the Howard Hughes Medical Institute. J.N. is supported by the National Institutes of Health MSTP grant GM07739. This work has been supported by grants from the National Institutes of Health and the Starr Foundation.

REFERENCES

Alonso, L. and Fuchs, E. 2006. The hair cycle. *J. Cell Sci.* **119:** 391–393.
Andl, T., Reddy, S.T., Gaddapara, T., and Millar, S.E. 2002. WNT signals are required for the initiation of hair follicle development. *Dev. Cell* **2:** 643–653.
Andl, T., Ahn, K., Kairo, A., Chu, E.Y., Wine-Lee, L., Reddy, S.T., Croft, N.J., Cebra-Thomas, J.A., Metzger, D., Chambon, P. et al. 2004. Epithelial Bmpr1a regulates differentiation and proliferation in postnatal hair follicles and is essential for tooth development. *Development* **131:** 2257–2268.
Arnold, I. and Watt, F.M. 2001. c-Myc activation in transgenic mouse epidermis results in mobilization of stem cells and dif-

ferentiation of their progeny. *Curr. Biol.* **11:** 558–568.

Balzac, F., Avolio, M., Degani, S., Kaverina, I., Torti, M., Silengo, L., Small, J.V., and Retta, S.F. 2005. E-cadherin endocytosis regulates the activity of Rap1: A traffic light GTPase at the crossroads between cadherin and integrin function. *J. Cell Sci.* **118:** 4765–4783.

Barker, N., van Es, J.H., Kuipers, J., Kujala, P., van den Born, M., Cozijnsen, M., Haegebarth, A., Korving, J., Begthel, H., Peters, P.J., and Clevers, H. 2007. Identification of stem cells in small intestine and colon by marker gene *Lgr5*. *Nature* **449:** 1003–1007.

Barrandon, Y. and Green, H. 1987a. Cell migration is essential for sustained growth of keratinocyte colonies: The roles of transforming growth factor-α and epidermal growth factor. *Cell* **50:** 1131–1137.

Barrandon, Y. and Green, H. 1987b. Three clonal types of keratinocyte with different capacities for multiplication. *Proc. Natl. Acad. Sci.* **84:** 2302–2306.

Benitah, S.A., Frye, M., Glogauer, M., and Watt, F.M. 2005. Stem cell depletion through epidermal deletion of Rac1. *Science* **309:** 933–935.

Bickenbach, J.R. and Mackenzie, I.C. 1984. Identification and localization of label-retaining cells in hamster epithelia. *J. Invest. Dermatol.* **82:** 618–622.

Blanpain, C. and Fuchs, E. 2006. Epidermal stem cells of the skin. *Annu. Rev. Cell Dev. Biol.* **22:** 339–373.

Blanpain, C., Lowry, W.E., Pasolli, H.A., and Fuchs, E. 2006. Canonical notch signaling functions as a commitment switch in the epidermal lineage. *Genes Dev.* **20:** 3022–3035.

Blanpain, C., Lowry, W.E., Geoghegan, A., Polak, L., and Fuchs, E. 2004. Self-renewal, multipotency, and the existence of two cell populations within an epithelial stem cell niche. *Cell* **118:** 635–648.

Botchkarev, V.A., Botchkareva, N.V., Nakamura, M., Huber, O., Funa, K., Lauster, R., Paus, R., and Gilchrest, B.A. 2001. Noggin is required for induction of the hair follicle growth phase in postnatal skin. *FASEB J.* **15:** 2205–2214.

Botchkarev, V.A., Botchkareva, N.V., Roth, W., Nakamura, M., Chen, L.H., Herzog, W., Lindner, G., McMahon, J.A., Peters, C., Lauster, R., McMahon, A.P., and Paus, R. 1999. Noggin is a mesenchymally derived stimulator of hair-follicle induction. *Nat. Cell Biol.* **1:** 158–164.

Cairns, J. 1975. Mutation selection and the natural history of cancer. *Nature* **255:** 197–200.

Chan, E.F., Gat, U., McNiff, J.M., and Fuchs, E. 1999. A common human skin tumour is caused by activating mutations in β-catenin. *Nat. Genet.* **21:** 410–413.

Chang, D.H., Angelin-Duclos, C., and Calame, K. 2000. BLIMP-1: Trigger for differentiation of myeloid lineage. *Nat. Immunol.* **1:** 169–176.

Chrostek, A., Wu, X., Quondamatteo, F., Hu, R., Sanecka, A., Niemann, C., Langbein, L., Haase, I., and Brakebusch, C. 2006. Rac1 is crucial for hair follicle integrity but is not essential for maintenance of the epidermis. *Mol. Cell. Biol.* **26:** 6957–6970.

Claudinot, S., Nicolas, M., Oshima, H., Rochat, A., and Barrandon, Y. 2005. Long-term renewal of hair follicles from clonogenic multipotent stem cells. *Proc. Natl. Acad. Sci.* **102:** 14677–14682.

Clayton, E., Doupe, D.P., Klein, A.M., Winton, D.J., Simons, B.D., and Jones, P.H. 2007. A single type of progenitor cell maintains normal epidermis. *Nature* **446:** 185–189.

Cotsarelis, G., Sun, T.-T., and Lavker, R.M. 1990. Label-retaining cells reside in the bulge area of pilosebaceous unit: Implications for follicular stem cells, hair cycle, and skin carcinogenesis. *Cell* **61:** 1329–1337.

DasGupta, R. and Fuchs, E. 1999. Multiple roles for activated LEF/TCF transcription complexes during hair follicle development and differentiation. *Development* **126:** 4557–4568.

de Guzman Strong, C., Wertz, P.W., Wang, C., Yang, F., Meltzer, P.S., Andl, T., Millar, S.E., Ho, I.C., Pai, S.Y., and Segre, J.A. 2006. Lipid defect underlies selective skin barrier impairment of an epidermal-specific deletion of Gata-3. *J. Cell Biol.* **175:** 661–670.

Di-Poi, N., Michalik, L., Desvergne, B., and Wahli, W. 2004. Functions of peroxisome proliferator-activated receptors (PPAR) in skin homeostasis. *Lipids* **39:** 1093–1099.

Dobereiner, H.G., Dubin-Thaler, B.J., Giannone, G., and Sheetz, M.P. 2005. Force sensing and generation in cell phases: Analyses of complex functions. *J. Appl. Physiol.* **98:** 1542–1546.

Elias, P.M. 2007. The skin barrier as an innate immune element. *Semin. Immunopathol.* **29:** 3–14.

Estrach, S., Ambler, C.A., Lo Celso, C., Hozumi, K., and Watt, F.M. 2006. *Jagged 1* is a β-catenin target gene required for ectopic hair follicle formation in adult epidermis. *Development* **133:** 4427–4438.

Fassler, R., Sasaki, T., Timpl, R., Chu, M.L., and Werner, S. 1996. Differential regulation of fibulin, tenascin-C, and nidogen expression during wound healing of normal and glucocorticoid-treated mice. *Exp. Cell Res.* **222:** 111–116.

Frame, M.C., Fincham, V.J., Carragher, N.O., and Wyke, J.A. 2002. v-Src's hold over actin and cell adhesions. *Nat. Rev. Mol. Cell Biol.* **3:** 233–245.

Frye, M., Gardner, C., Li, E.R., Arnold, I., and Watt, F.M. 2003. Evidence that Myc activation depletes the epidermal stem cell compartment by modulating adhesive interactions with the local microenvironment. *Development* **130:** 2793–2808.

Fuchs, E. 2007. Scratching the surface of skin development. *Nature* **445:** 834–842.

Fuchs, E. and Cleveland, D.W. 1998. A structural scaffolding of intermediate filaments in health and disease. *Science* **279:** 514–519.

Fuchs, E. and Green, H. 1980. Changes in keratin gene expression during terminal differentiation of the keratinocyte. *Cell* **19:** 1033–1042.

Fuller, M.T. and Spradling, A.C. 2007. Male and female *Drosophila* germline stem cells: Two versions of immortality. *Science* **316:** 402–404.

Gat, U., DasGupta, R., Degenstein, L., and Fuchs, E. 1998. De novo hair follicle morphogenesis and hair tumors in mice expressing a truncated β-catenin in skin *Cell* **95:** 605–614.

Gebhardt, A., Frye, M., Herold, S., Benitah, S.A., Braun, K., Samans, B., Watt, F.M., Elsasser, H.P., and Eilers, M. 2006. Myc regulates keratinocyte adhesion and differentiation via complex formation with Miz1. *J. Cell Biol.* **172:** 139–149.

Ghazizadeh, S. and Taichman, L.B. 2001. Multiple classes of stem cells in cutaneous epithelium: A lineage analysis of adult mouse skin. *EMBO J.* **20:** 1215–1222.

Gritli-Linde, A., Hallberg, K., Harfe, B.D., Reyahi, A., Kannius-Janson, M., Nilsson, J., Cobourne, M.T., Sharpe, P.T., McMahon, A.P., and Linde, A. 2007. Abnormal hair development and apparent follicular transformation to mammary gland in the absence of hedgehog signaling. *Dev. Cell* **12:** 99–112.

Guasch, G., Schober, M., Pasolli, H.A., Conn, E.B., Polak, L., and Fuchs, E. 2007. Loss of TGFβ signaling destabilizes homeostasis and promotes squamous cell carcinomas in stratified epithelia. *Cancer Cell* **12:** 313–327

Hardy, M.H. 1992. The secret life of the hair follicle. *Trends Genet.* **8:** 55–61.

Horsley, V., Aliprantis, A.O., Polak, L., Glimcher, L.H., and Fuchs, E. 2008. NFATc1 balances quiescence and proliferation of skin stem cells. *Cell* **132:** 299–310.

Horsley, V., O'Carroll, D., Tooze, R., Ohinata, Y., Saitou, M., Obukhanych, T., Nussenzweig, M., Tarakhovsky, A., and Fuchs, E. 2006. Blimp1 defines a progenitor population that governs cellular input to the sebaceous gland. *Cell* **126:** 597–609.

Huelsken, J. and Birchmeier, W. 2001. New aspects of Wnt signaling pathways in higher vertebrates. *Curr. Opin. Genet. Dev.* **11:** 547–553.

Huelsken, J., Vogel, R., Erdmann, B., Cotsarelis, G., and Birchmeier, W. 2001. β-Catenin controls hair follicle morphogenesis and stem cell differentiation in the skin. *Cell* **105:** 533–545.

Ichii, T. and Takeichi, M. 2007. p120-catenin regulates microtubule dynamics and cell migration in a cadherin-independent manner. *Genes Cells* **12:** 827–839.

Ito, M., Liu, Y., Yang, Z., Nguyen, J., Liang, F., Morris, R.J., and Cotsarelis, G. 2005. Stem cells in the hair follicle bulge contribute to wound repair but not to homeostasis of the epidermis. *Nat. Med.* **11:** 1351–1354.

Ito, M., Yang, Z., Andl, T., Cui, C., Kim, N., Millar, S.E., and Cotsarelis, G. 2007. Wnt-dependent de novo hair follicle regeneration in adult mouse skin after wounding. *Nature* **447:** 316–320.

Jamora, C., DasGupta, R., Kocieniewski, P., and Fuchs, E. 2003. Links between signal transduction, transcription and adhesion in epithelial bud development. *Nature* **422:** 317–322.

Janes, S.M. and Watt, F.M. 2006. New roles for integrins in squamous-cell carcinoma. *Nat. Rev. Cancer* **6:** 175–183.

Jensen, U.B., Lowell, S., and Watt, F.M. 1999. The spatial relationship between stem cells and their progeny in the basal layer of human epidermis: A new view based on whole-mount labelling and lineage analysis. *Development* **126:** 2409–2418.

Jones, P.H., Harper, S., and Watt, F.M. 1995. Stem cell patterning and fate in human epidermis. *Cell* **80:** 83–93.

Kaverina, I., Krylyshkina, O., and Small, J.V. 1999. Microtubule targeting of substrate contacts promotes their relaxation and dissociation. *J. Cell Biol.* **146:** 1033–1044.

Kee, S.H. and Steinert, P.M. 2001. Microtubule disruption in keratinocytes induces cell-cell adhesion through activation of endogenous E-cadherin. *Mol. Biol. Cell* **12:** 1983–1993.

Kobielak, K., Pasolli, H.A., Alonso, L., Polak, L., and Fuchs, E. 2003. Defining BMP functions in the hair follicle by conditional ablation of BMP receptor IA. *J. Cell Biol.* **163:** 609–623.

Kobielak, K., Stokes, N., de la Cruz, J., Polak, L., and Fuchs, E. 2007. Loss of a quiescent niche but not follicle stem cells in the absence of BMP signaling. *Proc. Natl. Acad. Sci.* **104:** 10063–10068.

Laurikkala, J., Mikkola, M.L., James, M., Tummers, M., Mills, A.A., and Thesleff, I. 2006. p63 regulates multiple signalling pathways required for ectodermal organogenesis and differentiation. *Development* **133:** 1553–1563.

Laurikkala, J., Mikkola, M., Mustonen, T., Aberg, T., Koppinen, P., Pispa, J., Nieminen, P., Galceran, J., Grosschedl, R., and Thesleff, I. 2001. TNF signaling via the ligand-receptor pair ectodysplasin and edar controls the function of epithelial signaling centers and is regulated by Wnt and activin during tooth organogenesis. *Dev. Biol.* **229:** 443–455.

Lavker, R.M. and Sun, T.T. 1982. Heterogeneity in epidermal basal keratinocytes: Morphological and functional correlations. *Science* **215:** 1239–1241.

Lechler, T. and Fuchs, E. 2005. Asymmetric cell divisions promote stratification and differentiation of mammalian skin. *Nature* **437:** 275–280.

Lee, J., Basak, J.M., Demehri, S., and Kopan, R. 2007. Bi-compartmental communication contributes to the opposite proliferative behavior of Notch1-deficient hair follicle and epidermal keratinocytes. *Development* **134:** 2795–2806.

Lee, J.W. and Juliano, R. 2004. Mitogenic signal transduction by integrin- and growth factor receptor-mediated pathways. *Mol. Cells* **17:** 188–202.

Levy, V., Lindon, C., Harfe, B.D., and Morgan, B.A. 2005. Distinct stem cell populations regenerate the follicle and interfollicular epidermis. *Dev. Cell* **9:** 855–861.

Levy, V., Lindon, C., Zheng, Y., Harfe, B.D., and Morgan, B.A. 2007. Epidermal stem cells arise from the hair follicle after wounding. *FASEB J.* **21:** 1–9.

Lew, D.J., Burke, D.J., and Dutta, A. 2008. The immortal strand hypothesis: How could it work? *Cell* **133:** 21–23.

Lo Celso, C., Prowse, D.M., and Watt, F.M. 2004. Transient activation of β-catenin signalling in adult mouse epidermis is sufficient to induce new hair follicles but continuous activation is required to maintain hair follicle tumours. *Development* **131:** 1787–1799.

Lorenz, K., Grashoff, C., Torka, R., Sakai, T., Langbein, L., Bloch, W., Aumailley, M., and Fassler, R. 2007. Integrin-linked kinase is required for epidermal and hair follicle morphogenesis. *J. Cell Biol.* **177:** 501–513.

Lowry, W.E., Blanpain, C., Nowak, J.A., Guasch, G., Lewis, L., and Fuchs, E. 2005. Defining the impact of β-catenin/Tcf

transactivation on epithelial stem cells. *Genes Dev.* **19:** 1596–1611.

Luger, K. and Hansen, J.C. 2005. Nucleosome and chromatin fiber dynamics. *Curr. Opin. Struct. Biol.* **15:** 188–196.

Mackenzie, I.C. 1997. Retroviral transduction of murine epidermal stem cells demonstrates clonal units of epidermal structure. *J. Invest. Dermatol.* **109:** 377–383.

Malanchi, I., Peinado, H., Kassen, D., Hussenet, T., Metzger, D., Chambon, P., Huber, M., Hohl, D., Cano, A., Birchmeier, W., and Huelsken, J. 2008. Cutaneous cancer stem cell maintenance is dependent on β-catenin signalling. *Nature* **452:** 650–653.

Manohar, A., Shome, S.G., Lamar, J., Stirling, L., Iyer, V., Pumiglia, K., and DiPersio, C.M. 2004. α3β1 integrin promotes keratinocyte cell survival through activation of a MEK/ERK signaling pathway. *J. Cell Sci.* **117:** 4043–4054.

Marinkovich, M.P. 2007. Tumour microenvironment: Laminin 332 in squamous-cell carcinoma. *Nat. Rev. Cancer* **7:** 370–380.

Maytin, E.V. and Habener, J.F. 1998. Transcription factors C/EBPα, C/EBPβ, and CHOP (Gadd153) expressed during the differentiation program of keratinocytes in vitro and in vivo. *J. Invest. Dermatol.* **110:** 238–246.

McLean, G.W., Komiyama, N.H., Serrels, B., Asano, H., Reynolds, L., Conti, F., Hodivala-Dilke, K., Metzger, D., Chambon, P., Grant, S.G., and Frame, M.C. 2004. Specific deletion of focal adhesion kinase suppresses tumor formation and blocks malignant progression. *Genes Dev.* **18:** 2998–3003.

Merrill, B.J., Gat, U., DasGupta, R., and Fuchs, E. 2001. Tcf3 and Lef1 regulate lineage differentiation of multipotent stem cells in skin. *Genes Dev.* **15:** 1688–1705.

Mikkola, M.L. 2007. Genetic basis of skin appendage development. *Semin. Cell Dev. Biol.* **18:** 225–236.

Millar, S.E. 2002. Molecular mechanisms regulating hair follicle development. *J. Invest. Dermatol.* **118:** 216–225.

Millar, S.E., Willert, K., Salinas, P.C., Roelink, H., Nusse, R., Sussman, D.J., and Barsh, G.S. 1999. WNT signaling in the control of hair growth and structure. *Dev. Biol.* **207:** 133–149.

Mills, A.A., Zheng, B., Wang, X.J., Vogel, H., Roop, D.R., and Bradley, A. 1999. p63 is a p53 homologue required for limb and epidermal morphogenesis. *Nature* **398:** 708–713.

Ming Kwan, K., Li, A.G., Wang, X.J., Wurst, W., and Behringer, R.R. 2004. Essential roles of BMPR-IA signaling in differentiation and growth of hair follicles and in skin tumorigenesis. *Genesis* **39:** 10–25.

Morgan, B.A., Orkin, R.W., Noramly, S., and Perez, A. 1998. Stage-specific effects of sonic hedgehog expression in the epidermis. *Dev. Biol.* **201:** 1–12.

Moriyama, M., Durham, A.D., Moriyama, H., Hasegawa, K., Nishikawa, S., Radtke, F., and Osawa, M. 2008. Multiple roles of Notch signaling in the regulation of epidermal development. *Dev. Cell* **14:** 594–604.

Morris, R.J., Liu, Y., Marles, L., Yang, Z., Trempus, C., Li, S., Lin, J.S., Sawicki, J.A., and Cotsarelis, G. 2004. Capturing and profiling adult hair follicle stem cells. *Nat. Biotechnol.* **22:** 411–417.

Morrison, S.J. and Kimble, J. 2006. Asymmetric and symmetric stem-cell divisions in development and cancer. *Nature* **441:** 1068–1074.

Muller-Rover, S., Handjiski, B., van der Veen, C., Eichmuller, S., Foitzik, K., McKay, I.A., Stenn, K.S., and Paus, R. 2001. A comprehensive guide for the accurate classification of murine hair follicles in distinct hair cycle stages. *J. Invest. Dermatol.* **117:** 3–15.

Nair, M., Teng, A., Bilanchone, V., Agrawal, A., Li, B., and Dai, X. 2006. Ovol1 regulates the growth arrest of embryonic epidermal progenitor cells and represses c-myc transcription. *J. Cell Biol.* **173:** 253–264.

Nguyen, B.C., Lefort, K., Mandinova, A., Antonini, D., Devgan, V., Della Gatta, G., Koster, M.I., Zhang, Z., Wang, J., Tommasi di Vignano, A., et al. 2006a. Cross-regulation between Notch and p63 in keratinocyte commitment to differentiation. *Genes Dev.* **20:** 1028–1042.

Nguyen, H., Rendl, M., and Fuchs, E. 2006b. Tcf3 governs stem cell features and represses cell fate determination in skin. *Cell* **127:** 171–183.

Niemann, C., Owens, D.M., Hulsken, J., Birchmeier, W., and Watt, F.M. 2002. Expression of ΔNLef1 in mouse epidermis results in differentiation of hair follicles into squamous epidermal cysts and formation of skin tumours. *Development* **129:** 95–109.

Nowak, J.A., Polak, L., Pasolli, H.A., and Fuchs, E. 2008. Hair follicle stem cells are specified and function in early skin morphogenesis. *Cell Stem Cell* **3:** 33–43.

Orkin, S.H. and Zon, L.I. 2008. SnapShot: Hematopoiesis. *Cell* **132:** 712.

Oro, A.E. and Higgins, K. 2003. Hair cycle regulation of Hedgehog signal reception. *Dev. Biol.* **255:** 238–248.

Oro, A.E., Higgins, K.M., Hu, Z., Bonifas, J.M., Epstein, Jr., E.H., and Scott, M.P. 1997. Basal cell carcinomas in mice overexpressing Sonic hedgehog. *Science* **276:** 817–821.

Oshima, H., Rochat, A., Kedzia, C., Kobayashi, K., and Barrandon, Y. 2001. Morphogenesis and renewal of hair follicles from adult multipotent stem cells. *Cell* **104:** 233–245.

Owens, D.M. and Watt, F.M. 2003. Contribution of stem cells and differentiated cells to epidermal tumours. *Nat. Rev. Cancer* **3:** 444–451.

Pan, Y., Lin, M.H., Tian, X., Cheng, H.T., Gridley, T., Shen, J., and Kopan, R. 2004. γ-Secretase functions through Notch signaling to maintain skin appendages but is not required for their patterning or initial morphogenesis. *Dev. Cell* **7:** 731–743.

Perez-Moreno, M. and Fuchs, E. 2006. Catenins: Keeping cells from getting their signals crossed. *Dev. Cell* **11:** 601–612.

Plikus, M.V., Mayer, J.A., de la Cruz, D., Baker, R.E., Maini, P.K., Maxson, R., and Chuong, C.M. 2008. Cyclic dermal BMP signalling regulates stem cell activation during hair regeneration. *Nature* **451:** 340–344.

Potten, C.S. 1974. The epidermal proliferative unit: The possible role of the central basal cell. *Cell Tissue Kinet.* **7:** 77–88.

Potten, C.S. 2004. Keratinocyte stem cells, label-retaining cells and possible genome protection mechanisms. *J. Investig. Dermatol. Symp. Proc.* **9:** 183–195.

Powell, S.K., Williams, C.C., Nomizu, M., Yamada, Y., and Kleinman, H.K. 1998. Laminin-like proteins are differentially regulated during cerebellar development and stimulate granule cell neurite outgrowth in vitro. *J. Neurosci. Res.* **54:** 233–247.

Pummila, M., Fliniaux, I., Jaatinen, R., James, M.J., Laurikkala, J., Schneider, P., Thesleff, I., and Mikkola, M.L. 2007. Ectodysplasin has a dual role in ectodermal organogenesis: Inhibition of Bmp activity and induction of Shh expression. *Development* **134:** 117–125.

Raghavan, S., Vaezi, A., and Fuchs, E. 2003. A role for αβ1 integrins in focal adhesion function and polarized cytoskeletal dynamics. *Dev. Cell* **5:** 415–427.

Rangarajan, A., Talora, C., Okuyama, R., Nicolas, M., Mammucari, C., Oh, H., Aster, J.C., Krishna, S., Metzger, D., Chambon, P., et al. 2001. Notch signaling is a direct determinant of keratinocyte growth arrest and entry into differentiation. *EMBO J.* **20:** 3427–3436.

Rendl, M., Lewis, L., and Fuchs, E. 2005. Molecular dissection of mesenchymal-epithelial interactions in the hair follicle. *PLoS Biol.* **3:** 1910–1924.

Rendl, M., Polak, L., and Fuchs, E. 2008. BMP signaling in dermal papilla cells is required for hair follicle formation. *Genes Dev.* **22:** 543–557.

Reynolds, A.B. 2007. p120-catenin: Past and present. *Biochim. Biophys. Acta* **1773:** 2–7.

Rhee, H., Polak, L., and Fuchs, E. 2006. Lhx2 maintains stem cells character in hair follicles. *Science* **312:** 1946–1949.

Schmidt-Ullrich, R. and Paus, R. 2005. Molecular principles of hair follicle induction and morphogenesis. *Bioessays* **27:** 247–261.

Schober, M., Raghavan, S., Nikolova, M., Polak, L., Pasolli, H.A., Beggs, H.E., Reichardt, L.F., and Fuchs, E. 2007. Focal adhesion kinase modulates tension signaling to control actin and focal adhesion dynamics. *J. Cell Biol.* **176:** 667–680.

Scholl, F.A., Dumesic, P.A., Barragan, D.I., Harada, K., Bissonauth, V., Charron, J., and Khavari, P.A. 2007. Mek1/2 MAPK kinases are essential for mammalian development, homeostasis, and Raf-induced hyperplasia. *Dev. Cell* **12:** 615–629.

Scott, J.A. and Yap, A.S. 2006. Cinderella no longer: α-Catenin steps out of cadherin's shadow. *J. Cell Sci.* **119:** 4599–4605.

Segre, J.A., Bauer, C., and Fuchs, E. 1999. Klf4 is a transcription factor required for establishing the barrier function of the skin. *Nat. Genet.* **22:** 356–360.

Senoo, M., Pinto, F., Crum, C.P., and McKeon, F. 2007. p63 is essential for the proliferative potential of stem cells in stratified epithelia. *Cell* **129:** 523–536.

Sharov, A.A., Sharova, T.Y., Mardaryev, A.N., Tommasi di Vignano, A., Atoyan, R., Weiner, L., Yang, S., Brissette, J.L., Dotto, G.P., and Botchkarev, V.A. 2006. Bone morphogenetic protein signaling regulates the size of hair follicles and modulates the expression of cell cycle-associated genes. *Proc. Natl. Acad. Sci.* **103:** 18166–18171.

Shi, Y. and Massagué, J. 2003. Mechanisms of TGF-β signaling from cell membrane to the nucleus. *Cell* **113:** 685–700.

Stenn, K.S. and Paus, R. 2001. Controls of hair follicle cycling. *Physiol. Rev.* **81:** 449–494.

St-Jacques, B., Dassule, H.R., Karavanova, I., Botchkarev, V.A., Li, J., Danielian, P.S., McMahon, J.A., Lewis, P.M., Paus, R., and McMahon, A.P. 1998. Sonic hedgehog signaling is essential for hair development. *Curr. Biol.* **8:** 1058–1068.

Taylor, G., Lehrer, M.S., Jensen, P.J., Sun, T.T., and Lavker, R.M. 2000. Involvement of follicular stem cells in forming not only the follicle but also the epidermis. *Cell* **102:** 451–461.

Trempus, C.S., Morris, R.J., Bortner, C.D., Cotsarelis, G., Faircloth, R.S., Reece, J.M., and Tennant, R.W. 2003. Enrichment for living murine keratinocytes from the hair follicle bulge with the cell surface marker CD34. *J. Invest. Dermatol.* **120:** 501–511.

Truong, A.B., Kretz, M., Ridky, T.W., Kimmel, R., and Khavari, P.A. 2006. p63 regulates proliferation and differentiation of developmentally mature keratinocytes. *Genes Dev.* **20:** 3185–3197.

Tumbar, T., Guasch, G., Greco, V., Blanpain, C., Lowry, W.E., Rendl, M., and Fuchs, E. 2004. Defining the epithelial stem cell niche in skin. *Science* **303:** 359–363.

van Genderen, C., Okamura, R.M., Farinas, I., Quo, R.G., Parslow, T.G., Bruhn, L., and Grosschedl, R. 1994. Development of several organs that require inductive epithelial-mesenchymal interactions is impaired in LEF-1-deficient mice. *Genes Dev.* **8:** 2691–2703.

Van Mater, D., Kolligs, F.T., Dlugosz, A.A., and Fearon, E.R. 2003. Transient activation of β-catenin signaling in cutaneous keratinocytes is sufficient to trigger the active growth phase of the hair cycle in mice. *Genes Dev.* **17:** 1219–1224.

Vauclair, S., Nicolas, M., Barrandon, Y., and Radtke, F. 2005. Notch1 is essential for postnatal hair follicle development and homeostasis. *Dev. Biol.* **284:** 184–193.

Vidal, V.P., Chaboissier, M.C., Lutzkendorf, S., Cotsarelis, G., Mill, P., Hui, C.C., Ortonne, N., Ortonne, J.P., and Schedl, A. 2005. Sox9 is essential for outer root sheath differentiation and the formation of the hair stem cell compartment. *Curr. Biol.* **15:** 1340–1351.

Waghmare, S.K., Bansal, R., Lee, J., Zhang, Y.V., McDermitt, D.J., and Tumbar, T. 2008. Quantitative proliferation dynamics and random chromosome segregation of hair follicle stem cells. *EMBO J.* **27:** 1309–1320.

Waikel, R.L., Kawachi, Y., Waikel, P.A., Wang, X.J., and Roop, D.R. 2001. Deregulated expression of c-Myc depletes epidermal stem cells. *Nat. Genet.* **28:** 165–168.

Wang, X., Bolotin, D., Chu, D.H., Polak, L., Williams, T., and Fuchs, E. 2006. AP-2α: A regulator of EGF receptor signaling and proliferation in skin epidermis. *J. Cell Biol.* **172:** 409–421.

Watt, F.M. 2002. Role of integrins in regulating epidermal adhesion, growth and differentiation. *EMBO J.* **21:** 3919–3926.

Wu, X., Tu, X., Joeng, K.S., Hilton, M.J., Williams, D.A., and Long, F. 2008. Rac1 activation controls nuclear localization of β-catenin during canonical Wnt signaling. *Cell* **133:** 340–353.

Wu, X., Quondamatteo, F., Lefever, T., Czuchra, A., Meyer, H., Chrostek, A., Paus, R., Langbein, L., and Brakebusch, C. 2006. Cdc42 controls progenitor cell differentiation and β-catenin turnover in skin. *Genes Dev.* **20:** 571–585.

Yamamoto, N., Tanigaki, K., Han, H., Hiai, H., and Honjo, T.

2003. Notch/RBP-J signaling regulates epidermis/hair fate determination of hair follicular stem cells. *Curr. Biol.* **13:** 333–338.

Yang, A., Schweitzer, R., Sun, D., Kaghad, M., Walker, N., Bronson, R.T., Tabin, C., Sharpe, A., Caput, D., Crum, C., and McKeon, F. 1999. p63 is essential for regenerative proliferation in limb, craniofacial and epithelial development. *Nature* **398:** 714–718.

Yang, L., Mao, C., Teng, Y., Li, W., Zhang, J., Cheng, X., Li, X., Han, X., Xia, Z., Deng, H., and Yang, X. 2005. Targeted disruption of Smad4 in mouse epidermis results in failure of hair follicle cycling and formation of skin tumors. *Cancer Res.* **65:** 8671–8678.

Yi, R., Poy, M.N., Stoffel, M., and Fuchs, E. 2008. A skin microRNA promotes differentiation by repressing 'stemness.' *Nature* **452:** 225–229.

Yuhki, M., Yamada, M., Kawano, M., Iwasato, T., Itohara, S., Yoshida, H., Ogawa, M., and Mishina, Y. 2004. BMPR1A signaling is necessary for hair follicle cycling and hair shaft differentiation in mice. *Development* **131:** 1825–1833.

Zenz, R. and Wagner, E.F. 2006. Jun signalling in the epidermis: From developmental defects to psoriasis and skin tumors. *Int. J. Biochem. Cell Biol.* **38:** 1043–1049.

Zhang, J., He, X.C., Tong, W.G., Johnson, T., Wiedemann, L.M., Mishina, Y., Feng, J.Q., and Li, L. 2006a. Bone morphogenetic protein signaling inhibits hair follicle anagen induction by restricting epithelial stem/progenitor cell activation and expansion. *Stem Cells* **24:** 2826–2839.

Zhang, J., Grindley, J.C., Yin, T., Jayasinghe, S., He, X.C., Ross, J.T., Haug, J.S., Rupp, D., Porter-Westpfahl, K.S., Wiedemann, L.M., Wu, H., and Li, L. 2006b. PTEN maintains haematopoietic stem cells and acts in lineage choice and leukaemia prevention. *Nature* **441:** 518–522.

Zhou, P., Byrne, C., Jacobs, J., and Fuchs, E. 1995. Lymphoid enhancer factor 1 directs hair follicle patterning and epithelial cell fate. *Genes Dev.* **9:** 700–713.

Very Long-term Self-renewal of Small Intestine, Colon, and Hair Follicles from Cycling $Lgr5^{+ve}$ Stem Cells

N. Barker,* J.H. van Es,* V. Jaks,[†] M. Kasper,[†] H. Snippert,* R. Toftgård,[†]
and H. Clevers*

*Hubrecht Institute, Uppsalalaan 8, 3584CT Utrecht, The Netherlands; [†]Karolinska Institutet, Department of
Biosciences and Nutrition, Novum, SE-141 57, Huddinge, Sweden

The intestinal epithelium and the hair follicle represent examples of rapidly self-renewing tissue in adult mammals. We have recently identified a novel stem cell gene $Lgr5$ expressed in multiple adult tissues. At the bottoms of crypts in small intestine and colon as well as in hair follicles, $Lgr5$ marks cycling cells with stem cell properties (Barker et al. 2007; Jaks et al. 2008). Using an inducible $Lgr5$-Cre knockin allele in conjunction with the $Rosa26$-$LacZ$ Cre reporter strain, long-term lineage-tracing experiments were performed in adult mice. The $Lgr5^{+ve}$ crypt-based cell generated all epithelial lineages during a 14-month period, implying that it represents the stem cell of the small intestine and colon. Similarly, lineage tracing during a 14-month period revealed that $Lgr5^{+ve}$ cells located in the bulge of the hair follicle sustained multiple rounds of hair growth. These observations support the counterintuitive notion that $Lgr5^{+ve}$ cells are actively cycling, yet represent long-term stem cells of these adult, self-renewing tissues.

$LGR5^+$ STEM CELLS IN SMALL INTESTINE

The self-renewing epithelium of the murine small intestine is organized into crypts and villi (Gregorieff and Clevers 2005). In the mouse, it turns over every 3–5 days. A massive rate of cell production occurs in the crypts and is balanced by apoptosis at the tips of the villi. The analysis of chimeric mice and of mutagen-induced somatic clones (Winton and Ponder 1990; Bjerknes and Cheng 1999) and the study of regeneration upon injury have allowed an operational definition of stem cell characteristics. Self-renewing stem cells cycle steadily to produce the rapidly proliferating transit-amplifying (TA) cells capable of differentiating toward all lineages. The estimated number of stem cells is between four and six per crypt (Bjerknes and Cheng 1999). Three differentiated cell types (enterocytes, goblet cells, and enteroendocrine cells) form from TA cells at the crypt–villus junction. These cells continue their migration in coherent bands along the crypt–villus axis. Each villus receives cells from multiple different crypts. The fourth major differentiated cell type, the Paneth cell, resides at the bottom of the crypt.

We have previously described the $Wnt/Tcf4$ target gene program in colorectal cancer cells. This program is physiologically expressed in intestinal crypts (van de Wetering et al. 2002; van der Flier et al. 2007). Because Wnt signals thus constitute the major driving force behind the biology of the crypt (Korinek et al. 1998), we hypothesized that some $Wnt/Tcf4(Tcf7l2)$ target genes may be specifically expressed in the stem cells. Indeed, the $Lgr5/Gpr49$ gene was expressed in a unique fashion at crypt bottoms. This Wnt target gene was expressed in the crypts, but not the villi, of mouse small intestine. In situ hybridization revealed expression in a limited number of cells located at all crypt bottoms as well as in adenomas in the small intestine of an APC^{min} mouse. This expression pattern clearly differed from that obtained with other Wnt target genes.

More than 30 years ago, Leblond and Cheng (1974) noted the presence of cycling cells between the Paneth cells and coined the name "Crypt Base Columnar" (CBC) cells. Cheng and Bjerknes (Bjerknes and Cheng 1981; Barker et al. 2007) and Gordon and colleagues (Stappenbeck et al. 2003) have proposed that these cells may harbor stem cell activity. The $Lgr5$ gene appeared to mark these cycling CBC cells, interspersed between Paneth cells (Barker et al. 2007). The $Lgr5$ gene encodes an orphan G-protein-coupled receptor (GPCR), characterized by a large leucine-rich extracellular domain (Hsu et al. 1998).

Heterozygous $Lgr5$-$LacZ$ knockin mice allowed us to detail the expression of $Lgr5$. Before birth, a dynamic and broad expression pattern was observed (N. Barker et al., in prep.). Around birth, $Lgr5$ expression was extinguished in most tissues. Expression in adult mice was restricted to rare, scattered cells in the eye, brain, hair follicle, mammary gland, stomach, adrenal gland, and intestinal tract. $Lgr5$-$LacZ$ expression was observed in slender CBC cells at the bottom of small intestinal and colon crypts. CBC cells typically expressed the Ki67 cell cycle marker. Bromodeoxyuridine (BrdU) labeling allowed us to define the average cycling time of CBC cells to be on the order of 1 day (Barker et al. 2007).

To document the potential "stemness" of $Lgr5^{+ve}$ CBC cells, we generated another knockin allele. An enhanced green fluorescent protein–internal ribosomal entry site (EGFP-IRES)-CreERT2 cassette was integrated at the first ATG codon of the $Lgr5$ gene (Fig. 1). GFP expression was specifically seen in CBC cells at crypt bottoms of the small intestine of these mice (Fig. 2, left panel). Typically, the GFP+ CBC cells were relatively broad at their base and contained a flat, wedge-shaped nucleus. In our original experiments (Barker et al. 2007), we crossed the $EGFP$-$IRES$-$CreERT2$ knockin allele with the Cre-activatable $Rosa26$-$LacZ$ reporter (Fig. 1) (Soriano 1999) and followed these mice during a 2-month period. Injection of

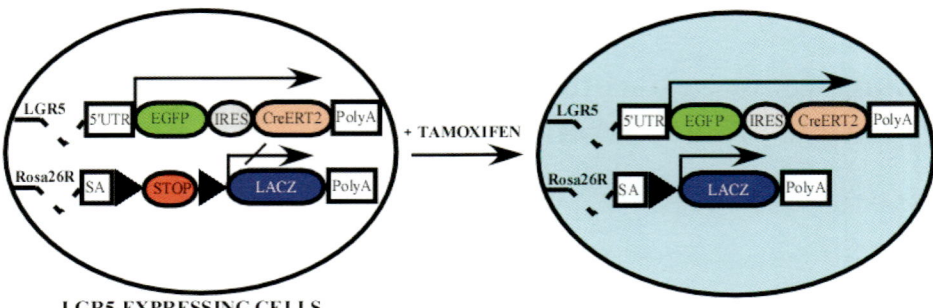

Figure 1. Expression of EGFP and CreERT2 from a single bicistronic message by gene knockin into the first exon of *Lgr5*. The resulting allele has been named *EGFP-IRES-CreERT2* and was crossed with *Rosa26-LacZ* mice (Soriano 1999). Injection of tamoxifen activates the Cre fusion protein, resulting in activation of the *LacZ* reporter (*right*).

tamoxifen activates the CreERT2 fusion enzyme in *Lgr5*-expressing cells. Cre-mediated excision of the roadblock sequence in the *Rosa26-LacZ* reporter should then irreversibly mark *Lgr5*[+ve] cells. Moreover, although potential progeny of these cells will no longer express GFP, the activated *LacZ* reporter should act as a genetic mark, facilitating lineage tracing. *LacZ* expression was not observed in noninduced mice (not shown). When adult mice were subjected to a very low-dose tamoxifen pulse and were sacrificed at 1, 5, 12, 35, and 60 days postinduction, the following observations were made. One day postinduction, occasional CBC cells in the small intestine and colon expressed *LacZ*. Parallel ribbons of cells emanated from the crypt bottoms and ran up the side of adjacent villi at later time points. The numbers of these ribbons did not significantly change during a 2-month period (Barker et al. 2007). Importantly, double-labeling of 60-day-induced intestine revealed the presence of goblet cells, Paneth cells, and enteroendocrine cells in the *LacZ*-stained clones, underscoring the multipotency of the stem cells.

To determine whether the *Lgr5*[+ve] CBC cells constitute a stem cell pool that sustains self-renewal over the life span of a mouse, we decided to follow induced *EGFP-IRES-CreERT2* × *Rosa26-LacZ* mice for 14 months. *LacZ*

staining of small intestines of these mice revealed that the numbers of clonal blue ribbons had remained essentially unchanged (Fig. 3). Histological analysis revealed that—like the situation at earlier time points—all cell types of the epithelium were represented in the blue clones, corroborating the multipotency of the cells. These observations implied that the *Lgr5*[+ve] CBC cells represent long-term stem cells of the small intestine.

LGR5[+] STEM CELLS IN THE COLON

The colon epithelium contains crypts, but it has a flat surface rather than carrying villi. The epithelium comprises two major differentiated cell types: the absorptive colonocytes and the goblet cells (Gregorieff and Clevers 2005). Before the discovery of the *Lgr5* marker, no stem cells had been identified in the colon. Analysis of the colon yielded essentially identical observations to what was seen in the small intestine. *Lgr5*[+ve] cells with an appearance very similar to that of the CBC cells of the small intestine were observed in the *EGFP-IRES-CreERT2* mice (Fig. 2, middle panel) (Barker et al. 2007). When we crossed the *EGFP-IRES-CreERT2* knockin allele with the Cre-activatable *Rosa26-LacZ* reporter (see

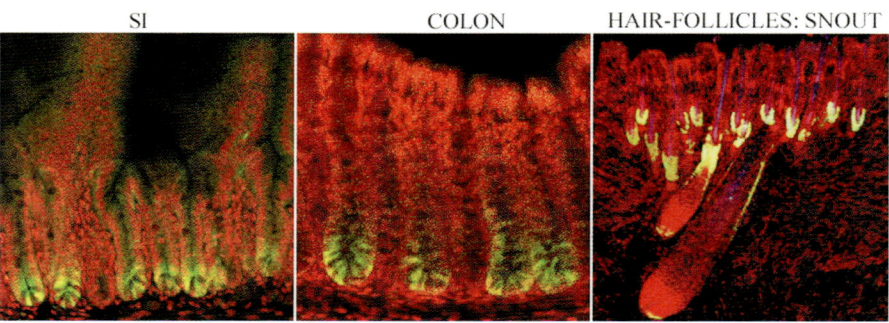

Figure 2. EGFP expression in an LGR5-EGFP-IRES-CreERT2 knockin mouse in the intestinal tract and hair follicle. Confocal GFP imaging was counterstained with the red DNA dye ToPro-3. (*Left*) *Lgr5* expression is restricted to the six to eight slender cells sandwiched between the Paneth cells at the crypt base of the small intestine. (*Middle*) Confocal imaging of EGFP expression in the colon confirms that *Lgr5* expression is restricted to a few cells located at the crypt base. (*Right*) Confocal imaging of EGFP expression in the skin reveals *Lgr5* expression in the bulge and the secondary germ of the telogen hair follicle, as well as in the lower outer root sheath (ORS) of the anagen hair follicle.

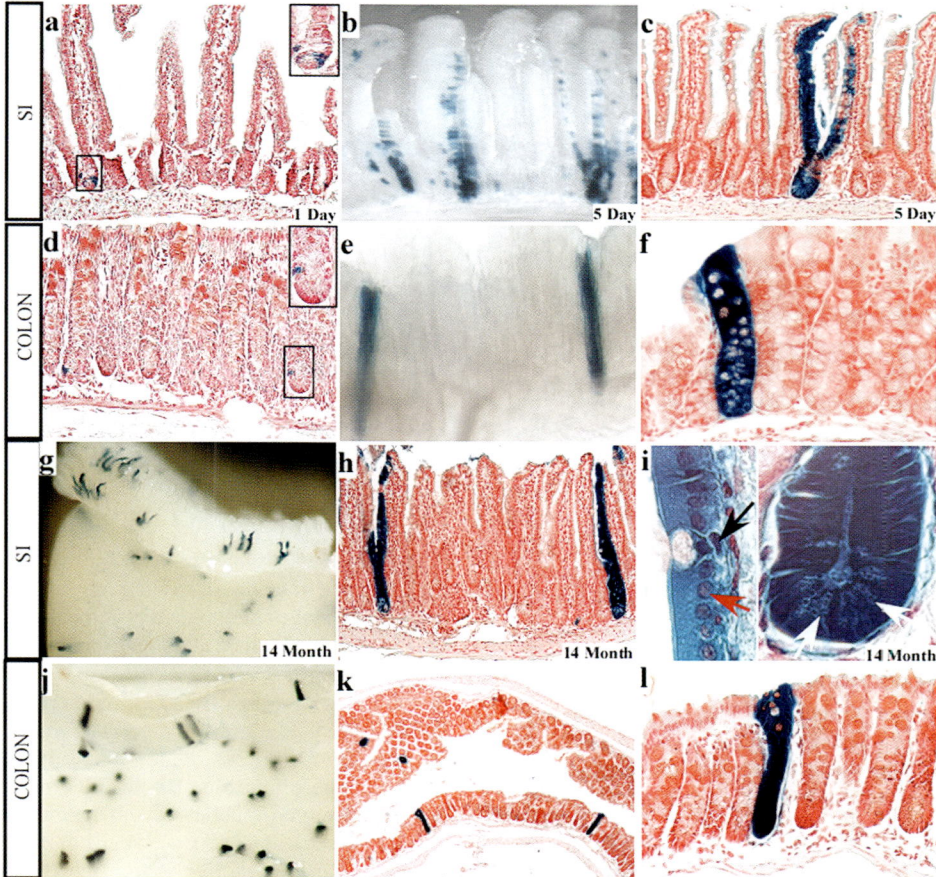

Figure 3. Very long-term lineage tracing in the small intestine and colon. (*a*) LGR5-EGFP-IRES-CreERT2 knockin mouse crossed with *Rosa26-LacZ* reporter mice 12 hours after tamoxifen injection; (*b*) whole-mount analysis at day 5 postinduction; (*c*) histological analysis of *b*; (*d–f*) same as *a–c* for colon epithelium; (*g–i*) whole-mount and histological analysis of tamoxifen-induced *LacZ* marking of small intestine after 14 months; (*j–l*) whole-mount and histological analysis of tamoxifen-induced *LacZ* marking of colon after 14 months.

Fig. 1) and followed these mice during a 2-month period, the *Lgr5*[+ve] cells yielded blue clones emanating from the bottoms of the colon crypts (Fig. 3) (Barker et al. 2007). These clones contained colonocytes as well as goblet cells and essentially remained unchanged during the 60 days of chase. One significant difference with the situation in the small intestine involved the kinetics of clone formation. At 5 days, blue staining in most crypts was still restricted to the bottom and entirely blue crypts were only rarely observed, implying that the colon stem cells were more often quiescent than their small intestinal counterparts.

To determine whether the *Lgr5*[+ve] cells of the colon epithelium constitute a stem cell pool that sustains self-renewal during the life span of a mouse, we analyzed the colon of induced *EGFP-IRES-CreERT2* × *Rosa26-LacZ* mice after 14 months. *LacZ* staining revealed that the numbers of clonal blue ribbons had remained essentially unchanged (Fig. 3). Histological analysis revealed that—like the situation at earlier time points—all cell types of the epithelium were represented in the blue clones, corroborating the multipotency of the cells. This is evidenced by the fact that no cells of labeled crypts were *LacZ*-neg-

ative (Fig. 3f). These observations implied that the *Lgr5*[+ve] cells represent long-term stem cells of the colon.

LGR5[+] STEM CELLS IN HAIR FOLLICLES

The skin is the largest organ in mammals. The interfollicular epidermis (IFE) is a stratified, keratinized epithelium that is constantly self-renewing and essentially produces a single differentiated cell type, the keratinocyte. In contrast, the hair follicles undergo cycles consisting of growth, involution, and resting phases (Alonso and Fuchs 2006). The stem cells of the hair follicle are thought to reside in the hair follicle bulge and have been defined by expression of the CD34 cell surface marker, expression of cytokeratin 15 (Cotsarelis et al.1990; Trempus et al. 2003; Blanpain et al. 2004; Morris et al. 2004), and retention of either DNA or histone labels during long time periods (Braun et al. 2003; Tumbar et al. 2004). Cells with stem cell properties have been isolated from other areas of the hair follicles (Ito et al. 2004; Nijhof et al. 2006). Of particular interest, at certain stages of its cycle, the lower parts of the murine vibrissae follicle contain cells that can reconstitute a complete hair follicle upon transplantation (Oshima et al. 2001). Thus, the bulge might not be the only reservoir of stem cells. Indeed, studying $Lgr5^{+ve}$ cells in the hair follicle, we have recently reported evidence for the notion that a $Lgr5^{+ve}$ stem cell pool exists that can sustain hair follicles for up to 6 months (Jaks et al. 2008). Specifically, we found that an actively cycling $Lgr5^{+ve}$ cell population exists in the bulge and the secondary germ of the telogen hair follicle, as well as in the lower outer root sheath (ORS) of the anagen hair follicle (see Fig. 2, right panel) (Jaks et al. 2008). Analysis of sorted cells in colony-forming and transplantation assays revealed that the $Lgr5^{+ve}$ cells comprise an actively proliferating and multi-potent stem cell population able to give rise to new hair follicles. We provided histological images of lineage tracing in induced *EGFP-IRES-CreERT2* × *Rosa26-LacZ* mice to demonstrate that $Lgr5^+$ cells maintained all cell lineages of the hair follicle during 6 months.

To determine whether the $Lgr5^{+ve}$ cells of the hair follicles constitute a stem cell pool that sustains self-renewal during the life span of a mouse, we analyzed the hair follicles of induced *EGFP-IRES-CreERT2* × *Rosa26-LacZ* mice after 14 months. *LacZ* staining revealed that the numbers of blue follicles had remained essentially unchanged (Fig. 4). Histological analysis revealed that—like the situation at earlier time points—all cell types of the follicle were represented in the blue clones, corroborating the multipotency of the cells (not shown). These observations implied that the $Lgr5^{+ve}$ cells represent long-term stem cells of the hair follicle.

METHODS

Mice. LGR5-EGFP-IRES-CreERT2 mice were generated by homologous recombination in embryonic stem cells targeting an EGFP-IRES-CreERT2 cassette to the ATG of LGR5. *Rosa26-LacZ* Cre reporter mice were obtained from Jackson Labs.

Tamoxifen induction. Mice of >8 weeks were injected intraperitoneally with 200 µl of tamoxifen in sunflower oil at 10 mg/ml.

BrdU injection. Mice were injected intraperitoneally at 4-hour intervals with 200 µl of a BrdU solution in phosphate-buffered saline (PBS) at 5 mg/ml.

Tissue preparation for immunohistochemistry, GFP confocal microscopy, and LacZ analysis. These were performed as described previously (Barker et al. 2007; Jaks et al. 2008).

CONCLUSIONS

Our previous observations have provided evidence for the existence of a pool of cycling $Lgr5^{+ve}$ stem cells in the epithelium of the small intestine and colon and in hair follicles. Of note, the first two organs are of endodermal origin, whereas the skin derives from ectoderm, implying that $Lgr5$ represents a stem cell marker across germ layers. The $Lgr5^{+ve}$ cells are generally not quiescent, but they are rapidly cycling, as evidenced by the expression of Ki67 and phosphohistone H3, the incorporation of BrdU, and the observed kinetics of self-renewal (Barker et al. 2007; Jaks et al. 2008). $Lgr5^{+ve}$ cells of the small intestine appear to be more actively dividing than their colonic counterparts, likely reflecting differences in the rate of epithelial turnover between the two organs. We have not analyzed cell cycle kinetics of the $Lgr5^{+ve}$ cells in the hair

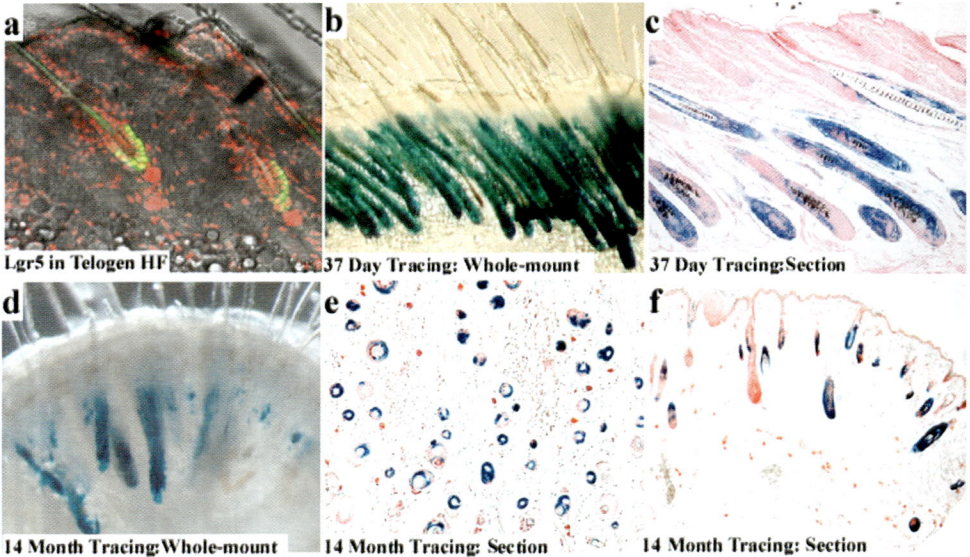

Figure 4. Very long-term lineage tracing in hair follicles. (*a*) LGR5-EGFP-IRES-CreERT2 knockin mouse. GFP expression in a telogen hair follicle to indicate where tracing starts; (*b*) entirely labeled hair follicles 37 days after tamoxifen induction; (*c*) histological analysis of *b*; (*d–f*) the same as in *b* and *c*, 14 months after tracing.

follicle. It appears somewhat counterintuitive that stem cells are actively dividing. This is, however, not unprecedented. Germ stem cells in the *Drosophila* testis and ovary of the fly, arguably the best-studied adult stem cells in animals, cycle throughout the lifetime of the adult fly (Ohlstein et al. 2004). Similarly, a recent elegant study based on a very similar tracing strategy as exploited in the current study, demonstrated that adult stem cells of mammalian skin are continuously cycling (Clayton et al. 2007). The observations in the current study imply that *Lgr5*[+ve] cells in the small intestine accomplish the surprising feat of predictedly completing more than 400 cell cycles during the duration of the experiment, without deleterious effects on their telomeres, their genomic content, or their capacity to serve as stem cells to the epithelium. In essence, we believe that the *Lgr5*[+ve] cells of the three organs studied here constitute stem cell pools that persist for the lifetime of an animal, given that a mouse rarely lives beyond 1.5–2 years of age. If these observations can be extrapolated to humans, our CBC cells complete an astounding number of cell cycles.

It is rather unusual that adult stem cells can be defined based on the expression of a single gene. This phenomenon may not be restricted to the intestine and hair follicle, because we observe highly restricted *Lgr5* expression in several other tissues. Although patterns of proliferation in stomach glands have indicated that the epithelial stem cells reside at the isthmus, halfway between the gland base and epithelial surface (Bjerknes and Cheng 2002), we find *Lgr5* expressed at gland bottoms. Ongoing lineage-tracing experiments imply that entire glands derive from these cells (N. Barker and H. Clevers, unpubl.). In the mammary gland, stem cells reside in the basal epithelial layer (Sleeman et al. 2007), where we observe *Lgr5* expression (not shown). *Lgr5* may thus represent a more general marker of adult stem cells. Two closely related genes exist in the mammalian genomes, *Lgr4* and *Lgr6*. Our ongoing studies using similar knockin strategies indicate that these two homologs are expressed in unique, yet overlapping, expression patterns when compared to *Lgr5*. At several sites, we believe that the two homologs mark stem cells, much like *Lgr5* (A. Haegebarth et al., unpubl.). The function of the three *Lgr* genes in stem cells remains to be determined, but it is likely redundant among the three homologs.

REFERENCES

Alonso, L. and Fuchs, E. 2006. The hair cycle. *J. Cell Sci.* **119:** 391–393.

Blanpain, C., Lowry, W.E., Geoghegan, A., Polak, L., and Fuchs, E. 2004. Self-renewal, multipotency, and the existence of two cell populations within an epithelial stem cell niche. *Cell* **118:** 635–648.

Barker, N., van Es, J.H., Kuipers, J., Kujala, P., van den Born, M., Cozijnsen, M., Haegebarth, A., Korving, J., Begthel, H., Peters, P.J., et al. 2007. Identification of stem cells in small intestine and colon by the marker gene *Lgr5*. *Nature* **449:** 1003–1007.

Bjerknes, M. and Cheng, H. 1981. The stem-cell zone of the small intestinal epithelium. III. Evidence from columnar, enteroendocrine, and mucous cells in the adult mouse. *Am. J. Anat.* **160:** 77–91.

Bjerknes, M. and Cheng, H. 1999. Clonal analysis of mouse

intestinal epithelial progenitors. *Gastroenterology* **116:** 7–14.

Bjerknes, M. and Cheng, H. 2002. Multipotential stem cells in adult mouse gastric epithelium. *Am. J. Physiol. Gastrointest. Liver Physiol.* **283:** G767–G777.

Braun, K.M., Niemann, C., Jensen, U.B., Sundberg, J.P., Silva-Vargas, V., and Watt, F.M. 2003. Manipulation of stem cell proliferation and lineage commitment: Visualisation of label-retaining cells in wholemounts of mouse epidermis. *Development* **130:** 5241–5255.

Cheng, H. and Leblond, C.P. 1974. Origin, differentiation and renewal of the four main epithelial cell types in the mouse small intestine. V. Unitarian Theory of the origin of the four epithelial cell types. *Am. J. Anat.* **141:** 537–561.

Clayton, E., Doupé, D.P., Klein, A.M., Winton, D.J., Simons, B.D., and Jones, P.H. 2007. A single type of progenitor cell maintains normal epidermis. *Nature* **446:** 185–189.

Cotsarelis, G., Sun, T.T., and Lavker, R.M. 1990. Label-retaining cells reside in the bulge area of pilosebaceous unit: Implications for follicular stem cells, hair cycle, and skin carcinogenesis. *Cell* **61:** 1329–1337.

Gregorieff, A. and Clevers, H. 2005. Wnt signaling in the intestinal epithelium: From endoderm to cancer. *Genes Dev.* **19:** 877–890.

Hsu, S.Y., Liang, S.G., and Hsueh, A.J. 1998. Characterization of two LGR genes homologous to gonadotropin and thyrotropin receptors with extracellular leucine-rich repeats and a G protein-coupled, seven-transmembrane region. *Mol. Endocrinol.* **12:** 1830–1845.

Ito, M., Kizawa, K., Hamada, K., and Cotsarelis, G. 2004. Hair follicle stem cells in the lower bulge form the secondary germ, a biochemically distinct but functionally equivalent progenitor cell population, at the termination of catagen. *Differentiation* **72:** 548–557.

Jaks, V., Barker, N., Kasper, N., van Es, J.H., Snippert, H., Clevers, H., and Toftgård, R. 2008. *Lgr5* marks cycling, yet long-lived, hair follicle stem cells. *Nature* (in press).

Korinek, V., Barker, N., Moerer, P., van Donselaar, E., Huls, G., Peters, P.J., and Clevers, H. 1998. Depletion of epithelial stem-cell compartments in the small intestine of mice lacking Tcf-4. *Nat. Genet.* **19:** 379–383.

Morris, R.J., Liu, Y., Marles, L., Yang, Z., Trempus, C., Li, S., Lin, J.S., Sawicki, J.A., and Cotsarelis, G. 2004. Capturing and profiling adult hair follicle stem cells. *Nat. Biotechnol.* **22:** 411–417.

Nijhof, J.G., Braun, K.M., Giangreco, A., van Pelt, C., Kawamoto, H., Boyd, R.L., Willemze, R., Mullenders, L.H., Watt, F.M., de Gruijl, F.R., et al. 2006. The cell-surface marker MTS24 identifies a novel population of follicular keratinocytes with characteristics of progenitor cells. *Development* **133:** 3027–3037.

Ohlstein, B., Kai, T., Decotto, E., and Spradling, A. 2004. The stem cell niche: Theme and variations. *Curr. Opin. Cell Biol.* **16:** 693–699.

Oshima, H., Rochat, A., Kedzia, C., Kobayashi, K., and Barrandon, Y. 2001. Morphogenesis and renewal of hair follicles from adult multipotent stem cells. *Cell* **104:** 233–245.

Sleeman, K.E., Kendrick, H., Robertson, D., Isacke, C.M., Ashworth, A., and Smalley, M.J. 2007. Dissociation of estrogen receptor expression and in vivo stem cell activity in the mammary gland. *J. Cell Biol.* **176:** 19–26.

Soriano, P. 1999. Generalized *lacZ* expression with the ROSA26 Cre reporter strain. *Nat. Genet.* **21:** 70–71.

Stappenbeck, T.S., Mills, J.C., and Gordon, J.I. 2003. Molecular features of adult mouse small intestinal epithelial progenitors. *Proc. Natl. Acad. Sci.* **100:** 1004–1009.

Trempus, C.S., Morris, R.J., Bortner, C.D., Cotsarelis, G., Faircloth, R.S., Reece, J.M., and Tennant, R.W. 2003. Enrichment for living murine keratinocytes from the hair follicle bulge with the cell surface marker CD34. *J. Invest. Dermatol.* **120:** 501–511.

Tumbar, T., Guasch, G., Greco, V., Blanpain, C., Lowry, W.E., Rendl, M., and Fuchs, E. 2004. Defining the epithelial stem cell niche in skin. *Science* **303:** 359–363.

van der Flier, L.G., Sabates-Bellver, J., Oving, I., Haegebarth, A.,

De Palo, M., Anti, M., van Gijn, M.E., Suijkerbuijk, S., van de Wetering, M., Marra, G., et al. 2007. The intestinal Wnt/TCF signature. *Gastroenterology* **132:** 628–632.

van de Wetering, M., Sancho, E., Verweij, C., de Lau, W., Oving, I., Hurlstone, A., van der Horn, K., Batlle, E., Coudreuse, D., Haramis, A.P., et al. 2002. The β-catenin/ TCF-4 complex imposes a crypt progenitor phenotype on colorectal cancer cells. *Cell* **111:** 241–250.

Winton, D.J. and Ponder, B.A. 1990. Stem-cell organization in mouse small intestine. *Proc. Biol. Sci.* **241:** 13–18.

The Heterogeneity of Adult Neural Stem Cells and the Emerging Complexity of Their Niche

A. Alvarez-Buylla,* M. Kohwi,[†] T.M. Nguyen,* and F.T. Merkle[‡]

*Department of Neurological Surgery and Institute for Regeneration Medicine, University of California, San Francisco, California 94143; [†]Department of Neuroscience, University of Oregon, Eugene, Oregon 97403; [‡]Department of Molecular and Cellular Biology, Harvard University, Cambridge, Massachusetts 02138

Neural stem cells persist in the adult mammalian brain in a neurogenic niche known as the subventricular zone (SVZ). SVZ neural stem cells (NSCs) can self-renew and are multipotent in culture. In rodents, adult NSCs correspond to SVZ astrocytes (type B cells) that are derived from radial glia, the NSCs of the embryonic and early postnatal brain. Type B cells generate transit-amplifying (type C) cells that give rise to young neurons (type A cells) and oligodendrocytes. Young neurons are born throughout the adult neurogenic niche and migrate tangentially through a complex network of chains that merge into the rostral migratory stream (RMS), a major pathway that leads into the olfactory bulb (OB). Within the OB, young neurons differentiate into multiple types of interneurons. The SVZ was thought to be limited to the lateral wall of the lateral ventricle, but recent work shows that the adult neurogenic niche is significantly more extensive and includes portions of the medial and dorsal walls of the lateral ventricle and the RMS itself. Furthermore, several recent studies explain why young OB neurons are generated in such an extensive region. Type B cells in different regions of the SVZ, although able to self-renew and generate both neurons and glial cells in vitro, are heterogeneous and committed to producing defined neuronal subtypes in vivo. The adult SVZ therefore provides a rich system to study not only neural replacement, but also the cellular and molecular mechanisms underlying regionalization and cell-fate specification.

The discovery of neural stem cells in the adult mammalian brain shattered the long-standing belief that neurogenesis is restricted to embryonic and early postnatal periods. The largest germinal region in the adult mammalian brain is the SVZ. The SVZ is classically described as a thin layer of proliferative cells lining the lateral wall of the lateral ventricle (LV) and separated from the ventricular lumen by a layer of ependymal cells (Smart and Leblond 1961; Altman 1969). The existence of NSCs in this region was first suggested by in vitro experiments in which SVZ cells were shown to self-renew and produce neurons, astrocytes, and oligodendrocytes (Reynolds and Weiss 1992; Morshead et al. 1994). In vivo, NSCs generate a large number of young neurons that migrate along the RMS to the OB where they replace multiple types of interneurons (Luskin 1993; Lois and Alvarez-Buylla 1994). NSCs in the SVZ also generate both parenchymal oligodendrocyte progenitors (OPCs) and myelinating oligodendrocytes, most of which migrate into the neighboring corpus callosum (Nait-Oumesmar et al. 1999; Picard-Riera et al. 2002; Menn et al. 2006).

Adult neurogenesis leads to the generation and replacement of specific types of neurons in restricted brain regions, including the OB and dentate gyrus of the hippocampus. Many processes of embryonic development are recapitulated during adult neurogenesis, such as neuronal differentiation, migration, maturation, and cell death. However, adult-born neurons confront an environment very different from those born in the developing brain. Adult-born neurons migrate through more complex and frequently extensive territories and must integrate into circuits that are already fully functional. Young neurons in the SVZ and RMS migrate along each other, forming long aggregates of cells called chains (Lois et al. 1996) and are able to migrate long distances in relatively short periods of time (Wichterle et al. 1997). Within 2–5 days from their time of birth in the SVZ, the majority of these young neurons have reached the OB. Once in the OB, young neurons move radially away from the RMS and begin their final differentiation and maturation, a process that takes 5–10 days (Petreanu and Alvarez-Buylla 2002). During this period, new neurons develop dendritic trees and synaptic spines and become functionally integrated into the OB circuitry (Carleton et al. 2003).

Initial studies in the neonatal rat brain suggested that new OB neurons originate from a restricted territory in the anterior SVZ, in a region close to the RMS (Luskin 1993; Lois and Alvarez-Buylla 1994). However, subsequent work uncovered an extensive network of chains of young neurons throughout most of the SVZ on the lateral wall of the LV (Doetsch and Alvarez-Buylla 1996), suggesting that migrating cells originate along the length of the lateral ventricular wall. Below, we review new experiments that suggest that the neurogenic SVZ covers regions of the lateral ventricular walls facing the pallium, subpallium, and septum, as well as the RMS. Furthermore, it was commonly assumed that adult NSCs would correspond to rare cells that divided infrequently but were highly plastic in their potential to generate multiple types of neurons and glial cells. We also review work that suggests that NSCs dedicated to producing specific cell types at specific developmental time points are restricted to different domains of the adult neurogenic niche. Thus, the adult neurogenic niche not only is more extensive than has been appreciated, but is also remarkably heterogeneous and complex.

These insights are of great biological interest and could have important therapeutic implications. For example, if we were to learn how mature neural circuits eliminate some neurons and recruit new ones, we might be able to develop effective strategies to replace neurons that are lost due to disease or trauma (Nottebohm 2004). Furthermore, the newfound heterogeneity of the neurogenic niche highlights the adult SVZ as a rich and unique experimental system to investigate the origin of neuronal diversity and to develop new strategies for cell replacement therapy.

ADULT NSCs ARE ASTROGLIAL CELLS THAT ORIGINATE FROM RADIAL GLIA

The primary precursors in the SVZ correspond to type B cells, a subpopulation of slowly dividing astroglial cells adjacent to a layer of ependymal cells (Doetsch et al. 1999; Laywell et al. 2000; Imura et al. 2003). Type B cells produce type C cells, a type of transit-amplifying cell that divides rapidly to produce young neurons, also known as neuroblasts or type A cells. It was thought that cells in this lineage (B→C→A) were separated from the ventricle by a layer of ependymal cells; therefore, the adult neurogenic region was referred to as an SVZ. However, recent work suggests that most, if not all, B cells actually contact the ventricle through small, specialized apical processes (Mirzadeh et al. 2008). These apical processes are tightly packed and surrounded by multiple ependymal cells, forming pinwheel-like structures that are only observed in the walls of the ventricle where neurogenesis continues throughout adult life. Therefore, it appears that the adult neurogenic niche is characterized by a proliferative ventricular zone (VZ) in addition to the SVZ. A similar neurogenic VZ has been described in the adult avian brain, where postmitotic ependymal cells are mixed with mitotic astroglial cells (Alvarez-Buylla et al. 1998).

This finding also illustrates the developmental history of adult NSCs. It is now clear that these astrocyte-like cells are derived from radial glial cells present in the embryonic and early postnatal brains (for review, see Merkle and Alvarez-Buylla 2006). Radial glia are now appreciated as the principal NSCs of the developing brain (Anthony et al. 2004; Miyata et al. 2004; Noctor et al. 2004). The cell bodies of radial glia form a VZ around the ventricles of the developing brain, which they contact via an apical process, much like B cells in the adult brain. Radial glia also contact the pial surface of the brain via a long, radially projecting basal process. These basal processes can be infected by Cre-expressing adenovirus. When this virus is injected into the brains of neonatal Cre-reporter mice, these basal processes become infected, and radial glia are specifically labeled in the VZ. This technique was used to demonstrate that adult NSCs are derived from radial glia (Merkle et al. 2004) and that the neurogenic niche is not restricted to the lateral wall of the LV, as discussed below.

THE ADULT GERMINAL LAYER IS MORE EXTENSIVE THAN PREVIOUSLY THOUGHT

The adult SVZ neurogenic niche was initially thought to be limited to the anterior lateral wall of the LV (Luskin 1993), but subsequent work shows that OB interneurons are born throughout the lateral ventricular wall (Doetsch and Alvarez-Buylla 1996). This conclusion has been confirmed by the labeling of primary progenitors in posterior regions of the lateral wall of the LV in neonates and adults

(Merkle et al. 2007). In vitro and in vivo studies in neonates and adults also show that progenitors in the RMS produce neurons (Gritti et al. 2002; Liu and Martin 2003; Hack et al. 2005). These progenitors continue to produce neurons for at least 4 weeks after being labeled with an adenovirus that expresses Cre under the GFAP (glial fibrillary acidic protein) promoter, suggesting that RMS progenitors are long-lived GFAP+ cells, much like SVZ stem cells (Merkle et al. 2007).

Perhaps more surprising is the recent demonstration that the dorsal wall of the LV facing the pallial VZ also generates interneurons destined for the OB (Kohwi et al. 2007; Merkle et al. 2007; Ventura and Goldman 2007). This pallial region extends laterally and caudally into the subcallosal zone (SCZ), a derivative of the pallial wall of the LV formed by the collapse of the LV due to an expanding hippocampus (Fig. 1). This region contains migratory neuroblasts (Seri et al. 2006) and produces neurons when its primary precursors are labeled in the neonate or adult (Merkle et al. 2007). Generation of OB interneurons from precursors of the dorsal wall of the LV is particularly unexpected given that during mouse development, the pallial VZ does not appear to generate its own inhibitory interneurons. Cortical interneurons are instead generated by the medial and caudal ganglionic eminences (Marin and Rubenstein 2003).

The adult germinal layer also includes regions of the anterior medial wall of the LV (SVZam). Embryonic and adult progenitors in this region can generate neurons in culture (Morshead et al. 1994) and produce OB interneurons after grafting (Kohwi et al. 2007) and labeling in vivo (Merkle et al. 2007). This medial neurogenic area appears to face a small region of the septum as well as the nucleus accumbens, a part of the ventral striatum. Therefore, the adult SVZ is a remarkably extensive germinal region covering most of the neuroepithelium-derived germinal compartments in the developing telencephalon (Fig. 1).

TYPE B CELLS GENERATE MULTIPLE TYPES OF INTERNEURONS IN THE OB

The rostral extension of the RMS forms the core of the OB. It is through this route that thousands of young neurons enter into the mouse OB and complete their tangential trajectory every day. The OB is a highly laminated structure; its neurons are organized into different layers based on their functions. Upon entering the bulb through its core, young neurons migrate radially into the granule and glomerular layers where they differentiate into granular and periglomerular local inhibitory interneurons (Kosaka et al. 1995; Carleton et al. 2003; Kohwi et al. 2005). These interneurons have important roles in modulating the activity of mitral and tufted cells, the primary projection neurons that relay sensory input from the olfactory epithelia directly to the cortex (Greer 1987; Shepherd and Greer 1998; Lledo et al. 2008).

Several subtypes of granular and periglomerular interneurons can be distinguished on the basis of morphology, connectivity, and expression of molecular markers. For example, within the granule cell layer, new neurons can be localized in the superficial or deep granule cell layers (Fig. 2) and can be further subdivided based on their expression

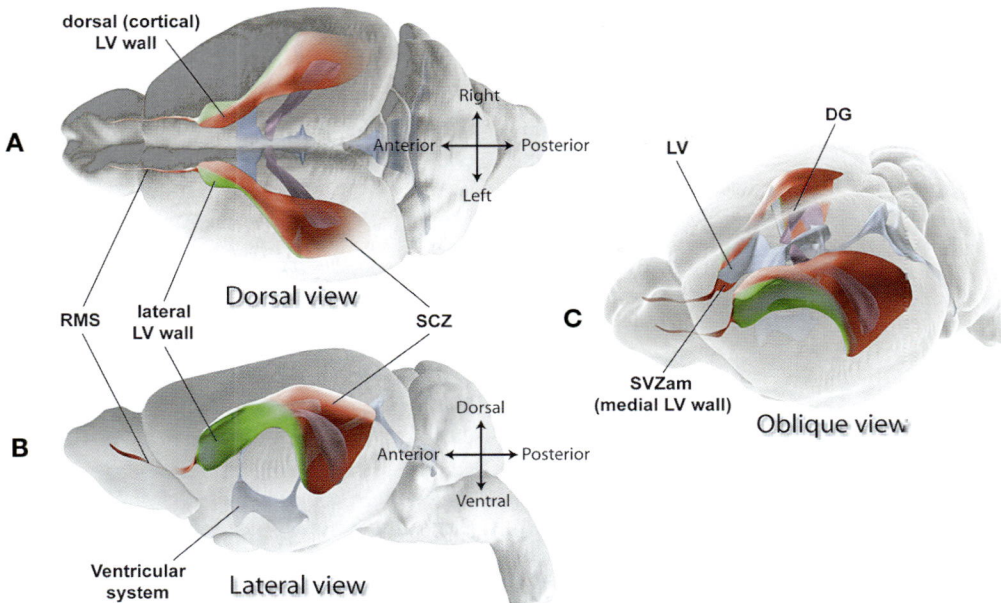

Figure 1. Spatial extent of the SVZ neurogenic niche in the adult mouse brain. The SVZ is traditionally thought to consist of a thin sheet of cells lining the lateral walls of the lateral ventricle, but recent work demonstrates that OB interneurons are born in a much more extensive neurogenic niche. To display its complex topography, the niche was digitally traced in serial sections of an adult mouse brain and combined to generate a three-dimensional model that is shown in dorsal (*A*), lateral (*B*), and oblique (*C*) views within a transparent model of the adult mouse brain surface that was reconstructed from serial magnetic resonance imaging (MRI) sections. The newly discovered neurogenic regions (*red*) include the rostral migratory stream (RMS), the anterior medial wall (SVZam) of the lateral ventricle (LV), and the dorsal (cortical) wall of the LV, which becomes the subcallosal zone (SCZ) in more posterior regions. Each region in the niche generates OB interneurons when its stem cells are targeted in the postnatal brain. For reference, the SVZ niche is shown relative to the ventricular system (*blue*) and the other major neurogenic region in the adult brain, the dentate gyrus (DG) of the hippocampus (*purple*). The borders of the niche are approximate and have been assembled from available stem-cell-targeting data.

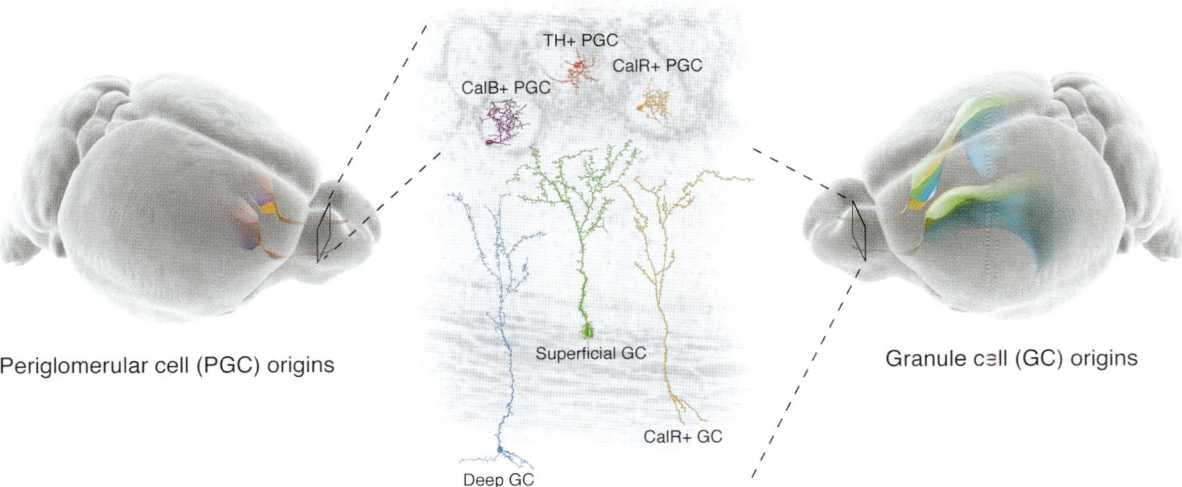

Figure 2. Region-specific production of OB interneuron subtypes in the SVZ. Colored and labeled camera lucida traces of different OB interneuron subtypes are shown in the central panel. To show their relative positions in the OB, the tracings are superimposed over a photomicrograph showing a partial cross section of the OB where superficial is up and the OB core is down. The side panels show oblique views of the adult neurogenic niche, colored to show the region of origin of periglomerular cells (PGCs, *left panel*) and granule cells (GCs, *right panel*). For example, superficial GCs (*green*) are largely produced by neural stem cells in the dorsal part of the neurogenic niche, which comprises the pallial SVZ (SCZ and dorsal wall of the lateral ventricle) and the dorsal portion of the subpallial SVZ, whereas deep GCs (*blue*) are produced in the ventral subpallial SVZ. CalR+ GCs (*yellow*) are produced primarily in the medial wall of the anterior SVZ but also at low levels in the pallium. PGCs are also produced in a region-specific pattern in the anterior SVZ but are rarely produced in the posterior SVZ, as indicated by a diminished intensity of coloring. It has been difficult to study regionalization within the RMS due to its small size, but because it is derived from both pallium and subpallium, it may give rise to different OB interneuron subtypes. The RMS is speckled with a mixture of colors to reflect this ambiguity.

of the calcium-binding protein calretinin (CalR). Superficial granule cells (GCs) primarily target the dendrites of tufted cells in the superficial external plexiform layer (EPL), whereas deep GCs primarily target the deep EPL and connect to mitral cells (Orona et al. 1983; Mori and Shepherd 1994). Because mitral and tufted cells project to different regions of the olfactory cortex, superficial and deep GCs likely subserve separate olfactory functions (Liu and Shipley 1994). Periglomerular cells (PGCs) can be subdivided into at least three different cell types: CalR$^+$, dopaminergic (tyrosine hydroxylase expressing [TH$^+$]), and calbindin (CalB$^+$) (Parrish-Aungst et al. 2007). Although in the mouse, all three PGC subpopulations are GABAergic (Kohwi et al. 2007), there appears to be a species-specific difference in the neurotransmitter characteristics of these cells because only the TH$^+$ PGCs are immunopositive for GABA among rat PGCs (Kosaka and Kosaka 2007).

Most, if not all, of these many cell types are continuously replaced throughout the life of the animal. We are now beginning to understand the origins of their heterogeneity and molecular factors involved in their specification, as described below.

MIGRATING NEUROBLASTS ARE HETEROGENEOUS

Many different types of OB interneurons are generated in the adult brain, but the origins of this diversity are not clear. One possibility is that neuroblasts migrating within the RMS are a homogeneous population of cells that differentiate into different subtypes only after reaching the OB, perhaps in response to local cues. Another possibility is that the migrating neuroblasts are already a heterogeneous group of cells destined to acquire specific fates. Staining for Pax6, a member of the paired-homeobox transcription factor family (Simpson and Price 2002), suggest that neuroblasts in the RMS are not homogeneous (Hack et al. 2005; Kohwi et al. 2005). Only a small subpopulation of neuroblasts expresses Pax6, whereas others are clearly negative. In the OB, Pax6 is expressed in a subpopulation of mature GCs and PGCs (Stoykova and Gruss 1994). Interestingly, Pax6 appears to be cell-autonomously required for the generation of dopaminergic PGCs and a subpopulation of superficial GCs (Hack et al. 2005; Kohwi et al. 2005). The deficit of even one copy of Pax6 results in profound reduction in TH$^+$ PGC production (Dellovade et al. 1998; Kohwi et al. 2005), indicating that Pax6 expression is essential for the generation of specific subpopulations of adult OB neurons. These studies raised the question of how diverse neuroblasts are generated. In addition to Pax6, other transcription factors have been associated with particular subpopulations of interneurons in the adult OB. In particular, the transcription factor SP8, which is highly expressed by the developing dorsolateral ganglionic eminence, cortex, and septum (Long et al. 2007), has been associated with the formation of CalR$^+$ interneurons in the OB (Waclaw et al. 2006). These studies demonstrate that neuroblasts are transcriptionally heterogeneous, suggesting that they are restricted to particular fates soon after they are born. Subsequent studies reveal that this heterogeneity arises in the SVZ and is influenced by both time and space.

TEMPORAL SPECIFICATION OF SVZ PROGENITORS

In the mouse, OB interneurons are first generated about embryonic day (E) 12–14 (Wichterle et al. 2001; Stenman et al. 2003; Tucker et al. 2006). However, the NSCs that produce these neurons in the embryo are morphologically very different from the NSCs that produce OB interneurons in the adult, likely due to the dramatic morphological changes that occur in the developing brain. A growing body of evidence suggests that in parallel with these morphological changes, NSCs produce different OB interneuron types at different times in development (De Marchis et al. 2007; Batista-Brito et al. 2008). This temporal code for cell-type production has been well described for cortical NSCs, which sequentially generate neurons that migrate to and occupy the deep to superficial layers of the cortex (for review, see McConnell 1992). Similarly, retinal progenitor cells sequentially generate different interneurons in a specific order (Cepko et al. 1996). Temporal specification of neural subtypes may be a theme that also applies to OB interneuron production as well. Earlier work suggests that early born neurons survive longer in the OB compared to those generated later in life (Kaplan et al. 1985). More recently, it has been shown that neurons born within the first week of life are more likely to differentiate into superficial GCs than cells born at later ages (Lemasson et al. 2005). Differences in the production of different types of PGCs have also been suggested on the basis of heterochronic and homochronic transplantation experiments (De Marchis et al. 2007). Compared to adults, neonate grafts derive a larger proportion of CalB$^+$ PGCs. Conversely, CalR$^+$ and TH$^+$ PGCs are more frequently observed from adult SVZ grafts, suggesting that progenitors at different developmental stages preferentially produce specific subtypes of OB interneurons. A more recent study used transgenic mice carrying a floxed reporter gene and Cre-ER under the control of the Dlx1/2 intragenic enhancer to label OB interneurons at different stages of development (Batista-Brito et al. 2008). This study confirms that CalB$^+$ cells are more likely to be produced at early developmental stages, whereas most CalR$^+$ cells are produced later, but in contrast to the earlier study, this genetic labeling technique suggests that TH$^+$ cells are produced in larger numbers at earlier time points rather than later. These inconsistent observations may be explained by technical differences. For example, SVZ transplants taken at different stages may not represent truly homologous regions because the morphology of germinal zones and transcription factor expression patterns change dramatically from the embryo to the adult. However, it is also possible that the Dlx1/2 intragenic enhancer may be differentially active at different time points and therefore label different subsets of SVZ cells.

These studies suggest that different OB interneuron types are produced in different numbers at different developmental time points, which should result in constantly shifting ratios of different OB interneuron subtypes. This would raise important functional questions about the maturation of olfactory function. At least among PGCs, however, the ratios of CalB$^+$, CalR$^+$, and TH$^+$ cells do not appear to change from P0 to adult (Kohwi et al. 2007), suggesting that factors besides temporal generation of these cells influence OB circuitry development and maintenance. For example, it is not clear that the rate of maturation of specific cell types remains constant from early development to adult. Indeed, many TH^+, $CalR^+$, and $CalB^+$ cells can already be detected in the P0 OB, soon after the onset of OB interneuron production in the embryo. In contrast, adult-born TH^+ neurons do not reach their peak of maturation until postnatal day 45 (Kohwi et al. 2007). Additionally, we do not yet know whether the turnover rate of different cell types changes over time. Half of adult-born OB interneurons undergo activity-dependent apoptosis between 15 and 45 days after birth (Petreanu and Alvarez-Buylla 2002), but it is not known whether this rate of apoptosis applies to all interneuron subtypes equally or whether this profile is consistent throughout early development to the adult. In the future, it will be important to understand the impact of development, maturation, and turnover of individual interneuron subtypes on OB neuronal circuitry. Because some markers such as TH are also modulated by activity, better cell-type-specific markers will help to clarify this issue.

SPATIAL SPECIFICATION OF SVZ PROGENITORS

Although it is clear that migrating neuroblasts are heterogeneous and that there is a temporal component to OB interneuron production, these studies do not address how multiple OB interneuron types are simultaneously produced in the developing and adult brain or why neurogenesis occurs in such an extensive niche. SVZ NSCs could be equivalent, each generating every different OB interneuron type, or they may be a heterogeneous population of progenitors, each restricted to producing just one cell type or a few cell types.

Regionalization of germinal zones is a general feature of neurogenesis in the developing brain where progenitors in different regions become specified for the production of different neuronal types. This parcellation has been extensively documented in the developing spinal cord where gradients of morphogens establish discrete dorsoventral territories of transcription factor expression, each associated—in time and space—with the production of different types of neurons and glial cells (Ericson et al. 1997; Hochstim et al. 2008). Similarly, segregation of progenitor zones based on distinct transcription factor expression profiles is also observed in the developing telencephalon (Fig. 3) (Campbell 2003; Puelles and Rubenstein 2003; Guillemot 2005; Long et al. 2007); some of these regions are retained in the walls of the adult LV. The lateral wall of the adult SVZ expresses Dlx1/2 (Stuhmer et al. 2002), ER81 (Stenman et al. 2003), Mash1 (Parras et al. 2004), Pax6 (Hack et al. 2005; Kohwi et al. 2005), Sp8 (Waclaw et al. 2006), and Gsh2 (Young et al. 2007), a set of transcription factors that are also expressed in the lateral ganglionic eminence (LGE) during embryonic development (Fig. 3). This common pattern of gene expression, together with transplantation experiments (Wichterle et al. 1999), led to the conclusion that the adult SVZ is largely equivalent to the developing LGE and is likely derived from this structure (Stenman et al. 2003; Waclaw et al. 2006).

The LGE is an important contributor to the adult SVZ as discussed above, but more recent work indicates that other domains of the developing telencephalon, including the pallium (cortex) and septum, contribute importantly to the adult germinal niche that produces OB interneurons (Fig. 3). Consistent with the view that a significant fraction of the adult SVZ is derived from the LGE, lineage-tracing experiments show that a large number of postnatally derived neurons are from progenitors that express Dlx5/6 (Kohwi et al. 2007) and Gsh2 (Young et al. 2007). However, several observations suggest that the embryonic pallium could contribute to OB interneurons as well. For example, lineage tracing of Emx1, a transcription factor highly expressed in the pallium during development, indicates that derived Emx1-expressing progenitor cells remain proliferative in the adult dorsal (cortical) SVZ and generate OB interneurons (Willaime-Morawek et al. 2006; Kohwi et al. 2007; Young et al. 2007). Interestingly, a significant subset of $CalR^+$ cells is derived from Emx1-expressing progenitors, whereas very few $CalR^+$ neurons are derived from the Gsh2 lineage (Fig. 3) (Kohwi et al. 2007; Young et al. 2007). Consistent with these observations, $CalR^+$ cells are generated when embryonic pallium, but not embryonic LGE, was grafted to the adult SVZ (Kohwi et al. 2007). These results suggest that the embryonic pallium also contributes to the adult neurogenic niche and may constitute a population of progenitors different from those derived from the LGE. However, because a small subpopulation of Emx1-lineage cells is also detected in the subpallium in the developing telencephalon of Emx1-Cre mice (Gorski et al. 2002; Willaime-Morawek et al. 2006), it is possible that these cells, and not the pallial progenitors, give rise to the subpopulation of Emx1-derived OB interneurons.

Direct evidence for the spatial specification of SVZ progenitors came from a study that used a modification of the technique originally used to demonstrate that adult NSCs were derived from radial glia (Merkle et al. 2004). A small volume of adenovirus-expressing Cre was injected in the neonatal brain in a region traversed by the radial processes of radial glial cells, the precursors of adult SVZ type B cells. The adenovirus was retrogradely transported up the radial process to the nucleus, where the virally encoded Cre recombines a conditional reporter gene, allowing the lineage of labeled NSCs to be traced. This procedure enables small groups of radial glia to be labeled in different discrete locations of the periventricular germinal niche (Merkle et al. 2007). This approach demonstrates that different types of OB interneurons are derived from unique locations of the SVZ (see Fig. 2). For example, neonatally targeted pallial radial glia, which project their radial processes through the corpus callosum into the cortex (Merkle et al. 2007; Ventura and Goldman 2007), generate many GCs and PGCs, confirming the conclusions of the Emx1 lineage-tracing studies. Interestingly, the majority of pallially derived PGCs expresses TH and a small subpopulation expresses CalR. In contrast, the majority of $CalB^+$ cells is derived from ventrolateral SVZ.

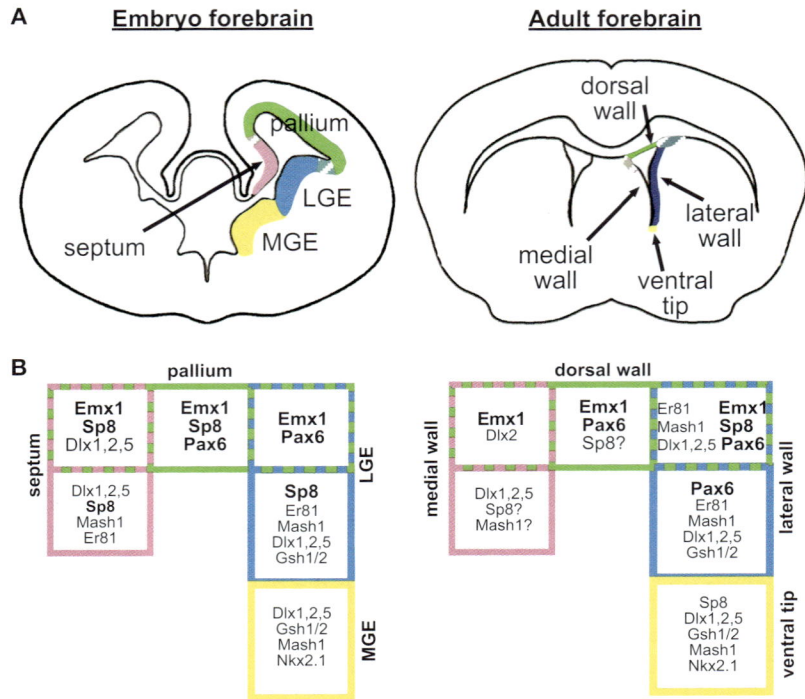

Figure 3. Transcription factor expression in the developing and adult forebrain germinal zones. Embryo: Different germinal zones are labeled with different colors. The pallial–septal and pallial–LGE borders express transcription factors associated with both neighboring progenitor zones and are shown by hatch marks in *A* and two-colored boxes in *B*, below the corresponding schematic. Transcription factors expressed in each region are indicated within each colored box. Some subtypes of OB interneurons have been associated with specific transcription factors; these are indicated in bold text (for references, see text). (MGE) Medial ganglionic eminence; (LGE) lateral ganglionic eminence. Adult: Adult germinal zones homologous to those in the embryo (based on anatomy and transcription factor expression) are shown in the same colors as those for the embryo. Note the increase in the number of transcription factors with overlapping expression at the dorsolateral boundary in the adult ventricle compared to the embryo. (*C*) Certain subtypes of OB interneurons that have been associated with specific transcription factors are indicated.

Different types of GCs are also derived from different SVZ domains. Targeting of NSCs in the pallial and dorsal subpallial SVZ leads to primarily superficial GC production, whereas ventral SVZ targeting leads to mostly deep GC production (Merkle et al. 2007). However, CalR+ GCs are not produced when NSCs in the dorsal subpallium were targeted; rather, they are derived in small numbers from the pallium, and in larger numbers from the medial wall and RMS (Fig. 2). Similarly, a subpopulation of CalR+ PGCs is also derived from this medial germinal zone. Although the ventral portion of this medial region faces the nucleus accumbens (an LGE derivative), more anterior and dorsal regions appear to face the septum, which is not derived from the LGE, but expresses the transcription factors Gsh2 and Dlx1/2. Interestingly the RMS generates nearly all OB interneuron subtypes, including a large percentage of CalR+ GCs and PGCs (Merkle et al. 2007). The RMS forms along what earlier in development correspond to the olfactory

ventricle. This ventricular wall is patterned and contains both pallial and subpallial components (Fig. 2) (Long et al. 2007).

The dendritic arbors of newborn OB interneurons also appear to be specified within different domains of the SVZ. A recent study used retroviruses to label proliferative cells in the anterior or posterior SVZ of neonatal rats (Kelsch et al. 2007). GCs with dendrites that branch into the superficial EPL are derived from cells labeled in the anterior SVZ of the neonatal rat, whereas GCs that branch deep within the EPL are derived from retroviral injections into the posterior SVZ. Because retroviruses can label all dividing cells in the SVZ, it is not clear whether this specification occurs in primary or secondary progenitors. Nonetheless, the work does suggest that differences in the rostrocaudal position of dividing SVZ progenitors affect the branching pattern of dendrites in young neurons that migrate to the OB.

To determine whether adult SVZ type B cells share the same regional specification as radial glial cells labeled in the neonatal brain, adult progenitors in different regions of the SVZ were targeted with an adenovirus-expressing Cre under the GFAP promoter. Type B cells labeled in this manner generate similar types of OB interneurons as radial glia targeted in the same region. This indicates that the spatial organization of SVZ progenitors is maintained during postnatal development (Merkle et al. 2007). Furthermore, it suggests that SVZ stem cells do not move tangentially during postnatal development. In summary, a growing body of evidence shows that the SVZ is organized into domains containing different types of progenitor cells. This conclusion parallels what has been described in the embryo, where the position of a progenitor in a germinal zone determines the types of neurons that it will generate.

When SVZ stem cells are cultured with high concentrations of growth factors, they can be passaged several times and generate neurons, astrocytes, and oligodendrocytes (Morshead et al. 1994), demonstrating self-renewal and multipotency under these conditions. To determine whether SVZ stem cells retain their regional identity in culture, they were cultured under adherent conditions developed by the Steindler laboratory (Scheffler et al. 2005). Under these conditions, SVZ stem cells produce OB interneuron types with regional specificity similar to that observed from in vivo targeting. For example, cultures derived from the medial wall produce many more CalR$^+$ neurons than cultures derived from the lateral wall of the SVZ. Within the lateral wall, ventral cultures give rise to higher numbers of CalB$^+$ cells than dorsal cultures. Although under the above culture conditions SVZ progenitors appear to retain a remarkable level of specification, it remains to be determined whether different culture conditions might alter this program or allow adult progenitor cells to be respecified.

Positional specification in the neonatal mouse brain could be inherited from early development. Alternatively, positional cues in the postnatal brain may continually instruct stem cells to generate specific types of neurons. Grafting experiments, where progenitors from one region are transferred to another location of the postnatal SVZ, have failed to demonstrate plasticity indicative of environmental cues reinstructing postnatal stem cells to acquire particular fates (Kelsch et al. 2007; Merkle et al. 2007). For example, radial glia in the ventral SVZ normally generate deep GCs continue to generate deep GCs when grafted into the dorsal SVZ, where superficial GCs are produced. This level of specification is maintained even after progenitors were cultured for multiple passages in vitro. This suggests that positional specification is an early event likely to occur before birth and is maintained postnatally. Although the data above strongly suggest that relocation within the SVZ is not sufficient to switch stem cell specification, this is not enough to conclude that cell fate cannot be modified. Environmental positional cues, which may have created positional patterns of transcription factor expression early in the embryo, may no longer be present in the postnatal brain. It will be interesting to determine whether some of the molecular mechanisms involved in the early specification remain operational in the adult.

CONCLUSION

The SVZ is the largest germinal region of the adult mammalian brain. Several studies suggest that the adult SVZ may be an important reservoir of precursor cells for brain repair (Lindvall and Kokaia 2006; Martino and Pluchino 2006). Although the extent to which the robust neurogenesis found in the rodent can be found in the adult human brain is unclear (Curtis et al. 2007; Sanai et al. 2007), NSCs and some level of SVZ proliferation has been described in autopsy material (Eriksson et al. 1998; Curtis et al. 2003; Sanai et al. 2005). Recent work in rodents indicates that this germinal region is much more extensive than previously thought and includes regions derived from all three main telencephalic walls in the embryo (Figs. 1–3). If a similar level of spatial specification exists in humans, different subtypes of neurons are likely to originate from SVZ regions that are separated from one another by long distances. This may be an important constraint in evolution. As brain size increases, there may be limits to the tangential migration required to transfer young neurons from unique sites of birth to their ultimate destination within the circuits where they will ultimately integrate. This is relevant not only to brain evolution, but also to its repair. The generation of specific neuronal subtypes for therapeutic applications may depend on whether the region-specific progenitors that give rise to them are still active in the adult. Stem cells in these different regions express unique sets of transcription factors likely derived from an early stage of embryonic development. An important challenge for future work is to decipher the mechanisms by which combinations of transcription factors expressed in different locations of the SVZ result in the generation of distinct subsets of olfactory interneurons. This new information will provide a fundamental basis for understanding how multiple variables, including time and space, regulate NSCs so that we can better harvest their therapeutic benefits.

ACKNOWLEDGMENTS

This work is supported by the John G. Bowes Research Fund and grants from the National Institutes of Health. We thank Kenneth Xavier Probst for assisting Florian Merkle with preparing the three-dimensional models shown in Figures 1 and 2.

REFERENCES

Altman, J. 1969. Autoradiographic and histological studies of postnatal neurogenesis. IV. Cell proliferation and migration in the anterior forebrain, with special reference to persisting neurogenesis in the olfactory bulb. *J. Comp. Neurol.* **137:** 433–458.

Alvarez-Buylla, A., Garcia-Verdugo, J.M., Mateo, A.S., and Merchant-Larios, H. 1998. Primary neural precursors and intermitotic nuclear migration in the ventricular zone of adult canaries. *J. Neurosci.* **18:** 1020–1037.

Anthony, T.E., Klein, C., Fishell, G., and Heintz, N. 2004. Radial glia serve as neuronal progenitors in all regions of the central nervous system. *Neuron* **41:** 881–890.

Batista-Brito, R., Close, J., Machold. R., and Fishell, G. 2008. The distinct temporal origins of olfactory bulb interneuron subtypes. *J. Neurosci.* **28:** 3966–3975.

Campbell, K. 2003. Dorsal-ventral patterning in the mammalian telencephalon. *Curr. Opin. Neurobiol.* **13:** 50–56.

Carleton, A., Petreanu, L.T., Lansford, R., Alvarez-Buylla, A., and Lledo, P.M. 2003. Becoming a new neuron in the adult olfactory bulb. *Nat. Neurosci.* **6:** 507–518.

Cepko, C.L., Austin, C.P., Yang, X.J., Alexiades, M.R., and Ezzeddine, D. 1996. Cell fate determination in the vertebrate retina. *Proc. Natl. Acad. Sci.* **93:** 589–595.

Curtis, M.A., Penney, E.B., Pearson, A.G., van Roon-Mom, W.M., Butterworth, N.J., Dragunow, M., Connor, B., and Faull, R.L. 2003. Increased cell proliferation and neurogenesis in the adult human Huntington's disease brain. *Proc. Natl. Acad. Sci.* **100:** 9023–9027.

Curtis, M.A., Kam, M., Nannmark, U., Anderson, M.F., Axell, M.Z., Wikkelso, C., Holtas, S., van Roon-Mom, W.M., Bjork-Eriksson, T., Nordborg, C., et al. 2007. Human neuroblasts migrate to the olfactory bulb via a lateral ventricular extension. *Science* **315:** 1243–1249.

Dellovade, T.L., Pfaff, D.W., and Schwanzel-Fukuda, M. 1998. Olfactory bulb development is altered in small-eye (*Sey*) mice. *J. Comp. Neurol.* **402:** 402–418.

De Marchis, S., Bovetti, S., Carletti, B., Hsieh, Y.C., Garzotto, D., Peretto, P., Fasolo, A., Puche, A.C., and Rossi, F. 2007. Generation of distinct types of periglomerular olfactory bulb interneurons during development and in adult mice: Implication for intrinsic properties of the subventricular zone progenitor population. *J. Neurosci.* **27:** 657–664.

Doetsch, F. and Alvarez-Buylla, A. 1996. Network of tangential pathways for neuronal migration in adult mammalian brain. *Proc. Natl. Acad. Sci.* **93:** 14895–14900.

Doetsch, F., Caille, I., Lim, D.A., García-Verdugo, J.M., and Alvarez-Buylla, A. 1999. Subventricular zone astrocytes are neural stem cells in the adult mammalian brain. *Cell* **97:** 1–20.

Ericson, J., Briscoe, J., Rashbass, P., van Heyningen, V., and Jessell, T.M. 1997. Graded sonic hedgehog signaling and the specification of cell fate in the ventral neural tube. *Cold Spring Harbor Symp. Quant. Biol.* **62:** 451–466.

Eriksson, P.S., Perfilieva, E., Bjork-Eriksson, T., Alborn, A., Nordborg, C., Peterson, D.A., and Gage, F.H. 1998. Neurogenesis in the adult human hippocampus. *Nat. Med.* **4:** 1313–1317.

Gorski, J.A., Talley, T., Qiu, M., Puelles, L., Rubenstein, J.L., and Jones, K.R. 2002. Cortical excitatory neurons and glia, but not GABAergic neurons, are produced in the Emx1-expressing lineage. *J. Neurosci.* **22:** 6309–6314.

Greer, C.A. 1987. Golgi analyses of dendritic organization among denervated olfactory bulb granule cells. *J. Comp. Neurol.* **257:** 442–452.

Gritti, A., Bonfanti, L., Doetsch, F., Caille, I., Alvarez-Buylla, A., Lim, D.A., Galli, R., Verdugo, J.M., Herrera, D.G., and Vescovi, A.L. 2002. Multipotent neural stem cells reside into the rostral extension and olfactory bulb of adult rodents. *J. Neurosci.* **22:** 437–445.

Guillemot, F. 2005. Cellular and molecular control of neurogenesis in the mammalian telencephalon. *Curr. Opin. Cell Biol.* **17:** 639–647.

Hack, M.A., Saghatelyan, A., de Chevigny, A., Pfeifer, A., Ashery-Padan, R., Lledo, P.M., and Gotz, M. 2005. Neuronal fate determinants of adult olfactory bulb neurogenesis. *Nat. Neurosci.* **8:** 865–872.

Hochstim, C., Deneen, B., Lukaszewicz, A., Zhou, Q., and Anderson, D.J. 2008. Identification of positionally distinct astrocyte subtypes whose identities are specified by a homeodomain code. *Cell* **133:** 510–522.

Imura, T., Kornblum, H.I., and Sofroniew, M.V. 2003. The predominant neural stem cell isolated from postnatal and adult forebrain but not early embryonic forebrain expresses GFAP. *J. Neurosci.* **23:** 2824–2832.

Kaplan, M.S., McNelly, N.A., and Hinds, J.W. 1985. Population dynamics of adult-formed granule neurons of the rat olfactory bulb. *J. Comp. Neurol.* **239:** 117–125.

Kelsch, W., Mosley, C.P., Lin, C.W., and Lois, C. 2007. Distinct mammalian precursors are committed to generate neurons with defined dendritic projection patterns. *PLoS Biol.* **5:** e300.

Kohwi, M., Osumi, N., Rubenstein, J.L., and Alvarez-Buylla, A. 2005. Pax6 is required for making specific subpopulations of granule and periglomerular neurons in the olfactory bulb. *J. Neurosci.* **25:** 6997–7003.

Kohwi, M., Petryniak, M.A., Long, J.E., Ekker, M., Obata, K., Yanagawa, Y., Rubenstein, J.L., and Alvarez-Buylla, A. 2007. A subpopulation of olfactory bulb GABAergic interneurons is derived from Emx1- and Dlx5/6-expressing progenitors. *J. Neurosci.* **27:** 6878–6891.

Kosaka, K. and Kosaka, T. 2007. Chemical properties of type 1 and type 2 periglomerular cells in the mouse olfactory bulb are different from those in the rat olfactory bulb. *Brain Res.* **1167:** 42–55.

Kosaka, K., Aika, Y., Toida, K., Heizmann, C.W., Hunziker, W., Jacobowitz, D.M., Nagatsu, I., Streit, P., Visser, T.J., and Kosaka, T. 1995. Chemically defined neuron groups and their subpopulations in the glomerular layer of the rat main olfactory bulb. *Neurosci. Res.* **23:** 73–88.

Laywell, E.D., Rakic, P., Kukekov, V.G., Holland, E.C., and Steindler, D.A. 2000. Identification of a multipotent astrocytic stem cell in the immature and adult mouse brain. *Proc. Natl. Acad. Sci.* **97:** 13883–13888.

Lemasson, M., Saghatelyan, A., Olivo-Marin, J.C., and Lledo, P.M. 2005. Neonatal and adult neurogenesis provide two distinct populations of newborn neurons to the mouse olfactory bulb. *J. Neurosci.* **25:** 6816–6825.

Lindvall, O. and Kokaia, Z. 2006. Stem cells for the treatment of neurological disorders. *Nature* **441:** 1094–1096.

Liu, Z. and Martin, L.J. 2003. Olfactory bulb core is a rich source of neural progenitor and stem cells in adult rodent and human. *J. Comp. Neurol.* **459:** 368–391.

Liu, W.-L. and Shipley, M.T. 1994. Intrabulbar associational system in the rat olfactory bulb comprises cholecystokinin-containing tufted cells that synapse onto the dendrites of GABAergic granule cells. *J. Comp. Neurol.* **346:** 541–558.

Lledo, P.M., Merkle, F.T., and Alvarez-Buylla, A. 2008. Origin and function of olfactory bulb interneuron diversity. *Trends Neurosci.* **31:** 392–400.

Lois, C. and Alvarez-Buylla, A. 1994. Long-distance neuronal migration in the adult mammalian brain. *Science* **264:** 1145–1148.

Lois, C., Garcia-Verdugo, J.M., and Alvarez-Buylla, A. 1996. Chain migration of neuronal precursors. *Science* **271:** 978–981.

Long, J.E., Garel, S., Alvarez-Dolado, M., Yoshikawa, K., Osumi, N., Alvarez-Buylla, A., and Rubenstein, J.L. 2007. Dlx-dependent and -independent regulation of olfactory bulb interneuron differentiation. *J. Neurosci.* **27:** 3230–3243.

Luskin, M.B. 1993. Restricted proliferation and migration of postnatally generated neurons derived from the forebrain subventricular zone. *Neuron* **11:** 173–189.

Marin, O. and Rubenstein, J.L. 2003. Cell migration in the forebrain. *Annu. Rev. Neurosci.* **26:** 441–483.

Martino, G. and Pluchino, S. 2006. The therapeutic potential of neural stem cells. *Nat. Rev. Neurosci.* **7:** 395–406.

McConnell, S.K. 1992. The control of neuronal identity in the developing cerebral cortex. *Curr. Opin. Neurobiol.* **2:** 23–27.

Menn, B., Garcia-Verdugo, J.M., Yaschine, C., Gonzalez-Perez, O., Rowitch, D., and Alvarez-Buylla, A. 2006. Origin of oligodendrocytes in the subventricular zone of the adult brain. *J. Neurosci.* **26:** 7907–7918.

Merkle, F.T. and Alvarez-Buylla, A. 2006. Neural stem cells in mammalian development. *Curr. Opin. Cell Biol.* **18:** 704–709.

Merkle, F.T., Mirzadeh, Z., and Alvarez-Buylla, A. 2007. Mosaic organization of neural stem cells in the adult brain. *Science* **317:** 381–384.

Merkle, F.T., Tramontin, A.D., Garcia-Verdugo, J.M., and Alvarez-Buylla, A. 2004. Radial glia give rise to adult neural stem cells in the subventricular zone. *Proc. Natl. Acad. Sci.* **101:** 17528–17532.

Mirzadeh, Z., Merkle, F.T., Soriano-Navarro, M., Garcia-Verdugo, J.M., and Alvarez-Buylla, A. 2008. Neural stem cells confer unique pinwheel architecture to the ventricular surface in neurogenic regions of the adult brain. *Cell Stem Cell* **3:** 265–278.

Miyata, T., Kawaguchi, A., Saito, K., Kawano, M., Muto, T., and Ogawa, M. 2004. Asymmetric production of surface-dividing and non-surface-dividing cortical progenitor cells. *Development* **131:** 3133–3145.

Mori, K. and Shepherd, G.M. 1994. Emerging principles of molecular signal processing by mitral/tufted cells in the olfactory bulb. *Semin. Cell Biol.* **5:** 65–74.

Morshead, C.M., Reynolds, B.A., Craig, C.G., McBurney, M.W., Staines, W.A., Morassutti, D., Weiss, S., and van der Kooy, D. 1994. Neural stem cells in the adult mammalian forebrain: A relatively quiescent subpopulation of subependymal cells. *Neuron* **13:** 1071–1082.

Nait-Oumesmar, B., Decker, L., Lachapelle, F., Avellana-Adalid, V., Bachelin, C., and Van Evercooren, A.B. 1999. Progenitor cells of the adult mouse subventricular zone proliferate, migrate and differentiate into oligodendrocytes after demyelination. *Eur. J. Neurosci.* **11:** 4357–4366.

Noctor, S.C., Martinez-Cerdeno, V., Ivic, L., and Kriegstein, A.R. 2004. Cortical neurons arise in symmetric and asymmetric division zones and migrate through specific phases. *Nat. Neurosci.* **7:** 136–144.

Nottebohm, F. 2004. The road we travelled: Discovery, choreography, and significance of brain replaceable neurons. *Ann. N.Y. Acad. Sci.* **1016:** 628–658.

Orona, E., Scott, J.W., and Rainer, E.C. 1983. Different granule cell populations innervate superficial and deep regions of the external plexiform layer in rat olfactory bulb. *J. Comp. Neurol.* **217:** 227–237.

Parras, C.M., Galli, R., Britz, O., Soares, S., Galichet, C., Battiste, J., Johnson, J.E., Nakafuku, M., Vescovi, A., and Guillemot, F. 2004. Mash1 specifies neurons and oligodendrocytes in the postnatal brain. *EMBO J.* **23:** 4495–4505.

Parrish-Aungst, S., Shipley, M.T., Erdelyi, F., Szabo, G., and Puche, A.C. 2007. Quantitative analysis of neuronal diversity in the mouse olfactory bulb. *J. Comp. Neurol.* **501:** 825–836.

Petreanu, L. and Alvarez-Buylla, A. 2002. Maturation and death of adult-born olfactory bulb granule neurons: Role of olfaction. *J. Neurosci.* **22:** 6106–6113.

Picard-Riera, N., Decker, L., Delarasse, C., Goude, K., Nait-Oumesmar, B., Liblau, R., Pham-Dinh, D., and Evercooren, A.B. 2002. Experimental autoimmune encephalomyelitis mobilizes neural progenitors from the subventricular zone to undergo oligodendrogenesis in adult mice. *Proc. Natl. Acad. Sci.* **99:** 13211–13216.

Puelles, L. and Rubenstein, J.L. 2003. Forebrain gene expression domains and the evolving prosomeric model. *Trends Neurosci.* **26:** 469–476.

Reynolds, B.A. and Weiss, S. 1992. Generation of neurons and astrocytes from isolated cells of the adult mammalian central nervous system. *Science* **255:** 1707–1710.

Sanai, N., Alvarez-Buylla, A., and Berger, M.S. 2005. Neural stem cells and the origin of gliomas. *N. Engl. J. Med.* **353:** 811–822.

Sanai, N., Berger, M.S., Garcia-Verdugo, J.M., and Alvarez-Buylla, A. 2007. Comment on "Human neuroblasts migrate to the olfactory bulb via a lateral ventricular extension." *Science* **318:** 393; author reply 393.

Scheffler, B., Walton, N.M., Lin, D.D., Goetz, A.K., Enikolopov, G., Roper, S.N., and Steindler, D.A. 2005. Phenotypic and functional characterization of adult brain neuropoiesis. *Proc. Natl. Acad. Sci.* **102:** 9353–9358.

Seri, B., Herrera, D.G., Gritti, A., Ferron, S., Collado, L., Vescovi, A., Garcia-Verdugo, J.M., and Alvarez-Buylla, A. 2006. Composition and organization of the SCZ: A large germinal layer containing neural stem cells in the adult mammalian brain. *Cereb. Cortex* (suppl. 1) **16:** i103–i111.

Shepherd, G.M. and Greer, C.A. 1998. Olfactory bulb. In *The synaptic organization of the brain* (ed. G.M. Shepherd). pp. 159–204. Oxford University Press, New York.

Simpson, T.I. and Price, D.J. 2002. Pax6: A pleiotropic player in development. *Bioessays* **24:** 1041–1051.

Smart, I. and Leblond, C.P. 1961. Evidence for division and transformations of neuroglia cells in the mouse brain, as derived from radioautography after injection of thymidine-H3. *J. Comp. Neurol.* **116:** 349–367.

Stenman, J., Toresson, H., and Campbell, K. 2003. Identification of two distinct progenitor populations in the lateral ganglionic eminence: Implications for striatal and olfactory bulb neurogenesis. *J. Neurosci.* **23:** 167–174.

Stoykova, A. and Gruss, P. 1994. Roles of Pax-genes in developing and adult brain as suggested by expression patterns. *J. Neurosci.* **14:** 1395–1412.

Stuhmer, T., Puelles, L., Ekker, M., and Rubenstein, J.L. 2002. Expression from a Dlx gene enhancer marks adult mouse cortical GABAergic neurons. *Cereb. Cortex* **12:** 75–85.

Tucker, E.S., Polleux, F., and LaMantia, A.S. 2006. Position and time specify the migration of a pioneering population of olfactory bulb interneurons. *Dev. Biol.* **297:** 387–401.

Ventura, R.E. and Goldman, J.E. 2007. Dorsal radial glia generate olfactory bulb interneurons in the postnatal murine brain. *J. Neurosci.* **27:** 4297–4302.

Waclaw, R.R., Allen II, Z.J., Bell, S.M., Erdelyi, F., Szabo, G., Potter, S.S., and Campbell, K. 2006. The zinc finger transcription factor Sp8 regulates the generation and diversity of olfactory bulb interneurons. *Neuron* **49:** 503–516.

Wichterle, H., Garcia-Verdugo, J.M., and Alvarez-Buylla, A. 1997. Direct evidence for homotypic, glia-independent neuronal migration. *Neuron* **18:** 779–791.

Wichterle, H., Garcia-Verdugo, J.M., Herrera, D.G., and Alvarez-Buylla, A. 1999. Young neurons from medial ganglionic eminence disperse in adult and embryonic brain. *Nat. Neurosci.* **2:** 461–466.

Wichterle, H., Turnbull, D.H., Nery, S., Fishell, G., and Alvarez-Buylla, A. 2001. In utero fate mapping reveals distinct migratory pathways and fates of neurons born in the mammalian basal forebrain. *Development* **128:** 3759–3771.

Willaime-Morawek, S., Seaberg, R.M., Batista, C., Labbe, E., Attisano, L., Gorski, J.A., Jones, K.R., Kam, A., Morshead, C.M., and van der Kooy, D. 2006. Embryonic cortical neural stem cells migrate ventrally and persist as postnatal striatal stem cells. *J. Cell Biol.* **175:** 159–168.

Young, K.M., Fogarty, M., Kessaris, N., and Richardson, W.D. 2007. Subventricular zone stem cells are heterogeneous with respect to their embryonic origins and neurogenic fates in the adult olfactory bulb. *J. Neurosci.* **27:** 8286–8296.

Notch, Neural Stem Cells, and Brain Tumors

T.J. Pierfelice,[*†] K.C. Schreck,[*‡†] C.G. Eberhart,[§¶#] and N. Gaiano[* *‡§]

*Institute for Cell Engineering, *Departments of Neurology, ‡Neuroscience, §Oncology, ¶Pathology, and #Ophthalmology, Johns Hopkins University School of Medicine, Baltimore, Maryland 21205

The Notch pathway has a fundamental role during cell-fate specification in the developing mammalian nervous system. During neocortical development, Notch signaling inhibits neuronal differentiation and maintains the neural stem/progenitor cell pool to permit successive waves of neurogenesis, which are followed by gliogenesis. In addition, recent evidence suggests that Notch signaling is not uniformly used among distinct proliferative neural cells types, with the canonical cascade functional in neural stem cells but attenuated in neurogenic progenitors. Although the role of Notch in neural development is increasingly well understood, it has recently become evident that Notch also has a role in brain tumor biology. Notch receptors are overexpressed in many different brain tumor types, and they may have an initiating role in some. Stem-like cells in brain tumors share many similarities with neural stem/progenitor cells and may require Notch for their survival and growth. Understanding the role of Notch signaling in neoplastic and non-neoplastic stem/progenitor populations will advance our understanding of basic principles regulating developmental and stem cell biology and may also lead to more effective therapies for brain tumors.

The Notch pathway regulates many different processes during mammalian development. Notch signaling has been particularly heavily studied in the developing nervous system, where Notch receptor activation inhibits neuronal differentiation and maintains the neural stem/progenitor cell pool. Notch signaling has also been found to promote glial character, and in light of findings that certain glial cells possess neural stem cell character, the ability of Notch to maintain progenitor character and to promote glial character is likely to be mechanistically related (Gaiano and Fishell 2002).

In addition to its role in development, Notch signaling has been implicated in many human cancers, including those in the brain. Because Notch pathway activation is well known for contributing to the maintenance of a proliferatively active cell state, the notion that aberrant Notch signaling could contribute to tumor formation is not surprising. The mechanisms by which Notch regulates neural stem cells in the developing nervous system are likely to be similar to those used during the regulation of putative brain tumor stem cells. As such, understanding parallels between both the signaling mechanisms and the cellular heterogeneity present in both settings (e.g., stem cell vs. transit-amplifying progenitor) will provide valuable insight.

In the first part of this chapter, we focus on Notch's role in embryonic neural stem/progenitor cells in the developing mammalian neocortex. Ongoing work in the field has focused on understanding how the Notch signal transduction cascade is regulated and also on the nature of cell–cell interactions that mediate signaling. In the second part of this chapter, we address Notch signaling in brain tumor formation and growth and how this signaling pathway may regulate the maintenance and behavior of brain tumor stem cells. We also consider parallels between Notch function in embryonic neural stem/progenitor cells and in brain tumor stem cells.

NOTCH AND NEURAL PROGENITORS

Neocortical Development

The mammalian neocortex is a highly organized six-layered structure that develops at the anterior end of the neural tube. The first neural stem/progenitor cells (simply referred to as progenitors hereafter, unless otherwise noted) line the ventricles in a germinal area referred to as the ventricular zone (VZ). A second germinal area, termed the subventricular zone (SVZ), forms just beneath the VZ after the onset of neurogenesis. The VZ and SVZ both contain highly proliferative neural progenitor cells that undergo symmetric and asymmetric cell divisions to either maintain the proliferative pool or produce the neurons of the different cortical layers (Noctor et al. 2004). In the mouse, neurogenesis begins around embryonic day 10.5 (E10.5) and lasts until around E17.5, when gliogenesis begins in the SVZ (Molyneaux et al. 2007). The newborn neurons produced in the VZ or SVZ exit these areas by radial migration along the processes of radial glial cells to their final destination in the neocortical plate, where neuronal differentiation and circuit formation take place (Noctor et al. 2001, 2004).

At least three types of well-defined neural progenitors exist in the developing neocortex: neuroepithelial progenitors (NEPs), radial glial cells (RGCs), and intermediate progenitor cells (IPCs) (Pontious et al. 2007). NEPs span the neural tube from the ventricular (inner) surface to the pial (outer) surface and comprise the VZ before the onset of neurogenesis. NEPs initially divide symmetrically to expand the progenitor pool but undergo asymmetric divi-

---

[†]These authors contributed equally to this work.

sions to yield the first neurons (Molyneaux et al. 2007). Early during neocortical development, many NEPs transform into RGCs, which also extend processes from the ventricular to pial surfaces, and in addition to a role as progenitors (see below), they serve as a migratory scaffold for newly generated neurons. RGCs differ from NEPs because they acquire aspects of astroglial character including expression of astrocyte-specific glutamate-aspartate transporter (GLAST) and brain lipid-binding protein (BLBP) (Götz and Barde 2005). Recent studies have shown that RGCs function as neuronal progenitors either by directly producing neurons (Noctor et al. 2001; Malatesta et al. 2003) or by producing IPCs, which give rise to neurons (Noctor et al. 2004). Time-lapse imaging in slice culture has shown that IPCs divide symmetrically to produce either two neurons or two IPCs (Noctor et al. 2004). IPCs are characterized by expression of Tbr2, Svet1, and Cux2 (Tarabykin et al. 2001; Nieto et al. 2004; Englund et al. 2005).

Several recent studies have suggested that a fourth neocortical progenitor cell type exists in addition to NEPs, RGCs, and IPCs. One study identified a short neural precursor (SNP) present during neurogenesis and located in the VZ (Gal et al. 2006). SNPs are morphologically distinct from RGCs in that their processes only span the VZ. In addition, SNPs are molecularly distinct from RGCs, with the former driving expression from the tubulin α1 (Tα1) promoter and the latter driving expression from the GLAST and BLBP promoters (Gal et al. 2006).

In support of the existence of two molecularly distinct VZ progenitor subtypes, we have recently shown that Notch signaling is not uniformly used in the VZ. Those cells that drive expression from the GLAST promoter (RGCs) are able to robustly activate the canonical Notch signaling cascade, whereas those VZ cells that drive expression from the Tα1 promoter (presumptive SNPs) have attenuated canonical Notch signaling (Mizutani et al. 2007; see below for further discussion). The relationship between the latter cell type, referred to as intermediate neural progenitor (INP), and the IPCs present primarily in the SVZ is unclear, although because both are neurogenic, it is reasonable to hypothesize that the former give rise in some part to the latter. If so, it may ultimately make more sense to call them both INPs but with designations that specify apical (aINPs or aIPCs in the VZ) or basal (bINPs or bIPCs in the SVZ) locations. We favor retention of the term "neural" to distinguish these cells from intermediate progenitors in other tissues. For a model of the lineage relationships among the four different neocortical progenitor subtypes described here (NEPs, RGCs, SNP/INPa, IPC/INPb), see Fig. 1.

Notch Signaling in Neural Progenitors

The Notch signaling pathway is evolutionarily conserved and has a fundamental role during animal development and in the adult, in particular by regulating cell-fate specification (Yoon and Gaiano 2005; Louvi and Artavanis-Tsakonas 2006). The function of Notch signaling was first characterized in the fly where heterozygous loss-of-function mutants displayed notching in the wing margin, and homozygous

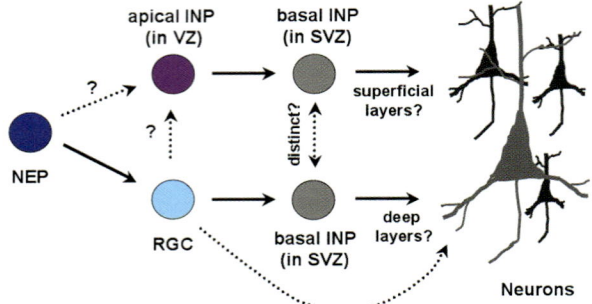

Figure 1. Schematic depicting potential lineage relationships between different neural progenitor subtypes present during neocortical neurogenesis. Note that the apical INP (aINP) is likely to be the same as the short neural precursor (SNP) described in the text, whereas the basal INP (bINP) is the same as the intermediate progenitor cell (IPC) described in the text, which is located primarily in the SVZ.

loss-of-function mutants contained supernumery neurons, in what was termed a "neurogenic" phenotype (Louvi and Artavanis-Tsakonas 2006). Since then, Notch function has been well documented in the developing nervous system, immune system, and the skin, muscle, and cardiovascular system, as well as in many other contexts.

The Notch signal is mediated through contact between adjacent cells. In mammals, there are four Notch receptors (Notch1–4) and multiple ligands of the Delta-like (Dll) and Jagged (Jag) families, all of which are single-pass transmembrane proteins. Upon ligand binding, the Notch receptors are cleaved by the γ-secretase complex, releasing the Notch intracellular domain (NICD). NICD translocates into the nucleus and associates with CBF1 and Mastermind-like (MAML) (Wu et al. 2000) to create a transcriptional activator complex (see Fig. 2). In the developing nervous systems, that complex activates the expression of the *Hes* and *Hey* genes (Iso et al. 2003), which are basic helix-loop-helix (bHLH) transcriptional repressors that antagonize the function of the proneural genes such as the *Neurogenins* and thereby inhibit neuronal differentiation (Bertrand et al. 2002).

The role of Notch signaling in the embryonic mammalian nervous system has been investigated using both loss- and gain-of-function approaches. Gain-of-function studies have revealed that Notch signaling inhibits neuronal and oligodendroglial differentiation while promoting progenitor and astroglial character (for review, see Gaiano and Fishell 2002). For example, in the mammalian neocortex, retroviral expression of activated forms of Notch1 or Notch3 in vivo promotes RGC identity embryonically and astrocyte fate postnatally (Gaiano et al. 2000; Dang et al. 2006b). Consistent with the gain-of-function findings, loss of Notch signaling results in an increase in neuronal differentiation and diminished progenitor maintenance (for review, see Yoon and Gaiano 2005; Yoon et al. 2008). Interestingly, although Notch signaling is required for neural progenitor maintenance, it does not appear to be necessary for the initial generation of neural stem cell/progenitors in vivo (Hitoshi et al. 2002).

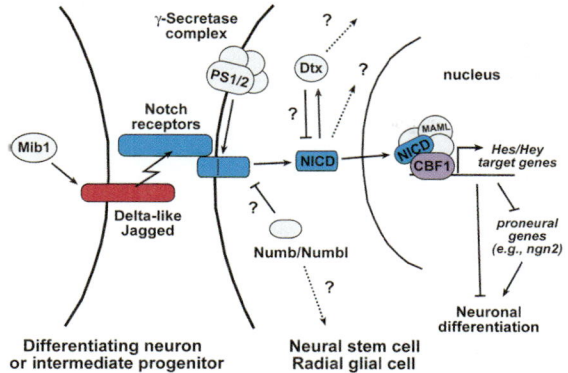

Figure 2. The Notch signaling pathway depicted between two cells in the neocortical germinal zone. The ligand-expressing cell may be either a neuron or an intermediate progenitor, whereas the signal-receiving cell is inhibited from undergoing neuronal differentiation. (PS1/2) Presenilin 1 and 2, the catalytic subunits of the γ-secretase complex; (Mib1) mindbomb1; (Dtx) Deltex; (NICD) Notch intracellular domain; (MAML) Mastermind-like; (CBF1) C-promoter-binding factor 1.

Notch Signaling: Puzzles and Progress

Although the core components of the Notch signaling cascade are known, both the nature of the cell–cell contacts that activate that signaling and the molecular mechanisms that modulate it remain areas of active investigation. Most prominent among pathway elements with uncertain function are the Deltex, Numb, and Numblike proteins. For example, some studies provide evidence that Deltex proteins (of which there are four in mammals) positively regulate Notch signaling (Xu and Artavanis-Tsakonas 1990; Matsuno et al. 1995; Ramain et al. 2001). However, other studies have suggested that Deltex can inhibit Notch/CBF1-dependent signaling, while positively transmitting a Notch signal that is CBF1-independent (Yamamoto et al. 2001; Patten et al. 2006).

Another ongoing area of some confusion in the Notch field relates to the role of the modulators *numb* and *numblike* (*numbl*). Numb was first identified in *Drosophila*, where it was shown to be an inhibitor of Notch during neural development (Spana and Doe 1996). Numb protein is localized asymmetrically during cell division such that one of the daughter cells receives most of the Numb protein and will therefore have Notch signaling inhibited, whereas the other daughter will not. The role of Numb with respect to Notch in the mammalian nervous system has been far less clear (see Yoon and Gaiano 2005; Petersen et al. 2006). Several *numb* and *numbl* knockout studies have been reported, but they show contradictory findings, with some indicating that Numb and Numblike promote differentiation (Zilian et al. 2001; Li et al. 2003) and others showing that they promote maintenance of the progenitor pool (Zhong et al. 2000; Petersen et al. 2002, 2004). Other functions have recently been ascribed to Numb and Numblike, including regulation of cell adhesion (Kuo et al. 2006; Rasin et al. 2007; Zhou et al. 2007), but how such functions relate to Notch signaling, if at all, remains to be determined.

Although confusion persists with respect to certain aspects of the Notch cascade, progress has recently been made in other areas, including cell–cell signaling during neocortical development. As indicated above, Notch signaling occurs between a signal-sending cell (expressing Notch ligands) and a signal-receiving cell (expressing Notch receptors). Traditionally, it has been assumed that the signal-sending cells were newborn neurons and this remains likely to be true to some extent. However, recent work examining the role of the protein mindbomb-1 (mib1), an E3 ubiquitin ligase required for Notch ligand endocytosis and consequently receptor activation, suggests that IPCs also have a role in activating Notch receptors to maintain RGC character in the VZ (Yoon et al. 2008). Conditional inactivation of *mib1* in the developing neocortex results in the loss of Notch activation, depletion of RGCs, and premature differentiation into IPCs and neurons. Of particular interest, this study provides strong evidence that ligands on one progenitor cell type (IPCs) can activate Notch receptors on another progenitor cell type (RGCs). In addition, this study details one of the best examples of Notch loss of function in the mammalian nervous system resulting in precocious neuronal differentiation at the expense of progenitor maintenance.

In another recent advance with respect to Notch signaling, real-time imaging has revealed that expression of the target gene *Hes1* oscillates in embryonic neocortical progenitors in a manner similar to what has been shown before and during somitogenesis (Shimojo et al. 2008). As a direct consequence of *Hes1* oscillations, the expression of *Ngn2* and *Dll1* also oscillate in neural progenitors but inversely with respect to *Hes1*. This work suggests that neocortical progenitors are maintained by a pulsatile inhibition of neurogenesis, but they are poised to be driven toward neurogenesis upon a sustained drop in *Hes1* and the resulting sustained increase in *Ngn2*. Many questions remain to be addressed, including whether the timing of the *Hes1* oscillations (2-hour periodicity) is important and how the decision to stabilize low (or potentially high) levels of *Hes1* is achieved.

Differential Notch Signaling in Neural Progenitors

Our group has recently shown that differential Notch use in neural progenitors exists and likely contributes to the generation of progenitor heterogeneity, in particular in the VZ (Mizutani et al. 2007). This was first observed using a transgenic Notch reporter (TNR) line that expresses enhanced green fluorescent protein (EGFP) as a readout of CBF1-dependent Notch signaling. The transgene includes four CBF1-binding sites and the SV40 basal promoter, together referred to as the CBF1-responsive element (CBFRE), upstream of EGFP. Examination of either transgene expression or reporter expression from constructs delivered into the embryonic brain transiently revealed VZ progenitor heterogeneity (Fig. 3). As indicated above, those VZ cells that contained canonical Notch signaling (and thus were EGFP+) were found to be RGCs, which drove expression from the GLAST promoter. In addition, those VZ cells that exhibited attenuated canonical Notch signaling (and thus expressed little to no detectable EGFP) drove expression from the Tα1 pro-

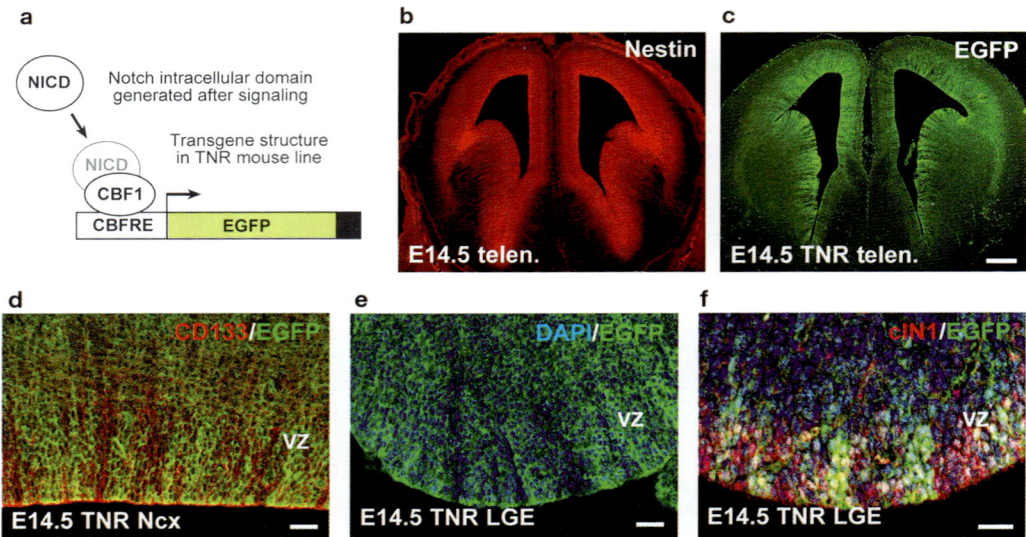

Figure 3. Notch signaling heterogeneity exists in the embryonic forebrain. Strategy to detect CBF1 activation in the TNR line (*a*). Nestin (*b*) is expressed in the telencephalic (telen.) VZ, as is EGFP in TNR embryos (*c*). EGFP^{hi} cells in the E14.5 neocortex (Ncx, *d*) and lateral ganglionic eminence (LGE, *e*) are interspersed with EGFP$^{lo/neg}$ cells. Double labeling reveals that cleaved (activated) Notch1 (*red*) is present in cells expressing high and low levels of EGFP (*f*). Bars: (*b,c*) 200 µm; (*d–f*) 25 µm. (Reprinted from Mizutani et al. 2007 [© Nature Publishing Group].).

moter and were likely neurogenic progenitors (Fig. 4). As suggested above, here we will call such cells aINPs for apical (VZ) intermediate neural progenitors.

Additional in vivo analysis revealed that aINPs did not activate the CBFRE even when NICD1 expression driven in those cells could inhibit their differentiation (Mizutani et al. 2007). Furthermore, the clustered organization of aINPs raised the interesting possibility that those cells possess a heritable block to CBF1 activation. We per-

formed additional analysis of RGCs and aINPs by using fluorescence-activated cell sorting (FACS) to isolate cells that were expressing the apical progenitor marker CD133 and were either EGFP$^+$ or EGFP$^-$. Consistent with the classification of those cells as RGCs and aINPs, respectively, the former were found to exhibit stem cell character in vitro and after in vivo transplantation, whereas the latter were biased primarily toward becoming neurons, but also oligodendrocytes in vivo.

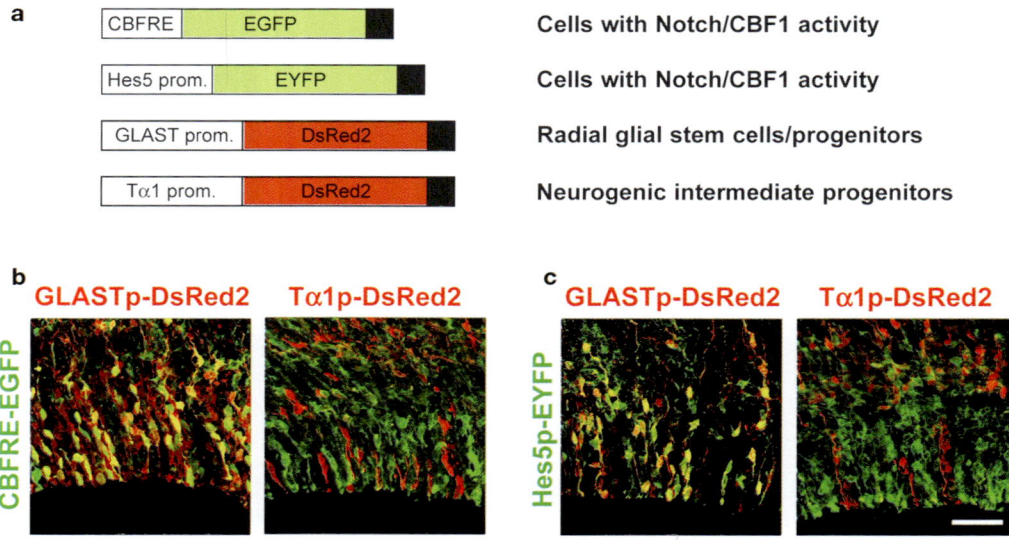

Figure 4. In utero electroporation was used to deliver plasmids designed to identify cells with CBF1 activity (CBFRE-EGFP, Hes5p-EYFP), radial glial NSC (GLASTp-DsRed2), or INP character (Tα1p-DsRed2). (*a*) After E12.5 coelectroporations, VZ cells with CBF1 activity (*green*) at E15.5 are predominantly GLASTp-DsRed2$^+$ and Tα1p-DsRed2$^-$ (*b,c*). Bar, 50 µm. (Modified from Mizutani et al. 2007 [© Nature Publishing Group].).

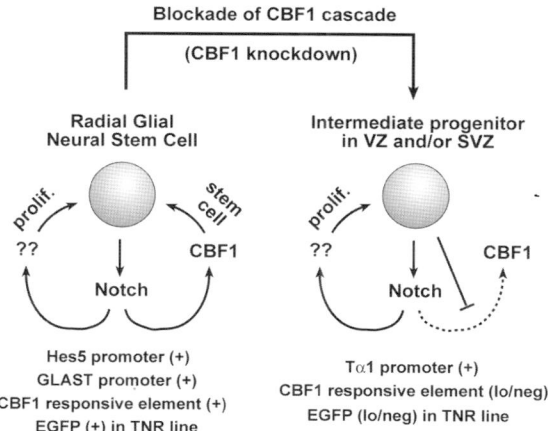

Figure 5. A model summarizing the molecular characteristics of radial glial neural stem cells (NSCs), and intermediate neural progenitors (INPs). Both cell types use Notch signaling to remain undifferentiated. However, the former utilize the canonical CBF1 signaling cascade, whereas the latter possess a heritable block to CBF1 signaling. (Reprinted from Mizutani et al. 2007 [© Nature Publishing Group].)

All told, our results indicate that the telencephalic VZ contains two progenitor subtypes that can be distinguished based on Notch/CBF1 signaling levels (Fig. 5). How Notch signaling in RGCs and aINPs is differentially regulated remains unknown, although it appears that aINPs possess a yet to be identified regulator of Notch signaling (a known or novel modulator) that can antagonize NICD signaling through CBF1 or might instead shunt NICD to noncanonical targets (CBF1-dependent or otherwise). Interestingly, we have shown that CBF1 knockdown, using short hairpin RNA (shRNA), promotes the conversion of RGCs to aINPs in vivo, suggesting that the differential Notch/CBF1 signaling observed between RGCs and aINPs is not just correlative, but has a causative role in the transition from one cell type to the other. Future studies will provide novel insight into how Notch signaling regulates different neural progenitor subtypes and into the general mechanisms of Notch pathway regulation.

NOTCH IN BRAIN CANCER

Given the significant role of the Notch pathway in development, it is not surprising that it would also have an important role in oncogenesis. Notch has been known to be active in cancers for some time. In 1991, it was shown that a translocation in T-cell acute lymphoblastic leukemia (T-ALL) causes the expression of a truncated, constitutively active form of Notch1 (Ellisen et al. 1991). Subsequently, it has been demonstrated that Notch pathway dysregulation is quite common in a wide range of hematopoietic and epithelial cancers (Bolos et al. 2007; Aster et al. 2008). In most tumors, Notch acts as an oncogene. In some cases, however, it appears that Notch has a protective effect against the formation of tumors, as has been shown for Notch1 in the skin (Nicolas et al. 2003).

Notch dysregulation was first demonstrated in brain can-

cers rather recently, when it was found that cell lines derived from malignant human gliomas overexpress the Notch ligands *Jag1* and *Dll1* (Ignatova et al. 2002). It has subsequently been shown that Notch pathway alterations are present in a wide range of brain tumor types, including neoplasms resembling stem-like embryonic neural cells, various glial cell types, and meningothelial cells. Many of these tumors are extremely aggressive, with no known effective therapy, and the possibility of targeting such cancers using agents that block Notch signaling has generated considerable excitement (Dirks 2008; Fan and Eberhart 2008). As such, the fact that Notch may have an important role in stem-like brain tumor cells is of great clinical interest.

Astrocytomas

The gliomas, a heterogeneous group of tumors, are classified based on their morphological resemblance to glial cell types, and they include multiple grades of astrocytomas, oligodendrogliomas, and ependymomas (Louis et al. 2007). Accumulating evidence suggests that Notch pathway components are expressed in gliomas and have an important functional role in their formation and growth. The most malignant astrocytic tumors, known as glioblastomas, overexpress Notch receptors and ligands (Ignatova et al. 2002; Purow et al. 2005). Inhibition of the pathway via *Notch1-*, *Jag1-* and *Dll1-*directed small interfering RNA (siRNA) leads to decreased glioblastoma proliferation in vitro, increased apoptosis, and decreased tumorgenicity in vivo (Purow et al. 2005). Similarly, pathway inhibition with a γ-secretase inhibitor (which prevents the activating cleavage of Notch receptors, thereby preventing NICD translocation to the nucleus) represses growth of glioma cell lines (Kanamori et al. 2007). Conversely, overexpression of activated Notch1 in a human glioma cell line results in increased colony (neurosphere)-forming ability and cell growth rate (Zhang et al. 2008). In further support of an oncogenic role for Notch, it has been shown that small genomic deletions including the *Notch2* locus correlate with better survival in human glioma patients (Boulay et al. 2007).

Expression profiles of Notch pathway members appear to correlate with tumor grade or clinical outcome in various ways. In one large study that included multiple sample sets, a microarray analysis of glioblastomas suggested that the tumors could be divided into three subtypes based on their gene expression profiles and that these subtypes correlated with age of onset and prognosis (Phillips et al. 2006). The subclass associated with the best prognosis was the "proneural" class, which was shown to have increased expression of genes associated with normal neurogenesis, as well as the Notch ligands *Dll1* and *Dll3*, the Notch target *Hey2*, and the proneural gene *Mash1* (Phillips et al. 2006). The other subclasses, associated with less favorably prognoses, did not have increased expression of Notch pathway components and instead showed a strong correlation with epidermal growth factor receptor (EGFR) activation.

At first glance, the above findings may seem to contradict the notion that Notch has an oncogenic role in gliomas. However, these data are in agreement with other recent observations on Notch component expression in gliomas of varying grades. For example, proportionately fewer glioblastomas overexpress NICD1 as compared with other less malignant astrocytic and oligodendroglial tumors (Cheung et al. 2006). Additionally, it has been shown that *Mash1*, which is inhibited by the Notch pathway, is up-regulated in some types of high-grade gliomas, possibly due to a decrease in Notch pathway activity (Somasundaram et al. 2005). These data suggest that Notch pathway activation may be important for low-grade gliomas but that the pathway is turned off in a portion of higher-grade tumors. Alternatively, some high-grade tumors may form independent of Notch pathway overexpression, whereas low-grade gliomas may be more commonly initiated via a Notch-dependent mechanism.

Other studies have yielded at least partially contradictory results with respect to the role of Notch in astrocytomas. In contrast to the decreased expression of active Notch1 reported by some researchers in more malignant astrocytomas, at least one study reported a positive correlation between increased levels of the Notch pathway target *Hey1* and increased tumor grade (Hulleman et al. 2008). Interactions of Notch with other oncogenic pathways are complicated and are likely to contribute to seemingly contradictory findings with respect to the role of Notch in particular. For example, one recent study suggested that EGFR is regulated by Notch1 via a p53-dependent mechanism (Purow et al. 2008), complicating the picture painted by Phillips and colleagues. All told, it is clear that there are many gaps in our understanding of the role Notch has signaling in gliomas, although progress is certainly being made.

Ependymomas

The Notch pathway also affects other types of glial brain tumors, including ependymomas, in which Notch pathway expression patterns have been used to infer the cell of origin. Taylor et al. (2005) have found that increased expression of the Notch pathway ligand Jag1 is present in supratentorial ependymomas but not in those arising in the posterior fossa or spinal cord. Because this pattern correlates with that seen in non-neoplastic radial glial cells, the authors have suggested that these act as cells of origin for supratentorial ependymomas, with radial glia identified using other markers serving similar roles in the more caudal neuroaxis. We have also detected high-level expression of Notch ligands, receptors, and targets in human ependymoma samples, and it appears that Notch receptors cluster at the tips of the distinctive tumor processes that rosette around blood vessels (C. Eberhart, unpubl.).

Medulloblastomas

Medulloblastomas are malignant embryonal brain tumors that arise in the cerebellum, usually in children. They are thought to develop from committed progenitor cells in the external germinal layer (EGL) or from stem/progenitor cells along the ventricles (Eberhart 2007). Notch2 activity is present in the developing cerebellar EGL and has a role in maintaining this proliferative pool

of progenitors, whereas Notch1 is expressed in differentiating cells (Solecki et al. 2001). Interestingly, we have observed that Notch2 promotes the growth of medulloblastoma cell lines, whereas Notch1 does not have this effect and in fact inhibits growth (Fan et al. 2004). These findings suggest that the two Notch receptors recapitulate their roles in normal cerebellar development in precursor-derived cerebellar tumors. Other groups have also shown that Notch pathway components, specifically *Notch2*, *Jag1*, and *Hes1*, show increased expression in mouse medulloblastoma models (Hallahan et al. 2004; Dakubo et al. 2006). It has also been observed that knockdown of Notch pathway activity in Sonic hedgehog-initiated tumors leads to increased cell death (Hallahan et al. 2004), indicating that the Notch pathway has an important role in medulloblastoma tumor maintenance.

Other Brain Tumors

Two final low-grade brain tumor types in which Notch has been implicated are meningioma and choroid plexus papilloma (CPP). In meningioma cell lines, introduction of activated Notch1 or Notch2 causes tetraploidy, and it has been speculated that Notch has a role in meningioma initiation (Baia et al. 2008). We have demonstrated more directly a tumor-initiating role for Notch3 in murine tumors. We found that constitutively active Notch3 (NICD3) is capable of causing CPP formation in mice when retrovirally expressed during neural development (Dang et al. 2006a). Strikingly, Notch1 does not have the same effect when injected at the same times during development, indicating that the various Notch receptors are not functionally equivalent with respect to tumor initiation in the mouse brain (Gaiano et al. 2000). We also found that Notch receptors and targets were expressed in CPPs from human patients (Dang et al. 2006a).

Brain Tumor Stem Cells

The idea that brain tumors contain a small population of stem-like cells capable of self-renewal, indefinite division, and "differentiation" into all of the cell types found in a given tumor is gaining acceptance in the field (Reya et al. 2001; Zhou and Zhang 2008). However, the identifying characteristics of cancer stem cells remain unclear. Relatively recently, it was shown that brain tumors contain subpopulations of cells that proliferate through numerous passages under neural stem cell culture conditions, give rise to a heterogeneous population of cells, and form xenografts recapitulating the phenotype of the tumors from which they were derived (Hemmati et al. 2003; Galli et al. 2004; Singh et al. 2004). Tumor-derived neurosphere colonies containing such stem-like cells have been derived from human glioblastomas (Galli et al. 2004), medulloblastomas, pilocytic astrocytomas, gangliogliomas (Singh et al. 2003), and ependymomas (Taylor et al. 2005).

The identification of robust molecular markers of cancer stem cells has been a significant challenge. One group used CD133, a cell surface marker previously shown to be present in neural progenitors (Uchida et al. 2000), to enrich for stem-like brain tumor cells. However, it has also been shown that not all brain tumors contain CD133$^+$ stem-like cells, and that CD133$^-$ cells can initiate xenografts from these tumors (Beier et al. 2007). In addition to CD133, stem-like brain tumor cells have also been prospectively identified on the basis of their expression of ABC-type transporters and resulting ability to efflux Hoechst dye 33342 faster than more differentiated cells in the population (Patrawala et al. 2005). This "side population" (SP, as defined by FACS) has also been demonstrated to be enriched for neural progenitors in fetal and adult brain (Murayama et al. 2002). In brain cancer, it was shown that the SP is enriched for cells able to form more clones in a neurospheres assay, suggesting that it coincides with cancer stem cells in brain tumors (Patrawala et al. 2005).

The intermediate filament Nestin can also serve as a marker of undifferentiated or poorly differentiated cells in brain tumors. Nestin was first found to be expressed in NEPs in the developing central nervous system (Lendahl et al. 1990) and is now known to be expressed in both stem cells and more differentiated neural progenitor cells (Wiese et al. 2004). Interestingly, it has been reported that Notch signaling is able to drive *nestin* gene transcription, and coexpression of the two proteins has been shown in Kras-induced tumors (Shih and Holland 2006). These data further support the idea that Notch may have a role in regulating cell fate and stem cell maintenance in brain tumors.

The role of Notch in maintaining neural progenitors has been well established, yet our understanding of the role it has in stem-like tumor cells is much less advanced. We have shown that inhibition of the Notch pathway by a γ-secretase inhibitor leads to an up to fivefold decrease in CD133$^+$ cells and a complete loss of side population within medulloblastoma cell lines (Fan et al. 2006). We have also demonstrated that the growth of these cell lines was dramatically decreased by the Notch blockade and that the treated cell lines were unable to form colonies in soft agar or xenografts in athymic mice (Fan et al. 2006). These results suggest that the stem-cell-like population in these tumors may have been eliminated as a consequence of the Notch blockade.

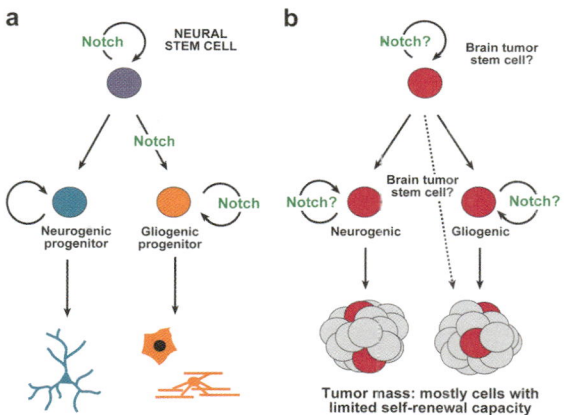

Figure 6. Schematic representation depicting the idea that there may be parallels between embryonic neural progenitor subtypes and brain tumor stem cells and that Notch signaling, well known to have a role in regulating the former (*a*), may also have an aberrant role in regulating the latter (*b*).

based on the cancer stem cell model, that Notch signaling contributes to brain cancer formation by regulating brain tumor stem cells in a manner akin to how the pathway normally regulates neural progenitors (see Fig. 6). Aberrant Notch signaling could maintain an undifferentiated and proliferative active state that would contribute to tumor formation and growth. Furthermore, our recent findings that (1) canonical Notch signaling is used in embryonic neural stem cells (i.e., RGCs) but not in more restricted neurogenic progenitors (INPs) (Mizutani et al. 2007) and (2) Notch inhibition preferentially ablates stem-like tumor cells while sparing other proliferating cell types (Fan et al. 2006) suggest that understanding the heterogeneous nature of neural stem/progenitor cells in normal and neoplastic contexts is likely to provide insight directly relevant to the eventual eradication of brain cancer.

CONCLUSIONS

The Notch signaling pathway has not only a vital role in the developing mammalian nervous system, but also an important role in oncogenesis. In neural development, molecular interactions among progenitors and/or more differentiated cells activate Notch receptors to maintain a neural stem/progenitor state and inhibit neuronal differentiation. Notch signaling has also been implicated in many cancers, including brain cancer, where pathway components are aberrantly expressed. The pathway appears in most cases to promote brain tumor formation and growth, but in some contexts, Notch signaling may suppress tumors or help to define a subset of patients with less aggressive disease.

Although the data we have reviewed summarize much of what is known about Notch in neural progenitors during development, and Notch in brain tumors, it is currently not clear how to connect these two aspects of Notch signaling directly. It is tempting to speculate,

REFERENCES

Aster, J.C., Pear, W.S., and Blacklow, S.C. 2008. Notch signaling in leukemia. *Annu. Rev. Pathol.* **3:** 587–613.

Baia, G., Stifani, S., Kimura, E.T., McDermott, M.W., Pieper, R.O., and Lal, A. 2008. Notch activation is associated with tetraploidy and enhanced chromosomal instability in meningiomas. *Neoplasia* **10:** 604–612.

Beier, D., Han, P., Proescholdt, M., Lohmeier, A., Wischhusen, J., Oefner, P.J., Aigner, L., Brawanski, A., Bogdahn, U., and Beier, C.P. 2007. CD133$^+$ and CD133$^-$ GBM-derived CaSCs show differential growth characteristics and molecular profiles. *Cancer Res.* **67:** 4010–4015.

Bertrand, N., Castro, D.S., and Guillemot, F. 2002. Proneural genes and the specification of neural cell types. *Nat. Rev. Neurosci.* **3:** 517–530.

Bolos, V., Grego-Bessa, J., and Pompa, J.L. 2007. Notch signaling in development and cancer. *Endocr. Rev.* **28:** 339–363.

Boulay, J.L., Miserez, A.R., Zweifel, C., Sivasankaran, B., Kana, V., Ghaffari, A., Luyken, C., Sabel, M., Zerrouqi, A., Wasner, M., Van Meirs, E., Tolnay, M., Reifenberger, G., and Merlo, A. 2007. Loss of *NOTCH2* positively predicts survival in subgroups of human glial brain tumors. *PLoS One* **2:** e576.

Cheung, H.C., Corley, L.J., Fuller, G.N., McCutcheon, I.E., and

Cote, G.J. 2006. Polypyramidine tract binding protein and Notch1 are independently re-expressed in glioma. *Mod. Pathol.* **19:** 1034–1041.

Dakubo, G.D., Mazerole, C.J., and Wallace, V.A. 2006. Expression of Notch and Wnt pathway components and activation of Notch signaling in medulloblastomas from heterozygous patched mice. *J. Neuro-Oncol.* **79:** 221–227.

Dang, L., Fan, X., Chaudhry, A., Wang, M., Gaiano, N., and Eberhart, C.G. 2006a. Notch3 signaling initiates choroid plexus tumor formation. *Oncogene* **25:** 487–491.

Dang, L., Yoon, K., Wang, M., and Gaiano, N. 2006b. Notch3 signaling promotes radial glial/progenitor character in the mammalian telencephlaon. *Dev. Neurosci.* **28:** 58–69.

Dirks, P.B. 2008. Brain tumor stem cells: Bringing order to the chaos of brain cancer. *J. Clin. Oncol.* **26:** 2916–2924.

Eberhart, C.E. 2007. In search of the medulloblast: Neural stem cells and embryonal brain tumors. *Neurosurg. Clin. N. Am.* **18:** 59–69.

Ellisen, L.W., Bird, J., West, D.C., Soreng, A.L., Reynolds, T.C., Smith, S.D., and Sklar, J. 1991. *TAN*-1, the human homolog of the *Drosophila Notch* gene, is broken by chromosomal translocations in T lymphoblastic neoplasms. *Cell* **66:** 649–661.

Englund, C., Fink, A., Lau, C., Pham, D., Daza, R.A., Bulfone, A., Kowalczyk, T., and Hevner, R.F. 2005. Pax6, Tbr2, and Tbr1 are expressed sequentially by radial glia, intermediate progenitor cells, and postmitotic neurons in developing neocortex. *J. Neurosci.* **25:** 247–251.

Fan, X. and Eberhart, C.G. 2008. Medulloblastoma stem cells. *J. Clin. Oncol.* **26:** 2821–2827.

Fan, X., Mikolaenko, I., Elassan, I., Ni, X.Z., Wang, Y., Ball, D., Brat, D.J., Perry, A., and Eberhart, C.G. 2004. *Notch1* and *Notch2* have opposite effects on embryonal brain tumor growth. *Cancer Res.* **64:** 7787–7793.

Fan, X., Matsui, W., Khaki, L., Stearns, D., Chun, J., Li, Y.M., and Eberhart, C.G. 2006. Notch pathway inhibition depletes stem-like cells and blocks engraftment in embryonal brain tumors. *Cancer Res.* **66:** 7445–7452.

Gaiano, N. and Fishell, G. 2002. The role of Notch in promoting glial and neural stem cell fates. *Annu. Rev. Neurosci.* **25:** 471–490.

Gaiano, N., Nye, J.S., and Fishell, G. 2000. Radial glial identity is promoted by Notch1 signaling in the murine forebrain. *Neuron* **26:** 395–404.

Gal, J.S., Morozov, Y.M., Ayoub, A.E., Chatterjee, M., Rakic, P., and Haydar, T.F. 2006. Molecular and morphological heterogeneity of neural precursors in the mouse neocortical proliferative zone. *J. Neurosci.* **26:** 1045–1056.

Galli, R., Binda, E., Orfanelli, U., Cipelletti, B., Gritti, A., De Vitis, S., Fiocco, R., Foroni, C., Dimeco, F., and Vescovi, A. 2004. Isolation and characterization of tumorgenic, stem-like neural precursors from human glioblastoma. *Cancer Res.* **64:** 7011–7021.

Götz, M. and Barde, Y. 2005. Radial glial cells defined and major intermediates between embryonic stem cells and CNS neurons. *Neuron* **46:** 369–372.

Hallahan, A.R., Pritchard, J.I., Hansen, S., Benson, M., Stoeck, J., Hatton, B.A., Russell, T.L., Ellenbogen, R.G., Bernstein, I.D., Beachy, P.A., and Olson, J. 2004. The SmoA1 mouse model reveals that Notch signaling is critical for the growth and survival of sonic hedgehog-induced medulloblastomas. *Cancer Res.* **64:** 7794–7800.

Hemmati, H.D., Nakano, I., Lazareff, J.A., Masterman-Smith, M., Geschwind, D.H., Bronner-Fraser, M., and Kornblum, H.I. 2003. Cancerous stem cells can arise from pediatric brain tumors. *Proc. Natl. Acad. Sci.* **100:** 15178–15183.

Hitoshi, S., Alexson, T., Tropepe, V., Donoviel, D., Elia, A.J., Nye, J.S., Conlon, R.A., Mak, T.W., Berstein, A., and van der Kooy, D. 2002. Notch pathway molecules are essential for the maintenance, but not the generation, of mammalian neural stem cells. *Genes Dev.* **16:** 846–858.

Hulleman, E., Quarto, M., Vernell, R., Masserdotti, G., Colli, E., Kros, J.M., Levi, D., Gaetani, P., Tunici, P., Finocchiaro, G., Rodriguez y Baena, R., Capra, M., and Helin, K. 2008. A role for the transcription factor HEY1 in glioblastoma. *J. Cell. Mol.*

Med. **13:** 136–146.

Ignatova, T.N., Kukekov, V.G., Laywell, E.D., Suslov, O.N., Vrionis, F.D., and Steindler, D.A. 2002. Human cortical glial tumors contain neural stem-like cells expressing astroglial and neuronal markers in vitro. *Glia* **39:** 193–206.

Iso, T., Kedes, L., and Hamamori, Y. 2003. HES and HERP families: Multiple effectors of the Notch signaling pathway. *J. Cell. Physiol.* **194:** 237–255.

Kanamori, M., Kawaguchi, T., Nigro, J.M., Feuerstein, B.G., Berger, M.S., Miele, L., and Pieper, R.O. 2007. Contribution of Notch signaling activation to human glioblastoma multiforme. *J. Neurosurg.* **106:** 417–427.

Kuo, C.T., Mirzadeh, Z., Soriano-Navarro, M., Rasin, M., Wang, D., Shen, J., Sestan, N., Garcia-Verdugo, J., Alvarez-Buylla, A., Jan, L.Y., and Jan, Y.N. 2006. Postnatal deletion of Numb/Numblike reveals repair and remodeling capacity in the subventricular neurogenic niche. *Cell* **127:** 1253–1264.

Lendahl, U., Zimmerman, L.B., and McKay, R.D. 1990. CNS stem cells express a new class of intermediate filament protein. *Cell* **60:** 585–595.

Li, H.S., Wang, D., Shen, Q., Schonemann, M.D., Gorski, J.A., Jones, K.R., Temple, S., Jan, L.Y., and Jan, Y.N. 2003. Inactivation of Numb and Numblike in embryonic dorsal forebrain impairs neurogenesis and disrupts cortical morphogenesis. *Neuron* **40:** 1105–1118.

Louis, D.N., Ohgaki, H., Wiestler, O.D., and Cavenee, W.A. 2007. *WHO classification of tumours of the central nervous system.* IARC Press, Lyon, France.

Louvi, A. and Artavanis-Tsakonas, S. 2006. Notch signaling in vertebrate neural development. *Nat. Rev. Neurosci.* **7:** 93–102.

Malatesta, P., Hack, M.A., Hartfuss, E., Kettenmann, H., Klinkert, W., Kirchhoff, F., and Götz, M. 2003. Neuronal or glial progeny: Regional differences in radial glia fate. *Neuron* **37:** 751–764.

Molyneaux, B.J., Arlotta, P., Menezes, J.R., and Macklis, J.D. 2007. Neuronal subtype specification in the cerebral cortex. *Nat. Rev. Neurosci.* **8:** 427–437.

Matsuno, K., Diederich, R.J., Go, M.J., Blaumueller, C.M., and Artavanis-Tsakonas, S. 1995. Deltex acts as a positive regulator of Notch signaling through interactions with the Notch ankyrin repeats. *Development* **121:** 2633–2644.

Mizutani, K., Yoon, K., Dang, L., Tokunaga, A., and Gaiano, N. 2007. Differential Notch signaling distinguishes neural stem cells from intermediate progenitors. *Nature* **449:** 351–355.

Murayama, A., Matsuzaki, Y., Kawaguchi, A., Shimazaki, T., and Okano, H. 2002. Flow cytometric analysis of neural stem cells in the developing and adult mouse brain. *J. Neurosci. Res.* **15:** 837–847.

Nicolas, M., Wolfer, A., Raj, K., Kummer, J., Mill, P., van Noort, M., Hui, C., Levers, H., Dotto, G.P., and Radtke, F. 2003. Notch1 functions as a tumor suppressor in mouse skin. *Nat. Genet.* **33:** 416–421.

Nieto, M., Monuki, E.S., Tang, H., Imitola, J., Haubst, N., Khoury, S.J., Cunningham, J., Gotz, M., and Walsh, C.A. 2004. Expression of Cux-1 and Cux-2 in the subventricular zone and upper layers II–IV of the cerebral cortex. *J. Comp. Neurol.* **479:** 168–180.

Noctor, S.C., Flint, A.C., Weissman, T.A., Dammerman, R.S., and Kriegstein, A.R. 2001. Neurons derived from radial glial cells establish radial units in neocortex. *Nature* **409:** 714–720.

Noctor, S.C., Martinez-Cerdeno, V., Ivic, L., and Kriegstein, A.R. 2004. Cortical neurons arise in symmetric and asymmetric division zones and migrate through specific phases. *Nat. Neurosci.* **7:** 136–144.

Patrawala, L., Calhoun, T., Schneider-Broussard, R., Zhou, J., Claypool, K., and Tang, D.G. 2005. Side population is enriched in tumorigenic, stem-like cancer cells, whereas ABCG2$^+$ and ABCG2$^-$ cancer cells are similarly tumorigenic. *Cancer Res.* **65:** 6207–6219.

Patten, B.A., Sardi, S.P., Koirala, S., Nakafuku, M., and Corfas, G. 2006. Notch1 signaling regulates radial glia differentiation through multiple transcriptional mechanisms. *J. Neurosci.* **26:** 3102–3108.

Petersen, P.H., Zou, K., Hwang, J.K., Jan, Y.N., and Zhong, W. 2002. Progenitor cell maintenance requires *numb* and *numb-*

like during mouse neurogenesis. *Nature* **419**: 929–934.

Petersen, P.H., Zou, K., Krauss, S., and Zhong, W. 2004. Continuing role for mouse *Numb* and *Numbl* in maintaining progenitor cells during cortical neurogenesis. *Nat. Neurosci.* **7**: 803–811.

Petersen, P.H., Tang, H., Zou, K., and Zhong, W. 2006. The enigma of the Numb-Notch relationship during mammalian embryogenesis. *Dev. Neurosci.* **28**: 156–168.

Phillips, H.S., Kharbanda, S., Chen, R., Forrest, W.F., Soriano, R.H., Wu, T.D., Misra, A., Nigro, J.M., Colman, H., Soroceanu, L., Williams, P.M., Modrusan, Z., Feuerstein, B.G., and Aldape, K. 2006. Molecular subclasses of high-grade glioma predict prognosis, delineate a pattern of disease progression, and resemble stages in neurogenesis. *Cancer Cell* **9**: 157–173.

Pontious, A., Kowalczyk, T., Englund, C., and Hevner, R.F. 2007. Role of intermediate progenitor cells in cerebral cortex development. *Dev. Neurosci.* **30**: 24–32.

Purow, B.W., Haque, R.M., Noel, M.W., Su, Q., Burdick, M.J., Lee, J., Sundaresan, T., Pastorino, S., Park, J.K., Mikolaenko, I., Maric, D., Eberhart, C.G., and Fine, H.A. 2005. Expression of Notch-1 and its ligands, Delta-like-1 and Jagged-1, is critical for glioma cell survival and proliferation. *Cancer Res.* **65**: 2353–2363.

Purow, B.W., Sundaresan, T.K., Burdick, M.J., Kefas, B.A., Comeau, L.D., Hawkinson, M.P., Su, Q., Kotliarov, Y., Lee, J., Zhang, W., and Fine, H.A. 2008. Notch-1 regulates transcription of the epidermal growth factor receptor through p53. *Carcinogenesis* **29**: 918–925.

Ramain, P., Khechumian, K., Seugnet, L., Arbogast, N., Ackermann, C., and Heitzler, P. 2001. Novel Notch alleles reveal a Deltex-dependent pathway repressing neural fate. *Curr. Biol.* **11**: 1729–1738.

Rasin, M.R., Gazula, V.R., Breunig, J.J., Kwan, K.Y., Johnson, M.B., Liu-Chen, S., Li, H.S., Jan, L.Y., Jan, Y.N., Rakic, P., and Sestan, N. 2007. Numb and Numbl are required for maintenance of cadherin-based adhesion and polarity of neural progenitors. *Nat. Neurosci.* **10**: 819–827.

Reya, T., Morrison, S.J., Clarke, M.F., and Weissman, I.L. 2001. Stem cells, cancer and cancer stem cells. *Nature* **414**: 105–111.

Shih, A.H. and Holland, E.C. 2006. Notch signaling enhances nestin expression in gliomas. *Neoplasia* **12**: 1072–1082.

Shimojo, H., Ohtsuka, T., and Kageyama, R. 2008. Oscillations in Notch signaling regulate maintenance of neural progenitors. *Neuron* **58**: 52–64.

Singh, S.K., Clarke, I.D., Terasaki, M., Bonn, V.E., Hawkins, C., Squire, J., and Dirks, P.B. 2003. Identification of a cancer stem cell in human brain tumors. *Cancer Res.* **63**: 5821–5828.

Singh, S.K., Hawkins, C., Clarke, I.D., Squire, J.A., Bayani, J., Hide, T., Henkelman, R.M., Cusimano, M.D., and Dirks, P.B. 2004. Identification of human brain tumor initiating cells. *Nature* **432**: 396–401.

Solecki, D.J., Liu, X.L., Tomoda, T., Fang, Y., and Hatten, M.E. 2001. Activated *Notch2* signaling inhibits differentiation of cerebellar granule neuron precursors by maintaining proliferation. *Neuron* **31**: 557–568.

Somasundaram, K., Reddy, S.P., Vinnakota, K., Britto, R., Subbarayan, M., Nambiar, S., Hebar, A., Samuel, C., Shetty, M., Sreepathi, H.K., Santosh, V., Hedge, A.S., Hedge, S.,

Kondaiah, P., and Rao, M.R. 2005. Upregulation of *ASCL1* and inhibition of Notch signaling pathway characterize progressive astrocytoma. *Oncogene* **24**: 7073–7083.

Spana, E.P. and Doe, C.Q. 1996. Numb antagonizes Notch signaling to specify sibling neuron cell fates. *Neuron* **17**: 21–26.

Tarabykin, V., Stoykova, A., Usman, N., and Gruss, P. 2001. Cortical upper layer neurons derive from the subventricular zone as indicated by *Svet1* gene. *Development* **128**: 1983–1993.

Taylor, M.D., Poppleton, H., Fuller, C., Su, X., Liu, Y., Jensen, P., Magdaleno, S., Dalton, J., Calabrese, C., Board, J., Macdonald, T., Rutka, J., Guha, A., Gajjar, A., Curran, T., and Gilbertson, R.J. 2005. Radial glia cells are candidate stem cells of ependymoma. *Cancer Cell* **8**: 323–335.

Uchida, N., Buck, D.W., He, D., Reitsma, M.J., Masek, M., Phan, T.V., Tsukamoto, A.S., Gage, F.H., and Weissman, I.L. 2000. Direct isolation of human central nervous system stem cells. *Proc. Natl. Acad. Sci.* **97**: 14720–14725.

Wiese, C., Rolletschek, A., Kania, G., Blyszczuk, P., Tarasov, K.V., Tarasova, Y., Wersto, R.P., Boheler, K.R., and Wobus, A.M. 2004. Nestin expression—A property of multi-lineage progenitor cells? *Cell. Mol. Life Sci.* **61**: 2510–2522.

Wu, L., Aster, J.C., Blacklow, S.C., Lake, R., Artavania-Tsakonas, S., and Griffin, J.D. 2000. MAML1, a human homologue of *Drosophila* Mastermind is a transcriptional coactivator for NOTCH receptors. *Nat. Genet.* **26**: 484–489.

Xu, T. and Artavanis-Tsakonas, S. 1990. *deltex*, a locus interacting with the neurogenic genes, *Notch, Delta* and *mastermind* in *Drosophila melanogaster. Genetics* **126**: 665–677.

Yamamoto, N., Yamamoto, S., Inagaki, F., Kawaichi, M., Fukamizu, A., and Kishi, N. 2001. Role of Deltex-1 as a transcriptional regulator downstream of the Notch receptor. *J. Biol. Chem.* **276**: 45031–45040.

Yoon, K. and Gaiano, N. 2005. Notch signaling the mammalian nervous system: Insights from mouse mutants. *Nat. Neurosci.* **8**: 709–715.

Yoon, K., Koo, B., Im, S., Jeong, H., Ghim, J., Kwon, M., Moon, J., Miyata, T., and Kong, Y. 2008. Mind bomb 1-expressing intermediate progenitors generate Notch signaling to maintain radial glial cells. *Neuron* **58**: 519–531.

Zhang, X.P., Zheng, G., Zou, L., Liu, H.L., Hou, L.H., Zhou, P., Yin, D.D., Zheng, Q.J., Liang, L., Zhang, S.Z., et al. 2008. Notch activation promotes cell proliferation and the formation of neural stem cell-like colonies in human glioma cells. *Mol. Cell. Biochem.* **307**: 101–108.

Zhong, W., Jiang, M.-M., Schonemann, M.D., Meneses, J.J., Pedersen, R.A., Jan, L.Y., and Jan, Y.N. 2000. Mouse *numb* is an essential gene involved in cortical neurogenesis. *Proc. Natl. Acad. Sci.* **97**: 6844–6849.

Zhou, J. and Zhang, Y. 2008. Cancer stem cells. *Cell Cycle* **7**: 1360–1370.

Zhou, Y., Atkins, J.B., Rompani, S.B., Bancescu, D.L., Petersen, P.H., Tang, H., Zou, K., Stewart, S.B., and Zhong, W. 2007. The mammalian Golgi regulates numb signaling in asymmetric cell division by releasing ACBD3 during mitosis. *Cell* **129**: 163–178.

Zilian, O., Saner, C., Hagedorn, L., Lee, H.-Y., Säuberli, E., Suter, U., Sommer, L., and Aguet, M. 2001. Multiple roles of mouse Numb in tuning developmental cell fates. *Curr. Biol.* **11**: 494–501.

Human ESC-derived Neural Rosettes and Neural Stem Cell Progression

Y. Elkabetz and L. Studer

Developmental Biology Program, Division of Neurosurgery,
Memorial Sloan-Kettering Cancer Center, New York, New York 10065

Neural stem cells (NSCs) are defined by their ability to self-renew while retaining differentiation potential toward the three main central nervous system (CNS) lineages: neurons, astrocytes, and oligodendrocytes. A less appreciated fact about isolated NSCs is their narrow repertoire for generating specific neuron types, which are generally limited to a few region-specific subtypes such as GABAergic and glutamatergic neurons. Recent studies in human embryonic stem cells have identified a novel neural stem cell stage at which cells exhibit plasticity toward generating a broad range of neuron types in response to appropriate developmental signals. Such rosette-stage NSCs (R-NSCs) are also distinct from other NSC populations by their specific cytoarchitecture, gene expression, and extrinsic growth requirements. Here, we discuss the properties of R-NSCs within the context of NSC biology and define some of the key questions for future investigation. R-NSCs may represent the first example of a NSC population capable of recreating the full cellular diversity of the developing CNS, with implications for both basic stem cell biology and translational applications in regenerative medicine and drug discovery.

The isolation of NSCs from the developing and adult brain has transformed our understanding of CNS development and presented new avenues for regenerative medicine (McKay 1997; Gage 2000). Much effort in the field has been devoted to the identification of factors directing the self-renewal and differentiation of NSCs in vitro and in vivo (Alvarez-Buylla and Lim 2004; Shen et al. 2004; Conti et al. 2005). Such studies demonstrated that single factors act instructively to specify neuronal versus glial fate choice (Johe et al. 1996). However, the ability to direct NSCs toward specific neuronal fates has remained remarkably poor. It is well established that multipotent NSCs from many CNS regions can clonally give rise to neurons, oligodendrocytes, and astrocytes in vitro. However, the neuronal subtype potential of in-vitro–expanded NSCs is largely restricted to GABAergic and, to a lesser extent, glutamatergic fates (Caldwell et al. 2001; Jain et al. 2003). Recent in vivo studies suggest that adult subventricular zone (SVZ) stem cells also exhibit region-specific biases toward specific neuronal subtypes such as $Pax6^+/TH^+$ cells or $calretinin^+/GABA^+$ cells (Merkle et al. 2007). Remarkably, restrictions in neuron subtype potential in the adult SVZ are cell autonomous and retained following extensive in vitro culture (Merkle et al. 2007). These studies suggest that both endogenous NSC populations and NSCs cultured in vitro exhibit developmental and regional bias in neuron subtype differentiation.

In contrast to NSC studies, work with neural plate-stage explants or studies with embryonic stem cells (ESCs) suggest that early neural cells can be specified towards a wide range of neuronal fates. For example, combined exposure to sonic hedgehog (SHH) and retinoic acid (RA) or SHH and FGF8 induces populations enriched in spinal motoneuron (MN) or midbrain dopamine neurons, respectively. These protocols work well in neural-plate-stage explants (Roelink et al. 1995; Ye et al. 1998) and in neural progeny of ESCs (Lee et al. 2000; Wichterle et al. 2002; Barberi et al. 2003) but not in cultured NSCs (Caldwell et al. 2001; Jain et al. 2003). The striking difference in neuron subtype fate potential of ESCs and NSCs suggests that neural patterning response becomes rapidly restricted during neural development. One of the major challenges in NSC biology is therefore the development of novel strategies to enhance the differentiation potential of NSCs. This could be achieved either by manipulating the genetic or epigenetic state of later-stage NSCs or by isolating and maintaining an earlier NSC stage with intact patterning potential.

We have recently reported the isolation of a novel NSC type from human ESCs (hESCs), termed rosette-stage NSCs (R-NSCs) (Elkabetz et al. 2008) based on their characteristic cytoarchitecture. R-NSCs are capable of extensive self-renewal and broad differentiation potential along CNS and peripheral nervous system (PNS) lineages and readily yield a wide range of neuron types inaccessible to later-stage NSCs (Lee et al. 2007a; Elkabetz et al. 2008). The differentiation potential of R-NSCs corresponds to that of neural-plate-stage cells, and R-NSC-like populations can indeed be isolated directly from neural-plate-stage embryos (Elkabetz et al. 2008). The broad differentiation potential of R-NSCs is reflected by the specific molecular signature and distinct extrinsic growth factor requirements. A major goal of future studies is the identification of the molecular mechanisms that enable broad patterning potential of R-NSCs. In this chapter, we summarize our current understanding of R-NSCs and discuss their properties in the context of ESC and NSC biology.

NEURAL STEM CELLS

The in vivo persistence of immature neural precursors in the adult rodent brain was first described more than four decades ago (Altman and Das 1965). However, these studies were largely ignored until studies in the 1980s confirmed a link between dividing precursors in the SVZ and postmitotic neuronal progeny in the adult canary brain (Goldman and Nottebohm 1983). Many subsequent studies confirmed and extended these results for both the developing and adult CNS, providing a quantitative description of endogenous murine precursor populations in vivo (Frederiksen and McKay 1988). These early studies revealed strong evidence for the presence of endogenous NSCs in vivo. However, efforts to manipulate NSC fate in a systematic manner required the establishment of relevant in vitro culture systems. Early attempts at isolating and manipulating the fate of neural precursors (Cattaneo and McKay 1990) got a significant boost with the availability of the neurosphere culture technique (Reynolds and Weiss 1992; Kilpatrick and Bartlett 1993; Gritti et al. 1995). Under these conditions, neural precursors are grown as free-floating aggregates on nonadherent plates in the presence of fibroblast growth factor-2 (FGF-2), epidermal growth factor (EGF), or both. For human neurospheres, supplementation with leukemia inhibitory factor (LIF) in addition to EGF and FGF-2 has been proposed (Galli et al. 2000). The demonstration of clonal neurosphere formation, self-renewal, and multilineage potential provided a valuable assay to probe NSC properties in vitro. Neurosphere assays were used for the prospective identification of NSC-like populations from primary tissue based on expression of CD133 (Uchida et al. 2000), SSEA-1 (Capela and Temple 2002), or combinations of positive/negative marker sets (Rietze et al. 2001). However, neurospheres are composed of heterogeneous cell types including many differentiated cells. Furthermore, in the adult SVZ, proliferating neurospheres are obtained more readily from transient-amplifying cells (type C cells) than from the SVZ stem cell compartment (type B cells) (Doetsch et al. 2002). Therefore, neurosphere data need to be interpreted cautiously, and neither primary nor secondary neurosphere formation is a conclusive assay of NSC identity.

An alternative approach for isolating NSCs in vitro is based on monolayer cultures in the presence of FGF-2 (Davis and Temple 1994; Palmer et al. 1995; Johe et al. 1996). Monolayer techniques allow direct observation of lineage relationships and the establishment of complete lineage trees in vitro (Qian et al. 2000). A recent variation of this technique is the neurosphere (NS) culture system yielding fairly homogeneous populations of radial glial-like NSC progeny in the presence of FGF-2 and EGF (Conti et al. 2005). NS cultures exhibit improved maintenance of neurogenic fate potential following long-term culture and represent one of the few examples of a symmetrically dividing stem cell population (Conti et al. 2005). However, the predominant neuron types generated from NS cells are GABAergic and glutamatergic neurons, similar to the neurosphere and FGF-2-expanded monolayer cultures.

DEVELOPMENTAL POTENCY

The two key criteria that define NSCs are capacity for extensive self-renewal and multilineage potential toward neurons, astrocytes, and oligodendrocytes (McKay 1997; Gage 2000; Temple 2001). Maintenance of NSCs in vitro or in vivo requires that at least one of the two daughter cells retains NSC potential. In the developing CNS, cells

undergo various phases of expansion and differentiation. At the earliest stages of neural development, cells symmetrically expand to yield the precursor pool subsequently generating the broad range of neural cell types. At later stages, cells undergo both asymmetric and symmetric divisions to yield transit-amplifying cells, neurons, and glia. There is evidence that cells at the neural-plate stage contain precursors with plasticity capable of changing anteroposterior (AP) and dorsoventral (DV) identity in response to inducing tissues (Yamada et al. 1993; Hynes et al. 1995) or defined patterning signals (Roelink et al. 1994; Ye et al. 1998). However, it is unclear whether cells at the neural-plate stage represent true NSC populations that undergo many self-renewal divisions before losing patterning potential. Accordingly, the identification of an "authentic" NSC in vivo with the ability to recreate the whole cellular diversity of the CNS has remained controversial (Mukouyama et al. 2006). The situation in the CNS is contrasted by the hematopoietic system, where a single transplanted hematopoietic stem cell (HSC) can successfully reconstitute hematopoiesis throughout the life span of the graft recipient (Morrison et al. 1995).

In the adult brain, there is ample evidence for continuous neurogenesis in specific regions of the CNS, such as in the SVZ along the lateral ventricle (Doetsch et al. 1999) and in the hippocampus (Palmer et al. 1997). However, adult NSC populations display an even more limited neuronal fate potential (Merkle et al. 2007). Although studies in the SVZ have reported broadened differentiation potential toward striatal neuron types upon noggin and BDNF (brain-derived neurotrophic factor) exposure (Chmielnicki et al. 2004), evidence for more general AP or DV respecification of adult NSCs is lacking.

One important question is whether restricted in vivo NSC potential can be overcome following in vitro culture. An expanded differentiation repertoire of isolated "glial-restricted" O2A progenitors toward neuronal fates has been reported after FGF-2 treatment (Kondo and Raff 2000). FGF-2 treatment has also been shown to alter expression of DV markers in neural precursors and to broaden differentiation from bipotent to tripotent fate characteristics in vitro (Gabay et al. 2003). Such studies raise the question of whether in vitro culture and FGF-2 exposure can dysregulate patterning and broaden fate potential. However, FGF-2 has complex roles during development and serves as a patterning factor in its own right (Crossley et al. 1996; Lumsden and Krumlauf 1996; Shimamura and Rubenstein 1997; Shimamura et al. 1997; Bertrand et al. 2000). Therefore, FGF signaling may have roles in both broadening and restricting potential depending on developmental context.

During normal development, there is evidence for regional specification even before neurulation (Quinlan et al. 1995), and early neural-plate cells appear to quickly adopt regional subtype identities (Chang and Hemmati-Brivanlou 1998) that impact their ability to respond to patterning morphogenetic cues. A number of studies isolated NSC-like populations from specific regions of the CNS including forebrain (He et al. 2001), midbrain (Ling et al. 1998; Studer et al. 1998), hypothalamus (Markakis et al. 2004), and spinal cord (Mayer-Proschel et al. 1997). These studies generally confirm that in-vitro–expanded NSC populations exhibit temporal region-specific neuronal differentiation potential. The issue of whether such restrictions can be overcome in an "instructive" environment remains controversial. Numerous studies have reported remarkable plasticity of grafted neural precursor populations upon heterotopic transplantation in the developing CNS (Brüstle et al. 1995; Campbell et al. 1995; Fishell 1995; Suhonen et al. 1996; Shihabuddin et al. 2000). However, these studies are limited largely to the analysis of morphological criteria and generic neuron marker expression; they did not carefully probe for reexpression of region-specific transcription factors. There is evidence that grafted cells can undergo morphological adaptation in the host region without undergoing respecification such as NSCs grafted into the retina (Takahashi et al. 1998). Future studies will need to revisit these questions using genetic lineage-marking strategies before and after transplantation.

Temporal fate restrictions have been investigated in heterochronic transplantation assays. Such studies have shown that early-stage NSCs have the ability to populate orthotopic regions in older-stage hosts but not vice versa (Frantz and McConnell 1996). Other examples of temporal restrictions are the increased bias of late NSC and neural precursor populations to undergo gliogenic differentiation (Quinn et al. 1999) and, conversely, the unresponsiveness of early-stage cortical NSCs to undergo glial differentiation in response to instructive gliogenic cues based on activating JAK-STAT signaling (Molne et al. 2000). The latter phenomenon has been linked to methylation of critical components of JAK-STAT signaling in early-stage NSCs (Fan et al. 2005). One interpretation of these data is that at any given stage of development, no single NSC possesses the capacity for recreating the full cellular diversity of the CNS (Mukouyama et al. 2006) and that NSCs progress through sequential stages of fate potential from the time of neural induction to adulthood.

A practical example for the difficulties of using NSCs to derive a specialized neuron type is the generation of midbrain dopamine neurons for application in models of Parkinson's disease. Functional midbrain dopamine neurons can be derived from short-term expanded precursors isolated from the early rodent or human midbrain (Studer et al. 1998; Sánchez-Pernaute et al. 2001; Parish et al. 2008). However, more extensive in vitro expansion results in a dramatic decrease in the efficiency to generate dopamine neurons (Yan et al. 2001). Many strategies have been developed in an effort to overcome this limitation, including treatments with extrinsic factors and genetic modification of key developmental genes (for review, see Kriks and Studer 2009). However, the goal of obtaining long-term expandable populations of NSCs yielding authentic midbrain dopamine neurons has remained elusive.

EMBRYONIC STEM CELLS AND THE DERIVATION OF NEURAL FATES

The challenges for obtaining patternable NSC populations have been addressed using two main strategies: the development of improved protocols for expansion and differentiation of NSC-like populations with defined regional

and temporal identity and the search for a novel NSC source with intact neural-plate-stage patterning potential. Mouse ESCs and the more recent isolation of human ESCs (Thomson et al. 1998) offer a powerful paradigm for generating unlimited numbers of early-stage neural progeny. Many protocols have been developed for inducing neural differentiation in ESCs. These include embryoid body (EB) differentiation in combination with retinoic acid treatment, conditioned medium, or neural survival factors (Bain et al. 1995; Okabe et al. 1996; Lee et al. 2000; Zhang et al. 2001; Rathjen et al. 2002); the use of neuron-inducing stromal feeders (Kawasaki et al. 2000, 2002; Barberi et al. 2003; Perrier et al. 2004); and the development of adherent feeder-free approaches (Reubinoff et al. 2001; Cibelli et al. 2002; Ying et al. 2003).

All three basic strategies yield neural progeny at stages suitable for neural patterning studies. Comparative analysis of developmental progression in vitro and in vivo showed a remarkable correlation from the stage of gastrulation to neurulation, patterning, neurogenesis, and gliogenesis (Barberi et al. 2003). Clonogenic experiments have demonstrated that ESC-derived neural progeny can self-renew extensively and clonally give rise to neurons, astrocytes, and oligodendrocytes (Barberi et al. 2003). These results in mouse ESCs provided access to early stages of neural development and yielded ESC-derived multipotent NSC populations in vitro. NSC-like populations have been also obtained from hESCs and demonstrated to engraft in the adult rodent SVZ in vivo and contribute human cells to the host rostral migratory stream and olfactory bulb (Tabar et al. 2005). However, the isolation of ESC-derived NSCs illustrated again the importance of developmental stage and neuronal fate potential. ESC-derived neural progeny readily respond to appropriate developmental cues that direct fate toward midbrain dopamine neurons (Lee et al. 2000; Barberi et al. 2003; Perrier et al. 2004), somatic motoneurons (Wichterle et al. 2002; Yan et al. 2005; Lee et al. 2007b), or forebrain neuron types (Barberi et al. 2003; Watanabe et al. 2005, 2007). However, patterning potential is restricted to narrow developmental windows and is lost progressively after in vitro expansion and isolation of FGF-2/EGF-expanded NSC populations (Tabar et al. 2005; Elkabetz et al. 2008). The broad differentiation potential of ESC-derived early-stage neural progeny raises the question of whether these cells reflect a novel, distinct NSC state that can be isolated in vitro and propagated long term under appropriate conditions.

NEURAL ROSETTES

Original Description of Neural Rosettes

A striking feature during neural differentiation of hESCs is the formation of neural rosettes. Neural rosettes are an early intermediate composed of radially organized columnar epithelial cells resembling the stage of neurulation. Neural rosette formation has been previously described under pathological conditions such as during tumor formation in teratomas or primitive neuroepithelial tumors (PNETs). Initial studies in hESCs suggested that neural rosettes represent abortive neural tube structures

(Zhang et al. 2001) expressing a broad range of neural precursor markers including Nestin, NCAM, Pax6, and Sox1. One critical feature of neural rosettes is the close temporal link between rosette formation and neural patterning potential (Perrier et al. 2004; Yan et al. 2005). Neural rosettes also exhibit a high proliferative capacity and a broad differentiation potential along both neuronal and glial lineages (Reubinoff et al. 2001; Zhang et al. 2001). There has been some controversy whether neural patterning is restricted to the earliest stages of neural rosette formation or whether patterning potential is maintained in later-stage rosettes. It has been suggested that expression of Sox1 marks later-stage rosettes with reduced patterning potential (Yan et al. 2005). However, other studies did not observe an obvious delay in the onset of Pax6 versus Sox1 expression and reported intact patterning potential toward midbrain dopamine neuron (Perrier et al. 2004) and spinal motoneuron fates (Lee et al. 2007b) in later-stage rosettes.

Recent work from our lab has characterized the patterning potential of neural rosette-stage cells in more detail and confirmed extensive AP and DV respecification in rosettes (Elkabetz et al. 2008), including the potential to yield neural crest stem cells (NCSCs) and precursor populations (Lazzari et al. 2006; Lee et al. 2007a). Interestingly, neural rosettes do not express distinct markers that define specific DV domains during neural tube formation, as best characterized for the developing spinal cord in vivo (Jessell 2000). The lack of defined DV domains within rosettes suggests that neural rosettes correspond to the neural-plate/neural-fold stage before DV specification, rather than representing abortive neural tube structures.

Stem Cell Nature, Cytoarchitecture, and Progression in Neural Rosettes

The broad differentiation potential and proliferative capacity of neural rosettes suggest that they may contain patternable NSC-like populations. The demonstration of a self-renewing rosette-stage NSC with intact differentiation potential has been challenging. Propagation of rosette-stage cells in FGF-2 and EGF, the growth factors most commonly used in NSC culture, results in the rapid loss of radial organization and the concomitant loss of patterning potential. These findings suggest that the extrinsic milieu (e.g., FGF-2/EGF exposure) suppresses rosette potential, or that proliferation in the presence of FGF-2 and EGF causes progression toward a later nonpatternable state.

The classic definition of NSCs is based on self-renewal potential and trilineage differentiation. In the case of rosette-stage stem cells, this definition may need to be adjusted to include maintenance of rosette cytoarchitecture and—most importantly—patterning potential. Therefore, NCSs at the rosette stage are expected to exhibit self-renewal and differentiation capacities broader than those observed in classic NSCs.

Neural rosettes and early-stage neuroepithelial cells in vivo share many morphological and functional characteristics including a pronounced apical-basal polarity. The apical side of neuroepithelial cells in vivo faces the amniotic fluid at the neural-plate stage and the neural tube lumen fol-

lowing neural tube closure. The basal side is linked to the basal lamina. Neural rosettes maintained in vitro lack equivalents of basal lamina or amniotic cavity. Therefore, apial-basal polarity must be defined based on marker expression and functional characteristics. Throughout the cell cycle, neuroepithelial cells are anchored to one another apically through junctional complexes (Hinds and Ruffett 1971; Aaku-Saraste et al. 1996) characterized by ZO1 expression. Toward mitosis, neuroepithelial cells round up at the apical side, concomitant with interkinetic migration of their nuclei along the apical-basal axis. In vitro, the expression of ZO1 and the location of mitotic nuclei marked by phosphohistone 3 (PH3) define the rosette lumen as the apical side of the rosette structure (Lazzari et al. 2006; Elkabetz et al. 2008). Early-stage neuroepithelial cells, similar to early rosettes in vitro, form a pseudostratified epithelium with all cells touching the lumen (Sauer 1935). As neural rosettes continue to grow in vitro, they mimic the processes occurring during neurulation and neural tube growth. Rosette cells gradually give rise to differentiated cells such as neurons migrating radially away from the rosette structure and a set of more restricted progenitors. The more complex organization of late-stage rosettes is reflected in their multilayered structure. Future studies must address whether the rosette lumen at these late stages continues to exert a niche function and retain a population with intact patterning and proliferation potential or whether all cells in late rosettes undergo progressive fate restrictions.

Early neural rosettes represent a symmetrically dividing undifferentiated neuroepithelial cell population with very small numbers of spontaneously differentiating neurons. This early expansion phase is followed by a neurogenic stage where rosette cells start to efficiently produce neurons. Concomitant to onset of neurogenesis, rosettes give rise to radial glial-like cells expressing classic markers such as brain lipid-binding protein (BLBP) and GLAST. These radial glial-like populations retain features of neuroepithelial cells such as interkinetic nuclear migration and apical endfeet. At later stages, there is a loss of epithelial polarity and increased EGF responsiveness associated with a loss of neurogenic bias and increased production of glial fates.

Forse-1: Anterior Bias and Neural Patterning in Rosettes

The broad differentiation potential of hESC-derived neural rosettes raises the question of whether cells with true NSC potential can be isolated at that stage. Such rosette-stage NSCs are predicted to combine self-renewal and multilineage differentiation potential with the ability to respond to patterning cues. Forse1 was previously characterized as a surface marker distinguishing anterior and posterior CNS precursor fates (Tole et al. 1995). The Forse1 antibody allows the separation of rosette-stage cells enriched for anterior markers such as BF-1 (FoxG1B). Forse1-negative rosette cells are enriched for precursors expressing posterior CNS and neural crest markers (Fig. 1). In addition, Forse1⁻ cells also include populations of more differentiated neurons and glia as well as nonneural progeny.

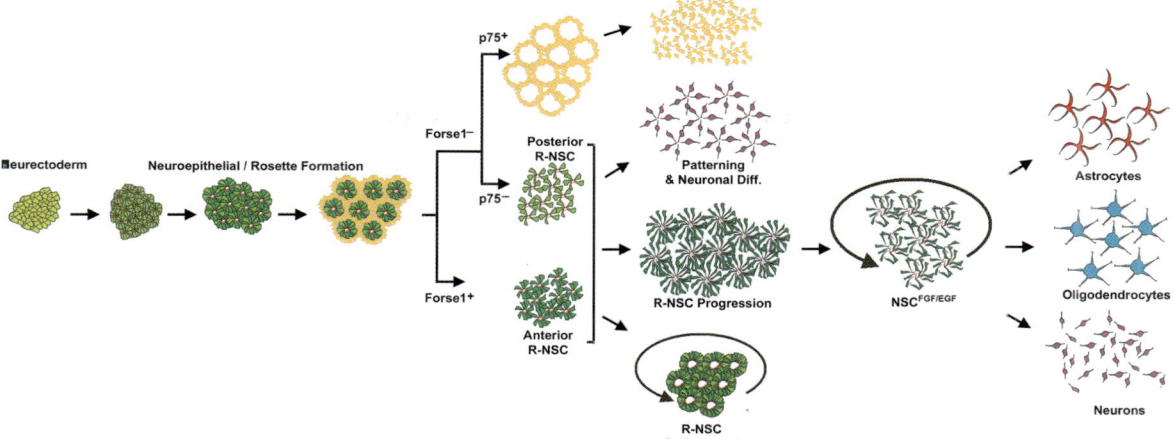

Figure 1. NSC progression model. During neural induction, hESCs form early neuroepithelial structures developing into neural rosette structures. At the rosette stage, cultures can be separated based on Forse1 expression into cells with anterior CNS character and into cells with more posterior markers including neural crest markers. Both anterior and posterior R-NSC–stage cells can be patterned toward a broad range of fates along the AP and DV axes. Depending on culture conditions, R-NSCs can be propagated and give rise to NSC[FGF/EGF] or yield self-renewing populations of R-NSCs. NSCs[FGF/EGF] also exhibit self-renewal potential and multilineage differentiation toward astrocytes, oligodendrocytes, and neurons but lose patterning potential.

Forse1[+] cells express classic NSC markers including nestin, BLBP, and vimentin while retaining expression of neuroepithelial markers such as Pax6, Sox1, and N-cadherin. Importantly, at the rosette stage, both Forse1[+] and Forse1[−] cells contain population of cells that can be respecified in response to appropriate AP and DV patterning cues (Elkabetz et al. 2008). Clonal studies have revealed that single Forse1[+] cells are capable of self-renewal, multilineage differentiation, and neural pattern-

ing, the key features that define NSCs at the rosette stage. These data indicate that rosettes contain regionally biased NSC populations that can be prospectively identified and that retain broad differentiation potential at the clonal level, termed R-NSCs. Figure 2 presents a summary of markers that distinguish R-NSCs from both undifferentiated hESCs- and hESC-derived NSCs[FGF/EGF].

Identification of an R-NSC–specific molecular profile was based on the hypothesis that independent of their

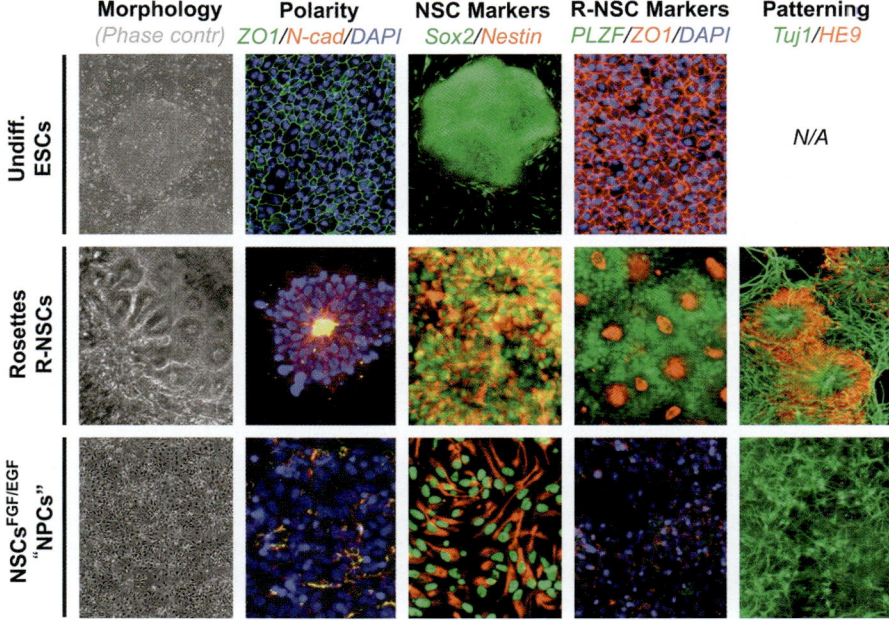

Figure 2. Marker and functional panel distinguishing ESCs, R-NSCs, and NSCs[FGF/EGF]. Representative markers show rosette cytoarchitecture ("morphology"), cell polarity, NSC identity, R-NSC state, and the ability of the cells to differentiate in region-specific neurons such as somatic motoneurons (patterning).

regional bias (Forse1$^+$ or Forse1$^-$), R-NSCs share a molecular profile distinct from undifferentiated hESCs and NSCs$^{FGF-2/EGF}$. R-NSC markers identified by global transcriptome analysis were highly enriched in nuclear proteins including transcription factors and transcriptional regulators (Elkabetz et al. 2008). Most R-NSC genes have not been previously associated with neural development and NSC biology, although most of them are expressed during early neural development in the mouse (www.informatics.jax.org). Null mutations for many of the R-NSC–specific markers do not result in major CNS defects, suggesting functional redundancy. However, these studies did not directly study CNS impact of these genes and may have missed more subtle neural phenotypes. On the basis of our studies, we predict—given the narrow developmental windows of expression—that more dramatic effects may be observed in gain-of-function rather than in loss-of-function studies. Gene ontology analysis revealed that gene expression in R-NSCs is enriched for components of the Notch, SHH, and Wnt signaling pathways and is distinct from NSC$^{FGF/EGF}$.

Growth Requirements and Intrinsic Growth Potential

A strong temporal correlation exists between patterning potential and rosette cytoarchitecture, although there is no evidence that rosette organization is directly required for mediating patterning response. When R-NSCs are first derived from hESCs, they rapidly expand in number and spontaneously form secondary rosettes, suggesting that a default intrinsic program drives R-NSC self-renewal. However, as R-NSCs progress in culture, this program is down-regulated as reflected by reduced proliferation potential, and the loss of rosette cytoarchitecture and patterning potential and becomes dependent on extrinsic factors. Therefore, a key question in R-NSC biology is the identification of extrinsic factors that promote R-NSC self-renewal.

Two well-known factors for propagating NSCs are FGF-2 and EGF. Both factors have been shown to promote in vitro proliferation of NSCs and neural precursors from many regions of the developing and adult CNS (Murphy et al. 1990; Reynolds and Weiss 1992; Kilpatrick and Bartlett 1993; Ray et al. 1993; Gritti et al. 1995; Johe et al. 1996; Kalyani et al. 1997; Studer et al. 1998). FGF-2–dependent proliferation in vitro is compatible with the early expression of FGF-R1 (FGF-receptor-1) and FGF-2 in the CNS (Orr-Urtreger et al. 1991; Wanaka et al. 1991), defects in cell proliferation associated with targeted deletion of these receptors (Yamaguchi et al. 1994; Ciruna et al. 1997; Raballo et al. 2000), and the proliferative effects following FGF-2 administration in vivo (Kuhn et al. 1997). EGF exposure also has well-known effects in the developing and adult SVZ (Craig et al. 1996; Doetsch et al. 2002) and is required for normal forebrain development at late embryonic and postnatal stages (Sibilia and Wagner 1995; Threadgill et al. 1995). Despite the important role of FGF-2 and EGF for NSCs in vitro and in vivo, at the rosette stage, FGF-2 and EGF treatments cause a rapid loss of rosette cytoarchitecture and patterning potential (Shin et

al. 2006; Elkabetz et al. 2008). Another problem regarding FGF-2 treatment at the R-NSC stage is the induction of non-CNS cells with neural crest and mesenchymal precursor features (Lee et al. 2007a; Elkabetz et al. 2008).

There are multiple lines of evidence for the importance of Notch signaling at the R-NSC stage. First, components of Notch signaling were selectively enriched in R-NSCs (Elkabetz et al. 2008). Second, high cell density promotes growth of R-NSCs and maintenance of rosette cytoarchitecture, and both of these responses can be blocked by Notch inhibition via the γ-secretase inhibitor DAPT (N-[N-(3,5-difluorophenacetyl)-L-alanyl]-S-phenylglycine t-butylester). Third, addition of recombinant proteins activating notch signaling including Dll4 and Jagged-1 promotes R-NSC maintenance (Elkabetz et al. 2008). However, Notch downstream effectors such as Hes1 and Hes5 are enriched both in R-NSCs and NSCs$^{FGF/EGF}$, and a role for Notch signaling is not restricted to the R-NSC state given its well-known effects in many NSC paradigms (Gaiano et al. 2000; Ohtsuka et al. 2001; Androutsellis-Theotokis et al. 2006; Mizutani et al. 2007). Therefore, it remains to be determined whether certain aspects of Notch signaling selectively impact the R-NSC state or whether Notch exerts a more generic self-renewal effect in R-NSCs, similar to its role in other NSC populations.

SHH has many fundamental roles in early neural development as both a morphogen and a proliferation factor. Ablation of SHH during mouse development causes many abnormalities in CNS patterning and beyond (Chiang et al. 1996). Conditional ablation studies in the CNS have revealed a role for SHH signaling in the maintenance of neural stem and precursor cells during development and adulthood (Ishibashi and McMahon 2002; Lai et al. 2003; Machold et al. 2003; Ahn and Joyner 2005; Palma et al. 2005). We observed that SHH is a potent mitogen for R-NSCs, inducing rapid proliferation over multiple passages without impacting rosette cytoarchitecture or gene expression profiles (see Fig. 3) (Elkabetz et al. 2008). However, it remains to be determined whether SHH acts by selectively promoting R-NSC proliferation, promoting R-NSC identity via inducing R-NSC–specific genes, repressing progression to NSC$^{FGF/EGF}$ state, or, in part, preventing differentiation toward dorsal fates including neural crest precursors.

Molecular Dissection of R-NSC Development

A striking illustration of the R-NSC state is the unique molecular signature as defined by global gene expression analysis comparing undifferentiated hESCs to R-NSC and NSC$^{FGF/EGF}$ progeny (Fig. 4). The identification of large sets of R-NSC–specific genes opens up new opportunities for defining the R-NSC state and the associated neural patterning potential at the functional levels. However, temporal analysis of gene expression also provides a better understanding for the molecular events involved in R-NSC induction and progression to the NSC$^{FGF/EGF}$ state.

At the level of cell adhesion, an early switch from E-cadherin expression in undifferentiated hESCs to N-cadherin expression is retained in R-NSCs and NSCs$^{FGF/EGF}$.

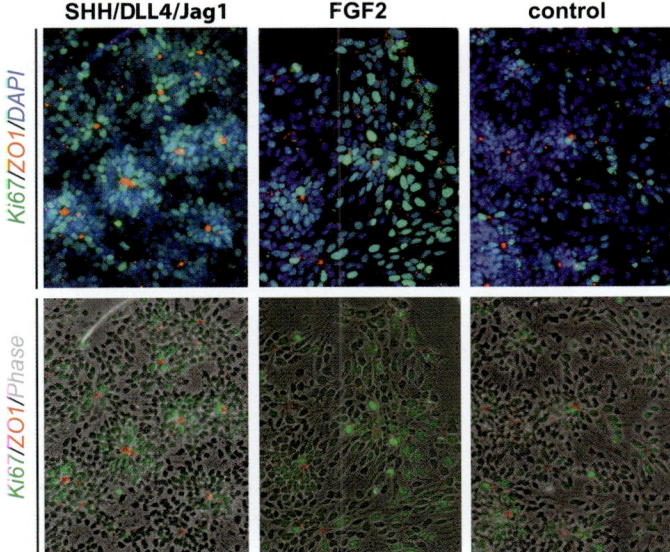

Figure 3. R-NSC maintenance and proliferation. Labeling of cells using Ki67 reveals that SHH/DLL4/Jag1 and FGF-2 treatments induce proliferation compared with untreated control cultures. However, only SHH/DLL4/Jag1 treatment efficiently maintains R-NSC cytoarchitecture, whereas FGF-2 treatment causes a progressive loss of cell polarity.

R-NSCs and NSCs[FGF/EGF] also share expression of classic neuroepithelial and radial glial markers such as Sox1, Sox2, Hes5, and Pax6. The increasing radial glial nature of R-NSCs and NSCs[FGF/EGF] is further illustrated by the temporal expression of FABP7 and SLC1A3 (GLAST) (Fig. 4). One striking feature of R-NSCs is their pronounced polarity, and several key polarity genes including CDC42 and Par6 show peak expression in R-NSCs. The greatest temporal specificity is observed for bona fide R-NSC markers (Elkabetz et al. 2008) such as Plagl1, Dach1, and Zbtb16 (PLZF).

The patterning potential of R-NSCs likely reflects an epigenetic state permissive for plasticity. Recent studies have demonstrated reprogramming of somatic cells back

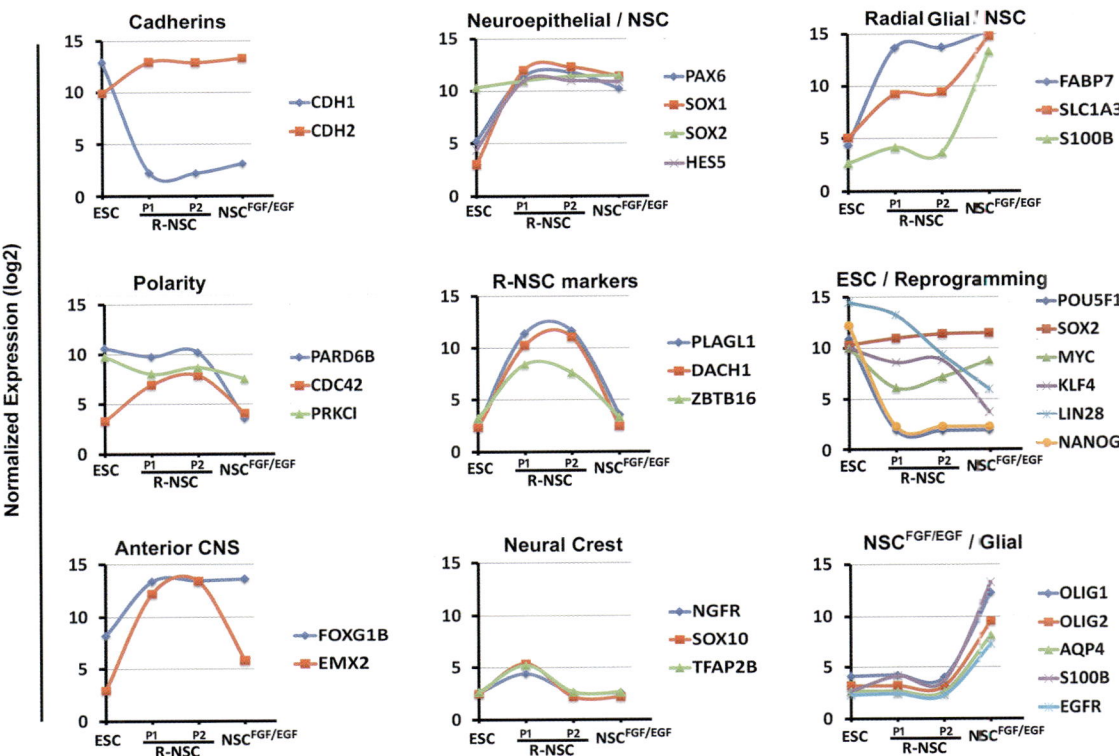

Figure 4. Expression of molecular markers: Time course analysis. Selected markers are shown grouped into specific functional categories (for details, see text).

to a pluripotent state using defined sets of transcription factors or RNA-binding proteins (Takahashi and Yamanaka 2006; Takahashi et al. 2007; Yu et al. 2007). We observed that expression of these reprogramming factors is rapidly abolished in the case of Oct4 (POU5F1) and Nanog. However, R-NSCs selectively retain high levels of LIN28, and levels of Sox2 and Klf4 were similar to those of NSCs$^{FGF/EGF}$. Although most anterior CNS markers such as FoxG1 are expressed highly in both R-NSCs and NSCs$^{FGF/EGF}$, Emx2 shows decreased expression at the NSCs$^{FGF/EGF}$ state. However, other anterior markers including Dlx1 and Dlx2 show increased expression in NSC$^{FGF/EGF}$ (data not shown). Finally, our data indicate that neural crest potential (P1 R-NSCs) and glial differentiation bias (NSCs$^{FGF/EGF}$) can be predicted based on global gene expression profiles.

In summary, molecular analysis offers insights into the sequential events controlling NSC progression. These studies also identified many potential markers for the prospective isolation of R-NSCs and NSCs$^{FGF/EGF}$ using genetic reporter lines. Importantly, molecular analysis also provides a high priority list of targets for functional interventions aimed at enhancing R-NSC self-renewal and preventing or reversing progression to the NSCs$^{FGF/EGF}$ state.

CONCLUSIONS AND QUESTIONS FOR THE FUTURE

The isolation of hESC-derived R-NSCs raises many key questions for future exploration. One obvious task is defining the role of R-NSC–specific genes in self-renewal and patterning potential and testing whether ectopic expression of key R-NSC genes in NSC$^{FGF/EGF}$ is sufficient to reverse the R-NSC state. R-NSC markers are also critical for defining cell heterogeneity and lineage relationships within R-NSCs. Another issue relevant for future translation is the development of improved protocols harnessing the patterning potential of R-NSCs while reducing the risk for neural overgrowth, as commonly observed in hESC-derived neural progeny-grafting paradigms (Sánchez-Pernaute et al. 2005; Ferrari et al. 2006; Roy et al. 2006; Aubry et al. 2008; Elkabetz et al. 2008).

Beyond the role of R-NSCs in hESC biology, it will be important to identify potential endogenous R-NSC populations. Initial studies suggest that R-NSC–like populations can be isolated from mouse embryos at the neural-plate stage (E8.25) (Elkabetz et al. 2008). The availability of genetic markers should allow prospective isolation of R-NSCs from primary tissue and in vivo lineage marking. Although current data suggest that R-NSCs represent a highly transient cell population, it will be critical to confirm whether R-NSC–like cells can persist into postnatal and adult stages.

The striking cytoarchitecture of rosettes raises the question of whether cell organization and cell-to-cell contact is critical for aspects of R-NSC biology such as self-renewal and symmetric division capacity or patterning potential. To this end, tools need to be developed that selectively disrupt polarity at the R-NSC stage. Given the emergence of rosette structures in many neural tumor types such as PNETs or medulloblastomas, understanding these struc-

ture/function relationships may provide insights beyond the neural stem cell field.

One fundamental idea of rosette biology is that similar to ESCs at the inner cell mass (ICM) stage, R-NSCs may capture a defined developmental state such as the neural-plate stage and perpetuate early cell potential in vitro. Such an "ES cell equivalent of the nervous system" could become a powerful tool for translational medicine and for applications in high-throughput screening assays requiring large numbers of homogeneous cell populations. Such a paradigm may for the first time provide access to a universal NSC stage capable of recreating the complete cellular diversity of the central and peripheral nervous systems.

ACKNOWLEDGMENTS

We thank members of the Studer lab for valuable discussions of the manuscript and Sonja Clairmont and Yael Strulovici for continued technical support. Our own work described in this review was supported by the National Institutes of Health (grants 1R01NS052671, 5R01NS044819), NYSTEM, the Starr Foundation, and Project ALS.

REFERENCES

Aaku-Saraste, E., Hellwig, A., and Huttner, W.B. 1996. Loss of occludin and functional tight junctions, but not ZO-1, during neural tube closure—Remodeling of the neuroepithelium prior to neurogenesis. *Dev. Biol.* **180:** 664–679.

Ahn, S. and Joyner, A.L. 2005. In vivo analysis of quiescent adult neural stem cells responding to Sonic hedgehog. *Nature* **437:** 894–897.

Altman, J. and Das, G.D. 1965. Autoradiographic and histological evidence of postnatal hippocampal neurogenesis in rats. *J. Comp. Neurol.* **124:** 319–335.

Alvarez-Buylla, A. and Lim, D.A. 2004. For the long run: Maintaining germinal niches in the adult brain. *Neuron* **41:** 683–686.

Androutsellis-Theotokis, A., Leker, R.R., Soldner, F., Hoeppner, D.J., Ravin, R., Poser, S.W., Rueger, M.A., Bae, S.K., Kittappa, R., and McKay, R.D. 2006. Notch signalling regulates stem cell numbers in vitro and in vivo. *Nature* **442:** 823–826.

Aubry, L., Bugi, A., Lefort, N., Rousseau, F., Peschanski, M., and Perrier, A.L. 2008. Striatal progenitors derived from human ES cells mature into DARPP32 neurons in vitro and in quinolinic acid-lesioned rats. *Proc. Natl. Acad. Sci.* **105:** 16707–16712.

Bain, G., Kitchens, D., Yao, M., Huettner, J.E., and Gottlieb, D.I. 1995. Embryonic stem cells express neuronal properties *in vitro. Dev. Biol.* **168:** 342–357.

Barberi, T., Klivenyi, P., Calingasan, N.Y., Lee, H., Kawamata, H., Loonam, K., Perrier, A.L., Bruses, J., Rubio, M.E., Topf, N., et al. 2003. Neural subtype specification of fertilization and nuclear transfer embryonic stem cells and application in parkinsonian mice. *Nat. Biotechnol.* **21:** 1200–1207.

Bertrand, N., Medevielle, F., and Pituello, F. 2000. FGF signalling controls the timing of Pax6 activation in the neural tube. *Development* **127:** 4837–4843.

Brüstle, O., Maskos, U., and McKay, R.D.G. 1995. Host-guided migration allows targeted introduction of neurons into the embryonic brain. *Neuron* **15:** 1275–1285.

Caldwell, M.A., He, X.L., Wilkie, N., Pollack, S., Marshall, G., Wafford, K.A., and Svendsen, C.N. 2001. Growth factors regulate the survival and fate of cells derived from human neurospheres. *Nat. Biotechnol.* **19:** 475–479.

Campbell, K., Olsson, M., and Björklund, A. 1995. Regional incorporation and site-specific differentiation of striatal precursors transplanted to the embryonic forebrain ventricle. *Neuron* **15:** 1259–1273.

Capela, A. and Temple, S. 2002. LeX/ssea-1 is expressed by adult mouse CNS stem cells, identifying them as nonependymal. *Neuron* **35**: 865–875.

Cattaneo, E. and McKay, R. 1990. Proliferation and differentiation of neuronal stem cells regulated by nerve growth factor. *Nature* **347**: 762–765.

Chang, C.B. and Hemmati-Brivanlou, A. 1998. Cell fate determination in embryonic ectoderm. *J. Neurobiol.* **36**: 128–151.

Chiang, C., Litingtung, Y., Lee, E., Young, K.E., Corden, J.L., Westphal, H., and Beachy, P.A. 1996. Cyclopia and defective axial patterning in mice lacking *Sonic hedgehog* gene function. *Nature* **383**: 407–413.

Chmielnicki, E., Benraiss, A., Economides, A.N., and Goldman, S.A. 2004. Adenovirally expressed noggin and brain-derived neurotrophic factor cooperate to induce new medium spiny neurons from resident progenitor cells in the adult striatal ventricular zone. *J. Neurosci.* **24**: 2133–2142.

Cibelli, J.B., Grant, K.A., Chapman, K.B., Cunniff, K., Worst, T., Green, H.L., Walker, S.J., Gutin, P.H., Vilner, L., Tabar, V., et al. 2002. Parthenogenetic stem cells in nonhuman primates. *Science* **295**: 819.

Ciruna, B.G., Schwartz, L., Harpal, K., Yamaguchi, T.P., and Rossant, J. 1997. Chimeric analysis of *fibroblast growth factor receptor-1 (Fgfr1)* function: A role for FGFR1 in morphogenetic movement through the primitive streak. *Development* **124**: 2829–2841.

Conti, L., Pollard, S.M., Gorba, T., Reitano, E., Toselli, M., Biella, G., Sun, Y.R., Sanzone, S., Ying, Q.L., Cattaneo, E., and Smith, A. 2005. Niche-independent symmetrical self-renewal of a mammalian tissue stem cell. *PLoS Biol.* **3**: e283

Craig, C.G., Tropepe, V., Morshead, C.M., Reynolds, B.A., Weiss, S., and van der Kooy, D. 1996. *In vivo* growth factor expansion of endogenous subependymal neural precursor cell populations in the adult mouse brain. *J. Neurosci.* **16**: 2649–2658.

Crossley, P.H., Minowada, G., MacArthur, C.A., and Martin, G.R. 1996. Roles for FGF8 in the induction, initiation, and maintenance of chick limb development. *Cell* **84**: 127–136.

Davis, A.A. and Temple, S. 1994. A self-renewing multipotential stem cell in embryonic rat cerebral cortex. *Nature* **372**: 263–266.

Doetsch, F., Caille, I., Lim, D.A., Garcia-Verdugo, J.M., and Alvarez-Buylla, A. 1999. Subventricular zone astrocytes are neural stem cells in the adult mammalian brain. *Cell* **97**: 703–716.

Doetsch, F., Petreanu, L., Caille, I., Garcia-Verdugo, J.M., and Alvarez-Buylla, A. 2002. EGF converts transit-amplifying neurogenic precursors in the adult brain into multipotent stem cells. *Neuron* **36**: 1021–1034.

Elkabetz, Y., Panagiotakos, G., Al Shamy, G., Socci, N.D., Tabar, V., and Studer, L. 2008. Human ES cell-derived neural rosettes reveal a functionally distinct early neural stem cell stage. *Genes Dev.* **22**: 152–165.

Fan, G., Martinowich, K., Chin, M.H., He, F., Fouse, S.D., Hutnick, L., Hattori, D., Ge, W., Shen, Y., Wu, H., et al. 2005. DNA methylation controls the timing of astrogliogenesis through regulation of JAK-STAT signaling. *Development* **132**: 3345–3356.

Ferrari, D., Sánchez-Pernaute, R., Lee, H., Studer, L., and Isacson, O. 2006. Transplanted dopamine neurons derived from primate ES cells preferentially innervate DARPP-32 striatal progenitors within the graft. *Eur. J. Neurosci.* **24**: 1885–1896.

Fishell, G. 1995. Striatal precursors adopt cortical identities in response to local cues. *Development* **121**: 803–812.

Frantz, G.D. and McConnell, S.K. 1996. Restriction of late cerebral cortical progenitors to an upper-layer fate. *Neuron* **17**: 55–61.

Frederiksen, K. and McKay, R.D. 1988. Proliferation and differentiation of rat neuroepithelial precursor cells in vivo. *J. Neurosci.* **8**: 1144–1151.

Gabay, L., Lowell, S., Rubin, L., and Anderson, D.J. 2003. Deregulation of dorsoventral patterning by FGF confers trilineage differentiation capacity on CNS stem cells in vitro. *Neuron* **40**: 485–499.

Gage, F.H. 2000. Mammalian neural stem cells. *Science* **287**: 1433–1438.

Gaiano, N., Nye, J.S., and Fishell, G. 2000. Radial glial identity is promoted by Notch1 signaling in the murine forebrain. *Neuron* **26**: 395–404.

Galli, R., Pagano, S.F., Gritti, A., and Vescovi, A.L. 2000. Regulation of neuronal differentiation in human CNS stem cell progeny by leukemia inhibitory factor. *Dev. Neurosci.* **22**: 86–95.

Goldman, S.A. and Nottebohm, F. 1983. Neuronal production, migration, and differentiation in a vocal control nucleus of the adult female canary brain. *Proc. Natl. Acad. Sci.* **80**: 2390–2394.

Gritti, A., Cova, L., Parati, E.A., Galli, R., and Vescovi, A.L. 1995. Basic fibroblast growth factor supports the proliferation of epidermal growth factor-generated neuronal precursor cells of the adult mouse CNS. *Neurosci. Lett.* **185**: 151–154.

He, W., Ingraham, C., Rising, L., Goderie, S., and Temple, S. 2001. Multipotent stem cells from the mouse basal forebrain contribute GABAergic neurons and oligodendrocytes to the cerebral cortex during embryogenesis. *J. Neurosci.* **21**: 8854–8862.

Hinds, J.W. and Ruffett, T.L. 1971. Cell proliferation in the neural tube: An electron microscopic and Golgi analysis in the mouse cerebral vesicle. *Z. Zellforsch. Mikrosk. Anat.* **115**: 226–264.

Hynes, M., Poulsen, K., Tessier-Lavigne, M., and Rosenthal, A. 1995. Control of neuronal diversity by the floor plate: Contact-mediated induction of midbrain dopaminergic neurons. *Cell* **80**: 95–101.

Ishibashi, M. and McMahon, A.P. 2002. A sonic hedgehog-dependent signaling relay regulates growth of diencephalic and mesencephalic primordia in the early mouse embryo. *Development* **129**: 4807–4819.

Jain, M., Armstrong, R.J., Tyers, P., Barker, R.A., and Rosser, A.E. 2003. GABAergic immunoreactivity is predominant in neurons derived from expanded human neural precursor cells in vitro. *Exp. Neurol.* **182**: 113–123.

Jessell, T.M. 2000. Neuronal specification in the spinal cord: Inductive signals and transcriptional codes. *Nat. Rev. Genet.* **1**: 20–29.

Johe, K.K., Hazel, T.G., Müller, T., Dugich-Djordjevic, M.M., and McKay, R.D.G. 1996. Single factors direct the differentiation of stem cells from the fetal and adult central nervous system. *Genes Dev.* **10**: 3129–3140.

Kalyani, A., Hobson, K., and Rao, M.S. 1997. Neuroepithelial stem cells from the embryonic spinal cord: Isolation, characterization, and clonal analysis. *Dev. Biol.* **186**: 202–223.

Kawasaki, H., Mizuseki, K., Nishikawa, S., Kaneko, S., Kuwana, Y., Nakanishi, S., Nishikawa, S., and Sasai, Y. 2000. Induction of midbrain dopaminergic neurons from ES cells by stromal cell-derived inducing activity. *Neuron* **28**: 31–40.

Kawasaki, H., Suemori, H., Mizuseki, K., Watanabe, K., Urano, F., Ichinose, H., Haruta, M., Takahashi, M., Yoshikawa, K., Nishikawa, S.I., Nakatsuji, N., and Sasai, Y. 2002. Generation of dopaminergic neurons and pigmented epithelia from primate ES cells by stromal cell-derived inducing activity. *Proc. Natl. Acad. Sci.* **99**: 1580–1585.

Kilpatrick, T.J. and Bartlett, P.F. 1993. Cloning and growth of multipotential neural precursors: Requirements for proliferation and differentiation. *Neuron* **10**: 255–265.

Kondo, T. and Raff, M. 2000. Oligodendrocyte precursor cells reprogrammed to become multipotential CNS stem cells. *Science* **289**: 1754–1757.

Kriks, S. and Studer, L. 2009. Protocols for generating ES cell-derived dopamine neurons. In *Development and engineering of dopamine neurons* (ed. R.J. Pasterkamp et al.). Landes Biosciences, Austin, Texas. (In press.)

Kuhn, H.G., Winkler, J., Kempermann, G., Thal, L.J., and Gage, F.H. 1997. Epidermal growth factor and fibroblast growth factor-2 have different effects on neural progenitors in the adult rat brain. *J. Neurosci.* **17**: 5820–5829.

Lai, K., Kaspar, B.K., Gage, F.H., and Schaffer, D.V. 2003. Sonic hedgehog regulates adult neural progenitor proliferation in vitro and in vivo. *Nat. Neurosci.* **6**: 21–27.

Lazzari, G., Colleoni, S., Giannelli, S.G., Brunetti, D., Colombo,

E., Lagutina, I., Galli, C., and Broccoli, V. 2006. Direct derivation of neural rosettes from cloned bovine blastocysts: A model of early neurulation events and neural crest specification in vitro. *Stem Cells* 24: 2514–2521.

Lee, G., Kim, H., Elkabetz, Y., Alshamy, G., Panagiotakos, G., Barberi, T., Tabar, V., and Studer, L. 2007a. Isolation and directed differentiation of neural crest stem cells derived from human embryonic stem cells. *Nat. Biotechnol.* 25: 1468–1475.

Lee, H.J., Al Shamy, G., Elkabetz, Y., Schoefield, C., Harrison, N.L., Panagiotakos, G., Tabar, V., and Studer, L. 2007b. Directed differentiation and transplantation of human embryonic stem cell derived motoneurons. *Stem Cells* 25: 1931–1939.

Lee, S.-H., Lumelsky, N., Studer, L., Auerbach, J.M., and McKay, R.D.G. 2000. Efficient generation of midbrain and hindbrain neurons from mouse embryonic stem cells. *Nat. Biotechnol.* 18: 675–679.

Ling, Z.D., Potter, E.D., Lipton, J.W., and Carvey, P.M. 1998. Differentiation of mesencephalic progenitor cells into dopaminergic neurons by cytokines. *Exp. Neurol.* 149: 411–423.

Lumsden, A. and Krumlauf, R. 1996. Patterning the vertebrate neuraxis. *Science* 274: 1109–1115.

Machold, R., Hayashi, S., Rutlin, M., Muzumdar, M.D., Nery, S., Corbin, J.G., Gritli-Linde, A., Dellovade, T., Porter, J.A., Rubin, L.L., et al. 2003. Sonic hedgehog is required for progenitor cell maintenance in telencephalic stem cell niches. *Neuron* 39: 937–950.

Markakis, E.A., Palmer, T.D., Randolph-Moore, L., Rakic, P., and Gage, F.H. 2004. Novel neuronal phenotypes from neural progenitor cells. *J. Neurosci.* 24: 2886–2897.

Mayer-Proschel, M., Kalyani, A.J., Mujtaba, T., and Rao, M.S. 1997. Isolation of lineage-restricted neuronal precursors from multipotent neuroepithelial stem cells. *Neuron* 19: 773–785.

McKay, R.D. 1997. Stem cells in the central nervous system. *Science* 276: 66–71.

Merkle, F.T., Mirzadeh, Z., and Alvarez-Buylla, A. 2007. Mosaic organization of neural stem cells in the adult brain. *Science* 317: 381–384.

Mizutani, K., Yoon, K., Dang, L., Tokunaga, A., and Gaiano, N. 2007. Differential Notch signalling distinguishes neural stem cells from intermediate progenitors. *Nature* 449: 351–355.

Molne, M., Studer, L., Tabar, V., Ting, Y.-T., Eiden, M.V., and McKay, R.D. 2000. Early cortical precursors do not undergo LIF-mediated astrocytic differentiation. *J. Neurosci. Res.* 59: 301–311.

Morrison, S.J., Uchida, N., and Weissman, I.L. 1995. The biology of hematopoietic stem cells. *Annu. Rev. Cell Dev. Biol.* 11: 35–71.

Mukouyama, Y.S., Deneen, B., Lukaszewicz, A., Novitch, B.G., Wichterle, H., Jessell, T.M., and Anderson, D.J. 2006. Olig2$^+$ neuroepithelial motoneuron progenitors are not multipotent stem cells in vivo. *Proc. Natl. Acad. Sci.* 103: 1551–1556.

Murphy, M., Drago, J., and Bartlett, P.F. 1990. Fibroblast growth factor stimulates the proliferation and differentiation of neural precursor cells in vitro. *J. Neurosci. Res.* 25: 463–475.

Ohtsuka, T., Sakamoto, M., Guillemot, F., and Kageyama, R. 2001. Roles of the basic helix-loop-helix genes *Hes1* and *Hes5* in expansion of neural stem cells of the developing brain. *J. Biol. Chem.* 276: 30467–30474.

Okabe, S., Forsberg-Nilsson, K., Spiro, A.C., Segal, M., and McKay, R.D.G. 1996. Development of neuronal precursor cells and functional postmitotic neurons from embryonic stem cells in vitro. *Mech. Dev.* 59: 89–102.

Orr-Utrreger, A., Givol, D., Yayon, A., Yarden, Y., and Lonai, P. 1991. Developmental expression of two murine fibroblast growth factor receptors, *flg* and *bek*. *Development* 113: 1419–1434.

Palma, V., Lim, D.A., Dahmane, N., Sánchez, P., Brionne, T.C., Herzberg, C.D., Gitton, Y., Carleton, A., Alvarez-Buylla, A., and Ruiz i Altaba, A. 2005. Sonic hedgehog controls stem cell behavior in the postnatal and adult brain. *Development* 132: 335–344.

Palmer, T.D., Ray, J., and Gage, F.H. 1995. FGF-2-responsive neuronal progenitors reside in proliferative and quiescent regions of the adult rodent brain. *Mol. Cell. Neurosci.* 6: 474–486.

Palmer, T.D., Takahashi, J., and Gage, F.H. 1997. The adult rat hippocampus contains primordial neural stem cells. *Mol. Cell. Neurosci.* 8: 389–404.

Parish, C.L., Castelo-Branco, G., Rawal, N., Tonnesen, J., Sorensen, A.T., Salto, C., Kokaia, M., Lindvall, O., and Arenas, E. 2008. Wnt5a-treated midbrain neural stem cells improve dopamine cell replacement therapy in parkinsonian mice. *J. Clin. Invest.* 118: 149–160.

Perrier, A.L., Tabar, V., Barberi, T., Rubio, M.E., Bruses, J., Topf, N., Harrison, N.L., and Studer, L. 2004. Derivation of midbrain dopamine neurons from human embryonic stem cells. *Proc. Natl. Acad. Sci.* 101: 12543–12548.

Qian, X.M., Shen, Q., Goderie, S.K., He, W.L., Capela, A., Davis, A.A., and Temple, S. 2000. Timing of CNS cell generation: A programmed sequence of neuron and glial cell production from isolated murine cortical stem cells. *Neuron* 28: 69–80.

Quinlan, G.A., Williams, E.A., Tan, S.S., and Tam, P.P. 1995. Neuroectodermal fate of epiblast cells in the distal region of the mouse egg cylinder: Implication for body plan organization during early embryogenesis. *Development* 121: 87–98.

Quinn, S.M., Walters, W.M., Vescovi, A.L., and Whittemore, S.R. 1999. Lineage restriction of neuroepithelial precursor cells from fetal human spinal cord. *J. Neurosci. Res.* 57: 590–602.

Raballo, R., Rhee, J., Lyn-Cook, R., Leckman, J.F., Schwartz, M.L., and Vaccarino, F.M. 2000. Basic fibroblast growth factor (Fgf2) is necessary for cell proliferation and neurogenesis in the developing cerebral cortex. *J. Neurosci.* 20: 5012–5023.

Rathjen, J., Haines, B.P., Hudson, K.M., Nesci, A., Dunn, S., and Rathjen, P.D. 2002. Directed differentiation of pluripotent cells to neural lineages: Homogeneous formation and differentiation of a neurectoderm population. *Development* 129: 2649–2661.

Ray, J., Peterson, D.A., Schinstine, M., and Gage, F.H. 1993. Proliferation, differentiation, and long-term culture of primary hippocampal neurons. *Proc. Natl. Acad. Sci.* 90: 3602–3606.

Reubinoff, B.E., Itsykson, P., Turetsky, T., Pera, M.F., Reinhartz, E., Itzik, A., and Ben Hur, T. 2001. Neural progenitors from human embryonic stem cells. *Nat. Biotechnol.* 19: 1134–1140.

Reynolds, B.A. and Weiss, S. 1992. Generation of neurons and astrocytes from isolated cells of the adult mammalian central nervous system (comments). *Science* 255: 1707–1710.

Rietze, R.L., Valcanis, H., Brooker, G.F., Thomas, T., Voss, A.K., and Bartlett, P.F. 2001. Purification of a pluripotent neural stem cell from the adult mouse brain. *Nature* 412: 736–739.

Roelink, H., Porter, J.A., Chiang, C., Tanabe, Y., Chang, D.T., Beachy, P.A., and Jessell, T.M. 1995. Floor plate and motor neuron induction by different concentrations of the amino-terminal cleavage product of sonic hedgehog autoproteolysis. *Cell* 81: 445–455.

Roelink, H., Augsburger, A., Heemskerk, J., Korzh, V., Norlin, S., Ruiz i Altaba, A., Tanabe,Y., Placzek, M., Edlund, T., and Jessell, T.M. 1994. Floor plate and motor neuron induction by *vhh-1*, a vertebrate homolog of hedgehog expressed by the notochord. *Cell* 76: 761–775.

Roy, N.S., Cleren, C., Singh, S.K., Yang, L., Beal, M.F., and Goldman, S.A. 2006. Functional engraftment of human ES cell-derived dopaminergic neurons enriched by coculture with telomerase-immortalized midbrain astrocytes. *Nat. Med.* 12: 1259–1268.

Sánchez-Pernaute, R., Studer, L., Bankiewicz, K.S., Major, E.O., and McKay, R.D. 2001. In vitro generation and transplantation of precursor-derived human dopamine neurons. *J. Neurosci. Res.* 65: 284–288.

Sánchez-Pernaute, R., Studer, L., Ferrari, D., Perrier, A.L., Lee, H., Viñuela, A., and Isacson, O. 2005. Long-term survival of dopamine neurons derived from parthenogenetic primate embryonic stem cells (Cyno-1) in rat and primate striatum. *Stem Cells* 23: 914–922.

Sauer, F.C. 1935. Mitosis in the neural tube. *J. Comp. Neurol.* 62: 377–405.

Shen, Q., Goderie, S.K., Jin, L., Karanth, N., Sun, Y., Abramova, N., Vincent, P., Pumiglia, K., and Temple, S. 2004. Endothelial cells stimulate self-renewal and expand neurogenesis of neural stem cells. *Science* **304:** 1338–1340.

Shihabuddin, L.S., Horner, P.J., Ray, J., and Gage, F.H. 2000. Adult spinal cord stem cells generate neurons after transplantation in the adult dentate gyrus. *J. Neurosci.* **20:** 8727–8735.

Shimamura, K. and Rubenstein, J.R. 1997. Inductive interactions direct early regionalization of the mouse forebrain. *Development* **124:** 2709–2718.

Shimamura, K., Martinez, S., Puelles, L., and Rubenstein, J.L. 1997. Patterns of gene expression in the neural plate and neural tube subdivide the embryonic forebrain into transverse and longitudinal domains. *Dev. Neurosci.* **19:** 88–96.

Shin, S., Mitalipova, M., Noggle, S., Tibbitts, D., Venable, A., Rao, R., and Stice, L. 2006. Long-term proliferation of human embryonic stem cell-derived neuroepithelial cells using defined adherent culture conditions. *Stem Cells* **24:** 125–138.

Sibilia, M. and Wagner, E.F. 1995. Strain-dependent epithelial defects in mice lacking the EGF receptor. *Science* **269:** 234–238.

Studer, L., Tabar, V., and McKay, R.D. 1998. Transplantation of expanded mesencephalic precursors leads to recovery in parkinsonian rats. *Nat. Neurosci.* **1:** 290–295.

Suhonen, J.O., Peterson, D.A., Ray, J., and Gage, F.H. 1996. Differentiation of adult hippocampus-derived progenitors into olfactory neurons in vivo. *Nature* **383:** 624–627.

Tabar, V., Panagiotakos, G., Greenberg, E.D., Chan, B.K., Sadelain, M., Gutin, P.H., and Studer, L. 2005. Migration and differentiation of neural precursors derived from human embryonic stem cells in the rat brain. *Nat. Biotechnol.* **23:** 601–606.

Takahashi, K. and Yamanaka, S. 2006. Induction of pluripotent stem cells from mouse embryonic and adult fibroblast cultures by defined factors. *Cell* **126:** 663–676.

Takahashi, M., Palmer, T.D., Takahashi, J., and Gage, F.H. 1998. Widespread integration and survival of adult-derived neural progenitor cells in the developing optic retina. *Mol. Cell. Neurosci.* **12:** 340–348.

Takahashi, K., Tanabe, K., Ohnuki, M., Narita, M., Ichisaka, T., Tomoda, K., and Yamanaka, S. 2007. Induction of pluripotent stem cells from adult human fibroblasts by defined factors. *Cell* **131:** 861–872.

Temple, S. 2001. The development of neural stem cells. *Nature* **414:** 112–117.

Thomson, J.A., Itskovitz-Eldor, J., Shapiro, S.S., Waknitz, M.A., Swiergiel, J.J., Marshall, V.S., and Jones, J.M. 1998. Embryonic stem cell lines derived from human blastocysts. *Science* **282:** 1145–1147.

Threadgill, D.W., Dlugosz, A.A., Hansen, L.A., Tennenbaum, T., Lichti, U., Yee, D., LaMantia, C., Mourton, T., Herrup, K., Harris, R.C., et al. 1995. Targeted disruption of mouse EGF receptor: Effect of genetic background on mutant phenotype. *Science* **269:** 230–234.

Tole, S., Kaprielian, Z., Ou, S.K., and Patterson, P.H. 1995. FORSE-1: A positionally regulated epitope in the developing rat central nervous system. *J. Neurosci.* **15:** 957–969.

Uchida, N., Buck, D.W., He, D.P., Reitsma, M.J., Masek, M., Phan, T.V., Tsukamoto, A.S., Gage, F.H , and Weissman, I.L. 2000. Direct isolation of human central nervous system stem cells. *Proc. Natl. Acad. Sci.* **97:** 14720–14725.

Wanaka, A., Milbrandt, J., and Johnson, Jr., E.M. 1991. Expression of FGF receptor gene in rat development. *Development* **111:** 455–468.

Watanabe, K., Kamiya, D., Nishiyama, A., Katayama, T., Nozaki, S., Kawasaki, H., Watanabe, Y., Mizuseki, K., and Sasai, Y. 2005. Directed differentiation of telencephalic precursors from embryonic stem cells. *Nat. Neurosci.* **8:** 288–296.

Watanabe, K., Ueno, M., Kamiya, D., Nishiyama, A., Matsumura, M., Wataya, T., Takahashi, J.B., Nishikawa, S., Nishikawa, S., Muguruma, K., and Sasai, Y. 2007. A ROCK inhibitor permits survival of dissociated human embryonic stem cells. *Nat. Biotechnol.* **25:** 681–686.

Wichterle, H., Lieberam, I., Porter, J.A., and Jessell, T.M. 2002. Directed differentiation of embryonic stem cells into motor neurons. *Cell* **110:** 385–397.

Yamada, T., Pfaff, S.L., Edlund, T., and Jessell, T.M. 1993. Control of cell pattern in the neural tube: Motor neuron induction by diffusible factors from notochord and floor plate. *Cell* **73:** 673–686.

Yamaguchi, T.P., Harpal, K., Henkemeyer, M., and Rossant, J. 1994. *fgfr-1* is required for embryonic growth and mesodermal patterning during mouse gastrulation. *Genes Dev.* **8:** 3032–3044.

Yan, J., Studer, L., and McKay, R.D.G. 2001. Ascorbic acid increases the yield of dopaminergic neurons derived from basic fibroblast growth factor expanded mesencephalic precursors. *J. Neurochem.* **76:** 307–311.

Yan, Y., Yang, D., Zarnowska, E.D., Du, Z., Werbel, B., Valliere, C., Pearce, R.A., Thomson, J.A., and Zhang, S.C. 2005. Directed differentiation of dopaminergic neuronal subtypes from human embryonic stem cells. *Stem Cells* **23:** 781–790.

Ye, W.L., Shimamura, K., Rubenstein, J.R., Hynes, M.A., and Rosenthal, A. 1998. FGF and Shh signals control dopaminergic and serotonergic cell fate in the anterior neural plate. *Cell* **93:** 755–766.

Ying, Q.L., Stavridis, M., Griffiths, D., Li, M., and Smith, A. 2003. Conversion of embryonic stem cells into neuroectodermal precursors in adherent monoculture. *Nat. Biotechnol.* **21:** 183–186.

Yu, J., Vodyanik, M.A., Smuga-Otto, K., Antosiewicz-Bourget, J., Frane, J.L., Tian, S., Nie, J., Jonsdottir, G.A., Ruotti, V., Stewart, R., Slukvin, I.I., and Thomson, J.A. 2007. Induced pluripotent stem cell lines derived from human somatic cells. *Science* **318:** 1917–1920.

Zhang, S.C., Wernig, M., Duncan, I.D., Brüstle, O., and Thomson, J.A. 2001. In vitro differentiation of transplantable neural precursors from human embryonic stem cells. *Nat. Biotechnol.* **19:** 1129–1133.

Metabolomics of Neural Progenitor Cells: A Novel Approach to Biomarker Discovery

M. Maletić-Savatić,[*][†] L.K. Vingara,[‡] L.N. Manganas,[*] Y. Li,[*] S. Zhang,[*]
A. Sierra,[*] R. Hazel,[*] D. Smith,[§] M.E. Wagshul,[*] F. Henn,[§] L. Krupp,[*]
G. Enikolopov,[†] H. Benveniste,[§] P.M. Djurić,[*] and I. Pelczer[‡]

[*]Departments of Neurology and Electrical and Computer Engineering, Stony Brook University,
Stony Brook, New York 11790; [†]Cold Spring Harbor Laboratory, Cold Spring Harbor, New York 11724;
[‡]Department of Chemistry, Princeton University, Princeton, New Jersey 08544; [§]Medical Department,
Brookhaven National Laboratory, Upton, New York 11973

Finding biomarkers of human neurological diseases is one of the most pressing goals of modern medicine. Most neurological disorders are recognized too late because of the lack of biomarkers that can identify early pathological processes in the living brain. Late diagnosis leads to late therapy and poor prognosis. Therefore, during the past decade, a major endeavor of clinical investigations in neurology has been the search for diagnostic and prognostic biomarkers of brain disease. Recently, a new field of metabolomics has emerged, aiming to investigate metabolites within the cell/tissue/organism as possible biomarkers. Similarly to other "omics" fields, metabolomics offers substantial information about the status of the organism at a given time point. However, metabolomics also provides functional insight into the biochemical status of a tissue, which results from the environmental effects on its genome background. Recently, we have adopted metabolomics techniques to develop an approach that combines both in vitro analysis of cellular samples and in vivo analysis of the mammalian brain. Using proton magnetic resonance spectroscopy, we have discovered a metabolic biomarker of neural stem/progenitor cells (NPCs) that allows the analysis of these cells in the live human brain. We have developed signal-processing algorithms that can detect metabolites present at very low concentration in the live human brain and can indicate possible pathways impaired in specific diseases. Herein, we present our strategy for both cellular and systems metabolomics, based on an integrative processing of the spectroscopy data that uses analytical tools from both metabolomic and spectroscopy fields. As an example of biomarker discovery using our approach, we present new data and discuss our previous findings on the NPC biomarker. Our studies link systems and cellular neuroscience through the functions of specific metabolites. Therefore, they provide a functional insight into the brain, which might eventually lead to discoveries of clinically useful biomarkers of the disease.

Early detection of disease is a pressing goal in clinical medicine. However, for most neurological disorders, specific biomarkers of the disease are not known. The diagnosis often relies on clinical or radiological criteria. The response to treatment is also qualitative. Therefore, during the past decade, the search for diagnostic biomarkers of specific disorders has escalated and has included genomics, proteomics, and, most recently, metabolomics analysis of a variety of disease states (Fiehn et al. 2001; German et al. 2005; Goodacre 2005; Hollywood et al. 2006).

Metabolomics, a nontargeted identification and quantification of a collection of metabolites in a biological system (the "metabolome"), is used to examine biochemical and metabolic processes in a given tissue or fluid (Fig. 1) (Fiehn 2002; Nicholson and Wilson 2003; Goodacre et al. 2004). The level of each metabolite within the metabolome depends on the specific physiological, developmental, and pathological state of a cell or tissue, and thus, the metabolome reflects tissue phenotype that results from both genetic and environmental influences (Gavaghan et al. 2000; Roessner et al. 2001; Phelps et al. 2002; Griffin 2004; Griffin and Nicholls 2006; Lindon and Holmes 2006). Therefore, by monitoring the biochemical status of an organism, one can indirectly assess its gene functions as well as the environmental effects on the genome expression (Fiehn et al. 2000; Raamsdonk et al. 2001; Harrigan and Goodacre 2003; Oksman-Caldentey and Saito 2005).

Such complex data collection has generally been accomplished using mass spectrometry or nuclear magnetic resonance (NMR) spectroscopy of body fluids or tissues in vitro, rather than of intact organs in a living organism (Beckonert et al. 2007). The complexity of the metabolome data sets has led to the development of different analytical strategies in order to discern its details, such as target analysis and metabolomic profiling (Goodacre 2004). Metabolomic fingerprinting, a subcategory of metabolomic profiling, represents a scan of a large number of intracellular metabolites with the aim of finding a specific signature of a certain tissue or certain state of the tissue. Metabolomic footprinting, on the other hand, identifies secreted metabolites in a given sample and is mostly used in microbial metabolomics (Kell et al. 2005).

Figure 1. Flow chart of main compartments within the cell. Genomics investigates the genome pool of the cell, which consists of about 25,000 genes. mRNA (~100,000) is investigated through transcriptomics, whereas proteins (~1,000,000) are investigated through proteomics. Metabolomics investigates the metabolome, which consists of about 2500 small molecules, such as amino acids, peptides, lipids, and sugars.

A correlation of unique metabolomic fingerprints with certain disease states not only indicates predictive biomarkers, but can also provide novel information related to the underlying disease biology. Although mostly investigated in cancer cell biology (Griffin and Shockcor 2004; Cheng and Pohl 2007), the feasibility of metabolomic profiling for assessment of a variety of conditions has been demonstrated in other diseases (Moolenaar et al. 2003; Prabakaran et al. 2004; Lindon and Holmes 2006; Holmes et al. 2008; Li et al. 2008). One of the most significant applications has been the investigation of the use of ^{1}H-NMR spectra of human sera for rapid, accurate, and noninvasive assessment of the severity of coronary artery disease (Brindle et al. 2002). More recent studies have indicated the existence of human metabolic phenotypes related to diet and hypertension (Brindle et al. 2003; Rezzi et al. 2007; Holmes et al. 2008). In addition, the use of metabolomics has been explored in drug-safety monitoring and toxicology (DeGraaf 1998; Waters et al. 2001; Nicholson et al. 2002; Griffin and Bollard 2004; Mortishire-Smith et al. 2004; Coen et al. 2008). Therefore, metabolomics represents a noninvasive and medically applicable set of techniques for detection of low quantities of known metabolites and identification of unknown compounds, and it provides a great opportunity for biomarker discovery and the prospect of an increasingly important role in medicine and biotechnology.

METABOLOMICS DATA ACQUISITION

To obtain the metabolomic profile of a desired sample, complex data collection has generally been accomplished using mass spectrometry or NMR spectroscopy of body fluids or tissues in vitro in order to detect low quantities of known metabolites and identify unknown compounds (Beckonert et al. 2007; Lenz and Wilson 2007). The NMR-based approach is high-throughput and allows rapid data acquisition of large data sets. In addition, the preparation of the samples is minimal, which increases the high-throughput nature of this form of metabolomics data acquisition. The principle of NMR is based on the fact that nuclei with odd atomic or mass numbers act like magnets oriented toward or against an external magnetic field—a property known as nuclear spin (Hatada and Kitayama 2004). The ^{1}H-NMR uses the external magnetic field to align the nuclear spins of the protons within the sample. To detect these spins, the nuclei are exposed to a rapidly varying electromagnetic field whereby, at one very specific frequency (denoted resonance), the spins absorb this electromagnetic radiation. After a pulse of electromagnetic radiation, the spins return to equilibrium at time constants T1 (longitudinal or spin–spin relaxation) and T2 (transversal or spin–lattice relaxation). During the nuclei relaxation process, there is an emission of radiation, and it is this emission that is detected as the ^{1}H-NMR signal. The exact resonance frequency, however, will vary from one proton to the next, depending on the exact chemical structure of the molecule, such as the molecular bonds that bind the protons and the number of carbon atoms in the molecule and their spatial relationship to the protons (Ross and

Bluml 2001; Hatada and Kitayama 2004). Therefore, ^{1}H-NMR records a unique signal for each chemically distinct hydrogen nucleus. Recently, an improved ^{1}H-NMR methodology has gained widespread use for (semi)heterogeneous samples, such as tissues. Magic angle spinning (MAS) NMR applies rotation along the magic angle of 54.7° at high speed, and this significantly enhances sample homogeneity and spectral resolution (Garrod et al. 1999; Griffin et al. 2003). However, none of these techniques can be used for the analysis of metabolomic profiles of intact organs in a living organism. Therefore, despite its great value for metabolomic analysis of isolated samples and fluids, NMR spectroscopy cannot provide in vivo data that would be useful for metabolic profiling of intact organs in health and disease.

A correlate of ^{1}H-NMR spectroscopy that provides information about the metabolic status of a tissue in vivo is proton magnetic resonance spectroscopy (^{1}H-MRS), a noninvasive magnetic resonance imaging (MRI) modality that investigates the functional and dynamic status of the tissue in real time. Therefore, complements the anatomically based MRI and other MRI modalities that provide information about a variety of physiological parameters, such as blood flow, tissue perfusion, and water mobility. In addition, ^{1}H-MRS surpasses conventional MRI imaging because it can detect metabolic abnormalities that might precede structural changes, therefore providing insights into the pattern of disease evolution and progression. Like NMR, MRS can also detect different nuclei, such as ^{1}H, ^{31}P, ^{13}C, ^{15}N, ^{19}F, and ^{23}Na, with ^{1}H and ^{31}P being the most frequently used. ^{31}P-MRS was the first to be applied in in vivo brain spectroscopy and has been used to evaluate energy metabolism (Burt et al. 1979; Bernard et al. 1983). However, ^{1}H-MRS has surpassed the use of ^{31}P-MRS because it allows the detection of different biomarkers that are specific for neurons (N-acetyl aspartate, NAA) and astrocytes (myoinositol, mI, and choline, Cho), the two most abundant cell types present in brain tissue (Ross and Bluml 2001). Thus, ^{1}H-MRS has become the mainstay of human brain spectroscopy.

The first in vivo ^{1}H-MRS of rat brain was acquired by Behar et al. (1983) and was followed by the first ^{1}H-MRS of human brain by Bottomley et al. (1985). As shown in Figure 2, the ^{1}H-MRS acquires data within a specific volume of the brain tissue, designated as volume or voxel of interest (VOI). Within the spectra, one can clearly discern metabolites that are present at high concentration in the brain tissue, such as NAA, Cho, mI, and creatine (Cr), the energy metabolite that is most stable (Fig. 2). However, many more metabolites exist within the scanned VOI (Table 1), but, due to a variety of data acquisition and processing constraints, only several more can be reliably detected in human brain spectra, such as glutamate, GABA, and lactate. The main data acquisition constraints come from the overabundant presence of water in brain tissue and magnetic field inhomogeneity. The quality of ^{1}H-MRS data greatly depends on the signal-to-noise ratio (SNR) because the method investigates not the water peak, but metabolites that are present in the brain tissue at much smaller concentrations. Therefore, historically, the first problem encountered in ^{1}H-MRS was the presence of

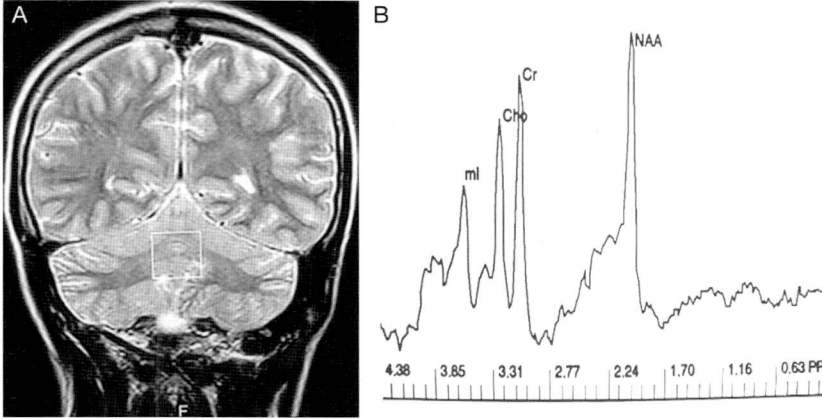

Figure 2. ¹H-MRS acquisition of a single voxel spectroscopy. (*A*) Coronal MRI scan with VOI in the cerebellum. Anatomical scan is done first to guide the positioning of the VOI within which the spectroscopy will be performed. (*B*) ¹H-MRS of the VOI in *A*, processed by FFT. Main metabolites are outlined: *N*-acetyl aspartate (NAA) is a biomarker of neurons, choline (Cho) and myoinositol (mI) are biomarkers of glial cells, and creatine (Cr) is a biomarker of energy metabolism. (VOI) Voxel of interest; (FFT) fast Fourier transform.

water, reaching up to 90 mol/liter concentration and creating a large background signal that overshadowed the signals of all other metabolites present. To overcome this problem, a variety of techniques had to be developed that either did not excite the background water signal or suppressed it substantially (Bottomley et al. 1985). Furthermore, although NMR of intact tissue and tissue extracts has provided high separation of individual spectral peaks, the field inhomogeneity intrinsic to in vivo ¹H-MRS causes broadening of the peaks and loss of significant detail about the individual metabolites. The analysis of such spectra would lead to unreliable results because metabolites of lower concentration may be hidden within the broad peaks and therefore pass undetected, and, if the peak area is used as a measure of metabolite concentration, broad peaks due to field inhomogeneity might lead

to misleading values. To resolve this problem, a variety of techniques have been developed, from generating simulated echoes to localize different regions of the brain (Frahm et al. 1989; Keevil 2006) to improved shimming (i.e., optimization of the field inhomogeneity). Finally, to be detected by ¹H-MRS, a metabolite must be small (<20 kDa), soluble, and mobile in the magnetic field (Ross and Bluml 2001). In an in vivo situation, a variety of conditions may change the solubility and/or mobility of a metabolite detected in vitro. Under such conditions, a metabolite visible by in vitro NMR might be invisible or erroneously increased by in vivo MRS. Fortunately, this problem has not been frequently encountered because major metabolites discovered by ¹H-NMR in vitro, such as NAA, Cr, Cho, glutamate, mI, and others, have all been detected by ¹H-MRS in vivo in the human brain (Ross and Bluml 2001; Ross et al. 2006).

Despite the above pitfalls, ¹H-MRS, since its approval by the Federal Drug Administration (FDA) in 1993, has been used to identify and characterize metabolic changes associated with a variety of neurological diseases (Di Costanzo et al. 2007). The list is long and includes neurodegenerative diseases, cerebrovascular diseases, dementias, brain tumors, and demyelinating disorders, as well as rare diseases such as Canavan disease, Rasmussen's encephalitis, creatine deficiency syndrome, and others (Bonavita et al. 1999; Lin et al. 2005). However, despite its value, ¹H-MRS has not become a part of a routine clinical assessment because of the limited number of metabolites that can be analyzed using traditional signal-processing methodologies. Fortunately, during the past few years, new analytical tools have emerged that allow for a more comprehensive analysis of ¹H-MRS data. The new approaches come from the field of metabolomics as well as from new developments in the ¹H-MRS field, where the search for more precise quantification methods applicable to data acquired by higher magnetic field strengths, such as 3T and 7T human MRI scanners, has been increasingly relevant.

Table 1. Metabolites detected by spectroscopy in the live human and rodent brain as well as in tissue biopsies

Metabolite	Concentration range (mmol/kg$_{ww}$)
Acetate	0.4–0.8
NAA	7.9–1.6
ATP	3.0
Alanine	0.2–1.4
GABA	1.3–1.9
Aspartate	1.0–1.4
Choline (total)	0.9–2.5
Creatine	5.1–10.6
Glucose	1.0
Glutamate	6.0–12.5
Glutamine	3.0–5.8
Glycine	0.4–1.0
Lactate	0.3–0.6
Myoinositol	3.8–8.1
Phosphocreatine	3.2–5.5
Phosphorylcholine	0.6
Phosphorylethanolamine	1.1–1.5
Pyruvate	0.2

Adapted from Govindaraju et al. (2000).
The list includes only the metabolites of >0.5mmol/kg$_{ww}$ concentration.

METABOLOMICS SIGNAL PROCESSING

Metabolomic signal processing consists of several steps: processing of raw data that allows for comparisons of different data sets, data mining that involves use of statistical tools for high-throughput analysis which identifies variables of interest, and quantitative measurements that provide both absolute and relative concentrations of selected metabolites which contribute mostly to the pathology investigated (Glassbrook et al. 2000; Dunn et al. 2005; Goodacre 2005). Herein, we describe some signal-processing algorithms used in metabolomics and ^{1}H-MRS.

The most widely used analytical methods for processing complex metabolomic data sets are based on multivariate statistical analysis. Principal component analysis (PCA) (Jolliffe 1986) is an unsupervised multivariate analysis that highlights the differences in a data set using a smaller number of factors, the principal components (PC) that capture the most variance in the data set. The first PC is the vector that accounts for the most variation in the data, and each observation is projected onto this line. The second PC is represented by another vector in K dimensional space, which is orthogonal to the first PC, and represents the second largest variation in the data set. Many PCs may be obtained from a single data set, which consists of a multitude of related spectra; however, the variation quickly tapers off within a few PCs. Most often, the first ten PCs explain more than 90% of the variance. The PCA data are presented as a scores plot that reveals clustering in the observations based on similarities in the spectra. Thus, PCA provides a rapid means of visualizing and comparing data sets, where different clustering indicates the true variance among data classes. Additionally, a loadings plot is obtained from PCA that assists in the interpretation of the scores plot by highlighting the spectral regions responsible for clustering. In ^{1}H-MRS data, it is expected that actual peaks, not noise, are the highest contributing factors (Goodacre et al. 2004).

Partial least squares—discriminant analysis (PLS-DA) is another analytical method widely used in multivariate analysis. Whereas PCA reduces the dimensionality of a multidimensional data set while retaining the characteristics of the data set that contribute most to its variance, PLS-DA is a regression extension of PCA that takes advantage of a priori class information to maximize the separation among groups of observations. Therefore, PLS-DA uses the basic principles of PCA but further minimizes the noise and maximizes the differences among classes. It is a compromise between the usual discriminant analysis and the discriminant analysis on the significant PC of the predictor variables. As such, PLS-DA has been used for predictive modeling of data sets and is especially suited to deal with a large number of predictors compared to the number of observations, which might pose a problem when analyzing complex metabolomic data sets (Perez-Enciso and Tenenhaus 2003). Although PCA and PLS-DA as well as other related multivariate analytical methods have been applied for analysis of the metabolomic data, they are not traditionally used for the analysis of ^{1}H-MRS data.

To specifically process ^{1}H-MRS data, one can use frequency- or time-domain-based methodologies. Herein, we outline the basic principles for ^{1}H-MRS signal processing and the most frequently used software. First, the signals that carry information in ^{1}H-MRS are exponentially decaying sinusoids, also known as free induction decay (FID) signals. They become smaller over time and are modeled as

$$y_n = \sum_{k=1}^{K} a_k e^{(-d_k + j2\pi f_k)t_n} + w_n \qquad (1)$$

where y_n are the measured signal samples; $n = 0, 1, 2, \cdots,$ $N - 1$; K is the number of decaying complex sinusoids in the model; $j = \sqrt{-1}$; a_k, d_k, and f_k are the amplitude, decay factor, and frequency of the kth complex sinusoid, respectively; $t_n = nT_s$, where T_s is the sampling interval; w_n represents noise sample. In addition, more general models that account for imperfections of the signal acquisition process include a Gaussian decay factor and a combination of Lorentzian and Gaussian decays (Slotboom et al. 1998; Bartha et al. 1999).

In ^{1}H-NMR, a significant amount of prior knowledge exists about the parameters used in Equation 1, often derived by imposing constraints on these parameters. For example, the prior knowledge in ^{1}H-NMR can be the information about frequencies and/or the decay factors of the metabolites, ratios of some of their amplitudes, or differences in their frequencies (DeGraaf 1998; Bartha et al. 1999). Obviously, with better prior knowledge, the methods for metabolite quantification yield more accurate results.

The most widely utilized method for quantification of ^{1}H-MRS data is based on the peak frequency domain and integration of the areas under the peaks of Fourier-transformed data (Mierisova and Ala-Korpela 2001; Zandt et al. 2001). The fast Fourier transform (FFT) is an algorithm that converts the data in the time domain (i.e., the time constants of decay of the sinusoids) into the frequency domain, generating peaks identified on the basis of the frequency at which they resonate, known as the chemical shift. The chemical shift is a dimensionless value measured in parts per million (ppm) when compared with a reference resonance frequency, usually that of 3-trimethylsilylpropionic acid (TSP) or tetramethylsilane (TMS). The area under a peak is, in principle, proportional to the concentration of the metabolite that generates the peak (Meyer et al. 1988; DeGraaf 1998). However, methods based on peak area computation have low estimation accuracy and most do not use the model (Eq. 1) directly (Vanhamme et al. 1997). This is primarily due to problems that arise from the tails of metabolite spectra that overlap with the spectral peaks of interest. One method, however, that fits a linear combination of single metabolite spectra with an arbitrary line shape under the constraints of a common baseline has shown good results and is known as the linear combination of model (LCModel) (Provencher 2001). To incorporate maximum prior knowledge into the analysis, LCModel uses complete spectra rather than individual resonances. It is noninteractive and nearly model-free, and it accounts automatically for the baseline and provides both absolute and relative quantifications of the detected compounds. As such, it has gained wide acceptance for in vivo ^{1}H-MRS processing (Jírů et al. 2003; Kanowski et al. 2004; Rosen and Lenkinski 2007).

Another group of methods for quantification of ^{1}H-MRS data are known as time-domain methods (Vanhamme et al. 1999, 2001). Typically, these methods rely on model assumptions as in Equation 1. One of the first widely used time-domain interactive methods was the variable projection method (VARPRO; van der Veen et al. 1988). A method that improves the robustness and flexibility of the analysis is known as AMARES (Barkhuijsen et al. 1985; Vanhamme et al. 1997). Some of the improvements result from the use of prior knowledge and the possibility of fitting echo signals. An interactive software package that combines VARPRO and AMARES and allows for frequency selective filtering, correction of *eddy* currents artifacts, enhancement of FID signals, and linear prediction is known as the magnetic resonance user interface (MRUI) package. An important drawback of all of the interactive methods is their reliance on human assistance.

In addition to these interactive methods, there are black-box methods that are also based on time-domain processing. They use the linear prediction principle or state–space representations. From theory, it is known that the observations represented by Equation 1 can be approximately modeled by a high-order autoregressive (AR) model, i.e.,

$$y_n = -\sum_{m=1}^{p} \alpha_m y_{n-m} + w_n \qquad (2)$$

where p is the order of the AR model, α_m are its parameters, and w_n is its driving noise. Because the information about the original parameters a_k, d_k, and f_k is in the AR parameters, the objective is to estimate the α_ms and then deduce the values of a_k, d_k, and f_k. In principle, the f_ks and d_ks are first obtained from the α_ms and then substituted back in Equation 1, to allow for the estimation of a_k. A popular method for processing decayed sinusoids based on Equation 2 exploits the concept of singular-value decomposition (SVD) (Haacke et al. 1989). A straightforward application of SVD to ^{1}H-MRS data first develops an overparametrized model such as that in Equation 2 and expresses it in a vector-matrix form, where the unknown vector is composed of the parameters α_m. This vector multiplies data matrix whose elements are various samples of the data, y_n. After performing SVD of the data matrix, the rank of that matrix can be reduced by preserving only a subset of the largest singular values and equating the remaining ones with zero. With the rank-reduced matrix, a significant noise contribution is removed, better solutions for α_m are obtained, and, as a result, better estimates of f_k and d_k are produced. Finally, novel approaches for better estimation of metabolite quantitation are still being developed (Poullet et al. 2007) with the aim to optimize processing and fitting of the acquired spectra, as reviewed in a number of papers (Mierisova and Ala-Korpela 2001; Provencher 2001; Vanhamme et al. 2001; Zandt et al. 2001).

Although the described methods are mostly used to quantify single-peak metabolites, a novel method has recently been developed for identifying metabolites with multiple peaks in the spectra (Cloarec et al. 2005). Statistical total correlation spectroscopy (STOCSY), the cross-correlation of correlation matrices, offers two major classes of correlations. One set is the resonance/peak pattern of individual components, which is charaterized by the highest correlation coefficient for any molecule, because all peak intensities vary together as the concentration of a particular molecule/metabolite varies across the data set. The added, important benefit of STOCSY is the next class of correlations that connects those components which vary positively and negatively in a concerted fashion. This is likely to happen if there are connections in the metabolic pathways or other reasons for coherences in pathophysiological processes reflected at the metabolic level. Therefore, STOCSY offers a unique analytical ability to provide information not only about the multipeak metabolites but also about metabolic pathways that might correlate during a certain cell/tissue state.

In summary, many signal-processing methodologies are now available for analysis of metabolomic data. Some of these processing methodologies are used for all "omics" data sets and some are used for quantification of single- and multipeak metabolites detected in the spectra. A combination of these methodologies provides comprehensive information not only about the patterns of differences among various data sets, but also about the most significant specific metabolites and specific pathways that correlate positively or negatively during the given state of the tissue. The outlined metabolomic methodologies have traditionally been applied to analyze NMR-acquired data sets in vitro. However, they can also be applied to analyze ^{1}H-MRS data sets in vivo, providing yet unexplored insights into the living brain metabolome.

CELLULAR AND SYSTEM METABOLOMICS: A NOVEL APPROACH FOR BIOMARKER DISCOVERY

Metabolomics studies have been limited by the almost exclusive systems biology approach and have not yet been able to translate systems data into concrete, clinically useful diagnostic and prognostic biomarkers of a disease. In contrast, we propose a more comprehensive use of metabolomics that aims to identify biomarkers through both cellular metabolomics (profiling of chosen cell types under in vitro conditions) and systems metabolomics (profiling of human tissue in vivo). Herein, we describe our strategy for identification of biomarkers of human brain disease(s), which combines the two methodologies. We also outline our signal-processing approach, which might provide qualitative and quantitative evaluation of individual metabolites and their contribution to the investigated condition in both healthy and diseased brain tissue.

Cellular metabolomics or the bottom-up approach (Fig. 3) is used to define both the whole metabolome and the unique metabolomic fingerprint of different cell types present in brain tissue in both normal and abnormal states. First, by using a 16T (700 MHz) ^{1}H-NMR, which provides

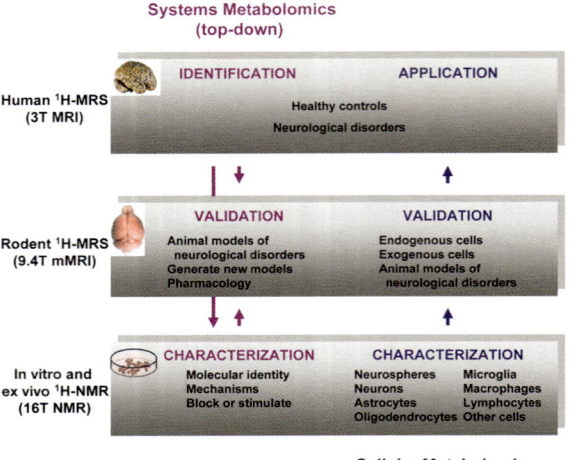

Figure 3. Flow chart of systems and cellular metabolomic strategies for biomarker discovery. A combination of such strategies readily translates clinically discovered biomarkers to cellular mechanistic and pharmacological studies, as well as in vitro gained knowledge to clinically applicable biomarkers. For example, systems metabolomics (top-down approach) uses ¹H-MRS of the human brain to identify biomarkers of the disease. Once they are discovered, they can be investigated on the cellular level, through NMR of cells and tissues, and validated in animal models using ¹H-MRS. On the other hand, cellular metabolomics (bottom-up approach) uses ¹H-NMR to characterize cellular and molecular biomarkers in vitro. Once they are discovered, they can be validated in animal models through ¹H-MRS spectroscopy and applied to human brain imaging for correlation with a specific disease.

the maximal spectral resolution, we can characterize very precisely the metabolomic profiles specific for different cell types. By analyzing the cellular metabolomic profiles, we can identify and characterize metabolites unique for each cell type in a healthy condition, thereby generating a database of cellular metabolomic fingerprints for any studied cell. To investigate whether a certain disease has an effect on the cellular metabolomic fingerprint, one can mimic a variety of disease conditions in culture and analyze both intracellular (fingerprint) and extracellular (footprint) changes of the metabolome. This is accomplished by the ¹H-NMR of whole cells and cell extracts, as well as the extracellular conditioned media obtained during the course of the investigated condition. Furthermore, in this system, we can both investigate the mechanisms that lead to specific metabolomic dysfunction and test whether any drug(s) reverses the effects of the disease and recovers the normal cellular metabolome in vitro.

We can then directly translate our findings to in vivo ¹H-MRS metabolomic profiling of the brain tissue. However, under the in vivo condition, relatively simple cellular and molecular metabolomic fingerprints are embedded in a complex tissue, where all cell types are present in a specific ratio. First, to test whether the cellular and molecular biomarkers retain their chemical shift in vivo, we can inject desired cells or molecules into the rodent cortex in vivo and then perform microMRI (¹H-MRS) spectroscopy of the injected regions. This step is

crucial for establishing the most optimal mMRI acquisition parameters for ¹H-MRS. To validate the specificity of the obtained results, we use several well-established methodologies to confirm that signals obtained from the ¹H-MRS spectroscopy indeed belong to the injected material. For example, we can perform ¹H-NMR of the homogenized tissue obtained from the same animal that underwent mMRI, fluorescence-activated cell sorting (FACS) or enzyme-linked immunosorbent assay (ELISA) analysis of the cell types and molecules injected, and/or immunohistochemical staining of the cortical sections to correlate histological findings with the quantitative mMRI data. Therefore, animal models provide an invaluable tool for validation of metabolomic data translated from the in vitro studies and for their correlation with the changes at the genome and proteome levels that occur in the same animal model.

If expected metabolomic fingerprints are obtained repetitively and reliably in vivo, we can then investigate the biomarker dynamics in a disease state by using animal models of human disease with pathology known to involve the respective cells and molecules of interest. To validate these data, the same methodologies used for validation of exogenous biomarkers can be used. Moreover, in animal models, one can manipulate the disease in a variety of ways and then analyze the changes of individual biomarkers and the whole metabolome both qualitatively and quantitatively. Finally, both spacial and temporal studies can be done to estimate not only the degree of disease process and the presence of repair molecules and cells, but also their dynamic changes over time.

The most significant contribution of this strategy lies in its direct application to the imaging and characterization of the human brain metabolome. Through our integrative bottom-up approach, we have discovered a metabolite that is enriched in NPCs (Manganas et al. 2007). We have detected this metabolite in the live human brain, a significant breakthrough that might have direct impact on our ability to test the role of NPCs in a variety of neurological and psychiatric diseases. We can now investigate NPC density in any age group, from neonates to elderly, and correlate it with a variety of clinical, systemic, cognitive, radiological, and other measures in any disease of interest. Similarly, we can investigate other cellular and molecular biomarkers, once characterized and validated in our in vitro and in vivo studies (see Fig. 3), emphasizing a significant potential of this experimental approach.

Overall, our studies have demonstrated that cellular metabolomics can provide invaluable insights into the metabolome of isolated cells in vitro, which can be directly translated to the metabolomics of human brain tissue in vivo. However, cellular metabolomics might not offer sufficient insight into the complex changes that occur in diseased tissue in vivo, because those changes involve interactions of many cell types and molecules present within the tissue. Therefore, to investigate specific disease biomarkers in a most comprehensive way, in addition to cellular metabolomics, we need to use systems metabolomics (top-down approach) and to translate those findings to animal models and cellular levels (see Fig. 3).

Using ^{1}H-MRS of the human brain in vivo to acquire systems metabolomic data, we first investigate the patterns that differentiate the diseased tissue from the normal tissue. By dissecting the specific patterns down to individual metabolites, as outlined below, we can identify the most significant contributors to the metabolic differences observed, as well as correlate them within metabolic pathways that might be impaired in a given disease (Fig. 3). Once identified, specific patterns can be further investigated in animal models of a chosen disease using mMRI, and specific individual biomarkers or identified pathways can be investigated on a cellular level using NMR. As in the cellular metabolomic strategy, the in vitro systems are useful for characterization of the metabolites and pathways, identification of the mechanisms involved, and creation of pharmaceutical agents that affect those specific compounds. Investigations of animal models in vivo in the system metabolomic approach also aim to validate the data acquired in both human and cellular studies. Therefore, in our research, both cellular and system metabolomics converge in the animal models used to validate the in vitro and human in vivo data and to study biomarker and pathway identity and mechanisms that control them.

The key to our proposed strategy lies in signal processing that combines traditional metabolomic tools with algorithms used for ^{1}H-MRS processing (Fig. 4). First, multivariate analysis, such as PCA, is used to process the complex data sets and to provide information about the principal components that contribute to the identified patterns of difference, whether these are detected in vitro or in vivo (Fig. 4). Such analysis of the metabolomic data is then faced with two key challenges: (1) It must determine whether the observed statistical patterns are meaningful, and (2) it must then translate these patterns into biologically relevant information in the form of metabolite identity and quantity (Weckwerth and Morgenthal 2005).

PCA provides the loadings plots that may be used to identify individual metabolites that contribute mostly to the observed pattern difference. To investigate the specific metabolites discovered by the multivariate analysis and loadings plots, we apply already developed algorithms used in ^{1}H-MRS processing.

Several signal-processing methods can be used to quantify individual metabolites with known frequency of oscillation, such as SVD and LCModel (Fig. 4). Our method of choice for quantitative analysis of metabolites with known frequency has been SVD-based, specifically designed to overcome the significant decline in biomarker concentration and spectroscopic resolution as we proceed from cells to live brain tissue. However, signal processing is very much limited by the quality of the data to be processed, particularly when the SNR is low. Spurious signals outside the region of interest can further contribute to the noise of the obtained spectra, and these effects may modify the final estimates of the cell and/or metabolite concentrations. Thus, to ensure reliable signal processing, many iterations within the selected frequency range must be performed to ascertain that signals are coming from the metabolite of interest. Another problem may arise if many cell types within the region of interest have the same biomarker and therefore contribute to the frequency band of interest. To minimize this confounder, the purity of the primary cultures must be ensured for best-quality data when investigating cellular biomarkers in vitro, and a combination of different cell types in vitro may be used to evaluate the anticipated in vivo conditions. Finally, a critical issue in data analysis is the probability of a false alarm, i.e., the probability of detecting biomarkers when there are none. Therefore, it is important to find the distribution of the estimation error from which we can deduce the sensitivity of the analytical method. These are all crucial performance specifications for any applied method of analysis, and they need to be precisely outlined to interpret the obtained results reliably.

Once we identify the specific cellular fingerprints through our in vitro analysis, we can use signal processing to isolate, amplify, and quantify single-cell signatures within complex metabolomic data of a chosen brain region. These signal-processing algorithms may enable the quantification of a variety of metabolites present in the tissue at low concentrations, such as those that indicate amino acid biochemistry; processing of proteins, sugars, and lipids; as well as the oxidoreductive status of the mitochondria. In addition, one can also apply correlative analysis such as custom-developed STOCSY to analyze metabolic pathways and metabolites that correlate in their expression pattern (up- or down-regulated) within the spectra. Such correlative analysis might provide useful information about the possible biochemical pathways that might be perturbed in the disease state. Finally, predictive modeling can be done using PLS-DA, which, in combination with all of the other analytical methodologies, might lead to the ultimate goal of our studies—a clinically useful application of specific metabolomic biomarker.

In summary, implementation of our experimental strategy and signal-processing methodologies might shed light on as yet undiscovered processes that may underline

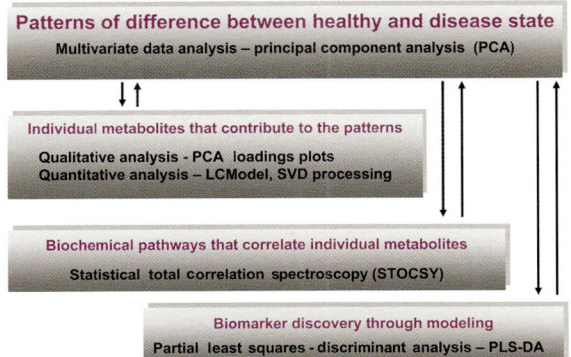

Figure 4. Flow chart of signal-processing tools used for metabolomic and spectroscopy signal processing. Such comprehensive analytical approach may lead to identification of patterns of difference between different samples as well as individual metabolites that contribute to the pattern difference and correlation between them. This analytical approach may ultimately identify and characterize specific diagnostic and prognostic biomarkers of a disease state. All of the data obtained may be used for predictive modeling. See main text for details.

the particular brain disease one aims to investigate. Our approach may indicate specific metabolic differences between healthy and disease states or between different parts of the diseased brain. Furthermore, it might reveal specific metabolites and metabolic pathways that significantly contribute to the observed metabolic difference. Clinically, such studies might lead to the identification of diagnostic and prognostic biomarkers for individual symptoms or for the whole phenotype of the disease. Finally, specific metabolites and pathways may be further investigated on the cellular level, for both mechanistic and pharmacological studies, which could further lead to identification of the rescue mechanisms that might lead to the establishment of the normal metabolomic profiles both in vitro and in vivo.

IMAGING OF NEURAL PROGENITOR CELLS: AN EXAMPLE OF METABOLOMICS BIOMARKER DISCOVERY

Since their discovery in the human brain (Eriksson et al. 1998; Roy et al. 2000), NPCs have been envisioned as a new hope for a variety of neurological diseases. Even though their significant therapeutic capability has been widely recognized, the absence of well-defined markers that would distinguish them from other neural cell types has hampered their investigation in the live human brain. Until recently, methodologies for identification of NPCs in the living human brain have not been available, and without the ability to identify NPCs, the analysis of their fate and function in the human brain has been virtually impossible.

During the past few years, molecular and cellular MRI imaging has emerged as a new field for visualization of macromolecules and cells in living organisms and has been attempted for identification and tracking of NPCs (Bulte et al. 2002; Modo et al. 2005). Among several paramagnetic reagents, superparamagnetic iron oxide (SPIO) particles have been used in both molecular and cellular MRI studies. Several methodologies for conjugation of SPIO particles to transfection agents have been tested in order to enable efficient intracellular magnetic labeling (Arbab et al. 2003; Bulte and Kraitchman 2004; Bulte 2006). Migration of SPIO-labeled NPCs has been studied by mMRI in several animal models of human disease, such as stroke, glioma, and multiple sclerosis (Hoehn et al. 2002; Modo et al. 2004; Anderson et al. 2005; Ben-Hur et al. 2007). Recently, a clinical study has reported tracking of SPIO-labeled pancreatic islet grafts used to treat diabetic patients (Toso et al. 2008). Although no immediate side effects were observed, the authors recognized that iron overload might be the major obstacle for use of SPIO-based compounds (Toso et al. 2008). In addition to SPIO-based MRI imaging, positron emission tomography (PET) and single-photon emission computerized tomography (SPECT) scanning of ex vivo radiolabeled cells have been attempted and have led to a semiquantitative method for evaluation of cell trafficking (Adonai et al. 2002; Chin et al. 2003). However, all of the above techniques are applicable only to exogenous NPCs and do not provide the means for tracking endogenous NPCs. In addition, they cannot be used for imaging of infants and young children because radiolabeled reagents

have not been approved by the FDA for children under 18 years of age. Therefore, there are currently no clinical high-resolution imaging techniques that would enable spatiotemporal localization and tracking of unlabeled NPCs.

We have approached NPC imaging from the metabolomics standpoint and have applied a cellular metabolomics strategy to investigate metabolic biomarkers of NPCs and their progeny. First, ^{1}H-NMR data were acquired for all individual cell types, grown as purified primary cultures in vitro (700-MHz Bruker NMR) (Manganas et al. 2007). As shown in Figure 5A, the raw spectra clearly indicate the differences in the metabolomes of NPCs and their progeny: neurons, astrocytes, and oligodendrocytes, the main cell types present in the brain. A major difference that can be deduced from the raw spectra is the presence of the 1.28 ppm metabolite in the NPC spectra and not in the spectra of other cell types. To investigate the metabolomic pattern of each cell type, we performed multivariate analysis (SIMCA-P, Umetrics). The PCA confirmed a clear difference in metabolomic patterns among the four cell types (Fig. 5). Interestingly, each cell type clustered separately, indicating that their metabolic profiles were specific. Whereas astrocytes and oligodendrocytes separated from neurons along the principal component 2 (PC2), NPCs clustered away from all of them along the PC1 (Fig. 5B). One of the main metabolites contributing to the separation of NPCs from other cell types was the 1.28 ppm metabolite, as seen in the PC1 loadings plot (Fig. 5D). Therefore, metabolomic profiling of NPCs indicates that they are clearly metabolically different from other neural cell types.

In addition to gaining the information on the NPC metabolomic profile, the first PCA indicated that clusters of astrocytes and oligodendrocytes are close together, suggesting that these cells share common metabolic compounds as compared to neurons and NPCs (Fig. 5B). However, the metabolic profiles of these cells have some differences, as indicated by further PCA processing. As shown in Figure 5C, astrocytes and oligodendrocytes cluster away from each other based on PC3. The metabolite that seems to contribute the most to this separation is choline, resonating at 3.23 ppm and seen in the PC3 loadings plot (Fig. 5F).

We have further characterized the 1.28-ppm metabolite in terms of its quantification, specificity, and basic molecular identity (Manganas et al. 2007). To exemplify our strategy described earlier in the text, we show an example of our translation from in vitro NMR data to animal and human brain imaging of NPCs.

Translation of the in vitro data to in vivo brain tissue imaging faces significant challenges, such as considerable decline in magnet strength (from 16T Bruker NMR scanner to 9.4T mMRI Bruker scanner to 3T Phillips MRI scanner) associated with decreased SNR; spectroscopy of tissue as compared to spectroscopy of isolated cell types; and presence of our cells of interest, NPCs, in a limited concentration within a certain brain region (hippocampus) that presents its own intrinsic challenges for ^{1}H-MRS acquisition (field inhomogeneity, close proximity to the cerebrospinal fluid, etc.). The data acquisition is essential for reliable processing because all signal-processing algorithms heavily depend on the SNR. Because an NPC

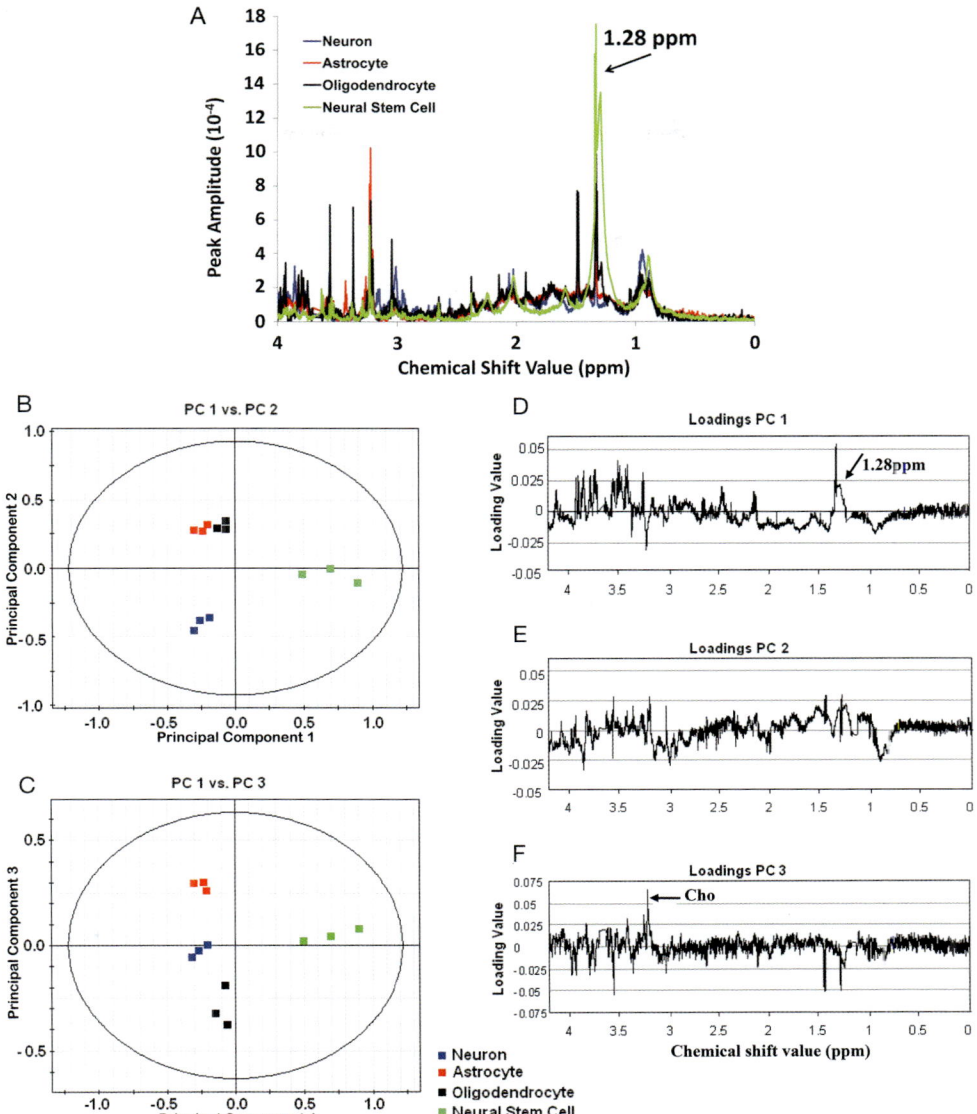

Figure 5. Multivariate analysis of the neural cell types in vitro. (*A*) NMR spectra of individual cell types indicate differences among them, as well as abundant presence of the 1.28 ppm metabolite that distinguishes NPCs from the other cell types. (*B*) PCA of the NMR spectra of neural cell types indicates tight clustering of individual cell types as well as clear separation of different cell types. Specifically, NPCs separate from others along PC1, whereas astrocytes and oligodendrocytes separate from neurons along PC2. (*C*) PCA of the NMR spectra of neural cell types identifies principal component 3 as the main contributor for the metabolic difference between astrocytes and oligodendrocytes. (*D,E,F*) Loadings plots for PC1, -2, and -3. Loadings plots for each PC indicate variations of individual metabolites in the spectra and can be used to identify those metabolites that mostly contribute to the observed pattern difference. Loadings plot for the PC1 (*D*) indicates that the 1.28 ppm metabolite is one of the major contributors for the separation of the NPCs from the other cell types. Loadings plot for the PC3 (*F*) indicates that choline (Cho) is one of the major contributors for the separation of the astrocytes from the oligodendrocytes.

metabolite, resonating at 1.28 ppm is embedded within the noise when in vivo spectra are processed by the FFT, we have exploited a signal-processing algorithm based on SVD that enabled us to detect NPCs in vivo using high-resolution ¹H-MRS of the rodent brain and the human brain (Manganas et al. 2007; Djurić et al. 2008).

Herein, we show two examples of detection of NPCs in vivo in an adult rat brain (Fig. 6) and in an adult human brain (Fig. 7). Meticulous placement of the VOI is crucial

for proper ¹H-MRS acquisition. For the mMRI of the rodent brain, the hippocampus (where NPCs reside) is close to the center of the magnetic field, and therefore shimming and ¹H-MRS acquisition are "easier" than shimming and acquisition of the spectra in the cortex, which serves as a negative control (Fig. 6). The opposite is true for the human brain ¹H-MRS—the hippocampal region is notoriously difficult for data acquisition, and this might lead to poor SNR and unreliable analysis. The examples of

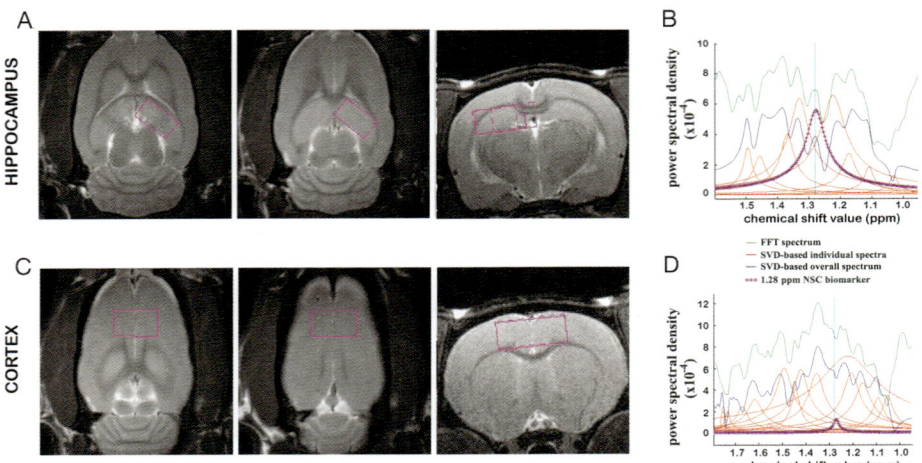

Figure 6. ¹H-MRS of the rodent brain in vivo. (*A*) Coronal and axial images of the rat brain with localization of the VOI in the hippocampus. (*B*) Singular-value decomposition (SVD)-based signal processing of the spectra obtained in *A*. Modeling of the individual peaks outlines the 1.28 ppm metabolite and enables its quantification. (*C*) Coronal and axial images of the rat brain with localization of the VOI in the cortex. (*D*) SVD-based signal processing of the spectra obtained in *C*. Modeling of the individual peaks within the similar range as in *B* indicates minimal presence of the 1.28 ppm metabolite in the cortex.

the SVD analysis of both rodent and human spectra are shown in Figures 6 and 7, respectively. Processing is done as described by Manganas et al. (2007). Briefly, the data are imported into MatLab and water is removed. In addition, signals whose tails might compromise the detection of the 1.28-ppm metabolite are also removed. We then search for the 1.28-ppm signal within a specific range (1.28 ± 0.02

ppm) and quantify it based on its amplitude. The processing is slightly different for rodent and human data because they differ in acquisition parameters and model order (Manganas et al. 2007). Both rodent and human in vivo data can be scrutinized by other analytical methods of multivariate analyses, which may yield more information about pattern differences among the different brain regions.

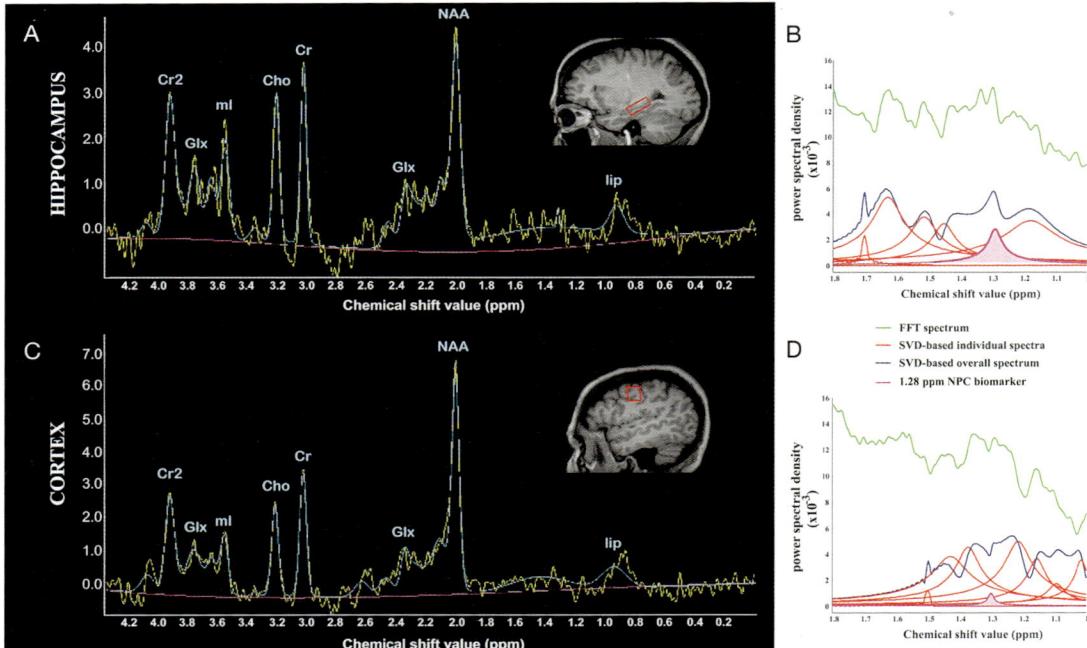

Figure 7. ¹H-MRS of the human brain in vivo. (*A*) Sagittal MRI scan of the human brain with localization of the VOI in the hippocampus. FFT spectra are shown with major metabolites indicated (Glx = glutamate and glutamine). (*B*) Singular-value decomposition (SVD)-based signal processing of the spectra obtained in *A*. Modeling of the individual peaks outlines the 1.28 ppm metabolite and enables its quantification. (*C*) Sagittal MRI scan of the human brain with localization of the VOI in the cortex. FFT spectra are shown. (*D*) SVD-based signal processing of the spectra obtained in *C*. Modeling of the individual peaks within the same range as in *B* indicates minimal presence of the 1.28 ppm metabolite in the cortex.

CONCLUSION

One of the biggest challenges in biomedical science is the translation of basic scientific accomplishments to clinical practice. Although drug or device developments have established principles of transition from in vitro to animal in vivo to human applications, other fields of medicine have not yet provided the means of clear transference. Our outlined strategy for experimental design and signal processing using metabolomics and spectroscopy tools might offer such means and a direct translation of biomarkers characterized in vitro to clinical applications in vivo. Similarly, our approach can provide an unprecedented insight into the functional status of live human brain tissue and can translate this information to more readily available validation models, such as animal models and studies at the cellular level. Furthermore, although a well-known correlate of high-resolution NMR, ^{1}H-MRS spectroscopy has never been scrutinized using complex data analysis paradigms well established for ^{1}H-NMR metabolomics. Application of metabolomics analytical methods to ^{1}H-MRS in conjunction with other spectroscopy signal-processing tools might therefore lead to a potential wealth of functional human brain ^{1}H-MRS tissue-profiling data. We hope that our introduction of the concept of cross-fertilization among different disciplines, such as metabolomics and ^{1}H-MRS, in terms of signal processing and data correlation, will prompt further research in these areas, which ultimately may lead to new discoveries that will improve our understanding of human brain diseases.

ACKNOWLEDGMENTS

This work was supported by NINDS R21NS05875-1 and 5K08 NS044276, and U.S. Army Medical Research grant DAMD170110754 (M.M.S.); T32DK07521-16 (L.N.M.); NINDS R01-NS32764, NARSAD, Seraph Foundation, Hartman Foundation, Hope for Depression Foundation, and Hazan Foundation (G.E.); Department of Energy FWP MO-065 (H.B.); and National Science Foundation CCF-0515246 and Office of Naval Research N00014-06-1-0012 (P.D.).

REFERENCES

Adonai, N., Nguyen, K.N., Walsh, J., Iyer, M., Toyokuni, T., Phelps, M.E., McCarthy, T., McCarthy, D.W., and Gambhir, S.S. 2002. Ex vivo cell labeling with 64Cu-pyruvaldehyde-bis(N4-methylthiosemicarbazone) for imaging cell trafficking in mice with positron-emission tomography. *Proc. Natl. Acad. Sci.* 99: 3030–3035.

Anderson, S.A., Glod, J., Arbab, A.S., Noel, M., Ashari, P., Fine, H.A., and Frank, J.A. 2005. Noninvasive MR imaging of magnetically labeled stem cells to directly identify neovasculature in a glioma model. *Blood* 105: 420–425.

Arbab, A.S., Bashaw, L.A., Miller, B.R., Jordan, E.K., Lewis, B.K., Kalish, H., and Frank, J.A. 2003. Characterization of biophysical and metabolic properties of cells labeled with superparamagnetic iron oxide nanoparticles and transfection agent for cellular MR imaging. *Radiology* 229: 838–846.

Barkhuijsen, H., de Beer, R., Bovee, W.M., Creyghton, J.H., and van Ormondt, D. 1985. Application of linear prediction and singular value decomposition (LPSVD) to determine NMR frequencies and intensities from the FID. *Magn. Reson. Med.* 2: 86–89.

Bartha, R., Drost, D.J., and Williamson, P.C. 1999. Factors affecting the quantification of short echo in-vivo 1H MR spectra: Prior knowledge, peak elimination, and filtering. *NMR Biomed.* 12: 205–216.

Beckonert, O., Keun, H.C., Ebbels, T.M., Bundy, J., Holmes, E., Lindon, J.C., and Nicholson, J.K. 2007. Metabolic profiling, metabolomic and metabonomic procedures for NMR spectroscopy of urine, plasma, serum and tissue extracts. *Nat. Protoc.* 2: 2692–2703.

Behar, K.L., den Hollander, J.A., Stromski, M.E., Ogino, T., Shulman, R.G., Petroff, O.A., and Prichard, J.W. 1983. High-resolution 1H nuclear magnetic resonance study of cerebral hypoxia in vivo. *Proc. Natl. Acad. Sci.* 80: 4945–4948.

Ben-Hur, T., van Heeswijk, R.B., Einstein, O., Aharonowiz, M., Xue, R., Frost, E.E., Mori, S., Reubinoff, B.E., and Bulte, J.W. 2007. Serial in vivo MR tracking of magnetically labeled neural spheres transplanted in chronic EAE mice. *Magn. Reson. Med.* 57: 164–171.

Bernard, M., Canioni, P., and Cozzone, P.J. 1983. Study of in vivo cellular metabolism by phosphorus 31 nuclear magnetic resonance (transl.). *Biochimie* 65: 449–470.

Bonavita, S., Di Salle, F., and Tedeschi, G. 1999. Proton MRS in neurological disorders. *Eur. J. Radiol.* 30: 125–131.

Bottomley, P.A., Edelstein, W.A., Foster, T.H., and Adams, W.A. 1985. In vivo solvent-suppressed localized hydrogen nuclear magnetic resonance spectroscopy: A window to metabolism? *Proc. Natl. Acad. Sci.* 82: 2148–2152.

Brindle, J.T., Antti, H., Holmes, E., Tranter, G., Nicholson, J.K., Bethell, H.W., Clarke, S., Schofield, P.M., McKilligin, E., Mosedale, D.E., and Grainger, D.J. 2002. Rapid and noninvasive diagnosis of the presence and severity of coronary heart disease using 1H-NMR-based metabonomics. *Nat. Med.* 8: 1439–1444.

Brindle, J.T., Nicholson, J.K., Schofield, P.M., Grainger, D.J., and Holmes, E. 2003. Application of chemometrics to 1H NMR spectroscopic data to investigate a relationship between human serum metabolic profiles and hypertension. *Analyst* 128: 32–36.

Bulte, J.W. 2006. Intracellular endosomal magnetic labeling of cells. *Methods Mol. Med.* 124: 419–439.

Bulte, J.W. and Kraitchman, D.L. 2004. Iron oxide MR contrast agents for molecular and cellular imaging. *NMR Biomed.* 17: 484–499.

Bulte, J.W., Duncan, I.D., and Frank, J.A. 2002. In vivo magnetic resonance tracking of magnetically labeled cells after transplantation. *J. Cereb. Blood Flow. Metab.* 22: 899–907.

Burt, C.T., Cohen, S.M., and Barany, M. 1979. Analysis with intact tissue with 31P NMR. *Annu. Rev. Biophys. Bioeng.* 8: 1–25.

Cheng, L.L. and Pohl, U. 2007. The role of NMR-based metabolomics in cancer. In *Handbook of metabonomics and metabolomics* (ed. J.C. Lindon et al.), pp. 345–374. Elsevier, Amsterdam.

Chin, B.B., Nakamoto, Y., Bulte, J.W., Pittenger, M.F., Wahl, R., and Kraitchman, D.L. 2003. In oxine labelled mesenchymal stem cell SPECT after intravenous administration in myocardial infarction. *Nucl. Med. Commun.* 24: 1149–1154.

Cloarec, O., Dumas, M.E., Craig, A., Barton, R.H., Trygg, J., Hudson, J., Blancher, C., Gauguier, D., Lindon, J.C., Holmes, E., and Nicholson, J. 2005. Statistical total correlation spectroscopy: An exploratory approach for latent biomarker identification from metabolic 1H NMR data sets. *Anal. Chem.* 77: 1282–1289.

Coen, M., Holmes, E., Lindon, J.C., and Nicholson, J.K. 2008. NMR-based metabolic profiling and metabonomic approaches to problems in molecular toxicology. *Chem. Res. Toxicol.* 21: 9–27.

DeGraaf, R.A. 1998. *In vivo NMR spectroscopy: Principles and techniques*. Wiley, New York.

Di Costanzo, A., Trojsi, F., Tosetti, M., Schirmer, T., Lechner, S.M., Popolizio, T., and Scarabino, T. 2007. Proton MR spectroscopy of the brain at 3 T: An update. *Eur. Radiol.* 17: 1651–1662.

Djurić, P.M., Benveniste, H., Wagshul, M.E., Henn, F., Enikolopov, G., and Maletić-Savatić, M. 2008. Response to comments on "Magnetic resonance spectroscopy identifies neural progenitor cells in the live human brain." *Science* **321:** 640.

Dunn, W.B., Bailey, N.J., and Johnson, H.E. 2005. Measuring the metabolome: Current analytical technologies. *Analyst* **130:** 606–625.

Eriksson, P.S., Perfilieva, E., Bjork-Eriksson, T., Alborn, A.M., Nordborg, C., Peterson, D.A., and Gage, F.H. 1998. Neurogenesis in the adult human hippocampus. *Nat. Med.* **4:** 1313–1317.

Fiehn, O. 2002. Metabolomics: The link between genotypes and phenotypes. *Plant Mol. Biol.* **48:** 155–171.

Fiehn, O., Kopka, J., Dormann, P., Altmann, T., Trethewey, R.N., and Willmitzer, L. 2000. Metabolite profiling for plant functional genomics. *Nat. Biotechnol.* **18:** 1157–1161.

Fiehn, O., Kloska, S., and Altmann, T. 2001. Integrated studies on plant biology using multiparallel techniques. *Curr. Opin. Biotechnol.* **12:** 82–86.

Frahm, J., Bruhn, H., Gyngell, M.L., Merboldt, K.D., Hanicke, W., and Sauter, R. 1989. Localized high-resolution proton NMR spectroscopy using stimulated echoes: Initial applications to human brain in vivo. *Magn. Reson. Med.* **9:** 79–93.

Garrod, S., Humpfer, E., Spraul, M., Connor, S.C., Polley, S., Connelly, J., Lindon, J.C., Nicholson, J.K., and Holmes, E. 1999. High-resolution magic angle spinning 1H NMR spectroscopic studies on intact rat renal cortex and medulla. *Magn. Reson. Med.* **41:** 1108–1118.

Gavaghan, C.L., Holmes, E., Lenz, E., Wilson, I.D., and Nicholson, J.K. 2000. An NMR-based metabonomic approach to investigate the biochemical consequences of genetic strain differences: Application to the C57BL10J and Alpk:ApfCD mouse. *FEBS Lett.* **484:** 169–174.

German, J.B., Hammock, B.D., and Watkins, S.M. 2005. Metabolomics: Building on a century of biochemistry to guide human health. *Metabolomics* **1:** 3–9.

Glassbrook, N., Beecher, C., and Ryals, J. 2000. Metabolic profiling on the right path. *Nat. Biotechnol.* **18:** 1142–1143.

Goodacre, R. 2004. Metabolic profiling: Pathways in discovery. *Drug Discov. Today* **9:** 260–261.

Goodacre, R. 2005. Metabolomics shows the way to new discoveries. *Genome Biol.* **6:** 354.

Goodacre, R., Vaidyanathan, S., Dunn, W.B., Harrigan, G.G., and Kell, D.B. 2004. Metabolomics by numbers: Acquiring and understanding global metabolite data. *Trends Biotechnol.* **22:** 245–252.

Govindaraju, V., Young, K., and Maudsley, A.A. 2000. Proton NMR chemical shifts and coupling constants for brain metabolites. *NMR Biomed.* **13:** 129–153.

Griffin, J.L. 2004. Metabolic profiles to define the genome: Can we hear the phenotypes? *Philos. Trans. R. Soc. Lond. B Biol. Sci.* **359:** 857–871.

Griffin, J.L. and Bollard, M.E. 2004. Metabonomics: Its potential as a tool in toxicology for safety assessment and data integration. *Curr. Drug Metab.* **5:** 389–398.

Griffin, J.L. and Nicholls, A.W. 2006. Metabolomics as a functional genomic tool for understanding lipid dysfunction in diabetes, obesity and related disorders. *Pharmacogenomics* **7:** 1095–1107.

Griffin, J.L. and Shockcor, J.P. 2004. Metabolic profiles of cancer cells. *Nat. Rev. Cancer* **4:** 551–561.

Griffin, J.L., Pole, J.C., Nicholson, J.K., and Carmichael, P.L. 2003. Cellular environment of metabolites and a metabonomic study of tamoxifen in endometrial cells using gradient high resolution magic angle spinning 1H NMR spectroscopy. *Biochim. Biophys. Acta* **1619:** 151–158.

Haacke, E.M., Liang, Z.P., and Izen, S.H. 1989. Constrained reconstruction: A superresolution, optimal signal-to-noise alternative to the Fourier transform in magnetic resonance imaging. *Med. Phys.* **16:** 388–397.

Harrigan, G.G. and Goodacre, R., eds. 2003. *Metabolic profiling: Its role in biomarker discovery and gene function analysis.* Kluwer, Boston.

Hatada, K. and Kitayama, T., eds. 2004. Basic principles of NMR. In *NMR spectroscopy of polymers,* pp. 1–34. Springer, New York.

Hoehn, M., Kustermann, E., Blunk, J., Wiedermann, D., Trapp, T., Wecker, S., Focking, M., Arnold, H., Hescheler, J., Fleischmann, B.K., Schwindt, W., and Buhrle, C. 2002. Monitoring of implanted stem cell migration in vivo: A highly resolved in vivo magnetic resonance imaging investigation of experimental stroke in rat. *Proc. Natl. Acad. Sci.* **99:** 16267–16272.

Hollywood, K., Brison, D.R., and Goodacre, R. 2006. Metabolomics: Current technologies and future trends. *Proteomics* **6:** 4716–4723.

Holmes, E., Loo, R.L., Stamler, J., Bictash, M., Yap, I.K., Chan, Q., Ebbels, T., De Iorio, M., Brown, I.J., Veselkov, K.A., et al. 2008. Human metabolic phenotype diversity and its association with diet and blood pressure. *Nature* **453:** 396–400.

Jírů, F., Dezortová, M., Burian, M., and Hájek, M. 2003. The role of relaxation time corrections for the evaluation of long and short echo time 1H MR spectra of the hippocampus by NUMARIS and LCModel techniques. *Magma* **16:** 135–143.

Jolliffe, I.T. 1986. *Principal component analysis.* Springer-Verlag, New York.

Kanowski, M., Kaufmann, J., Braun, J., Bernarding, J., and Tempelmann, C. 2004. Quantitation of simulated short echo time 1H human brain spectra by LCModel and AMARES. *Magn. Reson. Med.* **51:** 904–912.

Keevil, S.F. 2006. Spatial localization in nuclear magnetic resonance spectroscopy. *Phys. Med. Biol.* **51:** R579–R636.

Kell, D.B., Brown, M., Davey, H.M., Dunn, W.B., Spasic, I., and Oliver, S.G. 2005. Metabolic footprinting and systems biology: The medium is the message. *Nat. Rev. Microbiol.* **3:** 557–565.

Lenz, E.M. and Wilson, I.D. 2007. Analytical strategies in metabonomics. *J. Proteome Res.* **6:** 443–458.

Li, M., Wang, B., Zhang, M., Rantalainen, M., Wang, S., Zhou, H., Zhang, Y., Shen, J., Pang, X., Wei, H., et al. 2008. Symbiotic gut microbes modulate human metabolic phenotypes. *Proc. Natl. Acad. Sci.* **105:** 2117–2122.

Lin, A., Ross, B.D., Harris, K., and Wong, W. 2005. Efficacy of proton magnetic resonance spectroscopy in neurological diagnosis and neurotherapeutic decision making. *NeuroRx* **2:** 197–214.

Lindon, J.C. and Holmes, E. 2006. A survey of metabonomics approaches for disease characterization. In *Handbook of metabonomics and metabolomics* (ed. J.C. Lindon et al.), pp. 413–442. Elsevier, Amsterdam.

Manganas, L.N., Zhang, X., Li, Y., Hazel, R.D., Smith, S.D., Wagshul, M.E., Henn, F., Benveniste, H., Djurić, P.M., Enikolopov, G., and Maletić-Savatić, M. 2007. Magnetic resonance spectroscopy identifies neural progenitor cells in the live human brain. *Science* **318:** 980–985.

Meyer, R.A., Fisher, M.J., Nelson, S.J., and Brown, T.R. 1988. Evaluation of manual methods for integration of in vivo phosphorus NMR spectra. *NMR Biomed.* **1:** 131–135.

Mierisova, S. and Ala-Korpela, M. 2001. MR spectroscopy quantitation: A review of frequency domain methods. *NMR Biomed.* **14:** 247–259.

Modo, M., Mellodew, K., Cash, D., Fraser, S.E., Meade, T.J., Price, J., and Williams, S.C. 2004. Mapping transplanted stem cell migration after a stroke: A serial, in vivo magnetic resonance imaging study. *Neuroimage* **21:** 311–317.

Modo, M., Hoehn, M., and Bulte, J.W. 2005. Cellular MR imaging. *Mol. Imaging* **4:** 143–164.

Moolenaar, S.H., Engelke, U.F., and Wevers, R.A. 2003. Proton nuclear magnetic resonance spectroscopy of body fluids in the field of inborn errors of metabolism. *Ann. Clin. Biochem.* **40:** 16–24.

Mortishire-Smith, R.J., Skiles, G.L., Lawrence, J.W., Spence, S., Nicholls, A.W., Johnson, B.A., and Nicholson, J.K. 2004. Use of metabonomics to identify impaired fatty acid metabolism as the mechanism of a drug-induced toxicity. *Chem. Res. Toxicol.* **17:** 165–173.

Nicholson, J.K. and Wilson, I.D. 2003. Opinion: Understanding "global" systems biology: Metabonomics and the continuum of metabolism. *Nat. Rev. Drug. Discov.* **2:** 668–676.

Nicholson, J.K., Connelly, J., Lindon, J.C., and Holmes, E. 2002. Metabonomics: A platform for studying drug toxicity and gene function. *Nat. Rev. Drug Discov.* **1:** 153–161.

Oksman-Caldentey, K.M. and Saito, K. 2005. Integrating genomics and metabolomics for engineering plant metabolic pathways. *Curr. Opin. Biotechnol.* **16:** 174–179.

Perez-Enciso, M. and Tenenhaus, M. 2003. Prediction of clinical outcome with microarray data: A partial least squares discriminant analysis (PLS-DA) approach. *Hum. Genet.* **112:** 581–592.

Phelps, T.J., Palumbo, A.V., and Beliaev, A.S. 2002. Metabolomics and microarrays for improved understanding of phenotypic characteristics controlled by both genomics and environmental constraints. *Curr. Opin. Biotechnol.* **13:** 20–24.

Poullet, J.B., Sima, D.M., Simonetti, A.W., De Neuter, B., Vanhamme, L., Lemmerling, P., and Van Huffel, S. 2007. An automated quantitation of short echo time MRS spectra in an open source software environment: AQSES. *NMR Biomed.* **20:** 493–504.

Prabakaran, S., Swatton, J.E., Ryan, M.M., Huffaker, S.J., Huang, J.T., Griffin, J.L., Wayland, M., Freeman, T., Dudbridge, F., Lilley, K.S., et al. 2004. Mitochondrial dysfunction in schizophrenia: Evidence for compromised brain metabolism and oxidative stress. *Mol. Psychiatry* **9:** 684–697, 643.

Provencher, S.W. 2001. Automatic quantitation of localized in vivo 1H spectra with LCModel. *NMR Biomed.* **14:** 260–264.

Raamsdonk, L.M., Teusink, B., Broadhurst, D., Zhang, N., Hayes, A., Walsh, M.C., Berden, J.A., Brindle, K.M., Kell, D.B., Rowland, J.J., et al. 2001. A functional genomics strategy that uses metabolome data to reveal the phenotype of silent mutations. *Nat. Biotechnol.* **19:** 45–50.

Rezzi, S., Ramadan, Z., Martin, F.P., Fay, L.B., van Bladeren, P., Lindon, J.C., Nicholson, J.K., and Kochhar, S. 2007. Human metabolic phenotypes link directly to specific dietary preferences in healthy individuals. *J. Proteome Res.* **6:** 4469–4477.

Roessner, U., Luedemann, A., Brust, D., Fiehn, O., Linke, T., Willmitzer, L., and Fernie, A. 2001. Metabolic profiling allows comprehensive phenotyping of genetically or environmentally modified plant systems. *Plant Cell* **13:** 11–29.

Rosen, Y. and Lenkinski, R.E. 2007. Recent advances in magnetic resonance neurospectroscopy. *Neurotherapeutics* **4:** 330–345.

Ross, B. and Bluml, S. 2001. Magnetic resonance spectroscopy of the human brain. *Anat. Rec.* **265:** 54–84.

Ross, B.D., Coletti, P., and Lin, A. 2006. Magnetic resonance spectroscopy of the brain: Neurospectroscopy. In *Clinical magnetic resonance imaging* (ed. R.R. Edelman et al.), pp. 1840–1901. Saunders, Philadelphia.

Roy, N.S., Wang, S., Jiang, L., Kang, J., Benraiss, A., Harrison-Restelli, C., Fraser, R.A., Couldwell, W.T., Kawaguchi, A., Okano, H., Nedergaard, M., and Goldman, S.A. 2000. In vitro neurogenesis by progenitor cells isolated from the adult human hippocampus. *Nat. Med.* **6:** 271–277.

Slotboom, J., Boesch, C., and Kreis, R. 1998. Versatile frequency domain fitting using time domain models and prior knowledge. *Magn. Reson. Med.* **39:** 899–911.

Toso, C., Vallee, J.P., Morel, P., Ris, F., Demuylder-Mischler, S., Lepetit-Coiffe, M., Marangon, N., Saudek, F., James Shapiro, A.M., Bosco, D., and Berney, T. 2008. Clinical magnetic resonance imaging of pancreatic islet grafts after iron nanoparticle labeling. *Am. J. Transplant.* **8:** 701–706.

van der Veen, J.W., de Beer, R., Luyten, P.R., and van Ormondt, D. 1988. Accurate quantification of in vivo 31P NMR signals using the variable projection method and prior knowledge. *Magn. Reson. Med.* **6:** 92–98.

Vanhamme, L., van den Boogaart, A., and Van Huffel, S. 1997. Improved method for accurate and efficient quantification of MRS data with use of prior knowledge. *J. Magn. Reson.* **129:** 35–43.

Vanhamme, L., Van Huffel, S., Van Hecke, P., and van Ormondt, D. 1999. Time-domain quantification of series of biomedical magnetic resonance spectroscopy signals. *J. Magn. Reson.* **140:** 120–130.

Vanhamme, L., Sundin, T., Hecke, P.V., and Huffel, S.V. 2001. MR spectroscopy quantitation: A review of time-domain methods. *NMR Biomed.* **14:** 233–246.

Waters, N.J., Holmes, E., Williams, A., Waterfield, C.J., Farrant, R.D., and Nicholson, J.K. 2001. NMR and pattern recognition studies on the time-related metabolic effects of α-naphthylisothiocyanate on liver, urine, and plasma in the rat: An integrative metabonomic approach. *Chem. Res. Toxicol.* **14:** 1401–1412.

Weckwerth, W. and Morgenthal, K. 2005. Metabolomics: From pattern recognition to biological interpretation. *Drug Discov. Today* **10:** 1551–1558.

Zandt, H., van Der Graaf, M., and Heerschap, A. 2001. Common processing of in vivo MR spectra. *NMR Biomed.* **14:** 224–232.

Signaling Pathways Controlling Neural Stem Cells Slow Progressive Brain Disease

A. Androutsellis-Theotokis, M.A. Rueger, H. Mkhikian, E. Korb, and R.D.G. McKay

Laboratory of Molecular Biology, National Institute of Neurological Disorders and Stroke, National Institutes of Health, Bethesda, Maryland 20892

The identification and characterization of multipotent neural precursors open the possibility of transplant therapies, but this approach is complicated by the widespread pathology of many degenerative diseases. Activation of endogenous precursors that support regenerative mechanisms is a possible alternative. We have previously shown that Notch ligands promote stem cell survival in vitro. Here, we show that there is an intimate interaction between insulin and Notch receptor signaling. Notch ligands also expand stem cell numbers in vivo with correlated benefits in brain ischemia. We now show that insulin promotes recovery of injured dopamine neurons in the adult brain. This response suggests that activating survival mechanisms in neural stem cells will promote recovery from progressive degenerative disease.

It seems obvious that the nervous system is composed of many distinct components. The anatomical differences between neurons was known by 1900, but it took additional time before advances in molecular biology gave us techniques that rivaled the power of the Golgi method. Hybridoma technology was the first of these new methods to teach us that molecular differences between neurons are fundamental to the way the brain works (Zipser and McKay 1981). The sheer scale of the molecular diversity of the brain made the origin of these differences a compelling question. The first evidence for multipotent neural precursors in the central nervous system (CNS) was reported many decades ago (Nieuwkoop 1973; Ready et al. 1976). On the basis of this pioneering work, we established tools to identify and manipulate these cells. Neural precursors in the developing brain were first identified by a specific monoclonal antibody (Hockfield and McKay 1985; Frederiksen and McKay 1988). The antibody recognized an intermediate filament protein whose precise expression pattern is regulated by a conserved enhancer that directs gene expression to neural precursors (Lendahl et al. 1990; Zimmerman et al. 1994). These powerful regulatory elements in the *nestin* gene are now widely used to achieve specific genetic manipulation of the brain (Josephson et al. 1998; Lonigro et al. 2001; Tanaka et al. 2004). The initial identification of precursor cells in vivo was followed by evidence that proliferating *nestin*-positive cells would generate functional neurons in vitro and that these cells would incorporate into the brain (Cattaneo and McKay 1990; Renfranz et al. 1991; Vicario-Abejón et al. 1995, 2000). These experiments opened up the idea that large numbers of functional neurons could be generated in the laboratory.

Controlling the ex vivo production of neurons has become a widely pursued activity often justified as the basis for new cell therapies for degenerative disease. Transplants of fetal mesencephalic tissue into the striatum of patients with Parkinson's disease (PD) suggest that cell replacement therapies may be beneficial. The grafted neu-

rons can reinnervate the diseased brain and restore function, but problems limit this approach, including the scarcity of tissue suitable for transplantation (Lindvall and Björklund 2004). Stem cell technologies may provide large numbers of dopamine neurons that are specifically required for PD cell therapy. The most obvious source of these cells is the fetal midbrain that normally generates dopamine neurons during development. Midbrain precursors can divide in vitro and then differentiate into dopamine neurons, but the short proliferative phase limits the numbers of neurons that can be generated (Studer et al. 1998). We now know that dopamine neurons are derived from the midbrain floor plate, a region of the brain not previously thought to generate neurons (Kittappa et al. 2007). Distinct mechanisms may control proliferation of floor-plate precursors, but whatever the case, this limitation was overcome by deriving dopamine neurons from embryonic stem (ES) cells.

Some years after ES cells were first obtained, they were shown to incorporate into the inner cell mass of the blastocyst and generate chimeras with donor cells in all tissues including the germ line (Bradley et al. 1984; Thomas and Capecchi 1986). These results showed that ES cells were pluripotent, but it took additional experiments to demonstrate efficient production of somatic fates in vitro. Using our knowledge of neural precursors to optimize the differentiation protocol, we showed that neurons could be generated as efficiently from ES cells as from the developing brain (Okabe et al. 1996). This remarkable result led to further studies showing that ES-derived oligodendrocytes would myelinate axons in a model of human demyelinating disease (Brüstle et al. 1999). It is a measure of the enthusiasm for these ideas that oligodendrocytes derived by in vitro methods from human ES cells may soon be used in clinical trials (Keirstead et al. 2005). Endodermal fates can also be derived from ES cells. Induced insulin secretion, plasma glucose regulation, and weight control suggest that ES-derived cells conduct pancreatic endocrine functions (Lumelsky et al. 2001; Shim et al. 2007). Electrophysiological measurements, dopamine release, positron emission tomography (PET) chemistry, and behavioral responses show that functional dopamine neurons can be derived from ES cells (Kim et al. 2002; Rodríguez-Gómez et al. 2007). These grafting studies provide some of the clearest examples supporting the worldwide interest in using human pluripotent cells.

Despite this success, major difficulties inhibit implementing cell therapy for neurodegenerative disease. The intermediate steps between pluripotent precursors and mature functional neurons are still being defined (Tesar et al. 2007; Elkabetz et al. 2008). Even when we have solved the logistically complex task of generating consistently high-quality donor cells, a successful cell therapy will still demand strong knowledge of disease mechanisms. In PD, pathological evidence suggests that the deficits involve many neurons other than dopamine neurons (Braak and Del Tredici 2008). There is increasing evidence that widely disseminated pathology leads to the death of striatal medium spiny neurons in Huntington's disease (HD) and motor neurons in amytotrophic lateral sclerosis (ALS) and in the spinocerebellar atrophies (SCA) (Zuccato et al. 2001; Orr and Zoghbi 2007; Yamanaka et al. 2008). These insights into disease mechanisms raise concerns about the value of transplanting a specific neuron type in a degenerating brain where the disease may be initiated elsewhere and involve many other neuron types. Here, we suggest a different approach where injured neurons are not replaced but restored to health by treatments that stimulate endogenous precursors in large regions of the brain.

It might be valuable to consider for a moment what it takes to keep a cell alive. In 1890, embryology was a flourishing enterprise at Woods Hole where Thomas Hunt Morgan worked a few years after his graduation from Johns Hopkins University. In that same summer, Ross Harrison who was an undergraduate at Johns Hopkins also worked at Woods Hole. Ross Harrison showed that axons would grow out when brain explants were placed in a drop of clotted frog lymph. Subsequently, the technology for growing cells in culture was extended. HeLa cells were placed in culture in 1930, again at Johns Hopkins. Although neurons were among the first cells to be studied in cell culture, the clear demonstration that neurons could be derived from precursors that had expanded in the laboratory is surprisingly recent and depended on using defined growth factors (Barnes and Sato 1980). Access to pure growth factors became possible as techniques for cloning and expressing genes were commercialized. In Europe and America in 1982, insulin became the first recombinant protein approved for clinical use. Access to insulin and fibroblast growth factor-2 (FGF-2) allowed the first definitive demonstration that proliferating precursors would transform into neurons in cell culture (Cattaneo and McKay 1990).

In neural stem cells, we made the surprising observation that ligands for the Notch receptor also control cell survival through a transient activation of phosphoinositol-3 kinase (PI3K) Akt (Androutsellis-Theotokis et al. 2006). Notch encodes a *trans*-membrane receptor that is cleaved on activation to release an intracellular domain that is directly involved in transcriptional control and regulates cell fate (Heitzler and Simpson 1991; Ruohola et al. 1991; Greenwald and Rubin 1992; Spana and Doe 1996; Artavanis-Tsakonas et al. 1999). Transcriptional regulation has been the primary focus of contemporary Notch research. The mechanism that transfers information from the surface receptor to the nucleus was clarified when it was demonstrated that the Notch protein was cleaved and the intracellular domain was itself a transcriptional regulator (Kidd et al. 1986; Blaumueller et al. 1997). A detailed model is now available for the interaction between the Notch intracellular domain (NICD) and other transcriptional cofactors (Wallberg et al. 2002; Nam et al. 2006; Yoshimatsu et al. 2006).

The Notch receptor and ligands are expressed in the developing vertebrate CNS where gain- and loss-of-function experiments show that Notch inhibits the cascade of events required for the formation of neurons and promotes the differentiation of glia (Morrison et al. 2000; Tanigaki et al. 2001). In mice, the two homologs of the *Drosophila* Notch inhibitor numb are required for the maintenance of the neural precursor state in the developing and adult nervous system (Kuo et al. 2006).

In culture, the homogeneity of cells derived from the fetal telencephalon makes them ideal for signal transduction studies. Importantly, they can be grown in serum-free medium containing only three exogenous proteins: apotransferrin for iron transport, FGF-2, and insulin. Under these conditions, a proportion of the cells die. This death is abruptly blocked by soluble ligands that activate the Notch receptor, including Deltalike-4 (Dll4) and Jagged (Jag). The novel survival effect of Notch ligands involves many components of the phosphorylation cascade associated with surface trysoine kinase receptors (Androutsellis-Theotokis et al. 2006).

INSULIN AND DLL4 COOPERATE TO PROMOTE DOWNSTREAM SIGNALING

Neuroepithelial cells express both insulin-like growth factor-1R (IGF-1R) and Notch1, and their survival is sensitive to both the γ-secretase and PI3K inhibitors. The insulin superfamily contains insulin, IGF-1 and IGF-2, and the relaxins (INSL3–7). These growth factors bind to multiple receptors, but the activation of receptor tyrosine kinases dominates our current thinking on the cellular response to insulin and IGFs. There are about 60 receptor tyrosine kinases in the human genome and there is great interest in the molecular mechanisms that activate them (Weiss and Schlessinger 1998). Both IGFs act through the IGF-1R, and there is compelling evidence that IGF-2 controls cell survival during mouse development (Burns and Hassan 2001). IGF-1R activates an intracellular signaling cascade that includes acute increases in the phosphorylation of the PI3K and Akt kinases and this leads glucose uptake and cell survival (Dudek et al. 1997; Stokoe et al. 1997).

This idea that cell survival is mediated by both Notch and IGF-1R led us to investigate the interaction of Delta4 and insulin on receptor activation. Western blots with an antibody that recognizes the phosphorylated IGF-1R show that Dll4 induces receptor activation (Fig. 1a). The signal from insulin treatment is also shown (Fig. 1a,

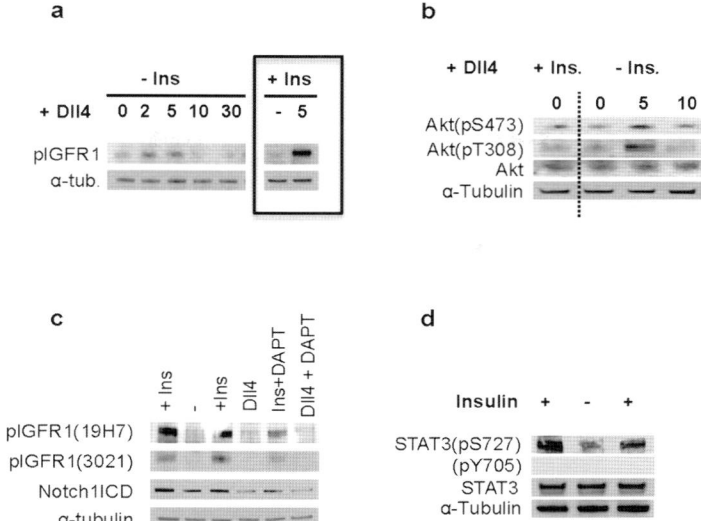

Figure 1. Notch and insulin signaling mutually facilitate receptor activation. (*a*) In the absence of insulin, Dll4 (500 ng/ml) induces the phosphorylation of the IGF-1/insulin receptor (the antibody used recognizes phosphorylated IGF-1 receptor and phosphorylated insulin receptor). (*Inset*) Acute (5 minutes) insulin (0.025 mg/ml) treatment of fetal NSC cultures following a 16-hour insulin starvation induces strong IGF-1R receptor phosphorylation. Cells were grown under conditions described by Johe et al. (1996) and Androutsellis-Theotokis et al. (2006). Antibodies against the following epitopes were used: pSer473-Akt (9271 5) and pThr308-Akt (9275) from Cell Signaling; STAT3 (482), pSer727-STAT3 (8001-R, pTyr705-STAT3 (7993), and Akt (5298); α-tubulin (Sigma-Aldrich T6074). (*b*) Dll4 activates Akt in the absence of insulin. Akt is phosphorylated on both the Ser-473 and Thr-308 residues; 16-hour insulin starvation was followed by acute Dll4 treatment (5 minutes). (*c*) Dll4 facilitates IGF-1/insulin receptor phosphorylation. Insulin starvation (16 hours) was followed by the indicated treatments for 1 hour. Insulin treatment (0.025 mg/ml) induces both IGF-1/insulin receptor phosphorylation and cleavage of the Notch receptor to generate the NICD. Dll4 (in the absence of insulin; 500 ng/ml) does not induce NICD cleavage or IGF-1/insulin receptor phosphorylation. Acute insulin treatment induces NICD cleavage and IGF-1R phosphorylation. Both IGF-1R phosphorylation and Notch cleavage were inhibited by the γ-secretase DAPT. (*d*) Insulin induces STAT3-S727 phosphorylation in fetal NSC cultures after 16 hours of insulin withdrawal.

inset). In the absence of insulin, Dll4 still stimulates rapid phosphorylation of both major activating sites on Akt (Fig. 1b). The phosphorylation of IGF-1R in response to Dll4 may explain the Akt activation.

In a further experiment, the effect of insulin on the level of the cleaved intracellular fragment of the Notch (NICD) protein was assessed. Neural stem cells (NSCs) were maintained for 16 hours without insulin and western blots were performed 1 hour after treating the cells (Fig. 1c). Addition of insulin gave strong phosphorylation of the IGF-1R. Consistent with the brief IGF-1R response to Dll4, no increase in activated IGF-1R was observed 1 hour after Dll4 treatment. Our previous report on the Notch survival pathway showed that a 1-hour treatment with Dll4 generates elevated levels of the NICD in the presence of insulin, but this was not observed in the absence of insulin (Fig. 1c). However, insulin treatment caused elevated levels of NICD and this was sensitive to γ-secretase inhibition. The reduced activation of IGF-1R in the presence of DAPT, the synergy between insulin and Dll4 on NICD levels, and the γ-secretase-dependent elevation of NICD by insulin all support a link between these two signals.

Following Dll4 treatment, NSCs show an activation of a serine residue (Ser-727) on STAT3 (signal transducer and activator of transcription 3) that is required for cell survival (Androutsellis-Theotokis et al. 2006). The action of JAK tyrosine kinases on Tyr-705 is considered to be the major regulator of STAT3 nuclear translocation and transcriptional activation (Levy and Darnell 2002). Elevated GFAP (glial fibrillary acidic protein) expression and astrocyte differentiation are seen when neural stem cells are treated with growth factors that induce JAK activity (Johe et al. 1996; Bonni et al. 1997). Undifferentiated NSCs show very low levels of Tyr-705 phosphorylation, and Ser-727 phosphorylation occurs after insulin treatment (Fig. 1d). This result shows that insulin and Notch activation have similar effects on this key downstream component of the signaling cascade that controls stem cell survival.

INSULIN AND DLL4 ACT THROUGH HES3 TO PROMOTE STEM CELL SURVIVAL

The effects of insulin and Dll4 on the survival of fetal and adult NSCs were assessed. Insulin and Dll4 alone promote the expansion of fetal mouse telencephalic cells in culture, but their effect is greater when combined (Fig. 2a). The concentration of insulin used far exceeds the half-maximal concentration for receptor phosphorylation, and dose-response curves for Dll4 also suggest that saturating levels of the ligand are present. The subventricular zone (SVZ) of the adult rat telencephalon unusually contains large numbers of precursor cells that generate proliferating cellular aggregates known as neurospheres (Reynolds and Weiss 1992; Luskin 1993; Doetsch and Alvarez-Buylla 1996). Here, we show that insulin and Dll4 allow the direct

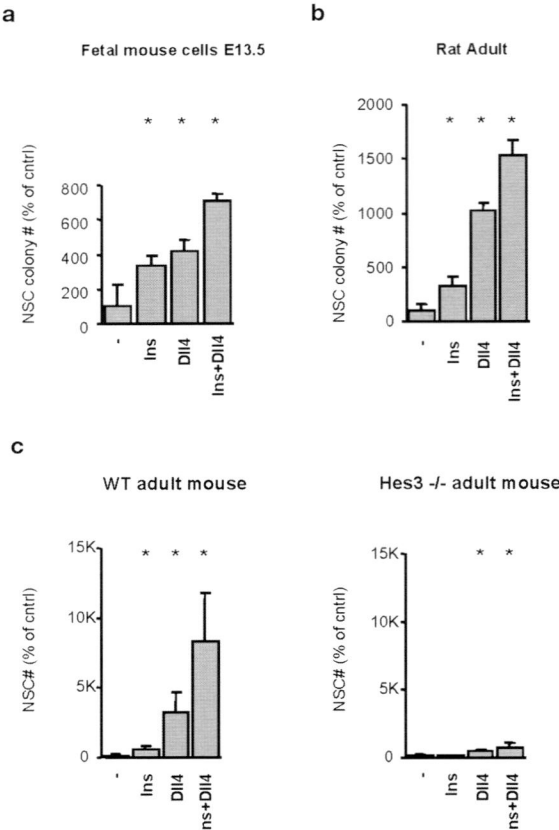

Figure 2. Hes3 mediates Notch and insulin survival functions. (*a,b*) Insulin and Dll4 increase the expansion of fetal NSC cultures and adult rat SVZ NSC cultures alone and when combined (colony numbers are shown). (*c*) Insulin and Dll4 promote the expansion of wild-type fetal NSC cultures and adult rat SVZ NSC cultures alone and when combined (cell numbers are shown). Their efficacy in Hes3 null cultures is significantly reduced.

isolation of highly enriched proliferating precursors when cells from the SVZ of the adult rat brain are dissociated and placed in attached culture (Fig. 2b).

Nuclear translocation of the NICD leads to the transcriptional regulation of *Hes* and *Hey* target genes. In the fetal brain, *Hes1, 3,* and *5* act cooperatively to sustain the neuroepithelial compartment (Hatakeyama et al. 2004). The survival signaling pathway downstream from Dll4 leads to a rapid and specific increase in mRNA for the *Hes3* gene (Androutsellis-Theotokis et al. 2006). Here, we show that *Hes3* is required for the survival of precursors from the adult SVZ in response to either insulin or Dll4 (Fig. 2c). This result strongly supports a critical role for this new signaling pathway in survival responses of adult stem cells to insulin.

INSULIN PROTECTS DOPAMINE NEURONS FROM 6-OHDA TOXICITY

As a consequence of the aging population, there are growing concerns about the financial and social costs of neurodegenerative disease. Growth factors that stimulate the recovery of the injured brain have been pursued since

Rita Levi-Montalcini made the observation of excess neuronal survival that led to the isolation of nerve growth factor (NGF) (Levi-Montalcini 1987). In the case of the degeneration of dopamine neurons seen in PD, new therapies using growth factors in the glial-cell-derived neurotrophic factor (GDNF) family are being pursued most avidly. This enthusiasm is based on the initial identification of GDNF as a survival factor for midbrain dopamine neurons in cell culture and the subsequent demonstration that lentiviral delivery of GDNF rescued injured midbrain dopamine neurons in primates (Kordower et al. 2000). The use of cytotoxic drugs currently provides the most widely used assay for developing therapies for PD. The specificity of these drugs takes advantage of the high-affinity systems that recover and recycle released dopamine. Dopamine neurons are specifically killed when the compound 6-hydroxy-dopamine is injected into the striatum of an adult rat (Przedborski et al. 1995). The striatum contains the axonal projections of dopamine neurons with cell bodies located in the ventral midbrain. These cell bodies degenerate with a predictable time course in the 4–6 weeks following injection of the toxin.

Single injections of insulin 2 weeks after the cytotoxic lesion was administered conferred significant protection on dopamine neurons; 13 weeks after insulin treatment (i.e., 15 weeks after 6-OHDA lesion), the insulin-treated brains showed a greater tyrosine hydroxylase (TH, a marker of dopamine neurons) signal in the ipsilateral to lesion hemisphere (Fig. 3a). When the retrogradely transported dye fluorogold was injected into the striatum ipsilateral to lesion 1 week before sacrifice, many TH[+] cell bodies were labeled in the insulin-treated animals (Fig. 3a). This result demonstrates that the rescued neurons in the substantia nigra had intact processes that extended into the striatum. In addition, the rescued neurons had cell bodies that were significantly larger in size than controls (Fig. 3a). This is in line with previous reports that the size of dopamine neuron bodies is a measure of their functional strength (Russo et al. 2007).

The behavioral impact of the rescued neurons was assessed by measuring the asymmetry in amphetamine-induced rotations. This assay is known to measure the deficit induced by the specific loss of dopamine neurons and is now widely used to monitor dopamine replacement or recovery (Dunnett et al. 1988). In dopamine neurons derived from the fetal brain, expanded fetal precursors, or ES cells obtained through an enriching step, the behavioral recovery occurs rapidly (Dunnett et al. 1988; Studer et al. 1998; Kim et al. 2002). A single insulin treatment induced a rapid reduction in the asymmetric rotations that stabilized after approximately 3 weeks and was sustained for a further 10 weeks (Fig. 3b). These anatomical data and behavioral kinetics establish that insulin rescues dopamine neurons with intact and functional nigrostriatal connections.

CONCLUSIONS

As we noted earlier, the use of defined growth factors was critical to the first production of neurons in vitro and to the identification of multipotent neural stem cells. In

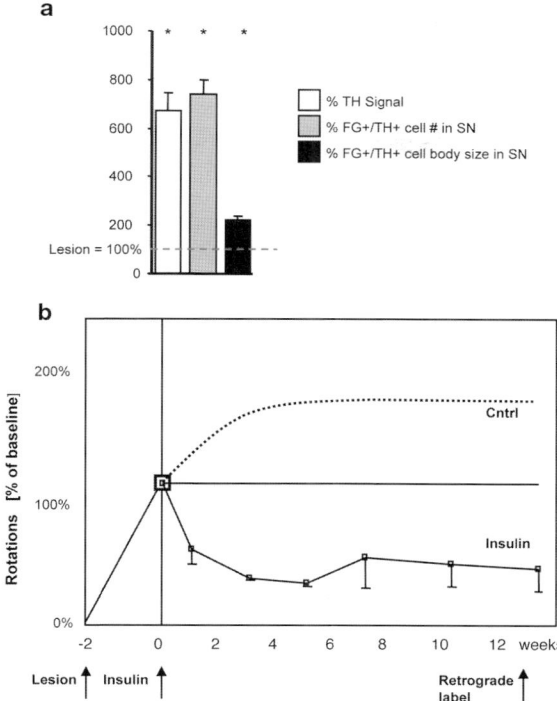

Figure 3. A single intracerebroventricular injection of insulin induces dopamine neuron survival and long-term motor behavior improvement in adult rats. (*a*) Insulin rescues dopamine neurons in vivo. Two weeks after the 6-OHDA lesion, animals were given a single intracerebroventricular injection (5 μl) with either artificial cerebrospinal fluid containing BSA (bovine serum albumin) (0.01%) as carrier (control) or insulin (8 mg/ml, in artificial cerebrospinal fluid containing 0.01% BSA). In control and insulin-treated animals, stereotaxic injections were placed into the right lateral ventricle using the following coordinates: bregma AP –0.9 mm, ML –1.4 mm, VD +3.8 mm; 12 weeks later, the rats were given a single injection of fluorogold in the ipsilateral to lesion and treatment striatum. One week after that, they were sacrificed and analyzed immunohistochemically. Relative to controls, insulin-treated brains showed a significantly increased TH signal in the striatum. Insulin-treated brains also had a greater number of TH⁺/fluorogold⁺ cell bodies. These cell bodies were also larger in size. Antityrosine hydroxylase antibodies P80101 and P40101 were obtained from Pel-Freez. (*b*) Amphetamine-induced rotometry shows an improvement in the motor behavior of rats. Asymmetric movement in control rats increased for approximately 4 weeks until it stabilized. In contrast, insulin-treated rats improved for approximately 3 weeks and stabilized. Unilateral lesion of the nigrostriatal dopamine pathway was achieved by stereotaxic injection of 50 μg of 6-hydroxydopamine (6-OHDA) into the right striatum, using the following coordinates: bregma AP +0.5 mm, ML –3.0 mm, VD +5.5 mm. The behavioral analysis was performed as described by Kim et al. (2002). Insulin (40 μg) was delivered as a single injection. Results shown are the mean ±S.D. or mean ±S.E.M. as indicated. Asterisks identify experimental groups that were significantly different (*p* value < 0.05) from control groups by the Student's *t*-test (Microsoft Excel), where applicable.

more recent work, we showed that Notch ligands control survival of neural precursors, fetal pancreatic precursors, and human ES cells (Androutsellis-Theotokis et al. 2006). This was an exciting finding because it identified Notch, STAT3, and Hes3 as novel components of the signaling cascade that regulates the survival and growth of neural stem cells. The work we present here shows that all three proteins are intimately associated with survival signaling downstream from insulin. In neural stem cells, insulin has a more important influence on cell survival than does FGF-2. FGF-2 in contrast supports cell growth but does not have a major effect on survival (Rao et al. 2008). The importance of insulin for the survival of vertebrate cells in culture was initially established empirically (Barnes and Sato 1980). The interaction between receptor tyrosine kinases and Notch illustrates how access to stem cells can define fundamental new features of the signaling logic controlling cell survival.

Recent work shows that insulin acts directly to regulate synapse and circuit function in the *Xenopus* visual system (Chiu et al. 2008). In the rat striatum, loss of pancreatic insulin reduces striatal dopamine function, and this loss can be restored when insulin is directly injected into the brain (Williams et al. 2007). These results suggest that the insulin/IGF receptors are a physiologically important stimulus in the vertebrate CNS and more specifically in the striatum. In the case of PD, the family of ligands related to GDNF has received the most attention, but the most prominent receptor (RET) for these ligands is not required for the maintenance of adult neurons (Jain et al. 2006). GDNF is normally synthesized by astrocytes, and recent studies developing a GDNF-based gene therapy suggests that expression in astrocytes has a better effect on dopamine neurons than expression in neurons (Do Thi et al. 2007). These results suggest that the neuroprotective responses of GDNF are indirect. Our data suggest that a general model that explains the survival of dopamine neurons must also include insulin and Notch receptors.

Although we focus on the relationship between stem cell survival and dopamine neurons in this chapter, it seems likely that these signaling pathways will have regenerative effects in other brain diseases and in other tissues. Two examples illustrate this point. Notch signaling controls muscle regeneration, and the efficacy of this process diminishes with age (Carlson et al. 2008). IGF-1 has a powerful role in the recovery of muscle after injury (Mourkioti and Rosenthal 2005). These results suggest that the Notch/IGF-1R system controls regeneration in many tissues.

Our data also show that these widely distributed prosurvival pathways can be readily activated by a single treatment. A simple strategy to activate a widely distributed prosurvival response has clear implications in medicine. Progressive degeneration of dopamine neurons occurs when humans are transiently exposed to a toxin that targets dopamine neurons (Langston et al. 1999). It is clear that insulin is a well-characterized and widely available drug that should be considered as a treatment for such cases of acute toxicity, but might we use insulin/IGF/Notch signaling more widely in PD?

Our understanding of PD has been advanced by the identification of α-synuclein as a familial cause of the disease and a component of Lewy bodies, protein aggregates found in the injured brain (Polymeropoulos et al. 1997; Spillantini et al. 1997). Postmortem histopathology, including analysis of Lewy bodies, shows that the major cortical, nigral, and local inputs to striatal neurons go through a characteristic series of changes in PD (Braak and Del Tredici 2008). These observations support a view that PD is a member of a wider group of diseases (Langston 2006). The evidence that Notch ligands and insulin have regenerative effects in vivo supports their use to delay the age-dependent damage that is a cardinal feature of PD and other neurodegenerative diseases. The pathology stresses the need to develop treatments that stimulate widely dispersed prosurvival responses. The insulin/IGF/Notch pathway may meet this need. It is important to note that the proliferation of endogenous neural precursors declines when dopamine neurons are absent, suggesting that stem cell functions will be depressed in PD (Höglinger et al. 2004). Mutually beneficial signals may pass between neural stem cells and dopamine neurons.

We have focused here on the role that endocrine hormones have in the survival of stem cells and adult dopamine neurons. We show that insulin and Notch receptors interact to control the signaling cascade that is fundamental to cell growth. In addition, we demonstrate pharmacological manipulation that promotes long-term recovery of dopamine neurons from injury. Recent work from our group also shows that the same core mechanism, involving the *foxa2* gene, specifies the early generation and the late survival of dopamine neurons in old age (Kittappa et al. 2007). FoxA2 is a member of the foxo group of forkhead genes that are central mediators in cell survival. We have defined an FGF receptor, FGF-R1, expressed in the adult dopamine neurons at most risk in PD (van der Walt et al. 2004; Murase and McKay 2006). These discoveries begin to define key cell-autonomous components of survival signaling that respond to the state of surrounding stem cells and astrocytes. A deeper understanding of how intrinsic and extrinsic factors interact to promote progressive positive change in dopamine neurons may provide a fundamentally new model to assist patients with PD and other neurodegenerative diseases.

ACKNOWLEDGMENTS

We thank R. Kageyama for the Hes3 null mice. This work was supported by the intramural research program of the National Institute of Neurological Disorders and Stroke.

REFERENCES

Androutsellis-Theotokis, A., Leker, R.R., Soldner, F., Hoeppner, D.J., Ravin, R., Poser, S.W., Rueger, M.A., Bae, S.K., Kittappa, R., and McKay, R.D. 2006. Notch signalling regulates stem cell numbers in vitro and in vivo. *Nature* **442:** 823–826.

Artavanis-Tsakonas, S., Rand, M.D., and Lake, R.J. 1999. Notch signaling: Cell fate control and signal integration in development. *Science* **284:** 770–776.

Barnes, D. and Sato, G. 1980. Serum-free cell culture: A unifying approach. *Cell* **22:** 649–655.

Blaumueller, C.M., Qi, H., Zagouras, P., and Artavanis-Tsakonas, S. 1997. Intracellular cleavage of Notch leads to a heterodimeric receptor on the plasma membrane. *Cell* **90:** 281–291.

Bonni, A., Sun, Y., Nadal-Vicens, M., Bhatt, A., Frank, D.A., Rozovsky, I., Stahl, N., Yancopoulos, G.D., and Greenberg, M.E. 1997. Regulation of gliogenesis in the central nervous system by the JAK-STAT signaling pathway. *Science* **278:** 477–483.

Braak, H. and Del Tredici, K. 2008. Assessing fetal nerve cell grafts in Parkinson's disease. *Nat. Med.* **14:** 483–485.

Bradley, A., Evans, M., Kaufman, M.H., and Robertson, E. 1984. Formation of germ-line chimaeras from embryo-derived teratocarcinoma cell lines. *Nature* **309:** 255–256.

Brüstle, O., Jones, K.N., Learish, R.D., Karram, K., Choudhary, K., Wiestler, O.D., Duncan, I.D., and McKay, R.D. 1999. Embryonic stem cell-derived glial precursors: A source of myelinating transplants. *Science* **285:** 754–756.

Burns, J.L. and Hassan, A.B. 2001. Cell survival and proliferation are modified by insulin-like growth factor 2 between days 9 and 10 of mouse gestation. *Development* **128:** 3819–3830.

Carlson, M., Hsu, M., and Conboy, I. 2008. Imbalance between pSmad3 and Notch induces CDK inhibitors in old muscle stem cells. *Nature* **454:** 528–532.

Cattaneo, E. and McKay, R. 1990. Proliferation and differentiation of neuronal stem cells regulated by nerve growth factor. *Nature* **347:** 762–765.

Chiu, S., Chen, C., and Cline, H. 2008. Insulin receptor signaling regulates synapse number, dendritic plasticity, and circuit function in vivo. *Neuron* **58:** 708–719.

Do Thi, N.A., Saillour, P., Ferrero, L., Paunio, T., and Mallet, J. 2007. Does neuronal expression of GDNF effectively protect dopaminergic neurons in a rat model of Parkinson's disease? *Gene Ther.* **14:** 441–450.

Doetsch, F. and Alvarez-Buylla, A. 1996. Network of tangential pathways for neuronal migration in adult mammalian brain. *Proc. Natl. Acad. Sci.* **93:** 14895–14900.

Dudek, H., Datta, S.R., Franke, T.F., Birnbaum, M.J., Yao, R., Cooper, G.M., Segal, R.A., Kaplan, D.R., and Greenberg, M.E. 1997. Regulation of neuronal survival by the serine-threonine protein kinase Akt. *Science* **275:** 661–665.

Dunnett, S.B., Hernandez, T.D., Summerfield, A., Jones, G.H., and Arbuthnott, G. 1988. Graft-derived recovery from 6-OHDA lesions: Specificity of ventral mesencephalic graft tissues. *Exp. Brain Res.* **71:** 411–424.

Elkabetz, Y., Panagiotakos, G., Al Shamy, G., Socci, N.D., Tabar, V., and Studer, L. 2008. Human ES cell-derived neural rosettes reveal a functionally distinct early neural stem cell stage. *Genes Dev.* **22:** 152–165.

Frederiksen, K. and McKay, R.D. 1988. Proliferation and differentiation of rat neuroepithelial precursor cells in vivo. *J. Neurosci.* **8:** 1144–1151.

Greenwald, I. and Rubin, G.M. 1992. Making a difference: The role of cell-cell interactions in establishing separate identities for equivalent cells. *Cell* **68:** 271–281.

Hatakeyama, J., Bessho, Y., Katoh, K., Ookawara, S., Fujioka, M., Guillemot, F., and Kageyama, R. 2004. Hes genes regulate size, shape and histogenesis of the nervous system by control of the timing of neural stem cell differentiation. *Development* **131:** 5539–5550.

Heitzler, P. and Simpson, P. 1991. The choice of cell fate in the epidermis of *Drosophila*. *Cell* **64:** 1083–1092.

Hockfield, S. and McKay, R.D. 1985. Identification of major cell classes in the developing mammalian nervous system. *J. Neurosci.* **5:** 3310–3328.

Höglinger, G.U., Rizk, P., Muriel, M.P., Duyckaerts, C., Oertel, W.H., Caille, I., and Hirsch, E.C. 2004. Dopamine depletion impairs precursor cell proliferation in Parkinson disease. *Nat. Neurosci.* **7:** 726–735.

Jain, S., Golden, J.P., Wozniak, D., Pehek, E., Johnson, Jr., E.M., and Milbrandt, J. 2006. RET is dispensable for maintenance of midbrain dopaminergic neurons in adult mice. *J. Neurosci.* **26:** 11230–11238.

Johe, K.K., Hazel, T.G., Muller, T., Dugich-Djordjevic, M.M., and McKay, R.D. 1996. Single factors direct the differentiation of stem cells from the fetal and adult central nervous system. *Genes Dev.* **10:** 3129–3140.

Josephson, R., Müller, T., Pickel, J., Okabe, S., Reynolds, K., Turner, P.A., Zimmer, A., and McKay, R.D. 1998. POU transcription factors control expression of CNS stem cell-specific genes. *Development* **125**: 3087–3100.

Keirstead, H.S., Nistor, G., Bernal, G., Totoiu, M., Cloutier, F., Sharp, K., and Steward, O. 2005. Human embryonic stem cell-derived oligodendrocyte progenitor cell transplants remyelinate and restore locomotion after spinal cord injury. *J. Neurosci.* **25**: 4694–4705.

Kidd, S., Kelley, M.R., and Young, M.W. 1986. Sequence of the Notch locus of *Drosophila melanogaster*: Relationship of the encoded protein to mammalian clotting and growth factors. *Mol. Cell. Biol.* **6**: 3094–3108.

Kim, J.H., Auerbach, J.M., Rodríguez-Gómez, J.A., Velasco, I., Gavin, D., Lumelsky, N., Lee, S.H., Nguyen, J., Sánchez-Pernaute, R., Bankiewicz, K., and McKay, R. 2002. Dopamine neurons derived from embryonic stem cells function in an animal model of Parkinson's disease. *Nature* **418**: 50–56.

Kittappa, R., Chang, W.W., Awatramani, R.B., and McKay, R.D. 2007. The *foxa2* gene controls the birth and spontaneous degeneration of dopamine neurons in old age. *PLoS Biol.* **5**: e325.

Kordower, J.H., Emborg, M.E., Bloch, J., Ma, S.Y., Chu, Y., Leventhal, L., McBride, J., Chen, E.Y., Palfi, S., Roitberg, B.Z., et al. 2000. Neurodegeneration prevented by lentiviral vector delivery of GDNF in primate models of Parkinson's disease. *Science* **290**: 767–773.

Kuo, C., Mirzadeh, Z., Sorianonavarro, M., Rasin, M., Wang, D., Shen, J., Sestan, N., Garciaverdugo, J., Alvarezbuylla, A., and Jan, L. 2006. Postnatal deletion of Numb/Numblike reveals repair and remodeling capacity in the subventricular neurogenic niche. *Cell* **127**: 1253–1264.

Langston, J.W. 2006. The Parkinson's complex: Parkinsonism is just the tip of the iceberg. *Ann. Neurol.* **59**: 591–596.

Langston, J.W., Forno, L.S., Tetrud, J., Reeves, A.G., Kaplan, J.A., and Karluk, D. 1999. Evidence of active nerve cell degeneration in the substantia nigra of humans years after 1-methyl-4-phenyl-1,2,3,6-tetrahydropyridine exposure. *Ann. Neurol.* **46**: 598–605.

Lendahl, U., Zimmerman, L.B., and McKay, R.D. 1990. CNS stem cells express a new class of intermediate filament protein. *Cell* **60**: 585–595.

Levi-Montalcini, R. 1987. The nerve growth factor: Thirty-five years later. *EMBO J.* **6**: 1145–1154.

Levy, D.E. and Darnell, Jr., J.E. 2002. Stats: Transcriptional control and biological impact. *Nat. Rev. Mol. Cell Biol.* **3**: 651–662.

Lindvall, O. and Björklund, A. 2004. Cell therapy in Parkinson's disease. *NeuroRx* **1**: 382–393.

Lonigro, R., Donnini, D., Zappia, E., Damante, G., Bianchi, M.E., and Guazzi, S. 2001. *Nestin* is a neuroepithelial target gene of thyroid transcription factor-1, a homeoprotein required for forebrain organogenesis. *J. Biol. Chem.* **276**: 47807–47813.

Lumelsky, N., Blondel, O., Laeng, P., Velasco, I., Ravin, R., and McKay, R. 2001. Differentiation of embryonic stem cells to insulin-secreting structures similar to pancreatic islets. *Science* **292**: 1389–1394.

Luskin, M.B. 1993. Restricted proliferation and migration of postnatally generated neurons derived from the forebrain subventricular zone. *Neuron* **11**: 173–189.

Morrison, S.J., Perez, S.E., Qiao, Z., Verdi, J.M., Hicks, C., Weinmaster, G., and Anderson, D.J. 2000. Transient Notch activation initiates an irreversible switch from neurogenesis to gliogenesis by neural crest stem cells. *Cell* **101**: 499–510.

Mourkioti, F. and Rosenthal, N. 2005. IGF-1, inflammation and stem cells: Interactions during muscle regeneration. *Trends Immunol.* **26**: 535–542.

Murase, S. and McKay, R.D. 2006. A specific survival response in dopamine neurons at most risk in Parkinson's disease. *J. Neurosci.* **26**: 9750–9760.

Nam, Y., Sliz, P., Song, L., Aster, J., and Blacklow, S. 2006. Structural basis for cooperativity in recruitment of MAML coactivators to Notch transcription complexes. *Cell* **124**: 973–983.

Nieuwkoop, P.D. 1973. The organization center of the amphibian embryo: Its origin, spatial organization, and morphogenetic action. *Adv. Morphog.* **10**: 1–39.

Okabe, S., Forsberg-Nilsson, K., Spiro, A.C., Segal, M., and McKay, R.D. 1996. Development of neuronal precursor cells and functional postmitotic neurons from embryonic stem cells in vitro. *Mech. Dev.* **59**: 89–102.

Orr, H.T. and Zoghbi, H.Y. 2007. Trinucleotide repeat disorders. *Annu. Rev. Neurosci.* **30**: 575–621.

Polymeropoulos, M.H., Lavedan, C., Leroy, E., Ide, S.E., Dehejia, A., Dutra, A., Pike, B., Root, H., Rubenstein, J., Boyer, R., et al. 1997. Mutation in the α-synuclein gene identified in families with Parkinson's disease. *Science* **276**: 2045–2047.

Przedborski, S., Levivier, M., Jiang, H., Ferreira, M., Jackson-Lewis, V., Donaldson, D., and Togasaki, D.M. 1995. Dose-dependent lesions of the dopaminergic nigrostriatal pathway induced by intrastriatal injection of 6-hydroxydopamine. *Neuroscience* **67**: 631–647.

Rao, R.C., Boyd, J., Padmanabhan, R., Chenoweth, J.G., and McKay, R.D. 2008. Efficient serum-free derivation of oligodendrocyte precursors from neural stem cell-enriched cultures. *Stem Cells* (in press).

Ready, D.F., Hanson, T.E., and Benzer, S. 1976. Development of the *Drosophila* retina, a neurocrystalline lattice. *Dev. Biol.* **53**: 217–240.

Renfranz, P.J., Cunningham, M.G., and McKay, R.D. 1991. Region-specific differentiation of the hippocampal stem cell line HiB5 upon implantation into the developing mammalian brain. *Cell* **66**: 713–729.

Reynolds, B.A. and Weiss, S. 1992. Generation of neurons and astrocytes from isolated cells of the adult mammalian central nervous system. *Science* **255**: 1707–1710.

Rodríguez-Gómez, J.A., Lu, J.Q., Velasco, I., Rivera, S., Zoghbi, S.S., Liow, J.S., Musachio, J.L., Chin, F.T., Toyama, H., Seidel, J., et al. 2007. Persistent dopamine functions of neurons derived from embryonic stem cells in a rodent model of Parkinson disease. *Stem Cells* **25**: 918–928.

Ruohola, H., Bremer, K.A., Baker, D., Swedlow, J.R., Jan, L.Y., and Jan, Y.N. 1991. Role of neurogenic genes in establishment of follicle cell fate and oocyte polarity during oogenesis in *Drosophila*. *Cell* **66**: 433–449.

Russo, S.J., Bolanos, C.A., Theobald, D.E., DeCarolis, N.A., Renthal, W., Kumar, A., Winstanley, C.A., Renthal, N.E., Wiley, M.D., Self, D.W., et al. 2007. IRS2-Akt pathway in midbrain dopamine neurons regulates behavioral and cellular responses to opiates. *Nat. Neurosci.* **10**: 93–99.

Shim, J.H., Kim, S.E., Woo, D.H., Kim, S.K., Oh, C.H., McKay, R., and Kim, J.H. 2007. Directed differentiation of human embryonic stem cells towards a pancreatic cell fate. *Diabetologia* **50**: 1228–1238.

Spana, E.P. and Doe, C.Q. 1996. Numb antagonizes Notch signaling to specify sibling neuron fates. *Neuron* **17**: 21–26.

Spillantini, M.G., Schmidt, M.L., Lee, V.M., Trojanowski, J.Q., Jakes, R., and Goedert, M. 1997. α-Synuclein in Lewy bodies. *Nature* **388**: 839–840.

Stokoe, D., Stephens, L.R., Copeland, T., Gaffney, P.R., Reese, C.B., Painter, G.F., Holmes, A.B., McCormick, F., and Hawkins, P.T. 1997. Dual role of phosphatidylinositol-3,4,5-trisphosphate in the activation of protein kinase B. *Science* **277**: 567–570.

Studer, L., Tabar, V., and McKay, R.D. 1998. Transplantation of expanded mesencephalic precursors leads to recovery in Parkinsonian rats. *Nat. Neurosci.* **1**: 290–295.

Tanaka, S., Kamachi, Y., Tanouchi, A., Hamada, H., Jing, N., and Kondoh, H. 2004. Interplay of SOX and POU factors in regulation of the *Nestin* gene in neural primordial cells. *Mol. Cell. Biol.* **24**: 8834–8846.

Tanigaki, K., Nogaki, F., Takahashi, J., Tashiro, K., Kurooka, H., and Honjo, T. 2001. Notch1 and Notch3 instructively restrict bFGF-responsive multipotent neural progenitor cells to an astroglial fate. *Neuron* **29**: 45–55.

Tesar, P.J., Chenoweth, J.G., Brook, F.A., Davies, T.J., Evans, E.P., Mack, D.L., Gardner, R.L., and McKay, R.D. 2007. New cell lines from mouse epiblast share defining features with human embryonic stem cells. *Nature* **448**: 196–199.

Thomas, K.R. and Capecchi, M.R. 1986. Introduction of homol-

ogous DNA sequences into mammalian cells induces mutations in the cognate gene. *Nature* **324:** 34–38.

van der Walt, J.M., Noureddine, M.A., Kittappa, R., Hauser,
M.A., Scott, W.K., McKay, R., Zhang, F., Stajich, J.M.,
Fujiwara, K., Scott, B.L., Pericak-Vance, M.A., Vance, J.M.,
and Martin, E.R. 2004. Fibroblast growth factor 20 polymorphisms and haplotypes strongly influence risk of Parkinson
disease. *Am. J. Hum. Genet.* **74:** 1121–1127.

Vicario-Abejón, C., Cunningham, M.G., and McKay, R.D.
1995. Cerebellar precursors transplanted to the neonatal dentate gyrus express features characteristic of hippocampal neurons. *J. Neurosci.* **15:** 6351–6363.

Vicario-Abejón, C., Collin, C., Tsoulfas, P., and McKay, R.D.
2000. Hippocampal stem cells differentiate into excitatory
and inhibitory neurons. *Eur. J. Neurosci.* **12:** 677–688.

Wallberg, A.E., Pedersen, K., Lendahl, U., and Roeder, R.G.
2002. p300 and PCAF act cooperatively to mediate transcriptional activation from chromatin templates by notch intracellular domains in vitro. *Mol. Cell. Biol.* **22:** 7812–7819.

Weiss, A. and Schlessinger, J. 1998. Switching signals on or off
by receptor dimerization. *Cell* **94:** 277–280.

Williams, J., Owens, W., Turner, G., Saunders, C., Dipace, C.,
Blakely, R., France, C., Gore, J., Daws, L., Avison, M., and
Galli, A. 2007. Hypoinsulinemia regulates amphetamine-
induced reverse transport of dopamine. *PLoS Biol.* **5:**
2369–2378.

Yamanaka, K., Boillee, S., Roberts, E.A., Garcia, M.L.,
McAlonis-Downes, M., Mikse, O.R., Cleveland, D., and
Goldstein, L.S. 2008. Mutant SOD1 in cell types other than
motor neurons and oligodendrocytes accelerates onset of disease in ALS mice. *Proc. Natl. Acad. Sci.* **105:** 7594–7599.

Yoshimatsu, T., Kawaguchi, D., Oishi, K., Takeda, K., Akira, S.,
Masuyama, N., and Gotoh, Y. 2006. Non-cell-autonomous
action of STAT3 in maintenance of neural precursor cells in
the mouse neocortex. *Development* **133:** 2553–2563.

Zimmerman, L., Parr, B., Lendahl, U., Cunningham, M.,
McKay, R., Gavin, B., Mann, J., Vassileva, G., and
McMahon, A. 1994. Independent regulatory elements in the
nestin gene direct transgene expression to neural stem cells or
muscle precursors. *Neuron* **12:** 11–24.

Zipser, B. and McKay, R. 1981. Monoclonal antibodies distinguish identifiable neurones in the leech. *Nature* **289:**
549–554.

Zuccato, C., Ciammola, A., Rigamonti, D., Leavitt, B.R.,
Goffredo, D., Conti, L., MacDonald, M.E., Friedlander, R.M.,
Silani, V., Hayden, M.R., Timmusk, T., Sipione, S., and
Cattaneo, E. 2001. Loss of huntingtin-mediated BDNF gene
transcription in Huntington's disease. *Science* **293:** 493–498.

Cancer Stem Cells in Brain Tumor Biology

J.N. RICH* AND C.E. EYLER*†

*Department of Stem Cell Biology and Regenerative Medicine, Cleveland Clinic, Cleveland,
Ohio 44195; †Department of Pharmacology and Cancer Biology and Medical Scientists Training Program,
Duke University, Durham, North Carolina 27710

Tumors are aberrant organ systems containing a complex interplay between the neoplastic compartment and recruited vascular, inflammatory, and stromal elements. Furthermore, most cancers display a hierarchy of differentiation states within the tumor cell population. Molecular signals that drive tumor formation and maintenance commonly overlap with those involved in normal development and wound responses—two processes in which normal stem cells function. It is therefore not surprising that cancers invoke stem cell programs that promote tumor malignancy. Stem-cell-like cancer cells (or cancer stem cells) need not be derived from normal stem cells but may be subjected to evolutionary pressures that select for the capacity to self-renew extensively or differentiate depending on conditions. Current cancer model systems may not fully recapitulate the cellular complexity of cancers, perhaps partially explaining the lack of power of these models in predicting clinical outcomes. New methods are enabling researchers to identify and characterize cancer stem cells. Our laboratory focuses on the roles of brain tumor stem cells in clinically relevant tumor biology, including therapeutic resistance, angiogenesis, and invasion/metastasis. We hope that these studies will translate into improved diagnostic, prognostic, and therapeutic approaches for these lethal cancers.

Primary brain tumors comprise a large family of cancers (>160 types according to the World Health Organization [WHO]) (Furnari et al. 2007). The most common primary intrinsic brain tumors are gliomas in adults and medulloblastomas in children. Gliomas are defined by their morphologic and marker similarities to the glia (supporting cells of the brain), which include astrocytes and oligodendrocytes, and they are named astrocytomas or oligodendrogliomas (note that ependymomas may be included as glial tumors, but they display very different biological behavior and thus are commonly considered separately). Gliomas are graded by histologic criteria that include the presence of mitoses, aberrant nuclear or cytoplasmic morphology, glomeruloid angiogenesis, and necrosis, according to a WHO system of grading from I to IV, with grades increasing with more severe malignancy. Grade III gliomas (anaplastic astrocytoma or anaplastic astrocyoma) and grade IV gliomas (glioblastoma multiforme) are the most common and lethal of the gliomas and are treated in a similar manner. Standard of care for malignant gliomas (grade III and IV gliomas) consists of maximal surgical resection, followed by external beam radiation with concurrent chemotherapy (the oral methylator temozolomide), and then adjuvant temozolomide chemotherapy (Stupp et al. 2005). Unfortunately, tumor recurrence is essentially universal and no therapies have clear benefit in improving the survival of patients experiencing tumor recurrence or progression. The median survival for glioblastoma patients remains to be only 15 months. The outcome for children diagnosed with medulloblastoma is relatively better than for adults with glioblastoma but even long-term survivors commonly suffer long-term disability, including decreased intelligence. In fact, since the recent improvements in treating childhood leukemias, brain tumors are now the most common cause of pediatric cancer deaths. Thus, brain tumors present a severe clinical challenge, and the overall survival rate of patients has changed little in 30 years.

This chapter serves to highlight the work of the Rich laboratory within the context of this field. Because a number of laboratories share a similar research focus, this discussion represents only a small fraction of the ongoing work in the field and contains opinions of the author that may differ from those of other researchers.

CANCER STEM CELLS IN BRAIN TUMORS

Cancers are not simple collections of homogeneous neoplastic cells. Instead, a tumor is an organ system comprised of a neoplastic compartment with associated vasculature, inflammatory cells, and reactive cellular and extracellular components (Reya et al. 2001). Bailey and Cushing (1926) long ago recognized that brain cancers display striking morphologic variation, as evidenced by the term glioblastoma multiforme. Glial tumors often contain mixed subpopulations that morphologically resemble astrocytes and oligodendrocytes, leading to an intermediate diagnosis of oligoastrocytomas in the WHO classification system. Genetic analysis has additionally demonstrated that chromosomal aberrations and gene expression vary regionally within the tumor (Fulci et al. 2002). Regional variance is also evident in the commonly observed mixed clinical responses detected for specific therapies, in which part of the tumor may be responsive to a therapy, whereas other areas of the tumor fail to respond (Pope et al. 2006). Differentiation markers have been assessed in human brain tumors and demonstrate that aberrant and multiple states of differentiation may be present in the same tumor.

Our understanding of the normal development of the nervous system has dramatically increased in recent years. The nervous system has a complex differentiation hierarchy ranging from a neural stem cell that can give rise to all of the major lineages in the brain parenchyma (primarily neurons, astrocytes, and oligodendrocytes) to lineage-committed progenitors that have a more restricted

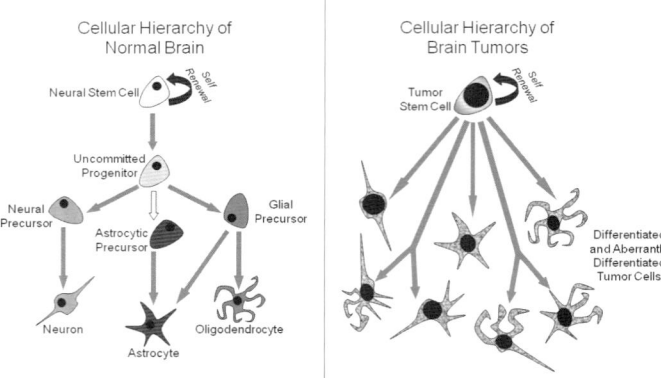

Figure 1. Brain tumors display a cellular hierarchy that resembles, but differs from, the normal neural cell hierarchy. Normal neural stem cells have the ability to self-renew while also dividing, to give rise to uncommitted progenitor cells that, in turn, give rise to lineage-restricted committed progenitors and finally to differentiated astrocytes, oligodendrocytes, and neurons. Similarly, tumors appear to have a cellular hierarchy, with self-renewing glioma stem cells able to generate a variety of more differentiated progeny, although patterns of differentiation appear to be less discrete than those in the normal brain. Many of the cancer stem cell–derived progeny display aberrant differentiation patterns, expressing more than one type of differentiation marker.

differentiation potential to terminally differentiated cells (Fig. 1) (Uchida et al. 2000; Rietze et al. 2001; Sanai et al. 2004). The recognition of the importance of differentiation state and the role of neural stem cells in development and wound responses (two processes that are recapitulated in carcinogenesis) have prompted the application of neural stem cell biology to neuro-oncology. Stem cell concepts can influence the understanding of brain cancer in two prominent areas: tumor origin and maintenance. The cell of origin for brain tumors is unresolved with genetic models supporting either a stem cell origin or a dedifferentiated committed cell of origin, with even more recent evidence suggesting a potential dual origin with a common final morphology (for review, see Furnari et al. 2007). No less controversial, the cancer stem cell hypothesis proposes that established tumors consist of a cellular hierarchy with a subpopulation of tumor cells able to maintain and propagate the tumor. Two competing models have been proposed: a stochastic model, in which any cell within the tumor has an equal chance of growth based on the genetic phenotype of the cell, and the hierarchical model, in which a subset of neoplastic cells can maintain tumor growth indefinitely (Reya et al. 2001). The initial

identification of a cancer stem cell occurred in leukemia (Lapidot et al. 1994; Bonnet and Dick 1997), but similar identifications in multiple systemic cancer types have followed (Al-Hajj et al. 2003; Hope et al. 2004; Dalerba et al. 2007; Li et al. 2007; O'Brien et al. 2007; Ricci-Vitiani et al. 2007). Cancer stem cells displaying these properties have been isolated from major types of brain tumors including gliomas, medulloblastomas, and ependymomas (Fig. 2) (Ignatova et al. 2002; Hemmati et al. 2003; Singh et al. 2003, 2004; Galli et al. 2004; Yuan et al. 2004; Taylor et al. 2005). Several issues have contributed to controversies surrounding the cancer stem cell hypothesis. These include (1) evidence that some cancers or some tumor models do not display a recognizable hierarchy (Quintana et al. 2008), (2) lack of universal markers that identify cancer stem cells, and (3) confusion regarding the implications of the cancer stem cell hypothesis in terms of the rarity of cancer stem cells (Kelly et al. 2007) and implications regarding the cell of origin. The cancer stem cell hypothesis does not require a rare cancer stem cell nor a stem cell origin for tumors (Clarke et al. 2006). It is unlikely that stem cell biology will explain the entirety of brain tumor biology, but it is increasingly evident that

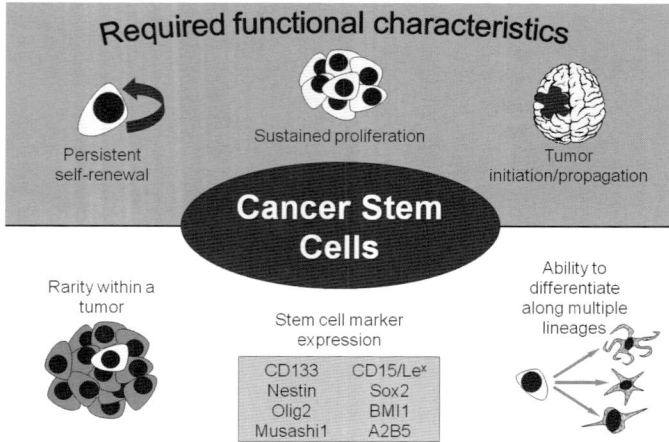

Figure 2. Cancer stem cells are defined by a capacity for sustained self-renewal, persistent proliferation, and tumor initiation or propagation. Some characteristics that are often, but not necessarily, associated with brain tumor stem cells include rarity within a tumor, expression of stem cell markers, and a capacity for multilineage differentiation.

stem cell signal transduction pathways are commonly dysregulated in brain tumors and that a tumor population can be commonly derived from human brain tumor surgical specimens that exhibit characteristics similar to those of normal stem cells. The acceptance of a cancer stem cell model is not mutually exclusive with a stochastic model of tumor initiation and maintenance that has been the leading paradigm in cancer biology for years, but it appears probable that these systems may be used in complement to inform research.

DEFINING BRAIN TUMOR STEM CELLS

No current agreement exists regarding the definition of a normal stem cell beyond long-term renewal and differentiation potential, so it is of little surprise that there is limited consensus regarding the defining characteristics of cancer stem cells. The current definition of a cancer stem cell requires self-renewal, sustained proliferation, and tumor initiation/propagation (Clarke et al. 2006). Because the stem-cell-like populations are defined in functional assays, some investigators have selected a nomenclature to represent the ability of cells to propagate tumors, but these terms fail to communicate that core characteristics may be shared between these tumor cells and normal stem cells (markers, signal transduction pathways, self-renewal capacity, etc.).

Normal stem cells commonly express specific antigens that permit prospective enrichment of cells that fulfill stem cell criteria, but no antigenic profile (the immunophenotype) is absolutely representative of a stem cell. Thus, we lack the ability to directly assess the creation of a perfect copy of a cell in real time. Rather, stem cell assays validate that a cell capable of self-renewal must have been present during an earlier step. Stem cells have developmentally regulated replication that can be either symmetric (yielding two identical cells—either two stem cells or two differentiated daughter cells) or asymmetric (giving rise to one differentiated and one undifferentiated daughter cell) in response to cell state and external cues. Measuring the differentiation status of a cell in a single division presents significant challenges to score a division as symmetric or asymmetric. To date, current techniques have included measurement of polarized proteins including Notch (Wu et al. 2007), but current techniques cannot verify in real time that a daughter cell has not undergone differentiation. The gold standard for defining a normal adult stem cell remains the generation of the full cellular constituents of the relevant organ from a single stem cell. For the nervous system, the ability to form neurons, astrocytes, and oligodendrocytes is required of normal neural stem cells. The differentiation cascade for the hematopoietic system is by far the best characterized, but the nervous system is increasingly well modeled. In cell culture systems without serum, neural progenitors may form three-dimensional structures called neurospheres that do not adhere to the culture surface. These complex structures tend to have the least differentiated cellular populations located on the surface of the sphere, with expression of differentiation markers more commonly occurring on cells in the interior. Complex neuronal processes may be formed in these structures, and

some correlation of similar structures in the breast (mammospheres) has permitted an assessment of self-renewal through the serial passage of the spheres from single cells and of proliferation rate from the size of the generated spheres. However, caution must be exercised in interpreting the significance of neurosphere generation (Singec et al. 2006). First, spheres may be generated from cells that were incompletely disaggregated with residual cohesive cells. Second, spheres may form and "grow" through the fusion of smaller structures that may be present at higher cellular densities. In addition, neurospheres are cell culture artifacts that do not have in vivo correlates. Finally, not all cells within a neurosphere have an undifferentiated state (in fact, stem-like cells are often a small minority) and even committed progenitors may be able to form a neurosphere. Thus, the presence of a neurosphere does not prove the presence of a stem cell in normal physiology.

Within the field of cancer stem cell biology, the current state of understanding has led to high variability in the rigor with which validation of the stem cell nature of a cellular population is approached. Many groups have simply used the presence of a stem cell marker or neurosphere generation as an indication of a cancer stem cell. This is inadequate, and these studies may create difficulty in a field that is already confusing and controversial. The requirements and challenges for the identification of a cancer stem cell are similar to those of normal stem cells. Two main approaches have been used to date. In the first, tumors are disaggregated and cultured in serum-free media until tumor spheres form (Ignatova et al. 2002; Hemmati et al. 2003; Galli et al. 2004; Yuan et al. 2004). These spheres are sequentially passaged to confirm sustained self-renewal. The advantage of this system is that it includes an important (if challenging) functional assay at the initial characterization. There are weaknesses, however, because the neurospheres are still mixed populations that represent only a small fraction of the original tumor. The underlying heterogeneity will, by definition, be lost. Although neurospheres can be subjected to differentiation conditions, it is unclear whether these conditions are fully representative of the diversity of cellular populations in the original tumor. In addition, serial neurosphere passage requires extended periods of cell culture that can rapidly (even in minutes) induce significant alterations in cellular biology and gene expression. Therefore, there has been a strong effort to identify cell surface antigens that can be used to prospectively enrich cancer stem cells from tumor populations immediately upon surgical resection (Singh et al. 2003, 2004). Although several markers may be informative in brain tumor stem cell identification (e.g., Prominin-1 [CD133], CD15 [SSEA-1, Lewis x structure (Le^x)], A2B5, BMI1, Nestin, Sox2, and Musashi1), there are significant deficiencies with the current available markers. In our own studies, markers are only reliably useful in segregating tumor-initiation potential after immediate derivation from an in vivo environment, suggesting that marker expression in cancer stem cells requires interactions with the microenvironment. In summary, the current methods for enrichment for cancer stem cells remain imperfect and will require improvements.

The gold standard for defining cancer stem cells remains tumor propagation. The current preferred assay is the in vivo limiting dilution assay, in which progressively smaller numbers of tumor cells are implanted in an orthotopic location to demonstrate the minimal number of cells required to form tumors (Singh et al. 2004). Presumably, the number of cells represents a surrogate for the frequency of true cancer stem cells. However, it is possible that for solid cancers, some tumors will require more than one cell to initiate tumor growth of even pure cancer stem cell populations. An ideal result would be to have single tumor cells demonstrate the capacity to form a tumor and give rise to daughter cells that share this characteristic (Quintana et al. 2008). To date, the brain tumor field has not seen reports of this efficiency but instead have a cellular requirement of 100–1000 from human surgical biopsy specimens (Singh et al. 2004; Bao et al. 2006a). Regardless, the requirement for in vivo tumor propagation is absolute. The many studies that solely assess tumor sphere formation or expression of a cancer stem cell marker cannot be considered to have demonstrated the presence of cancer stem cells and may detract from the field.

An unresolved question in the solid tumor cancer stem cell field revolves around the proliferative rate of cancer stem cells. Normal and leukemic stem cells share the ability to proliferate over the long term but are quiescent in normal conditions. In ex vivo studies, brain tumor stem cells are apparently proliferative, but this may represent a response to culture conditions. Preliminary studies of brain tumor specimens have suggested that tumors contain cells that coexpress cancer stem cell and proliferative markers, but the ability to distinguish a stem cell population from a transit-amplifying/committed progenitor population remains unreported.

The cancer stem cell hypothesis has engendered criticism due to a lack of clarity in the terminology. Many researchers and the lay public assume that a cancer stem cell is derived from a normal stem cell. Conceptually, brain tumors may be derived from neural stem cells, transit-amplifying cells, or terminally differentiated cells. A stem cell cell of origin is an attractive hypothesis because the long life of these cells would permit the accumulation of genetic and epigenetic alterations required for transformation. In addition, many characteristics of normal neural stem cells are similar to characteristics of cells in high-grade brain tumors (including diversity of cell populations, high migratory potential, and sustained proliferation). Several genetically engineered brain tumor models suggest that neural stem cells may be transformed with restricted oncogenic stimuli (for review, see Furnari et al. 2007). However, other models support a potential for dedifferentiation of more differentiated cells in tumor origination. Very recently, two groups in parallel have demonstrated that identical tumors (medulloblastomas) can be derived from different cells of origin with identical genetic alterations, strongly supporting a model in which cancers that display similar morphologies may be derived from different starting points (Schüller et al. 2008; Yang et al. 2008a). It is likely that no single rule may be applied to the originating cell for a single cancer.

DERIVATION OF BRAIN TUMOR STEM CELLS

Our approach in cancer stem cell derivation has been built on the seminal studies by the Dirks laboratory that were in turn based on protocols for neural stem cell derivation (Singh et al. 2003, 2004). We have used tumor sources of human surgical biopsy specimens immediately collected after resection and tumor xenografts maintained in an in vivo environment. We disaggregate these tumors and then prospectively sort for a single cell surface marker, Prominin-1 (cluster of differentiation 133 [CD133]), which is the most developed brain tumor stem cell marker. Prominin-1, a pentaspan transmembrane glycoprotein located on cellular protrusions, was originally identified separately through the development of antibodies against the mouse neuroepithelium (Weigmann et al. 1997) and CD34^{bright} hematopoietic stem and progenitor cells derived from human fetal liver, bone marrow, and blood (Miraglia et al. 1997; Yin et al. 1997). The function of Prominin-1 is unknown, but mutations in Prominin-1 are detected in patients with familiar macular degeneration, and they disrupt photoreceptor disk morphogenesis in a genetic mouse model (Yang et al. 2008b).

CD133 was first used to enrich tumor-repopulating cells in leukemia (Bühring et al. 1999; Horn et al. 1999) and informed prognosis (Lee et al. 2001). On the basis of its expression on neural stem/progenitor cells, CD133 was investigated as a brain tumor stem cell marker. In seminal studies first performed in vitro and then in vivo, brain tumor stem cells were exclusively detected in CD133^{+} cells from gliomas and medulloblastomas (Singh et al. 2003, 2004). Tumor-derived neurospheres from pediatric brain tumors also express CD133 and other stem cell markers (Sox2, Musashi-1, Bmi-1, maternal embryonic leucine zipper kinase, and phosphoserine phosphatase) (Hemmati et al. 2003). The CD133 marker is not absolute but has proven useful for segregating tumorigenic potential for the majority of tumors in our laboratory. The primary advantage to this approach is that the cellular heterogeneity that is the core of the cancer stem cell hypothesis is maintained. The cells that are collected are then cultured in appropriate media: Stem cell populations are grown in defined media with growth factors (Lee et al. 2006) but without serum, whereas non-stem-cell populations are maintained in serum. Under these conditions, cancer stem cells will tend to form neurosphere-like structures and nonstem cells will grow as adherent cells. We have found that the cells in both populations rapidly accumulate changes in gene expression and genetic markers when cultured, suggesting that characterization of cancer stem cells and their matched non-stem-cell brethren should be performed at low passage number. Even this caveat is likely inadequate because phosphorylated proteins become altered within minutes after culturing, but our current technologies do not permit a perfect system to maintain the original cellular phenotype. Even in the earliest reports using CD133, the investigators recognized variation among tumors in marker expression. CD133 is also informative in ependymomas in conjunction with other markers (Nestin and brain lipid-binding protein [BLBP]) (Taylor et al. 2005). Many reports have confirmed the utility of CD133 in prospective isolation of brain tumor stem cells (Bao et al. 2006a, b, 2008; Piccirillo et al. 2006; Calabrese et al. 2007), and CD133 has proven useful in a number of other solid cancers, including colorectal cancers (O'Brien et al. 2007; Ricci-Vitiani et al. 2007). However, challenges to the universal expression of CD133 have arisen, and some tumors have tumor-propagating potential without significant numbers of CD133$^+$ cells (Beier et al. 2007). Interestingly, primary glioblastomas have much higher levels of CD133$^+$ cells than do recurrent tumors. The difficulties with CD133 are multiple. For example, AC133 reagents (monoclonal antibodies against the CD133 glycoprotein) are challenging to use (Bidlingmaier et al. 2008). In flow cytometry assays, CD133$^+$ peaks are not fully separated from isotype antibody control peaks in most tumor preparations. Without a clear separation, CD133$^-$ and CD133$^+$ populations cannot be clearly delineated and require functional validation. The precise methodologies used to disaggregate tissues and purify cellular populations can have profound effects on CD133 fractions (Panchision et al. 2007). Cell culture conditions are important to maintain appropriate tumor stem cell populations (Lee et al. 2006), but direct transfer to an in vivo environment may be optimal for preservation of a CD133$^+$ tumor cell fraction (Shu et al. 2008). CD133 is not a static gene product but is a target of promoter methylation alterations in cancers (Tabu et al. 2008; Yi et al. 2008) and may be regulated during the cell cycle (Jaksch et al. 2008). The complexity of these conditions has translated into the common use of very small numbers of tumor specimens in even high-impact reports. It is almost certainly the case, however, that morphologically identical brain tumors have underlying complex cellular differences due to different cell-of-origin or oncogenic changes that are represented with different brain tumor stem cells that may express different marker immunophenotypes.

THE SIGNIFICANCE OF BRAIN TUMOR STEM CELLS IN NEURO-ONCOLOGY

Although it may appear that the cancer stem cell hypothesis is merely an academic exercise or a laboratory phenomenon, one cannot deny the near total failure in the development of therapies to improve the outcomes of brain cancer patients using traditional laboratory-investigative approaches. The use of temozolomide has been hailed as a tremendous advance in the treatment of malignant gliomas, but the benefit has been limited to less than 3 months of improved median survival for glioblastoma patients (Stupp et al. 2005). Not only has the genetic knowledge of brain tumor biology been inadequate to drive new effective therapies, but advanced imaging technologies are still unreliable in early tumor detection and prediction of the most important outcome: survival. The heterogeneity of brain cancers may be helpful in explaining many of our failures. To date, no direct proof of a role for cancer stem cells in brain tumor clinical trials has been reported, but several studies have examined the expression of cancer stem cell marker–positive cells in clinical brain tumor biopsy specimens.

CD133 immunohistochemistry of brain tumor specimens has shown variability in utility, likely due to the combination of tumor heterogeneity and reagent specificity. CD133$^+$ cells reside in a perivascular niche of tumors (Bao et al. 2006b; Calabrese et al. 2007; Christensen et al. 2008). Analysis of CD133 and proliferation has not demonstrated consistent relationships to date (Christensen et al. 2008; Ma et al. 2008), but CD133 may inform prognosis (Beier et al. 2008; Howard and Boockvar 2008; Thon et al. 2008; Zeppernick et al. 2008), although some studies have failed to demonstrate a link (Christensen et al. 2008). One study (Liu et al. 2006) found that CD133 mRNA increased after tumor recurrence. In summary, it is premature to consider CD133 to be a validated prognostic indicator. The validation of other potential markers remains less developed.

BRAIN TUMOR STEM CELLS IN THERAPEUTIC RESISTANCE

Unfortunately, patients afflicted with malignant gliomas suffer nearly universal treatment failure and death. As described above, surgical resection and cytotoxic modalities (radiation, chemotherapy) remain the mainstay of brain tumor therapy (of note, antiangiogenic therapy in the form of the humanized neutralizing antibody against vascular endothelial growth factor [VEGF] has shown initial promise). The mechanisms through which brain tumors become resistant to conventional therapy are poorly understood and are likely multifactorial. We examined a potential contribution of brain tumor stem cells to radiation resistance (Bao et al. 2006a). We found that ionizing radiation increased the relative frequency of tumor cells expressing cancer stem cell markers in treated xenografts. The relative enrichment of these cells was accompanied by maintained capacity for self-renewal and tumor propagation, whereas matched nonstem cancer cells were more likely to die. Cancer stem cell–enriched cultures treated with radiation demonstrate a lower apoptotic fraction than do nonstem cells in the same conditions, allowing for the outgrowth of cancer stem cells. To elucidate a potential mechanism, we studied the DNA-damage checkpoint response in which a cascade of proteins integrates signals from damage sensors to determine whether cells will initiate a cell cycle arrest with DNA repair or undergo apoptosis. Cancer stem cells treated with radiation or radio-mimetics displayed an increased activation of the DNA-damage checkpoint response compared to that of matched nonstem cancer cells. Although the activated proteins showed variability among samples, some proteins appeared to be activated at baseline (e.g., Rad17), as if the cancer stem cells were primed to respond to genotoxic stress, which may be an early event in cancer initiation. The role of the DNA-damage checkpoint response proved contributory because a pharmacologic inhibitor of the checkpoint sensitized the cancer stem cells to radiation. These results were supported by studies in genetically engineered medulloblastomas (Hambardzumyan et al. 2008). Other researchers have also found that neurosphere-forming brain tumor cells are more resistant to chemotherapy than are similar cells grown under differentiating conditions (Liu et al. 2006). In sum, although these studies suggest that cancer stem cells may contribute to the common therapeutic resistance of brain tumors and may be targetable with pharmacologic approaches, it also seems unlikely that the full extent of resistance in brain tumors derives from cancer stem cells.

BRAIN TUMOR STEM CELLS IN ANGIOGENESIS

Malignant gliomas are commonly angiogenic, with vascular proliferation serving as an informative histologic feature indicating a glioblastoma among the gliomas. Many growth factors are secreted by malignant gliomas to stimulate and maintain neoangiogenic vasculature. Targeted therapies have been developed against some of these pathways (for review, see Jain et al. 2007). Most clinical trials have demonstrated modest benefits from these agents, but a potential clinical efficacy has been seen in several trials of bevacizumab (Avastin), a neutralizing antibody against VEGF (Vredenburgh et al. 2007a,b). Interestingly, the activity of low-molecular-weight inhibitors against the VEGF receptors have been more modest in clinical trials, suggesting that the same molecular pathway may be targeted by different agents with different outcomes. During our studies of brain tumor stem cells, we noted that cancer stem cells form highly angiogenic tumors compared to the uncommon tumors that we detect in propagation studies with cancer stem cell–depleted cultures (Bao et al. 2006b). We found that conditioned media from cancer stem cells strongly induced endothelial cell migration, proliferation, and tube formation in contrast to nonstem cancer cell conditioned media. Characterization of angiogenic proteins in the conditioned media revealed a consistent up-regulation of VEGF. We were able to specifically block the effects of cancer stem cell conditioned media on endothelial cells using bevacizumab. In animal studies, bevacizumab strongly reduced the growth of tumors derived from cancer stem cells to a size and paucity of vascularity that is nearly identical to that of the uncommon tumors formed by nonstem cancer cells. Because nonstem cancer cells can survive implantation but rarely form tumors, the angiogenic drive may provide one explanation for the striking tumor propagation of cancer stem cells. In addition, cancer stem cells may provide an angiogenic drive to support the growth of nonstem cancer cells, suggesting that their effects in the tumor may not need to be limited solely to the direct production of progeny. In our studies, we noted that the cancer stem cells appeared to be located near the vasculature. These observations have been confirmed and extended in a seminal study that demonstrated that brain tumor stem cell growth is supported by endothelial cells and that tumor formation by the cancer stem cells requires support from a vascular niche (Calabrese et al. 2007). Additional studies further indicate the presence of cancer stem cell marker–positive cells located in the perivascular niche of patient specimens. In sum, these studies suggest that brain tumor stem cells have the ability to form their own tumor microenvironment through the elaboration of angiogenic factors, but at the same time, they remain dependent on that niche (Gilbertson and Rich 2007). These results may partially explain both the clinical activity of bevacizumab and the invasive phenotype in patients who suffer failure after bevacizumab treatment, because cancer stem cells display an invasive phenotype.

TARGETING BRAIN TUMOR STEM CELLS

There have been numerous recent reports of molecular targets that may be useful in ablating brain tumor stem cells. Several of these reports have focused on core stem cell/differentiation pathways, including BMI1 (Bruggeman et al. 2007; Godlewski et al. 2008), bone morphogenic protein (BMP) (Piccirillo et al. 2006), sonic hedgehog (Bar et al. 2007; Clement et al. 2007), Sox2 (Gangemi et al. 2008), Oct4 (Du et al. 2008), and Notch (Fan et al. 2006). Inhibitors of growth factor pathways, including epidermal growth factor (EGF) (Soeda et al. 2008) and platelet-derived growth factor (PDGF), may be also be useful against brain tumor stem cells. To discover new molecular targets, we have compared expression of gene products or activated signal transduction pathways between cancer stem cells and nonstem cancer cells. The rationale behind this approach is that previously unrecognized targets may be discovered in the small fraction of cells that we have found to be cancer stem cells in brain tumors (it is again important to note that cancer stem cells may not necessarily be uncommon).

In one study, we found that the cell surface protein L1 cell adhesion molecule (L1CAM, CD171) is preferentially expressed in brain tumor stem cell–enriched cultures (Bao et al. 2008). L1CAM cosegregates with CD133 in glioblastoma patient biopsy specimens and is expressed at higher levels than in human neural progenitors. L1CAM contributes to brain tumor stem cell survival as targeting L1CAM expression through lentiviral short hairpin RNA (shRNA) specifically induced apoptosis in brain tumor stem cell cultures and ablated neurosphere formation. We found that L1CAM mediates its effects on brain tumor stem cells at least in part through the regulation of the transcriptional regulator Olig2. Other studies have demonstrated that the targeted disruption of Olig2 in genetically engineered brain tumor models blocks tumor initiation (Ligon et al. 2007). In our studies, targeting L1CAM decreased Olig2 expression and increased the expression of the key Olig2 target, the p21[CIP1/WAF1] cyclin-dependent kinase inhibitor, and overexpression of Olig2 rescued the effects of L1CAM targeting. Most importantly, targeting L1CAM either before xenotransplantation or in established tumors reduced tumor growth and extended the life span of mice bearing tumor stem cell xenografts. These results demonstrated that analysis of brain tumor stem cells can identify novel molecular targets that may be useful for brain tumor therapy.

The phosphoinositol-3 kinase (PI3K) pathway is commonly dysregulated in malignant gliomas through mutations in either the subunits of PI3K or the phosphatase and tensin homolog (*PTEN*) tumor suppressor gene (Cancer Genome Atlas Research Network 2008). PI3K functions in part through regulation of the Akt/protein kinase B (PKB) survival pathway. We therefore examined the activation state of Akt in brain tumor stem cells compared to matched nonstem tumor cells (Eyler et al. 2008). The level of activating phosphorylation of Akt was lower at baseline in brain tumor stem cell cultures but was also more sensitive to inhibitory effects of low-molecular-weight Akt inhibitors. Whereas Akt inhibition of nonstem tumor cells was largely cytostatic, tumor stem cells displayed an apoptotic response with Akt or PI3K inhibitors. Akt inhibitors also reduced neurosphere formation and invasion. Finally, tumor initiation was impaired by Akt inhibition. These results and those of other laboratories studying BMP, Notch, sonic hedgehog (SHH), and epidermal growth factor (EGFR) (Fan et al. 2006; Piccirillo et al. 2006; Bar et al. 2007; Clement et al. 2007; Lee et al. 2008; Soeda et al. 2008) suggest that brain tumor stem cells may be particularly sensitive to targeted therapies against signal transduction pathways.

The c-*myc* oncogene is commonly involved in cancer initiation and maintenance, but the role of c-*myc* in glioma biology is poorly understood. We examined the potential role of c-*myc* in brain tumor stem cells as a result of its involvement in normal stem cell biology (Wang et al. 2008). Glioma stem cells derived from human surgical biopsies consistently expressed higher levels of c-*myc* mRNA and protein relative to the nonstem tumor cells. Targeting c-*myc* expression was cytostatic in nonstem tumor cells but potently induced apoptosis and blocked self-renewal in the glioma stem cells. Most importantly, targeting c-*myc* expression completely blocked tumor propagation in transplantation studies. These results are very similar to those of a genetically engineered glioma model in which *p53* and *Pten* are disrupted (Zheng et al. 2008).

These and several other studies have laid the foundation for novel insights into brain tumor biology through the analysis of molecular regulators of brain tumor stem cells. The extension of these studies into combination regimens with other therapies and potential clinical trial application may offer improved clinical outcomes.

PERSPECTIVE

Neuro-oncology has witnessed some important therapeutic advances, particularly in the treatment of pediatric brain tumors. Unfortunately, the outcome for adult patients with the most common intrinsic primary brain tumor, glioblastoma multiforme, continues to be extremely poor, with even the most exciting advances providing only minimal improvement in median survival in clinical trials. Fundamental changes in our paradigm in the development of prognostic markers, imaging, and therapy must occur for real change in patient outcome. It is potentially useful to take lessons from another area of medicine: infectious diseases. Mycobacterium tuberculosis (Mtb) is a major health burden in the developing world and in immunocompromised hosts. Few new effective antituberculosis agents have been developed. Recent studies suggest that traditional high-throughput Mtb drug development assays that essentially and nonspecifically target proliferation may not be useful for improving patient outcome because the model does not recapitulate in vivo conditions (Nathan et al. 2008). Rather, Mtb displays a cellular heterogeneity in a small fraction of the total population that is resistant to conventional therapies and is relatively quiescent. Nonreplicating bacte-

ria may be critical to the problem of persistent Mtb infection. The striking parallels to cancer stem cell biology cannot be ignored and are not surprising because nature tends to repeat patterns. It is probable that not all cancers display a clear cellular hierarchy of tumor growth but the heterogeneity of cancers is essential to incorporate in models. The concept of stem-cell-like cells within brain tumors is not new, but recent technologies have improved the ability to prospectively enrich for cancer stem cells, and the recent increase in genetic understanding of brain tumors has informed the development of genetic brain tumor models. Although it is unlikely that brain tumor stem cells will inform all of brain tumor biology, our current failure in clinical neuro-oncology demands the aggressive investigation of new areas of research. Our studies and those of other laboratories have suggested that brain tumor stem cells contribute to therapeutic resistance, tumor angiogenesis, and invasion. Further characterization of this cellular fraction may guide the development of biomarkers, imaging modalities, and treatments that will hopefully be more effective. However, this field is immature and progress will likely be made with stumbles and errors and will be a learning process. The current challenge in deriving and maintaining brain tumor stem cells is a major limitation in the field because most laboratories do not have access to viable clinical specimens and animal resources. However, it is important to find ways to adapt these techniques for widespread use because there is currently insufficient evidence to show that established cell lines maintained for long periods in serum are useful in cancer stem cell studies. Cell culture—particularly long-term cell culture in medium containing serum—is well recognized to induce genetic changes that were not present in the original tumor, limiting the utility of cell lines in modeling the original disease. The development of validated brain tumor models that can be shared in the field would be an important step forward. In addition, the functional assays for all brain tumor stem cell studies must be standardized with current use of serial neurosphere formation as a surrogate for self-renewal and tumor propagation. Available markers for brain tumor stem cells are imperfect and cannot be definitively linked to a stem cell phenotype, supporting an urgent need for improved markers. Because brain tumors are likely heterogeneous diseases, universal marker immunophenotypes may not be identifiable, but markers may assist in subcategorizing tumors. Molecular regulators of brain tumor stem cells may provide biomarkers, imaging targets, and therapeutic targets, but it is likely that molecules may be shared with normal somatic stem cells and thus their use may be complicated. Regardless of the outcome, the recognition of the potential importance of the cancer stem cell hypothesis has energized brain tumor research. The healthy debate between believers and skeptics will almost certainly lead to completely unforeseen directions in the field of brain tumor research and therapeutic development. In the end, we all hope to help those patients and families who are afflicted by brain tumors.

ACKNOWLEDGMENTS

Financial support was provided by the Childhood Brain Tumor Foundation, Pediatric Brain Tumor Foundation of the United States, Accelerate Brain Cancer Cure, Alexander and Margaret Stewart Trust, Brain Tumor Society, Goldhirsh Foundation, Duke Comprehensive Cancer Center Stem Cell Initiative Grant, and National Institutes of Health grants NS047409, NS054276, CA129958, and CA116659. J.R. is a Damon Runyon-Lilly Clinical Investigator supported by the Damon Runyon Cancer Research Foundation.

REFERENCES

Al-Hajj, M., Wicha, M.S., Benito-Hernandez, A., Morrison, S.J., and Clarke, M.F. 2003. Prospective identification of tumorigenic breast cancer cells. *Proc. Natl. Acad. Sci.* **100:** 3983–3988.

Bailey, P. and Cushing, H.A. 1926. A classification of the gliomata. In *Classification of the tumors of the glioma group on a histogenetic basis with a correlated study of prognosis*, pp. 53–95. Lippincott, Philadelphia.

Bao, S., Wu, Q., Li, Z., Sathornsumetee, S., Wang, H., McLendon, R.E., Hjelmeland, A.B., and Rich, J.N. 2008. Targeting cancer stem cells through L1CAM suppresses glioma growth. *Cancer Res.* **68:** 6043–6048.

Bao, S., Wu, Q., McLendon, R.E., Hao, Y., Shi, Q., Hjelmeland, A.B., Dewhirst, M.W., Bigner, D.D., and Rich J.N. 2006a. Glioma stem cells promote radioresistance by preferential activation of the DNA damage response. *Nature* **444:** 756–760.

Bao, S., Wu, Q., Sathornsumetee, S., Hao, Y., Li, Z., Hjelmeland, A.B., Shi, Q., McLendon, R.E., Bigner, D.D., and Rich, J.N. 2006b. Stem cell-like glioma cells promote tumor angiogenesis through vascular endothelial growth factor. *Cancer Res.* **66:** 7843–7848.

Bar, E.E., Chaudhry, A., Lin, A., Fan, X., Schreck, K., Matsui, W., Piccirillo, S., Vescovi, A.L., DiMeco, F., Olivi, A., and Eberhart, C.G. 2007. Cyclopamine-mediated Hedgehog pathway inhibition depletes stem-like cancer cells in glioblastoma. *Stem Cells* **25:** 2524–2533.

Beier, D., Wischhusen, J., Dietmaier, W., Proescholdt, M., Brawanski, A., Bogdahn, U., and Beier, C.P. 2008. CD133 expression and cancer stem cells predict prognosis in high-grade oligodendroglial tumors. *Brain Pathol.* **18:** 370–377.

Beier, D., Hau, P., Proescholdt, M., Lohmeier, A., Wischhusen, J., Oefner, P.J., Aigner, L., Brawanski, A., Bogdahn, U., and Beier, C.P. 2007. CD133⁺ and CD133⁻ glioblastoma-derived cancer stem cells show differential growth characteristics and molecular profiles. *Cancer Res.* **67:** 4010–4015.

Bidlingmaier, S., Zhu, X., and Liu, B. 2008. The utility and limitations of glycosylated human CD133 epitopes in defining cancer stem cells. *J. Mol. Med.* **86:** 1025–1032.

Bonnet, D. and Dick, J.E. 1997. Human acute myeloid leukemia is organized as a hierarchy that originates from a primitive hematopoietic cell. *Nat. Med.* **3:** 730–737.

Bruggeman, S.W., Hulsman, D., Tanger, E., Buckle, T., Blom, M., Zevenhoven, J., van Tellingen, O., and van Lohuizen, M. 2007. Bmi1 controls tumor development in an Ink4a/Arf-independent manner in a mouse model for glioma. *Cancer Cell* **12:** 328–341.

Bühring, H.J., Seiffert, M., Marxer, A., Weiss, B., Faul, C., Kanz, L., and Brugger, W. 1999. AC133 antigen expression is not restricted to acute myeloid leukemia blasts but is also found on acute lymphoid leukemia blasts and on a subset of CD34⁺ B-cell precursors. *Blood* **94:** 832–833.

Calabrese, C., Poppleton, H., Kocak, M., Hogg, T.L., Fuller, C., Hamner, B., Oh, E.Y., Gaber, M.W., Finklestein, D., Allen, M., et al. 2007. A perivascular niche for brain tumor stem cells. *Cancer Cell* **11:** 69–82.

Cancer Genome Atlas Research Network. 2008. Comprehensive genomic characterization defines human glioblastoma genes and core pathways. *Nature* **455:** 1061–1068.

Christensen, K., Schrøder, H.D., and Kristensen, B.W. 2008. CD133 identifies perivascular niches in grade II–IV astrocytomas. *J. Neurooncol.* **90:** 157–170.

Clarke, M.F., Dick, J.E., Dirks, P.B., Eaves, C.J., Jamieson, C.H., Jones, D.L., Visvader, J., Weissman, I.L., and Wahl, G.M. 2006. Cancer stem cells—Perspectives on current status and future directions: AACR Workshop on cancer stem cells. *Cancer Res.* **66:** 9339–9344.

Clement, V., Sanchez, P., de Tribolet, N., Radovanovic, I., and Ruiz i Altaba, A. 2007. HEDGEHOG-GLI1 signaling regulates human glioma growth, cancer stem cell self-renewal, and tumorigenicity. *Curr. Biol.* **17:** 165–172.

Dalerba, P., Dylla, S.J., Park, I.K., Liu, R., Wang, X., Cho, R.W., Hoey, T., Gurney, A., Huang, E.H., Simeone, D.M., et al. 2007. Phenotypic characterization of human colorectal cancer stem cells. *Proc. Natl. Acad. Sci.* **104:** 10158–10163.

Du, Z., Jia, D., Liu, S., Wang, F., Li, G., Zhang, Y., Cao, X., Ling, E.A., and Hao, A. 2008. Oct4 is expressed in human gliomas and promotes colony formation in glioma cells. *Glia* (in press).

Eyler, C.E., Foo, W.C., Lafiura, K.M., McLendon, R.E., Hjelmeland, A.B., and Rich J.N. 2008. Brain cancer stem cells display preferential sensitivity to Akt inhibition. *Stem Cells* **26:** 3027–3036.

Fan, X., Matsui, W., Khaki, L., Stearns, D., Chun, J., Li, Y.M., and Eberhart, C.G. 2006. Notch pathway inhibition depletes stem-like cells and blocks engraftment in embryonal brain tumors. *Cancer Res.* **66:** 7445–7452.

Fulci, G., Ishii, N., Maurici, D., Gernert, K., Hainaut, P., Kaur, B., and Van Meir, E.G. 2002. Initiation of human astrocytoma by clonal evolution of cells with progressive loss of p53 functions in a patient with a 283H TP53 germline mutation: Evidence for a precursor lesion. *Cancer Res.* **62:** 2897–2906.

Furnari, F.B., Fenton, T., Bachoo, R.M., Mukasa, A., Stommel, J.M., Stegh, A., Hahn, W.C., Ligon, K.L., Louis, D.N., Brennan, C., et al. 2007. Malignant astrocytic glioma: Genetics, biology, and paths to treatment. *Genes Dev.* **21:** 2683–2710.

Galli, R., Binda, E., Orfanelli, U., Cipelletti, B., Gritti, A., De Vitis, S., Fiocco, R., Foroni, C., Dimeco, F., and Vescovi, A. 2004. Isolation and characterization of tumorigenic, stem-like neural precursors from human glioblastoma. *Cancer Res.* **64:** 7011–7021.

Gangemi, R.M., Griffero, F., Marubbi, D., Perera, M., Capra, M.C., Malatesta, P., Ravetti, G.L., Zona, G.L., Daga, A., and Corte, G. 2009. *SOX2* silencing in glioblastoma tumor initiating cells causes stop of proliferation and loss of tumorigenicity. *Stem Cells* **27:** 40–48.

Gilbertson, R.J. and Rich, J.N. 2007. Making a tumour's bed: Glioblastoma stem cells and the vascular niche. *Nat. Rev. Cancer* **10:** 733–736.

Godlewski, J., Nowicki, M.O., Bronisz, A., Williams, S., Otsuki, A., Nuovo, G., Raychaudhury, A., Newton, H.B., Chiocca, E.A., and Lawler, S. 2008. Targeting of the Bmi-1 oncogene/stem cell renewal factor by microRNA-128 inhibits glioma proliferation and self-renewal. *Cancer Res.* **22:** 9125–9130.

Hambardzumyan, D., Becher, O.J., Rosenblum, M.K., Pandolfi, P.P., Manova-Todorova, K., and Holland, E.C. 2008. PI3K pathway regulates survival of cancer stem cells residing in the perivascular niche following radiation in medulloblastoma in vivo. *Genes Dev.* **22:** 436–448.

Hemmati, H.D., Nakano, I., Lazareff, J.A., Masterman-Smith, M., Geschwind, D.H., Bronner-Fraser, M., and Kornblum, H.I. 2003. Cancerous stem cells can arise from pediatric brain tumors. *Proc. Natl. Acad. Sci.* **100:** 15178–15183.

Hope, K.J., Jin, L., and Dick, J.E. 2004. Acute myeloid leukemia originates from a hierarchy of leukemic stem cell classes that differ in self-renewal capacity. *Nat. Immunol.* **5:** 738–743.

Horn, P.A., Tesch, H., Staib, P., Kube, D., Diehl, V., and Voliotis, D. 1999. Expression of AC133, a novel hematopoietic precursor antigen, on acute myeloid leukemia cells. *Blood* **93:** 1435–1437.

Howard, B.M. and Boockvar, J.A. 2008. Stem cell marker CD133 expression predicts outcome in glioma patients. *Neurosurgery* **62:** N8.

Ignatova, T.N., Kukekov, V.G., Laywell, E.D., Suslov, O.N., Vrionis, F.D., and Steindler, D.A. 2002. Human cortical glial tumors contain neural stem-like cells expressing astroglial and neuronal markers in vitro. *Glia* **39:** 193–206.

Jain, R.K., di Tomaso, E., Duda, D.G., Loeffler, J.S., Sorensen, A.G., and Batchelor, T.T. 2007. Angiogenesis in brain tumours. *Nat. Rev. Neurosci.* **8:** 610–622.

Jaksch, M., Múnera, J., Bajpai, R., Terskikh, A., and Oshima, R.G. 2008. Cell cycle-dependent variation of a CD133 epitope in human embryonic stem cell, colon cancer, and melanoma cell lines. *Cancer Res.* **68:** 7882–7886.

Kelly, P.N., Dakic, A., Adams, J.M., Nutt, S.L., and Strasser, A. 2007. Tumor growth need not be driven by rare cancer stem cells. *Science* **317:** 337.

Lapidot, T., Sirard, C., Vormoor, J., Murdoch, B., Hoang, T., Caceres-Cortes, J., Minden, M., Paterson, B., Caligiuri, M.A., and Dick, J.E. 1994. A cell initiating human acute myeloid leukaemia after transplantation into SCID mice. *Nature* **367:** 645–648.

Lee, J., Kotliarova, S., Kotliarov, Y., Li, A., Su, Q., Donin, N.M., Pastorino, S., Purow, B.W., Christopher, N., Zhang, W., Park, J.K., and Fine, H.A. 2006. Tumor stem cells derived from glioblastomas cultured in bFGF and EGF more closely mirror the phenotype and genotype of primary tumors than do serum-cultured cell lines. *Cancer Cell* **9:** 391–403.

Lee, J., Son, M.J., Woolard, K., Donin, N.M., Li, A., Cheng, C.H., Kotliarova, S., Kotliarov, Y., Walling, J., Ahn, S., et al. 2008. Epigenetic-mediated dysfunction of the bone morphogenetic protein pathway inhibits differentiation of glioblastoma-initiating cells. *Cancer Cell* **13:** 69–80.

Lee, S.T., Jang, J.H., Min, Y.H., Hahn, J.S., and Ko, Y.W. 2001. AC133 antigen as a prognostic factor in acute leukemia. *Leuk. Res.* **25:** 757–767.

Li, C., Heidt, D.G., Dalerba, P., Burant, C.F., Zhang, L., Adsay, V., Wicha, M., Clarke, M.F., and Simeone, D.M. 2007. Identification of pancreatic cancer stem cells. *Cancer Res.* **67:** 1030–1037.

Ligon, K.L., Huillard, E., Mehta, S., Kesari, S., Liu, H., Alberta, J.A., Bachoo, R.M., Kane, M., Louis, D.N., Depinho, R.A., et al. 2007. Olig2-regulated lineage-restricted pathway controls replication competence in neural stem cells and malignant glioma. *Neuron* **53:** 503–517.

Liu, G., Yuan, X., Zeng, Z., Tunici, P., Ng, H., Abdulkadir, I.R., Lu, L., Irvin, D., Black, K.L., and Yu, J.S. 2006. Analysis of gene expression and chemoresistance of CD133+ cancer stem cells in glioblastoma. *Mol. Cancer* **5:** 67.

Ma, Y.H., Mentlein, R., Knerlich, F., Kruse, M.L., Mehdorn, H.M., and Held-Feindt, J. 2008. Expression of stem cell markers in human astrocytomas of different WHO grades. *J. Neurooncol.* **86:** 31–45.

Miraglia, S., Godfrey, W., Yin, A.H., Atkins, K., Warnke, R., Holden, J.T., Bray, R.A., Waller, E.K., and Buck, D.W. 1997. A novel five-transmembrane hematopoietic stem cell antigen: Isolation, characterization, and molecular cloning. *Blood* **90:** 5013–5021.

Nathan, C., Gold, B., Lin, G., Stegman, M., de Carvalho, L.P., Vandal, O., Venugopal, A., and Bryk, R 2008. A philosophy of anti-infectives as a guide in the search for new drugs for tuberculosis. *Tuberculosis* (suppl. 1) **88:** S25–S33.

O'Brien, C.A., Pollett, A., Gallinger, S., and Dick, J.E. 2007. A human colon cancer cell capable of initiating tumour growth in immunodeficient mice. *Nature* **445:** 106–110.

Panchision, D.M., Chen, H.L., Pistollato, F., Papini, D., Ni, H.T., and Hawley, T.S. 2007. Optimized flow cytometric analysis of central nervous system tissue reveals novel functional relationships among cells expressing CD133, CD15, and CD24. *Stem Cells* **25:** 1560–1570.

Piccirillo, S.G., Reynolds, B.A., Zanetti, N., Lamorte, G., Binda, E., Broggi, G., Brem, H., Olivi, A., Dimeco, F., and Vescovi, A.L. 2006. Bone morphogenetic proteins inhibit the tumorigenic potential of human brain tumour-initiating cells. *Nature* **444:** 761–765.

Pope, W.B., Lai, A., Nghiemphu, P., Mischel, P., and Cloughesy, T.F. 2006. MRI in patients with high-grade gliomas treated with bevacizumab and chemotherapy. *Neurology* **8:** 1258–1260.

Quintana, E., Shackleton, M., Sabel, M.S., Fullen, D.R., Johnson, T.M., and Morrison, S.J. 2008. Efficient tumour formation by single human melanoma cells. *Nature* **456:** 593–598.

Reya, T., Morrison, S.J., Clarke, M.F., and Weissman, I.L. 2001. Stem cells, cancer, and cancer stem cells. *Nature* **414:** 105–111.

Ricci-Vitiani, L., Lombardi, D.G., Pilozzi, E., Biffoni, M., Todaro, M., Peschle, C., and De Maria, R. 2007. Identification and expansion of human colon-cancer initiating cells. *Nature* **445:** 111–115.

Rietze, R.L., Valcanis, H., Brooker, G.F., Thomas, T., Voss, A.K., and Bartlett, P.F. 2001. Purification of a pluripotent neural stem cell from the adult mouse brain. *Nature* **412:** 736–739.

Sanai, N., Tramontin, A.D., Quiñones-Hinojosa, A., Barbaro, N.M., Gupta, N., Kunwar, S., Lawton, M.T., McDermott, M.W., Parsa, A.T., Manuel-García Verdugo, J., Berger, M.S., and Alvarez-Buylla, A. 2004. Unique astrocyte ribbon in adult human brain contains neural stem cells but lacks chain migration. *Nature* **427:** 740–744.

Schüller, U., Heine, V.M., Mao, J., Kho, A.T., Dillon, A.K., Han, Y.G., Huillard, E., Sun, T., Ligon, A.H., Qian, Y., et al. 2008. Acquisition of granule neuron precursor identity is a critical determinant of progenitor cell competence to form Shh-induced medulloblastoma. *Cancer Cell* **2:** 123–134.

Shu, Q., Wong, K.K., Su, J.M., Adesina, A.M., Yu, L.T., Tsang, Y.T., Antalffy, B.C., Baxter, P., Perlaky, L., Yang, J., et al. 2008. Direct orthotopic transplantation of fresh surgical specimen preserves CD133+ tumor cells in clinically relevant mouse models of medulloblastoma and glioma. *Stem Cells* **26:** 1414–1424.

Singec, I., Knoth, R., Meyer, R.P., Maciaczyk, J., Volk, B., Nikkhah, G., Frotscher, M., and Snyder, E.Y. 2006. Defining the actual sensitivity and specificity of the neurosphere assay in stem cell biology. *Nat. Methods* **3:** 801–806.

Singh, S.K., Clarke, I.D., Terasaki, M., Bonn, V.E., Hawkins, C., Squire, J., and Dirks, P.B. 2003. Identification of a cancer stem cell in human brain tumors. *Cancer Res.* **63:** 5821–5828.

Singh, S.K., Hawkins, C., Clarke, I.D., Squire, J.A., Bayani, J., Hide, T., Henkelman, R.M., Cusimano, M.D., and Dirks, P.B. 2004. Identification of human brain tumour initiating cells. *Nature* **432:** 396–401.

Soeda, A., Inagaki, A., Oka, N., Ikegame, Y., Aoki, H., Yoshimura, S., Nakashima, S., Kunisada, T., and Iwama, T. 2008. Epidermal growth factor plays a crucial role in mitogenic regulation of human brain tumor stem cells. *J. Biol. Chem.* **283:** 10958–10966.

Stupp, R., Mason, W.P., van den Bent, M.J., Weller, M., Fisher, B., Taphoorn, M.J., Belanger, K., Brandes, A.A., Marosi, C., Bogdahn, U., et al. 2005. Radiotherapy plus concomitant and adjuvant temozolomide for glioblastoma. *N. Engl. J. Med.* **352:** 987–996.

Tabu, K., Sasai, K., Kimura, T., Wang, L., Aoyanagi, E., Kohsaka, S., Tanino, M., Nishihara, H., and Tanaka, S. 2008. Promoter hypomethylation regulates CD133 expression in human gliomas. *Cell Res.* **18:** 1037–1046.

Taylor, M.D., Poppleton, H., Fuller, C., Su, X., Liu, Y., Jensen, P., Magdaleno, S., Dalton, J., Calabrese, C., Board, J., et al. 2005. Radial glia cells are candidate stem cells of ependymoma. *Cancer Cell* **8:** 323–335.

Thon, N., Damianoff, K., Hegermann, J., Grau, S., Krebs, B., Schnell, O., Tonn, J.C., and Goldbrunner, R. 2008. Presence of pluripotent CD133+ cells correlates with malignancy of gliomas. *Mol. Cell. Neurosci.* (in press).

Uchida, N., Buck, D.W., He, D., Reitsma, M.J., Masek, M., Phan, T.V., Tsukamoto, A.S., Gage, F.H., and Weissman, I.L. 2000. Direct isolation of human central nervous system stem cells. *Proc. Natl. Acad. Sci.* **97:** 14720–14725.

Vredenburgh, J.J., Desjardins, A., Herndon II, J.E., Dowell, J.M., Reardon, D.A., Quinn, J.A., Rich, J.N., Sathornsumetee, S., Gururangan, S., Wagner, M., et al. 2007a. Phase II trial of bevacizumab and irinotecan in recurrent malignant glioma. *Clin. Cancer Res.* **13:** 1253–1259.

Vredenburgh, J.J., Desjardins, A., Herndon II, J.E., Marcello, J., Reardon, D.A., Quinn, J.A., Rich, J.N., Sathornsumetee, S.,

Gururangan, S., Sampson, J., et al. 2007b. Bevacizumab plus irinotecan in recurrent glioblastoma multiforme. *J. Clin. Oncol.* **25:** 4722–4729.

Wang, J., Wang, H., Li, Z., Wu, Q., Lathia, J.D., McLendon, R.E., Hjelmeland, A.B., and Rich, J.N. 2008. c-Myc is required for maintenance of glioma cancer stem cells. *PLoS One* **11:** e3769.

Weigmann, A., Corbeil, D., Hellwig, A., and Huttner, W.B. 1997. Prominin, a novel microvilli-specific polytopic membrane protein of the apical surface of epithelial cells, is targeted to plasmalemmal protrusions of non-epithelial cells. *Proc. Natl. Acad. Sci.* **94:** 12425–12430.

Wu, M., Kwon, H.Y., Rattis, F., Blum, J., Zhao, C., Ashkenazi, R., Jackson, T.L., Gaiano, N., Oliver, T., and Reya, T. 2007. Imaging hematopoietic precursor division in real time. *Cell Stem Cell* **5:** 541–554.

Yang, Z., Chen, Y., Lillo, C., Chien, J., Yu, Z., Michaelides, M., Klein, M., Howes, K.A., Li, Y., Kaminoh, Y., et al. 2008a. Mutant prominin 1 found in patients with macular degeneration disrupts photoreceptor disk morphogenesis in mice. *J. Clin. Invest.* **118:** 2908–2916.

Yang, Z.J., Ellis, T., Markant, S.L., Read, T.A., Kessler, J.D., Bourboulas, M., Schüller, U., Machold, R., Fishell, G., Rowitch, D.H., Wainwright, B.J., and Wechsler-Reya, R.J.

2008b. Medulloblastoma can be initiated by deletion of *Patched* in lineage-restricted progenitors or stem cells. *Cancer Cell* **2:** 135–145.

Yi, J.M., Tsai, H.C., Glöckner, S.C., Lin, S., Ohm, J.E., Easwaran, H., James, C.D., Costello, J.F., Riggins, G., Eberhart, C.G., et al. 2008. Abnormal DNA methylation of CD133 in colorectal and glioblastoma tumors. *Cancer Res.* **68:** 8094–8103.

Yin, A.H., Miraglia, S., Zanjani, E.D., Almeida-Porada, G., Ogawa, M., Leary, A.G., Olweus, J., Kearney, J., and Buck, D.W. 1997. AC133, a novel marker for human hematopoietic stem and progenitor cells. *Blood* **90:** 5002–5012.

Yuan, X., Curtin, J., Xiong, Y., Liu, G., Waschsmann-Hogiu, S., Farkas, D.L., Black, K.L., and Yu, J.S. 2004. Isolation of cancer stem cells from adult glioblastoma multiforme. *Oncogene* **23:** 9392–9400.

Zeppernick, F., Ahmadi, R., Campos, B., Dictus, C., Helmke, B.M., Becker, N., Lichter, P., Unterberg, A., Radlwimmer, B., and Herold-Mende, C.C. 2008. Stem cell marker CD133 affects clinical outcome in glioma patients. *Clin. Cancer Res.* **14:** 123–129.

Zheng, H., Ying, H., Yan, H., Kimmelman, A.C., Hiller, D.J., Chen, A.J., Perry, S.R., Tonon, G., Chu, G.C., Ding, Z., et al. 2008. p53 and Pten control neural and glioma stem/progenitor cell renewal and differentiation. *Nature* **7216:** 1129–1133.

Neural and Cancer Stem Cells in Tumor Suppressor Mouse Models of Malignant Astrocytoma

S. Alcantara Llaguno, J. Chen, C.-H. Kwon, and L.F. Parada

Department of Developmental Biology, The University of Texas Southwestern Medical Center, Dallas, Texas 75390

Malignant astrocytomas are highly invasive brain tumors that portend poor prognosis and dismal survival. Mouse models that genetically resemble the human malignancy provide insight into the nature and pathogenesis of these cancers. We previously reported tumor suppressor mouse models based on conditional inactivation of human astrocytoma-relevant genes *p53*, *Nf1*, and *Pten*. These mice develop, with full penetrance, varying grades of astrocytic malignancy that recapitulate the human condition histologically and molecularly. Our studies indicate a central role for neural stem cells and stem-cell-like cancer cells in tumor initiation and progression. These mouse models thus represent powerful tools for investigating various aspects of tumor development that otherwise cannot be explored in humans. Further studies will provide a better understanding of the biology of these tumors and will hopefully pave the way for more effective therapeutic approaches for these devastating diseases.

Gliomas are the most common primary intracranial neoplasms in both pediatric and adult populations. The most malignant form, glioblastoma multiforme (GBM), is the most prevalent and also the most lethal, with a median survival of less than 1 year (Maher et al. 2001; Zhu and Parada 2002). Despite numerous advances in the field, however, these tumors remain resistant to conventional therapies and portend dismal outcomes for patients (Stupp et al. 2005).

Gliomas are classified on the basis of predominant tumor cell type(s), and astrocytomas, which morphologically resemble astrocytes, comprise the majority of these tumors. These tumors exhibit cellular heterogeneity and extensive infiltration into adjacent normal structures, which precludes complete surgical resection. Radiotherapy and chemotherapy improve survival but are not curative (Maher et al. 2001; Zhu and Parada 2002).

Astrocytic tumors are further classified on the basis of histopathologic and clinical criteria into increasing degrees of malignancy: grades I through IV. Grade I tumors are benign, whereas grade II tumors are low-grade malignancies that undergo early diffuse infiltration, rendering them surgically incurable. Grade III (anaplastic) and grade IV (GBM) astrocytomas are highly malignant and invasive tumors that are lethal within years to months. GBMs can be further subtyped as secondary GBMs, with previous clinical history of a lower-grade lesion, whereas primary GBMs arise de novo. Unlike other malignancies, high-grade astrocytomas rarely metastasize outside the central nervous system (CNS); hence, tumor grade serves as the primary determinant of clinical outcome (Maher et al. 2001; Furnari et al. 2007).

At the molecular level, a variety of mutations have been described in human astrocytomas. The classical genetic alterations target pathways involved in cell cycle and apoptosis regulation, such as *P53*, *INK4A*, *CDK4, and RB,* as well as growth factor signaling through *EGF, PDGF*, *PTEN*, and *NF1*. Frequent mutations in these genes underscore the importance of mitogenic signaling through receptor tyrosine kinases coupled with inactivation of critical negative regulators of cell proliferation and senescence in the pathogenesis of these tumors (Zhu and Parada 2002; Louis 2006; Furnari et al. 2007).

NEURAL STEM CELLS IN DEVELOPMENT AND CANCER

Neural stem cells are self-renewing cells in the CNS that exhibit multipotent differentiation into all neural cell types in the brain, including neurons, astrocytes, and oligodendrocytes (Gage 2000). In the adult mammalian brain, the two major neural stem cell niches are the subventricular zone (SVZ) of the lateral ventricle and the subgranular layer of the dentate gyrus. The type-B neural stem cells in the SVZ give rise to type-C transient amplifying cells which then give rise to neuroblasts that migrate along a defined pathway, called the rostral migratory stream (RMS), and into the olfactory bulb (OB), where they differentiate into mature neurons (Doetsch et al. 1999; Alvarez-Buylla and Lim 2004). In the dentate gyrus, neurogenesis also occurs in the subgranular zone (SGZ), which produces local neurons that incorporate into the granule cell layer (Zhao et al. 2008).

Some cancer cells share important characteristics exhibited by stem cells, including unlimited replicative potential, diversity of progeny, telomere maintenance, and migratory properties (Sanai et al. 2005). In contrast to cancer cells, however, the function of stem cells at different stages of development is tightly regulated by diverse signaling pathways that impinge on various processes, including self-renewal, differentiation, and survival (Reya et al. 2001; Vescovi et al. 2006; Dalerba et al. 2007). Hence, it has been hypothesized that this population of undifferentiated cells that persist throughout the lifetime of an individual may have important roles in the natural progression of cancer.

Historically, the differentiated astrocyte has been thought to be the cell of origin of astrocytomas (Sanson et al. 2004; Sanai et al. 2005). On the other hand, numerous studies have suggested that these tumors may arise from the transformation of neural precursor cells (Holland et al. 2000) or dedifferentiation of mature astrocytes (Bachoo et al. 2002; Uhrbom et al. 2002). None of these studies have been conclusive so far. Given the recent in vivo identification of these progenitor populations in the adult brain, a previously thought of postmitotic organ, the proposal that neural stem cells can give rise to these tumors is an attractive hypothesis, but one that has yet to be experimentally verified.

Cancer cells with stem cell properties have been isolated from human cancers. These "cancer stem cells" have been operationally defined as a subpopulation of cells within tumors that maintain the self-renewing or propagating properties that confer the ability to initiate tumor formation in immunodeficient mice (Dalerba et al. 2007). These cells are thought to be responsible for the aggressive behavior, invasiveness, metastatic potential, and even resistance to conventional chemotherapy and radiotherapy of many tumors (Reya et al. 2001; Wang and Dick 2005; Dalerba et al. 2007). In human astrocytomas, the presence of stem-like cancer cells has been reported, and it has been suggested that the CD133[+] fraction of GBMs comprise the population of self-renewing stem-like cancer cells with enriched tumorigenic capacity (Singh et al. 2004). These cancer stem cells have also been shown to be sensitive to bone morphogenetic protein signaling inhibition while being resistant to radiation therapy (Bao et al. 2006; Piccirillo et al. 2006; Lee et al. 2008). Many questions remain, however, as the molecular mechanisms that regulate cancer and normal neural stem cell behavior are still being unraveled, and efforts are under way to exploit these cells as possible therapeutic targets.

TUMOR SUPPRESSORS IN NEURAL STEM CELL AND CANCER DEVELOPMENT

Various tumor suppressors have been implicated in cancer, and *Nf1, p53,* and *Pten* represent tumor suppressor pathways that are frequently involved in human malignant astrocytomas. In fact, these three tumor suppressors are among the most frequently mutated genes in sporadic human GBM (TCGA GBM Disease Working Group; http://cancergenome.nih.gov/dataportal/). *Nf1* encodes neurofibromin, a GTPase-activating protein that negatively regulates Ras, a downstream effector of receptor tyrosine kinase signaling (Le and Parada 2007). By virtue of epidermal growth factor receptor (EGFR) amplification and platelet-derived growth factor (PDGF) receptor overexpression, Ras signaling is hyperactivated (although *Ras* mutations are infrequent) in sporadic GBMs (Zhu and Parada 2002). On the other hand, *p53,* involved in apoptosis, cell cycle arrest, and DNA-damage repair, is frequently mutated early on in low- and high-grade astrocytomas, whereas mutations in *Pten*, which negatively regulates the phosphoinositide-3′-kinase (PI3K)-Akt signaling involved in cell proliferation, survival, and migration, are frequently found in high-grade astrocytomas (Furnari et al. 2007). Consistent with the above findings, individuals with germ-line mutations in

p53 (Li Fraumeni), *Nf1* (neurofibromatosis type 1), and *Pten* (Cowden disease) have increased incidence of developing astrocytomas compared to the general population (Rasmussen et al. 2001; Gutmann et al. 2002; Ichimura et al. 2004). These data underscore a central role for these tumor suppressors in the development of malignant astrocytomas.

Interestingly, *Nf1, p53,* and *Pten*, like other tumor suppressors, have lately been shown to function as negative regulators of neural stem cell function. *Nf1* deficiency was shown to promote neural stem cell proliferation and survival (Dasgupta and Gutmann 2005; Hegedus et al. 2007). Loss of *p53* increases proliferation in the SVZ neural stem/progenitors and provides a growth advantage compared to wild-type cells (Gil-Perotin et al. 2006; Meletis et al. 2006), and *Pten* loss increases neural stem cell proliferation and self-renewal (Groszer et al. 2001, 2006).

MOUSE MODELS OF MALIGNANT ASTROCYTOMA

Signature genetic lesions found in human tumors have been exploited in the mouse to generate genetically engineered animal models that have greatly enhanced our understanding of astrocytoma development. Strategies have included gain-of-function approaches, such as overexpression of active forms of Ras, Akt, EGFR, PDGF, and transforming antigens v-*src* and polyomavirus middle T antigen, often in combination with targeted deletions of Ink4A/Arf or Pten (Weissenberger et al. 1997; Holland et al. 2000; Ding et al. 2001; Bachoo et al. 2002; Uhrbom et al. 2002; Xiao et al. 2002; Fomchenko and Holland 2006; Furnari et al. 2007). These mutations were induced in the germ line or in specific cell populations, and tumor development was observed with variable penetrance. The first endogenous genetic tumor suppressor mouse model was based on heterozygous mice carrying *cis* germ-line mutations in *Nf1* and *p53*. These mice developed high-grade astrocytomas with varying penetrance depending on genetic background (Reilly et al. 2000).

TUMOR SUPPRESSOR MOUSE MODELS: DISSECTING THE ROLE OF NEURAL AND CANCER STEM CELLS

To better understand the biology of malignant astrocytomas, our lab previously developed mouse models based on tumor suppressor inactivation in specific cell types in the brain. We used Cre-loxP technology that permits more selective spatial and temporal ablation of tumor suppressors. We took advantage of a mouse transgenic expressing Cre recombinase under the control of the *hGFAP* promoter (*hGFAP-Cre*) (Zhuo et al. 2001) and combinations of tumor suppressor conditional alleles or germ-line mutations (Jacks et al. 1994; Groszer et al. 2001; Zhu et al. 2001; Lin et al. 2004). We first generated conditional mutant mice wherein *cis* heterozygous germ-line or somatic *p53* heterozygosity was combined with somatic *Nf1* heterozygosity driven by a Cre recombinase that is active in both neural stem cells and differentiated astrocytes (Zhu et al. 2005). These mice developed astrocytomas with 100% penetrance and were indistinguishable from the human malignancy based on known histologic and molecular criteria. This provided evidence that *Nf1* and *p53* loss of function is *sufficient* to initiate malignant astrocytoma formation. Variations in the genetic configurations of the tumor suppressor alleles also showed that *p53* inactivation concomitant or before *Nf1* inactivation is critical for tumor development. These are shown in Table 1 as *Mut1* or *Mut3* conditional mutant mice, which developed, with 100% penetrance, a spectrum of low- to high-grade malignant astrocytomas, whereas *Mut2* mutants very rarely developed tumors.

In an additional refinement, when we added somatic *Pten* heterozygosity to the *Nf1-p53* mouse models, mutant mice (*Mut4* and *Mut6*) were found to develop high-grade astrocytomas (Table 1) with decreased latency of tumor formation (Kwon et al. 2008). We also found that *Nf1* and *Pten* somatic heterozygosity alone was not sufficient for astrocytoma development (*Mut0*). This crucial role of *p53* in tumor initiation is consistent with the frequency of *p53* mutations observed in low-grade astrocytomas. These studies underscore the importance of the *Nf1, p53,* and *Pten* tumor suppressors in malignant astrocytoma formation and progression.

NEURAL STEM/PROGENITORS AS CANCER-INITIATING CELLS

Studies from these tumor-bearing mice show considerable evidence for the role of neural stem cells in the development of malignant astrocytomas. Analysis of conditional mutant mice at different stages of tumor development showed the earliest lesions in the neurogenic niche of the adult SVZ (Zhu et al. 2005). Presymptomatic mutant mice at young ages are histologically similar to littermate controls (Fig. 1). However, as these mutant mice aged further, we observed areas of hyperplasia as shown by hematoxylin and eosin (H&E) staining and confirmed by immunohistochemistry. A short-term pulse with bromodeoxyuridine (BrdU) showed an increase in proliferating cells in the SVZ, as compared to controls (Fig. 1). Longer-term pulses with BrdU in presymptomatic mutant mice uncovered migration defects, such that more BrdU$^+$ cells could be seen outside the SVZ (Kwon et al. 2008). Older mice subsequently developed tumors, the majority in areas near the SVZ, starting at 4 months of age (Zhu et al. 2005). Moreover,

we found growth changes using neurosphere cultures of SVZ cells isolated from presymptomatic young mice (data not shown). These data demonstrate that neural stem/progenitors show proliferation and migration defects before tumor formation, suggesting that these cells may be the cancer-initiating cells in our tumor suppressor mouse models.

The hypothesis that neural stem cells may be the cell of origin of astrocytomas is illustrated in Figure 2. We propose that tumor-initiating mutations target the reservoir of self-renewing stem cells, allowing for the accumulation of more mutations required for malignant transformation. The tumorigenic phenotype may manifest itself in dividing progenitor-like cells, which undergo deregulated cell divisions to create the tumor bulk. Other mechanistic scenarios may also be possible, given a different set of initiating mutations or environmental influences.

Direct examination of the role of neural stem/progenitors as the origin of these tumors will require directly targeting these cells for tumor suppressor inactivation. Whether more mature, differentiated CNS cells have the capacity to give rise to these tumors will also need to be addressed. This will require the use of more cell-type-specific promoters to drive Cre-mediated recombination. Cell ablation or depletion experiments will also determine whether these cells are required for tumor development.

Identification of the cell of origin will be crucial for investigating the various mechanisms of tumor initiation and progression. Susceptible cells for transformation can be targeted with genetic lesions found in human patients to discover other tumor-initiating mutations and determine cooperation among these genes in astrocytoma development. Identifying these cells will also allow mechanistic studies on signal transduction pathways involved early on during tumor development.

CANCER STEM CELLS IN MOUSE MODELS OF MALIGNANT ASTROCYTOMA

Malignant astrocytomas are very heterogeneous tumors. Although malignancies are predominantly astrocytic in character, a variety of tumor cell types can be seen in the tumor bulk, as shown by immunostaining (Fig. 3). Some tumor cells were found to express primitive neural stem/progenitor markers such as nestin. Other tumor cells were immunoreactive for markers of more differentiated

Table 1. Genetic configurations and phenotypes of *Nf1-p53-Pten* tumor suppressor mouse models

Mutant	Cre	Nf1	p53	Pten[a]	Tumor grade[a]
Mut0	hGfap-Cre	Nf1$^{f/+}$	p53$^{+/+}$	Pten$^{f/+}$	no tumors
Mut1	hGfap-Cre	Nf1$^{f/f}$	p53$^{-/-}$	Pten$^{+/+}$	low- to high-grade astrocytomas
Mut2	hGfap-Cre	Nf1$^{f/f}$	p53$^{-/+}$	Pten$^{+/+}$	very rare high-grade astrocytomas
Mut3	hGfap-Cre	Nf1$^{f/+}$	p53$^{-/+}$	Pten$^{+/+}$	low- to high-grade astrocytomas
Mut4	hGfap-Cre	Nf1$^{f/+}$	p53$^{-/+}$	Pten$^{f/+}$	high-grade astrocytomas
Mut5	hGfap-Cre	Nf1$^{f/+}$	p53$^{-/f}$	Pten$^{+/+}$	low- to high-grade astrocytomas
Mut6	hGfap-Cre	Nf1$^{f/+}$	p53$^{-/f}$	Pten$^{f/+}$	high-grade astrocytomas

Data from Zhu et al. (2005) and Kwon et al. (2008).
f indicates *flox/loxP*; + indicates wild type.
[a]Low grade = grade II; high grade = grade III or grade IV astrocytomas.

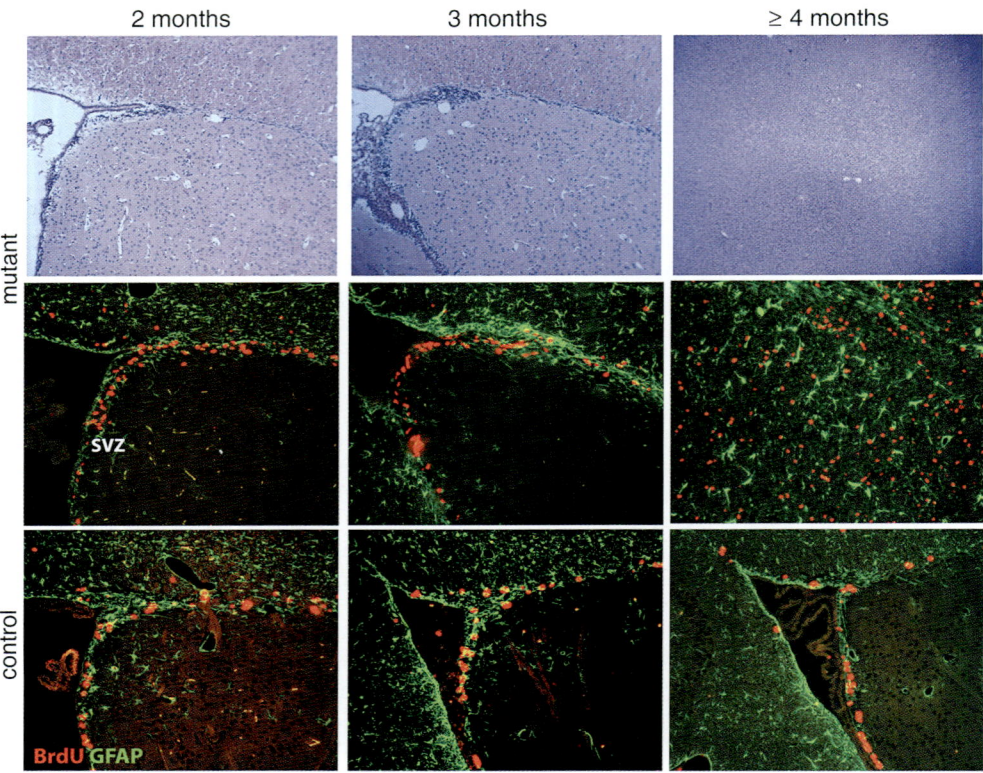

Figure 1. Tumor suppressor mouse models show early hyperplastic lesions in the SVZ. Time course analysis of Mut3 (*hGFAP-Cre;*
Nf1flox/+; p53−/+) conditional mutant mice at 2, 3, and 4 months of age shows progressive changes in the SVZ. H&E staining shows
increased cellularity in the SVZ by as early as 3 months of age in mutant mice. Short-term (1-hour) BrdU pulsing in these mice indicates an increase in proliferating BrdU+, as well as in GFAP+ cells in the SVZ before full-blown tumor formation. Controls show a
decrease in BrdU+ proliferating cells with age. Bar, 200 μm.

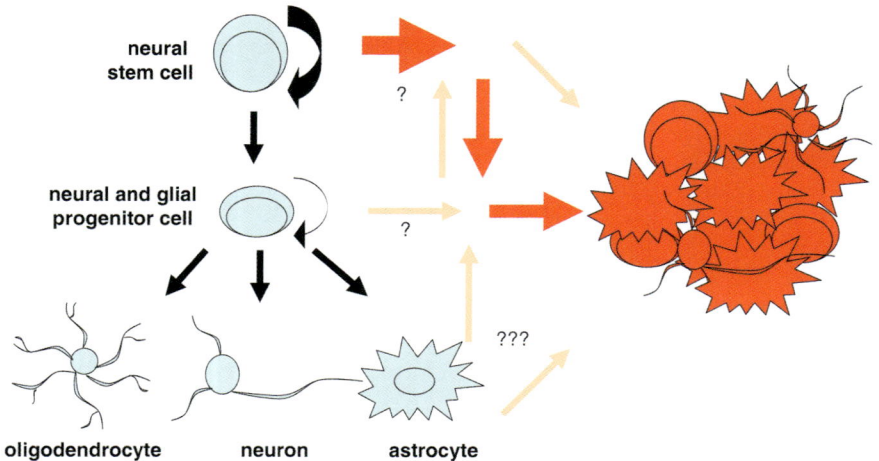

Figure 2. Cell of origin hypothesis of malignant astrocytomas. A neural stem cell origin of malignant astrocytomas proposes the lifelong self-renewing stem cells, as opposed to progenitor cells with limited self-renewal or differentiated astrocytes, as the path of least
resistance to tumorigenesis. These neural stem cells may transform into dividing progenitor-like cells that undergo uncontrollable
mitoses and give rise to tumors (*red arrows*). Alternative pathways, such as dedifferentiation of the more mature cell types into stem-
or progenitor-like cells (*light orange arrows*), are also shown.

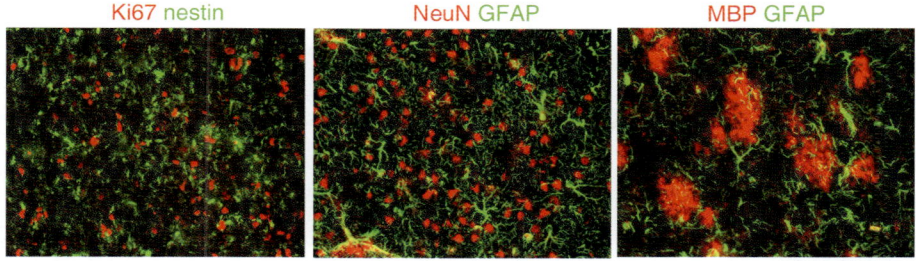

Figure 3. Tumor suppressor mouse models show the presence of cancer cells expressing markers of immature and multilineage differentiation. Full blown mouse tumors show heterogeneity in tumor cell composition, expressing markers of undifferentiated cells, such as nestin, and more mature astrocytic (Gfap), neuronal (NeuN), and oligodendrocytic (MBP) differentiation. Bar, 200 μm.

CNS cell types, including the astrocytic marker glial fibrillary acidic protein (GFAP), neuronal marker NeuN, and oligodendrocytic marker myelin basic protein (MBP), suggesting the capacity for multilineage differentiation (Fig. 3). These tumors can also give rise to self-renewing neurosphere-forming cells when cultured in serum-free media, similar to normal neural stem cell cultures (data not shown). These data suggest that a subpopulation of tumor cells may exhibit more stem-cell-like properties compared to the rest of the tumor bulk.

The crucial experiment will be to determine whether this rare subpopulation of cells has increased tumorigenic potential compared to other tumor cell types in serial transplantation assays. Compared to human "cancer stem cells," the use of *mouse* tumor cells orthotopically transplanted into immunodeficient *mice* may provide a better platform to determine the frequency of these "tumor-initiating" cells because of less interference of complex human cell–mouse microenvironment issues.

Identification of this rare subpopulation of cells will allow us to compare the molecular mechanisms that are operative in "cancer stem cells" vis à vis other tumor cells *and* normal neural stem cells. Exploiting their differences will be vital in designing novel therapeutic agents against these very aggressive cancers.

PERSPECTIVE

The use of genetic mouse models in studying human disease has greatly facilitated our understanding of cancer. Using tumor suppressor mouse models, we have validated mutations in human malignant astrocytomas as bona fide tumor-initiating mutations that lead to full-blown tumor development. We have also uncovered cooperativity between these tumor-initiating genes. Detailed studies of these tumor-bearing mice have also led to a greater understanding of the role of neural and stem-like cancer cells in tumor initiation and progression.

Further advances in the fields of neurodevelopment and cancer biology will provide more sophisticated tools for studying these malignancies. The interface between human and mouse cancers will need to be more fully exploited in order to rapidly translate basic science knowledge into practical clinical applications. The hope is that a more integrated understanding of the molecular and phys-

iological processes underlying this complex disease will provide us with the means of finally cracking open some of the many secrets of brain tumor formation, ultimately resulting in more targeted and effective therapies.

ACKNOWLEDGMENTS

The authors thank Linda McClellan, Shawna Kennedy, Steven McKinnon, Patsy Leake, and Alicia Deshaw for technical assistance, and Parada Lab members, especially Renee McKay, for helpful suggestions and discussion. This work is supported in part by the Children's Tumor Foundation Young Investigator Award to S.A.L., Basic Research Fellowships from American Brain Tumor Association (in memory of Daniel J. Martinelli and Geoffrey J. Cunningham) to C.-H.K., and by National Institutes of Health grant P50NS05260602 and American Cancer Society grant RP0408401 to L.F.P. L.F.P. is an American Cancer Society Research Professor.

REFERENCES

Alvarez-Buylla, A. and Lim, D.A. 2004. For the long run: Maintaining germinal niches in the adult brain. *Neuron* **41:** 683–686.

Bachoo, R.M., Maher, E.A., Ligon, K.L., Sharpless, N.E., Chan, S.S., You, M.J., Tang, Y., DeFrances, J., Stover, E., Weissleder, R., Rowitch, D.H., Louis, D.N., and DePinho, R.A. 2002. Epidermal growth factor receptor and *Ink4a/Arf:* Convergent mechanisms governing terminal differentiation and transformation along the neural stem cell to astrocyte axis. *Cancer Cell* **1:** 269–277.

Bao, S., Wu, Q., McLendon, R.E., Hao, Y., Shi, Q., Hjelmeland, A.B., Dewhirst, M.W., Bigner, D.D., and Rich, J.N. 2006. Glioma stem cells promote radioresistance by preferential activation of the DNA damage response. *Nature* **444:** 756–760.

Dalerba, P., Cho, R.W., and Clarke, M.F. 2007. Cancer stem cells: Models and concepts. *Annu. Rev. Med.* **58:** 267–284.

Dasgupta, B. and Gutmann, D.H. 2005. Neurofibromin regulates neural stem cell proliferation, survival, and astroglial differentiation in vitro and in vivo. *J. Neurosci.* **25:** 5584–5594.

Ding, H., Roncari, L., Shannon, P., Wu, X., Lau, N., Karaskova, J., Gutmann, D.H., Squire, J.A., Nagy, A., and Guha, A. 2001. Astrocyte-specific expression of activated p21-ras results in malignant astrocytoma formation in a transgenic mouse model of human gliomas. *Cancer Res.* **61:** 3826–3836.

Doetsch, F., Caille, I., Lim, D.A., Garcia-Verdugo, J.M., and Alvarez-Buylla, A. 1999. Subventricular zone astrocytes are neural stem cells in the adult mammalian brain. *Cell* **97:** 703–716.

Fomchenko, E.I. and Holland, E.C. 2006. Mouse models of brain tumors and their applications in preclinical trials. *Clin. Cancer Res.* **12:** 5288–5297.

Furnari, F.B., Fenton, T., Bachoo, R.M., Mukasa, A., Stommel, J.M., Stegh, A., Hahn, W.C., Ligon, K.L., Louis, D.N., Brennan, C., Chin, L., DePinho, R.A., and Cavenee, W.K. 2007. Malignant astrocytic glioma: Genetics, biology, and paths to treatment. *Genes Dev.* **21:** 2683–2710.

Gage, F.H. 2000. Mammalian neural stem cells. *Science* **287:** 1433–1438.

Gil-Perotin, S., Marin-Husstege, M., Li, J., Soriano-Navarro, M., Zindy, F., Roussel, M.F., Garcia-Verdugo, J.M., and Casaccia-Bonnefil, P. 2006. Loss of p53 induces changes in the behavior of subventricular zone cells: Implication for the genesis of glial tumors. *J. Neurosci.* **26:** 1107–1116.

Groszer, M., Erickson, R., Scripture-Adams, D.D., Lesche, R., Trumpp, A., Zack, J.A., Kornblum, H.I., Liu, X., and Wu, H. 2001. Negative regulation of neural stem/progenitor cell proliferation by the Pten tumor suppressor gene in vivo. *Science* **294:** 2186–2189.

Groszer, M., Erickson, R., Scripture-Adams, D.D., Dougherty, J.D., Le Belle, J., Zack, J.A., Geschwind, D.H., Liu, X., Kornblum, H.I., and Wu, H. 2006. PTEN negatively regulates neural stem cell self-renewal by modulating G0-G1 cell cycle entry. *Proc. Natl. Acad. Sci.* **103:** 111–116.

Gutmann, D.H., Rasmussen, S.A., Wolkenstein, P., MacCollin, M.M., Guha, A., Inskip, P.D., North, K.N., Poyhonen, M., Birch, P.H., and Friedman, J.M. 2002. Gliomas presenting after age 10 in individuals with neurofibromatosis type 1 (NF1). *Neurology* **59:** 759–761.

Hegedus, B., Dasgupta, B., Shin, J.E., Emnett, R.J., Hart-Mahon, E.K., Elghazi, L., Bernal-Mizrachi, E., and Gutmann, D.H. 2007. Neurofibromatosis-1 regulates neuronal and glial cell differentiation from neuroglial progenitors in vivo by both cAMP- and Ras-dependent mechanisms. *Cell Stem Cell* **1:** 443–457.

Holland, E.C., Celestino, J., Dai, C., Schaefer, L., Sawaya, R.E., and Fuller, G.N. 2000. Combined activation of Ras and Akt in neural progenitors induces glioblastoma formation in mice. *Nat. Genet.* **25:** 55–57.

Ichimura, K., Ohgaki, H., Kleihues, P., and Collins, V.P. 2004. Molecular pathogenesis of astrocytic tumours. *J. Neurooncol.* **70:** 137–160.

Jacks, T., Remington, L., Williams, B.O., Schmitt, E.M., Halachmi, S., Bronson, R.T., and Weinberg, R.A. 1994. Tumor spectrum analysis in p53-mutant mice. *Curr. Biol.* **4:** 1–7.

Kwon, C.H., Zhao, D., Chen, J., Alcantara, S., Li, Y., Burns, D.K., Mason, R.P., Lee, E.Y., Wu, H., and Parada, L.F. 2008. Pten haploinsufficiency accelerates formation of high-grade astrocytomas. *Cancer Res.* **68:** 3286–3294.

Le, L.Q. and Parada, L.F. 2007. Tumor microenvironment and neurofibromatosis type I: Connecting the GAPs. *Oncogene* **26:** 4609–4616.

Lee, J., Son, M.J., Woolard, K., Donin, N.M., Li, A., Cheng, C.H., Kotliarova, S., Kotliarov, Y., Walling, J., Ahn, S., et al. 2008. Epigenetic-mediated dysfunction of the bone morphogenetic protein pathway inhibits differentiation of glioblastoma-initiating cells. *Cancer Cell* **13:** 69–80.

Lin, S.C., Lee, K.F., Nikitin, A.Y., Hilsenbeck, S.G., Cardiff, R.D., Li, A., Kang, K.W., Frank, S.A., Lee, W.H., and Lee, E.Y. 2004. Somatic mutation of p53 leads to estrogen receptor α-positive and -negative mouse mammary tumors with high frequency of metastasis. *Cancer Res.* **64:** 3525–3532.

Louis, D.N. 2006. Molecular pathology of malignant gliomas. *Annu. Rev. Pathol.* **1:** 97–117.

Maher, E.A., Furnari, F.B., Bachoo, R.M., Rowitch, D.H., Louis, D.N., Cavenee, W.K., and DePinho, R.A. 2001.

Malignant glioma: Genetics and biology of a grave matter. *Genes Dev.* **15:** 1311–1333.

Meletis, K., Wirta, V., Hede, S.M., Nister, M., Lundeberg, J., and Frisen, J. 2006. p53 suppresses the self-renewal of adult neural stem cells. *Development* **133:** 363–369.

Piccirillo, S.G., Reynolds, B.A., Zanetti, N., Lamorte, G., Binda, E., Broggi, G., Brem, H., Olivi, A., Dimeco, F., and Vescovi, A.L. 2006. Bone morphogenetic proteins inhibit the tumorigenic potential of human brain tumour-initiating cells. *Nature* **444:** 761–765.

Rasmussen, S.A., Yang, Q., and Friedman, J.M. 2001. Mortality in neurofibromatosis 1: An analysis using U.S. death certificates. *Am. J. Hum. Genet.* **68:** 1110–1118.

Reilly, K.M., Loisel, D.A., Bronson, R.T., McLaughlin, M.E., and Jacks, T. 2000. *Nf1;Trp53* mutant mice develop glioblastoma with evidence of strain-specific effects. *Nat. Genet.* **26:** 109–113.

Reya, T., Morrison, S.J., Clarke, M.F., and Weissman, I.L. 2001. Stem cells, cancer, and cancer stem cells. *Nature* **414:** 105–111.

Sanai, N., Alvarez-Buylla, A., and Berger, M.S. 2005. Neural stem cells and the origin of gliomas. *N. Engl. J. Med.* **353:** 811–822.

Sanson, M., Thillet, J., and Hoang-Xuan, K. 2004. Molecular changes in gliomas. *Curr. Opin. Oncol.* **16:** 607–613.

Singh, S.K., Hawkins, C., Clarke, I.D., Squire, J.A., Bayani, J., Hide, T., Henkelman, R.M., Cusimano, M.D., and Dirks, P.B. 2004. Identification of human brain tumour initiating cells. *Nature* **432:** 396–401.

Stupp, R., Mason, W.P., van den Bent, M.J., Weller, M., Fisher, B., Taphoorn, M.J., Belanger, K., Brandes, A.A., Marosi, C., Bogdahn, U., et al. 2005. Radiotherapy plus concomitant and adjuvant temozolomide for glioblastoma. *N. Engl. J. Med.* **352:** 987–996.

Uhrbom, L., Dai, C., Celestino, J.C., Rosenblum, M.K., Fuller, G.N., and Holland, E.C. 2002. *Ink4a-Arf* loss cooperates with KRas activation in astrocytes and neural progenitors to generate glioblastomas of various morphologies depending on activated Akt. *Cancer Res.* **62:** 5551–5558.

Vescovi, A.L., Galli, R., and Reynolds, B.A. 2006. Brain tumour stem cells. *Nat. Rev. Cancer* **6:** 425–436.

Wang, J.C. and Dick, J.E. 2005. Cancer stem cells: Lessons from leukemia. *Trends Cell Biol.* **15:** 494–501.

Weissenberger, J., Steinbach, J.P., Malin, G., Spada, S., Rulicke, T., and Aguzzi, A. 1997. Development and malignant progression of astrocytomas in GFAP-v-*src* transgenic mice. *Oncogene* **14:** 2005–2013.

Xiao, A., Wu, H., Pandolfi, P.P., Louis, D.N., and Van Dyke, T. 2002. Astrocyte inactivation of the pRb pathway predisposes mice to malignant astrocytoma development that is accelerated by PTEN mutation. *Cancer Cell* **1:** 157–168.

Zhao, C., Deng, W., and Gage, F.H. 2008. Mechanisms and functional implications of adult neurogenesis. *Cell* **132:** 645–660.

Zhu, Y. and Parada, L.F. 2002. The molecular and genetic basis of neurological tumours. *Nat. Rev. Cancer* **2:** 616–626.

Zhu, Y., Romero, M.I., Ghosh, P., Ye, Z., Charnay, P., Rushing, E.J., Marth, J.D., and Parada, L.F. 2001. Ablation of NF1 function in neurons induces abnormal development of cerebral cortex and reactive gliosis in the brain. *Genes Dev.* **15:** 859–876.

Zhu, Y., Guignard, F., Zhao, D., Liu, L., Burns, D.K., Mason, R.P., Messing, A., and Parada, L.F. 2005. Early inactivation of p53 tumor suppressor gene cooperating with NF1 loss induces malignant astrocytoma. *Cancer Cell* **8:** 119–130.

Zhuo, L., Theis, M., Alvarez-Maya, I., Brenner, M., Willecke, K., and Messing, A. 2001. *hGFAP-cre* transgenic mice for manipulation of glial and neuronal function in vivo. *Genesis* **31:** 85–94.

Pten and p53 Converge on c-Myc to Control Differentiation, Self-renewal, and Transformation of Normal and Neoplastic Stem Cells in Glioblastoma

H. Zheng,[*ᵚ] H. Ying,[*ᵚ] H. Yan,[*] A.C. Kimmelman,[*†] D.J. Hiller,[#] A.-J. Chen,[*] S.R. Perry,[*¶]
G. Tonon,[*] G.C. Chu,[*‡¶] Z. Ding,[*] J.M. Stommel,[*] K.L. Dunn,[*] R. Wiedemeyer,[*] M.J. You,[*]
C. Brennan,[ᶥ] Y.A. Wang,[*¶] K.L. Ligon,[*‡§ɣ] W.H. Wong,[#] L. Chin,[*¶⁰] and R.A. DePinho[*¶]

*Department of Medical Oncology, ¶Center for Applied Cancer Science, Belfer Foundation Institute for Innovative
Cancer Science, ɣCenter for Molecular Oncologic Pathology, Dana-Farber Cancer Institute and Harvard Medical
School, †Harvard Radiation Oncology Program, ‡Department of Pathology, §Division of Neuropathology, ⁰Department
of Dermatology, Brigham and Women's Hospital, Harvard Medical School, Boston, Massachusetts; #Department of
Statistics, Stanford University, Stanford, California; ᶥDepartment of Neurosurgery, Memorial Sloan-Kettering Cancer
Center, New York; ᶥDepartment of Neurosurgery, Weill-Cornell Medical College, New York, New York

Glioblastoma (GBM) is a highly lethal primary brain cancer with hallmark features of diffuse invasion, intense apoptosis resistance and florid necrosis, robust angiogenesis, and an immature profile with developmental plasticity. In the course of assessing the developmental consequences of central nervous system (CNS)-specific deletion of p53 and Pten, we observed a penetrant acute-onset malignant glioma phenotype with striking clinical, pathological, and molecular resemblance to primary GBM in humans. This primary, as opposed to secondary, GBM presentation in the mouse prompted genetic analysis of human primary GBM samples that revealed combined p53 and Pten mutations as the most common tumor suppressor defects in primary GBM. On the mechanistic level, the "multiforme" histopathological presentation and immature differentiation marker profile of the murine tumors motivated transcriptomic promoter-binding element and functional studies of neural stem cells (NSCs), which revealed that dual, but not singular, inactivation of p53 and Pten promotes cellular c-Myc activation. This increased c-Myc activity is associated not only with impaired differentiation, enhanced self-renewal capacity of NSCs, and tumor-initiating cells (TICs), but also with maintenance of TIC tumorigenic potential. Together, these murine studies have provided a highly faithful model of primary GBM, revealed a common tumor suppressor mutational pattern in human disease, and established c-Myc as a key component of p53 and Pten cooperative actions in the regulation of normal and malignant stem/progenitor cell differentiation, self-renewal, and tumorigenic potential.

Malignant glioma, the most common intrinsic brain tumor in adults, is nearly uniformly fatal (Furnari et al. 2007). The most advanced form of malignant glioma, glioblastoma multiforme (GBM), carries a median survival of only 9–12 months, a statistical fact that has changed little despite innumerable clinical trials during the past several decades (Reardon et al. 2006). Several hallmark biological properties of malignant gliomas are the likely basis for the intractable nature of GBM, including an intense apoptosis resistance, rampant genome instability, and a propensity to disseminate early and widely throughout the brain—rendering surgical resection noncurable.

The identification of clonogenic TICs with stem-like properties across many different tumor types has revealed a level of plasticity in human cancer cells previously unrecognized (Bonnet and Dick 1997; Al-Hajj et al. 2003; Singh et al. 2004b; Taylor et al. 2005; O'Brien et al. 2007; Ricci-Vitiani et al. 2007). These clonogenic cells, regardless of their actual cell(s) of origin, constitute a reservoir of self-sustaining cells with the exclusive ability to self-renew and maintain the tumor. Moreover, the TIC or "cancer stem cell" hypothesis is predicated on the idea that not all glioma cells have equal proliferative potential and that cells with the greatest ability to proliferate and form new tumors have phenotypic and functional properties similar to those of NSCs (Singh et al. 2004a; Clarke et al. 2006). During the past several years, multiple reports have established the existence of TICs in human GBM that are endowed with superior tumorigenic potential in orthotopic models and possess self-renewal and multilineage cell differentiation potential—properties that are eerily reminiscent of NSCs (Singh et al. 2004b; Bao et al. 2006; Lee et al. 2006;).

The phenotypic similarities of GBM TICs and NSCs have fueled speculation that neural stem/progenitor cell compartments may be the preferred cell of origin for GBM, although the dedifferentiation and transformation of more differentiated glial cells have been demonstrated experimentally (Bachoo et al. 2002). The latter findings have implications for the cancer stem cell hypothesis because it suggests that although at any given time certain cells may have TIC properties, the ready capacity for dedifferentiation raises the possibility that "TIC differentiated progeny" could reacquire an immature tumorigenic state. If so, this plasticity would have implications for the treat-

---

ᵚThese authors contributed equally to this work.

ment of cancer because it would imply that complete erad-ication of the TIC compartment would not lead to durable cures, as is widely assumed, and that such eradication would need to be complemented with differentiation-inducing agents. In this light, it stands to reason that an improved understanding of the pathways governing processes of differentiation along the NSC-glial axis and, by extension, the TIC-progeny axis, may prove instrumen-tal in guiding more effective drug development efforts.

During the past 20 years, molecular and human genetic studies have revealed obligate pathways involved in GBM pathogenesis, and principal among these are the RB-p16/p18-CDK4/6, p53-MDM2-ARF, and PI3K/PTEN/AKT pathways (Zhu and Parada 2002; Furnari et al. 2007). A wealth of studies has established a central role for the phosphoinositol-3 kinase (PI3K) pathway in the pathogen-esis of glioma on the genomic, genetic, and epigenetic lev-els (Furnari et al. 2007). Historically, *p53* inactivation has been considered to be a more classical genetic lesion in low-grade astrocytomas that invariably evolve into second-ary GBM, whereas primary GBMs are thought to more commonly sustain loss of the CDKN2A locus, which con-tains both cyclin-dependent kinase inhibitor 2A (Ink4a) and an alternate open reading frame (ARF), a negative reg-ulator of the p53 pathway (Watanabe et al. 1996; Kleihues and Ohgaki 1999). On the other hand, recent population-based studies have observed *p53* mutations in primary GBM, raising the possibility that *p53* might also serve as a tumor suppressor in the primary GBM subtype (Ohgaki et al. 2004; Fukushima et al. 2006). Moreover, our recent effort on unbiased global copy-number analysis of human primary GBM has identified both *Pten* and *p53* deletion by genome topography scanning (Wiedemeyer et al. 2008).

In the course of assessing the effect of singular or com-bined impact of *p53* and *Pten* deficiencies on CNS devel-opment and gliomagenesis, our studies have generated a faithful model of primary GBM, prompted resequencing studies that uncovered a high frequency of *p53* mutations in human primary GBM (a lesion previously considered to be more classical for secondary GBM), and revealed cooperative interactions of *p53* and *Pten* deficiencies in the regulation of glial cell differentiation. Along these lines, we show that conditional *p53* and *Pten* inactivation in neural stem cells and resultant tumors promotes normal and malignant cell self-renewal, impedes their differentia-tion, and sustains potent malignant potential. An inte-grated genomic and functional analysis indicates that these two tumor suppressors converge on the regulation of Myc to control these cellular processes.

RESULTS

Combined Loss of *p53* and *Pten* in the Mouse Brain Leads to Highly Penetrant Malignant Glioma Development with Striking Similarity to Human Primary GBM

To genetically address the role and genetic interactions of *p53* and *Pten* in CNS development, NSC biology, and glioma pathogenesis, we used a constitutive *hGFAP-Cre* transgenic line (Zhuo et al. 2001) to target deletion of *p53*

alone or in combination with *Pten* in the NSC compart-ment and its derivative CNS lineages. Early and broad CNS *hGFAP-Cre* activity was evidenced by Cre-mediated activation of the *Rosa26-LacZ* reporter in the forebrain and hindbrain as well as in germinal zones of the embry-onic brain (E13.5). Consequently, Cre-mediated recombi-nation is present throughout neural stem/progenitor cells and mature lineages of the adult CNS (data not shown). Correspondingly, *hGFAP-Cre Pten^{lox/lox}* mice display the early postnatal lethal phenotype of hydrocephalus and macrocephaly, as previously reported in the Pten-deficient brain (Backman et al. 2001; Kwon et al. 2001; data not shown). Because broad CNS deletion in mice homozy-gous for *Pten^{lox/lox}* resulted in an early postnatal lethal phe-notype of hydrocephalus and macrocephaly, the *Pten^{lox/+}* genotype was used for glioma modeling in this study.

By 15–40 weeks of age, 42/57 (73%) mice in the *hGFAP-Cre p53^{lox/lox} Pten^{lox/+}* cohort presented with acute-onset neurological symptoms including seizure, ataxia, and/or paralysis, resulting in death with median survival of 28 weeks (Fig. 1A). Necropsy and histopathological analyses revealed that all 42 neurologically symptomatic mice harbored malignant gliomas that, by World Health Organization (WHO) 2007 classification criteria (Louis et al. 2007), were determined to be anaplastic astrocytomas (WHO Grade III, $n = 36$, 64%) or GBM (WHO Grade IV, $n = 14$, 25%) (Fig. 1B,C). All tumors exhibited marked cellular pleomorphism and diffuse infiltrative spread in a single-cell manner with formation of secondary structures of Sherer, including perineuronal and perivascular satelli-tosis and subpial collections in the cerebral cortex (Fig. 1D). The tumors classified as WHO Grade IV (GBM) dis-played the defining histopathological features of necrosis with pseudopalisading and, less frequently, necrosis com-bined with microvascular proliferation (Fig. 1D). These tumors had a striking resemblance to the human disease with increased mitotic indices as indicated by Ki-67 stain-ing and extensive immunohistochemical staining for the human glioma markers GFAP and Nestin (Fig. 1C). The remaining 15 (15/57) *hGFAP-Cre p53^{lox/lox} Pten^{lox/+}* mice succumbed to non-CNS-derived tumors, either sarcomas or breast adenocarcinomas that result from *hGFAP-Cre* activity outside of the CNS, as documented by low-level Cre-mediated deletion of the floxed alleles in non-CNS tissues (data not shown).

In addition to these end point analyses, we also con-ducted a serial histopathological analysis of 15 mice before the onset of neurological symptoms. Surprisingly, no mice were found to have developed tumors analogous to human low-grade gliomas. Instead, 8/15 mice exhibited high-grade (at least WHO Grade III) pathology including some very small lesions composed of highly anaplastic cells with features of nuclear atypia, multinucleated tumor cells, and/or dense cellularity (Fig. 1E); thus, even early lesions show highly aggressive malignant features from the outset. Together, the rapidly progressive clinical course and pathologic WHO Grade IV classification in a significant number of mice indicate that the *hGFAP-Cre p53^{lox/lox} Pten^{lox/+}* model most faithfully recapitulates the clinical and pathological features of the human primary GBM subtype.

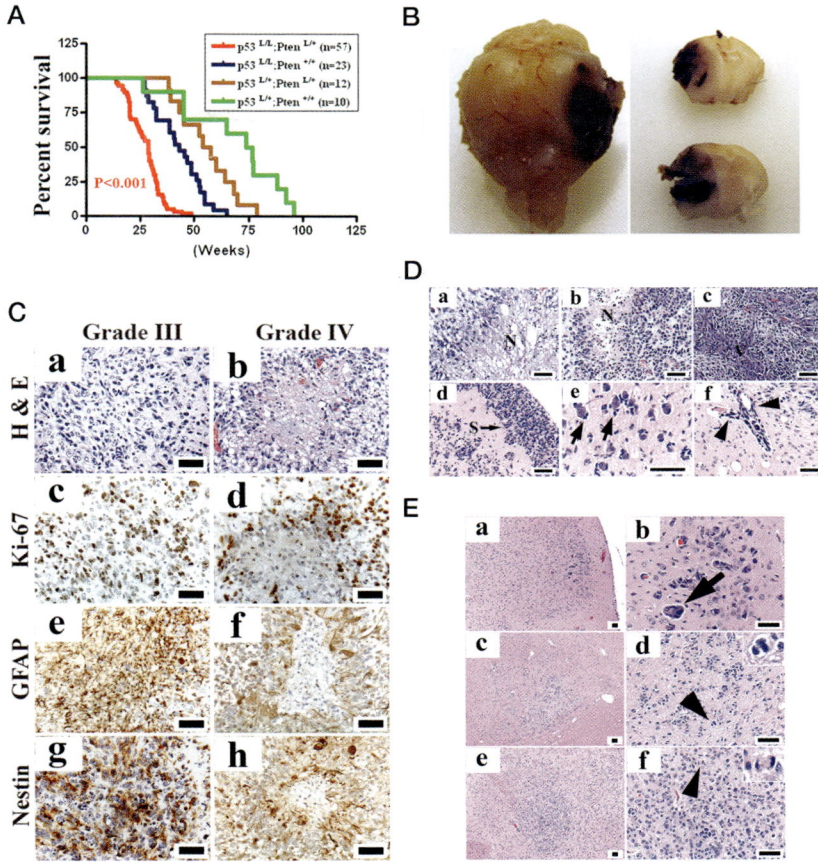

Figure 1. Inactivation of *p53* and *Pten* induces primary malignant gliomas. (*A*) Kaplan-Meier tumor-free survival curves for mice of indicated genotypes as a function of weeks. (*B*) Representative of malignant gliomas after biopsy. (*C*) H&E (hematoxylin and eosin) histology and immunohistochemical staining of sections of WHO grade-III and -IV malignant gliomas from *hGFAP-Cre;P53^{lox/lox};Pten^{lox/+}* mice. (*D*) Histological features of glioblastomas observed in *hGFAP-Cre;P53^{lox/lox};Pten^{lox/+}* mouse brains. "N" represents areas of the palisading with regional necrosis in the tumors (*a,b*); "V" regions suggest microvascular proliferation (*c*); "S," subpial spread (*d*); *arrows* in *e* point to perineuronal satellitosis and *arrowheads* in *f* point to perivascular satellitosis. (*E*) Small emerging tumors in asymptomatic mice show highly malignant histological features. *Arrow* in *b* points to multinucleated giant cell; *arrowheads* in *d* and *f* indicate mitotic figures amid the tumor cells. Bars, 50 µm. (Reprinted from Zheng et al. 2008 [© Nature Publishing Group].)

Acquired Molecular Profile of *hGFAP-Cre p53^{lox/lox} Pten^{lox/+}* Gliomas Mirrors Classical Alterations of the Human Disease

To further assess the human relevance of the *hGFAP-Cre p53^{lox/lox} Pten^{lox/+}* model, we examined the acquired molecular profile of the murine high-grade gliomas. In human high-grade glioma, loss of heterozygosity (LOH) of chromosome 10q encompassing the *Pten* locus is one of the most frequent genomic alterations, occurring in 60–70% of cases (Louis 2006; Furnari et al. 2007). Consistent with the critical need to extinguish Pten in advanced gliomas, the high-grade murine *hGFAP-Cre p53^{lox/lox} Pten^{lox/+}* glioma cells showed consistent loss of anti-Pten immunoreactivity as the result of losing the remaining wild-type Pten allele (Fig. 2A,B). Along with complete loss of Pten expression within the tumor cells, examination of key PI3K signaling surrogates revealed robust activation of AKT and S6-kinase in all samples tested (Fig. 2C). Moreover, the enhanced cyclin D1 expression and high level of vascular

endothelial growth factor (VEGF) observed in human high-grade disease are also present in these murine malignant gliomas, whereas much lower levels were detected in the adjacent normal brain tissues (Fig. 2C).

The highly penetrant GBM phenotype in the absence of an engineered receptor tyrosine kinase (RTK) allele was notable given the frequent amplification and overexpression of epidermal growth factor receptor (EGFR) in human primary GBM (Liberman et al. 1985) and the effectiveness of mutant activated RTK alleles in murine glioma models (Fomchenko and Holland 2006). At the same time, all human primary GBMs show coactivation of multiple RTKs, commonly EGFR and platelet-derived growth factor receptor-α (PDGFRα), and many of these activated RTKs do not show genomic or genetic alterations yet function cooperatively to maintain tumor cell viability (Stommel et al. 2007). Consistent with this notion, the robust anti-PDGFRα signal was detected in murine glioma cells relative to that in adjacent normal cells (Fig. 2D). Most tumors concurrently showed strong

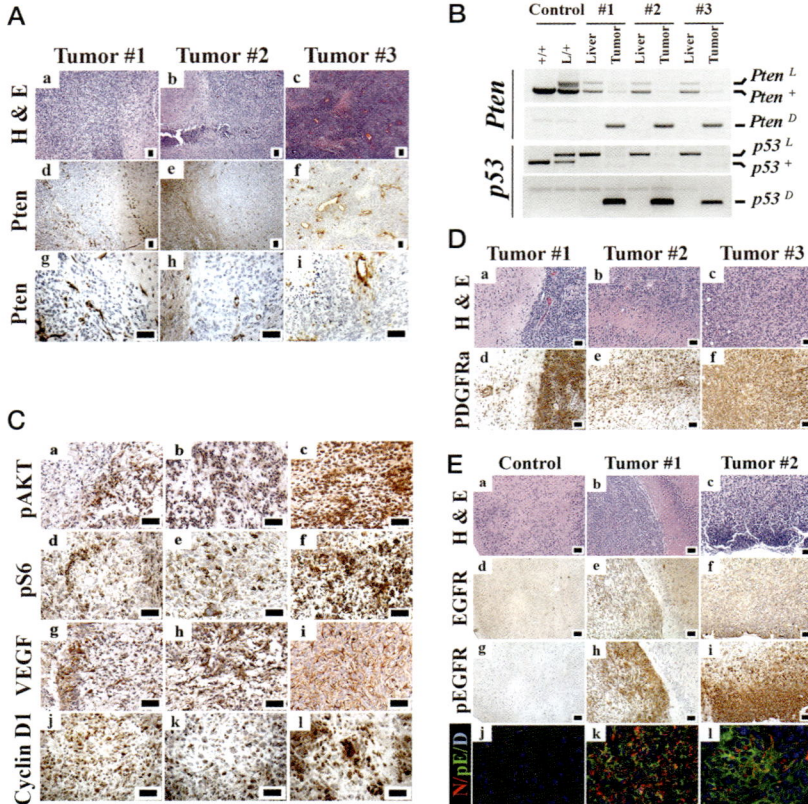

Figure 2. *hGFAP-Cre*;*P53*^{lox/lox};*Pten*^{lox/+} gliomas mirror key features of human malignant gliomas. (*A*) Pten expression is completely extinguished in tumor cells. (*B*) Wild-type *Pten* allele is lost in glioma cells. Genomic DNA isolated from liver tissues and brain tumor cells were subjected to polymerase chain reaction (PCR)-based assays for genotyping *Pten* and *p53* alleles. Note that "+" designates the *Pten* wild-type allele, "*L*" the conditional allele, and "*D*" the inactivated form of the conditional allele after Cre-mediated recombination. (*C*) Immunohistochemical staining of murine glioma sections with antibodies against activated p-AKT, p-S6 kinase, VEGF, and cyclin D1. Bars, 50 μm. (*D*) PDGFRα is specifically overexpressed in tumor cells as compared to normal brain cells. (*E*) EGFR is activated in malignant glioma cells. Ajacent sections from control normal brain and two independent high-grade gliomas were subjected to H&E, immunohistochemical staining for EGFR, and phospho-EGFR (pEGFR). Bars, 50 μm. (Reprinted from Zheng et al. 2008 [© Nature Publishing Group].)

regional activation of EGFR, which is less conspicuous in the relatively solid tumor centers and more robust in the subpial regions, as well as the more infiltrative fronts with high proliferative activity (Fig. 2E). Of interest, a fraction of strong phospho-EGFR tumor cells was intensely positive for Nestin, a finding that may relate to the prominent biological role of EGF signaling in less mature CNS cells including NSCs (Bachoo et al. 2002; Doetsch et al. 2002).

Pten and *p53* Are Frequently Mutated in Human Primary GBM

The striking clinical and histopathological resemblance of the p53/Pten model with human primary GBM prompted analysis of p53 mutational status in the primary GBM subtype in humans. To this end, we performed resequencing and copy-number analyses of *Pten* and *p53* genes in a collection of 35 clinically annotated and pathologically verified human primary GBM samples; 26 of these samples were previously profiled by array-CGH (comparative genome hybridization) (Wiedemeyer et al. 2008). Ten out of 35 (29%) tumors registered prototypical "hot spot"

tumor-relevant *p53* mutations, and 14/35 (40%) tumors possessed inactivating *Pten* mutations including missense mutations, insertions, deletions, and splicing mutations. Among the 10 samples harboring p53 mutations, six also showed *Pten* mutations or homozygous deletion. In addition, all five samples with wild-type *Pten* copy-number and sequence status were free of *p53* mutations. Together, these genomic studies indicate that combined inactivation of *p53* and *Pten* is an unexpectedly frequent genetic profile in human primary GBM.

Stem/Progenitor Cell Features

A classical feature of human primary GBM is its high degree of intertumoral and intratumoral morphological and lineage heterogeneity, hence the moniker glioblastoma "multiforme." These morphological characteristics were readily evident in *hGFAP-Cre p53*^{lox/lox} *Pten*^{lox/+} gliomas (Fig. 3A). The basis for the morphological variability is not known and may relate, among many possibilities, to the maintenance or acquisition of a developmental state with multipotency and/or differentiation plas-

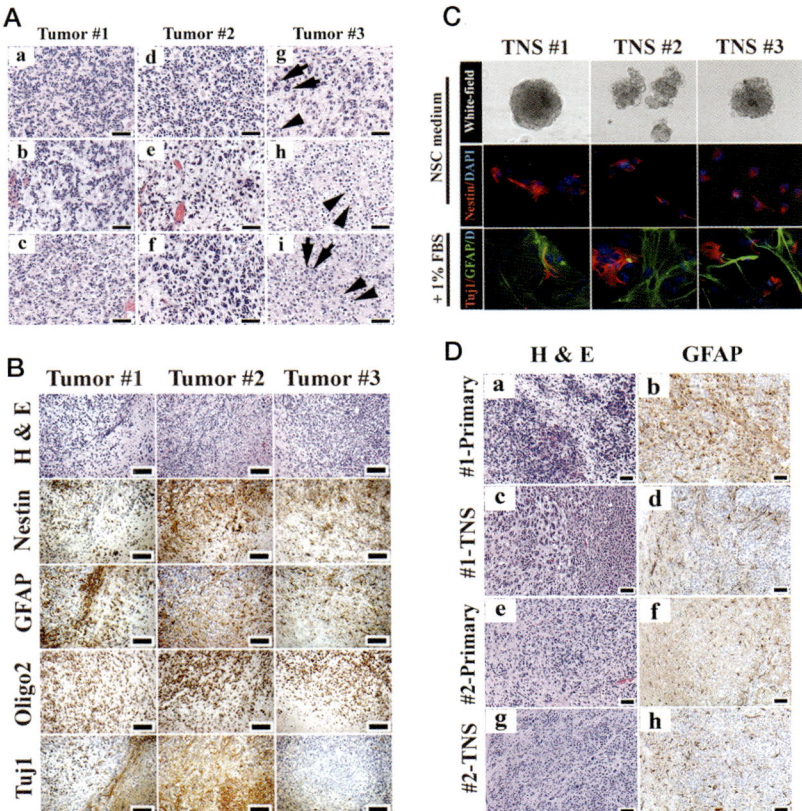

Figure 3. *hGFAP-Cre;P53^lox/lox;Pten^lox/+* murine malignant gliomas present stem/progenitor features. (*A*) Malignant gliomas possess high intertumoral and intratumoral histological heterogeneity. (*a–c*, *d–f*, and *g–i*) Three different regions of tumor 1, 2, and 3, respectively. Note the mixed conspicuously fibrillated matrix of astrocytoma and "fried egg" appearance of oligodendroglioma features in tumor 3 (*g–i*); *arrows* designate astrocytoma cells and *arrowheads* indicate oligodendroglioma cells. (*B*) Sections from three independent high-grade gliomas were subjected to H&E and immunohistochemical staining with antibodies against Nestin, GFAP, Olig-2, and Tuj1. Note the lack of Tuj1 staining in the no. 3 tumor cells. (*C*) TNS lines isolated from independent *hGFAP-Cre;P53^lox/lox;Pten^lox/+* gliomas were cultured in NSC medium or differentiation medium with 1% FBS and immunostained for lineage markers Nestin, GFAP, and Tuj1 as indicated. (*D*) Secondary tumors generated from orthotopic injection of TNS cells recapitulate the pathological features of primary tumors. Bars, 50 μm.

ticity resulting from cancer-relevant genetic and epigenetic alterations. Along these lines, it is intriguing that occasional tumors (5/50) presented with both astrocytic and oligodendroglial histopathological features within a single tumor (Fig. 3A), suggesting the existence of transformed cells in these tumors that retain the capacity to differentiate into different cell types. In line with this speculation, all murine tumors expressed stem or lineage progenitor markers commonly observed in human gliomas, including the stem/progenitor marker Nestin (N), the astrocytic lineage marker GFAP (G), and the oligodendroglial progenitor marker Olig2 (O) (Fig. 3B) (Ligon et al. 2007), and were negative for mature neuronal and oligodendrocyte markers NeuN and MBP (data not shown). Intriguingly, the murine tumors could be further classified into two subgroups on the basis of neuronal lineage marker Tuj1 expression that is present in about 80% of tumors and completely absent in the remainder, reminiscent of the proneural, proliferative, and mesenchymal subclasses found in the human disease (Phillips et al. 2006). In light of the developmental hierarchy and increasing evidence of the stem-like TICs residing within gliomas, these subsets

may reflect the TIC differentiation status, in which N+G+O+T+ tumors originate from TICs resembling NSCs or early progenitors and N+G+O+T– tumors from TICs resembling lineage-committed progenitors. Alternatively, the propensity for a proneural phenotype may reflect acquired alterations in the TICs that control lineage development.

This stem/progenitor profile is in accord with the ability of all tumors tested to readily generate tumor neurospheres (TNSs) from single-cell suspensions of freshly resected and dissociated malignant murine glioma tissues with (1) very robust NSC/progenitor marker Nestin expression, (2) limited capacity to differentiate into astrocytic and neuronal lineages upon exposure to differentiation agents (Fig. 3C), and (3) strong tumor-initiating potential in which as few as 500 cells orthotopically injected into the forebrain of SCID (severe combined immunodeficiency disease) mice could readily generate secondary tumors histopathologically resembling their primary tumors (Fig. 3D). These data provide further evidence for the existence of cells displaying the tentatively defined properties of TICs within murine gliomas.

Pten and *p53* Inactivation Impedes NSC Differentiation in a Myc-dependent Manner

Prompted by the immature marker profile of the murine tumor and the likelihood that NSCs/progenitor cells may be the preferred cell of origin for GBM (Sanai et al. 2005; Zhu et al. 2005; Gilbertson and Gutmann 2007), we hypothesized that the gliomagenic impact of *Pten* and *p53* deficiencies might be executed in part by affecting processes of NSC/progenitor cell self-renewal and/or differentiation. To test this hypothesis, we characterized the cellular and molecular properties of primary NSC cultures as a function of *p53* and/or *Pten* status. Consistent with previous reports that Pten and p53 negatively regulate the proliferation and self-renewal capacity of NSCs (GilPerotin et al. 2006; Groszer et al. 2006; Meletis et al. 2006), NSCs null for either Pten or p53 showed a modestly increased proliferative rate and neurosphere formation capacity after each passage. However, combined loss of p53 and Pten resulted in a marked increase in NSC proliferation and self-renewal capacity (Fig. 4A,B), pointing to strong cooperative interactions of these tumor suppressors in the regulation of NSC biology.

The enhanced NSC self-renewal, coupled with the aforementioned varied histological tumor presentations, raised the possibility that combined *p53* and *Pten* deficiency might impede the differentiation potential of the NSC/progenitor cultures. When continuously cultured in serum-free media supplemented with EGF and basic fibroblast growth factor (bFGF), NSCs of various genotypes showed similar morphology and robust expression of the NSC/progenitor marker Nestin and minimal expression of differentiated lineage markers, such as GFAP (glial fibrillary acidic protein) for astrocytes, O4 for oligodendrocytes, and Tuj1 for neuronal lineages (Fig. 4C). However, differentiation differences emerged among the genotypes. Specifically, upon exposure to differentiation-inducing medium containing 1% fetal bovine serum (FBS), the majority of wild-type and single-null NSCs assumed a flattened morphology, lost Nestin staining and differentiated into either GFAP-positive astrocytes, Tuj1-positive neurons, or O4-positive oligodendrocyte (Fig. 5A). In contrast, NSCs doubly null for both *p53* and *Pten* failed to respond fully to these differentiation cues and retained stem-cell-like morphology and lineage-marker staining after 7 days under differentiation-induction conditions. Immunofluorescence staining analysis confirmed that approximately 50–60% of cells retained Nestin expression, whereas the remaining cells showed a range of lineage marker profiles. Importantly, NSCs singly null for either *p53* or *Pten* did not exhibit such resistance to differentiation induction. Accordingly, inhibition of AKT by Triciribine (Yang et al. 2004) readily reversed differentiation resistance in the double-null NSCs (Fig. 5B). These findings point to the cooperative impact of *p53* and *Pten* deficiency in preserving the stem-cell-like state in the face of potent differentiation cues.

To understand the molecular basis for the impaired differentiation capacity of double-null NSCs, we performed transcriptome comparison of *p53*-null and *p53/Pten* double-null NSCs following exposure to a differentiation inducer. At day 1 of induction, 410 genes exhibited significant differential expression. Gene ontology (GO) analysis of these differentially expressed genes revealed enrichment of cell cycle and neural development functions, consistent with the observed differentiation defect. Intriguingly, promoter analysis on the 410 differentially expressed genes identified E2F and Myc motifs as two of the most enriched promoter-binding elements with 1.7- and 1.4-fold increases, respectively. The identification of c-Myc as a potential driving force in our system was encouraging given its established roles in cell cycle progression and apoptosis (Pelengaris et al. 2002; Patel et al. 2004) and the accumulating evidence of its functions as a key regulator of stem cell self-renewal and differentiation during development and oncogenic processes (Coppola and Cole 1986; Cartwright et al. 2005; Takahashi and Yamanaka 2006;

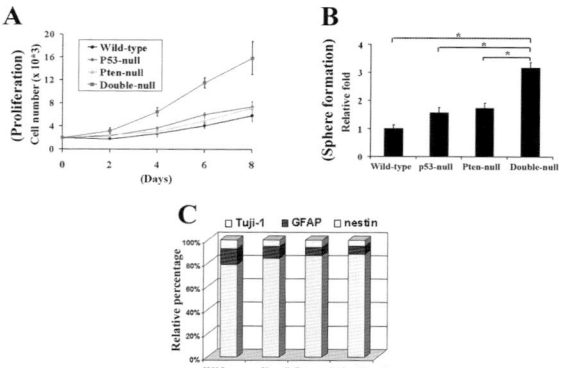

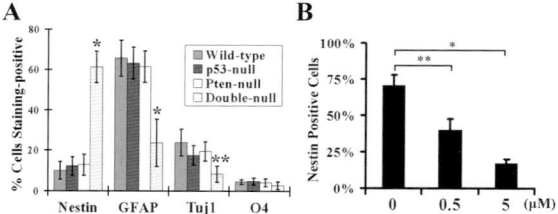

Figure 4. Mutant NSCs present enhanced proliferation and self-renewal. (*A*) Proliferation of *p53/Pten* double-null NSCs is enhanced as compared with wild-type and singly null NSCs ($n = 3$). (*B*) The number of multipotent neurospheres formed by *p53/Pten* double-null NSCs in culture is significantly increased as compared with neurospheres of wild-type or singly null NSC controls (*$P < 0.001$; $n = 3$). (*C*) Quantification histograms indicate that percentages of different lineage markers staining positive cells in wild-type and mutant NSCs are comparable when cultured in NSC medium ($n = 3$). Nestin, NSC/progenitor marker; GFAP, astrocyte lineage marker; Tuj1, neuronal lineage marker.

Figure 5. *p53* and *Pten* doubly inactivated NSCs are defective of their differentiation capacity. (*A*) Multilineage differentiation induced by 1% FBS was impaired in double-null NSCs as compared with wild-type and singly null NSCs (*$P < 0.005$; **$P < 0.05$; $n = 3$). (*B*) Inhibition of the PI3K/AKT pathway sensitizes *p53/Pten* double-null NSCs to differentiation. *p53/Pten* double-null NSCs cultured in 1% FBS in the absence or presence of AKT inhibitor (Triciribine) for 7 days were subjected to immunofluorescent staining with antibodies against Nestin and GFAP (**$P < 0.05$; *$P < 0.001$; $n = 3$).

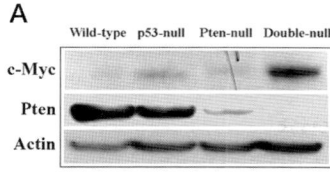

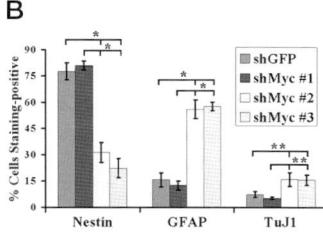

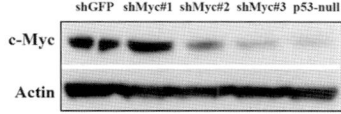

Figure 6. *p53* and *Pten* coordinately regulate cellular c-Myc protein levels and NSC differentiation. (*A*) Combined inactivation of *p53* and *Pten* in NSCs stimulates c-Myc protein expression. (*B*) Knockdown of c-Myc expression in *p53/Pten* double-null NSCs restores differentiation capacity in response to 1% FBS. (*Upper panel*) Histograms show percentage of indicated cells immuno-positive for different lineage markers (*$P < 0.005$; **$P < 0.05$; $n = 3$). (*Lower panel*) Western blot showing c-Myc protein expression of double-null NSCs infected with indicated lenti-shRNA. Note that both lenti-shRNA against GFP and shMyc no. 1 functioned as controls. (Reprinted from Zheng et al. 2008 [© Nature Publishing Group].)

Ben-Porath et al. 2008; Chang et al. 2008). Consistent with this, western blot analysis shows a substantial increase in c-Myc protein levels in the *p53* and *Pten* double-null NSCs and only marginally increased c-Myc expression in single *p53*-null or *Pten*-null NSCs when compared to wild-type cells (Fig. 6A). The prominent enrichment of Myc promoter-binding motifs in this unbiased integrated transcriptomic/promoter analysis, coupled with the impaired differentiation phenotype, raised the possibility that *p53* and *Pten* act coordinately to regulate c-Myc levels, which in turn control NSC self-renewal and differentiation.

To test this hypothesis, we examined the impact of c-Myc knockdown on the differentiation potential of *p53/Pten* double-null NSCs. We observed that two independent short hairpin RNAs (shRNAs) (nos. 2 and 3) against c-Myc, which attenuated c-Myc expression in double-null NSCs to the levels comparable to those levels seen in *p53*-null NSCs, largely restored the NSC differentiation capacity upon exposure to differentiation-inducing medium containing 1% FBS (Fig. 6B). Conversely, enforced c-Myc expression in *p53*-null NSCs, but not empty vector controls, repressed their differentiation potential, and the cells retained expression of the stem/progenitor markers of Nestin and Sox2 upon differentiation induction (data not shown). Together, these data indicate that concomitant loss of both *p53* and *Pten* tumor

suppressors impedes differentiation and enhances renewal of NSCs in a c-Myc-dependent manner.

c-Myc Regulates Tumor Neurosphere Differentiation, Renewal, and Tumorigenic Potential

The potent and pleiotropic activities of c-Myc requires tight control of its expression in developing and adult tissues, and its deregulation contributes to the genesis of various cancer types including gliomas (Nesbit et al. 1999; Bredel et al. 2005; Ben-Porath et al. 2008). Our finding that c-Myc impacts on NSC/progenitor cell biology prompted us to further assess the relevance of c-Myc on the malignant potential of murine-glioma-derived TICs; such TICs are enriched in TNS cultures. First, consistent with the importance of PI3K pathway activation, AKT inhibitor treatment of the murine glioma TNS cells strongly attenuated c-Myc protein expression and promoted their differentiation (Fig. 7A,B). Correspondingly, c-Myc knockdown using two independent shRNAs markedly reduced TNS cell proliferation and self-renewal capacity (Fig. 8A,B). Moreover, upon exposure to differentiation-induction conditions, c-Myc knockdown strongly sensitized the murine-glioma-derived TNS cells to differentiate, whereas control TNS cells retained an NSC/progenitor marker Nestin-positive phenotype (Fig. 8C). With respect to tumorigenic potential, intracranial injection of vector-transduced murine-derived TNS cells resulted in lethal infiltrating gliomas that resemble primary tumor pathology in 10/10 mice within 1 month (Fig. 8D), whereas 9/10 mice implanted with c-Myc knockdown TNSs continue to survive for several months. Together, these data support the view that c-Myc has a key

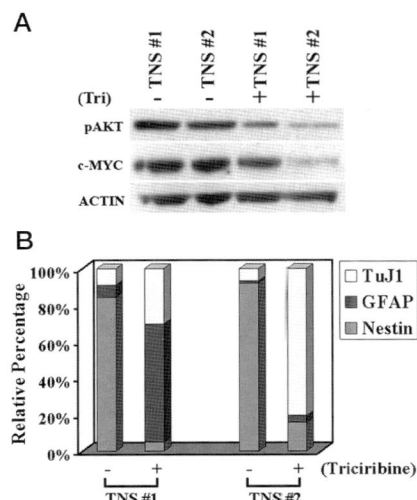

Figure 7. Inhibition of the AKT pathway induces TNS cell differentiation. (*A*) Inhibition of the AKT pathway in TNS cells with Triciribine attenuates their cellular c-Myc expression. (*B*) Inhibition of AKT pathway with Triciribine induces TNS cell differentiation ($n = 3$). Two independent TNS lines (TNS 1 and TNS 2) were cultured in 1% FBS in the absence or presence of Triciribine (5 μM) for 7 days before being subjected to immunostaining with antibodies against Nestin, GFAP, and Tuj1.

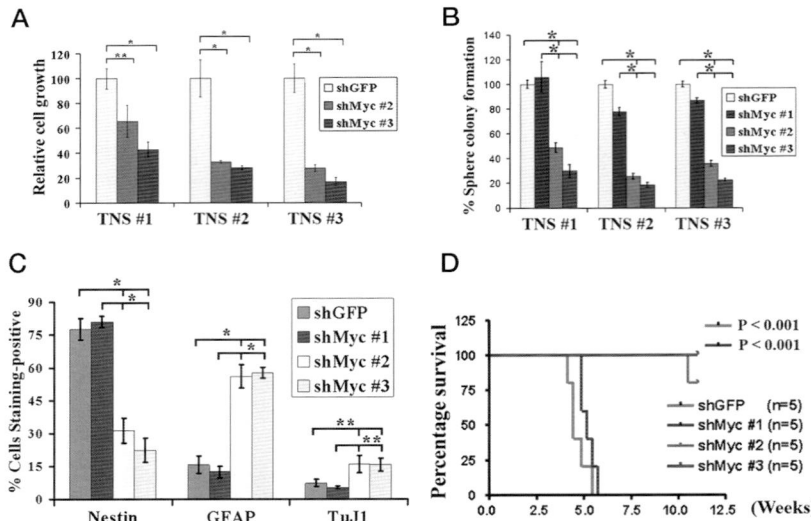

Figure 8. Attenuated c-Myc expression restores *hGFAP-Cre;P53^{lox/lox};Pten^{lox/+}* TNS differentiation potential and reduces their tumorigenic potential. (*A*) Knockdown of c-Myc expression reduces TNS cell growth. TNS cells infected with indicated lenti-shRNA virus were cultured in NSC medium and cell proliferation was measured 6 days after plating (**$P < 0.01$; *$P < 0.001$; $n = 3$). (*B*) Reduction of c-Myc expression in TNS cells reduces their self-renewal potential assessed by sphere formation (*$P < 0.001$, $n = 3$). (*C*) shRNA-mediated reduction of c-Myc expression in TNS cells sensitizes cells to differentiation stimuli (*$P < 0.005$; **$P < 0.05$; $n = 3$). For all graphs, values represent mean ± s.d. from at lease three experiments. (*D*) Knockdown of c-Myc expression represses TNS tumorigenic potency after orthotopic transplantation into SCID mice. (Reprinted from Zheng et al. 2008 [© Nature Publishing Group].)

role in impaired differentiation and potent tumorigenic potential of *p53/Pten*-null TNS cells.

CONCLUSION

In the present study, we demonstrate that dual inactivation of *p53* and *Pten* produces a primary GBM phenotype that shares many of the molecular and histopathological features of the human disease. Our genomic analysis of *p53* and *Pten* in human primary GBM revealed that they are the two most commonly mutated tumor suppressors, thereby supporting the relevance of the *hGFAP-Cre; p53^{lox/lox};Pten^{lox/+}* glioma model and the utility of genetically engineered mouse models to guide the analysis of human cancer genomics. More importantly, this mouse-modeling effort demonstrates how this approach affords an experimental framework to understand on a mechanistic level how specific genetic alterations cooperate to affect malignant transformation. Along these lines, using an integrative genomic and functional analysis, we have demonstrated c-Myc as a focal point for the cooperative actions of combined *p53* and *Pten* deficiencies, consistent with studies showing that p53 can repress *c-Myc* transcription through directly binding to the *c-Myc* promoter (Ho et al. 2005), and an activated PI3K can promote mammalian target of rapamycin (mTor)-dependent c-Myc translation as well as glucogen synthase kinase-3β (GSK-3β)-induced c-Myc protein stability (Sears et al. 2000; Gera et al. 2004).

p53 and PI3K/*Pten* pathways are universally altered in human GBM (Furnari et al. 2007). Loss of *p53* in mice has been shown to elevate self-renewal of adult NSCs (GilPerotin et al. 2006; Meletis et al. 2006), whereas murine brain-specific inactivation of *Pten* has been shown

to cause aberrant NSC and astrocyte proliferation and brain overgrowth during development (Backman et al. 2001; Kwon et al. 2001; Groszer et al. 2006), suggesting their relevance in the regulation of NSC biology. The role of *Pten* in stem cell biology gains further significance from recent hematopoietic studies establishing that acute somatic deletion of *Pten* leads to initial expansion but ultimate depletion of hematopoietic stem cells (HSCs) (Yilmaz et al. 2006; Zhang et al. 2006). This progressive depletion suggests either the activation of exhaustion pathways causing stem cell exhaustion and/or the absence of renewal-promoting pathways. The former may stem from decreased FoxO family activity, which has been shown to deplete long-term HSC reserves via loss of quiescence control, increased apoptosis, and increased intracellular reactive oxygen species (ROS) levels (Tothova et al. 2007). In the context of the findings of our current study, the latter may relate to insufficient activation of a c-Myc-driven stem cell renewal program.

The prominence of the c-Myc proto-oncogene in our model system was of particular interest because it represents one of a handful of factors known to both interfere with normal stem cell biology and enable malignant transformation (Shachaf et al. 2004; Wong et al. 2008). Our experimental data in premalignant NSCs are in line with the perspective that sustained c-Myc expression blocks cell differentiation and promotes self-renewal in embryonic stem cells (ESCs) and cancers (Lassman et al. 2004; Shachaf et al. 2004; Ben-Porath et al. 2008; Wong et al. 2008), a notion reinforced by the recent findings of c-Myc as one of the four factors sufficient to reprogram mouse or human fibroblasts into embryonic stem (ES) cells (Takahashi and Yamanaka 2006; Jaenisch and Young 2008). This function of c-Myc may relate to its ability to globally reprogram and maintain an active chromatin state (Knoepfler et al. 2006). The recent discovery by Wong et al. (2008) and Ben-Porath et al. (2008) that an ES-cell-like gene expression signature is enriched in poorly differentiated aggressive human tumors including malignant gliomas further supports the idea that c-Myc functions to promote tumorigenesis by inappropriately altering the machinery controlling the cellular differentiation state. Our finding that *p53* and *Pten* pathways converge on c-Myc and its downstream targets therefore provides a rational molecular explanation and model for the defective differential traits of the double-null NSCs as well as the TNSs derived from the resultant tumors (Fig. 9).

The remarkable similarities shared between normal stem cells and cancer cells, such as self-renewal and differentiation potential, have motivated efforts to compare these two cellular states (Reya et al. 2001). Identification of TICs with stem-like properties in diverse human cancers including GBM represents an important conceptual advancement in cancer biology, with important implications for the diagnosis and treatment of cancers (Clarke and Fuller 2006; Lobo et al. 2007). Like normal NSCs, glioma TICs, which appear to constitute a reservoir of self-sustaining cells with potent tumorigenic potential, reside in perivascular niches that maintain the stem-like properties of these cancer cells (Calabrese et al. 2007). Moreover, when orthotopically transplanted into the brains of immunodeficient mice, the isolated TICs form tumors that are histologically identical to the original tumor, suggesting that the TICs retain the capacity to generate all cell types found in the parent tumor and can fully recapitulate the neoplastic phenotype in vivo (Singh et al. 2004b; Lee et al. 2006). However, unlike normal NSCs that readily differentiate along a developmental hierarchy

into lineage-restricted progenitors and their differentiated progenies (Gage 2000; Alvarez-Buylla et al. 2001; Temple 2001), malignant glioma cells, which display phenotypic plasticity mirroring normal NSCs, often lack the terminal differentiation traits possessed by their normal counterparts. In support of this notion, the murine *p53/Pten* doubly null TNS cells show desensitization toward differentiation cues. Forced restoration of the differentiation capacity in the TNS cells through either AKT inhibitor treatment or shRNA-mediated c-Myc ablation greatly diminishes their tumorigenic potential. The dramatic attenuation of self-renewal and tumorigenic potential upon down-regulation of c-Myc in murine TNSs, combined with in silico evidence of its dysregulation in human primary GBM, encourages the identification and testing of novel or existing combinational therapies targeting differentiation-promoting pathways including c-Myc in the treatment of primary GBM.

ACKNOWLEDGMENTS

We thank A. Berns for providing *p53[L]* mice; S. Zhou and S. Jiang for excellent mouse husbandry and care; R.T. Bronson for helpful discussion on pathology analysis; K. Montgomery for discussion on sequencing; and Y.-H. Xiao, B. Feng, and J. Zhang for bioinformatic help. H.Z. was supported by the Helen Hay Whitney Foundation. H.Y. is a recipient of the Marsha Mae Moeslein fellowship from the American Brain Tumor Association. A.C.K. is a recipient of the Leonard B. Holman Research Pathway fellowship. Z.D is supported by the Damon Runyon Cancer Research Foundation. J.M.S is supported by a Ruth L. Kirschstein National Research Service Award fellowship. R.W. is supported by a Mildred Scheel Fellowship (Deutsche Kreb-

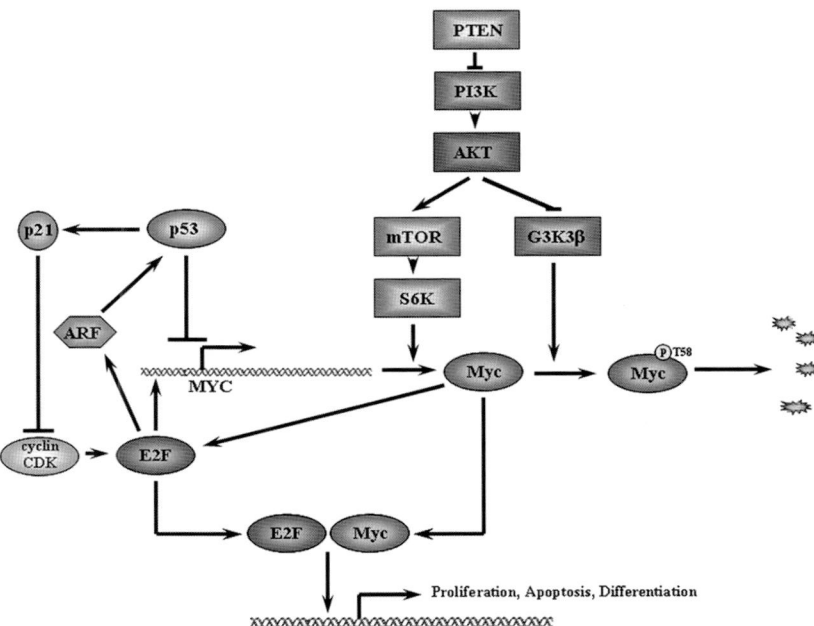

Figure 9. *p53* and *Pten* converge on c-Myc to regulate normal and malignant stem/progenitor self-renewal and differentiation. Combinatorial inactivation of both of the *p53* and *Pten* tumor suppressors collectively elevate c-Myc and E2F activity to enhance NSC self-renewal and impede their differentiation capacity.

shilfe). Grant support comes from the Goldhirsh Foundation (R.A.D.) and National Institutes of Health grants U01 CA84313 (R.A.D.), RO1CA99041 (L.C.), and 5P01CA95616 (R.A.D., L.C., W.H.W., C.B., and K.L.L.). R.A.D. is an American Cancer Society Research Professor supported by the Robert A. and Renee E. Belfer Foundation Institute for Innovative Cancer Science.

REFERENCES

Al-Hajj, M., Wicha, M.S., Benito-Hernandez, A., Morrison, S.J., and Clarke, M.F. 2003. Prospective identification of tumorigenic breast cancer cells. *Proc. Natl. Acad. Sci.* **100:** 3983–3988.

Alvarez-Buylla, A., Garcia-Verdugo, J.M., and Tramontin, A.D. 2001. A unified hypothesis on the lineage of neural stem cells. *Nat. Rev. Neurosci.* **2:** 287–293.

Bachoo, R.M., Maher, E.A., Ligon, K.L., Sharpless, N.E., Chan, S.S., You, M.J., Tang, Y., DeFrances, J., Stover, E., Weissleder, R., et al. 2002. Epidermal growth factor receptor and *Ink4a/Arf:* Convergent mechanisms governing terminal differentiation and transformation along the neural stem cell to astrocyte axis. *Cancer Cell* **1:** 269–277.

Backman, S.A., Stambolic, V., Suzuki, A., Haight, J., Elia, A., Pretorius, J., Tsao, M.S., Shannon, P., Bolon, B., Ivy, G.O., and Mak, T.W. 2001. Deletion of Pten in mouse brain causes seizures, ataxia and defects in soma size resembling Lhermitte-Duclos disease. *Nat. Genet.* **29:** 396–403.

Bao, S., Wu, Q., McLendon, R.E., Hao, Y., Shi, Q., Hjelmeland, A.B., Dewhirst, M.W., Bigner, D.D., and Rich, J.N. 2006. Glioma stem cells promote radioresistance by preferential activation of the DNA damage response. *Nature* **444:** 756–760.

Ben-Porath, I., Thomson, M.W., Carey, V.J., Ge, R., Bell, G.W., Regev, A., and Weinberg, R.A. 2008. An embryonic stem cell-like gene expression signature in poorly differentiated aggressive human tumors. *Nat. Genet.* **40:** 499–507.

Bonnet, D. and Dick, J.E. 1997. Human acute myeloid leukemia is organized as a hierarchy that originates from a primitive hematopoietic cell. *Nat. Med.* **3:** 730–737.

Bredel, M., Bredel, C., Juric, D., Harsh, G.R., Vogel, H., Recht, L.D., and Sikic, B.I. 2005. Functional network analysis reveals extended gliomagenesis pathway maps and three novel MYC-interacting genes in human gliomas. *Cancer Res.* **65:** 8679–8689.

Calabrese, C., Poppleton, H., Kocak, M., Hogg, T.L., Fuller, C., Hamner, B., Oh, E.Y., Gaber, M.W., Finklestein, D., Allen, M., et al. 2007. A perivascular niche for brain tumor stem cells. *Cancer Cell* **11:** 69–82.

Cartwright, P., McLean, C., Sheppard, A., Rivett, D., Jones, K., and Dalton, S. 2005. LIF/STAT3 controls ES cell self-renewal and pluripotency by a Myc-dependent mechanism. *Development* **132:** 885–896.

Chang, C.J., Mulholland, D.J., Valamehr, B., Mosessian, S., Sellers, W.R., and Wu, H. 2008. PTEN nuclear localization is regulated by oxidative stress and mediates p53-dependent tumor suppression. *Mol. Cell. Biol.* **28:** 3281–3289.

Clarke, M.F. and Fuller, M. 2006. Stem cells and cancer: Two faces of eve. *Cell* **124:** 1111–1115.

Clarke, M.F., Dick, J.E., Dirks, P.B., Eaves, C.J., Jamieson, C.H., Jones, D.L., Visvader, J., Weissman, I.L., and Wahl, G.M. 2006. Cancer stem cells—Perspectives on current status and future directions: AACR Workshop on cancer stem cells. *Cancer Res.* **66:** 9339–9344.

Coppola, J.A. and Cole, M.D. 1986. Constitutive c-*myc* oncogene expression blocks mouse erythroleukaemia cell differentiation but not commitment. *Nature* **320:** 760–763.

Doetsch, F., Petreanu, L., Caille, I., Garcia-Verdugo, J.M., and Alvarez-Buylla, A. 2002. EGF converts transit-amplifying neurogenic precursors in the adult brain into multipotent stem cells. *Neuron* **36:** 1021–1034.

Fomchenko, E.I. and Holland, E.C. 2006. Mouse models of brain tumors and their applications in preclinical trials. *Clin. Cancer Res.* **12:** 5288–5297.

Fukushima, T., Favereaux, A., Huang, H., Shimizu, T., Yonekawa, Y., Nakazato, Y., and Ohagki, H. 2006. Genetic alterations in primary glioblastomas in Japan. *J. Neuropathol. Exp. Neurol.* **65:** 12–18.

Furnari, F.B., Fenton, T., Bachoo, R.M., Mukasa, A., Stommel, J.M., Stegh, A., Hahn, W.C., Ligon, K.L., Louis, D.N., Brennan, C., et al. 2007. Malignant astrocytic glioma: Genetics, biology, and paths to treatment. *Genes Dev.* **21:** 2683–2710.

Gage, F.H. 2000. Mammalian neural stem cells. *Science* **287:** 1433–1438.

Gera, J.F., Mellinghoff, I.K., Shi, Y., Rettig, M.B., Tran, C., Hsu, J.H., Sawyers, C.L., and Lichtenstein, A.K. 2004. AKT activity determines sensitivity to mammalian target of rapamycin (mTOR) inhibitors by regulating cyclin D1 and c-*myc* expression. *J. Biol. Chem.* **279:** 2737–2746.

Gilbertson, R.J. and Gutmann, D.H. 2007. Tumorigenesis in the brain: Location, location, location. *Cancer Res.* **67:** 5579–5582.

GilPerotin, S., Marin-Husstege, M., Li, J., Soriano-Navarro, M., Zindy, F., Roussel, M.F., Garcia-Verdugo, J.M., and Casaccia-Bonnefil, P. 2006. Loss of p53 induces changes in the behavior of subventricular zone cells: Implication for the genesis of glial tumors. *J. Neurosci.* **26:** 1107–1116.

Groszer, M., Erickson, R., Scripture-Adams, D.D., Dougherty, J.D., Le Belle, J., Zack, J.A., Geschwind, D.H., Liu, X., Kornblum, H.I., and Wu, H. 2006. PTEN negatively regulates neural stem cell self-renewal by modulating G_0-G_1 cell cycle entry. *Proc. Natl. Acad. Sci.* **103:** 111–116.

Ho, J.S., Ma, W., Mao, D.Y., and Benchimol, S. 2005. p53-Dependent transcriptional repression of c-myc is required for G_1 cell cycle arrest. *Mol. Cell. Biol.* **25:** 7423–7431.

Jaenisch, R. and Young, R. 2008. Stem cells, the molecular circuitry of pluripotency and nuclear reprogramming. *Cell* **132:** 567–582.

Kleihues, P. and Ohgaki, H. 1999. Primary and secondary glioblastomas: From concept to clinical diagnosis. *Neuro-Oncology* **1:** 44–51.

Knoepfler, P.S., Zhang, X.Y., Cheng, P.F., Gafken, P.R., McMahon, S.B., and Eisenman, R.N. 2006. Myc influences global chromatin structure. *EMBO J.* **25:** 2723–2734.

Kwon, C.H., Zhu, X., Zhang, J., Knoop, L.L., Tharp, R., Smeyne, R.J., Eberhart, C.G., Burger, P.C., and Baker, S.J. 2001. Pten regulates neuronal soma size: A mouse model of Lhermitte-Duclos disease. *Nat. Genet.* **29:** 404–411.

Lassman, A.B., Dai, C., Fuller, G.N., Vickers, A.J., and Holland, E.C. 2004. Overexpression of c-MYC promotes an undifferentiated phenotype in cultured astrocytes and allows elevated Ras and Akt signaling to induce gliomas from GFAP-expressing cells in mice. *Neuron Glia Biol.* **1:** 157–163.

Lee, J., Kotliarova, S., Kotliarov, Y., Li, A., Su, Q., Donin, N.M., Pastorino, S., Purow, B.W., Christopher, N., Zhang, W., Park, J.K., and Fine, H.A. 2006. Tumor stem cells derived from glioblastomas cultured in bFGF and EGF more closely mirror the phenotype and genotype of primary tumors than do serum-cultured cell lines. *Cancer Cell* **9:** 391–403.

Libermann, T.A., Nusbaum, H.R., Razon, N., Kris, R., Lax, I., Soreq, H., Whittle, N., Waterfield, M.D., Ullrich, A., and Schlessinger, J. 1985. Amplification, enhanced expression and possible rearrangement of EGF receptor gene in primary human brain tumours of glial origin. *Nature* **313:** 144–147.

Ligon, K.L., Huillard, E., Mehta, S., Kesari, S., Liu, H., Alberta, J.A., Bachoo, R.M., Kane, M., Louis, D.N., DePinho, R.A., et al. 2007. Olig2-regulated lineage-restricted pathway controls replication competence in neural stem cells and malignant glioma. *Neuron* **53:** 503–517.

Lobo, N.A., Shimono, Y., Qian, D., and Clarke, M.F. 2007. The biology of cancer stem cells. *Annu. Rev. Cell Dev. Biol.* **23:** 675–699.

Louis, D.N. 2006. Molecular pathology of malignant gliomas. *Annu. Rev. Pathol.* **1:** 97–117.

Louis, D.N., Ohgaki, H., Wiestler, O.D., and Cavenee, W.K. 2007. *WHO classification of tumours of the central nervous system,* 4th ed. World Health Organization, Lyon, France.

Meletis, K., Wirta, V., Hede, S.M., Nister, M., Lundeberg, J., and

Frisen, J. 2006. p53 suppresses the self-renewal of adult neural stem cells. *Development* **133**: 363–369.

Nesbit, C.E., Tersak, J.M., and Prochownik, E.V. 1999. *MYC* oncogenes and human neoplastic disease. *Oncogene* **18**: 3004–3016.

O'Brien, C.A., Pollett, A., Gallinger, S., and Dick, J.E. 2007. A human colon cancer cell capable of initiating tumour growth in immunodeficient mice. *Nature* **445**: 106–110.

Ohgaki, H., Dessen, P., Jourde, B., Horstmann, S., Nishikawa, T., Di Patre, P.L., Burkhard, C., Schuler, D., Probst-Hensch, N.M., Maiorka, P.C., et al. 2004. Genetic pathways to glioblastoma: A population-based study. *Cancer Res.* **64**: 6892–6899.

Patel, J.H., Loboda, A.P., Showe, M.K., Showe, L.C., and McMahon, S.B. 2004. Analysis of genomic targets reveals complex functions of MYC. *Nat. Rev. Cancer* **4**: 562–568.

Pelengaris, S., Khan, M., and Evan, G. 2002. c-MYC: More than just a matter of life and death. *Nat. Rev. Cancer* **2**: 764–776.

Phillips, H.S., Kharbanda, S., Chen, R., Forrest, W.F., Soriano, R.H., Wu, T.D., Misra, A., Nigro, J.M., Colman, H., Soroceanu, L., et al. 2006. Molecular subclasses of high-grade glioma predict prognosis, delineate a pattern of disease progression, and resemble stages in neurogenesis. *Cancer Cell* **9**: 157–173.

Reardon, D.A., Rich, J.N., Friedman, H.S., and Bigner, D.D. 2006. Recent advances in the treatment of malignant astrocytoma. *J. Clin. Oncol.* **24**: 1253–1265.

Reya, T., Morrison, S.J., Clarke, M.F., and Weissman, I.L. 2001. Stem cells, cancer, and cancer stem cells. *Nature* **414**: 105–111.

Ricci-Vitiani, L., Lombardi, D.G., Pilozzi, E., Biffoni, M., Todaro, M., Peschle, C., and De Maria, R. 2007. Identification and expansion of human colon-cancer-initiating cells. *Nature* **445**: 111–115.

Sanai, N., Alvarez-Buylla, A., and Berger, M.S. 2005. Neural stem cells and the origin of gliomas. *N. Engl. J. Med.* **353**: 811–822.

Sears, R., Nuckolls, F., Haura, E., Taya, Y., Tamai, K., and Nevins, J.R. 2000. Multiple Ras-dependent phosphorylation pathways regulate Myc protein stability. *Genes Dev.* **14**: 2501–2514.

Shachaf, C.M., Kopelman, A.M., Arvanitis, C., Karlsson, A., Beer, S., Mandl, S., Bachmann, M.H., Borowsky, A.D., Ruebner, B., Cardiff, R.D., et al. 2004. MYC inactivation uncovers pluripotent differentiation and tumour dormancy in hepatocellular cancer. *Nature* **431**: 1112–1117.

Singh, S.K., Clarke, I.D., Hide, T., and Dirks, P.B. 2004a. Cancer stem cells in nervous system tumors. *Oncogene* **23**: 7267–7273.

Singh, S.K., Hawkins, C., Clarke, I.D., Squire, J.A., Bayani, J., Hide, T., Henkelman, R.M., Cusimano, M.D., and Dirks, P.B. 2004b. Identification of human brain tumour initiating cells. *Nature* **432**: 396–401.

Stommel, J.M., Kimmelman, A.C., Ying, H., Nabioullin, R., Ponugoti, A.H., Wiedemeyer, R., Stegh, A.H., Bradner, J.E., Ligon, K.L., Brennan, C., DePinho, R.A., and Chin, L. 2007. Coactivation of receptor tyrosine kinases affects the response of tumor cells to targeted therapies. *Science* **318**: 287–290.

Takahashi, K. and Yamanaka, S. 2006. Induction of pluripotent stem cells from mouse embryonic and adult fibroblast cultures by defined factors. *Cell* **126**: 663–676.

Taylor, M.D., Poppleton, H., Fuller, C., Su, X., Liu, Y., Jensen, P., Magdaleno, S., Dalton, J., Calabrese, C., Board, J., et al. 2005. Radial glia cells are candidate stem cells of ependymoma. *Cancer Cell* **8**: 323–335.

Temple, S. 2001. The development of neural stem cells. *Nature* **414**: 112–117.

Tothova, Z., Kollipara, R., Huntly, B.J., Lee, B.H., Castrillon, D.H., Cullen, D.E., McDowell, E.P., Lazo-Kallanian, S., Williams, I.R., Sears, C., et al. 2007. FoxOs are critical mediators of hematopoietic stem cell resistance to physiologic oxidative stress. *Cell* **128**: 325–339.

Watanabe, K., Tachibana, O., Sata, K., Yonekawa, Y., Kleihues, P., and Ohgaki, H. 1996. Overexpression of the EGF receptor and p53 mutations are mutually exclusive in the evolution of primary and secondary glioblastomas. *Brain Pathol.* **6**: 217–224.

Wiedemeyer, R., Brennan, C., Heffernan, T.P., Xiao, Y., Mahoney, J., Protopopov, A., Zheng, H., Bignell, G., Furnari, F., Cavenee, W.K., et al. 2008. Feedback circuit among INK4 tumor suppressors constrains human glioblastoma development. *Cancer Cell* **13**: 355–364.

Wong, D.J., Liu, H., Ridky, T.W., Cassarino, D., Segal, E., and Chang, H.Y. 2008. Module map of stem cell genes guides creation of epithelial cancer stem cells. *Cell Stem Cell* **2**: 333–344.

Yang, L., Dan, H.C., Sun, M., Liu, Q., Sun, X.M., Feldman, R.I., Hamilton, A.D., Polokoff, M., Nicosia, S.V., Herlyn, M., Sebti, S.M., and Cheng, J.Q. 2004. Akt/protein kinase B signaling inhibitor-2, a selective small molecule inhibitor of Akt signaling with antitumor activity in cancer cells overexpressing Akt. *Cancer Res.* **64**: 4394–4399.

Yilmaz, O.H., Valdez, R., Theisen, B.K. Guo, W., Ferguson, D.O., Wu, H., and Morrison, S.J. 2006. Pten dependence distinguishes haematopoietic stem cells from leukaemia-initiating cells. *Nature* **441**: 475–482.

Zhang, J., Grindley, J.C., Yin, T., Jayasinghe, S., He, X.C., Ross, J.T., Haug, J.S., Rupp, D., Porter-Westpfahl, K.S., Wiedemann, L.M., Wu, H., and Li, L. 2006. PTEN maintains haematopoietic stem cells and acts in lineage choice and leukaemia prevention. *Nature* **441**: 518–522.

Zheng, H., Ying, H., Yan, H., Kimmelman, A.C., Hiller, D.J., Chen, A.-J., Perry, S.R., Tonon, G., Chu, G.C., Ding, Z., et al. 2008. p53 and Pten control neural and glioma stem/progenitor cell renewal and differentiation. *Nature* **455**: 1129–1133.

Zhu, Y. and Parada, L.F. 2002. The molecular and genetic basis of neurological tumours. *Nat. Rev. Cancer* **2**: 616–626.

Zhu, Y., Guignard, F., Zhao, D., Liu, L., Burns, D.K., Mason, R.P., Messing, A., and Parada, L.F. 2005. Early inactivation of p53 tumor suppressor gene cooperating with NF1 loss induces malignant astrocytoma. *Cancer Cell* **8**: 119–130.

Zhuo, L., Theis, M., Alvarez-Maya, I., Brenner, M., Willecke, K., and Messing, A. 2001. hGFAP-cre transgenic mice for manipulation of glial and neuronal function in vivo. *Genesis* **31**: 85–94.

Establishment of a Normal Hematopoietic and Leukemia Stem Cell Hierarchy

M.P. CHAO, J. SEITA, AND I.L. WEISSMAN

Stanford Institute for Stem Cell Biology and Regenerative Medicine,
Stanford University School of Medicine, Stanford, California 94305

Many types of adult tissues, especially for high turnover tissues such as the blood and intestinal system, stand on a hierarchical tissue-specific stem cell system. Tissue-specific stem cells concurrently have self-renewal capacity and potential to give rise to all types of mature cells in their tissue. The differentiation process of the tissue-specific stem cell is successive restriction of these capacities. The first progeny of tissue-specific stem cells are multipotent progenitors (MPPs) that lose long-term self-renewal capacity yet have full lineage potential. MPPs in turn give rise to oligopotent progenitors, which then commit into lineage-restricted progenitors. This hierarchical system enables a lifelong supply of matured functional cells that generally have a short life span and a relatively high turnover rate. In this chapter, we review our findings and other key experiments that have led to the establishment of the current cellular stem and progenitor hierarchy in the blood-forming systems of mice and humans for both normal and leukemic hematopoiesis. We also review select signaling pathways intrinsic to normal hematopoietic and leukemic stem cell populations as well our recent findings elucidating the possible origin of the leukemia stem cell.

HIERARCHICAL STRUCTURE OF THE MOUSE HEMATOPOIETIC SYSTEM

The hematopoietic system is one of the most complicated but most investigated systems, which has centered on the identification of the hematopoietic stem cell (HSC). The concept of a tissue-specific stem cell in the blood system was first proposed by Till and McCulloch in 1961 when they found that rare cells in the mouse bone marrow could form myeloerythroid colonies in the spleens of irradiated mice transplants, in which a subset of these cells could self-renew (Till and McCulloch 1961; Becker et al. 1963; Siminovitch et al. 1963; Wu et al. 1968). This finding led to the eventual isolation of mouse HSCs in 1988 by prospective isolation using fluorescence-activated cell sorting (FACS) with combinations of monoclonal antibody cell surface determinants (Spangrude et al. 1988). The Thy-1^{lo}Lin-Sca$^+$ population identified had a frequency of 0.05% in the mouse adult bone marrow, and limited cell numbers could fully reconstitute the hematopoietic system in lethally irradiated mice long term while giving rise to progeny in all defined myelolymphoid lineages (Spangrude et al. 1988). Since that finding, the mouse functional HSC has been more extensively defined (Fig. 1) (Ikuta and Weissman 1992; Uchida and Weissman 1992; Christensen and Weissman 2001; Kiel et al. 2005). These mouse HSCs when transplanted at the single-cell level give rise to lifelong hematopoiesis, including a steady-state HSC pool between 20,000 and 100,000 that gives rise to more than 10^9 blood cells produced daily (Smith et al. 1991; Osawa et al. 1996; Morrison et al. 1997; Wagers et al. 2002b). Using a combination of cell surface marker expression and functional readout assays, the developmental cellular hierarchy of HSCs was then next achieved (Fig. 1). At the top of the hierarchy sits HSCs that possess lifelong self-renewal capacity. These cells then advance to the MPP phase (Morrison and Weissman 1994;

Christensen and Weissman 2001). Studies reveal that these MPPs are heterogeneous (Csawa et al. 1996; Christensen and Weissman 2001; Kiel et al. 2005) and are in the process of being further characterized (Adolfsson et al. 2005; Forsberg et al. 2006; Arinobu et al. 2007). Further downstream, MPPs give rise to two oligopotent progenitors: (1) the common lymphoid progenitor (CLP) (Kondo et al. 1997) and (2) the common myeloid progenitor (CMP) (Akashi et al. 2000), in which CMPs then give rise to megakaryocyte-erythrocyte progenitors (MEPs) and granulocyte-macrophage progenitors (GMPs). These downstream oligopotent progenitors then give rise to all of the lineage-committed effector cells of the hematopoietic system, with both CMPs and CLPs proposed to give rise to dendritic cells. The current hematopoietic stem cell hierarchy is outlined in Figure 1.

Under normal conditions, lineage-restricted progenitors do not detectably self-renew. They must therefore be derived continually from HSCs, in which the steady-state HSC pool is tightly regulated. Entry into the cell cycle is therefore restricted in the HSC population (<2% enter the cell cycle per day, whereas higher percentages of MPP, CLP, and CMP enter the cell cycle each day) (Ogawa 1993; Morrison and Weissman 1994; Kondo et al. 1997; Akashi et al. 2000; D. Bhattacharya et al., unpubl.). This multitiered scheme allows for an enormous amplification in the numbers of terminally differentiated cells and enables precise regulation of homeostasis exemplified by the observation that mature blood cells produce more than 1 million cells per second in adult humans (Ogawa 1993).

HIERARCHICAL STRUCTURE OF THE HUMAN HEMATOPOIETIC SYSTEM

Taking advantage of similar methods used to identify the mouse HSC hierarchy, the prospective isolation of

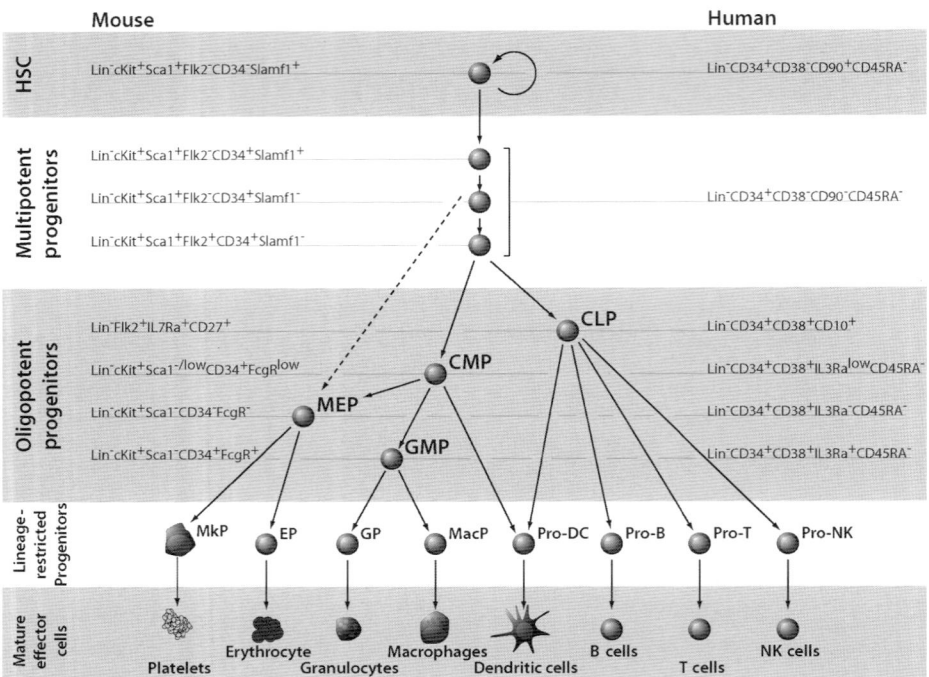

Figure 1. Model of the mouse and human hematopoietic hierarchy. The HSC resides at the top of the hierarchy and is defined as the cell that concurrently possesses both self-renewal capacity and potential to give rise to all hematopoietic cell types (multipotency). Throughout differentiation, an HSC first loses self-renewal capacity, then loses lineage potential step by step, and finally commits to matured functional cells of certain lineages. The cell surface phenotype of each population is shown for the murine and human systems. Intermediate precursors between the first lineage-committed progenitor and finally matured cell and subsets of matured B and T cells are omitted. In the mouse system, heterogeneity of multipotent progenitors by the difference of the cell surface marker phenotype has been revealed, and functional differences of these subsets have been discussed. For example, evidence suggests that some subset of MPPs directly give rise to MEPs without passing through CMPs (*dashed arrow*). (HSC) Hematopoietic stem cell; (CLP) common lymphoid progenitor; (CMP) common myeloid progenitor; (MEP) megakaryocyte-erythrocyte progenitor; (GMP) granulocyte-macrophage progenitors; (MkP) megakaryocyte progenitor; (EP) erythrocyte progenitor; (GP) granulocyte progenitor; (MacP) macrophage progenitor; (DC) dendritic cell; (NK) natural killer; (Lin) lineage markers.

human HSCs and downstream progenitors has also been achieved. The first positive marker used to identify human HSCs was identified two decades ago as the cell surface antigen CD34. CD34 is expressed in the nonhematopoietic tissues of endothelial cells and is a ligand for L-selectin (CD62L). Of hematopoietic cells in the human fetal liver, cord blood, and bone marrow, 0.5–5% express CD34 (Civin et al. 1984; Krause et al. 1996) with only CD34+ cells possessing in vitro clonogenic potential (Civin et al. 1984; DiGiusto et al. 1994; Krause et al. 1996). Although almost all CD34+ cells show stem cell potential in vitro, the CD34+ cell population is very heterogeneous. The majority of CD34+ cells (90–99%) coexpress the CD38 antigen. However, only 1–10% of CD34+ cells that do not express CD38 and mature lineage markers contain single cells with in vitro bilineage, lymphoid, and myeloid differentiation potential (Baum et al. 1992; Huang and Terstappen 1994; Hao et al. 1998; Miller et al. 1999). Furthermore, CD34+CD38− cells and not CD34+CD38+ cells are highly enriched for LTC-IC (Hao et al. 1996; Petzer et al. 1996) and contain severe combined immunodeficiency disease (SCID) human (hu) repopulating cells (Baum et al. 1992) and nonobase diabetic (NOD)-SCID repopulating cells (Larochelle et al. 1996; Bhatia et al. 1997) in vivo, with read-out in secondary NOD-SCID

transplants. Further elucidation of the human HSC phenotype occurred when the observation was made that single Lin−CD34+CD90+ and not Lin–CD34+CD90− cells generated lymphoid and myeloid progeny in both cell culture colony assays and in SCID-hu mice (Baum et al. 1992). Furthermore, virtually all CD34+CD90+Lin− cells reside in the CD38− fraction. In a more definitive demonstration of in vivo HSC function using methodology similar to mouse hematopoietic reconstitution assays, human clinical trials of autologous mobilized peripheral blood transplantation revealed that purified CD34+CD90+ cells provided long-term engraftment while being purged of unwanted cancer or T cells (Michallet et al. 2000; Negrin et al. 2000; Vose et al. 2001). All of these data combined support the idea that HSCs are contained in the Lin−CD34+CD38−CD90+ population.

Although the phenotype of human HSCs has been well-defined, the identification of the downstream progenitors has been less characterized compared to that in mouse. This specifically includes the immediate downstream MPP that has been comparably well characterized in the mouse (Morrison and Weissman 1994; Morrison et al. 1997; Christensen and Weissman 2001). Clues to the existence of this human MPP population was first demonstrated when retroviral or lentiviral marking of Lin−CD34+CD38− cells before xenotransplantation revealed that single Lin⁻CD34⁺ CD38⁻ cells had variable self-renewal and proliferation potential (Guenechea et al. 2001; Mazurier et al. 2004; McKenzie et al. 2006), suggesting that the human MPP lies in this population. Recently, we have identified this candidate MPP population in human cord blood as Lin⁻CD34⁺CD38⁻CD90⁻CD45RA⁻ given its ability for multipotency, incomplete self-renewal capacity, and downstream position from CD90⁺CD45RA⁻ cells (Majeti et al. 2007). Further downstream from this candidate MPP, early myeloid and lymphoid-committed progenitors have been identified in the human hematopoietic system that give rise to likely counterparts of murine CMP, GMP, and CLP (Fritsch et al. 1993; Ogawa 1993; Galy et al. 1995; Kimura et al. 1997; Manz et al. 2002). Using interleukin-3 (IL-3)Rα (a cytokine receptor that when activated supports proliferation and differentiation of primitive progenitors) and CD45RA (an isoform of CD45 that negatively regulates some select classes of cytokine signaling), CMP, GMP, and MEP populations can be identified that are verified by in vitro and in vivo clonality assays (Fig. 1).

PROPERTIES OF HEMATOPOIETIC STEM CELLS

Stem Cell Fate: Self-renewal versus Differentiation

HSCs under resting conditions are an asynchronously dividing cell population (Cheshier et al. 1999); when in the absence of injury, the total pool of HSCs remains approximately the same. Accordingly, half of all HSC cell divisions in the total HSC pool must be self-renewing to ensure that stem cell reserves are not exhausted throughout time. Thus, at every point of cell division, an HSC must decide between a process of self-renewal and one of differentiation down the committed progenitor pathway. Although by no means elucidated, important insights have been made to identify the molecular signals that govern this stem cell fate. This particularly includes identifying the genes that regulate HSC self-renewal.

Several developmental regulators of cell fate including Notch (Karanu et al. 2000; Varnum-Finney et al. 2000), HoxB4/HoxA9 (Sauvageau et al. 1995; Thorsteinsdottir et al. 1999, 2002), and Sonic hedgehog (Bhardwaj et al. 2001) have been demonstrated to have roles in self-renewal of HSCs. In addition, our lab has demonstrated that Wnt signaling is a crucial pathway in regulating self-renewal in HSCs (Reya et al. 2003). By transducing mouse HSCs with constitutively active β-catenin, we observed dramatic expansion and long-term growth in culture compared to control HSCs with a subsequent expansion of functional HSCs in vivo. Furthermore, ectopic expression of axin or a frizzled ligand-binding domain, inhibitors of the Wnt signaling pathway, lead to inhibition of HSC growth in vitro and reduced constitution in vivo (Reya et al. 2003).

In contrast to signals that promote self-renewal, signals that limit this potential have also been identified. We have hypothesized that telomere shortening could have a role given the notion that loss of telomere sequence is associated with cellular senescence. Telomerase, an RNA/protein complex responsible for extending telomeric DNA, is expressed by both mouse fetal liver and bone marrow HSCs (Morrison et al. 1996). In addition, telomerase activity is reduced as HSCs differentiate to multipotent progenitor populations, suggesting that such activity correlates with self-renewal capacity (Morrison et al. 1996). However, despite constitutive telomerase activity in HSCs, telomeres still shorten with HSC division in vivo (Notaro et al. 1997; Allsopp et al. 2001; Brummendorf et al. 2001), indicating that such activity is insufficient to maintain telomere length. In addition, serial transplantation of HSCs in mice is limited to approximately five to seven rounds (Harrison et al. 1978; Harrison and Astle 1982), perhaps indicating that HSCs cannot self-renew indefinitely. However, such observations should be interpreted with caution given that serial transplantation is not simply a measure of HSC life span but also a measure of homing to stem cell niches and establishment of appropriate cell–cell interactions within these niches. This caution is justified by the observation that HSCs that constitutively express high levels of TERT fail to have increased numbers of serial passages compared to wild-type HSCs (Allsopp et al. 2003). Nevertheless, the above data suggest an association with inhibition of cellular senescence and HSC self-renewal.

Migration

Since the emergence of the field of HSC biology, demonstration of stem cell activity has relied on transplantation of candidate stem cells into mouse or other species recipients. For the majority of these assays, cells are injected ectopically from the bone marrow or other classic hematopoietic niches. For successful engraftment to occur, these cells must find their way from the periphery past select tissue barriers to finally end up in hematopoietic microenvironments. This is best indicated by the success of bone marrow transplantation where cells harvested from bone marrow cavities of donors are intravenously injected into patients whose blood-forming system is compromised by radiation or chemotherapy. These cells successfully colonize the ablated bone marrow and reconstitute the entire hematopoietic system. The consistent accuracy of hematopoietic stem and progenitors homing to hematopoietic niches is quite amazing when one thinks about how many potential barriers HSCs must encounter before successful migration. However, it is unclear, a priori, why bone marrow transplantation should be successful. Why should HSCs and progenitors isolated from donor bone marrow accurately home to recipient hematopoietic niches? Is this process simply a random event or do HSCs take advantage of a normal physiologic process: cellular homing from blood to bone marrow?

To begin to examine this question, we first examined the fate of blood-borne HSCs using genetically marked parabiotic mice, which are surgically conjoined and share a common circulation. Through phenotypic and functional analyses, we determined that there are an estimated 100 steady-state HSCs in the blood of the mouse (Wright et al. 2001). When we isolated these HSCs and reintroduced them into the bloodstream, we found that there was rapid clearance from the circulation with a resident time between 1 and 5 minutes with subsequent reengraftment, demonstrating that HSCs rapidly migrate through the blood to engraft bone marrow (Wright et al. 2001). To determine whether these HSCs were trafficking to hematopoietic niches, we tracked them with luciferase and found that, indeed, HSCs homed to the bone marrow, liver, and spleen (Cao et al. 2004). In addition, single-cell HSCs that engrafted in the bone marrow exhibited engraftment of different bone marrow compartments (vertebrae, skull, femur, ribs) at different points in time, consistent with the hypothesis that HSCs are in a constant recirculatory flux (Cao et al. 2004). Thus, with respect to the success of bone marrow transplantation, the existence of a physiologic process by which HSCs and progenitors travel from bone marrow to blood and back provides a rationale for such success; the injected cells likely home to bone marrow niches in irradiated recipients because the mechanisms underlying this migration preexist in unmanipulated animals (Wright et al. 2001).

Given that HSCs appear to take advantage of a normal physiologic program of tissue homing, what are the cellular signals that govern this intrinsic mechanism of HSC migration to its respective niches? Although much of the mechanisms that influence HSC migration are not well understood, the extensive studies on lymphocyte homing guide the current hypothesis for HSCs because lymphocytes and HSCs share some similar cellular properties. Accordingly, this includes requiring homing receptor/addressin interactions in the vasculature, followed by chemokine/chemokine receptor interactions, integrin/receptor binding, and growth/survival factors.

The chemokine receptor CXCR4 and its respective ligand SDF-1α have been demonstrated to be signals important for HSC migration. Both mouse and human HSCs express CXCR4 (Mohle et al. 1998; Wright et al. 2002), and in cell culture chemotaxis assays, we show that out of a large panel of chemokines, murine HSCs show selective migration only to the CXCR4 ligand SDF-1α (Wright et al. 2002). Furthermore, mice deficient in CXCR4 or SDF1-α fail to establish bone marrow hematopoiesis (Nagasawa et al. 1996; Zou et al. 1998; Ara et al. 2003), whereas inhibitory antibodies to CXCR4 blocked engraftment of NOD-SCID mice by human HSCs, all consistent with the role for CXCR4/SDF-1α interactions in HSC homing to bone marrow (Peled et al. 1999).

With respect to the importance of integrin/receptor binding, β1 integrin has been shown to have a role in HSC homing. During normal development, colonization of fetal liver, bone marrow, and spleen by HSCs requires the expression and function of β1 integrin, which had been implicated in directing the homing of yolk sac, or aortagonad mesonephros (AGM)-derived HSCs, to these tissues (Hirsch et al. 1996; Potocnik et al. 2000). In the adult setting, blocking antibodies against VLA-4 (which is induced by SDF-1α), VLA-5 (α5β1), or their shared subunit β1 integrin prevent engraftment of NOD-SCID mice by human HSC-enriched CD34+ cord blood (Papayannopoulou et al. 1995; Peled et al. 2000). Furthermore, we have shown that inhibition of VLA-4 blocks bone marrow homing and engraftment by murine HSCs (Wagers et al.

2002a). As evidenced by the involvement of select chemokine pathways and integrin/receptor interactions, HSC migration is a dynamic and complex process that involves multiple interactions and signaling pathways.

ORIGIN OF LEUKEMIA STEM CELLS

Given the observation that the hematopoietic cell hierarchy is tightly regulated at every step from self-renewal of HSCs to differentiation down the progenitor pathway, deregulation of such homeostasis can lead to severe diseases, none the least of which is cancer. Given the observation that cancers are generally monoclonal in nature with heterogeneous cell populations at varying degrees of differentiation, it is not surprising that cancer would be organized in a cellular hierarchy similar to that of the hematopoietic and other tissue-forming systems with a cancer stem cell at the top of the hierarchy. Although the notion that tumors arise from stem cells was first proposed by Cohnheim (1875), it was not until the identification of the leukemia stem cell (LSC) in acute myelogenous leukemia (AML) that experimental evidence demonstrated the existence of a cancer stem cell.

Ever since the first enrichment of an LSC population in a hematological malignancy by Dick and colleagues, much research focus has been on elucidating the genetic and epigenetic events that cause transformation of a normal cell into an LSC as well as the stage of hematopoietic development from which LSCs arise. The first reported enrichment of an LSC was found in human AML when Dick and colleagues found that a small subset of AML cells, CD34+CD38− cells, were able to transplant AML into NOD-SCID and possessed key stem cell properties of differentiation and self-renewal (Lapidot et al. 1994; Bonnet and Dick 1997). Because of the phenotypic similarities between normal HSCs and LSCs, it has been postulated that LSCs arise from HSCs. Given the unique self-renewal property and subsequent long life span of HSCs, this cell population would be most ideal to acquire multiple mutations required for leukemic transformation. However, subsequent experiments demonstrated that AML LSCs, unlike HSCs, do not express CD90 (Thy1) (Blair et al. 1997; Miyamoto et al. 2000). These observations led to two hypotheses on the origin of the LSC: (1) AML LSCs originate from HSCs but aberrantly lose expression of CD90 during transformation or (2) AML LSC transformation occurs in a downstream committed progenitor cell. To test these hypotheses, we first investigated the cell surface phenotypes in AML-ETO1 translocation-associated AML patients who were atomic bomb survivors from Hiroshima. Consistent with previous studies, we found that the AML LSC was found in the Lin−CD34+CD38−CD90− (Fig. 2). However, when the bone marrow of long-term disease-free patients was analyzed, the AML1-ETO translocation was detected in Lin−CD34+CD38−CD90+ nonleukemic HSCs that gave rise to normal multilineage colonies in vitro. Interestingly, up to 40% of these HSCs possessed the AML-ETO1 translocation, whereas patients in remission retained approximately 1% of their HSCs as AML1-ETO-positive over several years posttreatment (Miyamoto et al. 2000). These observations led to an important conclusion: The AML1-ETO translocation is

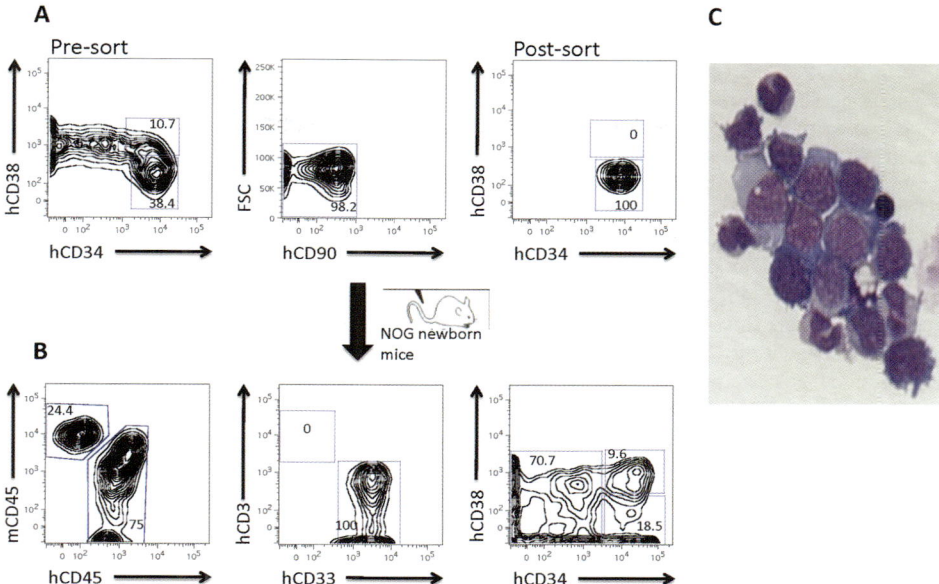

Figure 2. Cell surface phenotype of AML LSC. (*A*) Cell surface expression profile of bone marrow cells from a representative human AML patient. Lineage (CD3, CD19, CD20)-positive cells are gated out. The AML LSC (Lin⁻CD34⁺CD38⁻CD90⁻) population was prospectively isolated and injected via facial vein into NOD/SCID/IL2Rγ⁻/⁻ (NOG) pups. (*B*) Engraftment of human LSCs in mouse bone marrow after 10 weeks shows exclusively myeloid engraftment with recapitulation of CD34⁻, CD34⁺CD38⁺, and CD34⁺CD38⁻ populations. (*C*) Wright–Giemsa staining of purified human CD45⁺ cells indicates the presence of leukemia cells in recipient mouse bone marrow with blast morphology. (Image courtesy of Ravindra Majeti and Chris Park.)

necessary, but not sufficient, to induce leukemogenesis and occurs in an HSC population that is preleukemic. This conclusion was further substantiated by experiments involving transgenic mice expressing AML1-ETO driven from the MRP8 promoter. The MRP8 gene encodes a small calcium-binding protein of the S100 family that is only expressed in neutrophils, monocytes, and their immediate progenitors, CMPs and GMPs, but not HSCs (Lagasse and Weissman 1992). These hMRP8AML1-ETO transgenic mice were healthy with no evidence of leukemia; only after these mice were treated with a DNA alkylating mutagen did they develop leukemias (Yuan et al. 2001). To additionally address the possibility of a committed progenitor as the AML LSC, we investigated the role of the MLL-ENL translocation in the mouse hematopoietic system. The MLL-ENL translocation is frequently associated with the development of acute myeloid and lymphoid leukemias and has been shown to induce AML in mice (Lavau et al. 2000; Ayton and Cleary 2001; Cozzio et al. 2003). We retrovirally transduced the MLL-ENL fusion gene into prospectively sorted HSCs and downstream myeloid progenitors and found that transduction of HSCs, CMPs, or GMPs by MLL-ENL induced the exact same AML in vitro and in vivo, demonstrating that mutations in committed progenitors can initiate AML (Cozzio et al. 2003). To further elucidate the exact progenitor stage of LSC transformation, we recently showed that normal primitive hematopoietic cells (Lin⁻CD34⁺CD38⁻ CD90⁻CD45RA⁻) exhibiting a phenotype similar to that of LSC are multipotent progenitors, suggesting that AML stem cells could originate from this normally non-self-renewing population (Majeti et al. 2007). Therefore, in the case of AML, the LSC likely arises not at

the stage of HSC, but rather at a later stage of differentiation. Given these observations in AML, our lab then investigated other leukemia types to test this hypothesis.

Part of our investigation first turned to chronic myelogenous leukemia (CML). In 1977, Fialkow et al. (1977) first demonstrated by allelic X chromosomal markers that CML originates in a multipotent hematopoietic cell with expression of the canonical BCR-ABL translocation. Years later, our lab demonstrated that chronic phase CML HSCs harbor the BCR-ABL translocation (Jamieson et al. 2004b). These BCR-ABL⁺ HSCs outcompete their BCR-ABL⁻ HSC counterparts while simultaneously not increasing overall stem cell frequency. Correlating these findings in the mouse, we found that a chronic phase CML syndrome could be induced by blocking the expression of the transcription factor JunB with only HSCs and not progenitor populations capable of transplanting this CML-like disease (Passegué et al. 2004). Given these results, it is likely that in the chronic phase of CML, the transformation to LSC was at the HSC level. However, the pathogenesis of CML is triphasic, with transformation to a more acute leukemia phenotype occurring in the accelerated and blast crisis phases. Therefore, during the more aggressive accelerated and blast crisis phases of CML, is leukemic transformation at the level of the HSC or a committed progenitor? To first address this question, we determined whether other populations besides the HSC could initiate CML in a mouse model. We generated hMRP8p210^{BCR-ABL} transgenic mice in which the hMRP8 promoter expression of BCR-ABL is targeted exclusively to myeloid progenitors and progeny and is specifically absent on HSC. Interestingly, these mice developed a human CML-like disease, suggesting that a cell downstream from HSC could be the ultimate target of later leukemic transformation (Jaiswal et al. 2003). To support this notion in a more definitive matter, our lab examined CML patients during the chronic, accelerated, and myeloid blast crisis phases to determine population differences in stem cell potential. Interestingly, we found that in the blast crisis phase, the only cell population that was increased in frequency (besides the blast population) was the granulocyte-macrophage progenitor (GMP) population. In addition, BCR-ABL amplification occurred in the GMP population, whereas BCR-ABL transcript levels in HSCs remained constant during progression from chronic phase to blast crisis. In looking at β-catenin expression through CML disease progression, we found that GM progenitors in accelerated phase or blast crisis had increased β-catenin levels compared to GM progenitors in chronic phase or normal controls. Additionally, HSCs from chronic, accelerated, and blast crisis phase CML as well as normal HSCs all had similar levels of activated β-catenin, which was significantly lower than activated levels in blast crisis GMPs. Furthermore, these GMPs demonstrated aberrant self-renewal in vitro in a methylcellulose replating assay in which normal GM progenitors could not (Jamieson et al. 2004a,b). This GMP population, and not CML blasts, transplants into immunodeficient mice and can be serially transplanted into secondary mouse recipients, demonstrating in vivo self-renewal stem cell function (C.H. Jamieson, pers. comm.). Thus, it appears that in blast crisis CML, as with AML, the LSC arises from a committed progenitor population, the GMP, which has aberrantly acquired the ability to self-renew.

PROPERTIES OF LEUKEMIA STEM CELLS

Self-renewal

Like normal HSCs, a key requirement of LSCs is the ability to self-renew in order to propagate leukemic disease. Several signaling pathways including Wnt/β-catenin (Reya et al. 2003), Hox (Argiropoulos and Humphries 2007), and Bmi-1 (Raaphorst 2003) have been implicated in self-renewal roles for both normal and leukemic hematopoiesis. However, the normal homeostatic mechanisms regulating self-renewal in HSCs are likely disrupted in LSCs. First, we have demonstrated such abnormal self-renewal signaling in the Wnt/β-catenin pathway. As mentioned in CML, the candidate LSC in blast crisis, the GMP population, has significantly higher levels of nuclear active β-catenin compared to both CML HSCs and normal GMPs, which directly contributes to its ability to aberrantly self-renew in vitro and in vivo (Jamieson et al. 2004b). The abnormal self-renewing properties of the CML GMP can be subsequently inhibited by axin, an inhibitor of β-catenin.

Hox genes have also been implicated in self-renewal in both normal and abnormal hematopoiesis and have a role in stem and progenitor expansion. Hyperexpression of HoxB4 leads to expansion of the HSC pool (Sauvageau et al. 1995), whereas forced expression of Hoxa9, Hoxa10, Hoxb3, Hoxb6, and Hoxb8 confers a growth advantage in vitro and in vivo and leads to long latency leukemia (for review, see

Argiropoulos and Humphries 2007). Regulators of Hox gene expression are also aberrantly regulated in human leukemias. One such regulator is the mixed-lineage leukemia (MLL) transcription factor that is required for normal hematopoiesis, encoding chromatin modifiers that are required for proper maintenance of Hox gene expression. Rearrangements involving MLL fusion genes constitute 5% of all AML cases and 22% of those with acute lymphoblastic leukemia (Sauvageau et al. 1995; De Braekeleer et al. 2005). In our studies, we have shown that retroviral insertion of a leukemogenic MLL-ENL fusion gene into normal hematopoietic stem and progenitor cells induces an AML phenotype in vitro and in vivo, endowing the normally non-self-renewing committed progenitor populations with aberrant self-renewal properties evidenced by their ability for serial replating of methylcellulose colonies (Cozzio et al. 2003). We also observed similar acquisition of long-term self-renewal properties in multipotent progenitors when another MLL fusion gene, MLL-GAS7, was retrovirally inserted into HSC and early progenitors, which resulted in mixed-lineage leukemias (So et al. 2003).

Other molecular pathways involved in self-renewal in normal hematopoiesis that might also be implicated in leukemia include Bmi-1 (for review, see Raaphorst 2003). In 2003, we and other investigators demonstrated that Bmi-1 has a crucial role in the self-renewal of LSCs. Fetal liver cells from Bmi-1$^{-/-}$ mice were retrovirally transduced with Hox9 and Meis1a oncogenes (which generate a well-characterized myeloid leukemia; Kroon et al. 1998) and were transplanted into syngeneic mouse hosts. In the resulting leukemia that developed, Bmi-1$^{-/-}$ bone marrow leukemia cells were unable to form secondary transplants and were less efficient at generating leukemia colony-forming cells compared to Bmi-1 wild-type leukemia cells (Lessard and Sauvageau 2003). We then investigated the relative expression levels of Bmi-1 within the hematopoietic cell hierarchy by generating green fluorescent protein (GFP) knockin mice expressed under endogenous transcriptional regulatory elements of the Bmi-1 gene. We found that Bmi-1 is expressed in HSCs at its highest levels and down-regulated upon commitment to differentiation (Hosen et al. 2007). Looking at two murine leukemia models induced by p210BCR/ABL or TEL/PDGFβR + AML1/ETOm we found that expression of Bmi-1 was highest in leukemic HSCs compared to downstream progenitors, albeit at similar expression levels compared to normal HSCs (Hosen et al. 2007).

Programmed Cell Death

Given the general hallmark of tumor cells to escape homeostatic controls that allow unregulated proliferation, tumor cells must have aberrant abilities to escape apoptosis, a key regulator of proliferative homeostasis. Indeed, this is the case for acute leukemias and LSCs. Because the down-regulation of apoptosis is generally a key feature of tumor pathogenesis, it is not surprising that acute leukemias and LSCs have deregulated apoptotic signaling. The antiapoptotic gene Bcl-2 has long been shown to have a role in lymphoid malignancies, with a growing body of evidence suggesting that deregulation of Bcl-2 is important in the leukemic transformation of myeloid cells. Leukemia cells from most human AML subtypes have been found to express Bcl-2 at levels significantly higher than their normal cellular counterparts (Delia et al. 1992; Bensi et al. 1995) as well as highest levels of Bcl-2 expression found in early LSC progenitors (Konopleva et al. 2002). In a mouse model overexpressing the Bcl-2 gene driven by the major histocompatibility complex (MHC) class I promoter, HSCs were more resistant to radiation-induced apoptosis than wild-type HSCs, and these mice possessed a larger HSC compartment (Domen et al. 1998), suggesting that Bcl-2 has a specific role in the protection of HSCs from apoptosis with possibly similar behavior in the LSC population. Despite this, overexpression of Bcl-2 in transgenic murine models has been found to be relatively benign in terms of cellular transformation (Cory et al. 1994), suggesting that additional oncogenic mutations are needed for transformation. We showed that enforced expression of Bcl-2 in the myeloid lineage with the hMRP8 promoter allows monocytes from these mice to survive in culture in the absence of serum or exogenous growth factors compared to wild-type cells rapidly undergoing apoptosis, suggesting that Bcl-2 expression may promote a prolonged cellular survival that would allow a greater possibility of acquiring additional oncogenic mutations. In line with this hypothesis, we have shown that enforced Bcl-2 expression greatly increases the incidence of CML-like disease in hMRP8$^{BCR-ABL}$hMRP8^{Bcl-2} double transgenic mice (Jaiswal et al. 2003), as well as the incidence of acute promyelocytic leukemia in hMRP8$^{PML/RAR\alpha}$hMRP8^{Bcl-2} double transgenic mice (Kogan et al. 2001).

In addition to Bcl-2, we have shown that the proapoptotic Fas pathway might be involved in myeloid leukemia pathogenesis. Loss-of-function mutations in the Fas receptor (CD95) or in components of the Fas signaling pathway have been implicated in several human AMLs (Robertson et al. 1995; Bouscary et al. 1997). Furthermore, human CD34$^+$CD38$^-$ AML progenitors express lower levels of Fas/Fas-ligand and Fas-induced apoptosis compared to CD34$^+$CD38$^+$ blasts (Costello et al. 2000). To further assess the role of the Fas pathway in AML leukemogenesis, we crossed hMRP8Bcl-2 and Fas-deficient Faslpr/lpr mice and observed the development of AML in 15% of these crossed mice but not in either of the individual mice strains (Traver et al. 1998). These results suggest that Bcl-2 and the Fas receptor regulate two distinct apoptotic pathways in myeloid cells, with changes in both that are sufficient to induce AML at a low frequency.

Altogether, these observations demonstrate that the prevention of cell death through either up-regulation of antiapoptotic signals or down-regulation of proapoptotic signals is an important event in the cellular transformation to myeloid leukemias.

Immune Surveillance

Many tumors have mechanisms that function in the evasion of immune surveillance, mechanisms that contribute to disease pathogenesis. For example, tumors can downregulate components of the MHC class I antigen-processing pathway to avoid recognition by cytotoxic T lymphocytes, secrete immunosuppressive cytokines, or use methods to evade the innate immune system. We have recently identified CD47 as having a role in immune evasion in AML. CD47, a cell surface protein also known as integrin-associated protein, is known among other functions to inhibit phagocytosis by macrophages and other immune effector cells by binding its ligand signal regulatory protein α (Sirpα) on phagocytes. We have shown that CD47 is up-regulated on mouse and human bulk leukemia cells as well as LSCs in both CML and AML compared to normal cellular counterparts (S. Jaiswal et al.; R. Majeti and M.P. Chao; both in prep.). In addition, enforced CD47 expression facilitates engraftment of human leukemia cells in immunodeficient mice that otherwise cannot engraft (S. Jaiswal et al., in prep.). Furthermore, antibody blockade of the CD47-Sirpα interaction causes human leukemia cells that normally evade phagocytosis to become engulfed and eliminated by macrophages (R. Majeti and M.P. Chao, in prep.). Altogether, the identification of pathways that regulate evasion of immune surveillance in AML LSCs could provide insight into a new avenue of leukemic biology as well as represent a potential therapeutic target.

A MULTISTEP MODEL OF MYELOID LEUKEMIA PROGRESSION

In both AML and CML, we observed that initial leukemic mutations (AML1-ETO in AML and BCR-ABL in CML) are found in the HSC population and that leukemia initiated at the HSC level likely represents an initial leukemic phase that is chronic in nature. Rather, it appears that multiple independent genetic and epigenetic changes must occur to drive leukemia to a more aggressive phase, which is marked by LSC transformation at the committed progenitor level. For example, in the case of CML, studies have shown that 30–75% of normal, healthy individuals possess the BCR-ABL translocation in a minority of peripheral blood leukocytes (Biernaux et al. 1996; Bose et al. 1998). In addition, in our hMRP8p210$^{BCR-ABL}$ transgenic mouse model, these mice developed a disease similar to human chronic phase CML. However, hMRP8p210$^{BCR-ABL}$ mice crossed with apoptosis-resistant hMRP8Bcl-2 mice induced a blast crisis disease in 50% of the mice, again suggesting that additional mutations are necessary for the progression from chronic phase to blast crisis CML (Jaiswal et al. 2003). For perpetual propagation of leukemic mutations to occur, these mutations must occur in a cell population that can self-renew, otherwise the short life of a committed cell would not sustain these oncogenic events. Given that normally committed hematopoietic progenitors are targets for LSC transformation in AML and CML, it is probable that these downstream progenitors undergo several oncogenic events over time that ultimately lead to aberrant self-renewal, which then escape homeostatic controls and ultimately lead to proliferative leukemia (Fig. 3).

CONCLUSIONS

Both the mouse and human hematopoietic systems consist of a tightly regulated cellular hierarchy that is governed by step-by-step lineage restriction correlating with

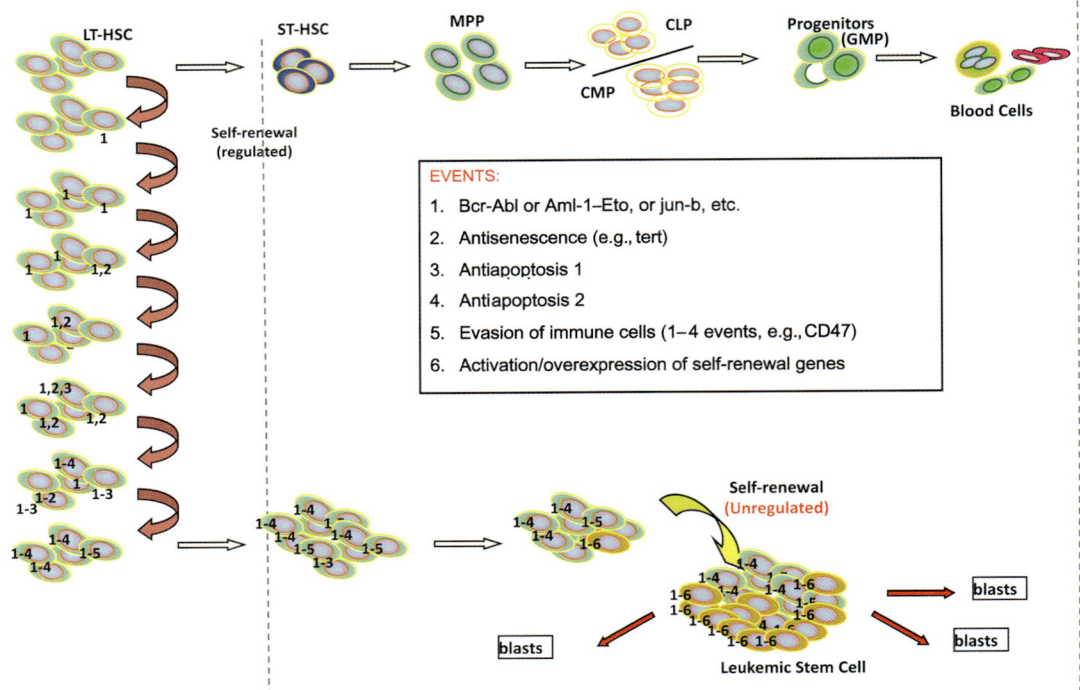

Figure 3. Multistep model of progression to AML or blast crisis CML. HSCs undergo differentiation to myeloid lineages in a quantal fashion (*top*). Given their long cellular life, HSCs acquire initial mutations that get passed down to self-renewing clonal HSCs. These HSC clones acquire further mutations or epigenetic events until the accumulation of multiple mutations and epigenesis leads to acquisition of aberrant self-renewal properties, with transformation of the HSC to the LSC. These LSCs then lead to a proliferation of leukemic blasts.

gradual loss of self-renewal. Identification and characterization of hematopoietic stem and progenitor cells have led to the success of bone marrow transplantation for a variety of human diseases and have provided precedence for the therapeutic potential of tissue-forming stem cells in other organ systems. However, the unique properties that separate HSCs from committed progenitors, including self-renewal, multipotency, cell survival, and migration, can also be aberrantly deregulated leading to the transformation of the HSCs into LSCs. Indeed, LSCs abnormally activate self-renewal pathways and antiapoptotic signals and possess metastatic features consistent with a loss in normal stem cell homeostasis. Although several genetic events have been identified in the transformation of normal hematopoietic stem and progenitor cells into LSCs, work must be done to identify all of the events that are necessary and sufficient for leukemogenesis.

Such elucidation can then lead to the identification of the sequential progression of these events and determine whether the progression is always similar across patients. One can then imagine being able to genetically profile a leukemia patient for these mutations and subsequently determine the stage of leukemia. In addition, knowledge of such events as well as determining the genetic and cellular differences between LSC and normal HSC will lead to novel therapies that eradicate the initiator of tumorigenesis, the cancer stem cell.

ACKNOWLEDGMENTS

We thank Siddhartha Jaiswal, Ravindra Majeti, Catriona Jamieson, Deepta Bhattacharya, Agnieszka Czechowicz, and Lisa Ooi for contributing data cited in this work. We also thank Derrick Rossi and David Bryder for manuscript assistance. M.P.C. is supported by a Howard Hughes Medical Institute Medical Fellowship and J.S. is supported by a fellowship from the California Institute for Regenerative Medicine. This research is supported by National Institutes of Health grant R01CA86017 to I.L.W.

REFERENCES

Adolfsson, J., Månsson, R., Buza-Vidas, N., Hultquist, A., Liuba, K., Jensen, C.T., Bryder, D., Yang, L., Borge, O.J., Thoren, L.A., et al. 2005. Identification of Flt3[+] lympho-myeloid stem cells lacking erythro-megakaryocytic potential a revised road map for adult blood lineage commitment. *Cell* **121:** 295–306.

Akashi, K., Traver, D., Miyamoto, T., and Weissman, I.L. 2000. A clonogenic common myeloid progenitor that gives rise to all myeloid lineages. *Nature* **404:** 193–197.

Allsopp, R.C., Cheshier, S., and Weissman, I.L. 2001. Telomere shortening accompanies increased cell cycle activity during serial transplantation of hematopoietic stem cells. *J. Exp. Med.* **193:** 917–924.

Allsopp, R.C., Morin, G.B., Horner, J.W., DePinho, R., Harley, C.B., and Weissman, I.L. 2003. Effect of TERT over-expression on the long-term transplantation capacity of hematopoietic stem cells. *Nat. Med.* **9:** 369–671.

Ara, T., Tokoyoda, K., Sugiyama, T., Egawa, T., Kawabata, K., and Nagasawa, T. 2003. Long-term hematopoietic stem cells require stromal cell-derived factor-1 for colonizing bone marrow during ontogeny. *Immunity* **19:** 257–267.

Argiropoulos, B. and Humphries, R.K. 2007. *Hox* genes in hematopoiesis and leukemogenesis. *Oncogene* **26:** 6766–6776.

Arinobu, Y., Mizuno, S., Chong, Y., Shigematsu, H., Iino, T., Iwasaki, H., Graf, T., Mayfield, R., Chan, S., Kastner, P., and Akashi, K. 2007. Reciprocal activation of GATA-1 and PU.1 marks initial specification of hematopoietic stem cells into myeloerythroid and myelolymphoid lineages. *Cell Stem Cell* **1:** 416–427.

Ayton, P.M. and Cleary, M.L. 2001. Molecular mechanisms of leukemogenesis mediated by MLL fusion proteins. *Oncogene* **20:** 5695–5707.

Baum, C.M., Weissman, I.L., Tsukamoto, A.S., Buckle, A.M., and Peault, B. 1992. Isolation of a candidate human hematopoietic stem-cell population. *Proc. Natl. Acad. Sci.* **89:** 2804–2808.

Becker, A.J., McCulloch, C.E., and Till, J.E. 1963. Cytological demonstration of the clonal nature of spleen colonies derived from transplanted mouse marrow cells. *Nature* **197:** 452–454.

Bensi, L., Longo, R., Vecchi, A., Messora, C., Garagnani, L., Bernardi, S., Tamassia, M.G., and Sacchi, S. 1995. Bcl-2 oncoprotein expression in acute myeloid leukemia. *Haematologica* **80:** 98–102.

Bhardwaj, G., Murdoch, B., Wu, D., Baker, D.P., Williams, K.P., Chadwick, K., Ling, L.E., Karanu, F.N., and Bhatia, M. 2001. Sonic hedgehog induces the proliferation of primitive human hematopoietic cells via BMP regulation. *Nat. Immunol.* **2:** 172–180.

Bhatia, M., Wang, J.C., Kapp, U., Bonnet, D., and Dick, J.E. 1997. Purification of primitive human hematopoietic cells capable of repopulating immune-deficient mice. *Proc. Natl. Acad. Sci.* **94:** 5320–5325.

Biernaux, C., Sels, A., Huez, G., and Stryckmans, P. 1996. Very low level of major BCR-ABL expression in blood of some healthy individuals. *Bone Marrow Transplant.* (suppl. 3) **17:** S45–S47.

Blair, A., Hogge, D.E., Ailles, L.E., Lansdorp, P.M., and Sutherland, H.J. 1997. Lack of expression of Thy-1 (CD90) on acute myeloid leukemia cells with long-term proliferative ability in vitro and in vivo. *Blood* **89:** 3104–3112.

Bonnet, D. and Dick, J.E. 1997. Human acute myeloid leukemia is organized as a hierarchy that originates from a primitive hematopoietic cell. *Nat. Med.* **3:** 730–737.

Bose, S., Deininger, M., Gora-Tybor, J., Goldman, J.M., and Melo, J.V. 1998. The presence of typical and atypical BCR-ABL fusion genes in leukocytes of normal individuals: Biologic significance and implications for the assessment of minimal residual disease. *Blood* **92:** 3362–3367.

Bouscary, D., De Vos, J., Guesnu, M., Jondeau, K., Viguier, F., Melle, J., Picard, F., Dreyfus, F., and Fontenay-Roupie, M. 1997. Fas/Apo-1 (CD95) expression and apoptosis in patients with myelodysplastic syndromes. *Leukemia* **11:** 839–845.

Brummendorf, T.H., Rufer, N., Baerlocher, G.M., Roosnek, E., and Lansdorp, P.M. 2001. Limited telomere shortening in hematopoietic stem cells after transplantation. *Ann. N.Y. Acad. Sci.* **938:** 1–8.

Cao, Y.A., Wagers, A.G., Beilhack, A., Dusich, J., Bachmann, M.H., Negrin, R.S., Weissman, I.L., and Contag, C.H. 2004. Shifting foci of hematopoiesis during reconstitution from single stem cells. *Proc. Natl. Acad. Sci.* **101:** 221–226.

Cheshier, S.H., Morrison, S.J., Liao, X., and Weissman, I.L. 1999. In vivo proliferation and cell cycle kinetics of long-term self-renewing hematopoietic stem cells. *Proc. Natl. Acad. Sci.* **96:** 3120–3125.

Christensen, J.L. and Weissman, I.L. 2001. Flk-2 is a marker in hematopoietic stem cell differentiation: A simple method to isolate long-term stem cells. *Proc. Natl. Acad. Sci.* **98:** 14541–14546.

Civin, C.I., Strauss, L.C., Brovall, C., Fackler, M.J., Schwartz, J.F., and Shaper, J.H. 1984. Antigenic analysis of hematopoiesis. III. A hematopoietic progenitor cell surface antigen defined by a monoclonal antibody raised against KG-1a cells. *J. Immunol.* **133:** 157–165.

Cohnheim, V. 1875. Congenitales, quergestreiftes muskelsarkom der nieren. *Virchows Arch. Pathol. Anat. Physiol. Klin. Med.* **65:** 64–69.

Cory, S., Harris, A.W., and Strasser, A. 1994. Insights from transgenic mice regarding the role of *bcl-2* in normal and neoplastic lymphoid cells. *Philos. Trans. R. Soc. Lond. B Biol. Sci.* **345:** 289–295.

Costello, R.T., Mallet, F., Gaugler, B., Sainty, D., Arnoulet, C., Gastaut, J.A., and Olive, D. 2000. Human acute myeloid leukemia CD34$^+$/CD38$^-$ progenitor cells have decreased sensitivity to chemotherapy and Fas-induced apoptosis, reduced immunogenicity, and impaired dendritic cell transformation capacities. *Cancer Res.* **60:** 4403–4411.

Cozzio, A., Passegué, E., Ayton, P.M., Karsunky, H., Cleary, M.L., and Weissman, I.L. 2003. Similar MLL-associated leukemias arising from self-renewing stem cells and short-lived myeloid progenitors. *Genes Dev.* **17:** 3029–3035.

De Braekeleer, M., Morel, F., Le Bris, M.J., Herry, A., and Douet-Guilbert, N. 2005. The MLL gene and translocations involving chromosomal band 11q23 in acute leukemia. *Anticancer Res.* **25:** 1931–1944.

Delia, D., Aiello, A., Soligo, D., Fontanella, E., Melani, C., Pezzella F., Pierotti, M.A., and Della Porta, G. 1992. bcl-2 proto-oncogene expression in normal and neoplastic human myeloid cells. *Blood* **79:** 1291–1298.

DiGiusto, D., Chen, S., Combs, J., Webb, S., Namikawa, R., Tsukamoto, A., Chen, B.P., and Galy, A.H. 1994. Human fetal bone marrow early progenitors for T, B, and myeloid cells are found exclusively in the population expressing high levels of CD34. *Blood* **84:** 421–432.

Domen, J., Gandy, K.L., and Weissman. I.L. 1998. Systemic overexpression of BCL-2 in the hematopoietic system protects transgenic mice from the consequences of lethal irradiation. *Blood* **91:** 2272–2282.

Fialkow, P.J., Jacobson, R.J., and Papayannopoulou, T. 1977. Chronic myelocytic leukemia: Clonal origin in a stem cell common to the granulocyte, erythrocyte, platelet and monocyte/macrophage. *Am. J. Med.* **63:** 125–130.

Forsberg, E.C., Serwold, T., Kogan, S., Weissman, I.L., and Passegué, E. 2006. New evidence supporting megakaryocyte-erythrocyte potential of flk2/flt3$^+$ multipotent hematopoietic progenitors. *Cell* **126:** 415–426.

Fritsch, G., Buchinger, P., Printz, D., Fink, F.M., Mann, G., Peters, C., Wagner, T., Adler, A., and Gadner, H. 1993. Rapid discrimination of early CD34$^+$ myeloid progenitors using CD45-RA analysis. *Blood* **81:** 2301–2309.

Galy, A., Travis, M., Cen, D., and Chen, B. 1995. Human T, B, natural killer, and dendritic cells arise from a common bone marrow progenitor cell subset. *Immunity* **3:** 459–473.

Guenechea, G., Gan, O.I., Dorrell, C., and Dick, J.E. 2001. Distinct classes of human stem cells that differ in proliferative and self-renewal potential. *Nat. Immunol.* **2:** 75–82.

Hao, Q.L., Smogorzewska, E.M., Barsky, L.W., and Crooks, G.M. 1998. In vitro identification of single CD34$^+$CD38$^-$ cells with both lymphoid and myeloid potential. *Blood* **91:** 4145–4151.

Hao, Q.L., Thiemann, F.T., Petersen, D., Smogorzewska, E.M., and Crooks, G.M. 1996. Extended long-term culture reveals a highly quiescent and primitive human hematopoietic progenitor population. *Blood* **88:** 3306–3313.

Harrison, D.E. and Astle, C.M. 1982. Loss of stem cell repopulating ability upon transplantation. Effects of donor age, cell number, and transplantation procedure. *J. Exp. Med.* **156:** 1767–1779.

Harrison, D.E., Astle, C.M., and Delaittre, J.A. 1978. Loss of proliferative capacity in immunohemopoietic stem cells caused by serial transplantation rather than aging. *J. Exp. Med.* **147:** 1526–1531.

Hirsch, E., Iglesias, A., Potocnik, A.J., Hartmann, U., and Fassler, R. 1996. Impaired migration but not differentiation of haematopoietic stem cells in the absence of β1 integrins. *Nature* **380:** 171–175.

Hosen, N., Yamane, T., Muijtjens, M., Pham, K., Clarke, M.F., and Weissman, I.L. 2007. Bmi-1-green fluorescent protein-knock-in mice reveal the dynamic regulation of bmi-1 expression in normal and leukemic hematopoietic cells. *Stem Cells* **25:** 1635–1644.

Huang, S. and Terstappen, L.W. 1994. Lymphoid and myeloid differentiation of single human CD34+, HLA-DR+, CD38− hematopoietic stem cells. *Blood* **83:** 1515–1526.

Ikuta, K. and Weissman, I.L. 1992. Evidence that hematopoietic stem cells express mouse *c-kit* but do not depend on steel factor for their generation. *Proc. Natl. Acad. Sci.* **89:** 1502–1506.

Jaiswal, S., Traver, D., Miyamoto, T., Akashi, K., Lagasse, E., and Weissman, I.L. 2003. Expression of *BCR/ABL* and *BCL-2* in myeloid progenitors leads to myeloid leukemias. *Proc. Natl. Acad. Sci.* **100:** 10002–10007.

Jamieson, C.H., Weissman, I.L., and Passegué, E. 2004a. Chronic versus acute myelogenous leukemia: A question of self-renewal. *Cancer Cell* **6:** 531–533.

Jamieson, C.H., Ailles, L.E., Dylla, S.J., Muijtjens, M., Jones, C., Zehnder, J.L., Gotlib, J., Li, K., Manz, M.G., Keating, A., Sawyers, C.L., and Weissman, I.L. 2004b. Granulocyte-macrophage progenitors as candidate leukemic stem cells in blast-crisis CML. *N. Engl. J. Med.* **351:** 657–667.

Karanu, F.N., Murdoch, B., Gallacher, L., Wu, D.M., Koremoto, M., Sakano, S., and Bhatia, M. 2000. The Notch ligand Jagged-1 represents a novel growth factor of human hematopoietic stem cells. *J. Exp. Med.* **192:** 1365–1372.

Kiel, M.J., Yilmaz, O.H., Iwashita, T., Yilmaz, O.H., Terhorst, C., and Morrison, S.J. 2005. SLAM family receptors distinguish hematopoietic stem and progenitor cells and reveal endothelial niches for stem cells. *Cell* **121:** 1109–1121.

Kimura, T., Sakabe, H., Tanimukai, S., Abe, T., Urata, Y., Yasukawa, K., Okano, A., Taga, T., Sugiyama, H., Kishimoto, T., and Sonoda, Y. 1997. Simultaneous activation of signals through gp130, c-kit, and interleukin-3 receptor promotes a trilineage blood cell production in the absence of terminally acting lineage-specific factors. *Blood* **90:** 4767–4778.

Kogan, S.C., Brown, D.E., Shultz, D.B., Truong, B.T., Lallemand-Breitenbach, V., Guillemin, M.C., Lagasse, E., Weissman, I.L., and Bishop, J.M. 2001. BCL-2 cooperates with promyelocytic leukemia retinoic acid receptor α chimeric protein (PMLRARα) to block neutrophil differentiation and initiate acute leukemia. *J. Exp. Med.* **193:** 531–543.

Kondo, M., Weissman, I.L., and Akashi, K. 1997. Identification of clonogenic common lymphoid progenitors in mouse bone marrow. *Cell* **91:** 661–672.

Konopleva, M., Zhao, S., Hu, W., Jiang, S., Snell, V., Weidner, D., Jackson, C.E., Zhang, X., Champlin, R., Estey, E., Reed, J.C., and Andreeff, M. 2002. The anti-apoptotic genes Bcl-X_L and Bcl-2 are over-expressed and contribute to chemoresistance of non-proliferating leukaemic CD34+ cells. *Br. J. Haematol.* **118:** 521–534.

Krause, D.S., Fackler, M.J., Civin, C.I., and May, W.S. 1996. CD34: Structure, biology, and clinical utility. *Blood* **87:** 1–13.

Kroon, E., Krosl, J., Thorsteinsdottir, U., Baban, S., Buchberg, A.M., and Sauvageau, G. 1998. *Hoxa9* transforms primary bone marrow cells through specific collaboration with *Meis1a* but not *Pbx1b*. *EMBO J.* **17:** 3714–3725.

Lagasse, E. and Weissman, I.L. 1992. Mouse MRP8 and MRP14, two intracellular calcium-binding proteins associated with the development of the myeloid lineage. *Blood* **79:** 1907–1915.

Lapidot, T., Sirard, C., Vormoor, J., Murdoch, B., Hoang, T., Caceres-Cortes, J., Minden, M., Paterson, B., Caligiuri, M.A., and Dick, J.E. 1994. A cell initiating human acute myeloid leukaemia after transplantation into SCID mice. *Nature* **367:** 645–648.

Larochelle, A., Vormoor, J., Hanenberg, H., Wang, J.C., Bhatia, M., Lapidot, T., Moritz, T., Murdoch, B., Xiao, X.L., Kato, I., Williams, D.A., and Dick, J.E. 1996. Identification of primitive human hematopoietic cells capable of repopulating NOD/SCID mouse bone marrow: Implications for gene therapy. *Nat. Med.* **2:** 1329–1337.

Lavau, C., Luo, R.T., Du, C., and Thirman, M.J. 2000. Retrovirus-mediated gene transfer of MLL-ELL transforms primary myeloid progenitors and causes acute myeloid

leukemias in mice. *Proc. Natl. Acad. Sci.* **97:** 10984–10989.

Lessard, J. and Sauvageau, G. 2003. *Bmi-1* determines the proliferative capacity of normal and leukaemic stem cells. *Nature* **423:** 255–260.

Majeti, R., Park, C., and Weissman, I. 2007. Identification of a hierarchy of multipotent hematopoietic progenitors in human cord blood. *Cell Stem Cell* **1:** 635–645.

Manz, M.G., Miyamoto, T., Akashi, K., and Weissman, I.L. 2002. Prospective isolation of human clonogenic common myeloid progenitors. *Proc. Natl. Acad. Sci.* **99:** 11872–11877.

Mazurier, F., Gan, O.I., McKenzie, J.L., Doedens, M., and Dick, J.E. 2004. Lentivector-mediated clonal tracking reveals intrinsic heterogeneity in the human hematopoietic stem cell compartment and culture-induced stem cell impairment. *Blood* **103:** 545–552.

McKenzie, J.L., Gan, O.I., Doedens, M., Wang, J.C., and Dick, J.E. 2006. Individual stem cells with highly variable proliferation and self-renewal properties comprise the human hematopoietic stem cell compartment. *Nat. Immunol.* **7:** 1225–1233.

Michallet, M., Philip, T., Philip, I., Godinot, H., Sebban, C., Salles, G., Thiebaut, A., Biron, P., Lopez, F., Mazars, P., et al. 2000. Transplantation with selected autologous peripheral blood CD34+Thy1+ hematopoietic stem cells (HSCs) in multiple myeloma: Impact of HSC dose on engraftment, safety, and immune reconstitution. *Exp. Hematol.* **28:** 858–870.

Miller, J.S., McCullar, V., Punzel, M., Lemischka, I.R., and Moore, K.A. 1999. Single adult human CD34+/Lin−/CD38− progenitors give rise to natural killer cells, B-lineage cells, dendritic cells, and myeloid cells. *Blood* **93:** 96–106.

Miyamoto, T., Weissman, I.L., and Akashi, K. 2000. AML1/ETO-expressing nonleukemic stem cells in acute myelogenous leukemia with 8;21 chromosomal translocation. *Proc. Natl. Acad. Sci.* **97:** 7521–7526.

Mohle, R., Bautz, F., Rafii, S., Moore, M.A., Brugger, W., and Kanz, L. 1998. The chemokine receptor CXCR-4 is expressed on CD34+ hematopoietic progenitors and leukemic cells and mediates transendothelial migration induced by stromal cell-derived factor-1. *Blood* **91:** 4523–4530.

Morrison, S.J. and Weissman, I.L. 1994. The long-term repopulating subset of hematopoietic stem cells is deterministic and isolatable by phenotype. *Immunity* **1:** 661–673.

Morrison, S.J., Prowse, K.R., Ho, P., and Weissman, I.L. 1996. Telomerase activity in hematopoietic cells is associated with self-renewal potential. *Immunity* **5:** 207–216.

Morrison, S.J., Wandycz, A.M., Hemmati, H.D., Wright, D.E., and Weissman, I.L. 1997. Identification of a lineage of multipotent hematopoietic progenitors. *Development* **124:** 1929–1939.

Nagasawa, T., Hirota, S., Tachibana, K., Takakura, N., Nishikawa, S., Kitamura, Y., Yoshida, N., Kikutani, H., and Kishimoto, T. 1996. Defects of B-cell lymphopoiesis and bone-marrow myelopoiesis in mice lacking the CXC chemokine PBSF/SDF-1. *Nature* **382:** 635–638.

Negrin, R.S., Atkinson, K., Leemhuis, T., Hanania, E., Juttner, C., Tierney, K., Hu, W.W., Johnston, L.J., Shizurn, J.A., Stockerl-Goldstein, K.E., et al. 2000. Transplantation of highly purified CD34+Thy-1+ hematopoietic stem cells in patients with metastatic breast cancer. *Biol. Blood Marrow Transplant.* **6:** 262–271.

Notaro, R., Cimmino, A., Tabarini, D., Rotoli, B., and Luzzatto, L. 1997. In vivo telomere dynamics of human hematopoietic stem cells. *Proc. Natl. Acad. Sci.* **94:** 13782–13785.

Ogawa, M. 1993. Differentiation and proliferation of hematopoietic stem cells. *Blood* **81:** 2844–2853.

Osawa, M., Hanada, K., Hamada, H., and Nakauchi, H. 1996. Long-term lymphohematopoietic reconstitution by a single CD34-low/negative hematopoietic stem cell. *Science* **273:** 242–245.

Papayannopoulou, T., Craddock, C., Nakamoto, B., Priestley, G.V., and Wolf, N.S. 1995. The VLA4/VCAM-1 adhesion pathway defines contrasting mechanisms of lodgement of transplanted murine hemopoietic progenitors between bone marrow and spleen. *Proc. Natl. Acad. Sci.* **92:** 9647–9651.

Passegué, E., Wagner, E.F., and Weissman, I.L. 2004. JunB deficiency leads to a myeloproliferative disorder arising from hematopoietic stem cells. *Cell* **119:** 431–443.

Peled, A., Petit, I., Kollet, O., Magid, M., Ponomaryov, T., Byk, T., Nagler, A., Ben-Hur, H., Many, A., Shultz, L., et al. 1999. Dependence of human stem cell engraftment and repopulation of NOD/SCID mice on CXCR4. *Science* **283:** 845–848.

Peled, A., Kollet, O., Ponomaryov, T., Petit, I., Franitza, S., Grabovsky, V., Slav, M.M., Nagler, A., Lider, O., Alon, R., Zipori, D., and Lapidot, T. 2000. The chemokine SDF-1 activates the integrins LFA-1, VLA-4, and VLA-5 on immature human CD34$^+$ cells: Role in transendothelial/stromal migration and engraftment of NOD/SCID mice. *Blood* **95:** 3289–3296.

Petzer, A.L., Hogge, D.E., Landsdorp, P.M., Reid, D.S., and Eaves, C.J. 1996. Self-renewal of primitive human hematopoietic cells (long-term-culture-initiating cells) in vitro and their expansion in defined medium. *Proc. Natl. Acad. Sci.* **93:** 1470–1474.

Potocnik, A.J., Brakebusch, C., and Fassler, R. 2000. Fetal and adult hematopoietic stem cells require β1 integrin function for colonizing fetal liver, spleen, and bone marrow. *Immunity* **12:** 653–663.

Raaphorst, F.M. 2003. Self-renewal of hematopoietic and leukemic stem cells: A central role for the Polycomb-group gene *Bmi-1. Trends Immunol.* **24:** 522–524.

Reya, T., Duncan, A.W., Ailles, L., Domen, J., Scherer, D.C., Willert, K., Hintz, L., Nusse, R., and Weissman, I.L. 2003. A role for Wnt signalling in self-renewal of haematopoietic stem cells. *Nature* **423:** 409–414.

Robertson, M.J., Manley, T.J., Pichert, G., Cameron, C., Cochran, K.J., Levine, H., and Ritz, J. 1995. Functional consequences of APO-1/Fas (CD95) antigen expression by normal and neoplastic hematopoietic cells. *Leuk. Lymphoma* **17:** 51–61.

Sauvageau, G., Thorsteinsdottir, U., Eaves, C.J., Lawrence, H.J., Largman, C., Lansdorp, P.M., and Humphries, R.K. 1995. Overexpression of *HOXB4* in hematopoietic cells causes the selective expansion of more primitive populations in vitro and in vivo. *Genes Dev.* **9:** 1753–1765.

Siminovitch, L., McCulloch, E.A., and Till, J.E. 1963. The distribution of colony-forming cells among spleen colonies. *J. Cell. Physiol.* **62:** 327–336.

Smith, L.G., Weissman, I.L., and Heimfeld, S. 1991. Clonal analysis of hematopoietic stem-cell differentiation in vivo. *Proc. Natl. Acad. Sci.* **88:** 2788–2792.

So, C.W., Karsunky, H., Passegué, E., Cozzio, A., Weissman, I.L., and Cleary, M.L. 2003. MLL-GAS7 transforms multipotent hematopoietic progenitors and induces mixed lineage leukemias in mice. *Cancer Cell* **3:** 161–171.

Spangrude, G.J., Heimfeld, S., and Weissman, I.L. 1988. Purification and characterization of mouse hematopoietic stem cells. *Science* **241:** 58–62.

Thorsteinsdottir, U., Sauvageau, G., and Humphries, R.K. 1999. Enhanced in vivo regenerative potential of *HOXB4*-transduced hematopoietic stem cells with regulation of their pool size. *Blood* **94:** 2605–2612.

Thorsteinsdottir, U., Mamo, A., Kroon, E., Jerome, L., Bijl, J., Lawrence, H.J., Humphries, K., and Sauvageau, G. 2002. Overexpression of the myeloid leukemia-associated *Hoxa9* gene in bone marrow cells induces stem cell expansion. *Blood* **99:** 121–129.

Till, J.E. and McCulloch, C.E. 1961. A direct measurement of the radiation sensitivity of normal mouse bone marrow cells. *Radiat. Res.* **14:** 213–222.

Traver, D., Akashi, K., Weissman, I.L., and Lagasse, E. 1998. Mice defective in two apoptosis pathways in the myeloid lineage develop acute myeloblastic leukemia. *Immunity* **9:** 47–57.

Uchida, N. and Weissman, I.L. 1992. Searching for hematopoietic stem cells: Evidence that Thy-1.1^{lo} Lin$^-$ Sca-1$^+$ cells are the only stem cells in C57BL/Ka-Thy-1.1 bone marrow. *J. Exp. Med.* **175:** 175–184.

Varnum-Finney, B., Xu, L., Brashem-Stein, C., Nourigat, C., Flowers, D., Bakkour, S., Pear, W.S., and Bernstein, I.D. 2000. Pluripotent, cytokine-dependent, hematopoietic stem cells are immortalized by constitutive Notch1 signaling. *Nat. Med.* **6:** 1278–1281.

Vose, J.M., Bierman, P.J., Lynch, J.C., Atkinson, K., Juttner, C., Hanania, C.E., Bociek, G., and Armitage, J.O. 2001. Transplantation of highly purified CD34$^+$Thy-1$^+$ hematopoietic stem cells in patients with recurrent indolent non-Hodgkin's lymphoma. *Biol. Blood Marrow Transplant.* **7:** 680–687.

Wagers, A.J., Allsopp, R.C., and Weissman, I.L. 2002a. Changes in integrin expression are associated with altered homing properties of Lin$^{-/lo}$Thy1.1^{lo}Sca-1$^+$c-kit$^+$ hematopoietic stem cells following mobilization by cyclophosphamide/granulocyte colony-stimulating factor. *Exp. Hematol.* **30:** 176–185.

Wagers, A.J., Sherwood, R.I., Christensen, J.L., and Weissman, I.L. 2002b. Little evidence for developmental plasticity of adult hematopoietic stem cells. *Science* **297:** 2256–2259.

Wright, D.E., Bowman, E.P., Wagers, A.J., Butcher, E.C., and Weissman, I.L. 2002. Hematopoietic stem cells are uniquely selective in their migratory response to chemokines. *J. Exp. Med.* **195:** 1145–1154.

Wright, D.E., Wagers, A.J., Gulati, A.P., Johnson, F.L., and Weissman, I.L. 2001. Physiological migration of hematopoietic stem and progenitor cells. *Science* **294:** 1933–1936.

Wu, A.M., Till, J.E., Siminovitch, L., and McCulloch, E.A. 1968. Cytological evidence for a relationship between normal hemotopoietic colony-forming cells and cells of the lymphoid system. *J. Exp. Med.* **127:** 455–464.

Yuan, Y., Zhou, L., Miyamoto, T., Iwasaki, H., Harakawa, N., Hetherington, C.J., Burel, S.A., Lagasse E., Weissman, I.L., Akashi, K., and Zhang, D.E. 2001. AML1-ETO expression is directly involved in the development of acute myeloid leukemia in the presence of additional mutations. *Proc. Natl. Acad. Sci.* **98:** 10398–10403.

Zou, Y.R., Kottmann, A.H., Kuroda, M., Taniuchi, I., and Littman, D.R. 1998. Function of the chemokine receptor CXCR4 in haematopoiesis and in cerebellar development. *Nature* **393:** 595–599.

Role of "Cancer Stem Cells" and Cell Survival in Tumor Development and Maintenance

J.M ADAMS, P.N. KELLY, A. DAKIC, S. CAROTTA, S.L. NUTT,
AND A. STRASSER

Walter & Eliza Hall Institute of Medical Research, Melbourne 3050, Australia

One critical issue for cancer biology is the nature of the cells that drive the inexorable growth of malignant tumors. Reports that only rare cell populations within human leukemias seeded leukemia in mice stimulated the now widely embraced hypothesis that only such "cancer stem cells" maintain all tumor growth. However, the mouse microenvironment might instead fail to support the dominant human tumor cell populations. Indeed, on syngeneic transplantation of mouse lymphomas and leukemias, we and other investigators have found that a substantial proportion (>10%) of their cells drive tumor growth. Thus, dominant clones rather than rare cancer stem cells appear to sustain many tumors. Another issue is the role of cell survival in tumorigenesis. Because tumor development can be promoted by the overexpression of prosurvival genes such as *bcl-2*, we are exploring the role of endogenous Bcl-2-like proteins in lymphomagenesis. The absence of endogenous Bcl-2 in mice expressing an Eµ-*myc* transgene reduced mature B-cell numbers and enhanced their apoptosis, but unexpectedly, lymphoma development was undiminished or even delayed. This suggests that these tumors originate in an earlier cell type, such as the pro-B or pre-B cell, and that the nascent neoplastic clones do not require Bcl-2 but may instead be protected by a Bcl-2 relative.

We address here two issues relevant to tumor development and maintenance. The first concerns the nature of the cells that perpetuate a tumor. It is now well established that normal tissues that turn over, such as the blood or gut, are sustained by rare tissue stem cells (see Weissman; Visvader et al.; both this volume). That principle stimulated the notion that the relentless growth of a tumor might be sustained not by most of its cells, but instead exclusively by a rare subpopulation, commonly termed the "cancer stem cells" (Wang and Dick 2005; Clarke et al. 2006). As reviewed recently (Clarke et al. 2006; Campbell and Polyak 2007; Adams and Strasser 2008; Vermeulen et al. 2008), this issue is attracting enormous interest, both because of its fundamental importance for tumor biology and its implications for therapy. We have investigated this issue for mouse hematopoietic tumors (Kelly et al. 2007a).

The second issue we address concerns the role of endogenous prosurvival members of the Bcl-2 family in tumorigenesis. Abatement of apoptosis is a key step in tumor development (Hanahan and Weinberg 2000; Cory and Adams 2002), and these proteins are the principal guardians against apoptosis (Adams and Cory 2007; Youle and Strasser 2008). Because overexpression of Bcl-2 can promote lymphoma development (Strasser et al. 1990), we have explored whether endogenous Bcl-2 is required for lymphomagenesis (Kelly et al. 2007b).

NATURE OF THE CELLS MAINTAINING THE INEXORABLE GROWTH OF TUMORS

Two Distinct Models for Tumor Propagation

As reviewed recently by us (Adams and Strasser 2008) and others (Wang and Dick 2005; Campbell and Polyak 2007; Vermeulen et al. 2008), two distinct models have been proposed to account for both the heterogeneity within a tumor and its inexorable growth. In the cancer stem cell model (Fig. 1A), tumor growth, like normal tissue development, relies exclusively on rare stem cells within it, and the vast majority of the cells, derived by differentiation from the cancer stem cells, lack self-renewal potential and hence do not contribute significantly to its perpetuation (Wang and Dick 2005; Clarke et al. 2006). Heterogeneity within the tumor is ascribed to somewhat aberrant differentiation from the cancer stem cell. In an alternative view (Fig. 1B), sometimes termed the "stochastic" (Wang and Dick 2005) or "clonal evolution model" (Campbell and Polyak 2007), most of the tumor cells contribute to tumor maintenance, albeit perhaps to varying degrees. This model ascribes tumor heterogeneity not only to differentiation, but also to intraclonal genetic and epigenetic variation plus microenvironmental influences. It envisions that a tumor is composed of subclones at different stages of neoplastic progression, each having a variable growth and survival advantage over normal cells.

The cancer stem cell model is thus highly hierarchical with a unique self-renewing cell type at the apex, whereas the clonal evolution model attributes much of the intratumor variation to subclonal differences in the mutational profile, and all except the terminally differentiated cells may well have some self-renewal capacity. Accordingly, in the cancer stem cell model, a phenotypically distinct and generally rare cell type maintains the tumor's growth, whereas in the clonal evolution model, the dominant subclone(s) sustains it.

Considerable confusion in the field has resulted because the term "cancer stem cell" is often also used to designate the normal cell in which the process of neoplastic transformation first began (the "cell of origin"). We follow here the current consensus that the term be restricted to the cell that maintains an established tumor

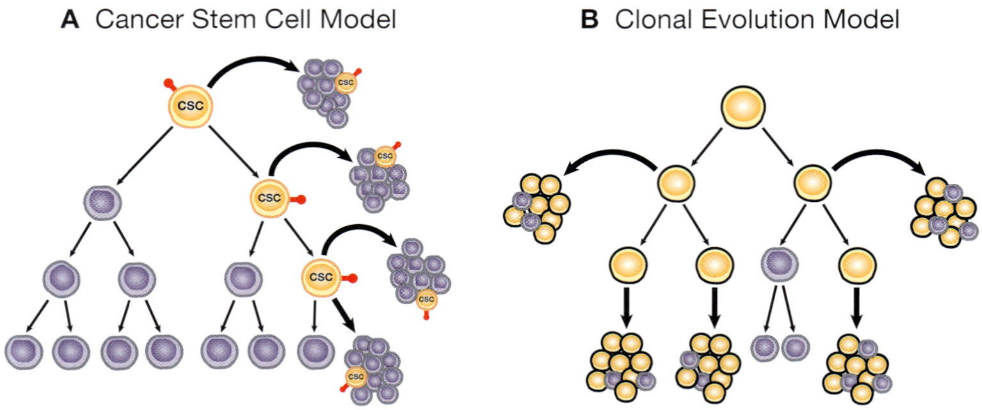

A Cancer Stem Cell Model   **B** Clonal Evolution Model

Figure 1. Models for the nature of sustained tumor growth. (*A*) In the cancer stem cell (CSC) model, only the CSC (*gold*), which can be isolated prospectively by surface markers (*red*), possesses self-renewal activity and hence represents the only relevant target for therapy. (*B*) In the clonal evolution model, a substantial proportion of the tumor cells (*gold*) can sustain its growth and hence therapy must attempt to eliminate all the cell types. (Modified from Adams and Strasser 2008.)

(Clarke et al. 2006). However, for clarity we often use more operational terms such as "tumor growth-sustaining," "tumor propagating," or "tumor perpetuating" cell.

Concerns about Xenotransplantation

The cancer stem cell model has arisen primarily from studies in which human tumor cells are transplanted at limit dilution into sublethally irradiated immunodeficient mice (Clarke et al. 2006). Support for this model was greatly stimulated by reports that only 1 in 10^4 to 10^7 of the cells in human acute myeloid leukemia (AML) could elicit leukemia in nonobese diabetic–severe combined immunodeficient (NOD-SCID) mice (Bonnet and Dick 1997; Wang and Dick 2005). As reviewed recently (Vermeulen et al. 2008; Visvader and Lindeman 2008), similar experiments have subsequently revealed putative cancer stem cell populations in diverse human solid tumors, including those of breast, colon, and brain origin (see, e.g., Al-Hajj et al. 2003; Singh et al. 2004; Ricci-Vitiani et al. 2007).

In our view, however, the interpretation of xenotransplantation experiments is problematic. First, it is now accepted that the growth of tumor cells requires an intricate network of interactions with different support cells, including fibroblasts, endothelial cells, macrophages, mast cells, and mesenchymal stem cells (Hanahan and Weinberg 2000), and many of the cytokines and receptors mediating these two-way interactions are incompatible between mice and humans (Arai et al. 1990). Second, whether many human tumor cells can home efficiently to an appropriate niche in the mouse is unknown. Third, the irradiation of the mice will kill may of the cells needed for an inflammatory response, which can aid tumor development (Lin and Karin 2007) and presumably also tumor engraftment. Finally, the natural killer cells remaining in NOD-SCID mice may eliminate some human tumor cell populations (Kong et al. 2008). Illustrating the limitations of xenotransplantation, 50% of human AML samples did not engraft irradiated NOD-SCID mice even when 10^7 or 10^8 cells were introduced (Pearce et al. 2006).

Proponents of the cancer stem cell hypothesis consider that the model is proven for human AML by evidence that cell populations prospectively isolated from the leukemia samples by surface markers (e.g., $CD34^+CD38^-$) seed leukemia in mice, whereas the majority cell population lacking that phenotype does not (Bonnet and Dick 1997; Wang and Dick 2005; Clarke et al. 2006). The hidden premise in this argument, however, is that the observed differences in engraftment must reflect differences in self-renewal ability within the patients. The nontransplantable human AML cell population might instead simply lack a feature needed for obtaining stromal support in the foreign microenvironment, such as a cytokine receptor responsive to mouse factors or a chemokine receptor that attracts the cells to a nurturing niche. Conversely, the transplantable population may simply have inadvertently acquired (perhaps by epigenetic changes) features that allow those cells to survive in the mouse milieu.

Abundant Transplantable Cells in Many Mouse Hematopoietic Tumors

To test the cancer stem cell hypothesis without the many complexities associated with xenotransplantation, we studied syngeneic transfers of cells from three types of primary mouse lymphomas or leukemias (Kelly et al. 2007a): the pre-B or B lymphomas arising in Eμ-*myc* transgenic mice (Adams et al. 1985), T lymphomas of Eμ-N-*ras* transgenic mice (Haupt et al. 1992), and the AML that develops in animals lacking PU.1 (Metcalf et al. 2006). These well-characterized models involve genes implicated in analogous human tumors, and the monoclonal tumors arise stochastically due to acquisition of mutations in other cancer-causing genes. Pertinently, cells from the preneoplastic animals do not seed tumors in recipients (Langdon et al. 1986).

The tumors in these models are relatively homogeneous, but small subpopulations of Eμ-*myc* lymphoma cells bore potential "stem cell" markers such as AA4.1 and Sca-1 (Fig. 2), so we included an AA4.1$^+$/Sca-1$^+$ subpopulation in

A

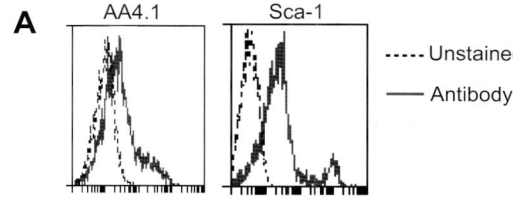

AA4.1   Sca-1

----- Unstained
—— Antibody

B

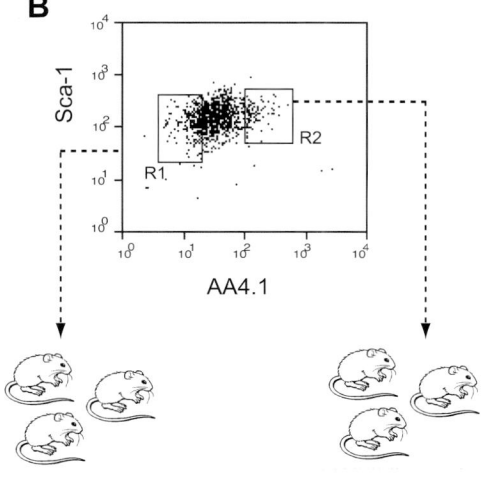

Sca-1

AA4.1

R1: Sca-1⁺ AA4.1 ^{lo} R2: Sca-1⁺ AA4.1 ^{hi}

Figure 2. Subpopulations of cells in Eμ-*myc* B lymphomas express progenitor markers. (*A*) Flow cytometry of cells stained with monoclonal antibodies to the surface markers AA4.1 and Sca-1, which are found on primitive hematopoietic cells (among others), revealed small subpopulations bearing these markers but not others examined (c-Kit, CD43, CD44, CD71). (*B*) Cells with the phenotype Sca-1⁺AA4.1^{hi} and Sca-1⁺AA4.1^{lo} were purified as indicated for transplantation tests.

one transplantation test. For the tests, we intravenously injected graded numbers of the lymphoma or leukemia cells into unmanipulated (e.g., nonirradiated) syngeneic mice and monitored tumor development (Kelly et al. 2007a).

Remarkably, ten cells from each B lymphoma sufficed to seed tumor growth, and the presence or absence of the presumptive stem cell marker made no difference (Table 1). Notably, with one B lymphoma (case 2), even transfer of a single cell (microscopically verified) succeeded in three of eight attempts. Similarly, ten cells sufficed with the T lymphoma, as well as with three of four of the AML cases (Table 1). The phenotypes of the tumors in the recipients mirrored those in the donors (Kelly et al. 2007a).

More recently, a model of pre-B acute lymphocytic leukemia (ALL) has been developed in animals whose B-lineage cells lack both PU.1 and IRF-8 (S. Carotta and S.L. Nutt, unpubl.). As few as ten of the ALL cells could seed leukemia in recipients. Thus, in all four types of primary uncultured murine hematopoietic tumors that we have studied, a substantial proportion of the tumor cells (>10%), rather than a rare subpopulation, drives tumor growth.

Several other recent studies with mouse leukemias have also demonstrated a high proportion of tumor-propagating cells. Pertinent to human AML, transplantation of colonies of mouse hematopoietic cells transformed by the MLL-AF9 oncogene, which has been generated by chromosome translocation in some human AML patients, revealed that a quarter of *all* the myeloid cells could seed leukemia in recipients (Somerville and Cleary 2006). Notably, the leukemia-propagating cells had a mature (Mac-1⁺ Gr-1⁺) phenotype, rather than that of a hematopoietic stem or early progenitor cell. Similarly, in another MLL-AF9 model, up to 50% of granulocyte-macrophage progenitors could initiate leukemia (Krivtsov et al. 2006). Furthermore, in pre-B ALL produced by the BCR-ABL translocation product in ARF

Table 1. A substantial proportion of tumor cells can sustain the growth of murine lymphoid and myeloid malignancies

Tumor model	Cell number injected (days to sacrifice)			
	10^5	10^3	10^{2a}	10
Eμ-*myc* B lymphoma				
Case 1	3/3 (25)	3/3 (25)	3/3 (32)	2/2 (35)
Case 2	3/3 (21)	3/3 (23)	3/3 (24)	3/3 (24)
Case 3 - Sca-1⁺ AA4.1^{hi}	3/3 (21)	3/3 (21)	n.d.	3/3 (17)
- Sca-1⁺ AA4.1/1^{lo}	2/2 (17)	2/2 (28)	2/2 (28)	2/2 (40)
Eμ-*N-ras* T lymphoma				
Case 1	3/3 (28)	3/3 (42)	3/3 (28)	3/3 (28)
PU.1$^{-/-}$ AML				
Case 1	1/1 (54)	2/2 (168)	1/2 (192)	0/2
Case 2	2/2 (84)	2/2 (85)	2/2 (224)	1/2 (114)
Case 3	1/1 (85)	2/2 (62)	2/2 (69)	2/2 (90)
Case 4	1/1 (30)	1/1 (37)	2/2 (79)	2/2 (88)

Cells from primary tumors of the indicated models, all on a C57BL/6 (Ly5.2) background, were mixed with 10^6 congenic spleen cells as carriers and injected into nonirradiated congenic (Ly5.1) recipients. Shown are the fraction of recipients that developed tumors and the average time (in days) from transplantation to tumor development. (Reprinted, with permission, from Kelly et al. 2007a [© AAAS].)
^{a}n.d. indicates not determined.

null mice, as few as 20 of the leukemia cells, and virtually all colonies generated by them, could seed leukemia in recipients (Williams et al. 2007). Thus, in all of these cases, the leukemia-propagating cells were abundant and displayed differentiated phenotypes, rather than resembling the hematopoietic stem cell.

A high frequency of tumor-propagating cells is not confined to genetically engineered models. Pioneering studies of spontaneous mouse leukemias and lymphomas of both lymphoid and myeloid origin revealed transplantable tumor cells that ranged from more than 1% to the majority of cells, and, in several striking examples, a single cell seeded a tumor (Furth and Kahn 1937; Hewitt et al. 1976). Thus, diverse monoclonal mouse hematopoietic malignancies, including those that closely match human tumors, are sustained by a substantial proportion of their cells. These results favor a model of tumor perpetuation by dominant clone(s) (Fig. 1B), perhaps by most of the cells that can form colonies in vitro under optimal conditions, rather than exclusively by a very minor subpopulation, as expected on the cancer stem cell model (Fig. 1A).

The disparity with the human AML results indicates to us that xenotransplantation greatly underestimates the proportion of cells, and range of cell types, within the human leukemias that drive neoplastic growth. Perhaps the rare human AML cells detected by xenotransplantation founded the original disease, which may have been akin to CML (chronic myelogenous leukemia), but subsequent mutations within the clone have created a dominant, more aggressive, and mature derivative that drives the AML in patients but cannot readily engraft mice (Fig. 3). If so, xenotransplantation might be telling us about the history of the disease, rather than the cell population that now maintains it.

Nature of the Cells Maintaining Solid Tumors

The cellular differentiation pathways in most organs are much less well understood than in hematopoiesis, and solid tumor development is more complex, with greater reliance on the microenvironment and angiogenesis. In addition, these tumors often eventually escape their tissue barriers and undergo the multiple changes required for metastasis. Accordingly, as reviewed recently (Vermeulen et al. 2008; Visvader and Lindeman 2008), the analysis of stem cells in most solid tumors is generally considered less advanced than that for AML (Clarke et al. 2006). None have yet been highly purified (Al-Hajj et al. 2003; Singh et al. 2004; Ricci-Vitiani et al. 2007), and because some fall within subpopulations (e.g., CD133+) that can contain up to 20% of the total cells, they need not be rare.

In some cases, the apparent rarity of human transplantable cells might reflect the need to cotransfer an essential support cell that happens to display similar cell surface markers. Notably, cotransfer of CD133+ support and tumor cells might explain the paradox that the colon cancer CD133+ population was estimated to contain 20 times more tumor-propagating cells than the unfractionated population (O'Brien et al. 2007). For example, CD133+ endothelial cells can enhance growth of transplanted human cancer cells (Calabrese et al. 2007). Although CD133 has been frequently used to isolate cancer stem cell populations, glioblastoma and metastatic colon carcinoma can be driven by either CD133+ or CD133− cells (Beier et al. 2007; Shmelkov et al. 2008). Moreover, the tumor-promoting cells within some Brca1-deficient mouse mammary tumors were CD44+/CD24−, whereas others were CD133+ (Wright et al. 2008). Thus, cancer stem cells may not have a consistent phenotype and need not be rare.

It is also unclear whether the markers used to isolate a cancer stem cell population are intrinsic to those cells or only transiently expressed. In breast cancer development, stem cell character has been linked to the epithelial-mesenchymal transition, a step essential for metastasis (Mani et al. 2008). Cells with the phenotype of breast cancer stem cells (CD44hi/CD24lo) (Al-Hajj et al. 2003) were generated from immortal mammary epithelial cells by inducing an epithelial-mesenchymal transition, even simply by treatment with transforming growth factor-β1 (TGF-β1) (Mani et al. 2008). Perhaps some differentiated cells can acquire stem cell character, which in normal tissues may be induced or maintained by signals from specialized niches (Morrison and Spradling 2008).

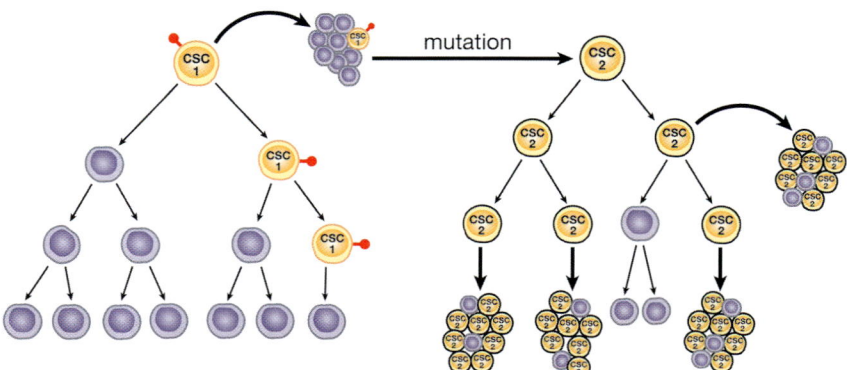

Figure 3. Clonal succession during tumor progression can create a dominant clone. At an early stage of development, a tumor might initially be driven primarily by rare cells of one phenotype (CSC1), but a mutation enhancing self-renewal in a differentiated derivative could create a dominant subclone driven by cells of a different phenotype (CSC2). In some human tumors (e.g., AML), CSC1 but not CSC2 might be able to engraft mice. (Modified from Adams and Strasser 2008.)

Pertinently, classical studies showed that engraftment of mouse solid tumors requires far more cells than hematopoietic ones, but those numbers fell markedly on coinjection of irradiated tumor tissue, suggesting that the solid tumors rely greatly on stromal support (Hewitt et al. 1976). Such studies with human tumors might reveal additional populations of tumor-propagating cells and far higher frequencies.

Implications for Tumor Propagation

The considerations above and others in recent reviews (Campbell and Polyak 2007; Hill and Perris 2007; Shipitsin and Polyak 2008; Vermeulen et al. 2008) raise many questions about the cancer stem cell model. Certainly, the evidence summarized above that a substantial proportion of the cells in many mouse leukemias are transplantable challenges its generality. Indeed, we suggest that nearly all tumors and leukemias of lymphoid origin must be driven by a dominant clone, because all of their cells exhibit a clonotypic rearrangement of their antigen receptor genes, and that distinctive hallmark of relatively mature differentiation invariably remains in their transplants. On the other hand, as reviewed elsewhere (Adams and Strasser 2008), some of the features expected from the cancer stem cell model have appeared in three types of mouse leukemias (Huntly et al. 2004; Deshpande et al. 2006; Neering et al. 2007), one type of murine breast cancer (Cho et al. 2008), and mouse skin carcinomas (Malanchi et al. 2008).

We therefore believe that tumors most likely fall on a spectrum spanning the two models in Figure 1. Indeed, the cells driving a tumor may well change during its progression (Fig. 3). The cancer stem cell model (Fig. 1A) may represent tumors at an early stage of development, such as the chronic phase of CML, whereas the clonal evolution model (Fig. 1B) may better describe the growth of more aggressive malignancies, such as CML in blast crisis or other acute leukemias, in which the dominant clones have acquired additional oncogenic mutations (Mullighan et al. 2008; Williams and Sherr, this volume).

More compelling tests of the cancer stem cell hypothesis might be provided by analysis of more mouse tumor models; by studies with human tumors that include cotransfer of human stromal cells or irradiated tumor tissue or exploit mice installed with human support cells; and by purification of the stem cells using more specific surface markers (Barker et al. 2007). Expression profiling and genomic sequence analysis of multiple subclones from the same tumor might reveal whether heterogeneity and differences in transplantability are simply due to differentiation, as in the cancer stem cell model (Fig. 1A) or instead often reflect a varied complement of mutations, as predicted by the clonal evolution model (Fig. 1B).

Why might tumors be propagated in two different ways? Perhaps a tumor tends to follow the cancer stem cell model if the key mutation occurred in a normal stem or primitive progenitor cell, as originally suggested (Bonnet and Dick 1997), whereas the clonal evolution model predominates among tumors that originate from more differentiated cells. Alternatively, or in addition, the nature of the mutations that create the tumor may be determinative. It is also conceivable that many tumors that initially follow the cancer stem cell paradigm progress on acquiring additional mutations to resemble the clonal evolution model (Fig. 3). For example, with metastatic neuroblastoma, as few as ten cells could engraft mice and no hierarchical organization was evident (Hansford et al. 2007).

Relevance to Therapy

Much of the excitement about the cancer stem cell model has arisen from the prospect that it might provide a new approach to therapy (Wang and Dick 2005; Clarke et al. 2006). If all self-renewal capacity resided in the cancer stem cells (Fig. 1A), they would be the critical therapeutic targets, and eliminating the bulk of the cells might have negligible impact on long-term patient survival. In addition, if the stem cell subpopulation, thought to be largely quiescent, were more refractory to therapeutic agents than other tumor cell populations, those cells might be primarily responsible for relapses. Hence, it is argued that targeting the cancer stem cells might yield more durable or even curative therapies, particularly if normal stem cells can be spared. For example, imatinib has revolutionized CML management but it is not curative, perhaps because the stem cells that drive this leukemia are refractory (O'Hare et al. 2006).

Despite the promise, to date there is only limited evidence that targeting putative cancer stem cells improves therapy. The cancer stem cells are reportedly more refractory to irradiation and chemotherapy (Bao et al. 2006; Liu et al. 2006) and administration to mice of an antibody to CD44, an antigen expressed on some human AML-initiating cells, markedly reduced leukemic repopulation (Jin et al. 2006). Approaches that force all tumor cells into cycle may hold promise. The quiescence of hematopoietic stem cells and some leukemia-initiating cells requires the PML (promyelocytic leukemia) protein, and arsenic trioxide, which promotes PML degradation, can force them into cycle (Ito et al. 2008). Accordingly, this well-tolerated drug markedly enhanced the sensitivity of mouse CML cells to chemotherapy (Ito et al. 2008).

Although treatment of some tumors may benefit from targeting the stem cells, if many (perhaps most) tumors are perpetuated by dominant clones, as we have argued above, curative therapy will usually require targeting *all* the cell populations within a tumor.

ROLE OF ENDOGENOUS BCL-2 IN MYC-INDUCED LYMPHOMA

Impaired apoptosis is a critical step toward malignancy (Hanahan and Weinberg 2000; Cory and Adams 2002). Its role in Myc-induced tumors is well established. Enforced Myc expression not only promotes proliferation and retards differentiation but also triggers apoptosis under suboptimal growth conditions, such as limiting cytokine (Green and Evan 2002). Accordingly, in Eμ-*myc* transgenic mice, in which *myc* is expressed throughout B-cell development (Adams et al. 1985), the premalignant animals exhibit an enlarged pre-B-cell population

(Langdon et al. 1986), but its expansion is limited by apoptosis, presumably due to consumption of the relevant cytokines. The pre-B or B-cell lymphomas that emerge stochastically have mutations that counter Myc-induced apoptosis, such as inactivation of the ARF-Mdm2-p53 pathway, which acts through the key apoptosis regulator, the Bcl-2 protein family (Adams and Cory 2007; Youle and Strasser 2008). Consequently, lymphomagenesis in Eμ-*myc* mice is accelerated by enforced expression of a prosurvival family member such as Bcl-2 (Strasser et al. 1990). Apoptosis ensues in such *myc-bcl-2* bitransgenic tumors if Bcl-2 expression is ablated (e.g., by Cre-lox-mediated elimination of the *bcl-2* transgene), underlining its crucial role (Letai et al. 2004).

These findings suggested that endogenous *bcl*-2 might be required for the development of Eμ-*myc* lymphomas, particularly because Bcl-2 is expressed in most stages of lymphopoiesis, including early progenitors (Li et al. 1993), and its overexpression enhances their survival (McDonnell et al. 1989; Strasser et al. 1991). We have therefore compared tumor development in the presence and absence of endogenous Bcl-2 (Kelly et al. 2007b). To bypass the complication that young Bcl-2-deficient mice succumb to polycystic kidney disease (Veis et al. 1993), we compared wild-type mice whose hematopoietic system was reconstituted with hematopoietic stem cell populations from either Eμ-*myc/bcl-2*$^{-/-}$ or Eμ-*myc/bcl-2*$^{+/+}$ (hereafter, Eμ-*myc*) embryos. We will denote the reconstituted animals by the genotype of their donor cells.

Preneoplastic Eμ-*myc/bcl-2*$^{-/-}$ Mice Have Much Fewer Mature B Cells

To determine how the absence of Bcl-2 affected B lymphopoiesis before the recipients developed a tumor, we enumerated the B-lymphoid cells at various stages of differentiation in their hematopoietic tissues by flow cytometry. The Eμ-*myc/bcl-2*$^{-/-}$ bone marrow contained pro-B-cell numbers similar to Eμ-*myc* recipients but about twofold to threefold less pre-B and sIg$^+$ B cells. Strikingly, the spleen (and lymph nodes) of Eμ-*myc/bcl-2*$^{-/-}$ recipients had less than 10% of the mature B cells in the Eμ-*myc* recipients (Fig. 4). Thus, endogenous Bcl-2 appears to be critical for the survival of mature Eμ-*myc* B cells but less important for the transgenic pro-B, pre-B, and immature B cells.

Bcl-2 Loss Accelerates Myc-induced Apoptosis of Mature B Cells

To assess whether the reduction in mature B cells in Eμ-*myc/bcl-2*$^{-/-}$ recipients reflected increased apoptosis, we purified donor-derived (Ly5.2$^+$) pro-B, pre-B, immature B, and mature B cells from bone marrow or spleen by flow cytometry and monitored their survival when cultured without cytokine. In the absence of cytokine, deregulated *myc* expression enhances apoptosis of B-lymphoid cells (Strasser et al. 1996). Significantly, apoptosis was accelerated markedly in the mature B cells but not the pre-B cells (or pro-B cells) from Eμ-*myc/bcl-2*$^{-/-}$ mice (Fig. 5). Thus, endogenous Bcl-2 is critical for countering the

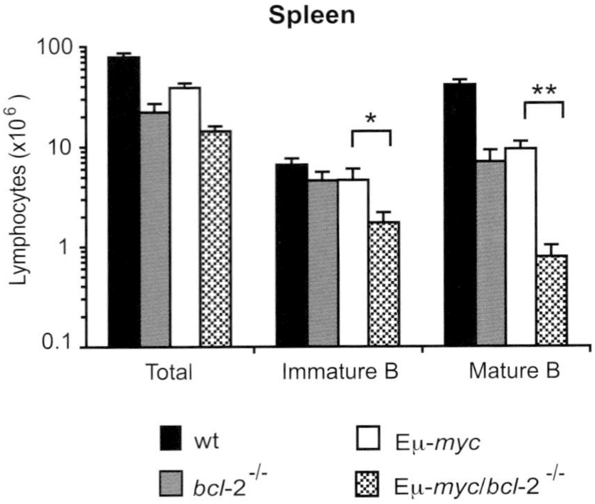

Figure 4. Preleukemic Eμ-*myc/bcl*-2$^{-/-}$ reconstituted mice have reduced numbers of mature B cells. Spleens were harvested from wild-type mice reconstituted with fetal liver cells of the indicated genotypes. Single-cell suspensions were stained with monoclonal antibodies to B-lineage surface markers, gated for donor-derived cells (Ly5.2), and analyzed by flow cytometry. Mean cell numbers ±S.E.M. are shown. (*) $p < 0.05$; (**) $p < 0.001$. (Reprinted from Kelly et al. 2007b.)

proapoptotic impact of deregulated Myc in mature B cells but appears dispensable for pro-B- and pre-B-cell survival.

Myc-induced Lymphomagenesis Is Unperturbed by Bcl-2 Loss

The dearth of B cells in Eμ-*myc/bcl*-2$^{-/-}$ mice and their accelerated apoptosis in culture led us to expect reduced or delayed lymphomagenesis. Remarkably, however, tumor incidence and latency in Eμ-*myc* and Eμ-*myc/bcl-*

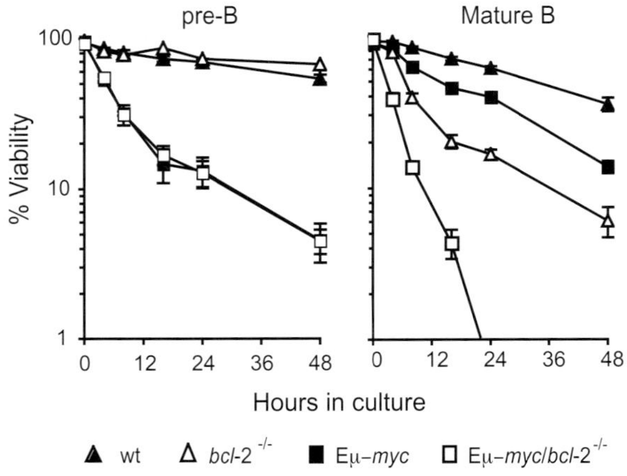

Figure 5. Accelerated apoptosis of Eμ-*myc/bcl*-2$^{-/-}$ mature B cells in culture. Donor-derived B-cell populations, preneoplastic pre-B cells, and mature B cells from the reconstituted mice were cultured without cytokines for the indicated periods, and cell viability was measured by staining with PI plus annexin V and flow cytometry (means ± S.E.M.). (Reprinted from Kelly et al. 2007b.)

2[-/-] reconstituted mice were indistinguishable (Fig. 6). The tumor phenotype was also the same (Kelly et al. 2007b): All were either pro/pre-B cells or immature B cells. The lymphomas were also just as aggressive, as judged from spleen enlargement, elevation of blood leukocytes, infiltration of the liver, lung, and kidney, and transplantability.

Interpretation

The failure of Bcl-2 loss to reduce the numbers of transgenic pro-B or pre-B cells in the bone marrow, or their survival when deprived of cytokine in vitro, probably reflects the low level of endogenous Bcl-2 in pre-B cells (Li et al. 1993). We surmise that other prosurvival Bcl-2 family members, such as Bcl-x_L, which is highly expressed in pre-B lymphocytes, are essential for inhibiting their Myc-induced apoptosis. We are therefore currently studying lymphomagenesis in the absence of Bcl-x_L. Intriguingly, our preliminary results suggest that its loss, unlike Bcl-2 loss, greatly inhibits the tumor development (P.N. Kelly, J.M. Adams, and A. Strasser, unpubl.).

The marked deficit of mature B cells in preleukemic Eμ-myc/bcl-2[-/-] mice presumably reflects their enhanced susceptibility to Myc-induced apoptosis. Despite this deficit, malignancy developed as rapidly as in the Eμ-myc animals, and the tumors were just as aggressive. These results demonstrate that the survival of the mature B cells is not essential for Myc-induced lymphomagenesis or sustained lymphoma growth. We surmise that neoplastic clones in Eμ-myc transgenic mice typically originate from a less mature cell type, one that does not require Bcl-2 for survival. An analogous observation is that preleukemic Eμ-myc/Eμ-max41 bitransgenic mice, which have less than 1% of the normal number of B lymphocytes, develop lymphoma at a rate comparable to Eμ-myc mice (Lindeman et al 1994). Conversely, Eμ-myc/Eμ-bcl-2 bitransgenic animals exhibit copious-cycling mature (sIg[+]) B-lymphoid cells, but all of the tumors that arise have a very primitive ("stem-cell"-like) phenotype, and the more mature bitrans-

genic B cells do not elicit tumors on transplantation (Strasser et al. 1990). These observations indicate that, even though many Eμ-myc lymphomas express surface immunoglobulin, indicative of a mature phenotype, the oncogenic mutations that cooperate with Myc most likely are acquired in a primitive lymphoid progenitor.

Our finding that loss of endogenous Bcl-2 accelerated apoptosis of preneoplastic mature Eμ-myc B cells, together with the demonstration that removing Bcl-2 decimated the tumors arising in Eμ-myc/Eμ-bcl-2 mice (Letai et al. 2004), supports the concept that inactivating the function (or expression) of Bcl-2, or its close relatives, will prove to be an important new approach to therapy (Fesik 2005; Adams and Cory 2007). Indeed, in preclinical studies, the compound ABT-737 and the closely related ABT-263, which both inactivate three Bcl-2 family members (Bcl-2, Bcl-x_L, and Bcl-w), have shown considerable promise with chronic lymphocytic leukemia and several other tumor types (Oltersdorf et al. 2005; Adams and Cory 2007; Tse et al. 2008). However, our finding that endogenous Bcl-2 is not required for the sustained growth of Eμ-myc lymphomas indicates that targeting Bcl-2 alone might not be effective for this class of tumor. Targeting several Bcl-2 family members (as ABT-737 does) represents one approach, but we surmise that antagonizing a single critical prosurvival family member, such as Bcl-x_L, offers the best prospect for selectively killing certain types of tumor cells while minimizing damage to normal tissues.

CONCLUSIONS

Our studies on transplantation of mouse hematopoietic tumors, and reports by several other groups, indicate that many such tumors are perpetuated by a substantial proportion of their neoplastic cells. These findings raise doubts about the interpretation of the xenotransplantation experiments that underpin the cancer stem cell model. We believe that model (Fig. 1A) is unlikely to appropriately represent many tumors, except perhaps at an early stage in their development (Fig. 3). Tumors that have accumulated the multiple mutations needed for full-fledged malignancy may be propagated instead by a dominant cell population (Fig. 1B). Thus, different tumors may fall on a spectrum between the cancer stem cell and clonal evolution models, or those models may represent early and late stages of tumor progression, respectively (Fig. 3). The important implication for treatment is that curative therapy is likely to require targeting not simply a rare cell population but most of the neoplastic cells found in the tumor.

Our investigation of the role of endogenous Bcl-2 in Myc-induced lymphomagenesis revealed that Bcl-2 loss caused a dramatic drop in mature B cells, presumably reflecting the Myc-induced apoptosis observed at that stage. Nevertheless, lymphoma development was unperturbed, and the tumors were just as aggressive. We conclude that the initiation, development, and continued growth of Eμ-myc lymphoma does not depend on endogenous Bcl-2, nor on the total number of B-lymphoid cells driven by the Eμ-myc transgene. We surmise that a Bcl-2

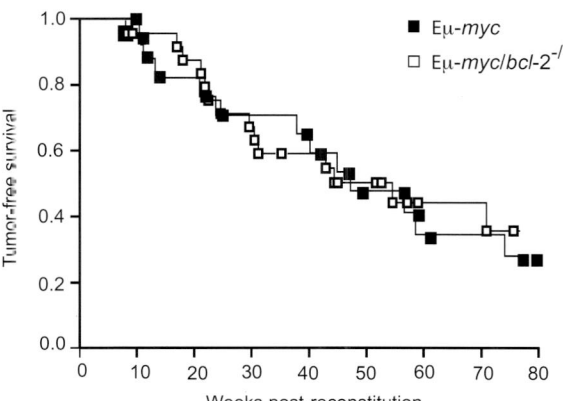

Figure 6. Loss of endogenous Bcl-2 does not delay Myc-induced lymphomagenesis. Kaplan–Meier analysis of tumorigenesis in mice reconstituted with Eμ-myc and Eμ-myc/bcl-2[-/-] embryos. (Reprinted from Kelly et al. 2007b.)

prosurvival relative, such as Bcl-x$_L$, has a more critical role. The implication for therapy is that it will be important to identify, for various types of tumor, the prosurvival protein(s) required for their maintenance, because those proteins will be critical targets for intervention.

ACKNOWLEDGMENTS

We are grateful to our colleagues, in particular Drs. S. Cory, J. Visvader, G. Lindeman, D.C.S. Huang, and P. Bouillet, for advice and discussions. This work was supported by fellowships and grants from the Australian National Health and Medical Research Council (Program Grant 257502), the Leukemia and Lymphoma Society (SCOR grant 7015), and the U.S. National Cancer Institute (CA 80188 and CA 43540).

REFERENCES

Adams, J.M. and Cory, S. 2007. The Bcl-2 apoptotic switch in cancer development and therapy. *Oncogene* **26:** 1324–1337.

Adams, J.M. and Strasser, A. 2008. Is tumour growth sustained by rare cancer stem cells or dominant clones? *Cancer Res.* **68:** 4018–4021.

Adams, J.M., Harris, A.W., Pinkert, C.A., Corcoran, L.M., Alexander, W.S., Cory, S., Palmiter, R.D., and Brinster, R.L. 1985. The c-*myc* oncogene driven by immunoglobulin enhancers induces lymphoid malignancy in transgenic mice. *Nature* **318:** 533–538.

Al-Hajj, M., Wicha, M.S., Benito-Hernandez, A., Morrison, S.J., and Clarke, M.F. 2003. Prospective identification of tumorigenic breast cancer cells. *Proc. Natl. Acad. Sci.* **100:** 3983–3988.

Arai, K.I., Lee, F., Miyajima, A., Miyatake, S., Arai, N., and Yokota, T. 1990. Cytokines: Coordinators of immune and inflammatory responses. *Annu. Rev. Biochem.* **59:** 783–836.

Bao, S., Wu, Q., McLendon, R.E., Hao, Y., Shi, Q., Hjelmeland, A.B., Dewhirst, M.W., Bigner, D.D., and Rich, J.N. 2006. Glioma stem cells promote radioresistance by preferential activation of the DNA damage response. *Nature* **444:** 756–760.

Barker, N., van Es, J.H., Kuipers, J., Kujala, P., van den Born, M., Cozijnsen, M., Haegebarth, A., Korving, J., Begthel, H., Peters, P.J., and Clevers, H. 2007. Identification of stem cells in small intestine and colon by marker gene *Lgr5*. *Nature* **449:** 1003–1007.

Beier, D., Hau, P., Proescholdt, M., Lohmeier, A., Wischhusen, J., Oefner, P.J., Aigner, L., Brawanski, A., Bogdahn, U., and Beier, C.P. 2007. CD133$^+$ and CD133$^-$ glioblastoma-derived cancer stem cells show differential growth characteristics and molecular profiles. *Cancer Res.* **67:** 4010–4015.

Bonnet, D. and Dick, J.E. 1997. Human acute myeloid leukemia is organized as a hierarchy that originates from a primitive hematopoietic cell. *Nat. Med.* **3:** 730–737.

Calabrese, C., Poppleton, H., Kocak, M., Hogg, T.L., Fuller, C., Hamner, B., Oh, E.Y., Gaber, M.W., Finklestein, D., Allen, M., et al. 2007. A perivascular niche for brain tumor stem cells. *Cancer Cell* **11:** 69–82.

Campbell, L.L. and Polyak, K. 2007. Breast tumor heterogeneity: Cancer stem cells or clonal evolution? *Cell Cycle* **6:** 2332–2338.

Cho, R.W., Wang, X., Diehn, M., Shedden, K., Chen, G.Y., Sherlock, G., Gurney, A., Lewicki, J., and Clarke, M.F. 2008. Isolation and molecular characterization of cancer stem cells in MMTV-*Wnt-1* murine breast tumors. *Stem Cells* **26:** 364–371.

Clarke, M.F., Dick, J.E., Dirks, P.B., Eaves, C.J., Jamieson, C.H., Jones, D.L., Visvader, J., Weissman, I.L., and Wahl, G.M. 2006. Cancer stem cells—Perspectives on current status and future directions: AACR Workshop on Cancer Stem Cells. *Cancer Res.* **66:** 9339–9344.

Cory, S. and Adams, J.M. 2002. The Bcl2 family: Regulators of the cellular life-or-death switch. *Nat. Rev. Cancer* **2:** 647–656.

Deshpande, A.J., Cusan, M., Rawat, V.P., Reuter, H., Krause, A., Pott, C., Quintanilla-Martinez, L., Kakadia, P., Kuchenbauer, F., Ahmed F., et al. 2006. Acute myeloid leukemia is propagated by a leukemic stem cell with lymphoid characteristics in a mouse model of *CALM/AF10*-positive leukemia. *Cancer Cell* **10:** 363–374.

Fesik, S.W. 2005. Promoting apoptosis as a strategy for cancer drug discovery. *Nat. Rev. Cancer* **5:** 876–885.

Furth, J. and Kahn, M.C. 1937. The transmission of leukemia in mice with a single cell. *Am. J. Cancer* **31:** 276–282.

Green, D.R. and Evan, G.I. 2002. A matter of life and death. *Cancer Cell* **1:** 19–30.

Hanahan, D. and Weinberg, R.A. 2000. The hallmarks of cancer. *Cell* **100:** 57–70.

Hansford, L.M., McKee, A.E., Zhang, L., George, R.E., Gerstle, J.T., Thorner, P.S., Smith, K.M., Look, A.T., Yeger, H., Miller, F.D., Irwin, M.S., Thiele, C.J., and Kaplan, D.R. 2007. Neuroblastoma cells isolated from bone marrow metastases contain a naturally enriched tumor-initiating cell. *Cancer Res.* **67:** 11234–11243.

Haupt, Y., Harris, A.W., and Adams J.M. 1992. Retroviral infection accelerates T lymphomagenesis in Eμ-N-*ras* transgenic mice by activating c-*myc* and N-*myc*. *Oncogene* **7:** 981–986.

Hewitt, H.B., Blake, E.R. and Walder, A.S. 1976. A critique of the evidence for active host defence against cancer, based on personal studies of 27 murine tumours of spontaneous origin. *Br. J. Cancer* **33:** 241–259.

Hill, R.P. and Perris, R. 2007. "Destemming" cancer stem cells. *J. Natl. Cancer Inst.* **99:** 1435–1440.

Huntly, B.J., Shigematsu, H., Deguchi, K., Lee, B.H., Mizuno, S., Duclos, N., Rowan, R., Amaral, S., Curley, D., Williams, I.R., Akashi, K., and Gilliland, D.G. 2004. MOZ-TIF2, but not BCR-ABL, confers properties of leukemic stem cells to committed murine hematopoietic progenitors. *Cancer Cell* **6:** 587–596.

Ito, K., Bernardi, R., Morotti, A., Matsuoka, S., Saglio, G., Ikeda, Y., Rosenblatt, J., Avigan, D.E., Teruya-Feldstein, J., and Pandolfi, P.P. 2008. PML targeting eradicates quiescent leukaemia-initiating cells. *Nature* **453:** 1072–1078.

Jin, L., Hope, K.J., Zhai, Q., Smadja-Joffe, F., and Dick, J.E. 2006. Targeting of CD44 eradicates human acute myeloid leukemic stem cells. *Nat. Med.* **12:** 1167–1174.

Kelly, P.N., Dakic, A., Adams, J.M., Nutt, S.L., and Strasser, A. 2007a. Tumor growth need not be driven by rare cancer stem cells. *Science* **317:** 337.

Kelly, P.N., Puthalakath, H., Adams, J.M., and Strasser, A. 2007b. Endogenous *bcl-2* is not required for the development of Eμ-*myc*-induced B-cell lymphoma. *Blood* **109:** 4907–4913.

Kong, Y., Yoshida, S., Saito, Y., Doi, T., Nagatoshi, Y., Fukata, M., Saito, N., Yang, S.M., Iwamoto, C., Okamura, J., et al. 2008. CD34$^+$CD38$^-$CD19$^+$ as well as CD34$^+$CD38$^-$CD19$^+$ cells are leukemia-initiating cells with self-renewal capacity in human B-precursor ALL. *Leukemia* **22:** 1207–1213.

Krivtsov, A.V., Twomey, D., Feng, Z., Stubbs, M.C., Wang, Y., Faber, J., Levine, J.E., Wang, J., Hahn, W.C., Gilliland, D.G., Golub, T.R., and Armstrong, S.A. 2006. Transformation from committed progenitor to leukaemia stem cell initiated by MLL-AF9. *Nature* **442:** 818–822.

Langdon, W.Y., Harris, A.W., Cory, S., and Adams, J.M. 1986. The c-*myc* oncogene perturbs B lymphocyte development in Eμ-*myc* transgenic mice. *Cell* **47:** 11–18.

Letai, A., Sorcinelli, M.D., Beard, C., and Korsmeyer, S.J. 2004. Antiapoptotic BCL-2 is required for maintenance of a model leukemia. *Cancer Cell* **6:** 241–249.

Li, Y.-S., Hayakawa, K., and Hardy, R.R. 1993. The regulated expression of B lineage associated genes during B cell differentiation in bone marrow and fetal liver. *J. Exp. Med.* **178:** 951–960.

Lin, W.W. and Karin, M. 2007. A cytokine-mediated link between innate immunity, inflammation, and cancer. *J. Clin. Invest.* **117:** 1175–1183.

Lindeman, G.J., Adams, J.M., Cory, S., and Harris, A.W. 1994. B-lymphoid to granulocytic switch during hematopoiesis in a transgenic mouse strain. *Immunity* **1:** 517–527.

Liu, G., Yuan, X., Zeng, Z., Tunici, P., Ng, H., Abdulkadir, I.R., Lu, L., Irvin, D., Black, K.L., and Yu, J.S. 2006. Analysis of gene expression and chemoresistance of CD133$^+$ cancer stem cells in glioblastoma. *Mol. Cancer* **5:** 67.

Malanchi, I., Peinado, H., Kassen, D., Hussenet, T., Metzger, D., Chambon, P., Huber, M., Hohl, D., Cano, A., Birchmeier, W., and Huelsken, J. 2008. Cutaneous cancer stem cell maintenance is dependent on β-catenin signalling. *Nature* **452:** 650–653.

Mani, S.A., Guo, W., Liao, M.J., Eaton, E.N., Ayyanan, A., Zhou, A.Y., Brooks, M., Reinhard, F., Zhang, C.C., Shipitsin, M., Campbell, L.L., Polyak, K., Brisken, C., Yang, J., and Weinberg, R.A. 2008. The epithelial-mesenchymal transition generates cells with properties of stem cells. *Cell* **133:** 704–715.

McDonnell, T.J., Deane, N., Platt, F.M., Nuñez, G., Jaeger, U., McKearn, J.P. and Korsmeyer, S.J. 1989. *bcl*-2-immunoglobulin transgenic mice demonstrate extended B cell survival and follicular lymphoproliferation. *Cell* **57:** 79–88.

Metcalf, D., Dakic, A., Mifsud, S., Di Rago, L., Wu, L., and Nutt, S. 2006. Inactivation of PU.1 in adult mice leads to the development of myeloid leukemia. *Proc. Natl. Acad. Sci.* **103:** 1486–1491.

Morrison, S.J. and Spradling, A.C. 2008. Stem cells and niches: Mechanisms that promote stem cell maintenance throughout life. *Cell* **132:** 598–611.

Mullighan, C.G., Miller, C.B., Radtke, I., Phillips, L.A., Dalton, J., Ma, J., White, D., Hughes, T.P., Le Beau, M.M., Pui, C.H., Relling, M.V., Shurtleff, S.A., and Downing, J.R. 2008. *BCR-ABL1* lymphoblastic leukaemia is characterized by the deletion of Ikaros. *Nature* **453:** 110–114.

Neering, S.J., Bushnell, T., Sozer, S., Ashton, J., Rossi, R.M., Wang, P.Y., Bell, D.R., Heinrich, D., Bottaro, A. and Jordan, C.T. 2007. Leukemia stem cells in a genetically defined murine model of blast-crisis CML. *Blood* **110:** 2578–2585.

O'Brien, C.A., Pollett, A., Gallinger, S., and Dick, J.E. 2007. A human colon cancer cell capable of initiating tumour growth in immunodeficient mice. *Nature* **445:** 106–110.

O'Hare, T., Corbin, A.S., and Druker, B.J. 2006. Targeted CML therapy: Controlling drug resistance, seeking cure. *Curr. Opin. Genet. Dev.* **16:** 92–99.

Oltersdorf, T., Elmore, S.W., Shoemaker, A.R., Armstrong, R.C., Augeri, D.J., Belli, B.A., Bruncko, M., Deckwerth, T.L., Dinges, J., Hajduk, P.J., et al. 2005. An inhibitor of Bcl-2 family proteins induces regression of solid tumours. *Nature* **435:** 677–681.

Pearce, D.J., Taussig, D., Zibara, K., Smith, L.L., Ridler, C.M., Preudhomme, C., Young, B.D., Rohatiner, A.Z., Lister, T.A., and Bonnet D. 2006. AML engraftment in the NOD/SCID assay reflects the outcome of AML: Implications for our understanding of the heterogeneity of AML. *Blood* **107:** 1166–1173.

Ricci-Vitiani, L., Lombardi, D.G., Pilozzi, E., Biffoni, M., Todaro, M., Peschle, C., and De Maria, R. 2007. Identification and expansion of human colon-cancer-initiating cells. *Nature* **445:** 111–115.

Shipitsin, M. and Polyak, K. 2008. The cancer stem cell hypothesis: In search of definitions, markers, and relevance. *Lab. Invest.* **88:** 459–463.

Shmelkov, S.V., Butler, J.M., Hooper, A.T., Hormigo, A., Kushner, J., Milde, T., St Clair, R., Baljevic, M., White, I., Jin, D.K., et al. 2008. CD133 expression is not restricted to stem cells, and both CD133$^+$ and CD133$^-$ metastatic colon cancer cells initiate tumors. *J. Clin. Invest.* **118:** 2111–2120.

Singh, S.K., Hawkins, C., Clarke, I.D., Squire, J.A., Bayani, J., Hide, T., Henkelman, R.M., Cusimano, M.D., and Dirks, P.B. 2004. Identification of human brain tumour initiating cells. *Nature* **432:** 396–401.

Somervaille, T.C. and Cleary, M.L. 2006. Identification and characterization of leukemia stem cells in murine MLL-AF9 acute myeloid leukemia. *Cancer Cell* **10:** 257–268.

Strasser, A., Harris, A.W., Bath, M.L., and Cory, S. 1990. Novel primitive lymphoid tumours induced in transgenic mice by cooperation between *myc* and *bcl*-2. *Nature* **348:** 331–333.

Strasser, A., Whittingham, S., Vaux, D.L., Bath, M.L., Adams, J.M., Cory, S., and Harris, A.W. 1991. Enforced *BCL2* expression in B-lymphoid cells prolongs antibody responses and elicits autoimmune disease. *Proc. Natl. Acad. Sci.* **88:** 8661–8665.

Strasser, A., Elefanty, A.G., Harris, A.W., and Cory, S. 1996. Progenitor tumours from Eμ-*bcl*-2-*myc* transgenic mice have lymphomyeloid differentiation potential and reveal developmental differences in cell survival. *EMBO J.* **15:** 3823–3834.

Tse, C., Shoemaker, A.R., Adickes, J., Anderson, M.G., Chen, J., Jin, S., Johnson, E.F., Marsh, K.C., Mitten, M.J., Nimmer, P., et al. 2008. ABT-263: A potent and orally bioavailable Bcl-2 family inhibitor. *Cancer Res.* **68:** 3421–3428.

Veis, D.J., Sorenson, C.M., Shutter, J.R., and Korsmeyer, S.J. 1993. Bcl-2-deficient mice demonstrate fulminant lymphoid apoptosis, polycystic kidneys, and hypopigmented hair. *Cell* **75:** 229–240.

Vermeulen, L., Sprick, M.R., Kemper, K., Stassi, G., and Medema, J.P. 2008. Cancer stem cells—Old concepts, new insights. *Cell Death Differ.* **15:** 947–958.

Visvader, J. and Lindeman, G.J. 2008. Cancer stem cells in solid tumours: Accumulating evidence and unresolved questions. *Nat. Rev. Cancer* (in press).

Wang, J.C. and Dick, J.E. 2005. Cancer stem cells: Lessons from leukemia. *Trends Cell Biol.* **15:** 494–501.

Williams, R.T., den Besten, W., and Sherr, C.J. 2007. Cytokine-dependent imatinib resistance in mouse BCR-ABL$^+$, *Arf*-null lymphoblastic leukemia. *Genes Dev.* **21:** 2283–2287.

Wright, M., Calcagno, A.M., Salcido, C.D., Carlson, M.D., Ambudkar, S.V., and Varticovski, L. 2008. *Brca1* breast tumors contain distinct CD44$^+$/CD24$^-$ and CD133$^+$ cells with cancer stem cell characteristics. *Breast Cancer Res.* **10:** R10.

Youle, R.J. and Strasser, A. 2008. The BCL-2 protein family: Opposing activities that mediate cell death. *Nat. Rev. Mol. Cell Biol.* **9:** 47–59.

The *INK4-ARF* (*CDKN2A/B*) Locus in Hematopoiesis and BCR-ABL–induced Leukemias

R.T. WILLIAMS* AND C.J. SHERR[†‡]

Departments of Oncology and [†]Genetics & Tumor Cell Biology and [‡]Howard Hughes Medical Institute, St. Jude Children's Research Hospital, Memphis, Tennessee 38105

Senescence and apoptosis programs governed by the Rb and p53 signaling networks can counter tissue stem cell self-renewal. A master regulator of Rb and p53 is the *INK4-ARF* (*CDKN2A/B*) locus that encodes two CDK inhibitors, p16[INK4A] and p15[INK4B], that maintain Rb in its active, hypophosphorylated form, and p14[ARF] (p19[Arf] in mice), that inhibits Mdm2 and activates p53. The *INK4-ARF* genes are epigenetically silenced in hematopoietic stem cells but become poised to respond to oncogenic stress as blood cells differentiate. Inactivation of *INK4-ARF* endows differentiated cells with an inappropriate self-renewal capacity, a defining feature of cancer cells. In BCR-ABL–induced (Philadelphia chromosome-positive [Ph[+]]) leukemias, *INK4-ARF* deletions frequently occur in clinically aggressive acute lymphoblastic leukemias (Ph[+] ALLs) but are not seen in more indolent Ph[+] chronic myelogenous leukemia (CML) or in CML myeloid blast crisis. Mouse modeling of Ph[+] ALL reveals that *Arf* inactivation attenuates responsiveness to targeted BCR-ABL kinase inhibitors, enhances the maintenance of leukemia-initiating cells within the hematopoietic microenvironment, and facilitates the emergence of malignant clones that harbor drug-resistant BCR-ABL kinase mutations. Thus, although BCR-ABL mutations typify drug resistance in both CML and Ph[+] ALL, loss of *INK4-ARF* in Ph[+] ALL enhances disease aggressiveness and undermines the salutary effects of targeted therapy.

The Ph[+], a balanced translocation between chromosomes 9 and 22, was the first identified cytogenetic anomaly linked to a specific form of human cancer (Nowell and Hungerford 1960; Rowley 1973). It is the founding genetic lesion of CML and a subset of Ph[+] ALLs. The t(9;22)(q34;q11) translocation joins the *ABL1* oncogene on chromosome 9 to a breakpoint cluster region (BCR) on chromosome 22 to generate the BCR-ABL fusion oncoprotein, a constitutively active tyrosine kinase that initiates both diseases. In CML, the p210[BCR-ABL] isoform is initially expressed in hematopoietic stem cells (HSCs) and in their derivative myeloid and lymphoid progeny, whereas in Ph[+] ALL, the synthesis of either of two alternative p185 or p210 isoforms is restricted to the B-cell lineage (Groffen et al. 1984; Chan et al. 1987; Clark et al. 1987).

Clinically, CML often presents as an indolent myeloproliferative disorder characterized by expansion within hematopoietic tissues of relatively mature myeloid cells that often spill over into the circulation. Such patients are said to be in chronic phase (CML-CP), but without effective therapy, they progress through an accelerated phase to lethal blast crisis (CML-BC) that resembles an acute leukemia and is distinguished by the rapid proliferation of primitive myeloid or lymphoid "blasts" in hematopoietic organs. De novo Ph[+] ALL resembles the lymphoid blast crisis of CML (CML-LBC) but without a clinically detectable chronic phase. Whole-genome single-nucleotide polymorphism (SNP) analysis indicates that CML-CP leukemic cells typically display no recurring gene amplifications or losses, consistent with the notion that BCR-ABL expression is necessary and perhaps sufficient to induce the early stages of disease (Mulligan et al. 2008b). In contrast, diagnostic blasts recovered from CML-LBC and Ph[+] ALL patients harbor several recurring genomic lesions, including frequent *INK4-ARF* deletions (Mulligan et al. 2008a)

that correlate with their decidedly poorer outcome, regardless of the therapeutic modalities used, as compared to CML-CP patients. Ph[+] ALL comprises a small fraction (5%) of ALL cases in children, but it represents about one third of adult ALL, the largest genetically defined subgroup (Armstrong and Look 2005); tragically, the outcome for patients of both age groups is equally poor (Arico et al. 2000; Gleissner et al. 2002).

Targeted therapy with BCR-ABL kinase inhibitors has revolutionized the treatment of CML. Imatinib (Gleevec), the first-generation FDA-approved kinase inhibitor, successfully maintains virtually all CML-CP patients in remission as long as they continue drug therapy (Wong and Witte 2004; Druker et al. 2006). Nonetheless, most if not all drug-treated patients harbor persisting leukemic stem cells that rapidly expand and contribute to clinical relapse after drug discontinuation. Furthermore, a small percentage of patients (5% in the first year and fewer thereafter) relapse despite continuous therapy, and the reappearance of leukemia is typically associated with development of drug-resistant mutations in the BCR-ABL kinase domain (KD) that impair drug binding (Shah et al. 2002). Second-generation BCR-ABL kinase inhibitors, including nilotinib (Tasigna) and dasatinib (Sprycel), were developed as more potent inhibitors that are capable of inhibiting many of the most frequent BCR-ABL drug-resistant KD mutants, and these were FDA approved for use in imatinib-resistant or -intolerant CML and for Ph[+] ALL (Kantarjian et al. 2006; Talpaz et al. 2006; Hochhaus et al. 2007). Importantly, all three targeted drugs display no efficacy against one specific KD "gatekeeper mutation," T315I, that frequently emerges at clinical relapse, particularly when patients are treated with the broader spectrum, more potent, second-generation inhibitors.

Although all three kinase inhibitors induce significant hematological and molecular responses in Ph[+] ALL patients, these are relatively short lived, and treated patients relapse despite continuous therapy within only 6 months of therapy initiation (Druker et al. 2001; Talpaz et al. 2006; Alvarado et al. 2007; de Labarthe et al. 2007); these patients frequently, but not invariably, harbor leukemic clones with drug-resistant KD mutations (Talpaz et al. 2006; Hochhaus et al. 2007; O'Hare et al. 2007). Similarly, the durable responses to kinase inhibition observed in CML-CP are not realized in CML-BC, suggesting that additional mutations "downstream" from the BCR-ABL kinase contribute to more aggressive disease and to the attenuated therapeutic response.

Here, we discuss the role that the *CDKN2A-CDKN2B (INK4-ARF)* tumor suppressors have in governing self-renewal in normal and BCR-ABL–transformed hematopoietic cells. We outline how robust, clinically relevant mouse modeling of Ph[+] ALL (which combines BCR-ABL expression and *Arf* inactivation) provides unique opportunities for implementing treatment strategies and for interrogating drug resistance mechanisms that counter targeted therapeutic agents.

THE *INK4-ARF* LOCUS AND CELLULAR SELF-RENEWAL

Current evidence suggests that the *INK4-ARF* locus can exist in four distinct states: epigenetically silenced, poised for activation, overtly activated, and inactivated (Fig. 1). Epigenetic silencing of *Ink4-Arf* by Bmi1-containing Polycomb complexes facilitates HSC self-renewal (Jacobs

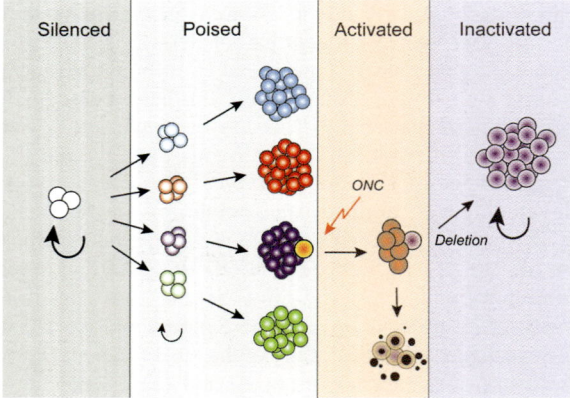

Figure 1. Four states of the *INK4-ARF* locus in hematopoietic cells. In HSCs that exhibit self-renewal capacity, the *INK4-ARF* locus is epigenetically silenced, whereas in progenitor cells and their progeny, the locus becomes poised to respond to various forms of stress, including sustained oncogenic signaling. Depicted is an oncogenic insult (ONC) that activates *INK4-ARF*, thereby eliminating the vast majority of cells through senescence or apoptosis. However, oncogene activation can also select for rare cells that have deleted the locus and acquire inappropriate self-renewal capacity. The latter cells have a greatly increased propensity to generate cancers.

et al. 1999; Lessard and Sauvageau 2003; Park et al. 2003; Iwama et al. 2004; Akala et al. 2008). Although *Bmi1* null mice are born with normal blood elements, they become aplastic within the first 2 months of life due to the loss of functional HSCs. The fact that this phenotype is manifested only after birth stems from the fact that the integrity of Bmi1-containing complexes is not required for HSCs within the fetal liver. However, during the first few weeks after birth, hematopoiesis becomes dependent on "definitive" HSCs that reside within the bone marrow where the requirement for Bmi1 becomes apparent. Although Bmi1-containing Polycomb complexes can regulate many genes, the maintenance of definitive adult HSCs depends strongly on *Ink4a* and *Arf*, because disruption of these genes significantly rescues blood cell formation in *Bmi1* null animals. Interestingly, mice that lack *Ink4a, Ink4b,* and *Arf* are even more highly prone to tumor development than mice lacking *Ink4a* and *Arf,* implicating an additional "back-up" role for *Ink4b* in tumor suppression (Krimpenfort et al. 2007) and raising the possibility that *Bmi1* null mice lacking the entire *Ink4-Arf* gene cluster might fare even better. Recent findings suggest that HSCs lacking *Ink4a-Arf* and *p53* lose their self-renewal potential and instead more rapidly generate transient amplifying cells that, although able to contribute to all blood cell lineages, exhibit diminished long-term repopulating activity after transplantation into irradiated recipients (Akala et al. 2008). Together, these findings point to the possibility that the *Ink4-Arf* locus acts as a gatekeeper in helping to establish the equilibrium between HSC self-renewal and the early differentiation steps that lead to lineage commitment.

As HSCs produce transiently amplifying progeny that undergo differentiation to form the various blood cell lineages, the *Ink4-Arf* locus is epigenetically remodeled and becomes "poised" to respond to abnormally increased and sustained thresholds of hyperproliferative signals (Fig. 1). Although the products of the locus are not detectably expressed under most normal circumstances, constitutively activated oncogenes can induce *Arf* expression and thereby trigger a p53-dependent transcriptional program that can eliminate oncogene-stressed cells through their senescence or apoptosis, the latter response being the more typical outcome for blood cells. Early B-cell progenitors are particularly sensitive to the oncogenic BCR-ABL kinase that efficiently induces an "activated" state of *Arf* expression. Although the accompanying p53-dependent apoptosis eliminates the vast majority of incipient leukemic cells, rare BCR-ABL–positive cells that sustain *Ink4-Arf* deletion can survive and acquire an abnormal self-renewing capacity that promotes leukemogenesis. Therefore, inactivation of the *Ink4-Arf* locus in cancer implies that the checkpoint had been activated at some earlier stage of tumor evolution (Fig. 1).

LINEAGE SPECIFICITY AND SELF-RENEWAL IN BCR-ABL–INDUCED LEUKEMIAS

CML-CP conforms to the cancer stem cell model in that the initiating BCR-ABL translocation develops in HSCs that can either self-renew or differentiate to more abundant mature progeny that are themselves nontumorigenic

(Clarke et al. 2006). We therefore anticipated that in Ph⁺ HSCs, like their nontransformed counterparts, epigenetic silencing would prevent *INK4-ARF* activation and thereby bypass any selection pressure for the subsequent deletion of the locus. However, because BCR-ABL–positive HSCs can generate more mature lymphoid and myeloid cells, these differentiated progeny should lose their capacity to silence *INK4-ARF*. Thus, in the accelerated phase of disease that leads to CML-LBC, we expected that the *INK4-ARF* locus would first be activated and then deleted, ultimately facilitating the acquisition of abnormal self-renewal capacity in lymphoid blasts that sustain the aggressive phase of the disease. The transition from CML-CP to CML-LBC is associated with a dramatic decline in patient survival despite the advent of therapies targeting BCR-ABL, implying that *INK4A-ARF* deletion might again be a deciding factor. Conceptually, de novo Ph⁺ ALL is like CML-LBC without a preceding chronic phase, consistent with the observation that the founding BCR-ABL translocation is restricted to lymphoid progenitors in this disease (Castor et al. 2005).

Recently, the frequencies of *INK4-ARF* deletions in Ph⁺ ALL and CML-LBC have been determined in an extended cohort of pediatric and adult patients by use of high-density microarray platforms that detect SNP arrays and by quantitative genomic polymerase chain reaction (PCR) analysis (Mullighan et al. 2008a,b). Whereas the *INK4-ARF* locus was intact in cells from CML-CP patients and those in myeloid blast crisis (CML-MBC), approximately two thirds of cases with Ph⁺ ALL or CML-LBC exhibited monoallelic or biallelic *INK4-ARF* deletions at the time of diagnosis (Table 1). In addition, these same patients also sustained frequent deletions of genes encoding two transcriptional regulators of the B-cell differentiation pathway—specifically, *IKFZ1* (*Ikaros*) and *PAX5*. Inactivation of these genes, frequently monoallelic, is expected to retard B-cell maturation and to expand the pool of Ph⁺-committed B-lymphoid cells that are susceptible to further mutational events, such as *INK4-ARF* loss. Aberrant recombination-activating gene (RAG)-mediated recombination events appear to underlie *IKFZ1* deletions and seem also to have a role in genomic deletions of the *INK4-ARF* locus in Ph⁺ ALL as well (Mullighan et al. 2008b), as has previously been described in T-cell ALL (Kohno and Yokota 2006). (In contrast, *INK4-ARF* deletions in many other forms of cancer seem to occur through RAG-independent recombination across regions of microhomology). The occurrence of RAG-dependent deletions in Ph⁺ lymphoid leukemias is likely to provide one explanation for the increased frequencies of *INK4-ARF* deletions in various forms of ALL versus their relative paucity in CML-MBC.

What molecular lesions contribute to acquisition of self-renewal in CML-MBC if the *INK4-ARF* locus is intact? The p53 tumor suppressor itself is mutated in a significant fraction of these patients (Calabretta and Perrotti 2004), reinforcing the idea that the p53-dependent tumor suppressor *pathway* is frequently targeted in blastic transformation of CML, but that the mechanisms differ between the lymphoid (*INK4-ARF* deletions) and myeloid (*p53* mutations) lineages. Additional lineage-specific mechanisms may also contribute to the self-renewal of myeloid blasts. For example, aberrant β-catenin signaling confers self-renewal to lineage-committed granulocyte-macrophage progenitors in CML-MBC (Jamieson et al. 2004) but is dispensable for BCR-ABL–induced ALL (Zhao et al. 2007).

MODELING Ph⁺ ALL IN THE MOUSE

We developed a highly efficient murine Ph⁺ ALL model system that entails the rapid ex vivo generation of primary, bone-marrow-derived populations of murine *Arf*⁻/⁻ pre-B cells that express p185^BCR-ABL. Introduction of these cells into healthy immunocompetent, syngeneic recipient host animals (hereafter "recipients") generates a fulminant ALL that quickly kills the mice. In direct contrast to the retarded leukemogenic potential of p185⁺;*Arf*⁺/⁺ cells in healthy animals, as few as 20 p185⁺;*Arf*⁻/⁻ cells are capable of producing a disseminated pre–B-cell ALL in recipients in less than 4 weeks (Williams et al. 2006, 2007). Abundant leukemic cells "marked" with green fluorescent protein (GFP) (expressed in tandem with BCR-ABL) can be recovered from all hematopoietic tissues of moribund animals and can be readily phenotyped, quantified, cultured in vitro, and efficiently transplanted serially and at the same limiting dilution into healthy secondary and tertiary recipients. Molecular assessments of immunoglobulin heavy (IgH) chain gene rearrangements and vector insertion sites, and measures of leukemogenic potential of multiple, independent single-cell-derived clones, revealed that virtually every p185⁺;*Arf*⁻/⁻ donor pre-B cell has leukemia-initiating cell (LIC) capacity. Thus, BCR-ABL expression combined with *Arf* inactivation is sufficient to guarantee ALL induction.

Although a variety of immunophenotypic markers have been identified that are associated with rare or infrequent "cancer stem cell" subpopulations in myeloid leukemias (Clarke et al. 2006), the *Arf* null status of p185⁺ pre-B cells represents a molecular determinant that confers at least a 4-log increase in their leukemogenic potential (Williams et al. 2007). The p185^BCR-ABL kinase confers growth factor (cytokine) independence to cells within the B-cell lineage; in contrast, *Arf* inactivation impairs BCR-ABL–induced apoptosis and dramatically facilitates the acquisition of limitless replicative potential by committed pre-B cells. The latency and phenotype of p185⁺;*Arf*⁻/⁻-

Table 1. Frequency of deletions in ALL

Subcategory	No. cases	Gene deletion frequency (%)		
		IKAROS	*PAX5*	*CDKN2A/B*
Ph⁺ B-ALL[a]	43	84	51	54
Non-Ph⁺ B-ALL	211	11	30	32
T-ALL	50	4	10	72

[a]No significant differences in gene deletion frequencies were observed between 21 pediatric and 22 adult cases subjected to analysis. Study of the *CDKN2A/B* gene cluster in 41 of these cases by quantitative PCR using primers directed to each of the *INK4A*, *ARF*, and *INK4B* exons indicated an overall deletion frequency of 64%.

induced ALL are mimicked in p185$^+$ cells that lack *p53*, suggesting that most, if not all, of *Arf*'s tumor-suppressive function is mediated via the *Arf-Mdm2-p53* axis.

One key aspect of this Ph$^+$ ALL model is that the donor animals, from which transformed p185$^+$ pre-B cells are derived, and the recipient mice are of the same genetic background, thereby preventing immune-mediated rejection of donor LICs. The system was developed with C57BL/6 mice, a strain that is ordinarily relatively resistant to BCR-ABL transformation. Although most experimental models of leukemia involve engraftment of modified donor bone marrow into lethally irradiated recipients, our model more closely recapitulates events in patients, in which somatic mutations initiate tumor development in the context of an otherwise healthy, immune competent host.

Implementation of luminescent imaging approaches in this Ph$^+$ ALL model system has now permitted spatial and temporal assessments of leukemic cell engraftment, expansion, and dissemination in vivo. Whereas previous studies relied on a murine stem cell virus (MSCV)-based p185-IRES (internal ribosome entry site)-GFP retroviral vector that encoded both p185 and a GFP marker gene in tandem, a luciferase reporter gene has now been substituted for GFP, thereby creating an analogous p185-IRES-Luc vector. Importantly, p185$^+$;*Arf*$^{-/-}$;GFP$^+$ donor cells and p185$^+$;*Arf*$^{-/-}$;Luc$^+$ cells produce equally aggressive disseminated pre–B-cell lymphoblastic leukemias on a per cell basis after inoculation into healthy recipients (Fig. 2). In the representative study illustrated here, 200,000 p185$^+$;*Arf*$^{-/-}$;Luc$^+$ cells were injected intravenously (by tail vein) into a cohort of healthy recipient mice on day 0, and mice were serially imaged every 3–4 days until their clinical deterioration necessitated their sacrifice at day 13. One noninjected animal was included as a negative imaging control, and all of the luminescent images presented in Figure 2 (top) were captured and analyzed under identical conditions.

Several important observations stem from these studies. First, the luminescent signal (i.e., disease) was barely detectable 3 days after inoculation of LICs, whereas by day 7, disseminated signals were appreciable in the limbs, sternum (bone marrow), and abdomen overlying the spleen. By days 10 and 13, further leukemic dissemination was observed as manifested by intense signals in the tail (likely reflecting circulating leukemic cells in the vasculature) and in the head region (reflecting leukemic infiltration into the meninges and into the calvarial bone marrow). Second, quantification of whole-animal luminescence revealed that there was significant consistency of signal among cohorts of animals analyzed at the same intervals and that there was an average log increase in signal intensity every 3 days (Fig. 2, bottom). This result is consistent with prior observations that leukemia onset is delayed by about 3 days for each log reduction in numbers of injected donor leukemogenic cells (Williams et al. 2007). Third, there is an approximately 3-log dynamic range of signal between barely detectable disease (i.e., just above background) at day 3 and that emanating from moribund recipient animals at day 13 (Fig. 2, bottom). These features now provide further opportunities to understand the biology of the disease in real time, test the efficacy of

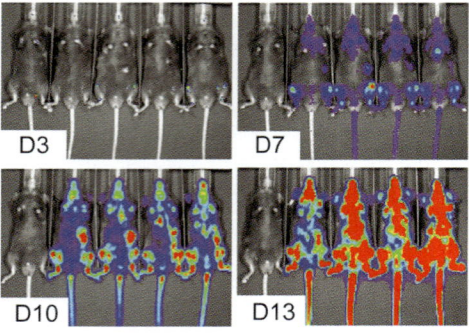

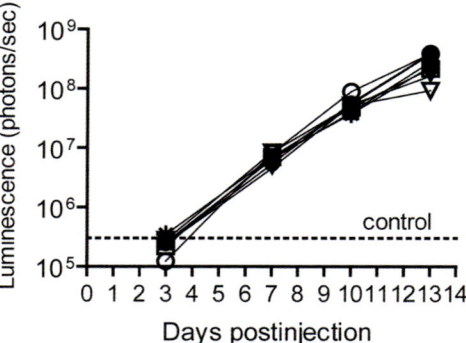

Figure 2. In vivo luminescent imaging of murine Ph$^+$ ALL. (*Top*) Serial luminescent images were captured every 3–4 days after intravenous injection of 200,000 p185$^+$;*Arf*$^{-/-}$ luciferase-expressing pre-B cells. (*Bottom*) Whole-body luminescent activity was quantified in these mice at the indicated times (days) thereafter. Each symbol and connecting line represents tumor progression in an individual animal.

therapeutic agents, develop novel treatment strategies, and interrogate mechanisms of antileukemic drug resistance.

RESISTANCE TO TARGETED THERAPY IN PH$^+$ ALL

In contrast to the success of BCR-ABL–targeted therapeutics in maintaining CML-CP patients in durable remission, patients with Ph$^+$ ALL typically experience significant early responses to single-agent kinase inhibition but then relapse, often with drug-resistant BCR-ABL KD mutations. Maximal intensity imatinib therapy (100 mg/kg twice per day by oral gavage) had only modest activity in our Ph$^+$ ALL mouse model (Williams et al. 2006), despite significant efficacy of the identical treatment regimen in murine CML models. Furthermore, these studies revealed that nontumor-cell-autonomous (cell-extrinsic) imatinib resistance, mediated in part through cytokine signaling within the hematopoietic microenvironment, contributed to therapeutic resistance in vivo (Williams et al. 2007).

We therefore initiated therapeutic trials with dasatinib, reasoning that its significantly enhanced potency relative to imatinib, coupled with its ability to inhibit the Src family of nonreceptor tyrosine kinases, would lead to an enhanced therapeutic response in vivo (Shah et al.

2004). Dasatinib is about 300-fold more potent than imatinib in in vitro measures of growth inhibition of p185+;*Arf*−/− cells. In contrast to earlier trials in which imatinib therapy was initiated 3 days after inoculation of p185+;*Arf*−/−;GFP+ cells, dasatinib treatment (10 mg/kg twice per day by gavage) was commenced 10 days after introduction of 200,000 p185+;*Arf*−/−;Luc+ cells. These day-10 recipients were estimated to have a 100-fold increase in leukemic burden compared to day-3 mice (see Fig. 2). They demonstrated clinical signs of leukemic cell expansion (lethargy, ruffled coat, reduced mobility) and typically had infiltration of their bone marrow compartment with 20–40% leukemic cells (a level of disease that would confer a diagnosis of ALL in a human patient). Luminescent imaging confirmed that cohorts of recipient animals had comparable disease burdens when therapy was initiated and allowed us to prospectively monitor the therapeutic response (Fig. 3, top). As expected, all vehicle-treated control mice succumbed to their leukemia within 3–4 more days (not shown). Remarkably, dasatinib therapy induced a dramatic reduction in disease burden within 1 week, and after 2 weeks of therapy, there was a 30–100-fold reduction in whole-animal luminescent signals (Fig. 3, bottom). Despite continuous twice-daily dasatinib therapy, however, all mice had persistent levels of measurable disease during the 10-week treatment window, and as therapy continued, virtually all animals relapsed with overt clinical signs of leukemia that closely correlated with dramatic increases in whole-body luminescent signals (Fig. 3, bottom). At relapse, many animals demonstrated foci of enhanced luminescent signals over their head and neck region, and often, these mice had clinical signs of central nervous system disease, including a bulging skull, irritability, and erratic behavior.

The significant initial response of murine Ph+ ALL to dasatinib therapy followed by therapeutic failure despite continuous therapy is reminiscent of the clinical experience with dasatinib and nilotinib in human Ph+ ALL (Kantarjian et al. 2006; Talpaz et al. 2006; Hochhaus et al. 2007). Molecular analysis of these dasatinib-resistant murine Ph+ leukemias has now revealed that a significant number of them individually harbor clones with BCR-ABL KD mutations that have been previously identified in human clinical trials. Indeed, the most frequently encountered KD mutation (T315I) in mice that relapsed on dasatinib dramatically impairs kinase inhibition by all three FDA-approved BCR-ABL inhibitors (imatinib, nilotinib, and dasatinib) and represents one of the key resistance determinants in treated Ph+ ALL patients. The acquisition of dasatinib-resistant KD mutants further reinforces the notion that the combination of BCR-ABL expression and *Arf* inactivation faithfully models the biology, genetics, and therapeutic responses of human Ph+ ALL and will provide unique opportunities to further interrogate the genesis of drug resistance in vivo.

Despite the considerable dasatinib-induced "debulking" of leukemic burden, there are tissue compartments that harbor functionally dasatinib-resistant cell populations in treated mice. What are the factors that contribute to the persistence of disease and facilitate the emergence

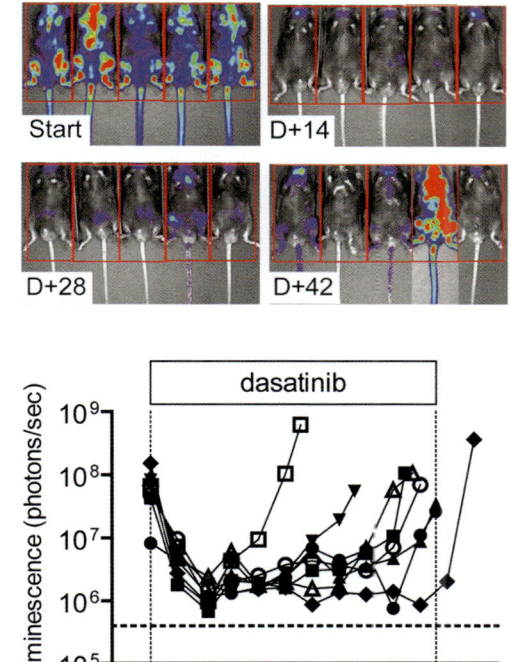

Figure 3. Dasatinib response and relapse of murine Ph+ ALL. (*Top*) Serial luminescent images were captured at the start of therapy (Start) 10 days after injection of 200,000 p185+;*Arf*−/− luciferase-expressing pre-B cells. Total body luminescence was quantified at the indicated days (D) thereafter as animals were maintained on continuous, twice-daily dasatinib therapy. An animal with elevated signal (abundant disease) at D+42 underwent relapse. (*Bottom*) Whole-body luminescent activity was quantified from these mice at the start of therapy and at weekly intervals during the course of dasatinib treatment and is plotted against time (days) after injection of donor cells. Twice-daily dasatinib therapy was continued for 10 weeks. Each symbol and connecting line represents an individual animal, all of which ultimately relapsed.

of drug-resistant mutants in the face of continuous therapy? A trivial explanation would be that mice are receiving suboptimal doses of dasatinib, but our experience suggests that the current dosing schedule is maximally intensive in animals that harbor significant disease burdens at the time of initiation of therapy. Alternatively, local drug concentrations in certain tissue microenvironments might fluctuate around cytostatic levels, but not cytotoxic levels, particularly when leukemic cells are protected by host factors produced through their interactions with stromal cells and extracellular matrix components. Moreover, the levels of dasatinib achieved in particular organ compartments might be suboptimal, a phenomenon that likely contributes to relapse within the central nervous system.

The unifying hypothesis is that in the context of BCR-ABL–driven lymphoid leukemias, *Arf* inactivation enhances the biological fitness of LICs, thereby promoting more efficient dissemination of leukemic cells, diminishing the efficacy of targeted therapy, and facilitating the emergence of cell-intrinsic drug resistance, most frequently manifested by BCR-ABL KD mutations. We reason that persistent BCR-ABL kinase activity initially selects for *INK4-ARF* deletions in vivo. Subsequently, nontumor-cell-autonomous drug resistance mechanisms operate to sustain a residual pool of viable p185[+]*;Arf* null leukemic cells during therapy. Persistent drug exposure then selects for the acquisition of drug-resistant BCR-ABL KD mutations in some cells that further expand. If *INK4-ARF* inactivation similarly facilitates acquisition of drug-resistant BCR-ABL mutations in humans, the third of Ph[+] ALL patients that retain the gene cluster at diagnosis might respond better than the majority to targeted therapy. Conversely, Ph[+] ALL patients that have already sustained *INK4-ARF* loss may need to be treated with the most potent kinase inhibitors together with conventional therapeutic agents or, ideally, with drugs that target the interaction of LICs with the host hematopoietic microenvironment.

CONCLUSIONS

In the hematopoietic system, the *INK4-ARF* locus functions as a master regulator of Rb and p53 function to prevent inappropriate progenitor cell self-renewal and to eliminate incipient cancer cells driven by sustained oncogenic signaling. Analysis of two distinct Ph[+] leukemias provides clear evidence that whereas the *INK4-ARF* locus is intact (and likely silenced) in CML, a disease that is very responsive to targeted therapies, frequent deletion of *INK4-ARF* occurs in BCR-ABL–dependent lymphoid leukemias (CML-LBC and Ph[+] ALL) and correlates with transient responses to therapy and poor survival. Our highly efficient murine Ph[+] ALL model has been further refined and has several desirable features that facilitate biological and therapeutic investigations. First, the leukemic phenotype is very robust and mimics the genetics and biology of human ALL, including frequent dissemination of leukemic cells to the central nervous system. Second, fixed numbers of tagged, homogeneous, and genetically defined (p185[+]*;Arf* [−/−]) donor cells, when engrafted into the hematopoietic system of an otherwise healthy syngeneic recipient, imitate the effects of the somatic mutation that initiates Ph[+] ALL in human patients. Third, disease burden can be monitored and quantified in real time before and in response to therapy, permitting the tracking of individual treatment responses and relapse. Fourth, in response to different targeted treatment protocols, we can now characterize the in vivo emergence and spectrum of drug-resistant BCR-ABL mutations. Collectively, these features underscore the value of modeling human cancer through innovative applications of mouse genetic systems and emphasize the utility of controlled preclinical "trials" for efficiently implementing and evaluating new therapeutic strategies.

ACKNOWLEDGMENTS

We thank members of the Sherr/Roussel and Williams laboratories for their contributions throughout the course

of these studies. This work was supported in part by Cancer Center Core grant CA21765 and by ALSAC of St. Jude Children's Research Hospital. C.J.S. is an Investigator of the Howard Hughes Medical Institute.

REFERENCES

Akala, O.O., Park, I.-K., Qian, D., Pihalja, M., Becker, M.W., and Clarke, M.F. 2008. Long-term haematopoietic reconstitution by Trp53[−/−], p16[Ink4a−/−], p19[Arf−/−] multipotent progenitors. *Nature* **453:** 228–232.

Alvarado, Y., Apostolidou, E., Swords, R., and Giles, F.J. 2007. Emerging therapeutic options for Philadelphia-positive acute lymphocytic leukemia. *Expert Opin. Emerg. Drugs* **12:** 165–179.

Arico, M., Valsecchi, M.G., Camitta, B., Schrappe, M., Chessells, J., Baruchel, A., Gaynon, P., Silverman, L., Janka-Schaub, G., Kamps, W., Pui, C.-H., and Masera, G. 2000. Outcome of treatment in children with Philadelphia chromosome-positive acute lymphoblastic leukemia. *N. Engl. J. Med.* **342:** 998–1006.

Armstrong, S.A. and Look, A.T. 2005. Molecular genetics of acute lymphoblastic leukemia. *J. Clin. Oncol.* **23:** 6306–6315.

Calabretta, B. and Perrotti, D. 2004. The biology of CML blast crisis. *Blood* **103:** 4010–4022.

Castor, A., Nilsson, L., Astrand-Grundstrom, I., Buitenhuis, M., Ramirez, C., Anderson, K., Strombeck, B., Garwicz, S., Bekassy, A.N., Schmiegelow, K., et al. 2005. Distinct patterns of hematopoietic stem cell involvement in acute lymphoblastic leukemia. *Nat. Med.* **11:** 630–637.

Chan, L.C., Karhi, K.K., Rayter, S.I., Heisterkamp, N., Eridani, S., Powles, R., Lawler, S.D., Groffen, J., Foulkes, J.G., Greaves, M.F., and Wiedemann, L.M. 1987. A novel *abl* protein expressed in Philadelphia chromosome positive acute lymphoblastic leukemia. *Nature* **325:** 635–637.

Clark, S.S., McLaughlin, J., Crist, W.M., Champlin, R., and Witte, O.N. 1987. Unique forms of the *abl* tyrosine kinase distinguish Ph[1]-positive CML from Ph[1]-positive ALL. *Science* **235:** 85–88.

Clarke, M.F., Dick, J.E., Dirks, P.B., Eaves, C.J., Jamieson, C.H.M., Jones, D.L., Visvader, J., Weissman, I.L., and Wahl, G.M. 2006. Cancer stem cells—Perspectives on current status and future directions: AACR workshop on cancer stem cells. *Cancer Res.* **66:** 9339–9344.

de Labarthe, A., Rousselot, P., Huguet-Rigal, F., Delabesse, E., Witz, F., Maury, S., Rea, D., Cayuela, J.-M., Vekemans, M.-C., Reman, O., et al. 2007. Imatinib combined with induction or consolidation chemotherapy in patients with de novo Philadelphia chromosome-positive acute lymphoblastic leukemia: Results of the GRAAPH-2003 study. *Blood* **109:** 1408–1413.

Druker, B.J., Sawyers, C.L., Kantarjian, H., Resta, D.J., Reese, S.F., Ford, J.M., Capdeville, R., and Talpaz, M. 2001. Activity of a specific inhibitor of the BCR-ABL tyrosine kinase in the blast crisis of chronic myeloid leukemia and acute lymphoblastic leukemia with the Philadelphia chromosome. *N. Engl. J. Med.* **344:** 1038–1042.

Druker, B.J., Guilhot, F., O'Brien, S.G., Gathmann, I., Kantarjian, H., Gattermann, N., Deininger, M.W., Silver, R.T., Goldman, J.M., Stone, R.M., et al. 2006. Five-year follow-up of patients receiving imatinib for chronic myeloid leukemia. *N. Engl. J. Med.* **355:** 2408–2417.

Gleissner, B., Gokbuget, N., Bartram, C.R., Janssen, B., Rieder, H., Janssen, J.W., Fonatsch, C., Heyll, A., Voliotis, D., Beck, J., et al. 2002. Leading prognostic relevance of the BCR-ABL translocation in adult B-lineage lymphoblastic leukemia: A prospective study of the German Multicenter Trial Group and confirmed polymerase chain reaction analysis. *Blood* **99:** 1536–1543.

Groffen, J., Stephenson, J.R., Heisterkamp, N., de Klein, A., Bartram, C.R., and Grosveld, G. 1984. Philadelphia chromosomal breakpoints are clustered within a limited region, *bcr*, on chromosome 22. *Cell* **36:** 93–99.

Hochhaus, A., Kantarjian, H.M., Baccarani, M., Lipton, J.H., Apperley, J.F., Druker, B.J., Facon, T., Goldberg, S.L., Cervantes, F., Niederwieser, D., et al. 2007. Dasatinib induces notable hematologic and cytogenetic responses in chronic-phase chronic myeloid leukemia after failure of imatinib therapy. *Blood* **109:** 2303–2309.

Iwama, A., Oguro, H., Negishi, M., Kato, Y., Morita, Y., Tsukui, H., Ema, H., Kamijo, T., Katoh-Fukui, Y., Koseki, H., van Lohuizen, M., and Nakauchi, H. 2004. Enhanced self-renewal of hematopoietic stem cells mediated by the polycomb gene product Bmi-1. *Immunity* **21:** 843–851.

Jacobs, J.J.L., Kieboom, K., Marino, S., DePinho, R.A., and van Lohuizen, M. 1999. The oncogene and Polycomb-group gene *bmi-1* regulates cell proliferation and senescence through the *ink4a* locus. *Nature* **397:** 164–168.

Jamieson, C.H., Ailles, L.E., Dylla, S.J., Muijtjens, M., Jones, C., Zehnder, J.L., Gotlib, J., Li, K., Manz, M.G., Keating, A., Sawyers, C.L., and Weissman, I.L. 2004. Granulocyte-macrophage progenitors as candidate leukemic stem cells in blast crisis CML. *N. Engl. J. Med.* **351:** 657–667.

Kantarjian, H., Giles, F., Wunderle, L., Bhalla, K., O'Brien, S., Wassmann, B., Tanaka, C., Manley, P., Rae, P., Mietlowski, W., et al. 2006. Nilotinib in imatinib-resistant CML and Philadelphia chromosome-positive ALL. *N. Engl. J. Med.* **354:** 2542–2551.

Kohno, T. and Yokota, J. 2006. Molecular processes of chromosome 9p21 deletions causing inactivation of the p16 tumor suppressor gene in human cancer: Deduction from structural analysis of breakpoints for deletions. *DNA Repair* **5:** 1273–1281.

Krimpenfort, P., Ijpenberg, A., Song, J.Y., van der Valk, M., Nawijn, M., Zevenhoven, J., and Berns, A. 2007. p15^Ink4b is a critical tumour suppressor in the absence of p16^Ink4a. *Nature* **448:** 943–946.

Lessard, J. and Sauvageau, G. 2003. Bmi-1 determines the proliferative capacity of normal and leukaemic stem cells. *Nature* **423:** 255–260.

Mullighan, C.G., Williams, R.T., Downing, J.R., and Sherr, C.J. 2008a. Failure of CDKN2A/B (INK4A/B-ARF)-mediated tumor suppression and resistance to targeted therapy in acute lymphoblastic leukemia induced by BCR-ABL. *Genes Dev.* **22:** 1411–1415.

Mullighan, C.G., Miller, C.B., Radtke, I., Phillips, L.A., Dalton, J., Ma, J., White, D., Hughes, T.P., Le Beau, M.M., Pui, C.-H., et al. 2008b. *BCR-ABL1* lymphoblastic leukemia is characterized by the deletion of Ikaros. *Nature* **453:** 110–114.

Nowell, P.C. and Hungerford, D.A. 1960. Chromosome studies on normal and leukemic human leukocytes. *J. Natl. Cancer Inst.* **25:** 85–109.

O'Hare, T., Eide, C.A., and Deininger, M.W. 2007. Bcr-Abl kinase domain mutations, drug resistance, and the road to a cure for chronic myeloid leukemia. *Blood* **110:** 2242–2249.

Park, I.-K., Qian, D., Kiel, M., Becker, M.W., Pihalja, M., Weissman, I.L., Morrison, S.J., and Clarke, M.F. 2003. Bmi-1 is required for maintenance of adult self-renewing haematopoietic stem cells. *Nature* **423:** 302–305.

Rowley, J.D. 1973. A new consistent chromosomal abnormality in chronic myelogenous leukemia. *Nature* **243:** 290–293.

Shah, N.P., Nicoll, J.M., Nagar, B., Gorre, M.E., Paquette, R.L., Kuriyan, J., and Sawyers, C.L. 2002. Multiple BCR-ABL kinase domain mutations confer polyclonal resistance to the tyrosine kinase inhibitor imatinib (STI571) in chronic phase and blast crisis chronic myeloid leukemia. *Cancer Cell* **2:** 117–125.

Shah, N.P., Tran, C., Lee, F.Y., Chen, P., Norris, D., and Sawyers, C.L. 2004. Overriding imatinib resistance with a novel ABL kinase inhibitor. *Science* **305:** 399–401.

Talpaz, M., Shah, N.P., Kantarjian, H., Donato, N., Nicoll, J., Paquette, R., Cortes, J., O'Brien, S., Nicaise, C., Bleickardt, E., et al. 2006. Dasatinib in imatinib-resistant Philadelphia chromosome-positive leukemias. *N. Engl. J. Med.* **354:** 2531–2541.

Williams, R.T., Roussel, M.F., and Sherr, C.J. 2006. *Arf* gene loss enhances oncogenicity and limits imatinib response in mouse models of Bcr-Abl-induced acute lymphoblastic leukemia. *Proc. Natl. Acad. Sci.* **103:** 6688–6693.

Williams, R.T., den Besten, W., and Sherr, C.J. 2007. Cytokine-dependent imatinib resistance in mouse BCR-ABL⁺, *Arf*-null lymphoblastic leukemia. *Genes Dev.* **21:** 2283–2287.

Wong, S. and Witte, O.N. 2004. The BCR-ABL story: Bench to bedside and back. *Annu. Rev. Immunol.* **22:** 247–306.

Zhao, C., Blum, J., Chen, A., Kwon, H.Y., Jung, S.H., Cook, J.M., Lagoo, A., and Reya, T. 2007. Loss of β-catenin impairs the renewal of normal and CML stem cells in vivo. *Cancer Cell* **12:** 528–541.

Delineating the Epithelial Hierarchy in the Mouse Mammary Gland

M.-L. Asselin-Labat,* F. Vaillant,* M. Shackleton,*† T. Bouras,*
G.J. Lindeman,* and J.E. Visvader*

*VBCRC Laboratory, The Walter and Eliza Hall Institute of Medical Research,
Melbourne, VIC 3050, Australia

Reconstitution assays have shown that mouse mammary stem cells reside within the mature mammary gland in vivo. Single cells could be prospectively isolated and shown to regenerate an entire mammary gland that exhibited full developmental capacity. The more recent identification of luminal progenitor populations has indicated that the mammary epithelium is organized in a hierarchical manner. Further definition of epithelial cell types in both mouse and human mammary glands will provide insight into the "cells of origin" in the different subtypes of breast cancer, as well as the nature of cancer-propagating cells. Here, we review the known characteristics of mammary stem and progenitor cells, their steroid receptor status, and the pathways that have thus far been implicated in regulating their self-renewal and differentiation.

Development of the mammary gland occurs predominantly after birth and is largely governed by the concerted action of steroid and peptide hormones together with locally produced cytokines (Hennighausen and Robinson 2005). During puberty, the mouse mammary gland undergoes extensive growth, resulting in the formation of a network of branching ducts that fills the mammary fat pad. In the adult, the mammary epithelium undergoes cycles of proliferation and differentiation with each estrus cycle, during which alveolar buds arise. These develop into alveolar structures during pregnancy and undergo terminal differentiation, thus allowing milk production in lactation. Involution represents the final stage of mammary gland morphogenesis and involves substantial structural remodeling of the gland, resulting in a near preparous (virginal) appearance. The massive expansion of mammary epithelium during puberty and pregnancy, together with the polyregenerative capacity apparent upon successive reproductive cycles, suggests the existence of long-lived stem cells in the mammary gland.

Pioneering studies in the mammary gland biology field led to the development of an in vivo transplantation assay in which the mammary fat pad (MFP) can be cleared of endogenous epithelium while retaining an intact stromal environment (De Ome et al. 1959). Small mammary tissue fragments (explants) (Hoshino and Gardner 1967; Daniel et al. 1968) or cell suspensions prepared from whole mammary tissue (Smith 1996) were shown to reconstitute functional mammary epithelial structures and were capable of serial transplantation, although regenerative senescence was induced after seven generations. Direct support for the stem cell concept came from transplantation studies using retrovirally marked donor tissue fragments, which suggested that a single stem cell was capable of repopulating the entire mammary epithelium (Kordon and Smith 1998). In human breast tissue, there is also evidence for a popula-

tion of stem cells. This includes the presence of large contiguous fields of cells within mammary ducts and lobules that exhibit clonality (Tsai et al. 1996) and the presence of distinct progenitor cell phenotypes in colony-forming assays (Stingl et al. 1998, 2001).

Three distinct epithelial cell types reside within the mammary gland: ductal, alveolar, and myoepithelial cells. The ductal and alveolar cells belong to the luminal epithelial cell lineage and line the lumina or constitute the alveolar units that emerge during pregnancy, respectively. The myoepithelium surrounds the luminal epithelium and contacts the basement membrane. In the lactational phase, these cells provide contractile properties to facilitate milk expulsion. It is presumed that a hierarchy of stem and committed progenitor cells give rise to mature luminal and myoepithelial cells, analogous to that in the hematopoietic compartment. Moreover, breast cancer is a very heterogeneous disease, and the observation of multiple subtypes based on gene expression profiling studies further suggests the existence of a hierarchy of epithelial cells. At least six distinct subtypes have been described (Perou et al. 2000; Sørlie et al. 2001; Sotiriou et al. 2003; Herschkowitz et al. 2007), and these are thought to originate in different breast epithelial cell types that serve as the "cell of origin." Elucidation of the molecular pathways involved in regulating the self-renewal and progressive differentiation of mammary stem and progenitor cells is critical for understanding the mechanisms that underpin mammary carcinogenesis.

PROSPECTIVE ISOLATION OF MAMMARY STEM CELLS

The mouse mammary stem cell (MaSC) has been prospectively isolated using the "gold-standard" transplantation assay developed for the mammary gland (De Ome et al. 1959). Using optimized methods of tissue dissociation to obtain a single cell suspension, we used a combination of cell surface markers and fluorescence-activated cell sorting

---

†Present address: Life Sciences Institute, University of Michigan, Ann Arbor, Michigan 48109.

(FACS) to resolve four distinct mammary populations. Cells were depleted of leukocytesand endothelial and erythroid precursor cells, using CD45, CD31, and TER-119, respectively, thus defining a Lin⁻ population. Cells were then sorted on the basis of expression of CD24/HSA (heat-stable antigen) and CD29/β1–integrin and transplanted into cleared mammary fat pads at limiting dilution to enable evaluation of the repopulating cell frequency (Bonnefoix et al. 1996). Only CD29^{hi}CD24^{+} cells were able to regenerate a mammary epithelial tree, at a frequency of 1 in 64 cells (Shackleton et al. 2006). Similarly, Stingl et al. (2006) showed that 1 in 62 CD49f^{hi}CD24$^{+/mod}$ cells could yield mammary outgrowths. It is relevant that CD49f (α6 integrin) is known to form heterodimers with CD29 in the mammary gland. Costaining revealed that the vast majority of CD29^{hi} cells were also CD49f^{hi} (Fig. 1, top right), although there were some CD29^{hi} cells that did not express high levels of CD49f. It seems probable that this integrin heterodimer mediates interactions between the basement membrane and stem cells, which likely reside in a basolateral position. CD24 appears to serve as a pan-epithelial cell surface marker in the mouse mammary gland (Shackleton et al. 2006; Sleeman et al. 2006; Stingl et al. 2006). Sleeman et al. (2006) have described three levels of CD24 expression in mouse mammary tissue: CD24^{neg}, CD24^{mod}, and CD24^{hi} subpopulations, corresponding to nonepithelial, myoepithelial/basal, and luminal cells, respectively. Only the CD24^{mod} cells, which are also negative for Sca-1, were capable of regenerating mammary outgrowths. It should be noted that different levels of fluorescence have been obtained with different anti-CD24 antibodies and with varying antibody concentrations, leading to differing CD24 nomenclature. However, there is full agreement on the emerging phenotype of the MaSC, which is CD29^{hi}

CD49f^{hi}CD24$^{+/mod}$Sca-1$^-$. This population, however, is not pure and contains MaSCs (< 10%) together with mature myoepithelial cells and likely basal progenitors.

Analogous to the hematopoietic stem cell, a single microscopically visualized cell was demonstrated to reconstitute an entire functional mammary gland. The isolated stem cells exhibited full developmental potential (Fig. 1, bottom left), giving rise to extensive outgrowths comprising milk-producing alveolar units in pregnant recipients (Fig. 1, bottom middle and right). Furthermore, the regenerated structures contained daughter cells with the same in vivo repopulating activity as the original stem cell (Shackleton et al. 2006). In vivo serial transplantation experiments demonstrated that clonal primary outgrowths can be passaged for up to 10 generations (Vaillant et al. 2008), revealing that the MaSC is capable of extensive self-renewal. Indeed, Stingl et al. (2006) estimated that the MaSC can execute 10 symmetrical self-renewing divisions. Thus, the prospectively isolated mammary repopulating cells fulfill the two defining features of a stem cell: multilineage differentiation and self-renewal. Although sorted MaSCs were shown to be cycling (Stingl et al. 2006), the CD29^{hi}CD24$^{+/mod}$ population is enriched for bromodeoxyuridine (BrdU)-label-retaining cells in vivo, suggesting that a pool of quiescent stem cells may reside in the mammary gland (Shackleton et al. 2006).

IDENTIFICATION OF A LUMINAL EPITHELIAL PROGENITOR CELL

In the mouse mammary gland, the CD29^{lo}CD49f^{lo}CD24$^{+/hi}$ population is enriched for luminal epithelial cells that are characterized by expression of cytokeratin 18 but lack smooth muscle actin, a myoepithelial marker. Using an

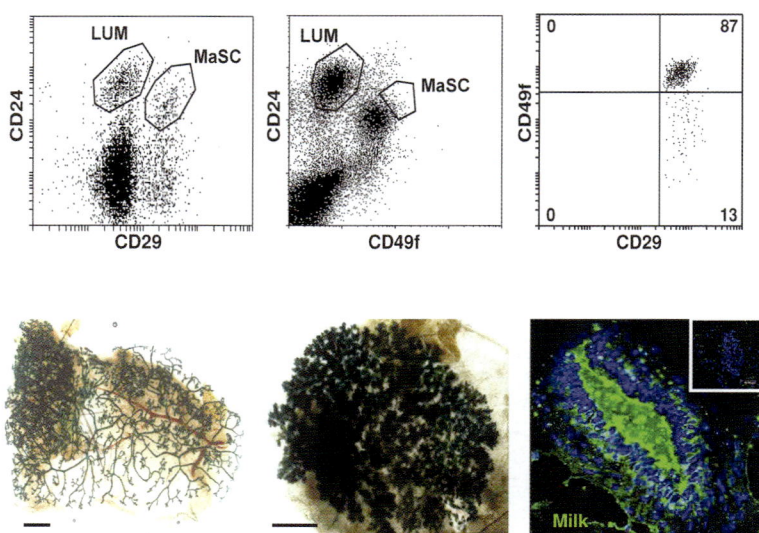

Figure 1. (*Top panels*) Expression of CD29, CD24, and CD49f in Lin⁻ mammary cells. Flow cytometric analysis of Lin⁻ cell suspensions costained with CD29 and CD24 (*left panel*) and CD49f and CD24 (*middle panel*). (*Right panel*) CD29 and CD49f expression in Lin⁻CD29^{hi}CD24^{+} cells. (LUM) Luminal epithelial cells; (MaSC) mammary stem cells. (*Bottom panels*) X-gal-stained outgrowths derived from single transplanted Rosa-26 cells in virgin (*left panel*) and pregnant (*middle panel*) recipients. Bar, 250 μm. (*Right panel*) Immunofluorescent staining for milk proteins in a mammary duct derived from a single cell and harvested during pregnancy.

additional cell surface marker, CD61/β3-integrin, the luminal CD29^{lo}CD24$^+$ population could be subdivided into two subpopulations (CD61$^+$ and CD61$^-$ fractions). The CD61$^+$ subset represents about 30% of the luminal population and has extensive colony-forming capacity in both two-dimensional and Matrigel cultures, yielding colonies that stain exclusively with luminal lineage markers. In contrast, the CD61$^-$ subset is highly enriched for mature luminal epithelial cells (Asselin-Labat et al. 2007). In this more mature population, a small fraction of cells can form colonies in vitro, indicating that CD61$^-$ cells also comprise a progenitor population (Asselin-Labat et al. 2007). In separate studies, CD133/prominin-1 was found to fractionate the luminal cell population into CD24$^{+/hi}$CD133$^-$ and CD24$^{+/hi}$CD133$^+$ cells enriched in progenitor and differentiated cells, respectively (Sleeman et al. 2007). CD61 expression, however, does not directly correlate with CD133 expression in the luminal cell population defined by CD29^{lo} and CD24$^+$ (Fig. 2, top). This observation suggests that these markers may be identifying distinct progenitor cells. During mammary gland ontogeny, the percentage of CD61$^+$ progenitor cells was highest in pubertal mammary glands, a stage involving extensive ductal branching and elongation to generate the mature ductal tree, and declined thereafter in the adult gland. Pregnancy was accompanied by a dramatic decrease in the proportion of CD61$^+$ progenitor cells, compatible with alveolar differentiation occurring during this stage. Taken together, these results suggest that CD29^{lo} CD24$^+$CD61$^+$ cells can act as progenitors for both ductal and alveolar

luminal cells, dependent on the hormonal milieu. The side population (SP) defined by Hoechst 33342 dye efflux, originally thought to be enriched in MaSCs (Welm et al. 2002), has subsequently been shown to be depleted of CD29^{hi}CD24$^+$ (MaSC-enriched) cells (Shackleton et al. 2006; Stingl et al. 2006). In fact, the SP contains sixfold more luminal progenitor cells than differentiated luminal cells, and 70% of SP cells are found in the luminal progenitor fraction (CD29^{lo}CD24$^+$CD61$^+$) (Fig. 2, bottom).

There are at least two potential hierarchical models for mammary stem cells giving rise to progressively restricted intermediate and differentiated epithelial cells in the mammary gland. In the model that we currently favor (Fig. 3), mature luminal and myoepithelial cells arise from unipotent progenitor cells. Although there appears to be a common luminal progenitor cell (CD61$^+$) for ductal and alveolar epithelial cells, this model does not exclude the existence of further intermediate cell types downstream from the common luminal progenitor cell, with distinct precursors for ductal and alveolar luminal cells. The generation of these will presumably be dependent on the hormonal environment and developmental stage. Markers of the myoepithelial progenitor cell are yet to be determined, but this cell is presumed to reside in the MaSC-enriched fraction, and the presence of myoepithelial cell-only colonies in cultures of human breast epithelial cells (Stingl et al. 1998) provides support for their existence. In an alternative model, bifurcation of the ductal and alveolar lineages may occur before luminal versus myoepithelial cell-fate decisions. Separate

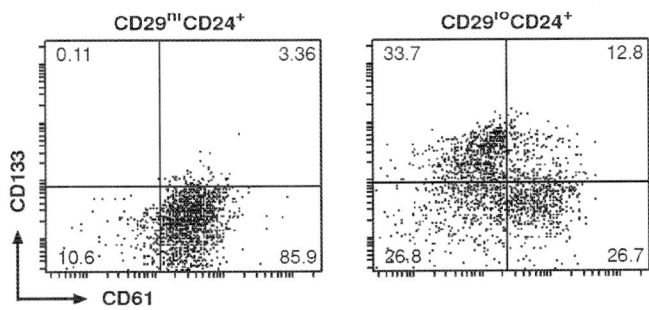

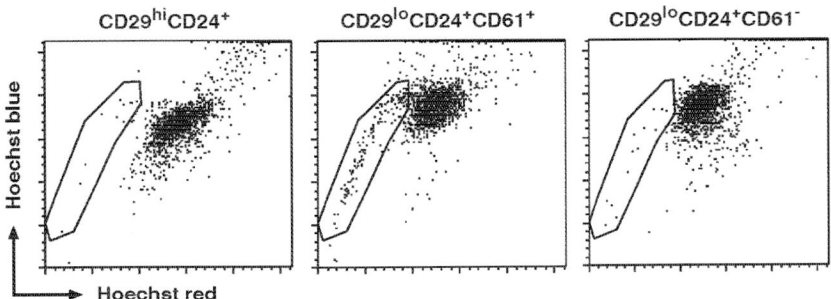

Figure 2. (*Top*) Prominin-1 (CD133) and CD61 costaining in MaSC-enriched (CD29^{hi}CD24$^+$) and luminal (CD29^{lo}CD24$^+$) populations. Cells were depleted of CD31$^+$CD45$^+$TER119$^+$ cells (Lin$^-$). (*Bottom*) Enrichment in the Hoechst side population (SP) cells of luminal progenitor cells (CD29^{lo}CD24$^+$CD61$^+$) but not MaSC-enriched (CD29^{hi}CD24$^+$) or mature luminal cells (CD29^{lo}CD24$^+$CD61$^-$).

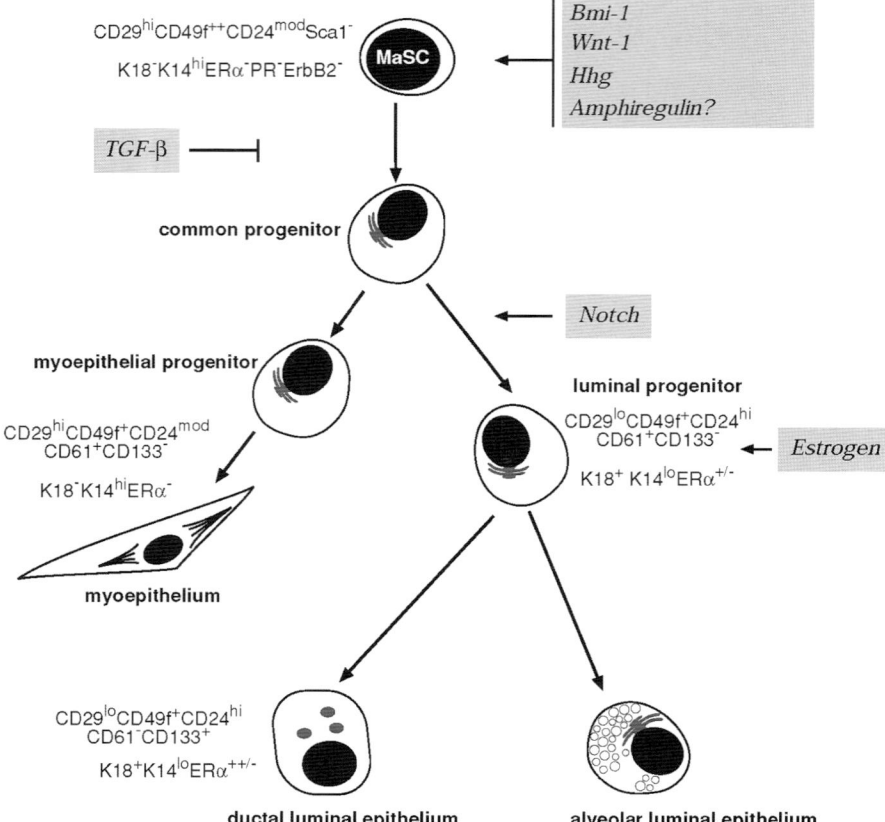

Figure 3. Schematic model of the mouse mammary epithelial hierarchy. The cell surface immunophenotypes and keratin expression profiles of the stem and intermediate cell types are shown. Genes and pathways that have been implicated in regulating cells in the hierarchy are indicated. (MaSC) Mammary stem cell.

progenitors for the ductal and alveolar lineages may exist as well as a novel alveolar myoepithelial cell. Some support for this hierarchical organization comes from the observation that transplantation of cultured mammary cells into cleared fat pads produced epithelial outgrowths with either a ductal-only or lobule-only morphology (Smith 1996). Further studies using additional cell surface markers and freshly isolated cells at clonal dilution will be required to unravel the epithelial hierarchy and to allow lineage tracing to determine the relationship between these cell types. One potential question that arises is Can the different hormonal influences impose a degree of plasticity on stem and progenitor cells in the mammary gland?

HUMAN BREAST STEM AND PROGENITOR CELLS

There is substantial evidence for the existence of a bipotent epithelial progenitor cell in human breast tissue. This includes a MUC1⁻CALLA$^{lo/hi}$ESA⁺(EpCam⁺)CD49f⁺ progenitor cell (Stingl et al. 1998, 2001), an immortalized bipotent human mammary epithelial cell line derived from MUC1⁻ESA⁺ cells (Gudjonsson et al. 2002), and multipotent epithelial cells that preferentially form mammospheres (Dontu et al. 2003). Using several postulated stem cell markers including stage-specific embryonal antigen-4

(SSEA-4); keratins 15, 5, and 6a; chondroitin sulfate; and Bcl-2, Villadsen et al. (2007) described putative stem cells that resided within ducts and were virtually absent from lobules. In vitro culture assays further showed that the cells with clonogenic capacity and bipotency were localized to ducts (Villadsen et al. 2007).

Different types of unipotent progenitors have also been described, including two types of luminal-restricted progenitors and a myoepithelial-restricted progenitor (Stingl et al. 2001; Villadsen et al. 2007). Luminal-restricted progenitors with the EpCAM⁺CD49f⁺ phenotype express the luminal marker MUC1 and can be further subdivided into two subsets, K19⁺K14⁻ and K19⁻K14⁻ cells (Villadsen et al. 2007), but the physiological relevance of these two populations is not yet understood. Differentiated luminal (EpCam⁺MUC1⁺CD49f⁻) and myoepithelial (EpCam^{lo} MUC1⁻CD49f⁺) cell populations have also been identified (Stingl et al. 2001; Stingl and Caldas 2007). Thus, there appear to be striking parallels in the cell types that reside in mouse and human mammary epithelium.

Studies on human mammary stem and progenitor cells have been hampered by the challenges of xenotransplantation into the mouse mammary fat pad. Toward this end, Kuperwasser et al. (2004) "humanized" the mouse mammary fat pad of immunocompromized NOD/SCID mice by injection of irradiated and unirradiated immortalized human fibroblasts in an attempt to recreate a stromal environment characteristic of that in human breast tissue. Breast organoids were transplanted into this microenvironment and were able to generate ducts and lobular units. More recently, Ginestier et al. (2007) identified a subpopulation of cells with stem/progenitor cell activity in normal human mammary tissue based on aldehyde dehydrogenase 1 (ALDH1) activity using the ALDEFLUOR assay. High ALDH1 activity is a feature of both stem and progenitor cells in the neural and hematopoietic systems (Armstrong et al. 2004; Hess et al. 2004; Matsui et al. 2004). ALDH1$^+$ breast epithelial cells were shown to be enriched for cells capable of forming mammary epithelial ducts in vivo. Although serial transplantation presents a major challenge for human stem cell work, it will ultimately be required to prove that these cells correspond to breast stem cells.

HORMONAL REGULATION OF MAMMARY STEM CELLS

The steroid hormones estrogen and progesterone have essential roles in mammary gland development, and both estrogen receptor α (ERα) and progesterone receptor (PR) are important prognostic markers in breast cancer. Estrogen exerts its activity by binding to nuclear ERα, and ERα knockout mice display profound defects in ductal development (Bocchinfuso and Korach 1997; Mueller et al. 2002). Interestingly, we have found that the mouse MaSC-enriched population CD29^{hi}CD49f^{hi}CD24$^{+/mod}$ does not express ERα or PR, which is also a target gene of ERα (Asselin-Labat et al. 2006). The CD24^{mod} population described by Sleeman et al. (2007) also lacked expression of these receptors. Notably, only a small proportion (6%) of the luminal progenitor population expressed ERα and PR, in contrast to the differentiated luminal cell population, in which 40% expressed these receptors (Asselin-Labat et al. 2006). Thus, the luminal progenitor cell may be the first cell type along the hierarchy that expresses ERα, and these cells may be the initial target of transformation in individuals that develop ERα$^+$ breast cancer. Other studies have reported that label-retaining mammary epithelial cells in the mouse mammary gland (Booth and Smith 2006) and side population cells in human breast tissue are enriched in ERα-positive cells (Clarke et al. 2005). These findings are compatible with the observation that some progenitor cells express ERα and lend support to the notion that a hierarchy of stem cells may exist, reminiscent of that occurring in the hematopoietic compartment.

Given the importance of estrogen to normal mammary development, how does a single ERα-negative mammary stem cell reconstitute an entire ductal tree? It seems likely that the MaSC creates its own epithelial niche through symmetrical and asymmetrical cell divisions, giving rise to both ER-positive and ER-negative progeny that respond to hormonal cues during puberty and coordinate the development of a branching ductal tree. It is known that estrogen mediates its effects in the mammary gland through paracrine signaling (Scully et al. 1997; Mallepell et al. 2006). It is therefore presumed that ERα-positive epithelial cells secrete a paracrine factor(s), in response to

activation by estrogen, which in turn activates MaSCs. Although these factors are yet to be fully established, amphiregulin, an agonist of the epidermal growth factor receptor, appears to be a critical mediator of estrogen effects in the mammary gland. Amphiregulin was recently found to be a direct target of estrogen through ERα and to regulate both mammary epithelial proliferation and ductal elongation (Ciarloni et al. 2007).

Although multiparity reduces the long-term risk of breast cancer, there is an increased risk of cancer for a number of years following pregnancy (Lambe et al. 1994). A recent meta-analysis highlighted that parity-associated risk reductions are limited to ER$^+$PR$^+$ breast cancers (Ma et al. 2006). It is not clear how pregnancy exerts this dual effect, but delineation of the cellular hierarchy may provide some insight. It is likely that pregnancy involves the expansion of a transit-amplifying cell population and that any cells harboring mutations would therefore be expanded, increasing the risk of developing breast cancer for a short term following pregnancy. Alternatively, the microenvironment might be transiently tumor-promoting after remodeling of the mammary gland to its prepregnant state (Schedin 2006). Of relevance, although the mammary gland returns to a near prepregnant state following weaning, it is not identical to the original gland (Medina and Smith 1999; D'Cruz et al. 2002; Master and Chodosh 2004). This finding suggests that there may be permanent changes in a specific cell population. For example, the number of MaSCs may be diminished, thus reducing the pool of long-lived cells that have a greater probability of accumulating mutations over time (Britt et al. 2007). The protective effect of a full-term pregnancy on breast carcinogenesis has also been observed in mice, indicating their utility to study the effects of pregnancy on cell populations (Sivaraman et al. 2002).

During mouse mammary gland development, the size of different epithelial subpopulations fluctuates. We have shown that the number of cells in the MaSC-enriched population CD29^{hi}CD24$^+$ varies during ontogeny, reaching its maximum at midpregnancy, and then declining to the level in virgin glands during involution (M.-L. Asselin-Labat, unpubl.). In aged virgin mice, the proportion of CD29^{hi}CD24$^+$ cells is increased by twofold to threefold compared to young adult mice, but multiparity only slightly reduces the number of CD29^{hi}CD24$^+$ cells compared to aged-matched controls. However, in vivo transplantation studies are required to assess whether the observed changes in the size of the CD29^{hi}CD24$^+$ population during aging and pregnancy reflect changes in MaSC number or function of MaSCs. Interestingly, a parity-identified (previously termed parity-induced) mammary epithelial cell population (PI-MECs) (Smith and Medina 2008) that persists following involution has been described using Wap-cre/Rosa-LacZ and Wap-cre/CAG-GFP double transgenic reporter systems (Wagner et al. 2002; Matulka et al. 2007). Transplantation of mammary fragments or large numbers of cultured cells showed that the PI-MECs were capable of regenerating a mammary epithelial tree, but it is not yet clear whether these correspond to definitive stem cells.

REGULATORS OF LINEAGE DIFFERENTIATION IN THE MAMMARY GLAND

Gata-3 has emerged as a key regulator of differentiation along the luminal lineage. This transcription factor was found to be an essential regulator of mammary gland morphogenesis at several different stages: It is required for the formation of primary mammary buds in the embryo, for ductal growth in the adult, and for alveolar development during pregnancy (Kouros-Mehr et al. 2006; Asselin-Labat et al. 2007). Significantly, Gata-3 is expressed exclusively in luminal epithelial cells, and its absence results in failed differentiation and a marked expansion of luminal progenitor cells. Interestingly, Gata-3 can induce milk protein gene expression in the MaSC-enriched population in the absence of a lactogenic stimulus. Its critical role in differentiation is consistent with the gradient of Gata-3 expression that is apparent along the hierarchy: The MaSC-enriched population expresses low levels of Gata-3, with moderate expression in luminal progenitor cells and highest levels apparent in differentiated luminal cells (Asselin-Labat et al. 2007). Expression of Gata-3 in the MaSC-enriched population implies a role in stem cell function, but this is yet to be determined. Costaining for ER and Gata-3 revealed that a substantial proportion of ER$^+$ cells in the mature luminal cell subset (CD61$^-$) coexpress Gata-3 (Asselin-Labat et al. 2007), compatible with recent findings that ERα and Gata-3 directly coregulate each other's expression (Eeckhoute et al. 2007).

The absolute level of Gata-3 in the mouse mammary gland appears to be critical, and heterozygous mice exhibit a profound defect in ductal morphogenesis. Moreover, in breast cancer, the level of Gata-3 is also a determinant of prognosis. The highest expression of this gene is observed in the "luminal" subtypes (A and B) of breast cancer (Perou et al. 2000; Sørlie et al. 2001; Sotiriou et al. 2003; Kouros-Mehr et al. 2008b), which express ERα and predict good prognosis (Voduc et al. 2008). Gata-3 mutations have also been identified in a subset of breast cancers (Usary et al. 2004), and Gata-3 polymorphisms are associated with breast cancer risk (Garcia-Closas et al. 2007). In the mouse mammary tumor virus–polyomavirus middle T (MMTV–PyMT) mouse model of tumorigenesis, Gata-3 is down-regulated during tumor progression from mammary adenomas to carcinomas (Kouros-Mehr et al. 2008a), consistent with Gata-3 acting as a tumor suppressor. Kouros-Mehr et al. (2008a) further showed that transduction of MMTV–PyMT carcinoma cells with a Gata-3-expressing lentivirus reduced tumor cell dissemination to the lung following transplantation. The role of Gata-3, however, in the initiation of mammary tumorigenesis has not yet been addressed.

Luminal cell differentiation is regulated by the Ets transcription factor Elf-5. Unlike Gata-3, Elf-5 deficiency does not lead to defects in ductal growth and morphogenesis (Zhou et al. 2005; Harris et al. 2006). However, both Elf-5$^{+/-}$ and Elf-5$^{-/-}$ mammary glands show defects in alveolar differentiation during pregnancy, whereas glands from transgenic Elf-5 mice exhibit precocious alveolar differentiation in the virgin state (Zhou et al. 2005; Oakes

et al. 2008). Accordingly, Elf-5$^{-/-}$ glands harbor an increased number of luminal progenitor cells during pregnancy, and transgenic mice show a decrease in mature luminal cells, indicating that Elf-5 regulates the differentiation of progenitor to alveolar cells. We speculate that Elf-5 lies downstream from Gata-3 in the differentiation pathway, because Gata-3 is expressed earlier in the epithelial hierarchy and has a critical role in ductal morphogenesis in the peripubertal mammary gland.

PATHWAYS THAT REGULATE MAMMARY STEM/PROGENITOR CELL PROLIFERATION

A number of pathways have now been identified that have a key role in normal mammary development and oncogenesis. There are accumulating data pointing to a critical role for the Wnt/β-catenin pathway in the mammary gland. Wnt-1 is a frequent site of integration by the MMTV (Nusse and Varmus 1982; Nusse et al. 1984) and overexpression of Wnt-1 causes mammary hyperplasia and tumors (Tsukamoto et al. 1988; Bocchinfuso et al. 1999). Evaluation of preneoplastic mammary tissue in MMTV–Wnt-1 mice revealed an expanded MaSC-enriched CD29^{hi}CD24$^+$ population and a fourfold to sixfold increase in the absolute number of MaSCs compared to age-matched wild-type control glands (Shackleton et al. 2006). Other studies have suggested that the oncogenic effects of Wnt-1 in mammary epithelium are initiated in a mammary progenitor cell population, with an expansion of the SP and Sca1$^+$ cell populations observed in hyperplastic MMTV–Wnt-1 glands (Li et al. 2003; Liu et al. 2004). Wnt-1 further appears to render these progenitor cells radioresistant (Woodward et al. 2007). Intriguingly, we have recently shown that luminal progenitor cells in preneoplastic Wnt-1 glands are capable of repopulating the fat pad in vivo, suggesting that Wnt-1 induces oncogenic activity in both stem and luminal progenitor cells. Alternatively, it is also possible that Wnt-1 predominantly affects stem cells but that their cell surface phenotype alters slightly upon commitment to a neoplastic state (Vaillant et al. 2008).

Hedgehog (Hh) signaling has been implicated in the self-renewal of multiple types of stem cells (Molofsky et al. 2004). Activation of human breast epithelial cells with the Hedgehog ligands Shh (sonic hedgehog) and Ihh (Indian hedgehog) enhanced mammosphere-initiating capacity (Liu et al. 2006). However, in vivo transplantation assays using mammary cells isolated from transgenic mice expressing the smoothened receptor (smo) showed a decrease in MaSC activity (Moraes et al. 2007). Because sphere formation appears to be more indicative of progenitor rather than stem cell activity, these data are compatible with Hedgehog signaling causing an increase in progenitor cell activity. In addition, recent findings have shown that loss of one copy of Patched-1, a negative regulator of Hedgehog, results in enhanced progenitor cell formation by stem cells in the mammary gland (Li et al. 2008). Hh can also mediate its effects via the polycomb gene *Bmi-1*. Indeed, overexpression of *Bmi-1* in human breast epithelial cells increased both the number and size of primary and secondary mammospheres, whereas knock down of *Bmi-1* levels reduced sphere formation (Liu et al. 2006), suggesting that *Bmi-1* promotes the proliferation of progenitor cells. The Notch pathway also has a central role in breast oncogenesis, and both Notch 1 and Notch 4 were identified as frequent sites of proviral insertion and activation in mouse mammary tumors (Kiaris et al. 2004; Raafat et al. 2004; Hu et al., 2006). We have recently shown that the Notch pathway promotes luminal cell commitment in the mammary gland and that deregulated Notch signaling leads to the uncontrolled expansion of luminal progenitor cells (Bouras et al. 2008). Consistent with increased progenitor activity, enhanced mammosphere formation was observed upon addition of Notch ligand to human mammary epithelial cell culture (Dontu et al. 2004).

Negative regulation of proliferation is also critical for normal development. Transforming growth factor-β1 (TGF-β1) regulates both mammary ductal morphogenesis and alveolar development (Daniel et al. 2001). Transgenic mice overexpressing an activated form of TGF-β1 displayed reduced lateral branching in their mammary glands, consistent with the proposal that TGF-β1 inhibits MaSC and/or progenitor cells located along the ducts (Pierce et al. 1993). In support of this, Smith and colleagues have reported that mammary expression of TGF-β1 results in senescence of MaSCs (Boulanger and Smith 2001; Boulanger et al. 2005; Kordon et al. 1995). TGF-β1 has also been implicated in tumorigenesis and metastasis in different organs including the mammary gland (Moses and Serra 1996; Siegel and Massagué 2003). Interestingly, the TGF-β1 pathway appears to be activated in a putative human breast cancer stem cell population (CD44+CD24−) isolated from pleural effusions of breast cancer patients (Shipitsin et al. 2007).

CELL OF ORIGIN OF BREAST CANCER

The MaSC-enriched population (CD29hiCD24+) expresses the basal markers keratin 14, p63, and epidermal growth factor receptor (EGFR) and lacks expression of ErbB2 (Asselin-Labat et al. 2006). These characteristics are reminiscent of the phenotype of the basal subtype of breast cancer and suggest that the stem cell may be the "cell of origin" for this cancer subtype. There is evidence indicating that ovariectomy prevents the formation of both ERα-positive and ERα-negative breast cancers, suggesting that ER-negative breast cancers are dependent on ovarian hormones for their formation (Nissen-Meyer 1964a,b; Gupta et al. 2007). Interestingly, despite no change in the size of the CD29hiCD24+ mammary population in ovariectomized mice, the repopulating capacity of these cells is reduced (M.-L. Asselin-Labat, unpubl.). This finding suggests that the MaSC-enriched population is sensitive to estrogen withdrawal and raises the question of whether cancer cells in the basal (ER-negative) subtype of breast cancer have lost their dependence on steroid hormones.

Breast carcinomas from *BRCA1* mutation carriers are ERα-negative and have a basal phenotype, suggesting that they may also originate in the basal cell compartment (Foulkes et al. 2003; Ribeiro-Silva et al. 2005). Interestingly, *BRCA1* seems to directly regulate ERα expression (Hosey et al. 2007). Knockdown of *BRCA1* expression in

primary human mammary epithelial cells results in the accumulation of undifferentiated cells with a basal phenotype, and in vivo transplantation of epithelial cells in which *BRCA1* levels have been reduced leads to abnormal outgrowths lacking luminal cells (Liu et al. 2008). In combination with studies showing that loss of heterozygosity (LOH) at the *BRCA1* locus is associated with high ALDH1 activity, these findings suggest a critical role for *BRCA1* in promoting the differentiation of mammary stem/progenitor cells. Furthermore, conditional deletion of *BRCA1* in mammary glands delayed development during puberty and impaired alveolar cell differentiation during pregnancy and lactation, eventually leading to tumor development (Xu et al. 1999). Similarly, deletion of both *BRCA1* and *p53* (frequently mutated in BRCA1 tumors) led to highly proliferative mammary tumors that were ERα-negative and expressed basal epithelial markers (Liu et al. 2007).

CANCER STEM CELLS

The "cancer stem cell" hypothesis postulates that a small subset of tumor cells with stem-cell-like properties drive tumor progression and recurrence. These cells lie at the apex of the hierarchy in tumors and give rise to more differentiated progeny with limited proliferative potential. Cancer stem cells (CSCs) do not necessarily arise from normal stem cells but may originate in progenitor cells or more differentiated cells that have acquired self-renewing properties. In human breast, Clarke and colleagues (Al-Hajj et al. 2003) first identified a population of cells with a CD44+CD24−/lo phenotype that had significantly higher tumor-initiating capacity than CD44−CD24+ cells when transplanted into immunocompromized mice. CD44+CD24−/lo tumorigenic cells isolated from a metastatic breast cancer xenograft model showed increased expression of Gli-1, Gli-2, and Bmi-1 mRNA compared to nontumorigenic cells (Liu et al. 2006), consistent with the cancer stem cell population having increased self-renewal capability.

In mouse models of mammary tumorigenesis, recent data have provided direct support for the cancer stem cell hypothesis. In MMTV–Wnt-1 mammary tumors, Cho et al. (2008) found that CD24+Thy-1+ cells have a higher tumor-generating capacity than "not-CD24+Thy-1+" cells when transplanted subcutaneously into syngeneic recipients. Recently, we have shown that CD29loCD24+CD61+ cells are enriched 20-fold in tumor-initiating cells compared to the CD29loCD24+CD61− population and can recapitulate a tumor with the same phenotypic characteristics as the parental tumor following serial transplantation (Vaillant et al. 2008). Similarly, CD61− cells isolated from some tumors that developed in p53+/− mice were also highly enriched for cancer stem cells. Conversely, in mammary tumors arising in doxycyline-inducible Myc bitransgenic mice, CD61 did not enrich for cancer stem cells, suggesting that this progenitor marker may not be relevant to all mouse tumor types or that the development of Myc mammary tumors follows a model distinct from that of the cancer stem cell paradigm (Vaillant et al. 2008). In MMTV-Neu mammary tumors, we were not able to identify a cancer stem cell subpopulation using a wide range of cell surface markers. These tumors are very homogeneous and exhibit luminal characteristics, consistent with their CD61[+] phenotype. This mouse strain seems more likely to follow a stochastic model of tumor formation. Thus, different mouse models of mammary tumors may follow distinct models of tumorigenesis.

CONCLUSIONS

The in vivo mammary reconstitution assay, together with the ability to sort different epithelial cell populations, has paved the way for understanding the molecular and cellular mechanisms that regulate both normal development and breast oncogenesis. With the discovery of additional markers to allow further purification of stem cells, it should be possible to answer many questions pertaining to mammary stem cells and their niche. The decision by an MaSC to self-renew rather than initiate differentiation is likely to be tightly regulated in order to maintain normal homeostasis and ensure continued production of differentiated cells. It is presumed that there is a small pool of quiescent stem cells within the mammary gland, but their location and the extracellular signals that activate them are presently unknown. The role of estrogen in regulating stem cell function is also unclear but almost certainly involves multiple paracrine factors that are yet to be identified. The earliest progenitor cell known to date to express ER and PR is the luminal CD61[+] cell, but it is possible that there is another upstream progenitor that expresses physiologically relevant levels of these receptors. Further elucidation of epithelial cells in the mammary gland and the pathways that regulate their activity is pivotal to understanding the "cells of origin" and cancer stem cells that underlie different breast tumors and for the design of more effective therapies.

ACKNOWLEDGMENTS

This work was supported by the Victorian Breast Cancer Research Consortium, the National Health and Medical Research Council (Australia), and the Australian Stem Cell Centre. M.A. was supported by an Australian Research Council Discovery Project postdoctoral fellowship.

REFERENCES

Al-Hajj, M., Wicha, M.S., Benito-Hernandez, A., Morrison, S.J., and Clarke, M.F. 2003. Prospective identification of tumorigenic breast cancer cells. *Proc. Natl. Acad. Sci.* **100:** 3983–3988.

Armstrong, L., Stojkovic, M., Dimmick, I., Ahmad, S., Stojkovic, P., Hole, N., and Lako, M. 2004. Phenotypic characterization of murine primitive hematopoietic progenitor cells isolated on basis of aldehyde dehydrogenase activity. *Stem Cells* **22:** 1142–1151.

Asselin-Labat, M.L., Shackleton, M., Stingl, J., Vaillant, F., Forrest, N.C., Eaves, C.J., Visvader, J.E., and Lindeman, G.J. 2006. Steroid hormone receptor status of mouse mammary stem cells. *J. Natl. Cancer Inst.* **98:** 1011–1014.

Asselin-Labat, M.L., Sutherland, K.D., Barker, H., Thomas, R., Shackleton, M., Forrest, N.C., Hartley, L., Robb, L., Grosveld, F.G., van der Wees, J., Lindeman, G.J., and Visvader, J.E. 2007. Gata-3 is an essential regulator of mammary-gland morphogenesis and luminal-cell differentiation. *Nat. Cell Biol.* **9:** 201–209.

Bocchinfuso, W.P. and Korach, K.S. 1997. Mammary gland development and tumorigenesis in estrogen receptor knockout mice. *J. Mammary Gland Biol. Neoplasia* **2:** 323–334.

Bocchinfuso, W.P., Hively, W.P., Couse, J.F., Varmus, H.E., and Korach, K.S. 1999. A mouse mammary tumor virus-*Wnt-1* transgene induces mammary gland hyperplasia and tumorigenesis in mice lacking estrogen receptor-α. *Cancer Res.* **59:** 1869–1876.

Bonnefoix, T., Bonnefoix, P., Verdiel, P., and Sotto, J.J. 1996. Fitting limiting dilution experiments with generalized linear models results in a test of the single-hit Poisson assumption. *J. Immunol. Methods* **194:** 113–119.

Booth, B.W. and Smith, G.H. 2006. Estrogen receptor-α and progesterone receptor are expressed in label-retaining mammary epithelial cells that divide asymmetrically and retain their template DNA strands. *Breast Cancer Res.* **8:** R49.

Boulanger, C.A. and Smith, G.H. 2001. Reducing mammary cancer risk through premature stem cell senescence. *Oncogene* **20:** 2264–2272.

Boulanger, C.A., Wagner, K.U., and Smith, G.H. 2005. Parity-induced mouse mammary epithelial cells are pluripotent, self-renewing and sensitive to TGF-β1 expression. *Oncogene* **24:** 552–560.

Bouras, T., Pal, B., Vaillant, F., Harburg, G., Asselin-Labat, M.-L., Oakes, S.R., Lindeman, G.J., and Visvader, J.E. 2008. Notch signaling regulates mammary stem cell function and luminal cell-fate commitment. *Cell Stem Cell* **9:** 429–441.

Britt, K., Ashworth, A., and Smalley, M. 2007. Pregnancy and the risk of breast cancer. *Endocr. Relat. Cancer* **14:** 907–933.

Cho, R.W., Wang, X., Diehn, M., Shedden, K., Chen, G.Y., Sherlock, G., Gurney, A., Lewicki, J., and Clarke, M.F. 2008. Isolation and molecular characterization of cancer stem cells in MMTV-*Wnt-1* murine breast tumors. *Stem Cells* **26:** 364–371.

Ciarloni, L., Mallepell, S., and Brisken, C. 2007. Amphiregulin is an essential mediator of estrogen receptor α function in mammary gland development. *Proc. Natl. Acad. Sci.* **104:** 5455–5460.

Clarke, R.B., Spence, K., Anderson, E., Howell, A., Okano, H., and Potten, C.S. 2005. A putative human breast stem cell population is enriched for steroid receptor-positive cells. *Dev. Biol.* **277:** 443–456.

D'Cruz, C.M., Moody, S.E., Master, S.R., Hartman, J.L., Keiper, E.A., Imielinski, M.B., Cox, J.D., Wang, J.Y., Ha, S.I., Keister, B.A., and Chodosh, L.A. 2002. Persistent parity-induced changes in growth factors, TGF-β3, and differentiation in the rodent mammary gland. *Mol. Endocrinol.* **16:** 2034–2051.

Daniel, C.W., De Ome, K.B., Young, J.T., Blair, P.B., and Faulkin Jr., L.J. 1968. The in vivo life span of normal and preneoplastic mouse mammary glands: A serial transplantation study. *Proc. Natl. Acad. Sci.* **61:** 53–60.

Daniel, C.W., Robinson, S., and Silberstein, G.B. 2001. The transforming growth factors β in development and functional differentiation of the mouse mammary gland. *Adv. Exp. Med. Biol.* **501:** 61–70.

De Ome, K.B., Faulkin, Jr., L.J., Bern, H.A., and Blair, P.B. 1959. Development of mammary tumors from hyperplastic alveolar nodules transplanted into gland-free mammary fat pads of female C3H mice. *Cancer Res.* **19:** 515–520.

Dontu, G., Abdallah, W.M., Foley, J.M., Jackson, K.W., Clarke, M.F., Kawamura, M.J., and Wicha, M.S. 2003. in vitro propagation and transcriptional profiling of human mammary stem/progenitor cells. *Genes Dev.* **17:** 1253–1270.

Dontu, G., Jackson, K.W., McNicholas, E., Kawamura, M.J., Abdallah, W.M., and Wicha, M.S. 2004. Role of Notch signaling in cell-fate determination of human mammary stem/progenitor cells. *Breast Cancer Res.* **6:** R605–R615.

Eeckhoute, J., Keeton, E.K., Lupien, M., Krum, S.A., Carroll, J.S., and Brown, M. 2007. Positive cross-regulatory loop ties GATA-3 to estrogen receptor α expression in breast cancer. *Cancer Res.* **67:** 6477–6483.

Foulkes, W.D., Stefansson, I.M., Chappuis, P.O., Bégin, L.R., Goffin, J.R., Wong, N., Trudel, M., and Akslen, L.A. 2003. Germline BRCA1 mutations and a basal epithelial phenotype in breast cancer. *J. Natl. Cancer Inst.* **95:** 1482–1485.

Garcia-Closas, M., Troester, M.A., Qi, Y., Langerod, A., Yeager, M., Lissowska, J., Brinton, L., Welch, R., Peplonska, B., Gerhard, D.S., et al. 2007. Common genetic variation in GATA-binding protein 3 and differential susceptibility to breast cancer by estrogen receptor α tumor status. *Cancer Epidemiol. Biomark. Prev.* **16:** 2269–2275.

Ginestier, C., Hur, M.H., Charafe-Jauffret, E., Monville, F., Dutcher, J., Brown, M., Jacquemier, J., Viens, P., Kleer, C.G., Liu, S., et al. 2007. ALDH1 is a marker of normal and malignant human mammary stem sells and a predictor of poor clinical outcome. *Cell Stem Cell* **1:** 555–567.

Gudjonsson, T., Villadsen, R., Nielsen, H.L., Ronnov-Jessen, L., Bissell, M.J., and Petersen, O.W. 2002. Isolation, immortalization, and characterization of a human breast epithelial cell line with stem cell properties. *Genes Dev.* **16:** 693–706.

Gupta, P.B., Proia, D., Cingoz, O., Weremowicz, J., Naber, S.P., Weinberg, R.A., and Kuperwasser, C. 2007. Systemic stromal effects of estrogen promote the growth of estrogen receptor-negative cancers. *Cancer Res.* **67:** 2062–2071.

Harris, J., Stanford, P.M., Sutherland, K., Oakes, S.R., Naylor, M.J., Robertson, F.G., Blazek, K.D., Kazlauskas, M., Hilton, H.N., Wittlin, S., et al. 2006. Socs2 and Elf5 mediate prolactin-induced mammary gland development. *Mol. Endocrinol.* **20:** 1177–1187.

Hennighausen, L. and Robinson, G.W. 2005. Information networks in the mammary gland. *Nat. Rev. Mol. Cell Biol.* **6:** 715–725.

Herschkowitz, J.I., Simin, K., Weigman, V.J., Mikaelian, I., Usary, J., Hu, Z., Rasmussen, K.E., Jones, L.P., Assefnia, S., Chandrasekharan, S., et al. 2007. Identification of conserved gene expression features between murine mammary carcinoma models and human breast tumors. *Genome Biol.* **8:** R76.

Hess, D.A., Meyerrose, T.E., Wirthlin, L., Craft, T.P., Herrbrich, P.E., Creer, M.H., and Nolta, J.A. 2004. Functional characterization of highly purified human hematopoietic repopulating cells isolated according to aldehyde dehydrogenase activity. *Blood* **104:** 1648–1655.

Hosey, A.M., Gorski, J.J., Murray, M.M., Quinn, J.E., Chung, W.Y., Stewart, G.E., James, C.R., Farragher, S.M., Mulligan, J.M., Scott, A.N., et al. 2007. Molecular basis for estrogen receptor α deficiency in BRCA1-linked breast cancer. *J. Natl. Cancer Inst.* **99:** 1683–1694.

Hoshino, K. and Gardner, W.U. 1967. Transplantability and life span of mammary gland during serial transplantation in mice. *Nature* **213:** 193–194.

Hu, C., Diévart, A., Lupien, M., Calvo, E., Tremblay, G., and Jolicoeur, P. 2006. Overexpression of activated murine Notch1 and Notch3 in transgenic mice blocks mammary gland development and induces mammary tumors. *Am. J. Pathol.* **168:** 973–990.

Kiaris, H., Politi, K., Grimm, L.M., Szabolcs, M., Fisher, P., Efstratiadis, A., and Artavanis-Tsakonas, S. 2004. Modulation of Notch signaling elicits signature tumors and inhibits Hras1-induced oncogenesis in the mouse mammary epithelium. *Am. J. Pathol.* **165:** 695–705.

Kordon, E.C. and Smith, G.H. 1998. An entire functional mammary gland may comprise the progeny from a single cell. *Development* **125:** 1921–1930.

Kordon, E.C., McKnight, R.A., Jhappan, C., Hennighausen, L., Merlino, G., and Smith, G.H. 1995. Ectopic TGFβ1 expression in the secretory mammary epithelium induces early senescence of the epithelial stem cell population. *Dev. Biol.* **168:** 47–61.

Kouros-Mehr, H., Slorach, E.M., Sternlicht, M.D., and Werb, Z. 2006. GATA-3 maintains the differentiation of the luminal cell fate in the mammary gland. *Cell* **127:** 1041–1055.

Kouros-Mehr, H., Bechis, S.K., Slorach, E.M., Littlepage, L.E., Egeblad, M., Ewald, A.J., Pai, S.Y., Ho, I.C., and Werb, Z. 2008a. GATA-3 links tumor differentiation and dissemination in a luminal breast cancer model. *Cancer Cell* **13:** 141–152.

Kouros-Mehr, H., Kim, J.W., Bechis, S.K., and Werb, Z. 2008b. GATA-3 and the regulation of the mammary luminal cell fate. *Curr. Opin. Cell Biol.* **20:** 164–170.

Kuperwasser, C., Chavarria, T., Wu, M., Magrane, G., Gray, J.W., Carey, L., Richardson, A., and Weinberg, R.A. 2004.

Reconstruction of functionally normal and malignant human breast tissues in mice. *Proc. Natl. Acad. Sci.* **101:** 4966–4971.

Lambe, M., Hsieh, C., Trichopoulos, D., Ekbom, A., Pavia, M., and Adami, H.O. 1994. Transient increase in the risk of breast cancer after giving birth. *N. Engl. J. Med.* **331:** 5–9.

Li, Y., Welm, B., Podsypanina, K., Huang S., Chamorro, M., Zhang, X., Rowlands, T., Egeblad, M., Cowin, P., Werb, Z., et al. 2003. Evidence that transgenes encoding components of the Wnt signaling pathway preferentially induce mammary cancers from progenitor cells. *Proc. Natl. Acad. Sci.* **100:** 15853–15858.

Li, N., Singh, S., Cherukuri, P., Li, H., Yuan, Z., Ellisen, L.W., Wang, B., Robbins, D., and DiRenzo, J. 2008. Reciprocal intraepithelial interactions between TP63 and hedgehog signaling regulate quiescence and activation of progenitor elaboration by mammary stem cells. *Stem Cells* **26:** 1253–1264.

Liu, B.Y., McDermott, S.P., Khwaja, S.S., and Alexander, C.M. 2004. The transforming activity of Wnt effectors correlates with their ability to induce the accumulation of mammary progenitor cells. *Proc. Natl. Acad. Sci.* **101:** 4158–4163.

Liu, S., Dontu, G., Mantle, I.D., Patel, S., Ahn, N.S., Jackson, K.W., Suri, P., and Wicha, M.S. 2006. Hedgehog signaling and Bmi-1 regulate self-renewal of normal and malignant human mammary stem cells. *Cancer Res* **66:** 6063–6071.

Liu, X., Holstege, H., van der Gulden, H., Treur-Mulder, M., Zevenhoven, J., Velds, A., Kerkhoven, R.M., van Vliet, M.H., Wessels, L.F.A., Peterse, J.L., Berns, A., and Jonkers, J. 2007. Somatic loss of BRCA1 and p53 in mice induces mammary tumors with features of human *BRCA1*-mutated basal-like breast cancer. *Proc. Natl. Acad. Sci.* **104:** 12111–12116.

Liu, S., Ginestier, C., Charafe-Jauffret, E., Foco, H., Kleer, C.G., Merajver, S.D., Dontu, G., and Wicha, M.S. 2008. BRCA1 regulates human mammary stem/progenitor cell fate. *Proc. Natl. Acad. Sci.* **105:** 1680–1685.

Ma, H., Bernstein, L., Pike, M.C., and Ursin, G. 2006. Reproductive factors and breast cancer risk according to joint estrogen and progesterone receptor status: A meta-analysis of epidemiological studies. *Breast Cancer Res.* **8:** R43.

Mallepell, S., Krust, A., Chambon, P., and Brisken, C. 2006. Paracrine signaling through the epithelial estrogen receptor α is required for proliferation and morphogenesis in the mammary gland. *Proc. Natl. Acad. Sci.* **103:** 2196–2201.

Master, S.R. and Chodosh, L.A. 2004. Evolving views of involution. *Breast Cancer Res.* **6:** 89–92.

Matsui, W., Huff, C.A., Wang, Q., Malehorn, M.T., Barber, J., Tanhehco, Y., Smith, B.D., Civin, C.I., and Jones, R.J. 2004. Characterization of clonogenic multiple myeloma cells. *Blood* **103:** 2332–2336.

Matulka, L.A., Triplett, A.A., and Wagner, K.U. 2007. Parity-induced mammary epithelial cells are multipotent and express cell surface markers associated with stem cells. *Dev. Biol.* **303:** 29–44.

Medina, D. and Smith, G.H. 1999. Chemical carcinogen-induced tumorigenesis in parous, involuted mouse mammary glands. *J. Natl. Cancer Inst.* **91:** 967–969.

Molofsky, A.V., Pardal, R., and Morrison, S.J. 2004. Diverse mechanisms regulate stem cell self-renewal. *Curr. Opin. Cell Biol.* **16:** 700–707.

Moraes, R.C., Zhang, X., Harrington, N., Fung, J.Y., Wu, M.-F., Hilsenbeck, S.G., Allred, D.C., and Lewis, M.T. 2007. Constitutive activation of smoothened (SMO) in mammary glands of transgenic mice leads to increased proliferation, altered differentiation and ductal dysplasia. *Development* **134:** 1231–1242.

Moses, H.L. and Serra, R. 1996. Regulation of differentiation by TGF-β. *Curr. Opin. Genet. Dev.* **6:** 581–586.

Mueller, S.O., Clark, J.A., Myers, P.H., and Korach, K.S. 2002. Mammary gland development in adult mice requires epithelial and stromal estrogen receptor α. *Endocrinology* **143:** 2357–2365.

Nissen-Meyer, R. 1964a. Prophylactic endocrine treatment in carcinoma of the breast. *Clin. Radiol.* **15:** 152–160.

Nissen-Meyer, R. 1964b. "Prophylactic" ovariectomy and ovarian irradiation in breast cancer. *Acta Unio Int. Contra Cancrum* **20:** 527–530.

Nusse, R. and Varmus, H.E. 1982. Many tumors induced by the mouse mammary tumor virus contain a provirus integrated in the same region of the host genome. *Cell* **31:** 99–109.

Nusse, R., van Ooyen, A., Cox, D., Fung, Y.K., and Varmus, H. 1984. Mode of proviral activation of a putative mammary oncogene (*int*-1) on mouse chromosome 15. *Nature* **307:** 131–136.

Oakes, S.R., Naylor, M.J., Asselin-Labat, M.L., Blazek, K.D., Gardiner-Garden, M., Hilton, H.N., Kazlauskas, M., Pritchard, M.A., Chodosh, L.A., Pfeffer, P.L., et al. 2008. The Ets transcription factor Elf5 specifies mammary alveolar cell fate. *Genes Dev.* **22:** 581–586.

Perou, C.M., Sørlie, T., Eisen, M.B., van de Rijn, M., Jeffrey, S.S., Rees, C.A., Pollack, J.R., Ross, D.T., Johnsen, H., Akslen, L.A., et al. 2000. Molecular portraits of human breast tumors. *Nature* **406:** 747–752.

Pierce, Jr., D.F., Johnson, M.D., Matsui, Y., Robinson, S.D., Gold, L.I., Purchio, A.F., Daniel, C.W., Hogan, B.L., and Moses, H.L. 1993. Inhibition of mammary duct development but not alveolar outgrowth during pregnancy in transgenic mice expressing active TGF-β1. *Genes Dev.* **7:** 2308–2317.

Raafat, A., Bargo, S., Anver, M.R., and Callahan, R. 2004. Mammary development and tumorigenesis in mice expressing a truncated human Notch4/Int3 intracellular domain (h-Int3sh). *Oncogene* **23:** 9401–9407.

Ribeiro-Silva, A., Ramalho, L.N., Garcia, S.B., Brandao, D.F., Chahud, F., and Zucoloto, S. 2005. p63 correlates with both BRCA1 and cytokeratin 5 in invasive breast carcinomas: Further evidence for the pathogenesis of the basal phenotype of breast cancer. *Histopathology* **47:** 458–466.

Schedin, P. 2006. Pregnancy-associated breast cancer and metastasis. *Nat. Rev. Cancer* **6:** 281–291.

Scully, K.M., Gleiberman, A.S., Lindzey, J., Lubahn, D.B., Korach, K.S., and Rosenfeld, M.G. 1997. Role of estrogen receptor-α in the anterior pituitary gland. *Mol. Endocrinol.* **11:** 674–681.

Shackleton, M., Vaillant, F., Simpson, K.J., Stingl, J., Smyth, G.K., Asselin-Labat, M.L., Wu, L., Lindeman, G.J., and Visvader, J.E. 2006. Generation of a functional mammary gland from a single stem cell. *Nature* **439:** 84–88.

Shipitsin, M., Campbell, L.L., Argani, P., Weremowicz, S., Bloushtain-Qimron, N., Yao, J., Nikolskaya, T., Serebryiskaya, T., Beroukhim, R., Hu, M., et al. 2007. Molecular definition of breast tumor heterogeneity. *Cancer Cell* **11:** 259–273.

Siegel, P.M. and Massagué, J. 2003. Cytostatic and apoptotic actions of TGF-β in homeostasis and cancer. *Nat. Rev. Cancer* **3:** 807–821.

Sivaraman, L., Gay, J., Hilsenbeck, S.G., Shine, H.D., Conneely, O.M., Medina, D., and O'Malley, B.W. 2002. Effect of selective ablation of proliferating mammary epithelial cells on MNU induced rat mammary tumorigenesis. *Breast Cancer Res. Treat.* **73:** 75–83.

Sleeman, K.E., Kendrick, H., Ashworth, A., Isacke, C.M., and Smalley, M.J. 2006. CD24 staining of mouse mammary gland cells defines luminal epithelial, myoepithelial/basal and non-epithelial cells. *Breast Cancer Res.* **8:** R7.

Sleeman, K.E., Kendrick, H., Robertson, D., Isacke, C.M., Ashworth, A., and Smalley, M.J. 2007. Dissociation of estrogen receptor expression and in vivo stem cell activity in the mammary gland. *J. Cell Biol.* **176:** 19–26.

Smith, G.H. 1996. Experimental mammary epithelial morphogenesis in an in vivo model: Evidence for distinct cellular progenitors of the ductal and lobular phenotype. *Breast Cancer Res. Treat.* **39:** 21–31.

Smith, G.H. and Medina, D. 2008. Re-evaluation of mammary stem cell biology based on in vivo transplantation. *Breast Cancer Res.* **10:** 203.

Sotiriou, C., Neo, S.Y., McShane, L.M., Korn, E.L., Long, P.M., Jazaeri, A., Martiat, P., Fox, S.B., Harris, A.L., and Liu, E.T. 2003. Breast cancer classification and prognosis based on gene expression profiles from a population-based study. *Proc. Natl. Acad. Sci.* **100:** 10393–10398.

Sørlie, T., Perou, C.M., Tibshirani, R., Aas, T., Geisler, S., Johnsen, H., Hastie, T., Eisen, M.B., van de Rijn, M., Jeffrey, S.S., et al. 2001. Gene expression patterns of breast carcinomas distinguish tumor subclasses with clinical implications. *Proc. Natl. Acad. Sci.* **98:** 10869–10874.

Stingl, J. and Caldas, C. 2007. Molecular heterogeneity of breast carcinomas and the cancer stem cell hypothesis. *Nat. Rev. Cancer* **7:** 791–799.

Stingl, J., Eaves, C.J., Kuusk, U., and Emerman, J.T. 1998. Phenotypic and functional characterization in vitro of a multipotent epithelial cell present in the normal adult human breast. *Differentiation* **63:** 201–213.

Stingl, J., Eaves, C.J., Zandieh, I., and Emerman, J.T. 2001. Characterization of bipotent mammary epithelial progenitor cells in normal adult human breast tissue. *Breast Cancer Res. Treat.* **67:** 93–109.

Stingl, J., Eirew, P., Ricketson, I., Shackleton, M., Vaillant, F., Choi, D., Li, H.I., and Eaves, C.J. 2006. Purification and unique properties of mammary epithelial stem cells. *Nature* **439:** 993–997.

Tsai, Y.C., Lu, Y., Nichols, P.W., Zlotnikov, G., Jones, P.A., and Smith, H.S. 1996. Contiguous patches of normal human mammary epithelium derived from a single stem cell: Implications for breast carcinogenesis. *Cancer Res.* **56:** 402–404.

Tsukamoto, A.S., Grosschedl, R., Guzman, R.C., Parslow, T., and Varmus, H.E. 1988. Expression of the *int*-1 gene in transgenic mice is associated with mammary gland hyperplasia and adenocarcinomas in male and female mice. *Cell* **55:** 619–625.

Usary, J., Llaca, V., Karaca, G., Presswala, S., Karaca, M., He, X., Langerod, A., Karesen, R., Oh, D.S., Dressler, L.G., et al. 2004. Mutation of GATA3 in human breast tumors. *Oncogene* **23:** 7669–7678.

Vaillant, F., Asselin-Labat, M.-L., Shackleton, M., Forrest, N., Lindeman, G.J., and Visvader, J.E. 2008. The mammary progenitor marker CD61/b3 integrin identifies cancer stem cells in mouse models of mammary tumorigenesis. *Cancer Res.* **68:** 7711–7717.

Villadsen, R., Fridriksdottir, A.J., Ronnov-Jessen, L., Gudjonsson, T., Rank, F., LaBarge, M.A., Bissell, M.J., and Petersen, O.W. 2007. Evidence for a stem cell hierarchy in the adult human breast. *J. Cell Biol.* **177:** 87–101.

Voduc, D., Cheang, M., and Nielsen, T. 2008. GATA-3 expression in breast cancer has a strong association with estrogen receptor but lacks independent prognostic value. *Cancer Epidemiol. Biomark. Prev.* **17:** 365–373.

Wagner, K.U., Boulanger, C.A., Henry, M.D., Sgagias, M., Hennighausen, L., and Smith, G.H. 2002. An adjunct mammary epithelial cell population in parous females: Its role in functional adaptation and tissue renewal. *Development* **129:** 1377–1386.

Welm, B.E., Tepera, S.B., Venezia, T., Graubert, T.A., Rosen, J.M., and Goodell, M.A. 2002. Sca-1^pos cells in the mouse mammary gland represent an enriched progenitor cell population. *Dev. Biol.* **245:** 42–56.

Woodward, W.A., Chen, M.S., Behbod, F., Alfaro, M.P., Buchholz, T.A., and Rosen, J.M. 2007. WNT/β-catenin mediates radiation resistance of mouse mammary progenitor cells. *Proc. Natl. Acad. Sci.* **104:** 618–623.

Xu, X., Wagner, K.U., Larson, D., Weaver, Z., Li, C., Ried, T., Hennighausen, L., Wynshaw-Boris, A., and Deng, C.X. 1999. Conditional mutation of *Brca1* in mammary epithelial cells results in blunted ductal morphogenesis and tumour formation. *Nat. Genet.* **22:** 37–43.

Zhou, J., Chehab, R., Tkalcevic, J., Naylor, M.J., Harris, J., Wilson, T.J., Tsao, S., Tellis, I., Zavarsek, S., Xu, D., et al. 2005. Elf5 is essential for early embryogenesis and mammary gland development during pregnancy and lactation. *EMBO J.* **24:** 635–644.

Stem Cell Biology in the Lung and Lung Cancers: Using Pulmonary Context and Classic Approaches

D.M. Raiser, S.J. Zacharek, R.R. Roach, S.J. Curtis, K.W. Sinkevicius, D.W. Gludish, and C.F. Kim

Stem Cell Program, Children's Hospital Boston, Department of Genetics, Harvard Medical School, and Harvard Stem Cell Institute, Boston, Massachusetts 02115

Classic stem cell biology approaches tailored specifically with lung biology in mind are needed to bring the field of lung stem cell biology up to speed with that in other tissues. The infrequent cellular turnover, the diversity of cell types, and the necessity of daily cell function in this organ must be considered in stem cell studies. Previous work has created a base from which to explore transplantation, label retention, and more sophisticated lineage-tracing schemes to identify and characterize stem cell populations in the normal lung. These approaches are also imperative for building on precedents set in other tissues in the exploration of the cancer stem cell hypothesis in lung cancers. Additionally, recent studies provide key leads to further explore the molecular mechanisms that regulate lung homeostasis. Here, we discuss strategies to advance the field of lung stem cell biology with an emphasis on developing new, lung-specific tools.

The application of stem cell biology approaches to the lung should reflect the impressively diverse composition of the respiratory epithelium. There are many epithelial cell populations in the lung, each with distinct functions and anatomical locations, that all cooperate to accomplish the vital task of gas exchange (Mason et al. 1997; Rawlins and Hogan 2006). Of particular emphasis for this chapter, the distal murine lung, found beyond the trachea and upper airways, consists of several types of epithelia (Fig. 1). The branching bronchioles and the terminal bronchioles to which they give way are lined with columnar Clara cells that secrete cytokine- and immunoregulatory molecule-rich surfactants and provide a protective barrier in the airways, and ciliated cells that propel secreted surfactant along the bronchiolar wall and function in mucociliary clearance (Massaro et al. 1994; Van der Schans 2007). In the alveolar space, cuboidal alveolar type II (AT2) cells produce the surfactant necessary for the flat alveolar type I (AT1) cells to perform their gas-exchange function and deliver oxygen to the blood (Dobbs et al. 1998; Fehrenbach 2001). Clara cells, ciliated cells, AT2 cells, and AT1 cells can be identified by their expression of Clara cell secretory protein (CCSP; also known as Scgb1a1, CCA, CC10, and uteroglobin), acetylated tubulin (Acet Tub), prosurfactant protein C (SPC), and aquaporin 5 (AQ5), respectively (Fig. 2). Notably, murine and human lungs have anatomical differences, such as the relative abundance of each of these cell types along the proximal-distal pulmonary axis (Liu et al. 2006; Rawlins and Hogan 2006).

Several key aspects of lung biology may set the lung apart from other tissues when considering the study of stem cells. First, unlike tissues such as the gut and skin, where the resident epithelium is renewed frequently, murine lung epithelium has little or no turnover in the absence of exogenous stimuli. For comparison, the lining of the small intestine turns over completely every 3–5 days; estimates of lung cell turnover time vary widely

(e.g., 6–17 months) and have not been carefully examined in the absence of injury (Rawlins and Hogan 2006). Second, the lung is more similar to the gut and skin than to other epithelial tissues, such as the mammary gland or prostate, in that the pulmonary system is constantly

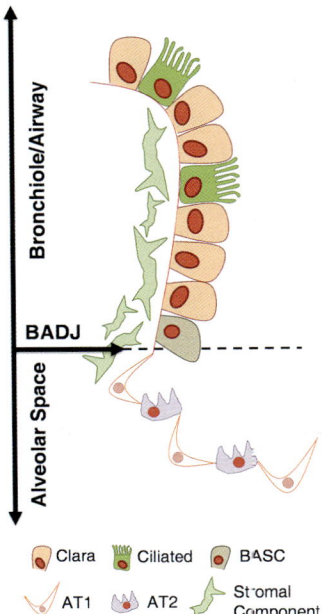

Figure 1. Epithelial organization of the murine distal lung. Clara cells and ciliated cells line the distal airways, with Clara cells being the prominent luminal cell population. The alveolar space is made up of alveolar type I (AT1) and type II (AT2) cells organized in sac-like structures. BASCs (bronchioalveolar stem cells) sit at the junction between the terminal bronchiole and the alveolar space (BADJ, *arrow*), and stromal components that may include mesenchymal cells and extracellular matrix molecules, underlie epithelia in both regions.

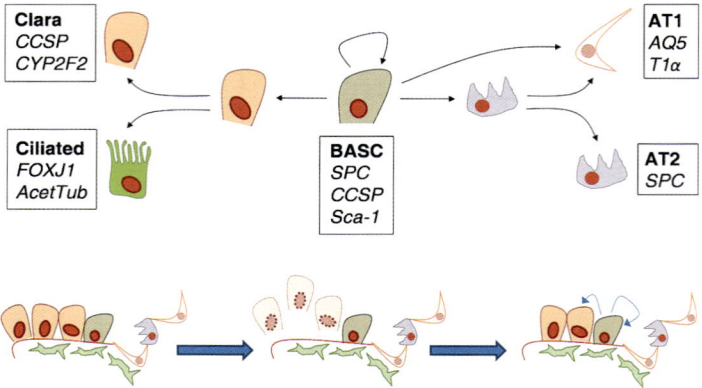

Figure 2. Model of BASC function and distal lung homeostasis. (*Top*) BASCs serve as stem cells for epithelial cells lining the airways and air space. They are capable of self-renewal (*arrow* back to BASCs) or differentiating to give rise to Clara, AT1, and AT2 cells in the appropriate physiological settings. In turn, a subset of Clara cells may serve as progenitor cells capable of differentiation to more specialized Clara and ciliated cells (*arrows*). Similarly, certain AT2 cells may be progenitor cells that give rise to AT2 and AT1 cells (*arrows*). Boxed near each cell type are markers used to identify each population. (*Bottom*) Example of proposed BASC function in vivo. BASCs and other lung epithelia are normally quiescent in vivo (*left*). When Clara cells are ablated by naphthalene (*middle*), BASCs are stimulated to proliferate, self-renew, and differentiate (*arrows*) to give rise to Clara cells to repair the damaged epithelium (*right*).

exposed to potential perturbagens from the outside environment. Together with this risk, the required function of the cells that comprise the lung epithelium for sustained viability makes repair of damage to the lung an urgent matter.

Putting together these unique aspects of lung cell biology, it follows that lung stem cell populations may be a subset of the same cells that function daily as contributors to the gas-exchange machinery, and their stem cell characteristics may only be gleaned in times when injury of more specialized, or more differentiated, cells occurs. An alternative hypothesis is that the majority of cells in the lung may have the potential to give rise to new epithelia, and thus the lung may not contain any stem or progenitor cells, just as has been proposed for the pancreas (Dor and Melton 2004). The concept of "facultative" stem or progenitor cells—defined as cells that differ from "dedicated" stem cells in that they are more differentiated, do not have long-term self-renewal capacity, and do not participate in tissue homeostasis but rather only respond to injury—has also been proposed for the lung and may include either of the models above (Rawlins and Hogan 2006; Stripp and Reynolds 2008). The sections below highlight studies suggesting that the distal lung contains resident stem or progenitor cells and describe the methods that are likely to yield an answer to which of these hypotheses is most correct.

STEM CELLS IDENTIFIED BY COMBINING FLOW CYTOMETRY AND LUNG CELL MARKER EXPRESSION

The isolation of BASCs added to the understanding of the epithelial diversity of the distal lung (see Fig. 1) (Kim et al. 2005). BASCs were initially identified based on their residence in the region between the bronchiolar and alveolar cells in terminal bronchioles, known as the bron-

chioalveolar duct junction (BADJ), and distinct coexpression of CCSP and SPC (Fig. 2). BASCs can be isolated from dissociated murine lung using a fluorescence-activated cell sorting (FACS)-based protocol wherein cells are sorted positively for expression of the cell surface marker Sca-1 and negatively for the endothelial marker CD31 and the hematopoietic marker CD45.

Isolated BASCs have the key stem cell properties of self-renewal and multipotency in that they can be passaged multiple times in culture, and, in clonal assays, they can differentiate into CCSP-positive, SPC-positive, or AQ5-positive cells (singly positive for each) when grown on Matrigel, a basement membrane matrix preparation. Additionally, BASCs are among the first cells to proliferate in vivo in response to naphthalene and bleomycin treatment injury models discussed below.

In addition to their response to lung injury, BASCs are implicated in the onset of lung adenocarcinoma initiated by oncogenic K-*ras* (Kim et al. 2005; Yang et al. 2008) or via other oncogenic aberrations (see below). Significant expansion of the BASC pool occurs following activation of oncogenic K-*ras* in vivo as well as in a cell-autonomous manner after K-*ras* is activated in isolated BASCs in culture, and many of the earliest lesions observed in this model arise at the BADJ where BASCs reside. We hypothesize that BASCs are the cell of origin for K-*ras*-induced lung adenocarcinoma, and we are currently creating the murine alleles needed to test this possibility.

MODELS OF LUNG INJURY IDENTIFIED PUTATIVE LUNG STEM AND PROGENITOR CELLS

Due to the low turnover rate of the adult lung epithelium, injury models are critical to the process of understanding which lung cell populations are required for epithelial regeneration. Stimulation and visualization of lung cell proliferation are facilitated by induction of lung injury; normal mouse lung has few to no epithelial cells that are capable of incorporating nucleotide analogs or staining for Ki-67, indicating that the lung epithelia is largely quiescent (Mason-Richie et al. 2008). Analyses of existing lung injury models, described below, have led to the identification of some putative lung stem cell populations. These models are also useful in formulating hypotheses regarding lung lineage relationships, given that each typically results in the perturbation of only one or a few particular types of epithelial cells.

Naphthalene

A compound typically found in trace amounts in cigarette smoke and an airborne pollutant, naphthalene causes acute toxicity to Clara cells (Stripp et al. 1995). Mice exposed to naphthalene are depleted of Clara cells, but the bronchiolar epithelium is reconstituted within 2–6 weeks (Van Winkle et al. 1995). A potential stem cell population for the Clara cell lineage—variant Clara cells—was identified based on their resistance to naphthalene, conferred by lack of expression of a cytochrome P450 enzyme necessary for naphthalene metabolism. Variant Clara cells (so-called because of their CCSP expression) were located in the proximal airways near neuroepithelial bodies and at the BADJ (Reynolds et al. 2000a; Hong et al. 2001; Giangreco et al. 2002). Variant Clara cells residing in the BADJ likely overlap with the BASC population in that they are both naphthalene-resistant and express CCSP. It also remains possible that the BADJ variant Clara cells contain a subset of cells that lack SPC, and therefore, some variant Clara cells may be distinct from BASCs. In short, naphthalene administration serves as one method to analyze BASCs, Clara cells, and the lineages to which they may give rise.

Bleomycin

Intranasal or intratracheal bleomycin treatment induces alveolar injury characterized by widespread AT1 cell loss, AT2 cell number reduction, and, in some animal strains, a deposition of fibrotic tissue in the pulmonary interstitium with associated inflammation (Adamson 1976; Aso et al. 1976). A return to normal numbers of AT1 and AT2 cells is seen when bleomycin is given at doses low enough to minimize the fibrotic response, and this repair was correlated with bromodeoxyuridine (BrdU) incorporation by AT2 cells (Kawamoto and Fukuda 1990). Homeostatic maintenance and reconstitution of the AT1 cell population have thus been classically attributed to AT2 cells acting as stem or progenitor cells in this setting. In culture, AT2 cells can be induced to exhibit AT1 cell characteristics (Isakson et al. 2001; Bhaskaran et al. 2007), yet our studies showed that AT2 cells lack self-renewal and differentiation ability (Kim et al. 2005); instead, they suggested that BASCs may be responsible for AT1 and AT2 cell maintenance and repair. Bleomycin is thus a tool for AT1, AT2, and BASC lineage analysis.

Cell Ablation Strategies

Cell ablation via engineered alleles is another classic technique useful to lung cell biology. In CCSP-HSVtk transgenic mice, expression of a herpes simplex virus thymidine kinase controlled by the CCSP promoter (Reynolds et al. 2000b, 2004; Hong et al. 2001) allows selective ablation of CCSP-expressing cells upon administration of the nucleoside analog ganciclovir. In contrast to the naphthalene injury model, CCSP-HSVtk mice are incapable of epithelial repair following ganciclovir treatment, further implicating variant Clara cells and BASCs as necessary regenerative populations following bronchiolar injury. In conjunction with the existing chemical injury models, an inducible means of ablating specific cell populations in the lung—such as the CCSP-HSVtk model or other systems such as doxycycline- and tamoxifen-inducibile alleles (e.g., rtTA/tet-O and CreERT²/Lox), which are also useful in lineage tracing (below) (Tichelaar et al. 2000; Rawlins et al. 2007)—will be invaluable in pinpointing the cell types that are necessary for lung repair in various contexts.

Partial Pneumonectomy

The murine lung undergoes regeneration after surgical removal of one or several lobes, referred to as partial pneumonectomy. This response is a compensatory overgrowth of the remaining lung attributed to cellular proliferation, yet the contribution of lung stem cells to this response is unknown. Using in vivo analyses after pneumonectomy and comparison of observed data to a mathematical model, we determined that both BASCs and AT2 cells are likely to be important in lung regrowth (Nolen-Walston et al. 2008). BASCs elicited the earliest and most significant proliferative response to partial pneumonectomy, before restoration of lung function capacity. Modeling data supported the hypothesis that BASCs may contribute to alveolar regeneration and showed that AT2 cells likely function as the cell type with the greater expansion potential to increase alveolar surface area; these studies support and contribute to our model of BASC function (below and Fig. 2). This work is the first demonstration of integrating approaches including physiology and systems biology to understand the adult lung and lung stem cell biology; such combined methodology is likely to be important in understanding the direct clinical implications of murine lung stem cell advances.

Although these and other injury models were crucial in the process of laying the groundwork for lung stem cell biology, it is of note that these models may have some limitations in precisely recapitulating how the lung is maintained and injured during aging or when lung injuries from exogenous agents, such as smoking, pollutants, and cold or flu viruses, are encountered. Most environmental insults faced by the lung epithelia contain many chemical components that likely impact many cell types at once: Cigarette smoke is likely to damage multiple cell types simultaneously. Therefore, chemicals that ablate only one cell type may not represent the entire potential of stem or progenitor cells when used for performing lineage tracing.

A STEM CELL MODEL FOR DISTAL LUNG HOMEOSTASIS

The in vivo lung injury and ex vivo culture experiments described above, together with new insights from the analysis of molecules that likely function in BASC control (see below), have led us to develop a working model for the function of BASCs and other distal lung epithelia in lung homeostasis (Fig. 2) (Kim et al. 2005; Kim 2007b). First, we hypothesize that BASCs sit atop a hierarchy of epithelial cells lining the airways and air space, aptly and uniquely positioned to contribute to epithelial repair in both compartments (Figs. 1 and 2). Thus, BASCs are capable of giving rise to Clara, AT1, and AT2 cells in the context of particular physiological settings. Subsets of Clara cells may serve as progenitor cells with the capacity to produce more differentiated Clara cells (e.g., variant Clara cells may be progenitors for more specialized Clara cells), as has been proposed by Stripp and colleagues (Hong et al. 2001), and Clara cells may also be progenitors for ciliated cells (Asmundsson et al. 1973; Evans et al. 1986; Rawlins et al. 2007; Reynolds et al. 2008). Similarly, AT2 cells may be progenitors that replenish subsets of specialized AT2 cells as well as AT1 cells. As depicted in Figure 2, there are no known molecular markers to functionally separate Clara cell or AT2 cell subsets in vivo. In this model, BASCs may be most similar to the multipotent hematopoietic stem cell, whereas Clara and AT2 cell subsets may be akin to lymphoid or myeloid progenitors, and specialized Clara cells and AT1 cells may be the terminally differentiated cells, as are T cells and red blood cells (Metcalf 2007). Importantly, in this model, evidence that Clara cells contribute to new Clara cells (e.g., via lineage tracing) would not rule out the potential contribution of BASCs to Clara cell lineages in particular physiological settings. Furthermore, a lack of BASC contribution to particular lineages in some experiments (e.g., to alveolar cells in lung development) does not exclude their ability to yield various lineage progeny in the right setting (e.g., to alveolar cells if both bronchiolar and alveolar cells are injured or absent).

Just as our model does not rule out the possibility of stem or progenitor cell activity in cell types other than BASCs, this hypothesis also does not require that BASCs contribute to developing or aging lung. In several independent studies, the BASC population does not become apparent—at least, as defined by SPC and CCSP coexpression—until after birth (Tyagi et al. 2007; Reynolds et al. 2008; Zhang et al. 2008; C. Kim, unpubl.). Thus, BASCs are likely not required for embryonic development of the lung. Wuenschell et al. (1996) reported that embryonic lung cells simultaneously exhibit staining for Clara, AT2, and neuroendocrine cell markers, whereas Whitsett and others have suggested that these cell types have distinct points of derivation (Perl et al. 2002). Because BASCs can differentiate into alveolar cells in culture and a significant percentage of alveoli are formed after birth, it remains possible that BASCs have a role in postnatal alveolar expansion and development; the developmental origins of BASCs remain to be defined.

LINEAGE TRACING IN THE LUNG

Lineage-tracing techniques (Stern and Fraser 2001; Barker et al. 2007) represent one approach needed to test models for lung homeostasis. To date, genetic lineage tracing of bronchiolar and alveolar cell types has been complicated by the expression of lineage markers in the embryonic lung epithelium and misexpression of transgenes driven by lung cell-type promoters (Whitsett and Perl 2006). To use lineage tracing in any tissue, it is necessary to have suitable markers that are exclusively expressed in the population of interest. This has been possible for analysis of the potential ciliated cell contribution to Clara cells using a tamoxifen-inducible ciliated cell-specific recombinase allele combined with a recombinase reporter allele (Rawlins et al. 2007). Because we have not yet identified a gene that is exclusively expressed in the BASC population for use in this manner, we are using genetic strategies that take advantage of our current knowledge of BASC markers, their known intersection or exclusion with regard to expression in other cell types, and multiple recombinase and reporter allele systems. We predict that as the more precise genetic tools we and others (see, e.g., Rawlins et al., this volume) are building become available, lung biology and lung stem cell biology will grow in its level of specificity and, more importantly, increase the knowledge of this complex organ.

Results from lineage tracing in the lung must be interpreted keeping in mind the biology of this tissue. A positive result in this scenario is sound evidence for lineage potential, but due to potential inconsistencies in recombination efficiency and injury efficacy, a negative result does not necessarily indicate that the cell population intended for marking is not a stem cell population. In addition, especially in the absence of injury in this relatively quiescent tissue, counting the percentage of labeled lung cells in a given pulse-chase experiment (e.g., tamoxifen to activate Cre recombinase, the pulse, followed by a waiting time for Cre reporter readout, the chase) is not necessarily related to self-renewal or differentiation without coincident evidence of cellular proliferation over the analyzed chase time.

LABEL RETENTION AS A LINEAGE TRACER

Traditionally, the definition of an adult stem cell has included the quiescent descriptor, indicating that the true stem cell divides infrequently. It follows that label retention can be used as a marker of stem cells when performing pulse-chase experiments. Indeed, many have used this technique successfully for this purpose (Potten et al. 1992; Tumbar et al. 2004), and yet, recent findings suggest that label retention does not necessarily identify the tissue stem cell (Barker et al. 2007; Kiel et al. 2007). It is equally important to note that the extent of stem cell quiescence is likely to vary from tissue to tissue. Thus, both label-retention and marker-based fate mapping can be useful in identifying stem cell populations, and the best approach may be a combinatorial one.

To test the utility of label retention in lung stem cell isolation, we are using a doxycycline-inducible H2B–green fluorescent protein (GFP) system in which chromatin-bound GFP can be expressed in a temporal fashion and lost over time with cell division (Kanda et al. 1998; Tumbar et al. 2004). We have determined that we can use this H2B-GFP allele with the reverse tetracycline activator driven by Rosa26 to label lung epithelia in vivo with temporal control (Fig. 3). With this approach, we will be able to analyze cell turnover in normal lung and in response to lung injury without the bias of known markers. Isolation and molecular characterization of GFP-positive lung cells with stem cell function could then lead to new markers for putative stem cells.

TRANSPLANTATION OF LUNG STEM CELLS

Upon isolation of putative adult stem cell populations, demonstration of their ability to effectively contribute to tissue-specific cell types and, in some cases, to repairing damaged or diseased tissue in a transplantation assay has proven to be compelling evidence for bona fide stem cell

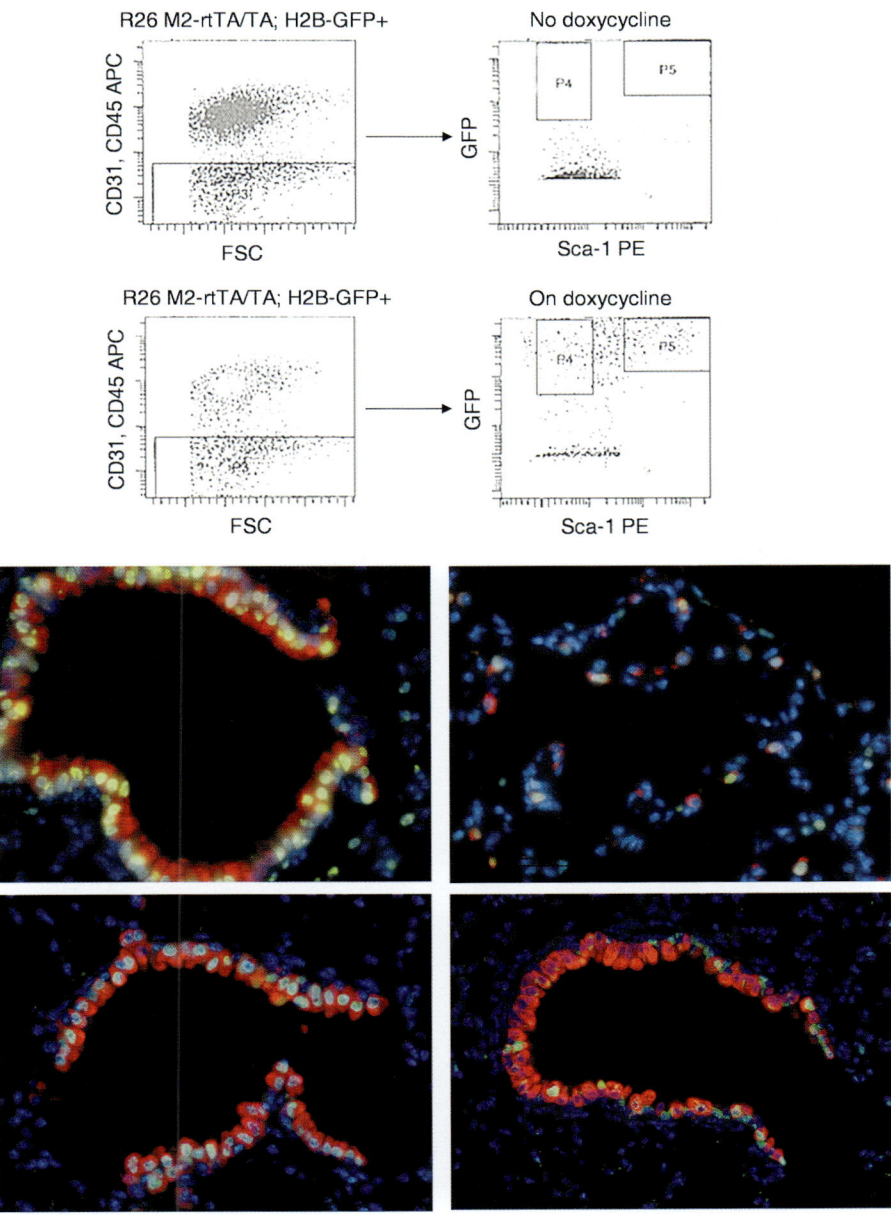

Figure 3. Inducible H2B-GFP expression in lung epithelial cells. FACS analysis shows that GFP expression in CD31- and CD45-negative lung cells, including BASCs (P5) and other epithelial cells (P4) from Rosa26-M2rtTA; H2B-GFP is absent in mice that did not receive doxycycline (*top graph*), whereas GFP is present in mice that received doxycycline treatment (*bottom graph*). In lung tissue sections, terminal bronchiole (TB) (*top left*) and alveolar space (*top right*) after a 1-week pulse of doxycycline and no chase period, showed GFP labeling of Clara cells and AT2 cells, respectively. (*Bottom, left and right*) TB 1 month after doxycycline pulse and mock-treated (*bottom left*) or naphthalene-treated (*bottom right*) mouse, showing loss of GFP label after repair. (*Red*) CCSP (*top left, bottom left* and *right*) or SPC (*top right*); (*green*) GFP; (*blue*) DAPI.

function (Till and McCullough 1961). Tissues in which this contribution has been observed include the brain, retina, skin, and muscle (Meissner et al. 2005; MacLaren et al. 2006; Mavilio et al. 2006; Cerletti et al. 2008). In some tissues, such as the blood and the mammary gland, it has even been possible to observe complete reconstitution of mature, fully differentiated tissues from the orthotopic transplantation of a single stem cell (Osawa et al. 1996; Shackelton et al. 2006). For other tissue-specific stem cells, subcutaneous injection of stem cells with supporting mesenchymal cells allowed the formation of structures that resemble the tissue of interest (Blanpain et al. 2004; Richardson et al. 2004; Lawson et al. 2007). The lung lags behind these tissues in the development of an orthotopic transplantation assay, due in no small part to the complex architecture of the lung and the unique air–epithelium interface that exists in the pulmonary lumen.

We are developing a transplantation protocol to assess the ability of putative lung stem cells to contribute to lung epithelial lineages. By using the injury models discussed above, we will be able to examine the contribution of transplanted stem cells in various contexts, both when the endogenous stem cell niche is intact and when the endogenous niche is cleared. We are also investigating the potential of purified adult lung cell populations to function in environments that support lung development. One such current approach is to inject normal putative lung stem cell populations from the mouse into the prelung mesenchymal field in the developing chicken embryo, similar to the example of how the developing chicken nervous system was used to test the potential of adult rat neural crest stem cells (Mosher et al. 2007). These studies may reveal a robust assay for lung stem cell self-renewal and differentiation potential in vivo. In combination with other genetic tools currently under development, transplantation studies will contribute to our understanding of the biological pool of cells with the potential to maintain lung epithelial tissue in vivo.

EXTENDING THE LUNG STEM CELL QUESTION TO CANCER

The cancer stem cell (CSC) hypothesis, which suggests that tumors are maintained by a population of cells possessing stem cell characteristics, has emerged as an attractive explanation for tumor growth, recurrence, and metastasis (Reya et al. 2001). Methods used in identification and characterization of stem cell populations from normal adult tissues have proven to be useful in uncovering cancer stem cell populations, yet this important concept in stem cell biology has not yet been carefully explored in the lung.

The results of multiple studies suggest that murine lung adenocarcinoma contains a CSC population. In a mouse model of lung adenocarcinoma initiated with oncogenic K-*ras*, the BASC population expands early and specifically after K-*ras* activation, and BASC-like cells can be found in the developed tumors (Jackson et al. 2001; Kim et al. 2005). Similar results have been reported in additional genetic models of lung adenocarcinoma, strongly implicating BASCs in tumor initiation and progression (Besson et al.

2007; Pei et al. 2007; Ventura et al. 2007; Yanagi et al. 2007; Yang et al. 2008). Development of a lung cancer cell transplantation assay has yielded further evidence suggesting the presence of a CSC population in murine lung adenocarcinoma, and studies to rigorously test the tumor-propagating potential of subpopulations of murine lung adenocarcinoma cells, including BASC-like cells, in this assay are under way in our lab (S. Curtis et al., unpubl.).

Several pieces of evidence suggest that human lung cancers also contain a population of cells with stem cell characteristics. First, the majority of newly diagnosed lung cancer patients have tumors that are already metastatic or refractory to chemotherapeutics or that quickly become resistant to chemotherapeutics. For example, the initial sensitivity of human adenocarcinomas with activating epidermal growth factor receptor (EGFR) mutations to EGFR tyrosine kinase inhibitors and the acquired resistance to these treatments suggest that drug-resistant CSCs may be present in these tumors (Kobayashi et al. 2005; Pao et al. 2005; Kosaka et al. 2006). Second, side population cells, isolated by their ability to efflux Hoechst dye, were identified in six human lung cancer cell lines; these cells exhibit increased drug-exporting transporter expression, enriched tumor-propagating capacity, and resistance to multiple chemotherapies (Ho et al. 2007). Third, CD133[+] cells from human lung tumors (and not CD133[–] cells) were recently shown to form self-renewing spheres in culture that could propagate tumors when transplanted subcutaneously into immunodeficient mice (Eramo et al. 2008). Importantly, although these studies support the likelihood of CSCs in lung cancers, the isolation and characterization of a population of human lung cancer cells that can serially propagate the lung tumor phenotype in the lung microenvironment have not been reported.

The most widely used technique for identification of CSCs uses a combination of cell surface markers that select cells with expression patterns similar to those of more primitive cells and distinct from those of differentiated cell lineages, followed by transplantation of sorted cancer cells into immunodeficient mice. CSCs in human leukemia, breast, brain, colon, pancreatic, and melanoma malignancies have been shown to be required for serial rounds of tumor propagation in transplantation assays that often use orthotopic strategies (Bonnet and Dick 1997; Al-Hajj et al. 2003; Singh et al. 2004; Dalerba et al. 2007; Li et al. 2007; O'Brien et al. 2007; Ricci-Vitiani et al. 2007; Schatton et al. 2008). Subcutaneous injections are commonly used to test tumor-propagating activity due to the ease of accessing the injection site and monitoring tumor progression, and the highly vascular environment under the kidney capsule that promotes cancer cell engraftment and proliferation is also used (Xin et al. 2005). However, such assays do not reflect the normal cellular milieu in which tumors arise, grow, and progress. Importantly, transplanting tumor cells into the normal microenvironment may most accurately reflect cancer progression, niche effects on CSCs, and cancer cell response to therapeutics (Li and Neaves 2006; Scadden 2006). Particularly important for future work to develop therapeutic intervention of lung CSCs, previous studies have shown that lung tumor cells growing in subcutaneous regions do not exhibit the same physiological response to chemotherapy as do tumors growing in the lung (Onn et al. 2003; Fujino et al. 2005). Therefore, injection of murine or human lung cancer cells into the lungs of recipient mice will be crucial in future studies to monitor CSCs in the lung microenvironment (McLemore et al. 1987).

ADDITIONAL MARKERS ARE NEEDED TO STUDY LUNG CANCER STEM CELLS

We have determined that markers of murine BASCs allow prospective identification of cancer stem cells from murine lung tumors in some settings (above), yet it will be crucial to find additional markers to isolate human lung CSCs. One key BASC marker, Sca-1, is not expressed on human cells, and a second positive BASC marker, CD34, has been shown to be present on other cell types found in human lung tissue (Gomperts and Strieter 2007). In addition, in our own recent studies, we found that the current commercially available antibodies for CD34 may overestimate BASC numbers and should be used with caution (R. Roach and C. Kim, unpubl.). For lung CSC studies, it may be useful to test markers that have been used to isolate CSCs from other types of primary solid human tumors. For example, human lung tumors could be sorted for CD133, which is expressed on brain, colon, and putative lung CSCs (Singh et al. 2004; O'Brien et al. 2007; Ricci-Vitiani et al. 2007; Eramo et al. 2008), and epithelial cell adhesion molecule (EpCAM), which is expressed on breast, pancreatic, and colon CSCs (Al-Hajj et al. 2003; Dalerba et al. 2007; Li et al. 2007), for use in serial transplantation experiments.

It will likely be important to also use bioinformatic and genetic approaches to identify murine and human lung CSCs. Bioinformatic analyses, which evaluate the differences between human normal lung and lung adenocarcinoma gene expression profiles, may reveal cell surface markers that correlate with tumor-propagation ability, because the number of CSCs is likely to be higher in lung tumors compared to normal tissues (Bhattacharjee et al. 2001; Beer et al. 2002; Garber et al. 2002; Powell et al. 2003). Direct comparison of normal stem cells to other lung cell types may also be valuable. For example, comparing gene expression data of BASCs versus other epithelial cells in normal and tumorigenic lungs may lead to the identification of novel lung CSC markers. In addition, culture-based assays may be useful surrogates for the identification of CSC markers. In parallel with studies to identify novel murine and human lung cancer cell or BASC surface markers, it is also critical to determine if the normal human lung has population(s) with stem cell functions.

GENETIC APPROACHES FOR CANCER STEM CELLS

Just as for testing models of lung homeostasis, the use of genetic strategies to identify lung CSCs will also be advantageous. One such approach would use cell-type-specific ablation to test the function of defined subsets of cancer cells in situ. Targeted ablation of putative cancer stem cells would further determine the necessity of that same cell population in the context of an endogenous tumor and thus directly interrogate the role of such cells in tumor maintenance. Targeted cell ablation can be achieved using lineage-restricted expression of suicide genes such as the diphtheria toxin A (DTA) chain or the herpes simplex virus-1 thymidine kinase (HSVtk) (above) (Breitman et al. 1987; Palmiter et al. 1987; Borrelli et al. 1988). Such strategies have been used in a variety of settings to understand the roles of cell populations in the context of development and niche effects (Lee et al. 2000; Visnjic et al. 2004; Ito et al. 2005; Sangiorgi and Capecchi 2008), and we anticipate a similar utility for this strategy in the study of lung cancer stem cell biology.

PATHWAYS IMPLICATED IN BASC FUNCTION

Just as the precise cellular mechanisms of lung homeostasis and lung cancer maintenance are poorly understood, the pathways that regulate lung stem cells are largely unknown. Several recent publications have described the identification of players that may be key in lung stem cell function, and assays more focused on particular lung cell subsets and stem cell activities are likely to lead to even more insight into lung homeostasis. Importantly, the factors that regulate features of stem cells, such as quiescence, self-renewal, and differentiation, may be distinct in the lung compared to other types of adult stem cells, given the unique attributes of lung biology discussed above. In addition, pathway abrogation can impact only one or several of the functions of lung stem cells, as seen in recent studies described below.

Regulation of stem cell quiescence, renewal, or terminal differentiation is ultimately dependent on modulating the cell cycle and cell growth machinery (Orford and Scadden 2008), and several components of the cell cycle regulation machinery have been implicated in BASC function. Deficiency of p18Ink4c, but not p27CIP/KIP, two different negative regulators of the Rb pathway, led to increased BASC numbers and a greater incidence of lung tumor development in corresponding knockout mice. Lungs of p18$^{-/-}$;Men$^{+/-}$ mice were more densely populated with BASCs, and these mice developed adenocarcinomas as well as neuroendocrine carcinomas. This finding raises the intriguing possibility that BASCs have the potential to contribute not only to adenocarcinomas, but also to tumors that are hypothesized to derive from neuroendocrine cells (Meuwissen and Berns 2005). Interestingly, although the tumor suppressor function of p27 is not required in BASC regulation (Pei et al. 2007), an alternative oncogenic function of p27 may be involved (Besson et al. 2007; Kossatz and Malek 2007; Sicinski et al. 2007). BASC numbers were elevated in mice bearing a *p27* mutant allele that is incapable of binding and inhibiting cyclins/CDKs. The *p27CK$^-$* allele also acted in a dominant fashion to promote tumorigenesis in multiple organs, including the lung (Besson et al. 2007). Because signaling directed by mitogen-activated protein kinase

kinase (MAP2K) or Ras and AKT may promote the cytoplasmic localization of this form of p27 (Chu et al. 2008), these effectors may contribute to BASC dysfunction in cancer.

Although oncogenic K-*ras* signaling was the first described driver of BASC expansion in tumorigenesis (Kim et al. 2005), the downstream effectors and pathways collaborating with K-*ras* in promoting this stem cell activity have been unknown. However, several recent publications now demonstrate the potential contribution of the phosphoinositol-3 kinase (PI3K)/PTEN pathway in BASC function and lung tumorigenesis (Yanagi et al. 2007; Dave et al. 2008; Iwanaga et al. 2008; Yang et al. 2008). Each study took a slightly different approach to conditionally inactivate PTEN within the lung, with varying outcomes; in some scenarios, PTEN disruption during postnatal stages led to BASC expansion, impaired differentiation into alveolar lineages, and increased lung adenocarcinoma development, whereas lung tumorigenesis did not occur in other studies of PTEN deficiency in the lung. Interestingly, elevated levels of activated AKT, Bcl-2, c-Myc, and Shh were detected in PTEN-disrupted lungs (Yanagi et al. 2007). BASC expansion in activated K-*ras*-induced tumors was diminished by inhibition of PI3K by the specific inhibitor PX-866 (Ihle et al. 2004). Given that PI3K is a master regulator of multiple growth signaling networks, it will be important to determine the specific downstream effectors important in this activity in the lung. Recent studies indicate that the kinase LKB1, a negative regulator of the PI3K/TOR (target of rapamycin) pathway, is frequently mutated in lung cancers, and its disruption is the most potent genetic event collaborating with oncogenic K-*ras* induction in lung tumorigenesis (Ji et al. 2007); whether mutation of LKB1 affects BASC activity is unknown.

Several other recent reports explored the role of lung developmental regulators on BASC function (Ventura et al. 2007; Reynolds et al. 2008; Zhang et al. 2008). The Wnt/β-catenin pathway is active in early lung development and declines upon terminal differentiation (Reynolds et al. 2008). Upon conditional expression of an activated form of β-catenin in the developing lung, differentiation of Clara cells was inhibited and an expansion of cells with BASC markers was observed (Reynolds et al. 2008; Zhang et al. 2008). When Gata6, a transcription factor known to be important in lung epithelial development (Yang et al. 2002), was conditionally depleted, severe differentiation defects resulted, coinciding with the premature detection of BASCs in embryonic lungs.

The Wnt receptor Fzd2 was determined to be an essential downstream effector of Gata6 function, indicating the potential importance of both canonical and noncanonical Wnt signaling in BASC function. MAPK14 (also known as p38α) has recently been implicated in acting downstream from Wnt5a and Fzd2 to suppress the Wnt/β-catenin noncanonical signaling pathway (Ma and Wang 2007), and the conditional knockout of MAPK14 in the postnatal lung also caused BASC expansion, disrupted epithelial differentiation, and collaborated with oncogenic K-*ras* to enhance lung tumorigenesis (Ventura et al. 2007). MAPK14 had been previously linked to affecting proliferation and differentiation through inhibition of EGFR and activation of c/EBPα, respectively, and were implicated in BASC function in these studies.

Our recent studies of Bmi1 deficiency within the lung indicate a spectrum of requirements for this Polycomb group protein in BASC function and lung cancer initiation and progression (Dovey et al. 2008), just as hematopoietic cells and neural cells may rely on Bmi1 for similar functions (Pietersen and van Lohuizen 2008). Although lung development occurs normally in Bmi1 null mice (van der Lugt et al. 1994), BASCs isolated from these mice are incapable of self-renewal in culture. Bmi1 null BASCs are capable of differentiation on matrigel, although preliminary studies indicate that they may differentiate more readily relative to wild-type BASCs (S. Zacharek and C. Kim, unpubl.). In response to lung injury, Bmi1 null mice show a reduced expansion of BASCs and inhibited ability to effect repair of the damaged epithelium (Fig. 4). The deficiencies observed in Bmi1 null BASCs in normal and injured lungs are mirrored in the oncogenic setting of K-*ras* activation: Lung tumorigenesis and BASC expansion initiated by oncogenic K-*ras* were inhibited by loss of Bmi1. Although the spontaneous activation of K-*ras*G12D from the K-*ras*LA2 allele normally results in rapid tumor progression and adenoma development within 9 weeks (Johnson et al. 2001), Bmi1 deficiency halted this progression at the earliest hyperplastic stage. The defects

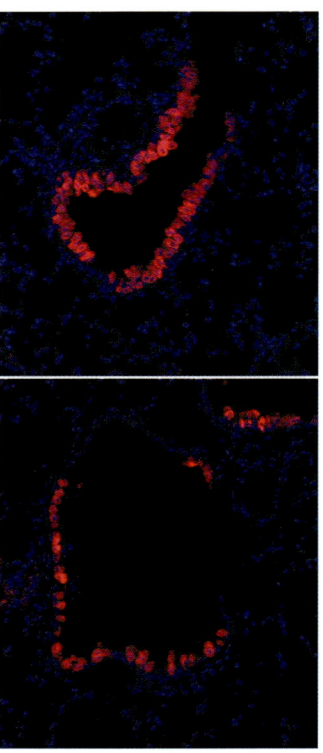

Figure 4. Bmi1 mutants are impaired in restoring the bronchiolar epithelium following naphthalene-induced lung injury. Lungs of Bmi1 mutant (*bottom*) or wild-type mice (*top*) were stained for CCSP (*red*) and the nuclear marker DAPI (*blue*) 1 month after naphthalene damage. Note the reduced incidence of Clara cells in the mutant.

exhibited by Bmi1 null BASCs in culture and in response to K-*ras* activation are partially due to up-regulated p19Arf function, although other effectors are clearly involved. Determining such additional Bmi1 target genes may point to other key pathways essential to BASC function.

Importantly, in the majority of the studies described above, the effects of these genetic abrogations on stem-cell-specific functions, such as self-renewal and differentiation on isolated cells, were not examined (Kim 2007a). This, coupled with the fact that the recombinase strains used in the studies discussed above were not BASC-specific, leave open the possibility that these pathways are not critical solely in BASCs, but are also crucial in other lung epithelial or stromal cells. Assessment of stem cell self-renewal, proliferation, and differentiation should be a minimum requirement in concluding that a gene regulates any adult stem cell population, including the lung. In addition, the development of new genetic tools, such as BASC-specific Cre lines coupled with conditional alleles of genes now suggested to promote (K-*ras*, p27CK, Bmi1, PI3K, Wnt, β-catenin) or restrain (p18, PTEN, MAPK14, GATA6) BASC function, will allow a more detailed evaluation of the cell-autonomous effects of these genetic alterations on BASC function.

CONCLUSIONS

Tools for studying stem cells and cancer biology that have proven their value in other tissues will have great utility in the elucidation of stem cell function in the lung, once we add a pulmonary-specific mindset to their use and interpretation. The most insightful work in this field is likely to come from combining the concepts described here: Chemical lung injury models should be combined with lineage-tracing techniques, cell sorting using molecular markers can be used with transplantation schemes, genetics should be applied to the study of cancer stem cells in endogenous tissues, and molecular screens should be performed with clonal assays for self-renewal and differentiation. With so many questions to answer, there is plenty of room for innovation to address the important question of how stem cell biology can be used to understand lung function and lung disease.

ACKNOWLEDGMENTS

We thank Jen Shepard Dovey, Jackie Lees, Andrew Hoffman, Rose Nolen-Walston, and all of our other collaborators for their efforts that contributed to the completed and ongoing studies described here. We also thank Sandra Ryeom for a critical reading of the manuscript. Our studies are supported by the Harvard Stem Cell Institute Seed Grant and Stem Cell Regulation Junior Faculty Program, RO1 HL090136, the Dana Farber Harvard Cancer Center Career Development Award, an American Cancer Society Research Scholar grant, the March of Dimes Basil O'Conner Award, the V Foundation for Cancer Research, the Joan Scarangello Lung Cancer Foundation, and the Children's Hospital Boston Stem Cell Program.

REFERENCES

Adamson, I.Y. 1976. Pulmonary toxicity of bleomycin. *Environ. Health Perspect.* **16:** 119–125.

Al-Hajj, M., Wicha, M.S., Bentio-Hernandez, A., Morrison, S.J., and Clarke M.F. 2003. Prospective identification of tumorigenic breast cancer cells. *Proc. Natl. Acad. Sci.* **100:** 3983–3988.

Asmundsson, T., Kilburn, K.H., and McKenzie, W.N. 1973. Injury and metaplasia of airway cells due to SO2. *Lab. Invest.* **29:** 41–53.

Aso, Y., Yoneda, K., and Kikkawa, Y. 1976. Morphologic and biochemical study of pulmonary changes induced by bleomycin in mice. *Lab. Invest.* **35:** 558–568.

Barker, N., van Es, J.H., Kuipers, J., Kujala, P., van den Born, M., Cozijnsen, M., Haegebarth, A., Korving, J., Begthel, H., Peters, P.J., and Clevers, H. 2007. Identification of stem cells in small intestine and colon by marker gene *Lgr5*. *Nature* **449:** 1003–1007.

Beer, D.G., Kardia, S.L., Huang, C.C., Giordano, T.J., Levin, A.M., Misek, D.E., Lin, L., Chen, G., Gharib, T.G., Thomas, D.G., et al. 2002. Gene-expression profiles predict survival of patients with lung adenocarcinoma. *Nat. Med.* **8:** 816–824.

Besson, A., Hwang, H.C., Cicero, S., Donovan, S.L., Gurian-West, M., Johnson, D., Clurman, B.E., Dyer, M.A., and Roberts, J.M. 2007. Discovery of an oncogenic activity in p27[Kip1] that causes stem cell expansion and a multiple tumor phenotype. *Genes Dev.* **21:** 1731–1746.

Bhaskaran, M., Kolliputi, N., Wang, Y., Gou, D., Chintagari, N.R., and Liu, L. 2007. Trans-differentiation of alveolar epithelial type II cells to type I cells involves autocrine signaling by transforming growth factor β1 through the Smad pathway. *J. Biol. Chem.* **282:** 3968–3976.

Bhattacharjee, A., Richards, W.G., Staunton, J., Li, C., Monti, S., Vasa, P., Ladd, C., Beheshti, J., Bueno, R., Gillette, M., et al. 2001. Classification of human lung carcinomas by mRNA expression profiling distinct adenocarcinoma subclasses. *Proc. Natl. Acad. Sci.* **98:** 13790–13795.

Blanpain, C., Lowry, W.E., Geoghegan, A., Polak, L., and Fuchs, E. 2004. Self-renewal, multipotency, and the existence of two cell populations within an epithelial stem cell niche. *Cell* **118:** 635–648.

Bonnet, D. and Dick J.E. 1997. Human acute myeloid leukemia is organized as a hierarchy that originates from a primitive hematopoietic cell. *Nat. Med.* **3:** 730–737.

Borrelli, E., Heyman, R., Hsi, M., and Evans, R.M. 1988. Targeting of an inducible toxic phenotype in animal cells. *Proc. Natl. Acad. Sci.* **85:** 7572–7576.

Breitman, M.L., Clapoff, S., Rossant, J., Tsui, L.C., Glode, L.M., Maxwell, I.H., and Bernstein, A. 1987. Genetic ablation: Targeted expression of a toxin gene causes microphthalmia in transgenic mice. *Science* **238:** 1563–1565.

Cerletti, M., Jurga, S., Witczak, C.A., Hirshman, M.F., Shadrach, J.L., Goodyear, L.J., and Wagers, A.J. 2008. Highly efficient, functional engraftment of skeletal muscle stem cells in dystrophic muscles. *Cell* **134:** 37–47.

Chu, I.M., Hengst, L., and Slingerland, J.M. 2008. The Cdk inhibitor p27 in human cancer: Prognostic potential and relevance to anticancer therapy. *Nat. Rev. Cancer.* **8:** 253–267.

Dalerba, P., Dylla, S.J., Park, I.K., Liu, R., Wang, X., Cho, R.W., Hoey, T., Gurney, A., Huang, E.H., Simeone, D.M., et al. 2007. Phenotypic characterization of human colorectal cancer stem cells. *Proc. Natl. Acad. Sci.* **104:** 10158–10163.

Dave, V., Wert, S.E., Tanner, T., Thitoff, A.R., Loudy, D.E., and Whitsett, J.A. 2008. Conditional deletion of Pten causes bronchiolar hyperplasia. *Am. J. Respir. Cell Mol. Biol.* **38:** 337–345.

Dobbs, L.G., Gonzalez, R., Matthay, M.A., Carter, E.P., Allen, L., and Verkman, A.S. 1998. Highly water-permeable type I alveolar epithelial cells confer high water permeability between the airspace and vasculature in rat lung. *Proc. Natl. Acad. Sci.* **95:** 2991–2996.

Dor, Y. and Melton, D.A. 2004. How important are adult stem cells for tissue maintenance? *Cell Cycle* **3:** 1104–1106.

Dovey, J.S., Zacharek, S.J., Kim, C.F., and Lees, J.A. 2008.

Bmi1 is critical for lung tumorigenesis and bronchioalveolar stem cell expansion. *Proc. Natl. Acad. Sci.* **105:** 11857–11862.

Eramo, A., Lotti, F., Sette, G., Pilozzi, E., Biffoni, M., Di Virgilio, A., Conticello, C., Ruco, L., Peschle, C., and De Maria, R. 2008. Identification and expansion of the tumorigenic lung cancer stem cell population. *Cell Death Differ.* **15:** 504–514.

Evans, M.J., Shami, S.G., Cabral-Anderson, L.J., and Dekker, N.P. 1986. Role of nonciliated cells in renewal of the bronchial epithelium of rats exposed to NO2. *Am. J. Pathol.* **123:** 126–133.

Fehrenbach, H. 2001. Alveolar epithelial type II cell: Defender of the alveolus revisited. *Respir. Res.* **2:** 33–46.

Fujino, H., Kondo, K., Ishikura, H., Maki, H., Kinoshita, H., Miyoshi, T., Takahashi, Y., Sawada, N., Takizawa, H., Nagao, T., Sakiyama, S., and Monden, Y. 2005. Matrix metalloproteinase inhibitor MMI-166 inhibits lymphogenous metastasis in an orthotopically implanted model of lung cancer. *Mol. Cancer Ther.* **4:** 1409–1416.

Garber, M.E., Troyanskaya, O.G., Schluens, K., Petersen, S., Thaesler, Z., Pacyna-Gengelbach, M., van de Rijn, M., Rosen, G.D., Perou, C.M., Whyte, R.I., et al. 2002. Diversity of gene expression in adenocarcinoma of the lung. *Proc. Natl. Acad. Sci.* **98:** 13784–13789.

Giangreco, A., Reynolds, S.D., and Stripp, B.R. 2002. Terminal bronchioles harbor a unique airway stem cell population that localizes to the bronchoalveolar duct junction. *Am. J. Pathol.* **161:** 173–182.

Gomperts, B.N. and Strieter, R.M. 2007. Fibrocytes in lung disease. *J. Leukoc. Biol.* **82:** 449–456.

Ho, M.M., Ng, A.V., Lam, S., and Hung, J.Y. 2007. Side population in human lung cancer cell lines and tumors is enriched with stem-like cancer cells. *Cancer Res.* **67:** 4827–4833.

Hong, K.U., Reynolds, S.D., Giangreco, A., Hurley, C.M., and Stripp, B.R. 2001. Clara cell secretory protein-expressing cells of the airway neuroepithelial body microenvironment include a label-retaining subset and are critical for epithelial renewal after progenitor cell depletion. *Am. J. Respir. Cell Mol. Biol.* **24:** 671–681.

Ihle, N.T., Williams, R., Chow, S., Chew, W., Berggren, M.I., Paine-Murrieta, G., Minion, D.J., Halter, R.J., Wipf, P., Abraham, R., Kirkpatrick, L., and Powis, G. 2004. Molecular pharmacology and antitumor activity of PX-866, a novel inhibitor of phosphoinositide-3-kinase signaling. *Mol. Cancer Ther.* **3:** 763–772.

Isakson, B.E., Lubman, R.L., Seedorf, G.J., and Boitano, S. 2001. Modulation of pulmonary alveolar type II cell phenotype and communication by extracellular matrix and KGF. *Am. J. Physiol. Cell Physiol.* **281:** C1291–C1299.

Ito, M., Liu, Y., Yang, Z., Nguyen, J., Liang, F., Morris, R.J., and Cotsarelis, G. 2005. Stem cells in the hair follicle bulge contribute to wound repair but not to homeostasis of the epidermis. *Nat. Med.* **11:** 1351–1354.

Iwanaga, K., Yang, Y., Raso, M.G., Ma, L., Hanna, A.E., Thilaganathan, N., Moghaddam, S., Evans, C.M., Li, H., Cai, W.W., et al. 2008. *Pten* inactivation accelerates oncogenic *K-ras*-initiated tumorigenesis in a mouse model of lung cancer. *Cancer Res.* **68:** 1119–1127.

Jackson, E.L., Willis, N., Mercer, K., Bronson, R.T., Crowley, D., Montoya, R., Jacks, T., and Tuveson, D.A. 2001. Analysis of lung tumor initiation and progression using conditional expression of oncogenic *K-ras*. *Genes Dev.* **15:** 3243–3248.

Ji, H., Ramsey, M.R., Hayes, D.N., Fan, C., McNamara, K., Kozlowski, P., Torrice, C., Wu, M.C., Shimamura, T., Perera, S.A., et al. 2007. LKB1 modulates lung cancer differentiation and metastasis. *Nature* **448:** 807–810.

Johnson, L., Mercer, K., Greenbaum, D., Bronson, R.T., Crowley, D., Tuveson, D.A., and Jacks, T. 2001. Somatic activation of the *K-ras* oncogene causes early onset lung cancer in mice. *Nature* **410:** 1111–1116.

Kanda, T., Sullivan, K.F., and Wahl, G.M. 1998. Histone-GFP fusion protein enables sensitive analysis of chromosome dynamics in living mammalian cells. *Curr. Biol.* **8:** 377–385.

Kawamoto, M. and Fukuda, Y. 1990. Cell proliferation during the process of bleomycin-induced pulmonary fibrosis in rats. *Acta Pathol. Jpn.* **40:** 227–238.

Kiel, M.J., He, S., Ashkenazi, R., Gentry, S.N., Teta, M., Kushner, J.A., Jackson, T.L., and Morrison, S.J. 2007. Haematopoietic stem cells do not asymmetrically segregate chromosomes or retain BrdU. *Nature* **449:** 238–242.

Kim, C.F. 2007a. MAPK-ing out the pathways in lung stem cell regulation. *Cell Stem Cell* **1:** 11–13.

Kim, C.F. 2007b. Paving the road for lung stem cell biology: Bronchioalveolar stem cells and other putative distal lung stem cells. *Am. J. Physiol. Lung Cell. Mol. Physiol.* **293:** L1092–L1098.

Kim, C.F., Jackson, E.L., Woolfenden, A.E., Lawrence, S., Babar, I., Vogel, S., Crowley, D., Bronson, R.T., and Jacks, T. 2005. Identification of bronchioalveolar stem cells in normal lung and lung cancer. *Cell* **121:** 823–835.

Kobayashi, S., Boggon, T.J., Dayaram, T., Janne, P.A., Kocher, O., Meyerson, M., Johnson, B.E., Eck, M.J., Tenen, D.G., and Halmos, B. 2005. *EGFR* mutation and resistance of non-small-cell lung cancer to gefitinib. *N. Engl. J. Med.* **352:** 786–792.

Kosaka, T., Yatabe, Y., Endoh, H., Yoshida, K., Hida, T., Tsuboi, M., Tada, H., Kuwano, H., and Mitsudomi, T. 2006. Analysis of epidermal growth factor receptor gene mutation in patients with non-small cell lung cancer and acquired resistance to gefitinib. *Clin. Cancer Res.* **12:** 5764–5769.

Kossatz, U. and Malek, N.P. 2007. p27: Tumor suppressor and oncogene...? *Cell Res.* **17:** 832–833.

Lawson, D.A., Xin, L., Lukacs, R.U., Cheng, D., and Witte, O.N. 2007. Isolation and functional characterization of murine prostate stem cells. *Proc. Natl. Acad. Sci.* **104:** 181–186.

Lee, K.J., Dietrich, P., and Jessell, T.M. 2000. Genetic ablation reveals that the roof plate is essential for dorsal interneuron specification. *Nature* **403:** 734–740.

Li, C., Heidt, D.G., Dalerba, P., Burant, C.F., Zhang, L., Adsay, V., Wicha, M., Clarke, M.F., and Simeone, D.M. 2007. Identification of pancreatic cancer stem cells. *Cancer Res.* **67:** 1030–1037.

Li, L. and Neaves, W.B. 2006. Normal stem cells and cancer stem cells: The niche matters. *Cancer Res.* **66:** 4553–4557.

Liu, X., Driskell, R.R., and Engelhardt, J.F. 2006. Stem cells in the lung. *Methods Enzymol.* **419:** 285–321.

Ma, L. and Wang, H.Y. 2007. Mitogen-activated protein kinase p38 regulates the Wnt/cyclic GMP/Ca^{2+} non-canonical pathway. *J. Biol. Chem.* **282:** 28980–28990.

MacLaren, R.E., Pearson, R.A., MacNeil, A., Douglas, R.H., Salt, T.E., Akimoto, M., Swaroop, A., Sowden, J.C., and Ali, R.R. 2006. Retinal repair by transplantation of photoreceptor precursors. *Nature* **444:** 203–207.

Mason, R.J., Williams, M.C., Moses, H.L., Mohla, S., and Berberich, M.A. 1997. Stem cells in lung development, disease, and therapy. *Am. J. Respir. Cell Mol. Biol.* **16:** 355–363.

Mason-Richie, N.A., Mistry, M.J., Gettler, C.A., Elayyadi, A., and Wikenheiser-Brokamp, K.A. 2008. Retinoblastoma function is essential for establishing lung epithelial quiescence after injury. *Cancer Res.* **68:** 4068–4076.

Massaro, G.D., Singh, G., Mason, R., Plopper, C.G., Malkinson, A.M., and Gail, D.B. 1994. Biology of the Clara cell. *Am. J. Physiol.* **266:** L101–L106.

Mavilio, F., Pellegrini, G., Ferrari, S., Di Nunzio, F., Di Iorio, E., Recchia, A., Maruggi, G., Ferrari, G., Provasi, E., Bonini, C., et al. 2006. Correction of junctional epidermolysis bullosa by transplantation of genetically modified epidermal stem cells. *Nat. Med.* **12:** 1397–1402.

McLemore, T.L., Liu, M.C., Blacker, P.C., Gregg, M., Alley, M.C., Abbott, B.J., Shoemaker, R.H., Bohlman, M.E., Litterst, C.C., Hubbard, W.C., et al. 1987. Novel intrapulmonary model for orthotopic propagation of human lung cancers in athymic nude mice. *Cancer Res.* **47:** 5132–5140.

Meissner, K.K., Kirkham, D.L., and Doering, L.C. 2005. Transplants of neurosphere cell suspensions from aged mice are functional in the mouse model of Parkinson's. *Brain Res.* **1057:** 105–112.

Metcalf, D. 2007. Concise review: Hematopoietic stem cells and tissue stem cells: Current concepts and unanswered questions. *Stem Cells* **25:** 2390–2395.

Meuwissen, R. and Berns, A. 2005. Mouse models for human lung cancer. *Genes Dev.* **19:** 643–664.

Mosher, J.T., Yeager, K.J., Kruger, G.M., Joseph, N.M., Hutchin, M.E., Dlugosz, A.A., and Morrison, S.J. 2007. Intrinsic differences among spatially distinct neural crest stem cells in terms of migratory properties, fate determination, and ability to colonize the enteric nervous system. *Dev. Biol.* **303:** 1–15.

Nolen-Walston, R.D., Kim, C.F., Mazan, M.R., Ingenito, E.P., Gruntman, A.M., Tsai, L., Boston, R., Woolfenden, A.E., Jacks, T., and Hoffman, A.M. 2008. Cellular kinetics and modeling of bronchioalveolar stem cell response during lung regeneration. *Am. J. Physiol. Lung Cell. Mol. Physiol.* **294:** L1158–L1165.

O'Brien, C.A., Pollett, A., Gallinger, S., and Dick, J.E. 2007. A human colon cancer cell capable of initiating tumour growth in immunodeficient mice. *Nature* **445:** 106–110.

Onn, A., Isobe, T., Itasaka, S., Wu, W., O'Reilly, M.S., Ki Hong, W., Fidler, I.J., and Herbst, R.S. 2003. Development of an orthotopic model to study the biology and therapy of primary human lung cancer in nude mice. *Clin. Cancer Res.* **9:** 5532–5539.

Orford, K.W. and Scadden, D.T. 2008. Deconstructing stem cell self-renewal: Genetic insights into cell-cycle regulation. *Nature Rev.* **9:** 115–128.

Osawa, M., Hanada, K., Hamada, H., and Nakauchi, H. 1996. Long-term lymphohematopoietic reconstitution by a single CD34-low/negative hematopoietic stem cell. *Science* **273:** 242–245.

Palmiter, R.D., Behringer, R.R., Quaife, C.J., Maxwell, F., Maxwell, I.H., and Brinster, R.L. 1987. Cell lineage ablation in transgenic mice by cell-specific expression of a toxin gene. *Cell* **50:** 435–443.

Pao, W., Miller, V.A., Politi, K.A., Riely, G.J., Somwar, R., Zakowski, M.F., Kris, M.G., and Varmus, H. 2005. Acquired resistance of lung adenocarcinomas to gefitinib or erlotinib is associated with a second mutation in the EGFR kinase domain. *PLoS Med.* **2:** e73.

Pei, X.H., Bai, F., Smith, M.D., and Xiong, Y. 2007. p18^{Ink4c} collaborates with *Men1* to constrain lung stem cell expansion and suppress non-small-cell lung cancers. *Cancer Res.* **67:** 3162–3170.

Perl, A.K., Wert, S.E., Nagy, A., Lobe, C.G., and Whitsett, J.A. 2002. Early restriction of peripheral and proximal cell lineages during formation of the lung. *Proc. Natl. Acad. Sci.* **99:** 10482–10487.

Pietersen, A.M. and van Lohuizen, M. 2008. Stem cell regulation by polycomb repressors: Postponing commitment. *Curr. Opin. Cell Biol.* **20:** 201–207.

Potten, C.S., Kellett, M., Roberts, S.A., Rew, D.A., and Wilson, G.D. 1992. Measurement of in vivo proliferation in human colorectal mucosa using bromodeoxyuridine. *Gut* **33:** 71–78.

Powell, C.A., Spira, A., Derti, A., DeLisi, C., Liu, G., Borczuk, A., Busch, S., Sahasrabudhe, S., Chen, Y., Sugarbaker, D., et al. 2003. Gene expression in lung adenocarcinomas of smokers and nonsmokers. *Am. J. Physiol. Cell. Mol. Biol.* **29:** 157–162.

Rawlins, E.L. and Hogan, B.L. 2006. Epithelial stem cells of the lung: Privileged few or opportunities for many? *Development* **133:** 2455–2465.

Rawlins, E.L., Ostrowski, L.E., Randell, S.H., and Hogan, B.L. 2007. Lung development and repair: Contribution of the ciliated lineage. *Proc. Natl. Acad. Sci.* **104:** 410–417.

Reya, T., Morrison, S.J., Clarke, M.F., and Weissman, I.L. 2001. Stem cells, cancer, and cancer stem cells. *Nature* **414:** 105–111.

Reynolds, S.D., Giangreco, A., Power, J.H., and Stripp, B.R. 2000a. Neuroepithelial bodies of pulmonary airways serve as a reservoir of progenitor cells capable of epithelial regeneration. *Am. J. Pathol.* **156:** 269–278.

Reynolds, S.D., Giangreco, A., Hong, K.U., McGrath, K.E., Ortiz, L.A., and Stripp, B.R. 2004. Airway injury in lung dis-

ease pathophysiology: Selective depletion of airway stem and progenitor cell pools potentiates lung inflammation and alveolar dysfunction. *Am. J. Physiol. Lung Cell. Mol. Physiol.* **287:** L1256–L1265.

Reynolds, S.D., Hong, K.U., Giangreco, A., Mango, G.W., Guron, C., Morimoto, Y., and Stripp, B.R. 2000b. Conditional Clara cell ablation reveals a self-renewing progenitor function of pulmonary neuroendocrine cells. *Am. J. Physiol. Lung Cell. Mol. Physiol.* **278:** L1256–L1263.

Reynolds, S.D., Zemke, A.C., Giangreco, A., Brockway, B.L., Teisanu, R.M., Drake, J.A., Mariani, T., Di, P.Y., Taketo, M.M., and Stripp, B.R. 2008. Conditional stabilization of β-catenin expands the pool of lung stem cells. *Stem Cells* **26:** 1337–1346.

Ricci-Vitiani, L., Lombardi, D.G., Pilozzi, E., Biffoni, M., Todaro, M., Peschle, C., and De Maria, R. 2007. Identification and expansion of human colon-cancer-initiating cells. *Nature* **445:** 111–115.

Richardson, G.D., Robson, C.N., Lang, S.H., Neal, D.E., Maitland, N.J., and Collins, A.T. 2004. CD133, a novel marker for human prostatic epithelial stem cells. *J. Cell Sci.* **117:** 3539–3545.

Sangiorgi, E. and Capecchi, M.R. 2008. Bmi1 is expressed in vivo in intestinal stem cells. *Nat. Genet.* **40:** 915–920.

Scadden, D.T. 2006. The stem-cell niche as an entity of action. *Nature* **441:** 1075–1079.

Schatton, T., Murphy, G.F., Frank, N.Y., Yamaura, K., Waaga-Gasser, A.M., Gasser, M., Zhan, Q., Jordan, S., Duncan, L.M., Weishaupt, C., et al. 2008. Identification of cells initiating human melanomas. *Nature* **451:** 345–349.

Shackleton, M., Vaillant, F., Simpson, K.J., Stingl, J., Smyth, G.K., Asselin-Labat, M.L., Wu, L., Lindeman, G.J., and Visvader, J.E. 2006. Generation of a functional mammary gland from a single stem cell. *Nature* **439:** 84–88.

Sicinski, P., Zacharek, S., and Kim, C. 2007. Duality of p27^{Kip1} function in tumorigenesis. *Genes Dev.* **21:** 1703–1706.

Singh, S.K., Hawkins, C., Clarke, I.D., Squire, J.A., Bayani, J., Hide, T., Henkelman, R.M., Cusimano, M.D., and Dirks, P.B. 2004. Identification of human brain tumour initiating cells. *Nature* **432:** 396–401.

Stern, C.D. and Fraser, S.E. 2001. Tracing the lineage of tracing cell lineages. *Nat. Cell Biol.* **3:** E216–E218.

Stripp, B.R. and Reynolds, S.D. 2008. Maintenance and repair of the bronchiolar epithelium. *Proc. Am. Thorac. Soc.* **5:** 328–333.

Stripp, B.R., Maxson, K., Mera, R., and Singh, G. 1995. Plasticity of airway cell proliferation and gene expression after acute naphthalene injury. *Am. J. Physiol.* **269:** L791–L799.

Tichelaar, J.W., Lu, W., and Whitsett, J.A. 2000. Conditional expression of fibroblast growth factor-7 in the developing and mature lung. *J. Biol. Chem.* **275:** 11858–11864.

Till, J.E. and McCulloch, E.A. 1961. A direct measurement of the radiation sensitivity of normal mouse bone marrow cells. *Radiat. Res.* **14:** 213–222.

Tumbar, T., Guasch, G., Greco, V., Blanpain, C., Lowry, W.E., Rendl, M., and Fuchs, E. 2004. Defining the epithelial stem cell niche in skin. *Science* **303:** 359–363.

Tyagi, S., Srisuma, S., Bhattacharya, S., and Mariani, T.J. 2007. Developmental ontogeny of bronchio-alveolar stem cells. *FASEB J.* **21:** 975–979.

van der Lugt, N.M., Domen, J., Linders, K., van Roon, M., Robanus-Maandag, E., te Riele, H., van der Valk, M., Deschamps, J., Sofroniew, M., van Lohuizen, M., et al. 1994. Posterior transformation, neurological abnormalities, and severe hematopoietic defects in mice with a targeted deletion of the *bmi-1* proto-oncogene. *Genes Dev.* **8:** 757–769.

Van der Schans, C.P. 2007. Bronchial mucus transport. *Respir. Care* **52:** 1150–1158.

Van Winkle, L.S., Buckpitt, A.R., Nishio, S.J., Isaac, J.M., and Plopper, C.G. 1995. Cellular response in naphthalene-induced Clara cell injury and bronchiolar epithelial repair in mice. *Am. J. Physiol.* **269:** L800–L818.

Ventura, J.J., Tenbaum, S., Perdiguero, E., Huth, M., Guerra, C., Barbacid, M., Pasparakis, M., and Nebreda, A.R. 2007. p38α MAP kinase is essential in lung stem and progenitor cell proliferation and differentiation. *Nat. Genet.* **39:** 750–758.

Visnjic, D., Kalajzic, Z., Rowe, D.W., Katavic, V., Lorenzo, J., and Aguila, H.L. 2004. Hematopoiesis is severely altered in mice with an induced osteoblast deficiency. *Blood* **103:** 3258–3264.

Whitsett, J.A. and Perl, A.K. 2006. Conditional control of gene expression in the respiratory epithelium: A cautionary note. *Am. J. Respir. Cell Mol. Biol.* **34:** 519–520.

Wuenschell, C.W., Sunday, M.E., Singh, G., Minoo, P., Slavkin, H.C., and Warburton, D. 1996. Embryonic mouse lung epithelial progenitor cells co-express immunohistochemical markers of diverse mature cell lineages. *J. Histochem. Cytochem.* **44:** 113–123.

Xin, L., Lawson, D.A., and Witte, O.N. 2005. The Sca-1 cell surface marker enriches for a prostate-regenerating cell subpopulation that can initiate prostate tumorigenesis. *Proc. Natl. Acad. Sci.* **102:** 6942–6947.

Yanagi, S., Kishimoto, H., Kawahara, K., Sasaki, T., Sasaki, M.,

Nishio, M., Yajima, N., Hamada, K., Horie, Y., Kubo, H., et al. 2007. Pten controls lung morphogenesis, bronchioalveolar stem cells, and onset of lung adenocarcinomas in mice. *J. Clin. Invest.* **117:** 2929–2940.

Yang, H., Lu, M.M., Zhang, L., Whitsett, J.A., and Morrisey, E.E. 2002. GATA6 regulates differentiation of distal lung epithelium. *Development* **129:** 2233–2246.

Yang, Y., Iwanaga, K., Raso, M.G., Wislez, M., Hanna, A.E., Wieder, E.D., Molldrem, J.J., Wistuba, II, Powis, G., Demayo, F.J., Kim, C.F., and Kurie, J.M. 2008. Phosphatidyl-inositol 3-kinase mediates bronchioalveolar stem cell expansion in mouse models of oncogenic *K-ras*-induced lung cancer. *PLoS ONE* **3:** e2220.

Zhang, Y., Goss, A.M., Cohen, E.D., Kadzik, R., Lepore, J.J., Muthukumaraswamy, K., Yang, J., DeMayo, F.J., Whitsett, J.A., Parmacek, M.S., and Morrisey, E.E. 2008. A Gata6-Wnt pathway required for epithelial stem cell development and airway regeneration. *Nat. Genet.* **40:** 862–870.

Epithelial Stem Cells of the Prostate and Their Role in Cancer Progression

R.U. LUKACS,* D.A. LAWSON,* L. XIN,‡ Y. ZONG,† I. GARRAWAY,#
A.S. GOLDSTEIN,ᵋ S. MEMARZADEH,# AND O.N. WITTE*†ᵋ§Δ

*Department of Microbiology, Immunology, and Molecular Genetics, University of California, Los Angeles;
‡Department of Molecular and Cellular Biology, Baylor College of Medicine; †Howard Hughes Medical
Institute at UCLA; #Department of Urology, David Geffen School of Medicine at UCLA; ᵋMolecular Biology
Institute, #Department of Obstetrics and Gynecology, §Broad Stem Cell Research Center at UCLA;
ΔDepartment of Molecular and Medical Pharmacology, David Geffen School of Medicine, University
of California, Los Angeles, California 90095

Prostate cancer is a leading cause of cancer-related death in adult men. It can regress dramatically upon antihormonal therapy, but it often recurs in a more aggressive, androgen-independent form. Defining the prostate tissue stem cells (PrSCs) and their involvement in cancer initiation and maintenance may lead to better therapeutics. Using a tissue-regeneration model in which dissociated prostate epithelial cells mixed with inductive mesenchyme give rise to prostatic tubules, we have identified a small population of prostate cells that contains multiple stem cell characteristics. In this system, prostate cancer can be initiated by autocrine or paracrine growth factor signaling and intracellular overexpression of genes often found mutated in human prostate cancer. Using an in vitro prostate sphere assay, we further defined the PrSC population and demonstrated their self-renewal and multilineage differentiation capabilities. Microarray analyses of the stem- and non-stem-cell populations have assisted us in finding and evaluating additional markers that can better define the PrSC population and further delineate the different cell types of the prostate, including those that serve as the target cell for tumor initiation.

Death rates from some leukemias, Hodgkin's disease, and testicular cancer have decreased significantly during past decades due to dramatic advances in their treatments. Although there have been improvements in prevention and early detection for the most common epithelial malignancies, there has not been significant improvement in the survival of patients with metastatic disease. It is widely accepted that transformation is caused by genetic changes leading to alterations in growth regulatory functions. New genome-wide screening methods have identified many of these potential oncogenes and tumor-specific mutations to target therapeutically. However, the fundamental question remains: Are we targeting the right cells?

Adult stem cells have the unique capability to self-renew and persist throughout a person's life span while retaining the potential to differentiate into any mature cell type of the tissue. Such mammalian tissue stem cells were first defined in the hematopoietic system but have since been identified in many solid organs as well, including the prostate (Till and McCulloch 1961; Isaacs 1987; Lois and Alvarez-Buylla 1993; Blanpain et al. 2007). The idea that primitive cells and cancer cells are related was postulated more than 150 years ago, when Rudolf Virchow (1855) noticed a histologic resemblance of teratocarcinomas to tissues of the developing fetus. Mounting evidence from more current studies suggests that stem cells and early progenitors are likely involved in cancer initiation as the cell of origin or in tumor maintenance acting as cancer stem cells.

Studies in leukemia suggest that some cell types may be more prone to transformation than others (Cozzio et al. 2003; Huntly et al. 2004). The theory that stem cells are the "cells of origin" for cancer is supported by their long-lived nature, which allows for multiple mutations to accumulate over time. Studies in blood and solid tissue cancers that show that stem cell pathways are present and often overactivated in tumors further support this belief (Reya et al. 2001). An alternate theory is that a mutated progenitor or differentiated cell can exploit and reactivate normal self-renewal pathways during carcinogenesis, potentially becoming the "cancer stem cell" (CSC) that maintains the tumor (Jamieson et al. 2004; Wicha et al. 2006). The idea of CSCs arose in the 1960s when two groups hypothesized that cancers are composed of a heterogeneous population of cells with different capacities for tumor propagation (Bruce and Van Der Gaag 1963; Southam et al. 1969). This theory is supported by studies in which distinct populations of cells isolated from acute myeloid leukemia (AML) (Lin⁻CD34⁺CD38⁻), brain tumors (CD133⁺), and breast cancers (Lin⁻CD44⁺CD24⁻/lo) were shown to contain all of the tumorigenic activity in their respective cancers (Bonnet and Dick 1997; Al-Hajj et al. 2003; Singh et al. 2004).

The CSC theory is an appealing concept based on elegantly designed experiments, but there are still many uncertainties, both theoretical and technical, about the interpretations of the data. Serial tumor transplantation experiments are the gold standard to measure the activity of cancer stem cells. However, it has been demonstrated that the growth of secondary tumors is highly dependent on the environment to which the cells are exposed. Factors such as the location of the transplantation, the presence or absence of highly irradiated feeder cells, and the inclusion of additional extracellular matrix components significantly affect the growth potential of implanted tumor cells (Hewitt et al. 1973; Peters and Hewitt 1974; Hill 2006).

Prostate cancer is the most commonly diagnosed cancer and the second highest cancer-related cause of death among men. The most frequently used treatment, androgen ablation, is based on the fact that the majority of prostate cancer cells are hormone-dependent (Huggins and Hodges 1941). This is a successful treatment initially, but the disease often recurs in an aggressive, androgen-independent form that is extremely difficult to treat. In the normal organ, androgen withdrawal through chemical or surgical castration causes approximately 90% of the differentiated cells to apoptose, leaving the primitive cell populations unaffected (Kyprianou and Isaacs 1988). This observation has led people to hypothesize that the subpopulation of remaining cancer cells after androgen ablation contains these primitive cells that can potentially cause the recurrence of the disease (Litvinov et al. 2003). Investigation of the potential roles of PrSCs in cancer initiation and maintenance may lead to more successful treatments for the disease.

THE PROSTATE GLAND AND ADULT PROSTATE STEM CELLS

The prostate is located at the base of the bladder in males, surrounding the urethra (Cunha et al. 1987). The gland is composed of tubules that have an epithelial compartment surrounded by stromal cells that include fibroblasts, smooth muscle, and myofibroblasts. The epithelium consists of two cellular compartments made up of three morphologically, functionally, and molecularly distinct cell types. Androgen-independent flat basal cells are attached to the basement membrane, where they maintain the homeostasis of the organ and express the high-molecular-weight cytokeratins (CK)5 and CK14. A subpopulation of basal cells also expresses the p53-family-related gene *p63*. Neuroendocrine cells reside largely in the basal compartment, where they secrete neuroendocrine peptides such as synaptophysin and chromogranin A that support epithelial viability (Bonkhoff 1998; Abrahamsson 1999). Luminal cells are CK8/18-positive androgen-dependent columnar cells that lay above the basal layer facing the lumen of each tubule, where they secrete prostatic proteins. Several groups have also reported a fourth population of epithelial cells, named transit-amplifying cells, that coexpress basal (CK5) and luminal (CK8) markers (Isaacs and Coffey 1989), as well as prostate stem cell antigen (PSCA) in later stages (Tran et al. 2002).

There are conflicting views of the lineage hierarchy of the prostate and the relationship among the different cell types. The existence of PrSCs was determined by the observation that the rodent prostate can undergo up to 30 cycles of involution and regeneration in response to androgen cycling (English et al. 1987). Two prevalent models have emerged to explain how these PrSCs give rise to the different cell types of the prostate. The linear model proposes that the PrSCs reside among the CK5-positive basal cells, where they can differentiate into the double-positive intermediate/transit-amplifying population and then finally into the CK8-positive luminal phenotype (Fig. 1A) (Isaacs and Coffey 1989; Hudson et al.

2001). This is partially based on the preferential survival of basal cells after castration and their ability to repopulate the luminal compartment upon androgen add-back (Isaacs 1987; Bonkhoff and Remberger 1996). In vitro studies have also shown that PSCA⁻ basal cells can give rise to PSCA⁺ intermediate cells coexpressing basal and luminal cytokeratins (Tran et al. 2002) that can differentiate into luminal cells (Hudson et al. 2000; Xin et al. 2007). Other investigators have proposed a branched model of differentiation where the luminal and basal cells are in separate lineages, maintained by separate progenitor cells (Fig. 1B). This model is supported by observations that after castration, residual quiescent luminal cells remain in the rodent prostate that have high proliferative activity upon androgen add-back (Evans and Chandler 1987; Tsujimura et al. 2002). Immunohistochemical (IHC) studies have also shown that fetal and prepubertal human epithelia appear positive for all cytokeratins, and no intermediate phenotype was seen in situ (Wernert et al. 1987; Wang et al. 2001). In this model, the PrSCs lose CK5/14 or CK8/18 as they differentiate into committed progenitors and then luminal or basal cells, respectively.

Several theories exist for the origin of the neuroendocrine cells as well. Due to their location among the basal cells during development and adulthood, and their expression of basal cell cytokeratins, many believe that neuroendocrine cells are derived from the PrSCs or basal cells (Fig. 1B) (Kellokumpu-Lehtinen et al. 1979; Bonkhoff and Remberger 1996). Other investigators have suggested that these cells are derived from neural crest cells, because they secrete neuropeptides such as sero-

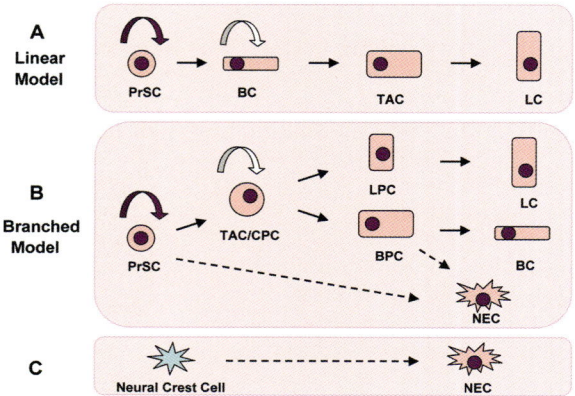

Figure 1. Models of prostate lineage hierarchy. (*A*) Schematic representation of the linear model of prostate differentiation where the basal, transit-amplifying, and luminal cells are in the same lineage. (*B*) Schematic representation of the branched model of prostate differentiation where the basal, luminal, and neuroendocrine cells are in separate lineages, but they are originally derived from one common stem cell. (*C*) An alternate theory for the origin of the prostate neuroendocrine cells suggests that they are derived from neural crest cells. (BC) Basal cell; (BPC) basal progenitor cell; (CPC) common progenitor cell; (LC) luminal cell; (LPC) luminal progenitor cell; (NEC) neuroendocrine cell; (TAC) transit-amplifying cell. (*Burgundy curved block arrow*) Self-renewal; (*white curved block arrow*) limited self-renewal.

tonin, thyroid-stimulating hormone (TSH)-like polypeptide, and somatostatin (Fig. 1C) (Aumuller et al. 1999). To study the PrSCs and better define their differentiation mechanism, recent efforts have focused on the development of efficient in vitro and in vivo assays.

APPROACHES FOR STUDYING PROSTATE STEM CELLS

Assays have been developed to grow multipotent cells outside their natural settings to identify stem-cell-specific markers for isolation, maintain their growth, and study their biology. Studies in the hematopoietic field have pioneered stem cell assay development, starting with the first demonstration of in vivo transplantation of hematopoietic stem cells (HSCs) into lethally irradiated mice and the first clonal growth of bone marrow cells in vitro (Ford et al. 1956; McCulloch and Till 1960; Bradley and Metcalf 1966). Many current neural and epithelial stem cell assays have been derived and adapted from these original methods.

In Vitro Colony-forming Unit Assay

Short-term in vitro assays are often preferred for studying differentiation and preliminary quantification of marker-based stem cell enrichments. Limiting dilution assays were used to measure frequencies of different hematopoietic cell subsets as they grew underneath a stromal layer as cobblestone area-forming cells (Ploemacher et al. 1989). Mammary and neural tissues can be cultured on solid substrate where they undergo differentiation in a hormone-, growth-factor-, or cell–cell interaction-regulated process (Reynolds and Weiss 1996; Dontu et al. 2003). Some of these methods were combined to develop in vitro prostate colony-forming unit (cfu) assays, in which dissociated adult prostate epithelial cells form colonies on top of feeder cells (Hudson et al. 2000; Uzgare et al. 2004; Lawson et al. 2007).

Primary mouse prostate cells have a colony-forming activity of 1 in 1000. Cells from castrated mice have a significant growth advantage over those from intact mice, indicating that stem or progenitor cells contribute to colony formation (Lawson et al. 2007). Mixing experiments using wild-type and transgenic green fluorescent protein (GFP) and *Discosoma* sp. red fluorescent protein (dsRed) prostate cells produce monochromatic colonies, suggesting that they are clonally derived. The majority of cells coexpress CK5 and CK8, indicating that they have an intermediate/transit-amplifying phenotype. One possibility for this observation is that the intermediate cells rather than the PrSCs from the prostate are responsible for giving rise to these colonies. Another explanation is that PrSCs give rise to the colonies, but without the proper niche signals, they differentiate into the intermediate phenotype. These colonies do not passage efficiently (Fig. 2A), further indicating that this assay promotes differentiation. Hudson et al. (2000) reported that two types of colonies can grow from primary human prostate cells: type I intermediate colonies and rare type II primitive basal cell colonies, suggesting that both stem and progenitor cells can thrive in this culture. These findings demonstrate that this assay could be used to quantitatively compare stem and progenitor cell enrichment techniques, but not to maintain them over time.

In Vitro Sphere Assay

A major advance in adult stem cell research was achieved by the discovery that undifferentiated multipotent neural cells can be grown and maintained in suspension using the neurosphere assay (Reynolds and Weiss 1996). Similar culture systems were developed for mammary stem cells, where it was shown that mammospheres can be serially passaged to demonstrate self-renewal activity in vitro (Dontu et al. 2003). Primary mouse prostate cells can also be maintained and passaged under floating conditions (Shi et al. 2007) where they exhibit increased expression of putative stem cell markers. To further replicate the stem cells' natural environment, a cell–matrix interaction component has also been introduced to this assay. Human prostate cancer cell lines and immortalized lines of normal cells can form spheroids in a semisolid matrix called Matrigel (Webber et al. 1997; Bello-DeOcampo et al. 2001; Lang et al. 2001). We have recently shown that a small fraction of primary prostate cells (1 in 1000) can also give rise to clonally derived spheres in Matrigel (Xin et al. 2007).

These spheres spontaneously develop a lineage hierarchy with the more immature p63+ basal cells localized to the outer rim and the PSCA+ transient-amplifying cells growing near the center. Sphere cells can be driven to terminal differentiation when allowed to adhere to a substratum and treated with testosterone, suggesting that the sphere-forming cells are multipotent. Spheres can be serially passaged for up to 12 generations, demonstrating their self-renewal activity in vitro (Fig. 2B). Dissociated cells from spheres can also be implanted in vivo to regenerate tubules containing both basal and luminal cells. This regeneration occurs at a low efficiency, suggesting that a part of the niche signaling is missing from the culture or that the spheres expand a progenitor population with less regenerative capacity.

In Vivo Regeneration Assay

Powerful in vivo engraftment and regeneration assays have been developed to characterize tissue stem cells from the blood (Harrison 1980) and the breast (Kordon and Smith 1998). The essence of these assays is that cells are put back into their natural environment already containing the appropriate structure and supportive stroma. Because the prostate cannot easily be cleared of epithelial cells, Cunha and Lung (1978) developed a prostate tissue fragment recombination assay that has been used to study prostate development and epithelial-mesenchymal interactions. In this procedure, fragments of rodent urogenital sinus mesenchyme (UGSM) are combined with adult prostate epithelial fragments and implanted under the kidney capsule of an immunodeficient mouse to generate prostatic tubules.

To use this assay for quantitative stem cell enrichment experiments, we modified the system to use single-cell-dissociated epithelial and UGSM cell preparations (Xin et

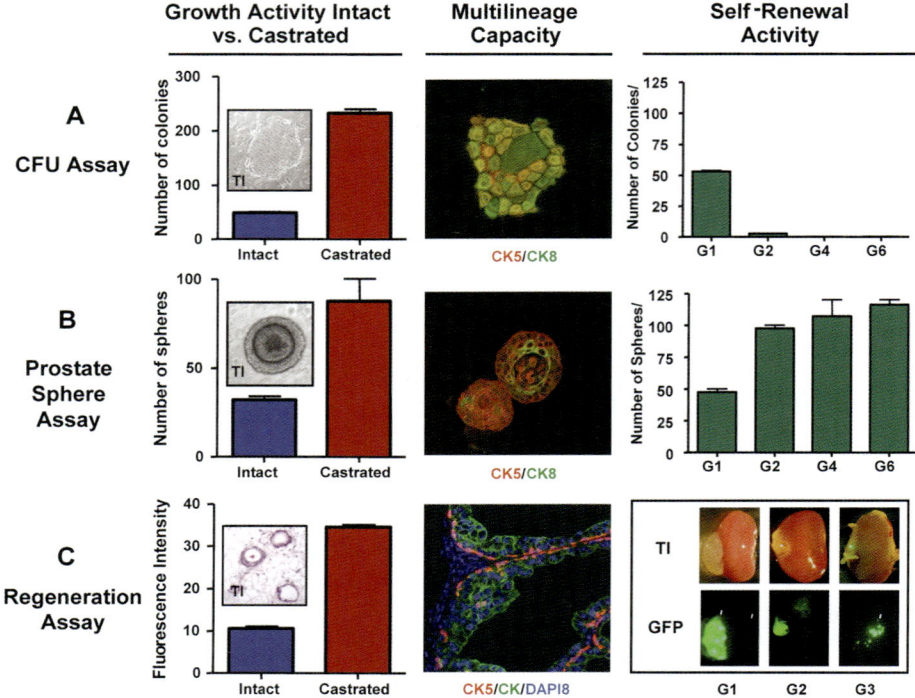

Figure 2. Summary of the prostate epithelial cell assays. (*A*) Graph represents the number of colonies that were formed after 5 × 10e4 prostate cells from castrated and intact C57BL/6 mice were plated on top of irradiated 3T3 cells in the cfu assay. Transillumination image (TI) shows a representative colony (*left*). Immunofluorescence stains show that colonies coexpress the luminal (CK8) and basal (CK5) cytokeratins (*middle*). Generation 1 (G1) colonies were dissociated and 5 × 10e4 cells were replated to test self-renewal. Rare colonies formed in generation 2 (G2), but did not passage further (*right panel*). (*B*) Graph represents the number of spheres that grew after 5 × 10e4 prostate cells from castrated and intact C57BL/6 mice were plated into Matrigel. TI image shows a representative sphere (*left*). Immunofluorescent image shows that the outer cells in spheres are CK5-positive, whereas the inner cells coexpress CK5 and CK8 (*middle*). Generation 1 (G1) spheres were dissociated and serially replated in equal numbers (5 × 10e4) for several generations (G2–G6) to test self-renewal (*right*). (*C*) Equal numbers (2 × 10e5) of prostate cells from castrated and intact transgenic mice harboring β-actin-driven GFPs were engrafted under the kidney capsule of SCID mice. Grafts were harvested after 8 weeks and the fluorescence intensities (× 10e6) of GFP signals were quantified by a CCD camera (*left*). Immunofluorescence staining shows CK5-positive cells located in the outer layer of the tubules and CK8-positive luminal cells facing the lumen (middle). Primary grafts were generated from 2 × 10e5 transgenic GFP⁺ cells plus 2 × 10e5 UGSM cells. After 8 weeks of regeneration, grafts were dissociated, and 2 × 10e5 cells were combined with fresh UGSM and transplanted into secondary and then tertiary recipients to show self-renewal in vivo. Transillumination (TI) and fluorescent (GFP) images show the grafts from generations 1, 2, and 3 (*right*).

al. 2003). Like Cunha's assay, the prostatic tubules derived from dissociated cell preparations display the same cellular composition as normal prostate tubules. Serial dilution experiments combining decreasing numbers of dissociated GFP⁺ prostate cells with noncolored wild-type cells revealed an average tubule-forming activity of approximately 1 in 2500. This number is lower than the 1 in 1000 cfu and sfu (sphere-forming unit) activity seen in the colony and sphere assays. The inconsistency could be an indicator that in addition to stem cells, progenitor cells can also give rise to colonies and spheres in vitro, whereas they cannot form tubules in vivo.

Mixing experiments produce only monochromatic tubules, suggesting that they are also clonally derived (Azuma et al. 2005; Xin et al. 2005). IHC analysis of these prostatic grafts reveals both CK5-positive basal cells in the outer layer of most tubules and CK8-positive luminal cells in the inner regions (Fig. 2C). Because all of the tubules are clonal, one can deduce that the cells giving rise to these ducts are multipotent. Comparison of the

regeneration activity of cells from castrated versus intact mice shows that the stem-cell-enriched fractions have higher growth capacity, just like in the in vitro assays (Fig. 2C) (D. Lawson, unpubl.).

Self-renewal activity has been demonstrated in HSCs (Lemischka et al. 1986) and mammary stem cells (Kordon and Smith 1998) using in vivo serial transplantation experiments. We took two different approaches to study PrSC self-renewal in our regeneration assay. Grafts generated from dissociated GFP⁺ prostate cells were harvested from primary recipients, dissociated into single cells, combined with fresh UGSM, and retransplanted into secondary and then tertiary recipients. Tubule formation was seen in all of the grafts, but the growth activity seemed to decrease with every passage (Fig. 2C). In a similar approach used to compare the in vivo self-renewal activity of the proximal versus distal prostate, Lynette Wilson's group saw more persistent self-renewal activity in cells from the proximal region, but they also observed a decrease in tubule formation over four rounds of transplantation (Goto et al. 2006). This decline could be interpreted as evidence that the cells giving rise to tubules do not have long-term self-renewal, but more likely, it is due to the technical shortcomings of this method.

In our second approach, we took advantage of the ability of the prostate to involute and regrow in the absence and presence of androgen. Dissociated cells from mice harboring a prostate-specific probasin promoter-driven luciferase (Pb-Luc) transgene (Xie et al. 2004) were used in our regeneration assay to allow visualization of the grafts throughout the experiment (Fig. 3A). The mice were imaged after 6 weeks using a charge-coupled device (CCD) camera to reveal a bright signal representing the outgrowth of the prostatic graft. The mice were then subjected to two rounds of androgen cycling, with CCD images taken at every involution and regrowth phase. The images show a cyclic decrease and increase in signal corresponding to the absence/presence of androgen (Fig. 3B). From sample mice taken at each stage, we confirmed that after androgen ablation, grafts had involuted and the luminal cells had undergone apoptosis, and after each regrowth, the tubules reformed to contain new luminal cells. This approach proves to be a unique alternative tool for the study of PrSC self-renewal in vivo.

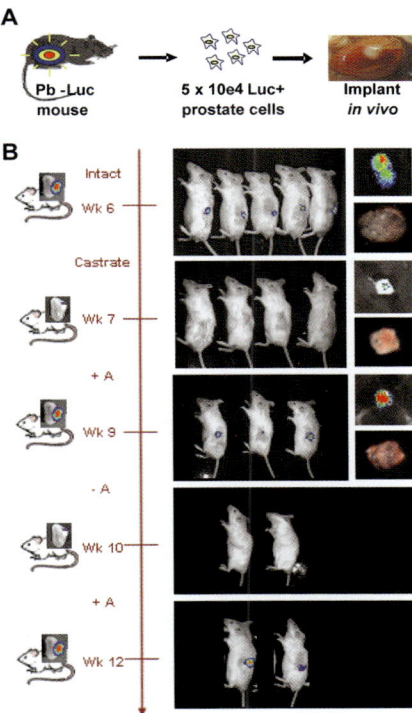

Figure 3. Development of in vivo prostate self-renewal assay. (*A*) Cells were dissociated from 8-week-old Pb-Luc mouse prostates; 5 × 10e4 cells were combined with 1 × 10e5 UGSM cells and implanted under the kidney capsule of SCID mice. (*B*) Luciferin was injected and CCD images were taken of the mice after 6 weeks of regeneration. CCD (*right top*) and transillumination (TI) (*right bottom*) images were also taken of a representative graft at this stage. Mice were then castrated and reimaged at 7 weeks, with a representative graft harvested for CCD and TI imaging. Mice were then subjected to two rounds of androgen cycling with CCD images taken after each regrowth and involution.

IDENTIFICATION AND ISOLATION OF THE PROSTATE STEM CELLS

Location

Initial studies showed that after androgen was given back to castrated rodents, DNA synthesis was only observed at the tips of the prostatic tubules during regeneration, suggesting that the stem cells might be in the region of the tubule distal to the urethra (Sugimura et al. 1986). It was later determined that those cells with high proliferation activity were actually transient-amplifying progenitor cells (Cunha et al. 1987; Tsujimura et al. 2002). Many stem cells have been identified by label-retention experiments, which exploit their slow-cycling, quiescent nature (Cotsarelis et al. 1990; Hong et al. 2001; Duvillie et al. 2003). [³H]thymidine or bromodeoxyuridine (BrdU) is incorporated into the DNA of all cells and gets diluted out of faster-cycling cells while remaining concentrated in slow-cycling cells. Because the prostate does not regenerate and involute under physiological conditions, Tsujimura et al. (2002) used androgen cycling in combination with BrdU labeling to identify a population of slow-cycling cells in the basal layer of prostate tubules proximal to the urethra. To functionally confirm that stem cells are enriched in this area, these authors showed that cells in the proximal region have a higher proliferation potential in vitro and are more capable of regenerating prostatic tubules in the dissociated prostate regeneration assay through several passages (Goto et al. 2006). In addition, stem-cell-associated telomerase activity is higher in the proximal region of the rodent prostate anterior lobe (Banerjee et al. 1998).

Isolation

The ability to isolate stem cells from tissue is crucial for the in depth study of their biology and involvement in development and disease. Weissman and colleagues developed methods to purify the HSCs by FACS (fluorescence-activated cell sorting) to identify the short-term repopulating HSCs based on defined cell surface marker expression (Spangrude et al. 1988). Since then, populations enriched for stem cell activity have been isolated from many adult tissues using various magnetic bead and FACS techniques (Shackleton et al. 2006; Stingl et al. 2006; Corti et al. 2007). We and other investigators first identified the glycosyl phosphatidylinositol–anchored cell surface protein Sca-1 as a marker that can enrich for prostate stem cell activity, using magnetic bead sorting methods combined with in vitro and in vivo readout assays (Burger et al. 2005; Xin et al. 2005).

Sca-1 had previously been identified on stem cells of various tissues, including the mammary, liver, skeletal muscle, and heart (Spangrude et al. 1988; Welm et al. 2002; Kim et al. 2005). Sca-1–positive cells are approximately threefold enriched in the proximal region of the prostate, where the PrSCs are thought to reside (Xin et al. 2005). Sca-1–enriched cell populations have a significantly higher tubule-forming activity in the dissociated prostate cell regeneration assay. Furthermore, Wilson and colleagues have demonstrated that the Sca-1–positive cells from the proximal region had even higher regenerative activity than the Sca-1–positive cells from the intermediate and distal regions (Burger et al. 2005). Using 7AAD-staining, DNA-staining, and pyronin-Y RNA-staining techniques, we have shown that the Sca-1–positive cell fraction is enriched for quiescent cells, as defined by low RNA and DNA staining (Xin et al. 2005).

In 2003, three different groups looked for molecular signatures of "stemness" by comparing the microarray expression profiles of embryonic, neural, and hematopoietic stem cells (Ivanova et al. 2002; Ramalho-Santos et al. 2002; Fortunel et al. 2003). The three independent lists of genes had only one in common: integrin α_6, also known as CD49f. A fraction of Sca-1–positive cells in the prostate express the stem cell markers CD49f and Bcl-2 (Burger et al. 2005; Lawson et al. 2007). Immunofluorescence analysis revealed that CD49f is strictly located on the basal surface of basal cells in the mouse prostate. Because Sca-1 is also present on some of the stromal and luminal cells, we predicted that combining this marker with CD49f would help to enrich for the basal cells of the proximal region. When the sphere-forming activity of FACS-sorted Lin(Ter119, CD45, CD31)⁻Sca-1⁺CD49f⁺ (LSC) cells were compared to unfractionated prostate cells subjected to the same conditions, there was a 50-fold enrichment for growth activity. In fact, only these LSC cells possess sphere-forming and in vitro self-renewal activity, and only LSC cells could give rise to prostatic tubules in vivo (Fig. 4) (Lawson et al. 2007).

Although using the LSC antigenic profile greatly enriches for PrSCs, only 1 in 44 LSC cells actually has sphere-forming and stem cell activity. To better study the

PrSCs, our efforts have focused on purifying the stem cell population further. Microarray-based experiments comparing the expression profiles of the LSC and non-LSC epithelial populations have produced several new marker candidates that are under evaluation for additional enrichment potential.

Human Prostate Stem Cells

Approaches used to identify human PrSCs have included similar functional evaluation of surface-marker-based fractionated cells as well as dye-efflux-based studies. Collins et al. (2001) isolated a subpopulation of basal cells expressing high levels of integrin $\alpha_2\beta_1$, based on their rapid adhesion to type I collagen. These cells have higher colony-forming activity and a greater ability to regenerate prostate-like glands in vivo than other basal cells. CD133 (prominin-1), a hematopoietic and neural stem cell marker, was found to be strictly expressed in the integrin $\alpha_2\beta_1^{hi}$ prostate cell fraction. The double-positive subpopulation has high in vitro growth activity as well as in vivo acinar structure regeneration potential, indicating that these markers enrich for stem cells (Richardson et al. 2004). The murine prostate stem cell marker CD49f has recently been found to also select for human prostate cells with high sphere-forming and self-renewal capabilities, which demonstrates that mouse and human stem cells share similarities (I. Garraway, unpubl.).

Several types of cancer cells and stem cells have been shown to express ATP-binding cassette (ABC) transporters associated with multidrug resistance. ABCG2 was

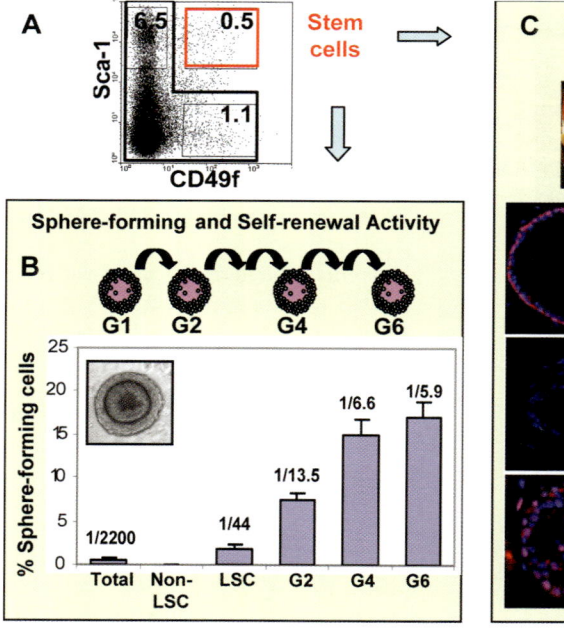

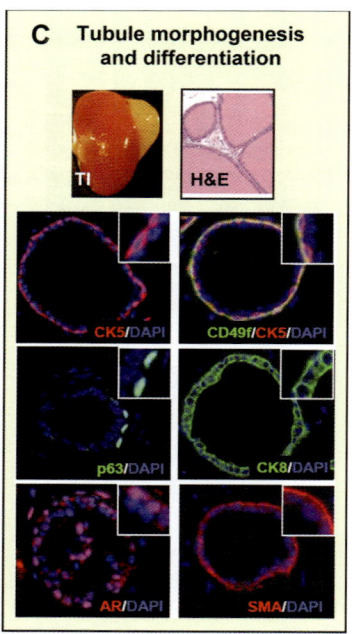

Figure 4. The LSC fraction enriches for stem-like activity. (*A*) FACS plot showing the LSC cell population (*red box*) and non-LSC cells of the prostate. (*B*) Graph shows the outgrowth of spheres from the total prostate, non-LSC, and LSC populations, followed by sphere growth from the second, fourth, and sixth generation (G2, G4, G6) of passaged LSC-derived spheres. (*C*) Transillumination image (TI) of a kidney with regenerated tissue (*top left*) and hematoxylin and eosin (H&E) stain of regenerated tubules (*top right*). Images show immunofluorescent staining of regenerated tubules for CK5, CD49f/CK5, p63, CK8, AR, and SMA. Sections are counterstained with DAPI. (Graph and stains reprinted, with permission, from Lawson et al. 2007 [© National Academy of Sciences].)

shown to be efficient at Hoechst dye efflux, creating a side population (SP) phenotype that has been used as an alternate approach to identify multipotent stem cells in bone marrow, skeletal muscle, and neural tissue (Bunting 2002) and more recently in the prostate (Bhatt et al. 2003). The SP cells of the prostate have slightly higher sphere-forming activity in vitro than the non-SP cells, although only by fourfold (Brown et al. 2007). Further characterization showed that only 1% of the basal cells were included in the SP fraction, and they expressed known stem cell genes such as *nestin, Bmi-1, hTeRT*, and *CD133* (Pascal et al. 2007). Although this evidence suggests that dye efflux methods have the potential to mark some primitive cells in the prostate and subdivide the basal fraction, these studies lack significant functional data.

HUMAN PROSTATE CANCER AND PROSTATE CANCER MODELS

Prostate cancer is predominantly a disease of aging, thought to progress in a step-like manner from prostatic intraepithelial neoplasia (PIN) lesions to local carcinoma and finally to metastatic disease. Many different genetic mutations have been identified in human prostate cancer over the years (Abate-Shen and Shen 2000). To recapitulate and study the disease, the most frequently found alterations have been used to create genetically engineered mouse models (Klein 2005). Our group has used the dissociated prostate regeneration assay for cancer studies, because a variety of individual or combinations of genetic alterations can be easily introduced into either the epithelial or mesenchymal compartment to drive transformation.

Alterations in the PTEN/AKT pathway are commonly seen in human prostate cancer as an activation of AKT or a loss of function in PTEN (Sun et al. 2001; Deocampo et al. 2003). Transgenic mice harboring a prostate-specific constitutively active AKT develop PIN lesions (Majumder et al. 2003), whereas conditional PTEN null mice develop prostate cancers that progress in a step-like fashion, as in the human disease (Wang et al. 2003; Ma et al. 2005). Using our regeneration assay, we also observe the growth of tubules containing PIN lesions upon the forced expression of myristoylated AKT1 and upon the inhibition of PTEN expression using short hairpin RNA (shRNA) (Xin et al. 2005). Alterations in androgen receptor signaling causing receptor hypersensitivity, promiscuity, or androgen-independent receptor *trans*-activation are also a common feature of prostate cancer (Feldman and Feldman 2001). When additional AR is introduced in addition to constitutively active AKT, regenerated grafts contain androgen-independent adenocarcinoma, resembling what is seen in the human disease (Xin et al. 2006).

Recently, there has been much excitement over Arul Chinnaiyan and colleagues' discovery of a class of translocations present in the majority of prostate cancers (Tomlins et al. 2005). This translocation is most commonly a result of the fusion of the androgen-responsive TMPRSS2 promoter and the coding region of the ETS family of transcription factors (ERG, ETV1, ETV4), which have previously been implicated in Ewing's sarcoma and other malignancies (Bohlander 2005; Kovar

2005). Other novel prostate-specific 5′ fusion partners have also been recently identified in prostate tumors with ETV1 outlier expression (Tomlins et al. 2007). Transgenic mice in which the overexpression of ERG is driven by the prostate-specific probasin promoter develop PIN lesions by 12–14 weeks of age (Tomlins et al. 2008). In recent experiments, we were able to recapitulate the PIN lesions seen in these transgenic mice by using lentivirus to introduce ERG1 into adult primary murine prostate cells and implanting these cells with UGSM under the kidney capsule of SCID (severe combined immunodeficiency disease) mice (Y. Zong, unpubl.).

Prostate cancer is unique in that it is often multifocal and heterogeneous, where several independent regions of the prostate display varying degrees of pathology. Heterogeneous genetic instability due to retroviral infection (Dong et al. 2007), perturbations in stroma, and "field effects" from global changes in the prostate (Harding and Theodorescu 2000) are three mechanisms by which this phenomenon can occur. Our dissociated prostate cell-regeneration model is a useful tool to recapitulate the multifocal heterogeneity of prostate cancer. We recently demonstrated that forced overexpression of fibroblast growth factor-10 (FGF-10) in the mesenchyme can drive the transformation of epithelial cells, resulting in pockets of high-grade PIN and carcinoma (Memarzadeh et al. 2007). A subset of these epithelial cancer cells remain tumorigenic upon transplantation despite removal of high FGF-10-expressing stromal cells. Paracrine FGF-10 signaling also leads to an increase in epithelial androgen receptor levels, which can synergize with AKT to advance the pathology.

THE CELL OF ORIGIN FOR PROSTATE CANCER

Cell of origin studies are fueled by the theory that not all cells are equally susceptible to transformation. Initial attempts to identify the cell(s) of origin for prostate cancer consisted of marker expression analysis of tumor samples, which only introduced controversy in the field. Most prostate cancers are adenocarcinomas that express markers associated with luminal epithelial cells such as CK8, AR, and prostate-specific antigen (PSA). Cells that express only basal markers such as CK5, CK14, and p63 are rarely observed in human cancers, leading to speculations that prostate cancer must initiate in a luminal progenitor or mature luminal cell that reacquired self-renewal (Nagle et al. 1987; Liu et al. 1999). Differentiated cells significantly outnumber primitive cells in all tissues. Therefore, the predominant presence of luminal cells in prostate cancer does not rule out a more immature cell as the original target for transformation.

Other studies have reported an expansion of intermediate cells in prostate cancer that coexpress both CK5 and CK8 (Verhagen et al. 1992). Reiter et al. (1998) have shown that a putative marker for late intermediate cells, PSCA, is often up-regulated in prostate cancer. These observations suggest that the disease may actually initiate in the transit-amplifying/intermediate progenitor cell population. An expansion of more primitive basal cells has also been observed in several cases. In the PTEN null mouse model where Cre deletes PTEN in both basal and luminal cells, there is a predominant expansion of the more primitive p63$^+$ basal cells, suggesting that PTEN deletion-mediated prostate cancer begins in the more stem-like cells (Wang et al. 2003). In this model, as well as in our AKT1-driven carcinoma model, there is a dramatic expansion of the Sca-1–positive stem-like cells, further suggesting that these cells are the preferred targets for transformation. In our FGF-10 paracrine model for multifocal prostate cancer, we also observe a dramatic response and expansion in the CK5 and p63$^+$ basal cells (Memarzadeh et al. 2007). All of these data suggest that the stem cell compartment can also serve as a responsive target for transformation.

One functional approach to identify the cell of origin is to control the expression of oncogenes or tumor suppressors by linking them to cell- or tissue-type-specific promoters in transgenic mice. When both p53 and RB (retinoblastoma) were conditionally knocked out in the prostate using probasin-driven Cre, adenocarcinomas were only observed in the proximal region of the prostate (Zhou et al. 2007). In the tips of the tubules, where the intermediate and more differentiated cells reside, only areas of mild dysplasia were observed. This observation supports the idea that the differentiation status of the target cell probably dictates the severity of the transformation, largely due to their self-renewal capacity. In a separate study, DiGiovanni et al. (2000) described a prostate cancer model where mice express human insulin-like growth factor-1 (IGF-1) in the basal cell compartment, under the bovine CK5 promoter, and develop a stepwise disease progressing from hyperplasia, through PIN lesions, to adenocarcinoma, as in human disease. Because the oncogene was driven exclusively by the CK5 promoter, this result clearly suggests that basal cells can serve as targets for transformation.

Another approach is to sort out defined cell populations of a given organ and subject them to known oncogenes to compare their relative susceptibility for transformation. Huntly et al. (2004) used this technique to show that *BCR-ABL* can only initiate chronic myelogenous leukemia (CML) in mice when expressed in HSCs and not in committed progenitors. In our initial attempts to investigate the tumor-forming capabilities of primitive and mature prostate cells, we compared the transforming capacity of Sca-1–positive and –negative cells after overexpressing AKT1 (Fig. 5) (Xin et al. 2005). Grafts from AKT-infected Sca-1–positive cells were ten times larger than the Sca-1–negative cell grafts. In addition, the Sca-1–enriched grafts contained tubules with PIN lesions, whereas the grafts from the Sca-1–negative cells contained mostly normal looking tubules. These data suggest that cells capable of initiating cancer exist within the PrSC-enriched Sca-1 fraction.

Since these early studies, identification of the LSC fraction of prostate cells has enabled us to get a much higher enrichment of PrSCs. Using the LSC-staining profile, we can also identify and isolate the more differentiated luminal (progenitor) cells and the stromal cells of the prostate. We have recently compared the susceptibility of these newly enriched prostate cell fractions to transformation from Akt1 overexpression, PTEN deletion, ERG1 overexpression, and enhanced stromal FGF-10 signaling. In all of these cases, the stem-cell-enriched LSC fraction was the most responsive target. Our preliminary data suggest that the stem or progenitor cell fraction of the prostate is a preferred target for several different types of oncogenes.

Studies in leukemia have shown that some fusion oncogenes (*MLL-ENL* and *MOZ-TIF2*) are more successful than others (*BCR-ABL*) at inducing transformation from committed myeloid progenitor cells, depending on their ability to confer self-renewal (Cozzio et al. 2003; Huntly et al. 2004). Several oncogenes and mutations found in prostate cancer, such as Bmi-1 and PTEN, have been linked to self-renewal (Stiles et al. 2004; Glinsky et al. 2005). One of our long-term goals is to test whether these specific oncogenes alone or in combination with others can induce transformation from prostate cells other than the stem cell fraction by inducing self-renewal.

Cancer Stem Cells in Prostate Cancer

Significant efforts have also been made to examine whether CSCs exist in prostate tumors and whether they are related to the PrSCs. Collins et al. (2005) isolated CD133$^+$/$\alpha_2\beta_1$ integrin^{hi}/CD44 cells from biopsies of human tumors and showed that these cells had higher growth capacity and invasiveness in vitro than their marker-negative counterparts. It remains to be determined whether these cells are truly CSCs, because their tumorigenic activity was not tested in vivo. Cells with varying tumorigenic activity have been isolated from prostate cancer cell lines and xenografts using CD44

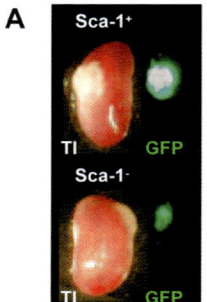
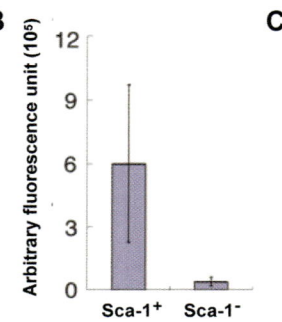
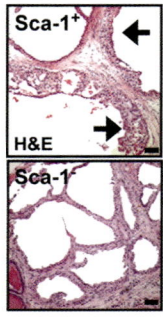

Figure 5. Prostate stem-cell-enriched fractions are efficient targets for transformation. (*A*) Transillumination (TI) and fluorescent (GFP) images of regenerated grafts from Sca-1$^+$ (*top*) and Sca-1$^-$ (*bottom*) cells infected with AKT-GFP lentivirus. (*B*) GFP fluorescent signals from the Sca-1-positive and -negative grafts were quantified. (*C*) Hematoxylin and eosin (H&E) staining of regenerated tissue from the AKT-infected Sca-1$^+$ (*top*) and Sca-1$^-$ (*bottom*) grafts. Arrows show areas of PIN lesion in the Sca-1$^+$ grafts. (Reprinted, with permission, from Xin et al. 2005 [© National Academy of Sciences].)

alone or in combination with the $\alpha_2\beta_1$ integrin[hi] profile, supporting the existence of a tumor cell hierarchy in the prostate (Patrawala et al. 2006, 2007). Although xenograft studies can be helpful, cell-line-based studies can be misleading due to cellular changes and selective pressures that can alter cancer cells in culture. Cells expressing embryonic stem cell (Oct4, Nanog, and Sox2) and early progenitor (CD44 and nestin) markers were found among immortalized primary human prostate cancer cells (Gu et al. 2007). These cells could clonally reconstitute the original tumor in mice to contain all three prostate cell lineages, demonstrating some CSC properties. Interestingly, all of the markers associated with putative prostate cancer stem cells so far are also known PrSC markers. Traditional cancer stem cell studies must still be performed, comparing the in vivo tumor-initiating potential of different cell populations sorted from primary tumors to identify CSCs in prostate cancer.

CONCLUSIONS

Characterizing the prostate cells that are more susceptible to transformation and capable of cancer maintenance represents an essential step to a better understanding and more efficient treatment of the disease. Conventional therapeutics target and kill proliferating cells, sparing the putative cancer stem cell fraction and allowing for the recurrence of the disease in many cases. To be successful, therapy must not only kill the proliferating tumor cells, but also eliminate or differentiate the CSCs. Elucidating the molecular circuitry that contributes to the maintenance of normal prostate stem cells can provide key insight into the molecular mechanisms of these unique cancer cells.

Elimination of CSCs could be approached by targeting therapies against specific self-renewal pathways. Notch and Sonic Hedgehog (Shh) pathways have been previously found to have important roles in the maintenance of several types of tissue, including prostate (Beachy et al. 2004; Karhadkar et al. 2004; Wang et al. 2006). Modulation of the Shh pathway with cyclopamine has also been shown to decrease the growth of medulloblastoma (Romer et al. 2004) and prostate cancer (Karhadkar et al. 2004) in mouse models. Now that we have the ability to isolate a highly enriched PrSC fraction, our immediate goal is to identify novel pathways that confer long-term self-renewal to these cells and test whether their modulation can control the progression of prostate cancer. We have recently identified *Bmi-1* as one such gene (R. Lukacs, unpubl.). Comparing the genetic expression profiles of tumorigenic cancer cells to normal PrSCs can aid in the identification of novel diagnostic markers and therapeutic targets that will spare the normal stem cells of the tissue. Alternatively, gaining a better understanding of the lineage hierarchy of the prostate and the mechanisms that control the differentiation process can reveal other potential targets with which differentiation of CSCs could be driven. The continual development of more precise and discriminative assays along with improvements in isolating highly purified stem cell fractions will be crucial for making progress in this field.

ACKNOWLEDGMENTS

We thank fellow lab members Stephanie Shelly and Houjin Cai for helpful suggestions. This work was supported by funds from the Prostate Cancer Foundation. O.N.W. is an investigator of the Howard Hughes Medical Institute. R.U.L. is supported by the California Institute for Regenerative Medicine training grant T1-00005. D.A.L was supported by the UCLA Graduate Student Dissertation Year fellowship grant. L.X. is supported by National Institutes of Health 1K99CA125937 Pathway to Independence. I.G. is supported by the Department of Defense Proposal (PC061088). S.M. has been supported by the NIH/National Cancer Institute, Clinical Scientist Training in Cancer Gene Medicine, K12-CA076905-09.

REFERENCES

Abate-Shen, C. and Shen, M.M. 2000. Molecular genetics of prostate cancer. *Genes Dev.* **14:** 2410–2434.

Abrahamsson, P.A. 1999. Neuroendocrine differentiation in prostatic carcinoma. *Prostate* **39:** 135–148.

Al-Hajj, M., Wicha, M.S., Benito-Hernandez, A., Morrison, S.J., and Clarke, M.F. 2003. Prospective identification of tumorigenic breast cancer cells. *Proc. Natl. Acad. Sci.* **100:** 3983–3988.

Aumuller, G., Leonhardt, M., Janssen, M., Konrad, L., Bjartell, A., and Abrahamsson, P.A. 1999. Neurogenic origin of human prostate endocrine cells. *Urology* **53:** 1041–1048.

Azuma, M., Hirao, A., Takubo, K., Hamaguchi, I., Kitamura, T., and Suda, T. 2005. A quantitative matrigel assay for assessing repopulating capacity of prostate stem cells. *Biochem. Biophys. Res. Commun.* **338:** 1164–1170.

Banerjee, P.P., Banerjee, S., Zirkin, B.R., and Brown, T.R. 1998. Lobe-specific telomerase activity in the intact adult brown Norway rat prostate and its regional distribution within the prostatic ducts. *Endocrinology* **139:** 513–519.

Beachy, P.A., Karhadkar, S.S., and Berman, D.M. 2004. Tissue repair and stem cell renewal in carcinogenesis. *Nature* **432:** 324–331.

Bello-DeOcampo, D., Kleinman, H.K., Deocampo, N.D., and Webber, M.M. 2001. Laminin-1 and $\alpha_6\beta_1$ integrin regulate acinar morphogenesis of normal and malignant human prostate epithelial cells. *Prostate* **46:** 142–153.

Bhatt, R.I., Brown, M.D., Hart, C.A., Gilmore, P., Ramani, V.A., George, N.J., and Clarke, N.W. 2003. Novel method for the isolation and characterisation of the putative prostatic stem cell. *Cytometry A* **54:** 89–99.

Blanpain, C., Horsley, V., and Fuchs, E. 2007. Epithelial stem cells: Turning over new leaves. *Cell* **128:** 445–458.

Bohlander, S.K. 2005. ETV6: A versatile player in leukemogenesis. *Semin. Cancer Biol.* **15:** 162–174.

Bonkhoff, H. 1998. Neuroendocrine cells in benign and malignant prostate tissue: Morphogenesis, proliferation, and androgen receptor status. *Prostate Suppl.* **8:** 18–22.

Bonkhoff, H. and Remberger, K. 1996. Differentiation pathways and histogenetic aspects of normal and abnormal prostatic growth: A stem cell model. *Prostate* **28:** 98–106.

Bonnet, D. and Dick, J.E. 1997. Human acute myeloid leukemia is organized as a hierarchy that originates from a primitive hematopoietic cell. *Nat. Med.* **3:** 730–737.

Bradley, T.R. and Metcalf, D. 1966. The growth of mouse bone marrow cells in vitro. *Aust. J. Exp. Biol. Med. Sci.* **44:** 287–299.

Brown, M.D., Gilmore, P.E., Hart, C.A., Samuel, J.D., Ramani, V.A., George, N.J., and Clarke, N.W. 2007. Characterization of benign and malignant prostate epithelial Hoechst 33342 side populations. *Prostate* **67:** 1384–1396.

Bruce, W.R. and Van Der Gaag, H. 1963. A quantitative assay for the number of murine lymphoma cells capable of proliferation in vivo. *Nature* **199:** 79–80.

Bunting, K.D. 2002. ABC transporters as phenotypic markers and functional regulators of stem cells. *Stem Cells* **20:** 11–20.

Burger, P.E., Xiong, X., Coetzee, S., Salm, S.N., Moscatelli, D., Goto, K., and Wilson, E.L. 2005. Sca-1 expression identifies stem cells in the proximal region of prostatic ducts with high capacity to reconstitute prostatic tissue. *Proc. Natl. Acad. Sci.* **102:** 7180–7185.

Collins, A.T., Habib, F.K., Maitland, N.J., and Neal, D.E. 2001. Identification and isolation of human prostate epithelial stem cells based on α(2)β(1)-integrin expression. *J. Cell Sci.* **114:** 3865–3872.

Collins, A.T., Berry, P.A., Hyde, C., Stower, M.J., and Maitland, N.J. 2005. Prospective identification of tumorigenic prostate cancer stem cells. *Cancer Res.* **65:** 10946–10951.

Corti, S., Nizzardo, M., Nardini, M., Donadoni, C., Locatelli, F., Papadimitriou, D., Salani, S., Del Bo, R., Ghezzi, S., Strazzer, S., et al. 2007. Isolation and characterization of murine neural stem/progenitor cells based on Prominin-1 expression. *Exp. Neurol.* **205:** 547–562.

Cotsarelis, G., Sun, T.T., and Lavker, R.M. 1990. Label-retaining cells reside in the bulge area of pilosebaceous unit: Implications for follicular stem cells, hair cycle, and skin carcinogenesis. *Cell* **61:** 1329–1337.

Cozzio, A., Passegue, E., Ayton, P.M., Karsunky, H., Cleary, M.L., and Weissman, I.L. 2003. Similar MLL-associated leukemias arising from self-renewing stem cells and short-lived myeloid progenitors. *Genes Dev.* **17:** 3029–3035.

Cunha, G.R. and Lung, B. 1978. The possible influence of temporal factors in androgenic responsiveness of urogenital tissue recombinants from wild-type and androgen-insensitive (Tfm) mice. *J. Exp. Zool.* **205:** 181–193.

Cunha, G.R., Donjacour, A.A., Cooke, P.S., Mee, S., Bigsby, R.M., Higgins, S.J., and Sugimura, Y. 1987. The endocrinology and developmental biology of the prostate. *Endocr. Rev.* **8:** 338–362.

Deocampo, N.D., Huang, H., and Tindall, D.J. 2003. The role of PTEN in the progression and survival of prostate cancer. *Minerva Endocrinol.* **28:** 145–153.

DiGiovanni, J., Bol, D.K., Wilker, E., Beltran, L., Carbajal, S., Moats, S., Ramirez, A., Jorcano, J., and Kiguchi, K. 2000. Constitutive expression of insulin-like growth factor-1 in epidermal basal cells of transgenic mice leads to spontaneous tumor promotion. *Cancer Res.* **60:** 1561–1570.

Dong, B., Kim, S., Hong, S., Das Gupta, J., Malathi, K., Klein, E.A., Ganem, D., Derisi, J.L., Chow, S.A., and Silverman, R.H. 2007. An infectious retrovirus susceptible to an IFN antiviral pathway from human prostate tumors. *Proc. Natl. Acad. Sci.* **104:** 1655–1660.

Dontu, G., Abdallah, W.M., Foley, J.M., Jackson, K.W., Clarke, M.F., Kawamura, M.J., and Wicha, M.S. 2003. In vitro propagation and transcriptional profiling of human mammary stem/progenitor cells. *Genes Dev.* **17:** 1253–1270.

Duvillie, B., Attali, M., Aiello, V., Quemeneur, E., and Scharfmann, R. 2003. Label-retaining cells in the rat pancreas: Location and differentiation potential in vitro. *Diabetes* **52:** 2035–2042.

English, H.F., Santen, R.J., and Isaacs, J.T. 1987. Response of glandular versus basal rat ventral prostatic epithelial cells to androgen withdrawal and replacement. *Prostate* **11:** 229–242.

Evans, G.S. and Chandler, J.A. 1987. Cell proliferation studies in the rat prostate. II. The effects of castration and androgen-induced regeneration upon basal and secretory cell proliferation. *Prostate* **11:** 339–351.

Feldman, B.J. and Feldman, D. 2001. The development of androgen-independent prostate cancer. *Nat. Rev. Cancer* **1:** 34–45.

Ford, C.E., Hamerton, J.L., Barnes, D.W., and Loutit, J.F. 1956. Cytological identification of radiation-chimaeras. *Nature* **177:** 452–454.

Fortunel, N.O., Out, H.H., Ng, H.H., Chen, J., Mu, X., Chevassut, T., Li, X., Joseph, M., Bailey, C., Hatzfeld, J.A., et al. 2003. Comment on " 'Stemness': Transcriptional profiling of embryonic and adult stem cells" and "a stem cell molecular signature." *Science* **302:** 393; author reply 393.

Glinsky, G.V., Berezovska, O., and Glinskii, A.B. 2005. Microarray analysis identifies a death-from-cancer signature predicting therapy failure in patients with multiple types of cancer. *J. Clin. Invest.* **115:** 1503–1521.

Goto, K., Salm, S.N., Coetzee, S., Xiong, X., Burger, P.E., Shapiro, E., Lepor, H., Moscatelli, D., and Wilson, E.L. 2006. Proximal prostatic stem cells are programmed to regenerate a proximal-distal ductal axis. *Stem Cells* **24:** 1859–1868.

Gu, G., Yuan, J., Wills, M., and Kasper, S. 2007. Prostate cancer cells with stem cell characteristics reconstitute the original human tumor in vivo. *Cancer Res.* **67:** 4807–4815.

Harding, M.A. and Theodorescu, D. 2000. Prostate tumor progression and prognosis. Interplay of tumor and host factors. *Urol. Oncol.* **5:** 258–264.

Harrison, D.E. 1980. Competitive repopulation: A new assay for long-term stem cell functional capacity. *Blood* **55:** 77–81.

Hewitt, H.B., Blake, E., and Proter, E.H. 1973. The effect of lethally irradiated cells on the transplantability of murine tumours. *Br. J. Cancer* **28:** 123–135.

Hill, R.P. 2006. Identifying cancer stem cells in solid tumors: Case not proven. *Cancer Res.* **66:** 1891–1895; discussion 1890.

Hong, K.U., Reynolds, S.D., Giangreco, A., Hurley, C.M., and Stripp, B.R. 2001. Clara cell secretory protein-expressing cells of the airway neuroepithelial body microenvironment include a label-retaining subset and are critical for epithelial renewal after progenitor cell depletion. *Am. J. Respir. Cell Mol. Biology* **24:** 671–681.

Hudson, D.L., O'Hare, M., Watt, F.M., and Masters, J.R. 2000. Proliferative heterogeneity in the human prostate: Evidence for epithelial stem cells. *Lab. Invest.* **80:** 1243–1250.

Hudson, D.L., Guy, A.T., Fry, P., O'Hare, M.J., Watt, F.M., and Masters, J.R. 2001. Epithelial cell differentiation pathways in the human prostate: Identification of intermediate phenotypes by keratin expression. *J. Histochem. Cytochem.* **49:** 271–278.

Huggins, C. and Hodges, C. 1941. Studies on prostatic cancer. I. The effects of castration, of estrogen and of androgen injection on serum phosphatases in metastatic carcinoma of the prostate. *Cancer Res.* **1:** 293–297.

Huntly, B.J., Shigematsu, H., Deguchi, K., Lee, B.H., Mizuno, S., Duclos, N., Rowan, R., Amaral, S., Curley, D., Williams, I.R., et al. 2004. MOZ-TIF2, but not BCR-ABL, confers properties of leukemic stem cells to committed murine hematopoietic progenitors. *Cancer Cell* **6:** 587–596.

Isaacs, J.T. 1987. Control of cell proliferation and cell death in the normal and neoplastic prostate: A stem cell model. In *Benign prostatic hyperplasia* (ed. C.H. Rogers et al.), pp. 85–94. National Institutes of Health, Bethesda, Maryland.

Isaacs, J.T. and Coffey, D.S. 1989. Etiology and disease process of benign prostatic hyperplasia. *Prostate* **2:** 33–50.

Ivanova, N.B., Dimos, J.T., Schaniel, C., Hackney, J.A., Moore, K.A., and Lemischka, I.R. 2002. A stem cell molecular signature. *Science* **298:** 601–604.

Jamieson, C.H., Ailles, L.E., Dylla, S.J., Muijtjens, M., Jones, C., Zehnder, J.L., Gotlib, J., Li, K., Manz, M.G., Keating, A., et al. 2004. Granulocyte-macrophage progenitors as candidate leukemic stem cells in blast-crisis CML. *N. Engl. J. Med.* **351:** 657–667.

Karhadkar, S.S., Bova, G.S., Abdallah, N., Dhara, S., Gardner, D., Maitra, A., Isaacs, J.T., Berman D.M., and Beachy, P.A. 2004. Hedgehog signalling in prostate regeneration, neoplasia and metastasis. *Nature* **431:** 707–712.

Kellokumpu-Lehtinen, P., Santti, R., and Pelliniemi, L.J. 1979. Early cytodifferentiation of human prostatic urethra and Leydig cells. *Anat. Rec.* **194:** 429–443.

Kim, C.F., Jackson, E.L., Woolfenden, A.E., Lawrence, S., Babar, I., Vogel, S., Crowley, D., Bronson, R.T., and Jacks, T. 2005. Identification of bronchioalveolar stem cells in normal lung and lung cancer. *Cell* **121:** 823–835.

Klein, R.D. 2005. The use of genetically engineered mouse models of prostate cancer for nutrition and cancer chemoprevention research. *Mutat. Res.* **576:** 111–119.

Kordon, E.C. and Smith, G.H. 1998. An entire functional mammary gland may comprise the progeny from a single cell. *Development* **125:** 1921–1930.

Kovar H. 2005. Context matters: The hen or egg problem in Ewing's sarcoma. *Semin. Cancer Biol.* **15:** 189–196.

Kyprianou, N. and Isaacs, J.T. 1988. Activation of programmed cell death in the rat ventral prostate after castration. *Endocrinology* **122:** 552–562.

Lang, S.H., Sharrard, R.M., Stark, M., Villette, J.M., and Maitland, N.J. 2001. Prostate epithelial cell lines form spheroids with evidence of glandular differentiation in three-dimensional Matrigel cultures. *Br. J. Cancer* **85:** 590–599.

Lawson, D.A., Xin, L., Lukacs, R.U., Cheng, D., and Witte, O.N. 2007. Isolation and functional characterization of murine prostate stem cells. *Proc. Natl. Acad. Sci.* **104:** 181–186.

Lemischka, I.R., Raulet, D.H., and Mulligan, R.C. 1986. Developmental potential and dynamic behavior of hematopoietic stem cells. *Cell* **45:** 917–927.

Litvinov, I.V., De Marzo, A.M., and Isaacs, J.T. 2003. Is the Achilles' heel for prostate cancer therapy a gain of function in androgen receptor signaling? *J. Clin. Endocrinol. Metab.* **88:** 2972–2982.

Liu, A.Y., True, L.D., LaTray, L., Ellis, W.J., Vessella, R.L., Lange, P.H., Higano, C.S., Hood, L., and van den Engh, G. 1999. Analysis and sorting of prostate cancer cell types by flow cytometry. *Prostate* **40:** 192–199.

Lois, C. and Alvarez-Buylla, A. 1993. Proliferating subventricular zone cells in the adult mammalian forebrain can differentiate into neurons and glia. *Proc. Natl. Acad. Sci.* **90:** 2074–2077.

Ma, X., Ziel-van der Made, A.C., Autar, B., van der Korput, H.A., Vermeij, M., van Duijn, P., Cleutjens, K.B., de Krijger, R., Krimpenfort, P., Berns, A. et al. 2005. Targeted biallelic inactivation of Pten in the mouse prostate leads to prostate cancer accompanied by increased epithelial cell proliferation but not by reduced apoptosis. *Cancer Res.* **65:** 5730–5739.

Majumder, P.K., Yeh, J.J., George, D.J., Febbo, P.G., Kum, J., Xue, Q., Bikoff, R., Ma, H., Kantoff, P.W., Golub, T.R., et al. 2003. Prostate intraepithelial neoplasia induced by prostate restricted Akt activation: The MPAKT model. *Proc. Natl. Acad. Sci.* **100:** 7841–7846.

McCulloch, E.A. and Till, J.E. 1960. The radiation sensitivity of normal mouse bone marrow cells, determined by quantitative marrow transplantation into irradiated mice. *Radiat. Res.* **13:** 115–125.

Memarzadeh, S., Xin, L., Mulholland, D.J., Mansukhani, A., Wu, H., Teitell, M.A., and Witte, O.N. 2007. Enhanced paracrine FGF10 expression promotes formation of multifocal prostate adenocarcinoma and an increase in epithelial androgen receptor. *Cancer Cell* **12:** 572–585.

Nagle, R.B., Ahmann, F.R., McDaniel, K.M., Paquin, M.L., Clark, V.A., and Celniker, A. 1987. Cytokeratin characterization of human prostatic carcinoma and its derived cell lines. *Cancer Res.* **47:** 281–286.

Pascal, L.E., Oudes, A.J., Petersen, T.W., Goo, Y.A., Walashek, L.S., True, L.D., and Liu, A.Y. 2007. Molecular and cellular characterization of ABCG2 in the prostate. *BMC Urol.* **7:** 6.

Patrawala, L., Calhoun, T., Schneider-Broussard, R., Li, H., Bhatia, B., Tang, S., Reilly, J.G., Chandra, D., Zhou, J., Claypool, K., et al. 2006. Highly purified CD44+ prostate cancer cells from xenograft human tumors are enriched in tumorigenic and metastatic progenitor cells. *Oncogene* **25:** 1696–1708.

Patrawala, L., Calhoun-Davis, T., Schneider-Broussard, R., and Tang, D.G. 2007. Hierarchical organization of prostate cancer cells in xenograft tumors: the CD44+α2β1+ cell population is enriched in tumor-initiating cells. *Cancer Res.* **67:** 6796–6805.

Peters, L.J. and Hewitt, H.B. 1974. The influence of fibrin formation on the transplantability of murine tumour cells: Implications for the mechanism of the Revesz effect. *Br. J. Cancer* **29:** 279–291.

Ploemacher, R.E., van der Sluijs, J.P., Voerman, J.S., and Brons, N.H. 1989. An in vitro limiting-dilution assay of long-term repopulating hematopoietic stem cells in the mouse. *Blood* **74:** 2755–2763.

Ramalho-Santos, M., Yoon, S., Matsuzaki, Y., Mulligan, R.C., and Melton, D.A. 2002. "Stemness": Transcriptional profiling of embryonic and adult stem cells. *Science* **298:** 597–600.

Reiter, R.E., Gu, Z., Watabe, T., Thomas, G., Szigeti, K., Davis, E., Wahl, M., Nisitani, S., Yamashiro, J., Le Beau, M.M., et al. 1998. Prostate stem cell antigen: A cell surface marker overexpressed in prostate cancer. *Proc. Natl. Acad. Sci.* **95:** 1735–1740.

Reya, T., Morrison, S.J., Clarke, M.F., and Weissman, I.L. 2001. Stem cells, cancer, and cancer stem cells. *Nature* **414:** 105–111.

Reynolds, B.A. and Weiss, S. 1996. Clonal and population analyses demonstrate that an EGF-responsive mammalian embryonic CNS precursor is a stem cell. *Dev. Biol.* **175:** 1–13.

Richardson, G.D., Robson, C.N., Lang, S.H., Neal, D.E., Maitland, N.J., and Collins, A.T. 2004. CD133, a novel marker for human prostatic epithelial stem cells. *J. Cell Sci.* **117:** 3539–3545.

Romer, J.T., Kimura, H., Magdaleno, S., Sasai, K., Fuller, C., Baines, H., Connelly, M., Stewart, C.F., Gould, S., Rubin, L.L., et al. 2004. Suppression of the Shh pathway using a small molecule inhibitor eliminates medulloblastoma in Ptc1+/−p53−/− mice. *Cancer Cell* **6:** 229–240.

Shackleton, M., Vaillant, F., Simpson, K.J., Stingl, J., Smyth, G.K., Asselin-Labat, M.L., Wu, L., Lindeman, G.J., and Visvader, J.E. 2006. Generation of a functional mammary gland from a single cell. *Nature* **439:** 84–88.

Shi, X., Gipp, J., and Bushman, W. 2007. Anchorage-independent culture maintains prostate stem cells. *Dev. Biol.* **312:** 396–406.

Singh, S.K., Hawkins, C., Clarke, I.D., Squire, J.A., Bayani, J., Hide, T., Henkelman, R.M., Cusimano, M.D., and Dirks, P.B. 2004. Identification of human brain tumour initiating cells. *Nature* **432:** 396–401.

Southam, C.M., Burchenal, J.H., Clarkson, B., Tanzi, A., Mackey, R., and McComb, V. 1969. Heterotransplantability of human cell lines derived from leukemia and lymphomas into immunologically tolerant rats. *Cancer* **24:** 211–222.

Spangrude, G.J., Heimfeld, S., and Weissman, I.L. 1988. Purification and characterization of mouse hematopoietic stem cells. *Science* **241:** 58–62.

Stiles, B., Groszer, M., Wang, S., Jiao, J., and Wu, H. 2004. PTENless means more. *Dev. Biol.* **273:** 175–184.

Stingl, J., Eirew, P., Ricketson, I., Shackleton, M., Vaillant, F., Choi, D., Li, H.I., and Eaves, C.J. 2006. Purification and unique properties of mammary epithelial stem cells. *Nature* **439:** 993–997.

Sugimura, Y., Cunha, G.R., and Donjacour, A.A. 1986. Morphogenesis of ductal networks in the mouse prostate. *Biol. Reprod.* **34:** 961–971.

Sun, M., Wang, G., Paciga, J.E., Feldman, R.I., Yuan, Z.Q., Ma, X.L., Shelley, S.A., Jove, R., Tsichlis, P.N., Nicosia, S.V., et al. 2001. AKT1/PKBα kinase is frequently elevated in human cancers and its constitutive activation is required for oncogenic transformation in NIH3T3 cells. *Am. J. Pathol.* **159:** 431–437.

Till, J.E. and McCulloch, E. 1961. A direct measurement of the radiation sensitivity of normal mouse bone marrow cells. *Radiat. Res.* **14:** 213–222.

Tomlins, S.A., Rhodes, D.R., Perner, S., Dhanasekaran, S.M., Mehra, R., Sun, X.W., Varambally, S., Cao, X., Tchinda, J., Kuefer, R., et al. 2005. Recurrent fusion of TMPRSS2 and ETS transcription factor genes in prostate cancer. *Science* **310:** 644–648.

Tomlins, S.A., Laxman, B., Dhanasekaran, S.M., Helgeson, B.E., Cao, X., Morris, D.S., Menon, A., Jing, X., Cao, Q., Han, B., et al. 2007. Distinct classes of chromosomal rearrangements create oncogenic ETS gene fusions in prostate cancer. *Nature* **448:** 595–599.

Tomlins, S.A., Laxman, B., Varambally, S., Cao, X., Yu, J., Helgeson, B.E., Cao, Q., Prensner, J.R., Rubin, M.A., Shah, R.B., et al. 2008. Role of the TMPRSS2-ERG gene fusion in prostate cancer. *Neoplasia* **10:** 177–188.

Tran, C.P., Lin, C., Yamashiro, J., and Reiter, R.E. 2002.

Prostate stem cell antigen is a marker of late intermediate prostate epithelial cells. *Mol. Cancer Res.* **1:** 113–121.

Tsujimura, A., Koikawa, Y., Salm, S., Takao, T., Coetzee, S., Moscatelli, D., Shapiro, E., Lepor, H., Sun, T.T., and Wilson, E.L. 2002. Proximal location of mouse prostate epithelial stem cells: A model of prostatic homeostasis. *J. Cell Biol.* **157:** 1257–1265.

Uzgare, A.R., Xu, Y., and Isaacs, J.T. 2004. In vitro culturing and characteristics of transit amplifying epithelial cells from human prostate tissue. *J. Cell Biochem.* **91:** 196–205.

Verhagen, A.P., Ramaekers, F.C., Aalders, T.W., Schaafsma, H.E., Debruyne, F.M., and Schalken, J.A. 1992. Colocalization of basal and luminal cell-type cytokeratins in human prostate cancer. *Cancer Res.* **52:** 6182–6187.

Virchow, R. 1855. Editorial. In *Archiv fuer pathologisce anatomie und physiologie und klinische medizin*, p. 23.

Wang, Y., Hayward, S., Cao, M., Thayer, K., and Cunha, G. 2001. Cell differentiation lineage in the prostate. *Differentiation* **68:** 270–279.

Wang, S., Gao, J., Lei, Q., Rozengurt, N., Pritchard, C., Jiao, J., Thomas, G.V., Li, G., Roy-Burman, P., Nelson, P.S., et al. 2003. Prostate-specific deletion of the murine Pten tumor suppressor gene leads to metastatic prostate cancer. *Cancer Cell* **4:** 209–221.

Wang, X.D., Leow, C.C., Zha, J., Tang, Z., Modrusan, Z., Radtke, F., Aguet, M., de Sauvage, F.J., and Gao, W.Q. 2006. Notch signaling is required for normal prostatic epithelial cell proliferation and differentiation. *Dev. Biol.* **290:** 66–80.

Webber, M.M., Bello, D., Kleinman, H.K., and Hoffman, M.P. 1997. Acinar differentiation by non-malignant immortalized human prostatic epithelial cells and its loss by malignant cells. *Carcinogenesis* **18:** 1225–1231.

Welm, B.E., Tepera, S.B., Venezia, T., Graubert, T.A., Rosen, J.M., and Goodell, M.A. 2002. Sca-1[pos] cells in the mouse mammary gland represent an enriched progenitor cell population. *Dev. Biol.* **245:** 42–56.

Wernert, N., Seitz, G., and Achtstatter, T. 1987. Immunohistochemical investigation of different cytokeratins and vimentin in the prostate from the fetal period up to adulthood and in prostate carcinoma. *Pathol. Res. Pract.* **182:** 617–626.

Wicha, M.S., Liu, S., and Dontu, G. 2006. Cancer stem cells: An old idea—A paradigm shift. *Cancer Res.* **66:** 1883–1890; discussion 1895–1896.

Xie, X., Luo, Z., Slawin, K.M., and Spencer, D.M. 2004. The EZC-prostate model: Noninvasive prostate imaging in living mice. *Mol. Endocrinol.* **18:** 722–732.

Xin, L., Ide, H., Kim, Y., Dubey, P., and Witte, O.N. 2003. In vivo regeneration of murine prostate from dissociated cell populations of postnatal epithelia and urogenital sinus mesenchyme. *Proc. Natl. Acad. Sci.* (suppl. 1) **100:** 11896–11903.

Xin, L., Lawson, D.A., and Witte, O.N. 2005. The Sca-1 cell surface marker enriches for a prostate-regenerating cell subpopulation that can initiate prostate tumorigenesis. *Proc. Natl. Acad. Sci.* **102:** 6942–6947.

Xin, L., Teitell, M.A., Lawson, D.A., Kwon, A., Mellinghoff, I.K., and Witte, O.N. 2006. Progression of prostate cancer by synergy of AKT with genotropic and nongenotropic actions of the androgen receptor. *Proc. Natl. Acad. Sci.* **103:** 7789–7794.

Xin, L., Lukacs, R.U., Lawson, D.A., Cheng, D., and Witte, O.N. 2007. Self-renewal and multilineage differentiation in vitro from murine prostate stem cells. *Stem Cells* **25:** 2760–2769.

Zhou, Z., Flesken-Nikitin, A., and Nikitin, A.Y. 2007. Prostate cancer associated with p53 and Rb deficiency arises from the stem/progenitor cell-enriched proximal region of prostatic ducts. *Cancer Res.* **67:** 5683–5690.

Role of β-catenin in Epidermal Stem Cell Expansion, Lineage Selection, and Cancer

F.M. WATT[*†] AND C.A. COLLINS[*]

*Wellcome Trust Centre for Stem Cell Research, University of Cambridge, Cambridge CB2 1QR, United Kingdom;
†Cancer Research UK Cambridge Research Institute, Li Ka Shing Centre, Cambridge, CB2 0RE, United Kingdom

The mammalian epidermis is an excellent model with which to analyze the factors that regulate adult stem cell renewal, lineage selection, and tumor formation. One of the key regulators of all three processes is β-catenin, the main cytoplasmic effector of the canonical Wnt signaling pathway. In this chapter, we review some of the ways in which β-catenin exerts its effects on cultured human epidermal cells and in genetically modified mice. We highlight the importance of the timing and level of activation and discuss some of the pathways activated downstream from β-catenin. Finally, we demonstrate the importance of Lef/Tcf-independent β-catenin signaling through interaction with the vitamin D receptor.

The epidermis provides a protective interface between the body and the environment. The main epidermal cell type is an epithelial cell, the keratinocyte. The major structures of the epidermis are the interfollicular epidermis (IFE), hair follicles (HFs), sweat glands, and sebaceous glands (SGs). In human epidermis, the sweat glands are present in both hair-bearing and hairless regions, whereas in mouse epidermis, the sweat glands are restricted to the paws. Because most in vivo studies are performed on the mouse, we will not refer to the sweat glands further in this chapter (Fig. 1).

To fulfill their functions, keratinocytes within the epidermis undergo terminal differentiation, resulting ultimately in cell death. Thus, the outermost, cornified, layers of the IFE, which form a physical barrier to penetration by microorganisms and prevent water loss, are anucleate cells filled with heavily cross-linked proteins and lipids. The cells of the hair shaft are also dead cells that are replaced during the hair growth cycle. Furthermore, during terminal differentiation, SG cells burst, releasing their lipid content onto the hairs and the surface of the skin to provide lubrication.

Because the terminally differentiated cells of the epidermis are dead cells and are continually shed from the skin surface, it was appreciated many decades ago that the tissue must be maintained by proliferation of less-differentiated cells that both replenish themselves and the terminally differentiated cells. Thus, the epidermis is one of the tissues in which it has long been acknowledged that there resides a stem cell compartment (Hall and Watt 1989). Our current view is that there are distinct pools of stem cells in discrete locations within the HF, IFE, and SG

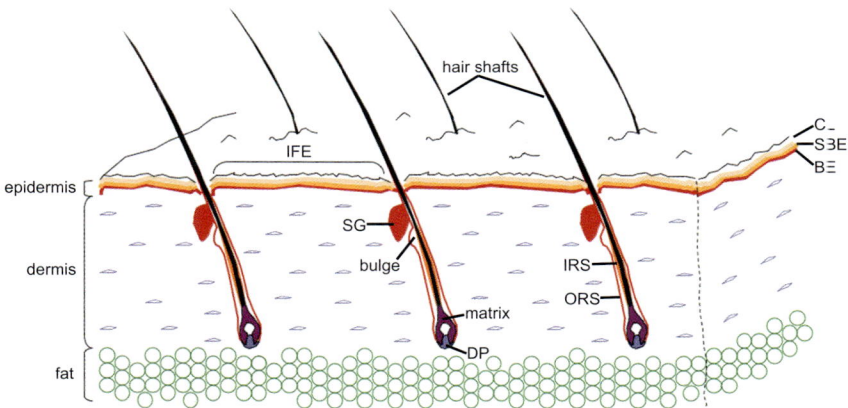

Figure 1. Organization of adult mouse epidermis. The epidermis is shown overlying the dermis and fat layers of the skin. (IFE) Interfollicular epidermis, comprising the BE, SBE, and CL. (BE) Basal epidermal layer; (SBE) suprabasal epidermal layers; (CL) cornified layers; (SG) sebaceous gland. Different regions of the HFs are shown, including the outer root sheath (ORS) and inner root sheath (IRS). (DP) Dermal papilla, the specialized mesenchymal cells at the base of the HFs.

(Owens and Watt 2003). When the epidermis is undamaged, each stem cell pool is responsible for maintaining a restricted number of differentiated lineages, such that IFE stem cells give rise to the differentiated lineage of the IFE, SG stem cells to the differentiated lineage of the SG, and HF stem cells to the eight lineages of the HF. However, when the epidermis is damaged or subject to genetic manipulation, stem cells in any location can give rise to any of the differentiated lineages of the epidermis (Watt et al. 2006; Jones et al. 2007).

The skin is an excellent tissue with which to investigate the factors that regulate adult stem cell renewal, lineage selection, and tissue assembly. In addition, it has long provided an experimental model for investigating cancer, from the earliest initiating events through to metastasis. In this chapter, we review the tools that are available to study these processes and some of the lessons that we have learned. One of the striking conclusions is that the principles of stem cell regulation in the epidermis can be applied much more widely to other adult and embryonic stem cells in a variety of organisms.

THE TOOL KIT

We have taken two complementary approaches to identify and manipulate epidermal stem cells: cultivation of human cells and generation of transgenic mice. The first approach is based on the discovery, in 1975, that cells isolated from human epidermis can be grown at clonal density and will form confluent multilayered sheets, in which proliferation occurs in the basal layer and IFE differentiation occurs in the suprabasal layers (Rheinwald and Green 1975; Watt 1998). Sheets of autologous cultured human epidermis can provide a permanent and functional IFE used to replace skin lost through burn injury. This led to the conclusion that clonal analysis of human epidermal cells could be used as a readout of stem cell activity (Barrandon and Green 1987; Jones and Watt 1993).

Different types of clonal growth assays are used, but in each case, stem cells are assigned to the category of cells capable of generating large numbers of both undifferentiated and differentiated progeny, whereas cells that show limited self-renewal ability and have a high probability of undergoing terminal differentiation are usually referred to as transit-amplifying cells (Jones et al. 2007). In vivo assays of function have included reconstitution of IFE following grafting of human keratinocytes into nude mice (Jones et al. 1995). The ability of the progeny of a single epidermal cell to reconstitute IFE, HF, and SG has been demonstrated for rodents, although not, so far, for humans (Blanpain et al. 2004; Claudinot et al. 2005). Retroviral vectors have been used to manipulate gene expression in human keratinocytes, making use of the high infection efficiency and minimal selection pressure involved (Gerrard et al. 1993; Levy et al. 1998).

One limitation of studies with cultured human keratinocytes has been that only differentiation along the IFE lineage can be examined. However, this is changing, with the discovery that cells immortalized from human SG can generate progeny that select the IFE and SG differentiation pathways in culture or after grafting into nude mice (Lo

Celso et al. 2008). Thus, in addition to investigating the factors that trigger exit from the stem cell compartment, we can begin to explore, at the single-cell level, the factors that regulate lineage selection (Lo Celso et al. 2008).

Studies of the stem cell compartment in mouse epidermis have been facilitated by characterization of promoters that drive transgene expression to different subpopulations of cells (Vassar et al. 1989; Carroll et al. 1993; Greenhalgh et al. 1993). The promoter that is used most frequently is the keratin 14 (K14) promoter, which targets basal layer cells, including stem cells, in the HF, SG, and IFE (Vassar et al. 1989). More recent refinements have been to selectively target the reservoir of stem cells in the HF bulge, via the K15 promoter (Liu et al. 2003) and to use inducible transgenes, so that the location and timing of transgene expression or deletion can be controlled, for example, by application of estrogen or progesterone analogs to the skin (Pelengaris et al. 1999; Vasioukhin et al. 1999; Wang et al. 1999). Recent studies show that it is possible to regulate the level of transgene expression by applying different concentrations of an inducing agent (Silva-Vargas et al. 2005).

Using keratinocyte cultures or transgenic mice as an assay system, a number of different markers for epidermal stem cells and the different differentiated lineages have been identified over the years. The first cell surface marker used to enrich for human epidermal stem cells was expression of high levels of integrin extracellular matrix receptors (Jones and Watt 1993; Jones et al. 1995); integrins now turn out to be a useful marker of stem cells in a range of other tissues, such as breast (Shackleton et al. 2006). The introduction of whole-mount preparations, for both human and mouse epidermis, has provided information about the spatial organization of cells in the different compartments and facilitates gathering of quantitative data (Jensen et al. 1999; Braun et al. 2003; Silva-Vargas et al. 2005). Finally, techniques for generating spontaneously immortalized mouse keratinocyte lines and for culturing primary adult mouse keratinocytes at clonal density have improved in recent years (Romero et al. 1999; Silva-Vargas et al. 2005; Wu and Morris 2005), providing the opportunity for direct comparisons between the behavior of mouse epidermal stem cells in culture and in vivo.

THE WNT PATHWAY

The pathway that is of central importance in regulating epidermal stem cell renewal and lineage selection is the Wnt pathway. The key components of the Wnt signaling pathway are summarized schematically in Figure 2 (Klaus and Birchmeier 2008). In the absence of a Wnt ligand, cytoplasmic β-catenin is rapidly degraded (Fig. 2a). However, when Wnt binds to cell surface receptors, the β-catenin destruction complex is inactivated. Thereafter, β-catenin accumulates in the cytoplasm and then translocates to the nucleus. In the nucleus, β-catenin forms a transcriptionally active complex with the Lef and Tcf transcription factors by displacing Groucho transcriptional repressors and interacting with coactivators such as CBP and Pygopous (Fig. 2b).

Within the epidermis, there are several ways in which the Wnt pathway can be regulated, one of which is via differential expression of pathway components. Different

a

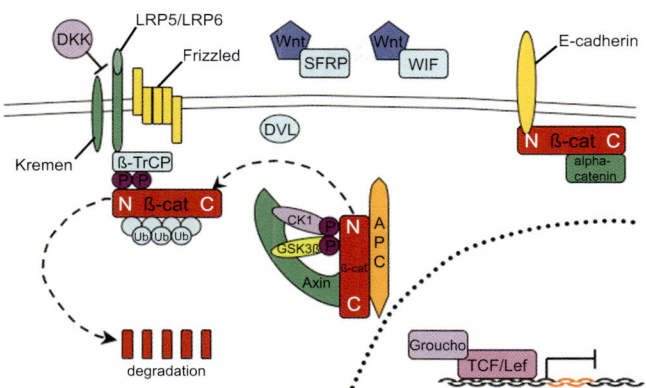

b

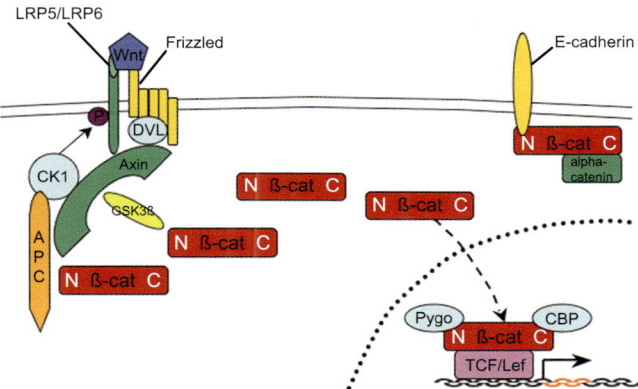

Figure 2. Diagram of the canonical Wnt signaling pathway. (*a*) In the absence of Wnt, the pathway is inactive; (*b*) the pathway is activated when Wnt binds to its receptor complex. (*Two parallel lines*) Plasma membrane; (*dotted line*) nucleus. (DKK) Dickkopf; (LRP) LDL-related receptor protein; (SFRP) secreted Frizzled-related protein; (WIF) Wnt inhibitory factor 1; (DVL) Dishevelled; (β-TrCP) β-transducin repeat-containing protein); (CK1) casein kinase 1; (GSK3β) glycogen synthase kinase 3β; (APC) adenomatous polyposis coli; (TCF) T-cell factor; (Lef) lymphoid enhancer factor; (Pygo) Pygopus; (CBP) CREB-binding protein; (P) phosphorylation; (Ub) ubiquitylation; (N) amino terminus; (C) carboxyl terminus. (Redrawn from Klaus and Birchmeier 2008.)

members of the Wnt family are expressed in specific subsets of cells in developing and adult epidermis; expression of *Frizzled* genes and Wnt antagonists is also dynamically regulated (Reddy et al. 2001, 2004; Sick et al. 2006). Lef1 and Tcf3, the main transcriptional effectors of the Wnt pathway in the epidermis, are expressed in different cell subpopulations (DasGupta and Fuchs 1999; Merrill et al. 2001). In the absence of Wnt, Tcf3 acts as a transcriptional repressor: Tcf3-repressed genes include transcriptional regulators of the IFE, SG, and HF lineages (Alonso and Fuchs 2003; Nguyen et al. 2006). Finally, in addition to its role as the key cytoplasmic effector of the Wnt pathway, β-catenin is a component of intercellular adhesive junctions, where it binds to the cytoplasmic domain of E-cadherin via the same amino acid residues as required for binding to adenomatous polyposis coli (APC) (Fig. 2) (Huelsken et al. 1994). Although β-catenin is not essential for adherens junction formation in the epidermis (Huelsken et al. 2001), due in part to coexpression with the related protein plakoglobin/γ-catenin, the ability of E-cadherin to bind β-catenin can lead to depletion of the pool of β-catenin available for Wnt signaling (Zhu and Watt 1999).

β-CATENIN IN CULTURED KERATINOCYTES

Our laboratory's interest in β-catenin originated not from its role in the Wnt pathway, but from its role in intercellular adhesion. We had previously established that integrin-mediated cell–extracellular matrix adhesion is a negative regulator of epidermal terminal differentiation (Adams and Watt 1989; Watt 2002) and found that cell–cell adhesion influences integrin expression and localization (Hodivala and Watt 1994). We therefore became interested in whether cell–cell adhesion regulates terminal differentiation. To study this, we infected human keratinocytes with a retroviral vector encoding the extracellular domain of H-2Kd and the transmembrane and cytoplasmic domains of E-cadherin (H-2Kd–E-cad), in order to inhibit adherens junction assembly. As a control, we deleted the binding site for β-catenin and plakoglobin (H-2Kd–E-cadΔC25) (Zhu and Watt 1996). As anticipated, cell–cell adhesion and stratification were inhibited. However, unexpectedly, the dominant-negative mutant had an inhibitory effect on keratinocyte proliferation and stimulated terminal differentiation. Terminal differentiation was stimulated even under conditions in which intercellular adhesion was prevented by disaggregating keratinocytes and holding them in suspension (Fig. 3a) (Zhu and Watt 1996).

In an attempt to discover why overexpression of the E-cadherin cytoplasmic domain could stimulate differentiation of disaggregated cells, we began to study β-catenin. We found that the subpopulation of cultured human keratinocytes with high proliferative potential, the putative epidermal stem cells, had a larger pool of noncadherin-associated β-catenin than transit-amplifying cells (Zhu and Watt 1999). Retroviral expression of stabilized amino-terminally truncated β-catenin (ΔNβ-catenin) increased the proportion of putative stem cells to almost 90% of the proliferative population in vitro, without blocking terminal differentiation or changing cell cycle kinetics (Zhu and Watt 1999). Furthermore, ΔNβ-catenin expression rescued keratinocytes from the differentiation stimulatory effect of overexpressing the E-cadherin cytoplasmic domain (Fig. 3b) (Zhu and Watt 1999). Conversely, β-catenin lacking armadillo repeats acted as a dominant-negative mutant and stimulated exit from the stem cell compartment in culture. The positive and negative effects of the β-catenin mutants on proliferative potential were independent of effects on intercellular adhesion.

The concepts that stem cells have a higher pool of β-catenin available for Wnt signaling than their more differentiated progeny and that β-catenin activation can expand the stem cell pool have subsequently been confirmed in work on hematopoietic stem cells (Reya et al. 2003) and other tissues.

β-CATENIN IN MOUSE EPIDERMIS

While our lab was examining intercellular adhesion of cultured keratinocytes, it was reported that expression of amino-terminally truncated β-catenin under the control of the K14 promoter led to formation of additional HFs in transgenic mice (Gat et al. 1998). This led us to wonder whether the phenotype was one of stem cell expansion, as would be predicted from our in vitro experiments (Zhu and Watt 1999). If so, inhibition of Wnt signaling should lead to a failure to maintain the epidermis through stem cell depletion. We tested this by using the K14 promoter to overexpress an amino-terminally truncated form of Lef1 that cannot bind β-catenin (ΔNLef1) and thereby acts as a dominant-negative inhibitor of canonical Wnt signaling (Niemann et al. 2002). The result was clear: Stem cell depletion did not occur; rather, the effect was to convert hair follicles into cysts of interfollicular epidermis with ectopic sebocytes. Our interpretation—that in vivo, β-catenin directed lineage selection rather than maintaining the stem cell compartment—was in agreement with other reports of the effects of overexpressing a different ΔNLef1 transgene (Merrill et al. 2001), of deleting β-catenin in the epidermis (Huelsken et al. 2001), or of inhibiting the pathway by overexpressing Dkk1 (Andl et al. 2002).

In subsequent work, we have examined, in some depth, the effect of selectively activating β-catenin in adult mouse epidermis. Our approach has been to overexpress ΔNβ-catenin fused at the carboxyl terminus to the ligand-binding domain of a mutant estrogen receptor (ΔNβ-cateninER) (Fig. 4) (Lo Celso et al. 2004). By topically applying the

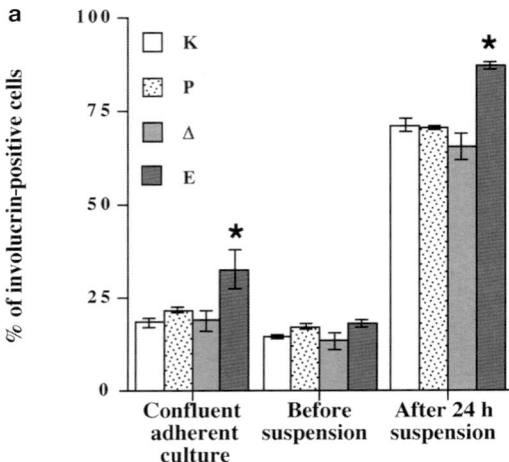

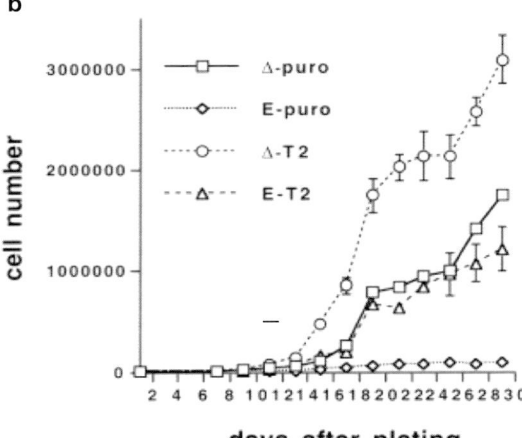

Figure 3. β-catenin-induced stem cell expansion in cultured human epidermis. (*a*) Stimulation of terminal differentiation by dominant-negative E-cadherin in adherent postconfluent cultures and in single-cell suspensions. The proportion of involucrin-positive keratinocytes in preconfluent cultures (Before suspension), confluent cultures and following 24 hours suspension in methyl cellulose (After suspension) is shown. Expression of the dominant-negative E-cadherin mutant H-2Kd–E-cad (E) stimulates differentiation relative to cells expressing a version of the construct in which the β-catenin-binding site is mutated, H-2Kd–E-cadΔC25 (Δ) (*P* < 0.005). (K) Uninfected keratinocytes; (P) keratinocytes transduced with empty retroviral vector. Error bars represent standard deviation. (*b*) Rescue of growth by β-catenin. Keratinocytes expressing H-2Kd–E-cad (E) or H-2Kd–E-cadΔC25 (Δ) were doubly infected with a retroviral vector encoding stabilized β-catenin (T2) or empty vector (puro) and plated at equal density in 35-mm dishes. Triplicate dishes were harvested on the days shown and cell numbers were determined. Error bars represent standard deviation of the mean. (*a*, Reprinted, with permission, from Zhu and Watt 1996 [© Company of Biologists]; *b*, reprinted, with permission, from Zhu and Watt 1999 [© Company of Biologists].)

inducing agent 4-hydroxy-Tamoxifen (4OHT), we can control when, where, and how long β-catenin is activated. In addition, we can induce different levels of β-catenin-dependent transcriptional activity with different doses of

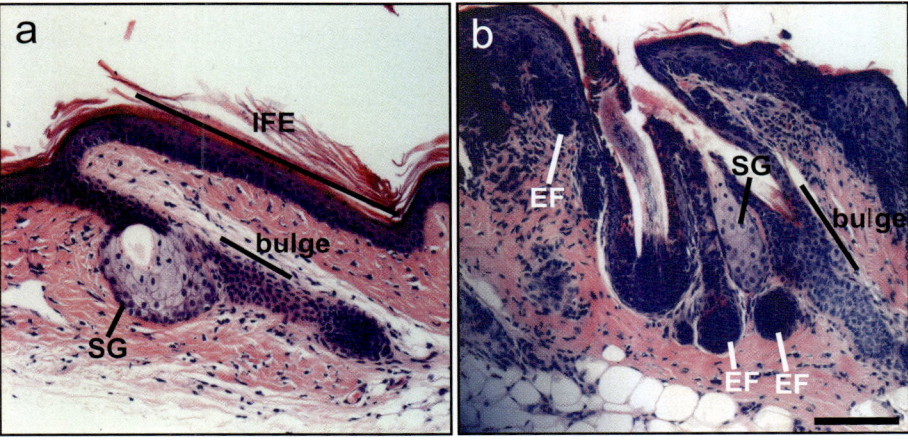

Figure 4. β-catenin-induced ectopic HF formation in adult mouse epidermis. (*a*) Nontransgenic tail skin treated with 4OHT; (*b*) K14ΔNβ-cateninER tail skin treated with 4OHT for 21 days. (SG) Sebaceous gland; (IFE) interfollicular epidermis; (EF) ectopic hair follicles. Bar, 100 μm.

4OHT. Finally, by examining epidermal whole-mount preparations, we can obtain quantitative data about the number and location of the ectopic HFs that form in response to β-catenin activation (Silva-Vargas et al. 2005). From our studies, we conclude that different regions of the epidermis exhibit differential sensitivity to a given level of β-catenin induction and that optimal HF morphogenesis is not achieved with the highest level of β-catenin signal, because under those circumstances, proliferation tends to predominate (Silva-Vargas et al. 2005).

The K14ΔNβ-cateninER model has allowed us to demonstrate that β-catenin-induced ectopic follicles contain cells with characteristics of HF stem cells (CD34, K15, and high levels of α6 integrin expression; clonal growth in vitro) and thus that, as in vitro (Zhu and Watt 1999), β-catenin activation can result in expansion of the stem cell compartment (Silva-Vargas et al. 2005). We have also shown, using lineage tracing, that ectopic follicles induced in the interfollicular epidermis are not obligatorily clonal in origin and are derived from the IFE, rather than the neighboring HF (Silva-Vargas et al. 2005). The latter finding has since been confirmed using a different experimental strategy (Ito et al. 2007).

The overall conclusion from our studies and those of other investigators is that β-catenin activation promotes differentiation of the HF lineages in embryonic and adult epidermis and can, under certain circumstances, expand the stem cell compartment (Alonso and Fuchs 2003; Ito et al. 2007; Närhi et al. 2008; Zhang et al. 2008). In addition to promoting the HF lineages, there are indications that β-catenin signaling negatively regulates SG differentiation (Huelsken et al. 2001; Merrill et al. 2001; Niemann et al. 2002; Lo Celso et al. 2004) and actively suppresses IFE differentiation in developing skin (Närhi et al. 2008; Zhang et al. 2008). This supports the concept that lineage selection involves both active transcription of the set of genes corresponding to the chosen lineage and repression of those in the lineage that has not been selected (Nguyen et al. 2006).

Although there is evidence that β-catenin signaling suppresses the IFE and SG lineages in vivo, this does not appear to be the case in culture. The expansion of the stem cell compartment that we observed in cultured human IFE keratinocytes was not correlated with impaired terminal differentiation, as evaluated by stratification and expression of involucrin, a marker of the differentiating layers of the IFE and the HF inner root sheath (Zhu and Watt 1999). Furthermore, activation of β-catenin in bipotential SG cells increases the proportion of cells that express involucrin, without decreasing the proportion that differentiate into mature sebocytes (Lo Celso et al. 2008). There are a number of potential reasons for the difference between the in vivo and in vitro results. One possibility is that the effects of β-catenin activation in vitro reflect the fact that the culture environment is not permissive for HF morphogenesis and that, as a result, the stem cell compartment expands. Another possibility is that the in vivo response to β-catenin is heavily dependent on reciprocal interactions with the underlying mesenchyme, which are not recreated in culture. These are interesting issues that remain to be explored.

HOW DOES β-CATENIN EXERT ITS EFFECTS ON ADULT EPIDERMIS?

To examine the pathways that act downstream from β-catenin to induce stem cell expansion and ectopic follicle formation, we performed Affymetrix microarray analysis using RNA isolated from total back skin of K14ΔNβ-cateninER mice treated for 7 days with 4OHT (Silva-Vargas et al. 2005). Bearing in mind that only the basal layer of the epidermis expresses the transgene, this search was definitely not restricted to direct β-catenin target genes. Nevertheless, in validation of the approach, we did identify a number of direct target genes, including many HF keratin genes (Zhou et al. 1995).

One of the pathways that was identified as up-regulated in response to β-catenin activation was the Hedgehog signaling pathway, with Sonic hedgehog (Shh) showing a 70-

fold increase in RNA level (Lo Celso et al. 2004; Silva-Vargas et al. 2005). To test the functional significance of this, we treated K14ΔNβ-cateninER epidermis with cyclopamine, a pharmacological inhibitor of Hedgehog signaling, at the same time as β-catenin was activated with 4OHT. The effect was to reduce the local increases in epidermal proliferation that occur on β-catenin activation. Cyclopamine treatment in combination with modest activation of β-catenin blocked ectopic HF formation and prevented existing follicles from entering anagen. However, in the presence of higher β-catenin activity, the effect of blocking Hedgehog signaling was to improve HF morphogenesis. The conclusion is that Shh signaling drives the local increases in proliferation required for ectopic HF formation but that excessive proliferation impairs morphogenesis (Silva-Vargas et al. 2005).

Another pathway that we identified as being induced on β-catenin activation is the Notch pathway. It has previously been well documented that this pathway is not required for HF development during embryogenesis; however, it is necessary for postnatal HF maintenance (Watt et al. 2008a). We identified one of the Notch ligands, Jagged1, as a direct β-catenin target gene. We further found that inhibition of the Notch pathway pharmacologically, with a γ-secretase inhibitor, or by genetic ablation of Jagged1, prevented β-catenin-induced ectopic HF formation (Estrach et al. 2006). By crossing K14ΔNβ-cateninER mice with mice in which Notch activation is inducible with 4OHT (K14N^{ICDΔOP}ERmice), we showed that the combined activation of both pathways did not give a sustained increase in HF formation relative to β-catenin alone. However, the follicles that did form in the double transgenics were more highly differentiated than in K14ΔNβ-cateninER single transgenics, as judged from their length and the markers of HF differentiation expressed (Estrach et al. 2006). As in the case of β-catenin activity, it appears that different levels of Notch activity result in different epidermal responses. In addition, non-cell-autonomous signaling between cells with different levels of Notch ligands can contribute both to exit from the stem cell compartment and to stem cell patterning (Watt et al. 2008a).

Several other important pathways that intersect with the Wnt pathway to regulate the epidermal stem cell compartment have been characterized (Fuchs 2008). However, one that stands out because it is so surprising involves the proto-oncogene c-Myc (Watt et al. 2008b). Myc is a β-catenin target gene and is the key effector of β-catenin-induced proliferation in intestine (Klaus and Birchmeier 2008). Nevertheless, in skin, β-catenin and Myc exert quite different effects. Whereas β-catenin can expand the stem cell compartment and induce ectopic follicle formation, Myc can deplete the stem cell compartment and promote IFE and SG differentiation (Watt et al. 2008b). In sebocyte cultures, β-catenin stimulates IFE differentiation, whereas Myc stimulates SG differentiation (Lo Celso et al. 2008). Myc, like β-catenin, can stimulate Hedgehog signaling; however, whereas β-catenin signaling increases Shh expression, Indian hedgehog (Ihh) is a direct Myc target gene (Lo Celso et al. 2008).

To examine whether activation of Myc and β-catenin are mutually antagonistic, we crossed K14ΔNβ-cateninER

transgenic mice with mice in which Myc is activated by 4OHT treatment (K14MycER; Arnold and Watt 2001). In these bitransgenic mice, there was inhibition of both β-catenin-induced ectopic follicle formation and Myc-induced SG differentiation (Lo Celso et al. 2008). Clearly, the key to understanding the different effects of β-catenin and Myc in the epidermis lies in defining the different effectors with which Myc and β-catenin interact. Nevertheless, our existing observations provide a very clear example of the context-dependent consequences of activating the same signaling pathways in different cell types.

TCF/LEF1-INDEPENDENT β-CATENIN SIGNALING

Although the classical transcriptional effector of the canonical Wnt pathway is the β-catenin/Tcf/Lef complex (see Fig. 2b), there is growing evidence that the Wnt pathway can activate Tcf/Lef-independent genes, as the following examples illustrate. Deletion of the Tcf/Lef-binding sites in the P-cadherin promoter does not prevent β-catenin from inducing P-cadherin gene transcription (Faraldo et al. 2007). β-catenin regulates the myogenic basic helix-loop-helix (bHLH) transcription factor MyoD via a Tcf/Lef-independent mechanism that involves β-catenin binding to MyoD and enhancing MyoD binding to E-box elements (Kim et al. 2008). During pituitary gland development, the paired-type homeodomain transcription factor Prop1 forms a complex with β-catenin, leading to both transcriptional activation of the lineage-determining transcription factor *Pit1* and repression of the lineage-inhibiting transcription factor *Hesx1* (Olson et al. 2006). Finally, β-catenin is known to bind to and activate the vitamin D receptor (VDR), which acts via vitamin D response elements (Pálmer et al. 2001; Shah et al. 2006).

The VDR, like β-catenin, is essential for postnatal maintenance of HFs. VDR null mice fail to undergo the first postnatal hair cycle; instead, the HFs convert to cysts of IFE, thereby resembling the phenotype of K14ΔNLef1 mice (Sakai et al. 2001; Niemann et al. 2002). When examining the epidermal genes that are up-regulated on 4OHT treatment of K14ΔNβ-cateninER mice, we found that many, including the bulge stem cell marker keratin 15, contain vitamin D response elements (VDREs) and that several are induced independently of Tcf/Lef (Pálmer et al. 2008a).

By crossing K14ΔNβ-cateninER mice with VDR null mice, we were able to show that the VDR is required for β-catenin-induced ectopic HF formation. Conversely, application of a vitamin D analog in combination with 4OHT stimulates the differentiation program within the HFs to a greater extent than activating β-catenin alone (Pálmer et al. 2008a).

Although it has been reported that VDR ablation results in gradual depletion of the HF stem cell pool (Cianferotti et al. 2007), in our hands, the degeneration of VDR null follicles does not reflect a loss of follicle stem cells (Pálmer et al. 2008b). Furthermore, in response to the phorbol ester TPA, (12O-tetradecanoylphorbol-13-acetate) VDR null bulge cells, like wild-type bulge cells, are competent to proliferate. We observed that in degenerating VDR null follicles, there was extensive internalization of the α6β4 integrin and that this correlated with reduced migration of VDR null keratinocytes in culture (Pálmer et al. 2008b). We therefore suggest that the failure of VDR null epidermis to maintain HFs in adult life is due, at least in part, to a failure of the cells to migrate along the follicle during anagen.

Our model for the interactions among β-catenin, Tcf/Lef, and VDR is shown schematically in Figure 5. In wild-type epidermis, initiation of the growth phase of the HF cycle (anagen) is dependent on activation of Wnt signaling, and in the absence of that stimulus, the HFs are in the resting (telogen) phase. However, in the absence of the VDR, the hair growth cycle cannot be maintained and the follicles degenerate. A high level of Wnt signaling, in the presence of endogenous vitamin D or a topically applied vitamin D analog, not only triggers anagen, but can also induce ectopic HFs. In the absence of the VDR, Wnt activation is unable to induce differentiation of ectopic follicles.

β-CATENIN IN TUMORS

A final issue to consider is how the information we obtain about normal stem cell renewal and differentiation impacts our understanding of disease mechanisms. One clear message is that tumor type within the epidermis is determined, at least in part, by the same pathways that regulate normal differentiation (Table 1) (Owens and Watt 2003). Prolonged activation of β-catenin in transgenic mice

Table 1. Different skin tumor types associated with different types of genetic alteration

Genetic change	Tumor type
β-catenin activation	pilomatricoma
β-catenin activation in absence of VDR	basal cell carcinoma
Ras activation	papilloma, squamous cell carcinoma
ΔNLef1	sebaceous adenoma, seboma
Ras and ΔNLef1	sebaceous adenoma, seboma

is sufficient to induce benign HF tumors (pilomatricomas; Fig. 5) (Gat et al. 1998), although these regress when the pathway is no longer active (Lo Celso et al. 2004). The importance of β-catenin activation extends to human HF tumors, because human pilomatricomas have been found to harbor activating β-catenin mutations (Chan et al. 1999).

Wnt signaling is frequently activated inappropriately in tumors (Klaus and Birchmeier 2008), and, consistent with this, deletion of β-catenin renders the epidermis resistant to developing chemically induced tumors (Malanchi et al. 2008). It is therefore surprising that the K14ΔNLef1 mice generated in our laboratory develop SG tumors with high frequency (Table 1) (Niemann et al. 2002). The tumor type is consistent with the ability of the ΔNLef1 transgene to promote sebocyte differentiation (Niemann et al. 2002). Furthermore, we have found that one third of human SG tumors that we examined has mutations in the amino terminus of LEF1, which block β-catenin binding (Takeda et al. 2006).

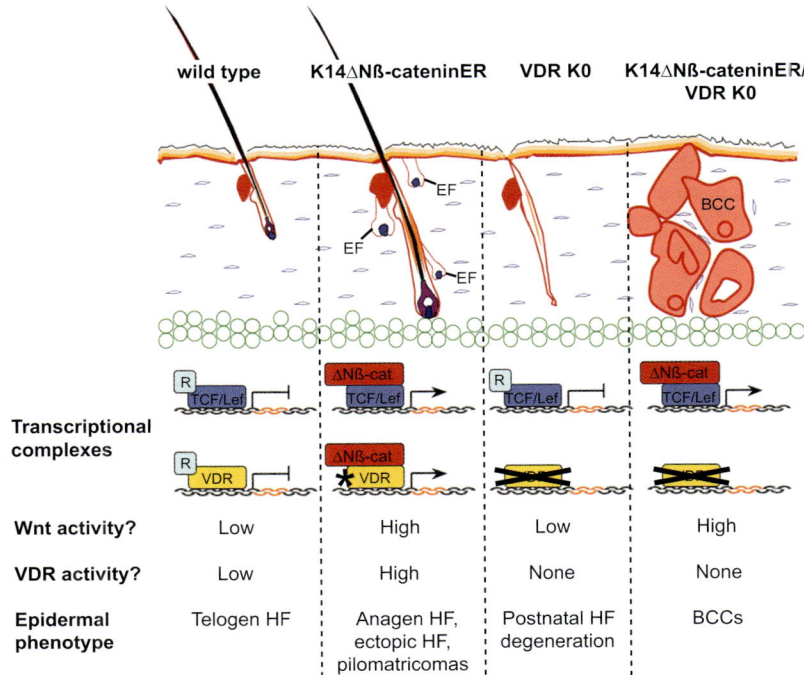

Figure 5. Interaction among β-catenin, Lef/Tcf transcription factors, and the VDR. The epidermal phenotypes and Wnt/VDR transcriptional activity of wild-type and VDR$^{-/-}$ mice are compared in the presence or absence of β-catenin activation in 4OHT-treated K14ΔNβ-cateninER transgenic mice. (EF) Ectopic hair follicle; (BCC) basal cell carcinoma. Tcf/Lef and VDR are shown bound to DNA in the presence of transcriptional corepressors (R) or β-catenin. Vitamin D ligand is indicated by an asterisk. (Based on Pálmer et al. 2008a.)

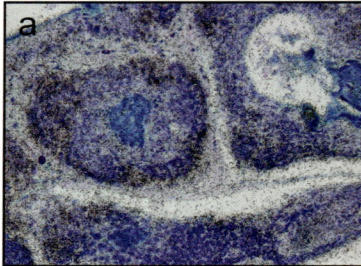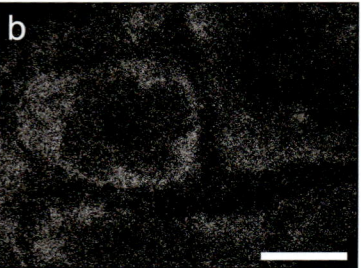

Figure 6. Patched1 is expressed in basal cell carcinoma-like lesions induced by β-catenin in the absence of the VDR. (*a*) Bright- and (*b*) dark-field images of epithelial mass induced by 4OHT treatment of K14ΔNβ-cateninER × VDR⁻/⁻ mouse skin. Radioactive in situ hybridization using the Ptch1 probe. Skin had been treated with 0.5 mg of 4OHT for 21 days. Bar, 100 μm.

There is evidence that ΔNLef1 not only determines tumor type, but also acts as a tumor promoter. In chemical carcinogenesis protocols, K14ΔNLef1 transgenics develop tumors in response to DMBA (dimethyl-benz[a]anthracene) treatment (tumor initiator) alone, whereas wild-type mice do not (Niemann et al. 2007). One of the ways in which ΔNLef1 acts is by blocking induction of the tumor suppressor gene *p53* (Niemann et al. 2007).

Evidence also exists that the intersection of the Wnt pathway with other pathways can determine tumor type (Table 1). During chemical carcinogenesis, DMBA is applied to induce Ha-Ras mutations, and this is followed by repeated applications of the phorbol ester tumor promoter TPA. DMBA/TPA treatment of wild-type mouse skin results in the development of papillomas and squamous cell carcinomas, which have the differentiation characteristics of IFE (Owens and Watt 2003). However, in the presence of the ΔNLef1 transgene, the type of tumors that develop on DMBA-mediated Ras mutation are SG tumors (Table 1) (Niemann et al. 2007).

A further example is that prolonged activation of β-catenin in the absence of the VDR results in the development not of pilomatricomas but of undifferentiated tumors resembling basal cell carcinomas (Table 1; Fig. 5) (Pálmer et al. 2008a). The tumors show evidence of up-regulated Ptch1 expression, a feature of basal cell carcinomas (Fig. 6). Conversely, activation of β-catenin in the presence of a vitamin D analog prevents β-catenin-induced formation of pilomatricomas. We have also found that human trichofolliculomas, another benign HF tumor, have cells with high levels of nuclear β-catenin and VDR, whereas infiltrative human basal cell carcinomas have high β-catenin levels and low VDR levels (Pálmer et al. 2008a). The observations suggest that the anticancer activity of vitamin D analogs may reflect, at least in part, inhibition of inappropriate Wnt signaling (Deeb et al. 2007).

CONCLUSIONS

Mammalian epidermis provides a powerful experimental model not only for studying adult stem cells, but also for examining the origin of different tumor types. Insights obtained from studies of the epidermis turn out to be widely applicable to other tissues. Technical advances in imaging cells in vivo will offer even more possibilities for unraveling the properties of epidermal stem cells, in particular their interactions with their neighbors.

ACKNOWLEDGMENTS

F.M.W. gratefully acknowledges financial support from Cancer Research U.K., the Wellcome Trust, and the Medical Research Council. C.A.C. is a Herchel Smith postdoctoral research fellow and holds an Evans-Freke Next Generation junior research fellowship. F.M.W. is grateful to all the members of her lab, past and present, who have contributed to the work described in this chapter. We thank Julia Turnock and Hector Pálmer for providing in situ hybridization data.

REFERENCES

Adams, J.C. and Watt, F.M. 1989. Fibronectin inhibits the terminal differentiation of human keratinocytes. *Nature* **340:** 307–309.

Alonso, L. and Fuchs, E. 2003. Stem cells in the skin: Waste not, Wnt not. *Genes Dev.* **17:** 1189–1200.

Andl, T., Reddy, S.T., Gaddapara, T., and Millar, S.E. 2002. WNT signals are required for the initiation of hair follicle development. *Dev. Cell* **2:** 643–653.

Arnold, I. and Watt, F.M. 2001. c-Myc activation in transgenic mouse epidermis results in mobilisation of stem cells and differentiation of their progeny. *Curr. Biol.* **11:** 558–568.

Barrandon, Y. and Green, H. 1987. Three clonal types of keratinocyte with different capacities for multiplication. *Proc. Natl. Acad. Sci.* **84:** 2302–2306.

Blanpain, C., Lowry, W.E., Geoghean, A., Polak, L., and Fuchs, E. 2004. Self-renewal, multipotency, and the existence of two cell populations within an epithelial stem cell niche. *Cell* **118:** 635–648.

Braun, K.M., Niemann, C., Jensen, U.B., Sundberg, J.P., Silva-Vargas, V., and Watt, F.M. 2003. Manipulation of stem cell proliferation and lineage commitment: Visualisation of label-retaining cells in wholemounts of mouse epidermis. *Development* **130:** 5241–5255.

Carroll, J.M., Albers, K.M., Garlick, J.A., Harrington, R., and Taichman, L.B. 1993. Tissue- and stratum-specific expression of the human involucrin promoter in transgenic mice. *Proc. Natl. Acad. Sci.* **90:** 10270–10274.

Chan, E.F., Gat, U., McNiff, J.M., and Fuchs, E. 1999. A common human skin tumour is caused by activating mutations in β-catenin. *Nat. Genet.* **21:** 410–413.

Cianferotti, L., Cox, M., Skorija, K., and Demay, M.B. 2007. Vitamin D receptor is essential for normal keratinocyte stem cell function. *Proc. Natl. Acad. Sci.* **104:** 9428–9433.

Claudinot, S., Nicolas, M., Oshima, H., Rochat, A., and Barrandon, Y. 2005. Long-term renewal of hair follicles from clonogenic multipotent stem cells. *Proc. Natl. Acad. Sci.* **102:** 14677–14682.

DasGupta, R. and Fuchs, E. 1999. Multiple roles for activated LEF/TCF transcription complexes during hair follicle development and differentiation. *Development* **126:** 4557–4568.

Deeb, K.K., Trump, D.L., and Johnson, C.S. 2007. Vitamin D signalling pathways in cancer: Potential for anticancer therapeutics. *Nat. Rev. Cancer* **7:** 684–700.

Estrach, S., Ambler, C.A., Lo Celso, C., Hozumi,, K., and Watt, F.M. 2006. *Jagged1* is a β-catenin target gene required for ectopic hair follicle formation in adult epidermis. *Development* **133:** 4427–4438.

Faraldo, M.M., Teulière, J., Deugnier, M.A., Birchmeier, W., Huelsken, J., Thiery, J.P., Cano, A., and Glukhova, M.A. 2007. β-Catenin regulates P-cadherin expression in mammary basal epithelial cells. *FEBS Lett.* **581:** 831–836.

Fuchs, E. 2008. Skin stem cells: Rising to the surface. *J. Cell Biol.* **180:** 273–284.

Gat, U., DasGupta, R., Degenstein, L., and Fuchs, E. 1998. De novo hair follicle morphogenesis and hair tumors in mice expressing a truncated β-catenin in skin. *Cell* **95:** 605–614.

Gerrard, A.J., Hudson, D.L., Brownlee, G.G., and Watt, F.M. 1993. Towards gene therapy for haemophilia B using primary human keratinocytes. *Nat. Genet.* **3:** 180–183.

Greenhalgh, D.A., Rothnagel, J.A., Quintanilla, M.I., Orengo, C.C., Gagne, T.A., Bundman, D.S., Longley, M.A., and Roop, D.R. 1993. Induction of epidermal hyperplasia, hyperkeratosis, and papillomas in transgenic mice by a targeted v-Ha-*ras* oncogene. *Mol. Carcinog.* **7:** 99–110.

Hall, P.A. and Watt, F.M. 1989. Stem cells: The generation and maintenance of cellular diversity. *Development* **106:** 619–633.

Hodivala, K.J. and Watt, F.M. 1994. Evidence that cadherins play a role in the downregulation of integrin expression that occurs during keratinocyte terminal differentiation. *J. Cell Biol.* **124:** 589–600.

Huelsken, J., Birchmeier, W., and Behrens, J. 1994. E-Cadherin and APC compete for the interaction with β-catenin and the cytoskeleton. *J. Cell Biol.* **127:** 2061–2069.

Huelsken, J., Vogel, R., Erdmann, B., Cotsarelis, G., and Birchmeier, W. 2001. β-Catenin controls hair follicle morphogenesis and stem cell differentiation in the skin. *Cell* **105:** 533–545.

Ito, M., Yang, Z., Andl, T., Cui, C., Kim, N., Millar, S.E., and Cotsarelis, G. 2007. Wnt-dependent de novo hair follicle regeneration in adult mouse skin after wounding. *Nature* **447:** 316–320.

Jensen, U.B., Lowell, S., and Watt, F.M. 1999. The spatial relationship between the stem cells and their progeny in the basal layer of human epidermis: A new view based on whole-mount labelling and lineage analysis. *Development* **126:** 2409–2418.

Jones, P.H. and Watt, F.M. 1993. Separation of human epidermal stem cells from transit amplifying cells on the basis of differences in integrin function and expression. *Cell* **73:** 713–724.

Jones, P.H., Harper, S., and Watt, F.M. 1995. Stem cell patterning and fate in human epidermis. *Cell* **80:** 83–93.

Jones, P.H., Simons, B.D., and Watt, F.M. 2007. Sic transit gloria: Farewell to the epidermal transit amplifying cell? *Cell Stem Cell* **1:** 371–381.

Kim, C.-H., Neiswender, H., Baik, E.J., Xiong, W.C., and Mei, L. 2008. β-Catenin interacts with MyoD and regulates its transcription activity. *Mol. Cell. Biol.* **28:** 2941–2951.

Klaus, A. and Birchmeier, W. 2008. Wnt signalling and its impact on development and cancer. *Nat. Rev. Cancer* **8:** 387–398.

Levy, L., Broad, S., Zhu, A.J., Carroll, J.M., Khazaal, I., Péault, B., and Watt, F.M. 1998. Optimised retroviral infection of human epidermal keratinocytes: Long-term expression of transduced integrin gene following grafting onto SCID mice. *Gene Ther.* **5:** 913–922.

Liu, Y., Lyle, S., Yang, Z., and Cotsarelis, G. 2003. Keratin 15 promoter targets putative epithelial stem cells in the hair follicle bulge. *J. Invest. Dermatol.* **121:** 963–968.

Lo Celso, C., Prowse, D.M., and Watt, F.M. 2004. Transient activation of β-catenin signalling in adult mouse epidermis is sufficient to induce new hair follicles but continuous activation is required to maintain hair follicle tumours. *Development* **131:** 1787–1799.

Lo Celso, C., Berta, M.A., Braun, K.M., Frye, M., Lyle, S., Zouboulis, C.C., and Watt, F.M. 2008. Characterization of bipotential progenitors derived from human sebaceous gland: Contrasting roles of c-Myc and β-catenin. *Stem Cells* **26:** 1241–1252.

Malanchi, I., Peinado, H., Kassen, D., Hussenet, T., Metzger, D., Chambon, P., Huber, M., Hohl, D., Cano, A., Birchmeier, W., and Huelsken, J. 2008. Cutaneous cancer stem cell maintenance is dependent on β-catenin signalling. *Nature* **452:** 650–653.

Merrill, B.J., Gat, U., DasGupta, R., and Fuchs, E. 2001. Tcf3 and Lef1 regulate lineage differentiation of multipotent stem cells in skin. *Genes Dev.* **15:** 1688–1705.

Närhi, K., Järvinen, E., Birchmeier, W., Taketo, M.M., Mikkola, M.L. and Thesleff, I. 2008. Sustained epithelial β-catenin activity induces precocious hair development but disrupts hair follicle down-growth and hair shaft formation. *Development.* **135:** 1019–1028.

Nguyen, H., Rendl, M., and Fuchs, E. 2006. Tcf3 governs stem cell features and represses cell fate determination in skin. *Cell* **127:** 171–183.

Niemann, C., Owens, D.M., Huelsken, J., Birchmeier, W., and Watt, F.M. 2002. Repression of β-catenin signaling in mouse epidermis results in transdifferentiation of hair follicles into squamous epidermal cysts and formation of skin tumours. *Development* **129:** 95–109.

Niemann, C., Owens, D.M., Schettina, P., and Watt, F.M. 2007. Dual role of inactivating Lef1 mutations in epidermis: Tumour promotion and specification of tumour type. *Cancer Res.* **67:** 2916–2921.

Olson, L.E., Tollkuhn, J., Scafoglio, C., Krones, A., Zhang, J., Ohgi, K.A., Wu, W., Taketo, M.M., Kemler, R., Grosschedl, R., Rose, D., Li, X., and Rosenfeld, M.G. 2006. Homeodomain-mediated β-catenin-dependent switching events dictate cell-lineage determination. *Cell* **125:** 593–605.

Owens, D.M. and Watt, F.M. 2003. Contribution of stem cells and differentiated cells to epidermal tumours. *Nat. Rev. Cancer* **3:** 444–451.

Pálmer, H.G., González-Sancho, J.M., Espada, J., Berciano, M.T., Puig, I., Baulida, J., Quintanilla, M., Cano, A., de Herreros, A.G., Lafarga, M., and Muñoz, A. 2001. Vitamin D(3) promotes the differentiation of colon carcinoma cells by the induction of E-cadherin and the inhibition of β-catenin signaling. *J. Cell Biol.* **154:** 369–387.

Pálmer, H.G., Anjos-Afonso, F., Carmeliet, G., Takeda, H., and Watt, F.M. 2008a. The vitamin D receptor is a Wnt effector that controls hair follicle differentiation and specifies tumor type in adult epidermis. *PLoS ONE* **3:** e1483.

Pálmer, H.G., Martinez, D., Carmeliet, G., and Watt, F.M. 2008b. The vitamin D receptor is required for mouse hair cycle progression but not for maintenance of the epidermal stem cell compartment. *J. Invest. Dermatol.* **128:** 2113–2117.

Pelengaris, S., Littlewood, T., Khan, M., Elia, G., and Evan, G. 1999. Reversible activation of c-Myc in skin: Induction of a complex neoplastic phenotype by a single oncogenic lesion. *Mol. Cell* **3:** 565–577.

Reddy, S., Andl, T., Bagasra, A., Lu, M.M., Epstein, D.J., Morrisey, E.E., and Millar, S.E. 2001. Characterization of *Wnt* gene expression in developing and postnatal hair follicles and identification of *Wnt5a* as a target of Sonic hedgehog in hair follicle morphogenesis. *Mech. Dev.* **107:** 69–82.

Reddy, S.T., Andl, T., Lu, M.M., Morrisey, E.E., and Millar, S.E. 2004. Expression of *Frizzled* genes in developing and postnatal hair follicles. *J. Invest. Dermatol.* **123:** 275–282.

Reya, T., Duncan, A.W., Ailles, L., Domen, J., Scherer, D.C., Willert, K., Hintz, L., Nusse, R., and Weissman, I.L. 2003. A role for Wnt signalling in self-renewal of haematopoietic stem cells. *Nature* **423:** 409–414.

Rheinwald, J.G. and Green, H. 1975. Serial cultivation of strains of human epidermal keratinocytes: The formation of keratinizing colonies from single cells. *Cell* **6:** 331–343.

Romero, M.R., Carroll, J.M., and Watt, F.M. 1999. Analysis of cultured keratinocytes from a transgenic mouse model of psoriasis: Effects of suprabasal integrin expression on keratinocyte adhesion, proliferation and terminal differentiation. *Exp. Dermatol.* **8:** 53–67.

Sakai, Y., Kishimoto, J., and Demay, M.B. 2001. Metabolic and cellular analysis of alopecia in vitamin D receptor knockout mice. *J. Clin. Invest.* **107:** 961–966.

Shackleton, M., Vaillant, F., Simpson, K.J., Stingl, J., Smyth, G.K., Asselin-Labat, M.L., Wu, L., Lindeman, G.J., and Visvader, J.E. 2006. Generation of a functional mammary gland from a single stem cell. *Nature* **439:** 84–88.

Shah, S., Islam, M.N., Dakshanamurthy, S., Rizvi, I., Rao, M., Herrell, R., Zinser, G., Valrance, M., Aranda, A., Moras, D.,

Norman, A., Welsh, J., and Byers, S.W. 2006. The molecular basis of vitamin D receptor and β-catenin crossregulation. *Mol. Cell* **21:** 799–809.

Sick, S., Reinker, S., Timmer, J., and Schlake, T. 2006. WNT and DKK determine hair follicle spacing through a reaction-diffusion mechanism. *Science* **314:** 1447–1450.

Silva-Vargas, V., Lo Celso, C., Giangreco, A., Ofstad, T., Prowse, D.M., Braun, K.M., and Watt F.M. 2005. β-Catenin and Hedgehog signal strength can specify number and location of hair follicles in adult epidermis without recruitment of bulge stem cells. *Dev. Cell* **9:** 121–131.

Takeda, H., Lyle, S., Lazar, A.J.F., Zouboulis, C.C., Smyth, I., and Watt, F.M. 2006. Human sebaceous tumours harbour inactivating mutations in LEF1. *Nat. Med.* **12:** 395–397.

Vasioukhin, V., Degenstein, L., Wise, B., and Fuchs, E. 1999. The magical touch: Genome targeting in epidermal stem cells induced by tamoxifen application to mouse skin. *Proc. Natl. Acad. Sci.* **96:** 8551–8556.

Vassar, R., Rosenberg, M., Ross, S., Tyner, A., and Fuchs, E. 1989. Tissue-specific and differentiation-specific expression of a human K14 keratin gene in transgenic mice. *Proc. Natl. Acad. Sci.* **86:** 1563–1567.

Wang, X.J., Liefer, K.M., Tsai, S., O'Malley, B.W., and Roop, D.R. 1999. Development of gene-switch transgenic mice that inducibly express transforming growth factor β1 in the epidermis. *Proc. Natl. Acad. Sci.* **96:** 8483–8488.

Watt, F.M. 1998. Epidermal stem cells: Markers, patterning and the control of stem cell fate. *Philos. Trans. R. Soc. Lond. B* **353:** 831–837.

Watt, F.M. 2002. Role of integrins in regulating epidermal adhesion, growth and differentiation. *EMBO J.* **12:** 3919–3926.

Watt, F.M., Lo Celso, C., and Silva-Vargas, V. 2006. Epidermal stem cells: An update *Curr. Opin. Genet. Dev.* **16:** 518–524.

Watt, F.M., Estrach, S., and Ambler, C.A. 2008a. Epidermal Notch signalling: Differentiation, cancer and adhesion. *Curr. Opin. Cell Biol.* **20:** 171–179.

Watt, F.M., Frye, M., and Benitah, S.A. 2008b. MYC in mammalian epidermis: How can an oncogene stimulate differentiation? *Nat. Rev. Cancer* **8:** 234–242.

Wu, W.Y. and Morris, R.J. 2005. Method for the harvest and assay of in vitro clonogenic keratinocytes stem cells from mice. *Methods Mol. Biol.* **289:** 79–86.

Zhang, Y., Andl, T., Yang, S.H., Teta, M., Liu, F., Seykora, J.T., Tobias, J.W., Piccolo, S., Schmidt-Ullrich, R., Nagy, A., Taketo, M.M., Dlugosz, A.A., and Millar, S.E. 2008. Activation of β-catenin signaling programs embryonic epidermis to hair follicle fate. *Development* **135:** 2161–2172.

Zhou, P., Byrne, C., Jacobs, J., and Fuchs, E. 1995. Lymphoid enhancer factor 1 directs hair follicle patterning and epithelial cell fate. *Genes Dev.* **9:** 700–713.

Zhu, A.J. and Watt, F.M. 1996. Expression of a dominant negative cadherin mutant inhibits proliferation and stimulates terminal differentiation of human epidermal keratinocytes. *J. Cell Sci.* **109:** 3013–3023.

Zhu, A.J. and Watt, F.M. 1999. β-Catenin signalling modulates proliferative potential of human epidermal keratinocytes independently of intercellular adhesion. *Development* **126:** 2285–2298.

Implications of Cellular Senescence in Tissue Damage Response, Tumor Suppression, and Stem Cell Biology

V. Krizhanovsky,* W. Xue,* L. Zender,*† M. Yon,*‡ E. Hernando,§ and S.W. Lowe*¶

*Cold Spring Harbor Laboratory, Cold Spring Harbor, New York 11724; ‡Departamento de Imunologia,
Instituto de Ciências Biomédicas, Universidade de São Paulo and Instituto de Investigação em Imunologia,
Instituto do Milênio, Brazil; §Department of Pathology, New York University School of Medicine, New York 10016;
¶Howard Hughes Medical Institute, Cold Spring Harbor, New York 11724

Cellular senescence is characterized by an irreversible cell cycle arrest that, when bypassed by mutation, contributes to cellular immortalization. Activated oncogenes induce a hyperproliferative response, which might be one of the senescence cues. We have found that expression of such an oncogene, Akt, causes senescence in primary mouse hepatoblasts in vitro. Additionally, AKT-driven tumors undergo senescence in vivo following p53 reactivation and show signs of differentiation. In another in vivo system, i.e., liver fibrosis, hyperproliferative signaling through AKT might be a driving force of the senescence in activated hepatic stellate cells. Senescent cells up-regulate and secrete molecules that, on the one hand, can reinforce the arrest and, on the other hand, can signal to an innate immune system to clear the senescent cells. The mechanisms governing senescence and immortalization are overlapping with those regulating self-renewal and differentiation. These respective control mechanisms, or their disregulation, are involved in multiple pathological conditions including fibrosis, wound healing, and cancer. Understanding extracellular cues that regulate these processes may enable new therapies for these conditions.

Cellular senescence is a stable form of a cell cycle arrest program that limits the proliferative potential of cells and thereby prevents immortalization. Initially defined by the phenotype of human fibroblasts undergoing replicative exhaustion in culture (Hayflick and Moorhead 1961), senescence can be triggered in many cell types in response to diverse forms of cellular damage or stress (Serrano et al. 1997; Campisi and d'Adda di Fagagna 2007). Senescent cells display a large flattened morphology and accumulate a senescence-associated β-galactosidase (SA-β-gal) activity distinguishing them from most quiescent cells (Campisi and d'Adda di Fagagna 2007). An additional feature of senescent cells is an accumulation of the senescence-associated heterochromatin that functions to repress the expression of certain genes responsible for the cell cycle progression (Narita et al. 2003). Consistent with its function in halting cellular proliferation, senescence also has an important role in tumor suppression. In addition, it has been suggested that senescence contributes to stem cell depletion or the decline of tissue regeneration capabilities during aging (Rossi et al. 2008).

One of the most potent triggers of senescence is DNA damage, which is induced by telomere attrition during replicative exhaustion (Vaziri et al. 1997). In the absence of the telomerase enzyme, mutations that overcome senescence allow continued proliferation in the presence of sustained DNA damage, leading to rampant genetic instability and eventually cancer progression. Consistent with the fact that many conventional chemotherapeutic agents can directly or indirectly damage DNA, senescent cells can accumulate in malignant tumors following chemotherapy (Schmitt et al. 2002; te Poele et al. 2002; Roninson 2003). Indeed, in a mouse lymphoma model, an intact senescence machinery contributes to a positive response to chemotherapy in vivo.

Several activated oncogenes can induce senescence, in part, through signals initiated by DNA damage or replicative stress. Although, on the one hand, paradoxically, this phenomenon can be rationalized by suggesting that normal cells possess cellular fail-safe mechanisms that counter the cancer-promoting effects of hyperproliferative mutations (Serrano et al. 1997). During the past several years, this "oncogene-induced senescence" has been established as an important tumor suppressor mechanism that constrains tumorigenesis in multiple forms of cancer in humans and mice. For example, nevi, premalignant lesions of melanoma in the human skin, contain senescent melanocytes bearing mutations in the BRAF gene (Michaloglou et al. 2005). Neurofibromas from NF1 mutant patients contain senescent cells in which a genetic defect in the NF1 gene leads to constitutively high levels of Ras activity (Courtois-Cox et al. 2006). Benign lesions of the prostate in human patients and mice lacking the tumor suppressor PTEN contain senescent cells, and disruption of the senescence program in mice leads to tumor progression (Chen et al. 2005). In mouse models of several cancers including lymphoma (Braig et al. 2005),

---

†Present address: Helmholtz Centre for Infection Research, Inhoffenstrasse 7, 38124 Braunschweig, Germany and Hannover Medical School, Department of Gastroenterology, Hepatology and Endocrinology, Carl-Neuberg-Strasse 1, 30625 Hannover, Germany.

hyperplasia of the pituitary gland (Lazzerini Denchi et al. 2005), skin carcinoma (Collado et al. 2005), and melanocytic lesions of UV-irradiated hepatocyte growth factor/scatter factor (HGF/SF) transgenic mice (Ha et al. 2007), premalignant lesions were shown to be limited by oncogene-induced senescence.

Consistent with the role of cellular senescence as a barrier to malignant transformation, senescent cells activate the p53 and p16/Rb (retinoblastoma) tumor suppressor pathways (Mooi and Peeper 2006; Campisi and d'Adda di Fagagna 2007; Collado et al. 2007). p53 promotes senescence by *trans*-activating genes that inhibit proliferation, including the p21/Cip1/Waf1 cyclin-dependent kinase inhibitor and miR-34 class of microRNAs (He et al. 2007a). In contrast, p16INK4a promotes senescence by inhibiting cyclin-dependent kinases 2 and 4, thereby preventing Rb phosphorylation and allowing Rb to promote a repressive heterochromatin environment (Narita et al. 2003). Both the Rb and p53 tumor suppressor pathways are negatively regulated by Bmi-1, CBX7, and other members of Polycomb repression complexes (Gil et al. 2005); these complexes act at the chromatin level of chromatin to repress the *INK4a* and *ARF* genes, both embedded in the same genomic locus on chromosome 9p.

In addition to up-regulation of the p53 and p16/Rb pathways, senescent cells often up-regulate inflammatory cytokines and other molecules known to modulate the microenvironment or immune response (Campisi and d'Adda di Fagagna 2007). Recently, high-throughput genetic approaches have identified several secreted molecules that functionally contribute to senescence (Acosta et al. 2008; Kuilman et al. 2008; Wajapeyee et al. 2008). In these studies, the secreted proteins insulin-like growth factor-binding protein-7 (IGFBP7) and interleukin-6 (IL-6) contribute to senescence induced by a mutant form of *BRAF*, suggesting that these factors might attenuate tumorigenesis. Another secreted cytokine, IL-8, and its receptor, CXCR2, appear to reinforce both replicative and oncogene-induced senescence (Acosta et al. 2008). The ability of secreted molecules to affect senescence is likely to have both autocrine and paracrine effects. Although the autocrine signaling might enforce senescence, its effect on the microenvironment might be coordinately different and depend on the nature of neighboring cells.

Our laboratory has been interested in the roles and regulation of cellular senescence for more than a decade. We previously established that oncogenes could trigger senescence (Serrano et al. 1997) and demonstrated that cellular senescence can contribute to the outcome of cancer chemotherapy (Schmitt et al. 2002). Later, we studied the molecular mechanisms of senescence and implicated chromatin-modulating factors and the miR-34 family of microRNAs as key regulators of the senescence program (Narita et al. 2003, 2006; He et al. 2007b). More recently, we studied the role of senescence in the biology of liver disease, including liver fibrosis and hepatocellular carcinoma (Xue et al. 2007; Krizhanovsky et al. 2008).

Most commonly, liver disease is caused by hepatitis viruses B and C and chronic alcohol abuse or is associated with obesity leading to nonalcoholic steatohepatitis, initially resulting in liver fibrosis (Bataller and Brenner 2005). Eventually, persistent liver damage leads to the development of liver cirrhosis, a major health problem worldwide (Bataller and Brenner 2005) and the 12th most common cause of death in the United States (NCHS 2004). In turn, liver fibrosis and cirrhosis are two of the main risk factors for developing hepatocellular carcinoma (HCC). Several reports suggest that up to 90% of HCC cases have a natural history of unresolved inflammation and severe fibrosis irrespective of the underlying cause of liver disease (Elsharkawy and Mann 2007). In biopsies, liver fibrosis is characterized by depositions of the extracellular matrix (ECM) known as fibrotic scars, which is deposited by activated hepatic stellate cells (HSCs), the main fibrogenic cell type in the liver. HSCs normally reside in proximity to blood vessels in the liver and are the main site of vitamin A storage in the body (Bataller and Brenner 2005). Upon liver injury, signaling from damaged hepatocytes leads to HSC activation, resulting in intensive proliferation of HSCs, migration from their normal location, ECM deposition, and formation of fibrotic scars. Recently, we demonstrated that senescence functions as both a tumor suppressive mechanism in liver carcinoma and a program that limits liver fibrosis.

SENESCENCE ACCOMPANIES P53 REACTIVATION IN A MOUSE MODEL OF HCC

We previously used a mouse model of liver carcinoma to determine the consequence of reactivating the p53 pathway in liver tumors (Xue et al. 2007). We used RNA interference (RNAi) to conditionally regulate endogenous *p53* expression using a p53 miR-30 design short hairpin RNA (shRNA) expressed from a conditional tetracycline-responsive promoter as described previously (Dickins et al. 2005). Embryonic hepatoblasts were transduced with retroviruses that express oncogenic *Hras*^{V12}, the tetracycline *trans*-activator protein tTA ("tet-off") and the p53 shRNA. In the absence of doxycycline (Dox), p53 shRNA knocked down p53 expression efficiently and led to hepatocarcinoma development when the triple-transduced cells were transplanted into recipient mice (Xue et al. 2007). Addition of Dox and *p53* reactivation resulted in tumor regression. Interestingly, we found that p53 reactivation in the tumor cells led to cellular senescence and not apoptosis, which was also associated with markers of cellular differentiation (see Figure S3 in Xue et al. 2007). Senescent cells triggered an immune response in vivo, resulting in infiltration of immune cells into the tumor and clearance of senescent cells. These results suggest that loss of a tumor suppressor may be required for the maintenance of some tumors; moreover, during cellular senescence in this system, cell-autonomous processes and non-cell-autonomous processes cooperate to eliminate the formerly malignant cell.

P53 REACTIVATION LEADS TO SENESCENCE IN CELLS THAT DEREGULATE PI3K SIGNALING

The phosphoinositol-3 kinase (PI3K) signaling pathway is an important component of the intracellular signaling cascade downstream from the Ras family of GTPases, which, when activated, are some of the most potent human oncogenes and triggers of senescence. Moreover, the PI3K pathway is frequently activated by genomic aberrations across many cancer lineages (Brugge et al. 2007). Hyperactivation of this pathway through loss of the *PTEN* tumor suppressor can promote cell proliferation and survival, but like oncogenic Ras, it can trigger senescence in some cell types (Chen et al. 2005).

We have previously shown that hyperactivation of the PI3K pathway, for example, by deregulation of *Akt*, together with p53 loss could promote hepatocellular carcinoma development (Zender et al. 2006). To determine whether tumors driven by deregulation of PI3K signaling respond to p53 reactivation by senescence, we studied the consequence of expression of a key PI3K target, *Akt*, in embryonic hepatoblasts (embryonic liver progenitor cells). Embryonic hepatoblasts were prepared from wild-type and *p53*$^{-/-}$ animals and were infected with retroviruses that expressed activated *Akt* (myr-*Akt*) (Wendel et al. 2004). Hepatoblasts derived from wild-type embryos infected with *Akt* exhibit flattened morphology and stain positive for SA-β-gal, whereas *p53*$^{-/-}$ hepatoblasts continue to proliferate and easily reach confluence when seeded at low density (Fig. 1A). These results suggest that *Akt* expression induces senescence in a manner that is dependent on p53.

To further understand the cooperation between *Akt* and p53 in senescence, we conditionally repressed p53 in *Akt*-transduced hepatoblasts. In addition to *Akt*, hepatoblasts were transduced with the tetracycline *trans*-activator protein tTA and a tet-responsive p53 shRNA and seeded subcutaneously into athymic nude mice. In the absence of Dox, expression of p53 was suppressed, leading to the development of aggressive liver carcinomas. Upon tumor formation, mice were left untreated or treated with Dox to suppress the p53 shRNA. Following Dox treatment, p53 was induced, and a substantial percentage of tumor cells became SA-β-gal positive (Fig. 1B); moreover, the tumors shrank and virtually disappeared after 20 days (Xue et al. 2007). In contrast, tumors present in untreated mice remained SA-β-gal negative and continued to grow. Therefore, like tumors expressing oncogenic *Hras*^{V12}, *Akt*-transformed hepatoblasts undergo senescence in vivo upon p53 reactivation.

In liver carcinomas expressing oncogenic *ras*, the onset of cellular senescence correlated with the appearance of differentiation markers (Xue et al. 2007). In these tumors, expression of differentiation markers cytokeratin 7 (CK7) and CK8 was lower in the absence of p53 than in its presence. Conversely, α-fetoprotein (AFP), which is typically expressed in progenitor cells and reduced upon differentiation, was detected in the tumors in the absence of p53 but

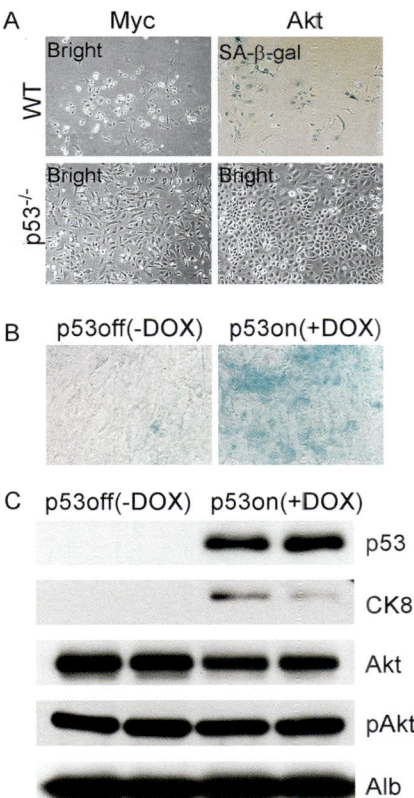

Figure 1. AKT induces cellular senescence and differentiation in cooperation with p53. (*A*) Wild-type (WT) but not *p53*$^{-/-}$ primary hepatoblasts senesce in response to AKT (*B*) AKT-driven tumor cells senesce in vivo in response to p53 restoration (+DOX), but not in the absence of p53 (-DOX). (*C*) Expression of the differentiation marker CK8 was increased in tumors following p53 restoration.

down-regulated upon p53 reactivation (Xue et al. 2007). Similarly, CK8 was induced following p53 reactivation in tumors expressing activated *Akt* (Fig. 1B). Therefore, along with induction of senescence, tumors show signs of differentiation following p53 reactivation.

SENESCENCE RESPONSE PROTECTS AGAINST LIVER FIBROSIS

Liver fibrosis caused by various types of liver damage can lead to cirrhosis, liver failure, or the development of hepatocellular carcinoma (Bataller and Brenner 2005). To investigate the role of cellular senescence in fibrosis progression, we induced fibrosis in mice by twice-weekly treatment with 1 ml/kg of CCl$_4$ for 6 weeks (Fig. 2A). This course of treatment is believed to induce mild cirrhosis in mouse liver (Rudolph et al. 2000), and, indeed, CCl$_4$-treated mice often displayed enlarged abdomens (Krizhanovsky et al. 2008), a phenotype that can be attributed to the ascites that often accompanies cirrhosis.

Livers from treated and untreated mice were dissected, and general liver architecture was evaluated by hematoxylin and eosin (H&E). Additionally, the presence of

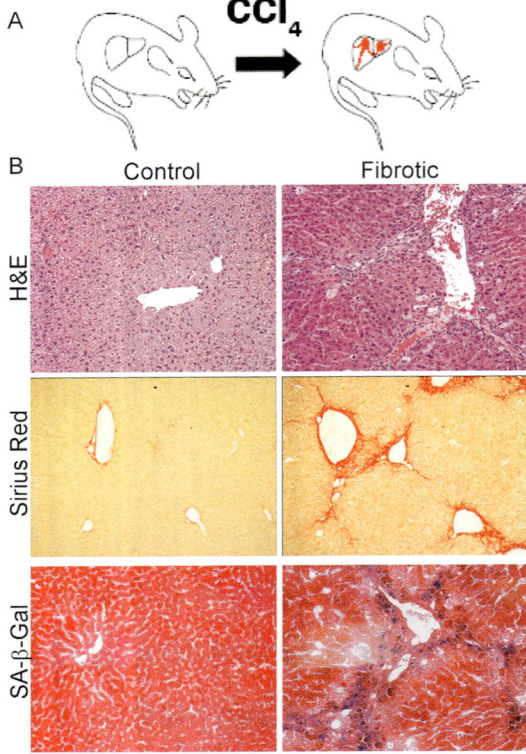

Figure 2. Senescent cells are present in mouse fibrotic liver. (*A*) Mice were treated with CCl$_4$ twice weekly for 6 weeks. (*B*) CCl$_4$-treated livers (fibrotic) but not control livers exhibit fibrotic scars (evaluated by H&E and Sirius Red staining). Multiple cells adjacent to the scar stain positively for the senescence marker SA-β-gal.

excessive ECM (Sirius Red) and the presence of senescent (SA-β-gal positive) cells were assessed. Livers from mice treated with CCl$_4$, but not vehicle, showed a distinctive fibrotic morphology and accumulated ECM in the fibrotic scars (Fig. 2B). Cells that stained positively for the senescence marker SA-β-gal were predominantly found in the areas adjoining fibrotic scars, the site of HSC proliferation, migration, and fibrotic ECM production (Fig. 2B). We find that cells in this region express senescence markers p16, p53, p21, and Hmga1 (Krizhanovsky et al. 2008).

On the basis of the observations above, we hypothesized that senescent cells might derive from activated stellate cells. Accordingly, the senescence markers p53 and Hmga1 were expressed in cells that were positive for αSMA and Desmin, markers of activated stellate cells (Krizhanovsky et al. 2008). Therefore, senescent activated stellate cells are present in chemically induced fibrotic livers in mice. Similar cells were also observed in livers from human patients with liver fibrosis (Krizhanovsky et al. 2008).

Both the p53 and p16/Rb pathways contribute to senescence in a cell-type-dependent manner, such that cells lacking either pathway alone sometimes retain a residual senescence response (Beausejour et al. 2003; Shay and Roninson 2004). To determine the impact of

disrupting both loci on senescence and fibrosis in the liver, we produced *p53$^{-/-}$;INK4a/ARF$^{-/-}$* compound mutant mice. Hepatic stellate cells were then prepared from wild-type and *p53$^{-/-}$;INK4a/ARF$^{-/-}$* animals. When plated at low density in standard media, these cells became activated and expressed αSMA within 2 weeks (Krizhanovsky et al. 2008).

We allowed wild-type and *p53$^{-/-}$;INK4a/ARF$^{-/-}$* stellate cells to become activated in vitro and examined their proliferative capacity 10 days thereafter in colony-formation assays. Whereas wild-type cells eventually undergo senescence in vitro and do not form colonies, *p53$^{-/-}$;INK4a/ARF$^{-/-}$* cells do not undergo senescence and continue to proliferate (Fig. 3A). Moreover, *p53$^{-/-}$;INK4a/ARF$^{-/-}$* cells display increased bromodeoxyuridine (BrdU) incorporation compared to wild-type cells (Krizhanovsky et al. 2008). These results demonstrate that *p53$^{-/-}$;INK4a/ARF$^{-/-}$*-activated stellate cells are immortalized in vitro.

To identify the consequences of impaired activated stellate cell senescence in vivo, we induced fibrosis in wild-type and *p53$^{-/-}$;INK4a/ARF$^{-/-}$* mice and tested their livers for fibrosis progression and amount of stellate cells. Liver fibrosis was significantly more pronounced in mutant *p53$^{-/-}$;INK4a/ARF$^{-/-}$* mice than in wild-type mice (Krizhanovsky et al. 2008) and was accompanied by a significant increase in the expression of the activated stellate cell marker αSMA (Fig. 3B) (see also Krizhanovsky et al. 2008). Therefore, senescence of activated stellate cells in wild-type mice protects the liver from unlimited expansion of these cells and more severe fibrosis.

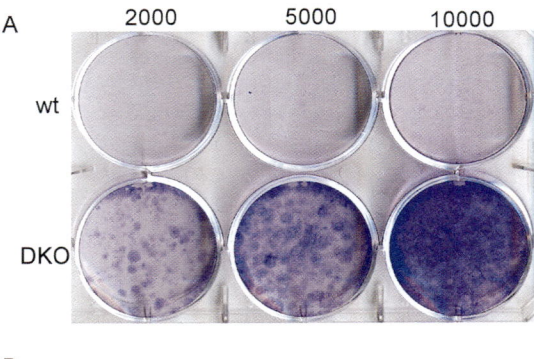

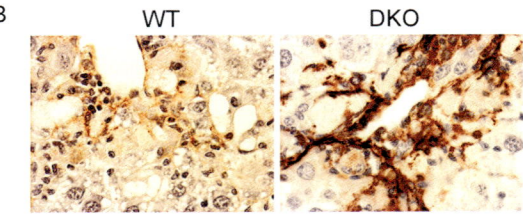

Figure 3. *p53$^{-/-}$;Ink4a/ARF$^{-/-}$* activated stellate cells are immortalized and contribute to fibrosis progression. (*A*) *p53$^{-/-}$;Ink4a/ARF$^{-/-}$* (DKO) but not wild-type-activated stellate cells form colonies following a 10-day colony-formation assay in vitro as evaluated by crystal violet staining. Numbers indicate amount of cells seeded per well. (*B*) Immunostaining identified higher expression of the activated stellate cell marker αSMA in fibrotic livers from DKO mice compared to wild-type mice.

EVIDENCE FOR HYPERPROLIFERATIVE SIGNALS IN ACTIVATED STELLATE CELLS

Several stimuli might be responsible for inducing senescence in activated stellate cells during the course of liver fibrosis. Telomere shortening is the driving force for senescence of normal human cells in culture (Campisi and d'Adda di Fagagna 2007); however, mouse cells have long telomeres, and it is therefore unlikely that in our model, telomeres in activated stellate cells will shorten sufficiently during the treatment period to lead to senescence. A boost in proliferation was observed before senescence in cells that undergo activated oncogene-induced senescence (Lin et al. 1998). We therefore examined fibrotic livers for the presence of activated *Akt*, which can mediate senescence in liver progenitor cells and other cell types (Chen et al. 2005). Indeed, elevated levels of an active phosphorylated *AKT* (Ser-473) were detected in cells positive for the stellate cell marker αSMA in fibrotic livers but not control livers (Fig. 4). Moreover, when we propagated human activated stellate cells in culture, phosphorylated *AKT* was detected in a proportion of late-passage cells (Krizhanovsky et al. 2008). These results are consistent with the possibility that the senescence of activated stellate cells is triggered by a hyperproliferative signal mediated, at least in part, by *AKT*.

SENESCENT ACTIVATED STELLATE CELLS ARE TARGETS OF THE IMMUNE SYSTEM IN VIVO

Interestingly, within 20 days of stopping CCl$_4$ treatment, liver fibrosis was almost resolved and senescent cells were not detected. Our previous work indicates that senescent cells can be cleared from tumors by the components of the innate immune system (Xue et al. 2007). Moreover, we have shown that senescent activated stellate cells up-regulate the expression of molecules involved in natural killer (NK) cell recognition and, furthermore, can be preferentially targeted by these cells in vitro (Krizhanovsky et al. 2008). In the fibrotic liver, immune cells (including NK cells) migrate into the fibrotic scar and create an inflammatory environment (Bataller and Brenner 2005; Muhanna et al. 2007). Using a combination of electron microscopy and immunophenotying by immunofluorescence, we noted that lymphocytes (including NK cells), macrophages, and neutrophils were detected adjacent to HSCs in fibrotic livers from CCl$_4$-treated mice (Fig. 5A) (Krizhanovsky et al. 2008). These immune cells were often in close proximity to cells expressing the senescent markers p53, p21, and Hmga1 (Krizhanovsky et al. 2008). Therefore, immune cells are localized adjacent to senescent activated stellate cells in fibrotic liver.

To evaluate an impact of the innate immune response on clearance of the senescent activated stellate cells and fibrosis resolution in vivo, we tested the impact of modulating NK cell activity on these processes in fibrotic livers. Following cessation of the fibrogenic stimulus, mice were treated with anti-NK neutralizing antibody to deplete NK cells or with polyI:C to induce an interferon-γ response and enhance NK cell activity. Livers from mice treated with the anti-NK antibody retained more senescent cells and fibrosis than controls, whereas senescent cells and fibrosis were more rapidly cleared in livers from mice treated with polyI:C (Krizhanovsky et al. 2008). We examined these livers for the presence of activated stellate cells by immunostaining. Numerous activated stellate cells were retained in the livers of anti-NK antibody-treated mice (Fig. 5B). In contrast, activated stellate cells were almost completely eliminated in the livers of polyI:C-treated mice. Therefore, the innate immune system can contribute to the elimination of senescent stellate cells from the fibrotic liver.

CELL-AUTONOMOUS AND NON-CELL-AUTONOMOUS EFFECTS OF SENESCENCE DURING TUMORIGENESIS AND WOUND HEALING

In cancer, senescence can serve as a tumor suppressor mechanism in multiple ways. First, oncogene-induced senescence provides an initial barrier to the development

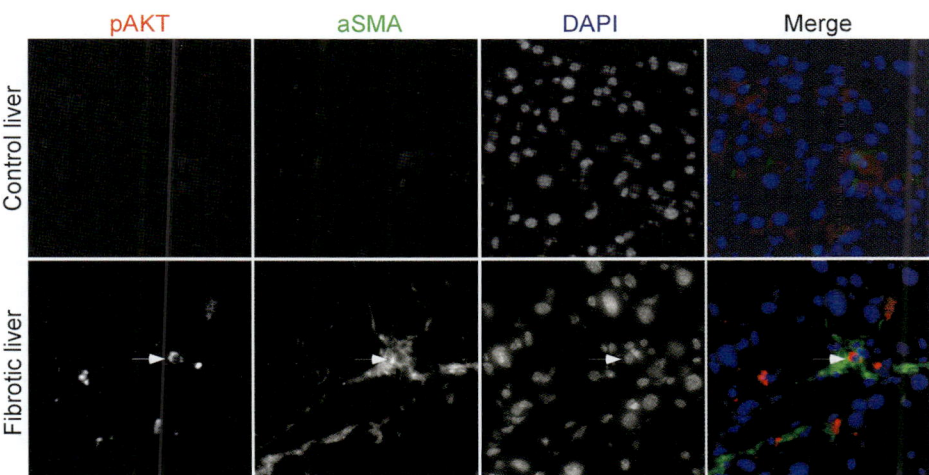

Figure 4. AKT signaling might contribute to the senescence of activated stellate cells. pAKT was detected in cells expressing the activated stellate cell marker αSMA in mouse fibrotic livers. Nuclei were identified by DAPI.

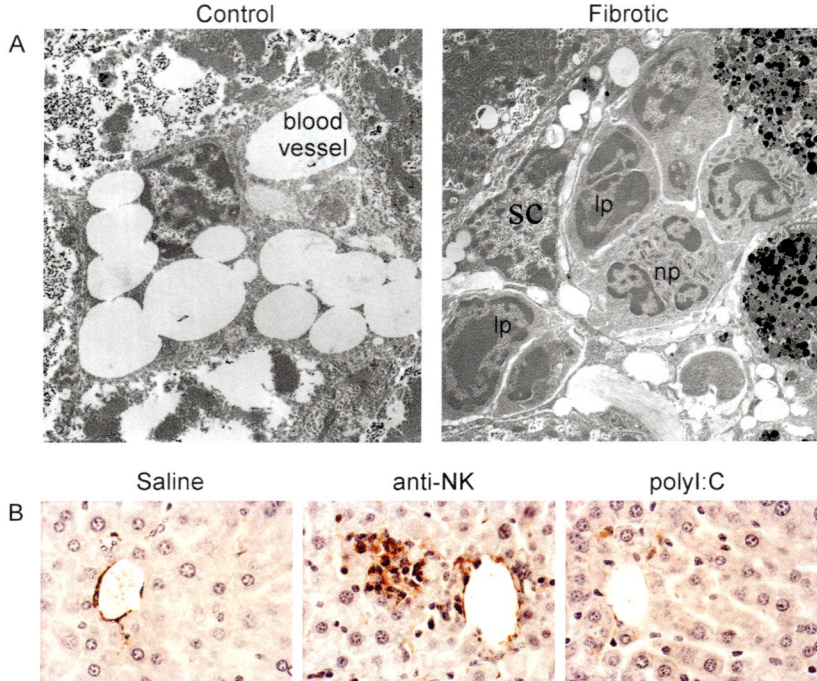

Figure 5. The immune system facilitates the clearance of senescent activated stellate cells in vivo. (*A*) Electron microscopy revealed that immune cells ([lp] lymphocytes; [np] neutrophil) are adjacent to activated HSCs in fibrotic mouse livers but not in normal mouse livers. (*B*) Mice treated with CCl₄ were treated with an anti-NK antibody (to deplete NK cells), polyI:C (as an interferon-γ activator), or saline (as a control) for 10 days. More activated stellate cells, identified by αSMA, are retained in mouse livers following depletion of NK cells.

of malignancies in skin, prostate, and hematopoietic and other systems, whereas telomere-based senescence may attenuate tumor progression (Narita and Lowe 2005). Second, upon tumor presentation, senescence can be induced in some tumor cells by chemotherapy agents (Schmitt et al. 2002). We have demonstrated that reactivation of the p53 pathway in tumors leads to activation of the senescence program and tumor clearance by the innate immune system in vivo (Xue et al. 2007). Therefore, the senescence machinery can remain intact in advanced cancers, where it remains capable of halting proliferation of tumor cells or cells carrying a strong oncogenic signal. If these processes occur in premalignant settings, senescence would limit tumorigenesis by preventing proliferation while at the same time exposing the damaged cells to a form of immune surveillance leading to their elimination.

Although the role of senescence in tumor suppression is now functionally established, its contribution to age-related disorders or other human pathologies is based on correlative data. However, our recent results show that senescence can act as a protective mechanism during tissue injury, in particular, by limiting the proliferation of activated stellate cells in response to acute injury and thus the associated fibrosis. Of note, tissue fibrosis contributes to a variety of pathologies beyond the liver, including those in the skin, lung, pancreas, kidney, and prostate. It therefore will be important to examine the role of senescence in other fibrotic conditions and wound-healing responses.

The potent cell cycle arrest that accompanies senescence is not the only mechanism by which the program limits fibrotic disease (Krizhanovsky et al. 2008). In addition, senescent cells decrease production of components of the ECM and up-regulate ECM degrading enzymes, thus decreasing the amount of deposited matrix. Moreover, senescent cells up-regulate a wide array of molecules that mediate interactions with the immune system that, in turn, may signal their subsequent elimination. Therefore, the senescence of activated stellate cells triggers a coordinated program that acts to inhibit the damage response by limiting the number of the cells and ECM production, on the one hand, and promoting tissue repair by signaling the elimination of senescence cells, on the other hand (Fig. 6).

Although our results suggest that senescence protects against liver fibrosis, in part, by stimulating clearance of these cells by the immune systems, it is noteworthy that chronic inflammation produced by viruses, alcohol abuse, or nonalcoholic steatohepatitis is a prerequisite for the development of hepatocellular carcinoma in humans. Although the pro-inflammatory signals that accompany liver damage are undoubtedly complex and arise from various sources, the contribution of senescent activated stellate cells in recruiting immune cells, although initially beneficial, may ultimately contribute to chronic inflammation that stimulates the emergence of malignant hepatocytes and cancer progression. In fact, xenotransplant studies mixing normal and senescent cells suggest that senescent cells can stimulate the transformation of premalignant epithelial cells (Krtolica et al. 2001).

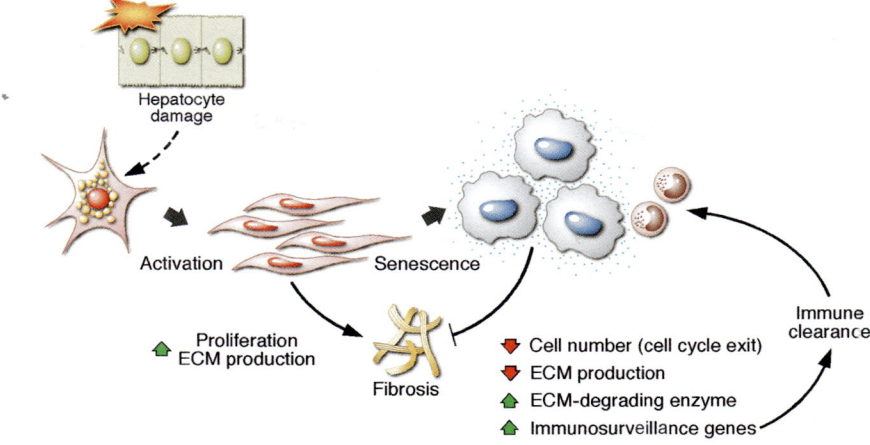

Figure 6. The eventual senescence of activated stellate cells limits fibrosis through a coordinated program involving cell cycle exit, down-regulation of ECM components, up-regulation of ECM-degrading enzymes, and enhanced immunosurveillance. (Reprinted, with permission, from Krizhanovsky et al. 2008 Supplemental Data [© Elsevier].)

Perhaps the role of senescence in limiting fibrosis evolved to protect the organism from acute tissue damage, but it can have a negative long-term role if the damage is chronic. It will be interesting to determine precisely how the senescence of mesenchymal cells involved in wound healing and fibrosis ultimately influences the initiation and progression of epithelial cancers.

ENDOGENOUS TRIGGERS OF SENESCENCE

In the context of tumor suppression, one of the key triggers of senescence is oncogene activation or other signals that produce a hyperproliferative response. The nature of this signal is the topic of much debate and likely involves activation of the *INK4a/ARF* locus as well as a DNA-damage response through signals produced by replication stress (Bartkova et al. 2006; Di Micco et al. 2006). One of the pathways that can lead to unrestricted proliferation and subsequent senescence in precancerous lesions is the PI3K pathway. Indeed, as presented here, hepatoblasts expressing hyperactive *Akt*—a key mediator of PI3K signaling—are prone to senescence in the presence of p53, and carcinomas induced by *Akt* and *p53* loss undergo senescence when p53 signaling is restored.

In principle, such hyperproliferative signals might drive senescence of activated stellate cells in liver fibrosis. Specifically, it is well established that upon liver injury, damaged hepatocytes send signals that trigger the massive expansion of stellate cells that are normally quiescent in the intact liver (Bataller and Brenner 2005). Although distinct from the situation produced in benign tumors, where hyperproliferative signals are produced by mutation, such strong signaling might lead to replicative stress and ultimately provide a built-in brake in the expansion of these cells. How this activation occurs and whether it is the definite driver of senescence during liver fibrosis remain to be determined. Nevertheless, we see up-regulation of PI3K signaling in senescent stellate cells and have noted DNA-

damage foci present in cells adjacent to the fibrotic scar (data not shown). Perhaps such signaling initially evolved to limit cell numbers during wound-healing responses and was co-opted as an anticancer mechanism later on.

EXPLOITING THE SENESCENCE-ASSOCIATED SECRETORY PHENOTYPE FOR SENESCENCE-MODULATING THERAPIES

It is now clear that the way in which senescence suppresses tumorigenesis or limits tissue damage extends beyond a cell-autonomous cell cycle arrest for which the program is best known. Hence, a non-cell-autonomous aspect of the program, the "senescence-associated secretory phenotype," clearly has an important role in limiting the accumulation of cells in tissues. On the one hand, this program stimulates the immune system to target senescent cells, and our studies show that this targeting has a key role in tumor suppression and tissue fibrosis (Xue et al. 2007; Krizhanovsky et al. 2008). Moreover, to at least some degree, these secreted proteins modulate the cell cycle exit program itself in an autocrine manner (Acosta et al. 2008; Kuilman et al. 2008; Wajapeyee et al. 2008).

The fact that the accumulation of senescent cells in tissues can be modulated externally suggests therapeutic strategies to modulate senescence-associated pathologies. In principle, this could be accomplished by biological therapies that interact with cell surface receptors to activate or inhibit senescence or by a variety of immune modulating drugs. In addition to their obvious anticancer potential, such strategies might be eventually used to induce senescence in fibrogenic cells from one side and enhance clearance of the senescent cells by the immune system from the other side, leading to better treatments for fibrotic conditions and possibly other wound-healing disorders. As examples, systemic delivery of the senescence inducer IGFBP7 can suppress the progression of melanoma xenografts in mice, and administration of polyI:C (which stimulates the immune system) accelerates the resolution of fibrosis (Krizhanovsky et al. 2008; Wajapeyee et al. 2008). Nevertheless, the success of such therapeutic applications will require a deeper understanding of the tissue-specific relationships between senescence and the microenvironment.

IMMORTALIZATION AND SELF-RENEWAL

Mutations that disable the senescence program—for example, loss of p53, Rb, and IGFBP7 or overexpression of Bmi-1 and CBX7—facilitate cellular immortalization and ultimately cancer development. In many ways, immortalization is similar to the process of cellular self-renewal, or the manner in which stem cells reproduce themselves, which is also a process of sustained cell division. It is striking that the normal process of stem cell self-renewal and the aberrant process of cellular immortalization share molecular pathways that have a central role in their regulation (Fig. 7). As discussed above, gene products of *INK4a/ARF* locus are established key elements in prevention of cell immortalization. Thus, p16/INK4a inhibits cyclin-dependent kinases 2 and 4, thereby preventing Rb phosphorylation and cell cycle progression. Both processes have been linked to self-renewal of stem cells in the hematopoietic and neuronal systems (Bruggeman et al. 2005; Molofsky et al. 2005), and in fact, p16 accumulates in certain stem cells with age and acts to limit self-renewal (Janzen et al. 2006; Krishnamurthy et al. 2006; Molofsky et al. 2006). Conversely, as discussed above, inactivation of the PTEN tumor suppressor can trigger senescence and limits the self-renewal of hematopoietic stem cells (Yilmaz et al. 2006).

Other convergent players that influence immortalization and self-renewal are the Polycomb proteins. In particular, the Polycomb repressor 2 complex protein Bmi-1 can promote cellular immortalization by suppressing the *INK4a/ARF* locus and appears to promote stem cell self-renewal in a variety of tissues (Bruggeman et al. 2005; Gil et al. 2005; Molofsky et al. 2005). Less understood are factors involved in Wnt signaling, which controls self-renewal and differen-

tiation of stem cells from multiple origins (Reya and Clevers 2005) and can facilitate immortalization in primary melanocytes (Delmas et al. 2007). Thus, the critical cellular decision to proliferate is regulated by multiple pathways, which can positively affect self-renewal or enable bypass of senescence and facilitate cellular immortalization.

The overlaps between immortalization (or escape from senescence) and self-renewal (or suppression of differentiation) can be extended to the settings of HCC progression and liver fibrosis studied here. For example, although reactivation of p53 in *Hras^{V12}* and *Akt*-expressing tumors drives a cell cycle arrest program with many hallmarks of senescence, this program is accompanied by the appearance of differentiation markers (see Fig. 1) (Xue et al. 2007). Conversely, in liver fibrosis following an initially proliferative burst, activated stellate cells eventually senesce, which might be viewed as a state of terminal differentiation. Interestingly, quiescent HSCs were recently reported to have stem cell properties, sharing the ability to proliferate and differentiate into multiple cell lineages (Yang et al. 2008). These cells might serve as regional multipotent progenitors that are unable to self-renew but are able to differentiate into multiple lineages.

CONCLUSIONS

In summary, the processes of senescence and immortalization, and differentiation and self-renewal, share overlapping signaling mechanisms and can be viewed as two sides of the same coin. What these similarities reveal, and the extent to which they apply to different contexts, remains to be determined. On the one hand, many of the processes currently linked to "self-renewal" have long been established to influence immortalization. As one example, there is much debate about the origin of cancer stem cells and how they acquire self-renewal capabilities; however, from the discussion above, it seems likely that in many cases, this reflects the process of immortalization studied for the last 30 years. Indeed, disruption of the ARF locus, well established to facilitate cellular immortalization, is sufficient to confer "cancer stem cell" properties to virtually all lymphoid cells expressing Bcr-Abl (Williams and Sherr 2008). On the other hand, these overlaps might have a deeper meaning, reflecting mechanisms that initially evolved to control cell numbers in normal tissues but were later incorporated into intrinsic tumor suppressive mechanisms to limit uncontrolled cell division. If this proves to be the case, it will be of interest to determine whether the self-renewal of stem cells is also influenced by autocrine and paracrine mechanisms that parallel those that influence differentiation. If so, such knowledge might stimulate the development of biological therapies to influence stem cell self-renewal and tissue regeneration.

ACKNOWLEDGMENTS

We thank L. Dow and P. Smith for editorial advice, members of the Lowe laboratory for stimulating discus-

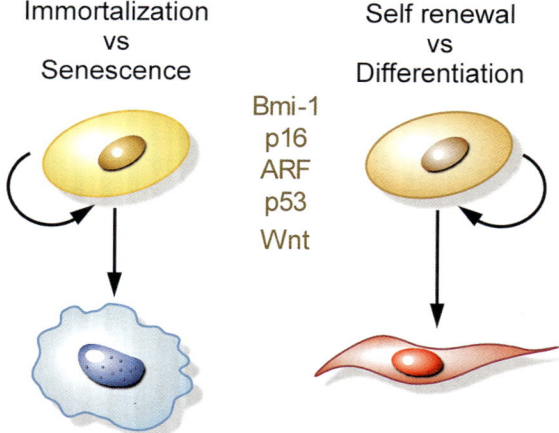

Figure 7. Molecular pathways driving cell fate decisions. Common regulators are illustrated.

sions, S. Hearn for electron microscopy, L. Chiriboga for immunohistochemistry, and J. Simon, K. Diggins-Lehet, L. Bianco, and the CSHL animal facility for help with animals. This work was supported by a postdoctoral fellowship from the Leukemia and Lymphoma Society (V.K.) and grant AG16379 from the National Institutes of Health (S.W.L.). M.Y. is supported by a doctoral fellowship from the National Council for Scientific and Technological Development (CNPq). L.Z. is a Seligson Clinical Fellow. S.W.L. is a Howard Hughes Medical Institute investigator.

REFERENCES

Acosta, J.C., O'Loghlen, A., Banito, A., Guijarro, M.V., Augert, A., Raguz, S., Fumagalli, M., Da Costa, M., Brown, C., Popov, N., et al. 2008. Chemokine signaling via the CXCR2 receptor reinforces senescence. *Cell* **133:** 1006–1018.

Bartkova, J., Rezaei, N., Liontos, M., Karakaidos, P., Kletsas, D., Issaeva, N., Vassiliou, L.V., Kolettas, E., Niforou, K., Zoumpourlis, V.C., et al. 2006. Oncogene-induced senescence is part of the tumorigenesis barrier imposed by DNA damage checkpoints. *Nature* **444:** 633–637.

Bataller, R. and Brenner, D.A. 2005. Liver fibrosis. *J. Clin. Invest.* **115:** 209–218.

Beausejour, C.M., Krtolica, A., Galimi, F., Narita, M., Lowe, S.W., Yaswen, P., and Campisi, J. 2003. Reversal of human cellular senescence: Roles of the p53 and p16 pathways. *EMBO J.* **22:** 4212–4222.

Braig, M., Lee, S., Loddenkemper, C., Rudolph, C., Peters, A.H., Schlegelberger, B., Stein, H., Dorken, B., Jenuwein, T., and Schmitt, C.A. 2005. Oncogene-induced senescence as an initial barrier in lymphoma development. *Nature* **436:** 660–665.

Brugge, J., Hung, M.C., and Mills, G.B. 2007. A new mutational AKTivation in the PI3K pathway. *Cancer Cell* **12:** 104–107.

Bruggeman, S.W., Valk-Lingbeek, M.E., van der Stoop, P.P., Jacobs, J.J., Kieboom, K., Tanger, E., Hulsman, D., Leung, C., Arsenijevic, Y., Marino, S., and van Lohuizen, M. 2005. *Ink4a* and *Arf* differentially affect cell proliferation and neural stem cell self-renewal in *Bmi1*-deficient mice. *Genes Dev.* **19:** 1438–1443.

Campisi, J. and d'Adda di Fagagna, F. 2007. Cellular senescence: When bad things happen to good cells. *Nat. Rev. Mol. Cell Biol.* **8:** 729–740.

Chen, Z., Trotman, L.C., Shaffer, D., Lin, H.K., Dotan, Z.A., Niki, M., Koutcher, J.A., Scher, H.I., Ludwig, T., Gerald, W., Cordon-Cardo, C., and Pandolfi, P.P. 2005. Crucial role of p53-dependent cellular senescence in suppression of Pten-deficient tumorigenesis. *Nature* **436:** 725–730.

Collado, M., Blasco, M.A., and Serrano, M. 2007. Cellular senescence in cancer and aging. *Cell* **130:** 223–233.

Collado, M., Gil, J., Efeyan, A., Guerra, C., Schuhmacher, A.J., Barradas, M., Benguria, A., Zaballos, A., Flores, J.M., Barbacid, M., Beach, D., and Serrano, M. 2005. Tumour biology: Senescence in premalignant tumours. *Nature* **436:** 642.

Courtois-Cox, S., Genther Williams, S.M., Reczek, E.E., Johnson, B.W., McGillicuddy, L.T., Johannessen, C.M., Hollstein, P.E., MacCollin, M., and Cichowski, K. 2006. A negative feedback signaling network underlies oncogene-induced senescence. *Cancer Cell* **10:** 459–472.

Delmas, V., Beermann, F., Martinozzi, S., Carreira, S., Ackermann, J., Kumasaka, M., Denat, L., Goodall, J., Luciani, F., Viros, A., et al. 2007. β-Catenin induces immortalization of melanocytes by suppressing *p16INK4a* expression and cooperates with N-Ras in melanoma development. *Genes Dev.* **21:** 2923–2935.

Di Micco, R., Fumagalli, M., Cicalese, A., Piccinin, S., Gasparini, P., Luise, C., Schurra, C., Garre, M., Nuciforo, P.G., Bensimon, A., et al. 2006. Oncogene-induced senescence is a DNA damage response triggered by DNA hyper-replication. *Nature* **444:** 638–642.

Dickins, R.A., Hemann, M.T., Zilfou, J.T., Simpson, D.R., Ibarra, I., Hannon, G.J., and Lowe, S.W. 2005. Probing tumor phenotypes using stable and regulated synthetic microRNA precursors. *Nat. Genet.* **37:** 1289–1295.

Elsharkawy, A.M. and Mann, D.A. 2007. Nuclear factor-κB and the hepatic inflammation-fibrosis-cancer axis. *Hepatology* **46:** 590–597.

Gil, J., Bernard, D., and Peters, G. 2005. Role of polycomb group proteins in stem cell self-renewal and cancer. *DNA Cell Biol.* **24:** 117–125.

Ha, L., Ichikawa, T., Anver, M., Dickins, R., Lowe, S., Sharpless, N.E., Krimpenfort, P., Depinho, R.A., Bennett, D.C., Sviderskaya, E.V., and Merlino, G. 2007. ARF functions as a melanoma tumor suppressor by inducing p53-independent senescence. *Proc. Natl. Acad. Sci.* **104:** 10968–10973.

Hayflick, L. and Moorhead, P.S. 1961. The serial cultivation of human diploid cell strains. *Exp. Cell Res.* **25:** 585–621.

He, L., He, X., Lowe, S.W., and Hannon, G.J. 2007a. microRNAs join the p53 network—Another piece in the tumour-suppression puzzle. *Nat. Rev. Cancer* **7:** 819–822.

He, L., He, X., Lim, L.P., de Stanchina, E., Xuan, Z., Liang, Y., Xue, W., Zender, L., Magnus, J., Ridzon, D., et al. 2007b. A microRNA component of the p53 tumour suppressor network. *Nature* **447:** 1130–1134.

Janzen, V., Forkert, R., Fleming, H.E., Saito, Y., Waring, M.T., Dombkowski, D.M., Cheng, T., DePinho, R.A., Sharpless, N.E., and Scadden, D.T. 2006. Stem-cell ageing modified by the cyclin-dependent kinase inhibitor p16INK4a. *Nature* **443:** 421–426.

Krishnamurthy, J., Ramsey, M.R., Ligon, K.L., Torrice, C., Koh, A., Bonner-Weir, S., and Sharpless, N.E. 2006. p16INK4a induces an age-dependent decline in islet regenerative potential. *Nature* **443:** 453–457.

Krizhanovsky, V., Yon, M., Dickins, R.A., Hearn, S., Simon, J., Miething, C., Yee, H., Zender, L., and Lowe, S.W. 2008. Senescence of activated stellate cells limits liver fibrosis. *Cell* **134:** 657–667.

Krtolica, A., Parrinello, S., Lockett, S., Desprez, P.Y., and Campisi, J. 2001. Senescent fibroblasts promote epithelial cell growth and tumorigenesis: A link between cancer and aging. *Proc. Natl. Acad. Sci.* **98:** 12072–12077.

Kuilman, T., Michaloglou, C., Vredeveld, L.C., Douma, S., van Doorn, R., Desmet, C.J., Aarden, L.A., Mooi, W.J., and Peeper, D.S. 2008. Oncogene-induced senescence relayed by an interleukin-dependent inflammatory network. *Cell* **133:** 1019–1031.

Lazzerini Denchi, E., Attwooll, C., Pasini, D., and Helin, K. 2005. Deregulated E2F activity induces hyperplasia and senescence-like features in the mouse pituitary gland. *Mol. Cell. Biol.* **25:** 2660–2672.

Lin, A.W., Barradas, M., Stone, J.C., Van Aelst, L., Serrano, M., and Lowe, S.W. 1998. Premature senescence involving p53 and p16 is activated in response to constitutive MEK/MAPK mitogenic signaling. *Genes Dev.* **12:** 3008–3019.

Michaloglou, C., Vredeveld, L.C., Soergas, M.S., Denoyelle, C., Kuilman, T., van der Horst, C.M., Majoor, D.M., Shay, J.W., Mooi, W.J., and Peeper, D.S. 2005. BRAFE600-associated senescence-like cell cycle arrest of human naevi. *Nature* **436:** 720–724.

Molofsky, A.V., He, S., Bydon, M., Morrison, S.J., and Pardal, R. 2005. Bmi-1 promotes neural stem cell self-renewal and neural development but not mouse growth and survival by repressing the p16Ink4a and p19Arf senescence pathways. *Genes Dev.* **19:** 1432–1437.

Molofsky, A.V., Slutsky, S.G., Joseph N.M., He, S., Pardal, R., Krishnamurthy, J., Sharpless, N.E., and Morrison, S.J. 2006. Increasing *p16INK4a* expression decreases forebrain progenitors and neurogenesis during ageing. *Nature* **443:** 448–452.

Mooi, W.J. and Peeper, D.S. 2006. Oncogene-induced cell senescence: Halting on the road to cancer. *N. Engl. J. Med.* **355:** 1037–1046.

Muhanna, N., Horani, A., Doron, S., and Safadi, R. 2007. Lymphocyte-hepatic stellate cell proximity suggests a direct interaction. *Clin. Exp. Immunol.* **148:** 338–347.

Narita, M. and Lowe, S.W. 2005. Senescence comes of age. *Nat. Med.* **11:** 920–922.

Narita, M., Narita, M., Krizhanovsky, V., Nuñez, S., Chicas, A., Hearn, S.A., Myers, M.P., and Lowe, S.W. 2006. A novel role for high-mobility group A proteins in cellular senescence and heterochromatin formation. *Cell* **126:** 503–514.

Narita, M., Nunez, S., Heard, E., Narita, M., Lin, A.W., Hearn, S.A., Spector, D.L., Hannon, G.J., and Lowe, S.W. 2003. Rb-mediated heterochromatin formation and silencing of E2F target genes during cellular senescence. *Cell* **113:** 703–716.

NCHS (National Center for Health Statistics). 2004. Chronic liver disease/cirrhosis. In *National Vital Statistics Report*. National Center for Health Statistics, Hyattsville, Maryland.

Reya, T. and Clevers, H. 2005. Wnt signalling in stem cells and cancer. *Nature* **434:** 843–850.

Roninson, I.B. 2003. Tumor cell senescence in cancer treatment. *Cancer Res.* **63:** 2705–2715.

Rossi, D.J., Jamieson, C.H., and Weissman, I.L. 2008. Stem cells and the pathways to aging and cancer. *Cell* **132:** 681–696.

Rudolph, K.L., Chang, S., Millard, M., Schreiber-Agus, N., and DePinho, R.A. 2000. Inhibition of experimental liver cirrhosis in mice by telomerase gene delivery. *Science* **287:** 1253–1258.

Schmitt, C.A., Fridman, J.S., Yang, M., Lee, S., Baranov, E., Hoffman, R.M., and Lowe, S.W. 2002. A senescence program controlled by p53 and p16^{INK4a} contributes to the outcome of cancer therapy. *Cell* **109:** 335–346.

Serrano, M., Lin, A.W., McCurrach, M.E., Beach, D., and Lowe, S.W. 1997. Oncogenic *ras* provokes premature cell senescence associated with accumulation of p53 and p16^{INK4a}. *Cell* **88:** 593–602.

Shay, J.W. and Roninson, I.B. 2004. Hallmarks of senescence in carcinogenesis and cancer therapy. *Oncogene* **23:** 2919–2933.

te Poele, R.H., Okorokov, A.L., Jardine, L., Cummings, J., and Joel, S.P. 2002. DNA damage is able to induce senescence in tumor cells in vitro and in vivo. *Cancer Res.* **62:** 1876–1883.

Vaziri, H., West, M.D., Allsopp, R.C., Davison, T.S., Wu, Y.S., Arrowsmith, C.H., Poirier, G.G., and Benchimol, S. 1997. ATM-dependent telomere loss in aging human diploid fibroblasts and DNA damage lead to the post-translational activation of p53 protein involving poly(ADP-ribose) polymerase. *EMBO J.* **16:** 6018–6033.

Wajapeyee, N., Serra, R.W., Zhu, X., Mahalingam, M., and Green, M.R. 2008. Oncogenic BRAF induces senescence and apoptosis through pathways mediated by the secreted protein IGFBP7. *Cell* **132:** 363–374.

Wendel, H.G., De Stanchina, E., Fridman, J.S., Malina, A., Ray, S., Kogan, S., Cordon-Cardo, C., Pelletier, J., and Lowe, S.W. 2004. Survival signalling by Akt and eIF4E in oncogenesis and cancer therapy. *Nature* **428:** 332–337.

Williams, R.T. and Sherr, C.J. 2008. BCR-ABL and CDKN2A: A dropped connection. *Nat. Rev. Cancer* **8:** 563.

Xue, W., Zender, L., Miething, C., Dickins, R.A., Hernando, E., Krizhanovsky, V., Cordon-Cardo, C., and Lowe, S.W. 2007. Senescence and tumour clearance is triggered by p53 restoration in murine liver carcinomas. *Nature* **445:** 656–660.

Yang, L., Jung, Y., Omenetti, A., Witek, R.P., Choi, S., Vandongen, H.M., Huang, J., Alpini, G.D., and Diehl, A.M. 2008. Fate-mapping evidence that hepatic stellate cells are epithelial progenitors in adult mouse livers. *Stem Cells* **26:** 2104–2113.

Yilmaz, O.H., Valdez, R., Theisen, B.K., Guo, W., Ferguson, D.O., Wu, H., and Morrison, S.J. 2006. Pten dependence distinguishes haematopoietic stem cells from leukaemia-initiating cells. *Nature* **441:** 475–482.

Zender, L., Spector, M.S., Xue, W., Flemming, P., Cordon-Cardo, C., Silke, J., Fan, S.T., Luk, J.M., Wigler, M., Hannon, G.J., et al. 2006. Identification and validation of oncogenes in liver cancer using an integrative oncogenomic approach. *Cell* **125:** 1253–1267.

Regulation of Self-renewal and Differentiation in the *Drosophila* Nervous System

T.D. SOUTHALL, B. EGGER, K.S. GOLD, AND A.H. BRAND

The Gurdon Institute and Department of Physiology, Development, and Neuroscience,
University of Cambridge, Cambridge CB2 1QN, United Kingdom

Stem cells can divide symmetrically to generate two similar daughter cells and expand the stem cell pool or asymmetrically to self-renew and generate differentiating daughter cells. The proper balance between symmetric and asymmetric division is critical for the generation and subsequent repair of tissues. Furthermore, unregulated stem cell division has been shown to result in tumorous overgrowth. The *Drosophila* nervous system has proved to be a fruitful model system for studying the biology of neural stem cell division and uncovering the molecular mechanisms that, when disrupted, can lead to tumor formation. We are using the *Drosophila* embryonic and larval nervous systems as models to study the regulation of symmetric and asymmetric stem cell division.

Stem cells have the capacity to renew themselves at each division while producing a continuous supply of differentiating daughters for the generation, and subsequent repair, of tissues (for review, see Weissman et al. 2001). These two aspects of stem cell biology must be carefully balanced during development to ensure that organs of the right size and tissue composition are formed. The spatially and temporally regulated differentiation of multipotent stem cells generates the enormous cellular diversity seen in the nervous system. A common strategy that stem cells use to generate cellular diversity is asymmetric cell division, where the unequal distribution of cell-fate determinants leads to the production of daughter cells with different fates (for review, see Morrison and Kimble 2006; Doe 2008; Gonczy 2008; Knoblich 2008). The *Drosophila* nervous system has proved to be a fertile ground not only for studying the asymmetric divisions of stem cells (Wu et al. 2008), but also, more recently, for understanding how unregulated division in a stem cell lineage can lead to tumorigenesis (Caussinus and Gonzalez 2005; Bello et al. 2006; Betschinger et al. 2006; Choksi et al. 2006; Lee et al. 2006a). Here, we focus on two key aspects of neural stem cell regulation: how the transition from proliferation to differentiation is achieved within a pool of stem cells and how the switch from self-renewal to differentiation is balanced within a single stem cell division.

SYMMETRIC VERSUS ASYMMETRIC DIVISION IN THE CENTRAL NERVOUS SYSTEM

Stem cells can proliferate through symmetric division, which leads to similar daughter cells, or they can divide asymmetrically to self-renew the mother cell and generate a differentiating daughter cell. The correct ratio between symmetric and asymmetric stem cell divisions is crucial for achieving the proper size of an organ and for the timely differentiation of a tissue. Failure to maintain this balance, or the incorrect activation of either division

mode, can lead to tumorous overgrowth or to premature differentiation (Morrison and Kimble 2006).

The importance of tight developmental controls over asymmetric division has been demonstrated in *Drosophila* by a number of studies showing that when the machinery that regulates asymmetric divisions is disrupted, neuroblasts proliferate aberrantly and form tumors (Albertson and Doe 2003; Caussinus and Gonzalez 2005; Bello et al. 2006; Betschinger et al. 2006; Choksi et al. 2006; Lee et al. 2006a,b). Less is known about the regulation of symmetric, proliferative stem cell division, which serves to increase a stem cell pool during normal development, nor how the transition from symmetric to asymmetric division is controlled. Such a transition can be seen in the developing mammalian cortex, where neuroepithelial cells initially proliferate symmetrically and then convert to radial glial cells (the main progenitor type) during neurogenesis (Gotz and Huttner 2005). The *Drosophila* larval optic lobe provides an elegant and tractable system for studying how the switch from one mode of division to another occurs.

The optic lobe generates the visual processing centers of the adult brain. It arises from two proliferative neuroepithelia in the larva, known as the inner and outer proliferation centers (IPC and OPC, respectively). During larval development, two morphologically and molecularly distinct cell populations can be identified in the OPC: lateral, symmetrically dividing neuroepithelial cells and medial, asymmetrically dividing medulla neuroblasts (Fig. 1) (White and Kankel 1978; Hofbauer and Campos-Ortega 1990; Ceron et al. 2001; Egger et al. 2007). Clonal analysis has shown that the symmetrically dividing neuroepithelial progenitors transform into asymmetrically dividing medulla neuroblasts at the medial edge of the OPC (Fig. 2) (Egger et al. 2007).

The transition from neuroepithelial cell to neuroblast occurs in a temporally regulated progression (Egger et al. 2007). The molecular mechanisms underlying this transition remain largely unknown, with the exception of one

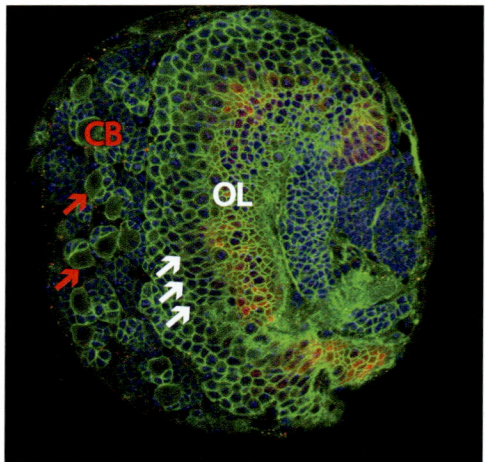

Figure 1. Neural stem cells in the larval brain. The symmetrically dividing neuroepithelial cells and asymmetrically dividing neuroblasts of the optic lobe (OL, *white arrows*) can be seen on the right-hand side of the brain lobe, whereas, on the left, the large neuroblasts of the central brain (CB, *red arrows*) are evident. (*Green*) Dlg; (*red*) l'sc-mRFP1; (*blue*) DNA.

recent study (Yasugi et al. 2008). Yasugi and colleagues have shown that the transition from symmetrically dividing neuroepithelial cells to asymmetrically dividing neuroblasts coincides with expression of the proneural gene *lethal of scute* (*l'sc*) in the most medial epithelial cells. The transformation of neuroepithelial cells to neuroblasts is driven by a "proneural wave" of L'sc expression that sweeps across the neuroepithelium from medial to lateral. Progession of this proneural wave is negatively regulated by the Janus kinase–signal transducer and activator of transcription (JAK/STAT) pathway, whose ligand, Unpaired (Upd), signals from the most lateral neuroepithelial cells (Yasugi et al. 2008).

To identify potential regulators of the symmetric-to-asymmetric transition, we characterized the transcriptional profiles of neuroepithelial cells and medulla neuroblasts. We devised a method by which to label and extract neuroepithelial cells and neuroblasts from the brains of live late third-instar larvae. Neuroepithelial and neuroblast cDNA samples were collected and directly compared on *Drosophila* whole transcriptome oligonucleotide microarrays. Statistical analysis of the data yielded nearly 200 genes that are significantly differentially expressed between neuroepithelial cells and neuroblasts. Several distinct classes of genes are strikingly enriched in the optic lobe neuroepithelium: the Delta/Notch and Hedgehog signaling pathways, proneural genes, and a number of genes with a role in retinal determination.

SELF-RENEWAL VERSUS DIFFERENTIATION IN THE CENTRAL NERVOUS SYSTEM

Drosophila neuroblasts divide in a regenerative fashion, producing a large daughter cell that self-renews and a smaller daughter cell (a ganglion mother cell [GMC]) that divides only once to give two postmitotic neurons or glial cells. This asymmetry in cell fate is achieved by the unequal partitioning of determinants (for review, see Morrison and Kimble 2006; Doe 2008; Gonczy 2008; Knoblich 2008; Zhong and Chia 2008). At each neuroblast division, cell-fate determinants, such as the phosphotyrosine-binding (PTB) domain protein Numb, the homeodomain transcription factor Prospero, and the NHL-domain protein Brain Tumour (Brat), are localized to the basal cortex by the adapter proteins Partner of Numb (Pon) and Miranda. When neuroblasts divide, only the basal daughter cell (the GMC) inherits Miranda, Pon, and their cargo.

We have been studying the role of Prospero in asymmetric division in regulating self-renewal and differentiation. Prospero is conserved in vertebrates, where the Prox (for Prospero-related homeobox) family of atypical homeodomain transcription factors appears to have a role in initiating the differentiation of progenitors in various

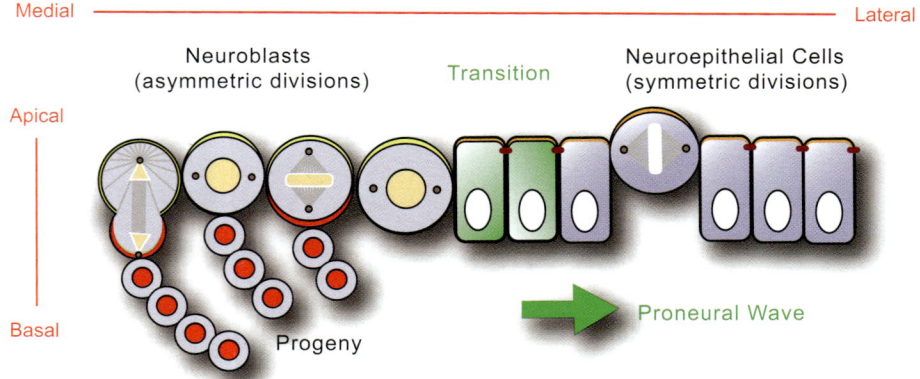

Figure 2. Neuroepithelial-to-neuroblast transition in the larval optic lobe. At the medial edge of the optic lobe, columnar neuroepithelial cells undergo a transition to neuroblasts. Neuroepithelial cells divide symmetrically, with horizontal spindle orientation, which results in the expansion of the progenitor pool. Medial neuroblasts divide asymmetrically, with vertical spindle orientation, and bud off smaller ganglion mother cells (GMCs) toward the medulla cortex. Most medial neuroepithelial cells (*green*) transiently express the proneural gene, *lethal of scute* (*l'sc*). This expression sweeps across the epithelium in the form of a proneural wave, from medial to lateral, as more neuroblasts are generated. (Adapted from Egger et al. 2007; Yasugi et al. 2008.)

tissues (Torii et al. 1999; Wigle et al. 2002; Dyer et al. 2003; Pistocchi et al. 2008). Prox1 is also associated with several reported cases of cancer progression (Versmold et al. 2007; Yoshimoto et al. 2007; Dudas et al. 2008; Petrova et al. 2008).

Prospero is transcribed in the neuroblast but not the GMC (Broadus et al. 1998). Prospero protein and its mRNA are anchored at the cortex of the neuroblast by the adapter protein Miranda (Ikeshima-Kataoka et al. 1997; Shen et al. 1997, 1998; Matsuzaki et al. 1998; Schuldt et al. 1998). Once segregated to the GMC by Miranda, Prospero is released from the cortex and enters the GMC nucleus. The nuclear localization of Prospero is one of the first molecular differences between a self-renewing neuroblast and a GMC (Hirata et al. 1995; Knoblich et al. 1995; Spana and Doe 1995). Therefore, identifying the genes regulated by Prospero should provide important insights into the genetic networks governing the switch from a stem cell to a differentiating cell.

To determine the sites to which Prospero binds throughout the genome, we used the in vivo chromatin profiling technique DamID (van Steensel and Henikoff 2000; van Steensel et al. 2001). In brief, DamID involves fusing any DNA or chromatin-associated protein of interest to a bacterial DNA adenine methyltransferase (Dam). When the fusion protein binds to its native site in the genome, the methylase leaves a unique methylation tag on the sequence GATC. Genomic DNA is then extracted and digested with DpnI, which cuts specifically at methylated GATC sites. The tagged sequences can then be isolated and identified on DNA tiling microarrays.

We generated transgenic embryos expressing a Dam-Prospero fusion protein, or Dam alone to control for nonspecific methylation, and collected genomic DNA material from stage-10–11 embryos. Following digestion with DpnI and ligation-mediated polymerase chain reaction (PCR) amplification of the methylated regions of DNA, the DNA was hybridized to full-genome tiling microarrays covering the whole euchromatic genome (Fig. 3A) (Choksi et al. 2006). The data from four biological replicates identified 1602 Prospero-binding sites. Of these, 1066 contain a previously published Prospero-consensus-binding site (Cook

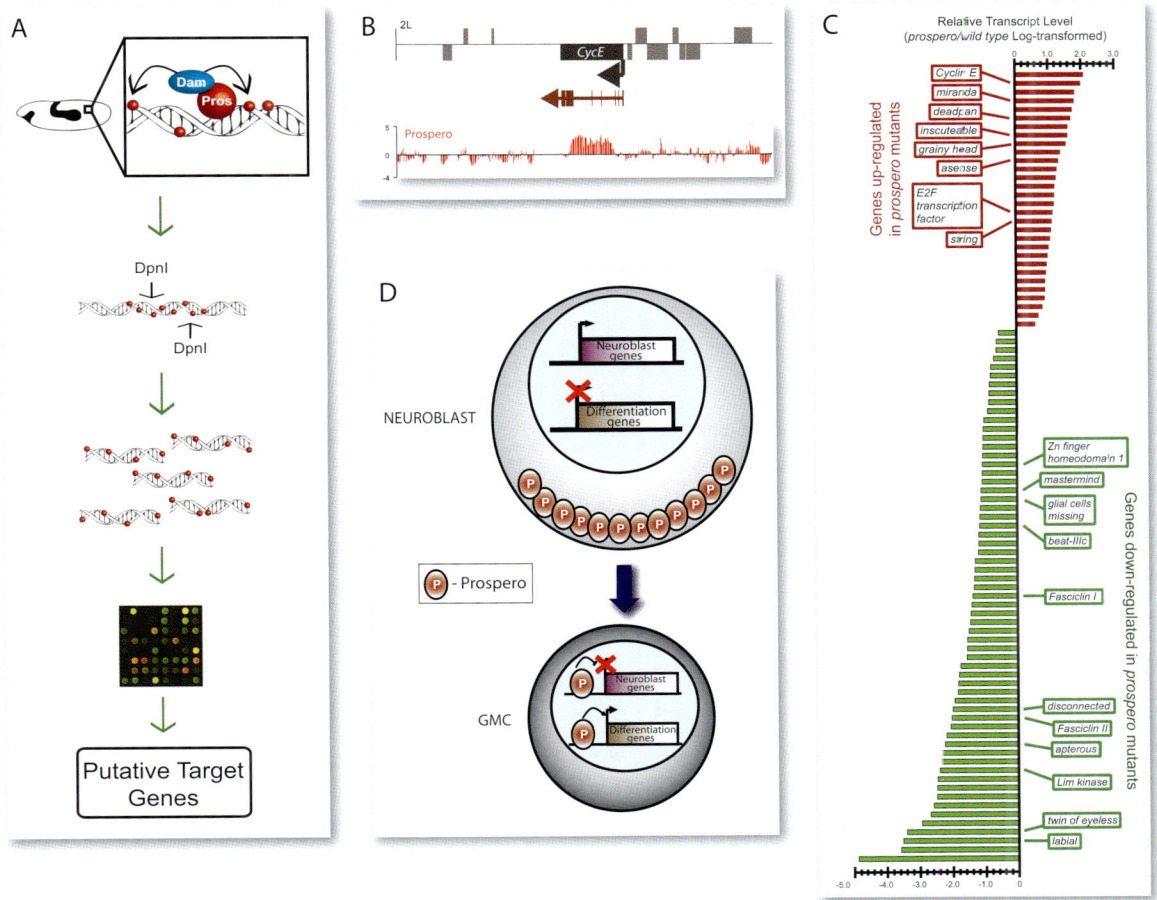

Figure 3. Genome-wide mapping of Prospero-binding sites. (*A*) Schematic diagram of the DamID technique. (*B*) Prospero binding at the *CycE* locus. (*Gray* and *black boxes*) Genes; (*brown boxes*) *CycE* exons. Bar heights are proportional to the average of normalized log-transformed ratio of intensities from four replicate DamID in vivo binding-site-mapping experiments. (*C*) High-resolution gene expression profiling of cells from the ventral nerve cord of *prospero* mutants. (*Red*) Genes that are up-regulated; (*green*) genes that are down-regulated. (*D*) Model showing Prospero's segregation from the neuroblast to the GMC, where it enters the nucleus to repress neuroblast genes and activate differentiation genes.

et al. 2003), corresponding to 730 genes (Choksi et al. 2006). An example of Prospero binding to the intronic regions of *Cyclin E* (*CycE*) is shown in Figure 3B.

An unbiased gene ontology analysis of these genes using GOtoolbox (Martin et al. 2004) highlighted cell-fate commitment, nervous system development, and regulation of transcription as the most enriched classes of biological function. This correlates with the known function and expression pattern of Prospero. Furthermore, Prospero binds to 41% of all annotated neuroblast genes and key regulators of the cell cycle, including *Cyclin E* (Fig. 3B), the *cdc25* homolog *string*, and *E2F*.

Expression of Prospero is maintained in glial cells but not in neurons, where the protein is absent (Spana and Doe 1995). Consistent with this, Prospero directly binds 45% of genes annotated to have a role in gliogenesis. However, to our surprise, we find that Prospero also binds to many genes involved in the terminal differentiation of neurons. These include 19% of all annotated neuronal differentiation genes, including the adhesion molecule FasciclinII (Lin et al. 1994) and the axon guidance molecule Netrin-B (Mitchell et al. 1996).

DamID provides information on whether a gene is bound by a transcription factor but not whether it is activated or repressed. To elucidate how Prospero regulates its target genes, we performed transcriptional profiling of neuroblasts and GMCs from wild-type and *prospero* mutant embryos. Approximately 100 cells were isolated from the ventral nerve cords of embryos using a glass capillary. The cells were expelled into lysis buffer, and cDNA was generated by reverse transcription and PCR amplification (Iscove et al. 2002; Choksi et al. 2006). These cDNA libraries were then compared on full-genome oligonucleotide microarrays.

We identified 91 genes that are directly bound by Prospero and show a significant change in transcription (Fig. 3C) (Choksi et al. 2006). Many genes involved in neuroblast cell-fate determination (*asense, deadpan, miranda, inscuteable*) and in cell cycle regulation (*Cyclin E* and *string*) show increased levels in a *prospero* mutant background, consistent with their being repressed by Prospero. Conversely, genes with a role in neuronal differentiation show decreased transcription in *prospero* mutants, indicating that Prospero is required for their transcription.

GMCs ARE TRANSFORMED INTO NEURAL STEM CELLS IN *PROSPERO* MUTANTS

If Prospero is key in promoting differentiation, then, in the absence of Prospero, neuroblasts should give rise to two self-renewing neuroblast-like cells. We followed the division pattern of individual neuroblasts in *prospero* mutant embryos by labeling with the lipophilic dye DiI (Choksi et al. 2006). Wild-type neuroblasts generate, on average, 16.2 cells that exhibit extensive axonal outgrowth (Fig. 4, left). However, *prospero* mutant neuroblasts produce an average of 31.8 cells with few, if any, projections (Fig. 4, right). Thus, *prospero* mutant neuroblasts produce much larger clones of cells with no axonal projections,

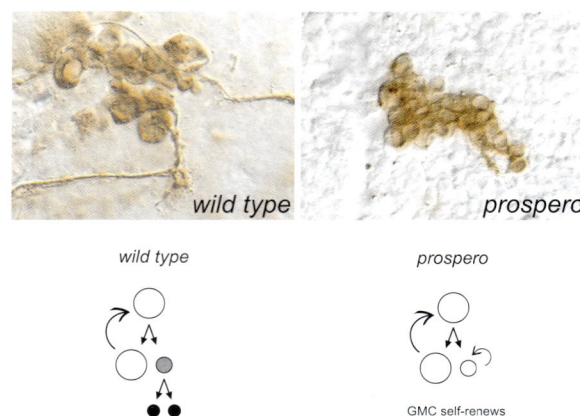

Figure 4. Neuroblasts overproliferate and fail to differentiate in *prospero* mutants. (*Left*) Wild-type clones, from a single S1 or S2 neuroblast labeled with DiI, extend axonal projections that often exit the central nervous system. (*Right*) *prospero* mutant clones give, on average, twice as many cells without axonal projections. (*Left schematic diagram*) Wild-type neuroblasts (*white*) divide in a self-renewing manner to produce a neuroblast and a GMC (*gray*). The GMC divides only once to produce two neurons or glial cells (*black*). (*Right schematic diagram*) In a *prospero* mutant, GMCs divide in a self-renewing manner.

suggesting that neural cells in *prospero* mutants undergo extra divisions and fail to differentiate. We followed the division pattern of individual GMCs in *prospero* mutant embryos by first labeling a single neuroblast with DiI. After the first cell division, we removed the neuroblast, leaving only a single labeled first-born GMC. Whereas wild-type GMCs give rise to just two daughter cells that extend axons, *prospero* mutant GMCs give up to seven cells that continue to express neuroblast markers and exhibit few, if any, signs of differentiation. We conclude that *prospero* mutant neuroblasts divide to give two stem-cell-like daughters. GMCs, which would normally terminate cell division and differentiate, are transformed into self-renewing neural stem cells that generate undifferentiated clones, or tumors. These results are similar to what has been observed in the larval brain, where clones of cells lacking Prospero or Brat undergo extensive cell division to generate undifferentiated tumors (Bello et al. 2006; Betschinger et al. 2006; Lee et al. 2006a).

Our data demonstrate that Prospero represses neural stem cell genes and is required for the activation of neuronal differentiation genes, suggesting that it acts as a binary switch between self-renewal and differentiation in neural stem cells. Consequently, when Prospero is absent, the daughter cell (to which Prospero is normally segregated) is no longer programmed to differentiate but continues to behave like a stem cell, undergoing further divisions and producing tumors.

Prospero is also present in the daughter cells of *Drosophila* midgut stem cells (Micchelli and Perrimon 2006; Ohlstein and Spradling 2006), suggesting that it may have a more universal role in regulating stem cell differentiation.

REFERENCES

Albertson, R. and Doe, C.Q. 2003. Dlg, Scrib and Lgl regulate neuroblast cell size and mitotic spindle asymmetry. *Nat. Cell Biol.* **5:** 166–170.

Bello, B., Reichert, H., and Hirth, F. 2006. The brain tumor gene negatively regulates neural progenitor cell proliferation in the larval central brain of *Drosophila*. *Development* **133:** 2639–2648.

Betschinger, J., Mechtler, K., and Knoblich, J.A. 2006. Asymmetric segregation of the tumor suppressor brat regulates self-renewal in *Drosophila* neural stem cells. *Cell* **124:** 1241–1253.

Broadus, J., Fuerstenberg, S., and Doe, C.Q. 1998. Staufen-dependent localization of *prospero* mRNA contributes to neuroblast daughter-cell fate. *Nature* **391:** 792–795.

Caussinus, E. and Gonzalez, C. 2005. Induction of tumor growth by altered stem-cell asymmetric division in *Drosophila melanogaster*. *Nat. Genet.* **37:** 1125–1129.

Ceron, J., Gonzalez, C., and Tejedor, F.J. 2001. Patterns of cell division and expression of asymmetric cell fate determinants in postembryonic neuroblast lineages of *Drosophila*. *Dev. Biol.* **230:** 125–138.

Choksi, S.P., Southall, T.D., Bossing, T., Edoff, K., de Wit, E., Fischer, B.E., van Steensel, B., Micklem, G., and Brand, A.H. 2006. Prospero acts as a binary switch between self-renewal and differentiation in *Drosophila* neural stem cells. *Dev. Cell* **11:** 775–789.

Cook, T., Pichaud, F., Sonneville, R., Papatsenko, D., and Desplan, C. 2003. Distinction between color photoreceptor cell fates is controlled by Prospero in *Drosophila*. *Dev. Cell* **4:** 853–864.

Doe, C.Q. 2008. Neural stem cells: Balancing self-renewal with differentiation. *Development* **135:** 1575–1587.

Dudas, J., Mansuroglu, T., Moriconi, F., Haller, F., Wilting, J., Lorf, T., Füzesi, L., and Ramadori, G. 2008. Altered regulation of Prox1-gene-expression in liver tumors. *BMC Cancer* **8:** 92.

Dyer, M.A., Livesey, F.J., Cepko, C.L., and Oliver, G. 2003. Prox1 function controls progenitor cell proliferation and horizontal cell genesis in the mammalian retina. *Nat. Genet.* **34:** 53–58.

Egger, B., Boone, J.Q., Stevens, N.R., Brand, A.H., and Doe, C.Q. 2007. Regulation of spindle orientation and neural stem cell fate in the *Drosophila* optic lobe. *Neural Dev.* **2:** 1.

Gonczy, P. 2008. Mechanisms of asymmetric cell division: Flies and worms pave the way. *Nat. Rev. Mol. Cell Biol.* **9:** 355–366.

Gotz, M. and Huttner, W.B. 2005. The cell biology of neurogenesis. *Nat. Rev. Mol. Cell Biol.* **6:** 777–788.

Hirata, J., Nakagoshi, H., Nabeshima, Y., and Matsuzaki, F. 1995. Asymmetric segregation of the homeodomain protein Prospero during *Drosophila* development. *Nature* **377:** 627–630.

Hofbauer, A. and Campos-Ortega, J.A. 1990. Proliferation pattern and early differentiation of the optic lobes in *Drosophila melanogaster*. *Roux's Arch. Dev. Biol.* **198:** 264–274.

Ikeshima-Kataoka, H., Skeath, J.B., Nabeshima, Y., Doe, C.Q., and Matsuzaki, F. 1997. Miranda directs Prospero to a daughter cell during *Drosophila* asymmetric divisions. *Nature* **390:** 625–629.

Iscove, N.N., Barbara, M., Gu, M., Gibson, M., Modi, C., and Winegarden, N. 2002. Representation is faithfully preserved in global cDNA amplified exponentially from sub-picogram quantities of mRNA. *Nat. Biotechnol.* **20:** 940–943.

Knoblich, J.A. 2008. Mechanisms of asymmetric stem cell division. *Cell* **132:** 583–597.

Knoblich, J.A., Jan, L.Y., and Jan, Y.N. 1995. Asymmetric segregation of Numb and Prospero during cell division. *Nature* **377:** 624–627.

Lee, C.Y., Wilkinson, B.D., Siegrist, S.E., Wharton, R.P., and Doe, C.Q. 2006a. Brat is a Miranda cargo protein that promotes neuronal differentiation and inhibits neuroblast self-renewal. *Dev. Cell* **10:** 441–449.

Lee, C.Y., Andersen, R.O., Cabernard, C., Manning, L., Tran, K.D., Lanskey, M.J., Bashirullah, A., and Doe, C.Q. 2006b. *Drosophila* Aurora-A kinase inhibits neuroblast self-renewal

by regulating aPKC/Numb cortical polarity and spindle orientation. *Genes Dev.* **20:** 3464–3474.

Lin, D.M., Fetter, R.D., Kopczynski, C., Grenningloh, G., and Goodman, C.S. 1994. Genetic analysis of fasciclin II in *Drosophila:* Defasciculation, refasciculation and altered fasciculation. *Neuron* **13:** 1055–1069.

Martin, D., Brun, C., Remy, E., Mouren, P., Thieffry, D., and Jacq, B. 2004. GOToolBox: Functional analysis of gene datasets based on Gene Ontology. *Genome Biol.* **5:** R101.

Matsuzaki, F., Ohshiro, T., Ikeshima-Kataoka, H., and Izumi, H. 1998. miranda localizes staufen and prospero asymmetrically in mitotic neuroblasts and epithelial cells in early *Drosophila* embryogenesis. *Development* **125:** 4089–4098.

Micchelli, C.A. and Perrimon, N. 2006. Evidence that stem cells reside in the adult *Drosophila* midgut epithelium. *Nature* **439:** 475–479.

Mitchell, K.J., Doyle, J.L., Serafini, T., Kennedy, T.E., Tessier-Lavigne, M., Goodman, C.S., and Dickson, B.J. 1996. Genetic analysis of *Netrin* genes in *Drosophila:* Netrins guide CNS commissural axons and peripheral motor axons. *Neuron* **17:** 203–215.

Morrison, S.J. and Kimble, J. 2006. Asymmetric and symmetric stem-cell divisions in development and cancer. *Nature* **441:** 1068–1074.

Ohlstein, B. and Spradling, A. 2006. The adult *Drosophila* posterior midgut is maintained by pluripotent stem cells. *Nature* **439:** 470–474.

Petrova, T.V., Nykänen, A., Norrmén, C., Ivanov, K.I., Andersson, L.C., Haglund, C., Puolakkainen, P., Wempe, F., von Melchner, H., Gradwohl, G., et al. 2008. Transcription factor PROX1 induces colon cancer progression by promoting the transition from benign to highly dysplastic phenotype. *Cancer Cell* **13:** 407–419.

Pistocchi, A., Gaudenzi, G., Carra, S., Bresciani, E., Del Giacco, L., and Cotelli, F. 2008. Crucial role of zebrafish *prox1* in hypothalamic catecholaminergic neurons development. *BMC Dev. Biol.* **8:** 27.

Schuldt, A.J., Adams, J.H., Davidson, C.M., Micklem, D.R., Haseloff, J., St Johnston, D., and Brand, A.H. 1998. Miranda mediates asymmetric protein and RNA localization in the developing nervous system. *Genes Dev.* **12:** 1847–1857.

Shen, C., Knoblich, J.A., Chan, Y., Jiang, M., Jan, L.Y., and Jan, Y.N. 1998. Miranda as a multidomain adapter linking apically localized Inscuteable and basally localized Staufen and Prospero during asymmetric cell division in *Drosophila*. *Genes Dev.* **12:** 1837–1846.

Shen, C.P., Jan, L.Y., and Jan, Y.N. 1997. Miranda is required for the asymmetric localization of Prospero during mitosis in *Drosophila*. *Cell* **90:** 449–458.

Spana, E.P. and Doe, C.Q. 1995. The prospero transcription factor is asymmetrically localized to the cell cortex during neuroblast mitosis in *Drosophila*. *Development* **121:** 3187–3195.

Torii, M., Matsuzaki, F., Osumi, N., Kaibuchi, K., Nakamura, S., Casarosa, S., Guillemot, F., and Nakafuku, M. 1999. Transcription factors Mash-1 and Prox-1 delineate early steps in differentiation of neural stem cells in the developing central nervous system. *Development* **126:** 443–456.

van Steensel, B. and Henikoff, S. 2000. Identification of in vivo DNA targets of chromatin proteins using tethered Dam methyltransferase. *Nat. Biotechnol.* **18:** 424–428.

van Steensel, B., Delrow, J., and Henikoff, S. 2001. Chromatin profiling using targeted DNA adenine methyltransferase. *Nat. Genet.* **27:** 304–308.

Versmold, B., Felsberg, J., Mikeska, T., Ehrentraut, D., Kohler, J., Hampl, J.A., Rohn, G., Niederacher, D., Betz, B., Hellmich, M., et al. 2007. Epigenetic silencing of the candidate tumor suppressor gene *Prox1* in sporadic breast cancer. *Int. J. Cancer* **121:** 547–554.

Weissman, I.L., Anderson, D.J., and Gage, F. 2001. Stem and progenitor cells: Origins, phenotypes, lineage commitments, and transdifferentiations. *Annu. Rev. Cell Dev. Biol.* **17:** 387–403.

White, K. and Kankel, D.R. 1978. Patterns of cell division and cell movement in the formation of the imaginal nervous system in *Drosophila melanogaster*. *Dev. Biol.* **65:** 296–321.

Wigle, J.T., Harvey, N., Detmar, M., Lagutina, I., Grosveld, G., Gunn, M.D., Jackson, D.G., and Oliver, G. 2002. An essential role for *Prox1* in the induction of the lymphatic endothelial cell phenotype. *EMBO J.* **21:** 1505–1513.

Wu, P.S., Egger, B., and Brand, A.H. 2008. Asymmetric stem cell division: Lessons from *Drosophila. Semin. Cell Dev. Biol.* **19:** 283–293.

Yasugi, T., Umetsu, D., Murakami, S., Sato, M., and Tabata, T. 2008. *Drosophila* optic lobe neuroblasts triggered by a wave of proneural gene expression that is negatively regulated by JAK/STAT. *Development* **135:** 1471–1480.

Yoshimoto, T., Takahashi, M., Nagayama, S., Watanabe, G., Shimada, Y., Sakasi, Y., and Kubo, H. 2007. RNA mutations of *prox1* detected in human esophageal cancer cells by the shifted termination assay. *Biochem. Biophys. Res. Commun.* **359:** 258–262.

Zhong, W. and Chia, W. 2008. Neurogenesis and asymmetric cell division. *Curr. Opin. Neurobiol.* **18:** 4–11.

Progenitor Cells for the Prostate Epithelium: Roles in Development, Regeneration, and Cancer

M.M. SHEN,[*†] X. WANG,[*†] K.D. ECONOMIDES,[¶] D. WALKER,[#] AND C. ABATE-SHEN[‡§]

*Departments of Medicine, †Genetics and Development, ‡Urology, and §Pathology, Herbert Irving Comprehensive Cancer Center, Columbia University College of Physicians & Surgeons, New York, New York 10032; ¶Department of Biological Sciences, Sanofi-Aventis, Bridgewater, New Jersey 08807; #Department of Molecular Biology, Bristol-Myers Squibb Research Institute, Princeton, New Jersey 08540

The identification of stem cell/progenitor populations represents a critical step for deducing the putative cell type(s) of origin for epithelial cancers and may provide important therapeutic insights. In the case of the prostate gland, recent studies have made significant progress in the identification of candidate stem cell populations, but they have left unresolved key questions about their tissue localization and functional properties. In our work, we have used genetic lineage marking in vivo to demonstrate that a rare epithelial cell population marked by expression of the Nkx3.1 homeobox gene in the androgen-deprived prostate contains bipotential progenitor cells that are capable of self-renewal. Inducible targeting of the *Pten* tumor suppressor in these castrate-resistant Nkx3.1-expressing cells demonstrates that this stem/progenitor population is also a potent cell of origin for prostate cancer in mouse models. These findings may help to explain several intriguing features of prostate cancer and its phenotypic progression.

The importance of the stem cell compartment as a potential target of oncogenic transformation was appreciated more than three decades ago (Park et al. 1971) and more recently has been highlighted due to substantial evidence supporting the existence of "cancer stem cells" (Reya et al. 2001; Pardal et al. 2003; Wicha et al. 2006). Such putative cancer stem cells can in principle arise through oncogenic transformation of an endogenous stem cell or a downstream lineally related progenitor. Therefore, to identify potential cell types of origin for cancer, it is essential to understand normal lineage relationships during tissue organogenesis and adult homeostasis. Here, we review current work on stem cells and epithelial lineages in the prostate epithelium and present our own studies identifying a stem/progenitor population that is a cell of origin for prostate cancer.

FEATURES OF THE PROSTATE EPITHELIUM

The rodent prostate is composed of four distinct lobes (anterior, dorsal, lateral, and ventral) that are arranged circumferentially around the urethra at the base of the bladder (Fig. 1A) (Abate-Shen and Shen 2000; Marker et al. 2003). These lobes display characteristic morphological and histological appearances but otherwise behave similarly in most assays for prostate growth and function. In contrast, the human prostate gland lacks a distinct lobular structure, but it does display a zonal architecture that is evident in histological sections (Fig. 1B). The embryological and molecular relationship between the individual prostate lobes of the rodent prostate and the zonal architecture of the human gland has been a subject of considerable debate. Although morphological criteria suggest that

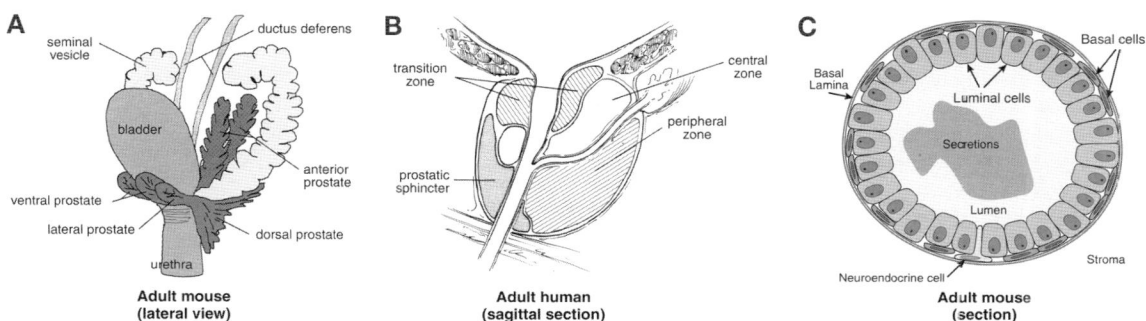

Figure 1. Anatomy of the prostate gland. (*A,B*) Schematic depiction of the anatomy of the mouse (*A*) and human (*B*) prostate gland. (*C*) Schematic illustration of the primary cell types within a section of a mouse prostatic duct. (Adapted from Abate-Shen and Shen 2000.)

the dorsal and anterior lobes are most closely related to the human prostate, all four lobes retain histopathological features of the human prostate epithelium.

There are three differentiated epithelial cell types in the mouse and human prostate (Fig. 1C) (Foster et al. 2002; van Leenders and Schalken 2003; Hudson 2004; Peehl 2005). Luminal cells are polarized secretory epithelial cells that produce the protein secretions that fill the lumen of the gland. The luminal epithelium corresponds to a continuous layer of tall columnar cells that express characteristic markers such as cytokeratins 8 and 18, the homeodomain protein Nkx3.1, and high levels of androgen receptor (AR). Basal cells are located beneath the luminal epithelium, and they express markers such as p63 and cytokeratins 5 and 14 as well as low levels of AR; interestingly, the basal layer is continuous in the human prostate, but it is discontinuous in the mouse prostate (El-Alfy et al. 2000). In addition, a subpopulation of basal cells coexpresses basal and luminal markers and has been termed "intermediate cells" (Verhagen et al. 1988; Bonkhoff et al. 1994; van Leenders et al. 2000; Hudson et al. 2001). Finally, neuroendocrine cells are rare cells that express endocrine markers but not AR. The relationship of neuroendocrine cells to the other two cell types has been unclear, although limited data have suggested that they represent a postmitotic cell type that is derived from luminal secretory cells (Bonkhoff et al. 1991, 1994, 1995).

DEVELOPMENT AND REGENERATION OF THE PROSTATE GLAND

Circulating testosterone synthesized by Leydig cells of the testis is required for most aspects of prostate induction, organogenesis, and tissue homeostasis. In the mouse, the prostate arises as epithelial buds from the urogenital sinus during late embryogenesis, initiating at 17.5 days postcoitum (dpc). During prostate organogenesis, epithelial cells in the prostatic buds express a combination of basal (p63, cytokeratin 5) and luminal (Nkx3.1) markers, whereas lineage-specific marker expression does not occur until neonatal stages. Following ductal canalization, distinct luminal and basal cell populations can be identified by morphological criteria as well as lineage-specific marker expression. Most ductal outgrowth occurs during before sexual maturity. During these stages, the prostatic ducts undergo extensive proximodistal outgrowth, with cell proliferation primarily occurring distally at the ductal tips (Sugimura et al. 1986a,c).

By early adulthood (8 weeks of age), the prostate epithelium is essentially growth-quiescent, with less than 1% of epithelial cells expressing the proliferation marker Ki-67 (Bhatia-Gaur et al. 1999). However, this tight control over epithelial proliferation deteriorates with age, because wild-type mice over 1 year of age frequently display epithelial hyperplasia and in rare cases can develop low-grade prostatic intraepithelial neoplasia (PIN), a precursor of prostate carcinoma (Kim et al. 2002b). In humans, renewed epithelial proliferation during aging is associated with benign prostatic hyperplasia (BPH) and is also evident in the emergence of PIN.

In the rodent and human prostate, androgen deprivation leads to rapid loss of most luminal cells through apoptosis, whereas the majority of basal cells survive (English et al. 1987; Evans and Chandler 1987b). The cells that undergo apoptosis are preferentially localized to the distal regions of prostatic ducts, whereas proximal ducts are less affected, a pattern that is opposite to the process of ductal outgrowth. Following administration of androgens to castrated rodents by implantation of hormone pellets or osmotic pumps, the prostate will regenerate during a span of approximately 2 weeks, with cellular proliferation occurring in both the basal and luminal compartments (Sugimura et al. 1986b,c; English et al. 1987; Evans and Chandler 1987b). If the source of androgens is subsequently removed, the regenerated prostate will once again fully regress, yet remain capable of androgen-induced regeneration, a process that can be repeated for more than 15 cycles of serial regression/regeneration (Tsujimura et al. 2002).

During prostate regression after androgen deprivation, most of the apoptosis occurs in the luminal population, whereas basal cells are largely unaffected. Thus, the 90% loss of cells during regression is almost entirely confined to the luminal population. However, substantial numbers of luminal cells still remain in the androgen-deprived state. These luminal cells, as well as the surviving basal cells, appear to be truly castration-resistant, because they persist after further removal of endogenous androgens following adrenalectomy or treatment with AR antagonists such as flutamide in the castrated state.

EVIDENCE FOR STEM CELLS IN THE PROSTATE EPITHELIUM

Although ample evidence exists for stem cell populations in many adult tissues, it is not evident that all tissues require stem cells for their maintenance and homeostasis. For example, substantial data suggest that tissue regeneration in the liver and pancreas appears to be driven by the proliferation of differentiated cells (Forbes et al. 2002; Dor et al. 2004). Given the growth quiescence of the adult prostate epithelium, it is unclear whether stem cell function is required for normal tissue homeostasis of the prostate gland. However, the observation that the prostate can undergo multiple rounds of serial regression/regeneration in response to androgen deprivation/administration implies the existence of a stem cell population in the prostate epithelium.

This ability of the prostate to undergo serial regression/regeneration implies that epithelial stem cells are castration-resistant, and do not require androgens for their survival. However, the ability of the prostate to regenerate rapidly in response to androgen treatment indicates that stem cells are directly or indirectly androgen-responsive. Notably, the prostate gland differs from other tissues (e.g., epidermis, intestinal epithelium, and hematopoietic system) in which there is a continual need for stem cell activity to maintain tissue homeostasis. Instead, the prostate appears to be more similar to growth-quiescent tissues such as the lung, in which stem cell activity is associated with wound repair and regeneration.

The proximodistal direction of ductal outgrowth and the predominant localization of proliferating cells to ductal tips during normal organogenesis (Sugimura et al. 1986a,c) suggest that stem cells should be preferentially localized to the proximal region of prostatic ducts (Kinbara et al. 1996). Moreover, explant experiments that involve microdissection of proximal versus medial and distal regions of the prostate ducts provide evidence that the proximal region is indeed enriched for cells with stem cell properties. In particular, cells from the proximal region are able to regenerate prostate ducts in tissue recombination/renal grafting experiments with a 17-fold-higher efficiency than cells from the distal region, and the proximal grafts can be serially grafted for at least four passages (Goto et al. 2006).

Additional evidence for the existence of stem cells in the prostate epithelium has been provided by assays that use ex vivo tissue recombination followed by renal grafting. When explanted sections of adult prostate ductal tips are combined with embryonic urogenital epithelium in renal grafts, massive proliferation of the epithelial cells occurs in these tissue recombinants (Norman et al. 1986; Kinbara et al. 1996). During 1 month's growth of such tissue recombinants, the number of epithelial cells can expand from 5000 to approximately 3×10^7, representing a 6000-fold increase (Norman et al. 1986; Kinbara et al. 1996; Kim et al. 2002b). This process can be repeated in serial transplantations of epithelium from the tissue recombinants for two additional rounds (Kim et al. 2002b). Interestingly, there appears to be a lower limit to the number of epithelial cells that can be used in these tissue recombinations, because approximately 3000–5000 cells are standardly used for successful grafts, but 600–1000 cells are usually unsuccessful (Norman et al. 1986), suggesting that the grafted epithelium requires the presence of a rare stem cell to proliferate significantly.

LOCALIZATION OF STEM CELLS IN THE PROSTATE EPITHELIUM

Evidence for a Basal Localization

Stem cells for the prostate epithelium have traditionally been thought to be located in the basal layer. This localization was originally proposed based on the relative androgen independence and higher proliferative index of basal cells, as well as their heterogeneous patterns of cytokeratin expression (Kyprianou and Isaacs 1988; Bonkhoff et al. 1994; Bonkhoff and Remberger 1996; Hudson et al. 2001). Consistent with this interpretation, studies of primary human prostate epithelial cells explanted into tissue culture have suggested that basal cells can generate luminal progeny, both in culture and following recombination with embryonic mesenchyme and renal grafting (Robinson et al. 1998; Collins et al. 2001; Richardson et al. 2004; Schmelz et al. 2005). Additional evidence has been provided by analysis of an immortalized rat basal cell line that can give rise to luminal cells (Danielpour 1999; Hayward et al. 1999). These and other studies have led to a model in which "intermediate" basal cells represent precursors of luminal cells (Verhagen et al. 1988; Bonkhoff et al. 1994;

van Leenders et al. 2000; Hudson et al. 2001). In general, the standard model proposes that a basal subpopulation of stem cells can generate transit-amplifying cells with an intermediate phenotype, and these in turn become precursors of luminal and possibly neuroendocrine cells (Fig. 2A) (Isaacs and Coffey 1989; Bonkhoff and Remberger 1996; Bui and Reiter 1998).

More recent work has used cell surface markers and flow cytometry/immunomagnetic bead approaches to identify candidate stem cell populations in normal mouse and human prostate epithelium, using markers such as CD133 (prominin), CD44, CD49f (α6 integrin), and Sca-1. For example, CD133$^+\alpha2\beta1^{hi}$ cells isolated from the human prostate have been reported to be enriched for clonogenic ability in culture and can form prostatic ducts in renal grafts (Richardson et al. 2004). However, despite the use of CD133 as a specific stem cell marker in a range of tissues, this identification is based on the use of a single monoclonal antibody (AC133) that detects a glycosylation-sensitive epitope that can yield distinct staining patterns under different immunostaining conditions. In particular, a recent study has shown that CD133 is in fact broadly expressed by luminal cells of the colon, lung, and pancreas (although the prostate was not examined) and has questioned the validity of CD133 as a marker for adult stem cells as well as colon cancer stem cells (Shmelkov et al. 2008).

In the mouse prostate, two groups originally reported enrichment for prostate epithelial stem cells among Sca-1$^+$ cells, which represent a heterogeneous population that comprises both luminal and basal cells and corresponds to between 15% and 20% of the total prostate epithelial population (Burger et al. 2005; Xin et al. 2005). A follow-up study further purified the Sca-1$^+$ population, showing that a Sca-1$^+$CD49f$^+$CD45$^-$CD31$^-$Ter119$^-$ population is 60-fold enriched for stem cells, as assessed by explant culture and tissue recombinant assays (Lawson et al. 2007). In this study, CD49f (α6 integrin) was reported to be specific for basal cells, although there is no direct evidence that the Sca-1$^+$CD49f$^+$CD45$^-$CD31$^-$Ter119$^-$ population is homogeneous. Most notably, a recent study has reported that 10% of single cells that are Lin$^-$Sca-1$^+$CD133$^+$CD44$^+$CD117$^+$ are able to reconstitute prostate duct formation following recombination with urogenital mesenchyme and renal grafting (Leong et al. 2008). Interestingly, this study has provided evidence that CD117 (c-*kit*) may have a critical role in stem cell maintenance and function during prostate regeneration. At present, however, it is unclear whether the Lin$^-$Sca-1$^+$CD133$^+$CD44$^+$CD117$^+$ population is restricted to basal cells, because CD117 is expressed by both basal and luminal cells (Leong et al. 2008).

Evidence for a Luminal Localization

In contrast, other work has suggested a luminal localization of stem cells, based on the proliferative activity of the luminal layer in the intact, castrated, and regenerating prostate (English et al. 1987; Evans and Chandler 1987a, b; van der Kwast et al. 1998). Interestingly, most cell proliferation in the regressed prostate occurs in the luminal

A

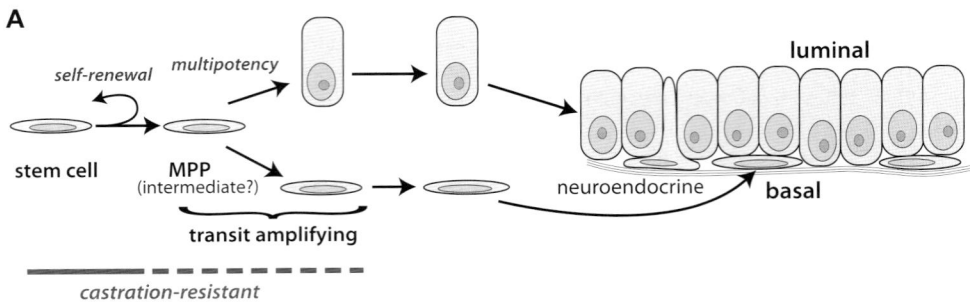

B

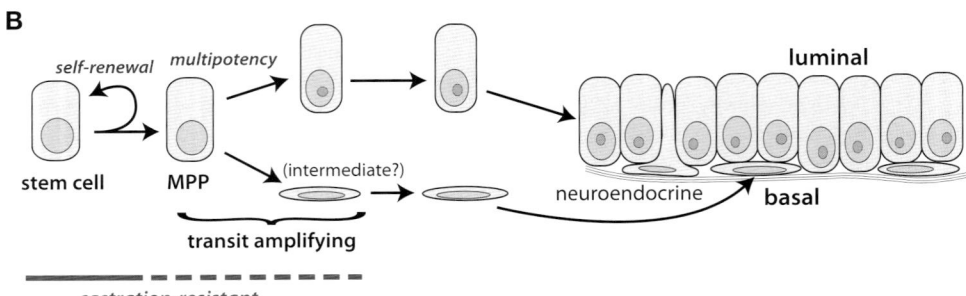

Figure 2. Models for lineage relationships in the prostate epithelium. (*A*) Conventional lineage model, in which a basal stem cell generates basal and luminal progeny. Intermediate cells correspond to a subset of basal cells that express luminal markers and may represent bipotential progenitors. The lineage relationship of neuroendocrine cells to other cell types is unclear. (*B*) Alternative model for lineage relationships, suggested by analyses of castration-resistant Nkx3.1-expressing cells (CARNs). In this model, a luminal stem cell may give rise to basal progeny, perhaps via intermediate cells. (MPP) Multipotent progenitor population.

epithelium, suggesting that at least some cells in the luminal compartment may have self-renewal capability (van der Kwast et al. 1998).

In particular, two recent lines of evidence support the notion that prostate stem cells may partially or exclusively reside in the luminal compartment. First, the tissue distribution of putative progenitors has been investigated by examining label-retaining cells (LRCs) generated during serial regression/regeneration of the mouse prostate (Tsujimura et al. 2002). This approach uses pulse-chase labeling with bromodeoxyuridine (BrdU) to identify cells that rarely proliferate and therefore retain BrdU over long periods of time, resulting in an LRC population that is in principle enriched for growth-quiescent stem cells (Bickenbach and Holbrook 1987; Cotsarelis et al. 1989; Morris and Potten 1999; Taylor et al. 2000). LRC populations in many tissues are believed to be enriched for stem cells, although recent work has shown that this is not the case in the hematopoietic system. Nonetheless, the analysis of LRCs generated during prostate regeneration indicated that these cells are preferentially localized to the proximal region of the prostatic ducts (closest to the urethra), consistent with an analysis of distribution of cells with stem cell properties in grafting experiments (Goto et al. 2006). Notably, LRCs are found in both luminal and basal compartments of the prostate epithelium (Tsujimura et al. 2002).

A second line of evidence has emerged from studies of *p63*-null mutant mice that display neonatal lethality with a complete absence of prostate tissue and lack all basal epithelial cells (Mills et al. 1999; Yang et al. 1999; Signoretti et al. 2000). All urogenital sinus epithelial cells are p63+ before prostate induction, and consistent with this observation, chimera experiments show that all prostate epithelial cells are derived from p63-expressing progenitors (Kurita et al. 2004; Signoretti et al. 2005). However, tissue-grafting experiments demonstrate that prostate tissue can form in explants from *p63*-null mutant embryos, despite the total lack of basal cells in the resulting tissue (Kurita et al. 2004; Signoretti et al. 2005). These grafted *p63* mutant prostate tissues display relatively normal prostate histology, apart from the absence of basal cells and the occasional formation of mucin-producing luminal cells that resemble intestinal epithelium (Kurita et al. 2004; Signoretti et al. 2005). Notably, these *p63* mutant prostate tissues can undergo multiple rounds of regression/regeneration in the absence of a basal cell layer (Kurita et al. 2004). These results indicate that the prostate stem cell compartment should at least partially reside within the luminal epithelium, although it remains conceivable that the *p63* mutation itself might alter stem cell localization.

In summary, considerable evidence exists in the literature that suggests the localization of prostate epithelial progenitor cells to the basal layer, although these studies do not necessarily identify the same cell populations. In addition, other substantive data support a luminal localization. To date, however, there has been no direct investigation of the lineage relationship between basal and luminal cells in vivo.

A PROGENITOR POPULATION
FOR PROSTATE REGENERATION

In recent studies, we have identified a novel cell population that contains prostate epithelial progenitors (X. Wang et al., in prep.). We identified this cell population through its unusual androgen-independent expression of the homeodomain protein Nkx3.1, which is required for proper ductal morphogenesis and secretory protein production (Bhatia-Gaur et al. 1999; Shen and Abate-Shen 2003; Abate-Shen et al. 2008). In the adult prostate, all luminal epithelial cells express Nkx3.1 and a small subpopulation of basal cells likely to correspond to intermediate cells (Fig. 3A,B). Although previous studies in cell culture and in vivo reported that Nkx3.1 expression in prostate epithelial cells is lost in the absence of androgens (Bieberich et al. 1996; Sciavolino et al. 1997; Prescott et al. 1998), we have observed that a rare population of epithelial cells within the castrated prostate expresses Nkx3.1 (Fig. 3C) (X. Wang et al., in prep.). These castration-resistant Nkx3.1-expressing cells (CARNs) are strictly luminal (Fig. 3D,E) and comprise less than 1% of total epithelial cells in the castrated prostate.

To determine whether these CARNs might correspond to prostate epithelial progenitor cells, we used Cre-recombinase-mediated genetic lineage marking to identify their progeny following prostate regeneration. In this lineage-marking approach, we used an inducible system to introduce a heritable genetic alteration in cells expressing a given promoter, as well as in all descendants of these cells. For this purpose, we generated a knockin allele that places an inducible Cre recombinase under the transcriptional control of the *Nkx3.1* promoter (X. Wang et al., in prep.). This inducible *Nkx3.1*[CreERT2] allele expresses a fusion protein of Cre with a mutated estrogen receptor (ER[T2]) that lacks the DNA-binding domain and fails to bind endogenous estrogens. The resulting CreER[T2] fusion protein is completely inactive in vivo, but it can be rapidly activated by administration of the synthetic ligand tamoxifen (Fig. 3F) (Feil et al. 1997; Indra et al. 1999; Kuhbandner et al. 2000).

To determine whether the CARN population in the castrated prostate contains progenitor cells, we induced Cre activity in castrated *Nkx3.1*[CreERT2/+];*R26R-YFP/+* or *Nkx3.1*[CreERT2/+];*R26R-LacZ/+* adult males with tamoxifen (Fig. 3G) and then initiated prostate regeneration by androgen treatment (X. Wang et al., in prep.). In these regenerated prostates, we observed frequent clusters of yellow fluorescent protein (YFP)-expressing or β-galactosidase-expressing cells, interspersed with occasional marked cells that had apparently remained solitary (Fig. 3H). The wide dispersion of the lineage-marked clusters suggests that many of these clusters are clonal, representing the progeny arising from a single progenitor during prostate regeneration. Notably, we observed that some lin-

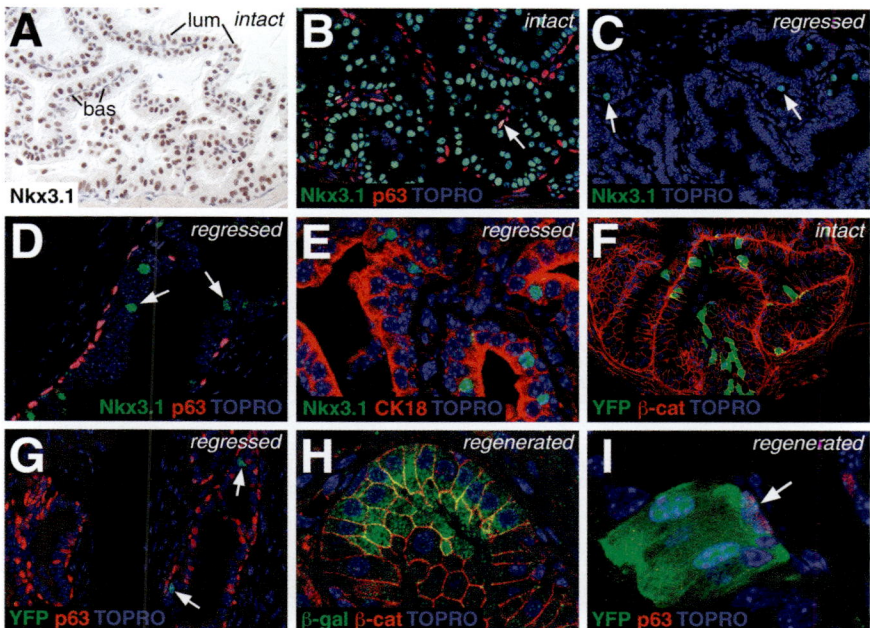

Figure 3. Identification and bipotentiality of castration-resistant Nkx3.1-expressing cells (CARNs). *(A)* Expression of Nkx3.1 in luminal cells of the anterior prostate from an intact adult male mouse detected by immunohistochemistry. (bas) Basal; (lum) luminal. *(B)* Expression of Nkx3.1 in luminal cells and a subpopulation of basal cells *(arrow)* in the intact prostate epithelium, as detected by immunofluorescence and confocal imaging. *(C)* Nkx3.1 is expressed in rare epithelial cells (CARNs, *arrows*) of the castrated anterior prostate. *(D,E)* CARNs are strictly luminal, because they never coexpress the basal marker p63 *(D)* and instead coexpress luminal cytokeratins, such as cytokeratin 18 (CK18) *(E)*. *(F)* Tamoxifen treatment of intact male mice carrying the inducible *Nkx3.1*[CreERT2] allele together with the *R26R-YFP* Cre reporter (Srinivas et al. 2001) results in specific yellow fluorescent protein (YFP) expression in the prostate epithelium. *(G)* Tamoxifen administration to castrated *Nkx3.1*[CreERT2/+];*R26R-YFP/+* male mice results in YFP expression in luminal cells (p63[-]) of the anterior prostate. *(H)* Androgen administration to castrated and tamoxifen-induced *Nkx3.1*[CreERT2/+];*R26R-LacZ/+* mice results in extensive proliferation of lineage-marked β-galactosidase[+] cells in the regenerated anterior prostate. *(I)* Detection of a lineage-marked YFP[+]p63[+] basal cell *(arrow)* following androgen-mediated prostate regeneration in a castrated, tamoxifen-induced *Nkx3.1*[CreERT2/+];*R26R-YFP/+* mouse.

eage-marked clusters contained both luminal and basal cells (Fig. 3I). Because the strictly luminal starting population can give rise to a mixed population of luminal and basal cells, we conclude that the CARN population contains bipotential progenitors.

In additional studies, we showed that the prostatic intraepithelial neoplasia phenotype of *Nkx3.1* mutants is partially suppressed in a serial regression/regeneration paradigm, suggesting that the number of prostate epithelial stem cells is reduced in *Nkx3.1* mutants (X. Wang et al., in prep.). Consistent with these results, we also found that the number of BrdU-label-retaining cells is reduced in *Nkx3.1* mutants relative to wild type following serial regression/regeneration. Moreover, we have previously observed that the ability of *Nkx3.1* mutant epithelial cells to proliferate is reduced in serial transplants of tissue recombinants with embryonic urogenital mesenchyme (Kim et al. 2002b; Abate-Shen et al. 2003). Consequently, stem cells are more likely to differentiate in the absence of *Nkx3.1,* leading to an expansion of the highly proliferative transit-amplifying population. Such a model is supported by the observation that the duration of epithelial proliferation during regeneration is significantly prolonged in castrated *Nkx3.1* heterozygous and homozygous males relative to wild type, resulting in epithelial hyperproliferation (Magee et al. 2003). Moreover, this interpretation is also consistent with the development of prostatic epithelial hyperplasia in young heterozygous and homozygous *Nkx3.1* mice (Bhatia-Gaur et al. 1999). Taken together, these results support a model in which *Nkx3.1* is required for prostate stem cell maintenance.

Finally, we have shown that the CARN population corresponds to a cell of origin for prostate cancer in mouse models (X. Wang et al., in prep.). Inducible deletion of the Pten tumor suppressor in CARNS in the regressed prostate of *Nkx3.1^{CreERT2/+};Pten^{flox/flox}* mice results in high-grade PIN/carcinoma lesions following androgen administration and prostate regeneration (Fig. 4). These lesions display membrane-localized phospho-Akt, greatly increased cellular proliferation, and evidence of microinvasion into the stroma (X. Wang et al., in prep.). These data indicate that the CARN population can represent an efficient target for oncogenic transformation.

In summary, our work differs in significant ways from earlier studies on prostate stem cells. First, we have used genetic lineage marking to identify a candidate stem/multipotent progenitor population (CARNs) that is both extremely rare and highly specific. Second, we have assayed progenitor properties in the context of prostate regeneration, which is known to be relevant to endogenous stem cell function. In contrast, the adult prostate epithelium is essentially growth quiescent during normal prostate homeostasis; thus, assays that rely on explants of adult epithelial cells may be less physiologically relevant. Third, our experiments have been performed in vivo, which may be relevant because many cell types can display greater plasticity when removed from their normal microenvironment and explanted in culture or in tissue recombination assays. Finally, we have shown that the CARN population represents a potential cell of origin for prostate cancer using an in vivo mouse model.

RECONCILING MODELS FOR PROSTATE STEM CELL LOCALIZATION AND LINEAGES

Taken together, our studies are compatible with a model in which the prostate epithelial stem cell displays a differentiated luminal phenotype and gives rise to both

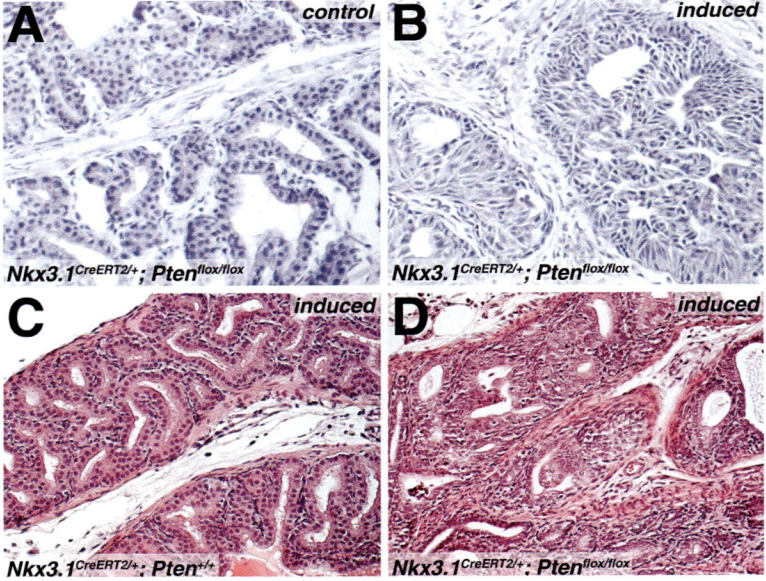

Figure 4. The CARN population is a cell of origin for prostate cancer. (*A,B*) High-grade prostatic intraepithelial neoplasia (PIN) and carcinoma in castrated, tamoxifen-induced, and regenerated prostate from a *Nkx3.1^{CreERT2/+};Pten^{flox/flox}* mouse but not in castrated and regenerated prostate from a control mouse of the identical genotype that was not induced with tamoxifen. (*C,D*) High-grade PIN and carcinoma in castrated, tamoxifen-treated, and regenerated prostate from a *Nkx3.1^{CreERT2/+};Pten^{flox/flox}* mouse but not from a control *Nkx3.1^{CreERT2/+};Pten^{+/+}* mouse.

luminal and basal progeny during androgen-mediated regeneration (see Fig. 2B). There may be significant technical reasons why previous studies have not identified luminal stem/progenitor cells in other assays. First, cell culture studies are performed in media that disfavor growth of luminal prostate epithelial cells, which are difficult to culture under conditions that maintain their differentiated phenotype (Peehl 2005). Moreover, flow cytometry experiments necessarily disrupt cell–cell interactions and potential regulation by a putative stem cell niche, which again may interfere with maintenance of a luminal phenotype in culture. Furthermore, the use of Matrigel to provide an extracellular matrix milieu for explant culture/renal grafting analyses may preferentially select for expression of CD44 and CD49f, which both mediate interactions with the extracellular matrix. Consequently, the results obtained for stem cell properties may be highly dependent on the methodologies used, and therefore, putative stem cells identified by one assay may differ from those identified by a distinct assay. In particular, it is not necessarily the case that stem cells for prostate organogenesis will be identical to those used during prostate regeneration.

Overall, our findings are not incompatible with a role for a basal stem cell in the prostate epithelium. Indeed, it is conceivable that there are independent stem cell lineages for basal and luminal cells within the prostate epithelium (Fig. 5). Alternatively, our study may have identified a "potential" stem cell within the prostate epithelium that is responsible for regeneration, perhaps analogous to facultative stem cell populations that have been described to participate in wound healing/regeneration responses in testis and pancreas (Nakagawa et al. 2007; Xu et al. 2008). Such a facultative stem cell population may express differentiation markers and would correspond to transit-amplifying cells that can display stem cell properties during wound healing/regeneration. Indeed, it is possible that two distinct stem cell populations may coexist within the prostate epithelium, one corresponding to an fetal/neonatal progenitor that functions during organogenesis and a second more differentiated population (CARNs) that functions during regeneration; such a model has been recently suggested for the *Drosophila* tracheal system (Weaver and Krasnow 2008).

CELL OF ORIGIN FOR PROSTATE CANCER

In principle, cancer can result from transformation of a rare stem cell, leading to the niche-independent dysregulated proliferation of cancer stem cells to generate a tumor or, alternatively, can result from transformation of a more restricted cell type (such as a transit/amplifying cell) and its "dedifferentiation" to acquire self-renewal properties characteristic of stem cells. Evidence for both modes of cancer stem cell generation has been provided by studies of hematopoietic cancers (see, e.g., Cozzio et al. 2003; Huntly et al. 2004; Passegué et al. 2004), but they are less well-studied for solid tumors. Furthermore, the activation of stem cells in chronic injury/wound healing may also serve as a stimulus for stem cell transformation (Beachy et al. 2004). However, the possible existence of cancer stem cells in solid tumors remains contentious (Hill 2006; Visvader and Lineman 2008).

The cancer stem cell model has several important translational implications. First, this model explains the difficulty in treating a relatively quiescent progenitor population with chemotherapeutic agents that target actively proliferating cells. Second, at least some stem cell populations may have a relatively drug-resistant phenotype due to expression of high levels of ABC family transporter proteins, which correlates with their dye efflux capability (Zhou et al. 2001, 2002; Kim et al. 2002a; Doyle and Ross 2003). Finally, the observation that normal prostate stem cells should be androgen independent raises the possibility that the emergence of androgen-independent prostate cancer might reflect a population shift in the percentage of cancer stem cells within the tumor.

With regard to prostate cancer, it has remained a long-standing conundrum that prostate carcinoma is effectively diagnosed by a complete absence of basal cells (Jiang et al. 2005; Humphrey 2007), whereas prostate epithelial stem cells have been thought to reside in the basal layer. This apparent discrepancy has led numerous studies and reviews to propose either that oncogenic transformation of the basal stem cell leads to its luminal differentiation in tumors or that transformation of a differentiated luminal cell results in the acquisition of stem cell properties. Our findings could potentially resolve this important conundrum because we have shown that the CARN population is luminal and contains stem/multipotent progenitors as

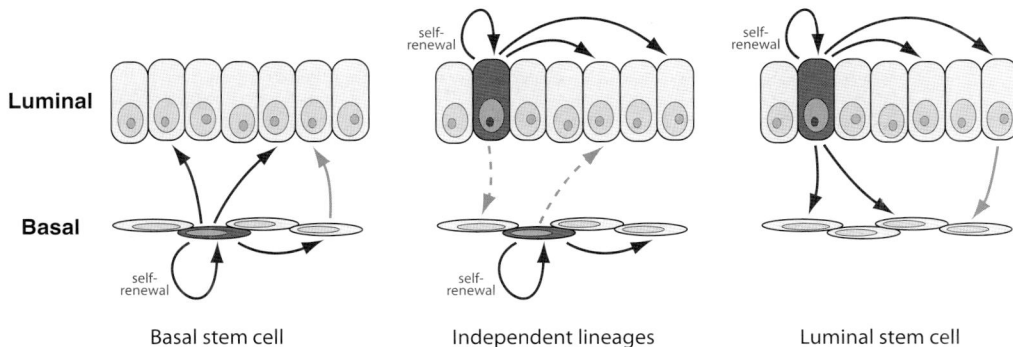

Figure 5. Alternative models for stem cell localization in the prostate epithelium. See text for discussion.

well as a potential cell of origin for prostate cancer. To the best of our knowledge, our work represents the first demonstration of a prostate epithelial stem cell/multipotent progenitor that can undergo oncogenic transformation to result in prostate cancer in vivo.

ACKNOWLEDGMENTS

This work was supported by grants from the National Institutes of Health (C.A.-S., M.M.S.), the DOD Prostate Cancer Research Program (K.E., C.A.-S., M.M.S.), and the National Cancer Institute Mouse Models of Human Cancer Consortium. Correspondence should be addressed to mshen@columbia.edu.

REFERENCES

Abate-Shen, C. and Shen, M.M. 2000. Molecular genetics of prostate cancer. *Genes Dev.* **14:** 2410–2434.

Abate-Shen, C., Shen, M.M., and Gelmann, E. 2008. Integrating differentiation and cancer: The *Nkx3.1* homeobox gene in prostate organogenesis and carcinogenesis. *Differentiation* **76:** 717–727.

Abate-Shen, C., Banach-Petrosky, W.A., Sun, X., Economides, K.D., Desai, N., Gregg, J.P., Borowsky, A.D., Cardiff, R.D., and Shen, M.M. 2003. *Nkx3.1; Pten* mutant mice develop invasive prostate adenocarcinoma and lymph node metastases. *Cancer Res.* **63:** 3886–3890.

Beachy, P.A., Karhadkar, S.S., and Berman, D.M. 2004. Tissue repair and stem cell renewal in carcinogenesis. *Nature* **432:** 324–331.

Bhatia-Gaur, R., Donjacour, A.A., Sciavolino, P.J., Kim, M., Desai, N., Young, P., Norton, C.R., Gridley, T., Cardiff, R.D., Cunha, G.R., Abate-Shen, C., and Shen, M.M. 1999. Roles for *Nkx3.1* in prostate development and cancer. *Genes Dev.* **13:** 966–977.

Bickenbach, J.R. and Holbrook, K.A. 1987. Label-retaining cells in human embryonic and fetal epidermis. *J. Invest. Dermatol.* **88:** 42–46.

Bieberich, C.J., Fujita, K., He, W.W., and Jay, G. 1996. Prostate-specific and androgen-dependent expression of a novel homeobox gene. *J. Biol. Chem.* **271:** 31779–31782.

Bonkhoff, H. and Remberger, K. 1996. Differentiation pathways and histogenetic aspects of normal and abnormal prostatic growth: A stem cell model. *Prostate* **28:** 98–106.

Bonkhoff, H., Stein, U., and Remberger, K. 1994. Multidirectional differentiation in the normal, hyperplastic, and neoplastic human prostate: Simultaneous demonstration of cell-specific epithelial markers. *Hum. Pathol.* **25:** 42–46.

Bonkhoff, H., Stein, U., and Remberger, K. 1995. Endocrine-paracrine cell types in the prostate and prostatic adenocarcinoma are postmitotic cells. *Hum. Pathol.* **26:** 167.

Bonkhoff, H., Wernert, N., Dhom, G., and Remberger, K. 1991. Relation of endocrine-paracrine cells to cell proliferation in normal, hyperplastic, and neoplastic human prostate. *Prostate* **19:** 91–98.

Bui, M. and Reiter, R.E. 1998. Stem cell genes in androgen-independent prostate cancer. *Cancer Metastasis Rev.* **17:** 391–399.

Burger, P.E., Xiong, X., Coetzee, S., Salm, S.N., Moscatelli, D., Goto, K., and Wilson, E.L. 2005. Sca-1 expression identifies stem cells in the proximal region of prostatic ducts with high capacity to reconstitute prostatic tissue. *Proc. Natl. Acad. Sci.* **102:** 7180–7185.

Collins, A.T., Habib, F.K., Maitland, N.J., and Neal, D.E. 2001. Identification and isolation of human prostate epithelial stem cells based on α(2)β(1)-integrin expression. *J. Cell Sci.* **114:** 3865–3872.

Cotsarelis, G., Cheng, S.Z., Dong, G., Sun, T.T., and Lavker, R.M. 1989. Existence of slow-cycling limbal epithelial basal cells that can be preferentially stimulated to proliferate: Implications on epithelial stem cells. *Cell* **57:** 201–209.

Cozzio, A., Passegué, E., Ayton, P.M., Karsunky, H., Cleary, M.L., and Weissman, I.L. 2003. Similar MLL-associated leukemias arising from self-renewing stem cells and short-lived myeloid progenitors. *Genes Dev.* **17:** 3029–3035.

Danielpour, D. 1999. Transdifferentiation of NRP-152 rat prostatic basal epithelial cells toward a luminal phenotype: Regulation by glucocorticoid, insulin-like growth factor-I and transforming growth factor-β. *J. Cell Sci.* **112:** 169–179.

Dor, Y., Brown, J., Martinez, O.I., and Melton, D.A. 2004. Adult pancreatic β-cells are formed by self-duplication rather than stem-cell differentiation. *Nature* **429:** 41–46.

Doyle, L.A. and Ross, D.D. 2003. Multidrug resistance mediated by the breast cancer resistance protein BCRP (ABCG2). *Oncogene* **22:** 7340–7358.

El-Alfy, M., Pelletier, G., Hermo, L.S., and Labrie, F. 2000. Unique features of the basal cells of human prostate epithelium. *Microsc. Res. Tech.* **51:** 436–446.

English, H.F., Santen, R.J., and Isaacs, J.T. 1987. Response of glandular versus basal rat ventral prostatic epithelial cells to androgen withdrawal and replacement. *Prostate* **11:** 229–242.

Evans, G.S. and Chandler, J.A. 1987a. Cell proliferation studies in rat prostate. I. The proliferative role of basal and secretory epithelial cells during normal growth. *Prostate* **10:** 163–178.

Evans, G.S. and Chandler, J.A. 1987b. Cell proliferation studies in the rat prostate. II. The effects of castration and androgen-induced regeneration upon basal and secretory cell proliferation. *Prostate* **11:** 339–351.

Feil, R., Wagner, J., Metzger, D., and Chambon, P. 1997. Regulation of Cre recombinase activity by mutated estrogen receptor ligand-binding domains. *Biochem. Biophys. Res. Commun.* **237:** 752–757.

Forbes, S., Vig, P., Poulsom, R., Thomas, H., and Alison, M. 2002. Hepatic stem cells. *J. Pathol.* **197:** 510–518.

Foster, C.S., Dodson, A., Karavana, V., Smith, P.H., and Ke, Y. 2002. Prostatic stem cells. *J. Pathol.* **197:** 551–565.

Goto, K., Salm, S.N., Coetzee, S., Xiong, X., Burger, P.E., Shapiro, E., Lepor, H., Moscatelli, D., and Wilson, E.L. 2006. Proximal prostate stem cells are programmed to regenerate a proximal-distal ductal axis. *Stem Cells* **24:** 1859–1868.

Hayward, S.W., Haughney, P.C., Lopes, E.S., Danielpour, D., and Cunha, G.R. 1999. The rat prostatic epithelial cell line NRP-152 can differentiate in vivo in response to its stromal environment. *Prostate* **39:** 205–212.

Hill, R.P. 2006. Identifying cancer stem cells in solid tumors: Case not proven. *Cancer Res.* **66:** 1891–1895.

Hudson, D.L. 2004. Epithelial stem cells in human prostate growth and disease. *Prostate Cancer Prostatic Dis.* **7:** 188–194.

Hudson, D.L., Guy, A.T., Fry, P., O'Hare, M.J., Watt, F.M., and Masters, J.R. 2001. Epithelial cell differentiation pathways in the human prostate: Identification of intermediate phenotypes by keratin expression. *J. Histochem. Cytochem.* **49:** 271–278.

Humphrey, P.A. 2007. Diagnosis of adenocarcinoma in prostate needle biopsy tissue. *J. Clin. Pathol.* **60:** 35–42.

Huntly, B.J., Shigematsu, H., Deguchi, K., Lee, B.H., Mizuno, S., Duclos, N., Rowan, R., Amaral, S., Curley, D., Williams, I.R., Akashi, K., and Gilliland, D.G. 2004. MOZ-TIF2, but not BCR-ABL, confers properties of leukemic stem cells to committed murine hematopoietic progenitors. *Cancer Cell* **6:** 587–596.

Indra, A.K., Warot, X., Brocard, J., Bornert, J.M., Xiao, J.H., Chambon, P., and Metzger, D. 1999. Temporally-controlled site-specific mutagenesis in the basal layer of the epidermis: Comparison of the recombinase activity of the tamoxifen-inducible Cre-ER(T) and Cre-ER(T2) recombinases. *Nucleic Acids Res.* **27:** 4324–4327.

Isaacs, J.T. and Coffey, D.S. 1989. Etiology and disease process of benign prostatic hyperplasia. *Prostate Suppl.* **2:** 33–50.

Jiang, Z., Li, C., Fischer, A., Dresser, K., and Woda, B.A. 2005. Using an AMACR (P504S)/34βE12/p63 cocktail for the detection of small focal prostate carcinoma in needle biopsy specimens. *Am. J. Clin. Pathol.* **123:** 231–236.

Kim, M., Turnquist, H., Jackson, J., Sgagias, M., Yan, Y., Gong, M., Dean, M., Sharp, J.G., and Cowan, K. 2002a. The multidrug resistance transporter ABCG2 (breast cancer resistance protein 1) effluxes Hoechst 33342 and is overexpressed in hematopoietic stem cells. *Clin. Cancer Res.* **8:** 22–28.

Kim, M.J., Bhatia-Gaur, R., Banach-Petrosky, W.A., Desai, N., Wang, Y., Hayward, S.W., Cunha, G.R., Cardiff, R.D., Shen, M.M., and Abate-Shen, C. 2002b. *Nkx3.1* mutant mice recapitulate early stages of prostate carcinogenesis. *Cancer Res.* **62:** 2999–3004.

Kinbara, H., Cunha, G.R., Boutin, E., Hayashi, N., and Kawamura, J. 1996. Evidence of stem cells in the adult prostatic epithelium based upon responsiveness to mesenchymal inductors. *Prostate* **29:** 107–116.

Kuhbandner, S., Brummer, S., Metzger, D., Chambon, P., Hofmann, F., and Feil, R. 2000. Temporally controlled somatic mutagenesis in smooth muscle. *Genesis* **28:** 15–22.

Kurita, T., Medina, R.T., Mills, A.A., and Cunha, G.R. 2004. Role of p63 and basal cells in the prostate. *Development* **131:** 4955–4964.

Kyprianou, N. and Isaacs, J.T. 1988. Activation of programmed cell death in the rat ventral prostate after castration. *Endocrinology* **122:** 552–562.

Lawson, D.A., Xin, L., Lukacs, R.U., Cheng, D., and Witte, O.N. 2007. Isolation and functional characterization of murine prostate stem cells. *Proc. Natl. Acad. Sci.* **104:** 181–186.

Leong, K.G., Wang, B.-E., Johnson, L., and Gao, W.-Q. 2008. Generation of a prostate from a single adult stem cell. *Nature* **456:** 804–808.

Magee, J.A., Abdulkadir, S.A., and Milbrandt, J. 2003. Haploinsufficiency at the *Nkx3.1* locus. A paradigm for stochastic, dosage-sensitive gene regulation during tumor initiation. *Cancer Cell* **3:** 273–283.

Marker, P.C., Donjacour, A.A., Dahiya, R., and Cunha, G.R. 2003. Hormonal, cellular, and molecular control of prostatic development. *Dev. Biol.* **253:** 165–174.

Mills, A.A., Zheng, B., Wang, X.J., Vogel, H., Roop, D.R., and Bradley, A. 1999. p63 is a p53 homologue required for limb and epidermal morphogenesis. *Nature* **398:** 708–713.

Morris, R.J. and Potten, C.S. 1999. Highly persistent label-retaining cells in the hair follicles of mice and their fate following induction of anagen. *J. Invest. Dermatol.* **112:** 470–475.

Nakagawa, T., Nabeshima, Y., and Yoshida, S. 2007. Functional identification of the actual and potential stem cell compartments in mouse spermatogenesis. *Dev. Cell* **12:** 195–206.

Norman, J.T., Cunha, G.R., and Sugimura, Y. 1986. The induction of new ductal growth in adult prostatic epithelium in response to an embryonic prostatic inductor. *Prostate* **8:** 209–220.

Pardal, R., Clarke, M.F., and Morrison, S.J. 2003. Applying the principles of stem-cell biology to cancer. *Nat. Rev. Cancer* **3:** 895–902.

Park, C.H., Bergsagel, D.E., and McCulloch, E.A. 1971. Mouse myeloma tumor stem cells: A primary cell culture assay. *J. Natl. Cancer Inst.* **46:** 411–422.

Passegué, E., Wagner, E.F., and Weissman, I.L. 2004. JunB deficiency leads to a myeloproliferative disorder arising from hematopoietic stem cells. *Cell* **119:** 431–443.

Peehl, D.M. 2005. Primary cell cultures as models of prostate cancer development. *Endocr. Relat. Cancer* **12:** 19–47.

Prescott, J.L., Blok, L., and Tindall, D.J. 1998. Isolation and androgen regulation of the human homeobox cDNA, *NKX3.1*. *Prostate* **35:** 71–80.

Reya, T., Morrison, S.J., Clarke, M.F., and Weissman, I.L. 2001. Stem cells, cancer, and cancer stem cells. *Nature* **414:** 105–111.

Richardson, G.D., Robson, C.N., Lang, S.H., Neal, D.E., Maitland, N.J., and Collins, A.T. 2004. CD133, a novel marker for human prostatic epithelial stem cells. *J. Cell Sci.* **117:** 3539–3545.

Robinson, E.J., Neal, D.E., and Collins, A.T. 1998. Basal cells are progenitors of luminal cells in primary cultures of differentiating human prostatic epithelium. *Prostate* **37:** 149–160.

Schmelz, M., Moll, R., Hesse, U., Prasad, A.R., Gandolfi, J.A., Hasan, S.R., Bartholdi, M., and Cress, A.E. 2005. Identification of a stem cell candidate in the normal human prostate gland. *Eur. J. Cell Biol.* **84:** 341–354.

Sciavolino, P.J., Abrams, E.W., Yang, L., Austenberg, L.P., Shen, M.M., and Abate-Shen, C. 1997. Tissue-specific expression of

murine *Nkx3.1* in the male urogenital system. *Dev. Dyn.* **209:** 127–138.

Shen, M.M. and Abate-Shen, C. 2003. Roles of the *Nkx3.1* homeobox gene in prostate organogenesis and carcinogenesis. *Dev. Dyn.* **228:** 767–778.

Shmelkov, S.V., Butler, J.M., Hooper, A.T., Hormigo, A., Kushner, J., Milde, T., St Clair, R., Baljevic, M., White, I., Jin, D.K., et al. 2008. CD133 expression is not restricted to stem cells, and both CD133⁺ and CD133⁻ metastatic colon cancer cells initiate tumors. *J. Clin. Invest.* **118:** 2111–2120.

Signoretti, S., Pires, M.M., Lindauer, M., Horner, J.W., Grisanzio, C., Dhar, S., Majumder, P., McKeon, F., Kantoff, P.W., Sellers, W.R., and Loda, M. 2005. p63 regulates commitment to the prostate cell lineage. *Proc. Natl. Acad. Sci.* **102:** 11355–11360.

Signoretti, S., Waltregny, D., Dilks, J., Isaac, B., Lin, D., Garraway, L., Yang, A., Montironi, R., McKeon, F., and Loda, M. 2000. p63 is a prostate basal cell marker and is required for prostate development. *Am. J. Pathol.* **157:** 1769–1775.

Srinivas, S., Watanabe, T., Lin, C.S., William, C.M., Tanabe, Y., Jessell, T.M., and Costantini, F. 2001. Cre reporter strains produced by targeted insertion of *EYFP* and *ECFP* into the *ROSA26* locus. *BMC Dev. Biol.* **1:** 4.

Sugimura, Y., Cunha, G.R., and Donjacour, A.A. 1986a. Morphogenesis of ductal networks in the mouse prostate. *Biol. Reprod.* **34:** 961–971.

Sugimura, Y., Cunha, G.R., and Donjacour, A.A. 1986b. Morphological and histological study of castration-induced degeneration and androgen-induced regeneration in the mouse prostate. *Biol. Reprod.* **34:** 973–983.

Sugimura, Y., Cunha, G.R., Donjacour, A.A., Bigsby, R.M., and Brody, J.R. 1986c. Whole-mount autoradiography study of DNA synthetic activity during postnatal development and androgen-induced regeneration in the mouse prostate. *Biol. Reprod.* **34:** 985–995.

Taylor, G., Lehrer, M.S., Jensen, P.J., Sun, T.T., and Lavker, R.M. 2000. Involvement of follicular stem cells in forming not only the follicle but also the epidermis. *Cell* **102:** 451–461.

Tsujimura, A., Koikawa, Y., Salm, S., Takao, T., Coetzee, S., Moscatelli, D., Shapiro, E., Lepor, H., Sun, T.T., and Wilson, E.L. 2002. Proximal location of mouse prostate epithelial stem cells: A model of prostatic homeostasis. *J. Cell Biol.* **157:** 1257–1265.

van der Kwast, T.H., Tetu, B., Suburu, E.R., Gomez, J., Lemay, M., and Labrie, F. 1998. Cycling activity of benign prostatic epithelial cells during long-term androgen blockade: Evidence for self-renewal of luminal cells. *J. Pathol.* **186:** 406–409.

van Leenders, G., Dijkman, H., Hulsbergen-van de Kaa, C., Ruiter, D., and Schalken, J. 2000. Demonstration of intermediate cells during human prostate epithelial differentiation in situ and in vitro using triple-staining confocal scanning microscopy. *Lab. Invest.* **80:** 1251–1258.

van Leenders, G.J. and Schalken, J.A. 2003. Epithelial cell differentiation in the human prostate epithelium: Implications for the pathogenesis and therapy of prostate cancer. *Crit. Rev. Oncol. Hematol.* (suppl.) **46:** S3–S10.

Verhagen, A.P., Aalders, T.W., Ramaekers, F.C., Debruyne, F.M., and Schalken, J.A. 1988. Differential expression of keratins in the basal and luminal compartments of rat prostatic epithelium during degeneration and regeneration. *Prostate* **13:** 25–38.

Visvader, J.E. and Lindeman, G.J. 2008. Cancer stem cells in solid tumours: Accumulating evidence and unresolved questions. *Nat. Rev. Cancer* **8:** 755.

Weaver, M. and Krasnow, M.A. 2008. Dual origin of tissue-specific progenitor cells in *Drosophila* tracheal remodeling. *Science* **321:** 1496–1499.

Wicha, M.S., Liu, S., and Dontu, G. 2006. Cancer stem cells: An old idea—A paradigm shift. *Cancer Res.* **66:** 1883–1896.

Xin, L., Lawson, D.A., and Witte, O.N. 2005. The Sca-1 cell surface marker enriches for a prostate-regenerating cell subpopulation that can initiate prostate tumorigenesis. *Proc. Natl. Acad. Sci.* **102:** 6942–6947.

Xu, X., D'Hoker, J., Stange, G., Bonné, S., De Leu, N., Xiao, X., Van de Casteele, M., Mellitzer, G., Ling, Z., Pipeleers, D., et al. 2008. β Cells can be generated from endogenous progenitors in injured adult mouse pancreas. *Cell* **132:** 197–207.

Yang, A., Schweitzer, R., Sun, D., Kaghad, M., Walker, N., Bronson, R.T., Tabin, C., Sharpe, A., Caput, D., Crum, C., and McKeon, F. 1999. p63 is essential for regenerative proliferation in limb, craniofacial and epithelial development. *Nature* **398:** 714–718.

Zhou, S., Morris, J.J., Barnes, Y., Lan, L., Schuetz, J.D., and Sorrentino, B.P. 2002. *Bcrp1* gene expression is required for normal numbers of side population stem cells in mice, and confers relative protection to mitoxantrone in hematopoietic cells in vivo. *Proc. Natl. Acad. Sci.* **99:** 12339–12344.

Zhou, S., Schuetz, J.D., Bunting, K.D., Colapietro, A.M., Sampath, J., Morris, J.J., Lagutina, I., Grosveld, G.C., Osawa, M., Nakauchi, H., and Sorrentino, B.P. 2001. The ABC transporter Bcrp1/ABCG2 is expressed in a wide variety of stem cells and is a molecular determinant of the side-population phenotype. *Nat. Med.* **7:** 1028–1034.

Stem Cells Use Distinct Self-renewal Programs at Different Ages

B.P. Levi and S.J. Morrison

Howard Hughes Medical Institute, Department of Internal Medicine, and Center for Stem Cell Biology,
Life Sciences Institute, University of Michigan, Ann Arbor, Michigan 48109-2216

Stem cells expand in number during development and persist throughout life by undergoing self-renewing divisions. The question of how stem cells self-renew throughout life is a fundamental problem in cell biology, with broad implications for understanding development, tissue regeneration, cancer, and aging. Recent insights demonstrate that self-renewal programs depend on key transcriptional regulators that are often shared among stem cells in different tissues but that often change between stem cells at different stages of life: Embryonic, fetal, young adult, and old adult stem cells are maintained by different self-renewal programs. Self-renewal programs change over time to contend with changes in tissue growth and repair demands as well as the increasing risk of malignant transformation during aging. The downstream mechanisms by which these programs regulate the cell cycle, developmental potential, and timing of differentiation are just starting to be elucidated. One key requirement for self-renewal is repression of the $p16^{Ink4a}$ and $p19^{Arf}$ tumor suppressors. This is accomplished by overlapping transcriptional regulators whose expression and function change with age, so as to maintain self-renewal potential throughout life while allowing increased expression of $p16^{Ink4a}$ and $p19^{Arf}$ in aging stem cells. This reduces stem cell function in aging tissues but also reduces cancer incidence.

Stem cells maintain themselves throughout life by undergoing self-renewing divisions in which a mother stem cell gives rise to one or two daughter stem cells that have a developmental potential indistinguishable from that of the mother cell (Molofsky et al. 2004). In the hematopoietic system, stem cell self-renewal can be measured in vivo by the ability of single hematopoietic stem cells (HSCs) to give long-term multilineage reconstitution of irradiated mice and for the progeny of these cells to give long-term multilineage reconstitution of multiple secondary recipients. Individual HSCs have extensive self-renewal capacity (Harrison and Astle 1982; Morrison et al. 1997), although this capacity declines with age (Chen et al. 2000; Janzen et al. 2006). In the nervous system, self-renewal is measured by subcloning multilineage colonies that arise from single stem cells to determine the number of multipotent daughter cells that are formed by individual multipotent mother cells (Bixby et al. 2002; Kruger et al. 2002; Molofsky et al. 2003, 2005, 2006). As in the hematopoietic system, neural stem cells can self-renew extensively, but this self-renewal potential declines with age (Kruger et al. 2002; Molofsky et al. 2006). Many of the same genes are required for hematopoietic and neural stem cell self-renewal, demonstrating that although self-renewal is measured differently in the two systems, both assays reflect the function of a core program of self-renewal regulators that is often shared between tissues.

Stem cells change their properties and their regulation over time, consistent with the changing demands of tissue growth. Stem cell numbers must expand rapidly during development to support tissue formation and growth. Both HSCs and neural stem cells are thought to undergo rapid, symmetric self-renewing divisions to enable this expansion during fetal development (Chenn and McConnell 1995; Morrison et al. 1995; Noctor et al. 2004). However,

as tissue growth slows postnatally, stem cell division also slows, and a higher proportion of the divisions are likely to be asymmetric (for review, see Morrison and Kimble 2006). Throughout adulthood, stem cells in many tissues appear to be quiescent most of the time but regularly enter cycle to maintain tissue homeostasis (Morrison and Kimble 2006). Self-renewal programs must therefore change between the fetal, neonatal, and young adult stages to accommodate developmental changes in the frequency and symmetry of stem cell division (Table 1).

Additional changes occur in stem cells between young adulthood and old adulthood as stem cells struggle to maintain homeostasis within aging tissues and to avoid neoplastic transformation. Genetic mutations (Rossi et al. 2007) and damage from reactive oxygen species (Ito et al. 2006) accumulate with age in stem cells, impairing stem cell function and increasing the risk of neoplastic transformation. There are also changes during aging in the rate at which stem cells divide, the fates they are likely to acquire, their frequency, and their self-renewal potential (Kuhn et al. 1996; Morrison et al. 1996; de Haan et al. 1997; de Haan and Van Zant 1999; Enwere et al. 2004; Rossi et al. 2005; Janzen et al. 2006; Molofsky et al. 2006). The overall effect of these changes is to reduce stem cell function and regenerative capacity in aging tissues. These changes likely reflect a combination of environmental changes (Conboy et al. 2003, 2005; Brack et al. 2007; Carlson et al. 2008), cell-intrinsic changes in stem cell regulation (Janzen et al. 2006; Krishnamurthy et al. 2006; Molofsky et al. 2006), and the accumulation of genetic and epigenetic damage within stem cells (Rossi et al. 2007). Stem cell self-renewal programs must therefore change during aging to cope with these environmental changes, to adjust to the changing demands of tissue homeostasis, and to avoid transformation (Table 1).

We review evidence that self-renewal programs differ among embryonic, fetal, young adult, and old adult stem cells. Our goal is not to comprehensively review self-renewal mechanisms but rather to focus on a subset of key age-specific mechanisms. To accomplish this, we focus primarily on embryonic stem (ES) cells, germ-line stem cells, HSCs, and neural stem cells.

EMBRYONIC STEM CELL SELF-RENEWAL

The zygote is totipotent and can form all lineages of the embryo including the extraembryonic trophectoderm and primitive endoderm. It is not clear whether totipotent stem cells self-renew in vivo as cells begin to specialize within the embryo by the 32-cell stage (3 days in mice) (Johnson and McConnell 2004). By implantation (4 days in mice), the blastocyst includes the extraembryonic trophectoderm as well as the embryonic inner cell mass (Yamanaka et al. 2006). The inner cell mass contains epiblast cells that when cultured give rise to pluripotent ES cell lines. ES cells self-renew indefinitely in culture and are pluripotent, giving rise to cells of all three germ layers, although they are not totipotent because they are unable to form extraembryonic tissues. Epiblast cells exist only in a limited window of development and although these cells must maintain pluripotency, it is not clear whether they actually execute self-renewing divisions in vivo. It is thus possible that the in vitro ES cell self-renewal program incorporates mechanisms that regulate pluripotency in vivo, whereas other aspects of the self-renewal program are acquired in culture.

Oct4, Nanog, and Sox2 are transcription factors that compose a core regulatory circuit required for ES cell pluripotency and self-renewal (Jaenisch and Young 2008). These three proteins bind and autoregulate their own promoters, regulate the expression of other pluripotency genes, and repress early differentiation genes (Nichols et al. 1998; Boyer et al. 2005; Loh et al. 2006; Wang et al. 2006; Masui et al. 2007). Nanog stabilizes the pluripotent state, reduces the propensity of ES cells to spontaneously differentiate, and is required for ES cells to contribute to the germ line (Chambers et al. 2007). Nanog, Oct4, and Sox2 induction also reprograms adult cells to pluripotency (Takahashi and Yamanaka 2006; Takahashi et al. 2007; Yu et al. 2007b; Park et al. 2008).

These core transcriptional regulators do not control self-renewal on their own. Caspase-3 is an intracellular protease that is activated during apoptosis and in ES cells upon differentiation. Caspase-3 is necessary and sufficient for ES cell differentiation, promoting this in part through the proteolytic cleavage of Nanog (Fujita et al. 2008). The discovery that caspase-3 regulates ES cell differentiation led to the discovery of Ronin. Ronin is a zinc finger DNA-binding protein that is also necessary and sufficient for ES cell self-renewal. Like Nanog, Ronin is cleaved by caspase-3 during differentiation (Dejosez et al. 2008). Ronin acts with other proteins to epigenetically repress gene expression independently of Oct4, Sox2, and Nanog. Thus, multiple pathways independently regulate ES cell self-renewal and pluripotency, and caspase-3 activity attenuates at least two of these to initiate differentiation.

Oct4, Nanog, Sox2, and Ronin are also required for the formation of the inner cell mass in vivo. Oct4 expression is restricted to the inner cell mass during early embryonic development and is rapidly shut off as differentiation occurs (Lengner et al. 2007). *Oct4*-deficient embryos fail to develop beyond the blastocyst stage and lack a pluripotent inner cell mass, which instead gives rise to extraembryonic trophoblast cells (Nichols et al. 1998). *Nanog* expression is largely restricted to the inner cell mass and early germ-line cells (Chambers et al. 2003; Mitsui et al. 2003). *Nanog*-deficient embryos fail to develop beyond the blastocyst stage and lack pluripotent epiblasts, although *Nanog*-deficient ES cells can contribute widely to somatic tissues outside of the germ line (Chambers et al. 2003; Mitsui et al. 2003). *Sox2*-deficient embryos also lack pluripotent epiblast cells (Avilion et al. 2003). Ronin expression is primarily detected within the inner cell mass. *Ronin* deficiency leads to peri-implantation lethality and prevents pluripotent cells from growing out of the

Table 1. Examples of genes that regulate stem cell self-renewal at distinct developmental stages

Gene	Gene product	Stem cells	References
Nanog	homeodomain transcription factor	ES cells	Mitsui et al. (2003); Chambers et al. (2007)
Pouf1/Oct4	Pou domain transcription factor	ES cells	Nichols et al. (1998); Lengner et al. (2007)
Thap11/Ronin	Thap domain epigenetic regulator	ES cells	Dejosez et al. (2008)
Caspase-3	protease	ES cells, HSCs	Fujita et al. (2008); Janzen et al. (2008)
Zfx	zinc finger transcription factor	ES cells, HSCs	Galan-Caridad et al. (2007)
Sox17	transcription factor	fetal HSCs	Kim et al. (2007)
Hmga2	chromatin regulator	fetal and young adult neural stem cells	Nishino et al. (2008)
Bmi-1	Polycomb epigenetic regulator	many adult stem cells	Lessard and Sauvageau (2003); Molofsky et al. (2003); Park et al. (2003)
Klotho	Wnt Inhibitor	old HSCs, satellite muscle cells	Liu et al. (2007)
Angiopoietin-1	cytokine	adult HSCs	Puri and Bernstein (2003); Arai et al. (2004)
thrombopoietin	cytokine	adult HSCs	Qian et al. (2007); Yoshihara et al. (2007)
Gfi-1	transcription factor	adult HSCs	Hock et al. (2004a)
Tel/Etv6	transcription factor	adult HSCs	Hock et al. (2004b)
let-7b	microRNA	old neural stem cells	Nishino et al. (2008)
$p16^{Ink4a}$	cyclin-dependent kinase inhibitor	many old stem cells	Janzen et al. (2006); Krishnamurthy et al. (2006); Molofsky et al. (2006)

inner cell mass in culture (Dejosez et al. 2008). These studies demonstrate that aspects of the ES cell self-renewal program that are required in culture are also required for the formation of the inner cell mass and for epiblast pluripotency in vivo.

Other genes that are required for the self-renewal of ES cells in culture are not necessary for the generation of the inner cell mass in vivo. Standard protocols for the derivation and propagation of ES cells use leukemia inhibitory factor (LIF), which promotes self-renewal by activating STAT3 (signal transducer and activator of transcription 3) via the gp130 receptor (Niwa et al. 1998; Ying et al. 2003). Although these genes are required for the self-renewal of ES cells in vitro, LIF and gp130 are dispensable for embryonic development in vivo (Li et al. 1995; Ware et al. 1995; Yoshida et al. 1996; Dani et al. 1998). The requirement for LIF signaling in culture may reflect the need to block differentiation-inducing cues in culture that do not exist in vivo; addition of chemical inhibitors of fibroblast growth factor-4 (FGF-4)-induced ERK (extracellular signal-regulated kinase) phosphorylation can maintain the pluripotency of ES cells in the absence of LIF and serum (Ying et al. 2008). A series of transcriptional regulators including Tbx3, Tcl1, and Esrrb (Ivanova et al. 2006) as well as Zfx (Galan-Caridad et al. 2007) are also required for ES cell self-renewal in culture but not early embryonic development in vivo. These studies illustrate the context dependence of self-renewal programs and the fact that mechanisms required for self-renewal in culture do not necessarily regulate pluripotency or self-renewal under physiological conditions in vivo.

Some key elements of the pluripotent self-renewal program are restricted to ES cells and epiblast cells, whereas others are not. Although Nanog and Oct4 regulate the maintenance of pluripotency in ES cells and epiblast cells, they are not required for the function of fetal or adult somatic stem cells (Chambers et al. 2007; Lengner et al. 2007). Sox2, on the other hand, is expressed by fetal and adult neural stem cells and is required for their self-renewal (Graham et al. 2003; Suh et al. 2007). Caspase-3 deficiency increases adult HSC numbers (Janzen et al. 2008) just as it promotes the maintenance of ES cells (Fujita et al. 2008). The Zfx transcription factor provides a particularly interesting example because it is required for the self-renewal of adult (but not fetal) HSCs in addition to ES cells (Galan-Caridad et al. 2007). It is not yet clear whether Ronin regulates somatic stem cell function. These studies demonstrate that although elements of the pluripotent stem cell self-renewal program can be used by somatic stem cells at later stages of development, other features seem to uniquely define the pluripotent state.

FETAL AND NEONATAL STEM CELL SELF-RENEWAL

Fetal development represents a distinct stage marked by tissue specialization and tremendous growth. Consistent with this, the pluripotent cells of the embryo are replaced by a series of tissue-specific multipotent stem cells with developmental potentials that match the demands of their growing tissues. As a result, fetal self-renewal programs acquire tissue-specific components that regulate multipotency and differentiation in response to tissue-specific cues. But although the developmental potential of fetal somatic stem cells is narrower than their embryonic progenitors, fetal stem cells remain highly mitotically active. HSCs from the fetal liver (Morrison et al. 1995) and neural crest stem cells from the developing peripheral nervous system (Morrison et al. 1999; Bixby et al. 2002) divide rapidly in vivo and appear to undergo frequent self-renewing divisions. The absolute numbers of stem cells in many tissues expand considerably during fetal development, and imaging of progenitor divisions suggests that this expansion is driven initially by symmetric self-renewing divisions, followed by asymmetric divisions that generate differentiated progeny (Chenn and McConnell 1995; Lechler and Fuchs 2005; Morrison and Kimble 2006). Fetal self-renewal programs must therefore change relative to embryonic self-renewal to account for tissue specialization as well as developmental changes in the symmetry of division.

Fetal Germ-line Stem Cells

The existence of fetal-specific self-renewal programs is illustrated by spermatogonial stem cells (Oatley and Brinster 2008). Primordial germ cells are specified from epiblast cells early in embryogenesis, then migrate to the genital ridges by E11.5. Primordial germ cells cycle rapidly and expand their numbers from 100 cells at E8.5 to 25,000 cells at E13.5. Once in the genital ridges, primordial germ cells form gonocytes that commit to a sex-specific fate. In male testis, fetal gonocytes briefly stop dividing, but they resume self-renewal by postnatal day 3 when they become spermatogonial stem cells.

This lineage illustrates the changes in fetal self-renewal programs that occur over time. Nanos3 and Nanos2 encode zinc finger proteins, thought to act as RNA-binding translational repressors, that are required for the maintenance/expansion of primordial germ cells and gonocytes, respectively (Tsuda et al. 2003). Nanos3 is expressed in primordial germ cells as early as E7.5, whereas Nanos2 is only detected in gonocytes after colonization of the genital ridges. In the absence of *Nanos3*, the number of primordial germ cells is normal at E7.5 but depleted by E15.5. In the absence of *Nanos2*, primordial germ cell numbers are normal at E14.5 but no spermatogonial stem cells are present in the testis by 4 weeks after birth. *Nanos3* is required to prevent cell death during primordial germ cell migration (Suzuki et al. 2008). *Nanos2* is required for specification of male-specific germ cell fate (Suzuki and Saga 2008) in addition to gonocyte survival (Tsuda et al. 2003). These results show that *Nanos3* and *Nanos2* are dispensable for the specification of primordial germ cells but are sequentially required for the maintenance of germ-line stem cells during development. Although *Nanos3* and *Nanos2* are required by germ-line stem cells during development, mice lacking these genes have no other overt phenotype, suggesting that *Nanos3* and *Nanos2* are not required by other somatic stem cells (Tsuda et al. 2003). The genes that Nanos3 and Nanos2 repress in order to ensure stem cell maintenance are not yet known.

Fetal and Neonatal Hematopoietic Stem Cells

Fetal and neonatal HSCs have a unique self-renewal program that depends on the Sox17 transcriptional regulator (Kim et al. 2007). *Sox17* is a Sox family high-mobility-group transcription factor that is expressed in fetal and neonatal HSCs but is absent from adult HSCs. Conditional deletion of *Sox17* in fetal mice using *Tie2-Cre* or in neonatal mice using *Mx-1-Cre* led to the rapid depletion of HSCs and hematopoietic failure (Fig. 1). However, when *Sox17* was conditionally deleted from HSCs using *Mx-1-Cre* 6 weeks after birth, no defect was observed in HSC frequency, function, or hematopoiesis. These findings demonstrate that *Sox17* is required for the maintenance of fetal and neonatal HSCs but not adult HSCs. This demonstrates that there is a distinct fetal/neonatal HSC self-renewal program and raises the question of what program takes over for the maintenance

of adult HSCs. Additional experiments are required to elucidate the mechanisms by which Sox17 promotes HSC maintenance.

Sox17 is also a marker of fetal identity in HSCs. Sox17 is expressed by all fetal and neonatal HSCs but by less than 1% of other cells in the fetal liver or neonatal bone marrow (Fig. 1A). At 3–4 weeks after birth, HSCs go through a concerted developmental change in which they transition from a rapidly dividing fetal phenotype to a slowly dividing adult phenotype (Bowie et al. 2006, 2007). This is also when Sox17 expression is extinguished by HSCs (Kim et al. 2007). Moreover, examination of HSCs in the middle of this transition shows that the Sox17[+] subset retains fetal characteristics, whereas the Sox17[−] subset in the same bones of the same mice have acquired adult characteristics (see Fig. 1D). A key transcriptional regulator of HSC identity and function thus changes between the fetal and young adult stages.

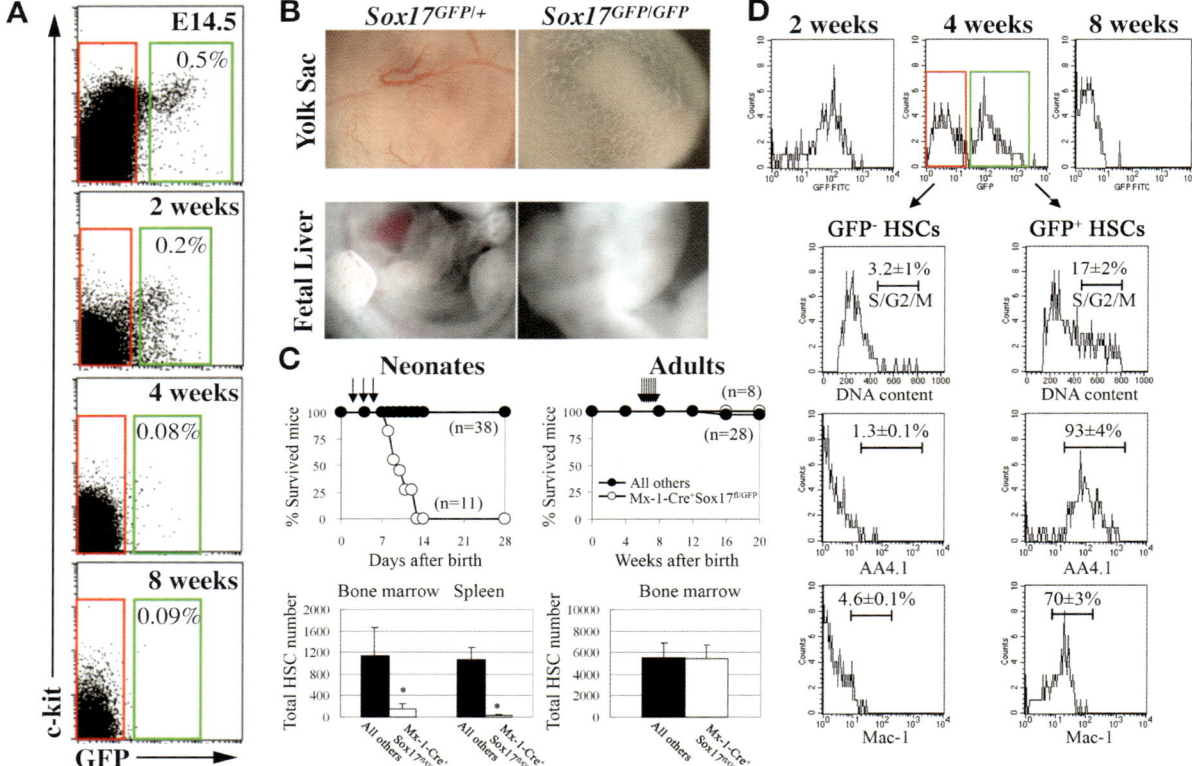

Figure 1. Sox17 is required for the self-renewal of fetal HSCs but not adult HSCs. (*A*) Green fluorescent protein (GFP) and c-Kit expression from unfractionated fetal liver or bone marrow cells analyzed by flow cytometry from *Sox17$^{GFP/+}$* mice. Boxes indicate GFP[−] (*red*) and GFP[+] (*green*) cells. GFP[+] cells represented fewer than 1% of cells in the fetal liver and were no longer detectable 8 weeks after birth in the bone marrow. (*B*) E11.5 embryos that lack Sox17 (*Sox17$^{GFP/GFP}$*) have a near-complete lack of hematopoiesis in the yolk sac or embryo. (*C*) *Sox17* was conditionally deleted (in *Mx-1-Cre^{+}Sox17$^{fl/GFP}$* mice) to test its role in hematopoiesis in neonatal (*left*) or young adult (*right*) mice. Although all neonatal mice died within 1 week of *Sox17* deletion, *Sox17* deficiency did not affect the viability of young adult mice. After *Sox17* deletion, HSCs were dramatically depleted in the bone marrow and spleen of neonatal mice, whereas HSCs were unaffected in young adult mice. (*D*) HSCs uniformly express *Sox17* 2 weeks after birth but uniformly lack expression by 8 weeks after birth (inferred based on GFP expression from the *Sox17^{GFP}* knockin allele). Four weeks after birth, when HSCs are making the transition from a fetal to an adult phenotype, half of the HSCs express *Sox17*. The *Sox17* expressing HSCs are rapidly dividing and express fetal HSC markers, whereas the subset of HSCs that lack *Sox17* expression are slowly dividing and express adult HSC markers. These results suggest that Sox17 is a marker of fetal/neonatal identity in HSCs. (Reprinted from Kim et al. 2007 [© Cell Press].)

Fetal Neural Stem Cells

Multipotent and self-renewing central nervous system (CNS) stem cells first arise in the context of the embryonic neuroepithelium. As neurogenesis begins, stem cells acquire a radial glial morphology with cell bodies in the ventricular zone and long processes that extend to the pial surface (Anthony et al. 2004). These radial glia often appear to divide asymmetrically, forming a radial glial daughter as well as a daughter cell fated to undergo neuronal differentiation (Noctor et al. 2004). Many of the daughter cells fated to form neurons amplify their progeny through subsequent symmetric divisions (Miyata et al. 2002; Haubensak et al. 2004; Noctor et al. 2004). Radial glial cells persist in the ventricle zone until the postnatal appearance of adult neural stem cells in a subventricular zone (Doetsch et al. 1999; Imura et al. 2003; Garcia et al. 2004). Lineage mapping has shown that adult neural cells are derived from fetal radial glial cells (Merkle et al. 2004). Neural stem cells thus undergo changes in their morphology and localization throughout fetal development.

Neural stem cells are fated to form an array of progeny at different stages of fetal development, starting with different types of neurons and ending with gliogenesis (Temple 2001). These changes in fate are likely regulated by both environmental and stem-cell-intrinsic changes that occur over time. In support of this, neural progenitors that are heterochronically transplanted into the CNS of older recipients form neurons appropriate to the age of their new environment, but older progenitors cannot form earlier subtypes of neurons upon transplantation into a younger developing forebrain (Frantz and McConnell 1996; Desai and McConnell 2000). This neuronal fate restriction is controlled by several components including FoxG1. FoxG1 is expressed by neural progenitors after the production of layer-1 neurons and represses layer-1 neuronal fate. Deletion of *FoxG1* causes layer-1 neurons to be continually produced during later developmental stages, where they are normally no longer produced (Hanashima et al. 2004; Shen et al. 2006).

The narrowing of neuronal subtype potential that occurs during cortical development suggests that multipotent forebrain cells do not generate identical progeny when they self-renew. Indeed, neural stem cells appear to undergo developmental restrictions throughout life but retain multipotency and proliferative potential. The observation that neural stem cells depend on many of the same self-renewal mechanisms that are used by HSCs (and that even HSCs undergo developmental changes in lymphocyte subtype potential over time) suggests that they depend on a core "self-renewal" program despite the fact that developmental potential changes over time. Existing data suggest that this core self-renewal program can be modified by factors that restrict neuronal subtype potential (or lymphocyte subtype potential in the case of HSCs). Self-renewal mechanisms are therefore not necessarily designed to generate identical progeny but rather to generate undifferentiated progeny that remain multipotent and self-renewing.

ADULT STEM CELL SELF-RENEWAL

Adult stem cells exhibit several differences relative to fetal stem cells that affect the regulation of self-renewal. In contrast to the rapid division of fetal stem cells, adult stem cells are often relatively quiescent. Adult mouse HSCs are quiescent most of the time, although all HSCs appear to enter the cycle on a regular basis and contribute to hematopoiesis (Cheshier et al. 1999; Kiel et al. 2007). As a result, adult self-renewal programs commonly feature mechanisms that promote quiescence. Adult stem cells often have less self-renewal potential than fetal stem cells from the same tissue. HSCs and neural crest stem cells from fetal mice have more self-renewal potential than those from adult mice (Chen et al. 2000; Kruger et al. 2002). Differences in self-renewal programs must therefore account for the difference in self-renewal potential. Finally, adult stem cells often have a narrower developmental potential than fetal stem cells in the same tissues. Adult HSCs remain multipotent and capable of generating lymphocytes but they lose the ability to make certain subtypes of lymphocytes that are only made by fetal HSCs (Hayakawa et al. 1985; Ikuta et al. 1990; Kantor et al. 1992). Neural crest stem cells from the adult gut remain multipotent and able to make neurons, but they lose the ability to make certain subtypes of neurons that are only made in the fetal gut (Kruger et al. 2002). As a result, multipotency of adult stem cells differs from that of fetal stem cells. For all of these reasons, self-renewal programs must be regulated differently in adult stem cells as compared to fetal stem cells.

Adult Germ-line Stem Cells

Cell signaling and transcriptional regulation distinguish the self-renewal of adult spermatogonial stem cells from their fetal counterparts. Glial-cell-line-derived neurotrophic factor (GDNF) is critical for the self-renewal of spermatogonial stem cells but not for the expansion of their fetal progenitors. Sertoli cells in the testis secrete GDNF (Tadokoro et al. 2002), whereas the spermatogonial stem cells express the GDNF coreceptors Ret and GFRA1 (Naughton et al. 2006). In the absence of *Gdnf*, *Ret*, or *Gfra1*, primordial germ cells colonize the testis normally and are present in normal numbers at birth, but spermatogonial stem cells fail to self-renew postnatally and become depleted (Naughton et al. 2006). Adult spermatogonial stem cells also require the zinc finger transcription factor Plzf (Buaas et al. 2004; Costoya et al. 2004) and the TFIID RNA polymerase complex component Taf4b (Falender et al. 2005). Mice lacking these genes exhibit normal spermatogonial development and initially produce mature sperm, but spermatogonial stem cells fail to self-renew and spermatogenesis is depleted (Costoya et al. 2004; Falender et al. 2005). Extrinsic and intrinsic factors that control spermatogonial stem cell self-renewal thus differ between adult and fetal stages.

Adult Hematopoietic Stem Cells

Many self-renewal mechanisms are conserved between fetal and adult HSCs. A number of transcriptional regulators including Rae28 (Ohta et al. 2002; Kim et al. 2004b), Meis1 (Hisa et al. 2004; Kirito et al. 2004; Azcoitia et al. 2005), c-Myb (Mucenski et al. 1991; Sandberg et al. 2005), Cbp (Rebel et al. 2002), Gata-2 (Tsai et al. 1994; Rodrigues et al. 2005), PU.1 (Kim et al. 2004a; Iwasaki et al. 2005), and Mll (McMahon et al. 2007) are required for the self-renewal of both fetal and adult HSCs. This demonstrates that self-renewal programs do not undergo wholesale changes between fetal development and adulthood but rather are modified to account for developmental differences in stem cell function.

Nonetheless, there are key differences in self-renewal regulators among fetal and adult HSCs. One prominent example is Bmi-1, which is consistently required for the self-renewal of all postnatal stem cell populations examined so far, including HSCs and neural stem cells (Lessard and Sauvageau 2003; Molofsky et al. 2003; Park et al. 2003). Bmi-1 is a proto-oncogene and a component of the Polycomb repressive complex 1 that regulates chromatin structure by recruiting epigenetic regulators to specific loci (Valk-Lingbeek et al. 2004). Bmi-1 is not required for the formation or maintenance of stem cells during fetal development because *Bmi-1*-deficient mice are born with normal numbers of stem cells. Rather, Bmi-1-deficient stem cells exhibit a postnatal self-renewal defect that leads to the depletion of stem cells by early adulthood. By postnatal day 30, HSCs are depleted from the bone marrow of *Bmi-1*-deficient mice, and bone marrow cells from these mice are unable to reconstitute irradiated recipient mice (Park et al. 2003). *Bmi-1*-deficient mice generally die by postnatal day 30, exhibiting growth retardation, hematopoietic failure, and neurological problems (ataxia, seizures) (van der Lugt et al. 1994).

One major mechanism by which *Bmi-1* promotes the self-renewal of HSCs and other postnatal stem cell populations is through repression of the p16[Ink4a] and p19[Arf] tumor suppressors (Fig. 2) (Jacobs et al. 1999; Bruggeman et al. 2005; Molofsky et al. 2005; Akala et al. 2008). p16[Ink4a] is a cyclin-dependent kinase inhibitor that promotes Rb (retinoblastoma) activation, slowing cell cycle progression or inducing cellular senescence (Lowe and Sherr 2003). p19[Arf] promotes p53 activity, also slowing cell cycle progression or inducing cellular senescence. p16[Ink4a] and p19[Arf] expression generally cannot be detected in fetal or young adult tissue, but these gene products are readily detected in postnatal stem cells in the absence of *Bmi-1* (Molofsky et al. 2003; Park et al. 2003). Deletion of *p16[Ink4a]* and/or *p19[Arf]* partially rescues the self-renewal defects observed in various stem cell populations, including HSCs and neural stem cells (Jacobs et al. 1999; Bruggeman et al. 2005; Molofsky et al. 2005; Oguro et al. 2006; Akala et al. 2008). These results demonstrate that repression of *p16[Ink4a]* and *p19[Arf]* is a fundamental requirement for the maintenance of adult stem cells.

Additional transcriptional regulators also distinguish adult from fetal HSC self-renewal. Gfi1 (Hock et al. 2004a), Tel/Etv6 (Hock et al. 2004b), and FoxO (Miyamoto et al. 2007; Tothova et al. 2007) gene products are all required for the maintenance of adult but not fetal HSCs. Gfi1 promotes the quiescence of adult HSCs, whereas Tel/Etv6 promotes their survival. FoxO tran-

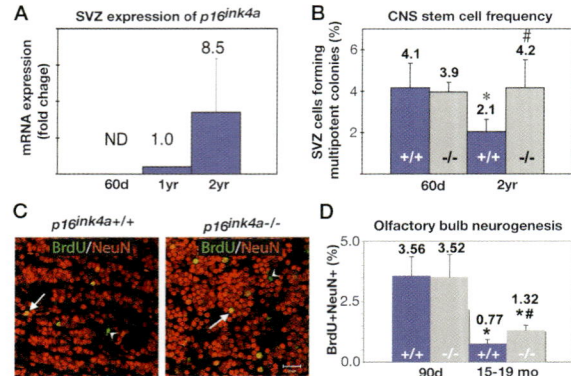

Figure 2. p16[Ink4a] expression increases with age and negatively regulates stem cell frequency and function. (*A*) *p16[Ink4a]* expression increases with age in uncultured subventricular zone cells. (*B*) The decline in stem cell frequency during aging is at least partially rescued by *p16[Ink4a]* deficiency. (*C,D*) The decline in olfactory bulb neurogenesis during aging is partially rescued by *p16[Ink4a]* deficiency (based on the frequency of bromodeoxyuridine [BrdU]-containing NeuN[+] neurons in the olfactory bulb). (Reprinted from Molofsky et al. 2006 [© Nature Publishing Group].)

scription factors promote the expression of enzymes that reduce levels of reactive oxygen species, such that in the absence of FoxO function, reactive oxygen species accumulate and HSCs are depleted. These results emphasize the importance of diverse transcription factors to regulate the maintenance of stem cell integrity in a developmentally regulated manner.

Fetal and adult HSCs also differ in their dependence on growth factors and signaling molecules. Thrombopoietin and its receptor Mpl promote the quiescence and maintenance of adult HSCs, but neither gene is required for the maintenance of fetal HSCs (Qian et al. 2007; Yoshihara et al. 2007). Angiopoietin-1 and its receptor Tie2 are similarly reported to promote the quiescence and maintenance of adult HSCs, but they are not required for the maintenance of fetal HSCs (Puri and Bernstein 2003; Arai et al. 2004). The ligands thrombopoietin and angiopoietin-1 are expressed in the bone marrow, and their corresponding receptors are expressed on HSCs. These results demonstrate that both cell-extrinsic and -intrinsic components of signaling networks regulate differences in stem cell maintenance between fetal and adult HSCs.

Adult Neural Stem Cells

Neural stem cells are maintained throughout adult life in two regions of the CNS and in at least two regions of the peripheral nervous system (PNS). Within the CNS, stem cells are maintained in the lateral wall of the lateral ventricle (Doetsch et al. 1999; Sanai et al. 2004) where they undergo neurogenesis throughout life, forming interneurons that migrate into the olfactory bulb (Sawamoto et al. 2006; Curtis et al. 2007). These new interneurons synapse on the projection neurons that connect the olfactory bulb to the cortex and likely regulate the ability to learn to discriminate chemically similar odorants (Magavi et al. 2005; Alonso et al. 2006). Stem cells also persist through adult life in the subgranular layer of the dentate gyrus (Palmer et al. 1997) where they form interneurons (Eriksson et al. 1998; van Praag et al. 2002). The new neurons that incorporate into the adult dentate gyrus appear to regulate certain forms of spatial learning and memory (Shors et al. 2001; Zhang et al. 2008). Within the PNS, multipotent and self-renewing neural crest stem cells persist throughout adult life in the wall of the gut (Kruger et al. 2002), although the physiological function of these stem cells remains unclear. A distinct neural-crest-derived stem cell population persists throughout life in the carotid body, where it gives rise to new glomus cells in response to oxygen stress (Pardal et al. 2007).

Just as in the hematopoietic system, some components of the neural stem cell self-renewal program are required by stem cells at all stages of development. For example, the Sox2 transcriptional regulator that is required for ES cell self-renewal is also required for the maintenance of undifferentiated neural stem cells at all stages from embryonic development through adulthood (Pevny and Placzek 2005). A loss of Sox2 function causes premature differentiation and depletion of neural stem cells at both fetal and adult stages, with a consequent impairment of neurogenesis (Bylund et al. 2003; Graham et al. 2003; Ferri et al. 2004). Sox2 is not alone in this regard because Numb and Numblike also appear to be required throughout fetal and adult life for the self-renewal of forebrain neural stem cells (Petersen et al. 2002, 2004; Shen et al. 2002; Kuo et al. 2006).

Some gene products are required throughout life by stem cells, but their precise function may change in concert with developmental changes in stem cells. For example, the orphan nuclear receptor Tlx is required to regulate cell cycle progression and fate determination within fetal and adult neural progenitors (Miyawaki et al. 2004; Roy et al. 2004; Shi et al. 2004); however, the precise ways in which these progenitors depend on Tlx varies depending on the location and time during development. In the retina, Tlx is required to regulate the proliferation of progenitors during fetal development and then to regulate glial differentiation postnatally (Miyawaki et al. 2004). In the developing cortex, Tlx is required to promote proliferation and to delay neurogenesis by ventricular zone progenitors, such that in the absence of Tlx, undifferentiated cells become depleted by premature neurogenesis and outer layers of cortical neurons are severely reduced (Roy et al. 2004). Within the adult subventricular zone, Tlx is required by stem cells to maintain self-renewal potential and to avoid premature glial differentiation (Shi et al. 2004). Thus, the way in which stem cells use Tlx to retain an undifferentiated state changes with location and time during development.

Other key transcriptional regulators are required by adult but not fetal stem cells. In the absence of Bmi-1, mice are born with normal numbers of neural stem cells, but p16[Ink4a] and p19[Arf] become derepressed postnatally, leading to the depletion of stem cells and neurogenesis from the CNS and PNS (Molofsky et al. 2003, 2005; Bruggeman et al. 2005). This demonstrates that neural stem cells share with HSCs a common dependence on transcriptional regulatory mechanisms that distinguish adult stem cell maintenance from fetal stem cell maintenance.

AGING STEM CELL SELF-RENEWAL

Stem cells must persist throughout life in numerous tissues to maintain tissue regenerative capacity. Yet, these cells must cope with ongoing change during adulthood. Genetic and epigenetic damage accumulates throughout life from reactive oxygen species, radiation, and chemical exposure. This damage impairs cellular function. For example, mitochondria accumulate damage with age, leading to a decline in mitochondrial function (Guarente 2008). Cells must compensate for this damage and declining function. In concert, the risk of neoplastic transformation increases with age as mutations accumulate. Cells must counteract this by augmenting tumor suppressor mechanisms during aging. Finally, there are systemic and local environmental changes within tissues during aging to which stem cells must adapt. For all of these reasons, stem cell self-renewal mechanisms change during aging.

Aging Hematopoietic Stem Cells

The properties of HSCs change in several ways as these cells age. The frequency of HSCs changes with age, although the nature of the change depends on the mouse strain: In some strains, such as C57BL, HSCs expand in number during aging (Morrison et al. 1996; Sudo et al. 2000; Yilmaz et al. 2006), whereas in other strains, HSCs become depleted during aging (de Haan and Van Zant 1999; Geiger and Van Zant 2002). HSC cell cycle status also changes during aging in a strain-dependent manner, with C57BL HSCs exhibiting increased cycling (Morrison et al. 1996; de Haan et al. 1997; Yilmaz et al. 2006). Aging HSCs also acquire an engraftment defect in which individual old HSCs are only one-third as likely as young HSCs to engraft in young irradiated recipient mice after transplantation (Morrison et al. 1996; Liang et al. 2005; Yilmaz et al. 2006). Nonetheless, individual reconstituting HSCs give similar overall levels of reconstitution, irrespective of whether they are from young or old donors. Finally, fate determination is affected because HSC differentiation appears skewed toward the production of myeloid cells at the expense of lymphoid cells in old mice (Morrison et al. 1996; Liang et al. 2005; Rossi et al. 2005). HSCs thus undergo many cell-intrinsic alterations in their functional properties during aging.

Despite ongoing Bmi-1 expression, p16[Ink4a] is induced in aging stem cells, reducing the cells' self-renewal potential (Janzen et al. 2006). p16[ink4a] is expressed by HSCs from old, but not young, mice. p16[ink4a] deficiency does not affect HSCs from young mice, but it increases the frequency, survival, self-renewal, and repopulating ability of HSCs from old mice. p16[ink4a] expression thus distinguishes old HSCs from young HSCs, demonstrating that the self-renewal of stem cells during aging is modified by increased tumor suppressor function. p19[Arf] expression also increases with age in a variety of tissues (Krishnamurthy et al. 2004), but whether p19[Arf] regulates stem cell aging has not yet been tested.

Aging Neural Stem Cells

Neural stem cell frequency, self-renewal capacity, and mitotic activity all decline with age, along with the rate of neurogenesis (Enwere et al. 2004; Maslov et al. 2004; Molofsky et al. 2006). These declines in neural stem cell frequency and function are partially caused by increasing p16^{Ink4a} expression. p16^{Ink4a} generally cannot be detected in fetal and young adult tissues but is induced in a variety of aging tissues (Fig. 2A) (Zindy et al. 1997; Krishnamurthy et al. 2004; Molofsky et al. 2006). p16^{Ink4a} deficiency partially rescues the decline in stem cell frequency (Fig. 2B), self-renewal potential, and neurogenesis in the aging CNS without affecting stem cell function or neurogenesis in the young adult CNS (Fig. 2C,D) (Molofsky et al. 2006). The decline in stem cell function at least partially reflects an active process in which tumor suppressors are induced during aging. Because p16^{Ink4a} deficiency also leads to early onset and increased incidence of malignancies, stem cells may induce p16^{Ink4a} expression during aging to avoid neoplastic transformation. This raises the possibility that stem cells balance regenerative (self-renewal) capacity with the risk of transformation during aging and that the induction of tumor suppressors during aging is a compromise designed to delay the onset of cancer.

The results described above raised two questions. First, why do old adult stem cells express higher levels of p16^{Ink4a} and p19^{Arf} than fetal and young adult stem cells? Second, given that Bmi-1 does not regulate p16^{Ink4a} and p19^{Arf} expression by fetal stem cells in vivo (Molofsky et al. 2003), are there other mechanisms that repress p16^{Ink4a} and p19^{Arf} in fetal stem cells? We have recently discovered that the high-mobility-group transcriptional regulator Hmga2 is highly expressed in fetal neural stem cells, but expression declines in young adult stem cells and is extinguished in old adult stem cells (Fig. 3A) (Nishino et al. 2008). This decrease in expression is partly caused by the increasing expression of let-7b microRNA, which is known to target Hmga2 (Fig. 3F) (Lee and Dutta 2007; Mayr et al. 2007; Yu et al. 2007a). Hmga2-deficient mice show reduced stem cell frequency and self-renewal potential throughout the CNS and PNS of fetal and young adult mice, but not old adult mice (Fig. 3D) (Nishino et al. 2008). Furthermore, p16^{Ink4a} and p19^{Arf} expression is increased in Hmga2-deficient fetal and young adult stem cells, and deletion of p16^{Ink4a} and/or p19^{Arf} partially restores their self-renewal capacity (Fig. 3B,C). Hmga2 thus promotes stem cell self-renewal during early life by decreasing p16^{Ink4a}/p19^{Arf} expression. These results identify a new pathway that regulates stem cell aging in which let-7b expression increases with age, reducing Hmga2 expression and increasing p16^{Ink4a}/p19^{Arf} expression, leading to reduced stem cell function (Fig. 3G).

The requirement for Hmga2 in fetal and young adult stem cells, along with the requirement for Bmi-1 in young and old adult stem cells, demonstrates that there are overlapping transcriptional mechanisms to prevent the expression of p16^{Ink4a} and p19^{Arf} in stem cells. Such pathways are temporally regulated by changes in let-7b and Hmga2 expression to allow p16^{Ink4a} and p19^{Arf} expression in aging stem cells (Fig. 4). In this way, self-renewal mechanisms can maintain stem cells throughout life while augmenting tumor suppression in aging stem cells.

p16^{Ink4a} and p19^{Arf} may be broadly important to diseases of aging beyond their function in stem cells. Several genome-wide association studies of large human populations revealed that polymorphisms at the p16^{Ink4a}/p19^{Arf} locus (chromosome 9p21) are associated with frailty (Melzer et al. 2007), atherosclerotic heart disease (Helgadottir et al. 2007, 2008; McPherson et al. 2007; Wellcome Trust Case Control Consortium 2007), and type-2 diabetes (Saxena et al. 2007; Wellcome Trust Case Control Consortium 2007; Zeggini et al. 2007). Although these studies do not pinpoint the polymorphisms that affect the risks of age-related diseases, there are only four genes in the vicinity of the mapped polymorphisms: p16^{Ink4a}, p19^{Arf}, p15^{Ink4b}, and ANRIL (a noncoding RNA) (Sharpless and DePinho 2007). p16^{Ink4a} expression increases with age in pancreatic β cells, and p16^{Ink4a} deficiency increases β-cell regenerative capacity (Krishnamurthy et al. 2006), providing a mechanism by which polymorphisms that affect p16^{Ink4a} expression or activity might affect risk for type-2 diabetes. It remains unclear whether these polymorphisms influence the risk of frailty and heart disease through their effects on tissue regenerative capacity or by mechanisms that are completely independent of stem/progenitor cells.

p16^{ink4a} may not be the only tumor suppressor that influences stem cell self-renewal, tissue regenerative capacity, age-related disease, and cancer. p53 is a well-known tumor suppressor that is frequently inactivated in human cancers. A polymorphism in p53 increases longevity and survival after cancer diagnosis in humans (Orsted et al. 2007). Overexpression of short isoforms of p53 in mice leads to early-onset aging phenotypes and decreased longevity (Tyner et al. 2002; Maier et al. 2004). These results raise the possibility that polymorphisms that increase p53 activity decrease cancer incidence but promote aging, whereas polymorphisms that decrease p53 activity increase cancer incidence but slow aging. However, bacterial artificial chromosome transgenic mice that bear a third copy of the p53 locus or the p15^{Ink4b}/p16^{Ink4a}/p19^{Arf} locus show a decreased cancer incidence but normal longevity and normal onset of aging phenotypes (Garcia-Cao et al. 2002; Matheu et al. 2004). Complicating matters further, mice that bear a third copy of the p53 locus and a third copy of the p15^{Ink4b}/p16^{Ink4a}/p19^{Arf} locus show increased longevity and delayed aging in a manner that cannot be explained by their reduced incidence of cancer (Matheu et al. 2007). These results suggest that the effects of p53 and p15^{Ink4b}/p16^{Ink4a}/p19^{Arf} expression on aging are context and dosage dependent and likely reflect a complex mix of mechanisms.

Insights from Other Systems

A series of studies conducted on aging muscle have shown that the function of aging stem cells is also affected by environmental change. Muscle satellite cells poorly regenerate injured muscles in aging animals, partly due to reduced Notch signaling in aging satellite cells (Conboy

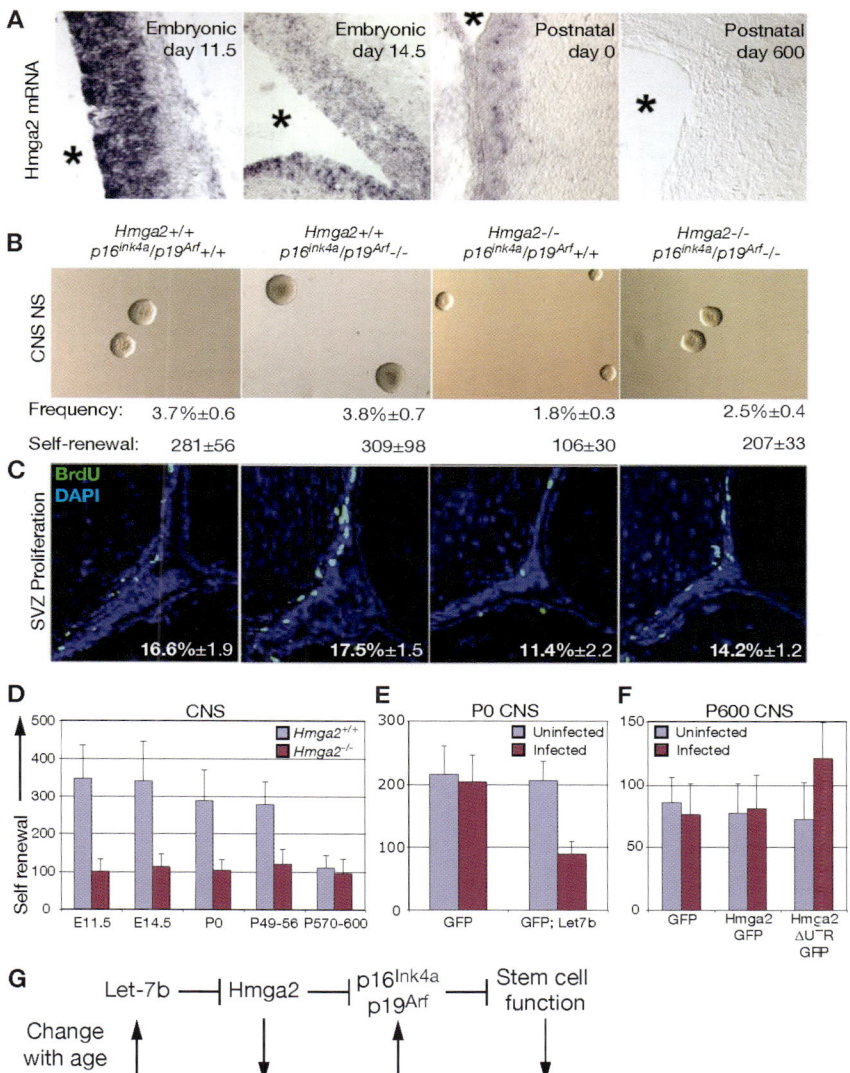

Figure 3. Hmga2 and *let-7b* regulate p16^Ink4a/p19^Arf expression and stem cell self-renewal in an age-related manner. (*A*) *Hmga2* is expressed by undifferentiated cells in the ventricular/subventricular zone and expression declines with age. Expression can no longer be detected in the subventricular zone of old adult mice. *Asterisk* indicates the ventricle. (*B*) *Hmga2* deficiency reduced the frequency (percentage of cells that formed stem cell colonies in culture) and self-renewal potential (number of secondary stem cell colonies generated per primary colony upon subcloning) of stem cells cultured from the fetal and young adult brain. *p16^Ink4a/p19^Arf* deficiency partially rescued these defects. Statistics reflect mean ± standard. (*C*) *Hmga2* deficiency reduced the rate of proliferation within the subventricular zone of young adult mice, but *p16^Ink4a/p19^Arf* deficiency partially rescued this defect. Numbers indicate the percentage of subventricular zone cells (mean ± S.D.) that incorporate a pulse of BrdU in vivo. (*D*) *Hmga2* deficiency reduced the self-renewal potential of stem cells cultured from the fetal and young adult brain but did not affect the self-renewal potential of stem cells from the old adult brain. (*E*) Overexpression of *let-7b*, but not *GFP* control, reduced the self-renewal of neural stem cells. (*F*) Deletion of the *Hmga2* 3′UTR (eliminating *let-7*-binding sites) was required to increase Hmga2 protein levels in old neural stem cells (data not shown). This *let-7*-insensitive form of *Hmga2*, but not wild-type *Hmga2*, significantly increased the self-renewal of neural stem cells from old mice. (*G*) Schematic of the mechanism by which changes in *let-7b* and *Hmga2* expression during aging regulate stem cell function. (Reprinted from Nishino et al. 2008. [© Cell Press].)

et al. 2003). However, when the circulatory systems of old and young mice are parabiotically fused, Notch signaling and muscle regeneration in the old mice are restored to young levels (Conboy et al. 2005). This demonstrates that age-related changes in stem cells are governed by both local and systemic environmental changes. Further mechanistic insights into these changes were provided by the observation that Wnt signaling is elevated in aging mice and that this contributes to the reduced regenerative capacity (Brack et al. 2007). This finding was supported by the discovery that *Klotho*-deficient mice, which exhibit elevated Wnt signaling and premature aging, lack a secreted Wnt antagonist (Liu et al. 2007). Other environmental changes also occur during aging, such as increasing TGF-β signaling. The increase in TGF-β signaling leads to increased expression of

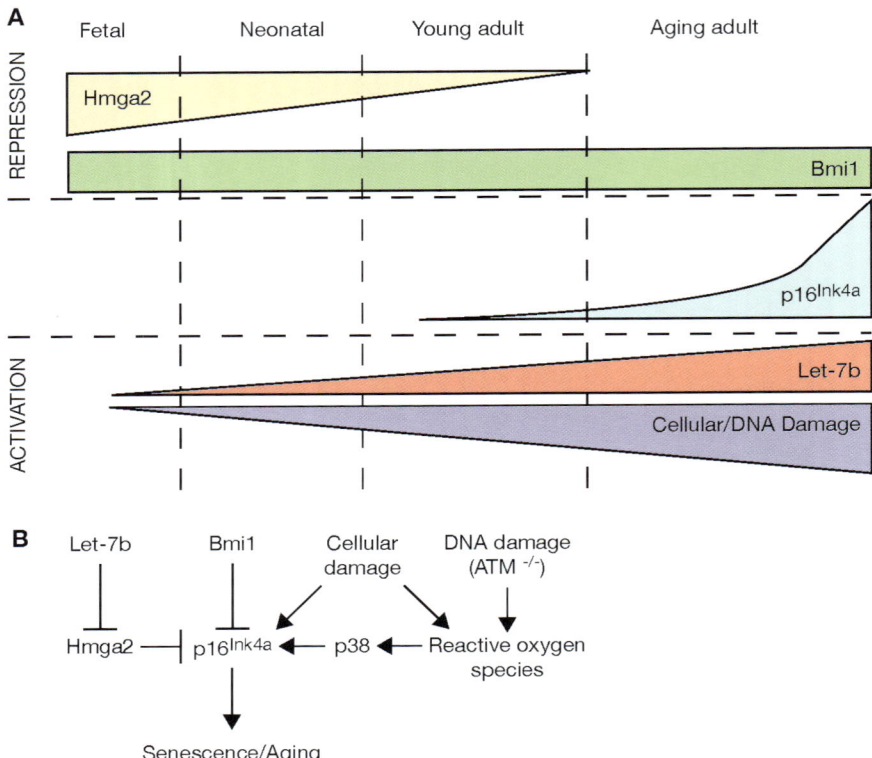

Figure 4. Age-related changes in p16^{Ink4a} regulation. (*A*) Schematic of genes and processes that affect p16^{Ink4a} expression during aging. (*B*) A summary of p16^{Ink4a} regulatory mechanisms within stem cells.

cyclin-dependent kinase inhibitors, including p16^{Ink4a} (Carlson et al. 2008). This provides a mechanism by which environmental changes that occur during aging can regulate cell-intrinsic factors identified in prior studies.

PERSPECTIVE

For a mammal to survive, stem cells must maintain their ability to self-renew throughout life. Tissue growth and regeneration demands change dramatically over the course of an animal's life, and the genetic programs regulating stem cell self-renewal change accordingly. Stem cells from embryonic, fetal, young adult, and old adult stages use somewhat different mechanisms to regulate their self-renewal (see Table 1).

Many key mechanisms that regulate stem cell self-renewal do not generically regulate all progenitor proliferation. Deletion of *Bmi-1*, *Hmga2*, or *p16^{Ink4a}* leads to self-renewal phenotypes in multipotent stem cells without affecting the proliferation of many restricted progenitors (Molofsky et al. 2003, 2006; Nishino et al. 2008). Sox17 is absolutely required for the maintenance of fetal HSCs but is not expressed by most restricted hematopoietic progenitors in the fetal liver (Kim et al. 2007). These results demonstrate that the self-renewal of stem cells is mechanistically distinct from the proliferation of many restricted progenitors. Nonetheless, this is not a black and white difference, because some restricted progenitors, such as lym-

phocytes and cerebellar granule precursor cells, also depend on Bmi-1 (van der Lugt et al. 1994; Leung et al. 2004). This suggests that some restricted progenitors use self-renewal mechanisms that are similar to those of stem cells.

A fundamental requirement for the maintenance of stem cells throughout life is the repression of p16^{ink4a} and p19^{Arf} tumor suppressors. This is often accomplished by stage-specific transcriptional regulatory mechanisms, such as Hmga2 and Bmi-1. The regulation of p16^{ink4a} and p19^{Arf} in a stage-specific manner confers the capacity for dynamic regulation. As the risk of cancer increases during aging, these genes are induced, reducing stem cell frequency and function in multiple tissues. Stem cell self-renewal is regulated by networks of proto-oncogenes and tumor suppressors that balance the need to maintain stem cells throughout life with the need to avoid neoplastic transformation. An important, although difficult, future question is whether the mechanisms that regulate stem cell aging also control life span and whether stem cell aging influences longevity.

The age-related changes in stem cell self-renewal programs are also likely to mean that different mutations are required for carcinogenesis at different stages of life. Cancers consistently arise from mutations that inappropriately activate self-renewal pathways (Reya et al. 2001; Pardal et al. 2003). Because self-renewal pathways change with age, we hypothesize that different mutations are required for neoplastic transformation at different ages, irrespective of whether cancers arise from stem cells or restricted progenitors/differentiated cells. This hypothesis remains to be tested directly, but it provides one explanation for the observation that adult cancers often have different mutation spectrums as compared to the childhood versions of the same cancers (Downing and Shannon 2002). A more sophisticated understanding of self-renewal mechanisms and how they change with age could provide new anticancer strategies.

ACKNOWLEDGMENTS

This work was supported by the Howard Hughes Medical Institute and the National Institutes of Health (1 RO1 AG024945-01). B.P.L was supported by an American Heart Association postdoctoral fellowship (0725726Z) and a Cancer Research Institute postdoctoral fellowship. We thank Ivan Maillard and Jeff Magee for reviewing the manuscript.

REFERENCES

Akala, O.O., Park, I.K., Qian, D., Pihalja, M., Becker, M.W., and Clarke, M.F. 2008. Long-term haematopoietic reconstitution by $Trp53^{-/-}p16^{Ink4a-/-}p19^{Arf-/-}$ multipotent progenitors. *Nature* **453:** 228–232.

Alonso, M., Viollet, C., Gabellec, M.M., Meas-Yedid, V., Olivo-Marin, J.C., and Lledo, P.M. 2006. Olfactory discrimination learning increases the survival of adult-born neurons in the olfactory bulb. *J. Neurosci.* **26:** 10508–10513.

Anthony, T.E., Klein, C., Fishell, G., and Heintz, N. 2004. Radial glia serve as neuronal progenitors in all regions of the central nervous system. *Neuron* **41:** 881–890.

Arai, F., Hirao, A., Ohmura, M., Sato, H., Matsuoka, S., Takubo, K., Ito, K., Koh, G.Y., and Suda, T. 2004. Tie2/angiopoietin-1 signaling regulates hematopoietic stem cell quiescence in the bone marrow niche. *Cell* **118:** 149–161.

Avilion, A.A., Nicolis, S.K., Pevny, L.H., Perez, L., Vivian, N., and Lovell-Badge, R. 2003. Multipotent cell lineages in early mouse development depend on SOX2 function. *Genes Dev.* **17:** 126–140.

Azcoitia, V., Aracil, M., Martinez, A.C., and Torres, M. 2005. The homeodomain protein Meis1 is essential for definitive hematopoiesis and vascular patterning in the mouse embryo. *Dev. Biol.* **280:** 307–320.

Bixby, S., Kruger, G.M., Mosher, J.T., Joseph, N.M., and Morrison, S.J. 2002. Cell-intrinsic differences between stem cells from different regions of the peripheral nervous system regulate the generation of neural diversity. *Neuron* **35:** 643–656.

Bowie, M.B., McKnight, K.D., Kent, D.G., McCaffrey, L., Hoodless, P.A., and Eaves, C.J. 2006. Hematopoietic stem cells proliferate until after birth and show a reversible phase-specific engraftment defect. *J. Clin. Invest.* **116:** 2808–2816.

Bowie, M.B., Kent, D.G., Dykstra, B., McKnight, K.D., McCaffrey, L., Hoodless, P.A., and Eaves, C.J. 2007. Identification of a new intrinsically timed developmental checkpoint that reprograms key hematopoietic stem cell properties. *Proc. Natl. Acad. Sci.* **104:** 5878–5882.

Boyer, L.A., Lee, T.I., Cole, M.F., Johnstone, S.E., Levine, S.S., Zucker, J.P., Guenther, M.G., Kumar, R.M., Murray, H.L., Jenner, R.G., et al. 2005. Core transcriptional regulatory circuitry in human embryonic stem cells. *Cell* **122:** 947–956.

Brack, A.S., Conboy, M.J., Roy, S., Lee, M., Kuo, C.J., Keller, C., and Rando, T.A. 2007. Increased Wnt signaling during aging alters muscle stem cell fate and increases fibrosis. *Science* **317:** 807–810.

Bruggeman, S.W., Valk-Lingbeek, M.E., van der Stoop, P.P.,

Jacobs, J.J., Kieboom, K., Tanger, E., Hulsman, D., Leung, C., Arsenijevic, Y., Marino, S., and van Lohuizen, M. 2005. *Ink4a* and *Arf* differentially affect cell proliferation and neural stem cell self-renewal in *Bmi1*-deficient mice. *Genes Dev.* **19:** 1438–1443.

Buaas, F.W., Kirsh, A.L., Sharma, M., McLean, D.J., Morris, J.L., Griswold, M.D., de Rooij, D.G., and Braun, R.E. 2004. Plzf is required in adult male germ cells for stem cell self-renewal. *Nat. Genet.* **36:** 647–652.

Bylund, M., Andersson, E., Novitch, B.G. and Muhr, J. 2003. Vertebrate neurogenesis is counteracted by Sox1-3 activity. *Nat. Neurosci.* **6:** 1162–1168.

Carlson, M.E., Hsu, M., and Conboy, I.M. 2008. Imbalance between pSmad3 and Notch induces CDK inhibitors in old muscle stem cells. *Nature* **454:** 528–532.

Chambers, I., Colby, D., Robertson, M., Nichols, J., Lee, S., Tweedie, S., and Smith, A. 2003. Functional expression cloning of Nanog, a pluripotency sustaining factor in embryonic stem cells. *Cell* **113:** 643–655.

Chambers, I., Silva, J., Colby, D., Nichols, J., Nijmeijer, B., Robertson, M., Vrana, J., Jones, K., Grotewold, L., and Smith, A. 2007. Nanog safeguards pluripotency and mediates germline development. *Nature* **450:** 1230–1234.

Chen, J., Astle, C.M., and Harrison, D.E. 2000. Genetic regulation of primitive hematopoietic stem cell senescence. *Exp. Hematol.* **28:** 442–450.

Chenn, A. and McConnell, S.K. 1995. Cleavage orientation and the asymmetric inheritance of Notch1 immunoreactivity in mammalian neurogenesis. *Cell* **82:** 631–641.

Cheshier, S., Morrison, S.J., Liao, X., and Weissman, I.L. 1999. In vivo proliferation and cell cycle kinetics of long-term self-renewing hematopoietic stem cells. *Proc. Natl. Acad. Sci.* **96:** 3120–3125.

Conboy, I.M., Conboy, M.J., Smythe, G.M., and Rando, T.A. 2003. Notch-mediated restoration of regenerative potential to aged muscle. *Science* **302:** 1575–1577

Conboy, I.M., Conboy, M.J., Wagers, A.J., Girma, E.R., Weissman, I.L., and Rando, T.A. 2005. Rejuvenation of aged progenitor cells by exposure to a young systemic environment. *Nature* **433:** 760–764.

Costoya, J.A., Hobbs, R.M., Barna, M., Cattoretti, G., Manova, K., Sukhwani, M., Orwig, K.E., Wolgemuth, D.J., and Pandolfi, P.P. 2004. Essential role of Plzf in maintenance of spermatogonial stem cells. *Nat. Genet.* **36:** 653–659.

Curtis, M.A., Kam, M., Nannmark, U., Anderson, M.F., Axell, M.Z., Wikkelso, C., Holtas, S., van Roon-Mom, W.M., Bjork-Eriksson, T., Nordborg, C., et al. 2007. Human neuroblasts migrate to the olfactory bulb via a lateral ventricular extension. *Science* **315:** 1243–1249.

Dani, C., Chambers, I., Johnstone, S., Robertson, M., Ebrahimi, B., Saito, M., Taga, T., Li, M., Burdon, T., Nichols, J., and Smith, A. 1998. Paracrine induction of stem cell renewal by LIF-deficient cells: A new ES cell regulatory pathway. *Dev. Biol.* **203:** 149–162.

de Haan, G. and Van Zant, G. 1999. Dynamic changes in mouse hematopoietic stem cell numbers during aging. *Blood* **93:** 3294–3301.

de Haan, G., Nijhof, W., and Van Zant G. 1997. Mouse strain-dependent changes in frequency and proliferation of hematopoietic stem cells during aging: Correlation between lifespan and cycling activity. *Blood* **89:** 1543–1550.

Dejosez, M., Krumenacker, J.S., Zitur, L.J., Passeri, M., Chu, L.F., Songyang, Z., Thomson, J.A., and Zwaka, T.P. 2008. Ronin is essential for embryogenesis and the pluripotency of mouse embryonic stem cells. *Cell* **133:** 1162–1174.

Desai, A.R. and McConnell, S.K. 2000. Progressive restriction in fate potential by neural progenitors during cerebral cortical development. *Development* **127:** 2863–2872.

Doetsch, F., Caille, I., Lim, D.A., Garcia-Verdugo, J.M., and Alvarez-Buylla, A. 1999. Subventricular zone astrocytes are neural stem cells in the adult mammalian brain. *Cell* **97:** 703–716.

Downing, J.R. and Shannon, K.M. 2002. Acute leukemia: A pediatric perspective. *Cancer Cell* **2:** 437–445.

Enwere, E., Shingo, T., Gregg, C., Fujikawa, H., Ohta, S., and Weiss, S. 2004. Aging results in reduced epidermal growth factor receptor signaling, diminished olfactory neurogenesis, and deficits in fine olfactory discrimination. *J. Neurosci.* **24:** 8354–8365.

Eriksson, P.S., Perfilieva, E., Bjork-Eriksson, T., Alborn, A.-M., Nordborg, C., Peterson, D.A., and Gage, F.H. 1998. Neurogenesis in the adult human hippocampus. *Nat. Med.* **4:** 1313–1317.

Falender, A.E., Freiman, R.N., Geles, K.G., Lo, K.C., Hwang, K., Lamb, D.J., Morris, P.L., Tjian, R., and Richards, J.S. 2005. Maintenance of spermatogenesis requires TAF4b, a gonad-specific subunit of TFIID. *Genes Dev.* **19:** 794–803.

Ferri, A.L., Cavallaro, M., Braida, D., Di Cristofano, A., Canta, A., Vezzani, A., Ottolenghi, S., Pandolfi, P.P., Sala, M., DeBiasi, S., and Nicolis, S.K. 2004. *Sox2* deficiency causes neurodegeneration and impaired neurogenesis in the adult mouse brain. *Development* **131:** 3805–3819.

Frantz, G.D. and McConnell, S.K. 1996. Restriction of late cerebral cortical progenitors to an upper-layer fate. *Neuron* **17:** 55–61.

Fujita, J., Crane, A.M., Souza, M.K., Dejosez, M., Kyba, M., Flavell, R.A., Thomson, J.A., and Zwaka, T.P. 2008. Caspase activity mediates the differentiation of embryonic stem cells. *Cell Stem Cell* **2:** 595–601.

Galan-Caridad, J.M., Harel, S., Arenzana, T.L., Hou, Z.E., Doetsch, F.K., Mirny, L.A., and Reizis, B. 2007. Zfx controls the self-renewal of embryonic and hematopoietic stem cells. *Cell* **129:** 345–357.

Garcia, A.D., Doan, N.B., Imura, T., Bush, T.G., and Sofroniew, M.V. 2004. GFAP-expressing progenitors are the principal source of constitutive neurogenesis in adult mouse forebrain. *Nat. Neurosci.* **7:** 1233–1241.

Garcia-Cao, I., Garcia-Cao, M., Martin-Caballero, J., Criado, L.M., Klatt, P., Flores, J.M., Weill, J.C., Blasco, M.A., and Serrano, M. 2002. "Super p53" mice exhibit enhanced DNA damage response, are tumor resistant and age normally. *EMBO J.* **21:** 6225–6235.

Geiger, H. and Van Zant, G. 2002. The aging of lympho-hematopoietic stem cells. *Nat. Immunol.* **3:** 329–333.

Graham, V., Khudyakov, J., Ellis, P., and Pevny, L. 2003. SOX2 functions to maintain neural progenitor identity. *Neuron* **39:** 749–765.

Guarente, L. 2008. Mitochondria—A nexus for aging, calorie restriction, and sirtuins? *Cell* **132:** 171–176.

Hanashima, C., Li, S.C., Shen, L., Lai, E., and Fishell, G. 2004. *Foxg1* suppresses early cortical cell fate. *Science* **303:** 56–59.

Harrison, D.E. and Astle, C.M. 1982. Loss of stem cell repopulating ability upon transplantation: Effects of donor age, cell number, and transplantation procedure. *J. Exp. Med.* **156:** 1767–1779.

Haubensak, W., Attardo, A., Denk, W., and Huttner, W.B. 2004. Neurons arise in the basal neuroepithelium of the early mammalian telencephalon: A major site of neurogenesis. *Proc. Natl. Acad. Sci.* **101:** 3196–3201.

Hayakawa, K., Hardy, R.R., Hertzenberg, L.A., and Herzenberg, L.A. 1985. Progenitors for Ly-1 B cells are distinct from progenitors for other B cells. *J. Exp. Med.* **161:** 1554–1568.

Helgadottir, A., Thorleifsson, G., Magnusson, K.P., Gretarsdottir, S., Steinthorsdottir, V., Manolescu, A., Jones, G.T., Rinkel, G.J., Blankensteijn, J.D., Ronkainen, A., et al. 2008. The same sequence variant on 9p21 associates with myocardial infarction, abdominal aortic aneurysm and intracranial aneurysm. *Nat. Genet.* **40:** 217–224.

Helgadottir, A., Thorleifsson, G., Manolescu, A., Gretarsdottir, S., Blondal, T., Jonasdottir, A., Jonasdottir, A., Sigurdsson, A., Baker, A., Palsson, A., et al. 2007. A common variant on chromosome 9p21 affects the risk of myocardial infarction. *Science* **316:** 1491–1493.

Hisa, T., Spence, S.E., Rachel, R.A., Fujita, M., Nakamura, T., Ward, J.M., Devor-Henneman, D.E., Saiki, Y., Kutsuna, H., Tessarollo, L., Jenkins, N.A., and Copeland, N.G. 2004. Hematopoietic, angiogenic and eye defects in *Meis1* mutant animals. *EMBO J.* **23:** 450–459.

Hock, H., Hamblen, M.J., Rooke, H.M., Schindler, J.W., Saleque, S., Fujiwara, Y., and Orkin, S.H. 2004a. Gfi-1 restricts proliferation and preserves functional integrity of haematopoietic stem cells. *Nature* **431:** 1002–1007.

Hock, H., Meade, E., Medeiros, S., Schindler, J.W., Valk, P.J., Fujiwara, Y., and Orkin, S.H. 2004b. Tel/Etv6 is an essential and selective regulator of adult hematopoietic stem cell survival. *Genes Dev.* **18:** 2336–2341.

Ikuta, K., Kina, T., Macneil, I., Uchida, N., Peault, B., Chien, Y.H., and Weissman, I.L. 1990. A developmental switch in thymic lymphocyte maturation potential occurs at the level of hematopoietic stem cells. *Cell* **62:** 863–874.

Imura, T., Kornblum, H.I., and Sofroniew, M.V. 2003. The predominant neural stem cell isolated from postnatal and adult forebrain but not early embryonic forebrain expresses GFAP. *J. Neurosci.* **23:** 2824–2832.

Ito, K., Hirao, A., Arai, F., Takubo, K., Matsuoka, S., Miyamoto, K., Ohmura, M., Naka, K., Hosokawa, K., Ikeda, Y., and Suda, T. 2006. Reactive oxygen species act through p38 MAPK to limit the lifespan of hematopoietic stem cells. *Nat. Med.* **12:** 446–451.

Ivanova, N., Dobrin, R., Lu, R., Kotenko, I., Levorse, J., DeCoste, C., Schafer, X., Lun, Y., and Lemischka, I.R. 2006. Dissecting self-renewal in stem cells with RNA interference. *Nature* **442:** 533–538.

Iwasaki, H., Somoza, C., Shigematsu, H., Duprez, E.A., Iwasaki-Arai, J., Mizuno, S., Arinobu, Y., Geary, K., Zhang, P., Dayaram, T., et al. 2005. Distinctive and indispensable roles of PU.1 in maintenance of hematopoietic stem cells and their differentiation. *Blood* **106:** 1590–1600.

Jacobs, J.J., Kieboom, K., Marino, S., DePinho, R.A., and van Lohuizen, M. 1999. The oncogene and Polycomb-group gene *bmi-1* regulates cell proliferation and senescence through the *ink4a* locus. *Nature* **397:** 164–168.

Jaenisch, R. and Young, R. 2008. Stem cells, the molecular circuitry of pluripotency and nuclear reprogramming. *Cell* **132:** 567–582.

Janzen, V., Forkert, R., Fleming, H.E., Saito, Y., Waring, M.T., Dombkowski, D.M., Cheng, T., DePinho, R.A., Sharpless, N.E., and Scadden, D.T. 2006. Stem-cell ageing modified by the cyclin-dependent kinase inhibitor p16[INK4a]. *Nature* **443:** 421–426.

Janzen, V., Fleming, H.E., Riedt, T., Karlsson, G., Riese, M.J., Lo Celso, C., Reynolds, G., Milne, C.D., Paige, C.J., Karlsson, S., et al. 2008. Hematopoietic stem cell responsiveness to exogenous signals is limited by caspase-3. *Cell Stem Cell* **2:** 584–594.

Johnson, M.H. and McConnell, J.M. 2004. Lineage allocation and cell polarity during mouse embryogenesis. *Semin. Cell Dev. Biol.* **15:** 583–597.

Kantor, A.B., Stall, A.M., Adams, S., Herzenberg, L.A., and Herzenberg, L.A. 1992. Differential development of progenitor activity for three B-cell lineages. *Proc. Natl. Acad. Sci.* **89:** 3320–3324.

Kiel, M.J., He, S., Ashkenazi, R., Gentry, S.N., Teta, M., Kushner, J.A., Jackson, T.L., and Morrison, S.J. 2007. Haematopoietic stem cells do not asymmetrically segregate chromosomes or retain BrdU. *Nature* **449:** 238–242.

Kim, H.G., de Guzman, C.G., Swindle, C.S., Cotta, C.V., Gartland, L., Scott, E.W., and Klug, C.A. 2004a. The ETS family transcription factor PU.1 is necessary for the maintenance of fetal liver hematopoietic stem cells. *Blood* **104:** 3894–3900.

Kim, I., Saunders, T.L., and Morrison, S.J. 2007. *Sox17* dependence distinguishes the transcriptional regulation of fetal from adult hematopoietic stem cells. *Cell* **130:** 470–483.

Kim, J.Y., Sawada, A., Tokimasa, S., Endo, H., Ozono, K., Hara, J., and Takihara, Y. 2004b. Defective long-term repopulating ability in hematopoietic stem cells lacking the *Polycomb*-group gene *rae28*. *Eur. J. Haematol.* **73:** 75–84.

Kirito, K., Fox, N., and Kaushansky, K. 2004. Thrombopoietin induces HOXA9 nuclear transport in immature hematopoietic cells: Potential mechanism by which the hormone favorably affects hematopoietic stem cells. *Mol. Cell. Biol.* **24:** 6751–6762.

Krishnamurthy, J., Ramsey, M.R., Ligon, K.L., Torrice, C., Koh, A., Bonner-Weir, S., and Sharpless, N.E. 2006. p16^{INK4a} induces an age-dependent decline in islet regenerative potential. *Nature* **443**: 453–457.

Krishnamurthy, J., Torrice, C., Ramsey, M.R., Kovalev, G.I., Al-Regaiey, K., Su, L., and Sharpless, N.E. 2004. *Ink4a/Arf* expression is a biomarker of aging. *J. Clin. Invest.* **114**: 1299–1307.

Kruger, G.M., Mosher, J.T., Bixby, S., Joseph, N., Iwashita, T., and Morrison, S.J. 2002. Neural crest stem cells persist in the adult gut but undergo changes in self-renewal, neuronal subtype potential, and factor responsiveness. *Neuron* **35**: 657–669.

Kuhn, H.G., Dickinson-Anson, H., and Gage, F.H. 1996. Neurogenesis in the dentate gyrus of the adult rat: Age-related decrease of neuronal progenitor proliferation. *J. Neurosci.* **16**: 2027–2033.

Kuo, C.T., Mirzadeh, Z., Soriano-Navarro, M., Rasin, M., Wang, D., Shen, J., Sestan, N., Garcia-Verdugo, J., Alvarez-Buylla, A., Jan, L.Y., and Jan, Y.N. 2006. Postnatal deletion of Numb/Numblike reveals repair and remodeling capacity in the subventricular neurogenic niche. *Cell* **127**: 1253–1264.

Lechler, T. and Fuchs, E. 2005. Asymmetric cell divisions promote stratification and differentiation of mammalian skin. *Nature* **437**: 275–280.

Lee, Y.S. and Dutta, A. 2007. The tumor suppressor microRNA *let-7* represses the HMGA2 oncogene. *Genes Dev.* **21**: 1025–1030.

Lengner, C.J., Camargo, F.D., Hochedlinger, K., Welstead, G.G., Zaidi, S., Gokhale, S., Scholer, H.R., Tomilin, A., and Jaenisch, R. 2007. Oct4 expression is not required for mouse somatic stem cell self-renewal. *Cell Stem Cell* **1**: 403–415.

Lessard, J. and Sauvageau, G. 2003. *Bmi-1* determines the proliferative capacity of normal and leukaemic stem cells. *Nature* **423**: 255–260.

Leung, C., Lingbeek, M., Shakhova, O., Liu, J., Tanger, E., Saremaslani, P., van Lohuizen, M., and Marino, S. 2004. Bmi1 is essential for cerebellar development and is overexpressed in human medulloblastomas. *Nature* **428**: 337–341.

Li, M., Sendtner, M., and Smith, A. 1995. Essential function of LIF receptor in motor neurons. *Nature* **378**: 724–727.

Liang, Y., Van Zant, G., and Szilvassy, S.J. 2005. Effects of aging on the homing and engraftment of murine hematopoietic stem and progenitor cells. *Blood* **106**: 1479–1487.

Liu, H., Fergusson, M.M., Castilho, R.M., Liu, J., Cao, L., Chen, J., Malide, D., Rovira, I.I., Schimel, D., Kuo, C.J., et al. 2007. Augmented Wnt signaling in a mammalian model of accelerated aging. *Science* **317**: 803–806.

Loh, Y.H., Wu, Q., Chew, J.L., Vega, V.B., Zhang, W., Chen, X., Bourque, G., George, J., Leong, B., Liu, J., et al. 2006. The Oct4 and Nanog transcription network regulates pluripotency in mouse embryonic stem cells. *Nat. Genet.* **38**: 431–440.

Lowe, S.W. and Sherr, C.J. 2003. Tumor suppression by *Ink4a-Arf*: Progress and puzzles. *Curr. Opin. Genet. Dev.* **13**: 77–83.

Magavi, S.S., Mitchell, B.D., Szentirmai, O., Carter, B.S., and Macklis, J.D. 2005. Adult-born and preexisting olfactory granule neurons undergo distinct experience-dependent modifications of their olfactory responses in vivo. *J. Neurosci.* **25**: 10729–10739.

Maier, B., Gluba, W., Bernier, B., Turner, T., Mohammad, K., Guise, T., Sutherland, A., Thorner, M., and Scrable, H. 2004. Modulation of mammalian life span by the short isoform of p53. *Genes Dev.* **18**: 306–319.

Maslov, A.Y., Barone, T.A., Plunkett, R.J., and Pruitt, S.C. 2004. Neural stem cell detection, characterization, and age-related changes in the subventricular zone of mice. *J. Neurosci.* **24**: 1726–1733.

Masui, S., Nakatake, Y., Toyooka, Y., Shimosato, D., Yagi, R., Takahashi, K., Okochi, H., Okuda, A., Matoba, R., Sharov, A.A., Ko, M.S., and Niwa, H. 2007. Pluripotency governed by *Sox2* via regulation of *Oct3/4* expression in mouse embryonic stem cells. *Nat. Cell Biol.* **9**: 625–635.

Matheu, A., Pantoja, C., Efeyan, A., Criado, L.M., Martin-Caballero, J., Flores, J.M., Klatt, P., and Serrano, M. 2004. Increased gene dosage of *Ink4a/Arf* results in cancer resistance and normal aging. *Genes Dev.* **18**: 2736–2746.

Matheu, A., Maraver, A., Klatt, P., Flores, I., Garcia-Cao, I., Borras, C., Flores, J.M., Viña, J., Blasco M.A., and Serrano, M. 2007. Delayed ageing through damage protection by the Arf/p53 pathway. *Nature* **448**: 375–379.

Mayr, C., Hemann, M.T., and Bartel, D.P. 2007. Disrupting the pairing between *let-7* and *Hmga2* enhances oncogenic transformation. *Science* **315**: 1576–1579.

McMahon, K.A., Hiew, S.Y., Hadjur, S., Veiga-Fernandes, H., Menzel, U., Price, A.J., Kioussis, D., Williams, O., and Brady, H.J. 2007. *Mll* has a critical role in fetal and adult hematopoietic stem cell self-renewal. *Cell Stem Cell* **1**: 338–345.

McPherson, R., Pertsemlidis, A., Kavaslar, N., Stewart, A., Roberts, R., Cox, D.R., Hinds, D.A., Pennacchio, L.A., Tybjaerg-Hansen, A., Folsom, A.R., et al. 2007. A common allele on chromosome 9 associated with coronary heart disease. *Science* **316**: 1488–1491.

Melzer, D., Frayling, T.M., Murray, A., Hurst, A.J., Harries, L.W., Song, H., Khaw, K., Luben, R., Surtees, P.G., Bandinelli, S.S., et al. 2007. A common variant of the p16^{INK4a} genetic region is associated with physical function in older people. *Mech. Ageing Dev.* **128**: 370–377.

Merkle, F.T., Tramontin, A.D., Garcia-Verdugo, J.M., and Alvarez-Buylla, A. 2004. Radial glia give rise to adult neural stem cells in the subventricular zone. *Proc. Natl. Acad. Sci.* **101**: 17528–17532.

Mitsui, K., Tokuzawa, Y., Itoh, H., Segawa, K., Murakami, M., Takahashi, K., Maruyama, M., Maeda, M., and Yamanaka, S. 2003. The homeoprotein Nanog is required for maintenance of pluripotency in mouse epiblast and ES cells. *Cell* **113**: 631–642.

Miyamoto, K., Araki, K.Y., Naka, K., Arai, F., Takubo, K., Yamazaki, S., Matsuoka, S., Miyamoto, T., Ito, K., Ohmura, M., et al. 2007. Foxo3a is essential for maintenance of the hematopoietic stem cell pool. *Cell Stem Cell* **1**: 101–112.

Miyata, T., Kawaguchi, A., Saito, K., Kuramochi, H., and Ogawa, M. 2002. Visualization of cell cycling by an improvement in slice culture methods. *J. Neurosci. Res.* **69**: 861–868.

Miyawaki, T., Uemura, A., Dezawa, M., Yu, R.T., Ide, C., Nishikawa, S., Honda, Y., Tanabe, Y., and Tanabe, T. 2004. Tlx, an orphan nuclear receptor, regulates cell numbers and astrocyte development in the developing retina. *J. Neurosci.* **24**: 8124–8134.

Molofsky, A.V., Pardal, R., and Morrison, S.J. 2004. Diverse mechanisms regulate stem cell self-renewal. *Curr. Opin. Cell Biol.* **16**: 700–707.

Molofsky, A.V., He, S., Bydon, M., Morrison, S.J., and Pardal, R. 2005. Bmi-1 promotes neural stem cell self-renewal and neural development but not mouse growth and survival by repressing the p16^{Ink4a} and p19^{Arf} senescence pathways. *Genes Dev.* **19**: 1432–1437.

Molofsky, A.V., Pardal, R., Iwashita, T., Park, I.K., Clarke, M.F., and Morrison, S.J. 2003. *Bmi-1* dependence distinguishes neural stem cell self-renewal from progenitor proliferation. *Nature* **425**: 962–967.

Molofsky, A.V., Slutsky, S.G., Joseph, N.M., He, S., Pardal, R., Krishnamurthy, J., Sharpless, N.E., and Morrison, S.J. 2006. Increasing *p16^{INK4a}* expression decreases forebrain progenitors and neurogenesis during ageing *Nature* **443**: 448–452.

Morrison, S.J. and Kimble, J. 2006. Asymmetric and symmetric stem-cell divisions in development and cancer. *Nature* **441**: 1068–1074.

Morrison, S.J., Hemmati, H.D., Wandycz, A.M., and Weissman, I.L. 1995. The purification and characterization of fetal liver hematopoietic stem cells. *Proc. Natl. Acad. Sci.* **92**: 10302–10306.

Morrison, S.J., White, P.M., Zock, C., and Anderson, D.J. 1999. Prospective identification, isolation by flow cytometry, and in vivo self-renewal of multipotent mammalian neural crest stem cells. *Cell* **96**: 737–749.

Morrison, S.J., Wandycz, A.M., Akashi, K., Globerson, A., and Weissman, I.L. 1996. The aging of hematopoietic stem cells. *Nat. Med.* **2**: 1011–1016.

Morrison, S.J., Wandycz, A.M., Hemmati, H.D., Wright, D.E., and

Weissman, I.L. 1997. Identification of a lineage of multipotent hematopoietic progenitors. *Development* **124:** 1929–1939.

Mucenski, M.L., McLain, K., Kier, A.B., Swerdlow, S.H., Schreiner, C.M., Miller, T.A., Pietryga, D.W., Scott, W.J., and Potter, S.S. 1991. A functional c-*myb* gene is required for normal murine fetal hepatic hematopoiesis. *Cell* **65:** 677–689.

Naughton, C.K., Jain, S., Strickland, A.M., Gupta, A., and Milbrandt, J. 2006. Glial cell-line derived neurotrophic factor-mediated RET signaling regulates spermatogonial stem cell fate. *Biol. Reprod.* **74:** 314–321.

Nichols, J., Zevnik, B., Anastassiadis, K., Niwa, H., Klewe-Nebenius, D., Chambers, I., Schöler, H., and Smith, A. 1998. Formation of pluripotent stem cells in the mammalian embryo depends on the POU transcription factor Oct4. *Cell* **95:** 379–391.

Nishino, J., Kim, I., Chada, K., and Morrison, S.J. 2008. Hmga2 promotes neural stem cell self-renewal in young but not old mice by reducing p16^{Ink4a} and p19^{Arf} expression. *Cell* **3:** 469–470.

Niwa, H., Burdon, T., Chambers, I., and Smith, A. 1998. Self-renewal of pluripotent embryonic stem cells is mediated via activation of STAT3. *Genes Dev.* **12:** 2048–2060.

Noctor, S.C., Martinez-Cerdeno, V., Ivic, L., and Kriegstein, A.R. 2004. Cortical neurons arise in symmetric and asymmetric division zones and migrate through specific phases. *Nat. Neurosci.* **7:** 136–144.

Oatley, J.M. and Brinster, R.L. 2008. Regulation of spermatogonial stem cell self-renewal in mammals. *Annu. Rev. Cell Dev. Biol.* **24:** 263–286.

Oguro, H., Iwama, A., Morita, Y., Kamijo, T., van Lohuizen, M., and Nakauchi, H. 2006. Differential impact of *Ink4a* and *Arf* on hematopoietic stem cells and their bone marrow microenvironment in *Bmi1*-deficient mice. *J. Exp. Med.* **203:** 2247–2253.

Ohta, H., Sawada, A., Kim, J.Y., Tokimasa, S., Nishiguchi, S., Humphries, R.K., Hara, J., and Takihara, Y. 2002. *Polycomb* group gene *rae28* is required for sustaining activity of hematopoietic stem cells. *J. Exp. Med.* **195:** 759–770.

Orsted, D.D., Bojesen, S.E., Tybjaerg-Hansen, A., and Nordestgaard, B.G. 2007. Tumor suppressor p53 Arg72Pro polymorphism and longevity, cancer survival, and risk of cancer in the general population. *J. Exp. Med.* **204:** 1295–1301.

Palmer, T.D., Takahashi, J., and Gage, F.H. 1997. The adult rat hippocampus contains primordial neural stem cells. *Mol. Cell. Neurosci.* **8:** 389–404.

Pardal, R., Clarke, M.F., and Morrison, S.J. 2003. Applying the principles of stem-cell biology to cancer. *Nat. Cancer Rev.* **3:** 895–902.

Pardal, R., Ortega-Saenz, P., Duran, R., and Lopez-Barneo, J. 2007. Glia-like stem cells sustain physiologic neurogenesis in the adult mammalian carotid body. *Cell* **131:** 364–377.

Park, I.H., Zhao, R., West, J.A., Yabuuchi, A., Huo, H., Ince, T.A., Lerou, P.H., Lensch, M.W., and Daley, G.Q. 2008. Reprogramming of human somatic cells to pluripotency with defined factors. *Nature* **451:** 141–146.

Park, I.-K., Qian, D., Kiel, M., Becker, M., Pihalja, M., Weissman, I.L., Morrison, S.J., and Clarke, M.F. 2003. Bmi-1 is required for the maintenance of adult self-renewing hematopoietic stem cells. *Nature* **423:** 302–305.

Petersen, P.H., Zou, K., Krauss, S., and Zhong, W. 2004. Continuing role for mouse *Numb* and *Numbl* in maintaining progenitor cells during cortical neurogenesis. *Nat. Neurosci.* **7:** 803–811.

Petersen, P.H., Zou, K., Hwang, J.K., Jan, Y.N., and Zhong, W. 2002. Progenitor cell maintenance requires *numb* and *numb-like* during mouse neurogenesis. *Nature* **419:** 929–934.

Pevny, L. and Placzek, M. 2005. *SOX* genes and neural progenitor identity. *Curr. Opin. Neurobiol.* **15:** 7–13.

Puri, M.C. and Bernstein, A. 2003. Requirement for the TIE family of receptor tyrosine kinases in adult but not fetal hematopoiesis. *Proc. Natl. Acad. Sci.* **100:** 12753–12758.

Qian, H., Buza-Vidas, N., Hyland, C.D., Jensen, C.T., Antonchuk, J., Mansson, R., Thoren, L.A., Ekblom, M., Alexander, W.S., and Jacobsen, S.E.W. 2007. Critical role of thrombopoietin in maintaining adult quiescent hematopoietic stem cells. *Cell Stem Cell* **1:** 671–684.

Rebel, V.I., Kung, A.L., Tanner, E.A., Yang, H., Bronson, R.T., and Livingston, D.M. 2002. Distinct roles for CREB-binding protein and p300 in hematopoietic stem cell self-renewal. *Proc. Natl. Acad. Sci.* **99:** 14789–14794.

Reya, T., Morrison, S.J., Clarke, M.F., and Weissman, I.L. 2001. Stem cells, cancer, and cancer stem cells. *Nature* **414:** 105–111.

Rodrigues, N.P., Janzen, V., Forkert, R., Dombkowski, D.M., Boyd, A.S., Orkin, S.H., Enver, T., Vyas, P., and Scadden, D.T. 2005. Haploinsufficiency of *GATA-2* perturbs adult hematopoietic stem-cell homeostasis. *Blood* **106:** 477–484.

Rossi, D.J., Bryder, D., Seita, J., Nussenzweig, A., Hoeijmakers, J., and Weissman, I.L. 2007. Deficiencies in DNA damage repair limit the function of haematopoietic stem cells with age. *Nature* **447:** 725–729.

Rossi, D.J., Bryder, D., Zahn, J.M., Ahlenius, H., Sonu, R., Wagers, A.J., and Weissman, I.L. 2005. Cell intrinsic alterations underlie hematopoietic stem cell aging. *Proc. Natl. Acad. Sci.* **102:** 9194–9199.

Roy, K., Kuznicki, K., Wu, Q., Sun, Z., Bock, D., Schutz, G., Vranich, N., and Monaghan, A.P. 2004. The *Tlx* gene regulates the timing of neurogenesis in the cortex. *J. Neurosci.* **24:** 8333–8345.

Sanai, N., Tramontin, A.D., Quinones-Hinojosa, A., Barbaro, N.M., Gupta, N., Kunwar, S., Lawton, M.T., McDermott, M.W., Parsa, A.T., Manuel-Garcia Verdugo, J., Berger, M.S., and Alvarez-Buylla, A. 2004. Unique astrocyte ribbon in adult human brain contains neural stem cells but lacks chain migration. *Nature* **427:** 740–744.

Sandberg, M.L., Sutton, S.E., Pletcher, M.T., Wiltshire, T., Tarantino, L.M., Hogenesch, J.B., and Cooke, M.P. 2005. c-Myb and p300 regulate hematopoietic stem cell proliferation and differentiation. *Dev. Cell* **8:** 153–166.

Sawamoto, K., Wichterle, H., Gonzalez-Perez, O., Cholfin, J.A., Yamada, M., Spassky, N., Murcia, N.S., Garcia-Verdugo, J.M., Marin, O., Rubenstein, J.L., et al. 2006. New neurons follow the flow of cerebrospinal fluid in the adult brain. *Science* **311:** 629–632.

Saxena, R., Voight, B.F., Lyssenko, V., Burtt, N.P., de Bakker, P.I., Chen, H., Roix, J.J., Kathiresan, S., Hirschhorn, J.N., Daly, M.J., et al. 2007. Genome-wide association analysis identifies loci for type 2 diabetes and triglyceride levels. *Science* **316:** 1331–1336.

Sharpless, N.E. and DePinho, R.A. 2007. How stem cells age and why this makes us grow old. *Nat. Rev. Mol. Cell Biol.* **8:** 703–713.

Shen, Q., Zhong, W., Jan, Y.N., and Temple, S. 2002. Asymmetric Numb distribution is critical for asymmetric cell division of mouse cerebral cortical stem cells and neuroblasts. *Development* **129:** 4843–4853.

Shen, Q., Wang, Y., Dimos, J.T., Fasano, C.A., Phoenix, T.N., Lemischka, I.R., Ivanova, N.B., Stifani, S., Morrisey, E.E., and Temple, S. 2006. The timing of cortical neurogenesis is encoded within lineages of individual progenitor cells. *Nat. Neurosci.* **9:** 743–751.

Shi, Y., Chichung Lie, D., Taupin, P., Nakashima, K., Ray, J., Yu, R.T., Gage, F.H., and Evans, R.M. 2004. Expression and function of orphan nuclear receptor TLX in adult neural stem cells. *Nature* **427:** 78–83.

Shors, T.J., Miesegaes, G., Beylin, A., Zhao, M., Rydel, T., and Gould, E. 2001. Neurogenesis in the adult is involved in the formation of trace memories. *Nature* **410:** 372–376.

Sudo, K., Ema, H., Morita, Y., and Nakauchi, H. 2000. Age-associated characteristics of murine hematopoietic stem cells. *J. Exp. Med.* **192:** 1273–1280.

Suh, H., Consiglio, A., Ray, J., Sawai, T., D'Amour, K.A., and Gage, F.H. 2007. In vivo fate analysis reveals the multipotent and self-renewal capacities of Sox2$^+$ neural stem cells in the adult hippocampus. *Cell Stem Cell* **1:** 515–528.

Suzuki, A. and Saga, Y. 2008. Nanos2 suppresses meiosis and promotes male germ cell differentiation. *Genes Dev.* **22:** 430–435.

Suzuki, H., Tsuda, M., Kiso, M., and Saga, Y. 2008. Nanos3 maintains the germ cell lineage in the mouse by suppressing both Bax-dependent and -independent apoptotic pathways. *Dev. Biol.* **318:** 133–142.

Tadokoro, Y., Yomogida, K., Ohta, H., Tohda, A., and Nishimune, Y. 2002. Homeostatic regulation of germinal stem cell proliferation by the GDNF/FSH pathway. *Mech. Dev.* **113:** 29–39.

Takahashi, K. and Yamanaka, S. 2006. Induction of pluripotent stem cells from mouse embryonic and adult fibroblast cultures by defined factors. *Cell* **126:** 663–676.

Takahashi, K., Tanabe, K., Ohnuki, M., Narita, M., Ichisaka, T., Tomoda, K., and Yamanaka, S. 2007. Induction of pluripotent stem cells from adult human fibroblasts by defined factors. *Cell* **131:** 861–872.

Temple, S. 2001. The development of neural stem cells. *Nature* **414:** 112–117.

Tothova, Z., Kollipara, R., Huntly, B.J., Lee, B.H., Castrillon, D.H., Cullen, D.E., McDowell, E.P., Lazo-Kallanian, S., Williams, I.R., Sears, C., et al. 2007. FoxOs are critical mediators of hematopoietic stem cell resistance to physiologic oxidative stress. *Cell* **128:** 325–339.

Tsai, F.-Y., Keller, G., Kuo, F.C., Weiss, M., Chen, J., Rosenblatt, M., Alt, F.W., and Orkin, S.H. 1994. An early haematopoietic defect in mice lacking the transcription factor GATA-2. *Nature* **371:** 221–226.

Tsuda, M., Sasaoka, Y., Kiso, M., Abe, K., Haraguchi, S., Kobayashi, S., and Saga, Y. 2003. Conserved role of nanos proteins in germ cell development. *Science* **301:** 1239–1241.

Tyner, S.D., Venkatachalam, S., Choi, J., Jones, S., Ghebranious, N., Igelmann, H., Lu, X., Soron, G., Cooper, B., Brayton, C., et al. 2002. p53 mutant mice display early ageing-associated phenotypes. *Nature* **415:** 45–53.

Valk-Lingbeek, M.E., Bruggeman, S.W., and van Lohuizen, M. 2004. Stem cells and cancer; the Polycomb connection. *Cell* **118:** 409–418.

van der Lugt, N.M., Domen, J., Linders, K., van Roon, M., Robanus-Maandag, E., te Riele, H., van der Valk, M., Deschamps, J., Sofroniew, M., van Lohuizen, M., et al. 1994. Posterior transformation, neurological abnormalities, and severe hematopoietic defects in mice with a targeted deletion of the *bmi-1* proto-oncogene. *Genes Dev.* **8:** 757–769.

van Praag, H., Schinder, A.F., Christie, B.R., Toni, N., Palmer, T.D., and Gage, F.H. 2002. Functional neurogenesis in the adult hippocampus. *Nature* **415:** 1030–1034.

Wang, J., Rao, S., Chu, J., Shen, X., Levasseur, D.N., Theunissen, T.W., and Orkin, S.H. 2006. A protein interaction network for pluripotency of embryonic stem cells. *Nature* **444:** 364–368.

Ware, C.B., Horowitz, M.C., Renshaw, B.R., Hunt, J.S., Liggitt, D., Koblar, S.A., Gliniak, B.C., McKenna, H.J., Papayannopoulou, T., Thoma, B., et al. 1995. Targeted disruption of the low-affinity leukemia inhibitory factor receptor gene causes placental, skeletal, neural and metabolic defects and results in perinatal death. *Development* **121:** 1283–1299.

Wellcome Trust Case Control Consortium. 2007. Genome-wide association study of 14,000 cases of seven common diseases and 3,000 shared controls. *Nature* **447:** 661–678.

Yamanaka, Y., Ralston, A., Stephenson, R.O., and Rossant, J. 2006. Cell and molecular regulation of the mouse blastocyst. *Dev. Dyn.* **235:** 2301–2314.

Yilmaz, O.H., Kiel, M.J., and Morrison, S.J. 2006. SLAM family markers are conserved among hematopoietic stem cells from old and reconstituted mice and markedly increase their purity. *Blood* **107:** 924–930.

Ying, Q.L., Nichols, J., Chambers, I., and Smith, A. 2003. BMP induction of Id proteins suppresses differentiation and sustains embryonic stem cell self-renewal in collaboration with STAT3. *Cell* **115:** 281–292.

Ying, Q.L., Wray, J., Nichols, J., Batlle-Morera, L., Doble, B., Woodgett, J., Cohen, P., and Smith, A. 2008. The ground state of embryonic stem cell self-renewal. *Nature* **453:** 519–523.

Yoshida, K., Taga, T., Saito, M., Suematsu, S., Kumanogoh, A., Tanaka, T., Fujiwara, H., Hirata, M., Yamagami, T., Nakahata, T., et al. 1996. Targeted disruption of gp130, a common signal transducer for the interleukin 6 family of cytokines, leads to myocardial and hematological disorders. *Proc. Natl. Acad. Sci.* **93:** 407–411.

Yoshihara, H., Arai, F., Hosokawa, K., Hagiwara, T., Takubo, K., Nakamura, Y., Gomei, Y., Iwasaki, H., Matsuoka, S., Miyamoto, K., et al. 2007. Thrombopoietin/MPL signaling regulates hematopoietic stem cell quiescence and interaction with in the osteoblastic niche. *Cell Stem Cell* **1:** 685–697.

Yu, F., Yao, H., Zhu, P., Zhang, X., Pan, Q., Gong, C., Huang, Y., Hu, X., Su, F., Lieberman, J., and Song, E. 2007a. *let-7* regulates self renewal and tumorigenicity of breast cancer cells. *Cell* **131:** 1109–1123.

Yu, J., Vodyanik, M.A., Smuga-Otto, K., Antosiewicz-Bourget, J., Frane, J.L., Tian, S., Nie, J., Jonsdottir, G.A., Ruotti, V., Stewart, R., Slukvin, I.I., and Thomson, J.A. 2007b. Induced pluripotent stem cell lines derived from human somatic cells. *Science* **318:** 1917–1920.

Zeggini, E., Weedon, M.N., Lindgren, C.M., Frayling, T.M., Elliott, K.S., Lango, H., Timpson, N.J., Perry, J.R., Rayner, N.W., Freathy, R.M., et al. 2007. Replication of genome-wide association signals in UK samples reveals risk loci for type 2 diabetes. *Science* **316:** 1336–1341.

Zhang, C.L., Zou, Y., He, W., Gage, F.H., and Evans, R.M. 2008. A role for adult TLX-positive neural stem cells in learning and behaviour. *Nature* **451:** 1004–1007.

Zindy, F., Quelle, D.E., Roussel, M.F., and Sherr, C.J. 1997. Expression of the p16^INK4a tumor suppressor versus other INK4 family members during mouse development and aging. *Oncogene* **15:** 203–211.

Complementary and Independent Function for *Hoxb4* and *Bmi1* in HSC Activity

A. FAUBERT,* J. CHAGRAOUI,* N. MAYOTTE,* M. FRÉCHETTE,* N.N. ISCOVE,[†]
R.K. HUMPHRIES,[‡] AND G. SAUVAGEAU*[§¶]

*Laboratory of Molecular Genetics of Hematopoietic Stem Cells, Institute for Research in Immunology and Cancer (IRIC), University of Montréal, Montréal, Québec, Canada H3C 3J7; [†]Terry Fox Laboratory, British Columbia Cancer Agency, Department of Medicine, University of British Columbia, Vancouver, British Columbia, Canada V5Z 1LS; [‡]The Ontario Cancer Institute, Department of Medical Biophysics, University of Toronto, Toronto, Canada M5G 2M9; [§]Department of Medicine, University of Montréal, Montréal, Québec, Canada H3C 3J7; [¶]Division of Haematology and Leukemia Cell Bank of Québec (BCLQ), Maisonneuve-Rosemont Hospital, Montréal, Québec, Canada H1T 2M2

Determinants regulating short- and long-term repopulating hematopoietic stem cell (STR-HSC and LTR-HSC) self-renewal remain largely uncharacterized. To gain further insights into HSC self-renewal, we investigated possible genetic interactions between two well-recognized regulators of this process: *Bmi1* and *Hoxb4*. Using complementation and overexpression strategies in mouse HSCs, we document that *Bmi1* is not required for the in vivo expansion of fetal HSCs but is essential for the long-term maintenance of adult HSCs. Importantly, we show that *Hoxb4* overexpression induces an expansion of *Bmi1[−/−]* STR-HSCs leading to a rescue of their repopulation defect. In contrast to *Hoxb4*, we also show that *Bmi1* fails to induce HSC expansion ex vivo. Consistent with these results, we report high levels of Angptl3 and Cbx7 in *Hoxb4*- and *Bmi1*-transduced cells, respectively. Together, these results support the emerging concept that fate and sustainability of this fate are two critical components of self-renewal in adult stem cells such as HSCs.

Self-renewal implies the coordination between cell division and fate such that for at least one progeny, cell differentiation, proliferative senescence, and apoptosis are excluded. In contrast to mouse embryonic stem cells that undergo symmetrical self-renewal divisions in response to leukemia inhibitory factor (LIF) and bone morphogenetic protein (BMP) signaling, most adult stem cells have the capacity to undergo two distinct types of self-renewal divisions: symmetrical and asymmetrical (Shen et al. 2002). In the hematopoietic system, HSCs undergo at least two rounds of mobilization that largely overlap with changes in proliferative status: Indeed, most HSCs are in cycle and symmetrically self-renew (thus expand) at a specific time in the mouse fetal liver (FL) and they then switch to a more quiescent program later in the bone marrow environment where the population is maintained for the entire life of the animal (no expansion, but maintenance of the population). Whether this maintenance phase is established by a balance between symmetrical self-renewal and stem cell loss, by asymmetric self-renewal division, or a combination of these processes remains unclear. In adult mice, HSC symmetrical self-renewal divisions can be experimentally triggered (reactivation of an embryonic program?) under specific conditions such as during the early phase following HSC transplantation (Pawliuk et al. 1996; Antonchuk et al. 2001), and this can be dramatically enhanced upon the engineered overexpression of the *Hoxb4* transcription factor or other homeobox genes (Sauvageau et al. 1995; Thorsteinsdottir et al. 1999, 2002; Antonchuk et al. 2001, 2002; Krosl et al. 2003; Cellot et al. 2007; Ohta et al. 2007).

Recent studies have documented that at least two different stem cell populations with more or less equal differentiation potential can be prospectively isolated. These include the long-term repopulating hematopoietic stem cells (LTR-HSCs) that supply hematopoiesis for the life span of the animal and the short-term repopulating HSCs (STR-HSCs) that contribute to repopulation for up to 3 months after transplantation (Morrison et al. 1997; Christensen and Weissman 2001; Wognum et al. 2003). Until recently, however, this property of short-term versus long-term repopulation was limited to phenotypically defined populations, but new studies have validated, at the single-cell level, the concept that the HSC compartment is highly heterogeneous and includes STR-HSCs and LTR-HSCs (Dykstra et al. 2007; N.N. Iscove, pers. comm.). In these studies, single STR-HSCs could provide multilineage reconstitution for up to 3–4 months post-transplantation, whereas LTR-HSCs contributed to life-long repopulation. This level of blood cell production, estimated at above 10^{12} progeny per STR-HSC, argues that these cells can extensively self-renew, thus introducing the notion (of N.N. Iscove) that it is the ability to sustain self-renewal rather than self-renewal itself that distinguishes LTR-HSCs from STR-HSCs.

Together, these data suggest the existence of two mandatory components that define self-renewal (SR) in LTR-HSCs: one that coordinates fate with cell division (referred to as "execution of SR") and one that ensures the sustainability of this execution through cell divisions (referred to as "sustainability of SR"). Several observations suggest that although *Hoxb4* overexpression enhances the execution of SR in defined conditions, the Polycomb group (PcG) gene *Bmi1* may be critical for regulating the sustainability of this process. Indeed, by phenotypical analysis, *Bmi1[−/−]* HSCs are present in normal numbers in FL, but they progressively decrease from embryonic to adult life (Park et al. 2003; Iwama et al. 2004). In addition, although *Bmi1* homozygous null HSCs fail to reconstitute myeloablated recipients for periods extending for more than a few weeks, this defect is complemented by the experimental reintroduction of *Bmi1* in these cells, thus indicating the independence on *Bmi1* for the generation of FL LTR-HSCs. Moreover, a single *Hoxa9 + Meis1* leukemia-initiating cell that lacks *Bmi1* expression expands sufficiently to induce lethality in mice, but this expansion is nevertheless limited in time (Lessard and Sauvageau 2003; Park et al. 2003; Iwama et al. 2004). Finally, *Bmi1* is not critical for stem cell homing and does not appear to regulate apoptosis (Park et al. 2003; Iwama et al. 2004). With this in mind, we now report a series of experiments suggesting that *Bmi1* ensures the sustainability of SR in LTR-HSCs and, in contrast to *Hoxb4*, fails to enhance the execution of SR under comparable experimental conditions.

Bmi1 IS REQUIRED FOR ACTIVITY OF ADULT HSC

Bmi1 homozygous null fetal liver cells (FLCs) were infected using retroviral vectors (Fig. 1a) and manipulated to assess *Bmi1* function on progenitor and stem cell activities (Fig. 1b). Gene transfer efficiencies for all experiments were between 14% and 100% (typically between 60% and 100%). As previously reported, progenitor activity is markedly reduced in *Bmi1*$^{-/-}$ FLCs kept in short-term culture (Fig. 1c). These cells also lack in vivo reconstitution ability (Fig. 1d) (Lessard and Sauvageau 2003). Retroviral transduction of *Bmi1* in *Bmi1*$^{-/-}$ FLCs rescues both their progenitor and in vivo reconstitution activities (Fig. 1c,d). Moreover, complemented *Bmi1*$^{-/-}$ cells (referred to hereafter as *Bmi1*$^{-/-}$–*Bmi1* cells) have the capacity to differentiate into myeloid (Gr-1$^+$ and CD11b$^+$) and lymphoid (CD45R$^+$ and CD4$^+$) lineages in vivo (Fig. 1e), and they can provide hematopoietic reconstitution when transferred into secondary recipients (data not shown). Because hematopoietic reconstitution is observed when wild-type HSCs are transplanted into *Bmi1*$^{-/-}$ mice (Park et al. 2003; Iwama et al. 2004), these results support a cell-autonomous dysfunction for *Bmi1* homozygous null HSC.

To further investigate the requirement for *Bmi1* in adult HSC activity, RNA interference (RNAi) studies were conducted using an approach that we previously validated (Chagraoui et al. 2006). *shBmi1*-transduced progenitor cells were significantly compromised in their ex vivo response to hematopoietic growth factors when compared to control cells (Fig. 1f). Interestingly, multipotent progenitors (i.e., CFU-GEMM) were most sensitive to the acute reduction in *Bmi1* levels (Fig. 1g). Consistent with these data, *shBmi1*-transduced cells were impaired in their in vivo reconstitution activity with only one mouse reconstituted in both the myeloid and the lymphoid compartments 16 weeks post-stem-cell transfer (SCT) (Fig. 1h, right panel, solid dot).

Together with previous reports confirming that *Bmi1*$^{-/-}$ HSCs are not impaired in their survival and homing potential (Park et al. 2003; Iwama et al. 2004), these results indi-

cate that *Bmi1* is not required for HSC specification but is essential for their long-term maintenance in vivo.

Bmi1$^{-/-}$ FL HSCS UNDERGO SR DIVISIONS

Taking into consideration that mouse HSC populations rapidly expand between embryonic days 10 (E10) and 14.5 (E14.5) (Rebel et al. 1996), experiments were designed to test whether *Bmi1* is required for HSC expansion (i.e., symmetrical self-renewal divisions). In these assays, freshly isolated FLCs were infected with a *Bmi1* retrovirus before their transplantation at limiting dilution into lethally irradiated hosts, and stem cell frequencies were evaluated using the CRU (competitive repopulating unit) assay (Szilvassy et al. 1990). Results from these experiments indicated that HSC frequencies (Fig. 1i,j) and absolute numbers (data not shown) are similar in *Bmi1*$^{-/-}$ or *Bmi1*$^{+/+}$ fetal livers, suggesting that although essential for HSC maintenance, *Bmi1* is not required for embryonic HSC expansion and therefore fetal-liver-derived HSCs can symmetrically self-renew in the absence of this gene.

Bmi1 OVEREXPRESSION DOES NOT INDUCE HSC EXPANSION

We next investigated whether *Bmi1* overexpression might trigger HSC expansion and regulate *Hoxb4* expression levels. For these experiments, bone marrow cells (BMCs) were transduced using control, *Hoxb4*, *Bmi1*, or both *Hoxb4* + *Bmi1* retroviruses (Figs. 1a and 2a). Western blot and quantitative reverse transcriptase–polymerase chain reaction (Q-RT-PCR) analyses of transduced cells showed that levels of *Hoxb4* are not affected in *Bmi1*-transduced cells and similarly that *Bmi1* is not regulated by *Hoxb4* (Fig. 2b and data not shown).

When compared to controls, *Hoxb4*-transduced KLS (c-Kit$^+$Lineage$^-$Sca-1$^+$) cells showed a modest expansion during a 7-day culture period (Fig. 2c). In contrast, expansion of *Bmi1*-transduced KLS cells was significantly higher (Fig. 2c). Accordingly, cell cycle distribution studies indicated a greater proportion of KLS cells in S phase in cultures initiated with *Bmi1*-transduced cells than in green fluorescent protein (GFP) control cultures (data not shown). In agreement with previous findings, both *Hoxb4* and *Bmi1* significantly enhanced the generation of myeloid progenitors in culture compared to controls (Fig. 2d) (Sauvageau et al. 1995; Iwama et al. 2004).

The effect of *Bmi1* overexpression on in vivo hematopoietic cell reconstitution and ex vivo HSC expansion was evaluated using the CRU assay. In contrast to results obtained within the same experiment with *Hoxb4*, *Bmi1*-overexpressing LTR-HSCs did not expand ex vivo and reconstituted primary recipients similarly to control cells (Fig. 2e,f). These experiments were also conducted with cells derived from E14.5 FL and likewise, *Bmi1* overexpression had no effect on LTR-HSC expansion (data not shown). Importantly, control and *Bmi1*-transduced HSCs showed similar expansion and differentiation properties, thus indicating that *Bmi1* overexpression is not detrimental for stem cell activity (Fig. 2f,g). Interestingly, *Bmi1* over-

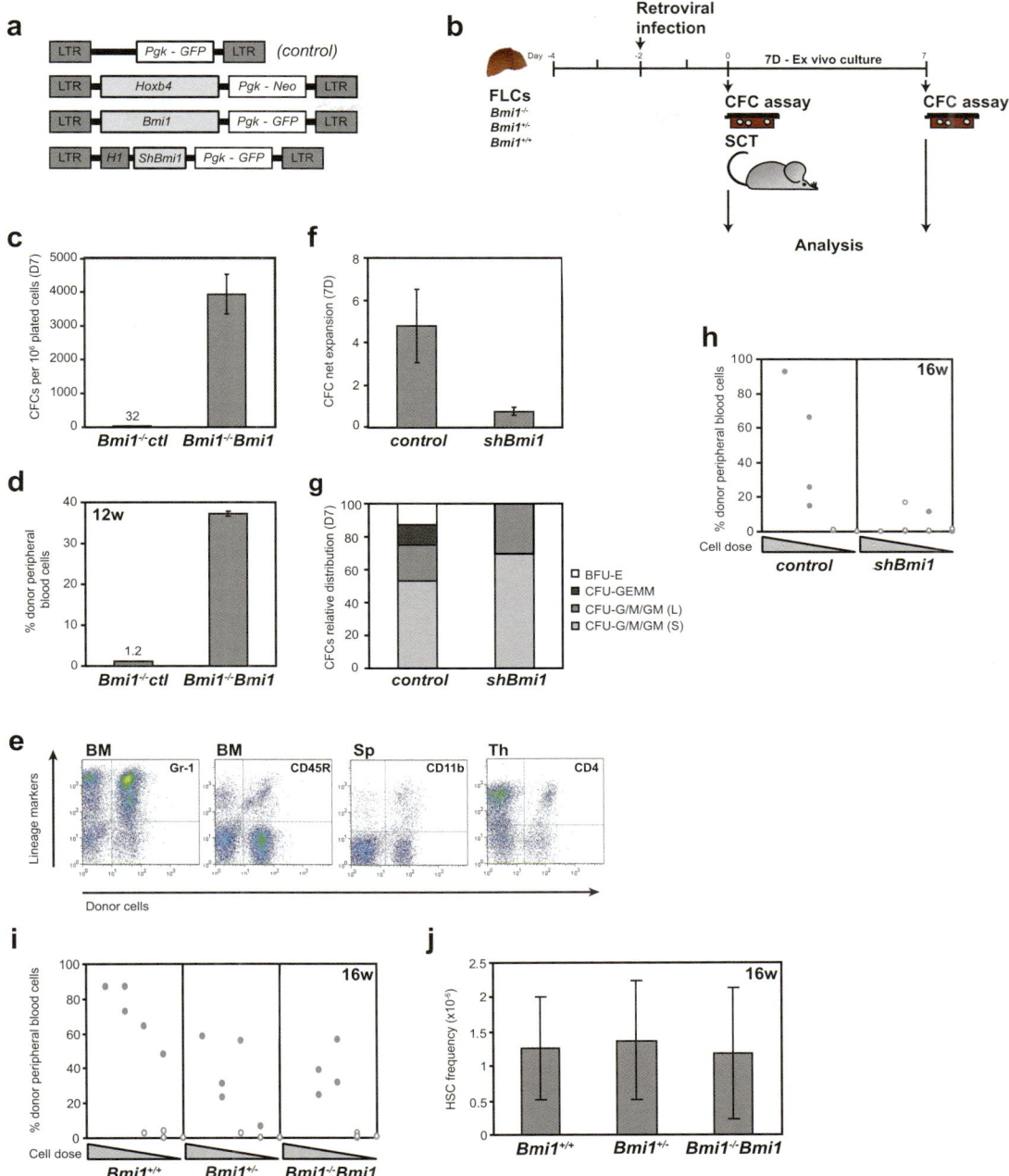

Figure 1. *Bmi1* is required for hematopoietic stem cell function. (*a*) Retroviral constructs used in this study. (Sequences are available upon request to the author.) (*b*) Experimental outline. *Bmi1*[+/+], *Bmi1*[+/−], or *Bmi1*[−/−] E14.5 fetal liver cells (FLCs) were transduced with a control, *Bmi1*, or *ShBmi1* retroviral vector. Stem cell transplantation (SCT) and colony-forming cell (CFC) assays were performed as described previously (Chagraoui et al. 2006). (*c*) CFC assays were performed 7 days postinfection (*n* = 50 colonies analyzed per condition) (mean ±S.D.). (*d*) Transduced *Bmi1*[−/−] FLCs (3.3 × 10⁶ cells) were transplanted (*n* = 3 mice per condition), and a proportion of donor-derived (Ly5.2[+]) peripheral blood cells was assessed 12 weeks later (mean ±S.D.). (*e*) Fluorescence-activated cell sorting (FACS) profiles of hematopoietic tissues from a representative *Bmi1*[−/−] *Bmi1* chimeric mouse to determine myeloid (Gr-1, CD11b) and lymphoid (CD45R and CD4) reconstitution by donor cells (*n* = 3 mice were analyzed). (*f*) Expansion of CFCs during a series of 7-day ex vivo cultures (compared CFC numbers were determined at D0) (mean ±S.D.). (*g*) Relative distribution of CFC subtypes following a 7-day ex vivo culture. (*h*) Donor-derived reconstitution of recipients of FLCs transduced with control versus *shBmi1* virus, as indicated in *a*. Reconstitution was performed with a cell dose of 10⁷ (highest dilution) to 40,000 (lowest dilution) cells per mouse. *n* = 10 mice per dilution. Each dot represents a mouse (*solid* vs. *open dots*: presence vs. absence of donor-derived lympho-myeloid reconstitution). Similar results were confirmed with bone marrow cells (BMCs) (data not shown). (*i*) Transduced FLCs were transplanted (*n* = 10 mice per condition) at limiting dilution (from 10⁶ to 20,000 cells) and peripheral blood cell reconstitution was assessed as in *h*. (*j*) HSC frequencies were evaluated using reconstitution results presented in *i* (error bars indicate a 95% confidence interval). Results (*i* and *j*) were confirmed in two independent experiments. (BFU-E) Burst-forming-unit erythroid; (BM) bone marrow; (CFU-GEMM) colony-forming-unit granulocyte/erythroid/monocyte/megakaryocyte; (L) large (CFU-G/M/GM colonies with >50 cells); (S) small (CFU-G/M/GM colonies with <50 cells); (Sp) spleen; (Th) thymus.

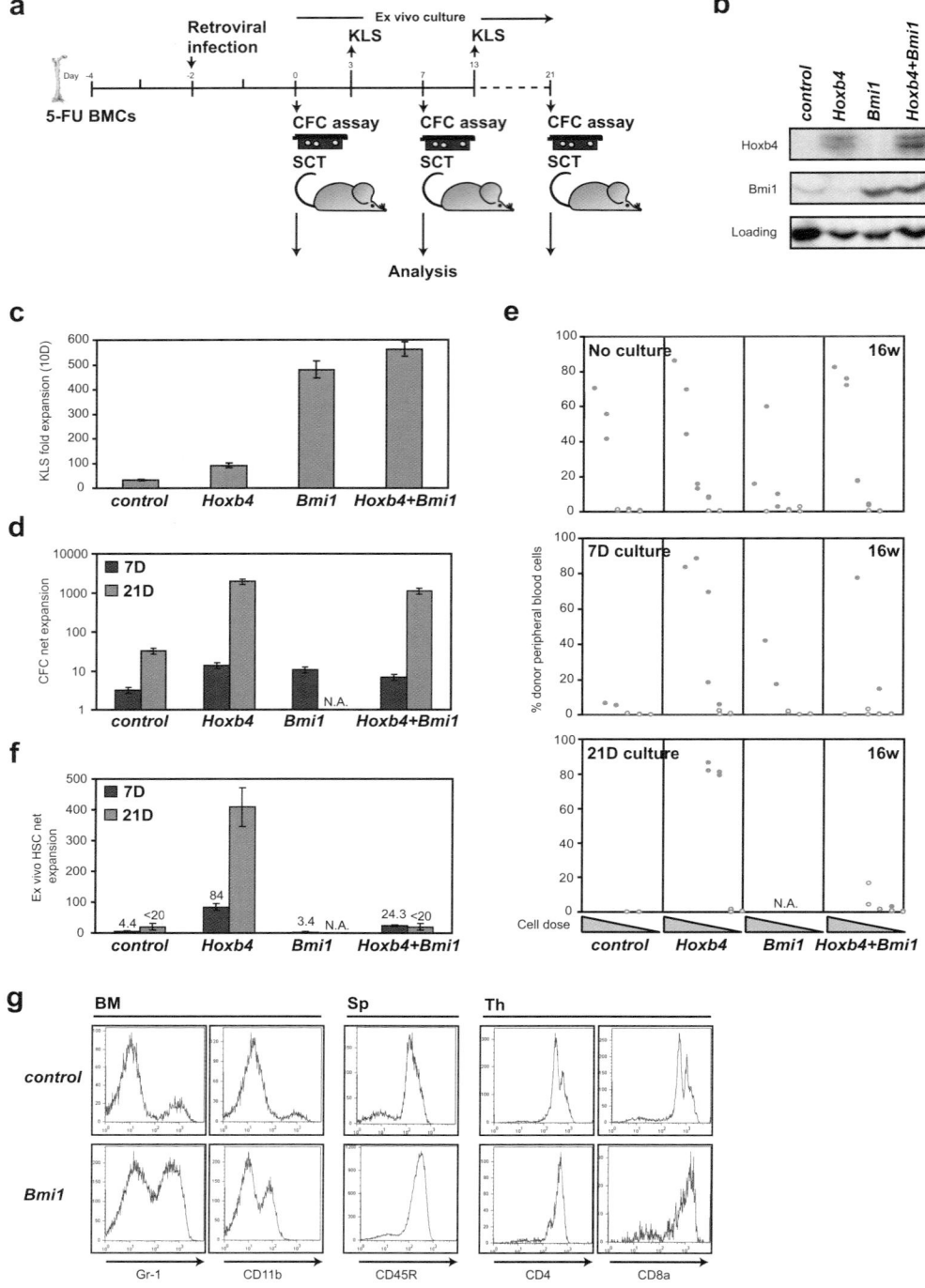

Figure 2. *Bmi1* overexpression is not associated with ex vivo stem cell expansion. (*a*) Experimental outline. 5-Fluorouracil-treated bone marrow cells (5-FU BMCs) were transduced with control, *Hoxb4*, *Bmi1*, or *Hoxb4 + Bmi1* retroviral vectors. CFC (*n* = 50–100 colonies analyzed per condition) and stem cell transplantation (*n* = 10 mice per condition) assays were performed on days 0, 7, and 21. (*b*) Western blot analysis of BMCs transduced with control, *Hoxb4*, *Bmi1*, or *Hoxb4 + Bmi1* retroviral vectors and immunoblotted with *Hoxb4*, *Bmi1*, or ribosomal protein S6 (loading) antibodies. (*c*) Expansion of c-Kit⁺Lineage⁻Sca-1⁺ (KLS) cells during a 10-day culture period (*n* = 3 experiments) (mean ±S.D.). (*d*) CFC expansion during 7 (*black*) or 21 days (*gray*) of ex vivo cultures (compared to D0 control CFC numbers) (mean ±S.D.). (*e*) Transduced BMCs were transplanted at limiting dilution (from 10⁶ to 500 cells), and reconstitution was assessed 16 weeks later. (*Upper, middle, lower panels*) Experiments performed with cells isolated at days 0, 7, and 21 after initiation of the cultures, respectively (see also *a*). Cell dose for the experiments is represented in the *upper panel:* 10⁶ to 500 cells. (*Middle, lower panels*) All cultures were seeded with 10⁶ cells and recipients were transplanted with a range of 200,000 to 20 (D7) or 2,000 to 20 (D21) cells. Results for D0 and D7 were confirmed in three independent experiments and also with FLCs (data not shown). (*f*) Seven-day (*black*) and 21-day (*gray*) ex vivo HSC expansions (in comparison to D0 control HSC numbers) were evaluated using reconstitution of recipients shown in *e* (≥95% confidence interval). (*g*) FACS profiles of hematopoietic tissues from selected recipients of *GFP* (control) versus *Bmi1*-transduced cells from *f*. Abbreviations are the same as those in Fig. 1.

expression conferred a small proliferative advantage to repopulating cells (i.e., higher reconstitution level per stem cell transplanted) (Fig. 2e, compare middle [*Bmi1*] and first [control] panels).

BMI1 IMPAIRS *HOXB4*-INDUCED HSC EXPANSION

As shown in the previous sections, *Bmi1* overexpression enhances proliferation of phenotypically primitive KLS hematopoietic cells, whereas *Hoxb4* mostly triggers HSC self-renewal divisions. On the basis of these results, we inferred that these two genes may display potent genetic interaction leading to enhanced HSC expansion (e.g., enhanced proliferation [*Bmi1*] of self-renewing [*Hoxb4*] HSCs). To verify this hypothesis, control, *Hoxb4*, and *Hoxb4* + *Bmi1* BMCs were kept in culture for up to 21 days and progenitor/stem cell contents were evaluated at various time points (Fig. 2a). As shown in Figure 2d, *Hoxb4* progenitor activity was not enhanced by the engineered cooverexpression of *Bmi1* after 7 or 21 days of culture. Importantly, expanded BMCs generated a large proportion of clones that expressed both *Hoxb4* and *Bmi1*, documenting the absence of negative interaction between these two genes in progenitors (data not shown).

In vivo reconstitution experiments performed with these same ex vivo expanded cells revealed that *Hoxb4*-induced HSC expansion (e.g., 400-fold at 21 days; Fig. 2f) is abrogated by the cooverexpression of *Bmi1* (Fig. 2f). Importantly, the loss of HSC activity in cultures initiated with doubly transduced cells is not due to nonspecific toxicity of *Bmi1* on HSC, because *Hoxb4* + *Bmi1* doubly transduced HSC are as competitive as *Hoxb4* cells in vivo when transplanted immediately after gene transfer (Fig. 2e, upper panel). Thus, ex vivo, where the HSC-expanding activity of *Hoxb4* is most prominent (Antonchuk et al. 2002), *Bmi1* dominantly interferes with *Hoxb4* activity while enhancing proliferation. As demonstrated using Annexin V staining, this effect was not consequential to a modulation of apoptosis, which was minimal in the KLS subfraction of all cultures analyzed (data not shown).

BMI1$^{-/-}$ STR-HSC FUNCTION IS RESCUED BY *HOXB4*

In an attempt to further investigate the genetic interaction between *Hoxb4* and *Bmi1*, experiments were performed to assess the possibility that *Hoxb4* might complement the *Bmi1*$^{-/-}$ progenitor and stem cell defects (Fig. 3a). As shown in Figure 3b and c, *Bmi1*$^{-/-}$ progenitor numbers, differentiation potential, and proliferative activity are fully rescued by the engineered overexpression of *Hoxb4*.

In vivo experiments further revealed that *Hoxb4* also complements the short-term hematopoietic reconstitution defects of *Bmi1*$^{-/-}$ cells to levels similar to those obtained for *Bmi1*$^{-/-}$*Bmi1* cells (Fig. 3d, upper panel). Confirmation that reconstitution was derived from *Hoxb4*-transduced *Bmi1*$^{-/-}$ cells was obtained by PCR-based analyses of peripheral blood samples isolated from several independent mice (Fig. 3d, middle [250-bp band] and lower panels) and from protein analysis of a representative specimen (Fig. 3e). Importantly, *Bmi1*$^{-/-}$*Hoxb4* cells were competent in myeloid (Gr-1$^+$), B lymphoid (CD45R$^+$), and T lymphoid (CD8a$^+$) differentiation in vivo (Fig. 3f).

Interestingly, the reconstitution potential of *Bmi1*$^{-/-}$*Hoxb4* cells progressively decreased, such that by 12 weeks posttransplantation, it became almost undetectable (Fig. 3g). The rare donor-derived cells that persisted at 12 weeks in recipients of *Bmi1*$^{-/-}$*Hoxb4* grafts were derived from myeloid (CD11b$^+$) and lymphoid (CD45R$^+$ and CD4$^+$CD8a$^+$) lineages (Fig. 3h). As might be expected from the results presented above, *Bmi1*$^{-/-}$ cells transduced with both *Hoxb4* and *Bmi1* persistently reconstituted all mice analyzed at 12 weeks post-SCT (Fig. 3g), and evaluation of the stem cell content in these grafts reveals that HSCs had expanded 20–35 times more than control HSCs (i.e., *Bmi1*$^{-/-}$*Bmi1*) (data not shown). Together, these results suggest that *Bmi1*$^{-/-}$ FL-derived HSCs were complemented by *Hoxb4* overexpression for reconstitution lasting up to 8–12 weeks. However, *Bmi1* is required to further sustain long-term stem cell activity (i.e., >12 weeks).

HOXB4 TRIGGERS EXPANSION OF *BMI1*$^{-/-}$ HSCS

A series of additional experiments were conducted to further understand the nature of *Hoxb4*-induced rescue of *Bmi1*$^{-/-}$ repopulating cells. We specifically designed these experiments to determine whether *Hoxb4* influenced the expansion (self-renewal) or the proliferative output of these *Bmi1*$^{-/-}$ cells (Fig. 4a). These experiments first showed that *Hoxb4* overexpression induced a significant expansion of Lin$^-$CD48$^-$CD150$^+$ populations in cultures initiated with *Bmi1*$^{-/-}$ FLCs to levels similar to those determined in cultures initiated with wild-type cells (Fig. 4b and data not shown). Furthermore, and consistent with results reported by Iwama et al. (2004), Lin$^-$CD48$^-$CD150$^-$ wild-type and *Bmi1*$^{-/-}$ cells, lacking long-term reconstitution potential (Kiel et al. 2005; Kim et al. 2006), were poorly responsive to *Hoxb4* overexpression (Fig. 4c).

The ex vivo expansion of *Hoxb4-Bmi1*$^{-/-}$ repopulating cells was assessed at an early time point (i.e., 4 weeks) using the CRU assay as described originally (Szilvassy et al. 1990). Results from these experiments showed that *Hoxb4* overexpression similarly triggered the net expansion of short-term repopulating (STR-HSC) cells from both genotypes (*Bmi1*$^{-/-}$ and *Bmi1*$^{+/+}$) to values representing 2.5- to 3.9-fold, whereas untransduced STR-HSCs decreased by approximately 10-fold in these cultures (Fig. 4d). Thus, independent of *Bmi1* genotype, cultures initiated with *Hoxb4*-transduced cells contained between 25 and 39 times more STR-HSCs than the control (Fig. 4d). Importantly, these cells maintained lymphoid and myeloid differentiation capacity but lacked the potential for long-term reconstitution (i.e., >12 weeks).

Confirmation of the lympho-myeloid potency of individual STR-HSCs was obtained for at least two clones where identical proviral integration profiles were identified in sorted myeloid (Gr-1$^+$) and lymphoid (CD45R$^+$)

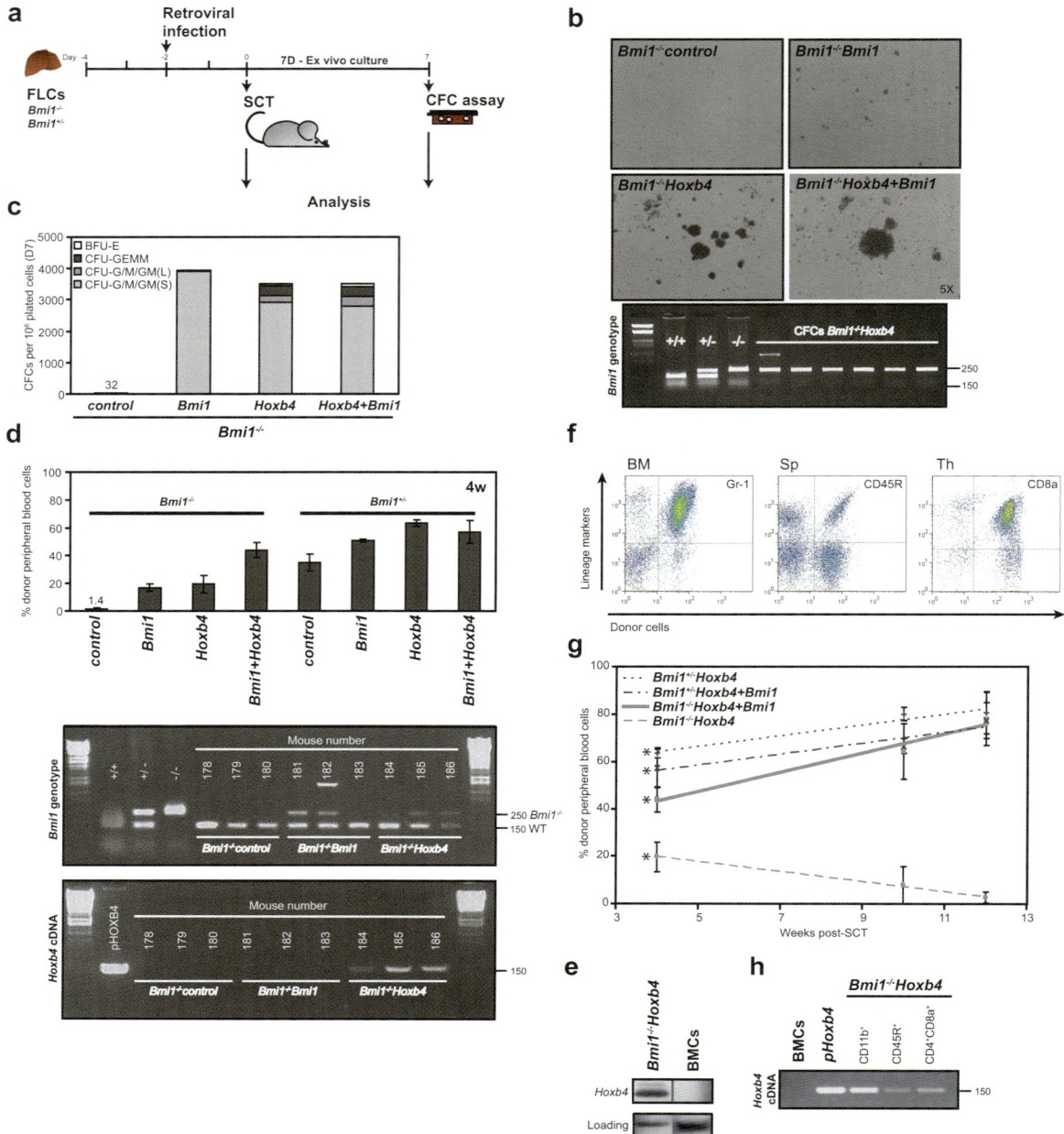

Figure 3. *Hoxb4* overexpression rescues *Bmi1⁻/⁻* STR-HSCs. (*a*) Experimental outline. *Bmi1⁻/⁻* and *Bmi1⁺/⁻* FLCs were transduced with control, *Bmi1*, *Hoxb4*, or *Hoxb4* + *Bmi1* retroviral vectors. CFC and SCT assays were performed at the indicated time points. (*b,c*) Transduced *Bmi1⁻/⁻* FLCs were kept in culture for 7 days and plated in methylcellulose-containing medium for 7 days for CFC evaluation as described by Chagraoui et al. (2006). Pictures were taken at day 7 after initiation of the methylcellulose cultures (*n* = 60 colonies analyzed per condition). (*b, Lower panel*) Genomic DNA was isolated from individual colonies picked from methylcellulose cultures initiated with *Hoxb4*-transduced *Bmi1⁻/⁻* (or *Bmi1⁻/⁻Hoxb4*) cells and subjected to PCR analysis to confirm that *Hoxb4*-rescued clones were of *Bmi1⁻/⁻* genotype (indicated by the ~250-bp fragment). (*d, Upper panel*) Transduced *Bmi1⁻/⁻* and *Bmi1⁺/⁻* FLCs were transplanted (3.3 × 10⁶ cells per mice, *n* = 3), as indicated in *a*, and donor-derived reconstitution was evaluated 4 weeks later (mean ±s.D.). (*d, Middle panel*) Same analysis as detailed in *b* (lower panel) except that DNA was extracted from peripheral blood isolated from recipients 178–186 at 4 weeks posttransplantation. (*d, Lower panel*) Detection of *Hoxb4* provirus by PCR in the same DNA. (*e*) Western blot analysis of protein extracts from sorted (Ly5.2⁺) BMCs isolated from recipients of wild-type control or *Bmi1⁻/⁻Hoxb4* cells. Membranes were immunoblotted with antibodies that specifically recognize *Hoxb4* or Tubulin β (loading). (*f*) Representative FACS profiles of hematopoietic tissues from a selected recipient of *Bmi1⁻/⁻Hoxb4*-transduced cells. Analysis was performed using cells isolated at 4 weeks posttransplantation (*n* = 3 mice analyzed). (*g*) Time course showing donor-derived reconstitution (mean ±s.D.) of several recipients of *Bmi1* heterozygous or homozymous null FLCs transduced with the indicated retroviruses. (*Asterisks*) Results shown at 4 weeks are derived from *d*. (*h*) The presence of *Hoxb4*-integrated provirus in the DNA of BMCs isolated from *Bmi1⁻/⁻Hoxb4* chimeric mice at 12 weeks posttransplantation. BMCs were sorted for myeloid (CD11b⁺), B-lymphoid (CD45R⁺), and T-lymphoid (CD4⁺CD8a⁺) populations, and amplification of the *Hoxb4* provirus was performed by PCR (*n* = 3 mice analyzed, one representative shown). Total BMCs were used as a negative control. Results were confirmed in three independent experiments.

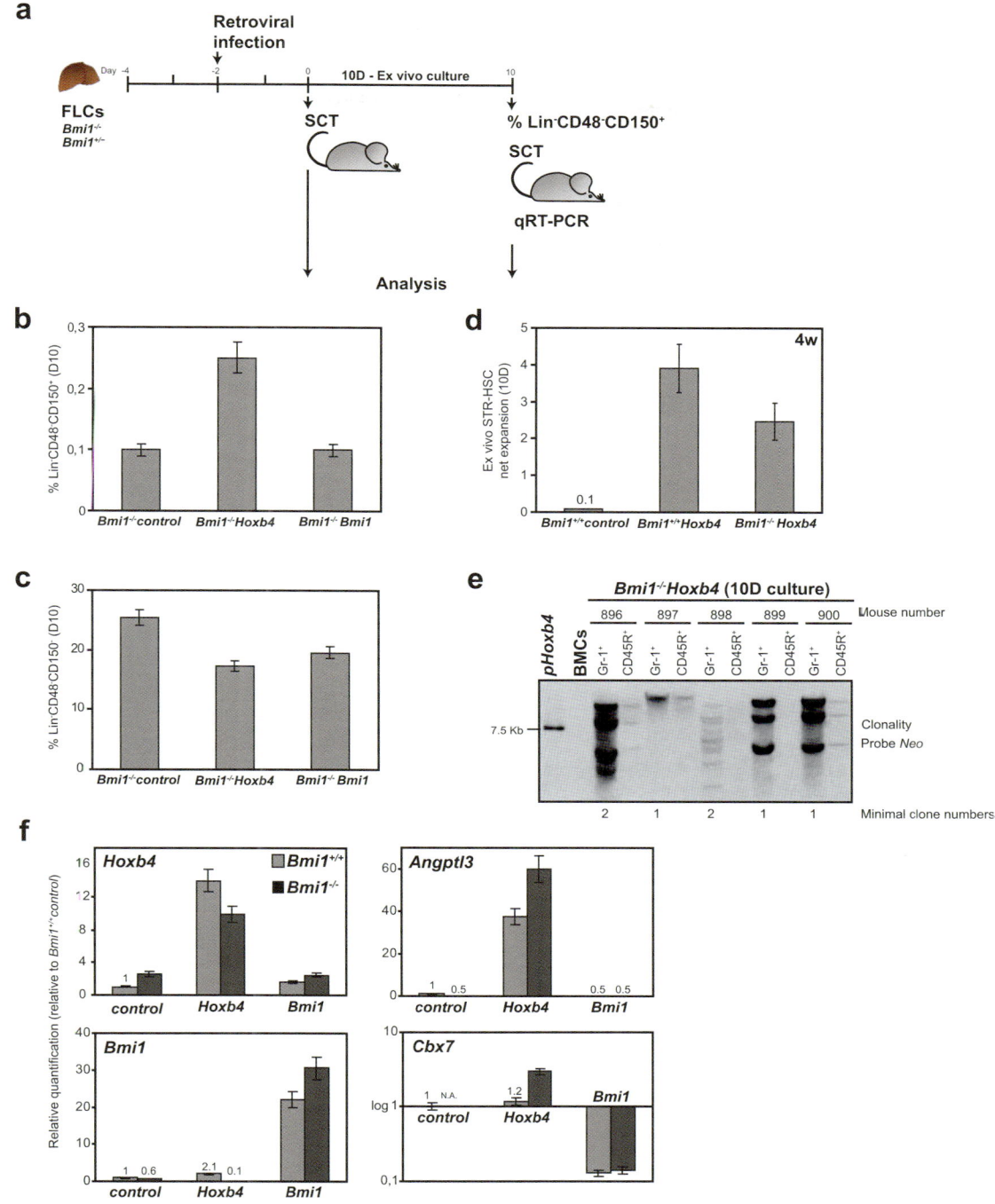

Figure 4. *Hoxb4* induces expansion of STR-HSCs in the absence of *Bmi1*. (*a*) Experimental outline. *Bmi1*$^{+/+}$ and *Bmi1*$^{-/-}$ FLCs were transduced with control; *Bmi1* or *Hoxb4* retroviral vectors and cells were analyzed as indicated. (*b,c*) Proportion of Lin⁻CD48⁻CD150⁺ and of Lin⁻CD48⁻CD150⁻ cells in cultures initiated with the indicated cell populations. FACS analyses were performed 10 days after the cultures were initiated (mean ±S.D.). (*d*) Determination of STR-HSC net expansion during a 10-day culture as described in *a–c*. STR-HSC determination was performed using the CRU assay with a minimum of ten recipients per dilution. Assessment of donor-derived reconstitution was performed at 4 weeks posttransplantation, hence the definition of STR-HSC in these experiments (error bar indicates a 95% confidence interval). (*e*) Southern blot analysis of genomic DNA isolated from *Hoxb4*-transduced *Bmi1*$^{-/-}$ BMCs and sorted based on the expression of Gr-1⁺ (myeloid) or CD45R⁺ (B-lymphoid) surface markers. Note that cells were collected at 4 weeks and 9 weeks posttransplantation for mouse 896 and mice 897–900, respectively. The DNA was digested with EcoRI, which cuts once in the integrated proviruses, and membranes were probed with a radiolabeled *neomycin* resistance gene fragment. Each fragment documents a unique insertion site for the *Hoxb4* provirus in the sorted populations. (*f*) Q-RT-PCR analyses of *Hoxb4* (*left upper panel*), *Bmi1* (*left lower panel*), *Angptl3* (*right upper panel*), and *Cbx7* (*right lower panel*) expression in sorted Lin⁻CD48⁻CD150⁺ *Bmi1*$^{+/+}$ and *Bmi1*$^{-/-}$ cells engineered to overexpress *GFP* (control) or *Hoxb4-GFP* or *Bmi1-GFP*. Results are presented as levels relative to those derived from *GFP*-transduced *Bmi1*$^{+/+}$ cells. Quality of RNA was evaluated using *Gapdh* primer sets. Primer sequences available upon request.

populations isolated from selected recipients 9 weeks post-SCT (Fig. 4e, mice 897 and 900). Further analysis of proviral integrations in bone-marrow-derived cells from these recipients revealed that at least six *Hoxb4*-transduced clones contributed to repopulation (see minimal clone numbers indicated at the bottom of Fig. 4e).

Most importantly, recipients 899 and 900 were repopulated by identical retrovirally marked clones, a typical signature for SR that occurred during the 10-day culture period before transplantation (Fig. 4e). The *Bmi1*$^{-/-}$ cell type (HSC vs. progenitor) that underwent SR as a result of *Hoxb4* overexpression is undefined. However, three arguments suggest that this cell type was an HSC. First, *Hoxb4* overexpression fails to recruit non-HSCs (progenitors) into this compartment (Park et al. 2003; Iwama et al. 2004). Second, the results presented in Figure 1i and j indicate that LTR-HSCs are present in normal numbers in *Bmi1*$^{-/-}$ fetal livers. Third, complementation experiments summarized in Figure 1j indicate that *Bmi1*$^{-/-}$ LTR-HSCs can expand normally in vivo during embryonic development.

DIFFERENT PATHWAYS ARE REGULATED BY *BMI1* AND *HOXB4*

The different impact of *Hoxb4* and *Bmi1* on HSC activity suggests that they affect distinct gene subsets. To gain insight into this possibility, we performed Q-RT-PCR studies on sorted Lin$^-$CD48$^-$CD150$^+$ cells engineered to overexpress *Hoxb4* or *Bmi1* and compared the expression of approximately 110 candidate gene targets in these cells. These genes included *Hox* members and cofactors, Polycomb group genes, and other stem cell regulators such as *Etv6* (Wang et al. 1998), *Gfi1* (Hock et al. 2004; Zeng et al. 2004), and *Lnk* (Ema et al. 2005). The expression of most of these genes was unaffected by *Hoxb4* or *Bmi1* (Table 1). However, we found that *Angptl3* (but not *Angptl2*) expression was strongly induced by *Hoxb4* in HSC-enriched cells and that this regulation was observed

in wild-type and *Bmi1*$^{-/-}$ cells (Fig. 4f). Conversely, the Polycomb group gene *Cbx7* is down-regulated by *Bmi1* and not by *Hoxb4* (Fig. 4f).

BMI1 INTEGRATES MULTIPLE PATHWAYS IN THE REGULATION OF HSC ACTIVITY

Hoxb4-induced complementation of *Bmi1*$^{-/-}$ repopulating cells is reminiscent of that recently described by our group for the loss of function of *E4F1* (Chagraoui et al. 2006) and by other groups for *arf/ink4a* and p53 (Jacobs et al. 1999; Bruggeman et al. 2005; Molofsky et al. 2006; Oguro et al. 2006; Akala et al. 2008). In fact, besides the recent results with *p53* (Akala et al. 2008), it remains challenging to identify a single gene whose gain or loss of function can fully rescue *Bmi1*$^{-/-}$ LTR-HSC activity. The recent identification of the *arf/ink4a* locus as a direct target for *Bmi1*, together with its direct interaction with E4F1, may suggest that both pathways are simultaneously regulated in self-renewing HSCs. Furthermore, the essential role for *Bmi1* as a cofactor to Ring1/2 ubiquitin E3 ligase in possibly promoting monoubiquitination of H2A (K119) suggests that other regulators may be involved (Fig. 5a).

CONCLUSIONS

Because HSC expansion occurs strictly as a result of symmetrical self-renewal divisions, we conclude that *Bmi1* is not essential for this function (at least before E14.5 in vivo and likely as a result of *Hoxb4* overexpression in vitro). This provides evidence to propose that execution and sustainability of SR are independently regulated in HSCs and that although the first appears to be regulated by *Hox* genes and their targets, the former is under the control of *Bmi1* (Fig. 5b). On the basis of this, it is tempting to speculate that in the absence of *Bmi1*, LTR-HSCs become functionally equivalent to STR-HSCs.

Table 1. Genes not regulated by *Hoxb4* and *Bmi1* overexpression in Lin$^-$CD48$^-$CD150$^+$ cells

Acin1	Egr1	Hoxa9	Nfat5	Rnf2
Angpti2	Epc2	Hoxb1	Notch1	Runx1
Asxl1	Etv6	Hoxb13	Numb	Runx2
Asxl2	Exh1	Hoxb2	Pard3	Scmh1
Atf4	Ezh2	Hoxb3	Pard6a	Scml2
Baz1b	Foxo3a	Hoxb5	Pard6b	Sf1
Bmp2	Gata2	Hoxb6	Pcgf2 (Mel18)	Sfpi1 (Pu.1)
Bmp4	Gfi1	Hoxb7	Pcgf6	Sh2b3 (Lnk)
Bmp7	Gli1	Hoxb8	Phc1 (Rae-28)	Smarcc1
Bteb1	Gli2	Hoxb9	Phc2	Stat3
Casp8ap2	Gli3	Irx5	Phc3	Stat5a
Cbfb	H3f3b	L3mbti1	Phf1	Suz12
Cbx2	Hoxa1	L3mbti2	Pknox1 (Prep1)	Tal1
Cbx4	Hoxa10	L3mbti3	Pknox2 (Prep2)	Tcf4
Cbx6	Hoxa11	Lef1	Prdm16	Tcfe2a (E2a)
Cbx8	Hoxa13	Lmo2	Prox1	Tgif
Cebpa	Hoxa2	Mef2c	Pten	Trim2
Crebbp	Hoxa3	Meis1	Rap1a	Trim28
Crem	Hoxa4	Mrg1 (Meis2)	Rarb	Xbp1
Dnmt3b	Hoxa5	Mrg2 (Meis3)	Rela	
Eed	Hoxa6	Myb	Rest	
Egfr	Hoxa7	Ncl	ring1	

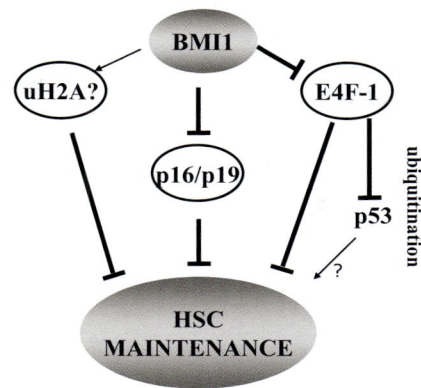

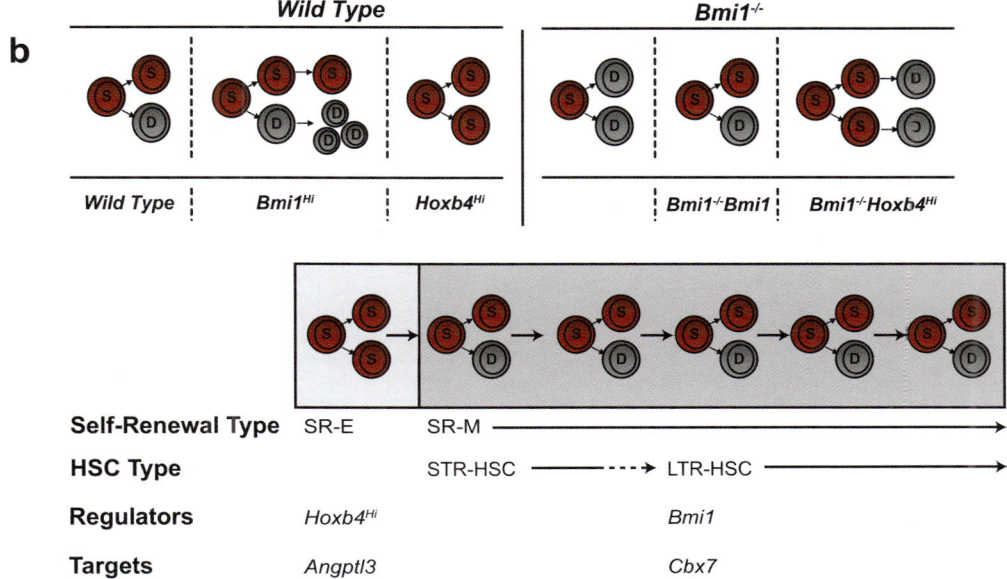

Figure 5. Complementary and independent function for *Hoxb4* and *Bmi1* in HSC activity.

ACKNOWLEDGMENTS

We thank Edlie St. Hilaire for her technical help regarding mouse manipulation and to Jana Krosl for critical reviews of this paper. We express our gratitude to Sonia Cellot, Simon Girard, Jana Krosl, and Martin Sauvageau for their good advice and to Christian Charbonneau, Danièle Gagné, and Pierre Chagnon for their help with the imaging, flow cytometry, and genomic core facilities of IRIC, respectively. This work was financed by the Canadian Institutes of Health Research and the National Institutes of Health. A.F. was a Fond de la Recherche en Santé du Québec (FRSQ) and National Cancer Institute of Canada (NCIC)/Terry Fox Foundation Studentships recipient. G.S. holds a Canada Research Chair in Molecular Genetics of Stem Cells.

REFERENCES

Akala, O.O., Park, I.K., Qian, D., Pihalja, M., Becker, M.W., and Clarke, M.F. 2008. Long-term haematopoietic reconstitution by *Trp53⁻/⁻p16Ink4a⁻/⁻p19Arf⁻/⁻* multipotent progenitors. *Nature* **453:** 228–232.

Antonchuk, J., Sauvageau, G., and Humphries, R.K. 2001. HOXB4 overexpression mediates very rapid stem cell regeneration and competitive hematopoietic repopulation. *Exp. Hematol.* **29:** 1125–1134.

Antonchuk, J., Sauvageau, G., and Humphries, R.K. 2002. *HOXB4*-induced expansion of adult hematopoietic stem cells ex vivo. *Cell* **109:** 39–45.

Bruggeman, S.W., Valk-Lingbeek, M.E., van der Stoop, P.P., Jacobs, J.J., Kieboom, K., Tanger, E., Hulsman, D., Leung, C., Arsenijevic, Y., Marino, S., and van Lohuizen, M. 2005. *Ink4a* and *Arf* differentially affect cell proliferation and neural stem cell self-renewal in *Bmi1*-deficient mice. *Genes Dev.* **19:** 1438–1443.

Cellot, S., Krosl, J., Chagraoui, J., Meloche, S., Humphries, R.K., and Sauvageau, G. 2007. Sustained in vitro trigger of self-renewal divisions in Hoxb4hiPbx1(10) hematopoietic stem cells. *Exp. Hematol.* **35:** 802–816.

Chagraoui, J., Niessen, S.L., Lessard, J., Girard, S., Coulombe, P., Sauvageau, M., Meloche, S., and Sauvageau, G. 2006. E4F1: A novel candidate factor for mediating BMI1 function in primitive hematopoietic cells. *Genes Dev.* **20:** 2110–2120.

Christensen, J.L. and Weissman, I.L. 2001. Flk-2 is a marker in hematopoietic stem cell differentiation: A simple method to isolate long-term stem cells. *Proc. Natl. Acad. Sci.* **98:** 14541–14546.

Dykstra, B., Kent, D., Bowie, M., McCaffrey, L., Hamilton, M., Lyons, K., Lee, S.J., Brinkman, R., and Eaves, C. 2007. Long-term propagation of distinct hematopoietic differentiation programs in vivo. *Cell Stem Cell* **1:** 218–229.

Ema, H., Sudo, K., Seita, J., Matsubara, A., Morita, Y., Osawa, M., Takatsu, K., Takaki, S., and Nakauchi, H. 2005. Quantification of self-renewal capacity in single hematopoietic stem cells from normal and Lnk-deficient mice. *Dev. Cell* **8:** 907–914.

Hock, H., Hamblen, M.J., Rooke, H.M., Schindler, J.W., Saleque, S., Fujiwara, Y., and Orkin, S.H. 2004. Gfi-1 restricts proliferation and preserves functional integrity of haematopoietic stem cells. *Nature* **431:** 1002–1007.

Iwama, A., Oguro, H., Negishi, M., Kato, Y., Morita, Y., Tsukui, H., Ema, H., Kamijo, T., Katoh-Fukui, Y., Koseki, H., van Lohuizen, M., and Nakauchi, H. 2004. Enhanced self-renewal of hematopoietic stem cells mediated by the Polycomb gene product Bmi-1. *Immunity* **21:** 843–851.

Jacobs, J.J.L., Kieboom, K., Marino, S., DePinho, R.A., and van Lohuizen, M. 1999. The oncogene and Polycomb-group gene *bmi-1* regulates cell proliferation and senescence through the *ink4a* locus. *Nature* **397:** 164–168.

Kiel, M.J., Yilmaz, O.H., Iwashita, T., Yilmaz, O.H., Terhorst, C., and Morrison, S.J. 2005. SLAM family receptors distinguish hematopoietic stem and progenitor cells and reveal endothelial niches for stem cells. *Cell* **121:** 1109–1121.

Kim, I., He, S., Yilmaz, O.H., Kiel, M.J., and Morrison, S.J. 2006. Enhanced purification of fetal liver hematopoietic stem cells using SLAM family receptors. *Blood* **108:** 737–744.

Krosl, J., Austin, P., Beslu, N., Kroon, E., Humphries, R.K., and Sauvageau, G. 2003. In vitro expansion of hematopoietic stem cells by recombinant TAT-HOXB4 protein. *Nat. Med.* **9:** 1428–1432.

Lessard, J. and Sauvageau, G. 2003. *Bmi-1* determines the proliferative capacity of normal and leukaemic stem cells. *Nature* **423:** 255–260.

Molofsky, A.V., Slutsky, S.G., Joseph, N.M., He, S., Pardal, R., Krishnamurthy, J., Sharpless, N.E., and Morrison, S.J. 2006. Increasing p16^{INK4a} expression decreases forebrain progenitors and neurogenesis during ageing. *Nature* **443:** 448–452.

Morrison, S.J., Wandycz, A.M., Hemmati, H.D., Wright, D.E., and Weissman, I.L. 1997. Identification of a lineage of multipotent hematopoietic progenitors. *Development* **124:** 1929–1939.

Oguro, H., Iwama, A., Morita, Y., Kamijo, T., van Lohuizen, M., and Nakauchi, H. 2006. Differential impact of *Ink4a* and *Arf* on hematopoietic stem cells and their bone marrow microenvironment in *Bmi1*-deficient mice. *J. Exp. Med.* **203:** 2247–2253.

Ohta, H., Sekulovic, S., Bakovic, S., Eaves, C.J., Pineault, N., Gasparetto, M., Smith, C., Sauvageau, G., and Humphries, R.K. 2007. Near-maximal expansions of hematopoietic stem cells in culture using NUP98-HOX fusions. *Exp. Hematol.* **35:** 817–830.

Park, I.K., Qian, D., Kiel, M., Becker, M.W., Pihalja, M., Weissman, I.L., Morrison, S.J., and Clarke, M.F. 2003. Bmi-1 is required for maintenance of adult self-renewing haematopoietic stem cells. *Nature* **423:** 302–305.

Pawliuk, R., Eaves, C., and Humphries, R.K. 1996. Evidence of both ontogeny and transplant dose-regulated expansion of hematopoietic stem cells in vivo. *Blood* **88:** 2852–2858.

Rebel, V.I., Miller, C.L., Thornbury, G.R., Dragowska, W.H., Eaves, C.J., and Lansdorp, P.M. 1996. A comparison of long-term repopulating hematopoietic stem cells in fetal liver and adult bone marrow from the mouse. *Exp. Hematol.* **24:** 638–648.

Sauvageau, G., Thorsteinsdottir, U., Eaves, C.J., Lawrence, H.J., Largman, C., Lansdorp, P.M., and Humphries, R.K. 1995. Overexpression of *HOXB4* in hematopoietic cells causes the selective expansion of more primitive populations in vitro and in vivo. *Genes Dev.* **9:** 1753–1765.

Shen, Q., Zhong, W., Jan, Y.N., and Temple, S. 2002. Asymmetric Numb distribution is critical for asymmetric cell division of mouse cerebral cortical stem cells and neuroblasts. *Development* **129:** 4843–4853.

Szilvassy, S.J., Humphries, R.K., Lansdorp, P.M., Eaves, A.C., and Eaves, C.J. 1990. Quantitative assay for totipotent reconstituting hematopoietic stem cells by a competitive repopulation strategy. *Proc. Natl. Acad. Sci.* **87:** 8736–8740.

Thorsteinsdottir, U., Sauvageau, G., and Humphries, R.K. 1999. Enhanced in vivo regenerative potential of *HOXB4*-transduced hematopoietic stem cells with regulation of their pool size. *Blood* **94:** 2605–2612.

Thorsteinsdottir, U., Mamo, A., Kroon, E., Jerome, L., Bijl, J., Lawrence, H.J., Humphries, R.K., and Sauvageau, G. 2002. Overexpression of the myeloid leukemia-associated *Hoxa9* gene in bone marrow cells induces stem cell expansion. *Blood* **99:** 121–129.

Wang, L.C., Swat, W., Fujiwara, Y., Davidson, L., Visvader, J., Kuo, F., Alt, F.W., Gilliland, D.G., Golub, T.R., and Orkin, S.H. 1998. The *TEL/ETV6* gene is required specifically for hematopoiesis in the bone marrow. *Genes Dev.* **12:** 2392–2402.

Wognum, A.W., Eaves, A.C., and Thomas, T.E. 2003. Identification and isolation of hematopoietic stem cells. *Arch. Med. Res.* **34:** 461–475.

Zeng, H., Yücel, R., Kosan, C., Klein-Hitpass, L., and Möröy, T. 2004. Transcription factor Gfi1 regulates self-renewal and engraftment of hematopoietic stem cells. *EMBO J.* **23:** 4116–4125.

Regeneration, Stem Cells, and the Evolution of Tumor Suppression

B.J. Pearson and A. Sánchez Alvarado

Howard Hughes Medical Institute, Department of Neurobiology and Anatomy,
University of Utah School of Medicine, Salt Lake City, Utah 84132

All multicellular organisms have requirements for tumor suppression to regulate cellular proliferation during either embryonic development or adult life. However, different organisms have vastly different requirements. Adult tumor suppression is probably not crucial to organisms possessing both short life spans and largely postmitotic soma. In contrast, animals with lifelong tissue turnover or those capable of regenerating body parts lost to injury must possess evolutionarily selected mechanisms to control rates of cell proliferation such that tissue homeostasis can be maintained or restored after injury. We hypothesize that these biological differences may help to explain why the lists of tumor suppressor genes in humans and *Drosopaila* are largely nonoverlapping. Here, we address this disparity by examining the tumor suppressor gene content of two outgroups to the vertebrates and flies/nematodes: the freshwater planarian and the single-celled choanoflagellate. Both of these organisms have recently had their genomes sequenced, giving us a first glimpse of which known tumor suppressor genes have been maintained during evolution. In addition, we attempt to resolve which genes may have had ancestral tumor suppressor function and which may have acquired this function de novo.

Of all the genes that can contribute to the exceedingly complex disease known as "cancer," perhaps the most amenable to functional dissection are the tumor suppressor genes (TSGs). Simply defined, a gene is a tumor suppressor if it can meet several criteria: (1) When it is removed from an organism, it results in a tumor or hyperproliferation phenotype, (2) mutations in these genes are mostly recessive and function cell-autonomously, and (3) the given gene must also be mutated or deleted in human cancers. A search of PubMed using the quoted phrase "tumor suppressor" retrieved 59,612 entries (as of October 7, 2008). Such a vast field of ongoing research into the functions of TSGs is driven by a desire to understand the ontogeny of human cancers and find new genes to target for therapy. However, it is important to remember that during evolution, TSGs were not selected because of their role in the disease of cancer. Instead, these genes likely evolved to function in indispensable cellular processes such as control of cell division and DNA-damage repair. Thus, the reason that these genes exist seems to be straightforward, but it does not explain the paradox that the major TSGs in human cancers are not TSGs in model laboratory invertebrates. Our collective lack of understanding of the evolution of TSGs may explain why a search of PubMed using the quoted phrases "tumor suppressor evolution," "evolution of tumor suppressors," or "evolution of tumor suppression" retrieved a total of zero entries. Thus, studies attempting to understand tumor suppression through the lens of evolutionary biology are uncommon.

There are approximately 50 million species of animals on our planet (Brusca and Brusca 2003). Of these, less than 10 have been used in the laboratory to understand the mechanisms of tumor suppression. Of these 10, only 2 species have been used outside of the vertebrates and both are from the super-phylum Ecdysozoa, i.e., *Drosophila melanogaster* and *Caenorhabditis elegans*. Remarkably, our understanding of tumor development and the disparities that are known to exist among species comes from this small, not statistically significant sample size. This notable shortcoming in our knowledge imposes serious limitations in our interrogation of tumor suppression mechanisms. First, the general absence of molecular studies in other phyla leaves unanswered the question of how tumor suppression evolved. Second, when these facts are taken into consideration, the difficulties of postulating bona fide conclusions about how tumor suppression works become readily apparent. Therefore, mechanistic studies of tumor suppression in underrepresented, but evolutionarily important organisms are not only warranted, but also essential to fill this lacunae of knowledge.

Because regulation of cellular proliferation is indispensable to the survival of multicellular organisms, it is likely that mechanisms for tumor suppression have ancestral evolutionary origins. Equally important, physiological and environmental changes often require adult organisms to modulate their rates of cell proliferation (e.g., regeneration of body parts lost to injury), and therefore mechanisms to relax tumor suppression in order to allow temporary hyperproliferation must have also evolved. To investigate these hypotheses, at least two questions need to be answered: Did the unicellular ancestor to animals already have tumor suppressor genes that were then recruited into tumor suppression? Or, did animals evolve tumor suppressor genes de novo to contend with survival and perpetuation issues raised by the emergence of multicellularity?

In this chapter, we investigate the evolutionary implications of tumor suppression. First, we explore the different types of tumor suppression requirements in various organisms with the objective of determining whether universal tumor suppressors exist in the Metazoa. Second, we examine the complement of known, well-studied tumor suppressor gene homologs across animals, including the newly sequenced genomes of the freshwater planarian *Schmidtea mediterranea* and the unicellular choanoflagellate *Monosiga brevicollis* (King et al. 2008). We then compare and contrast the functions of common TSGs in flies, *C. elegans*, and vertebrates and attempt to answer why differences exist while tracking these genes throughout the evolution of animals. Finally, we argue that the freshwater flatworm (planaria) can be used not only to reconcile evolutionary hypotheses, but also to offer a fresh view on tumor suppression mechanisms.

TYPES OF TUMOR SUPPRESSION VERSUS LIFE SPAN AND ADULT CELL PROLIFERATION

All animals have requirements for tumor suppression, but animal tumors can be generally cataloged as either embryonic/larval (arising during development) or late-onset (arising in adults). The type most prevalent in different organisms sharply correlates with life span and somatic cell proliferation during adulthood. For example, in animals possessing few adult somatic cell divisions and a short life span, adult tumor suppression is not nearly as important as tumor suppression during embryonic/larval development, where extensive cell proliferation is the norm.

In *Drosophila* and *C. elegans*, it is clear that their susceptibility to tumors is high during embryonic/larval development (e.g., *scribbled [scrib], lethal giant larvae [lgl], lethal giant discs [lgd], discs large [dlg], malignant brain tumor [mbt], brain tumor [brat]*) (Gateff and Schneiderman 1974; Xu et al. 1995; Woodhouse et al. 1998; Humbert et al. 2003; Klezovitch et al. 2004; Menut et al. 2007). This hypothesis is supported by the fact that no tumor suppressor mutation has yet been found in *Drosophila* or *C. elegans* to specifically affect adults (i.e., not arise during development). Even though no adult-onset tumors have been identified in these organisms, the TSGs that function during *Drosophila* development, for example, have extensively informed cell biological processes involved in vertebrate tumorigenesis and metastasis.

In principle, vertebrates have developmental tumor suppression requirements similar to those of flies and *C. elegans*. However, we hypothesize that vertebrates have fundamentally different requirements for adult tumor suppression due to notable differences in life span and the numbers of adult somatic cell divisions (Fig. 1A). For instance, flies and *C. elegans* have very short adult life spans (60 and 14 days, respectively) and therefore might not require robust adult tumor suppression. In contrast to these two invertebrates, the shortest captive life span documented for a vertebrate is 90 days (Terzibasi et al. 2007), but most vertebrates live 10–100 times longer than *Drosophila*. Furthermore, flies and *C. elegans* have relatively few adult somatic cell divisions (homeostasis) (0

and likely ~10^5, respectively), which may not be enough cell turnover to select for the maintenance of tumor suppressor function, especially when coupled to their short life spans. Humans, on the other hand, are predicted to turn over many of the approximately 10^{14} cells in our bodies throughout our lives, leading to homeostasis of about 10^{10} cells per day (Heemels 2000). That is 10^7 times higher than the number of cells present in an entire adult *C. elegans* and 10^5 more cells than a fly turns over in a lifetime. With these staggering differences in cell turnover, it is easy to hypothesize that natural selection may have selected for completely different tumor suppression mechanisms in vertebrates. The sheer number of constant cell turnover in humans, for example, must be under tight genetic control and simultaneously must require constant tumor suppression activity. Consistent with this observation is the fact that many adult TSGs in vertebrates do not appear to function as adult tumor suppressors in *Drosophila* or *C. elegans* (*phosphatase and tensin homolog deleted on chromosome 10 [PTEN], p53, retinoblastoma [Rb], neurofibromatosis 1 [NF1], p16^{ink4a}, p19^{ARF}, p21^{cip1}, p27^{Kip1}, breast cancer susceptibility gene 1 [Brca1], breast cancer susceptibility gene 2 [Brca2], deleted in lung cancer 1 [DLC1]*) (de Nooij et al. 1996; The et al. 1997; Huang et al. 1999; Sherr and McCormick 2002; Stevaux et al. 2002; Coqueret 2003; Lowe and Sherr 2003; Carroll and Stone-

A

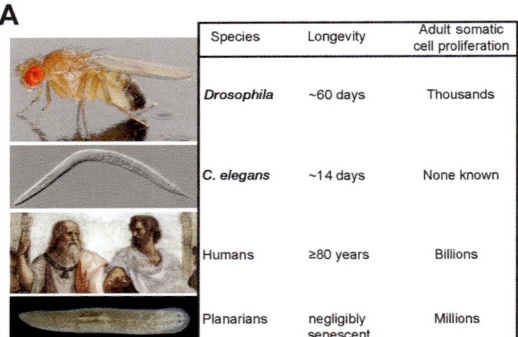

Species	Longevity	Adult somatic cell proliferation
Drosophila	~60 days	Thousands
C. elegans	~14 days	None known
Humans	≥80 years	Billions
Planarians	negligibly senescent	Millions

B

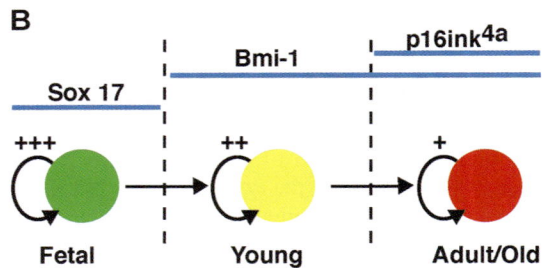

Figure 1. Cell proliferation and animal longevity. (*A*) Comparison of longevity and cell proliferative activities between humans and established (*C. elegans*, *Drosophila*) and emerging organisms (*S. mediterranea*). (*B*) Expression of known activators and suppressors of neuronal stem cells throughout mammalian life. Expression of the Polycomb group gene *Bmi-1* is continuously maintained after birth. As animals age, however, expression of the tumor suppressor gene *p16^{ink4a}* is observed, resulting in an inhibition of stem cell self-renewal (Molofsky et al. 2006). (*Drosophila*, nematode, and human images courtesy of http://commons.wikimedia.org.)

cypher 2005; Evers and Jonkers 2006; Lu and Abrams 2006; Rossi and Weissman 2006; Aoki and Taketo 2007; Moon et al. 2008; Xue et al. 2008).

Many of the old cells replaced during homeostasis in mammals are produced by tissue-specific adult stem cells. Remarkably, during all of these divisions, the adult stem cells strictly maintain their numbers. Implied in this is that there are genetic mechanisms to constantly prevent hypoproliferation (organismal senescence) or hyperproliferation (tumorigenesis) that exist specifically in adult stem cells themselves. Evidence for the existence of such mechanisms is provided by the age-dependent activation of TSGs as neuronal stem cells age in mammals (Fig. 1B). Young neuronal stem cells have been shown to depend on the function of the Polycomb group gene *Bmi-1* for their self-renewal. When this gene is eliminated, cell proliferation decreases and a marked increase in the expression of the tumor suppressor *p16^{Ink4a}* is observed. This observation suggests that *Bmi-1* may actually repress tumor suppression in order to promote stem cell proliferation in young individuals (Molofsky et al. 2003). Interestingly, as animals age and the proliferative capacities of neural stem cells in the subventricular zone and olfactory bulb all decline, a concomitant increase in the expression of *p16^{Ink4a}* is observed (Molofsky et al. 2006). These findings suggest that regulation of proliferation of somatic stem cells may have been selected through evolution to balance the longevity assured by stem-cell-dependent tissue renewal against the risk of developing hyperproliferative disorders such as cancer (Gatza et al. 2007).

PLANARIANS AND CHOANOFLAGELLATES AS NEW ORGANISMS TO UNDERSTAND TUMOR SUPPRESSION

In contrast to flies and nematodes, the adult freshwater planarian *S. mediterranea* possesses an abundant and experimentally accessible population of adult somatic stem cells known as neoblasts (Newmark and Sánchez Alvarado 2002). Many of these cells are continuously undergoing cell division (Newmark and Sánchez Alvarado 2000), and their progeny become determined and ultimately differentiate into all of the known cell types in planarians (Newmark and Sánchez Alvarado 2000; Eisenhoffer et al. 2008). In fact, planarians derive their well-known tissue homeostasis and regenerative properties from regulating the proliferation and functions of neoblasts and their division progeny (Reddien and Sánchez Alvarado 2004; Sánchez Alvarado and Tsonis 2006). After amputation, for instance, planarians develop at the cut surface a specialized structure known as a regeneration blastema, which is the result of extensive proliferation by neoblasts, and which will ultimately differentiate to replace the missing body parts lost to injury. The fact that planarians can temporarily allow their stem cells to hyperproliferate after injury, only to reign in this activity to normal levels, indicates the existence of molecular mechanisms that are constantly surveying changes in the environment and orchestrating the appropriate changes in proliferation required by such changes. Therefore, an investigation of tumor suppression mechanisms in this

invertebrate should help inform to what extent the molecular basis of tumor suppression may be evolutionarily conserved among multicellular organisms.

By the same token, the genetic composition of the closest unicellular relative of the multicellular animals, the choanoflagellate *M. brevicollis*, should provide us with a glimpse of the molecular tool kit available before the emergence of multicellularity. Analyses of the *Monosiga* genome indicate that a large number of transcription factor families previously thought to be restricted to metazoa are found in this organism, as well as proteins involved in multicellularity, such as cadherins, transmembrane C-type lectins, and other molecules in cell adhesion (King et al. 2008). Therefore, determining the presence or absence of known tumor suppressors in the genome of this organism should provide evidence for either the ancestry or novelty of known tumor suppressor genes.

EVOLUTIONARY SCENARIOS OF TUMOR SUPPRESSION

On the basis of the above information, we pose three different hypotheses that can explain the disparity in tumor development known to exist between vertebrates and flies/*C. elegans*. First, the genome of the ancestor of flies, *C. elegans*, and vertebrates contained all of the homologs of tumor suppressor genes, but only embryonic/larval tumor suppression function was maintained in flies and nematodes (or lost in vertebrates). Second, the genome of the ancestor of flies, *C. elegans*, and vertebrates contained all of the gene homologs for tumor suppression, but only in the vertebrate lineage did they acquire adult tumor suppressor function (or lost in flies/*C. elegans*). Third, the genome of the ancestor of flies and vertebrates did not have all of the homologs of TSGs, and these emerged de novo and independently in both clades of animals, giving us the disparate list of genes discussed above. These various scenarios may be resolved by examining the first three TSGs identified in flies, and the top three, most frequently mutated TSGs in human cancers. We can then search the genomes of outgroup organisms (planarians and *Monosiga*) to determine whether they contain these genes and examine their functions if they are known.

EVOLUTION OF TSGs IN *DROSOPHILA*: *SCRIB*, *LGL*, AND *DLG*

The embryonic/larval tumor suppressors in flies are referred to as neoplastic TSGs (nTSGs), and they were found in genetic screens in which larvae could not pupate or displayed imaginal disc overgrowth in conditional mutant screens (Gateff and Schneiderman 1974; Gateff and Mechler 1989; Gateff 1994; Xu et al. 1995; Pagliarini and Xu 2003; Menut et al. 2007). Analyses of these genes have shown that they function mainly in the establishment and maintenance of correct apical-basal cell polarity in proliferating epithelia and neuroblasts during fly development (Humbert et al. 2003; Vasioukhin 2006). When mutated, cells in developing epithelial structures hyperproliferate and can even metastasize when dissected and transplanted into the abdomen of another adult fly. Interestingly, if a developing tissue is only partially mutant for these nTSGs, the cells are often at a growth disadvantage and do not produce tumors nor metastasize. This observation is contrary to canonical mammalian tumorigenesis, in which the vast majority of human tumors are monoclonal, and the rare mutant cells for a TSG occur in an otherwise wild-type tissue (Hanahan and Weinberg 2000; Wang et al. 2007). Finally, it does not appear that vertebrate homologs of fly nTSGs function as tumor suppressors during development or in adults, although only several studies have been performed (see discussion). Because most human cancers also arise from epithelia and because human epithelia are also polarized, it seems reasonable to hypothesize that homologs of the fly nTSGs function in the later stages of cancer, when cellular polarity is commonly lost (Wodarz and Nathke 2007).

We traced the evolution of these nTSGs across the animal kingdom and our analyses led us to a few interesting conclusions (Fig. 2). First, there is similar gene redundancy in vertebrates, planaria, and *Monosiga* but, generally, single gene orthologs in flies and *C. elegans*. For instance, although flies have a (likely) single *dlg*, vertebrates, planaria, and *Monosiga* have at least four. This could explain the discrepancy of why individual vertebrate *dlg* homologs do not appear to have tumor suppressor function in vertebrates. Another interesting conclusion is that the genome of unicellular choanoflagellates already have clear homologs to *scrib* and *dlg*, but not *lgl*. Because *scrib*, *lgl*, and *dlg* are involved in the same structure of adherens and tight junctions in both invertebrates and vertebrates, it will be interesting to determine what their functions may be in choanoflagellates and what kind of complex *Monosiga* makes without the presence of *lgl*. The possibility exists that choanoflagellates use these genes in proliferation control in a manner similar to that of *Drosophila*. If shown to be true, this would mean that the

tumor suppressor activity of the nTSGs was either lost in the vertebrates or not sufficiently and rigorously tested. Alternatively, a similar conclusion suggesting that the nTSGs had ancestral tumor suppressor activity would also be reached if these molecules are found to also be tumor suppressors in planarians.

EVOLUTION OF TSGs IN VERTEBRATES: *p53, PTEN,* AND *Rb*

Although there are many TSGs in vertebrates, virtually every cancer has a mutation in at least one of three genes: *p53, PTEN,* and *Rb* (Fig. 3). Of these three, none more perfectly illustrate the differences between invertebrates and vertebrates to develop tumors than p53. p53 is mutated in more than 50% of all human cancers (Harris 1996), yet p53 mutant *Drosophila* or *C. elegans* develop normally and are tumor free (Brodsky et al. 2000; Ollmann et al. 2000; Derry et al. 2001; Rong et al. 2002; Wells et al. 2006).

On the basis of gene number and protein function, p53 possesses a complex, if not tortuous, evolutionary history. Although *Drosophila* and *C. elegans* only have a single p53 family member, there has been an expansion of the family in many other organisms (Nedelcu and Tan 2007). Vertebrates have three family members: p53, p63, and p73, of which the latter two are not robust tumor suppressors and have an additional carboxy-terminal sterile α motif (SAM) (Lu and Abrams 2006). Furthermore, p63 and p73 have many spliceoforms that can act as dominant negatives to p53 (often making them oncogenes) (Li and Prives 2007). The cross-regulation and redundant function of vertebrate p53 family members is exceedingly complex and can vary from tissue to tissue (Fei et al. 2002). Despite some differences in domain structure, the p53 family appears to have retained specific ancestral functions. For

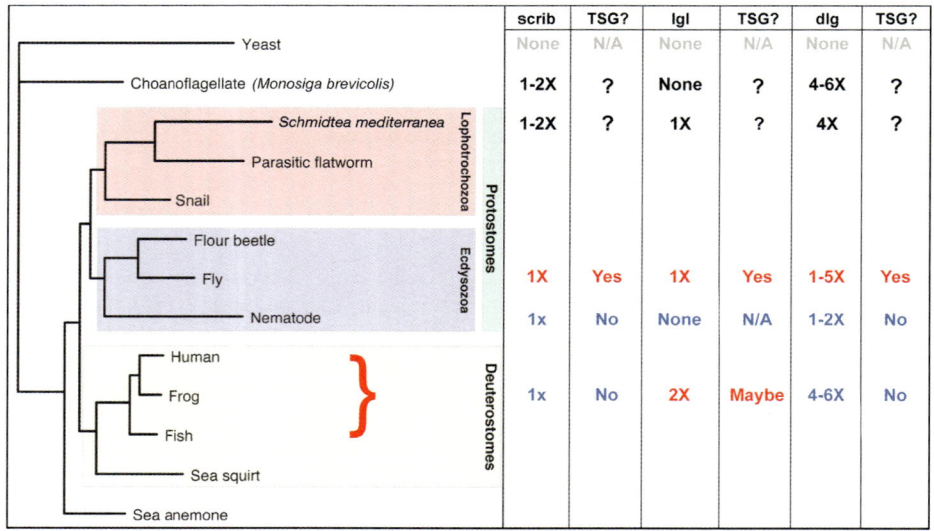

Figure 2. Evolution of the fly nTSGs scrib, lgl, and dlg. Reciprocal BLAST searches of the *Drosophila* neoplastic tumor suppressor genes *scrib*, *lgl*, and *dlg* were performed against the genomes of choanoflagellates (*M. brevicollis*), planaria (*S. mediterranea*), nematodes (*C. elegans*), and vertebrates (fish, frog, and human). Phylogenetic relationships are illustrated on the left and the number of homologs found for each of the genes is listed. (*Red text*) Evidence for TSG function; (*blue text*) evidence for no TSG function; (*black text*) function unknown.

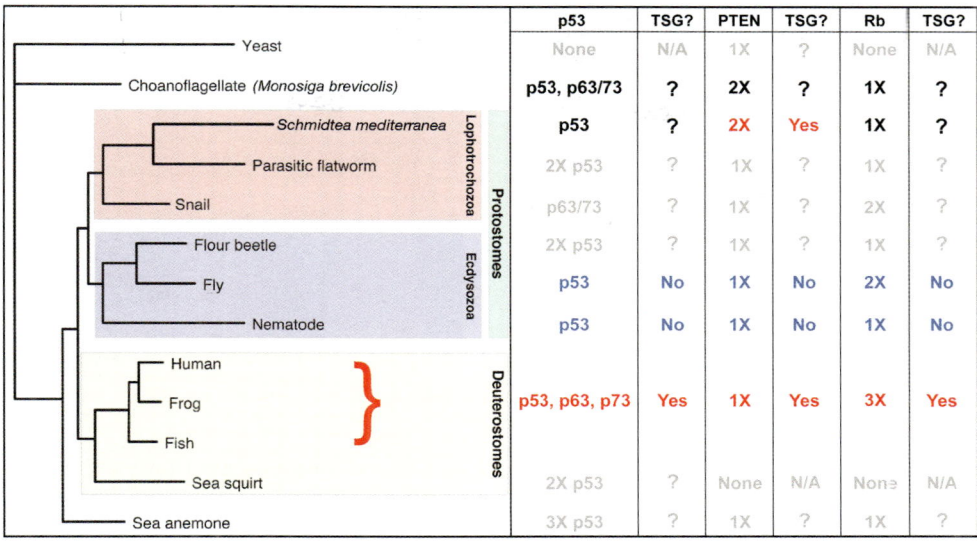

	p53	TSG?	PTEN	TSG?	Rb	TSG?
Yeast	None	N/A	1X	?	None	N/A
Choanoflagellate (*Monosiga brevicolis*)	p53, p63/73	?	2X	?	1X	?
Schmidtea mediterranea	p53	?	2X	Yes	1X	?
Parasitic flatworm	2X p53	?	1X	?	1X	?
Snail	p63/73	?	1X	?	2X	?
Flour beetle	2X p53	?	1X	?	1X	?
Fly	p53	No	1X	No	2X	No
Nematode	p53	No	1X	No	1X	No
Human / Frog / Fish	p53, p63, p73	Yes	1X	Yes	3X	Yes
Sea squirt	2X p53	?	None	N/A	None	N/A
Sea anemone	3X p53	?	1X	?	1X	?

Figure 3. Evolution of the vertebrate TSGs p53, PTEN, and Rb. Reciprocal BLAST searches of the vertebrate tumor suppressor genes *p53*, *PTEN*, and *Rb* were performed against the genomes of yeast (*Saccharomyces cerevisiae*, *Schizoaccharomyzes pombe*), choanoflagellates (*M. brevicollis*), sea anemone (*Nematostella vectensis*), planaria (*S. mediterranea*), parasitic flatworms (*Schistosoma mansoni*), snails (*Lottia gigantea*), flour beetle (*Tribollium castaneum*), nematodes (*C. elegans*), urochordates (*Ciona intestinalis*), and vertebrates (fish, frog, and human). Phylogenetic relationships are illustrated on the left and the number of homologs found for each of the genes is listed in the table. Question marks indicate instances in which orthology could not be conclusively established. Text color coding is the same as that in Figure 2.

example, all three p53 family members in vertebrates respond to genotoxic stresses and can induce apoptosis (Cuddihy and Bristow 2004; Irwin 2004; Murray-Zmijewski et al. 2006). Fly and *C. elegans p53* also have this function and as such can be assigned as ancestral (Lu and Abrams 2006). Apparent new functions for the p53 family in vertebrates are tumor suppression/cell cycle regulation (*p53*), epithelial stem cell self-renewal (*p63*) (Senoo et al. 2007), and neural differentiation and survival (*p73*) (Moll and Slade 2004; Rocco et al. 2006).

It is widely accepted that p53 has acquired tumor suppressor function in vertebrates from an ancestral role of DNA-damage sensing/repair. However, there has never been a functional study of a p53 family member in an invertebrate that has high requirements for adult tumor suppression, nor in an animal from the Lophotrochozoan superphylum. Planaria have a single homolog of p53, and once again, these organisms are in an advantageous position to resolve whether the "new" functions of the p53 family are actually new or ancestral (but lost in flies/*C. elegans*). Surprisingly, *Monosiga* has a homolog to both p53 and p63 (Nedelcu and Tan 2007). Determining how these genes function in a single-celled animal will almost certainly yield insights into the evolution of tumor suppressor activity of p53 family members.

Unlike p53, the Rb transcription factor has a clear evolutionary history and an ancestral function in cell cycle regulation and was the first tumor suppressor to be identified (van den Heuvel and Dyson 2008). One of the three *Rb* genes in humans is thought to be mutated in virtually all cancers (Weinberg 1995; Ajioka et al. 2007). In vertebrates, Rb functions by repressing the expression and activation of the E2F family of proteins, which are needed for multiple processes during the cell cycle (van den Heuvel

and Dyson 2008). Although no obvious homolog to *Rb* can be found in fungi, homologs of this gene can be found in plants and in unicellular (*Monosiga*) and multicellular animals. Flies have two Rb family members; however, one mutant shows only mild overgrowth defects (Du et al. 1996; Stevaux et al. 2002). Judging from the fact that *Monosiga*, *C. elegans*, planarians, and sea anemone (*Nematostella vectensis*) have a single Rb family member, this suggests that the ancestral condition in animals was a single gene. Interestingly, vertebrates, flies, and snails (*Lottia gigantea)* have expanded the family, and the urochordate *Ciona intestinalis* (sea squirt) has somehow lost this gene. Even though Rb is the closest example of a universal TSG that we cover in this chapter, its function has not been examined in planaria or *Monosiga*, so once again the sample size for making bona fide conclusions is small. We predict that Rb will function as a negative cell cycle regulator in planarians, *Monosiga*, and other metazoans.

The third most mutated gene in human cancers is *PTEN*. PTEN functions to balance levels of the intracellular signaling molecule phosphatidylinositol-3-phosphate (PIP_3) with the gene *phosphoinositol-3 kinase* (*PI3K*), and has crucial roles in cellular proliferation, differentiation, and migration in both vertebrates and invertebrates (Sulis and Parsons 2003). Mutations of this gene are found in a wide variety of cancers, including non-Hodgkin's lymphoma (Nakahara et al. 1998), breast (Kirkegaard et al. 2005), prostate (Li et al. 1997), and some types of pancreatic cancers (Asano et al. 2004). Altogether, this body of evidence suggests that PTEN may have a key role in regulating and misregulating tissue homeostasis in mammals (Marx 2007). Although the PTEN signaling pathway has many loops, intersections, and forks (Sulis and Parsons 2003), it is generally thought that in *PTEN* mutants, the canonical

PI3K-Akt-mTOR (mammalian target of rapamycin) pathway is the dominant route. This is supported by suppression of PTEN-mediated tumorigenesis by the small molecule rapamycin, which is thought to inhibit mTOR. In mice, for instance, rapamycin treatment compensates for *PTEN* loss and is effective in preventing leukemia-initiating cells, while keeping normal hematopoietic function in mice with a *PTEN* conditional deletion (Yilmaz et al. 2006).

Similar to p53, elimination of *PTEN* in *Drosophila* does not result in defects associated with tumor suppression (Huang et al. 1999). Instead, the *PTEN* mutant phenotype results in defects associated with the scale and growth of cells rather than with cellular proliferation. For example, a *PTEN* mutant eye imaginal disc in *Drosophila* does not show hyperproliferation, even though cells in the disc are enlarged and/or misshapen (Huang et al. 1999). On the basis of gene complement, it is likely that the ancestral condition was a single PTEN gene which was expanded to two genes in *Monosiga* and planarians and, again, surprisingly lost in the sea squirt (*C. intestinalis*). Unlike p53 and Rb, a PTEN-like gene does exist in the fungi. Vertebrates also have two closely related genes—*TPIP* and *TPTE*—that have no fly homologs. Interestingly, planaria have a single homolog to *TPIP/TPTE* that appears to be the ancestral condition. Therefore, the gene complement for the PTEN homologs in planaria appears to be much more similar to vertebrates.

Recently, we analyzed the function of two planarian PTEN orthologs (Oviedo et al. 2008). When both *Smed-PTEN1* and *Smed-PTEN2* are inhibited by RNA interference (RNAi), adult planaria show ectopic growths, hyperproliferation of stem cells, and breakdown of the subepithelial basement membrane that surrounds the worm (Fig. 4). Interestingly, rapamycin can suppress all three of these phenotypic outcomes (Oviedo et al. 2008). These results are the first to show that PTEN can function as a tumor suppressor in invertebrates. In addition, this discovery demonstrates that planaria can be used as a model to understand the early stages of tumorigenesis as well as to interrogate the function of tumor suppressor genes in adult animals. In the context of this review, these data suggest that the ancestral function of PTEN in multicellular animals was also in the tumor suppression that flies and *C. elegans* have lost.

DISCUSSION

Despite differences in tumor suppression between flies/*C. elegans* and vertebrates/planaria, it is clear that more experiments testing the ideas presented in this chapter are needed. First, it has recently been shown that *Drosophila* does indeed possess adult stem cells (Micchelli and Perrimon 2006). In addition, now that a cellular lineage has been at least partially elucidated for fly intestinal antibody-secreting cells (ASCs), it is possible to test what happens to these cells when *p53, PTEN,* or *Rb* are conditionally deleted in adults. It is formally possible that deletion of these TSGs will give rise to hyperproliferation and expansion of the ASC population, further cementing the ancestral role of these genes in tumor suppression. Because planarians have already been shown to be amenable to the

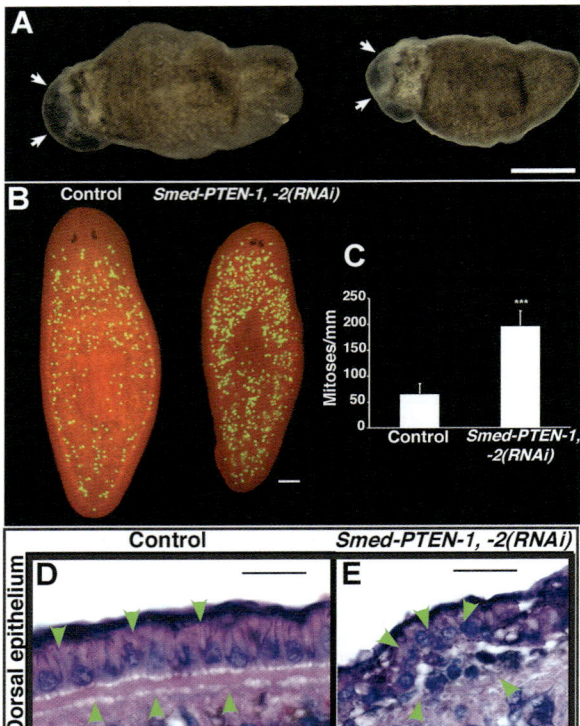

Figure 4. Tumor suppressor function of two PTEN homologs in planarians. (*A*) Abrogation of gene function via RNAi of the two planarian orthologs of PTEN (*Smed-PTEN-1* and *-2*) results in abnormal outgrowths (*white arrows*). Bar, 500 μm. (*B*) Whole-mount immunostaining using α-phosphorylated histone H3 (H3P) antibody in normal and *Smed-PTEN-1(RNAi);Smed-PTEN-2(RNAi)* worms indicates a noticeable increase of cellular proliferation in RNAi-treated animals. (*C*) H3P-signal quantitation from both control and simultaneous *Smed-PTEN-1(RNAi); Smed-PTEN-2(RNAi)* worms. A significant increase (*p* < 0.0001) in mitotic activity was observed in RNAi-treated animals. (*D*) Paraffin-tissue sections of normal animals stained with hematoxylin and eosin. (*Green arrowheads*) Columnar epithelium and basement membrane. (*E*) Hematoxylin and eosin stains of paraffin sections from *Smed-PTEN-1(RNAi);Smed-PTEN-2(RNAi)*. (*Green arrowheads*) Multiple cell in the epithelium and the disruption of basement membrane. (Data from Oviedo et al. 2008.)

study of TSGs in adult stem cell lineages and have been shown to have tumor suppressor function for *PTEN*, it will be interesting to test the function of *p53, Rb,* and other "vertebrate-specific" TSGs in these animals. Should the planarian homologs of these genes show tumor suppressor function, we can conclude that their ancestral roles were tumor suppression, which was most likely lost in flies and *C. elegans*. Finally, because we have shown that these genes evolved before multicellularity, testing gene function of these homologs in a choanoflagellate will also give perspective on their ancestral functions.

Contrary to the testing of p53/PTEN/Rb in invertebrates, fly nTSGs need to be further studied for tumor suppressor function in vertebrate ASCs. Only the tissue-specific deletion of the *lgl* ortholog has been performed and with mixed results (Klezovitch et al. 2004; Vasioukhin 2006). Interesting data have come from a few studies showing that the two mouse Dlgs can rescue the fly *dlg* mutant (Thomas et al. 1997; Humbert et al. 2003). Therefore, from the fly's perspective, the mammalian orthologs are tumor suppressors because they can suppress the tumor formation of a mutant fly. Additionally, planaria, *Monosiga*, and vertebrates have multiple family members of these proteins, and thus their tumor suppressor function may be hidden until combinations of mutants (or RNAi) are combined. To rigorously test whether the fly nTSGs have tumor suppression in vertebrates, multiple conditional alleles need to be deleted in adult animals and stem cell lineages followed—not an easy task in vertebrates. Because we have shown that planarians have a complement similar to that of vertebrates for these three fly nTSGs, testing their function in planarian stem cells is straightforward and would help to resolve whether these genes have ancestral function as tumor suppressors. We also showed that *Monosiga* has homologs to two of the three fly nTSGs examined in this chapter that can now be tested for tumor suppressor function. Should planarians or *Monosiga* show a tumor suppressor function for the fly nTSG homologs, this would lend support to the hypothesis that these genes had ancestral tumor suppressor function that was lost in vertebrates. In conclusion, the study of tumor suppression mechanisms in emerging model systems stands to broadly inform the function of TSGs not only in the ontogeny of animal cancers, but also in normal biological functions such as stem cell regulation, regeneration, and tissue homeostasis.

ACKNOWLEDGMENTS

B.J.P. was funded by Damon Runyon Postdoctoral Fellowship 1888-05. A.S.A. is a Howard Hughes Medical Institute Investigator.

REFERENCES

Ajioka, I., Martins, R.A., Bayazitov, I.T., Donovan, S., Johnson, D.A., Frase, S., Cicero, S.A., Boyd, K., Zakharenko, S.S., and Dyer, M.A. 2007. Differentiated horizontal interneurons clonally expand to form metastatic retinoblastoma in mice. *Cell* **131:** 378–390.

Aoki, K. and Taketo, M.M. 2007. Adenomatous polyposis coli (*APC*): A multi-functional tumor suppressor gene. *J. Cell Sci.* **120:** 3327–3335.

Asano, T., Yao, Y., Zhu, J., Li, D., Abbruzzese, J.L., and Reddy, S.A. 2004. The PI 3-kinase/Akt signaling pathway is activated due to aberrant Pten expression and targets transcription factors NF-κB and c-Myc in pancreatic cancer cells. *Oncogene* **23:** 8571–8580.

Brodsky, M.H., Nordstrom, W., Tsang, G., Kwan, E., Rubin, G.M., and Abrams, J.M. 2000. *Drosophila* p53 binds a damage response element at the *reaper* locus. *Cell* **101:** 103–113.

Brusca, R. and Brusca, G. 2003. *Invertebrates*. Sinauer, Sunderland, Massachusetts.

Carroll, S.L. and Stonecypher, M.S. 2005. Tumor suppressor mutations and growth factor signaling in the pathogenesis of NF1-associated peripheral nerve sheath tumors. II. The role of dysregulated growth factor signaling. *J. Neuropathol. Exp. Neurol.* **64:** 1–9.

Coqueret, O. 2003. New roles for p21 and p27 cell-cycle inhibitors: A function for each cell compartment? *Trends Cell Biol.* **13:** 65–70.

Cuddihy, A.R. and Bristow, R.G. 2004. The p53 protein family and radiation sensitivity: Yes or no? *Cancer Metastasis Rev.* **23:** 237–257.

de Nooij, J.C., Letendre, M.A., and Hariharan, I.K. 1996. A cyclin-dependent kinase inhibitor, Dacapo, is necessary for timely exit from the cell cycle during *Drosophila* embryogenesis. *Cell* **87:** 1237–1247.

Derry, W.B., Putzke, A.P., and Rothman, J.H. 2001. *Caenorhabditis elegans* p53: Role in apoptosis, meiosis, and stress resistance. *Science* **294:** 591–595.

Du, W., Vidal, M., Xie, J.-E., and Dyson, N. 1996. *RBF*, a novel RB-related gene that regulates E2F activity and interacts with *cyclin E* in *Drosophila. Genes Dev.* **10:** 1206–1218.

Eisenhoffer, G.T., Kang, H., and Sánchez Alvarado, A. 2008. Molecular analysis of stem cells and their descendants during cell turnover and regeneration in the planarian *Schmidtea mediterranea. Cell Stem Cell* **3:** 327–339.

Evers, B. and Jonkers, J. 2006. Mouse models of BRCA1 and BRCA2 deficiency: Past lessons, current understanding and future prospects. *Oncogene* **25:** 5885–5897.

Fei, P., Bernhard, E.J., and El-Deiry, W.S. 2002. Tissue-specific induction of p53 targets in vivo. *Cancer Res.* **62:** 7316–7327.

Gateff, E. 1994. Tumor suppressor and overgrowth suppressor genes of *Drosophila melanogaster:* Developmental aspects. *Int. J. Dev. Biol.* **38:** 565–590.

Gateff, E. and Mechler, B.M. 1989. Tumor-suppressor genes of *Drosophila melanogaster. Crit. Rev. Oncog.* **1:** 221–245.

Gateff, E. and Schneiderman, H.A. 1974. Developmental capacities of benign and malignant neoplasms of *Drosophila. Wilhelm Roux's Arch. Entwicklungsmech. Org.* **176:** 23–65.

Gatza, C., Moore, L., Dumble, M., and Donehower, L.A. 2007. Tumor suppressor dosage regulates stem cell dynamics during aging. *Cell Cycle* **6:** 52–55.

Hanahan, D. and Weinberg, R.A. 2000. The hallmarks of cancer. *Cell* **100:** 57–70.

Harris, C.C. 1996. p53 tumor suppressor gene: From the basic research laboratory to the clinic—An abridged historical perspective. *Carcinogenesis* **17:** 1187–1198.

Heemels, M.-T. 2000. Nature insight: Apoptosis. *Nature* **407:** 769.

Huang, H., Potter, C.J., Tao, W., Li, D.M., Brogioli, W., Hafen, E., Sun, H., and Xu, T. 1999. PTEN affects cell size, cell proliferation and apoptosis during *Drosophila* eye development. *Development* **126:** 5365–5372.

Humbert, P., Russell, S., and Richardson, H. 2003. Dlg, Scribble and Lgl in cell polarity, cell proliferation and cancer. *BioEssays* **25:** 542–553.

Irwin, M.S. 2004. Family feud in chemosensitvity: p73 and mutant p53. *Cell Cycle* **3:** 319–323.

King, N., Westbrook, M.J., Young, S.L., Kuo, A., Abedin, M., Chapman, J., Fairclough, S., Hellsten, U., Isogai, Y., Letunic, I., et al. 2008. The genome of the choanoflagellate *Monosiga brevicollis* and the origin of metazoans. *Nature* **451:** 783–788.

Kirkegaard, T., Witton, C.J., McGlynn, L.M., Tovey, S.M., Dunne, B., Lyon, A., and Bartlett, J.M. 2005. AKT activation predicts outcome in breast cancer patients treated with tamoxifen. *J. Pathol.* **207:** 139–146.

Klezovitch, O., Fernandez, T.E., Tapscott, S.J., and Vasioukhin, V. 2004. Loss of cell polarity causes severe brain dysplasia in *Lgl1* knockout mice. *Genes Dev.* **18:** 559–571.

Li, J., Yen, C., Liaw, D., Podsypanina, K., Bose, S., Wang, S.I., Puc, J., Miliaresis, C., Rodgers, L., McCombie, R., et al. 1997. *PTEN,* a putative protein tyrosine phosphatase gene mutated in human brain, breast, and prostate cancer. *Science* **275:** 1943–1947.

Li, Y. and Prives, C. 2007. Are interactions with p63 and p73 involved in mutant p53 gain of oncogenic function? *Oncogene* **26:** 2220–2225.

Lowe, S.W. and Sherr, C.J. 2003. Tumor suppression by *Ink4a-Arf:* Progress and puzzles. *Curr. Opin. Genet. Dev.* **13:** 77–83.

Lu, W.J. and Abrams, J.M. 2006. Lessons from p53 in non-mammalian models. *Cell Death Differ.* **13:** 909–912.

Marx, J. 2007. Molecular biology. Cancer's perpetual source? *Science* **317:** 1029–1031.

Menut, L., Vaccari, T., Dionne, H., Hill, J., Wu, G., and Bilder, D. 2007. A mosaic genetic screen for *Drosophila* neoplastic tumor suppressor genes based on defective pupation. *Genetics* **177:** 1667–1677.

Micchelli, C.A. and Perrimon, N. 2006. Evidence that stem cells reside in the adult *Drosophila* midgut epithelium. *Nature* **439:** 475–479.

Moll, U.M. and Slade, N. 2004. p63 and p73: Roles in development and tumor formation. *Mol. Cancer Res.* **2:** 371–386.

Molofsky, A.V., Pardal, R., Iwashita, T., Park, I.K., Clarke, M.F., and Morrison S.J. 2003. *Bmi-1* dependence distinguishes neural stem cell self-renewal from progenitor proliferation. *Nature* **425:** 962–967.

Molofsky, A.V., Slutsky, S.G., Joseph, N.M., He, S., Pardal, R., Krishnamurthy, J., Sharpless N.E., and Morrison, S.J. 2006. Increasing *p16^{INK4a}* expression decreases forebrain progenitors and neurogenesis during ageing. *Nature* **443:** 448–452.

Moon, N.S., Di Stefano, L., Morris, E.J., Patel, R., White, K., and Dyson, N.J. 2008. E2F and p53 induce apoptosis independently during *Drosophila* development but intersect in the context of DNA damage. *PLoS Genet.* **4:** e1000153.

Murray-Zmijewski, F., Lane, D.P., and Bourdon, J.C. 2006. p53/p63/p73 isoforms: An orchestra of isoforms to harmonise cell differentiation and response to stress. *Cell Death Differ.* **13:** 962–972.

Nakahara, Y., Nagai, H., Kinoshita, T., Uchida, T., Hatano, S., Murate, T., and Saito, H. 1998. Mutational analysis of the *PTEN/MMAC1* gene in non-Hodgkin's lymphoma. *Leukemia* **12:** 1277–1280.

Nedelcu, A.M. and Tan, C. 2007. Early diversification and complex evolutionary history of the p53 tumor suppressor gene family. *Dev. Genes Evol.* **217:** 801–806.

Newmark, P.A. and Sánchez Alvarado, A. 2000. Bromodeoxyuridine specifically labels the regenerative stem cells of planarians. *Dev. Biol.* **220:** 142–153.

Newmark, P.A. and Sánchez Alvarado, A. 2002. Not your father's planarian: A classic model enters the era of functional genomics. *Nat. Rev. Genet.* **3:** 210–219.

Ollmann, M., Young, L.M., Di Como, C.J., Karim, F., Belvin, M., Robertson, S., Whittaker K., Demsky, M., Fisher, W.W., Buchman, A., et al. 2000. *Drosophila* p53 is a structural and functional homolog of the tumor suppressor p53. *Cell* **101:** 91–101.

Oviedo, N.J., Pearson, B.J., Levin, M., and Sánchez Alvarado, A. 2008. Planarian PTEN homologs regulate stem cells and regeneration through TOR signalling. *Dis. Models Mech.* **1:** 131–143.

Pagliarini, R.A. and Xu, T. 2003. A genetic screen in *Drosophila* for metastatic behavior. *Science* **302:** 1227–1231.

Reddien, P.W. and Sánchez Alvarado, A. 2004. Fundamentals of planarian regeneration. *Annu. Rev. Cell Dev. Biol.* **20:** 725–757.

Rocco, J.W., Leong, C.O., Kuperwasser, N., DeYoung, M.P., and Ellisen, L.W. 2006. p63 mediates survival in squamous cell carcinoma by suppression of p73-dependent apoptosis. *Cancer Cell* **9:** 45–56.

Rong, Y.S., Titen, S.W., Xie, H.B., Golic, M.M., Bastiani, M., Bandyopadhyay, P., Olivera, B.M., Brodsky, M., Rubin, G.M., and Golic, K.G. 2002. Targeted mutagenesis by homologous recombination in *D. melanogaster*. *Genes Dev.* **16:** 1568–1581.

Rossi, D.J. and Weissman, I.L. 2006. Pten, tumorigenesis, and stem cell self-renewal. *Cell* **125:** 229–231.

Sánchez Alvarado, A. and Tsonis, P.A. 2006. Bridging the regeneration gap: Genetic insights from diverse animal models. *Nat. Rev. Genet.* **7:** 873–884.

Senoo, M., Pinto, F., Crum, C.P., and McKeon, F. 2007. p63 is essential for the proliferative potential of stem cells in stratified epithelia. *Cell* **129:** 523–536.

Sherr, C.J. and McCormick, F. 2002. The RB and p53 pathways in cancer. *Cancer Cell* **2:** 103–112.

Stevaux, O., Dimova, D., Frolov, M.V., Taylor-Harding, B., Morris, E., and Dyson, N. 2002. Distinct mechanisms of E2F regulation by *Drosophila* RBF1 and RBF2. *EMBO J.* **21:** 4927–4937.

Sulis, M.L. and Parsons, R. 2003. PTEN: From pathology to biology. *Trends Cell Biol.* **13:** 478–483.

Terzibasi, E., Valenzano, D.R., and Cellerino, A. 2007. The short-lived fish *Nothobranchius furzeri* as a new model system for aging studies. *Exp. Gerontol.* **42:** 81–89.

The, I., Hannigan, G.E., Cowley, G.S., Reginald, S., Zhong, Y., Gusella, J.F., Hariharan, I.K., and Bernards, A. 1997. Rescue of a *Drosophila NF1* mutant phenotype by protein kinase A. *Science* **276:** 791–794.

Thomas, U., Phannavong, B., Müller, B., Garner, C.C., and Gundelfinger, E.D. 1997. Functional expression of rat synapse-associated proteins SAP97 and SAP102 in *Drosophila dlg-1* mutants: Effects on tumor suppression and synaptic bouton structure. *Mech. Dev.* **62:** 161–174.

van den Heuvel, S. and Dyson, N.J. 2008. Conserved functions of the pRB and E2F families. *Nat. Rev. Mol. Cell Biol.* **9:** 713–724.

Vasioukhin, V. 2006. Lethal giant puzzle of Lgl. *Dev. Neurosci.* **28:** 13–24.

Wang, W., Warren, M., and Bradley, A. 2007. Induced mitotic recombination of p53 in vivo. *Proc. Natl. Acad. Sci.* **104:** 4501–4505.

Weinberg, R.A. 1995. The retinoblastoma protein and cell cycle control. *Cell* **81:** 323–330.

Wells, B.S., Yoshida, E., and Johnston, L.A. 2006. Compensatory proliferation in *Drosophila* imaginal discs requires Dronc-dependent p53 activity. *Curr. Biol.* **16:** 1606–1615.

Wodarz, A. and Nathke, I. 2007. Cell polarity in development and cancer. *Nat. Cell Biol.* **9:** 1016–1024.

Woodhouse, E., Hersperger, E., and Shearn, A. 1998. Growth, metastasis, and invasiveness of *Drosophila* tumors caused by mutations in specific tumor suppressor genes. *Dev. Genes Evol.* **207:** 542–550.

Xu, T., Wang, W., Zhang, S., Stewart, R.A., and Yu, W. 1995. Identifying tumor suppressors in genetic mosaics: The *Drosophila lats* gene encodes a putative protein kinase. *Development* **121:** 1053–1063.

Xue, W., Krasnitz, A., Lucito, R., Sordella, R., Van Aelst, L., Cordon-Cardo, C., Singer, S., Kuehnel, F., Wigler, M., Powers, S., Zender, L., and Lowe, S.W. 2008. *DLC1* is a chromosome 8p tumor suppressor whose loss promotes hepatocellular carcinoma. *Genes Dev.* **22:** 1439–1444.

Yilmaz, O.H., Valdez, R., Theisen, B.K., Guo, W., Ferguson, D.O., Wu, H., and Morrison, S.J. 2006. *Pten* dependence distinguishes haematopoietic stem cells from leukaemia-initiating cells. *Nature* **441:** 475–482.

Germ Cell Specification and Regeneration in Planarians

P.A. Newmark, Y. Wang, and T. Chong

Howard Hughes Medical Institute, Department of Cell and Developmental Biology,
University of Illinois at Urbana-Champaign, Urbana, Illinois 61801

In metazoans, two apparently distinct mechanisms specify germ cell fate: Determinate specification (observed in animals including *Drosophila, Caenorhabditis elegans*, zebra fish, and *Xenopus*) uses cytoplasmic factors localized to specific regions of the egg, whereas epigenetic specification (observed in many basal metazoans, urodeles, and mammals) involves inductive interactions between cells. Much of our understanding of germ cell specification has emerged from studies of model organisms displaying determinate specification. In contrast, our understanding of epigenetic/inductive specification is less advanced and would benefit from studies of additional organisms. Freshwater planarians—widely known for their remarkable powers of regeneration—are well suited for studying the mechanisms by which germ cells can be induced. Classic experiments showed that planarians can regenerate germ cells from body fragments entirely lacking reproductive structures, suggesting that planarian germ cells could be specified by inductive signals. Furthermore, the availability of the genome sequence of the planarian *Schmidtea mediterranea*, coupled with the animal's susceptibility to systemic RNA interference (RNAi), facilitates functional genomic analyses of germ cell development and regeneration. Here, we describe recent progress in studies of planarian germ cells and frame some of the critical unresolved questions for future work.

Since the late 19th century, when Weismann proposed that an immortal germ line was propagated from generation to generation, producing a mortal soma at each generation, developmental biologists have been fascinated by the question of how the germ cell lineage is established during embryogenesis (Weismann 1893; McLaren 1981; Extavour 2007). In addition to serving as a link between generations, the germ cells represent an intriguing example of cellular differentiation, in which highly differentiated cell types maintain their totipotency and are capable of reproducing themselves indefinitely (Rando 2006; Seydoux and Braun 2006; Cinalli et al. 2008).

Two apparently distinct modes of germ cell specification are typically observed in animals. Determinate specification (or preformation) refers to germ cell specification by localized maternal determinants early in embryogenesis (Nieuwkoop and Sutasurya 1981; Extavour and Akam 2003; Seydoux and Braun 2006; Extavour 2007). Determinate specification is observed in the most commonly studied invertebrates, *Drosophila* and *C. elegans* (for review, see Seydoux and Schedl 2001; Santos and Lehmann 2004), as well as in some vertebrates, including anuran amphibians (Houston and King 2000) and zebra fish (Raz 2003). In these animals, specialized cytoplasm (the germ plasm) is associated with germ cell formation (Eddy 1975). This cytoplasm contains granular inclusions consisting of RNA and protein; these granules are known as polar granules in flies, P granules in nematodes, and germinal granules in frogs, or more generally, as germ granules (Seydoux and Braun 2006).

In contrast to determinate specification, many organisms use inductive interactions to specify their germ cell lineage relatively late in embryogenesis. This epigenetic specification is observed in mammals as well as in a wide range of basal invertebrates and other vertebrate species (Nieuwkoop and Sutasurya 1979, 1981; Extavour and Akam 2003; Hayashi et al. 2007). Phylogenetic surveys

suggest that this mode is more widespread throughout the metazoa and is likely to be ancestral (Extavour and Akam 2003; Johnson et al. 2003; Extavour 2007). Although this usage of the term "epigenetic" accurately reflects the original meaning of the term "epigenesis," this word is now more commonly used to signify, "...a change in the state of expression of a gene that does not involve a mutation, but that is nevertheless inherited in the absence of the signal (or event) that initiated the change" (Ptashne 2007). To minimize confusion and to avoid the unintended mechanistic implications of the term "epigenetic specification," we refer to this mode of specification as "inductive specification" (Seydoux and Braun 2006).

Much of the progress made in understanding the mechanisms of germ cell determination has been driven by genetic analyses in *Drosophila* and *C. elegans*. Thus, numerous components of the *Drosophila* germ plasm have been identified, along with factors that are required for their localization to the posterior pole of the embryo (Starz-Gaiano and Lehmann 2001). Likewise, genetic analyses in *C. elegans* also have identified components of the germ plasm and revealed the importance of transcriptional repression mechanisms during the early stages of germ cell formation (Seydoux and Strome 1999). Many genes identified from these model invertebrates have been conserved evolutionarily. For example, homologs of *Drosophila vasa*, a gene encoding a DEAD-box RNA helicase that is a component of the polar granules (Hay et al. 1988; Lasko and Ashburner 1988), are expressed in germ cells in *Xenopus* (Komiya et al. 1994; Ikenishi and Tanaka 1997), zebra fish (Olsen et al. 1997; Yoon et al. 1997), and chick (Tsunekawa et al. 2000), as well as *C. elegans* (Gruidl et al. 1996) and numerous other organisms (Shibata et al. 1999; Mochizuki et al. 2001; Extavour and Akam 2003; Extavour et al. 2005). Similarly, homologs of *Drosphila nanos*, which encodes a CCHC zinc finger RNA-binding protein required for abdominal segmentation as well as germ cell differentiation and maintenance (Lehmann and Nüsslein-Volhard 1991; Wang and Lehmann 1991; Kobayashi et al. 1996; Forbes and Lehmann 1998; Deshpande et al. 1999; Hayashi et al. 2004), also serve as markers of germ cells throughout the metazoa (Extavour and Akam 2003; Extavour et al. 2005; Juliano et al. 2006).

Although mammalian germ cell formation requires inductive signaling rather than maternally supplied determinants (Hayashi et al. 2007), homologs of these and several other *Drosophila* genes also have roles in mammalian germ cells. In the mouse, *nanos3* is expressed in primordial germ cells soon after they are formed in both sexes, whereas *nanos2* is expressed in male primordial germ cells after they have migrated to the genital ridge (Tsuda et al. 2003). Knockout mutations of *nanos3* result in germ cell loss in both sexes and *nanos2* knockouts result in loss of spermatogonia (Tsuda et al. 2003; Suzuki et al. 2007). Like *nanos2*, the mouse *vasa* homolog is not expressed in primordial germ cells until they arrive at the genital ridge (Fujiwara et al. 1994; Toyooka et al. 2000) and knockouts of this gene do not result in defects in germ cell specification, but they cause male-specific defects in germ cell proliferation and/or survival (Tanaka et al. 2000). Additional mouse homologs of *Drosophila* polar granule components have been shown to be associated with chromatoid bodies, the germ granules of mammalian spermatocytes. These conserved molecules include tudor-domain proteins (Chuma et al. 2006; Hosokawa et al. 2007), components of the microRNA pathway, such as the products of *piwi* and *argonaute* genes, and various components of processing (P) bodies, sites at which untranslated mRNAs accumulate in somatic cells (Kotaja et al. 2006).

Despite the identification of many conserved components and the realization that general regulatory mechanisms are shared between inductive and determinate specification (e.g., transcriptional repression of somatic gene expression, chromatin remodeling to a generally repressive state, and an emphasis on posttranscriptional control mechanisms [Seydoux and Braun 2006; Cinalli et al. 2008]), several critical questions remain unanswered. What are the mechanisms that activate expression of germ cell–specific genes in the earliest stages of primordial germ cell formation? Are there conserved inductive signals that specify germ cell fate? Because mechanistic studies of inductive specification have been limited to the mouse, virtually nothing is known about the identity of germ cell–inducing signals in other organisms. In the mouse, bone morphogenetic protein (BMP) signaling from the extraembryonic ectoderm is required for primordial germ cell formation (Lawson et al. 1999; Ying et al. 2000). However, this signaling may be acting indirectly through the visceral endoderm, rather than directly upon cells of the proximal epiblast (de Sousa Lopes et al. 2004); thus, the nature of the inductive signal(s) is still not entirely clear. How is totipotency established and maintained in germ cells? The demonstration that germ-line stem cells from mouse neonatal testes (Kanatsu-Shinohara et al. 2004) and spermatogonial stem cells from adult testes (Guan et al. 2006; Seandel et al. 2007) can generate pluripotent cells with developmental potential similar to that of embryonic stem (ES) cells hints at the

therapeutic applications that could result from answering the above questions.

To answer such questions and expand our understanding of the inductive specification of germ cell fate, we are using the freshwater planarian as a model organism. Planarians are well known for regenerative abilities that enable them to produce a complete organism from a tiny fragment of the body. The fact that their germ cell lineage can also be regenerated is less well appreciated and forms the basis for this work. Before discussing the current state of our knowledge and outlining important areas for future research, it is first necessary to review briefly the salient aspects of planarian biology, as well as the recent technical advances that pave the way for a mechanistic analysis of germ cell development in these organisms.

INTRODUCTION TO PLANARIAN BIOLOGY

The remarkable regenerative abilities of planarians have led generations of biologists to study these organisms (Newmark and Sánchez Alvarado 2002; Reddien and Sánchez Alvarado 2004; Sánchez Alvarado 2006). Planarians are free-living, freshwater members of the phylum Platyhelminthes (the flatworms). Flatworms are among the simplest animals with three tissue layers (triploblasts) that display bilateral symmetry, an anterior concentration of neural tissue (cephalization), and the organization of specialized tissues into organs. Flatworms lack respiratory and circulatory systems and instead rely on diffusion to obtain oxygen. The planarian nervous system consists of bilobed cerebral ganglia at the anterior end and two longitudinal nerve cords that underlie the ventral body wall musculature and run the length of the animal (Agata et al. 1998; Okamoto et al. 2005; Cebria 2008). Sensory structures (photoreceptors, chemoreceptors) located at the anterior of the animal send projections to the cephalic ganglia, which then process external signals and direct the appropriate behavioral responses (MacRae 1967; Inoue et al. 2004).

Freshwater planarians reproduce either asexually, by transverse fission, or sexually, as cross-fertilizing hermaphrodites (Hyman 1951). Some planarians use exclusively one mode of reproduction, and others may alternate among them depending on the season. In the asexual mode of reproduction, the worm splits itself transversely into two fragments, each of which regenerates the missing tissue, thereby producing two planarians. This behavior can be regulated both by the physiological state of the animal and by external stimuli (e.g., population density and darkness) (Pigon et al. 1974; Morita and Best 1984). In the sexual mode, hermaphrodites cross-fertilize and then lay egg capsules that contain several embryos. In hermaphrodites, numerous testes are distributed dorsolaterally (Fig. 1a) and the ovaries are situated at the posterior region of the cephalic ganglia (Fig. 1b). Individual lobes of the testes (Fig.1c) possess an outer layer of spermatogonia that undergo three rounds of division with incomplete cytokinesis, generating eight primary spermatocytes that progress through meiosis to generate 32 spermatids (Franquinet and Lender 1973). Spermiogenesis packages the DNA in elongated sperm (~100 µm in length) with two 9 + 1 flagella at the anterior

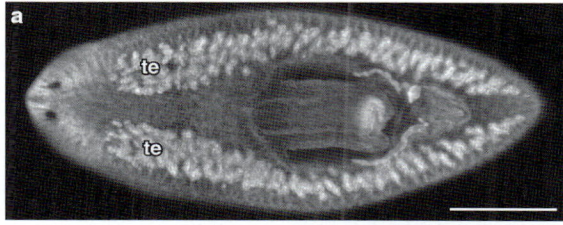

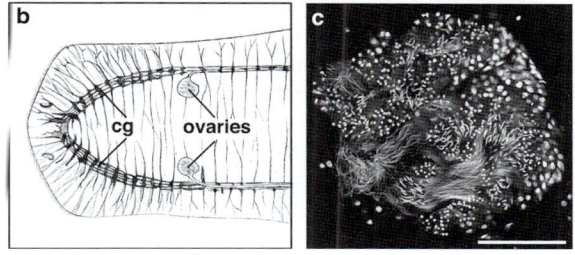

Figure 1. Reproductive organs in sexually reproducing planarians. (*a*) DAPI (4′-diamidino-2-phenylindole) labeling to visualize dorsolateral testes (te) clusters in a whole-mount specimen of *S. mediterranea*. Bar, 1 mm. (*b*) Position of the ovaries relative to the cephalic ganglia (cg) in *S. polychroa*. (*c*) Confocal projection of individual testis lobe from *S. mediterranea* labeled with Hoechst. Different stages of spermatogenesis and elongated spermatozoa are easily observed. Bar, 100 μm. (*b*, Adapted from Iijima [1884].)

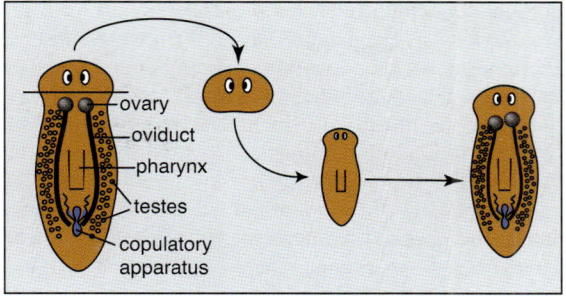

Figure 2. Regeneration of the planarian reproductive system from a fragment lacking all reproductive tissues. Amputation anterior to all of the reproductive structures produces a head fragment devoid of reproductive organs. This fragment will regenerate a complete animal; when it reaches the appropriate size, it will regenerate the gonads and the copulatory aparatus. (Adapted from an experiment described by Morgan [1902].)

end (Silveira and Porter 1964). Ciliated oviducts run along the nerve cords to transport fertilized eggs to the region posterior to the pharynx, where the copulatory apparatus and genital pore are located. Sperm ducts also project along the nerve cords, connecting individual testis lobes to the seminal vesicles and copulatory apparatus. In sexually reproducing planarians, the germ cell lineage does not appear to be segregated during embryogenesis; rather, when the planarian attains the appropriate size, gonads and the copulatory apparatus develop in the appropriate regions of the worm (Curtis 1902).

The developmental plasticity displayed by planarians is based on a population of stem cells that is maintained in the animal throughout adult life. These stem cells are referred to as neoblasts and they are the only proliferating somatic cells in the worm (Baguñà 1976; Newmark and Sánchez Alvarado 2000; Salvetti et al. 2000; Orii et al. 2005). In intact planarians, neoblasts are scattered throughout the parenchyma (mesenchyme) and their division progeny generate replacements for cells lost during the course of physiological cell turnover. When a planarian is transected, the neoblasts are stimulated to proliferate; as the neoblasts migrate toward the wound epithelium, they give rise to the regeneration blastema, the structure in which the missing parts will be regenerated (Agata and Watanabe 1999; Newmark and Sánchez Alvarado 2002; Reddien and Sánchez Alvarado 2004; Sánchez Alvarado 2006).

The regenerative abilities of planarians are not limited to their somatic tissues: They are also capable of regenerating their germ cells de novo. T.H. Morgan (1902) showed that a planarian head fragment, completely devoid of any reproductive structures, could regenerate functional gonads from the remaining somatic tissue (Fig. 2). Thus,

Morgan concluded that the germ cell lineage could be derived from somatic cells in these animals. During the degrowth (shrinkage) that is the planarian's response to starvation, the gonads and copulatory apparatus are resorbed (Schultz 1904; Berninger 1911), only to be regenerated when the animal reaches the appropriate size after feeding is resumed. Treatment with sublethal doses of γ-irradiation also leads to regression of the testes, followed by their regeneration (Fedecka-Bruner 1965, 1967). Following amputation of the head of sexually mature animals, the testes are resorbed and are only reformed after regeneration of the cephalic ganglia is complete (Ghirardelli 1965). Thus, inductive influences from the brain appear to be important for sexual differentiation.

THE PLANARIAN *SCHMIDTEA MEDITERRANEA*: AN EMERGING MODEL ORGANISM

Of the hundreds of species of free-living planarians, *S. mediterranea* has emerged as the model of choice for functional genomic analyses of regeneration (Newmark and Sánchez Alvarado 2002; Sánchez Alvarado 2006). It is a stable diploid, with a genome size of 8×10^8 bp (data from S. Johnston, Texas A&M University, and the Washington University Genome Sequencing Center [WUGSC]). In contrast, the common North American species often found in biology classrooms are mixoploids (mosaics of diploid and triploid cells) and have genomes that are about twice as large as that of *S. mediterranea*. Two strains of this species are found in nature: a sexual strain that reproduces as cross-fertilizing hermaphrodites and an asexual strain that reproduces via transverse fission. These strains can be distinguished genetically by a chromosomal translocation present in the asexual strain (Benazzi et al. 1975). Clonal lines of both strains have been used to produce expressed sequence tags (ESTs) from various tissues and regenerative stages of asexual animals (Sánchez Alvarado et al. 2002) and from developing juveniles and reproductively mature sexual animals (Zayas et al. 2005). The genome of the sexual strain has been sequenced to 11.6x coverage by whole-genome shotgun sequencing and a genome assembly is now available (WUGSC). The genome has been annotated using an automated annotation pipeline (Cantarel et al. 2008); the annotation data are available via the *Schmidtea mediterranea* genome database (SmedGD), maintained by the Sánchez Alvarado laboratory at the University of Utah (Robb et al. 2008).

The application of the tools of cellular and molecular biology has opened up a new era in studies of freshwater planarians (Newmark and Sánchez Alvarado 2002). It is now possible to identify and functionally characterize the genes required for regulating developmental processes in these animals. Gene expression in planarians can be inhibited specifically by RNA interference (RNAi), by either microinjection (Sánchez Alvarado and Newmark 1999) or feeding bacterially expressed double-stranded RNA (dsRNA) (Newmark et al. 2003; Reddien et al. 2005a). Because the neoblasts are the only proliferating somatic cells in the animal, they can be labeled specifically with bromodeoxyuridine (Newmark and Sánchez Alvarado 2000). Fluorescence-activated cell sorting (FACS) methodologies have been developed by Kiyokazu Agata's laboratory to isolate neoblasts and other planarian cell types (Asami et al. 2002; Hayashi et al. 2006). Automated whole-mount in situ hybridization techniques have been developed for high-throughput localization of gene expression patterns with single-cell resolution (Sánchez Alvarado et al. 2002; Zayas et al. 2005).

The functional genomic resources now at our disposal provide us with tremendous tools for discovering the genetic programs that underlie regeneration in planarians. The utility of these tools has already been borne out by the identification of several genes that are conserved between planarians and mammals and that are required for proper regulation of the planarian stem cell population (Cebria et al. 2002; Reddien et al. 2005a,b; Salvetti et al. 2005; Guo et al. 2006; Palakodeti et al. 2008).

NANOS FUNCTION IS REQUIRED FOR THE DEVELOPMENT AND REGENERATION OF GERM CELLS IN PLANARIANS

As described above, classic experiments suggested that planarians can regenerate their germ cell lineage; the neoblasts appear to be the source of newly regenerated germ cells. To understand the switch from somatic stem cell to germ cell, we characterized an *S. mediterranea* homolog of *nanos*, a gene required for germ cell development in widely divergent organisms. Homologs of *nanos* are expressed in the germ cell lineage of all animals studied to date; *nanos* expression is used routinely to identify early germ cells (Extavour and Akam 2003; Extavour et al. 2005; Juliano et al. 2006). In sexual *S. mediterranea*, *Smed-nanos* mRNA is detected in developing, regenerating, and mature ovaries and testes (Handberg-Thorsager and Saló 2007; Wang et al. 2007). A similar pattern was also reported for a *nanos* homolog from another planarian species, *Dugesia japonica* (Sato et al. 2006). However, consistent with an inductive origin of the germ cells, *nanos* RNA is not detected in the vast majority of newly hatched

planarians or in small tissue fragments that will ultimately regenerate germ cells (Sato et al. 2006; Wang et al. 2007).

To analyze the function of *nanos* in planarian germ cells, we assayed the effects of *nanos* RNAi upon regeneration of the reproductive organs. Mature animals were fed twice with *nanos* dsRNA and then amputated posterior to the ovaries to trigger regression of the testes; regeneration of the testes and ovaries could then be monitored. Feeding of dsRNA was resumed 2 weeks after amputation to allow the animals to grow and undergo sexual maturation. Control planarians fed bacteria containing vector alone showed proper regeneration of somatic tissues and reproductive organs. Worms fed *nanos* dsRNA displayed normal regeneration of their somatic tissues; however, they did not regenerate reproductive organs as assayed by several markers of germ cells at various stages of differentiation (Wang et al. 2007).

When newly hatched planarians were reared on food containing *nanos* dsRNA, they failed to develop testes or ovaries. Similarly, when sexually mature planarians were subjected to *nanos* RNAi, the testes and ovaries degenerated over the course of several weeks, resulting in animals that lacked gonads (Fig. 3). Together, our results showed that *nanos* is required for the postembryonic development, regeneration, and maintenance of planarian germ cells.

PRESUMPTIVE GERM CELLS IN ASEXUAL PLANARIANS

Surprisingly, *nanos* RNA was also detected in the asexual strain of *S. mediterranea*; *nanos*-positive cell clusters were found dorsolaterally, in the position of testes primordia in hermaphrodites (Handberg-Thorsager and Saló 2007; Wang et al. 2007). This pattern was also described in the asexual strain of the planarian *D. japonica*; however, no functional data were reported (Sato et al. 2006). Because *nanos* is expressed in germ cells from widely divergent metazoan phyla (Extavour and Akam 2003) and the *nanos*-positive cells observed in asexual planarians are located in positions at which germ cells are first observed postembryonically in sexual planarians, we refer to these cells as presumptive germ cells. These

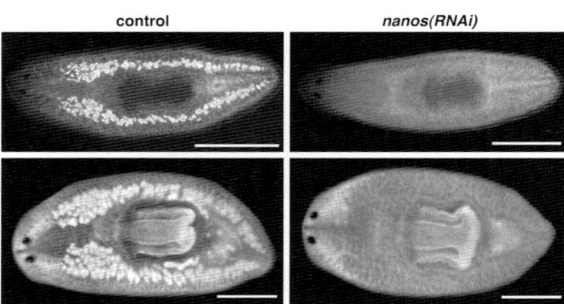

Figure 3. *nanos* function is required to maintain gonads in sexually mature planarians. (*Upper panels*) Fluorescent in situ hybridization to detect *germinal histone H4* RNA in control and *nanos* (*RNAi*) animals, 30 days after initiation of RNAi. (*Lower panels*) DAPI staining revealing testes in control animals but not in *nanos* (*RNAi*) animals. Bars, 1 mm.

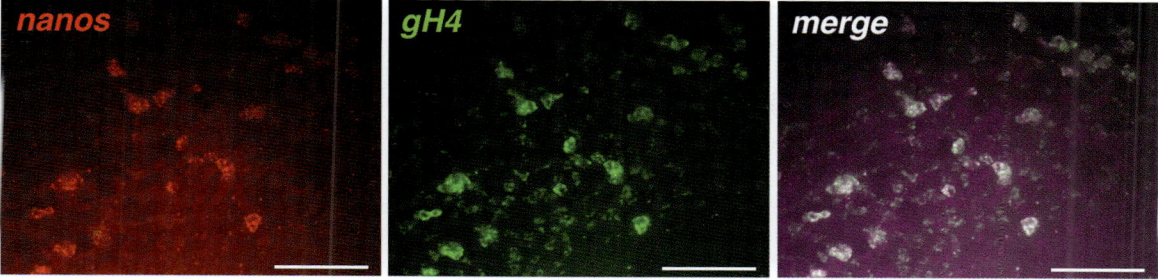

Figure 4. Coexpression of *nanos* and *germinal histone H4* (*gH4*) mRNAs in the presumptive germ cells of asexual planarians as visualized by whole-mount double fluorescent in situ hybridization. Individual fluorescent channels and the merged images are shown. Bars, 60 μm.

nanos-positive presumptive germ cells express high levels of *germinal histone H4* (*germinal H4*) (Fig. 4), a marker of both male and female germ cells in sexual planarians (Zayas et al. 2005; Wang et al. 2007). The *germinal H4* riboprobe also labels neoblasts, albeit more weakly (Fig. 4) (Wang et al. 2007), possibly due to cross-hybridization with somatic *histone H4* transcripts (Wolfe et al. 1989). In addition, these clusters express a planarian Piwi protein, Smedwi-1, which labels neoblasts as well as their committed progeny (Guo et al. 2006; Guo 2007). Expression of both *nanos* and *germinal H4* in asexual planarians is γ-radiation-sensitive, suggesting either that the presumptive germ cells are actively cycling or that they are derived from proliferating cells.

To test the role of *nanos* in asexual planarians, we performed RNAi. After *nanos* knockdown and amputation, somatic tissue regeneration proceeded normally: Head and tail regeneration were unaffected and cephalic ganglia and photoreceptors formed properly; similarly, *germinal H4* labeling of the neoblasts was unaffected. However, the dorsolateral *germinal-H4*-positive clusters were lost (Wang et al. 2007). Similarly, dorsolateral *germinal H4* clusters were lost after subjecting intact asexual planarians to *nanos* RNAi (Fig. 5). Thus, *nanos* function is required for the regeneration and maintenance of presumptive germ cells in asexual planarians. These results suggest that asexual planarians specify germ cells, but their differentiation is blocked at a step downstream from *nanos* function.

NEOBLASTS SHARE SEVERAL FEATURES WITH GERM CELLS

Transplantation experiments in which cell fractions enriched in neoblasts were introduced into lethally irradiated planarians suggested that neoblasts are capable of generating both somatic and germ cell lineages (Baguñà et al. 1989). This notion is supported by the regeneration of *nanos*-positive presumptive germ cells described above. Given the apparent ability of neoblasts to produce germ cells, it is reasonable to ask to what extent these somatic stem cells are themselves similar to germ cells. As viewed by transmission electron microscopy, the somatic neoblasts of *D. japonica* are morphologically indistinguishable from *nanos*-positive presumptive germ cells in asexual individuals (Sato et al. 2006). Ultrastructural analyses revealed that the cytoplasm of neoblasts contains

chromatoid bodies, which are electron-dense ribonucleoprotein particles that resemble the germ granules found in germ cells (Le Moigne 1967; Sauzin 1968; Morita et al. 1969; Hay and Coward 1975; Hori 1982; Auladell et al. 1993). Like germ granules, the chromatoid bodies can be found in association with mitochondria and the nuclear envelope. Chromatoid bodies diminish as neoblasts differentiate (Sauzin 1968; Morita et al. 1969; Hay and Coward 1975; Hori 1982). Furthermore, several genes with conserved roles in germ cell development are expressed in neoblasts, including homologs of *vasa*, *piwi*, *pumilio*, and *bruno* (Shibata et al. 1999; Reddien et al. 2005b; Salvetti et al. 2005; Guo et al. 2006; Rossi et al. 2006; Yoshida-Kashikawa et al. 2007; Palakodeti et al. 2008). Thus, the pluripotent neoblasts appear poised to adopt germ cell fate(s). As observed in more basal metazoans (e.g., sponges, cnidarians, and acoels), this pluripotent stem cell pool provides the source of the germ cell lineage, consistent with the notion that the germ line may have arisen initially from somatic stem cells (Extavour 2007).

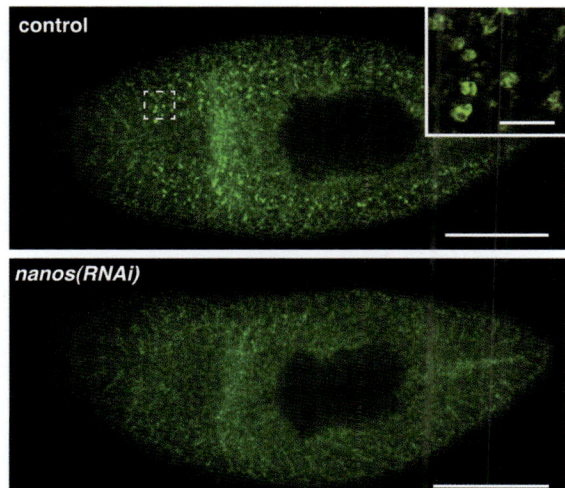

Figure 5. *nanos* function is required to maintain presumptive germ cells in asexual planarians. Fluorescent in situ hybridization to detect *germinal H4* RNA in control and *nanos* (*RNAi*) animals 35 days after initiation of RNAi. (*Inset*) Magnified image of boxed region showing *germinal-histone-H4*-positive clusters of presumptive germ cells observed in control animals but absent in *nanos* (*RNAi*) animals. Bars, 500 μm; (*Inset*) 50 μm.

PLASTICITY OF PLANARIAN GERM CELLS AND COMMITTED NEOBLASTS?

Given the relatively large numbers of cells in the planarian that can be defined morphologically as neoblasts, it is unlikely that all of these cells represent true stem cells (Baguñà et al. 1990). Rather, most of these cells may already be committed to specific fates, with only a small fraction representing true stem cells. Heterogeneity of this cell population is supported by electron microscopic analysis of fluorenscence-activated cell-sorted planarian cells (Higuchi et al. 2007).

To what extent are the committed progeny locked into their "assigned" fates? There are two lines of evidence for some degree of plasticity within the pool of neoblast progeny: One involves the germ cell lineage; the other, the somatic lineage. Experiments conducted by Gremigni and coworkers suggested that the planarian germ cell lineage was capable of contributing to the regeneration blastema and producing somatic cells (Gremigni and Puccinelli 1977; Gremigni and Miceli 1980; Gremigni et al. 1980a,b). These experiments used a mosaic strain of *Dugesia lugubris* (*Schmidtea polychroa*) in which the somatic cells are triploid. This pseudogamous strain produces triploid oocytes that are activated by haploid sperm; after activation, the sperm pronucleus is expelled from the egg. Thus, in these animals, the premeiotic female germ cells are hexaploid, whereas neoblasts that give rise to male germ cells undergo a round of chromosomal elimination, generating diploid premeiotic germ cells that produce haploid sperm (Benazzi Lentati 1970).

Using both karyological and cytophotometric analyses, Gremigni and his colleagues (1980a) reported that, depending on the level of amputation (i.e., whether the gonads had been transected before amputation), diploid and/or hexaploid cells were found to contribute to the blastema and could be observed among differentiated cell types (e.g, cells of the pharynx; Gremigni and Miceli 1980), suggesting that the germ cells could *trans*-differentiate to produce somatic cell types. Similarly, following amputation and regeneration of the ovaries, tetraploid oocytes were produced, albeit at low frequency (3.2%); uninjured, unamputated animals did not produce such tetraploid oocytes. The authors suggested that the tetraploid oocytes were derived from diploid male germ cells that had been mobilized by amputation and changed their fates to generate female germ cells (Gremigni et al. 1982).

Several recent observations suggest that this plasticity of committed germ cells is not limited to planarians. *Drosophila* spermatogonia (Brawley and Matunis 2004) and ovarian cystoblasts (Kai and Spradling 2004) are capable of reverting to germ-line stem cells after initiating differentiation and exiting their niches. Similarly, when cultured in vitro, germ-line stem cells from mouse neonatal testes (Kanatsu-Shinohara et al. 2004) and spermatogonial stem cells from adult testes (Guan et al. 2006; Seandel et al. 2007) can produce pluripotent cells similar to ES cells.

In addition to the apparent plasticity of committed germ cells, evidence for plasticity of somatic neoblast progeny comes from classic blastema explant culture experiments performed by Sengel (1960). When 3-day anterior blastemas were surgically isolated from the rest of the nonregenerating tissues and cultured in vitro, they were capable of differentiating to form pigment, muscles, and anterior structures, such as photoreceptors and cephalic ganglia. Thus, at this relatively early stage of regeneration, the cells of the anterior blastema were already specified to make head structures. When early posterior blastemas at a similar stage were cultured in vitro, they formed muscle and pigment but did not produce photoreceptors or cephalic ganglia. Intriguingly, when anterior blastemas were cocultured with posterior blastemas, they gave rise to small planarians in which the central body region (pharynx) and gut were formed. Thus, the juxtaposition of anterior and posterior blastemas led to the intercalary regeneration of central body structures. It is important to note that in planarians, proliferation does not occur in the blastema (Saló and Baguñà 1984, 1989); rather, neoblasts proliferate in the stump and postmitotic neoblast progeny migrate into the blastema. The explant experiments showed that these postmitotic blastemal cells are specified early in regeneration. However, this specification (and perhaps cell cycle exit) could be reversed by juxtaposing cells from anterior and posterior regions, suggesting that the fates of specified neoblast progeny can be altered by changes in extracellular signals. If this is the case, then it is possible that a relatively large population of neoblast progeny committed to specific developmental pathways—required for maintaining tissues during the course of cell turnover—may be able to reverse their state of commitment following amputation and generate additional pluripotent cells. As we refine the tools for labeling individual cells and following their fates in the animal, it will be worth reexploring this issue.

CONCLUSIONS

In the last decade, great strides have been made in developing freshwater planarians as models for studying stem cell biology, metazoan regeneration, and tissue homeostasis. In this brief overview, we have described many aspects of the biology of *S. mediterranea* that make it well suited for examining another fundamental problem in developmental biology: specification of the germ cell lineage. The demonstration that planarian germ cells can be regenerated de novo, combined with the functional genomic tools available for studying *S. mediterranea*, provides us with an exceptional opportunity to dissect the molecular mechanisms by which inductive signals can specify germ cell fate. Several important questions can now be addressed experimentally in this animal, including: What is the nature of the inductive signal and has it been conserved evolutionarily? What intrinsic changes in gene expression direct the transition from somatic stem cell to germ cell in response to the inductive signal(s)? Why are presumptive germ cells unable to complete their differentiation in asexual *S. mediterranea*? What mechanisms lead germ cells to follow male versus female differentiation pathways in these hermaphroditic organisms? How are physiological/metabolic signals coupled with meiotic entry, germ cell differentiation, and maintenance of the gonads? How plastic are the fates of committed neoblasts and differentiating germ cells in these animals? As the cohort of researchers interested in studying these fascinating animals continues to expand, we look forward to the answers that will emerge.

ACKNOWLEDGMENTS

We thank David Forsthoefel, Tingxia Guo, Joel Stary, and Ricardo Zayas for helpful discussions and David Forsthoefel and Ricardo Zayas for constructive suggestions on various drafts of the manuscript. Work in P.A.N.'s laboratory has been funded by grants from the National Institutes of Health, National Science Foundation, and the Damon Runyon Cancer Research Foundation. P.A.N. is an Investigator of the Howard Hughes Medical Institute.

REFERENCES

Agata, K. and Watanabe, K. 1999. Molecular and cellular aspects of planarian regeneration. *Semin. Cell Dev. Biol.* **10:** 377–383.

Agata, K., Soejima, Y., Kato, K., Kobayashi, C., Umesono, Y., and Watanabe, K. 1998. Structure of the planarian central nervous system (CNS) revealed by neuronal cell markers. *Zoolog. Sci.* **15:** 433–440.

Asami, M., Nakatsuka, T., Hayashi, T., Kou, K., Kagawa, H., and Agata, K. 2002. Cultivation and characterization of planarian neuronal cells isolated by fluorescence activated cell sorting (FACS). *Zoolog. Sci.* **19:** 1257–1265.

Auladell, C., Garcia-Valero, J., and Baguñà, J. 1993. Ultrastructural localization of RNA in the chromatoid bodies of undifferentiated cells (neoblasts) in planarians by the RNase-gold complex technique. *J. Morphol.* **216:** 319–326.

Baguñà, J. 1976. Mitosis in the intact and regenerating planarian *Dugesia mediterranea* n.sp. I. Mitotic studies during growth, feeding and starvation. *J. Exp. Zool.* **195:** 53–64.

Baguñà, J., Saló, E., and Auladell, C. 1989. Regeneration and pattern formation in planarians. III. Evidence that neoblasts are totipotent stem cells and the source of blastema cells. *Development* **107:** 77–86.

Baguñà, J., Romero, R., Saló, E., Collet, J., Auladell, C., Ribas, M., Riutort, M., Garcia-Fernàndez, J., Burgaya, F., and Bueno, D. 1990. Growth, degrowth and regeneration as developmental phenomena in adult freshwater planarians. In *Experimental embryology in aquatic plants and animals* (ed. H.-J. Marthy), pp. 129–162. Plenum, New York.

Benazzi, M., Baguñà, J., Ballester, R., Puccinelli, I., and Del Papa, R. 1975. Further contribution to the taxonomy of the "*Dugesia lugubris-polychroa* group" with description of *Dugesia mediterranea* n.sp. (Tricladida, Paludicola). *Boll. Zool.* **42:** 81–89.

Benazzi Lentati, G. 1970. Gametogenesis and egg fertilization in planarians. *Int. Rev. Cytol.* **27:** 101–179.

Berninger, J. 1911. Über die Einwirkung des Hungers auf Planarien. *Zool. Jahrb.* **30:** 181–216.

Brawley, C. and Matunis, E. 2004. Regeneration of male germline stem cells by spermatogonial dedifferentiation in vivo. *Science* **304:** 1331–1334.

Cantarel, B.L., Korf, I., Robb, S.M., Parra, G., Ross, E., Moore, B., Holt, C., Sánchez Alvarado, A., and Yandell, M. 2008. MAKER: An easy-to-use annotation pipeline designed for emerging model organism genomes. *Genome Res.* **18:** 188–196.

Cebria, F. 2008. Organization of the nervous system in the model planarian *Schmidtea mediterranea:* An immunocytochemical study. *Neurosci. Res.* **61:** 375–384.

Cebria, F., Kobayashi, C., Umesono, Y., Nakazawa, M., Mineta,

K., Ikeo, K., Gojobori, T., Itoh, M., Taira, M., Sánchez Alvarado, A., and Agata, K. 2002. FGFR-related gene *noudarake* restricts brain tissues to the head region of planarians. *Nature* **419:** 620–624.

Chuma, S., Hosokawa, M., Kitamura, K., Kasai, S., Fujioka, M., Hiyoshi, M., Takamune, K., Noce, T., and Nakatsuji, N. 2006. *Tdrd1/Mtr-1,* a *tudor*-related gene, is essential for male germ-cell differentiation and nuage/germinal granule formation in mice. *Proc. Natl. Acad. Sci.* **103:** 15894–15899.

Cinalli, R.M., Rangan, P., and Lehmann, R. 2008. Germ cells are forever. *Cell* **132:** 559–562.

Curtis, W.C. 1902. The life history, the normal fission, and the reproductive organs of *Planaria maculata. Proc. Boston Soc. Nat. Hist.* **30:** 515–560.

Deshpande, G., Calhoun, G., Yanowitz, J.L., and Schedl, P.D. 1999. Novel functions of *nanos* in downregulating mitosis and transcription during the development of the *Drosophila* germline. *Cell* **99:** 271–281.

de Sousa Lopes, S.M., Roelen, B.A., Monteiro, R.M., Emmens, R., Lin, H.Y., Li, E., Lawson, K.A., and Mummery, C.L. 2004. BMP signaling mediated by ALK2 in the visceral endoderm is necessary for the generation of primordial germ cells in the mouse embryo. *Genes Dev.* **18:** 1838–1849.

Eddy, E.M. 1975. Germ plasm and the differentiation of the germ cell line. *Int. Rev. Cytol.* **43:** 229–280.

Extavour, C.G.M. 2007. Evolution of the bilaterian germ line: Lineage origin and modulation of specification mechanisms. *Integr. Comp. Biol.* **47:** 770–785.

Extavour, C.G. and Akam, M. 2003. Mechanisms of germ cell specification across the metazoans: Epigenesis and preformation. *Development* **130:** 5869–5884.

Extavour, C.G., Pang, K., Matus, D.Q., and Martindale, M.Q. 2005. *vasa* and *nanos* expression patterns in a sea anemone and the evolution of bilaterian germ cell specification mechanisms. *Evol. Dev.* **7:** 201–215.

Fedecka-Bruner, B. 1965. Régénération des testicules des planaires après destruction par les rayons X. In *Regeneration in animals and related problems* (ed. V. Kiortsis and H.A.L. Trampusch), pp. 185–192. North-Holland, Amsterdam.

Fedecka-Bruner, B. 1967. Études sur la régénération des organes genitaux chez la planaire *Dugesia lugubris*. I. Régénération des testicules après destruction. *Bull. Biol. Fr. Belg.* **101:** 255–319.

Forbes, A. and Lehmann, R. 1998. Nanos and Pumilio have critical roles in the development and function of *Drosophila* germline stem cells. *Development* **125:** 679–690.

Franquinet, R. and Lender, T. 1973. Étude ultrastructurale des testicules de *Polycelis tenuis* et *Polycelis nigra* (Planaires). Evolution des cellules germinales mâles avant la spermiogenèse. *Z. Mikrosk. Anat. Forsch.* **87:** 4–22.

Fujiwara, Y., Komiya, T., Kawabata, H., Sato, M., Fujimoto, H., Furusawa, M., and Noce, T. 1994. Isolation of a DEAD-family protein gene that encodes a murine homolog of *Drosophila* vasa and its specific expression in germ cell lineage. *Proc. Natl. Acad. Sci.* **91:** 12258–12262.

Ghirardelli, E. 1965. Differentiation of the germ cells and regeneration of the gonads in planarians. In *Regeneration in animals and related problems* (ed. V. Kiortsis and H.A.L. Trampusch), pp. 177–184. North-Holland, Amsterdam.

Gremigni, V. and Miceli, C. 1980. Cytophotometric evidence for cell 'transdifferentiation' in planarian regeneration. *Wilhelm Roux's Arch. Dev. Biol.* **188:** 107–113.

Gremigni, V. and Puccinelli, I. 1977. A contribution to the problem of the origin of the blastema cells in planarians: A karyological and ultrastructural investigation. *J. Exp. Zool.* **199:** 57–72.

Gremigni, V., Miceli, C., and Picano, E. 1980a. On the role of germ cells in planarian regeneration. II. Cytophotometric analysis of the nuclear Feulgen-DNA content in cells of regenerated somatic tissues. *J. Embryol. Exp. Morphol.* **55:** 65–76.

Gremigni, V., Miceli, C., and Puccinelli, I. 1980b. On the role of germ cells in planarian regeneration. I. A karyological investigation. *J. Embryol. Exp. Morphol.* **55:** 53–63.

Gremigni, V., Nigro, M., and Puccinelli, I. 1982. Evidence of male germ cell redifferentiation into female germ cells in planarian regeneration. *J. Embryol. Exp. Morphol.* **70:** 29–36.

Gruidl, M.E., Smith, P.A., Kuznicki, K.A., McCrone, J.S., Kirchner, J., Roussell, D.L., Strome, S., and Bennett, K.L. 1996. Multiple potential germ-line helicases are components of the germ-line-specific P granules of *Caenorhabditis elegans. Proc. Natl. Acad. Sci.* **93:** 13837–13842.

Guan, K., Nayernia, K., Maier, L.S., Wagner, S., Dressel, R., Lee, J.H., Nolte, J., Wolf, F., Li, M., Engel, W., and Hasenfuss, G. 2006. Pluripotency of spermatogonial stem cells from adult mouse testis. *Nature* **440:** 1199–1203.

Guo, T. 2007. "Functional analysis of genes involved in regulating stem cell maintenance and differentiation in the freshwater planarian *Schmidtea mediterranea.*" Ph.D. thesis. University of Illinois, Urbana-Champaign.

Guo, T., Peters, A.H., and Newmark, P.A. 2006. A *bruno*-like gene is required for stem cell maintenance in planarians. *Dev. Cell* **11:** 159–169.

Handberg-Thorsager, M. and Saló, E. 2007. The planarian nanos-like gene *Smednos* is expressed in germline and eye precursor cells during development and regeneration. *Dev. Genes Evol.* **217:** 403–411.

Hay, E. and Coward, S. 1975. Fine structure studies on the planarian, *Dugesia.* I. Nature of the "neoblast" and other cell types in noninjured worms. *J. Ultrastruct. Res.* **50:** 1–21.

Hay, B., Jan, L.Y., and Jan, Y.N. 1988. A protein component of *Drosophila* polar granules is encoded by *vasa* and has extensive sequence similarity to ATP-dependent helicases. *Cell* **55:** 577–587.

Hayashi, Y., Hayashi, M., and Kobayashi, S. 2004. Nanos suppresses somatic cell fate in *Drosophila* germ line. *Proc. Natl. Acad. Sci.* **101:** 10338–10342.

Hayashi, T., Asami, M., Higuchi, S., Shibata, N., and Agata, K. 2006. Isolation of planarian X-ray-sensitive stem cells by fluorescence-activated cell sorting. *Dev. Growth Differ.* **48:** 371–380.

Hayashi, K., de Sousa Lopes, S.M., and Surani, M.A. 2007. Germ cell specification in mice. *Science* **316:** 394–396.

Higuchi, S., Hayashi, T., Hori, I., Shibata, N., Sakamoto, H., and Agata, K. 2007. Characterization and categorization of fluorescence activated cell sorted planarian stem cells by ultrastructural analysis. *Dev. Growth Differ.* **49:** 571–581.

Hori, I. 1982. An ultrastructural study of the chromatoid body in planarian regenerative cells. *J. Electron Microsc.* **31:** 63–72.

Hosokawa, M., Shoji, M., Kitamura, K., Tanaka, T., Noce, T., Chuma, S., and Nakatsuji, N. 2007. Tudor-related proteins TDRD1/MTR-1, TDRD6 and TDRD7/TRAP: Domain composition, intracellular localization, and function in male germ cells in mice. *Dev. Biol.* **301:** 38–52.

Houston, D.W. and King, M.L. 2000. Germ plasm and molecular determinants of germ cell fate. *Curr. Top. Dev. Biol.* **50:** 155–181.

Hyman, L.H. 1951. *The Invertebrates: Platyhelminthes and Rhynchocoela. The Acoelomate Bilateria.* McGraw Hill, New York.

Iijima, I. 1884. Untersuchungen über den Bau und die Entwicklungsgeschichte der Süsswasser-Dendrocoelen (Tricladen). *Z. Wiss. Zool.* **40:** 359–464.

Ikenishi, K. and Tanaka, T.S. 1997. Involvement of the protein of *Xenopus* vasa homolog (*Xenopus vasa*-like gene 1, *XVLG1*) in the differentiation of primordial germ cells. *Dev. Growth Differ.* **39:** 625–633.

Inoue, T., Kumamoto, H., Okamoto, K., Umesono, Y., Sakai, M., Sánchez Alvarado, A., and Agata, K. 2004. Morphological and functional recovery of the planarian photosensing system during head regeneration. *Zoolog. Sci.* **21:** 275–283.

Johnson, A.D., Drum, M., Bachvarova, R.F., Masi, T., White, M.E., and Crother, B.I. 2003. Evolution of predetermined germ cells in vertebrate embryos: Implications for macroevolution. *Evol. Dev.* **5:** 414–431.

Juliano, C.E., Voronina, E., Stack, C., Aldrich, M., Cameron, A.R., and Wessel, G.M. 2006. Germ line determinants are not localized early in sea urchin development, but do accumulate in the small micromere lineage. *Dev. Biol.* **300:** 406–415.

Kai, T. and Spradling, A. 2004. Differentiating germ cells can revert into functional stem cells in *Drosophila melanogaster* ovaries. *Nature* **428:** 564–569.

Kanatsu-Shinohara, M., Inoue, K., Lee, J., Yoshimoto, M., Ogonuki, N., Miki, H., Baba, S., Kato, T., Kazuki, Y., Toyokuni, S., et al. 2004. Generation of pluripotent stem cells from neonatal mouse testis. *Cell* **119:** 1001–1012.

Kobayashi, S., Yamada, M., Asaoka, M., and Kitamura, T. 1996. Essential role of the posterior morphogen nanos for germline development in *Drosophila. Nature* **380:** 708–711.

Komiya, T., Itoh, K., Ikenishi, K., and Furusawa, M. 1994. Isolation and characterization of a novel gene of the DEAD box protein family which is specifically expressed in germ cells of *Xenopus laevis. Dev. Biol.* **162:** 354–363.

Kotaja, N., Bhattacharyya, S.N., Jaskiewicz, L., Kimmins, S., Parvinen, M., Filipowicz, W., and Sassone-Corsi, P. 2006. The chromatoid body of male germ cells: Similarity with processing bodies and presence of Dicer and microRNA pathway components. *Proc. Natl. Acad. Sci.* **103:** 2647–2652.

Lasko, P.F. and Ashburner, M. 1988. The product of the *Drosophila* gene *vasa* is very similar to eukaryotic initiation factor-4A. *Nature* **335:** 611–617.

Lawson, K.A., Dunn, N.R., Roelen, B.A., Zeinstra, L.M., Davis, A.M., Wright, C.V., Korving, J.P., and Hogan, B.L. 1999. *Bmp4* is required for the generation of primordial germ cells in the mouse embryo. *Genes Dev.* **13:** 42436.

Le Moigne, A. 1967. Présence d'émissions nucléaires fréquemment associées à des mitochondries dans les cellules embryonnaires des planaires. *C.R. Soc. Biol.* **161:** 508–511.

Lehmann, R. and Nüsslein-Volhard, C. 1991. The maternal gene *nanos* has a central role in posterior pattern formation of the *Drosophila* embryo. *Development* **112:** 679–691.

MacRae, E.K. 1967. The fine structure of sensory receptor processes in the auricular epithelium of the planarian, *Dugesia tigrina. Z. Zellforsch. Mikrosk. Anat.* **82:** 479–494.

McLaren, A. 1981. *Germ cells and soma: A new look at an old problem.* Yale University Press, New Haven.

Mochizuki, K., Nishimiya-Fujisawa, C., and Fujisawa, T. 2001. Universal occurrence of the *vasa*-related genes among metazoans and their germline expression in *Hydra. Dev. Genes Evol.* **211:** 299–308.

Morgan, T.H. 1902. Growth and regeneration in *Planaria lugubris. Arch. Ent. Mech. Org.* **13:** 179–212.

Morita, M. and Best, J.B. 1984. Effects of photoperiods and melatonin on planarian asexual reproduction. *J. Exp. Zool.* **231:** 273–382.

Morita, M., Best, J., and Noel, J. 1969. Electron microscopic studies of planarian regeneration. I. Fine structure of neoblasts in *Dugesia dorotocephala. J. Ultrastruct. Res.* **27:** 7–23.

Newmark, P. and Sánchez Alvarado, A. 2000. Bromodeoxyuridine specifically labels the regenerative stem cells of planarians. *Dev. Biol.* **220:** 142–153.

Newmark, P.A. and Sánchez Alvarado, A. 2002. Not your father's planarian: A classic model enters the era of functional genomics. *Nat. Rev. Genet.* **3:** 210–219.

Newmark, P.A., Reddien, P.W., Cebria, F., and Sánchez Alvarado, A. 2003. Ingestion of bacterially expressed double-stranded RNA inhibits gene expression in planarians. *Proc. Natl. Acad. Sci.* (suppl. 1) **100:** 11861–11865.

Nieuwkoop, P.D. and Sutasurya, L.A. 1979. *Primordial germ cells in the chordates.* Cambridge University Press, London.

Nieuwkoop, P.D. and Sutasurya, L.A. 1981. *Primordial germ cells in the invertebrates: From epigenesis to preformation.* Cambridge University Press, London.

Okamoto, K., Takeuchi, K., and Agata, K. 2005. Neural projections in planarian brain revealed by fluorescent dye tracing. *Zoolog. Sci.* **22:** 535–546.

Olsen, L.C., Aasland, R., and Fjose, A. 1997. A *vasa*-like gene in zebrafish identifies putative primordial germ cells. *Mech. Dev.* **66:** 95–105.

Orii, H., Sakurai, T., and Watanabe, K. 2005. Distribution of the stem cells (neoblasts) in the planarian *Dugesia japonica. Dev.*

Genes Evol. **215:** 143–157.

Palakodeti, D., Smielewska, M., Lu, Y.C., Yeo, G.W., and Graveley, B.R. 2008. The PIWI proteins SMEDWI-2 and SMEDWI-3 are required for stem cell function and piRNA expression in planarians. *RNA* **14:** 1174–1186.

Pigon, A., Morita, M., and Best, J.B. 1974. Cephalic mechanism for social control of fissioning in planarians. II. Localization and identification of the receptors by electron micrographic and ablation studies. *J. Neurobiol.* **5:** 443–462.

Ptashne, M. 2007. On the use of the word 'epigenetic'. *Curr. Biol.* **17:** R233–R236.

Rando, T.A. 2006. Stem cells, ageing and the quest for immortality. *Nature* **441:** 1080–1086.

Raz, E. 2003. Primordial germ-cell development: The zebrafish perspective. *Nat. Rev. Genet.* **4:** 690–700.

Reddien, P.W. and Sánchez Alvarado, A. 2004. Fundamentals of planarian regeneration. *Annu. Rev. Cell Dev. Biol.* **20:** 725–757.

Reddien, P.W., Bermange, A.L., Murfitt, K.J., Jennings, J.R., and Sánchez Alvarado, A. 2005a. Identification of genes needed for regeneration, stem cell function, and tissue homeostasis by systematic gene perturbation in planaria. *Dev. Cell* **8:** 635–649.

Reddien, P.W., Oviedo, N.J., Jennings, J.R., Jenkin, J.C., and Sánchez Alvarado, A. 2005b. SMEDWI-2 is a PIWI-like protein that regulates planarian stem cells. *Science* **310:** 1327–1330.

Robb, S.M., Ross, E., and Sánchez Alvarado, A. 2008. SmedGD: The *Schmidtea mediterranea* genome database. *Nucleic Acids Res.* **36:** D599–D606.

Rossi, L., Salvetti, A., Lena, A., Batistoni, R., Deri, P., Pugliesi, C., Loreti, E., and Gremigni, V. 2006. *DjPiwi-1*, a member of the *PAZ-Piwi* gene family, defines a subpopulation of planarian stem cells. *Dev. Genes Evol.* **216:** 335–346.

Saló, E. and Baguñà, J. 1984. Regeneration and pattern formation in planarians. I. The pattern of mitosis in anterior and posterior regeneration in *Dugesia (G) tigrina*, and a new proposal for blastema formation. *J. Embryol. Exp. Morphol.* **83:** 63–80.

Saló, E. and Baguñà, J. 1989. Regeneration and pattern formation in planarians. II. Local origin and role of cell movements in blastema formation. *Development* **107:** 69–76.

Salvetti, A., Rossi, L., Deri, P., and Batistoni, R. 2000. An *MCM2*-related gene is expressed in proliferating cells of intact and regenerating planarians. *Dev. Dyn.* **218:** 603–614.

Salvetti, A., Rossi, L., Lena, A., Batistoni, R., Deri, P., Rainaldi, G., Locci, M.T., Evangelista, M., and Gremigni, V. 2005. *DjPum*, a homologue of *Drosophila Pumilio*, is essential to planarian stem cell maintenance. *Development* **132:** 1863–1874.

Sánchez Alvarado, A. 2006. Planarian regeneration: Its end is its beginning. *Cell* **124:** 241–245.

Sánchez Alvarado, A. and Newmark, P.A. 1999. Double-stranded RNA specifically disrupts gene expression during planarian regeneration. *Proc. Natl. Acad. Sci.* **96:** 5049–5054.

Sánchez Alvarado, A., Newmark, P.A., Robb, S.M., and Juste, R. 2002. The *Schmidtea mediterranea* database as a molecular resource for studying platyhelminthes, stem cells and regeneration. *Development* **129:** 5659–5665.

Santos, A.C. and Lehmann, R. 2004. Germ cell specification and migration in *Drosophila* and beyond. *Curr. Biol.* **14:** R578–R589.

Sato, K., Shibata, N., Orii, H., Amikura, R., Sakurai, T., Agata, K., Kobayashi, S., and Watanabe, K. 2006. Identification and origin of the germline stem cells as revealed by the expression of *nanos*-related gene in planarians. *Dev. Growth Differ.* **48:** 615–628.

Sauzin, M.-J. 1968. Présence d'émissions nucléaires dans les cellules différenciées et en différenciation de la planaire adulte *Dugesia gonocephala. C.R. Acad. Sci.* **267:** 1146–1148.

Schultz, E. 1904. Über Reduktionen. I. Über Hungerserscheinungen bei *Planaria lactea. Arch. Entwicklungsmech. Org.* **18:** 555–577.

Seandel, M., James, D., Shmelkov, S.V., Falciatori, I., Kim, J., Chavala, S., Scherr, D.S., Zhang, F., Torres, R., Gale, N.W., et al. 2007. Generation of functional multipotent adult stem cells from GPR125$^+$ germline progenitors. *Nature* **449:** 346–350.

Sengel, C. 1960. Culture *in vitro* de blastèmes de régénération de Planaires. *J. Embryol. Exp. Morphol.* **8:** 468–476.

Seydoux, G. and Braun, R.E. 2006. Pathway to totipotency: Lessons from germ cells. *Cell* **127:** 891–904.

Seydoux, G. and Schedl, T. 2001. The germline in *C. elegans:* Origins, proliferation, and silencing. *Int. Rev. Cytol.* **203:** 139–185.

Seydoux, G. and Strome, S. 1999. Launching the germline in *Caenorhabditis elegans:* Regulation of gene expression in early germ cells. *Development* **126:** 3275–3283.

Shibata, N., Umesono, Y., Orii, H., Sakurai, T., Watanabe, K., and Agata, K. 1999. Expression of *vasa (vas)*-related genes in germline cells and totipotent somatic stem cells of planarians. *Dev. Biol.* **206:** 73–87.

Silveira, M. and Porter, K.R. 1964. The spermatozoids of flatworms and their microtubular systems. *Protoplasma* **59:** 240–265.

Starz-Gaiano, M. and Lehmann, R. 2001. Moving towards the next generation. *Mech. Dev.* **105:** 5–18.

Suzuki, A., Tsuda, M., and Saga, Y. 2007. Functional redundancy among Nanos proteins and a distinct role of Nanos2 during male germ cell development. *Development* **134:** 77–83.

Tanaka, S.S., Toyooka, Y., Akasu, R., Katoh-Fukui, Y., Nakahara, Y., Suzuki, R., Yokoyama, M., and Noce, T. 2000. The mouse homolog of *Drosophila Vasa* is required for the development of male germ cells. *Genes Dev.* **14:** 841–853.

Toyooka, Y., Tsunekawa, N., Takahashi, Y., Matsui, Y., Satoh, M., and Noce, T. 2000. Expression and intracellular localization of mouse *Vasa*-homologue protein during germ cell development. *Mech. Dev.* **93:** 139–149.

Tsuda, M., Sasaoka, Y., Kiso, M., Abe, K., Haraguchi, S., Kobayashi, S., and Saga, Y. 2003. Conserved role of nanos proteins in germ cell development. *Science* **301:** 1239–1241.

Tsunekawa, N., Naito, M., Sakai, Y., Nishida, T., and Noce, T. 2000. Isolation of chicken *vasa* homolog gene and tracing the origin of primordial germ cells. *Development* **127:** 2741–2750.

Wang, C. and Lehmann, R. 1991. Nanos is the localized posterior determinant in *Drosophila. Cell* **66:** 637–647.

Wang, Y., Zayas, R.M., Guo, T., and Newmark, P.A. 2007. *nanos* function is essential for development and regeneration of planarian germ cells. *Proc. Natl. Acad. Sci.* **104:** 5901–5906.

Weismann, A. 1893. *The germ-plasm. A theory of heredity* (transl. W.N. Parker and H. Rönnfeldt). Charles Scribner's Sons, New York.

Wolfe, S.A., Anderson, J.V., Grimes, S.R., Stein, G.S., and Stein, J.S. 1989. Comparison of the structural organization and expression of germinal and somatic rat histone H4 genes. *Biochim. Biophys. Acta* **1007:** 140–150.

Ying, Y., Liu, X.M., Marble, A., Lawson, K.A., and Zhao, G.Q. 2000. Requirement of *Bmp8b* for the generation of primordial germ cells in the mouse. *Mol. Endocrinol.* **14:** 1053–1063.

Yoon, C., Kawakami, K., and Hopkins, N. 1997. Zebrafish *vasa* homologue RNA is localized to the cleavage planes of 2- and 4-cell-stage embryos and is expressed in the primordial germ cells. *Development* **124:** 3157–3165.

Yoshida-Kashikawa, M., Shibata, N., Takechi, K., and Agata, K. 2007. DjCBC-1, a conserved DEAD box RNA helicase of the RCK/p54/Me31B family, is a component of RNA-protein complexes in planarian stem cells and neurons. *Dev. Dyn.* **236:** 3436–3450.

Zayas, R.M., Hernandez, A., Habermann, B., Wang, Y., Stary, J.M., and Newmark, P.A. 2005. The planarian *Schmidtea mediterranea* as a model for epigenetic germ cell specification: Analysis of ESTs from the hermaphroditic strain. *Proc. Natl. Acad. Sci.* **102:** 18491–18496.

Novel Insights into the Flexibility of Cell and Positional Identity during Urodele Limb Regeneration

M. Kragl,*† D. Knapp,* E. Nacu,*† S. Khattak,* E. Schnapp,*‡
H.-H. Epperlein,§ AND E.M. Tanaka*†

*Max-Planck-Institute of Molecular Cell Biology and Genetics, 01307 Dresden,
Germany; †Center for Regenerative Therapies Dresden, 01307 Dresden, Germany;
§Department of Anatomy, TU Dresden, 01307 Dresden, Germany

The ability of diverse metazoans to regenerate whole-body structures was first described systematically by Spallanzani in 1768 and continues to fascinate biologists today. Given the current interest in stem cell biology and its therapeutic potential, examples of vertebrate regeneration garner strong interest. Among regeneration-competent vertebrates such as the fish, frog, and salamander, the salamander is particularly impressive because it can regenerate the entire limb and tail as well as various internal organs as an adult (Goss 1969). This spectacular natural phenomenon leads us to ask what cellular properties allow regeneration and what prevents this phenomenon in other vertebrates. From this perspective, it is imperative to know whether the stem cells in regenerating limbs harbor particularly special traits such as a higher plasticity in cell fate compared to tissue stem cells in other organisms. Flexibility in cell fate needs to be considered with respect not only to tissue identity, but also to patterning because limb amputation causes cells in a particular limb segment to form more distal limb elements. How positional identity is encoded in stem cells and how it is controlled to produce only the missing portion of the limb are also questions of fundamental importance.

Limb regeneration proceeds over the course of weeks and months and is a complex sequence of cellular events that unfold to produce the fully patterned and functional structure. We therefore discuss the problem of regeneration in four main parts: wound healing, blastema formation including the potency of cell fate, nerve dependence, and patterning.

EPITHELIAL WOUND HEALING

Reepithelialization during limb regeneration already shows distinctions to the mammalian situation. Whereas during mammalian wound healing, epithelial cell migration initiates only 48 hours after injury, during salamander limb regeneration, blood clotting is immediately followed by rapid migration of epithelial cells that close the wound within 8 hours. The cuboidal epithelial layer covering the end of the limb displays embryonic features where no basal lamina is formed between the epithelial and underlying layers. In subsequent days, this wound epidermis thickens to form the apical epidermal cap (AEC), which is molecularly and functionally similar to the apical ectodermal ridge (AER) of the developing limb bud (Saunders 1948; Christensen and Tassava 2000). For example, the AEC is essential for outgrowth of the regenerating limb because removal of the AEC or replacement of this specialized epithelium by mature skin blocks regeneration (Dearlove and Stocum 1974; Mescher 1976; Tassava and Garling 1979). Furthermore, the AEC expresses those factors found in the embryonic AER such as mito-

gens of the fibroblast growth factor (FGF) family, including FGF-1 and FGF-2 (Boilly et al. 1991; Mullen et al. 1996; Christensen et al. 2002; Dungan et al. 2002).

Lateral wounding does not induce regenerative events, whereas full amputation does. The AEC forms only after limb amputation and does not form at a wound on the side of the limb. In fact, lateral or internal wounds to the limb have more in common with mammalian wound healing than bona fide limb regeneration. For example, if a significant gap is made in the limb skeleton without amputating the limb (Fig. 1), the bone pieces do not grow together to replace the missing segment, whereas if the limb is amputated through the skeleton, that bone is regenerated (Weiss 1925; Bischler 1926). Thus, it is not even necessary for the bone to be present at the amputation site—removal of the bone before limb amputation still results in regeneration of a properly patterned skeletal element from the site of amputation, suggesting some plasticity in tissue cell fate, a topic that is addressed below.

Similar to the situation in bone, muscle regeneration exhibits distinct phenotypes when occurring along the side of the limb versus after full amputation. Carlson has extensively studied muscle regeneration in different vertebrate species and found that minced muscle preparations made along the limb are repaired in salamanders similar to repair in other vertebrates where the regenerated muscle is distinguishable from the original because of the central location of the nuclei in the regenerated fibers and the accumulation of fibrous tissue. In contrast, when the limb is amputated and the limb muscle regenerates in concert with other limb tissue, it is perfectly patterned and indistinguishable from the original (for review, see Carlson 2003).

---

‡Present address: Stem Cell Research Institute, DiBiT, San Raffaele Scientific Institute, 58 via Olgettina, 20132 Milan, Italy.

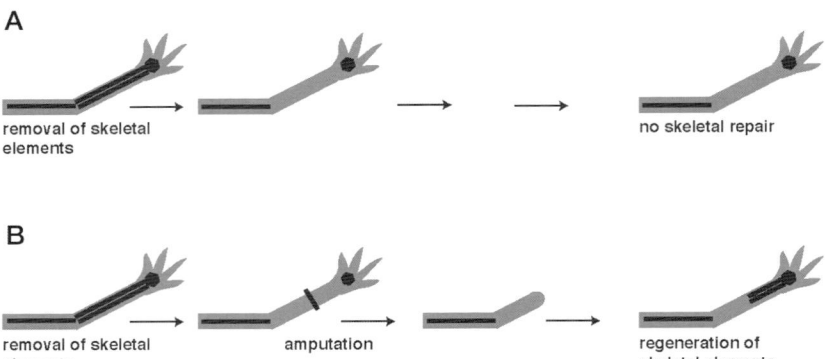

Figure 1. Plasticity during regeneration. (*A*) Skeletal elements do not grow back together when a gap is made, but the rest of the limb is kept intact. (*B*) If, however, a limb is amputated, even through a region lacking skeletal elements, the skeletal elements regenerate from the plane of amputation (Weiss 1925; Bischler 1926).

An exception to these cases is spinal cord injury where a gap lesion made in the spinal cord is repaired. The ependymoglial cells lining the canal at both cut ends heal over to make terminal bulbs. The ends then grow toward each other, with the posterior-facing end growing faster than the anterior-facing edge (Butler and Ward 1967). Eventually, the two tubes grow together and intercalate to make a seamless tube. The unique ability of the spinal cord to repair gap lesions may be related to its powerful properties as an organizer for tail regeneration. In contrast to the situation in bone, removal of spinal cord before tail amputation prevents tail regeneration. Furthermore, if the spinal cord is rotated 180° along the dorsoventral (DV) axis and the tail is then amputated through the inverted region, all tail tissues subsequently regenerate with an inverted DV orientation according to the orientation of the spinal cord (Holtzer 1956). This indicates that the spinal cord harbors positional information that is transmitted to the surrounding tissues and furthermore shows that the spinal cord seems to autonomously harbor the requisite positional information and growth information for tail regeneration. A molecule involved in these processes is sonic hedgehog, which is expressed in cells of the ventral spinal cord. Treatment of amputated tails with the inhibitor cyclopamine blocked regeneration, indicating that it is one of the factors from the spinal cord required for regeneration of the surrounding tissues (Schnapp et al. 2005).

BLASTEMA FORMATION

What may distinguish the stem or progenitor cells produced after limb amputation from those used in normal wound healing? The defining cellular structure of the regenerating limb is the blastema—a zone of mesenchymal progenitor cells that accumulates underneath the wound epidermis in the first week(s) of regeneration and represents the progenitor cells that will form all of the various limb tissues.

Which mature tissues contribute to the blastema and what are the properties of blastema cells? Selective X-irradiation of the limb stump before limb amputation demonstrated that the cells for the blastema arise locally

from the tissues at the amputation plane (Butler 1931). Tracking of triploid or tritiated thymidine-labeled tissue showed that multiple tissues including dermis, muscle, cartilage, and Schwann cells likely contribute to the blastema and thus the regenerating limb (Steen 1968; Wallace 1973; Namenwirth 1974; Dunis and Namenwirth 1977; Lheureux 1983; Holder 1989).

In terms of how mature tissues give rise to blastema cells, a concept long discussed in regeneration is the dedifferentiation of mature, postmitotic cells to produce blastema cells. Detailed histological examination of regenerating tissue argued that differentiated cells at the amputation plane lose their mature features and return to the cell cycle to produce blastema cells (Chalkley 1954; Hay 1959; Hay and Fischman 1961). Muscle represents the only tissue type where concrete experimental evidence consistent with the possibility of dedifferentiation exists. Implantation of lineage-labeled myotubes, as well as injection of labeled endogenous muscle fibers, suggests that multinucleated, postmitotic muscle cells return to the cell cycle and resolve into mononucleated cells to produce cycling blastema cells (Lo et al. 1993; Kumar et al. 2000; Echeverri et al. 2001). The details of such dedifferentiation studies have been reviewed elsewhere (see, e.g., Straube and Tanaka 2006) and are not discussed here. At the moment, it is not yet clear what proportion of the blastema forms from bona fide dedifferentiation of cells versus recruitment of resident stem cells, an issue that is important to resolve in the future. Furthermore, it will be necessary to determine whether regeneration involves activation of rare cells that expand enormously to form the blastema or whether a large proportion of cells in mature tissue enter into the blastema. Muscle tissue might serve as a good model because it harbors satellite cells, a pool of myogenic stem cells that are activated in response to muscle injury and can be identified using molecular markers (for review, see Buckingham et al. 2003).

Beyond the issue of dedifferentiation versus stem cell activation, it was unknown whether the various tissues all contribute a similar generalized pluripotent stem cell to the blastema or whether the blastema is composed of tissue-specific stem cells that respect their developmental origin. Indeed, the term blastema connotes a group of highly plastic progenitors, and it would be fair to say that the general expectation favored pluripotency of at least some blastema cells. For example, Wallace (1973) found that the implantation of healthy nerves into irradiated limbs was sufficient to rescue regeneration, implying that cells of neural origin could generate nonneural tissue. In a more recent study, satellite cells isolated from newt muscle tissue and labeled with bromodeoxyuridine (BrdU) were reimplanted into blastemas and contributed to cartilage and even epidermis of the regenerating limb, suggesting high plasticity of muscle-derived cells (Morrison et al. 2006). Such studies, however, were not definitive because cultivation and labeling of cells might have induced pluripotency, and the grafted cells may have been contaminated by cells from other tissues.

BLASTEMA CELLS RESPECT THEIR DEVELOPMENTAL ORIGIN

We wanted to determine if any tissue produces a highly pluripotent cell type in the blastema or whether blastema cells respect their developmental origin. To do this, we labeled every limb tissue separately in order to track its ultimate fate in regeneration (Fig. 2). Our labeling strategy was to graft the embryonic anlage of each limb tissue from a green fluorescent protein (GFP)-transgenic donor into a normal recipient. The donor embryos ubiquitously expressed GFP under the control of the CAGGS promoter (Sobkow et al. 2006). We implemented this method to successfully label limb muscle, dermis, cartilage, Schwann cells, and blood vessels. To achieve muscle and blood vessel labeling, we grafted mesoderm at somite level 3–4 at different DV locations; for dermis and cartilage, we transplanted lateral plate mesoderm. Because all lateral plate mesoderm transplants labeled both cartilage and dermis, we obtained specific cartilage or dermis labeling by performing a second graft of cartilage or dermis from the limbs of the labeled animals into unlabeled hosts. For Schwann cells, we obtained specific labeling by grafting the neural folds at embryonic stage 16. As expected, fluorescent cells migrated into the limb bud and contributed to tissues in the mature limb that developmentally derive from the respective grafted embryonic cells. Having confirmed integration into host tissue and specificity of each graft type, we amputated through the fluorescent region and tracked the fate of GFP[+] cells during regeneration.

The results from our experiments show for the first time that blastema cells do not acquire pluripotency because no type of graft generated progeny that contributed to all limb tissues. The final fate of GFP[+] blastema cells strongly reflected their tissue-specific origin. Labeled muscle fibers and satellite cells gave rise to muscle tissue in the regenerated limb but no cartilage; Schwann cells only generated new Schwann cells and blood vessels produced more blood vessels. Dermis-derived cells exhibited the highest flexibility, populating the regenerated skeleton and connective tissues at a significant frequency, but the lateral-plate-derived cells never contributed to muscle. The conversion between dermal and cartilage identity reflects

the close lineage relationship between these two tissue types, which both arise from lateral plate mesoderm.

Considering the overall lack of pluripotency that we observed, we were concerned enough to examine two potential confounding issues: (1) whether the transgene was silenced in cells that transited to another cell type and (2) whether cells are actually pluripotent but had been masked because during normal regeneration, cells stay close to their tissue of origin, and therefore most easily form their parent tissue type. To address these issues, we examined Wallace's paradigms to rescue irradiated limbs through implantation of nerve. In our experiments, the irradiated host was transgenically marked with constitutive nuclear-Cherry expression, and the nerve implant was derived from either the GFP–Schwann cell animals or the constitutive GFP–transgenic animal. When irradiated animals were implanted with GFP–Schwann-cell-labeled nerve, we observed GFP expression only in Schwann cells. Although these rescued appendages contained cartilage, the cartilage was unlabeled. In contrast, when nerve—where all cells were GFP[+]—was implanted, the rescued regenerate contained GFP[+] cartilage, and all cells in the appendage were either nuclear Cherry or GFP positive. These experiments allowed us to conclude that (1) the transgene is not silenced during regeneration and (2) Schwann cells maintain their cell identity even under conditions when the nerve is asked to rescue regeneration. This further shows that the cartilage cells of the rescued regenerate come from contaminating non-Schwann cells in the nerve sheaths, presumably connective fibroblasts. We observed no muscle in the rescued regenerates, suggesting that cells from the nerve do not form muscle. Taken together, our results show that pluripotent cells do not appear to be implemented during limb regeneration, because cells respect lineage restrictions defined during development. Gargioli and Slack (2004) had observed evidence of tissue-specific restriction in muscle, notochord, and spinal cord during anuran tadpole tail regeneration, although their results left open the possibility that other tissue types such as dermis or blood derivatives could have produced a pluripotent cell type. Furthermore, it was unclear whether tadpole tail regeneration in the frog represented a truly representative regeneration system because the tail is lost during metamorphosis, and the regeneration of neural structures is deficient in the frog tadpole. Interestingly, cell tracking during axolotl tail regeneration that robustly regenerates all tail structures showed that spinal cord progenitors may have a more generalized potential during regeneration (Echeverri and Tanaka 2002).

ROLE OF NERVES DURING REGENERATION

Despite their limited plasticity, blastema cells are undifferentiated progenitors that repeatedly divide and then progressively differentiate until the missing portion of the limb has formed. What signals support the proliferation of these cells? The wound epidermis (AEC), as mentioned above, is required, but the regenerating nerve is also required to maintain proliferation of blastema cells because denervation of the limb causes blastema cells to

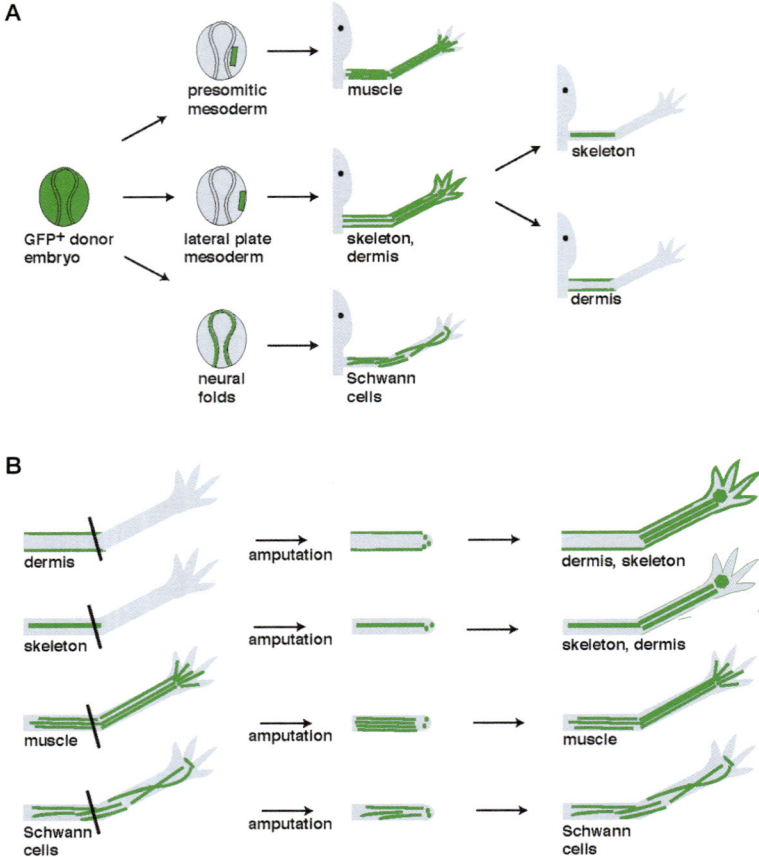

Figure 2. Blastema cells respect their developmental origin. (*A*) Specific labeling of different limb tissues was achieved by mapping and grafting the embryonic anlage of each limb tissue using GFP-transgenic donors. Lateral plate mesoderm ultimately contributed to dermis and skeletal elements, presomitic mesoderm to muscle tissue, and neural folds gave rise to Schwann cells. (*B*) These limbs were amputated through the fluorescent region, and GFP⁺ cells were tracked through the course of regeneration. Cells respected their developmental origin: Schwann cells only regenerated Schwann cells, muscle gave rise to muscle, and blood vessels formed new blood vessels. Dermis-derived cells gave rise to dermis and also contributed to skeletal elements. Likewise, skeleton regenerated new skeletal elements and also contributed to dermis. Both dermis and skeleton derive from the same embryonic anlage, the lateral plate mesoderm.

cease proliferating and thus prevents regeneration (Singer 1952). Interestingly, the nerve dependence of cell proliferation is specific to regeneration and is not found in the developing limb bud. Amazingly, when aneurogenic limbs are created by preventing innervation during development, these limbs can regenerate in the absence of nerve input, indicating that the limb becomes "addicted" to the nerve (Yntema 1959a,b). Singer (1952) demonstrated that either sensory or motor nerves can provide the crucial signal for blastema cell growth.

What factors may mediate the nerve dependence of regeneration? Transferrin, substance P, and rhGgf2 were suggested to be involved in this process because they were shown to be present in neural tissues and stimulated proliferation of blastema cells (Globus et al. 1983; Munaim and Mescher 1986; Globus 1988; Wang et al. 2000). However, FGFs and the recently identified nAG, a member of the anterior gradient family of secreted proteins, are of particular interest. Mullen and colleagues (1996) showed that FGF-2, like other members of the FGF family, is expressed in limb nerves and the AEC, and

its expression levels decrease after transection of the nerve. In denervated blastemas, FGF-2-soaked beads rescued blastema outgrowth. On a molecular level, FGF-2 was shown to be sufficient to restore the expression of the AEC-specific gene *dlx3* in denervated limbs. Although it remains to be resolved whether FGF-2 or other family members stimulate proliferation of blastema cells in vivo, these results strongly suggest that FGFs have an important role during nerve-dependent regeneration.

The recent discovery of nAG has provided new insight into the molecular cross-talk between nerve supply and blastema cell proliferation (Kumar et al. 2007). nAG is expressed in Schwann cells as soon as the first blastema cells divide and, later, in the specialized gland cells of the wound epidermis. Nerve transection causes loss of nAG expression in both nerve and wound epidermis, implying that the epidermal expression is nerve-dependent and may reflect a positive feedback between nerve and epidermis. To test the direct role of nAG in blastema cell proliferation, Kumar et al. (2007) showed that recombinant nAG induces the proliferation of cultured blastema cells.

Strikingly, when the nAG gene was ectopically expressed in denervated limb stumps 5 days postamputation, full limb regeneration occurred in these limbs. Therefore, this molecule appears to be an important contributing factor for nerve-dependent regeneration.

THE ALM DEFINES THE CONDITIONS REQUIRED FOR INDUCING ECTOPIC LIMB FORMATION

Although innervation is clearly an essential component of adult limb regeneration, nerve supply is clearly not sufficient to induce an entire regenerating limb. Deviation of nerves to a wound bed on the side of the limb induces an ectopic blastema-like structure to form but the outgrowth eventually regresses, indicating that the cut nerve endings are not the only component that directs formation of a blastema at the end of the amputated limb. If at the site of nerve deviation, a piece of full-thickness skin (including dermal fibroblasts) deriving from the opposite side of the contralateral limb were placed next to the wound, ectopic, fully patterned limbs would grow out (Lheureux 1977; Reynolds et al. 1983; Maden and Holder 1984; Egar 1988; Endo et al. 2004; Satoh et al. 2007, 2008). This showed that the combination of wound epidermis, nerve supply, and skin fibroblasts of different positional identities is sufficient to induce the formation of a fully patterned limb (Fig. 3). Endo et al. (2004) realized that this experimental system can be used to analyze the "minimal" essential conditions required to induce an ectopic limb on the lateral side of a limb. They developed the accessory limb model (ALM) that describes limb regeneration as a sequence of events starting with wound healing, followed by the induction of a blastema by nerves, and finally outgrowth and patterning in response to the juxtaposition of dermal fibroblasts coming from different positional identities. One advantage of using this system is that processes related to wound healing, nerve dependence, and fibroblasts can be analyzed separately (Satoh et al. 2007, 2008). For example, Satoh and colleagues found that nerve deviation, but not contralateral skin grafts, was required for the expression of marker genes such as *msx2, tbx5,* and *hoxa13*.

The ALM experiments as well as earlier blastema rotation experiments showed that juxtaposition of anterior and posterior limb cells is a crucial cue for growing out a properly patterned limb (Iten and Bryant 1975). These results indicate that mature salamander limb tissue retains spatial coordinate information that is used during regeneration to define where a limb should grow out. The molecular nature of the anteroposterior (AP) information has not been defined, but it is likely to be a conserved molecular network used to define these axes during embryonic limb development. For example, the retention of spatial coordinate information has been better studied during spinal cord regeneration, where continued expression of the embryonic morphogen sonic hedgehog from the floorplate in the mature and regenerating spinal cord controls DV neural progenitor cell identity (Schnapp et al. 2005). In contrast, the adult mammalian spinal cord does not display expression of the embryonic patterning markers seen in the urodele (Yamamoto et al. 2001).

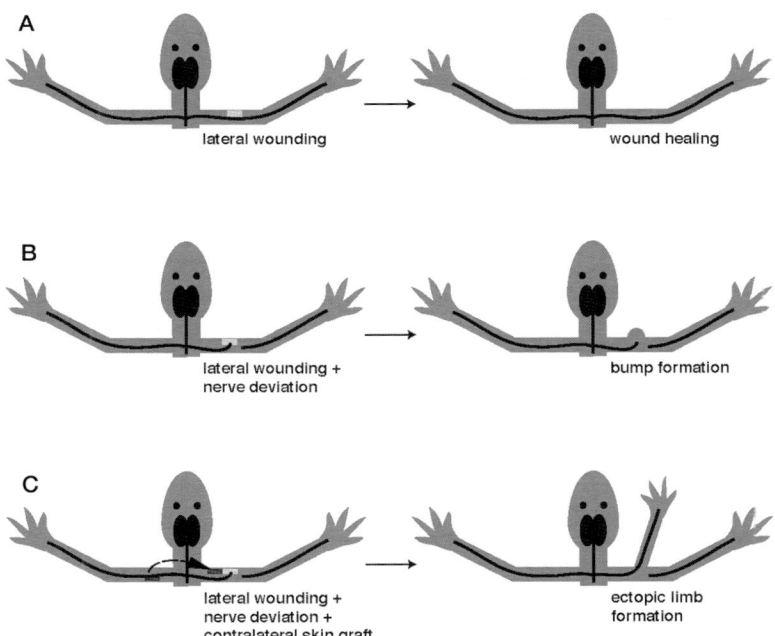

Figure 3. The accessory limb model (ALM). (*A*) A wound that is induced at the lateral site of the limb heals without scar formation. (*B*) If the limb nerve is transected and deviated to the wound, a bump forms that, however, does not possess the ability to regenerate an ectopic limb. (*C*) If, in addition, full-thickness skin from the opposite side of the contralateral limb is grafted next to a lateral wound with a deviated nerve, an ectopic limb forms (Endo et al. 2004).

PD INFORMATION EXISTS
AS A GRADED PROPERTY

Another dimension of spatial patterning is reformation of the correct number of limb segments along the proximodistal (PD) axis. Again, the results suggest that positional information for this axis is retained in adult salamander tissue. Under normal circumstances, salamander limb regeneration is unidirectional, meaning that cells at the amputation plane will always regenerate limb elements more distal to that location, a phenomenon called "the rule of distal transformation" (Rose 1962; Stocum 1975; Maden 1980). This was most clearly illustrated in an experiment where an "inverse" limb was created by suturing the wrist to the body (Fig. 4). After healing, amputation resulted in a limb where the wrist element lay in the proximal position and the upper arm was in the distal position. Such limbs regenerated lower-arm elements from the inverted upper arm, rather than upper-arm elements (Rose 1962). This result indicates that cells at the amputation plane have an identity associated with their position along the PD axis, a concept called positional memory (Wolpert 1969); however, amputation can then reprogram some cells or their progeny to a more distal identity. What cellular properties define positional memory and how is positional information manifested at the cellular level?

PD INFORMATION IS ENCODED BY
CELL SURFACE PROPERTIES

A number of grafting experiments have implicated local cell–cell interactions in patterning the regenerating PD axis. When a wrist blastema is transplanted onto an upper-arm stump, a normal regenerate forms in which the positions between upper arm and hand derive primarily from the upper-arm stump (Iten and Bryant 1975; Stocum 1975; Pescitelli and Stocum 1980). This phenomenon has been termed intercalary regeneration, where the upper-

arm stump intercalates the missing limb positions up to the transplanted distal blastema. Intercalary regeneration does not occur when a digit-stage distal regenerate or a mature hand is grafted to a proximal stump, implying that special conditions existing in the blastema are required to induce intercalary growth (Stocum 1975).

The ability of proximal and distal blastema tissue to respond to each other is also manifested in another manner that has been termed affinophoresis. When a distal blastema is grafted onto the side of a regenerating upper arm at the junction between mature and blastema tissue, the grafted distal blastema is carried along the regenerating proximal blastema so that it ultimately produces a second hand emanating from the wrist (Crawford and Stocum 1988), indicating that cells from the same PD region have a preferential affinity for like cells. This was further investigated by juxtaposing proximal and distal blastemas in hanging-drop cultures (Nardi and Stocum 1983). In such preparations, the proximal tissue engulfs the distal tissue, whereas distal-distal or proximal-proximal tissue confrontations maintain a straight border, indicating that distal and proximal cells have different adhesive properties.

It has been proposed that these observations reflect the cell surface recognition processes required for intercalary regeneration. Whether these recognition events are a cause or a consequence for PD patterning is one of the big challenges to address and mainly depends on elucidating the underlying molecules and genetic networks.

RETINOIC ACID RESPECIFIES CELLS
TO MORE PROXIMAL IDENTITIES AND
ALTERS CELL SURFACE PROPERTIES

A molecular handle for analyzing PD identity was provided by Niazi and Saxena's (1978) report that treatment of toad tadpole wrist blastemas with retinyl palmitate caused an entire arm to grow from this wrist. Maden (1983) further showed that retinoic-acid (RA)-induced proximalization in the salamander is dose-dependent, so that at low concentrations, wrist blastemas duplicate only the distal-most elements of the lower arm, and progressively increasing concentrations generate more arm elements (Fig. 5). These results showed that RA could force a distal cell to acquire proximal identity, subverting the normal distalization process that occurs during regeneration. The intercalation and affinophoresis assays showed that RA-transformed blastemas acquire the cell surface properties of a proximal blastema because they did not travel down to the wrist level during regeneration, but rather the ectopic regenerate emerged from the upper arm. Furthermore, in the hanging-drop assay, RA-treated distal blastemas were not engulfed by proximal blastemas (Fig. 6) (Nardi and Stocum 1983; Crawford and Stocum 1988).

RA-RESPONSIVE GENES ARE
INVOLVED IN PD IDENTITY

Considering the rule of distal transformation, it has remained unclear whether endogenous RA acts during normal limb regeneration to proximalize blastema cells. Nonetheless, RA has been used experimentally to identify

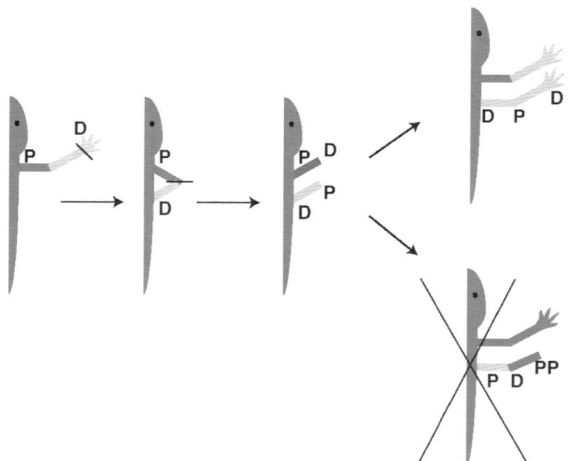

Figure 4. The rule of distal transformation. When an "inverse" limb is created, more distal (D) elements regenerate from the proximal (P) site. Regeneration of more proximal (PP) elements does not occur, demonstrating that distal-to-proximal changes of cell identity do not normally occur (Rose 1962).

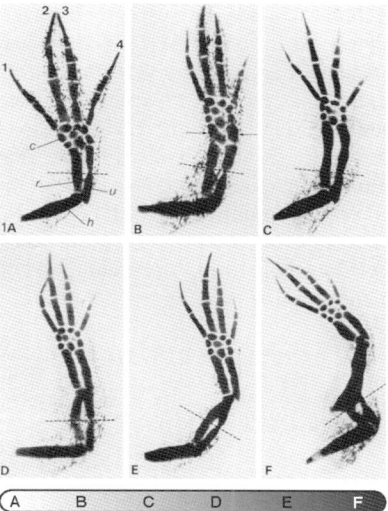

Figure 5. RA proximilizes the limb cell identity. RA respecifies cells from distal to proximal identity. Depending on the concentration (A–F), wrist blastemas produce limbs that contain duplications, with an entire arm growing out at the highest concentration (*F*). (Reprinted, with permission, from Maden 1983 [© Company of Biologists].)

sivity is a consequence of positional information. da Silva and colleagues (2002) proposed that Prod1 acts as a receptor that undergoes homophilic interactions that titrate the levels of Prod1 open to signaling via a heterotypic ligand. Thus, only the ligand-mediated interaction results in intracellular signaling to provoke proliferation and processes related to proximal identity. Because more Prod1 molecules are available on cells in the proximal domain, the putative ligand would bind and stimulate these cells. Interestingly, nAG was identified as a binding partner of Prod1 in a yeast two-hybrid screen, but whether an interaction between nAG and Prod1 mediates positional identity has not yet been resolved (Kumar et al. 2007). The nAG/Prod1 interaction brings up the exciting possibility that a link exists between nerve dependence and patterning, issues that were previously investigated separately.

Another class of RA-responsive molecules that dictate proximal identity are the Meis homeobox transcription factors (Mercader et al. 1999, 2005). During limb regeneration, Meis transcription factors are localized to the nucleus in the proximal domain of the blastema near the amputation plane, and overexpression of Meis1a or Meis2a plus their binding partner Pbx in distal blastema cells leads to their proximal translocation. Strikingly, electroporation of Meis antisense morpholinos into RA-treated wrist blastemas prevents limb duplications, demonstrating the requirement of the Meis proteins in RA-induced proximalization (Mercader et al. 2005). It is not yet known which genes become up-regulated by Meis transcription factors. Future experiments need to address whether Meis and Prod1 act in the same RA-induced pathway and, if yes, which one is upstream of the other.

two RA-responsive genes that are key regulators of PD identity: the GPI-linked protein Prod1 and the homeodomain protein Meis1/2. In a screen aiming to identify genes that are (1) up- or down-regulated by RA, (2) differentially expressed between proximal versus distal blastemas, and (3) coding for a cell surface molecule, da Silva et al. (2002) found that the only candidate meeting these three criteria was Prod1, a member of the Ly6 superfamily. Functional evidence supporting Prod1's role in PD recognition was first shown in the hanging-drop assay where phospholipase C and anti-Prod1 antibody treatment blocked proximal engulfment. Further evidence came from in vivo cell-labeling experiments where ectopic expression of Prod1 in distally fated blastema cells caused them to end up in a proximal limb location (Echeverri and Tanaka 2005). Still unresolved is whether Prod1 directly mediates proximal identity or whether its involvement in cell adhe-

TISSUE HETEROGENIETY OF PD IDENTITY

On the basis of our above studies tracking tissue cell fate during regeneration, it became clear that the blastema is a heterogeneous pool of progenitor cells. PD identity had so far been studied assuming that blastema cells behave homogeneously. Taking into account our recent results, we wanted to know if blastema cells deriving from the different tissues participate similarly in PD pat-

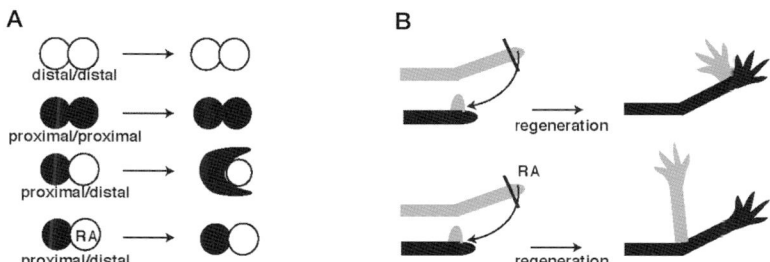

Figure 6. Two assays reveal that the cell surface properties of proximal and distal cells differ. (*A*) In the affinophoresis assay, a wrist blastema is transplanted onto an upper-arm blastema. The wrist blastema tracks along the regenerating limb and ultimately grows out at the wrist level (Crawford and Stocum 1988). An RA-treated wrist blastema behaves like an upper-arm blastema in the affinophoresis assay. (*B*) Hanging-drop cocultures show that of an upper-arm blastema engulfs the wrist blastema (Nardi and Stocum 1983). RA-treated wrist blastemas are no longer engulfed by an upper arm blastema, suggesting that it has acquired upper-arm properties.

terning or whether tissue-specific differences exist. We asked whether skeleton-, dermis-, muscle-, or Schwann-derived cells express markers of positional identity, namely, Meis1/2 and *hoxa13*, similarly or whether differences exist. Second, we asked whether distal Schwann- or cartilage-derived blastema cells displayed affinophoresis capabilities by transplanting GFP[+] Schwann or cartilage cells from the hand into the upper arm and asking if they ultimately came to reside in the hand or all along the limb after regeneration.

Our results revealed that cartilage- and Schwann-derived blastema cells display strikingly different behaviors with respect to PD identity assays. First, we found that 95% of bone-derived blastema cells harbored nuclear Meis in the proximal part of an upper-arm blastema, suggesting that these cells harbor a proximal identity. In contrast, no Schwann-derived blastema cells displayed nuclear Meis protein, suggesting that Schwann cells are neutral for PD positional identity. In addition, none of the Schwann-derived blastema cells expressed *hoxa13*, whereas some bone-derived cells expressed this marker. In concurrence with the molecular markers, we observed that cartilage and Schwann cells behaved differently in the affinophoresis assay. Hand-derived cartilage cells transplanted into the upper arm before limb amputation ultimately populated only the hand in the regenerate, indicating that they had maintained their distal identity. In contrast, hand-derived Schwann cells transplanted similarly spread throughout the PD axis of the regenerating limb, suggesting that they have no PD identity. These results clearly show that some blastema cells harbor positional identity whereas others do not. A second conclusion we can draw from this work is that hand-derived cartilage cells stably maintain their distal identity throughout the course of limb regeneration. Therefore, the proximal blastema is not able to efficiently convert distal tissue to a more proximal identity.

PERSPECTIVES

Our goal has been to understand which properties of the stem and progenitor cells of the regenerating limb blastema are important for regeneration and may distinguish them from mammalian cells that do not undergo regeneration. A surprising outcome of our cell-tracking experiments is that tissue-specific stem cells largely retain their tissue identity during the course of regeneration, and we see no evidence for the participation of a pluripotent stem cell. This means that with respect to tissue stem cell identity, the cells undertaking limb regeneration may not be astonishingly different from their mammalian counterparts. The ALM model has clarified that nerve signals and AP positional identity are important cues for initiating limb regeneration. Therefore, a key question is whether crucial differences exist between mammals and salamanders in the nerve/AEC growth stimulatory circuitry and/or maintenance of AP positional information in adult tissue required to induce the tissue-specific stem cells to undertake the complex program of regeneration. In addition to AP positional identity is information along the PD axis to define the limb segments. Interestingly, Rinn et al. (2006) have surveyed adult human fibroblasts for anatomically related differences and do find that fibroblasts along the PD axis of human appendages express different developmental genes reflecting their PD position. So far, it appears that the positional identity of these human fibroblasts is stable even under multiple rounds of division and even when cells are confronted with cells from different locations (Rinn et al. 2008). Therefore, the ability to reprogram PD identity after limb amputation could be an important contributing trait for regenerative ability.

ACKNOWLEDGMENTS

This work was supported by grants from the Volkswagen Foundation I/78766; DFG TA274/3-1 and funds from the Max-Planck Institute and the Center of Regenerative Therapies, Dresden. D.K. was a fellow of the Alexander von Humboldt Foundation. We wish to thank Andrea Merseburg for assistance on the nerve-dependence experiments and Heino Andreas, Tobias Richter, and Maritta Schuez for technical assistance.

REFERENCES

Bischler, V. 1926. L'influence du squellette dans la regeneration, et les potentialities des divers territoires du member chez *Triton cristatus*. *Rev. Suisse Zool.* **33:** 431–560.

Boilly, B., Cavanaugh, K.P., Thomas, D., Hondermarck, H., Bryant, S.V., and Bradshaw, R.A. 1991. Acidic fibroblast growth factor is present in regenerating limb blastemas of axolotls and binds specifically to blastema tissues. *Dev. Biol.* **145:** 302–310.

Buckingham, M., Bajard, L., Chang, T., Daubas, P., Hadchouel, J., Meilhac, S., Montarras, D., Rocancourt, D., and Relaix, F. 2003. The formation of skeletal muscle: From somite to limb. *J. Anat.* **202:** 59–68.

Butler, E.G. 1931. X-radiation and regeneration in *Amblystoma*. *Science* **74:** 100–101.

Butler, E.G. and Ward, M.B. 1967. Reconstitution of the spinal cord after ablation in adult *Triturus*. *Dev. Biol.* **15:** 464–486.

Carlson, B.M. 2003. Muscle regeneration in amphibians and mammals: Passing the torch. *Dev. Dyn.* **226:** 167–181.

Chalkley, D.T. 1954. A quantitative histological analysis of forelimb regeneration in *Triturus viridescens*. *J. Morphol.* **94:** 21–70.

Christensen, R.N. and Tassava, R.A. 2000. Apical epithelial cap morphology and fibronectin gene expression in regenerating axolotl limbs. *Dev. Dyn.* **217:** 216–224.

Christensen, R.N., Weinstein, M., and Tassava, R.A. 2002. Expression of fibroblast growth factors 4, 8, and 10 in limbs, flanks, and blastemas of *Ambystoma*. *Dev. Dyn.* **223:** 193–203.

Crawford, K. and Stocum, D.L. 1988. Retinoic acid coordinately proximalizes regenerate pattern and blastema differential affinity in axolotl limbs. *Development* **102:** 687–698.

da Silva, S.M., Gates, P.B., and Brockes, J.P. 2002. The newt ortholog of CD59 is implicated in proximodistal identity during amphibian limb regeneration. *Dev. Cell* **3:** 547–555.

Dearlove, G.E. and Stocum, D.L. 1974. Denervation-induced changes in soluble protein content during forelimb regeneration in the adult newt, *Notophthalmus viridescens*. *J. Exp. Zool.* **190:** 317–328.

Dungan, K.M., Wei, T.Y., Nace, J.D., Poulin, M.L., Chiu, I.M., Lang, J.C., and Tassava, R.A. 2002. Expression and biological effect of urodele fibroblast growth factor 1: Relationship to limb regeneration. *J. Exp. Zool.* **292:** 540–554.

Dunis, D.A. and Namenwirth, M. 1977. The role of grafted skin in the regeneration of X-irradiated axolotl limbs. *Dev. Biol.* **56:** 97–109.

Echeverri, K. and Tanaka, E.M. 2002. Ectoderm to mesoderm lineage switching during axolotl tail regeneration. *Science* **298:** 1993–1996.

Echeverri, K. and Tanaka, E.M. 2005. Proximodistal patterning during limb regeneration. *Dev. Biol.* **279:** 391–401.

Echeverri, K., Clarke, J.D., and Tanaka, E.M. 2001. In vivo imaging indicates muscle fiber dedifferentiation is a major contributor to the regenerating tail blastema. *Dev. Biol.* **236:** 151–164.

Egar, M.W. 1988. Accessory limb production by nerve-induced cell proliferation. *Anat. Rec.* **221:** 550–564.

Endo, T., Bryant, S.V., and Gardiner, D.M. 2004. A stepwise model system for limb regeneration. *Dev. Biol.* **270:** 135–145.

Gargioli, C. and Slack, J.M. 2004. Cell lineage tracing during *Xenopus* tail regeneration. *Development* **131:** 2669–2679.

Globus, M. 1988. A neuromitogenic role for substance P in urodele limb regeneration. In *Regeneration and development* (ed. S. Inoue et al.), pp. 675–685. Okada, Maebashi, Japan.

Globus, M., Vethamany-Globus, S., Kesik, A., and Milton, G. 1983. Roles of neural peptide substance P and calcium in blastema cell proliferation in the newt *Notophthalmus viridescens.* In *Limb development and regeneration* (ed. J.F. Fallon and A.I. Caplan), pp. 513–524. Liss, New York.

Goss, R.J. 1969. *Principles of regeneration.* Academic, New York.

Hay, E.D. 1959. Electron microscopic observations of muscle dedifferentiation in regenerating *Amblystoma* limbs. *Dev. Biol.* **3:** 26–59.

Hay, E.D. and Fischman, D.A. 1961. Origin of the blastema in regenerating limbs of the newt *Triturus viridescens.* An autoradiographic study using tritiated thymidine to follow cell proliferation and migration. *Dev. Biol.* **3:** 26–59.

Holder, N. 1989. Organization of connective tissue patterns by dermal fibroblasts in the regenerating axolotl limb. *Development* **105:** 585–593.

Holtzer, S. 1956. The inductive activity of the spinal cord in urodele regeneration. *J. Morphol.* **99:** 1–34.

Iten, L.E. and Bryant, S.V. 1975. The interaction between the blastema and stump in the establishment of the anterior-posterior and proximal-distal organization of the limb regenerate. *Dev. Biol.* **44:** 119–147.

Kumar, A., Velloso, C.P., Imokawa, Y., and Brockes, J.P. 2000. Plasticity of retrovirus-labelled myotubes in the newt limb regeneration blastema. *Dev. Biol.* **218:** 125–136.

Kumar, A., Godwin, J.W., Gates, P.B., Garza-Garcia, A.A., and Brockes, J.P. 2007. Molecular basis for the nerve dependence of limb regeneration in an adult vertebrate. *Science* **318:** 772–777.

Lheureux, E. 1977. Importance of limb tissue associations in the development of nerve-induced supernumerary limbs in the newt *Pleurodeles waltlii* Michah (author's transl.). *J. Embryol. Exp. Morphol.* **38:** 151–173.

Lheureux, E. 1983. The origin of tissues in the X-irradiated regenerating limb of the newt *Pleurodeles waltlii. Prog. Clin. Biol. Res.* **110A:** 455–465.

Lo, D.C., Allen, F., and Brockes, J.P. 1993. Reversal of muscle differentiation during urodele limb regeneration. *Proc. Natl. Acad. Sci.* **90:** 7230–7234.

Maden, M. 1980. Intercalary regeneration in the amphibian limb and the rule of distal transformation. *J. Embryol. Exp. Morphol.* **56:** 201–209.

Maden, M. 1983. The effect of vitamin A on the regenerating axolotl limb. *J. Embryol. Exp. Morphol.* **77:** 273–295.

Maden, M. and Holder, N. 1984. Axial characteristics of nerve induced supernumerary limbs in the axolotl. *Roux's Arch. Dev. Biol.* **193:** 394–401.

Mercader, N., Tanaka, E.M., and Torres, M. 2005. Proximodistal identity during vertebrate limb regeneration is regulated by Meis homeodomain proteins. *Development* **132:** 4131–4142.

Mercader, N., Leonardo, E., Azpiazu, N., Serrano, A., Morata, G., Martinez, C., and Torres, M. 1999. Conserved regulation of proximodistal limb axis development by Meis1/Hth. *Nature* **402:** 425–429.

Mescher, A.L. 1976. Effects on adult newt limb regeneration of partial and complete skin flaps over the amputation surface. *J. Exp. Zool.* **195:** 117–128.

Morrison, J.I., Loof, S., He, P., and Simon, A. 2006. Salamander limb regeneration involves the activation of a multipotent skeletal muscle satellite cell population. *J. Cell Biol.* **172:** 433–440.

Mullen, L.M., Bryant, S.V., Torok, M.A., Blumberg, B., and Gardiner, D.M. 1996. Nerve dependency of regeneration: The role of Distal-less and FGF signaling in amphibian limb regeneration. *Development* **122:** 3487–3497.

Munaim, S.I. and Mescher, A.L. 1986. Transferrin and the trophic effect of neural tissue on amphibian limb regeneration blastemas. *Dev. Biol.* **116:** 138–142.

Namenwirth, M. 1974. The inheritance of cell differentiation during limb regeneration in the axolotl. *Dev. Biol.* **41:** 42–56.

Nardi, J.B. and Stocum, D.L. 1983. Surface properties of regenerating limb cells: Evidence for gradation along the proximodistal axis. *Differentiation* **25:** 27–31.

Niazi, I.A. and Saxena, S. 1978. Abnormal hindlimb regeneration in tadpoles of the toad *Bufo andersoni*, exposed to vitamin A. *Folia Biol.* **26:** 3–11.

Pescitelli, Jr., M.J. and Stocum, D.L. 1980. The origin of skeletal structures during intercalary regeneration of larval *Ambystoma* limbs. *Dev. Biol.* **79:** 255–275.

Reynolds, S., Holder, N., and Fernandes, M. 1983. The form and structure of supernumerary hindlimbs formed following skin grafting and nerve deviation in the newt *Triturus cristatus. J. Embryol. Exp. Morphol.* **77:** 221–241.

Rinn, J.L., Bondre, C., Gladstone, H.B., Brown, P.O., and Chang, H.Y. 2006. Anatomic demarcation by positional variation in fibroblast gene expression programs. *PLoS Genet.* **2:** e119.

Rinn, J.L., Wang, J.K., Allen, N., Brugmann, S.A., Mikels, A.J., Liu, H., Ridky, T.W., Stadler, H.S., Nusse, R., Helms, J.A., and Chang, H.Y. 2008. A dermal HOX transcriptional program regulates site-specific epidermal fate. *Genes Dev.* **22:** 303–307.

Rose, S.M. 1962. Tissue-arc control of regeneration in the amphibian limb. *Symp. Soc. Study Dev. Growth* **20:** 153–176.

Satoh, A., Gardiner, D.M., Bryant, S.V., and Endo, T. 2007. Nerve-induced ectopic limb blastemas in the axolotl are equivalent to amputation-induced blastemas. *Dev. Biol.* **312:** 231–244.

Satoh, A., Graham, G.M., Bryant, S.V., and Gardiner, D.M. 2008. Neurotrophic regulation of epidermal dedifferentiation during wound healing and limb regeneration in the axolotl (*Ambystoma mexicanum*). *Dev. Biol.* **319:** 321–335.

Saunders, Jr., J.W. 1948. The proximo-distal sequence of the origin of the parts of the chick wing and the role of the ectoderm. *J. Exp. Zool.* **108:** 363–403.

Schnapp, E., Kragl, M., Rubin, L., and Tanaka, E.M. 2005. Hedgehog signaling controls dorsoventral patterning, blastema cell proliferation and cartilage induction during axolotl tail regeneration. *Development* **132:** 3243–3253.

Singer, M. 1952. The influence of the nerve in regeneration of the amphibian extremity. *Q. Rev. Biol.* **27:** 169–200.

Sobkow, L., Epperlein, H.H., Herklotz, S., Straube, W.L., and Tanaka, E.M. 2006. A germline GFP transgenic axolotl and its use to track cell fate: Dual origin of the fin mesenchyme during development and the fate of blood cells during regeneration. *Dev. Biol.* **290:** 386–397.

Steen, T.P. 1968. Stability of chondrocyte differentiation and contribution of muscle to cartilage during limb regeneration in the axolotl (*Siredon mexicanum*). *J. Exp. Zool.* **167:** 49–78.

Stocum, D.L. 1975. Regulation after proximal or distal transposition of limb regeneration blastemas and determination of the proximal boundary of the regenerate. *Dev. Biol.* **45:** 112–136.

Straube, W.L. and Tanaka, E.M. 2006. Reversibility of the differentiated state: Regeneration in amphibians. *Artif. Organs* **30:** 743–755.

Tassava, R.A. and Garling, D.J. 1979. Regenerative responses in larval axolotl limbs with skin grafts over the amputation surface. *J. Exp. Zool.* **208:** 97–110.

Wallace, H. 1973. Participation of grafted nerves in amphibian limb regeneration. *J. Embryol. Exp. Morphol.* **29:** 559–570.

Wang, L., Marchionni, M.A., and Tassava, R.A. 2000. Cloning and neuronal expression of a type III newt neuregulin and rescue of denervated nerve-dependent newt limb blastemas by rhGGF2. *J. Neurobiol.* **43:** 150–158.

Weiss, P. 1925. Unabhängigkeit der Extremitätenregeneration vom Skelett (bei *Triton cristatus). Wilhelm Roux' Arch. Entwicklungsmech. Org.* **104:** 359–394.

Wolpert, L. 1969. Positional information and the spatial pattern of cellular differentiation. *J. Theor. Biol.* **25:** 1–47.

Yamamoto, S., Nagao, M., Sugimori, M., Kosako, H., Nakatomi, H., Yamamoto, N., Takebayashi, H., Nabeshima, Y., Kitamura, T., Weinmaster, G., Nakamura, K., and Nakafuku,

M. 2001. Transcription factor expression and Notch-dependent regulation of neural progenitors in the adult rat spinal cord. *J. Neurosci.* **21:** 9814–9823.

Yntema, C.L. 1959a. Blastema formation in sparsely innervated and aneurogenic forelimbs of *Amblystoma* larvae. *J. Exp. Zool.* **142:** 423–439.

Yntema, C.L. 1959b. Regeneration in sparsely innervated and aneurogenic forelimbs of *Amblystoma* larvae. *J. Exp. Zool.* **140:** 101–123.

Summary: Present and Future Challenges for Stem Cell Research

B.L.M. HOGAN

Department of Cell Biology, Duke University Medical Center, Durham, North Carolina 27710

Stem cell research is being driven forward at an intense pace by creative interactions among scientists working in different fields. These include developmental and reproductive biology, regeneration, genomics, live cell imaging, RNA biology, and cancer biology, to name a few. Numerous model systems and techniques are being exploited, and lab scientists are teaming up with bioengineers and clinicians. The ferment of ideas that makes the field so exciting was in full evidence throughout the Symposium. However, many challenges still need to be overcome to translate basic discoveries into therapeutic outcomes that will save lives and fulfill the promises that have been made. This chapter summarizes some of the highlights of the Symposium and indicates future directions that are being taken by leaders in the field.

There are few topics in the biological sciences that have generated such an intense foment of ideas, legitimate therapeutic promise, irresponsible hyperbole, informed public debate, and political controversy as stem cell research. Perhaps only the human genome project has come close to capturing the imagination of such a broad swath of the scientific and lay community. This excitement, combined with the extraordinarily rapid progress of stem cell research during the past few years, has fueled many international meetings. However, there was nothing jaded about the 73rd Cold Spring Harbor Symposium entitled Control and Regulation of Stem Cells. This is because Cold Spring Harbor has a special cachet when it comes to meetings. Not only is the science of the highest quality, but the atmosphere is always special; the eclectic spirits watching from the walls, the natural and artistic beauty of the environment, and the well-orchestrated hospitality help to create an environment that brings out the best in participants. And so the meeting was a great success; new discoveries were unveiled, connections were made, and collaborations were initiated. The reports in this Symposium volume convey some of this energy, but the full excitement will have to be imagined. Likewise, this summary can only provide a brief overview of the important themes that were discussed and the questions that were raised for the coming years.

DEFINING STEM AND PROGENITOR CELLS: BREAKING DOWN OLD STEREOTYPES AND BUILDING A NEW CONSENSUS

One of the consequences of the increased breadth of stem cell research is that investigators are studying a much wider range of model organisms and tissues than previously. When studies were focused on just a few examples, such as the hematopoietic system, small intestine, hair follicle, and *Drosophila* male and female germ line, the definition of a stem cell was relatively straight-

forward. The properties of what might now be called "classical" stem cells are summarized in Figure 1. They are relatively less differentiated and quiescent cells that reside in a local microenvironment or "niche" that controls their behavior. During the lifetime of the organ, the stem cell population both self-renews and produces daughter progenitor cells that differentiate into one or more postmitotic specialized cell types. The progenitors can themselves self-renew and proliferate extensively, earning themselves the title of transit-amplifying (TA) cells. However, the time span for TA self-renewal is significantly shorter than for stem cells. Most importantly, the "classical" definition of a stem cell includes a stringent requirement for a functional activity: A single stem cell has the potential to maintain or regenerate an entire organ or tissue during the lifetime of the organism.

As new organs have been examined from a stem cell perspective, and greater scrutiny has been applied to old favorites, several new concepts have emerged, as discussed during the Symposium. One such concept is that organs, even relatively small ones, may contain more than one kind of stem cell, each controlled by a different regulatory mechanism. A good example is the *Drosophila* ovary. Once considered only as the home for the germline stem cell (GSC), it is now known to contain two other stem cell populations: the escort stem cell (ESC) and the follicular stem cell (FSC) (Xie et al.; Spradling et al.). Mammalian skeletal muscle has long been known to harbor a population of "satellite cells" that lie just underneath the basal lamina. Studies now show that there are at least two kinds of adult satellite cell: myogenic stem cells that only give rise to muscle and multipotent cells that can give rise to fat and fibroblasts as well (Cerletti et al.). Another example of a tissue containing multiple stem cell types is the mammalian epidermis. In this case, distinct pools of stem cells with different properties reside in the interfollicular epidermis, the hair follicles, and the sebaceous glands (Watt and Collins; Fuchs and Nowak). These pools are thought to be derived from different progenitors during the embryonic development of the epider-

---

All authors cited here without dates refer to papers in this volume.

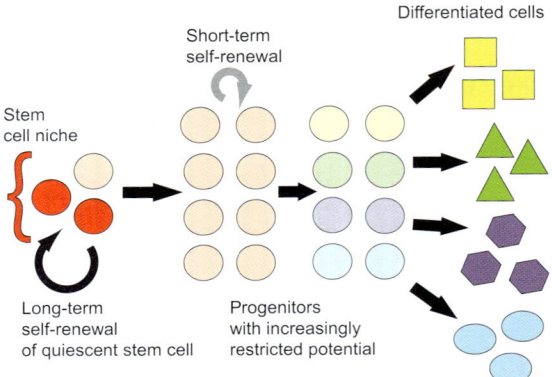

Stem
cell niche

Short-term
self-renewal

Differentiated cells

Long-term
self-renewal
of quiescent stem cell

Progenitors
with increasingly
restricted potential

Figure 1. Schematic representation of a tissue maintained through the activity of a "classical" stem cell and its descendants. This archetypal stem cell (*red*) is anchored in a three-dimensional niche (*red bracket*) and is quiescent, with a low proliferative rate. If division is very infrequent, BrdU incorporated into the DNA of the stem cell is retained during long chase periods. Upon division, individual stem cells may generate two identical stem cells, one stem cell and one multipotent progenitor (*pink*) that gives rise to differentiated progeny, or to two progenitors. The particular outcome may depend on intrinsic asymmetry in the segregation of determinants into the two products of a stem cell division, or on extrinsic differences in the environment inside and outside the niche. In either case, the capacity for self-renewal of the stem cell population persists for all, or a substantial fraction of, the lifetime of the organ (*curved black arrow*). Progenitors can divide rapidly as transit-amplifying (TA) cells, but their capacity for self-renewal is limited (*curved gray arrow*). Finally, progenitor cells give rise to terminally differentiated mature cell types (*multicolored shapes*).

mis. Likewise, there is now evidence that different populations of mesodermal cells in the mammalian heart originate in different embryonic heart fields and the epicardial organ (Nakano et al.).

Another emerging concept in stem cell biology is that even apparently homogeneous populations of stem cells are, when examined closely, heterogeneous in terms of their behavior and/or developmental potential. One example highlighted in the Symposium is the population of neural stem cells that resides in the subventricular zone (SVZ) of the lateral ventricle in the mammalian brain. It now appears that stem cells in different locations along the lateral wall express different transcription factor combinations. These impose different positional identities on the stem cells in the different spatial domains and regulate the fate of the daughter cells that arise from them (Alvarez-Buylla et al.). Another example of apparently homogenous multipotent cells having different positional identities was provided by the work of Elly Tanaka on blastema cells in the regenerating salamander limb (Kragl et al.). These relatively undifferentiated mesenchymal cells are located in the distal stump of the amputated limb and give rise to the replacement parts. Contrary to what was once thought, these blastema cells are not pluripotent. Rather, cells derived from one lineage (e.g., muscle) apparently give rise to the same lineage in the regenerated limb. Moreover, blastema cells still retain positional identity relative to the proximodistal (PD) axis of the original

limb and always give rise to cells with more distal identity. How positional memory is encoded in stem and progenitor cells, when this identity is acquired, and whether it can be changed experimentally are important questions for the future.

One preconceived idea about "classical" stem cells that has changed during the last few years is that they must be quiescent (Fig. 1). One example of stem cells breaking this mold was provided by the work of Hans Clevers on mouse small intestine (Barker et al.). His findings support the idea that the gene *Lgr5*, encoding a G-protein-coupled receptor, marks a population of relatively undifferentiated epithelial stem cells in the base of the crypt, intermingled with Paneth cells. The main evidence that these cells are stem cells comes from in vivo lineage-tracing studies using an *Lgr5-EGFP-ires-CreER^{T2}* "knockin" allele to drive recombination of a *Rosa26R^{lacZ}* reporter gene. Ribbons of cells expressing β-gal were seen running from the base of crypts to the top of the villi for as long as 14 months following activation of the reported allele by a pulse of tamoxifen. Moreover, the lineage label was seen in all differentiated cell types in the villus. Significantly, the *Lgr5*-expressing crypt cells are not quiescent, as expected if they are "classical" stem cells. Rather, they appear to actively divide about once every 24 hours. Previous studies of the dynamics of cell turnover in the small intestine have suggested that the stem cells are localized just above the base of the crypt, in what is known as the +4 position, and that these cells divide infrequently. In support of this idea, recent studies have used expression of the gene *Bmi1* as a marker of the +4 stem cells (Sangiorgi and Capecchi 2008). If a *Bmi1-CreER* knockin allele is used to drive recombination of a lineage reporter, then lines of cells are also seen running up a crypt for at least a year after induction. Bmi1$^+$ stem cells are not found throughout the small intestine but only in the most anterior region. Taken together, these results are compatible with several models. First, there may be two different stem cell populations in the anterior small intestine, one in the +4 position and another in the crypt base, with no functional difference between them under any conditions. Alternatively, the +4 cells, because of their relative quiescence, may be able to survive certain stressful physiological conditions under which the Lgr5$^+$ cells are damaged or lost. During recovery, the +4 cells would then give rise to new populations of Lgr5$^+$ cells. Under these conditions, the Lgr5$^+$ cells would be more like "long-term self-renewing progenitor cells" than classical stem cells (Fig. 1).

Given the complexities that are emerging from recent studies of adult tissues, it is clear that we need to be very precise when defining cells as stem cells, TA cells, or long-term self-renewing committed or differentiated progenitors. This is especially true for tissues that normally have a slow rate of turnover, such as the islets of the pancreas and the liver and the bronchioles of mouse lung (Rawlins et al.). There is a real need for more models and for new nomenclature to cover the different scenarios that may occur (Fig. 2).

In deriving models for the role of stem cells in adult organs, many different criteria must be taken into consid-

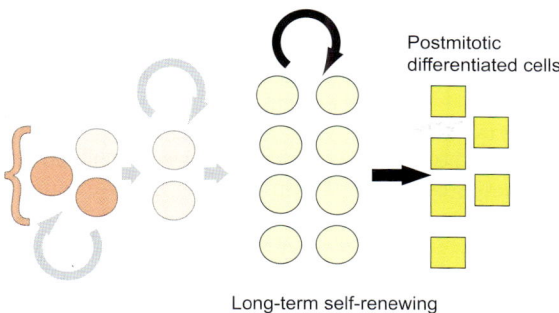

Postmitotic
differentiated cells

Long-term self-renewing
differentiated progenitors

Figure 2. Schematic representation of a tissue maintained primarily through the activity of long-term self-renewing differentiated cells. In this simplified model, a tissue is maintained by the long-term self-renewal (*black curved arrow*) of cells that express differentiated markers and may give rise to one or more postmitotic cell types (*yellow*). This strategy operates during the steady-state maintenance of a tissue that has a slow rate of turnover. However, in response to extensive tissue damage or certain physiological states, the tissue may draw upon a pool of quiescent classical stem cells (*red*) and their progeny (*peach*). These cells also have long-term self-renewal capacity (*gray curved arrows*) but do not normally contribute to tissue maintenance (*short gray arrows*). If the repair conditions are taken into consideration, the yellow cells could be classified as long-term self-renewing committed progenitors. This model can be applied to Clara epithelial cells in the bronchioles of mouse lung (Rawlins et al.). It is also relevant to β cells in the islets of mouse endocrine pancreas that can self-renew during long periods (Brennand et al. 2007; Teta et al. 2007). However, islet tissue can also be regenerated from pancreatic ducts in response to severe damage to the pancreas (Xu et al. 2008).

eration. First, it was emphasized again and again in the Symposium that studies must be based on quantitative, in vivo cell lineage-tracing studies. Ideally, lineage tracing needs to be performed at both the single-cell as well as the population level. Although single-cell lineage tracing is routine in *Drosophila*, new or more sophisticated tools are being developed for organisms such as the mouse and zebra fish. For example, we need the ability to (1) follow the fate of more than one cell type at the same time and (2) image stem cells and their divisions in real time in intact living tissues. A very impressive example of real-time imaging presented in the Symposium was the tracing of labeled spermatogonial cells in mouse testis (Yoshida). We also need more markers for stem cells and their progeny. In addition, a real challenge for the future is to devise models that reveal the full range of physiological conditions under which stem cells are expected to behave and for which they will have been selected during the long evolution of the species. Finally, we need more assays to test the ability of candidate stem cells to regenerate a complete tissue. This ability is the ultimate and perhaps only universal property of a stem cell. Moreover, it is also the most relevant regarding the quest for cell replacement therapy and for building replacement organs.

FUNCTIONAL ASSAYS FOR STEM CELLS

The gold-standard assay for stem cell function is the ability of a single hematopoietic stem cell (HSC) to recon-

stitute the hematopoietic system of an irradiated mouse (Chao et al.). HSC engraftment can also be used in the zebra fish, and Len Zon described the development of a transparent zebra fish to help investigators study how labeled cells home to and engraft into target tissues (Huang and Zon). A number of other transplantation assays have been reported over the years and the use of some of these assays to study stem cells was described in the Symposium. For example, muscle satellite cells can be transplanted into Mdx mutant mice, which are a model for Duchenne's muscular dystrophy (Cerletti et al.). In another example, spermatogonial stem cells can colonize the testis of an infertile recipient mouse (Brinster and Zimmermann 1994; Kanatsu-Shinohara et al.; Yoshida). A beautiful assay for reconstitution of mouse mammary gland is available. The assay involves introducing genetically marked mouse mammary cells sorted into different populations into the mammary fat pads of prepubertal mice that have been "cleared" of endogenous mammary cells. Lineage-tracing studies show that a single mouse mammary stem cell (MaSC) can reconstitute the entire mammary gland in this assay (Asselin-Labat et al.). Moreover, serial transplantation shows evidence for the long-term reconstitution ability of these cells. One challenge for the future is to "humanize" this assay so that it can be adapted for use with human mammary cells.

Two different assays were described for testing the ability of specific epithelial cells in mouse prostate to function as prostate tissue stem cells (PrSCs). The different conclusions regarding the identity of the PrSC reached using these assays highlight the need for multiple approaches to studying stem cell function. For example, Owen Witte's lab (Lukacs et al.) has focused on using an assay, initially developed by Cuhna, in which labeled cells from the prostate epithelium are combined with embryonic mesenchyme from the embryonic urogenital sinus and engrafted under the kidney capsule of immunodeficient mice. A small subfraction of the grafted prostate epithelial cells is able to give rise to tubules containing differentiated cells and resembling normal prostate. Taking an alternative approach, Michael Shen's lab used lineage labeling to follow the behavior of different epithelial cell populations during androgen-dependent serial regression and regeneration of normal mouse prostate. In this model, cells are first lineage labeled, the mice castrated to shrink the prostate, and then androgens are given back to promote regrowth. Using this model, the investigators have obtained evidence that a rare population of $NKX3.1^+$ luminal cells can give rise to both luminal and basal cells during regeneration. They speculate that mouse prostate, which normally turns over very slowly, contains more than one population of progenitor cell for regeneration—basal cells and $NKX3.1^+$ luminal cells (Shen et al.).

In the long run, it will be important to replace in vivo assays with culture methods in which the behavior of stem cells can be followed in real-time and high-throughput screens performed to test compounds for their effect on stem cell behavior. Hans Clevers described the development of an in vitro assay for culturing isolated crypts and associated villi in Matrigel with the goal of following the behavior of stem cells and their progeny in real time. The Witte lab reported using an assay in which dissociated epithelial cells from the mouse prostate are combined with mesenchyme and embedded in Matrigel for in vitro culture rather than grafting (Lukacs et al.). In the future, more of these in vitro assays must be developed to quickly identify and test mechanism regulating stem cell behavior.

THE STEM CELL NICHE: DYNAMIC PERSPECTIVES AND PRACTICAL APPLICATIONS

"Classical" stem cells are anchored in a highly regulated microenvironment or niche (Ohlstein et al. 2004; Xie and Li 2007). Polarized stem cells may be anchored in such a way that when they undergo asymmetric division, one cell (the mother stem cell) remains tightly bound to the niche, whereas the other daughter is displaced from the niche and behaves as a differentiating progenitor (Fig. 1). The niche functions to integrate the different signals regulating the behavior of stem cells. The critical word here is "integrate." In his Symposium talk, Alan Spradling beautifully articulated the concept of the niche and stem cell as a dynamic, integrative unit (Spradling et al.). He stressed that the interaction between the niche and the stem cells is two-way. The niche transmits up-to-date information (through local factors and cytokines, hormones, nutrients, mechanical stress, nerve activity, blood flow, etc.) about the needs of the organism. In response, the stem cells modify their behavior (proliferation rate, specification, fate of daughters, etc.) to maintain homeostasis, meet physiological demands, or repair the effects of injury. In return, the stem cells provide the niche with important information about their behavior so that both positive and negative signals can be ramped up or down accordingly. As stressed many times in the Symposium, the two-way nature of the conversation between stem cells and their niche has important implications in all organisms—not only in animals but also in plants.

The niche is composed of a variety of cells, including in some cases specialized cells that make direct contact with the stem cells, such as the cap cells of the *Drosophila* ovary (Xie et al.). In addition, the niche can include extracellular matrix molecules, blood vessels, and nerves. Two of the best-studied niches are those harboring the *Drosophila* male and female germ lines. The insights that can be gained from straightforward anatomical studies of the stem cell niche were highlighted by work on the organization of stem and ependymal cells of the mammalian SVZ (Alvarez-Buylla et al.). Future challenges will be to define the niche for stem cells in the hair follicle bulge, muscle satellite cells, crypt of the intestine, and plant root and shoot. Significant advances are being made in real-time imaging of stem cells in relation to their niche. For example, in mouse testis Yoshida provided evidence that the vasculature contributes to the niche in this tissue (Yoshida). A major challenge for the future is to identify the different extrinsic signals from the niche to the stem cells, their range of action and localization (Nusse et al.), and how they interact with intrinsic mechanisms functioning within the stem cells to control proliferation and

developmental potential. Two outstanding questions, for example, are whether conserved mechanisms such as the Wnt signaling pathway function in all niches (Nusse et al.), and what is the relative importance of positive signals versus inhibitors and antagonists in stem cell regulation.

During the past few years, the niche has been found to be a ruthlessly competitive environment as well as a nurturing one. Studies have shown that if stem cells are destroyed, daughter cells that have begun to differentiate can replace the stem cell and acquire their phenotypic characteristics, including the ability to self-renew over the long term. It was suggested that competition for access to the niche among stem cells as well as among their daughters serves an important quality control process during homeostasis (Xie et al.). In addition, several talks in the Symposium considered the aging of stem cells and whether this affects the niche, stem cells, or both (Xie et al.).

Understanding the stem cell niche and how it regulates stem cell behavior is likely to have important practical applications as stem cell research moves forward. For example, replicating or reconstructing critical aspects of the niche ex vivo may enable us to expand populations of rare and highly valuable stem cells that can be used for transplantation. In vitro culture will allow their detailed phenotypic analysis and the visualization of stem cell–niche interactions in real time. In the long-term, these studies will help us to bioengineer replacement organs. They may also help us to understand how stromal cells in epithelial tumors promote or restrict tumor growth and/or metastasis.

Finally, it is important to understand how the niche changes with age and how this influences the proliferation and differentiation of stem cells as they, too, age. Studies have shown that satellite cells in old mice are less able to regenerate muscle than those in young animals (Levi and Morrison). However, the activity of old satellite cells can be restored by factors circulating in young animals. Age-related changes in the behavior of other tissue stem cells were also described. Identification of the different pathways involved in aging is an exciting area of future research. Progress may allow us to promote the proliferation of young stem cells, if their numbers are rate limiting for tissue growth, for example, in premature babies. In addition, the possibility needs to be considered that changes in stem cells as they age may affect the class of oncogenic mutations that will promote their self-renewal (Levi and Morrison).

STEM CELLS AND ASYMMETRIC CELL DIVISION

As discussed earlier, some polarized stem cells are anchored to their niche in such as way that after division the mother cell remains in the niche and maintains the stem cell phenotype, whereas the differentiating daughter is displaced and proceeds to give rise to postmitotic progeny. There is intense interest in the mechanisms regulating such asymmetric division and, in particular, how specific determinants are segregated into one daughter versus the other. In the male germ line of *Drosophila*, the mother cell (the GSCs) is attached to the hub (niche) cell.

When asymmetric division takes place, the old centrosome always remains anchored to the hub–GSC interface and the differentiating daughter inherits the new centrosome. It has been argued that cell polarity and asymmetric division with the differential inheritance of cell intrinsic factors are ancient mechanisms that evolved to cope with the problem of aging in single-celled organisms (Macara and Mili 2008). In budding yeast, for example, aging factors remain in the mother cell and are excluded from the young bud. Consequently, there was considerable interest shown during the Symposium in modeling stem cell asymmetry in yeast, which is a superb tool for both cell biology and genetics (Thorpe et al.). In addition, basic mechanisms of asymmetric cell division are also being studied in single animal cells such as the T cell (Chang and Reiner). Naïve T cells are not polarized. However, they associate transiently with partners (the antigen-presenting cell) and use specialized contacts to initiate cell polarization. This leads to asymmetric cell division, with the segregation of determinants into the two daughter cells that then uncouple from their association.

MECHANISMS REGULATING THE SELF-RENEWAL AND POTENCY OF EMBRYONIC AND TISSUE STEM CELLS

Two distinguishing features of multipotent stem cells are their ability to self-renew and give rise to daughter cells that differentiate into specialized cell types. This potential to give rise to different lineages is most dramatic in pluripotential embryonic stem (ES) cells that can generate most cells of the embryo and adult. For this reason, there has been an enormous amount of hard work and ingenuity geared to defining the trancriptional network that controls the phenotypes of mouse and human ES cells (Jaenisch and Young 2008). The consensus that has emerged from several laboratories is that the pluripotential state is maintained by the combined activity of multiple components—transcription factors (TFs), chromatin regulatory factors, signaling pathways, and noncoding RNAs—that cooperate as part of a metastable self-regulating circuit (Cole and Young; Orkin et al.; Chen et al.; Zwaka; Kagalwala et al.) Work has focused on a small group of "key regulators," for example, Oct4 (Pou5f1), Sox2, nanog, and various chromatin-remodeling complexes, that form the core of the network. At present, it is unclear just how many components comprise this core group, but one goal for the future is to devise high-throughput screens to identify additional members. A recurring theme seems to be that the key TF regulators do not work individually. Rather, they bind in combinations, and sometimes even in association, to multi-input regulatory motifs ("hot spots") in the promoters of target genes. Moreover, among these target genes are the genes encoding the TFs themselves. Consequently, the circuits maintaining pluripotency have built into them feedforward and feedback loops, so that under steady state conditions they are self-sustaining.

One of the major challenges for the future is to build similar metastable transcriptional networks for multipotent cells that can give rise to a smaller range of cell types compared to ES cells, for example, endoderm or mesoderm. Hand in hand with this challenge is the need to make connections among the different circuits and identify the switches or gates that allow cells to pass from one state to the next. These gates function during both embryonic development and adulthood. In the embryo, they control how cells within an organ primordium become increasingly restricted in their developmental potential. In adult tissues, they control the flow of progenitor cells down the hierarchy from multipotent stem cell to differentiated cell type (Fig. 1). Examples were given of the exciting progress being made in defining the genetic circuitry of multipotent myogenic lineages and muscle satellite cells (Lagha et al.; Deato and Tjian) and the TF circuitry in early endoderm (Zaret et al.) and neural progenitors (Elkabetz and Studer). One quite unexpected revelation from the Symposium is that the strategies used by stem and progenitor cells to regulate pluripotency and the switch to other states are similar, in principle, between animals and plants (Kornet and Scheres; Lohda et al.).

TURNING LEAD INTO GOLD: REPROGRAMMING DIFFERENTIATED CELLS INTO THE PLURIPOTENT STATE

One of the most unexpected and energizing advances in stem cell research, and indeed in modern biology, has been the discovery that differentiated somatic cells can be induced to become pluripotential embryonic stem-cell-like cells (iPS cells). This process, known as "direct reprogramming," was first achieved by the forced expression in fibroblasts of four TFs (Sox2, Oct4, c-Myc, and Klf4) under the control of retroviral vectors (Takahashi and Yamanaka 2006). However, more recent studies have shown that the proto-oncogene c-*myc* can be omitted from the cocktail, although this reduces efficiency. Likewise, if cells already express *Sox2*, this gene does not need to be added. Even under the best conditions, the efficiency of direct reprogramming is very low—in the range of 0.01–0.1% of transfected fibroblasts—and requires 2–3 weeks of continuous culture, during which stepwise changes in gene expression and epigenetic modification occur. The generation of iPS cells opens up many exciting vistas including the possibility of generating patient-specific pluripotential cells for basic studies into disease mechanisms and, ultimately, cell therapy. Several talks in the Symposium described advances in the derivation of iPS cells. A recurring theme was the need for a careful analysis of the intermediate states that occur during the first 6–12 days of the complex reprogramming process (Hanna et al.; Maherali and Hochedinger). A future challenge is to circumvent the use of viral or DNA vectors to deliver reprogramming genes. The goal will be to replace the vectors with small molecules or drugs that will substitute for the proteins or induce transient coordinated up-regulation of the endogenous genes in order to start a reprogramming cascade.

Direct reprogramming of differentiated cells to pluripotency is a relatively new phenomenon. Other examples of induced global changes in the genetic program of cells that have been studied for longer amounts of time were well represented in the Symposium. It is critical that these models of extreme plasticity continue to be studied in parallel with direct reprogramming because each has significant advantages for studying different aspects of the rewiring process. Examples that were discussed include the reprogramming to pluripotency of the nuclei of differentiated cells transfered into the oocyte (de Vries et al.) and the conversion of primordial germ cells (PGCs) to embryonic germ (EG) cell lines (Surani et al.) and of spermatogonial stem cells to pluripotent stem cells (Kanatsu-Shinohara et al.).

Another process involving extensive and controlled reprogramming of the genome is the formation of totipotent germ cells during embryogenesis (Rangan et al.). Two basic mechanisms appear to be used within the animal kingdom. The first involves the specific allocation of maternally supplied "determinants" (RNA and proteins) to the cytoplasm of the future germ cells. These determinants are initially incorporated into granular inclusions known as germ plasm but are released to function at the transcriptional level and, in particular, the posttranscriptional level. This mechanism of "determinate specification" of the germ line has been extensively studied in *Drosophila* and *Caenorhabditis elegans*. The alternative mechanism, known as "inductive specification," requires signaling among cells within the early embryo and occurs in mammals and a wide range of other invertebrate and vertebrate species. It also occurs during a fascinating phenomenon that captured the intense interest of the Symposium audience: the regeneration of Planaria. Even very small fragments of Planaria can regenerate a complete gonad with germ cells. The germ cells are derived from multipotent neoblasts, or somatic stem cells, within the fragment in response to inductive signals (Newmark et al.). Intense effort is being expended on identifying the components of the germ plasm, how these components function, and the intercellular signals and downstream pathways involved in germ-line induction. From these studies, it appears that both mechanisms share many evolutionarily conserved components, for example, nanos, a zinc finger RNA-binding protein. Thus, the mechanisms differ largely in the initial localization of the conserved factors and the timing at which they become active. At the Symposium, investigators working with *Drosophila* presented evidence of a role for a whole new family of noncoding RNAs, known as Piwi RNAs, in regulating totipotency (Lin and Yin). Precisely how these RNAs regulate gene expression and chromatin modifications is a hot topic for the future.

REPROGRAMMING ADULT CELLS FROM ONE LINEAGE TO ANOTHER: THE THERAPEUTIC POTENTIAL OF TRANSDIFFERENTIATION

As we have seen, a major focus of the stem cell field has been on the mechanisms regulating pluripotency and the switching of differentiated cells to the pluripotent state by experimental manipulation of gene expression. One of the most provocative ideas highlighted during the Symposium was that adult cells can, under certain conditions, be switched from one differentiated state to another,

without going all the way back to pluripotency. This process is known as "transdifferentiation" or "transdetermination" if it occurs in embryonic tissues such as the imaginal disc of *Drosophila*. In fact, examples of transdifferentiation or metaplasia in adult tissues, usually under conditions of injury and repair, have been well documented, and the potential of harnessing this plasticity for therapeutic purposes has long been recognized (Slack 2007). Thus, it was very exciting to learn that Doug Melton's lab had been able to convert exocrine pancreas into cells with both the molecular and morphological phenotype of insulin-producing β cells (Zhou and Melton). This was achieved by forced expression of three TFs, each of which has a critical role in guiding the embryonic development of the endocrine pancreas. This unexpected but tremendously exciting finding has great clinical promise if it could be applied, for example, to deriving β cells from liver, which is more accessible than exocrine pancreas. This is because there are presumably fewer steps that could go wrong in the process of generating a β cell from a cell already committed to the endoderm lineage than from an iPS or ES cell. It would be premature, however, to abandon other strategies that are being used to generate more β cells for clinical use, for example, from a potential population of multipotent cells in pancreatic ducts (Xu et al. 2008).

CANCER STEM CELLS

About one third of the talks at the Symposium involved cancer-related research. In recent years, this field has been greatly influenced by the idea that stem cell biology can throw new light on the origin, progression, and treatment of human tumors (Bonnet and Dick 1997; Reya et al. 2001; Wang and Dick 2005). The idea that tumors are perpetuated and sustained by "cancer stem cells" is based on the finding that some cancers can be serially transplantated by grafting them into immunodeficient mice. Moreover, individual cells in the tumor differ in their efficiency to give rise to new tumors containing the same range of cell types as the first. A rare subpopulation of cells—the presumptive multipotent cancer stem cells—are significantly more efficient at giving rise to new tumors than are the majority. Because "classical" stem cells are relatively quiescent, it has also been argued that these tumor stem cells can escape the action of standard anticancer drugs designed to kill rapidly proliferating cells. In contrast, drugs designed to block pathways specifically required for stem cell self-renewal would, according to the model, be particularly effective in blocking the progression and recurrence of tumors.

The Symposium talks illustrated the fact that, over time, the "cancer stem cell model" has encompassed several very different ideas, often leading to confusion (Kelly et al. 2007; Visvader and Lindeman 2008). Thus, the original model has been extended to include the idea that cancers arise from stem cells in normal tissues. In other words, stem cells are likely to be the "cell of origin" of the original cancer because only individual tissue stem cells stay around long enough to accumulate the combination of oncogenic mutations that lead to malignant transformation. Defining the mechanisms regulating self-renewal and differentiation of normal tissue stem cells is therefore critical for understanding how cancers arise and progress (Williams and Sherr; Pierfelice et al.; Rich). However, strictly speaking, this model can only apply to tissues in which progenitor or TA populations (Fig. 1) normally have a strictly limited self-renewal capacity and a high probability of giving rise to terminally differentiated cells that are lost from the body. If the committed progenitor cells self-renew during long periods of time, presumably they too could accumulate multiple oncogenic mutations and become the "cell of origin" of a first cancer. An interesting idea that was brought up was that changes in the number or properties of tissue stem cells over time may account for changes in cancer risk in human populations, for example, the increased risk of breast cancer in nulliparous women (Asselin-Labat et al.).

Once a tumor has formed, there is likely to be intense competition within the tumor population, favoring the survival and expansion of cells with high self-renewal and low differentiation potential. According to what is known as the "clonal expansion model" of cancer progression, cells within the tumor that normally behave like short-lived progenitors will have a selective advantage if they acquire mutations that promote a more stem-cell-like phenotype. According to this idea, "cancer stem cells" or "cancer-sustaining cells" from early tumors will have different properties than cells with the same behavior that evolve from progenitor cells in more advanced tumors.

Identifying cancer stem cells depends heavily on the assay of grafting dissociated tumor cells into immunodeficient mice and obtaining serial transplantable tumors. The point was made that this assay may only measure the ability of human cells to grow in mouse, rather than a cancer stem cell phenotype per se (Adams et al.). This caveat gains credibility as we understand more about the complex reciprocal interactions between epithelial tumors and their support environment or niche that includes mesenchymal stroma, blood vessels, and immune cells. Several speakers addressed this issue. For example, a very elegant series of experiments was reported by Sean Morrison (Levi and Morrison), who took a systematic approach to developing an assay for propagating human mammary tumors in mice. In these and other assays, the tumor environment in severely immunocompromised mice is "humanized" by the addition of human mesenchymal cells that provide a "niche" for the tumor cells. In this way, the percentage of transplanted cells that can give rise to a tumor can be greatly increased (Quintana et al. 2008).

Other assays, most notably the "neurosphere assay" for neural stem cells, have been adapted to identify stem cells in cancers. This assay depends on neural stem cells grown in suspension that then form floating spheres containing undifferentiated and differentiated cell types that can be propagated over multiple rounds of dissociation and culture. Here again, the sphere-forming assay appears to work for some solid tumors but not for others. Owen Witte described the development of an in vitro sphere-forming assay for normal and tumor-derived prostate epithelial cells. In this assay, cells are combined with mesenchymal cells and grown embedded in Matrigel

where they give rise to spheres that can be propagated over multiple passages (Lukacs et al.).

Finally, speakers pointed out that testing the idea that stem cells are the cells of origin of certain cancers will require the identification of more promoters to drive genetic recombination and the expression of different oncogenes in specific cell populations (stem cells and TA cells) in normal tissue (Alcantara Llaguno et al.). In addition, more surface markers are needed to sort subpopulations of tumor cells. Of course, these challenges are not confined to cancer-related research but confront all investigators taking a genetic approach to stem cell biology with animal models. In conclusion, one take-home message from the Symposium for cancer researchers was that enthusiasm for the "cancer stem cell" model needs to be tempered by a rigorous and critical analysis of the data and methods.

CONCLUSIONS

In this summary, I have highlighted some of the topics covered during the Symposium and the challenges raised by the new ideas and data that were presented. I apologize to speakers whose work I did not refer to specifically. In the few months since the meeting, there have been many impressive advances and already some of the challenges I outlined have been met. Yet, stem cell research remains one of the most influential areas of science, in part because it brings together people from many different backgrounds—those working in cell biology, developmental biology, genomics, immunology, bioengineering, chemistry, drug design, and medicine—to work together to achieve goals benefiting us all. Although it may take longer than we hope for advances in basic research to reach the clinic, there is no doubt that they eventually will. We can only imagine the topics of future Cold Spring Harbor Symposia as regenerative medicine and stem cell research come to maturity and fruition.

ACKNOWLEDGMENTS

I would like to express my personal thanks to Bruce Stillman, David Stewart, and all of the wonderful faculty and staff at Cold Spring Harbor Laboratory for making the Symposium and this volume so successful. I also thank the National Institutes of Health and those in my laboratory for the opportunities and joy they provide.

REFERENCES

Bonnet, D. and Dick, J.E. 1997. Human acute myeloid leukemia is organized as a hierarchy that originates from a primitive hematopoietic cell. *Nat. Med.* **3:** 730–737.

Brennand, K., Huangfu, D., and Melton, D. 2007. All β cells contribute equally to islet growth and maintenance. *PLoS Biol.* **5:** e163.

Brinster, R.L. and Zimmermann, J.W. 1994. Spermatogenesis following male germ-cell transplantation. *Proc. Natl. Acad. Sci.* **91:** 11298–11302.

Jaenisch, R. and Young, R. 2008. Stem cells, the molecular circuitry of pluripotency and nuclear reprogramming. *Cell* **132:** 567–582.

Kelly, P.N., Dakic, A., Adams, J.M., Nutt, S.L., and Strasser, A.

2007. Tumor growth need not be driven by rare cancer stem cells. *Science* **317:** 337.

Macara, I.G. and Mili, S. 2008. Polarity and differential inheritance—Universal attributes of life? *Cell* **135:** 801–812.

Ohlstein, B., Kai, T., Decotto, E., and Spradling, A. 2004. The stem cell niche: Theme and variations. *Curr. Opin. Cell Biol.* **16:** 693–699.

Quintana, E., Shackleton, M., Sabel, M.S., Fullen, D.R., Johnson, T.M., and Morrison, S.J. 2008. Efficient tumour formation by single human melanoma cells. *Nature* **456:** 593–598.

Reya, T., Morrison, S.J., Clarke, M.F., and Weissman, I.L. 2001. Stem cells, cancer, and cancer stem cells. *Nature* **414:** 105–111.

Sangiorgi, E. and Capecchi, M.R. 2008. *Bmi1* is expressed in vivo in intestinal stem cells. *Nat. Genet.* **40:** 915–920.

Slack, J.M. 2007. Metaplasia and transdifferentiation: From pure biology to the clinic. *Nat. Rev. Mol. Cell Biol.* **8:** 369–378.

Takahashi, K. and Yamanaka, S. 2006. Induction of pluripotent stem cells from mouse embryonic and adult fibroblast cultures by defined factors. *Cell* **126:** 663–676.

Teta, M., Rankin, M.M., Long, S.Y., Stein, G.M., and Kushner, J.A. 2007. Growth and regeneration of adult β cells does not involve specialized progenitors. *Dev. Cell* **12:** 817–826.

Visvader, J.E. and Lindeman, G.J. 2008. Cancer stem cells in solid tumours: Accumulating evidence and unresolved questions. *Nat. Rev. Cancer* **8:** 755–768.

Wang, J.C. and Dick, J.E. 2005. Cancer stem cells: Lessons from leukemia. *Trends Cell Biol.* **15:** 494–501.

Xie, T. and Li, L. 2007. Stem cells and their niche: An inseparable relationship. *Development* **134:** 2001–2006.

Xu, X., D'Hoker, J., Stangé, G., Bonné, S., De Leu, N., Xiao, X., Van de Casteele, M., Mellitzer, G., Ling, Z., Pipeleers, D., et al. 2008. β Cells can be generated from endogenous progenitors in injured adult mouse pancreas. *Cell* **132:** 197–207.

Author Index

601

Subject Index

603